GENETICS MANUAL

CURRENT THEORY, CONCEPTS, TERMS

GENETICS MANUAL

CURRENT THEORY, CONCEPTS, TERMS

GEORGE P. RÉDEI
University of Missouri

World Scientific
Singapore • New Jersey • London • Hong Kong

Published by
World Scientific Publishing Co. Pte. Ltd.
P O Box 128, Farrer Road, Singapore 912805
USA office: Suite 1B, 1060 Main Street, River Edge, NJ 07661
UK office: 57 Shelton Street, Covent Garden, London WC2H 9HE

British Library Cataloguing-in-Publication Data
A catalogue record for this book is available from the British Library.

First published 1998
Reprinted 1999

GENETICS MANUAL: CURRENT THEORY, CONCEPTS, TERMS

Copyright © 1998 G.P. Rédei. All rights reserved. According to the United States of America copyright law, no part of this book may be translated, reproduced or transmitted in any form or by any means, electronic, mechanical, including photocopying, recording, or any information storage and retrieval system — except for brief quotations for critical reviews or limited number of definitions in research articles — without the permission of the author.

ISBN 981-02-2780-9

Printed in Singapore.

ACKNOWLEDGEMENTS

I thank my wife Magdi for her patience while writing and for critical reading of the text. My daughter Mari introduced me into word processing. Granddaughters, Grace and Paige are most inspirational.

Gratitude is due to Dr. Michael R. Speicher, Institut für Anthropologie und Humangenetik, LMU, München, Germany for the beautiful cover picture, showing a human male metaphase prepared by multiplex-fluorescence *in situ* hybridization, permitting the identification of each chromosome by a different color. I am grateful to countless number of colleagues on whose work this material is based and to whom I could not refer because of limitations of space. I appreciated the help of Dr. J. Perry Gustafson, University of Missouri, Columbia, MO for technical advice in the preparation of the typescript. I am indebted to colleagues, especially to Dr. Csaba Koncz, Max-Planck-Institut, Cologne, Germany for many useful discussions. My students at the University of Missouri, Columbia, MO, and students and colleagues, particularly Dr. András Fodor, at the Eötvös Lóránd University of Basic Sciences, Budapest, provided purpose for undertaking this project. Some of the chemical formulas are based on the Merck Index and on the Aldrich Catalog. I express my gratitude to Professor Jerry Hsueh, Fudan University, Shanghai for his interest, enthusiasm, and translation of my work into Chinese.

Editors Sook Cheng Lim and Cheryl Lim, World Scientific Publishing Company provided advice during the preparations. I am indebted also to my former Editor, Grace Tan for encouragement.

♥

TO PAIGE AND GRACE

♥

PREFACE

The primary goal of this Manual is the facilitation of communication and understanding across the wide range of biology that is now called genetics. The emphasis is on recent theoretical advances, new concepts, terms and their applications. The book includes about 18 thousand concepts and over 650 illustrations (graphs, tables, equations and formulas). Most of the computational procedures are illustrated by worked-out examples. A list of about 900, mainly recent, books is provided at the end of the volume, and additional references are located at many entries and illustrations. The most relevant databases are also listed.

The cross-references following the entries connect to a network within the book, so this is not just a dictionary or glossary. By a sequential search, comprehensive, integrated information can be obtained as you prepare for exams, or lectures, or develop or update a course, or need to review a manuscript, or just wish to clarify some problems. In contrast to encyclopedias, I have used relatively short but greater variety of entries in order to facilitate rapid access to specific topics.

This Manual was designed for students, teachers, scientists, physicians, reviewers, environmentalists, lawyers, administrators, and to all educated persons who are interested in modern biology. Concise technical information is available here on a broad range of topics without a need for browsing an entire library. This volume can always be at your fingertips without leaving the workbench or desk. Despite the brevity of the entries, the contents are clear even for the beginner.

Herbert Macgregor made the remarkable statement that in 1992 about 7,000 articles related just to chromosomes were scattered among 627 journals. Since then, the situation has become worse. Many publications — beyond a person's specialization — are almost unreadable because of the multitude of unfamiliar acronyms and undefined terms. Students and colleagues have encouraged me to undertake this effort to facilitate reading of scientific and popular articles and summarize briefly the current status of important topics.

According to Robert Graves (a good poem) "makes complete sense and says all that it has to say memorably and economically". I hope you will appreciate the sense and economy of this Manual.

I will be much indebted for any comment, suggestion and correction.

GPR
3005 Woodbine Ct.
Columbia, MO 65203-0906, USA
Telephone: (573) 442-7435
e-mail: plantgpr@showme.missouri.edu

"I almost forgot to say that genetics will disappear as a separate science because, in the 21st century, everything in biology will become gene-based, and every biologist will be a geneticist."

Sydney Brenner, 1993

FIRST TO READ

The material in this book is in alphabetical order. There are different styles of alphabetization, however. Numbers involved with the entries do not affect the order. Words standing alone precede the hyphenated or compounded terms, e.g. *in vivo* precedes *inactive*. Hyphenated words are ordered as if they would not be hyphenated but single words. The spelling of some terms varies because some are used frequently with or without hyphenation (as one word) any many words are spelled either with a *c* or *k*. In the literature some technical terms are spelled either with an *e* or *ae*. Here the most common usage is favored. Certain terms are in plural and that may affect the relative order of the entry. Some entries are qualified by another word added after and in others the qualifier comes first. An attempt was made to guide the reader to the entry sought by both ways when this appeared important. If the case sought after appears missing, try to use synonyms or related terms or concepts and you may find the one that was hiding at the first attempt. Thanks for your patience.

A

A: adenine, a purine base of nucleic acids. (See adenine)

2,5-A (adenine) oligonucleotides: are generated by 2,5-A synthethase from double-stranded RNA. These oligonucleotides activate RNase L which attacks infecting viruses of vertebrates. If the two genes encoding these two enzymes are transformed into plants, they provide resistance against RNA viruses. (See host - pathogen relationship. RNase)

Å (ångstrom): unit of length, 1/10 of a 1 nm; 10^{-7} mm.

A6: *Agrobacterium tumefaciens* strain with a Ti plasmid coding for octopine production in the plant cell. (See *Agrobacterium*, opines, octopine)

α: average inbreeding coefficient, $\alpha = \Sigma p_i F_i$ where p_i is the relative frequency of inbred individuals with F_i coefficient of inbreeding. This value in most human populations is less than 0.001 while in isolated human groups it may exceed 0.02 or 0.04. (See inbreeding coefficient)

A BOX: an internal control region of genes (5S ribosomal RNA and tRNA) transcribed by DNA-dependent RNA polymerase III; the consensus is 5'-TGGCNNAGTGG-3'. The tRNA genes have also an essential *intermediate segment* of about a dozen bases that has no consensus yet its length is necessary for function. Also there is nearby another regulatory consensus, the B box 5'-GGTTCGAANNC-3'. The matrix attachment region (MAR) is also an A box (with a consensus of AATTAAA/CAAA). (See MAR)

A CHROMOSOME: member of the regular chromosome set in contrast to a B or supernumerary chromosome. (See also B chromosome, accessory chromosome)

α COMPLEX: one of the alternate chromosome translocation complexes in *Oenothera*. (See β complex, translocation, *Oenothera*)

A DNA: see DNA

A MEDIUM for *E. coli*, g/L: K_2HPO_4 10.5, KH_2PO_4 4.5, $(NH_4)_2 SO_4$ 1.0, Na-citrate.2 H_2O, 0.5 plus glucose 0.4%, thiamin 1 mg/L, $MgSO_4$ 1 mM, and an appropriate antibiotic

α PARTICLES: see alpha particles

α SATELLITE: the centromeric DNA that is normally heterochromatic but it may have important role in controlling chromosome segregation and other centromere functions. (See centromere, satellite DNA, heterochromatin, segregation, meiosis)

A SITE: is a compartment on the ribosome; at the beginning of the translation process the first codon, Met or fMET lands at the P site and the next amino acid is delivered to the A site. Then the elongation of the peptide chain proceeds. (See protein synthesis)

α-AMANITIN: a protein synthesis inhibitor fungal octapeptide. It blocks RNA pol II (0.1 µg/mL); RNA polymerase III is also blocked by it but at much higher concentrations (20 µg/mL) but pol I is insensitive to it even at 200 µg/mL. LD_{50} in albino mice is 0.1 mg/kg. (See RNA polymerase, pol, LD_{50})

α-1,4-GLUCOSIDASE DEFICIENCY: see acid maltase

α-HELIX: a secondary structure of polypeptides with maximal intrachain hydrogen bonding. A most common conformation, when after 5.4 Å high, five right turns of an amino acid chain, every 18th amino acids occupy the same line as the first. (See pitch, protein structure)

A-KINASES: cAMP-dependent protein phosphorylating enzymes; the phosphorylation is dependent on sufficiently high level of cAMP. (See cAMP)

α-LACTOSE: milk sugar is converted into allolactose by the β-galactosidase gene of the *Lac* operon of *E. coli* and the latter then becomes the inducer of the operon. (See *Lac* operon)

α MATING TYPE FACTOR OF YEAST: is responsible for the secretion of the α factor (a pheromone), composed of 13 amino acids and it acts on *a* type cells. (See mating type determination)

AAA PROTEINS: are ATPases, enzymes cleaving off phosphates from ATP. (See ATP)

AAF: see alpha accessory factor

AARI: see TUP1

AARSKOG SYNDROME (Aarskog-Scott syndrome): autosomal dominant, autosomal recessive X-linked (Xq12) recessive short stature, hypertelorism (increased distance between organs or parts), scrotum (the testis bag) anomaly, pointed hairline (Widow's peak), broad upper lip, floppy ears, etc. The basic defect involves the RHO/RAC member of the RAS family of GTP-binding proteins. (See stature in humans, head/face/brain defects, RAS)

AATAAA: a consensus of 10 - 30 bp upstream from a CA dinucleotide at the site where cleavage, then polyadenylation of the mRNA commonly take place. This consensus may also be a signal for transcription termination although normally RNA polymerase II continues to work after passing it. (See polyadenylation signal, mRNA tail, transcription termination)

AATDB: *Arabidopsis thaliana* database provides general information on all aspects of the plant, including genes, scanned images of mutants, nucleotide sequences, genetic and physical map data, cosmid and YAC clones, bibliographical information, etc. Access is available without password through <http://www.weeds.mgh.harvard.edu/index.html> or by e-mail <curator@frodo.mgh.harvard.edu>. (See also AIMS, *Arabidopsis thaliana*, databases)

AAUAAA: is a consensus for polyadenylation of the mRNA. Apparently the poly-A RNA polymerase enzyme and associated protein attach to this sequence before cleavage of the transcript and post-transcriptional polyadenylation takes place. Yeast does not have this consensus. (See also AATAAA consensus's role in polyadenylation)

αβ T cells: recognize the major histocompatibility complex-bound peptide antigens. (See MHC, γδ T cells, T cell)

ABA (abscisic acid (3-methyl-5-[1'-hydroxy-4'-oxo-2'-cyclyhexen-1'-yl]-cis-2,4-pentadienoic acid): is a terpenoid, synthesized from mevalonate and xanthins, apparently through two pathways. It has multiple physiological functions in concert with other plant hormones, particularly with gibberellins and cytokinins by regulating seed dormancy, germination, leaf abscission, etc. Some *aba* genes have been cloned. In the ABA signal transduction farnesyl transferase may be involved. (See prenylation, farnesyl pyrophosphate, plant hormones)

ABASIC SITES: in the DNA where glycosylases have removed bases by cleaving the glycosylic bond. (See glycosylases, DNA repair)

ABAXIAL: not in the axis of body or of an organ.

ABC EXCINUCLEASE: is a 260,000 M_r protein complex containing the subunits coded for by the *uvrA*, *uvrB* and *uvrC* genes of *E. coli*. UvrA is an adenosine triphosphatase and also brings into position UvrB which after attaching to the DNA cuts it at the 3' position and that provides the opportunity for UvrC to incise at the 5' position. UvrD, a helicase releases the damaged oligomer along with UvrC. Following these events, DNA polymerase fills in the correct nucleotides. In yeast the RAD1, 2, 3, 4, 10, 14, carry out the same tasks as the ABC excinucleases of bacteria. In humans, the XPA (a damage recognition protein, comparable to UvrA), binds to the XPF-ERCC1 (**e**xcision **r**epair **c**ross-**c**omplementing protein) heterodimer and to the human single strand binding replication protein, HSSB. XPF (3' cut) and XPG (5' cut) are nucleases. The gap-filling polymerases are polδ and polε. XPB and XPD are helicase subunits of the TFIIH transcription factor. The excinuclease complex is released at the end of the process with the aid of the proliferating cell nuclear antigen (PCNA). This complex is capable of excision of cyclobutane pyrimidine dimers, 6 - 4 photoproducts (adjacent pyrimidines cross linked through C^6- C^4), nucleotide adducts (molecules with added groups) formed by mutagenic agents. (See excision repair, adduct, DNA polymerases, DNA ligase, helicase, transcription factors, PCNA, cyclobutane)

ABC TRANSPORTERS (ATP-binding cassette transporters): constitute a large family of proteins which hydrolyze ATP and mediate transfers through membranes. These are usually called now TAP. (See TAP, protein-conducting channel, TRAM, signal hypothesis, SRP, translocon, translocase)

ABCD MODEL: is an environmental matrix for the study of the performance of species. (See box)

The ABCD Model

	Environment 1	Environment 2
Genotype 1	A	B
Genotype 2	C	D

ABDOMEN IN *DROSOPHILA*: the body segment between the thorax and telson. (See *Drosophila*)

ABELSON MURINE LEUKEMIA VIRUS oncogene (*abl*): is the mammalian homolog of the avian Rous sarcoma virus. It codes for a plasma membrane tyrosine kinase. (See oncogenes, Rous sarcoma)

ABERRANT GENETIC RATIOS: occur when the chromosomes carrying the wild type or mutant allele of a gene have reduced transmission through meiosis or the viability of the gametes is diminished. Depending on the chromosomal location of the defect, either the one (wild type) or the other (recessive) allele may appear in excess of expectation of normal phenotypic ratios.

ABERRATION CHROMOSOMAL: see chromosome breakage

ABETALILIPOPROTEINEMIA: involves very low levels of the very low density (VLDL), low density (LDL) and high density (HDL) of these lipoproteins. The rare recessive anomaly is accompanied by excretion of lipoproteins, malabsorption of fat, acanthocytosis (thorny type erythrocytes), retinitis pigmentosa (sclerosis [hardening], pigmentation and atrophy [wasting away]) of the retina of the eye and irregular coordination of the nerves (ataxia). (See neuromuscular disease)

ABH ANTIGENS: in humans are secreted in the saliva and other glycoprotein-containing mucus in the presence of the *Se* (dominant allele, human chromosome 19cen-q13.11), and the gene codes for the α2L-fucosyltransferase enzyme. The secreted glycoproteins, A and B are about 85% carbohydrate and about 15% protein. Approximately 75-80% of Caucasoids are secretors (homozygous or heterozygous for *Se*). The precursors of the antigens are Galactose(β1-3)*N*-acetyl-D-glucosamine-R and Gal(β1-4)*N*-acetyl-glucosamine-R (where R stands for the extension of the carbohydrate chain).

Fucose
|

Antigen H has the critical structure of Galactose(β1-3,4)*N*-acetyl-D-glucosamine-R. Antigenic determinant A is formed by *N*-acetylgalactosamine, and the B antigen by galactose addition at non-terminal position to the H antigen. Thus the A, B, and H antigens are different from each other by these carbohydrates and in some variants by the number of fucose molecules. The A and B alleles are codominant. The recessive O blood group lacks a fucosidase activity that places a fucose, by an α1-2 linkage on a galactose. The Lewis blood group (Le [Les], 19p13.1-q13.11) is distinguished on the basis that its dominant allele Le places fucose in an α-1,4 linkage to the *N*-acetylglucosamine. Individuals that have no secretor activity but are Le belong to the Lewis blood group Le[a] whereas when both Se and Le are expressed they represent the Le[b] type. (See also ABO blood group, Bombay blood type).

ABIOGENESIS: spontaneous generation of life, origin of living cells from organic material during the early history of the earth. (See spontaneous generation, origin of life)

abl: B cell lymphoma (Abelson leukemia) oncogene encoding a non-receptor protein tyrosine kinase. This oncogene is activated by ionizing radiation and alkylating agents. In its absence the JNK/SAP kinases (Jun kinase) are not stimulated. (See leukemia, lymphoma. JUN, JNK/SAP)

ABL: see abetalipoproteinemia

ABL (Abelson murine leukemia virus oncogene): located to human chromosome 9q34.1 and mouse chromosome 2. When translocated to human chromosome 22 it may transcribe a fusion protein with an abnormal protein tyrosine kinase activity and this is probably the cause of

ABL continued
chronic myeloid leukemia. Acute lymphocytic leukemia is also associated with a similar translocation, the Philadelphia chromosome, but it appears that tyrosine kinase activation is different from that of the fusion protein. The ABL gene has an about 300 kb intron downstream from the first exon. This intron appears to be the target of the translocations and causes acute lymphocytic leukemia. Insertion of DNA sequence into the *abl* gene of mouse results in several morphological alterations and death. (See oncogenes, ARG, Philadelphia chromosome, leukemia)

ABLATION: mechanical removal of cells or tissues of stem cells or plant meristems to study the role of those cells in differentiation and development. The purpose can be achieved also by obtaining genetic deletions in these areas, heterozygous for appropriate marker genes. The deletion of the dominant allele reveals the function of the recessives and permits tracing cell lineages on the basis of the visible sectors formed. Familial retina ablation may occur in animals as a hereditary abnormality. (See also gene fusion, pseudodominance, deletion, cell lineages)

ABM PAPER: see diazotized paper

ABO BLOOD GROUP: is represented by three major type of alleles (human chromosome 9q34) displaying codominance. These blood types are extremely important because inappropriate mixing (in blood transfusion) results in agglutination that prevents the flow of blood through the veins and oxygen transfer, and it is potentially lethal. These antigens are actually carbohydrates (attached to polypeptides), and the genes A and B specify α-D-*N*-acetylgalactosaminyltransferase and α-D-galactosyltransferase enzymes, respectively. Gene O is not active as an enzyme. The A and B enzymes (M_r about 100,000) are dimeric and structurally similar to each other. The A and B molecules are identified as A and B antigens. The clinical characteristics

Blood Group (frequency in caucasoids*)	Genotype	Antigens Formed	Antibodies Formed	Clumping With	Blood Type Acceptable for Transfusion
O (0.45)	$i^O i^O$	neither	anti-A anti-B	A, B AB	O
A (0.44)	$i^A i^A$ or $i^A i^O$	A	anti-B	B, AB	A, O
B (0.08)	$i^B i^B$ $i^B i^O$	B	anti-A	A, AB	B, O
AB (0.03)	$i^A i^B$	A, B	neither	neither	A, B, O

*The frequency of these alleles vary in different populations. For the calculation of frequencies see gene frequencies. Actually the A type exists in A_1 and A_2 forms and in about 1 - 2 % of the A_2 and in about 25% of the A_2B individuals anti-A_1 antigens occur.

Occasionally maternal antibodies against the A and B antigens may enter, through the placenta, the fetal blood stream and affect adversely the erythrocytes causing anemia and hyperbilirubinemia. In such cases medical treatment may be required. The ABO system has also a limited use in forensic medicine in paternity suites, in typing blood stains, semen and saliva in criminal cases. Immunologically active forms may be recovered in old human remains and can be used also in archeological research. This blood group provided some correlative information in cancer research, e.g. in O individuals afflicted with carcinomas A antigen may be detected in 10-20% of the cases. It appears, changes in glycosyltransferase activity is not uncommon in several types of tumors. The frequency of the various ABO alleles varies a great deal in the world's population. It has been shown that the O blood type provided some protection against the most severe form of syphilis (*Treponema pallidum*) but somewhat higher susceptibility to diarrhea caused by some viral and bacterial infections. The B blood group may have afforded some protection against smallpox, plague and cholera. (See also ABH antigen, Lewis blood group, blood groups, forensic genetics)

ABORIGINE: the first group of inhabitants, humans, animals or plants.

ABORTION, SPONTANEOUS: is frequently caused by disease, chromosomal aberrations. Various types of chromosomal defects were cytologically detected in 30-50% of the aborted fetuses. About 15-20% of the verified human pregnancies are aborted spontaneously and an estimated 22% of the abortions occur before pregnancy is clinically detected. (See selective abortion, trisomy, chromosomal rearrangements, chromosome breakage)

ABORTIVE INFECTION: bacteria are infected with a phage capsule that carries bacterial rather than phage DNA and thus cannot result in the liberation of phage particles.

ABORTIVE TRANSDUCTION: the transduced DNA is not incorporated into the bacterial genome and in the absence of a replicational origin it can be transmitted but it cannot be propagated. Therefore the transduced fragment is contained in a decreasing proportion of the bacteria. (See transduction, transduction abortive [diagram])

ABRIN: agglutinin, a toxic lectin and hemagglutinin extracted from the seed of the tropical leguminous plant jequirity (*Abrus precarius*). Abrins A, B, C, D are glycoproteins of two polypeptide chains. The small A chain is an inhibitor of aminoacyl-tRNA binding and has nothing to do with agglutination. Abrin is more toxic to a variety of cancer cells (ascites, sarcomas) than to normal cells. (See aminoacyl tRNA synthetase, lectins, hemagglutinin)

ABRINE: N-methyl-L-tryptophan (α-methylamino-β-[3-indole]propionic acid), and is unrelated to abrin.

ABSCISIC ACID: a plant hormones regulating a variety physiological processes, including modification of the action of other plant hormones. Originally it was detected as a substance involved in the abscission of leaves. (See also ABA, plant hormones, formula below)

ABSCISSION ZONE: the thin-walled tissue layer (low in lignin and suberin) formed at the base of the plant organs before abscission takes place. (See abscisic acid)

ABSOLUTE LINKAGE: there is no recombination between (among) the genes in a chromosome. (See recombination, linkage)

ABSOLUTE WEIGHT: the mass of 1000 seeds or kernels after appropriate cleaning.

ABSORPTION: uptake of compounds through cell membranes or through the intestines into the bloodstream.

ABSORPTION SPECTRUM: the characteristic absorption peaks of a compound at various wavelengths of light; e.g. guanine has maximal absorption at about 278 nm at pH 9 but its maximum at pH 6.8 is at ca. 245 nm ultraviolet light; chlorophyll-a has an absorption maximum in benzene at ca. 680 and 420 nm visible light whereas chlorophyll-b maxima are at ca. 660 and 460 nm, respectively. These characteristics vary according to the pH and solvents used and are determined by spectrophotometers.)

ABUNDANCE: average number of molecules in cells.

ABUNDANT mRNAs: a small number of RNAs that occur with great numbers in the cells. (See mRNA)

ABZYMES: monoclonal antibodies with enzyme-like properties. If these antibodies can recognize the transition state analogs of enzyme-substrate reactions, they might have enzymatic properties. These abzymes would have numerous chemical and pharmaceutical applications. (See monoclonal antibody, antibody)

Ac—Ds (*Activator -Dissociator*): the first transposable element system recognized on the basis of its genetic behavior in maize. It contains 4563 bp and bordered by 11 bp imperfect, inverted repeats. The independently discovered *Mp* (*Modulator* of *p1* [pericarp color]) is basically the same transposon. *Ac* is an autonomous element and can move by its own transposase function. The *Ac/Mp* element makes a 3.5 kb transcript, initiated at several sites upstream, and a 2,421 base mRNA. A defective (deleted) version of it, *Ds* (*Dissociator*), is non-autonomous and requires the presence of *Ac* for transposition. *Ds* was originally discovered on the basis of frequ-

Ac - Ds continued

ent chromosome breakage associated with it. The *Ds* elements are quite varied in size but practically identical at the terminal sections to *Ac*. These elements have been identified first on the basis of mutation at known loci *(a, Adh, sh, wx, etc.)* upon insertion and reversions when the inserted element is evicted. More recently it has been shown that many of the insertions do not lead to observable change in the expression of the genes or their effect is minimal and their presence may be then revealed only by sequencing of the target loci.

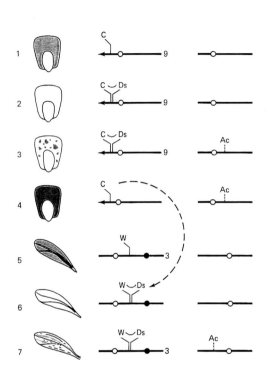

The possible phenotypic expression of genes in the presence of the *Ac - Ds* elements in maize. (1) The expression of the *C* allele in chromosome 9 in the absence of the transposable elements. (2) If *Ds* is introduced into the locus the function of *C* is disrupted and the kernel becomes colorless. (3) If *Ac* (transposase) is introduced into any other location of the genome, it may cause the movement of the transposable element and colored spots appear. (4) In case *Ds* is entirely dislodged from the germline, in the following generation full or partial function of the *C* gene is restored, depending whether the original site was completely restored or some modifications took place, and only diluted color appears. (5) The *W* allele in chromosome 3 controls the development of green leaf color. (6) If *Ds* moves into the gene it may disrupt its function and albinism is observed. (7) In case *Ac* is introduced by crossing, *Ds* may move as indicated by the green stripes. Remember, *Ds* lacks transposase function although it may be moved by *Ac* which carries the transposase.

The *Ac* element is transposed by a non-replicative manner and after meiosis only one of the sister chromatids displays *Ac/Mp* at the original site (called *donor site*). In the other chromatid the element may be at another location (*recipient site*) and the original location becomes "empty". The recipient sites are most commonly in the same chromosome and quite frequently within the vicinity of the donor site. The *Ds* element frequently initiates a series of events resulting in chromosomal breakage by the mechanism of *breakage-fusion-bridge cycles* and duplications between the original donor and recipient sites. The *Ds* element may move in an inverted manner to the vicinity of a locus and thus the revertants may still contain a *Ds* element.

In the control of transposition the 11 bp inverted terminal repeats and, in addition, sequences 0.05 to 0.18 kb have importance. The *Ac-Ds* target sites display 8 bp duplication which remains even after the removal of the element. The empty target sites may show internal deletions and rearrangements.

The transposition is mediated by a *transposase* enzyme that can mobilize the *Ac* element which codes for it but it may act on the *Ds* elements too (which are transposase-defective *Ac* elements). It appears that an increase in the number of *Ac* elements results in proportionally smaller revertant sectors, and the genetic background, developmental specificities (e.g., somatic or germline tissues) and physiological factors may influence the timing and frequency

Ac - Ds continued

of transposition. There is evidence in favor of methylation being one of the factor(s) affecting *Ac* expression. This family of transposable elements has additional members such *MITE* (miniature transposable element) that has the same termini but it is very short. The *Ds1* element is similar to *Ds* but it carries retrotransposons within its sequences.

Ac has been successfully transferred to other species such as tobacco and *Arabidopsis*. (See controlling elements, transposable elements, hybrid dysgenesis, insertional mutation)

ACANTHOCYTOSIS: see abetalilipoproteinemia, elliptocytosis

ACANTHOSIS NIGRICANS: a hyperkeratosis and hyperpigmentation of the skin that may accompany the Crouzon syndrome. (See Crouzon syndrome)

AcAP: an anticoagulant protein isolated from *Ancylostoma caninum* hookworm.

ACAT: see sterol

ACATALASEMIA: a rare autosomal recessive trait involving the deficiency of the enzyme catalase. This enzyme has a protective role in the tissues by removing the H_2O_2. Symptoms include small painful ulcers around the neck, gangrenes in the mouth and atrophy of the gum and very low catalase activity in the blood and other tissues. The heterozygotes have intermediate levels of catalase activity. Acatalasemia may be classified into different groups according to the clinical symptoms, both in humans and in animals.

ACATALASIA: the same as acatalasemia

ACC (1-aminocyclopropane-1-carboxylic acid): is a precursor of the plant hormone ethylene.

ACCEPTOR SPLICING SITE: the junction between the right end of one exon and the left end of the next exon. (See introns, splicing)

ACCEPTOR STEM: part of the tRNA, including the site (5'CCA3') where amino acids are attached. (See aminoacyl-tRNA)

ACCESS TIME: the time interval between calling in a piece of information from a storage source to the actual delivery of that information to the caller. (See also real time)

ACCESSIBILITY: the genetically determined ability of the genome to provide access for the V(D)J recombinase to rearrange the immunoglobulin genes. (See V(J)D recombinase, RAG, immunoglobulins, CDR, RSS)

ACCESSION NUMBER: see BankIt

ACCESSORY CELLS (called also companion cells): are epidermal cells next to the guard cells around the plant stomata, that appear different from the usual epidermal cells.

ACCESSORY CHROMOSOME: see B chromosome

ACCESSORY DNA: product of DNA amplification in the cell. (See amplification)

ACCESSORY PIGMENTS: complement chlorophylls in absorbing light (carotenoids, xanthophyll, phycobilins)

ACCESSORY PROTEINS: such as transcription factors that bind to upstream DNA elements for controlling transcription and other binding proteins that take part (not necessarily the main part) in a particular function. Accessory host proteins are involved also in the orientation or directionality of transposons. (See transcription factors, transposable elements, transposons)

ACCESSORY SEXUAL CHARACTERS: the structures and organs of the genital tract including accessory glands and external genitalia, but not the gonads, that are the primary sexual characters. (See sex determination, gonad, sex phenotypic)

ACCURACY: the percentage of correct identification of carcinogens and non-carcinogens on the basis of mutagenicity tests. The mutagenicity tests are much faster and much less expensive than direct carcinogenicity assays but it is important to know how well these simpler tests reveal the carcinogenic (or non-carcinogenic) properties of the chemicals tested. (See sensitivity, specificity of mutagen assays, predictability, bioassays in genetic toxicology)

ACCURACY OF DNA REPLICATION: see DNA replication error

ACE (affinity capillary electrophoresis): is a procedure to test the binding strength of ligands.

ACEDB: *Caenorhabditis elegans* (a nematode, useful for genetic analyses) database. (See

Caenorhabditis elegans)

ACENAPHTENE: a spindle fiber poison and thus polyploidization agent; it is also a fungicide and insecticide. (See polyploid, colchicine, spindle)

ACENTRIC (CHROMOSOME) FRAGMENT: lacks centromere and its distribution to daughter cells is at random, and it is frequently lost during meiosis. (See centromere, chromosome morphology)

ACENTRIC FRAGMENT: a broken off piece of a chromosome that lacks centromere and therefore its distribution to the poles during nuclear divisions is random and commonly it is lost.

ACERVULUS: a disk-like conidia-bearing reproductive structure of fungi. (See conidia)

ACETABULARIA: single-celled green alga that may reach the size of 2-3 cm and may be differentiated into rhizoids, stem and cap. It can survive enucleation for several months. The rhizoids, containing the nucleus, may regenerate into complete plants; x ≈ 10. (See enucleate)

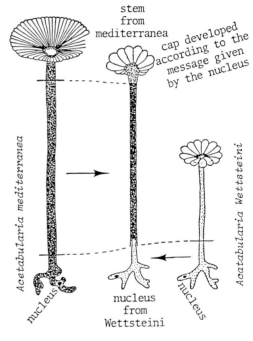

Acetabularia species are useful object for devclopmental genetic studies and show dramatically the role of the cell nucleus. Grafting of the nucleus-containing section of the cell of *A. wettsteinii* to *A. mediterrania* caused *A. mediterranea* to develop a cap according to the instructions of the nucleus donor species.
Experiment of J. Hämmerling during the 1940s. (Modified after Goldschmidt, R.B. 1958 Theoretical Genetics. Univ. California Press. Berkeley, CA, USA)

ACETO-CARMINE: see stains

ACETONITRILE (methyl cyanide): a highly poisonous liquid with ether-like odor, flash point 12.8° C (beware of the vapors) a polar solvent used (among others) for the separation of oligonucleotides by reverse-phase chromatography on silica gels.

ACETO-ORCEIN: see stains

ACETOSYRINGONE:(4-acetyl-2,6-dimethoxyphenol) and hydroxyacetosyringone are produced in plant cells (tobacco) and are one group of the compounds that induce the *vir* gene system of the *Agrobacterium* Ti plasmid. (See *Agrobacterium*, transformation [plants], virulence genes of agrobacteria)

ACETYL COENZYME A: see acetyl-CoA

ACETYL GROUP $(R\!-\!\overset{\overset{O}{\|}}{C}\!-\!CH_3)$: is derived from acetic acid CH_3COOH; the R stands for different chemical groups. (See acyl group)

ACETYL-CoA (acetyl coenzyme A, ACoA): is a heat-stable cofactor involved in the transfer of acetyl groups in many biological reactions (citric acid cycle, fatty acid metabolism, etc.). It has three major domains: the β-mercapto ethylamine unit, the panthothenate unit and adenylic acid. (See epinephrine)

ACETYL-CoA CARBOXYLASE DEFICIENCY (ACAC): recessive, in human chromosome 17q21. It causes multiple interference with gluconeogenesis, fatty acid and the branched-chain amino acid metabolism. (See branched-chain amino acids)

ACETYLCHOLINE (M_r 149): the acetylcholine receptor provides the connection between synapsing neurons and it is thus a signal transmitter. When acetyl choline binds to a receptor a Na^+/K^+ channel opens. The muscarinic acetylcholine receptors are activated by the fungal alkaloid, muscarine, whereas the nicotinic acetylcholine receptors are operating in the nerve and muscle cells. Acetylcholine receptors are diffusely distributed on the embryonic myotubes but become highly concentrated in a minute area in the post-synaptic membrane and they tether the synaptic cytoskeletal complex. (See ion channels, synapse, cytoskeleton, rapsyn, myotube, neuregulin, agrin, neurotransmitters, acetylcholine receptors, muscarinic acetylcholine receptors)

ACETYLCHOLINE RECEPTORS: are acetylcholine regulated cation (Na^+, K^+ and Ca^{++}) channels between the motor neurons and the skeletal muscles. The receptor in the skeletal muscle contains five transmembrane polypeptides, encoded by four separate yet similar genes. When acetylcholine attaches to the receptor a conformational change ensues resulting in a brief opening of the channel. They are easily isolated from the electric organs of some fishes. (See also muscarinic acetylcholine receptors, nicotinic acetylcholine receptors, ion channels, agrin)

ACETYLCHOLINESTERASE (ACHE) is encoded in human chromosome 3q25.2 by codominant alleles. (See acetylcholine, acetylcholine receptors, pseudocholinesterase deficiency)

achaete-scute **COMPLEX**: of *Drosophila* is a complex X-chromosomal locus regulating bristle formation and nerve differentiation. (See complex locus)

ACHENE: a single-seed dry fruit.

ACHIASMATE: nuclear division without the formation of chiasmata. (See meiosis, chiasma, distributive pairing)

ACHILLES' HEEL TECHNIQUE: is a technique applicable to systems where there is abundant sequence information, and it permits the cleavage of only a small set of restriction sites. It works this way: DNA sequences around the site or set of sites are synthesized and added to the genomic DNA along with RecA, and a methylase. After deproteinization a restriction enzyme is added. All the (methylated) restriction sites are protected from cleavage except those that were covered by the RecA-DNA complex. (See DNA sequencing, Rec, methylase, methylation of DNA, restriction enzyme)

ACHONDROGENESIS: has been described in two or more autosomal recessive forms involving deficiency in bone formation at the hip area and large head, short limbs, still-birth or neonatal death. The phenotypes show variations and clear-cut differentiation of the symptoms is difficult. (See achondroplasia, hypochondroplasia, stature in humans, collagen)

ACHONDROPLASIA: a rather common chromosome 4p16.3 dominant (homozygous perinatal lethal) type of human dwarfness that was observed e.g. in Denmark at a frequency of 1.1×10^{-4}. Its mutation frequency was estimated to be within the range 4.3 to 7×10^{-5}.

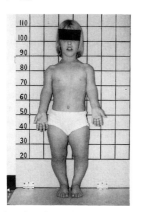

AN AUTOSOMAL DOMINANT TYPE ACHONDROPLASIAC ADOLESCENT. (Courtesy of Dr. D.L. Rimoin, Harbor General Hospital, Los Angeles, CA, and Dr. Judith Miles)

The proximal bones in the limbs are most reduced. Large head with disproportionally small mid-face, abnormal hip and hands are characteristic. The heterozygotes are generally plagued by heart, respiratory and other problems. Hypochondroplasia appears to be allelic to achondroplasia. The so-called Swiss type achondroplasia is recessive and the afflicted individuals show reduced amount of leukocytes (lymphopenia) and agammaglobulinemia. Pseudoachondroplastic dysplasias (spondyloepiphyseal dysplasia) are autosomal recessive but some ambiguities were noted regarding the pattern of inheritance because of apparent gonadal mosa-

Achondroplasia continued

icism. The different forms do not have clear phenotypic distinctions within the group and from the dominant achondroplasia. Some of the skeletal reductions and defects are aggravated by face, eye defects, cleft palate and muscle weakness. Achondroplasia is caused by defects in the fibroblast growth factor receptor 3 (FGFR-3), located in human chromosome 4p16.3. A recurrent missense mutation in a CpG doublet of the transmembrane domain of FGFR-3 caused an arginine substitution for glycine. Achondroplasiacs usually display normal intelligence. (See stature in humans, hypochondroplasia, pseudoachondroplasia, achondrogenesis, agammaglubulinemia, cleft palate, fibroblast growth factor, dwarfism)

ACHROMATIC: parts of the cell nucleus which are not stained by nuclear stains. A microscope lens that does not refract light into different colors.

ACID BLOB: a sequence of acid amino acids (negatively charged), responsible for activation of a transcription factor. (See transcription factors)

ACID FUCHSIN: a histological stain used to detect connective tissue and secretion granules (Mallory's acid fuchsin, orange G and aniline blue, and in the Van Gieson's solution of trinitrophenol staining of connective tissue of mammals). (See stains)

ACID MALTASE DEFICIENCY: is type II glycogen storage disease involving defect(s) in α-1,4-glucosidase activity. The disease causes accumulation of glycogen in most tissues, including the heart. The first symptoms appear by 2 months after birth and by 5-6 months death results due cardiorespiratory (heart and lung) failures. Although it is classified as an autosomal recessive trait in humans (GAA, 17q25.2-q25.3), the heterozygotes may be distinguished clinically. (See glucosidase, Gaucher diseases, glycogen storage diseases)

ACID PHOSPHATASE: cleaves phosphate linkages at low pH. Its levels are increased in most lysosomal storage diseases, particularly in Gaucher's diseases involving glucosyl ceramide lipidosis (defect in lipid metabolism involving cerebrosides, a complex of basic amino alcohols [sphingosine], fatty acids and glucose). Other diseases may also cause increase of acid phosphatase. In plants only acid phosphatases are found in appreciable quantities. Yeast has at least 4 genes with acid phosphatase function; one of them is constitutive, others are repressed by inorganic phosphate. ACP1 is in human chromosome 2p25, ACP2 in 11p12-p11. (See alkaline phosphatase)

ACIDIC DYES: stain basic residues.

ACIDIC SUGARS: see sialic acids

ACIDOSIS: a reduction of buffering capacity of the body resulting in lower pH of fluids.

ACINAR CELLS: are exocrine cell such as the mammary gland cells that secrete milk, lacrimal cells that secrete tears, etc.

AcMNPV (*Autographa californica* nuclear polyhedrosis virus): can be used for the construction of insect and mammalian transformation vector. (See baculovirus)

ACNE: inflammation of the sebaceous glands (that secrete oily stuff on the skin). It does not appear to be under strict genetic control but rather various environmental conditions, including bacterial infections, mechanical irritation, cosmetics, etc. cause it. It usually appears in puberty and disappears after but may leave behind permanent scars. Occasionally it occurs on infants. (See skin diseases)

ACONITASE: an enzyme controlling the dehydration of citrate to cis-aconitate and the hydration of the latter to isocitrate. This enzyme has also an important role in the transport of iron. Iron-containing proteins regulate many processes in both prokaryotes and eukaryotes. In eukaryotic cells the level of the storage protein ferritin increases when soluble iron level increases in the cytosol. The control of the process is mediated by a 30-nucleotide *iron-response element* to what aconitase binds and then blocks the downstream translation of RNA. Aconitase is an iron-binding protein, and the increasing level of iron within the cell dissociates it from the ferritin mRNA resulting in about two order of magnitude increase of ferritin by releasing the translation suppressor from the ferritin mRNA. The increased level of iron also decreases the stability of several mRNAs encoding the receptor that binds the iron-transporting transferrin

Aconitase continued

and thereby reduces the amount of the receptor. Aconitase also binds to the 3' untranslated tract of the transferrin receptor mRNA and enhances the production of the receptor, probably by stabilizing the mRNA. The human ACO1 gene is in chromosome 9p22-p14 and the mitochondrially located ACO2 is encoded in 22q11-q13. (See ferritin, translation repressor protein, rabbit reticulocyte *in vitro* translation system)

ACQUIRED: alteration occurred during the lifetime of an individual. (See constitutional)

ACQUIRED CHARACTERS, INHERITANCE OF: is an ancient idea supposing that the minor and major environmental effects may cause a long-lasting heritable changes in the genetic material. This view was proven incorrect by the advances of biology in the 19th century. It had been revived, however, in the Soviet Union by the poorly trained ideologues of marxism, the followers of Mitchurin and Lysenko. It seems to be resurrected periodically by modern biologists who claim the existence of environmentally inducible selective mutations. Most of these recent experiments also remain controversial because alternative explanations of the experimental data seem to be as good or even more satisfactory (see directed and local mutagenesis). Advantageous frameshift backmutations may take place, however, under selective conditions by recombination. (See also lamarckism, mitchurinism, lysenkoism, soviet genetics, transformation, recombination, frameshift, backmutation)

ACQUIRED IMMUNITY: is the consequence of natural infection or vaccination or direct transfer of antibodies or lymphocytes from an appropriate donor. (See innate immunity)

ACQUIRED IMMUNODEFICIENCY SYNDROME (AIDS): is apparently caused by the HIV-1 (HTLV-III) and HIV-2 (human immunodeficiency virus), retroviruses. The general structure of the virus includes three major structural proteins: *gag, pol* and *env*, and several regulatory and accessory proteins: *vif, vpr, vpu, vpt, tev/tnt*:

```
LTR----------gag--------------                      ----------------env-----------------LTR
                    --------------------pol----------------         ----------tat-----------
                                          --------vif-----        ----------rev---------    ----nef----
                                                      ---vpr--
    →                                                 ---vpu--
    ΔΔΔ TAR                                                                    TAR   ΔΔΔΔ
```

The gag proteins serve as structural elements, the pol is processed into reverse transcriptase, protease and integrase. The reverse transcriptase generates the enzymes that transcribes RNA into DNA. The protease processes the polyproteins into the various enzymes of the virus and integrase facilitates the entry of the virion into the host cells. The envelope proteins fo, vpr (15 kDa) accelerate replication and infection, rev (19 kDa) is transcribed from two exons regulates viral replication and its basic amino acid domain interacts with the target site (RRE), tat (16 kDa, two exons) is the primary regulator of the virus, vpu (15 - 20 kDa) facilitates the assembly of the virion, nef (25 - 27 kDa) is involved probably in signal transduction, it has similarity to G-proteins. The *tat* gene functions only through the left TAR (transactivation response element [Δ]) although such sequences are present also at the right LTR. The TATAA box is situated -24 to -28 positions from the GGT initiator codon. Further upstream in the enhancer region are the binding sites for the USF (upstream enhancer), Ets-1 (thymocyte-enriched protein), LEF (lymphocyte-specific high-mobility group protein), NK-κB (nuclear factor κ binding protein) and Sp1 (a mammalian transcription factor) binding proteins within the region -166 to -45. Around the transcription initiation site are the overlapping SSR (initiator) and IST (initiator of short transcripts) sequences. The virus does not have known genetic repair system and displays great antigenic variability, therefore it is difficult to develop an effective vaccine against it.

HIV1 and lentiviruses are suitable for the construction of transformation vectors that may integrate into non-dividing cells. Two of the viral proteins interact with nuclear import and mediate the active transport of the HIV preintegration complex into the nucleus through the nuclear pores. The infection begins when the virus penetrates the cell membrane and its own lipid membrane fuses with the cell membrane and the viral core is released into the cell. Inside the

Acquired immunodeficiency syndrome continued

cell, the viral reverse transcriptase synthesizes DNA copies of its RNA genome, and this DNA provirus integrates into the host genome with the aid of its terminal repeats, characteristic also for all types of insertion elements. The HIV contains genes for proteins and their regulation. HIV does not have a lytic phase so it does not kill the cells directly. Instead, it assembles its particles in the cytoplasm and then infects other cells. The primary targets are the helper T lymphocytes carrying the CD4 receptors. Impairing these cells the immune system is debilitated and that is the primary cause of the disease. The virus attacks also the macrophages and the dendritic cells of the nervous system. Generally the first sign of the disease is the susceptibility to *Pneumocystis carenii,* an opportunistic fungal pathogen causing influenza-like symptoms. This happens because the AIDS patients have only 200 CD4 helper cells per mL of blood versus 800 in normals. The other, most critical, diagnostic feature of AIDS is the development of Kaposi's sarcoma, a disease causing bluish eruptions all over the body that become cancerous. In tissue culture the infected and uninfected cells fuse into syncytia and this and immunological methods, are being used as a laboratory diagnostic procedures for the infections. AIDS is one of the most dreaded diseases of the 20th century is battled now with the most advanced techniques of molecular genetics yet no effective cure has been devised by 1997. Avoidance of infection through body fluids (blood, semen, saliva, etc.) is the only effective defense until immunization or cure can be developed. Although the majority of the specialists in medical virology and molecular biology maintains the view that the causative agent of the disease is HIV, some reject this assumption and others take a look at the mechanism with some reservations. They find likely or conceivable that AIDS is the result of the synergistic action of viral and other requisites such as the use of drugs (antibiotics, etc.), some types of autoimmune predisposition, etc., and thus has multifactorial origin. The AIDS disease has infected 21 million people world wide and their numbers are increasing daily by 8,500. There are now 10 known subtypes of the virus. The pharmaceutical industry is developing various drugs to combat the disease. None of the drugs so far provides a full cure or prevention yet definite progress is being made. The first and best known chemicals (AZT) attack the viral replication system, protease inhibitors are aimed at the assembly process of the viral coat protein in order to prevent multiplication of HIV. The virus depends on cutting and processing of host cellular protein and uses protease to this end. Unfortunately the protease inhibitors may cause very unpleasant side effects and HIV may develop resistance against the drugs. According to some estimates HIV may produce 10 billion virions daily. Since its genome contains 10^4 nucleotides, all possible mutational combinations can be readily tested by the virus. The estimated number of mutations/replication is 3×10^{-5}. In addition, recombination facilitates the production of new variants. The combination of two nucleoside analogs and protease inhibitor may reduce the level of the viral RNA copies from 20,000 and 1,000,000 copies/mL plasma below detectability (i.e. below 200 to 400 copies/mL). These figures are concerned with viral levels in the blood. The lymphnodes and other sanctuaries may regenerate the virus after the discontinuation of the therapy. Additional problems may arise from the irreversibility of the tissue (thymus) damage. HIV-1 replication requires the REV oncogene cofactor and eukaryotic peptide elongation factor EIF-5A. Some mutations in the elongation factor retained the ability to bind to the HIV-1 REV response element:REV complex and were expressed in human cells. When such T lymphocytes infected with replication-competent virus, replication was inhibited, however. Other current research attempts are focusing on the immune system to prevent infection. Although several interpretations are available it is still uncertain what is the cause that the period required for the development of full scale AIDS requires such a different length of latency after the initial infection. It had been assumed that the immune system is weakened by the ever-increasing viral diversity. Others believe that an immune dysregulation is responsible for the outbreak. Others suggest that the cellular immune system against AIDS is directed to both conserved and variable epitopes. It is assumed that the cytotoxic T lymphocytes alone cannot completely eliminate the virus and there is a need to achieve a balance between the viral load and the $CD4^+$ T lymphocytes. After a period of time

Acquired immunodeficiency syndrome continued
the increasing variations in the HIV-1 population deplete and foul up the immune system. In the so called non-progressor individuals, AIDS may not develop for more than 10 or even 20 years after the infection. In the cells of such a person there are high levels of $CD8^+$ $CD38^-$ cytotoxic lymphocytes, high peripheral blood $CD8^+$ major histocompatibility class I-restricted anti-HIV cytotoxic lymphocytes and those stay at an even level. Also there is a strong $CD8^+$ non-MHC-restricted HIV suppressor activity and high level of antibody against HIV. Recent studies (1996) indicate that in the non-progressors the gene coding for the chemokine receptor CCR-5 is mutated (contains deletion[s]), and thus it appears that CCR-5 assists infection by HIV-1. Heterozygosity for the mutant allele also appears to convey reduced susceptibility to infection. The frequency of the mutant alleles in Caucasian populations is about 0.092 and thus the predictable frequency of homozygosis ($0.092^2 \approx 0.0085$) is about 1%. This type of mutation seems to be much less common in African and Japanese populations. Also likely that some non-progressors were infected with less aggressive HIV variants. The residual genetic constitution of the infected individual may also affect the course of the disease. Recent studies revealed that homozygosity for a mutation in the chemokine receptor CCR5 (synonym CKR5) protects against HIV infection and heterozygosity may also be of some advantage. This is the receptor through which HIV infection takes place. The screening of the blood donations for possible HIV infection is based on the determination of the proportion of CD4/CD8 molecules. The normal ratio is about 2 and in infected blood it is below 1. (See retroviruses, CD4, $CD8^+$, HLA, fusin, telomere, T cells, RANTES, AZT, Nevirapine, TIBO, NF-κB, Sp1, HMG, enhancer)

ACRIDINE DYE: such as proflavine, acriflavine, acridine orange, etc. are potential frameshift mutagens by intercalating between the nucleotides of the DNA. Some acridines act by photosensitization of the DNA. It has been used to cure bacteria from plasmids (by selective removal), and to induce respiration-deficient mitochondrial mutations in yeast. (See mtDNA, fluorochromes, curing of plasmids, frameshift)

ACRIFLAVINE: see acridine dyes

ACROCENTRIC CHROMOSOME: has a near terminal centromere and one arm is very short; acrocentric chromosomes may fuse or become translocated (see Robertsonian translocations) and generate biarmed chromosomes.

ACROCEPHALOSYNDACTYLY: see Apert syndrome

ACRODERMATITIS ENTEROPATHICA: is an autosomal recessive blistering of the skin usually accompanied by lack of hairs on the head, eyebrows and eyelashes, and partial pancreatic hyperplasia and thymus hypoplasia. The deficiency in zinc-binding is characteristic and causes low levels of zinc and alkaline phosphatase (a zinc metalloenzyme) in the plasma. The treatment with zinc is very successful. (See alkaline phosphatase, zinc fingers, skin disease, Wilson disease, Menke disease, hemochromatosis)

ACROSOMAL PROCESS: a spike-like actin structure on the head of the sperms of several animals, and at its base is the acrosome which is a sac of hydrolytic enzymes destined to facilitate the penetration through the gelatinous coat of the egg. The process is enhanced by progesterone, probably by acting on a $GABA_A$ receptor. (See sperm, GABA, progesterone)

ACROSOME: see acrosomal process

ACROSOME REACTION: see acrosomal process

ACROSYNDESIS: a spurious end-to-end pairing of the chromosomes during meiosis.

ACRYLALDEHYDE: toxic compound made from allylalcohol by the enzyme alcohol dehydrogenase. Cells defective in this enzyme permit the selective survival on allylalcohol as it is not converted to acrylaldehyde. (See mutant isolation, alcoholdehydrogenase)

ACRYLAMIDE: in the presence of ammonium persulfate and TEMED (N,N,N',N'-tetramethylenediamine) it is polymerized into chains with various length depending on the concentration used. In the presence of N,N'-methylenebisacrylamide it becomes cross-linked and pores are formed depending on the length of the chains and the degree of crosslinking. It can be used to separate nucleotides by electrophoresis from 2000 to 6 bp, depending on the pore size of the

Acrylamide continued

gels. Acrylamide is a potent neurotoxin and can be absorbed through the skin. Although the polymerized form is considered non-toxic, it should be handled only with gloves because of the trace amounts of monomers. (See electrophoresis, gel electrophoresis)

ACTH (adrenocorticotropin): controls adrenocortical growth and steroidogenesis. The hypothalamus controls the ACTH releasing factor and in response to that the anterior pituitary releases this hormone. ACTH is encoded in human chromosome 2. (See animal hormones, adrenocorticotropin, cAMP, steroid hormones, hormone-response elements, pituitary gland, brain)

ACTIN: a protein of the cytoskeleton and the thin muscle fibers. Actin gene number varies in different organisms, yeast has only one, *Dictyostelium* 8, *Drosophila* 6, *Caenorhabditis* 4, humans about two dozen at dispersed locations. The cytoplasmic actins involved in cellular motility are similar in all eukaryotes; α-actins are located in the smooth, skeletal and cardiac muscles. The smooth muscles have in addition γ-actin. In the cytoplasm of mammals and birds there are β- and γ-actins. The amino acid sequence and composition of the actins is rather well conserved and differences exist mainly at the amino terminals. Actin genes have different numbers of introns and pseudogenes, permitting evolutionary inferences partly because the flanking sequences are much more variable than the genes. Some proteins bind actin in monomeric or filamentous form such as myosin (a contractile protein in muscles), α-actinin (involved in cross-linking actins), profilin (mediates the formation of actinin bundles), fimbrin (cross-linking parallel actin filaments), filamin (promotes the gel-formation by actins), tropomyosin (strengthens actin filaments), spectrin (attaches filaments to plasma membranes), gelsolin (fragments filaments), etc. (See cytoskeleton, myosin, filament, micofilament, myofibril)

ACTINOMORPHIC: a structure (flower) in multiple symmetry patterns (e.g. star-shaped).

ACTINOMYCETES: are actually prokaryotes rather than fungi as they once were assumed to be. (See *Streptomyces*, streptomycin, actinomycin)

ACTINOMYCIN D: is an antibiotic from *Streptomyces*; it is an inhibitor of transcription because it intercalates into the DNA between neighboring GC base pairs and hinders the movement of the transcriptase on the DNA template without interfering with replication; it is used in reverse transcriptase reactions to prevent self-primed second strand synthesis. Actinomycin D is a teratogen and carcinogen. (There are a number of other actinomycin antibiotics). (See *Actinomycetes, Streptomyces*, transcriptase, reverse transcription)

ACTION POTENTIAL: electrical excitation in muscle or neuron membranes. Mediates long-distance nerve signaling.

ACTION SPECTRUM: a representation of a degree of response to certain type of treatment(s), e.g. the course of photosynthesis in relation to wavelength of irradiation.

ACTIVATING ENZYME: see aminoacylation of tRNA

ACTIVATION ANALYSIS: is a nuclear technique used for the very sensitive detection of radionuclides for various purposes including forensic analysis. (See radionuclide)

ACTIVATION ENERGY: the energy required for converting 1 gram molecular weight of a compound from the ground state to the transition state. It is required, from outside, by molecules and atoms to undergo chemical reaction(s).

ACTIVATION OF MUTAGENS: many mutagens (and carcinogens) require chemical alterations to become biologically active. The mutagenic and carcinogenic properties of many agents overlap and thus active in mutagenesis and carcinogenesis. The activation generally requires enzymatic reactions. The most important enzymes are the mixed-function oxidases contained by the cytochrome P-450 cellular fraction. These reactions require NADPH and molecular oxygen and the general process is: **RH** (reduced reactant) + NADPH + H$^+$ + O$_2$ $\rightarrow$ **ROH** + NADP$^+$ + H$_2$O. These enzymes occur in multiple forms and can utilize a variety of substrates, hydrocarbons, amines and amids, hydrazines and triazines, nitroso compounds, etc. They occur in different tissues of animals, primarily in the endoplasmic reticulum of cells what is generally called microsomal fraction after isolation following grinding and centrifugal

Activation of mutagens continued
separation of the cellular fractions. These enzymes are subject to induction by phenobarbitals, methylcholanthrene and a variety of substrates. Other related activating enzymes are flavoprotein N-oxygenases, hydrolases, and reductases. Other enzymes of activating ability include various transferases that add glucuronyl, sulfuryl, glutathion, acetyl and other groups and either detoxify the compounds or further enhance their reactivity. These reactions are affected by the cellular and membrane transport, protein binding, excretion, genetic differences among the species and individuals, sex differences, age, circadian rhythm, nutritional status, etc. If the clearance of these compounds from the body is slow, the risk for the individuals increases. (See promutagen, proximal mutagen, ultimate mutagen, environmental mutagens, mutagen assays, microsomes, Ames test, bioassays of mutagenesis)

ACTIVATOR: *Ac*, the autonomous element of the *Ac - Ds* controlling element system of maize (see *Ac*). Also any DNA binding protein that enhances transcription, a positive modulator of an allosteric enzyme. (See allosteric control, modulation, transcriptional activator)

ACTIVATOR A: is synonymous with RF-C, a cellular replicator. (See RF-C)

ACTIVATOR I: same as RF-C

ACTIVATOR PROTEINS: stimulate transcription of genes by binding to TATA box binding protein (TBP) and the recruitment of TFIID complex to the promoter. Sometimes they require co-activator metabolites for function. The primary role of these is probably the remodeling of the nucleosomal structure so DNA-binding proteins can access their target. The DNA has multiple binding sites for activators in the promoter region. The potency of the activation domains of the activators may vary. (See also transcription factors, TBP, regulation of gene activity, promoter, co-activator, transcriptional activator, enhancer, chromatin, nucleosome, VDR, recruitement)

ACTIVE IMMUNITY: see immunity

ACTIVE SITE: a special part of an enzyme where its substrate can bind and where the catalytic function is performed, the catalytic site. (See substrate, enzymes, catalysis)

ACTIVE TELOMERIC EXPRESSION SITE: variable surface glycoprotein (VSG) genes are responsible for the diversity of antigenic variants in *Trypanosomas*. These generate different antigenic properties of the parasite. There are about thousand genes in this gene family and their activation is interpreted by their transposition to the vicinity of the centromere, the expression site of the silent copies. (See *Trypanosomas*, mating type determination in yeast, silent site)

ACTIVE TRANSPORT: passing solutes through membranes with the assistance of an energy donor required for the process. (See passive transport)

ACTIVIN: soluble protein that may contribute to the formation of dorsal and mesodermal tissues in the developing animal embryo; its activity may be blocked by follistatin. Activins belong to the family of *transforming growth factor-β* superfamily of proteins. They are serine/threonine protein kinases. (See protein kinases, TGF, organizer, follistatin)

ACTIVITY COEFFICIENT: is obtained by multiplying with it the concentration of a solute to obtain its thermodynamic activity.

ACUTE TRANSFORMING RETROVIRUS: contains a v-oncogene and it is an efficient oncogenic transformation agent. (See v oncogene, oncogenes, retrovirus)

ACYCLOVIR: see ganciclovir

ACYL GROUP (R—C(=O)—): where R can be a number of different chemical groups. (See acetyl group)

ACYLCYCLOHEXANEDIONE: an inhibitor of gibberellin biosynthesis. (See plant hormones)

Ad5 E1B: an adenovirus oncoprotein. (See adenovirus, oncogenes)

ADA: a Zn-containing protein that transfers methyl groups to its own cystein and thus repairs aberrantly methylated DNA (e.g. 6-O-methylguanine), and by binding to specific DNA sequences it activates genes involved in conveying resistance to methylation. (See also adenosine deaminase)

ADACTYLY: absence of digits in the fingers or toes. (See Holt-Oram syndrome, polydactily, ectrodactyly)

ADAM (**a d**isintegrin **a**nd **m**etalloprotease) family of enzymes such as KUZ (kuzbanian), responsible for partitioning of neural and non-neural cells during the development of the central and peripheral nervous system. (See neurogenesis)

ADAM COMPLEX (amputations): is the acronym for Amniotic deformity, Adhesions, Mutilations phenotype complexed with other anomalies caused by mechanical constriction of the amniotic sac but there is evidence also for the role of autosomal recessive inheritance. The phenotype may include bands on fingers, and loss of finger bones or even parts of legs (amputations), etc. (See limb defects)

ADAMS-OLIVER SYNDROME: autosomal dominant mutilations of the limbs, skin and skull lesions.

ADAPTATION: the organism has fitness in a special environment. Mutation provides the genetic variations from what the evolutionary process selects the genes that convey the best adaptation. In physiology, it defines adjustment to specific stimuli. (See also fitness)

ADAPTIN: is a major coat protein in a multisubunit complex on vesicles. These proteins bind the clathrin coat to the membrane and assist in trapping transmembrane receptor proteins that mediate the capture of cargo molecules and deliver them inside the vesicles. (See clathrin, cargo receptors, endocytosis)

ADAPTIVE CONVERGENCE: similarity in morphology and function among unrelated species within a particular environment, e.g. fins on fishes and the mammalian whales.

ADAPTIVE ENZYME: same as inducible enzyme. (See also e.g. *Lac* operon)

ADAPTIVE LANDSCAPE: represents the frequency distribution of alleles corresponding to the fitness of the genotypes, e.g. *AAbb* and *aaBB* means the fixation (*peak*) of the allelic pairs. The *pits* of fitness may mean the fixation of *AABB* or *aabb*, and the *saddle* corresponds to the polymorphic condition *AaBb*. The three conditions may be represented by a three-dimensional model. (See fitness)

ADAPTIVE MUTATION: see directed mutation

ADAPTIVE RADIATION: a phyletic line spreads over a variety of different ecological niches resulting in a rapid adaptation to these locales and appearing in strikingly different forms. (See phylogeny, niche)

ADAPTIVE RESPONSE: induction of (bacterial) repair enzymes which activate glycosylases or O^6-methylguanine methyltransferase and thereby mutated DNA is repaired. The name comes from the property of adaptation to higher doses of mutagens after an initial shorter exposure; it is mediated by the *Ada* gene product (37 kDa) of *E. coli*. (See alkylating agent, chemical mutagens, DNA repair, glycosylases, methylation of DNA)

ADAPTIVE TOPOGRAPHY: same as adaptive landscape.

ADAPTIVE VALUE: see fitness

ADAPTOR: tRNA is called an adaptor by the older literature because it adapts the genetic information in DNA through mRNA to protein synthesis. (See tRNA, aminoacyl-tRNA synthetase, protein synthesis)

ADAPTOR PROTEINS: adaptor proteins play key roles in cellular signaling such as phosphorylation, dephosphorylation, signal transduction, organization of the cytoskeleton, cell adhesion, regulation of gene expression, all distinct yet interacting systems. Proteins equipped with the Rous sarcoma oncogen (Src) homology domains SH2 and SH3 mediate the interactions between the phosphotyrosine kinase receptors of mitogenic signals and the RAS-like G proteins. The SH2 domain selects the phospho-Tyr-Glu-Glu-Ile sequences. Phospholipase C (PLC-γ1) and the protein-tyrosine phosphatase (PTPase) recognize several hydrophobic residues following pTyr on the ligand-binding molecule. The SHC homology proteins and the insulin-receptor substrate (IRS-1) recognize somewhat different sequences: Asn-Pro-X-pTyr.

Adaptor proteins continued

The SH3 binding sites involve about 10 proline-rich amino residues. The SH3-binding peptides can bind either in $NH_2 \rightarrow COOH$ or in the reverse orientation. The SOS (son of sevenless) adaptor protein binds to Grb2 (growth factor receptor-bound protein) are attached in $C \rightarrow N$ orientation. The pleckstrin domains are widespread in occurrence (serine/threonine and tyrosine kinases, and their substrates, phospholipases, small GTPases, dynamin, cytoskeletal proteins, etc.). Pleckstrin domains occur in cytoplasmic and membrane signaling molecules. The LIM domains facilitate binding of signaling molecules, transcription factors as well as the units of the cytoskeleton. These adaptor proteins may form partnerships with a variety of proteins and thus generate complex networks of signaling. (See separate entries mentioned)

ADAPTORS: see linker

ADDICTION: a complex phenotype relative to the abuse of drugs, alcohol, smoking or habituation to other non-natural behavior. It is generally controlled by multiple genes and deeply influenced by several social conditions and commonly associated with antisocial behavior. It is assumed that the long-term abuse of these substances causes molecular changes in the neuronal signaling. The adaptation may modify the autonomic somatic functions causing dependence and when the agent is withdrawn, resulting in withdrawal anomalies. The agent may alter the motivational control system resulting in craving. Chronic use of morphine upregulates components of the cAMP signal transduction pathway. In mice, with a deletion of the CREBα element the withdrawal symptoms were reduced indicating that CREB-dependent gene transcription is a factor of opiate dependence. The major receptor for opiates (morphin, heroine) is the trimeric G protein-linked μ. Long-term opiate use decreases the μ—opiate receptor signaling without reducing the number of receptors and leading to tolerance and dependence. In non-addicted individuals the opiate receptor opens an outward rectifying K^+ channel and reduces the phosphorylated state of a Na^+ channel. In an addicted individual the K^+ channel is shut off, however and the G protein—adenylate cyclase activates a protein kinase (PKA) and the phosphorylated Na^+ channel moves sodium inward the locus ceruleus, a pigmented structure at the floor of the brain. As a consequence the cyclic AMP response element (CRE) binding proteins (CREB) stimulate the transcription of RNA required for adaptation to the addictive drug. The psychoactive effects of cocaine can be superseded in rats by active immunization using a cocaine conjugate, GNC-KLH (a hapten and keyhole limpet haemocyanin). (See ion channels, G protein, cAMP, adenylate cyclase, CRE, CREB, behavioral genetics, keyhole-limpet hemocyanin)

ADDICTION MODULE: represents a prokaryotic system with resemblance to apoptosis in eukaryotes. The module includes the products of two genes: one is long-lasting and toxic, the other is short lived and protects against the toxic effect. The "addiction" is a dependence on the antagonist of the toxin. This system is usually controlled by plasmid elements. In the *hok-sok* module of the R1 plasmid the *sok* gene product is an antisense RNA, subject to degradation by a nuclease. Homologs of this plasmid system have been found also in the main bacterial chromosome. Encoded by the bacterial *rel* operon the MazE antitoxin protein is subject to degradation by the clpPA serine protease. It protects from the toxic effects of MazF toxin protein. MazE-MazF is regulated by the level of ppGpp which itself is toxic to the cells. MazE-MazF expression is regulated also by 3',5'-bispyrophosphate, synthesized by the RelA protein under amino acid starvation. (See apoptosis)

ADDISON DISEASE (adrenocortical hypofunction): autosomal dominant defect of the kidneys cortical layer resulting in excess potassium and sodium in the urine, decreased levels of cortisol and hyperpigmentation of the skin. (See pigmentation defects, kidney diseases, cortisol)

ADDISON-SCHILDER DISEASE: is an X-chromosome-linked adrenocortical (kidney outer layer) atrophy and diffuse cerebral sclerosis (hardening); the cerebral lesions resemble the symptoms of multiple sclerosis.

ADDITION LINES: carry an extra chromosome(s) coming from an other genome. (See alien addition)

ADDITIVE EFFECTS of genes means that each allele contributes quantitatively to the phenotype of an individual that carries it, i.e. there is no dominance. (See polygenic inheritance)

ADDITIVE GENES: each allele has a definite quantitative contribution to the phenotype without dominance within a locus and without epistasis between loci or overdominance between alleles. (See additive variance, quantitative genetics, heritability)

ADDITIVE VARIANCE: each allele contributes a special value (quantity) to the phenotype and there is no interallelic (overdominance) or interlocus (epistasis) effects of the variance. (See genetic variance, heritability, QTL)

ADDITIVITY OF GENETIC MAPS: ideally it means that the distance between genes A - C is equal to the sum of the distance between A - B and B - C if the order of genes is A B C. To this generally valid rule exceptions exist because of genetic interference. (See mapping genetic, recombination frequency)

ADDUCIN: a membrane protein mediating the binding of spectrin to actin. (See spectrin, actin)

ADDUCT as a verb: to draw to the median plane or axial line; as a chemical: the complex of two or more components such as the cyclobutane ring of pyrimidine dimers, benzo(a)pyrene-guanine, and other alkyl groups of mutagens added to nucleic acid bases. Lipid peroxidation generates various DNA adducts with mutagenic effects similar to that caused by exogenous carcinogens. This may be the cause by the "spontaneous" cases of carcinogenicity by high-fat diet. (See pyridine dimers, ethylmethane sulfonate, benzo(a)pyrene, excision repair, ABC excinuclease, malondialdehyde, pyrimidopurinones)

ADELPHOGAMY: sib-pollination of vegetatively propagated individual plants. (See sibling)

ADENINE: a purine base in either DNA or RNA. (See purines)

ADENOCARCINOMA: cancer of glandular tissues. (See pancreatic adenocarcinoma)

ADENOMA: usually benign gland-shape epithelial tumor. (See also endocrine neoplasia):

ADENOMATOSIS, ENDOCRINE, MULTIPLE (MEN): in the autosomal dominant MEN I (human chromosome 11q13) pancreatic adenonomas are prevalent, in MEN II pheochromocytoma and thyroid carcinoma (10q11) and in MEN III cancers of the nerve tissues are most common although the latter two conditions appear allelic in the pericentric region of human chromosome 10q. (See pheochromocytoma, SHC, adenoma, cancer)

ADENOSINE: adenine with a ribose added. (See adenine, nucleoside)

ADENOSINE 3', 5'-CYCLIC MONOPHOSPHATE (cAMP): is formed from ATP by adenylate cyclase enzyme. It has important regulatory functions as "second messenger" for microbial and animal cells. (See more also at cAMP, adenylate cyclase, G-proteins, signal transduction)

cAMP →

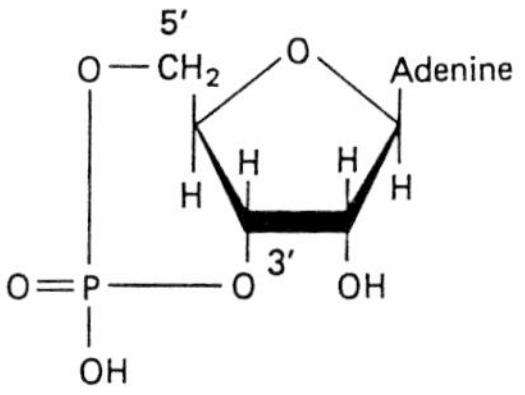

ADENOSINE DEAMINASE (ADA): an enzyme that hydrolyzes adenosine monophosphate to inosine. Its deficiency causes severe immunodeficiency and the patients lymphocytes are disabled to fight infections successfully. ADA is synthesized in the cells for the purpose of detoxification of excessive amounts of adenosine or its analogs; it inactivates 9-β-D-xylofuranosyl adenine, a DNA damaging chemical, and thus it can be used as a dominant selection agent in tissue culture. Its synthesis in the cells can be overproduced over ten thousand times by a strong inhibitor of ADA, 2'-deoxycoformicin (dCF), a transition-state analog for adenine nucleotide enzymes. It is encoded in human chromosome 20q12-q13. (See adenosine deaminase deficiency, mosaic, immunodeficiency)

ADENOSINE DEAMINASE DEFICIENCY (ADA deficiency): see adenosine deaminase; this is also called severe combined immunodeficiency disease (SCID), and it may be treated with gene therapy. Into T lymphocytes isolated from the patients, the normally functional human *ADA* gene is introduced by retroviral vectors and the lymphocytes injected into the afflicted children. Upon periodically renewed treatment, symptoms of the disease (chronic infections, diarrhea and muscle weakness) usually recede. Post-exercise cramping of the muscles may be caused by inadequate level of this enzyme. The ADA gene was located to human chromosome 20q13 area. (See adenosine deaminase, Lesch-Nyhan syndrome, gout, gene therapy, immunodeficiency, viral vectors, lymphocytes, deamination,)

ADENOSINE DIPHOSPHATE: see ADP

ADENOVIRUS: a large mammalian (1.8×10^8 Da), icosahedral, (diameter about 80 nm) double-stranded DNA virus with ca. 36 kbp genetic material. The human adenovirus DNA has a 55 kDa protein covalently bond to both 5'-ends. The initiation of replication depends on a viral 80 kDA protein reacting with the first deoxycytidylic residue and its 3'-OH group serves as the starting point. The complementary DNA strand is a template. After replication the 80 kDa protein is cleaved off but the 55 kDa protein stays on. The replication does not require the synthesis of Okazaki fragments because first one of the strands is completed with the aid of the protein-dCTP primer, then the other strand is replicated. Replication can start at either end because of protein primer is used. Both strands of the DNA are transcribed into overlapping transcripts. The integrated viral DNA is generally smaller than the genome of adenoviruses. After lytic infection, a cell may release about 100,000 virus particles. Upon infection it may produce a flu-type ailment, and upon integration into the genome it may cause cancer. The adenoviruses have broad host range and this makes them suitable for veterinary vaccine production. The adenovirus oncoprotein E1A induces progression of the cycle by binding to a protein complex p300/CBP. A histone acetylase (P/CAF) competes in this with E1A and inhibits its mitogenic activity. The main function of E1A to disrupt the association between p300/CBP and the histone acetylase. The viral E1B gene encoded 55 kDa protein inactivates the p53 tumor suppressor gene and the cancerous proliferation begins. However, a mutant form of adenovirus (dl1520) that cannot express this 55 K protein still can replicate in cells that are defective in the p53 suppressor and as a consequence can lyse these defective cells. This finding offers a promise for the destruction of p53-deficient cancer cells by injection with dl1520 mutants. Adenoviruses have been used as vectors for genetic transformation after some regulatory sequences have been deleted and replaced. The maximal carrying capacity is about 6-8 kbp. Since adenoviruses preferentially infect the respiratory tract, it may be used for somatic gene therapy of e.g. cystic fibrosis. It has been used also to transfer genes to skeletal muscles. Adenovirus vector and its load DNA can be taken up by the cell by a specific virus receptor and the $\alpha_v \beta_3$ or $\alpha_v \beta_5$ surface integrins. The adenoviral vector is not integrated into the human genome and thus does not lead to permanent genetic change and they have to be reapplied periodically (in weeks or months). Also, the current vectors may cause inflammation because of antivector cellular immunity. An advantage of this vector that it can be used in very high titers (up to 10^{13} particle/mL). Adenovirus is not known to induce human cancer. Adenoviral vector-mediated interleukin-12 gene therapy seems to protect mice with metastatic colon carcinoma. (See icosahedral, Okazaki fragment, replication, cancer, viral vectors, titer, gene therapy, p53, cystic fibrosis, antivector immunity)

ADENYLATE: salt of adenylic aci.

ADENYLATE CYCLASE (adenylyl cyclase): an integral membrane enzyme with an active site facing the cytosol, generating cAMP (cyclic AMP) from ATP and releases inorganic pyrophosphate (two phosphates). This enzyme is activated by the $G_{s\alpha}$ subunit-bound GTP. The enzyme has also a weak GTPase activity too that eventually breaks the $G_{s\alpha}$—GTP link and thus turns off the cyclase function. The activation of the cyclase function is initiated by the hormone epinephrine which binds to a membrane receptor and activates the G_s proteins. cAMP itself is degraded by cyclic nucleotide phosphodiesterase. Oscillation of its level is regulated by the cellular level of Ca^{2+}. (See adenosine 3', 5' cyclic monophosphate, G_s, G-protein, GTPase, GTPase activating protein, cyclic AMP-dependent protein kinase, animal hormones, epinephrine, calcium ion channel)

ADENYLATE KINASE DEFICIENCY (AKI1): dominant in human chromosome 9q32; it causes hemolytic problems.

ADENYLIC ACID: a phosphorylated adenosine

ADENYLYL CYCLASE: see adenylate cyclase

ADENYLYLATION: addition of adenine to an amino acid near the active site in a protein (by the enzyme adenylyl transferase); it may regulate activity of the target.

ADH (antidiuretic hormone): a short peptide (vasopressin). (See antidiuretic hormone)
ADH (alcohol dehydrogenase) : is an enzyme catalyzing the reversible reaction:
$$\text{acetaldehyde} + NADH + H^+ \Leftrightarrow \text{ethanol} + NAD^+$$
The ADH subunits (α, β, γ) are encoded in human chromosome 4q21. (See adh⁻, acetaldehyde dehydrogenase, mutant isolation, allylalcohol)

adh⁻: a mutant with a defective ADH enzyme. (See acrylaldehyde, mutation detection)
ADHALIN: see muscular dystrophy
ADHD: see attention-deficit hyperactivity
ADHERENS JUNCTION: cell surface where actin filaments attach. (See β-catenin)
ADHESION: sticking together, e.g., water molecules clinging to various surfaces. (See integrins, selectins, cadherins, plakoglobin, vinculin, talin, adherens junction)
ADHESION BELT: adherens belt connecting neighboring cells. (See adherens junction)
ADHESION PLAQUE (focal contact): the spot where a cell is anchored to the extracellular matrix by transmembrane proteins.
ADIPOCERE (grave wax): is hydrolyzation product of body fats after death; its formation may help the preservation of DNA of the brain in some ancient animal/human remains. (See ancient DNA)
ADIPOCYTE: fat storage cell; it is used as a depository of excess calorie intake or reserve when expenditure exceed intake of calories.
ADIPOSE: related to fat, e.g. adipose tissue = fat tissue
ADJACENT DISJUNCTION: neighboring members of translocation rings or chains move to the same pole; adjacent-1 when centromeres are non homologous, adjacent-2 when centromeres are homologous (nondisjunctional) at the poles. (See translocation chromosomal)
ADJACENT DISTRIBUTION: see adjacent disjunction
ADJUVANT, IMMUNOLOGICAL: if the immune response of an antigen is unsatisfactory because of the small amount present, the immune reaction may be enhanced by protecting the antigen from degradation and promoting slow release and increase its uptake by macrophages. For this purpose mineral oils, alum (a hydrated aluminium oxide), charcoal, Freund adjuvant, etc. can be used. (See immune response, antigen, Freund adjuvant)
aDNA: ancient DNA. (See ancient DNA)
ADNFLE: see epilepsy
AdoMet: S-adenosyl-L-methionine (current abbreviation is SAM) is a methyl donor for the enzymes guanosine 7-methyl transferase and the 2'-O-methyl transferase enzymes in the cap of pre-mRNAs and for other methylation reactions. (See SAM, cap, methylase, methylation of DNA, methylation of RNA)
ADOPTED CHILDREN: are frequently used in human genetics to determine the relative effects of genes and environment. These studies are frequently hampered, however, because either the families do not have biological children or the biological parents of the adopted children are not available for examination. According to civil law, adopted children may lose any legal ties to and identity with the natural parents. This loss of identity may carry some genetic caveats because of chances of inbreeding by the lack of information about descent. These problems may be similar to the ones encountered by artifcial insemination using anonymous sperm donors. (See twinning, artificial insemination)
ADOPTIVE CELLULAR THERAPY: infusion of immune effector cells (NK cells, macrophages, $\gamma\delta$ T cells, $\alpha\beta$ T cells, B cells, etc.) for the treatment or prevention of disease. (See gene therapy)
ADP: adenosine 5'-diphosphate, a phosphate group acceptor in various cellular processes. It is produced by hydrolyis of ATP; it can also regenerate ATP by oxidative phosphorylation.
ADP-RIBOSYLATION FACTOR: see ARF
ADRENAL: adjacent or pertinent to the kidney.
ADRENAL HYPERPLASIA, CONGENITAL (CAH): occurs in both X-linked and autosomal

Adrenal hyperplasia congenital continued

forms and apparently controled by several loci. The X-linked form is attributed to gonadotropin deficiency. The steroid hormone deficiency indicates a defect in steroid 21-hydroxylase, localized within the boundary of the HLA complex in human chromosome 6. The affected female babies are masculinized. It may be associated with Addison disease (hypotension, anorexia, weakness and pigmentation). One form of the disease is accompanied by difficulties in salt retention in the newborns. The prevalence is about 7×10^{-5}. An autosomal form in chromosome 8q21 is deficient in 11-β-hydroxylase and/or corticosteroid methyl oxidase II occurs at a frequency of 1×10^{-5}. Masculinization may be caused by several other genetic anomalies and also by various medications administered to the mother or maternal androgen-producing tumors. (See also HLA, hermaphroditism, genital anomaly syndromes, adrenal hypoplasia, Addison disease, STAR)

ADRENAL HYPOPLASIA, CONGENITAL (AHC): is characterized by abnormal underdevelopment (hypoplasia) of the genitalia and the gonads, insufficient function of the kidneys, hypoglycemia (reduced blood sugar), seizures, etc. Several forms of the disease (hypoadrenocorticism, polyglandular autoimmune syndrome) were reported with autosomal recessive inheritance. The X-chromosome-linked DAX1 locus encodes a dominant negative regulator of transcription, a nuclear hormone receptor protein with a DNA-binding domain. The DAX-1 transcription is mediated by the retinoic acid receptor. (See also hypogonadism, adrenal hyperplasia, epilepsy, retinoic acid, RAR, dominant negative, transcriptional activator)

ADRENALINE: see epinephrine, animal hormones

ADRENERGIC RECEPTORS: come in the forms of α_1, α_2, β_1, β_2 distinguished on the basis of their responses to agonists and antagonists and tissue-specificity. They all respond to the adrenal hormones, epinephrine and norepinephrine. (See epinephrine, membrane proteins, receptors, agonist, antagonist)

ADRENOCORTICAL: pertaining to the cortex (the outer layer) of the kidney.

ADRENOCORTICOTROPIN (ACTH): a pituitary peptide hormone that controls the secretion of steroid hormones of the kidney in response to cAMP. (See cAMP, glucocorticoids)

ADRENODOXIN: an electron carrier iron-sulphur protein in the kidney cortex mitochondria assisting cholesterol biosynthesis. (See cerebral cholesterinosis)

ADRENOLEUKODYSTROPHY: the X-linked neonatal form is a defect in peroxisome assembly. (See microbodies, Zellweger syndrome, Refsum disease)

ADROGENITAL SYNDROME: a complex genetic disorder based on anomalies of steroid biosynthesis and adrenal hyperplasia. Gene frequencies vary a great deal in different populations from 0.026 of Alaskan Eskimos to 0.004 in Maryland, USA. (See adrenal hyperplasia, allelic frequency, steroid hormones)

ADSORPTION: the tendency of molecules to adhere to a surface (different from absorption that is uptake through a membrane).

ADSORPTION CHROMATOGRAPHY: see column chromatography, thin layer chromatography

ADSORPTIVE ENDOCYTOSIS: see receptor-mediated endocytosis

ADVANTAGE OF HETEROZYGOTES: means that the fitness (the reproductive value) of the heterozygotes exceeds that of both types of homozygotes in a population and this may lead to balanced polymorphism. (See balanced polymorphism, polymorphism, fitness)

ADVANTAGEOUS MUTATIONS: are favored by a particular environment and they are expected to propagate under steady-state conditions by a rate per generation: $v = \sigma\sqrt{2s}$ where σ is the standard deviation caused by diffusion (migration) and s = selective advantage in the absence of dominance. E.g. if $\sigma = 10$ km, and s = 0.02 then the advance per generation in kilometers will be $10\sqrt{2 \times 0.02} = 2$; then it would take 250 generations to advance 500 km. (See mutation, mutation beneficial, migration, selection coefficient)

ADVENTITIA: the outer coating of organs by lose connective tissues composed mainly of fibrillin and elastin.

ADVENTIVE EMBRYOS: developing from the diploid tissues of the plant nucellus (without fertilization); they occur commonly in citruses. (See apomixia, nucellus)

AECIDIOSPORE: a dikaryon of plant rust fungi formed through a sexual process that did not involve nuclear fusion; the aecidiospores are products of the aecidium, a group of sporangia. (See aecidium, sporangium, fungal life cycle)

AECIDIUM or aecium a fruiting structure of fungi (Basidiomycetes-Uredinales) such as *Puccinia graminis tritici*. Aecidia are formed only on the intermediate host, barberry, but the spores infect only wheat.

Aegilops caudata: a diploid representative of the *Triticum* genus (2n = 14) carrying the 7-chromosome C genome (current name *Triticum dichasians*). *Aegilops cylindrica*: an allotetraploid of the wheat genus containing the CD genomes (C from *T. dichasians* and D from *T. tauschii*). Current name *Triticum cylindricum. Aegilops squarrosa: Triticum tauschii* by current name, is a diploid species in the wheat genus with the D genome. *Aegilops umbellulata* by current name *Triticum umbellulatum*, a diploid species of the wheat genus with the C^u genome. *Aegilops variabilis:* currently *Triticum peregrinum*, a species of the wheat genus; occurs in nature both as tetraploid (DM) and hexaploid (DDM) genomic constitution. (See *Triticum*)

AEGRICORPUS: a genetic-physiological complex determined by the host-pathogen interaction; the phenotype of the disease in plants. (See host-pathogen relations, Flor's model)

AEQUORIN: is a luminescent protein (green fluorescent protein) from jellyfish (*Aequoria victoria*); its activation is dependent on the level of the available Ca^{2+} and on this basis minute quantities and differences in this ion can be measured by optical means within the range of 0.5-10 mM. This may be of major importance because calcium may play regulatory functions in all eukaryotic cells. The chromophore results from the cyclization and oxidation of the Ser^{65} (Thr^{65})Tyr^{66} Gly^{67} amino acid sequence in the central helix of the 11-stranded β barrel. Aequorin has also the advantage that is non-invasive and non-destructive label in various organisms. (See calmodulin, Renilla GFP)

AER2: see TUP1

AEROBE: an organism that uses oxygen as the terminal electron acceptor in respiration. (See electron acceptor, respiration)

AEROBIC: a reaction (or organism) requires or takes place in the presence of oxygen.

AFFECTED-SIB-PAIR METHOD: is a non-parametric method for linkage analysis of susceptibility genes. For this purpose the risk ratio (λ_s) is determined, that is the risk of a sib of an affected proband compared to the average prevalence in the population. E.g. diabetes has a prevalence of 0.004 in the general population but its incidence among sibs of affected individuals is 0.06, hence $\lambda_s = 0.06/0.004 = 15$. This λ_s is for all loci responsible for the phenotype. The larger λ_s the higher is the genetic contribution. It is also affected by the interaction (can be multiplicative or additive) of the various factors contributing to the phenotype. The strength of the proof for linkage depends on the so called *maximal lod score* (MLS) symbolized by T and means the log odds in favor of linkage. Usually, the estimation is carried out in steps by selecting at each step linkage with markers increasing from T>1.0. Statistically valid linkage is expected when the T score reaches or exceeds 3. The T value may also increase by the use of larger populations. Recombination decreases MLS and the use of multiple loci increases the estimate. There is another advantage of this approach that both recessive and dominant alleles can be studied. It is applicable also for quantitative traits. (See recombination, frequency, maximum likelihood method applied to recombination, lod score, non-parametric tests)

AFFECTIVE DISORDERS: are psychological illnesses, psychoses. (See manic depression, autism, hyperactivity [ADHD], Tourette's syndrome)

AFFERENT: conducting or transferring toward the middle.

AFFINITY: unlinked genes segregate to the same gamete more frequently (quasi-linkage) or less frequently (reverse linkage) than expected on the basis of randomness. In immunogenetics: the intensity of interaction between a particular antigen receptor and its epitope. (See epitope)

AFFINITY CHROMATOGRAPHY: polyadenylated mRNA can be separated from other RNAs

Affinity chromatography continued

by adsorption on oligo T (thymine) cellulose or sepharose columns by virtue of the complementarity of the A and T bases. Similar procedures are used for the purification of antibodies on immobilized antigen media (antibody purification) and DNA-binding proteins can be isolated and enriched by a factor of 10^4 with the aid of affinity chromatography. (See also gel retardation assay, cDNA library screening)

AFFINITY LABELING: most commonly a photo-affinity hapten is used (i.e. one which is activated only upon illumination). The affinity label is bound to the antigen-binding site amino acids of the antibody and thus reveals where on the antibody the attachment is taking place. (See hapten, antigen, antibody)

AFFINITY MATURATION: selection of cells with high affinity for the antigen as clonal selection progresses. It takes place by accumulation of mutations in the *germinal center* of a lymphoid follicle in the paracortex of a lymphoid node. These alterations take place in response to the antigens arriving there through small capillary veins on the surface of the antigen-presenting cells and helper T cells. (See clonal selection, antibody, antigen-presenting cell, immune response, repertoire shift, immunoglobulins)

AFFINITY PURIFICATION: is required unless the antibody reacts with more than one antigen. If this is not the case, an affinity chromatography column is prepared by using pure antigen. Alternatively, monoclonal antibody must be used or the immunoglobulin library must be carefully analyzed for true or false positive immune reactions. (See antibody, antigen, monoclonal antibody)

AFFINITY-DIRECTED MASS SPECTROMETRY: detects interaction between proteins, receptors-ligands, proteins-nucleotides, etc. (See mass spectrum)

AFI: amaurotic familial idiocy; now called Tay-Sachs disease (TSD). (See Tay-Sachs disease)

AFIBRINOGENEMIA: is an autosomal recessive deficiency of fibrinogen (blood coagulation factor I). The afflicted individuals bleed very heavily after injury. Periodic blood accumulation under the skin (ecchymosis), nosebleeding (epistaxes), bloody tumors (hematomas), bloody cough (hemoptysis) or stomach-intestinal or genitourinary bleeding occur. Characteristically, for longer periods of time no symptoms appear. Therapy is intravenous injection of concentrated human fibrinogen. (See also antihemophilic factors, hemophilia, dysfibrinogenemia, fibrin-stabilizing factor)

AFLATOXINS: a group of heterocyclic mycotoxins produced under appropriate conditions by the *Aspergillus flavus* fungus. The aflatoxins are extremely carcinogenic because they affect DNA synthesis. The LD_{50} of aflatoxins orally administered to monkeys may be as low as 1750 µg per kg. Aflatoxin may be a contaminant on grains, peanuts, dry chili pepper, and on many other material humans and animals eat or are exposed to. Aflatoxins frequently cause mutations in *p53* tumor suppressant at codon 249 (AGG→AGT) resulting in hepatocarcinomas. (See environmental mutagens, p53, toxins)

AFLP: is a DNA fingerprinting technique involving restriction enzyme digestion and amplification of special fragments by PCR. (See DNA fingerprinting, PCR, restriction enzymes)

AFRICAN GREEN MONKEY: (*Cercopithecus aethiops*) kidney cells are the best laboratory host for the propagation of SV40 (Simian virus 40). (See SV40, cos)

AFTER MORNING PILL: see hormone receptors [RU-486]

AGAMEON: a species without sexual reproduction.

AGAMIC: a species reproducing asexually (without gametes).

AGAMMAGLOBULINEMIA: occurs as an X-chromosomal (congenital) and autosomal defect in the synthesis of γ-globulin, a component of the heavy chain of antibodies. The X-chromosome-linked (Xq21.33) is frequently called Bruton's agammaglobulinemia (XLA). The manifestation of XLA may differ in different families, indicating the involvement of several genes. It is conceivable that the defect is caused by rearrangement of the genes involved. Some of the individual have truncated V regions of the antibody. In the afflicted persons the IgG and IgM content is generally no more than 1% of the normal. The basic defect is also the absence of

Agammaglobobulinemia continued
plasma cells from the lymph nodes, spleen, intestine and bone marrow. The patients are very susceptible to pyogenic infectious bacteria (staphylococci, pneumococci, streptococci, and *Hemophilus influenzae*). Pus-forming inflammation of the sinuses, pneumonia, meningitis (inflammation of the brain), furunculosis (boils) are common but can be prevented by the use of antibiotics or raising the γ-globulin levels by regular injections. Without treatment these infections may become fatal. The afflicted children are not more susceptible to viral, enterococcal, gram-negative bacteria, protozoan or the majority of fungal infections. Another X-chromosome linked or autosomal agammaglobulinemia causes susceptibility to bacterial, fungal and viral infections and leukemia. This is generally accompanied by lymphopenia (decrease of lymphocytes in the blood). This disease is generally detected after the discontinuation of breast feeding of the babies or near the end of the first year of life. Agammaglobulinemia may occur also as an acquired disorder with onset at different ages, generally as a follow-up to other diseases. The prevalence is about 0.5 to 1×10^{-5}. (See gammaglobulin, immunoglobulins, immune system, antibody, hypogammaglobulinemia/common variable immunodeficiency, immunodeficiencies, achondroplasia, cancer, BTK)

AGAMOSPERMY: seed production without fertilization, apomixis. (See parthenogenesis, diplospory, apospory, adventitious embryos, apomixis)

AGANGLIONOSIS: congenital lack of intestinal ganglions. (See Hirschprung disease)

AGAR: gelling agent produced from marine algae with various degrees of purification (bacteriological agar, noble agar) and used for microbial and plant cell culture media. (See also gellan gum, agarose)

AGAROSE: a purified linear galactan hydrocolloid isolated from marine algae. In the crude form it is generally contaminated with salts and other substances, polysaccharides, proteins. Some commercial products are highly purified. It is used for electrophoretic separation of oligo and polynucleotides from 0.1 to 60 kb range, depending on the concentration of this matrix. The higher concentration (2%) separates the smallest molecules whereas the lowest concentration (0.3%) permits the separation of the largest fragments; 0.9 - 1.2% are the most commonly used concentration ranges separating 0.4 to 7 kb fragments. Contaminations of the agarose may interfere with further enzymatic handling of the eluted DNA. (See electrophoresis, gel electrophoresis)

Agave (sisal): basic chromosome number $x = 30$ and the various plant species may be diploid, triploid or pentaploid.

AGE (advanced-glycation endproduct): a sugar-derived carbonyl group added to a free amine that forms an adduct after rearrangment producing AGES. Age may cross-link amino groups in macromolecules and thus may promote aging, accelerate diabetes and may participate in other reactions. The cross-links may be broken by N-phenacetylthiazolium bromide and may have therapeutic application. (See Alzheimer disease, aging, diabetes, adduct)

AGE AND MUTATION IN HUMAN POPULATIONS: can be expressed by the formula $\mu_t = \alpha t + \mu_0$ where mutation rate at a given time is μ_t, α is the mutation rate per cell divisions and μ_0 the initial frequency of mutation. It is expected that mutation rate increases as the number of cell divisions increases in the spermatogonia and oogonia. The available data indicate that chondrodystrophy (achondroplasia, a dominant dwarfness) and acrocephalosyndactily (Apert's syndrome, pointed top of the head and syndactily [webbing in between or attachment of the fingers and toes]) increases at birth by about 2 - 4 fold with paternal age from 25 to 45 years. Other dominant mutations show similar tendencies but much less clear differences. The human eggs may be different because new egg cells are not formed in the female babies after birth; the oogenesis is almost complete in the newborn. Nevertheless some age differences are still expected. Chromosomal aberrations (trisomy) in the eggs may increase, however, from 1/2300 at age 20 to 1/46 after age 45, probably because of the prolonged meiotic dictyotene stage (diakinesis). Some of the eggs complete meiosis before each ovulation, a period extending over 30-40 years. Trisomy in sperm is much less common, partly because it is the product of new divisions, partly because the disomic sperm may be at a disadvantage in competition for fertilization. A normal human ejaculate may contain 250 - 400 million sperm cells. (See mutation rate, gonads, Apert syndrome, syndactily, gametogenesis, trisomy)

AGE CORRELATION BETWEEN MATES: is much higher in consanguineous marriages than in unrelated mates. On the average age correlation makes first-cousin marriages about twice, second cousin marriages about 1.7 times and third cousin marriages about 1.4 times as frequent as if there would be no correlation between the ages at marriage. Since some of the human hereditary diseases have late onset, the greater the age at marriage may reduce the reproduction of genes with late manifestation because the afflicted persons may not marry or chose not to have children. (See consanguinity)

AGE OF ONSET OF DISEASE: the probability can be calculated:
$(1 - \phi_1)(1 - \phi_2)...(1 - \phi_{x-1})$ where ϕ_x = the probability of onset between ages x and x + 1 the probability of surviving to age x before onset is $l_x = (1 - q_1)(1 - q_2)...(1 - q_{x-1})$ where q_x = the probability of dying at age x before the onset of the condition. (See aging)

AGE OF PARENTS AND SECONDARY SEX RATIO: is slightly decreasing from 0.517-0.516 at parental age group 15-19 to 0.512-0.511 at parental age 45-49. The decrease is approximately the same for both parents although it is slightly more affected by the age of the mother. (See sex ratio)

AGE-SPECIFIC BIRTH AND DEATH RATES: the probability that an individual of a certain age dies (or gives birth) within the following year is determined by population projection matrices. The numbers of giving birth and death rates in a time interval can be determined by 1- B/N for birth per women extracted from available census figures. One of such studies in 1966 found that in the age group of women 15 - 30 and 30 - 45 the mean number of children born per women was 1.37 and 0.465, respectively. Also, the study found that the average survival of age groups 0 - 15, 15 - 30 and 30 - 45 were 0.992, 0.988, and 0.964, respectively. Thus, if one takes a sample of 30 woman of age group 0 - 15 they will give birth to 30 x 1.37 = 41.1 children. In the age group 30 - 45 dealing with 20 human females the prediction will be 20 x 0.465 = 9.3, and so on. Similarly the survivors expected in the age group 0 - 15 will be 40 x 0.992 = 39.68, in the age group 15 - 30 the expectations are 30 x 0.988 = 29.64, by the end of the respective periods, and so on. The natural logarithm of the annual growth rate is called the *intrinsic rate of natural increase of the population* (r), and it means that once a stable equilibrium is reached for the various age groups it will increase by this intrinsic rate per year. Example: if the population growth at equilibrium at 15 years cycles is 1.307 then the annual $r = ln(1.307)^{1/15} \cong 0.0178$. The age-specific birth and death rates and *r* must be determined for each population because considerable variations may exist from time to time even in the same group, depending on cultural and economic conditions. (See human population growth)

AGENT ORANGE: a herbicide containing mainly the synthetic auxin 2,4,5-trichlorophenoxy acetic acid (2,4,5-T). It had been used as a defoliating agent and brush killer. The LD$_{50}$ of 2,4,5-T for mammals is 500 mg/kg, however, there are reports of much lower doses of high toxicity particularly at subcutaneous injection. It is frequently contaminated by dioxin, a carcinogen. The symptoms of agent orange exposure can be anorexia, hepatotoxicity, chloracne, gastric ulcers, porphyrinuria, porphyria, teratogenesis, etc. 2,4,5-T may be degraded by genes in the plasmids derived from *Pseudomonas ceparia*. (See LD$_{50}$, anorexia, hepatotoxicity, acne, chloracne, ulcer, porphyria, teratogenesis)

AGGA BOX: is an upstream transcriptional regulatory site. (See transcription factors, promoter)

AGGLUTINATION: clumping; it occurs when two different blood types are mixed or when bacteria are exposed to specific antisera. The basis of this phenomenon is a component of the complement on the antibody (C1$_q$) protein that binds to the Fc region of the IgG heavy chain and that is followed by a change in conformation of the antibody. This process is triggered by the binding of the epitope to the antibody. (See immunoglobulins, antibody, epitope, complement)

AGGLUTININ: is an antibody that causes agglutination of cognate antigen. (See abrin)

AGGREGATION CHIMERA: is produced *in vitro* by the assembly of genotypically different early embryonic cells. (See chimera, allopheny)

AGGREGULON: a protein complex involved in activation and repression of genes; the term reglomerate was used in the same sense.

AGGRESSION: a behavioral trait with great variance in animal and human populations and it may be an expression of innate self-assertion, frustration or a response to antisocial behavior encountered. It may be the consequence of affective disorders and mental illness (paranoia). Aggression was attributed by evolutionists to the means of survival in the struggle for life, and as such it is observable among the majority of animals. Accordingly in subhuman beings it is instinctive and largely depends on the species concerned. Among humans it has an animal component but it is determined also by the ethical and cultural factors of the individual and the standards of the population. While animals are not credited with conscientious value judgments, in human societies, the moral, ethical, religious and cultural principles may predominate. All human ethnic groups appear to have a condemning attitude toward violence. Yet mainly humans display violent aggression within species. It has been suggested that the human species lack the ability of submission, a widely common ability among other mammals. The genetic basis of aggressive behavior is generally not understood although it is known that a deficiency, e.g. in hypoxanthine-guanine phosphoribosyl transferase may result in hostile and self-mutilating behavior. The major problem is concerned with the large non-biological but cultural component of aggression. Unfortunately, human societies treat the cultural problem with double standards: killing and violent behavior is condemned yet even major religions approve patriotic or holy wars with the weapons of mass destruction. The questions remain unsettled whether capital punishment is appropriate for killers, is induced abortion an act of aggression, is euthanasia a merciful act or just another form of taking life? To what extent are criminals predestined by their genetic endowment to aggression and how much is the role of the social environment, and the free will? Obviously, some of the answers are beyond the scope of genetics. Mice deficient in α-calcium-calmodulin kinase II displayed reduced levels of serotonin and aggressive behavior. (See behavior genetics, submission signal, ethics, morality, instinct, Lesch-Nyhan syndrome, nitric oxide, calmodulin, serotonin, behavior in humans, personality, mental illness, paranoia)

AGING: is an exponential increase in mortality as a function of time or cell divisions. It is determined by some type of irreversible alterations in the DNA. In older cells the chromosomal telomeres are shortened, the frequency of nondisjunction dramatically increases by age, e.g. the incidence of Down syndrome may increase 200 fold in the offspring of just pre-menopausal mothers. The autosomal recessive Werner syndrome (gene frequency 1 to 5×10^{-3}) involves premature aging (graying of hair, atrophy of skin, osteoporosis, decreased libido, and increased risk of cancer) is characterized also by non-ketotic hyperglycinemia. Also, progeria (Hutchinson-Gilford syndrome), another autosomal recessive traits, causes very early senescence. Aging has been attributed to defects of the immune system and to diminished activity of superoxide dismutase, an enzyme normally destroying the highly reactive radicals that arise due to irradiation and aerobic metabolism. It has been suggested that aging is the result of degenerative changes in the mitochondria. Byproducts of oxidative phosphorylation, hydrogen peroxide and superoxide, may accumulate during senescence. This causes delays in mitochondrial replication. The slow replication leaves unprotected the D loop of mtDNA, possibly increasing the chances for deletions and mutations. Hereditary premature aging is also known in animals. Voltage-activated Ca^{2+} influx into the brain neurons is accelerated during aging. In *Caenorhabditis*, mutations are known that in combination may extend the life of the nematodes through two different pathways up to five fold. In yeast, activated GTPase (RAS), inactivation of the *LAG1* gene (encoding a membrane-spanning protein) and the *SIR* silencing complex extend life span. Aging in mammals seems to be associated with aging of the lymphocytes and their function. The telomerase enzyme has also been implicated in aging. Since aging usually occurs after the reproductive period, it is no longer the object of natural selection. Population geneticists entertain two genetic mechanisms for aging: the accumulation of deleterious mutations and increase in antagonistic pleiotropy among gene loci. In rodents, calorie restricted

Aging continued

(CR) diet has anti-aging effect. Aging has been attributed also to the gradual loss of the telomeric DNA repeats. (See senescence, killer plasmids, chromosome breakage, DNA repair, Werner syndrome, progeria, Cockayne syndrome, Bloom syndrome, longevity, ion channels, RAS, lymphocytes, telomere, telomerase, selection, silencer, RAS, mating type determination in yeast)

AGONIST: activates a receptor.

AGONISTIC BEHAVIOR: combative behavior. (See aggression)

AGOUTI HAIR PATTERN ⇒

AGOUTI: alternating light and dark bands on individual hairs of the fur in mammals such as mouse, rat, rabbit. The genes *agouti* and *extension* determine the relative amounts of eumelanin (brown-black) and phaeomelanin (yellow-red) pigments. *Extension* encodes the receptor of the melanocyte-stimulating hormone (MSH) and *agouti* is a signal sequence in the hair follicle, inhibiting eumelanin production and the melanocortin receptor, an MSH receptor. Agouti has been cloned and sequenced; it contains 5 exons but two of them are not translated. The secreted protein products has 131 amino acid residues. The alleles that produce increased amounts of pheomelanin makes the mice more prone to late-onset obesity and diabetes. The A^y and the A^{vy} increases the liability to neoplasias and others cause embryonic lethality. (See pigmentation of animals, melanin, melanocyte-stimulating hormone)

AGRETOPE: that part of an antigen that interacts with a desetope (antigen-binding site) of a MHC (major histocompatibility) molecule. (See desetope, epitope, histotop, antigen)

AGRICULTURAL PRODUCTIVITY: is affected by genetic improvement of plants and animals and the improved husbanding and culture practices. Between the years 1951-1980 the overall plant productivity in the USA has increased 166% and that of animals to 144%. The yield of maize after the introduction of hybrids increased to about 500%. (See heterosis, QTL)

AGRIN: a natural glycoprotein (200 kDa) causing the aggregation of acetylcholine receptors on muscle cells *in vitro* and *in vivo* is used for the formation of neuromuscular junctions. The process also requires a muscle-specific protein kinase (MuSK). Agrin deficient mutant mouse is inviable. (See acetylcholine receptors, laminin, neuregulins)

AGROBACTERIUM MINI-PLASMID: carries the T-DNA, including its borders, but it is free of other segments, including the *vir* genes. (*See Agrobacterium tumefaciens*, T-DNA)

Agrobacterium rhizogenes: a bacterium closely related to *A. tumefaciens*. It induces hairy roots rather than crown gall on the host plants. The genes responsible for the formation of hairy roots reside in the *Ri* plasmid. The hairy root tissues, unlike crown gall, readily regenerate into plants. The *Ri* plasmid has been used similarly to the *Ti* plasmid to construct plant transformation vectors. (See *Agrobacterium tumefaciens, Ri* plasmid)

Agrobacterium tumefaciens: a soil born plant pathogenic microorganism of the family of *Rhizbiaceae*. It is responsible for the crown gall disease of the majority of wounded dicotyledonous plants and it infects also a few monocots (*Liliaceae, Amaryllidaceae*). Several of its characteristics are similar to *Rhizobium, Bradyrhizobium* and *Phyllobacterium* species. The pathogenicity is coded in genes within the T-DNA of its Ti (tumor-inducing plasmid). T-DNA containing plasmids are the most important transformation vectors of plants. The T-DNA (transfered DNA) is an about 21 kb segment of the Ti plasmid with two direct repeat flanks bordering the oncogenes (responsible for tumorigenesis in the wild type plasmids), and some of the opine genes. Molecular biologists most widely use the Agrobacterial strains A6 and C58, containing octopine and nopaline encoding Ti plasmids, respectively. Certain *Agrobacteria* strains have *limited host range* (LHR) caused by an altered *virA* gene in the Ti plasmid. The *supervirulent* strains on the other hand overproduce the VirG protein. The infection of some species of plants is limited by a *hypersensitive reaction*. For the transfer of the T-DNA to other cells, including plant cells, requires the formation of a conjugation tube (pilus) controlled by virulence genes *virA, virG, virB1* to *virB11*. (See Ti plasmid, T-DNA, virulence genes of *Agrobacterium*, transformation [plants], host-pathogen relation, BIBAC)

AGROCINOPINE: is a phosphorylated sugar, an opine, produced in octopine plasmids from mannopine by the enzyme agrocinopine synthase. (See *Agrobacterium*, opines, octopine)

AGROINFECTION: is a method of plant transformation. More than one genome of the double-stranded DNA of Cauliflower mosaic virus is inserted in tandem within the T-DNA of *Agrobacterium tumefaciens*. Such a construction permits the escape of the viral DNA from the bacterial plasmid once it was introduced into plants. Geminiviruses can be introduced into plants in a similar way. (See cauliflower mosaic virus, geminiviruses, transformation genetic)

AGROPINE: is a bicyclic phosphorylated sugar derivative of glutamic acid; it is synthesized by *Agrobacterium* strain Ach5. (See opines)

AGROPYRON (x = 7): a genus of grasses; their chromosomes are homoeologous to that of several species within the genus of wheat and can be substituted for to introduce agronomically useful genes (e.g. disease resistance). Some hybrids are known as perennial wheat, a forage crop. (See chromosome substitution, alien transfer, homoeologous chromosomes)

AHONEN BLOOD GROUP: is a rare type, distinct from ABO, MNS, P, Rh, Duffy, Kidd and Dembrock. (See blood groups)

AHR: see arylhydrocarbon receptor

AICD: see memory immunological

AID: artificial insemination by donor. (See artificial insemination, AIH, ART, acquired immunodeficiency)

AIDS: see acquired immunodeficiency syndrome

AIG (anchorage independent growth): normal mammalian cells grow in monolayer anchored to a solid surface. Tumor cells grow independently of anchorage. (See anchorage, tumor, cancer, CATR1, oncogenes)

AIH: artificial insemination by husband. (See also AID, artificial insemination, ART)

AIMS: *Arabidopsis* information data base at Michigan State University; can be reached through Telnet (aims.msu.edu) with the user name guest1, without requiring a password. (See *Arabidopsis*)

AK1: adenylate kinase.

AKAP79 (A kinase anchoring protein): is a cytoplasmic protein binding to cyclic adenosine 3',5'monophosphate (cAMP)-dependent protein kinase (PKA, calcineurin [phosphatase 2B]) and protein kinase C (PKC) and appears to have a regulatory function as a scaffold for the cellular signaling system. (See signal transduction, T cell)

AKR MICE: a specially selected strain of the animal containing the genes *Akv-1* and *Akv-2* that code for ecotropic retroviruses causing thymic lymphosarcoma (leukemia). Ecotropic viruses replicate only in cells from what they have been isolated originally. (See replicase)

AKT ONCOGENE: was isolated from thymomas (cancer of the thymus) of AKR mice transformed by an ecotropic virus. In the mouse genome it is located in chromosome 12. A homolog of it is found in human chromosome 14-32.3 and it is frequently associated with chromosomal breakage. The Akt protein is a threonine/serine protein kinase (protein kinase B/PKB) and targeted by PI 3-kinase-generated signals. Akt is involved in the regulation of cellular proliferation/apoptosis, glycogen synthase kinase (GSK3) and protein synthesis. Akt is regulated also by an insulin-like growth factor (IGF) and the nerve growth factor (NGF). (See AKR, insulin-like growth factor, nerve growth factor, PKB, phosphoinositides, ecotropic retrovirus, apoptosis, glycogen, mice, oncogenes)

ALA: aminolevulinic acid, a first compound in the synthesis of porphyrins from glycine and succinyl CoA. (See heme)

ALAGILLE SYNDROME: autosomal dominant (human chromosome 20p11.2) involving obstruction of the bile duct (cholestasis) and jaundice, lung anomalies (pulmonary stenosis), deformed vertebrae, arterial narrowness, altered eye pigmentation, facial anomalies, etc. In some cases it is accompanied by translocations or deletions of the region. The multiplicity of the symptoms have been considered as a contiguous gene syndrome. (See contiguous gene syndrome, face/heart defects)

ALAND ISLAND EYE DISEASE: see albinism ocular

ALANINE: L-alanine [CH$_3$CH(NH$_2$)COOH] is a non-essential amino acid for mammals. The enantiomorph D-alanine may not be metabolized by some organisms and may even inhibit their growth. β-alanine [CH$_3$CH$_2$CH$_2$CO$_2$COOH] is synthesized by several microorganisms from aspartate but it occurs only in trace amounts in animal tissues, possibly through the action of intestinal microorganisms; γ-butyric acid [H$_2$NCH$_2$CH$_2$CH$_2$COOH] is structurally related. Its dipeptides with histidine are carnosine, homocarnosine and anserine. (See alaninuria, alanine aminotransferase, carnosinemia)

ALANINE AMINOTRANSFERASE (glutamate-pyruvate transaminase, GPT): autosomal dominant gene (8q24.2-qter) encodes the enzyme that catalyzes the reversible transamination of pyruvate and α-ketoglutarate to alanine. This enzyme exists in cytosolic and mitochondrial forms. (See amino acid metabolism, alaninuria, glutamate pyruvate transaminase, alanine)

ALANINE-SCANNING MUTAGENESIS: see homologue-scanning mutagenesis

ALANINURIA (with microcephaly, dwarfism, enamel hypoplasia and diabetes mellitus): the autosomal recessive condition is accompanied by the clinically demonstrable excessive amounts of alanine, pyruvate and lactate in the blood and urine. Both lactate and alanine are derived from pyruvate. (See alanine aminotransferase, amino acid metabolism, alanine, hypoplasia, diabetes)

ALBINISM: a pigment-free condition in plants and animals. The absence of skin and hair pigmentation in mammals is generally determined by homozygosity of recessive genes controling melanin synthesis. Melanocytes are the cells specialized for melanin synthesis. During embryonal development melanoblasts, precursor cells of melanocytes, move to surface areas. Melanin is synthesized in special cytoplasmic organelles, melanosomes. The precursor of melanin is the amino acid tyrosine and the conversion is catalyzed by the aerobic oxidase, tyrosinase (polyphenol oxydase). In one type of albinism tyrosinase activity in the hair follicle is still present, however. Albinism may involve the entire melanocyte system of the body or it may be limited to the eye. Albinism of the eye may occur in a sectorial manner in females heterozygous for the Xp 22.3-22.2- linked recessive gene. Albinism of the eye may involve problems of vision, involuntary eye movements (nystagmus), and head nodding. Males affected may be partially color blind (protonomalous). Albinism is controled by numerous single genes. The prevalence of albinism varies from 1/14,000 to 1/60,000 depending on the gene, and the ethnicity of the population; it is generally more frequent among negroids than among caucasoids. Ocular albinism may occur very frequently among some Indian tribes (1/150). Albinism involves hypersensitivity to light and increased susceptibility to some forms of cancer. It may be a component of complex syndromes and may be associated with deafness, neuropathy and bleeding disorders. Albina condition occurs with very high frequency in progenies of plants exposed to ionizing radiation. In plants over 100 genes can cause albinism when mutated. (See Himalayan rabbit, piebaldism, pigmentation of animals, Chédiak-Higashi syndrome, xeroderma pigmentosum, light-sensitivity diseases, tyrosinase, hair color)

↑ Cuna Indian albinos. (Courtesy of Dr. C. Keeler)

ALBINO animals defective in melanin synthesis. In plants more commonly *albina* is the designation of leaf-pigment-free individuals (because the word *planta* is of feminine gender)

ALBIZZIN (L-2-amino-ureidopropionic acid): a glutamine analog. (See asparagine synthetase)

ALBOMACULATA (or status albomaculatus): a green - yellowish white variegation caused by mutation in extranuclear genes in plants. (See chloroplast genetics)

ALBRIGHT HEREDITARY OSTEODYSTROPHY (pseudohypoparathyroidism): an autosomal recessive pseudoparathyroidism is based on a defect in an autosomal recessive G-protein mutation. An X-linked dominant form is based on a defect in the parathormone-adenylatecyclase-G$_s$-protein complex. (See hyperparathyroidism, parathormone, G-protein)

ALBRIGHT-McCUNE SYNDROME: is a pituitary neoplasia resulting from excessive secretion of growth hormone, caused by mutation and constitutive expression of the GTP-binding subunit (G_s) of a G-protein. (See pituitary gland, pituitary tumor, G-proteins)

ALBUMINS: are different proteins soluble in water and in dilute salt solutions, such as bovine serum albumin (BSA) used in chemical analyses. In the fetal serum of mammals the predominant protein is the albumin α-fetoprotein, transcribed from two genes in humans, and have about 35% homology and are immunologically cross-reactive despite their substantial divergence. Both serum albumin and α-fetoprotein, products of the same gene family, are synthesized in the liver and gut. After birth the production of the latter drops dramatically whereas the former is produced throughout life. Their tissue-specificity of expression resides within 150 bp from the beginning of transcription. The α-fetoprotein gene carries three enhancer elements 6.5, 5 and 2.5 kbp upstream that may increase the level of transcription up to 50-fold. The liver-specificity for the serum albumin gene is controled by the PE (proximal element) most important for promoter activity, located between the TATA and CCAAT boxes and three distal elements (DE I [around -100 base from the initiation of translation], most important for liver-specificity, DE II [around - 116], DE III around about -158]). The PE element (5'-GTTAATGATCTAC-3') is quite similar to sequences in the promoters of other liver- expressed genes. The PE binding protein is HNF-1 (88 kDa) is also shared by other liver genes, including the hepatitis virus promoter. The binding proteins associated with the DEs do not appear liver-specific inasmuch as they are used by a variety of other genes of ubiquitous expression, or the specificity is modified by so far unknown co-factors. (See serum, fetoprotein, transcription, enhancer, promoter, TATA box, CCAAT box, HNF)

ALCAPTONURIA: see alkaptonuria

ALCOHOL: an organic molecule formed from hydrocarbons by substituting - OH for H. The simplest representative is ethanol (ethylalcohol, CH_3—CH_2—OH, MW 46.07, boiling point 78.3°C). Ethanol usually contains 5% water; the absolute ethanol is very hygroscopic; for disinfection the 60-80% solutions are most effective.

ALCOHOL DEHYDROGENASE: see ADH, mutation detection

ALCOHOL FERMENTATION: the conversion of sugar into alcohol in the absence of air by glycolysis. (See glycolysis)

ALCOHOLISM: a chronic and addictive use of the chemical is a behavioral trait with some hereditary component of the manifestation. Alcoholism may involve fatal or very serious consequences in certain diseases, in pregnancy, and when certain types of medicines or drugs are used. The fetal alcohol syndrome includes microcephaly (small head), folded skin at the side of the nose, defective eyelids, upturned nose, etc. In adults it may cause cirrhosis of the liver leading to further (fatal) complications. Unfortunately, no association between alcoholism and any particular gene or chromosomal segment has been firmly established; it is apparently under polygenic control. Alcohol abuse during pregnancy may expose the fetus and the newborn to serious developmental harm (fetal alcohol syndrome, FAS) including physical and mental retardation that may seriously affect lifelong the health and function of the individuals. FAS is an increasingly serious social problem along with other abuse of drugs. About 0.001-0.002 fraction of the children are suffering from it. In mice the higher alcohol consumption appeared to be associated with defects in the $5-HT_{1b}$ serotonin receptor and genetic variations (Lys487Glu) of the aldehyde dehydrogenase 1 locus. The (ALDH1 and 2) aldehyde dehydrogenases are present in some far-East human populations and may increase the proclivity to alcohol consumption by a factor of 5 to 10. In some other populations low in ADH1 the alcohol consumption is moderate because of the poor tolerance. In mice the *Alp1* locus may be responsible for 14% and the *Alcp2* for 18% of the total alcohol preference. Interestingly the former gene is acting only in males and the latter only in females. Other loci controlling alcohol withdrawal sensitivity (chromosome 1), alcohol-induced hypothermia and amphetamine-induced hyperthermia, and another hyperthermia loci seem to be in chromosome 9 of mouse. These loci do not appear to be controlling general tendencies for substance abuse. Alcohol preference appears to be a quantitative trait. (See Dubowitz syndrome, polygenic inheritance, teratogenesis, serotonin, substance abuse, QTL, behavior in humans, aldehyde dehydrogenase)

ALDEHYDE: $R-\overset{\overset{H}{|}}{C}=O$

ALDEHYDE DEHYDROGENASE (ALDH, acetaldehyde dehydrogenase): form ALDH1 is encoded in human chromosome 9q21. Low level of this form of the enzyme is responsible for poor alcohol tolerance. ALDH2 functions the liver and it is encoded in 12q24.2 and ALDH3 is coded in human chromosome 17. (See alcohol dehydrogenase)

ALDOLASE-1 (fructose-1,6-bisphosphate aldolase): the ALDOA isozyme has been mapped to human chromosome 16q22-q24, ALDOB (fructose intolerance) in chromosome 9q22, ALDOC in human chromosome 17cen-q12. There is also a deoxyribose-5-phosphate aldolase but its deficiency is apparently harmless. ALDOA deficiency may be involved in a form of hemolytic anemia (γ-glutamylcysteine synthetase deficiency). (See fructose intolerance, anemia)

ALDOSE: a sugar that ends with a carbonyl group (= C=O)

ALDOSTERONE (18-aldo-corticosterone): the main electrolyte-regulating steroid hormone of the kidney cortex. (See steroid hormones, aldosteronism)

ALDOSTERONISM (hyperaldosteronism, glucocorticoid-remediable aldosteronism [GRA]): is controlled by two autosomal dominant genes. It is due to overactivity of aldosterone synthase and steroid 11β-hydroxylase, coded by human chromosome 8q. These two genes are quite similar in structure and frequently form also somatic recombinants. Increased aldosterone production and hypertension results by their activity. This hyperaldosteronism is suppressible by glucocorticoids and dexamethasone. The chimeric genes are under adrenocorticotropic hormone control. As a consequence aldosterone is secreted and causes water and salt reabsorption and high blood pressure. (See hypertension, aldosterone, glucocorticoid, dexamethasone, hypoaldosteronism, mineral corticoid syndrome)

ALEURON: a protein-rich outer layer of the endosperm of monocotyledonous kernels. Aleuron color genes have proven to be very useful chromosomal markers (such as loci *A, C, R, Bz, B* in maize and some of the functional homologs in other cereals) because the dominant alleles can be identified already among the seeds of the heterozygotes, and in case of maize, they can be classified on the cob in immobilized condition and in large numbers.

ALEXANDER's DISEASE: is an autosomal recessive anomaly of lipid metabolism accompanied by a megaencephaly (synonym macroencephaly), a pathological enlargement of the brain.

ALFALFA (*Medicago sativa*): a leguminous forage plant. Its closest wild relative *M. coerulea* (2n = 16) and the somewhat more distant *M. falcata* (2n =16). *M. sativa* autotetraploid (2n = 4x = 32. It is also called lucerne.

ALFALFA MOSAIC VIRUS: genetic material is four RNAs of 1.3, 1.0, 0.7 and 0.34 x 10^6 Da.

ALGA: can be prokaryotic such as the blue-green algae or eukaryotic such as *Chlamydomonas reinhardtii, Ch. eugametos, Euglena gracilis*, seaweeds etc.; they are photosynthetic microorganisms. (See *Chlamydomonas, Euglena*)

ALGENY: the genetic alteration of an organism by non-natural means such as genetic engineering, gene therapy, genetic surgery, transformation. (See items under separate entries)

ALGIN: a polymer of mannuronic acid in the cell wall of brown algae.

ALGOL (algorithmic oriented language): a computer language set by international procedure. (See algorithm)

ALGORITHM: a set of rules and procedures for solving problems in a finite number of sets; usually the repetitive calculation is aimed in finding the greatest common divisor of two members.

ALIEN ADDITION: chromosome(s) of an other species added to the genome of polyploids without seriously disturbing genic balance, in contrast to diploids where even small duplications or deletions may become quite deleterious. The procedure of addition is crossing the higher chromosome number species as pistillate parent with the lower chromosome number pollen donor. The F_1 is generally sterile but by doubling their number (with colchicine) may result in a fertile amphiploid. Upon recurrent back crossing the amphiploid with the recipient parent, monosomy results for the donor's chromosomes. After repeated backcrossing in

Alien addition continued

large populations, one may obtain plants with single monosomes for all chromosomes of the donor. These are called single monosomic addition lines. Upon selfing such individuals, disomic additions are obtained. These carry one pair of extra chromosomes. The purpose of addition is that occasionally the added chromosome, containing agronomically useful genes, may be substituted for its homoeolog and then a substitution line results. (See addition line, pistillate, pollen, amphidiploid, monosome, disomic, homoeologue, substitution line)

The general scheme for the generation alien addition lines in wheat. The same procedure is applicable to other polyploid plant species

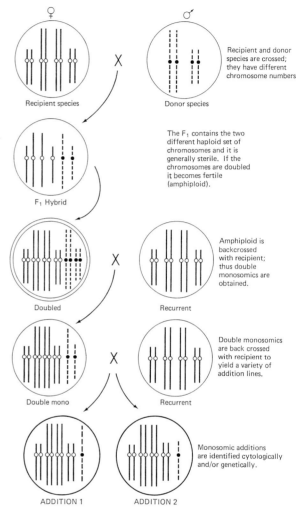

ALIEN SUBSTITUTION: chromosome(s) of an other species replacing own chromosome(s) of a species. Alien substitutions may be obtained from alien addition lines. Most commonly, however, monosomic lines are used. In polyploids monosomic lines can be maintained without too much difficulties because the genomes are better balanced. Mono somic plants can produce some nulli somic eggs. These recipients are then crossed with a donor species. In the F_1 the chromosome absent in the nullisomic will appear as a monosome of donor These monosomic individuals are repeatedly backrossed (6-8 times) with the recipient until in some individuals all the chromosomes of the donor are eliminated, except that particular monosome. Upon selfing this monosomic substitution line, a disomic substitution results that has the same chromosome number as the euploid line from what the nullisomic arose but one pair of the chromosomes will represent the donor. (See alien addition, monosomic, nullisomic, backcross, selfing, euploid, alien transfer lines, chromosome substitution)

ALIEN TRANSFER LINES: alien substitution lines are genetically interesting but the plantbreeder is rarely satisfied with the substitution of an entire chromosome because that may contain, besides the desirable gene(s), some undesirable ones. By homoeologous pairing and crossing over, short segment or even single genes can be borrowed from the alien chromosome. Also, induced translocations may achieve this goal, and the result of these operations is called a transfer line. (See alien substitution, translocation)

ALIGNMENT: finding the nucleotide or amino acid linear sequence matches in nucleic acids and polypeptides, respectively. Free aligning software can be obtained through the Internet:

<http://www.bozeman.mbt.washington.edu/phrap.docs/phred.html> or <http://www.bozeman.mbt. washington. edu/phrap.docs/phrap.html>. (See also BLAST)

AlignMaker: a computer program that produces a file resembling MapSearch output so it can be read by the AlterMap program.

ALK: see anaplastic lymphoma

ALKALINE CHROMATOGRAPHY: is used for the rapid separation of less than 150 base long DNA probes on Sepharose CL-4B (a beaded [60-14 μm pore size], cross-linked agarose). (See Sepharose)

ALKALINE LYSIS: is a procedure to extract plasmid DNA from bacterial cells by 0.2 N NaOH and 1% SDS. The plasmids may be further purified by CsCl-ethidium bromide gradient ultracentrifugation or by polyethylene glycol precipitation at 10,000 rpm. (See SDS)

ALKALINE PHOSPHATASE: cleaves phosphates at a pH optimum about 9; it is present in microorganisms and animal cells but absent from plant tissues. The alkaline phosphatase of *E. coli* is 86 kDA and contains 2 subunits. Human intestinal alkaline phosphatase (ALP1) is in chromosome 2q37, the placental enzyme (ALPPP/PLAP) enzymes are also in chromosome 2q37, the liver enzyme is coded by ALPL in chromosome 1p36.1-p34. (See acrodermatitis enteropathica, acid phosphatase)

ALKALOIDS: diverse (more than 2500), mainly heterocyclic, organic compounds containing nitrogen, generally alkalic nature and are secondary metabolites of plants; frequently of strong biological activity at higher concentrations (such as nicotine, caffeine, cocaine, morphine, strychnine, quinine, papaverine, atropine, hyosciamine, scopolamine codeine, capsaicine, lupinin, etc.). Of particular interest is colchicine (obtained from the lily, *Colchicum autumnale* that blocks the microtubules and thus causes doubling the chromosome number in cells. Vincristine and vinblastine (from *Vinca rosea*) are antineoplastic drugs. The "animal alkaloid", ptomain, found in decomposing cadavers, is actually a microbial product.

ALKALOSIS: diminished buffering capacity of tissues resulting in higher pH.

ALKANE: aliphatic molecules joined by a single covalent bond, e.g., $CH_3—CH_3$. (See alkene)

ALKAPTONURIA (alcaptonuria): is a recessive metabolic disorder (prevalence about 1/40,000) with a defect in the enzyme homogentisate 1,2-deoxygenase and therefore homogentisic acid is not metabolized into maleyl- and fumaryl-acetoacetic acids and eventually to acetoacetic acid and fumaric acid. The degradation of the aromatic amino acids follows the pathway:
PHENYLALANINE→TYROSINE→p-HYDROXYPHENYLPYRUVICACID→ **HOMOGENTISIC ACID→▪**
MALEYLACETOACETIC ACID⇒FUMARYLACETOACETIC ACID⇒FUMARIC & ACETOACETIC ACID
The accumulated homogentisic acid is excreted in the urine and it is readily oxidized into a dark compound, alkapton, staining dark the diapers of affected newborns. Also, it involves dark pigmented spots in the connective tissues and bones, and later during development arthritis may follow. This human hereditary biochemical defect was the first recognized in 1859 and its genetic control was identified by Sir Archibald Garrod in 1902. The gene (AKU) was located to human chromosome 3q21-q23. (See tyrosine, phenylketonuria, amino acid metabolism)

ALKENE: hydrocarbons with one or more double bonds, e.g., $H_2C=CH_2$. (See alkane)

ALKYLATING AGENT: alkylates other molecules; many of the chemical mutagens and carcinogens are alkylating agents. The mutationally most effective alkylation site in the DNA is the O^6 site of guanine. The alkyl may be removed from the DNA in *E. coli* by an alkyltransferase enzyme. The acceptor may be a cysteine residue of that protein. (See mutagens-carcinogens, mutagenic potency, environmental mutagens, carcinogen, chemical mutagens, alkyltransferase)

ALKYLATION: addition of a CH_3 group (or other member of the alkane series) to a molecule. Alkylation of DNA bases may lead to mutation through mispairing and base substitution. Also, alkylation may lead to disruption of the sugar-phosphate backbone of the DNA through depurination by AP nucleases. Thymine is alkylated at the O^4 posion and adenine at the N3 position. (See base pairing, hydrogen pairing, tautomeric shift, AP endonuclease, chemical mutagens, alkyltransferase, methyltransferase, diagram depicting the reactions of guanine)

Alkylation continued

THE MOST COMMON REACTIONS OF THE ALKYLATING AGENT, ETHYLMETHANE SULFONATE, WITH GUANINE →

ALKLYLTRANSFERASES: protect the DNA against alkyl adducts by transferring the methyl or ethyl (alkyl) groups to cystein and repairing the damage. (See DNA repair)

ALLANTOIS: a tubular part of the hindgut, forming later the umbilical cord of the fetus and it fuses with the chorion. It participates in the formation of the placenta. (See amnion, chorion)

ALLELE: alternative states of a gene (e.g., a^1 and a^2). Hybrids of a^1/a^2 are commonly of mutant phenotype although they may show incomplete (allelic) complementation. Two alleles are *identical* if their base sequence is identical although different from the wild type. *Non-identical alleles* are still in the same gene (and are non-complementary) yet their expression may be distinguishable. *Homoalleles* are affected in the same codon but at the same site a different nucleotide occurs, and therefore they cannot be separated by recombination in a heterozygote for the locus. *Heteroalleles* have their differences at non-identical sites within the codon or in another codon, therefore they can be separated by recombination. *Isoalleles* convey wild phenotype yet under special circumstances they can be recognized by appearance. *Multiple alleles* are alleles of the same locus but more than two alternatives exists. In some organisms alleles of the *a1* locus are symbolized as *a1-1*, *a1-2*, etc. Molecular geneticists involved in physical mapping of the DNA use this term for any DNA difference (e.g. restriction fragment) that displays Mendelian inheritance and occupies the same chromosomal site. (See gene symbols, RFLP, RAPD)

ALLELE-SHARING METHODS: are used for the detection of linkage by examining pedigrees whether a particular genetic locus (chromosomal fragment) is more common among individuals in a pedigree than expected by random segregation. It is basically a non-parametric method. (See non-parametric tests)

ALLELE-SPECIFIC PROBE FOR MUTATION (ASP): in principle, this would detect single base change mutations because under very high stringency of hybridization oligonucleotide probes (ASO) would hybridize only to that sequence which is exactly matching but not to another that has one base pair substitution. This would also identify heterozygotes because they would hybridize to both types of probes, mutant and normal. This procedure requires high skills but can be semi-automated. (See hybridization, mutation detection, probe, ASO)

ALLELIC COMBINATIONS in gametes (at independent loci) can be predicted by 2^n where n is the number of different allelic pairs, and it produces 4^n gametic combinations and the number of genotypes is 3^n. If the number of loci is n and each has a number of alleles, the number of zygotic genotypes can be calculated at 1 locus as $[a \times (a + 1)]/2$ and for n loci: $\left[\frac{a \times (a + 1)}{2}\right]^n$ thus e.g. 100 loci, each with three alleles $[(3 \times 4)/2]^{100} \geq 6.53 \times 10^{77}$ zygotic genotypes are possible. (See also multiple alleles, gametic array, Mendelian segregation)

ALLELIC COMPLEMENTATION: is partial or incomplete complementation among mutant alleles of a genes, representing different cistrons. If the alleles are defective when homozygous they do not contribute to the synthesis of functional proteins. Each of the two in a heterozygote has another non-overlapping defective polypeptide product. In the cytoplasm the correct polypeptide chains may combine in the heterodimeric or heteropolymeric proteins and due to the right assembly, the function of the protein may be restored. Since the available correct polypeptide chains are reduced in number relative to that in the wild type, thus only a reduced number of good protein molecules can be formed, therefore allelic complementation is incomplete. The beneficial effect of the non-defective peptide chains may be brought about also by conformation correction, i.e. the conformation of the defective chains is brought into line as an effect of the other polypeptide chain as long as there is no defect at the active site. The extent of allelic complementation can be best determined by *in vitro* enzyme assays when regulatory genes cannot modify the functions by higher intensity or prolonged transcription of the relevant cistrons. (See step allelomorphism, conformation, complementation mapping)

ALLELIC EXCLUSION: only one of the two alleles at a locus is expressed in immunoglobulin genes or one type of chain rearrangement is functional. (See immunoglobulins)

ALLELIC FIXATION: in a random mating population takes place when one allele completely replaces another and the process depends on the coefficient of selection and the size of the populations (see diagrams below).

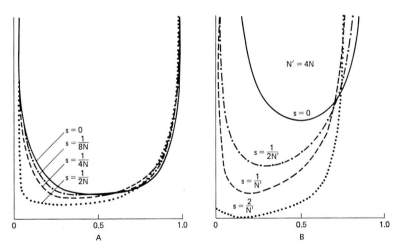

In very small populations only high selection coefficients can bring about changes in allelic distribution (**A.**). When the size of the population increases fourfold, relatively small selection coefficients effectively modify the direction of fixation (**B**). The ordinate represents allelic densities, the abscissa shows allelic frequencies from loss, (0) to complete fixation (1.0); s = selection coefficient, N = population size. (From Wright, S. 1931 Genetics 16:97)

ALLELIC FREQUENCIES: can be determined on the basis of the Hardy-Weinberg theorem.

Allelic frequencies continued

i.e. the genotypic composition of a random mating population is $p^2 + 2pq + q^2$, where p^2 and q^2 are the frequencies of the homozygous dominants and recessives, respectively. Thus if we consider a single allelic pair, A and a, and diploidy, the frequency of the A allele = double the number of homozygous dominants plus the number of heterozygotes. The frequency of the a allele = double the number of homozygotes plus the heterozygotes because the homozygotes have two copies of the same allele whereas the heterozygotes, each have one of each kind. The frequency of the recessive alleles in an equilibrium population is simply $1 - p$. In case of dominance the heterozygotes may not be directly recognized, therefore this procedure may not work. In case, however, the population is at equilibrium and the mating is at random the frequency of the recessive alleles is $q = \sqrt{q^2}$. If the size of the homozygous recessive class is very small, the vast majority of the recessive alleles are in the heterozygotes. In case of sex-linkage the male carries one dose (XY) whereas the female is XX. Thus the males display more frequently a recessive trait than the females which express it only when homozygous. Thus if the the expression of an X-linked recessive allele is 0.10 in the males, in the females it is expected to show up at $(0.10)^2 = 0.01$. The frequency of alleles in a population may change by selection, mutation and random drift and migration. (See selection, mutation, random genetic drift)

ALLELIC INTERACTION: see overdominance

ALLELIC RESCUE: is a procedure for cloning a mutant allele. The mutant cell is transformed by a vector carrying the wild type allele which has, however, an internal deletion overlapping the mutant site. When the cellular (gap) repair system fills in the deleted sequences using the mutant template, the plasmid vector carries a copy of the mutant gene and it can then be isolated along with the vector. (See DNA repair, transformation)

ALLELISM TEST: is carried out by complementation tests. If two recessive genes are allelic they fail to complement each other in the F_1 hybrids (i.e. the hybrid is of mutant phenotype). In case the hybrid of two recessive individuals is of wild phenotype (i.e. they complement each other) the two genes are not allelic. Thus the number of complementation groups reveal the number of different loci. In practice people talk about complementation groups, this is understood as different complementation groups. Allelic genes at the same multicistronic locus may show, however, partial or allelic complementation. (See allic complementation)

ALLELOMORPH: a historical term for allele. (See allele)

ALLERGY: sensitivity to a particular antigen(s) showing an immunological reaction. Common forms are the hay fever after exposure to pollen (ragweed), or drug, food, bacteria cold, etc. The allergic reaction may be a hereditary property (atopy). In asthma, hay fever, and various other allergic reactions the regulatory Re gene was implicated in the decrease of immunoglobulin E level, and in re/re individuals it seemed to be higher. The frequency of the re gene was estimated to be about 0.49. The IgE response in about 60% of the atopy cases was assigned to chromosome 11q13 or q12. The IgE response is apparently controlled by IgE receptor (FcεRI). It appears also that IgG Fc receptor, FcγRIII affects FcεRI assembly. The ragweed sensitivity was assigned to the HLA complex in human chromosome 6. Elevated levels of immunoglobulin E, controlled by an autosomal dominant gene with incomplete penetrance, were detected in the neutrophil chemotaxis defect, characterized by chronic eczema, repeated infections by staphylococci and eosinophilia (cytological structures readily stained with eosin stains). Asthma and other allergies are apparently under the control of multiple genetic loci. (See ragweed, HLA, immunoglobulins, hypersensitivity, atopy)

ALLIGATOR: *Alligator mississipiensis*, 2n = 36.

ALLIUM (onion, garlic): a monocot genus, 2n = 16 or 32; well suited for cytological analysis.

ALLOANTIBODY (isoantibody): is produced by an individual of a species against alloantigens within the species. These may be due to preceding transfusions or pregnancies and may cause hyperacute rejection in case of transplantation in another individual of the same species. (See alloantigen)

ALLOANTIGEN: genetically determined antigen variant within the species; it may be called

Alloantigen continued
also neoantigen when the epitope appears the first time. It is recognized within the same species by the lymphocytes with different haplotype. (See antigen, epitope, lymphocyte, haplotype, isoallogen, alloantibody)

ALLOANTISERA: are antibodies that can recognize a certain protein in a different individual.

ALLOCATALASIA: is characterized by normal catalase activity and stability yet the protein is a different variant. (See catalase)

ALLOCHRONIC SPECIES: does not occur at the same time level with others during evolution.

ALLOCYCLY: chromosomal regions or chromosomes or genomes may show cyclic variation in coiling and heteropycnosis. (See heterochromatin, heteropycnosis, Lyonization, Barr body)

ALLODIPLOID: a polyploid which has chromosome sets (genomes) derived from more than one ancestral organism, e.g. hexaploid bread wheat has A, B and D genomes. (See also autotetraploid, *Triticum*)

ALLOGAMOUS: the individuals are not pollinating the pistils of the same plant, they are cross-pollinating, and they are also called exogamous species.

ALLOGAMY: fertilization by gametes coming from different individual(s). (See also autogamy)

ALLOGENEIC: antigenic difference exists between two cells (in a chimera). (See antigen, isoallogen, allograft, also autologous)

ALLOGENEIC INHIBITION: in mice parental cells do no accept a graft from the F_1 but the reciprocal graft may succeed.

ALLOGENIC: same as allogeneic.

ALLOGRAFT: transplantation of tissues that carry cell surface antigens not present in the recipient. The result may be rejection and destruction of the graft and harm to the recipient. (See HLA, graft, isograft, heterograft)

ALLOHAPLOID: haploid cell derived from an allodiploid. (See allodiploid)

ALLOLACTOSE: the inducer of the *Lac* operon; it is the intermediate product of lactose (a disaccharide) digestion by β-galactosidase, and it is then converted to galactose and glucose by the same enzyme. (See galactosidase, lactose, *Lac* operon)

ALLOMETRIC DEVELOPMENT: different growth (development) rate of one part of the body relative to other parts.

ALLOMETRY: growth in different dimensions of space and time within an organism.

ALLOMIXIS: crossfertilization, allogamy.

ALLOPATRIC SPECIATION: is involved in geographic adaptation and sexual isolation of species inhabiting non-identical habitats. (See speciation)

ALLOPHENIC: the original meaning was that some genes may not be expressed in one cell type but can act as gene activators in other tissues. Also, the expression of genes in chimeric tissues of an embryo or adult that has been produced through the *in vitro* fusion of two or more genetically different (chimeric) blastomeres. These blastomeres developed upon the fusion of the gametes of two parents, each, and several different blastomeres can be fused, resulting in (tri-, quadri-, hexa-parental, etc.) multiparental offspring developing from these allophenic individuals. The fused blastomers are implanted into the uterus of pseudopregnant animals which carry the developing mosaic embryos to terms. (See biparental, chimera, blastomere)

ALLOPHYCOCYANIN: a fluorochrome; it is excited at wavelength 610 and 640 nm and emits bright red light at 650 nm. It is used for flowcytometry. (See fluorochromes, flow cytometry, phycobilins)

ALLOPLASMIC: the cytoplasm and the nucleus are of different origin.

ALLOPOLYPLOID: containing two or more kinds of genomes from different species, e.g. *Triticum turgidum* (macaroni wheat) is an allotetraploid containing the AABB genomes and *Triticum aestivum* (bread wheat) is an allohexaploid with AABBDD genomes or *Triticum crassum* (a wild grass) hexaploid DDDDMM. *Nicotiana tabacum* is an allotetraploid (2n = 48) containing the genomes of *N. tomentosiformis* (2n = 24) and *N. sylvestris* (2n = 24). When

Allopolyploid continued

N. tabacum is crossed with either of the progenitors in the F_1 there will be 12 bivalents (12") and 12 univalent (12') chromosomes. On the basis of chromosome pairing and chiasma frequency the degree of homology between genomes can cytologically be determined in meiosis. Allopolyploids generally acquired genes during evolution that suppress multivalent pairing of the chromosomes, therefore the gene segregation pattern resembles that of diploids with more than one pairs of alleles. A duplex autotetraploid may segregate in 35:1 to 19.3:1 (depending on the distance between gene and centromere) but an allotetraploid is expected to display a 15:1 ratio, and an allohexaploid a 63:1 proportion if there are 4 and 6 copies of the genes, respectively. Some genes (which have only two alleles) in hexaploids may display, however, a 3:1 segregation. (See also sesquidiploid, allopolyploid segmental, autopolyploid)

ALLOPOLYPLOID, SEGMENTAL: participating genomes have partial (segmental) homology yet are sufficiently different to cause some sterility. (See allopolyploid)

ALLOPURINOL (hydroxypyrazole pyrimidine) is an inhibitor of *de novo* pyrimidine synthesis and xanthine oxidase activity. It is a gout medicine. (See xanthine, gout)

ALLOREACTIVE (allorestrictive): T cell recognizing foreign antigen and mobilizes cellular defense against it, e.g. graft rejection. (See T cell, antigen)

ALL-OR-NONE TRAIT: either present or absent, and there are no intermediates.

ALLOSPECIFIC: different from the normal.

ALLOSTERIC CONTROL: modification of the activity of an enzyme by alteration at a site different from the active site by another molecule affecting its conformation without a covalent attachment. (See active site, conformation)

ALLOSTERIC EFFECTOR: is a molecule that is involved in bringing about allosteric control. (See allosteric control, allostery)

ALLOSTERY: a conformational change in a protein through the effect of a ligand molecule; the process is often called allosteric shift. (See allosteric control)

ALLOSYNDESIS: synapsis between non-entirely homologous chromosomes in an allopolyploid. (See homoeologous chromosomes, chromosome pairing)

ALLOTETRAPLOID: see amphidiploid

ALLOTOPIC EXPRESSION: a gene which is not organellar (mitochondrial or plastid by origin) is targeted and expressed in an organelle. (See also ectopic expression)

ALLOTYPE: difference in antibody caused presumably by allelic substitution mutation. (See isotype, immunoglobulins, antibody)

ALLOZYGOTE: an individual that at one or more loci possesses alleles which were not derived from the same common ancestor, i.e. are not identical by descent. (See inbreeding, coancestry, autozygous)

ALLOZYMES: different forms of an enzyme due to allelic differences of the genes.

ALL-WALKING APPROACH: a program used in physical mapping of DNA in connection with YACs. The STS (sequence-tagged sites) are derived from the ends of YAC inserts. Its advantages are that the position of the STS is defined compared to the case when the STS is internal, and it identifies chimeric YACs and the end-STS YACs tend to be larger than the others. (See YAC, STS)

ALLYL ALCOHOL: an eye-irritating liquid that permits positive selection of alcohol dehydrogenase mutations because the wild type cells (adh^+) in a suicidal manner convert this compound to acryl aldehyde and thus only the adh^- cells can survive. (See mutant isolation).

ALMOND (*Prunus amygdalus*): basic chromosome number x = 7 but 2x to 6x forms are known.

ALOPECIA: hair loss, baldness caused probably by an autoimmune condition occurring in different forms. In some cases it is accompanied by psychomotor epilepsy (involuntary movements) or palm and sole keratosis (callosity) and lower mental capacity. In humans it appears as autosomal dominant. (See autoimmune disease, keratosis)

ALPERT'S SYNDROME: acrocephalosyndactylia, pointed head top and fused fingers and toes, an autosomal dominant or recessive disease of humans involving the fibroblast growth factor receptor (FGFR2), defective also in the Pfeiffer syndrome. The estimated mutation rate is 3-4 x 10^{-6}. (See Pfeiffer syndrome)

ALPHA ACCESSORY FACTOR: enhances the affinity of pol α and primase for the DNA template. (See pol α, primase)

ALPHA COMPLEX: the translocation complex of chromosomes that transmits only through the females whereas the beta complex is transmitted only through the males (in *Oenothera*). (See translocation complex, complex heterozygotes)

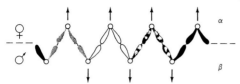

← Alternate distribution of a four-translocation complex at anaphase I (α and β)

ALPHA-FETOPROTEIN: see fetoprotein

ALPHA HELIX: is a hydrogen-bonded secondary structure of polypeptides when the polypeptide backbone is tightly wound around the longitudinal axis of the peptide bonds and the side groups of the amino acids are protruding along the generally right-handed helical structure. Most commonly each turn takes 3.6 amino acids. Frequently represented graphically also as a cylinder. (See protein structure, pitch)

ALPHAMERIC: symbols using letter identifications.

ALPHANUMERIC: a set of characters using letters and numerals and possibly other character symbols.

ALPHA PARAMETER: provides a combined estimate of the frequency of quadrivalent association (q), meiotic exchange (e), favorable anaphase distributions (d) and from these it predicts the frequency of double reduction, i.e. the production of *aa* gametes when the parental constitution is *AAAa* (see autotetraploids). The derivation of the α parameter is as follows. In a triplex the cytogenetic constitutions can be represented:

```
      A1              A3              A5              a1
  (W)======      (X)======      (Y)======      (Z)======
      A2              A4              A6              a2
```

The centromeres are represented by circled letters W, X, Y and Z, the chromatids are represented by dashed lines and the dominant and recessive alleles are numbered from *A1* to *A6* and *a1* to *a2*, respectively.

(i) In the **absence of recombination** the association of chromatids, allele and centromeres is: A1-A2, A3-A4, A5-A6 and a1-a2

In case of **recombination** between gene and centromere the following arrays are formed:

(ii) A1-A3, A1-A5, A3-A5 are the possible recombinant associations of dominant alleles which were attached originally to different centromeres

(iii) A1-a1, A3-a1, A5-a1 are the three possible dominant-recessive recombinant associations when only one chromatid originally attached to a centromere is considered in the quadrivalent

The total frequency of the gametes is 1 and the frequency of group (i) is designated as α, and the chance of each of the 4 types of associations within this group is α/4.

The combined frequency recombinant group (ii) and (iii) associations is 1 - α. Since groups (ii) and (iii) have 3 representatives, each, and combined 6, the frequency of each of the recombinant associations is (1 - α)/6.

Group (i) has 3 double dominants (A1-A2, A3-A4 and A5-A6) among the total of 4, and each has a frequency of α/4 and the combined frequency is 3 x (α/4). Group (ii) also has 3 double dominants (A1-A3, A1-A5, A3-A5) with individual frequencies (1 - α)/6 and their combined frequency is 3 x (1 - α)/6. Thus the total frequency of gametes with a dominant alleles in both chromatids is [3 x (α/4)] + [3 x (1 - α)/6] = [3α/4] + [(3 – 3α)/6]. After dividing both numerator and denominator of the second term by 1.5, we obtain [3α/4] + [(2 - 2α)/4] = (2 + α)/4 = the frequency of *AA* gametes. The frequency of the *Aa* gametes is obtained similarly.

Alpha parameter continued

Group (iii) has 3 *Aa* gametes with individual frequencies of $(1 - \alpha)/6$ and their combined frequency upon dividing numerator and denominator by 1.5 becomes $3 \times (1 - \alpha)/6 = (3 - 3\alpha)/6 = (2 - 2\alpha)/4$. The frequency of the double recessive gametes (*aa*) as shown above is $\alpha/4$.

Thus the total gametic output of the triplex $\frac{2 + \alpha}{4}$ *AA* : $\frac{2-2\alpha}{4}$ *Aa* : $\frac{\alpha}{4}$ *aa*

The practical meaning of the α parameter is best illustrated by an example. Let us assume that the cytological analysis indicates the value of $q = 0.7$, $e = 0.25$ and $d = 0.333$ and thus $\alpha = 0.7 \times 0.25 \times 0.333 \times 0.5 = 0.02914$. This value can then be substituted into the formulas of the Table below on the appropriate line.

Expected gametic frequencies of polyploids. Each term on the corresponding line has to be divided by the Divisor shown at the right column. (After Fisher, R.A. & Mather, K. 1943 Annals of Eugenics 12:1)

ZYGOTES	GAMETES			DIVISOR
	TETRASOMY			
	AA	**Aa**	**aa**	
Triplex	$2 + \alpha$	$2(1 - \alpha)$	α	4
Duplex	$1 + 2-\alpha$	$4(1 - \alpha)$	$1 + 2\alpha$	6
Simplex	α	$2(1 - \alpha)$	$2 + \alpha$	4
	HEXASOMY			
	AAA	**Aaa**	**Aaa**	**aaa**	
Pentaplex	$3 + \alpha$	$3 - 2\alpha$	α	0	6
Quadruplex	$3 + 3\alpha$	$9 - 5\alpha$	$3 + 3\alpha$	α	15
Triplex	$1 + 3\alpha$	$9 - 3\alpha$	$9 - 3\alpha$	$1 + 3\alpha$	20
Duplex	α	$3 + \alpha$	$9 - 5\alpha$	$3 + 3\alpha$	15
Simplex	0	α	$3 - 2\alpha$	$3 + \alpha$	6

Simplex : 1 *A*
Duplex: 2 *A*
Triplex: 3 *A*
Quadruplex: 4 *A*
Pentaplex: 5 *A*
Nulliplex: no *A*

The *A* value indicates the number of dominant alleles in the zygotes

Accordingly, for the simplex (line 3) AA becomes $\alpha/4 = 0.02914/4 = 0.007285$, Aa is obtained in frequency $2(1 - \alpha)/4 = 0.48543$, and $aa = (2 + \alpha)/4 = 0.507285$. Thus the double dominant gametes are expected below 1%, the *Aa* and *aa* in near equal frequency around 50%. The proportion of the double reduction (*aa* gametes) gives some indication on the relative distance of the gene from the centromere albeit not in precise units, directly convertible to map distances but approximating it. Theoretically, these calculations are elegant, unfortunately the determination of the variable components of α requires great experimental skills and very favorable conditions. Therefore the analysis of segregation in polyploids is very difficult. (See also autopolyploids, centromere mapping)

ALPHA PARTICLES: helium nuclei (contain 2 protons and 2 neutrons) emitted by radioactive decay. They release their high energy in a very short track; even in air they move only for a few centimeters. In living material they have minimal penetration yet because of their short track and high ionizing energy, they are very destructive (break chromosomes). (See ionizing radiation, linear energy transfer)

ALPHA TOCOPHEROL (vitamin E): tocopherols are plant products but required in mammals for the maintenance of fertility and for the prevention of muscle degeneration. They appear to be antioxidants for unsaturated lipids. Lipid peroxidation may result in cross-linking of proteins and may cause mutation, the appearance of age pigments (lipofuscin), etc. (See unsaturated fatty acids, fatty acids

ALPHAVIRUS: is a single, negative-strand RNA virus with 240 molecule basic capsid protein, surrounded by a lipid bilayer of 240 glycoprotein heteromeric envelope. It can infect a variety of cells and its genomic RNA is translated into non-structural proteins to begin then the replication of the viral RNA. Although it is a cytocidal virus it may be engineered into a vector for transient gene therapy or used for vaccination. This group of viruses includes the Sendai virus and the Semliki forest virus. (See Sendai virus)

ALPORT'S DISEASE: exists in different forms, autosomal dominant, recessive and X-linked

Alport's disease continued
type were described. The phenotypes vary but most commonly inflammation of the kidney(s), and deafness are present. Most probably the basic defect involves the basement membrane of the kidney glomerules and the Goodpasture antigen leading to kidney failure and hypertension. The basement membrane defect is an Xq22 coded collagen α-chain anomaly. Basically the Alport syndrome is the same as the Epstein syndrome. (See Goodpasture syndrome, collagen, basement membrane)

ALS: see amyotrophic lateral sclerosis

ALSTRÖM SYNDROME: is an autosomal recessive human defect involving obesity, retinitis pigmentosa, deafness and diabetes. Its frequency is elevated in some Louisiana and Nova Scotia populations of French origin. (See obesity, Bardet-Biedl syndrome)

AlterMap: a computer program replacing sections of the Kohara map of *E. coli* with the MapSearch alignments of the DNA fragments. (See Kohara map)

ALTERNATE DISJUNCTION: when in a translocation heterozygote each pole receives a complete set of the genetic material, and consequently it is genetically stable. (See adjacent distribution, translocations chromosomal, translocation complex)

ALTERNATION OF GENERATIONS: the cycles of haploid and diploid generations such as the gametophytic and sporophytic generations of plants. Also it is used for the cycles of sexual and asexual generations that coexist in some species. (See also life cycles, meiosis, mitosis, apomixis, parthenogenesis, fission)

ALTERNATIVE SPLICING: of mRNAs generates different protein molecules from the same genes. There are protein splicing factors to carry out these functions. Typically alternative splicing occurs in immunoglobulin synthesis (among other mechanisms) to generate a greater repertory of antibodies from a lesser number of genes. (See also splicing, introns)

ALTRUISTIC BEHAVIOR: is an evolutionary feature in animals where members of the species protect other members, especially the youngs even at their own peril. Through this behavior the survival of the population is promoted. Altruism may be manifested also in mating behavior. In a pride of animals some males may refrain from reproduction and allowing more powerful kins to mate with the females available. Apoptosis involves some elements of altruism of single cells that are suicidal to assure differentiation of a tissue or defend the organism against mechanical or biological injuries or attacks. (See kin selection, inclusive fitness, apoptosis)

Alu FAMILY: about 150-300 kb long nucleotide sequence monomers associated head-to-tail, and repeated about 300,000-500,000 times in the primate genome. These nucleotide sequences are cut by the Alu I restriction enzyme (recognition site AG↓CT) and hence the name of these gene families. Members of this family are considered also as transposable elements. The Alu sequences are specific for the human genomes but homologs appear in other mammals. The Alu sequences appear to have evolved from the 7SL RNA genes about 15-30 Myr ago. (See SINE, Myr)

Alu-EQUIVALENT: is a group of genomic sequences similar to the Alu family. (See Alu family)

ALZHEIMER'S DISEASE: a presenile-senile dementia (loss of memory and ability of judgment as well as general physical impairment) involves the accumulation of amyloid protein plaques in the brain resulting in degeneration of neurons. This protein, amyloid-β peptide (Aβ) comes from a larger amyloid precursor protein (βAPP) that is synthesized in normal brain and processed in a number of ways and encoded in chromosome 21q as an early onset dominant. The largest protein spans the cell membrane. One of the extracellular domains is a protease inhibitor. In normal cells this domain may be released but in the diseased cells the amyloid protein is processed incorrectly. Genes responsible for the amyloid protein synthesis have been cloned and others mapped to human chromosomes 1q31-42 (encoding STM2, a seven-transmembrane integral protein) and chromosome 14q24.3 encodes protein S182 which is 67% homologous to STM2). Human chromosome 19 encodes apolipoprotein E (APOE) that modifies the onset of Alzheimer disease. In chromosome 21q (early onset, dominant encodes βAPP), and is assumed to be involved with the disease. Aβ is the major component of the brain plaques and its ligand

Alzheimer disease continued

is a ≈ 50K relative molecular weight protein, identical to RAGE (receptor for advanced glycation endproduct) or AGE and for amphoterin controlling neurite outgrowth (an inflammatory process). RAGE mediates the interaction of Aβ with endothelial cells and neurons causing oxidative stress. Its interaction with microglia results in cytokine production, chemotaxis and binding movements. These processes can explain the dysfunction and death of neurons. The sporadic Alzheimer disease may be associated with the very low-density lipoprotein (VLDL) receptor gene. Alzheimer disease is common among Down syndrome (due to chromosome 21 trisomy) individuals. Some psychotropic drugs (affecting the nervous system) may alleviate some of the symptoms. At least four genes seem to be capable to cause Alzheimer symptoms in humans. The incidence of Alzheimer disease (AD) increases from 0.1% below 70 to double after age 80. The risk for first degree relatives varies from 24 to 50% by age 90. The concordance rate of monozygotic twins is 40-50% and among dizygotic twins 10-50%. (See β-amyloid, also behavior human, prion, Creutzfeldt-Jakob disease, scrapie, encephalopathy, mental retardation, Down's syndrome, corticotropin releasing factor, presenilins, AGE, microglia, NF-κB, presenilin, LDL, VLDL)

AMANITIN ($C_{39}H_{54}N_{10}O_{13}S$): see α-amanitin.

AMARANTHS: subtropical or tropical American seed plants, 2n = 2x = 32.

AMASTIGOTE: see *Trypanosoma*

AMATOXINS: bicyclic octapeptides (e.g. α-amanitin) produced by the fungus *Amanita phalloides* inhibit the function of RNA polymerase II (occasionally pol III) of eukaryotes but they do not affect the transcriptases of the prokaryotic type, e.g. those in the cytoplasmic organelles of eukaryotes. (See RNA polymerase, RNA replication, transcription, pol III)

AMAURIS: African species of butterflies are models to mimic by another butterfly *Papilo dardanus*. The purpose of the mimicking is that the models are distasteful to predators and thus the mimicking species improves its survival. (See Batesian mimicry)

AMAUROSIS CONGENITA: is an autosomal recessive blindness or near-blindness caused by a defect of the cornea (keratoconus). About 10% of the blinds are suffering from it. A dominant form of this disease is due to deficiency of the retina. The Leber's congenital amaurosis (LCA) has been mapped to human chromosome 17p13.1. (See eye diseases)

AMAUROTIC FAMILIAL IDIOCY (AFI): the old name of the Tay-Sachs disease. (See Tay-Sachs disease, Batten disease)

AMBER: a fossil tree resin up to millions of years old. It is hardened and became resistant to most environmental factors. Frequently it contains microbes, plants or animals or residues of them in usually well preserved state and thus may provide very valuable information on old organismal specimens, including the genetic material. (See ancient DNA)

AMBER: chain-terminator codon (UAG)

AMBER MUTATION: generates a chain-termination polar effect (the name has nothing to do with function rather it was named after Felix Bernstein whose German family name translates into amber). (See code genetic, polar mutation)

AMBER SUPPRESSOR: mutations in the anticodon triplet (3'-AUC-5') so that the amber mutation (5'-UAG-3') may be read as a tyrosine codon and thus the translation is not terminated. (See *supC, supD, supE, supF, supG, supU*)

AMBIDEXTROUS: see handedness

AMBIENT SIGNALS: the position of a particular cell determines its response to particular environmental stimuli.

AMBIGUITY IN TRANSLATION (mistranslation): may be brought about by antibiotics, or modification of the tRNA or the ribosomes and consequently an amino acid different from the correct one is incorporated into the nascent polypeptide. (See error in aminoacylation)

AMBIGUITY OF RESTRICTION ENZYMES: they can cut more than a single sequence, although with not equal efficiency, e.g. Hind I: GTT↓GAC, GTT↓AAC, GTC↓GAC.

AMELIA: see limb defects in humans

AMELIORATION OF GENES: DNA sequences incorporated into a genome by horizontal transfer tend to adapt to the codon usage of the recipient organisms during evolution. (See transmission, codon usage)

AMELOGENESIS IMPERFECTA: the autosomal dominant forms involve softness of the tooth enamel caused by lack of calcium. Similar symptoms but with calcium deposits in the kidneys and other variations indicated autosomal recessive inheritance. Two X-linked forms are distinguished. One of them is very similar in phenotype to the autosomal dominant form. In one, the thickness of the enamel is about normal but soft, in the other the enamel is hard but very thin. (See entries under tooth)

AMELOGENIN TEST: is a forensic sex typing tissue test.

AMENORRHEA: absence of menstruation may be caused by physiological factors (over- or underweight, pregnancy), hormonal, age, disease factors or by genetic causes such as pseudo-hermaphroditism, Turner syndrome, etc. (See hermaphroditism, Turner syndrome)

AMENSALISM: one organism is inhibited by another which is unaffected by this relationship.

AMERICAN TYPE CULTURE COLLECTION (ATCC): maintains and catalogues microbial stocks and other cultured cells.

AMES TEST: a bacterial assay based on backmutation of different histidine-requiring strains of *Salmonella typhimurium*. The reversions are capable of detecting various types of base substitutions and frameshift. A single plate generally detects mutations in 100,000 or more cells.

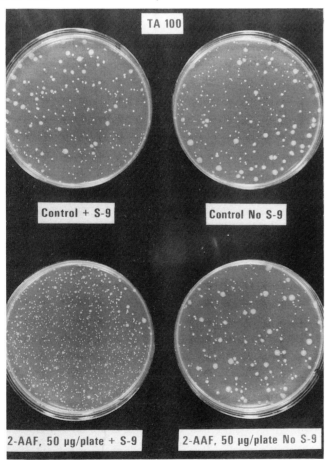

THE AMES TEST WITH AND WITHOUT THE USE OF THE MICROSOMAL (S-9) FRACTION. IT IS CLEAR THAT THE MICROSOMAL ENZYMES DO NOT AFFECT THE FREQUENCY OF REVERSIONS WITHOUT 2-AAF (2-ACETYL-AMINOFLUORENE). ALSO 2-AAF WITHOUT ACTIVATION IS NOT MUTAGENIC. THE S-9 FRACTION IS PREPARED, USUALLY, FROM RODENT LIVER HOMOGENATE AND IN ITS PRESENCE, THE PROMUTAGENS CAN ALSO BE ASSAYED. THE BACTERIA LACK THE ACTIVATING ENZYME COMPONENT.

In some strains the *his*⁻ genes are present in multicopy plasmids to further enhance its targets. The bacteria carry also mutations that interfere with genetic repair. The testing medium includes microsomal fractions of mammalian liver that can activate promutagens into ultimate mutagens. Thus the mutagenic effectiveness of the majority of chemicals may be increased by three orders of magnitude. The results of this assay are highly correlated with carcinogenecity of the compounds yet it requires only two days compared with the rodent tests, taking possibly several

Ames test continued

months for complete evaluation. It is also inexpensive and permits the evaluation of large number of compounds at low cost. (See bioassays in genetic toxicology, reversion studies in *Salmonella* and *E. coli* in genetic toxicology, microsomes, base substitution mutation, frameshift mutation, activation of mutagens)

AMETHOPTERIN: an inhibitor of dihydrofolate reductase, an important enzyme in the *de novo* biosynthetic pathway of purine and pyrimidine nucleotides. Synonymous with methotrexate, used as an antitumor drug and a selective agent in genetic transformation. (See aminopterin)

AMIDE BOND: a carbonyl group linked to an amine, e.g., $R-\overset{\overset{\displaystyle O}{\|}}{C}-NH_2$ (See peptide bond)

AMILORIDE ($C_6H_8ClN_7O$): a potassium-sparing diuretic, regulating K^+ and Na^+ balance in the cells. (See ion channels)

AMINO ACID: a protein building block. There are 20 natural amino acids. (See amino acids)

AMINO ACID ACTIVATION: see aminoacylation

AMINO ACID ANALYZER: an automated equipment similar to a high pressure liquid chromatography apparatus to separate and quantitate the amino acid composition of protein digests. (See chromatography)

AMINO ACID METABOLISM: amino acids are derived from the compounds in the glycolytic, the citric acid and the pentose phosphate pathways. The biosynthetic systems in different evolutionary categories may vary. Bacteria and plants are normally able to synthesize all the 20 primary amino acids whereas animals depend on the diet for the *essential amino acids*. Genetics of microorganisms played an important role in elucidating the pathways. Single gene mutations generate special requirement for all amino acids and these requirement can be met by feeding the appropriate precursor. In higher plants, auxotrophy only for very few amino acids exist, probably because amino acids may be synthesized by parallel pathways or functionally duplicated genes. In humans and other mammals, genetic defects are known that affect one way or another all the natural amino acids and many of their derivatives and thus cause inborn errors of metabolism. (See argininemia, citrullinemia, ornithine decarboxylase, ornithine aminotransferase, ornithine transcarbamylase, alanine aminotransferase (glutamate-pyruvate transaminase), alaninuria, aspartate aminotransferase [glutamate oxaloacetate transaminase], asparagine synthetase, aspartoacylase deficiency, cystinuria, cystinosis, cystathionuria, homocystinuria, cystin-lysinuria, glutamate synthesis, glutamate decarboxylase, glutamate dehydrogenase, glutamate formiminotransferase deficiency, glutamate pyruvate transaminase, glutamate oxaloacetate transaminase, glutaminase, glycine biosynthesis, glycinemia, methylmalonicaciduria, vitamin B_{12} defects, histidine operon, histidase, histidinemia, isoleucine-valine biosynthetic pathway, isovalericacidemia, 3-hydroxy-3-methylglutaryl CoA lyase deficiency, leucine metabolism, methylcrotonylglycinemia, methylglutaconicaciduria, hydroxymethylglutaricaciduria, lysine biosynthesis, hyperlysinemia, dibasicaminoaciduria, methionine biosynthesis, methionine adenosyl transferase deficiency, methionine malabsorption, phenylalanine, phenylketonuria, proline biosynthesis, hyperprolinemia, serine, threonine, tryptophan, tyrosine, alkaptonuria, valine, hypervalinemia, urea cycle, sarcosinemia, carnosinemia)

AMINO ACID REPLACEMENTS: take place by base substitution in the codons, e.g. a glutamic acid (GAA) residue may be replaced by glutamine (CAA), lysine (AAA), glycine (GGA), valine (GTA), alanine, (GCA), aspartic acid (GAT), and so on. The rate of amino acid substitution per site in protein had been estimated to be on the average 10^{-9}/year during evolution. This average may vary by 3-4 orders of magnitude among different proteins. (See also PAM)

AMINO ACID SEQUENCING: can be carried out in different ways. Today most frequently the putative amino acid sequence is deduced indirectly from the codon sequences in the DNA. Direct estimates can be obtained from polypeptides cleaved by proteolytic enzymes (trypsin, chymotrypsin, pepsin, and various other proteases, cyanogen bromide) to obtain manageable smaller fragments of proteins. These agents have preferences for certain cleavage points, represent-

Amino acid sequencing continued

ed by particular amino acids. Also, chemical breakage of disulphide bonds is utilized. The Edman degradation is then employed that is using end labeling and removal of single amino acids at a time. Eventually, the sequenced fragments must be ordered on the basis of overlapping ends. (See Edman degradation, amino acid analyzer, sequenator, DNA sequencing, databases)

AMINO ACID SYMBOLS IN PROTEIN SEQUENCES: alanine A, aspartic acid or asparagine B, cysteine C, aspartic acid D, glutamic acid E, phenylalanine F, glycine G, histidine H, isoleucine I, lysine K, leucine L, methionine M, asparagine N, proline P, glutamine Q, arginine R, serine S, threonine T, valine V, tryptophan W, unknown X, tyrosine Y, glutamic acid or glutamine Z. (See amino acids)

AMINO ACIDS: relatively simple yet diverse chemical compounds that all have at least one NH_2.

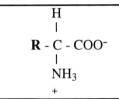

☛ Where **R** can be a *non-polar aliphatic* group: glycine (Gly), alanine (Ala), valine (Val), leucine (Leu), isoleucine (Ile), proline (Pro), *aromatic:* phenylalanine (Phe), tyrosine (Tyr), tryptophan (Trp), *polar uncharged:* serine (Ser), threonine (Thr), cysteine (Cys), methionine (Met), asparagine (Asn), glutamine (Gln), *negatively charged:* aspartic acid (Asp), glutamic (Glu), *positively charged:* lysine (Lys), arginine (Arg), histidine (His). About 20 natural amino acids are the building blocks of proteins. Some amino acids are modified in certain types of proteins. Cysteine and methionine always contain sulphur. The α-amino acids have both the amino and carboxyl group(s) attached to the same C atom. (See amino acid symbols in protein sequences, essential amino acids, nonessential amino acids)

AMINOACIDURIAS: diverse groups of hereditary diseases are characterized by urinal excretion of cystine (cystinosis), tyrosine (tyrosinemia), all kinds of amino acids (fructose intolerance), very large quantities of primarily threonine, tyrosine and histidine (Hartnup disease), hypervalinemia. Many diseases of the kidneys show excessive amino acid excretion. (See homocystinuria, Fanconi renotubular syndrome, Hartnup disease, also neuromuscular diseases, Rowley-Rosenberg syndrome)

AMINOACYLATION: to the acceptor arm (CCA-OH) of tRNA an amino acid is attached by its NH_2 end by an ATP-dependent enzymatic process. Rate of mischarging is 3×10^{-3}. Although this reaction requires a protein enzyme, a ribozyme may also be adapted to carry out aminoacylating function in a manner analogous to the ribozyme, peptidyl transferase. (See tRNA, amino acid-tRNA synthetase, ribozyme, aminoacyl-tRNA synthetase)

AMINOACYL- tRNA: an amino acid-charged tRNA at the 3' end. (See tRNA, protein synthesis, aminoacyl-tRNA synthetase, amino acylation)

AMINOACYL-tRNA SYNTHETASE: enzymes carrying out aminoacylation of tRNA. The first step is the attachment of the amino acid to the α–phosphate group of an ATP molecule, that is accompanied by the removal of an inorganic pyrophosphate group. The aminoacyl adenylate is bound then to the active site of one of the two types of aminoacyl-tRNA synthase enzymes. Class I enzyme handles Arg, Cys, Gln, Glu, Ile, Leu, Met, Trp, Tyr and Val. Class II enzyme is involved with Ala, Asn, Asp, Gly, His, Lys, Phe, Pro, Ser, Thr (for these abbreviations see amino acids). Class I synthase first attaches the aminoacyl-A to the 2'-OH of the terminal A of the amino arm of tRNA and subsequently it is moved to the 3'-OH by trans esterification. The class II enzymes bypass the 2'-OH transfer step. The enzymes recognize among the 40-80 or more tRNAs the appropriate carriers for each amino acid and this rather complex recognition process is directed by the so called *second genetic code*. The recognition of the proper tRNAs is determined by several sites on the tRNA, most importantly by the anticodon. In *E. coli* the anticodon is the most important for recognition for 17 of the 20 amino acids. For many of the isoaccepting tRNAs the 73. position of the amino acid accepting arm is also very important along with the anticodon. The enzyme capable also to correct errors in recognition, e.g. isoleucyl-tRNA cannot entirely prevent valine to attach to its binding site and may form a valyl-adenylate. This activated valine cannot, however, attach to either $tRNA^{Val}$ or

Aminoacyl-tRNA synthetase continued

tRNA^{Ile} rather it is hydrolyzed by tRNA^{Ile}, so no erroneous valyl-tRNA^{Ile} is formed. The overall mistakes in amino acid incorporation is about 1/10,000 residues. The aminoacyl-tRNA synthetases of the organelles are encoded by nuclear genes. The enzymes are organelle-specific, however. Some nuclear genes may encode both types of enzymes by differential transcription and processing. The reaction, catalyzed by the aminoacyl-tRNA synthetase can be catalyzed also by a reactive RNA. The aminoacyl-tRNA synthetases of higher eukaryotes form multiprotein complexes. (See arginyl t-RNA synthetase, glutamyl-tRNA synthetase, histidyl tRNA synthetase, leucine t-RNA synthetase, threonyl tRNA synthetase, methionyl tRNA synthetase, tryptophanyl tRNA synthetase, valyl tRNA synthetase, ribozyme)

AMINOBENZYLOXYMETHYL PAPER: is a diazotized (using 1-[(m-nitrobenzyloxy) methyl] pyridinium chloride [NBPC]) Whatman 540 or other comparable paper used for Northern blotting. (See Northern blotting)

AMINO END: of a protein is where the synthesis started on the ribosome. It is commonly a methionine residue although during processing of the protein the first amino acid(s) may be removed. The amino end of the polypeptide corresponds to the 5' end of the mRNA. (See also amino terminus)

AMINO GROUP: is derived from ammonia (NH_3) by replacing one of the hydrogens by another atom (H_2N—).

AMINOGLYCOSIDE PHOSPHOTRANSFERASES: (NPTII, aph(3')II): phosphorylate aminoglycoside antibiotics and cause resistance against these antibiotics. The genes for the two related enzymes were isolated from Tn*5* and Tn*60* bacterial transposons, respectively, and are used as dominant selectable markers (with appropriate promoters) in transformation of animal, and plant cells. (See kanamycine resistance, geneticin resistance, neo^R, neomycinphosphotransferase)

AMINOGLYCOSIDES: a group of antibiotics with a common characteristic of a cyclic alcohol in glycosidic linkage with amino-substituted sugars. They (streptomycin, kanamycin, etc.) affect the A site of the ribosomes and thus interfere with protein synthesis. (See ribosome, protein synthesis)

AMINOLEVULINIC ACID (δ-aminolevulinic acid, ALA): a precursor of porphyrin, required for the production of hemoglobin and chlorophylls. . The ALA dehydratase (ALAD) is coded in human chromosome 9q34. (See chlorophyll, hemoglobin, porphyrin)

AMINOPTERIN: inhibits the activity of dihydrofolate reductase at 10^{-8} to 10^{-9} M concentrations. This enzyme is required for the biosynthetic pathway of both pyrimidine and purines. This drug is also used in the HAT medium to shut down the *de novo* synthetic pathway of nucleotides when thymine kinase and hypoxanthine-guanine phosphorybosyl transferase mutations are screened for in mammalian cell cultures. (See also amethopterin, HAT medium)

2-AMINOPURINE (AP): an adenine analog that may incorporate into DNA in place of adenine and can form normal hydrogen bonds with thymine but it is prone to mispair with cytosine either with a single hydrogen bond in its normal state or after tautomeric shift with two hydrogen bonds. The mispairing may result in a replacement of an AT pair by a GC pair and thus result in mutation. AP may be highly mutagenic in some prokaryotes but not in eukaryotes. (See base analogs, base substitution, hydrogen pairing)

← 2-Aminopurine

AMINOTERMINAL: the only amino acid in a polypetide chain with a free α-amino group and it occurs at the end of the chain. (See also amino end)

AMINOTRANSFERASE: transaminase enzymes that transfer α-amino groups from amino acids to α–keto acids.

3-AMINO-1,2,4-TRIAZOLE: a carcinogenic standard (non-mutagenic in the Ames test); now a banned herbicide. (See Ames test)

AMISH: is a Mennonite religious group with strict and conservative principles and lifestyle. Their communities are relatively isolated from the surrounding populations. Because of this, consanguinous marriages are higher and certain genetically determined conditions are relatively frequent. The recessive Ellis-Van Creveld syndrome, pyruvate kinase deficiency, cartilage-hair hypoplasia, limb-girdle muscular dystrophy, and Christmas disease, are relatively common. The Amish brittle hair syndrome (also recessive) involving short stature, somewhat lower intelligence, brittle hair and reduced fertility, low sulfur content of the nails was first recognized in such a population. (See Ellis-Van Creveld syndrome, Christmas disease)

AMITOSIS: nuclear divisions without the characteristic features of the mitotic apparatus and involving the small (21 to 1,500 kb) acentric chromosomes in the macronucleus of some *Protists*. No mitotic spindle is evident and the nuclear membrane seems intact during the entire division. The distribution of the chromatin is not entirely random nevertheless. (See mitosis, *Paramecia*, fission, acentric, chromatin)

AMIXIS: term used in fungal genetics for apomixis. (See apomixis)

AML1: acute myeloid leukemia oncogene, a DNA-binding protein, encoded in human chromosome 21q22. (See leukemias)

AMMONIFICATION: release of ammonia upon decomposition of compounds (e.g. amino acids)

AMMUNITION: gene tagging with non-autonomous P elements of *Drosophila* that stays put after the removal of the helper (complete) element. (See hybrid dysgenesis, smart ammunition)

AMNIOCENTESIS: prenatal diagnosis of the genetic constitution of a fetus by withdrawing fluid or cells from the abdomen (amniotic sac) of a pregnant woman. This procedure is applicable after about 16 weeks of the pregnancy by the time the amount of the amniotic fluid is sufficient. The tests can be cytological, enzymological, immunological or molecular and may involve cell cultures to amplify the material. Amniocentesis can be used also for genetic counseling. Normally it entails minimal risk to either the fetus or the mother yet it should be used only in cases when it is warranted by other parts of the diagnoses. (See risk, counseling genetic, prenatal diagnosis, PCR, polymerase chain reaction)

AMNION: a strong membrane enveloping the mammalian embryo and fetus containing the am-

Amnion continued

niotic fluid that helps in protecting the embryo during the entire period of development until delivery. A similar membrane is found in other animals too. The amnion is closest to the embryo, covered by the allantoic mesoderm and the outer layer is the chorion. (See chorion, allantois)

AMOEBA: free-living or parasitic single-cell eukaryote. Some of the amoebae crawl by forming pseudopodia (leg-like extensions of the single cell). (See nuclear transplantation)

AMORPH ALLELE: is inactive (may be a deletion). (See allele)

amp: see ampere

Amp (ampicillin, 6-[D(-)α-aminophenyl acetamid]-penicillinic acid): member of the penicillin family antibiotics. The *Amp^R* genes are common in genetic vectors. (See antibiotics, vectors genetic)

AMP: adenosine 5'-monophosphate (adenylic acid); when additional 2 phosphates are added ATP is formed. (See also cAMP)

AMPA (α-amino-3-hydroxy-5-methyl-4-isoaxozolpropionate): a member of the glutamate receptor family of proteins, and it mediates the excitatory synaptic transmissions in the brain and the spinal cord. It also controls post-synaptic influx of Ca^{2+}, further regulating synaps. (See synaps, excitatory neurotransmitters)

AMPERE (A): 1 A = 1 C/sec. 1C (Coulomb) = 1 As (Amperesecund). (See Volt, Watt)

AMPHIBOLIC PATH: of metabolism involves both anabolic and catabolic reactions.

AMPHIDIPLOID: contains two genomes from at least two different species; it is obtained by doubling the number of chromosomes of amphiploids. (See amphiploid, chromosome doubling)

AMPHIGAMY: in the usual type of fertilization the gametic nuclei fuse. (See also dikaryon)

AMPHIHAPLOID: haploid cell of an amphidiploid, an allohaploid. (See haploid, amphidiploid)

AMPHIMIXIS: sexual reproduction. (See also apomixis)

AMPHIPATIC: has a charged and a neutral face (some proteins), structures that have hydrophilic and (polar) hydrophobic (non-polar) surfaces, e.g. lipids.

AMPHIPHYSIN: nerve protein of the synaptic vesicle bound to synaptotagmin. (See synaptotagmin, BIN1)

AMPHIPLOID: contains at least two genomes from more than one species. (See also amphidiploid, allopolyploid)

AMPHIPROTIC: can donate or accept protons and thus can behave weakly acid or alkalic, e.g. water or amino acids. (See proton, amino acids)

AMPHIREGULIN: regulator with both (+) and (-) effects. It regulates the proliferation of keratocytes and some fibroblasts and inhibits the proliferation of various tumor cells. It competes for the epidermal growth factor (EGF) receptor. (See EGFR, keratosis)

AMPHISTOMATOUS: leaves bearing stomata on both surfaces. (See stoma)

AMPHITHALLISM: homoheteromixis; both self-fertilization and outcrossing occurs in fungi.

AMPHITROPIC MOLECULE: carries out two different functions at different sites.

AMPHOLINE: is an ampholyte used for polyacrylamide, agarose and dextran gels, density gradient stabilizing in analytical and preparative electrofocusing. (See isoelectric focusing)

AMPHOLYTE: is an amphoteric electrolyte. (See amphoteric, electrolyte)

AMPHOTERIC: has dual, opposite characteristics such as behaving both as an acid and a base.

AMPHOTERINE: see Alzheimer disease

AMPHOTROPIC RETROVIRUS (polytropic retrovirus): replicates both in the cells from where it was isolated as well as in other types. (See also ecotropic and xenotropic retroviruses)

AMPICILLIN: an antibiotic that binds to bacterial cell membranes and inhibits the synthesis of the cell wall. The ampicillin resistance gene (*amp^r*) codes for a β-lactamase enzyme that detoxifies this antibiotic; the *Amp^r* gene is used also as marker for insertional inactivation and the concomitant ampicillin susceptibility. (See antibiotics. insertional mutation, pBR322)

AMPLICON: a DNA fragment produced by polymerase chain reaction (PCR) amplification. Also the amount of DNA present in an amplified gene or chromosomal segment. (See PCR)

AMPLIFICATION: temporary synthesis of extra, functional copies of some genes *in vivo*, or *in vitro* by the use of some forms of the polymerase chain reaction. Bacteriophage λ can be amplified by a series of nitrocellulose filter transfers after *in situ* hybridization. The addition of chloramphenicol (10-20 µg/mL) to pBR322 and pBR327 may amplify plasmid yield if the synthesis of protein is not completely prevented. Cosmid libraries may be amplified by starting on solid plates followed by liquid cultures. Replica plating can amplify animal cell cultures. Approximately 5×10^4 colonies can be accommodated on a 138 mm filter and this way about 30 filters are required to obtain a representative library of overlapping fragments. DNA amplification can occur in a *genetically programmed and predetermined* manner in eukaryotes. E.g. in the ovarian follicle of *Drosophila* large quantities of an egg shell protein is needed during oogenesis. The need is met by a disproportionately favorable replication of the chorion gene clusters in the X-chromosome and chromosome 3. DNA replication is initiated bidirectionally at a replicational origin and it generates multiple copies of the genes needed. A distance away the replication tapers off and the flanking regions are amplified less and less in proportion to the distance from the origin. Similar programmed amplification takes place in the ribosomal genes of amphibia during intense periods of protein synthesis in embryogenesis. The approximately 500 - 600 genomic copies of rRNA genes may be increased by a factor of 1000. The replication of detached DNA sequences follows a rolling circle type process, and the new DNAs (in about 100 rDNA repeats) are separately localized in micronuclei. The replicates of these nuclei are structurally similar to each other, indicating that they are the clonal products of a single replicating unit, but the new micronuclei generated in different cells may not be the same as judged by the length differences in the intergenic spacers. Ribosomal DNA amplification takes place during the amitotic divisions of the protozoon, *Tetrahymena*. Here again the macronuclear rDNA copies may be selectively amplified in the 10^4 range whereas in the micronuclear DNA there is only a single rDNA gene. A *genetically non-programmed amplification* takes place in several mutant cell lines to correct mutational defects. Enzyme deficiencies may be compensated for by producing multiple copies of the genes controlling low efficiency enzymes. Transfection of ADA genes to mammalian cells may be amplified in the presence dCF (see adenosine deaminase); mammalian cells can be amplified if they are co-transfected with the *dhfr* (conveying methotrexate resistance) gene and other desired sequences. In the presence of methotrexate the *dhfr* genes as well the flanking DNA may be amplified 1000 fold. The amplified DNA, in stable lines, is integrated into the chromosome in *homogeneously stained regions* (HSRs). In unstable cell lines, *dhfr* is in autonomously replicating elements, called double-minute chromosomes (DMs) that have no centromeres and can be maintained only in methotrexate containing cultures. Some general feature of amplifications are: (i) Expansion of a particular locus and flanking regions or the generation small supernumerary chromosomes called double minutes that contain the critical gene, (ii) the amplified unit may undergo rearrangements, (iii) the amplified sequences are not all identical and may change but these changes are somewhat unusual because a larger number of copies may be altered simultaneously in an identical manner. In vivo *amplification of genes during evolution* may account for the presence of gene families. Some of the amplified genes when the larger copy number was no longer advantageous, may have acquired new functions without entirely losing their structural similarity to the ancestral sequences. Other members of the amplified group lost their function(s) through deletions and mutations and became pseudogenes. (See PCR, nitrocellulose filter, *in situ* hybridization, chloramphenicol, pBR322, cosmid library, oogenesis, chorion, bidirectional replication, rolling circle, micronucleus, ADA, HSR, methotrexate, pseudogene)

AMPLIFICATION CONTROL ELEMENTS: amplification of genes in chromosome 3 and the X-chromosome of *Drosophila* is determined by less than 5 kbp DNA sequences that normally occur in the vicinity of the genes that are amplified under natural conditions of the genome

Amplification control elements continued
(e.g. the chorion protein gene). If these control elements are isolated, inserted into genetic vectors (P-elements) and reintroduced at random sites into the *Drosophila* genome, they may amplify other sequences in their new neighborhood. (See amplification, hybrid dysgenesis)

AMPLITAQ™: Taq DNA polymerase, a single polypeptide chain enzyme with minimal secondary structure, from the bacterium *Thermus aquaticus*. Its temperature optimum is about 75° C but can withstand ≤95° C without great loss of activity. It lacks intrinsic nuclease function but has a polymerization-dependent 5' → 3' exonuclease activity. It is a prefered enzyme for PCR. (See PCR, DNA polymerase, exonuclease)

AMPLITYPE: see DNA fingerprinting

AMPUTATIONS: see ADAM complex, limb defects

AMV ONCOGEN (*v-amv*): see MYB

α–AMYLASE: hydrolyzes α-1-4 glucosidic linkages of amylose, amylopectin and other carbohydrates and yields maltose, α-dextrin, maltotriose. β-amylase hydrolyzes starch into maltose. The human AMY genes are in chromosome 1p21.

AMYLOID ANGIOPATHY: see amyloidosis type VI

AMYLOIDOSIS: involves extracellular deposition of variable amounts of amyloids, a special fibrous glycoprotein of the connective tissues. Some of the familial nephropathies (kidney diseases), heart diseases, and neoplasias involve amyloidosis. Genetically these are inhomogeneous groups of diseases mainly with dominant, some with recessive pattern of inheritance. Amyloidosis type VI involves high incidence of hemorrhages due to accumulation of amyloids. The afflicted individuals (dominant) are low in cystein proteinase inhibitor, cystatin C, encoded in the region of human chromosome 20q13. (See cold hypersensitivity, amyloid, Mediterrenean fever, β-amyloid, Alzheimer disease)

AMYLOPECTIN: normally a minor variant of common starch. While starch (amylose) is an unbranched chain of D-glucose units of α1-4 glycosidic linkages, amylopectin contains also at every 24 to 30 residues branch points in α-1-6 linkages. Most commonly cereal grains contain amylose as the principal storage polysaccharide, recessive mutations may cause the predominance of amylopectin (dextrin). These two types of starches are easily distinguished *in situ* by a drop of iodine solution (I_2 0.12 g + KI 0.4 g in 100 mL H_2O); amylose stains blue-black while amylopectin appears red-brown. The amylose content of corn is desirable for the film and fiber manufacturing industry. Several genes (*ae, du*) may substantially increase the amylose content relative to that of amylopectin.

AMYLOPLAST: are plastids with a primary role of starch storage

AMYLOSE: see amylopectin

AMYOTROPHIC LATERAL SCLEROSIS (ALS, Lou Gehrig disease): hardening of the lateral columns of the spinal cord with concomitant muscular atrophy that may spread and may cause death in a few years after onset. According to a mouse model it is probably caused by a defect in the enzyme Cu, Zinc Superoxide Dismutase (SOD). This enzyme breaks down superoxide radicals (highly reactive compounds) to less reactive products although it may form other types of free radicals. The gene involved in the disease symptoms is a dominant "gain of function" mutation within the area 21q22.1-q22.2. The syndrome in different forms occurs at a frequency of about 1×10^{-5}. It was named after baseball infielder Henry Louis (Lou) Gehrig who was elected to the US National Hall of Fame in 1939, and suffered from this condition. ALS is sometimes associated with Parkinson's and Alzheimer disease-like phenotypes. This form may be caused or aggravated by nutritional factors (neurotoxins in the food, low calcium and magnesium uptake). A recessive autosomal type has an early onset from 3 to 20 years of age was assigned to human chromosome 2. (See neuromuscular diseases, Alzheimer's disease, Parkinson's disease)

ANABASINE (neonicotine): is an alkaloid in chenopods and solanaceous plants; it is highly toxic (LDlo orally 5 mg/kg for humans). (See LDlo)

ANABOLISM: energy-requiring synthetic processes of the cellular metabolism.

ANAEROB: organisms that live without atmospheric (free) oxygen.
ANAEROBIC: a process in the absence of (air) oxygen.
ANAGENESIS: an evolutionary change within a line of descent. (See also cladogenesis)
ANALBUMINEMIA: is a human chromosome 4 recessive absence or reduction of albumin from the blood serum that is not accompanied by very serious ailments although fatigue, mild anemia and mild diarrhea may be associated with it. (See albumins)
ANALGESIC: a medication alleviates pain without losing consciousness.
ANALOGOUS GENES: have similar function without common evolutionary descent. (See homologous genes)
ANALOGUE: a chemical compound similar to a natural one but it may or may not function in the metabolism or may even block the function of a normal metabolite or the enzyme involved.
ANALOGY: similarity is not based on common origin. (See homology, convergent evolution)
ANALYSIS OF VARIANCE: statistical method for detecting the components of variance. It is used for the evaluation of differences of experimental data involving different treatments. The square root of the quotient of the sum of squares of the variates and the mean square of the error variance is equal to t, and the corresponding probability, at each degree of freedom, can be read from a t-distribution table. Analysis of variance used also in calculating heritability by intraclass correlation. (See intraclass correlation, t-distribution)
ANAPHASE: in *mitosis*, at anaphase the centromere of the chromosomes split and that makes it possible for the spindle fibers to pull the two identical chromatids toward the opposite poles. This assures the genetic identity of the daughter cells. In *meiotic* anaphase I the centromeres do not split and the chromatids are held together as they move toward the poles. This is thus the means for the reduction of chromosome number. The anaphase II of meiosis essentially resembles anaphase in mitosis. The chromosome movements are mediated by microtubules and special motor proteins. (See also meiosis, mitosis, microtubules, motor protein)
ANAPHYLATOXINS: fragments released during activation of the serum complement (C) proteins of the antibodies. These activation peptides are called anaphylatoxins because they may elicit reactions similar to anaphylactic shock (violent reaction to antibodies and/or haptens that may be fatal). These fragments also cause contraction of the smooth muscles, release of histamine, other vasoactive amines, lysosomal enzymes, enhance vascular permeability. (See antibody, complement)
ANAPHYLAXIS: a rapid serological (antigen - antibody) reaction of an organism to a foreign protein. Either the crystalline fragment of the antibody (Fc) or the complement is involved. Prior sensitization may make the reaction quite violent and may cause death. (See immune system, antibody, complement)
ANAPLASIA: dedifferentiation.
ANAPLASTIC LYMPHOMA (large-cell non-Hodgkin lymphoma): is a lymphoma of children caused a 2p23:5q35 chromosomal translocation fusing a protein tyrosine kinase gene, ALK, to the nucleolar phosphoprotein genes (NPM). The resulting anomaly affects the small intestine, testis and brain but not the lymphocytes. Alk is related to the insulin receptor kinases and may eventually cause malignancies. (See leukemias, lymphoma, Hodgkin disease)
ANASTRAL SPINDLE: a mitotic spindle without asters, such as in higher plants. (See aster)
ANCHOR CELL: a gonadal cell of *Caenorhabditis* that induces neighboring cell's development into vulval opening. (See *Caenorhabditis*)
ANCHOR LOCUS: a gene with well known map position that can be used as a reference point for mapping new genes. (See anchoring)
ANCHORAGE DEPENDENCE: normal mammalian cells grow in culture in a monolayer attached to a solid surface; cancer cells are not contact-inhibited and pile up on each other. It appears that the suppression of cyclin E-CDK2 activity is required for cell anchorage. In transformed fibroblasts the cyclin E-CDK2 complex was active regardless of anchorage. (See CATR1, AIG, cyclins, cancer cells)
ANCHORING: the DNA fragments obtained during the initial stages of physical mapping must

Anchoring continued

be tied together by contigs. For the establishment of contigs large capacity YACs are used. These YACs must be correlated to molecular markers (anchors) along the length of the chromosome. Such anchors may be RFLPs, RAPDs, STSs and even the recombination maps obtained by strictly genetic methods. The relative position of two YACs is revealed when a YAC is found to bridge two anchors to one of which one of the YACs, and to the other anchor the other YAC is connected:

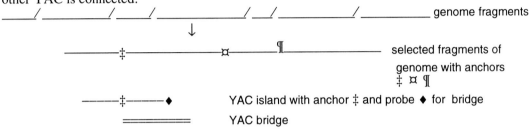

ANCIENT DNA: DNA of 50,000-100,000 years old may be still analyzed, however samples preserved in amber may last longer. Mitochondrial DNA extracted from 80-million years old bones and amplified by PCR had sequences different from any other sequences known so far. The conditions of preservation is critical. It is very important that during the PCR analysis the greatest caution is exercised to avoid contamination. It is advisable to test not just the sample but the immediate environment, the reagents themselves, verify that sample conforms to that of the species and be suspect if the fragments are too long. In case protein is present, the high ratio between D and L aspartic acid indicates that most likely the DNA had been degraded. The purpose of the analysis of ancient DNA is to obtain information on individuals, groups or to assess evolutionary relations. (See ancient organisms, PCR, ice man, mummies, Romanovs)

ANCIENT ORGANISMS: now extinct species recognized as paleontological reliques are difficult to study even by the most modern research techniques because the organic material has decayed. A 25-40 million years old bacterial spore, discovered in the digestive tract of a now extinct bee species, preserved as an amber inclosure was reported to be revived and its 16S ribosomal RNA was quite similar to the living species of *Bacillus sphericus*. Actually, the calculated rate of nucleotide substitution in the 16S RNA encoding DNA segment appeared to be 1.8 to 2.4×10^{-9} per site per year. Although, the isolation of the spore from the amber was carried out with extreme caution, some questions regarding possible contamination may be raised and newer studies failed to confirm DNA in amber. (ancient DNA, mummies, ice man, amber)

ANCIENT RNA: retrieved from extinct or very old specimens. (See ancient organisms)

ANCOVA: abbreviation for **a**nalysis **of cova**riance. (See correlation)

ANDALUSIAN FOWL: has been frequently used as an example for codominant segregation; when black and white fowl is crossed in the F_2 1 black: 2 blue: 1 white individuals are found; the "blue" has black and white (white-splashed) feathers. (See codominance)

ANDERSEN DISEASE: autosomal recessive deficiency of amylotransglucosidases(s) causing liver, heart and muscle disease because of the defect in glycogen storage. (See glycogen storage disease [Type IV])

ANDERSON DISEASE: involves lipid transport defects of the intestines and the retention of chylomicrons. (See lipids, chylomicron)

ANDERSON - FABRY DISEASE: is a human X-chromosome linked deficiency of α-galactosidase resulting in angiokeratoma (red or pink skin or mucous membrane lesions caused by dilation of veins). The gene is 12 kb with 7 exons encoding a 427 amino acid protein. (See galactosidase-β, angiokeratoma)

ANDROECIUM: the male region of a flower (the stamens).

ANDROGEN: a hormone that promotes virility but it is present at lower level also in females.

(See animal hormones, FGF)

ANDROGEN RECEPTOR: see hormone response elements (HRE), Kennedy disease, gynecomastia, testicular feminization

ANDROGENESIS: development of the male gamete into a paternal haploid or diploid embryo under natural conditions; it can be obtained by *in vitro* culturing and regeneration of plants from microspores. *In vitro* androgenesis can be *direct* when the microspores develop directly into plantlets or *indirect* when the microspores form first a callus and from that plantlet are regenerated in a second step. Androgenesis may result also when from a fertilized egg all the chromosomes of the female are lost and those of the male remain; again paternal offspring results. (See apomixis, embryo culture, anther culture, microspore culture, hydatidiform mole, gynogenesis)

ANDROGENITAL SYNDROME: see pseudohermaphroditism

ANDROGENOTE: diploid embryo with only paternal sets of chromosomes. (See androgenesis)

ANDROGENOUS: it is a pseudo- or true hermaphroditic stage in mammals or plants. (See hermaphrodite)

ANDROMEROGONY: development of an egg (or part of it) containing only the male pronucleus; the egg's own nucleus was removed prior to fusion with the male nucleus. (See androgenesis, pronucleus)

ANDROSOME: chromosome which normally occurs only in the males. (See sex chromosomes)

ANEMIA: a reduction of the red blood cells and hemoglobin below the normal level. It occurs when the production of erythrocytes does not keep up with losses. Several human diseases involve anemia, including some hereditary ones such as the thalassemias, sickle cell anemia, glucose-6-phosphate dehydrogenase deficiency, etc. Some anemias appear under autosomal dominant, autosomal recessive or X-linked control. (See Cooley's anemia, Fanconi's anemia, elliptocytosis, hemolytic anemia, pyruvate kinase deficiency, glutathione synthetase deficiency, thalassemia, siderocyte anemia, transcobalamin deficiency, magaloblastic anemia)

ANEMOPHILY: pollination by the wind.

ANENCEPHALY: a perinatal disorder of fetuses and newborns without brain (cerebrum and cerebellum); many of the afflicted die before birth, 1/16 survives birth but they rarely survive for a week. It may be due to a recessive mutation but some of the cases are due to non-genetic causes. Its prevalence is less than 1/1000. Prenatal test may be carried out if family history indicates genetic determination. (See neural-tube defects, prenatal diagnosis, genetic screening, MSAPF)

ANERGY: an unresponsiveness of the lymphocyte to an antigen because, e.g. a slightly modified peptide-MHC is attached to the T cell receptor or some of inducive factors are not functioning adequately. (See T cell, HLA)

ANESTHETICS: numb the nerve receptors; they are generally affecting the ligand-gated ion channels. (See ion channels)

ANEUGAMY: the chromosome number of the two gametes involved in the fertilization is different. (See also anisogamy, isogamy, heterogametic, homogametic)

ANEUHAPLOID: is a haploid which has incomplete set(s) of chromosomes. (See aneuploidy, haploid)

ANEUPLOIDY: chromosome number is either more or less than in a normal genome, e.g., $n \pm 1$ or $2n \pm 1$ or ± 2 or ± 3, etc. Aneuploids are trisomics or monosomics, single or multiple. Aneuploidy is frequent in cultured cells and in cancer cell. Hamerton (1971, after surveying 1291 spontaneous human abortions, found 5% monosomics, 11.9% trisomics, 4.1% triploids, and 1.2% tetraploids (note that the triploids and tetraploids are polyploids but not aneuploids and their frequency is included only for comparison.). Aneuploids are usually very deleterious yet sex-chromosomal aneuploidy e.g. Turner syndrome XO, Klinefelter syndrome XXY, etc. is not generally lethal in humans or other animals. Monosomics ($2n - 1$) have been very skillfully exploited for mapping genes to chromosomes in polyploid plants (wheat, oats, etc.). (See next page, hypoploid, hyperploid, triploid, pentaploid, monosomic analysis, MSAFP)

Aneuploidy continued

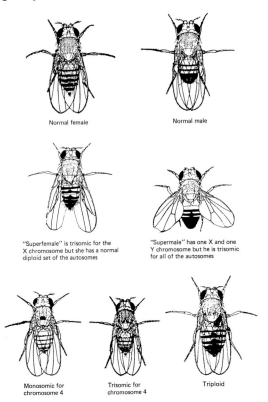

← NORMAL MALE AND FEMALE AS WELL AS VARIOUS RARE ANEUPLOIDS OF *DROSOPHILA*. (From Morgan *et al*. 1925 Bibliographia Genet. 2:3)

ANEURYSM: formation of small sacs of blood caused by the dilation of the veins. It is under autosomal dominant control in both the abdominal (more common in females) and in the brain aneurysm. (See collagen)

ANEUSOMATIC: the somatic chromosome number varies among the body cells because of the presence of supernumerary chromosomes and their frequent somatic nondisjunction. (See supernumerary chromosomes, nondisjunction, aneuploidy)

ANEUSOMY, SEGMENTAL: (See contiguous gene syndrome)

ANGELMAN SYNDROME (Happy Puppet Syndrome): is apparently an autosomal recessive human defect with somewhat irregular inheritance. Cytologically and molecularly detectable deletion in the 15q11-q13 region (similarly to the Prader-Willi syndrome) was observed. The unusual feature of this condition that the Angelman syndrome is transmitted only through the mother whereas in Prader-Willi syndrome the transmission is paternal. Imprinting has been suggested for the phenomenon. The affected individuals have motor function defects, mental retardation, epilepsy, speech defect or absence, and frequent protruding tongue accompanied by excessive laughter (hence the HPS term). It appears that the syndrome is due to an abnormal ubiquitin-mediated degradation of a brain ligase. (See disomic, Prader-Willi syndrome, mental retardation, imprinting, head/face/brain defects)

ANGIOEDEMA: dilation of the subcutaneous capillary veins leading to skin, respiratory tract and gastrointestinal fluid accumulations. The hereditary dominant form has been attributed to mutations in the serpin gene or to complement inhibitory factor deficiency. (See serpin, complement)

ANGIOGENESIS: the formation of blood vessels, and chronic inflammation. The vascular endothelial growth factor and its two receptors Flt-1 and Flk-1/KDR are required in rodents for angiogenesis. Angiogenesis factor (VEGF), peptide hormones, secreted by tumors to increase blood supply and ensure neoplasias and their further growth. There are two angiogenesis pathways. The fibroblast growth factor or tumor necrosis factor-α initiated path depends on integrin $\alpha_v\beta_3$, whereas angiogenesis initiated by vascular endothelial growth factor, transforming growth factor-α or phorbol ester use the $\alpha_v\beta_5$ path. The tumor necrosis factor-α induced angiogenesis uses the B61 cytokine-inducible ligand for the Eck protein tyrosine kinase receptor (RPTK). Angiogenesis is required also for the proliferation of tumors but the process can be restricted by the antibiotic minocycline, AGM and interferon α/β. (See tumor, cancer, VEGF [vascular endothelial growth factor], fibroblast growth factor [FGF], integrin, phorbol esters, tumor necrosis factor [TNF])

ANGIOKERATOMA: a recessive X-chromosome linked skin disease involving dilation of the small veins, warty growth, thickening of the epidermis primarily on fingers, toes and the scrotum. (See Anderson - Fabry disease, fucosidosis)

ANGIOMA: tumor of the blood or lymph vessels or a neoplasia which forms blood and lymph vessels. Many forms exist in humans; they are controled by autosomal dominant genes. (See also hemangioma)

ANGIOSPERM: plants that bear seeds within an ovary; the majority of higher plants belong to this taxonomic category.

ANGIOTENSIN: Asp - Arg - Val - Tyr - -Ile - His - Pro - Phe peptide stimulates the smooth muscles of the blood vessels, reduces the blood flow through the kidneys and decreases the excretion of fluid and salts, increases the secretion of aldosterone and stimulates the reabsorption of sodium. It is involved in the hereditary disorders of adrenocortical steroid biogenesis. Angiotensin I receptor AT_1 mediates the pressor (higher blood pressure) and angiotensin II receptor AT_2 has the opposite (depressor) effect. Angiotensin II cell surface receptor is directly stimulated by the Jak/STAT signal transduction pathway. (See aldosterone, hypertension, signal transduction, tachykinin, pseudoaldosteronism)

ÅNGSTRÖM (Å): 1 Å = 1/10 nanometer (nm)

ANGULAR TRANSFORMATION (arcsine transformation): is used with percentages and proportions. In a binomial distribution the variance is a function of the mean. The arcsine transformation prevents that, $\theta = \arcsin \sqrt{p}$, where p is a proportion which stands for an angle whose sine is the given quantity. The transformation stretches out both tails of a distribution of percentages and compresses the middle. It may be usefully applied to genetic data when the figures fall outside the 30% and 70% ranges. (See arcsine, sine)

ANHIDROSIS: reduction or lack of sweating. An X-linked hypo- or anhidrotic ectodermal dysplasia is caused by mutation in a transmembrane protein. Ectodermal dysplasia is part of about 150 syndromes. (See ectodermal dysplasia)

ANHIDROTIC ECTODERMAL DYSPLASIA: see ectodermal dysplasia

ANHYDRIDE: result of a condensation reaction where water was eliminated (between carboxyl and phosphate groups)

ANIMAL HORMONES: are the first chemical messengers that are secreted by some tissues and carried by the bloodstream to the specific sites of action where regulatory functions are carried out. They regulate either the synthesis or activity of enzymes or affect membrane transport in cooperation with the second messengers, cyclic adenosine monophosphate and cGMP. They are of three major types. *PEPTIDE HORMONES*: secreted by the hypophysis of the pituitary gland, are somatotropin (general growth hormone, GH), corticotropin (adrenocorticotropin, ACTH in the kidneys), thyrotropin (thyroid-stimulating hormone, TSH), follitropin (FSH, in gonads), lutrotropin (luteinizing hormone, LH, in gonads), prolactin (in mammary glands). Secreted by the neurohypophysis are: oxytocin (controls uterine contractions and milk production) vasopressin (antidiuretic hormone controls water reabsorption of the kidneys and blood pressure), secreted by the middle section of the hypophysis. Melanotropins (control melanin pigments). The pancreas secretes insulin (controls carbohydrate, fatty acid, cholesterol metabolism), and glucagon (stimulates glucose production by the liver). The ovary produces relaxin (controls pelvic ligaments, the uteral cervix, thereby labor), the thyroid gland is the source of parathyrin (involved in calcium and phosphorus metabolism), the kidneys release erythropoietin (a glucoprotein involved in erythrocyte production by the bone marrow), and renin (causes constriction of the blood vessels), the digestive tract secretes gastrin (promotes digestive enzymes), enterogastrone (controls the gastric secretion), cholecystokinin (regulates gall bladder), secretin (controls pancreatic fluids and bile production), pancreozymin (of duodenal origin stimulates pancreatic functions). *AMINO ACID HORMONES* are thyroxin and triiodothyronin secreted by the thyroid gland and affect many functions in the body. The kidney tissues secrete epinephrine (adrenaline) and norepinephrine (triiodothyronin) regulate blood pressure and heart rate, the pineal gland (a cone-shaped epithelial body at the base of the brain) produces melatonin affecting the pigment producing melanophore cells. The nerve cells produce serotonin (5-hydroxytryptamine) affecting contraction of the blood vessels and nerve

Animal hormones continued

function. *STEROID HORMONES* are produced in the testes (testosterone, regulates male reproductive capacities), in the ovaries (estrogen [estradiol-17β], involved in female reproductive functions), in the corpus luteum of the ovary and in the Schwann cells of the peripheral nervous system progesterone is made. It functions during menstrual cycles and pregnancy and in myelin formation. In the kidney cortex cortisol (corticosterone) is synthesized affecting glucose utilization and glucose levels in the blood. *EICOSANOID (HORMONELIKE) SUBSTANCES* are prostaglandins (triggering smooth muscle contraction, control fever and inflammations), leukotrienes (secreted by the white blood cells and affecting hypersensitivity reactions and pulmonary functions), thromboxanes are produced by the blood platelets and other cells and are involved in blood clotting, blood vessel constriction, etc. There are a large number of other hormones of important functions. (See also hormones, hormone receptors, hormone response elements, opiocortin)

ANIMAL HOST CELLS: are used for genetic transformation. *Xenopus* oocytes are well suited for such studies because they can propagate foreign genes in appropriate vectors quite efficiently. Similarly, COS cells of mice and other somatic cells have been used effectively. More recently techniques became available for the transformation of animal zygotes and embryos, and thus genetic information can be added or replaced in the germline, and transmitted to the sexual progeny. (See also transformation of animal cells, COS, vectors genetic, germline)

ANIMAL MODELS: certain biological phenomena cannot be studied in humans because mutants are not available and cannot be produced or manipulated effectively. In such cases animals such as *Caenorhabditis, Drosophila* and mice, are used for the experimentation (in behavioral genetics, neurobiology, etc.). Animal models may have important role in improving the techniques of gene therapy. The "shiverer" deletion of mice resulting in convulsions because of the loss of a gene coding for a myelin protein had been genetically cured by transfection of the wild type allele into the gamete. Similarly, the size of mice could be genetically increased by transformation using the rat somatotropin (RGH, growth hormone) gene fused to and regulated by a metallothionein promoter. The following monogenic human genetic disorders have models in mouse [abbreviations h. chr. = human chromosome, m. chr. = mouse chromosome]: *adenomatous polyposis* (protrusive growth in the mucous membranes, h. chr. 5q21-q22, mouse homolog *Apc^{Min}*, chr. 18), *androgen insensitivity* (sterility, h. chr. Xq11.2-q12, mouse gene *AR^{Tfm}*, m. chr. X), *X-linked agammaglobulinemia* (deficiency of γ globulin in blood, h. chr. Xq21.33-q22, mouse gene *Btk^{Xid}*, m. chr. X), *Duchenne muscular dystrophy* (an early muscular disability, h. chr. Xp21.3-p21.2, mouse *Dmd^{mdx}*, m. chr. X), *Greig cephalopolysyndactily* (multiple fusion of digits, h. chr. 7p13, mouse gene *Gli3^{Xt}* m. chr. 13), *mucopolysaccharidosis type VII* (a type of lysosomal storage disease, h. chr. 7q22, mouse gene *Gus^{mps}*, m. chr. 5), *α–thalassemia* (defect in the hemoglobin α chain, h. chr. 16p13.3, mouse gene *Hbath*, m. chr. 11), *β-thalassemia* (defect in the β–chain of hemoglobin, h. chr. 11p15.5, mouse gene *Hbbth*, m. chr. 7), *piebaldism* (color patches on the body, h. chr. 4p11-q22, mouse gene Kit^W, m. chr. 5), *ornithine transcarbamylase* (defect in the transfer of a carbamoyl group, $H_2N - C = O$, from ornithine to citrulline, h. chr. Xp21.1, mouse gene *Otc^{Spf}*, m. chr. X), *tyrosinase-positive type II* (occulocutaneous albinism, h. chr. 15q11-q12, mouse gene p^p, m. chr. 7), *phenylketonuria* (phenylalanine hydroxylase deficiency, h. chr. 12q22-q24.2, mouse gene *Pah^{enu2}*, m. chr. 10), *Waardenburg syndrome type 1* (h. chr. 2q35-q37, mouse gene *Pax3^{Sp}*, m. chr. 1), *aniridia* (absence of the iris, h. chr. 11p13, mouse gene *Pax6^{Sey}* m. chr. 2), *pituitary hormone deficiency* (h. chr. 3q, mouse gene *Pit1^{dw}*, m. chr. 16), *Pelizaeus-Merzbacher disease* (central brain sclerosis, h. chr. Xq21.33-q22, mouse gene *Plp^{jlp}*, m. chr. X), *Charcot-Marie-Tooth disease type 1A* (a progressive neuropathic muscular atrophy, h. chr. 17p12-p11.2, mouse gene *Pmp22^{Tr}*, m. chr. 11), *retinitis pigmentosa* (sclerosis and

Animal models continued

pigmentation of the retina, h. chr. 6p21.2-cen, mouse gene $RD2^{Rd2}$, m. chr. 17), *gonadal dysgenesis* (underdeveloped germcells in the testes, h. chr. Y11.2-pter, mouse gene Sry^{Sxr}, m. chr. Y), *tyrosinase negative oculocutaneous albinism* (see albinisms, h. chr. 11q14-q21, mouse gene Tyr_c, m. chr. 7). By disruption of hexosaminidase α subunit a model for the Tay-Sachs disease has been generated in mouse. Interestingly, these animals suffered no obvious behavioral or neurological deficit. Disrupting the hexoseaminidase β subunit (Sandhoff disease model) resulted in massive depletion of spinal cord axons and neuronal storage of ganglioside G_{M2}.

MOUSE POLYGENIC DISORDERS WITH SIMILARITIES TO HUMAN CONDITIONS [human problem - mouse strain]: alcoholism and opiate drug addictions - C57BL/6J, asthma - A/J, atherosclerosis - C57BL, audiogenic (sound-induced) seizures - DBA, cleft palate (fissure in the mouth) - A, deafness - LP, dental disease - C57BL, BALB/c, diabetes - NOD, epilepsy - EL, SWXL-4, granulosa cell tumors in the ovary - SWR, germ cell tumors in the ovary - LT, germ cell tumors in the testes - 129, hemolytic anemia - NZB, hepatitis - BALB/c, Hodgkin disease (pre-B cell lymphoma - SJL, hypertension - MA/My, kidney adenocarcinoma -BALB/cCd, leprosy (*Mycobacterium leprae*) - BALB/c, leukemia - AKR/J, C58/J, P/J, lung tumors - A, Ma/My, measles - BALB/c, osteoporosis - DBA, polygenic obesity - NZB, NZW, pulmonary tumors - A/J, rheumatoid arthritis - MRL/Mp, spina bifida (defect of the bones of the spinal cord) - CT, systemic lupus erythematosus (a skin degeneration) - NZB, NZW, whooping cough (pertussis) - BALB/c. (Some of the data by courtesy of GIBCO BRL CO.)

ANIMAL POLE: is dorsal end of the animal egg opposite the lower end, the vegetal pole, and where the sperm entry is located. After the entry, the egg cortex rotates slightly and in some species at the side opposite the entry a *gray crescent* is formed. (See vegetal pole)

ANIMAL SPECIES HYBRIDS: the most familiar example is the hybrids of the mare (*Equus caballus*, 2n = 64) and the jackass (*Equus asinus*, 2n = 62), and the stallion and the she-ass. The hybrid males do not produce viable sperm although they may show normal libido. The females may have estrus and ovulate but there is no proven cases of fertility. Zebras (2n = 44) also may form hybrids with both donkeys and horses. Buffalo (*Bison bison*, 2n = 60) may be crossed reciprocally with cattle (*Bos taurus*, 2n = 60) but their offspring (cattalo) has reduced fertility. The domesticated pig (*Sus crofa*, 2n = 38) forms fertile hybrids with several wild pigs with the same number of chromosomes. The sheep (*Ovis aries*, 2n = 54) interbreeds with the wild mouflons but the sheep x goat (*Capra hircus*, 2n = 60) hybrid embryo only rarely can be kept alive. Some monkeys can be interbred but primates are generally sexually isolated. There is no sexual barrier among the various human races, indicating close relationship but no hybrids are known between humans and any other species. These general rules do not hold for somatic cell hybrids because human cells can be fused with rodent or plant cells but they cannot be regenerated or even maintained successfully for indefinite periods of time. The hybridization barrier is not identical with other functional barriers.

HYBRID OF THE MALE GRANT'S ZEBRA AND THE FEMALE BLACK ARABIAN ASS, GLOUCESTER ZOO. (From Gray, A.P. Mammalian Hybrids. Commonwealth Agric Bureau. Farnham Roal, Slough, UK)

ANIMAL TRANSFORMATION VECTORS: most commonly Simian virus 40 (SV40) and Bovine papilloma virus (BPV) based vectors are used. The BPV vectors can be used for the

Animal transformation vectors continued

synthesis of large amounts of proteins specified by the gene(s) carried by the expression vectors. In addition, the BPV vectors can be maintained for long periods of time in cell cultures and may yield 10 mg of specific protein(s) per liter of culture/24 hr. The SV40 vectors can also be used for gene amplification in COS cells. Both of these vectors can serve as shuttles between animal and prokaryotic cells. (See also BPV and SV40 constructs, adenovirus, retroviral vectors, COS cells)

ANIMAL VIRUSES: include both invertebrates and vertebrate viruses. The Rhabdoviridae and the Bunyoviridae may infect also plants. The *double-stranded DNA* viruses may be *enveloped*: Baculoviridae, Poxviridae, Herpesviridae, Hepadnaviridae, Polydnaviridae, and double-stranded DNA viruses *without envelope*: Iridoviridae, Adenoviridae, Papovaviridae. The Parvoviridae have *single-stranded DNA* and they are *not enveloped*. The *single-stranded RNA* and *enveloped* group includes the Togaviridae, Bunyaviridae, Rhabdoviridae, Coronaviridae, Paramixoviridae, Toroviridae, Orthomyxoviridae, Arenaviridae, Flaviviridae, Retroviridae and Filoviridae. The *single-stranded RNA* and *non-enveloped* viruses are: Picornaviridae, Tetraviridae, Nodaviridae, Picornaviridae and Caliciviridae. The *double-stranded RNA* and *non-enveloped* viruses are Reoviridae and Birnaviridae. Their genetic material varies in size from 5 kb in the Parvoviridae to 375 kbp in the Poxviridae. The Polydnaviridae may have several copies of double-stranded circular DNAs. The Papovaviridae have only a single double-stranded DNA genetic material. The others may have two or more segments of linear nucleic acid genetic material. (See viruses)

ANIMALCULES: the pioneer microscopist Anthony Leuwenhoek (17th century) believed to see small encapsulated animals in the sperm of various animals and this observation supported his view that inheritance is only through the sperm and the females serve only as incubators. His observations lead to the notion of preformation, rather than epigenesis. (See preformation, epigenesis)

ANION EXCHANGE RESIN: a polymer with cationic groups and trap anionic groups and thus can be used in chromatographic separation.

ANIONS: negatively charged ions

ANIRIDIA: absence or reduction of the iris of the eye. It is freqently accompanied by cataract (opacity of the eye[s]), glaucoma (increased intraocular pressure causing deformation of the optic disk), nystagmus (involuntary movement of the eyeball), etc. The condition is caused by dominant defects in human chromosomes 2 and 11. In a Michigan population the rate of mutation appeared 4×10^{-6}. It may involve Wilms tumors and genital abnormalities due a deletion in human chromosome 11p13. The *Drosophila* locus *eyeless* and the mouse *Sey/Pax-6* are the corresponding homologs. (See Wilms tumor, WAGR, deletion, eyeless)

ANISOGAMY: the gametes are not identical, e.g. male and female (+ or -) are distinguishable. (See also isogamy)

ANISOMYCIN: an antibiotic isolated from *Streptomyces griseolus*. It inhibits peptidyl transferase during protein synthesis on the ribosomes. It also inhibits pathogenic fungi (e.g. mildew) in plants and it was found to be useful against infection by various species of the parasitic flagellate, *Trichomonas*, causing inflammation of the gum in the mouth, diarrhea, and vaginal discharge and irritation in humans and animals (particularly poultry and pigeons). (See antibiotics, protein synthesis)

ANISOTROPIC: the material varies in different directions, responds differently to external effects depending on directions.

ANKYLOSING SPONDYLITIS: an autosomal dominant rheumatism-type of disease with reduced penetrance. Onset is after age 20. A defect in the HLA B27 antigen is involved. (See HLA, immunodeficiency, connective tissue disorders, autoimmune disease, penetrance)

ANKYRIN: is a protein capable of binding fibrous proteins (e.g. spectrin) of the cytoskeleton and thus may be involved in some polar transports within the cell. (See cytoskeleton, spectrin, elliptocytosis, IκB, spherocytosis)

ANLAGE: group of cells of the embryo initiating specific biological structures. (See primordium)

ANNEALING: formation of double-stranded nucleic acid when two complementary single stranded chains meet (nucleic acid hybridization). The process is used to estimate DNA complexity, for identifying the presence of homologous sequences in the genome by radioactively labeled or fluorescent homologous and heterologous probes. (See c_0t curve, probe, chromosome painting, FISH, DNA hybridization)

ANNEXIN: are proteins composed of four or eight conserved 70-amino acid domains with variations mainly at the amino end. In mammals, there are at least 10 annexins and others exist in lower eukaryotes. Annexins bind to negatively charged phospholipids in the membranes. Annexins V and VII form voltage regulated ion channels for different cations whereas VII is specific for Ca^{2+}. Annexins II may assist exo- and endocytosis. An annexin-like protein may be involved in mitigating H_2O_2 stress. (See ion channels, endocytosis, exocytosis)

ANNULUS (a ring): e.g. specialized cells in a sporangium involved in opening.

ANOMALOUS GENETIC RATIOS are caused by many different mechanisms. Defective chromosomes or chromosomes carrying deleterious genes are transmitted at lower than normal frequencies and reduce the appearance of the genes residing in that chromosome (conversely the other allele may appear in excess). Monosomy, trisomy also modify segregation ratios. The genetic ratios may be altered by preferential segregation of certain chromosomes in meiosis, similarly segregation distorter genes can cause dysfunction of sperm carrying them. Meiotic drive in a population can work against the more fit alleles. (See Mendelian segregation, chromosomal breakage, aneuploidy, deletion, segregation distorter, gametophyte factor, certation, meiotic drive, preferential segregation, gene conversion, drift genetic, penetrance)

ANOMALOUS KILLER CELL (AK): is a T cell grown in the presence of IL-2 and acquires natural killer cell (NK)-like properties. (See killer cell)

ANOMER: stereoisomers of sugars differing only in the configuration of the carbonyl residue, e.g. α-D-(+)-glucose and β-D(+)-glucose.

ANONYMOUS DNA SEGMENT: a mapped DNA fragment without known gene content.

ANONYMOUS GENE: a mapped gene without information about its molecular mechanisms but known to affect the expression of a quantitative response such as a behavioral trait. If it displays two allelic states it can be used for (DNA) mapping. (See behavior)

ANONYMOUS PROBE: a DNA probe has no known gene(s) in it and its function is unknown. Nevertheless it provides information on the presence of sequences homologous to it and thus may be useful for taxonomic or evolutionary studies. (See physical mapping)

***ANOPHELES* MOSQUITO**: host and vector of the protozoan *Plasmodium falciparum*, the cause of malaria. (See thalassemia, sickle cell anemia)

ANOPHTHALMOS: autosomal recessive bilateral defect in the formation of the optic pit. It has been reported also as an Xq27-encoded fusion of the eyelids and other complications. (See microphthalmos, eye diseases)

ANOREXIA: lack of appetite or *anorexia nervosa* is a psychological disturbance of adolescents (primarily females) caused by an abnormal fear of gaining weight and therefore refusing to eat, habit of self-induced vomiting, unnecessary use of laxatives leading to emaciation, irregular or lack of ovulation, reduced interest in sex and other anomalies. Medical treatment may be required.

ANOVA: abbreviation for **an**alysis **of va**riance. (See analysis of variance)

ANOXIA: absence or deficiency of oxygen; reduces chromosomal damage during irradiation. (See radiation effects)

ANSERINE (β-alanine-1-methylhistidine): a dipeptide occurring in birds and some mammals but not in humans. (See carnosinemia)

ANT (*Formica sanguinea*): 2n = 48.

ANTAGONIST: blocks biological receptor activation. (See also agonist)

ANTELOPE (*Antilocapra americana*): 2n = 58.

ANTEATER (*Tamandua tetradactyla*): 2n = 54.

ANTENATAL DIAGNOSIS: determination of a particular condition before birth by amniocentesis or blood samplings or by other means. (See amniocentesis, prenatal diagnosis, fetoscopy)

ANTENNA: feeler organ on the head of insects. (See *Drosophila*)

ANTENNA PIGMENTS: in the chloroplasts collect light energy that is transmitted to the reaction centers for photochemical use. (See chloroplasts, chlorophyll, photosynthesis)

ANTENNAPEDIA: *Drosophila* gene (*Antp*; map location 3-47.5, salivary bands 84B1-2) with numerous alleles. The null alleles result in embryonic lethality. Initially the locus was recognized by mutations that transform the antennae into mesothoracic legs. Numerous other homeotic changes may accompany the mutations. The different alleles may involve various types of at the locus. The gene occupies about 100 kb, containing eight exons. These exons are transcribed from promoters P1 or P2 or from both. The transcripts may undergo alternate splicing. The homeobox motif is in exon 8. (See homeotic genes, morphogenesis)

ANTERIOR: indicates a direction in front of something or toward the head.

ANTEROGRAD: ahead or forward moving. (See also retrograd)

ANTHER: the pollen-containing parts of the male flowers. (See gametogenesis)

ANTHER CULTURE: is used for the isolation of haploid plants. The culture may start with microspores that are directly regenerated into plantlets (without an intermediate callus stage) or from anthers haploid tissues are isolated and first a callus is formed and then the calli are regenerated into plants. Both procedures are using tissue culture methods under aseptic conditions. The haploid cells may diploidize spontaneously or by induction and that results in perfect homozygosis of the plants. (See androgenesis, *Asparagus*, gametogenesis, embryo culture, YY plants)

ANTHERIDIUM: the male sex organ (gametangium) of lower plants and fungi.

ANTHESIS: the time of pollen shedding or receptivity of a flowering plant.

ANTHOCYANIN: plant flower pigments (delphinidine, cyanidin, pelargonidine, peonidine, petunidine, malvidine, etc.) from phenylalanine via trans-cinnamic acid and cinnamoyl-CoA, chalcones, and flavonones. CH_3 and OH groups on the B-ring determine the color produced; glycosylation (hexose or pentose) at the 3 and 5 positions (or at both) increases stability of the pigments, and these glycosides are called anthocyanidins. Each enzymatic step is controled by different genes and these original discoveries, beginning at the early years of the 20th century, prepared the way to biochemical genetics. By the use of antisense constructs of the gene chalcone synthase (CHS) the activity of this enzyme and chalcone flavonone isomerase (CHI) could also be reduced indicating that CHS regulates also the expression of CHI.

Delphinidin (purple)   Cyanidin (red)   Pelargonidin (salmon)

Pelargonidin displays an OH group at position 4', cyanidine has two OH groups at 3' and 4', delphinidine has three OH groups (3', 4' and 5'). Peonidine (not shown) has 3' OCH_3 and 4' OH. Petunidine: 3' OCH_3, and 5' OH. Malvidine: 3' and 5' OCH_3 and 4'OH. Further color variations may be brought about by glycosylation and acetylations of the A ring(s) [at left].

ANTHRANYLIC ACID: synthesis begins with the condensation of erythrose-4-phosphate + phosphoenolpyruvate, and from this shikimate and then chorismate are formed. Chorismate through prephenate contributes to phenylalanine and tyrosine and through another path it is a precursor of the amino acid tryptophan (actually indole-3-glycerol phosphate → indole and serine are converted to this amino acid). (See tyrosine, phenylalanine)

ANTHROPOMETRIC TRAITS: physical or physiological characters of humans (such as weight, head circumference, hair color, protein differences, behavior, etc.) that may be used for the characterization of human populations.

ANTI: a conformation of nucleotides, the CO and NH groups in the 2 and 3 positions of the pyrimidine ring (1, 2, 6 positions in the purine ring) are away from the glycosidic ring while in the SYN conformation they lie over the ring. The anti conformation is most common in nucleic acids and free nucleotides.

ANTIAUXIN: interferes with the action of auxins, e.g. 2,3,5-triiodobenzoic acid inhibits the growth promoting action of 2,4-D (dichlorophenoxyacetic acid) or the indoleacetic acid (IAA) analog 5'-azido-indole-3-acetic acid interferes with enzymes involved with IAA. (See plant hormones)

ANTIBIOTIC RESISTANCE: is brought about either by enzymatic inactivation of the antibiotic, or modification of the target, or active efflux of the substance or sequestration by binding to special proteins. It is generally determined by genes in bacterial plasmids and transposons. The mechanisms of resistance varies: penicillins and cephalosporins (β-lactamase hydrolysis), chloramphenicol (detoxification by chloramphenicol transacetylase that acetylates the hydroxyl groups or interferes with uptake), tetracyclines (interference with uptake or maintenance of the molecules), aminoglycosides (streptomycin, kanamycin, etc. enzymatic modification of the drug [phosphorylation] interferes with uptake or action), erythromycin, lincomycin (methylation of the small ribosomal subunit). Antibiotic resistance acquired through conjugative transfer of the resistance factors or mutation pose serious problems to medicine, e.g. the recent resistance of *Mycobacterium tuberculosis* to all known antibiotics. Antibiotic resistance genes are used generally to assure the removal (by carbenicillin or claphoran [cefotaxime]) of the carrier *Agrobacteria* after infection with plant transformation vectors. Also, the transformed bacterial, fungal, animal and plant cells are selectively isolated on the basis of antibiotic resistance. Insertional mutagenesis in bacteria is monitored by the inactivation of the resistance genes upon integration. (See antibiotics, pBR322, lactamase)

ANTIBIOTICS: a wide variety of chemicals produced by microorganisms and plants (also now by organic laboratory synthesis) that are toxic to other organisms. The major types of antibiotics are *penicillins, ampicillin* and *cephalosporins* (interfere with bacterial cell wall biosynthesis), *chloramphenicol* (binds to 50S ribosomal subunit and blocks the peptidyl transferase ribozyme function during protein synthesis of prokaryotes, *tetracyclines* inhibit the entry of the charged tRNA to the A site of the ribosome in prokaryotes. *Streptomycin* blocks the process of prokaryotic peptide chain elongation and causes also reading errors during translation. *Spectinomycin* inhibits the function of the 30S ribosomal subunit. *Kanamycin, geneticin (G418), neomycin, gentamycin, hygromycin* bind to 30S and 50S ribosomal subunits and prevent protein synthesis or cause misreading. *Erythromycin* inhibits the translocation of the nascent peptide chain on the prokaryotic ribosomes. *Lincomycin* inhibits chain elongation on the prokaryotic ribosome by its effect on peptidyl transferase but not in eukaryotes. *Rifampycin* interacts with the β subunits of the prokaryotic RNA polymerase. *Fusidic acid* interferes with the binding of aminoacylated tRNAs to the ribosomal A site by inhibiting the release of prokaryotic elongation factor EF-G and also eukaryotic elongation factor eEF-2. *Kasugamycin* blocks the attachment of tRNA^{fMet} to the P site of the prokaryotic ribosome. *Kirromycin* actually promotes the binding of elongation factor EF-TU-GTP complex to the prokaryotic ribosome but then inhibits the release of the elongation factor. *Thiosrepton*, from *Streptomyces azureus*, blocks prokaryotic peptide elongation from both prokaryotic and eukaryotic ribosomes. *Cycloheximide* interferes with peptide translocation on the eukaryotic ribosome. *Anysomycin* blocks the petidyl transferase on the eukaryotic ribosomes and is comparable in effect to that of chloramphenicol in prokaryotes. *Streptolydodigins* do not block RNA initiation but interfere with the elongation of the RNA chain in prokaryotes. *Cifrofloxacin* interacts with DNA gyrase. *Actinomycin D* inhibits primarily RNA polymerase II and to a lesser extent the other RNA polymerases but not DNA polymerase in either prokaryotes or eukaryotes; *α–amanitin* also inhibits eukaryotic RNA polymerase II and in very high concentration pol III but not pol I. *Pactamycin* blocks the eukaryotic initiator tRNA^{Met} to attach to

Antibiotics continued

the P site of the ribosome. *Showdowmycin* interferes with the formation of the eukaryotic eEF -tRNA^{Met} complex. *Sparsomycin* is a eukaryotic peptide chain translocation blocker. *Cefotaxime* (synonym *claforan*), *carbenicillin, vancomycin* are more effective as antibacterial agents than their toxicity to eukaryotic cells and are frequently used in plant tissue culture to prevent bacterial growth. Antibiotics which interfere with protein synthesis on prokaryotic ribosomes, cause similar damage to the ribosomes of eukaryotic organelles (mitochondria, plastids). The availability of antibiotics in the 1940s opened a new era in medicine and they became in the 1970s the most important selectable markers for the construction of vectors for genetic engineering, and are used for selective isolation of various genetic constructs in microbial, plant and animal cell genetics. The number of antibiotics is continuously increasing because of the need for effective new drugs since microorganism develop resistance to the old antibiotics. *Staphylococcus aureus* bacteria are now resistant to all antibiotics except vancomycin and it will be only a matter of time when resistance mutations will develop to this too. There are already *Enteroccus faecium* strains that are resistant to vancomycin. (See also antibiotic resistance, protein synthesis, selectable marker, cell genetics, vectors)

ANTIBODIES: are immunoglobulins that react — as a cellular defense — with foreign antigens. Antibodies contain two light chains, either κ or λ chains two one of the five heavy chains (μ, δ, γ, ε, α) and their variants. Both light and heavy chains contain variable and constant regions. The specificity resides in the variable regions. Antibodies have specificities to about a million different antigens. This specificity is achieved with the aid of a much smaller number of antibody genes by differential processing of the transcripts, mutation, recombination, gene conversion, and transposition within the families of immunoglobulin genes. Antibodies are made by the lymphocytes and my be attached to their membrane or may become humoral antibodies (secreted into the blood stream by the B lymphocytes). One particular B cell synthesizes only one type of antibody molecules. Each B cell deposits the first 100,000 antibodies it makes in its plasma membrane and serves there for antigen receptors. When a particular antigen binds to the B cell, it stimulates its clonal division and the production of more antibody. These series of the antibody are then made at the amazing rate of about 2,000 molecules/second then secreted into the blood plasma. An individual can make about 10,000 different heavy chain variants and about 1,000 different light chain variants. Since these chains can combine freely, the total number of different antibodies can be $10^4 \times 10^3 = 10^7$. IgM type antibodies (containing γ immunoglobulin chains) occur at the largest concentration in the blood serum and their half-life is the longest. The general structure of the antibody molecules is diagramed here. Each antibody molecule has two identical antigen-binding sites (diagram p67). The majority of the antigens have, however, several to many antigenic determinants (epitopes). Some of these antigens may be built of repeating units and in these cases they are *multivalent* because they have multiple copies of the epitope. The binding between epitopes (e) and antibody (a) is a concentration dependent, reversible process: $(a + e) \Leftrightarrow (ae)$. When the concentration of the epitope increases, the binding to the antibody is increasing and the intensity of the reaction is expressed by the *affinity constant*: $(k) = (ae)/(a)(e)$. When half of the (a) sites are filled $k = 1/e$. The values of (k) range from 5×10^4 to 10^{12} liters per mole. The *avidity* of an antibody for an antigenic determinant depends also how many binding sites are available. The affinity is increasing with time after immunization (affinity maturation). Antibodies are involved in the destruction of invaders either through stimulating the macrophage cells to phagocytosis, or by ions, using the complement enzymes or activating the killer cells. Usually their turnover is rapid; the half-life of antibodies is days to a few weeks. About 20% of the total plasma proteins represents a diverse set of antibodies. After the B lymphocytes respond to an antigen and differentiate into plasma cells, their rate of antibody production may reach 1,000 molecules/second after the immunization (affinity maturation). Receptors (FcRn) of the Fc domain (see diagram) contribute toward the phagocytotic functions, cytotoxicity and to neonate immunity. In the maternal uterus FcRn/IgG has been detected. The FcRn

Antibodies continued

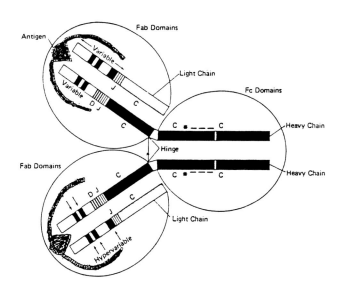

← ANTIBODY STRUCTURE
Antibodies have three main domains, the two Fab domains (Fragment antigen binding), including the light chains and parts of the heavy chains and one Fc (Fragment crystalline) domain The light chains have a size of about 23 kDA, the heavy chains vary from 53 to 70 kDa. X-ray crystallography revealed the domains as 2 x 4 nm oval or cylindrical in shape and the polypetide chain in each domain is folded in pleated β-sheets. The dimeric structure of the light and heavy chains are held together by disulfide bonds in variable numbers, depending on the particular molecules. The inter- and intra-chain disulfide bonds are not shown, except at the proline-rich hinge area that provides the molecules with some flexibility. At the amino end of the light and heavy chains are the variable and hypervariable regions that determine the specificity of the antibody. This region includes approximately 100 to 115 amino acids. The specificity is determined by complementarity between antibody and antigen in the antigen-binding "pocket" or surface as outlined by the hatched arms of the antigen around the area. The various specificities are determined by combinations of the variable (V), diversity (D) and junction (J) genes that account for about 25% of the amino acid residues and the remaining 75% is considered the framework. The complementarity determining regions are generally identified by CDR1, CDR2 and CDR3 (shown by the dark bands). The variable regions in the light and heavy chains are homologous. The constant regions (C) show very little variability within a species. There is a glycosylation site in the constant heavy chain region within the Fc domain (*). Also in the constant heavy chains there are sites for binding of the activator of the complement (---). The complement consists of about 30 different proteins of catabolic functions that are activated in a cascading manner after the binding of the antigen to the antibody and carry out the destruction of the foreign antigen.

receptors transfer maternal humoral immunoglobulins to the newborn before the immune system of the progeny is activated. During nursing the FcRn class receptors mediate the transfer of the IgG/FcRn complex by the milk. (See also immunoglobulins, immune system, complement, monoclonal antibodies, HLA, lymphocytes, T cell, TCR, killer cell, antigen, antigen presenting cell, MHC, neutralizing antibody, immunization alloantibody, natural antibody)

ANTIBODY DETECTION: is possible by several procedures: antibodies bound to proteins expressed in *E. coli* are detected by I^{125} (isotope) labeled antibodies that react to the species-specific determinants of the primary antibodies. Protein A labeled with I^{125} second antibody, conjugated to horseradish peroxidase (HRP) or HRP coupled to avidin, may be used to detect a second antibody coupled to biotin or by second antibody conjugated to alkaline phosphatase; using radio-labeled ligands. Antibodies can be detected also by agglutination and complement fixation. In agglutination a precipitate is formed upon the reaction. One of the procedure is the *Ouchterlony assay* by placing in neighboring wells of agar plates the antibody and the antigen and upon diffusion a visible precipitate is formed in about midway between the two wells

Antibody detection continued

if the antigen (e) and antibody (a) recognize each other:

ⓔ | ⓐ	ⓔ ⓐ ← wells
↑ precipitate between wells	no precipitate

The complementation fixing procedure has a unique feature inasmuch as the complement binds only to antibody that is complexed with the antigen. If the complement is fixed, adding red blood cells and cognate antibody to the reaction mix, resulting in no hemolysis, is the proof for fixation of the complement and the procedure can be quantitated by employing a series of dilutions. (See antibodies, complement, immunostaining)

ANTIBODY ENGINEERING: involves genetic modification of the immunoglobulin genes. (See antibody, humanized antibody)

ANTIBODY GENE SWITCHING: is preceded by pairing between members of the antibody

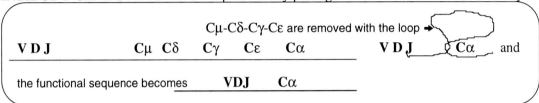

Cμ-Cδ-Cγ-Cε are removed with the loop →

V D J   Cμ   Cδ   Cγ   Cε   Cα   V D J   Cα   and

the functional sequence becomes   **VDJ**   **Cα**

genes families and the formation of loops that are then cut off at the stem and the deletion brings different heavy chain elements in the vicinity of the J (junction) genes. The site-specific switch then permits the expression of the genes moved to the vicinity of the J genes after the stem of the loop is cut off and the DNA strands are religated. The transcript is then further processed by removal of the introns. This is one of the mechanisms to generate greater diversity in the antibody proteins. (See immunoglobulins, antibodies, immune system)

ANTIBODY MIMIC: is a small synthetic polypeptide with specificity for a particular natural or synthetic epitope. (See epitope, antibody)

ANTIBODY, MONOCLONAL: see monoclonal antibody

ANTIBODY PREPARATION: an animal is injected with a pure antigenic mocule. After 2 to 3 weeks it develops antibodies against the epitope, then the animal is bled and from the serum the antibody is removed by precipitation with the cognate antigen and further purified. Hundreds of different antibody preparations are commercially available from biochemical supply companies.

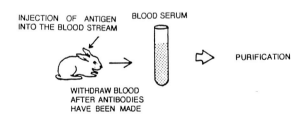

ANTIBODY PURIFICATION: the protein antigen may be coupled to a cyanogen-bromide-activated Sepharose. The epitope then retains the cognate antibodies while all other antibodies flow through. The antibody can then be retrieved by breaking the complex (with potassium thiocyanate, low-pH buffers, etc.). The methods must be adapted to the different proteins. Another procedure is to adsorb antibodies to protein antigens immobilized on diazotized paper or nitrocellulose filters following electrophoresis by SDS-polyacrylamide gels. The antibodies are then eluted with a suitable buffer. Antibodies can be used for qualitative and quantitative assays of antigens, including immunoprecipitation, western blotting and solid-phase radioimmunoassays (RIA). (See Sepharose, epitope, cyanogen bromide, diazotized paper, nitrocellulose filter, electrophoresis, SDS-polyacrylamide gels, immunoprecipitation, radioimmunoassay)

ANTIBODY, SECONDARY: a molecule, cell, or tissue may be labeled with the cognate antibody (primary antibody). Then to boost the level of recognition the primary antibody is reacted with another antibody (secondary antibody), labeled with an isotope (e.g., I^{125}) or a fluorochrome and this way a stronger signal can be obtained. (See antibody, fluorochromes)

ANTI-CHROMATIN: is a state of the chromatin not conducive for active transcription. (See chromatin, pro-chromatin)

ANTICIPATION: in successive generations it may appear as if the genetic trait (disease) would have occurred with an earlier onset in the more recent generations. This is, however, an artifact because when the investigator knows what is expected, the recognition becomes easier. There is also the possibility that individuals with early onset of the disease died early or failed to leave offspring. (See ascertainment test)

ANTICLINAL SELECTION: the selection takes different directions in different environments compared to the *synclinal selection* when the direction is the same. (See cline)

ANTICODING STRAND: is the transcribed strand of DNA. (See antisense RNA, coding strand, template strand, sense strand, plus strand)

ANTICODON: part of the tRNA which recognizes a mRNA code word by complementarity and it is one of the means of tRNA identity. In the mitochondria the "universal" genetic code does not entirely prevail but different eukaryotic mitochondria (except higher plants) use a somewhat different codon dictionary. In these systems the anticodons are also different inasmuch as there are no separate tRNAs for each of the synonymous codons. Rather the mtDNA codons recognized in pairs or in four-member sets of codons, and the anticodon - codon interaction is by G•U pairing or the 5'-terminal U of the anticodon of the four-member set can pair with any of the four bases in the mRNA codon. (See tRNA, genetic code, wobble)

ANTIDIURETIC HORMONE (vasopressin) is a small peptide hormone (ADH, M_r 1040) increases water reabsorption in the kidney and blood pressure; it affects a variety of functions, including learning and behavior. *Nephrogenic diabetes insipidus* an X-chromosomal human disease with problems of maintaining water balance, fails to respond to ADH. Vasopressin is very similar to oxytocin; only two amino acid difference exists between the two. The structure of vasopressin is:

$$\overset{\longleftarrow S-S \longrightarrow}{Cys - Tyr - Phe - Gln - Asn - Cys - Pro - Arg - GlyNH_2}$$

It binds to receptor molecules in the plasma membrane in the kidney and blood vessels and activates a specific membrane phospholipase. The phospholipase then breaks the bond between glycerol and phosphate in phosphatidylinositol-4,5-bisphosphate and releases inositol-1,4,5-triphosphate and diaglycerol. Vasopressin is encoded in the short arm of human chromosome 20 along with oxytocin. (See oxytocin, diabetes insipidus, phospholipase, diaglycerol; inositol, phosphoinositids, nocturnal enuresis)

ANTIFREEZE PROTEIN: is present in several species of fishes living in the northern regions. The protein binds, through free OH groups of amino acids, to the first ice crystals and thus prevents the expansion of the ice and thus the fish is protected. Somewhat similar proteins may play a role in other organisms too. (See hysteresis, cold hypersensitivity)

ANTIFUNGAL RESPONSE: insects defend themselves against fungi and microorganisms by the production of proteolytic enzymes, phagocytosis and by the production of antimicrobial peptides. In *Drosophila* antifungal drosomycin and several antimicrobial/antibacterial peptides cecropins, diptericin, drosocin, attacin and defensin are produced. The *spätzle, Toll, cactus,* and *dorsal* dorsoventral regulatory genes (corresponding to the mammalian NF-κB cascade) and the immunodeficiency gene, *imd* mediate these responses. (See antimicrobial peptides, morphogenesis in *Drosophila,* host - pathogen relationship, NF-κB)

ANTIGEN: substance (usually a protein) which alone or in combination with a protein elicits antibody formation. The protein antigen may be a large molecule with more than a single specificity due to its different subunits. A particular specificity of the antigen is determined by the epitope or a hapten conjugated with the protein molecule to form an antigen that reacts with the paratope of the antibody. (See antibody, epitope, paratope, superantigen, TI antigens)

ANTIGEN, MALE SPECIFIC: see grafting in medicine, H-Y antigen

ANTIGEN MIMIC: is a short polypeptide used for screening for specific paratope sites. (See paratope, antibody mimic)

ANTIGEN PRESENTING CELL (APC): binds antigens, internalize, process and express them on their surface in conjunction with class II type molecules (one of the two type of molecules coded for by the MHC genes). Helper T cells can be activated only in the presence of APC cells. Macrophages, dendritic (branched) cells and B lymphocyte cells express class II antigens and thus they can serve as APC *in vitro*; *in vivo* macrophages and dendritic cells are apparently the most important as APC. The activation of helper T cells requires that the T cells and the APC are derived from animals (mice) syngeneic in region *I* of the MHC, and the production of the lymphokine, interleukin-1 (IL-1) family member CD80. (See antigen, immune system, T cell, T cell receptor, HLA, syngeneic, lymphokines, interleukins, affinity maturation, CD80, proteasomes)

ANTIGEN PROCESSING AND PRESENTATION: antigen-presenting cells mediate the association of the native antigen with an MHC molecule and thereby the antigen is recognized by the T lymphocytes. The antigenic protein must be degraded to some extent and processed to be presented to the MHC molecules. The MHC I associated peptides are generally shorter than those associated with MHC II molecules that are derived from excreted proteins or other external proteins. Usually the peptides enter the endoplasmic reticulum before their epitope is presented to the MHC molecules. If the proteins lack the signal peptide to be transferred to the endoplasmic reticulum, their epitope may still be presented to the MHC molecules. The MHC Class II molecules are associated with invariant I_i polypeptide that mediates the folding of the MHC II molecules in the endoplasmic reticulum and compartmentalizing the MHC II molecules for special peptide binding in the endosomes. (See antigen presenting cell, HLA, lymphocytes, proteasomes, endosome, CLIP, major histocompatibility complex, TAP, T cell)

ANTIGEN RECEPTORS: are molecules on lymphocytes, responsible for recognition and binding of antigens and antigen-MHC. (See lymphocytes, HLA)

ANTIGENIC DETERMINANT: see epitope, antibody

ANTIGENIC DRIFT: the surface antigens of a pathogen may change by mutation. (See also antigenic variation, phase variation)

ANTIGENIC SIN: individuals who had been previously exposed to one virus and later encountering another virus variant of the same subtype, can make antibodies against the original viral hemagglutinin (HA) and also to the new one. This happens because the memory B cells were activated in a way specific for the progenitor virus. (See hemagglutinin, immune system)

ANTIGENIC VARIATION: is the property of prokaryotic and eukaryotic microorganisms to switch on the synthesis of different surface proteins to escape the immunological defense system of the host organisms. This goal is reached generally by transposition of genes relative to the promoter. The bacterium *Neisseria gonorrhoeae* (responsible for a venereal disease, manifested primarily in males but transmitted through both sexes and the various complications affecting both genders) relies on gene conversion for the purpose. (See phase variation, *Borrelia, Trypanosoma brucei*, cassette model of yeast, gene conversion)

ANTIGENOME: in the replicative form of the viral genetic material it serves as a template for the synthesis of the genome. (See RF)

ANTIHEMOPHILIC FACTORS: blood coagulation requires the formation of complexes between serine protease coagulation factors and membrane-bound cofactors. Tissue factor and proconvertin (VII) are required to activate factors IX and X. Plasma thromboplastin antecedent (XI) activates factor IX. Blood coagulation Factor (VIII) acts in concert with Factor IX_a, a proteolytic enzyme, to activate Factor X (Stuart Factor), and that in turn activates prothrombin (II) to thrombin that acts on fibrinogen (I) to convert it to fibrin (responsible for loose clot) and then the fibrin-stabilizing factor (XIII) generates the firm clots required for blood clotting. Hageman factor (XII) activates factor XI. Accelerin (V) stimulates the activation of II. In

Antihemophilic factors continued
classic recessive X-chromosomal hemophilia Factor VIII is defective, Factor IX deficiency, a partially dominant disorder of hemostasis (arrest of blood flow) is involved in Christmas disease (Xq26-q27). Blood clotting requires in addition calcium and thromboplastin (lipoprotein released into blood from injured tissues). A thromboplastin antecedent (XI) deficiency is responsible for hemophilia C. (See also hemophilia, Hageman trait, PTA deficiency, prothrombin deficiency, Stuart disease, vitamin K-dependent clotting factors, coumarin-like drug resistance, parahemophilia, afibrinogenemia, dysfibrinogenemia, fibrin-stabilizing factor, hypoproconvertinemia, von Willebrand's disease, platelet abnormalities, hemostasis)

ANTIHORMONES: are antagonists of hormones by altering the conformation of the hormones or by binding to the hormone receptor sites and thus preventing the attachment of hormones to the hormone responsive elements (HRE) in the DNA and thus blocking the transcription of the hormone-responsive genes. (See hormone responsive element, conformation)

ANTI-IDIOTYPE ANTIBODY: is a specific antibody that recognizes a particular antigen. Homologous anti-idiotypic antibodies are produced within the species whereas in different species heterologous anti-idiotypic antibodies are produced. (See antibody, idiotype)

ANTILOG: is the inverse logarithm and it is obtained if the base is raised to the power of the logarithm. The antilogarithm for $\log_{10} x$ is 10^x and for $\ln x$ the antilogarithm is e^x. (See logarithm).

ANTILOPE (blackbuck, *Antilope cervicapra*): the male is 2n = 31-33, the female 2n = 30-32.

ANTIMETABOLITE: compound that binds to an enzyme but is not generally utilized as a substrate, and thus interferes with normal metabolism. (See metabolism, metabolite)

ANTIMICROBIAL PEPTIDES: occur on or in animals and plants as a defense system. They can be linear molecules such as *cercopin* (moths, pig, *Drosophila*), making pores by lysis, *magainin* (frog skin-forming) pores, *bactenein* (bovine neutrophiles) affecting membrane permeability. Disulphides: *defensins* (in several organisms) making pores, *tachyplesins* (in horseshoe crab), affecting potassium efflux, *protegrins* (pig leukocytes). *Serprocidins* that are high molecular weight protease-like molecules: *protease 3* and *azurocidin* in mammals and *cathepsin G* in human neutrophils that inhibit metabolism. (See antifungal response)

ANTIMITOTIC AGENTS: block or inhibit mitosis such as ionizing radiation, radiomimetic-chemicals and inhibitors of the cell cycle. (See radiation effects, cancer therapy, cytostatic)

ANTIMONGOLISM CHROMOSOME: a chromosome 21 deletion in humans that compensates in some respects for the syndrome accompanied trisomy of complete chromosomes 21 (Down's syndrome; by old name mongoloid idiocy). This deletion G I causes the formation of large ears, prominent nasal bridges, an antimongoloid slant of the eyelids, long fingers and toes, micro- or dolicephaly and hypo-γ-globulinemia (rather than an excess as in Down's syndrome). (See Down's syndrome, dolicephaly, microcephaly, agammaglobulinemia)

ANTIMORPH: a (dominant) mutation which antagonizes the function of the wild type allele (by competing for the substrate).

ANTIMUTAGEN: protects against the mutagenic effect(s) of other agents. Generally hypoxia, reducing agents (such as dithiothreitol) lower the damage of ionizing radiation. Inhibitors of microsomal mutagen activating enzymes (such as 9-hydroxyellipticine, gallic and tannic acids, carbon monoxide, selenium, etc. may reduce the mutagenic effectiveness of chemicals. (See antimutator, mutagen, caffeic acid, methylguanine-O^6-methyltransferase)

ANTIMUTATOR: lowers mutation rate. Increased level of nuclease activity (editing function) and all other genetic repair mechanisms may act this way. Compounds that inactivate microsomal enzymes involved in conversion of promutagens into mutagens are also antimutagens. (See also AP nucleases, ABC excinucleases, DNA repair, mismatch repair, mutator)

ANTIONCOGENES: the normal alleles of some genes that in the mutant state incite tumors, e.g. the cloned normal allele of the human retinoblastoma gene codes for a DNA-binding protein and the cancer cells transformed by this gene are suppressed in proliferation. (See also tumor suppressor genes)

ANTIPAIN: a protease inhibitor (1 - 2 µg/mL) is effective against cathepsin A and B, papain and trypsin protease enzymes.

ANTIPARALLEL PAIRING of polynucleotide chains means that at the same end of the double helix one has 5' and the other has 3' ends of the paired nucleotides:

```
5'--------------------------------->3'
3'<---------------------------------5'
```

ANTIPODAL: haploid cells (nuclei) located in the plant embryosac at the end opposite to the place of the egg and the micropyle. (See embryosac)

ANTIPORT: membrane transport of substances in opposite directions.

ANTIRESTRICTION MECHANISMS: prevent the cleavage of the DNA by different mechanisms, e.g. methylation of critical bases (e.g. phages T2, T4, SPβ), inhibition of the endonuclease (e.g. T3, T7 phages), enhancing host-encoded methylase (phage λ), carrying hydroxymethyl cytosine in place of cytosine (T-even phages), 5-hydroxymethyluracil substitution for thymine (SPO1, SP8, φ25), reducing certain vulnerable restriction sites in their DNA (φ29), etc. (See restriction endonucleases, restriction-methylation)

Antirrhinum majus (snapdragon): a higher plant of the *Scrofulariaceae* family, (2n = 16). A favorite and attractive autogamous flower for genetic and cytogenetic studies. Large collection of mutants and transposable elements are available. (See TAM)

ANTISENSE DNA: library can be used for transformation and isolation of mutations or for other purposes of preventing gene expression. (See antisense RNA, aptamer, peptide nucleic acid)

ANTISENSE OLIGODEOXYNUCLEOTIDE (AS ODN): see antisense DNA, antisense RNA

ANTISENSE RNA: a transcript of a gene or transposon that may inhibit translation by pairing with the 5' end of the correct (sense) mRNA and thus prevents its ribosome binding and expression. In addition, some synthetic oligonucleotide analogs may block replication and transcription, interfere with splicing of exons, disrupt RNA structure, destabilize mRNA by interfering with 5' capping of mRNA, inhibit polyadenylation, activate ribonuclease H, when coupled to alkylating agents can cross-link nucleic acids at the recognized sequences, can be used as vehicles for targeted DNA cleavage, may inhibit receptors, etc. The various functions require a large variety of specific antisense constructs. Usually the antisense oligonucleotides are 12-50 nucleotide long. According to calculations in the human genome any 17 base sequence occurs only once, and in the mRNA populations; 13mer residues are unique. Shorter sequences do not have sufficient specificity. Long antisense sequences may have self-binding tracts that may cause lowered affinity for their target. Natural antisense RNA transcripts occur in all types of biological systems from viruses to higher eukaryotes. This fact indicates that in eukaryotes both strands of the DNA may be transcribed.

Antisense RNA (or DNA) was expected to become an important therapeutic tool fighting infections and cancer. This technology is still under development for cure against cytomegaloviruses, HIV1, Papilloma virus, autoimmune diseases (arthritis, etc.), leukemia (CMV) and blocking the immune system in case of organ transplants. To some surprise the antisense RNA may trigger an immune reaction because of the CpG blocks are unmethylated and the animal immune system responds to them as to bacterial molecules. In bacteria these bases are largely unmethylated in contrast to eukaryotes where a substantial fraction of the DNA is methylated. The phosphorothioate oligodeoxynucleotides are taken up by a variety of cell types, including some prokaryotes (*Vibrio*) and bind to both DNA and protein. (See host-pathogen relations, triplex, aptamer, pseudoknot, peptide nucleic acid, phosphorothioates, methylphosphonates, cap, fruit ripening, anthocyanin, co-suppression, RIP, Cytomegalovirus, Papilloma virus, autoimmune disease, leukemia, transplantation antigens, antisense technologies, sense strand, aptamer, AS ODN, anticoding strand, coding strand, triple strand formation)

ANTISENSE STRAND OF DNA: is the template strand of DNA from what the mRNA or other functional, natural RNAs are replicated as complementary copies. (See antisense RNA)

ANTISENSE TECHNOLOGIES: use RNA and DNA targets for the suppression or modification of gene expression. The antisense molecule then blocks the synthesis of RNA and protein.

Antisense technologies continued

Various forms of antisense molecules have been used (see antisense RNA); for antisense DNA technology the nucleotides are ligated by e.g. phosphorothioate linkage and not by the normal phosphodiesterase linkage in order to protect the antisense strand from nuclease attack.

$$\text{deoxyribose—O-P(=O)(O)-O—deoxyribose} \quad\quad \text{deoxyribose—O-P(=O)(S)-O—deoxyribose}$$

PHOSPHODIESTER   PHOSPHOROTHIOATE

Guanine-rich antisense sequences may have undesirable side effects on the telomerase enzyme and may form quadruplex structures and interfere with replication of the chromosomes and may bind to proteins and may affect their function. (See antisense RNA, fruit ripening, peptide nucleic acid, triple strand formation)

ANTISERUM: a blood serum that contains specific antibodies obtained from an animal after natural or artificial exposure to an antigen. Antisera are collected from the blood of fasted animals by centrifugation and allowed to clot at room temperature. The clot is then discarded and the straw-colored serum may be preserved either by lyophilization and stored at room temperature or at 4° with 0.02% sodium azide or deep frozen at -20° to -70° C.

ANTISERUM PURIFICATION: of polyclonal antibodies by affinity chromatography on protein A-Sepharose columns. Protein A binds the Fc domain of IgG of various sources but with not equal intensity. Further purification may be obtained by affinity chromatography with immobilized antigen of high purity. (See antibody purification, antibody, immunoglobulins)

ANTI-SHINE — DALGARNO SEQUENCE: CCUCC is complementary to the GGAGG Shine-Dalgarno consensus near the 3'-end of the 16S rRNA molecule. (See Shine - Dalgarno)

ANTISUPPRESSION: inactivates suppressor genes. (See suppressor gene, suppressor tRNA)

ANTITERMINATION permits the RNA polymerase to ignore transcription termination instructions such as bacterial *rho* and thus it proceeds through the termination signal. In phage λ after the transcription of two *immediate early genes* the RNA polymerase should stop. The switch to transcribe the next set of genes is controled by gene *N*, transcribed from the left promoter (P_L) and terminated by the rho-dependent tL_1 terminator and *cro*, transcribed from the right promoter (P_R) and terminated by the rho-dependent tR_1 terminator. The product of the *N* gene is protein N (pN), an antiterminator that permits readthrough to the delayed early genes in both tL_1 and tR_1. Although pN has a half-life of about 5 min, transcription is maintained because *N* is part of the delayed early transcripton. Gene *Q* is also part of the delayed early transcripton and its product pQ is also an antitermination protein that allows by readthrough the transcription at the late promoter P_R. The recognition site for pN is upstream at the N utilization sites, Nut_L and Nut_R, the former is near the promoter but the latter is near the terminator. pN can act on both rho-dependent and rho-independent systems. Different phages have different *nut* sites yet these all work in a similar manner. The *nut* elements include *boxA* and *boxB*, the former is required for binding the bacterial antitermination proteins, used by phages as well as by bacteria. The *boxB* is a phage-specific element.

Mutations in bacteria (*rpoB*) interact with pN. The *nus* loci (*A, B, G*) are involved with transcription termination; *nus E* codes for a protein in the 30S ribosomal subunit (p10). The product of *nusA* is a general transcription factor interacting with p10 and affect termination by binding to *boxA*. Gene *nusG* organizes the various Nus proteins that all together control rho-dependent termination whereas the *nusA* product combined with pN may interfere with termination where it normally is supposed to take place. In *E. coli* antitermination involves also the ribosomal *rrn* genes. This operon has in its leader sequence a *boxA* where the NusB-S10 protein dimer binds to the RNA polymerase as it passes through. This binding enables pol to continue transcription through the rho-dependent terminators of the transcripton. Protein

Antitermination continued

NusA does not bind to the bacterial RNA polymerase when it is associated with the σ factor but after pol attaches to the promoter σ may be released and that provides an opportunity for transcription and for the formation of the core polymerase-Nus complex. After termination of transcription the pol complex is released from the DNA and the separation of Nus from pol takes place. Thus the polymerase core enzyme may be in two alternative state one with σ for transcription and another with Nus with the potential for termination of transcription. Antitermination may then be mediated through pN after the polymerase binds Nus. Gene *Q* of phage λ has also a role in antitermination by permitting through its product the passage through the terminator signals. Transcription is modulated by preventing termination of transcription at T-rich sequences that occur at random within the gene but they dissociate RNA polymerase from the DNA when it arrives to the T-rich region at the end of the gene and where termination of transcription is expected. Other antitermination (attenuation) proteins are acting in the amino acid operons of bacteria, and allow the expression of the operon only after the protein that mediates attenuation of transcription is made, and thus attenuation is not dependent on the presence of an excess of charged specific tRNAs that slow down transcription when the supply of this particular amino acid is sufficient. (See also attenuation, RNA polymerases, rho, lambda phage, half-life, rrn, transcription, σ, transcription termination, tryptophan operon)

ANTITHROMBIN: an α-globulin neutralizing the blood clotting contribution of thrombin. (See blood clotting pathways, antihemophilic factors)

ANTITOXIN: see immunization

ANTITRYPSIN GENE (AAT or PI): in human chromosome 14q32.1 prevents the activity of the protease trypsin and elastase and the α-antitrypsin gene is supposed to be involved in pulmonary emphysema (increase of lung size because dilatation of the alveoli [the small sacs] of the lung), and liver disease. Different mutations may lead to one or the other or to both of these diseases. The so-called Z mutant group prevents the exit of the AAT protein from the liver, where it is synthesized, and as a consequence the liver disease (cirrhosis) appears. Smoking may increase the chances of the development of cirrhosis in the individuals of ZZ genotype by 3 orders of magnitude. The incidence of AAT deficiency is about 8×10^{-4} in the white population of the USA. The total length of the α-antitrypsin gene is 10.2 kb with coding sequences of 1,434 bp. (See emphysema, cirrhosis of the liver, liver cancer)

ANTIVECTOR CELLULAR IMMUNITY: see human gene transfer

ANTIZYME: protein binding to an enzyme and directs its degradation by proteasomes. (See proteasome)

ANUCLEATE: a cell after the nucleus was removed. (See cytochalasins, cytoplast, nuclear transplantation)

AORTA: the main arterial vein (carrying blood away from the heart) originating in the left heart ventricle and passing through the chest and abdomen. (See also coarctation of the aorta)

AORTIC STENOSIS: see coarctation of the aorta

AOTUS (owl monkey): see cebidae

AP: see amino purine base analog mutagen

AP1, AP2, AP3, AP4, AP5: (activated protein): is a group of transcription factors. AP1 is similar to the one coded for by the chicken virus oncogene *v-jun*. The human gene at chromosomal location 1p32-p31 shows 80% homology to the avian viral protein gene; their binding is greatly enhanced by the *fos* oncogen. In yeast AP1 has a homolog, GCN4 and the mammalian homolog is TFIID. The yeast and their mammalian factors can substitute for each other. The AP1 transcription factors are encoded by a family of genes where the binding motif is well conserved but other sequences may vary. These proteins bind to 5'-TGANTCA-3' consensus in DNA. AP2 binds only to TC-II but not to TC-I of the two identical and adjacent TC motifs (5'-TCCCCAG-3') upstream in the promoter of eukaryotic genes. AP2 binding affects enhancer activity. AP2 is an essential morphogenetic factor; in its deficiency head development is impaired. AP2 seems to have negative control on the cell cycle possibly by activation of p21 pro-

AP1, AP2, AP3, AP4, AP5 continued
tein. AP3 binds to TC-II and to the adjacent GT-I motif (5'-G[C/G]TGTGGA[A/T]TGT-3') and also to the so called core enhancer sequence (5' -GTGG[A/T][A/t][A/T]G-3') that is similar to parts of viral and prokaryotic enhancers but does not function by itself alone. AP4 binds to the 5'-CAGCTGTGG sequence that partially overlaps the GT-II motif (that is identical to GT-I, except two bases). AP5 binds to GT-II and adjacent sequences (5'-CTGTGGAATGT-3') and it is present in some cell types but not in others. The mouse *jun* genes (chromosomes 4 and 8) are inducible by serum and the phorbol ester, 12-o-tetradecanoyl phorbol 13-acetate (TPA). The *AP* loci of *Arabidopsis* are completely different and mean *apetala*, defective flower type. (See oncogenes, transcription factors)

AP ENDONUCLEASES: are basically repair enzymes in both prokaryotes (2 enzymes) and eukaryotes (encoded in humans by HAPIm BAP1, APE/APEX) that cut DNA 5' or 3' to modified (alkylated or otherwise mutated) DNA bases or at apurinic and apyrimidinic sites from where glycosylases already removed damaged purines or pyrimidines. Usually the first step is the recognition of the altered bases and the DNA sequence is cut in the vicinity. Then the exonuclease activity removes the damaged section and creates a gap. After that a repair synthesis adds the correct bases to the 3'-OH ends, using the undamaged strand of the double helix as a template. Ligation by covalent bonds restores the integrity of the DNA. The glycosylases have some specificities for deaminated cytosine residues, and uracil-N-glycosylase removes uracil residues, and hypoxanthine-N-glycosylase removes hypoxanthines formed by deamination of adenine. These endonucleases have thus antimutator activities. The eukaryotic DNA uses pol β or pol δ and pol ε for filling the gap. (See antimutator, DNA repair, glycosylases, DNA polymerases)

AP SITE: see apurinic site

APANDRY: development of diploid fruiting body of fungi by fusion of two female nuclei, without the involvement of any male gamete.

APC: see antigen presenting cell, see also Gardner syndrome, ASE1

APC: anaphase promoting complex; also called cyclosome; it is an 8 subunit protein complex containing CDC27, CDC16, CDC23, CDC26 and bimE. (See cell cycle, CDCs, bimE)

APERT or APERT-CROUZON SYNDROME: involves acrocephaly (top of the head pointed), syndactily (fingers fused) and mental retardation although some individuals have near normal intelligence. The symptoms vary. Many of the cases are sporadic, in others autosomal dominant inheritance is most likely; chromosomal rearrangement may also be present in some cases. This condition may be caused also by a defect in FGFR2 (fibroblast growth factor receptor), a protein tyrosine kinase, encoded at 10q25-q26. It is allelic to the Crouzon and Pfeiffer syndromes. (See syndactily, mental retardation, craniosynostosis syndromes, Crouzon syndrome, Pfeiffer syndrome, Jackson-Weiss syndrome)

APEX: the top part of a cell, organ or any structure. (See apical, meristem)

APH: aminoglycoside phosphotransferases, are enzymes phosphorylating aminoglycoside antibiotics, resulting in resistance to the antibiotics when the enzyme is present (introduced by transformation). (See also APH[3']II, antibiotics, aminoglycosides)

APH(3')II: aminoglycoside phosphotransferase enzyme inactivates kanamycin, neomycin and geneticin, commonly used antibiotic resistance markers for transformation in tissue culture; synonymous with NPTII. (See aminoglycoside, antibiotics)

APHERESIS: separation of certain component(s) of a patient's blood and reinfusion the remainder.

APHIDICOLIN: a tetracyclic diterpene of fungal (*Cephalosporium*) origin capable of blocking cell division and of antiviral activity; is an inhibitor of DNA polymerase α. (See pol, terpenes)

APHIDS: small sucking insects, parasites of almost all species of plants. At the site of the infestation the plants secrete a honeydew that may attract other types of insects. They reproduce sexually at the end of the growing season after males have differentiated. During the rest of the year only females are found that reproduce parthenogenetically and their ca. 20 generations

Aphids continued

produce daily three to seven nymphs. Thus the progeny of a single individual may run into billions during the year. Besides the direct damage by sucking, they spread viral diseases of plants. They can be controled by contact or systemic insecticides. (See parthenogenesis)

APICAL: indicates top position. (See apex)

APICAL DOMINANCE: the terminal bud of the main stem of a plant prevents or suppresses the formation of lateral buds or branches.

APICAL ECTODERMAL RIDGE (AER): is the group of cells at the tip of the limb bud, involved in the differentiation of the limbs of animals. (See ZPA, morphogenesis, organizer)

APIGENIN: is a flavone plant pigment.

Apis mellifera (honeybee): social insects with three types of individuals: diploid egg-laying queen (2n = 32), haploid drones and sexually undifferentiated diploid workers. The drones hatch from unfertilized eggs. The difference between the queen and the workers is due to different nutrition of the larvae. (See arrhenotoky)

APLASIA: failure of the development of an organ or a type of tissue.

APLASTIC ANEMIA: a condition of several blood diseases when the bone marrow may not produce the cellular elements of the blood. (See anemia)

APLYSIA: a sea mollusc, an invertebrate small animal, frequently used for behavioral studies.

APM: affected-pedigree-member. APM is used in determining identity by descent and as a nonparametric method to detect linkage. (See IBD)

APNEA: a breathing disorder, responsible for sudden infant death. The genetic basis is unclear.

APO1: see Fas

APOAEQUORIN: see aequorin

APOCYTOCHROME b GENE (cob): is located in mitochondrial DNA of yeast; cytochromes are heme-containing proteins involved in electron transport. (See mitochondrial genetics, mtDNA)

APOENZYME: the enzyme protein without the co-factors required for activity.

APOFERRETIN: the protein part (M_r 460,000) of ferritin, and contains ferric hydroxide clusters. About 20-24% of it is iron. Ferritin is the most readily available iron storage facility in the body. (See ferritin)

APOGAMETY (apogamy): embryo formation without fertilization from a cell of the embryosac, other than the egg cell. (See apomixis)

APOINDUCER: DNA binding protein that stimulates transcription. (See transcription)

APOLAR: molecules are generally insoluble in water because they do not have symmetrical positive and negative charges.

APOLIPOPROTEINS: lipid-binding proteins in the blood that transport triaglycerols, phospholipids, cholesterol and cholesteryl esters within the body. Apolipoproteins are the most important parts of the high-density lipoprotein (HDL) and they are prefered in order to reduce the risk of coronary heart disease. Different classes are distinguished (APOA1, human chromosome 11q23.2-qter and mouse chromosome 9), APOC3 and APOA4 are in the same region. Apolipoprotein B (human chromosome 2p24) exists in two lengths due to different editing of the transcript. APOC cluster is the short arm of human chromosome 19 and APOE appears to be linked to it. APOE deficiency causes hyperlipidemia and atherosclerosis. Some other apolipoproteins are genetically less well defined. Apolipoprotein A-IV may protect against atherosclerosis without an increase of HDL levels. Apolipoprotein B deficiency reduces male fertility in knockout mice. (See cholesterols, fatty acids, atherosclerosis, arteriosclerosis, HDL, hyperlipidemia, hyperlioproteinemia, lipoprotein lipase, cholesterol)

APOMICT: reproduces by apomixis. (See apomixis, *Rosa canina*)

APOMIXIA: parthenogenesis, common in *Caenorhabditis elegans*, bees, wasps, aphids, in some crustacea, isopoda, lepidoptera, etc.; it does not occur in humans. (See parthenogenesis, apomixis)

APOMIXIS: embryo (zygote) development without fertilization in plants and fungi. It occurs

Apomixis continued

regularly in certain species, e.g. in the polyploid *Festuca*, hawkweeds (*Hieracium*), etc. Some apomicts reproduce sexually after doubling the chromosome number. Apomicts may make possible the fixation of heterozygous condition. Apomixis may be genetically very different from somatic embryogenesis. If apomixis is preceded by meiosis, and the egg parent was heterozygous, segregation may occur among the apomictic progeny. (See parthenogenesis, apomixia, apogamety, agamospermy, androgenesis)

APOMORPHIC: a species trait evolved from a more primitive state of the same. (See plesiomorphic, symplesiomorphic, synapomorphic)

APOPAIN: see apoptosis

APOPLAST: intercellular material.

APOPTOSIS: programmed cell death (aging). The cells and the nuclei shrink and generally are absorbed after fragmentation. A generalized outline of the apoptotic cell death pathway is represented on the following page (modified after Thompson, C.B. 1995 Science 267: 1456). Ceramide is one of the regulatory molecules of the process. Tumor necrosis factor (TNF) is an inducer of apoptosis. The metabolites of ceramides, sphingosine and sphingosine-1-phosphate prevent the symptoms of apoptosis. These two molecules are supposedly second messengers for cell proliferation mediated by platelet-derived growth factor. The ceramide-mediated apoptosis is inhibited also by activation of protein kinase C brought about by sphingosine kinase and the increase of the level of sphingosine-1-phosphate. The latter molecules stimulate also the ERK-controlled reactions and inhibit the stress-activated kinases SAPK/JNK. In *Caenorhabditis* more than a dozen *ced* (cell death) genes have been identified; Ced3 protein is an interleukin-1 converting (ICE) cystein protease enzyme, involved in ceramide production. Ced9 is a suppressor of cell death. The lattter has 23% identity to the human oncogene BCL-2, controling follicular lymphoma. If this human gene is transfected to the nematode it suppresses apoptosis, indicating that the same function is controled over a wide evolutionary range. The *reaper* gene (*rpr*) of *Drosophila* is an activator of apoptosis and it is homologous to *ced-3* of *Caenorhabditis*. Another *Drosophila* gene, *hid* (head-involution defective), is linked to reaper. The 65-amino acid RPR protein has also similarity to the "death domain" of the tumor necrosis factor receptor (TNFR) family. TNFR1 and Fas induce cell death when activated by ligand binding or when overexpressed. The dead cells are disposed of generally by the macrophages without producing inflammation in the tissue. The apoptotic cells are recognized generally by the altered sugar groups or phosphatidylserine on their surface. The macrophage secretes an extracellular protein, thrombospondin which recognizes apoptotic cells. The *Alg-2* (apoptosis-linked gene) encodes a Ca^{2+}-binding protein that is required for T cell receptor-, Fas- and glucocorticoid-induced apoptosis. *Alg-3* is a homolog of the Alzheimer disease gene which is basically a senescence gene. Apoptosis may be a very natural response of the cells to be disposed when no longer needed. In some cancers the proliferation is out of hand because regulators of the process go awry. The baculovirus apoptosis inhibitor proteins (Cp-IAp and Op-IAP) as well as neuronal apoptosis inhibitor proteins (NAIP), located in human chromosome 5q13.1, are defective or deleted in spinal muscular atrophy. BAX is a heterodimeric protein that works in the opposite direction as BCL2 (chronic lymphocytic leukemia, B cell). The gain of function BAX mutations, knockouts were viable but in the lymphocytes cell lineages apoptosis was induced; in other cell lineages hyperplasia was observed. Thus the BAX expression depends on cellular context. The BCL2 gene functions similarly to *Ced-9* of *Caenorhabditis*, i. e. it suppresses apoptosis. The enzyme apopain, cleaving poly(ADP-ribose) polymerase (PAR) is also necessary for apoptosis to proceed. Apopain is generated from the proenzyme called CPP32, a protein related to ICE and CED-3. Lymphocyte apoptosis may be mediated by Type 3 inositol 1,4,5-trisphosphate receptor in the plasma membrane by promoting the influx of calcium. Apoptosis of neurons is mediated by the activation of JNK (JUN [oncogene] NH_2 terminal kinase) in a process opposing the effect ERK (extracellular signal-activated kinase) in the absence of the NGF, nerve growth factor.

Apoptosis continued

The p35 protein of the Baculovirus *Autographa californica* has similar antiapoptotic property for insects as well as for mammals as the Ced-9 gene product of *Caenorhabditis*. The Ced-9 gene product inhibits both the apoptosis promoting and protecting effects of the Ced-4 products.

THE PATHWAY OF APOPTOSIS AND ITS INHIBITION

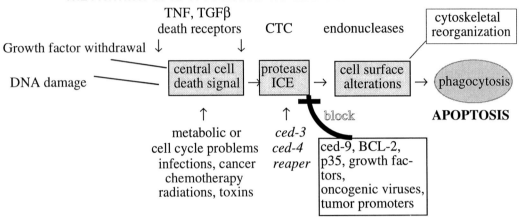

Ced-9 is homologous to the mammalian BTL-2. *Ced-3* and *Ced-4* are considered as promoters of apoptosis. Actually, *Ced-4* has two transcripts; the short transcript promotes apoptosis but the long transcript is somewhat protective. Overexpression of *ced-4L* actually can prevent programed cell death. It is interesting to note the structurally unrelated *BCL-x* and *Ich-1* genes are also involved in apoptosis, similarly to *Ced-4*, all have two alternative transcripts. This indicates that RNA splicing may have an important role in programed cell death. The basic leucine zipper proteins (bZIP) PAR (proline and acid rich) and other members of the protein family can also control apoptosis. Inappropriate activation of apoptosis may be involved in diseases such as AIDS, degeneration of the nervous system (Alzheimer disease, amyotrophic lateral sclerosis, Parkinson's disease, retinitis pigmentosa), constriction or obstruction of the blood vessels (ischemic stroke), anemias, liver diseases, auto-immune diseases, etc.

Recently for the ICE - ced protease enzyme system the generic name CASPASE has been suggested (c for cystein protease, aspase for cleaving aspartate). The individual enzymes would be designated also by numbers. The ced-3 enzymes are cysteine proteases, accompanied by nucleases that cut the DNA to approximately 180 bp fragments, the size of a nucleosomal unit. Apoptosis is used for multiple purposes besides aging: differentiation, homeostasis, cellular defense. Programed cell death occurs also in plants during the differentiation of the vascular system (xyleme), fruit ripening and senescence, the hypersensitive defense reaction against pathogens. (See aging, Hayflick's limit, necrosis, ceramides, sphingosine, ERK, SAPK, T cell, signal transduction, TNF, Fas, TGF, p53, interleukins, leukemia, cystein proteases, fragmentin-2, perforin, ICE, FADD/MORTI, FLICE, granzymes, Down's syndrome, acquired immunodeficiency, Alzheimer disease, amyotrophic lateral sclerosis, retinitis pigmentosa, aplastic anemia, CTC, apopain, addiction module, altruism)

APOREPRESSOR: repressor proteins that require another molecule, the co-repressor (frequently a late product of the metabolic pathway) to be active in controlling transcription. (See transcription, tryptophan operon, tryptophan repressor)

APOSPORY: seed formation without fertilization from diploid cells of the nucellus or integumentum. (See apomixis, agamospermy diplospermy, adventitious embryo, apomixis)

APOSTATIC SELECTION: predators often prefer the most abundant types of prey. (See frequency dependent selection)

APOTHECIUM: an open fruiting body of fungi on which the asci develop; it is similar to perithecium but the latter is a closed fruiting body. Among the genetically widely used organisms, *Ascobolus* develops apothecia. (See perithecium)

APOTRANSFERRIN: a transport protein without its ligand. (See transporters, ligand)

apo-VLDL: see apolipoprotein

AP-PCR: arbitrarily primed PCR. (See polymerase chain reaction)

APPLE (*Malus* spp.): about 25 species all with x = 17 and most of them are diploid although tetraploid and triploid varieties also occur. Apples are frequently self-incompatible but cross-fertile with other apples and do not easily hybridize with pears (*Pyrus*) but they hybridize with *Sorbus* (mountain ash). (See pears)

APPLICATION PROGRAMS (computer): programs serving special purposes such as word processing, graphics, telecommunication, DNA sequencing, data management, etc.

APRF (acute-phase response factor): is a transcription factor related to the p91 subunit of the interferon-stimulated gene factor-3α (ISGF-3α). It is phosphorylated by the mediation of cytokines, and along with Jak1 kinase, it is associated with gp130. Interleukin-6, leukemia inhibitory factor (LIF), oncostatin M (OSM) and ciliary neurotrophic factor (CNTF), cytokines, neurokines and neuronal differentiation factors are also involved. (See gp130, signal transduction, cytokines)

APRICOT (*Prunus armeniaca*): x = 7 but 2x and 3x forms are known.

APROTININ: inhibitor (at concentrations 1 - 2 µg/mL) of proteases kallikrein, trypsin, chymotrypsin, plasmin but not of papain. (See protease, kallikrein, trypsin, chymotrypsin, plasmin, papain)

APTAMER: is an oligo-RNA, oligo-DNA or a protein—oligo-RNA complex that can bind specifically a particular protein. E. g., a thrombin-binding aptamer will inhibit the action of thrombin in blood clotting and thus prevents the formation of blood clots. (See antisense RNA, SELEX)

APURINIC ENDONUCLEASE: see AP endonucleases

APURINIC SITE (AP): from where a purine has been removed from the nucleic acid.

APYRIMIDINIC ENDONUCLEASE: see AP endonucleases

APYRIMIDINIC SITE: from where a pyrimidine has been removed from a nucleic acid.

AQUAPORIN 1: a six-transmembrane protein domain and water channel in fluid absorbing and fluid secreting cells. It is activated by cAMP-dependent mechanisms or PKA. It is important for various types of cells, diabetes, kidney function, *Drosophila* neural development, nematodal infestation of plants, etc. (See CHIP, forskolin, cAMP, PKA, cell membranes, ion channels)

AQUEOUS: prepared with water (e.g., a solution) or watery in appearance.

ARABIDOPSIS **MUTAGEN ASSAYS**: in the mature embryo of the plants the inflorescence is represented by two diploid cells. Therefore if the seeds are exposed at such a stage to a mutagen, in the progeny of the emerging plants the segregation is about 7:1 for recessive mutations. This also indicates that one of the two apical cells became heterozygous for the new mutation induced. For mutagen assays it is sufficient to open up the immature fruits of the plants before the seed coat becomes opaque (about 10-14 days after fertilization) and albina or other color mutations in the cotyledons or embryo defects can be determined. The fruits on the plants emerging from the treated seed contain already the F_2 generation. Generally, two opposite fruits next to each other are examined because the phyllotaxy index assures that these are sufficient for complete sampling. Such a test permits the identification of about 80% of the spectrum of visible mutations. Since the plants are diploid and the "germline" consists of two cells, mutation rate on genome basis is calculated by counting all independent mutational events and dividing that number by the total number of plant tested x 4. *Arabidopsis* can activate many types of promutagens and therefore provides an efficient means to assess genotoxic effects of a wide variety of agents in a single culture at low cost. (See bioassays for genetic toxicology, *Arabidopsis thaliana*, phyllotaxy)

Arabidopsis thaliana: an autogamous plant of the crucifer family, 2n = 10, genome size has been estimated to be 9×10^7 to 1.5×10^8 bp. Its life cycle may be as short as 5-6 weeks, its seed

Arabidopsis thaliana continued

output may exceed 50,0000 per plant. Its mitochondrial DNA is 366,924 bp and apparently codes for 57 genes. Because of its small size thousands of individuals may be screened even on a Petri plate. This will probably be the first higher plant with the genome completely mapped and sequenced. Information is available through Arabidopsis Biological Resource Center, Ohio State University, 1735 Neil Ave., Columbus, OH 43210, USA. Tel.: 614-292-1982 (Scholl, seeds), 614-292-2115 (Davis, DNA). E-mail: <arabidopsis+@osu.edu>. Orders: by Fax 614-292-0603. E-mail: <seeds@ genesys.cps.msu.edu> (seed orders), <dna@genesys. cps.msu.edu>, on-line: inquire <aims@ genesys.cps.msu.edu>, Weeds World: < http://nasc. nott.ac.uk:8300>, AAtDB<http://weeds.mgh.harvard.edu/weedsworld/home.html> or <gopher://weeds.mgh.harvard.edu70/11arabidopsis/weedsworld>, Agricultural Library, USDA <http://www.nal.usda.gov/pgdic/>

ARABINOSE OPERON: consists of three juxtapositioned structural genes *araB* (L-ribulokinase), *araA*, (L-arabinose isomerase) and *araD* (L-ribulose-4-epimerase) transcribed in this order into a polycistronic araBAD mRNA starting at the O_{BAD} operator and initiated by the P_{BAD} promoter), The repressor-activator site is functioning by positive or negative control and it is transcribed in the opposite direction from the O_C operator and uses the P_C promoter. These genes are near the beginning of the *E. coli* map, while another gene *araF* is located at about map position 45. They form a common regulatory system: a regulon. Activation of the *ara* operon requires the presence of the substrate arabinose and that the catabolite activating protein - cyclic adenosine monophosphate complex be attached to the promoter. This operon is subject to catabolite suppression and as long as glucose is present in the medium (even when arabinose is also available) its transcription cannot begin. (See operon, polycistronic, operator, catabolite activating protein, cAMP, negative control)

ARABINOSURIA: was an early name of pentosuria but subsequently it turned out that L-xylulose was misidentified as arabinose; the current name of the recessive disorder is (essential) pentosuria. (See pentosuria, xylulose)

ARACHIDONIC ACID (arachidate): is an unsaturated fatty acid and it is called arachidonate when there are four double bonds in the molecule (it is synonymous with eicosatetraenoate). It occurs in lipids and plays a role in mediating signal transduction. Cyclooxygenase mediates the formation of proastaglandins, prostacyclins and thromboxanes whereas lipoxygenase catalyzes the synthesis of leukotrienes from arachidonic acid. (See fatty acids, cyclooxygenase, lipoxygenase signal transduction)

ARAF ONCOGENES: have been assigned to the mouse X-chromosome whereas in humans ARAF 1 is in chromosome Xp11-p11-2 and ARAF 2 was localized to either 7p11.4-q21 or 7p12-q11.21. These oncogenes are homologous to the RAF1 oncogene and are supposed to encode a serine/threonine kinase. (See RAF1)

ARBOVIRUSES: are parasites of blood-sucking insects. Their genetic material is RNA.

ARCHAEA: the third major group of living system besides bacteria and eukarya. *Methanococcus jannaschii* DNA (1.66 megabase) has been sequenced by 1996 and 1,738 predicted protein-coding genes have been identified. The organism has two, 58 kb and 16 kb, extrachromosomal elements. Only 38% of its genes appear similar to genes (by nucleotide sequences) of other fully sequenced bacteria or budding yeast. The metabolic genes bear similarities to bacterial genes whereas its genes involved in transcription, translation and replication resemble more eukaryotic genes. The complete nucleotide sequence is available through World Wide Web at <http://www.tigr.org/tdb/mdb/mjdb/mjdb.html>. (See life form domains)

ARCHAEBACTERIA: groups of prokaryotes that appear to have some similarities to eukaryotes inasmuch as displaying nucleosome-like structures in their DNA, introns in their genes, unlinked 5S RNA genes and their transcriptase enzyme is somewhat related antigenically to similar enzymes in lower eukaryotes.

ARCHEGONIUM: female sexual organ (gametangium) of lower plants (ferns) in what the eggs develop.

ARCHEOZOIC: geological period 400 to 100 million years ago when protists (unicellular organisms) evolved.

ARCHESPORE: the ancestral, enlarged cell that develops into the megasporocyte (megaspore mother cell) in plants. (See megagametophyte)

ARCHITECTURAL PROTEINS: modulate DNA structure in such a way that transcription factors would gain better access to the promoter area. (See UBF, high-mobility group proteins)

ARCHIVAL DNA: had been stored in museum, herbarium or other preserved samples of long dead cells; can be amplified with PCR techniques for analysis and find information on old populations or on extinct species. (See PCR, ancient DNA)

ARCHTYPE: a hypothetical ancestral form in evolution.

ARCSINE: the inverse of sine ($sine^{-1}$) and stands for the angle of whose sine is given. (See sine)

ARCSINE TRANSFORMATION: see angular transformation

ARCTOCEBUS: see Lorisidae

ARE (anoxia response element): DNA sequences regulating responses to anaerobiosis.

α-REPEAT: is a 171 bp abundant (up to 1,000,000) repeat in the human genome, localized primarily in the centromeric regions of the chromosomes. This and the Alu repeats constitute 5-10% of the human genome. (See Alu family)

ARF (ADP-ribosylation factor): is a GTP-binding protein involved in the transport between the endoplasmic reticulum and the Golgi apparatus and within the Golgi complex. Also, it is required for the physiological effect of cholera and pertussis toxins. (See signal transduction, cholera toxin, pertussis toxin, GTPase, SecA, SecB, translocase, translocon, ARNO, Golgi)

ARF1: a GTPase protein activating phospholipase D; it is activated by PtdIns (phosphatidyl inositol), and participates in the recruitment of coatomer and trans-Golgi network (TGN) clathrin coat. (See coatomer, clathrin, phosphatidyl inositol)

ARG: an oncogene related to ABL in human chromosome 1q24-q25 and in mouse chromosome 1. It encodes a tyrosine kinase, different from that of the ABL product. (See oncogenes, ABL)

ARGINASE: see argininemia

ARGININE (2-amino-5-guanidinovaleric acid): an essential amino acid. (See urea cycle)

ARGININEMIA (hyperargininemia): accumulation of high levels or arginine in the blood and urine caused by autosomal recessive arginase deficiency (ARG1 and ARG2 genes). ARG1 (6q23) coded enzyme represents 98% of the arginase activity in the liver and its deficiency is the common argininemia. Arginine accumulates in the blood because it is not degraded. It is a relatively rare disease. Treatment with benzoate and restriction of arginine intake may ameliorate the condition. Shope virus infection restores arginase activity in the cells. (See amino acid metabolism, citrullinemia, citrullinuria, urea cycle)

ARGININOSUCCINIC ACIDURIA: rare hereditary disorder (human chromosome 7q) involving mental retardation, seizures, hepatomegaly (enlargement of the liver that may become cancerous), intermittent ataxia, brittle and tufted hair, and accumulation of large quantities of arginosuccinic acid (an intermediate in the arginine-citrulline [urea] cycle) in the blood, urine, and the cerebrospinal (brain and spinal cord) fluid. Early and late onset types have been distinguished. (See urea cycle)

ARGINYL tRNA SYNTHETASE: the enzyme that charges the appropriate tRNA with arginine. The encoding gene was located to human chromosome 5. (See aminoacyl tRNA synthetase)

ARGOS: a secreted *Drosophila* protein containing a single EGF motif; it is a repressor of eye and wing determination and it acts against Spitz. (See Spitz, DER, EGF)

ARITHMETIC MEAN: $\bar{x} = \frac{\Sigma x}{N}$ is the sum of all measurements (x) divided by the number of measurements (N). (See mean)

ARITHMETIC PROGRESSION: a series with elements increasing by the same quantity, e.g., 1, 3, 5, 7, 9. (See geometric progression)

ARM RATIO: the relative length of the two arms of a eukaryotic nuclear chromosome. (See chromosome arm, chromosome morphology)

ARMADILLO: *Euphractus sexcinctus* 2n = 58; *Dasypus novemcinctus* 2n = 64; *Cabassous centralis* 2n = 62; *Chaetophractus villosus* 2n = 60.

Armadillo (*arm*, 1-1.2): homozygous lethal gene of *Drosophila*, involved in embryonic differentiation in connection with other genes. Its vertebrate homolog encodes is β-catenin. (See morphogenesis in *Drosophila*)

ARMS (amplification refractory mutation system): used with PCR, it may detect the strand that contains a known mutation or identify polymorphism of a particular DNA stretch. (See PCR)

ARMS OF BACTERIOPHAGE λ: when the stuffer segment is removed a left and a right segment (arms) of the genome remains and these are used for vector construction. (See stuffer DNA, lambda phage)

aRNA: ancient RNA. (See ancient organisms, ancient DNA)

ARNO (ARF nucleotide binding site opener): is a 399 amino acid human protein involved in the GDP⇔GTP exchange of ARF. It is a homolog of the yeast Gea1, and both are inhibited by brefeldin. (See ARF, GTP, brefeldin)

AROMATIC MOLECULE: a closed ring molecule with C in the ring, linked by alternating single and double bonds; they are frequently conjugated with other compounds.

ARPKD: see renal-hepatic-pancreatic kidney disease, polycystic kidney disease

ARRAY HYBRIDIZATION: is designed to identify single (or small number) of nucleotide changes in genomic DNA. The procedure requires a large array of oligonucleotide probes which are obtained by light-directed parallel chemical synthesis. Using in each oligonucleotide set four probes which differ only in one of the 4 bases, A, T, G, C, and the flanking bases are kept identical. Each sequence of the target is then queried by the complementary synthetic probes:

Target: 5'... TGAACTGTATCCGACAT...3'

Probes 3' TGACAT**A**GGCTGTA   match
            TGACAT**C**GGCTGTAG ⎤
            TGACAT**G**GGCTGTAG ⎬ mismatches
            TGACAT**T**GGCTGTAG ⎦

The single-base different hybridization probes are distinguished on the basis of the signals provided by the differences in the fluorescence labels and hybridization intensities of the probes. The procedure permits also the identification of more than one base and deletions. An entire genome can be thus scanned by a confocal device and its great merit is fastness. (See light-directed parallel synthesis, see also Chee, M. *et al.* 1996 Science 274:610)

ARRAYED LIBRARY: cloned DNA sequences are arranged on two-dimensional microtiter plates where they can be readily identified by row and column specifications. (See DNA library, microarray)

ARRESTIN: a 45 kDa protein regulating the phototransduction pathway in animals. (See phototransduction)

ARRHENOTOKY: a mechanism of sex determination; the males are haploid and the females are diploid for the sex genes (as in bees, wasps); the males develop from unfertilized eggs (a form of parthenogenesis) and display one or the other allele(s) for what the females (queens) are heterozygous for. (See chromosomal sex determination)

ARS: autonomously replicating sequences are about 100 bp long origins of replication of yeast chromosomal DNAs. The different ARS sequences share a consensus of 11 base pairs (5'-[A/T]TTTAT[A/G]TTT[A/G]-3'), and there are some additional elements around it that vary in the different chromosomes from where they were derived. Artificial yeast plasmids must contain ARS sequences to be maintained and they may remain stable as long as selective pressure exists for their maintenance, i.e. they carry essential genes for the survival of the yeast cell (missing from or inactive in the yeast nucleus). ARS elements occur also in organellar and other DNAs. (See YAC, yeast vectors, DUE, cell cycle, ORC, MCM)

ART (assisted reproductive technology): may be helpful to about 15% of the couples who are infertile. Various techniques are available to help conception. (See artificial insemination, intrauterine insemination, *in vitro* fertilization, oocyte donation, intrafallopian transfer of gamete and zygote, surrogate mother, sperm bank, preimplantation genetics, micromanipulation of the oocyte, ICSI, counseling genetic)

ARTERIOSCLEROSIS: thickening and hardening the walls of arterial veins; a common form of heart disease. (See atherosclerosis)

ARTHRITIS: an inflammation of the joints is caused by several factors with incomplete penetrance and expressivity. It is of a common occurrence in familial gout, a hyperuricemia (excessive uric acid production). Rheumatoid arthritis is generally considered as an autoimmune disease; it appears to be autosomal dominant but the genetic control is not entirely clear. TNFα, IFNγ, IL-2, IL-1 have been implicated in the inflammatory symptoms of rheumatoid arthritis. Anti-TNF antibody treatment may offer some promise. The prevalence is about 1% in general population. (See arthtropathy, arthropathy-camptodactyly, connective tissue disorders, TNF, IFN, IL, NF-κB)

ARTHROGRYPOSIS: unclear (autosomal) genetic determination of this malformation of low recurrence affecting limb deformations, hip dislocation, scoliosis (crooked spine), frequently short stature, amyoplasia (poor muscle formation), etc. X-linked forms have also been described. (See limb defects, connective tissue disorders)

ARTHROPATHY: any disease that affects the joints.

ARTHROPATHY-CAMPTODACTYLY (synovitis): is based on autosomal recessive inheritance. It involves inflammation of the joints (synovial membranes) resembling arthritis. It may have an early childhood onset. (See connective tissue disorders)

ARTHROOPHTALMOPATHY: see Stickler syndrome

ARTICHOKE (*Cynara scolymus*): vegetable crop; 2n = 2x = 34.

ARTIFICIAL CHROMOSOME: see YAC, BAC, PAC, human artificial chromosome

ARTIFICIAL INSEMINATION: may be used to overcome the consequences of male infertility in humans or to obtain larger number of offspring from male animals with economically desirable characters and high productivity. Generally the sperm is obtained from sperm banks where the semen is preserved at very low temperature. The cryopreservation may protect against sexually transmitted disease. (See sperm bank, ART, intrauterine insemination, surrogate mother, AID, AIH)

ARTIFICIAL INTELLIGENCE: a device (computer) with the ability to function similarly to human intelligence, i.e. capability of learning, reasoning and self improvement.

ARTIFICIAL SELECTION: see selection, selection conditions, selection index, gain, natural selection

ARYLESTERASE (ESA, paraoxonase): is encoded in human chromosome 7q22. The enzyme breaks down parathion and related insecticides. (See cholinesterase, pseudocholinesterase)

ARYL HYDROCARBON RECEPTOR (ARH): mediates the carcinogenic and teratogenic, immunosuppressive, etc. responses to arylhydrocarbons present in many environmental toxins (dioxin, benzo(a)pyrene, cigarette smoke, polychlorinated and polybrominated biphenyls, etc.) ARH regulated genes include cytochromes P450, uridine diphosphate - glucuronosyl transferase, growth factors and proteins. (See items under separate entries)

ARYLSULFATES: are aromatic molecules with bound sulfate. Deficiency of arylsulfatases is involved the lipidosis group of diseases, collectively designated as metachromatic leukodystrophy. (See lipidoses, Krabbe's leukodystrophy, metachromatic leukodystrophy)

α-SATELLITE DNA: is centromeric repetitive DNA. (See repetitious DNA, satellite DNA)

AS: see asparagine synthetase

AS ODN (antisense oligodeoxynucleotide): may bind oncogene mRNA and may inhibit cancer growth and regulate the formation of megakaryocytes. (See antisense technologies, cytofectin)

Ascaris megalocephala (horse threadworm): has very unusual chromosome behavior. It has only one pair of large chromosomes in the germline but during somatic cell divisions these large

Ascaris megalocephala continued
chromosomes are fragmented into a large number of small chromosomes. This organism made possible for Van Beneden in 1883 the discovery of reductional division in meiosis, a corner stone of the cytological basis of Mendelian segregation. *A. megalocephala univalens* has 2n = 2, *A. megalocephala bivalens* 2n = 4, and *A. lumbricoides* 2n = 43 chromosomes.

ASCERTAINMENT TEST: is generally required in larger mammals with few offspring to determine the segregation ratios on the basis of pooled data of several families. The problem involved in biased sampling (because those families where the parents are heterozygous for the recessive gene escape identification if no homozygotes are observed among the progeny) can be corrected for. The solution is mathematical. According to Mendelian expectation 3/4 of the single-child families have no affected children. Among the two-child families $(3/4)^2 = 9/16$ is the probability that neither will be of recessive phenotype. Of the remaining 7/16 of the families 6/7 will have one recessive and one dominant and 1/7 should have 2 recessives. Thus, the average expectation is $(1) \times (6/7) + (2) \times (1/7) = 8/7 = 1.143$. In the three-child families 27/64 will have no affected offspring, 9/37 will have 2, and 1/37 are expected with 3 recessives. Therefore the average expected is $(1) \times (27/37) + (2) \times (9/37) + (3) \times (1/37) = 1.294$. In the same manner the average expectation of recessives for various sizes of families can be determined:

Number of Children →	1	2	3	4	5	6	7	8
Average Homozygotes	1.000	1.143	1.297	1.463	1.639	1.825	2.020	2.223

With this information the number of observed and expected data for affected and unaffected of several size families can be analyzed with the chi square procedure and the goodness of fit can be evaluated:

Number of sibs/family	families	Number of affected sibs observed	expected	Number of unaffected sibs expected	observed
1	7	7	7 × 1 = 7.00	0	0
2	10	8	10 × 1.143 = 11.43	12	8.57
3	4	6	4 × 1.297 = 5.19	6	6.81
5	2	4	2 × 1.639 = 3.28	6	6.72
Total	25		26.90	24	22.10

$$\chi^2 = \frac{(25 - 26.9)^2}{26.9} + \frac{(24 - 22.1)^2}{22.1} = 0.298;$$ the degree of freedom = 1, and the probability of fit is >0.5 (for a χ^2 only values below 0.05 would have some ground for doubting the fit). A simpler (and less reliable) procedure for determining the average number of recessives:

$$\hat{q} = \frac{R - N}{T - N}$$

here R the number of recessive segregants observed, N = number of families showing recessives, T = the total number of children of these families. (See chi square)

ASCHHEIM-ZONDEK TEST (AZT) uses subcutaneous injection of the urine of human females into immature female mice to test for early pregnancy. Swelling, congestion and hemorrhages of the ovaries and precocious maturation of the follicles in the mice are positive indicators of pregnancy of the tested person. Today, either a hemagglutination or chorionic gonadotropin tests are used. Pregnancy immediately raises dramatically the level of this hormone.

ASCI: plural of ascus. (See ascus)

ASCITES: is a condition when ascitic fluid (may contain also cells) is excreted in response to cell proliferation in the abdominal cavity because of a neoplasia. The fluid is serum, containing polyclonal antibodies. Cirrhosis or hypoalbuminemia, and experimental injections may also cause ascites.

Ascobolus immersus: an ascomycete where the dissection of the ascospores is very simple, the spores spring off when touched and can be captured on microscope slides. This fungus has been extensively used for studies of recombination and gene conversion; x = 12, 16, 18.

ASCOCARPS: are the sites in fungi where the perithecia and apothecia (fruiting bodies) develop.

ASCOGENOUS HYPHAE: are diploid or bikaryotic hyphae leading to the formation of fruiting bodies in fungi. (See fruiting body, hypha)

ASCOGONIUM: the gametangium (oogonium) the female sexual organ of fungi (called also protoperithecium)

ASCOMYCETE: a large group of different fungi producing either asexual conidiospores and/or ascospores within asci as a consequence of meiosis. (See tetrad analysis)

ASCORBIC ACID: (vitamin C) is an anti scurvy (anti scorbutic) substance. It is required for proper hydroxylation of collagen and its deficiency leads to skin lesions and damage to the blood vessels, symptoms of scurvy. It is also a reducing compound and upon oxidation it is converted into dehydroascorbic acid. Together with Fe(II) and O_2 it is a hydroxylating agent for aromatics. In the process H_2O_2 is also formed. It has been claimed that in high daily doses it reduces the risks of the common cold and other ailments. Also, it has been found to be weakly mutagenic, probably because of its ability to generate free radicals. Primates and guinea pigs cannot synthesize this vitamin and they depend on dietary supplies.

ASCOSPORES: the haploid products of meiosis formed within an ascus. (See ascus, tetrad analysis)

ASCT1: is a zwitterionic amino acid transporter. (See transporters, zwitterion)

ASCUS: a sac-like structure in the *Ascomycete* fungi, containing the four products of meiosis (spores). In many fungi the number of ascospores may become eight due to a mitotic division following meiosis. The spores in the asci may be arranged in the same linear order as in the linear tetrad of meiosis (ordered tetrads such as *Neurospora, Ascobolus, Aspergillus*, etc.) or may be scrambled (unordered tetrad such as in yeast). Asci have been used very effectively to study the mechanics of recombination because the results of single meiotic events could be analyzed separately. (See tetrad analysis)

ASCUS-DOMINANT: a mutation or even a deletion affects (prevents) the expression of the dominant allele within an ascospore. It has been attributed to reduced dosage, defects in internuclear communication and transvection. (See transvection)

ASE1 (anaphase spindle elongation): is a gene encoding MAP, required for elongation of the mitotic spindle and separation of the spindle poles. It is degraded by the anaphase promoting complex (APC). (See MAP, spindle, cell cycle)

ASEPTIC: culture is free from contaminating microorganisms. (See axenic)

ASEXUAL REPRODUCTION: does not involve fusion of gametes of opposite sex or mating type. (See also reproduction)

AS-FISH (antisense fluorescent *in situ* hybridization): the sense strand of the DNA is labeled by the probe and thus it may make possible to label differentially the transcribed and non-transcribed heterologous DNA, introduced by transformation into the cell. (See FISH, antisense strand)

ASH: the mineral residue of tissues left after igniting the organic material.

ASH TREE: forest and ornamental trees (*Fraxinus excelsior*, 2n = 46; *F. americana*, 2n = 46, 92, 138)

ASHKENAZI(M): Jewish population that lived during the Middle Ages in German lands although migrated from there to Eastern Europe and other parts of the world. They preserved their ethnic identity and a special gene pool. Therefore certain hereditary conditions such as Tay-Sachs disease, Gaucher disease, Niemann-Pick's disease, Bloom's syndrome, higher I.Q, etc. occur at increased frequencies in the population compared to other ethnic groups. (See Sephardic, Jews and genetic diseases)

ASILOMAR CONFERENCE: in 1976, at the beginning of the more wide use of recombinant DNA, scientist convened at this California place to work out voluntary guidelines for protection against the potential hazards involving the application of the new techniques.

ASK1: is a member of the mitogen-activated MAP protein family; it is activated by TNF-α. It induces apoptosis but it may inhibit TNF-α-induced apoptosis. (See apoptosis, MAP, TNF)

ASLV (avian sarcoma-leukosis virus): see retroviruses
Asn-Pro-X-Tyr: amino acid sequence, responsible for the internalization of low-density lipoproteins (LDL) of the membranes. (See LDL)
ASO: allele-specific oligonucleotide probe. Screening can be carried by semi-automated procedures. (See allele-specific probe)
ASP ANALYSIS: is used to estimate linkage in cases when a particular trait is under polygenic control. The co-segregation of multiple markers is followed in individuals that express the particular trait and determine which of the markers are most consistently present in the individuals displaying the trait of primary interest. The analysis still requires multiple segregating families and the information obtained is evaluated by the MAPMAKER/SIBS computer program. (See MAPMAKER, QTL, interval analysis)
ASPARAGINE (α-aminosuccinamic acid): $NH_2COCH_2CH(NH_2)COOH$; its RNA codons are AAU, AAC.
ASPARAGINE SYNTHETASE (AS): of bacteria uses ammonia as an amide donor, rather than glutamine as the mammalian enzyme. Cells expressing the bacterial AS will grow on asparagine-free medium if the glutamine analog, albizzin is present. In AS transfected mammalian cells the gene can be amplified in the presence of β-aspartyl hydroxamate, an analog of aspartate, and thus AS can be used as a dominant amplifiable marker in mammalian cell cultures. The mammalian genes are present in human chromosomes 7q21-q31, 8pter-q21, 21pter-q22. The AS genes do not have TATA and CAAT boxes in the promoter. They are homologous to the hamster *ts11* gene that is required for passing the cell cycle through the G1 stage. (See amino acid metabolism, cell cycle, house keeping genes, CAAT box, TATA box)
ASPARAGINYL tRNA SYNTHETASE (ASNRS): charges the appropriate tRNA with asparagine In human cells it has been located in chromosome 18. (See aminoacyl tRNA synthetase)
Asparagus officinalis (a dioecious monocot, 2n = 20): sex determination by XX pistillate and XY staminate plants. By anther culture YY plants can be obtained that can be vegetatively propagated or by pollination they produce exclusively male progeny. The male plants are of special economic value because their yield/area of the edible spears is substantially higher. Almost half of the human populations excrete methanethiol in their urine after eating this vegetable. The excreter trait appears to be autosomal dominant. The ability to smell this particular odor may also be under dominant control. (See YY asparagus)
ASPARTAME (Nutra-Sweet): N-L-α-aspartyl-L-phenylalanine-1-methyl ester, an artificial low-calorie food and beverage sweetener; about 160 times as sweet as sucrose. Not recommended for phenylketonurics because it contains phenylalanine. (See saccharine, fructose, phenylketonuria)
ASPARTATE AMINOTRANSFERASE (glutamate oxaloacetate transaminase, GOT1, GOT2): one of the functional forms of this enzymes GOT1, is encoded in human chromosome 10q24.1-q25.1 and it is expressed in the cytosol. A homolog GOT2 is encoded in human chromosome 16q12-q21 and it is expressed in the mitochondria. Pseudogenes of the latter were located at 12p13.2-p13.1, 1p33-p32 and 1q25-q31. In liver, largely the mitochondrial enzyme is present whereas in the serum mainly the cytosolic enzyme is located. (See amino acid metabolism, asparagine synthetase)
ASPARTATE PHOSPHATASE: see two-component regulatory systems
ASPARTOACYLASE DEFICIENCY (aminoacylase-2 deficiency, Canavan disease, ACY2): this enzyme cleaves acylated amino L-acids into an acyl and amino acid group, whereas amino-acylase-1 (ACY-1) cleaves similarly all acylated L-amino acids, except L-aspartate. The autosomal recessive disorder has an early or late onset resulting in debilitating muscle, eye defects, mental retardation and spongy degeneration of the white matter of the brain. In the urine there may be a 200 fold increase of N-acetylaspartic acid. Its incidence is increased among Jews of Ashkenazy extraction and in Saudi Arabic populations. Chromsmal location 17pter-p13. (See amino acid metabolism, neuromuscular defects, mental retardation, eye diseases, Jews and genetic diseases)

ASPARTYLGLUCOSAMINURIA (AGA): a chromosome 4 recessive defect of this enzyme may eliminate an important S—S bridge of the protein resulting in neurological-mental and other defects. Its frequency is higher in populations of Finnish descent. (See amino acid metabolism, disulphide bridge)

AS-PCR: allele-specific PCR. (See polymerase chain reaction)

ASPERGILLUS : ascomycetes; *Aspergillus nidulans* (n =8, 2 x 10^7 bp) is a favorite organism for studies of recombination. One meiotic map unit is about 5-10 kbp. It has been extensively used for mitotic recombination. Asexual reproduction is by conidiospores (3-3.5 μm). This is a homothallic fungus and thus does not have different mating types. In the cleistothecium there are up to 10,000 binucleate ascospores in 8-cell linear, ordered asci. Transformation systems are available. It yields about 5 x 10^3 transformants/μg DNA. *Aspergillus flavus* is responsible for the of aflatoxin, an extremely poisonous toxin developing on infected plant residues, seeds, etc. (See aflatoxins, mitotic recombination, recombination, cleistothecium, conidia, tetrad analysis)

ASPLENIA: one form (Ivemark syndrome) is usually sporadic or autosomal recessive and it is associated with absence or enlargement of the spleen or multiple accessory spleens and cardiac and other organ malformations. Another form of asplenia involves most conspicuously cystic livers, kidneys and pancreas. (See spleen)

ASSAY: a test for mutagenic effectiveness or efficiency or the velocity of a chemical reaction catalyzed by enzymes or the test of function of any biological process.

ASSEMBLY INITIATION COMPLEX: the minimal elements required for the completion of the assembly of the viral components. (See bacteriophages)

ASSIGNMENT TEST: see somatic cell hybrids

ASSIMILATION: converting nutrients into the cell constituents. Also, blending of an initially different ethnic (cultural) group into the general population.

ASSOCIATION CONSTANT (K_a): indicates the association between components of a complex. The larger the K_a, the stronger is the association.

ASSOCIATION PHASE: coupling phase in linkage, a term used in fungal genetics. (See coupling phase, repulsion, crossing over, linkage)

ASSOCIATION SITE: periodically distributed, microscopically detectable multiple interstitial association points are called also nodules. The distance between the paired chromosomes is about 0.4 μm. (See zygotene stage, meiosis, synaptonemal complex, recombinational nodule)

ASSOCIATION TEST: is basically a 2 x 2 contingency chi square test based on a panel:

	First Variable	
	+	-
Second Variable +	a	b
-	c	d

where a, b, c, d represent the number of observations + +, - +, + - and - -, respectively; n = the total number of observations. If b = c = 0, there is no association. The significance of the association is tested by

$\chi^2 = \frac{n(|ad - bc| - 0.5)^2}{(a+c)(a+b)(b+d)(c+d)}$ chi square, and the probability of a greater chi square can be determined by a χ^2 table or χ^2 chart for 1 degree of freedom. The association test is most useful within a homogeneous population. A particular association may not be an indication of genetic linkage, physiological or cause-effect relationship but may provide useful information on relation between two diseases or whether the reciprocal crosses would be identical or not. (See chi square)

ASSORTATIVE MATING: mates are chosen on the basis of preference or avoidance (positive or negative assortative mating), rather than at random, e.g. tall people frequently chose tall spouses, educated, higher economic or social status individuals marry most commonly within their group. Traits unknown to the majority, like blood groups, usually do not come into consideration in mate selection. Assortative mating may have some effect on the expression of a quantitative trait and the heritability becomes $h^2 = \hat{h}^2[1 - (1 - \hat{h}^2)A]$ where A is the product of the average heritability and the phenotypic correlation, i.e. $r\hat{h}^2$. (See controled mating, mating system, inbreeding, correlation)

ASTER: see centrioles

ASTROCYTE: a type of branching cell, supporting the nervous system. (See glial cells)

ASV: avian sarcoma virus of birds is an oncogenic RNA virus that can induce sarcoma in rodents. (See sarcoma)

ASYMBIOTIC NITROGEN FIXATION: proceeds by a microorganism without dependence on cohabitation with other organisms such as by the members of the soil bacterial species *Azotobacter* and *Clostridium*. (See nitrogen fixation, symbiosis)

ASYMMETRIC CARBON atom has four different covalent attachments. (See covalent bond)

ASYMMETRIC CELL DIVISION: is a requisite for embryonal differentiation. Several protein factors specify the process. The orientation of the spindle in *Drosophila* involves the localization of the Numb and Prospero proteins in the basal cells and the polarity instructions may come from the product of the *inscrutable* locus. Yeast (Ash1p) and *Caenorhabditis* (SKN-1) also have controls similar to Numb and Prospero and *she* and *par* genes, respectively, are analogous to *inscrutable*. (See morphogenesis in *Drosophila*, spindle)

ASYMMETRIC HETERODUPLEX DNA: see Meselson - Radding model of recombination

ASYMMETRIC HYBRID: lost some of the chromosomes of one or the other parent. (See somatic hybrids)

ASYMMETRIC REPLICATION: at the replication fork DNA synthesis on the leading and lagging strands proceed in opposite direction relative to the base of the fork. (See replication, replication fork)

ASYNAPSIS: failure of chromosome pairing. (See also desynapsis, synapsis)

ATABRINE: a preparation of quinacrine, an antimalaria and antihelminthic (intestinal tapeworm) drug. The quinacrine mustard (ICR-100) is a radiomimetic mutagen. (See quinacrines, radiomimetic)

ATase: see Utase

ATAVISM: the recurrence of expression of traits of ancestors beyond great grandparents. It is based on either recessive, complementary recessive or recombination of genes or special environmental conditions. For a period of time in the 20th century it was no longer used in the genetic literature. Atavism may, however have real basis in the genetic material and may represent in an altered form of ancient genetic sequences that are expressed in an "atavistic" manner if appropriately activated by a developmental program shift. Under such circumstances from the rudimentary limb buds of the whales occasionally hind limb bones may develop. Hypertrichosis in humans, encoded in chromosome Xq24-q27.1, also represents such an atavistic reprograming. These atavistic changes may not be basically much different from expression of homeotic genes. (Atavus in Latin means great great grandfather). (See hypertrichosis, homeotic genes)

ATAXIA TELANGIECTASIA: is one of about a dozen human ailments involving ataxias: poor coordination of the muscles because of dilations in the brain blood vessels, reduced immunity, etc. Its appearance in human diseases is attributed to instability and breakage of chromosomes 14, 7, 2, 11 and 12 although the major locus appears to be at chromosome 11q22-q23. Leukemias and other malignancies are very common among the patients. Cultured cells of the individuals affected are highly sensitive to both X-ray and UV damage. Also standard radiation therapy for malignant tumors may be fatal to these individuals. The basic defect in AT is either in a DNA-dependent protein kinase that controls progression of the cell cycle or in DNA repair and recombination (its homologs are MEC1, SAD3, ESR1). Alternatively, it has been found that an inositol 1,4,5-trisphosphate receptor (IP_3R1) deficient mouse mutants either die *in utero* or when born display severe ataxia and die very shortly. Homozygosity for this recessive human gene has a frequency about 5×10^{-5} and the frequency of the carriers, prone to breast cancer and other malignancies, is about 1%. The spinocerebellar ataxia (SCA5) of human chromosome 11 is caused by an instability of the CAG trinucleotide repeats. SCA1 is in chromosome 6p23.5-p24.2 and has a CAG instability, SCA2 maps to 12q23-q24.1, SCA3 I

Ataxia telangiectasia continued
14q24.3-qter, SCA4 in 16q. The autosomal dominant cerebellar ataxia (ADCA type II) with pigmentary muscular dystrophy is coded in human chromosome 3p12-p21.1. (See Friedreich ataxia, DNA repair, excision repair, carcinogenesis, and a number of genetic diseases which may have ataxic symptoms such as neuromuscular diseases, gangliosidoses, β-galactosidase, Niemann-Pick disease, metachromatic leukodystrophy, neurofibromatosis, olivopontocerebellar atrophies, Refsum diseases, Usher syndrome, Hartnup disease, light-sensitivity diseases, cancer, cell cycle, DNA repair, trinucleotide repeats, AVED)

ATAXIN: proteins responsible for ataxias. (See ataxia)

ATCC: American Type Culture Collection maintains cell cultures of prokaryotes and lower and higher eukaryotes.

ATELES (spider monkey): see cebidae

ATF2: activating transcription factor, a family of proteins containing homologous basic/leucine-zipper (bZIP) binding domains; it is regulated by the JNK signal transduction pathway. Mutations in ATF2 interfere with the retinoblastoma and E1A oncogene's transcription suppressing activities. (See retinoblastoma, bZIP, JNK, adenovirus [E1A])

ATHEROSCLEROSIS: hardening and then degeneration of the walls of arteries by the deposition of fatty acid nodules on the inner wall and obstruction of blood circulation. In the first phase lipid-filled foam cells (macrophages) appear. Next fibrous plaques are formed of lipids and necrotic cells, covered by smooth muscle cells and collagen. The final phase lesion involves platelet and fibrous clots (thrombus). This group of vascular diseases are the most common causes of heart diseases From coronary heart disease 490,000 and from stroke 150,000 death results per year in the USA. The genetic mechanisms underlying varies and non-genetic factors have substantial share. (See cardiovascular diseases, sterol, HDL, CETPl)

α-THIOPHOSPHATE-dNTP: are point mutagens when incorporated into gapped DNA by DNA polymerase I. The thiophosphates are not effectively removed by the $3' \rightarrow 5'$ editing function of the DNA pol I enzyme. (See pol)

ATM (ataxia telangiectasia mutated): involves phosphatidylinositol kinase; in its presence ataxia develops. (See PIK, ataxia telangiectasia)

ATOM MICROSCOPY: is being developed for imaging atomic structures. The equipment is using mono-energetic sodium atoms ejected into a vacuum chamber and carried by noble gases such as argon. The beam is broken up into sub-components on a silicon nitride grid. The phase shift generated by two beams is then measured.

ATOMIC FORCE MICROSCOPE: an instrument that can image the surfaces of conductor and non-conductor molecules even in aqueous media. It can reveal molecular structure of surfaces, adhesion forces between ligands and receptors, and other biological processes in real time.

ATOMIC RADIATIONS: killed 100,000 and injured 60,000 in Hiroshima and Nagasaki at the end of World War II, and caused substantial increase of cancer but showed no significant increase in human mutation. The cause of this is not that these radiations were genetically ineffective, rather the human breeding system, avoiding marriage between relatives, did not favor homozygosity of the recessive mutations. Recent studies of the populations exposed to the radiation caused by the failure of the Chernobyl power plant indicate not only an increase in cancer but also of mutation. Most likely, some of the mutations induced will be maintained in the exposed populations and may contribute to an increase of the genetic load. The total radiation from natural sources (cosmic radiation, disintegration of terrestrial isotopes [uranium, thorium, potassium], etc.), reaching the human gonads was estimated to be 100-125 millirads per year. The atomic bomb tests conducted during the years 1956 to 1965 contributed an average of about 76 millirads and will exert their effect mainly up to the year 2000 through the short halflife radioactive elements (Cesium^{137}, Strontium^{90}) and will substantially decay by then. The long half-life Carbon^{14} will continue to pollute by an additional estimated 167 millirads even after year 2000. The meltdown of the atomic power plant in the Ukraine near the Byelorussian border in the spring of 1986 exposed nearby populations up to 75 rem whereas the exposure of the entire Byelorussia received about 3.3 rem. In 1986 in that country the total number of

Atomic radiations continued

thyroid cancer in children was 2, and by 1992 it reached to about 60 cases and the figures are still increasing. In the human minisatellite DNA the mutation rate doubled and in the feral populations of voles (*Microtus*) the base pair substitution frequency in the mitochondrial DNA was found to be in excess of 10^{-4}, an over two orders of magnitude increase above the appropriate control groups. Nevertheless the rodent populations appeared in good condition and their fertility was also good. The Hanford Nuclear Reservation in the state of Washington exposed nearby populations in excess of 33 rads over a period of three years. The official estimates place 0.025 rads per year as safe for airborne pollution by nuclear weapon plants for the civilians living in the neighboring area and for the workers in those plants 5 rad for the entire body per year. According to some estimates based on irradiation of mice 20-40 rad is the estimated doubling dose of mutation for ionizing radiation. It was estimated that the radioactive fallout from weapon testing may have increased by 2% the genetic risks over the natural background effects and by 8% for leukemia. The effects of atomic radiation on mutation rates in the minisatellite DNA remains controversial because of the difficulties of finding appropriate (concurrent) controls. The mutation rate in these very sensitive DNA sequences is much affected by environmental factors (pollution), age, etc.

When considering the harmful effects of radiation potentially released by atomic power plants, one must consider the harmful pollution generated by the coal-fired industry and the carcinogenic hydrocarbons released by the combustion in wood fireplaces, etc. Also, the shortage of energy may cause directly or indirectly substantial sufferings and even death to the genetically more vulnerable part of the populations, especially to children. (See also cosmic radiation, isotopes, radiation hazard assessment, doubling dose, plutonium, nuclear reactors, mutation in human populations, rad, rem, control)

ATOPY: hereditary allergy. (See allergy)

ATP: adenosine-5'-triphosphate is a universal carrier of metabolic energy by transferring the terminal phosphate to various acceptors and resulting in ADP (adenosine diphosphate) that is recycled to ATP by either the chemical energy of oxidative phosphorylation or the solar energy of photosynthesis. Besides the thermodynamic role, ATP has also catalytic activity e.g., in nitrogen fixation. ATP provides also binding energy through non-covalent interactions with various molecules in order to lower activation energy. It provides energy for charging tRNA with amino acids, for DNA synthesis, for bioluminiscence mediated by the firefly luciferase, it is indispensable in carbohydrate metabolism, it serves as a precursor of cyclic AMP that has major role in signal transduction and protein phosphorylation, etc. The major catabolic pathways (glycolysis, citric acid cycle, fatty acid and amino acid oxidation and oxidative phosphorylation) are coordinately regulated in the production of ATP. The relative abundance of ATP and ADP controls electron transfers in the cell. ATP is generated in the mitochondria and chloroplasts. ATP is the major link between anabolic and catabolic reactions mediated by enzymes. UTP (uridine triphosphate), GTP (guanosine triphosphate) and CTP (cytidine triphosphate) are also important in similar processes but have relatively minor role compared to ATP. (See ATP synthase, ATPase, cAMP)

ATP SYNTHASE: a protein complex forming ATP from ADP and phosphate (oxidative phosphorylation) on plasma membrane (bacterial, mitochondrial, chloroplast). (See also ATP, ATPases)

ATPase: enzymes are required for active transport of chemicals and other functions in the cells. The *P-type* ATPases maintain low Na^+, low Ca^{++} and high K^+ levels inside the cells, and generate low pH within cellular compartments and activate proteases and other hydrolytic enzymes of eukaryotes and generate transmembrane electric potentials. The *V-type* (vacuolar) ATPases secure low pH inside lysosomes and vacuoles of eukaryotes. The *F type* ATPases (energy coupling factors) are located in the plasma of prokaryotes and in the mitochondrial and thylakoid membranes of eukaryotes are actually ATP synthase enzymes generating ATP from ADP and inorganic phosphate. The *DNA-dependent* ATPases are type I restriction endonucleaves that depend on Mg++, ATP and SAM for cutting of DNA strands. After cleavage they function only as ATPases. (See ATP, SAM)

ATRAZINE (Lasso): see herbicides, photogenes

ATRESIA: closure of an organ (e.g. vagina, and it can be surgically corrected to permit procreation), parts of the digestive tract (pyloric atresia), etc. (See pyloric stenosis)

ATRIAL: adjectivization of atrium. (See atrium)

AT-RICH DNA: common in the repetitive sequences, and it is generally not transcribed. Some of the petite colony mutants of yeast mitochondrial DNA contain mainly AT sequences. (See mitochondrial genetics, mtDNA)

ATRIUM: entrance to an organ. (See atrial)

Atropa belladonna: plant of *the Solanaceae* family (n = 50, 72) is a source of alkaloids. (See henbane)

ATROPHY: under- or lack of nutrition, wasting away of cells and tissues. (See Kugelberg-Welander syndrome, Kennedy disease, dystrophy, muscular dystrophy, neuromuscular diseases, spinal muscular atrophy)

ATROPINE: a highly toxic alkaloid. (See henbane)

att SITES: lysogenic bacteria and temperate phage have consensus sequences at the position where site-specific integration and excision takes place. The *att* sites are about 150 nucleotides long in λ and 25 bp in the bacterium and 15 bp sequences are identical in both.

The **p**hage (**POP'**) sequence is   G C T T T T T T A T A C T A A
                                        C G A A A A A AT A T G A T T

the **b**acterial (**BOB'**) sequence is: G C T T T T T T A T A C T A A
                                              C G AA AA A A T A T G A T T

The underscored sequences are then reciprocally recombined and POB' and BOP' sequences are generated from the left (attL) and right (attR) sequences. The integration requires the phage-coded INT· and the bacterial coded HF proteins. The excision requires an additional protein XIS coded by the bacterial gene *xis* probably because it is not exactly the reverse type of process since the original attP and attB elements were not identical except the 15 bps. (See lambda phage)

ATTACHED X CHROMOSOMES: two X chromosomes fused at the centromere. They were exploited for cytogenetics. Among others, were first used to carry out half-tetrad analysis in *Drosophila*. Females with attached-X produce eggs but half of them have only autosomes and no X-chromosome. If the attached X-chromosomes carry different alleles of a locus, double dose of the same allele in the eggs can be achieved only if there is a recombination between that gene and the centromere because the first meiotic division is reductional and the second is equational. (See half-tetrad analysis)

ATTACHMENT POINT (*ap*): a mappable site in the chromosome of the chloroplast of *Chlamydomonas reinhardi* green alga, representing a hypothetical centromere-like element. It is called *ap* because it attaches to the chloroplast membrane and assists the disjunction of the ring DNA during division. In genetic recombination this is taken as the 0 coordinate of marker segregation. (See chloroplast genetics, mapping of chloroplast genes)

ATTACHMENT SITE: see *att* site

ATTENTION DEFICIT-HYPERACTIVITY (ADHD): is observed in 2 to 5% of elementary school children and causing learning disabilities and emotional problems. Boys have about 5-fold higher chance to be affected. It frequently goes into remission by progressive age but some personality disorders (hyperactivity, antisocial behavior, alcoholism, hysteria) may persist even in adulthood. About 25 to 30% of the parents of affected children had some of the symptoms in childhood. The genetic basis is unclear. (See also affective disorders, dyslexia)

ATTENUATION: a regulatory process in bacteria. (See attenuator region, host-pathogen relations tryptophan operon, antitermination)

ATTENUATION VIRAL: means a reduction in virulence that is achieved by subculturing in a new cell population where after a period of time numerous adaptive mutations occur that although permit them to grow well in the original cells but with diminished virulence. These mutations generally occur in the 5'-non-translated region and modify the translation of the viral RNA although attenuating mutations may occur all over the viral RNA genome. (See attenuator region)

ATTENUATOR REGION: where RNA polymerase may stop transcription when all the cognate tRNAs are charged, the mRNA assumes a special secondary structure and this leads to a temporary cessation of transcription, leading to a reduction of transcription by a factor of 8-10. It is one of the regulatory mechanisms of bacterial amino acid operons. The histidine operon does not even use the more common type of operator repressor/inducer system. (See diagram below and next page, tryptophan operon, TRAP, tryptophan, repressor, antitermination).

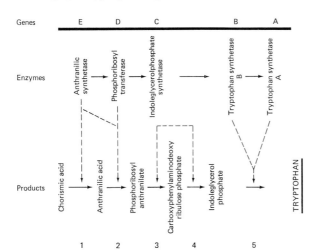

The biosynthesis of tryptophan from chorismic acid in *E. coli bacteria* requires five steps, mediated by five enzymes. The sequence of the encoding genes in the bacterial genetic map correspond to the sequence of the metabolic steps. This was the first case of recognition of such a co-ordinate regulation in bacteria. The five genes are controled primarily by a repressor, and attenuation provides an additional fine tuning. Some of the enzymes are composed from more than a single functional unit. Indoleglycerolphosphate synthetase catalyzes two synthetic steps as shown at left.

AUBERGER BLOOD GROUP: see Lutheran blood group

AUG CODON: in mRNA is the only one codon that specifies methionine yet there are two different tRNAs for methionine. In the majority of cases in prokaryotes one of the methionine-tRNAs is formylated at the amino group by N^{10}-formyltetrahydrofolate, and this formylmethionine tRNA initiates translation whereas the other methionine-tRNA carries methionine to all other sites in the polypetide. In eukaryotes the *initiator methionyl-tRNA* is not formylated, its initiator attribute is specified by the primary structure and conformation of the tRNA. Thus the overwhelming majority of the nascent proteins start at the NH_2 end with a methionine. In the mature protein this methionine may be absent because of processing. (See genetic code, protein synthesis)

AURICLES: small projections at the upper part of leaf sheath in cereals; they are of importance for taxonomic characterization.

AUSTRALOEPITHECUS: an extinct, fossil (5-1 million year old) of the bipedal Hominidae of the old world. Its brain size is intermediate between modern humans and apes. Exact relation to existing species is unclear. (Austral means Southern)

AUTISM: a human behavioral anomaly involving reticence, self-centered, subjective thoughts and actions, learning and communication difficulties. Its prevalence is 0.02 to 0.05% in the general population. The concordance between monozygotic twins appeared 36 to 96% while between dizygotic ones none to 24% was observed. Its incidence is higher in males than females. Generally it is associated with mental retardation and other psychological disorders. Although the incidence within affected families is 50 times above that in the general population, no single gene could be identified as a causative agent. (See continuation on next page.)

Autuism continued

The infantile autuism becomes apparent during the first year of life. Apparently it is under polygenic control. In autism with onset after an initial normalcy (Rett syndrome), the symptoms are shared but progressive dementia, uncoordination and deterioration of all mental functions follow. The latter type appears to be coded in the short arm of the X-chromosome. (See also affective disorders, mental retardation)

Attenuator region continued from preceding page

A genetic and molecular map ↓ of the tryptophan operon in *Escherichia coli* bacterium. Attenuation may dictate an early termination of transcription. The site of attenuation (*a*) is within the tryptophan leader sequence (*trpL*) and it is preceded by the site of the transcription pause (*tp*). Transcription is primarily under the control of the promoter - operator region and the process begins at the left end of the *trpL* site. The RNA polymerase pauses at the *tp* site before proceeding further. In case most of the tryptophanyl tRNAs (tRNA^{Trp}) are charged with tryptophan and therefore there is no need for additional molecules of this amino acid, transcription is momentarily terminated at the attenuator (*at*) site. If however the tRNA^{Trp} is largely uncharged because of shortage of tryptophan and active protein synthesis, the transcriptase RNA polymerase passes through the *at* site without interference. This passage is made possible by alterations in the secondary structure of the RNA transcript of the operon. The initial segment of the leader sequence encodes a short tryptophan-rich peptide. In case there is a scarcity in tryptophan, translation on the ribosome is stalled at the tryptophan codons in the leader sequence. During the pause (tp), the mRNA transcript assumes a hairpin-like structure by base pairing between segments marked by (2) and (3) and thus the passage through the *attenuator* (*a*) site is facilitated. In case, however, most of the cognate tRNAs are charged, the transcript shows base pairing between segments (3) and (4), resulting in stop-page of transcription until the oversupply is exhausted by protein synthesis. The tryptophan operon also relies on suppressive transcriptional controls. See base sequences of the attenuator on page 1076. (Modified after Yanofsky, C. 1981 Nature (Lond.) 289:751)

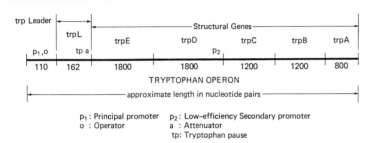

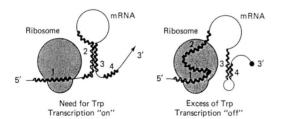

Need for Trp
Transcription "on"

Excess of Trp
Transcription "off"

AUTOALLOPOLYPLOID: is a polyploid in which the genome(s) is/are duplicated from one or more species, e.g. AAAABBBB or AAAABB. (See allopolyploid, sesquidiploid)

AUTOANTIBODY: antibody that is formed against the body's own antigens, such as in autoimmune disease. (See autoimmune disease)

AUTOANTIGEN (self-antigen): is a normal cellular protein yet it may be attacked by the cellular immune system similarly to what happens in autoimmune disease. (See immune system, immune reaction)

AUTOCATALYTIC FUNCTION of DNA is the process of replication. Also, any reaction that is promoted by its own product. (See replication)

AUTOCRINE: signal production within a cell in response to external stimuli.

AUTOCRINE STIMULATION: cells infected by proto-oncogene carrying virus secrete a growth factor that further stimulates the cell's proliferation. (See proto-oncogenes)

AUTOECIOUS: a parasite that completes its life cycle on the same host.

AUTOGAMY: self-fertilization common in hermaphroditic and monoecious plants; autogamy in the unicellular animals, *Paramecia*, is preceded by meiosis and one the four haploid products survive. This cell then divides into two cells by mitosis. These two identical cells then may fuse and a genetically homozygous diploid zygote is formed. (See also allogamy)

AUTOGENESIS: see Lamarckism

AUTOGENOUS CONTROL: the own product of genes regulates the coding gene either in a positive or negative way. (See negative control, positive control)

AUTOGENOUS EVOLUTION: structures and organelles evolved through differentiation of the cells own system. (See exogenous evolution)

AUTOGENOUS SUPPRESSION: the *Salmonella* RF2 translation termination protein occasionally fails to recognize or misreads the UGA stop codon resulting in readthrough by suppressing termination. (See translation termination, readthrough, recoding, stop codon)

AUTOGRAFT: the tissue transplantation is within one individual. (See also homograft)

AUTOIMMUNE DISEASE: the immune system fails to recognize the cell's own antigens and attacks them such as in lupus erythromatosus (a variety of skin and possibly visceral inflammations) based on making antibodies against its own DNA and RNA. In insulin-dependent diabetes the insulin producer β cells of the pancreas are attacked by the body's immune system, coded for by the major histocompatibility genes. The Rasmussen's encephalitis, a rare form of epilepsy and the paraneoplastic neurodegenerative syndrome (PNS) both are caused by autoantibodies against the glutamate receptors of the nervous system. Autoimmune diseases include a series of different diseases (p = prevalence, r = risk of siblings relative to risks in the general population): psoriasis (p: 2.8, r: 6), rheumatoid arthritis (p:1, r: 8), goiter (p: 0.5, r: 15), insulin-dependent diabetes (p: 0.4, r: 1.6), ankylosing spondilitis (p: 013, r: 54), multiple sclerosis (p: 0.1, r: 20), lupus erythematosus (p: 0.1, r: 20), Crohn disease (p: 0.06, r: 20), narcolepsy (p:0.06, r: 12), celiac disease (p: 0.05, r: 60), cirrhosis of the liver (p (0.008, r: 100) (See named diseases under separate entries, HLA, NF-κB, complement, Sjögren syndrome)

AUTOINDUCTION: is a type of cell to cell interaction in bacteria. The cells release small extracellular signaling molecules which are taken up again by the cells. It adjusts gene expression in the cells responding to a level appropriate for the local density of the signaling cells. The signals may be acylated homoserine lactones, Tra proteins, amino acids, short peptides and pheromones. (See also autoregulation, pheromones, quorum sensing, *tra*)

AUTOINTERFERNCE: defective viral particles may interfere the replication of intact ones.

AUTOLOGOUS: its origin is within the cell; it is a self-made molecule.

AUTOLYSIS: decomposition of cells and cell content by the action of the natural enzymes of the cells; it takes place generally in injured cells.

AUTOMIXIS: self-fertilization.

AUTOMUTAGEN: a metabolite of the organism may become mutagenic, e.g. tryptophan.

AUTONOMOUS CONTROLLING ELEMENT: a plant transposable element carries the transposase function and controls its own movement, e.g. *Ac* versus *Ds* in maize, the latter is a defective form of *Ac*, incapable of moving by its own power unless the autonomous (intact) *Ac* is present in the cell, (See transposable elements, *Ac - Ds, Spm*)

AUTONOMOUS DEVELOPMENTAL SPECIFICATION: maternal information or prelocalized morphogenetic information regulates the initiation of transcription of morphogenetic genes. (See morphogen)

AUTONOMOUSLY REPLICATING SEQUENCES: see ARS

AUTONOMY: cells transplanted into tissues of different genotype or forming parts of genetically different sectors, still maintain the expression encoded by their genotype, and not - or barely - affected by the genetically different tissue environment.

AUTOPHAGY: destruction of cytoplasmic particles within a cell.

AUTOPHENE: a genetically determined trait which is expressed independently of the position in case of transplantation. (See also allophenic)

AUTOPHOSPHORYLATION: upon binding a ligand to a receptor results in rapid phosphorylation of the receptor by its own subunits e.g., by members of a dimeric molecule generally at tyrosine sites. (See receptor tyrosine kinase)

AUTOPLOID (autopolyploid) more than two complete sets of an identical genomes per cell. If the number of chromosomes in the somatic cells is reduced to half it becomes a polyhaploid. Autopolyploids may be [auto]tetraploid (2n = 4x), hexaploid (2n = 6x), octaploid (2n = 8x), etc. Autotetraploids in meiosis may pair as quadrivalents, however, at a particular point of the chromosomes only two are synapsed. In autopolyploids pairing may be also as two bivalents, one trivalent and univalent, and may form four univalents. If all chromosomes pair as bivalents, it is called *selective pairing*, and the segregation of genes will resemble that of diploids with duplicate genes. Autotetraploids may carry a different allele in each of the four chromosomes, therefore they can produce a larger variety of gametes than diploids. The maximal number of gametic combinations can be determined by the formula:

$\begin{bmatrix} n \\ x \end{bmatrix}$ where n = the total number of alleles, and x = the number of alleles in a gamete,

thus in autotetraploids it becomes $\begin{bmatrix} 4 \\ 2 \end{bmatrix}$ for octaploids it is $\begin{bmatrix} 8 \\ 4 \end{bmatrix}$ and these can be rewritten as

$\frac{4 \times 3}{2 \times 1} = 6$ for autotetraploids and $\frac{8 \times 7 \times 6 \times 5}{4 \times 3 \times 2 \times 1} = 70$ for autooctaploids, and this means that the

GAMETIC OUTPUT OF AUTOTETRAPLOIDS						
	Absolute Linkage*			Independence from Centromere#		
Parent	AA	Aa	aa	AA	AA	aa
AAAa	1	1	0	13	10	1 (4.2%)
AAaa	1	4	1 (16.6%)	2	5	2 (22.2%)
Aaaa	0	1	1 (50.0%)	1	10	13 (54.2%)

Phenotypic segregation ratios in autotetraploids in case the dominance is complete in F$_2$				
	Absolute Linkage*		Independence from Centromere#	
Mating	Dominant	Recessive	Dominant	Recessive
AAAa selfed	1	0	575	1
AAaa selfed	35	1	19.3	1
Aaaa selfed	3	1	2.4	1
AAAa x AAaa	1	0	107	1
AAAa x Aaaa	1	0	43.3	1
AAAa x aaaa	1	0	23	1
AAaa x Aaaa	11	1	7.3	1
AAaa x aaaa	5	1	3.5	1
Aaaa x aaaa	1	1	1	1.2

* no recombination between gene and centromere (chromosome segregation)
\# the distance between gene and centromere is 50 map units or more and therefore recombination occurs freely as if they would not be syntenic (chromatid segregation or maximum equational segregation).

maximal number of allelic combinations with 4 different alleles 6 types of gametes are possible and with 8 different alleles in an octaploid the total number of gametic types is 70. In case all the four alleles are dominant, AAAA, the individual is a quadruplex, AAAa = triplex, AAaa = duplex, Aaaa = simplex and aaaa = nulliplex. The segregation ratios in F$_2$ depend on whether there is crossing over between the gene and the centromere, the type of pairing (as indicated

Autopolyploid continued
above) and the type of disjunction at anaphase II (alpha parameter). The phenotypic proportions in F_2 are determined by the gametic output of the parents or selfed individuals. The gametic output and F_2 segregation of autopolyploids is very difficult to generalize because the genes are rarely linked absolutely to the centromere and the frequency of recombination may vary from 0 to 50%. There are additional variables that may be estimated by the alpha parameter. Segregation ratios at higher level of polyploidy can be predicted only theoretically, the actual results may be quite different, however. (See synteny, bivalent, trivalent, univalent, synteny, alpha parameter, maximum equational segregation)

AUTOPROCESSING: a sequence of a protein (e.g. the C-terminal) is involved in its processing.

AUTORAD: the lab slang for autoradiogram. (See autoradiography)

AUTORADIOGRAPHY: labeling technique by which a radioactive substance reveals its own position in a cell or on a chromatogram when brought into contact with photographic film. For cytological analyses most commonly H^3-labeled thymidine is used because it gives the clearest resolution of chromosomal regions without serious DNA breakage whereas in molecular genetics the much higher energy P^{32}-labeled compounds are employed usually. (See also non-radioactive labels, immunoprobes)

AUTOREACTIVE: the lymphocytes recognize the individual's own molecules and develop an immune reaction to them. (See autoimmune disease)

AUTOREDUPLICATION: self-duplication

AUTOREGULATION: a compound (or system) controls the rate of its own synthesis, e.g. the

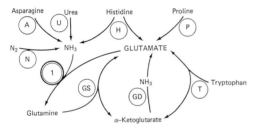

Metabolic steps involved in the regulation and autoregulation of glutamine synthetase
H (hut, histidine utilization), P (put, proline utilization) U (urease), A (asparaginase), N (nitrogenase, T (tryptophan transaminase), GS (glutamate synthase), GD (glutamate dehydrogenase)

bacterium *Klebsiella aerogenes* uses glutamine dehydrogenase to make glutamate from α–ketoglutarate and ammonia if the concentration of the latter exceeds 1 nM. If the concentration of ammonia is low, glutamate dehydrogenase cannot function to an appreciable extent. In this case the ammonia + glutamate are converted into glutamine by glutamate synthetase. The active form of glutamine synthetase is non-adenylylated. In the presence of high concentration of ammonia the enzyme is adenylylated and thus the activity is reduced by this mechanism of autoregulation. In its non-adenylylated states it represses glutamate dehydrogenase instead. (See regulation of gene activity)

AUTOSEGREGATION: may take place in an apomictic or vegetatively multiplied organism due to chromosomal loss or somatic mutation. (See apomixis, mutation)

AUTOSEXING: sex is identified by genetic markers rather than by the genitalia. This procedure has been exploited by silk worm breeders and poultry producers. Homozygosity for the *B* (barring) genes suppresses the appearance of colored spots on the head of the newly hatched chicks, controled by this sex-linked gene (remember that in birds the males are homogametic). The *B* gene is dominant yet it shows clear dosage effect. In the females that are heterogametic, the spot is evident. Thus they can separate early the hens from the roosters when the recognition of gender is very difficult by the anatomy. Because most of the roosters will be used for meat production while the hens will be used for egg production and they can be fed and managed accordingly. In the silk worm the male cocoons (chrysalis) produce 25 to 30% more silk than the females and therefore autosexing may have economic advantage. The silkworm eggs may be sorted by an electronic device according to color (sex). (See diagram on next page, chromosomal sex determination)

Autosexing continued

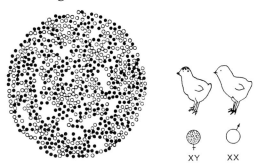

Left: AUTOSEXING IN THE SILKWORM. THE DOMINANT GENE IN THE Y CHROMOSOME PERMITS THE DISTINCTION BETWEEN THE EGGS WHICH WILL HATCH TO BECOME A MALE (PALE COLOR) AND FEMALE (DARK).
Right: THE HOMOZYGOUS MALE (B/B) CHICKS ARE LIGHT COLORED WHILE THE HEMIZYGOUS B/O FEMALES DEVELOP COLORED SPOTS ON THE HEAD (Remember that in the Lepidoptera and birds the males are WW and the females are WZ)

AUTOSOMAL DOMINANT MUTATION: is readily detected and identified in many instances because the novel type appears suddenly without precedence in the pedigree. Achondroplasia in humans is frequently cited as such an example. The homozygotes generally suffer perinatal death. Therefore most of the achondroplasiac dwarfs are heterozygotes and new mutants. These dwarfs are of normal and frequently of superior intelligence. One must not forget that over 70 gene loci are responsible for various types of dwarfing in humans. Autosomal dominant mutation rates (per gamete/generation) in human populations for ten diseases vary from 4 to 100×10^{-6}. (See mutation rate)

AUTOSOMAL RECESSIVE LETHAL ASSAY: a tester stock used for the detection of recessive second chromosomal lethals in *Drosophila* is of the following genetic constitution: *Cy L/Pm* where *Cy*, (*Curly*), *L* (*Lobe*) and *Pm*, (*Plum*) are heterozygous viable but homozygous lethal dominant genes. The *Cy* chromosome generally carries three inversions to prevent the recovery of cross overs. The heterozygotes of either sex are crossed with a mate that before the test carried no mutation in either of the two second chromosomes. Single F_1 male(s) are then backcrossed with the *Cy L/Pm* female tester. From their offspring *Cy L* individual sibs are mated. From this mating an F_2 is obtained. If all the survivors are *Cy L*, this indicates that a new lethal mutation occurred in the grandfathers' or grand-mothers' 2nd chromosome and therefore *non-Curly* and *non-Lobe* homozygous individuals could not live. The diagram does not show the genotypes in the F_2. (See sex-linked recessive lethal assays)

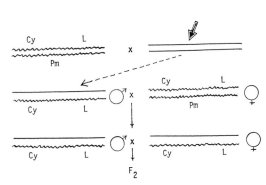

AUTOSOME: chromosome that is not a sex chromosome. (See also chromosomal sex determination)

AUTOSYNDESIS: may take place between homoeologous chromosomes in the absence of homoeologous pairing suppressor genes. (See homoeologous, chromosome 5B)

AUTOTOXIC ENTEROGENOUS CYANOSIS: an obsolete name for human familial NADH-methemoglobin deficiency. (See methemoglobin)

AUTOTROPH: can synthesize cellular C- and N- containing molecules from carbon dioxide and ammonia.

AUTOZYGOUS: a genotype which is not just homozygous but the alleles at the locus are *identical by descent*. (See also allozygous, inbreeding coeffcient, coancestry)

AUXANOGRAPHY: a method of mutant selection. Minimal medium is overlayered with auxotrophic spore or cell suspension and subsequently to different segments of the plate small quantities of various substances are added that the cells may need for growth. Where growth occurs, the cells utilize the compounds added and their nutritional requirement is identified.

AUXILIN: see clathrin

AUXINS: phytohormones produced by the plant metabolism such as indole-3-acetic acid or of synthetic origin such as α-naphthalene acetic acid or 2,4-dichlorophenoxyacetic acid. They play important role in cell elongation, signal transduction and required supplements for proliferation and regeneration in tissue culture. Genes (*iaaH, iaaM*) in the Ti plasmid of *Agrobacterium* have instructions for their production and regulation, and these play a role in crown gall formation (in cooperation with cytokinins). (See plant hormones, crown gall, Ti plasmid, embryogenesis somatic)

AUXONOGRAPHY: see auxanography

AUXOTROPH: mutant that requires nutritive(s) not needed by the wild type (prototroph). Auxotrophic mutations have been extensively used for the study of biochemical pathways and for the identification enzymes catalyzing particular metabolic steps. In genetic analysis auxotrophs facilitate selective techniques in backmutation, recombination, transformation, etc. A pyridoxine deficiency, causing seizure in humans, has been identified. True auxotrophic animal mutations are exceptional and they are rare also in higher plants. (See pyridoxine, autotroph)

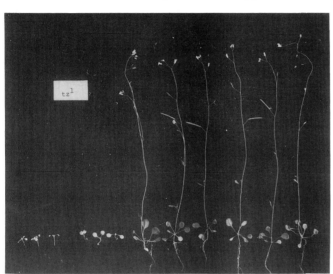

Left: THIAZOLE AUXOTROPHS OF ARABIDOPSIS ON BASAL MEDIUM; Right: ON THIAMINE MEDIUM

AVED (ataxia with vitamin E deficiency): is caused by mutation in the large subunit of microsomal triglycerid transfer protein (encoded in human chromosome 8q13). Although the intestinal absorption of α-tocopherol is normal, the hepatic secretion into the blood is defective. The condition is very similar to Friedreich ataxia. (See Friedreich ataxia)

AVENA: a genus of grasses with basic chromosome numbers x = 7 and form an allopolyploid series, 2n, 4n, 6n.

AVERAGE: is the arithmetic mean, i.e. the sum of all measurements (x) divided by the number of measurements (N); $\bar{x} = \frac{\Sigma x}{N}$. (See mean, median, mode)

AVERAGE INBREEDING COEFFICIENT: see α

AVIAN: pertaining to the taxonomic class of *Aves* (sing. *Avis*, bird(s)

AVIAN ERYTHROBLASTOSIS (*erbB*): viral oncogene (prevents maturation of the red blood cells in fowl) has its cellular homolog as a protooncogene in several eukaryotes. It is a protein kinase, phosphorylating primarily tyrosine residues. The normal allele specifies a plasma membrane receptor of epidermal growth factor (EGF). (See erythroblastosis fetalis)

AVIAN MC29 MYELOCYTOMATOSIS: viral oncogene (*myc*, causes carcinoma, sarcoma and myelocytoma [a kind of leukemia]); it is present as a cellular protooncogene in vertebrates and its homologs are present also in plant cells. (See oncogene, protooncogene, carcinoma, sarcoma, leukemias)

AVIAN MYELOBLASTOSIS: see MYB oncogene

AVIAN SARCOMA VIRUS: see ASV

AVIDIN: is a ca. 68,000 M_r protein of four subunits, each has strong affinity to biotin. It binds strongly to any molecule complexed with biotin such as nucleic acids, and biotin containing enzymes. It is widely used for non-isotopic labeling of nucleic acids. Originally it was found and isolated from raw egg white. Eating raw eggs may cause biotin deficiency (cooking inactivates it). It is isolated also from *Streptomyces avidinii* under the name streptavidin (See biotinylation, genomic subtraction)

AVIDITY: see antibody

AVIRULENCE: lack of competence for causing pathological effects by an infectious agent.

AVOGADRO NUMBER: number of molecules (= 6.02×10^{23}) in a gram molecular weight, a constant for all molecules.

AVOIDANCE LEARNING: is a classical test of animal behavior. In a two compartment box one is electrically wired to provide to the test animals an electric shock after a light turns on. Some of the animals, after a learning period, immediately move to the safe compartment as they see that light signals the coming shock in one compartment. The learning ability of inbred mice strains is genetically different. In some about half of the individuals "learn", in others only 10% associates the light signal with the shock. (See behavior genetics)

AXENIC: pure culture of organisms or cells without any contamination by other (micro) organisms. (See aseptic culture, tissue culture)

AXIAL ELEMENTS: are the lateral elements of the tripartite synaptonemal complex. (See synaptonemal complex)

AXILLARY: formed in the axil, the upper surface of the area between the leaf petiole and stem.

AXIS OF ASYMMETRY: in the majority of organisms three axes are recognized anterior - posterior (front - hind), dorsal - ventral (back - abdominal) and left - right. Recently genes have been identified controling asymmetry of the body. It had been known for a long time that changing the placement of internal organs has multiple deleterious consequences (see situs inversus viscerum). It has been shown now in the lefty mouse mutant that the expression of the transforming growth factor (TGFβ) family plays the role of a morphogen in controling asymmetry by expressing only in the left half of the gastrula. This asymmetry is transient and sets on before lateral asymmetry becomes visible. Similar genetically controled mechanism have been discovered also in chickens. (See morphogen, TGF, activin)

AXIS OF SYMMETRY: through which the objects or molecules form mirror images.

Axl: a receptor tyrosine kinase is human myeloid leukemia transforming protein. (See leukemias)

AXON: is a long nerve fiber that, generally in a bundle surrounded by a myelin sheath, communicates impulses between the central and the peripheral nervous system. Organelles and molecules can be transported along the nerve axons outward from the cell or back to the cell. (See neurogenesis)

AXONEME: microtubules and attached proteins that are the major part of cilia and flagella.

AXOPLASM: the cytoplasm of axons. (See axon)

5-AZACYTIDINE: a pyrimidine analog (and suspected carcinogen) that interferes with methylation of DNA bases. It may affect differentiation and development because hypomethylated genes are preferentially transcribed. It is noteworthy that some small eukaryotic genomes (yeast, *Drosophila*) do not contain methylcytosine yet their genomes are regulated during development. (See methylation of DNA, housekeeping genes)

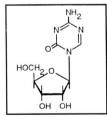

← 5-AZACYTIDINE

8-AZAGUANINE →

AZAGUANINE: a toxic analog of guanine and it is readily incorporated into RNA or DNA. (See HAT medium)

8-AZAGUANINE RESISTANCE: is a commonly used marker in mammalian cell cultures; the resistance is based on a deficiency of the enzyme azaguanine-hypoxanthine phosphotransferase and therefore this toxic purine cannot be processed by the metabolism. (See HAT medium)

AZASERINE (O-diazoacetyl-L-serine): an alkylating, antitumor, antifungal and mutagenic agent. The oral LD50 for rodents is 150-170 mg/kg. (See LD50)

6-AZAURACIL: an antineoplastic pyrimidine analog; its nucleotide is inhibitory to orotidylic-acid decarboxylase and may repress the synthesis of orotidylic acid pyrophosphorylase, key enzymes in the *de novo* pathway of nucleotide synthesis.

AZIDE: a compound with a NH_3; sodium azide, a respiratory inhibitor, is a strong mutagen for certain organisms at low pH but not for others. Nitrogenase enzymes may reduce azides to N_2 and NH_4. (See nitrogenase)

AZIDOTHYMIDINE: see AZT

AZOOSPERMIA: human gene AZF (azoospermia factor) appears to be the expression of the DAZ (deleted in azoospermia) site and it has been assigned to human chromosome Yq11. The frequency of the DAZ caused sterility is about 1.25×10^{-4} in men. This gene is substantially (42%) homologous to the *Drosophila* gene *boule* (*bol*) controlling meiotic G2 - M transition. Mouse gene *Dazla* is 33% homologous to DAZ. Both the mouse and the *Drosophila* genes are, however, autosomal yet they also involve male sterility. Recently it has been shown that the human AZF gene was originally in the short arm of human chromosome 3 (where highly homologous sequences still exist) and it was transposed to the Y chromosome, amplified and pruned. (See holandric genes, *twine, pelota, boule* [*bol*])

AZORHIZOBIUM: see nitrogen fixation

AZOTOBACTER: see nitrogen fixation

AZT (azidothymidine): thymidine analog with an azido (N_3) substitution of the 3'-OH group; it may slow down the reverse transcriptase activity of HIV virus by preferentially selecting this analog that has only minor effect on DNA polymerase of the mammalian cells. Unfortunately, some bone marrow damage is associated with the drug and this limits its usefulness in protecting against the full-scale development of AIDS. Should be remembered that the Aschheim-Zondek test for pregnancy is also abbreviated as AZT. (See acquired immunodeficiency, AIDS, HIV)

AZUROCIDIN: see antimicrobial peptides

B

B: back cross generation; the number of back crosses are indicated by subscripts, e.g. B_1. (See back cross)

B104: *Drosophila* retroposon similar to copia, gypsy and others. (See copia)

β BARREL: the polypeptide chain of a membrane protein forms a folded up β sheet arranged in the shape of a barrel. (See protein structure, membrane proteins)

B1 B CELL: are fetal and early infant B cells but they may exist in greater proportion than normal in leukemias and autoimmune diseases. (See B lymphocyte)

B BOX: part of the internal control region of tRNA genes; see also A box and internal control region of pol III genes. (See tRNA, pol III)

B CELL: see B lymphocyte

B CHROMOSOME: accessory (supernumerary) chromosome. They are generally heterochromatic and carry no major genes yet may be present in several copies in many species of plants. The B chromosomes have no homology to the regular chromosomal set (A chromo-somes) and are prone to nondisjunction be-cause their centromeres appear to be defective. If A-B translocations are constructed, the placement of genes to chromosomes, arms or even shorter regions may be facilitated. The principle of the use of A-B chromosome translocations is outlined at left. The A-chromosome or a translocated segment carries the dominant *A* allele and the B-chromosome has no counterpart to it therefore a null phenotype (*a*) appears in its absence. In the diagram the male has the translocation and the female is homozygous recessive for the *a* allele. In case there is no B-chromosomal nondisjunction—when the chromosomal con-stitution is as diagramed—both endosperm and embryo expresses the dominant gene. In case of nondisjunction the phenotypic effects de-pend on the constitution of the sperm which fertilizes the diploid polar nucleus of the endosperm or the embryo, respectively (see middle and bottom parts of the diagram). This type of difference between endosperm and embryo thus reveals the approximate physical and genetic position of the locus. Had the dominant allele been outside the translocated segment, the recessive allele would not have been unmasked. The example illustrated shows the most favorable case when the consequence of the translocation can be identified without tissue-specificity. (See mapping genetic, trisomic analysis, translocation genetic)

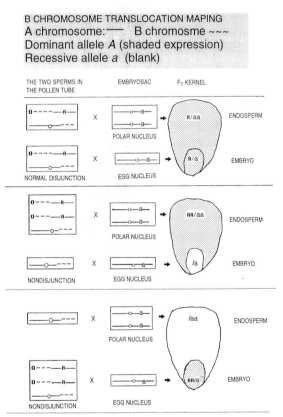

B LYMPHOCYTE (B cell): is responsible for humoral antibody synthesis. Its differentiation from hematopoietic cell depends on factors of the bone marrow, cytokines and antigens, T_H cells and a series of non-receptor and receptor tyrosine kinases and phosphatases, proteins mediating the pathway shown in italics in the diagram on next page outlining the developmental pathway of B lymphocytes. (See T cells, EBF, CpG motifs, CD40, TAPA-1, blood, bone marrow, thymus, immune system, immune reactions, germinal center)

B lymphocyte continued

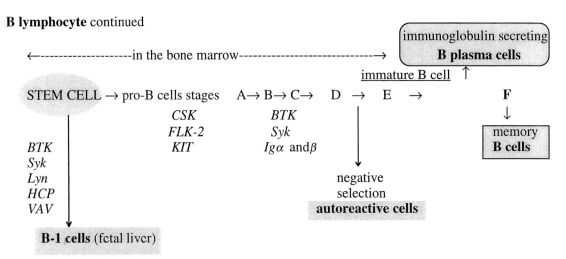

β **OXIDATION**: degradation of fatty acids at the β carbon into acetyl-coenzyme A. (See fatty acids, acetyl CoA)

B7 PROTEIN: is required for the activation of B cells. It is recognized by CD28 on surface of the antigen-presenting cells. The other required signal for the B cell activation is a foreign peptide antigen, associated with class II MHC molecule on the surface of an antigen-presenting cell. The same as BB-1 or CD80. (See also CD40, antigen presenting cells, MHC, B cell)

β **SHEET**: secondary structure of proteins where relaxed polypeptide chains are running in close parallel arrangement. (See protein structure)

BABOON (*Papio*): see Cercopithecidae

BAC (bacterial artificial chromosome): is a bacterial vector that can accommodate 350 kb DNA sequences and has much lower error rate than the still larger capacity yeast artificial chromosome (YAC). (See vectors, PAC, YAC, BIBAC)

BACH, SEBASTIAN (1685-1750): one of the greatest geniuses of classical music, had a family with over 50 more or less renowned organists, cantors and musicians. His first marriage to his second cousin Maria Barbara Bach included 3 musicians among the 4 surviving children (inbreeding coefficient 1/64 [3 offspring died in infancy]). From his second marriage to unrelated singer Anna Magdalena Wilcken (assortative mating) 13 babies were born and among the 5 survivors 3 were again musically talented. This family tree indicates that musical ability may be controled by relatively few genes, and the cultural environment also may have a major role. Recent studies demonstrated that musical talent is correlated with stronger development of the left planum temporale, increased leftward asymmetry of the brain cortex. (See dysmelodia)

BACILLUS: rod-shape bacterium. (See *Bacillus subtilis, Bacillus thüringiensis*)

Bacillus subtilis: is a gram-positive, rod-shaped soil bacterium and there it lives on decayed organic material and thus harmless. In starvation most of the cell content particularly the DNA moves to one end of the cell. This area, constituting about 10% of the cell is walled off and becomes a spore. The spore is extremely resistant to various environmental effects that would kill vegetative cells. Under favorable conditions the spore regenerates the bacterium. The size of its cells is similar to those of *E. coli*, and its DNA content is slightly higher. The DNA has different base composition from that in *E. coli*, A+T/G+C ratio in the former 1.38, in the latter 0.91 indicating that *B. subtilis* has more A+T than *E. coli*. For transcription it uses about 10 different σ factors although its major RNA polymerase is similar to that of *E. coli*. ($\alpha\alpha\beta\beta'\sigma$). The 43 kDa ($\sigma^{43}$) recognizes some of the consensus sequences of *E. coli* promoters. Its best known phage is SPO1 that is transcribed either by a phage RNA polymerase or by the host. (See <http://www.pasteur.fr/Bio/SubtilList.html>, *E. coli,* forespore, endospore)

Bacillus thüringiensis: a gram-positive bacterium that produces a the BT toxin. The toxin (delta

Bacillus thüringiensis continued

endotoxin) is within the crystalline inclusion bodies produced during sporulation. In alkaline environment (such as in the midgut of insects) the crystals dissolve and release proteins of M_r 65,000 to 160,000 that are cleaved by the proteolytic enzymes of the insects into highly toxic peptides. These toxins are most active against *Lepidopteran* larvae (caterpillars) but some bacteria produce toxins against *Diptera* and *Coleoptera* too. The extract of these bacteria can be directly used as a powder or suspension against the insects. The effectiveness is generally very good but the cost is substantially higher than that of chemical insecticides. A great advantage that these toxins are not harmful to mammals and to many other insects. The most economical solution is to transform plants (tobacco, cotton, etc., that can be transformed) with the *Bt2* gene that codes for the 1,115 amino acid residue pro-toxin protein. Actually, a smaller polypeptide, M_r 60K, and even smaller fragments are still fully active. The transgenic plants kill invading caterpillars within a couple of days and remain practically immune to any damage. The activity of the transgene has been further enhanced by the use of high efficiency promoters in the T-DNA constructs. In some instances transgenic cotton was still overpowered by bollworms. There are differences in the spectra of the bacterial toxins produced by different strains of the *Bacillus* and this provides an opportunity to extend the range of toxicity to other insect species. (See transformation of plants, promoter)

BACK CROSS: the F_1 is crossed (mated) by either of its two parents (see test cross). Each back crossing reduces by 50% the genetic contribution of the non-recurrent parent; thus after (r) back crosses it will be $(0.5)^r$. The percentage of individuals homozygous for the (n) loci of the recurrent parent in (r) number of back crosses $= [(2^r - 1)/2^r]^n$. The chance of eliminating a gene linked to a selected allele is determined by the intensity of linkage (p) and the number of back crosses (r) according to the formula $1 - (1 - p)^{r+1}$.

BACK MUTATION: a mutation (recessive) reverts to the wild type allele. (See reversion)

BACKGROUND, GENETIC: the (residual) genetic constitution not considering particular loci or genes under the special study.

BACKGROUND RADIATION: the natural radiation coming from cosmic or terrestrial sources. (See cosmic radiation, terrestrial radiation)

BACTENEIN: see antimicrobial peptides

BACTERIA: a broad taxonomic group of microscopically visible (prokaryotic) organisms with DNA as the genetic material (nucleoid) that is not enclosed by a distinct membrane within the cell and they may have various number of extrachromosomal elements, plasmids that constitute from about 2 to 20% of the DNA per cell. Bacteria are capable of protein synthesis and of independent metabolism even if they are parasitic or saprophytic. Their cell wall is mucopolysaccharide and protein. Their cells contain ribosomes (70S) but no mitochondria, plastids or endoplasmic reticulum or other compartments. They divide by fission in an exponential manner as long as nutrients and air are not in limiting supply. The division rate of bacteria during exponential growth can be expressed as $N = 2^g N_0$ where N is the number of cells after g generation of growth and N_0 is the initial cell number. After the exponential phase, unless their increasing need is met, the growth either declines or may become stationary. The generation time of *E. coli* under standard conditions may be 20-25 minutes.

Bacteria can be classified into three main groups *Archebacteriales*, *Eubacteriales* and *Actinomycetales*. *Eubacteriales* includes *Pseudomonadaceae* (*Pseudomonas aeruginosa*), *Azotobacteriaceae* (*Azotobacter vinelandii*), *Rhizobiaceae* (*Rhizobia, Agrobactria*), *Micrococcaceae* (*Mycrococcus pyogenes*), *Parvobacteriaceae* (*Haemophilus influenzae*), *Lactobacteriaceae* (*Diplococcus pneumoniae, Streptococcus faecalis*), *Enterobacteriaceae*, (*Aerobacter aerogenes, Escherichia coli, Salmonella typhimurium*), *Bacilliaceae* (*Bacillus subtilis, B. thüringiensis*). *Actinomycetales* includes *Mycobacteriaceae* (*Mycobacterium phlei, Mycobacterium tuberculosis*) and *Streptomycetaceae* (*Streptomyces coelicolor, S. griseus*). There are many other types of classification systems. It is common to identify bacteria as Gram-posi-

Bacteria continued

tive (indicating that they retain the deep red color of the Gram stain [crystal-violet and iodine] after treatment with ethanol) or Gram-negative that fail to retain it (and may appear colorless or just slightly pinkish). These properties depend on the composition and structure of the cell wall. The Gram-positive bacteria are surrounded by peptidoglycan outside of the plasma membrane, the Gram-negative cells have an outer membrane enveloping the peptidoglycan wall. The peptidoglycans are polymers of sugars and peptides and are cross-linked by pentaglycines that determine the shape of the cell wall and the bacterium. (See conjugation, bacterial recombination frequency, bacterial transformation, recombination molecular mechanisms in prokaryotes, bacteria counting)

BACTERIA COUNTING: is done either by counting the number of colonies formed or by determining cell density in a volume, using a photometer. By the first procedure an inoculum of a great dilution of a culture is seeded on a nutrient agar plate and incubated for a period of time (e.g. 2 days) so each colony formed represents the progeny of a single cell and the number of colonies indicate the number of *live* bacteria in the volume of the inoculum. By the second procedure the optical density indicates the cell density that becomes meaningful only if information is available on the correlation between light absorption and cell number, determined earlier by the plating technique described above. If the plate was seeded by 2 mL of the culture diluted 10^7 times and 100 colonies are observed then the number of live cells is $(100/2) \times 10^7 = 5 \times 10^8$ cells/mL. (See bacteria)

BACTERIAL ARTIFICIAL CHROMOSOME: See BAC

BACTERIAL RECOMBINATION FREQUENCY: in case of conjugation transfer of genes recombination frequency is determined by the time in minutes since the beginning of mating. This procedure is useful for genes that are more than 2 to 3 minutes apart. It takes about 90 - 100 minutes at 37° C to transfer from a Hfr donor bacterium to an F- recipient cell the entire genome (more than 4 million nucleotides). The efficiency of transfer depends also on the nature of the Hfr strain used. Thus approximately $5 - 6 \times 10^4$ nucleotides are transferred per minute. Bacterial recombination does not permit the recovery of the reciprocal products of recombination and all detected cross-over products are double cross-overs. If bacterial genes are closer than 2 to 3 minutes, then recombination mapping is used. For bacterial recombination generally selectable (auxotrophic) markers are used so the phenotypes can be easily recognized. The recipient strain carries genes, e.g. a and b^+, and the donor strain carries the alleles a^+ and b defining the interval where recombination is studied. In order to be able measure the number of successful matings, the donor strain carries also the prototrophic gene (c^+) and the recipient is marked by the auxotrophy allele (c) of the same locus. The c gene does not have to be very close to the interval studied:

$$\frac{a \quad b^+}{a^+ \quad b} \qquad \frac{c^+}{c}$$

The *frequency of recombination* (p) is than calculated:

$$p = \frac{\text{number of cells } a^+ b^+ \text{ constitution}}{\text{number of } c^+ \text{ cells}}$$

The a^+ b^+ recombinants are the result an exchange between a^+ and b^+ and also beyond the and c^+ site as shown by the arrows: $a^+ \uparrow b^+ \quad c^+ \downarrow$

To determine *gene order by recombination* one must use at least three loci in a reciprocal manner:

Would the gene order be a b c:

Donor	$a \downarrow b^+ \ c^+ \downarrow$	to obtain triple prototrophs double exchange is sufficient
Recipient	$a^+ \ b \ c$	
Donor	$\downarrow a^+ \downarrow b \ c$	in both of the reciprocal crosses
Recipient	$a \ b^+ \ c^+$	

Bacterial recombination frequency continued

Would the gene sequence be *b a c* in order to obtain
Donor	↓ b⁺ ↓	a	c⁺ ↓	triple prototroph recipients the number of exchanges
Recipient	b	a⁺ ↑	c	must be at least 4 as shown by the arrows
Donor	b	↓ a⁺ ↓	c	In the reciprocal cross only double recombination
Recipient	b⁺	a	c⁺	is required to produce b⁺ a⁺ c⁺

Thus depending whether the gene order is *abc* or *bac* we can tell from the frequency of protrotrophs in the reciprocal crosses.

Recombination frequency in bacteria within very short intervals, such as between alleles within a gene can be determined also by transduction. If the constitution of the donor DNA is a^+b^+ and the recipient is $a\ b$, *the frequency of transduction* (recombination) is:

$$\frac{[a^+\ b] + [a\ b^+]}{[a^+\ b] + [a\ b^+] + [a^+\ b^+]}$$

Gene order in bacteria can be determined also by a three-point transformation test as illustrated below in a hypothetical experiment when the donor DNA is $a^+\ b^+\ d^+$ and the recipient is $a^-\ b^-\ d^-$, and the reciprocal products of recombination are not recovered:

GENES			GENOTYPES OF TRANSFORMANTS				
a	+	−	−	−	+	+	+
b	+	+	−	+	−	−	+
d	+	+	+	−	−	+	−
Number of cells	12,000	3,400	700	400	2,500	100	1,200

Recombination for a particular interval is calculated by the number of recombinants in the interval(s) divided by the total number of cells transformed in that interval. We have here 7 classes of cells. Recombination is calculated in three steps: in the *ab*, *bc* and *ad* intervals.

In the ***ab*** interval we have here $\frac{[3400 + 400 + 2500 + 100]}{[1200 + 3400 + 400 + 2500 + 100 + 1200]} \approx 0.33$

in the ***bd*** interval $\frac{[700 + 400 + 100 + 1200]}{[12000 + 3400 + 700 + 400 + 100 + 1200]} \approx 0.14$

in the ***ad*** region $\frac{[3400 + 700 + 400 + 2500 + 100 + 1200]}{[12000 + 3400 + 700 + 400 + 2500 + 100 + 1200]} \approx 0.41$

Although the frequency of recombination in three-point transformation test is never exactly additive, it is clear that the *a-d* distance is the longest and thus the conclusion that the gene order is *abd* appears to be reasonable. (See also physical mapping, crossing over, mapping genetic, conjugation)

BACTERIAL TRANSFORMATION: genetic alteration brought about by the uptake and integration of exogenous DNA in the cell that is capable of expression. The exogenous DNA is generally supplied at a concentration of 5 to 10 µg/mL to transformation competent cells. *Competence* is a physiological state when the cells are ready to accept and integrate the exogenous DNA. Competence is maximal in the middle of the logarithmic growth phase. The donor DNA may synapse with the recipient bacterial chromosome and naked DNA generally replaces a segment of the bacterial genetic material rather than adding to it. The entire length of the donor DNA may not be integrated into the host and the superfluous material is degraded. The integrated DNA may form a permanent part of the bacterium's chromosomal genetic material. During integration only one or both strands of the donor DNA may be integrated. Transformation may be regarded as one of the mechanisms of recombination and can be used for determining gene order in the bacterial chromosome. The frequency of transformation in

Bacterial transformation continued

prokaryotes is generally less than 1% and most commonly it is within the range of 10^{-3} to 10^{-5}. Transformation of bacterial protoplasts (spheroplasts) may occur at much higher frequency. Transformation may mean also the transfer and expression of plasmid DNA in the cell. These plasmids may stay as autonomous elements within the bacteria. Transformation with the aid of plasmids is much more efficient. Also the competence can greatly be enhanced by some divalent cations and other means. (See transformation genetic, competence of bacteria)

BACTERIOCIN: natural bacterial products that may kill sensitive bacteria. (See colicins, pyocin, pesticin)

BACTERIOPHAGES: are viruses infecting bacteria The major types and characteristics of bacteriophages are:

PHAGE	TYPE	HOST	Da x 10^6	MORPHOLOGY
MS2, f2, R17	RNA, ss, virulent	E. coli	1	icosahedral
φ6	RNA, ds, virulent	Pseudomonas	3.3, 4.6, 7.5	icosahedral
φX174, G4, St-1	DNA, ss, virulent	E. coli	1.8	icosahedral-tail
M13, fd, f1	DNA, ss, virulent	E. coli	2.1	filamentous
P22	DNA, ds, temperate	Salmonella	26	icosahedral-tail
SPO1	DNA, ds, virulent	Bacillus subtilis	91	isosahedral-tail
T7	DNA, ds, virulent	E. coli	26	octahedral-tail
lambda	DNA, ds, temperate	E. coli	31	icosahedral-tail
P1, P7	DNA, ds, temperate	E. coli	59	head-tail
T5	DNA, ds, virulent	E. coli	75	octahedral-tail
T2, T4, T6	DNA, ds, virulent	E. coli	108	oblong head-tail

ss = single-stranded, ds = double-stranded
(See phage, phage lifecycle, phage morphogenesis, lambda phage, T4, T7, φX174, MS2, Mu bacteriophage, icosahedral, virulence, temperate phage, development)

BACTERIORHODOPSIN: a light receptor protein in the plasma membrane of some bacteria; it pumps protons upon illumination. (See rhodopsin)

BACTERIOSTASIS: preventing the reproduction of bacteria that may lead on the long run to their destruction but not immediate killing. Many antibiotics have such an effect. (See antibiotics)

BACTEROID: specialized, modified forms of bacteria such as the ones found in the root nodules where they act in the fashion of intracellular "organelles" in nitrogen fixation. (See nitrogen fixation.

BACTO YEAST EXTRACT: a water soluble fraction of autolyzed yeast, containing the vitamin B complex.

BACTO-TRYPTONE: a peptone rich in indole (tryptophan) used for bacterial cultures and classification of bacteria on the basis of activity.

BACULOVIRUSES: large (130 kbp) double-stranded DNA viruses, used for the construction of insect and mammalian transformation vectors. (See polyhedrosis, AcNBPV, transformation, viral vectors)

BADGER: (*Taxidea taxus*) 2n = 32

Bal 31: exonuclease removes simultaneously nucleotides from the 3' as well the 5' ends and thus it can be used for mapping functional sites in a DNA: The fragments can then be separated by electrophoresis and assayed after transformation. (See deletions unidirectional, exonuclease electrophoresis)

0 time	a b c d e	original DNA
after 1 time unit	b c d	digest
after 2 time units	c	digest

BALANCE OF ALLELES: this population model assumes that at the majority of loci several different alleles are present and these are kept in a dynamic equilibrium by the continuous but variable selective forces. (See balanced polymorphism, Hardy - Weinberg theorem, fitness, selection)

BALANCED LETHALS: are genetic stocks heterozygous for two or more non-allelic linked recessive lethal genes. Since both homozygotes die, only heterozygotes survive that are phenotypically wild type or in some cases exhibit mutant phenotype and keep on producing both types of the lethals. Such stocks can be maintained indefinitely as long as recombination between the linked loci can be prevented. For the balancing, generally spanning inversions are used that eliminate the cross-over gametes because of the duplications and deficiencies generated by recombination within the inverted region. The first balanced lethal of *Drosophila* contained the gene *Bd ^{1}* (*Beaded,* incised wing in heterozygotes, lethal in homozygotes, at map location 3-91.9 [slightly different in some other alleles]) and *l(3)a* a spontaneous lethal mutation (map position 3-81.6) within the inversion *In(3R)C*. In the multiple translocations of the plants of the different *Oenothera* species, there are also gametic and zygotic recessive lethal genes that are prevented from becoming homozygous and thus help to maintain these lethal genes by balanced heterozygosity. Besides the biological advantage of balanced lethals, they may be useful for various types of research. The *Bd* alleles have been extensively studied also at the molecular level and the developmental functions could be revealed by the availability of heterozygotes for the mutations. (See lethal factors, lethal equivalent, translocation ring)

BALANCED POLYMORPHISM: when the fitness (reproductive success) of heterozygotes exceeds both homozygotes at a locus, a stable genetic equilibrium may be established and the heterozygotes may reproduce the homozygotes in equal frequencies. This type of heterozygote advantage may lead to balanced polymorphism, i.e. the population may maintain several genotypes in stable proportions even if some of the homozygotes have low adaptive value. (See selection coefficient, fitness, balanced lethals, autosomal recessive lethal assay, balance of alleles, Hardy - Weinberg theorem, fitness, selection)

BALANCED TRANSLOCATION: is a reciprocal translocation where each of the interchanged chromosomes has a centromere. Unbalanced translocations have an acentric piece due to the interchange. (See translocation)

BALANCER CHROMOSOMES: are structurally modified (by inversions, translocation) so the recombinants (because of duplications or deficiencies in the meiotic products) are not recovered in the progeny, and facilitate the maintenance of certain chromosomal constitutions without recombination. (See *ClB* method, *Basc*, inversion, translocation chromosomal, Renner complex, balanced lethals, autosomal recessive lethal assay, *Oenothera*)

BALANCING SELECTION: includes heterozygote advantage (overdominance), or alleles differently selected by sex, season, niche in the habitat or in a frequency-dependent manner. (See selection, sexual selection, overdominance, frequency dependent selection)

BALB/c MICE: an albino inbred laboratory strain used frequently in immunoglobulin (antibody) and cancer research. (See mouse)

BALBIANI RING: is a puff (bloated segment) of the polytenic chromosome indicating special activity (intense RNA transcription) at the site, generating loosening up of the multiple elements of the chromosome. (See polytenic chromosomes, puff)

BALDNESS: is a sex-influenced trait in humans being more common in males than in females (particularly with later onset) probably depending to some extent on the level of androgen. It generally displays a developmentally manifested pattern starting from the front hairline toward the top of the head. This condition has a strong hereditary component but it may be caused by certain diseases (alopecia) and exposure to higher doses of ionizing radiation and to certain carcinostatic drugs. Early baldness may be determined by a dominant gene with better penetrance in the males than in the females. (See alopecia, hair, androgen, penetrance)

BALDWIN EFFECT: physiological homeostasis permits the survival of a species until mutation may genetically fix the adaptive trait in an originally inhospitable environment. (See homeostasis, canalization)

β-AMYLOID: exists as extracellular deposits of the brain plaques in Alzheimer disease. It is split off the amyloid precursor protein (APP). In the neuronal tissue APP_{695} is prevalent. If Val^{642} is replaced by Ile, Phe or Gly, the substitutions lead to fragmentation of nucleosomal DNA in the neurons and presumable contribute to neurotoxicity. (See Alzheimer's disease, amyloidosis)

BamH1: restriction enzyme with recognition sequence G↓GATCC.

BANANA (*Musa acuminata*, x = 11): is a fruit plant. The diploid fruits are loaded with seeds and have minimal edible pulp. The majority of the edible fruits are harvested from seedless triploid plants. When the triploids are crossed with diploids the progeny is partly tetraploid (2n = 44) and heptaploid (2n = 77) because of the high frequency of unreduced 3x and 6x gametes and their fruits are also seedless. Some of the related species have chromosome numbers 2n = 14, 18 and 20. (See seedless fruits, triploidy)

BAND: an element of the cross-striped chromosome. The banding may be due to condensation of chromomeres (e.g., in the salivary gland chromosomes of diptera) or to specific staining of the chromatin (see bands of polytenic chromosomes, chromosome banding). Electrophoretic separation of restriction enzyme digested DNA, or pulsed-field electrophoresis separated small chromosomes, as well as various proteins subjected to separation in electric field, generate bands in the substrate (gel) when visualized either by staining or by special illumination. (See chromosome banding, electrophoresis, pulsed field electrophoresis)

BAND CLONING: amplifying DNA bands, extracted from electrophoretic gels, in genetic cloning vectors for molecular analyses. (See cloning vectors)

BAND III PROTEIN: is a transmembrane protein consisting of about 800 amino acids. (See spectrin)

BAND-SHARING COEFFICIENT (S_{xy}): indicates the proportion of shared DNA fragments separated by electrophoresis; $S_{xy} = (2n_{xy})/(n_x + n_y)$ where n_x and n_y are the number of bands in *x* and *y* samples and n_{xy} is the number of shared bands. This coefficient may be used to determine the genetic composition of populations on the basis of DNA. (See DNA fingerprinting)

BAND SHIFTING: see gel retardation assay

BANDING PATTERN: the distribution of chromosome bands reflecting genetic differences or differences in expression of genes displaying more or less loose puffs. (See polytenic chromosomes, lampbrush chromosomes, chromosome banding, puff)

BANDS OF POLYTENIC CHROMOSOMES: deeply stained prominent cross bands on the chromosomes where the chromomeres of the elementary strands are appositioned; the salivary chromosomes of *Drosophila* display about 5,000 bands and for a period of time it was assumed that each corresponds to a gene locus. Today, it is known that the number of genes far exceeds the number of bands. (See polyteny, salivary gland chromosomes)

BankIt: is GenBank submission form for protein coding sequences. It generates then a GenBank accession number. Its address: <http://www3.ncbi.nlm.nih.gov/BankIt/>

BAP: 6-benzylaminopurine, a plant hormone. (See plant hormones)

***BAR* MUTATION**: of *Drosophila* (*B*, map position 1-57.0) reduces the eye to a vertical bar with about 90 facets in the male and to about 70 in the female compared with 740 in the normal males and 780 in the normal females; the heterozygous females have 360. The *B* mutation is actually a tandem duplication of salivary band 16A, arisen by unequal (oblique) crossing over. Thus the "normal allele" has 16A, the *Bar* 16A-16A, the *Ultrabar* 16A-16A-16A constitution. The phenotype is actually a position effect and not the cause of a dosage effect as the genetic analyses demonstrated. The process of unequal crossing over may be repeated and as many as 9 copies of band 16A can accumulate in a single X-chromosome. Also, the 16A band may be lost resulting in reversion by the loss of the *roo* transposable element. *B* mutations may be induced also by the

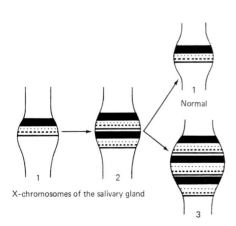

X-chromosomes of the salivary gland

Bar mutation continued

P hybrid dysgenesis element whereas chemical mutagens never produced this mutation. These facts indicate that, the *Bar* phenotype is caused by the breakage points in the duplications. The *Bar* phenotype may be the result of breakage in a regulatory element or within an intron and causes abnormal fusion of the exons. The *Bar* locus is very large, it spans at least 37 kb DNA. (See duplication, position effect, unequal crossing over, intron, exon, *ClB*)

BAR-CODE, GENETIC (molecular): bar codes generally represent two or four different width of vertical bars that correspond to digits 0 and 1, and which can in turn specify numbers of 0 to 9. The bar-coded information can be read by an optical laser and —through computer— the scanner can identify various types of information, including properties of a gene, phenotypic expression, etc. (See DNA chips, targeting genes)

BARDET-BIEDL SYNDROME: chromosome 16 recessive phenotype involving retinal dystrophy (retinitis pigmentosa), polydactily, and other anomalies of the limbs, obesity, underdeveloped genitalia and kidney malfunction, diabetes; mental retardation is also common. (See kidney diseases, eye diseases)

BARE LYMPHOCYTE SYNDROME (BLS): is a group of severe immunodeficiency diseases caused by defects in the major histocompatibility system. Some forms are due to defect(s) either in the HLA class I or class II genes involving lymphocyte differentiation. The current therapy is bone marrow transplantation. In the future gene therapy may be feasible. (See immunodeficiency, HLA, lymphocyte, MHC, gene therapy)

BARLEY (*Hordeum*): is cereal crop used for feed, food and the brewery industry. The cultivated *H. sativum* is diploid 2n = 14. Some of the wild barleys are polyploids. (See haploid [*H. bulbosum*])

BARNASE (*Bacillus amyloliquefaciens* ribonuclease): a ribonuclease that may be associated with chaperones. (See chaperones, barstar, ribonucleases)

BARR BODY: is a dark stained (heteropycnotic) structure visible at the periphery of the interphase nuclei of cells that have more than one X chromosome. XY cells do not have Barr body whereas normal XX female cells have one. The number of Barr bodies (named after M.L. Barr) is always one less than the number of X-chromosomes, indicating that the non-active X-chromosomes remain condensed (dosage compensation). Barr bodies are present also in XXY males. The Barr body is sometimes called sex chromatin. In the leukocytes the Barr body is enclosed in special nuclear appendage, called "drum-stick" because of its shape. (See lyonization, dosage compensation)

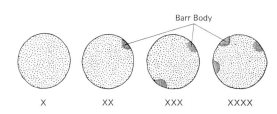

BARREL: protein β-sheets are closed by main chain hydrogen bonds. (See protein structure)

BARREN STALK: (*ba*) maize genes (in chromosomes 3, 2 and 9) affecting the tassel or ear or both and cause partial/full sterility. The tassel is male inflorescence. (See also tassel-seed)

BARRING GENE (*B*): see autosexing

BARSTAR: an inhibitor of barnase. (See barnase)

BARTH's SYNDROME: see endocardial fibroelastosis

BARTTER SYNDROME: is a dominant, human chromosome 11 encoded disease characterized by salt wasting and low blood pressure, accompanied by excessive amounts of calcium in the urine. The basic defect involves a potassium ion channel. (See ROMK, Gitelman syndrome, Liddle syndrome, hypoaldosteronism, ion channels)

β-ARK: β–adrenergic receptor kinase. (See adrenergic receptors)

BASAL: at or near the base.

BASAL BODY: a group of microtubules and proteins at the base of cilia and flagella of eukaryotes. (See microtubule, cilia, flagellum)

BASAL CELL CARCINOMA: see nevoid basal cell carcinoma

BASAL LAMINA: same as basement membrane

BASAL LEVEL ELEMENTS (BLE): have enhancer type functions in gene regulation and can occur at several positions. (See enhancer)

BASAL PROMOTER: generally situated in the region - 100 bp upstream of the transcription initiation site and contains various regulatory elements of transcription. (See promoter, transcription factors)

BASC: one of several similar *Drosophila* genetic stocks, containing the dominant *Bar* (*B*), the recessive eye color allele, *apricot* (w^a), and several *scute* inversions. The *B* and w^a markers identify the untreated chromosomes of the untreated females and eliminate the crossovers with the treated X-chromosomes of the males. The mutagenic effectiveness is determined on the basis of the reduced proportion of males in F_2 if a lethal or sublethal mutation was induced in the X-chromosome by the treatment. This type of analysis is called Muller-5 technique after H.J. Muller who designed the first stocks. The advantage of these stocks is that both males and homozygous females are completely fertile whereas the XO males are poorly viable and no crossovers appear along the X-chromosome. Variegation may occur in some unlinked genes. Rarely some exceptional females are also detected due to of unequal sisterchromatid exchange in the inversion heterozygote females. (See ClB, and also autosomal dominant mutation, autosomal recessive mutation)

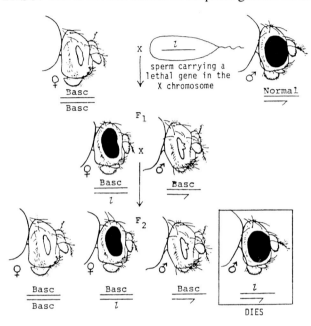

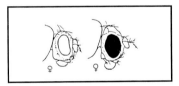

← EXCEPTIONAL FEMALES

BASE: the lowest part of a structure or a compound or ion which can combine with protons to yield salt; a *nitrogenous base* such as the pyrimidines and purines of nucleic acids. (See nucleic acid bases)

BASE ANALOGS: are nucleic acid bases or nucleosides similar to the normal compounds but cause mutation when incorporated into DNA either by incorporating in the wrong place or by mispairing with the incorrect base. The most commonly used mutagenic base analogs are 5-bromouracil (thymine analog) and 2-aminopurine (adenine analog). (See hydrogen bonding, base substitution, point mutation, universal base)

BASE MODIFYING AGENTS: nitrous acid causes oxidative deamination, hydroxylamine converts cytosine into hydroxylamino-cytosine (a thymine analog), alkylating agents place alkyl groups at several possible positions to purines and pyrimidines. (See chemical mutagens, base substitution, point mutation, hydrogen bonding)

BASE PAIR: hydrogen bonded A=T and G≡C in DNA or A=U in double-stranded RNA. (See mismatch, mispairing)

BASE PROMOTER: see core promoter

BASE SEQUENCING: see DNA sequencing

BASE STACKING: the nucleotides in parts of a polynucleotide chain may lay in such a way that the faces of the rings are appositioned. The stacking is most likely to occur near the chain termini where the bases move somewhat. It conveys some rigidity to the strand(s). The stacking is detectable by physical methods such as circular dichroism and optical rotatory dispersion. The stacking is eliminated by reagents which weaken hydrophobic reactions, and heating reduces stacking resulting in hyperchromicity. Destruction of hydrogen bonding also reduces stacking in double-stranded DNA. Base stacking may occur in double-stranded molecules where the pairing is weakened by deletions or mismatches. (See circular dichroism, optical rotatory dispersion, hyperchromicity)

BASE SUBSTITUTIONS: when a pyrimidine in the DNA is replaced by another pyrimidine or a purine is replaced another purine the change is a *transition*. When a purine is replaced by a pyrimidine, or vice versa, a *transversion* takes place. These changes cause mutation in the DNA and they may cause also amino acid replacement in the protein if the base substitution involves a non-synonymous codon. Many mutagens cause base substitutions, e.g. hydroxylamine (NH_2OH) targets cytosine (C) and hydroxyaminocytosine is formed that is a thymine analog. As a result a C≡G base pair is replaced by a T=A bp. Similarly 5-bromouracil may cause, through tautomeric shift, a T=A to be replaced by C≡G, and 2-aminopurine, an adenine analog, causes a tautomeric shift that may cause the replacement of an A=T pair by a G≡C pair. Some other chemicals, e.g. nitrous acid, by deamination, changes cytosine to uracil and converts adenine into guanine. These base substitutions may cause also reversions. A hydroxylamine-induced mutation may be reverted by bromouracil, etc. (See DNA, hydrogen bonding, point mutation, base analogs, evolution and base substitutions, incorporation error, replication error, chemical mutagens, substitution mutations)

BASEDOW DISEASE: see goiter

BASEMENT MEMBRANE: less than 500 nanometer thick laminated condensation of the extracellular matrix; several human diseases involve anomalies of basement membranes and/or associated proteins. (See extracellular matrix, Alport's disease)

BASIC CHROMOSOME NUMBER: is found in the gametes of diploid organisms and it is represented by x; in polyploids the haploid (n) number may be 2x, 3x and so on, depending on the number of genomes contained. The basic number is frequently called a genome. (See chromosome numbers, genome, polyploid)

BASIC COPY GENE: the silent copy of a *Trypanosoma* gene that is activated by transposition to an activation site in the telomeric region of the chromosome. (See *Trypanosomas*, telomere)

BASIC DYE: stains negatively charged molecules. (See stains)

BASIDIOMYCETES: a taxonomic group of fungi bearing the meiotic products in basidia. (See basidium)

BASIDIUM: is a fungal reproductive structure generally of club shape where meiosis takes place and then the haploid basidisopores are released for infection of host plants. (See stem rust)

BASOPHIL: a type of white blood cell, well stainable with basic cytological dyes. They contain conspicuous secretory granules and release histamine and serotonin in some immune reactions. Any other (acidic) structure or molecule with an affinity for positive charges. (See granulocytes, blood, immune system)

BASTA: see herbicides

BASTARD: a hybrid; an illegitimate or undesirable offspring.

BAT: *Carollia perpicillata* 2n = 21 in males, 20 in females; *Glossophaga soricina* 2n = 32; *Desmodus rotundus murinus* 2n = 28; *Atropzous pallidus* 2n = 46; *Eptesicus fuscus* 2n = 50; *Myotis velifer incautus* 2n = 44; *Nysticeius humeralis* 2n = 46.

BATEMAN's PRINCIPLE: the reproductive success of males shows greater variation than that of females because of the greater competition among males and larger number of male gametes.

BATESIAN MIMICRY: an adaptive evolutionary device. Certain species develop phenotypic characteristics of sympatric species (models) in order to secure better survival. The models are

Batesian mimicry continued

repugnant (distasteful) to certain predators and avoid them, and so the mimickers when mistaken for the models also escape destruction. The mimicry is more common among females than males (butterflies) because females are more subject to predation than males. (See adaptation, natural selection). EXAMPLES OF BATESIAN MIMICRY OF THE BUTTERFY *PAPILIO* IS SHOWN BELOW: (From Sheppard, P.M. 1959 Cold Spring Harbor Symp. Quant Biol. 24:131) ⇓

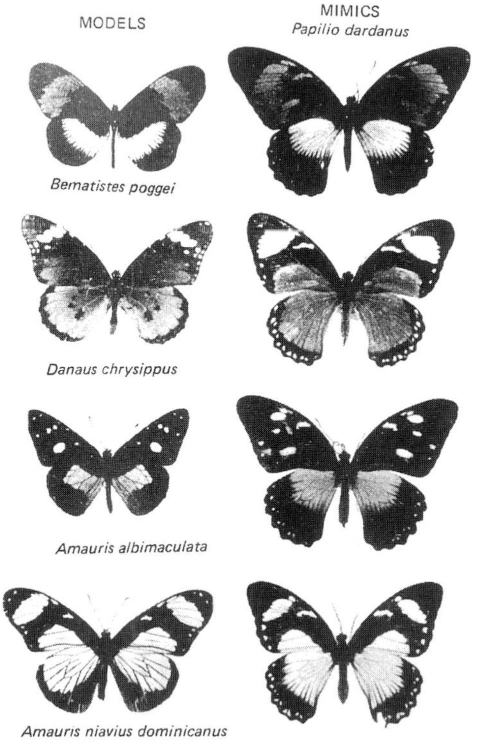

BATEN DISEASE: recessive human chromosome 16p12.1-p11.2. It is a juvenile onset familial amaurotic idiocy caused by lipid accumulation in the nerve tissues and vacuolization of the lymphocytes.

BATTEN-TURNER SYNDROME: see myopathy congenital, human chromosome 16.

BAYES' THEOREM: permits the estimation of various conditional probabilities and it is used for decision making processes. In the simplest general form:

$$P(A|B) = \frac{P[B|A]P[A]}{P[B|A]P[A] + P[B|A']P[A']}$$

Let us illustrate the use of it in genetics by assuming that we have three individuals, 2 homozygotes for a semi-lethal dominant factor and 1 heterozygote for the same semi-lethal and incompletely dominant gene. There are problems with visual classification at an early stage of the development. Assuming that the 2 homozygotes [A] have 60%, and the heterozygote [A'] is expected to have 80% viability.

Thus ($P[B|A] = 0.6$, and $P[B|A'] = 0.8$. The chance of selecting one individual of either genotype is $P[A] = 2/3$ and $P[A'] = 1/3$. The individual picked turns out to be very weak, and being uncertain about the choice, we would like to figure out — in view of the information available — what is the probabilty that we have chosen the heterozygote :

$$P(A|B) = \frac{[0.8][0.33]}{[0.8][0.33] + [0.6][0.67]} \cong 0.4$$

Basically the Bayesian method considers the classical population parameters as random variables with a specific *a priori* probability of distribution. Then the *conditional probability* is estimated on the basis of the a priori distribution. The conditional probability is thus a property of *a posteriori* distribution because the accepted or supposed a priori distribution is used for the estimation of an existing situation in a population. (See probability, conditional probability, risk)

B1, B2 REPEATS: are highly dispersed SINE elements in the mouse genome. (See SINE)

BB-1: the same as B7 or CD80

β-CATENIN: is a component of the cadherin-catenin cell adhesion complex. (See adherens junction)

B-CELL DIFFERENTIATION FACTOR: see interferon β-2 (IFNB2)

B-CELL GROWTH FACTOR: see IL-4

BCGF (B cell growth factor): 12 kDa cytokine produced by activated T cells. (See B cell, T cell, cytokine, lymphocytes)

BCIP: 5-bromo-4-chloro-3-indolyl phosphate is used in combination with nitro blue tetrazolium (NBT, reveals precipitated indoxyl groups) for the detection of antigen-antibody — antibody-AP (alkaline phosphatase) complexes.

BCL1, BCL2, BCL3, BCL5: are leukemia oncogenes. Bcl-1 is cyclin D1. Bcl-2 controls also apoptosis as a defense against malignant tumorigenesis; it also promotes regeneration of severed cells in the central nervous system. The Bcl-2 protein functions also as an ion channel and a docking protein. BCL-2 is located in the outer membrane of the mitochondria, nuclei and the endoplasmic reticulum. (See leukemia, apoptosis, malignant growth, cyclin)

β-CONFORMATION: an extended conformation of a peptide chain; a type of secondary structure. (See protein structure)

BCR: B-cell antigen receptor. (See B lymphocyte)

BCR (bcr): break point cluster region in the Philadelphia chromosome is an area where multiple chromosomal breakages have been observed leading to cancerous transformation (leukemia, ABL). The BCR polypeptide contains 1,271 amino acids and its normal function is a protein serine - threonine kinase. (See Philadelphia chromosome, leukemia, ABL)

B-DNA: a conformation of the DNA most common in hydrated living cells. (See DNA types)

BDNF (brain-derived neurotrophic factor): is an autocrine growth substance of neurons. (See autocrine, neuron)

BEADS-ON-A-STRING: originally this meant the (light-microscopically visible) chromosome structure in pachytene where the chromomeres could be seen as the term says. Today, it is used also for the DNA strands wrapped around the eight histones (nucleosomes) as detected by the electronmicroscope. (See also nucleosome, pachynema, chromomeres)

BEAGLE: see copia

BEANS (*Phaseolus* spp): pulse crops, all with $2n = 2x = 22$ chromosomes, including the most common *P. vulgaris* (French or navy beans) and the *P. lunatus* (Lima bean).

BEAR: *Ursus americanus* (black bear) $2n = 74$; *Tremarctos ornatus* (spectacled bear) $2n = 52$.

BEARE-STEVENSON SYNDROME: an autosomal dominant disorder involving furrowed skin, head bone fusions, facial anomalies, abnormal digits, umbilical, genital malformations and early death. The basic defect is in the fibroblast growth factor receptor 2. (See FGF)

BEAVER: *Castor canadensis,* $2n = 40$.

BECKER MUSCULAR DYSTROPHY (BMD): see muscular dystrophy

BECKWITH-WIEDEMANN SYNDROME (EMG syndrome): is caused by a dominant gene in human chromosome 11p15.5 area. The symptoms include enlarged tongue (detectable at birth) and generally umbilical anomalies (omphalocele = herniated intestines at the belly button area), hypoglycemia, enlargement of the internal organs (visceromegaly) and frequent concomitant kidney and liver anomalies, tumorous striated muscles (rhadomyosarcoma), etc. It may be associated with trisomy for chromosome 11 and it has been suggested that it is caused by paternal or maternal disomy when the normal (most commonly paternal) chromosome is lost from the trisomic cell lineage, or imprinting. A gene encoding a cyclin-dependent kinase inhibitor ($p57^{KIP2}$) is imprinted and preferentially expressed by the maternal allele may be responsible for some of the cases. Deletion, duplication and balanced translocation were also suggested for some cases. It has been shown recently that KVLQT1 gene, encoding a putative potassium channel, and mapped to the 1p15.5 region controls imprinting not just the Beckwith-Wiedemann gene but also the Jervell and Lange-Nielsen syndrome and the LQT heart arrhytmia. (See disomic, disomic uniparental, Wilms tumor, cancer, Jervell and Lange-Nielssen syndrome, rhabdomyosarcoma, Simpson-Golabi-Behmel syndrome, imprinting, ion channels, cyclin)

BEDWETTING: see nocturnal enuresis

BEECH (*Fagus*): a hardwood tree, *F. sylvestris*, $2n = 2x = 24$.

BEGONIA (*Begonia semperflorens*): ornamental plant, $2n = 34$.

BEHAVIOR GENETICS: analyzes the genetic determination and regulation of how organisms

Behavior genetics continued

behave. The majority of these traits (courtship, bird and frog songs) are under multigenic control and depend a great deal on the influence of the environment. In a few cases large effects of single genes have also been shown. In the honey bee a single gene controls the habit of uncapping the honeycombs containing dead larvae but another gene is required for the removal of the dead brood. If both genes are present the colony becomes resistant to the bacterial disease foulbrood because of the improved hygienic behavior. Alcoholism, criminality, etc. in humans may be determined by several genes and by the social environment. The Lesch-Nyhan syndrome is caused by a deficiency of the enzyme hypoxanthine-guanine phosphorybosyl transferase, and because of this, the salvage pathway of nucleic acid is inoperational. As a consequence purines accumulate and uric acid is overproduced leading to gout like symptoms but more importantly the nervous system is also affected, leading to antisocial behavior and self-mutilation. This gene has been isolated and cloned and may be transferred to the afflicted human body for gene therapy. Since behavioral traits are determined by the nervous system, neurogenetics may provide the answer for many serious conditions such as Alzheimer's disease (an amyloid accumulating pre-senile dementia), neurofibromatosis (a soft tumor of the nervous system affecting the entire body involving a protein resembling a GTPase activator), etc. Molecular analysis of memory and learning ability has made progress recent years through studies of simple organisms such as the slug *Aplysia* and also *Drosophila*. Several genes involved in the development of the nervous system of *Drosophila* have been cloned. Recent studies show that mice without the *fos* gene fail to nurse their pups presumably because of some brain lesions. For genetic and developmental analysis of nerv functions, the nematode, *Caenorhabditis* is particularly well suited because its entire nervous system consists of only 302 cells. (See courtship in *Drosophila*, personality, alcoholism, cognitive abilities, addiction, fate mapping, behavior in humans, aggression, ethics, instinct, morality, behavior in humans, avoidance learning, cross fostering, FOS)

BEHAVIOR IN HUMANS: has long been suspected to be genetically determined but with a few exceptions (Lesch-Nyhan syndrome, Tay-Sachs disease, Huntington's chorea, etc.) the genetic control is not completely understood. The majority of the behavioral traits are under the control of several genes. In such instances the tools of quantitative inheritance are needed, such as heritability, comparison of monozygotic-dizygotic twins with the general population, QTL mapping, etc. The approximate ratios of monozygotic:dizygotic concordance for alcoholism, females (1.1) versus males (1.7), dyslexia (1.7), Alzheimer's disease (2.1), major affective disorders (2.4), schizophrenia (2.7), autism (6.7). Heritabilities determined by intraclass correlation were: for memory (0.22), mental processing speed (0.22), scholastic achievement in adolescence (0.38), spatial reasoning (0.40), adolescent vocational interest (0.42), neuroticism (0.46), verbal reasoning (0.50), general intelligence (0.52). [Data based on Plomin, Owen & McGuffin, 1994 SCIENCE: 264:1733]. Cognitive abilities are studied also as part of the developmental genetic pattern (longitudinal genetic analysis). Multivariate genetic analysis determines the covariance (see correlation) among multiple traits. Although some genetic effects are specific to certain abilities, the majority of the genetic components have overlapping effects. The studies must also consider in assessing behavioral, cognitive genetic traits that form a continuum and the anomaly in a proband or several individuals may be just the extreme form of a normally existing behavioral pattern. Behavioral traits generally have about 50% or larger environmental components. These effects include family relationships and changes in that (e.g. divorce, death, accidents), social environment (economic status, schools, drug use, neighborhood), etc. The quantitative genetic approaches assumes that behavioral traits are complex and are the end result of cooperative action of individual genes, expressed as a phenotypic class rather than one gene - one disorder (OGOD). Some behavioral anomalies may show cosegregation with DNA markers such as used in QTL analysis. Although some of the mental anomalies, such as phenylketonuria (single recessive defect in phenylalanine hydroxylation) may account for about 1% of the affliction in mental asylums (see also the fragile-X syndrome). More than 100 single gene determined human diseases include mental retardation

Behavior in humans continued

as part of the syndromes. Defects in the X-chromosomally encoded gene product mitochondrial enzyme, monoamine oxidase A (MAOA) was attributed to violent behavior and also to schizophrenia. MAOA degrades serotonin, dopamine and norepinephrine. (See also behavior genetics, cognitive abilities, human intelligence, ethology, self-destructive behavior, personality, MAOA, homosexuality, aggression, autuism, dyslexia, morality, instinct, cocaine, serotonin, dopamine, norepinephrine, heritability, QTL, determinism)

BEHCET SYNDROME (TAP): a rare mouth and genital inflammation in humans; probably autosomal dominant.

BEHR SYNDROME: is a recessive infantile optical nerve atrophy. (See optic atrophy)

BEL: see *copia*

Bellevalia: subspecies of lilies (2n = 8 or 16), with large and well stainable chromosomes.

Bematistes pongei: an African butterfly mimicked by *Papilio dardanus*. (See Batesian mimicry)

BENCE-JONES PROTEIN: Some immunoglobulin heavy chain diseases (HCD) such as the lymphoproliferative neoplasms may contain only the antibody heavy chains (IgM, IgG, IgA), and even those are truncated and are deficient in most parts of the variable region. The γ and α type HCD cells synthesize no light chains but the μ HCD cells secrete an almost normal light chain that is detectable in the urine an it is called Bence-Jones protein. Some of the bone marrow cancer (myeloma) patients also produce Bence-Jones protein in their urine. These light chain immunoglobulins are generally homogeneous because they are the products of a clone of cancer cells and were very useful historically to gain information on antibody structure. (See myeloma, monoclonal antibody, immunoglobulins, antibody)

BENEFICIAL MUTATION: The majority of new mutations are less well adapted than the prevailing wild type allele in a particular environment or at best they may be neutral. Beneficial mutations are rare because during the long history of evolution the possible mutations at a locus had been tried and the good ones preserved. Nevertheless if a new mutation has 0.01 reproductive advantage, the odds against its survival in the first generation is $e^{-1.01} = 0.364$. Its chances to be eliminated by the 127th generation is reduced to 0.973 compared to a neutral mutation that would be eliminated by a chance of 0.985. Even mutations with an exceptionally high selective advantage may have a good chance to be lost ($e^{-2} \cong 0.1353$). Under normal conditions the selective advantage (s) is generally very small and the chance of ultimate survival $(y) = 2s$ and the chance of extinction $(l) = 1 - 2s$. In order that the mutation would have a better than 50% probability of survival, the requisites must be $(1 - 2s)^n < 0.5$ or $(1 - 2s) > 2$. Hence $-n \ln((1 - 2s) > \ln 2$ or approximately $- n(-2s) > \ln 2$, and therefore $n > (\ln 2)/2s$ or $\cong (0.6931)/2s$. If $(s) = 0.01$ and (n) = number of mutations, (n) must be larger than $0.6931/(2 \times 0.01) \cong 34.66$. In other words at least 35 mutational events must take place with at least 1% selective advantage of the mutants over the wild type that one would ultimately survive. If the rate of mutation is 10^{-6}, about a 35 million large population may provide such a mathematical chance. Under evolutionary conditions neutral or even deleterious mutations may make it however by random drift or chance in small populations. (See mutation neutral, mutation rate, mutation spontaneous, mutation in human populations)

BENTON - DAVIS PLAQUE HYBRIDIZATION: involves selection of recombinant bacteriophages on the basis of DNA hybridization with ^{32}P probes on an appropriate (nitrocellulose or nylon) membrane. For the screening of a mammalian or other large library, hundreds of thousands of recombinants need to be screened. In a 150 mm Petri dish 5×10^4 plaques may be used. (See DNA hybridization, Grunstein - Hogness screening, DNA library, recombination molecular mechanisms prokaryotes, plaque, plaque-forming unit)

BENZO(a)PYRENE: a highly carcinogenic polycyclic hydrocarbon generated by combustion at relatively lower temperature by polymerization of organic material. It occurs in automobile exhausts, by burning of coal, in cigarette smoke, fried and grilled meat (in char-broiled T-bone steaks more than 50 µg/kg had been detected, etc.) It has been estimated that 13,000 ton is annually released into the world's atmosphere by these processes. A single 0.2 mg intra-gastric dose per mouse, resulted in 14 tumors in five of the 11 animals treated. Exposure of the skin,

Benzo(a)pyrene continued
inhalation of the fumes proved the high and fast carcinogenicity. It is also a promutagen requiring metabolic activation in *E. coli*, yeast, *Drosophila*, various rodents and the plant *Arabidopsis*. Benzo(a)pyrene forms adducts with guanine by binding to the N2 position but it forms adducts also with deoxyadenosine. It causes also sister-chromatid exchange and the formation of micronuclei. (See environmental mutagens, carcinogens, Ames test, bioassays in genetic toxicology, sister chromatid exchange, micronucleus formation as a bioassay, adduct)

BENZYLADENINE (6-benzylaminopurine): see plant hormones

BER (base excision repair): see DNA repair, excision repair

BERNOULLI PROCESS: independent experiments which can provide only two outcomes, yes or no, success or failure are called Bernoulli trials and the two event classes and their probabilities are called Bernoulli process where p = probability of success and 1 - p = q = probability of failure. If in a sequence of 10 trials there are 4 successes, the probability of that sequence is $p^4 q^6$; if

$$p = \frac{2}{3} \text{ then the probability of the sequence becomes } \binom{2}{3}^4 \binom{1}{3}^6$$

The general formula becomes: $P(r \text{ successl } N, p) = \binom{N}{r} p^r q^{N-r}$ .... {1}

where p = probability of success, r = the exact number of successes and N = the number of independent trials. Let us assume that we observe a monogenic segregation where the penetrance of the mutant class is reduced from 25% to 20%, then the probability of finding a recessive mutant among 3 individuals will be according to {1}:

P (1 mutant among 3 individuals) = $\binom{3}{1}(0.20)^1 (0.80)^2 = \left(\frac{3!}{2!(2-1)!}\right) 0.2 \times 0.64 = 0.384$

If the population is increased to say 210, the chances increase to 70 for finding a mutant. (See also binomial probability)

BETA COMPLEX: one of the alternately distributed translocation complexes of the plant, *Oenothera* (See also the alpha complex alternative, multiple translocations, complex heterozygote)

BETA PARTICLES: are electrons emitted by radioactive isotopes; their mass is 1/1837 that of a proton. The negatively charged form of it is an electron whereas the positively charged is a proton. The beta particles have no independent existence; they are created at the instance of emission. In the biological laboratory the most commonly used isotopes emitting β radiation (with energy in MEV) are H^3 (0.018), C^{14} (0.155), P^{32} (1.718), S^{35} (0.167), I^{131} (0.600 and 0.300 but emits also γ radiations of various energy levels). The mean length of the path of H^3 is about 0.5 μm and that of P^{32} is about 2600 μm. (See linear energy transfer, isotopes)

BETA SHEETS: see protein structure

Beta vulgaris : beets (*Chenopodiaceae*), basic chromosome number 9. Includes sugar beets, fodder beets, mangold and chards are all important food and feed crops. The sugar beets represent a glowing example of the success of selective plant breeding by increasing sugar content (about 2% in the middle 18th century) by over 10-fold in some modern varieties. The most productive current varieties display triploid heterosis and improved disease resistance and the monogerm "seeds" facilitate mechanization of cultivation, etc. (See heterosis, triploid, monogerm seed)

BETEL NUT (*Arecia catechu*): a seed palm tree used as a stimulant; 2n = 4x = 32.

BETHLEM MYOPATHY: a dominant human disorder involving contractures of the joints, muscular weakness and wasting. It is associated with mutations of collagen type VII genes in human chromosome 21q22.3 and in 2q37. (See collagen, laminin)

BEV (Baculovirus expression vector): is a potential tool to control insect populations by biological means. (See Baculovirus, biological control, viral vectors)

bFGf: basic fibroblast growth factor

β-GALACTOSIDASE: an enzyme (lactase) that splits the disaccharide lactose into galactose and glucose. It can act on some lactose analogs too, e.g. on ONPG (o-nitrophenyl-β-D-galacto-

β-Galactosidase continued

pyranoside). This substrate (10^{-3} M or less) when exposed to the active enzymes (10^{10} molecules/mL) yield a yellow product (that has an absorption maximum at 420 nm) and can be used to measure the activity of the enzyme. In cells grown in A medium and with Z buffer the activity of galactosidase is determined by the formula: $1000 \times \dfrac{OD_{420} - (1.75 x OD_{550})}{tx(0.1 x OD_{660})}$ where OD is optical density at the wavelength indicated, and t is time of the reaction run in minutes. On Petri plates the activity of β-galactosidase is detected on EMB agar (containing eosin yellow, methylene blue and lactose) and in case the sugar is fermented, dark red color develops. Bacterial galactosidase is an inducible enzyme and induction takes place by allolactose that is formed upon the action of the residual few galactosidase molecules in the non-induced cells. A gratuitous inducer (induces the synthesis the enzyme although itself is not a substrate) is isopropyl β-D-thiogalactoside (IPTG). Constitutive mutants of the *E. coli z* gene can be identified on Xgal media containing 5-bromo-4-chloro-3-indolyl-β-D-galactoside dissolved generally in dimethylformamide (20 mg/mL). This compound is not an inducer of the enzyme but it is cleaved by it and thus a blue indolyl derivative is released. (See also *Lac* operon, galactosidase)

BGH: see bovine growth hormone

ß-GLUCURONIDASE: see GUS

BHK: baby hamster kidney cell; cultured fibroblasts of the Syrian hamster.

bHLH (basic helix-loop-helix protein): see helix-loop-helix

b/HLH/Z MOTIF: a basic amino acid sequence at the N terminus, required probably for DNA binding, helix-loop-helix structure, leucine zipper. This general structure is wide-spread among biologically active proteins involved in DNA binding. (See binding proteins, DNA-binding protein domains)

BI-ARMED CHROMOSOME: has two arms at the opposite sides of the centromere (O) such as:
----O------- (See also telochromosome, chromosome morphology)

BIALAPHOS: is an inhibitor is a glutamine synthetase normally produced by *Streptomyces hygroscopicus*. Upon splitting off two alanine residues it is activated into phosphinotricin. (See herbicides)

BIBAC (binary bacterial artificial chromosome): is a plant genetic expression vector that can be propagated in *Agrobacterium tumefaciens* and *E. coli* and it can deliver to plant chromosomes large (160 kb) foreign DNA sequences. (See BAC, *Agrobacterium*, transformation, vectors)

bicoid (bcd): a maternal effect mutation in *Drosophila*. The larvae have two abdomens and telsons, arranged in mirror image symmetry but no other body parts are affected. (See maternal effect, telson, Drosophila, morphogenesis in *Drosophila*)

BIDIRECTIONAL REPLICATION: from the replicational origin the replication moves in opposite directions in the DNA. (See replication bidirectional, replication fork)

BIDS: see hair-brain syndrome

BIFUNCTIONAL ENZYMES: have apparently evolved with the potential to adapt the amino and carboxy terminal tracts to fulfill the needs of metabolic requirements of different tissues. (See one gene—one enzyme theorem)

BigSeq: a computer program with similar purpose as Mask, and contains information on millions of contigs. (See contig, physical map)

BILAYER: of membranes consists of amphipathic lipids (and proteins) and the non-polar phase faces inward. The majority of cellular membranes are double membranes. (See amphipathic)

BILE SALTS: detergent type steroid derivatives involved in digestion and absorption of lipids.

BILINEALITY: more than a single locus determines a particular trait and they may segregate independently making chromosomal localization, by genetic techniques, very difficult.

BILIRUBIN: a bile pigment formed by the degradation of hemoglobin, and other heme containing molecules such as cytochromes. It is circulated in the blood as a complex with albumin and when deposited in the liver it forms bilirubin diglucoronide. It may arise from biliverdin, a breakdown product of heme, through reduction. (See bilirubinemia)

BimC: a family of motor proteins of the kinesin group, involved in the separation of the mitotic chromosomes by the spindle. (See motor proteins, spindle, mitosis)

BimE: is a subunit of the APC complex. (See APC, cell cycle)

BIMODAL DISTRIBUTION: when the population, represented graphically, displays two peaks or, in general, when the data are clustered in two modes, in two classes.

BIN: a group of markers (microsatellite DNA) mapped to the same location.

BIN1 (**b**ox-dependent Myc **in**teracting protein-**1**): is a tumor-suppressor protein (human chromosome 2q14), interacting with the Myc oncoprotein. It is related to amphiphysin that serves the similar purpose with breast cancer and to the RVS167 cell cycle control gene of yeast. (See MYC, tumor suppressor)

BINARY: any condition, choice or selection with two possibilities or a numeration system with a radix of 2. (See radix)

BINARY FISSION: splitting into two parts; bacteria, chloroplasts and mitochondria that do not have a mitotic mechanism reproduce this way after DNA replication has been completed.

BINARY VECTOR: of *Agrobacterium* carrying two plasmids, one has the T-DNA borders and other sequences that will integrate into the transformed cell's chromosomes, the other is a helper plasmid carrying the Ti plasmid virulence genes, required for transfer but no part of the latter plasmid is integrated into the host genome during transformation. (See T-DNA, virulence genes of *Agrobacterium*, transformation genetic)

BINASE: a 12 kDa dimeric ($\alpha\beta$) ribonuclease binding to N-1 of 3'-guanine monophosphate

BINDING ENERGY: is derived from non-covalent interaction between ligand and receptor, enzyme and substrate. (See ligand, receptor)

BINDING PROTEINS: are of a great variety and they control gene expression generally at the level of transcription (transcription factors, hormones, heat-shock proteins, etc.). The majority bind to upstream consensus sequences. The cap-binding proteins regulate the stability of the mRNA. Some of them are transcription termination factors, such as rho in bacteria or the Sal I box binding proteins in mouse. Their position and function at the DNA level is studied by foot-printing. Some proteins bind to the cellular membranes and control imports and exports, others mediate signal transduction. These proteins may have a combinatorial hierarchy and thus capable of influencing a multitude of processes in the cell, far in excess to their individual numbers. (See transcription factors, signal transduction, DNA-binding protein domains, single-strand binding protein, footprinting, affinity-directed mass spectrometry)

BINET TEST: see human intelligence

BINOMIAL COEFFICIENT: see binomial probability

BINOMIAL DISTRIBUTION: is useful in genetics for the direct estimation of segregation ratios in case of dominance by expansion of $(3 + 1)^n$ where n = is the number of heterozygous loci (note: the 3 + 1 must not be added). By expansion the binomial becomes:

$$1 \times 3^n + [n!]/1!(n-1)! \times 3^{n-1} + [n!/2!(n-2)!] \times 3^{n-2} + \cdots + [n!/(n-1)!] \times 3^{n-(n-1)} + 1 \times 3^{n-n}$$

The *exponent* of a base gives the number of loci with the dominant phenotype, the *power* identifies the frequency of that phenotype, and the *coefficients* show how many times quadruple, triple, etc. dominant phenotypic classes will be expected theoretically. The solution for four heterozygous pairs of alleles:

$$(1 \times 3^4) + \left(\frac{4 \times 3 \times 2 \times 1}{1 \times 3 \times 2 \times 1} \times 3^{4-1}\right) + \left(\frac{4 \times 3 \times 2 \times 1}{2 \times 1 \times 2 \times 1} \times 3^{4-2}\right) + \left(\frac{4 \times 3 \times 2 \times 1}{3 \times 2 \times 1 \times 1} \times 3^{4-3}\right) + (1 \times 3^{4-4}) =$$

$$(1 \times 3^4) + \left(\frac{24}{6} \times 3^3\right) + \left(\frac{24}{4} \times 3^2\right) + \left(\frac{24}{6} \times 3^1\right) + (1 \times 3^0) =$$

$$(1 \times 81) + (4 \times 27) + (6 \times 9) + (4 \times 3) + (1 \times 1)$$

Translated into genetic language in case of an Aa Bb Cc Dd heterozygote's F_2 progeny the phenotypic classes will be:

Binomial distribution continued

81 ABCD, [27 ABCd, 27 ABcD, 27 AbCD, 27, aBCD], [9 ABcd, 9 AbCd, 9 AbcD, 9 aBCd, 9 aBcD, 9 abCD], [3 Abcd, 3 aBcd, 3 abCd, 3 abcD], [1 abcd] or
81: 108 (4x27)*: 54* (6x9)*: 12* (4x3)*: 1*

For the calculation of genotypic classes among the segregants see trinomials, multinomials. (See Mendelian segregation, Pascal triangle)

BINOMIAL NOMENCLATURE: see taxonomy

BINOMIAL PROBABILITY: P is the complete binomial probability function whereas the $n!/(x!(n-x)!$ is the binomial coefficient, an integer that shows how many ways one can have x combinations of n, and $p=0.75$ and $q=0.25$ (because of the 3:1 segregation). In genetic experiments this shows—if we have n = independently segregating gene loci, and the inheritance is dominant—how many ways can have x combinations of n; e.g. if we deal with $n = 5$ loci and we wish to know the chance that at 3 (=x) loci the dominant phenotype would appear is then:

$$P = \binom{n}{x} p^x q^{(n-x)}$$

$$\binom{n}{x} = \frac{n!}{x!(n-x)!}$$

$$[5!/(3!2!)] = (0.75^3)(0.25^2) \cong 0.263672$$

The binomial distribution is obtained from the expansion of the binomial terms $(p + q)^n$; its standard deviation $\sigma = \sqrt{\dfrac{pq}{n}}$ and $p + q = 1$; n = is the exponent.

(See Pascal triangle, transmission disequilibrium, binomial distribution, trinomial distribution, Bernoulli process)

BINUCLEAR ZINC CLUSTER: is a domain of a DNA-binding transcriptional activator containing 2 Zinc ions about 3.5 Å apart and regulated by 6 cysteines. (See Zinc finger)

BIOASSAYS (biological assays): are used for determining the biological effect(s) of chemicals, drugs or any other factor on either live animals, or plants, or microorganisms or cells.

BIOASSAYS IN GENETIC TOXICOLOGY: have been designed to assess mutagenic (and indirectly carcinogenic) properties of factors human, animal, plant and microbial populations may be exposed to. Their range varies from testing chromosome breakage and point mutations in a wide variety of organisms using different endpoints. All the different procedures cannot be discussed or even enumerated here but the major types of tests include: (i) excision repair, (ii) reversion studies in *Salmonella* and *E. coli.*, (iii) sister chromatid exchange, (iv) mitotic recombination, (v) host-mediated assays, (vi) specific locus mutation assays, (vii) micronuclei formation, (viii) chromosome breakage, (ix) sex-linked lethal assays, (x) unscheduled DNA synthesis, (xi) sperm morphology studies, (xii) cell transformation assays, (xiii) dominant mutation, (xiv) somatic mutation detection, (xv) *Arabidopsis* mutagen assays, (xvi) human mutagenic assays. (See the essential features of these tests (i) to (xvi) under the separate entries, transgene mutation assay, mutation detection)

BIOCHEMICAL GENETICS: studies the genetic mechanisms involved in the determination and control of metabolic pathways. (See inborn errors of metabolism)

BIOCHEMICAL MUTANT: the chemical basis of the mutant function is identified. (See auxotroph)

BIOCHEMICAL PATHWAY: the chemical steps involved in a biological function are represented in a sequence. The individual steps are usually mediated by enzymes encoded by separate genes. (See one gene - one enzyme theorem)

BIOCOENOSIS (biocenosis): different organisms living together within the same environment; some of them may be dependent on others for survival or interact in various ways.

BIOFILM: bacterial aggregates surrounded by foamy substance and resistant to many types of disinfectants.

BIO-GEL: commercial ion exchange chromatography medium suitable for the separation of RNA from DNA, purification of oligonucleotides and linkers, etc.

BIODEGRADATION: decomposition, destruction of substances by bacterial or other organisms.

BIODIVERSITY: see species extant

BIOHAZARDS: working with pathogenic organisms or transgenic material containing potentially dangerous genes (coding for toxins). Containment (P1 to P4, the latter the most stringent) is necessary and the appropriate safety regulations must be complied with. Information for particular cases can be obtained from the local biohazard committees or from National Institute of Health, Building 31, Room A452, Bethesda, MD 20205, USA. (See biological containment, recombinant DNA and biohazards, <ftp//potency.berkeley.edu/pub/tables/hybrid.other.tab> or <cpdb@potency.berkeley.edu>)

BIOINFORMATICS: use of computers for information gathering from biological databases.

BIOLISTIC TRANSFORMATION: (biological-ballistic) introduces genes into the nuclei of cells (of the germline) by shooting DNA coated particles into the target cells, propelled by high- power guns. It is a most useful procedure when other methods of transformation are not sufficiently successful. It can accomplish transformation also in terminally differentiated cells. (See transformation, chloroplast genetics, mitochondrial genetics)

BIOLOGICAL CLOCK: is frequently called circadian rhythm; measures in various organisms daily periods and responses to alternation of daily light and dark cycles. The endogenous rhythms also influence gene activity and developmental patterns. (See circadian)

BIOLOGICAL CONTAINMENT: preventive measures to avoid the spread of potentially hazardous organism outside the laboratory. Recombinant DNA containing organisms with unknown biological impact in the environment may be prevented from accidental spreading using transformation vectors that lack the *bom* and *nic* sites facilitating plasmid mobilization. Also, the cloning bacteria may have an absolute requirement for diaminopimelic acid (lysine precursor), deficient in excision repair (*uvrB* deletion), auxotrophic for thymidine and *rec*⁻ (recombination and repair deficient) so even after accidental escape (assuming a mutation rate of 10^{-6} for each of the 5 loci they would require $(10^{-6})^5 = 10^{-30}$ chance to succeed in the environment. 10^{30} *E. coli* bacterial cell number has a mass of about 10^{11} metric tons. The mass of the Earth has been estimated to be 10^{20} tons. (See bio-hazards)

BIOLOGICAL CONTROL: pathogens or parasites are contained by propagation of their natural enemies, pathogens or parasites (e.g. *Aphelinus mali*) or genetically engineered organisms. *Colletotrichum truncatum* fungus is used as weed killer bioherbicide; *Sesbania exaltata* in various crops such as soybeans, rice and cotton. (See *Bacillus thüringiensis*, genetic sterilization, Dengue virus, BEV, antisense RNA)

BIOLOGICAL MUTAGENS: a large number of natural products present in different organisms may be mutagenic for others. Spontaneous mutation may be increased also by endogenous factors such as defective DNA polymerase or defects in the genetic repair system. (See also mutator genes, transposable elements, transposons)

BIOLOGICAL WEAPONS: may contain highly pathogenic microorganisms such as anthrax bacilli or *Corynebacterium diphteriae* (diphteria) or *Pasteurella pestis* (plague), etc.

BIOLUMINESCENCE: see luciferase, aequorin

BIOMETRY: mathematical statistical principles applicable to the study of genetic and non-genetic variation in biology. (See quantitative genetics, population genetics)

BIOMINING: certain bacteria obtain energy by oxidizing inorganic materials. This process may release acid which in turn can wash out metals from ores. *Thiobacillus ferrooxidans* can release copper and gold, *Pseudomonas cepacia* may assist phosphate mining. Eventually this biotechnology may become economical, especially for low grade ores. (See bioremediation)

BIOPHORE: a hypothesized hereditary unit of the pre-mendelian era. (See pangenesis)

BIOPOESIS: the evolution of living cells from chemical substances rather than from other cells. (See evolution prebiotic, origin of life)

BIOPTERIN: a pterin derived co-factor of enzymes functioning in oxidation - reduction processes.

BIOREMEDIATION: a procedure of adding organisms to an environment for the purpose of promoting degradation of harmful or undesirable properties of that environment. Some observations indicate that 44±18% of polycyclic hydrocarbons of the atmosphere are captured by the vegetation and eventually incorporated into the soil. Many polycyclic hydrocarbons are carcinogenic and mutagenic and pose serious health hazards to people and animals. Their removal from the atmosphre is desirable, however, it is not clear what is the consequence of eating plants that absorbed these semi-volatile compounds. Several organic compounds can be degraded by sequential exposure to anaerobic and aerobic bacteria. Bacterial mercuric ion reductase gene, in a reengineered form, has been introduced into *Arabidopsis* plants by transformation and the transgenic plants became resistant to $HgCl_2$ and to Au^{3+}. The transgenic plants evolved substantial amounts of Hg^0 (vapors). Plants can extract toxic substances from the soil (phytoextraction) and from water (rhizofiltration) and thus facilitate the cleaning up of the environment. A techniques of genetic engineering cytochrome P450 monooxygenase genes can be combined with toluene dioxygenase genes in e.g. *Pseudomonas*. Such bacteria then can degrade polyhalogenated compounds such as 1,1,1,2-tetrachloroethane (a powerful narcotic and liver poison) to 1,1-dichloroethylene and eventually to formic and glyoxylic acids which are still irritants but occur in natural products such as ants and fruits, respectively, but do not pose serious threat at low concentrations. (See environmental mutagens, biomining)

BIOSENSORS: analyze chemicals on the basis of molecular recognition. Among the different systems ligand-receptor binding and signal transduction pathways may be the most sensitive, especially when coupled to fluorescent stains. (See ligand, signal transduction, fluorochromes, aequorin)

BIOSPHERE: the range of habitat of organisms living in and on the soil, in bodies of water and the atmosphere.

BIOSYNTHESIS: synthesis of molecules by living cells.

BIOTA: the community of all living organisms in an environment.

BIOTECHNOLOGY: the purposeful application of biological principles to industrial and agricultural production such as molecular alteration of enzymes, cloned recombinant DNA and its translated products (e.g. human insulin produced by transgenic bacteria), replacement of defective genes by site-specific recombination, gene medicine (introducing transiently into cells genes capable of producing the medication required), transfer desirable genes into domestic animals and plants by genetic transformation to improve their economic value, clean up environmental pollutants by modified microorganism capable of digesting crude oil, etc. (See genetic engineering)

BIOTIC: related to living organisms

BIOTIDINASE DEFICIENCY (same as multiple carboxylase deficiency): is an autosomal recessive disease yet the heterozygotes may also be identified, however, by much less obvious symptoms. The biochemical basis is a deficiency of an enzyme (multiple carboxylase) that splits biocytin (biotin—ε-lysine) and thus generates free biotin from protein linkages. The symptoms that may have late onset or appear in neonates are hypotonia (reduced tension of muscles), ataxia (reduced coordination of the muscles), neurological deficiencies (hearing, vision), alopecia (baldness), skin rash, susceptibility to infections, etc. Generally, administration of biotin alleviates the symptoms and may restore normality. The prevalence varies within the 10^{-5} range. Simple procedure is available for the testing of blood by color on filterpaper, without purification. (See genetic screening, biotin)

BIOTIN: a vitamin, a mobile carrier of activated CO_2, its major biological role involves pyruvate carboxylase. It combines with avidin and thus used for non-radioactive labeling. (See non-radioactive labeling, fluorochromes, biotinylation)

BIOTINYLATION: is a very sensitive, non-radioactive labeling generated by incorporation into the DNA, with the aid of nick translation, biotinylated deoxyuridylic or deoxyadenylic acid The biotin in the DNA has great affinity for streptavidin carrying a dye marker and the labeled

DNA can thus be identified in light either cytologically or on membrane filters. (See also biotin, fluorochromes, labeling, FISH)

BIOTROPHPIC: a parasite living on live host. (See saprophytic)

BIOTYPE: physiologically distinct race within a species.

BiP: is a soluble heatshock protein 70, a chaperone. (See heatshock proteins)

BIPARENTAL INHERITANCE: nuclear genes are usually transmitted by the female and male parents, in contrast to cytoplasmic organelles (and their genetic material) are most commonly inherited only through the egg, and therefore the inheritance is uniparental (through the female). (See allophenic, mtDNA, mitochondrial genetics, chloroplast genetics)

BIPOLAR MOOD DISORDER: is a complex human disorder involving manic depression. Putative genetic determinants have been found in chromosomes 1, 2, 3, 4, 5, 7, 13, 15, 16, 17 and 18. (See affective disorders, manic depression, lithium)

BIPOLARITY: both strands of the DNA are transcribed in opposite directions.

BIRCH (*Betula*): the silver birch, hardwood tree *B. pubescens* is $2n = 28$ and the *B. verrucosa* is $2n = 56$; $x = 14$.

BIRTH CONTROL: see hormone receptors, sex hormones, menstruation

BIRTH RATES: see age-specific birth and death rates

BISEXUALITY: may be a case of hermaphroditism or just a behavioral anomaly. In the fruitfly some losses in the brain olfactory centers or receptors lead to a defect of the interpretation of pheromones causing anatomically male flies to court females as well as males. (See pheromones, hermaphrodite, homosexual, olfactory, olfactogenetics, sex determination)

BISON: American buffalo (*Bison bison*), $2n = 60$.

BISPECIFIC monoclonal **ANTIBODIES**: one of the two arms of the antibody has the recognition site for the surface antigens of a tumor cell, the other for the antigens of a killer lymphocyte. The bispecific antibody thus expected to bring these two cells together and destroy the tumor cells. (See also antibodies, monoclonal antibodies)

Biston betularia (peppered moth): is frequently used example for adaptive natural selection. The moth had predominantly overall greyish tones until the industrial revolution in the vicinity of Birmingham, England deposited black soot on the tree barks and favored the propagation of the dark colored (carbonaria) form of the moth that could hide better from predators. In unpolluted areas the light peppered form remained.

BIT: a binary digit with a two-way choice such as a value of 1 or 0, on or off, etc. The smallest unit of information a computer recognizes.

Bithorax: see morphogenesis in *Drosophila*

BITMAP: bits representing a graphic image in the memory of a computer.

BITNOTE: a message communicated through the computer (e-mail) using the Bitnet system, an IBM mainframe connection to the Internet. (See Internet)

BIVALENT: two homologous chromosomes, consisting altogether of 4 chromatids, paired in meiotic prophase. (See heteromorphic bivalent, interlocking bivalent, chromatid)

BIVARIATE FLOW CYTOMETRY: sorting chromosomes tagged with two fluorochromes (Hoechst 33258), specific for A=T and chloromycin A3, specific for G≡C, and excited by laser. (See flow cytometry, laser)

BK VIRUS: has 80% homology to Simian virus 40 with a somewhat different host range (human, monkey, hamster and other rodent cells). It may occur as an episomal element in two dozen to hundreds of copies. Its autonomous replicon may be useful for propagating DNA and genes in human cell cultures.

BKM SEQUENCES: tetranucleotide GATA and GACA repeats in the W chromosomes (comparable to the Y chromosome) of birds and reptiles, occasionally in other eukaryotic chromosomes. (See satellite DNA, tetranucleotide repeats)

BLACK BOX: a slang expression for an equipment that is too complicated inside to be generally understood. Figuratively, a living cell was considered to be a black box because some of its functions were observed yet all the mechanisms that drove these functions were not fully under-

Black box continued

stood. Geneticists knew segregation of genes and chromosomes but the molecular mechanisms underlying these processes were largely shut inside the "black box" until the discoveries of DNA replication, transcription, translation, gene regulation, cell cycle, etc.

BLACK LOCUST (*Robinia pseudoacacia*): leguminous tree; 2n = 20.

BLACK PEPPER (*Piper nigrum*): southeast Asian spice. Basic chromosome number probably 12, 13, or 16 and 2n = 46, 52, 104 and 128 have been reported.

β–LACTAMASE: an enzyme (synonym: penicillinase) capable of cleaving the β-lactam ring of antibiotics of the penicillin family. Their activity is determined by the R-group attached to the lactam ring. The majority of the synthetic penicillins are not susceptible to penicillinase action. The coding gene was originally detected in Tn*3*. The ampicillin resistance gene in the pBR322 plasmid codes for 263 amino acid residue pre-protein containing a 23 amino acid leader sequence which directs the secretion of the protein into the periplasmic space of the bacterium. The transcription of the gene starts counterclockwise at pBR322 coordinate 4146 and ends at 3297. Its mRNA *in vitro* contains a 5'-pppGpA terminus. The tetracycline resistance gene in pBR322 is transcribed from another promoter clockwise, starting at coordinate 244 or 245. The Tc^R gene encodes a polypeptide of 396 residues. Penicillinases occur naturally only in bacteria with peptidoglycan cell wall. The lack of the enzyme, in the absence of antibiotics, is inconsequential for the bacteria. The β-lactamase genes are used extensively in vector construction (to convey antibiotic resistance) and for the detection of insertional events that inactivate the enzymes. (See antibiotics, Tn*3*, periplasma, vectors)

BLANK ALLELE: is not expressed.

BLAST (basic local alignment search tool): is used for comparison of nucleotide sequences in apparently related (homologous) DNA [Altschul, S.F. *et al.* 1990 J. Mol. Biol. 215:403]. E-mail address <blast@ncbi.nlm.nih.gov>. (See DNA sequencing, homology, evolutionary tree, FASTA, BLOSUM, databases)

BLASTOCOELE: see blastula

BLASTOCYST: an early embryonal stage when the blastocoel is enveloped by a trophoblast cell layer, a pre-implantation stage of the animal zygote when the zona pellucida (the envelop of the egg) is still visible and the blastula begins to develop its inner cell mass. (See blastocoele, trophoblast, blastula)

BLASTOCYTE: an undifferentiated cell of an early zygote.

BLASTODERM: a single layer of cells at the embryonic stage surrounding a fluid-containing cavity (blastocoel) at the blastula stage of cell divisions.

BLASTOMA: a cell in the early stage of differentiation or a neoplastic tissue containing embryonic cells.

BLASTOMERES: the large fertilized egg through cleavage divisions produces small cells, the blastomeres. These divisions are extremely fast and during the short process RNA synthesis ceases and protein synthesis depends on reserve mRNAs. (See blastoderm, cleavage)

BLASTOPORE: near the site of the gray center of the animal pole where invagination of the blastula begins and eventually encompasses the vegetal pole. The *dorsal lip* of the blastopore organizes gastrulation. (See blastula, animal pole, vegetal pole, gastrulation, organizer)

BLASTULA: a product of the cleavage of the early zygote when it becomes a spherical structure in which the blastoderm envelops the blastocoele cavity. (See blastocoele)

BLASTX: computer program for gene searches. (See Nature Genet. 3:266 [1993])

BLEEDER DISEASE: see hemophilia

BLENDING INHERITANCE: was considered erroneously to account for some of the variations observed in nature up to the time of Mendel. A contemporary of Darwin, Fleeming Jenkins, an engineer, has pointed out that if this would be true, unique traits of single organisms should disappear in panmictic populations like a drop of ink in the sea.

BLEPHAROPHIMOSIS: a defect of the eyelids and nose, associated also with ovarian atrophy and small uterus. The dominant disorder was assigned to human chromosome 3q21-q23.

BLISTER: vesicle-like skin abnormality. Autosomal dominant, recessive and X-linked types exist. The autosomal dominant class: *bullous erythroderma* involves a hyperkeratosis with anomalies of the keratin tonofibrils. In one form the blisters are limited to hands and feet and appear only in warm weather after heavier exercises. Another form (*epidermolyis bullosa dystrophica*) may appear already at birth and the blisters may appear besides the extremities, on the ears and buttocks. The latter type accumulates and secretes sulfated glycosaminoglycans. In an early and transient form the blisters disappear generally by the end of the first year and do not return. In the *herpetiform epidermolysis bullosa* the larger vesicles appear in clusters on the palms, soles, neck and around the mouth apparently due to a mutation in the keratin 14 gene. Another epidermolysis bullosa appears to be due to deficiency of *galactosylhydroxylysyl glucosyltransferase*. In one form, a human chromosome 12-coded gelatin-specific *metalloprotease* deficiency may be involved. A *mottled* type (pigmented spots) displays recurrent blistering beginning at birth and premature aging. The *epidermolysis bullosa with absence of skin and deformity of nails* has a perfect penetrance. The autosomal recessive types: *epidermolysis bullosa dystrophica* (human chromosome 11q11-q13) is caused by excessive collagenase activity affecting primarily the hands, feet, elbows and knees at birth or infancy but may affect other organs too. The *epidermolysis bullosa letalis* may kill infants within about three weeks after birth but occasionally some survive to the first decade of life. In some forms the distal opening of the stomach (pylorus) may be constricted and atrophied, in other forms congenital deafness, muscular dystrophy may appear. X-linked epidermolysis with multiple complications (baldness, hyperpigmentation, dwarfism, microcephaly (small head), mental retardation, finger and nail malformation, and death before adult age is also known. (See keratosis, ichthyosis, skin diseasees)

BLK: see SRC oncogene family

BLOCH-SULZBERGER SYNDROME: see incontinentia pigmenti

BLOCK MUTATION: affects more than a single nucleotide in the cell, e.g. deletion; such mutations may not yield wild type recombinants if the defects overlap.

BLOCKED READING FRAME: translation is interrupted by nonsense codons. (See nonsense codons)

BLOCKING BUFFER: 3% BSA (bovine serum albumin) in phosphate buffered saline containing also 0.02% sodium azide. BSA blocks the binding sites on nitrocellulose filter that are not occupied by proteins transfered from (SDS polyacrylamide) gels. (See gel electrophoresis)

BLOCKS: an Internet tool for the search of functional DNA motifs: <blocks@howard.fhcrc.org>

BLOOD: the fluid that carries nutrients and oxygen by circulating through the blood vessels in the animal body. It is composed of red (non-nucleated mature erythrocytes) and white (nucleated leukocytes) cells. The white cells include granulocytes, neutrophils, eosinophils and basophils, lymphocytes and natural killer cells. The blood contains also platelets (thrombocytes) and the blood plasma, the non-corpusculate yellowish fraction. (See these cell types, hemolytic disease, blood groups, macrophages, immune system)

BLOOD CLOTTING PATHWAYS: are the *intrinsic pathway* involving the successive participation of the Hagemann factor (XII), plasma thromboplastin antecedent (PTA XI), Christmas factor (IX), antihemophilic factor (VIII), Stuart factor (X), phospholipid and proaccelerin and the *extrinsic clotting pathway* requiring proconvertin (VII), Stuart factor (X), proaccelerin and calcium ions. (See also antihemophilic factors, tissue factor)

BLOOD FORMATION (hematopoeisis): during early embryonic development the yolk and the aorta-gonad-mesonephros (AGM) region is involved, later the function in the embryo is switched to the liver and after birth the bone marrow is involved.

BLOOD GROUPS: an incomplete list of the types found in this volume: ABO, ABH, Ahonen, Colton, Diego, Dembrock, Duch, Duffy, En, Gerbich, I system, Kell-Cellano, Kidd, Lewis, Lutheran, LW, MN, Newfoundland, OK, P blood group, Radin, Rhesus, Scianna, Ss, Webb, Wright, Yt, Xg. These are distinguished mainly by the epitopes on the eryhtrocytes. (See epitope, erythrocyte)

BLOOD PRESSURE: is the pressure of the blood on the blood vessels (arteries). (See hypertension)

BLOOD **TRANSPOSABLE ELEMENT**: see *copia*

BLOOD TYPING: identification the blood group a person belongs to. (See blood groups)

BLOOM's SYNDROME: recessive human dwarfism; increases the frequency of chromosomal aberrations (particularly sister chromatid exchanges), and cancer (leukemia), sensitive to sunlight (red blotches over face) and usually shorter than normal life expectancy. It was attributed to a DNA ligase I deficiency but the cloning and sequencing of the gene indicates that this is not the primary defect rather a DNA helicase-like protein, encoded in 15q26.1, is involved. (See DNA repair, Xeroderma pigmentosum, Cockayne syndrome, carcinogenesis, light-sensitivity diseases)

BLOSUMs: are amino acid substitution matrices used to determine evolutionary changes in proteins [Henikoff, S. & Henikoff, J.G. 1992 Proc. Natl. Acad. Sci. USA 89:10915]. (See BLAST, FASTA)

BLOTTING: macromolecules separated by electrophoresis in agarose or polyacrylamide are transferred to a cellulose or nylon membrane and immobilized there for further study. (See Southern blot, Northern blot, Western blot, colony hybridization, immunoprobe)

BLOTTO (Bovine Lacto Transfer Technique Optimizer): is a 5% solution of non-fat, evaporated milk in 0.02% sodium azide (NaN_3). [It may contain RNase activity]. In 25-fold dilution it may be used for blocking background annealing in Grunstein - Hogness hybridization, Benton - Davis hybridization, dot blots, and non-single copy Southern hybridization. (See Denhardt reagent, heparin, Grunstein - Hogness screening, Benton - Davis plaque hybridization, dot blot)

BLUE GRASS (*Poa pratensis*): lawn and pasture plant; 2n = 36-123 in the polyploid series.

BLUE LIGHT RESPONSE: photomorphogenetic reaction (of plants) to illumination in the range of 400-500 nm wavelength. (See photomorphogenesis)

BLUEBERRY (*Vaccinium* spp): a fruit shrub with x = 12; *V. corymbosum* (high-bush blueberry) is tetraploid, the *V. angustifolium* (low-bush blueberry) is diploid.

BLUESSCRIPT M13: is a 2.96 kb genetic vector containing the bacteriophage M13 replication origin and a polycloning insertion site flanked by T7 and T3 phage promoters and useful for generation of single-stranded DNA or RNA complementary to the double-stranded DNA insert.

BLUNT END: of double-stranded DNA is generated by non-staggered cut and it terminates at the same base pair across both strands of the double helix. Bacterial DNA polymerase I (Klenow fragment) or phage T4 DNA polymerase can also generate 3' blunt ends of DNA by 5'→3' exonucleolytic activity. (See blunt end ligation)   BLUNT ENDS OF DNA → ─────────

BLUNT-END LIGATION: T4 phage DNA ligase joins non-staggered DNA ends or adds chemically synthesized duplexes to double-stranded blunt ends. (See DNA ligases, linker)

BLYM: chicken bursal lymphoma oncogene, located to human chromosome 1p32. It is homologous to transferrin, a glycoprotein with important role in the synthesis of ribonucleotide reductase, and thereby in DNA replication and mitosis. (See oncogenes, lymphoma, transferrin)

BMP: see bone morphogenetic protein

BMT: transformed monkey cell line expressing the T antigen of SV40 driven by a mouse metallothionein promoter. (See SV40, metallothionein)

BMYC: oncogene isolated from rat has extensive homology to the MYC oncogene but it maps to a different location than the other members of the MYC family LMYC, NMYC, PMYC, RMYC. (See MYC and other members of the family, oncogenes)

Bob: see OBF

BOB': see *att* sites

BODIPY: see fluorochromes

bol: see *boule*

bom: bacterial gene (basis of mobilization), required for the transfer of plasmids. (See plasmid mobilization, *mob, Hfr*)

Bombardia lunata: n = 7 is an ascomycete.

BOMBAY BLOOD TYPE: is relatively rare blood type discovered in India and subsequently on the Reunion Island in the Indian Ocean. This blood type has two main forms determined by the recessive alleles *h* (for H type red cell antigen). In the *h/h se/se* individuals also the enzyme fucosyltransferase 1 is inactive; in the *h/h Se/se* individuals a weak expression of the H antigen may be observed; *Se* is apparently coding for fuscosyltransferase 2. (See ABH antigen, Secretor, Lewis blood group)

Bombyx mori: see silk worm

BOND ENERGY: is required to break a chemical bond.

BONE MARROW: the red spongy tissue inside the bones gives rise to lymphocyte stem cells and erythrocytes; the yellow bone marrow is made of mainly fat cells. (See thymus)

BONE MORPHOGENETIC PROTEIN (BMP): is a maternally expressed factor in *Xenopus* embryos and in addition to bone differentiation it is involved in dorso-ventral organization of the embryo. BMP-1 is a procollagen protease (PCP) that assembles collagen within the extracellular matrix. The other BMPs belong to the transforming growth factor (TGF-β) family. BMP-4 regulates apoptosis in neural crest cells affecting skeletal bone and muscle formation. The growth/differentiation factors of mouse (GDF) belong to this family and their mutation shortens the limb bones (brachypodism) without affecting the axial skeleton. The CBFA-1 gene seems to be a major factor in ossification. (See also fibrodysplasia ossificans progressiva)

BOOK SYNDROME: autosomal dominant defect of tooth development, high degree of sweating and premature loss of hair color. (See hair color)

BOOKMARK: indicates an address on the Internet or other items in the computer where you wish to return.

BOOLEAN ALGEBRA: was developed by George Boole (1815-1864) for the use of formal logic. He supposed that in binary forms thinkable objects could be defined. Thus if x = horned and y = sheep then by selecting x and y the class of horned sheep is defined. Also 1 - x would define all things of the universe that are not horned and (1 - x)(1 - y) would identify all things that are neither horned nor sheep. This approach defines sets and subsets in discrete forms without intermediates yet capable of defining mutual relationships. Using simple symbols, syllogisms could be developed in mathematical forms. The switch gear of the telephone systems and the modern digital computers were developed on the basis of the Boolean binary logic.

BOOTSTRAP: is a statistical device that was introduced for computer operations (versus the classical type computations). The standard error by the classical method is computed as:

$$se(\bar{x}) = \left\{ \sum_{i=1}^{n}(x_i - \bar{x})^2/[n(n-1)] \right\}^{1/2}$$

in comparison, with the bootstrap procedure:

$$se[t(x)] = \left\{ \sum_{b=1}^{B} [t(x^{*b}) - \bar{t}]^2/(B-1) \right\}^{1/2}$$

where $se[t(x)]$ is the standard error of the bootstrap statistic, $t(x)$, B = bootstrap samples of size *n* from the data, $\bar{t}$ is the average of the B bootstrap replications ($t(x*^b)$).

The bootstrap algorithm can be applied to the majority of statistical problems and it is widely used for estimating the confidence level in evolutionary trees. The data points x_i need not be single numbers, they can be vectors, matrices or more general quantities, such as maps, graphs. The statistic t(x) can be anything as long $t(x^*)$ can be computed for every bootstrap data set x^*. Data set *x* does not have to be a random sample from a single distribution. Regression models, time series, or stratified samples can be accommodated by appropriate changes. For details and specific references see B. Efron & R. Tibshirani 1991 SCIENCE 253:390.

Bora Bora: 220 kb centromeric sequences in *Drosophila*. (See centromere)
BORDER SEQUENCES: see T-DNA
BORJESON SYNDROME (Borjeson-Forssman-Lehman syndrome): face, nervous system, endocrine defects assigned to human chromosome Xq26-q27. (See head/face/brain defects)
BORRELIA: spirochete bacteria; about 28 species (*B. burgdorferi, B. hermsii*, etc.) are responsible for relapsing fever or Lyme disease and other human ailments all over the world. The filamentous bacteria are 8 to 16 µm long flagellate cells infectious for birds and mammals. Their generation time is about 6 hours and in about 5 days within a single animal their population may exceed 10^6 cells and that coincides with the major symptoms (erythema migrans [enlarged red spots]) of the infection. Within weeks or months the bacteria may invade all major organs of the body, primarily the joints and cause arthritic symptoms. If untreated, Lyme disease may be fatal. Intravenous injection of rocephin or other antibiotics (also orally administered) may be the cures although some of the effects may persist for years. The vectors of the bacteria are the *Ixodes* arthropods (ticks) that live on grasses and low-growing bushes in wild-life (deer, mice, birds) frequented rural and suburban areas. Identification of the disease is difficult because of the complexity of the symptoms. Serological detection encounters problems because the outer membrane of the bacteria displays variable serotypes. *Borrelias* harbor several copies of approximately 23- 50 kb linear plasmids with genes for Vmps (variable major proteins). Transposition within and recombination between the plasmids assures great antigenic variation in these organisms. New serotypes appear at an estimated frequency of 10^{-4} to 10^{-3} per cell per generation. This fact accounts for the difficulties in developing effective immunsera. An attenuated strain of *Mycobacterium bovis*, the bacillus Calmette-Guerin (BCG) may serve as a suitable vector for the *B. burgdorferi* surface protein antigen A and may secure more than a year long protection by mucosal delivery. (See serotype, antigen, serum, mucosal immunity, *Ixodoidea*)
Bos taurus (cattle): 2n = 60.
BOSS: is a transmembrane protein product on the R8 photoreceptor in the eyes of *Drosophila*, encoded by gene *boss* (*bride of sevenless*, 3.90.5). It is the ligand for the receptor tyrosine kinase, encoded by the *sev* (*sevenless*, 1-33.38) gene. (See sevenless, son-of-sevenless, rhodopsin, receptor tyrosine kinase, daughter of sevenless)
BOTANY: a basic scientific field concerned with plants.
BOTTLENECK EFFECTS: if the size of the population is periodically reduced substantially, genetic drift may alter gene frequencies. Bottleneck effect is quite common in the transmission of mtDNA because only a small portion of it is passed through the germline and the heteroplasmy may be altered. (See genetic drift, mtDNA, heteroplasmy)
BOTTOM-UP ANALYSIS: see top-down analysis
BOTTOM-UP MAP: is relying on STS-based information. These are useful for relatively short chromosomal distances. Two STSs are 'singly linked' if they share at least one YAC and 'doubly linked' in case they share at least two YACs. Single linkage is generally not useful because of the high degree of chimerism among the YACs. In the first step STS are assembled into doubly linked contigs. Then, the doubly linked contigs are ordered either on the basis of radiation hybrids or traditional genetic recombination information. Finally single linkage can also be used to join contigs to the same short genetic region. (See mapping genetic, STS, contig, radiation hybrid, bottom-down mapping, YAC)
BOULE (*bol*): autosomal gene in *Drosophila* encoding a cell cycle protein regulating G2 - M transition. It is homologous to the human Y-chromosomal gene DAZ responsible for azoospermia. Its suspected function is translation and localization of mRNA. (See infertility, fertility, azoospermia, cell cycle, *twine, pelota, Dazla*)
BOUQUET: in leptotene the chromosomes are attached by their ends to a small area of the nuclear membrane while the rest of the chromosome length is looped across the nucleus. (See meiosis, leptotene stage)
BOVINE GROWTH HORMONE (BGH): is somatotropin; it has been commercially produced

by genetic engineering and it boosts significantly milk production. (See somatotropin)

BOVINE PAPILLOMA VIRAL VECTORS: By genetic manipulations, with pBR322 bacterial plasmid sequences added, they can be converted into a shuttle vector carrying genes between mouse and *E. coli*. (See transformation genetic, vectors, viral vectors)

BOVINE PAPILLOMA VIRUS (BPV): a papova virus (about 7.9 kbp DNA), responsible for wart in animals. The BPV_{69T} segment of their genome (5.5 kbp) has been used as a large capacity vector which can multiply into 10 to 200 copies. It can stay as an episome or can be integrated into the chromosomes of mammals.

BOVINE SPONGIFORM ENCEPHALOPATHY (BSE): see encephalopathies, Creutzfreldt-Jakob disease, Alzheimer disease, prions, scrapie

BOWEL DISEASE (chronic inflammatory bowel disease): see CIBD, Crohn disease

BOX: is generally used for a consensus sequence in the DNA, such a homeobox; domains of the internal control regions (box A, box, B, box C) that are the sequences where transcription factors bind.

BOX GENES: clustered mutations in exons or introns (in mosaic genes of mtDNA). (See mtDNA)

bp: base pair

BPV: see bovine papilloma virus

BRACHMANN-De LANGE SYNDROME: see De Lange syndrome

BRACHYDACTYLY: abnormally short fingers and toes controlled by autosomal dominant genes.

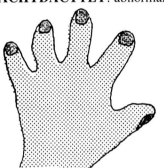

The expression may vary. Most commonly the middle bones (phalanx/phalanges) are affected (Type A); in some cases not all the fingers express the gene. In type B, in addition to the middle phalanges, the terminal ones are also short or absent. In type C more than 3 phalanges may appear. Type D involves short and flat terminal phalanges of the big toe and the thumb. In Type E the metacarpus and metatarsus (the bones between the wrist and fingers of the hand and the corresponding bones in the foot) are shortened. In still other types nervous defects, hypertension, shortening the bones of the arm accompany the hand and foot bone problems. In an autosomal recessive form the brachydactyly involves also small head (microcephaly). In another recessive form primarily the great toe is affected but the proximal (near the wrist) joints do not move. The dominant brachydactyly with severe hypertension gene has been assigned to human chromosome 12p. (See also polydactyly)

BRACHYMEIOSIS: the second meiotic division is missing. (See meiosis)

BRACHYURY: a homozygous (*TT*) dominant lethal (after 10 days of conception) gene in mice. The *Tt* heterozygotes are viable and have reduced tail (tailless), the homozygous *tt* also dies in 5 days. The different alleles of the complex locus have different effects of the development. The anomaly (chromosome 17) involves a genetic defect in the notochord development. The somites undergo differentiation but resorbed before birth. There are defects also in the posterior parts (limbs, allantois, umbilical vessels). The *t* alleles may display meiotic drive and from the male *t*/+ heterozygotes more than 99% of the progeny may receive the *t* allele. The distortion of the transmission (TRD) is controlled by at least six loci. In addition, there are at least 16 lethality loci within the *t* haplotype but these are not the primary causes of the distorted segregation. The *t* complex is also containing inversions which interfere with recombination between the wild type and the *t* complex. (See also Manx in cat, meiotic drive, killer spore, somite, notochord, haplotype)

BRACT: a small modified leaf from which flower may develop or a leaf on the floral axis subtending the flower.

BRADYKININ: see kininogen

BRADYTELIC EVOLUTION: has a very slow pace and it involves species whose adaptive environment extends over very long (geological) periods. In contrast, the tachytelic evolution is progressing at a fast pace whereas the horotelic evolution appears to show an average rate. (See evolution)

BRAIN, HUMAN: is a very complex structure and here only a few major landmarks are outlined as reference to several entries dealing with the central nervous system. The seven-layered hippocampus, consisting "gray matter" is not shown although this is the most important area at the basal-temporal region involved in memory and learning. (See figure below ↓)

BRAIN DISEASES: see Addison-Schilder syndrome, epiloia, mental retardation, affective disorders, see also craniofacial synostosis syndromes

B-RNA: see cowpea mosaic virus

BRANCH MIGRATION: during the process of molecular recombination the exchange point between two fixed sites of the DNA single strands can move left or right when the two single strands are separated they can simultaneously reassociate in an exchanged manner in both double helices. This strand invasion brings about heteroduplexes. In *E. coli* the RuvA (a specificity factor) and RuvB (an ATPase) proteins — induced by ultraviolet radiation damage to the DNA — bind to the Holliday junctions and increase the length of the heteroduplex. Mismatches in the synaptic strands may interfere with branch migration. (See Holliday model, Holliday juncture, heteroduplex, recombination molecular mechanisms, mismatch repair)

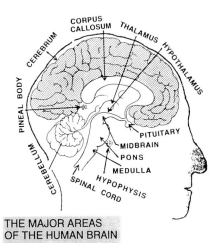

THE MAJOR AREAS OF THE HUMAN BRAIN

BRANCH POINT SEQUENCE: is short RNA tract (YNCUR$\boxed{A}$ Y, [Y stands for pyrimidine, R for purine and N can be either]) in the primary transcript near (18-38 base upstream) to the 3' end of an intron (**AG**) of mammals. After exon1-intron boundary is severed and the intron end is released with a 5'-**GU** pair at the end, it forms then a loop as it folds back by **G** making a 2'→5' bond with the A shown boxed above. Subsequently a cut is made at the 3' end of the intron, the intron is released, and the exon1 is attached to exon 2. (See introns)

BRANCHED CHAIN AMINO ACIDS: see isoleucine-valine biosynthetic steps

BRANCHED RNA: is an intermediate of RNA splicing. (See introns)

BRANCHIO-OTORENAL SYNDROME (BOR): is a human chromosome 8q13.3 dominant syndrome with incomplete penetrance and expressivity. It involves defects in the appearance of the ears, underdevelopment of the middle ear structures (malleus, incus, stapes) and the inner ear (cochlea) resulting in mild to severe hearing loss. This is accompanied also by underdevelopment of the kidney and the urinary tract. The gene bears homology to the *Drosophila* gene *eyes absent (eya)*.

Brassica oleracea (cabbage, kale): vegetable crops. Basic chromosome number is controversial 5 or 6 although cabbage is 2n = 18 and there are some indications of being an amphidiploid. (See turnip, swedes, rapes, mustards, radish, watercress)

BRASSINOLIDE: see brassinosteroids

BRASSINOSTEROIDS: are synthesized through the pathway campesterol→ campestanol→ cathasterone→ teasterone→ 3-dehydroteasterone→ typhasterol→ castasterone→ brassinolide. The latter compound has been shown to remedy de-etiolation, derepression of light induced genes, miniaturizing, male sterility and other symptoms of stress-regulated genes. These brassinosteroids bear close similarity to ecdysones, the animal molting hormones. The phytoecdysones were known for two decades in plants. All these plant hormones interact with each other in various ways and regulate signal transduction and gene activities. Unlike most animal steroid hormones, plant hormones are of small molecular size (except brassinosteroids) generally in the range of 28-350 Da. Brassinolide has a MW of about 480. (See plant hormones, de-etiolation, photomorphogenesis, hormones, steroid hormones)

BRCA1 (breast cancer antigen): is an exclusively nuclearly located protein. (See breast cancer)

BrdU: bromodeoxyuridine. (See bromouracil, hydrogen pairing, chemical mutagens)

BREAKAGE AND REUNION: the broken chromatids or DNA single strands are broken at the position of chiasmata, and reunited in an exchanged manner during genetic recombination. This process is a physical event not requiring (normally) DNA replication as it was one time hypothesized with the copy choice idea. Recently, it was found that recombination takes place also by replication. (See recombination molecular models, recombination by replication)

BREAKAGE-FUSION-BRIDGE CYCLES: may cause variegation in the tissues because some of the genes may not be present in one of the daughter cells whereas the other cell receives two copies. If this dominant gene determines color, its presence is immediately recognized in the cell lineages. (See diagram, *Ac - Ds*). (Photographs by courtesy of Barbara McClintock.)

Breakage - fusion-bridge cycles may occur in the kernels of maize plants if the end of the chromosomes is broken. The genetic and cytological consequences of such events are diagramed here. The relative size of the sectors (detectable when appropriate color markers are used) indicates the developmental time of the cycle. Early events involve large sectors, late events are indicated by small sectors. If the event occurs repeatedly, several sectors are observed.

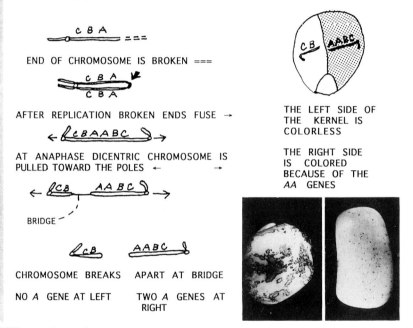

BREAKPOINT MAPPING: see inversions

BREAST CANCER (BRCA): breast cancer is one of the most common diseases of women. There is over 10% chance that the survivor to 90 or over will develop breast cancer and ≈5% of the cases are caused by either the BRCA1 or BRCA2 genes. BRCA1 (in 17q21) is responsible for 25% of the cases diagnosed before age 30. Predisdisposition to breast cancer is inherited as a dominant trait but the somatic expression (manifestation of cancer) requires that in these heterozygotes the normal allele would be lost or inactivated during the lifetime of the individual who inherited one BRCA susceptibility allele. The BRCA1 reading frame encodes 1863 amino acids (22 exons) with a Zn finger domain at the NH_2 end. It appears that amino acids 1,528 to 1863, i.e. the C-terminal domain is a transcriptional activator. The primary single transcript is 7.8 kb and is expressed primarily in the testis and the thymus but also in the breast and ovary. The transcript displays alternative splicings. The BRCA1 sequences are well conserved in mammals but absent from chicken. The defects in the gene varies in the different kindreds from 11 bp deletion to frame shifts, nonsense, missense mutations or other alterations causing instability. The phenotype of the patients varied between the kindreds indicating that the specific mutations at the locus may affect its expression. It was noteworthy that there were female carriers of the mutation(s) who by age 80 failed to develop breast or ovarian cancer.

Breast cancer continued

Apparently, the expression is affected to some extent by extraneous genetic and environmental factors. The BRCA1 protein may be aberrantly localized also in the cytoplasm and complicates the expression pattern. BRCA2 in human chromosome 13q12-q13 within a 6 cM region, is also dominant early onset disease, responsible for about 45% of all hereditary breast cancers. It does not create a substantial risk for ovarian and other cancers but the chance for breast cancer in males may be slightly elevated — in contrast to BRCA1. Deletions 185delAG in BRCA1 and deletion 617delT in BRCA2 occurs with carrier frequencies of 1.09% and 1.52%, respectively in Ashkenazy Jewish populations.

One of the remaining types of breast cancer is supposed to be due to a mutation in the *KRAS2* (Kirsten sarcoma) gene in human chromosome 6 when at codon 13 a G → A transition takes place resulting in Gly → Asp substitution. Ductal breast cancer was attributed to the loss of genes in human chromosome 1q21ter but also chromosomes 2, 14 and 20 were implicated. In families with high incidence of breast cancer a small secreted protein gene, expressed only in human breast cancer, was mapped to chromosome 21q22.3. A dominant gene product serologically reacts with murine monoclonal antibody DF3. In the region of chromosome 17p13.3, the *TP53* regulator of tumor protein p53 was found but it could not be ruled out that a regulator exists 20 megabases telomeric to *TP53*. Some forms of breast cancer affects also males and the incidence of breast cancer in Klinefelter (XXY) men is almost as high as that in women. Some types of breast cancers are associated with cancers in other organs. The risk of recurrence of breast cancer among first-degree relatives may be as high as 50%. The risk of breast cancer is increasing with age, unmarried status, obesity, radiation exposure, etc. In North-America the lifetime risk of breast cancer in females is about 10%. Of all breast cancer incidences about 5-10% is due to inherited causes. DNA repair defect seems to be involved. BIC (breast cancer information core) address:<http://www.nchgr.nih.gov/ dir/lab_transfer/bic>. (See p53, hormone receptors, tamoxifen, oncogenes, genetic screening, cancer, Li-Fraumeni syndrome, multiple hamartoma, granin, Klinefelter syndrome, tumor suppressor)

BREATHING OF DNA: a reversible, short-range strand-separation below the melting temperature. (See melting temperature)

BREEDING SYSTEM: Mating within a population may be random (each individual has an equal chance to mate with any member of the opposite sex) or it may be self fertilization (in monoecious species) or inbreeding (in dioecious species). Also random mating and inbreeding both take place within the group. The term breeding system denotes these alternatives. The predominant breeding system of a few plant species is tabulated (A: apomictic, D: dioecious, I: selfincompatible, M: monoecious, O: outbreeding, S: selfing)

ALfalfa O-S	Cherry O-I	Mulberry D	Rye grass O-I
Almond O-I	Chestnut O	Mustard O-I	Sorghum O-S
Alder M	Citrus O-I	Oak M	Soybean S
Antirrhinum O-S	Clover O-I	Oat S	Squash M
Apple O-I	Coffee O-I	Oenothera O-S	Spinach D
Apricot O	Collinsia S	Onion O	Spruce M
Arabidopsis S	Cotton O-S	Orchard grass O-I	Stock S
Ash M-D	Cucumber M	Osage orange D	Strawberry D-S
Asparagus D	Date palm D	Parsley O	Sugarcane O-I
Barley S	Datura S	Pea S	Sunflower O-I
Basswood O	Eggplant O	Peach O	Sweetclover O-S
Beach M	Elm M	Peanut S	Sweetpea S
Bean S	Fescue O-S	Pear O	Sycamore M
Belladonna O	Flax S	Petunia O-S	Teosinte M
Beet O-I	Grape O	Pine M	Timothy O
Birch M-I	Hemp D-M	Pineapple O-I	Tobacco S
Blue grass O-A-S	Hop D	Plum O-I	Tomato S
Broad bean S	Lentil S	Poplar D	Tripsacum M

Brome grass O	Lespedeza O	Potato O-S-I	Walnut M
Buckwheat O-I	Lettuce S	Radish O-I	Wheat S
Cabbage O-I	Lupine O-S	Rape seed O-I	
Carnation S	Maize M	Rice S	
Carrot O	Maple O	Rose O	
Castorbean M	Meadow foxtail O	Rubus O	
Celery O	Millet O-S	Rye O-I	

BREEDING VALUE: the quantitative value of a genotype, judged on the basis of the mean performance of the offspring. Actually, it is twice the mean deviation of the offspring from the mean of the parental population. The doubling is used here because each parent contributes a haploid gamete to the offspring and thus half of its genes. Breeders frequently call it the additive effect. The observed performance of individuals is called the *phenotypic value* (P) that is measured as the mean value of the population. The average value of two homozygotes, called the *midpoint* and it is equal to zero (only in case when the frequency of the two alleles is equal, 0.5) because the two parents deviate from it by a quantity of (+ a) and (- a), by definition, and their sum cancel out each other. The *genotypic value* of the heterozygote is designated by (d). In the absence of dominance d = 0, with complete dominance d = a, and with overdominance d > a:

```
aa                          0           Aa          AA
←         - a            →←         + a          →
                    ←      d        →
```

The mean value (x) is usually calculated as the weighted mean, i.e., multiplied by the genotypic frequencies in the population. If the population is in equilibrium, the mean is:

$\bar{x} = p^2(a) + 2pq(d) + q^2(-a) = (p^2 - q^2)(a) + 2pq(d) = (p+q)(p-q)(a) + 2pq(d)$ and because

$p + q = 1$, $\bar{x} = (a)(p - q) + 2pq(d)$ and if several loci are involved, $\bar{x} = \Sigma[(a)(p - q) + 2pq(d)]$
(See also gain, heritability, Hardy - Weinberg equilibrium, merit)

BREFELDIN A: (γ,4-dihydroxy-2-[6-hydroxy-1-heptenyl]-4-cyclopentanecrotonic acid λ lactone): is an inhibitor of passing peptides from the endoplasmic reticulum and Golgi complex. (See ARF, translocase, ARNO)

BRF: binding factor in RNA synthesis initiation.

BRIDGE: anaphase tie between separating centromeres in dicentric chromosomes. (See breakage-fusion-bridge, inversion).
Miotic anaphase single and double bridges and 1 or 2 chromatid fragments resulting from 2 and 4 strand recombination in a paracentric inversion heterozygote →

BRIDGING CROSS: if two genetically distant sexually incompatible species (A and B) are to be selected for gene transfer by sexual means, the problems can possibly be overcome by first mating one of the species with an intermediate compatible form (C) and then cross the hybrid (A x C) to the other (B). C serves as the bridge.

BRIGHT PARAMECIA: see symbionts hereditary

BRIGHT-FIELD MICROSCOPY: is ordinary light microscopy. (See also dark-filed microscopy, fluorescent microscopy, phase contrast microscopy, Nomarski, stereomicroscopy)

BRISTLE: see chaetae

BRITTEN & DAVIDSON MODEL: was suggested as a working hypothesis in the 1960s for interpreting the processes involved in the regulation of eukaryotic gene functions. The external stimuli were supposed to be directed to *sensor* genes that activated *integrator* genes that in turn transmitted the signals to *receptor* genes which affected than the *structural* genes, coding for protein. These systems might have operated than in series of interacting batteries.

Brn: eukaryotic transcription factors with POU domains controling terminal differentiation of sensorineural cells. They are homologous to unc-86 in *Caenorhabditis*. (See POU, unc-86)

BROAD BEAN (*Vicia faba*): 2n = 2x = 12 and large chromosomes has been used extensively for cytological research. It is an important crop in cool climates.

BROAD-BETALIPROTEINEMIA: is a hyperlipoproteinemia. (See apolipoproteins, hyperlipoproteinemia)

BROAD SENSE HERITABILITY: see heritability

BRODY DISEASE (ATP2A1): is a rare recessive disease encoded in human chromosome 16p12.1-6p12.2, and it involves impairment of muscle relaxation, stiffness and cramps in the muscles. The basic defect is associated with the muscle sarcoplasmic reticulum calcium ATPase (SERCA1). (See neuromuscular diseases, sarcoplasmic reticulum)

BROKEN TULIPS: a variegation (sectors) in the flowers caused by viral infection. These sectorial tulips have commercial value in floriculture. In the 17th - 18th century the bulbs were so highly valued that they fetched gold of equal weight. (See symbionts hereditary, infectious heredity)

(Courtesy of Stichting Laboratorium voor Bloembollenonderzoek, Lisse, The Netherlands) ➔

BROMOPHENOL BLUE: see tracking dyes

BROMOURACIL (BU): a pyrimidine base analog that is mutagenic in prokaryotes because it may lead to base substitution after tautomeric shift. When incorporated into eukaryotic chromosomes it may cause breakage upon exposure to light. On this basis it has been successfully used as a selective agent in animal cell cultures; the non-growing mutant cells failed to incorporate it and survived while the growing (wild type) cells were killed upon illumination. (See base substitution, tautomeric shift, hydrogen pairing, chemical mutagens)

BRONZE AGE: about 5,000 years ago marked the development of crafts and urbanization.

BRUSH BORDER: a dense lawn of microvilli on the intestinal and kidney epithelium that facilitates absorption by increasing the surface.

BRUTON's TYROSINE KINASE: see BTK, agammaglobulinemia

BRYOPHYTES: are mosses, liverworts and hornworts; they are green plants similar to algae but the organization of their body is more complex. Their gametangia is either unicellular or multicellular and show some cell differentiation. They usually have haploid and diploid life forms. The majority of them are terrestrial. (See also alga)

BSE (bovine spongiform encephalopathy): see encephalopathies, Creutzfeldt-Jakob disease

BTF2: same as TFIIH. (See transcription factors)

BTG: antiproliferative protein encoded in human chromosome 1q32. Its synthesis is regulated by p53. (See p53)

BTK (Bruton's tyrosine kinase): belongs to a family of non-receptor tyrosine kinases. Its deficiency results in immunodeficiency. (See agammaglobulinemia)

BUCCAL SMEAR: a sample of the epithelial cells from the inner surface of the cheek is spread onto microscope slides and used for rapid determination of the number of Barr bodies with 4-5% error rate. This procedure (mucosal swab) is non-invasive and the sampling is painless. The cells so sampled may also be used for DNA analysis by PCR. (See Barr body, PCR)

BUCKWHEAT (*Fagopyrum*): is a feed and food plant; all species have 2n = 2x = 16. Feeding the plants to animals or eaten by humans may increase sensitivity to light and skin rash.

BUD SPORT: genetically different sector (due to somatic mutation) in an individual plant (chimera).

BUDDING: an asexual reproduction by what the cell's cytoplasm does not divide into two equal halves, yet the bud after receiving a mitotically divided nucleus eventually reproduces all cytoplasmic elements and grows to normal size. (See *Saccharomyces cerevisiae*)

BUDDING YEAST: see *Saccharomyces cerevisiae*

BUdR (5-bromodeoxyuridine): animal cells deficient in thymidine kinase enzyme are resistant to this analog. It is a base analog mutagen, substituting for thymidine and may cause mutation.

BUdR continued

Its mutagenic efficiency is low, especially in eukaryotes although if cells with BUdR substituted chromosomes are exposed to visible light, chromosome breakage is induced. (See also bromouracil, base substitution mutation, sister chromatid exchange, hydrogen pairing)

BUERGER DISEASE: autosomal recessive predisposition to thromboangiitis (inflammation of the blood vessels); its frequency is relatively high in some oriental ethnic groups.

BUFFALO: Asiatic swamp buffalo (*Bubalus bubalis*) 2n = 48, the Murrah buffalo (*Bubalus bubalis*) 2n = 50, the African buffalos (*Syncerus caffer caffer*) 2n = 52, and the *Syncerus caffer nanus* is 2n = 54.

BUFFER: a chemical solution capable of maintaining a level of pH within a particular range depending on the components of an acid-base system. Also a special storage area in the memory of a computer from where the information can be utilized at different rates by different programs; e.g. the printer can store information faster than it can print it.

Bufo vulgaris (toad, 2n = 36): a primarily terrestrial small frog species and lives in water environment during the mating season. (See also *Rana, Xenopus*, frog, toad)

BULGE: unpaired stretches in the DNA. They are involved in binding of regulatory protein domains, in enzymatic repairs, slipped mispairing in the replication of microsatellite DNA, intermediates in frame shift mutations and essential elements for naturally occurring antisense RNA (See DNA repair, binding proteins, mispairing, frame shift mutation, microsatellite, antisense RNA)

BULKED SEGREGANT ANALYSIS: is used in mapping recombinant inbred lines. The individuals in the population are identical at a particular locus but unlinked regions in the chromosomes are represented at random. (See RAPD)

BULKING: a plant breeding procedure when selection of segregants after a cross is delayed to later generations when the majority of individuals become homozygous by continued inbreeding. The number of heterozygotes by F_n is expected to be $0.5^{(n-1)}$ for a particular locus where n stands for the number of generations selfed. Note the F_1 is produced by crossing therefore we use n - 1. (See also inbreeding progress of, inbreeding rate)

BULLER PHENOMENON: from dikaryotic mycelia nuclei may move into monokaryotic ones. (See di-mon)

BUNDLE: protein α-helices running along the same axis. (See protein structure)

BUNDLE SHEATH: are cells wrapped around the phloem bundles. Their chloroplasts do not show grana and synthesize carbohydrates through the C_3 pathway although the other chloroplasts in the same plant operate by the C_4 system. (See chloroplasts, C3 plants, C4 plants)

BUNGAROTOXIN: see toxins

BUOYANT DENSITY: a molecule (e.g. DNA) suspended in a salt density gradient (such as a CsCl solution, spun for 24 hr at 40,000 rpm in an ultracentrifuge tube) comes to rest in the salt gradient at the position where the medium (CsCl) density is identical to its own. The buoyant density of e.g. *Chlamydomonas* nuclear and chloroplast DNA is 1.724 and 1.695, respectively. Higher buoyant density reflects higher G + C content in the DNA. The refractive index determined by a refractometer, capable of 5 digit resolution, can be used to determine the density (ρ) of a cesium chloride solution in any sample withdrawn from the centrifuge tube. The relevant relationships at 25° C are the following:

CsCl % Weight	Density g/mL	Refractive Index	Molarity
50	1.5825	1.3885	4.700
55	1.6778	1.3973	5.481
56	1.699	1.3992	5.651
57	1.7200	1.4012	5.823
58	1.7410	1.4032	5.998

The density of *E. coli* DNA is about 1.710 and that of *Mycobacterium phlei* is 1.732 and a

Buoyant density continued

deoxy A-T polymer has a (ρ) value of 1.679. Since none of these values are directly readable from the tabulation above, interpolation is required that can be done by graphic representation. (See also density gradient centrifugation)

BUPHTHALMOS: see glaucoma

BURBANK, LUTHER (1849-1926): American breeder credited with the production of over 800 new varieties and strains of plants, mainly in Santa Rosa, California. In addition, his success had a stimulating effect on the development of plant breeding.

BURDO: a "graft hybrid" between tomato and nightshade (*Belladonna*), forming a periclinal chimera. The graft hybrids were assumed to be the result of fusion between the cells of different species combined at the site of the grafting. From this site then somatic hybrid cells regenerated into plants. (See periclinal, somatic cell hybrids)

BURKITT'S LYMPHOMA: is a human cancer caused by the Epstein-Barr virus is most common in central Africa but occurs in other parts of the world, involved in nasopharingeal (nose and throat) carcinoma and neoplasias of the jaws and the abdomen. It frequently involves a translocation between human chromosomes 8 to 14. (See Epstein-Barr virus)

BURSA: generally a sac like pouch; the bursae Fabricius are located in the intestinal tract of birds and produce the B lymphocytes. (See lymphocytes)

BURST SIZE: average number of phage particles released by the lysis of bacteria.

BUS: in a computer it is the circuit system that transmits information within the hardware or cables that link together various devices with the computer.

BUSHMEN: is a designation for nomadic people who live in the wilderness (bush), such as the Australian aborigines or people in the Kalahari Desert. The anthropological characteristics are shared within the group but there is no known evolutionary relationship among the bushmen inhabiting different geographical regions.

BvgS: *Bordatella pertussis* (bacterial) kinase affecting virulence regulatory protein BvgA. (See also pertussis toxin)

BYPASS REPLICATION: capable of bypassing a DNA defect and continues the process beyond it. (See DNA repair)

Byr 2: a serine-threonine kinase of *Schizosaccharomyces pombe*. An analog of RAF. (See raf, signal transduction, serine/threonine kinase)

BYTE: a computer unit of information consisting of a number of adjacent bits. Most frequently 1 byte is 8 bits that represent a letter or other characters that the computer uses.

bZIP: basic leucine zipper. (See leucine zipper, DNA-binding protein domains)

B-ZIP PROTEIN: a DNA binding, protein contains a basic amino acid zipper domain. (See binding proteins)

C

C: abbreviation of cytosine. (See cytosine)

C6: rat glioma cell line (tumor of tissues supporting the nerves). (See glioma)

c25: a prolactin/cycloheximide responsive transcription factor similar to IRF-1 (interferon regulator factor); controls proliferation and anti-proliferation responses in cells. (See transcription factors, IRF, interferon)

C AMOUNT OF DNA: content of DNA in the gametes is 1C; it is 2C in the zygotic cells before S phase, 4C after S phase. The 1C amount actually means that the chromosomes have only one chromatid containing a single DNA double helix. After replication each chromosome becomes double-stranded, i.e. composed of two chromatids held together at the centromere. The 4C stage usually is an indication of diploid or zygotic cell. (See cell cycle)

C BANDING: is a type of banding pattern near the centromere (and some other limited areas, such as telomeres) obtained after staining chromosomes with the Giemsa stain (a mixture of basic dyes), particularly when prior to staining the chromosomes are exposed to the protease trypsin. (See chromosome banding)

C GENES: genes coding for the constant region of the antibody molecule (such as Cμ, Cδ, Cγ, Cε, Cα). (See immunoglobulins, antibody)

C1 INHIBITOR: a proteinase inhibitor of the serpin family; it is a regulator of blood clotting. (See serpin)

C_3 PLANTS: produce three-carbon molecules as the first step in photosynthesis. (See Calvin cycle)

C_4 PLANTS: the first products of their carbon fixation are four-carbon molecules; their photosynthetic efficiency is greater than that of C_3 plants. (See C_3 plants)

C TERMINUS: the carboxyl end of a polypeptide chain. (See also collinearity, amino end)

C VALUE: the amount of DNA in a single chromatid or chromosome before the DNA has replicated, Thus the C value in the gametes is 1, in the diploid zygote it is 2 and after S phase before cell division takes place it may be 4 C.

C VALUE PARADOX: lack of relationship between evolutionary status (complexity of an organism) and its genome size, e.g. the size of the genome of the plant *Fritillaria* is 3×10^8 kbp whereas that of *Homo* is 3×10^6 kbp. (See copy number paradox, genome, pseudogenes, redundancy, repetitious DNA)

C_6 ZINC CLUSTER PROTEIN: a group of transcriptional activators. Their DNA binding sites have the conserved CGG...CCG triplets and in-between a number of other bases in the different members of the family. Furthermore, there is a 19 amino acid carboxy-terminal region at the zinc cluster side, containing the linker and the beginning of the dimerization element that directs the protein to its prefered site. (See transcription factors, transcriptional activator, zinc finger)

CAAT BOX (CCAAT): a consensus sequence in the untranslated promoter region of eukaryotic genes, recognized by transcription factors. House keeping genes may not have this box or a TATA box. (See C/EBP, AP, G box, asparagine synthetase, promoter, TATA box, house keeping genes)

CaaX BOX: is a membrane-binding protein motif.

CAB: see chlorophyll-binding protein

CABBAGE: see *Brassica*

CACAJO: see Cebidae

CACAO (*Theobroma cacao*): is a tropical plant, a source of chocolate and cocoa that is derived from the dried oily cotyledons of the seed. The economically useful species are 2n = 2x = 20.

CACHE: a memory in the computer that increases the speed and efficiency of the machine.

CACHECTIN: hormone-like protein product of macrophages releasing fat and lowering the concentration of fat synthetic and storage enzymes, encoded by a gene situated within the HLA cluster in human chromosome 6p21.3. (See HLA, macrophage)

CACHEXIA: a condition of emaciation, wasting away of muscles. I may be caused by a 24 kDa proteoglycan. It may be the consequence of cancer or other debilitating conditions. (See proteoglycan)

CACO: colon adenocarcinoma, a malignant adenoma. (See adenoma, carcinoma)

CACTUS: a protein product of the *cact* gene of *Drosophila* (2-52) controls dorso-ventral differention by maternal effect in the embryo. Its action is similar to Dorsal. Protein Toll dissociates Cactus from Dorsal and subsequently Dorsal moves into the nucleus. The signals from Toll to the Dorsal-Cactus complex is mediated by Pelle (serine/threonine kinase) and Tube. (See morphogenesis in *Drosophila*)

CAD (coronary artery disease): see coronary heart disease

CAD: a three enzyme complex trifunctional protein catalyzing the first three steps in the *de novo* pyrimidine pathways (carbamoyl phosphate synthetase II, aspartate transcarbamylase, and dihydroorotase). All the three identical polypeptide subunits (M_r 230,000, each) have active sites for the three reactions. In Syrian hamster the gene coding for it is 25 kbp, has 37 introns and the mRNA transcript is 7.9 kb. When amplified, this complex is 500 kbp. (See pyrimidine)

CADASIL (**c**erebral **a**utosomal **d**ominant **a**rteriopathy with **s**ubcortical **i**nfarcts and **l**eukoencephalopathy): is a complex syndrome encoded in a 800 kb region of human chromosome 19q13.1-13.2. It involves diffuse white matter in the brain, defects in the brain blood vessels, stroke, progressive mental illness, paralysis of the face, headaches, severe depression, etc. The basic defect is in a glycosylated transmembrane receptor protein, homologous to Notch of *Drosophila*. (See *Notch*, morphogenesis in *Drosophila*, Alzheimer disease, transmembrane proteins, stroke, brain human)

CADHERINS: cell adhesion molecules (glycoproteins) dependent on the presence of Ca^{++}. They have basic role in normal development and a reduction in their level increases the invasiveness of many types of cancerous growth. They have membrane-spanning and cytoplasmic domains. The former assures cell to cell contacts, the latter attaches to the cytoskeleton. Dominant negative cadherin mutants develop Crohn disease-like symptoms and adenomas. (See Crohn disease, adenoma, integrin)

Caenorhabditis elegans: a small nematode feeding on bacteria. It completes its lifecycle in about $3^1/_2$ days.

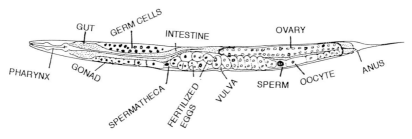

CAENORHABDITIS ELEGANS ADULT HERMAPHRODITE

Approximately 99.8% of the animals have the chromosomal constitution 5 pairs of autosomes and two X chromosomes and are hermaphrodites; 0.2% of the populations are XO males generated by nondisjunction. Its genome is about 1×10^8 bp. The nervous system includes only about 300 cells. The neurons represent 118 structural classes and the number of positions of identifiable chemical synapses appear to be 7600. More than 250 genes were identified by mutational analysis to be involved in behavior. Many of its genes have been cloned and sequenced. It is one of the best organisms for molecular developmental studies. The simplest developmental pathway of animals is seen in this nematode, with approximately 959 somatic cells, including the ca. 302 cells of its nervous system, in a thin 1.2 mm long body of the adult. The egg usually develops hermaphroditically or by fertilization by the rare males, into a 550-cell

Caenorhabditis elegans continued

embryo in the egg shell. Further divisions take place after hatching and passing, by moltings, through four larval stages. The entire process may be completed in about 3 days. Through the transparent body the migration of the embryonal cells and the formation of the organs can be traced under the microscope using Nomarski differential interference contrast optics. The pattern of differentiation and development displays very little variation. The somatic tissues generally have multicellular origin whereas the intestinal cells and the germline cells, each, are monoclonal. For differentiated functions, cytokinesis is not absolutely required. Differentiation is generally not controled by the cellular milieu but by the identity of the particular cell (demonstrated by ablation experiments). Nevertheless, signal transduction among cells may also be required for the formation of the egg discharging mechanisms. Through involvement of a single *anchor cell*, the *vulva* is formed and the *uterus* passes the egg through the anchor and vulva on the culture medium contained in a Petri plate where development is completed. A somatic "distal tip cell" controls the mitotic activity of the "germ cells" which undergo meiosis and produce the gametes. The hermaphroditic XX females are self-fertilized but by nondisjunction they produce also male gametes at a frequency of 0.1%. The XO males when mated with an XX female produce then males and females in equal proportion. More than 40% of it estimated 13,100 genes have some homology to genes of other organisms. Of 44 human disease genes analyzed 32 had significant similarities to its genes. Internet resources: Genome Sequencing Center, St. Louis, MO <http://genome.wustl.edu./gsc/gschmpg.html>, Sanger Centre, Cambridge, UK: <http://www.sanger.ac.uk>, Caenorhabditis Genetics Center: <gopher://elegans.cbs.umn.edu:70>, Leon Avery's World Wide Web Page: <http://eatworms.swm.ed.edu>. Gene Expression Pattern: <http://eatworms.swmed.edu/Worm_labs/Hope/ul33.jpeg> or <http://www.sanger.ac.uk/~sjj/C.elegans_blast_server.htm>. (See dauer larva, pou, unc-86, Nomarski differential interference contrast microscopy)

CAF (chromatin assembly factor): facilitates the structural organization of the chromosomal elements in association with acetylated histones. (See chromatin)

CAFÉ-AUX-LAIT SPOT: light brown skin macule; diagnostic signs of neurofibromatosis. (See neurofibromatosis)

CAFFEIC ACID (3,4-dihydroxycinnamic acid phenethyl ester, CAPE): is related to flavonoids present in honeybees' glue. It has antiviral, antiinflammatory, immunomodulatory effect and inhibits tumor growth, lipid peroxidation, lypoxygenase, ornithine decarboxylase, protein tyrosine kinase and the activation of NF-κB. (See named entries separately, antimutagen)

CAFFEINE: is a modified purine molecule, present in coffea, tea, cola nuts and other plants. It is a stimulant and diuretic. Caffeine interferes with the repair of spontaneously or otherwise caused chromosomal damage, and thus may enhance the mutagenic potential of irradiation and chemicals. Orally LDLo 192 mg/kg for humans. LD_{50} orally for male mice 127 mg/kg, for females 137 mg/kg. (See purine, theobromine)

CAGED COMPOUND: is rendered inactive by combining it with a photosensitive molecule and then introduced into the cell by microinjection, electroporesis or protoplast fusion and subsequently irradiated by UV light of a specific wavelength or by laser beam leading to liberation of the molecules (e.g. Ca^{++}, cAMP, GTP) from the sensitizer (e.g. nitrobenzyl side chain). This procedure permits the study of localized effects within the cytoplasm. Fluorochromes can also be caged for monitoring the behavior of subcellular structures, e.g. the function of tubulins. (See microinjection, electroporation, protoplast fusion, fluorochromes, UV, laser, tubulin)

CAIRNS STRUCTURE: a DNA molecule undergoing bidirectional θ replication. (See θ replication, bidirectional replication)

Cak (cyclin-dependent-activating kinase): is generally a component of the TFIIH transcription factor but in *Saccharomyces* it is not part of TFIIH. It activates CDC28. (See transcription factors, Cdk, MO15, cyclin, *CDC28*)

CAKβ: see CAM

CALCINEURIN (protein phosphatase-IIB): its action is dependent on calcium and calmodulin.

Calcineurin is the target of cyslosporin and FK506. (See calmodulin, immunosuppressant, cyslosporin, FK506, serine/threonine phosphoprotein phosphatases, T cell, NFAT)

CALCITONIN: an oligopetide hormone of the thyroid gland controling calcium and phosphate levels and an antagonist of parathyroid hormone. (See animal hormones, immune privilege)

CALCIUM ION CHANNEL: see ion channels, calmodulin, second messengers, dihydropyridine receptor, ω-agatoxin, ω-conotoxin, neurotransmitters

CALCIUM SIGNALING: calcium plays the role of a second messenger and activates many enzymes in the cell; it regulates synaptic activity of the neurons. When phospholipase C (PLC) is activated, the cells release intracellular Ca^{2+} through the calcium-release activated calcium channels (CRACs), receptor-operated calcium channels (ROCs) and store-operated calcium channels (SOCs). The incoming calcium then replenishes the Ca^{2+} store in the cell and it is also called CCE (capacitative Ca^{2+} entry). The CCE may be activated by hormones, by the immune reactions or by neural receptors. Light signal transduction in *Drosophila* requires PLC activation, synthesis of IP3 and the release of Ca^{2+}. For normal vision the flies must be able to maintain a Ca^{2+} homeostasis. Mutants are known (and cloned) that represent gene loci involved in calcium regulation. Nuclear gene expression may be regulated through the CRE element whereas the cytoplasmic signaling may be mediated by SRE. (See calmodulin, calcium ion channels, calcineurin, calcitonin, PLC, IP3, CRE, SRE, homeostasis)

CALCIUMPHOSPHATE PRECIPITATION: may significantly enhance the chances of DNA to be taken up by (*E. coli* or mammalian) cells to be transformed (transfected). (See transformation, cotransfection)

CALCOFLUOR WHITE (Tinopal): is a fluorescent brightener that can be used to monitor cell wall formation in protoplast suspensions. (See fluorochromes)

CALDESMON: 83 kDa actin and calmodulin-binding protein. During mitosis it dissociates from the microfilaments as a consequence of mitosis-specific phosphorylation of actomyosin ATPase. (See calmodulin, myosin, ATPase)

CALICO CAT: heterozygous *female* animals with black-yellow-white fur patches. The alternation of black and yellow fur is due to inactivation of the genes residing in one or the other of the two X-chromosomes. This pattern occurs only in the Klinefelter males (XXY). The white fur is an autosomal trait controled by the *S* (*spotted*) gene. (See also X-chromosome inactivation, tortoiseshell fur, lyonization)

CALLICEBUS: see Cebidae

CALLITHRICIDAE: are new world monkeys (marmosets and tamarins). *Callithrix argentata* 2n = 44; *Callithrix humeralifer* 2n = 44; *Callthrix jaccus* 2n = 46; *Callimico goldi* 2n = 48 some males 47; *Cebuella pygmaea* 2n = 44; *Leontocebus rosalia* 2n = 46; *Sanguinis fuscicollis illigeri* 2n = 46; *Sanguinus oedipus* 2n = 46; *Tamarinus mystax* 2n = 46; *Tamarinus nigricollis* 2n = 46. (See primates)

CALLOSE: a carbohydrate (glucan) forming on injured plant tissue.

CALLUS: a solid mass of plant cells (generally) on synthetic media; thickening of animal skin, unorganized bone growth.

CALMODULIN (CaM): is a 17,000 M_r acidic protein with 4 Ca^{2+} binding sites affecting (among others) membrane transport, chromosome movement and processes in fertilization; it functions as a regulatory unit of several enzymes although calmodulin itself is not an enzyme. Its binding results in conformational changes of the target proteins. The presence of Ca^{++}/ CaM is required for phosphorylation of several proteins (myosin light chain kinase, phosphorylase kinase). CaM-kinase II mediates the secretion of neurotransmitters. It activates tyrosine hydroxylase required in catecholamine biosynthesis. It is involved in the control of such brain functions as memory and learning. CaM-kinase II is capable of autophosphorylation even in the absence of Ca^{++}. CaM regulates adenylate cyclase (cAMP), and cAMP regulates CaM. A-kinase, regulated by CaM, phosphorylates the IP3 receptor, and cAMP and CaM-kinases control CREB. The delta subunit of phosphorylase kinase is CaM. (See cAMP, phosphorylases, autophosphorylation, signal transduction, cAMP, IP_3, CREB, neurotransmitters, aequorin)

CALNEXIN: is a membrane-bound chaperone glycoprotein; it may associate also with an antigen before entering the endoplasmic reticulum. (See chaperones, endoplasmic reticulum)

CALORIE: the amount of heat required to elevate the temperature of 1 gram of water from 14.5 to 15.5 C° (4.186 international joules). (See joule)

CALPHOSTIN C: an inhibitor of diaglycerol and C^{2+}-dependent phosphokinase CD (PKC). (See T cells)

CALRETICULIN: a calcium storage protein within the endoplasmic reticulum and also present in the nucleus. It may prevent the glucocorticoid receptor binding to its response element and it is thus a transcription factor. (See hormone response element, glucocorticoid)

CALVIN CYCLE: the pathway of fixation of CO_2 (1C) into 3-phosphoglycerate (3C), 1,3-bis-phosphoglycerate (3C) and glyceraldehyde-3-phosphate (3C). In this reaction 3 molecules of ATP and 2 molecules of NADPH are used for each CO_2 converted into carbohydrate. (See photosynthesis, Krebs—Szentgyörgyi cycle)

CALYPSO: *Drosophila* transposable element (7.2 kb) generally present in 10 to 20 copies per cell. (See transposable elements, hybrid dysgenesis)

CALYX: the collective name of sepals, the basal whorl of the flowers. (See flower differentiation)

CaM: see calmodulin

CAM (cell adhesion molecule): regulates monolayer formation in cultured mammalian cells and mediates neuronal connections with the aid of the fibroblast growth factor (FGF). Cell to cell adhesion is mediated by cadherins, immunoglobulins, selectins, and integrins. Cell-matrix adhesion is mediated by integrins and transmembrane proteoglycan. Focal cell adhesion, integrin-mediated contact between cells and the extracellular matrix is linked to the activation of $pp125^{FAK}$ protein kinase. This enzyme is a member of the family of FakB, PYK2/ CAKβ and RAFTK protein tyrosine kinases. PYK2 regulates calcium ion channels and the MAPK signaling pathway. The carboxy terminal of $pp125^{FAK}$ is expressed as a non-kinase $pp41/43^{FRNK}$. This latter protein is an inhibitor of $pp125^{FAK}$ and the phosphorylation of tensin and paxillin adhesion proteins. PYK2 activity is also coupled with the JNK signaling pathway. (See cadherins, immunoglobulins, selectins, integrins, proteoglycan, JNK, ion channels, extracellular matrix, MAPK, protein kinases)

CAM: see *Crassulacean* acid metabolism

cam (chloramphenicol transacetylase): conveys resistance to chloramphenicol which would block peptidyl transferase on the 70S ribosomes in prokaryotes and in eukaryotic organelles. (See chloramphenicol, antibiotics, protein synthesis)

CAMBIUM: a meristemic tissue layer around the stem of plants providing for growth in diameter. (See meristem)

CAMBRIAN: geological era 500-600 million years ago. The fossil plant relics are poorly preserved. All major animal groups, except the vertebrates, were already present. (See evolution, geological time periods)

CAMEL (*Camelus bactrianus*): 2n = 74 and its American relative, the vicuna (*Vicugna vicugna*) is also 2n = 74.

CaMK: calcium-calmodulin-dependent protein kinase. CaMK II is necessary for several physiological processes including learning and memory. (See calmodulin, kinases)

CaMO: see cauliflower mosaic virus

cAMP: adenosine 3':5' monophosphate (cyclic AMP, second messenger); has crucial role in signal transduction and general gene regulation in bacteria and animals. Some molecules can increase the level of cAMP (adenylate cyclase) by binding to certain transmembrane receptors whereas other cellular signal molecules are inhibitory. The stimulatory G_s protein activates adenylate cyclase. Actually the α_s subunit dissociates from the two other chains and rather binds to and hydrolyzes GTP, and then binds to adenylate cyclase and boosts the production of cAMP. Upon binding to adenylate cyclase GTPase activity will increase and an inactive G_s is formed again by recombining α_s with βγ. Cholera toxin (produced upon infection by the bacterium *Vibrio chol-*

cAMP continued

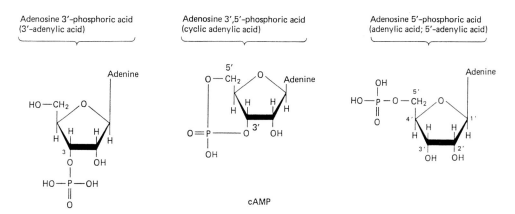

era) mediates the transfer of adenylate to α_s and this fact prevents the hydrolysis of its GTP and therefore the adenylate cyclase function stays on, resulting in increased levels of cAMP. This condition then opens a very active sodium and water efflux through the intestinal walls, causing debilitating diarrhea and dehydration of the entire body.

The inhibitory trimeric G_i has a special α_i subunit and it is activated by other types of cellular signals. In G_i the $\beta\gamma$ subunits are dissociated from the α_i subunit and these subunits then directly and indirectly interfere with adenylate cyclase. More importantly G_i opens K^+ ion channels in the plasma membrane. The pertussis toxin (due to *Bordetella* bacterial, whooping cough) —in contrast to the cholera toxin—mediates the adenylation of an α_i subunit that interferes with responding to the receptors and GDP remains bound to the G-protein and adenylate cyclase activity is not blocked and potassium channels are not opened.

One of the most important functions of cAMP is the activation of the four-subunit *protein kinase A*. This phosphorylase enzyme then adds phosphate groups from ATP to serine and threonine residues in certain proteins. Protein kinase A is activated by cAMP through forming a complex with two its regulatory subunits while the two other separated catalytic subunits are turned on. The enzyme exists in two forms, one is cytosolic whereas the other is bound to membranes and microtubules.

cAMP controls the transcription in a positive or negative manner of several other genes. The CREB (cAMP response elements)—located in the upstream region of genes—promote transcription upon phosphorylation of a specific serine residue in the binding proteins whereas the CREM α and β are repressors. cAMP controls also the secretion of cortisol by the kidney cortex, the thyroid hormone, the secretion of the luteinizing hormone (progesterone), adrenaline and glucagon production and glycogen and triaglyceride breakdown, heart muscle and other muscle functions, etc. (See adenosine-3',5' monophosphate, adenylate cyclase, G-protein, epinephrine, cAMP-dependent protein kinase, signal transduction, CREB, progesterone)

cAMP RECEPTOR PROTEIN (CRP): assists the binding of *E. coli* RNA polymerase to the operator of carbohydrate operons (positive regulation) but it may act also as an activator of the negative regulator CytR and as such becomes a corepressor. (See Lac operon, positive control, negative control, repressor, corepressor)

CAMPBELL MODEL: of recombination suggested first the mechanism of integration of the (pro)phage into bacteria by reciprocal recombination between the circular bacterial and temperate phage DNA molecules. (See bacteriophages, *E. coli*)

cAMP-DEPENDENT PROTEIN KINASE: see protein kinase A, phosphorylase *b* kinase

CAMPOMELIC DYSPLASIA (CD, CMD1: a most likely recessive bone-formation and testis-

Campomelic dysplasia continued
development defect due to mutation in the SOX9 gene in human chromosome 17q24.1-q25.1. The congenital bowing of the skeletal bones and malformation of the head bones usually causes death in early infancy, however, some individuals survive to early adulthood. This defect is often associated with chromosomal rearrangements and sex reversal. (See SRY, SOX, sex reversal)

CAMPTODACTYLY: autosomal dominant bent fingers.

CaMV: see cauliflower mosaic virus

CANALIZATION: is a genetic buffering mechanism that reduces the visible variations beyond what is expected on the basis of genetic diversity. Also, it eliminates from the populations those genotypes which cannot adjust to environmental fluctuations. Mutants generally have reduced buffering capacity compared to the wild type (that is best canalized for survival), and can be readily eliminated. (See also genetic homeostasis, homeostasis)

CANAVAN DISEASE: see aspartoacylase deficiency

CANAVANINE: a competitive inhibitor of arginine (natural plant product). (See competitive inhibitor)

CANCER: cells continue to divide when cell divisions are not expected, it is an uncontrolled growth, a malignant growth. Cells may spread through the blood stream to other locations in the body and initiate secondary foci of malignant growth (metastasis). It may occur in a variety of forms such as leukemia, adenoma, lymphoma, sarcoma but it is not exactly known how these different types are specified. The anomaly in malignant growth is that cells do not remain in a quiescent stage (G_0) and proceed either through terminal differentiation or death (apoptosis) but from G1 phase, and independently from normal cellular regulation continue indefinitely to S phase, mitosis and cell division. Normal cells pause at the G1 *restriction point* and respond to the instructions coming through cyclins. The cyclins (CLNs) coupled with the labile cyclin dependent kinases (CDK4, CDK6) are receptive to mitogens but they are kept in check by several INK (inhibitors of kinase) proteins. When normally tumor suppressors such as the retinoblastoma (RB) and other proteins (E2F family) become phosphorylated the cells exit from G1 and embark on DNA synthesis. The phosphorylated tumor suppressors transactivate CDKs and unless cell divisions are blocked at various checkpoints, divisions will continue and proceed in an accelerated and uncontroled manner. The G1 phase-specific cyclin E-CDK2 complex stimulates more phosphorylation of the tumor suppressor RB and E2F and the dependence on mitogens is diminished. CLN-A-CLN-B dependent kinase (CDK2) reinforces the phosphorylation process and dephosphorylation does not sets on until the completion of mitosis. CLN-E and CLN-A associated CDK2 assist the DNA replication machinery. CLN-dependent kinases are suppressed also by CDK inhibitors $p21^{CIP1}$, $p27^{KIP1}$ and $p57^{KIP2}$ but if the latter ones are deleted or mutated both cell number and size increases.

Cancerous growth may be initiated by mutations in structural genes, suppressor genes, by chromosomal rearrangements (translocations, inversions, transpositions, deletions) and viral insertions in the chromosomes. Although a single mutation may suffice for uncontroled cellular proliferation, several additional factors may be involved for full scale development. The initial mutation may take place in proto-oncogenes. Mutations or loss of the CLN-D1 locus (human chromosome 11q13) and the CDK4 gene (human chromosome 12q13), INK4a (CDK^{INK4a}, chromosome 9p21), p16, p53 are common in many types of cancers. p53 also regulates p21, p27 and p57 proteins. The gene encoding p53 is in human chromosome 17p13.15-p12 and it is altered in a large number of different cancers. The normal function of p53 is required for the development of the centrosome and proper segregation of the chromosomes. About 100 proto-oncogenes have now been discovered and their possible malfunction due to mutation, chromosomal rearrangement or loss are many fold (see diagram next page). These mutations, similarly to other mutations, are generally (not always) recessive and not expressed in diploid cells. When through another event(s) they become homozygous or hemizygous, cancerous growth may follow. This may be the cause why for the development of cancer a long period is required after the initial genetic lesion. These genes have normal cellular functions (some of the

Cancer continued

SOME PROTO-ONCOGENES AND THEIR ROLE IN CARCINOGENESIS

CHROMOSOME	5q		12p12	18q23	17p	
ALTERATION	del.		activat.	del	del	
GENE	APC	DDM-1	KRAS2	DCC	p53	various
PRODUCT		demethylation	oncoprot.	netrin	suppr.	metabolic regulators

| | | | | | |
| normal | abnormal | early | intermediate | late | |
| epithelium →proliferation → adenoma → adenoma → adenoma → carcinoma → metastasis |

del: deletion, activat = activation, oncoprot = oncoprotein, netrin = guidance proteins, suppr = tumor suppressor (Modified after Fearon, E.R. & Vogelstein, B. 1990 Cell 61: 757)

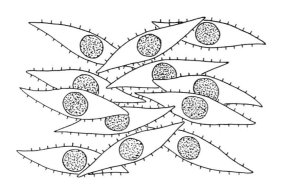

Cultured cancer cells lose their contact inhibition and proliferate in an overlapping fashion. (After Dulbecco, R. 1967. Sci. Amer. 216:[4] 28)

proto-oncogenes, e.g. *Ras*, encode G-proteins and play important role in healthy signal transduction) but mutation may alter their normal role and become (cellular) c-oncogenes. The v-oncogenes are viral counterparts of the c-oncogenes. When c-oncogenes are inserted into the chromosomes of animals they may provide efficient promoters for cellular genes involved in growth. In the development of cancer the products of more than a dozen tumor suppressor genes (e.g. p53, p16) play major role. Because of mutation, they no longer are capable of controling genes of the cell cycle. Normal mammalian cells grow in the cultures in an anchorage-dependent fashion, in monolayer. Cancerous cell differ in their growth habit by growing in a disarray, in layers. Since the human body contains about a billion cells per gram solid tissue, an average adult human body may have about 10^{14} to 10^{15} diploid cells according to the various estimates. Average mutation rates are within the range of 10^{-5} to 10^{-8} per genome, per generation, therefore all human individuals must have suffered numerous mutations with neoplastic potentials yet only a fraction of the human populations is afflicted by this group of diseases. The causes of this discrepancy are diploidy

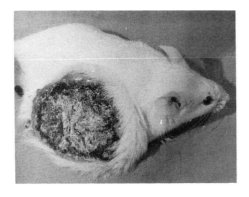

Benzo(a)pyrene induced cancer in the laboratory by painting the carcinogen on the skin of a rodent. (Courtesy of Drs. Nesnow, S. and Slaga, T.)

that masks recessive mutations and the immunological surveillance system recognizes mutant surface antigens, developed on cancer cells, and destroys them before they get out of control. When the immune system weakens by advancing age, by certain diseases (e.g. AIDS) or by taking immuno-suppressive drugs, the incidence of cancer increases. The cancerous growth itself is not necessarily lethal but it generally deprives the body from its normal metabolism and the patients succumb to secondary, opportunistic diseases. The general assumption is that cancer develops monoclonally, i.e. in the tumor each cell has a single common ancestor. This assumption does

Cancer continued

not preclude, however, that in the same cell lineage additional mutations occurred during the process of multiplication although experimental evidence does not show appreciable frequency of mutation during the progression of cancer. In very rare cases, it has been shown, however, that a particular cancer in an individual may not be monoclonal but more than a single founder cells contributed to its formation. (See genetic tumors, oncogenes, leukemias, melanoma, carcinogens, immunological surveillance, environmental mutagens, chromosome breakage, Wilms tumor, retinoblastoma, Bloom syndrome, xeroderma pigmentosum, neurofibromatosis, tuberous sclerosis, von Hippel-Lindau syndrome, nevoid basal cell carcinoma, multiple hamartomas, lentigens, leiomyoma, lipomatosis, Beckwith-Wiedemann syndrome, Werner syndrome, Rothmund-Thompson syndrome, focal dermal hypoplasia, Fanconi anemia, adenomatosis endocrine multiple, agammaglubulinemia, Wiskott-Aldrich syndrome, ataxia telangiectasia, Gardner syndrome, polyposis, colorectal cancer, polyposis hamartomatous, gonadal dysgenesis, Down syndrome, neuroblastoma, Li-Fraumeni syndrome, exostosis, breast cancer, Klinefelter syndrome, prostate cancer, liver cancer, phorbol esters, adenovirus, Epstein-Barr virus, gatekeeper gene, tumor suppressor, cell cycle and the named factors listed separately)

CANCER FAMILY SYNDROME: see Li-Fraumeni syndrome, Lynch cancer family syndrome

CANCER GENE THERAPY: transforming by (1) cytokine genes, (2) introducing genes encoding foreign antigens, (3) using antisense RNA constructs, (4) functional tumor suppressor alleles, (5) transfering the *Herpes simplex* virus thymidine kinase (HSVTK) gene making the tumor sensitive to ganciclovir (GCV) may be exploited. Phosphorylation of GCV (GCV-TP) inhibits DNA polymerase that stops cancer cell proliferation, (6) using the multiple drug resistance gene, MDT, (7) magic bullets have been attempted. Immunological defense has the difficulty that cancer cells are self origin and their immunological response to them is not good. Rejection of the malignant cells is very effective if the expression of a foreign gene is elicited by the tumor vaccines, e.g. after viral infection or transfection by a viral gene. The tumor cells so modified thus display foreign epitopes on their MHC molecules and thus recognized by the immune system causing the rejection of the parent-type tumor cells. New technologies are continuously developed. One of such cancer therapy design would convert the cancer genes (oncoprotein) into cancer killer genes by activating a toxin gene according to a construct outlined below. The oncoprotein has a DNA binding domain (DBD), transactivator domain (TAD) and the killer binding domain (KBD):

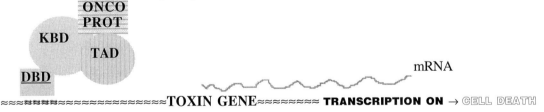

The toxin gene is transactivated and turned on, killing the cancer cell after it has been exposed to the transactivator by connecting the circuit through the oncoprotein (onc prot) to the upstream activator sequence (UAS). (Diagram modified after Da Costa *et al.* 1996 Proc. Natl. Acad. Sci. USA 93:4192).

Another possibility is to fuse the *Pseudomonas* exotoxin (PEA) to the extracellular domain of the HER2 transmembrane receptor protein. Transformed lymphocytes then can introduce into the cancer cells the toxin which inactivates elongation factor-2 (FE-2). Transfection of anti-sense RNA producing constructs has been proven successful in animal models. Human tissue factor (TF), an important receptor of the blood coagulation cascade, can be targeted to endothelial blood vessels of the tumor by appropriate antibody. TF complexed with antihemophilic factor VII activates the serine protease zymogen factors IX and X resulting in the formation of thrombin

Cancer gene therapy continuied

and blood clotting. Blocking specifically and locally the blood supply to the tumor may cause its substantial regression. (See gene therapy, cytokines, antisense technologies, tumor suppressor genes, multidrug resistance, magic bullet, epitope, MHC, transfection, ERBB2, exotoxin, EF-2, ganciclovir, plasmovirus, antihemophilic factors, tissue factor)

CANCER PROMOTER: see phorbol esters

CANCER THERAPY: see magic bullet, immunotoxin, adenovirus [dl1520], antimitotic agents, cytostatic, tamoxifen, cytokines, chemotherapy, radiation effects, cell cycle, Bcl, antisense technologies, multiple drug resistance, NF-κB, cancer gene therapy, plasmovirus

CANDELA: a unit of luminous intensity. A standard radiator produces 60 candela/cm^2 at the freezing temperature of platinum (-2,046° K); 1/60 candela = 1 candle (new unit). 1 foot candle = 10.76 lux; 1 lux = 1 lumen/m^2; 1 lumen = is the total visible energy emitted from one candle point luminous intensity.

"CANDIDATE GENE": is already mapped in a chromosomal region and possibly included in the DNA fragment to be mapped but it is not known (although hoped) that it is involved in a particular function or phenotype associated with a mutation.

CANDLE: see candela

CANNABINOIDS: about 60 related compounds synthesized by the hemp plants (*Cannabis*) and present in the psychoactive drug marihuana. Besides the psychoactive effects, they have various immunosuppressive properties. (See immunosuppressants, *Cannabis*)

Cannabis sativa (hemp): is a generally dioecious species (*Moraceae*) with a diploid chromosome number of 20, including either XX female or XY male flowers. The XXXY tetrasomics are also gynoecious. Hermaphroditic forms are also known that produce more seed and only somewhat reduced amount of fiber. Low-cannabinoid stocks are available. (See cannabinoids, hop)

CANOLA: is an erucic acid-free rape, *Brassica napus* (2n = 38, AACC genomes) oil crop. (See erucic acid)

CANONICAL SEQUENCE: a typical set of nucleotides in several genes (e.g. in a Pribnow box or Hogness box or other conserved elements).

CAP (catabolite activator protein): a homodimer with two subunits of M_r 22,000. It has a binding site for cAMP and DNA with a helix-turn-helix motif. As long as glucose is available in the nutrient medium the *Lac* operon of *E. coli* is inactive because glucose, a catabolite of lactose represses the operon (catabolite repression). In order to turn on the operon the CAP protein must be attached to the CAP site, in a process mediated by cAMP. The latter is formed by the adenyl cyclase enzyme from ATP and this process is inhibited by glucose. When glucose is used up, cAMP is formed and the latter binds to CAP and the complex binds to its palindromic site in the DNA (GTGAGTTAGCTCAC) near the promoter of the operon and transcription by the RNA polymerase is activated. Without CAP the promoter is very weak. In order to pursue normal transcription the grip of the repressor protein must also be lifted and this is mediated by lactose (allolactose). The cAMP - CAP complex regulates also the arabinose and galactose operons. (See negative control, positive control, transcriptional activation, regulation of gene activity, *lac* operon, arabinose operon, galactose operon, cAMP, palindrome)

CAP (ceramide-activated protein): CAP kinase mediates tumor necrosis factor and interleukin-1β functions. It phosphorylates also Raf1 on Thr 269 and increases Raf affinity for ERK kinases. (See RAF, ceramides, ERK, TNF, interleukin)

Cap: a methylated guanylic residue linked at transcription to the 5' end of the eukaryotic mRNA ($_{7ME}G_5$'ppp-mRNA) or 2,2,7-trimethyl guanosine in U RNA. The prokaryotic mRNA has only three phosphates at the 5' nucleotide end. The cap of the (long-life) eukaryotic mRN stabilizes it (makes it less sensitive to nucleases), assures the transport of the mRNA to the cytosol and facilitates its binding to the ribosome and initiation of transcription by lending itself as an anchor to eIF eukaryotic initiation factors. The cap also regulates splicing of the first intron. The effect of Cap on pre-mRNA splicing is mediated by a capping complex (CBC), composed of cap-binding nuclear proteins, CBP80 and CBP20. CBC then mediates the export of RNA. The U3 snRNA also has a trimethylated cap that is added in the nucleus and the RNA remains in the nucleus. (See also capping enzymes, regulation of gene activity, ribosome scanning)

CAP3: is the N-terminal domain of FLICE. (See FLICE)

CAPACITATION OF SPERM: see fertilization

Cap-BINDING PROTEIN: binds to the 5'-end of the mature mRNA. (See protein synthesis)

CAPILLARY: a structure resembling hair; usually bearing a small bore through which liquids can move such as in the capillary veins or capillary tubes, or soil capillary spaces

CAPILLARY TRANSFER: is used to draw a buffer by wicks from a reservoir to an electrophoretic gel containing separated DNA fragments. The gel is in contact with the absorbent papers. The moving stream elutes the DNA from the gel and deposits it onto a nitrocellulose or nylon filter in immediate contact with the gel, in between the gel and the stack of papers topped by a glass plate and weighted down. (See Southern blotting)

CAPPING ENZYMES generate the mRNA cap by using GTP and the diphosphate splits off from the first nucleotide triphosphate of the first residue in the pre-mRNA. The 5' terminal triphosphate is replaced by the guanyl group of GTP and it loses the γ and β PO_4 groups.

METHYL GUANINE IN THE CAP

THE CAP OF THE EUKARYOTIC mRNA

AT THE 3' P THE CHAIN CONTINUES

No capping takes place if at the terminal position there is a monophosphate. The capped G indicates the beginning of the transcript. The G is methylated by guanosine-7-methyl transferase at the 7 position. The 2'-OH is subsequently methylated by 2'-O-methyl transferase using SAM as a methyl donor. The capping reaction is associated with pol II because the triphosphate termini of U6 RNA, 5S RNA and the pre-tRNAs, are transcribed by pol III, and are not capped. The caps stabilize the mRNA and facilitates the ribosomal attachment. Initiation factor eIF-4E recognizes the cap and mediates its binding to the 40S ribosomal subunit. Picornaviruses do not need the cap and they inactivate the cap-binding proteins of the host and thus turn off the synthesis of host proteins. (See class II genes, cap, picornaviruses, pol II eukaryotic, transcription factors)

CAPS (cleaved amplified polymorphic sequences): are produced by digesting PCR products by restriction enzymes to find polymorphism in the DNA. (See PCR, restriction enzyme)

CAPSID: the protein shell of viruses; in complex viruses nucleic acids may also be found in the shell. (See viruses)

CAPSOMERE: the protein subunits of the viral capsids

CAPSULE: polysaccharide coat of bacterial cells or in general structure with content or a fungal sporangium, or a seed capsule of plants formed by fusion of two or more carpels. (See carpel)

CAR: cyclic AMP receptor. (See cAMP)

CARBACHOL (carbamylcholine chloride): is a cholinergic agonist and it is resistant to cholinesterase; carbachol may increase phosphorylation of RAS

CARBAMOYLPHOSPHATE SYNTHETASE DEFICIENCY (hyperammonemia): are autosomal defects in the urea cycle. Some of the enzymes responsible for the defect are mitochondrial, others are in several autosomes and also in the X chromosome. (See urea cycle, CAD, channeling)

CARBENICILLIN: semisynthetic antibiotic of the penicillin family; effective against gram negative and some gram positive bacteria. (See antibiotics)

CARBOCATION: see carbonium ion

CARBOHYDRATE: sugars and their polymers.

CARBON DATING: see evolutionary clock

CARBON FIXATION: photosynthetic organisms form sugars from atmospheric CO_2. (See C3, C4 plants)

CARBONIC ANHYDRASE (CA): catalyzes the reactions $H_2O_3 \rightleftharpoons H_2O + CO_2$ i.e. it provides an equilibrium between carbonic acid and carbon dioxide, and the addition reaction of $CO_2 + H_2O \rightarrow H^+ + HO-C(=O)-O^-$ i.e. the formation of carbonium ion from carbondioxide. This is a Zn enzyme of about M_r 30,000 is common in various eukaryotic tissue (1-2 g per L of mammalian blood). It is an extremely active enzyme but a much lower rate catalyzes also the hydration of acetaldehyde. In humans seven isozymic forms have been identified. CA 1, 2 and 3 were located to chromosome 8q13-q22. CA2 is about 20 kb apart from CA3 and it is transcribed in the same direction while CA1 and CA3 are separated by about 80 kb and their direction of transcription is opposite. CA1, 2, and 3 are common in the muscles. CA4 and CA7 genes are in another chromosome (16q21-q23) and are specific for the kidney, lung and liver mitochondria. CA6 gene is present in chromosome 1p37.33-p36.22 and the gene is expressed in the saliva. CA5 does not appear to be of specific significance. CA is found also in the mitochondria and this enzymes may participate in gluconeogenesis. CA2 deficiency is involved in osteopetrosis with renal tubular acidosis. (See osteopetrosis, mitochondrial genetics, mtDNA, gluconeogenesis)

CARBONIUM ION: a group of atoms containing only 6 electrons (rather than the normal octet); its is considered highly reactive and it is supposed to be involved in the reactions following alkylative processes in chemical mutagenesis and carcinogenesis. (See mutagen specificity, alkylating agents, carcinogen)

CARBONYL GROUP: —C=O such as occurs in aldehydes and ketones.

CARBOXYL GROUP: —C(=O)—OH

CARBOXYL TERMINUS: see C-terminus

CARBOXYPEPTIDASE: a Zinc-containing proteolytic enzyme cleaving the polypeptide chain at the carboxyl end after substrate binding caused an "induced fit", i.e. a conformational change in the enzyme protein. The "electronic strain" due to the presence of Zn accelerates catalysis.

CARCINOGEN: an agent that is capable of causing cancerous transformation of cells. Carcinogens include chemical, physical and viral agents. The chemical compounds may be genotoxic that act directly on the DNA (e.g. ethyleneimine, various epoxides, lactones, sulfate esters, mustard gas, 2-naphthylamine, nitroamides and nitrosoureas, etc.). They can be procarcinogens that require enzymatic activation (e.g., polycyclic or heterocyclic hydrocarbons such as benzo(a)pyrene, benz[a]anthracene, etc.) or can be inorganic compounds or elements that interfere with the fidelity of DNA replication. Some carcinogens act by some sort of physical means (e.g. various polymers, asbestos). Some hormones may also promote carcinogenesis in an indirect manner. Phorbol esters, n-dodecane, etc. may not be the primary cause of cancer but are considered to be promoters of cancerous growth. The latter group of chemicals not acting directly on DNA are frequently called *epigenetic carcinogens*. Carcinogens include most of the mutagens, many industrial and laboratory chemicals, pesticides, insecticides, fungicides, drugs, cigarette smoke, medicines, cosmetics, food preservatives, food additives, flame retardants, cross-linking agents, plastics, solvents, paints, adhesives, exhaust fumes, other products of combustion, tar, etc. Several natural plant products such as pyrrolyzidine alkaloids, safrole, mycotoxins such as aflatoxins, antibiotics, such as streptozotocin, viruses such as the Epstein-

Carcinogen continued

Barr virus, Simian virus 40, adenoviruses, etc. Many food products such as oxidized fats, overcooked meats, etc. are also potential carcinogens because of the carcinogens formed in them during their exposure to certain conditions. The direct assays of carcinogens involve testing the induction of skin or lung tumors in rodents, breast tumors in young female Wistar or Sprague-Dawley strains of rats, examination of rodent livers for carcinogenic response, etc. The determination of carcinogenecity at relatively low potency is extremely difficult by direct animal assays because tumorigenesis may occur only after a long delay (months or years) following exposure and because the required population size may be practically prohibitive. Also because all experiments must include an equal size concurrent control to obtain reliable information. The application of the carcinogen to the test animals may be by painting of the skin, subcutaneous or intravenous injection, feeding in the diet or drinking water, inhalation, etc. Since many of the carcinogens are also mutagens the preliminary tests are generally conducted by mutagenic assays that permit the evaluation in large populations, within short time and at a low cost. The mutagenic assays (see Ames test) usually try to substitute for the animal activation system by human or animal liver (microsomal) fractions. (See genetic tumors, neoplasia, cancer, oncogenes, mutagen assays, ionizing radiation, radiation effects, radiation hazard assessment, cocarcinogens, cigarette smoke, bioassays in genetic toxicology, <http://www.iarc.fr/monoeval/allmonos.htm>)

CARCINOGENESIS: the process of cancer induction and progression. (See cancer, carcinogen)

CARCINOMA: a malignant cancer tissue of epithelial origin.

CARCINOSTASIS: tumor growth inhibition.

CARDIO-AUDITORY SYNDROME: see Lange-Nielssen syndrome

CARDIOMYOPATHIES: a group of non-inflammatory heart diseases affecting about 25,000 persons annually in USA. (See for details cardiomyopathy hypertrophic, Duchenne muscular dystrophy, Becker muscular dystrophy, Barth syndrome [endocardial fibroelastosis], Acyl-CoA dehydrogenase deficiencies, cardiomyopathy dilated, superoxide dismutase)

CARDIOMYOPATHY, DILATED (DCM): involves thinner than normal ventricular heart wall, reduced contractility and heart failure. Prevalence in USA about 4×10^{-4} and about 1/4 of them are hereditary and may benefit from heart transplantation. Defects may involve recessive β oxidation of fatty acids, including acyl - CoA dehydrogenase, carnitin palmitoyl transferase II, and impaired mitochondrial oxidative phosphorylation. Also 4 dominant loci have been located in human chromosomes 1, 3, 9. (See cardiovascular diseases, acetyl coenzyme A, carnitin, fatty acids, mitochondrial diseases in humans)

CARDIOMYOPATHY, HYPERTROPHIC, FAMILIAL (FHC): has heterogeneous autosomal dominant symptoms (thickening of the heart's ventricular walls, shortness of breath, arrythmia and sudden death. Four genes are known to be involved controlling contractile heart proteins such as β myosin heavy chain, α tropomyosin, troponin T and cardiac binding protein C (human chromosome 11p11.2). (See cardiomyopathy dilated, cardiovascular diseases, myosin, troponin)

CARDIOVASCULAR DISEASES: affect the heart and the vein system. (See coronary heart disease, mucopolysaccharidosis, lipidoses, Tangier disease, lysosomal storage disease, hypertension, coarctation of the aorta, sickle cell disease, Ehlers-Danlos syndrome, Marfan syndrome, Norum disease, hyperthyroidism, LQT, cardiomyopathies, familial hypercholesterolemia, hypobetalipoproteinemia, homocystinuria, telangiectasia hereditary hemorrhagic, Williams syndrome, supravalvular stenosis)

CArG BOX: see SRE

CARGO RECEPTORS: are special molecules in or on the plasma membrane that are recognized by adaptins associated with clathrin coated vesicles and thus trap various molecules in the process of endocytosis. (See clathrin, endocytosis, receptors, adaptins)

CaRE: a cis-acting element responsible for induction of the c-fos protooncogene by calcium. (See FOS, cis-acting element)

CArG: a conserved promoter element [CC(A/T)$_6$GG] closely related to SRF.

CARL: see comparative anchor reference loci

CARNITIN (γ-trimethylamino-β-hydroxybutyrate): it facilitates the entry of fatty acids into mitochondria. It occurs in all organisms and it is most abundant in the muscles (0.1% of dry matter).

CARNOSINEMIA: carnosine is a neurotransmitter dipeptide of β-alanine and histidine. Enzymatically it may be split by carnosinase into two components or by the action of a methyltransferase it may be converted into anserine. An autosomal recessive defect in carnosinase may lead to the excretion of carnosine, anserine and homocarnosine and neurological disorders. (See neuromuscular diseases, anserine)

CAROLI DISEASE: see polycystic kidney disease

CAROTENOIDS: accessory light absorbing pigments of yellow, red or purple color, including carotene and xanthophylls.

CARP (*Cyprinus carpio*): 2n = 100-104.

CARPEL: the floral leaf forming the (enclosure) site for the ovules. (See ovule, flower differentiation)

CARRIER: human heterozygote for a recessive gene which is not expressed in that individual but may be transmitted to the progeny. *Obligate carrier* is identified on the basis of family history; the natural parents of a homozygous recessive individual must be carriers unless a rare mutational event happened in the heterozygous embryo after fertilization or the natural father is not identical with the legal one. The posterior probability that chromosome A is the carrier of the mutation in question is R/(R+1) where R= $\frac{P(Ma \mid A = D)}{P(Ma \mid A \neq D)}$ x $\frac{P(Mb \mid B = D)}{PMb \mid B \neq D)}$ and *Ma* and *Mb* represent the marker information for chromosomes A and B, respectively. D stands for the mutation-bearing chromosome so P(Ma|A=D) would indicate the probability that chromosome A would be carrying the mutation and P(Mb|B=D) would be the same for chromosome B. (See heterozygote, Bayes theorem)

CARRIER DNA: is a non-specific DNA that may be mixed with the specific DNA to facilitate transformation or other manipulations. (See transformation)

CARRIER PROTEIN: transports solutes through membranes while its conformation is altered. Some transport a single type of molecule (uniporter), others carry more in the same (symporter) or opposite directions (antiporter).

CARROT (*Daucus carota*): all cultivated forms are 2n = 2x = 18.

CARRYING CAPACITY: is a term of ecological genetics, and it means that in a particular environment only a certain number of species or individuals of a species can survive and their survival depends on their genetic adaptation. The carrying capacity of a vector is the size of DNA that it can accomodate.

CARTER-FALCONER MAPPING FUNCTION: is based on the assumption that there is substantial positive interference along the length of the chromosome:
map distance = 0.25 {0.5[ln(1 + 2r) - ln(1 - 2r)] + tan^{-1}(2r)} where r = the observed recombination fraction, ln = natural logarithm, tan = tangent. (See Haldane's mapping function, Kosambi's mapping function, mapping function)

CARTILAGE: a fibrous connective tissue; also it may be converted into bone tissues during postembryonic development. It is rich in collagen and chondroitin sulfate. (See collagen)

CARTILAGE-HAIR DYSPLASIA: see chondrodysplasia McKusick type

CARUNCLE: a small outgrowth on animal and plant tissues.

CARYONIDE: clonal derivative of a cell which after conjugation retains the original macronucleus in ciliates and all macronuclei in the subclones are derived from a single macronucleus. (See *Paramecium*)

CARYOPSIS: the "seed" (kernel) of some monocots containing the single embryo and endosperm and also tissues derived from the fruit (pericarp).

CAS: Chemical Abstract Service Registry that identifies chemicals by specific numbers.

CAS: is ≈100 kDA human microtubule-associated protein involved in the control of chromosome segregation. (See microtubule, spindle fibers)

CASAMINO ACIDS: hydrochloric acid hydrolysate of casein, containing amino acids (except tryptophan that is destroyed by the process). Total nitrogen content 8 to 10%, NaCl 14 to 38%.

CASCADE: (in genetics) is a sequence of events depending in specific consecutive steps such as feedback control, signal transduction and differentiation.

CASCADE HYBRIDIZATION: procedure for enriching a certain fraction of the DNA transcribed at a particular developmental stage. Total cDNA is hybridized in a cascade of events with 20, 50 and 100 in excess amounts of mRNAs synthesized at stage 2. The hybrid is then adsorbed to hydroxyapatite column. Unbound molecules are passed through. The unbound emanate is then hybridized with 100 excess of stage 1 mRNAs. Thus cDNA transcribed only in stage 1 is much enriched. (See mRNA, cDNA, hydroxyapatite)

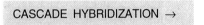
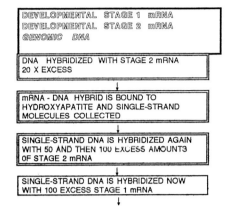

CASPASE: see apoptosis

CASSAVA (*Manihot esculenta*): belongs to the Euphorbiaceae. Originally an American perennial shrub and an important source of carbohydrate. The 98 species all have 2n = 36 chromosomes.

CASSETTE: see vector cassette

CASSETTE MUTAGENESIS: a synthetic DNA fragment is used to replace a short sequence of DNA and thus alter the genetic information at a site. (See mutation induction, localized mutagenesis, TAB mutagenesis)

CASSETTES MODEL OF SEX EXPRESSION: explains switches of the mating type in homothallic (*HO*) yeast by the transposition of either one or the other (*a* or α) silent, distant elements (cassettes) to the sex locus *MAT* where they can be expressed:

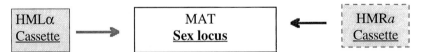

(See mating type determination in yeast)

CASTE: a specialized group within an insect society, e.g. workers among bees.

CASTOR BEAN (*Ricinus communis*): is an oil crop and ornamental plant adapted to a wide range of climates. The viscosity of its oil is rather constant at various temperatures and thus good lubricant for high-speed engines. Also, it is a very potent irritant and laxative and used for medical purposes. The plant also contains the lectin ricin which is one of the most toxic compounds known; 2-5 mg/kg is lethal for humans and in mice the intraperitonial minimal lethal dose is 0.001 µg ricin nitrogen/g body weight. All species are 2n = 20.

CASTRATION: surgical removal of the gonads or chemical prevention of testosterone production to prevent reproduction and reduce sex drive. A male deprived of testes is called also eunoch. The surgical removal of the female gonads is spaying (oophorectomy). (See also eunoch, ovariectomy)

CAT: chloramphenicol acetyltransferase; conveys resistance to the antibiotic chloramphenicol.

CAT (*Felis catus*): the domesticated cat, 2n = 38; the majority of wild cats have also 2n = 38 or 36.

CAT TRANSPORTERS: transport through membranes cationic amino acids (arginine, lysine, ornithine, histidine). (See transporters, ABC transporters, translocon)

CAT EYE SYNDROME: autosomal dominant; it involves vertical pupil, non-perforated anus, heart, kidney malformations, etc. The patients display tissue mosaicism and generally there is one or more copies of a modified human chromosome 22. (See coloboma, eye diseases).

CAT SCAN: see tomography

CATABOLIC: (an adjective) see catabolism

CATABOLISM: energy utilization from the degradative metabolism of chemical substances in the cells, mediated by enzymes.

CATABOLITE ACTIVATOR PROTEIN (CAP): see CAP

CATABOLITE REPRESSION: carbohydrate (glucose) represses the synthesis of enzymes involved in carbohydrate metabolism or other metabolic steps through decreasing the level of cyclic AMP. In some instances the glucose repression does not depend on cAMP because cAMP does not relieves its repression (e.g. pyrroline dehydrogenase, putrescine aminotransferase). Some bacteria which fail to synthesize sufficient amounts of cAMP (*Bacillus megaterium*) display intense glucose effect. Catabolite repression may be strongly modulated by the catabolite modulation protein factor (CMF). The catabolite repressor-cAMP system may also function as an activator of genes. Catabolite repression may switch to activation in case the gene carries overlapping promoters or when divergent dual promoters exist. (See also glucose effect, feedback controls, *Lac* operon, cAMP, DNA bending, cytosine repressor [CytR], FNR)

CATALASE: enzymes that mediate the reaction $2H_2O_2$ (hydrogen peroxide) $\rightarrow 2H_2O = O_2$. CAT is encoded in human chromosome 11p13. (See peroxide)

CATALYSIS: mediates chemical reactions without being used up in the process. Enzymes are biocatalysts. (See enzymes)

CATALYTIC ANTIBODY: contain catalytic and selective binding sites in one molecule and perform enzymatic reactions as enzyme mimics. Their catalytic efficiency is generally suboptimal.

CATALYTIC RNA: see ribozymes

CATALYTIC SITE: see active site

CATARACTS: are diseases of the eyes causing opacity of the lens. Cataracts are under X-chromosomal, autosomal dominant (1q21-q25, 2q99-q36, 16q22,1, 17q24) and autosomal recessive (galactosemia [17q21-q22], chondrodysplasia punctata, Zellweger syndrome [7q11, 12q11-13]) control. (See also Wilms tumor, Lowe disease, chondrodysplasia, eye diseases, microbodies)

CATCH: **c**ardiac defect, **a**bnormal face, **t**hymic hypoplasia, **c**left palate, **h**ypocalcemia are symptoms associated with a deletion in human chromosome 22q11.2. (See DiGeorge syndrome, velocardiofacial syndrome, cleft palate, hypocalcemia)

CATCRY SYNDROME: see cri du chat

CATECHOLAMINES: are neurotransmitters. (See neurotransmitters)

CATENANES: interlocked DNA circles: $\rightarrow$ ⟳⟲

CATENATED: attached like two links of a chain

CATENINS: are intracellular attachment proteins, connecting the inward-reaching carboxyl end of cadherins to actin filaments within the cell; β-catenin is involved in cell proliferation pathway and the development of adenomatous polyposis of the colon. (See cadherin, actin polyposis adenomatous intestinal, LEF)

CAT-EYE SYNDROME: the pupil appears vertical because a deformity of the iris, heart anomaly, imperforate anus, and various degrees of mental retardation occur. The anomaly is apparently caused by the presence of an extra copy of a very short metacentric chromosome 22 (partial trisomy or tetrasomy). The symptoms vary depending on the structure of this extra chromosome. (See mental retardation, trisomy, tetrasomy, duplication)

CATHEPSINS: intracellular proteolytic enzymes generally relegated to the lysosomes. (See antimicrobial peptides, lysosomes)

CATHODE RAYS: electromagnetic radiation emitted by the cathode toward an anode in a vacuum tube. After exposure of a metal target to this radiation, X-rays are generated. (See electromagnetic radiation, X-rays)

CATION: a positively charged atom or radical. (See ion, electrolyte)

CATION EXCHANGE: replacement of one positive ion by another on a negatively charged surface.
CATIONIC AMINO ACIDS: see Cat transporters
CATKIN: male inflorescence of some plants.
CATR1: a genetic element encoding 79 amino acids and a long untranslated region, expressed in malignant tumors. This element does not have homology with other oncogenes or tumor suppressor genes. It has been localized to human chromosome 7q31-q32. (See cancer, tumor oncogenes, tumor suppressor, AIG)
CATS (comparative anchor tagged sequences): see comparative maps, anchoring
CATTALO: the hybrid of buffalo (2n = 60) and cattle (2n = 60) with reduced male fertility.
CATTANACH TRANSLOCATION: involves the X-chromosome and autosome 7 of mouse. It includes several X-chromosomal fur color genes that may be subject to lyonization. (See lyonization, translocation)
CATTLE (*Bos taurus*): 2n = 60.
CAULESCENT: leaves are born separated by visible internodes on a stem. (See also rosette)
CAULIFLOWER MOSAIC VIRUS: (CaMV) is a (8 kbp) double-stranded DNA virus with limited potential use for genetic engineering. It infects cruciferous plants like turnip, broccoli, *Arabidopsis*, etc. The DNA of the virus shows three discontinuities ("gaps"), 1 in one of the strands and 2 in the other. The promoter of the 35S transcript of the virus drives high level constitutive expression of genes spliced to it and it is widely used; the promoter of the 19S peptide gene has not been proven nearly as good for biotechnology. (See agroinfection)
CAULOBACTER: is a dimorphic Gram-negative group of prokaryotes. The cells may have either a long flagellum or a thicker stalk of about 2/3 of the 2 µm length of the cell. The formation of these appendages have been extensively studied and they are regulated by the hierarchy of genes. (See Gram-negative/Gram positive)
CAULONEMA: in the gametophyte of some mosses a special type of cells are formed subapically and separated by oblique walls.
CAVITATION: after the formation of the 32-cell stage of the mammalian embryo a fluid secretion occurs during the blastocyte stage which initially accumulates between the cells then it is collected in the blastocoele. (See blastocyte, blastocoele)
C57BL: strains (B6 and B10) of black mouse commonly used for genetic studies.
CBAVD: see congenital bilateral aplasia of the vas deferens
CBF: CCAAT binding (factor) protein. (See CAAT)
CBL2 (oncogene): three cellular homologs of this viral oncogene is expressed in mammalian hematopoeitic (blood cell forming) systems. In humans, it was located to chromosome 11q23.3. Its translocations to chromosome 4 are associated with acute leukemia and B-cell lymphoma. The oncogene is present in the ecotropic Cas-Br-M virus and the form present in lymphomas is a recombinant between the virus and the cellular oncogene. The 100 kDa transforming fusion protein has sequence homology to the yeast transcription factor GCN4 and to *sli-1* regulator of vulval development in *Caenorhabditis*. Apparently, this family of genes modify receptor tyrosine-kinase mediated signal transduction. (See oncogenes, signal transduction, Jacobsen syndrome, Src)
CBP: is associated with CREB and mediates the induction of some promoters by cAMP. It interacts with p300, the nuclear hormone-receptors and the basic transcription machinery. In addition, it interacts with a range of transcriptional activators. The loss of CBP may be the basis of the Rubinstein-Taybi syndrome. (See cAMP, CREB, p300, E1A, transcriptional activators, histone acetyl transferase, Rubinstein syndrome)
CBP2: is a cellular binding protein required for splicing class I mitochondrial introns. The catalytic RNA domains must be in a folded form before CBP2 binds to them. Subsequently other proteins are attached to the 5' domain. In this ribozyme the catalytic component remains the RNA but the protein associated with increases the splicing rate by three orders of magnitude. (See intron, ribozyme, spliceosome, mtDNA)

CBP-II: similar to eIF-4F translation factor.

CBS (conserved sequence elements): are three short sequences in the mtDNA where the transition from DNA to RNA synthesis takes place. (See DNA replication mitochondria, mtDNA)

Cbs (chromosome breakage sequence): is a 15 bp nucleotide sequence at 50 to 200 sites in the micronuclear DNA of ciliates near the breakage sites where the genome rearrangement occurs in the macronuclear chromosomes. (See *Tetrahymena*)

CC CKR5: is a chemokine receptor. (See chemokines, RANTES)

CCAAT: see CAAT box

CCE (cell cycle element): has 11 bp that bind the histone nuclear factor (HiNF-M), required for the activation of Histone 4. (See HiNF, IFN, IRF-2)

CCG-1: hamster gene responsible for G1 phase cell cycle arrest.

CCL39: a hamster lung fibroblast cell line.

CCR: chemokine receptors; same as CKR. (See CC CKR5)

CCW: counterclockwise.

CD1: are antigen-presenting molecules—distantly related or unrelated to MHC proteins — for the T lymphocytes. CD1 can deliver to T cells endosomal lipoglycan antigens and the T cells are then stimulated to produce γ interferon and interleukin-4. CD1 is encoded in human chromosome 1q22-q23. (See T cell, cytokines, interferon, HLA)

CD2: is a T cell surface glycoprotein, mediating cell adhesion and the transduction of signals; it is encoded in human chromosome 1q22-q23.

CD3: immunoglobulins of γ (25 K), δ (20 K), ε (20 K) and ζ (16 K) molecular (weight) chains expressed on all T cells and assist in transducing signals when the major histocompatibility-antigen complex binds to the surface. They are encoded in human chromosome 11q23. (See immunoglobulin, HLA, signal transduction, T cell receptor)

CD4: cell surface proteins (M_r 55 K) with domain binding major histocompatibility (MHC II) molecules on helper T cells. CD4$^+$ T cells induce clonal proliferation of B cells along two paths. Foreign antigens switch on through the B cell antigen receptors the CD40L (CD40 ligand) and interleukin-4 (IL-4) promote the proliferation of B cells. CD95 (Fas) and Fas ligand (FasL) limit the access of mitogenic signals and mediate apoptosis in response to auto-antigens. For normal function both CD40 and Fas functions must be maintained. CD4 is encoded in human chromosome 11q13. (See HLA, T cells, B cells, CD8, CD40, LCK oncogen, autoantigen, apoptosis, interleukins)

CD5: is a transmembrane protein on the surface of T cells and some B cells. It may be a negative regulator of the T cell receptor (TCR) mediated signal transduction. Encoded at human chromosome 11q13. (See T cell receptor)

CD8: homo- or heterodimer proteins (M_r 70 K) on cytotoxic T cells, binding class I major histocompatibility molecules. CD8A and CD8B are encoded in human chromosome 2p12. CD8 is an active player in the T cell recognition complex. (See HLA, CD4, LCK oncogen)

CD11a/18: binds CD54 (ICAM) and mediates cell adhesion. CD11 is α integrin (human chromosome 16p12-p11), CD18 is β integrin (human chromosome 21q22.3). (See integrin)

CD19: most importantly the differentiation of B cells, it is a tyrosine kinase receptor. On mature B cells it is associated with CD81 and CD21. (See leukemia, BCP, B cell, genistein)

CD21: complement receptor 2. (See complement, TAPA-1)

CD22: is associated with immunoglobulin of B cell membranes. Tyrosine phosphorylated CD22 activates SHP protein tyrosine phosphatase and this down-regulates signaling through the immunoglobulin (Igμ). If CD22 is prevented from binding to μIg, the B cell may become 100 fold more receptive. (See lymphocytes, B cell, signal transduction)

CD25: is the interleukin-2 receptor α chain. (See interleukins)

CD28: is a homodimeric immunoglobulin (M_r 80 K, encoded in human chromosome 2q33-q34) present on the surface of helper T cells. The CD28/B7 system provides co-stimulation to T cell activation. (See T cell, B cell)

CD30: is a member of the tumor necrosis factor/nerve growth factor receptor family, including also TNF-R1, TNF-R2, CD40, CD27, Fas, etc. Their extracellular domain has cysteine-rich repeats. It interacts with TRAFs and indirectly with NF-κB. (See items separately)

CD34: is a protein in the bone marrow and is associated with hematopoietic function; it is encoded in human chromosome 1q32. (See hematopoiesis)

CD40: is a transmembrane protein expressed on the surface of B cells; their interaction with CD40 ligands is a requisite for the activation of B cells by helper T cells. This system appears critical for the development of humoral immunity. CD40 signaling may prevent Fas-induced apoptosis of B cells by cross-linking the immunoglobulin M complex. It induces B cell differentiation and Ig isotype switching and the expression of CD80. CD40 cytoplasmic tail interacts with CRAF1, a tumor necrosis receptor associated protein. CD40 is expressed also in dendritic cells, activated macrophages, epithelial cells and several tumors. In the absence of CD40, immunodeficiency arises and initiation of germinal centers suffers. (See immunodeficiency, B lymphocyte, T cells, CD40 ligand, germinal center, apoptosis, Fas, Igα, CRAF, CD80, B7, immunoglobulins, CRAF, dendritic cell, macrophage)

CD40 LIGAND (gp39/CD40L/TBAM): is a membrane-bound signaling molecule associated with CD40 transmembrane protein found on the lymphoid follicles of $CD4^+$ T lymphocytes. It may regulate adhesion and movement. (See CD40, T cell)

CD43: cell surface sialoglycoprotein of blood cells; it is deficient in the Wiskott-Aldrich im-munodeficiency. (See Wiskott-Aldrich syndrome)

CD44: is a family of cell surface receptors involved in adhesion and movement, The cytokine osteopontin (Eta-1) is one of its ligands, activating chemotaxis but not cell aggregation. Hyaluronate (a carbohydrate ligand) on the other hand affects growth. Thus metastasis of cancer cells may be controled by the state of CD44, encoded in human chromosome 11pter-p13. (See osteopontin)

CD45: is a protein tyrosine phosphatase, a transmembrane glycoprotein, activated by antigens on the surface of red blood cells and regulate T and B lymphocytes. Antibodies that react with this leukocyte antigen may prevent rejection of allografts. It is encoded in human chromosome 1q31. (See lymphocytes, allograft, antibody)

CD46: a complement regulatory protein. (See antibody, complement)

CD54: see ICAM

CD80: is a member of the interleukin family of proteins, same as B7, and BB-1. (See antigen-presenting cells, CD40, B7, interleukins)

CD95: see Fas

CD117: is the receptor for stem cell factors in thymic lymphocyte precursors. (See T cells)

CD ANTIGENS (cluster of differentiation antigens): are a large number of antigens of the leukocytes that can be classified and identified by monoclonal antibodies. (See CD proteins)

CD FRACTION (**c**onstant **d**osage): of the DNA in a chromosome including the functionally known genes. They are expected to be balanced with each other within a chromosome. These genes may, however, be amplified without serious detriment but changing the dosage of the syntenic genes (aneuploidy) may have very undesirable consequence. (See aneuploidy)

CD PROTEINS (**c**lusters of **d**ifferentiation): are accessory proteins on the surface of T cells. (See T cells, integrins, LFA, ICAM)

CDC (**c**ell **d**ivision **c**ycle): *cdc* genes encode cyclin-dependent kinase (CDK) proteins. (See cell cycle, CDK)

CDC2: protein when associated with mitotic cyclin proteins becomes the MPF (maturation promoting factor) of the oocytes, a serine/threonine protein kinase. Cdc2 controls the transition from the G_1 to the S phase and prevents the reinitiation of the cell cycle at G_2. Cdc2 also interacts with ORC. Phosphorylation of Cdc2 on threonine-14 (by Xenopus protein Myt1) and tyrosine-15 (by fission yeast protein Wee1) inhibits the activity of the Cdc2. For the activity of Cdc2, CAK (cyclin-dependent kinase) phosphorylates threonine-161. At the $G_2 \rightarrow M$ transition Cdk 25 protein dephosphorylates Thr^{14} and Tyr^{15} and consequently Cdc2 suppresses MPF during interphase. (See MPF, cell cycle, mitosis, ORC, Plx1)

CDC4: a protein required for the transition from the G1 to the S phase during the cell cycle in cooperation with CDC34, CDC53 and SKP1. CDC4 contains an F box which interacts with SKP1 and 8 WD-40 repeats common for proteins involved in protein-protein interactions. (See cell cycle, CDC34, WD-40, SKP1)

CDC6: a protein that appears in early S phase and late mitosis. Apparently it takes part in the formation of the pre-replicative complexes and ORC. (See cell cycle, ORC)

CDC7: a cell division cycle protein with histone (H1) specificity. (See cycle, histones)

CDC13: protein is a regulator of gene *cdc2* in cooperation with genes *wee1* and *cdc25*. (See *wee1*, *cdc25*, cell cycle)

CDC16: see CDC27

Cdc18: is a rate-limiting activator of replication and it interacts with ORC2. (See ORC2, cell cycle)

CDC25: the cyclin-dependent CDC25 phosphatases remove inhibitory phosphates from tyrosines and threonines. In human and mouse cells they represent a multigene family. In about a third of the breast cancers CDC25B is overexpressed. In *Saccharomyces cerevisiae CDC25* genes regulate RAS/cAMP pathway and their mutation causes defects in G_1 phase of the cell cycle. *Cdc25* is thus a proto-oncogene and when growth factors are exhausted it can induce apoptosis. (See cell cycle, RAS, budding yeast, MYC, Plx1)

CDC27: is a member of the tumor necrosis factor receptor family and restricts DNA replication to one round per cell cycle in cooperation with CDC16. (See TNF)

CDC28: (in *Saccharomyces cerevisiae*), *Cdc2* (in *S. pombe*) genes are responsible for the Start of mitosis in the cell divisional cycle. CDC28 protein is a kinase. (See cell cycle, CDK, CDC34)

CDC30: is a member of the tumor necrosis factor receptor family. (See TNF)

CDC34: is a ubiquitin-activating enzyme and it is required for the G1→S transition in the cell cycle. It facilitates the destruction of Cyclin 2 (CLN2) and Cyclin 3 (CLN3) and degrades CDK inhibitor, SIC1. CLN2 and CLN3 are phosphorylated by CDC28 before CDC34-dependent ubiquitination. (See cyclins, SIC1, ubiquitin)

CDC39: gene with a glutamine-rich repressor product affecting G1/S phase transition. (See cell cycle)

CDC40: is a member of the tumor necrosis factor receptor family. (See TNF)

CDC42: is RHO family GTPase protein; it affects in yeast the mating pheromone signaling and may be involved also in the Wiskott-Aldrich disease. (See RHO, mating type determination in yeast, GTPase, Wiskott-Aldrich disease)

CDC45/CDC46Mcm5: proteins are required for the initiation of chromosomal replication. (See MCM)

CDC53: see CDC4

CDK: cyclin-dependent kinases involved in cell division or apoptosis. These kinases do not operate without cyclin. CyclinD-CDK4, CDK6 are involved in G1 of the cell cycle, cyclinE-CDK2 drive G1 - S phase and DNA replication, cyclinA-CDK is active in S phase and cell division requires cyclinA-CDK1 and cyclinB in G2 and M phases. Full activity is achieved by phosphorylation by CAK. Some cells (myocytes) may be protected from apoptosis by $p21^{CIP1}$ and $p16^{INK4A}$ inhibitors of Cdk. Among the CDK inhibitors p15, p16, and p18 specifically inhibit CDK4 and CDK6 whereas p21, p27, p28 and p57 are inhibitors of a wide range of CDK cyclin complexes. CDK4 is tyrosine phosphorylated in G1 and dephosphorylation is required for the progression into S phase. UV irradiation may prevent dephosphorylation and cells are arrested in G1. If the CDK4 is not phosphorylated in G1, chromosomal breakage increases and a cell death may result. Cdk-activating kinase is a component of the Cak-complex and part of the carboxy-terminal of transcription factor TFIIH. The CDK proteins are similar in size (35-40K) and display >40% identity in the different organisms where the somewhat different enzymes are denoted differently. Cdc2 is the typical enzyme for fission yeast and in budding yeast the comparable protein is CDC28 whereas in human cells CDK1/CDK. The 300 amino residue catalytic subunit is inactive as a monomer or in the unphosphorylated form. In the inactive state

CDK continued

the substrate-binding site is blocked and the ATP-binding sites are not readily available for the phosphorylation required for activity. The binding of CDK to the ≈ 100 amino acid cyclin box is indispensable for function. CDK5 involved in neural development binds to protein p35 which apparently lacks a cyclin box domain. CDC2 may associate with a few different cyclins whereas CDC28 may be attached to nine different cyclins during the course of the cell cycle. The level and form of cyclins may vary during the cell cycle and their destruction is mediated by ubiquitins. In yeast, activation of CDC28 by the G1 cyclins stimulates then cyclins CLN1 and CLN2 and degrades mitotic cyclines (CLB) and after the G1 phase CLB levels may be elevated leading to the repression of G1 cyclins. When CDC28 is activated CLB decay begins. CDK-cyclin complex may be inhibited by phosphorylation near the amino end of CDC2 and CDK2 (at Thr 14 and Tyr 15). Phosphorylation at these two residues is followed by a rise of mitotic cyclins (CLB). At the end of the G2 phase Thr 14 and Tyr 15 are dephosphorylated by CDC25 phosphatase and CDC2 is activated. CDK may be inactivated also by protein CKI (a family of inhibitory proteins to the cyclin CDK complex by attaching to the complex). The inhibitory subunits include p21 and p27 and other proteins. Eventually the CKIs also decay and the cyclic events continue. (See cyclin, kinase, CAK, KIN28, PHO85, CDC28, cell cycle, apoptosis, UV, ubiquitin, CLB, p21, p27, p16, CKI)

cDNA: DNA complementary to mRNA (made through reverse transcription) and which does not normally contain introns. (See reverse transcriptases, transcription, mRNA)

cDNA LIBRARY: a collection of DNA sequences complementary to mRNA. (See mRNA, processed genes)

cDNA LIBRARY SCREENING: can be carried out by probing the DNA sequences either with special binding proteins for the purpose of high degree purification or they can be probed with any DNA or RNA in order to identify genes or gene products isolated from different organisms. (See also gel retardation assay, cloning, probe)

CDP: is a mammalian displacement protein at the CCAAT sequence of DNA and competes for binding of CP1, a CCAAT box binding protein. (See CP1, CAAT box)

CDP (cytidine diphosphate): a nucleotide involved the biosynthesis of phospholipid synthesis.

CDPK: Ca^{2+}-dependent protein kinases regulate signaling pathways. (See signal transduction)

CDR: **c**omplementarity **d**etermining **r**egion of the antibody's hypervariable region that binds the antigen. (See antibody)

CD-TAGGING METHOD: uses for insertion mutagenesis a DNA cassette that when inserted into an intron and transcribed and spliced, the mRNA will contain a special tag (guest tag). Upon translation the polypeptide will also carry a tag (guest peptide). The latter can be identified by monoclonal antibody, specially prepared for this epitope. The mRNA and the DNA sequences can be identified by PCR. Thus this method simultaneously labels DNA, RNA and the peptide; hence its name **c**entral **d**ogma tagging. (See insertional mutagenesis, monoclonal antibody, epitope, PCR)

CEBIDAE: families of new world monkeys. *Aotus trivirgatus trivirgatus* 2n = 54; *Aotus trivirgatus griseimembra* 2n = 52, 53, 54; *Ateles geoffroyi* 2n = 34; *Callicebus moloch* 2n = 46; *Callicebus torquatus* 2n = 20; *Cacajo* 2n = 46; *Cebus albifrons* 2n = 54; *Lagotrix ubericolor* 2n = 62; *Pithecia p. pithecia* 2n = 48; *Scaimii sciureus* 2n = 44. (See primates)

C/EBP: CAAT/Enhancer Binding Protein is transcription factor AP1, product of JUN and FOS oncogenes. (See AP, JUN, FOS)

CEBUS (capuchin monkey): see Cebidae

CEFOTAXIM (Claforan): a cephalosporin type general medical antibiotic with relatively mild toxicity to plant cells and it is thus widely used to free plant tissue from *Agrobacterium*, infected by for the purpose of genetic transformation. (See genetic transformation, *Agrobacterium*)

CEILING PRINCIPLE: a statistical procedure for the conservative estimation of the odds for the likelihood that DNA fingerprints would match. The odds against chance match is usually determined on the basis of the frequency of a particular genetic marker in a certain population (such

Ceiling principle continued
as Caucasians, Blacks, Hispanics, Orientals, etc.). The markers used for DNA fingerprint analysis are supposed to be of low frequency, below 10% or 5% but for the majority of subpopulations such information is not available yet. In such cases they take into account, say 0.1 as a maximal frequency (a ceiling). The chance for a particular person would have the same DNA marker as another individual in its group would be 0.1 x 0.1 = 0.01 and the probability that 8 markers would be identical by chance would be $(0.1)^8 = 1/100,000,000$ and if 0.05 is chosen as a ceiling it would be approximately 2.6×10^{-10}. The world's population of 6 billion is about 23% of 2.6×10^{10}. Some population geneticists disagree with the use of the rather arbitrary "ceilings" and advocate rather the theoretically more valid use of mean frequencies with the pertinent confidence intervals. Today, more information is being available on gene frequencies and therefore direct frequencies can be used for the majority of the genes involved. Abandoning the ceiling principle increases the accuracy of establishing individual liabilities and does not allow unwarranted advantage for the criminals. The recommended genetic markers for forensic comparison (in lieu of the ceiling principle) are variable number tandem repeats, VNTR (D2S44, 75 alleles and D1S80, 30 alleles), short tandem repeats, STRs (HUMTHO1, 8 alleles) simple sequence variations, SSV (DQA, 8 alleles, poly-marker [5 loci, 972 combinations]) and mtDNA D-loop with >95% diversity. (See DNA fingerprinting, confidence intervals, allelic frequency, Frye test)

CELERY (*Apium graveolens*): the stalks are used as food or the celeriac is a root vegetable; 2n = 2x = 22.

CELIAC DISEASE: in some individuals the intestinal enzymes are not digesting some water-insoluble proteins, such as the gliadin in wheat. This protein then causes inflammation of the intestinal lining and bloating of the abdomen. Its incidence in the general population is about 4/1000 and the genetic recurrence risk in the brothers and sisters of afflicted sibs is about 2 to 3%. The genes responsible for A2-gliadin synthesis were located to the long arm of chromosomes 6A, 6B and 6D of hexaploid wheat. Some gliadin genes are also in chromosome 1. There are substantial quantitative differences among the different chromosomes concerning the production of this protein. Some anthropologists suggested that the ancient Egyptians consumed high gliadin wheat varieties or some of the pharaohs were more susceptible to the disease (these families practiced high degree of inbreeding) because of the extended bellies of several royal mummies. This disease is under the control of the HLA genes in humans but one 6p locus, 30 cM from the telomere (thus outside HLA), has been identified for the predisposition. (See HLA, *Triticum*)

CELL: see cell structure, cell comparisons

CELL ADHESION MOLECULE: see CAM

CELL AUTONOMOUS: the product of the gene is limited to the cell expressesing it; it does not diffuse to other cells.

CELL BODY: the main part of the nerve cell containing the nucleus and excluding the axons and dendrites. (See neurogenesis)

CELL COMPARISONS: The cells of various organisms have common features yet differences exist that can be compared by the tabulation below.

CRITERIA	PROKARYOTES	PLANTS	ANIMALS
Cell wall	present	present	absent
Nucleus	non-enveloped nucleoid	enveloped	enveloped
Plastids	absent	present	absent
Mitochondria	absent	present	absent
Ribosomes	70S	80S	80S
(organellar)	not applicable	70S	70S
Endoplasmic Reticulum	absent	present	present
Centrioles	absent	absent	present

Cell comparisons continued

CRITERIA	PROKARYOTES	PLANTS	ANIMALS
Spindle Fibers	absent	present	present
Microtubules	absent	present	present
DNA Location	Nucleoid	Nucleus	Nucleus
	Plasmids	Mitochondria, Plastids	Mitochondria
Chromosomal	DNA	DNA	DNA
Composition	minimal protein	Protein	Protein
	or RNA	RNA	RNA
Division of the	Replication and	Replication	Replication
Genetic Material	Partition	Mitosis, Meiosis	Mitosis, Meiosis

CELL CORTEX: on the inner surface of the animal plasma membrane there is an actin-rich layer of the cytoplasm, mediating movement of the cell surface.

CELL CULTURE: generally the culture of isolated cells of higher eukaryotes is meant although growing bacteria or yeast is also cell culturing. (See also tissue culture)

CELL CYCLE: the phases of cell reproduction and growth are G1→ S→ G2→ M and cytokinesis. The duration of the cell cycle varies among different organisms, and it is influenced by several factors (temperature, nutrition, age, stage, etc.). In *Drosophila* embryos it may be completed within 8 minutes and in other early embryos the cycle may be completed within half an hour. In other cells the approximate numbers of hours required:

CELL TYPES	G1	S	G2	M	TOTAL	SOURCE
Onion roots	1.5	6.5	2.4	2.3	12.7	Van't Hoff
Mouse fibroblasts	9.1	9.9	2.2	0.7	22.0	John & Lewis
Xenopus early gastrula	3.5	4.5	8.0	0.5	16.5	John & Lewis
Xenopus late gastrula	2.0	2.0	3.5	0.5	8.0	John & Lewis
Saccharomyces	0.45	0.45	0.45	0.15	1.5	Fante
Schizosaccharomyces	0.26	0.24	1.85	0.16	2.5	Fante

The S and M phases are present in all dividing tissues, the G phases can be clearly distinguished only in cells where the divisions are less fast because of differentiation.

Before the cell enters the cell cycle the pre-replication complex (pre-CR) is assembled. This complex consists of the Origin Recognition Complex (ORC), the cell division cycle 6 protein (Cdc6p), Mcm (minichromosome maintenance proteins).

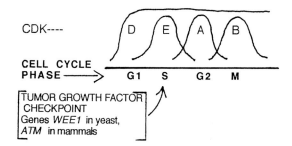

In the various phases of the cell cyclin-dependent kinases (CDKD, CDKE, CDKA and CDKB) reach their peak activity and at the same time other proteins (promoters and inhibitors) are also changing and balancing each other's effect on the progression of the cell cycle. The diagram on the right shows only the cyclin activities in relation to the mitotic phases and tumor formation checkpoint.

The cell cycle has been studied most frequently *in vitro* by the dividing egg of mammals manipulated by microinjection of various cellular components or by fusion of cells of different developmental stage. In plants, it is analyzed in somatic cells where division is triggered by the application of hormones and in yeast by accumulating conditional (temperature-sensitive) mutations and then by the introduction of the wild-type allele through transformation. Throughout the cell cycle (except mitosis) RNA and protein synthesis takes place. The G1 phase (Gap 1, named not very felicitously) is actually a phase of cell growth when a commitment is made for DNA replication. This point of commitment is called START by yeast cell biologists, and by ani-

Cell cycle continued

mal cell biologists RESTRICTION POINT. For START the cell requires the activity of a cyclin-dependent protein kinase composed of the catalytic subunit of protein Cdc28 and one of the three cyclins (Cln1, -2 or -3). After START, cyclin 5 B and cyclin 6 kinases are required. The latter two kinases are inhibited by protein Sic1 and the latter must be inactivated by proteolysis carried out by the ubiquitin-conjugated Cdc34. Some cells may stay very long at this preparatory stage and then it is called G_0 stage. In S phase DNA synthesis is completed. The cells cannot enter G2 (Gap 2, another cell growth phase) before the completion of S phase. Several genes were found to cooperate in *checkpoints* that provide clearance before the next phase can be entered. G2 is committed to the preparations of mitosis, a task that follows if during interphase all the nutritional requisites for carrying out mitosis build up. The two G phases and S phase combined are frequently named *interphase*, i.e., the phase between nuclear divisions (mitosis). In very rapidly dividing embryonic tissue G1 and G2, involved in cell growth, may be extremely brief or even absent and therefore the daughter cells become only half the size of the mother cells by each division. This is possible because the egg cell is generally very large at the time of fertilization. Some mature animal eggs may be thousands or tens of thousands times larger than an average body cell and it is loaded with nutritious material. In such cases the cell cycle may include only S (DNA synthetic) and M (mitosis, nuclear division) phases. The S phase can be identified by feeding labeled nucleotides which are then incorporated into the DNA. The fraction of the cells doing so, the *labeling index*, can be determined. From the fraction of cells undergoing mitosis in a tissue, the *mitotic index* is derived. The cell division is an extremely complex process, involving the cooperation of a very large fraction of all the genes and affecting the expression of many others. Cell division requires the presence of several *protein kinases* (phosphorylases), *phosphatases* and other activating proteins such as *cyclins*. The cyclins were named so because they are synthesized during the cycles of mitoses but not much is made during the intervening (interphase) periods. In fission yeast, gene *cdc2* *(cell division cycle)* is involved in the control through its 34 kDa phosphoprotein product (p34cdc2) that is a serine/threonine kinase, activated by cyclins and the complex becomes a cyclin-dependent protein kinase (Cdk). The cyclin-dependent protein kinase associated with cyclin B is also called MPF (*maturation protein factor*). In the phosphorylation a cyclic-AMP-dependent protein kinase (cAPK) also has a role. MPF dependent activation of cAMP-protein kinase A (PKA) and degradation of cyclin are required for the passage from mitosis into interphase. There are a large number (over 70) *cd* genes. The *cdc2* homolog in budding yeast is *cdc28*. Most of these genes have been well preserved during eukaryotic evolution and homologs are present in yeasts, in animals and higher plants. There are several different cyclins; the G1 cyclins are cyclin C and a number of different cyclin Ds and cyclin A when bound to the Cdk protein(s) control the onset of the S phase and the mitotic cyclin (cyclin B, encoded by *cdc13* in fission yeast) binds to Cdk before the onset of mitosis. Actually, similar genes and functions occur in all eukaryotes but they are named differently. The cyclin homolog genes in budding yeast are denoted by *CLN* and numbers. (Remember that in fission yeast the genes are symbolized with lower case letters, and + or - superscripts depending on whether wild type (+) or mutant genes (-) are represented. In budding yeast the wild type allele is capitalized and the mutant is in lower case. In both yeasts the genes are italicized whereas the protein symbols are not). Upon the binding of Cdc2 and cyclin the conformation of the former is altered allowing the phosphorylation of the threonine at position 161 of Cdc2 and the complex becomes a fully active promoter of mitosis. The Thr 161 phosphorylation is mediated by a Cdk (cyclin-dependent kinase), called also Cak (Cdk-activating kinase). cAPK is autophosphorylated at a Thr197 residue. Cyclin binding is also followed by dephosphorylation of the tyrosine 15 residue by the phosphatase, product of the *Wee1* gene. Before the cell could enter the M phase in the majority of organisms, phosphorylation of this protein decreases. In fission yeast, other proteins (encoded by genes *wee1, nim1, mik1*) exert negative control over the passage into the M phase. Eventually Cdc2 protein is dephosphorylated (gene *cdc25* regulates a phosphatase activity) and cyclin is degraded making the Cdc2 monomers

Cell cycle continued

available for another round of association with newly synthesized cyclins. The degradation of the cyclins is mediated by ubiquitin-dependent proteolytic cleavage pathways also controled by MPF. Exit from the M phase is regulated by genes that block entry into the M phase. The product of gene *suc1*, $p13^{Suc1}$ may be required (among other proteins) for the termination of mitosis. If S phase takes place but mitosis is not completed endopolyploidy may result, i.e. the chromosome number is multiplied (polyploidy). By this time the organellar material (plastids, mitochondria) are also readied for fission. There are some differences in the cell cycles of yeasts and other fungi from higher eukaryotes. In the former group the nuclear envelope is present throughout the cell cycle whereas in the latter it disappears from view from metaphase through telophase and is reformed after late telophase. In the fission yeast the cell division resembles that of higher eukaryotes by forming a cell plate in between the two daughter nuclei. In budding yeast one of the daughter nuclei moves into an extrusion of the cell, a bud, and eventually grows into a normal size cell. In fungi the spindle apparatus is located inside the nucleus rather than in the cytoplasm as in higher eukaryotes. These processes require also other regulatory mechanisms. It must be assured that DNA replication (S phase) produces complete sets of all essential genes and preferably *not* in multiple copies unless such amplification is required. In the regulation of cells cycle ubiquitin appears to have an important role. Three yeast proteins CD16, CD23 and CD27 (or their homologs in other organisms) mediate the attachment of ubiquitin to cyclin, resulting in its degradation at the end of mitosis. The initiation of a new cycle is hampered until an inhibitor (p27) is removed from the cyclin-CDK complex. This inhibitor is degraded by ubiquitin placed on the proteins of the spindle by CD16 before a new S cycle is entered. As the anaphase initiates the cell cycle becomes irreversible and as the cohesion between sister-chromatids breaks down the spindle fibers pull the new chromosomes (they were until now called chromatids) toward the poles. This process is mediated by the anaphase promoting protein complex (APC/NR/TSG24), also *cyclosome*. APC apparently manages that the chromosome would be properly arraigned in the metaphase plane and other proteins also would be functional.

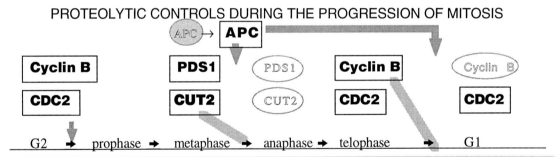

The heavy lines indicate activation the gray lines stand for blocking the transition. The symbols in outline indicate lack of activity and the gray circle stand for degradation. CDC2 = cell division cycle protein 2, PDS and CUT are non-cyclin proteins, APC = anaphase promoting complex.

The M phase cyclins are now lysed by the 26S proteosome after the telophase is completed. Other protein factors which are no longer needed are also ubiquitinated. APC operates through the proteolytic pathway shown above. Another proteolytic pathway during the cell cycle is mediated through protein CDC34 (see CDC34).

The $p34^{cdc2}$ and homologous proteins in cooperation with other factors (MPF) mediate the condensation of chromosomes through activation of H1 histone and control lamins to mediate the breakdown of the nuclear envelope (except in yeast). MPF controls tubulins and actins for the function of the mitotic spindle, etc. The cell cycle is intimately associated with signal transduction pathways, DNA topoisomerases, DNA polymerase, DNA ligase, RNA polymerases, tran-

Cell cycle continued

scription factors, etc. After all these events are successfully passed, the cell divides into two daughter cells and the cycle may be resumed depending also on environmental conditions. Aphidicolin blocks DNA synthesis, hydroxyurea interferes with the formation of DNA nucleotides and therefore DNA synthesis is halted. In case, along with hydroxyurea, caffeine is added an abortive DNA replication an mitosis results in cell death. Some of the mutations involving defective DNA repair inhibit cell cycle and cell divisions. The various growth factors are all directly or indirectly involved with the cell cycle. The p53, p16, p21 proteins are regulators of the cell cycle and some of their mutations no longer control the pace of orderly cell divisions and are thus instrumental in tumorigenesis. Breakdown of some cell cycle signals cause failures in attachment of the spindle fibers to the kinetochore resulting in nondisjunction and aneuploidy. An overview of events leading to overall fate of cells through the cell cycle can be represented:

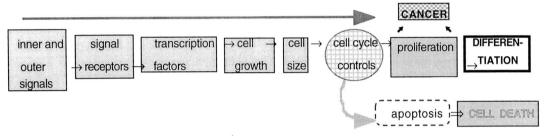

Meiosis has similar controls but the modulations are different. The primary oocyte is at an arrested G2 stage until a hormonal stimulation pushes the process to the first meiotic phase (reduction division) resulting in the formation of the first polar body in animals. In plants, this is the stage of the megaspore dyad. In mammalian oocyte the diplotene stage (called dictyotene) may then last from early embryonic development (about 3rd month in humans) to puberty. This is followed by the formation of the second polar body and the egg. In plants, from the four products of the "female" meiosis only one megaspore (most commonly the basal one) remains functional, and unlike in animals, undergoes 3 more divisions to form eventually the egg. Upon fertilization of the interphase egg, diploidy is restored, and cleavage divisions follow. (See mitosis, meiosis, cell division, CDC, CDK, endomitosis, CDC27 cyclin, CAK, $p15^{INK4B}$, $p16^{INK4B}$, ubiquitin, IFR, HiNF, licensing factor, SKP1, apoptosis, gametogenesis, amplification, polyteny, polyploidy growth factors, tumor suppressor gene, senescence, regulation of gene activity, asparagine synthetase, ataxia telangiectasia, proteasomes, apoptosis, cancer, differentiation, replication)

CELL DIVISION: in eukaryotes involving two steps, nuclear division (karyokinesis), followed by division of the rest of the cell (cytokinesis). A summary of the end results of the cell cycle, doubling of the chromatids as a result of the DNA replication during the S phase and changes in the C values of the nuclei. (See cell cycle, mitosis, cytokinesis)

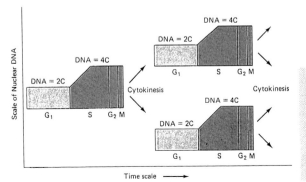

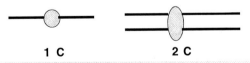

THE RELATION BETWEEN THE NUCLEAR CYCLES AND CELL DIVISIONS. DURING S PHASE THE AMOUNT OF DNA DOUBLES AND EACH 1 C CHROMOSOME WILL HAVE 2C AMOUNTS (2 CHROMATIDS), AND A DIPLOID CELL WILL HAVE A TOTAL OF 4 C AMOUNTS OF DNA BEFORE MITOSIS TAKES PLACE AND THEN CELL DIVISION (CYTOKINESIS) ENSUES.

CELL FATE: the program that determines the morphology and function of the undifferentiated cells in the embryo. (See fate map)

CELL FRACTIONATION: separation of the different subcellular organelles, generally by differential centrifugation, in variable density media. (See density gradient centrifugation, centrifuge)

CELL-FREE EXTRACT: prepared by grinding cells (tissues) in a buffer or other solutions and removal insoluble particulate material by filtration or centrifugation. Such extracts may be used for enzyme assays or for the purification of soluble cellular constituents.

CELL-FREE PROTEIN SYNTHESIS: *in vitro* protein synthesis in the presence of ribosomes, mRNA, tRNA, aminoacylating enzymes, amino acids and all the complex of translation factors and energy donor nucleotides. (See protein synthesis, rabbit reticulocyte assay, wheat germ assay)

CELL-FREE TRANSLATION: see cell-free protein synthesis

CELL FUSION: is the means to generate somatic cell hybrids. In contrast to hybridization by gametic fusion when the two nuclei fuse but usually only the maternal cytoplasm is preserved, in the fusion of somatic cells the entire content of the cells is combined in the somatic hybrid. For the fusion to take place polyethylene glycol, high concentration of calcium or higher pH medium have been used. (Inactivated Sendai virus addition also promotes the fusion of animal cells.) For the selective isolation of somatic cell hybrids both of the two types of cells generally carry recessive mutations that interfere with the survival of the cells on basic media. E.g. thymidine kinase deficient animal cells die because cannot synthesize thymidylic acid (DNA), hypoxanthine-guanine phosphoribosyl transferase deficient cells cannot make purine nucleotides (DNA). The fused cells being heterozygous at non-allelic loci are functional, however, and can selectively be isolated in large cell populations. Somatic cell fusion had important contributions to genetics involving human and other animals because it made possible to carry out allelism tests, facilitated the assignment of genes to chromosomes, identified the functional significance of chromosomal regions (in case of deletions). Somatic cells of very distantly related or entirely unrelated organisms, e.g. chicken and yeast, tobacco and human, human and rodent cells all can be fused although their further division may not usually be possible. (See also somatic hybridization, protoplast fusion, microfusion)

CELL GENETICS: actually nearly all genetics is cell genetics because geneticists generally think at the cellular level; in the narrow sense this term is applied to the genetic manipulations with isolated cells of multicellular organisms. (See cell fusion, somatic cell hybrids, fusion of somatic cells, mitotic crossing over, transformation)

CELL GROWTH: in any particular time $N = 2^g N$ where N is the final cell number, N_0 = the initial number of cells and g = the time required for a complete cell cycle This equation is valid as long there is no limitation on multiplication by nutrients, air, differentiation pattern, etc. In the absence of any limitation cell growth indicates cell doubling process. (See growth, growth retardation)

CELL HYBRIDIZATION: fusion of somatic cells. (See cell fusion, cell genetics)

CELL INTERACTION: the influence of cells on each other during differentiation and development. (See also contact inhibition, nurse cells, morphogenesis, maternal effect genes)

CELL JUNCTION: the area involved in the connection and communication between and among cells and extracellular matrices. (See extracellular matrix, morphogenesis, contact inhibition, transmembrane proteins)

CELL LETHAL: mutations may not be isolated or ascertained because of the cells involved cannot live.

CELL LINE: a (homogeneous) population of cells (of eukaryotes) that can be maintained in live (growing) conditions. (See also clone, clonal analysis)

CELL LINEAGES: the traces of the path of growth (multiplication) of the cells through several cell divisions, such as the study of the germline cells or the signs of visible mutations in the somatic tissues shown by the pattern of the sectors formed in chimeric organisms. (See clonal analysis, fate maps, nondisjuinction, phyllotaxis)

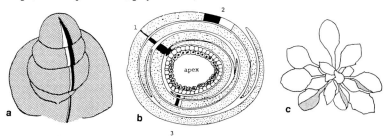

Mutation, deletion, nondisjunction during embryogenesis may potentially be identified during and after embryogenesis if the organism is heterozygous for a distinguishable somatic marker(s). (**a**) Organization of the apical dome of a wheat plant showing diagramatically the consequence of non-disjunction. (**b**) In the monocot apex the leaf initials wrap around the central axis and overlap each other; the humps at the leaves indicate the midribs. The oldest leaf initials are outside. Note that the outmost leaf and the one next below it have their midrib at opposite sides. Sector 1 is very narrow at the surface (old) leaf and it becomes wider in the (younger) ones below. Also the oldest sector is left from the midrib of the first leaf but it is at the right side of the one just below and again at the left side in the third. Sector (2) representing nondisjunction and twin sectors (black and white) occured only in one leaf because of a tangential event in a region of the embryonal apex. Sector (3) is a late occuring nondisjuntion indicated by the narrow twin sectors. (**c**) Somatic mutation in a single cell of the mature embryo of the dicot *Arabidopsis*. Three leaves displayed white sectors (sheded) because they differentiated from the same cell line of the apex. Non-sectorial leaves appeared in-between the mutant sectors because of phyllotaxis.

CELL-MEDIATED IMMUNITY: see immune system, T lymphocytes

CELL MEMBRANES: all cells are surrounded by membranes and inside the cells there are membrane enclosed bodies (nucleus, mitochondria, plastids, vacuoles, Golgi bodies, dictyosomes, lysosomes, peroxisomes). The endoplasmic reticulum, the mitochondrial crests, thylakoids are all membraneous structures. Cellular imports and exports pass through the membranes by active and passive mechanisms. The bulk of the plasma membranes consists of proteins and lipids (phospholipids, cholesterol, other sterols and glycolipids, triaglycerols, steryl esters, etc.). The composition varies in the different organisms and according to the particular membranes. The ultrastructure of the various membranes have common features and specificities. The basic structural element is the lipid bilayer of about 5 to 8 nm in thickness. In the double structure the polar head of the lipid face the aqueous environment and the tails inward are hydrophobic. Unsaturated lipids are concentrated in the inner layer of the structure. The outer surface of the membrane is also different from the inner surface that envelopes organelles or vesicles. The inner side carries on the surface charged groups, the outward surface may have a variety of peripheral proteins (glycoproteins) that determine the surface antigenicity of the cells. Some other proteins are integral parts of the membrane sunken in the fluid lipid bilayer. The fluidity is somewhat stabilized by the presence of sterols. The so called "seven membrane proteins" traverse the lipid bilayer and form within it a cluster of seven folds, the amino end at the outside and the carboxyl end inward. Other transbilayer polypeptides have hydrophilic domains both outside and inside, and outward are ports for communication (ion channels) with special proteins and lipids (transporters, permeases). Some of the peripheral proteins are attached to the membrane by electrostatic forces and H bonds. The proteins are regulators of membrane bound enzymes (e.g. phospholipase C) and mediate signal transduction. Membranes have a flexible structure to curl up into vesicles that ferry within the cell lipids and proteins in a protected manner. The membranes have the ability to fuse with another membrane at the delivery target. Exocytosis and endocytosis are the means of traffic. Fusion of the egg with the sperm, fusion between protoplasts of plant cells, somatic cell hybridization, protein synthesis within the endoplasmic reticulum, etc. are mediated by membrane functions. (See diagram below, cell structure)

Cell membranes continued

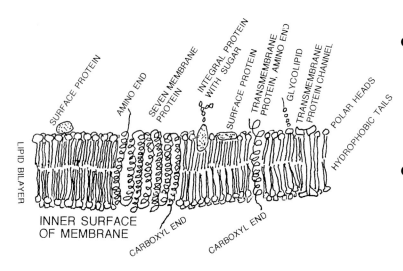

GENERALIZED DIAGRAM OF ←MEMBRANE STRUCTURE

- **CELL MEMORY**: the property of differentiated cells that they reproduce through numerous cell divisions similar specialized cells to what they have been committed. (See differentiation, morphogenesis)
- **CELL MIGRATION**: is common during animal development. Precursors of the blood cells, the germ cells, neurons, cells of the somites, etc. move through the embryo and are guided by cell surface receptor proteins and aided by the extracellular matrix of fibronectin whereas chondroitin sulfate proteoglycan interferes with the movement. The Kit protein in the membrane of the migrating cells and the ligand Steel factor produced by the cells which are contacted by the migrant, also control the movement. Gastrulation requires cell movement of mesodermal cells through the fibronectin-rich matrix in the blastocoele. (See proteoglycan, chondroitin sulfate, Steel factor, KIT oncogene, fibronectin, tenascin, integrin)
- **CELL NUMBERS**: in a small *Arabidopsis* plant about 5×10^6, in an *Arabidopsis* seed about 6 to 7 thousand, in a wheat embryo 10 days after fertilization about 40,000, and 150,000 at maturity (including the scutellum), at the surface of a maize endosperm about 1,400; in the human body about 6×10^{15} per 60 kg weight (ca. 1 billion per gram tissue) are found. The number of cells in a tissue depends on cell division, cell death and possibly on migration (in animals). The number of cells per particular structure is affected also by cell size. The local number of cells may be controlled by the tissue environment or extrinsic factors. The number of cells depends primarily on the cell divisional cycles, controled by a large number of hormones and other proteins; p27 in mouse appears to be a potent inhibitor of the growth of cell number and cell size by controlling cyclin and cyclin-dependent protein kinases. (See cell cycle).
- **CELL PLATE**: is the precursor of new cell wall in dividing plant cells.
- **CELL RECEPTORS**: see receptors, transmembrane proteins, signal transduction
- **CELL SAP**: the fluid, non-particulate cell content.
- **CELL SIZES**: vary a great deal depending on organisms and function. An *E. coli* cell is about 800 x 2,000 nm. Plant an animal cells generally have a diameter of 20 to 60 μm and their length is much more variable; some fibrous cells may be 20 cm long.
- **CELL SORTER**: specific cells can be labeled by cognate antibodies coupled with a fluorochrome (or any cell that incorporate labeled material). In a mixture where only one in a few thousands of cells carries this distinctive label, the latter ones can be separated using an electronic device. When a file of cells passes in front of a laser beam the fluorescent cells receive a different electric charge from the unstained ones. The high intensity electric field down in the path then separates the positively charged fluorescing cells from the negatively charged (unstained) ones. (See also labeling index, cell cycle, fluorochromes, antibody, laser)
- **CELL, STEM**: the stem cells of animals are capable of differentiation into various types of cells.
- **CELL STRUCTURE**: see animal and plant cells on page 165.

Idealized PLANT CELLS as viewed through

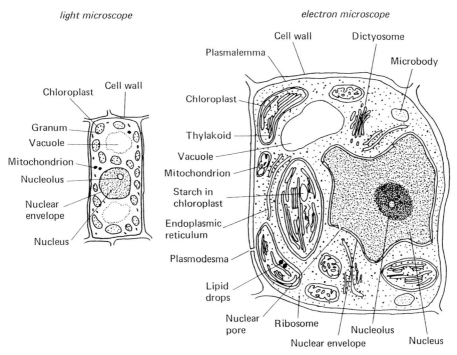

Idealized ANIMAL CELLS as viewed through

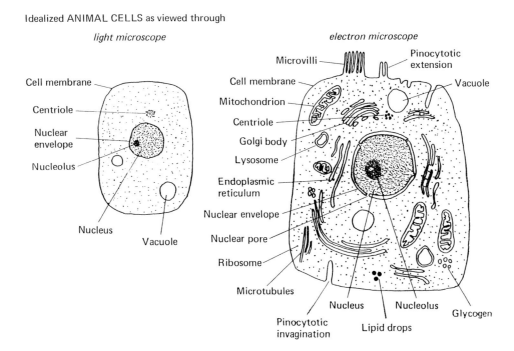

GENERALIZED STRUCTURE OF PLANT AND ANIMAL CELLS BY LIGHT AND ELECTRONMICROSCOPY

CELL THEORY: has been proposed in the 19th century. It states that the cells are the elementary units of life, and they can be produced only from pre-existing cells by mitosis (or meiosis). Abiogenetic reproduction of cells, assumed to exist in the 17th century has not been shown yet to exist in the present geological period of the earth. (See origin of life, spontaneous generation)

CELL TRANSFORMATION ASSAYS IN GENETIC TOXICOLOGY: generally hamster embryo cells or mouse prostate cells are exposed to chemicals and tumorigenicity is tested after introduction the cells into live animals (rodents). Also the activation of c-oncogenes (AKR, adenovirus) by chemicals is investigated. (See bioassay in genetic toxicology)

CELL WALL: exists in bacteria, fungi and plants; animal cells are surrounded only by membrane. The wall is made of polysaccharides, protein and lipids in bacteria. In plants it is mainly cellulose (polysaccharide) but lignin (a hard phenylanine and tyrosine polymer) suberin (a corky wax), cutin (a fatty acid polymer) also occur, and in fungi it may contain also chitin a linear polysaccharide differing from cellulose by a replacement at the C-2 OH group by acetylated amino group.

CELLULAR IMMUNITY: is mediated by the T cells. (See T cells)

CELLULASE: an enzyme digesting cellulose. Generally a collection of enzymes is used for removal of cell wall and gaining plant protoplasts, such as the *Onozuka R-10*. (See protoplast, macerozyme, cellulose)

CELLULOSE: a polysaccharide consisting of glucose subunits, it strengthens plant cell wall and forms the plant vascular system. (See cellulase)

CEN: the symbol of centromere DNA sequences. (See centromere)

CENOZOIC: geological period dating back to 75 million years ago when mammals and humans appeared. (See geological time periods)

CENP: is a protein that is diffusely located in the cytoplasm during G2 and prophase. During prometaphase it associates with the kinetochore until metaphase. At anaphase, it is located in the midzone of the spindle and degraded after cytokinesis. (See cell cycle, mitosis, spindle, kinetochore)

CENTIMORGAN: the unit of eukaryotic recombination; 1% meiotic recombination is one map unit (m.u.) = 1 centimorgan (cM). (See recombination frequency, mapping genetic, mapping function, CentiRay)

CentiRay (cR) : a chromosomal span within which a break can be induced with 1% probability by a specified dose of X-radiation. (See radiation hybrid)

CENTISOME: a quantitative unit of genomic sites.

CENTRAL DOGMA: the concept that the flow of genetic information follows the path DNA→RNA→Protein; it had to be slightly modified by the discovery of reverse transcriptases and ribozymes. (See ribozyme, reverse transcriptases)

CENTRAL NERVOUS SYSTEM: the brain and the spinal cord. (See brain human)

CENTRIC FUSION: fusion of two telocentric chromosomes into a biarmed single chromosome. (See also Robertsonian translocation, telocentric chromosome, acrocentric, misdivision)

CENTRIC SHIFT: changing the position of the centromere and thus the relative arm length of a chromosome by pericentric inversion or transposition. (See inversion, transposition, shift)

CENTRIFUGE: an instrument for sedimenting or separation of material by centrifugal force according to density. Low speed centrifuges generally spin the suspended material at less than 5,000-6,000 rpm (revolution per minute), table-top centrifuges usually reach a maximum speed of 10,000-12,000 rpm, high-speed centrifuges may reach about 20,000 rpm and usually are refrigerated so biological material would suffer minimal degradation. Ultracentrifuges, using refrigerated vacuum chambers may reach much higher speeds and may exceed 300,000 x g force and can separate even molecules. The conversion of revolution per minute into g force is generally done on the basis of tables provided by the manufacturers. The actual centrifugal force (F) = $\frac{\pi^2 S^2 M R}{900}$, where $\pi = 3.14159$, S = revolutions per minute, M = the mass in grams and R is the radius in centimeters. It is more convenient to express it as x g force where g = 980.665

Centrifuge continued

cm/sec/sec. Thus, e.g. when the maximal rpm is 20,000, the relative centrifugal force (RCF) may be 41,320 x g but this actually varies from the maximum at the bottom of the centrifuge tube to a lower value at the top, etc. (See ultracentrifuge, buoyant density centrifugation, density gradient centrifugation)

CENTRIOLE: hollow cylinders formed by nine microtubule triplets surrounded by a dense area in the centrosome; the two centrioles in each centrosome serve as the attachment point for the spindle fibers during nuclear divisions and along with the radial array of microtubules form the two *asters* in animal cells. (See also centrosome, centromere, spindle fibers)

CENTROMERE: region of the attachment of chromatids after chromosome replication and of spindle-fiber attachment at the kinetochore (localized within the centromere) during nuclear divisions. The centromere used to be called the primary constriction of the chromosomes because by the light microscopic techniques it frequently appears as a short slender region. (Secondary constrictions mark the juncture of the chromosomal satellites). In some organisms (*Juncaceae, Parascaris, Spirogyra, Scenedesmus*, etc.), the centromeres are "diffuse", i.e. their position spreads over the length of the chromosome (holocentric, polycentric chromosome). Under some conditions "neocentromeres" are visible, i.e. the spindle fibers may associate at more than one location within the chromosomes. The cloned centromeric region of yeast extends to a minimum of 150 bp. Although the base sequences in various centromeres of budding yeast are not identical, there is substantial homology. All contain a core (element II) of 83-84 base pairs that are 93 to 95% A=T. In addition, there are TCAC and TG identical stretches in flanking element I (11 bp) and on the other flank (element III, ~25 bp) there are GT and TG and T and CCGAA and TAAAA identical sequences that are separated by one to four different bases. The general structure of the yeast centromeres can be represented as:

CENTROMERE ↓

—|CONSERVED ELEMENT I|—A=T RICH CENTRAL ELEMENT II—|CONSERVED ELEMENT III|

The regions of the centromere elements (CDEI [8 bp], II [78-86 bp] and II [25 bp]I) are less susceptible to nuclease attack than the rest of the chromosomes and seems to indicate that they are associated with a different types of proteins. CDEIII is the site of the centromere binding factor 3 (CBF3), essential for segregation of the chromosomes. CDEI contains a nucleotide octamer (similar to the *octa* sequences present in the promoters of several genes where it binds a 39 kDa transcriptional activator protein (Cpf1p/CP1/CBF1). The arrangement of these 3 elements is shown above. A number of known proteins, with not entirely known functions, assure the formation of the centromere - kinetochore complex and the attachment of the spindle fiber.

The localized centromeres — in contrast to the holocentric ones — may be either *point* or *regional centromeres*. The point centromere contains about 250 bp, tightly packaged into a nuclease resistant structure and binds only a single microtubule. The point centromere may be arranged around a special nucleosome. The H2B and H4 histones may be altered and another highly variable protein, related to H-3 is present. The latter assists chromosome segregation. Such point centromeres are found in yeasts (*Saccharomyces cerevisiae, Schizosaccharomyces uvarum, Kluyveromyces lactii*).

The regional centromeres may consist of several kilobases and may be quite polymorphic. Actually, the *Schizosaccharomyces pombe* centromere is more similar to the mammalian centromeres than to that of the budding yeast. The *S. pombe* centromeres vary between 40 to 100 kb and the core of 4 to 7 kb may be surrounded by direct and inverted repeats. There is a larger array of proteins associated with the regional centromeres. In the majority of higher organisms the centromere is surrounded by heterochromatin and apparently not transcribed. The budding yeast centromeric DNA (CEN) in chromosome 3 (CEN3) appears to contain an open reading frame capable of coding for a peptide of 52 amino acids. There is no evidence for transcription to take place, however in the centromere. Actually this heterochromatic region suppresses the expression of open reading frames even if they are transposed or inserted in this

Centromere continued

region, indicating that "silencing" may be essential for the proper function of the centromere in chromosome disjunction. The centromeric Swi6p (yeast), HP1 (mammalian) proteins are repressors and *Drosophila* protein Pc (*polycomb*) is a negative regulator of the *Bithorax* (*BXC*) and the Antennapedia (*ANTX*) complexes. In the *Drosophila* centromeric region the 220 bp Bora Bora complex sequences, flanked either 5' or 3' by an about 200 bp "simple" sequence have been identified. The former is believed to contribute to the kinetochore formation, the latter may control sister chromatid association. The centromeres of mammals display considerable variations and larger number of proteins including centromere-binding proteins (CENP-A,-B,-C,-D, the kinesin-related MCAK and CENP-E, dynein, INCENPs [move to the micro-tubules in mitosis], etc). The centromeres may have functions also during interphase. In human cells the centromeres seem to be associated with the nucleolus during interphase. The fact that the centromeres are different yet functional when interchanged seems to indicate that the differences are not all functional specificities. The centromeric DNA in fission yeast is quite different from those of budding yeasts and shows more similarity to the centromeres of higher eukaryotes that are many times larger.

The availability of cloned centromeric DNAs permitted the construction of yeast artificial chromosomes (YACs). These YACs when properly constructed (see YACs and yeast centromeric vector) can be subjected to tetrad analysis. YACs are very useful tools for the physical mapping of larger DNAs. (See mitosis, meiosis, tetrad analysis, centromere mapping, YAC, kinetochore, aster, spindle fibers, microtubule, Roberts syndrome, octa, kinesin, dynein, α satellite, yeast centromeric vector [for nucleotide sequences])

CENTROMERE ACTIVATION: transposition of the centromere to a new position within the chromosome. (See also neocentromere, centromere, holocentric)

CENTROMERE INDEX: the length of the short arm divided by the length of the entire chromosome x 100. (See chromosome arm)

CENTROMERE MAPPING IN FUNGAL TETRADS: see tetrad analysis

CENTROMERE MAPPING IN HIGHER EUKARYOTES: in heterozygous autotetraploid (triplex: AAAa) or trisomic (duplex AAa) progenies the greater is the proportion of recessive individuals for a particular marker the further is that locus from the centromere because only crossing over between gene and centromere (maximal equational segregation) can produce double recessive gametes. Thus in very large population the relative distances can be estimated. More precise estimates of gene centromere distance can be obtained in allopolyploids by using telochromosomes that are usually not transmitted through the pollen. The experimental design may be as follows (only one pair of chromosomes is shown):

```
———a—Ⓞ—b—          X          ———A———-Ⓞ
———a—Ⓞ—b—                     ———a——Ⓞ—B—
       female                         male
```

The frequency of chromosomes of 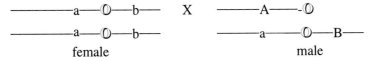 constitution indicates the frequency of recombination between locus A and the centromere (Ⓞ) in the male. The centromeres of the chromosomes of rice were mapped using RFLP markers in telo- and isotrisomics. The distance was calculated on the basis of linkage intensities of markers in the opposite sides of the centromere, infered by gene dosage. In mouse Robertsonian translocations can be exploited for centromere mapping. In most eukaryotes the centromeric region has repeated sequences and these can also be used for centromere mapping of RFLPs. (See allopolyploid, telochromosome, tetrad analysis, half-tetrad analysis, trisomic analysis, Robertsonian translocation, maximal equational segregation, RFLP)

CENTROMERIC VECTOR: see YAC

CENTROSOME: the center where spindle fibers (microtubules) originate (a microtubule organizing center) and develop from a pair of centrioles toward the centromeres during nuclear di-

Centrosome continued

visions in animals. The centrioles are surrounded by a pericentriolar material made of γ-tubulin and the fast-growing (plus) ends project into the cytoplasm whereas the slow-growing (minus) end of the microtubules are embedded into the γ-globulin ring. The division of the centrosome (also called centrosome cycle) is essential for the completion of the cell cycle in animals and it may be blocked by mutation in gene *cdc31*. The deficiency of the p53 tumor suppressor protein results in multiple centrosomes and unequal distribution of chromosomes. Plants do not have such distinct structures. Recently, mutations affecting the centrosomes have been isolated. The centrosome is paternally derived during fertilization in the majority of the animal species. Depending on the sperm donor, the size of the aster may vary. (See mitosis, centrioles, centromere, spindle, microtubule, kinesis, dynein, spindle pole body, aster, centrosomin)

CENTROSOMIN: one of the essential protein components of the centrosome. (See centrosome)

CEPH (Centre d'Étude du Polymorphism Humain): Paris, France-based research institute where human cells lines were collected and are maintained from four grandparents, two parents and their multiple children in order to map their genes and study their transmission.

CEPHALIC: involves the head or indicates the direction toward the head.

CEPHALOHEPATORENAL SYNDROME: see Zellweger syndrome

CEPHALOSPORIN TYPE ANTIBIOTICS: derived their name from *Cephalosporium acrimonium* and include a number of natural and semi-synthetic antibiotics that are resistant to the enzyme penicillinase. Their action involves interference with the cross-linking of the peptidoglycans of the bacterial cell wall. (See β-lactamase, antibiotics)

CERAMIDES: are structural units of sphingolipids, a fatty acid attached by -NH_2 linkage to a sphingosine molecule. (See sphingolipids, Farber's disease, CAP [ceramide activating protein])

CERBERUS: a factor expressed in the organizer of *Xenopus* embryos, causing the development of ectopic heads, duplicated hearts and livers. (It was named after the mythological three headed monster guarding the gate of the underworld). (See organizer, ectopic expression)

CERCOPIN: see antimicrobial peptides

CERCOPITHECIDAE (old world monkeys): *Allenopithecus nigroviridis* 2n = 48; *Cercocebus torquatus* 2n = 42; *Cercopithecus aethiops sabaceus* 2n = 60; *Cercopithecus ascanius* 2n = 66; *Cercopithecus cephus* 2n = 66; *Erythrocebus patas* 2n =54; *Macaca fascicularis* 2n = 42; *Macaca mulatta* 2n =42; *Miopithecus talapoin* 2n =54; *Papio* spp 2n = 42; *Presbytis melalophus* 2n = 44; *Presbytis senex* 2n = 44. (See primates)

CEREBELLUM: is a hind part of the brain supposed to be involved in the coordination of movements. Recent information indicates that the cerebellum acquires and discriminates among sensory informations rather than directly controlling movements. The Purkinje cells in the cortex are involved in the output and each Purkinje cell is innervated by the *mossy fiber system*, up to 200,000 parallel fibers originating from the deeper layers, and by a single *climbing fiber* originating from the oliva (a mass of cells) below the surface of the cortex. (See Purkinje cells, cerebrum, brain human)

CEREBRAL CHOLESTERINOSIS (CTX): human chromosome 2q33-ter recessive deficiency of sterol-27 hydroxylase, mitochondrial P-450 and other mitochondrial proteins as well as adrenodoxin reductase cause lipid (cholesterol) accumulation in tendons, brain, lung and other tissues. (See mitochondria, cholesterol, adrenodoxin)

CEREBRAL GIGANTISM: is a rare autosomal dominant (3p21 or 6p21) condition involving excessive bone growth and usually mental retardation. (See mental retardation)

CEREBRAL PALSY: see palsy

CEREBROSIDES: are sphingolipids, sugars linked to a ceramide. In the membranes of neural cells the sugar is generally galactose and in other cell membranes it is generally glucose. (See sphingolipids)

CEREBROTENDINOUS XANTHOMATOSIS: same as cerebral cholesterinosis.

CEREBRUM: the major part of the brain in two lobes, filling the upper part of the cranium. (See brain human, cerebellum)

CERENKOV RADIATION: when charged particles pass through an optically transparent material at speed exceeding that of light they cause emission of visible light. Cerenkov radiation is used in high energy nuclear physics for the detection of charged particles and measure their velocity.

CEROID LIPOFUSCINOSIS (NCL): is apparently autosomal recessive (assigned to three chromosomes) brown ceroid (wax-like) deposits in several internal organs, including the nervous system and causing spasms and mental retardation. Prevalence is about 1/12,500. The infantile subtype was located to chromosome 1p32 and it involves rapidly progressing mental deterioration due to a deficiency of palmitoylprotein thioesterase. The juvenile type was located to chromosome 16p. (See epilepsy, mental retardation, prevalence)

CERTATION: competition among elongating pollen tubes for fertilization of the egg. Genetically impaired pollentubes are at a disadvantage and this may cause a distortion of the phenotypic ratios because certain phenotypic classes may not appear or appear at a reduced frequency. (See gametophyte, gametophyte factor, male sterility, cytoplasmic male sterility, selection conditions, meiotic drive, segregation distortion)

CERULOPLASMIN: is a blue copper-transporting glycoprotein in the vertebrate blood. It is located in human chromosome 3q. (See Wilson disease)

CESIUM: alkali-metal element; its salts CsCl (MW 168.4) and Cs_2SO_4 (MW 361.9) are used as density gradient solutions for preparative and analytical ultracentrifugation, respectively. (See ultracentrifugation, buoyant density)

CETP (cholesterylester transfer protein): mediates the catabolism of HDL and the transfer of cholesterol to the liver and it may be thus anti-atherogenic. (See HDL, atherosclerosis)

CETYL TRIMETHYLAMMONIUM BROMIDE (CTAB): a detergent suitable for the precipitation of DNA. Generally used in a stock solution in 0.7 M NaCl.

CETYLPYRIDINIUM BROMIDE (CPB): a cationic detergent used for the precipitation of (radiolabeled) oligonucleotides.

CFTR (cystic fibrosis transmembrane conductance regulator): see cystic fibrosis

CG: dinucleotide is where the cytosine is most commonly methylated in vertebrates. The so called *maintenance methylase* enzyme acts on it when paired in a complementary manner in the DNA. The methylation may be inherited through DNA replication. (See methylation of DNA)

cGMP: cyclic guanosylmonophosphate, a second messenger. (See also cAMP, second messenger)

CHAETAE: bristles of insects, sensory organs of the peripheral nervous system. The large ones are called macrochaetae are mechanical sensory organs and the smaller ones of different types are microchaetae (a fraction of them are chemoreceptors). (See tormogen, trichome)

CHAGAS DISEASE: a non-hereditary infection by *Trypanosoma cruzi*. (See *Trypanosoma*)

CHAIN-SENSE PARADOX: expresses the problem of conformational switch from B to Z DNA. (See DNA types)

CHAIN TERMINATION: see transcription termination and nonsense codons. (See also initiator codon)

CHALAZA: the end of the plant seed where the funiculus unites with the ovule. The points in the bird eggs where the yolk is connected to the egg shell.

CHALCONES: phenylalanine is converted to *trans*-cinnamic acid and cinnamoyl-CoA that through a condensation reaction yield chalcone from what a series of other plant pigments (flavonones, flavones, flavonols, anthocyaniodins, etc.) are derived through single gene-controlled biochemical steps.

CHALONES: water-soluble glycoproteins which can inhibit mitosis.

CHAMBON'S RULE: splice points at the ends of intervening sequences are generally GT.....AG (except in tRNA genes and other minor classes of genes). (See introns, splicing)

CHAMELEON PROTEINS: contain a short amino acid sequence that may fold either as an α-helix or a β-sheet depending on its position. (See protein structure)

CHANCE: statistical probability or uncertainty. (See probability, likelihood)

CHANGE OF STATE: different levels of methylation of a genetic sequence. (See *Spm*)

CHANNEL: a path through what signals (molecules) can be transmitted.

CHANNELING: transfer of a common metabolite between two enzymes in a sequential and parallel function, e.g. a mutation may shut down the carbamyl phosphate pool leading to arginine synthesis but an overflow from the carbamyl phosphate precursor in the pyrimidine pathway may substitute for the defect and eliminates the dependence on exogenous arginine. This channeling goes both ways between the two metabolite pools. (See regulation of gene activity)

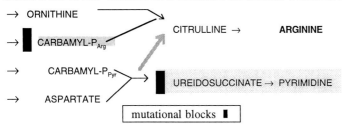

CHAOS: is a system extremely sensitive to minute perturbations, that are also called the 'butter-fly effect'. Although in common usage of the word chaotic conditions are meant to be uncontrolable and unpredictable because there are many degrees of freedom should be dealt with simultaneously. There are, however, high dimensional systems where the 'attractor' (the dynamics) is low dimensional. Among the infinite varieties in the system, one can be selected and stabilized by minute changes in one parameter. The task is to locate the critical points in a multitude of 'noise'. Various mathematical algorithms have been worked out to deal with the problems. The control of chaotic systems have great relevance to biology and genetics where a large number of genetic and environmental factors interact, such as the chaos in the heart, the neuronal information processing, epileptic seizures, etc. Chaos control (anticontrol) may then be applicable to harness the system.

CHAPERONE: protein mediating the conformational change or assembly of polypeptides (usually) without becoming a permanent part of the final product (e.g. the heatshock protein families Hsp70 and Hsp 60, rubisco, etc.) Their major function is probably the prevention of inappropriate conformation and interaction with incorrect ligands. The chaperones come in greatly different molecular sizes that is shown in their designation (in kDa). snRNAs may mediate the folding of rRNAs. The heatshock proteins may play a role also in signal transduction by modifying the folding of steroid hormone receptors. The chaperones may be classified into the Hsp70 (heatshock protein 70) and into the chaperonin families such as the Hsp60s. The chaperones recognize short extended polypeptides rich in hydrophobic residues that are released upon ATP hydrolysis. The chaperonins are large oligomeric ring complexes. The mitochondrial proteases (Lon, Afg3p, Rca1p) can carry out chaperone functions and assemble mitochondrial proteins. (See heat shock proteins, chaperonins, GroEL, flexer)

CHAPERONINS: ring proteins (chaperonin 60 and 10) mediate the assembly of 12 identical phage (λ, T4, T5)-encoded polypeptides and these serve as template for lining up the phage head precursor. Chaperonin 10 releases the phage proteins from chaperonin 60. Homologous proteins occur in the mitochondria and chloroplasts of eukaryotes. Their most essential function is ATP-dependent folding of proteins. The TF55/TCP1 chaperonins occur in the thermophilic archaea bacteria. The best known chaperonin proteins belong to two families the GroEL/GroES (include Hsp60 and rubisco binding proteins) and the TRiC. These form a porous cylinder of 14 subunits. The folding of proteins has numerous implications for normal function of proteins as well as for pathological conditions. (See chaperones, heatshock proteins, GreEL, TRiC, rubisco)

CHARACTER: in genetics it is a trait that may or may not be expressed through inheritance. Also any symbol used for conveying information, e.g., letters, numerals, punctuation marks, etc.

CHARACTER DISPLACEMENT: occurs when two species occupying similar but not identical habitat share a common area and in this shared zone each differ more from the other regarding a particular trait(s) than in the non-shared area.

CHARACTER MATRIX: is a device of classification of different groups regarding a trait as

Character matrix continued

have it (1) or not (0). On this basis then a *similarity index* can be obtained and the distinguished groups are called *operational taxonomic units* (OTU). If the differences are counted or measured a *distance matrix* is obtained. On the basis the similarities/differences branching *phenograms* or *dendrograms* can be constructed:

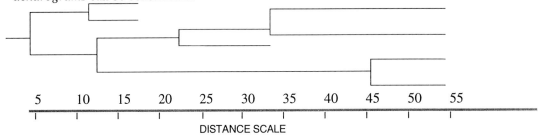

CHARCOT—MARIE—TOOTH DISEASE (hereditary motor and sensory neuropathy): it is known in multiple forms with autosomal recessive, dominant or even Xq13-linked types. The disease affects the nervous system and hearing; it is debilitating but not lethal. The average age of onset is in the early teens but is clearer manifested by the late 20s. The defect involves human chromosomes 1p21-23 or 17p12-p11.2 encoded myelin 22. Its presence can be detected by genetic screening in the affected families. In the basic defect connexin 32 encoded at Xq13.1 has also been implicated. This disease may be caused also by a duplication due to unequal crossing over in the peripheral myelin gene (chromosome 17) induced by a transposable element resembling *mariner* in *Drosophila*. An autosomal recessive form is determined by chromosome 8q and in 1p36 another locus has been implicated. (See neuropathy, connexins, myelin, genetic screening, MLE, MITE, *mariner*)

CHARGAFF's RULE: in double-stranded DNA nucleotides A = T and G ≡ C. (See hydrogen pairing, DNA). The recognition of this fact contributed significantly to the construction of the Watson - Crick model of DNA. (See Watson and Crick model)

CHARGED tRNA: the transfer RNA carries an amino acid. (See tRNA, protein synthesis, aminoacyl tRNA synthetase)

CHARGE-TO-ALANINE SCANNING MUTAGENESIS: see homologue scanning mutagenesis

CHAROMIDS: are specially constructed phage lambda-derived vectors. The particles must be of a minimum of 38 kb otherwise cannot be packaged into infectious particles. A conventional cosmid vector is 5 kb, therefore the minimal size of an insert would be 33 kb. When the fragment to be cloned is much smaller, charomids are used that carry repeating units of about 2 kb fragments of plasmid pBR322 in head-to-tail arrangement as space fillers. Thus, depending on the nature of charomids, they can be used for cloning DNAs from 2 to 45 kb length. In *recA*- bacteria (ED8767) these vectors are quite stable and can be used just like cosmids. (See cosmids, vectors, *Rec*)

CHARON VECTOR(S): modified phage λ plasmids (pronounce kharon; named after the ferryman of mythology who carried the dead souls through the infernal Styx river). The Charon vectors are primarily replacement vectors, i.e. they can place DNA into deleted parts of the stuffer region (between λ genes J and N, see diagram). There are a large number of Charon vectors with somewhat different features. *Charon 4* was used mainly to generate eukaryotic genomic libraries. In the replacement (stuffer) region there are E.coRI fragments 6.9 kb and 7.8 kb containing the β-galactosidase (*Lac Z*) and the biotin (*bio*) genes, respectively. Successful replacement is recognized by the inability of *lac*- bacteria carrying this vector to develop blue color on Xgal medium because the vector cannot provide the galactosidase function any more. Replacement of the *bio* gene results in biotin dependence in a similar way. The vector is also *Spi*-

Charon vectors continued

(Wild type lambda phages are unable to grow in bacteria containing prophage P2 and have the designation *Spi+* [sensitive to P2 interference]. The Spi selection is inoperable with Charon 4. Charon vectors that lost the *red* and *gam* functions [required for recombination] can grow in P2 lysogens and are called *Spi-* as long the bacteria have the *rec+* gene (mediating recombination) and the phage has its own *chi* element (a substrate for the *recBC* system of recombination) or the inserted foreign gene contains one (mammalian DNA has numerous *chi* elements). Also, *supE* and *SupF* (amber suppressors) must be present in the host to allow for selection of recombinant libraries by *in vivo* recombination because the vectors *A* and *B* genes carry amber mutations (chain termination codon UAG). Besides the EcoRI sites there is an XbaI site in the stuffer region that accepts an up to 6 kb insertion. *Charon 32* vector permits cloning DNA fragments at EcoRI (substitution up to 19 kb), HindIII (substitution up to 18 kb) and SacI sites (insertion up to 10 kb). In these cases *recA-* hosts suffice because the vector is *gam-*. When Charon 32 is used as a substitution vector for EcoRI-SalI, SalI-XhoI or EcoRI-BamHI fragments the host must be *recA+* because *gam* is lost. *Charon 34* and *Charon 35* are suitable for cloning fragments up to 21 kb at polycloning sites, using BamHi, EcoRI, HindIII, SacI, XbaI and SalI as well as their combinations. These vectors retain *gam* functions. *Charon 40* is useful for cloning fragments from 9.2 to 24.2 kb. In the stuffer region there are 16 restriction sites in opposite orientation near the ends. The poly-stuffer can be broken down into small fragments by the use of restriction enzyme NaeI (GCC↓GGC) and can be collected by precipitation with polyethylene glycol. The vector retains *gam*. (See vectors, lambda phage, *Spi+, supC, supD, SupE, supF, supG, supU,* restriction enzyme, Xgal, *Lac* operon)

CHASE METHOD: see haploids

CHASMOGAMY: pollination takes place after the flower opened. (See also cleistogamy)

CheA, CheB, CheR, CheW, CheY, CheZ: bacterial cytoplasmic proteins mediating the transduction signals of chemo-effectors through transducers to switch molecules. CheA is an autophosphorylating (histidine) kinase that phosphorylates also CheY and CheB (see below). CheY autophosphorylates also spontaneously in a few seconds and the process is accelerated by CheZ. CheA is central to information processing by the four transducers in cooperation with CheW. CheR, CheB: cytoplasmic bacterial proteins mediating the return of the chemotaxis excited cells to the normal state (adaptation). CheR is a methyltransferase, CheB is a methylesterase. (See effector, transducer proteins, autophosphorylation, esterases, transferases enzymes, chemotaxis, excitation, switch genetic)

	AB	Ab	aB	ab
AB	AABB	AABb	AaBB	AaBb
Ab	AAbB	AAbb	aAbB	Aabb
aB	aABB	aABb	aaBB	aaBb
ab	aAbB	aAbb	aabB	aabb

CHECKERBOARD: a representation of the genotypic array in the style of a checker board where the top line and the left-most column displays the gametic combinations. Note the 1:2:1 segregation in each of the four boxes. The left top individual is homozygous for both dominant alleles; the right bottom is homozygous for both recessives. At the diagonal from bottom left to top right all genotypes are heterozygous for the two genes. The diagonal from top left to bottom right are all homozygous at both genes. Also along the other diagonals the immediate neighbors or with one skipping are of identical constitution. This representation is called also Punnett square. (See also modified checkerboard, Mendelian segregation, Mendelian laws)

CHECKPOINT: critical phases in the progress of cell division where the cycle can be stopped and kept yet when conditions becomes appropriate progress may be resumed. The checkpoint indicates a step initiating a new direction in the progression; before the cell embarks on a new path, the preceding steps must be completed. Several proteins have now been identified in differ-

Checkpoint continued

ent organisms to mediate checkpoints. The checking is generally mediated by different cyclin-dependent kinases (CDKs). In yeasts there is only a single CDK (Cdk) whereas in animal cells different ones specialize for each checkpoint. The CDKs may be activated/deactivated by phosphatase/kinase action and in addition binding proteins, cyclin kinase inhibitors (CKI) and other modifiers may participate. These proteins are also subject to proteolytic destruction. In multicellular higher organisms apoptosis is also involved, and that does not stop the cell cycle at a stage but eliminates the cell with a defect. (See cell cycle, replication at the S phage in eukaryotes, $p56^{chk1}$, PIK, FK506, PIK, FK506.)

CHÉDIAK-HIGASHI SYNDROME (CHS): an autosomal recessive defect of the cytotoxic T cells in human chromosome 1q43. The afflicted individuals display reduced pigmentation of the hair and eyes, avoidance of light, reduction in the number of neutrophilic lymphocytes (neutropenia), high susceptibility to infections and lymphoma. In the heterozygotes the lymphocytes appear abnormally granular. Similar diseases occur in many mammals. Molecular investigations reveal a defect in a protein with carboxy-terminal prenylation and multiple potential phosphorylation sites (LYST). It may be involved as a relay integrating cellular signal response coupling. Its mouse homolog is the *beige* locus. (See albinism, lymphocytes, neutrophil, lymphoma, Hermansky-Pudlak syndrome)

CHEF: **c**ontour-clamped **h**omogeneous **e**lectric **f**ield that alternates between two orientations. The electric field is generated by multiple electrodes along a polygonal contour. The method applies the principles of electrostatics (statical electricity) to gel electrophoresis of very large molecules such as the entire DNA of small chromosomes. (See pulsed gel electrophoresis)

CHELATION: holding of a hydrogen or metal atom between two atoms of a single molecule. Hemin and chlorophyll are chelated. Chelation may improve solubility of metals and used also for relieving metal poisoning.

CHEMICAL MUTAGENS: include an extremely large number and wide variety of different chemical groups. Their effectiveness varies within a great range. About 80% of the chemical mutagens are also carcinogens. Some of these agents pose health hazards as industrial, agricultural and medical chemicals. A much smaller fraction is used for the experimental induction of mutation for research purposes. A not comprehensive classification includes (i) DNA base and nucleoside analogs (5-bromouracil or 5-bromodeoxyuridine (BuDR), 2-aminopurine (2-AP). These compounds are of relatively very low efficiency and can be utilized only under highly selective conditions in microbial populations. They act primarily either through incorporation at the wrong place in the DNA or by replacing their normal counterparts followed by mispairing (tautomeric shift) and in both cases causing base substitution. (ii) Chemical modifiers of nucleic acid bases such as nitrous acid (HNO_2) that oxidatively deaminates cytosine into uracil, adenine into hypoxanthine (resulting actually in A→G transitions) and guanine into xanthine (causing lethal effects because of interference with replication). Nitrous acid is not an effective mutagen for higher eukaryotic cells because of the high protein content of the chromosomes and nitrous acid has destructive effect also on proteins but it was used very advantageously for tobacco mosaic virus and other viruses. Hydroxylamine targets primarily cytosine and thus changes a G≡C pair into A=T. (iii) Alkylating agents of a diverse group such as sulfur- and nitrogen mustards, epoxides, ethyleneimins, unsaturated lactones (aflatoxin), alkyl and alkene sulfonates are powerful mutagens for different organisms. The nitrogen mustards induce primarily chromosome breakage like X-rays and frequently referred to as radiomimetic agents. The mustards are no longer used as laboratory mutagens because of the relatively low efficiency

Chemical mutagens continued

compared to their lethal effects and because the risks involved in routine handling. Among these, the most commonly used alkylating mutagens are methyl and ethyl methanesulfonate ($CH_3SO_2OCH_2CH_3$) that are highly effective in practically all organisms. They alkylate the guanine at the 7 position but the major mutagenic action results from the alkylation of the O^6 of guanine. To a lesser extent, they alkylate mutagenically other bases too. Alkylation may also break the linkage between the ribose and the N^9 of the purines resulting in depurination and breakage of the DNA strand. (iv) N-nitroso compounds such as nitrosamines, nitrosoureas, methyl-nitro-nitrosoguanidine (MNNG, $C_2H_5N_5O_3$), etc. are very potent mutagens. The latter one is inducing predominantly point mutations at pH 6 but it must be handled in dark because it rapidly decays in light. (v) A variety of compounds that produce free radicals such as hydrazines and hydrazides, hydrogen peroxide or organic peroxides, aldehydes and phenols are mutagenic but usually not employed for the induction of mutation, except perhaps maleic hydrazide that induces high frequency chromosomal breakage. (vi) Acridine dyes (proflavin, acriflavin, acridine orange, etc.) are intercalating agents and induce frameshift mutations at neutral pH. (See also physical mutagens, radiomimetic, alkylation, biological mutagenic agents, environmental mutagens, depurination, chromosome breakage, frameshift, tautomeric shift, laboratory safety, chemicals hazardous)

CHEMICAL MUTATION: alters the active site of an enzyme by modifying the (tertiary) structure of the protein.

CHEMICALS, HAZARDOUS: The total number of chemicals identified exceeds 6 million. The various industries use over 60,000 chemical compounds and an estimated 700 new chemicals are introduced annually for various uses. Since the biological effects (mutagenicity, carcinogenicity) of only a relatively small fraction is known with certainty, the majority of chemicals should be regarded with appropriate caution. (See environmental mutagens, chemical mutagens, carcinogens, laboratory safety)

CHEMIOSMOTIC COUPLING: uses a pH gradient through a membrane to drive energy requiring processes.

CHEMOAUTOTROPH: organism obtains energy from inorganic chemical reactions.

CHEMO-EFFECTORS: eliciting bacterial response by bacterial signal transducers in chemotaxis.

CHEMOHETEROTROPH: organism obtains energy from breakdown of organic molecules.

CHEMOKINES: chemicals (proteins) involved in defense systems or many activations by a chemical agents. Chemokines are participants in specific inflammatory responses and some of them are identical to specific lymphokines. The chemokine α is represented by IL-8 and others encoded in close vicinity to each other in human chromosome 4q12-q21. On the surface of human neutrophils 20,000 high affinity receptors have been found. They are encoded in human chromosome 2q34-q35. The β chemokine family includes MIP and the others are homologous to 28-73%, and are encoded in close linkage in human chromosome 17q11-q21. The MIP/RANTES receptor is encoded in 3q21. (See lymphokines, RANTES, MIP, fusin)

CHEMORECEPTOR: a gene product, activated by chemical signal(s).

CHEMOSMOSIS: chemical reaction across membrane.

CHEMOSTAT: is an apparatus for culturing bacteria at steady level of nutrients, aeration, temperature, etc. so continuous cell divisions are maintained.

CHEMOTAXIS: movement toward chemical attractants and away from chemical repellents. Macrophages, neutrophils, eosinophils and lymphocytes are attracted to a wide variety of substances causing inflammation. (See blood cells)

CHEMOTAXONOMY: study of evolutionary relatedness on the basis of chemical compounds.

CHEMOTHERAPY: curing a disease by chemical medication or fighting cancerous proliferation by anti-mitotic (cytostatic) chemicals. (See cancer, cancer therapy, cytostatic, multiple drug resistance)

CHENEY SYNDROME (acroosteolysis): is an autosomal dominant bone disease with similarities to pycnodysostosis but here rather than hardening of some of the bones (osteosclerosis), bone loss (osteoporosis) was accompanied with early loss of teeth, laxity of the joints, etc. In this syndrome the stature of the patients was not necessarily short. (See pycnodysostosis)

CHERNOBYL: see atomic radiation

CHERRY (*Prunus cerasus*): x = 7 but a wide variety of diploid and polyploid forms are known.

CHESTNUT (*Castanea* spp): monoecious tree, 2n = 2x = 24.

CHESTNUT COLOR in horses is due to homozygosity for the *d* allele; actually this color is expressed when the genetic constitution of the animals is *AAbbCCdd*.

CHETAH: *Acionyx jubatus*, 2n = 38.

chi ELEMENTS (χ): are crossingover hot spots-inciting DNA sequences in both prokaryotes and eukaryotes. The prokaryotic chi elements have the consensus 5'-GCTGGTGG-3' and nicking by the RecBCD complex takes place 4 to 6 bases downstream during recombination. In phage DNA chi sequences are required for the function of the bacterial genes *RecBC* coding for exonuclease V. The product of phage gene *gam* inhibits this enzyme. In *E. coli* there are about 500 to 600 chi elements. Chi promotes recombination in a region within 10 kb from its location. It may be activated by double-strand breaks within a range several kb downstream. Chi probably facilitates the production and availability of free 3' end for recombination. The hepatitis B virus encapsidation signal carries a 61 bp sequence (called 15AB) is a hot spot for recombination, and a cellular protein binding to this sequence appears to be a recombinogenic protein. The same nucleotide sequence (5'-CCAAG**CTGTG**CCTT**GGGTGGC**-3') has been identified with approximately 80% homologies also in the rat, mouse and human genomes. Note the pentanucleotides in bold, they are present also in the prokaryotic chi elements as shown above. It appears that this element is responsible for some of the chromosome rearrangements observed in hepatocarcinomas. (See crossing over, recombination, *rec*, hepatoma)

chi FORMS: recombinational intermediates representing consummated chiasmata. (See chiasma)

CHI SQUARE (χ^2): is a statistical device for testing the goodness of fit to a particular (null) hypothesis. $\chi^2 = \Sigma \dfrac{[\text{observed number - expected number}]^2}{\text{expected number}}$

where Σ stands for sum. When the degree of freedom is 1 the Yates correction may be justified, and the formula becomes:

$$\chi^2 = \Sigma \dfrac{[|\text{observed - expected}| - 0.5]^2}{\text{expected}}$$

The Yates correction may be applied for other degrees of freedom in case the size of any particular class is 5 or less. It may be a better practice, however, to avoid using this correction factor and keeping in mind that our chi square figure is conservative. Although the χ^2 is very useful it must be remembered that it has a general weakness because it was derived from the principles of normal distribution but it is actually applied for discrete classes. An alternative to the above χ^2 statistics the likelihood ratio criterion χ^2_L which is easier to compute in some cases and may give more realistic estimate: $\chi^2_L = 2\Sigma \text{ observed} \times ln\left(\dfrac{\text{observed}}{\text{expected}}\right)$; the degree of freedom is calculated the same way as in other cases.

A slightly different type of formula is used in tetrad analysis to determine whether the frequency of parental ditype (PD) tetrads really exceeds that of the non-parental ditypes (NPD). In linkage PD should exceed that of NPD.

$$\chi^2 = \dfrac{[PD - NPD]^2}{PD + NPD}$$

(See tetrad analysis, chi square, lod score)

When the population size exceeds 30, use $\sqrt{2\chi^2} - \sqrt{2n-1}$ as a normal deviate.

(See chi square table, homogeneity test, association test, degrees of freedom)

CHI SQUARE TABLE: displays the probability of a greater χ^2. Determine the value of the χ^2 by using the chi square formulas. Locate the nearest higher value in the body of the table (on next page) on the appropriate line of degrees of freedom (df). On the top line at the intersecting col-

Chi square table continued
umn you find the corresponding level of probability (**P**). For a particular degree of freedom, the smaller χ^2 value indicates a better fit to the theoretical expectation (null hypothesis). By statistical convention when P> 0.05, the fit is not questionable, when P is between 0.05 and 0.01 the fit is more or less in doubt. When P is below 0.01 the null hypothesis is no longer tenable. (In the majority of statistical books more extensive χ^2 tables can be found)

P→ df↓	0.99	0.90	0.75	0.50	0.25	0.10	0.05	0.01	0.005
1	0.00	0.02	0.10	0.45	1.32	2.71	3.84	6.64	7.90
2	0.02	0.21	0.58	1.39	2.77	4.60	5.99	9.92	10.59
3	0.11	0.58	1.21	2.37	4.11	6.25	7.82	11.32	12.82
4	0.30	1.06	1.92	3.36	5.39	7.78	9.49	13.28	14.82
5	0.55	1.61	2.67	4.35	6.63	9.24	11.07	15.09	16.76
6	0.87	2.20	3.45	5.35	7.84	10.65	12.60	16.81	18.55
7	1.24	2.83	4.25	6.35	9.04	12.02	14.07	18.47	20.27
8	1.64	3.49	5.07	7.34	10.22	13.36	15.51	20.08	21.97

CHIASMA (plural CHIASMATA): chromatid overlap in prophase (appearing through the light microscope like the Greek letter chi [χ], resulting in genetic exchange between bivalents.

Chiasmata in a male grasshopper chromosome. (Courtesy of Dr. B. John)

If the number of chiasmata truly represents the points of genetic exchange of chromosomes, from the cytological observation of chiasma frequency the length of the genetic map could be infered because each crossing over corresponds to 50% recombination. Thus the number of chiasmata multiplied by 50 should be equal to the sum of map distances. For example C.D. Darlington observed 3 chiasmata in chromosome 3 of maize, thus 3 x 50 = 150; actually according to the modern map the length of this chromosome is about 167 m.u. According to B. Lewin the number of chiasmata per meiocyte in *Drosophila melanogaster* was 6.6 and 6.6 x 50 = 300 and the new cytogenetic map appears to be 297 m.u. Recent analyses of plant recombination and chiasma data find that the frequency of recombination is higher than the chiasma frequency. This revelation may or may not affect the general validity of the correspondence suggested earlier because of the cytological difficulties obtaining very precise estimates on chiasmata. Also, the recombination data may have some inherent errors even if mapping functions are used. It is well known that interference varies along the length of the chromosome and according to species, etc., and no general mapping function can fully appreciate that fact. Chiasmata take place as the bivalents represent 4 strands (chromatids).

In about 6-10% of the X chromosomes there are no chiasmata, whereas 60-65% of the bivalents display one, and 30-35% shows two but higher number of chiasmata are very rare. In the majority of organisms chiasmata are rare at the tip and at the centromeric regions. Chiasmata are regulated also by special genes. In *Saccharomyces* the Rad50 protein (an ATP dependent DNA-binding protein, localized at interstitial sites) seems to be involved in the development of the axial structure of chromosomes, pairing and recombination. (See meiosis, crossing over, recombination frequency, mapping genetic, interference, chromatid interference, isochores, achiasmate, desynapsis, sister chromatid cohesion, mapping function, association point, recombination nodule)

CHIASMA INTERFERENCE: see chromatid interference, interference
CHIASMA TERMINALIZATION: see terminalization of chromosomes

CHIASMATA (plural of chiasma): see chiasma

CHICAGO CLSSIFICATION: of human chromosomes was based on banding and morphology and it has been refined since 1966. (See human chromosomes, Denver classification, Paris classification)

CHICK PEA (*Cicer arietinum*): grain legumes representing the pulses. The chromosome number is generally $2n = 2x = 16$.

CHICKEN: *Gallus domesticus* $2n$ = ca. 78.

CHIMERA: a mytological monster that had a serpent's tail, goat's body and lion's head and vomited flames through her mouth.

CHIMERA: mixture of genetically different tissues within an individual or other structures of two or more different fused elements; frequently it displays visible sectoring. (See also periclinal chimera)

CHIMERIC CLONES: two segments of DNA derived from non-contiguous regions of the chromosomes are joined together. This may be caused by a high level of homologous recombination. These joined areas are cloned together (co-cloning). Co-cloning is disadvantageous for the construction of physical maps. (See chimeric DNA, physical map)

CHIMERIC DNA: see recombinant DNA

CHIMERIC PLASMIDS: contain genetic sequences from other genomes along with their own DNA. (See plasmid, vector)

CHIMERIC PROTEINS: may be used to gain additional function(s) by the same molecule. That can be constructed by adding new domains. The desired domain may be provided with "sticky ends" through PCR procedures. The sticky ends are supposed to pair with homologous portions of a target in a single-strand vector and after replication one of the new strands of the double-stranded DNA will contain the sequences coding for a polypeptide corresponding to the donor molecule. Chimeric proteins may be obtained also by the use of translational fusion vectors. (See primer extension, translational gene fusion vector)

CHIMERIC YAC: is produced when more than one piece of DNA is ligated to the same vector arm. It is generally undesirable for chromosome walking toward a particular locus because it may direct toward different direction(s) than the region of interest. (See YAC, chromosome walking)

CHIMPANZEE: see Pongidae, primates

CHINESE RESTAURANT SYNDROME: is an adverse reaction (headache, stiffness of the neck and back, nausea, etc.) to the flavor enhancer monosodium glutamate in certain foods such as soy sauce, hot dog, etc. It may be controled by a recessive gene. (See monosodium glutamate)

CHIP (**ch**annel-forming **i**ntegral **p**rotein): a member of water transporters to various types of cells and tissues in cellular organisms. In humans it is encoded by the aquaporin-1 gene in chromosome 7p14.

CHIP: is an electronic circuit within a single piece of semiconducting material, e.g. silicon. (See semiconductor)

CHIPMUNK: *Eutamias amoenus* $2n = 38$; *Eutamias minimus* $2n = 38$; *Funambulus palmarum* $2n = 46$; *Funambulus pennanti* $2n = 54$; *Glaucomys volans* $2n = 48$.

CHIRAL COMPOUND: see enantiomorph

CHIRALITY: the dissimetry of a molecule, i.e its plane mirror image cannot be brought to coincide with itself.

CHIRONOMUS SPECIES ($2n = 8$): favorable organism (flies) for cytology because of the very conspicuous differences in the banding of salivary gland chromosome among species and within species, reflecting the activity of genes by the pattern of puff formation. (See giant chromosomes, puff, *Rhynchosciara*, *Sciara*)

CHITIN: poly-N-acetylglucosamine is part of the exoskeleton of insects, of other lower animals and of the cell wall of fungi. The chitinase enzyme plays a role in the protection of plants against fungal and insect damage. (See host—pathogen relations, exoskeleton)

Chlamydomonas eugametos (n = 7): unicellular green alga.

Chlamydomonas reinhardtii: unicellular green alga showing both haploid and diploid stages. The fusion of (+) and (-) mating type gametes (that are not identifiable by morphology) gives rise to diploid cells that immediately undergo meiosis and release haploid zoospores. This progeny can be subjected to unordered tetrad analysis. On solid media they produce only rudimentary flagellae but in liquid culture they are flagellated. The zoospores may readily divide by mitosis but they sexually differentiate under nitrogen starvation and gametic fusion follows. Their basic chromosome number, n = 8 (1×10^6 bp). They contain one large chloroplast (with about 196 kbp DNA). Chloroplast genes normally show uniparental inheritance in 99% of the progeny. When the male (mt^-)cells are irradiated with UV before mating, biparental plastids are formed. The heterozygotes for plastid genes (cytohets) display recombinations of plastid genes in their single-plastid progeny. Recombination is either reciprocal or non-reciprocal among the plastid genes. On this basis, strictly genetic maps could be constructed for the circular plastid DNA. Physical maps are also available and are used mainly for determining map positions. Molecular analysis revealed that one plastid gene *psa* (a photosystem protein) contain three exons; exon 1 is 50 kb away from exon 2 and from this exon 3 is 90 bp apart. In between, there are several other transcribed genes. Further complication is that exon 1 is in opposite orientation to the other two. It is supposed that for the expression of this gene transsplicing is used, i.e. (in contrast to the regular, common mechanism of splicing neighboring exons), here distant transcripts are brought together in the mRNA. The mitochondrial DNA has about 15 kb. Some alga mutants (*minutes*) are apparently mitochondrial and resemble *petites* in yeast. Since insertional mutagenesis became feasible, genetic and molecular analysis of the photosynthetic apparatus is greatly facilitated. [The name of this alga is often spelled as *Chlamydomonas reinhardi*] (See petite colony mutants, chloroplast DNA, chloroplast genetics, mitochondria)

CHLAMYDOSPORE: a thick-walled persistent asexual spore.

CHLORACNE: eruption on the skin caused by exposure to chlorine and related compounds.

CHLORAMBUCIL [p-(di-2-chloroethylamino)phenylbutyric acid]: a radiomimetic nitrogenmustard derivative causing primarily deletions and translocations; it had been used as an antineoplastic drug; it is also a carcinogen.

CHLORAMPHENICOL: is an antibiotic affecting peptidyl transferase in bacterial and mitochondrial protein synthesis. In human populations about 5×10^{-5} of the individuals may be very sensitive to the drug and develop anemia. (See cycloheximide, mitochondrial human disease)

CHLORAMPHENICOL ACETYLTRANSFERASE (CAT): gene has been extensively used as a reporter for transformation in cell culture by becoming resistant to the antibiotic chloramphenicol and thus selectable. (See antibiotics, transformation)

CHLORATE: has been used for the isolation of nitrate reductase deficient mutations that are not poisoned by chlorate. Some mutations, however, are hypersensitive to chlorate. Chlorate-sensitivity is apparently based also on uptake problems. (See nitrate reductase)

CHLORENCHYMA: a tissue with green plastids (chloroplasts).

CHLORIDE DIARRHEA, CONGENITAL: is a recessive defect involving ion transport, encoded in human chromosome 7q31.

CHLORONEMA: a gametophytic cell row in mosses. (See gametophyte)

CHLOROPHYLL: magnesium-porphyrin complexes (chlorophyll-a, -b, protochlorophyll, bactteriochlorophyll) in green plants and bacteria, receptors of light energy for carbon fixation and photosynthetic phosphorylation. (See Calvin cycle, chlorophyll-binding proteins)

CHLOROPHYLL-BINDING PROTEINS (CAB, chlorophyll A and B binding proteins, light-harvesting chlorophyll protein complex, LHCP): are situated in the membrane of the thylakoids. They modify the plane of orientation of chlorophylls. Due to this modification the chlorophyll does not fluoresce when excited by visible light (as it would do without CAB). The light energy absorbed by (an antenna) chlorophyll is rather transferred to a neighboring chlorophyll molecule and then excites this second chlorophyll while the first one returns to the ground state. This *resonance energy transfer* is continued to further neighbors until the *photochemical*

Chlorophyll-binding proteins continued

reaction center is reached. In this molecule an electron is raised to higher energy orbital, and this electron is transfered to the *electron transfer chain* of the chloroplast resulting in an electron hole (empty orbital). The electron acceptor gains thus a negative charge and the lost electron by the reaction center is replaced by another electron coming from a neighbor molecule which therefore becomes positively charged. As a consequence the light sets into motion an oxidation-reduction chain and the generation of ATP and NADPH. The CAB complex is encoded by about 16 genes, and a LHCPII system has also been identified. The red algae and cyanobacteria which have only chlorophyll a, use for light harvesting phycobilisomes. Several types of accessory pigments (carotenoids, pteridins, phycoerythrobilin, etc.) may be associated with the light-harvesting complex. (See photosynthesis, photosystems, photophosphorylation, Calvin cycle, phycobilins, Z scheme)

CHLOROPLAST: the green, chlorophyll-containing organelle of plant (algal) cells where photo-synthesis takes place. (See chloroplasts, chloroplast genetics, evolution of organelles)

CHLOROPLAST ENDOPLASMIC RETICULUM: see nucleomorph, endoplasmic reticulum

CHLOROPLAST ENVELOPE: a double membrane which surrounds this organelle and it con-trols the uptake of metabolites and the transport of proteins encoded by nuclear genes. Further-more, it participates in the biosynthesis of many plastid molecules; a few of the plastid envelope components are coded, however, by ctDNA. (See chloroplasts)

CHLOROPLAST GENETICS: was most successfully studied in the *Chlamydomonas* alga. *C. reinhardtii* has only one chloroplast with about 80 cpDNA molecules of 196 kbp each. The transmission of the cpDNA genome is largely uniparental, i.e. inherited most commonly through the mt^+ (comparable to egg) cytoplasm although in 1-10% of the cases biparental transmission may take place. Exceptionally, e.g. conifers, uniparental male transmission also may exist. The uniparental mt^+ transfer may be sometimes spurious in cases when the coding of the subunits of a particular protein is under the control of nuclear and organelle genes, respectively. In diploid vegetative zygotes of *Chlamydomonas* the biparental transmission of the extranuclear genes is most likely. Incubation in dark or postponing the meiosis by nitrogen starvation, however, favors the uniparental transmission.

In higher plants, the transmission of the plastid is usually through the female but biparental or only male transmission also occurs. The plastid nucleoids may be eliminated from or degraded in the sperms during the first or second nuclear division in the pollen or they are left behind when the generative nucleus enters the egg.

The most common algal mutations (at different loci) require acetate as carbon source because they cannot fix CO_2. Antibiotic resistance (or antibiotic dependent) mutations involve the rRNA or ribosomal proteins. Fluorodeoxyuridine is a specific mutagen for cpDNA. Arsenate and metronidazole are selective for non-photosynthetic mutations. Photosynthesis defective mutations (nuclear or cpDNA) can be screened also under long-wave-length UV. Also in higher plants streptomycin, spectinomycin (16S rRNA) lincomycin (23S rRNA), etc. resistance mutations could be induced and isolated.

The partial inactivation of the mt^+ cells by UV permits the transmission of cpDNA genes by the mt^- cells and this makes possible recombinational studies, however some progeny cells are heteroplasmic even after 3 divisions of the zygote. In the cpDNA only closely situated genes display linkage with about 1 kb/map unit. Mapping is practical by physical methods: either by deletional analysis or map-based sequencing or in interspecific crosses when RFLP exists by co-segregation with restriction endonuclease fragments (*C. eugametos* x *C. moewusii*). The nature of the recombination map was hotly debated until it turned out that genes within the inverted repeats map as if the area would be linear whereas recombination of genes outside this region reflects the circular nature of the cpDNA. A single recombination between two circular mole-cules may result in a cointegrate.

Recombination of cpDNA genes in higher plants also occurs albeit apparently quite rarely. The chloroplasts usually do not "synapse". Within cells of interspecific (intergeneric) somatic fusion in the *Solanaceae* recombination of antibiotic resistance markers has been demonstrated. Coseg-

Chloroplast genetics continued

regation of mitochondrial traits (cytoplasmic male sterility and chloroplast antibiotic markers was also shown).

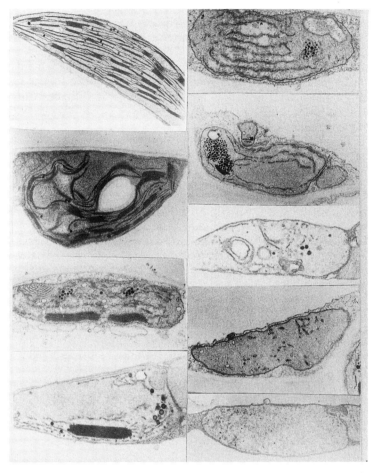

Mutant plastids in *Arabidopsis* induced by a nuclear mutator gene. At left in the top bloc, a normal chloroplast is shown. Because of the presence of the mutator, within single cells morphologically different plastids are visible. The mutator shows biparental recessive inheritance. The plastid mutations are transmitted only through the egg.

Chloroplast functions may be altered by transforming of appropriate genes (Atrazine resistance) into the cell nucleus, using T-DNA vectors. Transformation — using biolistic procedures — of cpDNA can be accomplished at high frequencies (up to 1×10^{-4}) in *Chlamydomonas* and higher plants. The insertion takes place around the passenger and vector junctions and the insert replaces the resident copy in the nucleoid. Co-transformation of antibiotic resistance genes situated in the inverted repeats and photosynthetic genes (situated in the unique regions) can be accomplished by the simultaneous use of two vectors. Alternatively, the bacterial *aadA* gene (that detoxifies antibiotics) is used. The gene is equipped by cpDNA transcription and translation elements and surrounded by the appropriate target sequences. Whether or not the transformation involves correction of a resident gene or insertion of a foreign gene (e.g. *uidA* [glucuronidase]), the integration requires homologous recombination within the flanking sequen-

Chloroplast genetics continued

ces. Transformation has been achieved also by direct transfer and integration of the cloned gene into protoplasts in the presence of polyethylene glycol. Under selective conditions sorting out of the transgene takes place rapidly.

The segregation and sorting out of chloroplast genes were subjected to analysis by the methods of population genetics and computer simulation. The biological observations do not seem to support the stochastic models of sorting out.

Mobile genetic elements as Group I introns, encoding the I-*Cre*I (or I-*Ceu*I) endonuclease and locating in the large ribosomal subunit gene or in the cytochrome b gene (I-*Csm*I endonuclease) in mt^+ (mating type) plastid nucleoids have been found in different *Chlamydomonas* chloroplasts. The I-*Cre*I endonuclease has a recognition site of 24 bp. These mobile elements resemble those of I-Sce in the mitochondria of yeast.

The promoter regions of the cpDNA genome is similar to that of prokaryotes. There are generally 10 nucleotides between the TATA box (TATAAT or longer) and the first translated codon and again 17-19 bases separate the TATA box from the 5'-TTGACA promoter consensus at about -35. Internal and further upstream promoters have also been identified. Many genes are transcribed into polycistronic RNAs. In spinach, 18 major RNAs were made from a single polycistronic transcript. The 3' termini of the transcripts generally contain inverted repeats that have probably only some processing and/or stabilizing functions along with some 5'-untranslated sequences. Binding proteins (3') seem to be involved in the processing of the RNA. Some observations indicate that polycistronic transcripts may bind to the ribosomes and translated without processing, perhaps with reduced efficiency. Translation of the chloroplast mRNA appears to be light-regulated. Apparently, an activator protein binds to the upstream untranslated region of the mRNA and the regulation is mediated through the redox state of this protein. Endonucleolytic processing of the transcripts may provide alternative leader sequences and binding sites for transcription factors. Activation may be exerted also by ribosomal proteins by induction and modulation of translation. The chloroplast mRNAs are not capped and the initiation of transcription is regulated in a manner similar to that in prokaryotes. In the untranslated upstream regions (UTR) there are binding sites for nuclearly encoded proteins that regulate transcription. Other proteins may bind to the 3' downstream sequences. Translational control is mediated by nuclear proteins. (See physical mapping, RFLP, cointegration, deletion mapping, biolistic transformation, β-glucuronidase, chloroplasts, mutation in cellular organelles, sorting out, introns, twintrons, maturase, mitochondrial genetics, transcription, σ, translation, mating type, polycistronic, promoter, nucleoid, binding protein, processing, redox reaction, RNA editing, endosymbiont theory, nucleomorph, metronidazole)

CHLOROPLAST IMPORT: the chloroplasts do not have structures for through-membrane traffic comparable to the nuclear pores. Four proteins are involved in import through the outer membrane of the chloroplast envelope and two with import through the inner membrane (IAP = import intermediate associated proteins). One of the four is immunologically related to heat-shock protein 70 and the other appears to be a channel protein. Two others (IAP34 and IAP85) are guanosine triphosphate binding proteins. Protein transport within the chloroplast and into the thylakoid lumen requires proteins SecA and SecY, homologous to translocation proteins present in bacteria. In addition plastocyanin and the 23-kDa and the 17 kDa subunits of the oxygen-evolving complex (OEC) are involved in thylakoid transport. (See mitochondrial import)

CHLOROPLAST MAPPING: in some algae the chloroplasts can be mapped by genetic means, a not completely natural process in most lower or higher plants. In addition, in higher plants the number of chloroplasts per cell may be quite high and recombination is a process involving simultaneously multiple events. Physical maps, based on molecular techniques are much more practical and provide more detailed information on the organization of the plastid genome. In the majority of higher plants (about 150 kbp) and *Chlamydomonas* algae (about 195 kbp) the 16, 23, 4.5 and 5 S ribosomal RNA genes occur in two repeats, separated by a long and a shorter sequence, coding for about 35 tRNAs and about 100 proteins. There are also different organizations. In *Euglena* alga with about 145 kbp genome the repeated rRNA clusters are adjacent. In

Chloroplast mapping continued

Pisum (pea) the genome is only about 120 kbp and the 16S and 23S rRNA genes are located in a single cluster. The first completely sequenced higher plant chloroplast genome of tobacco (155844 bp) the inverted rRNA repeats include 25,339 bp and the two single copy sequences contain 86,684 and 18,482 bps. (For more details see Sager, R. 1972 Cytoplasmic Genes and Organelles, Acad. Press. New York, Shinozaki, K. *et al.* 1986 EMBO J. 5:2043-2049, Palmer, J.D. 1991 in Molecular Biology of Plastids, Bogorad, L, & Vasil, I.K., eds. pp. 5-53, Acad. Press, San Diego, CA, USA, chloroplasts, chloroplast genetics)

CHLOROPLAST STROMA: the fluid phase of the chloroplast content the site of CO_2 fixation, RUBISCO, the Calvin cycle, chlorophyll synthesis, metabolism of amino acids, fatty acid synthesis, etc. Proteins can be exchanged among plastids. (See RUBISCO, Calvin cycle)

CHLOROPLASTS: chlorophyll-containing organelles (2-20 μm) in green plants with double-stranded circular DNA (cpDNA of 120-180 kbp) genetic material and a capacity of transcription and translation. The cpDNA codes for 16S, 23S, 5S and 4.5S ribosomal RNAs and has ribosomes of about 70S size, resembling those of prokaryotes. This genome codes for about 100 polypeptides and 35 different RNAs (rRNA, tRNA). The genes are frequently transcribed into polycistronic RNA, resembling bacterial gene clusters. Their rRNA genes (10-30 kb) are generally inversely repeated in land plants. Eighteen genera of the legumes *Fabaceae* do not have such inverted repeats. The unique sequences have a small (15 -25 kb) and a large (80-100 kb) tract. *Pelargonium hortorum* (geranium) displays a 76 kb repeat and this includes also some usually "unique" sequences (genes) here, however, duplicated. The cpDNA usually contains other inverted repeats. The cpDNA of *Chlamydomonas* algae is larger (195-294 kb) but has similar inverted repeats, and the unique sequences consist of two, about equal tracts with gene order differing from land plants. They have a recombination system acting among the different repeats. The *Euglena* cpDNA is similar in size (130-152 kbp) to that of land plants yet it lacks the two inverted repeats of rRNA but it has triple tandem repeats and two-fold tandem 16S rRNA genes. The tRNAs are clustered 2-5, each. Variations are found in several miscellaneous genes too. Other algae show additional variations. The colorless algae and plants (*Epifagus*) have smaller genomes (about 70 kbp) and lack about 95% of the genes encoding of the photosynthetic apparatus yet they have genes required for protein synthesis (rRNA, 17 tRNAs, 80% of the ribosomal proteins, etc.)

The major function of chloroplasts is photosynthesis. Actually, the chloroplasts reduce CO_2 and split water and release O_2. Enzymes of the chloroplast stroma converts CO_2 into carbohydrate. Some of the chloroplast genes have introns and some sequences in the cpDNA have homologies with both nuclear and mitochondrial and *E. coli* DNA. In bleached mutants or in antibiotic-sensitive plants callus growth can be maintained on carbohydrate-supplied culture media. The genetic code in the plastid DNA is the universal one, unlike in mitochondria. Several plastid functions are coded for by nuclear genes and the proteins or RNAs may be imported into the plastids from the cytosol with the assistance of transit peptides. Some of the plastid proteins are built by an assembly of nuclear and plastid gene coded subunits. The chloroplast genome has apparently evolved through endosymbiosis from an ancestral prokaryote(s). The size of the chloroplasts varies from to 1-3 μm in diameter and about 2 to 3 times as much in length. The double membrane enclosing its content has no pores. The internal flattened membrane vesicles called *thylakoids*, are stacked into *grana* and they harbor the photosynthetic apparatus. The grana are connected by the *stroma lamellae*. The stroma is the fluid phase of the chloroplasts. The photosynthetic apparatus of green bacteria and some lower algae is structurally simpler. (See chloroplast genetics, plastid male transmission, cpDNA, plastid number, compatibility of organelles and the nuclear genome, introns, spacers, chlorosome, chromatophore, chloroplast envelope, chloroplast endoplasmic reticulum, photorespiration, photosystems, evolution by base substitutions, organelle sequence transfer, differentiation of plastid nucleoids, endosymbiont theory, nucleomorph, nucleoid)

CHLOROQUINE: is an immunosuppressant compound that may be used for post-treatment of animal cells transfected by the calcium phosphate precitation method. It may increase the expression of the introduced DNA. (See transformation, calcium phosphate precipitation)

CHLORORESPIRATION: interaction between photosynthetic and respiratory electron transports. (See photosystems, photorespiration)

CHLOROSIS: a condition of plants when the chlorophyll content is reduced either by a nutritional deficiency (iron) or infection by viruses or other parasites. (See chlorophylls)

CHLOROSOMES: non-membraneous light-harvesting structures in green bacteria.

CHLOROSULFURON (2-chloro-N-[(4-methoxy-6-methyl-1-3,5-triazin-2-yl-amino carbonyl): is a herbicide to which resistant mutations have been isolated at the rate of about 1.2×10^{-7}. (See herbicides)

Ch-No38: a chicken protein involved in shuttle functions between nucleus and cytoplasm and assembly of ribosomes. (See ribosomes)

CHO (Chinese hamster ovary cells): frequently used for mutation studies in cell culture because of the hemizygosity permits the identification of recessive mutations in these diploid cells (2n = 44±).

CHOLERA TOXIN: causes the severely debilitating intestinal efflux of water and Na+ as a consequence of infection by the *Vibrio cholerae* bacterium. This enterotoxin enzyme mediates the transfer of adenylate from NAD+ to the Gs subunit of the G-proteins. As a consequence $G\alpha_s$ stops acting as a GTPase. Therefore adenylate cyclase continuously synthesizes cAMP resulting in the disturbance of the water and salt balance. Until recently, the etiology of the disease posed some hard problems because very often the *Vibrio* appeared entirely harmless. It has been shown the toxin production depends on the acquisition of the *ctx* gene complex from the filamentous phage CTX. The ctx complex then can be transfered from one bacterium to others. The information transfer requires that the bacterium has active pili. The new information explains the difficulties of immunization against this infection affecting hundreds of thousands in some years, particularly in South-East Asia. (See G-proteins, G_s, ARF, pilus)

CHOLESTEROLS: are amphipatic (lipid) molecules of the membranes with a polar head and a non-polar hydrocarbon tail (see cell membranes). Cholesterols are the principal sterols in animals, they occur also in plants (stigmasterols) and in fungi (ergosterol). Bacteria generally lack sterols. Cholesterols are synthesized from acetate through mevalonate, isoprenes, squalenes to a four-ring steroid nucleus. The principal regulatory enzyme is 3-hydroxy-3-methyl-glutaryl—coenzyme A (HMG-CoA) reductase. Most of the synthetic activity takes place in the liver and used as bile acids (cholesteryl esters). Cholesterols are parts of membranes, steroid hormones and precursor of vitamin D. Cholesterols are indispensable for numerous functions and are generally synthesized in sufficient amounts by the human body although dietary cholesterol may increase its level. Accumulation of excessive amounts of cholesterol bound to low-density lipoproteins (LDL) may lead to occlusion of the blood vessels (atherosclerotic plaques) leading to coronary heart disease. (See lipids, sphingolipids, prenylation, familial hypercholesterolemia, high-density lipoprotein, hypertension, CETP, Wolman disease)

CHOLESTERYL ESTER STORAGE DISEASE: see Wolman disease

CHOLINE: in the form of acetylcholine it is a neurotransmitter. (See acetylcholine)

CHOLINESTERASE: is a hydrolase that splits acyl groups from choline, and it is required for normal function of the nervous system. Organophosphorous insecticides (parathion, chlorpyrifos, [Dursban®], diazinon, and the nerve gas, sarin poison cholinesterase. Paraoxonase (PON1), a high density lipoprotein associated enzyme may detoxify organophosphate compounds but substantial human polymorphism exists in PON1. (See pseudocholinesterase)

CHONDROCYTE: a cartilage cell that can secrete collagen and glucosaminoglycans (polysaccharides with alternating sequence of more than one type of sugars). (See cartilage)

CHONDRODYSPLASIA (CD): in the autosomal *dominant* form (Conradi-Hünermann disease) extra bone formations takes place at the epiphysis (bone ends), and other abnormal bone formations but relatively rare (less than 30%) cataracts and skin anomalies are also found. The autosomal *recessive* forms were accompanied by excessive calcium deposits and hence the name

Chondrodysplasia continued

chondrodysplasia punctata because the cartilage appeared stippled reminding also to the Zellweger syndrome. About 2/3 of the cases had cataracts. The symptoms may be phenocopied by maternal exposure to the warfarin pesticide that depletes vitamin K-dependent blood coagulation although an autosomal CD punctata is accompanied by hereditary blood coagulation factor deficiencies that may be cured by vitamin K. Another *dominant X-chromosomal* dwarfism gene (supposedly at Xq28) also shows bald or scar-like spots ("punctata") apparently caused by lower peroxisomal functions. The rare autosomal dominant Murk Jansen type CD is characterized by short legs, extreme disorganization of the limb (metaphysis) and foot bones and defects of the spine, pelvis and fingers involving also accumulation of calcium but reduced phosphate levels in the blood. The basic defect was in the gene encoding the parathyroid hormone—parathyroid hormone-related peptide involving replacement of histidine^{223} by arginine. The autosomal recessive Hunter-Thompson acromesomelic chondrodysplasia (the shortening of the limbs is most pronounced in the distal bones) is caused by defects in the cartilage-derived morphogenetic protein (CDMP1), a member of the TGF-β superfamily of growth factors. The *metaphyseal chondrodysplasia* (McKusick type) is a recessive short-limb dwarfness accompanied by sparse blonde hair. (See dwarfness, vitamin K, vitamin K-dependent blood clotting factors, antihemophilic factors, peroxisome, collagen, Zellweger syndrome, TGF, microbodies, warfarin)

CHONDROITIN SULFATE: is a heteropolysaccharide composed of alternating units of glucuronic acid and acetyl-glucoseamine; with related compounds form the ground substance, an intracellular cement, of the connective tissue. (See mucopolysaccharidosis, spondyloepiphyseal dysplasia)

CHONDROME: the complete set of the mitochondrial genes. (See mtDNA)

CHOP: see GAGG153

'CHOPASE': recombination enzyme making double-strand break versus the 'nickase' which cut only one DNA strand. (See recombination molecular mechanisms)

CHOPROIDORETINAL DEGENERATION: is a type of X-linked retinitis pigmentosa, distinguished by a brilliant patch at the macula of the eye. (See retinitis pigmentosa, macula, macular degeneration, macular dystrophy, foveal dystrophy)

CHORDIN: see organizer

CHORION: the outermost envelope of the mammalian embryo (fetus), the non-cellular membrane around the egg of arthropods, fishes, etc. (See also amnion, allantois, chorionic villi)

CHORIONIC VILLI: thread-like protrusions, tufts on the surface of the chorion, an embryonic tissue. They are used for prenatal genetic examination by amniocentesis. (See amniocentesis)

CHORISMATE: is a precursor of the biosynthesis tryptophan, phenylalanine and tyrosine and their various derivatives. In the following pathway, in between the metabolites, the enzymes involved are shown in parenthesis. Phosphoenolpyruvate + Erythrose-4-phosphate -> (*2-keto-3-deoxy-D-arabinoheptulosonate-7-phosphate synthase*) -> 2-Keto-3-deoxy-D-arabino-heptulosonate-7-phosphate ->(*dehydroquinate synthase*) -> 3-Dehydroquinate -> (*3-dehydroquinate dehydratase*) -> 3-Dehydroshikimate -> (*shikimate dehydrogenase*) -> Shikimate -> (*shikimate kinase*) -> Shikimate-5-phosphate -> (*enolpyruvylshikimate-5-phosphate synthase*)->3-Enolpyruvyl-shikimate-5-phosphate -> (*chorismate synthase*) -> Chorismate. (See tryptophan, tyrosine, phenylalanine, phenylketonuria, alkaptonuria)

CHOROID: an inner layer of the eyeball supplying blood to nerves.

CHOROIDAL OSTEOMA: autosomal dominant eye neoplasias with bony-like cells. The choroidal sclerosis and choroid plexus calcification are autosomal recessive. (See choroid)

CHOROIDEREMIA: X-linked recessive atrophy of the choroid and the retina. (See choroide, retina)

CHOTZEN SYNDROME (acrocephalosyndactyly, ACS): dominant inheritance of syndactyly of fingers and toes, asymmetric and narrow head, etc., encoded in human chromosome 7p21-p22 by a gene appearing homologous to *Twist* of *Drosophila*, coding for a 490 amino acid acidic protein containing a basic helix-loop-helix motif. (See syndactily, Apert syndrome, Robinow

syndrome, limb defects, craniosynostosis syndromes, helix-loop-helix))
CHR: see cluster homology region
CHRISTMAS DISEASE: see antihemophilic factors
CHROMAFFIN CELL: is specifically receptive to staining by chromium salts.
CHROMATID: chromosomal strand containing one DNA double helix; after replication, each chromosome usually contains two chromatids, held together at the centromere. (See chromosome, centromere)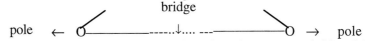
CHROMATID BRIDGE: whenever dicentric chromosomes are formed (paracentric inversion heterozygotes, breakage-bridge-fusion cycles, ring chromosomes) the spindle fibers are pulling the chromosomes to the opposite poles and at one region the chromosomal material is stretched (not unlike to a pulled rubber band) and this thin connecting tie is called chromosome bridge. The bridge eventually will break and this may lead to unequal distribution of genes to the poles, resulting in duplications and deficiencies in the daughter cells:

pole ← o⎯⎯bridge⎯⎯o → pole

(See breakage-bridge-fusion cycles, bridge, inversion, ring chromosome, bridge [photo])
CHROMATID INTERFERENCE: crossing over within a chromatid reduces the chance for another to occur. Chromatid interference can genetically be determined by tetrad analysis where in the absence of chromatid interference 2-strand, 3-strand and 4-strand double crossing over should occur in the proportion 1:2:1, and any deviation from this proportion is chromatid interference. (See double crossing overs, interference)
CHROMATID SEGREGATION: takes place after recombination and in case of polysomy it has detectable effect on the segregation ratios. (see autopolyploidy, trisomy)
CHROMATIN: material of the eukaryotic chromosome (DNA, RNA, histones and non-histone proteins). The DNA stretches in the chromatin of a chromosome from telomere to telomere. An "average" eukaryotic chromosome at metaphase may be 10 to 15 μm long but the DNA in it may 100,000 times longer when fully extended. The problem of accommodation is comparable to fitting a 2.5 km long thread into 2.5 cm long skein of 3 mm in diameter. The folding must be extremely orderly and stable to assure perfect synapsis preceding recombination and still flexible enough to make possible error-free replication within a few hours. The elementary DNA double helices form nucleosomal structures with histone proteins and attract various acidic proteins and RNA in the matrix. The DNA has short (80-100 bp) supercoiled stretches. When the majority of the protein is digested away from the DNA, the remaining structure appears in the form of a protein scaffold with DNA loops attached. The regions where the DNA is attached to the nuclear matrix is called MAR (matrix attachment region). The MAR sequences are generally rich in AT. They seem to contain transcriptional activators and recognition sites for topoisomerase II. Condensed chromosomes by various nuclear stains (Giemsa stain, a mixture of basic dyes) display characteristic C (centromeric region) and G bands that helps to identify several regions along the chromosomes. The nature of the specificity of the G staining is not known. That *rotational positioning* of the nucleosome in which the histone octamer is facing away from the minor grove of the DNA permits DNase I to cleave the chromatin into 10 bp sequences. *Translational positioning* of the nucleosomes in the chromatin defines the position of the nucleosomes relative to the site of transcription initiation of a gene about 300 to 150 bp away. In yeast the positioning is regulated by the MFα2 repressor. When a histone octamer is at the TATA box, transcription is hindered. Histone 1 is a repressor of all three eukaryotic RNA polymerases. During active transcription the nucleosomes are apparently not removed but only reconfigured. Genes in the chromatin may be attached to the nuclear scaffold at areas of their separation and thus are insulated from each other by these 'boundary elements'. The locus control element (LCR) in the chromatin seems to be regulating the activity of groups of genes. (See stains, FISH, chromosome painting, chromosome banding, nucleosome, chromosomal proteins, heterochromatin, euchromatin, high mobility group of proteins, nuclease hypersensitive sites, LCR, nuclease-sensitive sites, prochromatin, antichromatin, chromatin code)

CHROMATIN ASSEMBLY FACTOR (CAF): during genetic repair the chromatin and the nucleosomal organization has to be destabilized and after repair it must be reorganized. The nucleosomal structure, replication complex, the various transcription factors, etc. must be restored after excision of the DNA defects. This process requires a series of sequentially interacting proteins. (See chromatin)

CHROMATIN CODE: is the hypothesized system regulating the folding of chromatin fibers of eukaryotes. (See chromatin)

CHROMATIN DIMINUTION: occurs when pieces of chromosomes are excised or entire genomes are fragmented to generate minichromosomes. Such a phenomenon is normal during the 2nd to 8th cleavage divisions of *Ascaris* and related nematodes and during the formation of the macronuclei of ciliated protozoa. The fragments may be reintegrated again into much larger chromosome(s) in the generative cell nuclei or the germline may develop from a single cell that has not undergone chromosome diminution. The fragmentation is followed by the addition of 2-4 kb telomeric repeats (TTAGGC). Such repeats may be added also at other chromosomal breakage region (CBR) sites. AT-rich sequences near the CBR (approximately 1/4 of the germline DNA, including the two types of Tas retrotransposons) are eliminated during diminution. (See *Ascaris megalocephala*, *Paramecium*, telomere)

CHROMATIN FILAMENT: a 30 nm in diameter nucleosomal DNA fiber appearing as beads-on-string structure by electronmicroscopy; the beads are the nucleosomes. The folding of the fiber may be the consequence of H1 histone-induced contraction of the internucleosomal angle as the salt concentration approaches the physiological level. Decondensation required for transcription may be the consequence of depletion of the linker histone or acetylation of the core histone tail domain. (See nucleosome, histones, chromatin)

CHROMATIN REMODELING: is a change in the nucleosomal structure by establishing nuclease hypersensitive sites in active genes. (See DNase hypersensitive site, nucleosome, chromatin, histone acetyl transferase)

CHROMATIN STATE FIXATION: a hypothetical mechanism that stabilizes the function of certain genes and keeps others dormant (by heterochromatinization). (See heterochromatin)

CHROMATIN-NEGATIVE: cells do not display heterochromatic Barr bodies. (See sex chromatin, Barr body)

CHROMATOGRAPHY: in various forms partitions molecular mixtures between a stationary (sugar, cellulose, silica gel, sepharose, hydroxyapatite) and a water base or organic liquid phase.

CHROMATOPHORES: pigmented cells or the pigment-rich invaginations of the cell membrane.

CHROMATOSOME: is a part of the nucleosome, obtained as an intermediate during digestion with micrococcal nuclease; it contains a core particle of a histone octamer wrapped around by about 1 and 3/4 turn of about 146 bp plus 120 on each side, held together at the entry and exit points by H1 histone. (See nucleosome)

CHROMOCENTER: heterochromatin aggregates such as the common attachment point of the polytenic chromosomes in the salivary gland nuclei. (See salivary gland chromosomes)

CHROMOKINESIN: see NOD, Xklp1, spindle

CHROMOMERE: densely stained bead-like structures along the chromosomes at early prophase. The chromomeres represent increased coiling of the chromatin fibers. In the lampbrush chromosomes the loops seem to emanate from the chromomeres. In the salivary chromosome bands there are appositioned chromomeres of a large number of chromatids. Chromomeric structures were recognized and their constant number was observed by Balbiani in 1876. In the 1920s, John Belling counted their numbers in the genome and assumed that the chromomeres are physically identical with the genes, and the 2,193 chromomeres in *Lilium pardalinum* would be the number of genes in lilies. The chromomeric structure of the chromosomes was useful in identifying chromosomal aberrations by light microscopy. (See chromosome morphology, salivary gland chromosomes, lampbrush chromosomes, pachytene analysis, gene number, illustration on p. 188)

Chromomere continued

CHROMOMERES AS ILLUSTRATED BY BELLING [1928] (University of California Publ. Bot 14:307) →

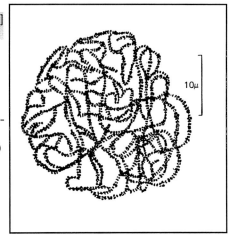

CHROMOMYCIN A3: is used as a stain, specific for GC nucleotides in the DNA.

CHROMONEMA: a term that had been used for the chromosomes of prokaryotes and for the smallest light-microscopically visible chromosome thread, chromatin filament, genophore. (See genophore, Mosolov model, chromosome structure, chromatin filament)

CHROMOPHORE (dye): a chemical substance that gives color to a structure upon binding to other compounds. (See stains, fluorochromes)

CHROMOPLAST: plastids in which red and yellow pigments (rather than chlorophylls) predominate, e.g., in the fruits of mature tomatoes and other plants, e.g. *Capsicum*. (See plastids)

CHROMOSOMAL ABERRATIONS: see chromosomal rearrangements, chromosome breakage, X-ray breakage of chromosomes

CHROMOSOMAL DNA: see chromatin

CHROMOSOMAL INHERITANCE: a somewhat outdated term for inheritance of nuclear genes since frequently organellar DNA molecules are also called chromosomes.

CHROMOSOMAL INSTABILITY: may occur spontaneously in cultured cells of animals and plants and their frequency depends on the age and composition of the nutrient media. Generally on liquid media the frequency of the anomalies is lower. Mitotic anomalies (nondisjunction) as well as polyploidy, aneuploidy and rearrangements occur. One of the major contributing factor may be the lack of coordination between nuclear divisions and cell divisions. Chromosomal instabilities may be found also in intact organisms caused primarily by natural insertion and transposable elements, and also by insertions introduced by transformation. (See also hybrid dysgenesis, RIP, Roberts syndrome, isochores, transposable elements)

CHROMOSOMAL INTERCHANGE, RECIPROCAL: see translocation

CHROMOSOMAL MOSAIC: not all the cells in the body have the same chromosomal constitution, i.e. patches of different chromosomal morphology or number co-exist. (See chimera)

CHROMOSOMAL MUTATION: is a term applied to mutations that involve defects detectable by the light microscope, and in general to mutations involving structural and numerical alterations of the chromosomes and mutations other than base substitutions. (See chromosomal rearrangements, chromosome breakage, deficiency, duplication, inversion, translocation)

CHROMOSOMAL POLYMOPHISM: in the population more than one type of chromosomal morphology or arrangement is present.

CHROMOSOMAL PROTEINS: include histones and a variety of non-histone proteins involved in the determination of the structure, replication, transcription and regulation of these processes in eukaryotes. In sperm, in place of histones protamines are found. In the majority of fungi histones are absent but eukaryotic viruses such e.g. SV40 chromosomes have nucleosomal structure. Prokaryotic chromosomes contain a minimal variety and quantity of proteins associated with the chromosomes in comparison with eukaryotes. (See chromatin, histones, non-histone proteins)

CHROMOSOMAL REARRANGEMENTS include (internal) deletions, (terminal) deficiencies, duplications, transpositions, inversions, translocations. The majority of the chromosomal rearrangements are deleterious because involve loss or altered regulations of functions (chromosomal aberrations are concomitant with several types of cancers) yet duplications and inversions play a role in evolution. Besides neoplasias, several hereditary human diseases involve chromo-

Chromosomal rearrangements continued

somal alterations. Chromosomal rearrangements play important role in the etiology of cancer. Deletions may eliminate genes controling important checkpoints in the cell cycle (tumor-suppressants). Inversions and translocations may result in gene fusions, and proliferative processes are activated (oncogenes, transcription factors, transcriptional activators, etc.). If the c-MYC oncogene is inserted to the immunoglobulin heavy chain or to an immuno-globulin κ or λ gene or T cell receptor gene, the oncogene may be activated. When the ABL oncogene (9q34.1) is translocated to the breakpoint cluster (BCR) in the Philadelphia chromosome (22q), myelogenous and acute leukemia may develop as a consequence of elevated protein tyrosine kinase activity. Inversion of human chromosome 14q(11;q32) involving the T cell receptor (14q11) and the immunoglobulin heavy chain (14q32.33) results in the fusion of the variable region of the immunoglobulin and the T cell receptor (TCRα) may cause lymphoma. Translocations may facilitate protein dimerization and changes in transcription. Site-directed chromosome rearrangements can be induced by the techniques of molecular biology. The *loxP* prokaryote gene can be inserted site-specifically by recombination into two selected chromosomal positions. Then the *Cre* prokaryotic recombinase is introduced into the same embryonic mouse stem cells by a transient expression transformation procedure. In such a manner thus translocation involving the MYC oncogene (chromosome 15) and an immunoglobulin (IgH) gene (chromosome 12) were reciprocally translocated. The selectable markers neomycin phosphotranferase (*neo*) and hypoxanthinephosphoribosyl transferase (*Hprt*) facilitated the selective isolation of the translocations. The frequency of this type of exchange is in the 10^{-6} to 10^{-8} range. (See chromosomal mutations, ataxia, Bloom's syndrome, Cockayne syndrome, Fanconi anemia, Lynch syndromes, Werner syndrome, Wiskott-Aldrich syndrome, xeroderma pigmentosum, RecQ, helicase, sterility, position effect, gene fusion, cancer, chromosome breakage, targeting genes, neo, HPRT, MYC, immunoglobulins, Cre/loxP, homing endonuclease)

CHROMOSOMAL SEX DETERMINATION: in the majority of eukaryotes the females have two X-chromosomes (XX) and the males have one X and one Y chromosome (XY), birds and butterflies have heterogametic females (WZ) and homogametic males (ZZ). In the nematode *Caenorhabditis* the males have only one X-chromosome (XO) and the females have two (XX); similar mechanism exists also in several fishes. In wasps and bees the females and workers hatch from fertilized eggs whereas the males are the products of unfertilized eggs, although in the body cells the males may double their DNA content during development. (See also sex determination, arrhenotoky, thelyotoky, deuterotoky, mealy bug)

CHROMOSOMAL VIRULENCE LOCI (*chv*): the majority of the virulence loci (*vir*, controling transfer of the T-DNA of the Ti plasmid) of *Agrobacterium* are situated in the plasmid but the *chv* genes are in the main DNA (chromosome) of the bacteria. (See *Agrobacteria*, Ti plasmid, virulence genes of *Agrobacterium*)

CHROMOSOME: the nuclear structures containing DNA, embedded in a protein and RNA matrix of eukaryotes; also the DNA strings of prokaryotic nucleoids and mitochondria and chloroplasts (sometimes called also genophores because the latter ones are associated with only small amounts of proteins in comparison to the eukaryotic nuclear chromosomes). The morphology (length, arm ratio, appendages [satellites] and banding pattern (natural or upon staining by special dyes) and number is characteristic for the species in eukaryotes. (See chromosome morphology, chromosome structure and additional items beginning with chromosomal or chromosome)

CHROMOSOME ABNORMALITY DATABASE: e-mail address <simon@bioch.ox.ac.uk>

CHROMOSOME ADDITION: see alien addition

CHROMOSOME ARM: the portion of a chromosome on either side of the centromere. (See chromosome morphology, isobrachial, heterobrachial, arm ratio)

CHROMOSOME ASSIGNMENT TEST: see somatic cell hybridization

CHROMOSOME 5B: a gene (*Ph*) in the B genome of wheat suppresses homoeologous pairing in this allohexaploid species. In case of loss of this chromosome or the pairing controlling region

Chromosome 5B continued

of it, synapsis may take place among all homoeologous chromosomes and multivalents are formed. Similar regulator genes occur in other chromosomes of wheat and other allopolyploid species. (See chromosome pairing, synapsis, multivalent, *Triticum*, allopolyploid)

CHROMOSOME BANDING: see C-banding, G-banding, Q-banding, R bands, T bands

CHROMOSOME BREAKAGE: chromosomes may break "spontaneously" or by the effects of chemical and physical agents (ionizing radiation). The breakage may involve only one of the chromatids or both (isochromatid breaks). Single breaks may lead to terminal losses of the chromosomes (chromosome deficiency) or double and multiple breaks may cause internal deletions and various rearrangements such as transposition, inversion and translocation. Deletions requiring only single breaks to occur at first order kinetics, chromosomal rearrangements generally follow second or multiple order kinetics. At first order kinetics the breakage occurs in a linear proportion to the dose of the agent causing it, at second order kinetics the number of breaks are proportional to the square of the dose (exponential response), i.e. at low doses the rise of the number of breaks is slow and at higher doses their numbers rise steeper. (See kinetics) The final outcome is modified also by the efficiency of repair mechanisms. Cancerous growth is frequently associated with chromosome breakage although it is unknown how many were caused by chromosome breakage and how often this occurred only during the process of abnormal cell proliferation. Several human syndromes are accompanied with an increased frequency of chromosome breakage and or deficiency of genetic repair: Fanconi's anemia, Bloom's syndrome, ataxia telangiectasia, xeroderma pigmentosum, Cockayne syndrome. (See Roberts syndrome, deletion, inversion, translocation, transposition, cancer, environmental mutagens, mutator genes, DNA repair, position effect, cancer, isochores, X-ray chromosome breakage)

CHROMOSOME BREAKAGE AS A BIOASSAY: many mutagenic and carcinogenic agents cause cytologically (by the light microscope) detectable chromosome breakage involving single chromatid lesions, double (isochromatid) breaks, deletions, transpositions, translocations, inversions, chromosome fusions, dicentric and acentric fragment formation, etc. The frequency of these alterations can be quantitated during mitotic and meiotic nuclear divisions of suitable plant (root tips, flower buds) and animal systems (cultured lymphocytes, fibroblasts, cells withdrawn from amniotic fluids during gestation, bone marrow cells and spermatocytes). The cytological assays can also reveal aneuploidy, polyploidy which do not involve chromosome breakage but non-disjunction may be the result of damage either to the centromere or to the spindle fibers. (See bioassays in genetic toxicology, heritable translocation assays)

CHROMOSOME BREAKAGE SYNDROMES: see fragile X syndrome, Bloom syndrome, Fanconi anemia, ataxia, trinucleotide repeats)

CHROMOSOME BRIDGE: see bridge, chromatid bridge

CHROMOSOME COILING: the status of condensation of chromosomes. During interphase the chromosomes are almost entirely stretched out but as the cell cycle proceeds the coiling increases reaching a maximum at metaphase. The two chromatids may be twisted around each other during prophase (relational coiling). This type of coiling does not permit the coiled strands to separate entirely unless the separation begins at one end and it is completed to the other end (plectonemic coiling). In case the coiling resembles pushing two spirals together, they can be separated in a single movement because they are not entangled (paranemic coiling). The coiling is not usually detectable in all cytological preparations unless special treatment (e.g ammonia vapor) is employed. (See also concatenate,

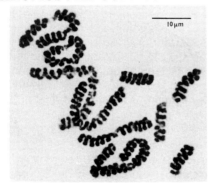

supercoiling). The *Tradescantia virginiana* photo is the courtesy of Vosa, C.G; from Clowes, F.A. & Juniper, B.E. 1968. Plant Cells. Blackwell, Oxford, UK

CHROMOSOME COMPACTION: the folding and packaging of the elementary chromosome fibers into the chromosomes, visible through light microscopy. (See chromatin)

CHROMOSOME COMPLEMENT: a haploid chromosome set. (See haploid)

CHROMOSOME CONDENSATION: the increasedly tight winding of the chromosomal coils from interphase to metaphase. (See chromosome coiling, mitosis, meiosis)

CHROMOSOME CONFIGURATION (meiotic configuration): the manner of pairing or assembly of chromosomes during meiosis. (See meiosis, translocation, inversion)

CHROMOSOME CONTAMINATION: see hybrid dysgenesis

CHROMOSOME CORE: the central axial part of the chromosome; well visible in lampbrush chromosomes. (See lampbrush chromosome, synaptonemal complex)

CHROMOSOME CRAWLING: is the same as inverse PCR. (See inverse PCR)

CHROMOSOME DIMINUTION: the fragmentation of the large (polycentric) meiotic chromosomes in the soma line of *Ascaris megalocephala univalens* with 1 pair of meiotic chromosomes and in *A.bivalens,* 2 meiotic pairs, into numerous, 52 to 72 and 62 to 144 small chromosomes in the soma, respectively. (See macronucleus, *Paramecia*, internally eliminated sequences)

CHROMOSOME DOUBLING: can be brought about by chemical or physical agents that block the function of the spindle fibers during meiosis or mitosis. Most commonly the alkaloid colchicine is used but others, e.g., acenaphtene (a petroleum product of industrial and pesticide, plastic manufacturing use), as well as other agents have also been employed. The purpose of the chromosome doubling is induction of polyploidy and in species hybrids to restore fertility of the hybrids which would be sterile without doubling the chromosome number because the distantly related chromosomes would not have homologs to pair with. (See colchicine, polyploid, amphidiploid)

CHROMOSOME ELIMINATION: during cleavage divisions of dipteran and hemipteral insects somatic chromosomes may be lost as a natural process but the germline cells retain the entire, intact genome. In the *Ascaridea* and some other species certain chromosomal segments may be lost as part of chromosomal differentiation during mitosis. In *Ascaris megalocephala univalens* there is only one pair of chromosomes during meiosis and that is fragmented into several smaller ones during somatic cell divisions. The macronuclei in *Paramecia* have only metabolic function, and disintegrate after the exconjugants are formed following fertilization and only the micronulei are retained. The macronuclei are reformed after mitoses of the diploid zygotes. Nondisjunction may also result in elimination because both of the non-disjoined chromosomes pass to one pole. The gene *polymitotic* (*pol,* map location 6S-4) of maize may eliminate several or even all the chromosomes after meiosis during successive divisions because nuclear divisions do not keep up with the rapid succession of cell divisions. In the pentaploid *Rosa canina* (2n = 35) 7 bivalents are formed both at male and female meiosis but in the male generally all the univalents are lost whereas in the female one set of the 7 chromosomes derived from the bivalents and the univalents are retained in the basal megaspore and in the egg. Thus upon fertilization the 35 chromosome number is restored in the zygotes. Chromosome elimination occurs regularly in species with supernumerary chromosomes that have unknown function and a defective centromere. Because of the nature of the centromere, the supernumerary chromosomes commonly display nondisjunction and additional losses after fragmentation. Chromosome elimination occurs when *Hordeum bulbosum* is crossed either with *H. vulgare* or hexaploid wheat. In somatic cell hybrids of human and mouse cells, the human chromosomes are gradually eliminated unless they carry genes essential for the survival of the cybrids. (See *Hordeum bulbosum*, haploids, cybrid, chromosome diminution, *Ascaris megalocephala*, *Paramecium*, *Rosa canina*, bivalent, univalent, B chromosomes, cybrid, assignment test)

CHROMOSOME ENGINEERING: generates rearrangements in the genomes, making alien additions, translocations, facilitate homoeologous pairing and recombination, alien transfers, alien substitutions, monosomics, chromosomal rearrangement, etc. with a primary goal, to improve the species for agronomic purposes. (See individual entries for these terms)

CHROMOSOME HOPPING: see chromosome jumping

CHROMOSOME INTERFERENCE: see interference

CHROMOSOME JUMPING: a special type of chromosome walking, taking advantage of the breakpoints of chromosomal rearrangements as guide posts; it permits the cloning of the two ends of a DNA sequence without the middle section. The procedure may take advantage of existing chromosomal rearrangements or the genomic DNA is partially digested with any restriction endonuclease or with an enzymes which cuts very rarely. The DNA fragments are circularized with the aid of DNA ligase and cloned in such a way that the cloning vector contains a known *E. coli* sequence between the ligation sites. The cloned product is then digested by a restriction enzyme that does not cut within the special *E. coli* sequence. Thus the fragments generated contain the *E. coli* sequence flanked by the cloned target DNA sequence that was originally far away (100-150 kb) in the chromosome. The *E. coli* sequence containing fragments are then recloned in a phage vector and the DNA is probed to a DNA library to identify the clones that contain sequences far away in the eukaryotic genome. This procedure thus facilitates the rapid movement toward the genetic cloning target. It may also be combined with chromosome walking to approach the desired gene. (See chromosome walking, jumping library)

CHROMOSOME KNOBS: dark-stained structures in the chromosomes best recognized during pachytene stage, representing local condensation of the chromatin. It is a characteristic feature of certain genomes within a species. The presence of knobs may affect recombination frequencies in their vicinity and may be involved in preferential segregation. (See chromosome morphology, karyotype, pachytene, preferential segregation)

CHROMOSOME LANDING: is an approach to gene isolation from large eukaryotic genomes by first identifying linkage to close physical markers. It is a substitute for chromosome walking which is frequently impractical in these organisms. (See chromosome walking)

CHROMOSOME LIBRARY: a collection of individual chromosomes, isolated by flow cytometric separation, or pulsed field gel electrophoresis. Such a library may facilitate the manipulation of large eukaryotic genomes. (See flow cytometry, pulsed field electrophoresis)

CHROMOSOME MAP: see mapping genetic, physical mapping, radiation mapping

CHROMOSOME MOBILIZATION: see *mob,* conjugation

CHROMOSOME MORPHOLOGY: is generally identified at metaphase and accordingly meta-, submeta-, acro-, and telocentric chromosomes are distinguished. Furthermore, secondary constrictions and appendages (satellites) are distinguished. The various banding techniques permit the analysis of individual chromosomes on the basis of differential staining. By the application of probes with fluorochromes (chromosome painting) different details (translocations, transpositions) can be identified. In interphase and prophase (pachytene) some chromosomes display chromomeres, cross bands (salivary gland chromosomes, giant chromosomes) or natural knobs. (See karyotype, chromosome banding, chromosome painting, fluorochromes, FISH, PRINS, pachytene analysis, chromosome knobs, nucleolar organizer, centromere, secondary constriction)

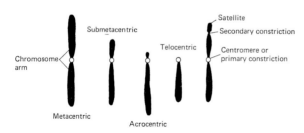

CHROMOSOME MUTATION: any change (beyond the size of a nucleotide or codon) involving the structure of the chromosome. (See chromosomal aberration, chromosomal rearrangement, point mutation, codon)

CHROMOSOME NUMBERS: is variable among the different species and may vary between species by polyploidy although it is an important, stable taxonomic feature. The basic number is represented by x, the gametic number by n and the somatic number by $2n$. Thus in a diploid species like *Arabidopsis* $x = 5 = n$. In hexaploid wheat $6x = 2n$. The chromosome number may vary between males and females as a mechanism of sex determination. Also, centromeric fusion may generate one bi-armed chromosome from two telocentrics (acrocentric) or the opposite may take place. In cultured cells of mice the chromosomes may become acrocentric and

Chromosome numbers continued
double in number. Numbers of genetically often studied organisms are shown below and additional numbers are found under the English (or scientific) name of the different organisms:

CHROMOSOME NUMBERS OF SOME REPRESENTATIVE ORGANISMS

MICROORGANISMS	x
Aspergillus nidulans	8
Chlamydomonas reinhardi	8
Dictyostelium discoides	7
Neurospora crassa	7
Saccharomyces cerevisiae	16
Saccharomyces pombe	2

PLANTS	
Arabidopsis thaliana	5
Barley (*Hordeum vulgare*)	7
Broad bean (*Vicia faba*)	6
Datura sp.	12
Epilobium sp.	18
Haplopappus gracilis	2
Lily (*Lilium sp.*)	12
Maize (*Zea mays*)	10
Oenothera lamarckiana	7
Petunia hybrida	7
Potato (*Solanum tuberosum*, 2n = 48 = 4x)	12+12
Tobacco (*Nicotiana tabacum*, 2n = 48 = 4x)	12+12
(*Nicotiana plumbaginifolia*, 2n = 24)	12
Tomato (*Lycopersicum sp.*)	12
Wheat (*Triticum aestivum*, 2n = 42 = 6x)	7+7+7
(*Triticum turgidum*, 2n = 24 = 4x)	7+7
(*Triticum monococcum*, 2n = 14)	7

ANIMALS	
Cattle (*Bos taurus*)	30
Caenorhabditis elegans (female 2n = 12, male 2n = 11)	6
Chimpanzee (*Pan troglodytes*)	24
Cricket (*Gryllus campestris*, female 2n = 30, male 2n = 29)	15
Drosophila melanogaster	4
Hamster (*Mesocricetus auratus*)	22
Honeybee (*Apis mellifera*, female 2n = 32, male 1n = 16)	16
Housefly (*Musca domestica*)	6
Homo sapiens	23
Mouse (*Mus musculus*)	20
Rat (*Rattus norvegicus*)	21
Sea urchin (*Strongylocetrotus purpuratus*)	18
Silkworm (*Bombyx mori*)	28
Swine (*Sus scrofa*)	19
Tetrahymena pyriformis	5
Toad (*Xenopus laevis*)	18
Wasp (*Habrobracon sp.*, female 2n = 20, male 1n =10)	10

(See genome, polyploid, haploid, acrocentric, telocentric, chromosome arm, sex chromosomes, Robertsonian translocation)

CHROMOSOME PAINTING: identification of chromosomes by *in situ* hybridization using fluorochrome-labeled probes. With recent refinements of these techniques each human chromosome can be distinctly identified by color and various rearrangements can be detected in an unprecedented manner. (See fluorochromes, fluorescence microscopy, chromosome morphology, *in situ* hybridization, FISH, WCPP, USP, spectral karyotyping, telomeric probes)

CHROMOSOME PAIRING (synapsis): is the ability of homologous (or under some circumtances homoeologous) eukaryotic chromosomes to associate intimately during the prophase of meiosis and form bivalents. The bivalents represent the essentially identical, homologous paternal and maternal chromosomes. In some organisms chromosomes may pair also during mitoses but this association is generally not considered equally intimate although in the salivary glands of *Drosophila* (and other dipterans) homologous chromosomes are tightly associated (somatic pairing). Synapsis — most commonly — begins at the termini of chromosomes at the zygotene stage and proceeds toward the centromere. By pachytene the pairing is complete and if it is not complete it is not going to completion later and some areas will remain unpaired. The pairing is genetically determined and single, specific genes may prevent pairing such as *␣␣␣␣*␣*1* (*asynaptic*, chromosomal location 1-56 in maize). Curiously, in *as1* maize crossing over may increase. In hexaploid wheat the *Ph* gene suppresses homoeologous association of chromosomes but when it mutates or deleted (monosomics and nullisomics for chromosome 5B) high degree of homoeologous pairing occurs even in hybrids of related species. Some *desynaptic* genes terminate pairing precociously. In polyploids (polysomics) the homologous chromosomes may display multivalent association but at any particular point the synapsis is only between two chromosomes. In the salivary glands of trisomic flies the three chromosomes may be paired all along their length. In diptera where the X and Y chromosomes share homologous euchromatic termini, a short-duration, delayed pairing occurs (touch-and-go pairing). Synapsis is facilitated by the formation, beginning in leptotene, of the synaptonemal complex, a tripartite protein structure, formed between the paired chromosomes. Synapsis provides the opportunity for the homologous chromosomes to experience crossing over and recombination. When chiasma and recombination takes place the distribution of the chromosomes at anaphase is orderly (*exchange pairing*) whereas in the absence of chiasmata (*distributive pairing*) the chance for nondisjunction increases. There are various types of pairings quite distinct from synapsis, these non-specific associations at the chromocenter of salivary gland chromosomes, association of telomeric heterochromatin in monosomes or self-pairing in certain univalents may be observed if they posses more or less homologous sequences. (See meiosis, distributive pairing, illegitimate pairing, crossing over, recombination, somatic pairing, parasexual mechanisms)

CHROMOSOME PARTIONING: in prokaryotes, after replication, one of the two chromosomes, each, is delivered to the daughter cells. (See also mitosis [for comparison])

CHROMOSOME POSITIONING: in prokaryotes the old and new (replicated) chromosomes tend to move to opposite poles of the cell. A defect in positioning may involve a condition analogous to nondisjunction in eukaryotes, i.e. the 0 - 2 distribution. (See also anaphase)

CHROMOSOME PUFFING: takes place in the polytenic chromosomes (of animals and plants) when genes are activated and begin synthesizing large amounts of RNA. When transcription is terminated the puffs recede. Puffing may be stimulated by the administration of hormones (e.g. ecdyson). (See illustration at puff)

CHROMOSOME ROSETTE: at the prometaphase stage (lasting about 5-10 min in human cells) the chromosomes are arranged like a wheel, centromeres oriented toward the hub and the arms assuming an arrangement like the spokes, If the chromosomes are painted by fluorochrome labels, the homologs appear at opposite positions of the rosette. (See mitosis, metaphase, FISH)

CHROMOSOME SCAFFOLD: the structurally preserved form of the chromosome freed from histone proteins. (See chromatin)

CHROMOSOME SEGREGATION: is the basis of mendelian inheritance (see mitosis and meiosis). In autopolyploids the chromosomes may segregate reductionally, e.g. at anaphase I in an autotetraploid chromosomes with A, A, A, A may move toward one pole and chromosomes

Chromosome segregation continued

a, a, a, a toward the other. This is called reductional segregation (R). Alternatively, the distribution may be *A a, a A,* and *a A, A a,* or *a A, a A* and *A a, A a,* (respectively) i.e. equational segregation (E) occurs. These two types of separations follow the proportion of 1R:2E. The term *chromosome segregation* is used also for cases in polyploids when genes are closely linked to the centromere and crossing over does not take place between them, in contrast to *maximal equational segregation* when one crossing over takes place between gene and centromere in each meiocyte. (See meiosis, autotetraploid)

CHROMOSOME SET: a group of chromosomes representing once all the chromosomes of the haploid set, the genome. It is represented by x. (See genome)

CHROMOSOME SORTING: see flow cytometry

CHROMOSOME STICKINESS: apparent adhesion of chromosomes that tend to stay together.

CHROMOSOME STRUCTURE: electronmicroscopic image of a chromosome reveals much more details of a chromosome than a lightmicroscopic one. The light microscope has, however, great adavantage for the study of chromosomal behavior in mitosis and meiosis such a chiasmata, nondisjunction, misdivision, etc. (See also chromatin, nucleosomes, Mosolov model).

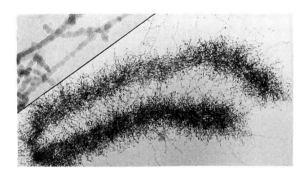

THE IMAGE AT LEFT REPRESENTS A LOOSELY PACKAGED ACROCENTRIC HUMAN CHROMOSOME ISOLATED FROM A BURKITT'S LYMPHOMA CELL. THE INDIVIDUAL FIBERS ARE SHOWN ON THE INSET AT APPROXIMATELY 85,000X MAGNIFICATION.
(Photomicrograph from Lampert, F., Bahr, G.F and DuPraw, E.J. 1969 Cancer 24:367)

CHROMOSOME SUBSTITUTION: can be alien substitution (see there) or inter-varietal substitution when one chromosome of a variety is replaced by the corresponding chromosome of another variety in polyploids where monosomic or nullisomic lines can be propagated Thus some desirable genes can be transfered without altering the genetic background. (See flowchart next page, linkage in breeding, genetic engineering)

CHROMOSOME SYMBOLS: see *Drosophila,* gene symbol

CHROMOSOME THEORY: developed at the turn of the 20 the century stated that the genetic material is contained in the chromosomes, and the Mendelian laws are based on the mechanisms of meiosis. (See Mendelian laws, Mendelian segregation)

CHROMOSOME UPTAKE: is outlined in a flowchart on page 197.

CHROMOSOME WALKING: is mapping the position of a DNA site or a gene by using overlapping restriction fragments. The principle is somewhat similar to classical cytogenetic mapping with overlapping deletions. It is used also for map-based isolation and then cloning of specific genes. The success of isolation of genes by this method is greatly affected by the size of the genome and even more importantly on the distance that must be "walked" from a known genomic position toward the desired gene. Some means must also be found for determining the function of the gene so its identity could be verified. In large eukaryotic genomes, the procedure is facilitated if physical maps are already available. (See cosmid vectors, YAC vectors, chromosome jumping, map based cloning, chromosome landing, position effect, see diagramatic outline on page 198).

CHROMOSOMIN: non-histone protein of the chromatin.

CHRONIC GRANULOMATOUS DISEASE (CGD): is a complex disease based on the inability of the phagocytizing neutrophils to destroy infectious microbes because defects in de-

Chronic granulomatous disease continued

livering high enough levels of oxygen to the neutrophil membranes. Laboratory diagnosis is generally based on the failure of the phagocytes to reduce nitroblue tetrazolium. Prenatal diagnosis exists for males and carriers. The afflicted persons are liable to infections. The human X-chromosomal gene was localized to Xp21 and it has apparently a defect in the cytochrome b system, probably most commonly in the b subunit (CYBB) whereas the autosomal form (human chromosome 16q24) is defective in the a subunit (CYBA). In some variants other functions may also be involved. (See immunodeficiency, cytochromes, neutrophil, phagocytosis)

CHRONIC LYMPHOCYTIC LEUKEMIA (CLL): see leukemia

CHRONIC RADIATION: radiation dose(s) is/are delivered continuously without interruptions. (See alfractionated dose)

CHRONOLOGY OF GENETICS: see genetics chronology of

CHRYSANTHEMUM (*C. indicum*): herbaceus, perennial ornamental; 2n = 20, 45-63

chv: chromosomal virulence loci in *Agrobacterium*. (See virulence genes of *Agrobacterium*)

CHYLOMICRIN: transporters of lipoproteins ingested or synthesized in the small intestines. (See hyperlipoproteinemia, Anderson disease)

AN OUTLINE OF INTERVARIETAL CHROMOSOME SUBSTITUTION. THE DISOMIC SUBSTITUTIONLINE HAS THE CHROMOSOMES OF THE ORIGINAL MONOSOMIC VARIETY OF WHEAT, EXCEPT THE BORROWED CHROMOSOME. IT IS EUPLOID. IF THE DONATED CHROMOSOME HAS ANY UNDESIRABLE GENE(S), RECOMBINATION MAY REMOVE IT/THEM.

CHYMASES: are similar to chymotrypsin and hydrolyze peptide bonds near the carboxyl end of hydrophobic amino acids. (See chymotrypsin)

CHYMOTRYPSINS: proteases (M_r 25 K) cleaving near aromatic amino acids, non-polar groups, ester bonds. They are targeted by nerve gas (DFP)

Ci (Curie): measure of radioactivity; 1 Ci= 3.7×10^{10} disintegrations/sec. (See radioactivity, isotopes, radioactive label)

cI: phage λ repressor. (See lambda phage)

CIBD (chronic inflammatory bowel disease): prevalence in the Western world 2×10^{-3}. (See Crohn disease)

CIGAR: CGGAAR (R=purine) enhancer motif of Herpes simplex virus. (See tat-garat)

CIGARETTE SMOKE: contains dozens of various combustion products including the most potent carcinogens and mutagens, e.g. benzo(a)pyrene about 20-40 ng/cigarette and it is responsible for the majority of cases of lung cancer cells carry mutations in the p53 tumor suppressor

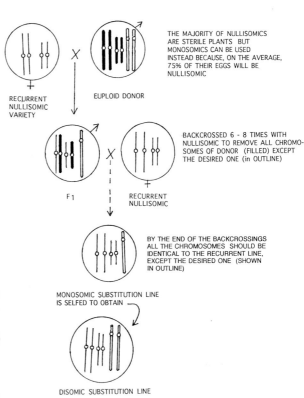

THE MAJORITY OF NULLISOMICS ARE STERILE PLANTS BUT MONOSOMICS CAN BE USED INSTEAD BECAUSE, ON THE AVERAGE, 75% OF THEIR EGGS WILL BE NULLISOMIC

RECURRENT NULLISOMIC VARIETY   EUPLOID DONOR

BACKCROSSED 6 - 8 TIMES WITH NULLISOMIC TO REMOVE ALL CHROMOSOMES OF DONOR (FILLED) EXCEPT THE DESIRED ONE (in OUTLINE)

F1   RECURRENT NULLISOMIC

BY THE END OF THE BACKCROSSINGS ALL THE CHROMOSOMES SHOULD BE IDENTICAL TO THE RECURRENT LINE, EXCEPT THE DESIRED ONE (SHOWN IN OUTLINE)

MONOSOMIC SUBSTITUTION LINE IS SELFED TO OBTAIN

DISOMIC SUBSTITUTION LINE

EUPLOID WHEAT EAR, VARIETY CHINESE SPRING

Cigarette smoke continued

gene, mainly G→T transversions, at codons 157, 248 and 273. The codon 157 host spot is absent from othert types of cancers. These hot spots are the sites of adduct formation by benzo(a) pyrene diol epoxide and guanine-N^2. (See p53, cancer, hot spot, benzo(a)pyrene, transversion)

Chromosome uptake by animal cells continued from page 195

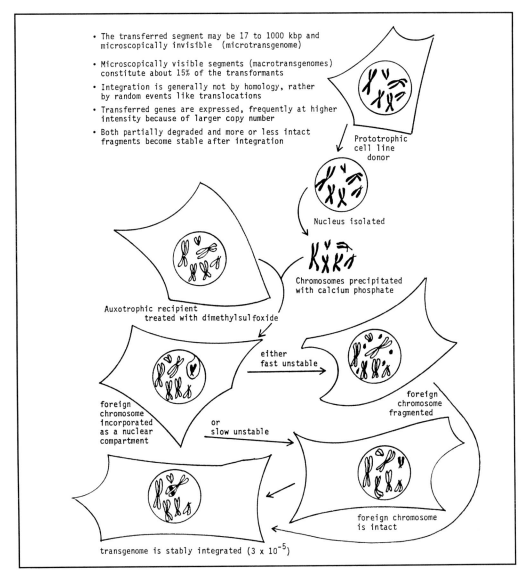

CIITA: is an apparently non-DNA-binding modulator of the synthesis of class II MHC molecules. (See MHC)

CILIA (singular cilium): hair-like structures formed from microtubules and are extensions of the basal bodies. They are used for locomotion (swimming) in watery media or on viscous films by a vibratory or lashing movement. (See microtubule)

CILIARY NEUROTROPHIC FACTOR (CNTF): is released by glial cells to repair damaged neurons. (See neuron)

Cin: is an invertase. (See invertases, Cin4)

Cin4: see non-viral retroposons, hybrid dysgenesis I - R

CIP (calf intestinal alkaline phosphatase): used for the removal of 5' phosphate from nucleic acids and nucleotides.

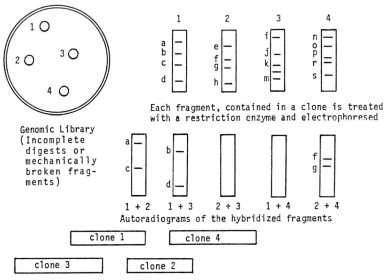

CHROMOSOME WALKING OUTLINE (to page 195)

CIR: cardiac inwardly rectifying ion channel. (See ion channels, I_{KAch})

CIRCADIAN RHYTHM: with a daily (24 h) periodicity; from the Latin *circa diem* [around the day]. Such periodicity (rhythm) may occur in plants in opening and closing flowers, translocation of metabolites or may affect various ways the daily behavioral and metabolic pattern of animals (such as sleep, social activities, etc.). Usually the circadian rhythm is more or less temperature independent. In *Drosophila* the *per* (period) mutation has, however a dimerization domain, PAS, and that is subject to temperature effects. In animals the control spot for the circadian periodicity resides in the suprachiasmatic nucleus of the hypothalamus. Glutamate, methyl-D-asparate or nitric oxide treatment injection into the brain produces the same effect as light in controlling the periodicity. (See also the *per* mutation in *Drosophila*, endogenous rhythm, hypothalmus, nitric oxide, clock genes, endogenous rhythm, brain human)

CIRCULAR DICHROISM: is the difference between the molar absorptivities for left-handed and right-handed polarized light and it is observed in chiral molecules (enantiomorphs). Circular dichroism detects Z DNA structures and the interaction of drugs and carcinogens/mutagens with the DNA. (See enantiomorph, chirality, Z DNA)

CIRCULAR DNA: covalently closed ring-shaped DNA molecules such as the genetic material of bacteria, eukaryotic organelles and the majority of the plasmids.

CIRCULARIZATION: forming a circle, like plasmids; unwanted circularization of DNA can be prevented by directional cloning or treatment with bacterial phosphatase to remove terminal phosphates needed for joining DNA ends. (See directional cloning)

CIRRHOSIS OF THE LIVER: autosomal recessive disease characterized by fibrous structure of the liver. It may be precipitated by various environmental factors such as copper toxicity, alcoholism, antitrypsin deficiency. It may occur as a component of other syndromes. (See Wilson's disease, galactosemia, antitrypsin, alcoholism, autoimmune diseases, syndrome)

CIS ARRANGEMENT: two genes (or two different mutant sites of a locus) are in the same chromosome strand. (See trans arrangement, cis-trans test)

CIS-ACTING ELEMENT: must be on the same DNA strand as its target to act on transcription. (See also trans-acting element)

CIS-DOMINANT: a dominance affecting only alleles in cis position but not those in trans. (See cis arrangement)

CIS-GOLGI: the side of the Golgi apparatus where molecules enter the complex. (See Golgi apparatus, trans-Golgi network)

CIS-IMMUNITY: the property of transposons to prevent integration of another element within the boundary of the insertion element. (See self-immunity, transposon)

CISPLATIN (cis-diamminedichloroplatinum, cis-DDP): a DNA cross-linking agent and an effective anticancer drug. (See cancer therapy)

CISS HYBRIDIZATION (chromosome *in situ* suppression): before DNA probes are applied to chromosomes for *in situ* hybridization, the chromosomes are treated with DNA to block the non-target sequences so these would not interfere with annealing of the specific labeled probes. (See *in situ* hybridization, probe)

CIS-TRANS TEST: is a procedure for determining allelism. If two independent *recessive* mutations are made heterozygous in a diploid (or by using an F' plasmid in bacteria) are in the opposite strands (in trans position, $\frac{m+}{m+}$) and fail to complement each other (i.e. the phenotype is mutant), then the two mutations are allelic (occupy the same cistron). When, however, the two recessive mutations both are in the same DNA strand (in cis $\frac{++}{mm}$ position) and are made heterozygous (or merozygous in prokaryotes) they are expected to be complementary, i.e. non-mutant in phenotype, because the strand containing the two non-mutant sites permits the transcription of a wild type (un-interrupted) mRNA. Molecular geneticists involved in physical DNA mapping use the term allele for any physical variation that is inherited by a Mendelian fashion and occupies the same chromosomal locus. (See allele, allelism test, cistron)

CISTERNA: membrane-enclosed space and frequently contains fluid.

CISTRON: segment of the DNA coding for one polypeptide chain or determining the base sequence in one tRNA or in one rRNA subunit. Mutant sites within a cistron generally do not fully complement, i.e. the heterozygotes are not wild type although in rare cases weak allelic complementation may be observed. (See allelic complementation)

CIS-VECTION: a position effect, operational only in case the genetic elements are syntenic, e.g. the promoter and structural gene, but not the regulator gene which may be effective also from trans position, i.e. if it is on the homologous or another DNA strand. (See position effect, synteny, operon, pseudoalleles, cis-acting element)

CITRIC ACID CYCLE: see Krebs - Szentgyörgyi cycle

CITRULLINEMIA (ASS): chromosome 9 recessive defect in the enzyme argininosuccinase. Normally citrulline is converted — via aspartate — into argininosuccinate. If the latter cannot be cleaved, citrulline and ammonia accumulate and as a consequence incontinence, insomnia, sweating, vomiting, diarrhea, convulsions, psychotic anomalies and even periods of coma may result. The disease has an early onset and may proceed progressively into adulthood; rarely the onset is during adult life. Craving for high arginine food (legumes) and avoidance of low arginine food and sweets is noticeable. The ASS gene may be present in 10 copies per human genome scattered over several chromosomes according to hybridization by a DNA probe. The multiple copies are presumably pseudogenes. (See argininemia, amino acid metabolism, pseudogene)

CITRULLINURIA: same as citrullinemia

CITRUS (*Citrus* spp): the taxonomy is unclear but several species are known, x = 9 and diploid as well as tetraploid forms exist among the lemons, oranges, mandarin, lime, grapefruit, etc.

CJD: see Creutzfeldt-Jakob disease

CJM: cell junction molecule. (See gap junction)

C-KINASE: a protein phosphorylase activated by Ca^{++} and diaglycerol.

CKI: an inhibitor of the CDK-cyclin complex. (See CDK)

CKR: chemokine receptor. (See chemokines, acquired immunodeficiency, CCR)

CLADE: a group of species descended from a common ancestral taxonomic entity.

CLADISTIC: representation of descent in the manner of a dendrogram, i.e. the divergence of taxonomic groups is shown by links and nodes. (See dendrogram, evolutionary tree, parsimony, character index)

CLADOGENESIS: an evolutionary change involving branching of lineage of descent. (See also anagenesis)

CLADOGRAM: see evolutionary tree, character index

CLAFORAN: see cefotaxim antibiotic

CLASS I GENES (eukaryotic): are transcribed by RNA polymerase I; these include 5.8S, 18S and the 28S ribosomal RNAs. (See pol I eukaryotic)

CLASS II GENES: are transcribed by eukaryotic RNA polymerase II; these include mRNA and snRNA (except U6 RNA). They carry in the mRNA transcript a 7-methyl guanine cap (with the exception of the picornaviruses) and a 2,2,7-trimethyl guanine in the U RNAs. In lower eukaryotes 75 to 125, in vertebrates 200 to 300 residues long poly-A tail is added posttranscriptionally. Histone mRNAs and U RNAs have no poly A tails. Some mRNAs include N^6-methylated adenine, U RNAs have modified uracils. They are regulated by cis- and trans-acting elements. The cap is associated with cap-binding protein, the capping enzyme. (See pol II, eukaryotic, transcription, transcription factors, transcription termination in eukaryotes, transcription unit, capping enzyme, polyA mRNA, U RNA, cis-acting, trans-acting)

CLASS III GENES: of eukaryotes are transcribed by RNA polymerase III; they include 5S ribosomal RNA and some small cytoplasmic RNAs. (See pol III eukaryotic)

CLASS SWITCHING: change in expression of immunoglobulin (antibody) heavy chain genes during cellular differentiation of an antibody-producing lymphocyte by changing the production from one immunoglobulin heavy chain to another. (See immunoglobulins, antibody gene switching)

CLASSICAL GENETICS: studies functions based on phenotype and genotype of the genetic material serve the primary guidance to the understanding of the mechanisms involved, in contrast to reversed genetics where the analysis begins with the molecules. (See reversed genetics)

CLASSICAL HEMOPHILIA: see hemophilia

CLSSSIFICATION: sorting out phenotypes (or genotypes) by groups. It may be difficult in case of continuous variation or when the penetrance or expressivity is low.

CLASTOGEN: any agent that can cause chromosomal breakage directly, or indirectly by affecting DNA replication. Clastogenic agents may be ionizing radiation, bleomycin, hydroxyurea, maleic hydrazide, etc. (See chromosome breakage)

CLATHRIN: 180 kDa proteins that in cooperation with smaller (35 kDa) polypeptides form the polyhedral coat on the surface of coated vesicles that are involved in intracellular transport between cellular organelles. Before fusing with the target the vesicle coats are stripped with the assistance of chaperones (hsp70) and another cofactor, auxilin. (See adaptin, cargo receptors, endocytosis, coatomer, chaperone)

CLAW-FOOT (Roussy-Levy hereditary areflexic dysplasia): an autosomal dominant anomaly usually involving paternal transmission. It bears resemblance to the Charcot-Marie-Tooth disease but it is accompanied by hand tremors. (See Charcot-Marie-Tooth disease)

CLAW-LIKE FINGERS AND TOES (curved nail of fourth toe): a rare apparently autosomal recessive nail deformity of the fourth (and fifth) toes and fingers.

CLB: a mitotic cyclin protein. (See cyclin, CDK)

***ClB* METHOD**: detects new sex-linked lethal mutations among the grandsons of *Drosophila* males on the basis of altered male: female ratios (C is an inversion, eliminating cross-over chromosomes *l* is a lethal gene: and *B* stands for *Bar eye*). If a new recessive lethal mutation (*m*) occurred in the X-chromosome of the grandfather, the grandson receiving this or the *ClB* chromosome will die. In F_2 the males carrying either the *Cl B* or the mutant X-chromosome will die; without the new mutation (*m*) the female : male ratio is 2:1. In case of a new lethal mutation no male grandsons may survive or in case the expressivity of the new lethal gene is reduced, some males may survive. (See also *Basc*, autosomal dominant, autosomal recessive mutation)

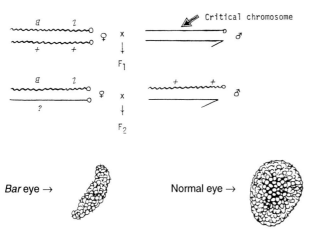

CLEAVAGE: early embryonal division (that gives rise to the blastomeres) by what the larger fertilized egg breaks up into several smaller cells without growth. In general, splitting cells, organelles and macromolecules into two.

CLEAVAGE NUCLEUS: the nucleus of the dividing egg.

CLEFT PALATE: an oral fissure frequently associated with harelip (lip fissure). Its incidence in the general population is about 4/10,000; its recurrence risk in sibs of an affected individual about 2% or more, in monozygotic twins 50%. Actually, these developmental anomalies may be parts of autosomal or X-linked dominant or autosomal recessive syndromes and some trisomies. In mouse deficiency of the β3 subunit of the A type γ-aminobutyric acid receptor results in cleft palate. (See recurrence risk, sib, trisomy, Van der Woulde syndrome, GABA, CATCH)

CLEIDOCRANIAL DYSOSTOSIS: autosomal dominant mutations may lead to deficiency of closure of the skull sutures, and chest, shoulder, hip and finger anomalies, short stature, etc. Autosomal dominant forms may show also reduced jaw size, absence of thumbs or toes, loss of distal digital bones. (See stature in humans, osteogenesis imperfecta, dysostosis, pycnodysostosis)

CLEISTOGAMY: shedding pollen before the flowers open, and thus results if self-fertilization in plants. (See autogamy, protandry, protogyny)

CLEISTOTHECIUM: a closed, spherical fruiting body of ascomycetes such as in *Aspergillus* and powdery mildew. The spores are released after the decay of the structure. (See perithecium, gymnothecium, ascogonium)

CLINE: a gradual change in the distribution of genotypes or phenotypes within a large population caused by the environment, population density or other factors. (See aclinal)

CLINICAL GENETICS: deals with practical genetic problems encountered during patient care. (See empirical risk, medical genetics, human genetics, genetic counseling, counseling genetic, informed consent)

CLINICAL TESTS FOR HETEROZYGOSITY: many genes fail to reveal their anomalous alleles by simple visual observation. Some of these are detectable by enzyme or serological assays in biopsies, by amniocentesis, and cell cultures. These tests may not be able to identify with certainty the genetic constitution because the same function may be affected by different gene loci. Some mutant alleles can be identified by RFLP or PCR analysis. (See carrier, PCR)

CLIP (class II-associated invariant-chain complex): see major histocompatibility complex

CliP: see sister chromatid cohesion

CLIPBOARD: of a computer is a storage area of the memory system from where information can be transfered to different documents created in the computer.

CLITORIS: an oval body (homologous to the male penis) in the inner side of the cleft between the opening of the urethra and the vagina of mammalian females. Its expansion followed by relaxing results in the sexual excitement (climax, orgasm) without ejaculation.

CLN: cyclin genes regulating cell division. (See cyclin, CDK)

CLOCK GENES: affect the biological clock such as the diurnal rhythm and endogenous rhythm or aging. (See circadian rhythm, endogenous rhythm, aging)

CLONAL ANALYSIS: is the study of pattern formation (genetic mosaics) as a consequence of mutation, deletion, recombination, nondisjunction. Such an analysis permits tracing the event to its origin, estimation of the number of cell divisions that have taken place since the event, the number of cells in the primordium involved, etc. Usually a requisite of such an analysis that appropriate genetic markers would be available. Cell autonomous genes are particularly useful for clonal analysis because genes of neighboring cells do not affect their expression. Heterozygotes for recessive color markers are frequently used because the loss or mutation of the dominant allele will result in the formation of recessive sector(s), pseudodominance. An unstable (ring) X-chromosome in *Drosophila* may lead to the formation of gynandromorphs. Nondisjunction or somatic recombination may lead to the formation of twin spots (sectors). In *Drosophila*, developmental compartments of polyclonal origin may be revealed on the basis what structures, organs are affected simultaneously or consequently. Mutations formed in the last three divisions of the wing imaginal disk involve changes in cuticular elements e.g., hairs and wing veins. The presence of sectors also reveals that a gene is cell autonomous and also whether it expression is modified by the product of other genes. Extracts of wild type lymph, mRNA or protein injected into mutants may correct genetic defects in a localized form, according to a gradient or in a particular pattern, depending on the nature of function of the corresponding mutant allele, being a specific transcription factor, a transmembrane protein, a signal receptor, etc. DNA labeled by fluorochromes, hybridized in the tissues, may identify the regions of mRNA distribution of a particular gene. The function of particular genes in regulation of others can be detected by immunostaining of particular loci with specific antibodies. (See also morphogenesis, cell lineage, fluorochromes, probe, immunostaining)

CLONAL RESTRICTION: the proliferating cells are propagated only within the preordained pattern of differentiation. (See clonal analysis, cell lineages)

CLONAL SELECTION: a few antibody producing lymphocytes which are committed to the production of a specific antibody are selected for clonal proliferation and stimulated for the synthesis of that specific antibody when infected by the pertinent antigen. This way immunity may build up. (See immune system, lymphocyte, antibody, immunoglobulins)

CLONE: progeny of asexual reproduction or molecular cloning yielding identical products. (See asexual reproduction, molecular cloning, cloning vectors)

CloneConvert: a computer program that converts a file into the 'miniset.dat' format to individual MapSearch Probes.

CLONIDINE: is an α-2-adrenoreceptor agonist. At low concentration it decreases presynaptic firing of noradrenergic cells. Its antagonist is idazoxan. (See synaps, adrenergic receptor)

CLONING: asexual reproduction in eukaryotes or replication of DNA (genes) with aid of plasmid vectors in appropriate host cells (molecular cloning). (See clone, cloning vectors, DNA library, plasmids, DNA, nuclear transplantation, transplantation of organelles)

CLONING BIAS: the deviation from randomness in the representation of fragments in a DNA library. The bias may be caused by rearrangement of direct repeats in *Rec*$^+$ bacterial strains, (but can be avoided by *recA* hosts); palindromic sequences may become unstable in some λ phage and plasmid vectors (can be avoided by the use of *recB* and/or *recC* and *sbcB* strains of *E. coli*, base modification of the DNA, host restriction enzymes, etc.). (See DNA library, restriction enzyme, cloning vectors)

CLONING SITES: are recognition sequences for restriction enzymes within genetic vectors where passenger DNA can be inserted. (See restriction enzymes, vectors, passenger DNA)

CLONING STRATEGY: the plan that will permit the identification of the cloned copy either by a suitable probe (DNA, RNA or antibody) or one may use a positional cloning or PCR based procedure. (See probe, PCR, positional cloning)

CLONING VECTORS: generally a plasmid, phage or eukaryotic virus-derived linear or circular DNA capable of reproduction (most commonly in bacteria or yeast) and producing (large number) molecular clones of the DNA inserted into it. Cloning vectors must have replicator mechanisms (replication drive unit) for self-propagation, multiple cloning sites (single or few recognition sites for several restriction enzymes), selectable markers (for verifying the success of molecular recombination and uncontaminated maintenance), regulatory elements for their copy number in the host (generally smaller plasmids can be present in larger number of copies), mechanisms for equal partition among the daughter cells and genetic stability to prevent rearrangement by host enzymes. It is frequently desirable to propagate in more than one host cell (shuttle vectors). Some of the cloning vectors are used only for propagation of DNA, others permit expression of the genes carried, others may be useful to isolate functional elements of the hosts (promoters, enhancers) by virtue of *in vivo* gene fusion. (See lambda vectors, cosmids, phagemids, YAC, SV40, retroviral vectors, agrobacterial vectors, plasmids, ColE1)

CLONING VEHICLE: cells suitable to propagate the cloning vectors, e.g. *E. coli*, yeast, *Agrobacterium*, etc. (See cloning vectors)

CLOSED PROMOTER COMPLEX: the transcriptase attached to the target promoter cannot start transcription because the DNA strands are not separated. (See also open promoter complex, pol prokaryotic RNA polymerase)

CLOSED READING FRAME: chain termination (nonsense) codons block its translation.

CLOSURE OF MAPPING: when approaching completion and two-genome-size DNAs had been mapped, the use of random clones is very inefficient, therefore non-random clones are employed. (See physical map)

CLOTTING FACTOR: see antihemophilic factors

CLOVERLEAF: a representation of the tRNA displaying the single strand stem (amino acid arm) and the D (dihydrouracil), AC (anticodon) and T (thymine) loops reminding to stem and the three leaflets of a leaf of a clover plant. (See tRNA)

CLOVERS (*Trifolium* spp) include about 250 species with the most prominent representatives: white clover (*T. repens*, 2n = 32)), the fragrant alsike (*T. hybridum*, 2n = 16), strawberry (*T. fragiferum*, 2n = 16), red clover (*T. pratense*, 2n = 14), *crimson clover* (T. incarnatuum, 2n = 14) and the subterranean clover (*T. subterraneum*, 2n = 16).

Clp: different prokaryotic proteins involved in chaperone or protease activities. (See protease, chaperone)

CLUBFOOT: a hereditary malformation of the foot with a prevalence of below 0.1% and with a recurrence risk among the sibs of an afflicted child is about 4%. Thus it appears to be under the control of more than one gene and with substantial environmental influences because even monozygotic twins may not be both afflicted. (See limb defects)

CLUSTER HOMOLOGY REGIONS (CHR): are homologous DNA sequences located in different chromosomes and probably are the results of gene duplication. In some cases the coded protein is specialized to tissue- or organelle-specific functions or its function has changed, assumed a new function. (See clustering of genes, duplication, evolution and duplication)

CLUSTERING OF GENES: bacteriophage genes are in a linear order of morphogenesis, several bacterial genes are clustered in the exact order of the biosynthetic pathway (tryptophan operon) others are only within a group but not in strict biosynthetic order (histidine operon) and they are under coordinated regulation. In the lower eukaryotes (fungi) some of the histidine genes and chorismic acid genes are in groups although they are not transcribed into a polycistronic RNA. The ribosomal and tRNA genes are in a linear array in prokaryotes and eukaryotes and processed

Clustering of genes continued

after transcription into individual molecules. The histone genes in *Drosophila* and sea urchin are in the same region but separated by spacers. Some of the antibody genes in mammals are clustered in genes families and are many times repeated. Some genes of the nematode *Caenorhabditis* are even polycistronic. (See operon, coordinate regulation, polycistronic mRNA, *His* operon, chorismate, histones, spacer DNA, tRNA, rRNA)

CLUSTERING OF PHENOTYPES: may be caused by exposure of similar environmental effect. Epidemiological factors such as carcinogens in the environment, viral infections may precipitate the expression of phenotypes in an unusual distribution.

CLUTCH: a cluster of eggs laid by a bird.

c.m. or c.M.: centi Morgan = 1% recombination; 1 map unit. (See mapping genetic, recombination)

C-MEIOSIS: meiosis is arrested because colchicine is poisoning the spindle fibers. (See meiosis, colchicine)

cMG1: a protein related to TIS11 (a 67 amino acid region is 72% identical), responds to epidermal growth factor plus cycloheximide.

CMI (cell-mediated immunity): see immunity

C-MITOSIS: mitotic anaphase is blocked by the poisonous effects of colchicine, and consequently the cell and its progeny may become polyploid. (See C-meiosis)

cms: cytoplasmically determined male sterility. (See cytoplasmic male sterility)

CMT: monkey cell line expressing SV40 T antigen. (See SV40)

CNS: central nervous system. (See brain human)

CNTF (ciliary neurotrophic factor): see APRF, neurotrphins, ciliary neurotrophic factor

C-ONCOGENES: cellular oncogenes. (See also v-oncogenes)

Co^{60}: see isotopes

CO$_2$ SENSITIVITY: of some strains of *Drosophila* is manifested as paralysis and death after anesthesia with the gas. The condition is caused by infection with rhabdovirus sigma. This RNA virus resembles the vesicular stomatitis virus (VSV) of horses (an acute febrile [fever causing] infection on the tongues, mouth membranes and lips) and to some fish viruses (PFR, SVC) that can also elicit carbondioxide sensitivity in *Drosophila* but the fly virus cannot infect vertebrates. In the non-stabilized state only the *Drosophila* females transmit the virus. In the stabilized state the transmission by eggs is 100% and it is transmitted also by the sperm although in the latter case stabilized infection does not ensue. The *ref* mutants (in chromosomes X, 2, 3) are refractory to this type of infection. (See *Rhabdoviridae, Drosophila*)

CoA: see acetyl coenzyme A

COACERVATE: colloidal aggregate of organic compounds. They probably have played a role in organic evolution. (See prebiotic evolution)

CO-ACTIVATOR: molecules required to activation of gene transcription in addition to TBP, TAF and the general transcription factors. (See TBP, TAF, transcription factors, TATA box, high mobility group of proteins [HMG], transactivator)

CO-ADAPTED GENES: represent genotypes capable of expression of a satisfactory (fit) phenotype. (See fitness)

COAGULATION FACTORS: see antihemophilic factors

COALESCENCE: the point or node of an evolutionary tree where two lineages merge (diverge) at a time or at any other scale. (See evolutionary tree)

COALESCENT: is a statistical parameter of the genealogical information of genetic data.

COANCESTRAL: gene(s) identical by descent in two individuals, e.g. uncle and niece, first cousin, etc. (See coefficient of coancestry, consanguinity, inbreeding coefficient)

COARCTATION OF THE AORTA: is an apparently autosomal polygenic narrowing of the blood vessels leading to congenital heart failures. (See cardiovascular disease, heart disease, supravalvular aortic stenosis)

COAT COLOR: see pigmentation of animals, fur color

COAT PROTEIN: the protein(s) of the viral capsid. (See capsid, capsomer)

COATED PITS: generated on the surface of coated vesicles by invagination and pinching off and thereby facilitating transport by losing the coat and fusing with other intracellular vesicles (lysosomes). (See lysosomes, dynamin)

COATED VESICLES: clathrin-coated vesicles. (See clathrin)

COATOMER (coat-protomer): is the large protein complex on the surface of vesicles (Golgi) mediating non-selective transport within cells. Their assembly requires ATP. After the transfer of the cargo, the coatomer still retained and docks with another membrane. (See clathrin)

cob: *apocytochrome b* gene in the mitochondrium (yeast) may exist without intron and with introns (called boxes). Its exons code for the apocytochrome 1 protein whereas the box 3 intron codes for a "maturase" protein that excises the introns from the long form gene and a box 7-coded protein splices the exons of the adjacent cytochrome oxidase (*oxi3*) gene. (See cytochromes, mtDNA)

COBALAMIN: (cyanocobalamin) vitamin B_{12}, a coenzyme for methylmalonyl CoA mutase; it has therapeutic use in anemia and acidosis. (See transcobalamin, transcobalamin deficiency)

cob-box **GENES**; encode mitochondrial cytochrome oxydase with introns. The intron boxes may have independent functions in processing the apocytochrome transcripts. (See mtDNA)

COBROTOXIN: the major protein toxin in the cobra venom. (See toxins)

COCA (*Erythroxylon coca*): is a source of cocaine; $2n = 2x = 24$.

CO-CARCINOGENS: may not be carcinogenic alone but they may act as tumor promoters such as phorbol esters. (See phorbol esters, carcinogens)

COCAIN (3-benzoyloxy)-8methyl-8-azabicyclo[3.2.1]octane-2-carboxylic acid methyl ester): is a topical anesthetic and a euphoriant. It is an addictive drug obtained either from *Erythroxylon* plants or produced by chemical synthesis. Cocaine-dependence and intense craving may return even after prolonged abstinence. The priming of the relapse seems to be mediated by the D_2-like receptor agonists of the dopamine system whereas the D_1-like receptor agonists prevented cocaine-seeking behavior. (See dopamine, agonist)

Cochliomya hominivorax (screwworm): a tropical and subtropical fly, an obligatory parasite of warm-blooded animals. Its infestations is causing myasis (weight loss and sometimes death). By puncturing the skin (hide) resulting in multimillion dollar annual loss to animal breeders and it is menacing also to people. For its biological control the mass release of genetically sterile (irradiated) males proved effective. (See genetic sterilization, myasis)

COCKAYNE SYNDROME: is characterized by very short stature, precocious aging, deafness, eye degeneration, mental retardation and sunlight sensitivity. The defect seems to be associated with a DNA helicase defect but unlike other DNA repair problems (ataxia telangiectasia, xeroderma pigmentosum, Bloom syndrome, Fanconi anemia) it does not involve increased proclivity to cancer. Some of the defects seem to be in the transcription-coupled nucleotide exchange repair. It may be controled by non-allelic autosomal recessive loci. Cockayne syndrome 2 has been located to human chromosome 10. The homologous yeast gene is *RAD28*. (See DNA repair, chromosome breakage, light-sensitivity diseases, DNA repair, xeroderma pigmentosum)

COCKROACH: *Blatta germanica*, $2n = 23$ male, 24 female.

CO-CLONING: see chimeric clones

COCONUT (*Cocos nucifera*): the source of copra (endosperm); used for food and the oils are processed for margarine, etc.; $2n = 2x = 32$. The coconut milk had been extensively used for plant tissue culture before synthetic plant hormones became available commercially.

CO-CONVERSION: two neighboring sites are simultaneously involved in gene conversion. (See gene conversion)

CO-CULTIVATION: permitting plant cell proliferation on agar plates in the presence of *Agrobacterial* suspension (equipped with a vector plasmid) for about 1 and 1/2 to 2 days. During this period the T-DNA is transfered to the plant cells. Then the bacteria are stopped or killed by an appropriate antibiotic (e.g. cefotaxim or carbenicillin) and the transformed cells are selectively grown further on a medium to what they are expected to be resistant. (See transformation genetic, vectors, carbenicillin, cefotaxim, antibiotics)

CO-DELETION ANALYSIS: genes that are situated very closely to each other are jointly lost more frequently than ones that are separated by larger distance and this fact permits the construction of deletion maps. (See deletion mapping)

CODE, GENETIC: specifies in adjacent nucleotide triplets (commaless code) each amino acid in the polypeptide chain. The triplets are generally called *codons* and are represented in mRNA nucleotides. The sequence of the amino acids is determined by the codon sequences in the mRNA that are recognized by the anticodons in the tRNAs. The genetic code is redundant because some amino acids have up to 6 codons. The redundancy is also called *degeneracy* because several codon are degenerated (reduced) to the meaning of one amino acid. The code is *almost universal* from prokaryotes to eukaryotes, i.e. identical codons are used for the same amino acids across taxonomic boundaries. Exceptions to these rules exist in some mycobacteria, ciliates and mitochondrial DNAs. The 4 regular nucleotides in all possible combinations of 4 (3^4) generate 64 triplets from which 61 are *sense codons*, i.e. specify amino acids and 3 are *nonsense* codons (stop codons) because at their position in the mRNA the translation into protein stops and the polypeptide chain is terminated. In the customary codon table the triplets are arranged into 16 families (see table of genetic code) where the first two nucleotides may be sufficient for specifications although all three must be present. In higher organisms usually only one of the DNA strands is transcribed into mRNA codons, in prokaryotes both strands may be coding but in opposite orientation (always 5' to 3'). Although individual codons are *not overlapping*, the genes may be read, however, in overlapping registers (see overlapping genes). This facilitates a better use of the relatively very small amount of DNA for a greater variety of functions in some bacteriophages. The *usage* of the redundant codons may vary from gene to gene (see codon usage). Some of the suppressor mutations in DNA may alter the anticodons of the tRNAs and the meaning of the stop or other codons may be altered although the mRNA codons remain the same. Changes in the codons require mutation in the DNA. (See also genetic code, amino acid symbols in proteins sequences, RNA editing)

CODEN: abbreviation of literature references as given by the Periodical Tables of the Chemical Abstracts Service.

CODING CAPACITY: of an organism or organelle is determined by the number and length of open reading frames. (See open reading frame)

CODING JOINTS: junctures of the VDJ segments of the immunoglobulin and T cell receptor genes. (See immunoglobulins, TCR)

CODING SEQUENCE: is transcribed into a functional RNA.

CODING STRAND: the DNA strand which has the same base sequence as the functional RNA within the cell, except that in DNA thymine occurs at the place of uracil. The terminology has some ambiguity because in some cases both strands of the DNA may be transcribed into RNA. (See also anticoding strand, antisense RNA, sense strand, template strand)

CODOMINANCE: both alleles at a locus are simultaneously expressed in the heterozygote. At the phenotypic level, observed without in depth analysis, codominance is not very common. At the protein level (using electrophoresis or serological tests) the majority of the heterozygotes display the products of both alleles at a locus. (See dominance)

CODON: a nucleotide triplet; 61 triplets specify the 20 natural amino acids, with 1 to 6 codons for each, and 3 codons (nonsense codons) signal the termination of the polypeptide chain. (See code genetic, genetic code, protein synthesis, anticodon)

CODON BIAS: see codon usage

CODON CHOICE: see codon usage

CODON FAMILY (synonymous codons): a group of codons encoding the same amino acid. (See genetic code)

CODON RECOGNITION: see protein synthesis, aminoacyl tRNA, anticodon

CODON USAGE: synonymous codons may not be selected at random but their usage varies from

Codon usage continued

gene to gene, e.g. in the MS2 RNA phage the UAC codon for tyrosine is used 3 times as frequently as the UAU codon. In the δ-chain of human hemoglobin (146 amino acid residues) the CUU codon for leucine and the GUA codon for valine were not used at all but the CUG and GUG codons were relied on 12 and 13 times, respectively. In plants the CUG codon for leucine was used on the average in 9% whereas the CUU codons were employed in 27-28%, and the AAG lysine codon was used almost twice as frequently as the AAA codon. Similar variations occur in the use of other codons in the majority of organisms and genes. The most highly expressed genes tend to use codons corresponding to the major tRNAs whereas the less frequently expressed genes rely mainly on rarer tRNAs. The highly expressed genes containing A and T chose for 3rd position C and if the first two bases are G and C, the third is preferentially T. The translation of highly expressed genes is stopped by TAA and those of lower expression by TAG and TGA nonsense signals. The rate and accuracy of the translation is lowered by substitution of rare codons into the mRNA and this may result also in frameshifts, skipping or termination. (See code genetic, genetic code, amino acids, Grantham's rule, antisense DNA, sense DNA, isochores, amelioration of genes, coding strand)

Co-eIF-2A: similar to eIF-2C

COEFFICIENT OF COANCESTRY (kinship, consanguinity): indicates the probability that one allele, derived from the same common ancestor, is identical by descent in two individuals.

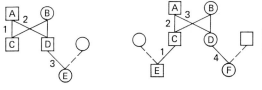

Uncle & Niece
$0.5^3 = 1/8$

First Cousins
$0.5^4 = 1/16$

The coefficient of consanguinity can be derived from the pedigrees. The solid line paths through which a particular allele is transmitted are numbered.

All diploid individuals have two alleles (paternal and maternal) at a locus and each parent has 50% chance for transmitting one or the other of these alleles to the offspring. Thus an allele of a grandmother has 0.5 chance to be transmitted to her daughter or son and that individual again has 0.5 chance for transmitting it to the grandchildren. Thus the probabilty that two first cousins would have the same allele of the grandmother 0.5^4. The degree of consanguinity varies a great deal in different cultures and within ethnic groups. Over-all in Europe, by the middle of this century, it remained below 1% (with rare exceptions). In North, Central and South America the frequency of consanguineous marriages was generally higher. In Asia it was significantly higher, especially in Southern India and Pakistan where in some areas the majority of the marriages were consanguineous. Similar situation existed in many African regions. The increased urbanization and expanding education result in reduction of inbreeding in human populations. (See inbreeding, inbreeding coefficient, relatedness degree, genetic load, lethal equivalent, incest)

COEFFICIENT OF COINCIDENCE: was designed by H.J. Muller in 1916 for estimating interference on the basis of dividing the *number* of double crossovers observed by the *number* of double crossovers expected by the probability of single crossover frequencies. In case the two crossing over events are independent the coefficient of coincidence is 1 and there is no interference. Genes far apart generally have higher coincidence than those which are closely linked. Example for computation. The size of the testcross progeny is 3,000 individuals. Frequency of single-recombinant individuals are 0.0687 and 0.2973. Thus expected frequency of double recombinants is 0.0687 x 0.2973 ≈ 0.0204. Hence the expected number of double recombinants is 0.0204 x 3,000 = 61.2. The number of double recombinants observed was 50. Hence the coefficient of coincidence is 50/61.2 ≈ 0.817. (Note that coincidence must be computed from integers and not from fractions). The *positive interference* is 1 - 0.817 = 0.183. Recent investigations indicate that coincidence is not related to the physical distance of genes (base pair or micrometers) but rather to the genetic distance expressed in map units. Accordingly, the coefficient of coincidence as a function of map distances for separated intervals can be calculated

Coefficient of coincidence continued
by the Foss equation (Genetics 133:681):

$$S_4 = (m+1)e^{-y}\sum_{i=0}^{\infty}\frac{y^{m(m+1)i}}{[m+(m+1)i]!} \qquad y = 2(m+1)x$$

where m = fixed number recombinational events resolved without crossing over between neighboring crossover events resolved with crossover, y = mean number of events per tetrad that can results in gene conversion disregarding accompanying crossovers in a test intervals, and S_4 = coefficient of coincidence for separated intervals, X = map distance in Morgan units (mean number of events resolved per tetrad in a given interval). (See interference, gene conversion, tetrad analysis, map unit)

COEFFICIENT OF CONSANGUINITY: see coefficient of coancestry

COEFFICIENT OF CROSSING OVER: is a term of Calvin Bridges (1937) stating the relation of the physical length relative to map distance. He found in *Drosophila* that 1 map unit corresponded to 4.2 μm length of the salivary gland chromosomes. Today, the genetic length of chromosomes can be expressed in nucleotide numbers. The estimates among the organisms may vary greatly depending on the amount of the DNA and resolution of the genetic markers.

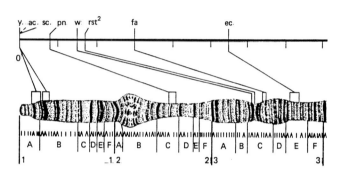

THE DISTAL END OF THE X-CHROMOSOME OF *DROSOPHILA*. (After Bridges, C.B. 1935 J. Herditry 26:69)

In *Arabidopsis* 1 map unit is about 140 kb whereas in maize it is about 240,000 kb. These ratios become important for efforts of map based isolation and cloning of genes. (See mapping genetic, mapping function, physical mapping, map based cloning)

COEFFICIENT OF INBREEDING: see F, inbreeding

COEFFICIENT OF KINSHIP: see coefficient of coancestry

COEFFICIENT OF SELECTION: see selection coefficient and fitness

COELOM: the inner cavity of the embryo.

COENOBIUM: a colony of unicellular organisms enclosed by a single membrane.

COENOCYTE: a multinucleate protoplast or cell aggregate without separation by walls.

COENZYME: a cofactor required for normal activity of enzymes; many coenzymes are vitamins; coenzyme A (acetyl coenzyme A) transfers acyl groups within cells.

COEVOLUTION: the evolution of two populations is concomitant because they have some mutual relationship, e.g. host and parasites, predators and preys or two gene loci evolve as a unit. (See selection types, frequency dependent selection)

COFACTOR: an inorganic or organic substance required for enzyme activity. (See enzymes)

COFFEE (*Coffea* spp): about 90 species with x = 11 and the plants are either diploid or tetraploid.

COFFIN-LOWRY SYNDROME (CLS): Xp22.3-p22.1 dominant mental retardation, face anomalies, tapered fingers, and lysosomal storage defects. The MAPK/RSK signaling pathway is responsible for the expression of the Rsk- kinase which controls this anomaly. (See mental retardation, head/face/brain defects, lysosomal storage diseases, MAPK, RSK)

COFFIN-SIRIS SYNDROME: does not have clear criteria but generally it involves mental retardation, broad nose, tapered fingers, sometimes the distal digits missing, poor nail development, hypertrichosis on the body but hypotrichosis on the scalp, etc. It is probably autosomal dominant with low penetrance although autosomal recessivity has also been claimed. (See hypertrichosis, Coffin-Lowry syndrome)

COFILIN: an actin- and phosphatidylinositol-binding protein involved in the regulation of the cytoskeleton. (See cytoskeleton, phosphoinositides)

COGNATE: has the ability of recognizing a particular molecule, like ligand and receptor, enzyme and substrate. (See ligand, receptor, substrate)

COGNATE tRNAs: are recognized by an aminoacyl synthetase. (See aminoacylation)

COGNITIVE ABILITIES: the major components are verbal and spatial abilities, memory, speed of perception, reasoning. These components may have subcategories such as verbal comprehension, verbal fluency, vocabulary, etc. (See behavior genetics, human intelligence)

COHEN & BOYER PATENTS: US # 4.237.224 and 4.468.464 were obtained by Stanford University on the construction of cloning plasmid chimeras (pSC101), and for the products commercially produced with the aid of them. Users must thus pay reasonable royalties for these key procedures. The fact of patenting became the subject of controversy because of its unusual nature, but it called the attention of institutions of learning to benefit directly from the commercial exploitation of the results of scientific discoveries. (See also patent)

COHEN SYNDROME: a highly complex autosomal recessive anomaly (prominent incisors, high nasal bridge, eye defects, mental retardation, hypothyroidism, etc.) with suggested chromosomal locations 15q11, 5q33, 7p, 8q22. (See eye diseases, mental retardation)

COHESIVE ENDS: two DNA molecules have base complementarity ends that can anneal.

```
─────────┐         THESE        ┌─────────
─────────┘     FIT TOGETHER     └─────────
```

COHORT: in population studies it is used to designate a particular group of individuals with common characteristics (age, treatment, ethnicity, education, taxonomic group etc.).

COILED COIL: two α helices of polypeptides wound around each other. (See α-helix)

COINCIDENCE: see coefficient of coincidence

CO-INDUCER: a chemical substance required, in addition to the inducer substrate, to activate the gene, e.g, in the arabinose operon besides the P protein, cAMP is also required for turning on the genes. (See *Arabinose* operon, *Lac* operon)

COINTEGRATION: two circular plasmid combine — without loss — into a double size plasmid.

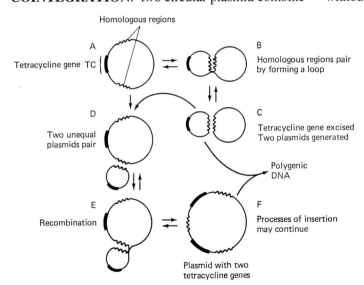

POLYGENIC PLASMIDS BY COINTEGRATION. (After Clewell, D.B., Yagi, Y and Bauer, B. 1975 Proc. Natl. Acad. Sci. USA 72:1720.)

Also, a circular donor plasmid carrying a transposable element fuses with a recipient circular plasmid and at each point of juncture there will be a copy of the transposable element. This process requires the breakage of the phosphodiester bond and duplicating the target site and the insertion element. This is replicative transposition. The cointegrate has thus, the two plasmids and two transposable elements. Upon resolution of the cointegrate two plasmids are produced, each carrying a transposable elements. (See plasmid, transposble element)

COISOGENIC: has identical genes with another strain, except at a single locus. (See isogenic, congenic)

COITUS: sexual intercourse

Col: see colicin

COLCEMID (demecolcine, N-deacetyl-N-methylcolchicine): a synthetic colchicine. (See colchicine)

COLCHICINE: an alkaloid produced by (*Colchicum autumnale* and other liliaceous plants) its tropolone ring specifically and strongly binds and disassembles microtubules of the mitotic spindle and thus the nuclear divisions are arrested in mitosis and cells skip at least one division resulting in doubling (or multiplying) the chromosome number (polyploidization). It is particularly useful for doubling the chromosome number of interspecific and intergeneric hybrids and thereby making the otherwise sterile individuals fertile by securing a pair for all chromosomes. Colchicine alleviates some of the symptoms of gout, a mammalian disease caused by uric acid overproduction. Colchicine is synthesized from phenylalanine and tyrosine. The LDLo oral dose for humans is 5 mg/kg and therefore it is a dangerous substance, taken up also through the skin. (See chromosome doubling, autotetraploid, allotetraploid, colcemid, gout, c mitosis)

COLD HYPERSENSITIVITY: autosomal dominant genes may be responsible for the development of urticaria (allergy-like rash), joint discomfort and fever after exposure to cold. It may be associated with amyloidosis (accumulation of fibrillar proteins) in various tissues. Amyloid nephropathy may be fatal. (See amyloidosis, hyperthermia, temperature-sensitive mutation, cold-regulated genes)

COLD-REGULATED GENES: are required for acclimation to low temperature. Some plant mutants may become very sensitive to cool temperature; one type of *Arabidopsis* mutant was killed at 18° C after exposure for a few days. The chilling may cause electrolyte leakage, and changes in the synthesis of steryl esters. Cold acclimation is frequently based on microsomal stearoyl coenzyme A desaturase activity. The unsaturated phosphoglycerides restore fluidity of cold-rigidified cell membranes. (See cold hypersensitivity, thermotolerance, antifreeze proteins, glycerophospholipid, fatty acids)

COLD SHOCK PROTEINS: are produced in response of abrupt change to low temperature. (See also heat shock proteins)

COLD SENSITIVE: mutant fails to grow normally at low temperature although it may be entirely normal at higher temperature. (See temperature-sensitive mutants)

COLD SPOTS: chromosomal areas where mutation, recombination or insertion are rare. (See also hot spot)

ColE1 replicon: is similar in nature to pMB1. The replicon produces 15-20 copies per cell and does not require any plasmid-encoded function for replication. It uses DNA polymerase I and III, DNA-dependent RNA polymerase and the products of bacterial genes *dnaB, dnaC, dnaD, dnaZ* that have very long life. In the presence of protein synthesis inhibitors (chloramphenicol, spectinomycin) when cell replication ceases this replicon can produce 2,000 to 3,000 plasmid copies per cell. The majority of modern bacterial vectors utilizes this replicon. The name is derived from natural plasmids coding for colicine production. The 4.2 megadalton plasmid is non-conjugative. (See colicins, plasmids [several entries], cloning vectors, phagemids, replicon)

COLEOPTILE: membrane-like first leaf in monocotyledones enclosing succeeding leaves at the stage of germination. (See embryogenesis in plants)

COLEORHIZA: an envelope of the root tip of germinating of grasses.

Coleus blumei: leafy, shade-tolerant ornamental plant; 2n = 24.

COLICINS: genetically (plasmid) controled, bacterial toxins that kill sensitive bacteria even at very low concentration of this substance, produced by killer *E. coli* and *Shigella sonnei* bacteria. Colicin E1 and colicin K inhibit active transport, colicin E2 may contribute to the degradation of DNA, colicin E3 interferes with protein synthesis of the sensitive bacteria by attacking the 16S rRNAs and removing a small fragment from their 3' terminus. Bacteria harboring Col plasmids are immune to each of these lethal effects of colicins and this property has been used for selection of Col-transformed bacterial cells. Each cell may normally carry about 20 copies of these plasmids. Colicin resistance mutations occur at appreciable frequencies in bacterial populations.

Colicins continued

The ColE1 plasmid has a molecular size of about 4.2 MDa and its is a non-conjugative. The ColE1 plasmid replicon has been used for the construction of genetic vectors and it has been engineered into cosmids and other plasmids. (See also ColE1, killer strains)

COLICINOGENIC BACTERIA: are killer strains producing colicines (bacterial toxins)

COLIFORM: enteric, gram-negative bacteria, related to *E. coli*. (See *E. coli*)

COLINEAR: see collinear

COLIPHAGE: bacteriophage, infectious for *E. coli* bacterium.

COLLAGEN: is a fibrous protein built mainly from hydrophobic amino acids (35% Gly, 11% Ala, 21% Pro and hydroxyproline). It forms a left handed helix with 3 residues per turn and made up of repeated units with glycine having every third positions. The collagens form triple helices from three polypeptide chains and they have structural roles in the cell. The 38 kbp chicken collagen gene (similar also in human or mouse) has 52 introns and short exons (54 to 108 bp) built of tandem repeats of 9 bases. There are about nine types of collagens, encoded by about 17 gene loci in humans. This protein is the major component of tendons and the cartilage. Several mammalian diseases are based on defects in collagen. (See osteogenesis imperfecta, osteoarthritis, Ehlers-Danlos syndrome, Alport syndrome, Stickler disease, Kniest dysplasia, achondrogenesis, hypochondrogenesis, aneurism aortic, epidermolysis, spondyloepiphyseal dysplasia, dermatoparaxis of cattle, metastasis)

COLLAPSIN: a member of the protein family of semaphorin; it seems to be inhibitory to axon outgrowth. (See semaphorin, CRMP-62, axon)

COLLATERAL: accessory.

COLLATERAL RELATIVES: animals in a breeding program that have one or more common ancestors but are not direct descendants of these ancestors.

COLLENCHYMA: parenchyma cells in the stem of plants that fit together closely; they have thickened wall at the corners.

Colletotrichum circinans: a fungal parasite producing smudge on sensitive onions. Generally white onions are susceptible because they lack the gene for the synthesis of parachatechuic acid that inhibits the parasite. Red onions having the *W* allele conveys resistance. Some varieties of onions have an *I* (inhibitor) allele that prevents the expression of *W* thus in the F_2 of the double heterozygotes there are 13 white susceptible and 3 red and resistant segregants.

COLLIE EYE: an eye anomaly of very high frequency among collie dogs that is detectable only by ophtalmoscope yet it frequently impairs the dogs vision.

COLLINEARITY (colinearity): amino acid sequences in the polypeptide correspond to the codon sequences in nucleic acids, the 5' end of the mRNA matches with the NH_2 end of the polypeptide chain. Some of the *Drosophila* genes, e.g. within the *ANTC* (*Antennapeadia* complex, chromosome 3-47.5) appear the same sequence in the map as the morphogenetic function they control (*lab, Pb, Dfd, Scr, Antp*). Similar collinearity has been shown in the *BXC* (*Bithorax* complex) *Ultrabithorax* segment. (The correct spelling is with two *l*s because it comes from theLatin *cum* +*linearis* (*collineo*) and thus requires the *ℓℓ*). In the homeotic complexes generally there is another "position effect"; the products of the more posterior acting genes appear to be more abundant than that of the anterior ones. (See *Lac* operon, lambda phage, morphogenesis in *Drosophila,* homeotic genes)

COLLINSIA (Scrophulariaceae): ca. 20 plant species 2n = 2x = 14; used for cytogenetic studies.

COLLOCHORES: are short heterochromatic sequences in the chromosomes supposed to be involved in their association, especially in the absence of chiasmata.

COLLOID: particles in the ranges of about 0.1 to 0.001 µm in diameter that can exist in fine suspensions (in gas, liquid or solids) or emulsions (in water).

COLOBIDAE (old world primates, langurs): *Nasalis larvatus* 2n = 48; *Presbytis crystatus* 2n = 44; *Presbytus entellis* 2n = 44; *Presbytis obscurus* 2n = 44; *Pygathrix nemaeus* 2n =44. (See primates)

COLOBOMA: appears as a missing or defective sector involving the iris, retina or the optic nerve. It may be associated with brachydactyly, abnormal movements, retardation. It may be controlled by autosomal dominant or recessive genes but X-linkage has also been suggested for some forms. (See eye diseases, cat eye syndrome, brachydactyly)

COLONY: a group of microbial cells grown at the same spot; they may have originated from a single cell or from several. The shape of the colony for a certain bacterial strain may vary according the nutrient content and diffusion, the movement and reproduction of the bacteria, and the local cell communication. Colony characteristics is used as a classification criterion.

COLONY HYBRIDIZATION: isolated DNA is cut into fragments with appropriate restriction endonucleases. A library of the fragments is established by cloning the fragments with the aid of a cloning vector (e.g. cosmid, YAC, etc). The bacteria presumably each carrying the DNA fragments in a chimeric plasmid are seeded at low density on agar plates so each colony would be separate. After the colonies are formed a replica plate is established from the master plate by pressing over it a membrane filter. After denaturation of the DNA on the nitrocellulose filter, it is hybridized with a labeled DNA or RNA probe. After washing off the unbound probe the filter is autoradiographed and the colonies containing the desired molecules of DNA are identified. Since the position of the colonies corresponding to the black dots on the photographic film (dot blot) can be identified, the bacteria containing the vector with that specific DNA can be further propagated. (See cloning vectors, cosmids, plaque lift, cloning vector, cosmid, YAC, denaturation, probe, autoradiography, DNA library, replica plating)

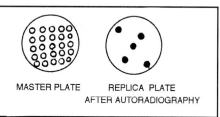

MASTER PLATE   REPLICA PLATE
                 AFTER AUTORADIOGRAPHY

COLONY STIMULATING FACTOR (CSF-1): protein activating RAS p21 protein by increasing the proportion of GTP-bound molecules. (See CSF, CSFR, RAS)

COLOR BLINDNESS: exists in different forms. Complete or nearly complete light sensing (monochromatism or achromatopsia) has a prevalence in the 10^{-5} range and it may not involve any alteration in the retina. The X-linked recessive gene is frequently modulated by adjacent chromosomal areas. One form of achromatopsia (autosomal recessive) involves also light sensitivity and the affected individuals are bothered by light but have better than average vision under dim conditions (day blindness). Partial color blindness (green color blindness, *deuteranopia*) is also X-linked recessive and it may affect 8% of the western European males. The green color vision is encoded by a cluster of genes which may recombine and undergo gene conversion, explaining the variations and the high frequency of this condition. The red color blindness (*protanopia*) appears to be determined by two X-chromosomal recessive loci with a frequency of 0.08 in males. Another type of color blindness involves loss of blue and yellow sensors but retains those for red and green (*tritanopia*) may exist in X-linked recessive or autosomal dominant forms and may occur at a very high frequency of 0.02 in some populations whereas in others it may be an order of magnitude or even less frequent. The blue cone pigment gene displays high homology to that of rhodopsin and substantial homology with the red and green pigments. In the latter form the rhodopsin receptor may be defective. The great chemist John Dalton suffered from "daltonism" or deuteranopia. (See hemeralopia, nyctalopia, rhodopsin, eye diseases, color vision, deuteranomaly)

COLOR VISION: in the human X-chromosome there is an array middle (2-7) to long-wavelength (2-4)-sensitive visual pigments. Humans with normal color vision have typically a single long-wavelength gene and two or three middle wavelength sensitivity genes. The multiple copies probably arose by unequal recombination and those with more copies may be able to see better the differences in hues. (See color blindness, rhodopsin, unequal crossing over)

COLORECTAL CANCER: may be controlled by a large number of genes involved in the "cancer family" syndrome. The most common types of cancers caused are adenocarcinoma of the colon and endometrial (inner mucous membrane) of the uterus but other types such as breast,

Colorectal cancer continued

ovarian and brain tumors as well as leukemia and other types may be under similar controls (Lynch type I). About 75% of the carcinomas show deletions of the short arm of human chromosome 17. This chromosomal segment may be responsible for the transition from the benign to the malignant state of the carcinomas. Nonpolyposis colorectal carcinoma causes 3.8 to 5.5 % of the colorectal cases whereas 0.2% were the contributions of adenomatous and 0.6% by ulcerative colitis. About 1/3 of the alterations involved the RAS oncogene (KIS = Kirsten murine sarcoma oncogen). Deletions involved also human chromosomes 22, 5, 6, 12q, 15, 17, 18. The human chromosome 18q21 locus encodes a TGFβ-regulated serine/threonine kinase receptor, MADR2. DCC (deleted in colorectal cancer) is based on a defect in a netrin receptor. Netrin is a human homolog of the *Caenorhabditis* gene products UNC-6 and UNC-40 involved in the guidance of neuronal axons. These findings indicate that besides the activation of the major oncogen, it is necessary to inactivate several tumor suppressors. UV light-induced cyclobutane pyrimidine dimers are more readily removed from the transcribed strand of the DNA than from the other. This called transcription-coupled repair. Such a repair system seems to be defective in several types of colorectal cancers. About 50% of the Western populations develop this type of cancer by age 70 and 10% of the cases become malignant. An estimated 15% of the cases have a strong dominant hereditary component. (See Gardner syndrome, polyposis, RAS, cancer, p53, p16, DNA repair, hereditary nonpolyposis colorectal cancer, mismatch repair, cyclobutane ring, neuron, Lynch cancer families)

COLOR-LESS TESTA: mutations occur in different plant species. These recessive mutations display delayed inheritance because the seed coat is maternal tissue but in F_3 usually about 1/4 of the individuals has uniformly the recessive seed-coat color. (See testa, delayed inheritance)

COLTON (Co): is relatively rare blood type encoded apparently in human chromosome 7. (See blood groups)

COLUMN CHROMATOGRAPHY (adsorption chromatography): separation of (organic) mixtures on sugar, resin, sephadex, silica gel or other columns established in glass tubes and eluting the components stepwise by different solvents or solvent mixtures. The eluates collected by fraction collectors can be monitored by spectrophotometry in the samples. (See chromatography, sephadex, spectrophotometry)

COMB TRAITS: of poultry are determined by two allele pairs and the interaction of their gene products specifies 4 comb types. *RRPP* and *RrPp* : walnut *RRpp* : rose, *rrPP*: pea, and *rrpp*: single comb in the proportion of 9:3:3:1. (See Mendelian segregation, epistasis, walnut comb)

COMBINATION: generating from **n** number of individuals all possible sets containing only x numbers in each set. Mathematically : $\binom{n}{x} = \frac{n!}{x!(n-x)!}$. (See binomial distribution)

COMBINATORIAL CHEMISTRY: is a method aiming at generating permutations of small molecular building blocks with the goal of finding the most effective pharmaceuticals.

COMBINATORIAL DIVERSIFICATION: the large number of various immunoglobulin genes may enter into different combinations and generate an enormous array of antibody molecules. (See immunoglobulins, antibody, junctional diversification, affinity maturation, somatic hypermutation)

COMBINATORIAL GENE CONTROL: the transcription of genes are regulated by the cooperative action of general and specific transcription proteins. It is not known how many such proteins exits but their number can be much smaller than the number of genes yet they can assure high degree of specificity. If we just assume that there is a total number (n) of 20 different inducible transcription factors (certainly an underestimated figure) and each gene requires 5 (x), the total number of specificities could be $\binom{n}{x} = \frac{n!}{x![n-x]!} = \frac{20!}{5!(15!)} = 15,504$ or if there would 27 inducible transcription factors and five would be used by each gene, the number of specificities $\binom{27}{5} = 80,730$ would be enough to regulate all the estimated human genes (ca. 75,000). (See regulation of gene activity)

COMBINATORIAL LABELING: is a cytogenetic method of chromosome analysis using simultaneously more than one fluorochrome-conjugated nucleotide. The number of useful combinations is $2^N - 1$. (See FISH, chromosome painting, fluorochromes)

COMBINATORIAL LIBRARY: antibody heavy and light chain cDNAs are amplified separately by PCR, then ligated and cloned in vectors. Thus a random combinatorial array of constructs are generated. *E. coli* cells infected with the vectors produce both antibody chains. But only the heavy chain contains the variable region and the first constant domain, the Fab region (see antibody diagram). This protein binds to the antigen but lacks the effector domain. The library can be screened by radioactively labeled antigen and after washing off the unbound radioactivity, the sought antigen-antibody can be spotted on the plate by the radioactivity fixed. This method permits a very efficient selection among millions of types of antibodies. This selection appears a thousand fold more efficient than the monoclonal method. A further improvement is provided by using filamentous phages (M13) that display antibodies on their surface. Screening can be done in liquid media and subjected to adsorption chromatographic purification. This procedure is called also epitope screening. (See epitope, antibody, antigen, filamentous phage, monoclonal antibody, chromatography)

COMBINING ABILITY: a term used in quantitative genetics and animal and plant breeding. *General combining ability* indicates that a particular stock has better than average performance in any hybrid combinations. *Specific combining ability* indicates a better than average performance only in certain hybrids. From (n) lines $\frac{n(n-1)}{2}$ single crosses and $\frac{[n-1][n-2][n-3]}{8}$ double crosses are possible. (See heterosis, hybrid vigor, double cross)

COMMENSALISM: species sharing the same natural resources without necessarily benefiting or suffering from the relationship.

COMMITMENT: see determination

COMMITMENT POINT: the determination (start) point of the cell division cycle. (See cell cycle)

ComP: *Bacillus subtilis* kinase affecting competence through regulator ComA.

COMPACTION: tight binding of cells to each other.

COMPANION CELLS: in plant conductive tissues are associated with sieve tubes. (See sieve tube)

COMPARATIVE ANCHOR REFERENCE LOCI (CARL): span the genomes and facilitate comparative genome mapping of different species of mammals. (See for data Lyons, L.A. *et al.* 1997 Nature Genetics 15: 47)

COMPARATIVE GENOMIC HYBRIDIZATION: is a method of cytological localization mutant DNA sequence. The normal sequence and the mutant sequence are labeled by different fluorochromes, and *in situ* hybridization is carried out by the mixture, The relative hybridization signals of the bands are monitored by fluorescence microscopy. (See fluorochromes, *in situ* hybridization)

COMPARATIVE MAPS: would reveal the evolutionary conservation of genes and nucleotide sequences across phylogenetic boundaries. For this purpose comparative anchor tagged sequences (CATS) are needed in the species where substantial amount of nucleotide sequence information is available. (See comparative anchor reference loci, unified genetic maps, evolution)

COMPARTMENTALIZATION: certain group of cells give rise to clones which are different by morphology and/or function from their surrounding tissues. The compartment may later further differentiate during development. Compartmentalization takes place also within the cells by assignment of special functions in a polar fashion or to special subcellular organs.

COMPATIBILITY: crossing or mating results in (normal) offspring. Also, tissue transplantation does not involve adverse immunological reaction or any other simultaneous treatments without undesirable consequence(s).

COMPATIBILITY GROUP: plasmids that may or may not co-exist in the same bacteria.

COMPATIBILITY OF ORGANELLE AND NUCLEAR GENOMES: can be determined in species where biparental transmission occurs and organelles between species can be transfered. Also on the basis of cybrids, it appears that compatibility differences are real and thus various cytoplasms can be classified. The reciprocal crosses are usually informative. (See plasmid male transmission, paternal leakage, ctDNA, mtDNA, cybrids)

COMPETENCE EMBRYONAL: a state of being receptive to stimuli for differentiation.

COMPETENCE OF BACTERIA: a physiological state of the (bacterial) cell when transformation (uptake and integration of DNA) is successful and it generally coincides with the second half of the generation time or its peak is near the end of the exponential growth phase. Competence can be induced by divalent cations and a combination of them. For more than four decades *E. coli* was refractory to transformation until in 1970 it was discovered that cold $CaCl_2$ makes possible the uptake of phage DNA. 1 mg supercoiled plasmid DNA yields 10^5 to 10^6 transformations. This frequency can further be increased by 2 to 3 orders of magnitude by the use of improved protocols involving also DMSO (dimethylsulfoxide, a wide-range solvent and penetrant). The bacteria so treated keep their transformation competence if stored at -70°C. Competence in *Bacillus subtilis* is regulated by the secreted competence stimulating factor (CSF, 520-720 Da peptide) and the pheromone ComX (the $\approx$ 10 amino acid C-terminal section of the 55 amino acid peptide). The process of competence development requires transcription factors and specific nutritional conditions. In the *Neisseria* bacteria competence is expressed constitutively. In *Haemophilus influenzae* arrest of cell division results in competence. (See transformation genetic, exponential growth, supercoiling)

COMPETITIVE EXCLUSION: two similar species generally do not coexist in the same niche indefinitely. They will either coalesce (be interbreeding) or one will be extinguished.

COMPETITIVE INHIBITION: the inhibitor competes with the substrate for the active site of the enzyme and its blocking effect can be relieved by an increase of the concentration of the substrate. (See regulation of enzyme activity)

COMPETITIVE REGULATION: besides the minimal enhancer, accessory cis elements are needed to ensure proper gene expression during the different developmental stages.

COMPETITIVE RELEASE: species or races are selected for general performance rather than to high adaptive specialization under conditions when there are limited or no competitive species.

COMPLEMENT: when antigen and antibody form a complex, in the Fc domain of the heavy chain, the activation of the complement takes place (see antibody). The complement consists of about thirty different proteins that have the role in the destruction of the foreign cells by lysis and activate the leukocytes to engulf the invaders by phagocytosis. Immunoglobulin M (IgM, 950 kDa) and the subclasses of immunoglobulin G (IgG, 150 kDa) bind complement C1q (410 kDa) components of C1 protein. The binding causes the sequential activation of C1r (83 kDa), C1s (83 kDa), C4 (205 kDa) and C2 (90 to 102 kDa) and the cleavage of C3 (185 kDa). There are a number other proteins that modulate these reactions. The cleavage of complement component C3 can be activated also by IgA, IgE, polysaccharides and endotoxins. Proteins B (95 kDa), D (24 kDa), P (properdin, 220 kDa) and a modulator H (150 kDa) are also involved in the cleavage of C3. The largest fragment of C3b then interacts with another protein group, properdin and participates in a positive feedback system to stimulate the cleavage of C3. Finally C3b initiates the cleavage of complement components C5 (210 kDa), C6 (104 to 128 kDa), C7 (92,4 to 121 kDa), C8 (152 kDa) and C9 (71 kDa). The complement components react with cell-specific receptors and some receptors react with more than one complement components. This reaction results then in enhanced phagocytosis, antibody dependent cellular cytotoxicity (ADCC), lysis, B-lymphocyte proliferation, etc. Components C1, C4 and C2 attack viral invaders, and their interaction increases permeability. C3a and C5a have anaphylatoxic effects, including the release of cellular histamine and regulate humoral immune reactions. C3b and C4b bind to surface receptors of several mammalian cells and to bacteria and cause their destruction. HIV and human T cell leukemia virus are, however, resistant to the human complement The Gal(α1-3)Gal terminal carbohydrates are present in the majority of mammals, except humans because hu-

Complement continued

man cells do not have a functional galactosyltransferase. If the porcine enzyme is transfected into human cells, the retroviruses become sensitive to the human serum.

The amount of the complement is measured by immunochemical and hemolytic properties. CH_{50} is a measure of the complement indicating that 50% of the antibody-sensitized erythrocytes released their hemoglobin. Genetically determined deficiencies are known for the complement proteins. Homozygotes may entirely miss a particular protein whereas heterozygotes may display only reduced amounts. *Lupus erythomatosus,* an autoimmune disease, is caused by C2 deficiency. Other deficiencies in the terminal components of the complement pathway may contribute to the symptoms of rheumatoid immune diseases. Susceptibility to various types of infections are also related to deficiencies in the components of the complement proteins. The complement is basically an innate component of the immune system and it is complementary to the function of the macrophages, mast cells, T cells, and B cells. Components of the complement system were localized to human chromosomes 6p21.3, 19p13.3-p132, 1p36.3-p34, 1p32, 9p34.1, 5p13. (See immunodeficiency, antibody, immunoglobulin, lymphocytes, macrophages, mast cells, humoral antibody, HLA, endotoxins, hemolysis, histamine, rheumatoid arthritis, immune system, T cells, convertase, TAPA-1, CD22, lupus erythematosus)

COMPLEMENT FIXATION: a serological measure of the degree of antigen-antibody reaction. (See antibody detection)

COMPLEMENTARITY: of nucleic acid bases means that Adenine pairs with Thymine and Guanine with Cytosine by two and three hydrogen bonds, respectively, and two complementary polynucleotide chains pair in an antiparallel manner. (See DNA structure, Chargaff' rule)

COMPLEMENTARITY DETERMINING REGION: see CDR

COMPLEMENTARY ALLELES: belong to different gene loci although partial complementation may occur between alleles that belong to different cistrons of the same gene locus. (See allele, allelism test, cistron, complementation mapping, complementation test *in vitro*)

COMPLEMENTARY BASE (nucleotide) **SEQUENCES**: the nucleotides can form hydrogen bonds according to the base pairing rules in double-stranded DNA or double-stranded RNA. (See hydrogen bonding, Chargaff's rule)

COMPLEMENTARY DNA: see cDNA

COMPLEMENTARY GENES: homozygosity of recessive alleles at different loci but controlling functions in the same biosynthetic pathway may be expressed by identical (or very similar) phenotype and in the F_2 of a double heterozygote and expected to display the phenotypic proportions 9 wild type and 7 mutants (homozygous for one [3] + for the other [3] + for both [1] of the recessive alleles = 7). This is a modification of the 9:3:3:1 (9/16, 3/16, 3/16, 1/16) ratio. (Modified Mendelian ratios)

COMPLEMENTARY SEGREGATION: see complementary genes

COMPLEMENTATION GROUPS: recessive mutations that complement in trans arrangement, belong to different complementation groups, i. e. they represent different gene loci whereas the non-complementary alleles belong to the same complementation group. (See trans, cis, allelism test, allelic complementation, complementation test, see diagrams next page)

COMPLEMENTATION MAPPING: is used for determining the pattern of the genetic differences among complementary alleles and the extent of the genetic lesions involved. (See diagrams next page)

COMPLEMENTATION TEST: a test of allelism; recessive alleles of the same cistron generally fail to complement in trans arrangement in heterozygotes, e.g: $\frac{A\,b}{a\,B}$ heterozygote is non-mutant if *a* and *b* are alleles belong to different loci. If two recessive mutations in the F_1 display mutant phenotype the two mutants are allelic. (See allelism)

COMPLEMENTATION TEST, *IN VITRO*: is carried out by cell extracts (enzyme assays). This is the most reliable test of complementation because *in vivo* the genes are regulated and that may boost complementation when the complementation may be actually quite week. A 5%

Complementation test *in vitro* continued

activity of an enzyme may appear as if it would be of wild phenotype *in vivo* but *in vitro*, in the absence of regulation, it can be readily distinguished from the 100% activity. (See allelism)

COMPLEMENTATION GROUPS (illustration to preceding page)

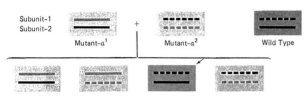

Random Association of Subunits in the Cytoplasm of the Compound

ALLELIC COMPLEMENTATION MAY BE BASED ON RANDOM CYTOPLASMIC ASSOCIATION OF PROTEIN SUBUNITS OF THE MULTIMERIC ENZYMES. THE MUTANT SUBUNITS MAY NOT BE FUNCTIONAL YET IF THEY ASSEMBLE FAVORABLY SOME ENZYME MOLECULES CAN SHOW ACTIVITY.

COMPLEMENTATION MAPPING (diagram for preceding page)

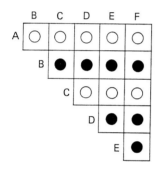
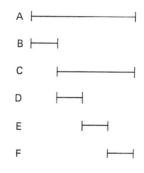

ON THE BASIS OF PARTIAL (ALLELIC) COMPLEMENTATION OF VARIOUS ALLELES OF A LOCUS CONTROLING A MULTIMERIC ENZYME, GENETIC MAPPING IS POSSIBLE. THIS IS NOT NECESSARILY A PHYSICAL MAP, RATHER A FUNCTIONAL ONE AND IT MAY NOT ALWAYS BE THEREFORE LINEAR. Open circles indicate lack of complementation, solid circles stand for complementation of recessive alleles identified here by A, B, C, D, E and F. (From Catcheside, D.G 1960 in Microbial Genetics. Hayes, W, and Clowes, R.C., eds., Cambridge Univ. Press, Cambridge, UK)

COMPLEMENTATION UNIT: is a cistron. (See cistron, complementation test)

COMPLETE DIGESTION: the reaction with a restriction endonuclease is continued until all potential cutting sites in the DNA are cleaved by the enzyme. (See restriction enzymes)

COMPLETE DOMINANCE: in the presence of a completely dominant allele, the phenotype controled by the recessive allele is not detectable under the conditions of study. (See Mendelian laws, Mendelian segregation, Modified Mendelian ratios, semidominance, codominance)

COMPLETE FLOWERS: have sepals, petals, stamens and carpels. (See flower differentiation)

COMPLETE LINKAGE: genes fail to recombine because of their extreme closeness in the chromosome or because of genetic factors interfering with crossing over (chromosomal aberrations), inviability of the recombinants (deficiency or duplication gametes in inversion heterozygotes) or absence of recombination in the heterogametic sex of male *Drosophila* or female silk worm. (See recombination, linkage)

COMPLETE MEDIUM: contains all the nutrients that potentially auxotrophic cells may require for growth. (See minimal medium, auxotroph)

COMPLEX: a functional aggregate of molecules without covalent bonds among them.

COMPLEX HETEROZYGOTE: viable heterozygotes for multiple reciprocal translocations when translocation homozygotes are lethal. It is actually a balanced lethal system and heterozygosity is permanent for the genes within the translocation complexes. If there are two multiple translocation complexes, it appears as if there would be only two groups of linked genes even when the basic number of chromosomes is larger. In meiosis the distribution of the chro-

Complex heterozygote continued

mosomes is alternate and paternal and maternal complexes always go to opposite poles. Recombination between the complexes is rare but if it occurs it may change the complexes. The size of the complexes may vary depending the number of chromosomes involved in the reciprocal translocation complexes. *Oenothera lamarckiana* carries the seven-chromosome *gaudens* complex (happy in Latin, i.e. the chromosomes do not contain dominant or semi-dominant deleterious genes) and the seven-chromosome *velans* translocation complex (concealing in Latin, since it contains semidominant genes that are responsible for paler color and narrower leaves). *Oenothera hookeri* has also 14 chromosomes but they are not involved in reciprocal translocations. When these two *Oenotheras* are crossed the F_1 is not uniform as in normal Mendelian crosses but they form "twin hybrids". The maternally transmitted *gaudens* complex makes the hybrids normal shape and color whereas the maternally transmitted *velans* complex containing hybrids are pale and have narrow leaves. Thus the F_1 reminds us to a testcross involving one pair of heterozygous alleles because of the two complexes. In *Oenothera* these complexes behave differently, some kill the male, others kill the female gametes and yet others cause zygotic lethality in the homozygotes but always only the complex heterozygotes are found in the sporophytic plants. The multiple translocations are identified by the light microscope as translocation rings. Such systems are common among *Oenothera* plants but occur also in other species. (See translocations, *Oenothera*, beta complex, zygotic lethal)

COMPLEX INHERITANCE: is found when multiple loci and interactions are involved in the expression of the gene products. N.E. Morton (Proc. Natl. Acad. Sci. USA 93:3471 [1996]) published parametric and non-parametric statistics for estimating linkage considering various such pedigrees. (See polygenic inheritance, multifactorial trait, QTL)

COMPLEX LOCUS: a generally large cluster of functionally related but not entirely similar cistrons, and alleles within the locus may show partial complementation (*scute* in *Drosophila*, *t* in mouse, *R* in maize). Some of the complex loci appeared to be more mutable than other genes because unequal crossing over between the repeated sequences generated new phenotypes. (See pseudoallelism, allelic complementation, step allelomorphism, Bar)

COMPLEX TRAIT: its inheritance is not based on single dominant or recessive alleles but multiple factors may be involved and various environmental effects contribute to their expression.

COMPLEX TRANSCRIPTION UNIT: the transcript of the gene may be processed in more than one way and the translated products vary according to cell- or tissue-specific functions.

COMPLEXIN: proteins of the nerve termini, binding syntaxin. (See syntaxin)

COMPLEXITY OF DNA: indicates the size of the DNA molecule as determined from the c_0t curve. (See c_0t curve, $c_0t_{1/2}$, kinetics, kinetic complexity)

COMPOSITE CROSS: individuals of various genetic constitution are hybridized in a mass for the purpose of studying the effect of natural selection or to obtain improved varieties.

COMPOSITE TRANSPOSON: carries genes (e.g. antibiotic resistance) beyond those required for transposition. (See transposon, insertion element)

COMPOUND CHROMOSOME: results from the fusion of telocentric chromosome into a bi-armed monocentric or by Robertsonian translocation between acrocentric chromosomes or by translocations or by interchromosomal transposition. (See telocentric, Robertsonian translocation, translocation, transposition)

COMPOUND EYE: in Arthropods (insects) the eye is composed of several, each structurally complete elements (ommatidium, about 800 in each eye of *Drosophila*). (See *Drosophila*)

COMPOUND, GENETIC: a heterozygote for two mutant alleles of the same gene and may display a phenotype that is intermediate between that of the two recessive homozygotes.

COMPOUND LEAF: is composed of several leaflets.

COMPRESSIONS IN GELS: in DNA sequencing gels abnormal intrastrand structures may form and cause anomalous pattern of migration, especially in DNAs with high G + C content. The compression may be avoided by using another DNA polymerase or by 2'-deoxyinosine-5'-triphosphate or 7-deaza-2'-deoxyguanosine-5'-triphosphate. (See DNA sequencing, deazanacleotides)

COMPTON EFFECT: as the energy of electromagnetic radiation increases above 0.5 MeV, the radiation of electrons may scatter and recoil depending on the surface hit and the angle of the incidence of the radiation. This may affect the effective dose absorbed by the object and may create hazards if the source is not effectively protected in all directions. (See X-ray)

CONALBUMIN: is an egg white iron-binding protein, encoded by a gene with 17 introns and regulated by estrogen and progesterone.

c-onc (cellular oncogene): a normal gene that may lose its ability to limit cell divisions and then initiates cancerous growth (see v-oncogene)

CONCANAVALINS: are agglutinin proteins which preferentially agglutinate cancer cells, used also as probes for cell surface membrane dynamics. Concanavalins are also mitogenic. (See lectins, cell adhesion)

CONCATAMER: repeated phage genomes associated in a linear array of DNA molecules in a head-to-head fashion, formed during normal replication, and must be cut to head capacity size for packaging into the capsid by a terminase gene product (endonuclease). (See headful rule, permuted redundancy, non-permuted redundancy, lambda phage, catenane)

CONCATENANE: interlocked DNA rings or chains→ ⟲⟳ (See also knotted DNA)

CONCEPTACLE: a cavity in the fern leaves (fronds) bearing gametangia. (See gametangia)

CONCERTED EVOLUTION: extensive homology within species and little homology between related species (such as found in intergenic spacer nucleotides and LINEs). (See molecular drive, LINE)

CONCORDANCE: identity of traits within twins or groups of individuals. (See discordance, twinning, zygosis, monozygotic twins, dizygotic twins, penetrance, expressivity)

CONCURRENT CONTROL: in an experiment involving certain type of treatment it is required that an adequate group of untreated individuals will be studied *simultaneously* to permit a reliable comparison and assessment of the effect of the treatment. (See control, historical control)

CONDENSATION REACTION: during the formation of a covalent bond water is released as a byproduct.

CONDITIONAL DOMINANCE: see dominance reversal

CONDITIONAL LETHAL: dies only under certain conditions (e.g. high, nonpermissive [restrictive] temperature) but is viable under others (e.g. low, permissive temperature). Many auxotrophs are conditional lethal because they survive only when the required nutrient is provided. (See temperature-sensitive mutation, auxotroph, conditional mutation)

CONDITIONAL MUTATION: is expressed only under the condition(s) required. Such mutations may be extremely useful for the study of conditional lethal genes. Conditional expression may be regulated also by agents that affect transcription and/or translation or conformation of the proteins (heatshock, heavy metals, hormones, repressors, DNA-binding proteins, dimerization of transactivator, signal transduction, etc). (See conditional lethal mutation, temperature-sensitive mutation, auxotrophs, and other terms mentioned under separate entries)

CONDITIONAL PROBABILITY: is not based on absolute frequencies but e.g. one group is fixed in a matrix. When we have 3 genotypes AA, Aa and aa, their frequencies add up to 1. If one of the groups is fixed then another must be of one of the 2 genotypes. In other words, one probability depends on what events have taken place previously. (See Bayes' theorem, risk)

CONDUCTANCE: a non-conjugative plasmid can be transmitted to a recipient cell by cointegration into a mobile, conjugative plasmid. (See conjugative plasmid, conjugation of bacteria, cointegration)

CONE: a fruiting structure bearing sporangia, e.g. a pine cone. (See sporangium). Also the retinal cone in the eye is a visual cell. (See also retinitis pigmentosa, retinoblastoma)

CONE PIGMENTS: are present in the retina and mediate color vision. (See cone)

CONFIDENCE BELTS: see confidence intervals

CONFIDENCE INTERVALS: population parameters can be estimated by *point estimates* and

Confidence intervals continued

by *interval estimates*. The former specifies the parameter itself, the latter defines the range of values within which the parameter is expected at a certain level of confidence (probability). If we obtained an average value in a population, we would like to know the probability within what range the real, true average may fluctuate by 95% or other confidence of choice. Confidence intervals (C.I.) for reasonably large populations can be determined with the formula: C.I. = $p \pm z\sqrt{pq/n}$ where p and q are the proportions observed ($p + q = 1$), z is the critical value for the normal distribution at a given level of confidence. $z = 1.96$ (for 95%), 2.58 (for 99%), 3.29 (for 99.9%), and it is not very useful to go for even higher z values because then the range will be so wide that it will become almost meaningless. n = the numbers in the population counted or measured. Example: a population of 140 consists of two groups represented by 60 ($p = 60/140 \approx 0.43$) and 80 ($q = 80/140 \approx 0.57$). C.I. for $p = 0.43 \pm 1.96\sqrt{[0.43][0.57]/140} = 0.43 \pm 1.96 \times 0.04184 = 0.43 \pm 0.8$. Therefore the frequencies and 0.51 and 0.35 are the 95% confidence limits of the experimentally observed p. Alternatively, the 95% confidence belts shown below can be used.

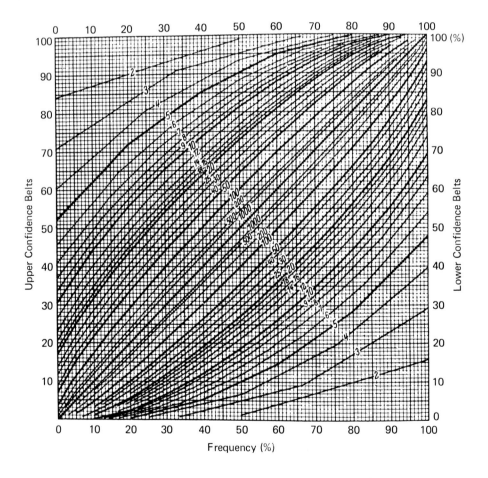

Use Of The Confidence Chart: calculate from the experimental data the frequencies p and q. At the intersections of the vertical line corresponding to "frequency" and the two belts (curves) numbered and

Confidence chart caption continued

indicating the size of the sample (population) one can read the percentage of upper and lower confidence. This chart facilitates also an estimate of the (1) repeatability of the fractions obtained, (2) the statistical range of the fractions, and (3) the size of the population where a chosen fraction will be represented by a number of individuals (counts) within the specified range. E.g. the frequency observed is 0.20. The confidence limits at probability 0.95 for a population of 100 will be 29 and 12. Or if the expected frequency is 0.25 and one is willing to accept a latitude of 10% (i.e. a frequency between 0.20 and 0.30), the minimum population size required is 200 at P = 0.95 because this latitude is bracketed by the 200 belts. (From Koller, S. 1956 in Biochem. Taschenbuch, Rauen, H.M, ed., Springer-Vlg., Berlin)

CONFIDENTIALITY: the code of ethics and many state laws in the US prohibit physicians to disclose any medical information to a third party even in court. Some courts permitted disclosure if the third party is in imminent danger by genetic or infectious disease. (See genetic pri-vacy, wrongful life, counseling genetic, paternity test, informed consent)

CONFLUENT: the cell culture extends over the entire surface of the medium.

CONFOCAL MICROSCOPY: permits the formation of a three-dimensional image by focusing laser or other light through a pinhole and with a dichroic mirror to a distinct small area of the object. The fluorescent light emitted from that small focal point is then reflected through another (confocal) pinhole (i.e. this second pinhole is exactly in the focus of that area of the object) and thus a sharp image is received on a detector. The image of the areas not in focus are thus obliterated. Subsequently, the light is focused at several other points and the image of each point is registered on the video screen and building up in a three-dimensional integrated picture. If the pinholes are moved in a synchronous manner, a scanning real-time imaging can be obtained. (See microscopy)

CONFORMATION: an arrangement of a molecule in space that does not require the severance of any bond because of their freedom of rotation.

CONFORMATION CORRECTION: multimeric proteins may be inactivated by mutation in a subunit that affects the conformation of that polypeptide chain. Another mutation may alter the conformation of another subunit in such a way that the first defect in conformation is corrected and the activity of the enzyme is at least partially restored. It was assumed that some of the allelic complementations observed are based on such a correction. (See also dominant negative, allelic complementation)

CONGENEIC: same as congenic; it is frequently used this way in immunogenetics.

CONGENER: related to something by origin and/or function, biologically or chemically.

CONGENIC RESISTANT LINES OF MICE (CR): contain a new histocompatibility gene locus introduced from another inbred (or any other) line. Earlier these were called IR (isogenic resistant) lines. Such CR lines are generated by mating inbred line #1 with known histoincompatible line #2. The hybrid accepts grafts of #1. In the progeny of hybrids mated *inter se* some segregants do not reject transplants of #1. These will be further backrossed with #1. After 7 backrosses the "hybrid" will have in over 99% ($1-0.5^7$) the same chromosomes as parent #1. Upon continued brother-sister matings (usually for 20 generations), an individual may show up that rejects grafts from #1. This is further backcrossed to #1 and the progeny will be selected to resistance against transplants to #1. Such a line is congenic resistant, i. e. almost identical (with the exception that histocompatibility gene) with #1. Some very closely linked genes to the histo-incompatibility locus of #2 may not be eliminated, however, from the CR line. Today the selection may be facilitated by serological assays rather than by expensive transplantation experiments. Also, the desired genes can be transferred by transfection without the need of repeated back-crosses. The development of a large number of CR lines permitted the genetic identification of allelism and complementation groups of histocompatibility genes. The serological assays permitted the identification of strong and weaker responses and on this basis major and minor histocompatibility genes could be classified. (See HLA, MHC)

CONGENIC STRAINS: they are identical, except at one locus or a very limited region of a chromosome; they are obtained by repeated backcrossses. These lines can be used to determine the effect of a particular gene on a selected genetic background. (See congenic resistant lines of mouse, coisogenic)

CONGENITAL: is a condition born with, regardless whether it is due to direct genetic or developmental causes. (See familial, hereditary)

CONGENITAL ADRENAL HYPERPLASIA: has a prevalence of about 0.0002-0.0001 but in some populations it may be several orders more frequent. It is based on the deficiency of one or another enzyme in cortisol (steroid) biosynthesis. The accumulation of the precursor results in virilization of female babies. The condition can be successfully treated by glucocorticoids given to the mother. The autosomal recessive gene(s) are closely linked to the HLA loci in human chromosome 6. Prenatal identification is feasible by linkage with appropriate DNA probes. There are, however, other types of steroid hydroxylase deficiencies located in other chromosomes. (See genetic screening)

CONGENITAL APLASIA of the VAS DEFERENS (CBAVD): autosomal recessive absence of the excretory channel(s) of the testes resulting in azoospermia. (See azoospermia, cystic fibrosis)

CONGENITAL HYPOTHYROIDISM: defective development of the thyroid gland causing goiter, mental retardation, deafness, etc. because of deficiency of the thyroid hormone.

CONGENITAL TRAIT: is evident at birth and is due to either hereditary or other causes. (See familial trait)

CONGRESSION: the assembly of chromosomes in the metaphase plane. (See mitosis, meiosis)

CONGRUENCE ANALYSIS: tests the appropriateness of conclusions reached by different methods. It is used in the study of evolutionary trees. (See homology, xenology)

CONIDIA (conidiospores; in singular conidium): asexual fungal spores that appear externally on a hypha by abstriction. Such a hypha is a conidiophore. (See also macroconidia, microconidia, fungal life cycle)

CONIDIOPHORE: see conidia

CONJUGATE REDOX PAIR: an electron donor and the corresponding electron acceptor, e.g. NADH and NAD^+.

CONJUGATED PROTEIN: contains prosthetic group(s); e.g. an iron or magnesium heme in hemoglobin and chlorophyll, respectively.

CONJUGATION: generally means mating; in bacteria the physical contact between F^+ donor and an F^- (recipient) cell and the unidirectional transfer of the (Hfr) chromosome by a rolling circle replication procedure through the conjugation tube. Approximately 12 µm DNA is transferred per minute. The standard genetic map of bacteria is based on the time in minutes required for the conjugational transfer of genes from donor to recipient. The conjugational transfer may be clockwise or counterclockwise depending on the orientation of the F element in the bacterial chromosome. For the transmission of plasmids from one cell to another requires a conjugational mechanism. The transforming plasmids generally lack the *mob* gene required for mobilization of a chromosome (genophore) or plasmid so recombinant DNA could be contained. For some of the plasmids the mobilization factor can be provided in trans by a helper plasmid (ColK). ColK plasmids code for a protein that opens up the circular DNA at the *nic* site close to *bom* (bacterial origin of mobilization). Some plasmids lack the *nic/bom* system and cannot be transferred through conjugation (non-conjugative plasmids). The latter type is favored for containment of recombinants. (See also Hfr, conjugational mapping, bacterial recombination frequency, relaxosome, ori_T, *tra* genes, non-plasmid conjugation, F plasmid, figure of conjugation on page 224)

CONJUGATION MAPPING: in bacteria the transfer and integration of genes of the Hfr chromosome to the recipient is measured by interrupted mating and the map is constructed on the basis of minutes required for the linear transfer of a particular marker(s). For the complete transfer ca. 90-100 minutes are required (40-45 kb/2-3 minute intervals). By conjugational recombination the complementary products of the exchange are not recovered, and the surviving products are the results of double recombination. The single strand transferred becomes double-

Conjugation mapping continued

stranded after entering the receipient (F⁻) cell. (See diagram below, Hfr, conjugation, bacterial recombination frequencies, F plasmid, conjugation bacterial)

Markers:	thr	leu	azi	T1	lac	T6	gal	λ	21	424
Hfr transfer (%)			>90	70	40	35	25	15	10	3
Minutes	8	8½	9	11	18	20	24	26	35	72

(On the basis of F. Jacob, and E. L. Wollman, 1961. Sexuality and Genetics of Bacteria, Academic Press, New York.

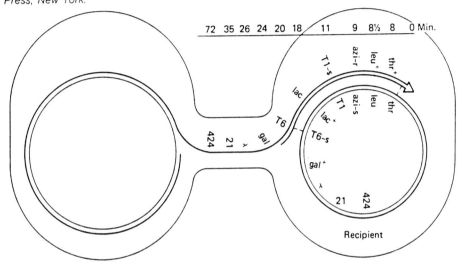

CONJUGATION, *PARAMECIA*: is the sexual process of these unicellular protozoa that most commonly reproduce asexually by fission. During conjugation two cells of opposite mating types appose and meiosis get on its way in the micronuclei. From the four meiotic products only one survives in both cells and that divides twice by mitosis generating for haploid nuclei in the two cells, each. One of the gametes is then passed over to the other cell through a conjugation bridge in a reciprocal manner resulting in mutual fertilization and in two diploid conjugants. Subsequently the pair separates and the *exconjugants* are formed. The macronucleus that has only metabolic but no known genetic role, disintegrates in both cells. Subsequently the diploid micronucleus undergoes two mitotic divisions, and two of the four mitotic products fuse to regenerate the macronucleus. The remaining two nuclei remain in the cells as diploid micronuclei. *Paramecia* may reproduce also by *autogamy* in the absence of conjugation. Meiosis may take place and four haploid nuclei arise from which again only one survives. The survivor divides again into two which after fusion generate a diploid homozygote. If the conjugation lasts for a longer periods, cytoplasmic elements are also transfered along with the gametes. (See *Paramecium*)

CONJUGATIVE PLASMIDS: see conjugation of bacteria, plasmid conjugative

CONNECTIN: same as titin. (See titin)

CONNECTIVE TISSUE DISORDERS: involve cells with substantial extracellular matrix that provide structural support, such as bone, cartilage, etc. In the majority of disorders collagen synthesis is affected. (See skin diseases, osteogenesis imperfecta, Marfan syndrome, Ehlers-Danlos syndrome, Stickler syndrome, Kniest dysplasia, arthritis, Reiter syndrome, lupus erythematosus, HLA)

Conjugation continued (to page 222)

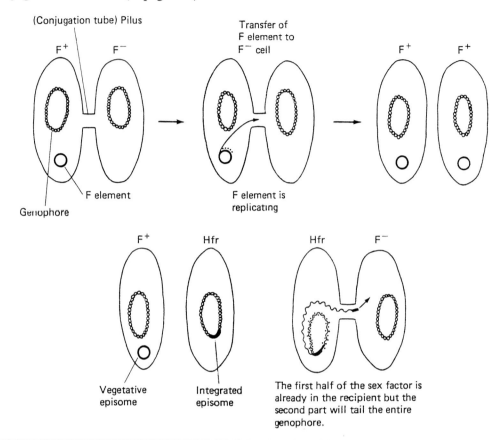

BACTERIAL CONJUGATION. TRANSFER OF THE SEX FACTOR F (plasmid) AND THE BACTERIAL CHROMOSOME (genophore) LEAD BY THE Hfr (high frequency recombination) ELEMENT. (See p. 222)

CONNEXINS: provide elements of gap junctions; they are built of six polypetides. All connexins have four membrane-spanning domains, two extracellular loops, a cytoplasmic loop and cytoplasmic amino and carboxyl termini. Connexins may form connexons, the pore of gap junctions. Connexins may have role in heart diseases, infertility, cataracts, etc. (See gap junctions, Charcot-Marie-Tooth disease)

CONPLASTIC: has the mitochondrial genome of one strain but the nuclear genome is derived from another. Conplastic strains are obtained by at least ten backcrossing of a female by a nuclear genome donor male. The probability that the nuclear genome is of the donor type is $1-0.5^n$; n = number of backcrosses. (See mtDNA, mitochondrial genetics)

CONRADI-HUNERMANN DISEASE: see chondrodysplasia

CONSANGUINITY: see coefficient of coancestry

CONSCIOUSNESS: awareness, sensory discrimination of events in the outside world and the body. Unfortunately neurophysiology does not have yet adequate tools to tackle most of the problems involved.

CONSENSUS: basically common (although generally not entirely identical) nucleotide sequence at certain positions among some DNAs or amino acids in proteins. Consensus is indicative of a common functionally important role such as the TATA box, transcription termination signals, etc. (See core sequences)

CONSERVATION GENETICS: is concerned with the population genetics principles involved in the maintenance of feral species. The maintenance of the populations depends on inbreeding, outbreeding, effective population size, deleterious mutation, reproductive success, adaptation to captivity, reintroduction, vagility, outbreeding depression, extinction, etc. (See the terms under separate entries, species extant)

CONSERVATIVE REPLICATION: an early idea of DNA replication suggesting that after replication each double-stranded molecule would contain either two old or two new single strands. (See semi-conservative replication, replication, DNA replication, replication fork)

CONSERVATIVE SUBSTITUTION: in a polypeptide chain an amino acid is replaced by another with similar properties.

CONSERVATIVE TRANSPOSITION: see transposition

CONSERVED SEQUENCES: see consensus

CONSPECIFIC: strains or varieties belonging to the same species.

CONSOMIC: after 10 or more backcrosses of a male to an inbred recipient female strain, all the chromosomes will belong to the recipient strain — at very high probability — except the Y chromosome. (See conplastic)

CONSTANT GENOME PARADIGM: visualized the genetic material in a stable form because insertion or transposable elements were not yet known. The Fluid Genome idea was developed after the discovery of the mobile genetic elements which are capable of restructuring the genome. (See transposons, transposable elements)

CONSTITUTIONAL: it was present in the individual at birth. (See acquired)

CONSTITUTIVE ENZYME: functions at a rather constant level in all cells, all the time. (See housekeeping genes, inducible enzymes)

CONSTITUTIVE GENE: the rate of its transcription is not subject to the effect of regulator gene(s). (See housekeeping genes)

CONSTITUTIVE HETEROCHROMATIN: is heterochromatic at all stages and in all cells, indicating that these sequences are never transcribed. (See heterochromatin, euchromatin)

CONSTITUTIVE MUTATIONS: lost their regulatory element(s) and they are in "on" position. (See constitutive gene)

CONSTITUTIVE SPLICING: of primary RNA transcripts occurs when the exons are spliced together in a single pattern consistent with their order in the gene. (See also alternative splicing, introns)

CONSTITUTIVE TRIPLE RESPONSE (CTR): is the reaction of plants to ethylene, namely the plumular hook is retained, prevention of geotropic response, reduction of stem elongation. (See plant hormones)

CONSTRICTION, CHROMOSOMAL: the primary constriction is the centromere and the secondary constriction may tie an appendage to the end of the chromosome by a relatively thin stalk. These secondary constrictions are frequently called *satellites* and they are associated with the nucleolus (nucleolar organizer region). The nucleolus, contains RNA and that region of the chromosome was believed not to contain DNA hence SAT for *sine acido thymonucleico* (without thymonucleic acid); that time (in the early 1930s) DNA was called thymonucleic acid). (See centromere, satellite)

CONSTRUCT: most commonly used for the designation of a specially built plasmid or engineered chromosome.

CONTACT GUIDANCE: the movement of axons may be guided by the physical environment. in the tissue. (See axon)

CONTACT INHIBITION: normal animal cells are anchorage dependent in culture and grow in monolayer because of inhibition by neighbor cells. Cancer cells lost the dependence on anchorage and constraints in growth by neighbors. This is associated also by changes in cell morphology. The oncogenic transformant cells thus can pile up in an apparently disorganized manner into tumors. (See saturation density, cancer, metastasis)

CONTAINMENT: a safe place from where hazardous material, including certain type of biological vectors, cannot presumably escape, and thus can be worked with safely. (See laboratory safety, biohazards)

CONTEXT, GENETIC: the contribution or effect of genes outside the locus of primary concern.

CONTIG: a set of partially overlapping DNA fragments that includes a complete region of the chromosome without gaps. For the determination of contigs large capacity vectors are used, e.g., YACs (up to 1-2 megabase), P1 plasmid (100 kb) and cosmids (40 kb). (See genome project, YAC, BAC, cosmid)

CONTIGUOUS GENE SYNDROME: deletions resulting in phenotypes consistent with overlapping functional sequences of more than one gene. Synonymous with segmental aneusomy. (See deletion, Prader-Willi syndrome, Wilms tumor, polyposis adenomatous, retinoblastoma, Miller-Dieker syndrome, Beckwith-Wiedemann syndrome, Alagille syndrome, di George syndrome, Lange-Giedion syndrome, Angelman syndrome)

CONTINGENCY TABLE: see association test

CONTINUITY, GENETIC: is assured by the mitotic nuclear divisions. The equal halves of each chromosome (derived by equational division) are shared between the two daughter chromosomes that are identical to those of the maternal cells (barring mutation, chromosomal breakage or accidents). (See mitosis)

CONTINUOUS TRAIT: displays a range of expression (such as weight, height, etc.) rather than an all-or-none appearance (such as white or red). Continuos traits are usually under polygenic control and subject to substantial environmental influence in expression. (See polygenes, QTL, continuous variation)

CONTINUOUS VARIATION: a trait shows a range of expression forming a quantitative series, without sharply separate classes. The majority of such quantitative traits are based on the collective effects of numerous genes (polygenic systems) that each may have only small effect but cooperatively they may bring about large variations. Traits subject to continuous variation are characterized by their fitting to the normal distribution or some classes of it. These variations are affected also by environmental factors and the outcome is a continuous or almost continuous series from relatively small to relatively large phenotypic expressions. Continuous variation cannot be classified into discrete categories such as black or white but they have a continuous spectrum. The characterization is made by counting or measurements or by systems of grading (such as learning scores, disease susceptibility, etc.). The study of continuous variation requires statistical tools such as mean, variance, standard deviation, tests of significance, correlation, regression, heritability, etc. (See topics mentioned under separate entries)

CONTRACEPTIVES: see sex hormones, hormone receptors

CONTRACTILE RING: is made of actin filaments and positioned in the cell equator, and it is instrumental in cutting apart the two daughter cells after the completion of mitosis. (See cell cycle, cytokinesis)

CONTRACTILE VACUOLES: occur in ciliated algae near the base of the flagella.

CONTROL (check): a standard to what the experimental data are compared. The standard must be identical and treated identically to the experimental material except of the special condition (genotype, developmental stage, time, chemicals, etc.) that is being studied. The *negative* control includes all elements except the genetic condition under investigation. The *positive* control is another similar sample, e.g. another DNA [female DNA extract if Y-chromosomal DNA is to be studied], just to see that he experimental system works. *Blind* control is used to test for possible contamination of the reagents, e.g. all the extraction, purification, steps are taken without the actual sample material. *No-template* control uses all the DNA amplification reagents except any DNA. (See also concurrent control, historical control, standard)

CONTROLED MATING: graphs next page show the percentage of homozygosis in successive generations of controlled mating systems and the limit of eventual homozygosity after an infinite number of generations. Controled indicates that the mating is not random but it is controlled by some plan. The different mating systems have important consequences for the genotypic composition of the population. (see chart on page 227).

Controled mating continued

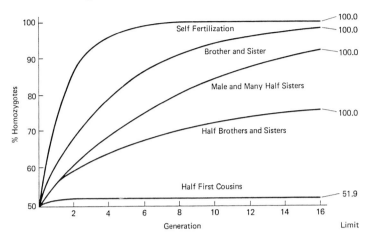

Controled mating is generally used by plant and animal breeders and in all types of genetic experiments. Controled mating of humans have been advocated by the discredited negative eugenics movements. The assortative matings based on ethnic, religious, cultural or other bases within smaller human groups may also be controled matings with the same biological consequences. Civil and religious laws impose some controled mating rules involving close relatives. Some state laws prohibited marriage among mentally defective individuals, including even sterilization. Most of the controled mating laws are no longer enforced in enlightened societies, except for the marriage between close relatives. Controled mating is very important in animal and plant breeding. [Chart was redrwan after Wright, S. 1921 Genetics 6:167.] (See breeding system, mating system)

CONTROLLING (Controing) **ELEMENTS**: a historical term for plant transposable elements that can occupy different positions in the plant chromosome and regulate the expression of various genes and can cause insertional mutations. (See Ac-Ds, Dt, Spm, Mu, TAM, transposable elements, insertion elements, insertional mutation)

CONVECTION: is an alternative term for non-oncogenic transformation of animal cells by exogenous DNA. (See transformation, transfection)

CONVERGENT EVOLUTION: similarity is based not on common ancestry but adaptive values; i.e. species from different lines of descent assume the forms, structure and function that is most valuable for their survival, e.g. sea mammals (whales, dolphins) are more similar in some traits to fishes than to terrestrial mammals. (See also divergence)

CONVERSION ASCI: see gene conversion

CONVERTANT: see gene conversion

CONVERTASE: a large enzyme complex that activates components of the complement. (See complement, properdin)

COOLIE's ANEMIA: see thalassemia

COOMASSIE BRILLIANT BLUE (acid blue, anazolene sodium): is a stain for proteins and a reagent for quantitative protein determination. Intravenal LD_{50} for mice is 450 mg/kg. (See LD_{50})

COOPERATIVITY: is a requisite for the formation of molecular complexes by one ligand promoting the binding of the following ones. (See ligand)

COORDINATE REGULATION: a group of genes are controlled by common regulatory elements such as the case in operons and regulons. The regulation may be induction or repression. (See *Lac* operon, *Arabinose* operon, regulon, SL1, SL2, homeotic genes)

COORIENTATION: by the end of metaphase the bivalent chromosomes or homologous chromatids tend to assume positions so the centromeres would face the opposite poles in anticipation of anaphase. (See meiosis, mitosis)

COP: a 12 subunit protein complex involved in the regulation of photomorphogenesis of plants. (See photomorphogenesis)

COPE's RULE: is a hypothesis that evolution usually involves increase in body size. The idea does not have general validity.

copia: RETROTRANSPOSON elements occur in all types of eukaryotes; in *Drosophila* they come generally in 5 - 7 kb sizes with about 300 bp direct and much shorter inverted terminal repeats. They frequently occur in 30-40 families and constitute 5-10% of the genome. The copia elements along with about 30 other insertion and transposable elements are involved in the mutability of the genomes by their movement. The frequency of their rearrangement is within the range of 10^{-3} to 10^{-4}. The long terminal repeats (LTR) encode a viral transposase-like protein. Retrotransposition cycles are initiated from an RNA copy of a transposable element. The transcription begins at the 5' region of the long terminal repeat (LTR) and this transcript is copied into a double-stranded DNA by a reverse transcriptase coded within the transposon. The synthesis of the first DNA strand is primed by the 3'- OH group of a host tRNA that immediately anneals to the LTR (this is called tRNA PBS [primer binding site]). Integration requires short inverted repeats (4-12 bp) at the end of the LTRs. Integration involves then a few bp duplications at the insertional target sites. These duplications are the consequences of the staggered cuts at the target that are filled in by complementary bases after integration. The length of the target site duplications reflects on the specificity of the transposon-encoded integrase. *Drosophila* retrotransposons have a number of features corresponding to retroviruses. They contain direct terminal repeats that have the sequences required for the initiation of transcription and polyadenylation. The majority of the *Drosophila* retrotransposons carry short inverted repeats at the end of LTRs. The LTR at one terminus contain TG and at the other CA bases. The majority of the retroposons have a purine-rich sequence immediately upstream to the 3'-LTR. These sequences are the priming sites for the second strand DNA (SSP, [second strand primer]). Most of the *Drosophila* retrotransposons have sequences starting with TGG immediately downstream to the 5'-LTRs that is complementary to the 3'-end of the tRNAs (the amino acid acceptor arm). The *Drosophila* retrotransposons have reverse transcriptase (*pol*), group-specific antigen (*gag*) polyproteins just like the retroviruses of vertebrates. Actually some plant retrotransposon-like elements are also organized in a similar manner but the majority of them are no longer able to transpose because some of their genes were reduced to pseudogenic forms. During retroviral life cycles extrachromosomal linear and circular DNA elements are formed with one or two LTRs. These features are also retained in some of the copia-like elements. These retroposons also produce viral-like elements in the eukaryotic cells quite similar to the real retroviruses. Also, some of the copia-like elements generate "strong-stop" DNA elements in the cells like retroviruses. These strong-stop DNAs and DNA-RNA heteroduplexes are left-overs of the first and second strand transcription by the reverse transcriptase enzyme. These copia-like elements transpose quite vigorously in both somatic and germline cells and induce mutations in the germline and the soma. Recombination between the multiple copies of the retroposons may cause all types of chromosomal aberrations in the host such as deletions, inversions, translocations. Perhaps the majority of the spontaneous mutations in the species that harbor transposable elements are caused by the movement (insertion) of these elements. The insertions may cause inactivation of exons but more frequently they insert into AT-rich sequences. The major representatives in *Drosophila* are *"17.6"* (7.4 kb) occurring in about 40 copies per genome and generate a 4-base duplication at the insertion target site, the LTR is 512 bp. *"297"* (7 kb) has approximately 30 copies. Its LTR is 415 bp, shows 1.7 kb homology between the right-hand ends with *17.6*. Its transposition-replication is primed by a $tRNA^{Ser}$. *"412"* (7.6 kb), copy number 40, LTR is 481 bp, target site duplication 4 bases. Its transposition is primed by $tRNA^{Arg}$. *"1731"* (4.6 kb) is present in about 10 copies and generates a target site replication of 5 bp. Its reverse transcription is primed probably by a fragment of the initiator $tRNA^{Met}$. *"3S18"* (6.5 kb) with target site duplication of 5 bp; occurs in about 15 copies. *"BEL"* (7.3 kb) is present in about 25 copies and the termini are very similar. *"blood"* (6 kb) occurs in 9 - 15 copies, LTR 400 bp, and generates a target site duplication of 4. The primer may be $tRNA^{Arg}$ similarly to elements *412* and *mdg1*. *"copia"* (5 kb) is present in 60 copies, LTR is 276 bp with 5 bp target site duplication. Its name came from copious amounts of its polyadenylated transcripts in the *copia* continued

cells. The initiator tRNA^{Met} primes for the reverse transcription of the copia RNA. Left end of the element: TGTTGGAATA TACTTATTCAA CCTACAAAG TAACGTTAAA, the right end: TATTAAAGAAA GGAAATATAA ACAACA. *"gypsy"* [synonymous with *mdg4*] (7.3 kb) with 10 copy number, LTR is 479 bp, and generates 4 bp target site duplication. These elements are associated with many mutations suppressed by *su(Hw)* [*suppressor of Hairy wing*], the product of this gene binds to an enhancer-like sequence within *gypsy* and affects the expression of adjacent genes including *gypsy*. The phenotype of some of the mutations caused by *gypsy* insertions are affected by *su(f)* [*suppressor of forked*]. The gypsy element is considered to be a retrovirus and its movement is controled primarily by the X-chromosomal mutation *flam* (flamingo). *"H.M.S.Beagle"* (7.3 kb) has 50 copies, LTR is 266 bp, and shows 4 base target site duplication. *"mdg1"* (7.3 kb), may be present in 25 copies, its LTR is 442 bp and creates 4 bp target site duplication; 14/18 of its primer binding sites are identical to that of *412* as well as the 27 bp adjacents to the left LTRs. tRNA^{Arg} is the most likely primer for their reverse transcription. *"mdg3"* (5.4 kb), has LTR of 267 kb; target site duplications are 4 bp. *"micropia"* (5.5 kb) hybridizes with *copia* in the Y chromosome of *D. hydei*. *"NEB"* (5.5 kb), *"opus"* (8 kb). *"roo"* (8,.7 kb, formerly called *B104*), LTR 429 bp and generates 5 bp target site duplication. *"springer"* (8.8 kb), LTR 405 bp and generates 6 bp target site duplication. (See retroviruses, retroposon, retrotransposon, transposable elements, transposase, reverse transcription, polyprotein, tRNA, suppressor, *Cin4*, hybrid dysgenesis, *Drosophila*)

COPOLYMER: is a molecule built from more than one type of components, e.g., a nucleic acid made of adenine and thymine units. If the sequence and quantity of the components do not follow a particular system it may be a random copolymer. If the units display periodic repetitions, it is called a repeating copolymer.

COPPER MALABSORPTION: see Menke's disease

COPROCESSOR: is an auxiliary processor to assist the main processor in special heavy tasks. Generally it also speeds up the processing of computers. (See processor)

COPROPORPHYRIA (CPO): autosomal deficiency of coproporphinogen oxidase resulting in excessive excretion of coproporphyrin III, an intermediate in porphyrin synthesis. (See porphyrin, porphyria)

COPULATION: sexual intercourse between female and male. Evolutionary selection of characteristics of the process would indicate that leaving more offspring even at the expense of decreased survival may be favored. Actually, this has been shown by the redback spider (*Latrodectus hasselty*) where hungry females frequently cannibalize their mates that copulate longer and thus transfer more sperm and thereby contribute more of their genes to the population.

COPY CHOICE: a hypothesis so named in the 1950s; it assumes that in recombination there may be no physical breakage and union between exchanged strands rather a replication and choice. This hypothesis bears similarity to Bateson's reduplication theory and gene conversion. Recently evidence has been presented that prokaryotic DNA polymerase III holoenzyme may slip frequently during replicating direct repeats. (See gene conversion, recombination in RNA viruses, breakage and reunion)

COPY NUMBER PARADOX: the genetically functional copies of organellar DNA appear much smaller than the physical copy number estimates. (See mtDNA, chloroplasts, C value paradox)

COPY-UP MUTATION: initiates runaway replication of the plasmid, i.e, the copy number is increasing beyond that in the wild type. (See runaway plasmids)

CORDYCEPIN: is a nucleoside analog (3'-deoxyadenosine) and it inhibits transcription and polyadenylation. (See polyA mRNA, polyA tail, polyadenylation signal)

CORE DNA: wraps around the histone octamer in the nucleosome core and does not include H1. (See nucleosome)

CORE ENZYME: prokaryotic transcriptase has two identical α and two β (ββ') subunits and to this core a fifth, σ subunit may be attached that is not essential for transcription but it seeks out the position of proper promoters. (See RNA polymerase)

CORE PARTICLE: see nucleosome

CORE POLYMERASE: is composed of only 3/10 subunits of prokaryotic DNA polymerase III (pol III). These subunits are α (polymerase), ε (proofreading 3'→5' exonuclease) and θ (enhances the activity of ε). (See DNA polymerases)

CORE PROMOTER: the most essential sequence within a promoter to carry out transcription. The core promoter of the eukaryotic RNA polymerase II usually includes the TATA box and may have also an initiator site, enhancer and the TATA box associated general transcription factors. In prokaryotes the core promoter includes only the RNA polymerase holoenzyme and the σ subunit. In eukaryotes instead of the σ, an assembly of the general transcription factors are found. In addition the TATA box binding proteins (TBP), TATA box associated proteins (TBA), initiator binding protein (IBP), and enhancer binding proteins (EBP) are usually present. The modulation and regulation of transcription requires a series activators, co-activators and different specific transcription factors. (See transcription factors, open promoter complex, transcription unit, base promoter, null promoter)

CORE, PROTEIN: the region common to the majority of structures in a superfamily or in a common fold.

CORE SEQUENCES: are usually invariable short tracts within more or less well preserved consensus sequences of the DNA. (See consensus)

CORECEPTOR: CD^8 and CD^4 T cell surface proteins along with the T cell receptor recognize MHC I and MHC II molecules. The CD cytoplasmic domains associate with a SRC protein tyrosine kinase. (See CD^8, CD^4, T cell receptor, MHC, SRC, protein tyrosine kinase)

COREMIUM: see hypha

COREPRESSOR: a metabolite which in combination with a repressor protein interferes with transcription and thereby with enzyme synthesis. (See repressor, aporepressor, repression, regulation of gene activity, cAMP receptor)

CO-RETENTION ANALYSIS: had been used in genetic mapping of mitochondrial genes. More or less large deletions frequently take place in the mitochondrial DNA resulting in simultaneous loss concomitant with simultaneous retention of antibiotic resistance markers. The co-deletion and co-retention frequencies provide converse estimates on linkage. Both co-deleted and co-retained markers must be present in groups indicating their physical position relative to each other. (See linkage, physical mapping, deletion mapping, mitochondrial genetics)

CORM: underground plant stem, modified for storage.

CORN: see maize

CORNEA PLANA: exists in autosomal dominant and recessive forms; most common in Finland. Characteristics: extreme farsightedness (hyperopia), opacity of the cornea especially at the margins. (See eye diseases)

CORNELIA de LANGE SYNDROME: see De Lange syndrome

COROLLA: the collective term for petals.

CORONARY HEART DISEASE: generally involves deposition of lipid plaques within the arteries surrounding the heart (atherosclerosis). Low density lipoprotein in the blood (above 200 mg/100 mL) increases the risk proportionally. On the other hand, high density lipoprotein content is favorable for avoidance. These account for the majority of all heart diseases and afflict 6-7% of the populations (predominantly males) in the Western industrialized countries. The underlying organic defects vary and many non-hereditary factors (diet, smoking, drug and alcohol consumption, age, etc.) and independent or concomitant diseases and conditions (blood pressure, diabetes, temperament, etc.) aggravate the liabilities. (See hypertension, cardiovascular disease, familial hypercholesterolemia, familial hyperlipidemia, hyperlipoproteinemia, lipoprotein, cholesterol, Tangier disease, diabetes, mucopolysaccharidosis, pseudoxanthoma elasticum, Marfan syndrome, homocysteinuria, lecithin:cholesterol acyltransferase deficiency)

CORPUS LUTEUM (yellow body): formed by luteinization of an ovarian follicle after the discharge of an ovum. In case of fertilization of the egg the corpus luteum increases in size and

Corpus luteum continued

persists for several months. If there is no fertilization the CL disintegrates. The CL secretes progesterone. (See ovary, Graafian follicle, ovum, luteinization, luteinizing hormone-release factor, corticotropin, animal hormones)

CORPUSCULAR RADIATION: is emitted by unstable radioisotopes (β particles [electrons] by ^3H, ^{14}C, ^{32}P, etc. and α particles [helium nuclei]) by uranium or fast and thermal neutrons during nuclear fission. (See also physical mutagens, isotopes, ionizing radiation, radiation effects, radiation measurement, radiation hazard assessment, electromagnetic radiation)

CORRELATION: interdependence of two variates. This relation may be statistical or physiological. The statistical correlation does not necessarily reveal any cause and effect link. The correlation may be positive (when the change of the variables follow the same direction) or negative (when the increase of one variable involves the decrease of the other). Thus the value of the correlation coefficient may vary between +1 and -1.

A GRAPHICAL REPRESENTATION OF CORRELATION. THE DOTS CORRESPOND TO THE POINTS OF THE MEASURED VALUES OF TWO QUANTITATIVE TRAITS

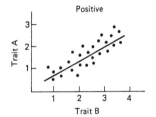
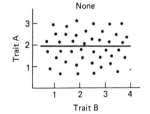
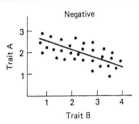

The correlation coefficient is independent from the scale of quantitations used, e.g. one can measure the correlation between intelligence and necktie color. For the calculation of the correlation coefficient, *covariance* has to be determined, i.e. the average product of the deviations of two variables from their respective means. It is estimated by dividing the sum of the products of the deviations from their means by the appropriate degrees of freedom. Thus **covariance** is:

$$(w) = \{\Sigma[(x_i - \bar{x})(y_i - \bar{y})]\}/(n-1)$$

for actual calculation the mathematically equivalent but computationally more convenient equation is used:

$$(w) = \{\Sigma(X_iY_i) - [\Sigma(X_i)\Sigma(Y_i)]/n\}/(n-1)$$

The analysis of covariance has many uses in biology and particularly in genetics. It may be used to separate the genotypic effects from treatment effects. It may reveal the relations among various types of variables. This type of analysis is useful to study the relationships among multiple classifications such as may occur if the experiments involve organisms of different genotypes, in different age groups and developmental stages and environments. In this Manual it is not possible to work out examples for all these different applications, only the basic procedure will be illustrated with step by step simple calculations.

The actual use of the formulas can be best shown by a hypothetical example. Let us assume that we measured (i) number of variates (in the following case *i* is 1 to 10) in two groups, each. In the columns x_i and y_i these measurements are listed from 1 to 10. To make the calculations easier (without changing the outcome) from each measurement a quantity (the same quantity, close to the means) is subtracted and we will name these X_i and Y_i, respectively as shown at the top of columns 3 and 4. Column 5 is the product of the lines in columns 3 and 4, X_iY_i. Columns 6 and 7 display the power of values in Columns 3 and 4, respectively: $(X_1)^2$ and $(Y_i)^2$. (See next page)

Correlation continued

COMPUTATION SCHEME FOR COVARIANCE

(1) x_i	(2) y_i	(3) $X_i = x_i - 150$	(4) $Y_i = y_i - 150$	(5) X_iY_i	(6) $(X_i)^2$	(7) $(Y_i)^2$
148	149	- 2	- 1	2	4	1
158	152	+ 8	+ 2	16	64	4
150	155	0	+ 5	0	0	25
143	142	- 7	- 8	56	49	64
162	160	+ 8	+ 10	80	64	100
150	160	0	+ 10	0	0	100
156	153	+ 6	+ 3	18	36	9
160	159	+10	+ 9	90	100	81
153	158	+ 3	+ 8	24	9	64
150	152	0	+ 2	0	0	4

$\bar{x} = 153$ $\bar{y} = 154$ $\Sigma X_i = + 26$ $\Sigma Y_i = + 40$ $\Sigma X_iY_i = 286$ $\Sigma(X_i)^2 = 326$ $\Sigma(Y_i)^2 = 452$

Substituting the values into the covariance formula:

$$W = \{\Sigma(X_iY_i) - [\Sigma(X_i)(\Sigma(X_i)]/n\}/(n-1) = \{286 - [(26) \times (40)]/10\}/9 = \mathbf{20.22}$$

Variances:

$$V_X = \{\Sigma(X_i)^2 - [(\Sigma X_i)^2]/n\}/(n-1) = \{(326) - [(26)^2]/10\}/9 = \mathbf{+28.71}$$

$$V_Y = \{\Sigma(Y_i)^2 - [(\Sigma Y_i)^2]/n\}/(n-1) = \{(452) - [(40)^2]/10\}/9 = \mathbf{+32.44}$$

The **coefficient of correlation**: $r = W/\sqrt{V_XV_Y} = 20.22/\sqrt{28.71 \times 32.44} = \mathbf{+0.663}$,
indicating that an increase in the values of X involved a positive change in the values of Y.
In genetic analysis the coefficient of regression is often used and that measures in quantitative units how much the dependent variable (Y) is changing as a function of the independent variable (X), e.g. we can determine how much the offspring's weight or height regresses to the weight

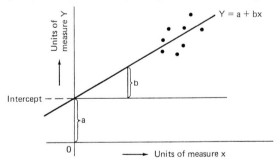

The linear regression line is specified by the equation given below: Y = a + bx. In this function, *a* is the value of the intercept, *b* defines the slope of the line, equal to the increase in *Y* per units of *x*. The solid circles correspond to the intersections of the values of the two variables. According to the table shown above on this page, we can consider the *x* axis as the units of measures of one of the variates and *Y* as the axis for the measurements of the other variates.

(in kg) or height (in cm) of the parents or to the mother or father. This is in contrast to correlation that could state only plus or minus and strong or weak correspondence but not in actual quantitative units. **The coefficient of regression**: $b = W/V_X =$ and in the example of the table $b = 20.22/28.71 = 0.704$. For a predictive value we use the linear regression equation : Y = a + bx from what a = Y - bx where (a) is the intercept of the straight line on the (Y) coordinate, (b) is the slope (indicating how much (y) changes by changes in (x). After substituting the

hypothetical data of our calculations into the equation $a = Y - bx$, since $(\bar{y} = Y = 154)$, and $\bar{x} = x = 153$, and $b = 0.704$, thus $a = 154 - (0.704 \times 153) = 46.29$. Therefore if the (x) independent variable is 158 kg, the dependent variable (Y) is expected to be $46.29 + (0.704 \times 158) =$

Correlation continued
157.52 kg. When the independent variable is 150 kg, the dependent variable (Y) is expected to be 46.29 + (0.704 x 150) = 151.89 kg. The example also testifies for the name of regression. Originally it was observed that large and small parents' children both tend to follow more the populations mean, i.e. they regress toward the mean. The offspring-parent regression is actually a measure of heritability. The linear regression is frequently represented graphically as shown on the preceding page. (See heritability, intraclass correlation)

CORTEX: outer layer of various tissues (egg, brain, kidney, tree bark, etc.)

CORTICAL GRANULES: small secretory vesicles under the egg membrane that by releasing some of their contents protect the egg from being fertilized by multiple sperms. (See fertilization, polyspermy)

CORTICAL REACTION: see fertilization, cortical granules

CORTICOSTEROID: 21-carbon steroids synthesized in the outer firm, yellowish layer of the adrenal (kidney) gland. The glucocorticoids regulate carbohydrate and protein metabolism whereas the mineralocorticoids regulate salt and water traffic. (See animal hormones, glucocorticoids, transcortin deficiency)

CORTICOTROPIN (adrenocorticotropin, ACTH): is a 39-amino acid peptide hormone of the anterior pituitary and regulates the synthesis of corticosteroid in the adrenal cortex. Its release — in response to stress, e.g.— is controlled by the releasing factors of the hypothalamus such as the thyrotropic releasing factor (TRF) and the luteinizing hormone releasing factor (LRF), corticotropin releasing factor. (See opiocortin, corticotropin releasing factor)

CORTICOTROPIN RELEASING FACTOR (CRF): it is supposed to be associated with the brain cognitive response. In Alzheimer's disease CRF is very low although CRF receptors accumulate. (See Alzheimer's disease, urocortin)

CORTISOL: is derived from progesterone (along with aldosterone) in the kidney cortex. It is chemically very similar to *cortisone*. The production of cortisone is controled by the pituitary hormone corticotropin and cAMP. The glucocorticoids promote gluconeogenesis and the deposition of glycogen in the liver, inhibit protein synthesis in the muscles and mediate fats and fatty acid breakdown in the adipose tissue, control inflammatory responses. (See glucocorticoids)

CORTISONE: see cortisol

COS: African green monkey cells with a chromosomally inserted simian virus (SV40) DNA, defective in the origin of replication but contain the intact T-antigen. (See SV40, African green monkey)

cos: cohesive ends (12 bp) of phage λ where linearization and cicularization of the DNA take place:
  pGGGCGGCGACCT------------
  -----------CCCGCCGCTGGAp   (See lambda phage)

CO-SEGREGATION: analysis for genetic linkage between two different genetic markers, e.g., a visible mutant phenotype and antibiotic resistance.

CO_2-SENSITIVITY: see symbionts hereditary

COSMIC RADIATION: strikes the earth from interstellar space, partly directly from the radiating bodies, partly (primary) or emanating from the collision of nuclei with cosmic (secondary) radiation in the upper atmosphere. The amount of cosmic radiation varies according to altitude and it is very high at the height of the space vehicles. The amount on the surface of the earth is about 0.028 to 0.045 rad per year. These figures should be corrected by a factor of about 0.8 because of protection by housing. For comparison, an average medical diagnostic chest X-ray delivers about 0.2 rad (but the gonads are protected). (See also isotopes; radiation natural, atomic radiation, radiation hazard assessment)

COSMIDS: approximately 5 kb cloning vectors are derived from phage λ DNA, containing one or two cohesive (*cos*) sites in the same orientation, an origin of replication (*ColE1*) and a selectable marker (*amp^R*). The cloning capacity of the plasmid is about 35-47 kbp because neither substantially smaller nor larger than the λ genome (≈ 49 kbp) can be packaged into the capsid

Cosmids continued
(see in-vitro packaging). For cloning smaller DNAs *charomids* must be used. Into the linearized cosmid the foreign DNA is ligated at appropriate cloning sites flanked by cosmid sequences. With the assistance of the phage terminase protein (an endonuclease with a specificity of cutting near *cos* sites), the *concatamer* is cut to head size to fill the capsid. After infection of *E. coli* , the cosmid recircularizes within the bacterial cell and due to the presence of the antibiotic resistance marker(s), the bacteria carrying the recombinant cosmid can be isolated. The cosmids can recombine with other plasmids (having homologous sequences) within *recA*$^+$ bacterial strains, and the cointegrates can be isolated if both carry independent selectable markers. There may be a problem, however, with rearrangements because of the active *recA*. A number of specially designed cosmids are available and they have been used for particular cloning needs. Cosmid **pJBS** had been used successfully for chromosome walking. After the recombinant cosmid is isolated from a library, the cloned foreign DNA segment can be cut up with restriction enzymes without destroying the vector that can be transformed into bacteria and yield small fragments that can be used as probes for overlapping fragments in the original library. Cosmid **c2RB** carries 2 *cos* sites and ampicillin and kanamycin resistance. Concatenation can be suppressed after insertion of the foreign DNA by phosphatase treatment (prevents the formation of phosphodiester bonds) or by directional cloning. There are two EcoRI sites in the vector flanking a single BamHI site and this makes it easier to walk from one recombinant cosmid to the next because the two small fragments so generated can be used as probes to rescreen the original library for overlapping clones. Cosmid pcos1EMBL carries tetracycline and kanamycin resistance, and the origin of replication of plasmid R6K (a replicator unrelated to Col1). With this plasmids the screening for recombinants can be carried out *in vivo*. Special cosmid vectors were developed for the transformation of animal cells. These carry antibiotic markers selectable in eukaryotic cells such as neomycin (*neo*), dihydrofolate (*dhfr*), hygromycin (*hph*) and other eukaryotic genes such a hypoxanthine-guanine phosphoribosyl transferase (*hgprt*), thymine kinase (*tk*$^-$), etc. The **pWE** series were used advantageously because of the ease of walking from one cosmid clone to another. They carry the phage T3 and T7 promoters on either side of NotI, an eight-base recognition site (GC↓GGCCGC) restriction endonuclease. The foreign DNA may be cloned into a BamHI site located between the phage promoters. After cloning, the cosmid is cut by restriction enzymes that do not affect the phage promoters. The cleavage products will contain small fragments adjacent to the promoters. The fragments that are downstream of the promoters are then transcribed into labeled RNA. These (radioactive) probes are used then to rescreen the library for overlapping fragments. Since NotI sites are rare in mammalian DNA, on the average about 1500 bp fragments are generated and frequently the NotI digests include the entire cloned fragment. These large fragments facilitate the construction of physical maps and may be useful in transformation because they may include the gene that is to be expressed. Some other pWE vectors may have polycloning sites rather than only the BamHI site. This may make the cloning easier but may create problems in the hybridization probes because of background hybridization of the labeled RNAs. Cosmids are used also with plant DNA. (See vectors, lambda phage, charomids, Rec, library, cosmid library, directional cloning, promoter, downstream)

COSMID LIBRARY: is a collection of DNA fragments of a genome cloned in cosmid vectors. (See cosmids, library)

COSMID MAPPING: physical mapping of cosmid-contained DNA sequences. (See cosmid, physical map)

COSMID WALKING: applies the "chromosome walking" procedure to DNA sequences in a cosmid library. (See genome projects)

CO-SPECIATION: indicates joint evolution of two organisms, e.g. host and parasite. The measure of co-speciaton is that the two phylogenies (on the basis of protein or DNA primary structure) are more similar than expected by chance. (See speciation, phylogeny)

COST OF EVOLUTION: evolution proceeds by replacing old alleles with new ones and as a

Cost of evolution continued

consequence some individuals are sacrificed; these pay the cost of evolution according to Haldane. The proportion of eliminated zygotes = sq^2 and the proportion of survivors = $\overline{w} = 1 - sq^2$ or in general term the *cost of natural selection*, $C = \int_0^t (sq^2/\overline{w})dt$ and hence $ln(p_t) - (ln(p_0)$ where s = selection coefficient, q = allelic proportion, $\overline{w}$ = fitness, t = time, dt = integral differentiation. (See evolution, natural selection, fitness, selection coefficient, allelic frequencies)

CO-SUPPRESSION: when a gene is introduced by transformation into a cell neither the resident nor the transgene copy of the same gene is expressed (repeat-induced gene silencing). The co-suppression may be complete or incomplete and it may be reversible during development. Co-suppression was discovered in plants but may occur in other organisms too. The cause of the phenomenon may be methylation of the genes, special allelic interaction or antisense RNA. (See methylation, antisense technology, transgene, RIP, transvection, silencer, paramutation, epigene suppression

$C_0t_{1/2}$: is the index of the reassociation of nucleic acid molecules when the reaction is half completed. It is proportional to the unique sequences in the reannealing molecules. (See c_0t curve)

c_0t CURVE (pronounce cot): when single-stranded DNA molecules are mixed they may reassociate, depending on the homology between the two types of strands. The facility of reassociation is determined by the degree of homology of the reactants, their concentration and the time allowed for the annealing. These three parameters are expressed by the c_0t curve. The reaction depends a great deal on the amount of redundant sequences in the DNA and the complexity of the DNAs. Redundant sequences and palindromic DNAs can anneal rapidly because of the similarities. For unique sequences it takes more time for the complementary sequences to

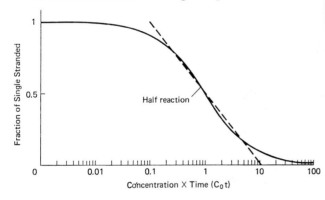

The time course of an ideal second-order reaction illustrates the features of the C_0t curve of nucleic acid reassociation. The ordinate corresponds to the fraction of single-stranded molecules. The abscissa denotes the mole value per liter of the material, multiplied by time (usually in seconds) on a logarithmic scale. (Modified after Britten, R.J. & Kohne, D.E. 1968 Science 161:529)

collide. For the characterization of DNAs of different types, generally the half c_0t values are used, i.e. the time when the reassociation half completed (see diagram). The reassociation of *E. coli* DNA generally takes place between c_0t values of about 0.1 and 10 (ca. 4.7 x 10^6 bp, mainly unique DNA) whereas a large and complex genome of rye containing about 80% redundancy in the 7.9 x 10^9 bp DNA, it is within 5 orders of magnitude of c_0t (between 0,001 to 100) in contrast to the two orders of magnitude in the bacterium. If single stranded homologous molecules are allowed to reanneal the process can be represented as $\dfrac{dc}{dt} = -kc^2$ where c represents the molecular concentration, t = time of the reaction, and k = constant that depends only on the length of the nucleic acid molecules, d means the differential integral. After integration we get:

$\int_t^0 \dfrac{dc}{c^2} = -kdt$ and hence $\left._{c}^{c_0}\left(-\dfrac{1}{c}\right)\right. = -k \left._t^0(t)\right.$ and $\dfrac{1}{c} - \dfrac{1}{c_0} = kt$ and if $t_{1/2}$ = is the time when

c₀t curve continued

$$c = c_0/2 \text{ and } \frac{2}{c_0} - \frac{1}{c_0} = \frac{1}{c_0} = kt_{1/2} \text{ and if the concentration is expressed as mass (g/L) then}$$

$c_0 = C_0/M_r$ where the molecular weight (M_r) is in Daltons and the complexity becomes

$$C_{0t_{1/2}} = \left[\frac{1}{k}\right] M_r$$

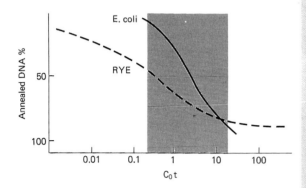

The reassociation kinetics of two genomes (7.9 x 10⁹ bp of rye and 4.7 x 10⁶ *E. coli*) of different complexities. The stippled area extends to two orders of C_0t values and covers most of the range of *E. coli* DNA. In contrast, 11% of the rye DNA reassociated at 0.001 to 100 C_0t. The half reassociation value of the non-repeated sequences of rye may be estimated by dividing 7.9 billion by 4.7 million is ca. 1,681 and after multiplying this figure by 2.5 (the C_0t values at half reaction of the *E. coli* DNA) we obtain 1681 x 2.5 ≈ 4,202 = the half reassociation value of the non-repeated rye. (Diagram and calculations were modified after Smith, D.B. and Flavell, R.B 1977 Biochim. Biophys. Acta 474:82)

C₀t VALUE: index of the rate of reassociation of single-stranded DNA molecules. (See C_0t curve, kinetics of reassociation)

COTRANSDUCTION: more than a single gene is transfered by transducing phage if the genes are closely linked. (See transduction, bacterial recombination)

COTRANSFECTION: transformation (genetic) by two or more genetically linked genes simultaneously. In the calciumphosphate precipitated donor DNA granules non-linked DNA molecules may also be included and integrated into the animal cell simultaneously. (See transformation genetic [transformation of animal cells], calciumphosphate precipitation)

COTRANSFORMATION: donor DNA molcules containing closely linked genes may be taken up and integrated simultaneously into the bacterial chromosome or into the chromosomes of other organisms. (See transformation mapping, cotransfection)

COTRANSLATIONAL TRANSPORT: the proteins synthesized on the ribosomes associated with the endoplasmic reticulum (rough endoplasmic reticulum) pass into the lumen of the endoplasmic reticulum during the process of translation (See protein synthesis, endoplasmic reticulum, signal sequence)

COTRANSPORT: two solutes passed through a membrane simultaneously by a transporter.

COTTON: see *Gossypium*

CO-TWIN: a pair of twins. (See twinning, zygosis)

COTYLEDON: seed leaf (an actually imprecise term because the seed before emergence may already have the initials of several leaves). At emergence the dicotyledonous plants have two cotyledons which have nutritive storage role before and for awhile after germination. The monocots have only a single very small cotyledon that has frequently a digestive role only. In dicots occasionally three or more cotyledons are formed as a developmental anomaly but tricotyledony is only exceptionally inherited. In zoology the cotyledons mean the tufts or subdivisions on the placental surface of the uterus.

COUMARIN: a plant metabolic product arising from 4-hydroxycinnamate through oxidation. It gives rise to dicoumarol, a vitamin K antagonist. These compounds are abundant in some varieties of sweet clover (*Melilotus*) and if the coumarin content is high, the forage value is much reduced because of the bitter taste the animals may not eat it. After fermentation (silage) di-

Coumarin continued

coumarol is formed that is very toxic due to its anti vitamin K and hemorrhagic effects. Plant breeding efforts have successfully reduced the coumarin content of some *Melilotus albus* varieties. (See vitamin K, prothrombin deficiency, coumarin-like drug resistance)

COUMARIN-LIKE DRUG RESISTANCE: due to an autosomal dominant gene some individuals are resistant to coumarin-like compounds such as warfarin, bishydroxycoumarin and phenindione (anticoagulants). These compounds are therapeutic agents in certain cases of surgery and in thromboembolic (blood clot forming diseases blocking arteries) diseases. (See blood clotting, antihemophilic factors, warfarin)

COUNSELING, GENETIC: human geneticists can make predictions about the probability of recurrence risks in families affected by hereditary anomalies and diseases such as birth defects, metabolic disorders, developmental problems, neurological or behavioral abnormalities, fertility problems, exposure to hazardous environment, consanguinity, pregnancy beyond age 35, etc. These predictions are based on family histories, cytological, biochemical and molecular analyses. The completion of the physical mapping of the human genome will greatly improve the potentials of accurate genetic counseling by the availability of appropriate probes and nucleotide sequence information. Genetic counseling should be sought before marriage or procreation in families where risk is indicated. The analyses may involve the prospective parents or the fetus may also be examined by amniocentesis. This procedure surgically withdraws with a syringe amniotic fluids 6-20 weeks following conception and analyzes the sloughed off, floating fetal cells by various methods. Under some circumstances of very high risks, termination of pregnancy may be opted for where there are no moral and/or legal objections. This, or any other decision, is made only by the family within the limits of the law. The physicians or geneticists provide only the facts but do not recommend the action(s) to be taken. From the viewpoint of genetics, selective abortion raises problems. If the families compensate for the aborted fetuses with new pregnancies, the frequency of the deleterious genes will rise in the population because the carriers may transmit the defective alleles and the problems are only postponed. If the carriers of serious hereditary defects refrain from reproduction, the frequency of these genes is supposed to decline eventually. The genetic counselor is a physician with thorough training in medicine, genetics, involving cytogenetic, molecular and statistical aspects of the field and is expected to be familiar with all the relevant techniques involved. In addition, he/she must have sufficient background in psychology and ethics. The genetic and medical facts must be explained in terms easily understandable by the families involved whose level of education may be quite variable from case to case. Genetic counseling is not supposed to follow any eugenic goals. (See risk, genetic risk, recurrence risk, empirical risk, utility index for genetic counseling, amniocentesis, supportive counseling, carrier, gene therapy, OMIM, prenatal diagnosis, ART, confidentiality, genetic privacy, wrongful life, informed consent)

COUNT PER MINUTE: see cpm

COUNTERCURRENT DISTRIBUTION: is partitioning relatively small molecules between two liquids that are polar to a different extent. Both liquids are moved in a special apparatus in opposite direction between steps of equilibration.

COUNTER-ION CONDENSATION: the association of ions with the polyelectrolyte DNA; in B DNA there is about 1 anionic charge per 1.7 Å distance. In case the line charge spacing changes the DNA may become unstable. (See DNA types)

COUPLED REACTION: the energy released by one reaction is utilized by the next reaction.

COUPLING PHASE: two or more recessive (or dominant) alleles in the same member of a bivalent chromosome (e.g. *A B* or *a b*). Some geneticists call this arrangement *cis*. (See recombination, repulsion)

COURTSHIP IN *DROSOPHILA*: is an extensively studied behavioral trait. Many genes are apparently involved but all appear to be pleiotropic and the mutations recovered affect more than one unrelated functions. Some are affected in sex-determination others are visual or olfactory mutants or affect female receptivity or male fertility, or circadian rhythm and others are involved in the "courtship song" of the flies (generated by the vibration of the wings). By a better understanding of neuronal functions and the availability of selection techniques progress is expected in this field. (See behavior genetics)

COUSIN: the child of the siblings of an individual's father or mother is a first cousin. The child of a first cousin is a second cousin or first cousin once removed. The cousins can be twice..... tenth, and so on removed depending on the steps in the relationship. (See cousin-german)

COUSIN-GERMAN: is a first cousin. (See german)

COUSIN MARRIAGES: may increase the chances of defective offspring depending on the coefficient of inbreeding. Infant death rate during the first year of life among children of first cousins approximately doubles relative to that in the general population. (See also coefficient of inbreeding, controled mating). Between 1959-1960 0.08% of the Roman Catholic marriages were inter first cousins, and among American Mormons during the period 1920-1949 it was 0.61%. In India the first cousin marriages in some societies was as high as 30% in the not too distant past. (See inbreeding coefficient, coefficient of coancestry, incest, genetic load)

COVALENT BOND: chemical linkage through shared electron pairs; it is typical for carbon compounds:

$$H \cdot + \cdot H \rightarrow H:H \quad \text{or} \quad 4H \cdot + \cdot C \cdot \rightarrow H:\overset{..}{\underset{..}{C}}:H$$

COVALENTLY CLOSED CIRCLE: a circular macromolecule (plasmid) in which all the building blocks are covalently linked and thus there are no open ends.

COVARIANCE: see correlation

COWDEN SYNDROME: multiple hamartomas. (See multiple hamartoma syndrome)

COWPEA (*Vigna unguiculata*): is a food and fodder crop of the tropics and subtropics primarily. All the 170 species have $2n = 2x = 22$ chromosomes.

COWPEA MOSAIC VIRUS: its genome contains B-RNA and M-RNA. The former codes for Vpg in a polyprotein complex at the carboxy terminus. (See VPg)

COYOTE: *Canis latrans* ($2n = 78$) North-American canine, can interbreed with domestic dogs ($2n = 78$) and the wolf.

CP1: is a mammalian transcription factor binding to the CCAAT box. (See CDP, CAAT box)

C11p11: a DNA probe used for the identification of certain cancer sequences.

CPB: see cetylpiridinium bromide

CPD: cross-linked adjacent pyrimidine dimers, photoproducts. (See pyrimidine dimers)

cpDNA (chloroplast DNA): its size varies generally between 8 to 13×10^7 Da, it is a circular double-stranded molecule with 20 to 40 copies per chloroplast of higher plants and 80 copies in the *Chlamydomonas reinhardtii* alga. (See chloroplasts, chloroplast genetics, *Chlamydomonas*)

CpG MOTIFS: of bacteria in unmethylated state induce murine B cells to proliferate and secrete antibodies. (See B cell)

CpG ISLANDS: regions of 500 bp or less 5' upstream of genes with these doublets. Their function is regulatory. (See isochores)

cPLA$_2$: catalyzes the hydrolysis of glycerophospholipid yielding arachidonic acid (see fatty acids) and lysophospholipid (membrane lipid). It is activated by MAP kinase in the presence of Ca^{2+}. (See integrin, MAP)

cpm (count per minute): is a measure of radioactivity; 1 µCurie ≅ 1,000,000 cpm. (see also dpm)

CPP32: a proenzyme of apopain. (See apoptosis, apopain, cysteine proteases, Yama)

CPSF: cleavage-polyadenylation protein factor.

cR: see CentiRay

CR LINES OF MICE: see congenic resistant

CRAF-1: is a protein factor interacting with the CD40 cytoplasmic tail by a region similar to the tumor necrosis factor receptor (TNF-α) associated factors (TRAF). CRAF is required for CD40-binding and dimerization. CRAF has five Zn-fingers and a Zn-ring finger. It participates in signal transduction. (See CD40, TRAF, TNF, signal transduction, zinc fingers, ring finger)

CRANIUM: the skeleton of the head, excluding the bones of the face, the container of the brain. (See brain human)

CRANIOORODIGITAL SYNDROME (otoplatodigital syndrome type II): is an X chromosomal head/face/brain defect which has overlapping symptoms with otopalatodigital syndrome and it is encoded in the same area of human chromosome Xq28. (See otopalatodigital syndrome, head/face/brain defects)

CRANIOSYNOSTOSIS SYNDROMES: occur in a great variety and involve premature closure of the sutures of the skull resulting in facial malformations and the defect is in one of the fibroblast growth factor receptors (FGFR). Most of them are under autosomal recessive control. (See Crouzon syndrome, Chotzen syndrome, Apert syndrome, Pfeiffer syndrome, Marfanoid, Shpritzen-Goldberg syndrome)

CRASSULACEAN ACID METABOLISM: some succulent plants can store large amounts of acids (malate) which are formed during the night and their level drops during the day.

CRE: cyclic AMP-response elements (TGANNTCA); DNA sequences upstream of transcription units of genes responding to cAMP). (See CREB)

CREATINE ([N-aminoiminomethyl]-N-methylglycine): in the form of phosphocreatine is the major source of energy generation (kcal/mol = 10.3). Creatine kinase catalyzes the reaction: phosphocreatine + ADP ⇔ ATP + creatine.

CREATIONISM: is the doctrine about how the universe came into existence based on oracle, and suggesting that organisms are as we see them at present because they have been so created and ordained. Another important aspect is "that the intellectual soul is created by God at the end of human generation, and this soul is the same time sensitive and nutritive" (Thomas Aquinas, 13th century). According to the Tertullian (2nd-3rd century) *transducianism* both soul and body are conceived and formed at exactly the same time. (See also evolution)

CREB: is a CRE (cyclic AMP-response element) binding protein requires phosphorylation at a serine residue at position 119 or 133 for transcriptional activity. CREB apparently activates T lymphocytes involving phosphorylation of CREB and that induces then transcription factor AP1 leading to the production of interleukin-2 and to the progression of the cell cycle. The CREB kinase is apparently identical with RSK2 a member of the RAS family. Mutations in CREB have been implicated in the Rubinstein-Taybi syndrome, in fusions with the MOZ (monocytic leukemia zinc finger protein) in case of acute amyloid leukemia (AML) and histone acetyl transferase displacement by AP1, leading to transformation. (See CRE, cAMP, calmodulin, T cell, AP1, cell cycle, immune system, RAS, signal transduction, Rubinstein-Taybi syndrome, leukemia [AML])

Cre/loxP: is a P1 phage recombinase system affecting specific target sites and it can be used also in various eukaryotes for mediating site-specific recombination or chromosomal breakage. The Cre protein (38 kDa) is a recombinase with specific recognition for the 34 bp *locus of crossing over of P1 (loxP)*, a pair of palindromic sequences: 5'-ATAACTTCGTATAG ~ CATACATTATACGAAGTTAT-3'
3'-TATTGAAGCATATCGTATGT ~ AATATGCTTCAATA-5'
and recombination takes place within the 8 base central core at the sites delineated by ~.

 recombination between the two loxP sites in the dimer yields the two monomers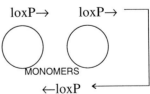

(See site-specific recombination Flp/FRT, chromosomal rearrangements, targeting genes, integrase, resolvase, homing endonuclease, ligand-activated site-specific recombination)

CREM a and b (modulator of CRE): are CREB-related transcriptional repressors in contrast to CREB which is a transcriptional activator. (See CRE)

CREMELLO: a color of horses of *AAbbCCDD* genetic constitution.

Crepis: composite flowers with chromosomes favorable for cytological studies. *C. capillaris* 2n

6; *C. tectorum* 2n = 8; *C. rubra* 2n = 10; *C. flexuosa* 2n = 14; *C. biennis* 2n = 40C.

CRETACEOUS PERIOD: 137 to 63 million years ago when the first human ancestors appeared. (See geological time periods)

CRETENISM: a hereditary or congenital deficiency of thyroid hormone, causing mental and physical retardation. (See goiter)

CREUTZFELDT - JAKOB DISEASE: is degenerative nerve disorder that begins with forgetfulness and nervousness – most commonly at middle age, sometimes earlier or later, and after a year or two progresses into jerky movement of the hands, insecure walk and expressionless face. These symptoms overlap with Gerstmann-Straussler disease and the two are probably identical basically although within both diseases different types of manifestations have been observed. The diseases are not limited to humans but in sheep (scrapie), goat and rodents highly similar nerve degenerations have been described. In the brain of the afflicted individuals amyloid protein deposits are found. The biochemical bases of the disorder was attributed to defects in the prion protein (PRP), a 27 to 30 kDa glycolipoprotein. The structural gene of PRP was assigned to mouse chromosome 2 and another gene in chromosome 17 was held responsible for the length of the incubation period of the disease. In humans the PRP gene is in chromosome 20p12-pter. It appears that PRP is a normal protein of the nervous system but proteolytic cleavage, amino acid replacements, insertions of 144 to 150 base pairs and insertions of 5 to 9 or more octapeptide repeats in-between the amino acids encoded by codons 51 and 91 may trigger the disease. The most commonly observed Pro→Leu replacement at codon 102 was attributed to the ataxia symptoms but changes at codon 117 (Ala→Val), 200 (Gln→Lys) and others were found to be associated with the degenerative phase of the PRP.

The injection of brain material into chimpanzees and other animals reproduces the disease. Since the infectious material does not contain any detectable amount of nucleic acid, scrapie – Creutzfeldt-Jakob – Gerstmann-Straussler diseases are considered to be the first infectious protein diseases. The Creutzfeldt-Jakob disease is attributed to autosomal dominant factors. These diseases occur in a familial manner, it is not exactly known when and how the genetic determination of the degeneration occurs. It has been detected among all ethnic groups; in some its frequency is much higher than in others. Among Jews of Lybian origin the incidence was reported to be 4×10^{-5}, nearly 50 times higher than in the general population. In addition, 41 to 47% of the cases observed were familial whereas in some other populations only 4 to 8% appeared familial. (See also prion, kuru, encephalopathies, Gerstmann-Straussler disease, bovine spongiform encephalopathy)

Cri du chat (cat cry): a deletion in the short arm of human chromosome 5 involves mewing like voice, mental and growth retardation and other disorders.

CRICKET(*Gryllus campestris*): 2n male = 29, female 2n = 30.

CRIGLER-NAJJAR SYNDROME (bilirubin glucuronosyl transferase deficiency, UDPGT): recessive, human chromosome 1q21-q23, causes a non-hemolytic jaundice and in case of total deficiency of UDPGT early infant death results. Partial deficiency of the enzyme is tolerated. (See uridine diphosphate glucuronyl transferase, Dubin-Johnson syndrome, Gilbert syndrome)

CRIMINAL BEHAVIOR: has some genetic component but the overwhelming motivation is provided by the family, social and economic conditions. The use of drugs, alcoholism, broken family ties, poverty, unemployment, etc. are the major factors. (See behavior in humans)

CRISIS IN ANIMAL CELL CULTURE: see Hayflick's limit

CRISS-CROSS INHERITANCE: is characteristic for X-chromosome linked recessive mutations. The recessive genes follow the pattern of inheritance of the X chromosome; they are expressed in the hemizygous male but only in the homozygous female:

Parents: XX *x* X^mY
 normal female mutant male

F_1 XX^m XY
 normal heterozygous female normal hemizygous male

Criss-cross inheritance continued

F_2	XX XX^m	XY X^{m}Y	
	females all normal phenotype	half of males mutant	

thus the X^m chromosome zig-zagged from grandfather to mother then from mother to grandson where it was expressed again

In case of Parents   X^{m}X^m x   XY
mutant female   normal male

F_1 XX^m XX^m X^{m}Y   X^{m}Y
normal females only   all males mutant
all the females are like the normal grandfather
and all the males are like the mutant grandmother

F_2 X^{m}X^m X X^m X^{m}Y   XY
mutant female   normal female   mutant male   normal male

In case the female is the heterogametic sex and the male is homogametic, like in the *lepidoptera* or birds, the inheritance follows the mirror image of what is shown above. (See sex linkage)

CRISTAE: invaginations of the inner membranes of the mitochondria. (See mitochondria)

CRITICAL POPULATION SIZE: although in a monogenic Mendelian F_2 generation 1/4 of the population is expected to be homozygous recessive, it rarely happens that every fourth individual meets this expectation. Therefore, it may be important to know how many individuals are needed in F_2 to find at least 1 of this desired phenotype. The statistical solution is a device that rules out the case when all the individuals would be the undesired type (3/4), e.g. $(3/4)^n = 1 - P$ where the (3/4) is the non-recessive class, (n) the number of individuals required in the population and P = probability. Thus, $(3/4)^n = 1 - 0.95$ must be solved for (n), $n(\log 3 - \log 4) = \log 0.05$, hence $n = (\log 0.05)/(\log 3 - \log 4) = -1.30/(0.477-0.602) = 10.4$, i.e. 11 (because fractions of individuals do not exist and the 0.95 probability is valid for 10.4 or more). Therefore, at 0.95 probability only 11 individuals give us an assurance to find at least 1 double recessive. The procedure is similar when we wish to determine the critical population size with a segregation ratio of 15:1 at 0.99 P: $(15/16)^n = 1 - 0.99 = 71.36$, i.e. 72. Similar calculations are useful for calculating the minimal population size required for the recovery of a mutant individual after mutagenic treatment if we know (or guess) the induced mutation rate. (See genetically effective cell number, mutation rate)

CRITICISMS ON GENETICS: essentially shares the same elements as those of other scientific fields. Most of the cases when genetics is criticized, the blame is attributable largely to perverse political systems but the biologists participating in it also deserve condemnation. Examples are the attempts to pursue negative eugenics, experimentation with biological warfare, inappropriate use of atomic energy, careless use of industrial, agricultural and medical chemicals, distortion of population genetic principles applied to environmental problems, cloning of animals and possibly humans. Some of the critiques argue that science cannot be left to scientists and the general public must be vigilant and reserve the decision-making. Some "scientists" also support this view. The problem is how to make decisions without being fully familiar with a particular field of science. A solution appears to be increasing continuous education on the progress of science. (See atomic radiation, radiation effects, chemicals hazardous, environmental mutagens, biological containment, genetic engineering, selection conditions, mutation, eugenics, nuclear transplantation)

Crk: is an adaptor protein in a signal transduction pathway, containing SH2-SH3-SH3 domains and it requires phosphorylation between the two SH3 domains (by the Abl tyrosine kinase). (See signal transduction, SH2, SH3, abl)

CRM: **c**ross-**r**eacting **m**aterial is a serologically identifiable protein (generally the product of a gene that fails to display enzyme activity); CRM⁻ designates a phenotype without an immunologically detectable protein. (See immune response)

CRMP-62 (collapsin response mediator protein): is a M_r 62 K protein required for axon extension in chickens and *Xenopus*. (See collapsin, UNC-33)

cRNA (complementary RNA): a DNA transcript. (See cDNA)

Cro **REPRESSOR**: of λ phage, cooperatively with *cI*, regulates lysogeny. (See lambda phage)

CROHN DISEASE (CD, regional enteritis): the autosomal recessive inflammation of the bowel is a familial condition because 10% of the affected individuals have relatives with the same affliction yet the genetic control is unclear. It is likely that more than a single genetic factor is involved in Crohn disease. CD is also called chronic inflammatory bowels disease, CIBD. TNF-α has been implicated in the symptoms. Genetic factors for CIBD and ulcerative colitis (UC) have been located to human chromosomes 3, 7 and 12. Some studies indicate the presence of *Mycobacterium paratuberculosis* RNA sequences in the clinical samples obtained from patients. The drinking water may be the source of infection. (See cadherin, TNF)

CRO-MAGNON MEN: see Neanderthal people

CROSS: mating between individuals of not-identical genetic constitution.

CROSS BREEDING: see cross fertilization

CROSS FEEDING: see channeling, syntrophic

CROSS FERTILIZATION: takes place when the sperm and the egg is produced by two different individuals of different genotypes (allogamy).

CROSS FOSTERING: is a procedure to test how much of a behavioral trait is hereditary and how much is the influence of the postnatal environment. Pups are separated from the natural mother and are given to foster mothers belonging to another inbred strain and then the behavioral differences compared with those individuals which were reared with the birth mother.

CROSS PROTECTION: see host - parasite relations

CROSS STERILITY: see incompatibility alleles

CROSSING: mating of two parental types of different genetic constitution.

CROSSING BARRIER: see incompatibility, incompatibility alleles

CROSSING OVER: process of reciprocal exchange between chromatids. Genes within a pair of homologous chromosomes may be linked in two fashions, either by coupling or by repulsion.

A B		A b	
a b	COUPLING	a B	REPULSION

Independent segregation observed by Mendel has the limitation of linkage and association of genes within a chromosome is not absolute either because crossing over and recombination may separate genes depending on their genetic (physical) proximity. Chiasmata during meiotic prophase is the physical basis of crossing over (the exchange between chromatids) and this is then detected in the genetic segregation as recombination. Crossing over takes place at the 4-strand stage, i.e. when each of the bivalents is composed of two chromatids associated at the centromere. This is the tetrad stage of the meiocyte. A single crossing over between two genes within a tetrad creates two reciprocally recombinant chromatids (see for exceptions gene conversion) whereas the other two chromatids remain unaltered (parental). Since 2/4 chromatids are crossovers, the frequency of recombination caused by a single crossing over event is 50% for that particular tetrad. Each individual heterozygous for linked genes has numerous meiocytes and crossing over does not take place in all of them at the same time therefore in a population of meiocytes the frequency of crossing over may vary from 0% to 50% depending on the genetic distance between the genes. The maximal frequency of recombination in meiosis is thus 50%. Occasionally higher than 50% recombination has also been observed in apparent contradiction to the principle described. This higher than 50% value does violate the principle because it is the result of selection during or after meiosis and gametogenesis or fertilization. One % recombination is considered, by convention, 1 genetic map unit (m.u) or 1 c.M (centi Morgan). Within a

Crossing over continued

single meiocyte simultaneously more than a single crossing over may take place. A single crossing over, however, always produces 50% recombination. If within the same genetic interval a second crossing over occurs, that may prevent the genetic detection of the first crossing over event because the second crossing over may restore the non-crossover type arrangement of the genes. The third crossing over within the same interval restores again the recombinant arrangement of the genes demarcating that interval:

PARENTAL	SINGLE CROSSOVER	SECOND CROSSOVER	THIRD CROSSOVER
A———B	A ↓ b	A ↓ B	A ↓ ↓b
a———b	a B	a ↑ b	a ↑ B

Thus each odd numbered crossovers generate detectable recombinants and the even numbered ones restore the original linkage phase of the alleles. Since multiple crossing overs are expected to occur at the product of the frequencies, double crossing within a meiocytes does not affect usually the other meiocytes.

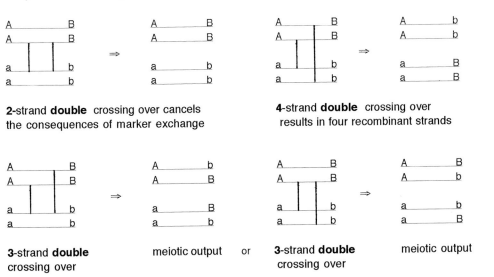

2-strand double crossing over cancels the consequences of marker exchange

4-strand double crossing over results in four recombinant strands

3-strand double crossing over   meiotic output   or   **3-strand double** crossing over   meiotic output

Two-strand, three-strand and four-strand crossing overs occur normally in the proportion of 1:2:1 (in the absence of chromatid interference). The frequency of recombination by all four types of double crossing over events combined is 50%.

Therefore if the frequency of a single crossing over is 0.30, double crossing over occurs by 0.30^2, and triple crossing over by the frequency of 0.30^3. Thus, if after the first crossing over the frequency of Ab + aB gametes is 0.30 after the second event it may be $0.30 - 0.30^2 = 0.21$ and after the third event $0.30 - 0.30^2 + 0.30^3 = 0.237$. The incidence of crossing overs may not occur as predicted by probability but the first crossing over may hinder the occurrence of a second one (positive interference) or it may stimulate it (negative interference). Double crossing over may involve two, three or four strands within a meiocyte (see illustration above). In yeast, each bivalent has at least one crossing over and usually not more than two. The probability of non-crossing over is commonly less than 0.1%. Usually crossing overs are not near to each other. Recombinational interactions are large but only a few of them progress to develop into crossovers (minimization). Non-crossover interactions exceed actual crossovers by a factor of 2 in *Neurospora*, 4 in *Drosophila* and based on the number of early recombination nodules in onions, the excess may be 30-40. It appears that the decision about crossing over is made before

Crossing over continued
the formation of the synaptonemal complex but the resolution of the Holliday junctions is delayed by or after pachytene.

Crossing over is generally limited to meiosis when the homologous pairs of chromosomes synapse. The chromosomes of some organisms, or under certain circumstance, my pair also during mitoses and this may result in somatic recombination. Mitotic crossing over resembles the meiotic event but the mechanism of exchange may not be identical. (See also coefficient of crossing over, cytological evidence for crossing over, oblique crossing over, time of crossing over, models of recombination, recombination frequency, mapping, mapping function, mitotic recombination, coincidence, interference, tetrad analysis, chromosome pairing, synapsis, chiasma, meiosis, mitotic crossing over, crossover, sperm typing, recombinational nodule, association point, pachytene, Holliday juncture, Holliday model, recombination mechanism eukaryotes)

CROSSING OVER MALE: see male recombination

CROSSING OVER MODIFIERS: see the various *Rec* alleles

CROSSING OVER, OBLIQUE: if the duplicated chromosomal region is obliquely paired recombination may take place in between the segments in the following way:

$$\frac{AAA}{aaa} \rightarrow \text{Aaa and aAA} \quad \text{or} \quad \rightarrow \text{AAa and aaA}$$

Such an event may also lead to increase or decrease of the chromatin and to cause position effect. (See *Bar*, position effect, unequal crossing over)

CROSS-LINKING: establishing bonds between two molecules or link together by covalent bond nucleotides in the same DNA or RNA. Chemical cross-linking agent is the bis-(2-chloroethyl)-methylamine and other mutagens and carcinogens (particularly the alkylating agents). The resulting quaternary structure leads to disruption of DNA functions. UV light can also produce cross-linking within and between nucleic acids and proteins. (See alkylating agents, DNA repair, pyrimidine dimers)

CROSSOVER: recombinant chromatid, chromosome or individual, originated by genetic exchange between homologous chromosomes through the process of crossing over. The crossovers can be single or multiple. (See crossing over, recombination, mapping function, coincidence, crossover fixation, mapping genetic)

CROSSOVER CONNECTION OF A PROTEIN: structural cores are connected by opposite ends of the cores across the surface of the domain.

CROSSOVER FIXATION: is a hypothesis to explain the relative homogeneity in repeat units of satellite DNA. According to this model the repeats can undergo frequent unequal crossing over during short evolutionary periods and then the same unit may be either propagated or eliminated after the recombinational event. If maintained, it can account for the homology of the sequences because there was not enough time yet to accumulate mutations even in the sequences that are not coding and are exempt of selection pressure. (See crossing over, crossover, satellite DNA, unequal crossing over)

CROSS-PATHWAY REGULATION: the biosynthesis of several different amino acids is coordinately regulated in fungi. (See tryptophan operon, channeling)

CROSS-REACTING MATERIAL: see crm

CROSS-REACTION: binding an antibody to an antigen; usually in the case when the formation of the antibody was not stimulated by it but by a very similar one. (See crm, antigen, antibody)

CROSS-REACTIVATION: see marker rescue, reactivation

CROSS-TALK: transmission of signals from receiver to sensor molecules between signal transduction pathways. (See signal transduction)

CROUZON SYNDROME (craniofacial dysostosis): autosomal dominant phenotype of ossified cartilages and various anomalies of the face, particularly protruding eyeballs. It is allelic to the Jackson-Weiss syndrome and to the Pfeiffer syndrome. The basic defect appears to be due to the fibroblast growth factor receptor 2 in human chromosome 10q25-q26. (See eye disease fibroblast growth factor)

CROWN GALL: a tumorous disease (mainly) of dicotyledonous plants caused by the soil-born *Agrobacterium tumefaciens*. The tumor development is initiated by the large (200 kb), Ti (tumor-inducing) plasmid containing oncogenes responsible for the production of auxins and cytokinin plant hormones. Removal of the bacteria about two days after infection does not stop the disease because the genes in the T-DNA part of the plasmids by that time have integrated into the plant genome. (See *Agrobacterium*, T-DNA)

CROZIER: the ascogenous, dikaryotic hyphae of ascomycetes form a three-cell hook type structure and in the cell at the tip the two nuclei fuse (karyogamy) to become diploid and after that an ascus is formed where meiosis takes place. (See ascus, fungal life cycle, *Neurospora*)

CRP: cAMP receptor protein. (See cAMP)

CRUCIFORM DNA: may be formed as a recombination intermediate between single strands of double-strand DNA (see Holliday model), and when inverted repeats, palindromes occur in both strands in a double-stranded DNA and within each strands these palindromes fold back and pair. (See Holliday model, palindrome, repeat inverted)

Cruciform recombination intermediate of DNA. (Courtesy of Drs. Potter, H. & Dressler, D.)

CRYOPRESERVATION: maintaining viability of biological tissues, enzymes, etc. at very low temperature, below -80° C or in liquid nitrogen (-195.8° C). (See sperm bank, artificial insemination)

CRYPTIC DOMINANCE: is a particular situation when recessive alleles of different loci fail to display complementation because of interaction of the gene products. (See epistasis)

CRYPTIC ELEMENT: such as a plasmid or transposon does not express a particular phenotype.

CRYPTIC PROPHAGE: can no longer exit from the chromosome of the bacterium and develop into a virion although some its genes are still transcribed. (See prophage)

CRYPTIC PLASMIDS: do not have known phenotype.

CRYPTIC SATELLITE: the satellite DNA is not displayed by ultracentrifugation as a separate (band) peak but it is masked within the main band of the DNA. (See satellite DNA, buoyant density, ultracentrifuge)

CRYPTIC SIMPLICITY: in the genome the repeated sequences are scrambled.

CRYPTIC SPLICE SITE: an unusual juncture where splicing of exons may take place in case the usual site is changed by mutation. (See splicing, introns)

CRYPTOBIOSIS: suspended life or reversible death.

CRYPTOGAMIC PLANTS: develop no flowers or seeds and multiply by spores such as ferns, mosses, algae. (See also spermatophytes)

CRYPTOGENE: mitochondrial DNA gene with primary transcript subject to pan editing that almost hides the original RNA sequences. (See kinetoplast, pan editing, *Trypanosoma*, RNA editing)

CRYPTOPOLYPLOIDY: increase in nuclear DNA content among related families with the same chromosome number. (See polyploid)

CRYPTORCHIDISM: failure of the testes to descend from the abdominal cavity into the scrotum. (See scrotum)

CRYSTAL LATTICE: see semiconductor

CRYSTALLIZATION: formation of crystals; a procedure preparatory to X-ray crystallographic analysis of macromolecular structures. (See X-ray diffraction analysis)

Cs (crossovers): recombinational events.

^{137}Cs: a cesium isotope emitting β and γ radiation; has a half-life of 33 years. (See isotopes)

CSAID (cytokine-suppressive anti-inflammatory drug): inhibitor of cytokine biosynthesis. (See cytokines, CSBP)

CSBP (CSAID-binding protein): are mitogen-activated protein kinases. (See CSAID)

CSF: cytostatic stability factor (probably the product of c-mos proto-oncogene) stabilizes MPF and thus prevents the exiting of the cell from M phase of mitosis. (See MOS, MPF, proto-oncogenes, mitosis, cell cycle)

CSF: colony stimulating factor, involved in autophosphorylation. The CSF receptor (product of the FMS oncogene) is homologous to the product of the KIT oncogene, a protein tyrosine kinase. (See KIT oncogene, FMS oncogene, protein tyrosine kinases, colony stimulating factor)

CSF1R: colony stimulating factor receptor, product of the FMS oncogene. (See FMS oncogene, CSF, KIT oncogene, colony stimulating factor)

Csk: a cytoplasmic protein tyrosine kinase; it has two amino-terminal protein-protein interaction domains (Src homology 2 and 3) and a carboxy-terminal catalytic domain. It is necessary for development of live mouse. In csk⁻ cells Src, Fyn and Lyn phosphotyrosine kinases activity is decreased. (See Src, B lymphocytes, Pyk)

CstF: is a cleavage-polyadenylation protein factor that may be associated with the CTD of RNA polymerase II. (See polI II, transcription factors, CPSF, CTD)

c_0t: find as if it would be cot.

CTAB: see cetyl trimethylammonium bromide

CTC: see cytotoxic T cell

CTD: **c**arboxy **t**erminal **d**omain repeat unit of the largest subunit of RNA polymerases (RNAP); can substitute for TATA box in genes that lack TATA sequences. This CTD of the α subunits is the contact site for transcription activator proteins and the upstream promoter element (UP) in *E. coli*. (See RNA polymerase, TATA box).

ctDNA: same as cpDNA or chloroplast DNA.

CTF: a CCAA motif binding proteins. (See binding proteins)

CTL (cytotoxic lymphocyte): see T cell

CTLA-4 (cytotoxic T lymphocyte antigen): is a T cell co-stimulatory molecule, it has signals opposite to protein CD28. (See T cell)

CTR: see constitutive triple response

CUCURBITS (*Cucurbitaceae*): *Cucumis sativus* (cucumber) 2n = 14; *Cucumis melo* (muskmelon) 2n = 24; *Citrullus lanatus* (water melon) 2n = 22; *Cucurbita pepo* (summer squash) and other squashes 2n = 40.

CULTIGENS: forms of cultivated plants. (See cultivar, variety)

CULTIVAR: a genetically distinct variety of crop plants, generally adapted to a region and has some agronomic value for the grower and the consumer. (See cultigens, variety)

CURARE: see toxins

CURIE: the basic unit of radioactivity contained in 1 g of radium, i.e 3.7×10^{10} disintegrations per second (dps). Most commonly 1/1,000th, the millicurie (mCi) or the 1/1,000,000th the micro curie (µCi, 2.2×10^6 desintegrations per minute [dpm]) are used in laboratory work. In most equipment only about half of the disintegrations are detectable and thus 1 µC corresponds to 1,000,000 counts per minute (cpm). Since the Ci unit defines the rate of disintegrations/time unit and the half-life of the different isotope may vary greatly depending on the species, the shorter half-life isotopes lose their isotopic atomes faster. (See isotopes)

CURING: a process of hardening by chemical or physical agents. Also, successful medication.

CURING OF PLASMIDS: removal of a plasmid or prophage from a bacterium, e.g. by the use of chemicals (acridine dye, Novobiocin, Coumeromycin) or irradiation. The [*PSI*⁺] form of the yeast prion can be "cured", i.e. converted into [*psi*⁻] form by treatment with guanidine hydrochloride or methanol, in a reversible process. (See prion, acridine dye)

CURRANTS (*Ribes* spp): all are 2n = 2x = 16.

CURVANS: see *rigens*

CUSHING SYNDROME (mixoma): also called by the acronym NAME (**n**evi, **a**trial mixoma, **m**ixoid neurofibromata, **e**ndocrine overactivity). Critical features are adrenal, hypothalmic, pituitary and lung tumors and hippocampal atrophy as a consequence of overproduction of glucocorticoids (hydrocortisone). (See nevus, neurofibromatosis, glucorticoids, adrenal, pituitary, hippocampus, atrophy)

CUT-AND-PASTE: is a mechanism of conservative transposition; transposase (e.g. Tn5) and host proteins bind the transposable element and target, double-strand DNA breaks at junction in a staggered manner, insertion follows, the gaps are filled with complementary nucleotides and (e.g. 9 bp [depending of the nature of the target cuts]) duplications are generated at the end of the "simple insert". (See transposons)

CUTICLE: a layer of substance over the surface of epidermal or epithelial cells.

CUTIN: a waxy material on the surface of plant cells that is slightly permeable to water or gaseous substances.

CUTIS LAXA: includes autosomal dominant, recessive or X-linked forms. The hereditary forms appear early in life and show loose skin and joints, folding skin. The recessive I form involves in addition lung, heart and digestive tract anomalies. Type II has bone malformation (dystrophy) and retarded development. Some other forms have the signs of early aging of the skin or emphysema and anemia. The X-linked form causes the formation of bony horns on the foramen magnum (the opening of the head to the vertebral canal) and deficiency of lysyl oxidase and disturbance in copper metabolism. (See skin diseases, collagen, Ehlers-Danlos syndrome, pseudoxanthoma elasticum)

CW: clockwise

Cxs : gene conversions accompanied by crossing over. (See gene conversion, crossing over)

CYANIDE (HCN): is a very powerful inhibitor of cellular oxidation by complexing with cytochrome oxidase and is thus an extremely dangerous poison. Sodiumthiosulfate and sodium nitrate are antidotes. Cyanides may be present in various foods (almond) and feed (Sudan grass, white clover).

CYANIDINE: see anthocyanin

CYANOBACTERIA (blue-green algae): are photosynthetic prokaryotes, and they contain phycobiliproteins and chlorophyll a. Their genetic material is DNA and can be manipulated in a way similar to other prokaryotes. (See phycobilins, phycocyanins, chlorophyll)

CYANOGEN BROMIDE (BrCN): a very toxic, lacrimant gas. It has been used to cleave polypeptides and fusion proteins and also as ligand for activated chromatography matrices for Western blotting. (See ligand, Western blotting)

CYANOGENIC PLANTS: see lotoaustralin

CYANOSIS: bluish discoloration in body areas caused by the accumulation of reduced hemoglobin and methemoglobin resulting from mutation or caused by other factors. (See methemoglobin)

CYBERNETICS: the theory of control and communication such as exist in the nervous system and also in mechanical devices such as the computer.

CYBRID: contains two different cytoplasms in fused cells. (See cell genetics, cell fusion)

CYCLIC AMP: see cAMP

CYCLIC GMP: see cGMP

CYCLIC ELECTRON FLOW: electrons emanating from *photosystem I* in light returning to their origin. (See photosynthesis, Z scheme)

CYCLIC PERMUTATION: see permuted redundancy

CYCLIC PHOTOPHOSPHORYLATION: cyclic electron flow drives ATP synthesis.

CYCLIN (CLN): proteins synthesized during cell divisional cycles and are responsible players in the process by activating protein kinases (CDKs) and control the cyclic sequence of divisional steps. (See cell cycle, CDK, CLB)

CYCLIN A: similar to cyclin B but appears earlier in the cell cycle. (See cell cycle, CDK)

CYCLIN B: has no known enzymatic activity; it is part of MPF and has a role accessory to the protein kinase (*cdc2*) subunit; it seems to determine the substrate specificity of MPF. (See cell cycle, MPF, cdc2, CDK, CAK)

CYCLIN D: is a protein involved in the cell cycle and it is also an oncogenic protein. (See cell cycle, CDK, CAK)

CYCLIN-DEPENDENT PROTEIN KINASE (CDK): in association with cyclin, phosphorylates proteins and thus promotes cell divisional events. (See cell cycle, CDK)

CYCLOBUTANE DIMER: cyclobutanes have 4 C atoms in a ring. Such a structure may also be formed by cross-linking adjacent pyrimidines in the DNA upon exposure to UV light. The most common, genetically effective, alteration in ultraviolet light exposed DNA is the formation of thymine dimers (as it is shown below). The formation of cyclobutane dimers is dependent on the wave-length of the radiation. At 280 nm it is induced almost five time as efficiently as at 240 nm. The dimer physically distorts the DNA and interferes with normal DNA replication and incorporation of nucleotides at wrong sites and the substitutions may lead to mutation. The dimers may be split by visible light-activated enzymes resulting in light repair. (See DNA repair, photolyase, pyrimidine dimer)

PYRIMIDINE (thymine dimers) FORM CYCLOBUTANE RING ↓

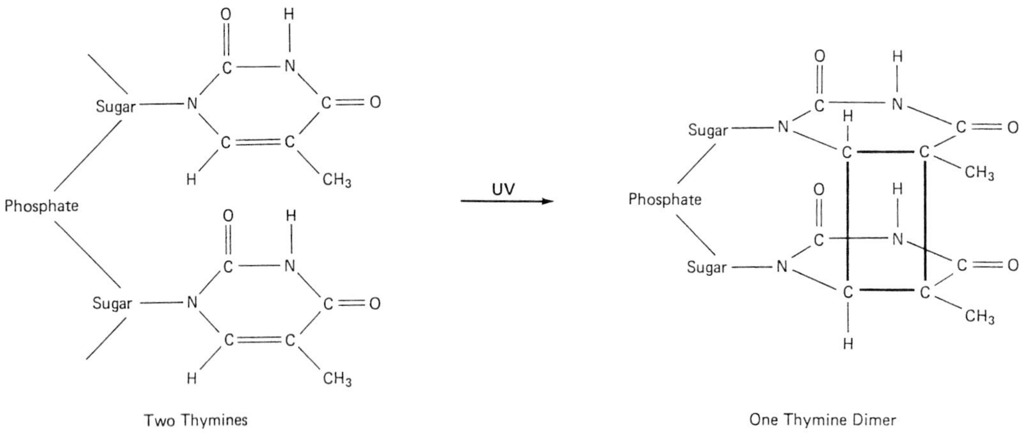

Two Thymines → UV → One Thymine Dimer

CYCLOHEXIMIDE: an antibiotic which blocks the peptidyl transferase on the 80S ribosomes but does not affect protein synthesis on the 70S ribosomes. (See antibiotics, protein synthesis, peptidyl transferase, signaling to translation, ribosome)   CYCLOHEXIMIDE →

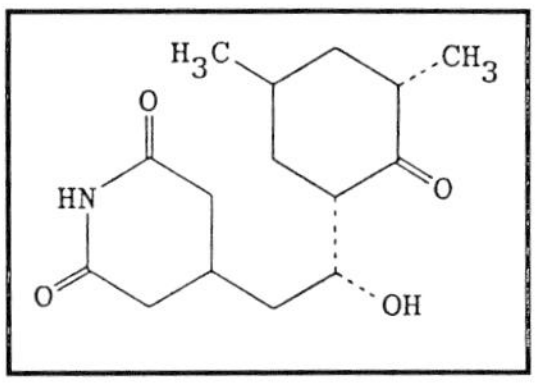

CYCLOIDEA ALLELES: determine the radial symmetry of the flowers.

CYCLOOXIGENASE (COX): an enzyme involved in the synthesis of the eicosanoids prostaglandins, prostacyclins and thromboxanes from arechidonic acid. This enzyme thus plays an important role in inflammation, pain, fever and suppresses immunosurveillance, stimulates tumorous growth. Resveratrol, a product of grapes and the legume *Cassia quinqueangulata*, acts as a chemopreventive of the tumorigenic effect of cyclooxygenases. (See arachidonic acid, lipoxygenase, prostaglandins, thromboxanes)

CYCLOPHILIN: is a peptidyl-prolyl isomerase and it catalyzes cis-trans isomerization of X-Pro peptide bonds. It is inhibited by the immunosuppressive antibiotic, cyclosporine. Cyclosporin bound to cyclophilin inactivates calcinuerin. The cyclophilins are usually heat-inducible proteins and seem to participate in heatshock protein-90 dependent signal transduction. (See cyclosporin, calcineurin, immunosuppressant, mitochondrial import, heatshock proteins, signal transduction, immunophilins)

CYCLOSOME: same as APC, see also cell cycle

CYCLOSPORIN: is a peptide antibiotic. It inhibits the activation of T lymphocytes and it is thus an immunosuppressant. (See immunosuppressant, antibiotics, calcineurin, cyclophilin, immunophilins)

CYCLOTRON: is an accelerator of electrically charged particles and atomic nuclei. It is used to cause transmutations in atomic nuclei, e.g. a normal magnesium into radioactive sodium.

CYLINDROMATOSIS (turban tumor): is a skin tumor caused by inactivation of a dominant gene in human chromosome 16q12-q13. It is a rare dominant disease.

CYME: an inflorescence where the growth of the apex ceases early to the benefit of the branches.

Cys$_4$ RECEPTOR: contain 4 cysteines—zinc domains in eukaryotic transcriptional regulators of α-helix-loop motif. The N-termini of the α helix contacts the nucleotide base, the N-terminal loop binds to the phosphate backbone and the C-end is the dimerization interface. (See zinc finger, binding proteins)

CYSTATHIONURIA: a recessive (human chromosome 16) deficiency of γ-cystathionase. Consequently cystathionine cannot be cleaved into cysteine and homoserine, resulting in benign defects. (See cystinosis, cystinuria, cystin-lysineuria, homocystinuria, amino acid metabolism)

CYSTEINE (HSCH$_2$CH[NH$_2$]COOH): upon hydrolyis of proteins in air it is converted to cystine.

CYSTEIN-HISTIDINE FINGER: see DNA-binding protein domains, Zinc finger

CYSTEINE PROTEASES: ICE, ICH1, NEDD2, ICH2/TX, CPP32/YAMA/apopain, MHC2 are implicated in apoptosis. (See apoptosis)

CYSTEINE STRING PROTEINS (CSP): contain palmitoylated cysteines on the peripheral membranes involved with the nerve synaptic vessels. (See fatty acids, synapse)

CYSTIC FIBROSIS: is apparently one of the most common serious recessive hereditary defects (human chromosome 7q31-q32), involving fibrous degeneration of the pancreas, bile ducts, the respiratory system, intestinal glands, sweat glands, male genital system, etc. The primary defect is apparently in chloride transport due to alteration of the gene regulating transmembrane conductance. It afflicts 1/2,000 white newborns (the frequency of the responsible alleles may exceed 0.02). In American blacks the prevalence is more than an order of magnitude less and it is even much less common among Orientals. The large CF gene (24 exons) has been cloned (250 kbp). More than 200 different point mutations and several deletions have been identified. Prenatal diagnosis is 98% effective on the basis of determining the high level of immunoreactive trypsin in the serum, characteristic for CF. Sweat test and DNA test can also be used for diagnosis. Pulmonary relief may be provided by inhaling DNase solutions and avoiding respiratory infections by the use of antibiotics. Pancreatic symptoms may be alleviated by enzyme replacement and proper diet. The intestinal mucilage may be removed by surgery. Cystic fibrosis can be detected on the basis of the pattern of heat inactivation of the enzymes acid phosphatase and α-mannosidase. Homozygotes display practically no activity, heterozygotes 40-60%, and the absence of the defective allele is indicated by 80-100% activity. The testing of γ-glutamyltranspeptidase, aminopeptidase M and alkaline phosphatase from the second trimester amniotic fluid permits prenatal diagnosis. Genetic screening can be carried out in newborns on the basis of immunoreactive trypsin. The normal allele of the cystic fibrosis gene is involved in the regulation of sodium and chloride absorption and therefore it has been called cystic fibrosis transmembrane conductance regulator (FTR), a protein kinase A and ATP regulated Cl$^-$ ion channel. A large deletion in the CFTR gene (ΔF508) conditions deficiency of internalization of *Pseudomonas aeruginosa* bacteria—to what cystic fibrosis patients are hyper-susceptible—and because of this, the epithelial cells are unable to clear from the lungs the mucosa by desquamation. The human CFTR$^+$ gene, equipped with the rat intestinal fatty acid-binding protein promoter, corrected, when transfected into mouse, a lethal intestinal defect in mouse. Mice heterozygous for the cystic fibrosis gene secreted 50% of the normal fluid and chloride ions in response to cholera toxin and thus CF may convey heterozygote advantage in natural selection. Many different mutations exist in the CF genes ranging from nil to only reduced synthesis of CFTR and various types of defects in regulation, processing and altered conductance. Defects in CF may be associated with defects in the excretory channel of the testis and azoospermia, asthma (breathing difficulties), nasal polyposis and hypertyrosinemia. Experiments indicated that gene therapy for cystic fibrosis is possible using primarily adenoviral vectors but the efficiency of local transformation is low <0.1%. 10% efficiency may be required for efficient remedy. The target should be the columnar cells lining the airways. Unfortunately these cells are rather refractory to gene transfer. It has been shown that DNase I treatment has beneficial effect on the removal of mucosa from the respiratory channel. Unfortunately DNase is inhibited by F-actin

Cystic fibrosis continued
secreted by the leukocytes that infiltrate the airways in response to infections. Actin-resistant DNase has been engineered that alleviates this problem. (See cystic fibrosis antigens, genetic screening, gene therapy, infertility, cholera toxin, azoospermia, polyp, tyrosinemia, CBAVD)

CYSTIC FIBROSIS ANTIGENS (calgranulin A and B): are determined by autosomal dominant genes in human chromosome 1q12-q22. Homozygosity for the cystic fibrosis gene (7q31-q32) is accompanied by absence of these proteins whereas the symptomless carriers have an intermediate level. It appears that these independent proteins track the basic defect in cystic fibrosis. Antigen A (also called calgranulin A because it is most abundant in the granulocytes) has a M_r 11,000 and antigen B (calgranulin B) has M_r 14,000. These two proteins are virtually identical with calcium-binding proteins in other sources. (See cystic fibrosis)

CYSTINE-KNOT: two disulphide bridges link adjacent antiparallel strands of a peptide chain and form a ring that is penetrated by the third one. Such cystine knots are found in NGF, TGF-β and PDGF-BB growth factors. (See TGF-β, platelet drived growth factor)

CYSTINE-LYSINURIA (diaminopentanuria): autosomal recessive increase of diamines (cadaverin) in the urine and causing ataxia and mental degeneration. (See amino acid metabolism, cystinuria, cystinosis)

CYSTINOSIS: a semi-recessive autosomal hereditary disorder under the control of more than one locus involving up to 100 fold amounts of cystine in the cells (lysosomes). Although it may not involve phenotypically obvious symptoms it may eventually cause kidney failure, eye defects, growth arrest and rickets. The heterozygotes can be identified by an increase in cystine in their cells. The onset may be early or late. A cystinosis gene has been assigned to the short arm of human chromosome 17. (See also Fanconi renotubular syndrome, amino acid metabolism, cystinuria, cystin-lysinuria, homocystinuria. (Photograph is the courtesy of Dr. D.L. Rimoin, Harbor General Hospital, Los Angeles, CA)

CYSTINURIA: is a recessive disease in several allelic forms causing variable degrees of cystin deposits in the kidney cysts and the bladder. Besides the increase of cysteine in the urine larger than normal amounts of other amino acids (lysine, arginine and ornithine) may also be excreted. High fluid intake may prevent or alleviate the amino acid deposits. It is encoded at human chromosome 2p21 at the same site as the rBAT transporter protein. (See amino acid metabolism, cystinosis, cystine-lysinuria, homocystinuria, rBAT/4F2hc)

CYSTOCARP: a spore-bearing structure formed in red algae after fertilization.

CYTIDINE: a pyrimidine base, cytosine, is associated with ribose or deoxyribose.

CYTIDYLATE: salt of cytidylic acid.   CYTIDINE (cytosine riboside) →

CYTIDYLIC ACID: cytidine plus phosphate (a DNA or RNA nucleotide).

CYTOBLAST: mitotic cell of the germarium in insects. (See germarium)

CYTOCHALASINS: are toxins that break cellular actin microfilaments, inhibit glucose transport, thyroid secretion, growth hormone release, phagocytosis, platelet aggregation and are used to evict nuclei from animal cells to produce cytoplasts and karyoplasts. (See toxins, cytoplast, karyoplast, actin, phagocytosis, nuclear transplantation)

CYTOCHEMISTRY: chemical analysis of isolated subcellular components, and employing also histochemical techniques for *in situ* tracing their action.

CYTOCHROMES: electron carrier heme proteins with important role in respiration, photosynthesis and other oxidation-reduction processes including the activation of promutagens and procarcinogens. (See promutagens, procarcinogen, activation of mutagens)

CYTODIFFERENTIATION: see differentiation, morphogenesis

CYTODUCTION: dominant cytoplasmic transmission of a particular hereditary state; heterokaryosis in yeast.

CYTOFECTIN GS2888 (dioleophosphatidylethanolamine): is a transfecting agent for mammalian cells. It is coupled with a fusogenic compound and a cationic lipid (GS2888). It carries plasmids, AS ODNs, etc. into cells efficiently and its toxicity is low. (See liposome, fusigenic liposome, AS ODN, transfection)

CYTOGENES: are located in the cellular organelles (plastids, mitochondria) and not in the nucleus. (See mitochondrial genetics, chloroplast genetics)

CYTOGENETICS: an area of genetics involving the study of chromosomal structure and behavior in connection with inheritance. The study of chromosomal anomalies and accompanied pathological conditions. Cytological analysis of the evolution of chromosomes.

CYTOHET: heterozygosity in cytoplasmic genes (in plastids and mitochondria) when the zygotes receive cytoplasmic material biparentally. (See chloroplast genetics, mtDNA)

CYTOKINES: peptides secreted in response to mitogenic stimulation and they participate in intercellular communication and cellular activation. Most of the cytokine receptors invoke tyrosine phosphorylation of cellular proteins including the cytokine receptors. The various cytokine receptors permit the action of different protein tyrosine kinases (PTK) in different cell types. They are instrumental in the induction and regulation of the immune system, cellular differentiation, blood cell formation, apoptosis, tumor inhibition, etc. The cytokine receptors belong to five superfamilies: (1) cytokine receptors, (2) interferon receptors, (3) TNF receptors, (4) interleukin-8 receptors and (5) TGF-β receptors. Cytokines (IL-2) have been administered locally as cancer therapeutic agents (in biodegradable polymer microspheres) or a more promising approach is transformation by cytokine genes. (See interleukins, interferons, colony stimulating factor, PTK,TGF, signal transduction)

CYTOKINESIS: division of the cytoplasm after nuclear division leading to the formation of two cells from one. (See mitosis, cell cycle)

CYTOKININS: see plant hormones

CYTOLOGICAL EVIDENCE FOR CROSSING OVER: was obtained first in 1931 in maize by using a line heterozygous for chromosome 9 bearing a large terminal knob and the *C/c*

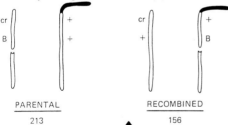

An outline of one of the *Drosophila* experiments on correspondence between genetic and physical exchange of chromosomes. (After Stern, C. 1931 Biol. Zbl. 51:547)

and *Wx/wx* genes. McClintock and Creighton found that the cytologically detected knob followed the syntenic genetic markers and corresponded to the expectation in the recombinants. The same time in *Drosophila* Curt Stern used a fragmented X-chromosome marked with *carnation* eye (*cr*) and *Bar* eye (*B*) and another X-chromosome with the wild type alleles and also a fragment of the Y-chromosome was attached to it. Again, the genetically observed recombination was associated with the physical exchange of the cytologically marked chromosomes. (See crossing over)

CYTOLOGICAL MAP: shows genetic sites in relation to microscopically visible structures such as chromosome bands, knobs, centromere, satellite(s). (See genetic map, physical map, RFLP, RAPD)

CYTOLOGICAL MARKER: a unique feature of the chromosome (e.g. knob, satellite, etc.), number of nucleoli in the nucleus, defective plastids, etc. visible by cytological analysis.

CYTOLOGY: the study of structure and related function of the cell and subcellular elements.

CYTOLYSIN: secreted by cells to dissolve other cells, e.g. during the immune reaction.

CYTOLYSIS: dissolving the cells into its chemical components.

CYTOMEGALOVIRUS (CMV): includes the herpes viruses, Epstein-Barr virus. These potentially tumorigenic viruses have DNA genetic material and great and diverse mammalian host

Cytomegalovirus continued
specificity. The name (megalo) comes from the observation that infection enlarges the host cells. The human cytomegalovirus can produce about 200 potentially antigenic polypeptides, however the CMV early-expressed genes may block antigen presentation by class I MHC molecules and thus the defense function of $CD8^+$ cytotoxic T lymphocytes. Phosphorylation of the immediate early (IE) proteins may prevent the interference with antigen presentation and the CTL defense system. The human CMV may cause serious disease to congenitally infected infants and adults with defective immune system. (See herpes, Epstein-Barr virus, MHC, antigen presenting cell, T cells, CTL)

CYTOPATHIC: causes pathological change to cell.

CYTOPHOTOMETRY: spectrophotometric chemical study of the content(s) of single cells. (See spectrophotometry, microscopy)

CYTOPLASM: all material enclosed by the cell membrane, including cellular organelles but the nucleus excluded. (See cell)

CYTOPLASMIC INCOMPATIBILITY: involves the disruption of fertilization or embryogenesis because of bacterial infection in insect species. (See infectious heredity)

CYTOPLASMIC INHERITANCE: is determined by non-nuclear genetic factors. (See chloroplast genetics, mtDNA, mitochondrial genetics)

CYTOPLASMIC MALE STERILITY (*cms*): in maize five different mitochondrial genomes have been identified and three of them (*T, S* and *C*) control cytoplasmically inherited male sterility. The *Rf* restorer genes restore fertility. An outline of the procedure for obtaining double-cross hybrid seed by the use of cytoplasmic male sterile lines of maize. *R* stands for a dominant fertility restorer gene; *r* does not affect cytoplasmic male sterility. The inner circles represent the cell nucleus and a band around it symbolizes the cytoplasm. Dots in the cytoplasm stand for determinants of mitochondrial male sterility. Plants with dotted cytoplasm can produce seed only when fertilized by *R* pollen.

Although *cms* has been identified in over 100 maize stocks all of them fall into these three groups. The various circular mtDNAs have been physically mapped and their sizes is different (T is 540 kb). The male sterility in the *Tcms* stock was attributed to a mitochondrial locus, *T-urf13*, near a 4.7 kb repeat absent from normal mtDNA. Fertility restoration requires specific nuclear genes or the loss of the *T-urf13* site caused by an intramolecular mtDNA recombination between two 4.6 kb repeats followed by an intermolecular recombination of a 127 bp repeat and resulting in a 0.4 kb deletion involving *T-urf13*.

The joint action of *Rf1* and *Rf 2* restorer genes is sporophytic, i.e. approximately all (95%) the pollen produced is fertile in the plants heterozygous for these dominant nuclear genes. In the *S* cytoplasm *Rf3* conveys gametophytic restoration, i.e. in the heterozygotes approximately half of the pollen is fully functional and half is aborted. *Rf4* acts sporophytically also in *S* cytoplasm. Maize mitochondria contain several smaller DNAs besides the main genome. The loss of mitochondrial plasmids *S-1* and *S-2* from the *S* cytoplasm results in reversion (fertility). The cytoplasmic male sterility of the *T* type is associated with susceptibility to *Helminthosporium maydis* (*Cochliobolus heterostrophus* Drechsler) blight and to other factors. The URF13 protein contains 115 amino acids residing in the inner membrane of the mitochondria. URF13 is a receptor of the toxin of the blight pathogens and also for the insecticide methomyl. When URF13

Cytoplasdmic male sterility continued

binds the toxin it forms a mitochondrial pore and uncouples oxidative phosphorylation. *Rf1* restorer gene alters the transcript of the *T-urf13* gene but *Rf2* does not do this although it too inhibits male sterility. The RF2 protein is very similar to mammalian mitochondrial aldehyde dehydrogenase. The elements of these systems are influenced also by the genetic background. In cultured cells *Helminthosporium*-resistant mitochondrial mutations occur that are also male fertile. Cytoplasmic male sterility systems were observed also in several other plant species. The 3.7 kb *pvs* mtDNA sequence in common bean plants is associated with cytoplasmic male sterility. This sequence includes two open reading frames, *orf239* and *orf98*. The 27 kDa protein product of *orf239* localizes in the callose layer and the primary cell wall of the pollen. Transforming *orf239* into tobacco (*N. plumbaginifolia*) without targeting it into the mitochondria caused pollen disruption. (See male sterility, mtDNA, symbionts hereditary, heterosis, hybrid vigor, mitochondrial plasmids, mitochondrial genetics, fertility restorer genes, RU maize)

CYTOPLASMIC TRANSFER: see conjugation *Paramecia*

CYTOPLAST: enucleated cell (cell that lost its nucleus). (See transplantation of organelles, karyoplast)

CYTOSINE (C): a pyrimidine base in RNA or DNA. (See pyrimidines)

CYTOSINE-REPRESSOR (CytR): regulates the transcription of at least 8 genes of *E. coli* involved in nucleoside uptake and catabolism. All CytR-repressed promoters have a catabolite repressor (CRP) binding site and repression is the outcome of the cooperation between the CytR and the CRP-cAMP complex. (See catabolite repression, cAMP)

CYTOSKELETON: bracing and scaffolding fibers (actin, microtubule and other filaments) in the cytoplasm. The cytoskeleton may affect different cellular functions, progression of the cell cycle, differentiation, movement, etc.

CYTOSOL: the soluble, non-particulate material of the cell that suspends all cellular structures within the plasma membrane.

CYTOSTATIC: hinders cell division.

CYTOSTATIC FACTOR: see CSF

CYTOSTATIC STABILITY FACTOR: maintains MPF at high level during metaphase II and prevents exit from metaphase. (See MPF, mitosis)

CYTOTACTIN: extracellular matrix protein affecting cell movement.

CYTOTAXONOMY: evolutionary studies based on the analysis of the chromosomes.

CYTOTOXIC T CELL (CTL, CTC): a type of T lymphocytes that destroys infecting cells or foreign antigens. They recognize target cells by the antigenic peptides presented. The natural killer cells (NK) lyse various types of cells without the cooperation of the MHC system. CTL and NK cells, however, may cooperate and induce the formation of cytolytic granules and lytic pores. In this process perforin expression on $CD8^+$ T cells and sometimes on $CD4^+$ cells has important role. Another mechanism of action was suggested relying on Fas-dependent apoptosis. Actually these two routes may be used simultaneously. $CD8^+$ T cells recognize cytopathic and noncytopathic viruses by presentation of their peptide antigens on MHC I molecules. The killing may involve also the secretion of antiviral lymphokines. Bacterial infections are handled also by activated macrophages, NK cells, T cell receptors, granulocytes, $CD4^+$ and $CD8^+$ T cells. IFNγ and perforin are the fighting molecules. In the autoimmunity disease, diabetes mellitus, perforin-dependent destruction of insulin-producing cells seems to be involved. Graft rejection is caused by the infiltration into the added tissue NK cells, macrophages, $CD4^+$ and $CD8^+$ T cells. The latter two recognize either MHC II or MHC I differences, respectively. (See lymphocytes, T cell, T cell receptor, perforin, granzyme, killer cells, apoptosis, CD4, CD8, MHC, Fas, macrophages, diabetes, graft rejection, cytomegaloviruses)

CYTOTOXICITY: the ability of any agent to harm, destroy or poison cells.

D

D = number of restriction enzyme recognition sites per DNA length. (See restriction enzymes)

Δ: universal symbol for deletions.

d (dalton): see also the now prefered Da and Dalton.

δ: yeast transposable element. (See Ty)

2,4-D: see dichlorophenoxyacetic acid

D ARM: see transfer RNA

δ DELETING ELEMENT (also called ΨJ_α): mediates the deletion of the δ gene from the α gene in the T cell receptor (TCR) site when there is no α gene rearrangement. (See T cell receptor)

D **ELEMENT**: see non-viral retrotransposable elements.

D-J: see immunoglobulins

D LOOP: is formed when in replicating small circular DNA (mtDNA) one of the strands is displaced while the other strand is copied or when in genetic recombination a single-strand DNA invades the RecA protein complex and displaces one strand of a duplex (See Meselson —Radding model, recombination molecular mechanism, prokaryotes, replication. mtDNA, strand displacement)

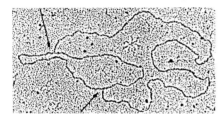

← D loop of mtDNA replication. (Courtesy of Dr. K Wolstenholme)

D2 PROTEIN: a histone-like protein in the nucleosomes of *Drosophila*. (See nucleosomes)

DA: TFIIA transcription factor associated with TFIID–DNA complex. (See transcription factors)

Da: see Dalton

DAB: transcription factor TFIIA associated with TFIIB–TFIID–DNA. (See transcription factors)

DACTYLOGRAM: fingerprint obtained by dactyloscopy (fingerprinting). (See fingerprints, dermatoglyphics)

DAF: DNA amplification fingerprinting. (See amplification, DNA fingerprinting)

DAF: see decay accelerating factor.

DAG: see diaglycerol

DAI: is an interferon-induced protein; a double-strand RNA activated inhibitor, and it is involved in kinase function regulating translation. When 25-30% of the factor is phosphorylated protein synthesis is severely inhibited. (See interferon, kinase, translation)

DALA: δ-aminolevulinic acid.

DALTON (Da): measurement unit of molecular mass (M_r), generally used to estimate the size of macromolecules, 1 Da = 1.661 x 10^{-24} g (1/12 of the MW of the C^{12} isotope). 1 megadalton is 10^6 Da. 1 pg of DNA is about 0.60205 x 10^{12} Da; the M_r of a nucleotide pair is about 650 Da; 1 kbp DNA (Na salt) is about 6.5 x 10^5 Da. The "average" molecular weight of an amino acid residue in a protein is about 110-120 Da.

DALY (disability-adjusted life year): is the sum of life years lost because of premature mortality and years of life with disabilities adjusted for the severity of the disability. On this basis congenital anomalies occupy the 10th rank among 17 leading disabilities in the world. (See genetic diseases)

DAM: female mammal. (See also sire)

dam: deoxyadenine methylation factor in a GATC sequence. (See methylation of DNA)

DAM METHYLASE: see methylation of DNA

DAMD (directed amplification of minisatellite DNA): uses PCR to produce probes for the determination of homologous variations among different species or genetic stocks for DNA fingerprinting, generally by RFLP. (See PCR, RFLP, minisatellite, DNA fingerprinting)

DANDY-WALKER SYNDROME: is a form of autosomal recessive of hydrocephalus with considerable variations. (See hydrocephalus)

DAPI: 4',6'-diamidino-2-phenylindole, a fluorochrome stain for chromosomes. When excited by UV light, blue light is emitted. (See chromosome morphology, fluorochromes, chromosome painting, FISH)

DARIER-WHITE DISEASE (keratosis follicularis): is an autosomal dominant keratosis, prevalent in areas where sebaceous glands (excreting fatty substances and cellular debris) are located, e.g. on the scalp, face, chest back, armpit and groin. (See keratosis)

DARK-FIELD MICROSCOPY: receives only the scattered light arriving from a side-ways illuminated object that appears light on a dark background. (See microscopy, fluorescent microscopy, phase-contrast microscopy, Nomarski)

DARK REACTION: light-independent enzymatic reactions, following the light reactions of photosynthesis, leading to the formation of monosaccharides. (See photosynthesis)

DARK REPAIR: repair of DNA not requiring light (excision repair). (See also DNA repair)

DARWINIAN FITNESS: see fitness, neutral mutation, beneficial mutation

DARWINISM: interpreting evolution as the outcome of natural selection, survival of the fittest. (See natural selection, neo-darwinian evolution)

DATABASES: (for some addresses and additional information see under special entries). provide information of different subjects by electronic means. The databases listed in this book were in operation in 1997. Addresses frequently change and many have overlapping information. Patience and experience is often required but they can be very helpful.

General Directories (Jump Stations)
<http://www.gen.emory.edu/MEDWEB/medweb.html >
Genetics Jump Station: < http://www.ifrn.bbsrc.ac.uk/gm/lab/docs/genetics. html>
Molecular Biology Jump Station: <http://www.ifrn.bbsrc.ac.uk/gm/lab/docs/molbiol.html>
WWW Worm <http://wwww.cs.colorado.edu/wwww>
National Center for Biotechnology Information: <http://www.ncbi.nlm.nih.gov>
ENTREZ: <info@ncbi.nlm.nih.gov> or <net-info@ncbi.nlm.nih.gov>
Infoseek: <http://www.cbi.nlm.nih.gov/Search/client.html>
Infoseek Yellow Pages: <http://www.infoseek.com/>
Gopher: <gopher://fly.bio.indiana.edu>
Netscape Net Search:<http://home.netscape.com/home/internet-search.html
FTP (file transfer protocol): for several databases and free software packages in molecular biology: <ftp://ftp.ebi.ac.uk> or <ftp://ncbi.nlm.nih.gov>
European Bioinformatics Institute: <http://www.embl-ebi.ac.uk/>
BioCatalog ("yellow pages": <http://www.ebi.ac.uk/biocat/biocat.html> or <http://www.bchs.uh.edu/Server>
Biopages list of resources: <http://www.golgi.harvard.edu/htbin/biopages>
Yahoo (www subject directory): <http://www.yahoo.com>
GenBank (Los Alamos Natl. Lab): <http://www.noc.lanl.gov>
NCBI(National Center for Biotechnology Information)GenBank: <http://www.ncbi.nlm.nih.gov/>
GenBank - general nucleotide sequence inquiries: <genbank%life@lanl.gov>
- sequence submission: <gb-sub%life@lanl.gov>
DDBJ - general nucleotide sequence inquiries: <ddjb@niguts.nig.junet>
- submission: <ddjbsub@niguts.nig.junet>
EMBL - general nucleotide sequence inquiries: <datalib@embl.earn>
- submission: <datasubs@embl.earn>. See European Bioinformatics
EST Sequence Information: <http://www.ncbi.nlm.nih.gov/dbEST/index.html>
<ftp://ncbi.nlm.nih.gov/repository/dbEST>

Databases continued
        STS, GenBank: <http://www.ncbi.nlm.nih.gov/dbSTS/index.html>
        SWISS-PROT (protein information): <http://expasy.hcuge.ch/sprot/sp-docu.html>
        BLOCKS (Fred Hutchinson Inst.'s protein blocks): <http://www.blocks.fhcrc.org>
        Protein structure images: <http://molbio.info.nih.gov/cgi-bin/pdb>
                <http:scop.mrc-lmb.cam.ac.uk/scop/>
        Homology search: <info@ncbi.nlm.nih.gov> human-mouse homology:
                <http://www3.ncbi.nlm.nih.gov/Homology>
        Cross-referencing model organism genes with human disease and other mammalian
                phenotypes: <http://www.ncbi.nlm.nih.gov/XREFdb/>
        Linkage analysis and software: <http://linkage.rockefeller.edu/>
        Genetics education (human): <http://www.kumc.edu/gec>
        Policy issues: <http://www.geneletter.org/>
ANIMAL GENETICS:
        USDA, ARS, National Agricultural Library (plants and animals):
                <http://www.nal.usda.gov/pgdic/>
        Poultry, Pig, Sheep and Cattle genome: <http://www.ri.bbsrc.ac.uk/>
        Caenorhabditis: <http:/www.sanger.ac.uk/~sjj/C.elegans_blast_server.html>
        Drosophila: <http://cbbridges.harvard.edu:7081/> or <http://flybase.bio.indiana.
                edu> or <http://www.embl-ebi.ac.uk>
        Mouse Informatics Database: <http://www.informatics.jax.org>
                Encyclopedia of the mouse genome: <http://www.informatics.jax.org>
                Mouse genome database <http://www-genome.wi.mit.edu>
                Recombinant inbred strain panels (mouse): Phone: 1-800-422 MICE or
                      207-288-3371. Fax: 207-288-3398
                Jackson Laboratory backross DNA panel map service (mouse):<lbr@aretha.
                      jax. org> or < meb@aretha.jax.org>
                TBASE (transgenic mice, knockouts):<http://www.gdb.org/Dan/tbase/tbase.
                      html>
        Human Gene Mutation Database: <http://www.cf.ac.uk/uwcm/mg/hgmd0.html>
                Mendelian Inheritance in Man, online (OMIM)
                          <http://www.ncbi.nlm.nih.gov/>
                          <http://gdbwww.gdb.org>
                          <ftp://ftp.gdb.org>
        Excite: <http://www.excite.com/Reviews/Science/Biology_and_Chemistry/
                Biology/Genetics/Human/index.html>
        WAIS Gateway. Human molec. genet (includes also abstracts) .:<http://
                www.informatik.uni-rostock.de/HUM-MOLGEN/>
        Human karyotypes:
                <http://www.pathology.washington.edu:80/Cytogallery/> or
                <http://gdbwww.gdb.org/gdb/ideo/docs/ideogram.html>
        Human Mapping Laboratories:
                <http://www.genethon.fr/> (information in French)
                <http://www.sanger.ac.uk/> (includes also *Caenorhabditis* and
                      other databases)
                <http://www-shgc.stanford.edu/> (radiation hybrid maps primarily)
                <http://www.well.ox.ac.uk/> (specific proteins, proteome and other
                      information)
                <http://www.genome.wi.mit.edu/> (human and mouse primarily)
    PLANTS:
        Mendel (The Commission on Plant Gene Nomenclature (CPGN) develops
                common nomenclature for sequenced plant genes): <http://probe.
                      nalusda.gov:8300/cgi-bin/browse/mendel/>

Databases continued

        Plant genenome database. USDA, Plant and Animal Genomes: <http://www.nal.usda.gov/pgdic/>

    EUKARYOTIC MICROBES:

        Dictyostelium: <http://worms.cmsbio.nwu.edu/dicty.html>
        Saccharomyces: <http://genome-www.stanford.edu/Saccharomyces/.>
        Saccharomyces Genome Database (SGD): <http://genome-www.stanford.edu>
        Yeast Protein Database (YPD): <http://www.proteome.com>
        Yeast, European Network Informatics (MIPS):<http://speedy.mips.biochem.mpg.de/mips/yeast/>
        Yeast, SCIENCE Magazine:
            <http://www.sciencemag.org/science/feature/data/genomebase.htm>

    PROKARYOTES:

        Haemophilus influenzae, Methanococcus, Mycoplasma, etc <http://www.tigr.org>
        E. coli: <http://cgsc.biology.yale.edu> or <http://genome4.aistnara.ac.jp/GTC/mori/research/dbservice/ecoli-e.html> or
        <ftp://ftp.pasteur.fr/pub/GenomeDB> or Encyclopedia of *E. coli* Genes and Metabolism<http://www.ai-sri.com/ecocyc/ecocyc.html>, <http://www.mbl.edui/html/ecoli.html>)
        Microbial Strain Data Network (MSDN): <http://www.bdt.org.br/bdt/msdn/>

    CARCINOGENS:

        <ftp//potency.berkeley.edu/pub/tables/hybrid.other.tab> or <cpdb@potency.berkeley.edu>, <http://www.iarc.fr/monoeval/allmonos.htm>)
    GENETIC TOXICOLOGY (TEHIP): <http://medlars.nlm.nih.gov>

Cautionary note: Many symbols have multiple synonyms or single symbols standing for different genes or spelling and capitalization may vary for the same word and various errors may be encountered. Some databases cannot be entered without permission; the conditions for access can usually be obtained from the URL addresses. Addresses frequently change.

DATE PALM (*Phoenix dactylifera*): one of the 12 species; dioecious, 2n = 2x = 36.

dATP: deoxyadenosine triphosphate.

DATURA: a members of the *Solanaceae* family (2n = 24) have been used for genetic studies involving cytological (trisomics) and cell culture methods, primarily. These species are the sources of the alkaloids atropin, hyoscine, hyoscyamine and scopolamine. (See also henbane, *Atropa*)

DAUER LARVA: represents an alternative form of larval development. At an early stage (e.g., at the second molt the larva (*Caenorhabditis*) becomes semi-dormant if feeding is inadequate or the culture is overcrowded. Such a larva does not feed and responds only to touching. It is a safety option for survival. Upon the appearance of new food supply, normal life cycle may be resumed. Dauer larva formation is under the control of a transforming growth factor (TGF-β) analog. (See *Caenorhabditis*, TGF)

DAUERMODIFICATION: induced modification of the phenotype that may be transmitted to the progeny but persists only for a few generations (therefore is not a mutation).

DAUGHTER CELLS: are formed after division from the parental one. (See cell division)

DAUGHTER CHROMOSOME: a replicated chromosome has two chromatids. When the chromatids are separated at the centromere during mitotic anaphase they become two single-stranded daughter chromosomes. (See chromatid, centromere, mitosis). Formation of Daughter Chromosomes→

Metaphase   Anaphase   METAPHASE   Anaphase   Metaphase

DAUGHTER OF SEVENLESS (DOS): the sevenless (SEV receptor tyrosine kinase) is a protein in the eye

Daughter of sevenless continued

developmental pathway. Corcscrew (CSW) is a phosphotyrosine phosphatase in the signaling path and its substrate is DOS, a pleckstrin homology domain protein transmitting light signals for the eye between sevenless and RAS1 in *Drosophila*. In this pathway Grb2 is an adaptor molecule with the son of sevenless (SOS) guanine exchange factor. The receptor sevenless tyrosine kinase (RTK) triggers neuronal differentiation in the single R7 cells of the ommatidia in response to the BOSS (bride of sevenless) ligand on the neighboring R8 photoreceptor cell. (See sevenless, son of sevenless, boss, ommatidia, RTK, pleckstrin domain, rhodopsin).

DAY BLINDNESS (hemeralopia): autosomal recessive, defective vision in bright light and total colorblindness. It is the result of defective cone-like bodies of the retina. (See night blindness, colorblindness)

DAY NEUTRAL: it does not respond to photoperiodic treatments. (See photoperiodism)

Dazla: see azoospermia

DBF4: a Cdc7 binding and activating protein and may be required for the initiation of DNA replication. (See Cdc7)

DBL ONCOGENE: same as MCF2

DBM PAPER: see diazotized paper

DBP: DNA **b**inding **p**roteins are histones, suppressors, activators, silencers and DNA and RNA polymerases, transcription factors. (See separate entries)

DCC: deleted colon carcinoma gene involved in cancerous growth. (See colorectal cancer, tumor suppressor genes, pancreatic adenocarcinoma, p16)

dCF (deoxycoformicin): see ADA

dcm: see methylation of DNA

DCMU (3[3,4-dichlorophenyl]-1,1-dimethylurea): an inhibitor of photosystem II. (See photosynthesis, Z scheme)

DctB: bacterial kinase, acting by phosphorylating protein DctD.

dCTP: deoxycytidine triphosphate

DDBJ: Data Submissions, Laboratory of Genetic Information Analysis, Center for Genetic Information Research, National Institute of Genetics, 111 Yata, Mishiama Shizuoka 411, Japan, General inquiries about nucleotide sequence database, e-mail: ddjb@niguts.nig.junet, submission forms: ddjbsub@niguts.nig.junet, telephone: 559 75 0771

DDB (aspartic, aspartic, glutamic acids): appears to be the core motif in various transposases. (See amino acid symbols in protein sequences)

DDT: see dichlorodiphenyltrichloroethane

DEAD-BOX PROTEINS: a family of ATP-dependent helicases, present in prokaryotes and eukaryotes and they can stabilize mRNA and facilitate translation with the involvement of the 43S complex containing eIF4A, eIF4B, eIF4F. The 4F has three subunits: eIF4A, eIF4E, and eIF4E. 4B and 4F form a helicase that binds the 5'-end of the untranslated RNA through the 4E subunit. The name DEAD comes from the single letter amino acid symbols of proteins: Asp (D)-Glu (E)-Ala (A)- Asp (D) identifying a sequence present in eIF4A. (See helicases, eIF-4A, DEAH box proteins, amino acid symbols in protein sequences, translation initiation)

DEAE-CELLULOSE: as a membrane it is used for the trapping of DNA from agarose gels. (See gel electrophoresis)

DEAE-DEXTRAN: is a polycationic diethylaminoethyl ether of dextran (a polysaccharide) that stimulates the uptake of proteins and polynucleotides into cell, promotes the infection of cells by viral RNA and DNA, may inhibit tumors in animals, and stimulates reactions to antibody. (See dextran, transformation genetic animal cells)

DEAE-SEPHACEL, DEAE-Sephadex, DEAE Sepharose are ion exchangers used for gel filtration, as ion exchangers and chromatographic media. (See Sephadex, Sepharose, ion-exchange resins)

DEAFMUTISM: hereditary loss of hearing and speech with a prevalence of about 0.03 to 0.04% and with a recurrence risk among afflicted sibs of about 12%. There is an estimated number of

Deafmutism continued

35 loci capable of causing this anomaly. The spontaneous rate of mutation was estimated to be about 4.5×10^{-4}. The incidence of deafness has a rather larger environmental component. (See Usher syndrome, deafness)

DEAFNESS: is a hearing deficit within a broad range from slight hearing difficulties to complete loss and this may be a progressive phenomenon. About 3/4 of the hearing problems have complete or partial genetic determination and the rest may be environmentally induced. About 10-5% of the population develops hearing problems by advancing age and about 0.001 fraction of the newborns are deaf or develop some kind of hearing loss by school age. The first indication of an infant's hearing loss is the lack of articulate, understandable talking. About 87% of the congenital deafmutism is caused by recessive factors. Dominant inheritance determines deafness to some low, middle and high tone sounds and at other frequencies the hearing may be normal. The Michel syndrome is responsible for a complete lack of the internal ear formation. The hearing problems may have a wide range of organic bases but usually are classified as *conductive* (transmission) *hearing deficit* that is caused by defects in the hearing canal or the middle ear. *Sensorineural* defects involve the inner ear and the associated nervous system. This classification may not be absolute because the types may overlap and further complicated in a number of syndromes. Conductive hearing problems occur in otopalatodigital syndrome, Treacher Collins syndrome, osteogenesis imperfecta, Crouzon syndrome, Turner syndrome. Sensorineural hearing defects are found in Alport syndrome, Jervel and Lange-Nielsen syndrome, Pendred syndrome, Usher syndrome, Stickler syndrome, Refsum syndrome, Wildervanck syndrome, Norrie's disease, albinism (coutaneous), Waardenburg syndrome, LEOPARD syndrome. Both conductive and sensorineural defects may be present in the Klippel-Feil syndrome and some other cases. Deletions or mutations in the Xq21 region may also cause deafness (DFN1, Mohr-Tranebjaerg syndrome, DFN3) in case of (surgical) injuries to the stapes (the stirrup-like bones in the ear) resulting in leakage of the fluid (perilymph) of the inner ear (gusher deafness). The molecular basis of this hearing impairment is in gene Brain Protein 4 (POU3F4), encoding a transcription factor with a pou domain. Approximately 6% of the hereditary deafnesses are X-linked and are brought about by different mechanisms such as defects in the iris, cornea or by ocular albinism or other types of albinisms. Thyroid hormone receptor β is essential for the normal development of the auditory function. Mitochondrially determined high sensitivity to aminoglycosides (streptomycin) may result in hearing loss. A non-syndromic recessive deafness was located to human chromosome 2p22-p23 and another to 21q22. (See the named syndromes under separate entries, mitochondrial diseases in humans, Charcot-Marie-Tooth disease, mucopolysaccharidosis [Hunter and Hurler syndromes], Wolfram syndrome, pou)

DEAH BOX PROTEINS: are involved in the processing of precursor RNAs. (See also deadbox)

DEAMINATION: removal of amino group(s) from a molecule. Cytosine is deaminated to uracil at the rate 3 to 7×10^{-13} sec^{-1} in double-stranded DNA, i.e. about 40 to 100 deaminations of this type occur daily in the human genome. Thus C≡G transitions to T=A are apparently of major significance for mutation and evolution. The deamination of 5-methyl-cytosine is 2 to 4 times higher than that of cytosine. In single-strand DNA the rate of deamination is about 140 times higher than in double strands. Mismatched Cs are deaminated 8 to 26 times the rate of normally paired ones. Transcribed strands are about 4 times more likely to show deamination than the non-template strands. Some of the deaminated nucleotides are, however, removed by uracil-DNA glycosylases. (See nitrous acid, transition, DNA repair)

DEATH: is an irreversible stop to vital functions, especially that of the brain and the genetic material. The genetic death is a population genetic term for lack of reproduction.

DEATH DOMAIN: regions of cytokine receptors that engage the apoptosis pathway. (See cytokine, apoptosis, TNFR)

DEATH RATES: see age-specific birth and death rates, apoptosis, Hayflick's limit

DEAZANUCLEOTIDES: are analogs of nucleotides, antiviral agents and used for compression of sequencing gels. (See compression in gels, DNA sequencing)

DEB: diepoxybutane an alkylating mutagen and carcinogen. (See mutagens)

DE-BRANCHING ENZYME: converts a nucleic acid loops (lariat) into a linear molecule. (See lariat RNA)

DEBRISOQUINE: is an adrenergic-blocking drug used for treatment of hypertension. The response to the drug (human chromosome 22, dominant) depends on cytochrome P450IID family of proteins. About 1 to 30% of the populations, depending on ethnicity, may be poor hydroxylators of this and other similar drugs and may suffer serious side effects upon treatment. (See cytochromes)

DEC-205: an integral membrane protein, homologous to the macrophage mannose receptor. It appears to have important role in antigen presentation and processing in antigen-capturing (dendritic) T cells. (See antigen presenting cell)

Decapentaplegic (dpp): see morphogenesis of *Drosophila*

DECAPPING: is a process of mRNA degradation by enzymatic decay in the 5'→3' direction. It is usually triggered by shortening of the poly(A) tail but it may be brought about also by other means. (See mRNA, polyA mRNA, polyadenylation signal)

DECARBOXYLATION: removal of COOH group(s) from a molecule.

DECAY ACCELERATING FACTOR (DAF): is an erythrocyte membrane protein. (See erythrocyte)

DECIDUA: membrane lining the uterus during pregnancy that is shed around delivery.

DECODING: although the genetic code intrigued biologists before the nature of the genetic material was firmly determined, the complete genetic code was deciphered between 1961 and 1966. From frame-shift mutations Crick's laboratory concluded that the code is written most likely in triplets of nucleotides. Since 1961 random co-polymers of RNA were used. e.g. in a 5A : 1 C copolymer the AAA triplet are expected to be more common the CCC triplets. By chance alone the AAA sequence is expected to have a frequency of $(5/6)^3 = 0.579$ (125/216). ACA is expected to have the frequency of $5/6 \times 1/6 \times 5/6 = 25/216$ (0.116) and CCC is expected in a frequency of $(1/6)^3 = 0.0046$ (1/216). Thus the 3 triplets' proportions were expected to be 125:25:1. In an *in vitro* protein synthesis assay this copolymer promoted the incorporation of lysine the most abundantly. Therefore the codon of AAA was expected for lysine. The 2A:1C codons could be AAC, ACA, CAA, therefore additional copolymers were needed for determining their meanings. Repetitive, ordered copolymers UUCUUCUUC permitted the incorporation of phenylalanine (UUC), serine (UCU) and leucine (CUU), depending in what register the sequence was read. The sequences GAUAGAUAG directed the incorporation Asp (GAU), Arg (AGA) and then translation was stopped. Thus UAG was identified as the amber stop codon. The most precise method used ribosome binding. A collection of charged tRNAs were allowed to bind single known sequence triplets. Then tRNAs were charged with radioactively labeled amino acids. Cognate anticodons of a specific charged tRNAs bound only one triplet and thus the codons were identified. The validity of the code was then confirmed also by recombination. At amino acid site 211 in the wild type tryptophan synthetase glycine was identified. One mutation at this site resulted in a replacement by arginine and in another mutant by valine. A transduction experiment between the mutants restored glycine at site 211. This could be achieved if in the wild type there was CCT, and in the mutants GCT and CAT, respectively. Recombination between the first bases $\frac{GCT}{CAT}$ could produce CCT (glycine), and thus verified the codons for glycine, arginine and valine. (See genetic code, code genetic, ribosome binding assay)

DEDIFFERENTIATION: loss of cellular differentiation frequently followed by new cell divisions. (See redifferentiation)

DEER: is chromosomally a very diverse group yet the American white-tail deer (*Odocoilus virginianus*) is 2n = 70 and the reindeer (*Rangifer tarandus*) is also 2n = 70.

DE-ETIOLATION: dark-grown plants usually elongate and fail to synthesize leaf pigments and thus show etiolation. Some mutations, however, show short hypocotyls and green pigment in the dark, i.e. they are de-etiolated. (See brassinosteroids)

DEFAULT of a computer is a preset instruction; it is followed until new instruction is given.

DEFECTIVE INTERFERING PARTICLE (DI): subgenome size mutants due to deletion(s) that require homologous virus for replication. They may have advantage in replication over the helper virus and thus secure their maintenance. (See helper virus)

DEFENSIN: see antimicrobial peptides

DEFICIENCY, CHROMOSOMAL: (terminal) loss of a piece of the chromosomes (see also deletion). Terminal losses of chromosomes can be readily induced by ionizing radiation at first order kinetics (see kinetics) and they frequently behave as null alleles. If in a heterozygote the wild type allele is destroyed or removed the remaining recessive may be expressed (pseudo-dominance). (See deletion, deletion mapping, duplication, duplication-deficiency)

DEFINED MEDIUM: contains chemically identified and characterized nutrients.

DeFinetti DIAGRAM: the genotype frequencies are represented as perpendicular lines from a point within an equilateral triangle in such a way as the length of the lines correspond to the frequencies.

DEGENERATE CODE: the same amino acid is coded for by more than one type of nucleic acid triplet (e.g. the RNA code word for phenylalanine can be either UUU or UUC; other amino acids may have a single or up to six codons (i.e. 6 codons degenerate [go down] into 1 amino acid). (See code genetic, genetic code)

DEGENERIN: is a member of a family of epithelial sodium channel proteins (interacting with collagen) in *Caenorhabditis* muscle contraction. Mutants uncoordinated (unc-105) have such a defect. Proteins MEC4 and MEC10, defective in mechanical signal transduction in the touch reception system, encode homologous proteins. Other homologs are the epithelial sodium channel (ENaC) genes of mammals. (See Liddle syndrome, ion channels, collagen)

DEGRADATIVE PLASMIDS: Pseudomonad plasmids that have the ability of degrading salicylate, camphor, octane, chlorobenzoate, 2,4,5-trichlorophenoxyacetic acid, etc.

DEGRADOSOME: a multienzyme complex of polynucleotide phosphorylase (exoribonuclease), RNase E (endoribonuclease) RhlB (helicase) and enolase (glycolytic enzyme) that has major role in processing mRNA. (See ribonuclease. helicase, glycolysis)

DEGREE OF FREEDOM: number of independent comparisons within numerical data; e.g. a 3:1 segregation has 1 degree of freedom because if one of the classes is specified within, say 4, the other can be either 3 or 1, i.e. there is one choice only. In cases where multiple comparisons can be made, e.g. in a segregation of 9:3:4 (recessive epistasis) the degree of freedom is 2 because if one class is specified as 4, the other two classes can be anything within 12/16, thus there is another choice left. In case of 9:3:3:1 segregation, degrees of freedom are 3 because if one class is chosen there are still three more to chose when one is fixed. In some experiments with multiple comparisons, e. g. in a variance analysis careful consideration must be given to the number of independent comparisons before the correct degrees of freedom can be determined. (See analysis of variance)

DEGREE OF RELATEDNESS: see relatedness degree of

DegS: *Bacillus subtilis* kinase regulating degradative enzymes thorough protein DegU.

DEHYDROGENASE: enzymes mediating removal of hydrogen from molecules.

De LANGE SYNDROME (Cornelia de Lange, Brachmann-de Lange syndrome): is most likely caused by new autosomal dominant mutations. The afflicted individuals do not reproduce. Its frequency by a Danish study appeared to be 6×10^{-6}. The gene is situated in the area 3q21-qter. About 30% of the cases are associated with various chromosomal anomalies, including duplication (and possibly deficiency) of the long arm of human chromosome 3. Some of the chromosomal anomalies may be unrelated to the syndrome characterized by the two eyebrows growing across the nose, hairy forehead and neck, long eyelashes, depressed nose bridge and uptilted nose tip, wide spacing of teeth, flat fingers and hands, altered palm print, mental retardation, etc. The physical anomalies are evident by the end of the second trimester. (See mental retardation, limb defects in humans, head/face/brain defects, hypertrichosis)

DELAYED EARLY GENES (DE): are turned on following the immediate early genes, about 2 minutes after phage infection. They use the early and new middle promoters. Their expression depends on protein synthesis. (See immediate early genes, late genes)

DELAYED INHERITANCE: the expression of some traits depend on the genotype of the diploid oocyte rather then the genetic constitution of the zygote. In such cases the reciprocal F_1 generation may be of two types (maternal or paternal), the F_2 may be uniform (because the genetic constitution of the F_1 is identical) and segregation is delayed to the F_3. Similar phenomenon is observed when the phenotype of the male gametophyte (pollen) is determined by the diploid microsporocyte rather than by the haploid nucleus of the microspore or pollen. (See testa, *Lathyrus odoratus*, *Limnaea*)

DELAYED-RESPONSE GENE: is activated by a growth factor after a lag period (about an hour). (See also early-response gene, early gene, late gene)

DELETION: loss of a (internal) chromosomal segment; it is generally symbolized with -, or d or Δ or *del*. Small deletions may appear as recessive null mutations or larger deletions may have a dominant phenotype. Deletions are distinguished from mutations by failing to revert to the normal allele and they may affect also the frequency of recombination. Specifically directed deletions can obtained in mice by taking advantage of the *loxP* and *Cre* factors of bacteriophage P1 (see targeting genes). Deletions may be responsible for various human hereditary anomalies. Deletions can be generated in isolated DNAs by cutting the double-stranded molecules with the aid of restriction endonucleases leaving behind complementary single-strand overhangs. These ends than can be sliced back by Bal 31 exonuclease or S1 nuclease before ligation of the free ends.

DELETION OF PHAGE λ DNA (b2)
(Westmoreland, B. *et al.* 1969
Science 163:1343)

DELETION IN THE SALIVARY GLAND
X CHROMOSOME OF *DROSOPHILA*
(Painter. T.S. 1934 J. Hered. 25:465)

Deletions may be detected with the aid of a light microscope if the chromosomes have clear landmarks such as knobs or when they display banding patterns such as the salivary gland chromosomes or by special staining techniques or when the deleted chromosome is paired either during mitosis or meiosis with the normal chromosome. Deletions of the DNA may be detected by electronmicroscopy if a deletion and a normal strand are hybridized *in vitro*. At the site of the deletion the normal chromosome or the normal DNA strand buckles out because it has no partner segment to pair with. (See also deficiency, cri du chat, Wolf-Hirschhorn syndrome, retinoblastoma, Prader-Willi syndrome, Angelman syndrome, Smith-Magenis syndrome, Beckwith-Wiedemann syndrome, Langer-Giedion syndrome, DiGeorge syndrome, Miller-Dieker syndrome, Wolf-Hirschhorn syndrome, contiguous gene syndrome, aging, deletion mapping, pseudodominance, duplication, nested genes, overlapping genes, knockout, Bal 31, S1)

DELETION ANALYSIS: involves a number of diverse procedures, including pseudodominant expression of a recessive allele in a heterozygote when the dominant allele is deleted, deletion mapping determines the extent of deleted segments on the basis of pseudodominance of linked genes, by genomic subtraction the normal DNA sequence extending over the gap can be isolated by molecular procedures, deletion of components of a gene (e.g. upstream regulatory elements) and their role can be identified, etc. (See deletion mapping, linker scanning, genomic subtraction, pseudodominance)

DELETION MAPPING: have been used in *Drosophila* and plants on the basis of pseudodominance, i.e. deletions of the normal sequences carrying the wild type alleles of syntenic genes identify the length of the deletions in the heterozygotes and determine their relative position.

```
a    b    c    d    e     ←RECESSIVE ALLELES IN HETEROZYGOTES
_____     deletion of the entire area unmasks the 5 genes
              _____     ←deletion unmasks d and e only
_____             ←deletion permits the expression of a, b, c
         _____       ←overlapping deletions define the sequence
                          c, d and so on
```

The fine structure of the *rII* gene of phage T4, about 300 sites, was mapped by about 2,000 mutants by the following principle:

```
          ═══════════════════════════════     long deletion
          ───────────────────────────────     wild type sequence
mutation→●─┼──────────────────────────        mutation carrying strand (otherwise normal)
          ──────                               shorter deletion
```

↖crossing over at the vertical line between the normal sections of two strands restores wild type strand in the recombinant DNA→ ────────────────────

Wild type was restored by recombination only between non-overlapping deletions and the crosses with the long deletion indicated that the mutation and the short deletion were both within the range of the long deletion. (See pseudodominance)

DELETION MUTATION: lacks a portion of the genetic material that may vary in extent from a single to millions of nucleotide pairs. Deletions may be generated by molecular biology techniques using restriction endonucleases and extension of the gaps by exonuclease. Since the restriction enzymes nick at specific locations, they can be used for site-directed deletions. (See deletion unidirectional, ionizing radiation, radiation effects, localized mutagenesis, genomic subtraction)

DELETION UNIDIRECTIONAL: DNA molecules blocked at the 3'-end by sulphur-containing nucleotides (thionucleotides) are protected from 3'→5' exonuclease action of T4 DNA polymerase are sliced back in the opposite direction in a time dependent extent. After removing the single strands left behind, nested sets of deletions of various lengths are obtained. (See Bal 31)

DELPHINIDINE: see anthocyanin

DELTA ENDOTOXIN: see *Bacillus thüringiensis*

DELTA 88: a 7 kb *Drosophila* transposable element. (See transposable elements)

DEME: an interbreeding population (a Mendelian population) without reproductive isolation; it is also used for denoting a natural breeding group with one male and several females.

DEMENTIA: deterioration of mental abilities, dullness of the mind due to innate defects of the brain or developmentally programmed disease (paranoia, schizophrenia, Alzheimer's disease, etc.) or caused by poisonous substances or drugs affecting the nervous system.

DEMETHYLATION: may take place by demethylase enzymes or by base excision repair. (See methylation of DNA)

DEMOGRAPHY: study of changes in the human populations by migration, birth, mortality, marriages, health, occupations, education, etc.

DE NOVO **SYNTHESIS**: formation of molecules through synthesis from (simple) precursors rather than by cannibalization (recycling) of more complex ones (salvage).

DEN: diethylnitrosamine an alkylating mutagen and carcinogen. (See mutagen, carcinogen)

DENATURATION: loss of native configuration of DNA (frequently separation of the complementary strands) or of proteins by damaging the non-covalent bonds using elevated temperature or chemicals, such as alkali, detergents and others. The forces between complementary

Denaturation continued

DNA strands can be measured by atomic force microscopy. Adhesive forces between complementary strands of 20, 16 and 12 base pairs was found to be 1.52, 1.11 and 0.83 nanonewtons, respectively (1 newton [N] = 1 kg/m/sec^{-2} = 10^5 din). (See atomic force microscope, c_0t curve)

DENATURATION MAPPING: electronmicroscopic localization of strand mismatches in DNA. (See mismatch, deletion)

DENATURING GRADIENT GEL ELECTROPHORESIS (DGGE): permits the separation of DNA fragments of identical size on the basis of different susceptibility to denaturation because of mutation even in a single nucleotide. The DNA fragments are electrophoresed in polyacrylamide gels in which there is an increasing gradient of a denaturing agent such as urea or formamide or both. The increased temperature also promotes separation of the strands of the DNA double helix. The partially denatured fragments migrate at a lower speed in the gel. (See electrophoresis)

DENDRITE: relatively short branch of neurons that receives information from other nerve cells. (See nerve cells)

DENDRITIC CELL: are leukocytes with the primary role to capture antigens and present them to T cells. (See antigen presenting cell, T cell)

DENDROGRAM: a chart showing relationship among entities in a form resembling branches of a tree. (See evolutionary tree, character matrix)

DENGUE VIRUS: a positive strand RNA virus translated into a single 350 kDa polyprotein and causes either the dengue fever or the hemorrhagic dengue. Both of them are debilitating diseases primarily in South-East Asia. The virus is transmitted by mosquitos. A way of protection was worked out involving the transformation of the mosquito (*Aedes egypti*) by a *Sindbis* virus vector carrying a 567-base antisense RNA of the premembrane coding region of the dengue-2 virus. The transduced mosquitos cannot support the replication and thus do not transmit the virus. (See biological control, antisense RNA)

DENHARDT REAGENT: is used to suppress background hybridization in Northern hybridization, RNA probing, single-copy Southern hybridization, annealing DNA immobilized on nylon membrane. It is made of Ficoll-400 (a non-ionic synthetic sucrose polymer), polyvinylpyrrolidone (an insoluble material removing phenolic impurities) and bovine serum albumin at a concentration of 2.5% (or less) each in water. (See DNA hybridization, Northern blot)

DENSITY GRADIENT CENTRIFUGATION: separation subcellular organelles or macromolecules on the basis of their density. The density of DNA (ρ) increases with increased GC content (ρ = 1.660 + [0.098 {G+C}]). During buoyant density centrifugation the DNA is positioned at the density of CsCl that corresponds to DNA density. The density of the CsCl solution is determined by refractometry. The nuclear DNA of most eukaryotes has an AT content of about 40%, corresponding to a density of about 1.7 g/mL. The DNA of cellular organelles may have a density different from that of the nucleus and thus form a satellite band in the ultracentrifuge. Single-stranded nucleic acids sediment faster and circular DNA molecules sediment slower. Among cellular organelles nuclei sediment fastest, then chloroplasts and mitochondria occupy the highest position in the centrifuge tube. For the separation of organelles either sucrose or percoll (polyvinylpyrrolidone-coated silica) are used most commonly. (See also centrifuge, buoyant density centrifugation, ultracentrifugation, density labeling)

DENSITY LABELING: growing cells first in dense (heavy isotope) medium, e.g. $^{15}NH_4Cl$ and then transfering to normal $^{14}NH_4Cl$ medium and the heavy and light DNA or recombinant strands can be separated by density gradient equilibrium centrifugation. (See density gradient centrifugation)

DENTAL NON-ERUPTION: autosomal dominant failure to cut teeth.

DENTATORUBRAL-PALLIDOLUYSAN ATROPHY (myoclonous epilepsy with choreoathetosis): involves involuntary jerky muscle movements, epilepsy, feeblemindedness, ataxia, brain degeneration due to a chromosome 12 dominant condition. The mutation causes an abnormal number (49-75) of CAG repeats in the atrophin protein compared to 7-23 under normal conditions. (See fragile sites, trinucleotide repeats, muscular atrophy)

DENTICLE: tooth-like extrusion (insects) or pulp stone in the mammalian tooth.

DENTIN DYSPLASIA: the pulp is absent or poorly developed, the root canal frequently empty and/or enlarged and the bluish teeth spontaneously lost. Several forms of autosomal dominant expression have been observed

DENTINOGESIS IMPERFECTA: the dominant condition is encoded in human chromosome 4q13-q21 causing blue-gray or brownish teeth due to dentin defect. The Brandywine type appears to be non-identical. (See dentin dysplasia, tooth)

DENVER CLASSIFICATION: of human chromosomes in 1960 arranged the chromosomes into 7 groups (A to G) on the basis of decreasing length and arm ratio. (See human chromosomes, Chicago classification, Paris classification)

DENYS-DRASH SYNDROME: is very severe mutation in the WT gene (Wilms tumor), affecting primarily the female gonads and genitalia, internal male genitalia and kidneys. (See Wilms tumor)

DEOXYRIBONUCLEIC ACID: see DNA

DEOXYRIBONUCLEASE: enzyme capable of breaking phosphodiester bonds in single- or double-stranded DNA. (See DNase free RNase, DNase hypersensitive sites, restriction enzymes, endonuclease, exonuclease)

DEOXYRIBONUCLEOTIDES: contain only a H rather than OH at the 2 position of the ribose, one of the nitrogenous bases (A, T, G, C) and phosphate; they are building blocks of DNA. (See ribonucleotide, nucleotide chain growth)

DEOXYRIBOSE: see ribose

DEOXYRIBOZYME: the hammerhead ribozymes contain deoxyribonucleotides yet they can cut RNA. Single-stranded DNAs may become Zn^{2+}/Cu^{2+} or Pb^{2+} metalloenzymes. The Zn^{2+}/Cu^{2+} enzymes may function as DNA ligase. (See ribozyme, DNA ligase)

DEPURINATION: loss of purine from nucleic acids

DEPRESSION: pschychological state of sadness, despair, low self-esteem generally accompanied by lack of appetite and sleeplessness. It is frequently associated with glucocorticoid overproduction that may lead to hippocampal atrophy. (See glucocorticoid, hippocampus)

DER: a transmembrane hormone receptor tyrosine kinase protein of the epidermal growth factor of *Drosophila*. It affects many phases of the development including photoreceptor determination, wing vein formation, etc. The activator ligand is the Spitz protein (a homolog of TGF-α; Argos is an inhibitor of DER. (See EGF, Argos, Spitz)

DEREPRESSION: removal of repression (so protein synthesis can go on). (See induction)

DERIVATIVE: mathematically is a function f' of a function f whose value at any point x_1 in the domain of f is: $f'(x_1) = \lim_{\Delta x \to 0} \frac{(x_1 + \Delta x) - f(x_1)}{\Delta x}$ if such a limit exists.

DERIVATIVE CHROMOSOME: has been modified by chromosomal rearrangement(s).

DERMATAN SULFATE: glucosaminoglucan; repeating units disaccharides, generally acetylgalactoseamines linked to a iuduronic acid. (See mucopolysaccharidosis)

DERMATITIS, ATOPIC: skin inflammation generally associated with itching, frequently evoked by allergic reactions to cosmetics, plants, animals, light, etc. (See skin diseases)

DERMATOGLYPHICS: examination of dermal ridges and creases on fingers, toes, palms and soles for the purpose of identification, diagnosis and forensic investigations. (See fingerprints)

DERMATOME: cell layer generates the mesenschymal connective layer of skin.

DERMATOPARAXIS: a hereditary disease in cattle. The procollagen peptidase that cleaves a peptide from the N-terminus of the chains is defective causing disorganized, poor fiber formation resulting in extreme brittleness of the hide. (See collagen)

DERMOMYOTOME: is the primordium of the vertebrate skeletal muscles.

DES (diethylsulfate): a potent alkylating mutagen and carcinogen.

DESATURASE: enzyme - on the endoplasmic reticulum of plants - introducing double bonds into the hydrocarbon portion (cis) of fatty acids in the presence of NADPH, light generated ferredoxin and O_2. (See fatty acids)

DESENSITIZATION: signal - response systems after prolonged stimulation display reduced responsiveness to stimulation by the same agent.

DESETOPE: the antigen binding site of the MHC molecule. (See antigen, MHC, agretope)

DESKTOP: of a computer that has the various menu bars and where the actual work is performed.

DESMIN: a filament protein.

DESMOCOLLINS: desmosome attached proteins. (See desmosome)

DESMOPLAKINS: proteins are involved in cell junctions. (See desmosome)

DESMOSOME: a protein plaque of cell junctions (between epithelial cells) into what desmin and keratin filaments of cells are tied. A defect of desmosomes is responsible for various skin diseases. (See junction complex, pemphigus, intermediate filaments, plakoglobin)

DESMOYOKIN: is a cell junction protein.

DESYNAPSIS: the loss of synapsis of the homologous chromosomes (after the completion of the recombination process). (See synapsis, asynapsis, sister chromatid cohesion, chiasma)

DETASSELING: removal the male inflorescence of maize. (See heterosis)

DETECTOR PROTEINS: sense environmental signals (in bacteria) in the periplasmic region. (See periplasma)

DETERGENT: the various types may have different chemical structures but they all have a large non-polar hydrocarbon end that is oil-soluble and a water-soluble polar end. The so called "soft detergents" are biodegradable. In the laboratory most commonly sodium lauryl sulfate (SDS, sodium dodecyl sulfate, $n-C_{11}H_{23}CH_2OSO_3^-Na^+$) is used in gel electrophoresis. Detergents are employed also for the solubilization of membrane proteins and lipid components. (See SDS, polyacrylamide gels, Tween 20, nonidet)

DETERMINATE INFLORESCENCE: the stem terminates in a flower rather than in an apical meristem. (See meristem, indeterminate inflorescence)

DETERMINATION: the establishment of a specific commitment to differentiation. It is a biochemical change within cells or tissues whereby they lose options for differentiation in all but one particular way. In plant cells the determination may be reversible, however. In the processes controled by homeotic genes transdetermination may overrule the regular pattern. (See transdetermination, homeotic genes, transplantation nuclear)

DETERMINISM, GENETIC: a tenuous social theory that crime, immorality, disease, poverty and all social ills are predetermined in families. Behavior genetics shows that inheritance plays a certain, variable role in behavior and by logical extension judgment of human values. Also, the question of the degree of responsibility is opened. If behavior is genetic how much is the role of free will? Ethicists and legal scholars would have difficulties to define "values". If it is measured can it be valued? The philosopher David Hume remarked: "Good sense and genius beget esteem: Wit and humor excite love." The problems certainly exceed the scope of this text although the moral dilemmas are inescapable for the geneticist once Pandora's box has been opened. (See human behavior)

DETERMINISTIC GENE: in its presence a certain condition or disease most likely to occur.

DETERMINISTIC MODEL: in a population the changes are based on fixed parameters such as the selection coefficient, mutation pressure, migration, etc. (See stochastic model)

DETOXIFICATION ENZYMES: see glutathione-S-transferase, paraoxonase

DETRIMENTAL MUTATION: has low fitness but their rate of survival is not below 10%. (See beneficial mutation, mutation, fitness)

DEUTERANOMALY: is a deficiency of the photopigment, sensitive to middle wave length yet they retain some trichromatic vision, depending on how much short and long wavelength receptors they have. The mildest cases have differences from normal in exon 5 of the X-chromo-

Deuteranomaly continued

somal gene. The next mildest anomaly involves exon 2 and either exon 3 or 4 but not both. In the most severe cases the anomaly was commonly in exon 2 but usually not in 3 or 4. Deuteranomaly affects about 5% of the US males but only 0.25% of the females because of X-linkage. (See colorblindness, night blindness)

DEUTERANOPIA: see color blindness

DEUTERIUM: ^2H, heavy hydrogen (atomic weight: 2.014725), a stable, non-radioactive isotope.

DEUTEROTOKY: unfertilized eggs develop into male or female. (See arrhenotoky, thelyotoky, chromosomal sex determination, sex determination)

DEVELOPMENT: is a sequence of events beginning with determination and through differentiation leads to the various stages of life of the organism. In multicellular organisms it begins with the fertilization of the egg and goes through epigenesis and terminates in death. The zygote grows and at an early stage polarity and segmentation becomes obvious. In animal embryos the cells move into organizing centers (gastrulation) and the development becomes rather rigidly determinate. The germline is laid down and is separated from somatic differentiation. Plant cells maintain high degree of totipotency through development and movement of cells is restricted by the rigid cell walls. In plants, *meristems* containing totipotent cells are organized and serve as the bases of growth, differentiation and development. In animals the *stem cells* assume the functions comparable to that of the meristem of plants. The development is based on a highly regulated cooperation of genes that are turned on, changed in the level of activity and turned off. Although development is potentiated primarily according to the genetic blueprint, the realization of the plan generally requires environmental cues. The motivation is provided by the signal transduction pathways.

Development is genetically controlled in all organisms. Many genes of the bacteriophages have been identified that control individual steps of differentiation of the structures and their assembly into a mature phage particle. The phage body is constructed according to the instructions of the viral genes but the cellular metabolism is used for execution of the project. Thus the cooperation of both host and of the viral genomes is responsible for the product. The simplest developmental pathway of animals is seen in the nematode, *Caenorhabditis elegans*. Developmental mechanisms are of focal interest of research using diverse organisms. (See *Caenorhabditis*, morphogenesis, segregation asymmetric, phage morphogenetic pathway on next page)

DEVELOPMENT, AUTONOMOUS: the process is independent of the (cellular, tissue) environment.

DEVELOPMENTAL CLOCK: the developmental fate of an organism or structure is determined as a function of time.

DEVELOPMENTAL CYCLE: the processes during ontogenetic developments in the alternating generations. (See alternation of generations, ontogeny)

DEVELOPMENTAL NOISE: a variation which cannot be attributed to any verified cause.

DEVELOPMENTAL THERAPEUTICS PROGRAM (DPT): involves information on drug discovery and it is accessible: <http://epnws1.ncfcrf.gov:2345/dis3d/DTP.HTML> and <http://www.nci.nih.gov/intra/lmpjnwbio.htm>

DEVIATION: difference from the usual type. (See also standard deviation, variance)

DEXAMETHASONE: see glucocorticoids

DEXTRAN: is an α-1,6 linked poly-D-glucose, sometimes with α-1,4 or α-1,3 linked branches. It is present in cross-linked form in the gel-filtration and anion exchanger agent Sephadex. It is synthesized by bacterial enzymes and is an important component (along with inorganic salts and lipids) of the dental plaques, responsible for tooth decay (caries) and gum disease. (See DEAE-dextran, Sephadex)

DEXTRIN: see amylopectin

DEXTROROTATORY: rotates clockwise the plane of polarized light.

DEXTROSE: same as glucose.

df or **d.f**. or DF (degrees of freedom): see degrees of freedom
DGGE: see denaturing gradient gel electrophoresis
dGTP: deoxyguanosine triphosphate. (See hypoxanthine for formula of guanosine)
Development continued from the middle of preceding page

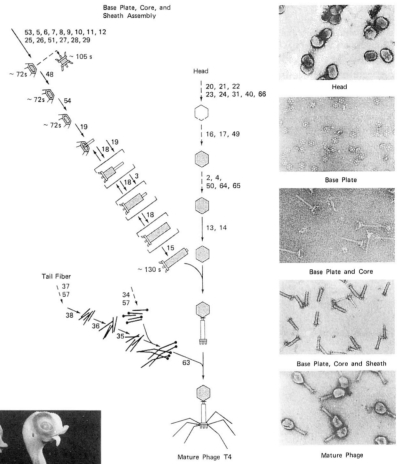

The morphogenetic and developmental pathway of phage T4. The numbers refer to genes. In case of mutation, incomplete phage parts may develop within the host bacteria. Many of these structures are true phage precursors and may form viable phage particles *in vitro* from appropriate mixtures of mutants blocked at different developmental steps. (After Wood, W. B. 1980 Quarterly Rev. Biol. 55:353, and King, J. 1971 J. Mol. Biol. 58:693)

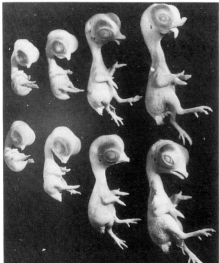

A clear example of the genetically programed development of simple traits of higher animals.
Bottom: normal and Top: "Donald Duck" chicken embryos after 8 to 11 days of incubation. The beak anomaly is the result of homozygosity of a recessive gene. The mutants cannot be identified until the eighth day of incubation, and within two days after the onset both lower and upper beaks display the developmental defect. By later stages the condition becomes more pronounced and the afflicted chicks cannot survive. Courtesy of Abbott, U.K. and Lantz, F.H. 1967 J. Heredity 58:240

DHAC 1 (RPD3): is a histone deacetylase. (See histone deacetylation, histone acetyltransferase)

DHFR: dihydrofolate reductase controls the reaction: dihydrofolate + NADPH + H+ →tetrahydrofolate + NADP+; *dhfr* deficient cells are resistant to methotrexate (amethopterin). The conversion of deoxyuridylic monophosphate to deoxythymidylic monophosphate requires tetrahydrofolate. (See folic acid, methotrexate)

DHR DOMAIN: same as PDZ domain.

DIABETES INSIPIDUS: autosomal dominant type, and like all diabetes involve an imbalance in electrolyte control resulting in excessive urination and as a consequence extreme thirst. In the case of the neurohypophysin type, the defect resides within the arginine V2 vasopressin receptor gene and the controling antidiuretic neurohypophyseal hormone. Large amounts of water cannot be reabsorbed by the kidneys and it is released, and may require a 10-15 fold higher than normal water intake (up to 20 L/day). In the autosomal dominant nephrogenic type, upon administration of antidiuretic hormone, cyclic adenosine monophosphate level increases in the urine remarkably. There is also an early onset juvenile insulin-dependent form. This may be recessive or polygenic; a problem hard to resolve because of the high frequency of the gene and the low penetrance. In another syndrome diabetes mellitus, diabetes insipidus and deafness occur together under probably autosomal recessive control. Some of the diabetes insipidus cases (nephrogenic and the neurohypophyseal) may be under X-linked control. (See diabetes mellitus, kidney disease, vasopressin, cAMP, insulin, neurophysin, MODY)

DIABETES MELLITUS: a recessive hereditary disease under the control of more than a single gene causing hyperglycemia due insulin deficiency and other factors, and by defects in glucose transport from the blood to cells. Diabetes of the mother may seriously affect the fetus. The prevalence of diabetes in most human population exceeds 3%, the recurrence risk among sibs varies depending on the type of genetic control or primarily non-genetic type of diabetes. The penetrance and onset are variable and it is affected by dietary and other environmental factors. Treatment involves dietary restrictions and insulin administration in the insulin-dependent diabetes. In common usage, without qualification, diabetes means the mellitus form although several other types of the condition have been identified. Diabetes is characterized by excessive amounts of sugar in the blood, and excretion of large amounts of urine which may or may not contain excessive amount of sugar. Diabetes can be medically characterized into insulin-dependent (IDDM) and insulin-independent (NIDDM) forms. In IDDM the loss of glucose may cause hunger and great thirst and frequent urination. The glucose loss may involve then increased catabolism of proteins and fat. As a consequence weight loss may follow. When the mobilization of fats increases and the oxidation of fatty acids becomes incomplete, ketone bodies and acetone may accumulate. The ketones appear in the urine and the acetone may be exhaled, giving the untreated patients a special odor. IDDM is treated with insulin and properly controled diet. The NIDDM may be caused by a defect in the glucagon receptor in human chromosome 17q25, near-telomeric region. Recently the NIDDM1 locus was assigned to chromosome 2 and NIDDM2 to chromosome 12. In NIDDM obesity is frequent sign but besides the increased blood sugar content, other symptoms may vary, depending on the cause. The latter type patients do not respond to insulin. The first clinical test for diabetes is the glucose-tolerance test. To fasted individuals about 100 g glucose is given in water orally. In healthy individuals the blood sugar level returns to normal after 2 - 3 hours but not in the diabetic ones. Various animals are also afflicted with this condition. Several forms of the insulin-dependent diabetes are associated with HLA antigen DR3/4 or 3 or 4. Diabetic and glucose intolerance symptoms are associated with a large number of syndromes. A gene involved in the control of IDDM7 has been located in human chromosome 2q34. An early onset insulin gene, IDDM2 (human chromosome 11p15.5) seems to be associated with tandemly repetitive DNA sequences that regulate insulin transcription. The principal gene for early onset (IDDM1) appears to be in human chromosome 6-21. IDDM3 was located to chromosome 15q. IDDM4 appears to be in 11q13, IDDM5 in 6q25 and chromosome18q harbors IDDM6, IDDM9 is in 3q21-q25, IDDM10 is near the centromere of chromosome 10. The glucokinase gene (GCK) was located to 7p15 and it is also involved in IDDM. The X-chromosomal IDDM

Diabetes mellitus continued

locus is DXS1068. This is basically a polygenic autoimmune disease brought about by the destruction of insulin-producing β cells in the Langerhans islets of the pancreas by the infiltration of T lymphocytes, B lymphocytes, macrophages and dendritic cells. T_H1 cells seem to play the primary role in the process whereas the T_H2 cells may not have either promoting or protecting effect. The concordance between monozygotic twins is only 50%. The MHC class II factors are most important in determining diabetes susceptibility. Diabetes may be responsible for blindness, kidney disease, heart attacks, strokes and for the amputational needs of the lower extremities. (See diabetes insipidus, insulin, HLA, Hirschprung disease, autoimmune disease, ion channels, blood cells, *VENTURE*, T cells, cytotoxic T cells, glucagon, superantigen, polygenic inheritance, Langerhans islet, MODY)

DIACYLGLYCEROL (DAG): a lipid in the cell membrane that activates protein kinase C. (See protein kinase)

DIAGENESIS: physical and chemical alterations that takes place in the geological sediments after deposition of organisms.

DIAKINESIS: is a nuclear division phase characteristic only for meiosis. It resembles very closely to the diplotene stage but the condensation of the chromosomes is further increased. The chiasmata tend to move toward the termini of the chromatids (terminalization) and the paired chromosomes (bivalents) begin their separation from each other as approaching metaphase. (See meiosis)

DIALLELE ANALYSIS: is used in quantitative genetics and (plant) breeding to assess the breeding value of genetic stocks by crossing them in all possible combinations. The data reveal both nuclear and extranuclear contributions. The great amount of work involved limits, however, the number of stocks that can be studied. Therefore some geneticists prefer heritability tests involving intraclass correlation. (See heritability, intraclass correlation)

DIALOG BOX: requests further information from the user of the computer or gives some warnings accompanied by a beep sound.

DIALYSIS: removal of small molecules through a semi-permeable membrane into water or into lower concentration solutes.

DIAMINOPIMELATE: a (polylysine) peptidoglycan component of the cell wall of many bacteria. *E. coli* strain EK2 (χ1776), is blocked in its synthesis and prevented the organism to survive in the mammalian gut, was used as a host for genetic vectors for recombinant DNA to assure containment. (See biohazards, biological containment)

2,6-DIAMINOPURINE: a mutagenic analogue of adenine.

DIAPAUSE: a relatively inactive period in the life cycle of an organism such as during pupation, hibernation, seed stage, etc.

DIAPEDESIS: exit of corpuscular elements through intact blood vessels. (See extravasation)

DIAPHORASE: see NAD, methemoglobin

DIAPHYSEAL ACLASIS: see exostoses

DIARRHEA: may have diverse origins. A rare autosomal dominant type responded to steroid hormone treatment. Autosomal recessive forms include a defect in chloride absorption and occurs at a frequency of about 7.6×10^{-5}. Another different recessive form was affected in the Na^+/H^+ exchange and a rare X-linked form was found to be very susceptible to infections and other complex problems, eczema, thyroid autoimmunity, diabetes, etc.

DIASTASE: an enzyme complex that can hydrolyze starch into sugar.

DIASTEMATOMYELIA: an autosomal recessive disease causing a split in the spinal cord by either fibrous or bony material and each half is wrapped. It is often associated with spina bifida, atrophy of the legs and other defects. (See neuromuscular diseases, spina bifida, atrophy)

DIASTROPHIC DYSPLASIA (DD): autosomal recessive (human chromosome 5q31-q33 or q34) anomaly with curved spine (scoliosis), clubbed foot, abnormal thumb, abnormal earlobes, premature calcification of rib cartilage, short stature, respiratory and cardiac insufficiencies, etc. (See stature in humans, limb defects in humans, dwarfism)

DIAUXY (diauxie): see glucose effect

DIAZOTIZED PAPER: is a modified filter paper (Whatman 540, Schleicher & Schuell 589 or equivalent) is used for nucleic acid transfers and Western blots. The paper is first treated with nitrobenzyloxymethyl pyridinium (NBPC) that leads to the formation of NBM paper that is then reduced with sodium bisulfite ($Na_2S_2O_4$) to aminobenzyloxymethyl (ABM) paper and through a reaction with nitrous acid (HNO_2) the amino group is converted into the diazo group of the DBM (diazotized) paper. NBM and ABM papers are commercially available. The stability of these modified papers depends on the conditions of storage. (See Western blot)

DIBASICAMINOACIDURIA: is an autosomal recessive defect involving accumulation of lysine, arginine and ornithine when protein-rich food is consumed without any increase in cystine. The anomaly may lead to liver and bone defects. (See hyperlysinemia, argininemia, citrullinemia, cystin-lysinemia, amino acid metabolism)

DICENTRIC BRIDGE: is formed when a dicentric chromosome is stretched because at anaphase the two centromeres are pulled in opposite direction. (See breakage - fusion - bridge cycles, paracentric inversions, dicentric chromosomes, bridge [for photomicrograms])

DICENTRIC CHROMOSOME: has two centromeres and is an unstable structure because at anaphase the two centromeres are pulled toward opposite poles and that generates a bridge that may be ruptured at any point, leading to unequal distribution of the genes of that chromosome. (See breakage-fusion-bridge)

DICENTRIC RING CHROMOSOMES: are formed when there is a sister-chromatid exchange in a ring chromosome. At mitotic anaphase the sister-centromeres are pulled toward the opposite poles causing the chromatids to break at potential points of stress and thus leading to unequal distribution of the genes (duplications and deficiencies in the cells). The broken ends may fuse and the process may continue with concomitant somatic instability revealed by sectors if appropriate genetic markers are present. (See ring bivalent, ring chromosomes, translocation chromosomal, diagram on next page)

DICHLORODIPHENYLTRICHLOROETHANE (DDT): an insecticide, now almost entirely avoided in the industrialized countries of the Northern Hemisphere because it weakened the egg shells of birds preying on insects and its accumulation in mammalian fat tissues, etc. Some insects developed resistance to it by eliminating HCl from the molecule with the aid of increased levels of DDT dechlorinase enzyme. It is toxic to humans: oral LDLo 6 mg/kg. Although it was suspected to be carcinogenic, the final tests did not confirm this. Recently it was discovered that main metabolite of DDT, p,p'-DDE is potent anti-androgen by binding to the androgen receptor. It is not in use for over 20 years, it still persists in the environment (its half-life being ≈100 years). It is blamed for reduced human sperm counts, increased testicular cancer, and other anomalies of the reproductive system. DDT thus became an anathema yet it is the most effective weapon against mosquitos, reponsible for malaria. (See LDLo, sperm)

2,4-DICHLOROPHENOXYACETICACID (2,4-D): a synthetic plant growth hormone. (See auxins)

DICHROIC MIRROR: transmits light in one color and when reflected passes it in another color. (See also circular dichroism)

DICISTRONIC TRANSLATION: transcripts carrying internal ribosomal entry sites (IRES) may be co-translated (expressed) in appropriate gene fusion vectors in the same sequence. (See IRES, ribosome scanning)

DICOT (dicotyledonous): a taxonomic category of plants with embryos having two cotyledons.

DICTYOSOMES: see Golgi apparatus

Dictyostelium discoideum: is a haploid (x = 7), unicellular amoeba with a genome size of 50,000 kbp. Diploid cells have also been observed. The cells form colonies that upon starvation differentiate into structures reminding to multicellular organisms. Actually, they form two types of cells, spores and stalk cells. At a stage when they are about 70% pre-spore and 30% pre-stalk they form aggregates that are called slugs, appearing somewhat similar to gastropod animals. They make during the growth phase about 3,000 to 5,000 different mRNAs and dur-

Dictyostelium discoides continued

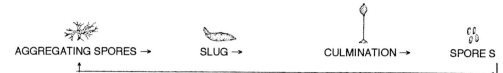

ing differentiation between 800 and 2,000, i.e., within the same order as lower animals. During the growth phase a *Pre-Starvation Protein Factor* (PSF) accumulates and when its level reaches the critical point it serves as a signal for development by activating adenylate cyclase, cAMP receptor and a Gα protein required for signal transduction although development may take place in the absence of cAMP. The steroid *Differentiation Inducing Factor* (DIF) is made by the end of aggregation. The interaction of these two factors leads to culmination and formation of the fruiting body. All the functions required for growth and development involve the regulated expression of about 2,000 genes in 7 linkage groups. (See cAMP, adenylate cyclase, signal transduction, Gα protein)

> **DICENTRIC RING CHROMOSOMES** may be a source of genetic instability. If sister chromatid exchange takes place in the following anaphase, double chromatid bridge is formed that may break equally, and in this case the distribution of the genes to the poles is normal. (The genes are represented by numbers). The chromatids may break also unequally resulting in unequal distribution of the chromatin (genes) that may be detected if appropriate markers are present. The broken ends usually heal and new dicentric ring chromosomes of different sizes are formed. (After B. McClintock 1941 Genetics 26:542) (See also preceding page)

DICTYOTENE STAGE: a late prophase stage of meiosis in human females. Meiosis in the oocyte begins by about the fourth month of the fetus but it is completed only before each ovulation. Thus, some cells remain in dictyotene for 30-40 years (until the end of ovulation). It has been hypothesized that this prolonged meiotic stage may be responsible for some of the chromosomal anomalies observed as a function of increasing age. (See meiosis, Down's syndrome)

DIDEOXY FINGERPRINTING: is a variation of the single-stranded conformation polymorphism (SSCP) technique that detects single-base or other subtle alterations in the DNA. One lane of the Sanger dideoxy termination reaction is electrophoresed in a non-denaturing gel. Mutations may show up as the presence of an extra segment or the absence of one or by altered mobility. (See single-strand conformation polymorphism, DNA sequencing [Sanger]).

DIDEOXY NUCLEOTIDES: see DNA sequencing

DIDEOXYRIBONUCLEOTIDE: lacks an O atom at both the 2' and 3' position of the ribose (see structural formulas below) and because of the latter feature it is incapable of supporting DNA chain elongation that is required for the DNA polymerase to add nucleotides to the chain. Therefore these molecules can be used for selective chain-terminators in the Sanger method of DNA sequencing. (See DNA sequencing, formula ↘)

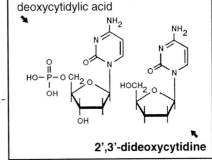

DIEGO BLOOD GROUP SYSTEM: is polymorphic in Mongolian, Chinese and American Indian populations. (See blood groups)

DIEPOXYBUTANE: see DEB

DIETHYLNITROSAMINE: see DEN

DiFERRANTE SYNDROME: see mucopolysaccharidosis

DIFFERENTIAL CENTRIFUGATION: partition of particles (cellular organelles) by their different rates of sedimentation in aqueous or organic liquid medium (sucrose, percoll, etc.) during centrifugation. (See centrifuge)

DIFFERENTIAL DISPLAY: see RNA fingerprinting

DIFFERENTIAL SEGMENT: the region of the X and Y chromosomes that lack homology and those chromosomal areas — in general — that fail to pair. Also, a particular borrowed chromosomal segment in a congenic strain. (See sex linkage, holandric genes, congenic)

DIFFERENTIATION: morphological and/or functional specialization of originally totipotent cells to meet particular needs of the organisms. Differentiation is determined by morphogen gradients, asymmetric distribution, cell position, etc. The pattern of differentiation is regulated in anterior/posterior, dorsal/ventral and by left/right gradients of the morphogens under genetic control. (See also determination, development, segregation asymmetric, morphogenesis, morphogenesis in *Drosophila*, primitive streak)

DIFFERENTIATION of PLASTID NUCLEOIDS: the SN (scattered nucleoids) in the small plastids are distributed between the thylakoids and grana (land plants and many algae). The CN (central nucleoids) are located at the center of the plastids (red alga, undifferentiated proplastids of higher plants). CL (circular lamellar) type occur as ring-shaped structures within the girdle lamella (brown algae). The PS (peripherally scattered) type lie near the inner plastid envelope and the SP type spreads around the pyrenoid. The proplastids may be of 0.2 µm diameter with 1- 2 cpDNA genomes or 2-3 µm in diameter with several cpDNA genomes. The smaller (proplastid precursors) may divide into structures like themselves or form the larger proplastids and change into colorless etioplasts in darkness and the latter upon illumination differentiate into green chloroplasts. In the *Chlamydomonas* alga the 5-6 nucleoids each may have 13-16 cpDNA while in *Euglena* the number of nucleoids may be 20-34 with 3-15 DNA rings each; in the chloroplasts of higher plants the 12 to 25 nucleoids each may contain 2-5 cpDNAs. (See chloroplasts, DNA replication in plastids, DNA replication in mitochondria)

DIFFUSION: movement of molecules from higher to lower concentration in a solution.

DIFFUSION, GENETIC: the distribution in a wave like manner of gene(s) in function of a variable such a geographic region, infestation by a parasite, etc. The rate of advance per generation under steady state condition is $v = \sigma\sqrt{2s}$ where σ = standard deviation and s = selection coefficient. The length of the wave advance: $\sigma([1/2]s)^{1/2}$. The rate of advance is obtained by $\sigma(2s)^{1/2}$ and if we arbitrarily substitute $\sigma = 20$ and $s = 0.01$ the km, the spread per generation would be $20(2 \times 0.01)^{1/2} \approx 2.83$ km. (See cline, legit)

DiGEORGE SYNDROME: is an autosomal dominant (human chromosome 22pter-22q11) defect of thymus and parathyroid formation and a variety of other anomalies (malformed ears, broad nasal bridge, far apart eyes, abnormal U-shaped mouth, small jaws, etc.), including immunodeficiency. Apparently autosomal recessive form has also been observed. This anomaly

DiGeorge syndrome continued
is frequently associated with deletions (1.5 megabase long) in chromosome 22 and some of the defects in this rather heterogeneous syndrome are due to a deletion in chromosome 10p13. This latter deletion accounts also for hypoparathyroidism associated with the velocardiofacial syndrome. In some instances transplantation of fetal thymus alleviated some of the symptoms. (See immunodeficiency, deletions, velocardiofacial syndrome, face/heart defects, CATCH)

DIGESTION: enzymatic hydrolysis of molecules *in vitro* or in the digestive tract *in vivo*.

DIGOXIGENINS: are aglycones of the steroid nucleus digoxine; they are used in non-radioactive forms to label amino acids and glycoconjugates.

DIGYNIC: has two sets of maternal chromosomes in a normally diploid cell.

DIHAPLOID: an individual having half the somatic chromosome number of that of a tetraploid from where it descended. Dihaploid plants are obtained generally by crossing tetraploids by diploids and among the offspring diploids are selected that carry some advantageous genes of the tetraploid that were not present in the parental diploid. (See haploid, autotetraploid)

DIHYBRYD: heterozygous for two pairs of alleles, e.g. *A/a* and *B/b*. (See Mendelian segregattion, modified Mendelian ratios)

DIHYDROFOLATE REDUCTASE (DHFR): enzyme synthesizes tetrahydrofolate, an essential precursor in the biosynthesis of thymine, purines and glycine. (See methotrexate)

DIHYDROPYRIDINE RECEPTOR (DHPR): is a voltage sensor in the L-type Ca^{2+} ion channels and ryanodine enhances the DHPR function. (See ion channels, ryanodine)

DEHYDROTESTOSTERONE: is a key hormone in maleness determination in mammals. It is made by steroid 5α-reductase enzyme from testosterone. Its deficiency causes defects in the development of the external male genitalia and the prostate but the epididymis, seminal vesicles and vas deferens are normal. The affected individuals are less prone to acne and baldness. The enzyme is encoded by two genes (SRD5A1, chromosome 5p15 and SRD5A2, 2p23) The dihydrotestosterone receptor deficiency leads to testicular feminization and Kennedy disease, both are located in human chromosome Xq11.1-q12. (See animal hormones, testicular feminization, Kennedy disease)

DIHYDROURIDINE (5,6-dihydro-2,4-dihydroxyuracil nucleoside): a postranscriptionally modified uridine in the tRNA. (See tRNA)

3,4-DIHYDROXYPHENYLALANINE: see DOPA   DIHYDROURACIL ⇨

DIHYDROZEATIN: a natural cytokinin. (See cytokinins)

DI-ISOSOMIC: in wheat, 20"+i", 2n=42, ["=bivalent, i=isosomic].

DI-ISOTRISOMIC: in wheat 20"+(i")1''', 2n=43, [i=isochromosome, "=disomic, 1'''=trisomic

DIKARYON: a single cell has two nuclei; the nuclei may be genetically identical (homokaryon) or genetically different (heterokaryon). These regularly occur in fungi and can be observed after fusion of somatic cells of other eukaryotes if the nuclei do not fuse. (See fungal life cycles)

DIKARYOTE: cells with two unfused nuclei. (See dikaryon)

DIMER: association of two units (e.g. two polypeptides).

DIMER 14-3-3: a protein factor that assists dimerization of Raf, an oncogene and a key player in the mitogen signal transduction pathways. (See raf, signal transduction)

DIMETHYLSULFATE: see DMS

DIMETHYLSULFOXIDE (DMSO): a very effective solvent of a wide range of organic chemicals with moderate toxicity; it also protects cells against low temperatures. DMSO may facilitate cellular DNA uptake and thus transformation of eukaryotic and prokaryotic cells.

Di-MON CROSS: involves dikaryotic and monokaryotic fungi. (See Buller phenomenon)

DIMORPHIC: displays two forms whether the two are chromosomes, cells or organisms.

2,4-DINITROPHENOL: is an indicator of pH, wood preservative, insecticide. It is highly toxic to animals by being an uncoupler of electron transport and oxidative phosphorylation. It blocks ATP formation by respiration. Uncoupling oxidative phosphorylation generates heat in hibernating animals, in those adapted to cold or in newborns.

DINUCLEOTIDE ABUNDANCE: see dinucleotide odds ratio

DINUCLEOTIDE ODDS RATIO: represent the frequency with which CpG or TpA sequences occur. Statistical survey of sequenced DNAs in different species indicates that the dinucleotide abundance within a species for different classes of DNAs (coding, intron, intergenic) tend to be similar but different in unrelated ones.

DIOECIOUS: the two sexes are represented by separate individuals like the majority of animals, fungi, and also some plants such as spinach, asparagus, date palm, poplar, osage orange, etc. (See breeding system)

DIOXIN: is a carcinogen and (frameshift) mutagen. It may occur as a contaminant in various industrial chemicals. LD_{50} orally for mice is 114 µg/kg. (See environmental mutagens, LD_{50})

DIPHTHERIA TOXIN: single-chain (535 residue) very potent toxin produced by *Corynebacterium diphteria* carrying a lysogenic phage; it has applied significance in cancer research because it blocks eukaryotic peptide chain elongation (eEF-2). Rat and mouse are resistant to this toxin because they lack its membrane surface receptor. Humans, guinea pigs and rabbits are however, sensitive to it. The human sensitivity is encoded in chromosome 5q23. (See toxins)

DIPLOCHROMOSOME: displays 4, (rather than) 2 chromatids in each arm because of replication without splitting of the centromere. (See endoreduplication, salivary gland chromosomes)

Diplococcus pneumoniae (old name *Pneumococcus*): is a member of the genus in the tribe *Streptococceae* and the family *Lactobacillaceae*. This is the most common cause of pneumonia. The lanceolate cells occur in doublets. The virulent forms are encapsulated and form shiny (smooth) colonies, the avirulent forms lack the protective coat (form rough colonies) and are destroyed by the enzymes of the host cells. Genetic transformation was discovered with this bacterium in 1928. (See transformation)

DIPLOID: has two complete basic sets of chromosomes, characteristic for the zygotes and for most of the body cells of the majority of animals and plants, and for the premeiotic phase of several other eukaryotes (fungi, algae, etc.)

DIPLONEMA: the structure of the chromosome thread at the diplotene stage of meiosis. (See meiosis, diplotene)

DIPLONT: a diploid individual (cell).

DIPLONTIC SELECTION: competition among diploid cells within a multicellular organisms. This can take place only when dominant mutation or other dominant-acting change(s) occur.

DIPLOPHASE: the diploid phase of an organism that exists also as a haploid.

DIPLOSPORY (apogamety): non-reduced embryosac develops without meiosis or by restitution. (See restitution nucleus)

DIPLOTENE STAGE: of meiosis is preceded by pachytene when the chromosomes begin to appear as clearly bipartite, doubled threads (diplonema) through the light microscope. The pairing (synapsis) of the bivalents appears somewhat relaxed but the four chromatids are held together by the chi-shaped (χ) structures of the overlapping chromatids, the chiasmata. The synaptonemal complex is generally the most conspicuous (by electronmicroscopy) when chiasmata are visible. (See meiosis, chromatid, chiasma, synaptonemal complex)

DIPOLE: molecules have two equal and opposite charges separated in space; a simple dipole molecule is H_2O.

DIQUAT: can serve as electron acceptor in photosystem I of photosynthesis. It is also a light-dependent herbicide. (See photosynthesis, herbicides)

DIRECT DNA TRANSFER: incorporation of DNA into plant protoplasts without bacteria using only plasmids or perhaps other "naked" DNA. (See transformation, naked DNA)

DIRECT REPEATS: adjacent or non-adjacent repeats of identical or similar nucleotide sequences of the same order such as ATG....ATG. These are common at the termini of insertion elements. (See insertion elements, transposons, transposable elements)

DIRECT SUPPRESSORS: modify the final product of the gene so the mutation is not (fully) expressed. (See suppressor mutations)

DIRECTED MUTATION: a controversial (Lamarckian) notion that mutations of advantageous phenotype are preferentially induced in an adaptive environment, i.e. mutations would not be random and preferentially selected as required by the ideas of neo-Darwinian theory. Although supportive evidence has been put forth even currently, none of them can be completely defended against the alternative, conservative, classical interpretations (see acquired characters). In depth analysis of adaptive mutation (directed mutation) in *E coli* revealed that these adaptive reversions of the *Lac* alleles required F' plasmid transfer replication and homologous recombination involving F' plasmid elements that may contain revertant alleles and may so account for the apparent directed mutations. In bacteria, the so called adaptive mutations took place only in the presence of a functional RecBCD recombinational pathway. The notion of the existence of true adaptive mutations is thus still controversial. Directed mutations can be obtained, however, by manipulating the DNA *in vitro* using the techniques of molecular biology and transformation. (See acquired characters, localized mutagenesis, PCR-based mutagenesis, TAB mutagenesis, linker scanning, Kunkel mutagenesis, homologue-scanning mutagenesis, lamarckism, lysenkoism, soviet genetics, neo-darwinian evolution)

DIRECTIONAL CLONING: vector DNA termini are ligated with different linkers that pro-duce different cohesive sequence at the two ends after restriction endonuclease cleavage. The passenger DNA (insert) can be ligated thus in a chosen orientation only. (See vectors)

PLASMID   INSERT   PLASMID

DIRECTIONAL SELECTION: alters the population mean in either plus or minus direction.

DIRECTIONALITY: phage λ *chi* sites may enhance recombination more on the left than on its right side or the transposase may be more functional in one orientation. (See orientation selectivity, lambda phage)

DIRECTORY: of the computer reveals the contents of the folders (the documents the user has generated).

DIRVISH (direct visual hybridization): is a mapping procedure for DNA sequences using fluorochrome-labeled samples hybridized to highly stretched DNA strands and their location is visualized with the aid of fluorescence microscopy. (See FISH)

DISACCHARIDE: covalently bound two monosaccharides, e.g. sucrose (glucose : fructose)

DISACCHARIDE INTOLERANCE: is a collective name for the inability of proper metabolism of sugars. The sucrose intolerance is caused by sucrase - isomaltase malfunction. The enzyme is generally present but does not function normally, resulting in diarrhea when sucrose is consumed. Sucrose, maltose and starch are well tolerated in lactase deficiency but lactose cannot be reabsorbed and cause bloating, diarrhea and general discomfort because of the high gas production by intestinal microbes. This defect is caused by an autosomal recessive gene in human chromosome 2. Generally two types of lactase-phlorizin hydrolase activity are distinguished. One type appears generally as a developmental defect after age 5. Its frequency may be quite variable in different populations; generally its frequency is low where dairy farming is prevalent and high where milk is absent from the diet. The gene is at human chromosome 3q22-q26. (See also fructose intolerance, lactose intolerance, glycosuria, phlorizin)

DISARMED VECTOR: agrobacterial genetic vector from which the oncogenes (encoding phytohormones) have been deleted in the T-DNA (and usually replaced by foreign, desirable genes). (See Ti plasmid, plant hormones)

DISASSOCIATION: separation of the two strands of DNA. (See C_0t curve, Watson and Crick model)

DISASSORTATIVE MATING: the mating partners are less similar in phenotype than expected by random choice. (See assortative mating, breeding systems)

DISC: see disk

DISC: is a death-inducing complex of FAS. (See FAS/CD95, apoptosis)

DISCONTINUOUS GENE: the translated exons are separated by introns. (See introns, exons)
DISCONTINUOUS REPLICATION: see Okazaki fragments, replication fork
DISCONTINUOUS VARIATION: is caused by qualitative genes and the expression can be classified into discrete groups with relatively low environmental effects. (See continuous traits)
DISCORDANCE: dissimilarity of a trait between individuals because of genetic difference. The term is used for gene loci carrying two different alleles in a diploid. (See dizygotic twins, twinning, zygosis, monozygotic twins, penetrance, expressivity)
DISCRIMINANT FUNCTION: estimates statistically the overlap between two populations and it has uses for classification and diagnosis, for the study of relations between populations and as a multivariate extension of the *t*-test. Example: a single variate X is distributed in two populations with means μ_1 and μ_2; the standard deviations (σ) are assumed to be the same. We wish to know to which of the two populations the specimen X belongs. We classify X to population 1 if $X < (\mu_1 + \mu_2)/2$ and to population 2 if $X > (\mu_1 + \mu_2)/2$. If X is from population 1, our decision is wrong if $X > (\mu_1 + \mu_2)/2$ or when δ (the distance between the two means) = $(\mu_2 - \mu_1)$. $(X - \mu_1)/\sigma$ follows the normal distribution and misclassification is probable in the tail from $\delta/2\sigma$ and ∞ in both cases. The δ/σ must exceed 3.0 to consider the classification accurate. The δ/σ is also called distance between two populations. Analysis of variance of the discriminant function can be used as follows:

	Sum of Squares	Degrees of Freedom
Between Groups	$(n/2)D^2$	2
Within Group	D	2n - 3
Total	$D(1+[n/2]D)$	2n-1

where D is the difference between $\bar{X}_1$ and $\bar{X}_2$. The significance of the difference can be determined by *z*. (See multivariate analysis, analysis of variance, z, Mather, K. 1965 Statistical Analysis in Biology. Methuen, London, UK)
DISCRIMINATOR REGION: is in the -10 to +1 region of prokaryotic promoters where in amino acid starving cells 5'-diphosphate 3'-diphosphate (ppGpp) binds. ppGpp is synthesized under starvation conditions and it represses rRNA protein genes and activates amino acid operons. The ppGpp repressed promoters are GC rich, the ppGpp activated promoters are richer in AT in the discriminator region. (See stringent control)
DISEASES IN HUMANS: see genetic diseases
DISEQUILIBRIUM: is a lack of equilibrium. (See also linkage disequilibrium and linkage equilibrium)
DISJUNCTION: separation of bivalents (chromosomes) during meiosis I or of chromatids in mitosis. (See meiosis, mitosis, nondisjunction)
DISK: a magnetic surface on what information can be recorded and later retrieved by the appropriate command to the computer. The capacity of amount of information stored is expressed in kilobytes (K) or megabytes (MB). The floppy disks (about 9 x 9 cm) can be low-density (400K, single-sided or 800K, double-sided) or high-density disks (1.4 MB). The hard disks hold 20 MB to several GB (gigabyte) or more information. 1 K = 1024 characters, about 170 words in English.
DISLOCATED HIP (hereditary): its prevalence in human population is about 0.075% and its recurrence risk among sibs is about 5%
DISOME: two homologous chromosomes
DISOMIC: an individual or cell with one specific or any of the chromosomes represented twice. *Maternal or paternal disomic* individuals arise from hybrids in case of nondisjunction for a particular chromosome in translocation heterozygotes (Robertsonian translocations). *Uniparental disomy* may be associated with trisomy when one of the three chromosomes, derived from one of the parents, is lost and the result is homozygosity (disomy) for the other parent's chromosomes. (See Beckwith-Weidemann syndrome, Prader-Willi syndrome, Angelmann

syndrome)

DISPERMIC FERTILIZATION: two sperms enter into one egg. (See also double fertilization)

DISPERSED GENES: are members of (multi) gene families. (See gene family)

DISPERSED REPEATS: see SINE, LINE

DISPLACEMENT LOOP: see D loop

DISRUPTIVE SELECTION: as a consequence of the process, the population breaks up into two (contiguous) groups with different mean values. (See selection types)

DISSECTION, GENETIC: determines the mechanism of genetic determination of a process.

DISSEMINATED SCLEROSIS: see multiple sclerosis

DISSOCIATION CONSTANT: K_d is the equilibrium constant for the dissociation of a complex of molecules (AB) into components (A) and (B); K_a is the dissociation constant of an acid into its conjugate base and a proton. The smaller the K_d value $\frac{(A)(B)}{(AB)}$ the tighter is the binding between the components.

DISTAL MARKER: is situated in a direction away from the centromere or another gene or in bacterial conjugation it is transfered after a particular site. (See centromere mappig, conjugational mapping)

DISTAL MUTAGEN: is formed when a promutagen through chemical modification is first converted into an intermediate (proximal mutagen) that finally becomes the ultimate or distal mutagen. (See mutagens, carcinogens)

DISTALIZATION RULE: in differentiation when new cells are generated (regenerated), each intercalary cell division produces progressively more distal cells until the circumferential filling of missing values are completed. (See polar coordinate model)

DISTANCE MATRIX: see character matrix

DISTORTED SEGREGATION: is seen when the transmission of one of the alleles is not the same as that of the other. The cause of this anomaly can be chromosomal defects, lethal or semilethal mutations or genes like the *Segregation distorter* (*Sd*, map location 2-54) in *Drosophila*. The distortion may reduce either the dominant or the recessive class depending on the linkage phase and map distance of the marker to the factor that disturbs normal phenotypic or genotypic proportions. (See segregation distorter, meiotic drive, gametophyte factor, certation, preferential segregation, megaspore competition, gene conversion)

DISTORTION IN CLONING: different DNA fragments, because of their nature, length, etc., may be replicated at different rates in the vectors and may bias the representation of the sequences in the library. (See library, vectors, replication)

DISTRIBUTIVE CIRCUIT IN SIGNAL TRANSDUCTION: one kinase may initiate multiple responses. (See signal transduction)

DISTRIBUTIVE PAIRING: the chromosomes may recombine during meiosis and are then distributed to the poles normally (*exchange pairing*). Chromosomes that are not involved in exchange display *distributive pairing* and may suffer nondisjunction. A *Drosophila* mutation *nod* (*no distributive disjunction*; map location 1-36) when homozygous may display high (800 fold) frequency of chromosome loss and nondisjunction of chromosome 4 during meiosis I. Recent information indicates that in the distribution of achiasmate chromosomes the heterochromatin adjacent to the centromere plays an important role. (See chromosome pairing, achiasmate, heterochromatin)

DISULPHIDE BRIDGE: a covalent bond between two sulphur atoms (—S—S—) of two cysteine residues of a polypeptide chain(s) affecting conformation. (See protein structure)

DITELO-MONOTELOSOMIC: in wheat, 20"+t"+t', 2n=43, ["=disomic, '=monosomic, t=telosomic]

DITELOSOMIC: in wheat, the chromosome constitution is 20"+t" (2n=42) [" indicates disomy, t=telosomic]

DITELOTRISOMIC: in wheat, 20"+(t")1"', 2n=43, ["=disomic, "'=trisomic, t=telosomic].

DITHIOERYTHRITOL (DTE): 2,3-dihydroxybutane, Cleland reagent; it protects SH groups.
DITHIOTHREITOL (DTT): DL-threo-1,4-dimercapto-2,3-butanediol; protects SH groups.
DITYPE: tetrad with two kinds of spores. (See tetrad analysis)
DIURON: herbicide interfering with photosynthesis.
DIVERGENCE: evolutionary differences in morphology, cytology and/or in the primary structure of nucleic acids and proteins that are believed to have descended from common ancestry. The divergence can be quantitatively estimated on the basis of average chiasma frequencies of the genomes if the species can be crossed. The average frequencies of amino acid substitutions in the proteins or base substitutions in nucleic acids can be used to estimate genetic distance and the time required to achieve it. Some caution may be required in interpreting evoluttionary divergence because similarity may be based also on evolutionary convergence. Divergence may be brought about by mutation, chromosomal rearrangements and recombination. The recombinational force in *E. coli* appears to be 50 times higher than that of mutation in clonal divergence. (See also convergent evolution, genetic distance, chiasma)
DIVERGENT DUAL PROMOTER: iuxtapositioned promoters may carry out transcription in opposite directions. (See promoter, divergent transcription, catabolite repression)
DIVERGENT TRANSCRIPTION: proceeds from two promoters in opposite orientation. (See promoter, transcription)
DIVERSIFICATION: see combinatorial diversification, junctional diversification
DIZYGOTIC TWINS: develop from two separate eggs fertilized by separate sperms. (See twinning, monozygotic twins, zygosis)
Δ(*lac - proAB*): a deletion for the bacterial *lac* (*lactose*; map position 8) and *prolineA* and *prolineB* (map position 6 for both; blocks before glutamate semialdehyde) genes. (See *Lac* operon)
D-LOOP: see displacement loop
DM: the HLA DMA and DMB gene products that are required to exchange CLIP from Class II major histocompatibility proteins so the T cell can be loaded with the antigen. The DM α and β subunits are similar to the α and β chains of the Class II molecules. (See MHC, HLA, CLIP)
DM (double minute) chromosome: has no centromere and can be maintained only under selective pressure for the gene(s) it carries and absent from the rest of the genome. (See YAC)
Dm: in front a gene or protein symbol indicates *Drosophila melanogaster* homolog.
DMBA (9,10-dimethyl-1,2-benzanthracene): a carcinogen standard, a mutagen; forms covalent DNA adducts. (See carcinogen, adduct)
DMC1: a meiotic protein in yeast controling recombination. (See RAD51)
Δ μH+: an energy donor, promotes translocation through cellular membranes.
DMS: dimethylsulfate, an alkylating agent, mutagenic and carcinogenic. (See alkylation)
DMSO: see dimethyl sulfoxide
dna: bacterial mutations involved in DNA replication.
DNA: deoxyribonucleic acid, the genetic material of all eukaryotes and bacteria and many viruses. The most direct proof for DNA being the genetic material was provided by genetic transformation. DNA is measured either in base pairs (bp) or spectrophotometrically: 1 unit of optical density (OD) of double-stranded DNA (at 260 nm wavelength) is about 50 μg/mL. One OD of single-strand DNA is about 40 μg/mL; 1 μg/mL DNA contains about 3.08 μM phosphate (see also DNA types); 3000 nucleotides are about 1 μm in length, 1 pg DNA is about 0.60205 x 10^{12} Da [10^{-12}/(1.6661 x 10^{-24} g)]. See also DNA types. (See Watson and Crick model, hydrogen-pairing, spectrophotometer, transformation)

Electromicrograph and interpretative drawing of single- and double-strand nucleic acids. (Courtesy of Prof. Y. Aloni).

DNA ALIGNMENT: see indel

DNA AMPLIFICATION: see amplification

DNA ANNEALING: reassociation of two complementary single strands into a double-stranded molecule. (See C_0t curve, DNA hybridization)

DNA BASE COMPOSITION: varies among different organisms:

SPECIES	A	T	G	C*
Humans	30.7	31.2	19.3	18.8
Wheat (*Triticum aestivum*)	25.6	26.0	23.8	24.6
Budding yeast (*Saccharomyces cerevisiae*)	31.3	32.9	18.7	17.1
Escherichia coli	26.0	25.2	24.9	23.9
Mycobacterium phlei	16.5	16.0	34.2	33.2
Bacteriophage T4	32.4	32.4	18.3	17.0
Bacteriophage φX174	24.3	32.3	24.5	18.2

*Methyl cytosine and hydroxymethylcytosine are combined.

The difference between the amounts of A and T, and G and C, respectively, were caused by analytical errors. Bacteriophage φX174 has single-stranded DNA, others are double-stranded.

DNA BENDING: two distant DNA sites are brought closer together because of non-straight run of the double helix caused by phosphate neutralization due to protein binding. The DNA may also bend by attaching a sequence-specific ligand to two sites separated by say 10 nucleotides. The lambda phage *Cro* repressor binds to three specific operator sites and causes DNA bending. Bending of chromatin may increase the chances of affinity of the TATA box-binding protein by over 100 fold and may also block gene expression. (See also looping of DNA, DNA kinking, lambda phage, TBP, flexer, high mobility group proteins)

DNA BINDING: N-methyl imidazole and N-methylpyrrole amino acids can recognize specific base pairs or short base sequences in the DNA and bind to them. There are also many DNA-binding proteins in the cells. (See DNA-binding protein domains)

DNA-BINDING PROTEIN DOMAINS: There are a variety of DNA-binding proteins and they have some common structural motifs belonging to four major groups: zinc fingers, helix-turn-helix, leucine zipper and helix-loop-helix.

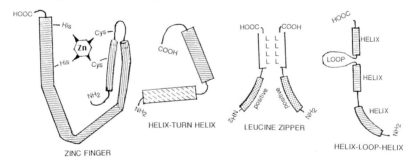

(See figures and more under the name of individual motifs, DBP, regulation of gene activity, hormone receptors, binding proteins, single-strand binding proteins, DNA bending, RNA-binding proteins)

DNA, BLUNTED: ═══════ lacks (single-stranded) cohesive termini such as ─────── and cannot circularize without modification of ends. (See cohesive ends)

DNA CHEMICAL SYNTHESIS: the 5'-end of the first nucleotide is protected by dimethoxytrityl (DMT) while the OH end is attached by a linker to silica. The reactive groups of all nucleotides are chemically protected. Afterwards DMT is removed by washing and the next nucleotide is activated and attached to the 3'-OH group. Using iodine, the 5' to 3' linkage is oxidized to generate a phosphotriester bond (one of the O of the phosphate group is methylated). The reaction is continued until the desired chain length is reached. About 70-80 residue polymers can be made this way. The process has also automated versions and routine synthetic services are commercially available.

DNA CHIMERA: a DNA molecule ligated together from originally different molecules.

DNA CHIPS: are a combinatorial array of oligonucleotides synthesized on a solid support (modified glass or polypropylene) *in situ* at specific "addresses" in a checkerboard like arrangement. The synthesis requires the use of masks to protect areas of the array. They may use of the 4 nucleotides (A, C, G, T) in any sequence length (4^s) up to many thousands. These procedures are expected to be exploited for mass, automated sequence analyses, mutation detection, etc. in a way analogous to the electronic microchip technology. This type of technology can be used to synthesize a vast array of nucleotide sequences at staggering speed. The procedure is a modification of the photolithography used by commercial printing for several decades. A mercury light is directed to a photolithographic mask and through it to a solid surface. There it activates at specific areas of nucleotides for chemical coupling. The 5' ends are protected. Subsequent exposure then removes the protection and potentiates the reaction of the end with another nucleotide. The cycle is then continued. The specificity of the nucleotides to be chosen is controled by the mask. An 1/6 cm^2 chip can accommodate hundreds of thousands of sequences of 20 µm. The space occupied by an oligonucleotide sequence of millions of molecules is called a *feature*. Each wafer contains a grid of 49-400 chips. The generation of the chips and the synthesis is mechanically controlled with minimal human labor and the whole process requires about half an hour. When the synthesis is complete the chips are separated and hybridized with an appropriate fluorochrome-labeled nucleic acid (DNA or RNA) probe. After removal the unhybridized probe sequences, the fluorescence is analyzed by a confocal microscope, requiring about half an hour. The resolution is going to be improved to 20 µm in the near future (1997), and each chip may permit the resolution of 400,000 probes/chip. The 3,000,000,000 nucleotide mammalian genome may require only 10 chips. Detection of single base pair mutations is a much more elaborate process than monitoring the expression of genes. The cost of the technology is comparable to an automatic sequencer. (For more see Southern E.M. 1996 Trends Genet. 12:110, Nature Genetics 14, No 4, Fodor, S.P.A. 1997 Science 277:393, see also SHOM, photolithography, <http://www.sciencemag.org/dmail.cgi?53241>)

DNA, CIRCULAR: the majority of plasmids, prokaryotic, and organellar (mitochondrial, plastid) DNAs form a covalently closed circle and a circular genetic map. (See genetic circularity)

DNA CLONING: see cloning

DNA COMPLEXITY: see c_0t curve

DNA COMPUTER: see Hamiltonian path

DNA CONSENSUS: see consensus

DNA DATA BASES: see databases

DNA DATING: see racemate

DNAs OF DIFFERENT DENSITIES SEPARATE IN DISTINCT BANDS IN CsCl IN THE TUBE OF THE ULTRACENTRIFUGE AND CAN BE VISUALIZED BY UV PHOTOGRAPHY OR BY ETHIDIUMBROMIDE STAINING AND THE BANDS CAN BE COLLECTED FOR FURTHER ANALYSIS.

DNA, DENATURED: hydrogen bonds are disrupted between the two strands; thus it is single stranded. (See DNA denaturation)

DNA DENATURATION: separation of the two strands of DNA duplexes by breaking the hydrogen bonds and the hydrophobic interactions among bases; it results also in reduced viscosity. This is accomplished by raising the pH or the temperature above 70° or 80° C. The "melting temperature" depends also on the base composition because G≡C binds by 3 and A=T by 2 hydrogen bonds. By lowering the temperature below 60° C renaturation may begin. (See also denaturation, hyperchromicity, hypochromicity, DNA, c_0t)

DNA DENSITY: see density gradient centrifugation, ultracentrifugation

DNA DENSITY SHIFT: may occur if heavy isotopes (^{13}C or ^{15}N) or 5-bromodeoxyuridine is incorporated into the DNA during replication, replacing the normal atoms or nucleotides.

DNA DEPENDENT PROTEIN KINASE: see DNA-PK

DNA DIAGNOSTICS: see prenatal analysis, genetic screening

DNA DRIVEN HYBRIDIZATION: in the molecular hybridization medium the DNA is in excess relative to RNA. (See DNA hybridization)

DNA EXTRACTION: there are a large number of procedures applicable to the major taxonomic

DNA extraction continued

groups. Only some basic outlines can be provided here. BACTERIA: *1.* grow bacteria in culture medium of choice to a high density (one to several days). *2.* Harvest cells by centrifugation. *3.* Lyse cell pellet in TE buffer (Tris-EDTA) containing SDS (sodium dodecyl sulfate) and proteinase K (free of DNase) at 37° C for an hour. *4.* Make it 0.5 M for NaCl and add CTAB (cetyl trimethylammonium chloride) to precipitate cell wall, polysaccharides, proteins, etc. but keep DNA in solution. *5.* Add equal volume of chloroform:isoamyl alcohol and centrifuge to remove CTAB and polysaccharides. *6.* Save supernatant. *7.* Remove protein by phenol:chloroform:isoamyl alcohol and centrifuge. *8.* From supernatant precipitate DNA by isopropanol. *9.* Wash the precipitated DNA with 70% ethanol. *10.* Take up DNA in TE buffer and store it in refrigerator. PLANTS: *1.* Use 10 - 50 g clean young tissue (from plants which have been kept in dark for 2 days to reduces starch). *2.* Grind it to powder in liquid nitrogen. *3.* Extract DNA in pH 8 Tris-EDTA buffer containing a detergent (SDS or Sarkosyl) for about 1 to 2 h at 55° C. *4.* Pellet debris by centrifugation at 4° C (6,000 rpm, 10 min) and collect supernatant DNA. *5.* Precipitate DNA from supernatant by 0.6 volume cold isopropanol (-20° C) by centrifugation (8,000 - 10,000 rpm, 4° C, 15 min). *6.* Pellet is taken up in TE buffer. MAMMALIAN TISSUES: *1.* Tissue or cell pellet (0.2 to 1 g) washed clean, and powdered in frozen liquid nitrogen. *2.* Lysis of cells (in NaCl, Tris buffer, EDTA, SDS, pH 8, proteinase K). *3.* Extraction of DNA in phenol:chloroform:isoamyl alcohol and centrifuge (10 min, 10,000 rpm). *4.* To supernatant add 0.5 vol 7.5 M ammonium acetate and 2 vol cold ethanol to precipitate DNA by centrifugation (2 min, 5,000 rpm). *5.* Wash DNA by 70% ethanol. *6.* Suspend DNA in TE buffer. All of the above procedures are basically very similar. Further purification may be necessary by ultracentrifugation in CsCl. Quantity may be determined on the basis of absorption of ultraviolet light of 260 nm (quartz cuvettes) in a spectrophotometer. In a reasonably pure DNA preparation the ratio of OD_{260}: OD_{280} is about 2:1. One OD is about 50 µg/mL for double-stranded DNA and 40 µg/mL for single-stranded DNA; 1 pg DNA is about 6.5×10^{11} Da. (See centrifuge, ultracentrifuge, spectrophotometer, SDS, Sarkosyl, TE buffer, EDTA, CTAB, Tris, proteinase K)

DNA FINGERPRINTING: has been developed since the 1980s as a molecular tool for the genetic specification of an individual, a taxonomic, or other related groups. DNA fingerprinting is an important tool for the study of evolution and also for establishing the degree of relatedness of genetic stocks and its application is relevant to civil and criminal court decisions requiring identifications with high precision. The perfect definition of identity would come from the complete sequencing of the genomes but that is not practical for routine purposes. The best approximation is to use Southern hybridization with some repetitive sequences as probes isolated from minisatellite DNAs of common occurrence in the eukaryotic genomes (see SINE). The first such probe was obtained from the four 33 basepair repeats in the intron I of the human myoglobin gene. These repeats have a core sequence very similar to the *chi* elements that are responsible for meiotic recombination hot spots. These elements generate unequal crossingover at high frequency but are silent in the somatic cells. Because of this nature they generate a high degree of specific genetic diversity. Actually, it appears that each and every individual is different from all others (except monozygotic twins or clones). These minisatellite DNA sequences differ from each other but share the highly conserved core:

CORE $\quad$ G G A G G T G G G C A G G A ^A_G
33.6 $\quad$ $\{(A G G G C T G G A G G)_3\}_{18}$ $\quad$ TRIMERIC SECTION REPEATED 18 TIMES
33.15 $\quad$ $\{A G A G G T G G G C A G G T G G\}_{29}$ $\quad$ REPEATED 29 TIMES IN HUMAN GENOME

The minisatellites are distributed widely in the human (and other eukaryotic genomes) within the chromosomes. In addition to the minisatellite probes, microsatellites can also be used. These are generally mono- to tetranucleotide repeats, occurring very frequently and dispersed in the eukaryotic genomes. When isolated genomic DNA is digested with one or more restric-

DNA fingerprinting continued

tion endonuclease(s), separated by agarose gel electrophoresis, Southern blotted and hybridized to minisatellite probes that are radioactively labeled, the autoradiograms appear similar to the diagram shown on next page. The total DNA without probes cannot be used because its quantity and complexity is too large and the individual bands cannot be distinguished due to the fact that many of them share the same or almost the same position in the gel. Such a gel does not appear cross-banded but it looks like a smear. Generally, the bottom even of probed gels contains a larger number of small fragments that are barely or not separated (not shown on the diagram). For the analyses fresh DNA samples are prefered because in older specimens degradation may take place and the larger fragments (closer to the start point [top]) become very weak or vanish. Some fingerprinting may be feasible, however, in frozen flesh or in over two thousand years old mummies. In the same gel, appropriate full range molecular size markers are also run at both sides to facilitate identification of fragment length with appropriate accuracy. The fragment length estimated on the basis of mobility in the gel compared to the corresponding size marker may have an error of 3-4 %.

In an experiment with 20 Englishmen, the following information was obtained. (Data and methods of calculations after A.J. Jeffreys *et al.* Nature 316:76):

DNA fragment size, kb	No. of fragments $M\pm$ standard deviation per individual		Probability x that fragment in A individual is present in B individual		Max. mean allelic frequency per heterozygosity	
Probes →	33.6	33.15	33.6	33.15	33.6	33.15
10-20	2.8±1.0	2.9±1.0	0.11	0.08	0.06	0.04
6-10	5.1±1.3	5.1±1.1	0.18	0.20	0.09	0.10
4-6	5.9±1.6	6.7±1.2	0.28	0.27	0.14	0.14

If one assumes that apparently co-migrating bands in the gels of A and B individuals represent identical alleles of the same minisatellite locus, the probability x that an allele in A is also present in B is proportional to the frequency of that allele according to $x = 2q - q^2$. If the frequency of the allele in the homozygotes is low then, the mean probability, $\hat{x} \approx 2\hat{q}$. Also if it is assumed that there is little variance among allelic frequencies, then the number of alleles $n \approx 1/\hat{q}$ and the mean homozygosity is approximately $\sum_1^n q_i^2 \approx n\hat{q}^2 = \hat{q}$. A certain proportion of the co-migrating bands in A and B belong to different minisatellite loci, and the estimates of mean allele frequency and homozygosity so estimated are maximal, and the true estimates will depend on the accuracy of the resolution and estimation of the size of the fragments in the gels. The mean probability that all fragments detected by probe 33.15 (see Table) are also present in individual B is $0.08^{2.9} \times 0.20^{5.1} \times 0.27^{6.7} \cong 2.78 \times 10^{-11}$. [Note that the probabilities were raised to the power of the corresponding mean fragment number]. Because the current population of the earth is about 6×10^9, an about more than 50 fold increase in the current population would be needed to find perhaps two individuals with identical genetic constitution. The precision of the analysis is further improved by the use of several probes. In the American forensic laboratory practice four or five probes are used that detect only two allelic variations, each, and making the precise reading of the bands easier because only two are shown by each run. This type of analysis still provides 99.9% accuracy for determining the identity of a person. The power of this procedure is also indicated by the fact that 0.5 to 5 μg DNA (present in a drop of blood, in a few thousands of cells) may be enough for an RFLP test. The procedure described above is the DNA fingerprinting or RFLP (restriction fragment length polymorphism). RFLP has superior analytical value to PCR because the former is based on the entire DNA whereas in the latter only fragments of the DNA are amplified. If RFLP is coupled

DNA fingerprinting continued

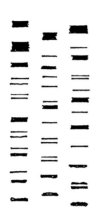

SCHEMATIC REPRESENTATION OF **DNA FINGERPRINTS**. The purified genomic DNA is digested with one or more restriction endonucleases and then separated by agarose gel electrophoresis and the run is Southern blotted onto a nitrocellulose or other substrate, denatured and hybridized with radioactively labeled probe, prepared from a minisatellite. After removal of the unbound label, it is autoradiographed at very low temperature for a few days. The film then reveals the restriction fragments that are homologous to the probe(s). The three code-bar-like patterns (at left) could be the DNA fingerprints of three different individuals or those of a single person probed with three different minisatellites. The largest fragments are at the top. The wide bands may indicate that several fragments of similar size did not separate well from each other.

with say 200,000 fold amplification of a sample using PCR, its efficiency may be further increased. The DNA sample prior to amplification may be as low as 500 pg (0.5 ng), however somewhat larger quantities (2 ng) may give better results. A single PCR amplification does not deal, however, with an entire genome-size DNA but only with an extremely small fragment, defined by the two 5'-terminal primers. By the use of Taq DNA polymerase, starting with pg quantities of DNA of 2 kb in length can be increased up to 1 µg in 30-35 cycles, and after dilution it can be subjected to even further amplification. For RFLP tests not just larger quantities of the DNA are required but it must be fresh or non-degraded. PCR analysis can be carried out not just with more minute samples, but some degradation may not prevent its suitability for the test. For forensic PCR tests, generally the DNA of the second exon of the human leukocyte antigen (HLA)-DQ-α is used most commonly. Actually one test kit, *Amplitype® HLA-DQ-alpha Forensic Kit*, marketed by the Perkin-Elmer Company is commercially available. This procedure has an accuracy of 98-99% but a newer PCR procedure, the *Poly-Marker Test* (PM) is supposed to have up to 99.7% precision. The PCR methods use dot blot hybridization and fluorochrome labels that eliminate the waiting period involved in exposing the Southern blot to X-ray film. The Amplitype relies on differences in six allelic types (n) that determine 21 different genotypes according to the general formula $[n(n+1)]/2$. The PCR method is less accurate than the RFLP test but its value can approach that of the latter by the use or larger number of probes. In both types of assays the proper statistical procedure is very important. The power of the statistics depends on databases revealing the frequencies of the allelic variations in the general or ethnic populations so the genotype of the individual to be tested can be compared to other similar or identical types. In other words the statistics will tell the probability of how many other individuals may have the same DNA fingerprint as the suspect. In court, then along with the DNA fingerprints, other evidence is considered that may further improve identification (such as dermatoglyphics, alibi, time frame, etc.).

For DNA fingerprinting a variable section of the mitochondrial DNA can also be successfully used. Since the size of this DNA is more limited than that of the nuclear DNA fragments, the potential information to be gained is less.

DNA fingerprinting can identify with great certainty members of a biological family. If the total number of identifiable bands of putative mother (M) and father (F) is T and of that n is shared by offspring (O), then the probability that (O) would share all the bands in common between (M) and (F) is Y^T where $Y = 1 - (1 - x)^2$ [for the definition of x see above]. The probability that (O) would share the specific maternal or the specific paternal fragments is x^{Mf} and x^{Ff} where Mf and Ff are the number of mother- or father-specific fragments. The use of DNA evidence in the criminal courts have been questioned on statistical grounds and because some of the crime or biological laboratory contractors may have based unwarranted conclusions on poorly performed and/or incompetently evaluated DNA tests. The statistical arguments brought forward are as follows. If the only evidence is the DNA profile and the probability of match for a particular individual is 1/1,000,000 but there may be 500,000 indi-

DNA fingerprinting continued

viduals who could be responsible for the case if the DNA evidence is not considered. There are then two possibilities: either the suspect is the criminal and he displays a perfect match to the DNA sample collected at the crime scene or another individual among the other 500,000 is the criminal and the DNA analyses match only by chance. The probability for the first case is $1/(500,001) \times 1 = 0.000001999$. The probability for the second case is $500,000 \times (1/500,001) \times (1/1,000,000) = 0.000,000999$. Since the ratio of these two fractions is about 0.5, and $1.5/0.5 = 3$. Thus, the statistical chance that the suspect is innocent is $1/3 \approx 0.33$. These methods are particularly useful to rule out identity, i.e. rule out that the blood, semen or tissue sample left at the crime scene did not come from a particular suspect or that a particular person may not be a relative to an individual. Positive identification is also likely at an extremely high probability. The technology and the experience of the technicians have improved since the 1980s. Also, the removal of contaminants from the specimens (sulfur, dyes of the clothing, etc.) may be facilitated by new procedures of purification of the DNA. Several of the commercial testing laboratories follow the guidelines recommended by the *Technical Working Group on DNA Analysis Methods* (TWIGDAM). The greatest care is still required in the use of these powerful techniques because in murder trials life or death of the suspects may depend on the correct identification. There are still some questions regarding the best procedures and the acceptance of DNA fingerprints and its statistical evaluation as evidence in the different courts of the USA. The best statistical procedures for the criminal justice system should be to determine the ratio of the conditional probabilities (Pr) of the claims of the prosecution (C) and the defense (C') based on the evidence (E) as $Pr(E|C)/Pr(E|C')$.

The mammalian mitochondrial DNA D loop is rich in variable base substitutions. In addition, the copy number of this DNA is very large making its analysis rewarding even from samples highly degraded such as in exhumed corpses or ancient DNA. In such instances DNA amplification is usually required. Essentially the same total DNA procedures have been used for the analysis of feral populations of animals, for anthropological studies, on ancient biological samples, to check the authenticity of cell cultures or plant varieties, etc. (See also fingerprinting, minisatellite, microsatellite, VNTR, ceiling principle, forensic genetics, Southern blotting, dot blot, nick translation, gel electrophoresis, PCR, RFLP, mtDNA, probe, SINE, myoglobin, intron, chi elements, labeling, fluorochrome, autoradiography, electrophoresis, molecular marker, allelic frequencies, HLA, feral, mtDNA, Bayes' theorem, conditional probability, Romanovs, heteroplasmy, utility index for genetic counseling, Frye test, The Evaluation of Forensic DNA Evidence. Natl. Acad. Press, 1996)

DNA FORMS: form I is superhelical and circular, II nicked circular, III linear. These conformational states affect electrophoretic mobility in gels. (See nick, electrophoresis)

DNA, GENOMIC: natural and complete nucleotide sequences of genes, introns included.

DNA GLYCOSYLASE: see glycosylase

DNA GROOVES: a B DNA double helix displays a 3.4 nm pitch including a wider (major) and a narrower (minor) groove (furrow) along its helical structure. (See DNA types, pitch, Watson and Crick model)

DNA GYRASE: DNA-binding protein, controlling coiling, site-specific nicking, and it is functional in replication, transcription, recombination and DNA repair. These enzymes are members of the DNA topoisomerase family. (See topoisomerase. nick)

DNA, HEAVY CHAIN: the chains of a DNA duplex can be separated and annealed with polyinosine-guanine (PI-G). The chains rich in cytosine bind more of PI-G and form a heavier band and sediment faster upon ultracentrifugation in CsCl. Therefore it is called heavy or C chain whereas the complementary chain is called light chain (W). The density of DNA can be increased also by substituting bromodeoxyuridine in place of thymidine. (See bromouracil, ultracentrifuge)

DNA HETERODUPLEX: see heteroduplex, models of recombination

DNA HYBRIDIZATION: annealing single-stranded DNA with complementary single-stranded DNA or RNA either for the purpose of measuring the degree of homology or for labeling it be-

DNA hybridization continued

fore selective isolation. Hybridization can be carried out either in a solution or with DNA immobilized on a membrane filter (Southern blots) or *in situ*, in denatured chromosomes or in microbial colonies (dot blots). The material intended for hybridization is labeled either radioactively (^{3}H, ^{14}C, ^{32}P) or with biotin and fluorescent dyes. (See chromosome painting, FISH, *in situ* hybridization, Southern blotting, Grunstein - Hogness screening, Benton - Davis plaque hybridization, Northern blot)

DNA IMMUNIZATION: see immunization genetic

DNA ISOLATION: see DNA extraction

DNA KINKING: regulation of transcription depends also on protein-induced localized modification of the DNA structure (bending). This modification depends on the nature of the proteins recruited for the tasks and the base sequences in the DNA. The bases stacked in the major and minor groves ('the rungs') of the helix may cause four major categories of architectural alterations relative to each other: (1). twist [slight rotation of the rung], (2) roll [slightly lifting the broader side of a rung], (3) tilt [uplifting the rung at one side], and (4). rise [slipping the rungs away from each other]. The elastic properties of the DNA determines how it is wrapped around histones in the nucleosomal structure, the packing of phage DNA into capsids, supercoiling, DNA looping, etc. These changes are reversible and depend on the physical environment of the DNA. (See nucleosome, supercoiling, DNA looping, triple helix formation, DNA bending)

DNA KNOTS: catenated (interlocking ring) DNA. (See catenated, catenene, knotted circle)

DNA LIBRARY: collection of restriction endonuclease-generated fragments, each containing different segments of the genome . (See restriction enzyme, fragment recovery probability)

DNA LIGASES: enzymes joining DNA termini in the cell nucleus. DNA ligase I (125 kDA) is the most important ligase in animal cells, DNA ligase II (72 kDA) and DNA ligase III (100 kDa) have minor role in mammalian cells. Other organisms have different DNA ligases. Bloom's syndrome and acute lymphoblastic leukemia are caused by DNA ligase deficiency. (See vectors, Bloom's syndrome, lymphoblastic leukemia, deoxyribozyme)

DNA, LIGHT CHAIN: see DNA, heavy chain

DNA LIKELIHOOD METHOD: is a type of maximum likelihood procedure for the calculation of branch length in evolutionary trees. (See evolutionary distance, evolutionary tree, unrooted evolutionary trees, least square methods, neighbor joining method, four-cluster analysis, minimum evolution methods)

DNA LOOPING: see looping of DNA, LCR

DNA MARKERS: special isolated or identified DNA sequences such as restriction fragments, RAPDS, microsatellites, minisatellites and any other sequences which can be used as probes. or followed by molecular or genetic/molecular analysis. (See terms under separate entries)

DNA METHYLATION: see methylation of DNA

DNA METHYLTRANSFERASE: see methyltransferase DNA, methylation of DNA

DNA MOBILITY SHIFT: see gel retardation assay

DNA MODIFICATION: see methylation of DNA, restriction enzymes, 5-azacytidine

DNA MUTATION: see mutation spontaneous

DNA, NATIVE: the hydrogen bonds between the two strands are intact. (See Watson and Crick)

DNA NICKING AND CLOSING: see topoisomerases

DNA, NONPERMUTED: is terminally redundant, e.g. **123456789012** and all molecules have the same sequence. (See also DNA permuted)

DNA OVERHANG: in a double-stranded DNA one of the chains protrudes at the end as shown

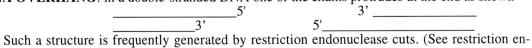

Such a structure is frequently generated by restriction endonuclease cuts. (See restriction enzyme)

DNA PACKAGING: the insertion of a piece of (concatameric) DNA into a phage capsid. (See also packaging of DNA, packing ratio)

DNA PAIRING: formation of hydrogen bonds between complementary bases of two single-stranded DNA. (See DNA, Watson and Crick model)

DNA PASSENGER: see passenger DNA

DNA, PERMUTED: phage DNA may be repetitious as the concatameric molecules cut off at different positions but to the same length. So a collections of these molecules may appear (with numbers substituted for nucleotides):

<u>123456789012</u>   <u>345678901234</u>   <u>567890123456</u>

(See also DNA non-permuted, redundant)

DNA PHOTOLYASE: splits pyrimidine dimers during direct repair. (See DNA repair, ultraviolet light, pyrimidine dimer, cyclobutane dimer)

DNA POLARITY: the first base in a DNA chain has a 5' phosphate (triphosphate) at the beginning and the subsequent bases are added to it at the 3'-OH position. From the nucleotide triphosphate two phosphates are removed and the third one (α group) forms a phosphodiester linkage with the first one. Thus the chain growth is $3' \leftarrow 5'$. The complementary DNA chain also has polarity and the two run antiparallel $\leftarrow\!\!\!\!\!\rightarrow$

DNA POLYMERASES: prokaryotes have 3 and eukaryotes have 5-6 major DNA polymerase enzymes with polymerase (and exonuclease) functions:

PROKARYOTES

pol I	**pol II**	**pol III**
109 kDa	90 kDa	≈1000 kDa
5' → 3' polymerase	5' → 3' polymerase	5' → 3' polymerase
3' → 5' exonuclease	3' → 5' exonuclease	3' → 5' exonuclease
5' → 3' exonuclease	none	none

DNA pol III holoenzyme forms a particle of nine subunits, and it is responsible for the replication of the prokaryotic chromosome.

EUKARYOTES

pol α	**pol β**	**pol γ**	**pol δ**	**pol ε**
3 subunits	1 subunit		1 subunit	1 subunit
180-300 kDa	300 kDa	40 kDa	170 - 230 kDa	250 kDa
nuclear	repair	polymerase	nuclear leading	3' → 5'
primase	nuclear	and	and lagging	exonuclease
3' → 5'		3' → 5' exonuclease	strand	nuclear
exonuclease		in mitochondria	replicase	

Recently in yeast **pol ζ** was discovered; it does not have 3'→5' exonuclease function although it is a repair enzyme that can bypass more efficiently thymine dimers than the other polymerases. (For more on functions see pol enzymes, DNA replication in eukaryotes and prokaryotes, recombination, DNA repair, ABC excinucleases, reverse transcription, PCNA)

DNA POLYMORPHISM: see DNA fingerprinting, band-sharing coefficient

DNA PRIMASE: see primase

DNA PROBE: a radioactively or non-radioactively labeled single-stranded DNA sequence that anneals specifically to its homologous, complementary DNA or RNA tract and thus by virtue of the label identifies the homologous strand (See Southern hybridization, dot blots, Northern hybridization, heterologous probe, insertional mutagenesis, nick translation, labeling)

DNA PROOFREADING: see proofreading, DNA repair

DNA REARRANGEMENTS: see phase variation, mating type determination in yeast, transposons, insertion elements, recombination, immunoglobulins, chromosome rearrangements.

DNA REASSOCIATION KINETICS: see c_0t curve

DNA RENATURATION: reannealing of denatured strands. (See renaturation, annealing)

DNA REPAIR: may be brought about by different mechanisms. The bulk of information is available from prokaryotic systems. The direct repair and the excision repair systems are error free because in the former no new DNA synthesis is required and in the latter the undamaged DNA

DNA repair continued

strand provides a normal template so there is no increase in errors beyond the normal level for the polymerase involved. The *SOS repair* is an error-prone repair mechanism that is operating when the path of the regular DNA polymerase is blocked by structural distortions and the enzyme is stalled and in the distress (lacking appropriate template) incorporates nucleotides in an unorthodox manner thus leading to mutation.

(I) DIRECT REPAIR systems remove pyrimidine dimers (cyclobutane dimers) by activating splitting enzymes, photolyases, upon exposure to visible light (Light repair). These enzymes contain chromophore cofactors, such as reduced flavin adenine dinucleotide ($FADH_2$) and folate, to absorb light. The *Bacillus subtilis* spore photoproduct lyase breaks between C-C bonds of 5-thyminyl-5,6-dihydrothymine, rather than cyclobutane dimers. Another direct repair enzyme is O^6-methylguanine methyltransferase that accepts at a cysteine site an O^6 alkyl group from mutated guanylic acid and thus restores the normal purine at the expense of its own inactivation. Direct repair does not involve any "unscheduled" DNA synthesis.

(II) EXCISION REPAIR SYSTEMS involve a various and limited amounts of unscheduled DNA synthesis. 1. *Mismatch repair*: MutS protein binds to various mismatched bases. MutH protein binds to a GATC tract in the DNA and MutL protein links the other proteins into a repair complex. The dam methylase (*dam*) enzyme recognizes the old DNA strand at replication and methylates it at a C site within the 5'-GATC tract. At that time the nascent DNA strand is still unmethylated. Mismatches are recognized (G—T well, C—C, poorly) and repaired within a 1 kb range of the GATC sequence as long as the new strand is not yet methylated. The MutH protein is an endonuclease that cuts 5' to GATC. The nicked strand then unwinds with assistance of helicase II (*uvrD* gene product). The site is braced by a single-strand binding protein (SSB, *ssb*), and the defective single strand is sliced off 3' to 5' by exonuclease I (*sbcB*) through the defect. On the 3' side of GATC the mismatch is removed either by exonuclease VII or *recJ* 5' to 3' exonuclease. Then DNA polymerases (*pol I, pol II* or *pol III*) repair the gap and the new ends are ligated (*dnaL*). 2. *Base excision repair:* defective bases are lifted out by cutting the N-glycosyl bonds by *DNA glycosylases* and leaving behind apurinic and apyrimidinic (AP) sites. The glycosylases remove deaminated cytosine (=uracil), deaminated adenine (=hypoxanthine) bases alkylated by alkylating mutagens, etc. Genes *ada* and *alkA* of prokaryotes are involved. After the base is removed *AP endonucleases* cut out the deoxyribose phosphates and possibly more nucleotides. The gap is filled in by pol I and ligation completes the repair process. 3. *Nucleotide excision repair:* damages to the DNA involving longer tracts are cut at two sites by the *ABC excinuclease* system and the intervening sequences are removed. This complex is encoded by genes *uvrA* (coding for the ATPase subunit of endonucleases), *uvrB* and *uvrC* (coding for the endonuclease subunits) of the excinuclease of *E. coli*. UvrD is a helicase that releases the excised oligomer. DNA pol I carries out repair synthesis and the M_r 75 ligase ties the ends. Excision repair may involve just a few and up to a few thousands of nucleotides. The gaps are filled in by any of the four DNA polymerases of prokaryotes. 4. *Transcription-coupled repair*: is a type of excision repair when the pyrimidine dimers are more readily removed from the transcribed strand than from the non-transcribed strand. This type of repair may be coupled with the more common excision repair system. Such a mechanism was first detected in *E. coli* but it turns out that the failure of this mechanism is the most common cause of base mismatches in colorectal human cancer. The vertebrate enzyme DNA polymerase β is involved in excision repair and in the release 5'-terminal deoxyribose phosphates from incised apurinic-apyrimidinic sites. The repair is confined to the transcribed strand and in mammals to genes transcribed by pol II. In humans, there is a requirement for the transcription-repair coupling protein (TRC). The Cockayne syndrome is a defect of two TRC factors. *E. coli* also requires a TRCF encoded by the *mfd* gene. The human diseases Xeroderma pigmentosum F and trichothiodystrophy are excision repair defects. Human DNA excision repair also requires XPA, a zinc finger protein and HSSB, a human single-strand binding protein, RPA another damage-recognition protein, the transcription factor

DNA repair continued
TFIIH with multiple subunits of helicase, zinc finger, kinase, and cyclin functions. Furthermore XPC, a ubiquitin, XPF (5' cut) and XPG (3' cut) nucleases, RFC an ATPase, PCNA, an auxiliary protein to DNA polymerase δ, repair polymerases ε/δ and DNA ligase are used.

III. SOS REPAIR is required when there is extensive damage to the DNA (presence of cyclobutane rings and adducts formed by UV light, cross-linking caused by chemical mutagens, etc.) that prevent the movement of the polymerase enzyme along the strands and the template cannot lend itself for accurate copying. The repair system is in distress and becomes error-prone. Shortly after the damage has taken place the activity of the RecA protein increases up to 50 times. Next the LexA repressor is cleaved that initiates the activity of a series of genes including the excinuclease system, *umuC* (with a repair function), *himA* (with multiple functions in replication, recombination and regulation). *LexA* repression targets an SOS box (containing a consensus $CTG-N_{10}-CAG$) in the vicinity of the promoter of its objects. Eukaryotic DNA polymerase ζ can bypass pyrimidine dimers. The SOS repair usually involves extensive DNA repair synthesis. SOS repair is considered to follow either the pathway of *damage avoidance* (DA), i.e. the complementary strand is used to rescue the blocked replication fork or by relies on *translesion synthesis* (TLS), i.e. the repair enzyme reads through the replication block. The block generated by the DNA-damaging agent determines the route of the SOS repair.

IV. RECOMBINATIONAL REPAIR makes corrections by exchanging the defects with correct sequences through recombination. RecA has here pivotal function. In *Drosophila* chromosome breakage, induced by the P transposable element is repaired by recombination and the efficiency of this repair is up to five times higher when the homologous sequence was syntenic within, at any position, of the X-chromosome rather than somewhere in trans position, in an autosome. (See recombination, molecular mechanisms in prokaryotes; DNA polymerases, P element). Retrotransposons or parts of them may be inserted into double-strand breaks and heal the defects. How frequent such events may be is not known but it is known that most eukaryotic genomes have a large number of retrotransposon elements and this repair function may justify their evolutionary maintenance. Specialized retroposons such as HeT-A and TART of *Drosophila* or the Ty5 yeast element may use their reverse transcriptase and RNA template to heal telomeres in case it is shortened by aging or if breaks damage it. The repair mechanisms of eukaryotes are much less well understood. The numerous *RAD* genes of yeast control radiation responses and are involved in repair. Short-patch DNA repair (such as damages caused by alkylating agents) requires the function of DNA polymerase β, a ubiquitous house-keeping enzyme. Repair of UV lesions involve DNA polymerase ε. e. g. xeroderma pigmentosum F, a human hereditary condition controlled by several dominant and recessive genes, involves a defect in excision repair of UV lesions that may be connected with polymerase ε. It has been estimated that each cell under normal conditions loses more than 10,000 bases from the spontaneous breakdown of DNA, and the damaged spots have to be repaired. The efficiency of repair in eukaryotes depends also on the time available before the onset of the S phase of the cell cycle. Proteins p53, p21, PCNA and other cellular proteins may have cell cycle-stalling functions. The efficiency of repair varies at different sites within a gene. Generally areas near active genes are more efficiently repaired than in inactive regions. The most rapid repair is found in the transcribed strands of active genes. The DNA repair systems play a role also in the cell cycle and tumorigenesis. In 1994 SCIENCE magazine declared DNA repair enzymes the molecules of the year. DNA repair in eukaryotes is complicated by the nucleosomal organization of the chromatin. (See also excision repair, X-ray repair, mismatch repair, radiation sensitivity, DNA-PK, PCNA, p53, p21, xeroderma pigmentosum, ataxia telangiectasia, Fanconi anemia, Bloom syndrome, Cockayne syndrome, colorectal cancer, aging, RAD28, TRCF, retrotransposons, retroposons, Ty, mating type determination in yeast, chromatin assembly factor, preferential repair, alkyltransferases, proofreading)

DNA REPAIR SYNTHESIS: see unscheduled DNA synthesis
DNA, REPETITIVE: see repetitious DNA, trinucleotide repeats

DNA REPLICATION ERROR: it has been estimated that in bacteria the individual nucleotide replacement error is about 10^{-9} to 10^{-10} per generation. (See replication error, error in replication, error-prone repair, proofreading)

DNA REPLICATION, EUKARYOTES: the mechanism of replication in prokaryotes is better understood because more simple *in vitro* assay systems are available for the smaller DNAs. In yeast the replicational origins are called ARS (*autonomously replicating sequences*). Although the genome of the eukaryote, *Saccharomyces cerevisiae* is only about 4 times larger than that of the prokaryote *E. coli,* the former has 400 replicational origins, each including about 300 bp. The yeast ARS are not identical with each other yet they have a short core sequence that is about 11 repeating A=T units. The single replicon of *E. coli* is over 4,000 kb but the yeast replicons are only about 40 kb. The speed of replication in yeast is about 3.6 kb/min but in *E. coli* it is almost 14 times as fast. In the large amphibian genome it proceeds at a speed of about 1/7 of that of the yeast. If the replication of the toad DNA proceeds only 500 bp/min, it would take more than 20 years to complete an S phase when actually at the gastrula stage it takes about 4 to 5 hours because the large genome relies on more than 15,000 replicons. Replication in eukaryotes just like in *E. coli,* proceeds bidirectionally and electron-microscopic examination detects a large number of replication "bubbles" or replication "eyes"

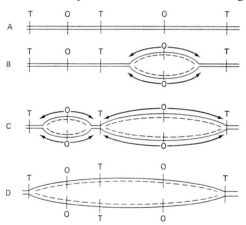

Multiple replicons are required for the large eukaryotic genomes. The horizontal lines represent the old DNA strands. The dashed lines show the newly, bidirectionally synthesized DNA within the "eyes". O = origin, T = terminus. Eventually the multiple replication bubbles coalesce. (Courtesy of Huberman, J.A. 1968 Cold Spring Harbor Symp. Quant. Biol. 33:509)

that are actually the replicational origins. As the replication within a bubble nears completion the neighboring bubbles coalesce as replication is completed along the length of the chromosome. The molecular structure and function of the eukaryotic replication fork is best understood in the eukaryotic SV40 virus that has a 5,243 bp chromosome with nucleosomal organization.

Human and other mammalian cellular fractions combine with the large T antigen of SV40 and in a plasmid replicate the viral genome, indicating the similarities of the interchangeable elements of the systems. In the presence of ATP, a double hexamer of the viral T antigen binds to the viral origin and modifies it. This protein - DNA complex then binds the *cellular replication protein A* (RPA, subunits 70, 34 and 11 kDa). RPA is a single-strand binding protein (SSB) similar in function to the helix-destabilizing proteins and in cooperation with the helicases assists in unwinding of the DNA duplex. The T antigen - RPA complex then binds eukaryotic *DNA polymerase* α, an over 300 kDa protein of three subunits, with RNA and DNA polymerase functions. This enzyme (pol α) then generates an RNA - DNA primer at the replicational origin. The *cellular replication factor C* (RFC, subunits 140, 41 and 37 kDa) binds to the 3'-end of the new DNA and brings to the growing point the *proliferating cell nuclear antigen* (PCNA, 29-36 kDa) and DNA-polymerase δ (pol δ, catalytic subunit 125 kDa, plus another 48 kDa subunit) then pol α is replaced by pol δ. Then the RCF/PCNA/pol δ aggregate proceeds in elongating the leading strand. On the lagging strand pol α/primase complex remains active at the initiation of the Okazaki fragments but after a switch the RCF/PCN/pol δ complex elongates also this strand. The elongation of the two strands is a coordinated process. After the Okazaki fragments are completed *RNase H* and 5'→3' *exonuclease MF1* (Fen1) remove the RNA primer and DNA ligase binds the ends of the Okazaki fragments into a continuous strand. Eukaryotic replication also utilizes topoisomerases in a

DNA replication eukaryotes continued

somewhat similar fashion to *E. coli*. When the replication fork encounters transcription initiation complexes, both in yeast (eukaryote) and prokaryotes (T4 page, *E. coli* bacterium) the progress of replication transiently slows down. (See also replication fork, DNA replication prokaryotes, DNA replication mitochondria, replication during the cell cycle, cell cycle, reverse transcription, DNA-PK, ARS, MCM, and diagrams)

DNA REPLICATION, MITOCHONDRIA: the mammalian mitochondrial DNA is a very short duplex circle (16.5 kbp). In mouse it has two origins of replication, *ori-H* and *ori-L*. Before replication at the *ori-H* site a displacement loop is formed of the H-strand, apparently to make room for the unidirectional replication. When the H strand loop is about 2/3 complete, replication of the L strand begins. The completion of the replication requires 2 hrs as it proceeds 270 base/min, i.e. more than two orders of magnitude slower than the process in the *E. coli* chromosome. Priming is initiated by an RNA transcribed on the L strand. Replication and transcription appear to be primed by the same process. The transition to L (light)-strand synthesis in human mtDNA is at the template GGCCG. In the larger mitochondrial DNAs (e.g. in maize the T mtDNA is 540 kbp and N mtDNA is 700 kbp) more *ori* sites are used. The 80 kbp yeast mtDNA has at least seven replicational origins. At the latter, there are 3 shorter G≡C rich regions, interrupted by 2 longer A=T rich sequences. In the DNA ribonucleotides are found, probably left-overs from the RNA primer. The principal replicase enzyme is DNA polymerase γ (180 to 300 kDa). [Note that this is not identical with the γ subunit (52 kDa) of prokaryotic DNA polymerase III coded for by *E. coli* gene *dnaX*]. The *Drosophila* mtDNA does not display D loops. The mtDNA may be divided into the daughter nucleoids either by pinching it off into two approximately equal pieces during the division of the nucleoid (*Physarum*) or dividing the nucleoid into two equal parts at the beginning of splitting of the organelle (*Paramecium*) or dividing the nucleoid prior to the beginning of division of the organelle (*Nitella*) or the approximately 30 small mitochondria fuse into a long mitochondrion during the G_1 phase of the cell cycle following mtDNA replication. Subsequently, the large mitochondrion divides into two and then fragmented further into spherical nucleoids (*Saccharomyces*). Within each nucleoid the number of mtDNA molecules may vary: 3-9 in yeast, 32 in the *Physarum* plasmodium or 2-6 in the mouse nucleoids. The number of mtDNA copies/cell may show substantial variations (≈1000 to ≥ 8000). In some fungi (*Candida*) the catenated mtDNA molecules may contain 7 units in a linear array. Most commonly the mtDNA appears as covalently closed circles yet in some cases (e.g. malignancy) two monomers may be catenated in a double circle or may be double-length chains. The 50 kbp mtDNA of *Tetrahymena* is also linear whereas in the *Paramecia* kinetoplasts mini and maxicircles occur. (See also mtDNA, DNA replication, MRP, CSB, kinetoplast, nucleoid, D loop, plasmodium, polymerase, catenane)

DNA REPLICATION, PROKARYOTES: the genetic material must be faithfully replicated to assure heredity. The replication takes place in a semi-conservative manner (see semi-conservative replication). Although the basic system of the process is very similar from viruses to eukaryotes, variations exist in the mode the final product is made. VIRAL DNA REPLICATION - Some viruses (φX174, G4, M13, fd) contain single-stranded viral DNA (V DNA) and thus the genetic material must be replicated via a complementary strand (C DNA) in the intermediate double-stranded step, the replicative form (RF). All types of DNAs require a replication *origin* (*ori*) to begin the process. A common feature of the origins is that they are rich in A=T to facilitate strand separation required for the semiconservative replication. The initiation of replication in the filamentous phage fd V DNA proceeds in opposite direction to the C DNA in the RF from closeby points. In the G4 phage the replicational origins of the V and C strands are widely separated and in its close relative, φX174 there is one origin for the V strand but multiple origins for the C strand. The replication of the about 15 times larger duplex DNA of phage T7 proceeds bidirectionally ↔ from the origin and so does the *E. coli* duplex DNA that is about 800 times larger. Since the DNA single strands run antiparallel, simultaneous

DNA replication, prokaryotes continued

semiconservative replication has some special requirements. This problem is met in a simple way by the 36 kb DNA of adenoviruses: first the replication starts at one of the 3'-ends and when it is completed then the other strand is copied using as template the other OH-end. Here each 5'-end cytidylic residue of the DNA is bound to a 55 kDa terminal protein. Then a 80 kDa viral protein displaces the end protein and uses the terminal deoxycytidylic acid as the beginning nucleotide template to lay down the first guanosinephosphate. The 80 kDA protein binds to nucleotides 9 to 18 at the end, and to the next 30 nucleotides host nuclear protein factor I is attached. Since the 80 kDA protein had formed already a complex with the polymerase, replication can now proceed.

REPLICATION IN *E. coli* - The size of the replicational unit, the *replicon*, in *E coli* is thus the whole genome (4.7×10^6 bp) whereas in mouse the average length of the about 2,500 replicons is about 150 kb. Despite of the large number of replicons in higher eukaryotes the pace of replication is much slower. In *E. coli* about 50 kb/min but in mouse about 2.2 bp/min.

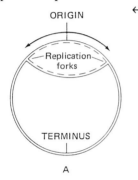

A

← The single origin in *E. coli* is in the *oriC* locus and the two forks must meet at about half way in the circular DNA. At about 100 kb from this meeting point there are the *ter* (termination) regions of about 23 bp length; *terD* and *terA* signal termination on one of the branches of the fork and *terC* and *terB* on the other. The protein coded by the *tus* gene recognizes the terminators and halts replication after the replication forks passed the *ter* sites. DNA synthesis can be started only if appropriate hydroxyl ends are available because this is a requirement of all DNA polymerases known. In *E. coli* the hexameric *helicase* proteins (330 kDa) are required to unwind the two strands and generate the **Y** form. This process uses energy liberated by hydrolysis of ATP to ADP. The unwinding is assisted by single-strand binding (SSB) *helix-destabilizing proteins* that make the bases free to be copied on both old strands. The unwinding removes the negative supercoiling and may even reverse the original coiling into positive supercoiling. This process has limitations and the tension must be relieved making single-strand cuts in the still unwound duplex. This nicking is carried out by topoisomerases (swivelases). DNA *topoisomerase type I* enzymes make a single cut in one strand of the helix and then permit the nicked strand to rotate once around the intact strand and then resealing the free ends and relieving negative supercoiling. *Topoisomerases type II* can cut and rejoin both strands of the duplex ahead of the fork and thus remove both negative and positive supercoiling. The latter group of enzymes include *gyrases,* and can make double cuts and convert positively supercoiled sections to negatively supercoiled ones. Type II topoisomerases can convert also catenated molecules ∞ into decatenated ones by permitting one duplex ring to slip out from the other ⇒ O O. At the bifurcation, one of the templates runs 5'→ 3' and the other 3' ← 5'. It is simple to proceed with replication of the old strand that faces the fork with its 5'-end because its complementary new strand can be elongated by adding bases to the 3'-OH terminus. This is called the *leading strand* (left branch of the fork on the diagram p. 293). While DNA synthesis (the new strand is shown by the heavier line) is going on, the strands at the fork must be kept separated by the *helix destabilizing proteins* (also called SSB, tetrameric *single-strand binding* proteins of 74 kDa, shown as small circles). The principal enzyme of replication is *DNA polymerase III* holoenzyme (pol III, 10 to 20 copies per cell). The α subunit of the enzyme (encoded by gene *polC*) carries out polymerization (chain extension) at the 3'-end of both *leading* and *lagging strands* (right branch of diagram). The ε subunit (gene *dnaQ*) is endowed with a 3' to 5' exonuclease function that reduces errors to 5×10^{-9} from 7×10^{-6}. The two (α and ε) together with θ, represent the *pol III core*.

DNA replication prokaryotes continued

DNA REPLICATION FORK OF PROKARYOTES ↳

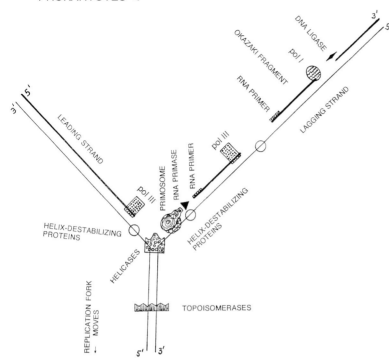

The β subunit (gene *dnaN*) apparently keeps the holoenzyme attached to the DNA. Additional polypeptide chains (γ, δ, δ', ψ, τ and χ) are also part of the huge holoenzyme ($\approx$ 1 MDa). The holoenzyme may be assembled in different ways under the control of gene *dnaX* products γ and τ. The lagging strand is synthesized in pieces (Okazaki fragments of 1 to 2 kb) because of the lack of a 3'-OH terminus to elongate a continuous strand starting from the base of the fork. The lagging strand requires a primer to be initiated. For primer a short RNA sequence is used. The *primosome* is a protein complex of helicases (*dnaB*, *dnaC* gene products), pre-priming protein (66 kDa, *dnaT* gene product), *priA* gene product (82 kDa monomer) recognizing primer assembly site and displaces the helix-destabilizing protein, and *primase* (60 kDa monomer product of gene *dnaG*). The primosome wraps around the DNA strands. Although this complex is shown only at the base of the lagging strand on the diagram it probably moves along to other locations as its role requires. Since the lagging strand is synthesized in pieces (Okazaki fragments) that carry RNA at the initiation region (5'), there is a need for other functions. *DNA polymerase I* (109 kDa monomer product of gene *polA*) serves triple roles. The larger C-terminal domain (68 kDa Klenow fragment, when cleaved by subtilisin) possesses a 5' to 3' polymerase and a 3' to 5' exonuclease functions. It can elongate the 3'-OH DNA or RNA termini by adding 5' nucleotidyl phosphates and can slice off nucleotides in the 3' to 5' direction. Its N-terminal domain (35 kDa) is a 5' to 3' exonuclease, an activity blocked during the polymerization reaction. Thus pol I can simultaneously extend the Okazaki fragments' 3'-OH end and remove the primer RNA at the 5'-end, and it is capable of a replacement replication (nick translation) that is just needed for making the lagging strand continuous. The 3' to 5' exonucleolytic activity has also the important function of "editing". If it finds "spelling errors" during synthesis, it removes the mismatched base and replaces it with the correct one. Its N-terminal domain reduces replicational errors from 10^{-5} to 10^{-7}, that is about the average spontaneous mutation rate in bacteria. After the Okazaki fragments reached full length they are joined by *DNA ligase*, an *E. coli* enzyme that attaches adjacent 5'-phosphoryl and 3'-OH ends of nucleotide chains in the presence of NAD^+ (nicotinamide adenine dinucleotide). Most of the details of the information on DNA replication was obtained in *in vitro* systems where the single components can be added or withdrawn and the role can be established without the complications of the *in vivo* analyses. In *E. coli* the transcriptionally most active genes are operating in the leading strand. (See also rolling circle replication and θ (theta) replication, replication fork, RNA replication in viruses, DNA chemical synthesis, replication speed)

DNA REPLICATION TYPES: are distinguished as type I and type II:

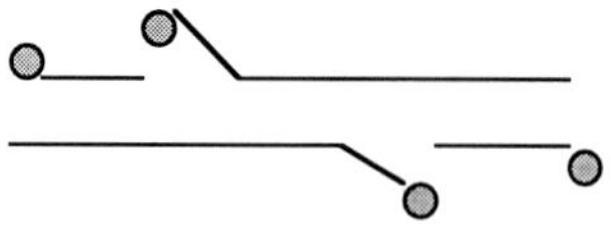

Type I is a linear duplex with one or more branches from the same DNA end

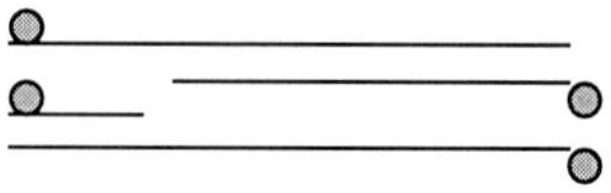

Type II molecules are partially single-partially double-stranded as shown on the left

DNA - RNA HYBRID: DNA is annealed with RNA in a double helix. (See annealing)

DNA, SELFISH: has little or no known use for the organism, yet it is of common occurrence such as the repetitive sequences. (See repetitious DNA, redundant DNA, SINE, LINE)

DNA SEQUENCE ALIGNMENT: see intel, DNA sequence information

DNA SEQUENCE INFORMATION: see databases

DNA SEQUENCING: determines the order of nucleotides in DNA fragments. Two procedures have been most widely used: the protocol of Maxam and Gilbert (M&G) and one or another modification of the method of Sanger and co-workers. By the *M&G method* both single and double-stranded DNA can be sequenced. The P^{32} end-labeled DNA sample is divided into five aliquots and by different, limited chemical breakage at one or two of the four bases, unique sets of labeled fragments are produced (G, G+A, A>C, C, C+T). From the five batches fragments are then separated according to length in five lanes in polyacrylamide-urea gels by electrophoresis. The gel is exposed to highly sensitive X-ray film to visualize the position the labeled bands. Since one end of each fragment in each batch terminates by the same nucleotide(s), according to the breakage mechanism indicated below, and in each batch these ends are specific, the sequence of the bases, in the different length fragments, can be read directly from the film of the gel (see illustrations below). The Maxam & Gilbert Methods is as follows:

```
                    ------------------------------------------
                    ------------------------------------------ DUPLEX DNA RESTRICTION FRAGMENT
        5'-END LABELED WITH  32P USING T4 KINASE
                            or
        3'-END LABELED WITH  32P USING T4 DNA POLYMERASE
USE EITHER A  SEPARATED SINGLE STRAND   32P------------------------------   OR

TWO PIECES OF THE DUPLEX                32P----------------    ----------------
                                                               ---------------- 32P
```

BREAK UP THE DNA CHEMICALLY TO GENERATE PIECES ENDING WITH EITHER OF THE FOUR NUCLEOTIDES

G	G + A	A > C	C	C + T
DIMETHYL SULFATE PIPERIDINE HEAT	DIMETHYL SULFATE PIPERIDINE ACID	NaOH PIPERIDINE	HYDRAZINE PIPERIDINE NaCl	HYDRAZINE PIPERIDINE

IN THE FIVE BATCHES THE LIMITED CHEMICAL BREAKAGE WILL YIELD FRAGMENTS OF DIFFERENT LENGTH THAT ARE ALL LABELED AT ONE OF THE ENDS BY ^{32}P AND THE OTHER END HAS EITHER ONE OR ONE OF THE TWO BASES ACCORDING TO THE SPECIFICITY OF THE CHEMICAL BREAKAGE.

THE REACTION PRODUCTS ARE THEN SEPARATED IN POLYACRYLAMIDE - UREA GELS BY ELECTROPHORESIS AND THE INDIVIDUAL FRAGMENTS WILL MOVE ACCORDING TO THEIR LENGTH AND

THEIR POSITION IS VISUALIZED AFTER EXPOSURE TO X-RAY FILM BECAUSE ONE OF THEIR ENDS CARRIES THE RADIOACTIVE LABEL. THUS THE FIVE GELS WILL APPEAR AS SHOWN BELOW WHEN RUN IN THE DIRECTION FROM BOTTOM UP AS SHOWN ON NEXT PAGE.

DNA sequencing continued

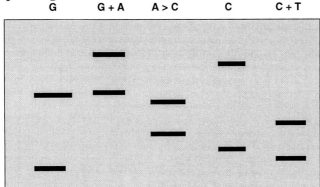

BECAUSE NO BANDS APPEAR IN THE **A >C** LANE CORRESPONDING IN POSITION IN THE **G + A** LANE OR IN THE **C** LANE CORRESPONDING TO THE **C + T** LANE THE SEQUENCE MUST BE

GTCATAGCA
(READ FROM BOTTOM TO TOP)

THE MAXAM AND GILBERT METHOD OF SEQUENCING SINGLE- OR DOUBLE-STRANDED DNA AND DIRECT READING OF THE NUCLEOTIDE SEQUENCE.

By the *Sanger method* single-stranded DNA is sequenced that has been cloned in M13 (or related) phage vectors. A chosen DNA fragment is then replicated in the presence of appropriate primer, α-P^{32}-dATP, limited supply of the four normal deoxyribonucleotide triphosphates and the large subunit of DNA polymerase I (Klenow fragment) or mainly Taq or phage T7 polymerases. In addition to these common components in four separate vessels the dideoxy analogs of the four nucleotides (dideoxyATP, dideoxyTTP, dideoxyGTP, dideoxyCTP) are added. These analogs when incorporated into the growing nucleotide chain, stop further elongation because they do not have OH at the 3' position where the new nucleotides normally attach during replication. As a result the chain growth stops and the position of this nucleotide in the chain is marked. The fragments generated by this stopped replication procedures are separated by electrophoresis as described in the M&G procedure and from the results of the sequencing gel the base sequences can be read directly. Recent years great technical advances were made in DNA sequencing by the use of laboratory equipment and computer technology and megabase size DNA stretches can be sequenced in much shorter time. A large number of viral, several prokaryotic, one eukaryote (*Saccharomyces*) genomes have been entirely sequenced and several others nearing completion. The average estimated error rate is 3×10^{-3} base.
(See DNA sequencing automated, BLAST, BLASTX, FASTA, BLOSUM, databases, evolutionary distance, array hybridization, DNA chips, an outline of the most widely used dideoxynucleotide method of Sanger *et al.* is outlined on next page).

DNA SEQUECING, AUTOMATED: the nucleotide sequences on the X-ray film obtained by autoradiography can be read directly or can be scanned with aid of computer programs The computer can be used also to search for particular sequences, transcription signals, homeoboxes, termination signals and any other type of conserved elements. In addition, through computer connections to DNA databases, homologies to previously determined sequences can be identified.

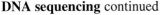

COMPUTER SCAN OF A SEQUENCING GEL EACH BASE IS LABELED BY A DIFFERENT FLUOROCHROME

SECTION OF A SEQUENCING GEL EACH LANE IDENTIFIES ANOTHER BASE

THE PEAKS (IN FOUR COLORS) CORRESPOND TO THE PATTERN OF THE SANGER SEQUENCING GEL. THE LASER SCANNER FEEDS TO THE COMPUTER THE INFORMATION ACCORDING TO COLOR THAT CORRESPONDS TO SPECIFIC BASES OF THE DNA. (Scan after a Promega advertisement; gel after a Pharmacia catalog)

GENETICS MANUAL

DNA sequencing continued from preceding page

INSERT INTO A CLONING SITE (CONTAINING THE *Lac* GENE) OF THE SINGLE-STRANDED M13 PHAGE. THE INSERTION INACTIVATES *Lac* IN THE PHAGE AND INFECT *Lac*⁻ BACTERIA. THE TRANSFORMANTS DO NOT PRODUCE COLORED PLAQUES ON Xgal MEDIUM.

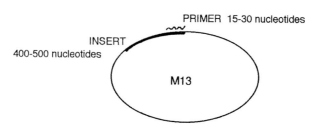

TO FOUR TUBES, EACH, ADD THE SEQUENCING VECTOR (M13), ^{32}P-DEOXYRIBOADENOSINE TRIPHOSPHATE, THE NORMAL DEOXYNUCLEOTIDE TRIPHOSPHATES AND A DNA POLYMERASE (EITHER THE KLENOW FRAGMENT OR SEQUENASE) AND IN EACH OF THE FOUR TUBES ADD ALSO ONE OF EACH OF THE FOUR TYPES OF DIDEOXYRIBONUCLEOTIDES:

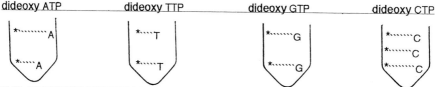

THE DIDEOXYNUCLEOTIDES ARE INCORPORATED IN PLACE OF THE NORMAL NUCLEOTIDES BUT LACKING THE 3'-OH END, THE CHAIN GROWTH STOPS. THEREFORE, THE SYNTHESIZED FRAGMENTS WILL TERMINATE EITHER WITH AN **A** or **T** or **G** or **C**, RESPECTIVELY. SINCE NORMAL NUCLEOTIDES ARE ALSO PRESENT, FRAGMENTS OF DIFFERENT LENGTH WILL BE GENERATED ON THE INSERT TEMPLATE ACCORDING TO THE CHANCE WHETHER NORMAL OR BLOCKING NUCLEOTIDES WILL BE INCORPORATED BY THE POLYMERASE. THE FRAGMENTS ARE THEN SEPARATED BY LENGTH BY POLYACRYLAMIDE GEL ELECTROPHORESIS. THE RADIOACTIVE dATP MARKS ONE END OF THE FRAGMENTS AND THEY CAN BE DETECTED IN THE GEL BY AUTORADIOGPAHY. THE RESULTS MAY BE AS DIAGRAMMED:

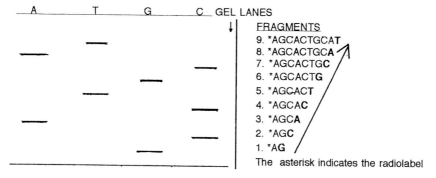

THUS THE **NUCLEOTIDE SEQUENCE** CAN BE READ FROM THE LAST BASE OF FRAGMENT NUMBER 1 TO FRAGMENT NUMBER 9: **G C A C T G C A T**

THE START OF THE RUN IS AT THE TOP AND THE LONGER IS A FRAGMENT THE SHORTER DISTANCE IT MOVES IN THE LECTRIC FIELD OF THE GEL

AN OUTLINE OF THE DIDEOXYNUCLEOTIDE SEQUENCING METHOD OF SANGER *et al.*

DNA SHUFFLING: *in vitro* homologous recombination in pools of randomly fragmented genes and reassembly by the polymerase chain reaction in order to test the consequences of the new sequences as evolutionary changes. (See localized mutagenesis, RCR-based mutagenesis).

DNA SILENCING: see silencer

DNA SPLICING: see splicing, splicing juncture

DNA STRUCTURE: see Watson and Crick model

DNA SUBTRACTION: see genomic subtraction

DNA SUPERCOILING: see supercoiling of DNA

DNA SYNTHESIS: see DNA replication, DNA chemical synthesis

DNA SYNTHESIS, MEIOSIS: during meiosis there is no DNA synthesis, except that may be needed for genetic repair. Before meiosis synthesis brings the level of DNA in the meiocytes to 4 C level. After the reductional division each cell has 2 C amount of DNA and at the end of meiosis there is 1 C amount in each of the four gametes. (See C amount DNA)

DNA THERMAL STABILITY: heat may disrupt the hydrogen bonds between the two strands of DNA. The stability depends on the base composition since G≡C are linked by 3 hydrogen bonds whereas A=T by only 2. Generally at temperatures exceeding 60° C denaturation may begin and upon prolonged exposure to 100° C it is completed. The resulting single strands have higher UV absorption at 260 nm wavelength then the duplexes (hyperchromicity). (See DNA denaturation, hyperchromicity)

DNA TOPOISOMERASES: are capable of nicking and reclosing single strands of (Type I) or causing scission in both strands and rejoining the cuts (Type II). These enzymes are ubiquitous from bacteria to eukaryotes and have roles in relaxing superhelical configuration, replication, transcription, recombination and repair. (See mentioned items under separate entries)

DNA TRANSPOSITION: see transposition

DNA TUMOR VIRUSES: may be integrated into the animal host DNA. The Papovavirus family consists of a variety of types and may be responsible for warts and carcinomas. The Hepadnaviruses (hepatitis-B) are responsible for liver carcinomas. The Herpes viruses (Epstein-Barr virus) may cause cancerous transformation of the lymphocytes (Burkitt's lymphoma, and nasopharyngeal cancer). Several RNA viruses belong to the retrovirus groups and may be trancribed into DNA and activate cancerous growth. (See mentioned terms under entries).

DNA TYPES: see in a table form below

Helix	bp/turn	Degree of rotation per base pairs	Rise/bp angström	Diameter angström	Conditions of existence
A	11	+32.7	2.9	23	75% rel. hum., Nam K or Ce
B	10	+36.0	3.38	19	92% rel. hum., low ions
C	9.7	38.6	3.34	19	66% rel. hum., Li ions
D	8		occurs only in guanine-free DNA		
E	7.5		occurs only in guanine-free DNA		
Z	12	-30.0	3.71	18	left-turned, alternating purines and pyrimidines, high salt

(bp = base pair, rel. hum.= relative humidity)

(See also H-DNA)

DNA TYPING: determination of the individual specificity of a DNA sample. (See DNA fingerprinting, DNA sequencing)

DNA, ULTRAVIOLET ABSORPTION: the maximal absorption is at about 260 nm but the maximum may be influenced by the solvent and pH. The maximal absorption of polycytidylic acid is between 270 and 280 nm whereas that of polyguanylic acid may vary between 250 - 290 nm, depending on the pH. Single-stranded molecules have increased absorption relative to double-stranded DNA. DNA isolated from plant or animal tissues generally have a 280/260 ≈ 2 absorption ratios. The OD_{260} of 1 corresponds to about 50 μg/mL double-stranded and 40 μg/mL single stranded DNA at about pH 8 in TE buffer. (See O.D., TE buffer, buffer, UV, cyclobutane dimer, physical mutagens)

DNA, UNIQUE: DNA sequences that occur only in single copies in the genetic material.

DNA VACCINES: see immunization genetic

DNAase: see DNase

Dnase: enzymes degrading DNA. (See restriction enzymes, DNA polymerases)

DNase-FREE RNase: may be prepared by different procedures: affinity chromatography on agarose 5'(4-aminophenylphosphoryl)-uridine-2'(3')-phosphate, by adsorption on Macaloid, by heating in the presence of iodoacetate. Commercially available preparations so labeled may not always be trustworthy. (See RNase-free DNase, DNase, RNase, Macaloid)

DNase HYPERSENSITIVE SITE (HSS): in DNA is very easily attacked by nucleases probably because they are unprotected by proteins (histones); their presence correlates with expression of adjacent genes. The generation of hypersensitive sites alters the nucleosomal structure of the chromatin. Hypersensitive sites are located in the vicinity of promoters or a promoter and enhancer. (See INI1, integrase, promoter, enhancer, nucleosome)

DnaJ: is a family of chaperone proteins. (See chaperones)

DNASIS: a computer program for DNA base sequence analysis.

DNA-PK (DNA-dependent protein kinase): has a role in transcription, replication, immunoglobulin gene switching and DNA repair. The sequence-specific enzyme works only when it is in cis-position to the site of phosphorylation. The Ku autoantigen, a transcription factor, attracts DNA-PK to specific DNA sequences. Ku binds to a specific negative regulatory element (NRE1) — in the long terminal repeat — and keeps in check inappropriate expression of the glucocorticoid-induced transcription of mouse mammary tumor virus (MMTV). Ku can bind also to other DNA sites although it displays preference to DNA ends and to the MMTV glucocorticoid receptor and octamer transcription factor 1 (Oct-1). DNA-PK phosphorylates *in vitro* glutathione S transferase-Oct-1 fusion protein and the glucocorticoid receptor. Phosphorylation by DNA-PK was contingent on the presence of the glucocorticoid receptor and Oct-1 binding sites. DNA-PK is required also for immunoglobulin (V[J]D) recombination. (See mouse mammary tumor virus, Oct-1, autoantigen, p350, Ku, immunoglobulins)

DNP: deoxyribonucleoprotein; also dinitrophenol.

dNTP: deoxyribonucleotide triphosphate of any of the nucleic acid bases.

Doc **ELEMENT**: see non-viral retrotransposable elements

DOCKING PROTEIN: a signal receptor protein in the cell membrane. (See signal transduction)

DODECYL SULFATE, SODIUM SALT (SDS): is an anionic detergent ($CH_3[CH_2]_{11}OSO_3Na$). It is used to separate membrane proteins from the lipid layers because the detergent replaces the lipids at the hydrophobic tract of the membrane protein. It is used also for SDS-acrylamide-gel electrophoresis where the unfolded proteins move according to their size rather than by charge. Synonym: lauryl sulfate Na salt. (See membrane proteins, electrophoresis)

DOG (*Canis familiaris*): 2n = 78; molecular data indicate its origin from wolves. (See wolf, fox)

DOG ROSE: see *Rosa canina*

DOGS (dioctadecylamidoglycylspermine): a lipopolyamine used for delivering exogenous DNA into cells. (See liposomes)

DOLPHINS: *Lissedelphis borealis* 2n = 44; *Orcinus orca* and other species are 2n = 44.

DOLYCEPHALY: the head being abnormally long.

DOMAIN: defined segments of proteins (with a characteristic tertiary structure) or cellular structures (chromosomes), e.g., DNA binding domain of transcription factors. (See also domains of antibodies)

DOMAINS OF ANTIBODIES: segments of light and heavy-chain polypeptides separable by chemical treatments. (See antibody)

DOMBROCK SYSTEM BLOOD (DO): about 64% of northern Europeans are Do(a^+) blood type; it is coded in the short arm of human chromosome 1. (See blood groups)

DOMESTICATION: taming and breeding animals under human supervision and for human use began in the neolithic (stone) age, about 10,000 years ago. Cultivation of plants started at about the same time.

DOMINANCE: property of an allele to be expressed in a heterozygote (merozygote) in the presence of any other allele at a gene locus. In the majority of species dominant mutations are much less common than the recessive ones. In humans, however, the known dominant mutations seem to exceed the recessives but many traits cannot be classified to either group. One must keep also in mind that the term dominance was conceived before biochemical characterization of gene expression became feasible. Usually, at a finer level of analysis both the "dominant" and the "recessive" alleles are expressed in the heterozygotes. In human or larger mammal pedigrees, because of the few offspring, it may be difficult to distinguish between dominant and recessive pattern of inheritance. The probabilities can, however, be analyzed statistically. Let us assume that 8 affected males have 6 affected female offspring and 5 unaffected male progeny. For non-dominant inheritance we would expect the probability to be $(0.5)^6 + (0.5)^5 = (0.5)^{11} \approx 0.00049$ (1/2048). One must consider also the problems of penetrance and environmental factors. (See also codominance, semidominance, dominance reversal, penetrance)

DOMINANCE, EVOLUTION OF: R.A Fisher suggested in the 1930s that dominance evolves through the acquisition of modifier mutations that convert gradually the original recessive mutations into dominant ones. The basis of this idea was that under feral conditions the majority of alleles are dominant while the majority of new mutations are recessive. Sewall Wright, J.B.S. Haldane and H.J. Muller hypothesized that dominance occurs by mutation as such and the rare advantageous dominant ones have the selective advantage of masking the deleterious recessive alleles in the populations. According to the molecular evidence that came to light since, the majority of the mutations (except the loss of genes) show codominance when their function can be determined with greater precision. The development of dominance has been attributed to the control of processes regulating metabolic pathways, to the "flux" by changes in activities of controled enzymes. (See dominance, co-dominance)

DOMINANCE HYPOTHESIS: attributes the superior value of hybrids (hybrid vigor, heterosis) to the accumulation of favorable dominant genes. (See hybrid vigor, overdominance, superdominance)

DOMINANCE REVERSAL: change of environment or developmental stage may alter and reverse the dominance - recessivity relationship of alleles. (See also epistasis, conditional lethal genes, temperature-sensitive mutation, dominance)

DOMINANCE THEORY: see dominance hypothesis

DOMINANCE VARIANCE: (in quantitative genetics) is due to the expression of dominant alleles. (See genetic variances, additive genes)

DOMINANT ALLELE: is expressed fully in a diploid (polyploid) even in the presence of recessive allele(s). (See dominance)

DOMINANT LETHAL ASSAYS IN GENETIC TOXICOLOGY: generally rodents are treated with mutagens during various stages of spermatogenesis. Subsequently they are mated with untreated females. After about two weeks of pregnancy the females are sacrificed and the number of lethal implantations are classified and counted. (See bioassays in genetic toxicology, implantation, autosomal dominant mutation)

DOMINANT NEGATIVE: the inhibition of the activity of a normal (wild type) protein when its subunits are mixed with mutant polypeptide subunits which alter the conformation and thus render it inactive. (See also conformation correction, allelic complementation)

DONKEY: *Equus asinus* 2n = 62.

DONOHUE SYNDROME (leprechaunism): is an autosomal defect in the insulin receptor. (See insulin, insulin receptor protein, dwarfism)

DONOR: provides genetic material (F+) to the recipient (F-) by bacterial conjugation or any other means of genetic transfer (including eukaryotes). Blood type O that is acceptable to any

Donor continued
individual with other alleles of the ABO blood group or other antigenic substances compatible to other serotype is also called a (universal) donor. (See conjugation, ABO blood group, serotype)

DONOR SITE: the original position in the DNA (map) of a non-replicative transposable element from where it may move to another position within the genome, to the *recipient site*. (See transposable elements)

DONOR SPLICING SITE: see introns

DOPA: 3,4-dihydroxyphenylalanine, an intermediate in melanin synthesis; the derivative dopamine is a neurotransmitter. (See neurotransmitter)

DOPAMINE: a catecholamine formed by decarboxylation of dopa and it is a precursor of epinephrine, norepinephrine and melanine. (See dopa, animal hormones)

DOPING NUCLEOTIDES: a short nucleotide string is synthesized where some codons in the first strand have any of the four natural nucleotides and at the third position they have either G or C. In the second strand at the positions corresponding to the above named codons there will be inosines. The rest of the codons correspond to the usual amino acids. When such an insert is added to a vector, a variety of random mutations may occur in the protein domain coded by the sequence and yields a library of proteins with different amino acids at critical regions. (See directed mutation, protein engineering, hypoxanthine [for inosine])

DORMANCY: a state of low metabolic activity.

DORSAL: relating to back position of a body or upper surface of a structure or organ.

DORSAL LIP: see blastopore

DORSALIZATION: the formation of dorsal elements is preferentially enhanced at the expense of ventral development during morphogenesis and embryo development. (See morphogenesis)

DOSAGE COMPENSATION: a single dose of a gene has the same phenotypic effect as two or more. Examples for this abound in X-chromosome-linked genes where the phenotype for such genes is practically identical in the XY males and XX females or in the WZ females and ZZ males or XO males and XX females. Such dosage compensation may occur with various doses of the X-chromosome and to some extent it may occur in various aneuploids. MSL and MLE proteins appear to be involved in dosage compensation in *Drosophila*. *Sex lethal* (*sxl*, 1-19.2, with many different alleles), controls both sexual dimorphism and dosage compensation through its 39 kDa protein product and promotes alternative splicing of *msl-2* RNA transcripts. Females with X:A (sex chromosome:autosome) ratio of 1 synthesize this protein and permits the transcription of both X-chromosomes. Males of X:A of 0.5 do not make the SXL protein and increase the expression of the majority of the X-linked genes. SXL permits the normal transcription of *Sxl* and *tra* in the females but in males it inserts stop codons into these two transcripts. There are binding sites for SXL in the 3' untranslated region (UTR) in the *msl-1* transcripts and these may cause down-regulation in the females. There are both 5' and 3' UTR binding sites in the *msl-2* transcripts in the females and thus other potentials for regulation. Msl-2 seems to be the main target and instrument of X-regulation. The SXL protein is suspected to have another function, down-regulating some X-chromosomal genes in the female flies. In the hermaphroditic XX females, X-chromosomal genes are expressed at about the same level as in the normal XO males. Some of the (*sdc*) genes affect both sex determination and dosage compensation. Other genes (dumpy series [*dpy*]) affect only dosage compensation. The DPY-27 protein is associated with the X-chromosomes in the hermaphrodites of *Caenorhabditis* but not in the XO males. DPY-27 like proteins are present in other organisms too and control chromosome condensation and segregation. The DPY-26 locus affects Dpy-27 and Dpy-30 proteins for dosage compensation and Sdc2 and Sdc3 for the coordination of dosage compensation and sex determination. SDC-1 is a 139 kDa protein with Zn fingers. This family of proteins has some structural features characteristic for motor proteins. SDC-3 regulates sex determination by its ATP-binding domain and a Zn finger-like domain seems to affect dosage compensation. Actually, both series are cooperating in reducing the level of X-chromosomal gene expression in the hermaphrodites. (See sex determination, lyonization, *Msl, Mle,* splicing, introns, Zn finger, motor protein, neo-X-chromosome, MSL)

DOSAGE EFFECT: the number of alleles of a certain type determines the degree of expression. (See gene titration, quantitative gene number, quantitative trait, mapping by dosage effect)

DOSE FRACTIONATION: the irradiation is delivered not in a single burst but there are intervals between the doses. This may permit intermittent repair. (See radiation effects)

DOSE RATE EFFECTS: usually at low dosage the correlation between the frequency of mutations and mutagen (ionizing radiation) is linear but at higher doses the response curve follows second or higher order kinetics because of the multiplicity of effects (hits) on the genetic material. (See radiation effects, kinetics)

DOSIMETER FILM (badge): contains photographic emulsion that can detect β, γ and X-radiation by blackening when developed. The films are used within the range of 20 keV for X rays and 200 keV for β radiation. (See radiation measurements)

DOSIMETER, POCKET: contains an ion chamber and detects ionizations due to X or γ radiation. The direct reading types are held against light to make a reading against a precharged value. Its useful range is generally 0 to 200 mR. Sometimes they may display false readings. (See radiation measurement)

DOT: Department of Transportation (USA) regulates shipment of certain substances. (See mutagens, carcinogens, environmental mutagens)

DOT BLOT: see colony hybridization

DOTTED: see *Dt* gene

DOUBLE BRIDGE: is visible at anaphase I if four-strand recombination occurs in paracentric inversion heterozygotes or in the case of sister-chromatid exchange in ring chromosomes. (See bridge, inversion)

DOUBLE CORTEX: human Xq21.3-q24 brain defect of the thinner cortex and disorganized neurons resulting in mental retardation and seizures. (See mental retardation, seizures)

DOUBLE CROSS HYBRIDS: were extensively used for the commercial production of hybrid corn. The seed companies produced the seed for the farmers according to the scheme:
 single cross (1)   [A x B] x [C x D]   single cross (2)
 DOUBLE CROSS SEED ↵
used by the farmers to produce corn for food, feed or industrial raw material. Currently, most of the corn hybrids are single-crosses. (See hybrid vigor, heterosis, cytoplasmic male sterility)

DOUBLE CROSSING OVER: a single recombination changes the synteny of linked markers:
 from   A———B   to   A———b
        a X b   a———B
whereas double recombination involving the same strand restores the original synteny of the two genes: A———B   to   A———B
        a X X b   a———b

The frequency of double crossing over is expected to equal the product of the frequency of the single crossing overs. 2-strand double crossing over within an inverted segment generally has no harmful effect in inversion heterozygotes and two normal and two inverted chromosomes are recovered in the gametes. 3-strand double crossing over within the inversion produces 1 acentric, 1 dicentric, 1 functional recombinant chromosome with a piece of the inverted segment and 1 functional (inverted or non-inverted chromosomes). 4-strand double crossing within a paracentric inversion heterozygote leads to the formation of double bridges (two dicentric chromosomes) and two acentric fragments and most likely no viable gametes. In pericentric inversions bridges do not occur but the gametes receiving the exchange strands (except the two strand double cross overs) will be duplicated or deficient and most likely non-viable. Double crossing over in translocation heterozygotes, depending on their site, may damage most of the gametes and reduces the fertility below 50%. (See crossing over, inversion)

DOUBLE EXCHANGE: see double crossing over

DOUBLE FERTILIZATION: occurs in plants when one of the generative nucleus fuses with

Double fertilization continued

the haploid egg and gives rise to the zygote whereas the other fuses with the two polar nuclei and initiates the development of the triploid endosperm. (See gametophyte, embryogenesis in plants, heterofertilization, semigamy, apomixis, polyembryony, adventive embryony, parthenocarpy, xenia, metaxenia)

DOUBLE HELIX: two deoxyribonucleotide chains formed by phosphodiester linkages are joined in a helical arrangement through hydrogen bonding between bases as determined essentially by the Watson and Crick model. The two DNA strands form plectonemic coils (wound around each other) that can be partially and locally released during replication by the helicase and topoisomerase enzymes. (See DNA types, Watson and Crick model, helicase, topoisomerase)

DOUBLE INFECTION: a bacterial cell may be infected with two different genotypes of the same or compatible bacteriophages and this may provide an opportunity for phage recombination. (See multiplicity of infection)

DOUBLE LYSOGENIC: a bacterium carrying a normal lambda phage, side-by-side with a λdg. This lambda, defective-galactose phage lost a piece of its genetic material but acquired the bacterial gal^+ gene and thus is capable of specialized transduction if another phage compensates for its defect. When such a phage is induced (to switch from prophage to vegetative phage) it produces a *high frequency lysate* and high *specialized transducing* ability. (See transduction)

DOUBLE MINUTES (DMs): extrachromosomally amplified chromatin containing a particular acentric chromosomal segment (gene). They occur frequently in cancer cells. (See cancer, YAC)

DOUBLE NEGATIVE CELL: lymphocytes in the thymus without CD4 or CD8 proteins.

DOUBLE POSITIVE CELL: expresses both CD4 and CD8 proteins but may die within the thymus. (See T cell, lymphocytes)

DOUBLE RECIPROCAL PLOT: see Lineweaver-Burk plot

DOUBLE RECOMBINATION: two recombination in between two stands (chromatids, DNA) within a determined interval (see double crossing over). It is genetically detectable only if there are multiple markers in the chromosome.

DOUBLE REDUCTION: is chromatid segregation which may occur when in a trisomic individual recombination takes place between a gene and the corresponding centromere. Consequently in a duplex, double recessive gametes are produced and e.g. an *AAa* individual produces *aa* gametes. (See trisomic analysis, autopolyploid, centromere mapping in higher eukaryotes)

DOUBLEISOTRISOMIC: in **wheat**, 20"+(i+i)1''', 2n=43, ["=disomic, '''=trisomic, i=isochromosome]. (See disomic, trisomic, isochromosome)

DOUBLEMONOISOSOMIC: in **wheat**, 20"+i'+i', (2n=42), ["=disomic, '=monosomic, i=isosomic]. (See disomic, monosomic, isosomic)

DOUBLEMONOTELOSOMIC: in **wheat**, 20"+t'+t', 2n=42, ["=disomic, '=monosomic, t=telochromosome. (See disomic, monosmic, telosome)

DOUBLE-STRAND-BREAK REPAIR MODEL of recombination: see Szostak model

DOUBLET: a double band in the salivary gland chromosome or a double band in an electrophoretic gel. Also, *Paramecia* that share a common endoplasm and a single macronucleus (See salivary gland chromosomes, gel electrophoresis, *Paramecium*)

DOUBLETELOTRISOMIC: in **wheat**, 20"+(t+t)+1''', 2n=43, ["=disomic, t=telochromosome, '''=trisomic]. (See disomic, telosome, trisomic)

DOUBLING DOSE: is the amount of a mutagen that doubles the spontaneous rate of mutation. Since mutation rate depends on the organisms, its genotype, developmental stage and environmental factors, the estimates arrived at by different investigators may not be identical. From mouse experiments it had been estimated that 1 R chronic ionizing radiation produces a mutation rate of 2.5×10^{-8} per locus per generation and the spontaneous rate was considered to be 1×10^{-6}, hence $(1 \times 10^{-6})/2.5 \times 10^{-8} = 40$ R was assumed to be the doubling dose of ionizing radiation for humans too. Other estimates considered the doubling dose for recessive

Doubling dose continued

mutations 32 R and for dominant ones 20 R. More recently the doubling dose for radiation has been estimated as 100 cGray (1 Rad). The doubling dose may have important meaning for the estimation of genetic radiation risks. Assuming that the average human dominant mutation rate is 10^{-5} to 10^{-6}, the chances for the occurrence of a human dominant mutations at 100 R dose (1 R = ca. 93 erg/wet tissue) exposure may be 32×10^{-5} to 32×10^{-6}, i.e. 1/3,125 to 1/31,250. Actually, these estimates may not be very accurate but direct data cannot be readily obtained in human populations. Another problem is that the mutant gametes may not compete successfully and the absolute genetic damage may be much larger than these estimates. An additional problem is that in human populations the mating is random (or almost random) therefore homozygous recessives may not appear most of the time unless the marriages are consanguineous. Atomic radiations in Hiroshima and Nagasaki substantially increased the incidence of cancer and teratogenesis but an increase in human mutations was not clearly detectable except by altered sex ratio. The Chernobyl atomic power plant accident in 1986 has shown, however genetic effects and increase in cancer caused by atomic radiation. A medical X-ray machine may deliver 0.04 to 1.0 rem (100 rem = 1 Sievert [Sv]) to the organ or structure routinely examined. Although the gonads are generally shielded during these examinations, some general risk remains, particularly for carcinogenic effects. (See radiation effects, mutation in human populations, radiation hazard assessment, relative mutation risk, coefficient of inbreeding, atomic radiation, teratogenesis, RBE, Rem, Rad, Sievert)

DOUBLING TIME: is the time required for the completion of a cell division, measured in minutes, hours or days for specific cell types under defined conditions. (See cell cycle)

DOUGLAS FIR (*Pseudotsuga menziesii*): is primarily a forest timber tree with 2n = 2x = 26.

DOWEX RESINS ®: are ion exchangers and used also for gel filtration and chromatography. (See ion exchange resins, gel filtration, chromatography)

DOWN PROMOTER: slows down the rate of transcription.

DOWN'S SYNDROME: is caused by primary trisomy or translocation of human chromosome 21 or 21q21-22.3. The incidence of this condition varies from less than 1/1,000 at maternal age around 20 years to close to 100/1,000 live birth when the mother approaches menopause. Trisomy 21 may affect only part of the body in a mosaic pattern. The transmission of disomic gametes is minimal through the males. The fertility of the afflicted persons is reduced (especially males) and the transmission risk by afflicted females is less than 50%. The most important characteristics involve a relaxation of the muscles controlling the eyefold, protruding tongue, flat face, shorter than expected height, two third displays the simian fold in the palm (see fig.), generally (but not always) reduced mental abilities (IQ 25 to 50 although with proper training they can learn to read and write and enjoy life and usually are quite sociable and affectionate. This trisomy is associated with a series of other anomalies such as heart disease, susceptibility to leukemia. Older literature calls this condition by the unfortunate term of mongoloid idiocy. Prenatal cytological examination of fetal cells in the amniotic fluid and sampling chorionic villi may reveal if the fetus has this condition. Unfortunately, amniotic cells may display anomalies that do not occur in the normal cells of the fetus. There is a correlation between the low level of maternal α-fetoprotein (MSAFP) and fetal Down syndrome. The interpret-

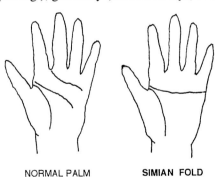

NORMAL PALM   SIMIAN FOLD

In a normal individual the palm creases do not go all the way from one side to the other (left) In Down's syndrome and some other congenital disorders the typical Simian fold is present. In addition, the hands may be shorter and the little finger crooked.

tation of the clinical analysis is further sharpened if unconjugated estriol (α-hydroxylation pro-

Down syndrome continued

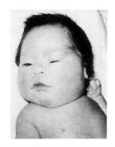
DOWN'S SYNDROME INFANT.
Courtesy of Dr. Judith Miles.

duct of estradiol) and gonadotropin levels are also measured. Ultrasonic analysis may also indicate Down syndrome prenatally if heavy skinfold on the neck, excessive fluid accumulation, narrow small intestines or short bones are indicated. The recurrence risk of trisomy is very low unless the family has a history of non-disjunctions and heterozygous for translocations involving chromosome 21. In the expression of the bone characteristics of this trisomy, the ETS oncogene and transcription factor (human chromosome 21) may play the major role.

The brain of DS individuals has fewer neurons and abnormal neuron differentiation. Also the neurons become apoptic earlier and predispose them to Alzheimer syndrome symptoms. These defects have been attributed to elevated levels of lipid peroxidation. The defects in the nervous system is attributed also to DYRK, a tyrosine phosphate regulated kinase. (See nondisjunction, trisomy, trisomic analysis, MSAFP, I.Q., amnion, estradiol, fetoprotein, translocation, prenatal diagnosis, amniocentesis, Alzheimer disease, apoptosis)

DOWNSTREAM: in the direction of the 3' terminus of the polynucleotide; the nucleotide chain grows downstream because the nucleotides are added to the 3' OH end of the preceding one.

DP1: is a transcription factor activated by MDM2. (See MDM2, E2F1)

DPA (DNA pairing activity): protein (120 kDa) promotes the formation of heteroduplexes in yeast . (See recombination, mechanisms, eukaryotes)

DPC4 (deleted in pancreatic cancer at human chromosome 18q21.1): is a gene missing in about 50% of the pancreatic cancer cells. It is homologous to the *Drosophila mad* (*many abnormal* [*imaginal*] *discs*, 3-78.6) gene, member of the *dpp* (*decapentaplegic*) family, encoding a transforming signal related to TGF-β. (See morphogenesis in *Drosophila*, tumor suppressor genes, TGF)

dpc (days post coitum): the time after mating has taken place.

dpm (disintegration per minute): is the number of disintegrations per 1 g radioactive radium = 1 μCi = 2.2×10^6 dpm but usually about half as many counts per minute (cpm) are shown in the equipments that generally work at about 50% efficiency.

DPN (diphosphopyridine nucleotide): current name NAD (nicotinamide adenine dinucletide).

dpp (decapentaplegic): see morphogenesis in *Drosophila* {17}

DPT: see developmental therapeutics program.

DQ ANTIGEN: is encoded by the DQ segment of the HLA complex. (See HLA)

DR ANTIGEN: is encoded by the DR alleles of HLA: (See HLA)

DRADA: a ds-RNA-dependent adenosine deaminase; an RNA editing enzyme in mammals; the same as dsRAD. (See RNA editing, dsRNA)

Dras: *Drosophila* homolog of *ras*. (see RAS oncogene)

DRES: *Drosophila*-related expressed sequences.

DRIFT GENETIC: is a change in gene frequencies caused by a sampling error in the gametic array of random mating populations. Such a change, due to chance, is most likely when the effective population size (the number of breeding individuals) is small. Genetic drift may occur repeatedly in a population and can cause substantial changes in the frequency of several genes. The change does not reflect the adaptive value of the alleles involved. (See effective population size, founder principle, genetic drift)

DRK: a downstream receptor kinase of the *Drosophila* light signal transduction pathway. Its vertebrate homolog of GRB2 and it is equipped with SH2 and SH3 domains and it is frequently called as a mediator protein in signal transduction. (See GRB2, SH2, SH3, signal transduction)

DROSOPHILA : species are dipteran flies. Genetically most thoroughly studied is the cosmolitan

Drosophila continued

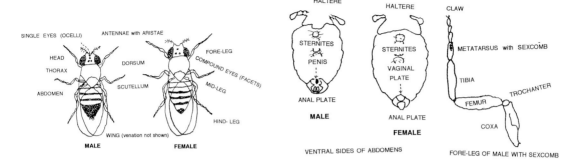

The major areas of the body of *Drosophila melanogaster*. Surface appendages (bristles, hairs) and wing venation are not shown here. (See more at figures in morphogenesis in *Drosophila*).

Sexing of the flies is by inspection of the sexcomb (present only on the metatarsus of the fore-leg of the male and the genitalia, visible on the posterior ventral part of the abdomen. For examination, the flies are anesthetized for 5-10 min with ether or other suitable fumes and placed under the dissecting microscope.

D. melanogaster, (n = 4). The species are reproductively isolated, i.e. the species may intermate if the opposite sex from the same species is not available but the F_1 offspring is sterile. The sterile eggs may be recognized by the lack of filaments (see diagrammatic life cycle). The size of the eggs is about 0.5 mm. Within less than a day it hatches into the first instar stage larva. (The instar is a larval growth stage in between moltings.) The larvae shed their cuticle by the process of molting. Three instar stages are distinguished. The larvae are very voracious feeders and reach a size of about 4.5 mm. The cuticle of the third instar darkens and hardens and becomes the puparium 4 days after hatching and in another 4 days the metamorphosis is completed and from the larva the imago (2 mm) emerges. (The imago is a fully differentiated adult form.) This adult type emerges by the process of eclosion through the anterior (fore) end of the pupa. Within a day its color darkens, the wings expand and the abdomen becomes rotund. There is no further growth after emergence. Within two days the imagos may be copulating after a mating courtship and may start laying eggs. The adult males live for about 33 days while the females die on the average after 26 days. See diagram of the life cycle and germline of *D. melanogaster*. Sexing (determining the gender) of *Drosophila* can be made by looking at the metatarsal segment of the fore leg that carries a structure called *sex comb* in the male only (see diagram above). Sexing is possible also on the basis of the genitalia viewed from the ventral side of the abdomen. The abdominal segmentation is also different in the two sexes. After copulation the female stores the sperm in the spermatheca (located above the uterus) and in the uterus. Once the female had been mated it may keep the sperm for a long time, therefore for each genetically controled mating generally virgin females are used. The eggs are fertilized just before laying. *Drosophila melanogaster* XX (female) or XY, XO (male) sex chromosomes and three pairs of autosomes. The X-chromosome (chromosome 1) is acrocentric (has one arm) has a genetic length of about 73 map units; chromosomes 2, 3, and 4 have genetic lengths of about 110, 111 and 3 map units, respectively. The Y chromosome has the *KL* male fertility complex on the long arm and the *KS* male fertility complex on the short arm and no other "visible" genes. The X-chromosome contains the nucleolar organizer region in a tandem array of repeats at the *bb* (*bobbed*) locus, coding for the 5.8S, 18S and the 28S ribosomal RNAs. The polytenic X-chromosome is divided into 20 regions; 1 at the tip and 20 at the centromere. It displays about 1000 bands. The Y chromosome does not display

Drosophila continued

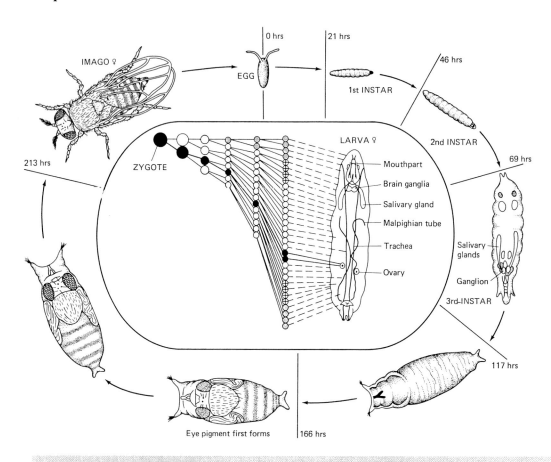

THE LIFE CYCLE OF *DROSOPHILA*. More than a hundred filamentous eggs (0.5 mm) are deposited by the female. The hatching maggots (instars) burrow into the food where they undergo two molts. The emerging larvae develop into pupae. The adults (imago) are about 2 mm long when they eclose from the pupal case. The average lifetime of the females is about 4 weeks; the males may live a little longer. The organs of the adults develop from the imaginal discs, present at the 1st instar stage. In the center (insert) the development of the germline is shown.

polyteny. The second chromosome contains 20 sections in the left arm (21 at the tip and 40th at the centromere) including about 869 to 927 bands and the right arm starts at the centromere with section 41 and ends with section 60 at the other telomere including 1009 to 1152 bands depending on the classification of doublets and singlets. Chromosome 3 appears the longest. The left arm begins with section 61 at the tip and section 80 at the centromere including 884 to 1032 bands. The right arm starts with section 81 at the centromere through section 100 at the other end, containing about 1147 to 1233 bands depending on the classification. Chromosome 4 is extremely short and does not display very clear polyteny and does not contain more than about 40 to 50 bands. It is divided into sections 101 and 102, proximally and distally to the centromere, respectively. Each section is further subdivided into A, B, C, D, E, F subsections by vertical lines positioned left to conspicuous bands. Within lettered subdivisions individual bands are numbered. The size of the nuclear genome is ca. 8×10^7 bp and the circular mito-

Drosophila continued

chondrial genome contains 19,500 bp. The origin of replication of the mtDNA is within an AT rich tract. The replication of the first strand is completed before the replication of the 2nd strand begins. It codes for ribosomal RNA and 22 tRNAs. It encodes also cytochrome b, cytochrome c oxidase, two subunits of ATPase and for 7 subunits of the NADH reductase complex. These genes are crowded next to each other with very few nucleotides keeping them apart. They contain no introns. Translation is initiated with ATA, ATT and ATG codons. TAA is the most common termination codon. Codon TGA spells Try rather than a stop like in the universal codon dictionary. ATA means Met rather than Ile, and AGA is a Ser codon rather than Arg. Codon usage is influenced by the predominance of AT pairs in this genome. "Infectious heredity" is also known in *Drosophilas,* caused by the (RNA) picornaviruses (DAV, DPV, DCV), and the vesicular stomatitis virus-like agent (sigma virus), and the sex-ratio (*SR*) condition (females do not produce male offspring) caused by spirochaete-like protozoa. In *Drosophila melanogaster* more than 4,000 genes and 9,000 chromosomal rearrangements are more or less well characterized. *Alleles* at a locus are generally distinguished by superscripts. Recessive alleles in a predominantly dominant allelic series are designated by capital letters but a superscript *r* is added, alternatively in a recessive series of alleles the dominant allele may be identified with lower case letters but with a *D* superscript their dominant expression is identified. The wild type alleles may be designated also by a (+) superscript, and an *rv* superscript and a number indicates revertants of recessive alleles.

Revertants of dominant mutations are considered deficiencies and are symbolized as *Df* with a distinguishing number, e.g *Df (1)1-D1* is a deficiency of chromosome 1 (X-chromosome) and the latter part of the symbol is explained by words. Alleles specifying the absence of a protein are symbolized with an *n* superscript and possibly by a number. *Proteins* are frequently designated by three Roman capital letters. Loci moved to new locations by transposable elements are enclosed in brackets followed by the new site in parenthesis, e.g. [*ry*$^+$] (*sd*) or [*Cp16*] (*52D*) indicates that the chorion protein gene was inserted by transformation into cytological site *52D*. *Mimics* (different loci but similar phenotype) are designated as e.g. *tu-1a, tu1b, tu-2* or by arbitrary numbers: *Sgs3, Sgs7* or by polytenic location *Act5C, Act42A* or by molecular weight *Hsp68, Hsp70* added to the letter symbols. *Modifier genes* such as suppressors or enhancers are symbolized with *e* or *E* and *Su* or *su* followed by the symbol of the gene acted upon in parenthesis, e.g. *su(lz^{34})*, suppressor of a *lozenge* allele. *Translocations* are symbolized with *T*, e.g., *T(1;Y;3)* indicates an X- and Y-chromosome interchange that may be followed by a number of other symbols. Chromosome rings are indicated as *R(1)2* where *R* stands for ring, *(1)* for the X-chromosome and *2* for ring #2. *Inversions* are identified by *In* and additional specifications, e.g. *In(2L)Cy* indicates an inversion in the left arm of chromosome 2 involving the dominant wing mutation *Curly* or *In(3RL)* stands for a pericentric inversion of chromosome 3. Additional specifications may also be applied. *Transpositions* are defined as three-break rearrangements (two are needed for excision and one at the target site). *Tp(3;1)ry* designates the movement of a third chromosomal *rosy* allele to the X-chromosome. *Duplications* are identified by *Dp* and additional information, e.g. *Dp(1;1)y^{bl}* designates a duplication in the X-chromosome containing the *yellow bristle* marker. When the duplication is a free fragment the designation is *Dp(1;f)* and it may further be specified by gene or band symbols added. *Dp(1;1;1)* indicates triplication. *Deficiencies* are defined by *Df*, irrespective whether a chromosomal segment or an entire chromosome (hypodiploid, 2n - 1) is involved, e.g., *Df(2R)vg* is a deficiency of the *vestigial* gene in the right arm of chromosome 2. Complex chromosomal rearrangements can also be symbolized but generally they require detailed descriptions. Deviations from the euploid, normal chromosomes change the phenotypic sex because sex in *Drosophila* is affected by the balance of the sex chromosomes and autosomes. The metafemales have the chromosomal constitution X/X/X;2/2;3/3;4/4, the triploid metafemales are X/X/X/X;2/2/2;3/3/3;4/4/4 (or 4/4 only). Metamales are X/Y;2/2/2;3/3/3;4/4/4. Inter-

***Drosophila* continued**
sexes may be X/X;2/2/2;3/3/3;4/4/4 with a Y chromosome either present or absent and the numbers of chromosome 4 may vary. Tetraploid females have been observed. Triploids are X/X/X;2/2/2;3/3/3;4/4/4 (or the sex chromosomes may be either (X/X/X/Y or attached XX/X or XX/X/Y). Haploids of X;2;3;4 are known. Aneuploids nullo-X (Y/Y;2/2,3/3;4/4), nullo-X nullo-Y (0/0;2/2;3/3;4/4), tetra-four (X/X;2/2;3/3;4/4/4/4), triplo-4 (X/X;2/2;3/3;4/4/4), haplo 4 (X/X;2/2;3/3;4), X0 male (X;2/2;3/3;4/4), XXY female (X/X/Y;2/2;3/3;4/4), XXYY female (X/X/Y/Y;2/2;3/3;4/4), XYY male (X/Y/Y;2/2;3/3;4/4) and XYYY male (either X/Y/Y/Y or attached XY/Y/Y plus normal autosomes) are also known. About 20% of the *Drosophila* genome consists of repeated sequences and display four satellite bands (1.672, 1.686, 1.688, 1.705) upon CsCl density gradient centrifugation [not amplified in the salivary glands]). Other repeated sequences, SINE (0.5 kbp) and LINE (5 - 7 kbp, include retroposons) are also part of the genome. *Drosophila* species may contain one or more of the over 30 different transposable elements. Detailed information on 4,000 gene loci and 9,000 chromosomal aberration (including references) are described in THE GENOME OF DROSOPHILA by D.L. Lindsley and G.G. Zimm, 1992, Acad. Press. San Diego, CA. (See salivary gland chromosomes, polytenic chromosomes, fruit fly, morphogenesis, imaginal disk, reproductive isolation, instar, molting, puparium, metamorphosis, eclosion, courtship, virgin, nucleolar organizer, infectious heredity, allele, suppressor, translocation, inversion, duplication, deficiency, triploid, aneuploid, tetraploid, SINE, LINE, <http://www.morgan.harvard.edu/> or gopher <flybase.bio.indiana.edu> or <FlyView:<http://www.pbio07unimuenster.de/html/About.html>.

DROSOPTERIN: is a bright red eye pigment in insects (*Drosophila*) and related pigments are common in other animals. The *se* (*sepia*) mutants of *Drosophila* (chromosome (3-26.0) accumulate a yellow pigment (a dihydropteridine) but unable to synthesize drosopterin or isodrosopterin because of the defect in PDA (2-amino-4-xo-6-acetyl-7,8-didydro-3H,9H pyrimido[4,5,6]-[1,4] diazepin) synthetase. (See pigmentation of animals)

DRPLA: see dentatorubral-pallidoluysan atrophy

DRUG DEVELOPMENT: the *Drosophila* hybrid dysgenesis factor, the P element is being explored to study transgene responses to drugs. (See hybrid dysgenesis, SAR by NMR, see also developmental therapeutics program)

DRUG RESISTANCE: see resistance transfer factors

DRUM STICK: see Barr body

Ds: (*Dissociator*): a defective transposable element of maize that can move only when *Ac* provides the transposase function. It name came from the observation that it was frequently associated with chromosome breakage. (See *Ac*, controlling elements, transposable elements)

DSB: double-strand break.

dsDNA: double-stranded DNA.

DSE: distal sequence element. (See Hogness box)

DSP (dual specificity protein phosphatase): one of its active sites dephosporylates serine, threonine, tyrosine in a protein whereas the deeper active site is specific only for tyrosine.

dsRAD: see DRAD

dsRNA: double-stranded RNA.

DSS (dosage-sensitive sex reversal): an approximately 160 kb duplication in the short arm of the X-chromosome upsets normal male gonad formation in XY individuals. Deletion of the same region, however does not affect gonadal differentiation. (See sex reversal, sex determination)

DST1: yeast gene encodes the 38 kDa strand transfer protein (STPα). (See recombination mechanism, eukaryotes)

DST2: yeast gene encodes the STPβ protein (identical to Sep 1). (See recombination mechanism eukaryotes)

dsx (*double sex*): abnormal location 3-48.1; regulates sexual differentiation in somatic cells of *Drosophila melanogaster*. The null allele converts males and females into intersexes. (See *Drosophila*, sex determination)

***Dt* GENE**: is actually a transposable element that may be situated in several locations in the maize genome. It causes the colorless *a1-m* allele to produce anthocyanin containing dots in the triploid aleurone tissue, depending on the dosage of the *Dt* elements. Also it may cause reversions to the *A* allele in the germline. *Dt*, like other transposable elements, induces also chromosome breakage and can newly arise through chromosomal breakage. The first *Dt* allele, *Dt1* was located to the initial position of the short arm of chromosome 9. *Dt2* (chromosome 6L-44), *Dt3* (7L), *Dt4* (4), *Dt5* (in the vicinity of *Dt1*) and *Dt6* (in the short arm of the chromosome 4) not far from the centromere were identified subsequently. (See transposable elements)

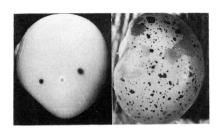

MAIZE KERNELS (LEFT) *aaa* and (RIGHT) *aaa^m Dtdtdt*).
(Courtesy of M.G. Neuffer, Genetics 46:625)

dTAF$_{II}$: see transcription factors

DTH (delayed types hypersensitivity): may cause of acute rejection of allografts and autoimmune reactions due to non-specific T cell responses. (See autoimmune, T cell)

DUBIN-JOHNSON SYNDROME (hyperbilirubinemia II): is an autosomal recessive disease with a relatively high frequency (8×10^{-4}) among Iranian Jews. It involves jaundice, hyperbilirubinemia, melanin-like deposits in the liver. It is a defect of the detoxification mechanism of the liver. Detoxification is mediated by conjugation with glutathione, glucuronides or sulfates, resulting in negatively charged amphiphilic compounds that can effectively be secreted into the urine. The excretion is carried out with the assistance of canalicular (tubular) multispecific organic anion transporter (cMOAT). Hyperbilirubinemia is caused also by the Gilbert syndrome (synonymous with Arias type hyperbilirubinemia and the Crigler-Najjar syndrome type II defects in the UDP-glucuronosyl transferase) in human chromosome 2 and deficiency of Factor VII in human chromosome 13q34. (See hyperbilirubinemia, glutathione-S-transferases, antihemophilic factors, Crigler-Najjar syndrome)

DUBININ EFFECT: change of dominance in alleles (variegation) because of transposition (of heterochromatin) in *Drosophila*. (See position effect, heterochromatin, dominance reversal)

DUBOWITZ SYNDROME: is a fetal-newborn physical and mental retardation complex under autosomal recessive control. The symptoms may resemble to those in fetal alcohol syndrome (non-genetic) and the autosomal recessive Fanconi anemia. (See alcoholism, Fanconi's anemia)

DUCH (DH blood group): is extremely rare; it has been identified in Aarhus, Denmark. (See blood groups)

DUCHENNE MUSCULAR DYSTROPHY: see muscular dystrophy

DUCK (*Anas platyrhynchos*): 2n = 78-80.

DUE (DNA unwinding element): a part of the autonomously replicating sequence (ARS). (See ARS)

DUFFY BLOOD GROUP, (*Fy*) ALLELES: in human chromosome 1q12-q21 control a red blood cell antigen, frequently used as genetic markers in population studies; 99% of the Chinese and 65% of the Caucasians are positive, whereas 95-99% of the black Africans are of the negative type (*Fy-/Fy-*). Individuals lacking these antigens [Fy (a⁻ b⁻)] are protected against the African malaria-causing *Plasmodium vivax* protozoon. (See blood groups)

'DUMPOSOME': protein Pdd1 of *Tetrahymena* that may be targeted to DNA degradation within cells of organisms. (See also apoptosis)

DUPLEX: in a polyploid (e.g. *AAaa*) or triploid or trisomic (e.g. *AAa*) carries two dominant alleles at a locus and the other allele(s) at the same locus are recessive. (See autopolyploid)

DUPLEX DNA: is double-stranded. (See Watson - Crick model)

DUPLICATE GENES: convey the same (or very similar phenotype) but segregate independently therefore in F_2 the phenotypic proportions are 15 dominant : 1 recessive. (See modified Mendelian ratios)

DUPLICATE RATIO: see duplicate genes

DUPLICATION: a eukaryotic chromosomal segment is repeated side-by-side (tandem duplication) or may be repeated at another location within the chromosome or in another chromosome. The duplication is cytologically detectable in duplication heterozygotes if it is of sufficient length. In prophase when paired with the non-duplicated strand it bulges out because the normal counterpart cannot pair with the extra chromosomal tract:

------------------------------ normal chromosome

------------ ∪ --------------- chromosome with duplication

The duplication may be direct →→ or inverted →← . After insertion of transposable elements duplications may occur at the target sites. Transposable elements generally carry in addition terminal direct or inverted repeats. Duplications and deficiencies are generated by crossing over within the inverted segments of para- and pericentric inversions, breakage-fusion-bridge cycles, crossing over between sister chromatids of ring chromosomes, unequal crossing over, adjacent I and adjacent II distribution of chromosomes in the meiosis of translocation heterozygotes, etc. Duplications may have evolutionary significance because the extra piece of DNA may be modified to carry out a new function(s). Molecular evolutionary studies provided many examples of duplications followed by differentiation and divergence. It has been estimated that the α and β chains of the human hemoglobin, after a duplication, separated about 500 million years ago. Subsequently, from the β chain the γ and δ chains evolved. (The rhesus monkey does not have, however the δ chain). Similarly, the protein superfamily of trypsin, chymotrypsin, elastase, thrombin, kallikrein. plasmin and bacterial trypsin all have structural similarities indicating common ancestry. (See also evolution and duplications, deficiency, chromosomal aberrations, duplication-deficiency, redundancy, deletion)

DUPLICATION-DEFICIENCY: cells are produced by adjacent-1 and adjacent-2 distribution of chromosomes in translocation heterozygotes, in para- and pericentric inversion heterozygotes when crossing over takes place within the inverted segment, in case there is a sister chromatid exchange in ring-chromosomes and all other cases when dicentric chromosomes are produced (breakage fusion-bridge cycles), etc. (See inversion, translocation, dicentric ring chromosome)

DURUM WHEAT: is a member of the allotetraploid series with AB genomes. (See *Triticum*)

DWARFISM: is caused by several *dominant* genes in apparently different autosomes. The distinguishing features include (i) narrow vertebrae, (ii) thickening of the tubular bone associated with far-sightedness, lower than normal level of blood calcium, (iii) low birth-weight, stiff joints and eye defects, (iv) Myhre syndrome involving pre- and postnatal growth retardation, anomalies of face morphology, poor mobility of joints, defects in sex organs, heart and hearing. Some *autosomal recessive* dwarfism, (v) like the Dubowitz syndrome involves intrauterine growth retardation, mental and hearing defects, sex organ anomalies, unusually high or low voice, no response to growth hormone, etc., (vi) and various other types accompanied by mental retardation, heart anomalies, hip dislocation, very short arms and digits and the absence of fibula (the smaller bone of the leg). (See also pituitary dwarfism, achondroplasia, hypochondroplasia, thanatophoric dysplasia, dyschondrosteosis, spondyloepiphyseal dsyplasia, chondrodysplasia, cleidocranial dysplasia, pycnodysostosis, osteogenesis imperfecta, Kniest dysplasia, Williams syndrome, animal hormone, somatotropin, corticotropin, growth retardation, Pygmy, SHORT syndrome, stature in humans, Mulbrey nanism, Donohue syndrome, limb defects, diastrophic dysplasia). Dwarfism occurs also in plants. This condition in plants may be remedied in some cases by the supply of gibberellic acid or ethylene. (See, plant hormones, brassinosteroids, de-etiolation)

DYAD: chromosome with two chromatids, also the two cells (dyads) formed by the first meiotic division. (See chromatid)

DYNACTIN: a dynein regulating protein with the role of moving organelles, and it may interact with the kinetochore and anchors there dynein. (See dynein)

DYNAMIC STATE: is the biochemical concept stemming from the realization that the constituents of the living body, irrespective whether they are metabolic or structural, are in a steady state of lux.

DYNAMIN: protein factors with guanosine triphosphatase activity and are associated with other proteins and play an essential role in coated vesicle formation and are localized at the plasma membrane around the neck of emerging coated pits. Several forms of dynamins are known and they display tissue-specificity. They are involved in vesicle recycling, nerve terminal depolarization, etc. (See coated pits, synaptojanin, synaptic vesicles, GTPase)

DYNE UNIT OF FORCE: is needed that a body of 1 gram can be accelerated one centimeter per second per second.

DYNEIN: is a protein associated with bundles of microtubules (axoneme) in cilia and flagella and assists their movement in an ATP-dependent process. Dynein is supposed to move the microtubules toward the centrosome. (See microtubules, tubulin, flagellum, cilia, axoneme, kinesin, dynactin, centrosome)

DYNIA FACTOR: a blood factor controlling the antihemophilic factor. (See blood clotting pathways, antihemophilic factors, hemostasis)

DYSAUTONOMIA: malfunction of the central nervous system. (See neuropathy, Riley-Days syndrome)

DYSBETALIPOROTEINEMIA: a defect in apolipoprotein E. (See apolipoproteins, hyperlipoproteinemia)

DYSCALCULIA: a brain defect manifested in difficulties handling numerical problems.

DYSCHONDROSTEOSIS: a dominant dwarfism of the forearms with about four times as high occurrence in females than in males. It may be caused by X - Y translocations. (See dwarfness, achondroplasia, hypochondroplasia, stature in humans)

DYSCRASIA: a pathological condition of different manifestations in the blood or plasma, plasma proteins.

DYSEQUILIBRIUM SYNDROME: autosomal recessive cerebral palsy, mental retardation, muscular insufficiency, poor motor control, etc. based on malfunction of dopamine-β-hydroxylase activity.

DYSERYTHROPOIETC ANEMIA (HEMPAS for **h**ereditary **e**rythroblastic **m**ultinuclearity with **p**ositive **a**cidified **s**erum test): an endopolyploidy apparently limited to the bone marrow cells. As a consequence anemia appears. The basic defect seems to be a deficiency in the enzyme N-acetyl-glucosaminyl transferase II affecting the biosynthesis of glycoproteins.

DYSFIBRINOGENEMIA: in contrast to fibrinogenemia where the synthesis of fibrinogen is reduced, these individuals fail to assemble the fibrinogen monomers into normally functional molecules. This chromosome 4 dominant condition most commonly is not associated with heavy bleeding but may suffer from periodic clot formation in the bloodvessels (thrombosis). (See also antihemophilic factors, afibrinogenemia, fibrin-stabilizing factor)

DYSGENIC: deleterious, undesirable genetic trait (See also hybrid dysgenesis)

DYSGENESIS: a mechanism or process producing dysgenic individuals. (See dysgenic, hybrid dysgenesis)

DYSKERATOSIS (DKC): the autosomal dominant forms involve hyperpigmentation of the skin, pre-cancerous skin lesions, poor bone and nail development, no dermal ridges (fingerprints). The benign form does not affect much the individual's well being. The autosomal recessive form is also similar. The Xq28 linked form afflicts primarily males who may be affected also by testicle atrophy, anemia, cancer, and lacrimation because of defects in the lacrimal ducts. (See keratosis, keratoma, skin diseases, fingerprints)

DYSLEXIA: a highly variable difference in the central nervous system causing difficulties in reading and understanding or tiredness of reading. In some forms it appeared to be associated with left-handedness and speech defects. Dyslexia may not necessarily be associated with deficiencies of the general intelligence (IQ) and can be compensated for by tutoring type of education. It has been suggested that some specific forms of it are determined by a dominant gene in human chromosome 6p21.3 with preferential penetrance in males. Chromosomes 15 also harbors a gene for dyslexia. (See mental retardation, IQ)

DYSMELODIA: an apparently autosomal dominant gene with low penetrance causes reduced musical ability. There is also another autosomal dominant gene (perfect pitch) enabling the individual to remember and play a tune. (See Bach)

DYSMORPHOLOGY: abnormal morphological change.

DYSOSTOSIS: defect in ossification (bone formation).

DYSPLASIA: abnormal organization of the cells within the tissue.

DYSPLOIDY: the basic chromosome number varies within a population either because of the presence of B-chromosomes or because of Robertsonian translocations or misdivision at the centromere. The change in number does not involve an increase or decrease of an integer of the basic chromosome set. (See polyploidy, aneuploidy, B chromosomes, misdivision, Robertsonian translocation)

DYSPNEA: difficulty in breathing due to various genetic or other causes.

DYSREPRODUCTIVE GENES: impair fertility of the individual because of lowering the adversely affect the reproductive system either by structural and developmental defects or as subvitals, semi-lethals and lethals.

DYSSEGMENTAL DWARFISM: autosomal recessive phenotype involving abnormal development of the vertebrae. It is accompanied with various other symptoms such as dwarfism, cleft palate, hydrocephalus, etc. A defect in the normal formation of collagen is suspected. (See dwarfism, cleft palate, hydrocephalus, collagen)

DYSTASIA (areflexic dystasia): see claw-foot

DYSTONIA: lack of muscle coordination caused either by a dominant factor at human chromosome 9q32-q34 in the region of the dopamine-β-hydroxylase locus. The prevalence of the disease (torsion dystonia) in Ashkenazy Jewish populations is high, $2\text{-}5 \times 10^{-4}$, and the penetrance is about 30%. Some dystonias respond favorably to dopa. Autosomal recessive expression (or possibly mitochondrial origin) may be accompanied by visual defects. In some X-linked forms the symptoms include also deafness. The hereditary progressive DOPA-responsive dystonia in human chromosome 14 is caused by a deficiency GTP cyclohydrase I. This enzyme controls biopterin biosynthesis required for dopamine. (See neuromuscular diseases, dopa, biopterin, idiopathic torsion dystonia)

DYSTROPHIN (DRP): is muscle protein that anchors muscle membranes to actin filaments in the myofibrils. The X-chromosomal gene contains 79 exons spanning 2,300 kb and the transcription requires about 16 hours. Utrophin is also a dystrophin type protein (DRP1) and DRP2 another dystrophin gene has been mapped to human chromosome Xq22. (See muscular dystrophy, gene size, exon, actin, myofibril)

DYSTROPHY: inadequate nutrition (of the muscles). (See muscular dystrophy, atrophy)

DYSZOOSPERMIA: an anomaly involving the formation of spermatozoa. (See spermatogonia, gametogenesis, spermatid, azoospermia)

DZ: dizygotic twin. (See dizygotic, twinning)

E.C: enzyme classification number. (See enzymes)
E. coli: (*Escherichia coli*, Colon bacillus; Enterobacteriaceae): is the most predominant bacterium in the intestinal flora of mammals. Normally it is not pathogenic but some strains may cause Winckel's disease (a possibly fatal jaundice of newborns), diarrhea and urinary infections. The size of single cells is about 1000 x 2400 nm, and its mass is about 2 pg. Its genome is about 4.7×10^6 bp. It can harbor a variety of plasmids, and it can be a lysogen. At 37° C its generation time is about 20 min. *E. coli* is a genetically and biochemically a most thoroughly studied organism. By 1997 its genome has been completely sequenced almost independently in the USA and Japan. (See conjugation, conjugation mapping, bacterial recombination frequency, recombination molecular mechanisms prokaryotes, databases <http://cgsc.biology.yale.edu> or <http://genome4.aist-nara.ac.jp/GTC/mori/research/dbservice/ecoli-e.html> or <ftp://ftp.pasteur.fr/pub/GenomeDB> or Encyclopedia of *E. coli* Genes and Metabolism <http://www.ai.sri.com/ecocyc/ecocyc.html>, <http://www.mbl.edui/html/ecoli.html>, Science 277: 1453)
E REGION OF GTP-BINDING PROTEINS: interacts with effectors (amino acids 32-42 in RAS). Mutation in this region may abolish oncogenic transforming ability without affecting GTP-binding. (See oncogenes, RAS, GTP)
E SITE (exit site): is on the ribosome to where the deacylated tRNA moves from the P site after the peptidyl-tRNA—mRNA complex moves from the A to the P site. The process may be reversible. Elongation factor EF-T mediates the regeneration of the active EF-Tu-GTP from EF-TU-GDP. Thus, besides the initially identified A and P sites this third, E sites is known now. In contrast to the A and P sites, the E site may not be a permanent structure rather just a transient intermediate stage of the peptide elongation process. (See EF-TU-GTP, protein synthesis, ribosome, amino acid activation)
E1A: is an adenovirus oncoprotein; it can reverse the growth-inhibitory effect of the transforming growth factor (TGF-β). (See TGF, T cells, adenovirus)
E2A: immunoglobulin enhancer-binding factor, a basic helix-loop-helix protein, encoded in human chromosome 19p13. (DNA binding domains, immunoglobulins)
EAA (excitatory amino acids): glutamate and aspartate particularly can activate neurotransmitters. (See neurotransmitter)
EADIE-HOFSTEE PLOT: in a coordinate system v/(S) is plotted against v, where v is the enzymatic reaction velocity and (S) is the substrate concentration. (see Michaelis-Menten equation, Linweaver-Burk plot)
EARLOBES ATTACHED: is generally a recessive human trait.
EARLY GENES: are transcribed early during development; they are involved in the infection process of the virus, and before replication begins. (See also early-response gene, late gene, delayed-response gene)
EARLY-RESPONSE GENE: is activated by growth factors within a few minutes without a prerequisite for protein synthesis. (See early gene, delayed-response gene, late gene)
EBF (early B cell factor): is a transcription factor specific for B cells and expressed at antigen-independent stages and it regulates the immunoglobulin α chain. (See B lymphocyte, T cell)
ebgA⁰: a gene of *E. coli* (map position 67) that through two mutations permits galactose utilization although this newly evolved galactosidase is immunologically distinct from the *LacZ* encoded β-galactosidase enzyme. (See lactose operon)
EBNA-2 and -5: Epstein-Barr virus antigens, and transactivating oncoproteins. (See oncogenes, Epstein-Barr virus, transactivator, transactivation response element)
E-BOX: the minimal DNA element (CANNTG, where N is any nucleotide) required for binding HLH and b/HLH/Z transcription factors. (See HLH, bHLH, helix-loop-helix)
EBP: enhancer binding protein. (See enhancer)
4E-BP1 (PHAS-1): is a binding protein of eIF-4E. After it is phosphorylated it stimulates the

4E-BP1 continued
activity of eIF-4F cap-binding protein, required for the beginning of translation. It appears that phosphorylation of 4E-BP1 releases eIF-4E from the inhibition by 4E-BP1. (See eIF-4F)

EBV: see Epstein-Barr virus

EC: embryonal carcinoma cells.

ECAF: endothelial attachment factor.

ECDYSONE ⇨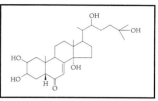

ECDYSONE: a steroid molting hormone of insects produced by the prothoracic gland. It stimulates apparently transcription and activates puffing at different locations in the salivary gland chromosomes of *Sciara*, *Drosophila* and other dipteran flies. This hormone and variants apparently regulate development also in crayfish, arthropods, schisostomes and nematodes. Similar compounds (β-ecdysones) occur also in plants. Ecdysone does not play a natural role in mammals yet the ecdysone receptor transgene controls ecdysone response. (See puffs, salivary gland chromosomes, animal hormones, steroid hormones, brassinosteroids)

ECERIFERUM LOCI: in various plant species determine the cuticular waxes or lack of them.

ECGF: endothelial growth factor

ECHOCARDIOGRAPHY: is an ultrasonic examination the structure of the fetal heart as part of a repertory of prenatal diagnosis of congenital defects in anatomy. (See prenatal diagnosis)

ECLIPSE: the (latent) period between viral infection of a bacterium and the burst of the new phage particles even if burst is induced. (See one-step growth, burst, induction of a lysogenic bacterium)

ECLOSION: the hatching of the adult form (imago) of insects from the puparium. (See pupa)

ECM: see extracellular matrix

ECODEME: a population adapted to a particular ecological condition.

ECOGENETICS: studies the genetically determined responses of organisms to environment(s).

ECOLOGICAL RACE: a distinctly adapted group of an organism without sexual isolation from the ancestral form. (See sexual isolation)

ECOLOGY: the study of the relation of living systems and the environment. (See ecogenetics)

EcoRI: type II restriction endonuclease with primary recognition site
$$5'\ pG{\downarrow}p\mathring{A}pApTpC$$
$$3'\ _{OH}CpTpTp\mathring{A}pAp{\uparrow}_{OH}G$$
Å indicates potential base to be methylated, arrows indicate sites of cut, the staggered cuts have a receding OH and a protruding p end -------OH
------------------p
Although the above 6 nucleotide sequence is the primary recognition site, alternate recognition sites with different preferences are also known. These secondary activities (named EcoRI* [star]) depend on the composition of the incubation mixture. (See restriction enzymes)

EcoSeq: a data base of a collection of DNA sequences of *E. coli* obtained from various sources.

ECOSYSTEM: the relation of living organisms to each other and to all environmental factors. (See species extant)

ECOTROPIC RETROVIRUS: replicates only in cells of the species from what it has been originally isolated. This specificity is determined by envelope glycoproteins that require specific receptors in the host. (See amphotropic)

ECOTYPE: a population adapted to a particular ecological condition. Besides the main genes determining the adaptive trait(s), the population may not be genetically homogeneous; usually it is not an isogenic line. (See adaptation, isogenic stocks)

ECTODERM: the surface layer of the embryo that develops into epidermis, skin, nerves, hair, nails, ears, eyes, enamel of teeth, internal mouth and anal tissues. (See gastrulation)

ECTODERMAL DYSPLASIA: several *autosomal dominant forms* involve complex skin alterations and may eliminate the dermal ridges and alters palmar prints (see fingerprints). In the ectodermal anhidrotic dysplasia the patients do not sweat and have cleft lip and palate or in the hypohidrotic form the sweating ability is only reduced whereas in the hidrotic form shows

Ectodermal dysplasia continued

normal sweating; the trichoodontochial form is accompanied by deficiency of tooth, hair, breast and nipple formation. The *EEC* (ectrodactyly [missing fingers]) *ectodermal dysplasia* has most of the symptoms mentioned above plus abnormal tear ducts. The *autosomal recessive* ectodermal dysplasias include also complex features such as sweat gland tumors (accrine tumors and ectodermal dysplasia), also anhidrotic types, dysplasia with neurosensory deafness, dysplasia with cleft lip and palate, mental retardation, syndactyly, hypohidrosis-hypothyroidism, hypohidrosis-hypothyroidism-lung disease. An *X-linked recessive* (Xq11-q21.1) ectodermal dysplasia is anhidrotic, shows reduction in hair and tooth development, hyperpigmentation around the eyes, short stature, etc. (See dermatoglyphics, fingerprints, ectoderm, anhidrosis, cleft palate, cyst, polydactily, EEC syndrome)

ECTOPIC EXPRESSION: the organ specific expression of a gene is altered by fusing the structural gene to a promoter with different or no organ-specificity of expression, or in more general meaning the displaced condition of an organ or function. (See promoter, transformation, allotopic expression)

ECTOPIC INTEGRATION: insertion of transforming DNA takes place at a target site different from that of the original map position of the transforming DNA. (See transformation)

ECTOPIC PAIRING: association between sites in non-homologous chromosomes. (See chromosome pairing)

ECTOPIC PREGNANCY: an abnormality when the fertilized ovum develops outside the uterus.

ECTOPLAST: the cytoplasmic membrane surrounding the protoplast.

ECTRODACTYLY: are split foot and hand, absence of fingers-polydactily. It is most likely under autosomal recessive control. (See adactyly, polydactyly)

ECZEMA: itching inflammation of the skin; it may be scaly or oozing and it may be caused by various factors even as simple as drying of the skin in winter, food allergy or infections or Kaposi sarcoma, etc. (See dermatitis, skin diseases, Wiskott-Aldrich syndrome, acquired immunodeficiency syndrome)

EDITING: some DNA polymerases, aminoacyl tRNA synthetases, aptamers have also nuclease functions and eliminate replicational mistakes or prevent translational errors. (See antimutator, error in aminoacylation, RNA editing, mtDNA)

EDMAN DEGRADATION: is a procedure for protein sequencing. The reagent phenylisothiocyanate causes the formation of phenylthiohydantoin and cleavage of the terminal residue. This amino acid can then be identified. The same process can be repeated many times and the amino acid sequences of the entire macromolecule can be determined step by step. The commercially available automatic protein sequenators use the same principle in an efficient way.

EDRF (endothelium-derived relaxing factor): it is nitric oxide. (See nitric oxide)

EDTA (ethylenediaminetetraaceticacid): is a chelating agent and as such an inhibitor of DNase, it is an anticoagulant, it is also used for plant nutrient media for improving the solubility of iron, etc. (See DNA extraction, embryogenesis somatic, DNA fingerprinting, versenes)

Edward: King of England, one of the 3 sons of Queen Victoria who did not inherit the hemophilia gene. (See hemophilias)

EDWARD'S SYNDROME: trisomy for human chromosome 18 with serious debilitating consequences or of prenatal, or post-natal death within a few months or may survive up to a few years. Generally the head is elongated the ears are set low, eyelids are abnormal, clenched fingers, hypoplasia of nails, and almost all organs are affected. Its incidence is about 1 in 7,500 to 10,000 births with a predominance of females among the afflicted. (See trisomy, trisomic analysis, hypoplasia)

EEC SYNDROME (ectrodactyly, ectodermal dysplasia, cleft lip/palate): autosomal dominant abnormality. (See ectrodactyly, ectodermal dysplasia)

eEF: eukaryotic elongation factor of peptide chain.

eEF-1a: translation factor, binds amino acid-tRNA, also a GTPase and regulates cytoskeletal (microtubule) rearrangements. (See protein synthesis)

eEF-1β: mediates GTP-GDP exchange on eEF-1α in translation. (See protein synthesis)
eEF-1γ: mediates GTP exchange with the aid of eEF-1β in translation. (See protein synthesis)
eEF-2: translation factor; stimulates peptide chain translocation on ribosome; when phosphorylated may slow elongation rate, it is also a GTPase. (See GTPase, translation, protein synthesis)
EEG (electroencephalogram): a record of the electric current developed in the brain, measured after electrodes were applied to the scalp, to the surface of the brain or into the brain material. It reveals the functional state of the central nervous system. The EEG pattern depends on a number of factors but has also a hereditary component indicating psychological responses of the family.
EF-2: translation elongation factor in eukaryotes. (See protein synthesis)
EF-1α: translation initiation factor in eukaryotes. (See protein synthesis)
E2F1: is a family of transcription factors activated by MDM2 oncogen. They are involved in the regulation of the cell cycle, apoptosis, neoplasia, etc. (See MDM2, DP1, pocket, cancer)
eIF-2B: translation factor involved in GTP -GDP exchange. (See protein synthesis, GTP)
EFFECTIVE MUTAGEN: causes all types of mutations (including chromosomal changes) at high frequency. (See also efficient mutagen)
EFFECTIVE NUMBER OF ALLELES: the number of alleles that are maintained in the population. (See effective population size)
EFFECTIVE NUMBER OF LOCI: is the number of loci involved in a quantitative trait, also called segregation index. (See gene number in quantitative traits)
EFFECTIVE POPULATION SIZE: is number of individuals that leave offspring in a population. Although from several viewpoints (economic, agricultural, ecological, demographic, insurance, welfare) the total number of individuals may be the most important, geneticists are concerned primarily with that fraction of the population that passes on genes to future generations. The *genetically effective size of the population* is represented as (N_e). If the effective population size is small (even if the mating is random), the allelic and genotypic frequencies may be biased. In cases when the two sexes are represented by unequal numbers the effective population size will be lower than the sum of the males and females. Each individual has 0.5 chance to contribute a particular allele to the offspring (through the egg and sperm) and the chance of contributing two of the same allele is 0.5 x 0.5 = 0.25. The probability that the same female contributes two alleles is:
$(1/N_f)(1/4)$ and for a male it is $(1/N_m)(1/4)$, where N_f and N_m are the number of females and males, respectively. Therefore, the probability that any two alleles of the population comes from the same individual is $\frac{1}{4N_m} + \frac{1}{4N_f} = \frac{1}{N_e}$ hence $N_e = \frac{4N_m N_f}{N_m + N_f}$. Fluctuations in population size from generation to generation as well as the non-random distribution of family size may affect the allelic sampling.

The sampling variance of the alleles is equal to:

$\sigma^2 = \frac{q(1-q)}{2N}$ and hence $\sigma = \sqrt{\frac{q(1-q)}{2N}}$ where q is the frequency of one of the alleles and N is the population size. If the frequency of the *a* allele is 0.5 and N = 25, the standard deviation of the frequency of the a allele becomes:

$$\sigma = \sqrt{\frac{0.5 \times 0.5}{2 \times 25}} = \sqrt{0.005} \approx 0.071$$

This indicates that chance alone can modify the frequency of the a allele in a small population and 31.74% of the loci (because according to the normal distribution 68,26% of the population is supposed to be within M ± 1σ) may carry the allele with frequencies more extreme than 0.429 to 0.571 rather than 0.5 (0.50 ± standard error). In other words, this also means if the population is broken up into smaller breeding units, 31.74% may have gene frequencies outside the range of 0.429 and 0.571. Any shift in the gene frequencies brought about by the random fluctuation in gametic sampling is called *random genetic drift*. The drift is completely

Effective population size continued
accidental. Such events may take place in the population in any breeding season if the number of breeding individuals is reduced. Once such a sampling bias of gametes (in mating) has taken place, the process may continue in either direction but there is a definite chance that the allelic composition of the small population permanently changes. A special case of the random drift is the *founder effect*. A small number of individuals (immigrants) introduced into a new habitat may not represent accurately the genetic constitution of the population of their origin. Their descendants then may form the basis of a divergent trend from the norm of their ancestors. Such a phenomenon is not uncommon if animal or plant species migrate to new parts of the world or to regions that is spatially isolated from their old homeland. (See speciation, isolation genetic, standard error, normal distribution)

EFFECTOR: small molecule which assists either in activating or deactivating a molecular event. Interaction of the RAS protein with effectors uses 32-40 amino acids, the effector region, varying in conformation in GTP- and GDP-bound RAS. (See GTP, RAS, signal transduction)

EFFECTOR CELL: see immune system

EFFECTOR DOMAIN: of antibody contains the region that recognizes specifically the cognate antigen. (See antibody, antigen)

EFFICIENT MUTAGEN: produces primarily point mutations and relatively low amount of chromosomal alterations. (See also effective mutagen, point mutation)

EFFLUX SYSTEMS: carry out energy requiring active pumping of harmful agents from the cells.

EF-G: translation factor and a motor protein that is also a GTPase; it mediates GTP-dependent transfer of peptidyl-tRNA-mRNA complex from ribosomal A site to P site. (See protein synthesis, ribosome, aminoacylation)

EF-Ts: prokaryotic translation factor protein involved in GTP - GDP exchange. (See protein synthesis)

EF-TU·GTP: active prokaryotic elongation factor of protein synthesis in which EF-Tu interacts with aminoacyl-tRNA and promotes the translocation of the peptidyl-tRNA from the ribosomal A to B site and the release from the ribosome the deacylated tRNA. (See aminoacylation, protein synthesis, tRNA)

E.GDP: bound (inactive) form of guanosine diphosphate in signal transduction. (See also E region of GTP-binding proteins)

EGF: epidermal growth factor binds to receptors such as the protein coded for by protooncogene ERBB, triggering growth-promoting signals. The *v-erbB* viral oncogene encodes a truncated EGF that continuously binds to a ligand and provides a constitutive supply of growth signals. EGF controls a wide range of cellular processes. (See growth factors, oncogenes)

EGFR (epidermal growth factor receptor): see ERBB1, LDL receptor

EGG (ovum): is the final haploid product of female meiosis. The eggs are huge cells (ca. 100 μn in diameter in humans) compared to the other ones in the body (about 20 μm). The egg cytoplasm contains yolk, a very condensed store of nutrients, especially in organisms that lay outside the eggs. In mammals, the yolk is comparatively minimal. In mammals the egg is surrounded by the *zona pellucida* membrane. In lower vertebrates there is a *vitelline layer* and also other ones. In birds there is the *egg white* and before laying the egg the *shell* is added in the oviduct. The vitelline layer of the insects is covered by the so called *chorion*, secreted by the *follicle cells* of the *ovary*. In the layer under the plasma membrane (*cortex*) are the *cortical granules*. Their content protects the egg from fertilization by more then one sperm. (See oocyte, ovary, gametogenesis, menopause, fertilization, polyspermic fertilization, RPTK)

EGG CYLINDER: a very early mammalian embryonal structure including the cells that will form the fetus and the embryonal sacs (amnion, allantois), some extra embryonic tissue that form the outer embryo sac (chorion) and the trophoblastic tissues of the placenta.

EGG NUMBER IN HUMANS: each of the two human ovaries contains more than 200,000 pri-

Egg number in humans continued
mary oocytes yet only about 400 eggs develop to maturity during the period from puberty to menopause. The size of a human egg is comparable to a period in this print. (See puberty, menopause, oogenesis)

EGG PLANT (*Solanum melongena*): is a tropical-subtropical vegetable with 2n = 2x = 24 chromosomes.

egr-1: mitogen-induced transcription factor, a serum-inducible nuclear phosphoprotein with Zinc finger. (See NGFI-A, TIS8, platelet derived growth factor, Zinc finger)

E.GTP: bound (active) form of guanosine triphosphate in signal transduction. (See also E region of GTP-binding proteins)

EHLERS-DANLOS SYNDROME (EDS): it is usually a dominant disorder involving loose joints that stretch excessively, excessive stretchability of the skin that is bruised easily and frequently pseudotumors appear after some trauma. The anomalies are caused either by a deficiency of procollagen peptidase or the lack of collagen type III. Low hydroxylysine content of the connective tissue prevents effective cross-linking. Lysyl oxidase level is reduced in the afflicted individuals by type VI and IX. The EDS has many different variations from mild to severe and its incidence may be estimated to be in the $1\text{-}2 \times 10^{-4}$ range. Type IV (chr. 2) suffers from mutation in type III pro-collagen and increases the risk of aneurysm. Type V disease has defect in the α chain of collagen mapped to human chromosome 9q34. (See collagen, skin diseases, connective tissue disorders, skin diseases, cardiovascular disease, Menkes syndrome, aneurysm)

EHRLICH ASCITES: see ascites

EI: ethylene imine (C_2H_5N), a powerful alkylating mutagen and carcinogen.

EICOSANOIDS: are mammalian autocrine signaling molecules affecting muscle contraction, platelet aggregation, pain, inflammation. They are produced at the expense of phospholipase degraded phospholipids. Eicosanoid is the systematic name of arachidonic acid. (See fatty acids, signal transduction, autocrine, platelet, phospholipase)

eIF: eukaryotic initiation factor of protein synthesis.

eIF-1: translation factor stimulating the 40S ribosomal subunit preinitiation complex. (See protein synthesis)

eIF-1A: ribosome dissociation factor and 40S preinitiation complex stimulating protein in translation.

eIF-2: a translation elongation factor mediating GTP-dependent Met-tRNA binding to 40S ribosomal subunit; it may repress protein synthesis if its α subunit is phosphorylated. (See protein synthesis, ribosome, GTP, aminoacylation)

eIF-2A: translation elongation factor controlling AUG-dependent Met-tRNA binding to the 40S ribosomal subunit. (See protein synthesis, aminoacyl-tRNA synthetase)

eIF-2C: translation initiation factor stabilizing ternary complexes. (See protein synthesis, ternary)

eIF-3: elongation initiation factor, activated by phosphorylation in translation. It stimulates the formation of the 40S ribosomal preinitiation complex. (See ribosome, protein synthesis)

eIF-3A: translation factor affecting ribosome dissociation by binding to the 60S ribosomal unit.

eIF-4A: eukaryotic translation (mRNA cap-binding protein) factor, RNA-dependent ATP-ase, helicase, stimulates mRNA binding to ribosome. (See protein synthesis, ATPase, helicase)

eIF-4B: eukaryotic translation initiation factor (mRNA cap-binding protein); elongation initiation of translation upon phosphorylation, a helicase, stimulates mRNA binding to ribosomes. (See protein synthesis, ribosome, helicase, mRNA, ribosome)

eIF-4C: function is similar to eIF-A.

eIF-4D: probably the same as eIF-5A.

eIF-4E: translation factor involved in binding of the mRNA cap to the 40S ribosomal subunit. (See cap, ribosome, protein synthesis)

eIF-4F: translation factor; recognizes mRNA cap; a helicase, elongation initiation factor and when phosphorylated may lead to overexpression and cancerous growth. (See protein synthesis, ribosome, helicase, cap, cancer)

eIF4G (eIF-4γ): is a translation initiation factor for the 40S eukaryotic ribosomes. It functions as an adaptor between the cap-binding proteins eIF-4A, eIF-4E, eIF-3, and also it may bind to the PolyA tail-binding protein. The latter has been shown to be involved in translation stimulation. (See protein synthesis, cap, IRES)

eIF-5: translation factor promoting joining of ribosomal units. (See ribosome, protein synthesis)

eIF-5A: assists in the formation of the first peptide bond during translation. (See protein synthesis, peptide bond)

eIF-6: similar to eIF-3A.

eIF-D: similar in function to eIF-5A.

EIGENVALUE: literally, this hybrid (German-English) word means proper value, and it is usually used (in physics) in the sense of a characteristic value.

EINCORN: see *Triticum* A genome

EJACULATE: see sperm

EK2: laboratory strain of *E. coli* defective in the synthesis of thymine and diaminopimelate. (See thymine, diaminopimelate, *E. coli*)

EKLF (erythroid Krüppel-like factor): is a Zn-finger homolog of the *Drosophila Krüppel* gene, involved in blood β-globin and blood non-globin gene transcription regulation. (See morphogenesis in *Drosophila*)

ELASTASE: a proteinase released by the lysosomes of blood granulocytes upon inflammation. It breaks down collagen if not inhibited by protease inhibitors. (See lysosome)

ELASTIN ($M_r \cong 70$ K): is a rubber-like, cross-linked, glycine- and proline-rich, fibrous protein present primarily in the blood vessels near the heart but also in the ligaments; modest amounts in the skin and tendons. (See supravalvular aortic stenosis)

ELC: expression linked copy. (See *Trypanosoma*)

ELECTROBLOTTING: transfer of macromolecules in an electric field from a gel onto a membrane.

ELECTROCARDIOGRAPHY: records the changes of the variations in the electric potentials of the heart muscles. The first deflections, denoted by P is due to the excitation of the atria. The QRS indicates the depolarization phase of the excitation of the ventricles. The T wave indicates the complete repolarization of the ventricles. These intervals are measured in fractions of a second. The long Q - T (LQT) interval may indicate abnormal function of either the potassium or sodium ion channel(s) or some other anomaly. (See Jervell and Lange-Nielsen syndrome, HERG, Ward-Romano syndrome)

ELECTROCHEMICAL GRADIENT: moves ions through biomembranes depending on their concentration and charge, at the two sides.

ELECTROELUTION OF DNA: separate DNA by electrophoresis in agarose gel containing 0.5 μg/mL ethidium bromide, locate band by long-wavelength U.V. light, cut out the band and place it into dialysis tube filled with 1 x TAE buffer. Let the slice sink to the bottom of the tube, seal, turn on current in 1 TAE buffer (4 to 5 V/cm) for 2 to 3 hours. Remove tube from apparatus and collect the buffer content of the tube containing the eluted DNA. (See electrophoresis, electrophoresis buffers)

ELECTROENCEPHALOGRAM: see EEG

ELECTROLYTE: a substance capable of dissociating into ions and then becoming a conductor of electricity. (See ion)

ELECTROMAGNETIC RADIATION: the genetically effective range includes UV light, X-rays and gamma rays. Under conditions of sensitization by various chromophores it may extend into the visible range. Some of the radiations may have contaminating components with potential genetic effects. Extremely low frequency electromagnetic fields (EMF) such as generated by power lines and household appliances have also been suspected to stimulate gene activity (MYC) and increase cancer risk. The experimental data are, however, insufficient to definitely rule in or out such an effect.

Electromagnetc radiation continued

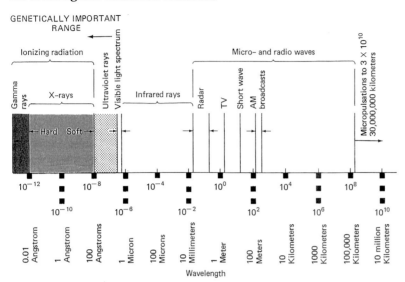

SPECTRUM OF ELECTRO-MAGNETIC RADIATIONS. (After Lerner & Libby, 1976 Heredity, Evolution and Society, Freeman, San Francisco)

(See radiation protection, radiation hazard assessment, illumination, light intensities, MYC)

ELECTROMORPH: a genetically determined variation that can be revealed by electrophoresis. (See electrophoretic polymorphism)

ELECTRON: is a small negatively [(4.80294 ±0.00008) x 10^{-10} absolute electrostatic units] charged particle; its mass is 1/1837 of the H nucleus and its diameter is 10^{-12} cm. All atoms contain one nucleus and one or more electrons. Cathode rays and Beta rays are electrons. (See cathode rays, beta particles, excitation)

ELECTRON ACCEPTOR: takes up electrons and therefore becomes reduced whereas the *electron donors* provide electrons to other molecules and become oxidized. (See electrons)

ELECTRON MICROSCOPY: in the light microscope the resolution of the objects is limited by the wave length (within the 350 to 750 nm range) of the light used for the illumination. The wavelength of electrons accelerated at high voltage (100 KV) may become only 0.004 nm. Because of aberrations of the lenses available, this minimum cannot be attained yet in the best equipment a resolution of about 0.1- 2 nm, i.e. at about 100 to 1,000 higher than with the light microscope may be possible. In the *transmission electronmicroscope*, instead of visible light the extremely thin (50-200 nm) specimen, specially fixed and stained, is placed on a grid is exposed to electron beams accelerated in a vacuum tube. The areas where the specimens reduce the focused electron flux, an image appears that can be viewed or photographed and enlarged by procedures similar to light microscopy. Electronmicroscopic specimens are usually prefixed in buffered 3% glutaraldehyde and postfixed in 1% buffered osmium tetroxide, then dehydrated and embedded in propylene oxide-resin mixture that requires polymerization before sectioning by glass or diamond microtomes. The sections are placed on a circular copper grid of 3 mm diameter, covered by a carbon or plastic film. Either before or after sectioning the biological specimen is treated usually by uranium or lead to assure contrast. Electronmicroscopic specimens may be exposed also to substrates of enzymes in order to localize an electron dense precipitate if the enzyme is active. The specimens may be coupled also with the specific fluorescent dyes or with colloidal gold which may be recognized by an antibody and thus its site is revealed. The material may also by *shadowed* — by spraying in an angle with other thin layers of metals (platinum, chromium) — for obtaining an apparent 3-dimensional image. Macromolecules may be studied also by *negative contrast*. The solution in 1% phosphotungstate (or uranyl acetate) is thinly spread on a carbon film and when dried an electron-dense layer is formed. At the position of the protein molecule there will be no tungstate. Thus a negative image is generated. A special technique in electronmicroscopy is *freeze-fracture* which can provide images of the internal organization of delicate structures such as biological double membranes. This freezing technique does not require fixation that usually denatures the material. The specimen is frozen in liquid nitrogen in the presence of a

Electron microscopy continued

cryoprotectant (glycerol, dimethylsulfoxide) to prevent the formation of ice crystals. The frozen material is then fractured with a knife to crack the double membrane between the two layers. The surface so generated is shadowed with a metal, and after the organic material is dissolved and removed, the remaining metal replica is examined. This replica is a negative image of the biological structure. A similar procedure is *freeze-etching*. Again the frozen sample is cracked, the moisture is removed by lyophilization and the inner and outer structures exposed are shadowed. (See also scanning electronmicroscopy, microscopy, freeze drying)

ELECTRON PARAMAGNETIC RESONANCE (EPR): can be used to measure the radiation exposure in human or mammalian populations. When paramagnetic substances are placed in a stationary magnetic field and exposed to electromagnetic radiation they register the field of strength and the frequency of the radiation. Paramagnetic substances contain molecules or atoms whose electrons move and produce weak magnetic fields. This method can then be used for the determination of the concentration of radiation-induced radicals in e.g. hydroxyapatite, present in teeth and bones. The tooth enamel is 97% hydroxyapatite and does not metabolize ^{90}Sr a common product of atomic fallout. Dentine which is made up of only 70% from hydroxyapatite, metabolizes this isotope. Therefore ^{90}Sr will be found mainly in the dentine. The isotope emits short range β-rays (0.6-1 mm path). One study showed that humans with chronic radiation disease the enamel received 1 Gy (gray unit) whereas the dentine showed 5.5 Gy γ-radiation. Thus it is possible to determine whether the radiation came internally or externally. (See radiation measurement, radiation hazard assessment, fallout)

ELECTRON TRANSPORT: movement of electrons toward the lower level of energy by electron carriers.

ELECTRON VOLT (eV): 1 eV = $1.60207 \pm 0.00007 \times 10^{-12}$ erg. (See erg)

ELECTROPHILE: eager to accept electrons. (See electron, nucleophile, nucleophilic attack)

ELECTROPHORESIS: separating charged molecules (such as polypeptides or polynucleotides) between the two poles of an electric field. The material is generally contained in a support medium (cellulose, starch, agarose polyacrylamide) and bathed in an appropriate buffer and may contain also specific stains. The basis of the separation of the components of a mixture is the relative charge, shape or molecular size of the components. Larger nucleic acid molecules are generally separated in agarose gels of various concentrations, depending the size of the molecules. Smaller nucleic acid fragments are separated in polyacrylamide gels. SDS-PAGE electrophoresis treats the proteins with the detergent sodium dodecyl sulfate (SDS) and separates them by polyacrylamide gel electrophoresis (PAGE). The detergent breaks the non-covalent bonds of the proteins and when β-mercaptoethanol added the disulfide bonds are also eliminated. The proteins so treated migrate in the gel according to their molecular weight rather than by their charge. Nucleic acids have uniform charge, thus their separation is based on their molecular size. Ultrathin (50 mm) gel electrophoresis permits the use of higher voltage without deleterious heat effect because of the increased heat transfer. Also, it increases the speed of separation by an order of magnitude although may decrease resolution and read length, and it is under experimental study. Another innovative approach is the matrix-assisted laser desorption and ionization method (MALDI) which permits single charged ions from proteins as large as 300 kDa to be produced and analyzed. Mass spectrophotometry MALDI (MALDI-MS) is capable of discriminating length difference in short oligonucleotides due to a single base. (See also pulsed field electrophoresis, isoelectric focusing, gel electrophoresis, two-dimensional gel electrophoresis, dodecyl sulfate sodium salt)

ELECTROPHORESIS BUFFERS: TAE (Tris-acetate EDTA), TPE (Tris-phosphate EDTA), TBE (Tris-borate EDTA), alkaline 50 mM NaOH-1 mM EDTA are commonly used.

ELECTROPHORETIC KARYOTYPING: small chromosomes are separated by pulsed field gel electrophoresis according to size. (See pulsed field)

ELECTROPHORETIC POLYMORPHISM: protein or DNA variations in a population affecting charge or size. In proteins, it may be detectable on the bases of amino acid (glutamic, and aspartic acid, lysine, histidine, arginine) substitutions by altered electrophoretic mobility or by size of polypeptides. In DNA, restriction enzyme digests provide such information after gel-electrophoresis. (See electrophoresis, isozyme, SDS-polyacrylamide gels, RFLP)

ELECTROPORATION: transport of DNA across cellular membrane with the aid of electric current pulses; the genes may display transient expression in the target cells; a means of genetic transformation. (See transformation, microfusion)

ELECTROSPRAY MS (mass spectrometry): a procedure for determining the molecular weight of unknown proteins and for the study of non-covalent interactions between proteins and other large molecules and thus facilitates their location and identification with the aid of databases. (See laser desorption mass spectrum, laser, mass spectrum, genome projects, STM)

ELEPHANTS: *Alphas maxis* 2n = 56; *Laxodonta africana* 2n = 56.

ELEPHANT MAN: a person with abnormally enlarged body parts. The cause of Joseph Merrick's anomaly has been attributed to neurofibromatosis and to Proteus syndrome but the real cause of it is unclear. It is different from elephantiasis, a tropical infection of the lymphatic nodes, limbs, genitalia, etc. causing inflammation and hypertrophy as the result the obstruction of the lymphatics by several species of nematodes. (See neurofibromatosis, Proteus syndrome)

ELICITOR: see host - pathogen relation

ELISA (enzyme-linked immunosorbent assay): is capable of detecting vg quantities of protein per gram tissue by the use of an antibody attached to a particular enzyme such as alkaline phosphatase (or peroxidase, urease). The enzyme + antibody complex binds to the protein antigen present in the reaction vessel and the cleavage of a chromogenic substrate (e.g. p-nitrophenyl-phosphate) by the enzyme results in the development of color (absorbance detected by spectrophotometer) that is proportional to the amount of the antigen protein. (See immunoglobulins, antibody, immune reaction, immunoprecipitation, immunolabeling)

ELK (oncogene): ELK1 (human chromosome Xp11.2) and ELK2 (14q32.3) are expressed in human lungs and testes. These oncogenes show homology with the ETS oncogenes. The Elk proteins bind to SRE element of the chromosomes and after phosphorylation may activate transcription. (See oncogenes, ETS, signal transduction, SRE)

ELL: a human RNA polymerase II elongation factor in chromosome 19. It increases the rate of transcription because it suppresses pausing of the polymerase along the DNA. Functionally it is similar to ELONGIN, another transcription elongation factor controlled by the von Hippel-Lindau tumor suppressor gene. (See von Hippel-Lindau syndrome)

ELLIPTOCYTOSIS (ovalocytosis): the shape of the erythrocytes (red blood cell) is not round as normal but elliptic, indicating an often fatal autosomal dominant hemolytic anemia. In the *Camelidae* elliptocytosis is a normal condition. Mutations in spectrin frequently lead to defects in forming the αβ~αβ tetrameric spectrin molecules is the basic cause and a variety of this conditions have been described. Some forms may be correlated with some protection against malaria. The α-spectrin gene was assigned to human chromosome 1q22-q25 (Rhesus-unlinked type elliptocytosis). The Rhesus-linked type was assigned to 1p34-p33. An atypical elliptosis is apparently recessive. (See anemia, spectrin, ankyrin, anemia, spherocytosis)

ELLIS-VAN CREVELD SYNDROME: human chromosome 4p16 recessive; it involves dwarfism, polydactily, short extremities, heart malformation, dystrophy of finger nails, 'partial hare lip', teething by birth, etc. It is quite common in some Amish populations. (See Amish, face/heart defects, polydactyly, hare lip, dystrophy)

ELOD: see interval mapping, lod score

ELONGATION FACTORS (EF): are proteins assisting translation on the ribosomes; see individual factors (eIF) separately listed. (See also protein synthesis, transcription shortening, transcription termination, translation termination)

ELONGIN: see ELL

ELUATE: that flows off a chromatographic column.

EM-ALGORITHM: is a principle that can be applied for the calculation of recombination frequencies in special cases.

EMASCULATION: see castration

EMB AGAR: see β-galactosidase

EMBDEN-MEYERHOF PATHWAY: an anaerobic process of glycolysis. The six-carbon sugars are converted into glucose-6-phosphate and through a number of steps to glyceraldehyde-3-phosphate and eventually to pyruvate. An oxidation step generates NADH and pyruvate is further oxidized to form the acetyl units of acetyl coenzyme A, that may be oxidized in the citric acid cycle, and become the core reactions of both carbohydrate and fat metabolism.

EMBEDDING: microscopic specimens requiring thin sections for examination may be surrounded (infiltrated) by paraffin wax, resins or other suitable material before sectioning by microtomes. (See microtomes, sectioning)

EMBEDDED GENE: a gene within the boundary of another. (See φX174)

EMBL: stores data bases for macromolecular sequences, European Molecular Biology Data Library, P.O.B. 10.2209, D-6900 Heidelberg, Germany (W), telephone: 011-49-6221-387-258 e-mail, general inquiries: datalib@embl.earn, for submissions and forms: datasubs@embl.earn

EMBL3: is a lambda DNA vector carrying polylinkers at both sides of the *red* and *gam* sites and if equipped with adjacent promoters, genes cloned into the "stuffer" region can be directly transcribed by RNA polymerase without subcloning. (See λDASH, λ FIX, GeneScribe, Lambda phage, stuffer region)

EMBRYO: the differentiated zygote, in mammals up to about a 1/5 of the normal time of gestation. In plants the embryo stage is considered to last up to the maturing and germination of the seed. (See embryogenesis, embryo culture, <http://visembryo.ucsf.edu>)

EMBRYO RESEARCH (human): NIH recommendations: embryos (14) 18 days or older are not acceptable. Unacceptable procedures: transfer human embryos to animals for gestation, transfer research embryos or parthenotes (unfertilized eggs) to humans, separation of blastomeres for generating twins, cloning by nuclear transplantation, creation of any type of chimeras (human-human, human-animal), creation of embryos in the lab from stem cells, cross-fertilization by human gametes with the exception of clinical laboratory sperm penetration tests with animal eggs, embryo transfer to cavities other than the uterus, sex selection with the exception to prevent sex-chromosome-linked disease, use sperm or eggs without consent, use of sperm or egg from donor who were paid more than reasonably expected. The embryo research policies vary in different countries, and may/may not permit the use of embryos/tissues for studying infertility, contraception, genetic screening, gene therapy, cloning, construction of chimeras, interspecific implantation, sex-selection, etc.

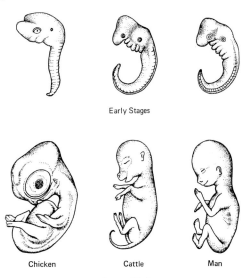

Early Stages

Chicken   Cattle   Man

Later Stages of Embryogeny

EMBRYOGENESIS IN VERTEBRATES
(This widely accepted Ernst Haeckel model has been seriously challenged recently.)

EMBRYO CULTURE: when developing plant embryos are lifted from the ovaries about 10-14 days after fertilization, they can survive and mature in aseptic cultures. This procedure may be applied for the rescue of hybrid embryos that may not survive in their natural environment because of the collapse of the endosperm, the nurturing tissue or the overgrowth of the nucellus. Although this technique itself does not overcome hybrid sterility, viable seedlings can be produced that may become fertile after doubling the chromosome number (with colchicine). The embryos (ear, fruit) require disinfection by calcium hypochlorite ($Ca[OCl]_2$ 5%, 8 to 10 min and several rinse in sterile H_2O). E2 culture medium mg/L H_2O, final conc.: NH_4NO_3 400, KNO_3 200, $MgSO_4 \cdot 7H_2O$ 100, $CaH_4(PO_4)_2$ 100, KH_2PO_4 100, K_2HPO_4 50. Before use

Embryo culture continued

add 2.5 mL/L chelated iron (FeSO$_4$.7H$_2$O 556 mg, diethylenetriamine pentaacetic acid 786.4 mg, H$_2$O 100 mL). Supplement it with 3% sucrose and add agar to solidify without making it too hard (about 0.6%) and autoclave. (See *in vitro*, embryo, ovary)

EMBRYOGENESIS: In vertebrates follows a rather similar pattern, especially at the early stages resembling also to lower forms. A large number of genes have been identified and the molecular mechanisms involved in their function are revealed. (See morphogenesis, embryonic induction, morphogenesis, gametogenesis, morphogenesis in *Drosophila*, figure on page 323).

EMBRYOGENESIS IN PLANTS: takes place after the egg has been fertilized by the sperm within the embryosac. The embryosac is located within the ovule. The pattern of embryogenesis of higher plant is shown below. The steps are genetically controled and numerous mutations have been identified that are responsible for the different morphogenic steps.

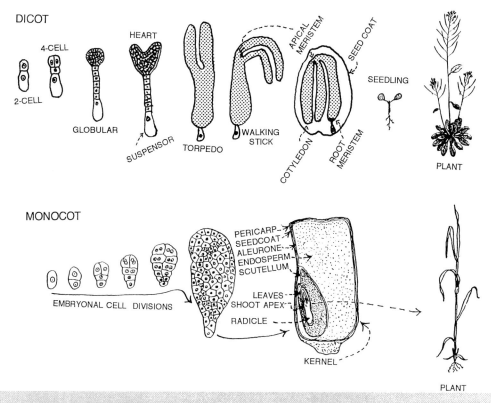

THE PATTERN OF EMBRYOGENESIS IN PLANTS IS DIFFERENT IN DICOTS FROM THAT IN MONOCOTS. ADDITIONAL VARIATIONS EXIST WITHIN THESE TWO GROUPS. THE DICOT PATTERN IS EXEMPLIFIED BY THE CRUCIFER *ARABIDOPSIS* AND THE MONOCOT PATTERN IS GENERALIZED ON THE BASIS OF MAIZE AND WHEAT.

The information regarding the molecular mechanisms involved in plants is lagging much behind to that in the best studied animals. The seed coat is entirely maternal tissue, and therefore the inheritance of seedcoat traits is "delayed". The endosperm develops from the fertilized fused "polar cells" and is generally triploid. In the majority of dicots, by the time the seed matures the endosperm tissue is generally insignificant in amount or entirely absorbed. In the majority of monocots the endosperm constitutes the bulk of the seed. In common language the monocot fruit is also called a seed but its proper designation is kernel. The outermost layer of the kernel is the maternal pericarp. Just under the pericarp there is the membrane-like seedcoat

Embryogenesis in plants continued

The outer layer of the endosperm is called aleurone. The scutellum of the monocots corresponds to the cotyledons of dicots. While the cotyledons emerge from the seed during germination, the scutellum does not. The "germline" of plants is not set aside during development as in animals yet the cells of the plant meristems form cell lineages that can be traced if visible genetic differences occur during ontogenesis. The size of the sectors formed permits an estimate on the time when the genetic alteration has taken place during development; large sectors indicate early event, small sectors show late mutations. (See megagametophyte, microgametophyte, phyllotaxy, morphogenesis, *Drosophila*, development, gametogenesis, embryo culture, gametophyte, gametogenesis)

EMBRYOGENESIS, SOMATIC: plant somatic embryos can be obtained by the techniques of tissue culture. The illustration shows embryogenesis in suspension culture and compares it with *in planta* embryogenesis. ↓ In most species it is not necessary to go to protoplasts and embryos may develop from other tissue cells on regeneration media. Several nutrient media have been developed in various laboratories (see Murashige and Skoog, Gamborg 5). For tobacco and *Arabidopsis* the following protocol is quite effective. *R4 Medium:* **I** Major mineral salts, mg/500 mL H_2O: NH_4NO_3 1,800, KNO_3 800, $MgSO_4.7H_2O$ 100, $CaH_4(PO_4)_2$ 100, KH_2PO_4 100, K_2HPO_4 50. **II** $CaCl_2$ 15g/100 mL. **III**. KI 75 mg/100 mL **IV**. Microelements mg/500 mL: H_3BO_3 6,200, $MnSO_4.H_2O$ 16,900, $ZnSO_4.7H_2O$ 8,600, $NaMoO_4.2H_2O$, $CuSO_4.5H_2O$ 25, $CoSO_4.7H_2O$ 29.54. **V**. Chelated iron: in 100 mL H_2O $FeSO_4.7H_2O$ 556 mg, diethylenetriamine pentaacetic acid 786.4 mg. **VI**. Vitamins, mg/50 mL: myo-inositol 5,000, thiamin 500, nicotinic acid amide 50, pyridoxine.HCl 50. Final solution: **I** 50 mL, **II** 290 µL, **III** 100 µL, **IV** 50 µL, **V** 500 µL, **VI** 100 µL, pH 5.6, sucrose 3 g, fill up to 100 mL, agar about 600 mg (varies according to batch) or Gellan gum 180 mg (must be separately dissolved in distilled water on hot plate with magnetic stirrer). The culture generally requires 5 stages with hormones added in µg/mL: R4-1 (callus initiation) 9RiP 1.5, 2,4-D 0.025 - 0.050, R4-2 (callus growth and regeneration) 9Rip 2.0, NAA 0.1, R4-3 (leafy callus) 9RiP 1.5, NAA 0.1, R4-4 (shoots appear) BAP 1, NAA 0.1, R4-5 (rooting) BAP 0.0005, NAA 0.05. After R4-4 *Arabidopsis* can be transferred to E2 medium *in vitro* (see embryo culture) where they produce seeds even in the absence of roots. Alternatively, after R4-5 the seedlings can be transferred to soil or commercial soil substitute. (RiP = isopentenyl adenosine, 2,4-D = dichlorophenoxy acetic acid, NAA: α-naphthalene acetic acid, BAP: benzylamino purine). Heat stable components are autoclaved, vitamins and hormones are filtered. (See embryo culture, cell genetics, tissue culture, ART)

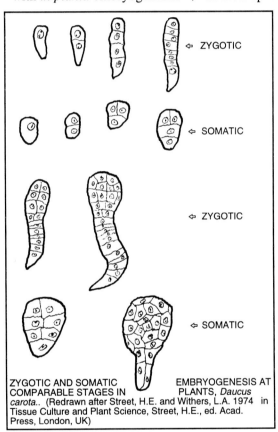

ZYGOTIC AND SOMATIC COMPARABLE STAGES IN EMBRYOGENESIS AT PLANTS, *Daucus carota*.. (Redrawn after Street, H.E. and Withers, L.A. 1974 in Tissue Culture and Plant Science, Street, H.E., ed. Acad. Press, London, UK)

EMBRYOID: develops through somatic embryogenesis. (See embryogenesis somatic)

EMBRYOID BODY: such as endoderm, ectoderm, neurons, yolk, cartilage, muscle cells formed from embryonic stem cells.

EMBRYONIC INDUCTION: interaction between and among tissues leading to tissue differentiation in the embryo. Vg1, activin, bone morphogenetic factor, fibroblast growth factor, etc. have been isolated from various sources and found to be effective in regulating embryonic differentiation in animals by binding to receptors, by being restricted in diffusion through the extracellular matrix, affecting signal transduction pathways, etc. (See embryogenesis, morphogenesis)

EMBRYONIC STEM CELL (ES): are cells of the early embryo that are capable of continuous growth and differentiation of animals. They are comparable to the meristems of plants.

EMBRYOSAC: of plants develops after meiosis from one of the (generally the basal) megaspores. After 3 divisions 8 cells are formed within a sac-like structure. The role of antipodals is unclear. The diploid polar cell is fertilized with one of the sperms and gives rise to the triploid endosperm. The egg is fertilized with the other sperm in the pollen tube and develops into the zygote and embryo, The synergids probably have an early role in nurturing the egg and zygote. (See gametophyte, embryogenesis, gametogenesis)

EMC: see enzyme mismatch cleavage, mutation detection

EMERIN: is a 254 amino acid serine-rich protein encoded by the STA gene, responsible for the Emery-Dreifuss muscular dystrophy. (Emery-Dreifuss muscular dystrophy)

EMERY-DREIFUSS MUSCULAR DYSTROPHY: see muscular dystrophy, emerin

EMF (extremely low frequency electromagnetic fields): see electromagnetic radiation

EMG SYNDROME: see Beckwith-Wiedemann syndrome

EMMER WHEAT: see *Triticum*

EMPHYSEMA, PULMONARY: is characterized by pathological inflation of the lungs and possible atrophy of the tissues. It appears to be due to autosomal dominant genes or to environmental causes (smoking). Emphysema may show up in infancy as a complication of other ailments. Emphysema develops in case of deficiency of α-antitrypsin. It can be medicated by regular supply of antitrypsin. (See antitrypsin, serpin)

EMPIRICAL RISK: the occurrence or recurrence of certain genetic defects cannot be predicted on the basis of theoretical principles or mathematical formulas because the genetic mechanisms and their control is not known. Some observations in larger populations indicate however a certain tendency of recurrence in families that had an afflicted individual in the pedigree. The calculation may take into consideration the empirical experience of onset of a disease with both genetic and age components. Example 100 families have 40 afflicted children older than 50 years and 170 unaffected who are still in the age group 20 to 50. Experience indicates that usually the age of onset begins at 50 and about half of the offspring of afflicted parents eventually expresses the condition. The empirical risk for the still healthy individuals then is (170 x 0.5)/210 $\approx$ 0.41. (See also genetic risk, risk, recurrence risk, clinical genetics, inbreeding coefficient)

EMS: a very potent alkylating mutagen and carcinogen. (See ethylmethane sulphonate)

En BLOOD GROUP: is limited apparently only or mainly to individuals of Finnish descent; it is very rare. (See blood groups)

ENaC (epithelial sodium channel): see ion channels, hypoaldosteronism

ENANTIOMORPHS: stereoisomers, molecules with structures that are mirror images of each other. In some instances only one of the enantiomorphs supports normal metabolism.

ENCEPHALITIS: brain inflammation. (See Rasmussen encephalitis, encephalopathies)

ENCEPHALOPATHIES: are usually recessive degenerative brain diseases. The childhood recurrent (autosomal dominant) form involves impaired muscular coordination, speech troubles. The Gerstmann-Straussler disease is also an autosomal dominant encephalopathy with late onset (around age 50) and involves feeble-mindedness. The central nervous system shows myeloid plaques, similar to those in the Creutzfeldt-Jakob disease. The bovine spongiform encephalopathy (BSE) is responsible for the fatal "mad cow disease" which bears similarities to

Encephalopathies continued

scrapie in sheep and it was believed that it is induced by a virus. Current views do not support the viral origin, rather it is attributed to an infectious protein (PrPc). In mice that have a null allele for the PrP^c gene, i.e. they are $PrnP^{0/0}$, even large doses of prions do not cause brain damage. There may be an undetermined chance that these encephalopathies are transmitted to humans from animals by consumption of the meat. Most of the encephalopathies have usually late onset. The incubation period for BSE may be several years. There is no direct evidence that humans are infected by BSE, however it is known that mice can be infected with scrapie. The major outbreak of BSE is attributed to the feeding of infected sheep offal to cattle. Infectivity can be abolished only by heating the animal products at high temperature and by some solvents. There are no sufficient data whether milk or eggs or poultry meat would transmit any source of infective material. (See prion, scrapie, Gerstmann-Straussler disease, Creutzfeldt-Jakob disease, kuru)

ENCYCLOPEDIA OF THE MOUSE GENOME: contains genetic and cytogenetic information on mouse, electronically obtainable through File Transfer Protocol (FTP), Gopher or World Wide Web Software (WWW) in Sun (UNIX) or in Macintosh versions. (See mouse, databases)

end : bacterial genes coding for DNA-specific endonucleases.

END LABELING: the *Klenow fragment* of bacterial DNA polymerase I can add α-P^{32} dNTP to the 3'-OH end of a nucleotide (polynucleotide) or by transfering the γ -phosphate of ATP to the 5'-OH end of DNA or RNA (forward reaction) or by exchanging the 5'-P of a DNA (in the presence of excess ADP) by the γ-P of radiolabeled ATP, using T4 bacteriophage *polynucleotide kinase*. This 5'-labeling is used for the Maxam & Gilbert method of DNA sequencing or in any other procedure when 5' labeling is required. (See Klenow fragment, DNA sequencing)

3'-END OF NUCLEIC ACIDS: the OH at the 3' C atom on the ribose or deoxyribose. (See nucleic acid chain growth)

5'-END OF NUCLEIC ACIDS: The first nucleotide in the chain retains the three phosphates whereas the subsequent ones form phosphodiester bonds with one phosphate between the 5' end and the 3' -OH position of another. (See phosphodiester bond, nucleic acid chain growth)

ENDANGERED SPECIES: see species extant

ENDEMIC: indigenous to a population rather then introduced, confined to a population or area.

ENDERGONIC REACTION: consumes energy.

ENDOCARDIAL FIBROELASTOSIS: is a human X-linked recessive thickening of the heart wall muscles due to collagen proliferation (Type I). In Type II (recessive Xq28, called also Barth syndrome) besides the defects in the heart muscles, neutropenia (decrease in the number of neutrophylic leukocytes) morphological anomalies of the mitochondria was detectable by electronmicroscopy. (See heart disease, see neutropenia, neutrophil, collagen)

ENDOCARP: the inner layer of the fruit wall of plants such as the stony pit of peaches, cherries.

ENDOCRINE: hormone producing glands in response to peptide activators within an organism secrete their product into the blood stream without the reliance on special duct.

ENDOCRINE NEOPLASIA, MULTIPLE (MEN): the autosomal dominant MEN I (11q13) involves endocrine adenomas in several tissues (stomach, lung, parathyroid, pituitary, colon, pancreas, etc.). The dominant MEN2 and MEN2A involving also phaeochromocytoma and amyloid-producing medullary thyroid carcinoma are apparently in chromosome 10q21.1. MEN2B is similar to MEN2A but frequently shows also neural hyperplasias of the mouth area and in the colon. In all of these cases involve receptor-like tyrosine kinases (RET). MEN3, also dominant, was located in vicinity of MEN2A, and also shares the same symptoms and displays neural tumors. (See adenoma, phaeochromocytoma, RET oncogene, protein-tyrosine kinase)

ENDOCYTOSIS: an uptake mechanism of cells involving invagination of the cell membrane and then cutting off the (clathrin coated) vesicle so formed inside the cell. The cargo receptor transmembrane proteins recognize various molecules by some sort of specificity. To one terminus

Endocytosis continued

of the cargo proteins an *adaptin* molecule is attached that in turn recognizes clathrins.

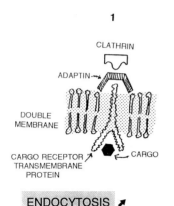
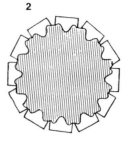

1. CLATHRIN / ADAPTIN / DOUBLE MEMBRANE / CARGO RECEPTOR TRANSMEMBRANE PROTEIN / CARGO / ENDOCYTOSIS
2. FORMATION OF A CLATHRIN-COATED VESICLE FROM SEVERAL INDIVIDUAL COMPLEXES AS SHOWN AT 1. THE CONTENTS ARE NOT SHOWN BUT INCLUDE THE LOADED CARGO RECEPTORS AND ATTACHED ADAPTINS
3. AFTER CLATHRINS AND ADAPTINS ARE SEPARATED AND REMOVED A MEMBRANE COATED VESICLE IS FORMED CONTAINING SEVERAL CARGO RECEPTORS AND CARGOS

Then clathrin coated vesicles are formed. When the clathrin and adaptin molecules separate from the membrane, a membrane-bound vesicle is formed enclosing the cargo molecules. The adaptins are multisubunit adaptor molecules that can recognize the (Tyr, X, Arg, Phe) peptide signals near the carboxyl end of the receptor, reaching into the cytosol. The carboxyl end of phosphorylated M6P proteins are also recognized by the adaptin molecules in the Golgi apparatus. The M6P proteins are named such by the mannose-6-phosphate groups that are linked to the amino ends of lysosomal enzymes. The M6P transmembrane proteins bind lysosomal hydrolases and assist in packaging them into transport vehicles that fuse with endosomes that transport molecules into lysosomes. The endosomes are membrane-enclosed transport vesicles, carrier of material to the lysosomes. An endocytotic compartment of the B cells may accumulate class II type MHC molecules and antigens. (See clathrin, lysosome, antigen processing, MHC)

ENDODERM: the internal cell layer of the embryo from what the lung, digestive tract, bladder and urethra are formed.

ENDODERMIS: in plants is composed of heavy-walled cells and without intercellular space among them; they are found around the vascular system particularly in roots.

ENDODUPLICATION: is basically similar to what is called endomitosis but it was first used to denote the diploidization of androgenetic or gynogenetic embryos. (See endoreduplication, androgenesis, gynogenesis)

ENDOGAMY: mating within a group, a sort of inbreeding. (See inbreeding)

ENDOGENOTE: is the tract of the recipient bacterial genome that is homologous to the donor DNA paired with it. In this merozygous condition homo- and heterogenotes can be distinguished, depending whether the donor sequences are identical or genetically different from that of the recipient. (See also exogenote, merozygote)

ENDOGENOUS RHYTHM: periodical (oscillatory) changes in cells that occur without external influences.

ENDOGENOUS VIRUS: is integrated into the host chromosome. (See provirus, temperate phage)

ENDOGLIN: the receptor protein on the vascular endothelial cells for the transforming growth factor β, encoded in human chromosome 9. (See telangiectasia familial hemorrhagic)

ENDOMITOSIS: chromosome replication is not followed by cell division and the result is polyploidy. In *Schizosaccharomyces* it was found that gene $rum1^+$ overexpression permits repeated Starts of the cell cycle without mitoses in between them, and deletion of the gene allows successive mitoses without an S phase in between them. It appears as if $rum1^+$ would regulate

Endomitosis continued
Cdc2 which apparently has two functional forms. Overexpression of *rum1⁺* would lock *Cdc2* in the form when it promotes the synthetic (S) phase and deletion of *rum1⁺* would switch *Cdc2* into the mitosis-promoting form. (See endoduplication, endoreduplication, polyploid, cell cycle, licensing factor, CDC27)

ENDONUCLEASE: enzyme that cuts DNA or RNA at internal positions. (See also exonuclease, restriction enzymes)

ENDOPLASMIC RETICULUM (ER): internal membrane system within the eukaryotic cell with secretory channels. Within these membrane-bound compartments proteins and lipids are synthesized. The rough endoplasmic reticulum has attached ribosomes causing the non-smooth appearance. Many nascent peptide chains are transferred to the ER where the synthesis of the proteins is completed. Within the ER the folding of the proteins takes place. If the folding would slow down the unfolded protein response (UPR) stimulates the transcription of ER-resident proteins. The Hac1 basic leucine zipper protein is a transcription factor that binds to the UPR element in the promoter of the UPR genes. The presence of Hac1 is regulated by splicing that bypasses the spliceosome and mediated by a tRNA ligase protein. (See chloroplast endoplasmic reticulum, signal peptide, signal sequence recognition particle, cell structure)

ENDOPOLYPLOID: cell with increased chromosome number because of endomitosis, i.e. the chromosomes have replicated but cell division was skipped. Such phenomena commonly occur repeatedly in cultured plant and animal cells. (See endomitosis, endoduplication)

ENDOPROTEASES: a family of enzymes involved in the processing of proteins, e.g. the large peptide hormone precursors, large secreted proteins, pheromones, etc.

ENDOREDUPLICATION: the chromosome replication is not followed by centromere and cell divisions and therefore multi-stranded (polytenic) chromosomes are formed. The mechanism of this phenomenon requires the prolongation of the S phase and suppression of the M phase. Polytenic chromosomes are present in the salivary gland cells of some flies (*Drosophila, Sciara*, etc.) but they occur in some plant tissues too such as the endosperm of cereals, in some cells (chalaza, antipodal) of *Allium ursinum, Aconitum ranunculifolium*, etc. (See polytenic chromosomes, endoduplication, chalaza, antipodal cells, replication during the cell cycle)

ENDORPHINS ("endogenous morphins"): are neuropeptide ligands of opiate receptors in the brain pituitary gland and peripheral tissues of vertebrates. Their physiological effects mimics morphines. They have potential therapeutic use in nervous disorders, pain perception, etc. (See opiocortin, pituitary, morphine, brain human)

ENDOSOME: see endocytosis

ENDOSPERM: the nutritive tissue within the seed developed from the fertilized polar cells in the embryosac. In some plants the endosperm develops to a substantial mass and persists as the bulk of the seed (the majority of the monocots) or it may be gradually consumed and only traces are visible by the time the embryo fully develops because that function is relegated to the growing cotyledons (majority of dicots). In monocots the surface layer of the endosperm is called *aleurone*. Normally the embryo is diploid and the endosperm is triploid. Some mutations may permit the development of the endosperm without fertilization of the fused polar nuclei. In species crosses the 2:3 proportion may be altered with deleterious consequences for the embryo. Crossing a diploid female with a tetraploid male, the embryo will be triploid and the endosperm will be tetraploid (3:4) but crossing a tetraploid female with a diploid male the number of genomes in the embryo remains the same (3n) but the endosperm will be pentaploid (3:5). Thus the gene dosage in the endosperm is usually different from that in the embryo. Some species of plants have different developmental patterns beginning with the embryosac. (See embryogenesis in plants, embryosac)

ENDOSPERM BALANCE NUMBER: term was suggested for the unusual observations that within some species the tetraploids do not effectively fertilize other tetraploids (xx) but may do so with diploids (2x). When their chromosome number is doubled (8x) they may be crossed with tetraploids, except the progenitor tetraploid. Thus a curious chromosome balance is observed. (See endosperm)

ENDOSPERM MOTHER CELL: the fused polar cells of the endosperm (2n) that when fertilized by a sperm gives rise to the triploid endosperm tissue in the seeds of plants. (See embryosac, gametogenesis)

ENDOSPORE: a dormant bacterial cell, resistant to most treatments that normally kill active cells. The process of sporulation has been extensively studied in *Bacillus subtilis*. Also a fungal spore which develops within the cytoplasm of a cell (sporangium). (See forespore)

ENDOSYMBIONT: an organism that lives within the cell of another. (See infectious heredity)

ENDOSYMBIONT THEORY: is an evolutionary idea suggesting that plastids and mitochondria were originally free-living microorganisms and later captured by nucleated cells and then they became cellular organelles in eukaryotes. This old hypothesis is now supported by DNA sequence analyses primarily of organellar and eubacterial and archebacterial rRNA genes. It appears that the plastid rRNA core sequences resemble that of gram-positive cyanobacteria and the mitochondrial rRNAs indicate descents from eubacteria. The dispute among evolutionist is still extant regarding monophyletic or polyphyletic endosymbiotic origin of these organellar genomes. Secondary endosymbiosis is believed to occur when a eukaryote ingested and retained other eukaryotes with chloroplasts. In these cases the organelles may be enclosed by four-layer membranes. (See chloroplasts, mitochondria, nucleomorph, rRNA)

ENDOSYMBIOSIS: see endosymbiont

ENDOTHELIN: is a peptide hormone secreted by the veins and inner tissues of the heart. It controls heart muscle functions, including myocardial infarction, protein kinase A (PKA) and chloride, potassium and calcium ion channel functions. Mutation in the endothelin may lead to hereditary hypoventilation. Endothelin mutations may account for the Hirschprung disease and the Shah-Waardenburg syndrome. (See PKA, ion channels, Hirschprung disease, Shah-Waardenburg syndrome)

ENDOTHELIUM: layer of epithelial cells of mesodermal origin, and lining organ cavities.

ENDOTHERMIC: a chemical reaction that takes up heat.

ENDOTOXINS: bacterial lypopolysaccharide toxins, attached to the outer membrane and secreted only inward the cell and released only when the cell disintegrates. They may present serious hazard to humans (diarrhea, bleeding, inflammation, increase of white blood cells, etc. (See toxins)

ENDPOINT: in chemistry it indicates the highest dilution during titration that still gives a detectable reaction with another substance. In genetic toxicology the method of identification of the lowest effective dose of a mutagen or carcinogen that causes point mutations, chromosome breakage, unscheduled DNA synthesis, etc. (See bioassays of genetic toxicology, unscheduled DNA synthesis)

END-PRODUCT INHIBITION: see feedback control

ENGRAFTING: see grafting

ENEDIYNES: are anticancer antibiotics. They cleave the DNA by generating benzenoid diradicals when activated. They cleave the DNA backbone by extracting from it hydrogen atoms. (See neocarzinostatin)

ENERGY CHARGE: $[(ATP) + 1/2(ADP)]/[(ATP) + (ADP) + (AMP)]$; i.e. it measures the phosphorylating capacity of the adenylate system. The energy charge is 0 if only AMP is available and it is 1 if all AMP is converted to ATP.

ENERGY COUPLING: transfer of energy from one reaction path to another.

ENGRAFTING: adding another tissue by surgical means, and if it is established it is engrafted.

engrailed (*en*, *Drosophila* gene, map location 2-62): among a variety of phenotypic consequences: half of the larval body segments are deleted and the remaining ones are duplicated. (See morphogenesis in *Drosophila*)

ENHANCER: cis-acting positive regulatory elements positioned (frequently near Z DNA) either upstream or downstream (by several kb) of the initiation of transcription. The enhancer may be located also within the transcription unit. It alters chromatin structure to facilitate transcrip

Enhancer continued

tion by binding special proteins. Enhancers are present in all eukaryotic genes and may increase the basal rate of transcription by two orders of magnitude. Prokaryotes usually do not use enhancers, however some σ subunits of *E. coli* RNA polymerase makes this enzyme enhancer-responsive. For more about enhancers see Simian virus 40. (See also G box, silencer, TAF, activator, regulation of gene activity, octa)

ENHANCER TRAPPING: see gene fusion

En-I (Enhancer-Inhibitor): system of transposable elements of maize. (See Spm)

ENKEPHALINS: are endogenous opioid peptides regulating pain, stress response, aggression and dominance behavior. (See opiates)

ENL: a serine- and proline-rich protein (encoded in human chromosome 19p13); in chromosomaltranslocations it may be respopnsible for acute lymphocytic leukemia. (See leukemias)

ENOL FORM: of a molecule contains an OH group where the keto form has C=O, e.g.

$$-C=C-OH \Leftrightarrow -C-C=O$$
$$|$$
$$H$$

enol - keto tautomerism; the enolic form may be subject to proton shift. (See tautomeric shift, hydrogen pairing)

ENOLASE (phosphopyruvate hydrolase, PPH, ENO): catalyzes the conversion of 2-phosphoglycerate into phosphoenolpyruvate; ENO1 is encoded in human chromosome 1pter-p36.13. The neuron-specific enzyme (ENO2) is encoded in human chromosome 12p13, and the muscle specific ENO3 is encoded in chromosome 17pter-p12.

ENUCLEATE: cell without a nucleus. (See nuclear transplantation, cytoplast)

ENTELECHY: the vitalists postulated inner force of organisms that is responsible for life and growth. In modern morphogenetic theory, the basis of the internal program of development.

ENTEROBACTERIA: a large group of gram negative bacteria with very wide distribution in insects, higher animals and plants and are represented by *Escherichia coli, Salmonella, Shigella, Serratia, Klebsiella, Proteus, Shigella, Erwinia, Yersinia*, etc. Many of them are pathogenic and some are saprophytes and others are facultative pathogens.

ENTEROKINASE DEFICIENCY: autosomal recessive defect in the activation of pancreatic proteolytic proenzymes (such as chymotrypsinogen, procarboxypeptidase, proelastase) by this intestinal enteropeptidase, and consequently hypoproteinemia and general weakness results. (See trypsinogen deficiency)

ENTEROTOXIN: is produced by enteric (intestinal) bacteria. (See toxins)

ENTHALPY (H): heat content of a system.

ENTHALPY CHANGE (ΔH): the difference between the energy required for disrupting a chemical bond and gained by forming new one(s).

ENTODERM: same as endoderm. (See endoderm)

ENTREZ: source of nucleotide and protein sequence information. Retrieval computer software and databases are distributed on CD-ROM. The software is public and available in Macintosh or IBM PC (Windows) formats. General questions through INTERNET: <info@ncbi.nlm.nih.gov> or <net-info@ncbi.nlm.nih.gov> (See databases)

ENTROPY: the measure of energy unavailable for use within a system. The entropy increases by natural processes of aging.

ENU (N-ethyl-N-nitrosourea): an ethylating agent; one of the most potent point mutagen in mouse (mutation rate/locus 6.6 to 15 x $10^{-4)}$, in *Arabidopsis* within the concentration range of 0.25 to 1.25 mM (18 h exposure of seeds) caused embryo lethals from 2.8 to 85.1% of the plants.

ENUCLEATION: removal of the cell nucleus. (See nuclear transplantation, cytochalasin)

Env Z: a phosphorylating enzyme (kinase) on the bacterial membrane envelop.

ENVELOPE: generally double-layer membranes surrounding organelles in the cell.

ENVIRONMENTAL DEVIATION: quantitative characters are manifested in a phenotypic value, called P that can be partitioned into a genotypic value G and an environmental deviation,

Environmental deviation continued
E, i.e., P = G + E where P is measured as the mean value of a population. The value of G can be determined by heritability estimates. (See heritability)

ENVIRONMENTAL EFFECTS: are of significant importance for the expression of genes. The genetic constitution provides the blueprint but the realization has frequently a variable environmental component. The genotype had been identified as providing a programed *reaction norm* but the temporal sequence and intensity of gene expression is determined by the internal (regulatory DNA elements) and the environment. Among the endless good examples, the expression of inducible enzymes, the immune reactions, the switching from fetal hemoglobin $\zeta\zeta\epsilon\epsilon$ in the early embryo to $\alpha\alpha G_\gamma A_\gamma$ by the 8th week then to $\beta\beta\delta\delta$ shortly before birth and by six months postnatal it switches to $\alpha\alpha\beta\beta$ with about 2% $\alpha\alpha\delta\delta$. In plants, the onset of the flowering in many species requires vernalization and/or appropriate photoperiodism. Sometimes environmental effects persist through meiotic events. (See dauermodification, delayed inheritance, maternal effects). The greatest effect the environment exercises on the expression of quantitative genes. (See quantitative traits, heritability). Many environmental factors are mutagenic, these may include food and feed additives, cosmetics, drugs, medicines, industrial and agricultural chemicals, natural food products, radioactive fallout, environmental mutagens.

ENVIRONMENTAL LOAD: some specific environmental conditions may favor and maintain alleles that convey inferior fitness to the individuals under usual, average conditions. The alleles so maintained represent the environmental load of the population. (See genetic load)

ENVIRONMENTAL MUTAGENS: carcinogens and teratogens occur in the laboratory, industry and agricultural environment. Many potentially mutagenic agents are present in every household. A very incomplete list is provided below:
* Acetaldehyde: an intermediate of organic solvents, preservative of certain food products; it may be present in small quantities in the sweet smell of ripening fruits
*Acridine dyes: are laboratory chemicals and may be present in some antimalaria drugs
*Acrylamides: used for gel electrophoresis and are absorbed also through the skin
*Actinomycin D: an antibiotic used to block transcription in the laboratory is a carcinogen
*Aflatoxins: about a dozen different natural products of *Aspergillus* fungi detectable on peanuts, grains and other feed and food
*Allylisothiocyanate: present in cruciferous vegetables, mustard condiments, horseradish
*3-Aminotriazole: herbicide that is largely undetectable by mutagen assays yet it is a powerful carcinogen
*Aramite: an insecticide used some time ago for fumigation of greenhouses
*Asbestos: a heat insulator and filtering agent, present in many buildings
*Atabrine: an antimalaria drug
*Atrazine: a herbicide that may be activated into a mutagen by the metabolism of plants
*Benzimidazole: may be present in pharmaceuticals, preservatives and insecticides
*Benzo(a)pyrene: regularly formed during combustion of many organic material (in fireplaces, automobile exhaust fumes, grilled-charcoal broiled meat, tobacco smoke) refined mineral oils, commercial wax products
*Bisphenol A: used for manufacturing resins and plastic coating of various containers
*β-Propiolactone: used as plasticizer, in wood processing, tobacco processing, additive to leaded gasoline, insecticides
*Bracken fern extracts
*Caffeine: is not mutagenic or carcinogenic itself but interferes with genetic repair mechanisms
*Captan: fungicide
*Carbontetrachloride: solvent, seed fumigant
*Chloral hydrate: microtechnical reagent, sedative
*2-Chloroethanol: polymerizing agent, insecticide, herbicide, it is present in some sedatives
*Chloroform: laboratory solvent, in some office supplies, may be formed in chlorinated drinking water

Environmental mutagens continued
* *Chloroprene: in elastomers and adhesives
* *Chloropromazine: in some tranquilizers
* *Colchicine: is used for polyploidization and it is very toxic
* *Cycasin: present in cycade plants
* *Diamines: contained by some hairdyes
* *Diazomethane: laboratory reagent
* *Dichlorvos: insecticide
* *Diepoxybutane: industrial polymerization agent, may be present in some pharmaceuticals
* *Diethylsulfate: a supermutagen
* *Dimethylsulfate: industrial methylating agent of cellulose, polymerizing compound, insecticide, stabilizer
* *Dithranol: in some antidermatosis and ringworm drugs
* *EDTA (ethylenediamine tetraacetic acid): chelator and antioxidant in laboratory and industry
* *Epichlorohydrine: present in some epoxy resins, gum, paint, varnish, nail polishes, manufacturing crease-resistant fabrics, paper processing, waterproofing
* *Epoxides: precursors of ethylene glycol, dioxane, carbowax, monoethanolamine, acetonitrile, plastics, gaseous sterilants. Epoxides may be trapped in the products or any material in contact with them may be contaminated
* *Estrogen (β-estradiol): fertility hormone used in ART is carcinogenic (See ART)
* *Ethidium bromide: used as nucleic acid stain and as a mutagen for mitochondrial DNA
* *Ethylmethane sulfonate: one of the most widely used and most potent laboratory mutagens. Some methanesulfonates are now banned ingredients of older prescription drugs
* *Ethyleneimine: used for the manufacture of flame-retardant clothing, crease-resistant and shrinkage controling of fabrics, in insecticides, soil-conditioners, synthetic fuels. It is a supermutagen
* *Ethylenedibromide: a fumigant for grain storage
* *Ethylmercuric chloride: a fungicide
* *Fats (after oxidation in rancy food) may become mutagenic and carcinogenic
* *Formaldehyde: disinfectant , preservative of museum specimens of animals and organs, fixative, in adhesives, crease-, crush- and flame resistance aid in automobile exhaust
* *Fumes: released by outdoor burning, defective wood-burning fireplaces, coal furnaces
* *Glycidol: may be present in glycerol; used for manufacturing water-repellent fabrics, food preservative. It is a very powerful mutagen
* *Hairdyes: certain types are mutagenic/carcinogenic
* *Hydrazines: in photographic materials, rocket fuels, preservatives, solvents, gasoline additives
* *Hydrogen peroxide: used for bleaching flour, starch, paper, tobacco, cosmetics (hair lightener), stabilizer, plasticizer
* *Hydroxylamine: used in nylon manufacturing, photographic materials, adhesive, paints. It converts specifically cytosine into a thymine analog
* *Hydroxyurea: breaks heterochromatin in the chromosomes and present is some antileukemic, dermatological drugs
* *Isotopic tracers for emitting β and γ radiations
* *Lindane: fungicide
* *Nicotine: an addictive component of tobacco; smoking is a major cause of human cancer
* *Nitrosoamines: extremely powerful mutagens but present in small quantities in nitrite treated meat products; formed from nitrites by the action of stomach acids and intestinal microbes
* *Nitrofurans (AF-2): onetime widely used preservative in Japan for fish and soybean products
* *Nitrogenmustards: were used as anticancer drugs, powerful radiomimetic agents
* *Nitrosoguanidines: are extremely potent laboratory mutagens
* *Nitrous acid: widely used preservatives for meat products are direct mutagens and can be converted by mammalian metabolism into nitrosoamines
* *Organomercurials: were widely used in fungicides

Environmental mutagens continued
  *Peroxides: bleaching agents, disinfectant; may be formed from tryptophan under UV
  *Phenol: used in DNA extraction; extremely toxic, causes vesicles on skin and possibly mutagenic
  *Phenylmethylsulfonyl fluoride (PMSF): used in pulsed-field gel electrophoresis is very toxic and absorbed through skin
  *Polychlorinated biphenyls: once used for various industrial processes and in pesticides
  *Propane sultone: may be present in detergents and dyes
  *Pyrrolizidine alkaloids: occur in several species of plants like *Senecio, Crotalaria,* etc.
  *Quinacrine: antimalarial drug
  *Sodium azide: laboratory reagent and powerful mutagen in some organisms not in others
  *Sodium bisulfite: preservative in fruit juices, wine and dried fruits
  *Sodium nitrite: common preservative of cold-cuts, fish and cheese and can be converted in the body into nitrosoamines
  *Streptozotocin: a prescription drug used against athletes' foot fungus, an alkylating agent
  *Thiotepa: used as flame-retardant, crease-resistant, water-proof fabrics, in manufacturing dyes, adhesives and drugs
  *Trichloroethylene: a weak mutagen that may be present in dry-cleaning agents, degreasing fluids, paint, varnish and food-processing solvents
  *Triethylenemelamine: a cross-linking agent, used for finishing rayon fabrics, waterproofing of cellophane, insect chemosterilant and anticancer drug
  *Trimethylphosphate: gasoline additive, insecticide, flame retardant, polymerizing agent, insect chemosterilant
  *Tri(2,-3-dibromopropyl) phosphate: is a flame retardant
  *Ultraviolet radiation: may cause skin cancer (melanoma) and mutation in tissues close to the surface; may be hazardous also through the peroxides generated. Wear protective goggles
  *Urethanes: used by the plastic industry, for manufacturing resins, fibers, textile finishes, herbicides, insecticides and drugs
  *Vinyl chloride: the monomer of polyvinyl plastics. May be liberated from the polymers by heat (in a locked-up car) and may contaminate the surface of plastic sheets, bags, water-hoses. The monomer is a powerful carcinogen and mutagen.
  Many of the mutagens and carcinogens require metabolic activation for effectiveness. There is an inter-individual variability of susceptibility particularly at low degrees of exposure. (See also Ames test, mutagen essays, bioassays for environmental mutagens, mutagenic potency, activation of mutagens, adduct, pseudo-cholinesterase deficiency, choline esterase)

ENVIRONMENTAL VARIANCE: see genetic variance, environmental deviation

ENZYME INDUCTION: requires the presence of the substrate or substrate analog. (See *Lac* operon, *Arabinose* operon)

ENZYME-LINKED IMMUNOSORBENT ASSAY: see ELISA

ENZYME MIMIC: catalytic antibody.

ENZYME MISMATCH CLEAVAGE (EMC): uses the resolvase enzyme of bacteriophages to recognize and cut at mismatches in double-stranded DNA that has been amplified by PCR. The PCR amplified DNA is expected to have matching and non-matching strands if there was mismatch in the original double-stranded DNA. This way heterozygosity or mutation may be detected after gel electrophoresis. (See resolvase, mismatch, mutation detection)

ENZYMES: are generally protein molecules. Some ribonucleic acids also have enzymatic functions (ribozymes). The protein part of the enzyme is the apoprotein or apoenzyme. Their catalytic function may require cofactors, such as metals (Cu^{2+}, Fe^{2+}, K^+, Mg^{2+}, Mn^{2+}, Mo, Zn^{2+}, etc.) or organic compounds (vitamins, nucleotides, etc.) called coenzymes, also called the prosthetic group of the enzymes. The complete enzyme (apoenzyme + prosthetic group) is the holoenzyme. Enzymes are organic catalysts; they mediate biochemical reactions without becoming parts of the reaction products. The site(s) of the enzyme molecule interacting with the substrate (the molecule to be acted on) are the active sites of the enzymes. Gene functions are

Enzymes continued
carried out by enzymes. RNA is transcribed on the DNA and the RNA is translated into protein. Enzymes are named by adding: ase to either the name of the substrate or to the name of the reaction, e.g., DNase, phosphorylase. The International Union of Biochemistry (*Enzyme Nomenclature*, 1972, Elsevier, Amsterdam) classified enzymes into six major groups and assigned code names to them. The six groups are 1. *oxidoreductases* (transfer electrons), 2. *transferases* (catalyze molecular group transfers), 3. *hydrolases* (cleave covalent bonds and transfer H and OH, respectively to the products), 4. *lyases* (form or remove double bonds), 5. *isomerases* (rearrange molecules internally), 6. *ligases* (mediate condensations by forming C-C, C-N, C-O and C-S bonds while cleaving ATP). In technical descriptions enzymes are identified by E.C. (enzyme classification) numbers where the first digit refers to the number of one of the above groups and the following digits specify more closely the nature of the reaction mediated, thus DNA polymerase I of *E. coli* has the E.C. number 2.7.7.7 whereas restriction endonuclease Eco RI is 3.1.21.4., bovine RNase is 3.1.27.5 and glucose-6-phosphate dehydrogenase is 1.1.1.49. (See also protein synthesis, inhibition, repression, induction, allosteric control, competitive inhibition, effector, Michaelis-Menten equation, Lineweaver-Burk equation, Eadie-Hofstee plot, regulation of enzyme activity, subunits, recombination mechanisms)

EOSINOPHIL: white blood cells (readily stainable by eosin dye), generally display bilobal large nucleus. They modulate allergic and inflammatory reactions of the animal body and mediate the destruction of parasites. (See granulocytes, blood)

EP: effector proteins stimulate the conversion of E.GTP into E.GDP. (See GTP, GDP)

EP (early pressure): in zebrafish gynogenetic embryos are produced by exposing early embryos, fertilized by UV-inactivated sperm, to high hydrostatic pressure immediately after fertilization to suppress the second meiotic anaphase. A heatshock, 15 minutes after fertilization, suppresses the first mitotic division and leads to gynogenesis. (See gynogenesis, zebrafish, UV)

EPH: a family of receptor tyrosine kinases involved in axon guidance, cell migration and embryo differentiation. (See tyrosine protein kinase)

EPIBLAST: precursor of the ectoderm and other embryonal tissues. (See ectoderm, organizer)

EPIBOLY: the first embryonic cell movement whereby the upper portion of the yolk bulges in the direction of the animal pole and resulting in the formation of a dome. The blastoderm cells begin spreading over the yolk. (See morphogenesis in *Drosophila*)

EPICANTHUS: a vertical skin fold near the eyelid; it is a normal characteristic for some oriental human races and occurs also as part of some syndromes. (See Down syndrome)

EPICHROMOSOMAL: outside the chromosomes; an epichromosomal vector does not integrate into the chromosome and thus does not cause insertional mutation but may not be able to replicate continuously. (See vectors, insertional mutation)

EPICOTYL: the stem section immediately above the cotyledons of plant embryos and seedlings.

EPIDEMIOLOGY: the study of the distribution, cause and modifying factors of diseases in human and other populations.

EPIDERMAL GROWTH FACTOR: see EGF

EPIDERMIS: the non-vascular outer cell layer, derived from the embryonal ectoderm of animals; the surface cell layer of plants.

EPIDERMOBLAST: gives rise to the epidermal cells.

EPIDERMOLYSIS: is the dissolution of the skin in spots resulting in blisters (bullae). It is a characteristic symptom of many hereditary skin diseases such as ichthyosis, keratosis, pemphigus, acrodermatitis, porphyria, etc. The epidermolysis bullosa simplex is encoded in human chromosome 8q24.13-qter and it is caused by mutation in plectin whereas a similar disease of mouse is associated with a defect in integrin. (See skin diseases, collagen, intermediate filaments, plectin, integrin)

EPIDIDYMIS: a narrow pouch-like structure attached to the testis serving for storage, maturation and forwarding the spermatozoa into the vas deferens during ejaculation.

EPIGENE CONVERSION: if the gene is introduced into the genome in an additional copy (by transformation of plants) the other copy is silenced by transcription. (See co-suppression)

EPIGENESIS: process by which differentiation takes place from undifferentiated cells as programed by the genome, without any mutational event. It is expected that during meiosis the epigenetic alterations would be erased by reversion to the embryonic stage. In fission yeast the mating type conversion genes (*mat*) are however transmitted not just mitotically but meiotically too. Similar examples in higher organism is the paramutant state and the position type variegation. (See preformation, paramutation, position effect, mating type determination)

EPIGYNY: the plant ovary being embedded in the flower receptacle and other flower organs are above it. (See flower differentiation, receptacle)

EPILEPSY: suddenly recurring impairment of consciousness, involuntary movements, nervous disturbances occurring simultaneously or separately, caused by acquired or genetic factors. Epilepsy is a very complex disorder. Some seizures may occur in early infancy others are triggered by light (photogenic epilepsy) or by reading. According to one study among 27 monozygotic twins 10 pairs were both affected (concordant) and 17 pairs were not, and among 100 dizygotic twins only 10 pair had it both. Thus some epilepsies appear to have a strong genetic component (heritability, $h^2 > 0.40$). *Grand mal* epilepsy is a severe form of seizures occur in 4-10% of the epileptics and it generally develops fully by adult age. *Petit mal* epilepsy is a milder form and it is less frequent. The *autosomal dominant nocturnal frontal lobe epilepsy* (ADNFLE, human chromosome 20q13.2-q13.3) is caused by mutation in the neural nicotinic acetylcholine receptor. Some seizures or spasms are associated with a number of genetic maladies (phenylketonuria, epiloia, Zellweger syndrome, Menke's syndrome, ceroid lipofuscinosis, adrenal hypoplasia, multiple sclerosis, glycogen storage diseases, lysosomal storage diseases, myoclonic epilepsy, porphyria, glutamate decarboxylase deficiency, rickets, galactosemias, Friedreich ataxia, West syndrome, Lesch-Nyhan syndrome.) A gene for partial epilepsy was located to human chromosome 10q. Epilepsy occurs also in animals; in chickens it is controled by a single recessive gene with reduced penetrance. The underproduction of γ-aminobutyrate (GABA), a molecule with important role in neurotransmission, may also lead to epileptic seizures. Rasmussen's encephalitis, a rare but severe form of childhood epilepsy is caused by auto-antibodies that turn against the brain's glutamate receptors. The *unc* (*uncoordinated*) mutants of the nematode, *Caenorhabditis* are also affected in GABA-mediated functions. Myoclonal epilepsy with ragged red fibers (MERRF) is determined by mitochondrial defects. (See disease mentioned under the specific entries, mitochondrial disease in humans, autoimmune diseases, seizures, glutamate decarboxylase deficiency disease)

EPILOBIUM: is a herbaceous plant species (2n = 36) in the family of *Onagraceae* distributed widely in Europe, Asia and Africa. It became a favorite organism for the study of non-nuclear inheritance since the early decades of this century. A variety of cytoplasmic mutations were discovered in the different species that were transmitted maternally. The identity of the cytoplasmic genetic material (plasmon) was preserved even when *Epilobium luteum x E. hirsutum* was backcrossed with *E. hirsutum* for 25 generations and less than 3×10^{-8} chance remained for the presence of *E. luteum* genes in the nucleus. Some of the information indicated, however, an interaction between nuclear genes and cytoplasmic genetic elements (plasmon-sensitive genes). Some of the cytoplasmic elements were assigned to the plastids and the others were presumably present in the mitochondria. These non-nuclear genes affected pigment variegation, male and female fertility and a number of morphological changes of leaf shape, plant height, etc. From the developmental patterns of the cell lineages (sector formation and sector size), the number of plasmon and plastome (chloroplast-coded elements) were estimated. It was a tragic fact that after the retirement and death of the primary research worker, Peter Michaelis, the majority of the *Epilobium* mutants were lost because of the lack of an organized system for their maintenance. (See also *Oenothera*)

EPILOIA (tuberous sclerosis, TS): is caused by more than one dominant gene; human chromosomes 9q32-q34, 3p26 and 12q23, 16p have been implicated. Its prevalence is about 2×10^{-4}.

Epiloia continued

Characteristic symptoms are seizures, mental retardation, papules on the face, low pigmented spots on the trunk and limbs, etc. The symptoms may vary substantially yet mental retardation is apparent in 50% of the afflictions. Brain nodules that may be revealed by tomography for diagnosis even in individuals who are otherwise non-symptomatic. The majority of the cases are due to new mutations thus sibs have no risk of recurrence. When the anomaly is present in the pedigree the recurrence risk may approach 50% although the penetrance may be incomplete. (See mental retardation, tomography)

EPIMASTIGOTE: see *Trypanosoma*

EPIMERS: stereoisomeric compounds that differ at one asymmetric configuration, e.g. glucose and mannose or ribulose 5'-phosphate and xylulose 5'-phosphate; they are reversibly interconverted by epimerase enzymes. (See also enantiomorph)

EPIMORPHOSIS: cell-division-requiring regeneration. (See also morphallaxis, regeneration in animals)

EPIMUTAGENIC: a physiological factor that causes epigenetic alteration. (See epigenesis)

EPINEPHRINE (adrenaline): is an animal hormone produced by the kidney (adrenal medulla). When it binds to specific transmembrane receptors of the G_s proteins, they may be phosphorylated to the $G_{s\alpha}$ — GTP form facilitating the activation of the adenylate cyclase enzyme which produces 3', 5' cyclic AMP. cAMP mediates the function of protein kinase A and setting into motion a cascade of protein phosphorylations and resulting in the activation of glycogen synthase, phosphorylase b kinase (causing glycogen breakdown), acetyl coenzyme A, carboxylase (required for the synthesis of fatty acids), pyruvate dehydrogenase (oxidation of pyruvate to acetyl-CoA), triacylglycerol lipase (resulting in mobilization of fatty acids in the mitochondria and plant peroxisomes), phosphofructokinase and fructose-2,6-bisphosphatase (involved in glycolysis and gluconeogenesis. (See G_s, adenylate cyclase, phosphorylase a, β-adrenergic receptor)

EPIPHYTE: a plant that grows on other plants without parasitizing it (mosses, lichens, orchids)

EPISOMAL VECTOR: see yeast episomal vector, transformation genetic

EPISOME: dispensable genetic element in bacteria that can exist in a free state within the cell or integrated into the main genetic material, comprising a couple of thousand to a few hundred thousands of nucleotides. A typical episome if the bacterial sex plasmid (F); the temperate phages also behave like episomes. Originally, the term was coined for a hypothetical chromatin attached to the *Drosophila* chromosome. (See mitochondrial plasmids)

EPISTASIS: interaction between products of non allelic genes, resulting in modification or masking of the phenotype expected without epistasis; the segregation ratios in F_2 may be altered depending on the type of epistasis involved. At recessive epistasis (a recessive allele of a locus is epistatic to another recessive allele of another independently segregating locus) the phenotypic classes in F_2 are 9:4:3, at dominant epistasis 12:3:1, rather than the common 9:3:3:1. Epistasis may be brought about by modification of gene function due to alterations in the signal transducing pathway. *Indirect epistasis*, in contrast, is not an intracellular phenomenon. A pregnant mother with some metabolic defect, may exert deleterious effect on the genetically normal (heterozygous) developing fetus by placental transfer of harmful metabolites. Similarly, breast feeding by such mothers may elicit the symptoms of the disease in the nursing babies. Such a situation may exist in case of phenylketonuric and myasthenic mothers. The problem of epistasis is of interest for quantitative genetic analysis and it may require sophisticated statistics to separate additive effects from interactions. Epistasis has thus significance for animal and plant breeding and also for human genetics to determine the role of genes in the development of cancer and other traits under multiple controls. Actually, the majority of genes do not operate independently from each other. (See signal transduction, phenylketonuria, myasthenia, modified Mendelian ratios, interaction deviation, synergism, recombinant congenic)

EPISTATIC SELECTION: consecutively occurring mutations my decrease fitness synergistically. (See fitness, selection)

EPISTEMOLOGY: the study of the nature and limitations of science.

EPITHELIAL CELL: cell of the surface layer of organs or bodies.

EPITHELIOMA: autosomal dominant benign or cancerous human skin lesions, encoded in the long arm of human chromosome 9. (See skin diseases)

EPITOPE: the binding site on the antigen for the paratope of the antibody; the antigenic determinant. The polymorphic (private) epitopes are specific for one MHC whereas the monomorphic (public) epitopes may be shared by more than a single MHC allele. The α-galactosyl epitope (Galα1-3Galβ1-4GlcNAc-R) is expressed on the surface of most mammalian cells (except humans and Old World primates) is the most common factor of tissue rejection. This epitope is present on several potentially pathogenic viruses, bacteria and protozoa. Retroviral vectors also carry this epitope and appears to be the major cause for triggering the complement cascade against them. The α-1,2-fucosyl transferase (H-transferase) may reduce the α-galactosyl residues on the viral and cellular surface and reduces thus the complement mediated immune reaction. The elimination of the α-galactosyl epitope either by gene knockout or by increasing the H-transferase expression may facilitate xenotransplantation. (See antibody, MHC, antigen, paratope, xenograft, gene therapy, complement, knockout)

EPITOPE SCREENING: if antibody is available for a protein it can be used to screen expression libraries of the antigen with the antibody. The antigen containing cells are plated, then transferred to nitrocellulose filter replicas and subsequently submerged in a solution of the antibody. The epitope containing colonies tightly bind the antibody and the position of the complex is revealed by incubating the filter by a second, radiolabeled antibody that binds to the first one. This procedure is of medical significance for the development of immunological defense systems. (See antibody, antigen, epitope, radioactive label)

EPONYM: the designation of a phenomenon or principle by the name of person(s) who discovered it or who was associated with it as a proband, e.g. Punnett square, centi Morgan, Hogness box, Abraham Lincoln hemoglobin, Lepore hemoglobin.

EPOXIDE: contain a three-membered (oxirane) ring e.g. $CH_2\!-\!\!-\!CH_2$ (ethylene oxide) and they are highly reactive. $\diagdown O \diagup$

EPPENDORF®: although this company merchandises other types of laboratory tools, generally the conical polypropylene microcentrifuge tubes of 1.5 mL capacity are meant. Such 0.5 to 1.5 mL tubes are marketed in equal quality by several companies.

EPS: exopolysaccharide.

EPSTEIN-BARR VIRUS (EBV): a 172 kbp linear DNA herpes virus with multiple direct repeats of 0.5 kb at both ends. Within the cells the EBV may make up to 100 circular, non-integrated copies or it may integrate into the chromosomes. Some of its RNA genes are transcribed by RNA polymerase III and use upstream elements for regulation. Herpes viruses causes infectious mononucleosis, Burkitt's lymphoma, nasopharingeal (nose-throat) carcinoma, Marek's disease (an avian tumor), etc. EBV infects and immortalizes B lymphocytes with the aid of the EBNA2 gene. EBNA2 is also a transcativator of other genes without binding to DNA. Responsive promoters are targeted through the CBF1 DNA binding transcriptional repressor protein. CBF1 binds to a consensus of GTGGGAA at a considerable distance from the EBNA2-responsive cellular promoters. EBNA2 counters transcriptional repression by CBF1. EBV can be used as a genetic vector. The vectors form only 2 to 4 copies in mammalian cells. They can carry 35 kb inserts and can be used as a shuttle vector. The EBV establishes latency in the target cells, making it a desirable genetic vector. (See Herpes, Burkitt's lymphoma, shuttle vector, carcinoma, cancer, oncogene, CBF)

EPSTEIN SYNDROME: macrothrombocytopathy. (See Alport syndrome)

EQUATIONAL DIVISION: mitosis.

EQUATIONAL SEPARATION OF CHROMOSOMES: takes place in mitosis when the sister chromatids go to opposite poles during anaphase. Similarly the separation of the chromosomes at anaphase I of meiosis may be equational in case of recombination or in the absence of muta-

Equational separation of chromosomes continued

tion. In heterozygous autopolyploids when crossing over takes place and the *Aa* and *aA* chromatids go to the same pole, the separation is equational. When the distance between a gene and its centromere is at least 50 map unit (or more) maximal equational segregation occurs. Such a separation may permit the formation of *aa* gametes in an *AAAa* (triplex) and increases the frequency of *aa* gametes in duplex (*AAaa*) and simplex (*Aaaa*) individuals. (See mitosis, meiosis, autopolyploidy, reductional separation)

EQUATORIAL PLANE: the middle region of a dividing cell nucleus where the chromosomes congregate before anaphase begins and where the nucleus will divide into two. N.b. sometimes it is called "plate" but there is no physical plate there at that stage. (See mitosis)

EQUATORIAL PLATE: see equatorial plane

EQUILIBRIUM: a state without a net change.

EQUILIBRIUM CENTRIFUGATION: the centrifugation is continued in a density gradient until each macromolecule (subcellular organelles) reaches a position corresponding to its density. (See density gradient centrifugation, ultracentrifugation)

EQUILIBRIUM CONSTANT (K_{eq}): the concentrations of all reactants and products at equilibrium under specified conditions. (See dissociation constant)

EQUILIBRIUM OF HETEROZYGOTES: both homozygotes in a random mating population are equally disadvantaged compared to the heterozygotes and heterozygote equilibrium takes place relative to homozygotes. (See Hardy - Weinberg theorem, genetic equilibrium)

EQUILIBRIUM OF MUTATIONS: the occurrence of a number of mutations in a random mating population has the same consequence as immigration, i.e. the frequency of the allele may increase. Unless the new mutation has substantial selective advantage, its survival (fixation) or death (extinction) are equally likely. If the mutation is frequent and random elimination (drift) is insignificant, maintenance of the mutant allele may be assured, however, even when it does not have a selective advantage. If a mutation from *A* to *a* is regular event, the frequency of *a* may increase at the expense of *A* :

$q_{n+1} = q_n + \mu(1 - q_n)$ {1} where q = frequency of the recessive allele, (1 - q) = frequency of the dominant allele, n = the number of generations and μ = mutation rate. After a number of generations (n), the initial frequency of the recessive allele (q_0) may increase to q_n by the acquisition of the same mutant alleles (A→a) as represented:

$e^{-n\mu} = (1 - q_n) / (1 - q_0)$ {2} and hence $-n\mu = \ln[(1 - q_n)/(1 - q_0)]$ {3}

If, e.g., $q_0 = 0.05$ (as we hypothesize for an example), the number of generations required to double its frequency to $q_n = 0.10$ can be computed if the value of μ (= mutation rate) is known, e.g., $\mu = 10^{-5}$. According to {3}, $\ln[(1 - 0.10)/(1 - 0.05)] = \ln[0.90/0.95] = \ln 0.94737 = -0.05407 = -n\mu$. When 0.05407 is divided by 0.00001 (the mutation rate given above, 10^{-5}), we get 5407, meaning that under the conditions specified by this hypothetical example, 5,407 generations are required to double the frequency of the recessive allele. Since this change depends not on the number of years but on the number of generations, species with many generations annually may change more rapidly than the ones with long times to sexual maturity and gestation. Therefore it is easier to test these mathematical models with bacteria or *Drosophila* or mice than with humans or elephants.

The rate of change of allelic frequencies in a random mating population can be expressed as:

$\Delta q_n = \mu(1 - q_n)$ {4} where Δ indicates change in the frequency of the recessive allele (q). Thus the rate of change {4} for the hypothetical experiment is $10^{-5}(0.90) = 0.000009$. The larger the number of alleles which can mutate, the larger is the chance for the change. One must take into account also the fact that mutations may revert. Accordingly: $\Delta q = \mu p - \rho q$ {5} where μp represents the mutation $A \to a$, and ρq stands for backmutation $A \leftarrow a$.

At mutational equilibrium $\hat{q} = \mu/(\mu + \rho)$ {6} and

$\hat{p} = \rho/(\mu + \rho)$ {7} where $\hat{q}$ and $\hat{p}$ represent the equilibrium

Equilibrium of mutations continued
frequencies of the recessive and dominant alleles, respectively.
It is evident if p is larger than q, a larger ρ is required to keep equilibrium with a smaller μ, e.g., if p = 0.8 and q = 0.2 (p + q is always 1) and μ = 0.00001, ρ must be 0.00004 to make 0.8 x 0.00001 = 0.2 x 0.00004 . Usually in nature the frequency μ (forward mutation) is larger because $A \to a$ change may occur by more mechanisms than base substitution alone (e.g. deletion, frame shift, etc.) whereas reversion is expected to take place mainly by nucleotide replacement. Therefore, in the maintenance of allelic equilibrium selection usually has a major role. (See also allelic frequencies)

ERBA: genes present in multiple copies in both the human (chromosome 17q22-q23 or 17q11-q12) and the mouse genomes. It was held responsible for the potentiation of ERBB1 and the protein is a receptor for thyroid hormones and functions in the nucleus as a transcription factor. It has the highest expression in the nervous system and, unlike other thyroid receptors, not in the liver. It is related to the avian erythroblastic leukemia virus oncogene and it is involved in leukemia and other cancers in humans including some translocations. (See also erbB, ERBB1, oncogenes, hormones, TRE, hormone response elements, regulation of gene activity)

erbA : retroviral oncogene produces a transcription factor (member of the steroid receptor family)

erbB : avian erythroleukemia oncogene (receptor for EGF). (See also avian erythroblastoma, ERBB1)

ERBB1: is an oncogene (human chromosome 7p12-p22, mouse chromosome 7) encodes the glycoprotein, epidermal growth factor receptor (EGFR), a transmembrane protein tyrosine kinase. The protein has two subunits, one contains phosphotyrosine and phosphothreonine, and the other contains, in addition, phosphoserine. ERBB2 has also been assigned to about the same chromosomal location as ERBB1 although it was considered to be distinct from the latter. The synonyms of ERBB2 are NEU and HER2. Probably ERBB2 is a normal growth factor but when its expression is enhanced it becomes a protooncogene. The amplification of its expression was detected in adenocarcinomas, and gastric cancer. ERBB2 product is present in neuroblastomas, breast and ovarian cancers. ERBB3 is a related (mammary) oncogene in human chromosome 12q13. (See also EGF, oncogenes, glycoprotein, EGF, protein kinase, protooncogene, breast cancer, neuroblastoma)

ERBBA: two human genes in chromosome 19 (locus EAR2, with preferential expression in the liver) and chromosome 5 (locus EAR3) encode steroid and thyroid hormone receptors. The homologous retroviral gene is *erb*. (See animal hormones, retrovirus)

ERCC: see excision repair, RAD25

eRF: recognizes stop codons in mRNA and terminates translation, stimulates peptidyl-tRNA cleavage and release.

erg: unit of energy or work in dyne acting through a distance of 1 cm. (See dyne unit of force)

ERG: a DNA-binding protein, oncogene, encoded in human chromosome 21q22.

ERGOSTEROL: the most common sterol in some fungi. It is D_2 provitamin and it is used as an antirachitic drug (for prevention of certain bone deficiency problems). (See also cholesterol)

ERGOT: dry sclerotia (compact dry mycelial mass) of *Claviceps purpurea* fungus on rye ears (and some other grasses) containing a large number of alkaloids and peptide alkaloids. Some of which promote the contraction of the uterus (oxytocic) and are highly toxic. Lysergic acid is one of the main alkaloids of ergot affects the nervous system, causes confusion and has been used as a psychomimetic and was supplied to the defenders of the soviet political "conception" trials to admit crimes they did not commit. (See alkaloids, psychomimetic)

ERK: a family of **e**xtracellular signal-**r**egulated protein **k**inases. (See signal transduction, SAPK)

ERP: is a transcription factor of the ETS family. (See transcription factors, ETS)

ERROR IN AMINOACYLATION: approximately 3×10^{-4} is the rate of charging a tRNA with the wrong amino acid. Several aminoacyl-tRNA synthetases carry out editing functions by hydrolyzing the mis-activated aminoacyl adenylates and aminoacyl-tRNAs. DNA aptamers are also involved in editing functions. (See ambiguity in translation, error in replication, aptamer, editing, aminoacyl tRNA synthetase)

ERROR IN REPLICATION: leads to base replacement and thus mutation. The rate of replicational errors of the different DNA polymerases varies but it is partly compensated for by the editing function of the (3'→5' exonuclease activity) of the polymerases. The error rate of α subunit of DNA polymerase III of *E. coli* is about 10^{-5} but it is reduced by about two orders of magnitude by exonuclease subunit ε. The repair polymerase, pol I of *E. coli* also has an error rate of about 1 x 10^{-5} but the 3'→5' exonuclease activity again reduces the errors by two orders of magnitude. The T7 DNA polymerase has an error rate of 10^{-3} to 10^{-4} but the repair system lowers it to 10^{-8} to 10^{-10}. The RNA polymerases of RNA viruses do not have proofreading and editing functions and their error rate may vary within the 10^{-3} to 10^{-4} range per nucleotide. (See editing, proofreading, reverse transcriptases)

ERROR IN TRANSLATION: see ambiguity in translation

ERROR-PRONE REPAIR: see SOS repair, DNA repair

ERROR TYPES: type I rejecting a true hypothesis, type II accepting a wrong hypothesis on the basis of statistical analysis.

ERROR VARIANCE: the variance arising from agents, conditions beyond the ability to control an experiment and with which the apparent effect of any controled factor must be compared to obtain meaningful evaluation. (See analysis of variance)

ERUCIC ACID (13-eicosenoic acid): a monoethenoid acid in the seeds of *Cruciferae* (rapeseed, mustard, horseradish, etc.) and *Tropaeolaceae* plants. It may constitute 50 to 80% of the fatty acids in these plants. It has poor digestibility but even worse, it accumulates in the muscles and livers of the animals (humans) and causes serious and irreversible pathological changes. The newly developed varieties of rapeseed (canola, *Brassica napus*) the erucic acid content has been reduced to 0.3% and thus make this plant a valuable oil-seed crop, particularly in cooler climates where soybeans do not fare well. The mutation blocks the conversion of oleic acid (18:1) to eicosenoic acid (20:1) and erucic acid (22:1). (See fatty acids, canola)

ERYTHERMALGIA: autosomal dominant disease with intense burning pain in the feet and hands, redness, heat-sensation and swelling. (See pigmentation defects)

ERYTHROBLAST: a nucleus-containing erythrocyte or in looser sense immature red bloods cell.

ERYTHROBLASTOMA: tumor-type mass of nucleated red blood cells.

ERYTHROBLASTOSIS: the circulating blood contains immature red blood cells.

ERYHTROBLASTOSIS FETALIS (neonatarum): when red blood cells of the developing fetus enter the maternal blood circulation of immunogenic mothers an immunization reaction may be generated. It may be particularly intense when the mother is Rh-negative and the fetus has Rh-positive blood. The antibodies formed by the mother then may enter the fetal blood supply and cause increased bilirubin production (causing nerve damage) and agglutination and lysis of the fetal blood resulting in intrauteral death or severe anemia unless the newborn's blood is replaced immediately after delivery. Today, the development of erythroblastosis may be prevented by monitoring, especially in the later stages of the pregnancy, for bilirubin or Rh-antibody accumulation. If the tests are positive, and before Rh-antibodies accumulate, the mother may be given anti-Rh γ-globulin that can be transferred through the placenta to the fetus and affords protection against erythroblastosis. Erythroblastosis may be caused also by infection by the avian retrovirus, AEV. (See Rh blood type, SU antigen in pigs)

ERYTHROCYTE (red blood cell): they are disk-shape, biconcave cells. Their red color is due to the red oxygen carrier molecules, hemoglobin they contain. The mature erythrocytes no longer contain nuclei or nuclear DNA. (See blood, hemolytic disease, sickle cell anemia)

ERYTHROLEUKEMIA: is based on an autosomal dominant gene causing neoplastic growth of the immature and mature red blood cells, increase in the size of liver and spleen and acute anemia. It is considered as a malignant disease. (See leukemia)

ERYTHROMYCIN: a group of three antibiotics produced by *Streptomyces erythreus;* they inhibit amino acid chain elongation on the ribosomes. (See antibiotics). The *ery* (erythromycin resistance) gene is the most distal marker to *ap* (attachment point) in the chloroplast

DNA of *Chlamydomonas reinhardi* according to recombination tests in cytohets (heterozygous for ctDNA). (See chloroplast genetics, chloroplast mapping)

ERYTHROPOIETIN (EPO): is a highly specific erythrocyte stimulating glycoprotein factor (M_r 51,000), produced mainly in the kidneys and acting in the bone marrow. It was isolated from anemic sheep but it is produced now also by recombinant DNA technology. It is used also as a medication to avoid the need for blood transfusion in patients prone to anemia. (See growth factors, megakaryocyte, hematopoiesis)

ES: embryonic stem cell; cells that can proliferate in an undifferentiated state but can give rise to differentiated cells as well and can be returned to an embryo to become part of it. These pluripotent cells are taken from blastocyst stage embryos. They are functionally similar to the meristem cells of plants. (See totipotency, pluripotency)

ESAG: expression site associated genes. (See *Trypanosoma*)

Escherichia coli: see *E. coli*

ESSENTIAL AMINO ACIDS: cannot be synthesized *de novo* by vertebrates and must be provided in the diet: arginine (in youngs), histidine, isoleucine, lysine, methionine, phenylalanine, threonine, tryptophan, valine. (See high-lysine corn, kwashiorkor)

ESSENTIAL ATHROMBIA: see thrombopathia

ESSENTIAL FATTY ACIDS: required by mammals although they cannot synthesize them (linoleate [18:2 cis-Δ^9, Δ^{12}], and α-linolenate [18:3 cis-Δ^9, Δ^{12}, Δ^{15}]) because they are unable to introduce additional double bonds beyond the C-9. (See fatty acids)

ESSENTIAL GENE: cannot be dispensed of without lethal consequences.

EST: see expressed-sequence tag

ESTABLISHED CELL LINE: of animals consists of fused cells (with one component being myeloma cancer cells) and is capable of indefinite growth without senescence. (See immortalization, HeLa)

ESTER: is formed when the OH end of an alcohol is combined with the COOH end of an acid leading to the removal of H_2 such as R—O—CO—R. Nucleotides may form ester linkages during chain elongation. (See phosphodiester bond, nucleotide chain growth)

ESTERASES: are enzymes involved in hydrolysis of ester bonds:

$$R_1-\overset{\overset{O}{\|}}{C}-R_2 + H_2O \rightleftharpoons R_1-\overset{\overset{O}{\|}}{\underset{\underset{O}{|}}{C}} + HO-R_2 + H^+$$

In biochemical systems the esters may be phosphate-, glycerol-esters and many other molecules. Esterase A-4 (ESA4) is encoded in human chromosome 11q13-q22, ESD was located to 13q14.11; other human variants exist.

ESTRADIOL: is a steroid hormone produced by the animal (human) ovaries and placenta and it regulates secondary sexual characters and female estrus and implantation. 17β-estradion and raloxifene indirectly activate the estrogen-response element. The β-estradiols are carcinogenic. (See animal hormones, estrogen response element)

ESTROGEN: see animal hormone

ESTROGEN RECEPTOR: to become functional it is phosphorylated at Ser^{118} by mitogen activated protein kinase (MAPK) in the presence of epidermal growth factor (EGF) and insulin-like growth factor (IGF). (See estrogen, MAPK, MAPKK, EGF, IGF)

ESTROGEN RESPONSE ELEMENTS: are generally located about 200-300 bp upstream of the transcription initiation site and despite their diversity they share generally a low homology consensus, e.g. the *Xenopus* vitellogenin and the chicken ovalbumin gene carry the dyad: GGTCANNNTG^A_TCC. (See hormone response elements, regulation of gene activity, animal hormones, raloxifene, tamoxifen)

ESTRUS: the recurrent sexually receptive period of female mammals (except humans) accompanied by a sexual urge (heat), it is induced by cyclic ovarian hormonal activity (oestrus).

ETHICS: the rules of animal and human behavior based on genetic and instinctive or social traditions. The Gene Letter URL (<http://www.geneletter.org>) is a source for current problems and information on genethics. (See behavior genetics, ethology, genetic privacy)

ETHIDIUM BROMIDE (EB): is a polycyclic fluorescent dye. Its fluorescence increases about 50 fold when it intercalates between DNA bases, and the sensitivity of the staining is so high that 2.5 ng DNA is detectable. It is used for revealing of DNA in agarose or polyacrylamide gels and buffers (0.5 mg/mL). The UV light absorbed by DNA at 254 nm is transferred to EB that itself has two absorption maxima at 303 and 366 nm and emits red-orange fluorescence at 590 nm maximum). DNA more than 250 ng/mL can quantitatively be measured spectrophotometrically. The detection of single-strand DNA or RNA is much less effective by EB. Intercalation of DNA into closed circular DNA is much less efficient than into linear DNA. Therefore EB-stained plasmid or organellar DNA can be separated by buoyant density centrifugation from linear DNA by becoming of lesser density. EB-stained gels can be destained with water or by 1 mM $MgSO_4$ for 20 min. From the DNA EB can be removed by washing several times with an equal volume of water-saturated 1-butanol or isoamylalcohol. EB is a powerful mutagen and must be handled carefully with gloves, and the used solutions after dilution to no more than 0.5 mg/L, require decontamination by 1 vol. 0.5 M $KMnO_4$ and careful mixing of 1 vol 2.5 N HCl for several hours, followed by neutralization with 1 vol. 2.5 N NaOH. This procedure reduces mutagenicity to 3×10^{-3} of the untreated solutions.

ETHNICITY: a human population related by biological features identifiable by anthropometric, cultural, linguistic, biochemical, serological and molecular characteristics. The identification may not use all of these methods, depending on the nature of the genetic distances in the comparisons. (See genetic distance)

ETHOLOGICAL ISOLATION: is a type of sexual isolation. Males or females refuse to mate for some "psychological" reason(s). (See sexual isolation)

ETHOLOGY: study of behavior. (See behavior genetics, ethics, aggression, instinct, morality)

ETHYL (—CH_2CH_3): is an alcohol radical.

ETHYLENE (C_2H_4): is the simplest plant hormone yet it controls a wide variety of morphogenetic and developmental processes from stem and root growth, seed germination, fruit ripening, senescence and sex determination. Ethylene is synthesized as a side branch of the Yang cycle. The key enzymes in ethylene synthesis are ACC synthase and ACC oxidase. ACC synthase is encoded by a multigene family and it is regulated by hormonal, physical and environmental signals. In *Arabidopsis* a series of mutants are available that determine constitutive ethylene response and are overproducers. The ethylene insensitive mutations include a histidine kinase (*ein1, etr1*), *ain* is ACC-insensitive, *ctr1* is a putative serine/threonine kinase similar to the members of the MAPK. Gene *ERS* (ethylene response sensor) has some structural similarity to *ETR* but its role is upstream in the pathway. The order of gene action has been determined for most of the numerous mutations. Ethylene plays a key role in plant signaling and, from practical points of view, in fruit ripening and disease resistance. Several of disease and stress resistance genes and other genes controlling differentiation contain GCC box repeats that are also ethylene response elements. The ethylene insensitive *ETR1* gene of *Arabidopsis* when transferred to yeast conveyed ethylene binding in this fungus too. The ethylene non-responding mutation *hookless1* seem to be a regulator of auxins. (See plant hormones, Yang cycle, fruit ripening, SAR, hypersensitive reaction, hormone-response elements, MAPK, MAPKK)

ETHYLENEIMINE: see EI

ETHYLENEOXIDE: (C_2H_4O) a colorless, explosive (!) gas used as a sterilizing agent for instruments, textiles, some food, soil, etc. It is hazardous to handle; irritant to mucous membranes and the lung and may cause pulmonary edema at higher concentrations.

ETHYLMETHANESULFONATE (EMS, $CH_3SO_2OCH_2CH_3$): an alkylating mutagen and carcinogen. Its prime target for mutagenesis is the alkylation of the O^6 position of guanine al-

Ethylmethane sulfonate continued
though it alkylates preferentially the 7 position of guanine but this rarely leads to base substitution. It alkylates all other nucleic acid bases too, and may cause strand scission by depurination. One of the most useful chemical mutagen for a wide range of organisms. (See alkylation, mutagens, carcinogens, depurination, hydrogen pairing)

ETIOLATED: plants grown in the absence of light display elongation and lack of the typical leaf pigments (chlorophyll and carotenoids). The chloroplasts have not developed and the plastids arrive only to the etioplast stage (thylakoid membrane system incomplete). The leaves are not fully expanded. (See photomorphogenesis, phytochrome, de-etiolation)

ETIOLOGY: the study of cause(s) of disease and its development.

ETIOPLAST: lacks the typical chloroplast pigments and display prolamellar body. (See etiolated, prolamellar body, proplastid, chloroplasts)

ETS: external transcribed spacers are located between pre-tRNA gene clusters:
$$ETS - 5' - 18S - ITS - 5.85 - ITS - 28S) - ETS$$
(See ITS, spacer DNA)

ETS ONCOGENES: ETS1 was located to human chromosome 11q23-q24 and it is involved in acute monocytic leukemia (AMol) when interferon β gene (IFB-1, chromosome 9p22), is translocated to it. ETS-1 and ETS-2 genes are physically contiguous in birds and the homologous two genes are coordinately expressed and the proteins are both in the cytoplasm and nucleus. In humans ETS-2 genes are in chromosomes 21q22, 1-q22.3 and their expression is not coordinate; the ETS-1 protein is cytoplasmic whereas the ETS-2 is nuclear. The translocation t(8;21)(q22;q22) is common in patients with acute myeloid leukemia (AML-M2). The breakage point in the latter cases is not within the ETS-2 sequences. The ETS-2 gene lacks TATA box and CAAT box and alternative elements are substituted for them. The ELK oncogenes display homology to the ETS oncogenes. This family of transcription factors has about 35 known members and regulate various aspects of growth, development and lymphocyte pool maintenance. (See leukemia, interferon β-1, oncogenes, ELK, SAP, ERP, lymphocytes)

EUBACTERIA: the majority of bacteria belong to this subkingdom of prokaryotes; the other subkingdom is *Archebacteria*.

EUCARYOTE: see eukaryote; (karyon [καρυον] is a Greek word and more appropriate to spell it with k. (See also prokaryote)

EUCHROMATIC: chromosomal regions containing euchromatin. (See euchromatin)

EUCHROMATIN: does not absorb the common nuclear stains during interphase and is normally transcribed into mRNA. (See also heterochromatin)

EUCIB: see European Collaborative Interspecies Backcross is a cooperative workgroup for plants.

EUGENICS: application of genetic principles to the breeding of the human race with the purpose of improvement. *Positive eugenics* wishes to enrich the frequency of "favorable" genes whereas *negative eugenics* wants to eliminate the "undesirable" genes by selective breeding. The word eugenics was coined by Sir Francis Galton in 1883, unaware of Mendel's discoveries, but his statistical data indicated that about 25% of the sons of eminent fathers were also eminent. From this he concluded that heredity has a role in talent, intellect and behavior. These fields are highly controversial because of the lack of scientific bases for the objective measurement of "desirable" and "undesirable" and the potentials of exploitation for political or unethical goals. The eugenics movements generally emerged during economic downturns in an effort to find scapegoats for the ills of the societies, and the support generally came from people without any understanding or training in genetics. By 1917, in 16 states of the USA laws were approved for compulsory sterilization of the feebleminded, insane, rapist, criminal and other "hereditary unfit" individuals. Actually, by 1935 about 30,000 sterilizations were carried out on the basis of these state laws. With the rise of scientific population genetics it became obvious that this type of negative selection is ineffective. According to the Hardy - Weinberg theo-

Eugenenics continued
rem if the frequency of the homozygous recessive "undesirable" individuals is 1/20,000 (5 x 10^{-5}) then at genetic equilibrium the frequency of that gene is $\sqrt{0.00005} \cong 0.0071 = q$ and the frequency of the carriers (heterozygotes) is $2 \times p \times q = 2 \times 0.9929 \times 0.0071 \cong 0.014099$, i. e. about 1/71. In other words, about 99.3% of this "undesirable" allele will be present in the heterozygotes of the population and about 1/71 individual will have a 50% chance to pass it on to their children. If all the "undesirable" individuals are prevented from reproduction by sterilization the frequency of their "bad" gene (q_n) will change in 100 generations to:

$$q_n = \frac{q_0}{1 + 100q_0} = \frac{0.0071}{1 + [100 \times 0.0071]} \cong 0.004152$$

Thus the initial gene frequency of 0.0071 will be reduced to about 0.004152 and the number of carriers by about 41% to 1/121 in about 3,000 years. Also, behavioral traits are under the control of multiple genes, scattered in the genome and each of them contributing partly to the phenotype. Furthermore characters as these are under polygenic control and are greatly affected by environmental influences. Thus, the negative eugenic measures are biologically ineffective and ethically unacceptable in enlightened societies. Nevertheless, in the name of eugenics, Hitler's regime exterminated 6 million Jews and millions of others of different ethnic groups as well as sterilized more than 250,000 people. Negative eugenics, although it is quite ineffective as shown above, still may have some justification in some forms chosen by the individuals at high genetic risk of disease. The simplest humane and intellectually and ethically correct solution is refraining from reproduction. Therapeutic abortion is counterproductive because it may increase the number of carriers although the "defective" individuals (homozygotes) are eliminated.

Positive eugenics has some biological and ethical problems too because human values cannot be adequately assessed. Certain measures may, however, be acceptable and are practiced in societies without naming them as eugenics, such as adequate scholarship to college students may facilitate their support of family and thus presumably intellectually better individuals may not be prevented from procreation because of economic hardship. Also, reproduction at younger age may reduce chromosomal defects (see Down's syndrome). The Nobel-laureate geneticist H. J. Muller advocated positive eugenics through his entire career. He considered a necessity to fight genetic load through *germinal choice*, meaning that spousal love should be separated from procreational role in marriage. He suggested the reliance of gene banks for artificial insemination of the women and recommended systematic screening the sperm donors on the basis of health, intellect and social consciousness. The sperm of the selected individuals were supposed to be stored frozen and used only some years after their death to make an objective and reliable assessment of their value. Although such a program may appear reasonable, problems in value judgment remain. Muller believed that in a true socialist political and social system these problems can be overcome. His own disenchantment with the marxist society of the USSR proved, however, otherwise. Some general fears of intervention in the human system of reproduction are still not completely dissolved. Will the controled insemination reduce the gene pool? Is it conceivable that the selection may foster the increase of some so far unforeseen genetic defects either by lack of recognition or by hitchhiking (linkage)? Is there a risk that political systems impose their selfish will upon the biologically desirable systems of reproduction? The problems of positive eugenics may not be ignored, however. The progress in medical technologies prevent the selection against formerly inferior traits. The use of medication, prosthetics, somatic gene therapy, etc. are devices of contraselection. Perhaps, the replacement of genes of the germline involved in clinically proven defects may offer a solution. The methodology appears now clear in principle but the consequences are still untested. Thus, the 'brave new world' is not yet at hand, primarily because our values cannot be defined in a simplistic manner. Eugenics must be separated from the racist views that discredited the field. Multiracial and multicultural societies offer unique advantages in the diversities for the betterment of life. Biologists must find the facts and the application of the principles discovered require the democratic decisions of the ethicists and the societies. Will ethicists know enough biology? (See gene therapy, Hardy-Weinberg theorem, sterilization in humans)

EUGLENA: is a green (has chloroplast) flagellate protozoon, n = 45.

EUKARYOTES: organisms with enveloped cell nucleus, mitosis, and meiosis such as fungi, plants and animals.

EUNUCH: a person whose testes were removed by castration that prevents sexual functions and if it takes place early in childhood reduces expression of the secondary sexual characters. They were used in the Middle-East and North Africa as guardians of the harems and chamberlains. The Chinese emperors employed hundreds of them in the forbidden city as counselors. In Italy, until 1871 the "castrati" dominated the operatic stages and church choirs because these male sopranos and contraltos had high-pitched voices of a great range. The removal of the testes contributed to the increased development of the vocal folds and combined with greater capacity of the chest and lungs facilitated finer expression then expected from women who were banned from such roles on "moral" grounds. (See castration)

EUPHENICS: corrective measures for genetically determined defects with the aid of non-genetic means (e.g., spectacles, prosthetics, insulin, etc.). (See eugenics)

EUTHERIANS: true placental mammals but not the marsupials and monotrenes. (See also marsupials, monotrene)

EUPLOID: cells or organisms with one or more complete genomes. (See also aneuploid)

EUSOCIAL: the altruistic behavior of the sterile worker class in social insects supporting the fitness of the colony by tending the needs of the reproductive casts. (See altruistic behavior, fitness, social insects)

eV: see electron Volt, Volt

EVAN'S BLUE: (direct blue) a strong oxidant and suspected carcinogen. It can be used for the isolation of intact protoplasts which do not stain by it whereas the damaged ones appear blue-green. (See protoplast)

EVE, FOREMOTHER OF MOLECULAR mtDNA: is an evolutionary hypothesis attempting to trace the origin of the human race on the basis of mitochondrial DNA. Mitochondrial DNA in humans is transmitted only through the females and because of the uniparental transmission recombination cannot reshuffle the base composition of the mtDNA. Because of technical and computational difficulties, the assumption — that a single woman's descendants populate the earth — could not unequivocally be placed to Africa, and the investigators remained divided concerning the notion that all humans descended from a single female or even from a single group of females. (See mtDNA, genetic drift, effective population size, out-of Africa, Y chromosome)

EVERSPORTING: organisms that carry unstable gene(s) and display somatic or germinal variegation. Many of the eversporting conditions are caused by the presence of transposable elements. (See transposable elements, transposons, retrotransposons, retroposons)

EVI ONCOGENES: integration sites for retroviral insertions in human chromosome 3q24-q28, translocated to chromosome 5q34 resulting in non-lymphocytic leukemia. The Evi-1 mouse gene may cause murine leukemias. Its protein product has numerous Zinc finger domains and is presumably a transcription factor. The EVI2A gene is in the proximal part of the long arm of human chromosome 17. The EVI2B is in the same chromosome about 15 kb apart. Its product is apparently a transmembrane protein with surface receptor function. A homolog of these genes is associated with murine myeloid leukemia. (See leukemia, oncogenes)

EVICTING PLASMID: is incompatible with another type of plasmid and if selection is favoring the evicting plasmid (e.g. carries a gene for antibiotic resistance) the other plasmid can be eliminated. These operations can be used for the construction of particular genetic vectors. (See vectors)

EVOCATION: induction of differentiation. (See differentiation)

EVOLUTION: is a process and theory of the biological and physical change that brings about the variety of the living world and its environment at the global and cosmic range. The general evolutionary theory integrates all ideas about the nature, origin and future of the universe. Or-

Evolution continued
ganic evolution is concerned with the origin, development, relationship of living organisms of past and present. Evolution is frequently contrasted with the views of creation as described in the Bible and other holy books of religions. The science of evolution is concerned with facts that can be experimentally studied with the available technological tools and has no room for faith while the primary criterion of religion is the faith in its teaching. Thus it is improper to compare these principles because they are not of the same nature and they are not required to support or exclude each other. The philosopher K.R. Popper defined the scientific method as "the criterion of potential satisfactoriness is thus testability, or improbability: only a highly testable or improbable theory is worth testing and is actually (and not merely potentially) satisfactory if it withstands severe tests — especially those tests to which we could point as crucial for the theory before they were ever undertaken." At another place he says "...I refuse to accept the view that there are statements in science which we have, resignedly, to accept as true merely because it does not seem possible, for logical reasons, to test them." Thus the testability and refutability are the most basic cornerstones of evolution. Evolution is using the methods of cytogenetics, population genetics, molecular biology and the geological fossil records of the past to establish the relationship, origin and variation in the living world. Ideas about evolution that are not supported by the methods refered to above, are just ideas but not science. (See Hardy—Weinberg theorem, mutation, migration, selection, fossil records, population genetics, evolutionary trees, evolutionary distance, PAUP, unified genetic map)

EVOLUTION AND BASE SUBSTITUTIONS: is generally estimated on the basis of number of changes per site per 10^9 years. Accordingly, the rates in various genomes: mammalian nuclear 2-8, angiosperm nuclear 5.4, mammalian mitochondrial 20-50, angiosperm mitochondrial 0.5, angiosperm cpDNA single copy regions 1.5, angisosperm cpDNA inverted repeats 0.3. The high mutation rate of mammalian mtDNA may be the result of the deficiency in excision, photoreactivation and recombinational repair as well as the low level of selection, and the ability of mtDNA's tRNAs to recognize all four synonymous codons within a family of amino acids. Base substitution rates in fungi as well as in plant mtDNAs are much smaller than in mammalian mtDNAs. (See base substitutions, evolution)

EVOLUTION AND DUPLICATIONS: molecular evidence indicates that duplications have played important roles in evolution. Duplication may arise by unequal crossing over between similar genes and may occur repeatedly. The example below shows the molecular consequences of an unequal crossing over in the DNA at the level of the protein products. Human haptoglobins have two α and two β chains and when an unequal crossing over takes place, duplication is detectable also in the protein product:

site   54
Hp^{1F} •━━━K━━━━━━━━━

             χ → Hp^2 •━━K━━━E━o
Hp^{1S} ━━━━━━━E━o   sites   54   113
site   54

these α1 chains have 84 amino acid residues and Hp^{1F} has at position 54 a lysine (K) whereas Hp^{1s} has glutamic acid (E) at the same site

the total length of Hp^2 (α2) chain originating by an unequal crossing over (χ) is 153 amino acid long containing both the K and the E sites

(See unequal crossing over, cluster homology region)

EVOLUTION, COST OF: evolution proceeds by replacing old alleles with new ones and the replacement may sacrifice also some individuals. This is the way the cost is paid. (See genetic load)

EVOLUTION OF AMOUNTS OF DNA: there is tendency toward increased amounts of genetic material in the more highly evolved organisms. Viruses have generally less DNA than

Evlution of the amount of DNA continued

bacteria and microorganisms have smaller genomes than higher eukaryotes (see genome). It is not entirely clear, however, why the total size of the genome varies by orders of magnitude among higher eukaryotes and e.g. amphibians and several plants have much more DNA than humans. This is frequently called the C-value paradox. (See C amount of DNA, C value paradox, genome)

EVOLUTION OF CHEMICALS: see evolution prebiotic

EVOLUTION OF ORGANELLES: the cytoplasmic organelles of eukaryotes, mitochondria and plastids contain DNA but the base composition of this DNA displays very little similarity to that of the nucleus. The organellar DNA is usually circular versus the nuclear that is linear. The ribosomes in the organelles bear more similarities to prokaryotic ribosomes, being also 70S, in contrast to the ribosomes in the cytosol that are 80S. Several mitochondrial codons do not conform with the universal code words (see mtDNA) and no organisms are known that would have the same nuclear codon dictionary. Plastids use the universal codons. The mitochondrial genome has probably polyphyletic origin with major similarity to purple non-sulphur photosynthetic bacteria. The plastid genome may have derived from non-cyano-bacterial oxygenic bacteria and it may be closest to *Prochloron*. The plastid genome also most likely descended from more than a single species. It is assumed that these organelles were captured by means of endocytosis of some lower organisms, perhaps repeatedly during evolution. The double membranes wrapping the cytoplasmic organelles may have their origin part from the ancestral donor part from the plasmalemma of the recipient cells. In the Euglenoids the plastids have triple membranes, suggesting that this is the relic of double endocytotic events. The size of the mitochondrial genome is very variable among the various organisms. In mammals, it is about 16 kbp but in the muskmelon it may be up to 2,400 kbp and in the majority of other plants varying generally between 200 to 500 kbp and thus exceeding those of the plastids which are rather conservatively between 120 and 270 kbp. The transcriptional machinery in plastids and mitochondria resembles that of prokaryotes. Some of the chloroplast genes have upstream elements similar to the promoter sequences of bacteria. The plastid RNA polymerase is not inhibited by rifampicin whereas this antibiotic blocks transcription in prokaryotes. Both plastids and mitochondria have genes for rRNAs and tRNAs and these also resemble more the prokaryotic ones than that of eukaryotes in the cytosol. More than a dozen of the plastid genes and about three of the mitochondrial genes have introns, an uncommon feature of prokaryotes. Less than one third of the ribosomal proteins in the plastids are coded within this organelle, the rest is of cytosolic origin. This and other facts indicate that the acquisition of the organelles was followed by loss of some functions that could be taken care of by genes in the nucleus. There is also a regular import of proteins into the organelles from the cytosol, aided by the transit peptides of the nuclear coded proteins. The photosynthetic rubisco protein small subunits are coded by nuclear genes whereas the large subunits are transcribed and translated of plastid DNA. Also, organellar gene functions, including mutability, is regulated by nuclear genes. It appears that some DNA sequences have homologs among the nucleus, plastids and mitochondria, indicating the availability of some DNA transfer mechanism(s). (See mitochondria, mtDNA, chloroplast, chloroplast genetics, ribulose1,6-bisphosphate carboxylase - oxydase)

EVOLUTION OF PROTEINS: the amino acid sequence in the proteins reflects the coding sequences in the DNA or RNA genetic material. Although the genetic code has synonymous codons, the primary, secondary, tertiary and quaternary structure reveals important information how proteins function. Adaptive evolution relies primarily on the expression of genes rather than on their structure although the latter is important for tracing the path of this evolution. Certain proteins such as cytochrome c (coded by the nucleus but present in the mitochondria) reveals similarities and differences in the amino acid sequence in widest range of eukaryotes and displays similarities even to bacterial cytochrome c2. Such analyses carried out 2 to 3 decades ago have shown that similar functions require similar structures. Furthermore, the similarities within apparently related groups is greater than among those that are more dis-

Evolution of proteins continued

tant by any type of classification. These studies brought to the recognition that some gene loci are *orthologous,* i.e. they seem to be directly connected by descent across phylogenetic groups whereas other genes, arising by duplication, (*paralogous loci*) may show greater difference in the primary structure because the ancestral copy of the gene could continue providing the needed function while the duplication was more free for (adaptive) evolutionary experimentation. (See homology)

EVOLUTION OF THE GENETIC CODE: the genetic code is practically universal with the exception of a few mitochondrial codons. Furthermore, the structures of tRNAs reveal much greater similarities than expectable on the basis of random assembly of the nucleotides. The most plausible interpretation of these facts is that these nucleotide sequences developed by an evolutionary process from common ancestral sequences and also indicates a relation of descent of eukaryotes from prokaryotes. It seems that the most archaic genetic code existed in RNA. The original sequences were probably relatively short although contained probably meaningless tracts mixed with useful ones. For the development of well-organized nucleotide sequences specifying an ancestral gene, mechanisms were required for the recognition and correction of errors. This ability probably evolved when the first DNA sequences appeared. Modern DNA polymerases are generally endowed with synthetic, proof-reading and editing (endonuclease) functions. This was indispensable for the development of larger and conservative genetic molecules. The first codon probably specified those amino acids which were most abundant among the molecules formed under abiotic conditions. It was suggested that the earliest codons were of the GNC type (where N is any base). This idea is supported by the fact that G≡C base pairs having three hydrogen bonds are more stable than the A=T pairs and under simulated prebiotic conditions glycine (GGC), alanine (GCC), aspartic acid (GAC) and valine (GUC) are produced in a relative abundance. It is conceivable that these four amino acids formed the first protein involved in replication. The next three amino acids might have been glutamic acid, serine, and phenylalanine. These seven early amino acids could later have served as precursors of others, first through abiotic pathways, later the synthesis might have been facilitated by enzymes. The earliest codons might have been somewhat ambiguous aggregates of nucleotides. Later each amino acid probably shared its codon with its daughter amino acid(s). The successive subdivision of codon domains are still reflected in the similarities of the codons among structurally related amino acids. Apparently, the expansion of the amino acid repertory took place in parallel with the evolution of the code. Eventually the assignment of all 64 codons were completed. In the primordial dictionary, probably only the first two bases were required for the specification of an amino acid. Today, in most of the degenerate codon domains still the first two positions are identical. With the acquisition of new amino acids and new codons, the stability, specificity and efficiency of the primordial proteins adaptively changed. The archaic, ambiguous codons with lesser precision must have been then eliminated, culminating in the development of a codon dictionary common to prokaryotes and eukaryotes as we know it today. Through functionally improved enzymes, faster and more reliable replicational processes emerged. As the efficient and precise replication, transcription and translation began, the prebiotic synthesis was no longer needed in this era, about 3 billion years ago. (See spontaneous generation, origin of life, genetic code, code genetic, transcription, translation, prebiotic)

EVOLUTION OF THE KARYOTYPE: chromosome number is a rather good characteristic of related species. Cytotaxonomists define homologies among taxonomic groups on the basis of the karyotype, the morphology of the chromosomes at (generally) metaphase. Some important landmarks of the chromosomes (chromomeres, band, knobs) can be better visualized at prophase (pachytene). Some chromosomal aberrations (inversions) are identified also at anaphase. Salivary gland (polytenic) chromosomes reveal a great deal of morphological information on the chromosomes. With the introduction of the "banding techniques" (Giemsa, C-banding) various taxonomic groups can be rather well defined. Various chromosomes or chromosomal segments can be identified by *in situ* hybridization with DNA probes "painted" with

Evolution of the karyotype continued

fluorochromes of various colors (such as fluorescein isothiocyanate [FITC, green], SpectrumOrange [red], 4',6'-diamidino-2-phenyl indole [DAPI, blue fluorescence in UV], biotin derivatives, etc.), and the genomes in amphiploids can be distinguished. Although the number of chromosomes may vary even in closely related taxonomic groups because centromeric fusion or fission of telochromosomes and bi-armed inversions:

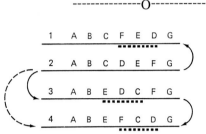

By analysis of the natural banding pattern of salivary gland chromosomes, inversions could be traced among related species. Acrocentric chromosomes may undergo Robertsonian translocations with changes in chromosome num-

Method of tracing the origin of chromosomal inversions

bers by making from two acrocentrics one bi-armed chromosome (quite common in mice). If, e.g. chromosome arms A and B as well as A and D, and similarly D and C as well as C and B arc fused in diploids cell, at meiotic metaphase I, the configuration shown at left and below is displayed. Such a situation may lead to sterility, reproductive isolation and eventually speciation.

Although inversion heterozygosity may lead to sterility depending of the frequency of crossing over in the inverted segment, many wild populations contain various types of inversions. Paracentric inversions do not affect the fertility of the females (either in plants or animals) because the inversion bridge prevents the incorporation of the crossover chromosomes into the eggs (see inversions paracentric) Chromosomal inversions can frequently be traced in the polytenic chromosomes. Inversions may occur repeatedly and may partially involve the same chromosomal segment each time. The serial origin of such inversions can be detected. (See inversions, chromosome painting, telochromosome, Robertsonian translocation, acrocentric, reproductive isolation, polyteny)

EVOLUTION, NON-DARWINIAN: is supposed to take place by the fixation of neutral mutations in contrast to Darwinian evolution that postulates evolution by survival of the fittest. Because of the degeneracy of the genetic code, many mutations will leave the phenotype or function unaltered. There is no reason why these neutral mutations would not be fixed. Although it must be remembered that the codon usage in different genes and/or organisms is not identical, and the codon selection may have an adaptive nature. Also, there is no reason to doubt that certain amino acid substitutions have no affect on protein structure and function and thus appear without selective effect. Rapid evolution is favored by the fixation of neutral changes. If neutral mutations are fixed, there is no need of concomitant elimination of the old genotypes. Random genetic drift may also lead to changes in gene frequencies and possibly to evolution. (See neutral mutation, beneficial mutation, drift genetic)

EVOLUTION, PREBIOTIC: before living cells appeared on Earth, organic molecules must have been formed. In 1953 Stanley Miller and Harold Urey have shown that when mixtures of ammonia, methane, hydrogen and water were exposed to electric sparks (simulating stormy conditions of the early atmosphere), carbon monoxide, carbon dioxide, amino acids, aldehydes and hydrogen cyanide were generated during the experiments lasting for several weeks. Later more extensive experiments revealed hundreds of organic molecules under such simulated abiotic conditions. It was of particular significance that more than ten different amino acids, various carboxylic acids, fatty acids, adenine, sugars, etc. could be obtained in abiotic experiments with increased sophistication. HCN and heavy metal ions present in the early earth environment and in the laboratory simulation could promote polymerization. Thus the conclusion appeared logical that primitive macromolecules could be formed *de novo* and the macromolecules required for life could eventually evolve. (See also origin of life, spontaneous generation unique and repeated, abiogenesis)

EVOLUTIONARY CLOCK: measures the time required in 1% amino acid replacement in a protein in a million years (MY) is based on the assumption that mutations are neutral and occur at random and the amino acid changes reflect the time elapsed since a particular event. A better procedure is to determine the rate of nucleotide substitution in homologous genes because mutation in synonymous codons do not lead to amino acid replacement. The rate of protein evolution is frequently expressed as the number of point mutations per 100 residues in the protein. This parameter is generally symbolized with the acronym PAM. The rate of amino acid substitution per site per year was estimated to be on average 10^{-9} which is frequently referred to as the "pauling" unit of molecular evolution (named after the Nobel-Laureate chemist, Linus Pauling). Individual proteins evolve at a very different rates (see table at left). Perhaps

PROTEINS	AMINO ACID SUBSTITUTIONS PER RESIDUES PER MILLION YEARS
FIBRINOPEPTIDES	90
PANCREATIC RIBONUCLEASE	33
HEMOGLOBINS	14
CYTOCHROME C	3
HISTONE 4	0.06

better comparisons can be made if enzymes rather than proteins are used. The data may be biased if orthologous and paralogous loci cannot be safely distinguished. From the comparison of enzymes it appears that eukaryotes and eubacteria shared common ancestors about 2 billion years ago. The divergence of plants and animals took place about 1 billion years ago. The similarity of fungi to animals is somewhat better than to plants. (See racemate, orthologous, paralogous)

EVOLUTIONARY DISTANCE: can be numerically estimated on the basis of allelic differences (amino acid differences in proteins, RFLPs, nucleic acid base sequences, etc.). The larger number of loci are used the greater is the accuracy of the estimate. There are several procedures in the literature. Here the method of Nei (Molecular Population Genetics, Elsevier/North Holland, 1975) will be illustrated. The normalized identity of alleles in populations is determined by dividing the arithmetic means of the products of allelic frequencies by the geometric means of the sum of squares of the homozygote frequencies in each population to be compared.

$$I = \frac{\{[p_1 \times p_2] + [q_1 \times q_2]\}/L}{\sqrt{\{[p_1^2 + q_1^2] \times [p_2^2 + q_2^2]\}/L^2}}$$

where I is the index of identity, p and q are allelic frequencies in the two populations and L is the number of loci studied (including even those with single allele representation). The evolutionary (genetic) distance is then calculated from the natural logarithm of I, i.e., $D = -\log_e I$. This type of calculation is meaningful and reliable if a minimum of 25 loci are compared. It can be used with any type of alleles, including even proteins of different electrophoretic mobilities. Example for the procedure is in the table below. By using the procedure outlined, the human racial distance and the number of years of divergence have been estimated (after M. Nei & A.K. Roychoudhury 1974, Amer. J. Hum. Genet. 26:421) and it turned out to be minimal:

HUMAN RACES	GENETIC DISTANCE	YEARS OF DIVERGENCE
Caucasoid - African Negroid	0.023	115,000
Caucasoid - Oriental Mongoloid	0.011	55,000
African Negroid - Oriental Mongoloid	0.024	120,000

The evolutionary distance can be estimated more precisely on the basis of nucleotide sequences and replacements (M. Kimura 1980, J. Mol. Evol. 16:111): The pyrimidine ⇄ pyrimidine and the purine ⇄ purine substitutions are designated as P and the pyrimidine ⇄ purine or purine ⇄ pyrimidine transversions are represented by Q. The evolutionary distance K is:

$$K = -(0.5) \log_e[(1 - 2P - Q)\sqrt{1 - 2Q}]$$

Evolutionary distance continued
The standard error of K:

$$s_k = \frac{1}{\sqrt{n}} \left\{ \sqrt{[a^2P + b^2Q] - [aP + bQ]} \right\}$$

where $a = \frac{1}{[1 - 2P - Q]}$ and $b = 0.5\left[\frac{1}{1 - 2P - Q} + \frac{1}{1 - 2Q} \right]$

In a sequencing study of the 438 nucleotide β – globin genes of chicken and rabbit the P value was $58/438 = 0.132$ and $Q = 63/438 = 0.144$, and after the appropriate substitutions the evolutionary distance thus appeared to be 0.347 ± 0.0329.

The time of divergence (in million years) varies according to the specific gene and taxonomic category considered, e.g. in Echinodermata—Chordata, Annelida—Chordata for ATPase: 786 and 1059, for cytochrome: 883 and 1078, for cytochrome oxidase I: 1160 and 1465, for cytochrome oxidase II 608 and 773, for 18S RNA: 1288 and 1214, respectively. (See DNA sequencing, array hybridization, amino acid sequencing, evolutionary tree, genetic distance, minimum evolution method, four-cluster analysis, neighbor joining method, evolutionary tree, least square methods, transformed distance, Fitch-Margoliash test, DNA likelihood method, protein-likelihood method)

EVOLUTIONARY SUBSTITUTION RATE: estimates the number of mutations leading to an evolutionary divergence in million years: $\sum \frac{(d_{ij} - 2d_{ik})}{n_i n_j} x \frac{1}{t_j}$ where d_{ij} and d_{ik} are the number of the DNA base substitutions per number of sites in genotypes i and j and between genotype i and evolutionary tree node k, respectively; $n_i n_j$ are the number of pairwise comparisons, t is the time in million years when the i genotype was deposited in the rock stratum and the k is the node furthest from the common root. (See evolutionary tree, evolutionary distance, PAUP)

EVOLUTIONARY TREE: displays the descent of organisms from one another. Initially trees

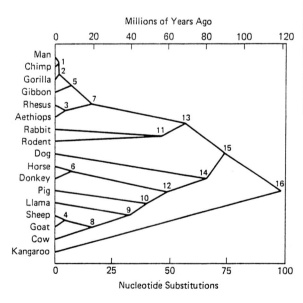

An evolutionary tree of mammals constructed on the basis of molecular clocks of seven proteins. The right end of the tree was fixed by paleontological information on marsupial divergence from the placental mammals, an estimated 120 million years ago. The nodes of divergence are based on the putative nucleotide subsdtitutions in the seven coding genes. (After Fitch, W.M. & Langley, C.H. 1976 Fed. Proc. 35:2092

were constructed on the bases of morphological traits and later on the basis of chromosome numbers and pairing affinities in hybrids and chiasma frequencies. The former criteria are not well quantifiable, the latter features are applicable to only relatively closely related forms because of the sexual isolation among distant species. The numbers of substitutions at a large or all comparable sites of macromolecules provides the best quantitive measures. Evolutionary trees based on nucleotide sequences are more reliable than those based on proteins because silent genes are not included in the proteins. Since DNA sequences are available in several organisms, nucleotide

Evolutionary tree continued

sequence alignment and percentage of homologies are frequently used. Caution may be necessary to interpret evolutionary trends because the similarities or differences can be brought about either by convergence or divergence. From the nucleotide sequences in the DNA, the amino acid sequence in the protein can be relatively easily infered on the basis of the genetic code with some caution. Before nucleotide sequencing became practical in the late 1970s, amino acid sequencing was available and many of the evolutionary trees constructed were based on protein primary structure. Since amino acids have up to six synonymous codons, there is no simple way from the amino acid positions in the protein to determine the nucleotide sequence in the gene. Furthermore, a present nucleotide or amino acid sequence may have evolved in different ways. Molecular evolutionists rely on the principle of *maximal parsimony*, i.e. they suppose that the actual evolution took place in the simplest possible way (what may not be true). Let us assume that the nucleotide sequence A C C A is the result of evolution. Through two mutational events (indicated by →) the original sequence may have changed the following ways:

```
A - A - A - A        A → C - C → A        C - C - C → A
|   ↓   ↓   |        |   |   |   |        ↓   |   |   |
A   C   C   A        A   C   C   A        A   C   C   A
```

Let us assume that in three different taxons (or orthologous loci) at a particular site valine, leucine and serine are found. Valine has four code words (GTT, GTC, GTA, GTG), leucine has six (TTA, TTG, CTT, CTC, CTA, CTG) and serine has six as well (TCT, TCC, TCA, TCG, AGT, AGC). A single mutation may replace the first G of the valine codon by T and a second mutation may lead to a T→C change at the second nucleotide and thus in place of a valine residue serine may occur. From the four valine triplets only two may result in serine codons via two mutational events.

```
GTA  GTG   VALINE
 ↓    ↓
TTA  TTG   LEUCINE
 ↓    ↓
TCA  TCG   SERINE
```

There are many ways to construct evolutionary trees and each has advantages and disadvantages under a particular condition. Evolutionary trees of genes may not be the same as phylogenetic trees because individual genes evolve at different rates. (See also evolutionary distance, indel, rooted evolutionary tree, unrooted evolutionary tree, DNA sequencing, amino acid sequencing, METREE, patristic distance, PAUP, heuristic search, exhaustive search, four-cluster analysis, least square methods, transformed distance, Fitch-Margoliash test, DNA likelihood method, protein likelihood method; see also Miyamoto, M.M. & Ctacraft, J. 1992 Phylogenetic Analysis of DNA Sequences, Oxford Univ. Press for detailed discussions)

EWING SARCOMA: see EWS, sarcoma, neuroepithelioma

EWS: Ewing sarcoma oncogene; a binding protein, encoded at human chromosome 22q12.

EXAGERATION: a recessive mutation in one chromosome and a deficiency for the same site in the homologous chromosome results in more extreme mutant phenotype than homozygosity for the same recessive mutation.

EXAPTATION: an adapted function is modified to become useful for a different biological function.

EXCHANGE PAIRING: chromosomes that have paired and recombined in meiosis are distributed normally to the pole. (See distributive pairing, chromosome pairing)

EXCHANGE PROMOTING PROTEIN (EP): increases the separation of GDP from the bound form (E.GDP) in signal transduction. (See GDP, E.GDP)

EXCINUCLEASES: see DNA repair

EXCISION: the release of a phage, insertion element, episome or any other element or DNA sequence from a nucleic acid chain. In *precise excision*, for the affected or involved sequence the "wild type" DNA is copied from the homologous DNA strands and thus substituted for the

Excision continued

stretch. Any interruption of this repair copying may result in retaining some parts of the inserted piece and thus resulting in *imprecise excision*. (See transposable elements, transposons, T-DNA, insertion elements, episome)

EXCISION REPAIR (dark repair): removal of damaged DNA segments followed by localized removal and replacement (unscheduled DNA synthesis). In the first step in the base excision repair a DNA glycosylase excises the damaged base and then the abasic deoxyribose is removed by AP endonucleases. In nucleotide excision repair an enzyme system hydrolyzes the phosphodiester bonds on both sides of the damaged single-strand tract and the sequence including the defect (12-13 nucleotides in prokaryotes, and 27-29 nucleotides in eukaryotes) is removed by excinucleases. Subsequently the gap is refilled by repair polymerase and the ends are religated. Excision repair genes have been identified in many organisms, including mammals. XPA (human single-strand biding protein [HSSB]) and its homologs (Rpa in yeast) are essential binding proteins for nucleotide exchange repair. The damage-binding proteins (DDB or UV-DDB) generally recognize defective areas caused by photoadducts. These proteins may recruit others such as XPCC (human) and some similar proteins of yeast (Rad4 and Rad23). The transcription complex TFIIH is composed of several subunits required for repair (XPB and XPD in humans) and in yeast (Ss12 and Rad3). TFIIH contains also kinases (Cdk2). The human excision repair genes were identified by genetic transformation of UV-sensitive Chinese hamster cells with human DNA (**e**xcision **r**epair **c**ross-complementing genes, ERCC). Gene ERCC3 (human chromosome 2q21) appears to be at the locus responsible for xeroderma pigmentosum type II (B) and it controls an early step of excision repair involving a DNA helicase function. ERCC1 (*RAD10*) appears to be in human chromosome 3, ERCC4 in chromosome 16, and ERCC5 (*RAD2*) in chromosome 13q22. Excision repair gene UV-135 (ERCM2) appears to be in human chromosome 13 and UV-24 in chromosome 2. Transcription factor TFIIH is also involved in nucleotide excision repair in transcriptionally not active DNA in cooperation with Rad 2 and 4. The repair synthesis is carried out by pol δ and pol ϵ in cooperation with PCNA (proliferating cell nuclear antigen) and RFC (replication factor) auxiliary proteins. (See also DNA repair, mismatch repair, AP endonucleases, ABC excinucleases, excision repair bioassays, X-ray repair, UVRBC, ultraviolet light, pyrimidine dimer, glycosylases, photolyase, phosphodiester bond, DNA ligase, DNA polymerases, DNA ligase, cyclobutane dimer, adducts, xeroderma pigmentosum, Cockayne syndrome, Bloom's syndrome, trichothiodystrophy, light-sensitivity diseases)

EXCISION REPAIR BIOASSAYS: are essentially biochemical procedures. The mutagen may cause the formation of cyclobutane rings, alkylation of bases, formation of covalent adducts between bases and the reactive chemical, depurination of the DNA, interstrand cross-links in the DNA, cross-links between DNA and protein, breakage of the phosphodiester bonds, intercalation of the mutagen between DNA bases, etc. The defects are detected in the extracted DNA cleaved by acids (formic acid or trifluoroacetic acid). The acid hydrolysis does not give as reliable results as the use of DNase or phosphodiesterase enzymes. The free bases and the pyrimidine dimers are analyzed by one-dimensional or two-dimensional paper or thin-layer or column (Sephadex or Dowex) chromatography. By similar procedures the alkylated bases can be identified. Identification of adducts have been attempted by using radiolabeled mutagens and detecting the sites of the adduct formation. This procedure, however, is not completely reliable. The use of single-strand specific nucleases may detect the consequences of strand breakage resulting in strand separation and then liable to digestion by these nucleases. DNA breakage has been evaluated also by centrifugation of the extracted DNA in 5-20% alkaline sucrose-gradients where the intact DNA peaks closer to the bottom of the centrifuge tube while the damaged one floats at the lower sucrose concentration areas. A different type of approach is to provide radioactively labeled bases after the mutagenic treatment and monitor the extent of repair replication that is supposed to be increased if the mutagen damages the DNA. The extent of repair is reflected by the increase in radioactivity in the DNA. Extensive repair replication can be assessed by adding bromodeoxyuridine (BrDU) to the mutagen-treated

Excision repair bioassays continued

DNA. Incorporation of BrDU increases the buoyant density of the DNA and that is detectable upon separation by ultracentrifugation. Another variation is the exposure of BrDU containing DNA tracts to irradiation by 313 nm UV-B. The patches containig the analog are cleaved by this and thus reveal by photolysis the original sites of damage. (See also unscheduled DNA synthesis, ultracentrifuge, ultraviolet light, buoyant density, bioassays in genetic toxicology, mutation detection, site-specific mutation)

EXCISION VECTOR: carries the prokaryotic *Cre* gene. If a mouse is transformed by Cre and its mate carries a gene closely flanked by loxP, then their F_1 offspring (or any heterozygote) may expell the targeted gene:

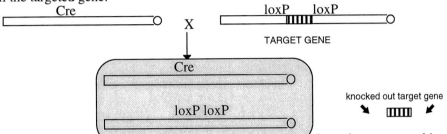

(See *Cre*, targeting genes, homologous recombination)

EXCITATION: illumination with ultraviolet light may raise transiently the orbital electrons of an atom to a higher level of energy. Such excitation may lead to electron loss and may be sufficient to cause mutation although the level of energy may drop back to ground level. (See ultraviolet light, electron)

EXCITED STATE: energy rich atom or molecule after the absorption of light energy.

EXCITATORY NEUROTRANSMITTERS: open cation channels and facilitate the influx of Na^+ to depolarize post-synaptic membranes. (See acetylcholine, glutamate, serotonin, neurotransmitter)

EXCONJUGANT: ciliates (*Paramecia*) may pair (conjugate) and the paired cells mutually fertilize each other that is followed by separation of the two, now diploid cells, the exconjugants. (See *Paramecium*)

EXENCEPHALY: an overgrowth of mid-brain neural tissue caused by a defect in protein p53. (See p53, neural-tube defects)

EXERGONIC REACTION: releases free energy.

EXHAUSTIVE SEARCH: is a procedure of constructing an evolutionary tree by analyzing several individuals to calculate the length of the branches of evolutionary trees. (See heuristic search, PAUP, evolutionary tree)

EXINE: outer layer (of pollen grains)

EXOBIOLOGY: makes inquiries about the possibilities of extraterrestrial life forms. (See extraterrestrial)

EXOCARP: outer layer of the fruit wall, the peel of citrus fruits, the skin of peaches.

EXOCRINE: secretory glands that release substances (enzymes, hormones, etc.) through ducts in outward direction. (See endocrine)

EXOCYTOSIS: secretion of molecules from the cell into the medium or transfer of molecules into membrane-enclosed storage compartments in the cell. (See exocytotic vesicles)

EXOCYTOTIC VESICLES: serve as vehicles for transport by exocytosis. (See exocytosis)

EXOGAMY: cross fertilization.

EXOGENIC HEREDITY: is based not on biological inheritance but on cultural transmission of traditions, scientific information, laws, ethics, values, etc.

EXOGENOTE: when the bacterial genetic material recombines, the recipient genome is called endogenote and the corresponding segment of the donor genome is an exogenote. (See conjugation bacterial)

EXOGENOUS: influences coming from outside.

EXOGENOUS EVOLUTION: evolution of cells of higher organisms by inclusion of prokaryotic cells such as plastids and mitochondria. (See autogenous evolution)

EXON: segments of eukaryotic mosaic genes that are represented in the mature mRNA and they are translated into protein. The exon intron boundaries are usually conserved, e.g.:

5'-ACTGCA**gt**aagg...tttcctctctct**ag**TGGGCG-3'   DNA
exon | intron | exon

All coding genes have exons but introns are exceptional in prokaryotes yet they occur also in the plastid and mitochondrial genes. The parts of RNA genes that are retained in the mature transcripts are also called exons. The length of individual exons is usually short, 10 to 400 nucleotides. The total average exon size per genes in mouse is about 2,300 and in humans ca. 3,400 although large differences exist among the different genes. (See also introns, spliceosome, branch point sequence, mosaic genes)

EXON CONNECTION: a method for isolation of genes by following the steps: 1. isolate RNA from a cell line, 2. obtain cDNA, 3. use PCR with primers of suspected adjacent exons, 4. clone the product of PCR, 5. sequence is tested for connection of exons, 6. Southern blot is performed with the probe generated, 7. identify and isolate the whole gene. (See exon, cDNA, PCR, DNA sequencing, gene isolation, introns)

EXON SCRAMBLING: in eukaryotes the exons may be spliced correctly but in different order from that in the genomic DNA. This scrambling is believed to be mediated by loops in the pre-mRNA and makes possible the alternative processing of the transcripts. These exons are not polyadenylated. (See exons, introns)

EXON SHUFFLING: exons of the same gene may be processed and thus expressed in more than one pattern and may be recruited for the synthesis of more than one protein. It has been suggested that the role of introns in the earliest cells during evolution was to facilitate the assembly of new genes by exon shuffling. Exon shuffling is most common in vertebrates but it has been detected also in the cytosolic glyceraldehyde-3-phosphate dehydrogenase gene of plants. (See exon, intron)

EXON THEORY: suggested that split genes arose by aggregation of exons during prebiotic or early evolution. Convincing experimental evidence is still lacking.

EXON TRAPPING: in physical mapping identifies "candidate genes" in transcribed sequences. (See candidate gene, physical mapping)

EXONUCLEASE: enzyme that digests polynucleotide chain beginning at either the 5' or the 3' termini. (See Bal1, recombination mechanisms, endonuclease, restriction enzymes, mismatch repair)

EXONUCLEASE III: is a multifunctional enzyme; in *E. coli* it works as a DNA-repair endonuclease, as an exonuclease 3'→5', phosphomonoesterase, and ribonuclease. (See endonuclease, exonuclease, phosphomonoesterase, ribonuclease)

EXONUCLEASE V: digests double-stranded DNA and displays DNA-dependent ATPase activity; it is encoded by the *recBCD* complex of *E. coli*.

EXONUCLEASE VII: cuts single-stranded DNA from both 3' and 5' ends.

EXOSKELETON: the hard shell of the body (insects, crustaceans, etc.) including the vertebrate nails, hoofs, hair and other epidermal structures are considered as exoskeleton. (See chitin)

EXOSOME: a segment of DNA associated with but not integrated into the chromosome, yet it can express its genetic information. (See also episome)

EXOSTOSIS (EXT, diaphyseal aclasis): autosomal dominant (human chromosome 8q24.1-p13, 11p11.1, 19p) phenotypes involving growth of extra cartilage or bony projections at the end of bones (mainly on hands and fingers but rarely on head, except the ear). Exostosis is accompanied generally by short stature. The incidence of such defects in Western populations is about $1-2 \times 10^{-5}$ and the estimated rate of mutations is $6 - 9 \times 10^{-6}$. Their chance for bone cancer is increased to 0.5-2% of the cases. (See stature in humans, limb defects in humans, Langer-Giedion syndrome, Ehlers-Danlos syndrome, fibrodysplasia, metachondromatosis, cancer)

EXOTHERMIC REACTION: releases heat.

EXOTOXIN: toxins secreted outside the cells (body) such as the protein toxins of several bacteria, e.g. *Clostridium botulinum* nerve toxin, etc. (See toxins)

EXPANSION CARD: a circuit board that can be inserted into some computers and permits the user to perform some special, additional functions.

ExPaSy (Expert Protein Analysis System): links to the Glaxo Institute of Molecular Biology, Geneva 2-dimensional electrophoretic data base to the SWISS-PROT protein sequence database. (See laser desorption MS, electrospray MS, MELANIE II, SWISS-PROT, databases)

EXPECTED PHENOTYPIC SUPERIORITY: is $i_p \sigma p_1$ where i is the selection intensity when fraction p is selected in a breeding program. (See selection intensity)

EXPERIMENTS: are conducted to test a hypothesis (deductive method) or to generalize from empirical data (inductive method). In either case the work has clearly defined objective(s). Genetics is basically an experimental science. (See science, genetics)

EXPLANT: a cut-out piece of (plant) tissue, used for *in vitro* culture. (See tissue culture)

EXPONENTIAL GROWTH: is taking place when nutrients and other factors required for growth are optimal. The multiplication of the cells is determined by the exponent of 2; e.g., after 10 divisions of a cell the number of cells becomes $2^{10} = 1024$ and if the number of initial cells were 8, then after 10 divisions the number of cells becomes $2^{10} \times 8 = 8{,}192$. This type of growth is called also logarithmic growth because $\log_2[1024] = 10$. [The conversion to $\log_2$ from $\log_{10}$ is carried out as follows :
$\log_2[1024] = \log_{10}[1024](1/\log_{10}[2]) = 3.010299957 \times \log_{10}(1/0.301029995) = 3.010299957 \times 3.321928095 \approx 10$. (See growth curve)

EXPRESSED-SEQUENCE TAG (EST): is a probe of short nucleotide sequences for genes that are expressed in a particular tissue although no information may be provided regarding their function or role. The use of ESTs greatly facilitated the analysis of the functional fraction of the eukaryotic genomes. Although patenting for them was sought, the ethical justification for their patenting has been criticized. (See physical mapping, patent)

EXPRESSION: means phenotype in genetics, the phenotype may be any morphological trait or a (protein) product of a gene. (See phenotype, genotype)

EXPRESSION CASSETTE: contains all the sequences required for the expression of a gene and can be inserted into various expression vectors for transcription, also the structural gene may be replaced by another structural gene. (See expression vector, structural gene)

EXPRESSION LIBRARY: includes cDNAs in expression vectors permitting their ready use for transformation and expression in several hosts.

EXPRESSION VECTOR: carries to the target cell by transformation the structural gene and all the regulatory signals required for expression. (See vectors)

EXPRESSION-LINKED COPY: in the *Trypanosomes* for the expression of a different type of antigen gene transposition is required. (see *Trypanosoma brucei*)

EXPRESSIVITY: degree of expression of a gene. (See also penetrance)

EXTENSINS: plant cell wall glycoproteins.

EXTERNAL TRANSCRIBED SPACERS: see ETS

EXTERNAL GUIDE SEQUENCES (EGS): are RNA oligonucleotides that can bind to specific RNA sites in an RNA molecule and guides to this location ribonuclease P that cleaves at specific sites. (See Ribonuclease P)

EXTINCTION: loss of a genotype (phenotype) from a population; disappearance of a conditioned response; suppression of cell-type-specific function in fused somatic cells. In spectrophotometry it means the intensity of the absorption (ε_{max}) at the absorption peak (λ_{max}). (See species extant, O.D.)

EXTRACELLULAR MATRIX (ECM): a complex of polysaccharides and proteins, secreted by cells and its role in the tissues is structural and physiological. ECM includes collagens, a variety of fibrous proteins of triple-helical coiled coil structure, glycoproteins (laminin, fibronectin), proteoglycans, etc. (See fibronectin, collagen, laminin, proteoglycans, basement membrane, integrin, ICAM, elastin, fibronectin)

EXTRACHROMOSOMAL INHERITANCE: determined by non-nuclear elements. (See mtDNA, mitochondrial genetics, chloroplast genetics, plasmids)

EXTRAGENIC: the genetic factor is outside the boundary of the gene that it affects, e.g., suppressor tRNA. (See suppressor tRNA, suppressor gene)

EXTRANUCLEAR GENES: are in the cellular organelles (mitochondria, plastids), except the cell nucleus.

EXTRANUCLEAR INHERITANCE: is determined by genes in cellular organelles other than the nucleus (See ctDNA, mtDNA, symbionts hereditary)

EXTRAPOLATION: to calculate values beyond an interval on the basis of the knowledge of values known within the interval. This is usually done with the assistance of the linear regression equation, $Y = a + bx$. Extrapolation may involve some risks of error if the value of x is increasing beyond the range known and even more seriously if the linear regression equation (Y) values does not represent the values properly but in effect may be a curved line beyond the interval. (See correlation, interpolation linear)

EXTRATERRESTRIAL LIFE: living creatures outside the earth have been claimed on the basis of various observations but so far no unchallenged evidence has been available.

EXTRAVASATION: passing cells from the blood into tissues. Such events occur during metastasis. (See diapedesis, metastasis)

EXTREME-VALUE DISTRIBUTION THEORY: the probability of an alignment score is caused by chance in physical mapping of DNA. (See physical mapping)

EXULES: successive generations of insects (aphids).

EX-VIVO: an operation carried out outside the body but the manipulated organ or gene is returned to the body. (See *in vivo*, *in vitro*)

EYE COLOR: in humans is polygenically controled trait. It is frequently assumed that one gene locus (BEY) brown/blue and another is responsible for green/blue (GEY) color. Brown (dark) tends to be dominant over blue, and green over blue. It is not too uncommon that blue-eyed parents have both blue-eyed and brown-eyed offspring. This is no basis for suspecting illegitimacy. Brown sectors in blue eyes (sometimes almost invisibly small) are attributed to somatic mutation. Such individuals are more likely to have brown-eyed children even if their spouses have blue eyes. The eye color is determined by pigmentation of the retinal layer of the iris and the nature of the semi-opaque layers in front of the iris. A blue layer may be present in all individuals but it may be masked by another layer of melanin in the front part of the iris. Therefore blue-eyed babies may develop brown eyes when melanin is accumulating in the front part of the iris at later stages. Green and hazel eyes indicate that melanin partially masks the reflections from the deeper layers of the iris. Gray eyes are variants of the blue color. Black eyes indicate very deep brown melanin layer. When all pigments are absent in albinos, the eyes appear pink or very pale blue because of the reflections of the blood vessels. Some eye diseases involving coloboma may change eye color. Phenylketonuriacs may display light eye color because of the defect in pigment synthesis. It is not too uncommon that the color of the two eyes of person does not match (heterochromia iridis) due to a developmental anomaly. (See pigmentation in animals, eye diseases, coloboma, phenylketonuria)

EYE DISEASES (ophtalmologic diseases): are concerned with anatomical, and pathological conditions related to vision. More than 0.5% of the population of the USA is afflicted with some serious eye problems and almost 0.02% is legally blind. In other parts of the world the eye-related diseases may be even more prevalent. The genetic component of these ailments is variable and it is usually part of complex syndromes affecting anatomical and physiological disorders. *Clouding of the cornea* occurs in the Hurler, Marquio, Maroteaux-Lamy syndromes, Fabry's disease. *Cataracts* may be present in Wilms tumors, Lowe syndrome, galactosemia, Werner syndrome, myoonic dystrophy. *Cherry-red spot* may appear on the retina in sphingolipidoses, gangliosidoses and mucolipidoses. *Reduced vision* occurs in neuroaminidase de-

Eye diseases continued

ficiency. Retinitis pigmentosa (most commonly clumped pigments and atrophy of the retina, contraction of the field of vision), Usher, Laurence-Moon, Bardet-Biedl syndromes, Refsum disease may involve eye pigment disorders. *Amaurosis congenita, retinoschisis* are malformations or deficiency of the retina. *Loss of eye pigments* and reduced vision in albinism, *Blue sclera* (the normally white outer surface of the eye) in osteogenesis imperfecta are characteristic. *Variegation of the eye color* (iris) is found in the Waardenburg syndrome. *Retinal neoplasia* is the most critical feature of Norrie disease. *Hypoplasia of the iris* occurs in Rieger syndrome and *Reduction in eye size* is caused by microphtalmos. *Myopia* is common in the Stickler syndrome, Marfan syndrome, Kniest dysplasia. *Tumors of the eye tissues* develop in retinoblastoma and von Hippel-Lindau syndrome. *Tumorous eyes* are found also in the Crouzon syndrome. In human trisomy 21 (*Down syndrome*) slanted eyelids and white spots around the iris appear. *Autoimmune eye disease*: may involve ocular cicatricial (scar-like) pemphigoid. (See under separate entries the named conditions; see also mitochondrial diseases in humans, color blindness, ophthalmoplegia, oculudentodigital dysplasia, Michel syndrome, focal dermal hypoplasia, Hirschprung disease, glaucoma, strabismus, nystagmus, coloboma, aspartoacylase deficiency, cat eye syndrome, galactosemia, myopia, Rothmund-Thompson syndrome, Rieger syndrome, Cohen syndrome, Zellweger syndrome, cataracts, ferritin, foveal dystrophy, macular dystrophy, macular degeneration, cornea plana, anophthalmos, microphthalmos, pemphigus)

EYELASHES LONG (trichomegaly): may be an autosomal dominant anomaly associated with cataract and spherocytosis and other ailments. An autosomal recessive form was also studied, and the latter involved mental retardation, retinal degeneration and a number of other symptoms with unclear relevance to the formation of the eyelashes. Multiple rows of eyelashes were also observed to follow apparently either recessive or dominant inheritance or being only a non-genetic anomaly. (See spherocytosis)

eyeless: is a *Drosophila* mutation controling eye differentiation but not the ocelli. The *Small eye* (*Sey/Pax-6*) of mice and the human Aniridia genes are homologous. (See Aniridia)

EYEPIECE: a microscope lens through which the eye directly views the object through the tube. The eyepiece only enlarges the image but does not afford better resolution unlike the lens of the objective. (See microscopy, objective)

EYESPOT: a light-perception structure in algae involved in the movement of the flagella (singular flagellum).

Eyk: a receptor tyrosine kinase encoded by a chicken protooncogene. (See protooncogene, tyrosine kinase)

EZRIN: a cytoskeletal protein. (See ICAM, T cell receptor)

F

F: coefficient of inbreeding expresses the probability for homozygosity of alleles, identical by descent, at a locus. It can simply be determined from the pedigree of an individual in a family.

Each gamete has 0.5 chance to transmit a particular allele through the paths available. Thus in the progeny of a half-brother and half-sister mating, being three relevant ancestors involved, the coefficient of inbreeding, $F = 0.5^3$. Similarly, the F value of the offspring of first cousins is 1/16 because through two loops, represented by E-C-B-D-F and F-D-A-C-E, from the great grand-parental gametes the same allele can be transmitted to the I offspring is $0.5^5 + 0.5^5 = 0.03225 + 0.03125 = 0.0625 = 1/16$. Note that the looping is done backward from the I offspring. (See coefficient of coancestry, relatedness)

F' (F prime): fertility factor that carries regularly bacterial chromosomal gene(s).

F⁺: bacterium carrying a sex element (F, fertility factor). (See F plasmid)

$F_1, F_2, ... F_n$: subsequent filial generations; the parents of the F_1 are homozygous for different alleles of the corresponding gene(s)

F DUCTION: a gene is transferred to the F⁻ bacterial cell by an F plasmid, called F'; it is the same process as that of sexduction. (See sexduction)

***F* ELEMENTS:** see non-viral retrotransposable elements. (See hybrid dysgenesis I - R, FB, F factor)

F FACTOR: bacterial fertility elements (F plasmid, sex plasmid). (See also Hfr, episome)

F_2 LINKAGE ESTIMATION: In some species where controled mating or crossing is difficult linkage intensities (recombination frequencies) can be estimated in the selfed progeny of F_1 or in the offspring of brother - sister matings. In test crosses the recombination frequencies can be determined in a straightforward manner by dividing the number of recombinants by the sums of the parental plus recombinant individuals in the progeny. In F_2 the computations rely on indirect means, using statistical tables (see table on page 362). The product ratio tables were constructed on the basis of the formula:

$$\frac{ad}{bc} = \frac{T(2+T)}{(1-T)^2}$$ where the meaning of a, b, c, d is given below, and $T = (1 - p)^2$ for coupling, and $T = (p)^2$ for repulsion data; p is recombination frequency. The procedure can be best described by an example. Let us consider a hypothetical F_2 population with four phenotypic classes (a, b, c, d):

AB (**a**)	*Ab* (**b**)	*aB* (**c**)	*ab* (**d**)	Total
660	38	40	183	921

Linkage is obvious, because the frequency of the double homozygous recessive class (*ab* or briefly **d**) exceeds that of the single heterozygotes (*Ab* and *aB*, classes **b** and **c**), the linkage phase is coupling in the two parents (*AB* and *ab*).

We can use the product ratio method (formula R for repulsion) and formula C in coupling:

$$\frac{a \times d}{b \times c} = R, \text{ and } \frac{b \times c}{a \times d} = C.$$

(b x c)/(a x d) = (38 x 40)/(660 x 183) ≅ 0.012584865 = C. In the table on page 362, exactly this fraction is not found therefore interpolation is necessary to find the right value. Interpolation is expected to provide a more accurate estimate than values in the product ratio table where the recombination fraction is shown in increments of 0.005. Corresponding to y = 0.085 and 0.090 in the C column, the table shows 0.01116 and 0.01262. The procedure of **linear interpolation** is illustrated on the following page.

F_2 linkage estimation continued

	1	2	3
C values	0.01116	0.012584865	0.01262
y values	0.085	?	0.090

Calculation: $\dfrac{C2 - C1}{C3 - C1} = \dfrac{0.012584865 - 0.01116}{0.01262 - 0.01116} = \dfrac{0.001424865}{0.00146} = 0.988004109 = \alpha$

$1 - \alpha = 0.0199589$; **y2** (the recombination fraction sought) = $(y1)(1 - \alpha) + (y3)(\alpha)$ = $0.085(0.0199589) + 0.090(0.988004109) = 0.08899204$ or ≈ **0.08899**

The interpolated SeC value is obtained: $0.2339(0.0199589) + 0.3025(0.988004109)$ = 0.305236136

The standard error of the y2 is calculated by $\dfrac{305236136}{\sqrt{921}} = 0.010057872$

Thus the interpolated recombination fraction, (y2) = 0.08899 ± 0.01006

The efficiency of the product ratio method in F_2 coupling phase equals that of the testcrosses when the recombination frequency is low (below 8%) but approaching independent segregation the efficiency is decreasing and maximally in F_2 2.5 times larger populations may be required to obtain equally dependable results. In repulsion, especially at close linkage, (below 8% recombination) the F_2 and product ratio procedures are very inefficient. The product ratio method is very useful when both recessive alleles reduce gametic or zygotic viability but not so when only one of the alleles is causing differential mortality. Generally, the product ratio method is equal in efficiency to the maximum likelihood procedure. The dependability of the results is determined primarily by the data collected and not by the statistical procedures. (See recombination, maximum likelihood method applied to recombination frequencies, mapping functions, mapping genetic, table on next page)

F_3 LINKAGE ESTIMATION: may be used in both repulsion and with coupling phases. In repulsion the information is not very trustworthy unless the recombination frequency is about 0.1 or less. Recombination can be estimated by combining F_2 and F_3 data as follows:

$x = \dfrac{a/a \; B/b}{n} = \dfrac{\text{Number of segregating lines}}{\text{Number of a/a lines}}$ then p (recombination fraction) = $\dfrac{x}{2-x}$ and

the standard error of p, $s_p = \dfrac{2\sqrt{x[1-x]}}{[2-x]^2 \sqrt{n}}$

F⁻ PHENOCOPY: a bacterial cell that has no F pilus although it carries an F element yet it is not in a conjugative state. (See bacterial conjugation, conjugation mapping) a conjugative state.

F PLASMID: is called also F factor or fertility plasmid of *E. coli*; it is about 100 kb and has four major regions. The (1.) *inc* and *rep* tracts control its vegetative replication. When only this is retained it is called a *miniplasmid*. The (2.) *IS* sequences are insertion elements and Tn*1000* is a transposon (called also γδ) is similar to Tn*3*. In the (3.) *silent region* few functions are known. The (4.) *tra* sites control transmission of the plasmid which originates at the *oriT* site. The *tra* region includes more than two dozen genes. The *tra M* is regulated by the *traJ* gene which in turn is negatively regulated by *fin* (fertility inhibition) gene products. The *tra* operon includes also genes for the formation of the sex pilus through which plasmid and/or chromosomal DNA is transfered to the recipient cells. The total number of genes in this plasmid is about 30. The plasmid may be present in 1 or 2 copies per F⁺ bacterial cells. It is an episome and can integrate clockwise or counterclockwise at various sites into the bacterial chromosome. When excised it may become an F' plasmid. (See conjugation, pilus, mapping, transposon, episome, F' plasmid)

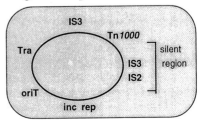

F₂ linkage estimation continued

RECOMBINATION FRACTIONS DETERMINED BY THE PRODUCT RATIO METHOD USING F$_2$ DATA. MODIFIED AFTER F.R. IMMER, GENETICS 15:81. RECOMBINATION FRACTION = y, PRODUCT RATIO IN REPULSION: R, PRODUCT RATIO COUPLING: C. FACTOR TO BE DIVIDED BY $\sqrt{N}$, TO OBTAIN STANDARD ERROR OF Y, FOR REPULSION SeR AND FOR COUPLING SeC. N = TOTAL NUMBER OF INDIVIDUALS IN F$_2$. WHEN THE VALUE OF THE PRODUCT RATIO DOES NOT CORRESPOND TO A CLOSE ENOUGH VALUE OF y, INTERPOLATION IS REQUIRED. MORE PRECISE ESTIMATES CAN BE OBTAINED BY THE TABLE OF W.L. STEVENS (J. GENETICS 39:171), USING 5 DECIMALS, RATHER THAN 3 AS IN THIS TABLE. SINCE RECOMBINATION FRACTIONS ARE GENERALLY VARIABLE, FOR MOST PURPOSES THIS ACCURACY MAY BE ENTIRELY SATISFACTORY. SEE PAGES 360-361.

y	R	C	SeR	SeC	y	R	C	SeR	SeC
0.005	0.00005000	0.00003361	1.0000	0.0707	0.255	0.1536	0.1396	0.9243	0.5191
0.010	0.00020005	0.0001356	0.9999	0.1001	0.260	0.1608	0.1467	0.9214	0.5244
0.015	0.0004503	0.0003076	0.9997	0.1226	0.265	0.1682	0.1540	0.9186	0.5297
0.020	0.0008008	0.000516	0.9996	0.1417	0.270	0.1758	0.1616	0.9158	0.5351
0.025	0.001252	0.0008692	0.9993	0.1585	0.275	0.1837	0.1695	0.9128	0.5404
0.030	0.001804	0.001262	0.9988	0.1736	0.280	0.1919	0.1717	0.9099	0.5456
0.035	0.002458	0.002733	0.9985	0.1877	0.285	0.2003	0.1861	0.9069	0.5509
0.040	0.003213	0.002283	0.9979	0.2007	0.290	0.2089	0.1948	0.9039	0.5560
0.045	0.004070	0.002914	0.9975	0.2129	0.295	0.2179	0.2038	0.9008	0.5612
0.050	0.005031	0.003629	0.9969	0.2246	0.300	0.2271	0.2132	0.8977	0.5663
0.055	0.006096	0.04429	0.9962	0.2357	0.305	0.2367	0.2228	0.8946	0.5714
0.060	0.007265	0.005318	0.9956	0.2463	0.310	0.2465	0.2328	0.8913	0.5764
0.065	0.008540	0.006296	0.9947	0.2565	0.315	0.2567	0.2432	0.8882	0.5815
0.070	0.009921	0.007366	0.9939	0.2663	0.320	0.2672	0.2538	0.8850	0.5864
0.075	0.01141	0.008531	0.9930	0.2758	0.325	0.2780	0.2649	0.8817	0.5914
0.080	0.01301	0.009793	0.9920	0.2850	0.330	0.2892	0.2763	0.8784	0.5963
0.085	0.01471	0.01116	0.9910	0.2339	0.335	0.3008	0.2881	0.8750	0.6012
0.090	0.01653	0.01262	0.9899	0.3025	0.340	0.3127	0.3003	0.8716	0.6061
0.095	0.01846	0.01419	0.9889	0.3109	0.345	0.3250	0.3128	0.8684	0.6110
0.100	0.02051	0.01586	0.9977	0.1392	0.350	0.3377	0.3259	0.8648	0.6157
0.105	0.02267	0.01765	0.9864	0.3272	0.355	0.3508	0.3393	0.8614	0.6205
0.110	0.02495	0.01954	0.9850	0.3351	0.360	0.3643	0.3532	0.8580	0.6254
0.115	0.02734	0.02156	0.9837	0.3428	0.365	0.3783	0.3675	0.8544	0.6301
0.120	0.02986	0.02369	0.9822	0.3503	0.370	0.3927	0.3823	0.8509	0.6347
0.125	0.03250	0.02594	0.9809	0.3578	0.375	0.4076	0.3977	0.8473	0.6395
0.130	0.03527	0.02832	0.9793	0.3652	0.380	0.4230	0.4135	0.8437	0.6442
0.135	0.03816	0.03083	0.9776	0.3723	0.385	0.4389	0.4298	0.8400	0.6488
0.140	0.04118	0.03347	0.9760	0.3792	0.390	0.4553	0.4467	0.8363	0.6534
0.145	0.04434	0.03624	0.9744	0.3862	0.395	0.4723	0.4641	0.8328	0.6580
0.150	0.04763	0.03915	0.9726	0.3930	0.400	0.4898	0.4821	0.8291	0.6626
0.155	0.05105	0.04220	0.9708	0.3999	0.405	0.5079	0.5007	0.8252	0.6672
0.160	0.05462	0.04540	0.9689	0.4064	0.410	0.5266	0.5199	0.8215	0.6718
0.165	0.05832	0.04875	0.9670	0.4129	0.415	0.5460	0.5398	0.8178	0.6762
0.170	0.06218	0.05225	0.9650	0.4194	0.420	0.5660	0.5603	0.8139	0.6808
0.175	0.06618	0.05591	0.9629	0.4258	0.425	0.5867	0.5815	0.8101	0.6853
0.180	0.07043	0.05973	0.9610	0.4320	0.430	0.6081	0.6034	0.8062	0.6897
0.185	0.07464	0.06371	0.9588	0.4383	0.435	0.6302	0.6260	0.8024	0.6941
0.190	0.07911	0.06787	0.9567	0.4445	0.440	0.6531	0.6494	0.7985	0.6986
0.195	0.08374	0.07220	0.9545	0.4506	0.445	0.6768	0.6735	0.7945	0.7029
0.200	0.08854	0.07671	0.9521	0.4565	0.450	0.7013	0.6985	0.7907	0.7073
0.205	0.09351	0.08140	0.9499	0.4624	0.455	0.7766	0.7243	0.7867	0.7116
0.210	0.09865	0.08628	0.9475	0.4683	0.460	0.7529	0.7510	0.7827	0.7159
0.215	0.1040	0.09136	0.9452	0.4741	0.465	0.7801	0.7786	0.7787	0.7204
0.220	0.1095	0.09663	0.9426	0.4799	0.470	0.8082	0.8071	0.7747	0.7247
0.225	0.1152	0.1021	0.9401	0.4857	0.475	0.8374	0.8366	0.7705	0.7288
0.230	01211	0.1078	0.9376	0.4913	0.480	0.8676	0.8671	0.7665	0.7331
0.235	0.1272	0.1137	0.9351	0.4970	0.485	0.8990	0.8986	0.7623	0.7374
0.240	0.2334	0.1198	0.9324	0.5026	0.490	0.9314	0.9313	0.7583	0.7416
0.245	0.1400	0.1262	0.9297	0.5081	0.495	0.9651	0.9651	0.7542	0.7459
0.250	0.1467	0.1328	0.9271	0.5136	0.500	1.0000	1.0000	0.7500	0.7500

See description of the product ratio method on page 360-361.

F₂ SEGREGATION: see Mendelian segregation

FAA: is a histolological fixative containing formalin, acetic acid, ethanol. (See fixatives)

Fab DOMAIN: of the antibody includes a whole light chain and part of the heavy chain, (can be split off by digestion with papain) including the paratope (the antigen recognition site). (See antibody)

FABRY's DISEASE: is an X-chromosome-linked disorder caused by a deficiency of α-glycosidase enzymes. The hemizygous males have skin lesions, opacity of the eye, periodic fevers, burning sensation in the extremities, edema (fluid accumulation) due to kidney malfunction. The heterozygous females have much milder symptoms yet they also develop opacity of the eyes. The afflicted individuals survive to adulthood when death is caused by heart, kidney and brain problems. The heterozygotes can be identified and the condition is detectable by amniocentesis and thus genetic counseling is effective. Galactosylgalactosyl-glucosyl ceramide accumulates in the endothelium (the tissue lining the heart, blood and lymph vessels and other cavities), in the nerve system, the epithelium (surface cell layers) of the kidney and cornea (the transparent outer coat in front of the eyes). Attempts have been made to correct the condition by retroviral vectors with α-glycosidase A expression. (See galactosidases, sphingolipidoses, sphingolipids, viral vectors, gene therapy)

FACE/HEART DEFECTS: see Alagille syndrome, Williams syndrome, Noonan syndrome, Ellis-van Creveld syndrome, DiGeorge syndrome, velocardiofacial syndrome

FACET: small flat surface on a structure or component of the insect eye. (See also ommatidium)

FACILITATION: by repeated impulses to the nerve cells, the amount of neurotransmitter is increased but eventually this process may lead to exhaustion of the neurotransmitter supply.

FACS (fluorescence-activated cell sorting): see cell sorter

FACTOR D: synonymous with TIF-1B

FACTORIAL: $1 \times 2 \times 3 ... \times (n-1) \times n = n!$, the products of integers from 1 to n, the factorial of n numbers. Note: $0! = 1$.

FACTORIAL EXPERIMENT: is analyzing simultaneously the effects of several factors, e.g. $a_1b_1, a_2b_1, a_1b_2, a_2b_2$. The comparison $a_2b_2 - a_1b_2$ assesses the " main effect" of A when B remains constant and $a_2b_1 - a_2b_2$ determines the B "main effect" when A is constant. Such experiments are very economical for testing simultaneously the consequences of several factors and their interactions (AB), and are frequently used to determine dose responses by using different levels of the individual elements such a_1 and a_2.

FACULTATIVE: no absolute single determination but it may show alternative forms or functions.

FACULTATIVE HETEROCHROMATIN: behaves as heterochromatin only at certain conditions, like in the mammalian X-chromosome when present in more than a single copy. (See heterochromatin)

FAD: flavin adenine dinucleotide, a riboflavin containing coenzyme in some oxidative-reductive processes; also an acronym of familial Alzheimer disease. (See Alzheimer disease)

FADD/MORT1 (FAS-associated death domain): the FLIP proteins interact with FADD in the presence of FLICE and inhibit apoptosis. (See FAS, FLICE, death domain)

FAK (focal adhesion tyrosine kinase): a protein tyrosine kinase occurring at high level in SRC transformed cells. Its role is in mediating cell adhesion and migration. FAK is localized at the cell membrane integrin receptors. Association of integrin with fibronectin results in binding of the Gerb2 adaptor to SRC and FAK protein tyrosine kinases and in the activation of MAPK (mitogen-activated protein kinase) in signal transduction pathway. (See CAM, Grb, SRC, MAPK, signal transduction)

FALLOPIAN TUBE (uterine tube): connects the mammalian ovary to the uterus. Upon maturation of the egg it is released into the fallopian tube and fertilization may take place after the sperm injected into the vagina travels to the egg by the cervical canal. (See fertilization)

FALLOT's TETRALOGY: pulmonary stenosis, atrial septal defect and right ventricular hypertrophy (constituting the *trilogy*), the *tetralogy* = trilogy + the right shifting of the aorta over-

Fallot's tetralogy continued

riding the interventricular septum and thus receiving both arterial and venous blood, the *pentalogy* = the tetralogy + open foramen ovale or atrial septal defect. (See meaning of these terms under separate entries, heart disease)

FALLOUT: radiation emitted by isotopes released into the atmosphere by applying or testing nuclear weapons, by nuclear explosions and atomic power plants. (See atomic radiations)

FALSE ALLELISM: a series of overlapping deficiencies may mimic an allelic series at a gene locus. For an illustration see the consequences of overlapping deficiencies observed at the tip of the short arm of chromosome 9 of maize. Note particularly that the *yg/pyd* heterozygotes are normal green demonstrating that in the *pyd* deficiency the wild type allele is present despite the phenotype of the homozygotes, and the allelism is only apparent. (See allelism, position effect)

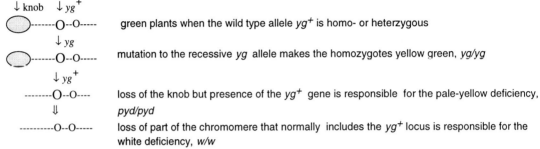

In the F$_1$ all *yg$^+$* homo- and heterozygotes as well as the *yg/pyd* plants are **normal green** but the *yg/w* plants are **yellow-green** and the *pyd/w* individuals are **pale yellow**. (Modified after B. McClintock, Genetics 29:478)

FALSE NEGATIVE: the test performed indicates the absence of a condition but in fact it is an incorrect observation. (See false positive)

FALSE POSITIVE: the test results indicate the presence of a particular condition but the observation is incorrect. (See false negative)

FAMILIAL ADENOMATOUS POLYPOSIS: see FAP

FAMILIAL HYPERCHOLESTROLEMIA (FHC): involves increase in low-density lipoprotein (LDL) bound cholesterol, and as a consequence at birth the homozygotes develop surface tumors filled with lipids (xanthomas) between the first two digits. Later on these plasias appear at the tendons, the cornea of the eye may also display gray lipid-containing little sacks and by adulthood coronary heart disease is expected. The basic defect affects the cell membrane or its receptor of LDL. Under normal condition the LDL passes into the lysosomes where upon degradation the liberated cholesterol feedback-represses hydroxy-methylglutaryl Coenzyme A reductase (HMGR), an enzyme responsible for cholesterol biosynthesis. The fungal products lovastatin and compactin are inhibitors of this enzyme and may be considered for treatment of FHC. The LDL receptor with 18 exons code for 13 polypeptides with homologies to different proteins of a large superfamily. The dominant gene was located to human chromosome 19p13. 2-p13.11 has various base substitutions, frameshift and deletion mutations. Some of the deletions are apparently caused by unequal recombination in the Alu sequences of introns 4 and 5 or others and missplicing. The heterozygotes may show the same symptoms but less evidently. The serum cholesterol and low-density lipoprotein levels in the homozygotes may exceed up to twice or more to that of the normal (75-175 mg/100 mL). About 5% of the myocardial infarction patients are heterozygous for this condition. The prevalence of FHC homozygotes is in the 10^{-6} range and that of the heterozygotes is higher than 0.002. Animals transplanted with autologous but genetically engineered hepatocytes developed a persisting 30-50% decrease of serum cholesterol levels. FHC may be identified at birth. This condition may not result in obesity, high blood pressure or some other common indicators of susceptibility to heart ailments. Using retroviral vectors carrying the LDL receptor gene, gene therapy is promising although

Familial hypercholesterolemia continued
improvement of the efficiency of the transformation is desirable. (See coronary heart disease, lipoprotein, lipids, cholesterols, sphingolipidoses, unequal crossing over, Alu, genetic screening, missplicing, gene therapy, hypertension)

FAMILIAL HYPERTRIGLYCERIDEMIA: is an autosomal dominant hyperlipoproteinemia leading to coronary heart disease. (See coronary heart disease, hyperlipoproteinemia)

FAMILIAL MEDULLARY THYROID CARCINOMA (FMTC): see RET oncogene

FAMILIAL TRAIT: appears in certain families; its cause may be hereditary or other, e.g., deafness may result from about two dozen genetic conditions, but ear infection, diabetes of the mother, birth injury, or senescence may be responsible for it. (See congenital trait)

FAMILY: offspring descended from a common mother (and father); in taxonomy a group of genera. (See also nuclear family)

FAMILY BOX: a group of RNA nucleotide triplets with identical nucleotides near the 5'-end but different at the 3'-end yet coding for the same amino acid. (See code genetic)

FAMILY HISTORY: is an essential source for the genetic counselors to determine recurrence risks. (See risk, genetic risk, recurrence risk, empirical risk, genetic counseling)

FAMILY OF GENES: a similar set of genes (coding for basically similar polypeptides); they probably arose during evolution by successive duplications and modifications. (See duplication)

FANCONI'S ANEMIA (FA, aplastic anemia, Fanconi's syndrome, Fanconi's disease): is a recessive genetic disorder assigned to at least five complementation groups. The A complementation group (FA-A) was localized to human chromosome 16q24.3, the FA-C group to 9q22.3, the FA-D to 3q22-q26. The earlier assignment to chromosome 20q could not be confirmed by newer data. The disease shows a wide range of symptoms: leukopenia (less than 5,000 leukocytes/mL blood), thrombocytopenia (reduced platelet count), pigmentation of the skin and various malformations at different degrees of expression. The homozygous cells suffer high frequency chromosome defects. Both red and white blood cells appear normal. It is generally fatal by the teen years. The genetic basis is chromosomal instability and associated leukemia and other malignancies. It is accompanied by genito-urinary anomalies, cystinuria, heart disease, dwarfism, skeletal problems, microcephaly (small head), deafness, etc. The genetic repair system appears to be defective in the complementation group C (FAC) protein (163 kDa) which seems to posses a nuclear localization signal. The A complementation group coded protein is cytoplasmic and its precise function is unclear, however general regulatory role is suspected. Gene therapy for the C protein appears promising although does not remedy all the problems of the disease. (See also hemostasis, platelet anomalies, DNA repair, stature in humans, limb anomalies, light-sensitivity diseases, cancer, Dubowitz syndrome, Bloom syndrome, gene therapy)

FANCONI RENOTUBULAR SYNDROME: an autosomal dominant, late onset kidney disease resulting in aminoaciduria, glycosuria; sometimes it is associated with bone problems. There are types with and without cystinosis. (See aminoacidurias, cystinosis, glycosuria)

FAO: Food and Agricultural Organization of the United Nations, Rome, Italy.

FAP: familial adenomatous polyposis; genetically controlled polyps in the colon that may develop into colorectal cancer after additional somatic mutations. (See polyp, Gardner syndrome)

FAP-1: is a protein tyrosine phosphatase regulating the activity of the Fas cell surface receptor. (See Fas)

FAR: is a protein required for pheromone-induced cell cycle arrest at G1. (See cell cycle)

FARBER's DISEASE: is concerned with a defect of the enzyme ceramidase. It is manifested in early infancy as irritability, swelling of the joints, motor abnormalities, mental retardation and early death. (See sphingolipids, ceramide)

FARNESYL PYROPHOSPHATE: is an intermediate in the synthesis of cholesterol and has a role in mediating binding of proteins to membranes. (See prenylation)

FARNESYLATION: see prenylation

Fas (CD95, APO1): a member of the tumor necrosis factor and nerve growth factor receptor protein family including also Cd40, CD27, CD30, OX40, SFV-2. FasL: Fas ligand, encoded in human chromosome 1. FasL expression is induced by PMA. Fas plays a key role also in apoptosis and autoimmune disease by binding the FasL. FasL can be rapidly induced by PMA but protein tyrosine kinase inhibitors (herbimycin, genistein) and cyclosporin inhibit the induction of FasL. The antibody to Fas is an immunoglobulin M (IgM) whereas the antibody to APO-1 is IgG3. Human Fas is a transmembrane protein containing 325 amino acids encoded in chromosome 10q. Fas is upregulated by interferon-γ and tumor necrosis factor-α in human B cells. Fas signaling leads to apoptosis and it is triggered by cross-linking of Fas with Fas antibodies, and by cells expressing FasL. Fas-induced apoptosis is faster than that induced by TNF-receptor. The Fas and the TNF induce apoptosis in the cells of the immune system. (See apoptosis, TNF, autoimmune disease, FAP-1, PMA, cyclosporin, immune system)

FASCICLE (fasciculus): a bundle

FASCICLINS: are cell adhesion molecules of the immunoglobulin superfamily and are expressed in motor neurons. (See titin)

FAST BLUE: a fluorescent neuronal tracer dye.

FAST COMPONENT: of the nucleic acid reassociation reaction represents the repetitive sequences. (See c_0t curve, c_0t value)

FAST GREEN: is a histochemical stain for basic proteins; generally used in combination with other stains, e.g. safranin or pyronin.

FASTLINK: a computer program for genetic linkage analysis.

FAST NEUTRONS: are particulate, ionizing radiations released at atomic nuclear fission. They are highly effective mutagens. (See radiation, physical mutagens)

FASTA: a program used for sequence comparisons in DNA, RNA and proteins [Pearson, W.R. 1991 Genomics 11:635]. E-mail address <fasta@ebi.ac.uk>. (See BLAST, BLOSUM)

FAT: fatty acid - glycerol ester. Its caloric value (9.3 kcal/g) is approximately 2.3 of that of carbohydrates and proteins. Oxidation of fats (rancidity) may result in mutagenic compounds in natural products. (See fatty acids, cholesterol, triaglycerols, atherosclerosis)

FATAL FAMILIAL INSOMNIA (insomnia-dysautonomia thalamic syndrome): autosomal dominant degenerative disease of the thalamic nuclei (the basal; part of the brain involved in transmission of sensory impulses). A progressive insomnia, defect in the autonomous nervous system, speech defect, tremor, seizures that eventually may lead to death. The pathological symptoms may resemble to that of the Creutzfeldt-Jakob syndrome with the difference that the spongy transformation of the cells is limited here to the thalamus. In mice devoid of prion protein (PrP^C) because of mutation, the circadian activity rhythm, including the sleeping pattern is altered. (See prion, encephalopathies, Creutzfeldt-Jakob syndrome)

FATE MAP, MORPHOGENETIC: indicates the positions in the blastoderm from which adult structures (legs, eyes, nerves, etc.) develop. In basic principles it has some resemblance to the genetic mapping of genes to chromosomes (both ideas were developed by A. Sturtevant). The procedure is generally as follows. The investigator constructs a diploid individual with an unstable chromosome with the wild type allele and the homologous chromosome carries the recessive gene whose expression is to be traced to developmental origin or control center. When the unstable chromosome is lost, the critical recessive allele can display pseudodominance in the sector which no longer carries the wild type allele. Most commonly the defective chromosome in *Drosophila* is an X-chromosome. Thus its loss generates also a gynandromorphic sector. The association between the gynander sector and the critical gene expression can be classified and the developmental origin of the function determined. The developmental distance calculated from fate mapping are expressed in *sturt* units; 1 sturt is the fate mapping quotient multiplied by 100. In other words, 1 sturt means that two cell clusters are different in 1% of the mosaics.

The results of a fate mapping experiment using an unstable ring X-chromosome and the *drop dead* mutation of *Drosophila* are shown on next page. The homo- and hemizygous indi-

Fate map, morphogenetic continued

viduals walk in an uncoordinated manner and suddenly die about ten days after eclosion. (Abridged from Y. Hotta & S. Benzer, Nature 240:527)

Behavior ↓	Number of individuals with the constitution indicated in the head cuticle and in the abdominal cuticle			
	Head		Abdomen	
	XX	XO	XX	XO
Normal	91	8	54	28
drop-dead	6	72	23	51
Total	97	80	77	79

The frequency of *drop-dead* gynander and non-gynander (normal) heads and abdomens can be calculated similarly to recombination frequencies. Thus,

$$\text{HEAD:} \quad \frac{6+8}{97+80} = 0.079 \qquad \text{ABDOMEN:} \quad \frac{23+28}{77+79} = 0.327$$

The fractions above indicate that the *drop-dead* phenotype is much closer associated with the head cuticle gynandromorphy than with that of the abdomen showing that this behavior is determined by the development of the head. Anatomical studies have shown perforations in the brain of the flies, confirming the conclusions of the fate-mapping experiments. (See gynandromorphs, deletion mapping, mapping, physical mapping)

FATTY ACIDS: aliphatic carboxylic acids of long chain structure, parts of the membrane phospho- and glycolipids, cholesterols, also in fats and oils. Saturated fatty acids are formic, acetic, propionic, butyric (4:0), lauric (12:0), myristic (14:0), palmitic (16:0), stearic (18:0), arachidic (20:0), behenic (22:0) and lignoseric (24:0) acids (Within the parenthesis the number of carbon atoms in the chain and after the colon the number of double bonds are indicated). Unsaturated fatty acids are crotonic acid (4:12), plamitoleic (16:1), oleic (18:1), vaccenic (18:1), linoleic (18:2), linolenic (13:3), and arachidonic (20:4) acids. Triacylglycerides (glycerol esters with three fatty acids) are one of the major source of energy in mammals, particularly during hibernation (when nearly all the energy comes from these compounds). Linoleate and linolenate are essential fatty acids for mammals. Lipids are complexes of fatty acids with phosphates, sterols or sugars and have indispensable functions in cellular membranes. Lipids associated with proteins have the role of transporting fatty acids. When fatty acid nodules are deposited on the inner walls of the blood vessels the disease atherosclerosis results, a major cause of heart disease. Fatty acid biosynthesis is important not just from the viewpoint of human disease but its knowledge may help in the developing plant varieties for better diet but also for the production of special raw material for industrial purposes. (See apolipoproteins, fat, triaglycerols, lipids, lipidoses, sphingolipids, sphingolipidoses, atherosclerosis)

FAVISM: a hemolytic anemia caused by eating even extremely small quantities of *Vicia faba* (broad bean) or inhaling its pollen. The initial reaction (headache, dizziness, nausea, chills, etc.) may occur within seconds and it may be followed within a day by jaundice and blood in the urine. People who are deficient in glucose-6-phosphate dehydrogenase are susceptible to this condition. The light sensitivity of the skin after eating the seeds of *Fagopyrum vulgare* (buckwheat) is different; the latter is caused by a photodynamic substance. (See glucose-6-phosphate dehydrogenase deficiency)

FB: see hybrid dysgenesis

FBI SITE (fold back inhibition): the transcript of Tn*10* transposase folds back a region (UGGUC) complementary to the Shine-Dalgarno sequence (AUCAG) and prevents ribosome binding and thus reduces transposition. (See Tn*10*)

FBS: fetal bovine serum.

Fc: crystallizable fragment of immunoglobin containing the C end of the heavy chains. The Fc receptors are expressed in monocytes and macrophages and some other cells of the immune system. (See antibody, immune system)

FcRn: crystallizable fragment receptor of the antibody. (See antibody)

F-DISTRIBUTION: is used for testing the significance of the difference between variances. The *F-test* is based on $F = (s_1)^2/(s_2)^2$ [s = standard deviation] that are computed from two different populations or samples. In this procedure the null hypothesis is that the two samples are identical. If, however, the variance ratio exceeds the values, at the appropriate degrees of freedom, in the body of the table, the differences are significant. (The name F is in honor of R.A. Fisher; see table on page 369)

F-DUCTION: transfer of genes from donor to the recipient bacteria with the aid of an F' plasmid.

FEATURE: see DNA chips

FECUNDITY: the reproductive ability determined by the quantity of gametes produced per time units. (See fertility)

FEEDBACK CONTROL: a late metabolite of a synthetic pathway regulates synthesis at earlier step(s). The feedback control is mediated by two specific sites on an early enzyme in the biosynthetic pathway. One site serves for recognition of the substrate, the other site recognizes a later product in the biosynthetic path. When this late product (end-product) accumulates because it is not utilized in proportion to its synthesis, it may combine with the early enzyme's feedback recognition site, resulting in a reversible conformational change and cessation or lowering the activity of this enzyme. The feedback is an economy device of the cell; the production of a metabolite is slowed down in the absence of a need. Feedback systems may operate in a number of self-explanatory ways (see at left); for the compensating feedback a note is needed. E inhibits the path between C and D but product F may alleviate the inhibition by reducing the activity of enzyme A thus E never accumulates excessively. If through mutation feedback-sensitivity is eliminated, the system may be overproducing the end-product and this may be beneficial from economic viewpoint in industrial or agronomic organisms. These are the most common regulatory mechanisms in eukaryotes. (See regulation of enzyme activity, inhibition, repression, attenuation negative control, positive control)

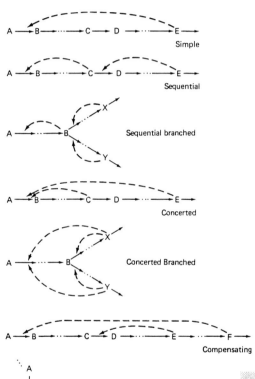

THE MAJOR TYPES OF FEEDBACK CONTROLS. THE BROKEN LINES INDICATE THE FEEDBACK LOOPS. THE SOLID LINES SHOW THE BIOCHEMICAL SYNTHETIC PATHWAY. (Redrawn after Savageau, M.A. Currt. Top. Cell Reg. 6:63)

FELINE: pertaining to cat or cats.

FEMALE: the individual producing the larger usually non-motile gamete (egg). Its symbol is the "Venus mirror" ♀ or in pedigree charts a circle ○ This is the most ancient symbol of genetics, used already three thousand years ago on stone tablets in Asia Minor. (See pedigree analysis, Mars shield)

F-DISTRIBUTION at 5% and 1% probability levels. At significance the F value computed must exceed the numbers in the table the intersection of the degrees of freedom (Condensed from S. Koller, Biochemisches Taschenbuch, H.M. Rauen, ed., Springer-Vlg, Berlin.)

df DE-NOMI-NATOR	\multicolumn{11}{c}{P = 5%}

df DENOMINATOR	1	2	3	4	5	6	8	10	15	20	30	
1	161	200	216	225	230	234	239	242	246	248	250	254
2	19	19	19	19	19	19	19	19	19	19	20	20
3	10.1	9.6	9.3	9.1	9.0	8.9	8.8	8.8	8.8	8.7	8.6	8.5
4	7.7	6.9	6.6	6.4	6.3	6.2	6.0	6.0	5.9	5.8	5.8	5.6
5	6.6	5.1	4.8	4.5	4.4	4.3	4.2	4.1	3.9	3.9	3.8	3.7
6	6.0	5.1	4.8	4.5	4.4	4.3	4.2	4.1	3.9	3.9	3.8	3.7
8	5.3	4.5	4.1	3.8	3.7	3.6	3.4	3.4	3.2	3.2	3.1	2.9
10	5.0	4.1	3.7	3.5	3.3	3.2	3.1	3.0	2.9	2.8	2.7	2.5
12	4.8	3.9	3.5	3.3	3.1	3.0	2.9	2.8	2.6	2.5	2.5	2.3
14	4.6	3.7	3.3	3.1	3.0	2.9	2.7	2.6	2.5	2.4	2.3	2.1
16	4.5	3.6	3.2	3.0	2.9	2.7	2.6	2.5	2.4	2.3	2.2	2.0
18	4.4	3.6	3.2	2.9	2.8	2.7	2.5	2.4	2.3	2.2	2.1	1.9
20	4.4	3.5	3.1	2.9	2.7	2.6	2.5	2.4	2.2	2.1	2.0	1.8
25	4.2	3.4	3.0	2.8	2.6	2.5	2.3	2.2	2.1	2.0	1.9	1.7
30	4.2	3.3	2.9	2.7	2.5	2.4	2.3	2.2	2.0	1.9	1.8	1.6
40	4.1	3.2	2.8	2.6	2.5	2.3	2.2	2.1	1.9	1.8	1.7	1.5
50	4.0	3.2	2.8	2.6	2.4	2.3	2.1	2.0	1.9	1.8	1.7	1.4
60	4.0	3.2	2.8	2.5	2.4	2.3	2.1	2.0	1.8	1.8	1.7	1.4
80	4.0	3.1	2.7	2.5	2.3	2.2	2.1	2.0	1.8	1.7	1.6	1.3
100	3.9	3.1	2.7	2.5	2.3	2.2	2.0	1.9	1.8	1.7	1.6	1.3
∞	3.84	3.00	2.60	2.37	2.21	2.10	1.94	1.83	1.67	1.57	1.46	1.00

					P = 1%						
	1	2	3	4	5	6	8	10	15	20	30	∞
1	4100	5000	5400	5600	5800	5900	6000	6000	6200	6200	6200	6400
2	98	99	99	99	99	99	99	99	99	99	99	100
3	34	31	29	29	28	27	27	27	27	27	27	26
4	21	18	17	16	16	15	15	15	14	14	14	13
5	16	13	12	11	11	11	10	10	9.7	9.6	9.4	9
6	14	11	9.8	9.2	8.8	8.5	8.1	7.9	7.6	7.4	7.2	6.9
7	12	9.6	8.5	7.8	7.5	7.2	6.8	6.6	6.3	6.2	6.0	5.7
8	11	8.7	7.6	7.0	6.6	6.4	6.0	5.8	5.5	5.4	5.2	4.9
10	10	7.6	6.6	6.0	5.6	5.4	5.1	4.9	4.7	4.4	4.3	3.9
12	9.3	6.9	6.0	5.4	5.1	4.8	4.5	4.3	4.0	3.9	3.7	3.4
16	8.5	6.2	5.3	4.8	4.4	4.2	3.9	3.7	3.4	3.3	3.1	2.8
18	8.3	6.0	5.1	4.6	4.3	4.0	3.7	3.5	3.2	3.1	2.9	2.6
20	8.1	5.9	4.9	4.4	4.1	3.9	3.6	3.4	3.1	2.9	2.8	2.4
25	7.8	5.6	4.7	4.2	3.9	3.6	3.3	3.1	2.9	2.7	2.5	2.2
30	7.6	5.4	4.5	4.0	3.7	3.5	3.2	3.0	2.7	2.6	2.4	2.3
40	7.3	5.2	4.3	3.8	3.5	3.3	3.0	2.8	2.5	2.4	2.2	1.8
50	7.2	5.1	4.2	3.7	3.4	3.2	2.9	2.7	2.4	2.3	2.1	1.7
60	7.1	5.0	4.1	3.7	3.3	3.1	2.8	2.6	2.4	2.2	2.0	1.6
80	7.0	4.9	4.0	3.6	3.3	3.0	2.7	2.6	2.3	2.1	1.9	1.5
100	6.9	4.8	4.0	3.5	3.2	3.0	2.7	2.5	2.2	2.1	1.9	1.4
∞	6.64	4.60	3.78	3.32	3.02	2.80	2.51	2.32	2.04	1.88	1.70	1.00

FEMALE GAMETOPHYTE: see gametophyte

FERAL POPULATION: inhabits nature rather than living under laboratory or domesticated conditions (e.g. feral mice, feral *Drosophila*, feral pigs, etc.)

FERMENTATION: energy-producing anaerobic degradation of carbohydrates.

FERMENTORS: precisely controlled mass culture vessels that may be used for industrial manufacturing of biological products (such as alcohol, antibiotics, proteins, etc.) by bacterial, fungal, animal or plant cells.

FERNS: are lower plants of the *Pterophyta* taxonomic group. They exist in two generations: sporophyte and gametophyte. The sporophytes develop roots, rhizomes and leaves. These diploid plants form generally one type of sporangium on the lower surface of the leaves in aggregates, called sori. Meiosis takes place within the sporangia and the released haploid products develop into generally heart-shaped gametophytes that (unlike in higher plants) form an independent organism on soil (thallus). When mature, on the lower surface of the gametophyte, eggs develop within the female archegonia and sperms in the male antheridia (6-10 cells). After a swimming spore fertilizes an egg, the diploid zygote grows on the gametophyte until the young sporophyte becomes ready (rooted) for independent existence. The gametophytes then die and through the sporophytes the life cycles are repeated. Ferns have a wide range of chromosome numbers of relatively large size, and many species are well suited for cytological analyses also because of the long haploid phase.

FERREDOXIN: an iron-containing protein involved in electron transport.

FERRITIN: ubiquitous, iron-storing cellular proteins, and thus protect the cells from the toxic effects of this heavy metal. Also, it transports iron to the sites of need. Ferritin consists of 24 subunits around an iron core that are encoded in human chromosomes at 11q13 and 19q13.1. The synthesis is determined by an iron-regulatory protein (IRP) and an iron-responsive element (IRE) CAGUGU. Mutation in the latter results in the dominant hyperferritinemia that may cause cataracts. (See aconitase, apoferritin, eye diseases)

FERTILIN: is a protein mediating sperm - egg fusion. (See fertilization)

FERTILITY: the production of viable offspring (gamete) per individual in the reproductive period. *Effective fertility* denotes the reproductive rate of individuals afflicted by a disease compared to that of normal, healthy individuals. The ability to form fertile hybrids has been generally used to define species. Different species were not expected to produce fertile hybrid progeny. This distinction has some problems because, e.g. the female hybrids of *Drosophila melanogaster* female and *D. simulans* male are viable and fertile whereas the males of the same cross die at the larval/pupal stage of development. In the reciprocal crosses, *D. simulans* female x *D. melanogaster* male most of the females also die as embryos, and only a few escape death. (See fecundity, reproductive isolation, species, speciation, infertility, spermiogenesis)

FERTILITY FACTOR: see F factor

FERTILITY INHIBITION: the conjugal function of the bacterial cell is prevented by an inhibitory plasmid. (See bacterial conjugation)

FERTILITY RESTORER GENES: may assist to overcome the cytoplasmically determined male sterility. Their ability to function will depend also on the nature of the cytoplasm. In the S cytoplasm of maize the restoration of fertility by *Rf 3* is gametophytic, i.e., only 50% of the pollen is fertile in the heterozygous plants, whereas in the T cytoplasm the restoration of fertility is under sporophytic control and the heterozygotes for the two complementary genes fertility genes *Rf1* and *Rf2* produce nearly 100% viable pollen. The *Rf4* gene is a sporophytically expressed restorer in the C cytoplasm. Fertility restoration may involve also some additional minor factors. The cytoplasmic male sterility in these cytoplasms is controlled by mitochondrial plasmid genes. The fertility restorer genes have applied significance in the commercial production of hybrids. The T (Texas) cytoplasmic male sterility is associated with susceptibility to *Helminthosporium* blight. The fertility restorer genes do not alleviate, however, the symptoms of the disease, indicating that the latter is controlled by other means than the fertility. Cytoplasmic male sterility occurs in several plant species. (See cytoplasmic male sterility, pollen sterility, mtDNA)

FERTILIZATION: involves the fusion of gametes of opposite sexuality. It may take place in dif-

Fertilization continued

ferent forms in hermaphroditic organisms such as the majority of plants and some of the animals: self-fertilization (autogamy), cross-fertilization (allogamy) or a mixture of the two. In the dioecious species of the majority of animals and a few plants only cross-fertilization can happen. The fertilization results in the formation of zygotes and the restoration of the chromosome number to the 2n level. In many plant species double fertilization takes place: the generative sperm fertilizes the egg and the vegetative sperm unites with the fused polar nuclei and as a result the triploid endosperm mother cell is generated. In some plant species the sperm and egg nuclei fail to fuse and each contributes independently to the development of the embryo: *semigamy*. In some plants the generative sperm and the vegetative sperm may come from a different pollen grain and thus *heterofertilization* results. If the genetic constitution of these two sperms is not identical, there is a discordance (non-correspondence) between the genetic constitution of the embryo and the endosperm. Embryo development without fertilization is called *apomixis* in plants and *parthenogenesis* in animals. In the majority of plants and animals only the nucleus of the sperm enters the embryosac and egg, respectively, and the cytoplasmic genes are transmitted only through the female in these cases. In other plants, cytoplasmic elements are also transmitted by the sperm during fertilization resulting in complete biparental inheritance. In plants only a limited number of pollen tubes grow down the pistil. In animals, more than a thousand sperms may attach to the surface of the egg but only one succeeds in fertilization although normal ejaculates may contain 200 million sperms. In autogamous plants the pollen count is much more limited than in the allogamous species. A large quantity of sperm is essential for normal fertilization of plants and animals. Human males with sperm counts per ejaculate below 20,000 are generally sterile. The number of eggs per ovule in plants or number of ova released during ovulation in animals may also be only one or a few. *Drosophila* females can store about 700 sperms after a single injection by the males for a prolonged period of time and fertilization may take place gradually. After a second mating within a period of two weeks, the leftover sperms may be destroyed. Usually the fertilization takes place within the female body, in others the fertilization is an external process or the eggs may be deposited in a pouch of the female where fertilization and further development follows. Even in cases of internal fertilization a variable amount of time may elapse between pollination or copulation and the actual penetration and fusion of the sperm and egg nuclei. In mammalian fertilization the first step after the sperm reaches the follicular cells is binding to the egg membrane. The sperm is "capacitated" to this task by changes in the sperm plasma membrane through the secreted products of the female. The sperm then binds to the zona pellucida membrane. This membrane is made up of cross-linked network of glycoproteins ZP1, ZP2, ZP3. The latter ones are *sperm-receptors*. In mouse the sp56 spermatid-specific protein recognizes ZP3. After this step, the protease and hyaluronidase content of the acrosome is released, facilitating the penetration and passing through the zona pellucida. After one sperm has fused with the egg plasma membrane, the so called cortical reaction blocks the entry of other sperms as the zona hardens. The *cortical reaction* is mediated by an increase in cytosolic calcium through an inositol phospholipid signaling pathway. On the surface of the egg coat microvilli develop in such a manner that they firmly hold the sperm. After fertilization, the egg becomes a zygote and the process is completed when the male and female pronuclei fuse into a *zygote nucleus*. The sperm donates to the egg not just the male pronucleus, containing a complete haploid set of chromosomes but also the *centriole*, that is not available in the egg. The centrioles are then associated with the female *centrosome*. This makes possible the mitotic divisions and embryogenesis. (See also apomixis, androgenesis, gametogenesis, embryogenesis, selective fertilization, polyspermic fertilization, certation, megaspore competition, sperm, IVF, artificial insemination, RPTK, centrosome)

FES: feline fibrosarcoma viral oncogene (*fes*). It is in the long arm of human chromosome 15, and in most cancer cells it is translocated to chromosome 17, causing leukemia. Its protein product is a protein tyrosine kinase. (See also ABL, ERBA, PTK)

FET: genes involved in iron uptake and metabolism.

FETAL TISSUE RESEARCH: see embryo research

α–FETOPROTEIN (AFP): is expressed in the embryonic yolk and liver of mammals. Serum albumin and α-fetoprotein genes are linked at about 15 kb apart, and each encodes about 580 amino acids. The two proteins are immunologically cross-reactive and display about 35% homology. The rate of transcription of AFP drops four orders of magnitude after birth. Regulatory sequences are positioned within 150 bp 5' and there are also enhancers 6.5, 5 and 2.5 kbp upstream from the transcription initiation site. The gene (human chromosome 4q11-q22) is a classical example for tissue-specific and developmental regulation. (See MSAFP)

FETOSCOPY: viewing (or possibly sampling of tissues) of the fetus within the womb to detect probable developmental or biochemical anomalies in case there is a good indication for them. Fetoscopy may involve up to 10% risk to the fetus and if other less invasive (e.g. sonography) methods are available it should be avoided. It is used also for the detection of fetal heartbeat. (See prenatal diagnosis, amniocentesis, sonography, echocardiography)

FETUS (foetus): the unborn child of viviparous animals at the stages following the embryonal state after substantial differentiation has been completed. In humans the fetal period is from nine weeks after conception to birth. (See vivipary)

FEULGEN: a microtechnical staining method detecting deoxyribose and thus DNA in warm acid-hydrolyzed tissue followed by staining with leucobasic fuchsin. (See stains)

Ffh: is a GTPase of the signal recognition particle of bacteria. (See FtsY, SRP)

FGF: fibroblast growth factor. It is also a FGF-related oncogene. FGF-B (the bovine form, FGF2) was assigned to human chromosome 4q25. FGF-6 and FGF-4 (human chromosome 12q13) bear similarity to NT2 oncogene. FGF may have critical role also in organ differentiation. The fibroblast growth factor receptor (FGFR) has a cytoplasmic tyrosine kinase domain. Mutations in FGFR2 has been associated with the Crouzon, Jackson-Weiss, Pfeiffer, Beare-Stevenson syndromes. FGF1 (in human chromosome 5q31) is implicated in angiogenesis, FGF3 11q13) is homologous to a mouse mammary carcinoma gene, FGF5 (4q21), FGF 7 is a keratocyte growth factor, FGF8 is an androgen-induced growth factor, FGF9 is the glia-activating factor. (See growth factors, signal transduction, HST oncogene, INT oncogene, organizer, tyrosine kinase, achondroplasia, hypochondroplasia, thanatophoric dysplasia, Jackson-Weiss syndrome, Pfeiffer syndrome, Apert syndrome, Crouzon syndrome, craniosynostosis, angiogenesis, glial cell, keratosis)

FGR ONCOGENE: located to human chromosome 1p36.2-p361 encodes an actin like and a tyrosine kinase sequence that is homologous to the viral gene *fgr*.

fi PLASMID: inhibits the fertility factor of bacteria. (See F element)

FIALOURIDINE (FIAU, [1-(2-deoxy-2-fluoro-D-arabinofuranosyl)-5-iodouracil]): is an inhibitor of DNA polymerase γ, responsible for the synthesis of mtDNA. (See DNA polymerases, mtDNA)

FIBRILLARIN: is a fibrillar protein present in many snRNPs within the nucleolus; it is also an autoantigen. (See snRNP, autoantigen)

FIBRILLIN: a large glycoprotein molecule involved in the formation of microfibrils along with elastin. The fibrillin microfibrils are encoded by two genes in human chromosomes 15q15-21.3 (FBN1) and 5q23-q31 (FBN2), respectively. (See Marfan syndrome, Marfanoid syndrome)

FIBRIN: an insoluble protein formed during blood clotting from fibrinogen by the action of thrombin. (See thrombin, fibrinogen, fibrin-stabilizing factor deficiency)

FIBRIN-STABILIZING FACTOR DEFICIENCY: is controlled by X-chromosomal recessive factors causing a deficiency of blood coagulation factor XIII that normally stabilizes clot formation by cross-linking fibrin. The wounds heals slow, the umbilicus bleeds days after birth, bleeding may occur inside the joints (hemarthrosis), in the genito-urinary tract (hematuria), abortion with bleeding, intracranial bleeding (generally lethal) may occur. Apparently, the condition is caused not by the lack of factor XIII rather than by the formation of defective molecules. (See also afibrinogenemia, defibrinogenemia, antihemophilic factors, hemophilia,

von Willebrand disease)
FIBRINOGEN: see fibrin
FIBRINOLYSIN: a serine endopeptidase. (See plasmin)
FIBROBLAST: elongated connective tissue cells (e.g., tendons and other supportive structures) which may be able to produce collagen; it is relatively easy to culture *in vitro*. (See collagen)
FIBROBLAST GROWTH FACTOR (FGF): stimulates blood vessel growth. The gene was located in human chromosome 4q25. The FGF receptors (FGFR) proteins have an extracellular immunoglobulin-like domain and the cytoplasmic domains are tyrosine kinases. FGFR3 (human chromosome 3) is located near the achondroplasia mutation site. FGFR2 in chromosome 10 seems to be associated with the Crouzon syndrome (identical with the Jackson-Weiss syndrome). FGFR1 in chromosome 8 co-segregates with the Pfeiffer syndrome. FGF3 deficiency accounts for achondroplasia, hypochondroplasia and thanatophoric dysplasia. (See signal transduction, angiogenesis, achondroplasia, Crouzon syndrome, Pfeiffer syndrome, achondroplasia, hypochondroplasia, thanatophoric dysplasia)
FIBRODYSPLASIA OSSIFICANS PROGRESSIVA (FOB): rare autosomal dominant progressive ectopic (malplaced) bone formation. The basic defects appears to be the inappropriate production of bone morphogenetic protein (BMP-4) by the lymphocytes. Sometimes accompanied by short fingers, anomalous vertebrae, deafness, baldness and mental retardation. (See exostosis, metachondromatosis, bone morphogenetic protein)
FIBROIN: silk protein. (See silk fibroin, silk worm)
FIBRONECTIN: extracellular matrix protein involved in development, wound healing and tumorigenesis. (See also integrin, RGD)
FICOLL®: a synthetic polymer of sucrose. It is used in the Denhardt reagent and in gel-loading buffers to increase the density of the sample applied to the wells in the agarose gel. (See gel electrophoresis, Denhardt reagent)
FIDELITY IN GENE CONVERSION: the converted allele is identical to the converter.
FIDELITY OF REPLICATION: see error in replication
FIDELITY OF TRANSCRIPTION: may be mediated by the very low level 3'→5' ribonuclease activity of eukaryotic RNA polymerase II. (See RNA polymerase)
FIDELITY OF TRANSLATION: errors occur approximately in the 10^{-4} range. (See ambiguity in translation)
FIDUCIAL LIMITS: synonymous with confidence limits, confidence intervals. (See confidence intervals)
FIG (*Ficus carica*): the genus includes about 2,000 species with x = 13, and the majority of them are diploid (2n = 26) although triploid and tetraploid forms are also known. It is a freeze-sensitive fruit tree.
FIGURE 8: is the configuration of cointegration of two ring DNAs before completion of the process. ⊙⊙ (See cointegration)
FIELD-FLOW FRACTIONATION (FFF): is a size-based fractionation of large molecules through thin channels, and from the retention time hydrodynamic properties are estimated.
FIELD-INVERSION GEL ELECTROPHORESIs: see FIGE
FIGE: field-inversion (10-0.02 Hertz) gel electrophoresis; it has been used to separate DNAs in the range of 15 - 700 kb, similar in size to the genetic material in the chromosomes of lower eukaryotes. Its advantage over pulsed-field gels is that the runs are straight. (See pulsed field electrophoresis. Hertz)
FILAMENT: myosin and actin fibers in animal tissue. (See also stamen)
FILAMENTATION: bacteria grow as long filaments during SOS repair. (See SOS repair)
FILAMENTOUS PHAGES: are frequently used as cloning vectors in genetic engineering: M13 (6408 b), f1 and fd (6408 b) single-stranded DNA phages. They have many advantages among others they are easy to purify, have little constraint on packaging and accept relatively large inserts (up 5 times their original DNA). M13 and its derivatives are generally used in DNA sequencing by the Sanger method. They contain multiple cloning sites within a truncated *E.*

Filamentous phages continued
coli lacZ gene. The vectors containing a successful insertion produce white plaques whereas the plaques of the phage without the insertion are blue on X-gal medium because *LacZ* is not interrupted and thus expressed. (See also phagemids, DNA sequencing Sanger, cloning vector, cloning site, Xgal)

FILE: a unit of related records in the computer.

FILE SERVER: permits additional users to share files and application programs on a computer.

FILIAL: offspring (generation). (See F_1)

FILOPODIUM: slender amoeba-like cell involved in movement and/or growth.

FILTER HYBRIDIZATION: one of the nucleic acid components is immobilized on a (nitrocellulose) filter and the other labeled, liquid component is allowed to anneal with it. (See DNA hybridization)

FILTER MATING: the bacterial cells at high density conjugate on the surface of a filter.

FILTERABLE AGENT: was the name of tobacco mosaic virus, and other viruses, before their nature was revealed. The name comes from the fact that they passed through bacterial filters and retained their biological activity. (See TMV)

FILTRATION ENRICHMENT: a method of selective isolation of all types of nutritional mutants used primarily in fungi. The wild type mycelia grow on minimal media but cells that have any type of special nutritional requirement most likely will not. A filter will retain the wild type mycelia while the non-growing spores pass through and thus separated from the bulk. This mutant-enriched filtrate can then be analyzed for the specific nutritional requirement. (See mutant isolation, mutation detection)

FIMBRIAE: flagellum-like appendages in the surface of bacteria. (See also pilus)

FINE STRUCTURE OF GENES: initially this meant recombinational analysis at high resolution in large populations, frequently based on selective identification of the recombinants that permitted mapping of the sites of different alleles within the locus. The ultimate fine structure analysis is using DNA (RNA) sequencing of cloned genes. (See mapping genetic, physical mapping, deletion mapping)

FINGER: see Zinc finger, Ring finger

FINGERPRINTING OF MACROMOLECULES: two-dimensional separation (by chromatography and/or electrophoresis) of digests of proteins or electrophoretic separation pattern of restriction enzyme digests of DNA for the purpose of characterization. (See chromatography, electrophoresis, RNA fingerprinting, DNA fingerprinting)

FINGERPRINTS: are analyzed by forensic and police investigations to determine personal iden-

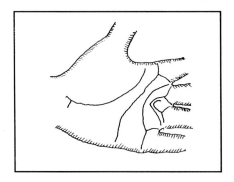
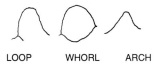

Fingerprints and palm-print patterns. The main types of the fingerprints are "loop", "whorl" and "arch" as shown but the additional dermal ridges within and outside are omitted here. The average number of ridges for women is about 127 and for men about 144. The far-left palmprint illustrates the main angles of the palmar triradial ridges. These are separate from the palmar creases affected in certain hereditary anomalies. (See Down's syndrome)

tity through dactyloscopy, a branch of dermatoglyphics. Ink is applied to the fingertips and impressions are made on paper. There may be a slight difference between the fingerprint pattern between the corresponding left and right hand fingers yet the overall pattern is characteristic for an individual. Fingerprints are determined by a relatively small number of genes yet each person has a unique fingerprint pattern. There is about 0.43 correlation between pa-

Fingerprints continued

ternal and offspring fingerprints. The pattern is determined by the end of the third months of the fetus. Fingerprints are influenced very little by environmental effects. Even if the epidermis is destroyed, regrowth reestablishes the original pattern. Toe and palm prints may provide additional confirmatory evidence for identity. There are three major types of fingerprints *loops*, *whorls* and *arches* with 65, 30 and 5% of the fingerprints falling into these broad categories (see illustration on preceding page). Autosomal dominant genes are known that either eliminate the dermal ridges or alter their pattern. There are elaborate systems for their classification into up to a million subgroups within which still individual differences are detectable. The uniqueness of the fingerprints became known by the second half of the 16th century at about the same time as sexuality of plants was identified by the same anatomist, Nehemiah Grew. Francis Galton, the founder of quantitative genetics, already used it for forensic purposes in the 19th century. Although lip prints have not been used for the same purpose as fingerprints, they also may permit individual identifications. (See ectodermal dysplasia, dyskeratosis, DNA fingerprinting, DNA fingerprinting)

FINE STRUCTURE MAPPING: genetic mapping the position of mutations (alleles) within a gene locus. (See allele, locus)

FIR (*Abies* spp.): timber trees, 2n = 24.

FIRE: intragenic (negative) regulatory element in the c-fos proto-oncogene and it must be relieved before fos can be expressed. (See FOS)

FIRST DIVISION SEGREGATION: see tetrad analysis

FIRST MESSENGER: peptide hormones. (See animal hormones)

FIRST STRAND DNA: the immediate product of reverse transcription. (See reverse transcription, cDNA)

FIS (factor for inversion stimulation): is a protein involved in the movement of inversion and transposable elements. (See invertases, insertion elements, transposable elements)

FISH (fluorescence *in situ* hybridization): detects eukaryotic chromosomal sites by the use of non-radioactive, fluorescence-labeled probes. In human chromosomes it permits the resolution of about 0.05-10 million base pairs as a band. The multiplex FISH technique using several fluorochromes simultaneously permits the identification of quite complex chromosomal rearrangements in amazing colors. Using DNA fibers, cloned probes can be mapped with a resolution down to 1 kb. The effectiveness of the FISH approach has greatly advanced by the use of epifluorescent filters and computer evaluation, permitting the simultaneous identification of 27 different DNA probes. This *multiplex FISH* can be seen in color on the cover of this book and described by Speicher, M.R, Gwyn, S. & Ward, D.C. 1996 NATURE GENETICS 12:368. (See chromosome painting, spectral karyotyping, non-radioactive labels, fluorochromes, *in situ* hybridization, probe, WCPP, telomeric probes, stretching chromosomes, FRET, DIRVISH)

FISH-EYE DISEASE: see lecithin:cholesterol acyltransferase deficiency

FISSION: a mode of asexual reproduction involving the division of a single cells or organelles by cleaving into two (generally) equal progeny (daughter) cells or organelles. (See alternation of generations, life cycles) → →

FISSION YEAST: see *Schizosaccharomyces*

FITC: a fluorochrome, a conjugate of fluorescein with isothiocyanate, used for cellular, chromosomal labeling. (See chromosome morphology, fluorochromes, chromosome painting, FISH)

FITCH—MARGOLIASH METHOD for TD: using *n* number of species, an evolutionary distance is computed by minimization of the branches of the evolutionary tree:

$$S_{\text{Fitch-Margoliash}} = \left[\frac{2\Sigma i < j\{(d_{ij} - e_{ij})/d_{ij}\}^2}{n(n-1)} \right]^{1/2} \times 100$$

where d_{ij} is the observed distance between species *i* and *j* and e_{ij} is the patristic distance. (See evolutionary tree, evolutionary distance, patristic distance, transformed distance)

FITNESS: the reproductive value in a population under specified conditions, determined by a genotype.

Estimation of Fitness and Equilibrium Frequencies of the Alleles at Heterozygote Advantage in a Hypothetical Population	
OBSERVED (N = 2000):	AA = 900, Aa = 1000, aa = 100
REPRESENTATION OF ALLELES (Σ = 4000):	A: (2 x 900) + 1000 = 2800, a : (2 x 100) + 1000 = 1200
ALLELIC FREQUENCIES:	A: 2800/4000 = 0.7 = p, a: 1200/4000 = 0.3 = q
EXPECTED FREQUENCY OF GENOTYPES ON THE BASIS OF $p^2 + 2pq + q^2$	AA: $(0.7)^2$ = 0.49, Aa: 2 x 0.7 x 0.3 = 0.42, aa : 0.3^2 = 0.09
EXPECTED NO. OF GENOTYPES $N(p^2 + 2pq + q^2)$	AA: 0.49 x 2000, Aa: 0.42 x 2000, aa: 0.09 x 2000 AA: 980 Aa: 840 aa: 180
FITNESS $\left(\dfrac{\text{observed number of genotypes}}{\text{expected number of genotypes}}\right)$	AA: $\dfrac{900}{980}$ = 0.92 Aa: $\dfrac{1000}{840}$ = 1.19 aa: $\dfrac{100}{180}$ = 0.56
STANDARDIZED FITNESS (rel. to w_2)	AA: $\dfrac{0.92}{1.19}$ = 0.77 Aa: $\dfrac{1.19}{1.19}$ = 1.00 aa: $\dfrac{0.56}{1.19}$ = 0.47
SELECTION COEFFICIENTS	s = 1 - 0.77 = 0.23 t = 1 - 0.47 = 0.53
ALLELIC FREQUENCIES AT EQUILIBRIUM	A: $\dfrac{t}{s+t} = \dfrac{0.53}{0.23+0.53}$ = 0.7, a: $\dfrac{s}{s+t} = \dfrac{0.23}{0.23+0.53}$ = 0.3

(See also selection coefficient, hybrid vigor, mutation rate in fitness, inclusive fitness)

FITNESS, STANDARDIZED: see selection coefficient

fix: see nitrogen fixation

FIXATION: the allele becomes homozygous in every individual of the population.

FIXATION INDEX: is the average coefficient of inbreeding in a population. In case of random mating the probability that an offspring would have exactly the same two ancestral alleles at a locus is (1/2)N where N is the number of diploid individuals in the population. The probability of having two different alleles at the same locus is 1 - (1/2)N. The coefficient of inbreeding of the first generation of this population is also (1/2)N by definition of inbreeding. In each succeeding generation, the non-inbred part of the population will have a chance to produce offspring with an allele pair of identical by descent. Therefore the coefficient of inbreeding in the next generations will be (1/2)N + [(1 - (1/2)N] x F, where F is the inbreeding coefficient of the preceding generation. After the gth generation the coefficient of inbreeding of this population will be: $F_g = (1/2)N + [1 - (1/2)N]F_{g-1}$

and this is called the *index of fixation*. Its complement is the *panmictic index* (P_g) that represents the average non-inbred fraction of the population:

$P_g = 1 - F_g$

The probability for the offspring to have two identical *A* or *a* alleles is Fp_{AA} and Fq_{aa}, respectively. Also, the probability of two alleles of a locus being non-identical by descent is 1 -F and the proportions of *AA*, *Aa* and *aa* are p^2, 2pq and q^2 (according to the Hardy-Weinberg theorem). Because the population will have both inbred and non-inbred components, its genetic structure will be: $F(p_{AA} + q_{aa})$ and $(1 - F)(p^2_{AA} + 2pq_{Aa} + q^2_{aa})$

When a population is completely inbred only homozygotes are found that is a change in genotypes but may not be a change in allelic frequencies if both alleles have equal fitness. The change may actually be from $AA + 2Aa + aa \rightarrow AA + AA + aa + aa$, that is mathematically

Fixation index continued

the same. The ultimate probability of fixation $P_f = \dfrac{1-e^{-N_e sp}}{1-e^{-N_e s}}$ may be estimated also on the basis of the *initial frequency of the gene* (= p), the *selection advantage* (= s) and the *effective population size* (N_e). [The base of the natural logarithm = $e \approx 2.718$]. (See also inbreeding, panmixis, inbreeding rate, Hardy - Weinberg theorem, mutation neutral, mutation beneficial, hybrid vigor)

FIXATIVE: is an agent(s) required to treat biological materials before staining is applied for microscopic examination. The fixative rapidly kills the cells, immobilizes the structures and assures better staining. Many different types of fixatives have been developed since the introduction of microscopic techniques. Some of the most widely used (especially for cytology) with composition in parts (volume):

DESIGNATION	ETHANOL	PROPIONIC ACID	ACETIC ACID	CHLOROFORM	
Farmer's	3	0	1	0	
Farmer's modif.	3	1	0	0	
Carnoy A*	6	0	1	3	*fresh
Carnoy B*	6	0	3	1	*fresh

Newer fixatives may contain formalin 5%, glacial acetic acid 5%, 90% or 70% ethanol, glutaraldehyde (4% in 0.025 M phosphate buffer pH 6.8), etc., and detergents (Tween 20) may be added for facilitating penetration or aspiration is applied for a few minutes. Fixation time is a day or two. The fixed material can be stored for months in 70% ethanol in refrigerator. For electron microscopy different fixatives are required. The fixatives used in photographic processing are also different. (See light microscopy, electron microscopy, stains)

FixL: *Rhizobium* kinase regulating N_2 fixation by FixJ. (See nitrogen fixation)

FFU: focus forming unit, the measure of focus formation in transformed (cancer) cells.

FK506 (Tacrolimus): an immunosuppressive protein in combination with rapamycin binds to cellular protein FKB12. FK506 intercepts the signal of the T lymphocyte receptor while rapamycin interferes with the signal of cytokines and growth factors. The FKBP12 - FK506 complex inhibits the serine - threonine phosphatase, calcineurin. The rapamycin - FKBP12 complex binds to FRAP and regulates p70 ribosomal protein S6 kinase that is required for the progression from G_1 phase of the cell cycle. Also called TOR, RAFT, FRAP. Members of this protein family are common in animals and occur also in plants. (See immunosuppression, cell cycle, calcineurin, immunophilins, ataxia telangiectasia, signaling to translation, checkpoint, cell cycle)

FK1012: is a lipid-soluble ligand, a dimeric form of FK506 and it directs the interaction between proteins linked to the FKBP12 receptor. (See FK506)

FKB12: see FK506

FKH (forkhead): embryonic lethal *Drosophila* gene at 3-95; human homolog (13q14) DNA-binding protein; translocations to 2q35 (PAX) may result in rhabdosarcoma (epithelial tumor).

FLAGELLAR ANTIGEN: of *Salmonella* is encoded by the *H1* and *H2* genes. At a particular time either one or the other flagellin is expressed. The expression is controlled by "phase variation", i.e. a transposition of the 970 bp long DNA segment. In one of the positions the *rh1* regulatory element represses the *H1* after the inversion the other, *H2* is expressed. (See *Trypanosoma*, mating type determination in yeast)

FLAGELLIN: the protein material of the (bacterial) flagellum. (See flagellum, phase variation)

FLAGELLUM (plural flagella): a cell appendage used for back and forth movement of microbial cells; in bacteria it is controlled by about 50 genes. (See flagellin)

FLANKING DNA (flanking gene): nucleotide sequences adjacent to a gene, or adjacent genes.

FLATWORM (*Planaria torva*): 2n = 16.

FLAX (*Linum usitatissimum* (2n = 30, 32): includes the fiber crop and the seed crop, linseed,

Flax continued

altogether with about 200 species. In the highly variable *Linum* genus the basic chromosome numbers vary: x = 8, 9, 12, 14, 15, 16 but the most common 2n numbers are 18 and 30. Some strains show also non-nuclear inheritance.

FLAVIN NUCLEOTIDES: riboflavin-containing coenzymes. (see FMN and FAD)

FLAVONE: along with, flavonones, flavonols are plant pigments derived from chalcones. (See also anthocyanidins and chalcone)

FLAVONOIDS: pigments with trimeric heterocyclic nucleus and frequently glycosylated. (See favone)

FLAVOPROTEINS: are tightly bound with a flavin nucleotide prosthetic group.

flea: see *copia*

FLETCHER FACTOR: causing an asymptomatic hereditary anomaly involved in the early regulation of the intrinsic blood clotting pathway. (See blood clotting pathways, antihemophilic factors, hemostasis)

FLEXER: proteins that determine DNA folding and bending similarly to the chaperones of proteins. (See chaperone, DNA bending)

FLICE: is a member of the ICE family proteases, apparently the same as MACH (mort-associated Ced-3 homolog). (See ICE, apoptosis)

FliG, FliM, FliN: bacterial switch proteins controling flagella rotation. (See phase variation)

FLIP-FLOP RECOMBINATION: occurs between two inverted repeats (in organelle genomes) and produces equal mixtures of two isomeric forms. (See mtDNA, chloroplast genetics)

Flk-1: a receptor tyrosine kinase with supposed role in endothelial and B lymphocyte differentiation, angiogenesis and formation of solid tumors; its ligand is VEGF (vascular endothelial growth factor). (See vascular endothelial growth factor, Flt-1,Tie-1, Tie-2, KIT, FLT)

FLOOD FACTOR: is a blood protein resembling blood factor VII; it shortens the slightly long prothrombin action time in asymptomatic individuals. (See antihemophilic factors, blood clotting pathways, hemostasis)

FLOPPY DISK: see disk

FLORA: a description of higher plant communities growing in a particular area or the vegetation present in an area.

FLORAL EVOCATION: is the process of commitment to flower differentiation of plants. (See commitment, determination, flower differentiation)

FLORAL INDUCTION: internal and external factors bringing about floral evocation. (See flower differentiation)

FLORET: an individual small flower being a part of an inflorescence.

FLOR's MODEL: developed by the plant pathologist, H.H. Flor in the 1950s, claiming that for each virulence gene of the pathogen a gene exists in the host and thus corresponding gene pairs (GCP) exist. These can be expressed in four different categories. Actually, the models were further elaborated and only pathogenicity and host reaction alleles were recognized and the terms virulence and resistance were deemed unnecessary. The expression of the disease symptoms (aegricorpus), depending also on the environmental factors, thus required an interorganismal genetic system. This rather complicated interacting system was considered by many plant pathologists a great idea to explain the genetic control of host - pathogen relationships. Many geneticists dismissed it as commonplace. They argued that in all organisms, all functions are genetically controlled and the interaction of gene products determine the phenotypes within organisms and also between organisms, and disease is no exception. Thus the model—they argued—in the absence of biochemical or molecular facts is not helpful. Since several plant genes have been cloned, there will be new opportunities to study host - pathogen relationships in physico-chemical terms. The resistance of maize against *Cochliobolus carbonum* (*Helminthosporium*) seems to support the latter argument inasmuch as a plant enzyme degrades the fungal toxin. The discoveries that some resistance genes encode elicitors of the signal transduction pathways in one way or another still do not invalidate the simple in-

Flor's model continued

terpretation. Recent experiments using the yeast two-hybrid system with the *Lac* reporter indicate that the *AvrPto* bacterial virulence gene interacts with the tomato resistance gene *Pto*. This reaction was evident only in this genic combination. Analysis of the Pto protein also revealed that a 95 amino acid stretch (129 to 224) of the protein was alone responsible for the specific interaction between the pathogen and the host. (See aegricorpus, host - pathogen relations, immune response, quadratic check, two-hybrid system)

floury (fl-2): gene of maize improves the nutritional value of the kernels by reduction the contents of prolamine and zein resulting in an increase in lysine, tryptophan and methionine. (See *opaque, kwashiorkor*)

FLOW CYTOMETRY: in the flow cytometer, suspended particles (cells, chromosome, etc.) are stained with fluorochromes and the dyes are excited by a laser beam. The particles then sorted with the aid of a computer according to their special properties. (See bivariate flow cytometry)

FLOWCHART: graphical display of procedures of analyses for the solution of a particular problem or subset of that problem.

FLOWER DIFFERENTIATION: is under very strict genetic control. Numerous genes have been identified which change the basic pattern of this morphogenesis. An idealized wild type dicot flower is shown at left, and the most common three types of homeotic conversions are indicated on the diagram. The homeotic transformations may not be complete and thus e.g., carpelloid sepals, stamenoid petals or petaloid stamina, etc. were identified. Usually, an entire

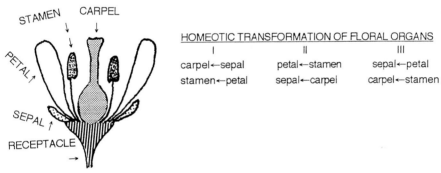

whorl of the flower ([1] sepals, [2] petals, [3] stamina, [4] carpels) is affected. The conversions generally involve adjacent whorls and simultaneously more then two whorls may be affected. On the basis of mutational evidence it has been suggested that the four whorls belong to A (1 + 2), B (2 + 3) and C (3 + 4) identity groups in the floral meristem. The actual phenotype of the mutants is determined which identity groups are affected by the mutation(s). E.g., a type II change is the result of interaction between domains A (*Apetala 1* [*AP1*] and 2 [*AP2*] genes) and B domain (*Apetala 3* [*AP3*] and *Pistillata* [*PI*] genes). Mutations in the C and B domains results in the *agamous* (*AG*) mutation in *Arabidopsis* and in either the *plena, pleniforma* or *petaloida* alterations in *Anthirrinum*. AP1, AP3, PI and the AG proteins all contain a MADS box, but AP2 represents another DNA binding protein. The plant MADS box proteins contain another conserved element, the K box that may form amphipathic α helices. AG binds also to another consensus ($[CC(A/T)_6GG]$, the CArG-box. These proteins interact with each other and may form AP1/AP1, AP3/PI and AG/AG dimers including the truncated AG/PI heterodimer. The association of these proteins is mediated to a large degree by the so called L (linker) region of these proteins that involves amino acids 31-35 situated between the MADS box and the K (keratin homology) box: N-terminus < MADS DOMAIN > < L REGION >< K BOX > C terminus

The MADS box appears to represent a large family of genes with critical functions in plant development. Flower differentiation seems to be controlled at the level of transcription and the specific RNA transcripts appear in regions of the flower primordia which are affected by the

Flower differentiation continued

specific genes. These genes are studied under the control of different promoters introduced by transformation. Recent studies indicate also that at least some of the genes are also regulated post-transcriptionally. (See MADS box, morphogenesis, homeotic genes)

FLOWER PIGMENTS: see anthocyanin, flavone, chalcones

FLOXING: see targeting genes

FLP/FRT: is a *Saccharomyces* recombinase system that can be expressed also in *Drosophila*. FLP is a recombinase and FRT define target sites of transposable elements. The system is very similar to the Cre-loxP of bacteriophage P1. The 43 kDa "flip" (FLP) is encoded by the 2μ circular yeast plasmid. Recombination (x) takes place within the core sequence and between FLP binding sites. This as well as the *Cre/loxP* system can be used for site-specific integration of transgenes and also for engineering chromosomal rearrangements in higher eukaryotes. (See recombinase, *Cre/loxP*, targeting genes, transposable elements, chromosomal rearrangements, homing endonuclease)

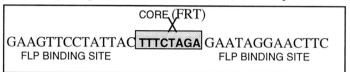

Flpter: fractional length from the hybridization signal to the terminus of the short arm of the chromosome; it is used for the localization of FISH labels on the chromosome. (See FISH)

FLT ONCOGENE: is in human chromosome 13q12 and it encodes a protein tyrosine kinase. It shows homology to FMS and ROS. Flt-1 is essential for animal embryonal vasculature but not for endothelial differentiation. The FLT3/FLK2 receptor tyrosine kinase is closely related to KIT and FMS. (See vascular endothelial growth factor, Flk-1, KIT, FLK, oncogenes, Tie-1, Tie-2, receptor tyrosine kinase, FMS, ROS)

FLUCTUATION TEST: was designed originally to determine whether the mutations observed were induced by a particular treatment or the treatment merely was a means of screening for that particular class of mutations that preexisted in the cultures. The design of the fluctuation test by Luria and Delbrück in 1943 initiated the experimental bacterial genetics. It can be used also in other microorganisms, in animal and plant cell cultures and the principles can be adapted also to most types of mutation experiments. The principle is as follows: from the same original culture two series are generated. In series I the culture is continued until a particular cell density is reached. In series II, small inocula are placed in say 10 (or more) vessels and the culture is continued until in all vessels of series I and II the same cell density is reached. Then plating is made in say 10 Petri plates from the single vessel culture I, and one Petri plate from each of the 10 test tubes of series II. After incubation for a period of time the number of mutations are scored separately on each the total of 20 Petri plates. The abridged data of an experiment of M. Demerec (J. Bacteriol. 56:63) are tabulated on next page. The averages ($\bar{x}$) and the variances $(\Sigma x^2 - nx^2)/(n-1)$ are determined and if it appears that the average number of mutational events in both series I and II are practically the same but the variance in series II significantly exceeds that of series I, the conclusion is that the treatment was not the cause of the mutations. In a fluctuation test of bacterial mutation for streptomycin resistance series I showed an average of 127.6 mutations, and the variance was 145.4. In series II the average was almost the same 123.3, and the variance was 4,392.0. Thus streptomycin only revealed the presence of preexisting mutations for streptomycin resistance. The logic of this argument is that in the single vessel #I the preexisting mutants were distributed uniformly and therefore upon plating the variance was low. In the 10 vessels of series #II mutations occurred or did not before plating. If mutations occurred early, many mutants appeared on that Petri plate, if mutation occured late in the test tube, few mutations were detectable after plating and if no mutations occurred in a particular test tube, the Petri plate failed to display any. The propagation of the pre-existing mutations caused the large number of the mutant colonies on

Fluctuation test continued

	SERIES I			SERIES II	
Culture No.	Resistant Cells (X)	X^2	Culture No.	Resistant Cells (X)	X^2
1	146	21,316	1	67	4,489
2	141	19,881	2	159	25,281
3	137	18,169	3	135	18,225
4	128	18,384	4	291	84,681
5	121	14,641	5	75	5,625
6	110	12,100	6	117	13,689
7	125	15,625	7	73	5,329
8	135	18,225	8	129	16,641
9	121	14,641	9	86	7,396
10	112	12,544	10	101	10,201
SUMS	1,279	164,126	SUMS	1,233	191,557
Average $\bar{x}$	127.6		Average $\bar{x}$	123.3	
Variance	154		Variance	4,392.0	

some plates and when there were no mutations prior to the treatment, none appeared after plating in series II. In series one in the single batch the pre-existing mutations were uniformly dispersed without fluctuation in the variation. Thus the different fluctuations on the identical series of Petri plates was the critical factor for the conclusion that the treatment was ineffective. (See mean, variance)

FLUID GENOME PARADIGM: see constant genome paradigm.

FLUIDITY: lipids can move within membranes. (See cell membranes, membranes)

FLUORESCEIN: a fluorochrome with excitation at 490 nm wavelength of light and emission at 525 nm. Generally used in conjugates with avidin, isothiocyanate or antibodies.

FLUORESCEIN DIACETATE: is a vital stain for protoplasts.

FLUORESCENCE: a property of chemical compounds to emit radiation (light) upon absorption of radiation from another source. The fluorescent radiation generally has longer wavelength than the absorbed one, e.g. nucleotides irradiated by 260 nm UV light display a visible purple color. The fluorescence lasts only as long as the exposure to the irradiation. (See UV)

FLUORESCENCE MICROSCOPY: uses a microscope with an illuminator equipped with a light filter that assures that the stage receives only that narrow spectral band of light that is required for the excitation of the fluorochrome used for staining the specimen. In front of the ocular there is another filter that transmits only the fluorescence emitted but shuts out the exciting wavelength light. By staining with more than one dye and changing filters, in the same specimen different structures (molecules) may be distinguished by different bright colors. Some of the fluorochromes permit viewing also living cells. (See fluorochromes, chromosome painting, FISH, microscopy)

FLUORESCENT DYES: see fluorochromes

FLUOROCHROMES: non-radioactive labels such as DAPI, fluorescein, rhodamine B, FITC, Texas Red, R-Phycoerythrin, RED613, RED670, allophycocyanin, isothiocyanate, etc. The dipyrromethene boron difluoride dyes may have special advantages for automatic sequencing of DNA. (See under individual names, fluorescent microscopy, biotinylation, FRET, FISH)

FLUOROGRAPHY: enhances the sensitivity of autoradiographic detection by adding to the sample scintillants such as 2,5-diphenyloxazole (PPO) or sodium salicylate. (See autoradiography, scintillation counters)

FLUOROSCOPE: is a medical apparatus for X-ray examination of deep-seated tissues. The image is seen on a fluorescent screen, coated with calcium tungstate or zinc cadmium sulfide and other materials. It was favored in the past because it could detect also motions. It has became obsolete with introduction of television cameras and because it used about ten fold higher doses of radiation than diagnostic X-ray machines and posed radiation hazards to both operator and patient. (See also X-rays, tomography, sonography)

FLUOROURACIL (FU): a uracil analog that is incorporated into RNA. It is metabolized into deoxyfluorouridine monophosphate, a potent antineoplastic agent. ➔

FLUSH-CRASH CYCLES: the size of a natural population fluctuate greatly.

FLUSH END: see blunt end (of DNA)

FLUX: the spread of a label from a metabolic precursor to other molcules in the cell.

FLUX, GENETIC: alterations in the cell in response to internal and external stimuli.

fMet: formylmethionine is generally the beginning (modified) amino acid in the translation on the ribosome of prokaryotes. (See also formylmethionine, protein synthesis)

FMN: flavin mononucleotide is composed of flavin (riboflavin) phosphate and dimethylisoalloxazine base (yellow enzyme); its a coenzyme for oxidation-reduction mediating proteins.

FMR1: see trinucleotide repeats, fragile sites, KH module

FMR1 MUTATION: involves expansion of the CGG trinucleotides which are freqently methylated on C and thus silenced. Hypermethylation of FMR1 is called *full mutation* whereas the *premutation* is an intermediate range of expansion (50-200) of these repeats. (See fragile sites, trinucleotide repeats, human intelligence, KH module)

FMS: feline sarcoma virus protooncogene is a receptor of colony-stimulating factor-1 [CSF], and macrophage colony stimulating factor. This gene is located in human chromosome 5q33.2-q33.3. It is involved in different types of leukemias. Its sequence is homologous to oncogene FLT, a tyrosine kinase. (See leukemias, colony stimulating factor, FGR, FLT. protooncogene)

FNR: is ferrous/ferric binding bacterial aerobic/anaerobic regulator protein with some similarity to the catabolite repressor protein. (See catabolite repression)

FOCAL CONTACT: an adhesion plaque on the surface of a cell that is attached to the extracellular matrix by transmembrane proteins (integrin).

FOCAL DERMAL HYPLOASIA (Goltz syndrome, FDOF): X-linked dominant lethal in males involving atrophy, skin pigmentation in a linear pattern, papillomas (epithelial neoplasms), polydactyly, underdevelopment of teeth, small defective eyes, mental retardation, etc. (See skin diseases, cancer, pigmentation defects, polydactyly, eye diseases, mental retardation)

FOCUS FORMATION: neoplastic (cancerous) cells grow up in dense clusters. (See cancer)

FOCUS FORMING UNIT: see FFU

FOETUS: see fetus

FokI: endonuclease recognizes the 5'-GGATG-3' sequence and cleave DNA 9 and 13 base away.

FOLATE: salt of folic acid. (See folic acid)

FOLD BACK DNA: palindromic (inverted repeat sequences) in a single strand of polynucleotides pair within the same strand:

FOLD BACK INHIBITION SITE: see FBI site

FOLDED LEAF: protein α-helices wrapped around a hydrophobic core. (See hydrophobic, α-helix

FOLDON: the protein domain involved in folding.

FOLIC ACID (pteroyl glutamic acid): a water-soluble vitamin; it is required for *de novo* nucleotide synthesis, amino acid conversions. As a derivative of pteridines it is involved in many oxidation reactions, mediating animal coloring, various light reactions, tetrahydrofolic acid is an important coenzyme, etc. (See fragile sites, methotrexate, aminopterin, phosphoribosylglycinamide formyltransferase, formiminotransferase deficiency, 5,10-methylenetetrahydrofolate dehydrogenase, formyltetrahydrofolate synthetase)

FOLLICLE: a small secretory sac or gland. The cell layer covering the ovary, in botany a simple dry fruit, dehiscing along one suture and formed of a single carpel.

FOLLICLE STIMULATING HORMONE: see FSH

FOLLISTATIN: a maternally expressed protein that by binding activin may interfere with activin-induced mesoderm formation. Follistatin-deficient mice die within hours after birth because of multiple defects indicating its requirement for several proteins of the transforming growth factor family. (See activin, mesoderm)

FONTS: the various styles of characters and scripts that the computer can use.

FOOTPRINTING: fragments of 5'-labeled double-stranded DNA are partially degraded by a DNase in the presence and also in the absence of a protein that is expected to bind to certain sequences in the DNA. Subsequently both samples (with and without the binding protein) are sequenced (Maxam and Gilbert method) and a comparison reveals the nucleotide sequences protected from DNase by the binding protein: thus from the path (footprints) of the DNase, the position of the binding protein is revealed. These sites may have importance for transcription initiation. *In vitro footprinting* is basically very similar to the methylation interference technique but it is performed in living cells and followed by DNA extraction and electrophoresis. (See regulation of gene activity, methylation interference assay)

FOOTPRINTING GENETIC: the role of sequenced genes without known function can be determined by inserting transposable elements (Ty) and disrupting their function in a sequential manner. This process can screen yeast cell populations in 10^{11} range and detects a wide range of mutations of variable severity or fitness. (See Ty, transposon footprint, mutation detection)

FOOTPRINTS, TRANSPOSABLE ELEMENT: after a mobile genetic element (insertion- or transposable element) moves it usually leaves behind some nucleotides (as a footprint) at the original target site. (See insertional mutation, transposon mutagenesis)

FORBES DISEASE: is a defect or deficiency of amylo-1,6-glucosidase and/or oligo-1,4-1,4-glucantransferase. (See under alternative name of glycogen storage disease III)

FORE TRIBE: of New Guinea is most affected by the kuru disease because of a behavioral tradition. (See kuru)

FORESPORE: after DNA replication a *Bacillus subtilis* cell is partitioned into two. One is small the other is larger (also called mother spore), and the smaller is the forespore. Before the forespore becomes a spore, the mother spore engulfs it and it is just pinched off at a later developmental stage and eventually matures into a spore. (See endospore, *Bacillus subtilis*)

FORELOCK, WHITE: is a dominant autosomal human trait (a lock of white hairs in the front part of the scalp). (See Waardenburg syndrome)

FORENSIC GENETICS: genetic studies used for legal purposes or in the judiciary courts relying on fingerprints, blood groups, other antigens, isozymes, VNTR (variable number tandem repeats of DNA) by employing RFLP (restriction fragment length polymorphism) or other heritable criteria to identify biological relationship, paternity or criminals, etc. Most of the chemical analyses require small samples of blood (60 µL) or semen (5 µL) or hair roots. For some of the tests dried blood or semen spots are useful even if they are weeks, months or years old. On the basis of polymorphic proteins identity can be defined to higher than 99% probability and for exclusion of an individual requires much less effort. Generally the spectrum of the protein components (separated by electrophoresis) in the sample obtained from the person to be identified is compared with the spectrum of the same protein components within the same (ethnic) population. The product of the frequencies expected by chance is compared with the actually observed data. The first test is generally for the ABO blood group but this alone rarely suffices because of the limited variations and the failure to identify all heterozygotes. Other proteins assayed for polymorphism are adenylate kinase (AK), adenylate deaminase (ADA, must be examined within 6 months), carbonic anhydrase (CA-II, within a week), erythrocyte acid phosphatase (EAP, within 6 months), esterase (EsD, within 1 month), glyoxylase (GLO), hemoglobin (Hb), peptidase A (pepA), phosphoglucomutase (PGM, within 6 months), gammaglobulin (Gm, displays about a dozen antigens determined by very closely linked genes and the clusters have different frequencies in different ethnic groups), Lewis anti-

Forensic genetics continued

gens (Le^a, Le^b), and the rhesus antigens (47 Rh determinants, within 6 months). In semen samples generally ABO, GLO, Pep A, PGM, Le are used. If in a case of rape, the semen sample is obtained by vaginal swabs it may be contaminated by vaginal fluids (may obscure semen fluids), proteolytic enzymes (that may degrade the proteins), or the presence of bacteria my interfere with blood typing. Actually female cells may be separated from sperm by digestion with sodium dodecylsulfate/proteinase K that does not destroy sperm. Later the sperm can also be digested in the same reaction mixture with (dithiothreitol) DTT added. The greatest specificity of identification can be obtained by analysis of the DNA in body fluids or skin or other tissue samples. (See fingerprinting, DNA fingerprinting, VNTR, RFLP, blood groups, ADA, PGM, hemoglobin, gammaglobulin, sodium dodecil sulfate, Frye test, forensic index)

FORENSIC INDEX: provides statistical information regarding the probability that the evidence (E) collected at a crime scene or from other potentially incriminating object would belong to the perpetrator (P) or to a suspect (S). Let us assume that the perpetrator and the suspect have DNA (VNTR or STR) or protein profile (blood group or enzymes) A and we must find what is the probability for the suspect being liable for that event or fact (C) or the suspect and the perpetrator are different individuals (event C'). These conditional probabilities are called L

$$(\text{forensic index}) \rightarrow \quad L = \frac{\Pr(E \mid C)}{\Pr(E \mid C')} = \frac{\Pr(S = A \mid P = A, C)}{\Pr(S = A \mid P = A, C')}$$

S = A indicates that the profile of S is A, i.e. the match is L times more probable if S and P are the same persons. Actually the decisions are more complex because in the match the relatedness within the population (population structure) must be considered. (See inbreeding coefficient, inbreeding and population size, fixation index, Bayes theorem, conditional probability, DNA fingerprinting, VNTR, STR, paternity index)

FORKED TONGUE: enables snakes to assess different signals (pheromones) simultaneously.

FORMAL GENETICS: see classical genetics

FORMIMINOTRANSFERASE DEFICIENCY: autosomal recessive physical retardation without mental retardation, anemia, etc. caused by oversupply of folate. (See folic acid)

FORMYLMETHIONINE: is the translation initiation amino acid in prokaryotes and cytoplasmic organelles (plastids and mitochondria) of eukaryotes but it is not used in the cytosol of eukaryotes. It is carried to the 70S ribosomes by a forrnylmethionine tRNA (tRNA$_i^{Met}$ or tRNA^{fMet}) that is distinct from the regular tRNA^{Met}. (See protein synthesis)

10-FORMYLTETRAHYDROFOLATE SYNTHETASE: is a key enzyme of folic acid metabolism. (See folic acid)

FORSKOLIN: is a diterpene isolated from the plant *Coleus forskohlii*. It is an activator of adenylate cyclase and some other mechanisms that depend on cAMP. (See also signal transduction, adenylate cyclase, cAMP)

FORTRAN (formula translating system): computer languages with specific problem orientations.

FORWARD MUTATION: mutation from wild-type to mutant allele. (See also reversion)

fos: murine osteosarcoma (chondrosarcoma) proto-oncogene, general transcription factor (AP1). The fos—jun heterodimers bind to the 5'-TGAGTCAA-3' sequence. Fos controls also complex behavioral traits such as nurturing. (See proto-oncogene, AP1, sarcoma, jun, apoptosis, FOS oncogene, apoptosis)

FOS ONCOGENE: in human chromosome 14q21-31 is homologous to the v-oncogene *fos*. The human FOS has a normal expression in fetal membranes almost as high as that is detectable in osteosarcomas of mouse. Products of FOS and Jun contribute to the formation of the AP1 transcription factor and participate in multiple ways in tissue differentiation. Recent studies indicate that Fos knockouts in mice fail to nurse their pups. (See JUN, AP1, oncogenes knockout, behavior genetics, apoptosis)

FOSSIL: petrified remains or impression of an organism of past geological ages, preserved in the earth or rocky layers. (See fossil record)

FOSSIL RECORD: is used to reveal the pattern of macroevolution (evolution of taxonomic categories above the species level) at the geological scale. Macroevolution was traditionally infered from the paleontological data, the appearance of petrified taxa in the successive geological strata. The age of the remains is estimated by *radioisotope dating*. If the fossils are less than 40 thousand years old their age is infered from the amount of carbon-14 (^{14}C) contained. This isotope is produced at a relatively constant rate from nitrogen-14 (^{14}N) under the bombardment of cosmic radiation through the ages. This ^{14}C is utilized by the organisms the same way as the more common ^{12}C. The former is unstable, however, and it decays to half in each 5,730 year cycle. The amount of ^{14}C in the organic material serves as a clock with an accuracy of ± 1 to 2%. The age of fossils over 40,000 years old is infered from the age of the sedimentary rocks where the organism died (if that is the site of the fossil and it was not moved by geological changes in the strata). The age of the rocks is estimated by the decay of other isotopes. Uranium-238, e.g., decays into lead-206 with a half-life of 4,510 million years. Therefore the proportion of these two elements in the rocks indicates their geological age. The evolutionary relation of the fossils can be better defined if protein and DNA analyses are also feasible. (See also half-life, isotope)

FOULBROOD: a disease of the honey bees caused by *Bacillus alvei*. Resistance against it is based on homozygosity of two non-allelic recessive genes determining behavior. One gene is responsible for uncapping the honeycombs when the larvae die, the other gene is responsible for the removal of the dead. (See behavior genetics)

FOUNDER EFFECT: same as founder principle.

FOUNDER MOUSE: is a chimeric animal obtained after transformation that may or may not involve the germline. (See chimera, germline, microinjection)

FOUNDER PRINCIPLE: a new population descends from a limited number of immigrants (because of sampling error), resulting in genetic drift. It is called also founder effect. (See effective population size, drift genetic, porphyria variegata)

FOUR-CLUSTER ANALYSIS: is a procedure for determining the evolutionary relationships among four large groups of organisms such as animals, plants fungi and protists without consideration to the variation within each of these groups. If we designate the four monophyletic groups as A, B, C and D, three unrooted evolutionary trees can be generated such as $T_1 = [(AB)(CD)]$, $T_2 = [(AC)(BD)]$ and $T_3 = [(AD)(BC)]$ from which one is expected to be correct on the basis that the correct construct would have the shortest sum of tree branch length. The three sums of branch lengths may be designated as S1, S2 and S3 and the differences S1 - S2, S1 - S3 and S2 - S3 are determined by an appropriate algorithm. (See evolutionary distance, evolutionary tree, least square methods, neighbor joining method, unrooted evolutionary trees)

FOUR-O'CLOCK: see *Mirabilis jalapa*

FOVEAL DYSTROPHY: is an autosomal dominant lesion of the macula in the eye fundus with aminoaciduria. (See macula, eye diseases)

FOX: these canid species are quite variable genetically and by chromosome number. *Vulpes velox* (kit fox) 2n = 50; *Vulpes vulpes* (red fox) 2n = 36; *Vulpes fulva* (American red fox) 2n = 34; *Urocyon cinereoargentus* (gray fox) 2n = 66; *Otocyon megalotis* (bat-eared fox) 2n = 72; *Lyalopex vetulus* (hoary fox) 2n = 74; *Cerdocyon thous* (crab-eating fox) 2n = 74. (See wolf)

F-PILI (or F-pilus or sex pilus): bacterial cell appendage that forms the conjugation tube through which the F element, conjugative plasmids and the Hfr bacterial chromosome is mobilized into the F- cells. (See F factor, sex factor, Hfr, conjugation)

FPS (fixed pairing segment): is hypothesis that in eukaryotes recombination is not entirely random but occurs in tracts which have either fixed at both, or one end or at the middle. During a single meiosis only a fraction of these segments pair and there is positive interference in their vicinities. (See interference)

fps: chicken sarcoma oncogene.

FRACTION 1 PROTEIN: the old name of ribulose bisphosphate carboxylase/oxygenase enzyme that is the largest single protein encoded by the plastid and forms about 50% of the proteins in the chloroplasts. (See rubisco)

FRACTIONAL MUTATION: displaying mosaicism (variegation) in the tissues of the body. If the mutation was induced in the germ cells, it indicates that the mutagenic agent associated with only one strand of the DNA and therefore mutant and non-mutant sectors arose and DNA repair occurred during the post-fertilization stage. Fractionals may be due also to unstable genes, mitotic recombination, nondisjunction, transposable elements, etc. (See unstable genes, mitotic recombination, gene conversion, nondisjunction, transposable elements

FRACTIONATED DOSE: irradiation is provided not in a chronic manner but with interruptions between each exposure although the doses are summed up. (See chronic radiation)

FRAGILE SITES: occur in several human chromosomes. The overall frequency of autosomal fragile sites is about 2×10^{-3}. Generally three types are distinguished: (i) folate sensitive [shows up if the cell culture medium is deficient in folate], (ii) elevated pH triggers their appearance and (iii) 5-bromodeoxyuridine (BdUR) is required for expression. The best studied is the fragile X syndrome. Fragile sites were identified also at 2q11, 3p14.2, 6p23, 9p21, 9q32, 10q23 (folic acid sensitive), 10q 25 (BdUR sensitive), 11q23, 12q13, ,16 p12, 16q22 (appeared only in the presence of Epstein-Barr virus [EBV[) antigen), 17p12, 20p11. The fragile sites are generally dominant and involve overlapping syndromes frequently mental retardation, cancer susceptibility and other symptoms. In the fragile X syndrome the number of CGG repeats may run into hundreds whereas under normal conditions only about 30 repeats are found. In Friedreich's ataxia GAA repeats are found in the first exon. Five neurological disorders, spinal and bulbar muscular atrophy (Kennedy disease), spinocerebellar ataxia (olivopontocerebellar atrophy) Type 1, Huntington's chorea, dentatorubral-pallidoluysian atrophy, Machado Joseph disease display poly CAG sequences within their genes and encode polyglutamine. Myotonic dystrophy is accompanied by CTG repeats in the untranslated last exon of the protein kinase gene. (See fragile X chromosome, Kennedy disease, glutamine-repeat diseases, anticipation, Huntington's chorea, ataxia, Machado-Joseph disease, dentatorubral-pallidolyusian atrophy, FMR1 mutation, human intelligence, mental retardation, trinucleotide repeats)

FRAGILE X CHROMOSOME (FRAXA): displays poorly stainable sites under the light microscope that are liable to breakage and may result in mental retardation and cancer. The affected individuals have also macrocephaly (large head), prominent jaws, macroorchidims (enlarged testes) and high pitched funny voice. The condition is caused by folate deficiency leading to low levels of thymidylate. It involves amplification of a CCG repeat in the FMR-1(fragile site mental retardation) gene (at Xq27.3) that occurs in human populations by a frequency of $\approx$ 1/1,250 in males and 1/2,500 of females. The males are predominantly affected (80%); about a 1/3 of the carrier females are also mentally retarded. Usually, 20% of the males with this X-chromosome are phenotypically normal and their daughters are also normal but their grandsons display the chromosome and the phenotype. Prenatal diagnosis is feasible. On folate-deficient cell culture media the critical X-chromosome displays a constriction at the Xq27-p28 site. Other fragile sites may account for some of the less common forms of the disease. (See fragile sites, mental retardation, Jacobsen syndrome, trinucleotide repeats, head/face/brain defects)

FRAGILE X SYNDROME: see fragile X chromosome

FRAGMENT RECOVERY PROBABILITY: indicates the number of genomic fragments to be screened in order to recover a desirable one with a chosen probability. Probability = $1 - (1 - f)^n$, where f = the size of an average fragment divided by the size of the genome, n = the required number of fragments to be cloned. Example: P = 0.95, average fragment size 1×10^6 Da, and the size of the genome is 2.6×10^9 Da. Then:
$n = \ell n(1 - P)/\ell n(1-f) = \ell n(0.05)/\ell n[1- (1000000/2600000000)] \approx 7787$, i.e. about 7,787 clones will include the wanted one by a probability of 95%. (See restriction enzymes, DNA library)

FRAGMENTIN-2: a cytotoxic serine protease; it can trigger apoptosis in combination with perforin. (See perforin, apoptosis, ICE, granzyme, RNKP-1)

FRAGRANCES: occur in all types of organisms and serve various adaptive purposes. Among animals, the pheromones are means of communication and are used both as attractants and repellents. In plants, the fragrances may be the means of aiding pollination or dispersal but also

Fragrances continued

as insect repellents. Chemically the fragrances are diverse. Many plant fragrances are monoterpenes such as citral, thymole (in thyme), linaleol, 1,8-cineol (in levander). In the *Menthas* carvon, piperitones, menthone, menthole terpene rings represent the scents. In the *Eucalyptus* species geraniols and cineols occur. In the majority of the species of plants the fragrances represent chemical complexes. Their inheritance is usually complex and frequently in the hybrids the parental fragrances are hard to recover. (See pheromones)

FRAMESHIFT MUTATION: insertion or deletion of bases changing the reading frame of the code words, leading to new amino acid sequences from the site toward the carboxyl end of the polypeptide. If one or two bases are either lost or gained, the genetic message from that site on is generally garbled whereas if the loss or gain involves triplets there is a possibility to continue reading in a normal manner. Frame shift mutations are caused frequently by acridine dyes and cross-linking mutagens. The discovery of frameshift mutagens contributed to the recognition that the genetic code relies on nucleotide triplets. Frame shift mutation can be represented by the following folly: ↓

normal text: JOE AND BOB ATE THE BIG HOT DOG AND DID NOT SIP ICE TEA
deletion ↓ and shift   JOE AND BOB ATE THE BIG HOT DOG AN**D IDN OTS IPI CET** EA
an addition ⇊ restores   ⇊
the meaning behind it   JOE AND BOB ATE THE BIG HOT DOG AND **IDN OTS** SIP ICE TEA

(See mutation, base substitution, deletion mutation, duplication, insertional mutation, genetic code, codon, reading frame)

FRAMESHIFT SUPPRESSOR: are generally insertion(s) or deletion(s) of nucleotides that are capable of restoring the normal reading frame within the gene. (See frameshift mutation, suppressor mutation, reading frame)

FRAMESHIFT TRANSLATIONAL: see overlapping genes

FRAMEWORK AMINO ACIDS: of antibodies secure the scaffolding of the hypervariable region but do not involve the CDR sequence. (See antibody, immunoglobulins, CDR)

FRAMEWORK MAP: includes several (or only one) collection(s) of genes or DNA sequence groups that are used to position loci or sequences (STS) relative to these panels. (See STS, radiation mapping)

FRAP (TOR, RAFT1): see FK506

FRATERNAL: involving brothers; it is used also for describing dizygotic twins as fraternal twins even when girls are involved (in the latter case the biologically correct usage should be "sororal twins" but it is not used). (See also twins, twinning)

FRASER SYNDROME: autosomal recessive (probably human chromosome 9) malformations involving facial anomalies (hypertelorism), underdeveloped kidneys, fusion of the labia pudendi (the fleshy borders at the mons pubis of the external female genitalia), enlargement of the clitoris (the female erectile body [homologous to the penis of males]), defective fallopian tubes (connecting the ovaries with the uterus), and ovaries, etc. (See kidney disease, genital anomaly syndromes, hypertelorism)

FRAX: fragile X chromosome . (See fragile X syndrome)

FREE ENERGY (G): the energy that can be obtained from a system for other purposes. (See also entropy)

FREE RADICAL: an atom or group of atoms with an unpaired electron, and it is therefore extremely reactive. Free radicals may be produced by exposure of wet tissues to ionizing radiation and thus leading to physiological and genetic damage of the cells.

FREEMAN-SHELDON SYNDROME (arthrogryposis): is characterized by deformities of the limbs and other structures. One form has been mapped to human chromosome 5.5-pter.15p1.

FREEMARTIN: somewhat masculinized sterile bovine (cattle, sheep, goat, pig, etc.) female born as twin with a male. The sterility is attributed to the circulation of blood containing male-specific antigens and hormones. Freemartins do not occur in humans although women treated with male hormones to prevent miscarriage have been reported to deliver female babies that after

Freemartin continued

puberty might have shown some secondary virile characteristics. The exact origin of the term is unclear. In old English a spayed heifer (neutered bovine female) was called martin. Also, St. Martin has been regarded as a protector of rogues (off-type creatures). In Scottish, ferry-cow means a cow [temporarily] barren. (See hormones in sex determination, puberty, spaying)

FREEZE DRYING (lyophilization): is a procedure for the preservation of biological specimens, bacteria, enzymes frozen at about -50 C° and dehydrating under high vacuum in a specially constructed equiment. The preserved samples are usually sealed in glass under vacuum for further storage. Generally, the activity of the enzymes is well maintained and the bacterial cells can be revived even after years of storage.

FREEZE ETCHING: is different from freeze fracture inasmuch as it allows the electronmicroscopic study of membrane surfaces rather than internal structures. The specimens are frozen in liquid nitrogen, the material is cracked and the water is removed by sublimation in a freeze dryer. The etched parts are shadowed and viewed in the electron microscope. An improved version of this procedure involves *rapid freezing* with a copper block (-269 C°, liquid helium) after being slammed against it and then lyophilized. This way the internal cell parts, filaments can be well visualized. (See also freeze etching, freeze drying, electronmicroscopy, membranes)

FREEZE FRACTURE: is technique for preparation membrane-containing specimens for electronmicroscopic examinations. The specimen is frozen in liquid nitrogen under the protection of an antifreeze (cryoprotectant) to prevent ice crystal formation and concomitant distortion. After cracking the frozen blocks some surfaces of the broken pieces expose the interior of cellular membrane bilayers. The membrane faces are then shadowed with platinum, and after the organic material is removed it can be viewed by electronmicroscopy. (See electronmicroscopy, shadowing, membranes)

FREQUENCY-DEPENDENT SELECTION: see selection types

FREQUENCY DISTRIBUTION: representation of a population in classes according to the frequency of individuals in each class.

FRET (fluorescent energy transfer): a fluorophore donor molecule which has an absorption maximum at a shorter wavelength can be excited and then can transfer the energy of an adsorbed photon no-radiatively to an acceptor molecule which has an excitation maximum at a longer wavelength. The distance over which FRET can be measured is about 40 to 100 Å and in general it depends on the 1/6 power of the distance but it is modified by several factors. (See fluorochromes)

FREUND ADJUVANT: is a water — light weight mineral oil emulsion containing an emulsifier and antigen; sometimes dry, dead *Mycobacterium butyricum* is also added. This preparation boosts the immune reaction in case of weak or small amounts of the antigen. (See antigen)

FRIEDREICH ATAXIA (FRDA): with optic nerve atrophy and deafness it is an autosomal dominant disease. An autosomal recessive form (9q12-q13) is a rare brain-spinal chord degenerative malfunction, characterized by hypoactive knee and ankle jerks, poor coordination of the limbs, spasms, etc. Most of the cases are point mutation in the gene encoding the 210 amino acid frataxin protein but this condition frequently involves GAA repeats in the first exon. The defect is concerned with a phosphatidylinositol-4-phosphate kinase and mitochondrial iron homeostasis. The prevalence of Friedreich ataxia is about 2×10^{-5}. (See fragile sites, ataxia telangiectasia, epilepsy, phosphatidylinositol, trinucleotide repeats, AVED)

FRIEND MURINE LEUKEMIA: see FMS oncogene

FROG: *Rana pipiens* 2n = 26, *Rana temporaria* 2n = 26.

FROND: a leaf-like thallus of lichens or leaves of ferns.

FRP: is a human phosphatidylinositol kinase. (See PIK)

FRS: Fos-regulating kinase. (See FOS)

FRUCTOSE: is a mono keto-hexose, present in the disaccharide saccharose. It plays key roles in metabolism through phosphorylated derivatives (fructose-1-phosphate, fructose-6-phosphate,

Fructose continued

fructose-1,6-bisphosphate, etc.). Fructose utilization is not impaired in diabetes. Fructose is about twice as sweet-tasting as glucose and its use may reduce the caloric intake. The "corn sweeteners" contain fructose, industrially produced from starch. Fructose and fructose-containing food and beverages, especially at acid pH and heating or just by long storage may liberate furans, furaldehyde and levulinic acid that may be toxic. Furfural is actually an insecticide. Levulinic acid LD50 intraperitonially is 450 mg/kg for mouse. (See fructose intolerance, fructosuria, aspartame, saccharin)

FRUCTOSE-2,6-BISPHOSPHATASE: breaks down fructose-2,6-bisphosphate.

FRUCTOSE INTOLERANCE (hereditary fructose intolerance): a human chromosome 9q21.3 recessive disorder caused by a deficiency of the enzyme fructose-1-phosphate aldolase. The patients begin sweating, trembling, feel dizzy and nauseating 20 minutes after ingesting fructose. The immediate clinical findings are fructosuria, hypophosphatemia (abnormally low amounts of phosphate in the blood), aminoaciduria (amino acids in urine), fructosuria (moderate amounts of fructose in urine), hyperbilirubinemia (excess of bilirubin [red bile pigment] in blood), etc. The chronic symptoms include jaundice, enlargement of the liver (hepatomegaly), vomiting, dehydration, edema (excessive fluid in the tissues), ascites (fluids in the abdominal cavity), seizures, fructose accumulation in the urine (fructosuria) and in the blood (fructosemia), cirrhosis (destruction of cells and increase of connective tissues) in the liver, etc. On a diet low on fruits, honey or fructose-containing sweeteners, the patients may be quite normal but consuming fructose may make them sick and infants may even die if the formula contains fructose. The apparent toxic effects of fructose in plant cell cultures is caused by the breakdown products (mainly furfural) of autoclaving. (See fructosuria, aldolase)

FRUCTOSURIA (essential fructosuria): is a rare autosomal disorder (prevalence is about 8 x 10^6). The biochemical basis of this non-debilitating anomaly is a deficiency of fructokinase.

FRUIT: mature ovary of plants (may include also other parts of the flower); with the exception of the seed within, it is genetically maternal tissue. (See flower differentiation, gametogenesis in plants)

FRUIT FLY: *Drosophila melanogaster* 2n = 8; *D. ananssae* 2n = 10; *D. melanica* 2n = 10; *D. obscura* 2n = 10; *D. pseudoobscura* 2n = 10; *D. virilis* 2n = 10; *D. willistonii* 2n = 6. (See *Drosophila*)

FRUIT RIPENING: is the consequence of changes in the composition and softening of the cell walls. The increase in respiration involves an increase of the production of the plant hormone ethylene which affects the expression of a number of genes, notably polygalacturonase (PG), 1-aminocyclopropane 1-carboxylic acid synthase (ACCS), and 1-amino cyclopropane 1-carboxylic acid oxidase (ACCO, tomato gene pTOM13). By transforming tomatoes with a single PG antisense construct, PG activity could be reduced to 1% but for a more efficient control of ripening the amount of the other enzymes had also been reduced to decrease ripening that is under polygenic control. For commercially effective storage without softening ripe-harvested tomato fruits, the use of the antisense RNA of the LE-ACS2 and LE-ACS4 loci was required. The berries stored at 20° C remained firm for months but could be fully matured when exposed either to ethylene (C_2H_4) or its analog (C_3H_6). The fruits so handled were practically indistinguishable by color, scent and consistency from vine-ripened fresh fruits. (See antisense technology, plant hormones)

FRUITING BODY: a collective name of fungal organs (perithecium, cleistothecium, apothecium, locule) containing the haploid reproductive spores.

FRYE TEST: is a legal ruling concerning admissible scientific evidence to the court. According to the Frye v. United States (D.C. Cir. 1923] 293 Fed. 1013) states that new scientific methods must be generally accepted by the scientific community before evidence from such methods is admissible to the courts. This became a highly controversial issue because the 'general acceptance' is difficult to define. Criminal defense lawyers frequently argued that electrophoretic pattern of proteins and DNA fingerprints, and their statistical evaluation should not be presented to the jury because some scientists may dissent about some aspects of the data. (See forensic genetics, DNA fingerprinting, ceiling principle)

FrzE: *Myxococcus xanthus* kinase affecting FrzE and FrzG proteins regulating bacterial motility and development.

FSH (follicle stimulating hormone or follitropin): controls ovarian follicles, estrogen secretion, menstrual cycles, spermatogenesis and D2 cyclin. Ovarian and testicular tumors have high levels of cyclin D mRNA. FSH deficient males have small testes yet fertile to a variable degree. (See animal hormone, follicle, ovary, menstruation, gametogenesis, differentiation, NGFI-A, cell cycle)

F-TABLE: see F distribution

F-TEST: see F distribution

FT-IR (Fourier transform infrared spectroscopy): a very sensitive method for the detection of intramolecular changes as function of time is converted into function of angular frequencies. (See also Raman spectroscopy, tumorigenesis)

FtsY: a prokaryotic transport protein, structurally related to mammalian SRP54. The related bacterial Ffh is ribonucleoprotein binding to FtsY in a GTP-dependent manner. (See SRP, GTP)

FUCHSIA: the ornamental plant what Gregor Mendel was examining while photographed in 1862 with his monasterial colleagues. This plant has 2n chromosome numbers 22, 56, 66 and 77 and it was one of Mendel's lucks that he did not experiment further with this erratic material which would not have permitted the type of studies he conducted with the stable peas.

FUCHSIN (triaminotrimethylmethane): a red cytological stain.

FUCOSE: is 6-deoxy sugar. It is present in several antigenic glycoproteins, it may be associated with several immunoglobulins and present in plant cell walls.

FUCOSIDOSIS: recessive, human chromosome 1p34 (and a pseudogene at 2q31) defect of α-fucosidase resulting in severe neurodegeneration and angiokeratoma although some forms are less detrimental. The accumulation of the fucose is detectable in the amniotic fluid. (See amniocentesis, angiokeratoma)

FUMARATE HYDRATASE (FH): is encoded in human chromosome 1q42.1 but both cytoplasmic and mitochondrial forms of the enzyme exist due probably to alternative processing of the transcript. Its deficiency results in mental and physical impairment. (See mitochondrial diseases in humans)

FUN GENES (functions **un**known genes): see orphan genes

Funaria hygrometrica (Bryophyte): chromosome numbers vary, 14, 28, 56.

FUNCTIONAL REDUNDANCY: according to this hypothesis functions my be carried out by gene products with highly similar activity but the genes being members of separate regulatory circuits would make the system more flexible than a combinatorial control of development. (See combinatorial gene control)

FUNCTIONALITY OF MUTAGENS: the number of chemical groups reacting in mutagenesis. (See nitrogen mustard, sulfur mustard)

FUNDAMENTAL THEOREM OF NATURAL SELECTION: see natural selection

FUNDAMENTALISM: the religious beliefs that regards as the only absolute right what is written in the holy books (Bible, Koran) or in the works of the basic ideologues (e.g. Marx, Engels, Lenin). Accordingly, only creationism explains properly the genesis of life, provides specific guide-lines to human ethics and ideology or political or economical theory, respectively. (See creationism, lysenkoism)

FUNGAL INCOMPATIBILITY: is based on interaction of the products of non-allelic genes and thus prevents self-fertilization similarly to the outcome of S alleles of plants. In *Ustilago maydis* (a pathogen of maize) stable dikaryons can be formed only between different mating type alleles of a multiallelic *b* locus — recognized by pheromones —. The same *b* locus is responsible also for plant pathogenicity. The *bE* and *bW* alleles encode different homeodomain proteins. The E variants and W variants differ primarily in the N-terminal amino acids. Activ-

Fungal incompatibility continued

ity requires that the E and W allele products would dimerize and this can happen only in appropriate allelic combinations; the majority of the over 300 combinations are active. In other fungi several multiallelic gene pairs coding for interacting homeodomains have been discovered; in yeast only two mating types *a* and *α* exist. *Vegetative incompatibility* prevents the fusion of hyphae in case of *het* alleles or non-allelic *het* genes are carried in their nuclei. (See fungal life cycle, incompatibility).

FUNGAL LIFE CYCLES: display an enormous variety of specializations in the various taxonomic groups and cannot be represented here. The general scheme is, however, relatively simple and shared by all fungi (See hypha).

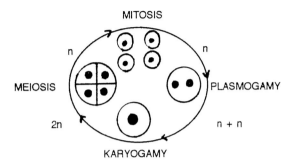

The approximately 2,000 genera of fungi have a variety of modes of reproductions that share some basic similarities. Between 9 o'clock and 3 o'clock are the haploid phases (n). Haploid cells fuse at 3 o'clock (plasmogamy) but the nuclei are still separate and a dikaryon (n + n) is formed. In a following step, at 6 o'clock nuclear fusion takes place (karyogamy) and the cell becomes diploid (2n). This is followed by meiosis (9 o'clock) yielding four haploid sexual spores that may divide again by mitosis (12 o'clock) before the spores are released. These sexual spores may differentiate and the cycle is reinitiated.

FUNGUS (plural fungi): eukaryotic thallophytes yet separate subkingdom from plants and bacteria. Includes many saprophytic, parasitic and pathogenic species of enormous variety in structure and function. Fungal genetics provided and is providing understanding for basic genetic phenomena such as recombination, biochemical pathways, cell cycle, etc. The yeasts and other ascomycetes are among the most important tools of modern genetic research.

FUNICULUS: a vascular stalk of the plant ovule, a cord-like structure, and in animals it includes also the umbilical cord, etc.

FUR COLOR: of animals is determined by the melanin pigments formed in the melanocytes and the migration of the melanoblasts. The various genes involved then modify, reduce or intensify pigmentation. In addition, the actual visible color depends on the dorso-ventral distribution of the phaeomelanin and eumelanin pigments. Superimposed on these are species- or genotype-specific striping, spotting, lyonization, temperature-sensitivity, etc. (See melanin, pigmentation of animals, hair color)

FURIN: a Golgi-associated proteinase; it may activate stromelysin. (See Golgi, stromelysin)

FUS3: a protein kinase of the MAPK family. (See signal transduction, MAPK)

fushi tarazu (*ftz*): *Drosophila* mutation of the pair-rule class, every other body segment is missing. (See morphogenesis in *Drosophila*)

FUSICOCCIN: toxin of the fungus *Fusicoccus amygdali*; an activator of plasma membrane H^+ ATP-ases, causes H^+ secretion, and K^+ influx into guard cells and thus opening of stomata of plants.

FUSIDIC ACID ($C_{31}H_{48}O_6$): is an antibiotic (ramycin) isolated from *Fusidium coccineum*. This compound mimics the effect of the *rel$^-$* (relaxed control) mutations and also permits the synthesis of guanosine polyphosphates (ppGpp, pppGpp) and ribosomal RNA and ribosomal protein. Tetracycline has similar effect. (See antibiotics, stringent control, relaxed control)

FUSIGENIC LIPOSOME: is similar to the liposome vectors, except they are engineered to carry on their surface hemaglutinating neuroaminidase (HNA) and a fusion protein. The hemagglutinating Japanese virus (HVJ) and the Sendai virus produce these proteins and makes them capable of fusing with the cell membrane at neutral pH. HNA is required for binding to cell recep-

Fusigenic liposome continued

tors containing sialoglycoproteins or sialolipids. The fusion protein is in an inactive form until it is hydrolyzed to two polypeptides, F1 and F2; F1 interacts with cholesterol to facilitate fusion and then the DNA carried by the liposome is delivered into the cell. (See liposome, sialic acid, Sendai virus, hemagglutinin, cholesterol, cytofectin)

FUSIN (LESTR, HUMSTR, CXCR4): a co-receptor of the CD4 antigen required for fusion with the membrane and entry of a virus (HIV) into a cell. It is a heterotrimeric GTP-binding protein. (See CD4, CD8, HIV, acquired immunodeficiency, RANTES, MIP)

FUSION OF SOMATIC CELLS: see cell fusion

FUSION PROTEIN: is synthesized when neighboring genes are transcribed and translated together. It contains full or incomplete parts of two normal proteins. Also, a group of proteins mediating membrane fusion in cells. (See also gene fusion, transcriptional gene fusion vectors, translational gene fusion vectors, read-through proteins)

FUSOME: a structure that is formed in the germline cell of insects during the mitotic divisions. It anchors the mitotic divisions leading to the formation of the nurse cells and the oocyte. After the divisions are completed it fades away. Its existence can be detected by antibodies to spectrin, a filamentous membrane protein. (See nurse cell, morphogenesis, spectrin)

FUTILE CYCLE: it is an apparently useless chemical reaction in the cell, e.g. fructose-6-phosphate is phosphorylated to fructose diphosphate and simultaneously it is hydrolyzed back to fructose-6-phosphate and resulting in cleavage of ATP into ADP and P_i (unnecessary ATPase activity). (See also substrate cycle)

FUZZY INHERITANCE: is a statistical term used for cases when linkage information is computed by allele sets and using set recoding. (See Nature Genetics 11:402[1995])

Fv (fragment variable): is a functional antibody molecule, composed of light and heavy chain antigen-binding sites of one region of the antibody. (See antibody)

φX174: is a single-stranded DNA bacteriophage with overlapping genes. (See bacteriophages)

THE PHYSICAL MAP OF BACTARIOPHAGE φX174 WITH 10 GENES A to H. ALTHOUGH SOME OF THE GENES OVERLAP EACH OTHER (OR INCLUDED), THERE ARE NON-GENIC (INTERGENIC) SEQUENCES (SHOWN IN SOLID BLACK). THE RELATED G4 PHAGE IS SIMILAR. (Courtesy of Godson, N.G. Stadler Symp. 12:143) ↓

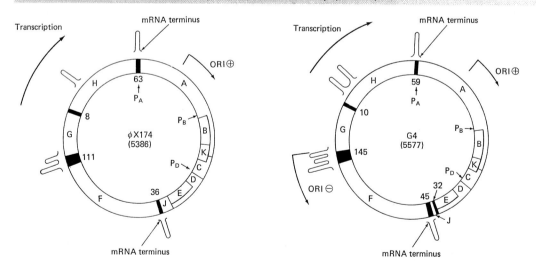

FYN: oncogene is a member of the Src non-receptor protein tyrosine kinase gene family. (See Src, Csk, protein tyrosine kinase)

G: see guanine; G is also used to denote generations after mutagenic treatment of mice; G_0, G_1, G_2, etc.; this designation is somewhat confusing with the preempted cell cycle symbols. (See cell cycle)

g: general intelligence. (See intelligence quotient, human intelligence)

G418 ($C_{20}H_{40}N_4O_{10} \cdot 2H_2SO_4$): an aminoglycoside antibiotic. (See geneticin)

G_α: G-protein involved in hormonal stimulation of adenylate cyclase and may regulate ion channels or phospholipase C. The human G_S 1α gene contains 13 exons and 12 introns in a total size of 20 kbp. Gα types: $G_i\alpha, G_o\alpha, G_x\alpha, G_t\alpha$. These are very highly conserved proteins across phylogenetic ranges. In $G_S\alpha$ only 1/394 amino acid difference was found between man and rat, and the protein is entirely identical between humans and bovines. (See G-proteins)

G BANDING: a chromosome staining methods, using Giemsa stain (a complex basic dyes, containing azures, eosin, glycerol and methanol) after pretreatment with the proteolytic enzyme, trypsin, it permits the identification of dark cross-bands that vary among the individual eukaryotic chromosomes and usually facilitates their identification even when their length and arm ratio is similar. The darkly stained bands represent heterochromatin. (See chromosome banding, stains, rye) G-BANDED HUMAN CHROMOSOME →

G BOX: a guanine-rich upstream cis element regulating transcription. Commonly it has the struct-ure GA|CAACGTG|GC. The G-box activator protein binds to the framed core sequence. It is commonly found in environmentally sensitive genes. (See Simian virus 40, CAAT box, regu-lation of gene activity)

***G* ELEMENT**: see non-viral retrotransposable element. (See retrotransposons, retroposons)

g FORCE: see centrifuge

G4 PHAGE: is single-stranded DNA phage, closely related to φX174. (See bacteriophages, φX174)

G_0 PHASE: the state of a pause for the cell before it enters the G_1 phase and until divisional activities would start again after mitosis. (See cell cycle)

G_1 PHASE: the first phase of the cell cycle following mitosis (C value = 2). (See cell cycle)

G_2 PHASE: the phase following DNA replication during the cell cycle (C value = 4). (See cell cycle)

G PROTEINS: are guanine nucleotide binding proteins that serve as intermediaries in biological signaling pathways. The signal is received by *receptors* and the *G-proteins* forward it by mediation of different number of intermediaries to the *effectors* that regulate genes in response to the signals. G-proteins are activated aluminumfluoride and the α subunit can be ADP-ribosylation mediated by bacterial toxins (cholera, pertussis).

G-proteins are heterotrimeric (α, β, γ) proteins that control the opening and closing of the signal transduction pathways by changing the attached GDP ⇌ GTP. In the GTP-associated form they have key role in signal transduction from receptors to effectors. There are also low molecular weight small G-proteins with a single (α) subunit. G-protein (G_s) is involved in the regulation of the level of the enzyme adenylyl cyclase and thus cAMP and cAMP-dependent protein kinase. The G_i form is involved in the inhibition of adenylate cyclase; G_{iia2} is required for insulin function. The light-activated GTPase activity is mediated by the G_t-protein, also called transducin. G-proteins stimulate the hydrolysis of phosphoinositides with the aid of phospholipase C. cAMP degradation by cyclic nucleotide phosphodiesterase is also mediated by G-proteins and indirectly by Ca^{2+}. G-proteins regulate also ion channels. When a proper ligand binds to a transmembrane receptor, the trimeric G-protein dissociates into a βγ and an α subunit. The α subunit stimulates adenylyl cyclase, and first the transition of GDP to GTP and

G proteins continued

later the transition of GTP to GDP through the mild GTPase activity. In the G-GDP state, reassociation of the three subunits follows. G-proteins regulate also Ca^{2+} metabolism and indirectly control allosteric effector proteins. Several human diseases are associated with defects in the G proteins (pituitary tumors, McCune-Albright syndrome, Albright hereditary osteodystrophy, puberty precocious) or with defects in the G protein receptors (hypercalcemia, hypercalciuria, hyperparathyroidism, diabetes insipidus, retinitis pigmentosa, color blindness, glucocorticoid deficiency). (See G region, G' region, G" region, G_α-, G_i-, G_q-, G_s-, G_o-, G_t-protein, GTPase, cAMP, adenylate cyclase, rhodopsin, signal transduction, calmoduline, receptor, effector, cholera toxin, pertussis toxin, GTP, cAMP, phosphodiesterase, ion channel, adenylate cyclase, see also the mentioned diseases under separate entries. Internet information source <http://receptor.mgh.harvard.edu/GCRDBHOME.html>)

G QUARTET: guanine-rich nucleotide sequences may form, four-stranded complexes. (See antisense technologies)

G REGION: consensus N-K-X-D (see Amino Acid Symbols) in GTP-binding proteins that interacts with guanine in GTP. (See G-proteins)

G' REGION OF GTP BINDING PROTEINS: with highly conserved consensus D-X-X-G-Q (see Amino Acid Symbols) involve GTP-ase function and may affect oncogenicity. (See signal transduction, G-proteins)

G" REGION: in RAS interacts with GTP through the E-T-S-A-K (see Amino Acid Symbols) consensus. In some G-proteins H-(F/M)-T-C-A(T/V)-D-T (see Amino Acid Symbols) may be the corresponding functional area. (See G-proteins)

G8 RNA: see thermal tolerance

GA: see gibberellic acid, plant hormones

GABA: γ-aminobutyric acid plays an important role in neurotransmission of vertebrates and invertebrates; in the nematode *Caenorhabditis* a series of *unc* (uncoordinated movement) genes respond to GABAergic neuronal effects. $GABA_A$ receptors mediate synaptic inhibition but upon intense activation they may excite rather than inhibit neurons. (See glutamate decarboxylase deficiency disease, epilepsy, *Caenorhabditi*s, neuron, cleft palate)

Gαβγ: heterotrimeric G-proteins and the three subunits are of 39-52, 35-36 and 7-10 kDa size, respectively. In mammalian cells several genes are known for each subunit and their cDNAs may generate additional variations by alternative splicing. (See G proteins, splicing)

GADD153: is a cellular enhancer-binding protein mediating stress of growth and differentiation. Under stress it may be activated by phosphorylation of Ser^{78} and Ser^{81} residues and consequently enhanced transcription and inhibited adipose cell differentiation. It is the same as CHOP. Gadd (**g**rowth **a**rrest and **D**NA **d**amage) are activated under varied stress conditions. (See enhancer, DNA repair)

GAG: see glycosaminoglycan

gag: group-specific antigen, a viral coat protein. (See retroviruses)

GAGA: is a multipurpose transcriptional activator binding to the GA/CT sites in the promoter. Its major function may be to rearrange the chromatin to facilitate transcription. GAGA activates chaperones and binds to the promoter of the *Ultrabithorax* and other *Drosophila* genes. (See position effect, morphogenesis in *Drosophila*, heatshock proteins, heterochromatin)

GAIN: is a practical measure of heritability, frequently used by animal breeders. By this criterion heritability, h^2 = (gain)/(selection differential). See graphical representation below (After Lerner, I.M. & Libby, W.J. Heredity, Evolution and Society, Freeman, San Francisco). The selection differential is the difference between the mean of the parental population and the mean of a portion of the parents selected for further reproduction to improve the herd. The gain/selection differential is frequently called *realized heritability*. The breeder may improve the gain either by increased heritability or by enhanced intensity of selection. Heritability estimates improve if environmental variation is kept at low level by proper feeding and health care of animals or appropriate tillage, fertilization, weed and pest control in plants. The inten-

Gain continued

sity of selection is increased if the proportion of the individuals selected for parents is reduced. Although this may appear to be an easy approach to improve selection gains, the small populations may increase inbreeding and becomes counterproductive. In large mammals the males generally have more offsprings than the females. By the use of artificial insemination, the breeding value of the males can be determined even more precisely than that of the dams. Generally the estimates improve with the age of the animals because larger number of offspring is available for evaluation. In practice, the selection is aimed simultaneously at several traits. Often these traits are negatively correlated because high performance may make the animals (plants) more susceptible to disease. Thus the gain in one trait may mean a loss in the others. Therefore breeders frequently use a *selection index* that weighs each trait by a score and the total of the scores becomes the basis of the selection value. There are statistical methods for predicting

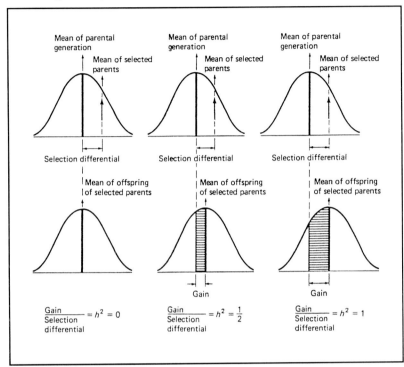

the quantitative performance in a selective breeding program: $Y_O = \bar{Y} + H_n(Y_p - \bar{Y})$ where Y_O is the predicted average performance of the progeny, Y_p is the average of the two parental families selected, $\bar{Y}$ = the average of the original population, H_n is heritability in the narrow sense. Example: the average number of eggs laid per year in a flock of chickens is 250, the heritability is 0.25, the average of the selected family of parents is 274, then the expectation for the offspring $Y_O = 250 + 0.25(274 - 250) = 256$. The genetic gain from mass selection is computed from the covariance: $(XY) = w = \frac{1}{2} \sigma^2_A$. For determining the covariance see correlation; σ^2_A = additive variance (see genetic variances); the plot-to-plot environmental variance is σ_e^2 and the plant-to-plant environmental variance $= \sigma^2_{we} = \sigma^2_{wf} + \sigma^2_{me}$ and the genotype environmental variance, $\sigma^2_{G \times E} = \sigma^2_{A \times E} + \sigma^2_{D \times E}$. If we consider the within-family variance = 0, then the gain for mass selection,

$$\Delta G_m = \frac{\frac{1}{2} i \sigma^2_A}{\sqrt{\sigma^2_A + \sigma^2_D + \sigma^2_{A \times E} + \sigma^2_{D \times E} + \sigma_e^2 + \sigma^2_{me}}}$$

and i = selection intensity, σ^2_D = the dominance, $\sigma^2_{A \times E}$ = additive × environment, and

Gain continued

$\sigma^2_{D \times E}$ = dominance x environment variances. This procedure is applicable to large populations at relative ease. In case of phenotypic recurrent selection, the equation for gain in mass selection above for cycle needs to be multiplied by 2 because the selection is applied to both parents. Additional formulas for other type of selections are to be found in Moreno-González, J. & Cubero, J. *Plant Breeding*, pp. 281-313, Hayward, M.D. *et al.*, eds., 1993, Chapman & Hall, London, New York. (See also polygenes, breeding value, selection index, quantitative genes, heritability, intraclass correlation, correlation)

GAIN-OF-FUNCTION MUTATIONS: generally mutations lead to loss of structures, e.g. hairs or bristles or certain function, e.g. auxotrophy. Some of the homeotic mutants, however gain additional structures such as extra petals or stamens in the flowers or legs on the head in *Drosophila*. These "gains" are the result of homeotic transdetermination regulated by altered transcription and/or transcript processing. (See transdetermination, transcription, processing)

GAL: see galactose utilization

GAL4: is a positive regulatory protein of the yeast galactose genes and it binds to a specific upstream regulatory DNA sequence. (See galactose utilization)

GALACTANS: are polymers of galactose. (See galactose)

GALACTOKINASE DEFICIENCY: may be due to defects at GALK1 (human chromosome 17q24) or GALK2 (chr. 15). (See galactosemias, galactose)

GALACTOSE: one of the most common six-carbon monosaccharides differing from glucose only sterically at carbon-4 chiral center (an epimer of glucose). It can be converted to glucose by an epimerase enzyme (UDP-Gal →UDP-Glc). Galactosyl groups are present in some anthocyanins, collagens, and immunoglobulins. Lactose (the milk sugar) is a disaccharide of galactose + glucose, split by the enzyme lactase. (See galactosemias, galactose utilization, chirality, epimers, galactosidase, galactose [formula on p. 397], galactose utilization, epilepsy, eye diseases, genetic screening)

GALACTOSE OPERON: see galactose utilization

GALACTOSE UTILIZATION: is coordinately regulated in prokaryotes and eukaryotes. The galactose genes of *E. coli* are either clustered (*galE* [UDP-galactose-4-epimerase, *galEo* [operator], *galE1p* [promoter], *galE2p* [promoter of galEK], *galK* [galactokinase], *galT* [galactose-1-phosphate uridyltransferase]) all at map position 17 or galR [galactose regulator] at map position 61, *galP* [galactose permease] at map position 63, and *galU* [glucose-1-phosphate uridyltransferase] at map position 27. In yeast the uptake of galactose is mediated by galactose permease (gene *GAL2*). In the presence of ATP, galactose is phosphorylated (Gal-1-P) by galactokinase (gene *GAL*). Galactose-1-phosphate (Gal-1-P) + uridine-diphosphoglucose (UDP-glucose) generate UDP-galactose + glucose-1-phosphate by the action of galactose-1-phosphate uridyl transferase (gene *GAL7*) while the UDP-galactose-4-epimerase (gene *GAL10*) mediates the formation of UDP-glucose from UDP-galactose. Genes *GAL1, GAL7* and *GAL10* form a cluster in yeast chromosome 2 and *GAL2* is in chromosomes 12. These yeast genes are coordinately inducible up to a 1000 fold by the presence of galactose although they are transcribed from separate promoters. The GAL enzymes are regulated by gene *GAL4* (linkage group 16) and gene *GAL80* (linkage group 13). *GAL4* is apparently a positive regulator of genes *1, 2, 7,* and *10* whereas some *GAL80* mutations abolish the need for induction of the same genes and convert them either to constitutive forms or make them non-inducible. It is assumed that the normal role of the product of gene *GAL80* is to prevent the transcriptional activation by *GAL4* but the combination of the GAL1 protein with GAL80 protein inactivates the latter and then GAL1 activates GAL4, the activator of the system. This GAL1 protein is an enzymes as well as a regulator of transcription. The activation by *GAL4* depends on *upstream activating sequences* (UAS) located 200 to 400 base pair upstream of the genes, *1, 2, 7, 10* and *GAL80*. The presence of two UAS is sufficient for full expression. The consensus within the 17 bp palindromic (↔) UAS is:

Galactose utilization continued

$$5'\text{-C G G A}^{\text{G}}_{\text{C}}\text{ G A C A G T C}^{\text{G}}_{\text{C}}\text{ T C C G -3'}$$
←――――――   ――――――→

The protein product of gene *GAL4* is about 100 kDA and it contains three essential domains. Amino acids from 1 - 65 are involved in DNA binding, residues 65 - 94 are concerned with dimerization. Amino acids 148 - 196 and 768 - 881 mediate activation of transcription (activation domain) and at the C-terminus the sequence 851 - 881 also bind the *GAL80* gene. At the N-terminus amino acid residues 10 - 32 display a Zinc-finger motif, common to binding proteins. At the C-terminus there is a high density of acidic amino acids, a characteristics of regulatory proteins. The presence of and inactivation by insertion elements in bacterial genomes were first recognized by a study of the *gal* operon in *E. coli*. (See also galactose, operon, coordinated regulation, palindrome, IS elements, regulation of gene activity, Zinc fingers, binding proteins, two-hybrid method, galactosemia)

GALACTOSEMIAS: are autosomal hereditary diseases in humans caused by the deficiency of either the enzyme galactokinase or more commonly galactose-1-phosphate uridyltransferase (human chromosome 9p13). As a consequence galactose cannot be transformed into glucose. Since the milk sugar is a disaccharide of galactose and glucose, galactose accumulates in the blood and excreted in the urine. The accumulating galactose causes severe intestinal problems and the accumulating galactose-1-phosphate may damage the liver, brain, eyelens (cataracts) and other organs. Unless this anomaly is detected right after birth, infant death may result. By a diet free of any source of galactose, damage may be prevented. This condition is quite common, about 4×10^{-4}. A human galactokinase gene has been mapped to chromosome 17q24 and it is responsible also for cataracts. Deficiency of the enzyme that converts UDP- galactose ⇔ UDP-glucose, galactose epimerase (GALE, chromosome 1p35-p36) also leads to galactosemia.

β–**GALACTOSIDASE**: Probably the best studied bacterial gene, *lac* involves the determination and control of the enzyme β-galactosidase (see *lac* operon). The enzyme *α-galactosidase* is an α-galactosyl hydrolase (melibiase, α-galactoside galactohydrolase, ceramide trihexosidase) is deficient in patients suffering from *Fabry's disease*. In the plasma and in most of the tissues the trihexosyl ceramides Gal-Gal-Glc-Cer or Gal-Gal-Cer (Gal = galactose, Glc = glucose, Cer = ceramide) accumulate in the tissues. In various organs extensive deposition of lipids occur and the patients suffer skin lesions, pain, paresthesia (burning, prickling sensation) in the extremities, ectasia (dilation, distention) in the skin vessels, edema (accumulation of fluids) in the legs, hypohidrosis (diminished sweating), albuminuria (protein accumulation in the urine), hyposthenuria (lowered amounts of solids in the urine). Death may result from renal failure. The disease is X-chromosome linked (q22). Heterozygous females have the same symptoms as hemizygous males but at reduced level. β-Galactosidase (a group of enzymes splitting galactosides, galactose linkages) activity is greatly reduced in patients affected by a group of human diseases called *gangliosidoses*. The general symptoms involve deterioration of psychomotor (brain and movement) activities, severe bony deformities and generally death by 2 years of age. These diseases occur in all ethnic groups as incurable autosomal recessive defects. Heterozygotes may be detected by β-galactosidase assays and the diseases can be identified by amniocentesis. (See also galactose utilization, Fabry's disease sphingolipidoses, gangliosidosis general, Krabbe's leukodystrophy, lactosyl ceramidosis, *Lac* operon)

GALACTOSYL CERAMIDE LIPIDOSIS: see Krabbe's leukodystrophy

GALAGO: see Lorisidae

GALECTIN: a β-galatoside binding protein regulating growth and immunological responses. It may induce apoptosis in activated human T cells. (See apoptosis, T cell)

GALL: generally undifferentiated tissue growth in plants, caused by infection. (See crown gall)

GALT (gut-associated lymphoid tissue): see Peyer's patches *gam*: see lambda phage, Charon vectors

GAMBORG MEDIUM (B5): for plant tissue culture is suitable for growing callus and different plant organs. Composition mg/L: KNO_3 2500, $CaCl_2.2H_2O$ 150, $MgSO_4.7H_2O$ 250, $(NH_4)_2SO_4$ 134, $NaH_2PO_4.H_2O$ 150, KI 0.75, H_3BO_3 3.0, $MnSO_4.H_2O$ 10, $ZnSO_4.7H_2O$ 2.0, $Na_2MoO_4.2H_2O$ 0.25, $CuSO_4.5H_2O$ 0.025, $CoCl_2.6H_2O$ 0.025, Ferric-EDTA 43, sucrose 2%, pH 5.5, inositol 100, nicotinic acid 1.0, pyridoxine.HCl 1.0, thiamine.HCl 10, kinetin 0.1, 2,4-D 0.1 - 1.0. The microelements, vitamins and hormones may be prepared in a stock solution and added before use. For kinetin other cytokinins may be substituted such as 6-benzylamino purine (BA) or isopentenyl adenine (or its nucleoside), for 2,4-D (dichlorophenoxy acetic acid), naphthalene acetic acid (NAA) or indole acetic acid (IAA) may be substituted or a combination of the hormones may used in concentrations that is best suited for the plant and the purpose of the culture. For solid media use agar or gellan gum. Heat labile components are sterilized by filtering through 0.45 µm syringe filters. This medium may be purchased from commercial suppliers in a dry mix ready to dissolve. (See Murashige & Skoog medium, embryo culture, cell culture, cell fusion, agar, gellan gum, plant hormones)

GAME THEORY: is dealing with decision making under uncertainty. Before a decision is made, the probabilities of a set of actions, e.g. $p(\theta_1)$ and $p(\theta_2)$ must be assessed, generally by a subjective manner. Such a procedure is most widely used in the business world (marketing) under competitive conditions. It may be applied also to natural sciences where exact statistical methods are not practical due to the variability and uncertainty of the conditions.

GAMETANGIA: sex organs of fungi; oogonium in the "female" and antheridium in the "male".

GAMETE: haploid male or female generative cell (egg, sperm). Gametic fusion (formation of the zygote) takes place during sexual reproduction. The zygote (2n) has twice the number of chromosomes of the haploid (n) gametes. (See gametogenesis)

GAMETE COMPETITION: if multiple gametes are available, their success in fertilization may be determined by genetically controlled viability or vigor. It occurs commonly among sperms of animals and plants, pollen tubes (certation) and also among eggs in multiparous animals or in plants where more than one megaspore of the tetrad may produce the egg. (See also certation, meiotic drive, selective fertilization, preferential segregation, segregation distorter)

GAMETIC ARRAY: of diploids, in case of independent segregation, can be determined by different procedures:

in a dihybrid: $(A + a) \times (B + b) = $ AB, Ab, aB, ab

in a trihybrid: $(A + a) \times (B + b) \times (C + c) = $ ABC, ABc, AbC, Abc, aBC, aBc, abC, abc or using any type of gene symbols such as I/i, R/r A/a the combinations can be read from left to right by following the paths of the arrows and at right we obtain the gametic arrays. In general, in diploids, in case of independent segregation, the gametic output can be determined by 2^n where n corresponds to the number of allelic pairs, e.g., in a trihybrid cross $2^n = 2^3 = 8$ as derived above. For gametic array in autopolyploids and trisomics see autopolyploidy and trisomy, respectively. (See Mendelian segregation)

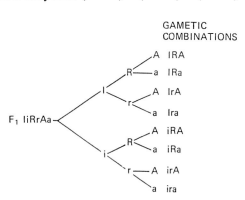

← DERIVATION OF THE GAMETIC ARRAYS

GAMETIC LETHAL: death at the egg or sperm stage. (See zygotic lethal)

GAMETOCIDE: any chemical which causes male sterility. They may be used in plant breeding to spare the efforts of emasculation in large-scale hybridization or may be used as birth-control agents when applied shortly before copulation. (See male sterility)

GAMETOCYTE: cell that produces gametes. (See also *Plasmodium*)

GAMETOGENESIS: the process of the formation of gametes. See diagrams below.

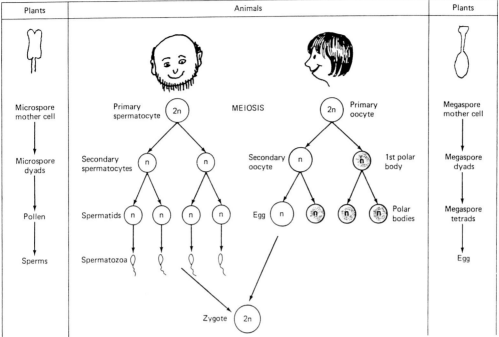

▲ A COMPARATIVE VIEW OF ANIMAL AND PLANT GAMETOGENESIS ▲

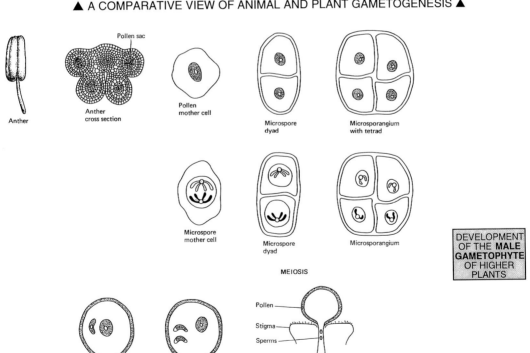

DEVELOPMENT OF THE **MALE GAMETOPHYTE** OF HIGHER PLANTS

GENETICS MANUAL

DEVELOPMENT OF THE **FEMALE GAMETOPHYTE** OF HIGHER PLANTS. THE MEIOTIC STAGES (showing only one bivalent) AT THE DEVELOPMENTAL PHASES ARE INDICATED. NOTE THE EMBRYOSAC AT STAGE 14.

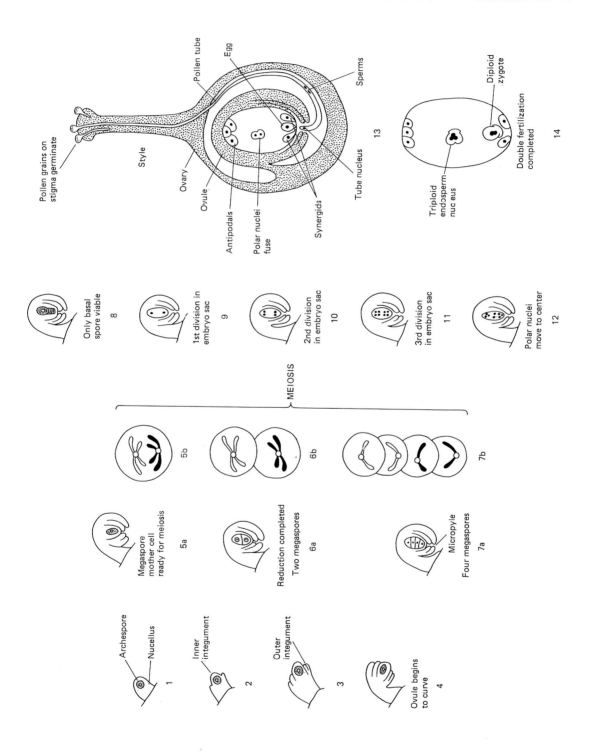

Gametogenesis continued

The animal egg is formed by differentiation without further cell division from a haploid product of meiosis and so do the spermatozoa from the spermatids. The development of the female and male gametes of higher animals is represented on page 399. Basically, gametogenesis in animals and plants shows substantial similarities because in both cases it is based on meiosis. (See also spermiogenesis, gametophytes)

GAMETOPHYTE: the cells resulting from the meiosis of plants that have half the chromosome number of the zygotes. The gametophytes (megaspores, microspores) will form the gametes (egg and sperm). Selection at the gametophyte level is much more effective than in the sporophytic generation when the intended target of the selection is expressed at this developmental stage. The effectiveness of selection is particularly needed when the frequency of the gene selected for is low. Selection at the haploid level was apparently successful for tolerance to herbicides, toxins secreted by pathogens, alcohol dehydrogenase mutations, and possibly against certain stress effects. (See diagrams of the male [page 399] and female [p. 400)] gametophyte development, see also sporophyte, cytoplasmic male sterility, male sterility, pollen competition, certation, gametophyte factor, incompatibility alleles)

GAMETOPHYTE FACTOR: affects the haploid gametophyte and may be responsible for reduced transmission of the chromosome (gamete) that carries it in a heterozygote. Gametophyte factors generally have more detrimental effect on the male but in rare cases the female is also influenced to various degrees. (See certation, gametophyte, meiotic drive, preferential segregation, selective fertilization, zygotic lethal)

GAMMA FIELD: is an area or space where usually chronic exposure is provided from a source of electromagnetic radiation (e.g. ^{60}Co). Such a field may help in studies assessing the effects of longterm exposures on various biological material (mutation, chromosome breakage, physiological changes) in case of nuclear accidents. (See electromagnetic radiation, radiation effects, radiation hazard assessment, radiation protection, gamma rays)

GAMMA INTERFERON ACTIVATION SITE (GAS): TTNCNNNAAA. (See signal transduction Jak-STAT, ISRE, STAT)

GAMMA RAYS: are ionizing radiations (photons, electromagnetic radiation) emitted by isotopes (such as ^{137}Cs, ^{60}Co, and others). They are similar to X-rays but have much higher energy and have an ability to traverse even several centimeters of lead. Gamma-rays from ^{60}Co (1.2 - 1.3 MeV) have a linear energy transfer 0.3 LET compared to hard X-rays (250 keV). [LET measures ionizing radiation in keV/nm path]. (See also physical mutagens, electromagnetic radiations, Volt, eV)

GAMMAGLOBULIN: is an immunoglobulin (IgG) consisting of either the light chains κ or λ and the heavy chains have one of the $C_{\gamma 3}$, $C_{\gamma 1}$, $C_{\gamma 2}$, $C_{\gamma 4}$ coded constant regions. (See antibody, immunoglobulins, agammaglobulinemia, immunodeficiency)

GAMMOPATHY: a condition of defective immunoglobulin (gammaglobulin) synthesis.

GAMODEME: the same as deme. (See deme)

GANCICLOVIR (GCV): a derivative of acyclovir (2-amino-1,9-dihydro-9-[{2-hydroxyethoxy} methyl]-6-H-purine-6-one); both are antiviral (herpes) drugs.

GANGLION: a group of nerve cells outside the central nervous system.

GANGLIOSIDES: are sphingolipids containing several units of acidic sugars attached to the fatty acid chain; they are common in nerve tissues. Their synthetic pathway is: Uridine-diphosphate[UDP]-glucose + ceramide -> glucosyl ceramide + UDP-galactose -> galactosyl-glucosyl ceramide. Galactosyl-glucosyl ceramide + cytidine monophosphate-N-acetyl-neuraminic acid (CMP-NANA) -> ganglioside G_{M3}. Ganglioside GM_3 + UDP-N acetyl-galactoseamine -> ganglioside G_{M2} + UDP. Ganglioside G_{M2} + UDP galactose -> ganglioside G_{M1} + UDP. Ganglioside G_{M1} + nCMP-NANA -> higher gangliosides. If the sialic acid group (acetyl neuraminic acid, glucosyl neuraminic acid) is removed asialogangliosides are generated. Several diseases, sphingolipidoses, are involved in their accumulation and breakdown. (See sphingolipids, sphingolipidoses, gangliosidoses, Tay-Sachs disease)

GANGLIOSIDOSES: include a variety of forms. The general gangliosidosis Type I is β-galactosidase deficiency disease leading to severe, progressive degeneration of the brain and death by the age of 2. The overall symptoms resemble to those the Tay-Sachs disease caused by hexoseaminidase A deficiency and the Niemann-Pick disease brought about by sphingomyelinase deficiency. The newborns already show abnormally low activity accompanied by facial and other edemas (fluid accumulation), the distance between the upper lip and nose is enlarged, the ears are set low, light hairiness on the front and neck, the spinal column is deformed, the fingers are short, poor appetite and lethargy and general weakness. The liver and spleen become enlarged. Type II juvenile gangliosidosis has a later onset and death is delayed to age 4 to 5. This form has also very low β-galactosidase levels yet another enzyme seems to be involved. In contrast to type I disease, in type II liver and spleen enlargement as well as bone deformities are absent. The heterozygotes can be identified by β-galactosidase assay and the recurrence may be avoided by genetic counseling. Type III is an adult form and it is controlled by a locus different from type I. Besides the autosomal Type 3, there is an X-linked GM3-gangliosidosis and the latter affects young children. The classification of gangliosidoses is quite complicated. (See also GM-gangliosidoses, galactosidase, sphingolipidoses, sphingolipids, Tay-Sachs disease, Sandhoff's disease, spleen)

GAP: GTPase activating protein are encoded in the long arm of human chromosome 5. Tyrosine-phosphorylated GAP is in the cell membrane whereas the unphosphorylated is mainly in the cytosol. RHO and RAS-related GTPases are abundant in the cells and they regulate signal transduction and the cytoskeleton. (See RAS, RHO, GTP, signal transduction)

GAP GENES: in *Drosophila* are missing some segments or have fused segments. (See morphogenesis, morphogenesis in *Drosophila*)

GAP JUNCTIONS: connecting channels between apposed cells permitting the transfer of molecules between cells; the same task in plants is assigned to the plasmodesmata. (See connexin, plasmodesma)

GAP PENALTY: when similarities are sought in nucleotide alignments for the sake of construction of evolutionary trees (or determine relatedness) and gaps are encountered, these are subtracted from the matches to avoid unwarranted conclusions regarding homologies.

GAPO SYNDROME: an autosomal recessive defect characterized by retarded **g**rowth, reduced hair development (**a**lopecia) toothlessness (**p**seudoanodontia) and progressive wasting of the **o**ptical nerves. It bears similarity to progeria. (See progeria, growth retardation, Gombo syndrome)

GARDNER SYNDROME (APC): is an autosomal dominant (human chromosome 5q21-q22) mutation or deletion occurring at a frequency of about 2-3 x 10^{-5} and it is causing adenomatous intestinal polyposis (a cancer). In addition, the syndrome includes several other symptoms, especially if the genetic lesion extends to a larger segment in the region of several genes nearby. In some cases the polyps are limited only to the colon but in some cases other parts of the intestinal tract, the stomach, the forehead, soft bony tissues, epidermal cysts may also become tumorous. Some forms were associated with increased ornithine decarboxylase activity. Presymptomatic diagnosis may detect deletions by the use of appropriate DNA markers. (See colorectal cancer, skin diseases, cancer, polyposis, Turcot syndrome, Muir-Torre syndrome)

GARGOYLISM: is a defect in L-iduronidase such as in the Hurler and Hunter syndromes. (See mucupolysaccharidosis)

GARLIC (*Allium sativum*): spice, 2n = 16. (See also onion)

GAS: see **g**amma interferon **a**ctivation **s**ite, interferons, signal transduction

GAS CHROMATOGRAPHY: see chromatography, gas-liquid chromatography

GAS-LIQUID CHROMATOGRAPHY (GLC): is suited for the separation of volatile compounds according to their ability to be dissolve in the material of the column bed. An inert gas (helium) driven through the column carries the volatile compounds, and they are sequentially eluted and collected. Some material must be converted to more volatile derivatives before applied to the columns. This method has been extensively used to separate and isolate fatty acids and other compounds. (See chromatography)

GASTRIN: hormones of 14 to 34 amino acid residues, released in the stomach and regulate stomach acid secretion, other enzymes and esophagal and gall bladder contraction.

GASTRULA: is an early stage of embryonic development following the blastula. Gastrulation patterns vary in different animal taxa. The general pattern is an invagination of the epithelial layer into the blastocoel (at the so called vegetal pole) forming the endoderm that gives rise later to the gut. The outer layer, the epithelium, becomes the ectoderm that will form the epidermis and the nervous system. The cells in-between these two layers develop into a mesoderm that will differentiate into the notochord (a vertebral column or its substitute), into the connective tissues, bones, cartilages, fibers, muscles, the urogenital system and the vascular system, including the heart and bloodvessels. Gastrulation of the human embryo takes place during the third week of embryonal development. In arthropods, gastrulation is followed by anterior-posterior segmentation and dorsal-ventral, medial-lateral identification of embryonal regions. (See also blastula, morphogenesis)

GATA: mammalian transcription factors. GATA-1 recognizes the $\frac{TGATAG}{ACTATC}$ or very similar upstream DNA sequences. Several other GATA factors have been identified also in other vertebrates. GATA-3 is a hematopoietic factor responsible also for the differentiation of T cells of the immune system. (See transcription factors, hematopoiesis, T cell)

GATEKEEPER GENE: is supposed to be acting in the pathway of carcinogenesis by representing a certain threshold that must be passed before mutation of the tumor suppressor or activator gene(s) can mediate the development of the recognizable oncogenic transformation. (See oncogenic transformation, oncogenes, transformation oncogenic, progression, cancer, phorbol esters)

GAUCHER's DISEASE: is a chromosome 1q21 recessive complex of glucosyl ceramide lipidoses. Glucosyl ceramide sphingolipids accumulate in the reticuloendothelial "Gaucher cells" because of deficiency of a β-glucosidase. These Gaucher cells occur in the lymphoid tissues, spleen, bone marrow, inside the veins, lung alveoli and other tissues. The type I disease occurs in various age groups and the most characteristic symptoms are enlargement of the spleen and bone anomalies. The neuronopathic or malignant type II form appears before age of six months and results in death by age of 2 years. The cranial nerves and the brain stem is attacked although there is not much lipid accumulation in these tissues. The less severe juvenile form (Type III) may permit survival to the age of 30 years. Gaucher's disease is of relatively common occurrence. Cure can be provided by enzyme replacement therapy. Prenatal diagnosis is feasible by the use of RFLP and enzyme assays. (See sphingolipid, sphingolipidoses, glucosidase, RFLP, prenatal diagnosis, lysosomal storage disease)

GAUDENS: *Oenothera lamarckiana* contains a ring of 12 translocation chromosomes and one bivalent. The translocations contain two complexes, *gaudens* (happy) conveys green color and *velans* (concealing) determines narrow leaves, pale color and disease susceptibility. The (complex) heterozygotes appear normal. Because of the translocations and the recessive lethal genes they carry half of the progeny is inviable (*gaudens/gaudens* and *velans/velans* homozygotes) and the other half (the balanced lethal translocation heterozygotes) breeds true and is of normal phenotype. (See multiple translocations, complex heterozygote, *Oenothera*)

GAUSSIAN DISTRIBUTION: see normal distribution

GAZELLA: in the Dorcas gazella (*Gazella dorcas*) and the Grant's gazella (*Gazella granti*) the male has 31 chromosomes the female 30. In the *Gazella leptoceros* the males are 2n = 33, the females 2n = 32. The Thomson's gazella (*Gazella thomsoni*) is 2n = 58.

G-BOX ELEMENT: is an upstream binding site (GACAACGTGGC) in plants of which the critical part is the CAACGTG core sequence that binds the G-box factor protein, a transcriptional activator.

G-BASE: genomic data base of mouse, for access see Mouse Genome Database, Encyclopedia of the Mouse Genome. (See mouse, databases)

GC BOX: in eukaryotic promoters generally contain the 5'-GGGCGG-3' motif, a binding site for transcriptional regulator proteins.

GC (guanine-cytosine) **CONTENT**: of DNA is contributing to the higher buoyant density of the molecules. In the DNA of the majority of eukaryotes the G-C content is about 40%. (See buoyant density, density gradient centrifugation, ultracentrifugation, DNA base composition)

GCN4: yeast transcription factor of a leucine zipper structure controling the transcription of several genes. Its transcription is triggered by amino acid starvation when eukaryotic peptide elongation factor eIF-2 becomes phosphorylated. The DNA binding site consensus is for GCN4: ATGACTCAT / TACTGAGTA (See leucine zipper)

Gcn5p: see p300

GCSF (G-CSF): see granulocyte colony stimulating factor, lymphokine

GD: see hybrid dysgenesis

γδ ELEMENT (Tn*1000*): is an insertion element (IE) of the bacterial F plasmid that may produce various Hfr bacterial strains either by cointegration or recombination. The pDUAL/pDelta vector series of the γδ family vectors have been successfully used for generating (nested) deletions in both strands of a cloned insertion sequence. The plasmid replication origin and some selectable marker(s) are located in both strands in such a way that none of the essential information would be outside the transposon. (See cointegration, Hfr, F factor, F plasmid, Tn3 family, nested)

γδ T CELLS: express Vγ2 and Vδ2 immunoglobulin genes, recognize non-peptidic antigens, i.e. the antigen does not require processing in order to be recognized by them. The γδ T cells are very different from the αβ T cells and they can be stimulated also by non-peptide antigens such as phosphocarbohydrates, X-uridine and X-thymidine-5'-triphosphates (TUBBag3 and TUBBag4, respectively) and isopentenyl pyrophosphate. The molecules may be the product of nucleic acid salvage pathways and intermediates of the lipid metabolism. (See immunoglobulins, T cells, αβ T cells, salwage pathway)

GDB: genome database, the official depository of information of the human genome project. It can be accessed by Internet <http://gdbwww.gdb.org/>.

GDNF (glial-cell-line-derived neurotrophic factor): assists the maintenance of central dopaminergic, noradrenergic and motor neurons and peripheral and sympathetic neurons. This protein is structurally related to the transforming growth factor (TGF-β) family and GDNF is a receptor tyrosine kinase. GDNF function requires a glycosyl-phosphatidyinositol-linked protein (GDNFR-α) and RET. It has been considered as a potential drug for Parkinson's diseases, amyotrophic lateral sclerosis and Alzheimer's disease. (See dopamine, neuron, receptor tyrosine kinase, Parkinson's disease, amyotrophic lateral sclerosis, Alzheimer's disease, kinase, TGF, RET oncogene)

GDRDA: see **g**enetically **d**irected **r**epresentational **d**ifference **a**nalysis

GDT: guanosine 5'-diphosphate.

GECN (genetically effective cell number): see genetically effective cells

GEF: translation factor similar in function to eIF-2B. (See eI factors)

GEIGER COUNTER (Geiger - Müller counter): registers the rate of disintegration of radioactive isotopes. They are necessary in all isotope laboratories, also for monitoring contamination and spillage; they detect also environmental pollution of radioactivity in case of fallout. It measures β radiation with an efficiency of 30-45% and for γ radiation (shield closed) 5,000 counts per minute per milliröntgen (mR). A typical full-scale reading is 0.2 to 20 mR/hr. It is a very useful equipment for surveying because of the good sensitivity and response in seconds. Its shortcomings are energy dependence, saturation at high rates and interference by ultraviolet and microwave radiations. A special adaptation of the Geiger counter is the strip counter that detects radiation in chromatograms, membrane filters, blots, etc. (See also scintillation counters, ionization chambers, radiation hazard assessment)

GEITONOGAMY: pollination by neighboring plants that have basically the same genetic constitution.

GEL ELECTROPHORESIS: nucleic acid fragments are electrophoresed in agarose and poly-

Gel electrophoresis continued
acrylamide gels, depending on the size of the fragment. In agarose larger, in polyacrylamide smaller fragments can be separated, e.g. in 0.3% agarose 5-60 kb, in 0.7% 0.8-10 kb fragments can be analyzed, in 5 % polyacrylamide 0.5-0.8 kb, in 12% 0.04-0.1 kb fragments can be resolved (see DNA fingerprinting). Proteins can be electrophoresed in various media (paper, starch, polyacrylamide) by charge or by size in polyacrylamide-sodium dodecylsulfate (SDS) gels. (See electrophoresis)

GEL FILTRATION: porous polymers such as Sephadex, Bio-Gel (commercially available in various pore sizes) can be used to separate high molecular weight DNA from smaller molecules (unincorporated dNTPs, linkers, etc.) The large DNA is excluded while the smaller fragments are retained on the gel during chromatography. For rapid purification it can be used in syringes. (See Sephadex, linker)

GEL MOBILITY ASSAY: see gel retardation assay

GEL RETARDATION ASSAY: compared to DNA alone, DNA bound protein retards the movement of the complex in the electrophoretic field (band shifting), and this way DNA binding proteins can be isolated and analyzed. To the protein bound to DNA other protein(s) may also bind making the complex increasingly slow moving from the start site; this process is called *supershift*. A more specific test uses DNA affinity chromatography. (See DNA-binding domains, DNA binding proteins, affinity chromatography)

GELDING: castrated male horse.

GELLAN GUM: is a synthetic polysaccharide, used for solidifying plant tissue and bacterial culture media (instead of agar). (See agar, embryo culture)

GELSOLIN: an actin binding protein that regulates the cytoskeleton. (See actin, cytoskeleton)

Gem: a GTP-binding protein, induced by mitogens; it is related to RAS. (See GTP, mitogen, RAS)

GEMINIVIRUSES: contain single-stranded small DNA genomes (2.7 kb). Some can infect monocots others infect dicots. Their capsules may be geminate (doubled) or their DNA may exist in two partially identical (200 bases common) rings. They may be used for plant vector construction. (See agroinfection)

GEMMULES: an ancient term for the hereditary units.

GenBank: data bank for information on nucleic acids, Los Alamos Natl. Laboratory, Group T-10, Mail Stop K710, Los Alamos, NM 87545, USA, Tel: (505) 665-2177, e-mail: general inquiries genbank%life@lanl.gov, sequence submission and forms: gb-sub%life@lanl.gov. (See also EMBL, DDBJ, sperm bank, databases)

GENDER: the sexual type, e.g. female or male in societal or lexicographic context; in physiology or genetics the word sex is more appropriate.

GENDER PRESELECTION: separation of X and Y bearing sperms before fertilization. Success in such a procedure may prevent the transmission of X-chromosome linked genetic defects. It may become an alternative to ethically or morally objectionable procedures of negative eugenenics. [Anyhow it should be called sex preselection.] (See eugenics, abortion)

GENE: a specific functional unit of the DNA (or RNA) coding for RNA or protein. A common structural organization of protein encoding genes in eukaryotes:

_{enhancer - promoter - leader -exons - introns - termination signal - polyadenylation signal - downstream regulators}

GENE ACTION: the type and mechanism of expression of genes.

GENE ACTIVATION: turning genes on; initiate expression of genes.

GENE ACTIVATOR PROTEINS: see transcriptional activators

GENE ALIGNEMENT: arranging the nucleotide sequences of functionally or by evolution related genes in such a manner that the homologous and non-homologous stretches of nucleotides can be assessed. (See homology)

GENE AMPLIFICATION : see amplification

GENE ASSIGNMENT: locating genes to chromosomes.

GENE BANK: see GenBank, DDBJ, EMBL, sperm bank, databases

GENE BLOCK: a group of syntenic genes. Gene blocks can be preserved in their original linkage phase if they are within inversions because the single recombinants are generally inviable and the double recombinants are very rare within short regions. Paracentric inversion testers have been used to locate advantageous gene blocks for utilization in plant breeding projects. Inversion homozygotes are backcrossed with inbred stocks, and the F_1 is backcrossed with the inversion-homozygote tester. This progeny is half homozygous for the inversion and half is heterozygous for it. The two groups can be easily distinguished by genetic markers or by semisterility of the heterozygotes (if any crossing over takes place). If the heterozygotes surpass the parental forms in quantitative traits, the good performance is attributed to the inverted segment tested. Favorable gene blocks (quantitative gene loci, QTL) may also be identified by linkage to RFLP markers. (See inversions, QTL, operon, gene cluster)

GENE CENTER: geographical area where the greatest genetic diversity within a species is found, and therefore it is considered as the evolutionary cradle of that species. (See evolution)

GENE-CENTROMERE DISTANCE: see centromere mapping, tetrad analysis, alpha parameter.

GENE CHIPS: see DNA chips

GENE CLONING: propagation of a piece of DNA, in identical copies, in a bacterial, viral or yeast (or other) vector in order to increase its quantity; the same as molecular cloning.

GENE CLUSTER: juxtapositioned genes sometimes with related function. (See operon, regulon, transcripton, immunoglobulin genes)

GENE CONVERSION: a biological event that results in the change of one allele to another present in the homologous chromosome. It is a specific type of non-reciprocal recombination. As a consequence the meiotic output is changing from 2:2 to 3:1 or if the conversion takes place in the opposite direction, to 1:3. In case octads are formed by an additional mitotic division following meiosis, other types of conversion asci can be identified as shown on the illustration at left (the left octad is normal, the other five indicate gene conversions). Gene conversion within a locus proceeds in a polarized fashion, following the direction of DNA replication. Gene conversion may involve *map expansion* because within very short distances the neighboring sites may be co-converted and thus reducing the chance of their separation. This is in contrast to classical recombination when

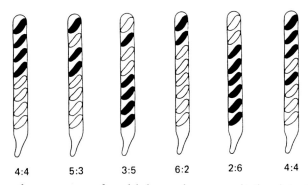

4:4   5:3   3:5   6:2   2:6   4:4

the presence of multiple markers permit the detection of higher number of recombinational events. The conversion is characterized by *fidelity*, i.e. the converter and the converted alleles are identical. The 3:1 and 1:3 spore ratios in the tetrads generally occur with equal frequency, and this was named *parity*. Mitotic gene conversion may produce somatic sector(s). Although gene conversion is not a classical recombinational event itself, it is accompanied by exchange of markers at the flanking region in half of the cases. Originally gene conversion was discovered in ascomycetes, now it is believed to occur in other eukaryotes too but the identification of gene conversion is still the easiest on the basis of tetrad data (or half tetrads in *Drosophila* with attached X-chromosomes). The frequency of gene conversion may be highly variable from gene to gene ranging from less than 0.5% to a few percent in yeast. The average size of the converted segment may extend to a 1000 bp. Not all aberrant spore output can be attributed to gene conversion, polysomy, polyploidy, nondisjunction, premeiotic mitotic recombination, suppressor mutations, etc. may all produce aberrant asci similar to the results of gene conversion. The different types of gene conversions are best identified in the asci where meiosis is followed by an additional mitosis. The detection of gene conversion is relatively

Gene conversion continued

simple in organisms where the (pollen) tetrads are preserved (*Salpiglossis, Arabidopsis*). The use of PCR technology may permit the detection of converted sequences within any eukaryotic gene. (See recombination mechanisms, recombination models, sex circle model, PCR)

GENE COPY NUMBERS: in lower organisms the majority of the genes occur in single copies per genome yet even in bacteria, ribosomal genes may be present in seven copies. In the amphibian oocytes, during the great need for protein synthesis ribosomal genes may be amplified 1,000 to 1,500 fold and form more than a thousand nucleoli. After meiosis this excessive amount of rRNA genes are discarded. In maize plants, there may be 10,000 to 20,000 copies of ribosomal genes per diploid cells. *Xenopus* may have 24,000 copies of the rRNA genes and 200 copies of each of the tRNA genes. *Drosophila* has about 500 copies of 5S RNA genes in the right arm of chromosome 2. Some particular sequences (SINE and LINE) occur in all eukaryotes in high numbers. In many higher eukaryotes repetitive sequences may exceed 80% of the genome. (See amplification, SINE, LINE, gene number)

GENE DISRUPTION: see insertional mutation, targeting genes, insertion elements, transposons

GENE DOSAGE: the number of identical and repeated genes in the genome.

GENE DUPLICATION: see duplication

GENE EDITING: parts of natural genes are replaced or completed by synthetic DNA chains or a natural repair process eliminates gaps or mismatches in the DNA (called also proofreading). (See editing, DNA polymerases, mtDNA, RNA editing)

GENE EVOLUTION: the process by what once similar genes diverged or different genes assumed similar structure and function. (See divergence, convergence)

GENE EXPRESSION: the realization of the genetic blueprint encoded in the nucleic acids. Gene expression may be modified, enhanced, silenced, timed by the regulatory mechanism of the cell, responding to internal and external factors. (See regulation of gene expression, protein synthesis, SAGE)

GENE FAMILY: the number of closely related genes; probably originated through duplications (and some divergence) during evolution. The members of these gene families are frequently (closely) linked but may also be dispersed in the genome. (See evolution of proteins, immunoglobulins, protein isomorphs)

GENE FARMING: cloning, transformation and propagation of genes in another species.

GENE FLOW: spread of genes in a population by migration of individuals. Gene flow, depending on its intensity, may rapidly alter gene frequencies in a population. Gene flow may be hindered or prevented by geographic isolation, physiological factors (differences in sexual maturity and breeding seasons), genetically (by chromosomal rearrangements causing hybrid sterility, incompatibility alleles and differences in chromosome number [polyploidy]). In neighboring populations, at the overlapping borders, repeated back crosses may occur resulting in *introgressive hybridization* and permanent inclusion of new allele into the gene pool in one or more populations. The availability of transformation techniques may overcome the natural gene flow and transfect genes among taxonomic groups that were earlier unable to exchange genetic information because of complete sexual isolation. The wave of advance of an advantageous gene was calculated by R.A. Fisher: $r = \sqrt{2gm}$ where g = the initial growth rate and m = the migration rate per time and space. (See migration, introgressive hybridization, Wahlund's principle, transformation)

GENE-FOR-GENE: relationship between host and pathogen. (See Flor's model, host - pathogen relation)

GENE FREQUENCY: frequency of a certain allele relative to all alleles at a locus within a particular population (See allelic frequencies, Hardy-Weinberg theorem, selection, drift, genetic equilibrium, forensic genetics, DNA fingerprinting, ceiling principle)

GENE FUSION: attaching to a structural gene by *in vitro* genetic engineering a selected promoter or other element(s), or a promoterless structural gene is transformed into a host cell and it

Gene fusion continued

is expressed only if it can trap *in vivo* an appropriate host promoter (enhancer). The procedure permits a study of the nature of the fused heterologous genetic element. If gene fusion occurs between coding regions of two genes, the translation product becomes a *fusion protein* that contains amino acid sequences from two structural genes. The function of the fusion protein may be modified by this process. Fusing ablation factors, such as ricin or diphteria toxin to site- or tissue-specific promoters may facilitate the study of differentiation and development because certain cell types can be eliminated during critical periods. (See transcriptional gene fusion vectors, translational gene fusion vectors, trapping promoters, fusion protein, read-through proteins, ablation, diphteria toxin)

GENE GUN: see biolistic transformation

GENE IDENTIFICATION: may be required after a particular DNA tract has been sequenced but its function is unknown. The simplest approach is checking the DNA databases for homologous sequences among genes with known function. Extensive amounts of redundant sequences may make the comparisons difficult but computer programs are available to identify repeats in human genes(<pythia@anl.gov> or <ftp:ncbi.nlm.nih.gov>) or BASTX for other organisms: (<ftp://ncbi.nlm.nih.gov;pub/jmc>). BLAST, FASTA can search databases. (See databases, Blast, Fasta, BLOCKS, and Fickett, J.W. 1996 Trends in Genetics 12:316)

GENE INTERACTION is a common misnomer; actually in most cases the products of the genes interact — with a few exceptions such as gene insertion, gene fusion, etc.— (See gene product interaction, epistasis, modified Mendelian ratios, morphogenesis in *Drosophila*)

GENE ISOLATION: the first gene isolation was reported in 1969. The *Lac* gene of *E. coli* was inserted in reverse orientation in bacteriophages λ and $\phi 80$ by a modification of specialized transduction. The DNA of these phages was extracted, denatured and the heavy chain of λ was combined with the heavy chain of $\phi 80$. Since the base sequences of the two phage strands were not complementary, only the *Lac* sequences annealed, and the phage DNA sequences remained single-stranded. S_1 nuclease degraded the single strands but the double-stranded *Lac* gene was preserved in pure form. Somewhat similarly, genes from F' plasmids could also be isolated. These ingenious methods did not have general applicability. A more general procedure was developed by isolation cDNA from mRNA. The problem was that a eukaryotic cell may contain over 40,000 mRNA molecules at a time. To be reasonably certain (say at 99% probability) that the desired molecule is really included, a very large number of molecules had to be isolated in order that the desired one be included:

$$[1 - (1/40,000)]^n = 1 - P = 1 - 0.99 = 0.01 \text{ hence}$$

$$n = \frac{\log 0.01}{\log[1 - (1/40,000)]} \cong 184,213$$

where n is the number of molecules to be screened to find at least 1 at P (=0.99) probability.

The desired mRNA may be enriched by "cascade hybridization". The mRNAs can be extracted from cells at different developmental stages. Also, substrate induction, heat shock, drug, hormone or pathogen caused induction can be used for enrichment of the mRNA. If a *DNA library* is available and the gene can be probed by colony hybridization, the fragments containing the gene or parts of it can be identified by the use of DNA probes. The simplest method of isolation of genes uses *heterologous probes*. Such a probe contains a homologous sequence of the gene to be isolated. The probe is labeled by *nick translation* using either radioactive nucleotides or biotinylation or any other non-radioactive fluorochromes or immunoprobes. The probe permits the selective isolation of the DNA fragment annealed with. If the amino acid sequence of at least part of the gene product is known, synthetic probes can also be generated.

Genes can be labeled also by transposon mutagenesis or by insertional inactivation using transformation (transfection). In case close genetic or physical mapping information is available, the gene may be isolated by the use of overlapping YAC clones and "chromosome walking" (*map-based gene isolation*). Essential regulatory elements of the gene can be identified by

Gene isolation continued

linker scanning. The identity of the gene generally requires confirmation by *in vitro* translation and testing the function of the protein so obtained. (See also biotinylation, chromosome walking, cloning, colony hybridization, cosmids, DNA library, DNA probes, fluorochromes, heterologous probes, immunoprobes, insertional mutation, linker scanning, nick translation, plasmid rescue, radioactive labeling, synthetic probes, transfection, transformation, transposon mutagenesis, YAC vectors)

GENE KNOCKOUT: see knockout, gene disruption, targeting genes, excision vector, *Cre/loxP*

GENE LIBRARY: a collection of cloned genes. (See cloning)

GENE LOCUS: see locus

GENE MANIPULATION: see genetic engineering

GENE MAPPING: see mapping genetic, mapping function, physical mapping

GENE MUTATION: molecular alteration within a gene (base substitution or frameshift, point mutation, substitution mutation). (See mutation)

GENE NOMENCLATURE ASSISTANCE: for human genes: <http://www.gene.ucl.ac.uk/nomenclature/> or <nome@galton.ucl.ac.uk>. (See also gene symbols, databases [Plants])

GENE NUMBER: the number of genes per genome of an organism can be estimated on the basis of mRNA complexity, or by total sequencing of the genome. The estimates based on mRNA complexity may be biased because not all genes are expressed simultaneously. The gene number can be best determined when the entire genome is sequenced by this method the single-stranded RNA phage, MS2 was found to have 4 genes. The gene number has been estimated also from mutations frequencies. If the overall induced mutation rate e.g. is 0.5 and the average mutation rate at selected loci is 1×10^{-5} then the number of genes is $0.5/(1 \times 10^{-5}) = 50,000$. Although this method is loaded with some errors, the estimates so obtained appear reasonable. On the basis of mutation in *Arabidopsis* the number of genes was estimated to be about 28,000, in *Drosophila* 17,000 genes were claimed on the basis of complexity, in humans 75,000-100,000 genes are expected on the basis of physical mapping, of these about 4,000 may involve hereditary illness or cancer. In *Saccharomyces* in the 5885 open reading frames 140 genes encode rRNA, 40 snRNA and 270 tRNA. About 11% of the total protein produced by the yeast cells (proteosome) has metabolic function, 3% each, is involved in DNA replication and energy production, respectively; 7% is dedicated to transcription, 6% to translation and 3% (200) are different transcription factors. About 7% is concerned with transporting molecules. About 4% are structural proteins. Many proteins are involved with membranes. The minimal essential gene number has also been estimated by comparing presumably identical genes in the smallest free-living cells *Mycoplasma genitalium* and *Haemophilus influenzae*, both completely sequenced. Such an estimation indicated the minimal number to be 250. (See gene number in quantitative traits, Cell 86:521 [1996], Science 276:1962 [1997])

GENE NUMBER IN QUANTITATIVE TRAITS: has been estimated by various complex statistical procedures (Mather & Junks 1971 Biometrical Genetics, Chapman & Hall, London, UK) but none of the estimates are entirely reliable because the number of genes with minor contribution or greatly influenced by environmental effects, genetic linkage, etc. confound the picture. Perhaps the number of polygenes controlling one quantitative trait may not be more than 5 or 6 major genes rather than hundreds, postulated by some authors. A very simple formula was provided by Sewall Wright in 1913:

$$\text{gene number (n)} = \frac{R^2}{8([s_1]^2 - [s_2]^2)}$$

where R is the difference between parental means, $[s_1]^2$ is the variance of the F_1 and $[s_2]^2$ is the variance of the F_2 generations. (See also polygenes, QTL, gene number)

GENE ORDER IN THE CHROMOSOME: can be determined by three-point or multipoint test-crosses in eukaryotes or by similar principles in prokaryotes. (See mapping genetic, bacterial recombination, chromosome walking, physical map)

GENE POOL: sum of alleles that can be shared by members of an interbreeding population. (See population genetics)

GENE PRODUCT: the transcript(s) of a gene and by extension the processed transcripts and even the translated polypetides or RNAs. (See processing, transcript, polypeptide, RNA)

GENE PRODUCT INTERACTION: is responsible for epistasis, additive, complementary and suppressor type of modifications of Mendelian segregation ratios. These are frequently called gene interactions but actually the gene products interact. (See modified Mendelian ratios, see examples under morphogenesis in *Drosophila*)

GENE REARRANGEMENT: see immunoglobulins, phase variation, sex determination in yeast, transposons, chromosomal rearrangements, gene replacement, targeting genes

GENE REGULATION: see regulation of gene activity

GENE RELIC: is usually a member of a multigene family that does not have all the elements necessary for function, it is actually a pseudogene. Their existence is explained by losses during evolution. (See pseudogene, processed pseudogene)

GENE REPLACEMENT: can be accomplished with the aid of genetic vectors that carry a different allele and the flanking sequences of a chromosomal locus. This constitution permits intimate homologous pairing in the area. If double crossing over or gene conversion takes place, the allele in the vector may replace the one in the chromosome. Because the frequency of such an event is very low, selectable markers must be used to screen out the replacement in a large population. For the selection one may use an antibiotic resistance gene with a defect in the upstream area and in the vector the same but with a defect downstream may restore antibiotic resistance and that can selectively be isolated on media containing the antibiotic. (See also targeting genes, localized mutagenesis, site-specific mutation, transformation, Cre/LoxP, FLP/FRT, knockout, homologous recombination, site-specific recombination)

GENE RESCUE: see plasmid rescue, marker rescue

GENE SCANNING: see linker scanning

GENE SIZE: can be measured in different ways. If only the translated number of codons are considered, the smallest genes appear to be the 21 bp *mccA* coding for the antibiotic heptapeptide, microcin C7 (MW 1,177 Da) of *Enterobacteria*. The other is the pentapeptide encoded within the 23S ribosomal subunit by only 15 nucleotide pairs. The largest mammalian genes, including introns and upstream and downstream regulatory sequences, may be in the range of hundreds of kbp. The human dystrophin gene with 2.34×10^6 bp includes 79 exons and it is probably the longest gene known. The "average" gene may have 400 codons and thus encode 46 to 48 kDa proteins. The human genome of about 3×10^9 bp contains an estimated 75,000 genes. *Haemophilus influenzae* bacterium has 1,749 sequenced genes whereas budding yeast in its 1.8×10^7 genome encodes 5,885 genes by 12,068 kb DNA; thus its "average" gene is about 2,050 nucleotides long. The smallest gene numbers, four, are found in some viruses. (See introns, exons, genomic DNA, *Enterobacteria*, Mbp, dystrophin, ribosomal RNA, *Haemophilus influenzae*, *Saccharomyces cervisiae*)

GENE SYMBOLS: the abbreviated representation of the function of the genes or it designates it in a unique manner using a single or more letters. Very frequently the allele that fails to carry out the normal function provides the name for the locus, e.g. the white eye locus in *Drosophila* is symbolized as *w* although the normal color of the eye is red. The symbols used vary in different organisms. Generally the symbols begin with the first letter of the name and it is usually italicized. The recessive alleles in eukaryotes are symbolized with lower case letters whereas the wild type alleles either begin with a capital letter or all the letters are capitalized. Symbols of genes in the same chromosome strand are usually separated by a space in between them. Genes in the homologous strands customarily have a slash in between them (*a/b*). Genes in non-homologous chromosomes are separated by a semi-colon and one space (*a; d*). Multiple alleles in *Drosophila* are designated by the same letter(s) representing the locus and further identified by superscripts, e.g. w^a, w^{a2}, w^{aM}, w^{a79i} or other additional distinctive signs. Recessive or dominant alleles in a series of mutant alleles in a series are frequently symbolized

Gene symbols continued

as a^R or a^D, respectively. The common dominant allele may be designated also as a^+ or A^+. Absence of a gene or lack of its function may be symbolized by a lower case letter such as a^-. Isoenzyme determining alleles may be designated as Adh^F, Adh^S, the superscript indicating fast or slow run in the electrophoretic field. A^n means null allele, a^l may be used for a lethal allele, if necessary with additional specifications. Non-allelic loci with similar phenotypes may be symbolized with the same letter(s) and subscripts: a_1 and a_2. Also, non-allelic loci with similar phenotypes (mimics) may be symbolized as *tu-1a, tu-1b, tu-2*. Different loci encoding similar proteins may be designated with the same letters but attaching to the letters different numbers or by the addition of an abbreviated form of the molecular weight of the protein (*Hsp68, Hsp83*). Transpositions are symbolized with the designation of the original symbol followed in parenthesis the new location: *[ry⁺](sd)*, indicating that the *rosy* gene was moved from chromosome 3-52 location to the *scalloped* locus in chromosome 1-51.5. The designation of transformants follows that of transpositions. Modifier genes, such as suppressors may be designated as the symbol of the modifier, followed in parenthesis the gene modified: $su(lz^{34})$. Some symbols may carry also the name of the discoverer or the location of discovery of the mutation or the mutagenic agent used. Reversion may be indicated by *rv* in superscript. Chromosomal aberrations are designated by capital letters and additional specifications. Translocations (reciprocal interchanges between/among non-homologous chromosomes) are represented as *T(1;Y;3)* indicating that chromosomes 1 (X-chromosome), Y and 3 are involved. Each chromosome may be further specified using a capital letter superscript indicating the approximate position of the break-point as P (proximal to the centromere), D (distal), or M (median). An X-chromosomal ring (of *Drosophila*) may be symbolized as *R(1)1*. Paracentric inversions are represented as *In(2L)* or *In(2R)*, depending whether the left or right arm of chromosome 2 is involved. *In(2L,R)* indicates pericentric inversion of chromosome 2. To this symbol, genes closest to the break-points may be attached. For transposition (non-reciprocal transfer of chromosomal segments) the symbol is *Tn* and in parenthesis first the donor, followed by the recipient chromosome, e.g., *Tn(2;3)*. Again, the gene(s) involved may be included in the symbols. Deficiencies are symbolized by *Df* followed by the indication of the chromosome (arm) and locus involved: *Def(2R)vg*. Duplications are symbolized with *Dp* such as *Dp(3;1)* indicating that duplicated segment of chromosome 3 is located in the X-chromosome. When the duplicated segment has a centromere and it is a free element, it is symbolized with a letter *f*, e.g., *Dp(1;f)*. In case there are multiple repeats: *Dp(1;1;1)*. When a combination of multiple chromosomal rearrangements occur, they are indicated one after the other with a "+" sign between them. The location of break-points may be designated by the euchromatic (1 to 102) or heterochromatic (h1 to h61) segment numbers. The older symbols in plants followed generally the customs in *Drosophila*. Recently, largely for convenience of typing or printing, the subscripts are substituted with a number written together with the gene symbol and the allelic number is attached hyphenated: *a2-5* the second *a* locus and allele 5 (rather than superscript 5). Mouse geneticists identify loci with three or four italicized letters, the first is capitalized. Human geneticists also use three or four (commonly not italicized) all-capital letter symbols with additional numbers. In Yeast and *Arabidopsis* the new gene symbols use three italicized capital letters for the wild type and three italicized lower case letters for the recessive alleles. In the majority of fungi the wild type alleles are designated with a superscript "+". Allelic designation frequently follows the locus designation in parenthesis: *ilv(STL6)* or *pyr-3 (KS43)*. Suppressor mutation symbols may include also the gene they modify: *su(met-7)-1*. The symbol *ssp* means supersuppressors. Mitochondrial mutations are designated as *mi*, and additional numbers. RFLP fragments are identified with an italicized three-letter symbol of the laboratory and a serial number. Transposable elements are symbolized similarly to the genes. In human genetics only capital letter symbols (no more than 4-5 letters) are used without sub- or superscripts. Hyphens or punctuations in the symbols

Gene symbols continued

are exceptional. Different loci by the same symbol are numbered e.g. BPAG1, BPAG2. Alleles may be indicated by an asterisk after the symbol and followed by other designation e.g. ACY1*2. A slash between two symbols stand for the diploid genotype, hetero- or homozygous. Lack of synteny is indicated by semicolons(s) between the symbols. If linkage is unknown comma is used. Gene order is usually started from the short arm down.

Bacterial geneticists designate the loci with italicized three lower case letters followed by a capital letter: *lacI, lacZo, lacZ* indicating the lactose utilization operon regulatory (inhibitor), operator and the β-galactosidase genes, respectively. The letters *p, o, a,* stand for promoter, operator and attenuator, respectively.

Protein products of the genes are generally symbolized with the abbreviations of the genes but they are all in capitals and not italicized.

Gene symbols have been periodically revised in some organisms and this may make reading the literature difficult. Creating new symbols is a cheap attempt to gain citations. If new symbolism is warranted that should not be used retroactively to published and used symbols. (See *Drosophila,* databases [plants - Mendel], gene nomenclature assistance)

GENE SYNTHESIS: generating nucleotide sequences by the methods of organic chemistry. These sequences—and their variations—can then be tested for function after transformation into suitable host cells. The first entirely synthetic genes coded for tRNAs. (See genes synthetic, synthetic genes)

GENE TAGGING: place into a gene an insertion or transposable element or any other DNA sequence with the aid of genetic transformation (transfection). The inserted sequence is known and can be probed and the gene can be identified by molecular hybridization and/or genetically by inactivation or altering the function of the gene (insertional mutation). Some insertions going into intergenic or untranslated (intron) regions may not affect the expression of the gene involved. (See also labeling, probe, insertional mutation, targeting genes)

GENE TARGETING: see gene replacement, targeting genes, targeted gene transfers, knockout, localized mutagenesis, excision vector, *Cre/loxP, FLP/FRT*

GENE THERAPY: insertion of a functional normal gene into an organism for the purpose of correcting or compensating for its genetic defect. It can be carried out either in somatic cells or in the germline (gametes, zygotes). The methods potentially available are microinjection of (foreign) DNA or transformation with retroviral or adenovirus vectors, liposomes (see transformation of animals, human gene transfer) and gene replacement by homologous recombination. Another possibility is "knockout" when the function of a deleterious gene is eliminated by insertional inactivation or deletion. The technology is available for transformation or *in vitro* mutation that can be followed by injecting embryonic stem cells into blastocytes that are introduced into the uterus of a females to develop genetically modified embryos and eventually viable offspring. Recently laboratory progress were made by introducing into mice with induced tyrosinemia, hepatic cells that could proliferate in the defective liver. The transformation technology needs refinements before it can be widely applied to humans. The techniques of embryo implantation are widely used to overcome female inability of conceive without surgical assistance. Before implantation, these fertilized embryos may be then tested for expression of transferred remedial genes. These procedures may become potentially useful for preventing the expression of genetic diseases under the control of single genes such as the Lesch-Nyhan syndrome, Tay-Sachs disease, cystic fibrosis, muscular dystrophys, Gaucher's disease, β-thalassemia, ADA, melanoma, neuroblastoma, multiple myeloma, lymphoma, breast cancer, colorectal cancer and several others. ADA patients treated by gene therapy appeared safe and had no diminishing effects even 4 years after the use of retroviral transformation. Liposomal vectors carrying the human leukocyte antigen (HLA)-B7 and the β_2 microglobulin cDNA to tumors can express these genes. Using adenovirus vector with the human cystic fibrosis transmembrane conductance regulator (CFTR) resulted in expression of the gene for 9 days in the nasal or bronchial membranes. With a retroviral vector the low density lipoprotein (LDL) receptor, important for familial hypercholesterolemia, has been successfully transform-

Gene therapy continued

ed and functioned. Promising initiative were made by treatment of patients with retroviral vectors carrying interleukin-4 (IL-4) to fibroblasts resulting in infiltration of the tumor with $CD3^+$ and $CD4^+$ T cells and cell adhesion molecules. The treatment resulted in some trial in the increase of $CD8^+$ tumor-specific cytotoxic T lymphocytes (CTL) and eosinophils as well as $CD16^+$ killer cells. The somatic cell genetic therapy may target cancer cells with interleukins, tumor necrosis factor, granulocyte macrophage colony stimulating factor to reinforce the immune system or use monoclonal antibodies against cancer cells equipped with toxins or sources of radiation (see lymphocytes, magic bullet). Some of the genes that are suitable for germline modification may be targeted to specific organs for alleviating or to overcome the symptoms of the disease. In some cases, e.g. neurological disorders, *ex vivo* methods have been sought of for the restoration of the normal function (in Alzheimer disease) of nerve growth factor (NGF) or transplanting dopamin producing tissue (in Parkinson disease). Bone marrow transplantation may alleviate or reverse the course of lysosomal storage diseases.

Gene therapy is opposed by some people on biological and/or ethical ground. The arguments against stem from the fears of unforeseable damage to the human gene pool and the possibility of using these procedures for "genetic enhancement". Genetic enhancement would have similar goals as eugenics and eventually may be exploited to create "supersoldiers" or other antisocial individuals with "uniform" genetic makeup. These fears are frequently fanned by political agenda or by unfounded speculations. The arguments in favor of gene therapy is that it provides means to prevent the perpetuation of "disease genes" by specifically targeting the single defects. It may result not only in elimination of suffering but may also reduce health maintenance cost on the long run. It is true that all the possible consequences of gene therapy have not been seen in an evolutionary history. The same criticism may also apply also to several drugs that are part of the current medical practice. Many of the medicines have physiological and genetic side effects (e.g. diagnostic and therapeutic X-rays, several antibiotics, anticancer drugs, etc.) yet the benefits are supposed to outweigh their risks. In human gene transfer there are some potential risks of new constructs to develop by recombination with the viral vector. In some cases the decision is very difficult, e.g. human dwarfism can be cured by the application of growth hormones or by functional growth hormone genes. Dwarfism is not an acute life-threatening anomaly yet it interferes in many ways with the normal fulfillment of life. The question arises then how far social philosophy should be permitted to affect the life of an individual. Animal models can be successfully applied for the testing of the physiological and biochemical consequences of gene therapy but may not detect all the consequences for human behavior and mental abilities. Thus gene therapy still have to face not just biological, technical problems but ethical ones as well. The freedom of scientific inquiry and the innate human striving for knowledge should not be prevented, however, for any unknown reasons. Although the same caution may be necessary as it was applied with the techniques of "recombinant DNA". Some of the problems to be solved include the development of more effective vectors and extrapolating successfully from animal models to humans. Somatic gene therapy involves apparently small risks. The possible harmful consequences of germline alterations is much more difficult to assess. (See diseases and terms under specific entries, transformation genetic, human gene transfer, immunostimulatory DNA, *ex vivo*, viral vectors, liposome, T cells, immune system, adoptive cell therapy, sickle cell anemia, antisense technologies, mosaic, epitope, cancer gene therapy, ribozyme)

GENE TITRATION: determining the quantitive expression of gene as a function of dosage. (See dosage effect, titration)

GENE-TOX: genetic toxicology; study of factors (physical and chemical agents) that are reponsible for mutation and cancer or both.

GENE TRANSFER: see transformation, human gene transfer

GENE TRANSFER BY MICROINJECTION: was the principal means of transformation of animals in the 1980s. Today gene targeting and other procedure are prefered. (See diagram on page 414, transformation genetic [animals], gene replacement, targeting genes).

Gene transfer by microinjection continued

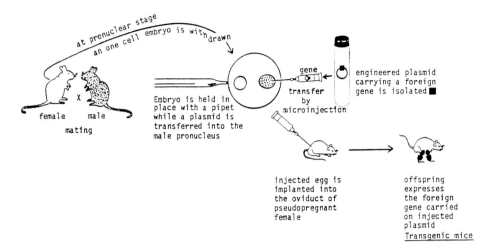
Transgenic mice

GENE TRAP VECTORS: are equipped with a reporter gene that can insert at a splice acceptor site. The resulting gene fusion may facilitate the transcription of the reporter gene. It may be used with (mouse) embryonic stems cells to detect genes expressed during early development. (See ES, gene fusion, reporter gene)

GENEALOGY: list and description of successive ancestors in a family. (See pedigree)

GENERAL ACID-BASE CATALYSIS: proton transfer from and to a molecule, water excepted.

GENERAL RECOMBINATION: recombination between homologous sequences. (See also illegitimate recombination, recombination genetic, gene conversion)

GENERAL TRANSCRIPTION FACTORS: see transcription factors

GENERALIZED TRANSDUCTION: can be mediated by either temperate or virulent bacteriophages when the empty phage capsid scoops up at random DNA fragments of a bacterium and transduces it into another cell. For the completion of the process the transferred DNA must be integrated into the genetic material of recipient bacterium. (See marker effect, *pac* sites, also specialized transduction, abortive transduction)

GENES, SPLIT: contain introns; the vast majority of eukaryotic genes are therefore split into segments of exons. (See introns, exon)

GENES, SYNTHETIC: were produced since the 1970s. In 1976 Khorana synthesized the tyrosine suppressor tRNA genes of *E. coli* by classical methods or organic chemistry. *In vitro* synthesis of 26 oligonucleotide tracts were ligated into a 207 bp DNA containing the 86 nucleotide sequence of the gene plus leader, promoter and the terminator sequence. This gene turned out to be biologically active and suppressed an amber mutation when transformed into bacterial cells. (See tRNA, suppressor tRNA, synthetic genes [for a diagram])

GeneScape: a computer program that seeks out miniset clones, DNA sequences, gene alignments to restriction maps and allows to zoom from a display to the entire map of *E. coli*.

GENE-SCRIBE®: a commercial transcriptional kit with a T7 phage RNA polymerase. It can be used to transcribe cloned genes without subcloning. (See EMBL3, λDASH, λFIX)

GENETHICS: see ethics

GENETIC ASSIMILATION: an adaptive mechanism in a population to fix genes as permanent parts of the genome by selection. (See adaptation, fixation, fixation index)

GENETIC BACKGROUND: all residual genes, besides the one(s) of special interest.

GENETIC BALANCE: see balance of alleles, balanced lethals, balanced polymorphism

GENETIC BAR-CODE: see DNA chip, bar code

GENETIC BLOCK: mutation in a gene may prevent or slow down the flow of a metabolic pathway. (See null allele, leaky mutant)
GENETIC BURDEN: same as genetic load.
GENETIC CASCADE: genes of a developmental pathway are activated in successive waves; the expression of "earlier" genes activate the next ones.
GENETIC CIRCULARITY: is a consequence of circular DNA genetic material, i.e. the genetic map has no ends although one point is generally designated as origin. (See DNA circular)
GENETIC CODE: consists of 64 contiguous nucleotide triplets from which 61 specify 20 amino acids and three serves as signals for termination of translation on the ribosomes sSee table below). The number of triplet codons for a particular amino acid varies from one to six. (See code genetic, amino acid symbols in proteins sequences, genetic code second)

THE GENETIC CODE IN RNA TRIPLETS

5' NUCLEOTIDE	SECOND NUCLEOTIDE				3' NUCLEOTIDE
	U	C	A	G	
U	Phe	Ser	Tyr	Cys	U
	Phe	Ser	Tyr	Cys	C
	Leu	Ser	ochre	opal	A
	Leu	Ser	amber	Trp	G
C	Leu	Pro	His	Arg	U
	Leu	Pro	His	Arg	C
	Leu	Pro	Gln	Arg	A
	Leu	Pro	Gln	Arg	G
A	Ile	Thr	Asn	Ser	U
	Ile	Thr	Asn	Ser	C
	Ile	Thr	Lys	Arg	A
	Met	Thr	Lys	Arg	G
G	Val	Ala	Asp	Gly	U
	Val	Ala	Asp	Gly	C
	Val	Ala	Glu	Gly	A
	Val	Ala	Glu	Gly	G

Ala = alanine (4)
Arg = arginine (6)
Asp = aspartic acid (2)
Asn = asparagine (2)
Cys = cysteine (2)
Glu = glutamic acid (2)
Gln = glutamine (2)
Gly = glycine (4)
His = histidine (2)
Ile = isoleucine (3)
Leu = leucine (6)
Lys = lysine (2)
Met = methionine (1)
Phe = phenylalanine (2)
Pro = proline (4)
Ser = serine (6)
Thr = threonine (4)
Trp = tryptophan (1)
Tyr = tyrosine (2)
Val = valine (4)

The universal genetic code for amino acids in RNA codons. The table shows in outline the three nonsense codons (chain-termination codon) and 61 sense codons coding for aminoacids. The numbers after the amino acids (right-most column) indicates the number of synonymous codons for each amino acids.

    methionine and tryptophan each have only.......................... 1
    asparagine, aspartic acid, cysteine, glutamic acid,
    glutamine, histidine, lysine, phenylalanine, tyrosine, each has 2
    isoleucine has.. 3
    alanine, proline, threonine, and valine have........................ 4
    arginine, leucine, serine all have....................................... 6 codons.

The codon usage is not random, it varies among organisms and genes.

Exceptional codon meanings:

Mycoplasma capricolum	*Tetrahymena thermophila*	*Euplotes octacarinatus*	Mitochondria			
			Mammal	*Drosophila*	Yeast	*Neurospora*
UGA: Trp	UAA: Gln	UGA: Cys	AUA: Met	AUA: Met	AUA: Met	CUN: Thr
	CAG: Gln	UAA: stop	AUU: Met	AUU: Met	CUA: Thr	
	UAG: Gln	UAG: absent	AUG: Met	AUG: Met	CUC: Met	
			AUC: Met	....	CUU: Met	
			UGA: Trp	UGA: Trp	CUG: Met	
			AGA: stop	AGA: Ser		
			AGG: stop	The UGA "universal" stop codon means		

Trp in the mitochondria of vertebrates, insects, molluscs, echinoderms, nematodes, plathyhelminthes, fungi and ciliates. Selenocysteine is also coded by UGA in *E. coli* and mammals.

GENETIC CODE, SECOND: see aminoacyl-tRNA synthetase

GENETIC COLONIZATION: infection of plants by agrobacteria results in the expression of bacterial genes in the plant cells and the gene products, such as opines, are utilized only by the bacteria. (See *Agrobacterium*, transformation, opines)

GENETIC COMPLEMENTATION: see complementary alleles, complementation groups, allelic complementation, complementation maps

GENETIC CONSERVATION: preservation of species and subspecific genetic variations in protected areas, national parks, game reserves, botanical gardens, zoos, seed depositories, microbial culture collections, sperm banks, etc.

GENETIC CORRELATION: linked genes are expected to segregate together depending on the frequency of recombination. Members of the same family display correlation, and even assortative mating shows correlations although the latter may not be genetic. From the correlation between certain phenotypes the chromosomal location of genes can be predicted. The term genetic correlation in animal breeding is defined as a measure of the ratios of additive variances: Cov $(X,Y)/\sqrt{Var(X) x Var(Y)}$. (See correlation)

GENETIC COUNSELING: provides information, medical diagnosis on hereditary bases, recurrence risk and possible management of genetic anomalies for the benefit of the family. It has no eugenic purpose and it does not make recommendations for decision; it merely informs the concerned individual(s). (See counseling genetic, risk, recurrence risk, utility index for genetic counseling)

GENETIC DEATH: occurs if an organism leaves no offspring or a gene is not transmitted to subsequent generations. (See mutation beneficial, mutation neutral, fitness, selection coefficient, selection conditions)

GENETIC DETERMINATION: same as heritability.

GENETIC DISCRIMINATION: prejudicial treatment on the basis of phenotypic or genotypic constitution by employers, insurance companies or any other person or institution.

GENETIC DISEASES: an estimated 4,000 human genes are directly or indirectly involved in the determination of human malformations and physical and mental disabilities. According to some estimates 15 to 20 % of newborns are afflicted by some hereditary problems and a large fraction of the abortions are caused by chromosomal anomalies and/or recessive or dominant lethal genes. Approximately 25% of the hospitalizations are due to maladies with a genetic component. Very often genes are not the absolute cause of the disease because many of them can be prevented by proper life-style and preventive medication if disposition exists. The occurrence of genetic disease sometimes can be avoided or the risks reduced by proper education, premarital genetic counseling. (See also eugenics, gene therapy, selection coefficient, inbreeding, consanguinity, risk, recurrence risk, DALY, prevalence)

GENETIC DISSECTION: analyzes the mechanism(s) of genetic determination and control of biological traits, morphogenesis and/or function. (See one gene — one enzyme theorem, morphogenesis in *Drosophila*, metabolic pathways)

GENETIC DISTANCE: genetic distance (d) can be measured by different procedures. One simple solution is based on a geometric model is $d^2 = 1 - \sqrt{p_1 p_2} - \sqrt{q_1 q_2}$ where *p* and *q* represent the frequencies of the two alleles of a locus in populations 1 and 2, respectively. For actual determination of the distance between two populations more than one allelic pair must be considered. Genetic distance, F_{ST} is calculated also as $V_p/\bar{p}(1-p)$ where V_p is the variance between gene frequencies in a set of *n* populations and $\bar{p}$ = their average gene frequencies. (See evolutionary distance, evolutionary tree)

GENETIC DIVERGENCE: see divergence

GENETIC DIVERSITY: the variations in the gene pool of a population or the genetic variations in the populations. (See gene pool, genetic variation, genetic conservation)

GENETIC DRIFT: is a change in gene frequencies by sampling error(s) of the gametic array so

Genetic drift continued

the genes are not maintained on the basis of their fitness or selective advantage they may convey but the selection is the outcome of chance. In case of two alleles selection by chance alone is determined by the frequency of the alleles, the binomial distribution and population size. Thus, if the frequency of allele A is p and that of a is q the frequency of alleles by chance alone will follow the binomial distribution of $(p + q)^n$ where n = the number of individuals that leaves offspring surviving to the reproductive age; e.g., in case the allelic frequencies are equal and four individuals survive, the probability that all 4 will be homozygous recessives is 0.0625. (See effective population size, founder principle, binomial distribution, Pascal triangle, Eve foremother)

GENETIC ENDPOINT: classification of the types of genetic lesions such as mutation, chromosomal aberration, unscheduled DNA synthesis, etc. that are detected in mutagen testing. (See bioassays in genetic toxicology)

GENETIC ENGINEERING: construction of a special chromosomes by cytogenetic manipulations, somatic cell fusions, or introduction of organelles into cells by mechanical means (genetic microsurgery). Isolation and propagation of DNA molecules in suitable hosts, molecular modification of genes and regulatory elements for special purposes, and transfer genes among diverse organism by by-passing the constraints of sexual reproduction and manipulate them for medical, industrial and agricultural use. (See transformation, cloning vectors, chromosome substitution, alien addition, alien substitution, alien transfer lines, metabolite engineering, protein engineering, pathogen identification)

GENETIC ENHANCEMENT: see gene therapy, plant breeding, animal breeding, eugenenics

GENETIC EQUILIBRIUM: exists when gene frequencies are stable for generations. (See also mutations and genetic equilibrium). In a panmictic diploid equilibrium population the frequency of heterozygotes is twice the square root of the product of the frequencies of the two homozygous classes: $2 = H/\sqrt{D \times R}$ where H, D, R and stand for heterozygotes, homozygous dominants, and homozygous recessives, respectively. This is derived from the middle term of the Hardy - Weinberg formula, $2pq = H = 2\sqrt{p^2q^2} = 2\sqrt{D \times R}$ and hence $2 = H/\sqrt{D \times R}$.

The four populations represented in the body of the table below all have identical gene frequencies, p = 0.8, q = 0.2 yet the genotypic proportions are quite different. According to the definition in the text only population 4 is in equilibrium.

Populations	Genotypic Frequencies		
	AA	Aa	aa
1	0.80	0.00	0.20
2	0.70	0.20	0.10
3	0.60	0.45	0.00
4	0.64	0.32	0.04

This principle can graphically be represented in the figure at right. In an equilibrium population the frequency of heterozygotes is represented by a parabola as the proportion of the alleles

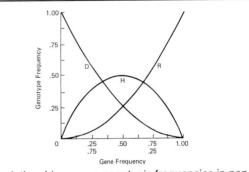

The relationship among genotypic frequencies in panmictic populations when D, H and R have equal fitness. The heterozygotes reach a maximum of 50% when D and R, each are at 25%. When p = 0.33 and q = 0.66 the heterozygotes will be more common than the homozygotes. When H = 0, either D or R equals 1

vary from 0 to 1 to 0 as long the three genotypes have equal fitness. With respect to an individual locus, equilibrium is attained within one generation of random mating. As long as random mating prevails and there is no selection; gene and genotypic frequencies do not change and the Hardy-Weinberg principle prevails. Multiple loci require more generations to attain equilibrium. Also, equilibrium depends on the intensity of linkage among the loci.

Genetic equilibrium continued

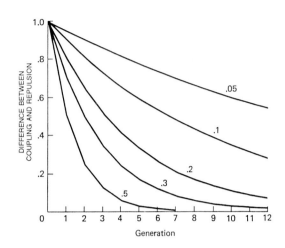

Progress toward genetic equilibrium in case two genes are linked in repulsion at zero generation time. At equilibrium the repulsion and coupling phases are equal. Four of the curves (0.05 to 0.3) represent the courses to equilibria at various intensities of recombination; 0.5 indicates independent segregation. (Modified after Falconer, D.S. Introduction to Quantitative Genetics. Longman, London, UK).

Progress toward equilibrium is delayed if the genes are sex-linked. If in the original mating the homogametic sex is homozygous for a recessive allele (X^aX^a) and the heterogametic sex carry the other allele (X^AY) the allelic frequencies in the two sexes will follow an oscillatory path during the generations because of the zig-zag pattern of inheritance of the X chromosome. In equilibrium the allelic proportions in the two sexes will be represented by the proportions of the X chromosomes. Somewhat similar situation exists in hermaphrodites carrying self-sterility alleles. The mating of plants self-incompatibility alleles (S) produce the following offspring:

FEMALE	MALE	OFFSPRING
S_1S_2	either S_1S_3 or S_2S_3	0.5 S_1S_3 and 0.5 S_2S_3
S_1S_3	either S_1S_2 or S_2S_3	0.5 S_1S_2 and 0.5 S_2S_3
S_2S_3	either S_1S_2 or S_1S_3	0.5 S_1S_2 and 0.5 S_1S_3

Half of the progeny is the same as the male whereas the other half has a different constitution. The genetic constitution of the females do not reappear in the immediate progeny because of the self-sterility. Therefore, if the frequencies of the alleles are not identical, the genotypes most common among the parents will be the least frequent among their offspring although will reappear in advanced generations. Equilibrium is reached however if the frequencies of the alleles are equal.

In polyploids the progress toward equilibrium is quite complicated and can be determined according to C.C. Li (First Course in Population Genetics, Boxwood Press, CA.). If the gametic output of an autotetraploid population is $G_0 \equiv x(AA) + 2y(Aa) + z(aa) = 1$, the frequency (p) of A = x + y and the frequency (q) of a = y + z. The gametic proportions in the course of generations (n) is expressed as $d = (y^2 - xz) = y^2 - (p - y)(q - y) = y - pq$, and *d* is the index of divergence from the equilibrium condition. This index is reduced by 2/3 during each generation of random mating. The gametic proportions and gene frequencies can thus be obtained as:

$$y_n = pq + d_n = pq + \left(\frac{1}{3}\right)^n d \to pq, \quad x_n = p - y_n = p^2 - \left(\frac{1}{3}\right)^n d \to p^2, \quad z_n = q - y_n = q^2 - \left(\frac{1}{3}\right)^n d \to q^2$$

(See Hardy-Weinberg, linkage disequilibrium, autopolyploidy, sex-linkage, self-sterility, Wahlund's principle)

GENETIC ESSENTIALISM: is a criticism of modern genetics for "equating" human (and other) beings with a molecular entity (DNA), including social, historical and moral complexities and responsibilities. (See vitalism)

GENETIC FINE STRUCTURE: analysis involves recombination within the boundaries of individual genes.

GENETIC FINGERPRINTING: see DNA fingerprinting

GENETIC HAZARDS: see risk, genetic risk, λ_s, recurrence risk, empirical risk, genotypic risk ratio, radiation hazard assessment, radiation effects, environmental mutagens

GENETIC HOMEOSTASIS: the property of a population to maintain its genetic composition and resist changes in gene frequencies by phenotypic regulation under variable environmental conditions. (See also homeostasis, canalization)

GENETIC HOMOLOGY: the degree of similarity in the base sequences of DNA and RNA or the amino acid sequences in the proteins. (See DNA sequencing, RNA sequencing, protein structure, amino acid sequencing, homology, databases)

GENETIC INFORMATION: the instructions in the nucleic acids for the cellular machinery.

GENETIC ISOLATION: lack of ability to interbreed (incompatibility) and/or hybrid inviability or sterility between/among different taxonomic groups. (See isolation genetic, speciation)

GENETIC LOAD: sum of deleterious genes in the genome. Recessive alleles cannot generally be detected in the heterozygotes. These heterozygotes may continuously contribute homozygotes to the population and if the recessives are deleterious, they may adversely affect the fitness of the population and thus constituting a genetic load. The amount of hidden genetic variation is revealed by the coefficient of inbreeding. In F_1 100% of the population is heterozygous. In successive generations of selfing the heterozygosity is decreasing by $(0.5)^n$ where n = the number of selfed generations (e.g. by F_5 there are 4 selfings). Thus, the sum of the heterozygotes = $1 - (0.5)^n$. The coefficient of inbreeding F, in the offspring of first cousins is 0.0625 whereas among unrelated individuals it is presumed to be 0. Thus, if the mortality range in a certain age group is say 11% in the general population, and 16% among the children of first cousins, the difference is 5%. Therefore, 16 x 0.05 = 0.80, (80%) would be the average mortality if the coefficient of inbreeding would reach 100% . Since recessive zygotic lethality requires homozygosity at the same locus (present in both parental gametes), and according to the Hardy - Weinberg theorem the frequency of the double recessive genotypes is expected to be q^2, and the frequency of at least one lethal equivalent gene is then $\sqrt{0.80} \cong 0.89$, indicating that almost 90% of the gametes carried a lethal gene or a combination of genes that cause lethality at homozygosity. On this basis, the genetic load of this population is close to 1 lethal equivalent factor per gamete. Some other investigations estimated the genetic load to be twice as high in some populations. The amount of the genetic load may vary in different populations and is affected by exposure to environmental mutagens, drugs, exposure to chemicals in the food chain (natural toxins or insecticides, pesticides) or in industrial pollutants or occupational hazards, presence of mutator genes (transposable elements), natural or other types of radiations (X-rays, UV, etc.). Completely dominant lethal mutations do not contribute to the genetic load because they may eliminate the carriers of the genetic defect and thus no load is passed on to successive generations. Of course, some of the mutations may show intermediate types or conditional expression and may or may not contribute to the load. Some deleterious genes are closely linked to advantageous genes and thus transmitted beyond their merit by this "hitchhiking" effect. In such a situation a *recombinational load* may exist. *Environmental load* is generated in a highly variable environment where under certain condition genes are selected that normally convey inferior fitness. *Incompatibility load* arises in cases of deleterious maternal - fetal interactions, such as that may arise if the mother is Rh negative for this blood antigen and the fetus is positive or if the mother expresses phenylketonuria but the fetus is heterozygous (maternal epistasis). The frequency of deleterious recessive alleles ($\hat{q}$) in the population is proportional to the mutation rate (μ) and the coefficient of selection (s): $\hat{q}^2 = \mu/s$. By rewriting the formula, the mutational load becomes $\mu = s\hat{q}^2$ and $\hat{q} = \sqrt{\mu/s}$. In the absence of dominance in a random-mating population the mutational load, $L = 2\mu/(1 + \mu)$. The mutational load of completely dominant alleles is 2μ. (See also allelic frequencies, Hardy - Weinberg theorem, mutation neutral, mutation beneficial, selection coefficient, fitness, coefficient of inbreeding, consanguinity, incest, genetic risk, lethal equivalent, incompatibility, epistasis)

GENETIC MANIPULATION: application of genetic, cytological or molecular techniques for constructing altered organisms. (See genetic engineering, chromosome engineering)

GENETIC MAP: the relative position of genes or other chromosomal markers represented in a linear manner on the basis of recombination frequencies. (See mapping genetic, physical map, mapping function, deletion maps, linkage group)

GENETIC MARKERS: help to identify nuclear chromosomes, cytoplasmic organelles and isolated cells on the basis of their inherited behavior and facilitate the identification of the genetic mechanisms involved in special phenomena, such as recombination, gene conversion, mutation, chromosomal rearrangements, genetic transformation, cell fusion, selection, etc.

GENETIC MATERIAL: either DNA (in eukaryotes and the majority of prokaryotes) or RNA (in some viruses). These nucleic acids can occur in either double of single-stranded form. (See RNA, mtDNA, ctDNA, prion, Watson and Crick model)

GENETIC MILIEU: see genetic background

GENETIC MOSAIC: is an individual with cell patches of different genetic constitution. It may come about by somatic mutation, movement of insertion- or transposable elements, somatic recombination, nondisjunction, deletion, etc. (See individual entries, chimera)

GENETIC NETWORK: the connections between DNA, RNA, protein, cis- and transacting regulators, operons, epistasis, signals and the signal transducing systems, feedback, involving a large number of genetic and environmental inputs. (See separate entries mentioned)

GENETIC NOMENCLATURE: see gene symbols

GENETIC POLYMORPHISM: gene loci in a population are represented by more than one allelic form. (See allele)

GENETIC PRIVACY: the right of an individual to keep his/her genetic record closed to the public. There are two aspects of this right: protection from discrimination by employers, insurance companies, etc., and second, it may hinder research on genetic disorders and development of new drugs. In the USA law recognizes now "protected medical information". (See genetic testing, wrongful life, ethics)

GENETIC RECOMBINATION: see recombination, recombination frequency, crossing over, bacterial recombination frequency, intragenic recombination, mapping genetic, illegitimate recombination, unequal crossing over, site-specific recombination, sister chromatid exchange, recombination variations of, recombination molecular mechanism prokaryotes, mtDNA, mitochondrial genetics, chloroplast genetics, recombination mechanisms eukaryotes, recombination models

GENETIC REPAIR: see DNA repair.

GENETIC RISK: is the chance that an offspring will be affected by a hereditary defect. The risk can be infered from the heritability of a particular gene or gene complex in a population. In case of simple Mendelian inheritance such as for example cystic fibrosis. In some Caucasian populations in genetic equilibrium, the frequency of this anomaly is $\approx 1/2{,}000 = 0.0005$. Thus if the frequency of the recessive allele is $\sqrt{0.0005} \approx 0.022 = q$. At genetic equilibrium the frequency of carriers (heterozygotes) is $H = 2pq = 2 \times (1 - q) \times q \cong 0.043 \cong 1/23$. If a person is heterozygous for such a deleterious gene ($q = 0.5$) marries a spouse by random choice ($q = 0.022$), the chance that they will have an afflicted offspring is $0.5 \times 0.022 = 0.011$, i. e. approximately 1/91. If the same heterozygous person marries a first cousin who may have a 0.25 chance carrying the same allele, the probability that they will have an afflicted child may be as high as $0.5 \times 0.25 = 0.125$, i.e. 1/8. If, however, an average Caucasian will have an offspring with an average Japanese spouse ($q = 0.004$) the probability that this child would be afflicted by cystic fibrosis is only $0.022 \times 0.004 = 0.000088$ or 1/11363. The genetic risks will slowly rise with the application of medical care that compensates for the hereditary defects by medicine, e.g. administration of insulin to diabetics or by the use of gene therapy without replacing the defective gene(s). The remedial treatments will not much affect the incidence of rare diseases on the shorter term. If the incidence of a dominant human anomaly is 1×10^{-5} at the present time, it may take 3,000 years (100 generations) to increase its prevalence to 1×10^{-3} whereas the incidence of recessive anomalies will rise much slower because the alleles are already sheltered from selection in the heterozygotes. The genetic risk can now be estimated

Genetic risk continued

with good precision if molecular information is available on the nucleotid sequences of a gene. E.g. in familial hypercholesterolemia if in the gene encoding cardiac β-myosin a substitution of Glu for Gly at position 256 involves only 0.56 chance for the penetrance of the disease whereas a Gln→ Arg change at position 403 predicts a 100% penetrance and thus sudden death. (See also genetic load, genetic counseling, empirical risk, risk, genotypic risk ratio, Hardy - Weinberg theorem, allelic frequencies, amniocentesis, clinical tests for heterozygosity, mutation rate, genetic screening, prenatal tests)

GENETIC SCREENING: is applied as (i) *prenatal tests* during pregnancy, to (ii) *newborns* whether afflicted by autosomal recessive disorders that require immediate medical attention to prevent severe later consequences. Most frequently the tests include biotidinase deficiency, congenital hypothyroidism, galactosemia, hereditary tyrosinemia, homocystinuria, maple syrup urine disease, phenylketonuria, and sickle-cell anemia. These test are mandated by law in the USA. Congenital adrenal hyperplasia, cystic fibrosis and others may also be involved. The tests may be performed on blood withdrawn from the neonates by specialized laboratories using standard and reliable procedures such as ELISA, enzyme assays, immuno essays, Guthrie test, etc. (iii) *Carrier testing* detects heterozygotes for "recessive" disorders in order to facilitate informed decisions by prospective parents regarding risks, especially in populations where the frequency of the deleterious genes is expected to be high (Tay-Sachs disease among Ashkenazi Jews [0.02], thalassemia in people of Mediterranean ethnicity may occur at frequencies exceeding 0.1 in high malaria areas, cystic fibrosis with variable (generally about 0.02) frequency but much higher in ethnic populations of high degree of consanguinity. About 70% of those afflicted by cystic fibrosis had a CTT (Phe) deletion of codon 508 in exon 10 (ΔF508). This assay is not yet used widely. (iv) *Presymptomatic* and susceptibility screening may be applied at younger age individuals with liability to late onset genetic anomalies such as autosomal polycystic kidney disease. Charcot—Marie—Tooth disease, Huntington's disease, familial hypercholesterolemia, retinitis pigmentosa. Some tests provide predictions regarding susceptibility to diabetes mellitus, coronary heart disease, breast cancer, etc. Testing for such predispositions must require confidentiality because of the obvious relevance to finding jobs or health insurance. (See terms used under specific entries, prenatal diagnosis, GMS, RDA, polymerase chain reaction, sperm typing, preimplantation genetics, ART)

GENETIC SEGREGATION: see Mendelian segregation, modified segregation ratios, meiosis, preferential segregation, somatic segregation.

GENETIC SIMILARITY INDEX: expresses the similarities between different strains on the basis of the number of shared restriction fragments identified by probes such as DNA minisatellite sequences, etc. (See minisatellite, microsatellite, probe, genetic distance)

GENETIC STABILITY: the gene and chromosomal mutabilities are relatively rare, transposable elements are absent, and the population is in genetic equilibrium. (See mutability, genetic equilibrium, transposable elements)

GENETIC STERILIZATION: heavy doses of ionizing radiation (X-rays) break the chromosomes but do not necessarily kill the irradiated animals that are even capable of mating. In their progeny, because of the chromosomal rearrangements that follow, sterility or lethality occur or although the irradiated males copulate, cannot fertilize the eggs of the females and they leave no offspring. This basic genetic knowledge was successfully applied to insect eradication. The screwworm (*Cochliomya hominivorax*) a tropical and subtropical parasite of warm-blooded animals produces larvae that hatch in the wounds of livestock and cause great damage to the hide and make it inferior for the leather industry. Additional damage results to agriculture by weight loss in cattle and sheep and to game animals but the fly may pose hazards also to people. The chemical control of this insect is difficult on live animals and not without danger of pollution and health effects. Therefore, pupae were reared in a large laboratory and treated by about 7,500 R X-radiation, and every weeks 2 million irradiated males were released in the areas with heavy infestation. The monogamous females so mated either failed to produce offspring or when more sophisticated chromosomal constructs were used "genetic

Genetic sterilization continued

time bombs" were generated that kept on killing the offspring due to the chromosomal or genic defects (temperature-sensitive alleles). In some areas and some years this pest control was so effective that the screw worm population was reduced to 1% of the years before the initiation of the program. Similar procedure was successfully applied also to mosquito control. Particularly good result were observed in the control of lepidopteran insects with holocentric chromosomes where the delayed and sustained lethal effects could be best exploited. (See holocentric chromosomes, translocations, radiation effects)

GENETIC SURGERY: replacing single or a few (defective) genes of an organism with the aid of (plasmid) vectors or introducing into cells foreign genetic material (organelles, chromosomes) with microsyringes or microcapillaries controlled by micromanipulators under microscopes. (See gene replacement, gene therapy, genetic engineering, gene transfer by microinjection, targeting genes)

GENETIC SWITCH: mechanisms based on interaction between specific DNA and protein sequences to turn genes on and off. (See regulation of gene activity, DNA binding proteins, DNA-binding protein domains, immunoglobulins, transposition, mating type determination in yeast, phase variation, *Trypanosomas, Borrelia*, serotype)

GENETIC SYSTEMS: means primarily the prevalent mode of reproduction (selfing, inbreeding, random mating, assortative mating etc.). Generally the mechanisms affecting variability (recombination, mutational mechanisms, etc.) are also included in this term. (See mating systems)

GENETIC TESTING: may reveal the liabilities of an individual to certain diseases and genetic anomalies. An individual may benefit from it because glucose-6-phosphate dehydrogenase deficiency may make a person more susceptible to environmental oxidants (ozone, nitrogen dioxide). Thalassemias may increase the dangers of exposure to lead and benzene, porphyrias to chloroquine, barbiturates, pseudocholinesterase deficiency to organophosphate and carbamate insecticides, etc. Molecular tests may reveal non-symptomatic heterozygosity for genetic diseases and may predict the risk for having various disorders in the offspring. On the other hand, employers and health insurance companies may discriminate against individuals on the basis of the genetic records. (See genetic privacy)

GENETIC TOXICOLOGY: see gene-tox

GENETIC TRANSFER: may be mediated by the gametes during sexual reproduction, by cytoplasmic organelles, plasmids, episomes, infectious heredity, bacteriophages (transduction) or plasmids, fusion of somatic cells, transfer of isolated organelles, transformation, vectors, viruses, retroviruses, prions, microinjection, electroporesis, targeting genes)

GENETIC TRANSFORMATION: see transformation genetic, transformation oncogenic, transfection

GENETIC TRANSFUSION: transfer of organelles and cellular inclusions by protoplast fusion.

GENETIC TRANSLATION: see protein synthesis, regulation of gene activity

GENETIC TUMORS: more than two dozens tumor genes have been assigned to *Drosophila melanogaster* chromosomes 1, 2 and 3. The majority of these are not malignant and occur freely or attached to internal organs in the thorax and in the abdomen. They are distinguished already at the third instar larva stage and persist through the life of the individuals. The majority of the tumors become melanotic. The melanotic tumors determined by genes *mbn* and *Tum* have malignant characteristics. In several inbred mice strains ovary tumors, testis tumors, B cell lymphoma, kidney adenocarcinoma, leukemia, pulmonary tumors are under polygenic control. *Bilateral retinoblastoma* (an eye tumor) of humans is controled by a dominant gene. Deficiencies involving the long arm of chromosome 13 may also induce retinoblastoma. Genes involved in the skin disease, *xeroderma pigmentosum* is based on deficiency in the genetic repair mechanism. Initially the disease involves excessive freckle formation and may become tumorous. Exposure to ultraviolet light (sun light) enhances the formation of skin tumors, particularly in fair skinned and albino individuals. The incidence of leukemia may increase in cases of trisomy or partial deficiency for chromosome 21. Both DNA (SV40, adeno-

Genetic tumors continued

virus, bovine papilloma virus, etc.) and RNA viruses (Epstein-Barr virus, retroviruses) can cause tumorigenesis in mammals. The loss or mutation in a gene controling protein p53 may lead to tumorigenesis presumably due to loss of this suppressor gene. Genetic hybrids between the species of the platyfishes (*Xiphophorus*) are prone to develop melanoma. Approximately 30 species-crosses of tobaccos may produce tumorous offspring that form callus *in vitro* cultures without a requirement for phytohormones. In the *Nicotiana glauca* (2n = 24) x *N. langsdorffii* (2n = 18) hybrids more than one locus is involved in tumor development. In the hybrids of *N. longiflora* (2n = 20) x *N. tabacum* (2n = 48) one chromosomal segment appears to be responsible for tumorigenesis. *N. saunderae* may inhibit the expression of tumors. In the majority of dicotyledonous and some monocotyledonous plants agrobacterial infection and the insertion of the T-DNA of the Ti plasmid may lead to tumor formation by genetic transformation. Certain viral infections also result in tumorous growth in plants. Several insects stimulate the formation of gall tumors in plant tissues through their metabolic products. For the *in vitro* development of plant tumors, the additions of phytohormones (primarily natural or synthetic auxins) are required. Some cultures, however, become "habituated" after a course of culture and the exogenous auxin supply may no longer be required. The plant tumors are non-malignant. (See also cancer, tumor, carcinogens, SV40, adenoviruses, retroviruses, reverse transcription, *Agrobacterium*)

GENETIC VARIABILITY: the ability or proneness (proclivity) to hereditary change. (See mutation, mutator genes, transposable elements, homeostasis, genetic homeostasis)

GENETIC VARIANCE: is caused by the various effects of the genotype (V_g). The variance observed is, usually the phenotypic variance (V_p) that is the outcome of the mutual action of the genotype and the environmental variance (V_e). The genetic variance itself has three components: $V_g = V_a + V_d + V_i$ where V_a is the additive genetic variance or breeding value, V_d is the dominance variance and V_i = interactions. The interactions can be epistatic effects among the individual quantitative traits and the effect of the environment on gene expression. (See also variance, midpoint value, breeding value, polygenes, gain, heritability)

GENETIC VARIATION: hereditary differences within or between populations.

GENETICALLY DIRECTED REPRESENTATIONAL DIFFERENCE ANALYSIS (acronym GDRDA): targets and identifies traits that differ between congenic lines without prior knowledge concerning their biochemical function. It determines linkage to known genes or to polymorphic DNA markers. (See congenic, DNA markers, RDA)

GENETICALLY EFFECTIVE CELLS: are those cells of the germline that actually contribute to the formation of the gametes and thus to the offspring. The number of genetically effective cells can be determined in autogamous species on the basis of the segregation ratios after mutation. In case the genetically effective cell number (GECN) is 1 the segregation is either 3:1 or 1:2:1. If the GECN is 2 the segregation of dominant:recessives is 7:1, in case of GECN = 4, the expected ratio is 15:1 because only one of the cells of the germline segregates while the other cell(s) provide only non-mutant offspring. Thus, the pooled phenotypic numbers yield the 7:1 (4 + 3):1, and 15:1, (4 + 4 + 4 + 3):1 proportions. These ratios may be (slightly) altered if the transmission of the gametes carrying the recessive alleles is impaired or if the viability of the recessive homozygotes is reduced. For the determination of GECN the aberrant progenies should be left out of consideration. (See also planning of mutation experiments, critical population size, mutation rate)

GENETICALLY EFFECTIVE POPULATION SIZE: see effective population size

GENETICIN (G418): an aminoglycoside antibiotic. (See antibiotics, aminoglycoside phosphotransferase, G418))

GENETICS: the study of inheritance, variation and the physical nature and function of the genetic material. (The term was suggested by William Bateson in 1906 for the then entirely new discipline). Genetics may be pursued as a basic science where only the discovery of new principle(s) and their integration into the store of knowledge are the goals. Alternatively, ap-

Genetics continued

plied branches of genetics rely on the established genetic principles and are used for agricultural (plant and animal breeding), industrial (biotechnology) purposes or for the improvement of human health (medical genetics). Applications of genetics are expanding into paleontology, archeology and forensic areas. The tools of genetics are integrated today into all biological disciplines from taxonomy, evolution, cytology, development, behavior, physiology, to biochemistry, biophysics and molecular studies. Thus, genetics escaped from its classical boundaries of heredity and cytology and became the core and unifying element of biology. (See heredity, inheritance, reversed genetics, population genetics, human genetics, medical genetics, clinical genetics, quantitative genetics, experiments, science)

GENETICS, CHRONOLOGY OF: a very broad overview includes only the most important milestones of basic genetics compiled somewhat subjectively. Paraphrasing G.B. Shaw, who would dare to say who is greater than Shakespeare? (To keep the length minimal, applied aspects of genetics are not included.)

200-300 B.C. Greek Philosophers discuss heredity
1694 Camerarius recognizes sex in plants
1761 Kölreuter reports thousands of attempted and some successful plant hybridizations
1839 Schleiden (plants) and Schwann (animals) discover cellular organization
1865 Mendel recognizes the basic principles of inheritance
1866 Haeckel points out the role of nucleus in heredity
1869 Galton lays down the foundations of statistics-based inheritance
1871 Miescher reports about nuclein
1873- on Mitosis, meiosis, chromosome numbers, supremacy and continuity of chromosomes are recognized
1900 Mendel's work is rediscovered
1902 Sutton proposes the chromosomal theory of inheritance
1902 Benda recognizes mitochondria
1902 Garrod reports on alkaptonuria as an inherited biochemical trait
1906 Bateson suggests the term genetics
1909 Johannsen coins the terms gene, genotype, phenotype and explains pure lines
1909 Correns and Baur discover non-Mendelian inheritance of chloroplasts
1910-11 Morgan discovers sex-linkage and crossing over
1910 von Dungern and Hirschfeld show that blood groups are inherited
1913 Sturtevant constructs the first linear map of 6 genes of the *Drosophila* X chromosome
1913 - 1925 Bridges and Sturtevant discover deficiency, nondisjunction, duplication, inversion, translocation
1926 Chetverikoff and Helena Timoféeff-Ressovskaya found experimental population genetics
1926 D'Hérelle describes bacteriophages
1927 Landsteiner and Levine lay the foundations of immunogenetics
1927 Muller and then Stadler induce mutations by X-rays
1928 Griffiths observes bacterial transformation
1930-on Fisher, Wright and Haldane, working independently, lay down the foundations of theoretical population genetics
1939-on Delbrück and Luria initiate phage genetics
1940 Beadle and Tatum conduct experiments leading to biochemical genetics and to the gene - polypeptide theory
1944 Auerbach and Robson discover chemical mutagenesis
1944 Avery, MacLeod and McCarty demonstrate that the transforming principle is DNA
1946 Lederberg and Tatum show bacterial recombination
1949 Chargaff discovers the variable base comoposition and A=T, G≡C relations in different DNAs
1951 McClintock discovers transposable elements
1952 Lederberg reports transduction
1953 Watson and Crick construct a valid DNA model
1955 Fraenkel-Conrat and Williams prove that RNA can also be a genetic material
1956 Kornberg shows *in vitro* replication of DNA
1957 Taylor in plants, and in 1958 Meselsohn and Stahl show that DNA replication is semiconservative
1957-on Beginning of the understanding of the machinery of protein synthesis
1960 Marmur and Lane hybridize nucleic acids
1960 Barski makes somatic cell hybrids
1961 Nirenberg and Ochoa laboratories independently

	demonstrate the nature of the genetic code
1961	Jacob and Monod propose the operon concept
1965	Southerland discovers cAMP and opens the ways into inquiries on signal transduction and transcription factors
1969	Shapiro *et al.* isolate the *lac* operon
1970	Temin and also Baltimore discover reverse transcription
1970	Khorana synthesizes *in vitro* a tRNA gene
1972	Transformation by recombinant DNA begins in Cohen, Berg and Lobban laboratories using plasmid vectors
1977	Development of efficient DNA sequencing by Gilbert's and by Sanger's laboratories
1978	Shortle and Nathans make localized mutagenesis
1980	Capecchi *et al.*, Ruddle *et al.* transform mice
1981	Schell *et al.* transform plants by *Agrobacterium*
1981	Cech discovers ribozymes
1983	Varmus and others identify c-oncogenes
1985	Mullis *et al.* develop the PCR procedure
1995-on	Sequencing of complete DNA genomes of prokaryotes and by 1996 also yeast
1996	Beginning of the mass identification of the function of the sequenced genes

GENETICS OF BEHAVIOR: see behavior genetics

GENETICS OF CANCER: see cancer, genetic tumors

GENETICS AND PRIVACY: the information rapidly accumulating on risks based on various screening techniques and DNA sequencing may result in discrimination by insurance companies, potential employers and possibly by the society in general. Therefore there is considerable concern that such information should not be divulged without the consent of the individual and the privacy should be legally protected.

GENIC BALANCE: in some organisms (e.g. *Drosophila*) in sex determination the proportion of the sex chromosomes and autosomes has crucial role. In *Drosophila* (1 X) and (2 sets of autosomes) means male, whereas (2X):(2 set of autosomes) = females (1:1). In general, all individuals with chromosomal ratios above 1 are females and those between 0.5 and <1 are intersexes. In humans and mice the XO is female while in *Drosophila* XO is male. In trisomics and nullisomics the nature of the individual chromosome(s) present or absent makes a great difference in the phenotype. (See sex determination, trisomy, nullisomic, nullisomic compensation)

GENICULATE BODY (geniculate nucleus): part of the brain where optic and auditory fibers are received.

GENISTEIN (4,5,7-trihydroxyflavone): is an inhibitor of protein tyrosine kinase. (See leukemia BCP)

GENITAL ANOMALY SYNDROMES: in the genetical males, these syndromes may involve hypospadias or cryptorchidism or micropenis. In hypospadias the urethra may open at the lower side of the penis or between the anus and the scrotum. Cryptorchidism indicates that the testes does not descend from the abdominal cavity into the scrotum (the testicular bag). In females most commonly either the ovaries, uterus or the fallopian tubes (connecting the ovaries with the uterus) or the vagina fail to develop normally, clitoromegaly, fusion of the labia occur. (See hermaphroditism, pseudohermaphroditism, gonadal dysgenesis, Smith-Lemli-Opitz syndrome, Opitz syndrome, Wilms tumors, Robinow syndrome, Fraser syndrome, Wolf-Hirschhorn syndrome, Bardet-Biedel syndrome, adrenal hyperplasia, trisomy, testicular feminization)

GENIUS: according to the Roman meaning of the word, it is a guarding spirit influencing a person for better or worse. C.D. Darlington (1964) defined it who "changes the environment of others for his own and even for succeeding generations, for his own species and even for the whole living world". Francis Galton in his book Hereditary Genius (1869) came to the conclusion that eminence is biologically inherited. Darlington had a less asserting view "the sons of great men are given the best chances with the worst results". Human intelligence cannot be exempted from the general biological laws although environment has great influence on the development of the hereditary qualities. A good example is Marie Curie who became the first woman who received, with her husband Pierre, the Nobel Prize for physics in 1903 and she also a second time in 1911 for chemistry. In 1935 their daughter Irene and her husband Frédérick Joliot were awarded the Nobel Prize for chemistry. (See human intelligence, Bach Sebastian)

GENMAP: is a computer program for mapping genetic data based on least squares. (See least squares)

GENOMATRON: is a gene-mapping machine.

GENOME: a complete single set of genes of an organism (taxonomic unit) or organelle, also the basic haploid chromosome set. The size of genomes in rounded nucleotide numbers varies in the different taxonomic categories:

Human mitochondrion	1.7×10^3	bp
MS2 (single-stranded RNA bacteriophage)	3.5×10^3	bases
φ X174 (single-stranded DNA bacteriophage)	5.4×10^3	bases
SV40 (double-stranded animal DNA virus)	5.2×10^3	bp
Tobacco mosaic virus (single-strand RNA)	6.4×10^3	bases
Influenza virus (single-strand RNA, animals)	1.4×10^4	bases
λ (double-stranded DNA bacteriophage)	4.9×10^4	bp
Vaccinia virus (double-stranded DNA, animals)	1.9×10^5	bp
T2 (double-stranded DNA phage)	2.1×10^5	bp
Chlamydia (bacteria)	6.0×10^5	bp
Escherichia coli bacterium	4.6×10^6	bp
Calotrix (bacteria)	1.3×10^7	bp
Saccharomyces cerevisiae (fungus, eukaryote)	1.2×10^7	bp
Drosophila (insect)	8.0×10^7	bp
Caenorhabditis elegans (nematode)	1.0×10^8	bp
Arabidopsis thaliana (higher plant)	1.5×10^8	bp
Mouse	2.5×10^9	bp
Homo sapiens	3.0×10^9	bp
Toad (*Bufo bufo*)	6.0×10^9	bp
Maize (higher plant)	2.0×10^9	bp
Trillium luteum (higher plant)	6.5×10^{10}	bp
Fritillaria davisii (higher plant)	1.5×10^{11}	bp

The average genome size of birds is about 1/3 of that of the mammals, mainly because the avian introns are shorter. (See mtDNA, chloroplasts, plants Bennett & Leitch Ann Bot. 76:113)

GENOME ANALYSIS: determining the origin of the component genomes in allopolyploid species on the basis of chromosome pairing, univalent(s) and multivalent associations, chiasma frequencies, chromosome substitution, chromosome morphology, chromosome banding, hemizygous ineffective alleles, DNA probes, etc. (See terms under separate entries)

GENOME EQUIVALENT: the mass of the DNA/RNA is the same as that of a genome.

GENOME-LINKED VIRAL PROTEIN: see VPg

GENOME MUTATION: affects chromosome numbers. (See aneuploid, polyploid)

GENOME ORGANIZATION: see genome, chromatin, euchromatin, heterochromatin, repetitive DNA

GENOME PROJECTS: are focused on the physical mapping and sequencing of entire genomes of humans and other higher and lower eukaryotes as well as of prokaryotes. Upon completion of these projects a detailed inventory of all genes will become available and this in turn will facilitate new generalization of organization and function of the cells and will permit the application of the new principles and the new technologies to human economic fields as well as for preventing and curing diseases. The complete nucleotide sequence of the 4 genes of the MS2 RNA virus has been determined by 1976, and by 1995 all the 1749 genes of *Haemophilus influenzae* bacterium had been sequenced. The genome of *Saccharomyces cerevisiae* yeast has

Genome projects continued

also been completely sequenced. The large eukaryotic genomes such as that of humans, containing over 3 billion bp is ordered first into sequential stretches by the use overlapping fragments. The first step is breaking up the human chromosomal DNAs (average of 250 Mb) into 100-2,000 kb fragments and cloned them in YACs. The YACs are cleaved into an average of 40 kb fragments and cloned by cosmids. The contents of the cosmids are then cloned in 5-10 kb capacity double-stranded DNA plasmid vectors or into the single-stranded filamentous phage M13 vector of 1 kb load. The fragments at each step can be tied into contigs by "chromosome walking". The nucleotide sequences of the smaller clones can be analyzed. The entire human genome requires a minimum of about 3,000 YAC or 20,000 BAC or 75,000 cosmid or 600,000 plasmid or 3,000,000 M13 phage clones. An alternative approach the complete sequencing is to proceed from sequence-tagged connectors (STC). The human chromosomes would be cloned in BAC vectors and sequence 500 nucleotides at the ends. The 600,000 BAC end sequences represent 10% of the genome and scattered at every 5 kb across the genome. They are called *sequence-tagged connectors* because they allow each BAC clone to be connected to about 30 others (150 kb insert/5 kb $\cong$ 30). The BAC inserts are digested by a restriction enzyme to determine its size. A "seed" BAC is sequenced and checked against the data of sequence-tagged connectors to identify the overlapping clones. In a following step, two BACs that show internal homology by the restriction enzyme digests and minimal overlap at their end, are completely sequenced. By such a procedure the entire human genome could be sequenced in 20,000 clones. The advantage of this proposal (Venter, J.C, Smit, H.O & Hood, L. 1996 Nature *381*:364) is that some of the low-resolution mapping (YAC and cosmid steps) could be eliminated and automatic sequencing procedures could be applied, reducing cost and labor. The BAC clone sequencing could be done by many groups world-wide. The already known sequence-tagged sites (STS) and expressed sequence tags (EST) could be readily located and additional genes could be easier placed. The procedure suggested would greatly facilitate the sequencing other smaller genomes of interest. (See physical mapping, YAC, cosmid, vectors, restriction enzyme, DNA sequencing, STS, EST, BAC, YAC, cosmid, DNA chips)

GENOME SCANNING: is cutting up the genome first by 8-bp-recognizing restriction endonuclease(s) into large fragments and followed by using more frequent cutter enzymes to generate physical information on the entire genome. These fragments can then be used to establish a physical map. (See physical map, restriction enzyme)

GENOME SEQUENCE SAMPLING (GSS): a chromosomal DNA, digested with several restriction enzymes, is cloned into cosmids. Hybridization with YAC clones of the same chromosomal DNA identifies all the cosmides that contain sequences present within the YAC. The cosmids are then broken down into contigs and their ends are identified by hybridization to pure cosmid DNA. The 300-500 bps of the ends are sequenced and aligned in sequence, permitting the generation of a rather high density physical map.

GENOMIC CLONE: is prepared from chromosomal DNA, rather than from cDNA. (See genomic DNA, cDNA)

GENOMIC DNA: the native DNA including exons, introns and spacer sequences (versus the processed genes which are transcribed from mRNA to DNA by reverse transcription and have only the coding sequences). (See processed genes, reverse transcription)

GENOMIC EXCLUSION: takes place in the ciliate *Tetrahymena pyriformis* in case one of the two mates has a defective genome (micronucleus) that is therefore not included in the meiotic progeny. The first progeny becomes heterokaryotic having only the normal diploid micronucleus and an old macronucleus that is genetically not concordant with the micronucleus. After a subsequent matings the normal micronucleus forms a macronucleus concordant with its own genetic constitution. As a result the strain is purged from the defect. (See conjugation *Paramecia*)

GENOMIC FORMULAS: n = haploid, 2n = diploid, 3n= triploid etc. and n -1 or 2n-2 = nullisomic, 2n-1 monosomic, 2n+1= trisomic, 2n+2=tetrasomic, etc. where n = haploid chromosome number. The basic chromosome number is however x and the diploids are 2x = 2n. (See polyploids, aneuploids)

GENOMIC LIBRARY: a set of cloned genomic DNAs; it expected that a good library includes at least one copy of all the genes of a particular genome. (See cloning, genome, fragment recovery probability)

GENOMIC MISTMATCH SCANNING: see GMS

GENOMIC STRESS: such as dissimilar genetic backgrounds in hybrids, *in vitro* cell culture, etc. may activate dormant transposable elements and cause a genetic instability. (See transposable elements, somaclonal variation)

GENOMIC SUBTRACTION: a method that removes from wild type DNA all the sequences that are present in a deletion mutant but retains the wild type DNA sequences corresponding to the deletions by denaturing a mixture of wild type and biotinylated mutant DNA. Allowing the mix to reassociate, the biotinylated sequences are subtracted by several repeated cycles of binding to avidin-coated polystyrene beads (that have great affinity for biotin). The remaining (non-biotinylated) DNA is wild type and, contains only the sequences that were deleted in the mutant but present in the wild type and can then be amplified by PCR and studied by standard techniques of sequencing. This method also permits the isolation of genes affected by the deletion (caused by e.g. ionizing radiation). (See physical mutagens, gene isolation, biotinylation, avidin, PCR, DNA sequencing, RDA, RFLP subtraction)

GENOMICS: study of the molecular organization of genomic DNA and physical mapping. (See genomic DNA, DNA sequencing, DNA chips, physical mapping, proteome)

GENOPHORE: gene string not associated with large amounts of protein (bacterial chromosome).

GENOTOXIC CHEMICALS: cause gene mutation, chromosomal aberration and cancer. (See gene-tox, databases)

GENOTYPE: the genetic constitution; the full set of genes.

GENOTYPIC FREQUENCIES: see Hardy-Weinberg theorem

GENOTYPIC MIXING: after infecting a cell with viruses of different genotypes, in a single viral capsid more than one type of viral DNA may be included. (See also rounds of matings)

GENOTYPIC RISK RATIO (GRR): the total number of offspring affected/twice the number of affected homozygotes. (See genetic hazards, risk, genetic risk, empirical risk)

GENOTYPIC SEGREGATION: see trinomial distribution

GENOTYPIC VALUE: is a quantitative genetics term indicating the genetically determined component (G) of the phenotypic variation; phenotypic value (P) = G + E, where (E) stands for environmental variation. (See also midpoint, breeding value, additive effects)

GENOTYPING: identifying the genotypic constitution at one or more loci by genetic, molecular, immunological or any other means using cells, tissues or whole organisms. (See genotype)

GENUS: a taxonomic category including usually several species of common descent. Some genera are monotypic, however, inasmuch as they consist of a single species. (See species)

GEOGRAPHIC ISOLATION: populations cannot exchange genes because of physical distance or other physical factors (mountain ranges, lakes, etc.) keep them apart. (See speciation)

GEOLOGICAL AGE: see evolutionary clock

GEOLOGICAL TIME PERIODS: Archeozoic, Pterozoic, Paleozoic, Mesozoic, Cenozoic. (See individual entries, origin of life)

GEOMETRIC MEAN: see mean

GEOMETRIC PROGRESSION: a series of elements increasing by the same factor, e.g. 2, 6, 18, 54 (i.e by a factor of 3 in this example). (See arithmetic progression)

GEORGE III: king of England (1738-1820) might have been a victim of porphyria. (See porphyria)

GEOTROPISM: growth influenced by gravity; positive geotropism directed toward (+) and negative (-) away from gravity. Plant roots grow downward (+) and the shoots upward (-).

GERANYL PYROPHOSPHATE: is a precursor of farnesyl pyrophosphate. Two molecules of farnesyl pyrophosphate then join by the pyrophosphate end and through the elimination of both pyrophosphates, squalene is formed. Squalene is then cyclicized to form lanosterol before being converted into cholesterol. (See prenylation, cholesterols)

GERBICH (Ge blood group): is distinguished by its encoding β and γ sialoglycoproteins (glycophorins). These red blood cell membrane proteins are suspected of being the receptors of the *Plasmodium falciparum* merozoite (malaria-causing protozoon). (See blood group, malaria)

GERBIL: *Gerbillus cheesmani* 2n = 38; *Gerbillus gerbillus* 2n = 43 male, 42 female.

GERM: (pathogenic) microorganism or an initial cellular structure capable of differentiation and development into a special organ or organism.

GERM CELLS: the reproductive (sex) cells of eukaryotes such as spores, eggs and sperms. The spores frequently come about by non-sexual processes such as the conidia of fungi and may not function similarly to sex cells. The egg and sperm are direct or indirect products of meiosis that have undergone a process of differentiation without division, e.g., the spermatozoa of animals arise from the spermatids or the sperms of plants are formed by post-meiotic division of the microspore nuclei. The eggs of animals arise by an additional division of the haploid secondary oocytes. The egg of plants is formed through three divisions from one of the haploid megaspores. (See gametogenesis, conidia)

GERM LAYERS: gastrulation forms the most inner layer, *endoderm*, the surface layer, *ectoderm* (epithelium), and the in-between mesenchyme cell layer the *mesoderm*. (See gastrula)

GERM PLASM: development (in *Drosophila*) begins with the formation of the primordial germ cells also called pole cells. The syncytial nuclei congregate at the posterior segment of the pole. Cellularization begins after about 2 hours. During gastrulation the germ cells move to the embryonic gonad and form the germline stem cells. In both males and females after 4 rounds of cell divisions 16 cell are formed. In the male, all 16 contribute to sperm formation. In the female these 16 cell remain interconnected but only one becomes an oocyte and the other 15 become polyploid nurse cells and nourish the oocyte. The oocyte proceeds with meiosis. The oocyte and nurse cells become surrounded by about 80 maternal (somatic) follicle cells. The development of the germ plasm is controled by the interacting products of a series of genes. (See also morphogenesis in *Drosophila*, germline, *Drosophila*)

GERMAN (germen): closely related, such as having the same parents. (See cousin german)

GERMAN MEASLES: see rubella virus

GERMARIUM: is the location of the pro-oocytes which through mitotic divisions gives rise to the oocysts and one of them becomes the oocyte. (Seen oocyte primary, karyosome)

GERMINAL CENTER: is a group of naive (uncommitted) B lymphocytes when activated by a specific antigen. The naive cells may develop into either memory B cells after antigen selection or become plasma cells. In the presence of interleukin-1,-10 and CD40 ligands they become memory B cells. By removal of CD40 ligand, the cells differentiate into plasma cells. A rapidly growing center includes also antigen-specific helper T cells. (See T lymphocyte, clonal selection, antigen, CD40, plasma cell, memory immunological, OBF, somatic hypermutation, B lymphocyte)

GERMINAL CHOICE: is the idea that parents should not necessariliy rely on their own gametes for producing offspring but adopt eggs or sperms or even fertilized eggs from superior gene pools as a practical measure of positive eugenics. (See also sperm bank, *in vitro* fertilization, eugenenics, ART)

GERMINAL MUTATION: occurs in the germline, gonads or in the gametes. (See germline, gonad, gamete)

GERMINAL VESICLE: the large nucleus of the amphibian oocyte. This nucleus contains the three eukaryotic RNA polymerases and can transcribe also exogenous (microinjected) DNA. The oocyte then translates the mRNAs into a variety of proteins. (See *in vitro* translation systems)

GERMLINE: the cell lineage that contributes to the formation of the gametes; in animals the germline is determined very early in the zygote. According to some views, plants do not have germline, certainly not in the sense of animals, because the generative cell lineage is not set aside definitely in early development and plant cells may retain totipotency for almost the en-

Germline continued

tire life of the individuals. Nevertheless, by "fate maps" the cell lineage giving rise to megaspores and microspores of plants can be traced to origin. In *Drosophila* for the development of the germline the product of the *nanos* (*na*) gene locus is essential. (See cell lineages, *Drosophila* life cycle, genetically effective cell number, morphogenesis in *Drosophila*, germ plasm)

GERMPLASM (Keimplasma): the sum of the genetic determinants transmitted through the gametes to the progeny. (See genotype, germ plasm)

GERONTOLOGY: clinical, biological and sociological study of aging. (See aging, apoptosis, Hayflick's limit)

GERSTMANN-STRAUSSLER DISEASE (GSD): is a chromosome 20 dominant brain disease with substantial similarities to the Creutzfeldt-Jakob disease. There are some apparent differences inasmuch as in GSD there are numerous multicentric tuft-like plaques in the cerebral and cerebellar cortex, in the basal ganglia and the white matter of the brain. GSD appears to have greater recurrence risk than the Creutzfeldt-Jakob disease. (See Creutzfeldt-Jakob disease, scrapie, prion, encephalopathies, encephalopathy bovine spongiform, brain human)

GESTATION: the time from fertilization of the ovum (ova) to the delivery of the newborns in viviparous animals. The average days of gestation: mouse 19, rats 21, rabbits 31, cats 63, dogs 63, sows 114, sheep 151, cows 284, mares 366, chimpanzees 238, women 267, elephants 624.

GESTATIONAL DRIVE (green beard effect): maternal genes recognizing and favoring special genes of the offspring, already during gestation, and favoring or disfavoring a genetic constitution may lead to consequences somewhat similar to meiotic drive. The concept is not generally accepted by population geneticists. (See meiotic drive)

GFP: see green fluorescing protein

GH: growth hormone such as the hGH (human, encoded in 17q22-q24) or rGH (rat) growth hormones. (See hormone response elements, hormones, pituitary dwarfness)

GHOST: an empty phage capsid without its genetic material. Also, electronic noise.

GHRH (growth hormone release hormone): stimulates the release of growth hormones from the pituitary. Somatostatin inhibits growth hormone secretion. (See animal hormones, pituitary, somatostatin, brain human)

G_h: G protein with GTP-binding signaling function and transglutaminase activity. (See G-protein)

G_I PROTEIN: is a member of the trimeric G-protein family; it activates adenylate cyclase and thus opens K+ channels. (See G-proteins, signal transduction, adenylate cyclase)

GI_{50}: a chemical dose that provides 50% growth inhibition, e.g. for a certain cancer cell line.

GIANT CHROMOSOMES: are the polytenic chromosomes and the lampbrush chromosomes. (See lampbrush chromosomes, polytenic chromosomes, salivary gland chromosomes)

Gibberella fujikuroi: a plant-pathogenic fungus that produces by its normal metabolism the plant hormones gibberellins. (See plant hormones) GIBBERELLIC ACID ➤

GIBBERELLINS: see plant hormones, *Gibberella fujikuroi*

GIBBON: see Pongidae, primates

GIEMSA STAIN: contains azure II, azure-eosin, glycerol, and methanol. (See G-banding, chromosome banding, rye)

GIERKE'S DISEASE: see glycogen storage disease type I

GIFT (**g**amete **i**ntra**f**allopian **t**ransfer): a method of artificial insemination. (See artificial insemination, ART)

GIGA: prefix for 10^9 size or quantity.

GILBERT SYNDROME: a very common autosomal dominant hyperbilirubinemia, similar to the Crigler-Najjar syndrome and probably controled by genes allelic to it. (See Crigler-Najjar syndrome, Dubin-Johnson syndrome, hyperbilirubinemia).

Gin: is an invertase. (See invertases)

GINGER (*Zingiber officinale*): perennial rhizome spice. It dilates blood vessels, relieves pain, reduces flatulence, increases perspiration, and it is a stimulant; 2n = 2x = 22.

GIP: is a G protein subunit, and a potential oncoprotein. (See G protein, oncoprotein)

GIRAFFE (*Giraffa camelopardalis*): is 2n = 30; the *Okapia johstoni* is 2n = 45.

GIRDLE BANDS: concentric rings of thylakoids. (See chloroplasts, thylakoids)

GIRK (G-protein-gated inwardly rectifying K$^+$ channel): is a heterotrimeric guanine nucleotide-binding protein. (See ion channels, G proteins)

GITELMAN SYNDROME: hypocalciuria and hypomagnesemia. (See Bartter syndrome, Liddle syndrome, hypoaldosteronism)

GLANZMANN'S DISEASE: is a variety of blood platelet anomalies determined by autosomal recessive genes. The overall symptoms include bleeding under the skin (ecchimosis), tiny, round and flat purplish (later yellow or blue) spots under the skin caused by blood release (petechia), bleeding of the tooth gum (gingiva), nosebleeds (epistaxes), gastrointestinal bleeding, excessive uterine bleeding (menorrhagia) or bleeding from the uterus at irregular intervals (metrorrhagia). The platelets may appear normal yet their number is reduced (thrombocytopenia). Sometimes the size of the platelets increases and their shape becomes abnormal and they appear isolated rather than aggregated. (See also platelet abnormalities, hemophilias, and other terms under separate entries)

GLAST: Na$^+$ - dependent transporters of glutamate and aspartate; it has 68% homology with another glutamate transporter GLT. (See transporters)

GLAUCOMA: may be controled by autosomal dominant or recessive genes and may be manifested at birth, during the juvenile years or in adults. The incidence of the different forms may vary from 10^{-4} to a couple of percent in the general population, usually at higher risk in the adult life. The most general features are opacity of the eye lens caused by a gray gleam on the iris and an increased intraocular pressure which distorts eventually the vision. In the early stages or in any mild forms the anterior chamber of the eye is open (open angle glaucoma). This stage may pass into an intermittent form that may be transient but can last for several months and eventually the angle becomes closed resulting in great pressure and swelling of the cornea accompanied by substantial pain. Eventually, if untreated, total blindness may follow. Glau-coma may be monitored by testing the eye pressure before the visible onset of the condition. The penetrance and expressivity of this disease is highly variable. The gene (GLC1A) coding for juvenile open angle glaucoma (JOAG) was assigned to human chromosome 1q21-q31. The adult onset open angle glaucoma gene is at 3q. (See eye diseases)

GLC: see gas liquid chromatography

Gle1: see RNA export

GLGF REPEATS: same as DHR domain or PDZ domain

GLI ONCOGENE: has been located to human chromosome 12q13. It is highly amplified in gliomas. GLI appears homologous to the *Krüppel* gene of *Drosophila* encoding a DNA-binding protein, regulating embryo morphogenesis. Similarly, GLI is also expressed in embryonal carcinomas but not in late developing ones. Other homologous genes were also found in the human genome, altogether 6 loci in five different chromosomes. (See oncogenes, Kr (34) in morphogenesis of *Drosophila*, *Greig's* cephalopolysyndactyly syndrome, Rubinstein-Taybi syndrome, *hedgehog*, syndactyly, polydactyly)

GLIADIN: see zein, glutenin

GLIAL CELL (neuroglia): can be either astrocytes or oligodendrocytes or microcytes; the first two have supportive roles, the latter phagocytizes the waste products of the nerves. (See FGF)

GLIOMA: tumor of the tissues supporting the nerve cells.

glnA: bacterial glutamine synthase

glnAp2, glnAp1: major and minor glutamine synthase promoters, respectively, in bacteria.

GLOBAL GENETIC EFFECTS: involve most or all of the genome.

GLOBINS: are ancestral protein molecules that diverged over a billion years ago into the oxygen-carrying muscle protein myoglobin and into the respiratory hemoglobins of the red blood cells. (See also myoglobin, hemoglobin, leghemoglobin, haptoglobin)

GLOBOID CELL LEUKODYSTROPHY: see Krabbe's leukodystrophy

GLOBOSIDE: glycosphingolipid with the most common structure: acetylgalactoseamine-galactose-galactose-glucose-ceramide

GLOBULIN: salt-soluble proteins with many diverse cellular functions.

GLOMERULONEPHROTIS: is an autosomal dominant kidney disease associated with very spare hairs and red lesions due to dilation of the blood vessels (telangiectasis). (See hair, kidney diseases, skin diseases, telangiectasis)

GLOVES: frequently recommended for laboratory work when handling hazardous material or when contamination by hands must be avoided. Remember that surgical latex gloves easily develop invisible holes and permit unseen contamination of the hands. Some plastic gloves are damaged by organic solvents. For most operations neoprene gloves provide the greatest safety. (See laboratory safety)

GLT: see GAST

GLUCAGON: is a polypeptide hormone, secreted by the α cells of the pancreas when the level of blood glucose sinks below a certain level. The hormone then increases the concentration of blood sugar by breaking down glycogen with the cooperation of epinephrine. (See epinephrine, animal hormones, cAMP, diabetes mellitus)

GLUCAN: a polymer (repeating units) of glucose.

GLUCANASE: glucan-digesting enzyme. (See glucan, host-pathogen relation)

GLUCOCORTICOID: is kidney cortex hormone which regulates carbohydrate, lipid and protein metabolism, muscle tone, blood pressure, the nervous system, etc. It inhibits the release of adrenocorticotropin, slows down cartilage synthesis, mitigates inflammation, allergy and various immunological responses. Cortisol (hydroxycortisone) is an important natural glucocorticoid whereas dexamethasone is a synthetic product that is two orders of magnitude more pottent than cortisol. The glucocorticoid-mediated immunosuppression involves the activation of the IκBα gene and an increase of its cytoplasmic protein product. When the nuclear regulator factor NF-κB is active (because of the expression of TNF) its inhibitor, the IκBα protein is degraded and NF-κB moves into the nucleus and activates the immune system. Dexamethasone — in contrast to the natural glucocorticoids — causes an increased transcription of IκBα and the NF-κB translocation to the nucleus is inhibited, leading to less nuclear NF-κB and reduction of inflammation because the immune system is suppressed. The familial and sporadic glucocorticoid deficiencies are caused by defective adrenocorticotropic hormone receptors. The deficiency of the glucocorticoid receptor (94 kDa encoded in 5q31) causes cortizol and dexamethasone resistance. The glucocorticoid receptor is an indispensable transcription factor. (See adrenocorticotropin, NF-κB, IκB, cortisol, dexamethasone, opiocortin, immunosuppression, Cushing syndrome, calreticulin, immunophilins, GRE)

GLUCOCORTICOID RESPONSE ELEMENTS (GRE): are located generally about 100 to 2,000 nucleotide pairs upstream from the transcription initiation site (the human growth hormone response element is within the transcribed region). These elements, such as the mammary tumor virus (MTV), metallothionein (MTIIA), tyrosine oxidase (TO), tyrosine amino transferase receptor element, respond to different activating proteins as their name indicates that despite their differences in structure they share a consensus: CGTACANNNTGTTCT. (See hormone response elements, regulation of gene activity, DNA looping, mammary tumor virus, metallothionein, tyrosine aminotransferase)

GLUCOGENIC AMINO ACIDS: can be converted into glucose or glycogen through pyruvate (alanine, cysteine, glycine, serine, tryptophan), α–ketoglutarate (arginine, glutamine, histidine proline), succinyl CoA (isoleucine, methionine, threonine, valine), fumarate, (phenylalanine, tyrosine) and oxaloacetate (asparagine, aspartate). (See amino acids)

GLUCONEOGENESIS: synthesis of sugars from non-carbohydrate precursors (such as oxaloacetate, pyruvate).

GLUCOSE (glycose): a 6-carbon sugar (dextrose), an aldohexose. (See galactose [for formula])

GLUCOSE EFFECT: is a form of catabolite repression when as long as glucose is available in

Glucose effect continued

the nutrient medium, the synthesis of enzymes involved in the utilization of other carbohydrates is prevented. Glucose may act at three levels: (1) inhibits the uptake of inducer molecules by relying on the dephosphorylated component of the phosphoenolpyruvate-dependent glucose phosphotransferase. (2) lowers the level of cAMP and its receptor and activates indirectly adenylate cyclase. (3) Increases the level of catabolites that repress the synthesis of inducible enzymes. In fungi, the mechanism of glucose effect may be mediated through the function of hexokinase. The Mig1/CREA Zinc-finger protein and the Tup1 general suppressor and Snf1 a transactivator protein have also been implicated. (See also feedback control, repression, catabolite repression, Zinc finger, Tup1, SW1, transactivator)

GLUCOSE TOLERANCE TEST: see diabetes

GLUCOSE-GALACTOSE MALABSORPTION (GGM): see SGLT

GLUCOSE-6-PHOSPHATE DEHYDROGENASE: the first enzyme in the pentose phosphate pathway that converts G-6-P into 6-phosphoglucone-δ-lactone. The final product of the pathway is D-ribose-5-phosphate, and NADPH is also generated. Although about 90% of the cellular glucose in mammals is converted to lactate by glycolysis, 10% is driven through the pentose phosphate path and this is the principal reaction to provide the erythrocytes with NADPH for the reduction of glutathion. The deficiency of the enzyme caused by Xq28-chromosomal genes was first identified as a hemolytic anemia caused by the antimalarial drug 8-aminoquinoline. Most of the afflicted individuals are essentially asymptomatic until exposed to drugs such as certain analgesics, sulfonamides, antimalarial drugs (atabrine), quinine, etc. or certain other diseases. G-6-P dehydrogenase deficiency is widespread in human populations probably because the heterozygotes and hemizygous males are protected against falciparum malaria by a 46-58% reduction of the infectious disease. Heterozygotes (XX) may display lyonization. In the Jewish populations of Kurdistan, Caucasus and Iraq the frequency of the defect reached 58.2, 28.0 and 24.8%, respectively, whereas in geographical areas free of malaria it was generally less than 2%. Cavalli-Sforza and Bodmer estimated that G-6-P dehydrogenase deficiency conveyed an extremely high 0.15% selective advantage against malaria. (See analgesic, malaria, selection coefficient, selection conditions, pentose phosphate pathway, glycolysis, glutathion)

GLUCOSE-PHOSPHATE ISOMERASE: see phosphohexose isomerase

GLUCOSIDASE (GCS1): enzyme digests 1,2-N-linked glycoproteins; it is encoded in human chromosome 2p13-p12. (See acid maltase, Pompe diseases, Gaucher disease)

GLUCOSIDES: when D-(+) glucose is treated with an alcohol (methanol) and HCl, methyl D-(+)glucoside is formed that still has one methyl group yet its properties resemble that of an acetal. Acetals may be formed from aldehydes and they are common in different plants. Cardiac glucosides present in plants such as *Digitalis, Scilla*, etc. have cardiotonic effect (strengthen heart function) and used as medicine. Many of the plant glucosides are highly toxic and cause anorexia (loss of appetite), nausea, vomiting, salivation, diarrhea, headache, drowsiness, delirium, hallucinations and possibly death. Glycosides linked to cyanides occur also in common food plants such as beans, apricot and almond seed, etc. Forage plants such as Sudan grass, white clover, etc. may contain enough cyanide to kill a 50 kg animal if it eats 1 to 2 kg fresh plant material. Through plant breeding efforts the synthesis of the glucoside (lotoaustralin) may be blocked or the production of the enzyme linamarase may reduce the toxicity. (See lotoaustralin, cyanide)

GLUCOSYLATION: attaching glucose to another molecule.

GLUCURONIC ACID: is a derivative of uronic acid (a derivative of glucose) and it is present in glucosaminoglycans. (See mucopolysaccharidosis, GUS)

GLUME: the lower-most bract of the grass florets. The glume is generally free from the fruit, in some case, however, it may be firmly associated with the kernels.

GluR: see glutamate receptor

GLUT4: an insulin-dependent glucose transporter. (See insulin)

GLUTAMATE (HOOCCH[NH$_2$]CH$_2$CH$_2$CONH$_2$): is an uncharged derivative of glutamic acid and has key role as a nitrogen donor in the cell. (See amino acids, glutamine, glutamate synthase, glutamate synthetase)

GLUTAMATE DECARBOXYLASE DEFICIENCY DISEASE (GAD): is pyroxine-dependent epilepsy. The two enzymes require the cofactor pyridoxal phosphate. These enzymes convert glutamic acid into γ-aminobutyric acid (GABA) that controls neurotransmission in vertebrates and invertebrates. The phenotype is autosomal recessive (2q). (See epilepsy, GABA, amino acid metabolism)

GLUTAMATE DEHYDGROGENASE (M$_r$ 330,000): catalyzes oxidative deamination of glutamate in the mitochondria resulting in the formation of α–ketoglutarate. The reaction requires NAD$^+$ or NADP$^+$ as cofactors and it is regulated allosterically by GTP and ADP. Then in turn α–ketoglutarate and ammonia may form again glutamate. If the concentration of NH$_3$ is low glutamate dehydrogenase cannot function to an appreciable extent. In such a case NH$_3$ plus glutamate are converted to glutamine by non-adenylylated glutamine synthetase. In the presence of high amount of NH$_3$, *glutamine synthetase* is adenylylated and becomes inactive and in this form it represses its own synthesis (autoregulation). In its non-adenylylated state (when the level of ammonia is low) it represses *glutamate dehydrogenase* instead. From glutamine and α–ketoglutarate, glutamate can be synthesized by *glutamate synthase* in the presence of NADPH + H$^+$. Glutamine synthase also serves as an inducer for tryptophan permease which together with tryptophan transaminase also may contribute to glutamate synthesis. In its non-adenylylated state glutamine synthetase activates also the histidine utilization operon (*hut*). This operon also yields glutamate and ammonia. In humans this enzyme (*GLUD*) is coded by a small multienzyme family; its level is relatively high in the brain. The principal and functional *GLUD* is in human chromosome 10q23. This gene is homologous to mouse locus *Glud-2* in chromosome 14. (See also UTase, glutamate synthase, olivopontocerebellar atrophy, autoregulation)

GLUTAMATE FORMIMINOTRANSFERASE: an autosomal recessive deficiency of this enzyme leads to the accumulation of formiminoglutamate $^-$OOC-CH-(CH$_2$)$_2$COO$^-$
$$|$$
$$NH$$
$$|$$
$$HC=NH$$
and folic acid in the urine and in the serum causing physical and mental retardation. (See amino acid metabolism, mental retardation)

GLUTAMATE OXALOACETATE TRANSAMINASE (GOT2): is encoded in human chromosome 16q21 but the protein is mitochondrially located. In many plants and lower animals the enzyme is mitochondrially coded. Pseudogenes were found at two locations in chromosome 1 and in chromosome 12. (See mtDNA, aspartate aminotransferase mitochondrial, tyrosine aminotransferase)

GLUTAMATE RECEPTORS (GluR): are cation channels, mediating the postsynaptic current in the central neurons. Certain mutations in GluR-B subunits lead to increased calcium uptake and concomitant seizures if e.g. the position 586 arginine prevent editing of pre-mRNA.

GLUTAMATE SYNTHASE: catalyzes the reaction that leads to: α-ketoglutarate + glutamine + NADPH + H$^+$ → 2 glutamate + NADP$^+$. The result of the combined action of glutamate synthetase and glutamate synthase in bacteria is:
α-ketoglutarate + NH$_4^+$ + NADPH + ATP → L-glutamate + NADP$^+$ + ADP + P$_i$
(See glutamate dehydrogenase, glutamine, autoregulation)

GLUTAMATE SYNTHETASE: in *E. coli* is a ca. 800,000 M$_r$ protein containing flavin, iron and S^{2-}. (See glutamate synthase, glutamate dehydrogenase, glutamic acid, glutamine, autoregulation)

GLUTAMATE TRANSPORTER: see GLAST

GLUTAMATE-PYRUVATE TRANSAMINASE (GPT1): catalyzes the reversible reaction:
$$HOOCCH_2CH_2COCOOH + CH_3CH(NH_2)COOH \leftrightarrows HOOCCH_2CH_2CH(NH_2)COOH + COOHCOCH_3$$
α-ketoglutaric acid   L-alanine   L-glutamic acid   pyruvic acid

The soluble enzyme is encoded in human chromosome 8q24.2-qter. Cytosolic and mitochondrial forms exist. It is also called alanine aminotransferase (AAT1). (See amino acid metabolism, glutamine, alanine aminotransferase)

GLUTAMIC ACID: $HOOC-C(H)(NH_2)-(CH_2)_2-COOH$. (See glutamine)

GLUTAMINASE (GLS): is an enzyme converting glutamine into glutamic acid and it has been mapped to human chromosome 2q32-q34. It is activated by phosphate and it may affect the neurotransmitter role of glutamate. (See amino acid metabolism, glutamine, glutamic acid)

GLUTAMINE: $HOOC-C(H)(NH_2)-(CH_2)_2-C(O)NH_2$ (See glutamic acid)

GLUTAMINE AMIDOTRANSFERASES: are a group of enzymes with two domains, one binds glutamine and the other binds another molecule. After cleaving ammonia from glutamine they transfer it to the other substrate, generally in the presence of ATP.

GLUTAMINE-REPEAT DISEASES: see Huntington's chorea, Kennedy disease, dentatorubral-pallidoluysian atrophy, olivopontocerebral atrophy, Macho-Joseph disease. (See also fragile sites, trinucleotide repeat)

γ-GLUTAMYL CARBOXYLASE (GGC): the enzyme required for the post-translational modification of vitamin K dependent proteins used for blood clotting and bone proteins. (See vitamin K-dependent blood clotting factors)

GLUTAMYL-tRNA SYNTHETASE (QARS): the enzyme charging the cognate tRNA with glutamic acid; it is encoded in human chromosome 1q32-q34. (See aminoacyl tRNA synthetase)

GLUTARALDEHYDE: see fixatives

GLUTARICACIDEMIA (GA): GAI autosomal recessive (19p13.2) glutaryl-CoA dehydrogenase deficiency results in increase in glutaric acid in the blood and in the urine resulting in neurodegenerative disorders. GAIC encoded in human chromosome 4q32-qter involves deficiency in the electron transfer flavoprotein oxidoreductase. GAIIA (15q23-q25) causes the excretion, besides glutaric acid, also lactic, ethylmalonic, isovaleric and different forms of butyric acids. Similarly an X-linked (Xq26-q28) acyl-CoA dehydrogenase deficiency results in the abnormal excretion of glutaric and other organic acids. (See glutaricaciduria, aminoacidurias)

GLUTARICACIDURIA: an autosomal recessive glutaryl-CoA dehydrogenase deficiency leading to accumulation of glutaric acid in the urine and degeneration of the nervous system and impairment of muscle functions. Limiting of amino acid intake may alleviate the symptoms. An autosomal dominant form was identified as a defect in an electron-transfer flavoprotein. Some glutaricacidemias are also called glutaricaciduria. Glutaricaciduria II (GAIIA) was assigned to 4q32-qter. (See neuromuscular diseases, aminoacidurias, glutaricacidemia)

GLUTATHIONE: is γ-Lglutamyl-L-cysteinylglycine, is a reducing agent and protects SH groups in proteins. About 10% of the blood glucose is oxidized to 6-phosphogluconate by glucose-6-phosphate dehydrogenase G6PD using $NADP^+$ and the reducer NADPH keeps glutathion reduced. Deficiency of G6PD results in destruction of red cells and thus anemia. (See glucose-6-phospate dehydrogenase)

GLUTATHIONE PEROXIDASE (GPX1): was assigned to human chromosome 3q11-q12. Its deficiency causes hemolysis and jaundice. The ailment may be caused also by selenium-deficient diet. (See hemolytic anemia, gluthatione synthase deficiency)

GLUTATHIONE REDUCTASE (GSR): the gene was located to human chromosome 8p21. Its deficiency results in hemolytic anemia. (See hemolytic anemia)

GLUTATHIONE SYNTHETASE DEFICIENCY: is a form of human chromosome 20q11.2 recessive hemolytic anemia and/or 5-oxyprolinuria. It may result also in excess metabolic pyroglutamic acid in the urine and to a variety of ailments. GST2 (γ-glutamylcystein synthetase) gene was assigned to human chromosome 6p12 also causes hemolytic anemia. (See hemolytic anemia, glutathione, anemia)

GLUTATHIONE-S-TRANSFERASES (GST): is a family of enzymes metabolizing, and detoxifying mutagens and carcinogens. GST 3 was assigned to 11q13, GST2 to 6p12, GST1, GST4, GST5 all at 1p13.3. GSTPL (glutathione transferase-like enzyme) is encoded in 12q13-q14. These enzymes, despite of different location of the coding units, show homology.

GLUTATHIONURIA (GGT): a recessive defect (human chromosome 22q11.1-q11.2) in γ-glutamyl transpeptidase enzyme and accumulation of glutathione in the urine.

GLUTEN: a mixture of several seed proteins in cereals. The main fractions are the alcohol-soluble gliadin and the alkali-soluble glutenin. The proportion of the components is genetically determined and define nutritional value and baking quality. (See glutenin, zein)

GLUTENIN: about half of the seed storage protein in wheat; soluble in 70% ethanol and alkali but insoluble in water. It is a polymer of extremely large molecular weight, up to the tens of millions. Its composition bears similarity to the muscle protein titin, comprising about 27,000 amino acid residues. The similarities based on (PEVK) proline, glutamate, valine and lysine sequences may be attribuited to the fact that both proteins require great elasticity in the bread dough. It was (indirectly selected by humans) to retain gas bubbles in the dough to return to the original position after extension. In wheat, gliadin occurs with glutenin. The former conveys resistance to extension while the latter provides the softness and viscosity of the dough. (See gluten, gliadin, *Triticum*)

GLYCAN: is a general old term for polysaccharides.

GLYCEROL ($CH_2OH-CHOH-CH_2OH$): is an intermediate in carbohydrate and lipid biosynthesis.

GLYCEROPHOSPHOLIPID: fatty acids are esterified to glycerol and a polar alcohol is linked to it by phosphodiester bond; parts of cell membranes (synonymous with phosphoglycerides).

GLYCINE BIOSYNTHESIS: glycine (NH_2CH_2COOH) is synthesized by hydroxymethyltransferase from serine ($HOCH_2CH(NH_2)COOH$) while tetrahydrofolate is converted to N^5,N^{10}-methylene tetrahydrofolate. The transferase gene has been located to human chromosome 12q12-q14 whereas the tetrahydrofolate cyclases are in human chromosomes 8q21, 18-qter. Glycine is synthesized alternatively from CO_2 and NH_4^+ by glycine synthase in the liver of vertebrates. (See glycinemia ketotic)

Glycine max (soybean): a leguminous plant (basic chromosome number 20). The seed contains 20 to 23% oil and its protein content (meal) may exceed 40% and it is one of the most important source for vegetable oil products and textured proteins for human food. Also, it is used as supplements to animal feed mixtures.

GLYCINEMIA, KETOTIC (PCC): is caused by two genes at two human chromosomal locations (PCCB at 3q13.3-q22 and PCCA at 13q32). The biochemical defect is propionyl-CoA carboxylase deficiency. This enzyme's primary known role is the generation of D-methylmalonyl-CoA which is epimerized into the L form and subsequently by a mutase—with vitamin B_{12} cofactor— to succinyl-CoA. These processes concomitantly somehow produce ketosis, hypoglycemia and hyperglycinemia. The symptoms are growth retardation, vomiting, lethargy, protein intolerance, low level of neutrophilic leukocytes, reduction in platelet number, etc. (See ketoacidosis, amino acid metabolism, glycine biosynthesis, methylmalonicaciduria)

GLYCOCALYX: a carbohydrate-rich membrane glycoprotein-lipid layer of the eukaryotic cell surface.

GLYCOGEN: is the main storage polysaccharide in animal cells; about 7% of the wet weight of

Glycogen continued

the liver is glycogen and it is present in the muscle cells. It is branched at every 8 to 12 residues. As needed, it is hydrolyzed into glucose to supply energy with the aid of enzymes that are associated with its granular form. Glycogen is synthesized from glucose-6-phosphate by first changing it into glucose-1-phosphate by phosphoglucomutase. Then UDP-glucose pyrophosphorylase converts G-1-P and UTP into UDP-glucose and pyrophosphate (PP_i). Glycogen synthase then converts UDP-glucose into glycogen. *Glycogen synthase a* is the dephosphorylated active form of the enzyme whereas the phosphorylated *glycogen synthase b* is inactive. The reaction requires a primer of α1-4 polyglucose and the protein glycogenin. The branching is generated by branching enzymes amylo-(1→4) to (1→6) transglycolase or glycosyl-(4→6) transferase. The glycogen metabolism is regulated by glucagon and insulin in the liver and mainly by epinephrine and insulin in the muscles. The level of glucagon is regulated by cAMP. *Glycogen synthase kinase-3* regulates glycogen and protein synthesis by insulin, modulates transcription factor AP-1 and CREB, dorso-ventral patterning of embryogenesis. (See epinephrine, insulin, glycogen storage diseases, AP, CREB, Akt, PTG)

GLYCOGEN STORAGE DISEASES: Several hereditary defects have been identified in the synthesis and catabolism of glycogen: 1. *Gierke's disease* (type I glycogen storage disease) involves a deficiency of glucose-6-phosphatase, determined by an autosomal recessive gene. The patients develop liver enlargement (hepatomegaly) and subnormal level of blood sugar content (hypoglycemia), increased levels of ketone bodies (acetone) in tissues and fluids (ketosis), as well as high amounts of lactic and uric acids in the blood. 2. *Type II glycogen disease* (chromosome 17) is determined by an autosomal recessive condition causing a deficiency of lysosomal α-1,4-glucosidase (acid maltase). Infants develop excessive enlargement of the heart (cardiomegaly) because of the deposition of glycogen in the heart. By age 2 they succumb to cardiorespiratory failure. The defect can be diagnosed prenatally after amniocentesis. A milder form of the disease is also known with prolonged survival. 3. *Type III glycogen disease* (see Forbes disease) is also caused by autosomal recessive mutations. The basic physiological defects involve in variable forms the glycogen debranching process. The symptoms are not as severe as in Type II disease and the patients may survive longer and by age some of the symptoms may be somewhat alleviated. 4. *Type IV disease* involves an autosomal recessive defect of the glycogen branching enzymes. Progressive destruction of the liver cells is accompanied by an increase of the connective tissues and the liver substance (cirrhosis) and increase of the size of the liver and spleen and accumulation of fluids in the abdominal cavity (ascites) results in death before age two. 5. In *McArdle's disease* (chromosome 11) the homozygosity of autosomal recessive genes causes variable symptoms accompanied by glycogen accumulation. Phosphorylating activity in the muscle tissues is deficient. Painful cramps on physical exercise is the first symptom with an onset around age 20. There is no hypoglycemia or increase of lactate in the blood but some patients excrete myoglobins in the urine. 6. *Type VI* (chromosome 14) patients accumulate glycogen and some have reduced phosphorylating activity. 7. *Type VII disease*, determined by chromosome 1 recessive genes, resembles Type V disease but the patients have reduced phosphofructokinase activity. 8. *Type VIII* glycogen storage disease is caused by an Xp-chromosomal recessive gene and thus affects primarily males. It is based on a leukocyte phosphorylase b activation deficiency. Some glycogen diseases involve multiple enzyme defects. Type II glycogen storage disease is also called Pompe's disease. These diseases are frequently associated with muscle weakness and various other adverse effects. (See also glycogen, epilepsy, acid maltase deficiency, neuromuscular disease, Pompe disease)

GLYCOGEN SYNTHASE: see glycogen

GLYCOGENOSIS: is used to designate glycogen storage diseases. (See glycogen storage)

GLYCOLIPID: a lipid with a carbohydrate group.

GLYCOLYSIS: the catabolic pathway from carbohydrates to pyruvate (see Embden-Meyerhof pathway, pentose monophosphate shunt)

GLYCOPHORIN: a 131 amino acid transmembrane glycoprotein.

GLYCOPROTEIN: proteins with covalently linked carbohydrate(s).

GLYCOSAMINOGLYCAN (synonym mucopolysaccharide): is heteropolysaccharide alternating N-acetylglucosamine + uronic acid and N-acetylgalactosamine + uronic acid (glucuronic acid). This family of compounds include chitins, chondroitin sulfate, heparan, heparin, hyaluronic acid, keratans, keratin.

GLYCOSE: is the generic name of monosaccharides, e.g. glucose, fructose, mannose, etc.

GLYCOSIDASE: digests glycosidic bonds.

GLYCOSIDIC BOND: sugar linked to either alcohol or purine, or pyrimidine or sugar through an oxygen or nitrogen atom.

GLYCOSOME: peroxisomes (microbodies) filled with glycolytic enzymes. (See glycolysis, microbody)

GLYCOSURIA: an incompletely recessive defect in glycose reabsorption by the kidney resulting in high sugar level in the urine. (See phlorizin, disaccharide intolerance)

GLYCOSYLASES: are enzymes involved in excision of damaged purines and pyrimidines from the sugar-phosphate backbone of DNA. Different enzymes work on different bases. The uracil-DNA glycosylases (human gene UNG, chromosome 12) remove uracils that are formed by spontaneous or induced deamination of cytosine to avoid U-G mispairing potentially leading to GC→AT transitions. The uracil-DNA glycosylase (UDG) is one of the most efficient of these repair enzymes. The enzyme pushes and pulls out the improper uracil nucleotide from the major groove of the DNA. Subsequently UDG excises U, AP endonuclease cleaves the DNA backbone, deoxyribophosphodiesterase removes the 5'-phosphate group and DNA polymerase β replaces the correct nucleotide and ligase finishes the job. It has been estimated that in a human cell 100 to 500 cytosine residues are deaminated daily. The 3-methyladenine-DNA glycosylase (gene AAG/MPG, chromosome 16) works on N-3- and N-7 methylation adducts of purines (including hypoxanthine) and cyclic adducts. The pyrimidine hydrate-DNA glycosylase removes damaged or altered pyrimidines. The formamidopyrimidine-DNA glycosylase excises oxydatively damaged purines such as 8-oxyguanine and 8-hydroxyguanine and formamidopyrimidines. DNA glycosylase remove also deaminated 5-methylcytosines that are common in eukaryotic DNA. All these excision repair enzymes maintain the working conditions of the human cells that each suffer more than 10,000 damages each day. The yeast or *E. coli* glycosylases have similar functions but the proteins involved are different in size. (See excinucleases, AP endonucleases, DNA repair, transition, adduct)

GLYCOSYLATION: the attachment of sugars to proteins either through a hydroxyl group of serine or threonine (O-glycosylation) or to the amide group of an asparagine (N-glycosylation). Glycosylated proteins have many different types of cellular functions. Some antibiotics (tunicamycin) interfere with the process.

GLYCOSYLTRANSFERASES: are enzymes adding glucose to proteins and lipids involved in the formation of lipopolysaccharides used for bacterial cell wall. The ABO blood group alleles also encode glycosyltransferases (also the B gene product adds galactose). These enzymes shape the cell surface, determine cell to cell contacts, play some role in cancer, and have important role in various sphingolipidoses. (See sphingolipidoses, ABO blood group)

GLYOXYLATE CYCLE: converts acetate into succinate and finally to carbohydrate.

GLYOXYSOME: vesicles in plant seeds, special type of microbodies (peroxisomes) in plants mediate the conversion of fatty acids to succinic acid to produce through the glyoxylate cycle peroxiacetyl CoA and glucose. (See microbody, glyoxylate cycle)

GLYPICAN: see Simpson-Golabi-Behmel syndrome

GLYT: glycine-specific transporters to the nervous system; they may be inhibitory neurotransmitters through a ligand-gated Cl⁻ channels, activated by glycine or may modulate glutamate-mediated neurotransmission. (See transporters, ion channels)

GMENDEL: computer program for analysis of segregation and linkage. (See J. Hered. 81:407)

GM1-GANGLIOSIDOSIS: see gangliosidosis type I

GM2-GANGLIOSIDOSIS: see Tay-Sach disease, Sandhoff disease

GM3-GANGLIOSIDOSIS: see gangliosidosis type III, Sandhoff disease

GM-CSF: granulocyte—macrophage colony stimulating growth factor, a lymphokine. (See lymphokines, macrophage, granulocyte, MCSF)

GMO: genetically modified organisms; a name actually used for transgenic plants and animals.

GMS (genomic mismatch scanning): is a method designed to scan large genomic DNA samples for differences in order to identify alterations, e.g. those responsible for hereditary disease. The principles are as follows: two DNA samples (diseased and healthy) are digested with restriction endonuclease. Fragments of one of the samples is methylated. Then both samples are denatured and allowed to hybridize. From reannealed DNA only those strands are subjected to further study which are hybrids (i.e. one of the two strands is methylated but other is not). These hybrids are exposed to bacterial mismatch repair enzymes that recognize mismatches and at that site nick the unmethylated strand. The nicked strands are then removed and the intact duplexes retained. These would be expected to include the desired marker(s). The method is very elegant in principles but cannot yet be applied to the very complex human genome with great amount of redundancy and complexity. (See also RDA, mismatch repair, genetic screening)

GnRHA (gonadotropin-releasing hormone agonist): when administered at a constant rate, it shuts down mammalian reproductive functions and induces a condition resembling the menopause. It can be employed as a fertility controling agent but must be supplemented with periodic treatments with other hormones to prevent menopause-like side effects. It can be used also to save an implanted ovum or zygote by preventing ovulation. (See gonadotropin releasing factor, *in vitro* fertilization, ART, menopause, menstruation)

GNRP (guanine nucleotide releasing protein): when activated by receptor tyrosine kinase in the signal transduction pathway, a RAS protein switch is turned on. (See RAS, signal transduction)

GO: a dormant stage of cell divisions in fission yeast. (See cell cycle, *Schizosaccharomyces pombe*)

G_o PROTEIN: a subunit of the trimeric G-protein; it activates K^+ channels and shuts down Ca^{++} ion channels. Mutations in the gene encoding it causes behavioral anomalies in *Caenorhabitis* similar to those caused by defect in the serotonin receptor. The main symptoms are hyperactivity, premature egg laying and male impotence due to defects in neuronal and muscle functions. (See signal transduction, ion channels, serotonin, G_α)

GOAT (*Capra hircus*): 2n = 60.

GOAT-SHEEP HYBRIDS: the domesticated sheep (*Ovies aries*, 2n = 54) can be fertilized with the domesticated goat (*Capra hircus*, 2n = 60) but the hybrid embryo rarely develops normally although occasionally some hybrids grew up.

GOGAT: glutamine — 2-oxoglutarate transferase.

GOITER, FAMILIAL: is actually a collection of various metabolic anomalies involving enlargement of the thyroid gland that may become obvious by viewing the neck. The defect may involve various dominant or recessive mutations in the thyroglobulin gene. The thyroglobulin gene (TG) is in human chromosome 8q24 extending to about 300 kb genomic DNA, containing 37 exons and large introns. The dimeric thyroglobulin protein has a molecular weight of ca. 660,000. This protein is iodinated at tyrosine residues to form mono- and diiodotyrosines. *Thyroxine* is a tetraiodothyronine but also *triiodothyronine* is formed upon activation by peroxidase. The iodinated proteins are transported by the blood and increase the metabolism and regulate the function of the nervous system, kidney, liver and heart functions. *Hyperthyroidism* occurs by overproducing iodinated thyroglobulin hormones, resulting in goiter, fast heart rate, fatigue, muscular weakness, heat intolerance and sweating, tremor and emotional instability. Excessive secretion of thyroid hormones is called by the synonymous *Graves* or *Basedow* disease. The latter condition may or may not be genetic although its frequency may be quite high (0.008). The basic defect may involve autoimmunity of the receptor of the horm-

Goiter familial continued

one. *Hypothyroidism* is the consequence of underproduction of the thyroid hormone, resulting in fatigue, lethargy, low metabolism, cold-sensitivity and menstrual problems in females. This condition may lead to *cretenism* which is most commonly caused by failure of releasing *thyrotropin*, the glycoprotein thyroid-stimulating hormone of the anterior pituitary. Cretenism also means an arrest of physical and mental development caused by this hormonal deficiency. Hypothyroidism may also lead to deafness. Defects in deiodination of iodotyrosines also may cause hypothyroidism. Goiter type diseases are known in the majority of mammals. Thyroxine binding globulin is encoded in human chromosome Xq28 and a thyroxin binding serum globulin is autosomal. (See hyper thyroidism, animal hormones, tyrosine)

GOLDBERG-HOGNESS BOX: see Hogness box

GOLDENHAR SYNDROME: autosomal dominant and recessive forms with different expressions of facial and other developmental deformities.

GOLGI APPARATUS: flat vesicles containing cellular storage and transport material involved

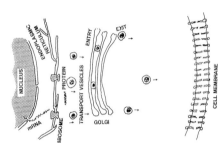

TRANSPORT FUNCTION OF THE GOLGI APPARATUS. FROM THE NUCLEUS, THROUGH THE NUCLEAR PORES, mRNA IS REACHING THE RIBOSOMES SITTING ON THE ENDOPLASDMIC RETICULUM (ER). THE PROTEIN SYNTHESIZED MAY ENTER THE LUMEN OF THE ER WITH THE ASSISTANCE OF A SIGNAL PEPTIDE TRANSFER PARTICLE-MEDIATED SYSTEM. THE PROTEINS MAY EMERGE THEN IN TRANSPORT VESICLES TO ENTER THE GOLGI AT THE CIS SIDE END EXIT AT THE " BULBOUS" ENDS OF THE STACKED MEMBRANE VESICLES. IN THE GOLGI THE PROTEINS ARE GLYCOSYLATED AND MODIFIED POST-TRANSLATIONALLY.

in glycosylation, sulfation, proteolysis, etc. in animals. The homologous structures in plants, are frequently called dictyosomes. Some of the Golgi structures are located next to endoplasmic reticulum and are called *cis Golgi* others are at a distance (*trans Golgi*). In these vesicles some proteins are modified after the completion of their synthesis in the endoplasmic reticulum. (See endoplasmic reticulum, dictyosome, cell structure, trans-Golgi network, cis-Golgi)

G_{olf}-**PROTEIN**: is a trimeric G-protein; stimulating cAMP in the control of olfactory neurons. (See G-proteins, olfactory, olfactogenetics)

GOMBO SYNDROME: is an autosomal recessive growth retardation with eye, brain, skeletal and mental defects. (See growth retardation)

GOMORI'S (Gömöri's) **STAIN**: is used primarily for histological localization of phosphatases and lipases in sectioned specimens by the light microscope. (See stains, histochemistry)

GONADAL DYSGENESIS: failure of normal differentiation of the gonads (ovary, testis). It is a common cause of sterility in aneuploids. Gonadal dysgenesis of XY chromosomal constitution occurs in mammalian females. They have "streak gonads" and fail to develop the secondary sexual characteristics. Gonadal neoplasias are frequent in these individuals. It has been shown that the testis-determining factor resides in a Y-chromosomal segment and either deletion or base substitution may lead to an inactive human SRY products, a DNA binding protein involved in testis determination. Transfection of the TDY (the mouse homolog of SRY) DNA into XX mouse induced male development. Gonadal dysgenesis may occur also in XX females which had higher than normal level of gonadotropins and underdeveloped male gonads. The cause of the dysgenesis may reside either in autosomal recessive genes or in the sex chromosomes. (See H - Y antigen, testicular feminization, hermaphroditism, pseudohermaphroditism, SRY, SOX, campomelic dysplasia, gonad)

GONADOTROPIN: a group of hormones that regulate gonadal and placental functions. (See MSAFP, GnRHA)

GONADOTROPIN-RELEASING FACTOR: see luteinizing hormone-releasing factor.

GONADOTROPIN-RELEASING HORMONE AGONIST: see GnRHA

GONADS: the organs of gametogenesis, such as ovary and testis. (See gametogenesis)

GONOCHORISM: normal development of both types of sexual organs. (See hermaphrodite)

GONOSOMIC: occurring in the somatic cell of the gonads.

GONZALO OF SPAIN: a great grandson of Queen Victoria of England who inherited from her grandmother, Beatrice, the classic X-chromosomal hemophilia gene and died by hemorrhage in an automobile accident at age 20. (See hemophilia, anticoagulation factors, Queen Victoria)

GOODNESS OF FIT TEST: see chi square

GOOSE: *Anser anser*, 2n = 80.

GOOSEBERRY (*Ribes* spp): tart berry fruits; 2n = 2x = 16.

GOPHER: genetic information databases, accessible through INTERNET electronic networks. Software for gopher is free and can be obtained with FTP (file transfer protocol) by <gopher @boombox.micro.umn.edu> (The name comes from a ground squirrel)

GORDON SYNDROME: involves hypertension and high salt concentration in the blood with normal filtration rate in the kidneys. (See hypertension)

GORILLA: see Pongidae, primates

GORLIN-CHAUDHRY-MOSS SYNDROME: is a very rare autosomal recessive craniofacial dysostosis (head malformation), excessive hairiness, heart and lung defect (patent ductus), hypoplasia (reduced growth) of the female external genitalia. (See hypertrichosis)

GORLIN-GOLTZ SYNDROME: see nevoid basal cell carcinoma

GOSSYPIUM (cotton): a member of the malvaceae family of plants. Economically the most important are the long staple upland species, *G. hirsutum* (2n = 4x = 52) that produces 95% of the cotton fibers, and *G. barbadense* an extra long staple (Sea Island, Egyptian) cotton (also a tetraploid), contributes about 5% of the world fiber. There are 30 diploid species. *G. herbaceum* and *G. arboreum* carry the *A* genome and are the only diploids with spinnable lint. The *B* genome is represented by North-African and Cape Verde Islands species. The *C* genome occurs in Australian diploids. *D* genome plants occur in Mexico, Peru, Galopagos Islands and the USA. The *E* genome species occur in North Africa, Arabia and Pakistan. The *F* genome is represented by a single African species. The new world tetraploids contain the *A* and *D* genomes. Most cottons are naturally cross-pollinated but tolerate inbreeding. The various genomes are distinguished primarily on the basis of chiasma frequencies and the number of univalents in the species hybrids, although some chromosome morphological differences also exist.

GOUT: is a complex hereditary disorder of the joints leading to arthritis caused by overproduction and/or underexcretion of uric acid. In an autosomal recessive gout glucose-6-phosphatase is deficient. In the X-linked gout hypoxanthine-guanine phosphoribosyl transferase deficiency exists. Some gouts are associated with increased turnover of nucleic acids. Autosomal dominant and polygenic forms also are known. Gout may be asymptomatic initially but at later stages the joints and the kidney may become permanently injured. The first signs are pain in the great toe but it may spread to other parts of the foot and also to the wrists and other body parts. The prevalence may vary in different populations from 0.2 up to 10%. The serum urate level may vary from 6 to over 9 mg/100 mL serum. Generally fewer women than men suffer from it but in women the gout may be more sever and destructive. If the diet is very low in proteins, gout may not appear. The uric acid crystals (urate) activate the Hageman factor in the viscous fluid (synovia) of the joints that in turn set into motion a series of events leading to inflammation. Chickens that lack the Hageman factor or in dogs with suppressed number of leukocytes (leukopenia), the inflammatory reaction fails, indicating the role of these factors in gout attack. After the first attack the gout symptoms apparently disappear for weeks or many months yet to return with greater strength. The chronic arthritic gout may produce ulcerating tophi (a chalky urate deposit) in the joint and may cause severe deformation of the affected area. Urates may be deposited also in kidneys, cartilage and bone tissues. Tophi may be present at the fingertips, palms, soles, eyelids, nasal cartilages, and in the eye. Rarely it is observ-

Gout continued

ed in and on the penis, the aorta, on the heart wall (myocardium), valves, tongue, the entrance of the larynx (epiglottis), and vocal cords. Urate deposits occur in between the vertebral disks and cartilages. There is very little or any urate in the spinal cord or in the nervous system. In the kidney medulla urate crystals may accumulate and kidney stones may be formed (lithiasis) in 20 to 40% of the affected persons. Gout is frequently associated with obesity, and hypeuricemia is common in case of diabetes mellitus. Hyperlipoprotein and high triglyceride levels are common in gout. Alcoholism may aggravate hyperuricemia. Serum urate levels are about the same in people of European origin, in North-American Indians, Hawaiians, Japanese, and Chinese. In some Polynesians and Australian aborigines and South-American Indians the urate level may be higher. Overproduction of uric acid is correlated with the availability of L-glutamine and phosphoribosyl-1-pyrophosphate that are rate-limiting precursors in purine biosynthesis. Uric acid is dramatically overproduced in case of (partial) deficiency of hypoxanthine-guanine-phosphoribosyltransferase (HPRT). Glucose-6-phosphatase and glutathione reductase also increase uric acid synthesis. Exposure to lead may increase the occurrence of gouty arthritis due to inflammation of the kidney (saturnine gout). Starvation, Down's syndrome, psoriasis (a skin disease causing silvery scaling and plaques) may increase uricemia. Acute attacks of gout may be successfully treated with colchicine and allopurinol, an inhibitor of xanthine oxidase. Both of these compounds may be highly toxic. Gout due to genetically determined factors is called primary gout, the secondary gout is the result of ingestion of certain chemicals and drugs. Many famous historical persons were afflicted by gout: Medici, Newton, Darwin, Luther, Calvin, Benjamin Franklin, Cotton Mather (who reported plant hybrids in America in 1721), and others. (See also Lesch-Nyhan syndrome, colchicine, antihemophilic factors, Hageman factor)

gp: in general, the abbreviation for glycoprotein; the gp is usually followed by a number.

gp: gene of phage, e.g. the first gene of λ phage entering the capsid is *gpNu*. (See lambda phage)

GP32 PROTEIN (of phage T4): is required for (i) configuration of the single-strand DNA (ssDNA) to accommodate the replisome, including DNA polymerase, (ii) melting adventitious secondary structures, (iii) protect ssDNA from nucleases and (iv) facilitating homologous recombination. (See replication fork, replisome)

gp39: same as CD40 ligand.

gp130: is a subunit of the interleukin 6 family receptors. (See interleukins, APRF)

GPA: genes of yeast homologous to Gα cDNAs involved in mammalian G-protein coding. GPA1 protein is 110, and GPA2 is 83 amino acid longer at the N termini than the mammalian proteins. GPA1 may be involved in mating signal transduction, GPA2 controls cAMP level. GPA1 (α subunit of G-protein) plays a negative role (growth arrest) in mating signal transduction whereas the *STE4/STE18* (β, γ subunits) are responsible for a positive transducing signal (enhancement) for mating. (See G proteins, cAMP, mating type determination in yeast)

G6PD: glucose-6-phosphate dehydrogenase deficiency is responsible for one type of hemolytic anemia in humans; it is controled by a sex-linked recessive gene (map location X28). It catalyzes the reaction G6P + TPN$^+$ + H$_2$O ⇔ 6-phosphogluconic acid + TPNH + H$^+$. (See Zwischenferment, glutathione, glucose-6-phosphate dehydrogenase, malaria)

GPI ANCHORS: glycosyl-phosphatidylinositol cell surface-membrane proteins.

G1ps: G1 (gap 1) pre-synthetic phase (preceding S phase) of mitosis. (See cell cycle, mitosis)

G$_q$-PROTEIN: is a member of the trimeric G-protein family; activates phospholipase C-β and responds to acetylcholine. (See G-proteins, signal transduction, phospholipase, acetylcholine, acetylcholine receptors)

GRAAFIAN FOLLICLE: small sac-like structures on the ovary of mammals containing a mature egg (secondary oocyte). The release of the egg is called ovulation and afterwards the follicle is transformed into a corpus luteum. (See luteinization, corpus luteum)

GRADIENT CENTRIFUGATION: is a technique of separation of cells, subcellular organelles and macromolecules on the basis of their density and shape by centrifugation. The larger part-

Gradient centrifugation continued

icles may be separated at high speed centrifuges while for macromolecules ultracentrifuges are used. The medium of separation may be sucrose, percol, cesium salts, etc. The material is placed either on the top of the medium which is made in various concentrations in steps; i.e., first we place in the centrifuge tube 60% sucrose, layer on top of it 40%, then 20% solutions. Alternatively, cesium salts may be used at an average density of the macromolecule. In the latter case during high-speed centrifugation the medium forms a continuous density gradient. In either case, the material will accumulate either at the top of the step (layer) which has higher density than the substance to be separated or it will accumulate as a band in the medium that corresponds to the density of the macromolecule (DNA, ribosomes, viral particles. (See also ultracentrifuge, buoyant density, DNA density)

GRADING UP: means that an animal breed is repeatedly backcrossed with males of another, more desirable livestock to improve its productivity and/or quality. (See gain)

GRADUALISM: evolution is supposed to proceed by slow acquisition of adaptive mutations in the Darwinian sense. (See Darwinian evolution)

GRAFT: transplantation of plant or animal tissues by surgical means.

GRAFT HYBRID: chimera produced by fusion of two genetically different cells in tissues. Non-chimeric type graft hybrids were postulated by the followers of the Mitchurin, Lysenko and Glushchenko's group of Soviet ideologues. They called them also vegetative hybrids, and claimed that grafting alters the hereditary material of both graft and scion. These claims were not reproducible by appropriate methods of experimentation, and several of the results were due either to ignorance or deliberate deception. (See also lysenkoism, Mitchurin, Soviet genetics, acquired characters)

GRAFT REJECTION: the manifestation and result of histoincompatibility between transplanted tissues. (See cytotoxic T cells, HLA, MHC)

GRAFTING: transfer one piece of tissue or an organ from one place to another within the body or to another body. Among plants grafting has been practiced by horticulturists as a means of propagation. Grafted roses and other ornamentals, as well as fruit trees assure the maintenance of genetic uniformity in the grafts where multiplication by seed would produce a heterogeneous offspring because of heterozygosity at multiple loci. Some grafts are horticulturally advantageous because the root stock may be resistant to soil-born pests and can secure crops in more valuable varieties of *Vitis vinifera* grapes. *Vitis rotundifolia*, a wild grape stocks may be 20 fold more resistant to the *Dactylosphaera vitifolii* root parasite than the standard varieties. Grafting may be used to propagate inviable plants on appropriate stocks and study the physiological interactions between scion and stock in such complex processes as flowering response, etc. (See also graft hybrids)

GRAFTING IN MEDICINE: grafting is practiced in modern medicine by transplanting skin, kidneys, liver, heart and other organs. Allografts are generally incompatible with the host immune system. The immune response is controled by a large number of genes that are part of the major histocompatibility gene families. Experimental studies on tissue transplantation are carried out with inbred strains of mice (see congenic resistant). The histocompatibility genes are codominant and F_1 hybrids between different inbred lines may accept graft from both parents whereas the two parental lines may be incompatible with each other. F_1 hybrids generally accept grafts also from their later generation offspring. The incompatibility is inherited in a Mendelian manner and 3/4 of the F_2 individuals are compatible with one or the other parent and 1/4 is not. If the number of independent histocompatibility loci is (n), $(3/4)^n$ = the number of histocompatible individuals in F_2 and in a backcross it is $(1/2)^n$. There are some confounding factors, however. Within some highly inbred lines skin grafts from male to female may be rejected but not from female to male. This may be due to "male-specific" antigens encoded by the Y chromosome. Also, tumor tissues may be rejected when skin grafts are accepted because of tumor-specific antigens. Heterotopically (placed to a non-regular position) transferred hearts and kidneys may be accepted when skin grafts are rejected. Even ap-

Grafting in medicine continued

parently accepted graft may produce very low level of antibodies. Also allogeneic inhibition may occur, i.e. parental animals fail to accept transplants from their offspring but the offspring may accept the transfer from that parent. Most of these principles of grafting were derived from studies on inbred mice strains. (See HLA, allogeneic)

GRAM MOLECULAR WEIGHT: grams of a compound equal to its molecular weight: mole.

GRAM NEGATIVE/GRAM POSITIVE: classification of bacteria depending on retention of the Gram stain (gentiana violet after an iodine stain, and then extracted by acetone or alcohol).

GRAM STAIN: see bacteria, Gram negative/Gram positive

GRANA (sing. granum): dark green pile of flattened membrane vesicles (thylakoids) in the chloroplasts. (See chloroplast, chloroplast genetics)

GRANDCHILDLESS-KNIRPS SYNDROME: in *Drosophila*: maternal effect genes cause embryonic lethality by eliminating pole cells and one or more abdominal segments. (See morphogenesis, pole cells)

GRANDE: the wild type cells of yeast in comparison with the petite mitochondrial mutants deficient in respiration. (See petite, mtDNA)

GRANIN: calcium-binding acidic proteins (21 to 76 kDa) in the Golgi network. Their function is processing secreted proteins and they are subject to processing by converting into biologically active peptides. The granin consensus bears similarity to breast cancer gene proteins, BRCA1 and 2. (See Golgi, breast cancer)

GRANTHAM's RULE: highly expressed genes preferentially use, from the synonymous codons, those that have a pyrimidine at the third position of the triplets. (See genetic code)

GRANULE EXOCYTOSIS: see cytotoxic T cells

GRANULOCYTE COLONY STIMULATING GROWTH FACTOR: G-CSF (See lymphokines)

GRANULOCYTES (polymorphonuclear leukocyte): specialized white blood cells such as neutrophils, eosinophils, basophils; they contain numerous lysosomes and secretory vesicles (granules) and play important role in the defense system of the animal body. (See lysosome)

GRANULOMATOUS DISEASE, CHRONIC (CGD): is a group of X-chromosomal recessive conditions involving chronic infections, based on defects in NADPH-oxidase subunits of the neutrophils and other phagocytotic leukocytes. If the normal enzyme is activated it generates superoxide that is converted to antimicrobial hydrogen peroxide. The 91 kDa membrane glycoprotein, a phagocyte oxidase ($gp91^{phox}$) is a part of the cytochrome b system. (See neutrophil, leukocyte, superoxide dismutase, hydrogen peroxide, McLeod syndrome)

GRANUM IN CHLOROPLASTS (plural grana): the multilayered thylakoids appear as dark "grains" in the chloroplasts viewed by the light microscope. (See chloroplasts, chloroplast genetics)

GRANZYMES: cytotoxic T cell serine proteases, responsible for apoptosis by activating the precursor (CPP32) of the protease cleaving poly(ADP-ribose) polymerase. (See CTL, ICE, apoptosis, perforin)

GRAPEFRUIT: *Citrus paradisi*, 2n = 18, 27, 36.

GRAPES: *Vitis vinifera*, 2n = 38; the *Muscadina* species, 2n = 2x = 40. (See resveratrol)

GRAPES, SEEDLESS: are normal diploids but a gene prevents the division of the embryo, presumably because of shortage of hormones (stenospermocarpy). (See seedless fruits)

GRASSES (cultivated for herbage): blue grass (*Poa pratensis*) 2n = 36-123; Italian ryegrass (*Lolium multiflorum*) 2x = 14; meadow fescue (*Festuca pratensis*) 2x = 14; orchardgrass (*Dactylis glomerata*) 4x = 28; perennial ryegrass (*Lolium perenne*) 2x = 14; smooth brome (*Bromus inermis*) 8x = 56; tall fescue (*Festuca arundinacea*) x = 7, 2n = 42; timothy (*Phleum pratense*) 6x = 42.

GRASSHOPPERS (*Orthoptera*): are suitable objects of cytological and evolutionary investigations because of the large size of chromosomes and in numbers (n = 13 to 57 in males that are of XO constitution); *Melanopus differentialis* 2n = 24. The variation in chromosome number is supposed to be due to chromatin reorganization rather than to polyploidy.

GRATUITOUS INDUCER: a substrate analog of an inducible enzyme that may trigger transcription of the gene concerned, such as IPTG (isopropyl thiogalactoside) for the *Lac* operon, although it is not metabolized by the *z* gene of the operon. (See inducer, *Lac* operon)

GRAVES DISEASE: see goiter

GRAVITROPISM: a tendency of plant organs such a roots to grow in the direction of the terrestrial gravitation. The mechanism of this response is unclear although amyloplasts and other cytoplasmic characteristics have been suggested as possible receptor sites. (See statolith, phototropism)

GRAY CRESCENT: a pale area in some amphibian eggs, opposite to the sperm entry; at this point will the dorsal parts be initiated. (See dorsal)

GRAY MATTER: is a tissue of the hippocampus. Its anomalies may lead to psychomotor retardation, seizures that are resistant to anticonvulsant therapy. Some of the hereditary infantile seizures respond dramatically to large doses of vitamin B_6 (pyridoxine). (See brain human)

GRAY UNITS: of ionizing radiations 1 Gy = 100 rad. (See R, rad, rem, Sievert)

GRB2 (growth factor receptor-bound protein): is a vertebrate adaptor protein with SH2 and SH3 binding domains; it is a downstream receptor kinase. It mediates the activation of guanine nucleotide exchange (GTP⇔GDP) on RAS, a homolog of the *Drosophila* protein DRK. (See DRK, signal transduction, SH2, SH3)

GRE (glucocorticoid receptor element): situated upstream from the TATA box gene regulatory tract. (See glucocorticoid, glucocorticoid response element)

GREEK-KEY: a protein configuration where β-sheets are connected across the end of a barrel. (See barrel)

"GREEN BEARD EFFECT": an idea that some unique traits are specially favored by the parents altruistic behavior during evolution. (See gestational drive)

GREEN FLUORESCENT PROTEIN: see aequorin, Renilla GFP

GREEN REVOLUTION: the development of new plant (cereal crop) varieties that because of the shorter and stronger stems and improved disease resistance permitted more intensive agricultural practices (use of higher doses of fertilizers, irrigation, etc.) and resulted in 2-3 fold increases in grain yield.

GREIG's CEPHALOPOLYSYNDACTYLY SYNDROME (GCPS): dominant, in the short arm of human chromosome 7p13, involves polysyndactyly and malformation of the head without mental defects. Molecular analysis indicates that the anomaly is concerned with GLI3 oncogen, a CREB-binding protein. The protein product of the *cubitus inrruptus locus* of *Drosophila*, involved in the regulation of limb development is also a homolog. (See hedgehog, Rubinstein-Taybi syndrome, GLI oncogene, morphogenesis in *Drosophila*, polydactyly)

GRM: general regulator of mating type in yeast in cooperation with PRTF. (See mating type determination in yeast, PRTF)

gRNA (guide RNA): in the kinetoplast, mitochondrial DNA of some protozoa the pan edited primary RNA transcripts substantially modified by U additions or deletions but some short sequences (50-100 bases) may remain homologous to the primary transcript. These sequences are apparently anchored to the 3'-end and thus pairing may get started. Additional homology may occur in the middle (20-30 bases) and at the 5'-end (ca. 10 bases). These gRNAs may serve as templates for editing. The process requires a series of enzymatic steps. Free uridine triphosphates are the source of the Us inserted and they are added to the 3' ends generated by the enzymatic cleavage. (See kinetoplast, *Trypanosoma*, *Leischmania*, RNA editing, pan editing, mtDNA)

gro: hamster gene activated by mitogens (see KC, N51, MGSA)

GroEL: is a homo-tetradecameric chaperon, composed of 57 kDa subunits of three functional domains each, arranged as a hollow cylinder of two stacked rings with a seven fold symmetry of

E. coli.. It binds to the smaller GroES molecule. (See chaperone, chaperonin)

GROUNDNUT (*Arachis hypogea*): about 40 - 70 species, 2n = 2x = 20; some have higher ploidy.

GROUND STATE: stable, normal, not excited form of an atom or molecule.

GROUP TRANSFER POTENTIAL: ability of a compound to donate an activated group (e.g. phosphate or acyl).

GROWTH: see cell growth, growth curve, exponential growth

GROWTH CURVE: cell multiplication may start at an exponential rate under ideal conditions for proliferation then it either reaches stationary phase (growth flattens) and only maintenance of the cell population takes place (S curve). The exponential growth is named so because the growth mathematically can be defined by an exponent of the base 2. Thus after 10 divisions of a cell the expected number is 2^{10}. In case the initial cell number was 100, after exponential growth, the number of cells 10 generations later would be 2^{10} x 100 = 1024 x 100. Alternatively, the growth may decline and the level of the population decreases. In higher organisms such growth curves can be observed only in isolated cell cultures. In differentiated tissues the growth has structural limitations.

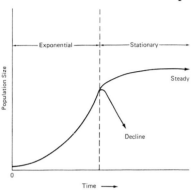

GROWTH-ASSOCIATED KINASE: is an M-phase histone-1 kinase, functions at its peak in mitotic M phase but its activity ebbs at other phases; it is active also during meiosis. (See histones, mitosis)

GROWTH FACTORS: see FGF, PDGF, EGF, IGF-I, IL-2, IL-3, NGF, TGF, erythropoietin, cell cycle, growth hormone pituitary.

GROWTH HORMONE, PITUITARY: gene complex is in human chromosome 17q23-q24 and encodes a 190 amino acid protein, human growth hormone (hGH), and also the somatotropin, a chorionic growth hormone (191 amino residue with ≈85% homology to hGH) and a growth hormone-like protein (GHL, 22 kDa). The expression of the hormone gene complex is regulated by the 33 kDA growth hormone transcription factor (GHF1) and a growth factor response protein (GFRP1). The hGH is released by a 44-amino acid growth hormone-release factor (GHRF) encoded at chromosome 20p12. (See pituitay dwarfness, pituitary, brain human, Rowley-Rosenberg syndrome, Gapo syndrome)

GROWTH HORMONES: see animal hormones, plant hormones

GROWTH RETARDATION: reduction in the rate of increase in size, cell multiplication, differentiation and development. (See retardation, Rowley-Rosenberg syndrome, GAPO syndrome, Gombo syndrome)

GRUNSTEIN - HOGNESS SCREENING: involves *in situ* lysis of bacterial colonies on nitrocellulose filters (or other membranes) and non-covalent attachment of the probe DNA to that support medium. (See also Benton - Davis plaque hybridization, probe)

G_s-PROTEIN: is a stimulatory G protein; when bound to GTP it stimulates the activity of adenylate cyclase, the membrane bound enzyme which generates cAMP. G_s has α, β and γ subunits, the GTP/GDP binding site being on the α subunit. When GDP is at the nucleotide-binding site, adenylate cyclase activity ceases. Displacement of GDP and replacement by GTP (mediated by the hormone, epinephrine) restores the active form. At this stage the α subunit with bound GTP, dissociates from β and γ. (See G-proteins, adenylate cyclase, signal transduction)

GSH: reduced glutathione. (See glutathione)

GSK3β (glycogen synthase kinase 3β): is a protein encoded by *Drosophila* gene zeste white (z^{W3}, chromosome 1-1); homologous proteins are present in other animals. It is assumed that GSK is mediating a step in the intestinal polyposis carcinogenic pathway. (See polyposis adenomatous intestinal, translation initiation)

gsp : c-oncogene, its product is the α-subunit of G-proteins. (See G-, c oncogene)

GSP: gene-specific primer. (See directed mutation, c- oncogene)

GST: glutathione-S-transferase. (See glutathione)

GT - AG RULE (Chambon's rule): the first two and the last two nucleotides of introns are GT and AG, respectively; some exceptions are known. (See intron)

G$_t$-PROTEIN (transducin): is a member of the trimeric G-protein family; it activates cGMP phosphodiesterase in photoreceptors. (See G-proteins, rhodopsin)

GTBP (G/T mismatch-binding protein): is encoded in human chromosome 2 and its mutations lead to genetic instability at single nucleotide sites. (See mismatch , DNA repair)

GTF (general transcription factors): proteins that are required for the initiation of transcription by RNA polymerase I (TFIs), RNA polymerase II (TFIIs) and RNA polymerase III (TFIIIs). (See transcription factors)

GTP: guanosine triphosphate.

GTPase: proteins mediating the conversion of GTP (guanosine triphosphate into GDP (guanosine diphosphate). These enzymes regulate translation, signal transduction, cytoskeletal organization, vesicle transport, nuclear import, protein translocation across membranes, etc. Two different GTPases may modulate each others's activity. (See RAS, RAB, RAN, RASA, dynamins, Arf, GAP, SAR)

GTPase-ACTIVATING PROTEIN (GAP): increases GTP hydrolysis to GDP by several orders of magnitude in signal transduction. (See signal transduction)

GTP BINDING PROTEIN SUPERFAMILY: includes transitional factors, transmembrane signaling proteins, Ras proteins and tubulins. (See signal transduction)

GUANIDINIUM CHLORIDE: used in molecular genetics similarly to guanidinium isothiocyanate for isolation of undegraded RNA. (See RNA extraction)

GUANIDINIUM ISOTHIOCYANATE: is used for the isolation of RNA. It breaks up cells, dissociates nucleoproteins and inactivates tough RNase enzymes (at 4 M solutions) in the presence of the reducing agent β-mercaptoethanol. (See RNA extraction)

GUANINE: purine base in DNA and RNA

GUANINE NUCLEOTIDE RELEASING PROTEIN (GNRP): hydrolyzes GTP, bound to G proteins, into GDP. (See signal transduction)

GUANOSINE: the nucleoside of guanine.

GUANYLATE CYCLASE: mediates the formation of cyclic guanosine monophosphate, cGMP.

GUANYLIC ACID: guanine nucleotide

GUAVA (*Psidium guajava*): subtropical, tropical, small, allogamous fruit tree; 2n = 2x = 22.

GUEST PEPTIDE: see CD tagging

GUEST RNA: see CD-tagging

GUEST TAG: see CD-tagging

GUESSMER: usually 30-7- base long synthetic oligonucleotides representing limited degeneracy and using neutral bases (inosine) at sites of ambiguity. The nucleotide sequence is generated on the basis of information of amino acid sequences in the protein. This label can be used for screening for specific coding sequences (genes). If the codons would be picked at random, the synthetic sequence would represent at least 76% homology by chance but by considering codon usage of the organism, the homology may be over 90%. The probes are labeled with the aid of polynucleotide kinase or primer extension with the Klenow fragment. (See probe, primer extension)

GUIDE RNA: chaperones the alignment of splicing by attaching either to the intron or to the exon sequences of the transcript. (See also RNA editing, gRNA, intron, splicing)

GUINEA PIG: *Cavia porcellus*, 2n = 64.

GUS: ß-glucuronidase enzyme (gene) is frequently used as a reporter for the *in vivo* testing of promoters, identifying site-specific expression or monitoring the excision of transposable ele-

GUS continued

ments. Several substrates of the enzyme are useful for releasing a blue color upon activity of GUS. Deficiency of the enzyme in mammals leads to lysosomal storage diseases. (See lysosomal storage diseases, gene fusion, reporter gene)

GUSTDUCIN: see taste

GUTHRIE TEST: detects phenylketonuria because if the blood contains phenylanine, the analog β-2-thienylalanine does not interfere with the growth of *Bacillus subtilis*. (See phenylketonuria, genetic screening)

Gy: Gray units of radiation; 1 Gy = 100 rad. (See rad)

Gy: billion years of geological time.

GYMNOSPERM (Coniferophyta): plants with seeds which are not enclosed in an ovary. Typical representatives are the pine trees (2n = 24).

GYMNOTHECIUM: fruiting body of some ascomycetes fungi; it may cause skin infections. (See perithecium, cleistothecium, ascogonium)

GYNANDER: same as gynandromorph

GYNANDROMORPH: sex mosaic (part male/part female); same as gynander. They are the result of the loss one of the X-chromosome during development of *Drosophila* and other organisms where the XO chromosomal constitution leads to the development of male phenotypic characteristics. The loss of the X-chromosome reveals the recessive alleles present in the remaining homolog. These sex mosaic individuals can be exploited for fate mapping. The right side of the fly shown at left shows the male body characteristics and has a ruby eye because the left sector is X0. The left side appears like females (XX). (See fate mapping, lyonization, variegation)

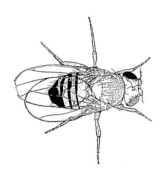

← GYNADROMORPH OF *DROSOPHILA*. From Morgan, T.H. *et al.* Bibliographia Genet. 2:1.

GYNECOMASTIA: increased development of the mammary gland of males involving estrogen accumulation and reduction of testosterone. A transient mild form may not be abnormal during puberty. (See pseudohermaphroditism male, Klinefelter syndrome, Kennedy syndrome, animal hormones)

GYNOECIUM: the carpels, and enclosed structures by it, in the flowers. (See fruits)

GYNOGENESIS: reproduction by parthenogenesis, i. e. the sperm does not fertilize the egg but stimulates the cleavage of the unreduced egg (pseudogamy). Also, embryos developed by transfer of male pronuclei into the egg, and thus diploid, are called *gynogenones*, in contrast to *parthenogenones* (gynogenotes) which arise by parthenogenesis. (See apomixis, parthenogenesis, androgenesis, EP)

GYPSY: is a somewhat diverse ethnic group migrating from the Indian subcontinent north and southward, presumably before the 9th century, to Asia and to Egypt and from there to most of the northern hemisphere although they are now found all over the world. Their Indian origin is asserted by orally transmitted legends. Linguistic evidence indicates Sanskrit roots. Their ethnic identity has been preserved by cultural and genetic isolation. J.B.S. Haldane (1935) used ABO blood type frequencies to show that the Hungarian Gypsies are more closely related to some Eastern Indian populations than to that of Hungary although some of them lived in that country since the early 15th century. They prefer to be called Roma. (See ethnicity)

gypsy RETROPOSON: see copia

GYRASE: a DNA topoisomerase that reverses the direction of coiling in DNA, resulting in negative supercoiling. (See DNA replication, transcription, supercoiling, topoisomerase)

H

H: heavy chain

H1, H2A, H2B, H3, H4: H1° and H5 are variants of H1. (See histones)

h (Planck constant): an energy quantum of radiation that relates it to the frequency of the oscillator that emitted it $E = h\nu$ where E is the energy quantum, ν is its frequency, numerically 6.624×10^{-27} erg/sec

h: also indicates human homolog of a gene or protein standing in front of the symbol.

H-2: the major histocompatibility gene cluster in the mouse is located to chromosome 17, proximal to the centromere within a segment of about 1.3 cM consisting of about 2,000 kb DNA. They encode cell surface glycoproteins that have major role in recognition and immune response to foreign antigens. The gene order in this cluster encoding class I, class II and Class III and Class I-like polypeptides is:
K- A- E - C2 - Bf - SLP - OH - C4 - TNF - D - D2 - D3 - D4 - L - Q - T - T1a - centromere
The transplantation antigens are the class I proteins coded for by genes K, D and L. The class II proteins, encoded by genes A and E, occupy the surface of B and T lymphocytes and the macrophages and participate in cell immune responses. The Class III proteins coded for by genes C2, Bf, SLP, OH and C4 are the complement proteins of the serum involved in the lysis of foreign material after the recognition by the antibody. The Q and T loci determine the so called differentiation antigens present in the blood cells. TNF is the tumor necrosis factor gene. (See also HLA, antibody, lymphocytes, TNF)

h^2: is the symbol of heritability; it is derived historically from Sewall Wright's definition of heritability as the ratio of the standard deviations of the additive and phenotypic variances, $h^2 = V_A/V_P$. Heritability is not a squared entity, and h^2 stands for heritability and not for its square. (See more at heritability, correlation, offspring-parent regression, intraclass correlation, heritability estimation in humans)

HA: see hemagglutinin

HABITUATION (accoutumance à l'auxine, anergie à l'auxine): plant tissues after prolonged culture may dispense of the continued reliance on exogenous auxins for proliferation. This alteration does not involve somatic mutation yet it bears similarity to oncogenic transformation. In animal cells the SV40 T antigen loss after a period of time still may not cease proliferation in the absence of the oncoprotein. (See tissue culture, somatic mutation, transformation oncogenic, tumor, oncoprotein, SV40)

HAC (human artificial chromosome): critical elements (telomeres, centromeres, replicator), and can eventually be used to ferry desirable genes to human cells for medical purposes. (See YAC, BAC, PAC, human artificial chromosome, HAEC)

Hae **II**: restriction enzyme with recognition site $\overset{\downarrow}{\text{GCGC}}\overset{\uparrow}{}$.

Hae **III**: restriction enzyme with recognition site GG↓CC.

HAEC (human artificial episomal chromosome): was constructed by using the replicational origin of the Epstein-Barr virus. Such a construct may carry over 300 kb inserts and may be maintained in human cells without integration (as an episome) in the genome. (Epstein-Barr virus)

HAEM: iron-porphyrin; occurs in different forms. (See also heme)

HAEMOCHROMATOSIS: see hemochromatosis

HAEMATOPOIETIC GROWTH FACTOR: see IL-3

HAEMATOXILIN: see hematoxilin

HAILEY-HAILEY DISEASE (benign pemphigus): is an autosomal dominant skin disease involving vesicle formation generally on the neck, groin and armpits. The benign disease is precipitated by infection by the fungus *Candida albicans* but antifungal, antibacterial drugs may also initiate it. (See skin diseases, pemphigus)

HAIR: the human hair has a high- and a low-sulphur protein; the former is about 40% of the hair

Hair continued

proteins. In some animal hairs a high-tyrosine protein also occurs. (See hair color, hair whorl, alopecia, hypotrichosis, hypertrichosis, baldness, De Lange syndrome, glomerulonephritis with spare hairs, hairy ears, hairy elbows, hairy nose, hairy palms and soles, hair-brain syndrome)

HAIR COLOR IN HUMANS: strikingly blond and red hair colors (prevalence about 2% or less) appear to be autosomal recessive but the latter may have some expression in the heterozygotes. Red hair is hypostatic to brown and black. Brown hair appears to be autosomal dominant and it is closely linked to green eye color. Dark hair appears to be dominant. The babies' first hair may not be concordant with that of later years. The hair color is determined by the relative proportions of the reddish phaeomelanin and the black eumelanin pigments. Their level is controled by the melanocyte stimulating hormone (MSH) and its receptor (MC1R). Environmental factors (temperature, sunshine, diseases) may also effect transiently the color. Graying of the hair proceeds usually by aging, however, precocious graying may be determined by dominant genes and it may be a symptom shared by several syndromes such as the Book, the Waardenburg, the Werner syndromes and pernicious anemia (a vitamin B12 deficiency) may also cause it. Actually, the inheritance of human hair is determined by many loci. In human chromosome 19 alone there are 6 loci homologous to fur color genes of the mouse. (See pigmentation in animals, forelock white, albinism, aging, hypostasis)

HAIR WHORL: may be clockwise that is autosomal dominant over counterclockwise rotation. (See hair)

HAIR-BRAIN SYNDROME (trichothiodystrophy, BIDS): autosomal recessive brittle hair, low intelligence, short stature, reduced fertility and reduced cystine-rich protein in the hair and nails. (See hair, stature, trichothiodystrophy)

HAIRPIN: double-stranded structure in nucleic acids brought about by folding back of palindromes like a hairpin. At the end where the arrow is pointed → ⊂ the unpaired structure may be digested by S_1 single-strand-specific nuclease. (See S_1 nuclease, palindrome)

HAIRY EARS: occur primarily in the male humans and has been thought to be due to Y-chromosomal (holandric) gene(s) or to two genes in the homologous segments of the X and Y chromosomes. Most likely it is determined by autosomal dominant factors and sex-influenced inheritance. (See sex-influenced, holandric genes)

HAIRY ELBOW: autosomal dominant hairiness on the elbows associated with short stature. (See hair)

HAIRY NOSE: autosomal dominant (?) hairs on the nose tip; onset after puberty in the male only (See hair)

HAIRY PALMS and SOLES: apparently autosomal dominant, male transmitted, site-specific hairiness. (See hair)

HAGEMAN FACTOR: is a protein (M_r ca. 80,000) present in the blood plasma and serum of the majority of the mammals but absent in dolphins, killer whales and birds. It is involved in the pathway of coagulation; it also affects vascular permeability, dilates blood vessels, contracts smooth muscles, provokes pain, promotes the migration of leukocytes, induces fibrinolysis (dissolution of fibrin), etc. (See also antihemophilic factors, Hageman trait)

HAGEMAN TRAIT: is controlled by an autosomal recessive gene causing deficiency of the Hageman factor in the blood. Normally the individuals lacking this factor do not show any disease symptom although blood coagulation is slow in the laboratory. No therapy is required yet in case of surgery it is advisable to keep at hand appropriate blood or plasma. (See Hageman factor)

HALDANE's MAPPING FUNCTION: $(1 - e^{-m}) 0.5 = y$ (recombination fraction), where m is the number of exchanges, e is the base of the natural logarithm. Hence $(1 - e^{-m}) = 2y$ and $e^{-m} = 1 - 2y$ and $m = -\ln(1 - 2y)$ and the corrected map distance estimate is $x = m/2$ because each exchange produces maximally 50% recombination. This formula does not take into account interference and thus frequently overestimates map distances. The graph permits reading directly the recombination frequencies corrected by Haldane's mapping function. Different

Haldane's mapping function continued

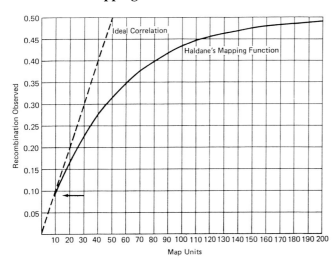

organisms may require the use of different mapping functions (Stahl, F. W. Genetic Recombination, Freeman, San Francisco, CA). (See also Kosambi's mapping function, recombination frequencies, mapping genetic, coefficient of coincidence)

READ THE POINT OF INTERSECTION BETWEEN THE RECOMBINATION FRACTION OBSERVED (ORDINATE) LINE AND THE SOLID LINE CURVE CORRESPONDING TO THE MAPPING FUNCTION AND PROJECT THE POINT TO THE BOTTOM LINE REPRESENTING MAP UNITS. FOR EXAMPLE 0.3 RECOMBINATION CORRESPONDS TO ABOUT 45-46 MAP UNITS. (After Haldane, J.B.S. 1919 J. Genet. 8:299)

HALDANE'S RULE: when in the F_1 offspring of two different animal races one sex is absent, or sterile, that sex is the heterogametic one. The cause is not an imbalance between autosomes and sex-chromosomes, rather it is due to the rather general fact that lethality is usually completely recessive whereas deleterious mutations may show a series of less debilitating effects that may be expressed also in the heterozygotes. (See sex-chromosomes, sex determination)

HALF CHROMATID: involves only one of the strands of the DNA double helix in the chromatid. The expression of mutation is delayed by one division if it occurs in a half-chromatid and it may result in somatic sectors. (See chromatid)

HALF LIFE: in general, the time required for the decay of one half of a compound; the decay for example of a promutagen or procarcinogen may lead to the formation of even more active mutagen and/or carcinogen. The time required for 50% decay of a radioactive isotope. The half life of H^3 and C^{14} is 12 and 5,700 years, respectively, and that of P^{32} 14.3 and I^{131} 8 days, respectively. (See also evolutionary clock, isotopes)

HALF-MUTANT: is Hugo de Vries' term for the new types that arose relatively frequently in *Oenotheras* with translocation rings. The half-mutants produced normal (translocation-ring) and sterile progenies due to recombination between the differential segment and its homolog in another chromosome within the ring resulting in smaller rings and other chromosomal changes. When the non-crossover original translocation ring was recovered in the egg and sperm, the offspring was the same as the original complex heterozygotes. (See complex heterozygote)

HALF-SIBS: share only one biological parent; they are half-sisters or half-brothers. (See sibling)

HALF-TETRAD ANALYSIS: tetrad analysis is feasible in a limited number of organisms where the four products of meiosis can separately be recovered either as a tetrad or after a post-meiotic division as an octad, a phenomenon common is the ascomycetous fungi. In *Drosophila* half-tetrad analysis can be carried out in the presence of attached X-chromosomes. In this case the products of meiosis have either two attached X-chromosomes or are nullisomic for the X-chromosome. X-chromosomal genes are inherited as a block unless recombination takes place. Flies heterozygous for attached X-chromosomal genes can become homozygous only after crossing over and produce two different non-parental gametes from a single meiocyte. This type of recombination provided the first direct evidence that genetic recombination takes place at the 4-strand stage of meiosis. Half-tetrad analysis has been adopted to a plant (*Medicago*) case using four RFLP markers in situations when restitution nuclei were observed at a high frequency. Three of the markers were linked (chromosome 1) and the fourth was independent

Half-tetrad analysis continued

(chromosome 6). The analysis permitted the localization of the centromere. (See illustration, tetrad analysis, gene—centromere distance, attached X-chromosomes, restitution nucleus)

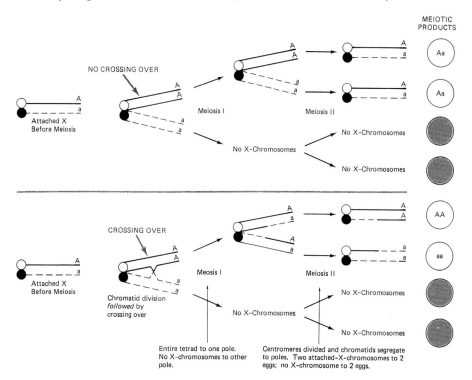

BECAUSE THE CENTROMERES OF THE ATTACHED-X CHROMOSOMES ARE STABLY FUSED, THEY CANNOT PASS TO OPPOSITE POLES DURING ANAPHASE I. SEGREGATION OF THE CHROMATIDS IS EQUATIONAL AT ANAPHASE II. INSTEAD OF FOUR VIABLE MEIOTIC PRODUCTS, ONLY TWO X-DISOMIC AND TWO LETHAL, NULLISOMIC GAMETES ARE FORMED. IF THE TWO ATTACHED CHROMOSOMES CARRY DIFFERENT ALLELES, ALL THE NON-CROSSOVER GAMETES WILL BE THE SAME AND ONLY THE CROSS-OVERS WILL BE OF TWO NON-PARENTAL TYPES. (After Shult, E.E. and Lindegren, C.C. 1959 Can. J. Genet. Cytol. 1:189)

HALF-TRANSLOCATION: is the half of an original reciprocal translocational event. Such a situation occurs, e.g. when a presumably reciprocal interchange takes place between the proximal region of an X-chromosome and the small 4th chromosome of *Drosophila* but one of the translocated strands is not recovered because it gets lost during segregation of chromosomes into the functional gamete. (See translocation chromosomal)

HALF-VALUE LAYER: is reducing the transmission of radiation to half.

HALLERMANN-STREIFF SYNDROME: most likely autosomal recessive bird-like face, long, narrow nose, cataracts, sparse hairiness, occasionally teeth by birth and proportionate dwarfism. (See dwarfs, tooth)

HALLERVORDEN-SPATZ DISEASE (neuroaxonal dystrophy): human chromosome 20p12.3-p13 located recessive brain atrophy accompanied by involuntary movement and early death. The brain accumulates iron in the basal ganglia.

HALOPHYTE: salt-tolerant or -resistant plant species.

HALOTHANE GENE: controls malignant hyperthermia syndrome in pigs (in chromosome 6). [Halothane is a 2-bromo-2-chloro-1,1,1-trifluoroethane, an anesthetic]. (See also PSE)

HALTER (singular), **HALTERES** (plural): see morphogenesis in *Drosophila* (balancing organs).

HALVORSON 5: nutrient concentrate for yeast g/L $(NH_4)_2SO_4$ 20, K_2HPO_4 43.5, succinic acid 29, $CaCl_2 \cdot 2H_2O$ 1.99, $MgSO_4 \cdot 7H_2O$ 5.11, *trace elements* 5 mL [containing mg/0.5 L $Fe_2(SO_4)_3$ 307, $MnSO_4$ 280, $ZnSO_4 \cdot H_2O$], pH 4.7. This medium is usually diluted 5 times.

HAMARTOMA: a tissue overgrowth (of darker color).

HAMILTONIAN PATH: is a directed graph if and only if there is a sequence of one way path beginning with v_{in} (vertex in) and ending at v_{out} and enters every other vertex exactly once. This mathematical method may serve the basis for designing a molecular (DNA) computer that may assist in developing very complex combinatorial approaches to manipulating macromolecules, e.g. designer enzymes.

HAMILTON's RULE: see inclusive fitness

HAMMERHEAD: see ribozymes

HAMSTER: *Cricetulus griseus*, 2n = 22; *Mesocricetus auratus* 2n = 44.

HAND CLASPING: some authors claimed that it is genetically determined whether the left or right hand fingers are on top when the hands are clasped. The latter is supposed to be more common among females. The control may be autosomal dominant or polygenic. (See handedness)

HANDCUFFING: the products of plasmid replication associated incA and incC (incompatibility factors of replication) that cause antiparallel pairing of the two DNAs preventing replication until the hadcuffs are disrupted.

HANDEDNESS: about 93% of human population is right-handed; left-handedness is higher in the younger than in the older age groups indicating greater longevity of the right handed individuals. The inheritance of handedness is unclear; it was attributed to polygenic control and it has been suggested that it is due to a homozygous recessive state and the heterozygotes being ambidextrous (able to use both hands with equal facility).

HANGING-DROP SLIDE: the material to be microscopically examined hangs from the cover slip into the concavity of the slide in a drop of a solution. ⇨

HAPLO-INSUFFICIENT: the gene in a single dose (such as in hemizygotes) cannot assure its normal function and may be even lethal. (See Turner syndrome, hemizygous ineffective)

HAPLODIPLOIDY: in some social insects the males (drones) are haploid (because they develop from unfertilized eggs) whereas the females (queen and workers) are diploid (because they hatch from fertilized eggs). Consequently and usually full sisters (progeny of a single mating) are more closely related to each other and less closely related to their brothers than to their daughters and sons. (See sex determination)

HAPLOID: contains only single complete set(s) of chromosomes (see also monoploid). The gametes are haploid; the meiotic products of polyploids are often called *polyhaploid*. Many of the fungi are haploid during most of their life cycle, except after fusion of the nuclei in dikaryotes preceding meiosis. (See meiosis, fungal life cycle). Haploid individuals among diploids show up spontaneously at low frequency. Haploid lines are very rare in animals although haploid frog cell lines have been used in cultures. The spontaneous frequency of haploids varies among different lines of the same species. High frequency of haploids occur in crosses involving the wild barley, *Hordeum bulbosum*, and certain wheat varieties crossed with the grass *Aegilops caudata (Triticum dichasians)*. In these crosses one of the genomes (e.g. *H. bulbosum*) is eliminated at high frequency. Haploid plants have been obtained in many species in artificial culture of immature pollen or microspores and anthers. The microspores can be cultured without separation from the anthers, although separation of the pollen sacs may improve their development because the anther walls may contain growth inhibitors. From these haploid cells haploid plants can be regenerated. Successful embryogenesis is generally easier if the cultures are initiated at the uninuclear stage of the microsopore. Haploid plants may be obtained actually through two different routes from microspores: either by so called *direct androgenesis* when the haploid cells guided through embryo without delay. Alternatively, the haploid initial cells can be converted to callus and from the callus then plants are generated (indirect androgenesis). The latter procedure is less desirable because during callus

Haploid continued

growth spontaneous chromosome doubling and—even worse—polyploidization and other types of chromosomal aberrations frequently occur. If the microspores are exposed to colchicine or other agents that block the mitotic spindle, chromosome doubling may be induced. Haploid cell cultures have great advantages for mutation studies because all the recessive mutations are detectable without the masking effect of the dominant alleles at the loci. Similar advantages are available in the use of hypoploid animal cell cultures, such as derived from XY males where X-chromosomal genes are present in a single dose or *e.g.* in the culture of Chinese hamster ovary cells (CHO) when individual chromosomes are spontaneously be eliminated. Although effective screening methods are available in animal cells (BUdR, antibiotic, temperature), regeneration of animals from isolated cells cannot be accomplished in culture, except stem cells. Haploids may have great advantage in plant breeding work because by doubling the chromosome numbers, in single step 100% homozygosity results. Ordinarily, by self-fertilization or inbreeding 6 to 8 cycles result in only 98 to 99% homozygosity ($0.5^6 \cong 0.0156$ heterozygosity). S.S. Chase developed a successful method for selective isolation of haploid maize. Females recessive for several markers easily recognizable in the kernels, are pollinated by the corresponding dominant stocks (*A, B, Pl, R*). Kernels that fail to show the paternal markers in the endosperm are discarded because they originated by unintended selfing. The seedlings displaying the dominant paternal markers are also discarded because they are most likely diploid. The true haploids arise thus from fertilization of the endosperm nucleus by the dominant sperm but the egg that develops into an embryo without fertilization yields seedlings with maternal traits only (thus haploid), can be recognized. After doubling their chromosome number, homozygotes are obtained. Some caution may be required in critical studies because through spontaneous mutation some variations may arise in these otherwise completely homozygous lines. Haploidy may be induced also by pollination with heavily irradiated or by other means damaged pollen. Some genotypes display a proclivity for spontaneous androgenesis. (See androgenesis, apomixis, anther culture, embryo culture, embryogenesis somatic, selective medium, antibiotic resistance, bromouracil)

HAPLOIDIZATION: reduction of the chromosome number to the haploid level.

HAPLOID-SPECIFIC GENES: are turned on in response to mating factors (yeast). The responding consensus is 50 bp upstream of the translation initiation site. (See mating type determination in yeast, consensus, upstream, translation)

HAPLONTIC: during most of its life the organism is haploid. (See haploid)

Haplopappus gracilis: a composite plant with only two pairs (n = 2) of good size chromosomes. Due to self incompatibility genes, its culture is somewhat inefficient.

HAPLOTYPE: set of genes in each chromosome of the genome inherited ordinarily as a bloc. (this term is most commonly used in immunogenetics). Originally it was used for representing the haploid set of genes of the MHC (multiple histocompatibility) antigens. (See HLA)

HAPLOTYPE ANALYSIS: infers the relative position of genes and DNA markers by assuming a minimum number of crossing overs along the chromosome. (See crossing over)

HAPPINESS: is most likely genetically determined because the correlation between monozygotic and dizygotic twins was found 0.44 and 0.08, respectively. It has been suggested that happiness may be determined by the D4 dopamine receptor and unhappiness is related to the control of serotonin metabolism. (See dopamine, serotonin)

HAPPY PUPPET SYNDROME: is an abandoned, derogatory name of the Angelman syndrome. (See Angelman syndrome)

HAPTEN: a small molecule that in association with protein (carrier) can act as an antigen. Alone, they are only antigenic but not immunogenic. The hapten-carrier is basis of the immune response to the complex. (See antigen, immune system, affinity labeling)

HAPTOGLOBIN: is a mammalian serum protein composed of two α and two β chains. The α_1 chain contains 84 amino acids, and differs from the α_2 chain that is of nearly of double size in

Haptoglobin continued

the presence of the Hp^2 allele due to a duplication in an intercalary segment, presumably brought about by an unequal crossing over event during its evolution. The α_2 chain occurs only in humans and it is thus most likely of relatively recent origin. This protein is attached to hemoglobin and has a role in recycling heme. The HP gene was located to human chromosome 16q22. A number of electrophoretic variants have been identified. The frequency of the genes responsible for the α_1 chain vary a great deal in ethnic populations. (See globin, hemoglobin, plasma proteins, unequal crossing over)

HARD DISK: is a permanently sealed disk of the computer and operates faster than the floppy disks and it has large capacity. (See disk)

HARD HEREDITY: inheritance determined by a permanent genetic material such as DNA and RNA, rather than the diffuse hypothetical gemmules, pangenes, etc. hypothesized before the acceptance of Mendelian heredity. (See Mendelian laws, pangenesis, gemmules)

HARDWARE: a physical equipment such as the computer machine. (See software)

HARDY - WEINBERG EQUILIBRIUM: the genotype and gene frequencies remain constant from generation to generation because there is random mating between the individuals and neither selection nor mutation or migration affect the composition of the population. (See Hardy - Weinberg theorem)

HARDY - WEINBERG THEOREM: for one allelic pair $p^2 + 2pq + q^2$ where p is the frequency of the dominant and q is the frequency of the recessive allele (1 - p). If the genotypic frequencies are available, the allelic frequencies can be derived because the two types of homozygotes have two copies of the alleles concerned whereas the heterozygotes have one of each. In the case of three alleles at a locus such as in the ABO blood group then the frequency of the i^O recessive allele is $r = \sqrt{\frac{O}{N}}$. Since the combined frequencies of i^O and I^A is $r^2 + 2pr + p^2$ the frequency of the i^A allele is $p = \sqrt{\frac{A+O}{N}} - r$, and therefore the frequency of the allele, by subtraction, becomes $q = 1 - p - r$. The O, A, and B are the actually observed numbers of the representatives of the blood groups and N is the population size. In the case of trisomy the various possible genotypes will be given by $(p_1A_1 + p_2A_2)^3$ after the expansion of this binomial. (See also allelic frequencies, ABO blood group

HARE: *Lepus americanus* 2n = 48; *Lepus towsendii* 2n = 48.

HARELIP: a hereditary cleft of the upper lip. The maxilla (upper jaws) and palate (the partition of the oral and nasal cavity) may also be affected. The incidence in the general population varies between 0.04 to 0.08% and the recurrence risk among brothers and sisters is about 0.2%. It is somewhat more common among males than females; it is a sex-influenced trait. (See cleft palate, sex-influenced, Van der Woulde syndrome)

HARLEQUIN STAINING OF CHROMOSOMES: results in different coloration of sister chromatids as the result of one or two cycles of replication in the presence of the nucleoside analog 5-bromodeoxyuridine (BrDU). The chromatids replicated in the presence of BrDU absorb less of fluorescent stain Hoechst 33258 than the chromatid that has replicated in the presence of thymidine. On a black and white photo negative, the chromatid free of BrDU appears lighter and in the print darker than the one that incorporated BrDU. Thus chromatids containing one, two or no BrdU can be distinguished. Therefore such a staining permits the detection of sister-chromatid exchange. (See sister chromatid exchange, bromouracil)

HARTNUP DISEASE: is an autosomal recessive disorder involving photosensitivity, rash, cerebellar ataxia (impaired muscle coordination by the brain) and aminoaciduria. The uptake of methionine and tryptophan and to some extent of other neutral amino acids by the intestines is reduced. It is diagnosed by urine analysis for increase in neutral amino acids. (See tryptophan, light-sensitivity diseases, ataxia telangiectasia)

HARVEST INDEX: the proportion of the economically directly usable productivity of crops, e.g. grain versus straw.

HARVEY MURINE SARCOMA VIRUS (transformation gene): was originally derived from rats and it is found to encode a 21 kDa oncoprotein (p21) or RAS. The human homolog was mapped to chromosome 11p14.1. (See RAS, p21)

HARWEY: *Drosophila* transposable element (7.2 kb). (See transposable elements)

HAT MEDIUM: of animal cell culture contains **h**ypoxanthine, **a**minopterin and **t**hymidine has been extensively used for the isolation of bromodeoxyuridine and azaguanine resistant mutants and complementary fused cell lines by the rationale outlined in the figure below. (See also selective isolation, bromodeoxyuridine, azaguanine, aminopterin)

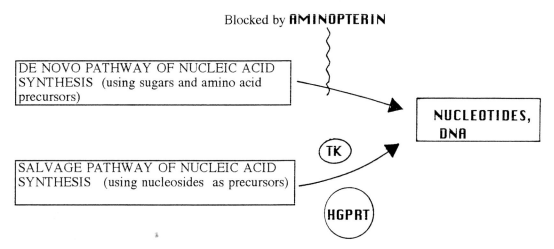

ALTERNATIVE PATHWAYS OF NUCLEIC ACID BIOSYNTHESIS. THE UPPER, *DE NOVO*, PATHWAY CAN BE SELECTIVELY BLOCKED BY AMINOPTERIN AN INHIBITOR OF DEHYDROFOLATE REDUCTASE (AN ENZYME ESSENTIAL FOR THE BIOSYNTHESIS OF THYMIDYLATE). IN CASE OF SUCH A BLOCK, PYRIMIDINES AND PURINES MAY BE STILL SYNTHESIZED THROUGH THE SALVAGE PATHWAY FROM NUCLEOSIDES. TK (THYMIDINE KINASE) CAN MAKE THYMIDYLIC ACID AND HGPRT (HYPOXANTHINE GUANINE PHOSPHORIBOSYL TRANSFERASE) CAN SUPPLY GUANYLIC ACID. IN CASE TK IS INACTIVATED BY MUTATION (*TK*-), THE CELLS BECOME RESISTANT TO 5-BROMOXYURIDINE BECAUSE THE CELLS CANNOT CONVERT IT INTO A NUCLEOTIDE ANALOG. SIMILARLY, IF HGPRT IS INACTIVATED (*HGPRT*-) BY MUTATION THE GUANINE ANALOG, 8-AZAGUANINE CANNOT BE INCORPORATED INTO DNA AND THUS THE CELL WILL BE RESISTANT TO IT. THEREFORE ON HAT MEDIUM BOTH BROMODEOXYURIDINE AND AZXAGUANINE RESISTANT MUTANT CELLS CAN SELECTIVELY BE ISOLATED.

HATCHING TIME OF POULTRY: the eggs of chicken hatch in three weeks, turkey, goose and duck eggs require four weeks incubation.

HAUSTORIUM (plural haustoria) organs of parasites (e.g. fungi) that penetrate the host cells for the purpose of absorbing nutrients.

HAYFLICK's LIMIT: human cells usually die in culture after 50 to 60 or fewer cell cycles. Cancer cells are ordinarily not subjected to such a limit by senescence. (See immortalization, hybridoma, senescence, apoptosis)

Haynaldia villosa: a diploid wild grass (2n = 14) carrying the V genome. It can directly be crossed with tetraploid wheats and the AABBVV hybrids can be backcrossed with hexaploid wheat (AABBDD) and AABBDD + 7V additions can thus be generated. (See addition lines, *Triticum*)

HAZARDOUS CHEMICALS: see chemicals hazardous, environmental mutagens, biohazards

HAZELNUT (*Corylus* spp): a monoecious shrub with edible nuts; 2n = 2x = 28.

HBV: hepatitis B virus.

HCK: see SRC kinase family of oncogenes

HCP: a non-receptor tyrosine phosphatase. (See B lymphocyte)

HDA1, HDA2, HDA3: are histone 3, histone 4 deacylating enzymes of yeast. (See DHAC, histone deacetylation)

HDL: see high-density lipoprotein, Tangier disease

H-DNA: a protonated molecule; apparently it does not have natural biological role. (See DNA types)

HEAD/FACE/BRAIN DEFECTS: see Prader-Willi syndrome, holoprosencephaly, de Lange syndrome, Angelman syndrome, Miller-Dieker lissencephaly, Walker-Wagner syndrome, Opitz syndrome, Smith-Lemli-Opitz syndrome, Opitz-Kaveggia syndrome, Borjeson syndrome, fragile X chromosome, Langer-Giedion syndrome, Coffin-Lowry syndrome, Aarskog syndrome, cranioorodigital syndrome, otopalatodigital syndrome

HEADFUL RULE: bacteriophages replicate their DNA in concatamers but the phage head (capsid) has a limited storage capacity, therefore the molecules have to be cut to "headful" length. (See concatamer, lambda phage)

HEARING DEFICITS: see deafness

HEART DISEASE: affects about 1% of the population and only a little less of the newborns; for about 15% it is lethal. Congenital heart disease frequently occurs as component of various syndromes. About 2% of the cases are caused or precipitated by environmental effects and various diseases (alcoholism, lithium, thalidomide, retinoic acid [a derivative of vitamin A], trimethadione [anticonvulsant drug], viral infections (rubella), maternal diabetes and phenylketonuria, trisomies (21, 8, 13, Turner syndrome), deletions, by the DiGeorge syndrome, asplenia, patent ductus, Holt-Oram syndrome, Ellis-van Creveld syndrome, mucopolysaccharidosis, Pompe's disease, endocardial fibroelastosis, coarctation of the aorta, Noonan syndrome, LEOPARD syndrome, Fallot's teratology, mitral prolapse, myotonic dystrophy, Jervell and Lange-Nielsen syndrome, Opitz-Kaveggia syndrome, etc. (See coronary heart disease, cardiovascular disease, hypertension, and the syndromes named above)

Heartbreaker (*hbr*): a non-translated, short insertion element in several grasses (cereals). (See transposable elements plants)

HEAT REPEATS: a repeated element first discovered in the proteins: **h**untingtin, translation **e**longation factor EF3, the **A** subunit in protein phosphatase 2A (PP2A) and **T**OR1, a target for rapamycin. Similar repeats have since been found in different proteins of several species. (See huntingtin, EF3, TOR, rapamycin)

HEAT SHOCK ELEMENTS: see HSE

HEAT SHOCK PROTEINS (hsp): are most commonly molecular chaperones that under heat stress inhibit or prevent denaturation of other proteins by binding to interactive surfaces and thus securing that these chaperoned proteins maintain their structure required under normal, non-stress conditions. Some hsp proteins may have other functions as well. Hsp110 is present in prokaryotes and eukaryotic organelles in response to stress. Hsp90 family members bind ATP, and are involved in autophosphorylation but in animals they are immunofilins (binding to immune-suppressive antibiotics such as cyclosporin). Hsp70 members are made in all types of organisms and their main role is mediating the folding of nascent proteins. The heptameric hsp60 molecules participate in the molding of proteins using ATP $\rightleftarrows$ ADP for energy. The hsp40 proteins may be involved in sorting polypeptides. The smaller hsp proteins may form aggregates before acting on other molecules. Prokaryotes do not contain the very smallest hsp proteins. The *hsp* genes are classified into families on the basis of the molecular weight of the proteins (in kDa) encoded by them. (See HSE, transcription factors inducible, chaperone, thermal tolerance, cold shock)

HEAVY CHAIN: see antibody

HEAVY CHAIN OF DNA: see DNA heavy chain

HEDGEHOG: *Erinaceus europaeus*, 2n = 48.

hedgehog (*hh*, 3-81): is a segment polarity type embryonic lethal mutation in *Drosophila*. The product of the wild type allele of *hh* encodes a signaling molecule that is processed by autoproteolysis into two active species. These two proteins display differences in tissue distribu-

hedgehog continued

tion and instruct adjacent cells to express the organizing signal encoded by the *decapentaplegic* (*dpp*) locus. The Hh protein is secreted under the control of the En (engrailed) protein. The *en* locus is continuously expressed in the posterior part of the embryo. The anterior part of the embryo expresses the *ci* (*cubitus interruptus*) locus which encodes a Zinc-finger binding protein of the Gli family of transcription factors. If *ci* is not expressed the *hh* gene product shows up and posterior compartment properties appear without the expression of the *en* signal. Increased levels of *ci* products induce the expression of *dpp* independently from *hh*. Expression of the normal *ci* product in the anterior cells results in limb development by limiting the expression of *hh* to posterior cells and mediating the ability to respond to the protein signal of the *hh* locus. Ci transduces the Hh signal by activating the *dpp* and *ptc* (*patch*) genes. The Patch gene product and the cyclic AMP-dependent protein kinase A interfere with inappropriate expression of *dpp* if Hh product is not available. The Ci product at low level represses *dpp* and at higher concentration appears to be an activator of the same gene. Hh induces both decapentaplegic (dorsal compartment) and wingless (ventral compartment) and these two modulate each other's function to assure normal axial development. The product of *patched* (*ptc*) gene is the receptor of *Hh* and *smoothened* (*smo*) is a signaling component of *Ptc*. The *hh* gene has at least three homologs in humans and other vertebrates: the sonic hedgehog, Indian hedgehog and desert hedgehog. (See more about *ptc, decapentaplegic* under nevoid basal cell carcinoma, Gli in Greig's cephalopolysyndactyly syndrome, Rubinstein syndrome, *engrailed* in morphogenesis in *Drosophila, sonic hedgehog*, tumor suppressor pathway)

HeLa cell line: immortalized human cancer cell line originated in 1951 from a highly malignant cervical carcinoma of the patient, Henrietta Lacks, who died the same year but the line is maintained all over the world indefinitely. (See immortalization)

HELICASE: enzyme unwinding the double helix of a DNA for replication and transcription, repair, recombination and in reactions associated with binding and hydrolysis of DNA. About 60 helicases have been identified; some unwind DNA and RNA hybrids. *E. coli* bacterium encodes about 12 different helicases, and all organisms have several types. The activity of helicase may be coordinated with DNA polymerase and the complex may travel at about 1,000 nucleotide/second. The RecQ family of helicases are implicated in mutator functions and chromosomal rearrangements such as occurring in the Bloom's syndrome, Werner's syndrome, Cockayne syndrome, Xeroderma pigmentosum B and D, and others. (See DNA replication in prokaryotes, chromosomal rearrangements, DEAD-box, Bloom syndrome, Werner syndrome, Cockayne syndrome, ABC excinuclease, Rep, recA, recombination molecular mechanism in prokaryotes, branch migration)

HELICAL TWIST: is the angle between neighboring DNA base pairs; it is within the range of $24°$ and $51°$ with a mean of 36.1 ± 5.9.

HELIX: (of macromolecular structure) is a coil of a three-dimensional ribbon ; the DNA helix is similar to a staircase with the steps corresponding to the bases and the strings of the staircase representing the sugar-phosphate backbone of the two polynucleotide chains; thus it is not a simple spiral. (See DNA, Watson and Crick model)

HELIX DESTABILIZING PROTEIN (RF-A): is a single-strand binding protein instrumental in DNA replication. (See DNA replication, binding proteins, RF-A)

HELIX-LOOP-HELIX: are polypeptides each with three-partite structures; two of the helices are connected through a loop in each component of the dimer; the third helix of the components is rich in basic amino acids and this end binds to DNA. The monomers thus appear as: HOOC-helix-loop-helix positively charged helix-NH_2. They recognize the CANNTG (E-box) sequence in the DNA. The H-T-H proteins may form dimers and recognize two different neighboring DNA binding sites. The proteins may saddle into the major groove of the DNA through their positively charged amino acids in the α helix of the NH_2 terminus. (See DNA-binding protein domains, regulation of gene activity, helix-turn-helix motif)

HELIX-TURN-HELIX MOTIF: are parts of regulatory proteins of prokaryotes and homeodomains of eukaryotes. One α-helix (*recognition helix*) fits into the major groove of the DNA and the other is positioned at right angle above it and it allows interaction with other proteins. The other monomer of the dimeric structure is binding to the next major groove along the DNA. There are large varieties of these proteins in bacteria and eukaryotes yet they contain a conspicuous symmetrical structure formed of antiparallel β sheets or α helices separated by a turn of several amino acids. H-T-H motifs occur in the homeodomain proteins, in various repressors (λ *cI* and *cro* proteins, in the *E. coli* catabolite activator [CAP] proteins, etc.). (See DNA-binding protein domains, regulation of gene activity, lac operon, lambda phage, DBP, monomeric, DNA grooves, protein structure)

Helminthosporium maydis: fungus is the causative agent of the maize disease, southern corn leaf blight. Plants with the T (Texas) cytoplasmic male sterility are extremely susceptible to the disease and suffer serious damage. (See also cytoplasmic male sterility, cms)

HELPER PLASMID: see binary vector

HELPER T CELL: see T cell

HELPER VIRUS: provides the functions that a defective virus particle lacks.

HEMAGGLUTININ: substances that cause agglutination of the blood such as antibodies, lectins, certain viruses (influenza, mumps, etc.). (See fusigenic liposome)

HEMANGIOMA: several forms of neoplasias of blood vessels under autosomal recessive gene control in humans. (See also angioma)

HEMATOPOIESIS: blood formation.

HEMATOPOIETIC RECEPTORS: bind hormone (GH, PRL, EPO, GCSF) or cytokine (interleukin) ligands and are involved in cellular signaling. The receptors may be homodimers (α-chain) and or heterooligodimers (α and β chains). (See mentioned items under separate entries)

HEMATOXYLIN: is a histological and cytological stain in different formulations. (Delafield's hematoxylin, Heidenehain hematoxylin, iron hematoxylin, etc.). (See stains)

HEME: iron- or magnesium-porphyrin prosthetic group in hemoglobin, cytochromes and chlorophyll, respectively. (See haeme [for structure].

HEMERALOPIA: see day blindness

HEMICELLULOSE: a polymer of neutral polysaccharides present in the plant cell wall matrix.

HEMIGLOBIN: see methemoglobin

HEMIMETHYLATED: only one of the two DNA strands is methylated. (See methylation of DNA)

HEMIN (ferriporphyrin chloride, $C_{34}H_{32}ClFeN_4O_4$): it is used for the treatment of porphyria and the preparation of rabbit reticulocyte lysate. (See porphyria, haem, rabbit reticulocyte *in vitro* translation)

HEMIN REGULATED INHIBITOR: see HRI

HEMIZYGOUS: gene(s) present in a single dose in an otherwise diploid cell or organism (e.g., X-chromosome-linked genes in an XY male or XO cells). (See dosage compensation)

HEMIZYGOUS INEFFECTIVE: are special class of recessive mutations in allopolyploids. These recessive alleles are not expressed in monosomics (hemizygotes) and in case of nullisomics the dominant wild type is expressed for these loci. Their expression requires two doses of these recessive alleles and never display pseudodominance. Thus, the hemizygous ineffective alleles resemble recessive suppressor mutations in diploids. Recessive mutations in allopolyploids are generally of this type because the homoelogous loci cover up the mutations at the other corresponding genes. (See allopolyploid, monosomic, nullisomic, pseudodominance, suppressor gene, homoeologous alleles)

HEMOCHROMATOSIS (HFE, haemochromatosis): is disease of iron accumulation, accompanied by cirrhosis (fibrous condition) of the liver, diabetes, dark pigmentation of the skin, heart abnormality and cancer. If diagnosed early by determining plasma iron and ferritin (a red 80,000 M_r serum protein with 2 iron-binding sites to transport iron, also called siderophilin). It

Hemochromatosis continued

is curable by lowering the iron level through venesection ([phlebotomy], letting out blood by cutting the vein. The recessive gene is located within the HLA-H complex in human chromosome 6. The frequency of the gene is about 6 to 10% with an incidence of homozygosity of 2-3 per 1,000 birth. (See also HLA, Menke's disease, Wilson disease, liver cancer)

HEMOCYANIN: copper containing protein in several invertebrates that is used to bind O_2 in a way similar to hemoglobin. (See keyhole limpet hemocyanin, hemoglobin)

HEMOCYTOMETER: is a special microscope slide with compartments of known exact volume when a coverglass is laid over. It is used for the microscopic counting of the number of blood cells, or any other suspended cells, or protoplasts in a certain volume.

HEMOGLOBIN: oxygen-transporting heme proteins (M_r 64,500) in the red blood cells. The four polypeptide chains are attached to four heme prosthetic groups (Fe^{2+} state). The human adult hemoglobin consists of two α (141 amino acids) and two β (146 amino acids) polypeptide chains whereas in the early embryo the polypeptide composition is ζζεε. The ζ (zeta) is α–like, the ε (epsilon) is β-like. By the 8th week of gestation the embryonal hemoglobin is replaced by fetal hemoglobin with a structure ααG$_γ$A$_γ$ (the latter ones are β-like). Just before birth these two γ chains are replaced by β– and δ-globin chains. By the age of six months after birth 97-98% of hemoglobin A (HbA) is ααββ and about 2% is ααδδ (HbA$_2$) and a very small amount is still fetal hemoglobin (HbF). Thus, there are two gradual developmental switches in the β-like genes but only one in the α–like genes. These changes are programed as the sites of the synthesis shift from the yolk sac of the embryo to the liver, spleen and bone marrow of the fetus and finally to the bone marrow in adults. The two gene families, α and β, of hemoglobin include α, γ, ζ and ∂ (θ, theta, unknown function) were located to human chromosome 16p13, and β, δ, ε and γ are at chromosome 11p15.5. Both the α family (→ ζ, ψζ, ψα, ψα, α2, α1, ∂) and the β family (→ ε, G$_γ$, A$_γ$, ψβ, δ, β) include pseudogenes (ψ) and transcribed in the order shown by the arrow. Hundreds of different amino acid substitutions and deletions are known in the globin chains and some of them lead to hemoglobinopathies, blood diseases. One of the most famous among them is a substitution of valine for glutamic acid at residue 6 of the β chain (Hemoglobin S). This is responsible for a hydrophobic change on its surface, resulting in an abnormal quaternary association of the subunits. When the oxygen level is reduced, the subunits polymerize into a linear array of fibers that alters the normally doughnut-shaped red blood cells to assume a sickle shape and therefore it is called *sickle cell anemia* in homozygous condition. In heterozygous state it is called the *sickle cell trait*. The sickling erythrocytes are inefficient in oxygen transport. In the arteries, normal hemoglobin is 95% saturated with oxygen, in the venous blood — on the return to the lung — the saturation is only about a third less. Over 3 million Americans carry this abnormal HbB gene and there are many millions more all over the tropical and subtropical regions of the world where malaria is common. The heterozygotes are apparently at a selective advantage when infested with the *Anopheles* mosquito that spreads the causative protozoon, *Plasmodium falciparum*. A milder form of sickling, expressed only in the homozygotes, is caused by another mutation, called hemoglobin C (HbC). Other variants are HbD and HbE. Many other amino acid substitutions that have been identified on the basis of altered electrophoretic mobility or amino acid substitutions at precisely identified residue sites in either of the chains, may not involve a disease. In the M hemoglobins, the oxygenated molecule has hereditary deficiencies of NADH-methemoglobin reductase activity and remain largely charged with oxygen. The stability of the hemoglobin molecules depends mainly on the close fit of the hemes to the globin chains that would be thermolabile at 50 C°. Mutations affecting the tight fit of the heme pocket may cause the formation of methemoglobin or the loss of the heme. Such unstable hemoglobins are e.g. the Hb Zürich in which at β-chain residue 63 His changed to Arg. Hb Köln β 98 has a Val → Met substitution. The increase in positive charge (e.g. Hb Zürich) favors the loss of heme. In case of Hb Hammersmith β 42, Phe → Ser substitution, there is no change in charge yet instability occurs. Instabilities were associated with α chain substitutions too, e.g. in Hb Boston 58 His →

Hemoglobin continued

Tyr. Other anomalies include *hereditary persistence of high fetal hemoglobin*. In the latter case numerous types of variations exist. It appears that deletions involving a regulatory region of 3.5 kb at 5' to the δ gene may be a very important factor for the high expression of fetal hemoglobin. In one form all the cells are equally affected whereas in the *heterocellular hereditary persistence of fetal hemoglobin (HHPFH)* only some of the cells display this anomaly. In the Lepore hemoglobins δβ or βδ fusion products were observed, indicating the possibility of unequal crossing over between these DNA sites. Another type of hemoglobin anomalies involve the slow rate or lack of synthesis of one or the other hemoglobin chain resulting in thalassemias. The HBA gene is located to human chromosome 16p13.33-p13.11 whereas the HBB, HBD and HBG genes are in 11p15.5. (See globin, methemoglobin, thalassemia, anemia)

HEMOGLOBIN EVOLUTION: molecular genetics evolved from the early studies (late 1940s) showed that the globin genes can be obtained in pure forms relatively easily. Over 90% of the soluble proteins in the blood plasma are globins. The still nucleated erythrocytes and the reticulocytes synthesize mainly globin mRNA. Amino acid sequence variants were detected in these proteins at the beginning of protein sequencing. Since globin genes occur both in animals and in plants, evolutionary studies became attractive. The number of amino acids in globins of plants as well in animals are fairly close: soybean leghemoglobin 143, seal myoglobin 153, human α 141, human β 146. Also there are sequence homologies in the primary structure of

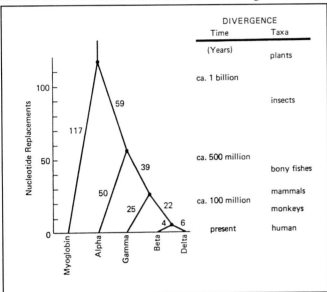

EVOLUTION OF HEMOGLOBINS. THE NUMBERS OF THE PRESUMED NUCLEOTIDE REPLACEMENT IN THE GENETIC CODE DURING EVOLUTION ARE INDICATED BY THE NUMBERS AT EACH NODE. THE TIME OF DIVERGENCE IS ESTIMATED ON AVERAGE AMINO ACID REPLACEMENT DATA AND EVOLUTIONARY CLOCKS. (After Fitch, W.M. & Margoliash, E. 1970 Evol. Biol. 46:67)

the chains that with aid of estimated average replacement data, the evolutionary time could also be quantitated. In vertebrates the globin genes have two introns at around the 30th amino acid-coding region and another around the 100th. In plants there are 3 introns two at about the same location and a third near the middle in between. The size of the introns vary substantially from 99 triplets middle intron of the leghemoglobin, to 4800 of the first intron in seal myoglobin. In vertebrates the heme binding is in the second exon, closer to the 3' end; the residues required for tetramer association are in exon 3. (See fig above, globin, hemoglobin, leghemoglobin, myoglobin, evolutionary clock, evolutionary tree)

HEMOGLOBIN SWITCHING: during development various different hemoglobin genes are activated, resulting in the formation of the different fetal and eventually adult types. (See hemoglobins)

HEMOGLOBINOPATHIES: diseases involving hemoglobins. (See hemoglobins, methemoglobin, thalassemia, anemia, hemophilia)

HEMOLYSIN: proteins secreted by pathogenic bacteria and causes the formation of pores in the mammalian cell membranes and dissolves red blood cell membranes. (See erythrocyte)

HEMOLYSIS: disruption of the membranes of erythrocytes by antibodies, hemolysin enzyme and chemicals. (See hemolytic disease)

HEMOLYTIC ANEMIA: disease may occur due to several causes such as emphysema, sensitivity to high temperature, Rh blood type, etc. (See anemia, hemolysis, phosphohexose isomerase, glutathione synthetase deficiency, glutathione peroxidase, glutathione reductase, pyruvate kinase deficiency)

HEMOLYTIC DISEASE: see erythroblastosis fetalis, Rh blood type, Su blood type

HEMOPHILIAS: are hereditary bleeding diseases caused by a deficiency of one or another antihemophilic blood-clotting protein factors. The classic *hemophilia A* has a prevalence of about 1/10,000. The estimated mutation rate is about $2\text{-}3 \times 10^{-5}$. The most famous case of hemophilia involved the descendants of Queen Victoria of England. She was heterozygous for the gene (probably through a new mutation) and transmitted it through marriage to the Russian, German and Spanish royal families (see pedigree chart).

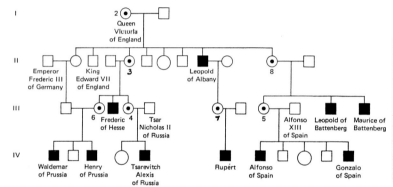

THE FAMILY OF QUEEN VICTORIA OF ENGLAMD. HEMOPHILIAC MALE DESCENDANTS ARE REPRESENTED BY BLACK SQUARES. THE CIRCLES, WITH A DOT INSIDE, STAND FOR THE CARRIER FEMALES, 8: BEATRICE, 7: ALICE, 6: IRENE, 5: VICTORIA, 4: ALEXANDRA, 3: GRANDDAUGHTER ALICE)

The inept handling of the social problems by Tsar Nicholas II of Russia has been attributed partly to his preoccupation with and worries about the affliction of his son Tsarevitch Alexis by hemophilia. Thus, a single recessive gene (Xq28) might have had contributed indirectly to the largest upheaval in the social order of the world. The so called *Christmas disease* (named after the family Christmas) is called hemophilia B is also an X-chromosome linked disease (Xq26-q27). The latter defect is about 1/5 to 1/10 as frequent as the classic hemophilia. These diseases are generally expressed in the males (because of X-chromosomal hemizygosity). Heterozygotes also have generally lower amounts of the proteins responsible for the condition, and thus the carriers can be identified in most cases. The temporarily effective therapy involves treatment with antihemophilic factor concentrates. In certain cases, the therapy is difficult because the patients may produce significant amounts of the so called circulating anticoagulant (IgG type antigen) that counteracts the beneficial effects of the clinically transfused blood or the supplied blood factors. Prenatal diagnosis is potentially feasible after 10 weeks by DNA analysis and physical mapping or blood samples can be studied for factor VIII. (See also antihemophilic factors, blood clotting pathways, X-chromosomal linkage, Hageman trait, PTA deficiency, prothrombin deficiency, Stauart factor, vitamin K-dependent clotting factors, coumarin-like drug resistance, parahemophilia, afibrinogenemia, dysfibrinogenemia, fibrin-stabilizing factors, von Willebrand's disease, Glanzmann's disease, thrombopathic purpura, thrombopathia, hemostasis, platelet abnormalities, pseudohemophilia)

HEMOPOIESIS: same as hematopoiesis.

HEMORRHAGE: bleeding.

HEMP: see *Cannabis sativa*

HEMOSTASIS: checking or arrest of blood flow. Hereditary diseases involved in hemostasis are the hemophilias, platelet abnormalities, the Nishimina factor, Tatsumi factor, Fletcher factor, Dynia factor, Flood factor, thrombocytopenia, May-Hegglin anomaly and another number

Hemostasis continued
of bleeding anomalies that are characterized by abnormal bleeding such as the Fanconi's dissease, Meekrin-Ehlers-Danlos syndrome, von Willebrand disease, osteogenesis imperfecta, hemorrhagic telangiectasis, Osler-Rendu-Weber syndrome, pseudohemophilia, stroke. (See the mentioned items under separate entries)

HENBANE: *Hyoscyamus niger* (2n = 33, 34) member of the *Solanaceae* family of plants requiring both vernalization and long photoperiods for flowering. It is a source of the hyoscyamine and scopolamine alkaloids (ca. 0.04% in the leaves). The Egyptian henbane *Hyoscyamus muticus* has higher alkaloid content (0.5%) and contains also hyoscipicrin and choline. These alkaloids are used as smooth muscle relaxant and sedative. Hyoscyamine (synonym atropine) has a low lethal oral dose of 5 mg/kg in humans, LD_{50} in mice intravenously 95 mg/kg. (See alkaloids, vernalization)

HENDERSON-HASSELBALCH EQUATION: $pH = pK + \log[(A^-)/(HA)]$, where pH is the hydrogen ion concentration, pK = is an equilibrium constant (1/K), (A^-) is a proton acceptor and HA = is a proton donor. This equation expresses the relations between pH and ratio of acid to base in a solution.

HENNIGIAN CLADISTICS: its is basically a version of the parsimony method of analyzing evolutionary pathways. (See maximum parsimony, parsimony, cladistics)

HENRIETTA LACKS: see HeLa cell line

HENRY OF PRUSSIA: great-grandson of Queen Victoria of England, and afflicted with hemophilia. (See hemophilia)

HENSEN's NODE: is group of cells in the primitive streak, contributing to the formation of head, notochord and endoderm of the embryo. (See primitive streak, notochord, endoderm)

HEPARAN SULFATE: are repeated disaccharide units composed of glucoseamine linked to uronic acid or either to (sulfated) glucuronic acid or (sulfated) L-iduronic acid. Heparan sulfates may be present in mucopolysaccharides. (See iduronic acid, glucuronic acid, mucopolysaccharidosis)

HEPARIN: is a mucopolysaccharide consisting of repeated units of uronic acid (glucosamine) and glucuronic acid disulfate. It is secreted into the bloodstream primarily by the liver and it has anticoagulant effects. It is employed as an anticoagulant medicine, and in the molecular biology laboratory it is used (with or without dextran) for *in situ* hybridization and in Southern hybridization as a blocking agent on nitrocellulose (but usually not on nylon) membranes. (See *in situ* hybridization, Southern hybridization)

HEPARINEMIA: accumulation of heparin in the blood often causing problems in blood clotting and thus bleeding. (See hemostasis, platelet anomalies, blood clotting pathways)

HEPATITIS B VIRUS: is a member of the hepadnavirus (DNA) family. The virus does not kill the cell upon integration rather it causes a stable transformation and replicates along with the chromosomes. It is endemic in South-East Asia, tropical Africa and along the Amazon river in South America. Transmission is by body fluids (oral ingestion, blood transfusion, sexual contacts, nursing) of the infected persons. Incubation period is 1-6 months. The disease caused involves fever, vomiting, jaundice, arthritis, etc. Some individuals completely recover but remain carrier and may have a high chance of developing cirrhosis or cancer of the liver. (See liver carcinoma).

HEPATITIS C VIRUS: a member of the Flaviviridae family, has a positive-strand RNA genome of 9,400 nucleotides, encoding a polyprotein of 3010-3033 residues. A serine protease processes its polyprotein and this is required for its replication. An estimated 1% of the world population is infected by it and chronic infection may lead to liver diseases and cancer. Immunization is not available. (See positive-strand virus)

HEPATIS DELTA VIRUS (HDV): has a closed circular RNA genome of 1.7 kb, packaged into folded and base-paired rods. It codes for two proteins that are edited from a single RNA transcript. The editing is carried out by a *double-stranded-RNA-adenosine deaminase*. This virus cannot be packaged without the hepatitis B virus. (See hepatitis B virus)

HEPATOCYTE: a type of liver cell; they are arranged in folded sheets, facing blood-filled spaces (sinusoids). The hepatocytes are responsible for synthesis, degradation and storage of many substances. The detoxification in the hepatocytes, with the assistance of cytochrome P450, may also produce ultimate mutagens and carcinogens as part of the process. Hepatocytes secrete also the bile that mediates the absorption of fats. (See hepatoma, scatter factor)

HEPATOCYTE GROWTH FACTOR (HGF): is a mitogen and morphogen and controls processes in liver and placental development. Its receptor is a heterodimeric transmembrane protein tyrosine kinase. It has numerous functions such as organ regeneration, angiogenesis, metastasis of tumors. Also it is a "scatter factor" because it dissociates epithelial cells and stimulates cell motility. (See also scatter factor, mitogen, morphogen, metastasis)

HEPATOMA: liver tumor; originally the transition stage between the generally benign adenoma and the malignant carcinoma of the liver. Ovarian hormones suppress hepatoma development and therefore male mice and men show five fold higher incidence of hepatocarcinomas. (See liver cancer, chi elements)

HEPATOTOXICITY: toxicity to the liver.

HEPES (N--2-hydroxyethypiperazine-N'-2-ethanesulfonic acid): is used for the preparation of buffers in the pH range 7.2-8.2.

HEPTAMER: see immunoglobulins

her: mutation converts X0 *Caenorhabditis* males into females.

HER2: synonymous with ERBB2, see ERBB1.

HERBACEOUS: plant tissue without woody components.

HERBAL: old books with description of plants considered mainly for spice or medicinal use.

HERBARIUM: museum collection of dried plant specimens, classified, identified and described

HERBICIDES: are plant growth regulating chemicals. The first herbicides were synthetic auxins (dichlorodiphenyl trichloroacetic acid, 2,4-D). Some of the general type weed killers such as 3-amino triazole, were very potent human carcinogens although not responding as positive in the majority of short-term mutagen-carcinogen assay systems. Some herbicides kill plants as germination inhibitors (pre-emergence weed killers such as atrazine). Atrazine interferes with electron transport in photosystem II (photosynthesis). About 20 species of weeds have developed resistance to atrazine since its introduction to agricultural practice in the 1950s. Recently atrazine was replaced by glyphosat, an inhibitor of the biosynthesis of aromatic amino acids by blocking enol-pyruvylshikimate-3-phosphate synthase (EPSP). No plant species developed yet resistance to glyphosat. Sulfonylureas are selective herbicides of dicotyledonous plants and inhibit the acetolactate synthase enzyme (ALS) in the branched pathway (leucine, valine, isoleucine). By highly selective isolation system resistant mutants were obtained. The non-selective herbicide gluphosinate-ammonium (phosphinothricin, BASTA) specifically inhibits the glutamine synthethase enzyme (GS). The plants accumulate highly toxic levels of ammonia Herbicide research took advantage of transformation techniques by developing bacterial EPSP transgenic crop plants (cotton, sorghum, maize, alfalfa, canola, tomato, sugarbeet) to make them resistant to glyphosat while the weeds can be eliminated. The resistance of cultivated plants against sulfonylureas could also be enhanced four fold by insertion of the ALS gene. The high level of activity of these enzymes permits the transgenic plants to escape death. BASTA resistance was engineered into plants by transformation with a gene from *Streptomyces* bacteria that inactivates the herbicide by acetylation, and toxic levels of ammonia do not accumulate. Hydantocidin, a spironucleoside, binds to the regulation site of adenylosuccinate synthetase and thus blocks purine biosynthesis in the cells. (See bialaphos, spironucleoside)

HEREDITARY: it is being biologically inherited because of the genetic constitution of the family. (See also familial, congenital)

HEREDITARY NONPOLYPOSIS COLORECTAL CANCER (HNPCC, chromosome 2p): is accompanied by high rate of mutation in microsatellite sequences. The mutability is caused by deficiency in mismatch repair controled by four loci MSH2, MLH1 and to lesser extent by

PMS1 and PMS2, homologous to microbial enzymes, MutS and MutL. Its prevalence is about 2×10^{-3}. (See colorectal cancer, microsatellite, polyposis, mismatch repair, trinucleotide)

HEREDITARY TYROSINEMIA: see tyrosinemia

HEREDITY: study of storage, transmission and faithful expression of the genetic information. (See inheritance, genetics, reverse genetics)

HEREGULINS: are transmembrane protein tyrosine kinase receptor ligands present in breast carcinoma and fibrosarcoma cell lines.

HERG: a human inward rectifying K^+ ion channel is responsible for the LQT2 syndrome and it is encoded in chromosome 7q35-36. (See ion channels, LQT1, LQT3, Jervell and Lange-Nielson syndrome)

HERITABILITY: can be estimated by offspring-parent regression, intraclass correlation and by special methods in humans. Heritability below 0.25 is considered low and above 0.75 it is high. Generally heritability of genes barely affecting fitness (e.g., spotting of the fur coat) is higher than four those that are important for reproductive success (fertility). The estimates of heritability are valid only for the population that provided the information. In different populations different sets of polygenes may exist and they may provide different heritability estimates. In experimental animals and plants generally the narrow sense heritability is used because that is much more predictive.

SELECTED HERITABILITY ESTIMATES IN VARIOUS ANIMALS AND PLANTS

HUMANS		CATTLE		MAIZE	
schizophrenia	0.75	white spots	0.95	plant height	0.51
epilepsy	0.50	milk production	0.43	kernel number	0.40
MOUSE		conception rate	0.03	yield	0.29
tail length	0.60	SWINE		ear number	0.20
litter size	0.15	litter number	0.20	SOYBEAN	
DROSOPHILA		weaning weight	0.10	maturity	0.75
abdom. bristles	0.50	CHICKEN		plant height	0.62
egg number	0.20	egg weight	0.75	oil percent	0.55
SHEEP		egg number	0.25	seed weight	0.54
wool length	0.55	viability	0.10		

(See h^2, heritability in the broad sense, heritability in the narrow sense, correlation, intraclass correlation, heritability estimation in humans, realized heritability, QTL, behavior genetics)

HERITABILITY, BROAD SENSE: total genetically determined fraction of the phenotypic variance. (See variance, genetic variances)

HERITABILITY ESTIMATION IN HUMANS: may be important in the study of various anthropometric traits, and other polygenically determined conditions such as weight, height, behavior, various types of mental illness, epilepsy, heart disease, diabetes, etc. Since monozygotic twins are expected to be identical genetically, any variation between them is supposed to be environmental. Dizygotic twins are of the same age exactly but genetically as different as any other sibs. Thus, heritability (h^2) is calculated frequently as:

$$\frac{\text{variance of dizygotics} - \text{variance of monozygotics}}{\text{variance of dizygotics}} = h^2$$

or

$$\frac{\text{percent monozygotic concordance} - \text{percent dizygotic concordance}}{100 - \text{dizygotic concordance}} = h^2$$

or using the correlation coefficients (r) of monozygotic twins reared together (r_{mzt}) and reared apart (r_{mza}):

$$\frac{(r_{mzt}) - (r_{mza})}{1 - (r_{mza})} = h^2$$

The interpretation in all of these cases have some limitations. The estimate is valid only if the

Heritability estimation in humans continued

variance is of the additive type because direct estimation, in the absence of controled matings, cannot be carried out. In the presence of dominance variance, the genetic determination will be underestimated in upper two formulas. Because of the common assortative matings in human populations leads to positive correlation between the parents; it thus overestimates heritability. Furthermore environmental variation may not be the same for monozygotic (identical) twins as for dizygotic ones. Generally, it has been difficult to find large enough number of twins for precise comparisons. Also, it must be kept in mind that in humans, just as in other organisms, heritability measured in one population, even for the same trait, may not be valid for another population. (See also correlation, intraclass correlation, h^2, monozygotic twins, heritability in the broad sense, heritability in the narrow sense, variance, confidence interval, standard error)

HERITABILITY, NARROW SENSE: genetically determined fraction of the phenotypic variance, excluding nonfixable interactions such as due to overdominance. (See heritability, heritability broad sense)

HERITABLE TRANSLOCATION TESTS AS BIOASSAYS IN GENETIC TOXICOLOGY: animals exposed to mutagens are tested for sterility or semi-sterility and examined also cytologically for multivalent association during diakinesis to metaphase I in the spermatocytes. Since may of these translocations are transmitted to the progeny, this procedure permits an assessment how much particular mutagenic agents increase the genetic load of mammals, potentially also of humans. (See chromosome breakage in bioassays for genetic toxicology)

HERMANSKY-PUDLAK SYNDROME: rare, recessive, human chromosome 10q23 disease although in some endemic populations (Puerto Rico, Switzerland) its prevalence is $5-6 \times 10^{-4}$. It involves pigmentation defects (ocular albinism, freckles but inability to get tanned), predisposition to bruising and bleeding, ceroid storage defects, large and abnormal melanocytes, lower platelet count and defective lysosomes. Survival is usually limited to 20-25 years. The defect seems to involve a transmembrane protein. This syndrome bears similarities to the Chédiak-Higashi syndrome. (See Chédiak-Higashi syndrome, albinism, melanocyte, platelet, lysosome, ceroid, lysosomal storage diseases)

HERMAPHRODITE: both male and female sex organs are present in the same individual. If the same plant bears both male and female flowers but individual flowers are either male (pollen producer) or female (egg producer), it is called monoecious. Hermaphroditism is the most common form of sexual differentiation in plants but normally it is very limited in the animal kingdom (to flatworms, nematodes, some annelids and crustaceans, etc.). *True hermaphroditism*: the same individual develops both ovarian and testicular structures. Its frequency is low in mammals, including humans. The majority of human true hermaphrodites has 46XX constitution and about 1/4 is 46XY, the remaining groups have sex chromosomal mosaicism, and are considered to be males by appearance until puberty. In the majority of cases *pseudohermaphroditism* is observed, i.e., the sex chromosomal constitution does not match the gonadal phenotype. The chromosomally XY individual appears feminine in many ways or the XX individual appears virile. Pseudohermaphroditism may arise by mutation in or translocation of the SRY gene, mutation in the Müllerian duct inhibitor substance gene, defects in the androgen receptor or deficiency in the testosterone 5α-reductase (See also intersex, pseudohermaphrodite male, testicular feminizaton, freemartins, sex ratio, sex reversal, sex determination, gonadal dysgenesis, Müllerian ducts, Wolffian ducts)

HERPES: a family of double-stranded (linear) DNA viruses with 120-200 kbp genetic material; the capsid is made of about 30 polypeptides. The viruses replicate in the cell nucleus that protrudes through the nuclear membrane into the endoplasmic reticulum. There are many types of herpes viruses and some benign forms are common in humans. Type 1 is responsible for cold sores and Type 2 for obnoxious, painful eruptions in the genital area. A member of the family, the Epstein-Barr virus is held responsible for nasopharingeal (nose-throat) carcinoma. The cytomegalovirus causing usually mild symptoms is also a herpes virus as well the equine and gallid herpes (the latter is responsible for the Marek's disease of poultry). Herpes simplex virus 1 (HSV-1) is used as a therapeutic genetic vector to deliver and express genes to the central nervous system, selective destruction of cancer cells and for prophylaxis against HSV and other infectious agents. (See Epstein-Barr virus, Marek's disease, cytomegalovirus)

HERS DISEASE: see glycogen storage diseases Type VI

HERSHEY CIRCLE: nicked phage λ DNA circle in the bacterial host. (See nick)

HERTZ (Hz): the unit of the number of cycles in alternating electric current. In the USA the standard is 60 Hz. 1 megahertz (MHz) is 1 million pulses/second. Modern personal computers employ 100-200 MHz microprocessors.

He-T SEQUENCES: repetitive DNA in the heterochromatin and near the telomeres of *Drosophila*. (See heterochromatin, telomere)

HeT-A: polyadenylated 6 kb retroposons (no LTR) specific for the telomeres of *Drosophila*, belonging to the family of LINE elements. The 2.8 kb open reading frame encodes a Zinc finger protein but no reverse transcriptase; thus for transposition it is expected to depend on this function, coded elsewhere in the genome. (See telomere, retroposon, LINE, Zinc finger, TART)

HETERADELPHIAN: conjoined twins or other rare teratological anomalies such as extra limbs without hereditary basis.

HETEROALLELES: nonidentical alleles which may recombine (involve different nucleotide sites in a cistron). (See allele, cistron)

HETEROBRACHIAL: the two arms of the chromosomes are not identical in length. (See chromosome arm, isobrachial, chromosome morphology)

HETEROCARYON: see heterokaryon (the desirable spelling)

HETEROCELLULAR: a body or tissue mixed of different cells.

HETEROCHROMATIN: parts of the chromosome that stain dark even during interphase (heteropycnotic). The *constitutive heterochromatin* remains in highly condensed state in all cells of an organism. It contains repetitive DNA that is not transcribed and translated; it is assumed to contain monotonous satellite DNA that is generally rich in AT sequences. It is frequently localized to both sides of the centromeres and to the telomeres. Certain chromosomal regions, such as most of the Y-chromosomes and the B-chromosomes are heterochromatic. The constitutive heterochromatin can be seen as distinctly and characteristically located cross bands of the chromosomes after stained with Giemsa or other stains. *Facultative heterochromatin* may be in a relaxed state when it is expressing some information in some cells at certain developmental stages but not under other conditions. The mammalian X-chromosome, displaying dosage compensation, becomes heteropycnotic in the additional copies, e.g. in the normal females one of the two X-chromosomes is heterochromatic although both of the X-chromosomes can be expressed in the males when they are present in a single dose (XY) or in the XO females the single X-chromosome is not heteropycnotic. In XXX trisomic individuals two are heteropycnotic and in XXXX superfemales three are heteropycnotic and show two or three Barr bodies, respectively when in the normal females (XX) there is only one Barr body. The facultative heterochromatinization in the two X-chromosomes permits the expression one or the other alleles of a gene associated with these chromosomes and this results in a variegation called *lyonization*. The function of heterochromatin has been investigated since the 1930s but no general function could be assigned to it, probably because there are different functional properties of this material, collectively identified as heterochromatin. Most commonly regulatory and structural roles were attributed to heterochromatin. Lyonization proves some regulatory role. Position effect shows that genes transfered to heterochromatic regions can be silenced. This silencing may be permanent in stable position effect, presumably the transposed gene lacks the appropriate environment (promoter) for transcription. In case of variegation type position effect (PEV), it was assumed that some binding proteins functioning between the homologous chromosomes have a trans-sensing role in the phenomenon. The role of heterochromatin in crossing over remains confusing because both enhancing and suppressing effects on the phenomenon have been observed. Heterochromatin being the site of polygenes has been suggested but not demonstrated conclusively, unless it is accepted on the basis that some of the

Heterochromatin continued

redundant genes are heterochromatinized. It came to light recently that centromeric heterochromatin (but not the telomeric) has a role in the disjunction of achiasmate chromosomes, The GAGA transcription factor has been implicated in stimulating variegation type position effect of heterochromatin. The *kl, k* and *cry* fertility genes of *Drosophila* seem to be in the Y-chromosomal heterochromatin and the ribosomal gene repeats, *bb* (bobbed) can be found in both the X and Y chromosomes. The enhancer of the second chromosomal (2-54) segregation distorter *(Sd)* locus (Esd) and also the activator *Sd*, responder *(Rp*, 2-56.61) were localized in the heterochromatin. The euchromatic abnormal ovule *(abo*, 2-44) mutations can be normalized by the *ABO* heterochromatic repeats in the X and Y chromosomes. About 13 autosomal genes were associated with heterochromatin. (See satellite DNA, Barr body, lyonization, banding, position effect, silencer, RAP polygenes, achiasmate, locus control region, GAGA, position effect, PEV, histone deacetylase)

HETEROCHRONIC MUTANT: expresses its genetic information with a timing or spatial pattern different from that of the wild type or it regulates development by a (+) or (-)control.

HETEROCLITIC ANTIBODY: binds to an antigen that is different but similar to the one which induced its formation.

HETEROCYST: is a terminally differentiated cell.

HETERODIMER: a protein with two different polypeptide subunits.

HETERODISOMY: see uniparental disomy

HETERODOX CHROMOSOMES: are special structures such a polytenic, lampbrush, sex, etc. chromosomes.

HETERODUPLEX: base paired polynucleotide chains with the two strands of different origins and may not be entirely complementary. In case the heteroduplex area includes different alleles, postmeiotic segregation may occur resulting in aberrant ascospore octads or sectorial colonies from single (haploid) spores. (See Meselson—Radding model of recombination)

HETEROENCAPSIDATION (transencapsidation): the viral coat and the enclosing genetic material does not match by origin.

HETEROFERTILIZATION: the polar nucleus and egg of plants fertilized by genetically different sperms, and therefore the egg and the endosperm become non-concordant (See gametophyte)

HETEROGAMETIC SEX: produces both X (Z) and Y (W) chromosome-containing gametes, or gametes with X and without X if the diploid has the chromosomal constitution XO. (See sex determination, MR)

HETEROGENEOUS NUCLEAR RNA: see hnRNA

HETEROGENOTE: see endogenote

HETEROGRAFT: the donor and recipient of the graft are different species. (See graft, allograft, isograft, grafting in medicine)

HETEROHYBRID DNA: one of the annealed strands is methylated, the other is not.

HETEROIMMUNE: when a lysogenic bacterium has a normal prophage and the λdgal transductant, and the two have different repressors, the transduction is heteroimmune. One species of animals was immunized by another species or the antigen (of any type) used evokes a pathological change. (See lambda,,λdgal, specialized transduction, HFT)

HETEROKARYON: more than one nucleus per cell of more than one type. This is of common occurrence in fungi when two different haploid cells undergo plasmogamy or when somatic cells of other organisms fuse and the cell fusion is not followed by fusion of the nuclei. (See also dikaryon, homokaryon, fungal life cycle, somatic cell hybrids)

HETEROKARYON TEST: two marked cells are fused and subsequently uninucleate cells are selected. If the nuclei are not fused and recombined, yet the uninucleate progeny still carries both parental markers, the markers must be carried by the cytoplasm.

HETEROLABEL: sisterchromatid exchange can be identified if the two sister chromatids are different by harlequin staining. (See sister chromatid exchange, harlequin staining of chromosomes)

HETEROLOGOUS PROBE: is used for the identification, localization, isolation and cloning of specific genes in an organism by employing a labeled (radioactive, fluorescent) nucleotide sequence of presumably similar structure and function, obtained from another species. (See probe)

HETEROMORPHIC BIVALENT: two morphologically distinguishable members of a homologous chromosome pair. (See bivalent) ────── or ──────●

HETEROMULTIMERIC: proteins with non-identical subunits, encoded by different genes.

HETEROPLASMY: the extranuclear genetic material within a eukaryotic cell is not homogeneous but contains genetically different components in analogy to heterozygosis. The exhumed remains of Tsar Nicholas of Russia and his brother Georgij Romanov's mitochondria displayed two populations of mtDNA with both C and T at position 16,169, respectively. (See mitochondria, chloroplast, Romanovs, DNA fingerprinting)

HETEROPLASTIDIC: within the same cell more than one type of chloroplasts exist as a consequence of mutation in the ctDNA or cell fusion. (See chloroplast genetics)

HETEROPLOID: the chromosome number deviates from the normal.

HETEROPOLYMER: a synthetic polynucleotide containing more than a single type of bases.

HETEROPOLYSACCHARIDE: contains more than one kind of monosaccharides.

HETEROPYCNOSIS: one chromosome or part of it is darkly stained when the rest of the chromosome(s) absorb(s) little or no stain at a particular stage. Dark staining at interphase is an indication of tight coiling and being in an inactive state. (See heterochromatin, Barr body)

HETEROSELECTION: in a population selection for heterozygotes may prevail and improving its ability to respond to new environmental challenges. (See selection, homoselection)

HETEROSIS: heterosis is frequently defined as the superiority of the F_1 hybrids over the midparent value. This definition expresses heterosis as Σdy^2 where d is the dominance effect at all loci and y is the difference of gene frequency between the two parental populations. This mathematical definition assumes that only dominant genes are responsible for heterosis. In agricultural practice the breeders and growers require that the hybrids surpass in performance any known parental types and overdominance and epistasis may also be involved (see also hybrid vigor). Heterosis was known for ages and was systematically exploited in agricultural practice by cross pollination between open pollinated, distinct varieties of maize. Since the second decade of this century inbred lines of maize were used for the production of single-cross hybrids. A single-cross is the F_1 generation of two inbreds. Inbreeding reduces vigor of the inbreds but the single-cross may surpass not just the parental inbred lines but also the open-pollinated varieties from what the inbreds were isolated by several years of selfing in plants or brother-sister mating in animals. During the process of inbreeding, a progressive increase toward homozygosity takes place. Also the different inbreds become homozygous for different alleles of the population. Therefore, not all inbreds are capable to contribute valuable genes to show heterosis. The breeders must select for inbred lines that produce superior hybrids, i.e. they have good combining ability. The *specific combining ability* means that a certain inbred contributes to valuable hybrids only with a particular other inbred line. *General combining ability* implies that an inbred yields high in combination with many other inbreds. Heterosis was most successfully exploited in maize breeding where it was practically difficult and expensive to use F_1 hybrids for commercial production. In the late 1910s it was discovered that *double-crosses* were also very productive and it is much more economical to generate seeds in large quantities for agronomic production. A double-cross is the intercross of two F_1 lines such as (A x B) x (C x D). In order to obtain 100% hybrid seeds, first the plants had to be detasseled, the male inflorescence, the tassel, had to be cut off before shedding pollen and when the female inflorescence (silks) were receptive, they had to be pollinated. This manual operation was very inefficient and costly. When cytoplasmic male sterility and fertility restorer genes were discovered, cross-pollination could be carried out economically at large scale. Before hybrid corn was introduced into commercial production the yield per hectare in

Heterosis continued

the USA was about 1,400 kg. By the time the use of hybrid corn became practically universal, the yield increased five to ten fold. Heterosis is now exploited in many other plant species, including some autogamous species such as tomatoes. Hybrid vigor is commercially utilized also in the poultry and pork industry. (See also inbreeding, maize, cytoplasmic male sterility, hybrid vigor, overdominance and fitness)

HETEROSPECIFIC: belonging to another species.

HETEROSPORIC: produces both micro- and macrospores. (See microspore, macrospore)

HETEROSTYLY: the anthers in a flower are at a height different (generally lower) from that of the stigma in order to avoid self-pollination (inbreeding). (See incompatibility alleles)

HETEROTHALLISM: in lower eukaryotes (fungi) is comparable to dioecy in higher plants, or bisexuality in animals, i.e. the plus and minus mating type spores are carried by different individuals (thalli, colonies). This definition does not require that the species would have sexual organs (that many fungi lack), nor are these spores immediate meiotic products. (See also homothallism)

HETEROTOPIC GRAFT: a piece of tissue is transplanted to a location different from its original site. (See graft)

HETEROTROPHS: cannot meet all their requirements from inorganic nutrients.

HETEROTROPIC: an allosteric enzyme needs more than its substrate for modulation. (See allostery, modulation)

HETEROZYGOTE: an individual with different alleles at one or more gene loci. (See also homozygote)

HETEROZYGOTE ADVANTAGE: see hybrid vigor, heterosis, selection coefficient, fitness, inbreeding coefficient

HETEROZYGOTE PROPORTIONS: in the F_2 of a monofactorial cross the genotypic proportions are 1AA:2Aa:1aa. If propagation is continued by selfing the frequency of homozygotes will increase (inbreeding) and that of the heterozygotes will decrease as discovered by Mendel. The proportion of each homozygotes (AA or aa) relative to the heterozygotes (Aa) will be in compliance with the formula: $[(2^{n-1}) - 1]:[2]$ where n stands for the number of filial generations. Applied this formula to F_2: $[(2^{2-1}) - 1]:[2] = 1 : 2$ and by F_6: $[(2^{6-1}) - 1]:[2] = 31 : 2$. (Note that the F_1 was not produced by selfing. The proportion of both types of homozygotes combined relative to the heterozygotes will be 2 : 64 at one locus. The same results can be obtained also at another way: 0.5^{n-1}, thus the frequency of heterozygotes at a single locus, in F_6 will be $0.5^{6-1} = 0.03125$ that is the same as 2/64 obtained above. (See inbreeding, inbreeding progress of)

HETEROZYGOUS: in a diploid or polyploid the alleles at a locus are not identical; similar condition may exist in haploids (merozygotes) carrying a duplication or a plasmid with the same locus. (See homozygous, Mendelian segregation)

HETEROZYGOSITY, AVERAGE: varies in an outbreeding population in *Drosophila* as well as in man; many genes are homozygous whereas heterozygosity at some other loci may vary from 0.10 to over 0.70. In a random mating population heterozygosity (H) for a locus can be estimated: $H = 1 - \sum_{i=1}^{k} x_i^2$ where x_i = frequency of the ith allele, k = the number of alleles, e.g. if the frequencies of the alleles A1 (0.4), A2 (0.2), A3 (0.3) and A4 (0.1) then $H = 1 - (0.4^2 + 0.2^2 + 0.3^2 + 0.1^2) = 1 - 0.3 = 0.70$. In case the size of the population is very small the following formula may give somewhat better estimate: $H = 2n/(2n - 1)(1 - \Sigma \hat{x}_i^2)$ where $\hat{x}_i$ = the frequency of the ith allele and n = the number of alleles studied. (See utility index of polymorphic loci in genetic counseling, counseling genetic)

HETS: heterozygotes.

HEURISTC SEARCH: is a sequential estimation the shortest branches of an evolutionary tree starting with a single individual and choosing then others. It is carried out usually with the PAUP computer program. (See PAUP, evolutionary tree, exhaustive search)

Hevea brasiliensis (rubber plant): is a major source of latex in plants grown mainly in Asia and West Africa for the production of natural rubber (a polymer of cis-1,4 polyisoprene $[C_5H_8]_n$). The origin of the plant is South America. The basic chromosome number x = 18 but 9 has also been suggested. Some of the species are either diploid, 2n = 36 or triploid, 2n = 54.

HEXAPLOID: has six basic sets of chromosomes (genomes) in its cells (chromosome number = 6x). Hexaploids are generally allohexaploids, i.e. each pair of the sets are of different evolutionary origin, e.g., in the somatic cells of the common bread wheat (*Triticum aestivum*) 2n = 42 contain 6 x 7 chromosomes of the AABBDD genomic composition. Autohexaploids would have problems with multivalent association, unequal disjunction consequently with sterility. (See polyploidy, allopolyploid)

HEXASOMIC: one particular chromosome is present in six copies within a cell. (See aneuploid)

HEXOKINASE: phosphorylates glucose to glucose-6-phosphate using ATP donor. HK1, HK2 and HK3 genes are in human chromosomes 10q12, 2p13 and 5q35.2, respectively.

HEXOSAMINIDASE A and B: are lysosomal enzymes involved in the breakdown of ganglioside sphingolipids forming about 6% of the membrane lipids in the gray matter of the brain and present in smaller amounts in other tissues. (See gangliosides, Tay-Sachs disease, Sandhoff's disease, lysosome)

HEXOSE: a sugar with 6 carbon backbone, e.g. glucose or fructose.

HEXOSE MONOPHOSPHATE SHUNT: the same as pentose phosphate pathway.

Hfr: high-frequency recombination strain of a bacterium; the sex element (F plasmid) is integrated into the bacterial chromosome resulting in about a thousand fold more efficient transfer of the bacterial chromosome with the integrated F+ element into the F- recipient cell. The transfer of the chromosome makes possible recombination but unlike in eukaryotic crossing over, only one of the recombinant strands is recovered. The sex factor and the bacterial chromosome are transferred in a unidirectional manner.

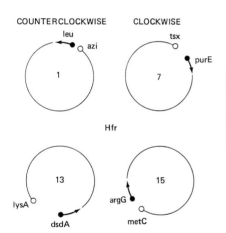

FOUR DIFFERENT Hfr GROUPS, 1, 7, 13 AND 15. THE ARROWS ON THE DOTS INDICATE THE DIRECTION OF THE TRANSFER AND THE GENE FIRST TRANSFERRED. THE OPEN CIRCLES SHOW THE POSITION OF THE MARKERS LAST TRANSFERRED. IN *E. coli* MORE THAN TWO DOZEN Hfr STRAINS ARE KNOWN.

The F element can be integrated into the bacterial chromosome at different map positions and it may be transferred either clockwise or counterclockwise (←diagram). The transfer involves a process of a rolling circle type of replication. The genes closest to the point of the initiation of transfer are transferred first in proportion to their distance from that point. The transfer point is generally in the middle of the sex element, so that the recipient cell can receive the intact F requires the transfer of the entire Hfr plasmid. The rate of transfer depends also on the nature of a particular Hfr element. The complete transfer of the entire bacterial genome, plus the F element is about 4,700,000 nucleotides and requires about 90-100 minutes. (See also rolling circle replication, *mob*, conjugation, conjugation mapping, bacterial recombination frequency)

HFS CELLS: are human fibroblast cells transformed by replication/transcription origin deficient SV40 DNA and produce T antigen (Tag) and thus can provide helper functions for replication to high copy number minireplicons containing only viral replicational origins. (See SV40, replicon)

HFT: high frequency transducing lysate. When bacteria are transduced by normal lambda phage and the defective λdgal phage, upon induction both λdgal and normal phages can be released at

HFT continued
about equal frequencies because the wild type phage can provide the information missing in the defective λ DNA. The two types of particles can be separated by CsCl density centrifugation because the density of the wild type particles is about 1.7 g/cm^3 whereas that of the defective may be as low as 1.3. (See specialized transduction, LFT, lambda phage)

HGF: hybridoma growth factor; hepatocyte growth factor. Its receptor is the product of the Met oncogene. The hepatocyte growth factor may interfere with the neoplastic activity of the c-myc oncogene in mouse. (See signal transduction, hybridoma, hepatocyte, macrophage-stimulating factor, Met, hepatocyte growth factor, Ron, HNF)

HGPRT: see hypoxanthine-guanine phosphoribosyltransferase; the same as HPRT

HhaI: restriction enzyme has the same recognition site as CfoI: GCG↓C.

HHV 8: human herpes virus 8; it is consistently associated with Kaposi sarcoma. (See Kaposi sarcoma)

HICKORY (*Carya* spp.): hardwood forest and shade trees, 2n =32, 36.

Hieracium: a genus of the *Compositae* family with many weedy species in America and Europe. Several of them have euploid chromosome number (e.g. *H. japonicum* 2n = 18). The apomicts have chromosome numbers 2n = 27, 36 and 45. These apomicts, studied by Gregor Mendel, upon the recommendation of Carl Nägeli, Professor of Botany at the University of Munich, caused great worries to Mendel concerning the general validity of his discoveries about inheritance because the apomicts failed to segregate as expected. Thus Nägeli, despite his great fame, almost thwarted the development of genetics with plants as he also vigorously argued against the existence of hard heredity in bacteria by advocating his theory of *pleomorphism*, meaning that these organisms lack fixed genetic material and vary freely. If the theory would have been accepted, it would have foiled also the development of bacteriology and medicine. (See Mendel's laws)

HIGH-DENSITY LIPOPROTEIN (HDL): is located in particles rich in protein and relatively low in cholesterol and cholesteryl esters. HDL is the benign lipoprotein; atherosclerosis is inversely related to the concentration of HDL. Lipoprotein lipase mutations may result in lower level of HDL. (See cholesterol, lipids, Tangier disease, LDL, hypertension, atherosclerosis)

HIGH-ENERGY BOND: upon hydrolysis the covalent linkage liberates large amounts of free energy, e.g., phosphodiester bond of ATP, thioester linkage in acetyl Coenzyme A. (See phosphodiester, thioester)

HIGH-FREQUENCY LYSATE: when a temperate bacteriophage excises from the bacterial chromosome it may pick up an adjacent site. E.g., bacteriophage λ may gain from *E. coli* map position 17 a *galactose* gene but because of the reciprocal recombination, a piece of its own genetic material, required for lysogeny, may be left behind. Thus, the new phage becomes λdgal (lambda-deficient-galactose). Such a phage particle may multiply vegetatively but its infecting ability is reduced. When a helper phage is inserted into the bacterial chromosome next to λdgal, the wild type phage gene may compensate for the defect and a *double lysogenic bacterium* is produced. Upon induction (to liberate phage), a *high-frequency lysate*, containing the two types of the phages is produced with high transducing ability. (See specialized transduction, lambda phage)

HIGH-FREQUENCY TRANSDUCING LYSATE: HFT

HIGH-LYSINE CORN: the seed proteins of cereals is composed of four major fractions in variable but frequently in closely equal amounts: prolamine (zein in maize, gliadin in wheat), glutenin, albumin and globulin. In maize, the genetically determined lysine content is highly variable and subject to increase and decrease by selection for high or low zein content, respectively. Two families of genes *opaq* (*o*) and *floury* (*fl*) scattered over the maize map are particularly influential on reducing the low-lysine protein fractions with the concomitant increase (doubling or more) of the percentage of this essential amino acid and that of tryptophan as well. In some *fl2* lines the level of methionine is also substantially higher. Particularly beneficial effects are due to genes *o2, o7, fl2* and *fl3*. In wheat in mol/10^5 g pro-

High-lysine corn continued

tein the lysine contents are ca. : gliadin 5.0, glutenin 17.6, albumin 78.4 and globulin 98.0. Cereals with improved nutritional values are desirable for feeding of animals and more importantly for the production of cereal food for human populations suffering from malnutrition as a result of protein deficiency in the diet (kwashiorkor). (See also essential amino acids, kwashiorkor)

HIGH-MOBILITY GROUP OF PROTEINS (HMG): are associated with functionally active chromatin and render the genes more sensitive to DNase and probably to RNase II. They are regulated by cell-cycle-dependent phosphorylation, affecting their ability to bind to DNA. HMG proteins are important for growth and development. A large number of transcription factors contain HMG-like domains. The specificity of these proteins varies but a common feature is that they distort DNA structure and have an affinity for distorted DNA structures. The change in DNA electrophoretic mobility is correlated with this altered structure. Recurrent rearrangements of the HMGI-C group was detected in some benign tumors. (See chromatin, nonhistone proteins, cell cycle, coactivator, transcription factors, SOX, DNA bending)

HIGH-PERFORMANCE LIQUID CHROMATOGRAPHY (HPLC): a mixture of compounds are applied to chromatographic columns with strong ion exchange resins. The solvent is forced through the resin under pressure for rapid and sharp separation of the components of the mixture. The eluates are electronically scanned and identified. (See chromatography)

HIGHLY REPETITIVE DNA: contains high degree of redundancy and reassociates very rapidly after denaturation. (See SINE, LINE, annealing, c_0t value)

HILL REACTION: illuminated chloroplasts evolve oxygen and reduce an artificial electron acceptor (ferricyanide → ferrocyanide). It was an important tool to study the mechanism of photosynthesis, namely that the evolved oxygen comes from water rather than from CO_2 and demonstrated that isolated chloroplasts can perform part of the reactions, and revealed the light-activated transfer of an electron from one substance to another against a chemical-potential gradient. (See photosynthesis)

HILUM: a depression or pit where vessels and nerves enter an organ; the place where the plant seed is connected to its stalk in the fruit.

him: high incidence of males mutation in *Caenorhabditis* have a high level of nondisjunction in XX hermaphrodites and thus produce X0 males. (See nondisjunction, chromosomal sex determination, *Caenorhabditis*)

HIMALAYAN RABBIT: carries temperature-sensitive tyrosinase genes controling pigmentation. The extremities, paws, ears and tail having lower blood circulation and concomitant lower body temperature develop darker pigmentation. Similar pattern of pigmentation occurs in other rodents and in the Siamese cats. Tyrosinase is a copper enzyme (also called polyphenol oxydase) is involved in the formation of 3,4-dihydroxyphenylalanine (DOPA) that is responsible for the production of melanin in the hair and skin and darkening of wounded fruits and other plant tissues. (See albinism, piebaldism, Siamese cat, pigmentation of animals, temperature-sensitive mutation)

*Hind*III: restriction enzyme with recognition site A↓AGCTT.

HiNF (histone nuclear factor): a 48 K M_r protein, identical to interferon regulatory factor IRF-2. (See IRF-2, histones)

HINGE: see antibody

HINNY: she-ass (2n = 62) x stallion (2n = 64) hybrid. The reciprocal (mare x jackass) is called mule. Mules are easier to produce because the jackass willingly mates with the mare but the stallion mates with the she-ass only under special circumstances (blindfolded). The hybrids' body resembles closer the female parent as an apparent cytoplasmic influence. These sterile hybrids—known since the beginning of human civilization—may retain some sexual drive and occasionally fertility has been reported in backcrosses with either the jackass or the stallion. The jackass backcrosses are entirely sterile but the backcrosses with stallions appear more nor-

Hinny continued

mal. The advantages of the mule and hinny is that they are stronger yet as resistant to stressful conditions as the donkey, and they can thrive on much less feed than the horse. The adults are generally healthier than the horse and well adapted to work under primitive conditions. These hybrids do not require iron hoof plates (horshoe). They are more obstinate than the horse although they are shrewd and better self-disciplined under conditions scary to horses. (See mule, animal species hybrids)

HIPPOCAMPUS: is ventral part of the brain and the presumed site of memory-amnesia-learning deficit, etc. (See brain human)

HIPPOPOTAMUS (*Hippopotamus amphibius*): 2n = 36.

HIR: genes involved in histone regulation. (See histones)

HIRSCHPRUNG's DISEASE (megacolon): occurs in several forms involving megacolon (large and dilated terminal section of the intestine), microcephaly (reduced head size), short stature, coloboma of the iris (the iris has two distinct colors) and unequal development of the two sides of the face, obstructed anus (the terminal end of the intestine) and bladder were observed. The symptoms may not all occur in each individual and were attributed most commonly to autosomal recessive (chromosomal deletion 13q1-q32.1), less frequently to autosomal dominant mutation (human chromosome 10q11.2) in the RET oncogene (receptor tyrosine kinase), or even to multifactorial inheritance. Mutations in the glial cell-derived neurotrophic factor (GDNF), encoded in human chromosome 5p may also have a minor role in the disease. Prevalence is about 1/5000 but among sibs about 1/25. The chance for males to be affected is 3 to 5 times higher than in females. Its incidence is high in Down's syndrome (6%) and in piebaldism. This disease occurs also in horses and mice and probably in pigs. Endothelin-3 mutations may be also responsible factors. (See eye diseases, stature in humans, diabetes mellitus, RET oncogene, psoriasis, hypogammaglobulinemia, piebaldism, endothelin, Shah-Waardenburg syn-drome, aganglionosis)

HIRSUTE: the word means hairy.

HIS (HS2) **ONCOGENES**: were identified in mouse chromosomes 2 and 19 from leukemia cells. The human homolog was assigned to 2q14-q21. (See oncogenes, leukemias)

His **OPERON**: is a coordinately regulated gene cluster. In *Salmonella typhimurium*

gene order on the map: ***E*** (2) ***I*** (3) ***F*** (6) ***A*** (4) ***H*** (5) ***B*** (7,9) ***C*** (8) ***D*** (10) ***G*** (1)

and the numbers in parenthesis after each gene indicate the order of the biosynthetic steps the genes mediate. The substrates and products in the pathways are ordered below:

(1) PHOSPHORIBOSYL PYROPHOSPHATE + ATP → **(2)** PHOSPHORIBOSYL-ATP → **(3)** PHOSPHORIBOSYL-AMP → **(4)** POSPHORIBOSYL-FORMIMINOAMINOIMIDAZOLE-4-CARBOXAMIDE RIBONUCLEOTIDE + GLUTAMINE → **(5)** > ? → **(6)** IMIDAZOLE GLYCEROL PHOSPHATE + AMINOIMIDAZOLE-4-CARBOXIAMIDE RIBONUCLEOTIDE → **(7)** IMIDAZOLEACETOL PHOSPHATE → **(8)** L-HISTIDINOL PHOSPHATE → **(9)** L-HISTIDINOL + PHOSPHATE → **(10) L-HISTIDINE**

HISTAMINE: a decarboxylated histidine. A vasodilator (expands blood vessels), mediates the contraction of smooth muscles, it is released in large amounts as part of the allergic response. In allergic persons an immunoglobulin E (IgE)-dependent histamine-releasing factor (HRF) is produced by lymphocytes. (See also basophils, mast cell)

HISTIDASE (histidine ammonia-lyase): see histidinemia

HISTIDINE KINASE: a signal transducing molecule of two components. One of the components, in response to environmental cues, autophosphorylates a histidine in the *catalytic* domain. Then the phosphate is transfered to an aspartate within the second component which is *a response regulator*, binding to another cellular protein to induce a cellular response. The histidine kinase may be either a transmembrane signal receptor or a cytoplasmic protein receiving the signal indirectly from another transmembrane receptor. In yeast the histidine kinase, SLN1 seems to be a transmembrane protein with the kinase domain in the NH_2 end region and it is joined to a response regulator domain at the COOH end. It appears that the sensor part is outside the cell and the kinase and the response regulator are within the cytoplasm. In yeast

Histidine kinase continued

downstream of the histidine kinase (Sln1) the Ssk1 protein, normally phosphorylated by Sln1, and is then inactive but when it is not phosphorylated it controls other protein(s) in the signal transduction pathway. Besides *SSK1* the mitogen-activated *HOG1* gene also encodes another *SLN1* suppressor. Hog1 appears to be a homolog of MAP (mitogen activated protein kinase). and also Pbs2 which seems similar to the MAP kinase kinase (MAPKK). The *ETR1* (ethylene response) histidine kinase gene of the plant *Arabidopsis* is organized after the yeast pattern and this histidine kinase gene is a signal transducer for ethylene. Downstream in the pathway acts the *CTR1* (constitutive triple response) gene encoding a serine-threonine kinase negative regulator of *ETR1*. (See histidine, histidine operon, signal transduction, ethylene, phytohormones, MAP)

HISTIDINE OPERON: see *His* operon

HISTIDINEMIA: is caused by a deficiency of histidase enzyme (histidine ammonia-lyase) involved in the removal of a NH_3 from histidine. It frequently causes mental retardation and neurological abnormalities. Identification, by enzyme assays, is feasible prenatally. Its pattern of inheritance is not entirely clear, however an autosomal recessive gene was assigned to human chromosome 12q22-q23. Its prevalence is 1 in 20,000 to 40,000, however in some isolated populations the heterozygote frequency was estimated to be as high as 3%. Histidinemia symptoms may be caused also by non-genetic factors. Histidase alleles were described in chromosome 10 of the mouse. (See amino acid metabolism)

HISTIDYL tRNA SYNTHETASE (HARS): charges $tRNA^{His}$ by histidine. Its gene is in human chromosome 5. (See aminoacyl-tRNA synthetase)

HISTOBLASTS: cell groups that give rise to dorsal epidermis (tergites) and ventral epidermis (sternites), respectively. (See epidermis)

HISTOCHEMISTRY: determines tissue components on the basis of *in situ* analysis of their chemical reactions by color and/or by specific antibodies. (See Gomori's stain)

HISTOCOMPATIBILITY: see HLA (human leukocyte antigen), leukocytes

HISTONE ACETYLTRANSFERASE: acylates amino acid (serine) residues in the nucleosomes; in the active chromatin histone 2a is less acylated than in the inactive one. (See p300, TF_{II} 230/ 250, PCAF, nucleosomes)

HISTONE DEACETYLASE: may activate silenced genes such as those showing position effect by centromeric or telomeric heterochromatin. (See DHAC1, HDA1, HOS, RPD, histone, heterochromatin)

HISTONES: five basic (rich in arginine and lysine) DNA binding proteins; H2A, H2B, H3 and H4 are parts of the nucleosome core and H1 is generally a linker between nucleosomes. The histone genes were highly conserved during evolution. The N terminal domains of the core histones and the C terminal domain of H2A protrude from the nucleosomes and provide means for interacting with other proteins involved in genetic regulation. Phosphorylation of serine and acetylation of lysine residues in the N terminal domain of core histones contributes to modulation of transcription by the nucleosomal structure. The H2A tail contains a leucine zipper which may be involved in interaction with the TATA box associated proteins. The 'deviant' histones may associate with the nucleosomal structure and exercise regulatory roles. The centromeric nucleosomes bind specific proteins (CENP-A, CENP-B, CENP-C) that are essential for proper chromosome segregation. The H1 and H5 linker histone in the nucleosomal structure contain a "winged helix", a bundle of three α-helices attached to a three-stranded antiparallel β-sheet. This structure is found in sequence-specific regulator proteins such as the catabolite activator protein and the hepatocyte-activating factor HNF3. The HNF3 protein replaces the H1 linker within chromatin containing the serum albumin enhancer. Similarly, H3 may be replaced by CENP-A at the centromere. Histones regulate also telomeric sites and the silent mating type locus. The yeast nucleosomes are compact and may not have H1 although a H1 gene was discovered in chromosome XVI. In *Tetrahymena* H1 may act as positive or negative regulator for some genes but does not have a general regulatory role in transcription.

Histones continued

These pieces of information indicate the special role of histones and nucleosomes in gene regulation through chromatin structure. The histone genes lack introns, their mRNA is not polyadenylated. (See nucleosome, H1TF2, HU, HiNF, CCE, chromatin, RPD)

HISTORICAL CONTROL: the concurrent controls are run along the experimental series. In some instances this is impossible to use, (e.g., some epidemiological studies, effects of environmental pollution effects, etc.) because there is no means to exempt the control from the overall consequences of the factors studied. For example if we want to assess the effect of a particular compulsive vaccination system, there is no contemporaneous non-vaccinated cohort group for the comparison. In such cases, similar data preceding the experiment or control data collected elsewhere under similar conditions are used as standard for the comparisons. (See control, concurrent control, cohort)

HISTOTOPE: the site of the MHC molecule recognized by the T cell receptor. (See MHC, TCR, agretope, desetope, epitope)

'HIT AND RUN TECHNIQUE': expected to target for mutation or replacement of only a single nucleotide or a very minute segment of a gene. (See targeting genes)

HITCHHIKING: because of close linkage to genes of selective advantage, non-advantageous genes may be maintained in a population. (See linkage drag, genetic load)

HIT-THEORY: see target theory

HIV-1, HIV-2: see acquired immunodeficiency (AIDS), viral vectors, lentivirus

Hix: is an invertase. (See invertases)

HKR ONCOGENES: were renamed GLI: see GLI

HKT (high-affinity kalium transport): a is a membrane protein confering the ability of potassium uptake. (See ion channels)

HL60: is a human granulocyte line. (See granulocyte)

HLA: human leukocyte antigen. (The corresponding functions in mouse are determined by the *H2* gene cluster). The development of congenic resistant lines in mice permitted eventually the determination of histocompatibility by serological tests rather than by tissue transplantation. The genes concerned with the determination of specificity are generally designated as the major histocompatibility complex (MHC). The *HLA class I* molecules have two polypeptide subunits. The 44 kDa highly variable heavy chain is coded within the *MHC* cluster in the short arm of human chromosome 6p21.3 whereas the invariable 12 kDa β_2-microglobulin is encoded at another location in the genome. The heavy chain has three domains of about 90 amino acids each, and one of them interacts with the β_2-microglobulin. These proteins are transmembranic with about 30 amino acids extending within the cell and a larger portion is outside the cell. The heavy chain consists of three domains $\alpha 1$, $\alpha 2$ and $\alpha 3$, encoded by genes HLA-A, HLA-B and HLA-C. The class I polypeptides were originally designated as transplantation antigens. *HLA class II* antigens are heterodimeric composed of a 33 to 34 kDa α chain and a 28 to 29 kDa β chain. The human MHC class II genes are HLA-DR, HLA-DQ and HLA-DP. All of the HLA genes are interrupted by several introns. The α_3 and the β_2-microglobulin of the class I HLA genes and the β_2 and α_2 polypeptides of the class II polypeptides are homologous with the heavy chain constant region of immunoglobulin M. The β_2-microglobulin bears similarity to the V gene of immunoglobulins. The HLA cluster occupies about 3,800 kb in the DNA, including interspersed other genes not all shown here:

centromere ← DP - DQ - DR - C4 - C2 - TNF - B - C - A → telomere

The DP cluster contain alternating 2 α and 2 β genes in opposite orientation; DQ also has the same arrangement of the 2 α and 2 β genes; DR has β_1, β_2, β_3 in the same left to right sequence and 1 α genes in opposite orientation. Both Class I and Class II genes appear to have pseudogenes. The HLA genes have large number of alleles and certain alleles of some genes, e.g., A1 and B8, are in linkage disequilibrium, i.e. they are syntenic much more often than expected by random recombination.

The HLA genes display high polymorphism that may be correlated with the specificity of the

HLA continued

immune reaction. In the N-termini of the A proteins there may be 7% differences. In exons 2 and 3 of the A genes the base substitutions generally result in amino acid replacements whereas in exon 4 half of the base substitutions are silent at the protein level. The β chains of DP, DQ and DR are highly polymorphic but the α chains of DR and DP are conserved. The variations were suspected to be the result of gene conversion but some of the variation implicate interallelic recombination.

The amount of HLA gene products vary substantially in different cells. B lymphocytes display much A, B and somewhat less C antigens whereas in T lymphocytes all three proteins are much less active. Interferons and tumor necrosis factors stimulate the expression of the Class I genes. Class II genes are expressed primarily on B lymphocytes, myelocyte resembling cells, and macrophages. When activated, T lymphocytes as well as skin fibroblasts and some endothelial cells may express them. Class II genes' transcription is induced only by γ interferon. If mouse lymphocytes are transformed with the HLA heavy chain genes of Class I, the heavy chain can function with the murine $β_2$-microglobulin and it is detectable by monoclonal antibodies against HLA determinants while in transgenic cells cytotoxic lymphocytes did not attack consistently the HLA determinants. Genetically engineered HLA class I indicated that the first two external regions of the heavy chain ($α_1$ and $α_2$) are critical for encoding serological determination. Class II genes are expressed on the surface of mouse lymphocytes and the mouse cell can provide the invariant antigen chain.

Within the long DNA tract of the HLA cluster, several genes involved in disease susceptibility have been identified such as 21-hydroxylase deficiency (in between DR and A, causing congenital adrenal hyperplasia, CAH1), hemochromatosis (HFE, either between B and A or distal to A) causing cirrhosis of the liver, diabetes, dark pigmentation and heart failure, the gene for juvenile myoclonic epilepsy, JME, (proximally to A) causing convulsions limited to certain areas, and is possibly involved in ragweed sensitivity, etc. The tumor necrosis factors TNF-A and TNB-B, cachetin (hormone-like protein product of macrophages releasing fat and lowering the concentration of fat synthetic and storage enzymes), lymphotoxin (lyses cultured fibroblasts) are also coded within the HLA cluster.

The most important functions of the complex are in immune recognition, defense against bacteria and viruses and they have also a number of other not entirely clarified roles. The glycoproteins encoded by the HLA genes are deposited on the surface of the majority of cells. These surface antigens are the ID cards of the individuals. The identity is determined by the inheritance of the specificity of the MHC alleles and not processed further as it is the case with the antibody transcripts. The class I gene products are located on the cytotoxic T cells (CTC) that destroy the cells of the body when infected. The class II proteins are on the surface of B lymphocytes and macrophages that are involved in binding humoral (circulating antigens). Antigens are separated into polypetides by cellular proteases and the polypeptides associate with specific compartments of the specific HLA protein molecule. The CD8 (**c**luster of **d**ifferentiation 8 protein, 34 kDa, encoded by a gene in human chromosome 2, product of differential processing of the transcript) becomes part of the T lymphocyte cytotoxic or suppressor molecule in association with Class I HLA proteins recognized by the T lymphocyte receptor (TCR). [The TCR protein is a complex of different polypeptide chains encoded at 14q11.2 (TCRα), at 7q35 (TCR β), at 14q11.2 (TCR δ), at 11q23 (TCRε) and at 7p15-p14 (TCRγ)]. CD4, a 55 kDa protein (encoded in human chromosome 12) is associated with TCR and with Class II HLA protein on specific T lymphocytes. The CD8 and CD4 proteins act also as signal transducers in association with CD3 antigens (multiple cistrons encoded at 11q23) and the helper T lymphocytes secrete lymphokines that activate B cells to secrete antibodies. Also they turn on a protein tyrosine kinase to initiate signal transfers to the cell nucleus. The human MR1 gene locus (1q25.3), called also HLALS, encodes a polypeptide that has similarities to the Class I major histocompatibility antigens. (See also major histocompatibility complex, *H-2*, congenic resistant lines, antibody, T cells, TCR, CD8, CD4, CD3, TAP, adrenal hyperplasia, hemochromatosis, myoclonic epilepsy juvenile, ragweed sensitivity, ankylosing spondylitis,

Reiter syndrome, lupus erythematosus, psoriasis)

HLH: amphipathic (both hydrophobic and hydrophilic sides) α-helices of proteins connected by a loop.

HLM ONCOGENE: is a mosaic of various types of retroviral elements found in avian and mammalian viruses. Oncogene HLM2 was assigned to human chromosome 1. (See retrovirus)

HLP: histone-like protein. (See histones)

HMBA (hexamethylene bisacetamide): cause genome-wide transient demethylation in mouse eryhtroleukemia cells while the DNA is not replicating. (See methylation of DNA)

HMG: see high mobility group of proteins

HMG BOX: high mobility group domain of the SRY gene product and other proteins with high affinity DNA-binding but differing in sequence specificity. (See SRY)

HMG—CoA REDUCTASE: see cholesterol

HML and **HMR**: the left (α) and right (a) mating type gene cassettes, respectively, in chromosome 3 of yeast. These elements are expressed only when transposed to the mating type, MAT site. (See mating type, mating type determination in yeast)

hMLH1: a human DNA repair locus in chromosome 3p21 where mutation may lead to colorectal cancer. (See hereditary nonpolyposis colorectal cancer, DNA repair)

HMS: see copia

HNGFR: high molecular nerve growth factor receptor. (See LNGFR)

HNF (hepatocyte nuclear factor): proteins are transcription factors of liver-specific genes.

HNNP (hereditary neuropathy with liability to pressure palsy): see palsy, neuropathy, MLE, MITE

HNPCC (human non-polyposis colon cancer): hereditary nonpolyposis colorectal cancer. (See colorectal cancer)

hnRNA: heterogeneous nuclear RNA; all the thousands of diverse species of RNA found in the eukaryotic nucleus, including primary transcripts and pre-mRNA in various stages of processing. The hnRNA includes introns and other transcribed but not translated RNAs. These RNAs may be associated with proteins of 34 kDA to 120 kDA size. There are about 20 different proteins within the particles. The six most common core proteins, A12, A2, B1, B2, C1, C2 occur in multiple copies within each globular aggregate. The complex takes a beads-on-string like structure with each globular structure (about 20 nm in diameter) contain 100 to 800 nucleotides and have a sedimentation coefficient of about 40S. The U1AP protein is coded by a gene in human chromosome 19q13.3. The U1-70K snRNP is the major antigen recognized by anti-(U1)RNP sera in autoimmune diseases. (See snRNA, S, RNP, U1-RNA, autoimmune disease, KH domain)

hnRNP: heterogeneous ribonucleoprotein. (See hnRNA)

hobo: see hybrid dysgenesis

HODGKIN DISEASE: is a malignant lymphoma with an unknown genetic determination because the familial nature of the condition is not sufficiently clear. W.F. Bodmer argued in favor of a gene linked with a HLA complex. In a two-marker case (A and A'), the distribution of 1AA:2AA':1A'A' would be expected in the progeny of heterozygotes without linkage. In a survey of 32 afflicted sib pairs the actual proportions were 16:11:2 that is significantly different from 1:2:1 at the level of 0.005 probability. If it is assumed that the frequency of the gene *a* (Hodgkin) is 0.01 and all *aa* individuals become afflicted and only 0.05 of the *aA* individuals develop the disease and none of the *AA* do, then only about 0.01 of the Hodgkin patients will be homozygotes for *aa*. In a case of 32 two-offspring families with two afflicted children the expected frequency of *aa* was about 2.5×10^{-6}. Also, assuming that the two-sib families may have *aA* x *AA* parents (about 4×10^{-3}) and the frequency of two-child-afflicted families is about 2.5×10^{-3}. These latter figures thus point to the possibility that in some populations the frequency of the recessive homozygotes may indicate high familial expression of the disease. (See ascertainment test, Hardy - Weinberg theorem, HLA, lymphoma, leukemia, anaplastic lymphoma)

HOECHST STAIN 33258: used for banding of AT-rich minor groove DNA sequences in the chromosomes and it is also an antibiotic. (See Harlequin staining of chromosomes, sister-chromatid exchange)

HOG-1: a protein kinase of the MAPK family. (See signal transduction, osmosis)

HOGNESS (-Goldberg) **BOX** (TATA box): a 7- 8 base pair region of conserved homology rich in TA, preceding the transcription initiation of the mRNA by about 19-31 residues in the promoter region of eukaryotic genes:

```
                                    TATA BOX
GC box---- CCAT box in animals -- T A T A A A A -- PY A PY (transcription start)
   or AGGA                        82 97 93 85 63 83 50
   in some plant genes         approx. % of conservation
```

Exceptionally, some promoter regions lack the TATA box, these are called *TATA-less promoters* such as U1, U2, U4 and U5 RNA promoters. The TATA box is generally surrounded by GC rich sequences (proximal and distal sequence elements, PSE and DSE). In these U promoters PSE and DSE are still present. Transcription by RNA polymerase II requires the association of the TATA box with a TATA binding protein (TBP) or additional TATA associated factors (TAF). (See also Pribnow box, TBP, TAF, open promoter complex)

HOGSTEEN PAIRS: refers to nucleotide pairs, by an association different from that proposed by Watson and Crick. The A—T pairs have an 80° angle between the glycosylic bonds and a 8.6 Å distance between anomeric carbons (differing in configuration about C). In the *reversed Hogsteen* pairs one base is rotated 180° relative to the other. (See Watson and Crick model)

HOLANDRIC GENE: is Y-chromosome linked. The mammalian Y chromosome appears largely heterochromatic under the light microscope and it carries very few genes. The H-Y antigen gene has been assigned to the proximal region of the long arm of Y and the testis-determining factor, formerly called TDF, now SRF is proximal to the centromere in the same arm in humans. The long arm contains also the pseudoautosomal region (PAS); this DNA sequence Yp (SMCY) is homologous to an X-chromosomal tract, Xp (SCX), the region where X and Y crossing over can occur. The gene for surface antigen MIC2Y was assigned to the euchromatic region Ypter - q1 of the Y chromosome. The homolog was assigned to an X-chromosomal band between Xp22.3 and Xpter. The azoospermia factor (AZF) Sp3 or HGM9 maps at the site of H-Y and may be identical with it. Genes controlling body height and tooth length were suspected to be in the Y-chromosome. An arginosuccinate and an actin pseudogene were located to the human Y chromosome. A gene for hairy ears was suspected to be in the Y chromosome but its status is not resolved with certainty. (See Y chromosome, sex determination, differential segment, H-Y antigen, SRF, pseudoautosomal, azoospermia, pseudogene, actin, heterochromatin, surface antigen)

HOLANDRIC INHERITANCE: genetic transmission (only) through the male.

HOLISM: the view that organisms represent an integrated system of elements and mechanisms and that the integrated system is more than a collection of the parts and therefore cannot properly be understood on the basis of the separated components. (See organismal genetics)

HOLLIDAY JUNCTURE: the points were the polynucleotides forming the Holliday structure are exchanged during recombination. This juncture may be bound by Rad1 protein in the presence of Mg^{2+} and cut by this endonuclease. Rad1 appears to be the catalytic subunit of the Rad1/Rad10 endonuclease. (See Holliday model, Holliday structure, Holliday model RAD, recombination molecular mechanism of)

← ELECTRONMICROPGAPH (Courtesy of Drs. H. Potter & D. Drechsler)

HOLLIDAY MODEL: of general recombination is best explained by the figure, p. 480. (After Potter, H. and Dressler, D. 1976, Proc. Natl. Acad. Sci. USA :73:3000; originally proposed by Holliday, R. 1974 Genetics 78:273). This is the most widely accepted model of both prokaryotic and eukaryotic recombination. (See also Holliday juncture, Holliday structure)

Holliday model of recombination continued

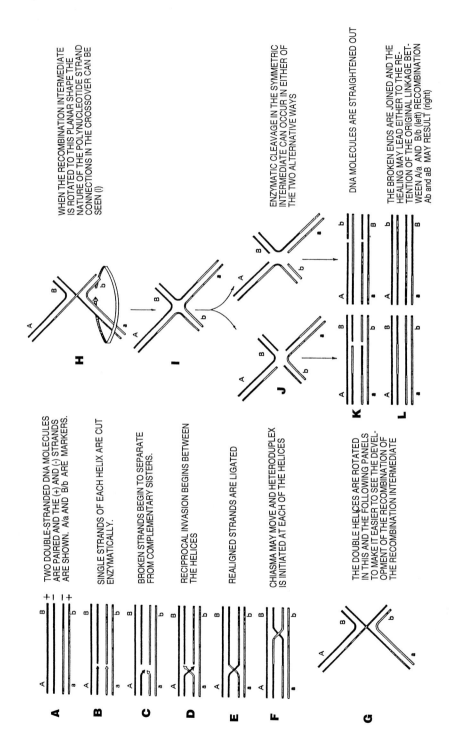

HOLLIDAY STRUCTURE: a recombinational intermediate of DNA displaying a four-strand cruciform arrangement. (See Holliday model step I). Its resolution requires a specific endonuclease and depending on the manner this takes place either by crossing over (flanking marker exchange) or gene conversion (non-crossover) without outside marker exchange results (steps L). (See model on preceding page, cruciform DNA)

HOLOCENTRIC: in several species of insects (*Lepidoptera, Hemiptera, Homoptera*) and in certain plants (*Luzula*) the spindle fiber attachment is not limited to the centromere (kinetochore) but the microtubules can be attached to many points along the chromosome, appearing as if the centromere would be diffuse. In some genotypes of maize and rye additional spindle fiber attachment sites (neocentromeres) accompany the major centromere. This characteristics may influence the distribution of the chromosomes to the poles (preferential segregation, non-disjunction). Species with holocentric chromosomes may be subjected to high doses of chromosome-breaking agents and still a more or less orderly anaphase distribution may take place. This feature has been exploited for biological control *lepidopteran* agricultural pests. The heavily irradiated individuals with broken chromosomes may survive and mate but in their progeny the chromosomal fragments may fuse leading to multiple translocations and lethal offspring. (See screwworm control, genetic sterilization, centromere, neocentromere, centromere activation, translocation)

HOLOENZYME: a functionally complete enzyme, including co-factors. (See apoenzyme)

HOLOGENESIS: a view that claims that humans originated at many locations of the globe.

HOLOGRAPHY: a three-dimensional photography with the aid of split laser beams.

HOLOPROSENCEPHALY: is dominant with human chromosomal locations 2p21 but possibly other locations also exist (7q36, 21q22.3, 18p11.3, 2q35, 12q13). The anomalies involve cleft lip, hypotelorism (abnormal distance between organs), defective head and face, mental defects due to neural tube anomalies. These human genes appear to be homologous to the *Drosophila hedgehog* gene. (See head/face/brain defects, *hedgehog*, tumor suppressor pathway)

HOLT-ORAM SYNDROME: is a dominant or sporadic arm, thumb and heart malformation associated either human chromosome 12q24.1, the gene is homologous with the *Drosophila* gene *Serrate* and the *Brachyury* mouse gene family. The prevalence of the syndrome is 1×10^{-5}. (See heart disease, polydactyly, adactyly, *Brachyury*, thrombocytopenia, sporadic, Tabatznik syndrome)

HOMEOALLELE: alleles in the homoeologous chromosomes. (See homoeologous chromosome)

HOMEOBOX: a conserved (183 bp) sequence within homoeotic genes. (See homeotic genes)

HOMEODOMAIN: the part of the protein that is coded for by the homeobox; it contains a DNA-binding helix-turn-helix motif protein domain. The homeodomain contains three α helices and a flexible N-terminal arm. The third or recognition helix takes position in the major groove of the DNA. The N-terminus keeps contact with several bases in the minor groove of the DNA. The best conserved part of the homeobox is the TAAT motif. The homeodomain genes determine the anterior posterior pattern of development and they are usually clustered in the genome. (See homeotic genes, DNA-binding protein domain, helix-turn-helix motif, homeobox, pseudogenes, morphogenesis in *Drosophila*, anterior, posterior)

HOMEOGENE: see homeotic gene

HOMEOGENETIC INDUCTION: cells or tissues are started on a certain path of development by induced cell(s).

HOMEOLOGOUS: see homoeologous

HOMEOLOGOUS RECOMBINATION: may take place between DNA (chromosome) strands that are similar but not entirely homologous. Recombination between *E. coli* and *Salmonella* with about 16% non-homology is about 10^{-5} of that of recombination within the species. Mismatches within the species affected recombination in a less dramatic extent. In mouse 2/232 mismatches reduced recombination to about 5%. Mitotic recombination in budding yeast at a difference of about 17-27% resulted in reduced exchange by a factor of 13-180. Meiotic recombination is also much reduced in higher eukaryotes in case of sequence differences. (See illegitimate recombination)

HOMEOPATHY: administering small doses of a medicine to a sick person that in larger doses given to a healthy individual would produce the same disease symptoms as the one it is intended to be cured.

HOMEOSTASIS: the property of a system to maintain its composition by a flexible adjustment of the function of its genes (genetic homeostasis), or a physiological, or developmental buffering capacity of cells or developing organisms under a range of conditions. (See logarithmic stability factor, stress)

HOMEOSTASIS, GENETIC: the property of a population to equilibrate its genetic composition and resist mutational changes.

HOMEOTIC GENES: specify an alternative competence for differentiation of a part of the body, e.g. in *Drosophila* legs in place of antennae, in plants petals in place of stamens, etc.; they contain a homeobox. Homoeosis (term coined by Bateson in 1894) was recorded by the ancient world (King Midas, 7th century B.C. grew 60-petalled roses). The discovery of the first homeotic gene, *bithorax* (*bx*, 3.58.8) by Calvin Bridges in 1915 stimulated more interest in these genetically determined developmental anomalies. Subsequently, other such developmental genes were discovered both in *Drosophila* and all types of higher organisms. In the plant *Arabidopsis* 35-70 homeogenes were estimated to exist. The homeotic genes are generally large complexes, the *BXC* complex occupies more than 300 kb and less than 1/10 of it codes for mRNA. They are interspersed with introns and intergenic DNAs required for the developmental regulation of the complex. In *Drosophila melanogaster* the homeotic genes are basically continuous (only introns are wedged within), in *Drosophila viridis*, the *Ultrabithorax* complex is mapped to two different salivary chromosome bands in chromosome 2. The first molecular analysis conducted (in D.S. Hogness laboratory, 1983) on the *antennapedia* complex, *ANTC* (3-47.5), revealed that this complex spans 335 kb and includes several transcription units. All homeotic genes have a 180 bp conserved consensus sequence, the so called *homeobox* that specifies the *homeodomains* of regulatory proteins. The organization of the homeoproteins is represented below by the structural features of the *Ant* gene of *Drosophila*:

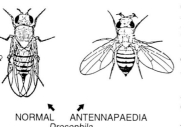

NORMAL   ANTENNAPAEDIA
Drosophila

[MSSLYY?N]-variable region-[IYPWM]-intron-[RKRGRQTYTRYQTLELEKQ⏎ acidic tail-COOH

NH$_2$ tract   homeopeptide   1   10→   helix I

FHFNRYLTRRRRIEIAHALC⏎

←22   28→   helix II   ←38

LTERQIKIWFQNRRMKWKK⏎

42→   helix III   ←58

EN] 60 **basic** amino-acid-rich
60   HOMEODOMAIN

(For the identity of amino acids see amino acid symbols in protein sequences). The 7-8 amino acids at the amino termini are rather well conserved across taxonomic groups, except the one represented by (?). The homeodomain of this protein may form 3 helical units between amino acids 1 (Ser) and 22 (Glu), between 28 (Arg) and 38 (Leu) and between 42 (Glu) and 58 (Lys). The best conserved amino residues are in outline. This homeodomain region has homology to the helix-turn-helix motif of prokaryotic repressor proteins and to the MAT α2 protein with repressor function at the mating type site in yeast. These three helical regions may fold into a helix-turn-helix DNA binding motif. The homeoprotein may make (unspecific) surface contact with the phosphate backbone of the DNA in the major groove. The homeodomain's conserved residues, preceding the helixes, attach to the minor groove of the DNA and for the more speci-

Homeotic genes continued

fic contacts probably helix III is responsible. These homeoproteins regulate processes of differentiation and either point or frameshift mutations may abolish their DNA-binding abilities and alter the pattern of differentiation. Binding can take place also at more than one DNA sequence as long as some basic similarities are shared in the base sequences. Therefore, one homeoprotein may regulate more than a single gene although at a different degree. Homeoboxes display great sequence similarities among different organisms. Homeobox, *HOX2* of the mouse is in chromosome 11 but its human homolog is in human chromosome 17 (containing in a 180 kb region altogether 9 homeobox genes separated by a few units of recombination), while *HOX1* of the mouse is in chromosome 6 but its homolog is in human chromosome 7p along with other seven homeoboxes. *HOX1* is also homologous to the *ANTC* homeobox of *Drosophila*. Hox genes in the mouse occur also in chromosome 11, 15 and 2. Each Hox/Hom contains a cluster 9-11 genes with an average length of about 10 kb and thus the clusters are about 100 kb, each. The genes within the cluster show paralogy and most likely arose by serial duplications during evolution. The genes within a paralogous group also refered to as cognate. Some similarities are apparent among the homeotic loci in different organisms and these are called orthologous because they appear to have common evolutionary origin. Statistical surveys of *Drosophila*, mouse, *Caenorhabditis*, humans, etc. indicates that the flanking areas of the homeotic gene clusters were conserved during evolution. Homeoboxes are regulated by cis- and trans-acting elements directly or indirectly. (See also POU, developmental pattern formation, helix-turn-helix, mating type yeast, morphogenesis, Wolf-Hirschhorn syndrome, syndactyly, coordinate regulation, operon, paralogous loci, collinearity)

HOMING ENDONUCLEASES: are encoded within mitochondrial and nuclear introns and catalyze the movement of introns. The best studied representatives of these enzymes is I-Sce. It is mitochondrially encoded within the ω^+ yeast factor (21S rRNA). After being spliced out it is translated into Sce protein. The VDE endonuclease is within the intron of a vacuolar membrane ATP-ase. They are transcribed and translated together and when the protein is processed the VMA1 ATPase and the VDE endonuclease are produced by evicting VDE and splicing the amino and carboxy-terminal amino acid sequences of VMA1.

The *Sce* system can be used for site-specific chromosome breakage and repair studies after introduction with appropriate vectors into the cells. The event can be monitored in a *S1neo* construct where the *neo* antibiotic resistance gene is interrupted by a 18 bp *I-Sce I* sequence and therefore *neo* is not expressed. The *I-Sce I* sequence is flanked by a CATG duplication:

S1neo≈≈≈CATG≈≈CATG. After I-Sce I cleavage two types of *neo* genes are found (1) *neo*≈≈≈CATG expressed and *neo*≈≈≈Δ≈ not expressed because of the Δ deletion. Other type of constructs can also be produced that can restore *neo* activity by cleavage and repair. *I-Sce I* can be introduced into the cells also by electroporation and can bring about non-homologous repair. The double-strand breaks generated by the homing endonuclease increased gene targeting up to a 1000 fold. Thus these systems — although basically different from the *Cre/loxP* or *FLP/FRT* — can be used for similar purposes. (See mitochondrial genetics, *Cre/loxP*, *FLP/FRT*, intron homing, site-specific recombinases, chromosomal rearrangements)

HOMINIDAE: the family of humans, *Homo sapiens, Homo erectus, Homo habilis* and other most closely related, now extinct, genera that are more highly evolved than the closest family of the *Pongidae* that includes the orangutan, chimpanzee and gorilla. The brain size in modern humans varies, the Australian aborigines have a brain volume over 1,200 cm^3 whereas that in *Homo erectus* was over 1,000, *Homo habilis* above 700, *Australopithecus* 400 to 500, gorilla 500 and orangutan and chimpanzee just below 400. Apparently the brain size changed very little in the last 300,000 years. The taxonomic classification and evolutionary descent of *Primates* is not entirely clear. The majority of anthropologists favor the idea that *Homo sapiens* evolved in Africa and then spread to Asia and Europe in reletively recent time. One tree of

Hominidae continued
hominid descent is shown below.

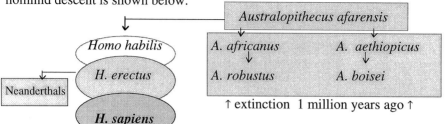

New (1997), skeletal mtDNA analysis of Neanderthal remains supports the view that these people represent an evolutionary dead end as shown at left

(See hologenesis, Primates, Eve foremother of molecular DNA, out-of-Africa)

Homo sapiens (man): 2n = 46. (See hominidae, human races, primates)

HOMOALLELES: differ at the same nucleotide site of a codon, and therefore only four different alleles can be produced, containing at a particular site either A or T or C or G and within the nucleotides recombination is impossible. Such an allele can be changed by intracodon (between nucleotides) recombination or base substitution. (See homoeoalleles, heteroalleles)

HOMOCARYON: see homokaryon

HOMOCYSTINURIA: may be due to different causes: (i) recessive deficiency of cystathionine synthetase [human chromosome 21q22], (ii) defects in vitamin B12 metabolism, (iii) poor intestinal absorption of B12, (iv) deficiency of methylenetetrahydrofolate reductase. Within group (i) different forms were also found; some responded also to vitamin B6 (pyridoxine). The general symptoms involve dislocation of the eye lens, thromboembolism (obstruction of the blood vessels), bone abnormalities, mental retardation, psychological disorders, etc. The diagnostic test for homocystein and methionine in the urine is carried out by the cyanide-nitroprusside reaction. Prenatal and carrier identification are practical. Types (ii) and (iii) respond favorably to vitamin B12 (hydroxycobalamin) and cultured cells required methionine. Type (iv) individuals respond favorably to pyridoxine and folic acid. (See coronary heart disease, hypertension, genetic screening, amino acid metabolism, cystinuria, cystinosis, cystathionuria, vitamin B_{12} deficiency, methionine biosynthesis, hyperhomocysteinemia)

HOMOEOALLELES: evolutionarily and functionally closely related genes in polyploid species. (See also homoalleles)

HOMOEOBOX: see homeobox, homeotic genes

HOMOEOLOGOUS ALLELES: occur in the homoeologous chromosomes of allopolyploid species; also called homoeoalleles.

HOMOEOLOGOUS CHROMOSOMES: nonidentical yet related chromosomes derived from a common ancestor and despite some evolutionary divergence, show partial homology. (See also homologous chromosomes)

HOMOEOTIC: related by evolutionary descent but modified during the course of evolution; similar but not entirely identical (recent usage is generally homeotic although homoeotic is the etymologically correct spelling)

HOMOGAMETIC SEX: (XX) produces only X-chromosome-containing gametes, in contrast to the heterogametic (XY) individual, which can have both X- and Y-chromosome-bearing gametes. The homogametic sex can be either female (XX) or male (ZZ). (See chromosomal sex determination, heterogametic)

HOMOGAMY: mating between similar types.

HOMOGENEITY TEST: can be used to determine whether different sets of data are statistically homogeneous enough to consider them to be part of the same population and whether the information is homogeneous enough to permit pooling. This test is basically a chi square procedure but the use of the Yates correction is not allowed. Without testing the homogeneity of separate sets of experiments, the information should not be pooled even when the combined data fit well to a null hypothesis. The use of this test is best explained by a tabulated example:

Homogeneity test continued

HOMOGENEITY TEST (using pea data of Mendel)

Family 1 (n = 36)	D	R	df⊗	chi²	Family 3 (n = 97)	D	R	df⊗	chi²
(1). observed numbers	25	11			(1) observed numbers	70	27		
(2) expected (3:1)	27	9			(2) expected (3:1)	72.25	24.25		
(3) difference (1)-(2)	2	2			(3) difference (1)-(2)	2.75	2.75		
(4) square of difference	4	4			(4) square of difference	7.563	7.563		
(5) divide (4) by (2)	0.148	0.444	1	0.592	(5) divide (4) by (2)	0.105	0.312	1	0.417

Family 2 (n = 39)	D	R	df⊗	chi²	Families Combined (n = 172)	D	R	df⊗	chi²
(1). observed numbers	32	7			(1) observed numbers	127	45		
(2). expected (3:1)	29.25	9.75			(2). expected (3:1)	129	43		
(3) difference (1)-(2)	2.75	2.75			(3) difference (1)-(2)	2	2		
(4) square of difference	7.563	7.563			(4) square of difference	4	4		
(5) divide (4) by (2)	0.259	0.776	1	1.035	(5) divide (4) by (2)	0.031	0.093	1	0.124

HOMOGENEITY TEST	Chi Squares	Degrees of Freedom	Probability*
Total of the 3 families	2.044	3	
Combined data	0.124	1	0.72
HOMOGENEITY (difference of the above lines)	1.92	2	0.38

* Use chi square table (here chi square charts were used). ⊗ df = degrees of freedom

(See also chi square, null hypothesis, Yates correction)

HOMOGENEOUSLY STAINED REGION: due to amplification in cancer cells, extended chromosomal bands are detectable by light microscopy.

HOMOGENOTE: see endogenote

HOMOGENTISIC ACID: see alkaptonuria

HOMOGRAFT: there is no known genetic difference between the transplanted and host tissues.

HOMOHISTONT: a non-chimeric tissue derived from a chimera. (See chimera)

HOMOHYBRID DNA: annealed product of two methylated or unmethylated DNA sequences.

HOMOKARYON: see dikaryon

HOMOLOGOUS CHROMOSOMES: contain the same gene loci and form bivalents in meiosis.

HOMOLOGOUS GENES: carry out basically the same function, and descended from common ancestors yet their primary structure may not be entirely the same. (See analogous genes)

HOMOLOGOUS PROTEINS: occur in different species and display similar structure and function such as the various globins, the majority of metabolic enzymes, etc.

HOMOLOGOUS RECOMBINATION: genetic exchange between essentially identical chromosomes (polynucleotide chains). Homologous recombination can be used for gene disruption and studying the consequence of lack of function of a particular locus. For this a plasmid is constructed that carries in between the flanking sequences of the target gene other nucleotide sequences such as an antibiotic resistance gene. The flanks permit homologous pairing and recombination and may take place within the boundary an interrupting DNA stretch (e.g. neomycin):

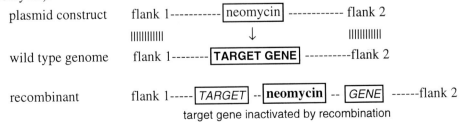

target gene inactivated by recombination

Homologous recombination continued

Such a manipulation is useful in animal cell cultures and the recombinant cells can be selected by the resistance to the antibiotics. Extremely rare double cross-overs can thus be recovered. Homologous recombination can be enhanced by about 10 fold by cutting within the target by an appropriate restriction endonuclease. The frequency of somatic homologous recombination in mammalian cells may vary within the range of 10^{-8} to 10^{-6} per cell per generation. One may note that this term "homologous recombination" gained a new meaning because classical genetics recognized recombination always as homologous. (See also illegitimate recombination, targeting genes, Cre/loxP, FLP/FRT, excision vector)

HOMOLOGUE-SCANNING MUTAGENESIS: is applicable to gene families coding for structurally related proteins. Homologous domains of the proteins (7 to 30 amino acids) are substituted for each other and in the substituted domain then amino acids are replaced by another amino acid, e.g. alanine. Then the binding of the substituted domains to the protein receptor (e.g. growth hormone, protein kinase) is analyzed. Such an analysis may reveal any change in receptor binding and function. A simpler procedure is the "charged-to-alanine scanning mutagenesis" in which blocks of amino acids (4 to 8) are replaced by alanine. (See site-specific mutation, directed mutagenesis, alanine-scanning mutagenesis)

HOMOLOGY: similarity based on nucleotide sequences in the DNA and RNA or amino acid sequences in the protein. It indicates also evolutionary relationship. Information on homology of human and yeast genes can be obtained by <http://www.ncbi.nlm.nih.gov/XREFdb>. The discovery of homology sometimes appears puzzling. *Saccharomyces* yeast contains a gene homologous to the *NifS* gene of nitrogen fixing *Azotobacter* although yeast does not fix nitrogen. Further studies revealed that this gene actually inserts sulfur into metal — sulfur centers of metalloenzymes using pyridoxal phosphate as cofactor. The reliance on homology for understanding genetic functions is of great importance. Yeast, *Drosophila*, *Arabidopsis* are relatively simple organisms and can genetically be manipulated by mutation, recombination, transformation, etc. and can help to shed light how series of genes function in a genomic context in other organism, e.g. humans, where the application of these laboratory techniques (e.g. controled mating) are limited or impossible. Evolutionists distinguish between *repetitive homology* such as the multiplicity of legs in the millipedes and *non-serial homology*, e.g. the leaves of a plant. On next page the amino acid sequence homology of cytochrome c protein is shown in 35 different organisms. The shading indicates complete identity at the alignments. (Courtesy of Margaret Dayhoff, ed. Atlas of Protein Sequence and Structure, 5. Natl. Biomed. Res. Found. Georgetown Univ. Washington, DC) Alignment of nucleotides sequences provides more critical information because the amino acid sequences do not distinguish among synonymous codons (See also analogy, DNA sequence alignment, indel, orthology, paralogy, xenology, congruence analysis, phylogenetic weighting, page 487 illustration)

HOMOMERIC PROTEIN: built from identical subunits.

HOMOMULTIMERIC PROTEIN: consists of more than two identical subunits.

HOMOPLASTIDIC: the chloroplasts within a cell are genetically identical. (See chloroplast genetics)

HOMOPLASY: parallel evolution (similarity is not based of common ancestry).

HOMOPOLYMER: a synthetic polynucleotide chain built from only one type of nucleotides, e.g. AAAA or CCCC, etc.

HOMOPOLYSACCHARIDES: are built of one type of sugar subunits.

HOMOSELECTION: in small populations selection may favor homozygotes and this increases the specialization to the unique environmental niche. (See heteroselection, selection)

HOMOSEXUAL: an individual is attracted to the same sex, that identifies him-/herself as homosexual, develops sexual fantasies about the same sex and practices homosexuality. Various studies estimated the prevalence of homosexuality from 2 to 10% in human populations. Homosexual orientation occurs in also other mammals, especially under conditions when the opposite sex is not available. (Continued on page 488)

Continued from preceding page.
HOMOLOGY OF THE AMINO ACID SEQUENCE OF CYTOCHROME *C* IN 35 ORGANISMS

Homosexual continued from page 486.

In some primitive human societies magical powers are attributed to homosexuals, however in the majority of societies and religions homosexuality is disapproved. The ancient Greeks accepted it as abnormal condition. The majority of the individuals may begin same-sex orientation between age 5 to 30, most commonly by puberty. The bases of homosexuality are not entirely clear. It had been attributed to hormonal, psychological, anthropological, genetic and moral causes or to a combination of part or all of these factors. The exact scientific study of homosexuality is difficult because generally a multitude of factors influence the development of all behavioral traits and homosexuality may also have different categories from obligate homosexuality to bisexuality and primarily heterosexuality. Recent anatomical studies attributed brain differences concomitant with homosexual orientation. Twin studies suggested higher concordance of this type of sexual orientation between monozygotic twins than between dizygotic ones. The observation that male homosexuals have more homosexual males than lesbian females in their kindred may indicate either that the genetic bases of female homosexuality may not be identical to that of males but may also point to some environmental causes. One study involving 114 families of homosexual index cases found an increase of same-sex orientation among maternal uncles and cousins in comparison to paternal relatives, suggesting the possibility of sex-linked transmission of the gene(s) concerned. In 40 families where at least two homosexual males occurred, in approximately 64% of the sib-pairs tested an apparent linkage was observed to the DNA marker Xq28 with a multipoint lod score of 4, indicating a higher than 99% probability for synteny. The heritability of male homosexuality was reported to be 0.50 but the inheritance of female homosexuality seems more complex. Some of this human information is debated by research workers. In *Drosophila* the *satori* (*sat*) mutants of males do not court or copulate with females but have sexual interest in males. The locus co-maps and allelic with *fru* at 91B chromosomal band. The *fru*^{sat} flies lack the male-specific Lawrence muscle (MOL). The fru^{sat} protein, expressed in some brain cells is probably a transcription factor with two zinc-finger domains. The social status of homosexuals is an ethical, rather than a biological problem yet the ethical solution may be facilitated by better biological information. (See lod score, sibling, twinning, kindred, ethics, behavioral genetics, sex determination)

HOMOSEXUAL CROSS: the mitochondrial genome of yeast appears to have sex-factor-like elements ω^+ and ω^- and the crosses between $\omega^+ \times \omega^-$ was called heterosexual while the crosses $\omega^+ \times \omega^+$ or $\omega^- \times \omega^-$ are homosexual crosses. (See mtDNA, mitochondrial genetics)

HOMOTHALLISM: the same individual (thallus) of lower eukaryotes produces both *plus* and *minus* or *A* and *a* mating type spores that can fuse into a zygospore. The homothallism bear similarity both to monoecy and autogamy in higher plants. These spores are not necessarily immediate meiotic products. (See heterothallism)

HOMOTROPIC ENZYMES: allosteric enzymes regulated by their substrate. (See allostery)

HOMOZYGOSITY IN A RANDOMLY SELECTED INDIVIDUAL: at a locus for an allele, under equilibrium conditions between mutation and genetic drift, is approximately $F = \dfrac{1}{1 + 4N\mu}$ where N = population size, µ = mutation rate. (See genetic equilibrium, genetic drift)

HOMOZYGOUS: in a diploid or polyploid the alleles at a locus are identical; being a homozygote.g. *aa* or *aaaa* or *AA* or *AAAA*.

HONEY BEE: *Apis mellifera* 2n = female 32, male n = 16.

HOOGSTEEN PAIRING: takes place between nucleotides of a third strand of DNA in the major groove of the duplex. This happens when in the duplex one strand is polypurine and the other is polypyrimidine; the third strand is most commonly polypyrimidine in the triplex DNA. (See triple helix formation)

HOP (*Humulus lupulus*): is a climbing dioecious plant with 2n = 20 chromosomes although forms (*H. japonicus*) with 2n = 16 females and 2n = 17 males have also been reported. Its main use is for brewing beer but it may be mixed to breads is some areas of the world. The soft resins and essential oils of the flowers add a typical bitter flavor to the brew and have some preservative effect. It is related to hemp (*Moraceae*). Some extracts have pharmaceutical value.

HOPEFUL MONSTER: neo-Darwinism assumes that evolutionary changes take place by accumulation of minor mutations that favor fitness of an organism. An alternative suggestion raised the possibility that some major mutations conveyed distinctive alterations to an individual transcending conventional taxonomic boundaries, and such an individual was called a hopeful monster. (See saltation, salatatory replication, gradualism)

HOPPING: movement from one location to another by a transposable or insertion element. Also in translation when the peptidyl-tRNA passes further downstream to a similar or same codon. (See transposable elements, insertion elements, recoding)

Hordeum bulbosum: a wild barley; it exists in diploid and tetraploid forms ($x = 7$). It gained particular attention because when crossed with the common barley (*Hordeum vulgare*, $2n = 14$) or with wheat ($2n = 42$), its chromosomes are eliminated from the zygote and thus haploids are produced at a high frequency. (See haploids, chromosome elimination)

Hordeum vulgare: cultivated barley ($2n = 14$). It is used for animal feed, human food and for the industry to produce malt and beer. The world production is about half of that of maize and about 40% of wheat. It evolved probably from the wild *H. spontaneum* ($2n = 14$) a species with what it is readily crossable and forms fertile hybrids. *H. spontaneum* has 2 dominant genes (*Bt* and *Bt$_1$*) that makes the ear brittle. The two-rowed barley develops fertile flowers in the central part of the spikelets whereas in the six-row barley all three flowers are female-fertile under the control of gene *v*. In the naked barley the glumes (husks) are not attached to the kernel because of gene. It is an autogamous species. (See databases, USDA)

HORIZONTAL TRANSMISSION: see transmission

HORMESIS: increased growth by irradiation at low doses or by other stress factors. (See radiation)

HORMONAL EFFECTS ON SEX EXPRESSION: although sex-determination is under the control of genes within the sex-chromosomes, the expression of sex characteristics may be influenced by hormones administered through medical treatment. Human females treated by steroid hormones to prevent miscarriage may give birth to females that may become somewhat masculinized after puberty. The bovine freemartins also display some virile features presumably caused by intrauterine exposure to male sex hormones. Genetically determined subnormal production of the pituitary growth hormone (human gene assigned to chromosome 17q23-q24) may involve recessive sexual anomalies. Castration and ovariectomy lead to intersex phenotype that can be further enhanced by grafting ovaries into male or testes into female chickens. The plant hormone gibberellin may affect sex ratio in some species, and can restore fertility in some genetic dwarfs. (See hormones, testicular feminization, animal hormones, growth hormones, freemartins, plant hormones, sex determination)

HORMONE RECEPTORS: are located on the surface or within target cells and transmit their signals to the genes that are set into action when the signals reach them. The general structure of he Zn-fingers of the steroid and thyroid (peptide) hormone receptors is shown on next page. Steroid hormones are not readily soluble in aqueous media like the blood and require special carrier serum proteins to be carried to the plasma membrane. The glucocorticoid receptor is usually situated in the cytoplasm and moves to the nucleus after binding with a ligand. The estrogen and progesterone receptors are mainly in the nucleus whereas the thyroid receptors are present only in the nucleus. The hormones combine with hormone receptor proteins. These complexes then may bind also to other complexes of hormones and navigate to the hormone response elements (HRE) in the upstream regulatory regions of the genes. The HREs are found in either the promoter or in the enhancer regions where they regulate the transcription of particular genes. These HRE elements vary from hormone to hormone specificities but all have a consensus sequence of dyad symmetry, sometimes palindromic and frequently separated by 3 or more non-conserved bases. The most important functional parts of the hormone receptors is the 66 to 68 conserved and basic amino acid rich tract that binds to the

Hormone receptors continued

DNA. The binding may require that a ligand be associated with the receptor or the removal of the ligand-binding domain. The DNA binding region of the hormone receptor forms two zinc fingers by cross-linked to Zn. The two fingers are coded for by two separate exons and the entire DNA-binding region is separated by introns from other coding regions of the receptor gene. The structure shown below represents the estrogen receptor but it is characteristic for many transcription factors. Thus the hormone receptors are hormone-inducible transcription factors. For the DNA binding the 3 amino acids following the last Cys residue of the right finger appears most important. The palindrome-like structure of the HRE in the DNA indicates that the hormone receptor protein works as a dimer. The function of the receptor as a transcription factor requires hormone binding. The hormone binding is determined primarily by the residues near the C terminus. This hormone

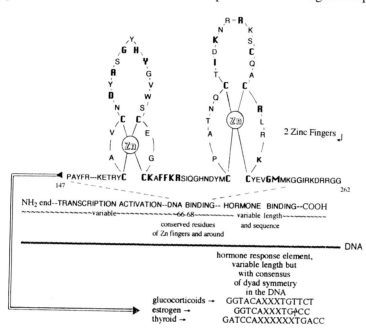

ZINC FINGERS OF THE ESTROGEN RECEPTOR DNA-BINDING PROTEIN. THE MOST CONSERVED RESIDUES ARE SHOWN IN BOLD. NUCLEAR LOCALIZATION IS SPECIFIED BY THE CARBOXY TERMINUS. THE AMINO ACID SYMBOLS IN PROTEIN SEQUENCES ARE GIVEN BY THE SINGLE-LETTER CODE. THREE DIFFERENT HORMONE RESPONSE CONSENSUS SEQUENCES ARE SHOWN AT THE LOWER PART OF THE DIAGRAM. THE HORMONE RESPONSE ELEMENT IS GENERALLY ABOUT 200 BASES UPSTREAM FROM THE TRANSCRIPTION INITIATION SITE.

binding region is critical for specificity. If a chimeric protein is constructed that replaces this binding region with the amino acid sequences of another hormone receptor, then the chimeric protein activates transcription of genes normally receptive to the first segments of the chimera in response to the latter hormone but not to the hormone that normally activates these genes. The receptor protein does not bind to the HRE unless it is linked to the hormone. Yet if the hormone binding domain is deleted, the receptor can bind to the cognate HRE. When the human estrogen response element (HRE) is transfected into yeast along with a functional estrogen receptor gene, estrogen can promote the expression of yeast genes even if their normal UAP (upstream activating sequence) had been deleted. In some instances all the components of transcriptional activation by hormones is present yet no activation is observed because an inhibitor may tie up the system and may prevent, e.g., dimerization of the receptor. These basic molecular genetic studies found medical applicability. The proliferation of breast cancer depends on a continuous supply of estrogen. The drug tamoxifen can compete for the estrogen binding site but this complex is not capable to activate transcription. Thus it may be used as an antineoplastic drug in combination with surgery, irradiation and chemotherapy if clinical evidence indicates that the tumor has estrogen and progesterone receptors. The steroid analog, RU486, developed in France, is a progesterone analog and can block the implantation of the

Hormone receptors continued

fertilized ova because of the depletion of normally functional receptors. Thus it is used — in countries where it is approved — as an "after morning pill" for prevention of pregnancy. The most commonly used birth control pills contain a combination of estrogen and progesterone and their elevated level shuts down the production of the pituitary hormones and thus preventing ovulation. (See hormones, transcription factors, signal transduction, breast cancer, testicular feminization, hormone-response elements, hormones, tamoxifen)

HORMONE-RESPONSE ELEMENTS (HRE): are short DNA sequences flanking genes that respond rapidly to activation by steroid or peptide hormones. Steroid hormones, retinoic acid, thyroid hormone and vitamin D_3 interact with ligand activated transcription factors. The receptors for these steroid/nuclear receptor superfamily are bound to the interspaces of the "half-sites" (n) between the tandem repeats of six base pairs:

| AGGTCA | n | AGGTCA | 9-CIS RETINOIC ACID RECEPTOR SITE
| TCCAGT | | TCCAGT |

| AGGTCA | nnnnn | AGGTCA | ALL-TRANS RETINOIC ACID RECEPTOR SITE
| TCCAGT | nnnnn | TCCAGT |

| AGGTCA | nnn | AGGTCA | VITAMIN D_3 RECEPTOR SITE
| TCCAGT | nnn | TCCAGT |

These six nucleotide pairs are unchanged but the number of bases (n) between the boxes varies. These are recognized by heterodimers of receptor proteins. The binding domains may include Zn fingers. (See glucocorticoid response elements, estrogen response elements, thyroid hormone response elements, RAR, RAX, Zn fingers, regulation of gene activity, ethylene, retinoic acid)

HORMONES: are peptides (polypeptides), amino acid-derivatives and steroids, synthesized in a gland of animals and carried by the blood to the site of their action(s) where they control the local function (first messengers). Animal hormones are e.g. somatotropin (growth hormone), corticotropin (in adipose and kidney tissues), thyrotropin (stimulates thyroids), follitropin and lutropin (in gonads), prolactin and lipotropin (in mammary glands), insulin (regulates sugar metabolism), serotonin (in nervous system), testosterone, estrogen (in most cells), progesterone (in uterus), prostaglandins (in smooth muscles), etc. Plants synthesize 5 different groups of hormones that are quite different (except the brassinosteroids) in chemical nature from animal hormones. These control primarily cell elongation (auxins), cell divisions (cytokinins), elongation, germination (gibberellins), abscission of leaves and fruits, dormancy, germination (abscisic acid), elongation, ripening, morphogenesis (ethylene). These plant hormones display numerous types of interactions and are involved in complex manner in signal transduction. (See also, steroid hormones, animal hormones, hormone receptors, hormone-response elements, plant hormones, brassinosteroids)

HOROTELIC EVOLUTION: see bradytelic evolution

HORSE: *Equus caballus*, 2n = 64. The Mongolian wild horse (*Equus przewalskii*) 2n = 66. (See mule, hinny)

HORSERADISH (*Armoracia rusticana*): is a perennial pungent condiment; 2n = 4x = 32.

HORSE THREADWORM: see *Ascaris*

HOS1, HOS2, HOS3: histone 3 and 4 deacetylase enzymes of yeast. (See histone deacetylation)

HOST CELL REACTIVATION: DNA repair mechanisms operating in bacterial cells harboring defective bacteriophages. (See DNA repair)

HOST-CONTROLED DNA MODIFICATION: see restriction modification

HOST-MEDIATED ASSAY: tests whether the host metabolism can convert a compound (a promutagen) into a mutagen, detectable by the cells passed through the host e.g., yeast, bacterial or animals cells injected into and then withdrawn from the abdominal cavity of a mouse previously exposed to a potential mutagen. Forward or backmutation frequency in the cells is assessed after withrawal from the abdominal cavity of the test animal and then plating the cells onto selective solid media or growing them in liquid media. Such a rapid assay was designed for substituting it to the slow and expensive direct mammalian assays of promuta-

Host-mediated assay continued
gens. (See figure below, bioassays in genetic toxicology)

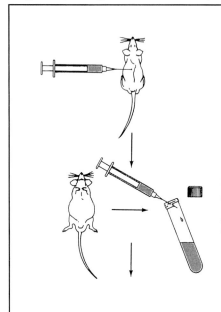

Microbial or cultured animal cells are injected intraperitoneally into a mammalian host for proliferation.

After a period of 1 to 3 days a test chemical dissolved in saline or in saline plus a solubilizer is injected subcutaneously once or more into the mouse or rat. The control animals are treated with the solvent solution only. This mutagenic exposure lasts for a few hours or a few days.

A sample of the injected cell population is withdrawn from the abdominal cavity into centrifuge tubes in preparation for assaying the mutagenic effectiveness of the test chemical in comparison with the control.

The recovered cells are plated on Petri dishes or incorporated into soft agar culture tubes to screen the mutants produced within the host as a consequence of the treatment.

A GENERAL OUTLINE OF THE HOST-MEDIATED MUTAGEN TESTing. (See text on page 491.) Modified after Fischer, G.A., Lee, S.Y. & Calabresi, P. 1974 Mutation Res. 26: 501

HOST - PATHOGEN RELATIONS: are based on susceptibility or tolerance/resistance genes in the host and virulence and avirulence genes in the pathogen, respectively. The development of the disease depends on the appropriate genetic determination and expression of genes in the two organisms. In plant breeding successful efforts were made to transfer disease resistance genes from wild relatives to the cultivated varieties (e.g., the *Ry* dominant genes of the wild potato, *Solanum stoloniferum* conveys resistance against the potato Y virus (PYV) or leaf rust (*Puccinia triticina*) from *Aegilops umbellulata* (new name *Triticum umbellulatum*) to the cultivated bread wheat, etc. These classical methods of breeding for resistance had to overcome problems of linkage of the resistance to agronomically undesirable genes present in the primitive species. Some viral plant pathogens have lines with extragenomic *satellite RNA* and the presence of the latter may suppress the expression of the full scale disease (*attenuation*). The disease symptoms may be prevented in some cases also by the simultaneous presence of another related plant virus (*cross protection*). This phenomenon, also called *preimmunity*, can be exploited by transferring through transformation into the plant species a coat protein gene of some of the RNA viruses. Another molecular plant breeding approach is to introduce into the plants antisense mRNA of the viral protein that would prevent the synthesis of viral proteins within the plants. The latter method may be coupled with a ribozyme expression system that would degrade the viral RNA.

Various genetic mechanisms are known to be involved in resistance against bacterial and fungal pathogens. The *hypersensitivity reaction* of the plant tissues localizes the infection by death of the surrounding host cells resulting, in addition, in the liberation of antimicrobial cellular substances. A collection of a chemically diverse substances, called phytoalexins regulate the synthesis of plant cellular compounds that are involved in the defense mechanisms or are the consequences of the infection process. Various genes involved in the synthesis of phenol-derivatives may be the regulatory targets. The activation or enhancement of the transcription of genes determining cell wall components (polysaccharides, lignin, suberin, saponin, etc.) may provide a barrier to infection. The activation of plant enzymes (*pathogen-*

Host-pathogen relations

esis-related proteins) such as various glucanases, chitinases, proteases may lead to continued breakdown of the cell walls of the pathogens and thus facilitating the escape of potential hosts from microbial infection. Plants do not have an immune reaction to invading agents yet some type of systematic resistance may be induced by exposure to non-pathogenic microorganisms, with certain chemicals such as nicotinic acid-derivatives, etc. Some microorganisms secrete or contain in their cell walls organic molecules (proteins, glucans, glucosamins, fatty acid-derivatives, etc.) that provoke the defense mechanisms of plants or the plants themselves may produce such molecules upon contact with the pathogens. These compounds are called *elicitors*. Elicitors come in many different chemical composition and display substantial specificities for different hosts and pathogens. Many of these substances seem to reach the plant cell nucleus through the various ion channels or the signals are transduced through the membrane systems to activate the appropriate plant cells and functions. Recently several plant genes have been cloned that are involved in the disease-resistance-tolerance mechanisms (RPS1, RPS2, Pto, [*Pseudomonas syringae*], Cf-9 [*Cladosporium fulvum*], L^6[*Melampsora lini*], N [tobacco mosaic virus], Hm1 [*Cochliobolus carbonum*], etc. It is believed that these genes act by activating the plant defense system though responding to the elicitors of the avirulent pathogens or by producing enzymes that degrade the fungal toxin. In some instances inorganic antimicrobial compounds accumulate; in *Theobroma cacao* cyclooctasulfur accumulation in the xylem walls was observed in *Verticillium dahliae* resistant plants. Salicylic acid is generally considered as an agent of improving plant resistance probably through its stimulating action of hydrogen peroxide release from the cells. Plant mutants with reduced ability to synthesize salicylic acid, phytoalexins become more susceptible to pathogens. These latter compounds are credited by systemic acquired resistance (SAR). The Xa21 (*Xanthomonas oryzea*) gene of rice conveys resistance to this plant pathogen. DNA sequence analyses indicate a leucine-rich repeat, characteristic for serine/threonine kinases. It has been suggested that this protein serves as signal transducer to alert the plant cell defense system. In some pathogenic bacteria (*Pseudomonas*) the same virulence factors mediate pathogenicity in plants (*Arabidopsis*) and in animals (mouse). It is assumed that plant disease resistance in general is mediated by a signal transduction path. A pathogen-generated ligand, produced by an avirulence gene, is recognized by an extra- or intracellular receptor, encoded by a plant receptor gene and that sets into motion the process of defense. Different plant species and different resistance reactions indicate the involvement of a leucine-rich repeat (LRR) and leucin zipper binding sites shared by many resistance genes. The majority of isolated disease resistance genes display a nucleotide binding domain, a common feature of many ATP- and GTP-activated protein families involved in signal transduction. The *Pto* protein kinase of tomato and the closely linked *Prf* gene seem to be binding directly to the *Pseudomonas syringae* avirulence gene. Many of the putative resistance genes cloned may lack the stringent criteria to be the principal determinants of resistance to a particular pathogen. Some plant pathogenic fungi (*Helminthosporium victoriae*) can be infected with double-stranded RNA viruses (totivirus) and causes lytic disease to the fungus. Recently in a wild relative of the cultivated beets a membrane protein was indentified that conveys resistance against a nematode (a lower eukaryotic animal). (See also immune system, antisense RNA, signal transduction, alien substitution, transformation, Floor's model, ribozyme, hypersensitive reaction, SAR, biological control, antimicrobial peptides, plant defense, plantibody, 2,5-A, salicylic acid)

HOST-RANGE RESTRICTION: the same oncoprotein may not be transforming different species because the host cellular machinery is not favoring its function.

HOST-RESISTANCE GENES: in mammals is generally quantitated on the basis of the length of survival after infection by viruses, bacteria or protozoa or on the basis of the density of the pathogen in the infected foci. The inheritance is assessed on the basis of crosses with recombinant inbred strains in mouse. Most of the resistance genes are linked to the H-2 complex (MHC) but

Host-resistance genes continued

other locations as source of resistance have also been identified. Several resistance genes have been cloned. Comparative analyses assist in the use of the information for human and livestock research. (See host - pathogen relations)

HOT SPOT: highly mutable site within a gene or high-frequency recombination site in chromosomes. Recombinational hot-spots are generally expressed as cM/kb (map unit/kilobase). Within the a^1 gene of maize the recombination frequencies varied between 5×10^{-3} to 8×10^{-2} cM/kb whereas in the 140 kb distance between a^1 and $sh2$ 6×10^{-4} cM/kb was observed. Within the pseudoautosomal site (*PAR*) of *Drosophila* 1 cM/53 kb was found. The frequency of recombination shows great variation along the length of the chromosomes and there is no equivalence between genetic and physical length. (See chi elements, cold spot, coefficient of crossing over, mariner, MITE, coefficient of crossing over)

HOUSEFLY: *Musca domestica*, 2n = 12.

HOUSEKEEPING GENES: are functional throughout the life of a cell and in the majority of cells and tissues. In the promoters the TATA and CAAT boxes are not present. Their 5' CpG islands are not methylated. (See asparagine synthetase, TATA box, CAAT box, transcription illegitimate, methylation of DNA, capping enzymes)

HOX: see homeobox, PAX

*Hpa*II: restriction enzyme recognition site is C↓CGG.

HPBO: human platelet basic protein is a member of a large cytokine family. (See cytokines)

HPLC (high performance liquid chromatography): is suitable for the separation (among other molecules) of oligonucleotides up to 20 residues by partitioning them between a stationary phase (such as a chromatography column) and a mobile phase (such as solvents) forming a gradient pumped through a system, carefully monitored. (See sequenator, electrophoresis. high-performance liquid chromatography)

HPFH: see thalassemia

HPFN: human plasma fibronectin is a cell adhesion molecule. (See CAM)

HPRT: hypoxanthine phosphoribosyl transferase is controlled in humans by a recessive gene (map location X26). Its deficiency causes the Lesh-Nyhan syndrome involving choreoathetoses (involuntary, uncoordinated movements), spasticity (high muscular tension), mental retardation, tendency to self-mutilation, overproduction of uric acid, renal damage, etc. It can be detected prenatally. The molecular basis of the defect is an interruption of the salvage pathway of nucleotides because the guanosine and hypoxanthantosine cannot be phosphorylated. Defect of this enzyme has been extensively exploited for *in vitro*, selective isolation of fused mammalian cells and for the isolation of mutations. (See salvage pathway, HAT medium, Lesh-Nyhan syndrome, HGPRT, epilepsy)

HPV: human papilloma virus. (See papilloma virus, bovine papilloma virus)

HR: hypersensitive reaction. (See host-pathogen relationships)

hr: human recombinant in abbreviation, e.g. hrEGF, human recombinant epidermal growth factor

hRAS (HRAS): RAS proto-oncogene of humans. (see RAS)

HRE: see hormone-response elements.

HRI: hemin regulated inhibitor; hemin apparently interferes with the function of a specific protein kinase involved in translational activation; in the absence of hemin the protein kinase is active. (See hemin, protein kinase)

hRP-A: a human homolog of the single-strand binding protein involved in genetic recombination. It is required for *in vitro* replication of SV40 DNA and growth. (See SV40)

HS SITE: see DNase hypersensitive site

hsdR: some mutations in *E. coli* that eliminate restriction function but not the modification in the restriction-modification system. (See restriction enzyme, restriction-modification)

HSE: heatshock elements located in the upstream region of eukaryotic heatshock responding genes and recognized by heat shock protein transcription factors. A typical heatshock element

HSE continued
is an about 14 bp palindromic sequence with a GAA core: 5'-CTN**GAA**NN**TTC**NAG-3'
3'-GAN**CTT**NN**AAG**NTC-5'
(See heat shock proteins, transcription factors inducible, HSTF)

Hsp: see heatshock proteins

Hsp70: a heatshock protein gene (activated by elevated temperature); it is a chaperone. Male mice lacking Hsp70 is unable to produce spermatids and mature sperm and showed dramatic increase of apoptosis of the spermatocytes. Female mice of the same constitution did not show meiotic disturbances or infertility. (See heatshock proteins, HSE, HSTF, apoptosis)

HST ONCOGENE: was assigned to human chromosome 11q13; originally identified in stomach cancer but it was found also in Kaposi sarcoma. Its product is a heparin-binding protein with homology to the fibroblast growth factor (FGF). It is called also K-FGF. (See FGF, oncogenes, growth factors, Kaposi sarcoma)

HSTF: heatshock transcription factor regulates about 20 heatshock genes by binding cooperatively to more than one heat shock element. It is activated by phosphorylation. (See HSE)

HTDV (human teratocarcinoma derived virus): is an endogenous human retrovirus family occurring in about 25-50 copies per genome, and also about 10,000 solitary LTRs are present. They are well expressed in different organs. (See retroviruses, solitary LTR)

HTF ISLANDS: 1-2 kb sequences around the 5' region of genes where the Hpa II restriction endonuclease cleaves Tiny Fragments because the cytidines are not methylated. At most other regions, because of methylation, only larger fragments are cut. The mammalian genome may display 30,000 HTFs.

H1TF2: a histone transcription factor binding to CCAAT box. (See histones)

HTLV (human T cell leukemia virus): is the causative agent of some adult leukemias and partial paralysis (HTLV-1-associated myelopathy). (See leukemia, myelopathy)

HTML: hypertext markup language permits the display what an internet (www) address and content looks like. HTTP (hypertext transfer protocol) is the program to use with the www. (See www)

HU: "histone-like" prokaryotic protein involved in maintaining the DNA structure and transposon integration. (See histones)

hUBF: human upstream binding factor is a transcription factor. (See transcription complex, transcription factors)

HUGO (Human Genome Organization): source of mammalian genomic data, available in Macintosh Hypercard disks from HUGO Europe, One Park Square West, London NW1 4IJ, UK, Phone: 44 71 935 8085. Fax: 44 71 706 3272. INTERNET: s.brown@sm.ic.ac.uk

HUMAN ARTIFICIAL CHROMOSOME: are not yet equivalent to similar constructs in *Saccharomyces* or *Schizosaccharomyces* and the closest are the minichromosomes that may be less than 1/10 of the size of the chromosome before truncation yet mitotically quite stable. (See minichromosome, YAC, BAC, HAC)

HUMAN CHROMOSOME MAPS: linkage and mapping information until the 1960s were obtained mainly by studying family pedigrees. This procedure was very inefficient because in human populations controled mating is not feasible and because of the small populations available, crossing over frequencies could not be used the same manner as in e.g., *Drosophila*. Some of the gene locations were determined by deletion mapping but until 1956 even the number of human chromosomes was uncertain. In the late 1960s and early 1970s new chromosomal staining techniques made possible the recognition of chromosome bands, permitting the more precise location of genes relative to these beacons. During the 1960s the culture of isolated mammalian cells has been improved and in 1960 fusion of somatic cells was accomplished following in 1965 by the production of mammalian somatic cell hybrids. It was discovered that in human-mouse somatic cell hybrids human chromosomes are preferentially lost, facilitating the assignment of particular genes to specific human chromosomes. Only those human genes were expressed in the cultures that were present and their expression ceased

Human chromosome maps continued

when the critical chromosome was eliminated. By induced breakage, the location of genes could be further specified when it was seen that the retention of particular cross bands was essential for the expression of a particular gene. In 1960 nucleic acid hybridization was discovered and by the mid 1960s nucleic acid hybridization was used for localizing genes in cultured human HeLa cancer cells. Somatic cell hybridization permitted the chromosomal assignment of three times as many genes within a few years as family analysis accomplished over half a century. By 1970 methods were developed to assign mouse DNA fragments to chromosomal location by *in situ* hybridization. *In situ* hybridization also facilitated the location of about twice as many genes than the pedigree analysis. The discovery of restriction endonuclease enzymes in the late 1960s and 1970's opened a new era in genetics of physical mapping, localizing cloned genes or just "anonymous DNA" sequences to chromosomal positions. Isolated DNA sequences could be ordered on the basis of the overlapping fragments by the use of chromosome walking and jumping first employed in *Drosophila* in 1983. The use of isolated genes from other species (*Drosophila*, mouse, yeast, etc.) as probes, permits now the identification of gene position of homologous genes and DNA sequences. By the extension of these techniques, individual human genes can be isolated, cloned and sequenced, thus the technology is in use for the ultimate mapping of the human genome at the nucleotide level, to recognize the structural bases of functions, regulation and organization of the genome. The detailed mapping is assisted now by sophisticated computer programs to determine the order of several gene loci on the basis of the maximum likelihood principle. This information, sought by the human genome projects, will contribute to evolutionary information as well to medical applications of gene therapy. The 1966 (Nature [Lond.] 380:14 March) comprehensive genetic map included 5,264 microsatellites (AC/TG). The average interval size was 1.6 cM and 59% of the map was covered by 2 cM intervals and 1% was in intervals larger tan 10 cM. (See reference point chart next page, pedigree analysis, mapping, deletion mapping, chromosome banding, radiation hybrids, somatic cell hybridization, nucleic acid hybridization, *in situ* hybridization, probes, restriction endonucleases, RFLP, chromosome walking, DNA sequencing, maximum likelihood, gene therapy, mouse)

REFERENCE POINTS FOR MAPPING GENES OR DNA SEQUENCES TO HUMAN CHROMOSOMES. EACH CHROMOSOME IS NUMBERED; 1 IS THE LONGEST BY CYTOLOGICAL EVIDENCE. SINCE THE METAPHASE LENGTH IS SOMEWHAT AMBIGUOUS BECAUSE THE CONDENSATION, DUE TO GENETIC AND OTHER CAUSES, MAY VARY AND LENGTH IS NOT AN ABSOLUTELY CONSISTENT FEATURE OF THE KARYOTYPE. THE APPROXIMATE MINIMAL DNA CONTENT (Mb) OF EACH CHROMOSOME IS SHOWN IN PARENTHESIS UNDER EACH DIAGRAM. THE CENTROMERE IS REPRESENTED BY CONSTRICTIONS AND THE PERICENTROMERIC REGION IS DIAGONALLY HATCHED. THE SHORT ARMS (SHOWN ABOVE THE CENTROMERE) ARE DESIGNATED BY p (for petit) AND THE LONG ARMS ARE DESIGNATED BY q. EACH ARM HAS CROSS-BANDS CORRESPONDING TO OBSERVATIONS BY GIEMSA, Q AND C BANDING TECHNIQUES. THE FINE HATCHING INDICATES HETEROCHROMATIC REGIONS. EACH CHROMOSOME, STARTING AT THE CENTROMERE, IS DIVIDED INTO DOMAINS AND SUBDOMAINS DESIGNATED BY NUMBERS. e.g. IF WE SEE THAT THE LOCATION OF TNFR-2 (encoding tumor necrosis factor 2, a 75 kDa protein) IS AT 1P36.3-P36.2, WE KNOW THAT IT IS IN THE TERMINAL REGION OF CHROMOSOME, EITHER IN THE FIRST BLACK BAND OR ABOVE. THE TELOMERIC REGION IS USUALLY DESINATED BY ter. ACROCENTRIC CHROMOSOMES (13, 14, 15, 21, 22) AND THEIR HETEROCHROMATIC AREAS AND SATELLITES ARE CROSS-HATCHED FINELY. THE LARGE NUMBER OF GENES ARE IMPOSSIBLE TO REPRESENT ON A SINGLE PAGE. THE LOCATION OF THE PHYSICAL MARKERS OCCUPIED, by 1996 (Nature 380, suppl.), ABOUT 138 PAGES IN FINE PRINT. (Illustration is based partly on the New Haven Human Gene Mapping Library Chromosome Plots. Number 4. HGM9.5, constructed by Spence, M.A. & Spurr, N.K)

HUMAN GENE TRANSFER: the vectors can be replication-deficient retroviral or adenoviral vectors. The retroviral vectors may integrate 9 kb passenger DNAs into the chromosomes and may destroy tumor suppressor genes or activate oncogenes if inserted nearby and thus may permanently alter the genome either favorably by inserting the desired genes or undesirably as mentioned above. They integrate only into replicating cells only. (continued on page 498)

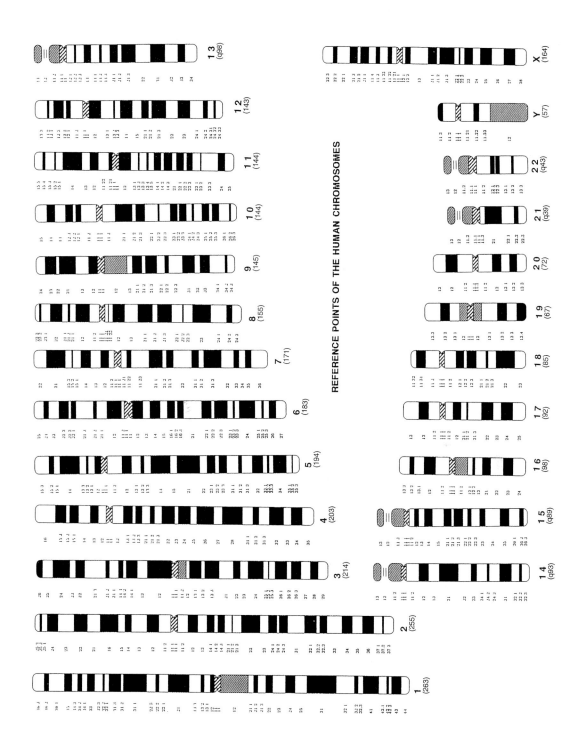

Human gene transfer continued from page 496

These vectors have been used only in *ex vivo* experiments.

Adenoviral vectors have a carrying capacity of about 7.5 kb DNA. These enter the cells by special receptors. Adenoviral vectors do not integrate into the chromosomes and are not replicating indefinitely but they have to be reapplied after a few weeks or months. They are well suited for *in vivo* use because they may be efficient in replicating and non-replicating cells and may have high titer (10^{13} virion/mL). The adenoviral vectors may involve inflammation of the tissues and may encounter antivector cellular immunity.

Plasmid-liposome carriers also have been explored under *in vivo* conditions. Human gene transfer may be used to insert selectable markers into T cells, to stem cells, to tumor-infiltrating lymphocytes, neoplastic cells in hematopoietic lines and to sarcoma cells, etc. Gene transfer may have also therapeutic goals. Most of the human gene transfer experiments were plagued with inconsistent results and the expectations based on animal model experiments were not entirely fulfilled. The vectors need further improvement regarding "homing-specificity" (targeting), side effects (inflammation), elimination of the possibility of insertional mutations, eliminating possible immunological reactions against the vectors. (See transformation genetic, vectors, gene therapy, adenovirus)

HUMAN GENETICS: basic genetics using human (cells or individuals) as the subject of study. The human genome contains approximately 75,000 genes and their physical mapping is progressing rapidly. In general the function of one or more physically identified gene is discovered weekly. (See clinical genetics, medical genetics)

HUMAN GENOME NEWS: monthly publication on the human genome project by the National Institute of Health and the U.S. Department of Energy. Information: Human Genome Management Information System, Betty K. Mansfield, Oak Ridge National Laboratory, 1060 Commerce Park, MS-6480, Oak Ridge, TN 37831, USA. Phone: 615-576-6669. Fax: 615-574-9888. INTERNET: <bkq@ornl.gov>

HUMAN IMMUNODEFICIENCY VIRUS: see acquired immunodeficiency syndrome

HUMAN INTELLIGENCE (IQ): is not a simple qualitative trait but it is generally defined as the composite index of a variety of scores that each is expected to have different genetic determination. The intelligent quotients (IQ) also vary, depending on the variety of test batteries used for the analysis, such as cognitive, verbal, mathematical, logical, etc. performance. Since intelligence is such a complex trait, it must be under the control of many genes. All polygenic traits have a genetic and environmental components, so does intelligence. As a consequence it is impossible to separate clearly ability from achievement, and the tests involve generally the latter. Nevertheless, it is obvious that all traits are expressed on the basis of a genetic blueprint in the DNA, and "intelligence" cannot be an exception. Shortly before the end of the 19th century, Francis Galton, the father of statistical genetics, came to the conclusion that the apparent mental abilities of parents and children are correlated. Galton found that 36% of the sons of the 100 most distinguished men were still eminent but only 9.5% of their grandsons and 1.5% of their great grandsons were such. Also, 23% of the brothers of eminent men were also eminent. Since then several studies confirmed the existence of such general correlations. These correlations may be biased to a great degree because the environmental conditions of parental and offspring population is also highly correlated. Before World War I, intelligence quotients were introduced for standardized quantitative measurements of intelligence. The Binet test had been widely used for determining scholastic performance and predictions. According to these quotients children scoring according to the average of their age group were classified with a score of 100 and those whose performance corresponded to 2 years behind or 2 years ahead of their peers were assigned IQ values 80 or 120, respectively. Within similar socioeconomic and educational groups, these figures were meaningful. The inheritance of IQ was studied by mono- and dizygotic twins, reared together and apart as well as by comparing adopted children with foster and birth parents. These studies have proven by objective measures the existence of a significant hereditary component of the IQ indexes. These IQ values should be very cautiously considered if different ethnic or socio-

Human intelligence continued

economic groups are compared. Also, the developmental rate of individuals may vary a great deal; some of the children are early or late "bloomers" and this condition limits the predictive value of the indexes. There can be no question that certain genes, concerned with the nervous system, are responsible for mental retardation. The IQ indexes should not be used, however, for ethnic or social group-discrimination because the statistical ranges of individual IQ values are highly overlapping in human populations. Humans can be judged only individually. Also, the diverse individual values of athletes, business people, laborers, physicians, geneticists, theoretical physicists, etc. are all important for the good function of the human societies. Sex differences in mental test scores have been examined repeatedly in large representative populations. Although the overall differences between the sexes appear small, the males' abilities in mathematical and mechanical performance appeared higher whereas in associative memory, reading comprehension and perceptual speed the females were at advantage. The genetic meaning of these data are not clear. Males and females share the same chromosomal complement, except the Y-chromosome which by current knowledge contains minimal information beyond what is located in the pseudoautosomal segment and the genes related to sex determination and male fertility. Fragile sites in several human chromosomes are associated with trinucleotide repeats. The C residues of these variable length repeats may be methylated to a variable extent and the increase of methylation is correlated with a decrease of intellectual abilities of the individuals. (See polygenes, heritability, behavior genetics, behavior in humans, genius, mental retardation, cognitive abilities, myopia, fragile sites, FMR1 mutation, trinucleotide repeats, autuism, dyslexia)

HUMAN MUTAGENIC ASSAYS: humans cannot be subjected to mutagenic treatments and mutagenic risks to human populations is determined largely by indirect means, using microbial, animal and plant bioassays of genetic toxicology. Other methods may involve epidemiological efforts involving dominant mutations, survey of chromosomal aberrations in blood samples, testing cell cultures for mutability, and biochemical and molecular methods to assay genetic repair in cell culture, monitoring changes in DNA by restriction fragment polymorphism using appropriate probes, etc. Increase in recessive mutations are difficult to detect because of the human mating system does not favor homozygosis. Although the populations of Hiroshima and Nagasaki were exposed to very high doses of ionizing radiation as a consequence of the atomic explosion during world war II, no statistically significant increase in gene mutation could be detected although developmental defects and incidence of neoplasia increased. Similarly, the meltdown the nuclear reactor in Chernobyl in 1986 caused a significant increase in birthdefects and cancer, it is too early to tell whether these were only teratological effects or genetic causes are also involved. More recent data seems to indicate that genetic alterations were also caused. (See also atomic radiations, bioassays in genetic toxicology, host-mediated assays, substitution mutation)

HUMAN POPULATION GROWTH (Modified after Biraben, J.N. 1980 Population 4:1):

Years	400	1	500	1000	1500	2000
Millions of people	160	250	200	250	460	≈ 6000

HUMAN RACES: are distinguished by anthropologists on the basis of anthropometric traits. Geneticists delineate the races on the basis of gene frequencies shared within the group and different from other "racial" populations. The main human races caucasoids, mongoloids (including Chinese, Japanese, Koreans and American Indians, etc.) and Negroids. Khoisanoids or capoids (Bushmen and Hottentots) and Pacific races (Australian aborigines, Polynesians, Melanesians, Indonesians) may also be distinguished. Many other subgroups within the larger ethnic groups may be classified. There is no genetic incompatibility among the various human races and there is no well-founded scientific evidence that interracial marriage would lead to the disruption of co-adapted gene blocks resulting in biological or mental deterioration in the offspring. The three major human races are genetically closely related as indicated by determination of evolutionary genetic distance. The human race is extremely closely related on the basis of DNA sequences to primates, primarily to chimpanzees, yet the genetic isolation is complete. (See allelic frequencies, evolutionary distance, genetic isolation, racism, miscegenation, interracial human hybrids, primates, *Homo sapiens*)

HUMANIZED ANTIBODY: a chimeric molecule with the variable region of the mouse fused to the constant region of the human antibody. This molecular chimera retains its binding specificity and some characteristics of the human antibody. Some humanized antibodies have been tried as immunosuppressors of lymphatic and breast tumors. (See antibody, immune system)

HUMORAL ANTIBODY: is made by the B lymphocytes and is circulating in the bloodstream rather than being attached to the surface of T lymphocytes. (See immune system, antibody)

HUMORAL ANTIGEN: is secreted into the bloodstream. (See B cell, T cell, immune system)

HUMUS: decaying organic matter in soil.

HUNTER SYNDROME: see mucopolysaccharidosis (MPS II)

HUNTER-THOMPSON CHONDRODYSPLASIA: see chondrodysplasia

HUNTINGTIN: see Huntington's chorea

HUNTINGTON'S CHOREA: is a dominant genetic disorder (chromosome 4p16.3) with a prevalence of 4 to 7 x 10^{-5} live births. It is a progressive degeneration of the basal ganglia (spinal cord) and the brain cortex causing uncoordinated (choreic) movements and loss of mental abilities (dementia). It exists in the juvenile (akinetic-rigid) form with an onset in the teen years or with the late onset at age 30 to 50. Expectancy of survival, after the first symptoms appear, is about 20 years. The late onset form is generally inherited through the mother and the early onset through the father. Although this difference of this gene with a near-perfect penetrance was attributed initially to a mitochondrial co-factor but it appears to be determined by differential methylation (imprinting). Prenatal testing is feasible because of a very closely linked DNA marker, D4S10 (G8). Because of the onset is frequently delayed beyond the reproductive period, early analysis of risk on the basis of linkage to this tight molecular marker is desirable. Various biochemical alterations (GABA, glutamic acid decarboxylase, choline acetylase deficiency) may be associated with the disease. It has been shown that the 'huntingtin' protein (≈ 350 kDa) in the basal ganglia and cerebral cortex displays 37 to 121 glutamine (CAG) repeats vs. the normal protein which has only 11-34 repeats. The huntingtin protein resembles neuronal nitric oxide synthase. The normal Huntington gene is essential for embryo survival. (See genetic screening, DNA marker, mental retardation, epilepsy, fragile sites, nitric oxide)

HURLER SYNDROME (mucopolysaccharidosis I, MPS I): it is also called Scheie syndrome or Hurler-Scheie phenotype. (See mucopolysaccharidosis, Hunter syndrome, hypertrichosis)

hut **OPERON**: histidine utilization genes. Histidine is synthesized from three precursors ATP and phosphoribosyl-pyrophosphate (PRPP) and glutamine, and it is involved in the regulation of other amino acids by yielding both glutamate and ammonia. (See also histidine operon)

HUTCHINSON-GILFORD SYNDROME: see aging

HUVEC: human umbilical vein endothelial cells.

HVR: hypervariable regions in the DNA. (See DNA fingerprinting, somatic hypermutation)

H-Y ANTIGEN: is a histocompatibility antigen controled by a Y-chromosome gene. It was recognized by rejection of skin grafts of male donors by female recipients but the male recipients did not reject the grafts donated by females. The mice involved in these studies were inbred for many generations and were supposed to be isogenic, except for factors in the Y-chromosome. Therefore the rejection was attributed to a male-specific antigen. Thereupon it was hypothesized that the H-Y male-specific cell surface antigen is also a male (testes) determining factor. The H-Y gene was located near the centromere in the long arm of the Y-chromosome encoding a 11-peptide residue of the SMCY protein of almost 1500 amino acids. The X-chromosome has a homolog SMCX with about 200 amino acid site differences scattered along the entire length, except in the 11 residue H-Y antigen where there is only a single amino acid difference between the SMCY and SMCX. In the mouse the H-Y controling gene (*Hya*) in the short arm of the Y chromosome and the epitope is defined by the octamer Thr-Glu-Asn-Ser-Gly-Lys-Asp-Ile. It was also found that exceptional individuals with XY chromosomal constitution displayed female phenotype and cytological analyses revealed a deletion at the tip

H-Y antigen continued

of the short arm of the Y-chromosome. Also chromosomally XX exceptional males carried a translocated terminal segment of the short arm of the Y-chromosome in one of the X-chromosomes. This terminal segment was thus identified as the testis-determining factor (TDF) and thus the H-Y antigen have a true histocompatibility role but it is apparently not responsible for testis differentiation. The H-Y antigen is the product of several Y chromosomal genes. The current name of *TDF* is *SRY* (**s**ex-determining **r**egion **Y**). (See chromosomal sex determination, sex determination, sex reversal, SRY, gonadal dysgenesis, testicular feminization, freemartins)

HYACINTH (*Hyacinthus orientalis*): bulbous fragrant spring flower, 2n = 16.

HYALOPLASM: the very finely granulated part of the cytoplasm. (See cytoplasm)

HYBRID: progeny of two genetically not identical parent. (See F_1)

HYBRID ANTIBODY: have more than one epitope-binding sites because they are produced by genetic or chemical modification. (See hybrid hybridoma, humanized antibody, antibody)

HYBRID ARRESTED TRANSLATION (HART): when a mRNA is hybridized with a cDNA, only those sequences can be translated *in vitro* that are not base-paired. Translation of the paired sequences is prevented by the pairing; on this basis the coding sequence for a particular polypetide can be identified. Hybridization of mRNA by antisense RNA also prevents translation. (See antisense RNA, cascade hybridization, hybrid-released translation)

HYBRID, ASYMMETRIC: after somatic cell fusion some chromosomes of one of the "parental" cells were eliminated. (See alien addition, cell fusion, somatic cell genetics)

HYBRID DEPLETION: identifies any cDNA that encodes a subunit of a multimeric protein. It is managed by preparing a mRNA pool coding for the protein of interest. The mRNA is hybridized to a cDNA cloned in a single-stranded vector and the hybrids are fractionated by CsCl equilibrium centrifugation providing at the bottom of the centrifuge tube the unhybridized, i.e. the antisense RNA. When it is injected into *Xenopus* oocytes along with the coding mRNA of the original pool it may block the translation of the complementary mRNA. (See mRNA, density gradient centrifugation, antisense technology)

HYBRID DNA: a heteroduplex. (See heteroduplex)

HYBRID DYSGENESIS: is a historical term for various genetic phenomena caused by transposable elements in *Drosophila*. It entails mutation and chromosomal rearrangements in hybrids

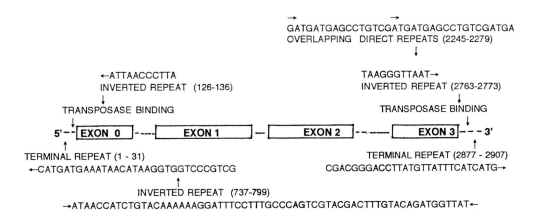

THE STRUCTURE OF A COMPLETE *P* ELEMENT OF *DROSOPHILA* (2,907 bp) FLANKED IN THE GENOMIC DNA TARGET SITE BY 8 bp DIRECT REPEATS. THE NUMBERS IN PARENTHESIS INDICATE THE NUCLEOTIDE POSITIONS (beginning with the 5' end). THE HORIZONTAL ARROWS INDICATE THE DIRECTION OF THE REPEATS; THE VERTICAL ARROWS POINT TOWARD THE POSITIONS IN THE LINEAR SEQUENCE OF THE *P* ELEMENT. (After Engels, W.R. 1989 in Mobile DNA, pp. 437-484. Berg,D.E & Howe, M.M., eds. Amer Soc. Microbiol. Washington, DC, USA

Hybrid dysgenesis continued

of two genetic stocks. *P - M* SYSTEM: The *P-strains* carry a transposable element and a suppressor of transposition. In the other strains (*M*) there is a genetic factor that derepresses the transposase and thus the hybrid becomes genetically unstable while both of the parental forms are stable. (The P and M originally indicated paternal and maternal conditions, respectively.). In *Drosophila* over 30 different hybrid dysgenesis systems have been identified. The best known among them is the *P - M* system. The physical structure of a complete *P* element is shown in the diagram on the preceding page.

Some *P* elements have internal deletion, duplications and substitutions (such as π_2, Pc[ry]). The *P* element can be *autonomous* (transpose by their own power) and non-autonomous (require a more complete [helper] element to move it). The autonomy of an element has been successfully tested by introduction through transformation an *in vitro* engineered *P* element (*Pc[ry]*) carried on a plasmid along with the *rosy* gene (*rosy* is the structural gene for xanthine dehydrogenase [map position 3-52] and mutants have rosy eyes). The wild type allele introduced into ry^- homozygotes have normal eyes and are capable of moving into the *sn* (*singed*, map location 1-21, responsible for bristle [microchaetae] deformations) and this fact can easily be monitored. The movement of *P* is controled by the transposase function that is encoded by the four exons that extend to almost the entire length of the element (see p. 501). The transposase begins transcription at base 85 and terminates at 2696, thus the transcript includes about 2.5 kb. The transposase enzyme is a 86.8 kDa protein. The *P* element can excise almost completely and leave behind the original genomic sequence or it may delete internal sequences, including sometimes even flanking nucleotides, involving known genes with a total length of rarely exceeding 7 kb. The imprecise excisions generate the defective elements. The frequencies of these excisions vary, ranging from 0.4 to nearly 2% per generation of the dysgenic flies. The mutability at the *sn* locus varies from 20 to 60% but may reach up to 90% when two reverse oriented *P* elements *(double P)* were at the target site. The targets for insertion are not distributed at random and *P* is inserted by several orders of magnitude more frequently in the *sn* locus than into the alcohol dehydrogenase (*Adh*) gene. Also, there is a tendency for *P* elements to become clustered. For insertion the non-translated upstream regions of genes is favored compared to the coding regions. Also, transposition in the germline is much more frequent than in the somatic cells. The suppression of transposition of the *P* chromosomes may only partially be relieved in the strains designated as *M'*. These transposable elements induce a variety of genetic events, including recombination in the male *Drosophila,* mutation and chromosomal rearrangements. The *P - M* system slightly boosts the effect of other mutagens. The frequency of X-chromosomal rearrangements was estimated to be 10% per generation and the second breakpoint tends to stay within the same chromosome. The active transposable elements cause also segregation distortion because the transmission of the *P* chromosomes is reduced compared to the *M* chromosomes. The transposons are also associated with gonadal abnormality and (GD) sterility at temperature particularly above 27° C. The activity of the *P* element my be affected also by the cytological sites and *cis*-effects. Various mutations induced by *P* are subject to suppressors. This transposon, similarly to others, can be used for gene tagging and isolation, particularly its special constructs carrying selectable markers such as neomycin resistance so they can be screened efficiently (smart ammunition). The element *pogo* (about 2.2 kb) is somewhat similar to *P*, and it has either 23 bp inverted terminal repeats and no target site duplication or a 21 bp inverted terminal repeats flanked by duplication of TA. The transposable element *hobo* (variable up to 3 kb) with up to 50 copies and 8 bp target site duplication. Some (H) *hobo* elements are located in euchromatic regions and others are empty (E) sites. *Hobo* is not activated by reciprocal crosses but its presence is associated with high degree of instabilities. *"HB"* is a small (1.6 kb) element with 20 copies and 8 bp target site duplication. *HB* contains one reading frame of 444 bp that shares 25% homology with the amino acids of the *Tc* element of *Caenorhabditis elegans.* The other

Hybrid dysgenesis continued
best studied transposable system of *Drosophila* causing hybrid dysgenesis is the *I - R* SYSTEM. The complete *I* element is 5.4 kb and has many feature of a retrotransposon, except that it does not have long terminal repeats. Thus it is structurally quite different from *P* yet some of its functions warrant its description. The counterpart of the *M* cytotype of the *P - M* system is the *R* (responsive) cytotype. Hybrid dysgenesis is observed in the crosses of *R* females with *I* males. These sterile females are called SF (stérilité femelle) whereas the reciprocal non-dysgenic ones are RSF. The female sterility of *I-R* does not involve gonadal anomalies (in contrast to GD in *P-M*) but hatching of the eggs is reduced. Eventually the *R* strains may be converted to *I* by "chromosome contamination", i.e. accumulation of chromosomes derived from an *I* strain by crossing and segregation. *I* factor activity involves mutation (recessive and dominant) that are frequently clustered, indicating their occurrence shortly before or at meiosis. The frequency of mutation varies at different loci and does not follow the same pattern as with *P*. The molecular structure of the complete I element of 5,371 bp is known. It does not have terminal repeats, however four TAA reiterations are near the 3'-end of one strand and in the genomic DNA there are 12 bp duplications at the target site. One of the strands of *I* has open reading frames (ORF) I (1,278 bp) and II (3,258 bp), separated by 471 bases. There is probably another ORF of 228 bp. The base sequences in ORF II are similar to viral and virus-like transposases' reverse transcriptases. There are apparent coding sequences at the COOH end for RNase H (ribonuclease digesting RNA in RNA-DNA hybrid molecules as required in reverse transcription). Elements similar to *I* have been detected in the mammalian *L1*, *Drosophila* non-viral retrotransposable elements (*F* family), *R2* ribosomal insertions in silk worm, *Cin4* element in maize, and in the *ingi* elements of *Trypanosoma brucei*. Near the 3' end of ORF II and the longest ORFs of L1 and Cin4 code the amino acid sequence: Cys-Pro-Phe-Cys-Gln-Gly-Asp-Ile-Ser-Leu-Asn-His-Ile-Phe-Asn-Ser-Cys that resembles the metal-binding domain of general transcription factor TFIIIa. ORF I has a sequence with some homology to the DNA-binding viral *gag* polypeptides (group-specific antigen). The *I* elements are most common near the centromeric regions. Mutations induced by *I* are stable and do not revert (unlike to *P*). Many of the *I* elements are truncated and show internal deletions and rearrangement. The *R* factor is quite complex, and it is determined by both nuclear and cytoplasmic regulatory components. Its role is to release the *I* elements' expression. The *FB* family of transposable elements are complex ca. 6.5 kb or smaller size transposons causing a variety of genetic effects. They seem wide-spread in *Drosophila* but present in small copy number. They cause chromosomal rearrangement in 1/1,000 chromosomes. (See transposable elements, retroposons)

HYBRID HISTOCOMPATIBILITY PHENOMENON: allogeneic inhibition.

HYBRID HYBRIDOMA: originates by fusion of two different hybridomas. (See hybridoma)

HYBRID INVIABILITY: is a post-mating or zygotic mechanism of sexual isolation. The hybrids either die before sexual maturity or the offspring is sterile. (See hybrid lethality, hybrid sterility, zygotic lethal)

HYBRID LETHALITY: the parental forms are normal yet hybrid embryos are aborted; this phenomenon is not uncommon when different species (with different chromosome numbers) are crossed. (See hybrid inviability, hybrid sterility, zygotic lethal)

HYBRID NUCLEIC ACID: double-stranded structure from two strands of different origin. (See DNA, DNA hybridization)

HYBRID PCR PRODUCTS: can occur when the amplified sample is heterozygous or when related sequences are amplified with the same primer. (See PCR)

HYBRID RELEASED TRANSLATION: cloned DNA is mobilized of a membrane filter and annealed with mRNAs. The hybridized mRNA is eluted and then translated in an *in vitro* system (wheat germ or rabbit reticulocytes). The labeled polypeptides are then analyzed by electrophoresis. (See hybrid arrested translation)

HYBRID RESISTANCE: see allogeneic inhibition

HYBRID STERILITY: the hybrid gonads or gametes are abnormal and incapable of normal sexual union. More commonly the sterility affects the males or the male gametes and the females may be successfully backcrossed with normal males. (See hybrid dysgenesis, hybrid inviability, hybrid lethality, zygotic lethal)

HYBRID SWARMS: arise when the habitats of two related species are adjacent and mass outcrossing takes place. In such cases the parental and the hybrid forms may not be easily recognized in the zone.

HYBRID VIGOR: the superior performance (growth, fitness) of hybrids has been observed for centuries before the birth of modern genetics. In its simplest case, hybrid vigor may be attributed to the presence of *complementary dominant* genes. If one parent is *AAbb* and the other is *aaBB*, their offspring is *AaBb* that has now the favorable dominant alleles at both loci (*dominance theory of hybrid vigor*). Geneticists measure vigor by reproductive advantage, fitness. The fitness of a homozygous recessive (R) class may be $w_{aa} = 1 - s$, where s is the coefficient of selection. The frequency of the homozygous recessives in a population may be determined by the rate of mutation (μ) from allele A to allele a. The average proportion of the recessive class is expected to be:

$\hat{p}^2 = \mu/s$ and the average frequency of recessive alleles is expected to be $\hat{q} = \sqrt{\mu/s}$. The average fitness of the population then becomes $\hat{w} = 1 - s\hat{q} = 1 - s\sqrt{\mu/s} = 1 - \sqrt{s^2\mu/s} = 1 - \sqrt{s\mu}$.

If each of the n alleles contribute equally to the performance of the individual (additive effect), the average reduction of fitness caused by homozygosity of the recessive alleles in the populations is $n\sqrt{s\mu}$. For example, assuming that an organism has 10,000 gene loci and an average mutation rate of 10^{-5}, and the selection coefficient against the recessive alleles is 0.01. After substitution we obtain $n\sqrt{s\mu} = 10,000\sqrt{0.01 \times 0.00001} \cong 3.16$. This indicates that inbreeding may reduce fitness of the population by a factor of about 3 compared to the situation when each locus has at least one dominant allele at all loci of a diploid organism. On the average this hypothetical example is in agreement with experimental observations on inbred and hybrid populations and lends support to the dominance theory of hybrid vigor (heterosis). If the mid-parent value, $m = 0.5 (P_1 + P_2)$ and $0.5(P_1 - P_2) = d$ and $h = F_1 - m$, it is possible to predict — in case of perfect additivity of all genes and the F_1 displays hybrid vigor—the best performance to be expected by accumulating all the favorable alleles in an inbred:

$$P_{mx} = m + \frac{h}{\sqrt{H/D}}$$ where H are the heterozygotes and D = the homozygous dominants

Increased vigor of hybrids may be caused also by *overdominance* (superdominance), i.e., the heterozygote *Aa* is surpassing both *AA* and *aa*. Let us assume that the selection coefficient of each of the two classes of homozygotes (*AA* and *aa*) is -0.05. In such a case the three genotypic classes (according to the Hardy - Weinberg theorem) would occur in one generation of reproduction: *AA* (95):*Aa* (200):*aa* (95) if the population is mating at random, the allelic frequencies are equal (0.5) and the fitness of both types of homozygotes is also equal. Since the size of the population (the three classes combined) is 390 rather than 400 as expected without selection, the proportion of the surviving zygotes is 390/400 = 0.975, indicating 2.5% (10/400) reduction by a single cycle of reproduction. This reduction may not be significant for the majority of species that produce more offspring than the number the habitat can maintain. The reduction in size may become, however quite serious if overdominance occurs at not only one but at several or many loci (see table on next page). The table indicates that for most populations even a 0.5% disadvantage of the homozygotes at larger number of loci may have very serious adverse consequences and would be hardly acceptable in herds of domesticated animals or in crop plants. The contribution of overdominance at single loci may be large enough to be of selective advantage in feral conditions or improve the performance of agricultural species. (See also inbreeding, selection coefficient, superdominance, overdominance)

Hybrid vigor continued

REDUCTION IN POPULATION SIZE AT THREE DIFFERENT PERCENTAGE OF DISADVANTAGE OF THE HOMOZYGOTES

Number of over-dominant loci	5%	1%	0.5%
10	0.77633	0.95111	0.97528
100	0.07952	0.60577	0.77856
1000	1.01×10^{-11}	0.00665	0.08183

HYBRID ZONE: at the geographical boundary of two races natural hybrids occur.

HYBRIDIZATION: crossing (mating) of genetically different individuals; annealing DNA single strands with RNA or a single-stranded DNAs of different origin (probe). (See Mendelian laws, c_0t curve, Southern hybridization, Northern hybridization, Western hybridization, South-Western method, *in situ* hybridization, dot blot)

HYBRIDIZATION PROBE: a radioactively or by fluorochrome [biotin] labeled nucleotide sequence that will hybridize with the complementary nucleotide sequences and identifies the homologous tract(s) either on an extracted DNA or *in situ* in a chromosome. (See probe, radioactive label, biotinylation)

HYBRIDOGENETIC: a species hybrid containing say the A and B genomes and normally mates with one of the parental species (say B) but its gametes transmit only one of its genomes (say A), thus the A genome is reproduced clonally while the B genome is added newly by each mating

HYBRIDOMA: myeloma (cancer) cell fused with spleen lymphocyte, producing monoclonal (identical by origin) antibodies. The cancer cell assures the rapid proliferation and indefinite growth while the other component determines the specificity. Hybridomas have many applications in basic research and are very useful for the production of monoclonal antibodies and for the generation of lymphokines. (See senescence, immortalization, monoclonal antibody, lymphokines, magic bullet)

HYBRIDOMA GROWTH FACTOR (HGF): see interferon β-2

HYDATIDIFORM MOLE: is a human hyperplasia resulting from an abnormal fertilization when the epithelial layer of the ovum is induced to proliferate into a tuft of cysts resembling a bunch of grapes. The karyotype of such structures is 46 (XX) and all the chromosomes are derived from a diploidized sperm of 23 (X) constitution. (See androgenesis)

HYDRA: an about 25-30 mm freshwater animal of about 100,000 cells. The freshwater hydra (*Hydra vulgaris attenuata*) is 2n = 32. Propagates mainly by budding; a new clonal offspring is produced every 1.5 to 2 days. Sexual reproduction has minor role; the production of a few eggs — predominantly by hermaphroditic means — takes a few weeks. It produces a variety of cell types and differentiated body under the control of morphogens similar to those in other more complex animals. It has excellent abilities for regeneration.

HYDROCARBONS: organic compounds containing only hydrogen and carbon; can be aliphatic (alkanes [paraffin], alkenes, alkynes, cyclic aliphatic) or aromatic. Although some of them are chemically rather inert (paraffin), the complex polynuclear hydrocarbons (benzo-a-pyrene, benzanthracenes, methylcholantrenes) are highly toxic, carcinogenic and mutagenic. They are present in combustion products. (See environmental mutagens, cigarette smoke)

HYDROCEPHALUS: is a disease of cerebrospinal fluid accumulation in the brain as a result of defect in its secretion and absorption. It can be the symptom of physiological or mechanical lesions and it may be due by several autosomal recessive (Dandy-Walker syndrome of occlusion of the openings of the fourth ventricle of the brain, Albers-Schönberg osteopetrosis [extreme bone density and bone proliferation]), autosomal dominant (achondroplasia, osteogenesis imperfecta congenital Type II, defects in bone formation), X-linked recessive (narrowing of a brain fluid channel following inflammation or bleeding) and X-linked dominant (orofacial-digital syndrome I, involving oral, digital and mental abnormalities) genetic causes. Its preval-

Hydrocephalus continued
ence is 0.01 to 1 x 10^{-3} births. It may be entirely sporadic or the recurrence risk when it is hereditary may vary from 15 to 50% within families, depending on the type of inheritance involved. Prenatal diagnosis may be feasible using ultrasound detection or brain tomography or magnetic resonance imaging. (See prenatal diagnosis, mental retardation, Walker-Wagner syndrome, orofacial-digital syndrome, Dandy-Walker syndrome, osteopetrosis, sonography)

HYDROCORTISONE: a glucucorticoid hormone, an antinflammatory drug.

HYDROGEN BOND: a weak bond between one electronegative atom and a hydrogen atom that is covalently linked to another electronegative atom C or N. Hydrogen bonds tie together the polynucleotide chains of a DNA double helix (A=T and G≡C), and also affect conformation of proteins. (See hydrogen pairing, Watson and Crick model, protein structure)

HYDROGEN PAIRING: secures the double-stranded form of DNA or RNA by establishing hydrogen bonds between the O atom attached to the C atom at position 4 (or position 6, depending whether the American or the Beilstein numbering system is used for the pyrimidine ring) of thymine (uracil) and the NH_2 group at the C of position 6 in adenine. The second hydrogen bond is formed between the hydrogen attached to the N at position 3 (or according to the Beilstein system 1) of the pyrimidine ring of thymine. Cytosine pairs with three hydrogen bonds with guanine between positions 4 - 6, 3 - 1 and 2 - 2 as shown by the figure on the opposite page. Cytosine and adenine cannot form hydrogen bonds unless a tautomeric shift occurs. Similarly, the pairing of thymine (or analogs) with guanine requires another tautomeric shift. The tautomeric shift is an isomerization of the bases, changing the position of a hydrogen from position 3N to position O^4 on the thymidine molecule or moving one hydrogen from N^6 position to 1N position in adenine.

As the diagram shows the bases may undergo keto - enol transformations according to the scheme below:

According to organic chemistry —C=C—OH ⇌ —C—C=O is keto-enol tautomerism
enol   keto
                H

(See DNA, base substitution, point mutation, base analogs, tautomeric shift, imino transformation, diagram on page 507)

HYDROGEN PEROXIDE (H_2O_2): may act as intracellular messenger in both plant and animal tissues. In plants salicylic acid inactivates catalase and that may be a factor in activation of plant defense mechanisms to pathogens. In animal tissues it may be involved in the activation of NF-κB transcription factor, involved in the regulation of the immune reaction and immunosuppression. H_2O_2 may initiate apoptosis. When cells are stimulated by cytokines, phorbol esters and growth factors H_2O_2 may be released into the extracellular space. (See NF-κB, host-pathogen relationship, immunosuppression, glucocorticoids, granulomatous disease, superoxide dismutase)

HYDROGEN PUMP: see proton pump

HYDROGENOSOMES: mitochondria-like organelles in *Trichomonads*. They are surrounded by double membrane and produce ATP from pyruvate but have no DNA. (See mitochondria)

HYDROLASES: enzymes carrying out hydrolysis reactions. (See hydrolysis)

HYDROLYSIS: splitting a molecule by inserting a molecule of water and one of their parts will obtain OH and the other H from the H_2O

HYDROPATHY INDEX: indicates the relative hydrophilic and hydrophobic properties of chemicals.

HYDROPATHY PLOT: is used to determine the hydrophobic amino acid tracts in (membrane) proteins on the basis of the energy requirement for transfer into water from a non-polar solvent. (See membrane proteins)

HYDROPHILIC: readily miscible with water; it has polar groups.

HYDROPHOBIC: insoluble in water, or poorly (if at all) soluble in water; it lacks polar groups.

.Also, (unjustified) fear of (drinking) water such as occurs in the viral disease called *rabies* which is also called hydrophobia.

Hydrogen pairing continued from page 506

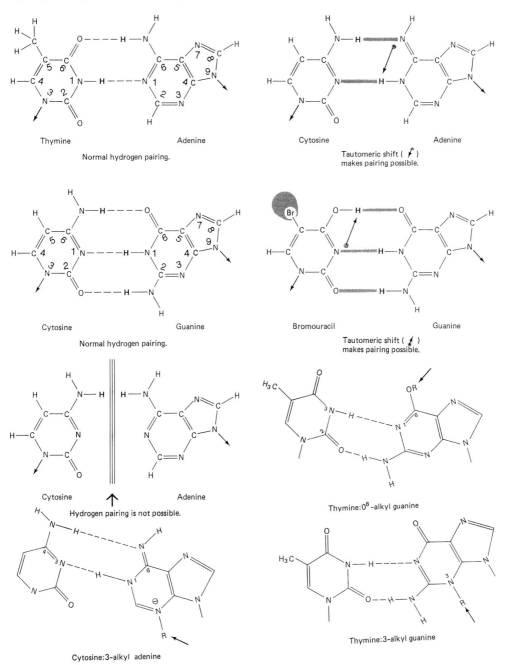

HYDROGEN BONDING BETWEEN DNA BASES. NORMAL BONDS: DASHED LINES; AFTER TAUTOMERIC SHIFT (↗ or ↙ respectively) THE HYDROGEN BOND SHIFTS AND IS THEN REPRESENTED BY STIPPLED LINES.

HYDROPONIC CULTURE: growing plants in salt solutions, rather than in soil, under semi-axenic conditions.

HYDROPS: an edema (accumulation of fluids) may occur in kidney diseases or in erythroblastosis fetalis and in many other diseases. (See erythroblastosis fetalis).

HYDROPS FETALIS: a prenatal anemia accompanied by fluid accumulation in the fetal body caused by failure to synthesize the α chain of hemoglobin (an extreme form of thalassemia major) or by destruction of hemoglobin in other hemolytic anemias (rh). (See thalassemia, hemoglobin, Rh, erythroblastosis fetalis)

HYDROXYAPATITE: is calcium phosphate hydroxide (may contain also silica gel). The phosphate residues of nucleic acids bind to calcium and thus double-stranded DNA binds to it stronger than the single-stranded molecules. On this basis, by adsorption chromatography, the two types of DNA can be separated from each other and DNA-RNA hybrids from RNA. The single-stranded molecules come off the columns by low molarity buffer and at higher molarity the double-stranded molecules can be eluted.

HYDROXYL RADICAL (—OH): is widespread in biological molecules.

HYDROXYLAMINE (NH_2OH): is an antagonist of pyridoxalphosphate (PLP)-requiring enzymes, cleaves Asp-Gly linkages at high pH, blocks oxidation of H_2O but permits electron transfer through photosystems I and II from artificial donors. Its poisonous effect (α effect) may be based also on its high nucleophilic reactivity. For genetics it is important that hydroxylamine reacts with carbonyl groups (C=O) of pyrimidines and targets specifically cytosine residues in nucleic acids and generates hydroxylaminocytosine (a thymine analog) and causes the transition of a G≡C base pair into a T=A pair, resulting in base-specific mutations. It is however a weak mutagen, effective in prokaryotes but without much effect in higher eukaryotes. (See transition mutation, base analog, base substitution, hydrogen pairing)

21-HYDROXYLASE DEFICIENCY: see adrenal hyperplasia

5-HYDROXYMETHYL CYTOSINE: the most common form of cytosine in T-even phage DNA. (See DNA methylation, 5-azacytidine, DNA base composition)

3-HYDROXY-3-METHYLGLUTARYL CoA LYASE DEFICIENCY: leucine is degraded in six enzymatic steps into acetoacetic acid and in the process acetyl-CoA is generated. The last enzyme in this path is 3-hydroxy-3-methylglutaryl CoA lyase. Deficiency of this enzyme causes acidosis (accumulation of acids) and reduction of sugar in the blood (hypoglycemia). This potentially fatal disease is controled by autosomal recessive mutation. (See isoleucine-valine metabolic pathway, isovalericacidemia, methylcrotonylglycinemia, methylglutaconicaciduria, amino acid metabolism)

HYDROXYTRYPTAMINE: see serotonin

HYDROXYUREA: an inhibitor of ribonucleoside reductases and thereby prevents the formation of deoxyribonucleotides from ribonucleotides. Consequently it blocks DNA synthesis in the S-phase of the cell cycle and thus it is also an antineoplastic agent. (See neoplasia, S phase)

HYENA: *Crocuta crocuta*, 2n = 40; *Hyena brunnea*, 2n = 40.

HYGROMYCIN B (Hyg,(5-deoxy-5-[[3-[4-[6-deoxy-β-D-arabino-hexofuranos-5-ulos-1-yl]oxy]-3-hydroxyphenyl]-2-methyl-1-oxo-2-propenyl]amino]-1-2-O-methylene-D-neoinositol): an antibiotic commonly used for screening transformed plant and animal cells transformed by the expression of hygromycin phosphotransferase gene (*hph*) as a selectable marker (confering resistance). It is also an antihelminthic drug. (See antibiotics)

Hylandra suecica (formerly *Arabidopsis suecica*): a putative amphidiploid (2n = 26) of *Arabidopsis thaliana* (2n = 10) and *Cardaminopsis arenosa* (2n = 32).

HYMENIUM: a fruitingbody-forming tissue in fungi.

HYOSCYAMINE: an anticholinergic alkaloid that blocks neurotransmission through the parasympathetic system (originating in the brain and controlling the heart, head, neck, chest, abdomen and pelvic organs). It is an alkaloid of the solanaceous species of plants *Hyoscyamus*, *Datura* and *Atropa*. (See alkaloids)

HYOSCYAMUS: see henbane

HYPERACTIVITY: see attention deficit-hyperactivity
HYPERAMMONEMIA: see carbamoylphosphate synthetase deficiency
HYPERARGININEMIA: see argininemia
HYPERBILIRUBINEMIA: excessive amounts of bilirubin in the blood causing jaundice, common symptom of several diseases involving destruction of red blood cells. It may be involved in indirect epistasis. (See Dubin-Johnson syndrome, Crigler-Najjar syndrome, Gilbert syndrome, bilirubin, epistasis)
HYPERCALCEMIA-HYPERCALCIURIA: two forms are distinguished, the familial idiopathic and infantile idiopathic. Both are autosomal dominant, and the differences are not entirely clear. The infantile form affects primarily females. The calcium level in the blood and urine is elevated. In some cases it involves bone and blood vessel disorders. The basic defect seems to be in a G-protein receptor. (See G-proteins, hypercalciuric hypercalcemia)
HYPERCALCIURIC HYPERCALCEMIA (FHH): is a usually mild autosomal syndrome involving increase in calcium and magnesium level in the blood and urine, and frequently gallstones. Occasionally, there is an increase in parathyroid activity. The basic defect is attributed to G-protein receptors and the connected Ca^{2+} sensors that are involved in parathyroid hormone release from the cell. (See hypercalcemia-hypercalciuria, G-proteins, parathormone)
HYPERCHOLESTEROLEMIA: see familial hypercholesterolemia
HYPERCHROMICITY: single-stranded nucleic acids have increased UV absorption relative to the double-stranded molecules. (See DNA, DNA denaturation, DNA thermal stability)
HYPERGLYCEMIA: increase in blood glucose level.
HYPERHOMOCYSTEINEMIA: is caused by mutation in the 5,10-methylenetetrahydrofolate reductase (MTHFR) enzyme, and homozygotes for the condition have increased levels of homocysteine and an increased risk of cerebrovascular (brain) and peripheral (vein) and coronary heart disease. (See homocystinuria, tetrahydrofolate)
HYPERINSULINEMIA: see hypoglycemia
HYPERKALEMIC PARALYSIS: a group dominant periodic paralysis diseases induced by high level of potassium caused by human chromosome 17q23-q25 defects. (See periodic paralysis)
HYPERLIPIDEMIA: the autosomal dominant condition has similarities to familial hypercholesterolemia but in this case hypercholesterolemia is not found in the offspring. Generally very low and low density lipoproteins accumulate. Early onset is relatively rare and results in hypertriglyceridemia. Its incidence exceeds 5 times that of hypercholesterolemia. (See coronary heart disease, cholesterol, lipid, apolipoprotein, familial hypercholesterolemia, familial triglyceridemia, sterol, HLP)
HYPERLIPOPROTEINEMIA: caused by a dominant gene in human chromosome 19q, coding normally for apolipoprotein E-d., a 299-amino acid polypeptide. A defect in this protein involves the accumulation of chylomicron (a small lipoprotein which normally transports dietary cholesterols and triaglycerides [triglycerides] from the intestines to the blood stream). As a consequence of the defect, the conditions for coronary heart disease may develop. Several forms of this disease are usually distinguished. Some types (IV) are induced by high carbohydrate diet, alcohol, uremia, glycogen storage diseases and steroid contraceptives. Recessive form of the disease is associated with human chromosome 8p22. The latter type involves large amounts of chylomicron accumulation even on normal diet but disappears on fat-free diet. It may not lead to early atherosclerosis. Lipoprotein lipase activity is deficient in this recessive form. The lipase gene encodes a 475 amino acid protein, including a 27 residue leader peptide. (See coronary heart disease, lipoprotein, triaglyceride, cholesterol, sex hormones, chylomicron, atherosclerosis, lipase, apolipoprotein)
HYPERLYSINEMIA: is an autosomal recessive phenotype caused by a defect in the enzyme lysine:α-ketoglutarate reductase. This enzyme is a bifunctional complex also of saccharopine dehydrogenase, controlling a step subsequent to α-ketoglutarate reduction in lysine degradation. The defect in this enzyme causes the accumulation of lysine in the blood resulting in physical and mental retardation. Hyperlysinemia may occur also if the transport of lysine into the mito-

chondria fails, and dibasic aminoaciduria also involves excessive urinary excretion of lysine. (See dibasicaminoaciduria, lysine biosynthesis, amino acid metabolism)

HYPERMORPHIC MUTANT: overexpresses a particular trait. (See also hypomorphic, gain-of-function mutation)

HYPERMUTATION (somatic hypermutation): is a common phenomenon in the variable region-coding sequences of immunoglobulin genes. Their rate of mutation has been estimated to be six orders of magnitude higher than the average somatic rate of mutation in other genes. If the variable κ immunoglobulin gene segment was replaced by heterologous sequences (prokaryotic *neo*, *gpt*, or β-globin) the rate of hypermutations did not decrease. (See antibody, affinity maturation, immune response, immunoglobulins, junctional diversification, somatic hypermutation, transposon, transposable elements)

HYPERNEPHROMA (adenocarcinoma of kidney): most commonly it involves rearrangement (translocations) of the short arm of human chromosome 3 or loss of 3p14.2. Trisomy or tetrasomy 7 may also be associated with kidney carcinomas. (See also renal cell carcinoma, von Hippel-Lindau syndrome)

HYPERPARATHYROIDISM: is an autosomal dominant adenoma of the parathyroid gland. The condition is caused by a defect in thyrotropin receptor A, G-protein receptor. Another dominant form is characterized by multiple bone tumors on the jaws. An autosomal recessive form affecting newborns is also known. Mutations in the calcium-sensing receptor (Casr) may upset parathyroid hormone production and may cause familial hypocalcemia or hypocalciuric hypercalcemia. (See adenoma, goiter, G-protein, parathormone, hypocalcemia)

HYPERPLASIA: abnormal increase of normal cells in a normal tissue. (See also neoplasia)

HYPERPLOID: contains extra chromosome(s) beyond the normal number.

HYPERPROLINEMIA: Type II is caused by autosomal deficiency of the enzyme Δ'- pyrroline carboxylate dehydrogenase and consequently accumulation of proline and also glycine in the blood (the mechanism is unclear). The patients generally show some degree of mental retardation and convulsions. In Type I disease proline oxidase is deficient and renal problems occur with or without mental defects. (See proline biosynthesis, amino acid metabolism)

HYPERSENSITIVE REACTION (HR): at the place of infection by fungi, bacteria and viruses the plant cells may suddenly die in a limited area and thus stop the spread of the infection and convey resistance to the host. The hypersensitivity reaction in case of fungi is elicited by hyphal cell membrane and in particular by the fatty acids eicosapentaenoic (EPA) and arachidonic (AA) acids and it is inhibited by β 1,3 and β 1,6 glucans. The availability of EPA and AA is determined by lipoxigenase activity. Peroxides may also have a role in the development of HR and may be a signal transducer. The development of the HR may be the cause or consequence of the alteration of the ion channel functions. In bacteria *hrp* (hypersensitivity and pathogenesis) genes and mutants were identified and these interact with various plant products, e.g. the plant flavonoids (acetosyringone) initiates the activation of the virulence cascade of *Agrobacterium*. The hairpin 44 kDa regulatory protein has been implicated in the HR expression. The HR may be suppressed by the bacterial avirulence (*avr*) genes, depending on particular host species. In animals the hypersensitivity is called allergy. (See host-pathogen relation, hydrogen peroxide, allergy)

HYPERSENSITIVE SITE: includes nucleotide sequences which are readily cut by endonuclease because these tracts are (relatively) free of chromosomal proteins. These sites generally found in front of transcribed genes and it is supposed that these facilitate the attachment of the transcriptase enzyme. (See regulation of gene activity)

HYPERSENSITIVITY: is a term used in mutagen (carcinogen) testing, indicating the percentage of compounds classified as carcinogens (mutagens) among all compounds tested by a system.

HYPERTELORISM: abnormally long distance between two organs or organ parts in the body.

HYPERTENSION: is probably under the control of a few genes. In rats, genetic analysis identified at least two genes with major effects BP/SP-1 and BP/SP-2 (blood pressure/sodium pump) in rat chromosome 10 (human chromosome 17?). BP/SP-1 is probably linked closely to gene

Hypertension continued

ACE1 (angiotensin converting enzyme). Hypertension has also an important physiological component associated with the lithium-sodium countertransport and thus depending on environmental (dietary) factors. Hypertension occurs in about of a third of the human populations and is frequently evoked by kidney diseases. Hypertension is the most common cause of heart disease. (See coronary heart disease, cardiovascular disease). Various familial disorders are frequently associated with hypertension: coarctation of the aorta, polycystic kidney disease, Alport syndrome, pheochromocytoma, neurofibromatosis, aldosteronism, hypoaldosteronism, hyperthyroidism, homocystinuria, Wilms tumor, familial hypertriglyceridemia, hyperlipidemia, hydroxysteroid dehydrogenase deficiency (11βHSD). Blood pressure may vary from 80 mm (Hg) or less at the diastole (contraction of the heart ventricles) to the very high 200 mm at systole (expansion of the ventricles). The average normal blood pressure in adults is 80 at diastole and 120 at systole. In children it is lower, in older people it is usually higher. Elevated salt diet is normally conducive to hypertension as the heart releases an atrial natriuretic peptide. The formation of this peptide is regulated by a guanyl cyclase A receptor (GC-A). Disruption of the GC-A gene in mice results in high blood pressure irrespective of the amount of salt in the diet. Mutations in the 11-β-hydroxysteroid dehydrogenase may lead to hypertension. (See aldosteronism, mineral corticoid syndrome, Liddle syndrome, Gordon syndrome, brachydactyly, hypotension, angiotensin, LDL, nitric oxide, stroke, debrisoquine)

HYPERTHERMIA (malignant hyperthermia): may involve only an increase of the skin temperature or a general elevation of the body temperature (hyperpyrexia) occurs after anesthesia. The susceptibility is under dominant control in humans (one of the genes is in human chromosome 19q13.1) but it is recessive in the light skinned pigs. The basic cause may be a defect in the regulation of the voltage-gated Ca^{2+} channels. (See temperature-sensitive mutation, cold hypersensitivity, ion channels)

HYPERTHYROIDISM: an autosomal dominant defect may be caused by inadequate response of the thyrotropin secreting cells to the pituitary thyroid-stimulating hormone (TSH). As a consequence high levels thyroid hormones appear and goiter, increased pulse rate, fatigue, nervousness, sweating and heat intolerance and other symptoms develop. In a recessive form (human chromosome 14), a deficiency of the TSH receptor is caused by an insert of a 8-amino acid sequence near the NH_2 end of the protein. A human chromosomal site (22q11-q13) is coding for a thyroid autoantigen but that is not the receptor as once assumed. (See thyroid hormone response element, TRE, hormones, cardiovascular disease, goiter)

HYPERTONIC: the salt concentration of this type of solutes are high enough to draw out water from a cell. (See also isotonic, hypotonic)

HYPERTRICHOSIS: excessive hairiness. In the autosomal dominant hypertrichosis universalis hair covers abundantly the entire body until the end of infancy, in another form gum disease was also present. In an autosomal recessive form the excessive hairiness was accompanied by nerv disease (neuropathy). An Xq24-q27.1-linked hypertrichosis affects more the males than the females where lyonization causes patchy appearance of the hairs. The condition is very rare; only about 50 cases were described. Excessive hairiness does occur, however as part of several syndromes such as the Hurler syndrome, de Lange syndrome, Coffin-Sirius syndrome, Lawrence-Seip syndrome, Schinzel-Geidion syndrome, Gorlin-Chaudhary-Moss syndrome. (See hair, hirsute, atavism, lyonization, and the syndromes listed above)

HYPERTROPHY: an overgrowth generally with larger than normal cells.

HYPERVALINEMIA (valinemia): accumulation of valine in the urine and blood plasma because of a defect in the enzyme valine transaminase caused by an autosomal recessive factor. Symptoms include drowsiness and vomiting. Prenatal diagnosis is feasible. (See isoleucine-valine biosynthetic pathway)

HYPERVARIABLE SITES: in the light and heavy chains of antibody molecules are responsible for their high specificity in recognizing different antigens. They are short polypeptide loops and called also the complementarity determining regions (CDRs). (See antibody)

HYPHA (plural hyphae): fungal filaments, cylindrical structural units of the mycelia. They are surrounded by a wall and filled with cytoplasm unless they are vacuolated. They elongate by growing at the tips (apex). When branching hyphae aggregate, they may form a standing-up mycelium, called *coremium,* or horizontal strands may form *rhizomorphs*. The spherical or irregular aggregates functioning as enduring (dormant) bodies are *sclerotia*. Hyphae aggregating in pseudoparenchyma tissue is the *stroma*. Hyphae may be involved in plasmogamy and function in some sort of sexual function (*somatogamy*) although they may not be sexually specialized. (See fungal life cycles)

HYPOALDOSTERONISM (adrenal hyperplasia): the two genes responsible for the recessive 11-β-hydroxylase deficiency have been located to human chromosome 8q21, and a defect of these is responsible for the accumulation of 11-deoxycorticosterone and consequently for hypertension, and other hormonal defects. The same enzymes are involved also in the hydroxylation of 18-hydroxysteroids and 17-hydroxysteroids. These genes are members of the P450 (cytochrome) enzyme coding family. Affected females show masculinization and the males precocious puberty because of the accumulation of steroids. Some of the aldosterone deficiency mutations are allelic to the aldosterone overproducing defect (aldosteronism) indicating the presence of a multifunctional protein. The pseudohypoaldosteronism which is caused not by aldosterone deficiency but by the recessive deficiency of the mineral corticoid receptor, encoded at human chromosome 4q31.1 or q31.2. This condition is characterized by salt wasting in the urine and respond favorably to salt administration. The hypoaldosteronism defects appear quite common among oriental (Persian) Jews. Pseudohypoaldosteronism type 1 is due to a defect in an epithelial sodium channel (ENaC) resulting in hyperkalaemic (high in potassium) acidosis and salt wasting. The α subunit of the channel is encoded in human chromosome 12p13.1-ter whereas the β and γ subunits are coded by chromosome 16p12.2-p13.11. (See aldosteronism, P-450, Jews and disease, Liddle syndrome, Bartter syndrome, Gitelman syndrome, mineral corticoid syndrome, cardiovascular diseases, ion channels)

HYPOBETALIPOROTEINEMIA: involves reduced levels of apolipoprotein B that may result in increased levels of blood cholesterol and atherosclerosis. (See apolipoproteins, atherosclerosis, hyperlipoproteinemia, cardiovascular diseases)

HYPOBLAST: precursor of the mesoderm and endoderm. (See mesoderm, endoderm)

HYPOCALCEMIA: reduced amount of calcium in the blood in several diseases involving defects in ion channels. (See ion channels)

HYPOCHONDRIASIS: a type of affective disorder when illness is imagined on the basis of irrelevant signs. (See affective disorders)

HYPOCHONDROGENESIS: a connective tissue disorder caused by mutation in collagen.

HYPOCHONDROPLASIA: this autosomal phenotype reminds to achondroplasia but the tibia (shin bone) and the head are rather normal. The fingers are short but the hand is not three-pronged. This gene appears to be allelic to that responsible for achondroplasia. It may be caused by a defect in the function of fibroblast growth factor 3, FBG3. (See stature in humans, achondroplasia, achondrogenesis, dwarfism, fibroblast growth factor, dwarfism)

HYPOCHROMICITY OF NUCLEIC ACIDS: in double-stranded molecules the free rotation of the bases is hindered, resulting in reduced optical density in UV light compared to single-stranded molecules (hyperchromicity). (See DNA, hyperchromicity)

HYPOCOTYL: the section of a plant embryo or plant situated between the cotyledon attachment point and the radicle or root, respectively.

HYPODONTIA (adontia): autosomal dominant condition of lack or underdevelopment of teeth. (See also teeth, dental non-eruption, dentinogenesis imperfecta, dental ankylosis)

HYPOGAMMAGLOBULINEMIA/COMMON VARIABLE IMMUNODEFICIENCY: is a complex immunodeficiency with uncertain familial determination. (See gammaglobulinemia, immunodeficiency, Hirschprung disease)

HYPOGLYCEMIA: involves lower than normal blood sugar content and may result in shaking, cold sweat, low body temperature, headache, irritability and eventually even coma. An auto-

Hypoglycemia continued
somal recessive gene may cause it by a deficiency of glycogen syntethase in the liver. Another recessive gene in human chromosome 1p31 may cause acyl-coenzyme A-dehydrogenase deficiency and hypoglycemia. Mutations in the sulfonylurea receptor gene may cause hyperinsulinemia (human chromosome 11p14-p15.1) and consequently hypoglycemia. Hypoglycemia may results indirectly from different ailments. (See epilepsy, ion channels, sulfonylurea)

HYPOGONADISM: less than normal function of the gonads (ovary and testes); it is frequently associated with retardation in growth and mental abilities. Mutation in the DAX-1 human gene may result in adrenal defects such as adrenal hypoplasia and hypogonadotropic hypogonadism (deficiency in the hormone gonadotropin and gonadal hypofunction). (See adrenal hypoplasia)

HYPOHIDROSIS: reduced ability to sweat caused by autosomal recessive defect of the sweat glands.

HYPOKALEMIA: autosomal recessive low potassium level in the body causing neuromuscular abnormalities. (See periodic paralysis, hypoaldosteronism, pseudoaldosteronism)

HYPOLACTASIA: see disaccharide intolerance, lactose intolerance

HYPOMELANOSIS of ITO: pale skin spots in whorls or patches of tan associated with diverse other anomalies indicating that this syndrome is a mixed bag, often associated with breakage of different chromosomes. (See incontinentia pigmenti, pigmentation defects, skin diseases)

HYPOMORPHIC ALLELE: has reduced activity compared to the wild-type allele.

HYPOPHOSPHATASIA: is either a *dominant* mutation expressed in adults as a deficiency of the liver (general) alkaline phosphatase gene in human chromosome 1p36.1-p34 or the *recessive* mutation in the same locus appears as the infantile hypophosphatasia. The adult phenotype involves early loss of teeth, bowed legs like in rickets. The level of intestinal alkaline phosphatase is normal. In some forms the stature is somewhat shorter. The infantile type is manifested already before birth, involves severe skeletal anomalies, increased levels of phosphoethanolamine and inorganic pyrophosphate in the urine and in the serum higher levels of pyridoxal phosphate. Death may occur within a year. In some instances infusion of normal blood plasma resulted in prolonged normalization. It was suggested that a cofactor of the enzyme is missing. (See stature in humans, dwarfism, hypophosphatemia, rickets)

HYPOPHOSPHATEMIA: a dominant bone disease coded in human chromosome Xp22 region resulting in low level of phosphate in the blood and no or minimal response to vitamin D. The defect may involve abnormal phosphate absorption too. Its prevalence is 2×10^{-4}. (See hypophosphatasia, exostosis)

HYPOPLASIA: underdevelopment of an organ. (See aplasia)

HYPOPLOID: contains less than the full set of chromosomes. (See autopolyploid, nullisomic)

HYPOPROCONVERTINEMIA: a recessive bleeding disease caused by deficiency of blood clotting factor VII in human chromosome 13q34-qter. The afflicted individuals may be deficient also in antihemophilic factor X. (See antihemophilic factors, proconvertin)

HYPOSPADIAS: the urethra opens at the lower part of the penis; it has autosomal dominant and recessive transmission. Prevalence is close to 0.3% with a heritability of about 0.57. (See penis)

HYPOSTASIS: a condition when a gene expression is masked by another, the epistatic one. (See epistasis)

HYPOTENSION: in contrast to hypertension involves low blood pressure. Some of the cases may be determined by identical loci just different alleles. The autosomal dominant orthostatic form (*Shy-Drager syndrome*) is characterized by incontinence, anhidrosis (absence of sweating), ataxia, tremor, and low norepinephrine level in the plasma. The *pseudohypoaldosteronism (PHA-1)* is autosomal dominant and involves serious dehydration, salt wasting, high level of potassium, and a form of hyperaldosteronism. The mutation seems to affect genes controlling the same ion channel as in the Liddle syndrome. *Gitelman syndrome* is a human chromosome 16 recessive that involves a defect in the Na-Cl co-transporter and consequently salt wasting. (See hypertension, Liddle syndrome, aldosteronism, mineral cortical syndrome)

HYPOTHALMUS: part of the brain where vision, visceral activities, water balance, temperature, sleep, etc. control centers are located. (See brain human)

HYPOTHESIS: is a supposition or multiple alternative suppositions (hypotheses) are used to explain certain experimental data and generally statistical approaches are used to decide which of the alternatives have the greatest probabilty to be true. One must keep in mind that the statistical methods do not prove cause-effect relationships or mechanisms, only the degree of chance or likelihood is indicated. The working hypothesis is used as guidance to design experimental procedures to test the most likely mechanism involved. In experimental science, such as genetics, only the testable hypotheses have any value. (See null hypothesis, probability, likelihood, maximum likelihood, chi square, genetic risk)

HYPOTHYROIDISM: see goiter

HYPOTONIC: the concentration of salts in this type of media is lower than the osmolality of the cell, therefore water from the cells may be drown out into the medium. (See hypertonic, hypotonic, osmolality, osmotic pressure)

HYPOTRICHOSIS: rare autosomal recessive reduction of hairs on the face and absence of pubic hairs even after puberty. Baby hair is shed shortly before or after birth. (See hair, hypertrichosis)

HYPOTROPHY: less than normal growth of a tissue or organ caused by inadequate nourishment.

HYPOVIRUSES: are persistent and hard to transmit infectious agents in fungi and cause no symptoms. However the synthetic transcripts of such viruses when introduced into the chestnut blight fungus (*Cryphonectria parasitica*) by electroporation can reduce the virulence of the fungus and thus indicates a means of biological control. (See symbionts hereditary)

HYPOXANTHINE: is a purine base very similar to guanine, except that from the C2 position NH_2 is removed. Its nucleoside is (confusingly) inosine and its nucleotide is called inosinic acid. It may occur in some anticodons of tRNA. When guanine is deaminated into hypoxanthine through chemical mutagens (nitrous acid), lethal mutation occurs because hypoxanthine cannot support nucleic acid replication. The enzyme hypoxanthine-guanine phosphoribosyl transferase (deficient in the Lesch-Nyhan syndrome) can phosphorylate both of these purines in the salvage pathway of nucleic acid. (See HPRT, salvage pathway, HAT medium, nitrous acid mutagenesis)

HYPOXANTHINE-GUANINE PHOSPHORIBOSYL TRANSFERASE (HGPRT): is the enzyme that donates the ribose phosphate moiety to the purine bases hypoxanthine and guanine from 5'-phosphoribosyl-1-pyrophosphate and thus forms the corresponding nucleotides. Another enzyme, adenosine phosphoribosyl transferase synthesizes adenylic acid from adenine. HGPRT is the key enzyme in the salvage pathway of nucleic acid synthesis. (See salvage pathway, HAT medium, Lesch-Nyhan syndrome)

HYPOXIA: low oxygen concentration in the environment of cells or tissues. The effectiveness of ionizing radiation in induction of mutation or chromosome breakage is reduced under low oxygen concentration. (See oxygen effect)

HYSTERECTOMY: surgical removal of the uterus. (See uterus)

HYSTERESIS: a time lag between two processes. Also, lowering of the point of freezing without lowering the temperature required for melting; it may be regulated by antifreeze proteins. (See antifreeze proteins, temperature-sensitive mutation)

I BLOOD GROUP (Ii system): I and i are universal erythrocyte antigens that exhibit alteration during development but minimal polymorphism. The synthesis of I/i antigens results from the cooperation of glycosyltransferases on common substrates and there is not a single diagnostic immunodeterminant sugar specific for the blood type; it may be associated with autoimmune hemolytic anemias. (See blood types, ABO blood typehemolytic disease, hemolytic anemia)

I **ELEMENT**: see non-viral retrotransposable elements. (See retroposon, retrotransposon)

I.U.: (international unit) is a quantity of various vitamins, hormones, enzymes, etc. that bring about a standard response as determined by the International Conference for Unification of Formulas.

IAA: indole acetic acid. (See plant hormones)   INDOLE-3-ACETIC ACID →

IAP: see chloroplast import

IARC: International Agency for Research on Cancer, Lyon, France; publishes reviews on carcinogens as Scientific Publications of IARC. (See cancer, environmental mutagens, databases)

IATROGENIC: (adjective for) any adverse effect resulting from or concomitant to medical treatment.

IBD: identical by descent, i.e. an allele inherited from an ancestor is exactly the same as that of the particular ancestor and not the result of a new mutation identical by state (IBS). IBD-APM is identical by descent for **a**ffected-**p**edigree-**m**ember. (See inbreeding coefficient)

IBIDS: see ichthyosis

IBS: see IBD

ICAM (CD54, intercellular adhesion molecule): proteins which bind together (epithelial) cells through integrins. The concentration of ICAM-2 in uropod projections (rather than evenly distributed, are special targets for cytotoxic lymphocytes (CTL). The targeting is assisted also by the cytoskeletal - membrane linker protein ezrin. (See N-CAM, cadherin, integrin, CD proteins, LFA, uropod)

ICE (interleukin-1β converting enzyme): is a cystein protease which is activated during Fas-mediated apoptosis. Its family includes CPP32 and Ich-1 proteases. ICE is a mammalian homolog of *ced-3*. (See Fas, apoptosis, interleukins, CPP32, granzyme B, RNKP-1)

ICE MAN (Ötzi): is a mummy discovered frozen in the Tyrolean Alps in 1991, and its age was estimated to be 5,100 to 5,300 years. Some claims have been made that it may be a scientific hoax but detailed analysis negates this possibility. Apparently 10 genome equivalent quantity of DNA was recovered per gram of its tissue, constituting only of redundant sequences. One DNA sequence of the hypervariable region of the mitochondrial DNA indicated close similarities to central and northern European populations. (See mtDNA, ancient DNA, mummies)

I-CELL DISEASE: see mucopolysaccharidosis

ICH: a protease implicated in apoptosis. (See apoptosis)

ICHTHYOSIS: is a non-inflammatory keratosis of the skin appearing in different forms and under different genetic controls, often associated with syndromes. In the *autosomal dominant* ichthyosis vulgaris the scaling of palms and soles appears during the first three months after birth and it may be caused by a deficit in keratohyalin, a precursor of filagrin, an element of keratin fibers. A number of different variations of the dominant types exist. The *autosomal recessive* forms have also great variations involving redness of the skin, brittle hair, blisters and liver disease, physical and mental retardation, etc. The recessive forms are generally more serious than the dominant ones. The harlequin type (so named because of the diamond-shape 4-5 cm diameter scaly, horny spots, like the pattern on the robes of harlequins) may cause death within the first week after birth. Some recessive forms involve hair loss, progressive neural defects, enlarged liver, kidney defects. In the Sjögren-Larsson syndrome the frequency of the recessive gene responsible for the disease in a county of Sweden appeared to be 0.01 and that of the carriers 0.02 while the prevalence was 8.3×10^{-5}. This latter form of ichthyosis was

Ichthyosis continued
accompanied by fatty aldehyde dehydrogenase deficiency, neurological disorders and keratosis. Another form of recessive ichthyosis is connected to triglyceride storage anomalies. In the autosomal recessive ichthyosis (14q11) the activity of keratinocyte transglutamase activity is much reduced. X-linked ichthyosis may involve asymmetric malformations of lung, thyroid, several nerves, etc. and hypoplasia at the same side as the ichthyosis. Another X-linked (distal part of Xp) group showed deficiency of placental steroid sulfatase. Also called IBIDS. (See keratosis, skin disease, collagen, lyonization)

ICOSAHEDRAL: a body with 20 facets like the capsids of some viruses. (See isometric phage)

ICR: compounds synthesized by the International Cancer Research Institute, Philadelphia, PA, are acridines with alkylating side chains and cause frameshift mutations. (See mutagens)

ICR: internal control regions through the zona pellucida. (See A box)

ICSI (intracytoplasmic sperm injection): some people's sperm unable to penetrate the egg. In such cases the spermatozoon may need to be injected mechanically into the ooplasm. (See ART)

ID: see idant

IDANT: a higher hierarchial genetic structure as conceived during the 19th century. Biophores ($\approx$ alleles) aggregated into ids ($\approx$ loci) and the ids formed idants ($\approx$ chromosomes).

IDAXOZAN: see clonidine

IDDM: insulin-dependent diabetes mellitus. (See diabetes mellitus)

IDENTICAL TWINS: see twinning

IDENTIFIER: sequences within introns or non-coding 3' regions are involved in genetic regulation. The identifier sequences are transcribed by RNA polymerase III and may define chromatin regions and facilitate the transcription by RNA polymerase II as promoters are made more accessible to soluble trans-acting molecules. (See introns, regulation of gene activity, RNA polymerase, trans-acting element)

IDENTITY INDEX: see evolutionary distance

IDENTITY OF STATE: when the coefficient of inbreeding is determined, the identity of an allele must be specified as *identity by descent* (i.e. passed on by a common ancestor) not by coincidentally occurring mutation. Mutation may generate only *identity by state*. (See coefficient of inbreeding, consanguinity)

IDIOGRAM: diagram of the chromosome set, including all essential morphological features of the chromosomes such as arm ratio, satellites, banding pattern, etc. (See human chromosome maps)

IDIOMORPHS: the two mating type determining gene loci in fungi.

IDIOPATHIC TORSION DYSTONIA (ITD): is a dominant (human chromosome 9q) neurological disorder ($\approx 30\%$ penetrance) with onset before the late 20s, affecting first the limb muscles and later the effects may spread. Its frequency ($1.6 - 5 \times 10^{-4}$) is higher among Ashkenazy Jews than in other populations. (See dystonia)

IDIOPLASM: a 19th century hypothetical concept about the physical basis of heredity.

IDIOTOPE: the determinant of the antigenic specificity. (See antigen, antibody)

IDIOTYPE: originally (in 1884) it was coined for identifying the entire complex of genetic determinants in a cell but in that sense it is no longer in use. In modern immunogenetics it means the idiotopes distinguishing one group of antibody-producing cells from other groups of immunoglobulin producing cells. Thus the idiotype represents the specificity of all the idiotopes because the antigens bind within or at the idiotopes. The recurrent (also called public, major or cross-reactive) idiotypes regularly appear during immune response but the private (or minor) idiotypes may or may not appear even in genetically identical individuals. (See epitope, paratope, allotype, isotype, antibody, immunoglobulins, HLA, anti-idiotype)

IDIOTYPE EXCLUSION: in a family of genes only one gene product is expressed. (See allelic exclusion, locus exclusion)

IDLIING REACTION: in amino acid starved cells synthetic activity is reduced (see stringent control), and the ribosomes are uncharged with aminoacylated tRNAs. Under such conditions uncharged tRNAs may be attached to the ribosomes and block any residual protein synthetic activity as the idling reaction. (See protein synthesis)

IDURONIC ACID: is an epimer [a diastereomer (stereoisomers without being mirror images]) of glucuronic acid and dermatan sulfate, heparan sulfate and heparin. (See mucopolysaccharidosis)

IES: see internally eliminated sequences

IF-1, IF-2, IF-3: protein synthesis initiation factors in prokaryotes and eukaryotic organelles. (See protein synthesis, eIF)

IFN: see interferons, IFR

IFN -gamma: same as MAF, a lymphokine.

IFR (interferon regulatory factor): is a negative regulator of interferon and when it binds to CCE it activates Histone 4 transcription, involved in the progression from G1 to S phase of the cell cycle. (See histones, interferons, HiNF, CCE, cell cycle)

Ig: means immunoglobulin (with additional letters to identify types). (See antibody, immunoglobulins, immune system)

Igα and Igβ: are immunoglobulin-associated proteins that form disulfide linked heterodimers (Igα—Igβ). They belong to a large family of antigen receptor-associated signal transducers and they have a tyrosine-containing cytoplasmic motif. B cell activation requires this dimer for triggering the Src and Sky family kinases for receptor phosphorylation. These proteins have additional regulatory role too. The heterodimer is essential for the differentiation of B cells. (See lymphocytes, immunoglobulins, Src, Sky, CD40)

IGF: see insulin-like growth factor

IGNORANT DNA: is a molecule amplified as a by-product of coincidental amplification and may be slightly harmful but may be also advantageous under certain circumstances, e.g. rRNA, satellite sequences, etc. (See also selfish DNA, junk DNA, satellite)

IGS: see intergenic spacer

IHF (integration host factor): a host protein required for site-specific recombination. It is a member of the high mobility group proteins. (See high mobility group proteins, site-specific recombination, recombination, intasome)

Ii GENES: see ABO blood group

iIF-3P: translation elongation factor stimulating the preinitiation complex on the 40S ribosomal subunit. (See protein synthesis, eIF)

IIH: is shorthand for TFIIH: (See transcription factors)

I_{KAch}: inwardly rectifying acetylcholine regulated K^+ channel, activated by G_i protein; inhibits the opening of the voltage-gated Ca^{2+} channels and thus regulates nerve and muscle functions. (See ion channels, G_i protein)

IκB: a regulatory protein binding to transcription factor NF-κB and keeps it in the cytoplasm; until it is degraded by proteolysis after being phosphorylated, it does not release the DNA binding transcriptional activator NF-κB to migrate into the nucleus. Inactivation of IκB usually requires stress signals or pathogenic attack. The various types of IκB proteins are characterized by ankyrin repeats and these are present also in several morphogenetic factors of *Drosophila* (e.g. *cactus*). (See signal transduction, regulation of gene activity, NF-κB, PEST, glucocorticoid, immunosuppression, ankyrin, morphogenesis in *Drosophila*)

IL-1: lymphocyte activating factor. The IL-1α (17 kDa) and β (also about 17 kDa) chains are encoded in human chromosome 2q13 at very close linkage. IL-1ra is an a cytokine antagonist of IL-1. (See interleukins)

IL-2: T-cell growth factor, TCGF. It promotes the proliferation and differentiation of lymphocytes. The IL-2Rβ and the IL-2Rγ chain subunits are members of the cytokine receptor superfamily. The IL-2Rβ (human chromosome 22q11.2-12) and γ (Xq13) chains are shared by the other IL receptors. The human Il-2Rα (chromosome 10p14-p15) has only 13 amino acid in its cytoplasmic domain and it may not have significant role in signal transduction. The IL-R receptor complex has prominent role in various signal transduction pathways. Mice with IL-2

Il-2 continued
receptor chain β (IL-Rβ) shows excessive differentiation of B cells into plasma cells and excessive amounts of immunoglobulins G1 and E and the production of autoantibodies resulting in hemolytic anemia. Defects in the widely shared IL-2Rγ causes severe combined immunodeficiency (SCID) in humans. (See interleukins, B cells, T cells, immunoglobulins, autoantibody, anemia, SCID, cytokines, NF-AT, AP-1, Oct-1, NF-κB)

IL-3: haematopoietic growth factor is a 28 kDa glycoprotein, encoded in human chromosome 5q 23-q31. (See interleukins)

IL-4: B-cell growth factor, 18 kDa glycoprotein, encoded in human chromosome 5q31. (See interleukins)

IL-5 (eosinophil differentiation factor): encoded in human chromosome 5q23.3-q31, near IL-4.

IL-6: an interleukin group of proteins (21-28 kDa), encoded in human chromosome 7p21; it binds to upstream DNA sequences in some genes. Its receptor, IL-6R, is a transmembrane protein. IL-6 formation is induced by infection and mediates primarily humoral immune reactions. (See interleukins)

IL-7: stimulates lymphocyte precursors in the bone marrow.

IL-8: activates neutrophils, cell migration, adhesion, inflammation.

IL-9: promotes cell proliferation and differentiation; it is the erythroid colony stimulating factor.

IL-10 (CSIF): is produced primarily the T_H cells and acts as **c**ytokine **s**ynthesis **i**nhibitors, yet it may stimulate the growth of mast cells and the CD8$^+$ T cells. (See mast cell, T cell)

IL-11: controls hematopoiesis in the bone marrow. (See hematopoiesis)

IL-12 (interleukin-12): is one of the most potent stimulator of T helper cells (T_H, natural killer cells (NK), B lymphocytes and thus increases the production of interferon γ (IFN-γ). Il-12 is a very promising tool in controlling HIV, Leischmaniasis, malaria, tuberculosis, schisostomiasis and other infectious diseases although some side effects (toxic shock syndrome, atherosclerosis) may be limiting its application. (See also separate entries)

IL-14: is produced by activated T cells and B cell lymphoma. (See T cell, lymphoma)

IL-15: is T cell growth factor, produced by monocytes and epithelial cells. (See T cell, monocyte)

IL-16: 130 amino acid proteins (M_r 13,500), secreted by activated CD8$^+$ cells and bind to T cells by the CD4 receptor. They suppress HIV and SIV. (See HIV, SIV, T cell)

ILLEGITIMATE CHILD: is begotten by another male than the legal father. This status can now be identified with great certainty by DNA fingerprinting. The mother seems always certain. Illegitimacy, contrary to some myths, does not endow the "love baby" with better abilities, despite some famous examples (Leonardo da Vinci, Francis Bacon, etc.). Actually, the out-of-wedlock children generally suffer physical and emotional stress. Illegitimate offspring in old pedigrees may confound recombination frequencies, especially at tight linkage, it may cause problems in prenatal diagnosis of some disorders, and may frustrate genetic counseling. (See DNA finger-printing, prenatal diagnosis, genetic counseling)

ILLEGITIMATE INSERTION: insertion elements may be incorporated into the genome by a process that does not require complete or even substantial homology unlike the general recombination events that usually have the requisite of homology between the sites of recombination. (See illegitimate recombination)

ILLEGITIMATE PAIRING: synapsis between not entirely identical strands, which may lead to aberrant crossover products. (See also oblique crossing over, unequal crossing over)

ILLEGITIMATE RECOMBINATION: recombination when the synaptic strands are not or not entirely homologous. (See homeologous recombination, recombination, synapsis)

ILLUMINATION: various types of units of measurement are in use. Most common is Lux = 1 standard new candela from 1 meter distance per 1 meter^{-2} or it is measured in Watt or Joule units. (See joule, watt, candela)

IMAGE ANALYZER (image processor): uses video cameras with high light-sensitivity, attached to the microscope. The camera is connected to a computer that can further enhance the image going to the camera. It thus can electronically enhance (by digitalization) and process the im-

Image analyzer continued

age and can detect details that cannot be perceived by the human eye by direct viewing through the microscope. The electronic system may show the picture in "false colors", i.e. selected by the operator rather than the natural color that the object has. The use of such devices can also free the image from background "noise" and enhance greatly the clarity. In addition, the operator's eyes are not stressed. The tissue-specific distribution and expression of the bacterial luciferase transgene can be monitored by microchannel plate enhanced photon counting analyzers. This setup consists of a microscope, image processor, TV monitor and a computer. The heart of the equipment can be outlined:

MICROSCOPE→ photocathode→ microchannel plates→ phosphor screen→ DISCRIMINATING VIDICON

This system can resolve a single photon. By comparison a single native bacterial cell releases about 10,000 photons. Image analyzer (e.g. VAX Station II/GP4) is used also for scanning DNA fingerprint autoradiograms. (See microscopy, luciferase)

IMAGE PROCESSING: see image analyzer

IMAGINAL DISKS: after fertilization of the *Drosophila* egg, the zygote nucleus begins dividing within a common cytoplasm and forms a syncytium (a multinucleate protoplasm) without actual separation by cell membranes. After about 2 hours and about 13 divisions the cellular blastoderm stage is reached (about 6,000 cells) and by then the cells move to the periphery of the embryo and the central space is occupied by the yolk. Within less than a day the larva is hatched and this developmental stage (3 instar steps) lasts for about 4 days. Inside the larva some groups of cells, distinguishable by location shape and size, the *imaginal disks* (50,000 cells) are set aside for serving as initials for the various organs and structures of the adult organism. The imaginal disks are not used by the larva and removal of them does not kill the organism. As the larva is metamorphosed into pupa most of the larval tissues disintegrate to support the development of the differentiation from the disks. During pupation, from these "prefabricated elements", the imago is assembled by cell and tissue fusions. From the anterior (front) part the head, from the middle region the thorax and from the posterior (hind) part the abdomen is formed. If the imaginal disks are lifted from their original position after the 3rd instar stage (about 5 days) and grafted into a new location of another larva, these disks develop into extra eye, wing, leg or other structures, depending on which imaginal disk has been chosen for the surgery and not on the host tissue. Thus, in the imaginal disks the developmental determination has been completed much before the onset of morphological and functional differentiation. For the correct differentiation it is required that the transplant would be placed in a larva and not into an adult. In the abdomen of an adult the disks only proliferate. After a series of transfers through adults, he original determination of the disks may change, and when the proliferated disk tissue is transplanted into a larva, an antennal disk may give rise to a leg instead. This altered course of differentiation is called *transdetermination*. The path of the transdetermination is not accidental or random. E.g., a genital disk may develop into a proboscis (mouthpart) or may pass through some indirect steps in a certain order such as genitalia → proboscis → antenna → leg. A wing from the genital disk cannot be formed directly but may arise through antenna or leg. The process of transdetermination is only a change in competence for differentiation without a mutational change whereas in the homeotic mutants one structure is replaced or altered after a change in the DNA. (See *Drosophila*, morphogenesis, homeotic genes, development, fate maps, determination, proboscis, antenna, blastoderm, homeotic mutants)

IMAGO: the adult form of an insect, male and female. (See *Drosophila*, imaginal disk)

IMBIBITION: absorption of water or other liquids.

IMINO FORM: of an amino acid arises from an amino form according to the reaction below:

$NH_2-C-COOH$ + flavin → **$NH=C-COOH$** + flavin H_2
AMINO   α-IMINO

Similar transformation can take place on nucleotides and can facilitate unusual hydrogen pairing between bromouracil and guanine or cytosine and adenine. (See hydrogen pairing)

IMMEDIATE EARLY GENES: of a virus are first turned on after infection without any requirement for synthesizing virally encoded proteins. (See delayed early genes, late genes)

IMMORTALIZATION: means that a cell in culture would not senesce but would proliferate indefinitely. Normally, cultured cells will die but cancerous transformation or by fusing normal cells with a cancer cells makes them "immortal". Immortalization and tumorous growth are, however, under separate genetic control. (See hybridoma, cancer, telomeres)

IMMUNE PRIVILEGE: grafts to certain sites (eyes, testis, brain) may be protected from rejection caused by the constitutive expression of the Fas ligand (FasL). Other factors may be blood tissue barriers, direct drainage of the tissue fluid into the blood, absence of efferent lymphatics, potent immunosuppressive environment (TGFβ), neuropeptides, α-melanocyte-stimulating hormone, vasoactive intestinal peptide, calcitonin gene-related peptide [CGRP], and membrane-bound inhibitors of the complement activation and fixation. Immune privilege may be necessary for pregnancy, for avoiding certain diseases and graft rejection, etc. (See TGF, Fas, calcitonin, complement)

IMMUNE RESPONSE: see MHC, immune system

IMMUNE SUPPRESSION: removal of the antigen. This can be achieved by destroying the target cells by CTL, B cell, macrophages or killer cells or by cytokines and other immunosuppressive agents. Contrasuppression prevents suppression and allows for the development of immunity. (See immune tolerance, veto cell)

IMMUNE SYSTEM: is a complex defense organization of vertebrate animals.

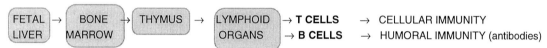

DEVELOPMENT OF THE IMMUNE SYSTEM AND ITS MAJOR COMPONENTS

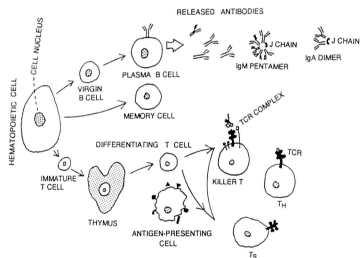

THE B AND T LYMPHOCYTES IN MATURE INDIVIDUALS ARE GENERATED BY THE HEMATOPOIETIC CELLS IN THE BONE MARROW. THE VIRGIN B CELLS BECOME EITHER PLASMA CELLS OR MEMORY CELLS UPON THE ENCOUNTER OF A FOREIGN ANTIGEN AND LYMPHOKINES. AS A RESPONSE THE PLASMA CELLS MANUFACTURE ANTIBODIES THAT ARE FIRST EMBEDDED IN THE PLASMA MEMBRANE OF THE B CELLS, AND WHEN THE MEMBRANE IS 'SATURATED', THEY ARE RELEASED AS CIRCULATING ANTIBODIES. ANTIBODY IgM POLYMERIZES INTO A PENTAMER AND JOINS A J PEPTIDE CHAIN.

THE IgA ANTIBODY FORMS A DIMER AND IT IS BRACED ALSO BY A J POLYPEPTIDE CHAIN. THE MEMORY CELLS DO NOT BECOME ACTIVE UNTIL FOLLOW-UP EXPOSURE TO THE PARTICULAR ANTIGEN. EACH OF THE ANTIBODY-PRODUCING CELLS MAKES ONE SINGLE TYPE OF ANTIBODY AND ITS CLONAL PROGENY ALSO MAKES THE SAME ANTIBODY. DIFFERENT ANTIBODIES ARE MADE BY DIFFERENT B CELL CLONES. THE IMMATURE T CELLS DIFFERENTIATE INTO T CELLS WITHIN THE THYMUS. THE THYMUS IS A BILOBATE ORGAN THAT DEVELOPS IN THE EMBRYO AND IT REACHES ITS MAXIMAL FUNCTION DURING PUBERTY AND THEN BEGINS A GRADUAL, SLOW DECLINE, ACCOMPANIED BY A DECREASE OF T LYMPHOCYTE PRODUCTION AND A WEAKENED IMMUNE RESPONSE. THE MATURE T CELLS ARE OF THREE MAIN TYPES; *KILLER CELLS* (CYTOTOXIC T LYMPHOCYTES, CTL), *HELPER T CELLS* (T_H) AND *SUPPRESSOR T CELLS* (T_S). THE SPECIFICITY OF THE T CELL FUNCTION DEPENDS ON THE SUR-

Immune system continued

> FACE ANTIGENS CARRIED BY THE ANTIGEN-PRESENTING CELLS. THE TCR (T CELL RECEPTOR) HAS SUBSTANTIAL STRUCTURAL AND FUNCTIONAL HOMOLOGY WITH THE IMMUNOGLOBULIN PROTEINS. THE TCR COMPLEX IS FORMED BY THE PARTICIPATION OF THE MHC PROTEIN, ENCODED BY THE HLA COMPLEX, AND A SERIES OF CELL MEMBRANE PROTEINS. THE KILLER CELLS DESTROY THE MEBRANES OF THE INVADING CELLS, FOLLOWED BY LYSIS OF THEIR CONTENTS.

The main cellular components of the immune system are the lymphocytes and macrophages. The B lymphocytes synthesize the antibodies that react and destroy the foreign antigens. The T lymphocytes generate antigen receptor-MHC protein complex that specifically recognizes individual antibodies and are involved in the destruction of foreign antigens. Lower animals may use as defense phagocytosis. Plants do not have an immune system. The immune system protects the body from microbial infections and also from non-infectious foreign macromolecules such as proteins, polysaccharides, and tissue grafts. Smaller foreign molecules (*haptens*) may also trigger the immune system primarily when associated with proteins. The molecules that elicit the immune response are called antigens. The antigens stimulate the formation of antibodies and the antibodies are the actual defense molecules. The immune reaction is extremely specific and minute differences between antigens, such as single amino acid substitutions or isomers may be specifically recognized. The immune system distinguishes between the body's own antigens from those of extraneous origin. It appears that this absence of immune response to the body's self antigens is an *acquired immunological tolerance*. The immature immune system may learn how to ignore even a foreign antigen if exposed to it during a very early stage. Immunological tolerance may be produced in later stages of the development with immuno-suppressive drugs, by very high or repeated exposure to very low amounts of the foreign antigen. Also, altering the so called *antigen-presenting cells* may affect tolerance. These antigen-presenting cells bind, process and combine foreign antigens with Class I and II proteins of the HLA complex (MHC). The T lymphocytes can be made tolerant to foreign antigens more readily than the B cells. In some rare *autoimmune diseases* even self-discrimination may break down with very serious consequences for the individual because the body's defense system may destroy its own tissue. The key players in mounting an immune response are the approximately two trillion lymphocytes (white blood cells) of the human body. The *B lymphocytes* are involved in the *humoral* (circulating in the blood) immune response. The B cells make antibodies. These are called B cells because they are made in the bursa Fabricius (intestinal pouches) in birds. In humans, the B cells originate from the stem cells of the fetal yolk sac, then in the liver and finally in the bone marrow (humans do not have bursa). The pre-B cells appear in the fetal liver by the 8th week of gestation and contain μ immunoglobulin chains in their cytoplasm but not on their surface. About 2 weeks later B cells appear. By the 13th week B cells have μ immunoglobulin on their surface. By the 12th week the B cell production shifts to the bone marrow. IgD (immunoglobulin δ) appears by 14th week in the spleen, lymph nodes and blood. After the 14th week the HLA antigens begin to appear. As the antibody gene rearrangements proceed all other immunoglobulin genes may be expressed. The proliferation and differentiation of B cells requires the presence of antigens on their surface and some soluble factors obtained from T cells. The T_H (helper) and T_S (suppressor) cells regulate the development of B cells. Activation of helper T cells requires the presence of APC cells (*antigen-presenting cells*), HLA Class I and II proteins and interleukin-1. The T_H cells secrete some mediators (e.g. lymphokines). The activated lymphocytes and mediators then induce the formation of receptors that is followed by proliferation and differentiation of both T and B lymphocytes. The suppressor lymphocytes may prevent the induction of the immune response by exposure to a wide variety of molecules. The other important representatives of the immune system are the *T lymphocytes* (shaped in the thymus), their primordia develop by the 5 to 6th week of human gestation and the first mature T cells appear by 9 to 10th week. The T cells are involved in the *cell-mediated* immune response, i.e., they react with antigens bound to their surface. As the maturation of the T cell proceeds the CD antigens (CD1, Cd2, CD4, etc.) ap-

Immune system continued

pear on their surface and identify T cell subsets. By the 16 to 20th week the spectrum of subsets reaches that of the adults. The T cells usually respond to antigens in association of the HLA complex (MHC, major histocompatibility complex). T cells with the CD8 antigens respond to the Class I proteins of the HLA complex whereas the CD4 T cells are limited to the HLA Class II proteins (DP, DQ, DR). This phenomenon is called *MHC restriction*. Maternal IgG may cross the placenta after the 16th week and provides early immune protection for the fetus. Fetuses and newborns may not respond much to foreign cells because the lymphocyte development may be arrested when contacted by antigens and anti-idiotype antibodies. IgM and IgA are normally (in the absence of infection) not present in substantial amounts in the newborns. Although after birth the maternal IgG tends to be eliminated, it is gradually replaced by the infant's own antibodies. By age year 3, IgG levels become sufficient. Before the end of the first year IgM levels reach that of adults but IgA levels rise very slow, reaching adult levels only by age 9 to 12 years. The appearance of the immune reaction is the result of a developmental process. The *virgin T cells* (that have never been exposed earlier to a specific antigen) become *effector cells,* i.e., they begin to proliferate and produce either the cell-mediated (T cell) or the humoral (B cell) response. Some other lymphocytes are also induced to proliferation and differentiation but this time they do not participate in the immune response but become *memory cells* that will "remember" the same antigen by virtue of their differentiation, and assume the effector cell role at a later similar exposure (*clonal mechanism of secondary immune response*). The immune response is mounted relatively slow (in days or weeks) when first time exposed to a particular antigen (*primary immune response*) but on subsequent exposures the response is much faster because of he availability of the memory cells. This phenomenon is the basis of immunization (vaccination) as it is used in medicine. (See antibody, immunoglobulins, HLA, T cells, TCR, B cells, autoimmune disease, complement, lymphokines, phagocytosis, hapten, antigen, MHC, antigen presenting cell, blood cells, lymphocyte, lymphocyte homing, interleukin, immune system diseases, innate immunity, memory immunological, leukalexins, host-pathogen relations)

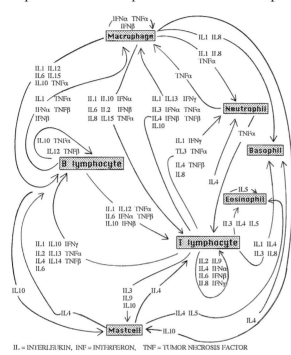

IL = INTERLEUKIN, INF = INTERFERON, TNF = TUMOR NECROSIS FACTOR

COMMUNICATIONS IN THE IMMUNE SYSTEM

IMMUNE SYSTEM DISEASES: see agammaglobulinemia, DiGeorge syndrome, Wiskott-Aldrich syndrome, Reiter syndrome, arthritis, lupus erythematosus, celiac disease, immunodeficiency, Sjögren syndrome, autoimmune diseases

IMMUNE TOLERANCE: failure of the immune system to respond to antigen(s). Neonates were supposed to display immune tolerance, however newer studies indicate that their immune response may be only different. Immune tolerance is regulated either by anergy or by suppression. The suppression may be mediated by orally supplied low doses of antigens or by clonal anergy when high doses of the antigen is provided. Oral tolerance may extend to tolerance of the gut immune system and thus can present systemic immune reaction to ingested proteins.

Immune tolerance continued

The brain-blood barrier prevents lymphocytes from entering the central nervous system. The peripheral immune tolerance may include T-cell activated repression by the secretion of TGF-β. (See antigen, anergy, T cell, TGF, mucosal vaccines, T cell, T cell receptor, thymus, immune system)

IMMUNITY: is a protected state against biological (microbial) or other agents. *Acquired immunity* develops as a reaction of the immune system of the organism. *Passive immunity* is conveyed by the transfer of antibodies or activated lymphocytes from another body. *Natural immunity* means that the cells are not susceptible to a particular organism, e.g., humans are not infected by wheat rust. *Genetic or familial immunity* indicates that a group of individuals are resistant to or free of a particular type of infection. *Maternal or intrauterine immunity* results when humoral antibodies are passed through the placenta into the fetus. *Cell mediated* immunity is the result of T lymphocyte activation and *humoral immunity* is provided by the antibodies secreted by the B lymphocytes. *Neonate immunity* is due to passing IgG immunoglobulins by the mother's milk with the assistance of the FcRn receptor. *Phage immunity* designates the condition when a lysogenic bacterium cannot be superinfected by another bacteriophage. (See immune response, antibody, humoral antibody, MHC, FcRn, lymphocyte, superinfection, mucosal immunity)

IMMUNIZATION: induction of immunity. It can be active immunization by introducing specific antigens into the body to promote the formation of a specific antibody or passive immunization by introducing antibodies into the body that may convey temporary immunity. During vaccination live (attenuated) or killed microbes are introduced into the body either through the blood stream or by oral means. The bonding subunit of *E. coli* heat-labile enterotoxin (LT-B) is a very effective immunogen. Transgenic plants encoding LT-B, express the foreign peptides, and able to bind the natural ligand after oligomerization. Mice fed by antigens produced by transgenic plants could orally be immunized. Also, tobacco plants transgenic for murine antibody κ chain, hybrid immunoglobulin A - G heavy chain, an immunoglobulin J chain and a rabbit secretory component, respectively, after intercrossing produced segregant progeny that simultaneously expressed all four polypeptides. The polypeptide chains were successfully assembled within the transgenic hybrid into a functional high molecular weight secretory immunoglobulin that could recognize streptococcal surface adhesion molecules. Interestingly, in the plants all this could take place in a single cell whereas in animals the process requires two cell types. The results indicate the possibility of using immunoglobulins manufactured by plants for oral vaccination. In 'reactive immunization' the antigen is so highly reactive that a chemical reaction takes place at the site of the combining antibody. This mechanism may enhance catalytic antibody chemistry. The common type of immunization does not result in covalent interaction between antigen and antibody. (See antibody, immunoglobulins, immunization genetic, vaccines, transformation genetic [transformation of plants])

IMMUNIZATION, GENETIC: involves introducing into an animal a gene or expression gene library encoding a particular protein(s) by biolistic transformation or vectors resulting in antibody production. This method does not involve the risk of live or attenuated pathogen vaccines. (See immunization, antibody, biolistic transformation, viral vectors, library)

IMMUNOADSORBTION: binding cognate antibody to antigen in order to facilitate the separation of a specific type of antigen or antibody from a mixture. (See immunofiltration).

IMMUNOBLOTTING: separate proteins by gel electrophoresis and identify an appropriate component by the specific monoclonal antibody labeled by fluorochrome or radioactivity. (See Western blot, electrophoresis)

IMMUNODEFICIENCY: may have several different causes; some of them involve milder effects whereas others are lethal. Immunodeficiency is frequently caused by defects in the lymphocyte differentiation system (thymus). Some of the immunodeficiencies are parts of other syndromes such as Down's syndrome, sickle-cell anemia, ataxia telangiectasia, glycoprotein

Immunodeficiency continued

deficiency, glucose-6-phosphate dehydrogenase deficiency, immunoglobulin imbalance [deficiency of IgA and IgG but increased IgM], defects of the HLA histocompatibility system and some caused by non-genetic causes, such as infection [HIV] or drugs and various allergies that may have a genetic component. X-linked immunodeficiency with hyper immunoglobulin M (IgM) but absence of IgG, IGA and IgE is caused by a defect in the interaction between CD40 and CD40L. (See agammaglobulinemia, hypoglobulinemia, severe combined immunodefficiency, bare lymphocyte syndrome, Nezlof syndrome, DiGeorge syndrome, chronic granulomatous disease, Wiskott-Aldrich syndrome, HLA, CD40, ZAP-70, bare lymphocyte syndrome, ADA).

IMMUNODEFICIENCY, VIRAL: see acquired immunodeficiency (AIDS)

IMMUNODOMINANCE: of the many possible epitopes of the antigen the cytotoxic lymphocytes (CTL) recognize only one or a few. (See epitope, CTL)

IMMUNO-ELECTRONMICROSCOPY: cell constituents are labeled by fluorescent antibodies and thus the location of proteins can be identified with electronmicroscopic resolution.

IMMUNOELECTROPHORESIS: can be carried out by different procedures. *Diffusion*: over the electrophoretically separated proteins, antisera (antibody or antibodies) are placed in a trough. The protein bands are permitted to diffuse in a radial manner from their original position in the gel while the antibody is diffusing vertically and at their position of reaction with each other an elliptical arc of precipitation is formed. *Rocket electrophoresis*: the antigen is placed in the wells of the gel that contain antibody. After turning on the electric current the antigen moves and forms a rocket-shape trail of precipitation as it reacts with the antibody. The length of the rocket indicates the amount of antigen applied to a well. (See *ELISA*)

IMMUNOFLUORESCENCE: identification of a protein (antigen) by *direct* adsorption to specific antibody that has an attached fluorochrome. Alternatively, to the first (unlabeled) antibody a second cognate, fluorochrome-labeled antibody is added for the sake of immunostaining (*indirect* method). (See fluorochromes)

IMMUNOGEN: substance which elicits immune response. (See immune system)

IMMUNOGENETICS: is concerned with the hereditary and molecular aspects of antigen and antibody systems and the immune response.

IMMUNOGLOBULINS: are the structural units of the antibody molecules. Each antibody is a heterotetramer consisting of two identical light immunoglobulin chain (either of two λ or two κ) and two identical heavy chains. There are five classes (*isotypes*) of immunoglobulins IgM, IgG, IgA, IgD and IgE, identified according to the heavy chain subunits: $\mu, \gamma, \alpha, \delta$ and ε. The Ig heavy chains are glycoproteins, containing about 15, 4, 10, 18, 18% sugars in a molecular mass of 70, 50, 55, 62 and 70 (kDa), for $\mu, \gamma, \alpha, \delta,$ and ε, respectively. IgM is a pentamer containing five μ heavy chains and five light chains and one J chain; its molecular mass is about 900 kDa. The J polypeptide ($\approx$20 kDa) is synthesized within IgM secreting cells is covalently attached between two Fc domains (see antibody) and it is supposed to initiate the oligomerization. (The J polypetide is not the product of the J genes located between the variable and constant gene clusters). IgA antibody is either a monomer, dimer or trimer with molecular weights 153, 325 and 580, respectively, and may have a J chain. IgA besides the J polypeptide may have also a secretory component (SC) of 400 kDa. The SC is picked up by the secretory IgA dimers from the surface of epithelial cells and it braces the IgA dimers and protects them from proteolysis. IgG, IgD, IgE antibody monomers are of 150, 180 and 190 kDa, respectively. Each of the 5 major types of heavy chains can be associated with either one of the light chains. The five classes of antibodies have somewhat specialized roles. IgG is the humoral, circulating antibody. IgE deals with the allergic reactions and IgA has the major role in defense against microbial infections. IgM and IgA have 2 subclasses, IgG and the λ chains have 4, designated as e.g., IgM1, IgM2, and so on. Within these subclasses *allotypes* are distinguished, 25 for human IgG, 2 for IgA and 3 for the kappa (κ) chain. These allotypes represent antigenic markers on the immunoglobulin chains and are designated as e.g., Gm1, Gm2... Am1...Km1... and so on

Immunoglobulins continued

for IgG, IgA and κ chain markers, respectively. The allotype variants represent amino acid substitutions at one or more sites. The characteristic series of allotype markers are inherited as gene blocs and are called *haplotypes*.

The genes encoding immunoglobulins are clustered in three supergene families. The heavy chain genes are clustered at human chromosomal location 14q32.33. The κ genes are at human chromosome 2p12, whereas the λ gene family is in human chromosome 22q11.12.

Within the approximately 7,000 kb heavy chain region in the long arm of chromosome 14 the genes L (encodes the signal that leads the polypeptide to the endoplasmic reticulum), V_H (variable heavy), D (diversity), J (juncture) and the various constant heavy chain genes C_μ to C_α, are arranged in groups separated by long base sequences (~~) and spacers (■) with switch signals (S) in front of the constant heavy chain genes. The general organization of the immunoglobulin genes are very similar in all mammals although they are located in different chromosomes:

5'—L—V_n—V_1—D_1 • D_{20}—JJJJ~~~■μ ■ δ ■ $γ_3$ ■ $γ_1$ ■ $ε_ψ$ ■ $α_1$ ■ $γ_ψ$ ■ $γ_2$ ■ $γ_4$ ■ ε ■ $α_2$ —3'

Up front of this sequence, the basal promoter of the Ig heavy chains contain non-translated regulatory transcriptional elements, such as the dispensable *heptamer* consensus: (5'-CTCATGA-3') and 10 to 40 bp downstream the indispensable *octamer* consensus: (5'-ATGCAAAT-3'). The latter is 30 to 60 bp upstream from the TATA box, that is followed within about 20 to 30 bp the transcriptional initiation site for the LVDJ and constant heavy chain sequences, μ to $α_2$. (The orientation of the octamer is opposite to the direction of transcription). Actually, between the LVDJ region and constant heavy chain genes, there are enhancer elements of the 5'-CAGGTGGC-3' motif and three core repeats of multiple GC sequences.

The κ cluster is similarly organized in human chromosome 2:

5' —L—V_S — — V_n ———JJJJJ— 1 κ constant gene group— 3'

The λ genes are in human chromosome 22. The variable genes occur in six groups. Here the J genes are not clustered separately but situated in front of the six constant λ gene groups. Some of the λ genes are outside the clusters and may not be functional. The individual λ segments are quite variable in size. The V_S is one of the switching sequences explained below.

The base promoter of the κ light chain contains at about 100 bp upstream from the transcription initiation site a pentanucleotide consensus, then within -90 to -60 bp the octamer consensus (oriented in the direction of the transcription) follows. This does not have the heptamer shown at the heavy genes. The TATA box is at about the same distance from the transcription initiation site as in the heavy chains. There are enhancer elements, designated as κB (5'-GGAAAGTCCCC-3') and Eκ1 to Eκ3 (variants of the enhancer motif shown at the heavy chains), between the LVJ genes and the constant κ gene group. The strongest enhancer is the κB. Both the heavy and light gene enhancers act preferentially in the B lymphocytes. The heavy chain enhancers seem to be constitutive whereas the light gene enhancers become active after the rearrangement of the genes (discussed below). Besides the enhancers, the immunoglobulin genes seem to have silencers of expression for non-lymphocyte chromosomes. The turning on the promoters requires transcription factors, one of them the 60 kDa OTF-2, has specficity for the immunoglobulin enhancer consensus 5'-CAGGTGGC-3'. The 90 kDA OTF-1a general mammalian transcription factor, is also present in the lymphocytes. The DNA-binding domains of these two factors are very similar but their other domains are different. These enhancers bind also other type of proteins and the only lymphocyte-specific enhancer appears to be the octamer. The light-chain specific transcription factor is protein NF-κB that binds to the 5'-GGGPu(C/T)TPyPy(C/T)C-3' motif. After the immunoglobulin light chain has undergone rearrangement, preparatory to transcription, the pattern of the *nuclease-sensitive sites* in the promoter region is altered.

The antibody specificity is determined by the variable-diversity regions of the light and heavy chains. The antigen has to fit, be complementary to the NH_2 end of the antibody. The hyper-

Immunoglobulins continued

variable region of the antibody (that has the highest specificity for the antigen) is represented by the three complementarity determining regions (CDR1, CDR2, CDR3).
Although in the germline the complete array of all immunoglobulin genes is present, during development various rearrangements and elimination of genes take place. Thus the mRNA does not represent all the genes all the time in the somatic cells but different ones may be re-presented, depending on their transcription, stimulated as the immune response unfolds. The heavy chain genes can generate an enormous variety of polypeptides. The variable region is put together from a menu of hundreds of V_H, about 20 diversity (D) segments and 5 or more joining segments (J_H). The V, J and constant heavy gene clusters are separated by sequences containing 12 or 23 spacers, flanked by conserved heptamer (7mer) and nonamer (9mer) nucleotide tracts that serve apparently the purpose of rearrangements. The general organization of the switching sequences are diagrammed below. The heavy chain genes are in mouse chromosome 12 and the corresponding human gene cluster is in chromosome 14. The light κ genes of the mouse are in chromosome 6 whereas the homologous human genes are in chromosome 2. The λ chain genes are in mouse chromosome 16 and in human chromosome 22. The 12 and 23 bp spacers are shown in bold numbers:

```
        9MER              7MER           7MER           9MER          7MER
VH-<GGTTTTTGT -23 - CACATGT>--<CACAGTG -12 - ACAAAAACC>-D -<CACAGTG-12--⌐

 → -ACAAAAACC>--<GGTTTTTGT -23 - CACATGT>--J--CH   (mouse chromosome 12)
        9MER              9MER           7MER
```

Recombination is limited to the 12 and 23 base spacers (12/23 rule). The RAG (recombination activating) gene products provide signals for the recombinational signal sequences (RSS) for double-strand breaks. The 12/23 and the RAG signals are concerted and mutation in one signal prevents cleavage at both. The blunt signal ends and the coding ends then form a hairpin-like structure by transesterification with the assistance of a score of repair enzymes. The hairpin intermediates lead to the assembly of the antigen-receptor sequences of the genes. This process has similarities to retroviral integration.

The switching recognition sites in the two mouse light chain gene series are as follows:

$U_κ$-<CACAGTG -12 - ACAAAAACC>-----<GGTTTTTGT - 23 - CACATGT>---J--$C_κ$

$U_λ$-<GGTTTTTGT -23 - CACATGT>------<CACAGTG -12 - ACAAAAACC>-J--$C_λ$

The consensus sequences shown above are present in between each of the variable and joining genes in the clusters. Thus rearrangements alone can generate an enormous variability, particularly within the heavy and κ clusters which have, each, hundreds of variable region genes. The λ cluster has only a very small number of variability genes, thus itself it contributes minimally to the antibody arrays, yet in combination with the highly variable heavy chains it may produce about 100,000 different antibodies. The κ — heavy chain assemblies are capable to generate more than a million types of antibodies. This assembly process is limited to the lymphoid cells. The T cell receptor assembly occurs only in the T cells and the complete assembly of the Ig genes takes place in the B cells. The accessibility of the V(J)D recombinase to the complementarity region is regulated by the transcription system. The enhancer motif called E-box µE3 and other upstream regulatory elements control also accessibility of the recombinase.

The V and J consensus sequences display opposite orientation. The opposite orientation means that the same nonamer (9MER) can be either $\frac{ACAAAAACC}{TGTTTTTGG}$ or $\frac{GGTTTTTGT}{CCAAAAACA}$ in the DNA doubles-strands as the sequences are inverted horizontally and vertically (→↓).

The types of sequences shown insure that V genes and the J genes do not recombine within the V or J group but the V genes recombine with the J genes because one type of spacing can recombine only with another type of spacing. Also in the heavy gene cluster a V gene is obligated to transpose to a D gene and that in turn can be relocated to a J gene. A V gene and the J gene have the same type of spacers (different from the D gene) thus the relocation of a V gene

Immunoglobulins continued

to a J gene must involve a D gene. Although these types of relocation mechanisms are frequently called recombination and a *recombinase system*. One must keep in mind that these events are not crossing overs. The 12 and 23 base spacers apparently represent one or two turns, respectively, of the DNA double helix and thus seem to indicate the mechanics of the rejoining. The V and J genes can recombine also by a breakage - reunion mechanism that may take place at the heptamers (7mers) at the ends of the two coding units and this results in the elimination of the interjacent segment with the *signal ends*, and the ends of the coding sequences are the *coding ends*. The signal ends may be joined to form a circular DNA structure and this circularization, followed by elimination, is apparently a major source of the rearrangements. In some instances there is an inversion of a V gene relative to the J gene. In such a case the intervening material may not be deleted although the V gene may be inactivated. Some of the rearrangements are *non-productive* because they occur at random and are in the wrong register. Since codons are triplets, two third of the rearrangements may result in garbled sequences. Also some nucleotides may be lost at the coding ends and also some may be added (*N nucleotides*) by the enzyme deoxynucleotidyl transferase. At the coding ends, the 5' terminus of one of the DNA strands may covalently fuse with the 3' terminus of its complementary strand. When the resulting hairpin structure breaks a protruding end of nucleotides may be formed that can serve as a template to generate an inverted terminal repeat of a few nucleotides (*P nucleotides*). The 96th codon at the end of the V_H gene is actually generated by a fusion between V and J genes. This is a critical point because the 96th amino acid is part of the antigen-binding region as well as the connection of the light and heavy chains. Both deletions and additions increase the variability of the antibody genes. The V_H regions than can combine with any of the 5 constant heavy chains (μ, δ, γ, ε, α and their subclasses, *isotypes*) and this is the source of another variation.

The expression of the immunoglobulin genes requires that the promoter would be transposed to the vicinity of an enhancer (see base promoter structure above). The enhancer becomes normally active only in the B lymphocytes. Before any antigen is encountered, IgM and IgD class antibody production starts. In such virgin B cells, both immunoglobulins have identical variable regions but they may differ in constant regions of the μ and the δ chains. Since B cells are diploid, genes only in one of the two homologous chromosomes can be expressed at a time and only one type of rearrangement can function in a cell; this is *allelic exclusion*.

The virgin B cells can then differentiate either into *plasma cells* or into *memory cells* upon exposure to an antigen. The former becomes an immediate producer of antibody, the latter will be activated into plasma cells only upon a subsequent exposure to the same antigen. The differentiation is aided by *lymphokines* (a variety growth regulating proteins), and various T lymphocytes (T_H, T_S) and macrophages. Association of the virgin lymphocytes with an antigen triggers the mechanism of *isotype switching*. Isotype switching brings about the selection of the proper heavy chain constant region by a process of transcription although DNA deletions may also be involved. In the undifferentiated B lymphocyte transcription begins at the heavy chain gene leader sequence upstream and continues through the variable and diversity regions to the end of the δ gene, passing through exons and introns. Polyadenylation signal follows the last exon of the transcribed constant region of the μ and δ genes:

5' Promoter- Enhancers--V_n···V_1---D_{20}···D_1-J $_n$···J_1--**S**-μ --**S**- δ -**S**-γ --**S** ε -**S**- α 3' **DNA**

    virgin B cell   ···V_2---D_5-J_5----μ-----δ·· **PRIMARY RNA TRANSCRIPTS**

processing   ···V_2 D_5 J_5 μ polyA and V_2 D_5 δ polyA **two mRNAS**

translation V_2 D_5 J_5 μ and V_2 D_5 δ **IgM and IgD POLYPEPTIDES**

Actually two types of μ chains exist at the early stage of the B lymphocyte. The μ_m chain (with

Immunoglobulins continued

a hydrophopbic C terminus and alternative processing event) and μ_Σ. The former is included in the lymphocyte membrane as a monomeric IgM antibody (2 light -2 μ chains) whereas the secreted IgM becomes pentameric and adopts also a ca. 20 kDa J peptide with the composition of $(\mu_2\lambda_2)_5$J, (the J is synthesized in the B lymphocyte but not coded in the constant heavy chain cluster).

After recombination and deletion occurred at one of the **S** switch points, in the non-coding, ca. 2 kb upstream tracts, at the $(GAGCT)_n GGGGT$ motifs, of the constant heavy chain genes, the rearranged gene is transcribed. The introns are eliminated and the transcript is processed into polyadenylated mRNA. The mRNA is translated into the individual heavy chain monomers:

switching at γ $\quad\quad V_2$---D_5-J_5----γ $\quad\quad$ **PRIMARY TRANSCRIPT**
$$\downarrow$$
IgG 50 kDA polypeptide

switching at α $\quad\quad V_2$---D_5-J_5----α $\quad\quad$ **PRIMARY TRANSCRIPT**
$$\downarrow$$
IgA 55 kDa polypeptide

In the examples above the same VD and J genes were shown, actually any of the VDJ genes can be selected before switching. Additional variation is generated by somatic mutation during the proliferation of the lymphocyte clones. The frequency of these mutational events (about 10^{-3} per base) appears to be higher than the usual mutation rate of other types of genes. After the heavy chain is finished it may combine with any of the two light chain polypeptides. IgG and IgA are further polymerized to form the final antibody. The activity of B cells is terminated partly by binding the antigen to the secreted antibody. This event then prevents the binding of the antigens to the B lymphocyte receptors and thus the stimulation of immunoglobulin synthesis ceases. Some of birds (ducks, geese, swans) produce immunoglobulin Y (IgY) which does not occur in mammals or in some other avian species but it bears similarities to IgG and IgE. IgY is produced in a larger and smaller forms; the latter is deficient in the crystalline fragment. (See antibody, immune system, immunization, lymphocytes, complement, HLA, TCR, T cell, T cell receptors, B cell, DNA-PK, RAG, RSS, V(J)D recombinase, SCID, terminal nucleotidyl transferase, repertoire shift, affinity maturation, CDR, somatic hypermutation, hypermutation, accessibility, translin, transposons [Tn*3*, Tn*5*, Tn*7*, Tn*10*])

IMMUNOLABELING: see immuno-staining, ELISA, immunofluorescence, RIA, immunoscintography, monoclonal antibody

IMMUNOLOGICAL LEARNING: the quality of the antibody improves as the clonal selection progresses. (See clonal selection)

IMMUNOLOGICAL MEMORY: survivors of a cell (individual) that mounted an immune response are more effectively protected in case of a following infection. The mechanism of this protection may be based on maintenance of specific T or B cells even in the absence of the antigen or some types of lymphocytes have the ability to remember the antigen. Some lymphocytes may be maintained by recurrent low levels of infection or some regulatory networks of cytokines respond in case of second infection. (See immune system, lymphocytes, T cell, B cells)

IMMUNOLOGICAL SURVEILLANCE: is considered one of defense mechanism against cancerous body cells. The surface antigens of the transformed (cancer) cells are different from their normal counterparts. The immune system that continuously monitors the body for invading microorganisms, and other foreign antigenic material (macromolecules, grafts, etc.) recognizes also the cancer cells at an incipient stage and with the aid of the immune system, disposes of them. Experimental data in support of this idea comes from the fact that antibodies

Immunological surveillance continued

produced against mammary cancer preferentially recognized the metastatic cells without reacting with the normal cells. Since cancerous transformation may be initiated by a single mutation (although for the development of neoplasias additional events are required), and mutation rates per cell are in the range of 10^{-9}, yet because the human body may have 4 to 5 orders higher numbers of cells, cancer mutations may affect each person numerous time during the human life. Nevertheless, the incidence of cancer death is about 0.2 fraction of all deaths. If no biological protection would be available, all individuals should have had cancerous transformation(s). (See immune system, cancer, chromosome breakage, genetic tumors)

IMMUNOLOGICAL TESTS: are extremely sensitive for the detection of the presence of a particular protein or other molecules which can form cross-reacting material (crm) with specific antibodies. Frequently, in case of the quantity of the material is very low and standard biochemical assays are not sensitive enough for the identification, immunoprobes are the choice for testing. (See antibody preparation, immunoprobe, immunoscreening, immunofluorescence)

IMMUNOLOGICAL TOLERANCE: see immune system

IMMUNOMODULATORS: viral encoded proteins regulating antigen presentation, regulators of cytokines, cytokine antagonists, inhibitors of apoptosis and interfering with the functions of the complement. (See antigen presenting cell, cytokines, apoptosis, complement)

IMMUNOPHILINS: are two classes of proteins which bind immunosuppressants such as either rapamycin and FK506 (FKBP) or cyclosporin. All known immunophilins display rotamase (peptidyl-prolyl *cis-trans* isomerase) activity *in vitro*. The 59-kDa member of the FKBP family is a component of the inactive glucocorticoid receptor. The immunophilins apparently interact with protein kinases of the signal-transducing paths and with the heatshock protein 90, a chaperone. (See FK506, cyclosporin, cyclophilin, T cell, immunosuppressants)

IMMUNOPRECIPITATION: the reaction of antigen with cognate antibody may lead to blood coagulation or to the selective precipitation of a protein. (See also Western blotting)

IMMUNOPROBE: or immunoblot; bacterial colonies are immobilized on a filter and a specific antibody is added. This antibody can bind the epitope of a second antibody or antibody plus protein A that may be labeled by a radioactive isotope (I^{135}) or a biotinylated molecule. The complex can then be detected by autoradiography on the dot blot or separated in SDS-polyacrylamide gel and the labeling identifies the substance of interest. (See Western blot, colony hybridization, protein A, probe, DNA probe)

IMMUNOSCINTIGRAPHY: by using a scintillation camera (capable of detecting the flashes emitted by radioactive isotopes) radioactively labeled monoclonal antibodies can be localized in the body or tissues by even a three-dimensional image.

IMMUNOSCREENING: the product of a gene is identified on the basis of a cognate antibody.

IMMUNOSTAINING: purified antibodies can be labeled by fluorochromes and their specific recognition sites can be visualized *in situ* with aid of fluorescence light microscopy. Also, antibodies labeled by colloidal gold permits their analysis by electronmicroscopy.

IMMUNOSTIMULATORY DNA (ISS): contains within short stretches of plasmid vehicles CpG dinucleotides: 5'GACGTC-3', 5'-AGCGCT-3' or 5'AACG%%-3'. Such sequences promoted the production of interferon-α and -β and interleukin-12. The significance of this finding for gene replacement therapy is that ISS may cause the production of proinflammatory cytokines and thereby down-regulate gene expression. (See therapy, cytokines, T cells)

IMMUNOSUPPRESSANT: blocks or reduces the immune response by irradiation, specific antimetabolites or specific antibodies. (See cyclosporin, cannabinoids, calcineurin, immunophilins)

IMMUNOSUPPRESSION: activation of the immune system generally involves the activation of cytokines and cell adhesion. Repression of this process requires the inhibition of the transcription factors required to develop the key elements of the immune system. Glucocorticoids, prednisone (also a glucocorticoid), the fungal cyclic oligopeptides, cyclosporin, cyclophosphamide (carcinogen), azathioprine (an arthritis drug), cytarabin (cytosine analog), mercaptopurine (a purine analog), methotrexate (a folic acid antagonist), muromonab-CD3 (a murine monoclonal antibody [IgG$_{2\alpha}$] targeted to the lymphocyte membranes), etc. are used. (See glucocorticoids)

IMMUNOTHERAPY: includes immunization, use of immunopotentiators, immunosuppressants, hyposensitization to allergens, transplantation of bone marrow or thymus.

IMMUNOTOXIN: may be an antitoxin, or specific antibody equipped with a bacterial, fungal or plant toxin. The monoclonal antibody provides the means of homing on the special target cell(s) of cancer (lymphoma, melanoma, breast and colorectal carcinomas) or graft rejection (bone marrow transplant) or T cells responsible for autoimmune disease (arthritis, lupus or HIV infected T cells.) In addition it may carry cytokine and soluble receptors to assist targeting. The toxin (*Pseudomonas* exotoxin, diphteria toxin, ricin, abrin, α-sarcin) then specifically destroys the target by inhibiting local protein synthesis without affecting other cells. Lysosome targeting (lysosomotropic) amines (NH$_4$Cl), chloroquine and carboxylic ionophores (monesin) protect the cells from some immunotoxins (e.g. diphteria toxin) but make them more sensitive to others (e.g. to ricin). The clinical applicability of this therapy is still very limited. (See magic bullet antitoxin, monoclonal antibody)

IMPALA (*Aepyceros melampus melampus*): 2n − 60.

IMPATERNATE: originated by parthenogenetic reproduction. (See parthenogenesis)

IMPDH (inosine-5'-monophosphate dehydrogenase): is involved in lymphocyte replication. Mycophenolic acid (MPA), an approved immunosuppressive drug is its potent inhibitor.

IMPERFECT FLOWER: has either the male or the female sexual apparatus, it is monoecious or dioecious but not hermaphroditic. (See flower differentiation, hermaphrodite)

IMPERFECT FUNGI: have no known sexual mechanism of reproduction. (See fungal life cycles)

IMPETIGO: a pus-forming skin infection caused by the plasmid-carrying *Staphylococcus aureus* bacteria.

IMPLANT: a grafted addition to the body or an inserted artificial objec or an implanted zygote.

IMPLANTATION: attachment of the blastocyst to the lining of the uterus after about a week of the fertilization and embedding into the endometrium (in humans). (See blastocyst, uterus)

IMPORTIN α and β: are protein factors mediating passage through the nuclear pore by binding the α subunit to a nuclear localization sequence of a protein. (See nuclear pore, nuclear localization sequence)

IMPOTENCE: inability to initiate or maintain erection of the penis caused by organic or psychological causes. (See nitric oxide)

IMPRINTING: the expression of behavioral or other traits may be influenced by the parental source of chromosome, i.e. the paternal and maternal genomes may have different effect (imprinting) on the developing offspring because of the modification of an allele by a cis-element or different methylation of the sequence. The insulin-like growth factor gene (*IGF-2*) of mice, transmitted through females is not transcribed (imprinted) in most of the tissues and only the one transmitted through the male is active. If the offspring receives a mutant copy of the gene through the male and a normal copy through the female, the heterozygote is crippled. The choroid plexus (the brain tissue secreting the cerebrospinal fluid) and the leptomeninges (the innermost of the three membranes covering the brain and the spinal chord) were not subject, however, to *IGF-2* gene imprinting in mice. The *IGF-2* is a single chain polypeptide and it is an autocrine regulator of hormone response and growth. It appears that the methylation takes place in CG-rich islands of 200 to 1,500 base pairs and conspicuously, several of the imprinted genes are either in chromosomes 11p or 15q in humans, and in the mouse in chromosome 7. In human chromosome 15q11-q13 an imprinting center (IC) was that is involved in epigenetic resetting of this 2 Mb domain. When the untranslated *H19* mouse gene was disrupted and the *Ins-2* and *Igf-2* genes— 100 kb upstream of H19—were transmitted by the female. In the offspring, 27% higher weight was observed compared to animals that received the same chromosome from their father. In this mouse chromosome 11 *mash-2* en-

Imprinting continued

codes a helix-loop-helix protein and it is maternally expressed only in the placenta. Normally, *Ins-2* (insulin) is expressed paternally in the embryo yolk but biparentally in the pancreas. The tissue-specificity of imprinting of the insulin-like growth factor is determined by which of the four promoters of the gene was used. In some cases the demonstration of true imprinting is difficult because in human diseases the penetrance and expressivity of the genes may widely vary. Imprinted genes frequently carry special repeats, display unusual sex-specific rates of recombination and the size of their introns are relatively short. Parthenogenesis may cause embryonic lethality in mouse if the imprinted paternal genes are not expressed. During tumorigenesis both the paternal and maternal copies of the IGF-2 gene are expressed. (See methylation of DNA, regulation of gene activity, IGF, Angelman syndrome, Prader-Willi syndrome, Beckwith-Weidemann syndrome, insulin, insulin-like growth factor, myotonic dystrophy, ataxia, MYF-3, KIP2, polar overdominance, *VENTURE*, diabetes mellitus, insulin-like growth factor, lyonization, parthenogenesis)

IN SITU **HYBRIDIZATION** (ISH): the DNA double strands within the cells (chromosomes) are separated (denaturation) in cytological preparations on microscope slides and labeled (radioactively or by fluorochromes or by immunoprobes) and complementary DNA or RNA strands (probes) are annealed. The cells are then visualized by microscopy as usual for cytological microtechniques and can either be autoradiographed or viewed through fluorescence microscopy to ascertain the position of the hybridized probe. Thus chromosomal location of molecular markers can be determined. (See DNA hybridization, somatic cell hybrids, probes, fluorochromes, chromosome painting, FISH, immunoprobe, nick translation, *in situ* PCR, PRINS)

IN SITU **PCR**: employs PCR technology within single cells. The DNA is amplified and then *in situ* hybridization is used. Its utility is enhanced by treating the cells by reverse transcriptase before PCR. Such a procedure permits the detection of cellular genic activity, viral infection, expression of just introduced transgenes for gene therapy. (See PCR, RT-PCR, PRINS, FISH)

IN VITRO: the reaction or culture is carried out within a glass vessel rather than in an intact cell or in natural culture conditions, respectively such as *in vitro* culture, *in vitro* fertilization or *in vitro* enzyme assay. (See ART)

IN VITRO **FERTILIZATION** (IVF): extracted mammalian eggs can be fertilized by competent sperms (AID or AIH) outside the body and then surgically implanted into the uterus. Such procedure may have a higher than 40% chance for success and helps to overcome sterility barriers in many women with infertility problem caused by factors other than the eggs. *In vitro* fertilization produces normal babies but the chance of having non-identical twins is greatly increased because the gynecologist generally implants more than a single fertilized egg to assure a reasonable success. IVF has applied significance in animal breeding. From the ovaries of cows eggs can be retrieved in the slaughterhouses and after *in vitro* fertilization can be re-implanted into any (even some sterile) cows. By this procedure more beef can be produced. There is an opportunity to obtain more offspring from genetically superior individuals of domestic animals by the use of surrogate females. It has applications also in wild life maintenance in endangered species (in zoos) or in species were the natural rate of reproduction is unsatisfactory. (See twinning, artificial insemination, preimplantation genetic, GnRFA, ART, ICSI, test tube baby)

IN VITRO **MUTAGENESIS**: mutation is produced in isolated DNA sequences that is then reintroduced into the cells by transformation. (See localized mutagenesis, site-specific mutagenesis, gene replacement TAB mutagenesis)

IN VITRO **PACKAGING**: recombinant DNA equipped with the phage λ *cos* sites and genes required for packaging (*origin* of replication and other sequences of about 4-6 kb at both *cos* neighborhoods) can accept inserts so that the total length will remain in the range of 37-52 kb, can be packaged into phage capsids that may infect *E. coli* and yeast cells and bring about transformation. (See cosmid vector)

IN VITRO **PROTEIN SYNTHESIS**: see *in vitro* translation systems

IN VITRO **TRANSLATION**: see rabbit reticulocyte translation assay, wheat germ translation, translation *in vitro*, oocyte translation.

IN VIVO: the process is taking place in intact cells or in tissues of a live organism or cell. (See also *in vitro*)

INACTIVE-X HYPOTHESIS: see Lyon's hypothesis

INBORN ERROR OF METABOLISM: a historical term for biochemical genetic defects. Generally mutation in single genes block or change the metabolic pathways at a single specific step. The consequences of the mutation may be alleviated either by providing the missing compound or by avoiding the supply of the accumulated precursors that cannot be further processed because of the defect in the enzymatic step. (See auxotrophy, one gene — one enzyme theorem, biochemical genetics)

INBRED: a line developed by continued inbreeding until the majority of the genes become homozygous. In mice generally 20 generations of brother sister mating is employed to produce such lines. In species where self-fertilization is feasible (e.g. plants) ten generations of inbreeding results in more than 0.999 % homozygosity ($1 - 0.5^{10}$), unless the genes are very tightly linked in repulsion. (See coefficient of inbreeding)

INBREEDING: mating among biological relatives, including self-fertilization, brother sister mating and mating with ancestors or offspring. (See coefficient of inbreeding, inbreeding in autopolyploids, inbreeding and death rates, inbreeding and population size, inbreeding progress)

INBREEDING AND DEATH RATES: inbreeding results in homozygosity of deleterious and lethal genes and as a consequence spontaneous abortions, infant mortality and the frequency of hereditary diseases will increase with this type of mating system. The frequency of the adverse consequences depends upon the frequency of these undesirable genes in the population concerned and the degree of inbreeding. When infant mortality of first cousin marriages and that of general population marriages are compared the frequency in the former is about double. Some of the variation within columns may be also due to random statistical error (After Fraser, F.C. & Biddle, C.J. Amer. J. Human Genet. 28:522) :

ETHNICITY	FIRST-COUSIN MARRIAGES %	GENERAL POPULATION %
Canadian	8.8	4.1
French	9.4	4.4
Japanese	6.2	3.9
Swedish	8.5	4.0

In a more recent analysis the excess death rate — up to age 10 — of the progeny of first cousin marriages in Japan, Pakistan, India and Brazil combined appeared 4.4%. (See coefficient of coancestry, incest, genetic load, lethal equivalent)

INBREEDING AND POPULATION SIZE: inbreeding increases more in smaller populations than in large one. The increase can be reduced by controled mating, i.e., when the mating pairs are selected from different families or if an equal number of mates are selected from each family of the herd. The rate of inbreeding (Δ_F) = $1/2N_e$. If the actual size of the population is say 10, then $\Delta_F = 1/(2 \times 10) = 0.05$. In case the effective population size is only say 0.75 of the total population then $\Delta_F = 1/2N_e = 1/(2 \times 10 \times 0.75) = 1/15 = 0.066$. The coefficient of inbreeding becomes $F_g = 1 - 1 - \Delta_F)^g$ where F_g is the inbreeding coefficient of the gth generation and Δ_F is the rate of inbreeding. (See inbreeding coefficient, inbreeding depression, inbreeding rate, population size effective)

INBREEDING AUTOPOLYPLOIDS: since autopolyploids have more alleles present per locus, homozygosity at a locus is achieved only after a larger number or generations. The table on p. 533 provides a comparison on the proportion of homozygotes in diploids and tetraploids of the initial genetic constitutions *Aa, Aaaa* and *AAaa*, respectively, after 5 generations of self-fertil-

Inbreeding in autopolyploids continued
ization (After Burnham, C.R. 1962, *Discussions in Cytogenetics*, Burgess, MN):

GENERATION	*Aa* (diploid)	*Aaaa* Chrom. segr[1]	*Aaaa* Max. equat.[2]	*AAaa* Chrom. segr[1]	*AAaa* Max. equat.[2]
F_2	0.500	0.250	0.295	0.050	0.099
F_3	0.750	0.380	0.460	0.194	0.285
F_4	0.875	0.493	0.581	0.326	0.442
F_5	0.938	0.558	0.674	0.438	0.566
F_6	0.969	0.648	0.747	0.531	0.662

[1] Chromosome segregation indicates that gene is absolutely linked to the centromere
[2] Maximal equational segregation occurs when the gene segregates independently from the centromere

INBREEDING COEFFICIENT: the probability that *two alleles at a locus in an individual* are identical by *descent* from a common ancestor, i.e. the chance that an individual is homozygous for an ancestral allele by inheritance (not by mutation). *Consanguinity* (coancestry) is a similar concept but the coefficient of coancestry indicates the chances that *one allele in two individuals* would be identical by descent. The coefficient of inbreeding is symbolized by F. The calculation of F is based on the fact that in a diploid at each locus there are two alleles and only one is contained in any gamete (either one in a particular egg or sperm). Thus each individual has 0.5 chance for passing on a particular allele to a particular offspring. Examples make simpler to illustrate the method of calculation required. Brother (X) and sister (Y) have two common parents (W) and (V). An offspring of the mating (X) x (Y) → (I) has a chance to inherit a gene from grandparent (W) through two routes (W)→(X)→(I) or (W)→(Y)→(I), therefore its chances for homzygosity for one allele derived from (W) is $0.5^2 = 0.25$. In other

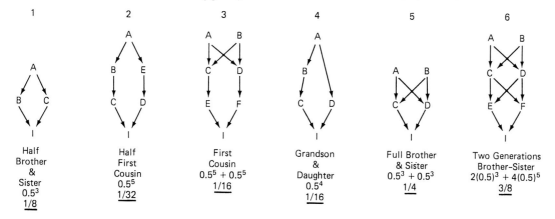

CALCULATION OF THE COEFFIOCIENT OF INBREEDING ON THE BASIS OF THE PATHS OF ALLELE TRANSMISSION

words, in the F_2 the chance is 1/4 for homozygosity for any allele according to the Mendelian law. In a half brother and half sister progeny grandparent (A) can transmit a particular allele to (I) either through (B) or (C) parents and the inbreeding coefficient of (I) is $0.5^3 = 1/8$ because three individuals are involved in the transmission route (A), (B) and (C). Similarly, the inbreeding coefficient of other types of matings can be calculated as indicated on the chart. In half first cousin mating individuals (C), (B), (A), (E), and (D) are involved in the transmission path, each with a 0.5 chance thus the coefficient of inbreeding becomes $0.5^5 = 0.03125 = 1/32$. In two generations of brother-sister matings (see scheme 6), the transmission of alleles may follow the routes [E-C-F, F-D-E], {E-C-A-D-F, F-D-B-C-E, E-D-A-C-F and F-C-B-D-E}, i.e. [2] and {4} paths of $[0.5]^3$ and $\{0.5\}^5$, respectively. Thus the coefficient of inbreeding

Inbreeding coefficient continued
is $2[0.5]^3 + 4\{0.5\}^5 = 0.375 = 3/8$.

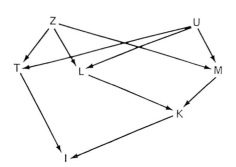

If there are multiple paths through the same ancestor, all the paths through the shared ancestors must be included in the calculation with the precaution that within the same paths the same ancestor must be counted only once. Let us illustrate the method with another example where (Z) and (U) are the common ancestors and again the inbreeding coefficient of individual (I) is sought. There are 2 routes through (Z): T-Z-L-K and T-Z-M-K and also 2 paths through (U): K-M-U-T and K-L-U-T. Each of the 4 paths involve four ancestors contributing genes to (I). Therefore the coefficient of inbreeding of (I) is $4(0.5)^4 = 0.25 = 1/4$. Under practical conditions of breeding much more complicated schemes may be encountered yet their solution can be sought on the basis of these much simpler examples. It is easier to determine the loops of gamete contribution by working backward from the critical individual, (I) in this case. It is conceivable that the common ancestors are not completely unrelated, contrary the assumption used in the calculations above, but they may have some degree of relatedness and their inbreeding coefficient, F_A (ancestral coefficient of inbreeding) must also be taken into account. Therefore the general formula for the coefficient of inbreeding $F = \Sigma[(0.5)^n(1 + F_A)]$ where Σ is the sum of the paths through which an individual can derive identical alleles from the ancestors and n = the number of individuals in the paths. The $1 + F_A$ is correction factor for the inbreeding coefficient of the common ancestor in the path. The knowledge of the coefficient of inbreeding may be very important in a breeding project but it may be quite relevant also to human families. Let us assume that the frequency of a recessive genetic disorder is $q^2 = 1 \times 10^{-6}$ and if the population is in a genetic equilibrium the frequency of that allele is $q = \sqrt{q^2} = \sqrt{0.000001} = 0.001$. Then related parents have the risk of an afflicted child is: $q^2(1 - F) + q(F)$. Since the inbreeding coefficient of the offspring of first cousins is F= 0.0625 (see chart), after substitutions we get: 0.0000001 x 0.9375) +(0.001 x 0.0625) = 0.000063. Since 0.000063 is 63 fold higher than 0.000001 (the frequency of individuals with this affliction in the general population), the first cousin parents take a 63 fold greater risk than unrelated parents to have an offspring afflicted with any hereditary disease that has a gene frequency of 0.001. (See F, coefficient of coancestry, consanguinity, relatedness degree, inbreeding progress, inbreeding rate, fixation index, genetic load)

INBREEDING DEPRESSION: when heterozygotes of normally outcrossing or dioecious or monoecious species are propagated by inbreeding and the deleterious recessive genes become ho-homozygous, the viability, vigor and fitness of the individuals and the population declines. (See inbreeding progress, inbreeding coefficient, inbreeding and death rates, heterosis, controled mating)

INBREEDING, PROGRESS OF: the proportion of homozygosity of any selfed or inbred generation for any number of allelic pairs can be computed by the formula:
$[1 + (2^g - 1)]^n$ where g is the number of generations selfed (note that, e.g. by F_5 there is 4 selfings (because the F_1 is the result of crossing) and n is the allelic pairs involved. Example of expansion of the binomial in case of 3 pairs of alleles in F_5:
$[1+ (2^4 - 1)]^3 = [1 + (16 - 1)]^3 = 1^3 + 3(1)^2(15) + 3(1)(15)^2 + 1(15)^3$ or rewritten
$$1^3(15)^0 + 3(1)^2(15)^1 + 3(1)^1(15)^2 + (1)^0(15)^3$$
where the first exponent in each term indicates the number of heterozygous genes and the second exponent shows the number of homozygous genes in each class of individuals. In this example 1 will be heterozygous for all 3 loci and homozygous for none (3 x 15), i.e. 45

Inbreeding progress of, continued

will be heterozygous for 2 loci and homozygous for 1, (3 x 15^2), i.e. 675 will be heterozygous for 1 locus and homozygous for 2, and 15^3, i. e. 3375 will be heterozygous for none of the 3 loci but will be homozygous for all 3 in a total population of 4096 (1 + 45 + 675 + 3375 = 4096). (See also heterozygote proportions, binomial, coefficient of inbreeding, fixation index)

INBREEDING RATE: is determined as $\Delta_F = 1/(2N_e)$, where N_e = effective population size. If the actual size of a population size is 10 then $\Delta_F = 1/(2 \times 10) = 0.05$. In this computation the effective population number was considered to be equal to the total number. Under practical circumstances this is rarely the case. If we suppose that the effective size is only 3/4 of the actual number then $\Delta_F = 1/(20 \times 0.75) \cong 0.066$, a higher fraction. The coefficient of inbreeding can be also calculated as: $F_g = 1 - (1 - \Delta_F)^g$ where g = the number of generations of inbreeding and F = inbreeding coefficient.
Accordingly, after 25 generations, $F_{25} = 1 - (1 - 0.066)^{25} = 1 - (0.934)^{25} \cong 0.819$ whereas if 10 would be the effective size, $F_{25} = 0.723$ in this hypothetical case. (See also effective population size, inbreeding progress, inbreeding coefficient, fixation index)

INCENP: see sister chromatid cohesion

INCEST: legally prohibited sexual intercourse between close (biological) relatives. Thus, incest is primarily a legal, ethical and moral concept but from the viewpoint of genetics the consanguinity of the partners is considered and the legal restrictions may not be sufficient to avoid the deleterious consequences for the progeny of such matings. In ancient Egypt the pharaohs frequently (and legally) married their sisters and that may explain the large number of child mummies at the burial places. Relatively scarce data are available on children of first degree relatives (parent x child, brother x sister) yet it is clear that about 40% suffers more or less severe physical and/or mental defects. Interestingly, in populations where uncle x niece and cousin marriages are practiced for centuries the number of birth defects is not as high as would be expected. Apparently, the inbreeding continued for many generations purged the gene pool from the most deleterious alleles that are maintained at random mate selection. (See coefficient of coancestry, coefficient of inbreeding, genetic load)

INCHWORM MODEL: during transcription the RNA polymerase is flexibly connected to the template. The front end domain is tightly associated with the DNA and that is followed by a loose association including the catalytic domain and that is followed by another tight association to the transcript. (See transcription, protein synthesis)

INCIDENCE: the frequency of occurrence of a genetic alteration or disease in a population. (See also prevalence)

INCIPIENT SPECIES: a group of organisms that is in the process speciation. (See speciation, evolution)

INCLUSION BODY: is a protein aggregate in *E. coli* cells expressing at high rate some foreign gene(s). These proteins are visible by phase contrast microscopy and can be concentrated by centrifugation of lysed or sonicated cells. (See phase-contrast microscope, sonicator)

INCLUSIVE FITNESS: is a type of altruistic behavior. Parents defend their children — at their own risk—for the sake of maintenance of their genes that the offspring shares with them. This altruistic behavior is positively correlated with the degree of relationship. (See fitness, altruistic behavior, kin selection)

INCOMPATIBILITY: see for plants S alleles, incompatibility alleles for mammals. (See also ABO blood group, Rh, erythroblastoma, histocompatibility [in tissue and organ transplantation, plasmid incompatibility, fungal incompatibility)

INCOMPATIBILITY ALLELES: self-incompatibility in a large number of plant species (tobacco, clovers, crucifers, fescue, beets, cherry, etc.) may prevent self-fertilization or zygote formation. The number of different incompatibility alleles may run into the hundreds in some species. A plant pollen carrying a particular incompatibility allele may not successfully develop a pollen tube in the stylar tissue of the same genetic constitution but may successfully fertilize another plant of different allelic constitution. A sperm with an S1 allele thus incom-

Incompatibility alleles continued

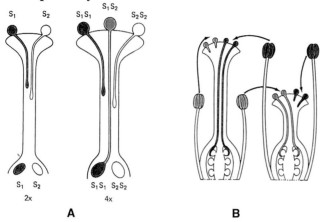

(**A**) THE COMPATIBILITY OF THE DIPLOID POLLEN IN A DUPLEX TETRAPLOID PLANT. (**B**) INCOMPATIBILITY ASSOCIATED WITH DIMORPHISM OF STYLE AND STAMEN LENGTH. (After Linskens, H.F. & Kroh, M. 1967. Encyclopedia of Plant Physiology. Vol. 18. Springer. Berlin, Germany)

patible with a stylus of S1S1 type but may fertilize an S2 egg. In an S1S2 heterozygote neither S1 nor S2 sperm may be successful but they have no barriers in S3S3 plants. In some species compatibility may be determined by the sporophytic tissue and in an S1S2 stylus, if the S2 allele is dominant, the S1 pollen may be successful and produce S1S1 seed. In some cases the S1S2 pollen of a tetraploid plant may be compatible with an S1S1 egg if the S2 allele is dominant. Incompatibility may be based also on different timing of pollen release and receptivity of the stigma. Compatible combinations may arise through induced mutations. Heterostyly (different height of the stylus and stamen) may also prevent self-fertilization. The selfincompatibility is caused by S-specific glycoproteins and ribonuclease enzymes. In the Brassicaceae two genes mediating incompatibility are encoding the S locus receptor kinase (SRK) and the secreted glycoprotein (SLG). It is assumed that a pollen-born ligand ties SLR and SLG into a signaling complex that prevents the germination or the growth of the pollen tube on the stigma or in the style. It is an interresting observation that an SRK incompatibility protein provides also protection against *Pseudomonas syringae* infection. In the Solanaceae the S locus encodes another type of glycoproteins with ribonuclease activity (S-RNase). (See mentor pollen effect, male sterility, apomixis, gametophyte, genetic load, fungal incompatibility, Rh blood group, ABO blood group, self-incompatibility alleles)

INCOMPATIBILITY LOAD: see genetic load

INCOMPATIBILITY MOTHER-FETUS: see erythroblastosis fetalis, genetic load, killer cells

INCOMPATIBILITY PLASMIDS: utilize the same system of replication and cannot co-exist. When they introduced into the bacterial cell they have to compete with each other and one is eliminated as a results. Plasmids (phages) that carry the same replicon belong to the same incompatibility group. The incompatibility is determined by RNA I, RNA II and the Rop protein. There are more than 30 plasmid compatibility groups. (See RNA I)

INCOMPLETE DIGESTION: reaction with restriction enzymes is terminated before all potential cleavage sites are cut. A larger variety and some large size fragments are cut because some neighboring sequences are not cleaved apart. (See restriction enzymes)

INCOMPLETE DOMINANCE: or semidominance is observed when the expression of the gene does not entirely mask or prevent the expression of the recessive allele at the same locus in a hybrid. (See dominance, epistasis)

INCOMPLETE LINKAGE: recombination takes place between or among the syntenic genes in question. (See recombination, crossing over, synteny, linkage)

INCONTINENTIA PIGMENTI (Bloch-Sulzberger syndrome): X-linked (Xp11) "marble-cake-like" dark pigmentation on skin of the trunk, preceded generally by an inflammation. It may begin soon after birth and may fade by the age of 20. The anomaly may be associated with eye, tooth, bone and heart anomalies. (See hypomelanosis of Ito, pigmentation defects)

INCORPORATION ERROR: is a mechanism of mutation when a nucleic acid base analog or a wrong base is inserted into the nucleic acid during replication. As a consequence a base pair is replaced by another. Thus the meaning of the codon may change and appears as a visible mut-

Incorporation error continued

ation if the codon change may lead to an amino acid substitution at a critical site in the protein. Example: during replication a 5-bromouracil is inserted into the DNA at a C site resulting in a BrU—G base pair. During next replication the a BrU—A pair is formed and at a subsequent replication a T=A pair is substituted at a site where originally a C≡G pair existed. (See replication error, base substitution mutations, hydrogen pairing)

INCROSS: hybridization between two strains that have the same genetic background.

INDEL: an **in**sertion or **del**etion in the DNA nucleotide sequences, e.g. By the simple example below the alignment score can be illustrated. $Pr = p^3 q^2 r^1$ where p is the probability of identity (match), q is the probability of substitution (mismatch) and r is the probability of an indel. The alignment score can be derived as follows:

```
AAGTTC
 |  |  |  ←match  |  |
A-GCCC  ←mismatch CC
↑
INDEL
```

$S' + \log Pr = 3(\log p) + 2(\log q) + (1(\log r)$ and $S = S' - n\log s = S' - 6(\log s)S =$ a constant satisfying $\log(p/s) = 1$). And $S = 3 - 2\mu - 1\delta$ where $\mu = \log(q/s)$ and $\delta = \log(r/s)$ and $S = \#$ identities $- \mu\#$ substitutions $- \delta\#$ indels. The task can be resolved by computer programs based on high level mathematics that cannot be presented here. (See Waterman, M.S., Joyce, J. & Eggert, M. 1991 Phylogenetic Analysis of DNA Sequences, pp. 59-89, Miyamoto, M. M. & Cracraft, J., eds., Oxford Univ. Press, New York, DNA sequence information, data bases)

INDEPENDENCE: two events are independent when the occurrence of one does not affect the chance of occurrence of the other. Genes at a distance of 50 map units or more segregate independently, the sex of the first child is (normally) independent from the sex of the next one, if two pennies are tossed they can land on head or tail independently from each other unless they are defective or biased.

INDEPENDENCE TEST: see association test (contingency chi square)

INDEPENDENT ASSORTMENT: alleles of different loci (non-allelic genes) may reassort freely in the gametes and therefore segregate independently in zygotes in the absence of linkage. The independent assortment of the alleles is one of the most essential discoveries of Mendel and it is frequently called Mendel's third law. (See Mendelian laws)

INDEPENDENT EVENTS: do not affect or influence each other.

INDETERMINATE INFLORESCENCE (raceme): the main axis can elongate indefinitely but the branches terminate in a flower bud. (See raceme)

INDEX: an alphabetical or other ordered set of files or symbols or numbers distinguishing particular things in an array, e.g. allele a^1, the "1" distinguishes this allele among all other *a* alleles, or *NK3* homeobox 3 of *Drosophila* or *adp^{fs}* an *adipose* allele conveying female sterility, or the second asymmetric leaf locus as_2 (*As-2*) of *Arabidopsis*.

INDEX CASE: see proband (propositus, proposita)

INDEX LOCUS: see polymorphism information content (PIC)

INDEX VALUE: is a concept used for selection in animal breeding. Each trait is weighted by a score and these scores summed in an index value. The use of this index alleviates the danger of selecting for one particular trait only and thus jeopardizing the overall success of the program because frequently high performance is accompanied by disease susceptibility or low fertility, etc. (See gain, selection)

INDIRECT DIAGNOSIS: identification of a gene on the basis of linkage rather than by direct evidence.

INDIRECT END-LABELING: determines the distance of a DNase-hypersensitive site from a restriction enzyme cleavage site. The chromatin is first digested with a Klenow fragment of DNAase I, then isolated and treated with a restriction endonuclease. The double digest is then separated by electrophoresis and probed with a sequence adjacent to the restriction site. The size of the fragment generated by the double cuts indicates the distance of the DNAase hypersensitive site from the site of restriction. This procedure can thus localize in the DNA the sites

Indirect end labeling continued

where transcription may be initiated because the hypersensitive sites are correlated with the position of transcriptionally active genes. (See DNase hypersensitive site, Klenow fragment, restriction endonuclease)

INDIRECT SUPPRESSION: some suppressor mutations do not correct the primary change in the gene, rather they modify the translation process and thereby suppress the expression of the mutation. (See suppressor mutation)

INDOLE: a heterocyclic compound, present in many organic associations in biological materials. When excited by ultraviolet light, it displays a characteristic fluorescence spectrum, suitable for its rapid detection. Among other roles it is a precursor of tryptophan synthesis: anthranylate → indole glycerol phosphate → indole + serine → tryptophan.

INDOPHENOLOXIDASE: see under the new name superoxide oxidase

INDUCED FIT: of enzymes happens when the conformation is so modified that the activity improves; this may be caused by binding to a ligand or substrate. (See ligand, DNA binding proteins, *Lac* operon)

INDUCED HELICAL FORK: after the binding protein contacted a few bases of the DNA, it keeps apart the double helix. (See Watson and Crick model, binding proteins)

INDUCED MUTATION: was obtained by exposure to a mutagen, and presumably generated by the mutagen rather than by an incidental spontaneous event. (See spontaneous mutation, mutation)

INDUCED REPLISOME REACTIVATION: see replication-restart

INDUCER: a substrate or an analog of a substrate of an enzyme prevents a repressor protein from attaching to the promoter (operator) of a gene and thus facilitates its expression. (See induction, gratuitous inducer, repression)

INDUCIBLE ENZYME: presence of substrate or substrate analog is required for their synthesis. (See as examples *Lac* operon, *Ara* operon)

INDUCTION: has several meanings; phage induction indicates the facilitation of the transition from the prophage stage to the lytic phase. Induction of enzymes set into motion the catalytic activity. Gene expression is induced by assembly of the pre-initiation complex of transcription. Embryonic development is induced by the transmission of the various exogenous and endogenous signals. (See prophage, enzyme induction, regulation of gene activity, transcription, signal transduction. morphogenesis in *Drosophila*, photomorphogenesis)

INDUCTION DEVELOPMENTAL: the fate of a cell or tissue is affected by the interaction of the embryonic cell or tissue with its neighbors. (See embryonic induction)

INDUCTION OF A LYSOGENIC BACTERIUM: to liberate phage particles by first inducing a change from prophage to vegetative phage state. (See zygotic induction, prophage)

INDUCTION OF AN ENZYME: initiates the synthesis of new enzyme molecules by the presence of an inducer that may be the substrate or an analog of the substrate of that enzyme. (See derepression)

INDUSIUM: a membrane-type layer over the sorus (sporangial cluster) of ferns.

INDUSTRIAL MELANISM: as industrialization (coal-burning pollution) increased (in Britain) the dark variants (dominant) of the black peppered moth (*Biston betularia*) increased as a selective trend to camouflage the insect on the soot covered tree barks. (See natural selection, adaptation)

INFANTILE AMAUROTIC IDIOCY: See Tay-Sachs disease

INFECTIOUS CENTER: a spot from where infectious phage or bacteria can be produced.

INFECTIOUS DRUG RESISTANCE: drug resistance genes are carried on conjugative plasmids of bacteria. (See conjugation bacterial, plasmid)

INFECTIOUS HEREDITY: see symbionts hereditary, segregation distorter, plasmid, prions

INFECTIOUS NUCLEIC ACID: may be a purified viral DNA or RNA that may propagate in the host cell and code subsequently for viral particles.

INFECTIOUS PROTEIN: see prion

INFERTILITY: may have various causes such as anatomical abnormalities of the sexual organs (hermaphroditism, dysgenesis, testicular feminization, polycystic ovary disease), infectious diseases, hormonal abnormalities, malfunction of CREM, psychological factors, organic diseases, medications, alcoholism or other substance abuse, malnutrition, chromosomal defects (trisomy, translocations, inversions, deletion, duplications, aneuploidy), hereditary abnormalities (cystic fibrosis, mental retardations, Kallman's syndrome, Kartagener syndrome, myotonic dystrophy). In the U.S. about 15% of the couples are involuntarily infertile. (See fertility, fecundity, spermiogenesis, sterility, azoospermia, cell cycle, CREM, and the other entries)

INFILTRATION: introduction various substances into biological tissues by diffusion, frequently facilitated by evacuation (under negative pressure).

INFLORESCENCE: a cluster of flowers, characteristic for taxonomic classification of plants.

INFLUENZA: is a viral infection of the respiratory tract with possible secondary infection by *Streptococcus, Staphylococcus* and *Haemophilus* bacteria. The RNA virus has several types, designated by origin such the Spanish, Hong Kong, Russian strains or as type A (most common and reoccurring in 2-3 year cycles), Type B (causes epidemics in 4-5 year cycles), Type C is a sporadically occurring one. Birds, horses, swines, cats also have influenza type infections by different viruses.

INFORMATICS: a system of databases and electronic retrieval.

INFORMATION: obviously the greater amount of information is available about a population the easier and more reliable is the decision of the geneticist about a parameter of that population or populations. Statistically, the information $I_p = \dfrac{1}{Vp}$ indicating that the total amount of information is inversely proportional to the variance (V) of the statistic employed. The calculation of the information for a particular set of data can be carried out by:

$I = \Sigma \left(\dfrac{1}{m} \left[\dfrac{dm}{d\theta} \right]^2 \right)$ where m is the expectation in terms of parameter θ, and Σ is the sum of all classes. R.A. Fisher pointed out that maximizing the likelihood function provides an estimate of T, which has the limiting value of $1/nV_T = I$. The reciprocal of the variance of the maximum likelihood estimate permits assessing the value of other estimates. If the variances obtained by other methods are not $1/nI$, they do not give us the full possible information and thus inferior to the maximum likelihood statistics. (See maximum likelihood, variance)

INFORMATION (in statistics): it is called sometimes "support" or "lod-score". (See lod score)

INFORMATION RETRIEVAL: the procedures to obtain information from a set of stored data, such as a specific nucleotide sequence in the databases. (See databases)

INFORMATIONAL MACROMOLECULES: DNA, RNA, proteins that can convey genetic, developmental, biochemical and evolutionary instructions to a cell or organism.

INFORMED CONSENT: may be a dilemma of a genetic counselor regarding the information he/she may wish to withhold from the counseled because of the psychological impact. Legally, all the dangers should be exposed that the professional evaluation indicates. Action to be pursued requires informed consent. (See counseling genetic, genetic privacy, confidentiality)

INFORMOSOME: mRNA complexed with protein and thus acquiring very low turn-over rate and stability.

ingi: see hybrid dysgenesis, I - R

INGRESSION: the movement of cells from the surface into the inner region. (See gastrulation)

INH: a protein complex containing at least six species, isolated from oocytes; it inhibits the activation of pre-MPF. (See maturation protein factor)

INHERITANCE: receiving genes from ancestors and passing them on to offspring. These genes are coded by DNA in eukaryotes and prokaryotes whereas in some viruses the transmitted genetic material is RNA. The genetic material may be located in the nucleus (nuclear inheritance) or carried by the nucleoid in prokaryotes. Extranuclear inheritance is mediated by the

Inheritance continued
  genetic material in mitochondria and chloroplasts. Prokaryotes and cytoplasmic organelles may have also plasmid vehicles of heredity. Contrary to some common loose wording, traits are not inherited, only the genes, which determine their expression, are transmitted. (See genotype, phenotype, heredity, genetics, reverse genetics, DNA, RNA, prions, genealogy, pedigree, acquired characters inheritance)

INHERITANCE, DELAYED: see delayed inheritance

INHIBIN: an antagonist of activin. Inhibins are glycoproteins (A and B) in the seminal and follicular fluids and inhibit the production of follicle-stimulating hormone and regulate gametogenesis, embryonic and fetal development as well as blood formation (hematopoiesis). (See activin, FSH)

INHIBITION: see inhibitor

INHIBITOR: a substance that interferes with the *activity* of an enzyme versus a repressor that prevents the *synthesis* of the enzyme. (See regulation of enzyme activity)

INI1 (integrase interacting protein): tethers the retroviral (HIV) integrase enzyme and facilitates the integration at or near the DNase hypersensitive site of the eukaryotic chromosome. (See integrase, DNase hypersensitive site, Cre/loxP, FLP, resolvase)

INITIATION CODON: is the first translated codon. In prokaryotes it is most commonly AUG translated into formylmethionine but UUG and GUG can also be used. In eukaryotes the AUG does not code for formylmethionine but for methionine. The initiation codon is charged to a specific initiator tRNA. (See aminoacyl-tRNA synthetase)

INITIATION COMPLEX: contains the small subunit of the ribosome with associated mRNA, aminoacylated tRNA and the various initiation protein factors and energy donor nucleotide triphosphates. (See protein synthesis)

INITIATION FACTOR FOR TRANSCRIPTION: see initiation complex, IF, eIF

INITIATION FACTORS OF PROTEIN SYNTHESIS: are involved in initiation of translation. (See also eiF, IF and iIF)

INITIATOR (Inr): see promoter

INITIATOR CODON: the starting site of translation in the mRNA; generally 5'-AUG-3' but can be less frequently 5'-GAG-3', 5'-GUG-3' or 5'-GUA-3'. (See protein synthesis, translation)

INITIATOR tRNA: carries formylmethionine (prokaryotes) and initiation methionine (eukaryotes) to the P site of the ribosome to begin translation. This tRNA has a structure distinguished from the rest of the transfer RNAs. (See protein synthesis, ribosome, tRNA)

INK (p^{INK}): polypeptide inhibitors of cyclin-dependent kinases involved and cause cell cycle G1 phase arrest. (See cancer, cell cycle)

INNATE: inherited, congenital. (See congenital)

INNATE IMMUNITY (natural immunity): is based on cell surface receptors and other protein molecules encoded by the germline. These systems generally recognize carbohydrate structures and then stimulate the synthesis of various molecules such cytokines, interleukins and tumor necrosis factor. Some natural killer cells recognize inimical cells by lectin-like membrane receptors. This innate immunity is thus a rather fixed, rigid system in comparison to the acquired immunity mediated by immunoglobulins which are greatly adaptable and variable. Innate immunity can shape the development of the acquired immunity by interacting with it. It may guide the selection of antigens by lymphocytes and the secretion of cytokines by the helper T lymphocytes. Innate immunity is the first line of defense by initiation of inflammation through recruiting phagocytic and bactericidal neutrophils. The complement is part of innate immunity but it also cooperates with the acquired immunity system. Innate immunity is present also in insects (*Drosophila*) although they lack the adaptive immunity system of the vertebrates. (See immune system, complement, acquired immunity, lymphocytes, antibody)

INORGANIC PYROPHOSPHATASE: cleaves off 2 molecules of phosphates from molecules.

INOSINE: see hypoxanthine

INOSINIC ACID: see hypoxanthine

INOSITIDES: see phosphoinositides

INOSITOL: in cells it occurs generally as myo-inositol as part of the vitamin B-complex. It is formed through cyclization from glucose-6-phosphate. It is an indispensable constituent of some lipids (phosphoinositides). In some forms of diabetes it may accumulate in the urine. (See signal transduction, diabetes)

INOSITOL TRIPHOSPHATE: see phosphoinositides

IN-PLANTA TRANSFORMATION: see transformation genetic

Inr: see promoter

INSECT CONTROL, GENETIC: see genetic sterilization, holocentric chromosomes

INSECT RESISTANCE IN PLANTS: some plant species contain genes for insect tolerance and these are being incorporated into plant breeding material by conventional techniques. The most successful insect resistance gene, the *Bacillus thüriengiensis* toxin gene, has been transformed into several species of dicots and provides almost complete defense when it is expressed under the control of efficient promoters. (See *Bacillus thüringiensis*)

INSERTION ELEMENTS: DNA sequences generally shorter than 2000 bases which can insert into any part of a genome (see also transposons) are common in all organisms from prokaryotes to eukaryotes. Some of the bacterial insertion elements have the characteristics shown:

NAME	SIZE (bp)	INVERTED TERMINAL REPEATS (bp)	TARGET DUPLICATION
*IS*1	768	18/23 (*E. coli*)	8-11
*IS*2	1,324	32/41 "	5
*IS*3	1,258	29/40 "	3
*IS*5	1,250	16 "	4
*IS*10	1,329	17/22 (*Tn*10)	9
*IS*66	2,548	18/20 (*Agrobacterium tumefaciens*)	8

They are not carrying any genetic information beyond that needed for insertion. Viruses also can act as insertion elements in cells. Insertion elements are the major factors of "spontaneous" mutability. According to some estimates 5-15% of the spontaneous mutations in bacteria are caused by *IS* elements. The presence of *IS*1 may increase deletion frequency of the *gal* operon by 30 to 2,000 fold. *IS* elements cause also chromosomal rearrangements. Many *IS* elements have a large number of potential target sites, others display clear preferences. The mechanism of transposition is either dependent on replication and the new copy is transposed or it involves simply a relocation of the existing element. The movement of the *IS* elements is affected also by host genetic factors (DNA polymerase, gyrase, histone-like proteins, dam methylase, DnaA protein, and proteins mediating recombination). The presence of *IS* elements may alter not just mutation but also gene expression by their presence in control regions of genes. In *Drosophila melanogaster* the average insertion density (average number of large insertions/kb/chromosome) is about 0.004 and the average frequency of their movement is 0.023. Some of the historical insertions were modified and became permanent regulatory elements of the genes. (See also hybrid dysgenesis, copia elements, retroposons, Ti plasmids, insertional mutagenesis, Tn, transposable elements, Ty)

INSERTION VECTORS: have restriction enzyme site(s) where foreign DNA can be inserted.

INSERTIONAL INACTIVATION: insertion any type of DNA into an antibiotic resistance or other gene, inactivates the function (e.g. becomes antibiotic sensitive) because of the disruption of the gene's continuity. (See insertional mutation, pBR322)

INSERTIONAL MUTATION: insertion of a transposable genetic element within a gene disrupts its function and as a consequence lethal effects or altered functions are displayed. Recently it has been shown that many of the insertions do not lead to observable change in the expression of the genes or their effect is minimal and their presence is revealed only by sequencing of the target loci. The gene carrying the insertion can be selectively isolated by labeled probes for the insert. (See insertion elements, transposons, T-DNA, gene isolation, REMI)

INSOMNIA: see fatal familial insomnia

INSTAR: insect larvae between the processes of molting (shedding the outer cover layer). In *Drosophila* three moltings take place after hatching and before pupation. The first and second instars last for about one day each, whereas the third instar for about two days. (See *Drosophila*)

INSTINCTS: a variety of innate behavioral patterns that are developing without learning although learning may reinforce them, e.g. motherly love, nursing, fearing for life, conform to some standards, etc. (See behavior genetics, ethics, ethology, aggression)

INSULATOR: DNA element(s) that prevent(s) interactions between enhancers and inappropriate target promoters and they may permit the expression of stably integrated transgenes.

INSULIN: is one of the most important peptide hormone (MW≈5,700) in the body and it is synthesized in the pancreas. The pancreas is a large gland behind the stomach. Inside, the Langerhans islets produce and secrete this hormone into the blood. Insulin regulates glucose uptake into the muscles by glucose transporters, and into the liver by glucokinase. In the muscles and the liver, with the assistance of glycogen synthase, glycogen is made. The breakdown of glycogen is mediated by the insulin-regulated glycogen phosphorylase. Glycolysis and acetyl coenzymeA synthesis is boosted by insulin through the enzymes phosphofructokinase and the pyruvate dehydrogenase complex. Fatty acid synthesis in liver is promoted by acetyl-CoA carboxylase and neutral fat (triaglycerol) is stimulated by insulin with the aid of lipoprotein lipase. The deficiency of insulin leads to the hereditary diabetes mellitus.

Diabetes is a very complicated disease, controled by several genes involved either in the differentiation of the Langerhans islets or at various steps in the synthesis and regulation of the hormones. About 5 to 10% of the population in western countries is involved in one or another form of this disease. Of these, however, only 1/10 is insulin-dependent. The insulin-dependent diabetes (IDDM) is largely familial and occurs early in life. Those who become diabetic after age 40 are generally affected by the non-insulin-dependent form of the disease (NIDdM). The latter has a relatively small hereditary component, and it is most common among obese individuals between age 50 to 60. One of the most important clinical manifestation of the disease is hyperglycemia (excessive amount of sugar in the blood), but there are many other complex characteristics. Actually, about 60 known human diseases are accompanied with diabetes symptoms. The early onset diabetes is managed by a well-controlled diet and continuous supply of insulin. The treatment of the other forms (NIDDM) varies according to the causative metabolic defect.

The primary structure of insulin was determined first by the double Nobel-laureate, F. Sanger in 1953, the same year as the Watson - Crick model of DNA was published. These two events signaled the beginning of molecular biology. Insulin is synthesized as a preproinsulin equipped with a "signal sequence" that direct the molecules into the secretory vesicles. After removal of this signal peptide, proinsulin is formed that is stored in the β cells. When the blood glucose level increases insulin-specific peptidases process the protein into the functional final form. The A chain of bovine insulin contains 21 amino acids and the B chain has 30, the two chains are joined by disulfide bridges between the A7 and B7 as well as A20 and B19 residues and a third disulfide bridge is formed within the A chain between residues 6 and 11 (see protein structure). The A chain is the same in humans, dogs, pigs, rabbits, and the bovine B chain is identical with that of pig, goat, dog and horse. The function of insulin requires the presence of a receptor.

The insulin receptor protein is made of two α and two β chains. The identical α chains bind insulin above the surface of the plasma membrane, whereas the two β chains reach inside the cell through the membrane with their carboxy termini. Upon binding of insulin, the β chains become a specific protein tyrosine kinase that first autophosphorylates and then phosphorylates other proteins in a cascading series of events. As a consequence a number of enzyme activities are altered by phosphorylation of tyrosine and serine residues. These multiple reactions explain why diabetes occurs in many forms and is part of another dozens of syndromes under the

Insulin continued

control of a variety of genes and manifested also in aneuploidies (Turner syndrome, Klinefelter syndrome, Down syndrome). Insulin receptor substrate-1 tend also to the integrin family of surface receptors and this way integrin and growth factor signaling are connected.

The various forms may be under the control of autosomal dominant genes (diabetes insipidous nephrogenic Type II, diabetes mellitus juvenile with early onset). The latter one was attributed in one case to a substitution of serine for phenylalanine at the 24th position of the β chain. Only a single locus seems to encode the human insulin structural gene in chromosome 11p15.5 The human INS and the mouse *Ins* (chromosome 7) are expressed only from paternal allele at some stages of the development (imprinting). In rats and mice two distinct loci are involved. Some mutations interfere with the proteolytic processing of proinsulin, resulting in its accumulation (hyperproinsulinemia) and without diabetes symptoms.

Non-insulin-dependent diabetes was observed to be associated with a mutation of the glucose transporter-4 protein at residue 383 (Val [GTC]→ Ile [ATC]) in some cases but not in others. Vasopressin and oxytocin deficiency may also be involved in diabetes. These two octapeptides regulate muscle contraction and renal secretion. Diabetes is generally associated with abnormalities of kidney functions.

Autosomal recessive inheritance appeared to account for the juvenile-onset diabetes mellitus Type I and this locus was assigned to human chromosome 6 in close proximity to the HLA Class II genes. This locus is probably a regulatory one, determining insulin susceptibility with a penetrance of 71% for the homozygotes and 6.5% for the heterozygotes. It was reported that if at position 57 of the HLA DQ-β chain has Asp, most likely diabetes was absent whereas among diabetics there was a high chance that this 57th residue was non-Asp. Other studies indicated that HLA genes DR3 and DR4 and HLA-DQ have predisposing effects on the expression of diabetes. Actually, it appeared that HLA-DQw1.2 allele was protective but HLA-DQw8 increased the chances of developing insulin-dependent diabetes. Population genetic studies also confirmed that having a non-polar amino acid at position 57, the chance to develop diabetes may increase by a factor of about 30.

X-chromosomal linkage was observed for nephrogenic diabetes insipidus Type I, and it was suggested that this form of the disease is associated with a defect in the vasopressin 2 (V2) receptor.

Insulin for therapeutic purposes, up to recent times, was obtained exclusively from animal pancreas collected at the slaughterhouses. By the use of recombinant DNA human insulin can industrially be produced with aid of bacterial cultures and further processing. The industrial production of human insulin by genetic engineering is more successful than that of other proteins because it does not require glycosylation. In yeast, proinsulin mRNA is normally produced but the translation of the message is very inefficient. The human insulin has a definite advantage for some patients who may be allergic to the slightly different animal protein. Lower than normal insulin resistance is frequently associated by infections, diabetes and some forms of cancer. Insulin resistance is mediated by tumor necrosis factor α and it decreases tyrosine kinase activity in the insulin receptor by activating serine phosphorylation. (See diabetes, HLA, TNF, obesity, insulin-like growth factor, insulin receptor)

INSULIN-LIKE GROWTH FACTORS (IGF-1, IGF-2): are required for passage of the cell cycle from G1 to S phase; promote cell maintenance, metabolism, cell division. The human gene for IGF-1 and IGF-2 are located in chromosome 11p15-p11, separated by about 12-13 kbp. The human IGF-2 gene is expressed only from the paternal chromosome. In mice, the IGF genes (chromosome 7) display tissue-specific imprinting and also expressed from the paternal chromosome. Somatomedin, a mammalian second messenger, also has insulin-like functions in regulating bone and muscle growth in conjunction with the pituitary hormone receptor. A minor 4.8 kb mRNA generates the prepro-IGF2 whereas the major 6 kb mRNA encodes a post-transcriptionally regulated IGF. (See signal transduction, insulin, imprinting, growth factors, pituitary dwarfness, achondroplasia, somatomedin, mannose-6-phosphate receptor, Simpson-Golabi-Behmel syndrome)

INSULINOMA: a relatively benign pancreatic cancer leading to excessive secretion of insulin

INSULIN RECEPTOR PROTEIN: is a tyrosine kinase; its deficiency is the Donohue syndrome. (See insulin, Donohue syndrome)

INSULIN-RECEPTOR SUBSTRATES (IRS1, IRS2): after multiple-site tyrosine phosphorylation they bind to and activate phosphatidylinositol-3'-OH kinase and other proteins with SH2 domains. (See insulin, insulin-like growth factor, SH2)

INT ONCOGENE: INT1 was assigned to human chromosome 12q13 and to mouse chromosome 15. The product of *Drosophila* gene *wingless* (*wg*) is a homolog and therefore the gene is sometimes mentioned as WNT. INT4 (human chromosome 17q21-q22, mouse chromosome 11) is also homologous to the *Drosophila wg* gene. In mammary carcinomas INT1 is frequently the target site for insertion and inactivation. INT2 is also a mammary tumor gene, and it was assigned to human chromosome 11q13 and its only relation to INT1 is its presence in mammary tumors. INT2 product shows some relationship to fibroblast growth factor 6. INT3, another mammary tumor gene, encodes a transmembrane protein. (See oncogenes, FGF, transmembrane proteins, morphogenesis in *Drosophila*)

INTASOME: a bacterial nucleoprotein complex, including a negative supercoiled phage DNA, wrapped around by several copies of the phage-encoded integrase and the bacterial integration host factor protein (IHF). (See integrase, IHF)

INTEGRAL MEMBRANE PROTEIN: a protein in the membrane without covalent linkage to it and it is bound there by about two dozen uncharged and/or hydrophobic amino acids.

INTEGRASE: a protein that specifically recognizes transposon or phage integration sites and opens them up for insertion to take place. (See integration, transposable elements, lambda phage [site 27815], INI1, IHF, intasome)

INTEGRATED CIRCUIT (chip): is an electronic circuit within a single piece of semiconducting material. (See semiconductors)

INTEGRATED MAP: based on the combined information on genetic linkage and physical mapping. (See mapping genetic, physical map)

INTEGRATING VECTOR: see yeast integrating vector

INTEGRATION: a DNA sequence is inserted at both termini by covalent linkage into the host DNA. DNA structure (bends) affects the site of integration. HIV integrates non-randomly either into purified naked DNA or preferentially into CpG stretches modified by cytosine methylation and within distorted nucleosomal DNA. The murine leukemia virus (MLV) integrates non-randomly with preference into the major groove of the nucleosomal DNA. Retroviruses most commonly select DNase hypersensitive sites and stretches involved in active transcription. Agrobacterial vectors are most commonly integrated into the non-translated regions of potentially expressed genes and similar predilections were observed in the bacterial transposons. The different yeast Ty elements also display preferences. (See Ty, T-DNA, DNase hypersensitive sites, MLV, HIV, transposons)

INTEGRATION: in mathematics is the finding of a function of which the integrand is a derivative or finding an equation among finite variables that is the equivalent of the differential equation integrated. $F(x) = \int f(x)dx$ and $\frac{dF(x)}{dx} = f(x)$. (See derivative)

INTEGRATION HOST FACTOR: see IHF

INTEGRATION PLASMID: has homologous sequences to the chromosome and after transformation, by recombination, it may be inserted in multiple copies into the yeast chromosome. Linearization of the plasmid my increases its efficiency of integration by 10-1000 fold.

INTEGRATIVE CIRCUITS IN SIGNAL TRANSDUCTION: multiple kinases initiate a common response. (See signal transduction)

INTEGRATIVE SUPPRESSION: elimination of the manifestation a genetic defect by insertion of a normal copy of the gene.

INTEGRATOR GENE: a hypothetical regulator and coordinator of eukaryotic genes.

INTEGRIN (ITG): heterodimeric family of integral membrane proteins that in connection with

Integrin continued
fibronectin and ICAM control cell adhesion and cell shape, signaling, intracellular Ca^{2+}, inositol lipid metabolism. Leukocytes express a variety of integrins that participate in inflammatory and immune responses. Integrins have an α and a β subunit that are translated separately and their association is not by covalent linkage. Some of the integrin α and β subunit genes were assigned to human chromosome 2, whereas β -7 integrin appears to be in chromosome 12. The α-6 integrin forms 991 residue extracellular, 23 amino trans-membrane and a 36 amino acid intracellular domains. Integrin is associated in the cells with a 59K serine/threonine protein kinase, containing four ankyrin-like repeats. This integrin-linked kinase (ILK) regulates probably integrin-mediated signal transduction. Integrins, by binding to ligands, may control substratum adhesiveness and thereby cell migration. The integrin-mediated signal transduction pathway may be represented (After Clark, E.A. & Brugge, J.S. 1995 Science 268:233):

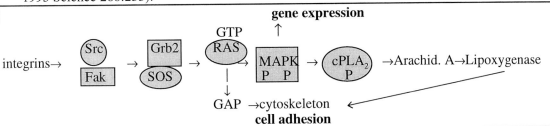

(For symbols see individual entries, Arachid. A = arachidonic acid, see fibronectin, cadherin, ICAM, ankyrin, signal transduction, angiogenesis, epidermolysis, atresia, cell migration)

INTEGRON: is a mobile DNA element, a transposon with a cassette, flanked by a 5' and a 3' conserved sequence. The internal cassette can accommodate antibiotic resistance genes by site-specific recombinase (integrase) located in the 5' element. It contains several rightward (P2, P2 , P4 and P5) and one leftward (P3) promoter sites and ribosome-binding site. The 3' element includes a conserved sulfonamide resistance gene. The origin of the sulfonamide resistance is based on the lack of sensitivity of dihydropteroate synthase to sulfonamide inhibition. Its origin is unclear since sulfonamides are synthetic antibiotics. Dihydropteroate is a precursor of folic acid. The integron is the vehicle of antibiotic resitance genes among various types of bacteria. (See integrase, folic acid, transposon, R plasmid)

INTEGUMENT: the maternal somatic tissue layers that surround the ovule of plants and give rise to the seed coat; thus it may show delayed inheritance. (See megagametophyte development, delayed inheritance)

INTEIN: elements inserted into proteins before completion of its sequence and removed after. They are spliced out post-translationally by autocatalytic proteolysis and ligation. Some inteins are site-specific endonucleases. They may also function as transposons and insert their coding sequences into intein-less genes. Inteins show different structures and they occur in prokaryotes, algal chloroplasts and other lower eukaryotes. (See also intron)

INTELLIGENCE QUOTIENT (IQ): see human intelligence

INTENSIFIER: an animal or plant gene that intesifies (darkens) color.

INTERACTION VARIANCE: is due to epistasis between quantitative traits and the effect of the environment on gene expression. (See analysis of variance, epistasis)

INTERALLELIC COMPLEMENTATION: see allelic complementation

INTERBAND REGION: in the polytenic chromosomes the relatively lighter stained space between the characteristic darkly stained bands. (See figure at coefficient of crossing over)

INTERBREEDING: within a population individuals of different genotypes may mate. (See random mating, mating systems)

INTERCALATING MUTAGENS (such as acridines, some nitrogen mustards, etc.): can insert within nucleotide sequences and cause frameshift mutations, short insertions and deletions. Intercalation causes also separation of the base pairs, lengthening and untwisting of the double helix. (See acridine dye, nitrogen mustards)

INTERCELLULAR: situated in between cells. (See also intracellular)

INTERCEPT: see correlation [in a linear regression a = Y - bx]).

INTERCHANGE: reciprocal translocation of chromosomes. (See translocations)

INTERCHANGE TRISOMIC: see trisomic tertiary, trisomic analysis

INTERCISTRONIC REGION: the number of nucleotides between the end of one gene and the beginning of the next one.

INTERCROSS: mating between individuals (siblings) of the same parentage.

INTERFACIAL ENZYMOLOGY: the enzymes act on substrates located on a surface and thus their activity is regulated also by the concentration of the surface. Such enzymes may reside within or on cellular membranes.

INTERFERENCE, CHROMATID: see chromatid interference

INTERFERENCE, CHROMOSOME: one crossing over may either reduce (negative interference) or increase (positive interference) the occurrence of additional ones. (See coincidence, mapping function, chromatid interference)

INTERFERON: specific glycoproteins developed after viral infection or as a reaction to RNA or other compounds, and have antiviral activity and possibly also antitumor activity. Interferons have three major forms, IFN-α, -β and -γ. Interferons can be produced after transformation in a variety of cells (yeast, silkworm, mouse, hamster, etc.). Interferons produced by leukocytes contain predominantly the α type. Lymphoblastoid cells (lymphocytes stimulated by an antigen) have 90% α and 10% β interferons. Induced fibroblasts contain mainly the β chain. The γ interferon is produced by antigen- or mitogen-stimulated T lymphocytes. Interferon induced protein genes have been located to human chromosomes 21, 10, 4, and 1. Intereferons may interfere with protein synthesis by causing phosphorylation of eukaryotic peptide chain initiation factor eIF-2.

Interferon α (leukocytic interferon, IFNA) genes (up to 30) were located to human chromosome 9p21-p13. An *interferon α receptor* (antiviral protein, AVP) was assigned to human chromosome 21q22. Translocations of INFA have been identified in leukemia patients. It has been claimed that intranasal use of interferon α may prevent the common cold.

Interferon β-1 (IFB1, human chromosome 9p21-pter), is structurally homologous to α interferon. Patients with acute monocytic leukemia displayed translocations of IFB1 to chromosome 21 and the breakpoint was within this gene at about 17 cM from the site of the ETS-2 oncogene (chromosome 21q22.1-q22.3). Actually, in chromosome 9 several interferon genes occur. *Interferon β-2* (IFNB2, human chromosome 7p21-p15) is induced by tumor necrosis factor (TNF) and interleukin (IL1) when interferon β-1 (IFNB1) is not induced. It is identical with the B-cell differentiation factor (BSF2) and hybridoma growth factor. *Interferon γ* (IFNG0, human chromosome 9) induces the expression of HLA Class II genes. The induction is modulated by a factor in human chromosome 16 (probably a receptor) and by another in chromosome 21 that may control the transduction of the γ interferon signal. In RAG (recombination activating) cells a human chromosome 6 factor was also required in human-rodent cell hybrids. IFNG induces the production of a 98-residue polypeptide (IP-10, chromosome 4q21) and other activating polypetides. The interferon regulatory factor (IRF) family includes the interferon consensus sequence binding protein (ICSBP, expressed constitutively in B lymphocytes) and other transcription factors (IRFs). Their N-terminal region binds to DNA (IFN-stimulated responsive element, ISRE) and the C-terminal contains the regulatory sequences. (See also lymphokines, leukemia, interferon response element, B cell, hybridoma, leukemia, modulation, signal transduction, receptor protein, IRF, interleukin)

INTERFERON RECEPTORS: IFNα and β share the same receptor (63.5 kDa), encoded in human chromosome 21. The IFNγ receptor binds its ligand at the 245 amino acid extracellular domain whereas the 222 amino acid intracellular domain is involved in signal transduction. (See interferon, interferon regulatory factors, signal transduction [Jak-Stat]).

INTERFERON REGULATORY FACTORS (IFR): bind to the upstream regions of both α and

Interferon regulatory factors continued
β interferon genes and serve as a transcription activator (IRF-1) or IRF-2 hve an antagonistic effect. They compete for the same cis element. Both factors were assigned to human chromosome 5p23-q31. The IFR-2 protein is identical with HiNF. (See interferon, upstream, cis-acting element, leukemia, macrophage, cytokines, transcription factors, CCE, histones, HiNF)

INTERFERON SEQUENCE RESPONSE ELEMENT (ISRE): AGTTTCNNTTTCN[C/T]. (See GAS, signal transduction Jak-Stat, interferon, interferon receptors)

INTERFERON-INDUCED PROTEINS: are located in different human chromosomes 1, 2, 4, 10, 12, 21. Interferon-inducible cytokine IP-10 (human chromosome 4q21) is located close to the break-point that is associated with monocytic leukemia. (Monocytes are phagocytic mono-nuclear leukocytes that develop into macrophages in lung and liver). IP-10 has substantial homology to several activating peptides and it may control inflammatory responses. (See cytokines, leukemia, macrophage)

INTERGENIC COMPLEMENTATION: is an evidence that the two genes are not allelic and belong to separate loci. (See also allelic complementation, allelism test)

INTERGENIC SPACERS (IGS): are sequences (↓) between rRNA genes that have very short transcripts (appear like a feathers) but may be contributing to initiation of DNA replication before the beginning of embryonic transcription. (See ITS, tRNA, pol III)

(Courtesy of Spring *et al.* J. Microsc. Biol. 25:107)
INTERGENIC SPACERS ⟋

INTERGENIC SUPPRESSOR: one mutation suppresses the expression of another situated in a different locus. Most commonly the suppressor encodes a mutant tRNA. (See suppression)

INTERKINESIS: see interphase

INTERLEUKINS: proteins secreted by white blood cells, phagocytes, B lymphocytes and are involved in stimulating growth and differentiation of lymphocytes concerned with the natural defense system of the body. (See lymphokines, IL-1, IL-2, IL-3, IL-4, IL-16)

INTERLOCKING BIVALENTS: when another chromosome passes through the terminalizing chiasmata (ring bivalents), the non-homologous chromosome may be trapped (interlocked) within the ring. (See ring bivalent)   ⇐ centromere of the chromosome

INTERMEDIARY METABOLISM: enzymes within the cells produce energy from nutrients and use it to synthesize other compounds or organize cellular components.

INTERMEDIATE FILAMENTS: are ubiquitous 10 nm protein filaments abundant in the eukaryotic cells and are encoded by at least 50 genes. They are composed of keratins, desmin, vimentin, neurofilament proteins, glial fibrillary acidic protein (GFAP), lamins, etc. Their anomalies may result in epidermolysis, keratosis and possibly other skin diseases. (See keratin, keratosis, epidermolysis, desmosome, vimentin, lamins, skin diseases)

INTERMEDIN: a melanocyte stimulating protein factor. (See melanocyte)

INTERNAL CONTROL REGIONS (ICR): see A box

INTERNAL MEMBRANE: membranes within the cell excluding the plasma membrane.

INTERNAL PROMOTER: see promoter

INTERNAL TRANSCRIBED SPACERS: see ITS, ETS, tRNA

INTERNALLY ELIMINATED SEQUENCES (IES): during the formation of the macronuclei of *Ciliates* chromosome diminution and fragmentation occurs. The DNA deleted involves repetitive sequences but the IES are not parts of the repetitive sequences, and could include even functional genes. (See chromosome diminution, *Paramecia*, *Ascaris*, macronucleus)

INTERNET: a complex system of interconnected electronic communication networks.

INTERNODE: a segment of plant stem between nodes (somewhat enlarged structures from which leaves originate).

INTERORGANISMAL GENETICS: see Flor's model

INTERPHASE: is the part of the cell cycle  G1, S, G2, M G1, S, G2, M when mitosis (M) or meiosis is not in progress. During interphase between mitoses the cells are actively

Interphase continued

synthesizing DNA (S phase) and during the G phases other molecules are made and the cellular organelles are dividing. In meiotic divisions DNA synthesis is limited only to repair and all the DNA is made during the interphase preceding meiosis. (See cell cycle, mitosis, meiosis)

INTERPOLATION, LINEAR: see the procedure at entry F_2 *linkage estimation*

INTERRACIAL HUMAN HYBRIDS: the genetic distance based on gene frequencies is relatively small (see evolutionary distance) among the various human races and despite the theoretical expectation of deleterious effects of breaking up co-adapted genetic sequences, interracial hybrids suffer apparently no physical harm. Problems may arise, however, in cultural adaptation because the hybrids generally are classified socially with the minority race (whatever the majority is) and discrimination against minorities is not uncommon in all racial, ethnic and cultural groups. Despite the great similarities of the genetic structure of all primates, no hybrids between humans and other primates have been verified. (See human races)

INTERRUPTED MATING: stopping bacterial conjugation in definite intervals (by stirring the culture) in order to determine the order of gene transfer and establish map position on the basis of minutes required for transfer of particular gene(s) from donor to recipient (See conjugation mapping)

INTERSEX: the true intersex types have both male and female gonads, a very rare condition. More common are those that have either female or male gonads and chromosomal constitution but they express, to various degrees, the secondary sexual characteristics normally unexpected for their chromosomal constitution. The intersex phenotype is determined by autosomal genes and in species where the sex chromosome:autosome ratio determines sexuality, the aneuploids appear as intersex types. In *Drosophila* the *tra* (*transformer*, chromosome 3-45) homozygotes of XX chromosomal constitution are sterile males. The *tra2* (chromosome 2-70) mutation in XX background has similar effects as *tra*; in XY background the males look normal and behave normal yet their sperm is not motile. The *dsx* (*doublesex*, 3-48.1) locus has numerous alleles. The homozygotes for their null alleles (in either XX or XY background) and the heterozygotes for the dominant alleles (in XX background only) are intersexes. The *ix* gene (*intersex*, 2-60.5) makes females intersex with reduced male and irregular female external genitalia. The *ix/ix* males look normal but they are homosexual largely and the *ix/+* heterozygotes court females and young males but not adult ones. The *tra* and *dsx* genes regulate sex expression by alternative splicing of the RNA transcript involved in normal sexuality. In some insects, like the Gypsy moth (*Lymantria dispar*) the sex-determining genes have "strong" and "weak" alleles in some populations and the crosses yield regularly intersex individuals. The crosses between head lice (*Pediculus capitis*) and body lice (*P. vestimenti*) produce intersexes in F_2 and F_3. 98 to 96% of angiosperm plants are hermaphroditic and among the 2 to 4% dioecious species intersexes occur, depending on the number of each of the sex chromosome they carry. E.g., in *Melandrium* the XXY and XXXY males produce occasionally intersex flowers but the XXXXY individuals are hermaphroditic. (See hermaphroditism, sex determination, introns, gynandromorphs, homosexuality)

INTERSPERSED REPETITIOUS DNA: see SINE, LINE

INTERSTITIAL SEGMENT: the chromosomal region between the centromere and the translocation break point. (See translocation)

INTERVAL MAPPING: considers pairs of adjacent markers and maximizing the likelihood for quantitatively expressed gene loci (QTL) being in between. ELOD is the expected lod score, θ = recombination fraction, p = the proportion of variance contributed by QTL, n = sample size. The calculation may be difficult under practical conditions because there may be more than one QTL per linkage group. An interval mapping based on the least squares method may be better. (See QTL, lod score, least squares, ASP analysis)

$$\text{ELOD} = \left(\frac{1-2\theta}{1-\theta}\right) \frac{n}{2} \log\left(\frac{1}{1-p}\right)$$

INTERVARIETAL SUBSTITUTION: is basically similar to alien substitution except the chromosomes belong to different varieties of the same species. (See alien substitution)

INTERVENING SEQUENCE (IVS): see intron

INTIMATE PAIRING (synapsis): the very close apposition of the chromosomes at the meiotic prophase that makes possible crossing over (gene conversion) and recombination. (See recombination mechanisms eukaryotes, recombination molecular mechanisms prokaryotes, recombination models, synapsis)

INTINE: the inner layer of the wall of the pollen grain.

INTRACELLULAR: being within cell(s).

INTRACELLULAR CLOCK: differentiation of particular cells into a particular type of tissue during embryonal development is controled by the timing of the signal for differentiation received. E.g., an animal epithelium excised at the gastrula stage grafted into the eye disc of an embryo may differentiate into a neural tube but if the same tissue if grafted to the same position a few hours later it may differentiate into an eye lens.

INTRACHROMOSOMAL RECOMBINATION: see sister chromatid exchange, transposition

INTRACLASS CORRELATION: a form of analysis of variance used for the estimation of heritability on the basis of variances between and within classes, e.g. in the progeny of a larger number of different males mated to a smaller number of females, each of which has several offspring. Thus a comparison can be made how much is the variance within the litter of a female mated (sired) to the same male (within "sires") and what is the variance among the total offspring of individual males mated to different females (between "sires"). The intraclass correlation of the sires is 1/4 of the heritability for the trait considered because each male contributes half of the genetic material to the offspring quantitatively analyzed, and these again will contribute only half of their chromosomes by their haploid sperm or egg. The procedure of calculation is illustrated by a hypothetical example below. (See correlation, heritability, variance, variance analysis)

PROCEDURE FOR CACULATING HERITABILITY BASED ON **INTRACLASS CORRELATION**. THE MEAN SCORES OF THE OFFSPRING ARE REPRESENTED IN THE BODY OF THE TABLE. (Y STANDS FOR INDIVIDUAL OR GROUP MEASUREMENTS)

Males → Females ↓	A Y_i $(Y_i)^2$	B Y_i $(Y_i)^2$	C Y_i $(Y_i)^2$	D Y_i $(Y_i)^2$
E	2 (4)	3 (9)	3 (9)	4 (16)
F	3 (9)	2 (4)	4 (16)	3 (9)
G	3 (9)	3 (9)	4 (16)	2 (4)
H	4 (16)	2 (16)	3 (9)	4 (16)
I	3 (9)	4 (16)	4 (16)	6 (36)
J	5 (25)	2 (4)	5 (25)	5 (25)
Sum Y_i	20	16	23	24
Sum Y_i^2	72	46	91	106

Sum (Σ) all Y_i = $Y_{..}$ = 83,   Sum ($\Sigma\Sigma$) all Y_i^2 = 315,   n (all measurements) = 24,
n_i (no. of families of females) = 6

Correction factor (C) = $Y^2_{..}/n$ = $83^2/24$ = 6889/24 = 287

Uncorrected Sum of Squares ($\Sigma Y_i^2/n_i$) = $(20^2 + 16^2 + 23^2 + 24^2)/6$ = 1761/6 = 293.5

Sum of Squares "Between Males" = SS_S = $(\Sigma Y_i^2) - C$ = 293.5 - 287 = 6.5

Sum of Squares "Within Males" = SS_P = $\Sigma\Sigma Y_i - (\Sigma Y_i^2/n_i)$ = 21.5

Mean Square (MS) is SS/df
(The lower index S stands for "Sires", the lower index P is for progenies)

ANALYSIS OF VARIANCE

	df	SS	MS
Between Males (SS_S)	3	6.5	2.17
Within Males (SS_P)	20	21.5	1.075

Intraclass correlation continued

$$\hat{\sigma}^2_S = \frac{MSS - MSP}{n_i} = \frac{2.17 - 1.075}{6} = 0.1825 \text{ (the male's variance component)}$$

$$\hat{\sigma}^2_P = MSP = 1.075 \text{ (the progeny variance component)}$$

The Males Intraclass Correlation (r_1), $\hat{\sigma}^2_S/(\hat{\sigma}^2_S + \hat{\sigma}^2_P)$ is equal to 1/4 of the heritability, hence $h^2 = 4\hat{\sigma}^2_S/(\hat{\sigma}^2_S + \hat{\sigma}^2_P) = 4[0.1825/(0.1825 + 1.075)] = 0.58$

(heritability is denoted by h^2 for historical reasons but it is not the second power of an entity)

INTRACYTOPLASMIC SPERM INJECTION: see ICSI

INTRAFALLOPIAN TRANSFER: of gametes (GIFT) or zygotes (ZIFT) are infertility treatment procedures whereby the spermatozoa and the mature oocytes or *in vitro* generated zygotes, respectively are placed surgically into the fallopian tube of the female where fertilization and/or segmentation may proceed. These procedures have higher rate of success for conception than the *in vitro* fertilization but these too lead frequently to twinning if more than single egg or zygote is used. (See ART, *in vitro* fertilization, TET, PROST)

INTRAGENIC RECOMBINATION: is rare event because alleles of a locus are very close to each other. Intragenic reciprocal recombination is expected to yield wild type and double mutant recombinants that can be verified only if flanking non-allelic markers are available. These outside markers — ideally — should not be more than 5-10 map units at both sides of the locus. When representing the mutant alleles by m^1, m^2 and m^3, respectively. The p, t and a are outside markers. The + sign indicates non-mutant. The following crosses are required to determine linear order of the m alleles in a simple case:

TESTCROSSES	RECIPROCAL RECOMBINANT PHENOTYPES
$\dfrac{p\ m1\ t+\ a}{p+\ m2\ t\ a+}$ X $\dfrac{p\ m\ t\ a}{p\ m\ t\ a}$	$p\ m\ t\ a^+$ and $p^+\ m^+\ t^+a$
$\dfrac{p^+\ m2\ t\ a^+}{p\ m3\ t^+\ a}$ X $\dfrac{p\ m\ t\ a}{p\ m\ t\ a}$	$p^+m^+\ t^+a$ and $p\ m\ t\ a^+$
$\dfrac{p\ m1\ t+\ a}{p+\ m3\ t\ a+}$ X $\dfrac{p\ m\ t\ a}{p\ m\ t\ a}$	$p\ m^+\ t\ a^+$ and $p^+\ m\ t^+\ a$

Among the recombinants, the *m* phenotype indicates double recessive alleles in the same strand whereas the *m*⁺ phenotype is an indication of recombination between the two recessive alleles present in the heterozygous parent which is testcrossed. According to these results the order of the mutant alleles and the markers must be $\underline{p - m^3 - m^1 - m^2 - t - a}$ and none of the recombinant classes are supposed to be contaminants because the markers are consistent with the recombination events suggested. The double mutant recombinants may be further tested by recombination to yield the two mutant classes.

INTRAGENIC SUPPRESSOR: see suppressor gene

INTRAUTERINE FERTILIZATION (IUI): the sperm is deposited directly into the uterus, bypassing the cervix (the anterior part of the uterus which may form a barrier to the passage of the sperm). This type of medical intervention may overcome human infertility. (See ART)

INTRINSIC RATE OF NATURAL INCREASE: see ages-specific birth and death rates.

INTRINSIC TERMINATORS: of transcription in prokaryotes require no cofactors (are rho-independent), for the termination of transcription. (See transcription termination in prokaryotes)

INTROGRESSION: the transfer of genes from one group of the species into another. The two populations may be inhabiting the same area (sympatric) or they have only occasional contacts because they live in a different area (allopatric). After the initial crossing the mating continues generally only within the group therefore only a few of the "borrowed" genes are maintained.

INTROGRESSIVE HYBRIDIZATION: aimed to accomplish introgression. (See introgression)

INTRON HOMING: the process of insertion of an intron at a particular site within an intron-less

Intron homing continued

homologous site. Accordingly, some introns of yeast mitochondria serve also as mobile genetic elements besides having ribozyme function. The homing introns, mitochondrial or nuclear, are endonucleases with similarities to restriction enzymes but their recognition sequence is much longer. (See introns, ribozyme, mtDNA)

INTRONS: are nucleotide sequences within a gene that are not represented in the mature RNA transcripts of that gene (intervening sequences). Introns are transcribed but not translated into protein that would be part of the products of the exon-coded genes or will not be included into the final RNA encoded by the gene. Introns separate the coding sequences of the exons. Introns occur in the majority of the eukaryotic genes, including genes of mitochondria and chloroplasts and also in viruses of eukaryotes but they are exceptional in prokaryotic genes. In land plant plastid DNA ca. 20 introns are present and their size varies from 400 to over 1000 bp. The average intron size in mice is about $1,800 \pm 300$, in humans about $3,000 \pm 550$. In the X chromosome the intron size is substantially longer. In imprinted genes the intron number, and particularly size, is much smaller. The RNA maturases in mammals are actually intron sequences within protein genes and assist the splicing of the pre-mRNA transcripts. In algae the cpDNA introns are quite variable in number and size; in *Euglena* 149 introns were detected. In the red alga *Porphyra* there were no introns in the cpDNA. In *Chlamydomonas* the insertion site of an intron corresponds to an *E. coli* rRNA surface domain. Introns may appear as late-comers to the eukaryotic organelle genes since in multiple evolutionary lines the introns are non-homologous either by sequence or location. The thymidylate synthetase gene of T4 bacteriophage has an intron however, and the archebacterial $tRNA^{Leu}$ and the large ribosomal subunit genes have introns. One of the largest introns (64 kb) was found in the human thyroglobulin gene, involved in the regulation of energy metabolism and in the various forms of the disease goiter. The overall profile of a gene can be represented as :

5'- enhancer-promoter - transcription initiation site - leader - exon ♦**intron**♦ exon-termination signal - 3'

The origin of introns is unclear. Some arguments favor their ancient evolutionary presence, preceding the divergence of eukaryotes from prokaryotes. Some introns may be located at critical regions of the gene and dividing it into functional domains or separating α-helices from β-sheets. It has been suggested that increased replication rate is inversely proportional with the number of introns. Indeed, introns are rare in prokaryotes and reduced in number in yeast. Introns may regulate genes controling complex developmental pathways. True, large homeotic genes of *Drosophila* seem to have more introns than other, simple genes. The greater density of introns within genes may be promoted by sexual reproduction that enables them to propagate in a selfish manner through gametes of both parents. Introns could protect against the deleterious consequences of recombination because if such events occur within introns rather than within exons, no harms would ensue to the coding capacity. Introns may also protect against illegitimate recombination by interrupting homoelogous tracts at random sites. Although these problems are not yet resolved the mechanics of transactions concerning introns have made substantial progress.

The number of introns vary a great deal. Some genes e.g., the histone, human α and β interferon genes have no introns, whereas the γ interferon gene has several. The large (2 Mbp) human dystrophin gene has more than 60 introns. (Dystrophin is a muscle protein, deficient in muscular dystrphy). The average size of an intron generally varies between 75 and 2,000 bp but some introns are several times as long or no more than about a dozen nucleotides. No introns occur in the 5S, 5.8S, U RNA, 7SL and 7SK RNA genes. The number of introns is generally small in organisms with compact genomes like *Drosophila* and budding yeast but fission yeast has a larger number of them. Interestingly, the mitochondrial genes of budding yeast have relatively more introns than the nuclear genes.

Introns are removed from the primary transcripts of the genes during processing. Removal of introns is essential for the expression of genes. Group II introns have six domains, and domain 1 (D1) and D5 are essential for splicing. Thus the mRNA becomes much shorter than the primary transcript. When the mRNA is hybridized to the genomic coding strand of DNA, the

Introns continued

latter reveals loops at the positions where from the mRNA the introns were removed. The chicken ovalbumin gene contains seven introns and loops corresponding to their location and length can be detected by electron microscopy (map by Dugaiczyk *et al.*, Stadler Symp. 11:57, and interpretative drawing by Rédei, 1982 ↓). Alternative mRNAs can be obtained from the same DNA sequence by controlling the length, number or pattern of exons used, depending on the site of (i) initiation of transcription, or (ii) on the alternative sites of polyadenylation signals, or (iii) on the selective retention of particular exons in the mRNA. By these mechanisms the products of single genes can be diversified during development or differentiation.

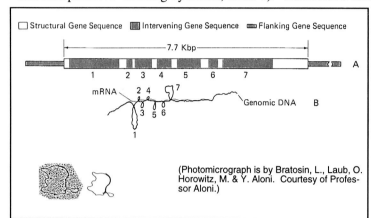

(Photomicrograph is by Bratosin, L., Laub, O. Horowitz, M. & Y. Aloni. Courtesy of Professor Aloni.)

Alternative splicing may be accomplished by using more than a single promoter, resulting in alternative long or short transcripts, depending on the site of transcription initiation. Alternative polyadenylation signals, at more than a single location downstream, may truncate the transcript at the 3'-end.
An example for mechanism (i), alternative promoters; the exons are bracketed, introns are in bold and in parenthesis:

TATA*1*--[1]--**(1)**--*TATA2*--[2]-- **(2)**--[3]-- **(3)**--[4]--**(4)**--[5]--**(5)**--[6] **DNA**

→

mRNA 1: --[1]--[4]--[5]--[6]-- alternative transcripts   **mRNA 2**: --[2]--[3]--[4]--[5]--[6]--

Example for alternative truncation (mechanism ii):

TATA--[1]--**(1)**--[2]--**(2)**--[3]--**(3)**--[4]--**(4)**--*poly A signal I*--[5]--[6]--**(6)**- *poly A signal II*--**DNA**

mRNA 1: -- [1]--[2]--[3]--[4]--AAAAA   or   **mRNA 2** :- -[1]--[2]--[3]--[5]--[6]--AAAAA

Example for mechanism (iii), alternative splicing between identical promoter and polyA signal:

TATA 1--[1]--**(1)**--[2]--**(2)**--[3]--**(3)**--[4]--**(4)**--[5]--**(5)**--[6]--**(6)**--- **DNA**

mRNAs: --[1]--[2]--[3]--[4]--[6]--AAA,--[1]--[3]--[4]--[5]--[6]--AAA,--[1]--[2]--[4]--[5--[6]--AAA
↑   ↑   ↑   ↑   ↑   ↑

Alternative splicing occurs in various eukaryotic genes and various viruses of eukaryotes. In the small genomes of the latter systems a single transcript may permit the production of several proteins. In the determination of correct splicing both cis- and trans-acting proteins cooperate. When the introns are removed the exons are *spliced* together and the continuity of the mRNA is restored. There are two splice sites, the *upstream donor site* and the *downstream acceptor site*. When the invariant bases (shown in bold below) are altered, splicing generally fails. Other neighboring bases may also effect the efficiency of splicing. In animal nuclear genes an A residue in the vicinity of the splice site is also required but its position does not have to be absolutely fixed relative to the splice site. In yeast there is an absolute requirement for a con-

Introns continued

THE INTRON - EXON BORDER SEQUENCES ARE CONSERVED WITHIN GROUPS OF DIFFERENT CLASSES OF RNA TRANSCRIPTS (AFTER CECH, CELL 44: 207) THE BOLD LETTERS INDICATE CONSTANT BASES, THE ↓ INDICATES SPLICE SITES, PY STANDS FOR ANY PYRIMIDINE AND PU FOR ANY PURINE AND X MEANS ANY BASE.

INTRON	5' SPLICE JUNCTION	3' SPLICE JUNCTION
Common nuclear pre-mRNA	CPuG↓**GU**^U_G AGU	(PY)$_n$**AG**↓X
Yeast nuclear pre-mRNA	↓**GU**AUGU	(PY)$_n$**AG**↓X
tRNA	X↓X	X↓X
[1] Introns Group I	U↓	G↓
[2] Introns Group II	↓**GUGCG**	(Py)$_n$**AU**↓
Euglena plastid mRNA	↓**GUG**^C_U **G**	(Py)$_n$**A**^U_C ↓

[1] Nuclear rRNA genes of in some lower eukaryotes, mitochondrial and plastid rRNA genes
[2] Yeast mitochondrial genes for cytochrome oxidase and cytochrome b

served UACUAAC tract within 6 to 59 nucleotides upstream from the 3' splice signal. Some genes have more than a single splice site pairs and that facilitates alternative splicing and the assembly of different mRNAs. Although originally introns were regarded as junk DNA, it is known now that some introns are translated but have independent functions from the exons although some have roles in the processing of the transcript shared by the neighboring exons, others seem to have regulatory functions.

On the basis of the structural information about the splice sites, some special introns have been classified into group I and group II (see table above). The principal characteristics of group I introns are (i) their splicing does not always require a protein enzyme, rather (ii) a short internal sequence facilitates their folding and splicing that is initiated by (iii) an extraneous guanosine or a phosphorylated form of it. The group II introns do not have the conserved internal sequences yet they are capable of fold-back pairing and the splicing requires an intrinsic signal rather than an extraneous guanine. The spliced group II introns form a *lariat* (similar to that of a tethering rope), the 5' P end forms a phosphodiester bond with the 2'-OH group of a nucleotide within the chain at some distance. The loop itself may have three nucleotides (GpApA):

```
                    2' OH
                     |
 5' P-X-O-p-X-O-p-X-O-p-X-O-p-X-O-p-X-O-p-X-OH   intron in the RNA
```

```
            ⌒5' P
lariat →   /      \   2'-5' phosphodiester bond in the lariat
          |  2' O  |
          |   |   /  branch point
           \ X-O-p-X-Op-X-O-p-X-O-p
            ⌣⌣⌣⌣⌣⌣⌣⌣
```

The splicing in group I introns requires *transesterification* (phosphodiester linkage exchanges). The transesterification takes place without severing the bonds first. The self-splicing is mediated by ribozymes, RNAs that have enzyme-like catalytic functions. The ribozymes facilitate the formation of the proper configuration of the RNA transcript as well as may function similarly to endonucleases. Actually the a!2 ribonucleoprotein of the *COX1* yeast

Introns continued

mitochondrial gene catalyzes the cleavage of the DNA target, recognized by complementarity of the sequences within the intron RNA. After cleaving the DNA the aI2 protein reverse-transcribes the intron RNA using the 3' end of the DNA as a primer. Actually, the aI2 RNA cleaves the sense strand of the DNA whereas the aI2 protein cuts the antisense strand and the latter also boosts the activity of the aI2 ribozyme.

SELF-SPLICING OF A GROUP I INTRON (1) A GUANOSINE (OR GUANYLIC ACID) CONDUCTS A NUCLEOPHILIC ATTACK AGAINST A PHOSPHATE GROUP NEAR THE POINT OF JUNCTURE, AND TRANSESTERIFICATION FOLLOWS LATER AT BOTH 5' AND THE 3' JUNCTIONS. THIS ACCOMPANIED BY THE RECOGNITION OF INTERNAL SEQUENCES ↓↓↓ THAT GUIDE THE FOLDING. (2) EXON TERMINI IN BOLD AND THE ENDS OF THE INTRON IS IN OUTLINE LETTERS. (3) SPLICED EXON AND INTRON ARE DISPLAYED.

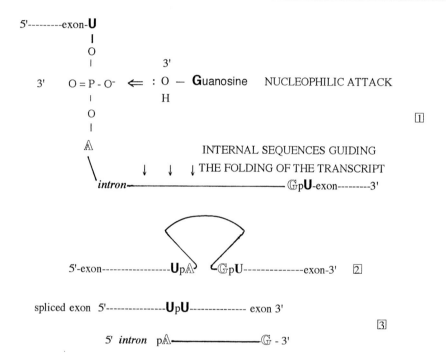

The intron assumes a complex secondary structure by base pairing of some complementary sequences separated by non-pairing tracts. The extraneous guanosine makes then a nucleophilic attack (an electron-rich molecule reacts with an electron poor that is willing to accept electrons) at the exon-intron boundaries resulting in a nucleotide chain breakage at the site and an exclusion of the intron. Not all the group I introns are able to self-splicing without the help of *maturase* proteins encoded by some introns within some yeast mitochondrial genes. The self-splicing of group II introns is somewhat similar only to that of group I. In group II there is no guanine participation. The nucleophilic attack is initiated by the 2'-OH group of an adenine within the intron. Then, after appropriate folding to bring the 5' and 3' junctures to each others vicinity, the exon is spliced by transesterification and the intron is released as a lariat (see lariat). The primary RNA transcript of nuclear pre-mRNAs is capped. Cleavage takes place at the 5' end of the intron upstream of a ↓pGU pair as indicated by the arrow. The free 3' end of the intron then forms a lariat with an internal A residue and in another step the intron with the lariat is released and the 3' end of the first exon is ligated to the 5' end to the next exon. This outcome is just the same as shown above earlier. The splicing of the nuclear pre-mRNAs re-

Introns continued

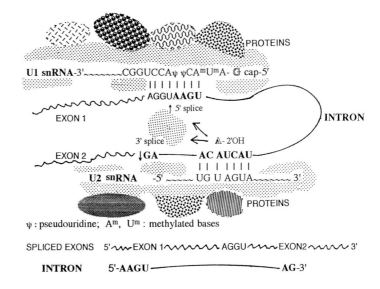

THE U1 AND U2 snRNAs FACILITATE THE PROPER CONFIGURATION OF THE FOLDING OF THE TRANSCRIPT AND THUS IDENTIFY THE CORRECT SPLICING SITES. THE A-2'OH NUCLEOTIDE LEADS THEN TO NUCLEOPHILIC ATTACK THAT IS FOLLOWED BY TRANSESTERIFICATION AND LIGATION OF THE EXON

quires, however, a more complex machinery, involving spliceosomes, complexes of small ribonucleoproteins (snRNP).
The spliceosomes are assembled from short U-rich RNAs, ranging in size from 65 (U7) to 217 (U3) nucleotides and a set of different proteins. The most common U RNAs in the mammalian nucleus are U1, U2, U4 and U6. The U3 RNA is involved in RNA processing in the nucleolus. The U7 RNA assists in the formation of the 3' end of the histone mRNAs that have no introns. The corresponding U RNAs are highly homologous even among taxonomically very different organisms such as humans and dinoflagellates. The homologous U RNAs among vertebrates are almost identical (ca. 95% similarity). The spliceosomes contain a common set of seven proteins (generally designated by capital letters [B, A', etc]) or by molecular weight (e.g. 59K, 25K) and they may have a few specific proteins. About 1% or less of the introns use as terminal nucleotides AU and AC (rather than the most common GU and AG). In plants, (cruciferes) exceptionally AU-AA introns may occur. Some of these introns use also the U12 snRNA for processing. The intron-encoded U22 small nucleolar RNA facilitates the processing of the 18S ribosomal RNA in *Xenopus* oocytes. The U1 snRNA specifically binds to the 5' splice site of the intron and protects this region from RNase attack but it does not cleave at the splice site. The U2 snRNA complex does the same near the 3' end of the intron around the adenosine residue whose 2'-OH is involved in the formation of the lariat. The spliceosomes are ellipsoid complexes of about 25 x 50 μm dimensions that actually bring together the splicing sites to make the cut and paste process possible. These bindings generally require some degree of complementarity but that is not complete. Besides these two spliceosomes, others (U2, U4, U6), and a number of binding proteins may be involved in facilitating and stabilizing the binding.
The position of *introns in tRNA* genes are located next to the third (3') base of the anticodon triplet, with a single base in-between. Yeast has about 40 nuclear-coded tRNAs and 10% of them has a single intron containing from 14 to 46 bases. This regular location is somewhat enigmatic because tRNA^{Tyr} genes of yeast with deleted intron sequences are still transcribed, processed normally and function. In the primary transcript, the intron has, however a triplet that is complementary to the anticodon. For the correct splicing it appears that the various loops and arms of the tRNA cloverleaf are important. The first step in splicing of tRNA transcript is a cleavage by an endonuclease at the ends of the intron. In yeast the 5' terminus of the

Introns continued

exon is then represented by an OH group and the 3' terminus by a 2',3'-cyclic phosphate. Then a cyclic phosphodiesterase breaks up the ring and generates a 3'-OH and a 2'-O-(PO_3) terminus. The 5'-OH end is then phosphorylated by a kinase, requiring ATP. Thus, the termini become ready for joining by an *RNA ligase* enzyme. Essentially similar reactions take place in splicing nuclear tRNAs in mammals and plants.

The splicing mechanisms shown and discussed above involve splicing exons within the same primary RNA transcripts. In some instances exons located originally in different transcripts, even from different chromosomes, may be spliced by *transsplicing*.

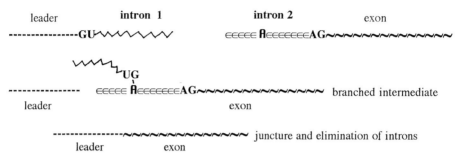

A TRANSSPLICING PATHWAY

Transsplicing can generate mRNAs from leader sequences remote to the exons. Transsplicing is common in the protozoa, *Trypanosomes* where to a single 35 bp leader sequence different exons can be joined.
(The exons are scattered among the approximately 100 chromosomes). The 3' end of the leader and the 5' end of the coding exons can be joined since the 5' end of the intron carries the generally conserved 5' GU sequence, the remote other intron, preceding the remotely transcribed coding exon, terminates at 3' with the conserved AG bases and has also an adenosine nucleophile somewhere in the vicinity of its 5' terminus. Then the two originally remote introns can be brought together by a 5' to 2' bond as a branched molecule, and subsequently both introns are eliminated and the leader is joined to the coding exon as outlined in the diagram just above. (Transsplicing occurs also in other eukaryotes, e.g. in the nematode *Caenorhabditis elegans*, and also in the chloroplast genes of plants. The open reading frames of class II introns have sequences reminding to reverse transcriptases of the type of retroposons. There is no evidence, however, that these would function as reverse transcriptases. (See transcription, ribozymes, endonuclease, ligase, branch point sequence, intron group III, mt DNA, RNA maturase, speckles intranuclear, spliceosome, *Neurospora* mitochondrial plasmids, intein)

INTRONS GROUP I: self-splicing introns of some ribosomal genes, using mechanisms different from the pre-mRNA or the majority of tRNA genes. (See introns for characterization, mitochondrial genetics for introns as mobile elements and plasmids)

INTRONS GROUP II: introns in a few mitochondrial, chloroplast genes differing in structure and splicing mechanisms from the common introns of eukaryotes. They have ribosomal activities. (See introns, mitochondrial genetics)

INTRONS GROUP III (twintron): are relatively short, group II introns inserted within the boundary of group II introns in the cpDNA of *Euglena*. (See ctDNA)

INVAGINATION: folding inward.

INVARIANCE: the reciprocal value of the variance. (See variance)

INVASION OF DNA STRANDS: see branch migration, Holliday model

INVASIVE: penetrating cells (and commonly causing their destruction).

INVERSE POLYMERASE CHAIN REACTION: can be used for molecular analysis of flanking regions of a target: ◊◊◊◊◊-----------ΔΔΔ ← RESTRICTION FRAGMENT INCLUDING FLANKS AND TARGET
the fragment shown above is circularized by its cohesive ends and cut with a restriction enzyme within the target. As a consequence a head-to-tail association of the flanks is produced:

------◊◊◊◊◊ΔΔΔ------
→ ←

This head-to-tail sequence of the flanks is amplified by PCR for nucleotide sequencing. The two flanks may be separated if appropriate restriction endonuclease site is known at or near their boundary. (See PCR, target)

INVERSION: chromosomal segment turned around by 180°;(๏) centromere.
Thus if the NORMAL ARRANGEMENT of the genes is: A B C D E ๏ F G H
after PARACENTRIC INVERSION the order becomes: A **D C B** E ๏ F G H
after PERICENTRIC INVERSION the sequence is: A B ᗡ Ǝ ๏ ƎᗡƆH

Such chromosomal rearrangements require two breaks, both of them in one chromosome arm in case of *paracentric inversions*, and one in each arm across the centromere in case of *pericentric inversions*. Pericentric inversion may alter the arm ratio of the affected chromosome as seen above. Within a single chromosome more than one inversion may be present. These multiple inversions may be independent or may be overlapping or a shorter inversion may be included within a longer one. If the chromosomes are well suited for cytological analysis, the various types can be identified with aid of the light microscope. Inversions as such may not have much phenotypic consequence, unless they cause "position effect", influence the expression of the gene because they interrupt either the coding or the non-translated regulatory sequences. These effects may be serious in the relatively rare inversion homozygotes. During the early years of genetics inversions were thought to be "C factors", crossing over inhibitors. Actually, inversions inhibit crossing over in inversion heterozygotes only in the vicinity of the breakage points where the rearrangement prevents intimate pairing of the chromatids. Crossovers are not observed primarily because the strands involved in recombination are usually not transmitted through the paracentric inversion females thus they do not generate defective gametes. The consequences for the sperms are not the same for paracentric inversion heterozygotes because the microspore tetrad is not linear and the crossover chromatids are not tied by the inversion bridge in such a way that they would not be included in any microspore. There fore in the males all four products of meiosis could potentially be transmitted, however 50% of the gametes are still defective. In pericentric inversion heterozygotes duplication-deficiency gametes are formed (without a bridge) in both females and males if crossing over takes place within the inverted segments during meiosis. Thus crossing over may occur but the crossover gametes or zygotes may not be viable. In animals the consequence of duplication-deficiencies may be different from that in plants. In plants the defective gametes are usually prevented from fertilization because of inviability, whereas in animals the defective sperm may be capable of fertilization but the offspring resulting from such a mating may not survive.

The cytological consequences of crossing over within inverted chromosome segments is diagramed on next 2 pages. Crossing over within pericentric inversions has the same genetic consequence as that of in paracentric inversions, namely 50% of the gametes formed by meiocytes which have suffered recombination are duplication-deficient. I.e., some of the genes are present in the same strand twice and others are entirely absent, and therefore generally inviable. Inviability may be gametic or may be zygotic. In the gametophytes of plants generally the former is the case whereas in animals the latter is prevalent. Crossing over in pericentric inversion heterozygotes cytologically differs from that in paracentric inversions. In the former, dicentric chromosomes and acentric fragments are not generated by recombination. Double crossing over within the same inverted paracentric segment, involving two homologous chromatids does not produce defective gametes. Three-strand double crossing overs yield,

Inversion continued

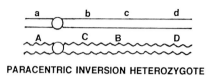

PARACENTRIC INVERSION HETEROZYGOTE

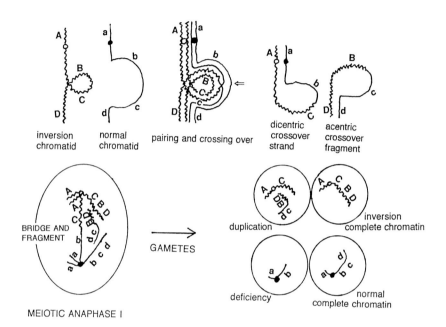

PARACENTRIC INVERSION HETEROZYGOTES PRODUCE 50% DEFECTIVE GAMETES WHEN CROSSING OVER TAKES PLACE WITHIN THE INVERTED CHROMOSOMAL SEGMENT. SINCE THE FREQUENCY OF CROSSING OVER DEPENDS ON THE LENGTH OF THE SEGMENT INVERTED. THE FREQUENCY OF THE DEFECTIVE GAMETES VARIES FROM 0 TO 50%. AT THE TOP, AN INVERTED AND A NORMAL CHROMOSOME ARE SHOWN. SUCH CHROMOSOMES CAN PAIR ONLY IF THE INVERTED STRANDS FORM A LOOP AND THUS GENE-BY-GENE ALIGNMENT BECOMES POSSIBLE. THE CONFIGURATION OF ONE INVERTED AND ONE NORMAL STRANDS ARE SHOWN ON THE SECOND LINE AT LEFT. IN THE CENTER OF THAT LINE THE PAIRING AND CROSSING OVER (⇐) ARE DIAGRAMED. AS A RESULT OF RECOMBINATION, ONE OF THE CROSSOVER STRANDS BECOMES DICENTRIC AND DEFICIENT FOR MARKER (D) AND DUPLICATED FOR MARKER (A/a). THE OTHER CROSSOVER STRAND BECOMES ACENTRIC, AND DEFICIENT FOR A/A AND DUPLICATED FOR D/d. AT MEIOTIC ANAPHASE I THE CROSSING OVER RESULTS IN A CHROMATID BRIDGE BECAUSE OF THE CHROMATIDS ARE TIED TOGETHER EVEN WHEN ANAPHASE SEPARATION PROCEEDS. THE TIE MAY EITHER HOLD TOGETHER THE CROSSOVER CHROMATIDS AND THEY CANNOT GET TO EITHER POLE, RATHER THEY REMAIN IN THE MIDDLE OF THE METAPHASE PLANE OR IF THE TIE BREAKS, DEPENDING ON SITE OF THE BREAKAGE, ADDITIONAL DUPLICATION AND SIMULTANEOUS DEFICIENCY MAY BE GENERATED. IN ANIMALS AND PLANTS, AFTER RECOMBINATION OF PARACENTRIC INVERSION HETEROZYGOTES GENERALLY ONLY VIABLE (NORMAL OR INVERTED) CHROMOSOMES ENTER THE EGG AND THUS FEMALE STERILITY USUALLY NOT OBSERVED. IN THE MALES (WHERE POLARITY DOES NOT OCCUR DURING GAMETOGENESIS), IF RECOMBINATION TAKES PLACE WITHIN THE INVERSION, 25% OF THE GAMETES CONTAIN ONLY NORMAL CHROMOSOMES, 25% CARRY INVERTED CHROMOSOMES CONTAINING COMPLETE CHROMATIN. BUT THE OTHER 50% RESULTS IN DUPLICATION-DEFICIENCY OR DEFICIENCY OF THE GENES, AND SUCH GAMETES ARE GENERALLY STERILE OR RESULT IN EMBRYO LETHALITY. RECOMBINATION IN PERICENTRIC INVERSION HETEROZYGOTES DOES NOT RESULT IN DICENTRIC BRIDGE BUT HALF OF BOTH MALE AND FEMALE GAMETES MAY BECOME DEFECTIVE.

Inversion continued

however both acentric fragments and dicentric chromosomes besides the two parental ones. Four-strand double crossing over results in the formation of two acentric fragments and two dicentric chromosomes and generally all gametes become thus defective (see figure below).

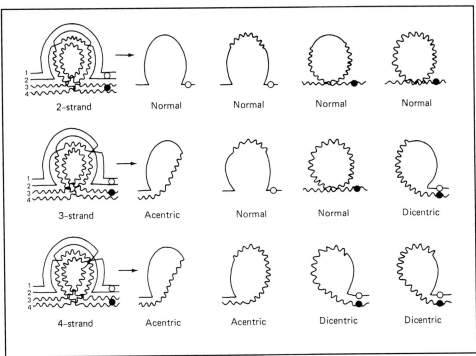

DOUBLE CROSSING OVER WITHIN PARACENTRIC INVERSIONS. ONLY 2-STRAND DOUBLE CROSSING OVER RESULTS IN TWO COMPLETE CHROMOSOMES. PERICENTRIC INVERSION HETEROZYGOTES DO NOT YIELD DICENTRIC STRANDS YET THE OVERALL GENETIC CONSEQUENCES OF DOUBLE RECOMBINATION ARE THE SAME AS IN PERICENTRIC INVERSION HETEROZYGOTES.

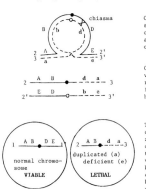

Inverted chromosome segment is marked with lower case bold letters (chromatid ends are numbered).

Chromosome pairing at the 4-strand stage displays a loop. Only two of the chromatids (involved in a chiasma) are shown for the sake of simplicity.

Crossing over in pericentric inversion heterozygotes does not result in a chromatid bridge and fragment yet the recombinants have duplications and deficiencies.

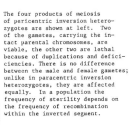

The four products of meiosis of pericentric inversion heterozygotes are shown at left. Two of the gametes, carrying the intact parental chromosomes, are viable, the other two are lethal because of duplications and deficiencies. There is no difference between the male and female gametes; unlike in paracentric inversion heterozygotes, they are affected equally. In a population the frequency of sterility depends on the frequency of recombination within the inverted segment.

TWO-STRAND PARACENTRIC INVERSION HETEROZYGOTE DISPLAYS DOUBLE CHROMATID BRIDGE AND TWO CHROMATID FRAGMENTS. BECAUSE OF THE CHROMATID TIE KEEPS THE CROSSOVER CHROMATIDS IN THE MIDDLE OF THE CELL, THEY FAIL TO BE INCORPORATED INTO THE EGG AND THUS USUALLY FEMALE STERILITY IS PREVENTED.

RECOMBINATION IN PERICENTRIC INVERSION HETEROZYGOTES

Inversions continued

Paracentric inversions may play an important role in speciation because they may cause hybrid male sterility and thus partial sexual isolation. W.S. Stone has estimated that during the evolution of the approximately 2,000 species of *Drosophila*, about 350 million paracentric inversions might have occurred, and from these tens of thousands have been permanently fixed, and the fate of another similar in number is still undecided. Pericentric inversion may also play a role in speciation because of the complete sexual isolation, and the preservation of the inverted segment as supergenes, refractory to recombination. The descendence of inverted chromosomes in natural populations can be determined by analyzing banded chromosomes or restriction fragment patterns. Inversions occur also in human populations at a frequency of less than 1% per birth. The actual rate may be much higher because the afflicted fetuses are spontaneously aborted. Inversions have been added to various mutation tester strains of *Drosophila* so the new mutations could be identified within a particular chromosome without confounding the results by recombination because inversions eliminate the crossovers. Short inverted nucleotide repeats are found at the termini of insertion (transposable) elements. Inverted duplication of nucleotide sequences form palindromes and facilitate the formation of double-stranded sequences in single-stranded nuclcic acid chains and have determining roles in conformation of the molecules, such as in tRNAs. (See recessive lethal tests in *Drosophila*, speciation, mutation chromosomal, unbalanced chromosomal constitution, palindrome, conformation)

INVERSIONS IN PHAGES AND BACTERIA: one group of enzymes are invertases (transposases) mediating inversions between inverted terminal repeats of transposable elements. The other classes are the integrases. The inversion may affect an invertible promoter or the entire gene. The invertase of phage MU (*gin*) is positioned outside the inversible 3000 bp G region flanked by 34 bp inverted repeats. In one orientation the host specificity genes Sv and U ($\rightarrow$) permit the attachment of Mu to *E. coli* and when inversion ($\leftarrow$) takes place the phage can be adsorbed to other host bacteria. A similar system is found in the phase variation of *Salmonella*. (See phase variation, transposase, resolvase)

INVERSION LOOP: see inversion

INVERSION, PARACENTRIC: involves only one arm of a chromosome. (See inversion)

INVERSION, PERICENTRIC: inversion spanning across the centromere. (See inversion)

INVERTASES: are proteins (Cin, Hin, Gin) involved in viral transposition and bear extensive homology with the N-terminal domains of the resolvases of transposons. In biochemistry the enzyme sucrase which hydrolyzes sucrose to fructose and glucose is also called invertase. (See Cin,Tn, resolvase)

INVERTED REPEAT: see repeat inverted, direct repeat

INVERTRON: a model suggesting the role of 5'-linked proteins in DNA replication and integration of agrobacterial plasmids. (See *Agrobacterium*, transformation genetic)

INVOLUCRE: a whorl of bracts around an inflorescence. (See bract)

IODINE STAIN: for coloring starch (amylose) blue-black, while amylopectin (dextrin) is colored reddish-brown (iodine120 mg and potassium iodide 400 mg in 100 mL water)

iojap (ij): a nuclear mutations in maize located in chromosome 3L-90 (*ij1*) and in chromosome 1L (*ij2*), respectively, causing leaf striping because of defects in the development of plastids. The *ij* gene appears to be a specific mutator of extranuclear DNA and displays normal Mendelian inheritance whereas the striping itself is maternally transmitted. Defects in plastid ribosome development was implicated in this variegation.

ION: a positively (cation) or negatively (anion) charged atom or radical. (See electrolyte)

ION CHANNELS: pores with special passage specificity, e.g. a membrane protein when bound to acetylcholine permits the influx sodium (sodium channel); a variety of ion channels exist (mechanically gated, voltage-gated, ligand-gated, cAMP-gated, etc.). The anion-gated ion channels appear structurally different from most of the cationic channels inasmuch as they may have different subunit associations creating double or triple pores. The thousands of different

Ion channels continued

odors activate in the nose the trimeric G protein, G_{olf}, which in turn activates adenylate cyclase and the cAMP-gated cation channels open and transmit the signal to the brain. Some olfactory receptors utilize the IP_3-gated ion channels. Vision perception is mediated by cyclic guanosine monophosphate (cGMP)-gated channels. Light rapidly induces the formation of guanylate cyclase, generating cGMP and it is degraded by cGMP phosphodiesterase. The photoreceptors (rhodopsin pigment) are in the retina of the eye. Voltage-gated Ca^{2+} ion channels regulate the influx of calcium through the plasma membrane. The L-type ion channels of the neurons may be shut off when the intracellular level of calcium increases beyond a point. The Ca^{2+} ions then serve as wide-spread intracellular messengers and regulate many diverse cellular functions, particularly the secretion of neurotransmitters. Their modulation is due to the βγ subunits of the trimeric G protein. The autosomal dominant human disease, *periodic paralysis*, appears to be due to an amino acid substitution in the α subunit of a sodium channel transmembrane protein. Cystic fibrosis is due to a defect in transmembrane conductance regulator protein kinase A and ATP-regulated chloride ion channel. In pancreatic β cells ATP-dependent K^+ channels are important for the glucose-induced insulin secretion and targets of sulfonylureas, used for oral treatment of non-insulin-dependent diabetes (NIDDM). Truncation of the sulfonylurea receptor (SUR) causes persistent hyperinsulinemic hypoglycemia of infancy and unregulation of insulin secretion in severe hypoglycemia. SUR and BIR (β inward regulator) are coded at a human chromosome position 11p15.1. The ion channels can be inward or outward rectifying, depending on the predominant direction of the flow of the ions. The rectification is not always an intrinsic property of the channel protein but accessory substances, spermidine, spermine and other polyamines, may control it. (See signal transduction, G-proteins, neurotransmitters, calmodulin, rhodopsin, periodic paralysis, cystic fibrosis, IP_3, dihydropyridine receptor, ryanodine, diabetes, hypoglycemia, sulfonylurea, pyrethrin, LQT, HERG, Jervell and Lange-Nielson syndrome, myokymia)

ION-EXCHANGE RESINS: can be cation or anion exchangers also their cross linkage determines their use for separation of molecules of different sizes. There is a variety ion exchange resins; they are made by the copolymerization of styrene and divinylbenzene and various other substances and are combined to produce the phosphocelluloses, diethylaminoethyl-cellulose (DEAE), carboxymethylcellulose (CMC), etc. They can be used for separation and purification of monovalent ions or polyelectrolytes of high molecular weight.

ION PUMP: mediates ion transport through membranes by the use of energy (ATP).

IONIC BOND: is a non-covalent attachment between a positively and a negatively charged atom.

IONIZATION CHAMBER: is an equipment to measure the radioactivity of gases by the ionizations generated through molecular collisions. Electrodes collect the ions and the current (amplified and registered) is proportional to the radioactivity. (See also scintillation counter, Geiger counter, radiation measurement)

IONIZING RADIATION: high-energy electromagnetic radiation causing intramolecular alterations (ion pairs) in organic material and thereby capable of inducing mutation and cancerous transformation in living cells. The maximum legal permissible occupational limits for human exposure: during a life time should not exceed 0.5 mSv per year, the legal limit actually should be 0.2 mSv. 1 Sv (Sievert) = 100 rem (röntgen equivalent man); 1 rem = 1 rad of 250 kVp X-rays. (See also radiation effects, radiation measurement, radiation hazard assessment, cosmic radiation, physical mutagens, Gray units, radiation protection, electromagnetic radiation)

IONOPHORES: hydrophobic molecules involved in ion transport through cell membranes.

IP₃ (inositol triphosphate): derived from inositol phospholipid PIP_2; in response to external signals it releases Ca^{++} from the endoplasmic reticulum. (See phosphoinositides, olfactogenetics)

IPCR: see inverse polymerase chain reaction

IPTG: isopropyl-β-D-thiogalactoside, a gratuitous inducer analog of the *Lac* operon. (See also β-galactosidase, *Lac* operon, gratuitous inducer)

IQ: see human intelligence

I - R: see hybrid dysgenesis

Ir: immune response genes is the old name of the HLA genes that constitute the MHC complex. (See HLA)

IRA1, IRA2: negative regulators of RAS in yeast, antagonistic to CDC25; they are structurally related to GAP. (See RAS, GAP)

IRAK: interleukin receptor-associated kinase. (See NK-κB, interleukins)

IRE: iron responsive element, a 28-nucleotide 5'-UTR sequence in the ferritin mRNA is necessary for iron regulation. IRE sequences in the 3'UTR of the transferrin receptor mRNA protect the mRNA from degradation if the iron level is low. (See ferritin, transferrin, UTR)

IRES (internal ribosome entry site): may be present in circular viral RNAs (picornaviruses) and they can be translated on eukaryotic ribosomes. These viral RNAs are not capped at the 5' untranslated region and carry several AUG codons, and show specific sequences serving as ribosome landing pads. The interaction with ribosomes requires the cellular eIF-4F eukaryotic translation initiation factor. Recently IRES type elements have been found in other viruses and in eukaryotes. The presence of IRES elements permits the dicistronic transcription of genes and facilitates gene targeting, homologous recombination and modification of genes. The use of promoterless vector constructs will position the IRES carrying sequences within transcribed regions rather than into non-translated regions of the genome. IRES elements are frequently borrowed from the family of encephalomyocarditis virus (EMCV) family. The advantage of this system is that it does not interfere with the regular Cap-mediated ribosome scanning. (See ribosome scanning, dicistronic transcription, targeting genes, gene fusion. eIF4F, eIF4G)

IRF: interferon-regulatory transcription factor; homolog of c25 rat factor. (See c25)

IRF4: see LSIRF

IRIS: the circular pigmented membrane in front of the lens of the eye and in its center the pupil is located. The monocot genus of perennial flowers (2n = 44, *Iris germanica*)

IRON AGE: about 3,000 to 4,000 years ago; the beginning of recorded history.

IRRADIATION: see radiation

IRS: interferon-response factor. (See interferon)

IS: see insertion elements

ISADORA: is a 8.3 kb transposable element of *Drosophila*, generally present in 8 copies.

ISCHEMIC: involving restriction of the blood vessels.

ISGF-3α: see APRF

ISOACCEPTOR tRNAs: group of different tRNAs that accept the same amino acid. (See tRNA, aminoacyl-tRNA synthetase, wobble, protein synthesis)

ISOALLELES: wild-type alleles which are distinguishable only by special techniques.

ISOALLOTYPE: variable antigens of one immunoglobulin (IgG) molecule subclass but invariant on the molecules of another subclass or subclasses. (See allotype)

ISOANTIGEN: an allelic variant of an antigen within the species. (See also alloantigen)

ISOBRACHIAL: the two arms of a chromosomes are identical. (See isochromosome, chromosome arm, chromosome morphology)

ISOCHORES: segments (generally larger than 200 kb) of the DNA of higher eukaryotes of rather homogeneous GC or AT content. Usually one rather light and two or three different rather heavy components can be separated by buoyant density centrifugation in the presence of certain ligands, e.g., silver ions (Ag^+). In the mammalian genome the GC poor isochores represent about 62% and the GC-rich and GC-very-rich isochores are 22% and 3-4% of the genome. Genes within isochores have base content characteristic for the isochores. In the high GC isochores the gene concentration is much higher than in the AT-rich areas. The isochores may affect codon usage and replication pattern. The cytologically identifiable Giemsa-stained bands are low in GC, the T bands are high and the R bands occupy a position in between. Because of the high GC content of the T bands, their codon usage is biased. R bands and G/R borders are characterized by higher frequency of chromosomal exchanges (breakage and chiasmata). Viral and transposon integration sites are correlated with the base composition of the elements; sequences similar to their own are the prefered targets. (See CpG islands, chromosome banding, codon usage, transposons)

ISOCHROMATID BREAKS: damage simultaneously inflicted to both chromatids of a single chromosome. (See radiation effects)

ISOCHROMOSOME: has two identical arms and generally comes about by misdivision of a telocentric chromosome. (See misdivision, trisomic secondary)

ISODIAMETRIC: its diameters (lines passing through it center) are the same in all directions.

ISODISOMY: see uniparental disomy

ISOELECTRIC FOCUSING: an electrophoretic separation technique based on the isoelectric point of the molecules to be separated. The isoelectric point is at a pH where a solute has no electric charge and thus does not move in the electric field. When, e.g., denatured protein mixture is placed in an electrophoretic gel that contains a pH gradient established by different buffers, the polypeptides are migrating to their isoelectric zones. (See also electrophoresis, two-dimensional electrophoresis)

ISOELECTRIC POINT: when a charged molecule in a solution—because of the pH—shows no net electric potential and consequently does not move in an electric field. (See isoelectric focusing)

ISOENZYMES: multiple, distinguishable forms (by primary structure [electrophoretic mobility], substrate affinity, reaction velocity and/or regulation) of enzymes that catalyze the same reaction. Same as isozyme.

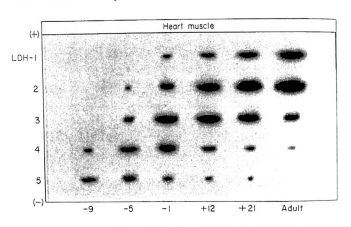

LACTATEDEHYDROGENASE ISOENZYME PROFILE IN THE HEART MUSCLES IN THE MOUSE SHOWS DRAMATIC CHANGES DURING DE-DEVELOPEMENT. NINE DAYS BEFORE BIRTH, ISOZYME LDH-5 IS PREDOMINANT, AND LDH-1, LDH-2 OR LDH-3 ARE NOT DETECTABLE. IN THE ADULT ANIMAL ALMOST THE OPPOSITE IS TRUE. LDH IS A KEY ENZYME IN THE PATHWAY OF CONVERTING SUGARS INTO AMINO ACIDS, LIPIDS, etc. THE NUMBERS AT THE BOTTOM INDICATE DAYS BEFORE (-) OR AFTER (+) BIRTH. (From Markerts, C.L. and Ursprung, U. 1971 Developmental Genetics. Prentice-Hall, Englewood Cliffs, NJ.

ISOFORM: due to some difference in amino acid composition, caused by mutation or alternative splicing of the RNA transcript, essentially the same protein is somewhat altered. Polypeptides in the cellular organelles may have basically similar catalytic function as those residing in the cytosol or organ- or tissue-specific forms of enzymes. (See alternative splicing, splicing)

ISOGAMY: the gametes forming a sexual union appear to be the same, e.g. in protozoa. (See anisogamy, heterogametic, homogametic)

ISOGENIC STOCKS: their genes are represented by the same alleles at all loci.

ISOGRAFT: the genotype of the donor and recipient tissues match. (See also allograft, heterograft, grafting in medicine)

ISOLABELING: ^{3}H label (or other) in both daughter chromatids after one replication in ^{3}H-thymidine medium (or other labeling medium) at a certain tract(s) as the consequence of sister chromatid exchange. (See sister chromatid exchange) → ⇨

ISOLATION GENETIC: may be based on the presence of inversions in the population that in case of recombination yield defective gametes. Also, since recombination within inverted segments produces defective gametes, advantageous gene blocks may be preserved as "supergenes". Genetic isolation may be the first step in speciation. (See also incompatibility)

ISOLEUCINE - VALINE BIOSYNTHETIC PATHWAY:

ketohydroxy-methylvalerate ↔ dihydroxy-methylvalerate ↔ keto-methylvalerate ↔ **isoleucine**
$\quad\quad\quad\quad\quad\quad\quad$ **1** $\quad\quad\quad\quad\quad\quad\quad\quad\quad$ **2** $\quad\quad\quad\quad\quad\quad\quad\quad$ **3**
ketohydroxy-isovalerate $\quad$ ↔ $\quad$ dihydroxy-isovalerate $\quad$ ↔ $\quad$ keto-isovalerate $\quad$ ↔ $\quad$ **valine**

Steps 1 and 2 are controlled by identical enzymes in both pathways in several organisms and therefore genetic defects in either may generate nutritional requirement for both valine [$CH_3CH(CH_3)CH(NH_2)COOH$] and isoleucine [$CH_3CH(CH_2CH_3)CH(NH_2)COOH$] or the accumulation of the intermediates. The *maple syrup urine disease* (MSUD) exists in different forms in humans and other mammals is caused by a block in the degradation, the decarboxylation (step 2) and accumulation of leucine, isoleucine and valine. The keto-methyl-valerate accumulation then causes the characteristic maple syrup odor of the urine. One of the genes was assigned to human chromosome 1p31 and a pseudogene to 3q24. The more serious aspect of the disease is physical and mental retardation, potential coma and death. The disease is controled also by two non-allelic recessive genes, (BCKDHA [branched chain keto acid decarboxylase and dehydrogenase] locus is at 19q13.1-q13.2, the BCKDHB is at 6p22-p21). Gene A is responsible for the biosynthesis of the α chain of the enzyme (BCKDHA) and locus B for the β chain (BCKDHB). In one form of the A type disease (10 mg/day) thiamin has reduced the hyper amino acidemia without dietary limitations. Apparently the larger subunit (M_r 46,500) of the enzyme is part of a mitochondrial protein complex. The enzyme complex contains also component E2 (M_r 52,000) that transfers the acyl group of the keto acid from the E1 component (protein A) to coenzyme A.

The disease may be due to single base pair substitution resulting in a tyrosine substitution at asparagine site 394 of protein A or to deletions of several nucleotides. The defect is detectable prenatally but carriers cannot be identified because of recessivity. The prevalence varies greatly in different ethnic groups from 3×10^{-5} to 6×10^{-3}. Defects in the transaminase reaction (step 3) also may be controlled by a common glutamic-branched-chain-amino acid transaminase. However in humans, hypervalinemia (valinemia) is caused by a defect in a specific transaminase, resulting in lack of ability to catabolize valine into keto-isovalerate without affecting the level of leucine and isoleucine in the blood. (See genetic screening, hypervalinemia, 3-hydroxy-3-methyl-glutaryl CoA lyase deficiency, leucine metabolism, methylcrotonylglycinemia, methylglutaconicaciduria, hydroxymethylglutaricaciduria, methacrylaciduria)

ISOMERASE: interconverts enantiomorphs. (See enantiomorph, chirality)

ISOMERIZATION OF STRANDS: one crossed over DNA strand changes into a two strand crossover through rotation of the molecules. (See Meselson-Radding model of recombination)

ISOMERS: are different compounds that have the same molecular formula but their structure is different. Their differences my be relatively subtle (D- and L-glucose) or may be as large as that between ethyl alcohol and methyl ether.

ISOMETRIC PHAGE: is enclosed in an isosahedral capsid. (See icosahedral)

ISONYMY: individuals having the same family name have a probability of kinship. According to some suggestions the frequency of isonymous couples multiplied by a factor of 1/4 may provide information on the coefficient of inbreeding in the population. One fourth of the married pairs have identical family names before marriage because of the inheritance of the grandfather's name through two sibs and and 3/4 have different surnames. The probability of isonymy for the first cousin is 1/4, for second cousins 1/16 and for third cousins 1/64. These fractions when multiplied by 1/4 may provide the coefficient of inbreeding, i.e. (1/4 x 1/4) = 1/16, (1/16 x 1/4)=1/64, and 1/64 x 1/4) =1/256 for the three types of matings, respectively.

ISOPENTENYLADENINE (6-[γ,γ-dimethylallylamino]purine): a post-transcriptionally modified base in tRNA; it is also a cytokinin plant hormone. (See tRNA, plant hormones)

ISOPRENE: 2-methyl-1,3-butadiene, $CH_2 = C(CH_3) - CH=CH_2$, unit of terpenoids, fragrances, rubber, etc. Isoprenylated proteins are anchored to the cell membranes. The role of the oncogene product, RAS, functions in carcinogenesis and signal transduction in isoprenylated form, attached to a farnesyl pyrophosphate. (See RAS, signal transduction, membrane proteins, prenylation, plastoquinone)

ISOPROPYL THIOGALACTOSIDE: see IPTG

ISOPYCNIC: molecules with equal density. (See buoyant density)

ISOPYCNOTIC CHROMOSOME: does not display heterochromatic regions. (See heterochromatin)

ISOSCHIZOMERS: restriction endonucleases with very similar recognition sites and their cleaved ends being identical (cohesive) and thus capable of joining each other, e.g.

      Bgl II   5'   A↓pGpApTpCpT
                 3'   TpCpTpApGp↑
      Bam HI   5'   G↓pGpApTpCpC
                 3'   CpCpTpApGp↑G

Their termini after ligation are :

                GGATCT
                CCTAGA

This juncture is no longer recognized however by either of the two enzymes because the bases at the left and right ends are incompatible with both Bgl II or Bam HI.

Other isoschizomers are Hpa II   5'   C↓ÇGG   3'   does not cut when Ç is methylated
                                   Msp I   5'   C↓CGG   3'   is indifferent to methylation

Isoschizomers SmaI (CCC↓GGG) and Xma I (C↓CCGGG) recognize the same sequence but cut it (↓) at different position. (See restriction enzymes, restriction-modification)

ISOSTERES: have similar electron arrangement but different chemical structures and used for substrate analog designs.

ISOTHERMAL: being at identical temperature.

ISOTONIC: the active salt concentration of this medium is the same as in the cell. A NaCl solution of 0.9% is isotonic with the human blood and can be used to maintain temporarily good osmotic conditions after substantial bleeding by injecting sterile *sol. natr.-chlor. isotonica* into the (blood vessels) venae. (See hypotonic, isotonic)

ISOTOPE DISCRIMINATION: the heavier atoms may be used less effectively then the lighter ones. (See isotopes)

ISOTOPES: two or more nuclides have the same atomic number thus are the same elements but differ either in mass (stable isotopes, such as Hydrogen and Deuterium) or radioactive isotopes (atom) that disintegrate by emission of corpuscular or electromagnetic radiation (α, β or γ rays). The latter ones are particularly useful in biology as radioactive tracers of minute amounts of labeled (H^3, C^{14}, P^{32}) nucleotides or (C^{14}, S^{35}, I^{125}) proteins. Stable isotopes can (N^{15}, C^{15}, H^2) be used also as density labels for distinguishing old and newly synthesized molecules. The type of radiations and energies in million electron volts are: H^3: β, 0.017-0.019, C^{14}: β 0.155, P^{32}: β 1.71, S^{35}: β 0.167, I^{131}: β 0.605, 0.250, and γ 0.164, 0.177, 0.284, 0.364, 0.625, I^{125}: γ 0.0355, Y^{90}: β 2.24, K^{42}: β 3.6, 2.4 and γ 1.5, Cs^{137}: β 0.518, 1.17 and γ 0.663, Co^{60}: β 1.56, γ 2.33, U^{238}: α 4.180, γ 0.045, Ra^{226} is generally understood to be radium because it has a long half-life (1,620 years) and thus quite stable. It emits primarily α radiation (helium nuclei) 4.750 MeV and γ electromagnetic radiation (0.188 MeV). Radium is converted to a number of other isotopes. Among those radon, the commonly present gas diffusing from the rocky soils, has a short half-life (3.825 days) yet it is usually replenished from the source. It emits α radiation and may pose an environmental health hazard if its concentration reaches higher levels. Radium is no longer used for medical or in very limited extent for industrial purposes because of the hazards in handling. Luminous watch dials are made now from fluorochromes. (See Curie, μC, fluorochromes, biotin, non-isotopic labeling)

ISOTOPIC GRAFT: a group of cells or a piece of tissue is transplanted to equivalent position of another animal.

ISOTYPE: closely similar immunoglobulins.

ISOTYPE SWITCHING: a process of immunoglobulin gene rearrangement. (See immunoglobulins)

ISOTYPIC EXCLUSION: either only λ or κ light chains are used with any of the heavy chains for the formation of a particular antibody (both light chains cannot be used simultaneously). (See antibody, immunoglobulins)

ISOVALERICACIDEMIA (IVD): is controlled by a recessive gene in human chromosome 15q13-q15. The disorder is characterized by sweaty-feet like odor (butyric and hexanoic acid) of the urine, dislike of protein, vomiting, anemia, ketoacidosis, high isovaleric acid content of the blood leading to injury of the nervous system. The biochemical lesion is in isovaleric acid CoA dehydrogenase deficiency. Several types of mRNAs of this gene have been identified due to differences in transcription and different mutations or deletions. (See amino acid metabolism, isoleucine-valine biosynthetic pathway, Jamaican vomiting sickeness)

ISOZYME: see isoenzymes

ISOZYGOTIC: homozygous for all genes.

ISP45: see mitochondrial import

ISRE (interferon sequence response element): see signal transduction [Jak-STAT]

ITAM (immuno receptor tyrosine-based activation motif): see lymphocytes, T cell receptor

ITD: see idiopathic torsion dystonia

ITG: see integrin

ITERATION: involves repetitions; a commonly used form is the iterated integral when we differentiate first with respect to one of the variables while holding the other constant and then differentiate the result with respect to the other variable. For biological experiments the distribution may be fitted to a negative binomial, requiring a similar procedures in the calculation of the maximum likelihood of the values. For complex numerical iterations computer programs may be required. (See negative binomial, maximum likelihood)

ITERATIVE CROSSES: are used in breeding programs such as the three-way and double-crosses employed for the testing or production of hybrid maize. (See heterosis)

ITERONS: are short DNA repeats which may bind to the plasmid replication protein and inhibit replication.

ITK: is a non-receptor tyrosine kinase. (See TCR)

ITP: inosine triphosphate. (See hypoxanthine)

ITR: inverted terminal repetition such as occurring in human adenovirus DNA. (See adenovirus)

ITS: internal transcribed spacers are short sequences within eukaryotic pre-tRNA transcription units (5' 18S - ITS - 5.8S - ITS - 28S 3'), and these clusters are separated by external transcribed spacers. (See ETS, tRNA)

IUCD: intrauterine contraceptive device that prevents implantation of the egg. Although it is an effective method of birth control, it may promote infection and may cause some discomfort.

IUI: see intrauterine fertilization, ART

IVF: see *in vitro* fertilization, ART

IVS: intervening sequences, same as introns. (See introns)

IXODOIDEA: group of insects (ticks) common carriers of *Borrelia* infection, causing Lyme disease and viral infections resulting in encephalitis. (See *Borrelia*)

Dermatocentor marginatus ➡ (tick)

J CHAIN: a 15 kDa polypeptide participating in the formation of the pentameric IgM and dimeric IgA antibody molecules. (See immunoglobulins)

J CHROMOSOME: a chromosome moving toward the pole appearing like⇨

J GENE: see immunoglobulins, J chain

JAB: a co-activator of AP1 transcription factor by transactivating c-Jun and JunD. (See AP, Jun, transactivator)

JACOBSEN SYNDROME: is a dominant fragile site involving human chromosome 11q23.3 is also within a distance of 100 kb to the CBL2 oncogene and CCG repeats. This trinucleotide repeat is also called FRA11B. The CpG repeats are liable to methylation. It involves growth and psychomotor retardation, anomalies of the face, finger and toe development. (See fragile sites, trinucleotide repeats, Huntington's chorea, ataxia, Machado-Joseph disease, Kennedy disease, dentatorubral-pallidolyusian atrophy)

JACKKNIFING: is a statistical device for the estimation of bias and variance of genetic parameters without providing essential estimates on the distribution of the estimates.

JACKPOT VESSEL: in a series of dilutions or in a fluctuation test one vessel has more than the expected number of cells, caused either by a clump of cells or a preexisting mutation.

JACKSON LABORATORY BACKCROSS DNA PANEL MAP SERVICE: makes available DNA from the reciprocal mouse crosses (C57BL/6J x *Mus spretus*), characterized by SSLP markers, proviral loci and several other sequences. Information: Lucy Rowe or Mary Barter, Jackson Laboratory, 600 Main Str., Bar Harbor, ME o4608, USA. Phone: 207-288-3371 ext. 1687. Fax: 207-288-5079. Internet: <lbr@aretha.jax.org> (L. R.) or <meb@aretha.jax.org> (M. B.)

JACKSON-LAWLER SYNDROME: is a keratosis of the skin and involves teeth at birth. (See keratosis)

JACKSON-WEISS SYNDROME: see Crouzon syndrome, Pfeiffer syndrome, Apert syndrome

JAK KINASES: Jak3 is required for the progression of the development of B lymphocytes. (See signal transduction by interferon signaling, Janus kinases)

JAK-STAT PATHWAY: several Jak kinases and signal transducers and activators of transcription (STATs) regulate the signal transduction of interleukins and interleukin-mediated transcription. The pathway may be activated by interferons, phospholipase C (PLC), growth hormones, epidermal growth factor (EGF), platelet-derived growth factor (PDGF, CSF)

JAMAICAN VOMITING SICKNESS: is caused by the consumption unripe ackee fruit, a common food for people of the island. The obnoxious component of the fruit is hypoglycin A which may reduce blood glucose content to the level of 10 mg/100 mL and may cause even death. The compound is a specific inhibitor of isovaleryl-CoA dehydrogenase, and isovaleric acid accumulates in the blood causing depression of the central nervous system. The poisoning has similar effect as human isovalericacidemia. (See isovalericacidemia)

JANUS KINASES: include the Jak kinases and Tyk2; they are non-receptor tyrosine phosphorylating enzymes. (See Jak)

JAROVIZATION (yarowization): see vernalization

JASMONIC ACID ([±]-1α,2β-[Z]-3-oxo-2-[2-pentenyl]cyclopentanacetic acid): is a fatty acid derivative protease inhibitor in plants and an activator of stress response genes in case of infection or wounding. (See plant defense)

JAUNDICE: may be caused by hyperbilirubinemia and is characteristic also for several hereditary syndromes. (See hyperbilirubinemia)

JAVA MAN: a representative of *Homo erectus* with small cranium (brain ≈ 815-1067 cm^2) and robust jaws who lived about 100,000 years ago. (See hominids)

JE: PDGF (platelet-derived growth factor) and serum-inducible cDNA. (See PDGF)

JERVELL and LANGE-NIELSEN SYNDROME: is an autosomal recessive heart and auditory (deafness) syndrome. In the electrocardiograms the interval Q - T is prolonged. In this method the excitation of the heart atrium is denoted by the P wave, followed by the QRS complex of deflections and excitations (depolarization) of the ventricles, and the T waves indicate the repolarization of the ventricles. Fibrillations (uncoordinated arrhytmia) of the heart atrial muscles are also observed as a consequence of inadequacy of potassium and/or sodium ion channels. Sudden death may occur. (See heart disease, deafness, electrocardiography, LQT, HERG, Ward-Romano syndrome, Beckwith-Wiedemann syndrome, ion channels)

JESUIT MODEL: there are more potential replicational origins than actually selected in eukaryotes. (See replication bubble)

JEWS and GENETIC DISEASES: *Askenazi Jews*: - COMMON: Riley-Day syndrome, Tay-Sachs disease, Gaucher's disease, Niemann-Pick syndrome, Diabetes mellitus, Pentosuria, Dystonia, about 1% of the women carries deletions at various positions in the BRCA1 and BRCA2 breast cancer genes, Cohen syndrome; Canavan disease, pentosuria, PTA deficiency disease. RARE OCCURRENCE: juvenile form of Gaucher's disease, Glucose-6-phosphate dehydrogenase deficiency, Bloom's syndrome. *Sephardic Jews*: - COMMON: Mediterranean fever, UNCOMMON: Tay-Sachs disease. Among *Oriental Jews* of Persian origin: hypoaldosteronisms, Dubin-Johnson syndromes appear relatively common. In Lybian Jewish populations the Creutzfeldt-Jakob disease is disproportionally frequent. For these differences of diseases (gene frequencies) there is no generally valid explanation. It has been suggested that genetic drift in small isolated populations may be the cause. The fact that most of these diseases are based on mutations at different sites within the respective loci, is at variance with this argument. The high frequency of Tay-Sachs, Gaucher, and Niemann-Pick diseases involve lysosomes but how this could be the cause is unclear. Selective advantage of the heterozygotes, specific for these particular populations has also been considered. (See diseases at separate entries, Ashkenazim, Sephardic, human intelligence, Amish, founder principle, evolutionary distance, aspartoacylase deficiency)

JIMPY MICE: is a special strain of these animals with reduced rate of cerebroside synthesis resulting in neurological defects. (See cerebroside)

JIMSON WEED: see *Datura*

SEED CAPSULE OF DATURA →

Jockey : see non-viral retrotransposable elements

JNK (Jun amino terminal kinase): a kinase that acts on the amino terminal of Jun oncogenes and other transcription factors. It is the same as SAPK. They belong to the MAK family of protein kinases that are activated by stress (heat shock, tumor necrosis factor). SAPK appears to be inhibited by p21, a transforming protein. Activated JNK stimulates the transcriptional activity of AP1 (See SAPK, MAK, p21, T cell, AP1, Pyk, JUN, ATF2)

JOINING OF DNA: see ligase, blunt-end ligation, cohesive ends

JOINT PROBABILITY: when two events are independent from each other, the probability of their joint occurrence can be obtained by multiplication of the independent probabilities. The same rule applies also to more than two independent frequencies. Independence means that the occurrence of one has no bearing on the occurrence of the other(s). (See probability)

JOUBERT SYNDROME: autosomal recessive developmental defect of the human brain (cerebelloparenchymal disorder, cerebellar vermis agenesis)

JOULE: 1 joule = 10^7, the energy expended per 1 second by an electric current of one ampere in a resistance of 1 ohm; approximately 0.24 calorie.

JUBERG-MARSIDI SYNDROME: Xq12-q21 mental retardation, growth and developmental anomaly is based on mutation in a helicase. The X-linked α-thalassemia and mental retardation seems to involve the same protein. (See thalassemia)

JUDASSOHN-LEWANDOWSKY SYNDROME: is a hereditary keratosis of the nails (onchyogryposis), palm, sole and mouth. (See keratosis)

JUMP STATIONS: collections of links for genetic and biological information regarding data

Jump stations continued
bases, journals newsgroups, etc. Genetics Jump Station: <http://www.ifrn.bbsrc.ac.uk/gm/lab/docs/genetics.html>, Molecular Biology Jump Station <http://www.ifrn.bbsrc. ac.uk/gm/lab/docs/molbiol.html>. (See also databases [general directories])

JUMPING FRENCHMAN OF MAINE: is a rare and obscure apparently autosomal recessive anomaly characterized by very rapid emotional reactions.

JUMPING GENES: move in the genome because they are within transposons. (See transposable elements)

JUMPING LIBRARY: is generated by circularizing large eukaryotic DNA fragments and cloning the junctions of the circle. The large fragments are obtained by using restriction enzymes that cut the DNA very rarely. (See chromosome jumping, DNA library)

JUN (*jun*): avian fibrosarcoma oncogene homolog JUN-A is in human chromosome 1p32-p31 and in mouse chromosome 4. Its homologs are present in other vertebrate species too and it appears to be identical to transcription factor AP-1 along with the product of oncogene FOS; they activate several genes. The products of JUN and FOS are bound together with a leucine zipper and at their carboxyl end they have a DNA-binding domain (5'-TGAGTCA-3'). They apparently form the C/EBP protein. The JUN-B and JUN-D oncogenes are closely linked in mouse chromosome 8. The JUN-B human homolog is in human chromosome 19p13.2. (See AP1, C/EBP, oncogenes, fos, JNK, bZIP)

JUNCTION COMPLEX: the assembly of the various types of junctions (tight junctions, adhesion belt, desmosome) within cells. (See gap junctions, desmosome)

JUNCTION SEQUENCE: see introns

JUNCTIONAL DIVERSIFICATION: when immunoglobulin genes are recombined to generate specific antibodies a few nucleotides may be lost or added to the recombining ends. (See immunoglobulins, antibody, RAG, combinatorial diversification)

JUNIPER (*Juniperus communis*): woody species, 2n = 22.

JUNK DNA: DNA that appeared without any obvious function such as some introns and spacers when the term was coined in 1980. Today, several introns are known with maturase and other functions. Some of the non-coding DNA is interspecifically conserved, indicating some kind of biological function. (See also selfish DNA, trinucleotide repeats, SINE, LINE)

JURASSIC PERIOD: about 190,000,000 to 137,000,000 years ago; an era dominated by the dinosaures and other reptiles although the ancestral forms of most vertebrates were also present and even primitive mammals began to appear.

JUVEBIONE: see juvenile hormone

JUVENILE HORMONE: secreted in the larval state and prevent precocious metamorphosis into the pupal stage of the insect. They include ethyl-polyprenyl components. Similar terpenes and terpene-related substances, e.g. juvebione (e.g. in balsam fir), gossypol (in cotton) occur in plants and also effect the feeding insects. Synthetic hormones have been produced with similar physiological effects. (See metamorphosis, molting, pupa, abscisic acid)

JUVENILE MORTALITY: is frequently the function of the consanguinity of the parents, e.g. stillbirth and neonatal death if the parents are unrelated was found in one study to be 0.044 whereas if the parents were first cousins (consanguinity 1/16) it was 0.111, similarly infant and juvenile death rates were 0.089 and 0.156, respectively. (See coancestry, inbreeding, mortality)

JUVENILE ONSET: a hereditary condition appearing in childhood. (See diabetes mellitus)

JUXTACRINE SIGNALING: the membrane-anchored growth factors and cell adhesion molecules are signaled through the juxtacrine mediators. (See signal transduction)

K1: non-nucleoidal methylation sites of *E. coli* transducer proteins, spaced seven amino acid residues apart.

κ : see symbionts hereditary

κ **CHAIN**: see immunoglobulins

K_a: see dissociation constant

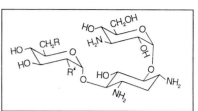

KANAMYCIN → (R = NH_2; R = OH)

K_A/K_S: this is the ratio of non-synonymous and synonymous mutations. The former leads to amino acid replacement in the protein. The ratio indicates thus adaptive change and it has been used to measure molecular evolution. (See also Grantham rule, molecular evolution)

KALANCHOE: a bryophyllum of several species with chromosome numbers varying from 34 to nearly 300 among them. They have been used for studies of development and differentiation and had been a favorite subject for investigations of the effects of agrobacteria on plants. (See *Agrobacterium*)

KALILO: see killer plasmids

KALLIKREIN: a serine proteinase in the pancreas, saliva, urine and blood plasma that cleaves kallidin (a kind of kinin) from globulin and has a vasodilator and possibly some type of skin-irritating effect.

KALLMAN SYNDROME: an autosomal recessive malfunction of the gonads resulting in infertility, lack of ability to smell and cleft palate and cleft lip. It is apparently caused by defects in the steroid hormone receptor(s). There is also an X-linked (Xp22.3) form of the disease. It has been suggested the X-linked gene codes for a cell adhesion molecule. (See infertility)

KAPPA DELETING ELEMENT: assists in the rearrangement of the immunoglobulin κ chain. (See immunoglobulins)

KANAMYCIN: is an aminoglycoside antibiotic frequently used as a selectable marker in genetic transformation. (See antibiotics, antibiotic resistance, aminoglycoside phosphotransferase, transformation genetic, vector, Geneticin; see structural formula at the top of this page)

KANGAROO: *Macropus rufus,* 2n = 20; the rat kangaroo (*Potorous tridactylus apicalis*), 2n = 13 in the male and 12 in the female.

KAP: human phosphatase with specificity for Thr^{160} in Cdk2. (See Cdk2)

KAPOK (*Ceiba*): a south-east Asian bamboo fiber tree, 2n = 72-88 with unknown basic chromosome number.

KAPOSI SARCOMA (hemangiosarcoma): shows red-purple nodules and plaques that become tumorous. It is due to autosomal dominant genes but it is also an opportunistic tumor because it is expressed mainly when some types of infections, such as *Pneumocystis* microorganisms are present as in the case in the AIDS and other immunodeficiencies. The higher frequency and aggressiveness of Kaposi sarcoma in AIDS is explained by the synergism between the cellular basic fibroblast growth factor (FGF) and the Tat enhancer protein of the virus. A novel herpes-virus has also been blamed for causing the disease. (See acquired immunodeficiency, FGF)

KAPPA PARTICLES: lysogenic bacterial symbionts that kill the sensitive hosts of *Paramecium aurelia*. (See symbionts hereditary, *Paramecium*)

KAR3: is a 84 kDa motor protein of the kinesin family in yeast. (See kinesin)

KARTAGENER SYNDROME (dextrocardia): a complex syndrome of left-right inverted location of major visceral organs, involving also lack of ciliary movement and sperm motility. (See situs inversus viscerum)

KARYOGAMY: nuclear fusion in fungi following fusion of the cytoplasms of two cells, plasmogamy. (See fungal life cycles)

KARYOKINESIS: division of the cell nucleus. (See cell cycle, mitosis)

KARYOLYMPH: the fluid fraction of the cell nucleus in contrast to the particulate ones, e.g. chromosomes.

KARYOPHERIN (α and β): are cytosolic proteins that mediate nuclear traffic in cooperation with nucleoporin, with a GTPase and RAN. (See RNA export, nuclear pore, nuclear localization , RAN, GTPase)

KARYOPLAST: nucleus surrounded by only a thin layer of cytoplasm and membrane. (See cytoplast, transplantation of organelles)

KARYOSOME: the mass of chromatin of the oocytes before metaphase of animal cells.

KARYOTYPE: characteristics of a (mitotic) metaphase chromosome set by number, morphology, arm ratio, secondary constrictions, banding pattern, etc. (See chromosome morphology, electrophoretic karyotyping)

KARYOTYPE EVOLUTION: Chromosome number is a rather good characteristic of a species although identical number of chromosomes may not indicate any relationship.

Subobscura

Pseudoobscura

Melanogaster

Ananassae   Willistoni

THE HAPLOID CHROMOSOME SET IN THE SUBGENUS SOPHOPHORA OF *DROSOPHILA*. THE LOWEST CHROMOSOMES ARE THE X. NOTE THE GRADUAL FUSION OF THE ARMS AND CHROMOSOMES FROM LEFT TO RIGHT. THE X CHROMOSOME OF *D. ANASSAE* PROBABLY EVOLVED FROM THAT OF *D. MELANOGASTER* BY A PERICENTRIC INVERSION. IN *D. WILLSTONII*, THE SMALL DOT-LIKE CHROMOSOME SEEMS TO BE INCORPORATED INTO THE BI-ARMED X CHROMOSOME. (After Sturtevant, A.H. 1940 Genetics 25:337)

The morphology of the chromosomes and arm ratio are also frequently used to assess similarities but substantial differences may exist within related groups and pericentric inversions may alter the relative position of the centromere. Centromere fusion and Robertsonian translocations may convert telocentric chromosomes into bi-armed ones. Misdivision may generate telocentrics from bi-armed chromosomes and change the same time the chromosome numbers. Paracentric inversions may serve the purpose of speciation and the sequence of change in the pattern of chromosome bands (in polytenic chromosomes or specially banded chromosomes) may be traced by the techniques of light microscopy. Translocations and other types of chromosomal aberrations may also be followed in related species. The similarities of the karyotypes can be assessed also on the basis of chiasma frequencies if the species are closely related enough to permit meiotic analyses. Chromosomal mapping of classical genetic markers or restriction fragments, sequence-tagged sites, RAPDS may also reveal the order of genes and nucleotide sequences and their evolutionary path. Karyotic changes (chromosome number and chromosome arm number) per evolutionary lineage per MY varies a great deal from 1.395 in horses to 0.025 in whales to 0.029 in other vertebrates (lizards, teleosts). (See chromosomal aberrations, inversion, misdivision, polytenic chromosomes, banding techniques, RFLP, RAPDs, sequence-tagged sites, polyploidy, evolution of the karyotype)

kb: kilobase, i.e. 1,000 bases.

kbp: kilobase pairs, thousand pairs of nucleotides in double-stranded nucleic acids.

KC: a cytokine protein, homologous to N51, MGSA and gro. (See cytokines, N51, MGSA, gro)

KCNA: potassium voltage-gated ion channel caused diseases encoded at several loci in the short arm of human chromosome 12 and 19, involving neurological disorders. (See ion channels, LQT)

K$_d$: see dissociation constant

kDa: kilo Dalton (1,000 Da). (See Dalton)

kDNA (kinetosome DNA): see kinetosome

KEARNS-SAYRE SYNDROME: see mitochondrial disease in humans, optic atrophy

KEEL: two petals associated along the edge.

KELL-CELLANO BLOOD GROUP (KEL): the KEL antigen is a 93 kDa membrane glycopro-

Kell-Cellano blood group continued
tein, associated with the cytoskeleton and it is encoded in human chromosome 7q32 area. Its precursor substance (Kx) is coded in the X chromosome (McLeod syndrome). Its mutations may cause "horny" appearance of the erythrocytes (acanthocytosis) and granulous inflammations in response to infectious and other factors. The frequencies of the KEL and Kx alleles in England were found to be between 0.0457 and 0.9543, respectively. (See blood groups, McLeod syndrome)

KELP: large brown algae.

KELVIN: temperature scale is used primarily in thermodynamics; 0° C = 273° K; the conversion between the C° and K° is: C° = K° - 273 , e.g. 100 C° = 373 K° or 0 K° = -273 C°

KENAF (*Hibiscus cannabinus*): warm-climate fiber crop; 2n = 2x =36.

KENNEDY DISEASE: is a non-lethal spinal and bulbar muscular atrophy, sensory deficiency, frequently with gynecomastia and impotence expressed primarily in adults. The recessive gene was mapped to Xq12. The basic defect is a CAG repeat (22-52x) in the first exon of the androgen receptor in the spinal cord, brains stem and sensory neurons. Longer repeats and not the number of repeats increase the severity of the symptoms. [This syndrome is not named after President Kennedy's back ailment.]. (See atrophy, neuromuscular disease, androgen receptor, gynecomastia, fragile sites, trinucleotide repeats, spinal muscular atrophy, testicular feminization, dihydrotestosterone)

KERATIN: a protein of the surface layer of skin, hair, nails, hoofs, wool, feather and porcupine quills, etc. Keratins may be high-sulphur, acidic matrix proteins or low-sulphur, basic fibrous proteins. These two types usually appear in pairs and are controled in humans by autosomal dominant genes. Point mutation in keratin genes in human chromosomes 12 and 17 may lead to various epithelial anomalies. (See keratosis)

KERATITIS: autosomal dominant inflammation of the cornea.

KERATOMA (hyperkeratosis): formation of keratoses on the palms and other parts of the body.

KERATOSIS: either a wart-like flat or emerging (scaly) spot(s) that may become cancerous or soft friable (sometimes colored), non-invasive benign skin lesion. Both may have a number of different forms and are under the control of autosomal dominant genes. These skin lesions generally appear during adulthood but some start in very early childhood and develop progressively and may be the signals of more serious conditions. Sunburn may lead to keratosis and squamous cell carcinoma if mutation occurs in p53. (See skin diseases, psoriasis, ichthyosis, Darier-White disease, Judassohn-Lewandowsky syndrome, Jackson-Lawler syndrome, pachyonychia, intermediate filaments, FGF, dyskeratosis, p53)

KERMIT: *Drosophila* transposable element (4.8 kb). (See also copia)

KERMIT: a computer program for telecommunication through CMS.

WHEAT KERNEL WITH THE EMBRYO AT BASE

KERNEL: is a "seed" (grain) covered by the pericarp (fruit) and not just with the → seed coat, such as a wheat or maize kernel; in barley it may have also glumes (husks) attached.

KETOACIDOSIS: occurs in several human diseases when ketones accumulate due to a defect in succinyl-CoA; 3-ketoacid CoA-transferase, in diabetes mellitus, in Gierke's disease (type I glycogen storage disease), in glycinemia, in methylmalonic aciduria and in lactic aciduria. (See individual entries)

KETOGENIC AMINO ACIDS: (tryptophan, phenylalanine, tyrosine, isoleucine, leucine, lysine) can serve as precursors of ketone bodies (acetoacetate, D-3-hydroxybutyrate, acetone) formed primarily from acetyl coenzyme A if fats are degraded . (See amino acid metabolism)

KETONE: $\begin{array}{c} R \\ | \\ R-C=O \end{array}$

KETOSE: monosaccharide with the carbonyl group being a ketone, such as fructose.

KETOSIS: a condition with accumulation of ketone bodies, e.g. in diabetes mellitus.

KETOTIC: showing ketosis. (See ketosis)

KeV (kilo electron volt): 1,000 electron volt. (See electron volt)

KEYHOLE LIMPET HEMOCYANIN (KLH): a carrier protein (from *Megathura crenulata*) that can be linked to synthetic amino acid sequences with 12 to 15 hydrophobic residues through an NN_2 or COOH-terminal cysteine. The synthetic peptides can be used to raise antisera against them and these may be advantageous because their recognition is independent of the conformation of the whole protein. (See antiserum, hemocyanin)

KGK (keratocyte growth factor): is involved in epithelial cell development in wound repair.

KH MODULE: the K homology motif is present in the heterogeneous nuclear ribonuclear protein (RNP) K, protein, ribonuclease P, the Mer splicing modulator of yeast and the fragile X product. A common feature of these domains is the binding of single-stranded RNA. (See FMR1, hnRNA, ribonuclease P, splicing, fragile X, splicing)

KID: kinase-inducible domain in transcription factors. (See transcription factors)

KIDD BLOOD GROUP (Jk): is apparently associated with human chromosome 18q11-q12. The frequency of the Jk^a allele in the world appeared to be 0.5142 and that of the Jk^b 0.4858. (See blood groups).

KIDNEY DISEASES: may be caused by various environmental and genetic factors. (See urogenital dysplasia, renal dysplasia and retinal aplasia, renal dysplasia and limb defects, renal hepatic-pancreatic dysplasia, renal tubular acidosis, nephrosis, nephrosialidosis, nephritis, kidney stones, diabetes insipidus, Bardet-Biedl syndrome, oncocytoma, Addison disease, glomerulonephrotis, kidney cell carcinoma, hypernephroma)

KIDNEY STONES (nephrolithiasis): accumulation of primarily oxalate crystals in the kidneys caused by intestinal malabsorption of dietary salts. Polygenic and autosomal dominant control have been claimed to be responsible although one form appears to be X-linked recessive. In the X-linked (Xp11.22) form the defect is caused by various mutations in an outwardly rectifying chloride ion channel. (See kidney disease, ion channel).

KIF: a motor protein superfamily. (See kinesin)

KILLER CELLS (NK, natural killer cell): are part of the immune system; they are large granular lymphocytes and they cooperate with other elements of the defense system to eliminate foreign organisms. The natural killer cells mount a cytotoxic reaction against invaders without prior sensitization. Similarly to other lymphocytes, they kill some virus-infected cells and tumor cells. NK cells are controled (inhibited) by MHC receptors, specific for Class I molecules of the major histocompatibility complex. The NK cells do not express conventional receptors for antigens. Actually, they kill allogeneic cells because of the absence of the autologous MHC molecules and cells that display different antigens because of viral infection. Human NK Class I receptors consist of an immunoglobulin domain containing lysine in the transmembrane section whereas in mice the transmembrane portion of receptors resemble C type lectins. The activity of the NK cells is increased by interferons, especially by γ interferon but IFNγ provides protection for normal cells against NKs. The NK cells usually cooperate with the cytotoxic T cells (CTL). They are often refered to as 'non-MHC-restricted' cells because they can attack even cells which do not express MHC. The outermost layer of the human placenta lacks class I and class II MHC proteins and this is sufficient to protect the hemiallogeneic fetal cells against T cells but this would not protect, however, from NK killer cells. The trophoblast cells (extraembryonic ectodermal tissue) in contact with the placenta (the tissue connecting the maternal and fetal system) express the HLA-G class I molecules and this is sufficient to protect the pregnancy from some adverse immunological reaction. There are also mononuclear killer cells that have antibody-dependent cellular cytotoxic ability. (See antibody, immune system, lymphocytes, T cells, blood cells, allogeneic, autologous, lectin, cytotoxic T cells, HLA, MHC, anomalous killer cell, monocyte)

KILLER PLASMIDS: the mitochondrial *kalilo* plasmid (8.6 kb) has 1,338 bp inverted terminal repeats in *Neurospora intermedia* from Hawaii. At the 5'-end it is covalently linked to a 120 kDa protein. The plasmid has two, non-overlapping, opposite orientation open reading frames. ORF1 codes for a RNA polymerase (homologous to that of phage T7) and ORF2 is a DNA polymerase gene. Integration of this kalDNA into the mtDNA causes senescence and death

Killer plasmids continued

because it builds up at the expense of the normal mtDNA. It is transmitted in heterokaryons. The *maranhar* plasmid is prevalent in *Neurospora crassa* from south Asia. In function, *maranhar* is similar to *kalilo* although the proteins encoded are substantially different. (See killer strains, mitochondrial genetics, mitochondrial plasmids, senescence, mitochondrial disease in humans, pollen killer)

KILLER STRAINS: of *Paramecium* harbor symbionts which release toxins lethal to sensitive strains. Bacteria harboring colicinogenic plasmids may destroy colicin-sensitive strains. In yeasts the double-stranded RNA viruses, L (4.6 kb) and M (two 1.8 kb dsRNA) result in the production of a killer toxin, affecting sensitive strains. These two viruses occur together because only the L strain encodes the capsid protein whereas the actual toxin is encoded by the M genome. L and M do not exist outside the cells and are transmitted during mating. In some insects *Wolbachia pipientis* bacterial infection of the males kills all the offspring sired by these males. (See *Paramecia* , symbionts hereditary, colicins)

KILLER TOXIN: is secreted by many yeast cells. The producer cells themselves are immune to the toxin but on the sensitive strains a pore is bored and that kills the cells. The K1 killer strains contain two doublestranded RNA viral genomes: the M_1 dsRNA is 1.8 kb encodes the toxin (42 kDa) and the immunity substance precursors and the larger L-A dsRNA (4.6 kb) replicates and maintains M_1. These virus-like particles are transmitted during cell divisions and mating. Several genes are required for the maintenance of MAK (maintenance of killer), SKI (superkiller), KEX (killer expression, endopeptidase and subtilisin like proteins), KRE (killer resistance) affects cell wall receptors and function of the killer state. (See colicins, *Paramecium*, subtilisin)

KILOBASE (kb): 1,000 bases in nucleic acids.

KIN SELECTION: generally, natural selection favors the survival of the fittest individuals. In some instances this principle may not be so obvious because of altruism supports individuals for the benefit of the population that share their genes. Selfless females may sacrifice themselves to predators in attempts to rescue their multiple offspring. In social insects (bees, ants, termites) only the queens and selected males reproduce yet the workers and soldiers of the colonies protect the reproductive individuals even at the cost of their life. The fitness of these non-reproducing castes is measured by the success of their mating sibs. Actually, the survival of their genes is assured indirectly through these reproducing individuals. It is a nepotism, motivated by natural selection. (See altruistic behavior, selection, fitness, inclusive fitness)

KIN28: is a cyclin-dependent kinase of *Saccharomyces cerevisiae;* it phosphorylates the C terminal domain of RNAP II to facilitate transcription. (See CDK, RNAP)

KinA, KinB: *Bacillus subtilis* kinases affecting sporulation regulatory proteins SpoA and SpoF, respectively. (See sporulation)

KINASE: enzyme that joins phosphate to a molecule. (See protein kinases)

KINDRED: a group of biological relatives with a determined pedigree. (See pedigree analysis)

KINECTIN: membrane protein, binding intracellular vesicles to kinesin. (See kinesin)

KINESIN: a cytoplasmic protein involved in moving vesicles and particles along the microtubules; it assists segregation of chromosomes and transport of organelles by using energy derived from ATP hydrolysis. The ca. 340 residue NH_2 domain has the motor function whereas the COOH terminus probably binds to organelles or microtubules. It belongs to the KIF family of motor associated proteins. Kinesin was assumed to move the microtubules away from the centrosome. (See myosin, dynein, kinectin, centrosome)

KINETIC COMPLEXITY: is measured by the reassociation kinetics of denatured DNA; the increase in complexity (large number of unique diverse sequences) requires longer time for reassociation. (See c_0t curve, kinetics)

KINETICS: analysis of reaction rates. The reactions may run to completion or may remain incomplete. *0 order* of the reaction kinetics indicates that the velocity of the process is constant and independent from the initial concentration of the substrate. *First order* kinetics indicates

Kinetics continued
that only one substrate is involved (monomolecular reaction). Mutation rates, terminal chromosomal deletions below a certain dose of mutagens or clastogens also follow first order kinetics.

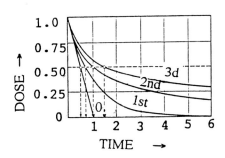

2nd order kinetics indicates bimolecular reactions, and ionizing radiation-caused chromosomal rearrangements (inversion, transposition, translocation) may also be in this category. *Multiple order* kinetics may be involved with more than two reacting factors. (See radiation effects, LET, clastogen, DNA repair)

REACTION KINETICS: DOSE (CONCENTRATION) PLOTTED AGAINST TIME. THE 0, 1st, 2nd, 3rd ORDER KINETICS ARE MARKED ON THE CURVES.

KINETIN (6-furfurylaminopurine): a cytokinin plant hormone. (See plant hormones, below)
KINETOCHORE: the structural protein part of the centromere where spindle fibers are attached. (See Roberts syndrome, CENP, SKP1)
KINETOPLAST: the mitochondrial DNA in some protozoa (*Trypanosoma*) is organized into kDNA (kinetoplast DNA), concatenated circular molecules. The ca. 20-50 copies of the *maxicircular* genome (22 kbp) resembles the mtDNA of other species, the heterogeneous 5,000 to 10,000 copies of the *minicircles* are 0.5-2.8 kbp and represent about 95% of the kDNA. The sum of kDNA constitutes 7-30% of the total cellular DNA in the different *Trypanosoma* and *Leishmania* species. The maxicircle transcripts are subject to editing by adding or (rarely) by subtracting uridine from the primary transcript. Such editing, called also pan editing, may cause some of the edited transcripts to have more than 50% uridine. The gene which encodes the transcript subject to pan editing is called a *cryptogene* because the sequences of the original transcripts are almost concealed by this process. In the minicircles short (50-100 base) sequences (guide RNA [gRNA]) remain complementary to the edited RNA. The kDNA may have major role in the life cycle of the protozoon while it is in the insect gut but not in the mammalian blood. (See *Trypanosoma*, *Leishmania*, mtDNA, gRNA, RNA editing)
KINETOSOME: see kinetoplast
KING GEORGE (III, 1762-1830) **MADNESS**: a neurotic behavior caused by porphyria. (See porphyria)
KININ: a group of endogenous peptides acting on blood vessels, smooth muscles and (injury-sensing) nerve endings. (See kininogen)
KININOGEN: kinin precursor α_2-globulins of either 100,000-250,000 M_r or 50,000-75,000 M_r, are split by kallikrein to bradykinin and lasyl-bradykinin (kallidin), respectively. The bradykinins regulate blood vessel contraction, inflammation, pain and blood clotting. (See Williams factor, kinin)
KINK: a distortion in the DNA structure.
KINSHIP: see coefficient of coancestry
KIP2 (p57KIP2): is a cyclin-dependent kinase inhibitor protein, encoded in human chromosome 11p and in mouse chromosome 7. At some developmental stages it is expressed from the ma-ternal chromosome. KIP1 (p27^{KIP1}) has also closely related function in the cell cycle. (See imprinting, CDK, cancer)
Kirsten-RAS: oncogene of a rat sarcoma virus. (See RAS, oncogenes)
KIS: Kirsten murine sarcoma virus oncogene.
KIT ONCOGENE: is a homolog to the viral *v-kit* gene of a feline sarcoma ; it is in human chromosome 4q11-q12, and chromosome 5 of mouse. This protooncogene codes for transmem-

KIT oncogene continued

brane tyrosine kinase with homology to CSF1R and to PDGF. The KIT gene was assumed to be responsible also for piebaldism in humans and mouse. The receptor for the KIT product appeared to be encoded by the *W* (*white fur*) and the PDGF genes of mice, located to about the same or identical chromosomal site as the KIT homolog. Another mouse locus, *Sl* (*Steel*) encodes MGF (mast cell growth factor), a ligand for the growth factor receptor. The KIT oncogene product is required also for the phasic contraction of the mammalian gut and the control of hematopoietic cells. (See also piebaldism, CSF1R, PDGF, oncogenes, transmembrane proteins, tyrosine kinase, FLT, hematopoiesis, mast cells)

KJELDAHL METHOD: determines total nitrogen content in organic or inorganic material after hot sulfuric acid digestion in the presence of a catalyzator (Se, Hg). After titration of the distilled ammonia in the presence of phenolphtalein, 1 mL 0.1 N H_2SO_4 bound by the ammonia corresponds to 0.0014 g nitrogen, and the amount of nitrogen multiplied by 6.25 estimates protein content.

KL: a male fertility complex in the long arm of the Y chromosome of *Drosophila melanogaster*. (See sex determination, *KS, Drosophila*)

KLEBSIELLA: a gram-negative, facultative anaerobic enterobacterial genus (closely related to *E. coli*) with several species, widely present in nature (including hospitals) and capable of causing urinary and pulmonary (lung) and wound infections but it has the desirable feature of fixing atmospheric nitrogen. (See nitrogen fixation)

KLENOW FRAGMENT: the large fragment of bacterial DNA polymerase I lacking 5'-to-3' exonuclease activity (located in the small fragment of the enzyme) but retaining the 5' → 3' polymerase and the 3'→ 5' exonuclease functions. It is generated by cleavage with subtilisin and other proteolytic enzyme. It has been used for nick translation and the Sanger's dideoxy method of DNA sequencing. (See nick translation, DNA sequencing, DNA replication in prokaryotes, DNA polymerase I)

KLH: se keyhole limpet hemocyanin

KLINEFELTER SYNDROME: is caused most commonly by XXY chromosomal constitution, although similar are the consequences of XYY, XXXY, XXXXY and some of the mosaicisms involving more than one X and Y chromosomes. The XXY condition affects about 1 to 2 boys among 1,000 births. XYY is somewhat less frequent, and the more complex types are much more rare. The XYY males may be fertile although the other symptoms are common with those of XXY. The Klinefelter syndrome is characterized by underdeveloped testes (hypogonadism) and seminiferous ducts.

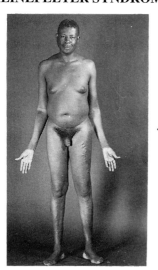

← KLINEFELTER MALE. Courtesy of Dr. K.L. Becker. By permission of the American Fertility Society.

The afflicted individuals are generally sterile, although effeminate yet heterosexual in behavior. Their height is usually above average and the limbs appear longer than normal. About half of them show increased breast size and they are about as likely to develop breast cancer as women. They are more likely to develop insulin-dependent diabetes and heart failure (mitral valve prolapse) than normal males. Some develop speech problems. Their intelligence is generally low, particularly in those having a higher number of sex chromosomes. Klinefelter individuals are frequently slow in development and have learning disabilities. Although they tend to be shy and immature in behavior, they were considered to be prone to violence but this latter classification turned out to be based on false statistics. True, there is a higher incidence of Klinefelter syndrome in prison populations but this due to their mental deficiency. Klinefelter symptom occurs also in various other mammals. (See sex chromosomes, sex determination, sex mosaics, sex chromosomal anomalies in humans, XX males, gynecomastia, trisomy, polysomic cells)

KLIPPEL-FEIL SYNDROME: in the autosomal dominant form it is associated with fusion of cervical (neck) vertebrae and malformation of head bones conducive to conductive and/or sensorineural hearing loss. The autosomal recessive forms did not involve hearing deficit. (See deafness)

Kluyveromyces lacti: a yeast species in which the first time linear plasmid was found.

KNIEST DYSPLASIA (metatropic dwarfism): is an autosomal dominant disease concerned with the locus of collagen II α-1 polypetide (human chromosome 12q13.11-q13.2). The defect involves a deficiency of the C propeptide that is required for normal fibril formation. The urine contains increased amounts of keratan sulfate. There are general problems with the cartilage. Specifically, the afflicted individuals cannot close tightly their fist, their palm is purplish, severe myopia (nearsightedness) appears, the retina is detached and various types of bone defects, including dwarfism, may be evident. (See Stickler syndrome, eye diseases, collagen, connective tissue disorders, dwarfism)

knirps (kni, map location 3-46): zygotic gap mutation in *Drosophila*. The first seven abdominal segments are abnormal but head, thorax, the eighth abdominal segments and tail are normal.

KNOB: heterochromatic cytological landmark of a chromosome. Knobs have been used to identi-

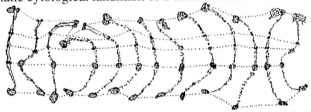

CHROMOSOME KNOBS OF THE PACHYTENE CHROMOSOMES OF 13 DIFFERENT HOUSECRICKETS. The dotted lines connect homologies. (From Wenrich, D.H. 1916 Bull. Mus. Comp. Zool. Harvard 60:58)

fy the fate of particular chromosomes during meiosis and evolution. Knobs were implicated in preferential segregation in maize and affecting the frequency of recombination of syntenic markers. Knobs (altogether about 22) were observed on all chromosomes of maize. These are composed of tandem arrays of about 180 bp elements and may function as neocentromers. (See preferential segregation, neocentromere)

KNOCKIN (Ki): insertion a functional copy within an inactive gene. (See knockout)

KNOCKOUT: inactivation of a gene by any means (e.g. deletion, insertion, targeted gene transfer) to determine the phenotypic, metabolic, behavioral or other consequences, and to draw conclusions concerning its normal function. Removal of genes by site-specific recombination is expected to specifically knock out a discrete genic sequence; it has been learned recently that the consequence of the knockout may depend also on the neighboring genes. The general procedure involves first growing embryonic stem cell (ES) of mice. The cells are then electroporated with a gene construct carrying a selectable marker (e.g. neomycin resistance) within the coding sequence and thus it is inactivated. The disrupted gene is flanked by sequences homologous to the target to facilitate homologous double crossing over. The cells are grown out on selective media and recombinants are isolated. These recombinants will not be able to carry out the normal function of the target gene because of the insert in the coding sequence. Eventually the ES cells are injected into host blastocytes with a micromanipulator and transferred into the blastocoele. In case of success, the developing embryo becomes chimeric and some of the germline cells or the entire germline will carry the knockout and transmits it to some of the progeny. In 17 days after the transfer knockout offspring may be obtained. The rate of success varies depending on the technical skills of the investigator and the mice concerned but it may be also quite high, over 50%. (See site-specific mutagenesis, targeting genes, excision vector, Internet: <http://www.gdb.org/Dan/tbase/tbase.html>)

KNOTTED CIRCLE OF DNA: see concatenane   ← A KNOTTED CIRCLE

KOALA: *Phascolarctos cinereus*, 2n = 16.

KOHARA MAP: of *E. coli* is a restriction fragment map, based on partial digests of 3400 phage lambda clones with eight restriction endonucleases. (See E. coli, RFLP, vectors, cosmids)

KOLMOGOROV-SMIRNOV TEST: is a non-parametric statistical procedure for the analysis of frequency distributions. It may be used in genetics for different problems, e.g. analysis of gene or physical marker frequency distributions in different populations. The method usually applied to expected and observed cumulative distributions The differences are expressed as differences between relative cumulative frequencies and usually statistical tables are used to ascertain the maximum difference between observed and expected cumulative frequency at certain level of significance. The largest absolute vertical deviation (D) = maximum $| F_s(x) - T_T(x) |$ where $F_s(x)$ is the sample cumulative distribution frequency (CDF) and $T_T(x)$ = the theoretical CDF. The procedure is simple to carry it out and computer programs are also available. Because of space limitation, I cannot include an illustrative sample. Details can be found in Hays, W.L & Winkler, R.L 1971 Statistics: Probability, Inference and Decision, Holt, Rinehart and Winston, New York or in Sokal, R. R. & Rohlf, F.J. 1969 Biometry. W.H. Freeman & Co, San Francisco, CA, USA)

KOSAMBI's FUNCTION: $x = 0.25 \ln[(1 + 2y)/(1 - 2y)]$, where x is the recombination frequency corrected by the mapping function and y is the observed recombination fraction. Kosambi's formula considers "average" interference, and it has been used in different species. (See also Haldane's mapping function, coefficient of coincidence, mapping genetic, recombination frequency, mapping function)

k-RNA (kinetoplast RNA): is mitochondrial RNA in *Trypanosomas*. (See *Trypanosoma*, RNA editing)

KRABBE'S LEUKODYSTROPHY (globoid cell leukodystrophy): is a rare chromosome 14 recessive disease, called also galactosyl ceramide lipidosis. The enzymatic basis of the condition is the deficiency of galactocerebroside β-galactosidase. The onset of the disease is expected within the first half year of life and it is generally fatal by around age 2. Exceptionally its onset may be delayed to late childhood. The early symptoms are irritability, hyperactivity that is followed by lethargy, degeneration of the nervous system resulting in blindness and deafness. Cerebrospinal fluid (CSF) proteins accumulate. In the white matter of the brain myelin is reduced and infiltrated with globoid cells that are rich in galactosyl ceramides. (See also galactosidase, sphingolipidoses, sphingolipids, metachromatic leukodystrophy, Addison disease)

KRAS: see RAS oncogene

KREBS-SZENTGYÖRGYI CYCLE: (1) **Oxaloacetate** → (2) Citrate ⇔ cis-Aconitate (3) cis-Aconitate ⇔ Isocitrate (4) Isocitrate ⇔ α-Ketoglutarate (5) α-Ketoglutarate⇔ Succinyl Coenzyme A (6) Succinyl CoA ⇔ Succinate (7) Succinate ⇔ Fumarate (8) Fumarate ⇔ Malate (9) Malate ⇔ **Oxaloacetate.** This simplified outline does not show the energy donors and cofactors. This cycle is the most efficient path to generate energy. Also called tricarboxylic acid cycle and citric acid cycle.

Krev-1: probably the same as Rap1. (See Rap)

KROX20: a serum inducible primary-response gene with Zn-finger, originally from *Drosophila*. It controls the myelination of the peripheral neurons. (See serum response element, Zinc finger, myelin, neuron)

Krox-24: see NGKI-A

KS: a male fertility complex in the short arm of the Y chromosome of *Drosophila melanogaster*. (See sex determination, *KL*, *Drosophila*)

KSS1: a protein kinase of the MAPK family. (See signal transduction, MAPK)

Ku: a heterodimeric serine/threonine protein kinase that binds to DNA in cooperation with transcription factors. Ku also interacts with the termini of DNA double strand breaks and targets DNA-activated protein kinase (DNA-PK). Ku has a role in DNA repair and recombination, including immunoglobulin V(D)J rearrangements, and it is encoded by the XRCC5 human gene. (See transcription factors, p350, immunoglobulins)

KUGELBERG-WELANDER SYNDROME: is a muscular atrophy expressed at infancy and it is determined either by a dominant or recessive factor in human chromosome 5q11.2-q13.3. (See neuromuscular disease, atrophy)

KUNKEL MUTAGENESIS: template DNA is generated in a bacterial strain of *dut ung* constitution. Such are defective in dUTPase (*dut*) and uracil-N-glycosylase (*ung*). Consequently several uracil residues are incorporated into the single-stranded M13mp19 phage DNA and these cannot be removed by DNA repair because of the defective glycosylase. To this DNA a mutagenic nucleotide primer is added and then in the presence of all 4 deoxyribonucleotides a new strand is synthesized using the U-containing single-strand DNA template of M13. After the new M13 DNA synthesis is completed, the molecule is transfected into wild type *E. coli* which gets rid of the U-containing strand and synthesizes DNA containing the mutant nucleo tide and the resulting phage plaques contain this localized, directed mutation. (See localized mutagenesis, bacteriophages, glycosylases)

KURTOSIS: is a departure from the symmetrical frequency distribution by displaying excess or deficiency at the shoulders compared to the tails and the highest point of the curve (peakedness). At normal distribution $(x - \mu)^4/\sigma^4 \cong 3$. A higher ratio indicates kurtosis. (See also normal distribution, skewness, moments)

KURU: infectious human chronic degenerative disease characterized by tremors, ataxia (lack of muscular coordination), strabismus (contorted visual axis in the two eyes), dysarthria (stemmering and stuttering), dysphagia (difficulties in swallowing), fasciculation (bundling of nerve and muscle tissues), etc. Generally, within a year after onset death results. It affects about 1% of the Fore women and their daughters and sons. It shows vertical transmission because in the Fore tribe New Guinian populations as a cannibalistic, religious ritual the females and their youngsters consume some flesh of dead relatives. It occurs sporadically also in neighboring tribes due to stenosis, LEOPARD syndrome, or it may be viral or it is due to prions. (See prion, Creutzfeldt-Jakob disease, LEOPARD syndrome, stenosis, encephalopathies).

KUZ: see ADAM

kV: kilovolt, 1,000 V. (See Volt)

KVLQT1: see Beckwith-Wiedemann syndrome

kW: kilowatt, 1,000 Watt. (See Watt)

KWASHIORKOR: a condition caused by malnutrition of humans on diets low in essential amino acids, particularly lysine, tryptophan and methionine. The symptoms are emaciation with altered pigmentation in patches on the hair and skin. On dark spots of the limbs and back the epithelial layer may be shed showing as pink blotches the raw flesh. If it is coupled also with a deficiency in caloric intake, the condition is further aggravated by losing both flesh and fat and generally coupled with dehydration as well (marasmic kwashiorkor). It is a wide-spread anathema particularly in the tropical and subtropical areas of the world with underdeveloped agriculture and political turmoil. The term's origin is an African Gold Coast (Ghanaian) language meaning pink boys, descriptive of the syndrome. This severe malnutrition affects primarily children. The therapy is a gradual return to balanced, nutritious diet. (See essential amino acids, high-lysine corn)

KYNURENINE: an intermediate in tryptophan metabolism. In humans an autosomal recessive deficiency of the enzyme kynurinase results in excessive amounts in xanthurenic acid in the urine. Xanthurenic acid is a tryptophan metabolite that accumulates also in case of pyridoxal phosphate (vitamin B_6) shortage in the diet. (See tryptophan)

← KYNURENIC ACID

L

λ: lambda bacteriophage. (See lambda phage)

L1: LINE1 (long interspersed repeat). (See LINE)

L1: see hybrid dysgenesis I - R

L amino acids (levorotatory amino acids): the natural amino acids.

L VIRUS: see killer strains

La: is an autoimmune antigen, transiently associated with pre-tRNAs and 5S rRNAs. It mediates both transcription initiation and termination by polIII. (See pol III, ribosomal RNA)

LABELING: attaching or incorporating into a molecule a radioactive or fluorescent compound that permits the recognition of the molecule itself in the cell or in the extract of the cells or any molecule with what it is hybridized, attached to or into what the label is inserted. (See also probe, nick translation, radioactive labeling, immunoprobe, fluorochromes, radioactive tracer, gene tagging, non-radioactive labels, biotinylation, FISH, immunofluorescence, aequorin)

LABELING INDEX: shows the fraction of cells that incorporate labeled nucleotides, i.e., the percent of S phase cells in a tissue. The fraction of labeled cells in relation to DNA content permits a convenient estimate of the cells in G1 phase (1 DNA unit), G2 + M (2 DNA units) and in between 1 and 2, indicating S phase. The fractions of cells can be analyzed with the aid of an automatic cell sorter. (See labeling, cell cycle, cell sorter)

LABOR: the process of child delivery.

LABORATORY SAFETY: the most important requisite is to know the potential hazards of equipment and the biological an chemical materials to be employed. Develop plans how to cope with possible accidents and how to dispose of spillage, fumes, fire, etc. Most of the commercial suppliers provide safety information for chemicals ordered. Use non-porous (neoprene) gloves. Sometimes using bare hands may be justified because accidental contact can be immediately sensed and proper washing can decontaminate the body. Fume hoods (with proper air exchange) must be used with chemicals that evaporate or sublimate. Biological hazards can be minimized by appropriate sterilization and the use of certified laminar flowhoods. Laboratory waste must be segregated for solids and liquids. Do not dump any chemical (mutagens and carcinogens) into drains that may hurt plumbers and cause problems at the waste water treatment. Radiation hazards may be prevented by monitoring with radiation counters and appropriate shielding, and by keeping workbenches clean. All laboratory personnel must be properly instructed about safety and checked regularly for compliance. Remember the admonitions of Paracelsus the 15th century physician and scientist that "Poison is everything and no thing is without poison. The dosage makes it either a poison or a remedy." (See environmental mutagens, chemical mutagens, ionizing radiation, radiation hazard assessment, gloves)

LABRUM: an anterior-most structure of the head of arthropods; in general morphology edges; lips are designated as such.

***Lac* OPERON**: *E. coli* can utilize the milk sugar lactose by splitting it into glucose and galactose with the aid of the β-galactosidase enzyme encoded by the *Lac z* gene. This enzyme is not made unless lactose or one of its analogs is present in the culture medium (inducible enzyme) and even then not until there is glucose available. This particular metabolic response is under a very precise and complex genetic regulation in prokaryotes. When transcription begins actually 3 genes are transcribed into a 3-cistronic RNA. The *Lac y* gene encodes a galactoside permease, a membrane protein that facilitates the uptake of the substrate of the galactosidase. The *Lac a* gene is transacetylase that acylates the galactoside with the assistance of acetyl-coenzyme-A. The complete nucleotide sequence of this regulatory upstream region is shown below. For the sake of brevity the map (8 min) is shown on top of next page. When galactosidase is not synthesized, a repressor protein is blocking a definite tract within a section of the promoter region, a part of the operator gene, the repressor-binding sequences. The product of the *Lac i* (1040 bp) gene is 152 kDa, 4 subunit repressor protein. The transcription of the *Lac i*

***Lac* operon** continued

gene is separately regulated from the genes that it controls by suppression (negative control). Since the repressor binding site is within the operator gene where the transcriptase enzyme

```
Lac i -⌈ --|CAP-cAMP|- ⌈operator-
           {repressor binding sequences-
                     transcription initiation}⌋ ⌉ – leader - z - y - a
        <--------------------PROMOTER-------------------->
```

(RNA polymerase) is attached to carry out its function, it prevents the expression (transcription) of the down-stream genes (z, 3510 bp; y, 780 bp; a, 825 bp) of the operon. The Lac Z protein is 125 kDA, the Lac Y and the Lac A are both about 30 kDa. Because of the coordinated operation of iuxta-positioned genes, the system was named *operon*. The *Lac i* gene is transacting because its product can flow to the operator irrespective where the gene is located within the cell, it can be in the vicinity of the operon or it can be carried by a plasmid. Furthermore, if there is an inactive i^- gene next to the operon within the bacterial chromosome and the z, y and a genes are transcribed, introduction of a i^+ (wild type) suppressor gene in a plasmid, the transcription is blocked from this transposition. Also it shows that the active form of i is dominant. The tetrameric repressor protein is normally a homotetramer, i.e., the four subunits are identical. If the cell has two different i genes that code for differently altered repressor monomers, the aggregate becomes a heterotetramer and may show *allelic complementation*. The i^{-d} gene product in the presence of the wild type polypeptide imposes a conformational change on the repressor tetramer and renders it inactive by "dominant negative complementation". This phenomenon also indicates that the monomers alone are not functional but their aggregate (quaternary structure) is the functional repressor. Base substitution mutations within the *Lac i* may abolish the ability of the protein to bind to the operator and then protein synthesis can go on constitutively (without a need for induction). Mutations may also just reduce its binding and then without induction still a reduced level of transcription can go on. If the mutations affect the inducer-binding sites of the repressor, it may no longer bind the inducer and it becomes a super-repressor mutation, i^s. The number of repressor molecules per cell is about 5 to 10.

Similarly, mutations within the operator region may alter the binding of the wild type suppressor. If the repressor is bound to it rather than the RNA polymerase, the transcription of the three structural genes (z, y, a) cannot proceed. The repressor does not prevent the binding of the RNA polymerase to the promoter, it may even enhance the binding of the polymerase but it blocks the initiation of transcription. Since the RNA polymerase is at its site before induction, as soon as the inducer makes contact with the repressor transcription is initiated almost immediately. The operator gene is unique because it does not have any product; it merely serves as the starter site for transcription if the RNA polymerase can attach to it. Therefore the operator must always be in cis position, in front of the structural genes. The binding sequences within the operator shown below have inverted repeats (**BOLD**) and the inactivating mutations are shown below in OUTLINE (modified after Watson, J.D. *et al.* 1987 Molecular Biology of the Gene, Benjamin/Cummings, Menlo Park, CA):

```
         TGG AATTGT GAGCGGATA ACAATT
         ACC TTAACT CT CGCCTAT TGTTAA
                       ⇔
            A - TGTTA --- C ----- T
            T - ACAAT --- G ----- A
```

The left side of the operator is more likely to render it unreceptive to repressor binding. If the repressor cannot bind to the operator, an operator-constitutive system emerges that is functional without induction (o^c). When the cells are grown without galactoside the number of

Lac operon continued

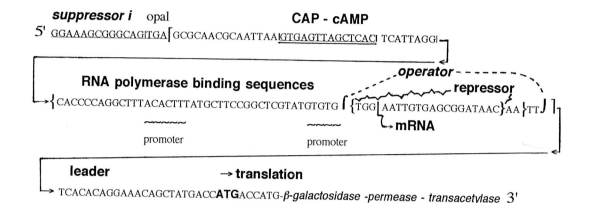

THE STRUCTURAL AND FUNCTIONAL ELEMENTS OF THE *LAC* OPERON OF *E. coli*. THE PROMOTER REGION IS ENCLOSED WITHIN THE SIGNS [ ] THE I CAT AND cAMP BINDING SITE I IS WITHIN THE PROMOTER AND IS SHOWN UNDERSCORED. THE SEQUENCES OCCUPIED BY THE DNA-DEPENDENT RNA POLYMERASE { ARE DELIMITED } WITHIN THIS REGION THE TWO SPECIFIC PROMOTER ELEMENTS ARE UNDERLINED ~~~~. THE OPERATOR REGION ⌠ ⌡ INCLUDES THE REPRESSOR-BINDING-SITE { } AND TRANSCRIPTION BEGINS WITHIN IT ↳ . THE LEADER SEQUENCE IS FOLLOWED BY THE ATG FORMYLMETHIONINE CODON AND CONTINUES INTO THE TRANSCRIPTS OF THE β-GALACTOSIDASE (z) - PERMEASE (p) - TRANSACETYLASE (a) GENES

galactosidase molecules may be less than 5 per cell. After supplying galactose, within a couple of minutes, the number of galactosidase molecules increases by about a thousand fold. If the inducer is used up, the system reverts to the uninduced state very rapidly because the half-life of the mRNA is only about 3 minutes. Although, the already synthesized proteins may linger on for a little longer. Induction can take place without β-galactoside if thiogalactosides, particularly the often used isopropylthiogalactoside (IPTG), are provided. These thiogalactosides induce the synthesis of the galactosidase enzyme although the enzyme cannot use these analogs as substrates. Therefore these are *gratuitous inducers*. The presence of the slight amounts of the permease protein in the non-induced cells is required to initiate the uptake that eventually jump-starts the system.

When the inducer is added to the system, it combines with the repressor and mediates a conformational change in it and thereby prevents it from attaching to the repressor-binding sites in the operator region. The repressor protein has two essential sites; one that binds inducer and the other that binds the operator. Although the binding of the repressor is not exclusively limited to the operator, it binds to this region by about ten million fold more

Lac **operon** continued

effectively than to other sequences of the genome. When the inducer is taken up by the cells, the unspecific binding to all over the genome does not change but the binding to the operator is almost completely relieved.

The mechanisms discussed above (*negative control*) are only parts of the total regulatory system and we will have to look at the role of the CAP - cAMP site upstream in the promoter. CAP stands for *catabolite activator protein*. It has been called also cyclic AMP receptor protein (CRP, encoded by gene *crp* at map position 73 min). CAP is a 22.5 kDa dimeric protein and the subunits contain a DNA-binding and another, transcription-activating sites. Thus, CAP interacts both with DNA at the evolutionarily conserved upstream site and with α-subunits of the RNA polymerase. The CAP protein becomes active upon forming a complex with cyclic adenosine monophosphate (cAMP). cAMP is formed from ATP by the enzyme adenylate cyclase (the encoding *cya* gene is at map position 84 min) and its formation is reduced by glucose. We have discussed earlier that as long as glucose is available for the cells β-galactosidase and the companion enzymes are not formed in appreciable amounts. This phenomenon is *glucose effect* or by another name *catabolite repression*. The basis of this repression is that there is not enough cAMP to activate transcription by the CAP-cAMP complex. If the CAP-cAMP system is defective the *lac* operon cannot function even if the repressor is not formed and the operator is constitutive. Therefore the CAP-cAMP complex constitutes a *positive regulatory* element of the *lac* operon in contrast to the *negative regulatory* i - o system seen above. We can conclude also that this system uses great wisdom in managing cellular energies: it calls to duty the enzymes only when they are needed and turns off the synthesis as soon as they are no longer needed. One of the products of the galactosidase, glucose reduces the synthesis of the enzyme that splits it off from lactose. (See also suppression, helix-turn-helix motif, transcription, translation, allelic complementation, galactose, galactose utilization, galactosidase, β-galactosidase [for assay], cAMP receptor protein, *Lac* repressor)

Lac **REPRESSOR**: a protein that is negatively regulating the *Lac* operon in the absence of lactose in the medium. When the repressor is combined with lactose (inducer) or isopropyl-β-D-thiogalactoside (gratuitous inducer), cAMP and CAP, the transcription of the structural genes may be started. The bound repressor either inhibits binding of the RNA polymerase or prevents elongation of the transcript. The Lac repressor (LacR) have a small headpiece (the binding domain) and a large core (regulatory domain). The intact 38 kDa repressor forms a homotetramer of 152 kDa. Each of the two LacR core dimers arranges the headpieces in such a manner that they bind maximally to the operator. Headpiece monomers bind only weakly. IPTG and the binding is reduced by three orders of magnitude. The LacR also mediates DNA looping that it would make contacts at multiple sites. There are about 20 members of the family of Lac repressor proteins. The lactose, fructose and raffinose repressors are tetramers whereas the others are dimeric. The majority of the Lac repressor (*LacI*) family members are most effective if no other proteins binds to them. The PurR (purine repressor), however, requires the presence of a co-repressor (hypoxanthine and guanine) ligand. (See *Lac* operon, purine repressor, cAMP receptor protein, hypoxanthine)

LacI^q : is a mutant *lacI* that synthesizes about 10 times more repressor than the wild type allele.

LACTACYSTIN: is a *Streptomyces*-produced inhibitor of proteasome by affecting the amino-terminal threonine. It inhibits several proteases, the cell cycle and causes neurite outgrowth in neuroblastoma cells. (See ubiquitin, neurite, neuroblastoma, proteasome)

LACTAMASE: see β-lactamase

LACTASE: β-galactosidase

LACTASE DEFICIENCY: see disaccharide intolerance

LACTATE DEHYDROGENASE (LDH): catalyzes the reaction ⇨ pyruvate + NADH + H$^+$ ⇌ lactate + NAD$^+$. Lactate is not utilized as such but it is converted back to pyruvate. The rationale of the reaction is to regenerate NADH for glycolysis in the skeletal muscles and shifting the metabolic burden from muscle to liver. In yeast the L- lactate dehydrogenase

Lactate dehydrogenase continued

is also called cytochrome b_2. The mammalian enzyme is a tetramer consisting of four subunits, each with an approximate M_r of 33,500. These subunits are encoded by two separate genes and can be combined into 5 different isozymic forms such as $A_4B_0, A_3B_1, A_2B_2, A_1B_3, A_0B_4$. In the skeletal muscles the A chains (synonymous with M) predominate whereas in the heart the B chains (synonymous with H). The isozymic forms also vary during development. In cancer cells the LDH isozymes are present in fetal rather than in the adult form. In humans the LDHA gene was assigned to chromosome 11p15. 4 and LDHB to 12p12.1 - 12p12.2. The testes-specific LDHC protein is also encoded in the close vicinity of LDHA. The nucleotide and amino acid homology between human LDHA and LDHB varies between 68 and 75%. In the mouse LDHC displayed 72-73% homology with LDHA and in humans the similarity was just slightly higher between the two amino acid chains. In Japanese population LDHA and LDHB deficiencies occurred at frequencies of 0.19 and 0.16, respectively. These figures are supposed to be too high for other populations. LDHA deficiency in humans was associated with skin lesions, myoglobinuria and fatigue. LDHA deficiency was found to be caused by 20 bp deletions in exon 6 but nonsense mutation GAG →TAG at codon 328 was implicated at another human deficiency of LDHA. LDHB deficiency was found to be due to a replacement of the highly conserved arginine 173 by a histidine. (See isozymes, code genetic, dehydrogenase)

LACTOFERRIN: an iron-binding glycoprotein in milk, in other secretions and in the neutrophils. It defends the cells against infections and it is a growth regulator and modulates killer cells of the immune system. It is also a peptide messenger, binding to conserved DNA sequences and activates transcription. (See messenger polypeptide, killer cells)

LACTIC ACID ($CH_3CHOHCOOH$, MW 90.08): is formed from pyruvate through glycolysis or by fermentation of lactose (yogurt). Lactic acid is the main endproduct of sugar metabolism of lactic acid bacteria. Lactic acid is formed in the skeletal muscles and oxidized in the heart for providing energy or it is converted again into glucose by gluconeogenesis. It is the most important preservative in silage fermented at moderate temperature.

LACTIC ACIDOSIS: see mitochondrial disease in humans

LACTICACIDURIA: due to an autosomal recessive condition blood lactate and pyruvate level is increased. In the different forms of the disease pyruvate carboxylase, pyruvate dehydrogenase, and phosphoenolpyruvate carboxykinase enzymes may be defective. Mental retardation, loss of hair, lack of muscle coordination and infant death may occur. (See ketoacidosis)

LACTOSE (milk sugar): a disaccharide of galactose + glucose. (See formula on p. 582)

LACTOSE INTOLERANCE: see disaccharide intolerance

LACTOSE PERMEASE: see *Lac* operon

LACTOSYL CERAMIDOSIS: is a deficiency of β-galactosyl hydrolase involving a hereditary sphingolipidosis type of disease. (See also galactosidase, sphingolipids)

LACUNA: a hole in the bacterial lawn caused by the production of bacteriocin.

LacZ: gene for β-galactosidase enzyme (map position 8 min). Another locus of *E. coli* (*ebgA^0*, map position 67 min) may evolve into a β-galactosidase gene if the locus at position 8 is deleted. At least two different mutations are required to acquire this enzyme activity but it may evolve to this state through different paths. The new activity is based on an immunologically different protein from that of the *lacZ* product. (See *Lac* operon)

lacZ ΔM15: in the bacterium the amino terminal of the β-galactosidase gene is deleted. (See *Lac* operon)

LADDER: a collection of precisely known length oligonucleotides that can be used as standards for identifying the length of DNA fragments separated by electrophoresis. The synthetic ladders generally have a certain number of progressive base increments, e.g. 123, 246, 369.

LaFLORA DISEASE: see myoclonic epilepsy

LAG PHASE: of a culture is when growth is minimal and under favorable conditions it may be followed by exponential growth. (See exponential growth)

LAGGING STRAND: of DNA facing the replication fork by its 5' end and the elongation can be accomplished only by adding 5'-P ends of the nucleotides to the 3'-OH ends by phosphodiester linkage and therefore it must be synthesized in pieces, in Okazaki fragments. (See DNA replication, replication fork)

LAGOTRIX (woolly monkey): see Cepidae

LAMARCKISM: is a largely discredited evolutionary theory embodying the ideas of the French biologist J.B. de Lamarck (1774-1829). He proposed a comprehensive theory claiming that evolution proceeds by the inheritance of gradually acquired characters. He supposed that the use or lack of use of a body structure eventually leads to reinforcement or lapse of that trait by strive and direct environmental influence. The giraffes stretched their neck to reach the tree tops, thus their *inner drive* and the circumstances contributed to their familiar shape. Contrarily, the current concepts, the neodarwinian theory believes that the longer-neck animals could feed better and thus through continuous selection for increased neck length, their progeny had a selective advantage and facilitated the propagation of cumulative mutations that assured the survival of the best adapted genotypes. Neo-Lamarckism is basically an identical dogma to Lamarckism, except it emphasizes the use or disuse idea rather than the inner drive or autogenesis aspects. (See also soviet genetics, lysenkoism, transformation, directed mutation)

LAMBDA CHAIN (λ chain): is an immunoglobulin light chain. (See immunoglobulins)

LAMBDA PHAGE (λ): is a temperate bacteriophage, an obligate parasite. It belongs to the family of *lambdoid* phages, that includes phages $\phi 21$, $\phi 80$, $\phi 81$, etc. Lambdoid phages are characterized by cohesive ends, ability to recombine, and inducibility by UV. Each λ particle contains in its icosahedral head (0.05 μm in diameter) one double-stranded DNA molecule of ca. 48,500 bp that has been completely sequenced. In its *E. coli* host it can replicate either autonomously and produce hundred progeny particles in 50 min at 37° C. Alternatively, it may insert into the host chromosome as a *prophage*, generally near the map position of the galactose (*gal*) operon. Phage λ usually does not kill all bacterial cells and therefore its plaques are turbid.

Gene *cI* is the principal phage gene enforcing the lysogenic state (being a provirus in the host chromosome). In addition, *cI* assures immunity to infection by other λ phages. When *cI* is inactivated, the phage may enter *productive growth,* i.e., it is liberated from the host chromosome (*induction*) and begins autonomous replication resulting in lysis of the host. Spontaneous induction occurs at frequency of 10^{-3} per generation. Exposure to UV light may cause induction in most cells. Mutations *recA-* and *cI ind-* prevent induction by radiation as well as by spontaneous means. The *N* gene of the phage regulates both *cI* (the repressor of autonomous replication) and the morphogenetic genes involved in operations during the lytic lifestyle. The λ genome may assume either a linear or circular form because at the ends (12 bp) of the DNA strands the complementary cohesive sites (*cos*) may open up to a linear form or circularize: 5' pGGGCGGCGACCT   CCCGCCGCTGGAp 5'. In-between these ends the genes (encoding about 50 proteins) are in functional units corresponding to their temporal sequence of expression. This relatively small genome transcribes genes in both strands of the DNA double helix. Because of this, it has both left- and rightward promoters (*transcrpitons*). The multiple promoters serve the purpose of most efficient regulation of expression of the genes. The immunity region contains the major regulatory genes of the phage. After infection of the host cell or induction, transcription begins at either leftward or at the rightward promoters. The leftward promoter, p_L mediates the transcription of genes involved in recombination (integration and excision) under the control of the (12.2 kDa) *N* gene product that may prevent transcription termination of genes using the p_L or p_R promoters. Some mutations (*ninL* and *ninR*) on both side of *N* may make the system insensitive to the N protein. The fate of the behavior of the phage is determined by a balance of the action of its own and some host genes. For lysogeny, genes *cII* and *cIII* (in opposite promoters, p_R and p_L, respectively) activate promoters p_E (that includes the *cI*) and p_L (that includes gene *int*). The *cI*, the λ repressor

Lambda phage continued

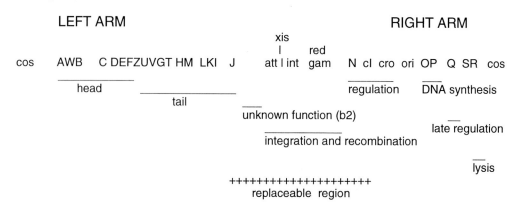

The major functional regions of phage λ. The head and tail genes are essential for morphogenesis and for the packaging of foreign DNA into phage vectors. The region between J and N (the stuffer) can be removed and replaced by foreign genes in the replacement vectors. Just left of the DNA replication genes is the phage immunity region. Before gene N is the left *nin* and after gene P there is right *nin* region controling transcription termination. The *ori* gene is the replicational origin and transcription may proceed left- or right-ward from p_L and p_R respectively. There are additional promoters for the delayed early genes (p_I, internal), for the transcription of genes located in the replaceable region and p_E and p_M for the establishement and maintenance of *cI* repressor controlling the transition from lysogeny to morphogenesis in cooperation with other genes. The total length of the DNA in the wild type phage is about 49.5 kbp. In order to assure packaging, the total DNA in the vectors cannot be less than 78% and no more than 105% of the wild type genome. The region of the A to J genes (ca. 20 kb) and that from p_R to *the cos* site (8 to 10 kb) must be retained for the viability of the particles.

produces then a 236 amino acid protein that prevents the expression of genes involved in phage DNA synthesis and morphogenesis and activates the synthesis of the Int and Xis proteins that in turn mediate recombination and thus the phage can integrate into the host chromosome as a prophage. In that state practically all λ functions cease, except *rex A* and *cI* that exclude superinfection by other phages (*immunity*) and replication of phage DNA becomes synchronized with that of the host. Actually, the prophage appears as an integral part of the host chromosome. At this state the concentration of the *cI* repressor is regulated by the p_M maintenance promoter. The recognition site of the lambda repressor is: TATCACCGCCAGAGGTA / ATAGTGGCGGTCTCCAT The ability of λ to have a lysogenic state, makes it a temperate phage. If for any reason *cI* is not functional the bacterial plate shows clear plaques, indicating that the phages are in the reproductive growth phase. The first gene turned on at the p_R promoter is *cro* that is followed by *cII* that has the product of λ gene P, a DNA pre-priming protein, with host proteins. The expression of gene Q prevents transcription termination of the genes involved in lysis and morphogenesis. Lysis of the host cell is mediated by the products of the S and R genes and this permits the liberation of the phage progeny. Under the influence of protein Q the $p_{R'}$ promoter assists in the transcription of all late genes and its role ends somewhere in the recombination-control region around the area of *b*. In the integrated state the prophage is replicated under the control of the host as if it would be an intrinsic part of the bacterial genome. Autonomous replication of phage DNA requires the function of the replicational origin (*ori*) where the O protein binds and replication originates. The replication is preceded by circularization of the phage DNA with the mechanical assistance of the cohesive (*cos*) ends

Lambda phage continued

then sealed by a DNA ligase. When the DNA injected into the host cell first replicates by the bidirectional theta (θ) replicational form and generates monomeric DNA circles. When the function of the *gam* gene is turned on, replication is switched to the bacterial recBC protein-dependent rolling circle type (sigma replication, σ). This replication results in the formation of linear concatamers attached by the *cos* ends. These concatamers are then stuffed into the preformed phage heads (encapsidation) as shown below and by the diagram of page 588.

The $p_{R'}$ promoter mediates the transcription of the head and tail genes. The head and tail components are made and assembled separately before joining into the mature phage particle (see illustration on next page). In this process, the cooperation of host proteins (groE) are needed. Into the prehead then the concatenated λ DNA molecules are filled in. The DNA was synthesized in head-to-tail molecules, involving more than a single genome (concatamer). The protein products of the *Nu1* and *A* genes bind to the DNA not far from the left *cos* sites and are brought to the prehead. The *FI* gene product is then reels it into the head until the terminal *cos* site is found where it is cut off by the Nu1-A protein complex (*terminase*). The product of the *D* gene stabilizes the head shell and the FII protein completes the head that is attached now to the pre-fabricated tail assembly. The mature phage particle is equipped with injector mechanisms that facilitate the adsorption of the particles to the host cell and introduction of its genome through the host cell membrane.

The phage genes under the control of the p_L promoter are often called *early genes*, the ones under p_R are *delayed early* and those regulated through $p_{R'}$ are refered to a *late genes*. The two major promoters, p_L and p_R may not necessarily stop at the termination signals represented by the arrowheads but may continue transcription and thus represent "read-through". The regulation of the *N* gene product is accomplished by acting at the *nut* sites rather then at the terminator sequences (t_L and t_R). The nut_L is located before the terminator less than 60 bp from p_L and the nut_R is about 250 bp from p_R. The *nut* regions have a dyad symmetry, including *box A* and *boxB*. The region of binding of the *Q* gene product is designated *qut*. The activities of the N and Q proteins frequently called *antitermination* function. The function of the λ N gene is regulated also by bacterial (N utilization) genes *nusA*, *nusB*, *nusE* and *nusG*. *NusA* is a transcription factor *nusG* encodes ribosomal protein 10. NusB protein and S10 dimer bind to the *boxA* sequences in bacteria. Nus G assembles the other Nus proteins for binding to RNA polymerase. NusA can facilitate transcription termination by the N protein at intrinsic terminators but all Nus proteins are required for stopping transcription in rho-dependent termination. When the Nus complex and other proteins associate with RNA polymerase, it becomes modified at the *nut* site and can pass through the terminator sites without stopping transcription. The polymerase core can associate either with the σ subunit of the transcriptase or with the Nus N complex. When the N protein replaces the σ subunit of the RNA polymerase the termination signals are ignored. The *qut* sites permit a change in transcriptase to work faster and avoid stoppage at the terminator sites and proceed through the lytic phase into the vegetative growth stage.

Recombination of λ may be mediated by the closely linked genes *exo* (also called *redX*, *redα* or *redA*) and the *bet* complex (also called *β*, *redβ*, *redB*). These two *red* complexes are involved in general recombination and are not dependent on the function of the host *recA* function. The *exo* gene codes for a 5'-exonuclease (M_r 24,000) that can convert a branched DNA structure to an unbranched nicked duplex by the process called *strand assimilation*. The *bet* gene product (subunit M_r 28,000) binds to the exonuclease and promotes reannealing with the complementary DNA strand. The *gam* gene product (M_r 16,500) in a dimeric form binds to the bacterial host enzyme recBC and inhibits its activities. General recombination is affected by *Chi* (χ) sites near the *A*, between *I* and *J*, *xis* and *exo*, within *cII*, between *Q* and *S* genes. *Chi*s do not have a gene product, rather they offer a suitable site for an enhancement of recombination (about five fold, measured by burst size) through the RecBC pathway.

Lambda phage continued

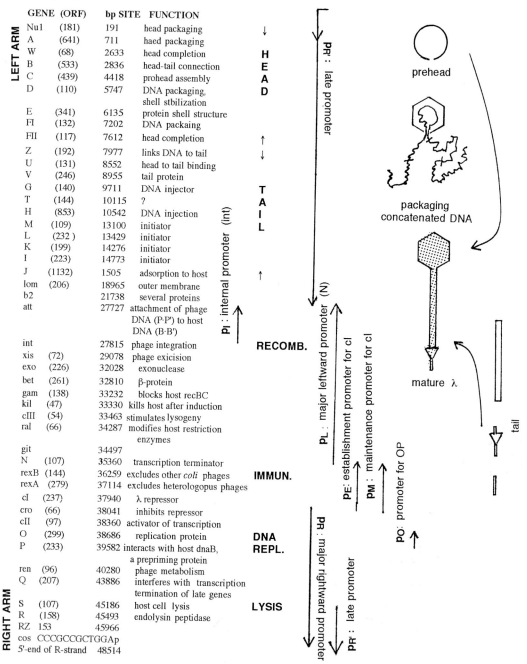

	GENE	(ORF)	bp SITE	FUNCTION		
LEFT ARM	Nu1	(181)	191	head packaging	↓	
	A	(641)	711	haed packaging		
	W	(68)	2633	head completion		**H**
	B	(533)	2836	head-tail connection		**E**
	C	(439)	4418	prohead assembly		**A**
	D	(110)	5747	DNA packaging, shell stbilization		**D**
	E	(341)	6135	protein shell structure		
	FI	(132)	7202	DNA packaing		
	FII	(117)	7612	head completion	↑	
	Z	(192)	7977	links DNA to tail	↓	
	U	(131)	8552	head to tail binding		
	V	(246)	8955	tail protein		
	G	(140)	9711	DNA injector		**T**
	T	(144)	10115	?		**A**
	H	(853)	10542	DNA injection		**I**
	M	(109)	13100	initiator		**L**
	L	(232)	13429	initiator		
	K	(199)	14276	initiator		
	I	(223)	14773	initiator		
	J	(1132)	1505	adsorption to host	↑	
	lom	(206)	18965	outer membrane		
	b2		21738	several proteins		
	att		27727	attachment of phage DNA (P·P') to host DNA (B·B')		
	int		27815	phage integration		**RECOMB.**
	xis	(72)	29078	phage exicision		
	exo	(226)	32028	exonuclease		
	bet	(261)	32810	β-protein		
	gam	(138)	33232	blocks host recBC		
	kil	(47)	33330	kills host after induction		
	cIII	(54)	33463	stimulates lysogeny		
	ral	(66)	34287	modifies host restriction enzymes		
	git		34497			
	N	(107)	35360	transcription terminator		
	rexB	(144)	36259	excludes other *coli* phages		**IMMUN.**
	rexA	(279)	37114	excludes heterologopus phages		
	cI	(237)	37940	λ repressor		
	cro	(66)	38041	inhibits repressor		
	cII	(97)	38360	activator of transcription		
	O	(299)	38686	replication protein		**DNA**
	P	(233)	39582	interacts with host dnaB, a prepriming protein		**REPL.**
	ren	(96)	40280	phage metabolism		
	Q	(207)	43886	interferes with transcription termination of late genes		
RIGHT ARM	S	(107)	45186	host cell lysis		**LYSIS**
	R	(158)	45493	endolysin peptidase		
	RZ	153	45966			
	cos	CCCGCCGCTGGAp				
	5'-end of R-strand		48514			

P_I : internal promoter (int)
P_L : major leftward promoter (N)
P_E : establishment promoter for cI
P_M : maintenance promoter for cI
P_O : promoter for OP
P_R : major rightward promoter
P_R' : late promoter
P_R' : late promoter

THE MAP OF BACETRIOPHAGE LAMBDA SHOWING THE ORDER OF THE MAJOR GENES. ORF INDICATES THE SIZE OF THE OPEN READING FRAMES AND bp SITE STANDS FOR THE NUMBER OF NUCLEOTIDES STARTING AT THE LEFT ARM. THE BASE SITES MAY VARY IN DIFFERENT λ PHAGES. THE MORPHOGENETIC PATHWAY IS ALSO OUTLINED. (Data from Hendrix, R.W. et al., eds 1983 Lambda II. Cold Spring Harbor Laboratory Press, Cold Spring Harbor, NY)

Lambda phage continued

Near the *chi* sites an octamer consensus exists: 5'-GCTGGTGG-3'. The stimulation of recombination by *chi* is not limited to its location but it is effective over a considerable distance. *Chis* act "dominant" because it is enough to be present in one of the recombining partners. When *Chi* is present in only one of the partners, recombinants without active *chi* are much more frequent than the reciprocals (*non-reciprocality*). Enhancement of recombination to its left is greater than to its right side (*directionality*).

Recombination may be also *site-specific*, requiring a special nucleotide sequence, the *primary att* sites of high efficiency and the much lower efficiency secondary *att* sites. The *att* sites have common 15 bp core sequences that are 70 to 80% AT. In the phage the primary *attP* sites are represented as ---POP'--- and there is a corresponding site of homology in the bacterium *attB* (~~~BOB'~~~). After recombination (insertion) at the crossing over region (*O*), by the means of pairing as an inversion loop, in the bacterial cell there will be:

~~~ BOP' ----- POB' ~~~

In excision the process is reversed. The insertion is mediated by λ protein Int and the excision uses the products of λ genes *int* and *xis*. Both processes require also the bacterial integration protein factor (IHF, M_r 20,000) and other host gene products. The coding region of *int* extends 84 to 1151 bp right from the crossing over site. The M_r of the Int protein is about 40,000 and at the amino terminus it is rich in basic residues and it works as a topoisomerase. The Xis protein shares amino acid sequences with Int. The *xis* product contains 72 amino acids with 25% Lys and Arg. Because of the site-specific recombination, λ phage can mediate *specialized transduction*.

Phage λ has been used for the development of a number of genetic transformation vectors. Genes not exceeding 5% of the genome can be inserted at a single target site into the λ DNA and can be propagated (*insertional vectors*). The nucleotide sequences between genes *J* and *N* can be deleted (representing about 25% of the genome) and replaced, and the DNA still can produce nearly normal size plaques as long as the total length of the DNA remains no less than 37 kb and not more than 52 kb (*replacement vectors*). In *cosmid vectors* it is essential to retain the *cos* sites plus 4-6 kb at both termini, including also the origin of replication. The vectors so constructed can propagate foreign DNA exceeding 30 kb. (See Charon vectors, cosmid vectors, theta replication, rolling circle, rho terminator, specialized transduction, burst size, chi elements, recombination, helix-turn-helix, DNA-binding protein domains)

LAMBDOID PHAGES: are closely related to λ phage such as P22 of *Salmonella*. (See P22)

LAMBERT: a unit of luminous intensity; 1 lumen per cm^2. The lumen is the unit of luminous flux emitted in a unit solid angle (steradian) by a uniform point source of one candela. (See candela [candle]).

LAMELLA: a thin plate or membrane sheet.

LAMELLIPODIUM: sheet-like cellular extension involving actin and assisting cell movement.

LAMINS: intermediate filament proteins; upon their polymerization the nuclear lamina are formed; during interphase lamins support the nuclear membrane. (See intermediate filaments)

LAMININ: is a protein localized in the synaptic cleft between the neuronal basal lamina of the muscle cell sheath and the acetylcholine receptors. (See synapse, acetylcholine, agrin)

LAMPBRUSH CHROMOSOME: giant chromosome in oocytes (mainly in amphibia), with conspicuous loops extending about 40 μm on a core resembling brushes used to clean the glass chimney of kerosene lamps. These loops are highly active in RNA synthesis, and although not polytenic, they are well visible under the light chromosome. A set of amphibian lampbrush chromosome may display 10,000 loops alternating with condensed chromomeres. On the surface of the DNA loops the nascent RNA transcripts may be visible. This high activity may precede meiosis and continues during it in order to secure a good supply to meet the needs later of the rapidly developing

zygote. (See giant chromosomes, newt)

LANDRACE: is a locally adapted variety of plants or breed of animals that may include a number of different genotypes; it was produced by intuitive selection of the growers or husbandryman, respectively.

LANGE-NIELSSEN SYNDROME (cardio-auditory syndrome): involves heart arrhythmia and deafness, and sudden death. The gene responsible KVLQT1 spans the region 1p15. This area includes also the Beckwith-Wiedemann syndrome. (See Beckwith-Wiedemann syndrome, LQT)

LANGER-GIEDION SYNDROME: generally involves mental retardation, small brain (microcephaly), bulbous nose, sparse hair, emergences on the bones (exostosis), etc. The autosomal dominant phenotype is based on deletions in the 8p22-8q24.13 chromosomal region. (See deletion, exostosis, trichorhinophalangeal syndrome, mental retardation, head/face/brain defects)

LANGERHANS ISLETS: cells with dentate nucleus in the pancreas, arranged in groups. They appear to have antigen-presenting abilities. (See diabetes mellitus)

LANGUAGE, COMPUTER: a set of representations used by a computer program.

LANGURS: see Colobidae

LANOSTEROL: is a precursor of cholesterol. (See geranyl pyrophosphate, cholesterol)

LARGE T ANTIGEN: see SV40

LARIAT RNA: is formed during the splicing reaction of the primary transcript of eukaryotic genes. In the first step the pre-mRNA is cut at the junction of exon 1 and intron resulting in a piece of RNA containing intron - exon 2 and this "2/3 molecule" immediately forms a loop like lariat (reminding to the tethering rope of cowboys). In the second step intron - exon 2 junction is cut and exons 1 and 2 are ligated (spliced). Afterwards intron is released and the procedure follows for the other junctions. (See introns, splicing)

LARON TYPE DWARFISM: see pituitary dwarfness. (See dwarfness)

LARVA: an insect at the developmental stage after hatching from the egg. It resembles more to a worm than to an adult insect and after pupation the imago emerges from it. During the larval stage the germline and the imaginal discs are laid down. (See *Drosophila*)

LASCAUX CAVE: located in the caves of Dordogne, southwestern France, reveals the remark-

able art of the Cro-Magnon men, originated about 15,000 years ago. These colored paintings disclose a great deal about the types of animals present in the area in the paleolithic age and attests the intelligence and artistic abilities of the early European ancestors. More recently, even older (35,000-40,000 years) ancient artwork and artifacts have been discovered both in Europe and Africa. The animal shown at left displays great similarity to a present day longhorn bull. A comparison with the modern breeds of cattle, the success of animal breeding can be assessed. (Courtesy of the Caisse Nat. Mon. Hist. Sites, France)

LASER (light **a**mplification by **s**timulated **e**mission of **r**adiation): equipment produces electromagnetic radiation in the infrared and visible spectrum by stimulation of atoms. The laser beam does not diffuse like that of an electric light. The radiation travels in the same direction, at the same wavelength of very narrow frequency band. Thus it focuses all the energy to a fine point. Lasers can produce radiation at many wavelengths and frequencies. Intense light sources and electron currents can activate gases, semiconductors and ions in solid material to produce coherent laser light. The lasers can be pulsed or continuous. The former produces extremely high peaks of power, the latter highly stable and gives pure emissions. Laser radiation is applicable to many scientific instruments (e.g. laser scanners), photochemistry (e.g., laser photolysis) and also to surgery because of the properties outlined above.

LASER DESORPTION MASS SPECTRUM: measures the molecular weight of proteins and thus assists rapid identification and comparison with information in data bases. (See laser, mass spectrum, electrospray MS, matrix-assisted laser desorption ionization/mass spectrometry)

LATE GENES: are transcribed only later during the life cycle of an organism. The viral genes involved in replication and in the synthesis of structural (coat) proteins (e.g. during the lytic phase of a bacteriophage). These genes use late promoters, different from the early ones. (See also early genes, delayed-response genes)

LATE PERIOD: of phage development starts with the beginning of DNA replication. (See development, lambda phage)

LATENT PERIOD: the infection has taken place but the symptoms are not manifested yet; in phage biology the time between injection of the phage DNA into the bacterium and the beginning of lysis. (See burst)

LATE-REPLICATING CHROMOSOME: or chromosomal region is heterochromatic and genetically not active. (See Barr body, lyonization, heterochromatin)

LATERAL TRANSMISSION: is the same as horizontal transmission. Lateral transmission is suggested generally when similar nucleotide sequences occur among species that may not be related by orthologous descent. A common objection for such cases is that convergent evolution of these genes or sequences may not be ruled out with great certainty. (See transmission, transposable elements, transformation, infectious heredity, orthologous genes, convergent evolution)

LATHYRISM: is caused by the presence of β-cyano-alanine and its decarboxylation product, β-aminopropionitrile in the seeds of the food and forage legume, *Lathyrus sativus*. Some other contaminating weed legumes may also be major culprits in lathyrism. In some countries the ground seed is used for a filler in bread making. As a feed its amount should be kept below 10% of the ration. Cooking substantially reduces its adverse effect. These compounds affect the cross-linking of collagen and resulting is spasms, pain, paralysis of the lower extremities (paraplegia), abnormal sensitivity (hyperesthesia), burning sensation of the skin (paresthesia), curvature of the spine, rupture of the aorta, etc. (See collagen)

Lathyrus odoratus (sweet pea): an ornamental legume (2n = 14), a favorite object of early studies on the gene-controled anthocyanin synthetic pathways. The shape of the pollen is determined by the dominant *L* allele (long ▢) and the recessive *ℓ* allele (disc-shaped ◯). Though this is a gametophytic trait, the phenotype is determined by the genotype of the anther (sporophytic) tissue and thus is an example of delayed inheritance. (See delayed inheritance)

LATIN SQUARE: a square array of items in parallel columns and rows in such a manner that each must occur once but only once in each row and column. Such a design lends itself for the evaluation of the experiments by analysis of variance and it can be employed in agricultural field experiments but it is useful for pharmacological, microbial and other research where the partition of the data into relevant variances is very important. (See analysis of variance)

LAURYL SULFATE: see dodecyl sulfate sodium salt

LAW: both civil and criminal law considers genetic principles and methods. Examples are: regulating marriage, social policy, demographic factors, using dermatoglyphycs (fingerprinting), serological methods, DNA fingerprinting, patents, etc. The branch of genetics involved in legal matters is called forensic genetics. (See forensic genetics, patents, genetic privacy)

LAW OF LARGE NUMBERS: the increase in the size of a population assures the closer fit of the observed data to a valid null hypothesis or that the experimentally observed mean represents the true mean of the population dispersed according to the normal distribution. (See normal distribution, mean, null hypothesis)

LAWN, BACTERIAL: a Petri plate containing nutrient agar and inoculated with bacterial cells.

LAWRENCE-MOON SYNDROME: is an autosomal recessive mental retardation involving retinal defects, underdeveloped genitalia and partial paralysis.

LAWRENCE-SEIP SYNDROME: a lipoatrophy (loss of fat substance from under the skin); it is accompanied by enlargement of the liver, excessive bone growth and insulin-resistant diabetes.

LB AGAR: add 1.5 to 2% agar to LB medium. (See LB bacterial medium)

LB (Luria - Bertani) **BACTERIAL MEDIUM**: H_2O, deionized, 950 mL, bactotryptone 10 g, bacto yeast extract 5 g, NaCl 10 g, adjust pH to 7 with 5 N NaOH, fill up to 1 L.

LC50: lethal concentration 50 of a chemical is expected to cause death in 50% of the treated cells or individuals of a population. (See also LD_{50}, LCLo, LDLo)

LCA ONCOGENE: isolated from human liver carcinomas was assigned to human chromosome 2q14-q21. (See oncogenes)

LCK ONCOGENE: is a lymphocyte non-receptor protein tyrosine kinase of the SRC family. It is located to human chromosome 1p35-p32. (See oncogenes, tyrosine kinase, signal transduction, T cell)

LCLo: lethal concentration low. The lowest concentration of a substance, in air that causes death to mammals in acute (<24h) or subacute or chronic (>24 h) exposure. (See LDlo)

LCR (locus control region): is a nuclease-hypersensitive region far upstream or downstream or other sites apart from the structural gene. Its presence is required for the expression of a particular gene locus and its role is probably opening of the chromatin for transcription. All the genes are in competition for the assistance of the LCR and those which are closer to it have an advantage. LCR may be at many kb distance from the locus it controls. The multiple genes appear to be transcribed alternately as the chromatin loops back and forth. (See nuclease sensitive site, regulation of gene activity, heterochromatin, chromatin, position effect, looping of DNA)

λDASH: a replacement vector for the stuffer DNA fragment which carries multiple cloning sites on both sides of *red* and *gam* genes and appropriate promoter(s) specific for T3 or Ty RNA polymerase and so the insertions can be transcribed without a need for recloning. (See λFIX, EMBL3, stuffer DNA, lambda phage, cloning)

LDL: see low-density lipoproteins, familial hypercholesterolemia, sterol

LDL RECEPTOR: mediates LDL endocytosis; it is encoded in human chromosome 19-p13.1-13.3. It is homologous with the EGF receptor. (See low-density lipoproteins, EGFR, lysosomes, LDL)

λdgal: lambda phage deficient but carries the galactose gene. (See specialized transduction)

LDLo (lethal dose low): the lowest dose of a substance introduced by any route, except inhalation, over a period of time in one or more portions, and has caused death to mammals. (See LClo, LD50)

LD50: a calculated dose of a substance expected to kill 50% of the experimental population exposed to. (See LDLo)

LDP (long-day plant): see photoperiodism

LEADER PEPTIDE: directs the translocation of proteins. In bacteria it consists of 16-26 amino acids involving a basic amino terminal, a polar central domain and a non-helical carboxy domain. The latter is essential for recognition by leader peptidase to cut it off. For mitochondrial import, the leader has 10-70 residues and it is rich in positively charged and hydroxylated residues. Similar leader peptides mediate the import of proteins from the cytosol to the chloroplasts. The leader sequences are not entering the target organelles. (See also signal peptide)

LEADER SEQUENCE: non-translated stretch of nucleotides at the 5'-end of the mRNA. (See mRNA)

LEADING STRAND: of DNA is facing the replication fork by the 3'-OH end extended by adding directly 5'-deoxynucleotidephosphate to that end (after removal of the γ and β position phosphates from the nucleotide triphosphate precursors. (See also lagging strand, replication fork, replication, DNA replication)

LEAF SKELETON HYBRIDIZATION: is a procedure carried out on plant leaves infected by cauliflower mosaic virus vector with the same purpose as colony hybridization of bacteria, i.e. to detect foreign gene sequences in the CaMV vectors. (See CaMV, colony hybridization)

LEAKY MUTANT: has an incomplete genetic block in a synthetic step. (See genetic block)

LEARNING BEHAVIOR: is partly genetically controlled and it has been found that slow-learning mice have reduced amount of protein kinase C in the hippocampus of the brain, thus it appears that the PKC gene is a part of the polygenic system affecting this complex trait. (See brain human)

LEARNING DISABILITY: see Tourett's syndrome, dyslexia

LEAST SQUARES: formula is the basis for the theory of regression: ⟶ $$b = \frac{S[y\{x - \bar{x}\}]}{S[x - \bar{x}]}$$

where S means sum, y = one of the variates, x = other variate, $\bar{x}$ = mean of x. Least square methods used also for the estimation of evolutionary distance. The smallest minimum sum of squared differences computed from paired data indicates the best topology for an evolutionary tree. (See minimum evolution method, four-cluster analysis, neighbor joining method

LEAVING GROUP: the displaced molecular group in a chemical reaction.

LEBER OPTIC ATROPHY: see mitochondrial disease in humans, optic atrophy, amaurosis congenita, eye diseases

LECITHIN: is a glycerol phospholipid, also called phosphatidyl choline.

LECITHIN:CHOLESTEROL ACYLTRANSFERASE DEFICIENCY (Norum disease, Fish-eye disease): is either a relatively rare recessive Norum disease coded in human chromosome 16q22 and in mouse chromosome 8 or by a dominant mutation at the same locus in the fish-eye disease. This defect in lipid metabolism causes proteinuria, anemia, renal and heart defects. In the Norum disease there is a general failure in esterification of cholesterol in high-density lipo-protein whereas in the fish-eye diseases the deficiency is more specific. The name comes from the opacity of the eye resembling that of boiled fish. (See lipoprotein, apolipoprotein, cardio-vascular disease)

LECTINS: were first identified as plant proteins that agglutinate erythrocytes by virtue of binding to surface sugars. Lectins occur also in invertebrate and vertebrate animals and may serve as ligands also in the natural killer cells of mouse. They also play a role in general cell adhesion, surface recognition and protection against bacteria and viruses. (See concavalin, selectin, cell adhesion, killer cells)

LEECH: a hirundinaceous (blood-sucking) lower animal with only ≈350 nerve cells per ganglion.

LEF (lymphoid-enhancer binding factor): a transcription factor regulating lymphocyte different-iation; it is a member of the high mobility group proteins. It interacts with β-catenin and regulates signal transmission to the nucleus. (See high mobility group proteins, lymphocytes)

LEFT-HANDEDNESS: see handedness

LEGHEMOGLOBIN: is a hemoglobin protein coded for by genes of leguminous plants and this protein accumulates in the root nodules formed by the presence of nitrogen fixing bacteria. The leghemoglobin transfers oxygen to the electron-transport system of the bacteria and pre-vents the accumulation of toxic amounts of oxygen that would interfere with nitrogen fixation. (See also globin, myoglobin, hemoglobin)

LEGHORN WHITE: the poultry breeds White Plymouth Rock and White Wyandotte have plumage determined by recessive genes *ii cc*. The White Leghorn on the other hand has the *IICC* genotype where *I* is a dominant suppressor gene of color (*C*). Thus when it crossed to another white breed, e.g. White Wyandotte, in the F_2 of their hybrids the segregation is 13 white or speckled and 3 black. The blacks have the genotype either *iiCC* or *iiCc*.

LEGIONNAIR's DISEASE: is caused by the bacterium *Legionella pneumophila*.

LEGIT: a statistical concept worked out by R.A. Fisher (Biometrics 6:353) showing the change in allelic frequencies in a cline. If the allelic frequency is known, the legit can be read from the table published in the Biometrics article cited above. (See cline, diffusion genetic)

LEGUME: a taxonomic group of plants (e.g. pea, beans) with an ability to accumulate atmos-pheric nitrogen in symbiosis with some nitrogen-fixing bacteria. (Rhizobia)

LEIGH's ENCEPHALOPATHY: the autosomal recessive disorders occur in multiple forms and are characterized by high pyruvate and lactate concentrations in the serum and in the urine. The biochemical findings may be caused either by a defect of pyruvate carboxylase or a necessary co-factor of the enzyme, thiamin triphosphate (TTP). The latter deficiency may be brought about by the absence of thiamin pyrophosphate-adenosine triphosphate phosphoribosyl trans-

Leigh's encephalopathy continued

ferase. The pyruvate carboxylase gene appears to be in human chromosome 11. The nuclear-encoded (3q29 and 5p15) flavoprotein subunit of succinate dehydrogenase is controlling a mitochondrial enzyme complex II, including succinate dehydrogenase. Gluconeogenesis may become defective and necrotic brain lesions, heart and respiratory disorders may be present. Prenatal diagnosis is successful in some cases. (See neuromuscular diseases, gluconeogenesis, mitochondrial diseases in humans)

LEIOMYOMA: generally benign tumors of the smooth muscles occurring in the uterus, genitalia, gullet, etc., controled by autosomal dominant genes. (See cancer)

LEISHMANIA : is a protozoon related closely to *Trypanosoma*. In mammals, including humans, it causes skin and visceral ailments ranging from relatively mild to life-threatening types depending on the infectious species. The amastigote form occurs in the mammalian blood and the promastigote form in the gut of the sandfly which spreads the infection. The virulence and transmission is controled by lipophosphoglycan on the surface of its cells. (See *Trypanosoma*, kinetoplast, *mariner*)

LEK: is a social mating group of animals.

LEMMA: the upper cover bract of the grass flower, frequently bearing an awn at the tip. (See palea)

LEMON (*Citrus limon*): 2n = 18 or 36. (See orange, grapefruit)

LEMUR: *Cheirogaleus major* and *C. minor*, 2n = 66; *Hapalemur griseus griseus*, 2n = 54; *Hapalemur griseus olivaceus* 2n = 58; *Lemur catta* 2n = 56; *Lemur coronatus* 2n =46; *Lemur fulvus albifrons* 2n = 60; *Lemur fulvus fulvus* 2n = 48; *Lemur macaco* 2n = 44; *Lemur variegatus subcinctus* 2n = 46. (See prosimii, primates)

LENGTH MUTATION: involves either deletion, duplication, insertion or any other alteration of the chromosome or nucleic acid sequence affecting the size of a tract. (See chromosomal aberrations)

LENTICEL: structure on tree barks permitting passage of gaseous substances; lens-shaped glands at the base of the animal tongue.

LENTIGINS (singular lentigo): freckles. (See LEOAPRD syndrome, xeroderma pigmentosum)

LENTIL (*Lens culinaris*): a pulse, popular in the old world because of its high protein ($\approx 25\%$), 2n = 2x = 14.

LENTIVIRUSES: cytopathic retroviruses with similarity to HIV. They are suitable for vector construction by pseudotyping. An advantage of these vectors is being suitable for transformation of non-dividing cells. (See HIV, pseudotyping)

LENTZ - HOGBEN TEST: the same as the ascertainment test.

LEOPARD CAT (*Felis bengalensis*): 2n = 38.

LEOPARD SYNDROME: it is an acronym for **L**entigines, **E**lectrocardiographic conduction abnormalities, **O**cular hypertelorism, **P**ulmonary stenosis, **A**bnormalities of the genitalia, **R**etardation of growth, sensorineural **D**eafness, and autosomal dominant inheritance. The lentigens may turn neoplastic. (See lentigines, hypertelorism, stenosis, sensorineural, cancer)

LEOPOLD OF ALBANY: hemophilic son of Queen Victoria, who died at age 31 of hemorrhage after a fall, and who transmitted this gene through her daughter, Alice to his grandson Lord Trematon who also died from hemorrhage after an auto accident. (See hemophilia)

LEOPOLD OF BATTENBERG: hemophilic grandson of Queen Victoria who died of hemorrhage after surgery at age 33. (See hemophilia)

LEPORE: see thalassemia

LEPRECHAUNISM (Donahue syndrome): dwarfism caused by a defect in the insulin receptor.

LEPTIN: is assumed to be a hormonal feedback signal acting on the hypothalamus - and regulating food intake and metabolic rate. A deficiency or resistance to leptin causes increased food intake in mouse and humans whereas injection of leptin, the hormone, encoded by the *ob* gene of mice, reduces body weight. Recombinant human leptin corrects sterility in *ob/ob* female mice. Leptin is signaling through the STAT proteins and modulates the activity of insulin. (See obesity, STAT, insulin)

LEPTONEMA: see leptotene stage, meiosis

LEPTOTENE STAGE: is distinguished in meiosis when the chromosomes appear under the light microscope as single strands although they have been doubled during the preceding S stage. The chromosomes are much relaxed and stretched out. The threads (leptonema) are usually quite tangled and their termini cannot be distinguished by light microscopy. Bead-like structures of localized condensations, the chromomeres, are usually visible. (See meiosis, chromomere)   LEPTOTENE STAGE →

LESBIAN: homosexual female. (See homosexuality)

LESCH-NYHAN SYNDROME: see HPRT

Lespedeza bicolor: forage legume, 2n = 18, 20, 22.

LESTR (leukocyte-expressed seven-transmembrane-domain receptor): same as fusin

LET: linear energy transfer is a measure of energy delivered by ionizing radiation in keV(kilo electron volts) per nanometer path within the target. Approximate LET values in biological material: gamma rays from ^{60}Co source (1.2 to 1.3 MeV): 0.3, hard x-rays (250 keV): 3.0, β-rays from ^{3}H (0.6 keV): 5.5, recoil proton from fast neutrons: 45, heavy nuclei from fission (α-particles): 5,000. (See also ionizing radiation, kinetics, DNA repair)

let-23: a gene of *Caenorhabditis* coding for a transmembrane receptor protein. (See signal transduction)

LETHAL EQUIVALENT: is the designation of genetic factors that in an individual, in combinations, are responsible for death, e.g., if three genes each are expected to reduce life expectancy by a chance of 0.33, their combined lethal equivalent value is ≈1, i.e. approximately the same as one that causes early death with 100% probability. The actual calculation must be carried out however by a different procedure because no information can be obtained on the number of the individual recessive sublethal genes. Thus if it is observed that the infant mortality among the offspring of totally unrelated parents is (F = 0) is 8%, and among the offspring of first cousins (F = 1/16) of the same population is 13%. [F is the coefficient of inbreeding]. The difference 0.08 - 0.13 = 0.05, 5%. Then 5% x 16 = 0.8, i.e. 80%. Since among diploids two recessive alleles at a locus are required for expression, hence the frequency of lethal equivalent recessive genes is $\sqrt{0.8}$ < 0.89, thus the number of lethal equivalent recessive genes per gamete in this particular population is ≈ 0.89, close to 1 per gamete. (See coefficient of inbreeding, coefficient of coancestry, genetic load, incest)

LETHAL FACTORS: have been extensively used for the genetic analyses of developmental pathways because they permit studies on the consequences of arrest of differentiation at particular stages. Conditional lethal mutations which survive under certain temperature regimes or on supplemented nutritive media offer particular advantages because they can easier be analyzed by biochemical methods. The presence of lethal factors may modify segregation ratios. Dominant and recessive lethal factors may not permit the expression of syntenic genes or only at low frequency depending on the extent of recombination. Therefore genes on the homologous strand may appear in excess. Recessive lethal mutations may cause 2:1 segregation instead of 3:1 in F_2. (See also segregation distorter, meiotic drive, certation)

LETHAL MUTATION: fails to survive under normal condition. (See also conditional lethal mutations)

LETTUCE (*Lactuca sativa*): composite salad crop with chromosome numbers n = 8, 9 and 17.

LEUCINE METABOLISM: leucine [(CH3)2CHCH2CH(NH2)COOH] may be formed from isoleucine [CH2CH(CH3)CH(NH2)COOH]. (See isoleucine-valine biosynthetic pathway, 3-hydroxy-3-methylglutaryl CoA lyase deficiency, methylcrotonylglycinemia, methylglutaconicaciduria, hydroxymethylglutaricaciduria)

LEUCINE-RICH REPEATS: 15-29 leucines assist ligand recognition in processes such as signal transduction, cell development, DNA repair and RNA processing. (See leucine zipper. ligand, DNA repair, introns)

LEUCINE ZIPPER: dimeric regulatory proteins of two subunits; at the COOH terminus leucine occurs at every seventh position; at the amino terminal domain positively charged amino acids are found and this region is bound to DNA in a zipper-like manner. The leucine zippers may be homo- or heterodimeric and thus increase their specificity by a combinatorial control mechanism. The basic leucine zipper (bZIP) contains 25 conserved amino acids of these 9 substitutions disrupt function. The binding domain is usually 100 amino acid long (See DNA-binding protein domains, regulation of gene activity, Max, induced helical fork)

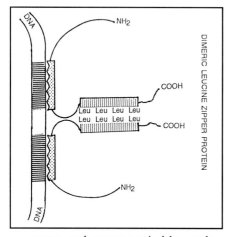

LEUCOCYTE (leukocyte): white blood cell. (See blood)

LEUCOPLASTS: plastids without carotenoids and chlorophylls.

LEUCYL tRNA SYNTHETASE (LARS): charges tRNA^{Leu} by leucine. Its gene is in human chromosome 5. (See aminoacyl-tRNA synthetase)

LEUKALEXINS: are released by effector cells and they slowly degrade DNA of the target cells. (See immune system)

LEUKEMIA: a cancer of the blood forming organs, characterized by an increase of the number of leukocytes and accompanied by enlargement or proliferation of the lymphoid tissues. It involves anemia and increasing general weakness and tiredness. Two main types are generally distinguished: *acute myelogenous* (AML) that affects primarily adults and *chronic lymphocytic* leukemia (CLL) where the circulating malignant cells are differentiated B lymphocytes. The *acute lymphoblastic leukemia* affects mainly children. The acute myeloid form frequently involves a translocation between human chromosome 8 and 21 [t(8;21)(q22;q22)]. In chromosome 21 the breakpoints are generally within the same intron of the receptor for the granulocyte-macrophage colony stimulating factor (CSF2R). Patients in relapse usually have activated Jak-2 protein tyrosine kinase. Tyrphostins (AG-490) effectively block leukemic cell growth in these cases by inducing apoptosis in the cancer cells without affecting normal blood-forming cells. The *chronic myeloid leukemia* (CML) is associated generally with a translocation between chromosomes 22q11.21 and 9q34.1. The translocation is within the BCR (break point cluster region) of chromosome 22 and in chromosome 9; the ABL gene (Abelson murine leukemia virus oncogene) is affected. The translocation causes a fusion between the 5' proximal region of one of the BCR spanned loci and the ABL gene. The fused BCR-ABL region encodes a 210 kDa protein. Transplantation experiments in mice demonstrated hematological malignancies in the recipients of this protein. *Chronic lymphocytic leukemia Type2* (BCL2) is a follicular lymphoma involving translocations between chromosome 18q21 - chromosome 14. It has been suggested that first chromosomal segment 13q32 (site of the immunoglobulin heavy chain gene J) moved to gene BCL in chromosome 18 and placing BCL under the influence of the IgJ$_H$ enhancer, leads to a low-grade malignancy. In another step a translocation with chromosome 8 brought in the MYC c-oncogene (myelocytomatosis v-oncogene homolog in chromosome 8q24) and the double translocation caused a high-grade malignancy. Transfection with a BCL-2 construct resulted in oncogenic potential. BCL-2 product is an integral mitochondrial membrane protein (M_r 25,000). *B cell leukemia/lymphoma* (BCL3 or BCL4) is apparently caused by a human chromosome 19q13 oncogene. Usually it involves translocations. *Chronic B cell leukemia* (BCL5) involves also a translocation between BCL-2 and a truncated MYC resulting in high-grade expression. *Chronic lymphatic leukemia* (CLL or BCL1) is a clearly familial disease involving translocations (11;14)(q13;q32). Again the translocation involves the joining segments 3 and 4 of the heavy chain Ig in chromosome 14. As a consequence the immune system is weakened and the leukemia is often associated with other disturbances of the immune system (hyperthyroidism, pernicious anemia, rheumatoid ar-

Leukemia continued

thritis, auto-immune disease). *Acute T cell leukemia* (ATL) is due to the activation of the α-chain of the T cell receptor (TCRA, chromosome 11p13) by translocations t(11;14)(p13;q11). The 14q11 band contains the variable region of TCRA and the 11p15.5 is the location of the HRAS (Harvey rat sarcoma) oncogene. *Lymphoid leukemia*, LYL1, involves translocations between chromosome 7 at the T cell receptor β-chain gene (TCRB, 7q35) and the LYL1 gene, chromosome 19p13. *Myeloid/lymphoid mixed lineage leukemia* (MLL) involves translocations of 11q23 and displays lymphoid myeloid phenotypes and it is present in infancy. The MLL gene is an apparent regulator of a homeobox, shared by other animals. The very common B-cell precursor (BCP) leukemia respond favorably to the CD19-associated tyrosine kinase inhibitor, genistein. CD19 is B cell-specific TK receptor and its destruction leads to apoptosis of 99.999 percent of the leukemia cells. *Stem cell leukemia* (SCL) gene is normally involved in the differentiation of the hematopoietic cells but it may form a complex with the oncogene product LMO-2 and GATA proteins and the tetramer may act in the development of acute T cell leukemia. (See interferon, T cell, TCR, RAS, immunoglobulins, ABL, Philadelphia chromosome, BCR, MYC, colony stimulating factor, apoptosis, leukemia inhibitory factor, homeobox, Hodgkin disease, CREB, neurofibromatosis, myelodysplasia, GATA, genistein)

LEUKEMIA INHIBITORY FACTOR (LIF): a cytokine involved also in the development of motor neurons. (See neurons, cytokines)

LEUKOCYTE: same as leucocyte. (See blood)

LEUKOPENIA: a condition with reduced number of leukocytes and neutrophils in the blood.

LEUKOSIS: is the organic basis of leukemia. (See leukemia)

LEUKOTRIENES: 20 carbon carboxylic acids with one or more conjugated double bonds and some oxygen substitutions. They are formed from arachidonic acid by lipoxygenase and control inflammatory and allergic reactions such as asthma, and indirectly obesity. Lekotrienes may activate the transcription factor peroxisome proliferator-activated receptor α that is involved in the regulation of enzymes of fatty acid oxidation and indirectly in inflammatory reactions. Leukotriene C4 synthase (LTC4S) is encoded in human chromosome 5q35. (See lipoxygenase, fatty acids, peroxisome, integrin, anemia, hemolytic anemia)

LEVOROTATORY: the plane of polarized light is rotated counterclockwise.

LEWIS BLOOD GROUP: is characterized by the expression of the Le gene coding for α4-L-fucosyltransferase. Four different types can be characterized with the frequencies indicated: (*1.*) Le^{a-b+}, 0.7, (genotype Le/le Se/se), (*2.*) Le^{a+b-}, 0.20 (genotype Le, se), (*3.*), 0.09 (genotype le, Se/se), (*4.*). Le^{a-b-}, 0.01 (genotype le, se). The ABH blood group may cause less severe reaction in incompatible application in blood transfusion than the ABO yet it may cause serious hemolytic problems. (See ABH antigen, ABO blood group, Bombay blood group, Secretor)

lexA: gene of *E. coli* (map position 91) controls resistance/sensitivity to X-rays and UV, it is a repressor of all SOS repair operons. Its protein product is autoregulated. The LexA repressor protein (22 kDa) is cleaved by the RecA protein and then the genes that it repressed become activated. (See DNA repair)

LEYDIG CELLS: see Wolffian ducts

λFIX: is a replacement vector very similar to λDASH. (See λDASH)

LFA (lymphocyte-associated proteins): belong to the integrin family present on white blood cells and T lymphocytes and promote cell adhesion. (See CD proteins)

LFT: low frequency transducing lysate is generated when aberrant excision of a phage particle takes place from a single lysogen. (See HFT, lysogen)

LH: luteinizing hormone: (See luteinizing hormone, animal hormones)

LHCP (light-harvesting chlorophyll protein complex): see chlorophyll-binding protein complex

LHERMITTE-DUCLOS DISEASE: see multiple hamartoma syndromes

LIABILITY: the appearance of many human traits cannot be explained by simple genetic mechanisms because they show up more frequently in certain human families than in others or in the

GENETICS MANUAL

Liablility continued

general population yet the proportions of afflicted individuals varies a great deal. The manifestation of these traits is commonly explained by polygenic systems. Polygenic systems are generally influenced by various environmental effects that makes predictability of the manifestation quite difficult. The frequency of diabetes, schizophrenia, hypertension, dental caries, peptic ulcer, various forms of cancers, etc. belong to these categories. Until a certain threshold is reached the person is considered healthy; beyond that point medical attention is necessary. The passing of the threshold requires special unidentified environmental conditions. Naturally, it is important to have some predictive ability regarding the liability of individuals to succumb to such diseases. A relatively simple statistical procedure exists to assess the heritability of the liability based on the normal distribution of these traits. In order to proceed with the empirical calculation, the table below may help:

1	2	3	4	5	6	7	8	← COLUMNS
$Q \rightarrow$	0.001	0.005	0.010	0.050	0.100	0.150	0.200	
$t \rightarrow$	3.090	2.576	2.326	1.645	1.282	1.036	0.842	
$z \rightarrow$	0.0037	0.1446	0.02665	0.10314	0.17550	0.23316	0.27996	

Q = the incidence of the trait, t = Student's t distribution with one-tail of the normal distribution, z = is the truncation point at the t value.

Let us assume that the incidence of the trait in the general population is $Q_p = 0.001$ and the incidence among the offspring of an afflicted individual is $Q_a = 0.1$. The truncation point in the general population is thus $t_p = 3.090$ (first column) and that in the afflicted family $t_a = 1.282$ (6th column). The mean liability of the affected parent $\mu^1 \cong z_a/Q_a = 0.0037/0.001 = 3.7$ The mean liability of the offspring of the affected parent is $\mu^2 \cong t_p - t_a = 3.090 - 1.282 = 1.808$. The heritability of the liability of the trait determined by offspring-parent regression is $h^2 = 2b_{O\bar{P}}$ (where regression is on one parent rather than on the midparent value) and $b_{O\bar{P}} = \mu^2/\mu^1 = 1.808/3.7 \cong 0.49$; hence heritability, $h^2 = 2 \times 0.49 = 0.98$. This hypothetical value is obviously very high for heritability. We must keep in mind, however that the trait has low penetrance for being controled by a single gene and most likely it is under the control of multiple genes and is not expected to occur frequently because the threshold event needed for its manifestation. Would have been the incidence of the trait in the general population $Q_p = 0.005$ and that of the afflicted family $Q_a = 0.500$, the heritability would have come out by this calculation as 0.65. The liability is increasing with consanguinity of the parents. (See also normal distribution, Student's t distribution, z, correlation, heritability, polygenic inheritance, GTL, consanguinity, penetrance, recurrence risk, risk, threshold trait)

LIBIDO: sexual drive, the motive for sexual contact although it may exist even in the absence of sexual potency such as in the sterile mules.

LIBRARY: a collection of cloned fragments, representing the entire genome (at least once). The construction of the library requires cutting up the genomic library by one or more restriction enzymes or mechanical shearing of the DNA. Subsequently the DNA fragments are ligated into appropriate cloning vectors and transformation is carried out in a suitable host (most commonly into *E. coli*). The screening of the cloned colonies requires either nucleic acid hybridization, south-western analysis, immunochemical procedures, recombinational assays or genetic analysis and identification through one or more of these final steps. (See gene bank, vectors, transformation, restriction endonucleases, south-western method)

LICENSING FACTOR (replicational licensing factor): see MCM

LIDDLE SYNDROME (pseudoaldosteronism): is an autosomal dominant moderate hypertension encoded in human chromosome 16. It is involved in the control of a Na$^+$ ion channel and renal sodium reabsortion. (See aldosteronism, mineral cortical syndrome, hypertension, degenerin, Bartter syndrome, Gitelman syndrome)

LI-FRAUMENI SYNDROME (LFS): in some families several types of cancers (breast, sarcomas, brain, lung, laryngeal, adrenal cancers, melanomas, prostate cancer, gonadal germ cell tu-

Li-Fraumeni syndrome continued
mor, and leukemias) are found with about 50 fold increased incidence, compared to the general population. The assumption is that in these families alleles of the dominant p53 tumor-suppressor gene are segregating. LFS is encoded in human chromosome 17p13.1. (See tumor suppressor genes, *p53*, malignancy, cancer, Lynch cancer families, substitution mutation)

LIF: see APRF

LIFE BEGINNING on EARTH: latest evidence indicates ≈ 3,800 Myr before present. (See Myr)

LIFE CYCLE: the successive changes in generations of organisms, including the modes of reproductions. In higher organisms this includes the generation of gametes, fertilization and other modes of propagation and the development of the adult forms. (See gametophyte, sporophyte, germline, alternation of generations, and see also the name of individual organisms)

LIFE FORM DOMAINS: bacteria, archaea and eukarya. (See archea, prokaryotes, eukaryotes)

LIFTING: see phasmid

LIGAND: molecule that can bind (by non-covalent bonds) to a receptor by virtue of special affinity. Ligands are transcription factors, growth factors, hormones, neurotransmitters, antigens, morphogens, membrane receptors, etc.

LIGAND-ACTIVATED SITE-SPECIFIC RECOMBINATION: the Cre recombinase is fused to a ligand-binding domain of the estrogen receptor (ER), and thus the recombinase may be activated by tamoxifen (but not by estradiol). A DNA sequence flanked by loxP sites can then be excised. (See Cre/loxP, site-specific recombination, tamoxifen, estrogen, estradiol)

LIGASE CHAIN REACTION: to both strands of denatured DNA two pairs of complementary oligonucleotides are added at immediate vicinity of each other and amplified. The ligation products are useful only for the signal amplification but not for amplifying other DNA copies.

LIGASE, DNA: enzymes that tie together ends of single polynucleotide chains. The *E. coli* enzyme requires 5' phosphate and 3' OH termini and NAD, the T4 enzymes requires ATP for the reaction. The T4 enzyme can ligate both cohesive and blunt ends although the latter reaction is much slower but its efficiency can be increased by monovalent cations (NaCl) and polyethylene glycol (PEG). Human cells contain at least four ligase isozymes all are ATP-dependent. The efficiency of these ligases depends on the total concentration of the substrates and also on the closeness of the ends to be ligated. DNA ligases have essential natural role in replication, recombination and in the laboratory for the construction of vectors or joining any DNA molecules. (See replication, recombination, vectors)

LIGASE, RNA: mediate the joining of RNA 5' and 3' ends in the presence of ATP; they have important role in the final processing of RNA transcripts. There are a very large number of RNA ligase ribozymes that structurally belong to three classes and within the classes to several families. (See ligase DNA)

LIGATION: covalent joining of nucleotide ends by DNA or RNA ligase, respectively.

LIGATION OF EMBRYOS: imposing a constriction on an e.g., *Drosophila* embryo. The development of the thoracic (central) structures requires the cooperation of both poles. Ligation at the blastoderm stage interferes with the development of only one segment. (See morphogenesis)

LIGHT: see electromagnetic radiations

LIGHT CHAIN: see DNA heavy chain, antibody

LIGHT DIRECTED PARALLEL SYNTHESIS: light directs the simultaneous synthesis on solid support. The pattern of masking of the light activates different regions of the solid state support and facilitates the chemical coupling reactions. The activation is the result of the removal of different photolabile protecting groups. After deprotection the first set of building blocks (amino acids or nucleotides bearing the photolabile compound), the entire surface is exposed to light and the reaction takes place only in the areas which were exposed to light in the preceding step. Subsequently the substrate is illuminated by using another mask and then a second block is activated. The masking and illumination pattern determines the location and the ultimate product. This way complex molecules can be fabricated in a combinatorial manner and the product may be useful for industrial and molecular biology studies. (See array hybridization, DNA chips)

LIGHT HARVESTING CHLOROPHYLL PROTEIN COMPLEX: see LHCP, CAB

LIGHT INTENSITIES: of various emitters (Lambert units [1 Lambert = 1 new candela/cm^2/π]): sunshine at noon 519,000, sun at the horizon 1,885, moonlight, 0.8, Tungsten bulb of 750 watt 7,500, mercury vapor light of 1,000 watt 94,000. These are approximate average data. (See illumination, electromagnetic radiation)

LIGHT MICROSCOPY: is used for the study of small biological specimens, biological tissues, cells and subcellular organs down to a resolution of 0.2 µm. The most essential elements of the light microscope are the objective lenses for viewing details of different sizes. The nosepieces generally contain objectives with numerical aperture 0.1 (4X), 0.25 (10X), 0.65 (high dry, 40X) and 1.25 (oil-immersion, 100X) lenses. Eyepieces are usually 10X and permit the viewing of the objects at about 1,000 fold larger, at maximum. The condensor focuses the light, coming from a special low voltage illuminator. The microscope stand includes a stage for the slides on what the specimens are moved and various adjustment knobs for focusing and adjusting light intensities. The light microscopes may be equipped with other devices, e.g. camera stand or built in camera, filters, etc. The common light microscopes require special handling of the specimens before viewing, e.g., fixation, staining, sectioning, etc. On the slides, the specimen is generally covered by a very thin (about 0.13-0.25 mm) cover glass (slips). For viewing natural specimens in three dimensions, stereomicroscopes are used that may permit magnifications within the range of 2X to 160X, generally with zooming capabilities. (See resolution optical, fluorescence microscopy, phase-contrast microscopy, Nomarski, confocal microscopy, electron microscopy, fixatives, stains, sectioning)

LIGHT REACTIONS: can be carried out in light only. (See DNA repair)

LIGHT REPAIR (of DNA): splitting of pyrimidine (thymine) dimers by visible light-inducible enzymes. (See DNA repair)

LIGHT RESPONSE ELEMENTS (LRE): are nucleotide sequences binding transcriptional regulators of light receptor protein (e.g. phytochrome) genes. (See phytochrome, hormone response elements)

LIGHT-SENSITIVITY DISEASES: see albinism, xeroderma pigmentosum, Bloom syndrome, trichothiodystrophy, ataxia telengiectasia, Fanconi anemia, Cockayne syndrome, Hartnup disease, protoporphyria, Rothmund-Thompson syndrome, RAD, DNA repair, excision repair

LIGNIN: a rigid woody polymer of coniferyl alcohol (derived from phenylalanine and tyrosine) and related compounds, occurring in plants along with cellulose. It is the industrial source of vanillin, dimethyl sulphoxide. Lignin is used in manufacturing certain plastics, rubber, precipitation of protein, etc.

LIGULE: is a "tongue-like" outgrowth on the upper and inner side of the leaf blade (at the leaf sheath) of grasses.

LIKELIHOOD: is a statistical concept dealing with a hypothesis based on experimental data and can be expressed as L(H/R) where L is the likelihood, H is the hypothesis and R the results obtained experimentally. In contrast, probability is fixed by the fit of the data to a pre-conceived null hypothesis. The likelihood ratio LR = probability of data model/probability data of null hypothesis or the log likelihood ratio, i.e. log (LR). (See maximum likelihood, lod)

LILAC: the plant *Syringa vulgaris* is an ornamental shrub (2n = 46, 47, 48); lilac is also a rodent fur color determined by epistatic action of certain gene combinations.

LILIACEAE: a family of monocotyledonous plants; many of them have been exploited for cytological studies because of the large chromosomes, *Lilium* subspecies (2n = 24), onion (*Allium cepa* , 2n = 16, 32), hyacinth (2 = 16), *Trillium* subspecies, (2n = 10), *Bellevalia* 2n = 8, 16). In 1928 John Belling counted 2193 chromomeres in the pachytene chromosomes of *Lilium pardalinum* and believed this number represented the number of genes in the species. In *Fritillaria davisii* (2n =24) the DNA content in the somatic nuclei was estimated to be 295 pg (approximately 50 times the amount in human somatic cells or 1,000 times that *Arabidopsis* or *Drosophila*). (See DNA, genome, Dalton)

LIM DOMAIN: a cystin-rich zinc-binding unit facilitating protein-protein interactions in signaling molecules, transcription factors, cytoskeletal proteins. LIM has the properties of an organizer. (See adaptor proteins, organizer, Williams syndrome)

LIMB BUD: the embryonal cell group initiating limb development. (See AER, ZPA, organizer)

LIMB DEFECTS IN HUMANS: can be isolated but most commonly associated with particular syndromes. In some cases entire limb(s) are absent (amelia) or there is a partial reduction (meromelia) frequently missing some parts between the extremities (hand or foot) and the trunk (phocomelia). Not uncommonly, extra fingers, toes are involved or fusion or deformation(s) are observed. The defects may be caused by teratogenic factors, chromosomal defects or mutations and may be accompanied by a variety of other symptoms of complex syndromes. (See adactyly, polydactyly, syndactyly, Holt-Oran syndrome, Mebiuss syndrome, Roberts syndrome, De Lange syndrome, Poland syndrome, ADAM complex, Chotzen syndrome, thrombocytopenia, orofacial-digital syndromes, Majewski syndrome, Pätau's syndrome, exostosis, arthrogryposis, clubfoot, renal dysplasia and limb defects)

Limnaea peregra: is a freshwater snail displaying delayed inheritance of its shell shape. (See delayed inheritance, maternal effect genes)

LINCOMYCIN: (antibiotic): 6-[1-methyl-4-propyl-2-pyrrolidinecarboxamidol]-1-thio-o-erythro-D-galactooctopyramide

LINCOLN, ABRAHAM, PRESIDENT (1809-1865): is suspected to have had the relatively mild form of spinocerebellar ataxia although earlier it was also suggested that he might have had the Marfan or the Marfanoid syndrome. (See ataxia, Marfan syndrome, ataxia)

LINE: a genetic stock with defined characteristics.

LINE: long ($\approx$ 5-7 kbp) interspersed repetitive DNA sequences, including retroposons generated by reverse transcription; may occur in 10^4 to 10^5 copies in eukaryotic genome. (See SINE)

LINE1 (L1): the major mouse LINE that can be used for DNA fingerprinting of YAC sequences from the mouse genome. (See DNA fingerprinting, YAC)

LINEAGE: the line of descent from an ancestral cell or ancestral individual(s). (See cell lineage)

LINEAR ENERGY TRANSFER: see LET

LINEAR PROGRAMMING: analysis or solution of problems when linear functions are maximized or minimized.

LINEAR QUARTET: in the embryosac of plants the 4 megaspores and in the asci of some fungi 4 spores arranged in the order they were formed during meiosis. (See megaspore, embryosac, tetrad analysis)   LINEAR SPORE QUARTET →

LINEAR REGRESSION: see correlation

LINEARIZATION: conversion of the covalently closed circular nucleic acid into an open linear form.

LINEWEAVER- BURK EQUATION: is the reciprocal of the Michaelis-Menten equation: $\frac{1}{v} = \frac{K_m + (S)}{V(S)} = \frac{K_m}{V}\left[\frac{1}{S}\right] + \frac{1}{V}$. It provides a simpler solution for representing enzyme kinetics; 1/v is plotted against 1/(S) or (S)/V against (S). (See also Michaelis-Menten equation and the Eadie-Hofstee plot for explanation of the symbols)

LINKAGE: association of genes within the same chromosome but can be separated by recombination. In test cross four phenotypic classes are expected in equal numbers if the segregation is independent (absence of linkage). These four classes can be designated *AB* (a), *Ab* (b), *aB* (c) and *ab* (d) for the *n* individuals. For the verification of linkage chi square tests (χ^2) can be used. First the segregation for *A* can be examined, $\chi^2_1 = \frac{[a+b-c-d]^2}{n}$ then in a second step the χ^2_2 (for *B*) $= \frac{[a-b+c-d]^2}{n}$ is determined. The linkage $\chi^2_L = \frac{[a-b-c+d]^2}{n}$.

Example: *AB*: 191, *Ab* : 37, *aB* : 36, *ab* : 203, Total 467. Without linkage in each class 116.7

Linkage continued
is expected in a testcross.

	Chi square	Degree of Freedom	Probability of Greater χ^2
Segregation for A - a,	0.259	1	>0.5
Segregation for B - b	0.362	1	>0.5
Linkage	220.645	1	<10^{-10}
Total	221.266	3	

The segregation of each marker is very normal but the linkage chi square is extremely high indicating that the segregation is not independent. Another simpler formula that does not take into account the transmission of the markers (p = parental, r = recombinant) is shown below:

$$\chi^2 = \frac{(observed_p - expected_r)^2}{(observed_p - expected_r)}$$

Similarly, for F$_2$ the chi squares for the *A* and the *B* locus and for linkage can be determined for the *AB* ($\frac{9}{16}$ n), *Ab* ($\frac{3}{16}$ n), *aB* ($\frac{3}{16}$ n), ab ($\frac{1}{16}$ n) classes designated as a, b, c and d, respectively:

$$\chi^2_A = \frac{[a+b-3c-3d]^2}{3n}, \quad \chi^2_B = \frac{[a-3b+c-3d]^2}{3n} \quad \text{and} \quad \chi^2_L = \frac{[a-3b-3c+9d]^2}{9n}$$

When we want estimate the probability of finding linkage all over the genome, the chance is relatively very small of finding one. The P value obtained by the χ^2 procedure can be converted to an actual estimate of linkage based on a Bayesian estimate:

Probability of linkage = $1 - \left(\dfrac{P}{P + f_{swept}}\right)$ where P is the value obtained by χ^2 and f_{swept} is the fraction of the genome that appears to be linked around the marker within the so called *swept radius*. The swept radius is the maximal distance on both sides of the marker between two loci with a certain probability. Usually a probability limit of 0.95 is chosen to be significant. Thus the swept radius is (shown in parenthesis) for 10 cM (21%), 20 cM (35%) and 30 cM (44%), respectively, for this critical probability. (See recombination frequency, F$_2$ linkage estimation, tetrad analysis, recombination mechanisms, mapping genetic, mapping function, interference, coincidence, linkage group, chi square, sex linkage, sex-linked lethal mutation, lod score, affected-sib-pair method, maximum likelihood method applied to recombination, Bayesian theorem)

LINKAGE, COMPLETE: linkage is practically complete in the heterogametic sex in insects, e.g., in the majority of the species of *Drosophila* male (XY) and in the silkworm female (WZ). Recombination in male flies have been observed, however, in the presence of transposable elements causing hybrid dysgenesis. Linkage is practically complete within inversions because the recombinant gametes or zygotes are lethal (balanced lethals). In case of multiple translocations involving all or most of the chromosomes, genes situated even in different chromosomes do not segregate independently and this is characteristic for the complex heterozygotes. (See inversions, translocations, hybrid dysgenesis, complex heterozygote)

LINKAGE DETECTION, TETRADS: the linear spore tetrads can be arranged as PD, TT and NPD. At linkage the frequency of the double recombinant class (NPD) must be smaller than the single recombinant (TT) class. If the deviation between TT and NPD is very small a chi square test may be necessary by using the formula: $\chi^2 = \dfrac{(PD - NPD)^2}{PD + NPD}$. (See tetrad analysis, chi square, linkage)

LINKAGE DISEQUILIBRIUM: certain groups of genes are syntenic in higher frequency than it would be expected on the basis of unhindered genetic recombination. In other words, the tighter the linkage between genes, the less the chance to reach linkage equilibrium. Another

Linkage disequilibrium continued

cause of this persistent association may be based on their selective advantage as a group or recent introgression. With these possibilities in mind the linkage disequilibrium determined for several loci may provide information on the chromosomal location of a gene. Disequilibrium is expected to be the greatest for a group of markers that are closest to the gene what needs to be assigned to a position. The quantitative measure of linkage disequilibrium:

$$\delta = (p_D - p_N)/(1 - p_N)$$

where p_D and p_N represent the two different linkage phase chromosomes. The decline (Δ_n) of the linkage disequilibrium depends on the recombination frequency (unless a particular gene block has high selective advantage and linkage drag exists): $\Delta_n = \Delta_0(1 - r)^n$ where r = recombination frequency, n = number of generations Δ_0 is the beginning disequilibrium. (See Hardy - Weinberg equilibrium, recombination genetic, coupling, repulsion, mutation dating)

LINKAGE DRAG: undesirable genes are preserved in a population on the basis of hitchhiking if no recombination occurs between the selected desirable and the undesirable genes. (See hitchhiking, genetic load)

LINKAGE EQUILIBRIUM: see genetic equilibrium

LINKAGE GROUP: syntenic genes (situated in one chromosome) belong to a linkage group. Therefore the number of linkage groups equals the number of chromosomes in the basic set. When the linkage information about a chromosome is incomplete, it is possible to see two or more linkage groups for a particular chromosome because gene clusters may recombine freely when the distance between them is 50 map units or more. If trisomic analysis is used all syntenic genes must appear as a linkage group, irrespective of their dispersal along the chromosome. In multiple translocations (complex heterozygotes) several interchanged chromosomes are transmitted together and form "super linkage groups" such as in the plants *Oenotheras*. Similar translocation complexes may exist also in animals. In the termite *Kalotermes approximatus* meiotic chains of 11 to 19 chromosomes have been observed, forming a single male-determining linkage group. (See mapping, synteny, trisomic analysis)

LINKAGE IN AUTOTETRAPLOIDS: see recombination in autotetraploids

LINKAGE IN BREEDING: in applied genetics linkage may be very useful if advantageous traits are controled by closely linked genes. The opposite is true if disadvantageous characters are syntenic. The breeder may benefit by knowing linkage with neutral, visible or easily detectable chromosome markers that permits the monitoring of the inheritance of sometimes hard-to-recognize quantitative traits. The male silkworm produces 25-30% more silk than the female. By having a dominant color (dark) gene in the Y chromosome, the less productive female-producing eggs (XY) can be separated by an electronic sorter before hatching, and eliminated. In the baby chickens it is difficult to identify the males by genitalia but the presence of the *B* (*Barring*) X-chromosome-linked gene reduces the color of head spots more effectively in two (XX, males) than in single dose (XY, females) and thus by autosexing the males can early be assigned to meat, whereas the females to egg production regimes. The use of RFLP techniques facilitates the recognition of linkage of quantitative trait loci with DNA markers. The knowledge of recombination frequency may facilitate the planning of size of the populations for selecting a combination of desirable genes. The yellow seedcoat color gene in the flax is very tightly linked to high quantity and high quality of oil. Unfortunately, low yields and susceptibility to disease are also linked in coupling. Therefore, it is practically impossible to produce a commercially acceptable flax variety with yellow seed coat. The availability of appropriate linkage information may facilitate also the positional cloning of agronomically desirable genes. If the frequency of recombination is known the expected frequency of double homozygous dominant offspring can be predicted by the the following formulas, for coupling $0.25(1- p)^2$ and for repulsion $0.25(p)^2$ where p = recombination frequency. Genetic transformation by cloned genes has an enormous advantage over the classical breeding methods because it does not have to go through the tedious and long-lasting period of selection after crossing if the linkage is tight. (See QTL, autosexing, gain, chromosome substitution, positional cloning, transformation)

Linkage in breeding continued

The frequency of sixfold double homozygous dominants for desirable genes located in two chromosomes. The numbers in the body of the table indicate the number of individuals carrying 12 advantageous alleles (+) in an F_2 population of 10^6 size. (After Power, L. 1952 In: Heterosis, Gowen, J.W., ed., Iowa State College Press, Ames, IA, p. 315).

Recombination Frequency	Genetic Constitution of the F_1		
	$(+++/---)$ $(---/+++)$	$(++-/--+)$ $(--+/++-)$	$(-+/-+-)$ $(-+-/+-+)$
0.000	62,500	0.000	0.000000
0.075	33,498	1.448	0.000063
0.225	8,134	57.787	0.410526
0.375	1,455	188.787	24.441630
0.450	523	234.520	105.094534
0.500	244	244.141	244.140625

LINKAGE MAP: see genetic map, mapping genetic, physical map

LINKED GENES: are in the same chromosome and their frequency of recombination is less than 0.5. (See linkage, recombination)

LINKER: (commercially available) very short DNA sequences that can be added by blunt-end-ligation to DNA termini to generate particular cohesive ends for insertion of passenger DNA fragments in molecular cloning or transformation. (See blunt-end ligation)

LINKER INSERTION: insertion of a DNA sequence in place of deleted sequences to test for specific functional units in the region deleted. (See linker scanning)

LINKER (NUCLEOSOME): the 40-60 of nucleotides (generally associated with H1, sometimes, with H5 histone) that connect the cores of nucleosomes. (See nucleosome, histones)

LINKER SCANNING: a molecular method for the identification of upstream regulatory elements.

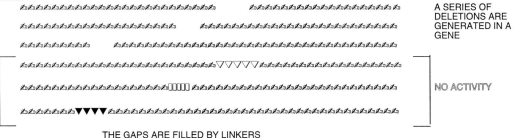

LINKER SCANNING FOR THE IDENTIFICATION OF THE POSITION OF ESSENTIAL REGULATORY ELEMENTS

The procedure is as follows: in defined lengths of upstream DNA fragments deletions are induced at different positions. The gaps are filled in such a way that the total original length of all fragments is neither reduced nor extended even by a single base pair. If the linker falls into

Linker scanning continued
the gaps that was normally the site of a regulatory element in front of the gene, expression is reduced or abolished, serving as an evidence that the site of the deletion (now the linker) involved an essential upstream element. (See linker, promoter)

LINKING NUMBER: indicates how many times one strand of DNA double helix crosses over the other strand in space. The linking number may reflect the degree of supercoiling in a closed DNA molecule. The linking number has two components the writhing number (W) and the twisting number (T); a change in linking number $\Delta L = \Delta W + \Delta T$. A change in linking number requires the breakage of at least one strand. Molecules that are the same as their linking number are called topological isomers. After a break, one strand can be rotated about the other and such a reaction may change one isomer into the other by DNA topoisomerases. These topological processes are the requisites for several functional activities of DNA such as replication, recombination, transcription and others. (See writhing number, isomer, topoisomerase)

LINKING NUMBER PARADOX: in a nucleosome the DNA is coiled by about 1 and 3/4 turns over the histone octamer and forms close to -2 superhelical turns. Yet when the histones are removed from the DNA-histone fiber, only -1 negative supercoiled turn is found. The explanation of this paradox may be that the DNA released from the histones becomes more tightly coiled than in the nucleosome-restricted form, i.e. the bound DNA has different structural periodicity than the free form. (See nucleosome, supercoiled DNA, histones)

LINSEED: see flax

LION (*Panthera leo*): 2n = 38; lion x panther [leopard] (*Panthera pardus*) are sterile.

LIPASE: enzymes that digests triaglycerols into fatty acids. (See triaglycerol, lipids, fatty acids)

LIPASE DEFICIENCY: *hepatic lipase* (HL) dominant 15q21, the *liposome lipase* (8p22) result in triglyceride-rich high- and low-density lipoproteins occur also in a recessive form at the same locus and their deficiencies are called then *chylomicronemia* or *hyperlipoproteinemia*, respectively. There is also an autosomal dominant *hormone-sensitive lipase* (HSL). The pancreatic lipase deficiency and the lysosomal acid lipase are autosomal recessive.

LIPID BILAYER: is the main structural element of membranes with two layers of hydrophobic lipid tails facing each other and the hydrophilic heads exposed at the outer parts. Within the lipids proteins occur and some transmembrane proteins pass through the lipid bilayer. (See cell membrane, seven membrane proteins)

LIPID STORAGE DISEASE: see lysosomal storage disease

LIPIDOSES: see lysosomal storage diseases, sphingolipidoses, lipid storage disease

LIPIDS: usually large molecules polymerized from fatty acids and associated frequently with sugar, protein and inorganic components. Lipids are the principal material of membranes; they are insoluble in water but readily soluble in non-polar organic solvents and detergents. (See fatty acids, sphingolipids, sphingosine, steroids, cell membranes, low-density lipoproteins, high-density lipoproteins, liposome, <http://www.lipidat.chemistry.ohio-state.edu>)

LIPOATE: carrier of H atoms and acyl groups in α–keto acids.

LIPO-CHITOOLIGOSACCHARIDE: see nodule

LIPOFECTION: transformation with the aid of liposome transfer. (See liposomes)

LIPOFUSCINOSIS: see ceroid lipofuscinosis

LIPOIC ACIDS: see lipoate

LIPOMATOSIS: development of usually benign tumors in fat tissues. They may be under autosomal dominant control and are frequently associated with break-points in human chromosome 12q13-q14. An autosomal recessive lipomatosis of the pancreas (Schwachman-Bodian syndrome) involves frequent chromosomal breakage. (See cancer)

LIPOSOMES: are delivery vehicles of macromolecules to cells. Within a protective coat of unilamellar vesicles consisting of phosphatidyl serine and cholesterol (1:1), they can carry DNA first to cell membrane, then after fusion, to the nuclear membrane, and eventually into the nucleus. The loading of the liposomes requires DNA in a buffer and thoroughly mixed with the ether-lipid mixture. The ether is then removed. The delivery may be facilitated by poly-

Liposomes continued
ethylene glycol and dimethyl sulfoxide. With appropriate coating (e.g. monoclonal antibody, DNA, viral vectors, drugs, etc.), they can be targeted to specific sites within the cell. Guanidium-cholesterol cationic lipids (BGSC, BGTC) can be used for construction of eukaryotic vectors. (See lipids, DOGS, fusigenic liposome, cytofectin)

LIPOPROTEIN: lipid-protein (apolipoprotein) complex; transports hydrophobic lipids.

LIPOPROTEIN LIPASE DEFICIENCY: see hyperlipoproteinemia

LIPOTROPIN: adenocortical peptides controling lipolysis. (See opiocortin)

LIPOXYGENASE: oxygenates arachidonic acid into leukotrienes. (See leukotriene, arachidonic acid, cyclooxygenase)

LIQUID-HOLDING RECOVERY: is a type of genetic repair which takes place in irradiated cells if kept in liquid medium (and allowed to utilize energy) for two days before plating.

LIQUID SCINTILLATION COUNTER: see scintillation counters

LISA (localized *in situ* **a**mplification): see PRINS, amplification, PCR

LISCH NODULE: bloody spots (hamartomas) on the iris, characteristic sign of neurofibromatosis. (See neurofibromatosis, hamartoma)

LISSENCEPHALY: see Miller-Dieker syndrome

LITHIUM: is a drug used for the treatment of manic depression. It causes various developmental anomalies in diverse organism. It appears that it antagonizes the enzyme glycogen synthase kinase-3β. (See manic depression)

LITTER: youngsters given birth to at a time by a multiparous animal.

LIVER CANCER (liver cell carcinoma, LCC): is generally initiated upon the integration of the hepatitis B virus in one or more chromosomal location. Antitrypsin deficiency, hemochromatosis and tyrosinemia may also be conditioning factors. When the viral infection takes place in an individual genetically susceptible to the liver carcinoma he/she may have an eventual incidence of 0.84 (males) and 0.46 (females) of the cancer. Without the viral integration the chances for the cancer are 0.09 and 0.01, respectively, even when the genetic constitution is permissible for its development. One of the viral integration site (17p12-p11.2) is near the p53 tumor suppressor gene. The flanking sequences are homologous to the autonomously replicating sequence (ARS1) which is required for the replication of the *Saccharomyces cerevisiae* integrative plasmids. Other integration sites identified are 11p13 and 18q11.1-q11.2. (See cancer, antitrypsin, hemochromatosis, tyrosinemia, ARS, yeast vectors, hepatitis B virus, tumor suppressor gene, chi elements)

LIVERWORT: see bryophytes

L19 IVS (intervening sequence lacking 19 nucleotides): is the 395 nucleotide linear RNA left after 19 nucleotides were consumed by the *Tetrahymena* ribozyme intron originally 414 nucleotide long. This RNA molecule is capable of nucleotidyl transferase activity and can elongate various nucleotide sequences in a protein enzyme type manner. Prefered substrates are the nucleotide sequences that can pair with guanylate rich sequences in L19IVS, i.e., have C residues. (See ribozyme, intron, *Tetrahymena*)

LMP-2, LMP-7: proteasomal subunits. (See proteasome)

LMYC: see MYC

LNGFR: low molecular weight nerve growth factor receptor. (See HNGFR, nerve growth factor [NGF])

LOAD: see genetic load

LOCALIZED DETERMINANTS: in some species, different parts of the egg determine the formation of different blastomeres depending on what cellular portions they received during cleavage. (See blastomere, cleavage)

LOCALIZED MUTAGENESIS: involves *in vitro* and *in vivo* manipulations. An amber mutation in a phage can be corrected by synthesizing a short DNA sequence complementary to a section of ϕX174 DNA single strand having a single mismatch at one base (A) corresponding to the amber codon. The synthetic sequence will have correctly C there. The short synthetic

Localized mutagenesis continued

DNA is then extended to a double-stranded form. After a cycle of replication the corrected strand makes a correct complementary copy of itself, and its progeny is thus permanently cor-

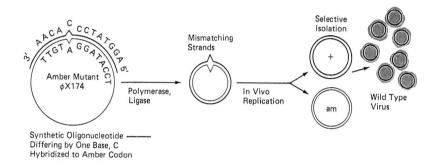

AN AMBER MUTATION IS CORRECTED BY A SHORT DNA SEQUENCE CONTAINING A CYTOSINE RATHER THAN AN ADENINE IN THE CODON. THE SYNTHETIC NUCLEOTIDE STRAND IS COMPLETED BY ADDING THE APPROPRIATE NUCLEOTIDES THROUGH THE USE OF A DNA POLYMERASE AND THE STRANDS ARE TIED TOGETHER BY DNA LIGASE. THE NEW SINGLE-STRANDED DNA HAS A SINGLE MISMATCH. AFTER TRANSFECTION OF THE PHAGE TO *E. coli* BACTERIA, EACH STRAND REPLICATES AND BOTH AMBER AND WILD TYPE φX174 GENOMES ARE MADE. BY SELECTIVE SCREENING THE WILD-TYPE PHAGES CAN BE ISOLATED AS THE RESULT OF THIS DIRECTED, LOCAL MUTATION. (After Itakura, K. & Riggs, A.D. 1980 Science 209:1401)

rected by localized mutagenesis. The other old single strand of the DNA will produce another amber mutant phage (see figure above).

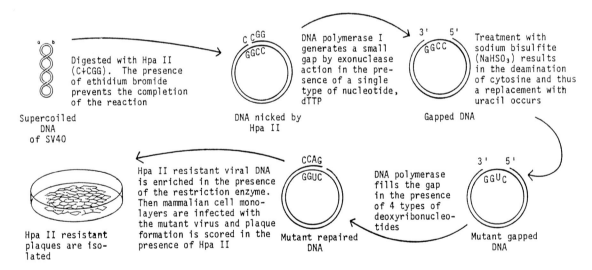

LOCAL MUTAGENESIS USING RESTRICTION ENDONUCLEASE GAPPING AND *IN VIVO* MUTAGENESIS FOLLOWED BY DNA POLYMERASE CORRECTION OF THE GAP. (After Shortle & Nathans, Proc. Natl. Acad. Sci. USA 75:2170):

Similar methods can be applied to any DNA that can be transfered to cells by transformation.

Localized mutagenesis continued

Site-directed transposon mutagenesis is feasible by identifying genes, which contain a P element within or near a particular gene. P elements are preferentially inserted at the 5'-ends of genes and thus gene-specific promoters can be placed there. (See insertional mutagenesis, site-specific mutagenesis, hybrid dysgenesis, gene replacement, PCR-based mutagenesis, TAB mutagenesis, Kunkel mutagenesis, homologue-scanning mutagenesis, targeting genes)

LOCULE: fruiting body of fungi or the cavity of the ovary of plants occupied by the ovule or the place of the uterus of some mammals where the embryo is attached.

LOCUS: the site of a gene in the DNA (or RNA) or the site of a gene in the chromosome.

LOCUS CONTENT MAP: displays the observations on the presence or absence of genes in chromosome fragments produced by ionizing radiation, This information may not be suitable for genetic mapping if contrasting alleles are not available. The data obtained can be confirmed by physical mapping. (See radiation hybrids)

LOCUS CONTROL REGION: see LCR

LOCUS EXCLUSION: both alleles at a locus are prevented from expression in deference to allele(s) at another locus. (See allelic exclusion, idiotype exclusion)

LOCUS HETEROGENEITY: the same phenotype is produced by more than one combination of the genetic constitution.

LOCUST: *Locusta migratoria*, 2n = 23.

LOD SCORE: expresses the relative probability of linkage compared to the odds against linkage (log odds) because the score is obtained by dividing the sum of logarithms indicating linkage by the sum of logarithms suggesting independent segregation. The principle briefly is as follows:

$$\text{Probability of linkage (P)} = \frac{\text{Recombination Frequency } \theta}{\text{Recombination Frequency } 1/2}$$

Here θ stands for any value of recombination, and 1/2 recombination indicates independent segregation. In case several linkage probabilities exist, the formula above can be rewritten as:

$$\text{Relative linkage (P)} = \frac{p_1(\theta_1)p_2(\theta_2)...p_n(\theta_n)}{p_1(1/2)p_2(1/2)...p_n(1/2)}$$

hence the relative odds of linkage is expressed:

$$\text{Log (odds of linkage)} = \frac{\log p_1(\theta_1) + \log p_2(\theta_2) + \log p_n(\theta_n)}{\log p_1(1/2) + \log p_2(1/2) + \log_n(1/2)} = \text{lod score}$$

The lod scores for all θ (recombination frequency) values summed up is expressed as $Z(\theta)$. This value may be empirically defined at various values of θ:

θ	θ^2	$(1 - \theta)^2$	$2[\theta^2 + (1 - \theta)^2]$	$Z(\theta)$
0	0	1	2.00	0.30303
0.05	0.0025	0.9025	1.81	0.25768
0.10	0.0100	0.8100	1.64	0.21484
↓	↓	↓	↓	↓
0.50	0.2500	0.2500	1.00	0

Thus a lod score if $\theta = 1/2$ is 0 because log 1 = 0. A lod score >3 indicates approximately 95% probability of linkage but at lod score of >4 gives a probability of about 99.5%. It is most useful for sequential data as available from (human) pedigrees. It is widely used also for mapping restriction fragments (RFLP). The probability of two genes being independent may also be obtained by the formula based on Bayes' Theorem, e.g., for the 22 human autosomes:

$$P(\theta = 1/2) = \frac{21}{\lambda + 21}$$ where λ is the average valuee of $\lambda(\theta)$ and it is equal to antilog $Z(\theta)$.

The chance that any particular unknown gene would be in a human autosome is 21/22 and the chance for another gene being in the same chromosome is 1/22. The computation, based on maximum likelihood is somewhat complicated when multipoint crosses are evaluated but computer programs (e.g. Mapmaker, Joinmap, VITESSE, FAST-LINK) are available for linkage estimation. (See mapping genetic, linkage, antilog; see also O'Connel, J.R. & Weeks, D.E. 1995 Nature Genetics 11:402)

LODICULE: scale like structures at the base of the ovary of grasses; they mediate the opening of the flower when the pollen is shed in allogamous species.

LOG PHASE: see exponential growth

LOGANBERRY (*Rubus loganobaccus*): fruit shrub, 2n = 42.

LOGARITHM: can be defined as $x = b^y$ and from this $y = \log_b x$ or in words y is the logarithm of x to the b base. The common (Briggs) logarithm uses base 10, the base of the natural logarithm is called e and it is 2.71828 (to five digits). Logarithms can be determined to any base and can be converted to each other, e.g., the natural logarithm, $\log_e$ (commonly designated also as ln) can be converted to common logarithms as $\log_{10} x = \log_e x \, (1/\log_e 10)$. (See also antilogarithm)

LOGARITHMIC GROWTH: see exponential growth

LOGARITHMIC STABILITY FACTOR: measures the developmental homeostatic values of individuals of certain genotypes under different environmental conditions. LSF is calculated as the absolute difference between two logarithmic means. If LSF = O the developmental homeostasis is maximal. (See homeostasis)

LOGIC: study of the methods of reasoning; it deals with propositions, their implications, possible contradictions, conversions, etc. *Symbolic logic* employs mathematical symbols for the propositions, quantifiers, mutual relations among propositions, etc. Logic is one of the most important tools of science for interpreting the experimental data. (See syllogisms)

LOH: loss of heterozygosity. (See also pseudodominance)

LOLLIPOP STRUCTURE: when single strands of DNA carrying palindromic terminal repeats is allowed to fold back the complementary ends pair while the non-complementary sequences in-between remain single-stranded and thus form a stem and loop configuration as shown at the left. Denatured insertion elements may also display lollipop structures if allowed to pair within the single strand. (See palindrome, insertion element)

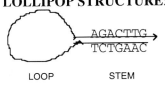

LONG-DAY PLANTS: require usually more than 12 hr daily illumination to be able to flower. (See photoperiodism, short-day plants)

LONG INTERSPERSED NUCLEOTIDE ELEMENT: see LINE

LONG-PATCH REPAIR: is mismatch repair; in contrast base excision and nucleotide excision repairs are short-patch repairs. (See DNA repair, excision repair, mismatch repair)

LONG TERMINAL REPEAT (LTR): retroviruses at the two flanks may have 2 - 8 kb repeated nucleotide sequences; similarly a few hundred repeats are found at the ends of retrotransposons. (See retroviruses, transposable elements, retroposons, solitary LTR)

LONGEVITY: represents the length of life. The average life expectancy in years in the U.S.A. in 1850 was 38.3 for males and 40.5 for females and by 1950 it became 66.5 and 73.0, respectively and it has increased since by approximately seven years. Raymond Pearl, an American pioneer of human population genetics found (during the early years of the 20th century) the life expectancy at birth of sons of fathers who's death was before age 50, 50-79, and over 80 to be 47, 50.5 and 57.2 years, respectively. Despite of regression, approximately similar life expectancy differences continued after age of 40. In *Drosophila* increased egg-production, receipt of male accessory fluid and courtship shorten lifespan. The lifespan of males is also shortened by mating and courtship. The hermaphrodite *Caenorhabditis* has shorter life if mated but this is independent from egg production or receipt of sperm. The males of this species are not affected by mating. Some suggestions were made regarding correlation between longevity and MHC allels, apolipoprotein E and B, angiotensin-converting enzyme, etc. (See mortality, aging, age and mutation, age-specific birth and death rates, copulation, apolipoproteins, MHC, angiotensin)

LONGHORN CATTLE: is a primitive form, used extensively as draft farm animal before the era of mechanization; it resembles the cattle depicted (15,000 years ago) by neolithic humans in European caves. (See Lascaux cave)

LONG-PERIOD INTERSPERSION: long sequences of repetitive DNA are alternating with long unique sequences.

LONG-TERM DEPRESSION: see memory

LONG-TERM POTENTIATION (LTP): activation of synaptic junctions in the brain relating to memory. The non-associative LTP is induced by high-frequency stimulation of the presynaptic termini. The associative type LTP increases the probability that neighboring synapses on the same neuron will be strengthened if they activated within milliseconds. This provides an opportunity to encode associations between different events if they occur concurrently. (See memory)

LOOP: single stranded (unpaired) nucleic acids sequences (loops) alternating with double-stranded, paired regions in isolated molecules. Paired regions of the meiotic chromosomes of inversion heterozygotes, unpaired regions of normal chromosomes "paired" with deletions. (See lollipop, inversion, tRNA, deletion, looping of DNA, lampbrush chromosome)

LOOP DOMAINS MODEL: the DNA fibers form 5 to 100 kb loops that are attached to the nuclear matrix (protein) at the *matrix attachment region* (MAR). These domains of the chromatin are supposed to be transcriptional and replicational units. The matrix is also called nuclear scaffold. The MAR regions are expected to regulate the expression/silencing of genes (See chromatin, A box, T box)

LOOPING OF DNA: is brought about by protein binding to DNA at two different positions from 10 to thousands of nucleotides apart. The association of proteins with specific DNA sequences permit the DNA to fold back from a longer distance to the location of the promoter of the gene and thus regulates its expression, recombination and replication of the genetic material. (See hormone response elements, transcription factors inducible, regulation of gene activity, LCR)

LOOP-OUT - LOOP-IN MUTAGENESIS: generation of deletions and replacing the deleted segments with other sequences.

LORDOSIS (curvature of the spine): it is generally applied to the condition when the curvature is expressed above the normal level such e.g. in some types of muscular dystrophy. (See muscular dystrophy)

LORISIDAE: a family of prosimii. *Arctocebus calabrensis* 2n = 52; *Galago crassicaudatus* 2n = 62; *Galago demidovii* 2n = 58; *Galago senegaliensis braccatus* 2n = 36, 37, 38; *Perodicticus potto* 2n = 62. (See prosimii)

LOS: loss of heterozygosity mutation. In a heterozygote one of the alleles (infrequently both) are lost. If the remaining allele is non-functional or suffers a secondary (somatic) mutation, the normal function of the gene is discontinued. (See LOH)

LOTKA—VOLTERRA FORMULA: quantifies the competitive interactions between populations:

$$\frac{dN_1}{dt} = r_1 N_1 \left[1 - \frac{(N_1 + \alpha_{12} N_2)}{K_1} \right] \quad \text{and} \quad \frac{dN_2}{dt} = r_2 N_2 \left[1 - \frac{(N_2 + \alpha_{12} N_1)}{K_2} \right]$$

Subscripts *1* and *2* stand for the two populations (N), *dN₁/dt* and *dN₂/dt* are the calculus symbols for the rate of change; *1/K* is the effect of an individual of species 1 on the growth of species 1 and α_{21}/K_1 is the effect of an individual of species 2 on species 1 and similarly $1/K_2$ and α_{12}/K_2 stand for the influence of one individual of species 1 on species 2, and *r* is the constant of intrinsic rate of growth.

LOTOAUSTRALIN: a glucoside linked with cyanide and making some white clover varieties toxic or lethal (if 1 to 2 kg is consumed) to animals (50 kg). This toxic effect has been circumvented by breeding white clovers deficient in the synthesis of the glucoside and in linamarase. Several other plant species (Sudan grass, flax, some beans, almond, apricot) produce potentially toxic cyanides.

LOU GEHRIG'S DISEASE: see amyotrophic lateral sclerosis

LOUIS-BAR SYNDROME: ataxia telangiectasia. (See ataxia)

LOW FREQUENCY TRANSDUCING LYSATE: see LFT

LOW-DENSITY LIPOPROTEINS (LDL): form 22 nm diameter particles containing bi-layered vesicles about 1,500 cholesterol molecules esterified to long-chain fatty acids and form a vehicle of delivery of lysosomes from where the cholesterol is made available for membrane synthesis. If there is a defect in the synthesis of the receptor proteins, the cell may not draw cholesterol from the blood resulting in atherosclerosis and eventually artery disease. LDL receptors are involved in high affinity and broad specificity endocytosis such as the macrophage scavenger receptors that are supposed to mediate cell adhesion, host defense and atherosclerosis. (See atherosclerosis, cholesterol, lysosome, hypertension, HDL, Alzheimer disease)

LOW-ENERGY PHOSPHATE: a compound that can release relatively low energy upon hydrolysis, e.g. glucose-6-phosphate.

LOWE's OCULOCEREBRORENAL SYNDROME: is an X-chromosome linked (Xq25-q26) eye defect, mental retardation, aminoaciduria, kidney anomaly, etc. The defect is in an inositol-5-phosphatase, synaptojanin. (See mental retardation, synaptojanin)

Lox (loxP): see *Cre/Lox*

L-PHASE: due to various shocks (temperature, osmotic, antibiotics, etc.) the bacteria may lose their walls in possibly reversible manner yet they can multiply.

LPS: lipopolysaccharide, lipid-carbohydrate complexes such as exist on the walls of some bacteria.

LQT: is a cardiac (heart) disease involving left ventricular arrhytmia. Several genetic defects results in the same symptoms (SCN5A, LQT3 [chromosome 3], HERG [LQT2, chromosome 7], KVLQT1 [LQT1, chr. 11], and another in chr. 4). All of these loci are involved in the control or regulation of muscle cell K^+ or Na^+ ion channels. (See cardiovascular diseases, ion channels, Jervell and Lange-Nielson syndrome, Ward-Romano syndrome, electrocardiogram for LQT, KCNA, Beckwith-Wiedemann syndrome)

LRE: see light response elements

LRF (luteinizing hormone releasing factor): see luteinization, corticotropin

λ_S: sibling risk/population sibling risk prevalence. (See sibling, risk, prevalence)

LSIRF (IRF4): is a transcription factor, specific for mature B and T cells.

LTD: see memory

LTP: see memory

LTR: long terminal repeats in movable genetic elements and oncogenic viruses. (See retroviruses)

LUC: the firefly luciferase gene.

LUCIFERASE: can be effectively used as a tissue- or developmental stage-specific reporter of gene expression. The luciferase of the firefly (*Photinus pyralis*) is a single polypeptide of 550 amino acids. In the presence of oxygen, and ATP (or coenzyme A) it oxidizes luciferin, resulting in the emission of yellow-green light flashes which can be monitored by luminometers, scintillation counters or by extended exposure to highly sensitive photographic film. Other insects also produce luciferases which produce somewhat different light emission. The bacterial luciferases such as the one produced by *Vibrio harveyi* or other similar strains consist of two subunits A and B. The substrate of the bacterial luciferase enzymes is n-decyl aldehyde (decanal) and for the emission of light reduced flavin mononucleotide ($FMNH_2$) is required. While the aldehyde (RCHO) is converted to acid (RCOOH) in the presence of oxygen (O_2), water (H_2O), FMN and *light* are generated. About 20 additional genes cooperate with and modulate the expression of the bacterial luciferase. The light can be monitored in the tissues without destruction by the use of microchannel plate enhanced photon counting image analyzer that uses a video camera, an image processor, a TV processor and computerized controls. Such an equipment may detect even a single photon. A much less expensive luminometer can also be used with ground tissues. (See image analyzer, GUS)

BACTERIAL LUCIFERASE EXPRESSED IN AN *ARABIDOPSIS* LEAF →

LUCIFERIN: upon activation cleaves a pyrophosphate off ATP and luciferyl adenylate is formed. Upon the action of firefly luciferase—in the presence of O_2— light is emitted. Oxyluciferin is generated then in the presence of CO_2 and AMP, and subsequently through some steps luciferin is reformed. These reactions are suitable for monitoring gene expression and for the quantitation of ATP. (See luciferase, ATP)

LUCY: a 3-million years old *Australopithecus afarensis* skeleton of about 1 m tall, (discovered in 1974 in Ethiopia) believed to be a representative of the origin of the human family tree. Although — as the name implies — the skeleton was assumed to be that of a female (because of the small size), some paleoanthropologist argue that she was rather a male because the pelvis would not have been sufficient for child bearing. (See hominids)

LUDWIG EFFECT: biological habitats generally divided into numerous microniches and the species respond to these by selection.

LUMEN: the interior compartment of a membrane-enveloped structure in the cell. Also a unit of luminous intensity. (See also Lambert)

LUMINESCENCE: light emission from cool sources such as excited gases (neon light), fluorescent tubes, television and computer screens and bioluminescence of fireflies, glowworms and certain fishes and bacterial luciferase. (See luciferase)

LUNG: see small cell lung carcinoma

LUPINES (leguminous plants): yellow lupine (*Lupinus luteus*, 2n = 46, 48, 52); blue lupine (*L. angustifolius*, 2n =40); white lupine (*L. albus*, 2n = 30, 40, 50) are crop plants used for "green manure" or in alkaloid-free forms as forage crops in acid soils. The perennial lupine (*L. polyphyllus* (2n = 48) is an attractive ornamental. The alkaloid-free lupines are one of the best examples of scientific plant breeding.

LUPUS ERYTHEMATOSUS: is a variety of skin and subcutaneous inflammations, each with a specific medical name. The genetic basis is not clear because some of the common symptoms are caused apparently by certain drugs (e.g. procaine anesthetics) and in other cases viral infection was suspected because of the similar symptoms observed on family dogs. Some indications are for the involvement of steroid hormone problems because of its incidence with the Klinefelter syndrome. In some cases along with the normal DNA, a low molecular weight DNA was associated with the disease. In other individuals anti-RNA or antinuclear antibodies were identified. Lymphocyte defects were also shown in some afflicted individuals. Females are 8 fold more likely to be afflicted by it than males in the age group 15 to 50 but later or earlier the relative risks differences are much less. The life time risk for a US Caucasian female is about 0.15% and much less for Blacks or Hispanics. (See also autoimmune disease, complement, HLA, procaine anesthetics)

LURIA - BERTANI MEDIUM: see LB

LUTEINIZATION: after ovulation the ovarian follicle is converted into a corpus luteum (yellow body). (See corpus luteum, Graafian follicle, luteinizing hormone-releasing factor)

LUTEINIZING HORMONE: see animal hormones

LUTEINIZING HORMONE-RELEASING FACTOR (LH-RF or LRF, gonadotropin-releasing factor): is a hypothalmic neurohormone stimulating the secretion of pituitary hormones. It is involved in the control of mammalian fertility. (See GnRHA, *in vitro* fertilization, ART, corticotropin, puberty precocious)

LUTHERAN BLOOD GROUP: (so named after the first patient identified,1945). The Lu(a⁻b⁻) phenotype is determined by either dominant or recessive alleles of the gene in human chromosome 19cen-q13, distinguishable by hereditary pattern and by serotype. The frequency of Lu(a) allele is about 0.04 and that of Lu(b) is about 0.96. The Lu phenotype may be caused by dominant inhibitors identified as mice antigens A3D8 and A1G3 under the control of a locus in human chromosome 11p. This 80 kDa antigen apparently has different epitopes on the same protein molecule. The Lu system appears to be identical to the Auberger blood group. (See blood groups, epitope)

Lux: see luciferase genes of bacteria, LUC

LUX: unit of illumination, 1 Lux = one lumen/m². (See lambert)

LUXURIANCE: is a hybrid vigor displayed by the somatic tissues of reduced-fertility or sterile hybrids between different species or genera. (See hybrid vigor)

LUXURY GENES: are not involved in functions indispensable for the viability of a cell and common basic function but only in differentiated, special cells.

Luzula purpurea (Juncaceae, n=3): and other related plant species have holocentric chromosomes and polyploid as well as aneuploid series. (See holocentric, polyploid, aneuploid)

LW BLOOD GROUP: produces the rodent anti-rhesus antibody. (See Rh blood group)

LYASE: enzymes catalyze additions or removal of double bonds.

Lychnis: synonyms *Melandrium* and *Silene*. (See *Melandrium*) MALE KARYOTYPE →

LYCOPENE ($C_{40}H_{56}$): a carotenoid pigment. (See carotenoids)

LYL: lymphoid leukemia oncogene, encoded in human chromosome 19p13.

LYMANTRIA: *Lymantria dispar*, 2n = 62. It has been used for studies on sex determination.

LYME DISEASE (borreliosis): see *Borrelia, Ixodoidea*

LYMPH: the transparent (yellowish or sometimes pinkish) fluid filtered through the capillary blood vessels from blood.

LYMPH NODES: are small (1 to 25 mm) nodules in the body where lymphatic vessels, lymphocytes and antigen-presenting cell accumulate. They have defensive roles by removing toxic and infectious agents. (See thymus, lymph, lymphocytes, lymphoma, Hodgkin disease, thymus, T cell, T cell receptor, antigen-presenting cell, spleen)

LYMPHOBLASTIC LYMPHOMA: a leukemia, a neoplasia of the lymph-producing cells (such as Hodgkin's disease). Its basis is a DNA-ligase deficiency. (See Hodgkin's disease)

LYMPHOCYTE HOMING: the immune processes are systematically distributed through movement of lymphocytes from their origin (bone marrow for B cells and thymus and bone marrow for T cells) to lymph nodes, glands, tonsils, spleen. In these organs microbial antigens are deposited and the naïve B and T cells are exposed to them. That propels their differentiation into memory cells and effector cells, respectively. These differentiated cells then home on the bone marrow (B cells) or to extralymphoid effector sites (T cells). These are complicated processes and may go through multiple and reversible substeps. The passage from the blood vessel requires an interaction of the lymphocyte receptors with vascular ligands, activation by G-protein-linked receptors and finally passage through the vessel wall (diapedesis). After stepping out and entering special tissues they home on e.g. B cell follicles or T zones during memory cell formation in germinal centers. They also seek up sites of inflammation. In each of these niches they are subject to various regulatory forces by attaching e.g. to TGFβ-regulated integrin that directs them to intraepithelial sites by attachment to cadherin. They are exchanging signals with antigen presenting cells, surface receptors, chemoattractant receptors, immunoglobulins, antigen receptors, chemokines, etc. Probably the RHO homolog of the RAS oncogene plays a central role in the signaling paths and the antigens also affect the trafficking. There is also some sort of competition and also and orderly assignment of the various lymphocytes to specific sites. (See memory immunological, germinal center, TGFβ, integrin, cadherin, RHO, antigen presenting cell, chemokines, immunoglobulins, antigen, signal transduction, receptors, ligand)

LYMPHOCYTES: are cells produced in the fetal hepatocytes and in adults by bone marrow and may differentiate into T lymphocytes and B lymphocytes. In the thymus the T cell receptor immunoglobulins undergo rearrangement and the terminal nucleotidyl transferase generates additional diversity. The T lymphocytes are involved in cell-mediated immunity; the B cell are responsible for the humoral antibodies. During the development of B cells the secreted and the membrane-bound immunoglobulin (Ig) undergo a series of events. The B cell antigen receptor (BCR) is formed by complexing a membrane Ig with the heterodimer of Ig-α and Ig-β (*pro-B stage*). The complex contains now an extracellular Ig-like domain, a trans-membrane domain and a short cytoplasmic domain. The signal transducing part of the receptor contains a two-tyrosine motif in the α and β cytoplasmic domains. When the hematopoietic stem cell of the

Lymphocytes continued

bone marrow becomes committed to B cell development the Ig gene undergoes a series of rearrangements. In the first step a D_H (diversity heavy chain; see immunoglobulins) joins a J_H (juncture heavy chain) gene. The one of the V_H (variable heavy) is attached to the $D_H J_H$ segment in order to form a functional heavy chain. Under normal conditions a functional μ heavy chain is expressed (pre-BI stage). This transition from *pro-B* to *pre-BI* constitutes an important checkpoint, contingent of the formation of Ig-β. The next step (*pre-BII*) is reached when temporary light chain immunoglobulin attaches to the VpreBI globulin. After this Ig-κ chain undergoes a rearrangement and it joins a V-$J_κ$ (variable kappa light chain gene) and the development thus arrives to the *Immature B Stage*. The λ5 - VpreB chain is replaced by a κ-μ-BCR (B Cell Receptor) complex and the so formed *Mature B Cell* emerges from the bone marrow in the presence of a functional Ig-α cytoplasmic domain. This B cell is then activated when it comes in contact with an antigen and secretes Ig. This requires the Ig-α cytoplasmic domain and the expression of a cytoplasmic (Btk) tyrosine kinase. For T cell - dependent antigen response Ig-α and Btk are not needed. The tumor infiltrating lymphocytes have potential use in cancer therapy. The differentiation and maintenance of the lymphocytes requires interleukine-7 (IL-7) and its receptor (IL-7R). The transcription factor Ikaros (a Zn-finger protein) is specific for lymphoid cell differentiation. This protein has binding sites in several other genes involved in lymphocyte development such as CD3-T cell receptor, CD4, CD2, terminal nucleotidyl transferase, interleukine-2Rα, NF-κB. (See T cells, B cell, TCR, immune system, immunoglobulins, antibody, TIL, memory immunological, lymphocyte homing, monocyte, germinal centers, Cd3, CDE4, CD8, NF-κB, dendritic cell, macrophage, monocyte, neutrophil, basophil, eosinophils, immunoglobulins)

LYMPHOKINES: are proteins secreted by lymphocytes in response to specific antigens and are involved in multiple ways in serological defense mechanisms of the body and in differentiation. Interferons (IF), interleukins (IL), granulocyte colony stimulating factors (G-CSF), granulocyte-macrophage colony stimulating factors (GM-CSF), macrophage activity factor (MAF or IFN gamma), T-cell replicating factor (TRF), migration inhibition factor (MIF), tumor necrosis factor (TNF) belong to this large group. (See immune system, chemokines, interleukins, interferons, tumor necrosis factor, cytotoxic T cells, rel)

LYMPHOMA: cancer of the lymphocytes, neoplasia of lymphoid tissues. (See Chédiak-Higashi syndrome, lymphoblastic lymphoma, leukemia, anaplastic lymphoma, translin)

LYMPHOPENIA: recessive a deficiency of lymphocytes, an autoimmune anemia. (See lymphocyte, autoimmune disease)

LYMPHOTACTIN: a cytokine but lacks 2 of the 4 cystin residues characteristic for chemokines. It is expressed in $CD4^+$ T cells and $CD4^-$ and $CD8^-$ T cell receptor $αβ^+$ thymocytes. It has no chemotactic activity for monocytes and neutrophils. (See blood, lymphocytes, monocytes, cytokines, chemokine, T cells)

LYMPHOTOXINS: are forms of lymphokines of the tumor necrosis factor family that cause the lysis of some cells, e.g., cultured fibroblasts and are required for growth, differentiation and regulation of the immune system. They are encoded within the human HLA gene cluster. (See lymphokines, HLA, TNF, immune system)

LYN: a tyrosine protein kinase of the SRC family involved in the differentiation of B lymphocytes. (See SRC, Csk, B lymphocytes)

LYNCH CANCER FAMILIES: are groups of dominant genes responsible for the development of certain cancer syndromes such as endometrial cancer, adenocarcinoma of the colon and others. The instability may be caused by defects of the genetic repair system. (See colorectal cancer, Li-Fraumeni syndrome)

LYON HYPOTHESIS: in mammalian cells with more than one X-chromosomes, usually all but one are heteropycnotic (highly condensed and thus dark-stained at all stages) and form n - 1 Barr bodies. The heteropycnosis may not equally affect the entire length of the chromosome. Mary Lyon, British geneticist suggested that the frequently observed mosaicism in these ani-

Lyon hypothesis continued

mals is the phenotypic consequence of the chromosomal observation. The heteropycnotic chromosomes carry their genes in an inactivated state (possibly key nucleotides are methylated) and depending which allele of a heterozygote is expressed, sectorial mosaicism is displayed. In order to carry out the inactivating switch the X-chromosome must have an intact *inactivation center*. During development the X-chromosomes may switch between their inactive - active states. These decisions are not made entirely by the X-chromosome(s) because in human triploids with XXY constitution both X-chromosomes remain active. Typical example is the tortoiseshell cat color that is observed almost exclusively in females heterozygous for yellow - black fur color. The exceptional (less than 1/500) tortoiseshell male cats are of XXY constitution (Klinefelter syndrome). Similar genetic heterogeneity has been observed also in humans when in heterozygous females two classes of lymphocytes were observed, one with normal and the other without testosterone binding capacity. (See Barr body, heteropycnosis, methylation of DNA, regulation of gene activity, lyonization, dosage effect)

LYONIZATION: variegation in mammalian females as predicted by the Lyon hypothesis. In an XX mammalian female one of the X-chromosomes remains in a condensed state, replicates its DNA asynchronously and its genes are not transcribed. The other X-chromosome displays a more open structure and its gene content is expressed. One of the X-chromosome is selected for inactivation by a chromosomal locus, *Xic* (X-chromosome inactivating center). This *Xic* locus is responsible for *counting* the number of X-chromosomes to be inactivated. In case of more than 2 X-chromosomes only one remains active normally. In addition, the *Xic* must be in cis position to the genes to come under its influence (*spreading effect*). Thus *Xic* also selects the chromosome to be inactivated (*choice function*). If the *XIC* locus is lost after inactivation has been initiated, the inactive state is maintained. There is another functional site within the *XIC* locus, *Xce* (X control element). *Xce/Xce* homozygotes undergo normal, random inactivation. In *Xce* heterozygotes there is a higher chance that that chromosome remains active which carries a strong *Xce* allele. The transcribed product of *Xic* is XIST (X inactive specific transcript) that is transcribed only from the inactive X-chromosome. In *XIST/Xist* heterozygotes there is a large RNA transcript in the nucleus, apparently associated with the inactive X, and it cannot be translated. Thus it appears that the regulation is mediated by RNA. During early embryogenesis *Xist* expression precedes the inactivation of the X-chromosome and in XY males it is expressed only in the male germ tissues and only before meiosis. In marsupials the inactive X-chromosome is the paternally transmitted one. Similar

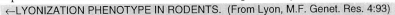
←LYONIZATION PHENOTYPE IN RODENTS. (From Lyon, M.F. Genet. Res. 4:93)

observations were also made in the extraembryonic endoderm cells of some rodents and to some extent in humans. Some human X-chromosomal genes escape inactivation, e.g. Xg blood group and the closely linked steroid sulfatase locus [ichthyosis]. In the inactivation differential methylation between the active X and the inactive X chromosome is implicated although the overall methylation appears to be the same. In the inactive X chromosome CpG islands in 5' region are mostly methylated whereas in the active ones they are not. (See Lyon hypothesis, dosage effect, ectodermal dysplasia anhidrotic, Wiskott-Aldrich syndrome, Xg, ichthyosis, pseudoautosomal, Barr body)

LYOPHILIZATION: see freeze drying

LYSATE: when a cell is lysed (dissolved) its contents are released; the lysis of bacterial cells may produce bacteriophage particles as lysate. (See lysogeny)

LYSENKOISM: T.D. Lysenko (1898-1976) presided over the greatest scandal in cultural history. Beginning in the 1930s and with the culmination of his almost complete victory in 1948, he destroyed genetics and most of biology in the Soviet Union. His group of charlatans, misled ignorants, psychopaths and common criminals brought havoc to biology in the Soviet Union

Lysenkoism continued
and other countries in her political interest sphere. The lysenkoists rejected the Mendelian inheritance, cytogenetics, biochemical genetics and statistics as "capitalist fraud". The movement traced its origin to Mitchurin, an uneducated railwayman who successfully practiced empirical plant breeding and attempted to interpret his observations without being familiar with scientific principles and facts. The lysenkoists denied the existence of genes, the role of "hard heredity", rather they claimed the inheritance of the phenotype and continuous change of heredity under the influence of nutrition, temperature, day-length and other environmental conditions. This was a revival of the ancient myths of pangenesis, lamarckism, inheritance of acquired characters, directed genetic change by agrotechnical methods (fertilizers, irrigation, etc.), grafting, vernalization (yarowisation), blood transfusion, etc. Although these ideas themselves were disgraceful and appalling, the major problem was that lysenkoism became a state political doctrine, imposed on and substituted for all scientific activities in biology, totally subjugating agriculture and making inroads even into medicine. Marx and Engels were supporters of lamarckism in the hope that the future of the human race may be improved by these doctrines. The soviet dictator Stalin embraced the lysenkoist ideas because they promised instant increased agricultural productivity by the application of vernalization, summer culture of potatoes, species transformation, graft hybridization, supplementary pollination, etc. at a period of famines, caused by the senseless social experimentation of the soviet bureaucracy. The promised results failed to come true, the agricultural productivity further declined partly because of the application of lysenkoism. The failures were then attributed to sabotage by the scientists and technicians. Therefore after 1948 Stalin granted virtually unlimited power to Lysenko to purge the universities and research institutes from his and the "people's enemies". Thousands of scientists actually lost their life either by execution or imprisonment. Genetics that had enjoyed reasonably good support before 1948, and world wide renowned scientists of the Soviet Union were persecuted, forced underground or physically eliminated. Resurrection of genetics in Russia could come only after his death. (See lamarckism, soviet genetics, acquired characters, directed mutation, hard heredity, pangenesis, vernalization)

LYSERGIC ACID: a psychomimetic alkaloid-derivative. (See ergot, psychomimetic)

LYSIDINE: a cytidine with a lysine residue qualifying it for recognition of the AUA codon by the tRNA. (See tRNA, aminoacylation, cytidine, genetic code)

LYSIN: a complement-dependent lytic protein such as hemolysin, bacteriolysin. (See complement, antibody)

LYSINE BIOSYNTHESIS: some fungi and algae synthesize lysine ($NH_2[CH_2]_4CH[NH_2]COOH$) from α-aminoadipic acid ($HO_2CCH[NH_2]CH_2CH_2CH_2COOH$). Bacteria, some fungi and higher plants make lysine from diaminopimelic acids ($HOOCCH(NH_2)(CH_2)_3CH(NH_2)COOH$). Asparate semialdehyde is also a precursor of lysine. For mammals it is an essential amino acid and they rely on the diet. Some food or feed may have too low levels of this amino acid to meet the requirements and then human malnutrition or low weight gain in animals results. Some mutant plants, e.g. high lysine maize may be used to avoid the nutritional problems. The degradation of lysine may take place by number of different ways, and a common intermediate is α-ketoglutarate. (See high-lysine corn, kwashiorkor, lysine intolerance, lysine malabsorbtion, hyperlysinemia)

LYSINE INTOLERANCE: an autosomal recessive human disorder resulting in periodic vomiting and coma based on a defect in the degradation of lysine. (See lysine biosynthesis, dibasic aminoaciduria)

LYSINE MALABSORPTION: causes excessive amounts of lysine in the urine and low levels in the serum due to a recessive autosomal mutation. (See lysine biosynthesis)

LYSINURIA: see lysine intolerance

LYSIS: disruption of the (bacterial) cell (before the release of virus or plasmid from a cell) by spontaneous or any other means such as exposure to lysozyme, by boiling or alkaline treatment or the dissolution of eukaryotic cells or cellular organelles by lysozyme or detergents.

LYSIS FROM WITHOUT: when too many phages (>20) attack a single bacterial cell, the bacterium may disintegrate by the perforations suffered.

LYSIS INHIBITION: in case of shortage of bacterial cells (that he phage senses by some means) it postpones lysis for several hours and produces the maximal possible number of virions from that cell (e.g. 400 copies of T4).

LYSOGEN: bacterium with phage integrated into its genome. In such a state the replication of the phage DNA is under the control of the host and the phage's own replication system is repressed (stable lysogen). Rarely, the suppression of the phage's own replication fails after insertion (abortive lysogeny) and the prophage cannot be replicated and it is lost by dilution during subsequent divisions of the host. (See lambda phage)

LYSOGENIC BACTERIUM: can harbor temperate phage in an integrated state in the chromosome; after induction, phage particles can be formed and released. (See prophage, lysis)

LYSOGENIC IMMUNITY: see immunity in phage

LYSOGENIC REPRESSOR: prevents the lysis of a lysogen.. (See lamda phage)

LYSOGENY: the phage coexists with the bacterium as a prophage. (See also symbionts hereditary)

LYSOPHOSPHATIDIC ACID (oleoyl-sn-glycero-3-phosphate): is derived from L-α-phosphatidic acid, dioleyl by the action of phospholipase A. It stimulates smooth muscle contraction and may affect blood pressure. (See also Pyk-2)

LYSOPIN: an octopine type opine. (See opines)

LYSOSOMAL STORAGE DISEASES: involve defects in lysosomal hydrolase enzymes and accumulation of the products of these hydrolases affecting in a variable manner various organs. Glycosphingolipid storage diseases are the Tay-Sachs disease and the Gaucher disease, and according to a mouse model N-butyldeoxynojirimycin may hinder the accumulation of the substrate for β-hexoseaminidase A and may alleviate the symptoms of the Tay-Sachs disease. (See mucolipidoses, mucopolysaccharidoses, lipidosis, neuroaminidase deficiency, glycoprotein storage disease, lipase deficiency, Tay-Sachs disease, Gaucher's disease, Farber's disease, Fabry's disease, Hurler's syndrome, Hunter syndrome, Sanfilippo syndrome, Maroteaux-Lamy syndrome, Hermansky-Pudlak syndrome, Coffin-Lowry syndrome, gangliosidoses, neuromuscular disease, lysosomes)

LYSOSOMES: cytoplasmic organelles (in eukaryotes) containing hydrolytic enzymes. The digestion is either by phagocytosis-type mechanism or by selective ingestion and lysis of molecules. The latter type process requires Mg-ATP, molecular chaperones (heatschock proteins 73 and 70). The receptor for the uptake may be a 96 kDa (Lamp2) lysosomal membrane protein. (See lysosomal storage diseases, lysozymes, LDL receptor, Wolman disease)

LYSOZYMES: enzymes capable of digesting bacterial cell walls; have important role in phage infection and liberation. (See lysosomes)

LYST: see Chediak-Higashi syndrome

LYT ONCOGENE: is associated with lymphoblastic leukemias and non-Hodgkin type lymphomas when the gene (in human chromosome 10q24) is translocated to chromosome 14q11 or 14q32. The normal product of LYT is similar to that of NFKB (NF-κB) products that are transcription factors. (See NFKB, REL oncogene, Hodgkin's disease, leukemias)

LYTIC: capable of initiating and performing lysis.

LYTIC CYCLE: the life cycle of the phage involving infection, growth (one-step-growth) and lysis of the host to liberate the infectious phage particles. (See lambda phage, one-step growth)

LYTIC INFECTION: see lysis of bacteria

λ**Zap**: carries the *lacIq* and tetracycline resistance in an F' plasmid. This vector is very effective for cDNA cloning. Its cloning capacity is about 10 kb at multiple cloning sites. In the presence of helper M13 of f1 helper phages the cloned DNA is excised and placed into a small plasmid to facilitate the restriction mapping or sequencing of the insert. Suitable also for making RNA transcripts of either strand using T3 or T7 transcriptases. (See lambda phage, F' plasmid, filamentous phages, cloning site, helper phage, restriction enzyme, sequencing of DNA)

M

M: abbreviation for mitosis or in statistics for the mean. (See mitosis, mean)

M9 BACTERIAL MINIMAL MEDIUM: 5 concentrated salt solution g/L H_2O, Na_2HPO_4. 7 H_2O 64, KH_2PO_4 15, NaCl 2.5, NH_4Cl 5 → 200 µL, 1 M $MgSO_4$ 2 mL, glucose 4 g, fill up to 1 L after any supplements (e.g. amino acids) added.

M COMPONENT: same as paraprotein.

M CYTOTYPE: see hybrid dysgenesis

M13 PHAGE: see bacteriophages, DNA sequencing

M PHASE: part of the cell cycle when the structure, movement and separation of chromosomes or chromatids are visible and the nuclear divisions are completed. (See mitosis and meiosis)

M PHASE HISTONE-1 KINASE: see growth-associated kinase

M PHASE PROMOTING FACTOR (MPF): same as maturation promoting factor.

M RNA: see cowpea mosaic virus (CPMV)

M VIRUS: see killer strains

Ma (mille mille annus): million years. (See My)

Mab: monoclonal antibody.

Mab: mouse antibody.

MAC: mammalian artificial chromosome. (See also HAC, HAEC, YAC, BAC, PAC, SATAC)

MACADAMIA NUT (*Macadamia* spp): is a delicious nut, 2n = 2x = 28.

MACALOID: is a clay colloid capable of adsorbing ribonuclease during disruption of cells and can be centrifugally removed from the RNA preparation after extraction by phenol. (See RNA extraction)

MACARONI WHEAT (hard wheat): *Triticum durum* (2n = 4x = 28) varieties containing the AB genome and taxonomically classified as a subgroup of *T. turgidum*. The endosperm of the commercially used varieties have more β-carotenes and therefore the milled products show yellowish color, desirable for pastas. It is milled to a more grainy product (semolina) that favors cooking quality. The durum wheats have higher protein and mineral content than the soft wheats. The turgidum wheats, except the durums, are not used for food. (See *Triticum*)

MACEROZYME: is a pectinase enzyme of fungal origin; it is used for the preparation of plant protoplasts in connection with a cellulase. (See cellulase, protoplast)

MACH: see FLICE

MACHADO-JOSEPH SYNDROME (Azorean neurologic disease): a 14q chromosome dominant defect of the central nervous system involving ataxia, and other anomalies of motor control. In the gene locus abnormally increased number of CAG repeats occur. If the CAG repeats are translated into polyglutamine, cell death results. (See ataxia, muscular atrophy, fragile sites, trinucleotide repeats)

MACROCONIDIUM: a large multinucleate conidium. (See conidia, microconidia, fungal life cycle)

MACROEVOLUTION: major genetic alterations which produced the taxonomic categories above the species level.

MACROGAMETOPHYTE: megaspore. (See gametogenesis in plants)

MACROLIDE: a type of antibiotic (including more than one keto and hydroxyl groups such as erythromycin, troleandomycin) associated with glycoses. *Streptomyces* bacteria produce these compounds that inhibit the function of the 70S ribosomes by binding to the large (50S) subunit. (See antibiotics, glycoses, *Streptomyces*, protein synthesis, ribosome)

MACROMOLECULE: molecules with molecular weight of several thousands to several millions such as DNA, RNA, protein and other polymers.

MACROMUTATION: a genetic alteration involving large, discrete phenotypic change.

MACRONUCLEUS: a larger type of polyploid nucleus in Protozoa. While inheritance is mediated through the small micronucleus, the macronucleus is directing metabolic functions by being transcriptionally active and the latter is responsible for the phenotype of the cells. After the internally eliminated sequences (IES) are removed from the micronuclear DNA, the left-over tracts are joined into macronuclear-destined sequences (MDS) to form the macronucleus. The macronucleus divides by fission rather than by mitosis. The macronuclei are formed by the fusion of two diploid micronuclei of the zygotes. (See *Paramecium*, chromosome diminution, IES)

MACRONUTRIENT: required in relatively large quantities by cells.

MACROORCHIDISM: a condition of larger than normal testes. (See fragile X)

MACROPHAGE (Mφ): phagocytic cell of mammals with accessory role in immunity. Macrophage cells are produced by the bone marrow stem cells as monocytes which enter the blood stream, and within two days of circulation they enter the various tissues of the body and develop into macrophages. The macrophages may differ in shape but generally they have single large round or indented nuclei, extended Golgi apparatus, lysosomes and vacuoles for the ample storage of digestive enzymes. The cell membrane is forming microvilli of various sizes, suited for phagocytosis (engulfing particles) and pinocytosis (uptake of fluid droplets) and thus are important as part of the cellular defense mechanism against foreign antigens, including even tumor antigens. The macrophages have a wide array of receptors, including the Fc, complement, carbohydrate, chemotactic peptide and extracellular matrix receptors. (See also granulocytes, immune system, antibody)

MACROPHAGE ACTIVITY FACTOR: MAF, a lymphokine, identical to IFN-gamma. (See lymphokine, interferon)

MACROPHAGE COLONY STIMULATING FACTOR (M-CSF): same as FMS oncogene, receptor of colony stimulating factor. Its receptor (MCSFR) is a protein tyrosine kinase. (See FMS, colony stimulating factor, RAS, signal transduction)

MACROPHAGE-STIMULATING PROTEIN (MSP): is 80 kDa serum protein stimulating responsiveness to chemoattractants in mice, induces of ingestion of complement-coated erythrocytes and inhibits nitric oxide synthase in endotoxin- or cytokine-stimulated macrophages. MSP is structurally homologous to hepatocyte growth factor, scatter factor (HGF-SF) but their targets are different. MSP has been located to human chromosome 3p21, a site frequently deleted in lung and renal carcinomas. (See HGF-SF, Met, Ron, macrophage, scatter factor)

MACROSPORE: see megaspore

MACULA: spot or thicker area, particularly on the retina; it is often colored. (See retina, choroidoretinal degeneration, macular degeneration, macular dystrophy, foveal dystrophy)

MACULAR DEGENERATION: a dominant-acting 6q25 deletion causing degeneration of the vitelline layer of the eye. It manifests with some variations in expression and onset. (See macula, eye diseases)

MACULAR DYSTROPHY: autosomal recessive (8q24) and X-linked forms with symptoms similar to macular degeneration. (See macula, eye diseases)

Mad (Mothers against decapentaplegic): is a *Drosophila* gene controlling several developmental events in the fly. Its human homolog is called SMAD1 (Sma in *Caenorhabditis*). SMAD is a TGF (transforming growth factor)/BPM (bone-morphogenetic protein) cytokine family-regulated transcription factor involving serine/threonine kinase receptors. SMAD is also called hMAD; MADR is its receptor. (See tumor growth factor, cytokines, *dpp*, serine/threonine kinase, MYC)

mad (many abnormal discs; 3-78.6): is a homozygous disc and cell autonomous lethal affecting several morphogenetic functions in *Drosophila*.

MAD COW DISEASE: see encephalopathy

MADS BOX: a conserved motif of 56 residues in a DNA-binding protein of transcription factors involved in the regulation of *MCM1* (yeast mating type), *AG* (agamous homoeotic gene of *Arabidopsis*), *ARG80* (arginine regulator in yeast), *DEF A* (deficient-flower) mutation of

MADS box continued
Antirrhinum) and SRF (serum response factor in mammals) regulating the expression of the c-fos protooncogene. The MADS box has amino terminal sequence specificity and carboxyl dimerization domains. In addition the SRF recruites accessory proteins such as ELK1, SAP-1 and MCM1 relies on MATα1 and MATα2, STE12 and SFF. (See separate entries, mating type determination in yeast, MEF)

MAF: macrophage activity factor, a lymphokine, an oncogene and a regulator of NF-E2 transcription factor. Heterodimers of Maf protooncogene family members promote the association with and the expression of NF-E2, whereas the homodimers are inhibitors of it. (See macrophage, lymphokine 4, oncogenes, NF-E2, homodimer, transcription factors)

MAGAININ: see antimicrobial peptides.

MAGIC BULLET: a specific monoclonal antibody is supposed to recognize only one type of cell surface antigen (e.g. one on a cancer cell) or growth factor receptors or differentiation antigens, and carries either the *Pseudomonas* exotoxin, or the diphteria toxin, or ricin or maytansinoids (an extract of tropical shrubs or trees) or a radioactive element (such as Y^{90}) capable of selectively unloading these harmful agents at the target cell and thus killing the cancer cell without much harm to any other cell. (See hybridoma, monoclonal antibody, receptor, antigen, diphteria toxin, ricin, immunotoxin, maytansinoid, isotopes)

MAGIC SPOT: pppGpp and ppGpp nucleotides that serve as effectors of the stringent control. (See stringent control)

MAGNETIC RESONANCE: see nuclear magnetic resonance spectroscopy.

MAGNETORECEPTION: the response of organisms to the magnetic field of the Earth for behavior and orientation.

MAGNIFICATION: increase in the units of ribosomal genes, hypothesized to occur by extra rounds of limited replication or unequal sister-chromatid exchange. At the *bobbed* (*bb*, 1-66.0) locus present in both sex-chromosomes of *Drosophila* in about 225 copies organized as large tandem arrays, separated by non-transcribed spacers. The copy numbers of *bb* in wild population Y-chromosomes may vary 6-fold. (See ribosomes, microscopy

MAIZE (*Zea mays*): belongs to the family of Gramineae (grasses) and the tribe Maydeae along with teosinte (*Euchlena mexicana*) and the genus *Tripsacum* (with a large number of species). The male inflorescence of maize is more similar to that of teosinte than the female inflorescence. The ear or maize carries generally 8 to 24 rows of kernels whereas teosinte has only two rows. The teosinte ear is fragile, the maize ear is not. The seed as is commonly named, is really a karyopse (caryopse), a single-seed fruit (kernel). The basic chromosome number x = 10. *Tripsacum* resembles more to some members of the *Andropogonaceae* family than to these two closer relatives and its basic chromosome number x = 18. Teosinte can be crossed readily with maize and the offspring is fertile whereas *Tripsacum* is more or less strongly isolated sexually and their hybrids are not fully fertile. Teosinte genes display the same chromosomal and gene arrangements as maize in contrast to *Tripsacum* that is more dissimilar. All three genera have evolved in the western hemisphere. Maize is one of the genetically most thoroughly studied plants. It is a monoecious species; under natural conditions it is allogamous. Approximately 500 kernels may be fixed on the cob for easy Mendelian analysis for a good number of endosperm, pericarp and embryo and seedling characters. Mendelian genes have been identified at about 1000 loci and more than 500 have been mapped. RFLP maps are also available. Cytogenetic analyses are facilitated by the characteristic pachytene chromosomes. Several genes controling meiosis have been identified. Cytogenetic studies with maize contributed significantly to the understanding of chromosomal rearrangements. Transposable genetic elements were first recognized in maize. Transformation is possible but requires either electroporation or the biolistic technique. The discovery and the commercial production of hybrid corn (heterosis) made an unprecedented increase in the food and feed supply. For genetic database contact by e-mail: <dhancock@teosinte.agron.missouri.edu> or <http://www.nal.usda.gov/pgdic>

MAJEWSKI SYNDROME: the autosomal recessive phenotype has lots of similarities with oral-facial-digital syndromes, particularly with the Mohr syndrome. It is distinguished from the latter on the basis of laryngeal (throat) anomalies and polysyndactyly of the feet. (See orofacial-digital syndromes, polydactily)

MAJOR GENE: determines clear, qualitative phenotypic trait(s). (See minor gene)

MAJOR GROOVE: the DNA helix as it turns displays two grooves in a 3.46 nm pitch; the wider one (almost 2/3 of the pitch) is the major groove. (See DNA, hydrogen pairing, Watson and Crick model)

MAJOR HISTOCOMPATIBILITY COMPLEX (MHC in humans, H-2 in mice): triggers the defense reactions of the cells against foreign proteins (invaders). Their existence was first recognized by incompatibilities of tissue grafts. The MHC molecules are transmembrane glycoproteins encoded by the HLA complex in humans and by the H-2 in mouse. The class I molecules have three extracellular domains at the NH_2 end and the COOH terminus reaches into the cytosol. The extracellular domains are associated with β_2 microglobulin. The class II MHC molecules are formed from α and β chains without the microglobulin. The class I heavy chain and the β_2 microglobulin are translocated to the endoplasmic reticulum before their translation is completed and their assembly takes place there with the assistance of the chaperons, BiP (a heatshock protein) and calnexin (glycoprotein). The MHC I molecules then associate with TAP and is ready for picking up a foreign antigenic peptides. After a peptide is bound, the system is released and after passing through the Golgi where their attached carbohydrates may be modified, exocytosis moves them to the surface of the cell membrane and this provides a chance for the $CD8^+$ T cells (CTL) to react to them. At this stage the cells that recognize self-antigens are eliminated by apoptosis. The MHC molecules resemble immunoglobulins, and the class II molecules, especially the β chains are highly polymorphic (The HLA-DRB 1 locus has more than 100 alleles.). The MHC molecules bind foreign antigens and present them to the lymphocytes. MHC Class I proteins accumulate the cut pieces of peptides from inside the cells whereas Class II MHC proteins are attached to the pieces of antigens from outside the cell. Before a Class II protein is loaded with an invader peptide it carries a neutral (dummy) peptide called CLIP which is then replaced by foreign pieces regulated by acidic conditions in the presence of DA. The Class I molecules are expressed on practically all cells that T cells recognize. Class II molecules are found on $CD4^+$ B cells and other antigen presenting cells. The T_H cells destroy any cell that present the antigen with MHC II molecules. The helper T cells do not attack directly the invaders but stimulate the action of macrophages. The MHC genes rely on transactivation by CIITA (non-DNA-binding) and RFX5, NF-X, NF-Y and other DNA-binding transcription factor proteins. There is a great diversity among MHC molecules; gene conversion may create variations in 10^{-4} range in sperms. The histocompatibility system of rabbits is called RLA, in rats RT1, in guinea pigs GPLA. (See HLA, immune system, blood cells, T cells, CTL, microglobulin, DA, proteasome, TAP, antigen presenting cell, antigen processing and presentation, bare lymphocyte syndrome, apoptosis, self-antigen, RFX)

MAJORITY CLASS SPORES: see polarized recombination

MALARIA: an infectious disease caused by one or another species of the *Plasmodium* protozoa. Transmission is mediated by *Anopheles* mosquito bites but it may be transferred by transfusion or transplacental infection with contaminated blood. This disease is prevalent in the subtropical and tropical areas of the world. The symptoms are chills, fever, sweating, anemia, etc. The attacks are recurring according to the major reproductive cycles of the parasite. Heterozygotes of sickle cell anemia are somewhat resistant to the disease and that explains the higher than expected frequency of this otherwise deleterious gene. Plasmodium degrades hemoglobin by two plasmalepsin proteases. Two minor and one major quantitative trait loci have been identified in *Anopheles gambiae* that inhibit the development of *Plasmodium cynomolgi* B in the midgut of the mosquito. (See hemoglobin, sickle cell anemia, *Plasmodium*, QTL, glucose-6-phosphate dehydrogenase)

MALATE DEHYDROGENASE: the product of the MDH1 (human chromosome 2p23) is cytosolic whereas the MDH2 enzyme (encoded in 7p13) is located in the mitochondria. (See Krebs-

Szentgyörgyi cycle, mitochondria)

MALDI: see electrophoresis

MALDI/TOF/MS: matrix-assisted laser desorption ionization/time of flight/mass spectrometry

MALE GAMETOCIDE: a chemical that destroys male gametes and thus may facilitate cross-pollination in plants or may serve as a human birth-control agent. (See crossing, birth control)

MALE GAMETOPHYTE: see gametophyte

MALE RECOMBINATION: is normally absent in *Drosophila* (except when transposable elements are present) and in the heterogametic sex (females) of the silk worm. Mitotic recombination may occur, however, even in the absence of the meiotic one. (See *MR*, hybrid dysgenesis). In maize and *Arabidopsis* certain cases indicate reduced frequency of recombination in the megasporocytes compared to the microsporocytes. (See recombination frequency, *MR*, hybrid dysgenesis)

MALE-SPECIFIC PHAGE: infects only those bacterial cells that carry a conjugative plasmid.

MALE STERILITY: may be caused by various chromosomal aberrations, such as inversions, translocations, deficiencies, duplications, aneuploidy, polyploidy, etc., and by cytoplasmic factors. Generally, in plants male sterility is more common than female sterility because the pollen, the male gametophyte, has a more independent life phase than the megaspore and cannot rely well on support by sporophytic tissues. In plants the male sterility may be caused by a mutation in the male gametophyte or it may be the result of the abnormal development of the anthers or other parts of the flower, e.g. failure to release the normally developed pollen. Common cause of male sterility is an alteration of the mitochondrial DNA, cytoplasmic male sterility (*cms*). Incompatibility between nuclear genes and certain cytoplasms as well as viral infection may also cause male sterility. Certain chemicals (mutagens, chromosome-breaking agents (maleic hydrazide) may have gametocidic effects. Male sterility may result in transgenic plants when a ribonuclease gene is driven by a tapetum-specificic promoter within the anthers and thus destroying the pollen. The fertility can be restored if the sterile plants are employed as female in crossing with a male carrying the transgene *barstar* inhibiting the ribonuclease. Other self-destroying — restoring combinations have also been considered. A healthy human male ejaculates 250-400 million sperms each time, and if for any reason this number is reduced to 20,000 or below, male sterility may result although only a single sperm functions in fertilization. (See cytoplasmic male sterility, chromosomal defects, gametocides)

MALIGNANT GROWTH: a defect in the regulation of cell division that may lead to cancer and may cause the spreading of the abnormal cells. (See metastasis, cancer, oncogenes, cell cycle)

MALONDIALDEHYDE: a mutagenic carbonyl compound, generated by lipid peroxidation. (See adduct, pyrimidopurinone)

MALTHUSIAN THEOREM: the population growth exceeds the rate of food production and thus jeopardizes the survival of humans. This 19th century principle was contradicted by the fact that between 1961 and 1983 the available food calories per capita have increased from 2320 to 2660 (over 7% in two decades). Unfortunately, this growth is slowing down again, and between 1984 to 1990 the increase was only about 1%. For about 2,000 years the global human population grew by an annual rate of about 0.04%. Between 1965 to 1970 the rate increased to 2.1% and in 1995 it was about 1.6% per year. (See human population growth)

MAMA (monoallelic mutation analysis): mutations are screened in somatic cell hybrids of lymphocytes and hamster cells with aid of DNA techniques.

MAMMALIAN COMPARATIVE MAPPING DATABASE: is part of the Mouse Genome Database URL: <http://www.informatics.jax.org>; questions, problems, etc. can be addressed to Mouse Genome Informatics Project: <mgi-help@informatics.jax.org>; tel: (207) 288-3371 ext. 1900, fax (207) 288-2516.

MAMMALIAN GENOME: a monthly publication of the Mammalian Genome Society. Contact V.M. Chapman, Dept. Molecular and Cellular Biology, Roswell Park Cancer Institute, Elm & Carlton, Buffalo, NY 1463, USA.

MAMMARY TUMOR VIRUS (of the mouse, MTV): is causing an estrogen-stimulated adenocarcinoma. The virus is transmitted through the milk. (See glucocorticoid response element)

MANATEE (*Trichechus manatus latirostris*): large herbivorous aquatic mammals, 2n = 48.

MANGO (*Mangifera*): tropical fruit tree, 2n = 2x = 40.

MANIC DEPRESSION: is psychological condition characterized by recurrent periods of excessive anguish (unipolar) or by manic depression (bipolar) the latter form is accompanied, in addition, by hyperactivity, obsessive preoccupation with certain things or events. Depression and other affective disorders may involve 2 to 6% of the human populations. It appears that the recurrence in the families is higher with the early onset types. Also, the bipolar types appear to have higher hereditary components. The genetic control of depression is unclear, X-linked recessive, autosomal dominant (chromosome 11p) genes have been implicated but the majority of assignments were not well reproducible. These may be major genes but other genes are also involved. The physiological bases may also vary from defects in neurotransmitters to electrolyte abnormalities, etc. Commonly recommended therapy involves monoamine oxidase inhibitors, tranquilizers (prozac), lithium, etc. (See also affective disorders, bipolar mood disorder, tyrosine hydroxylase, lithium)

MANNOPINE: N2-(1'-deoxy-D-mannitol-1'-yl)-L-glutamine. (See opines, Ti plasmid)

MANNOSEPHOSPHATE ISOMERASE (MPI): a Zn^{2+} monomeric enzyme converts mannose-6-phosphate into fructose-6-phosphate. It is coded in human chromosome 15q22.

MANNOSE-6-PHOSPHATE RECEPTOR (MPR): same as insulin-like growth factor; it plays a role in signal transduction, growth and lysosomal targeting. LOH mutation of this receptor (in human chromosome 6q26-q27) causes liver carcinoma. (See insulin-like growth factor, LOH)

MANNOSIDOSIS (MANB): the recessive (α-mannosidase B) deficiency has been located to human chromosome 19p13.2-q12 involves large increase of mannose in the liver causing susceptibility to infection, vomiting, facial malformations, etc. Another mannosidase defect at another autosome caused excessive mannosyl-1-4-N-acetylglucosamine and heparan in the urine, involved apparently glycoprotein abnormalities and a variety of physical and mental defects.

MANN-WHITNEY TEST: is a powerful non-parametric method for determining the significance of difference between two normal-distributed populations. This method is useful for evaluating scores of samples even if the size of them is not identical. The procedure is illustrated by small samples, however samples of n > 20 are prefered. The scores are ranked (T) as follows (in case of ties the average ranks are assigned to the two):

Populations	I	II	II	I	I	II	II	Sum of I = T_I = 10
Scores	1	4	5	7	8	9	10	Sum of II = T_{II} = 18
Rank	1	2	3	4	5	6	7	n_I = 3, n_{II} = 4

The null hypothesis to be tested is that the distributions of the two populations (I and II) are identical. Then the U is determined for sample I:

$U_I = n_I \, n_{II} + \{[n_I(n_I + 1)]/2\} - T_I = 3 \times 4 + (12/2) - 10 = (12 + 6) - 10 = 8$. In case the resulting U value is larger than $(n_I \, n_{II})/2$ then calculate $U' = n_I \, n_{II} - U_I$. For large populations the sampling distribution for U is approximately normal and $E(U) = (n_I \, n_{II})/2$ and the variance is determined $\sigma^2 = [n_I \, n_{II}(n_I + n_{II} + 1)]/12$; hence the z value (the standard normal variate) = $[U - E(U)]/\sigma_U$ and the probabilities corresponding to z can be read from statistical tables of the cumulative normal probabilities and a few commonly used corresponding values are as follows for z = 1.65, 1.96, 2.58 and 3.29, the *P* values are 0.90, 0.95. 0.99 and 0.999, respectively. (See Wilcoxon's signed rank test, standard deviation, probability, null hypothesis)

MANX CAT: is tailless because of fusion, asymmetry and reduction in size of one or more caudal vertebrae. This phenotype is caused by a dominant gene that is lethal in homozygotes. The Manx protein is essential also for the development of the notochord of lower animals. (See also brachyury, notochord)

MAOA (monoaminoxidase A): enzyme is involved in the biosynthetic path of neurotransmitters from amino acids. Mutation in the 8th exon (amino acid position 936) converted the glutamine codon CAG to TAG (chain termination codon) and resulted in mild mental retardation,

MAOA continued
continued and impulsive aggression, arson, attempted rape, and exhibitionism in human males with this X-chromosomal recessive defect. The block of MAOA resulted in accumulation of normetanephrine (a derivative of the adrenal hormone epinephrine), and tyramine (an adrenergic decarboxylation product of tyrosine) and a decrease in 5-hydroxyindole-3-acetone. The heterozygous women were not affected behaviorally or metabolically; monoamine oxidase B level remained normal. Both enzymes are in the short arm of the human X-chromosome and the enzymes are located in the mitochondrial membrane. (See mitochondria)

MAP: see microtubule associated proteins. (See *ASE1*, see also map genetic)

MAP-BASED CLONING (positional cloning): isolation of gene(s) on the basis of chromosome walking and propagation usually by YAC and/or cosmid clones. (See chromosome walking, chromosome landing, position effect)

MAP DISTANCE: indicates how far syntenic genes are located from each other in the chromosome as estimated by their frequency of recombination; 1 map unit = 1% recombination = 1 centi Morgan. The greater the distance between two genes, the higher is the chance that they are separated by recombination. A single recombination between two genes in a meiocyte produces maximally 50% recombination that is 50 map units. The distance between syntenic genes may exceed 50 map units several times; these longer distance are determined then in a staggered manner, proceeding step-wise from left to right and right to left. In prokaryotes, using conjugational transfer and recombination, map distances are measured in minutes of transfer (see conjugational mapping). (See also recombination frequencies for map units in kbp DNA, radiation hybrids, mapping function)

MAP EXPANSION: the distance between two distant markers exceeds the sum of the distances of markers in between; it is commonly observed in gene conversion. (See gene conversion)

MAP, GENETIC: the order of genes in chromosomes determined on the basis of recombination frequencies. (See also mapping, recombination frequencies, physical map, radiation hybrids, RFLP, RAPD, mapping function)

MAP KINASE: a family of serine/threonine protein kinases associated with mitogen activation. They have key role in signal transduction pathways. (See cell cycle, signal transduction)

Map Manager v 2.5: software for storing, organizing genetic recombination data and data base for RI strains of mouse. Information: K.F. Manly, Roswell Park Cancer Institute, Elm & Carlton, Buffalo, NY 14263, USA. Phone: 716-845-3372. Fax: 716-845-8169. INTERNET: <kmanly@mcbio.med.buffalo.edu>.

MAP, PHYSICAL: see physical map

MAP UNIT: 1% recombination = 1 map unit (m.u. or 1 centiMorgan, c.m. or cM). In approximate kilobase pairs equivalent to one centiMorgans in a few species: *Arabidopsis* ≈140; tomato ≈ 510; human ≈ 1,108; maize ≈ 2,140.

MAPK: mitogen-activated protein kinase. The family includes KSS1, HOG-1, FUS3, SLT-2, sapk-1, erk-1, etc. (See signal transduction, MAPKK, MAPKKK)

MAPKK: mitogen-activated (MAP) protein kinase kinase. This protein mediates signal transduction pathways by phosphorylating RAS, Src, Raf and MOS oncogenes. When such an active kinase was introduced into mammalian cells the AP-1 transcription factor was activated and the cells formed cancerous foci and became highly tumorigenic in nude mice, indicating that MAPKK is sufficient for tumorigenesis. (See signal transduction, tumor)

MAPKKK: is mitogen activated protein kinase kinase kinase. (See also TAK1)

MAPLE (*Acer* spp.): hardwood trees, and the sugar maple is used for collecting syrup, 2n = 26.

MAPLE SYRUP URINE DISEASE: see isoleucine-valine biosynthetic pathway

MAPMAKER 3.0: software for constructing linkage maps using multipoint analysis in testcross and F_2. MAPMAKER/QTL is for quantitative trait loci. Available for Sun (Unix), PC (DOS) and Macintosh. Contact: Eric Lander, Whitehead Institute, 9 Cambridge Center, Cambridge, MA 02142, USA. Fax: 617-258-6505. INTERNET: <mapmaker@genome.wi.mit.edu>

MAPPING BY DOSAGE EFFECT: if the activity of enzymes is proportional to their dosage

Mapping by dosage effect continued

and disomics can be distinguished from critical trisomics, genes (for the enzymes) located in a specific trisome can be identified and assigned to that specific chromosome or in the case of telotrisomic to a particular chromosome arm. Theoretically the enzyme activity is expected as follows if the locus is situated in the long arm of a particular chromosome:

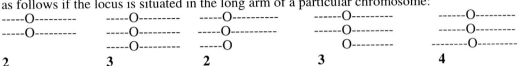

```
-----O---------   -----O---------   -----O----------   ------O---------   ------O---------
-----O---------   -----O---------   -----O----------   ------O---------   ------O---------
                  -----O---------   -----O             O---------         --------O---------
     2                  3                 2                  3                   4
```

In human trisomy 21 several genes show increased expression ranging from 1.21 to 1.61 relative to the normal disomic condition. Thus in practice the dosage effect may not be perfectly additive yet it may be clear enough for classification. (See trisomics)

MAPPING, GENETIC: mapping of chromosomes can be carried out on the basis of recombination frequencies of chromosomal markers, either genes or DNA markers such as RFLPs, RAPDs, etc., or molecular methods are used in physical mapping (chromosome walking). As a hypothetical example let us assume that genes *a, b, c, d* and *e* are syntenic and the recombination frequencies between them was found to be:

a   0.06   *b*   0.04   *c*   0.06   *d*   0.16   *e*   0.18   *g*

The sum of the recombination frequencies is 0.06 + 0.04 + 0.06 + 0.16 + 0.18 = 0.50 indicating that the recombination between *a* and *g* is independent. The results shown could be obtained between two genes at a time but such a two-point cross would not have permitted the determination of the order of the genes relative to each other. For the determination of genes at "left" and genes at "right" a three-point cross is required as a minimum, and multipoint crosses are even more helpful (see also recombination frequencies). Let us see the results of a hypothetical three-point testcross:

Phenotypic Classes →	ABD	abd	Abd	aBD	ABd	abD	AbD	aBd
Number of Individuals	34	34	5	5	10	10	1	1
		Parental		Recombinants Interval I		Recombinants Interval II		Recombinants Intervals I + II
Total of 100		68		10		20		2

According to the data above, the number of recombinants in interval I was 10 + 2 and in interval II 20 + 2 (the double recombinants had recombination in both interval I and II and their number must be added to the numbers observed). Thus the frequency of recombination between *A* and *B* is 12/100 = 0.12 and between *B* and *D* 22/100 = 0.22. The number of recombinations between *A* and *D* is 10 + 20 + 4 = 34 (the 4 is the double of the number of recombinants in intervals I and II because these represented double recombination events). Thus the relative map positions are *A - B - D*. Had we found that the combined parental numbers were 68 but the recombinants *Abd* plus *aBD* 10 and *ABd* plus *abD* 2, and *AbD* plus *aBd* 20, we had to conclude that the gene order was *A - D - B*, because the lowest frequency class (0.10 x 0.20 = 0.02) must have been the double recombinants, and thus the gene order would have been *A - D - B*. The observed recombination frequencies may have to be corrected by mapping functions because not all double-crossovers might have been detected (see mapping functions). The recombination frequencies may be biased also by interference when the frequency of double-crossovers are either higher or lower than expected on the basis of the product of the two single-crossovers (see coincidence, interference). The (corrected) recombination frequencies can be converted to map units by multplication with 100 and 1 map unit (m.u. or centi-Morgan (cM]) is 0.01 frequency of recombination. Recombination frequencies can be estimated also in F_2 by using the product ratio method (see there). In the latter case recombination frequencies can be calculated only between pairs of loci yet from the data of two pairs

Mapping genetic continued

involving 3 loci, the gene order can be determined. (See also recombination frequencies, crossing over, QTL, deletion mapping, chromosome walking, physical maps, Mapmaker, Joinmap, maximum likelihood method applied to recombination, mapping functions, radiation hybrids, F_2 linkage estimation)

MAPPING FUNCTIONS: correct map-distance estimates from recombination frequencies when the recombination frequency in an interval exceeds 15-20 % and double crossing overs are undetectable because of the lack of more densely positioned markers. (See Haldane's mapping function, Kosambi's mapping function, Carter-Falconer mapping function, mapping, recombination frequency, coefficient of coincidence)

MapSearch: locates regions of a genomic restriction map that resemble best a local restriction the so called "probe".

MapShow: computer program displaying MapSearch alignments and draws Probe-to-Map alignments in Sun Workstations. (See mapping)

MAR (matrix attachment region): the attachment region has a consensus of so called A box (AA TAAATCAA) or a T box (TTA/TAA/TTTA/TTT). (See chromatin, loop domains model)

MARANHAR: see killer plasmids, *Neurospora*

MARCKS: protein substrates of the protein kinase C (PKC), involved in differentiation; they bind actin filaments.

MAREK's DISEASE: see Herpes, Epstein-Barr virus

MARFAN SYNDROME (MFS): includes tall thin stature, long limbs and fingers, chest deformations are also common. The three most consistent defects are skeletal, heart-vein (cardiovascular) and eye (ectopia lentis) abnormalities. The disease may affect the development of the fetus and may be recognized in early development and the life expectancy in serious cases may not much exceed 30. The cause of death is generally heart failure but the defect may surgically be corrected in some cases. The penetrance appears very good but the expressivity is highly variable. The symptoms frequently overlap with other anomalies, particularly with those of the Ehlers-Dunlop syndrome. The latter involves a defect in collagen. The primary defect in MFS involves the elastic fiber system glycoprotein, fibrillin. This protein of the connective tissue contains repeats resembling sequences in the epidermal growth factor (EGF) where the lesion observed leads to the identification of the basic molecular cause. Formerly a collagen defect was suspected. Several investigators confirmed a transversion mutation at codon 293 leading to CGC (Arg) → CCC (Pro) replacement. Similar molecular defects have been identified in *Drosophila, Caenorhabditis* and cattle. This dominant gene has been assigned to human chromosome 15q15-q21.3. Interestingly, mosaicism for trisomy 8 causes similar symptoms. The prevalence of MFS is about 1×10^{-4} but this figure may not be entirely reliable because of the wide range of manifestation of the symptoms. The recurrence risk is about 50%; 15-30% of the cases may be due to new mutation that is the cause of the most severe cases, whereas the familial incidence generally entails milder symptoms. The estimated mutation rate is $4 - 5 \times 10^{-6}$. (See also Marfanoid syndromes, coronary heart disease, cardiovascular disease, connective tissue disorders, penetrance, expressivity, transversion mutation, Ehlers-Dunlop syndrome, fibrillin)

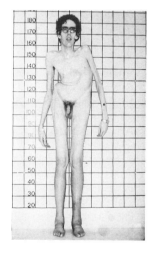

(Courtesy of Dr. D.L. Rimoin, Los Angeles)

MARFANOID SYNDROMES: may resemble the Marfan syndrome but one of the forms has no ectopia lentis (displacement of the crystalline lens of the eye). Another form does not involve cardiovascular defects. The marfanoid-craniosynostosis is called Shprintzen-Goldberg syndrome and the anomaly is caused by mutation in the fibrillin-1 gene. (See Marfan syndrome, eye diseases, craniosynostosis syndromes, fibrillin)

mariner: probably the smallest transposable element in eukaryotes (1,286 bp). It has not been observed in *Drosophila melanogaster* but it has been detected in African species of the *D. melanogaster* subgroup, *D. sechellia* (1-2 copies, *D. simulans* (usually 2 copies), in *D. yakuba* (about 4 copies), in *D. teissieri* (10 copies), *D. mauritiana* (20 to 30 copies). The element contains 28 bp inverted terminal repeats and a single open reading frame (1.038 bp) beginning with an ATG codon at position 172 and termination with an ochre (TAA). Overlapping AATAA bases may serve as polyadenylation signal:

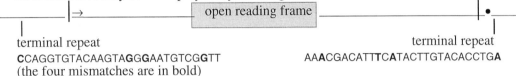

terminal repeat   terminal repeat
CCAGGTGTACAAG**T**A**GGG**AATGTCG**G**TT   AAA**C**GACATTT**C**A**T**ACTTGTACACCTGA
(the four mismatches are in bold)

The target in the untranslated leader of the w^{pch} is 5'- TGGCGTA↓TAAACCG-3'. The arrow marks the insertion and the TATA indicate probably the target site duplication. The mariner is different from other transposable elements inasmuch as inducing a high frequency of somatic sectors (4×10^{-3}) at the w^{pch} (white *peach*) locus. Germline mutation is about 2 to 4×10^{-3}, with about twice as high in the males than females (no sex difference in somatic mutation). Before the transposable element was recognized, the somatic instability was attributed to the factor named *Mos* in chromosome 3. This element also causes dysgenesis but it does not display the reciprocal difference observed in the *P - M* and *I - R* systems, however mariner transmitted through the egg shows higher rates of somatic excisions. Mariner homologs occur in other species too, including humans but the (*human*) sequences are pseudogenic although increase unequal crossing over in human chromosome 17p11.2-p12. The mariner sequences are expressed also in *Leishmania*. (See also hybrid dysgenesis, transposable elements, Charcot-Marie-Tooth syndrome, neuropathy, HNPP, unequal crossingover, MLE, MITE, *Leishmania*)

MARKER: genetic marker is any gene in a chromosome or cytoplasmic organelle, used as a special label for that chromosome or chromosomal area. Molecular marker is a macromolecule (nucleic acid or protein) of known size and electrophoretic mobility to be used as a reference point in estimating the size of unknown fragments and molecules. (See RFLP, RAPD, ladder)

MARKER-ASSISTED SELECTION: is used for animal and plant breeding once linkage has been established between physical markers (RFLP, microsatellite loci) and other, economically desirable traits (e.g. disease resistance, productivity) that have low expressivity and/or poor penetrance because of the substantial environmental influences. The physical markers should not have ambiguity of expression and thus may facilitate much faster progress in breeding. (See RFLP, microsatellite, expressivity, penetrance, QTL)

MARKER EFFECT: by generalized transduction — theoretically — any bacterial gene should be transferred by the transducing phage. As a matter of fact some genes are transduced 1,000 fold better than others. The differences have been attributed to the distribution of *pac* sites in the bacterial chromosome. Also recombination by transduction may vary along the length of the bacterial chromosome. (See generalized transduction, *pac* site)

MARKER EXCHANGE MUTAGENESIS: see targeting genes

MARKER PANELS: are DNA probes or genes that cover reasonably well portions of the genome regarding linkage.

MARKER RESCUE: integration of markers into normal phage from mutagen-treated (irradiated) phage during mixed infection of the host; it is similar to cross reactivation where from defective phages by recombination normal phages can be obtained. (See reactivation, multiplicity reactivation, Weigle reactivation)

MARKOV CHAIN STATISTICS: a sequence x_1, x_2...of mutually dependent random variables constitutes a Markov chain if there is any prediction about x_{n+1}. Knowing $x_1...x_n$ may without loss be based on x_n alone. Among others, it is used in physical mapping of genomes, for ascertaining frequency distributions in populations.

MARMOSETS: new world monkeys. (See Callithrichicidae)

MARMOTA: *Marmota marmota* 2n = 38; *Marmota monax* 2n = 38.

MAROTEAUX-LAMY SYNDROME: see mucopolysaccharidosis

MARQUIO SYNDROME: see mucopolysaccharidosis

MARRIAGE: in every state of the United States and many other countries marriages between parent and child, grandparent-grandchild, aunt-nephew, uncle-niece and brother-sister are illegal. In about half of the states first cousin and half sibling unions are also prohibited. In some societies consanguineous marriages are legal. (See consanguinity, miscegenation) ♂

MARS SHIELD: symbol of male; in pedigrees usually □ is used. (See Venus mirror) ○

MARSUPIAL: mammalian group of animals that carry their undeveloped offspring in a pouch, e.g. the kangaroos and other Australian species, also the North-American opossum. The organization of the genetic material of marsupials differs in several ways from other placental mammals. Ohno's law is not entirely complied with inasmuch as genes in the short arm of the eutherian X chromosomes are dispersed in three autosomes. Marsupial Y chromosomes are extremely small yet they contain testis-determining and other sequences homologous to other mammals although the Y chromosome is not critical for sex determination. Their sex chromosomes also carry a pseudoautosomal region at their tip but do not form synaptonemal complex and do not recombine. (See eutheria, monotrene, Ohno's law, SRY, pseudoautosomal)

MARTEN (*Martes americana*): 2n = 38.

MAS: see marker-assisted selection

MAS ONCOGENE: MAS1 was assigned to human chromosome 6q27-q27; it encodes a transmembrane protein. (See transmembrane protein, oncogenes)

MASH2: mammalian helix-loop-helix transcription factor controlling extraembryonic trophoblast development but not that of the mouse embryo. (See MYF-3, DNA-binding protein domains)

Mask: a computer program that produces ambiguous files in order to protect confidential DNA sequences in EcoSeq programs.

MASKED mRNA: present in eukaryotic cells in such a form and cannot be translated until a special condition is met. (See mRNA)

MASKED SEQUENCES: of nucleic acids are associated either with proteins or other molecules to protect them from degradation.

MASPIN: a protease inhibitor of the serpin family; maspin sequences are frequently lost from advanced cancer cells. (See serpin)

Mas6p: see mitochondrial import

MASS SPECTRUM: when in the mass spectrometer molecules are exposed to energetic electrons they are ionized and fragmented. Each ion has a characteristic mass to charge ratio, m/e. The m/e values are characteristic for particular compounds and provide the mass spectrum for chemical analysis. (See laser desorption mass spectrum, electrospray mass spectrum, affinity-directed mass spectrometry)

MAST CELLS: reside in the connective or hemopoietic tissues and play important role in natural and acquired immunity. They release TNF-α (tumor necrosis factor), histamines and attract eosinophils (special white blood cells) and destroy invading microbes especially if they have IgE or IgG (immunoglobulins) on their surface. They are responsible for the inflammation reactions in allergies. (See IL-10, immune system, TNF, immunoglobulins)

MASTER CHROMOSOME: the large circular genome within the mitochondria. (See mtDNA)

MASTER MOLECULE: regulates series of reactions in differentiation, involving several genes. (See morphogenesis in *Drosophila*, neuron-restrictive silencer factor)

MASTER-SLAVE HYPOTHESIS: interpreted redundancy in the genomes by multiple copying 'the slaves' from the original 'master' sequences. (See redundancy)

MAT CASSETTE: see mating type determination in yeast.

***MAT* LOCUS**: in yeast is involved in the determination of mating type in yeast. (See mating type determination in yeast)

MATE KILLER: mu particle. (See symbionts hereditary)

MATERNAL EFFECT GENES: display delayed inheritance because only the offspring of the homozygous or heterozygous dominant females is affected; these females themselves may appear normal. Also, genes with products (RNA, protein) in the follicle and nurse cells that may diffuse into the oocytes and the embryo, and thus not just the zygotic genes, affect early development. (See morphogenesis, delayed inheritance, *Limnaea*, indirect epistasis)

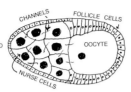
THE DROSOPHILA EGG CHAMBER CYSTS ARE SURROUNDED BY FOLLICLE CELLS AND INSIDE THE DIPLOID PRIMARY OOCYTE IS SHOWN WITH THE OFTEN POLYPLOID NURSE CELLS. THESE ARE CONNECTED BY CYTOPLASMIC BRIDGES. THE MATERNAL GENES AFFECT OOGENESIS AND THE EARLY DEVELOPMENT OF THE EMBRYO

MATERNAL EMBRYO: develops from an unfertilized egg. (See apomixis, parthenogenesis)

MATERNAL GENES: see maternal effect genes

MATERNAL INHERITANCE: genetic elements (generally extranuclear) are transmitted only through the female. (See mtDNA, mitochondrial genetics, chloroplast genetics)

MATERNITY VERIFICATION: is rarely needed as the Romans held "mater certa"; in case of legal disputes, the methods of forensic genetics are applicable. (See forensic genetics, paternity testing)

MATING ASSORTATIVE: see assortative mating

MATING BACTERIAL: see conjugation

MATING CONTROLED: see controled mating

MATING INTERRUPTED: see interrupted mating, conjugation mapping

MATING NONRANDOM: see inbreeding, assortative mating, autogamy (self-fertilization), controled mating

MATING RANDOM: see random mating, mating systems

MATING SUCCESS (K): $\dfrac{\text{No. of females mated by mutants/ No. of mutant males}}{\text{No. of females mated by wild types/ No. of wild type males}}$

MATING SYSTEMS: can be random mating, self fertilization, inbreeding, assortative mating or a combinations of these. (See also individual entries, Hardy-Weinberg theorem)

MATING TYPE: the designation of individuals with plus or *a* ("male") and minus or *α* ("female") labels when sex like in higher eukaryotes cannot be recognized yet genetically two types exist that do not "mate" within group but in between groups and the diploid zygote subsequently undergoes meiosis and reproduces the mating types in 2:2 proportion.

MATING TYPE DETERMINATION IN YEAST: *Saccharomyces cerevisiae* yeast can exist in homothallic and heterothallic forms. The heterothallic yeast cells are haploids and are either of a or α mating type. The homothallic yeast cells are diploid and heterozygous for the mating type genes (a/α). The diploid cells arise by fusion of haploid cells. Recognition of the opposite mating type cells is mediated by the pheromones, the α factor (13 amino acids) and the a factor (12 amino acids) peptide hormones, respectively. The two types of cells are equipped also by surface receptors for the opposite mating type pheromones. The diploid cells lack these factors and receptors and do not fuse but may undergo meiosis and sporulate by releasing both α and a haploid cells.

The two mating types are coded by genes in chromosome 3 on the left and right sides of the centromere, respectively. These genes are clustered within the so called mating type cassettes and are silent at their positions named *HMLα* and *HMRa* locations. They are expressed when transposed to the mating type site, *MAT*. At *MAT* either the left *HMLα* or the right *HMRa* cassette can be expressed within a particular homologous chromosome. The *MAT* site is approximately 2 kb and it is about 187 kb from *HMLα* and at about 93 kb from *HMRa*. The overall structure of the regions is shown on the following page. In the outline the *MAT* site is shown empty but in reality it is occupied by either the *HMLα* or by the *HMRa* cassette.

GENETICS MANUAL

Mating type determination in yeast continued

```
         HMLα         ←      187 kb      →  MAT  ←   93 kb  →    HMRa
    W  X  Yα  Z1  Z2                                           X  Ya  Z1
----------------------------------------O-----------------------------------------
    ↑                                                                    ↑
                                       SIR
```

When one or the other silent complex is unidirectionally transposed to the *MAT* site, the expressed *MAT*α and *MAT*a, respectively, are generated. At the original (left and right) locations the *Y*α (747 nucleotides) and the *Y*a (642 nucleotides) are kept in place and silent by the product of the *SIR1-4* gene (**S**ilent **I**nformation **R**egulator). Repression is mediated also by the autonomous replication sequences (ARS) and the pertinent enhancer (E) elements. The transposition activity of a *MAT* cassettes requires also the presence of nuclease hypersensitive sites in the flanking regions. The transposition at the *MAT* sites is initiated when the HO endonuclease makes a staggered cut at the *Y Z1* junction generating strands with four base overhangs:

5'-GCTTT↓CGGCAACAGTATA-3' *MAT*a	5'-ACTTCGCGC AACA↓GTATA-3' *MAT*α
3'-CGAAAGGCG↑TTGTCATAT-5'	3'-TGAAGCGCG↑TTGTCATAT -5'

The *HO* gene is expressed only in the haploid cells but not in the diploids and only at the end of the G1 phase of the cell cycle and both new cells have the same mating type and can transpose their mating type gene only after the completion of the next cell cycle. The function of *HO* is repressed by the product of gene *SIN1-5* and the expression of gene *SW1-5* is required for the expression of *HO*. Mother cells selectively transcribe *HO*. In daughter cells, Ash1p suppressor of *HO* may accumulate as a result of asymmetric mRNA distribution. For the actual mating the function of other (sterility, *STE*) genes are also needed. *STE2* and *STE3* are cell type-specific receptors of G proteins. *GPA1* encodes the Gα subunit, *STE4* the Gβ subunit and *STE18* the Gγ subunit. The GβGγ subunits, after activation, regulate further downstream units of the pheromone signaling pathway, including the kinases STE20, a series of MAP kinases (encoded by *STE11*, *STE7* and *FUS/KSS* genes). The Ste5p protein (encoded by *STE5*) is presumed to be a scaffold for organizing all the kinases.

Transposition involves pairing with the homologous *Z* sequences, followed by an invasion of a double-stranded receiving *Y* site and degradation of these *Y* sequences. Subsequently the *X* site is invaded. New DNA synthesis takes then place using as template the sequences of the invader molecule that is replacing the old DNA tract. Integration is mediated in the pattern of a gene conversion mechanism as suggested by the Holliday model of recombination.

The α mating type gene has two elements α1 (transcribed from right to left) that induces the α mating functions and α2 (transcribed in left to right similarly to a) that keeps in check the expression of the a mating type expression. The simultaneous expression of a1 and α2 represses also *SIR* and other haploid-specific genes. These processes recruite also proteins PRTF (Pheromone Receptor Transcription Factor) and GRM (General Regulator of Mating type) that recognize specific nucleotide sequences within the haploids and diploids. The α2 protein also interacts the MCM1 DNA binding and the MADS box proteins.

Whether *HML* or *HMR* switching occurs depends on the surrounding sequences and it is not intrinsic to the elements. *Mat a* cells recombine with *HML* almost an order of magnitude more frequently than with *HMR*. *MAT*α cells recombine in 80-90% of the cases with *HMR*. The switching is controled by 700 kb element 17 kb proximal to *HML*.

The expression of the mating type genes involves a complex cascade of events. The mating type protein factors interact with G protein-like receptors situated in the cell membrane. These G proteins then transduce the mating signals through a series of phosphorylation reactions to transcription factors that control the turning on/off genes mediating the cell cycle, cell fusion

Mating type determination in yeast continued

and conjugation. The industrial strains of yeast are frequently diploid or polyploid and may be heterozygous for mating type. In such a case they fail to mate. If they are plated on solid medium they may sporulate and the chromosome number will be reduced. (See Holliday model, signal transduction, cell cycle, silencer, sex determination, pheromones, MCM1, MADS box, homothallic, heterothallic, rare-mating, regulation of gene activity, Ty, *Schizosaccharomyces pombe*, ORC)

MATRIX: solutes in cells, organelles or chromosomes, etc.; the *extracellular matrix* fills the space among the animal cells and it is composed of a meshwork of proteins and polysaccharides, secreted by the cells.

MATRIX ALGEBRA: deals with elements that are arranged as shown at the left.

$$\begin{bmatrix} 1 & 4 & 7 \\ 2 & 5 & 8 \\ 3 & 6 & 9 \end{bmatrix} \begin{bmatrix} 2 & 1 \\ 4 & 6 \end{bmatrix}$$

A matrix with r rows and c columns is of *order* $r \times c$ or an $r \times c$ *matrix*. If $r = c$, it is a *square matrix*. A matrix with 1 row is *row vector* and a matrix with only 1 column is a *column vector*.

MATRIX-ASSISTED LASER DESORPTION IONIZATION/TIME of FLIGHT/MASS SPECTROMETRY: is a procedure for separation of DNA fragments mixed with a carrier that is painted subsequently on the surface of a solid face target. Laser desorbs and ionizes the fragments and acceleration in a mass spectrometer is used to determine fragment length. (See laser, mass spectrum, laser desorption mass spectrum)

MATROCLINOUS: the offspring resembles the mother because it developed either from an unfertilized egg, or from an egg that underwent nondisjunction and carries two X chromosomes or from a female with attached X-chromosomes or by imprinting or dauermodification or due to set of dominant genes or the failure of transmission of a particular chromosome through the sperm or caused by non-nuclear (mitochondrial, plastid) genes. (See *Rosa canina*, chloroplast genes, mtDNA)

MATTHIOLA (garden stock, wallflower): a cruciferous ornamental (2n = 14), and an early object of genetic research. In floricultural use the presence of a lethal factor, tightly linked to the simple flower character (*S*), has been of special interest. This causes the appearance of the "ever-segregating" full-flower trait (*s*) in 1:1 proportions rather than in 3:1. As a result of sophisticated breeding techniques and seed selection in trisomic offspring, the commercially available seed germinates and develops into nearly 100% full-flower plants that are, however, completely sterile due to the recessive lethal factor. Thus, the seed supply is entirely dependent on commercial sources.

MATURASES: proteins that mediate a conformational change in the pre-mRNA transcript and cooperate in the splicing reactions. (See introns, mitochondrial genetics)

MATURATION DIVISIONS: same as meiosis

MATURATION OF DNA: phage proteins cut the linear, continuous DNA into pieces that can be accommodated by the phage capsids.

MATURATION PROMOTING FACTOR: see MPF

MAURICE OF BATTENBERG: a hemophiliac grandson of Queen Victoria of England; son of carrier daughter Beatrice. (See hemophilias)

MAURICEVILLE PLASMID: see *Neurospora* mitochondrial plasmids

MAX: a b/HLH/Z (basic helix-loop-helix) protein hetero-oligomerizes with the Myc oncoproteins and this state is required for malignant transformation by Myc. Its DNA recognition site is CA CGTG. MAX may be orchestrating the biological activities of b/HLH/Z transcription factors. The basic α helices follow the major groove of the DNA in a *scissors grip*. (See Myc, helix-loop-helix, RFX, leucine zipper)

MAXAM-GILBERT METHOD: see DNA sequencing

MAXICELLS: are bacterial cells that lost most or all of their chromosomal DNA because of

Maxicells continued

heavy irradiation by UV light. Therefore they do not replicate their DNA. The plasmid they contain may have escaped the irradiation, and represents an appropriate replicon, and can carry on replication of that plasmid and direct the synthesis of plasmid-coded proteins. This makes such cells ideal for the expression of the plasmid-born protein without a background of cellular proteins. Especially useful are those maxi cells containing lambda vectors which have sufficient expression of the phage repressor and thus do not permit λ protein expression. (See minicells, lambda phage, replicon, plasmids)

MAXICIRCLE: the large mitochondrial genome. (See mtDNA)

MAXIMAL EQUATIONAL SEGREGATION: takes place when a gene is segregating independently from the syntenic centromere. It has particular significance in polyploids because it facilitates an increase of double (or multiple) recessive gametes and thus affects segregation ratios as a function of the map distance between gene and centromere. (See autopolyploids, trisomic analysis, synteny, polyploidy)

MAXIMAL PARSIMONY: see evolutionary tree

MAXIMAL PERMISSIVE DOSE: see radiation hazard assessment

MAXIMIZATION of GENE EXPRESSION: can be achieved by the selection or modification of optimal promoters or in prokaryotes by varying the bases immediately after the Shine-Dalgarno sequence or manipulation of the triplet preceding the first methionine codon. (See regulation of gene expression, regulation of protein synthesis, Shine-Dalgarno)

MAXIMUM LIKELIHOOD METHOD APPLIED TO RECOMBINATION frequencies: the justification for the use of the maximum likelihood principle in estimating recombination frequencies is that the value obtained has the smallest variance among all procedures. The estimation is based on the maximization of:

$$\frac{n!}{a_1! \, a_2! \ldots a_t!} (m_1)^{a_1} (m_2)^{a_2} \ldots (m_t)^{a_t}$$

where n is the population size, $a_1 \ldots a_t$ stand for the number of individuals in the different phenotypic or genotypic classes, $m_1 \ldots m_t$ represent the expected proportions of individuals in classes $1 \ldots t$. After maximizing the logarithm of the likelihood (L) expression with respect to the recombination fraction (p), we have:

$L = C + a_1 \log m_1 + a_2 \log m_2 + \ldots a_t \log m_t$; where C is a constant of the maximum likelihood that is eliminated upon differentiation:

$$\frac{dL}{dp} = a_1 \frac{d \log m_1}{dp} + a_2 \frac{d \log m_2}{dp} + a_t \frac{d \log m_t}{dp} = 0$$

Test cross examples:

	PARENTAL	RECOMBINANT	RECOMBINANT	PARENTAL	
Gametic Genotypes →	AB	Ab	aB	ab	Σ
Observed in **Coupling**	4032	149	152	4035	8368
Expected Coupling	$\frac{1}{2} n(1-p)$	$\frac{1}{2} n(p)$	$\frac{1}{2} n(p)$	$\frac{1}{2} n(1-p)$	n
	RECOMBINANT	PARENTAL	PARENTAL	RECOMBINANT	
Gametic Genotypes →	AB	Ab	aB	ab	Σ
Observed **Repulsion**	638	21,379	21,096	672	43,785
Expected Repulsion	$\frac{1}{2} n(p)$	$\frac{1}{2} n(1-p)$	$\frac{1}{2} n(1-p)$	$\frac{1}{2} n(p)$	n

For the coupling experiment above:

$$L = 4032 \log\left(\frac{1}{2} - \frac{1}{2}p\right) + 149 \log\left(\frac{1}{2} p\right) + 152 \log\left(\frac{1}{2} p\right) + 4035 \log\left(\frac{1}{2} - \frac{1}{2}p\right)$$

After maximization and differentiation:

$$\frac{dL}{dp} = \frac{4032}{1-p} + \frac{149}{p} + \frac{152}{p} - \frac{4035}{1-p} = 0 \quad \text{and} \quad p = \frac{149 + 152}{8368} \cong 0.03597$$

The standard error s_p is calculated:

Maximum likelihood method applied to recombination continued

$-\frac{1}{V_p} = S\left(m\,n\,\frac{d^2\log m}{dp^2}\right)$. Since a $\frac{d\log m}{dp}$ was defined earlier, after a second differentiation and substitution (mn) for (a), we obtain:

$-\frac{1}{V_p} = -\frac{n}{2}\left(\frac{1}{1-p} + \frac{1}{p} + \frac{1}{p} + \frac{1}{1-p}\right) = \frac{n}{(1-p)} \cong \frac{8368}{0.034676} \cong 241{,}319$ and hence

$V_p = 0.000004143$ and $s_p = \sqrt{V_p} \cong 0.00204$ or by the general formula $s_p = \sqrt{\frac{p[1-p]}{n}}$.

Recombination in F_2 can also be estimated with the aid of the maximum likelihood principle and it will be exemplified by a coupling phase progeny:

	PARENTAL	RECOMBINANT	RECOMBINANT	PARENTAL		
Phenotypic classes	AB	aB	Ab	ab	Σ	
Expectation	$\frac{n}{4}(2+P)$	$\frac{n}{4}(1-P)$	$\frac{n}{4}(1-P)$	$\frac{n}{4}P$	n	(1)
Observed	663	36	40	196	935	

$L = 663 \log\left(\frac{1}{2} + \frac{1}{4}P\right) + 36 \log\left(\frac{1}{4} - \frac{1}{4}P\right) + 40 \log\left(\frac{1}{4} - \frac{1}{4}P\right) + 196 \log\left(\frac{1}{4}P\right)$ (2)

Upon maximization: $\frac{dL}{dP} = \frac{663}{2+P} - \frac{36}{1-P} - \frac{40}{1-P} + \frac{196}{P} = 0$ (3)

This can be reduced: $\frac{663(1-P)(P)}{2P - P^2 - P^3} - \frac{76(2+P)(P)}{2P - P^2 - P^3} + \frac{196(2+P-2P-P^2)}{2P - P^2 - P^3}$ (4)

Common denominator omitted and multiply

$663(P - P^2) - 76(2P + P^2) + 196(2 + P - 2P - P^2)$ (5)

Multiplication completed:

$663P - 663P^2 - 152P - 76P^2 + 392 + 196P - 392P - 196P^2$ (6)

Terms summed up: $392 + 315P - 935P^2 = 0.0001585$ (close to zero) (7)

                        ↑   ↑   ↑

Designate terms   c   b   a

The right side of eq. (7) can be determined only after solving the quadratic equation below:

$P = -b \pm \frac{\sqrt{b^2 - 4ac}}{2a} = \frac{315 \pm \sqrt{99225 + 1466080}}{1870} = \frac{-315 \pm \sqrt{1565305}}{1870} = \frac{-315 \pm 1251.1215}{1870} = \frac{-1566.1215}{1870} = -0.837498$; after changing sign, $P = 0.837498$

Thus $P = 0.837498$, and $\sqrt{P} = 0.9151492 = 1 - p$ and hence the recombination fraction $p = 1 - 0.9151492 = 0.0848508$

The variance of P, $V_P = \frac{2P[1-P][2+P]}{n[1+2P]} = 0.0004495$, where $n(=\Sigma) = 935$ and

the variance of p, $V_p = \frac{V_P}{4P} = 0.0001342$ and

the standard error $s_p = \sqrt{V_p} = \sqrt{0.0001342} = 0.01158$

Thus the frequency of recombination between the two genes is 0.085 ± 0.012
Data may be entered at step (6) to expedite routine calculations. (See also maximum likelihood principle, recombination frequency, F_2 linkage estimation, information)

MAXIMUM LIKELIHOOD PRINCIPLE: provides a statistical method for estimating the optimal parameters from experimental data. The best statistics for the computations is that provides

Maximum likelihood principle continued

the smallest variance. E.g. the variance of the median of a sample is $\frac{\pi\sigma^2}{2n}$ which is $\frac{\pi}{2} = 1.57$ times the size of the variance of the mean $(\bar{x})$. Therefore the mean is a much better characteristic of the population than the median. The binomial probability is expressed as:

$\binom{n}{r} p^r(1-p)^{n-r}$ giving the probabilty (p) that (r) events occur in a sample of (n). The relative probability of r/n events for different values of (p) is called the *likelihood*. The procedure that facilitates finding a population parameter (θ) that maximizes the likelihood of a particular observation is a *maximum likelihood* procedure. If the dispersion of a population follows the normal distribution, the variance $V = \sigma^2$, is a maximum likelihood estimator of the distribution of that population. All other methods need to be compared with and tested against this method before their results can be accepted and used.

Naturally, all statistics provide only predictions and not direct proof regarding the biological mechanism concerned. Therefore careful collection of data, replications, sufficient sample sizes, etc. are indispensable for accuracy and predictability. The maximum likelihood mandates that the choice of the parameter (θ) makes the likelihood, $L(X_1, X_2...X_n|\theta)$ the largest value. Example: a random sample of 20 is obtained and among them, say 12 belongs to a particular class. We can hypothesize that the true frequency of this class is either (I): $p = 0.6$ or (II): 0.5 or (III): 0.7. According to the normal distribution then:

(I) $\binom{20}{12} (0.6)^{12} (0.4)^8 = \frac{20!}{12!8!}(0.6)^{12}(0.4)^8 = 125{,}970 \times 0.002176782 \times 0.00065536 \cong 0.17971$

(II) $\binom{20}{12} (0.5)^{12} (0.5)^8 = \frac{20!}{12!8!}(0.5)^{12}(0.5)^8 = 125{,}970 \times 0.000244140 \times 0.00390625 \cong 0.12013$

(III) $\binom{20}{12} (0.4)^{12} (0.6)^8 = \frac{20!}{12!8!}(0.4)^{12}(0.6)^8 = 125{,}970 \times 0.013841287 \times 0.00006561 \cong 0.11440$

Obviously, hypothesis (I) has the maximum likelihood to be applicable to this case. After this simple demonstration we can generalize the likelihood function as:

$L(X_1, X_2...X_N|p) = \binom{N}{r} p^r(1-p)^{N-r}$ where X are the samples, N = population size, p = probability, and r = 0, 1,...N. The maximized likelihood is conveniently expressed by the logarithm of the likelihood function: $\log L = \log\binom{N}{r} + (r)\log(p) + (N-r)\log(1-p)$

After differentiation to (p) and equating it to zero:

$\frac{d}{dp}\log L = \frac{r}{p} - \frac{N-r}{1-p} = 0.$ After bringing it to the common denominator:

$\frac{r(1-p) - (N-r)p}{p(1-p)} = 0$ The denominator omitted:

$r(1-p) - (N-r)p = 0 = r - rp - Np + rp$, and hence

$p = r/N$ and this is the *maximum likelihood estimator* of p. Similarly it can be shown that for a population in normal distribution the arithmetic mean of the sample $\left(\frac{\sum x_i}{N}\right)$ is the maximum likelihood estimator of the μ. The probability P for a multinomial distribution is: $\frac{N!}{X!Y!Z!...} p^X q^Y r^Z$ where p, q, r... are the probabilities of X,Y,Z...classes. Although we may not know these probabilities but we may have experimentally observed the classes (genotypes, alleles, etc.), and we can derive the likelihood function which permits the estimation of the parameters of p, q, etc. If in a random mating population the proportion of A is p^2, that of B is $2pq$ and that of C is q^2, we can write the likelihood function as:

$L = \frac{N!}{A!B!C!}(p^2)^A(2pq)^B(q^2)^C$ from which after logarithmic conversion and differentiation we

Maximum likelihood principle continued

can obtain the value of $p = \frac{2A + B}{2N}$ and $q = \frac{2C + B}{2N}$ and the variance $V_p = \frac{pq}{2N}$. For an in-depth treatment of maximum likelihood, mathematical statistics monographs should be consulted. The maximum likelihood method is widely used in decision-making theory. In genetics, it is most commonly used for the estimation of recombination and allelic frequencies. (See maximum likelihood method applied to recombination frequencies, probability, information)

MAXIMUM PARSIMONY: same as maximal parsimony

MAY-HEGGLIN ANOMALY: is an asymptomatic granulocyte and platelet disorder resulting often in thromobocytopenia. (See hemostasis, platelet anomalies)

MAYTANSINOIDS: extract of tropical trees or shrubs with an LDLo of 190 µg/kg as intravenous dose for humans. (See magic bullet, LDLo)

Mbp: megabase pair, 1 million base pair.

µC (microcurie): 3.7×10^4 dps. (See Curie, isotopes)

McARDLE's DISEASE: see glycogen storage disease Type V

MCF ONCOGENE (synonymous with DBL, ROS): the human mammary carcinoma protooncogene was assigned to human chromosome Xq27. It encodes a serine-phosphoprotein (p66). (See oncogenes, ROS)

Mch: ICE-related proteases. (See ICE, apoptosis)

MCK: is a muscle-specific kinase. (See MyoD)

McLEOD SYNDROME: is a recessive human Xp21 region deficiency of the Kx blood antigen. The symptoms vary because of overlapping defect with closely linked genes, particularly CGD (chronic granulomatous disease). It may be associated with acanthocytosis, characteristic for abetaloliproteinemia. (See abetalipoproteinemia, granulomatous disease chronic, Kell-Cellano blood group)

MCM1: a yeast DNA binding protein, product of the **mi**nichromosome **m**aintenance gene involved also in the regulation of mating type and it controls the entry into mitosis. (See mating type determination in yeast, MCM3, ARS, Cdc45/Cdc46/Mcm5)

MCM3: is apparently the same as the replicational licensing factor (RLF), that appears in tight binding to DNA during interphase but released during S phase. This factor assures that within a cell just one cycle of DNA replication occurs. This protein belongs to the family of MCM1 to MCM5 factors detected in yeast. (See MCM1, replication licensing factor)

mcr: see methylation of DNA

MCS: multiple cloning sites, see polylinker

M-CSF: see macrophage colony stimulating factor

mdg: see copia

MDM2: is a cellular oncoprotein that can bind and downregulate p53 tumor suppressor, and attach to the retinoblastoma suppressor, and can stimulate transcription factors E2F1 and DP1. (See oncogenes, transcription factors, retinoblastoma)

MDR: see multidrug resistance

MDS (macronuclear destined sequences): from the germline DNA during vegetative development of ciliates internal sequences are eliminated by the process of chromosome diminution and only the MDS is retained in the macronucleus. (See chromosome diminution, *Paramecium*, macronucleus)

MEAL WORM (*Tenebrio molitor*): sex is determined by a larger X and a smaller Y chromosome in this insect.

MEALY BUG: is a member of the coccidea taxonomic group of animals with the name reflecting the "mealy" appearance of the wax coat of the body of the insects. They received attention by the peculiarity of their chromosome behavior. During the cleavage divisions immediately after fertilization, all the chromosomes are euchromatic. After blastula one half of the chromosomes (2n = 10) becomes heterochromatic in the embryos which develop into males. At interphase these heterochromatic chromosomes clump into a chromocenter. By metaphase the hetero

Mealy bug continued
chromatic and euchromatic sets are no longer distinguishable. In the males the first meiotic division is equational and during the second division the two types of chromosomes go to opposite poles. Two of the four nuclei are heterochromatic and two euchromatic. The heterochromatic nuclei then disintegrate and the euchromatic cells proceed to spermiogenesis. The euchromatic set of the fathers becomes later the heterochromatic chromosomes of the sons. This was verified by X-raying the females and males. Only 3% of daughters of males irradiated by 16,000 R survived but the sons were unaffected even after 30,000 R. Some sons survived even after 90,000 R exposure of the fathers. Thus, sex appears to be determined developmentally in these insects. (See chromosomal sex determination)

MEAN: the *arithmetic mean* $\bar{x}$ is equal to the sum (Σ) of all measurements (x) divided by the number (n) of all measurements, or $\bar{x} = \frac{\Sigma x}{n}$. The *geometric mean* (G) is the nth root of the product of all measurements: $G = \sqrt[n]{x_1 \cdot x_2 \ldots x_n}$. The *harmonic mean* (H) is the inverse average of the reciprocals of the measurements $H = \frac{n}{\Sigma[1/x]}$.

Examples: $\bar{x} = \frac{2+8}{2} = 5$, $G = \sqrt[3]{2 \times 8 \times 4} = \sqrt[3]{64} = 4$, $H = \frac{2}{[1/2] + (1/8)} = 3.2$. The *weighted mean* is the calculated mean multiplied by the pertinent frequency of the groups in a population. (See variance)

MEAN LETHAL DOSE: of a mutagen or toxic agent is represented by LD_{50}. (See LD_{50}, LDLo, LC50)

MEAN SQUARES: the average of the squared deviations from the mean; it is obtained by dividing the sum of the squared deviations by the pertinent degrees of freedom. Basically, this is the estimated variance. (See variance, variance analysis, intraclass correlation)

MEANDER: two consecutive β-sheets of a protein are adjacent and antiparallel. (See protein structure)

MEASUREMENT UNITS: <u>Length</u>: 10 ångström (Å) = 1 nanometer (nm), 1000 nm = 1 micrometer (μm), 1000 μm = 1 millimeter (mm), 10 mm = 1 centimeter (cm), 100 cm = 1 m <u>Volume</u>: 1000 microliter (μL or λ = 1 milliliter (mL), 1000 mL = 1 liter (L); <u>Weight</u>: 1000 picogram (pg) = 1 nanogram (ng), 1000 ng = 1 microgram (μg), 1000 μg = 1 milligram (mg), 1000 mg = 1 gram (g), 10 g = 1 dekagram (dg), 100 dg = 1 kilogram (kg). <u>Generally</u>: milli = 10^{-3}, micro = 10^{-6}, nano = 10^{-9}, pico (p) = 10^{-12}, fempto (f) = 10^{-15}, atto (a) = 10^{-18} and kilo (k) = 10^3, mega (M) = 10^6, giga (G) = 10^9 and tera (T) = 10^{12}. (See M_r, Dalton)

MEC: see degenerin, ion channels

MEC1: is a kinase locus of yeast; its phosphorylates RAD53, a signal transducer of DNA damage. The homologs are *SAD3, ESR1,* and the human gene responsible for ataxia telangiectasia. (See ataxia, RAD, signal transduction)

MECHANISM-BASED INHIBITION: see regulation of enzyme activity

MECHANOSENSORY GENES: are involved in the neurobiological control of touch.

MECKEL SYNDROME: is a rare complex recessive syndrome (17q21-q24) with most specific characteristics are the brain defects, cystic kidneys and polydactyly. In Finnish populations it may occur however in the near 10^{-4} range. (See neural tube defects, polydactily)

MED-1: null promoter.

MEDIAN: a statistical concept that indicates that equal numbers of (variates) observations are on its sides at both minus and plus directions. (See mean, mode)

MEDICAL GENETICS: genetics applied to medical problems. (See clinical genetics, human genetics)

MEDICINAL CHEMISTRY: is involved with drug design and development.

MEDITERRANEAN FEVER: a human chromosome 17 recessive disease with recurrent spells

Mediterranean fever continued

of fever, pain in the abdomen, chest and joints and red skin spots (erythema). It is a type of amyloidosis. In some populations the prevalence, gene frequency and carrier frequency may be 0.00034, 0.019, and 0.038, respectively. (See amyloidosis)

MEDLINE: a medical bibliographic system of the National Library of Medicine USA. It can be reached on-line as part of the MEDLARS database. (See databases)

medRNA: mini-exon-dependent RNA. (See *Trypanosoma brucei*)

MEDULLA: the inner part of organs, the basal part of the brain connecting with the spinal chord. (See brain human)

MEEKRIN-EHLERS-DANLOS SYNDROME: is a connective tissue disorder.

MEF: a series of myocyte enhancer binding factors that specifically potentiate the transcription of muscle genes and thus differentiation of various types of muscles and myoblasts. The MEF group belongs to the family of MADS domain protein. (See MADS box, MyoD, MYF5, MRF4)

MEGADALTON (Mda): is 10^6 dalton; 1 da (or Da) = 1.661×10^{-24} g.

MEGAGAMETOPHYTE: is one of the four, the functional, haploid products of meiosis (megaspores) in the female sexual process of plants, and it develops into embryo sac. Its origin and most prevalent developmental paths is outlined in a figure at gametophyte. (See gametophyte [female])

MEGALOBLAST: large, nucleated, immature cells giving rise to abnormal red blood cells.

MEGALOBLASTIC ANEMIA: the autosomal dominant (human chromosome 5q11-q22) deficiency of dehydrofolate reductase (involved in the biosynthetic path of purines and pyrimidines) resulting in hematological and neurological anomalies. The symptoms may be alleviated by 5-formyltetrahydrofolic acid. An autosomal recessive type can be completely remedied as long as thiamin is provided. (See phosphatase [ACP1], megaloblast, folic acid, thiamin, anemia, transcobalamine deficiency)

MEGAKARYOCYTES: large cells in the bone marrow with large lobed nuclei; their cytoplasm produces the platelets. Megakaryocyte formation from stem cells is regulated by the cytokine receptor cMpl and its ligand the megakaryocyte lineage-specific growth factor (meg-CSF) which is homologous to erythropoietin and has both meg-CSF and thrombopoietin-like activities. (See erythropoietin, thrombopoietin, platelet)

MEGASPORE: see megagametophyte, gametogenesis

MEGASPORE COMPETITION: determines which of the four products of meiosis (megaspores) in the female (plant) becomes functional. It occurs only in a few species such as *Oenotheras*. (See also certation, pollen, gametogenesis)

Megaspore Tetrad — TOP Megaspore Develops — BASAL Megaspore Develops

MEGASPORE COMPETITION IN THE *OENOTHERAS* WHERE NORMALLY THE TOP SPORE OF THE TETRAD IS FUNCTIONAL BUT IN CASE THERE IS A DELETERIOUS GENE IN THE TOP SPORE, THE BASAL SPORE MAY COMPETE WITH IT SUCCESSFULLY. (Ater Renner O. from Goldschmidt, R. 1928 Einführung in die Verebungswissenschaft. Springer-Vlg. Berlin, Germany)

MEGASPORE MOTHER CELL: a diploid cell that produces the haploid megaspores through meiosis in the female plant. (See gametophyte [female])

MEGASPOROCYTE: the same as megaspore mother cell. (See gametogenesis, gamatophyte)

MEI41: is a 270 kDa *Drosophila* phosphatidylinositol kinase. When inactivated meiotic recombination is reduced. (See PIK)

MEIOCYTE: cell that undergoes meiosis. (See meiosis)

MEIOSIS: two step-nuclear divisions which reduce somatic chromosome number (2n) to half (n).

Meiosis continued

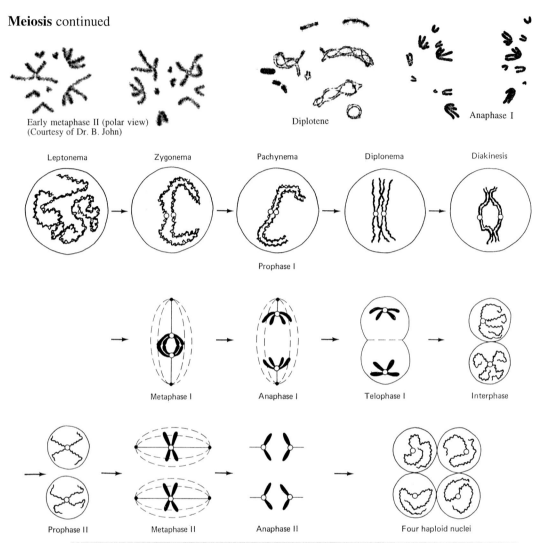

A GENERALIZED COURSE OF MEIOSIS REPRESENTED BY ONLY ONE PAIR OF CHROMOSOMES.

and is usually followed by gamete formation. Meiosis is genetically the most important step in the life cycle of eukaryotes. Meiosis procedes from the 4C sporocytes (in diploids) and includes one numerically reductional and one numerically equational chromosome divisions. Synapsis takes place at prophase I to metaphase I. Chiasmata may be visible by the light microscope during prophase I. Centromeres do not split at anaphase I and the sister chromatids are held together at the centromeres during the separation of the bivalents at anaphase I. At the completion of meiosis I the chromosome number is reduced to half (yet 2C) and by the end of meiosis II, the C-value of each of the 4 haploid daughter cells is 1. The major stages of meiosis are shown in the diagram above. These stages are rather transitional than absolutely distinct. The nucleolus is not shown. The nuclear membrane is generally not discernible by the light microscope from metaphase to anaphase but reappears at telophase. The spindle fibers are represented by dashed and solid thin lines. The genetic consequences of the meiotic behavior of the chromosomes is best detected in ascomycete fungi with linear tetrads. The duration of meiosis generally much exceeds that of mitosis and the longest is the prophase

Meiosis continued

stage. In the majority of plants the completion of mitosis requires 1 - 3 hours whereas meiosis may need from 1 - 8 days. In yeast meiosis takes place in about 7 hours. The stages most revealing for the cytogeneticist, pachytene (2 - 8), diplotene (0.5 - 1), metaphase I (1.5 - 2) and anaphase (0.5 - 1) require generally the number of hours indicated in parenthesis. In human females meiosis may stall at the late prophase stage, at dictyotene and the subsequent divisions take place only before the onset of ovulations (and following fertilization), a period repeated approximately 13 times annually during about 40 years. The activation of meiosis requires C_{29} sterols in both male and female. (See also leptotene, zygotene, pachytene, diplotene, diakinesis, metaphase, anaphase, interphase, mitosis, dictyotene, nucleolus, cell cycle, gametophyte, C amount of DNA)

MEIOSIS I: is the first stage of meiosis when through reduction of the chromosome number the 4C amount of DNA in the meiocyte takes place and each of the two daughter nuclei has 2C amounts of DNA. (See meiosis, C amount of DNA)

MEIOSIS II: follows meiosis I and it is basically an equational division of the chromosomes resulting in four daughter nuclei of the meiocyte that each have only 1C amount of DNA. (See meiosis, C amount of DNA)

MEIOTIC DRIVE: results in unequal proportions of two alleles of a heterozygote among the gametes in a population because certain meiotic products are not or less functional and consequently the proportion of other gametes increases. Meiotic drive may become a microevolutionary factor because it may alter gene frequencies. Meiotic drive may favor selectively disadvantageous gene combinations and thus may contribute to the genetic load of a population. Meiotic drive operates in the males or in the females but not in both. (See segregation distorter, preferential segregation, polarized segregation, certation, megaspore competition genetic load, transmission, polycystic ovarian disease, symbionts hereditary, killer strains, brachyury)

MEK: a member of the extracellular signal-regulated kinase (ERK) family. (See signal transduction)

MEKK: MEK kinase (i.e. a kinase kinase). (See MEK, signal transduction)

MEL ONCOGENE: was isolated from human melanoma although its role in melanoma is unclear; it was assigned to the broad area of human chromosome 19p13.2-q13.2). (See melanoma, oncogenes, $p16^{INK4}$)

MELANDRIUM (synonymous with Lychni*s, Silene*): dioecious plant (2n = 22+XX or 22+ XY, Caryophyllaceae. (See intersex)

MELANIN: see albinism, pigmentation of animals, piebaldism, agouti

MELANIE II: a computer software package that can match two-dimensional protein gel data to information in database. (See databases)

MELANISM: increased production of the dark melanin pigment. (See melanin, industrial melanism)

MELANOCYTE: produces melanin. (See melanin)

MELANOCYTE STIMULATING HORMONE (MSH): exists in forms α-MSH, β-MSH and γ-MSH. These adenocortical hormones regulate melanization. (See opiocortin, agouti, pigmentation of animals)

MELANOMA: a form of cancer arising in the melanocytes or other tissues. The most prevalent form of it appears as a mole of radial growth of reddish, brown and pink color with irregular edges that penetrate, as they progress, into deeper layers. It may originate in dark freckles on the head or other parts of the body. Excessive exposure to sun light may condition its development although susceptibility to melanoma is determined by autosomal dominant genes. Melanoma may be one of the most aggressively metastatic cancer. The development of melanoma is regulated by the melanoma mitogenic polypeptide, encoded by GRO human gene in chromosome 4q21. The melanoma-associated antigen, ME49 appears in the early stages of this cancer, and it is coded by an autosomal dominant locus (MLA1) in human chromosome 12q12-

Melanoma continued

q14. Melanoma-associated antigen, MZ2-E, is coded for by another autosomal dominant locus. The melanoma-associated antigen p97 is a member of the iron-binding transferrin protein family is coded by an autosomal dominant gene, MAP97 (MF12) at human chromosome 3q28-q29. The melanoma-specific chondroitin sulfate proteoglycan, expressed in melanoma cells, is encoded by an autosomal dominant gene in human chromosome 15. In cultured melanoma and metastatic tissues a mutant CDK4 protein was found that was unable to bind the p16^{INK4a} protein and thereby interferes with normal regulation of the cell cycle inhibitor. The p16 seems to be identical with MTS1 (multiple tumor suppressor) in human chromosome 9p21. Approximately 5 to 10% of the melanoma patients have at least one afflicted family member. Genetic hybrids between the species of the platyfishes (*Xiphophorus*) are prone to develop melanoma. (See MEL oncogene, melanocyte, cancer)

MELANOSOME: see albinism

MELAS SYNDROME: see mitochondrial disease in humans

MELATONIN: a hormone, synthesized in the pineal gland and controling reactions to light, diurnal changes, seasonal adjustment in fur color, aging, sleep, reproduction, etc. Melatonin is synthesized from serotonin. (See serotonin)

MELT-AND-SLIDE MODEL: DNA polymerase pol I, its dual functions are carried out by pol I occupying the duplex primer-template site for the polymerase action, and for the editing reaction the DNA melts (strands separate), unwind and the single-strand DNA is transferred to the exonuclease site of the polymerase enzyme. (See DNA repair, DNA replication)

MELTING: breakdown of the hydrogen bonds between two paired nucleic acid strands (denaturation, breathing of DNA, Watson and Crick model)

MELTING CURVE OF DNA: higher temperatures cause progressively higher disruption of the hydrogen bonds between DNA strands; it is affected also by the base composition of the DNA because there are three hydrogen bonds between G≡C and two between A=T. (See also renaturation, c_0t curve, hydrogen pairing)

MELTING TEMPERATURE: the temperature where 50% of the molecules is denatured (T_m); the DNA strands may be separated (depending on the origin of the DNA, G≡C content, solvent, and homology of the strands) generally above 80° C and melting may be completed below 100°. (See C_0t curve, hyperchromicity, melting curve of DNA)

MELTRINS: are metalloproteinase proteins mediating the fusion of myoblasts into myotubes. (See myotubes)

MEMBRANE: lipid protein complexes surrounding cells and cellular organelles and forming intracellular vesicles. (See cell membranes)

MEMBRANE CHANNELS: permit passive passing of ions and small molecules through membranes. (See cell membrane, ion channel)

MEMBRANE FILTERS: may be used for clarification biological or other liquids, trapping macromolecules, for exclusion of contaminating microbes, for Southern and Northern blotting, etc. The filters may be cellulose, fiberglass, nylon and might have been specially treated to best fit for the purpose.

MEMBRANE POTENTIAL: electromotive force difference across cell membranes. In an average animal cell inside it is 60 mV relative to the outside milieu. It is caused by the positive and negative ion differences between the two compartments.

MEMBRANE PROTEINS: may be *integral* parts of the membrane structure and cannot be released. The *transmembrane* proteins are single amino acid chains folded into (seven) helices spanning across the membrane containing lipid layers. The latter have a hydrophobic tract that passes through the lipid double layer of the membrane and their two tails, one pointing outward the membrane and the other reaching into the cytosol are hydrophilic. Some of the membrane proteins are attached only the outer or to the inner layer of the membrane lipids and they are called *peripheral membrane proteins*. The membrane proteins regulate not just the cell membranes but cell morphology by anchoring to the cytoskeleton, pH, ion channels and gener-

Membrane proteins continued

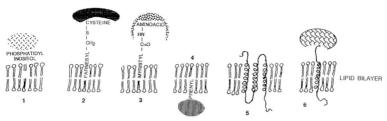

1. GLOBULAR PROTEIN ATTACHED TO THE OUTER SURFACE BY A GLYCOSYL-PHOSPHATIDYL INSOSITOL ANCHOR. 2. THE PROTEIN IS ATTCHED TO THE OUTER SURFACE BY A THIOETHER LINKAGE BETWEEN THE SULFUR OF CYSTEINE AND A FARNESYL MOLECULE. 3. AMINO ACID AND MYRISTIL ANCHOR JOINS A PROTEIN TO THE OUTER SURFACE OF THE MEMBRANE. 4. PRENYL RESIDUE ANCHORS THE PROTEIN TO THE INNER PART OF THE MEMBRANE. 5. TRANSMEMBRANE PROTEIN CHAIN PASSES THROUGH THE MEMBRANE THREE OR SEVEN TIMES. 6. TRANSMEMBRANE PROTEIN ANCHORS ANOTHER PROTEIN WITHOUT COVALENT LINKAGE.

al physiology of the cell. The structure of membranes can now be analyzed with the aid of membrane mutants, available in several organisms. (See diagram, prenylation, farnesyl pryphosphate, myristic acid, cell membranes, cytoskeleton)

MEMBRANE-SPANNING HELICES: see membrane proteins, seven membrane proteins

MEMBRANE TRANSPORT: movement of polar solutes with the aid of a transporter protein through cell membranes. (See cell membrane)

MEMORY: information storage in the brain or in a computer. The mammalian brain deals with synaptic strength as memory. If synapses are used repeatedly the strength is improved and long-term potentiation (LTP) takes place. The opposite of LTP is LTD (long-term depression). The latter may erase the effect of LTP. Both of these mechanism are triggered by the inflow of Ca^{2+}, regulated by an ion channel (NMDAR), glutamate-activated N-methyl-D-aspartate receptor channel. It is not entirely clear how discrimination between LTP and LTD is accomplished. Nerve growth factor (NGF) gene transfer to basal forebrain of rats resulted in recovery from age-related memory loss. *Explicit memory*, the remembrance of facts, resides in the hippocampal region of the brain. *Implicit memory* involves perceptual and motor skills that may be more widely distributed; memory of conditional fear appears to be in the amygdala. (See brain human, ion channels, long-term potentiation)

MEMORY IMMUNOLOGICAL: immunological memory rests with the lymphocytes. There are three phases in its development: (i) activation and expansion of the CD4 and CD8 T cells, (ii) apoptosis, and (iii) stability (memory). During phase (i), lasting for about a week the antigen selects the appropriate cells and an up to 5,000-fold expansion of the specific cells takes place. They also differentiate into effector cells. Between a week and a month time, as the antigen level subsides, the T cells die and effector function declines. This is called activation-induced cell death (AICD). The memory stage may then last for many years and they respond to low exposure of the reintroduced antigen and respond very rapidly. A certain level of antigen maintenance seems to be required to keep memory cell at high level but it appears that memory cells can be saved even in the absence of the antigen. The conditions for CD4 and CD8 cell maintenance appear different. Memory B and T cells do not provide immediate protection against infection but after infection they start rapid formation of effectors. The duration of the memory depends on the strength of immunization. The protection against peripheral reinfection is antigen-dependent. (See immune system, apoptosis, CD4, CD8, lymphocytes, germinal center, T cell, B cell)

MEN: MEN1 was located to human chromosome 11q13, responsible for the production of a 610 amino acid protein, encoded by 10 exons. (See endocrine neoplasia)

MENAQUINONE (vitamin K_2): is synthesized by bacteria and it is an electron carrier. (See also vitamin K)

MENARCHE: the first menstruational event, followed by a period of about 3 years of no ovulations before the regular menstruation begins and continues until menopause. (See menstruation, menopause)

MENDELIAN LAWS: the term was first used by Carl Correns (1900), one of the rediscoverers of these principles. *First law*: uniformity of the F_1 (if the parents are homozygous) and the reciprocal hybrids are identical (in the absence of cytoplasmic differences). *Second Law*: independent segregation of the genes to F_2. (in the absence of linkage). *Third law*: independent assortment of alleles in the gametes of diploids. Thomas Hunt Morgan (1919) also recognized three laws of heredity: (1) free assortment of the alleles in the formation of gametes, (2). independent segregation of the determinants for different characters, (3). linkage - recombination. Mendel himself never claimed any rules as such to his credit. He did not observe any linkage among the 7 factors he studied in peas although he had less than 1% chance for all factors segregating independently. This was called Mendel' luck. If he would have found linkage that would have been probably recorded in his notes. Unfortunately after his death, his successor at the abbey, Anselm Rambousek, disposed most of the records. After he experimented with *Hieracium*, an apomict (unknown that time), he developed doubt about the general validity of his discoveries. Yet, before ending his career he stated: "My scientific work has brought me a great deal of satisfaction, and I am convinced that I will be appreciated before long by the whole world". That appreciation began in 1900, 16 years after his death and continues since. (See Mendelian segregation, epistasis, modified Mendelian ratios)

MENDELIAN POPULATION: is a collection of individuals which can share alleles through interbreeding. (See population genetics)

MENDELIAN SEGREGATION: for independent loci can be predicted on the basis of the table below:

NUMBER OF DIFFERENT ALLELIC PAIRS	1	2	3	4	n
KINDS OF GAMETES AND NUMBER OF PHENOTYPES IN CASE OF DOMINANCE	2	4	8	16	2^n
NUMBER OF PHENOTYPES (IN CASE OF NO DOMINANCE), AND NUMBER OF GENOTYPES	3	9	27	81	3^n
NUMBER OF GAMETIC COMBINATIONS	4	16	64	256	4^n

Mendelian segregation ratios may show only apparent deviations in case of epistasis. Reduced penetrance or expressivity may also confuse the segregation patterns and in such cases it may be necessary to determine the difference between male and female transmission. (See epistasis, penetrance, expressivity, segregation distorter, certation, modified Mendelian ratios)

MENDELIZING: segregation corresponds to the expectations by Mendelian laws. (See Mendelian laws, Mendelian segregation)

MENINGIOMA: usually slow proliferating brain neoplasias classified into different groups on the basis of anatomical features. Generally meningiomas involve the loss of human chromosome 22 (hemizygosity) or part of its long arm or some lesions at 22q12.3-q13 where the SIS oncogene, responsible for a deficit of the platelet derived growth factor (PDGF) is located. Chromosomes 1p, 14q, and 17 have also been implicated. (See SIS, cancer)

MENINGOCELE: see spina bifida

MENKE's SYNDROME (MNK, kinky hair disease): the gene is situated in the centromeric area of the human X-chromosome. The phenotype involves hair abnormalities, mental retardation and short life span. Apparently, the defect is in the malabsorption of copper through the intestines. The prevalence is in the 10^{-5} range. It is detectable prenatally and the heterozygotes can be identified although its inheritance is apparently recessive. Lysyl oxidase level is reduced in the afflicted individuals. In the mouse homolog, Mottled-Bridled, the non-exported copper is tied up by metallothionenin and the afflicted individual dies within a few weeks after birth. (See mental retardation, Wilson disease, acrodermatitis, hemochromatosis, Ehlers-Danlos syndrome, collagen, metallothionein)

MENOPAUSE: the end of the periodic ovulation (menstruation) around age 50 in human females. (See menstruation, menarche)

MENSES: same as menstruation

MENSTRUATION: the monthly discharge of blood from the human (primate) uterus in the absence of pregnancy. If the egg is not fertilized, it dies and the endometrial tissue of the uterus is removed amids the bleeding. Fertilization takes place within the oviduct. About 3 days are required for the egg to reach the uterus through the oviduct where it is implanted within a day or two, and about a week after being fertilized. Fertilization may occur if the coitus takes place in period about two weeks after the beginning of the last monthly menstruation. The calendar rhythm method of birth control relies on knowledge of this receptive period. Unfortunately, its effectiveness is not very high. (See hormone receptors, sex hormones, ovulation, menarche)

MENTAL RETARDATION: is a collection of human disabilities caused by direct or indirect genetic defects and acquired factors such as diverse types of infections (syphilis, toxoplasma coccidian protozoa) viruses [rubella, human immunodeficiency virus, cytomegalovirus, herpes simplex, coxsackie viruses], bacteria [*Haemophilus influenzae*, meningococci, pneumococci], etc., mechanical injuries to the brain pre-, peri- and postnatally, exposure to lead, mercury, addictive drugs, alcoholism or deprivation of oxygen during birth, severe malnutrition, deficiency of thyroid activity, social and psychological stress, etc. An estimated 2 to 3% of the population is suffering from mild (IQ 50 - 70%) or more or less severe (IQ below 50%) forms of it. An estimated 90% of the cases can be helped by special education programs. Approximately 10% of the human hereditary disorders have some mental-psychological debilitating effects.
Autosomal dominant type hemoglobin H disease associated mental retardation due to a lesion in the α-globin gene cluster with chromosomal deletion and without it have been observed. Other cases of mental retardation were also observed involving autosomal dominant inheritance caused by breakage in several chromosomes. Autosomal recessive inheritance was involved in mental retardation associated with head, face, eye, and lip abnormalities, hypogonadism, diabetes, epilepsy, heart and kidney malformations, phenylketonuria. X-chromosome linked mental retardation was observed as part of the syndromes involving the development of large heads, intestinal defects, including anal obstructions, seizures, short statures, weakness of muscles, obesity, marfanoid appearance, etc. In some cases the "kinky hair" syndrome (Menkes syndrome), caused apparently by abnormal metabolism of copper and zinc, also involved mental retardation. A fragile site in the X chromosome (Xq27-q28) apparently based on a deficiency of thymidine monophosphate caused by insufficient folate supply is associated with testicular enlargement (macroorchidism), big head, large ears, etc. The transmission of the fragile X sites (FRAX) is generally through normal males. The carrier daughters are not mentally retarded and generally do not show fragile sites. In the following generation, about a third of the heterozygous females display fragile sites and become mentally retarded. This unusual genetic pattern was called the Sherman paradox and it is interpreted by some type of a pre-mutational lesion. The pre-mutation ends up in a genuine mutation only after transmitted by a female which had already a microscopically undetectable rearrangment. The risk of the sons was estimated to be 50% from mentally retarded heterozygous females, 38% from normal heterozygous mothers, and 0% from normal transmitting fathers. The probabilty of these sons being a mentally sound carrier was estimated as 12, 0 and 0%, respectively. The risk of the daughters of the same mothers to become a mentally affected carrier was calculated to be 28, 16 and 0%, and being a mentally normal carrier was estimated 22, 34 and 1%, respectively. The chance of mental retardation for the brother of a proband whose mother has no detectable fragile X site, may vary from 9 - 27% and among first cousins this is reduced to 1-5%. It was proposed (Laird 1987 Genetics 117:587) that the expression of the fragile X syndrome is mediated by chromosomal imprinting. The imprinting can, however, be erased by transmission through the parent of the other sex. The fragile X syndrome is apparently caused by localized breakage and methylation of CpG islands at the site. Currently the most reliable diagnosis ofthis condition is based on DNA probing. Besides the fragile X syndrome autosomal and sex-

Mental retardation continued

chromosomal trisomy and chromosome breakage associated with translocations may be contributing factors of mental retardation. Also, mutations causing metabolic disorders (phenylketonuria, homocystinuria, defects in the branched-chain amino acid pathway [maple syrup urine disease], anomalies in amino acid up-take (Hartnup disease), defects involving mucopolysaccharids (Hunter, Hurler and Sanfilippo syndromes), gangliosidoses and sphingolipidoses (most notably the Tay-Sachs disease, Farber's disease, Gaucher's disease, Niemann-Pick disease, etc.), galactosemias, failure of removal of fucose residues from carbohydrates (fucosidosis), defects in acetyl-glucosamine phosphotransferase (I-cell disease), defects in HGPRT (Lesch-Nyhan syndrome), hypothyroidism, a variety of defects of the central nervous system, and other genetically determined conditions may be responsible for mental retardation. The incidence, establishment of genetic risks and possible therapies are as variable as the underlying causes.

Mental retardation is defined as borderline: IQ $\approx$70-85, mild in case of IQ $\approx$50-70, moderate: IQ $\approx$35-50, severe: IQ $\approx$25-35 and profound: IQ $\leq$20.

(For more specific details see separate entries of the conditions here named, Huntington's disease, biotidinase deficiency, myotonic dystrophy, muscular dystrophy, hydrocephalus, cranofacial dysostosis, spina bifida, tuberous sclerosis, neurosfibromatosis, Menke's syndrome, Smith-Lemli-Opitz syndrome, Smith-Magenis syndrome, Seckel's dwarfism, Laurence-Moon syndrome, Noonan syndrome, Lowe's syndrome, mental retardation X linked, Apert syndrome or Apert-Crouzon disease, Prader-Willi syndrome, Rubinstein syndrome, cerebral gigantism, Langer-Giedion syndrome, Miller-Dieker syndrome, Walker-Wagner syndrome, Wilms tumor, Roberts syndrome, West syndrome, Russel-Silver syndrome, Opitz-Kaveggia syndrome, De Lange syndrome, Bardet-Biedl syndrome, focal dermal hypoplasia, ceroid lipofuscinosis, autism, dyslexia, human intelligence[IQ], psychoses, aspartoacylase deficiency, glutamate formiminotransferase deficiency, CADASIL, Cohen syndrome, Coffin-Lowry syndrome, fragile sites, FMR1 mutation, trinucleotide repeats, human intelligence, human behavior, head/face/ brain defects, craniosynostosis, double cortex, periventricular heterotopia, heritability, QTL)

MENTAL RETARDATION X-LINKED: these hereditary defects occur in a variety of forms: (i) MRXS with diplegia (paralysis), (ii) associated with psoriasis (skin lesions), (iii) with lip deformities, obesity and hypogonadism, (iv) Renpenning type with short stature and microcephaly, (v) with seizures, (vi) with Marfan syndrome-like habitus, (vii) with fragile X-chromosome sites, etc. (See Marfan syndrome, fragile X-chromosome, Lowe's syndrome, mental retardation)

MENTHAS: are a group of dicotyledonous species of plants of various (frequently aneuploid) chromosome numbers. *M. arvensis:* 2n = 12, 54, 60, 64, 72, 92; *M. sylvestris* : 2n = 24, 48; *M. piperita*: 2n = 34, 64. They were the source of menthol (peppermint camphor) and other oils used in cough drops, nasal medication, anti-itching ointments, candy, liquors, etc. Menthol appeared non-carcinogenic although doses (above 1 g/kg) may cause 50% death in laboratory rodents when administered subcutaneously or orally.

MENTOR POLLEN EFFECT: the simultaneous application of dead or radiation damaged compatible (mentor) pollen with incompatible pollen may in some instances helps to overcome the incompatibility of the latter and fertilization may result. (See incompatibility alleles)

MENU: of a computer lists the various functions you can choose. The menu bar on top of the screen of the monitor displays the titles of the menus available.

Mer: a human T cell protooncogene encoded receptor tyrosine kinase. (See TCR)

Mer⁻ PHENOTYPE: mammalian cell defective in methylguanine-O^6-methyltransferase.

MERCAPTOETHANOL: keeps SH groups in reduced state and disrupts disulphide bonds while proteins are manipulated *in vitro*.

MERCAPTOPURINE: a purine analog inhibiting DNA synthesis and it is therefore cytotoxic.

MERICLINAL CHIMERA: the surface cell layers are different from the ones underneath just like in the periclinal chimeras but the difference is that the different surface layer does not cover the entire structure but only a segment of it. (See chimera, periclinal chimera)

MERISTEM: undifferentiated plant cells capable of production of various differentiated cells and tissues; functionally similar to the *stem cells* of animals. (See diagram below, stem cells)

MERISTIC TRAITS: are quantitative traits that can be represented only by integers, e.g., the number of kernels in a wheat ear or the number of bristles on a *Drosophila* body. (See quantitative traits)

7mer-9mer: seven or nine base long conserved sequences in the vicinity of the V-D-J (variable, diversity, junction) segments of immunoglobulin genes in the germline DNA. (See immunoglobulins)

MERIT, ADDITIVE GENETIC: the same as the breeding value of an individual.

MERODIPLOID: see merozygous

MEROGENOTE: see merozygous

MEROGONE: fragment of an egg.

MEROMELIA: see limb defects in humans

MEROSIN: see muscular dystrophy

MEROZOITE: see *Plasmodium*

MEROZYGOUS: a prokaryote, diploid for part of its genome (merogenote). Prokaryotes are functionally haploid but transduction or plasmid may add another gene copy into the cell.

MERRY: see mitochondrial diseases in humans

MESELSON - RADDING MODEL OF RECOMBINATION: explains gene conversion (occurring by asymmetric heteroduplex, symmetric heteroduplex DNA) and crossing over occurring from one initiation event as indicated by the data of *Ascobolus* spore octads. In yeast the aberrant conversion tetrads arise mainly from asymmetric heteroduplexes as suggested by the Holliday model (see there). Symmetric heteroduplex covers the same region of two chromatids whereas asymmetric heteroduplex means that the heteroduplex DNA is present in only one chromatid. The heteroduplexes can be genetically detected very easily in asci containing spore octads. In the absence of heteroduplexes, the adjacent (haploid) spores are identical genetically. If the heteroduplex area carries different alleles the two neighboring spores may become different after post-meiotic mitosis. Actually heteroduplexes may be detectable also in yeast (that forms only 4 ascospores) by sectorial colonies arising from single spores. Branch migration indicates that the exchange points between two DNA molecules can move and eventually they can reassociate in an exchanged manner in both DNA double helices involved in the recombination event. Rotary diffusion indicates that the joining between single strands can take place by movement of the juncture in either direction and thus making the heteroduplex shorter or longer. (See diagram on the following page recombination models, recombination molecular mechanisms)

MESENCHYMA: unspecialized early connective tissue of animals that may give rise also to blood and lymphatic vessels. (See mesoderm)

MESOCARP: the middle part of the fruit wall. (See exocarp, endocarp)

MESODERM: middle cell layer of the embryo developing into connective tissue, cartilage, bone, lymphoid tissues, blood vessels, blood, notochord, lung, heart, abdominal tissues, kidney, and gonads. (See morphogenesis)

MESOKARYOTE: organism(s) that occupy somekind of a middle position between prokaryotes and eukaryotes. They are endowed with cytoplasmic organelles like plant cells but their nuclear structure reminds to prokaryotes. The amount of chromosomal basic proteins in the nucleus is low, the chromosomes are attached to the membrane yet they develop a nuclear spindle apparatus. Several microtubules pass through the nuclear membranes and in the majority of the species then pull the chromosomes to the poles with the membrane and without being attached to the chromosomes. In the dinoflagellate *Cryptecodinium cohnii* 37% of the thymidylate is replaced by 5-hydroxymethyluracil. More than half of the DNA is repetitious. The vegetative cells appear to be haploid and thus mutations can be readily detected. Both homo-

and heterothallic species are known. (See prokaryote, eukaryote)
Meselson-Radding model of recombination continued from preceding page

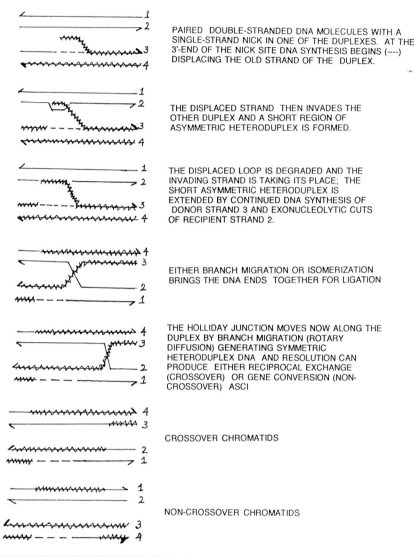

THE MESELSON - RADDING MODEL OF RECOMBINATION (See also preceding page)

MESOLITHIC AGE: about 12,000 year ago when domestication of animals started. (See paleolithic, neolithic)

MESOPHYLL: the parenchyma layers of the leaf blade.

MESOPHYTE: plants avoiding extreme environments such as wet, dry, cold or warm.

MESOSOME: an invaginated membrane within a bacterial cell.

MESOTHORAX: the middle thoracic segment of insects, bearing legs and possibly wings. (See *Drosophila*)

MESOZOIC: geological period in the range of 225 to 65 million years ago; age of the life and extinction of dinosaurs and several other reptiles. (See geological time periods)

MESSENGER POLYPEPTIDES: are extracellular signaling molecules that can pass the cell membrane, enter the cell nucleus and recognize in the DNA a special sequence motif and activate then transcription. These messengers are different from hormones or other signaling molecules because they do not need membrane receptors or special ligands or series of adaptors or phosphorylation to be activated and becoming nuclear co-activators of gene expression. (See lactoferrin)

MESSENGER RNA: see mRNA

MESTIZO: offspring of Hispanic and American Indian parentage. (See miscegenation, mulatto)

MET ONCOGENE: a hepatocytic growth factor receptor gene in human chromosome 7q31. Its α subunit is extracellular, its β subunit is an extra- and transmembrane protein. It is a tyrosine kinase as well as a subject for tyrosine phosphorylation. The receptor of HG-SF is the product of Met. (See tyrosine kinase, oncogenes, hepatocyte growth factor)

METABOLIC BLOCK: a non-functional enzyme (due to mutation in a gene, or to an inhibitor) prevents the normal flow of metabolites through a biochemical pathway.

METABOLIC PATHWAY: a series of sequential biochemical reactions mediated by enzymes under the control of genes. Genetic studies greatly contributed to their understanding along with the use of radioactive tracers. (See radioactive tracer, auxotrophy)

METABOLISM: enzyme-mediated anabolic and catabolic reactions in cells.

METABOLITE: product of metabolism. (See metabolism)

METABOLITE ENGINEERING: transformation of certain genes into a new host may lead to production of substances that the organisms never produced before or larger quantities or different qualities of certain proteins. Examples are: production of indigo or human insulin in *E. coli* or novel antibiotics or expressing antigens of mammalian pathogens in plants or changing the pathway of diacetyl formation in yeast to acetoin production and thus shortening the lagering process in brewing beers, etc. (See genetic engineering, protein engineering)

METACENTRIC CHROMOSOME: the two arms are nearly equal in length. (See chromosome morphology)

METACHONDROMATOSIS: autosomal dominant multiple exostoses particularly on hands, feet, knees and limb bones, without deforming the bones or joints and may disappear by time. (See exostosis, fibrodysplasia)

METACHROMASIA: the same stain colors different tissues in different hues.

METACHROMATIC LEUKODYSTROPHY (MLD): is a sulfatide lipidosis. Two distinct forms have been identified that are due to two recessive alleles of a gene, mutations in *A* and *I*, in human chromosome 22. Due to the deficiency arylsulfatase A, cerebroside sulfate accumulates in the lysosomes. The accumulation of galactoside-sulfate-cerebrosides in the plasma membrane and particularly in the neural tissues (myelin) causes a progressive and fatal degeneration in the peripheral nerves, liver and kidneys. As a consequence failure of muscular coordination (ataxia), involuntary partial paralysis, hearing and visual defects as well as lack of normal brain function arise after 18 to 24 months of age and usually causes death in early childhood. In the juvenile form of the disease the symptoms appear between age 4 and 10 years. There is also an adult type of the disease with an onset after age 16 and involves schizophrenic symptoms. The mutations involve either a substitution of tryptophan at amino acid residue 193, or at threonine 391 by serine or a defect at the splice donor site at the border of exon 2. MLD has been observed also in animals. The reduced activity of the enzyme can be identified also in cultured skin fibroblasts of heterozygotes and prenatally in cultured amniotic fluid cells of fetuses. (See also sphingolipidoses, sphingolipids, Krabbe's leukodystrophy, prenatal diagnosis)

METACLINE HYBRIDS: were called in *Oenotheras* the exceptional progeny that occurred only in the reciprocal crosses due to the difference in transmission through egg and sperm the different complex translocations. (See complex heterozygotes, certation, megaspore competition)

METACYCLIC *TRYPANOSOMA*: lives in the salivary gland of the insects which spread sleeping sickness. (See *Trypanosoma*)

METAFEMALE: has more than the usual dose of female determiners; its chromosomal constitution may be XXX. (See *Drosophila*, triplo-X; illustration at aneuploids)

METAGENESIS: sexual and asexual generations alternate.

METALLOPROTEIN: the prosthetic group of the protein is a metal, e.g. hemoglobin.

METALLOPROTEINASES: may be cell surface enzymes mediating the release of tumor necrosis factor α, a cytokine involved in inflammatory reactions. (See TACE)

METALLOTHIONEIN: is an SH-rich metal-binding protein in mammals. Its main function is detoxification of heavy metals. The promoter is activated by the same heavy metal the protein product binds. This promoter is very useful for experimental purposes because structural genes attached to it can be turned on and off by regulating the amount of heavy metal in the drinking water of the transgenic animals. It is encoded in human chromosome 16q13. (See transgenic, Menke's disease)

METAMALE: see supermale, *Drosophila*

METAMERISM OF THE BODY

METAMERISM: anterior - posterior segmentation of the body of annelids and arthropods. In chemistry it is rarely used for a type of structural isomerism when different radicals of the same type are attached to the same polyvalent element and give rise to compounds possessing identical formulas.

METAMORPHOSIS: distinct change from one developmental stage to another, such as from larva to adult or from tadpole to toad. (See *Drosophila*)

METAPHASE: stage in mitosis and meiosis when the eukaryotic chromosomes have reached maximal condensation and spread out on the equator of the cell (metaphase plane) and their arm ratios and some other morphological features can be well recognized. In meiosis I the bivalent chromosomes may be associated at their ends if chiasmata had taken place during prophase. The ring bivalents indicate crossing over between both arms whereas rod bivalents are visible when crossing over was limited to only one of the two arms. (See meiosis, mitosis, chromosome rosette, ring bivalent)

METAPHASE ARREST: may take place if the nuclear division is blocked by toxic agents or when two separate kinetochores are joined by translocation. Homologous recombination does not lead to metaphase arrest indicating kinetochore tension may be responsible for the event. (See meiosis)

METAPHASE PLANE: the central region of the cell where the chromosomes are located during metaphase. Often incorrectly called metaphase plate (but no plate is involved). (See mitosis, meiosis)

METASTABLE: a potentially transitory state; it can change to more or less stable form.

METASTASIS: the spread of cancer cells through the blood stream and thus establishing new foci of malignancy in any part of the body. The invasiveness of cancer cells requires an active state of the cell surface gelatinases (collagenases) so they could penetrate the extracellular matrix of the target cells. On the cell surface actually precursors of the gelatinases are found that are proteolytically activated by metalloproteinases. (See CD44, cancer, oncogenes, malignant growth, contact inhibition, saturation density, collagen, extracellular matrix, metalloprotein)

METATARSUS: the middle bones beyond the ankle but preceding the toes or in insects the basal part of the foreleg distal to the tibia but proximal to the tarsal segments and the claw. It carries the sexcombs in the male *Drosophila*. (See *Drosophila*)

METAXENIA: a physiological modification of maternal tissues of the fruit in plants by the genetically different embryo, e.g., in green hybrid apples the exocarp (skin) may become reddish if the pollen carries a dominant gene for red color. (See xenia, fruit)

METAZOA: all animals with differentiated tissues; thus protozoa are excluded.

METHACRYLATEACIDURIA: β-hydroxylisobutyryl CoA deacylase deficiency — involving the catabolism of valine — leads to urinary excretion of cysteine and cysteinamine conjugates of methacrylic acid, and to teratogenic effects. Methacrylic acid [$CH_2=C(CH_3)COOH$] is a degradation product of isobutyric acid [$(CH_3)_2CHCOOH$] and the amino acid valine [$(CH_3)_2CH$

(NH$_2$)CHCOOH]. (See isoleucine-valine biosynthetic pathway)

METHEMOGLOBIN (ferrihemoglobin, hemiglobin): is a hemoglobin with the ferroheme oxidized to ferriheme and has impaired reversible oxygen binding ability. Mutation and certain chemicals (ferricyanide, methylene blue, nitrites) may increase the normally slow oxidation of hemoglobin. Reduction of methemoglobin may be accomplished either through the Embden-Meyerhof pathway or through the oxidative glycolytic pathway (pentose phosphate pathway). The principal methemoglobin reducing enzyme is NADH-methemoglobin reductase that may be adversely affected by autosomal recessive genetic defects. As a consequence, lifelong cyanosis (bluish discoloration of the skin and mucous membranes) results following the accumulation of reduced hemoglobin in the blood. Males are apparently more affected. This anomaly may be corrected with the reducing agents, methylene blue or ascorbic acid. NADH-methemoglobin reductase is often called diaphorase. (See hemoglobin, NADH, Embden-Meyerhof pathway, pentose phosphate pathway, hexose monophosphate shunt)

METHIONINE ADENOSYLTRANSFERASE DEFICIENCY: homocystinuria and tyrosinemia may cause hypermethioninemia. The most direct cause of hypermethioninemia is an autosomal recessive defect in methionine adenosyltransferase. Besides these genetically determined causes methionine accumulation may be due to prematurity at birth or the overactivity of cystathionase when cow milk (rich in methionine) is fed, etc. Hypermethioninemia itself may not have very serious consequences. (See methionine biosynthesis, homocystinuria, tyrosinemia, amino acid metabolism)

METHIONINE BIOSYNTHESIS: methionine is biosynthesized from homocysteine by methionine synthase using N^5-methyltetrahydrofolate as a methyl donor:
$$HSCH_2CH_2CH(NH_2)COOH \rightarrow CH_3SCH_2CH_2CH(NH_2)COOH$$
      homocysteine   methionine

Methionine then can be used by methionine adenosyltransferase to generate S-adenosylmethionine and subsequently S-adenosylhomocysteine which after hydrolyzing off adenosine yields again homocysteine. (See homocystinuria, amino acid metabolism, methionine adenosyltransferase deficiency)

METHIONINE MALABSORBTION: is under the control of an autosomal recessive gene, and it results in the excretion of α-hydroxybutyric acid (with its characteristic odor) in the urine. Mental retardation, convulsions, diarrhea, respiratory problems and white eye accompany the condition. (See methionine biosynthesis)

METHIONYL tRNA SYNTHETASE (MARS): charges tRNA^{Met} by the amino acid methionine. The MARS gene is in human chromosome 12. (See aminoacyl-tRNA synthetase)

METHOTREXATE (amethopterin): a folic acid antagonist and an inhibitor of dihydrofolate reductase. It is an inhibitor of thymidylic acid synthesis. It is used also as selectable agent in genetic transformation. (See folic acid, transformation, selectable)

METHYLASE: enzymes in bacteria protect the cell's own DNA from type II restriction endonucleases by transfering methyl groups from S-adenosyl methionine to specific cytosine or adenine sites within the endonuclease recognition sequence (cognate methylases). When eukaryotic DNAs are transfered to *E. coli* cells by transformation for cloning, their methylation pattern may be lost because the methylation system of the prokaryotic cell is different from that of the eukaryote. *E. coli* does not methylate the C in a 5'-CG-3' but the dam methyltransferase methylates A in the 5'-GATC-3' sequence and the dcm methyl transferase methylates the boxed C in the 5'-C[C]AGG-3' group. Such methylations may change the restriction pattern of cloned DNAs depending whether the particular restriction enzyme can or cannot digest methylated DNA. (See methylation of DNA, methylation-specific PCR)

METHYLATION INTERFERENCE ASSAY: detects whether a binding protein can attach to the specific DNA sites and thus provides information on the binding site and on the protein. The analysis is carried out by combining DNA and binding proteins and followed by treatment with methylating enzyme. If the protein binds to a specific guanine site(s), that base will not be methylated. Piperidine breaks DNA at bases modified by methylation, and sites protected from methylation by bound protein are not cleaved by peperidine. (See methylation of DNA)

METHYLATION OF DNA: in many eukaryotes 1-6% or more of the bases in DNA is methylcytosine. In T2, T4 and T6 bacteriophages' DNA 5-hydroxy-methylcytosine occurs in place of cytosine. Methylation of other bases thymine (= 5-methyluracil), adenine and guanine may also occur. Some of the alkylations of DNA bases leads to mutations by base substitutions. Methylation protects DNA from most of the restriction endonucleases (see restriction endonuclease types). In the majority of *E. coli* strains two enzymes are responsible for DNA methylation, *dam* and *dcm* methylase; *dam* methylates adenine at the N^6 position within the sequence 5'-GATC-3'. This sequence occurs at the recognition sites of a number of frequently employed restriction enzymes (Pvu I, Bam HI, Bcl I, Bgl II, Xho II, Mbo I, Sau 3AI, etc). Mbo I (↓GATC) and HpaII (C↓CGG) are sensitive to methylation but Sau 3AI and MspI, respectively, are not, and their recognition sites are identical (isoschizomers), therefore when the DNA is methylated, the latter ones still can be used. For several restriction enzymes to work, the DNA must be cloned in bacterial strains that do not have the *dam* methylase. Mammalian DNA is not methylated at the N^6 position of adenine, therefore Mbo I is always supposed to work, as well as Sau 3AI. The DNAs of eukaryotes are most commonly methylated on C nucleotides, in $_{GC}^{CG}$ sequences. The dcm methylase methylates the internal C positions in the sequences 5'-CCAGG-3' and 5'-CCTGG-3'; this methylation interferes with cutting by EcoRII [↓CC(A/T)GG] but not by BstN I, although at another position [CC↓(A/T)GG] of the same sequence. *E. coli* strain K also has methylation-dependent restriction systems that recognize only methylated DNA: *mrr* (6-methyladenine), *mcrA* [5-methyl-C(G)], *mcrB* [(A/G)5-methyl C]. Mammalian DNA with extensive methylation at 5-methyl C(G) is e.g., restricted by *mcrA*. Once DNA is methylated this feature may be transmitted to the following cell generation(s) by an enzyme, *maintenance methylase* although methylation is usually lost through the meiotic cycle. In bacteria the expression of methylated genes may be reduced by a factor of 1,000 but in mammals the reduction may be of six orders of magnitude. Methylation in the promoter region usually prevents transcription of that gene. It has been suggested that genomic imprinting is caused by differential methylation. In mice demethylation may prevent the completion of the embryonic development. In *Arabidopsis* plants demethylation to about 1/3 of the normal level caused either by a DNA (*ddm1*) demethylation mutation or by introducing by transformation an antisense RNA of the cytosine methyltransferase (MET1) caused alterations in the morphogenesis and developmental time of the plants. It is assumed that methylation of infective (inserted) DNA is part of the eukaryotic defense system. The epigenetic state of methylation can be transfered in the ascomycete, *Ascobolus* by a mechanisms resembling or related to recombination. After fertilization, the methyl moieties are generally removed from the CpGs and an unmethylated state is maintained through blastula stage. Housekeeping genes stay unmethylated whereas the methylation of tissue-specific genes varies by tissues. Reduced methylation causes developmental anomalies in plants and animals. (See also transposition, silencing, cross linking, chemical mutagens, alkylation, paramutation, imprinting, regulation of gene activity, Sp1, methylation resistance, ascomycete, 5-azacytidine, HMBA, methylation-specific PCR, methyltransferase)

METHYLATION OF RNA: 2'-O-methyladenosine, 2'-O-methylcytidine, 2'-O-methylguanosine, 2'-O-methyluridine, 2'-O-methylpseudouridine are minor nucleosides in RNA. The cap of mRNA is a 7-methyl guanine, and a N^6-methyladenosine occurs near the polyadenylated tract in mRNA. (See capping enzyme)

METHYLATION RESISTANCE: see MNNG

METHYLATION-SPECIFIC PCR (MSP): methylation of DNA is usually detected by the inability of the majority of restriction endonucleases to cleave methylated sites. MSP detects methylated CpG sites in minute amounts of DNA. The DNA is treated with sodium bisulfite to convert all non-methylated cytosines to uracil. The methylated and unmethylated DNAs are amplified then by two different primers designed to represent original CpG rich sequences. The amplified DNA can then be sequenced to compare the differences between the two samples. (See methylation of DNA, polymerase chain reaction)

3-METHYLCROTONYL GLYCINEMIA: is caused by the deficiency of a mitochondric enzyme involved in the degradation of leucine, β-methylcrotonyl-CoA-carboxylase. As a consequence of this autosomal recessive condition, muscle defects, and in some cases urinary overexcretion of 3-methylcrotonyl glycine and 3-hydroxyisovaleric acid occur. Some patients respond favorably to biotin because this vitamin is a cofactor of the enzyme. (See isoleucine-valine metabolic pathway, isovaleric acidemia, 3-methylglutaconaciduria, 3-hydroxy-3-methyl-glutaricaciduria, amino acid metabolism)

5-METHYLCYTOSINE: is common in tRNA and DNA of eukaryotes. (See 5-hydroxymethyl cytosine, 5-azacytidine)

METHYL-DIRECTED REPAIR: see mismatch repair

METHYLENE BLUE: aniline dye, for microscopic specimens, an indicator of oxidation-reduction, an antiseptic, an antidote for cyanide and nitrate poisoning.

5,10-METHYLENETETRAHYDROFOLATE DEHYDROGENASE: an enzyme in folate biosynthes is encoded in human chromosome 14q24. (See folic acid)

METHYLGLUTACONICACIDURIA: the autosomal recessive condition is caused by a deficiency of 3-methylglutaconyl-CoA hydratase, an enzyme mediating one of the steps in the degradation of leucine. The patients may develop nerv disorders such as partial paralysis, involuntary movements, eye defects, etc. In some cases there is a marked increase of methylglutaric and methylglutaconic acid (an unsaturated dicarbonic acid) in the body fluids. Leucine administration may exagerate the symptoms. (See isoleucine-valine metabolic pathway, isovalericacidemia, 3-methylglutaconaciduria, 3-hydroxy-3-methylglutaricaciduria, amino acid metabolism)

METHYLGREEN-PYRONIN: a histological stain; coloring DNA blue-green and RNA red. (See stains)

METHYLGUANINE-O^6-METHYLTRANSFERASE: an enzyme that reverses the alkylation of this base, and it is thus antimutagenic. Cells defective in the enzyme (Mer$^-$, Mex$^-$) are extremely sensitive to DNA-alkylating agents. (See methylation of DNA, antimutator)

METHYLJASMONATE: fragrance of jasmine and rosemary plants; it is a proteinase inhibitor.

METHYLMALONICACIDURIA: there are several forms of the metabolic disorder, methylmalonic-CoA mutase deficiency (MUT), in human chromosome 6p21.2-p12, and another is caused by a defect in the synthesis of adenosyl-cobalamin (cblA, vitamin B_{12}) a necessary cofactor in the biosynthesis of succinyl-CoA from L-methylmalonyl-CoA by MUT. A third type of methylmalonic aciduria is due to a defect in the enzyme epimerase (racemase) that converts D-methylmalonyl-CoA to the L form. This pathway can be represented as:

Propionyl-CoA -> D-methylmalonyl-CoA -> L-methylmalonyl-CoA -> Succinyl-CoA
CARBOXYLASE   EPIMERASE   MUT
                                                 B_{12}

In these disorders methylmalonic acid and glycine may accumulate in the body fluids, and the affected individuals may show serious (growth and mental retardation, acidosis [keto acids in the blood]) or almost no adverse effects. Administration of vitamin B12 may alleviate the problem in some cases. (See methylcrotonylglycemia, amino acid metabolism, vitamin B_{12} defects, ketoaciduria)

METHYLMERCURIC HYDROXIDE: may be added to the electrophoretic agarose running gel of RNA, and when it is stained with ethidium bromide in 0.1 M ammonium acetate the color of the RNA is enhanced. It is also used to treat mRNA for preventing the formation of secondary structure during the synthesis of the first strand of cDNA. Note that MMH is an extremely toxic volatile compound. (See cDNA, electrophoresis)

METHYLMETHANESULFONATE: a powerful alkylating agent and mutagen/carcinogen. (See ethylmethanesulfonate)

METHYLPHOSPHONATES: are oligonucleotide analogs used for antisense operations. They

Methylphosphonates continued
are readily soluble in water and resistant nucleases. The oligonucleoside methylphosphonates form stable complexes with both single- and double-stranded DNA and RNA. They have both antiviral and anticarcinogenic effects. (See antisense RNA)

METHYLTETRAHYDROFOLATE CYCLOHYDROLASE DEFICIENCY: phosphoribosylglycinamide formyltransferase. (See folic acid)

METHYLTHIOADENOSINE PHOSPHORYLASE (MTAP): is encoded in human chromosome 9p21 area, and its defect or deletion is characteristic for many malignant tumors, and it may be associated (linked) to a tumor suppressor activity of CDK-4. (See CDK)

METHYLTRANSFERASE, DNA: is responsible for the methylation of CpG sites. The activity of this enzyme is increased during the initiation and progression of carcinogenesis. (See cancer, methylation of DNA, methylguanine-O^6-methyltransferase)

METHYLVIOLET: an aniline dye for bacterial microscopic examination.

METREE: is a computer program package for infering and testing minimum evolutionary trees; designed by the Institute of Molecular Evolution, Pennsylvania State University, Philadephia, PA, USA. (See evolutionary tree)

METRONIDAZOLE (2-methyl-5-nitro-1-imidazole ethanol): a radiosensitizing agent and mutagen for chloroplast DNA, and a suspected carcinogen. (See chloroplast genetics, formula ↓)

MeV (mega electron volt): million electron volt. (See electron volt)

MEVALONICACIDURIA (MVK): is a recessive (human chromosome 12) defect with huge increase of mevalonic acid in the urine, caused by a defect in mevalonic acid kinase.

MF1: is a 5' to 3' exonuclease. (See DNA replication eukaryotes)

MGD: see mouse genome data base.

MGMT: see methylguanine-O^6-methyltransferase

MGSA (melanoma growth stimulating activity): see KC, N51, gro

MGT: member of the MGMT group of enzymes; it is encoded in human chromosome 10q. (See MGMT, Mer⁻)

MHC: major histocompatibility complex is involved in immunological reactions; it is controled by linked multigene families (HLA) determining cell surface antigens and thus cellular recognition. The acronym is used also to the myosin heavy chain. (See HLA, immune system)

MHC RESTRICTION: see immune system

MIC: minimal inhibitory concentration. (See micRNA)

MICELLE: a round body of (protein) substances surrounded by lipids.

MICHAELIS - MENTEN EQUATION: measures enzyme kinetics in a process:

$$(E) + (S) \underset{k_2}{\overset{k_1}{\Leftrightarrow}} (E) \overset{k_3}{\to} \text{Product(s)} + (E),$$

where k_1, k_2, k_3 are constants of the reactions, (E) = enzyme, (S) = substrate concentration and

$$v = \frac{V(S)}{K_m + (S)} \quad \text{or} \quad K_m = (S)\left[\frac{V}{v} - 1\right]$$

where v is the velocity of the reaction when half of the substrate molecules is combined with the enzyme, V = the maximum velocity, and

$$K_m = \frac{[(E) - (ES)](S)}{(S)}$$

is the Michaelis - Menten constant. (see also Linweaver -Burk plot)

MICHEL SYNDROME (oculopalatoskeletal syndrome): is an autosomal recessive multiple defect involving the eyelid, opacity of the cornea, cleft lip and palate, defects of the inner ear and spine column, etc. It causes complete deafness. (See deafness, eye diseases)

micRNA: (messenger RNA-interfering RNA). This RNA is transcribed on short sections of the complementary strand DNA and it prevents gene expression. (See antisense RNA)

MICROARRAY: is a microtiter tray with wells containing cDNAs to what mRNAs (or fluorochrome-labeled-cDNAs) can be hybridized and after incubation and scanning the amounts of the mRNAs derived from the same tissue can be quantitated (on the basis fluorescence intensity). If the tissues are, e.g. from healthy and diseased sources, on the basis of the level of ex-

Microarray continued
pression, the genetic cause of the disease can be infered. The primary importance of this techniques is to monitor the expression pattern of many genes on a single plate from minute amounts of material. (See also SAGE, laser desorption MS, electrospray MS)

MICROBE: small organisms like the eukaryotic fungi, algae and protozoa and the prokaryotic blue-green algae, bacteria and viruses. They are frequently associated with disease.

MICROBODIES: also called peroxisome; 0.15 - 0.5 μm diameter bodies in eukaryotic cells containing oxidase and catalase enzymes. The proxisomes have indispensable roles in fatty acid β oxidation, phospholipid and cholesterol metabolism. Fatty acid granules are also named microbodies. The peroxisome biogenesis disorders (PBD) are recessive lethal diseases in variable forms. The most extreme form is the Zellweger syndrome, the Refsum disease and the adrenoleukodystrophy are milder and the rhizomelic chondrodysplasia punctata involves also bone defects. Peroxisome mutations have been identified also in yeast (PAS). (See also glyoxisome, peroxisomes, Zellweger syndrome, Refsum disease, adrenoleukodystrophy, chondrodysplasia, oxalosis, peroxidase and phospholipid deficiency, peroxisomal 3-oxo-acylcoenzyme A thiolase deficiency)

MICROCELL: a micronucleus, a piece of chromatin, a chromosome or a few chromosomes surrounded by a membrane. (See micronucleus)

MICROCEPHALY: abnormal smallness of the head; generally involves mental retardation. It is a condition due to various genetic and environmental causes (e.g. X-ray, heat [febrile] exposure of the fetus). The incidence of the autosomal recessive form is about 2.5×10^{-5}. A deletion of chromosomal segment 1q25-32 or mutation in 1q31-32 may be the cause of severe cases. The recurrence risk among sibs was estimated to be 0.19 but the risks may vary depending on the cause of the defect. It is generally accompanied by other abnormalities. Microcephaly with normal intelligence characterizes the autosomal recessive Nijmegen breakage syndrome. The latter is associated with chromosomal instability, immunodeficiency and radiation sensitivity. (See mental retardation, hydrocephalus, cranofacial dysostosis, cerebral gigantism)

MICROCHAETAE (pl.): hairs of insects.

MICROCHROMOSOMES: are uniformly very small chromosomes of the avian genome and are rich in GC content. They replicate ahead of the larger (macro)chromosomes. Their density of genes appears very high probably because their introns are short. (See intron)

MICROCIN: an enterobacterial heptapeptide inhibiting protein synthesis: Acetyl-Met-Arg-Thr-Gly-Asn-Ala-Asp-X where the first amino acid is acetylated and X is an acid labile group. It is encoded by 21 bp and thus appears to be one of the smallest translated genes. (See gene size)

MICROCINEMATOGRAPHY: is a time-lapse motion photorecording of living material. Successive frames delayed in real time, are projected at normal speed giving the sensation as if the events, movements of cells or chromosomes, etc., would have taken place at an accelerated time sequence and thus, e.g., the progress of mitosis requiring 1 - 2 hrs can be seen in motion, in minutes. (See phasecontrast microscopy, Nomarski)

MICROCOCCAL NUCLEASE: from *Staphylococcus aureus* degrades DNA (with preference to heat-denatured molecules) and RNA and generates mono- and oligonucleotides with 3'-phosphate termini.

MICROCONIDIA: small uninucleate conidia. (See macroconidium, conidia, fungal life cycles)

MICRODELETION: loss of a chromosomal segment, too short to be visualized by light microscopy.

MICROENCEPHALY: abnormal smallness of the brain caused by developmental genetic blocks or degenerative diseases. (See microcephaly)

MICROEVOLUTION: variation within species which may lead to speciation.

MICROFILAMENTS: actin and myosin containing fibers in the cells serving as part of the cytoskeleton and mediate cell contraction, amoeboid movements etc.

MICROFUSION: fusion of protoplast fragments with intact protoplasts to generate cybrids (generally by electroporation). (See cell fusion, somatic hybridization, cybrid)

MICROGLIA: mesodermal cells supporting the central nervous system. Microglia are a class of monocytes capable of phagocytosis. They can bind, through scavanger receptors, β-amyloid fibrils (present in Alzheimer plaques) resulting in the production of reactive oxygen species and leading to cell immobilization, and cytotoxicity toward neurons. (See monocyte, Alzheimer disease)

β_2-**MICROGLOBULIN**: the class I MHC α chain is non-covalently associated with this polypeptide which is not encoded by the HLA complex. The α_3 domain and the microglobulin are similar to immunoglobulins and they are rather well conserved, in contrast to the α_1 and α_2 domains which are highly variable. Its defect results in renal amyloidosis. (See TAP, HLA, amyloidosis)

MICROGNATHIA: abnormally small jaws.

MICROGRAPH (photomicrograph, electronmicrograph): photograph taken through a microscope (light- or electronmicroscope). (See microscopy)

MICROINJECTION: is a method of delivery of transforming DNA or other molecules into animal or other cells by a microsyringe. This procedure is not considered as a highly efficient method of transformation in plants. An advantage is, however, that the delivery can be targeted to cells but not into chromosomal location unless gene targeting is used. (See transformation animals, caged compounds, targeting genes, gene transfer by microinjection, targeting genes).

MICROMANIPULATION OF THE OOCYTE: the penetration of some types of disadvantaged sperms can be facilitated by mechanically opening an entry point through the zona pellucida (a non-cellular envelope of the oocyte) and thus facilitating the penetration of the sperm (PZD). Alternatively, with the aid of a microneedle the sperm can directly deposited under the zona pellucida (SZI) into the space before the vitellus (egg yolk). (See ART, preimplantation genetics, *in vitro* fertilization)

MICROMANIPULATOR: a mechanical device usually employing glass needles or microsyringes to carry out dissections or injections of cells, while viewed under the microscope.

MICROMERE (small micromeres): small cells in the vegetal pole arising from the 8-cell blastomere and giving rise to the coelom. (See vegetal pole, blastomere, coelom)

MICRONUCLEUS: the reproductive nucleus of *Infusoria*, as distinguished from their vegetative macronucleus; also a small additional nucleus containing only one or a few chromosomes in other taxonomic groups. In other organism broken chromosomes may be visible as micronuclei. (See *Paramecium*, microcell)

MICRONUCLEUS FORMATION AS A BIOASSAY: micronuclei are formed when broken chromosomes, chromosomal fragments fail to be incorporated into the daughter nuclei during cell division. Also, the damage to the spindle apparatus may result in the appearance of micronuclei. These phenomena have been exploited for testing mutagenic agents specifically causing these types of genetic damage to animal and plant cells. Such assays can be done in cultured cells but *in vivo* assay of meiotic plant cells or mammalian bone marrow polychromatic erythrocytes have also been used. (See bioassays in genetic toxicology)

MICRONUTRIENTS: are required for nutrition in small or trace amounts.

MICROPHTHALMOS (nanophthalmos): genetically determined (dominant and recessive) forms involve extreme reduction of the eye(s). In some instances it does not involve additional defects. Its frequency in the general population is low, about 0.004%, and the incidence among caucasian sibs is about 12-14% in the recessive form. Transformation of mice with diphteria toxin genes attached to eye-specific (γ-crystalline, a globulin of the lens) or the pancreas-specific Elastase I, a collagen digesting enzyme, promoted this developmental condition (ablation). The microphthalmia-associated transcription factor (MITF) mutations transform fibroblasts into melanocytes. (See eye diseases, anophthalmos, Waardenburg syndrome, ablation)

micropia: see copia

MICROPLASMID (miniplasmid): see πVX, recombinational probe

MICROPROJECTILE: see biolistic transformation

MICROPROPAGATION: regeneration plants from somatic, usually apical meristem cells, by *in vitro* techniques. It may be useful for rapid propagation of rare plants or non-segregating hybrids or secure virus free stocks. (See synthetic seed, embryo culture)

MICROPYLE: a pore of the plant ovule, between the ends of the integuments, through which the pollen tube (sperms) reaches the embryosac. The pore on the ovules of arthropods (and some other invertebrates) serving the penetration of the sperm.

MICROSATELLITE: mono- or tetranucleotide repeats distributed at random (10-50 copies) in the eukaryotic chromosome and can be used for constructing high-density physical maps and for the rapid screening for genetic instability, evolutionary relationships, detection of bladder cancer in cells found in the urine, etc. The microsatellite region may expand (most commonly) or contract and thus result in genetic instabilities. Microsatellite loci may mutate at a frequency of 0.8% per gamete, more frequently than in somatic cells. Some estimates are in the 10^{-4} to 10^{-5} range and makes them useful for linkage analysis using PIC. In the majority of the cases the expansion of the loci does not change the linkage phase of the flanking markers and therefore gene conversion is implicated. In humans they estimated one (≥4 bp) microsatellite per 6 kbp genomic DNA. Three quarters of the repeats are A, AC, AAAN or AG. The AC repeats are distributed at random in the genome and used most commonly for linkage studies. In the human X chromosome 3 and 4 base repeats occur in every 300-500 bp. Large scale survey indicates that most of the microsatellites were generated as a 3' extension of retrotranscripts and may serve as pilots to direct integration of retrotransposons. (See hereditary non-polyposis colorectal cancer, mismatch repair, trinucleotide repeats, minisatellite, VNTR, PIC, DNA fingerprinting)

MICROSATELLITE MUTATOR (MMP): increases frameshift and other mutations in G-rich microsatellites. (See microsatellite, minisatellite)

MICROSATELLITE TYPING: in case of linkage disequilibrium the inheritance of microsatellite markers can be followed on PCR amplified DNA. Example the CAAT repeat in the human tyrosine hydroxylase genes (chromosome 11p15.5) using the Z alleles displayed the pattern of segregation below in the ethidium bromide stained non-denaturing gel (from Hearne, C.M. Ghosh, S & Todd, J.A. 1992 Trends Genet. 8: 288):

father	mother	sib-1	sib-2	sib-3	paternal grandfather	paternal grandmother	ALLELES
▮▮▮	▮▮▮	▯▯	▮▮▮	▮▮▮			Z
				▮▮▮			4
	▮▮▮	▮▮▮				▮▮▮	12
▮▮▮	▮▮▮		▮▮▮			▮▮▮	16

The ▮ symbolize gel bands and the ▯ stand for homozygosity of the Z allele)
(See DNA fingerprinting, microsatellite, PIC)

MICROSCOPY: is using a special optical device for viewing objects that are not discernible clearly by the naked eye. (See light microscopy, stereomicroscopy, resolution, fluorescence microscopy, dark-field microscopy, phase-contrast microscopy, Nomarski, confocal microscopy, electronmicroscopy, atomic force microscopy, atom microscopy, scanning tunneling)

MICROSOMES: membrane fragments with ribosomes and enzymes obtained after grinding up eukaryotic cells and separating the cellular fractions by centrifugation. After about 9,000 x g force (10 min) the microsomes (S9) remain floating while other cellular particulates sediment. For the Ames genotoxicity bioassay generally Sprague-Dawley rat livers are used. The rats are previously fed in the drinking water with polychlorinated biphenyl (PCB), Araoclor 1254 a highly carcinogenic substance (requires special caution of disposal!) in order to induce the formation the P-450 monooxygenase activating enzyme system associated with the endoplasmic reticulum. (See Ames test, activation of mutagens, carcinogen)

MICROSPECTROPHOTOMETER: a spectrophotometer which can measure monochromatic light absorption of microscopical objects. (See spectrophotometer)

MICROSPORANGIUM: the sac that contains the microsopore tetrad of plants and the microspores of fungi and some protozoa.

MICROSPORE (small spore): the immature male spore of plants that develops into a pollen grain. (See microspore mother cell, microsporocyte, meiosis, gametogenesis)

MICROSPORE CULTURE: *in vitro* method for the production of haploid plants by direct or indirect androgenesis. (See diagram) →

MICROSPORE DYAD: the microspore mother cell after the end of the first meiotic division is divided into two haploid cells (the microspore dyad). (See meiosis)

MICROSOPORE MOTHER CELL: see microsporocyte

MICROSPORES: haploid products of male meiosis in plants that develop into pollen grains, and the smaller generative cells of lower organisms. (See also microsporocyte, megaspore)

MICROSPOROCYTE: the cell within which meiosis takes place and the microspores develop. (See microspore, gametogenesis)

MICROSOPOROGENESIS: see meiosis in plants.

MICROSPOROPHYLL: a leaf on which microsporangia develop in lower plants.

MICROTECHNIQUE: the procedures used for the preparation of biological specimens for microscopic examination (involving fixation, staining, sectioning, squashing, etc.)

MICROSURGERY: dissection or other surgical operations carried out under the microscope, generally with the aid of a micromanipulator. (See micromanipulator)

MICROTOME: an instrument that by means of various (sliding or rotating or rocking) motions cuts serial thin (usually within the range of 1-20 μm) sections of the embedded or frozen specimens to be examined by light or electronmicroscopy. The cutting edge may be steel or glass. Electronmicroscopy also requires sectioning of the specimens, usually employing a diamond knife. (See embedding, sectioning, stains, smear)

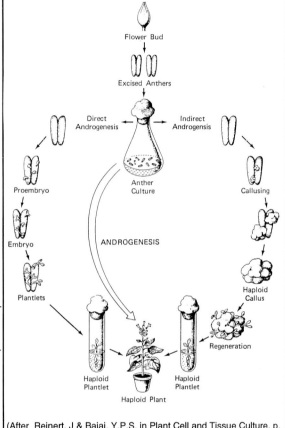

(After Reinert, J & Bajaj, Y.P.S. in Plant Cell and Tissue Culture, p. 251. Springer Vlg., New York

MICROTUBULE: various types of long cylindrical filaments of about 25 nm in diameter within cells built by polymerization of tubulin proteins. Microtubules are hollow tubulin filaments of the spindle apparatus of the dividing nuclei, elements of the cytoskeleton, cilia, flagella, etc. The energy for polymerization is provided by hydrolysis of GTP to GDP. The beginning of the polymerization is called nucleation and in animal cells that begins at the centrosomes. Microtubule elongation is a polarized process. Microtubules are somewhat unstable molecules and their growth may be blocked by antimitotic drugs such as colchicine or colcemid. Taxol (an anticancer extract of yew plants) stabilizes the microtubules and arrests the cell cycle in mitosis. After polymerization the microtubules undergo modification, e.g., a particular lysine of α-tubulin may be acylated and tyrosine residues removed from the carboxyl end. Microtubule-associated proteins (MAP) mediate the "maturation" of microtubules. MAPs aid the differentiation of nerve axons and dendrites that are loaded with microtubules. Microtubules move

Microtubule continued
various organelles such as the chromosomes during nuclear divisions in the cells with assistance of the proteins kinesin and dynein. The cilia and flagella involved in movements are built of bundles of microbules. The microbule protein complexes bind ATP at their "head" and associate with organelles by their "tail". The "plus" end of microtubules where tubulin subunits are added rapidly and the "minus" end where the addition is more slow. There are several protein factors that move the tubulins and associated structures. In prophase the duplicated centrosomes are separated by the bimC, KAR3 and cytoplasmic dyneins (450-550 kDa, motility 75 µm/minute). The bimC family members (120-135 kDa and with 1-2 µm/min motility) have different names in the different species (KLP61F and KRP_{130} in *Drosophila*, Cin8 in *Saccharomyces*, cut7 in *Schizosaccharomyces*). The KAR3 family (65-85 kDa, motility 1-15 µm/min) is involved also in spindle stabilization during metaphase. During prometaphase the microtubules are captured by the kinetochore with the assistance of the MCAK (mitotic centromere associated kinesin). The chromosomes congregating at the metaphase plane are moved by KIF4 (140 kDa) and related proteins that are involved also during metaphase in the chromosome alignment. During anaphase the movement of the chromosomes toward the poles is propelled by CENP-E and the KAR3 family of proteins. The elongation of the spindle fibers is mediated by the MKLP, bimC and cytoplasmic dynein proteins. (See tubulin, mitosis, meiosis, centrosome, spindle, centromere)

MICROTUBULE ASSOCIATED PROTEINS (MAP): control stability and organization of microtubules. (See microtubule, kinesin, dynein)

MICROTUBULE ORGANIZING CENTER: areas in the eukaryotic cells from where the microtubules emanate and grow such as the mitotic centers (poles) that give rise to the mitotic spindle. (See spindle, mitosis, also MTOC)

MICROVILLI: are small emergences on the surface of various cells in order to increase their surface. The microvilli of the chorion may be sampled for amniocentesis in prenatal genetic examinations. (See amniocentesis)

MIDBODY: after division microtubule fragments (midbodies) may be detected in animal cells.

MIDBRAIN: is the middle part of the brain. Degeneration of the motor neurons in this region may be responsible for Parkinson disease. (See brain human, Parkinson disease)

MIDDLE LAMELLA: the material (pectin mainly) that fills the intercellular space in between plant cells.

MIDDLE REPETITIVE DNA: is made up of relatively short repeats dispersed throughout the genome of (higher) eukaryotes. (See SINE, LINE, redundancy)

MIDGUT: the middle portion of the alimentary tract of insects and other invertebrates.

MIDPARENT VALUE: see breeding value [midpoint]

MIDPOINT: see breeding value

MIF (macrophage inhibitory factor): a pro-inflammatory pituitary factor; it may override glucocorticoid-mediated inhibition of cytokine secretion. (See cytokines, glucocorticoid)

MIFEPRISTONE (11-[44-(dimethylamino)phenyl-17-hydroxy-17-(1-propynyl)-(11β,17β)-estra-4,9-dien-3-one): is an antiprogestin, targeted to the progesterone receptors. This receptor family includes the glucocorticoid, mineralocorticoid, androgen, estrogen and vitamin D receptors. (See RU-486 and the other mentioned)

MIGRAINE: a neurological anomaly causing recurrent attacks of headaches, nausea, light and sound-avoidance. It may affect more frequently females (24%) than males (12%) to a very variable degree. It is generally attributed to multiple genes. Familial hemiplegic migraine (FHM) is however a rare autosomal dominant condition coded in human chromosome 19p13 as is episodic ataxia type 2. It appears that the basic defect is in a brain-specific Ca^{2+} ion channel α-1 subunit translated from 47 exons of CACNL1A4 gene. (See ion channels, ataxia)

MIGRATION: see gene flow

MIGRATION OF DNA IN GELS: is affected by molecular size, configuration of the macromolecule, concentration of the support medium (agarose, polyacrylamide), voltage, changing direction in the electric field, base composition, presence of intercalating dyes, buffer, etc.

MIGRATION INHIBITION FACTOR: MIF is a lymphokine. (See lymphokines, lymphocytes, immune system)

MIGRATION IN POPULATIONS: see gene flow

MIL ONCOGENE: is the avian representative of the RAF murine oncogene, a protein serine kinase; it is also related to the murine leukemia virus (MOS). (See oncogenes, RAF, MOS)

MILLER-DIEKER SYNDROME: characterized by smooth brain (lissencephaly I), more like in the early fetus, defects in other internal and external organs, mental retardation and death before age 20. Apparently deletions in area 17p13.3 are responsible for the recessive phenotype. Lissencephaly II may be caused by other genes involving also hydrocephaly and severe brain malformations (Walker-Warburg syndrome). The gene encodes a subunit of brain platelet-activating acetylhydrolase. (See deletion, head/face/brain defects)

MILLETS (*Eleusine, Pennisetum*): are arid climate grain crops. The cultivated *E. coracana* is $2n = 4x = 36$, tetraploid. *P. americanum* is $2n = 2x = 14$ diploid. The common millet (*Panicum miliaceum*) is an old grain crop; $2n = 4x = 36$. The foxtail millet (*Setaria italica*) is $2n = 2x = 18$ is mainly a hay crop.

MIM: see mitochondrial import

MIMICRY: the process and result of protective change in the appearance of an organism that makes it resemble the immediate environment for better hiding or imitating the features of other organisms that are distasteful or threatening to the common predators. (See also Batesian mimicry)

MIMICS: individuals that develop mimicry as a form of adaptation and evasion of predators. Also genes which are controling practically the same phenotype yet they are not allelic.

MIMOTOPE: a conformational mimic of an epitope without great similarity in amino acid residues. (See epitope)

MINERAL CORTICOID SYNDROME (AME): causes hypertension without overproduction of aldosterone. This syndrome is activated by cortisol. 11β-hydroxysteroid dehydrogenase converts cortisol to cortisone and activates thus the mineralcorticoid receptor; the patients are deficient in this enzyme which is inhibited by glycyrrhetinic acid (present in licorice). (See hypertension, aldosteronism, Liddle syndrome)

MINERAL REQUIREMENTS OF PLANTS: 9 macro elements (H, C, O, N, Ca, Mg, P, S) and 7 micro elements (Cl, B, Fe, Mn, Zn, Cu, Mo). Under some circumstances other elements may also be beneficial. (See embryo culture)

MINI Mu: deletion variants of phage Mu cloning vehicles that still carry the phage ends, a selectable marker and replicational origin. (See Mu bacteriophage)

MINICELL: DNA-deficient bodies surrounded by cell wall (in bacteria). Because they have no DNA, they cannot incorporate labeled precursors either into RNA or protein. In case the minicells descended from parents with plasmids, they may contain DNA and thus can make RNA and direct protein synthesis depending on the nature of the DNA fragment they carry. (See maxicells)

MINICHROMOSOME: in eukaryotic viruses (SV40, polyoma virus) the histone-containing, small nucleosome-like structure of genetic material; also in eukaryotes, an extra chromosome with extensive deletions. Such minichromosomes can be generated by the insertion of human telomeric sequences $(TTAGGGG)_n$ between the centromere and the natural telomere and eliminating the sequences distal to the insertion point. Human mini Y chromosomes have about 32.5 to 9 Mb compared with the normal Y chromosome of 50-75 Mb. These minichromosomes permit the analysis of the role of different sections of the chromosomes and possibly it may become feasible to extend the analysis of mammalian chromosomes in yeast cells. (See human artificial chromosome)

MINIGELS: are used for the separation of small quantities (10 to 100 ng DNA), in small fragments (< 3 kb) on about 5 x 7.5 cm slides (10 to 12 mL agarose), in a small gel box (ca. 6 x 12 cm) for 30 to 60 min at 5 to 20 V/cm. (See electrophoresis)

MINIGENE: some of its internal sequences are deleted *in vitro* before transfection.
MINIMAL MEDIUM: provides only the minimal (basal) menu of nutrients required for maintanence and growth of the wild type of the species. (See complete medium)
MINIMAL TILING PATH: a tightly overlapping set of bacterial vector clones, suitable for sequencing eukaryotic genomes.
MINIMIZATION: see crossing over
MINIMUM EVOLUTION METHODS: use an estimate of a branch length of an evolutionary tree construct on the basis of pairwise distance data calculated by various mathematical algorithms. The most plausible tree should be that provides the smallest sum of total branch length. (See evolutionary distance, evolutionary tree, least square methods, four-cluster analysis, neighbor joining method, algorithm)
MINIPLASMID: see πVX microplasmid, recombinational probe
MINIPREP: a small-scale quick preparation of DNA from plasmids or from other sources.
MINIREPLICON: is a vector consisting of a pBR322 replicon, a eukaryotic viral replicational origin (SV40, Polyoma) and a transcriptional unit. These vectors can be shuttled between *E. coli* and permissive mammalian cells. Also, deficient replicons containing only the replicational origin. (See replicon, shuttle vector)
MINISATELLITE: in eukaryotic genomes short (14-100 bp) tandem highly polymorphic repeats occur at many locations. They are supposed to be products of replicational errors (slippage) and localized amplifications. Their high variability is probably due to frequent unequal crossing over and duplication-deficiency events. These sequences are highly variable; the variations are associated with diabetes mellitus and various types of cancers. They are used as RFLP probes in physical mapping. (See also microsatellite, SINE, VNTR, DNA fingerprinting, RFLP, unequal crossing over)
MINISEGREGANT: budlike extrusions of animal cells with pinched-off DNA.
MINOR HISTOCOMPATIBILITY ANTIGEN: has some role in immune reaction but it is not coded in the HLA region. (See HLA, MHC)
MINOS: a 1,775 base transposable element of *Drosophila hydei* with 255 bp inverted terminal repeats and with two non-overlapping open reading frames. (See transposable elements animals)
MINUS END: of microtubules or actin filaments is less liable for elongation. (See plus end)
MINUS POSITION OF NUCLEOTIDES: indicate the distance from first translated triplet of the transcript of a genes.
MINUTES: approximately 60 dominant mutations in *Drosophila* that slow down the development of heterozygotes and lethal when homozygous.
MIP-1α (macrophage inflammatory protein): is a chemokine mediating virus induced inflammation and it is related to RANTES. (See RANTES, blood cells)
Mirabilis jalapa (four-o'clock): an ornamental plant and early object of inheritance, 2n = 58.
MISCEGENATION: sexual relations between partners of different human races. In the majority of human tribes marriage was generally limited within the tribes, however marriage by "capture" existed in ancient societies where the conquerors in war abducted females. Miscegenation was applied particularly to marriage between whites and blacks in the United States and in some South-American societies. There is no genetic justification against interracial marriage. Although marriage between Blacks and Orientals was not prohibited by any law, marriage between Caucasians and Blacks was unlawful in about 15 states of the U.S. until 1967 when the Federal Court ruled that the choice to marry resides with the individual. Racial and social (cast) discrimination in marriage still exists in many underdeveloped countries and in backward communities. (See mulatto, mestizo, racism, marriage)
MISCHARGED tRNA: linked to a wrong amino acids. (See aminoacyl- tRNA synthetase, protein synthesis)
MISDIVISION OF THE CENTROMERE: is a vertical rather than longitudinal division and it generates two telochromosomes from one bi-armed chromosome. The telochromosomes may open up to isochromosomes and may undergo misdivision again forming telochromosomes:

Misdivison continued

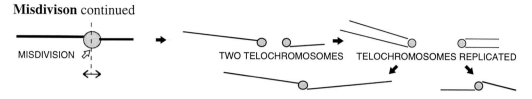

The cycles of bi-armed → telochromosome → isochromosomes → telochromosomes may continue.

MISINSERTION: DNA polymerase binds to a correctly matched 3' end of a primer where right and wrong dNTP substrates compete for insertion and occasionally the wrong may succeed resulting in misinsertion. (See also DNA editing, dNTP)

MISMATCH: there is one or more wrong (non-complementary) base(s) in the paired nucleic acid strands. (See transition mismatch, transversion mismatch)   ATCG A GTCA
TAGC C CAGT

MISMATCH EXTENSION: DNA polymerase binds to either a matched or mismatched primer end, and the mismatch is extended in replication. (See mismatch)

MISMATCH OF BASES: non-complementary bases in a (hetero)duplex DNA.

MISMATCH REPAIR: is an excision repair that removes unpaired bases and replaces them through unscheduled DNA synthesis with correct base pairs. Mutations in *E. coli* gene *mutL* and *mutS* increase genetic instabilities and in yeast defects in *PMS1, MLH1* and *MLH2* may increase genetic instability 100-700 fold because of deficient mismatch repair. Similar are the consequences of deletions of RTH1, encoding a 3'→5' exonuclease. The G/T binding protein (GTBP, 100K) and hMSH2 (160K, homolog of bacterial mismatch-binding protein) are essential for mismatch recognition in human cells. In fission yeast the mismatch repair enzyme, exonuclease I reduces mutation rate. Defects in the bacterial methyl-directed mismatch repair system greatly enhances mutability. The repair system is directed to DNA strands by methylation of adenine in the d(GATC) sequences. Since newly synthesized strands are not methylated, this is the criterion for recognition by the repair system (MutH, MutL, MutS, ATP). The mismatch may be located kilobases away from the d(GATC) tract. Interaction of this complex with heteroduplexes activates the MutH-associated endonuclease that responds to an initial sequence discontinuity or nick. Excision from the 5' side of the mismatch requires the RecJ exonuclease or exonuclease VII and digestion from the 3' end needs exonuclease I and also a helicase to unwind the strands because these proteins hydrolyze only single strands. RecJ and and exo VII require that the unmethylated (GATC) would be downstream from the defect. Exo I usually does not have this precondition. The replacement synthesis requires DNA polymerase III and other polymerases do not work in this system. In the eukaryotic systems the MutL homologs are PMS1 in yeast and PMS1, PMS2 in mammals. The yeast PMS1 system corrects G-T, A-C, G-G, A-A, T-T, T-C mismatches but C-C or long insertion/deletion sequences were barely repaired if at all. The human repair system fixes 8-12 pair mismatches and also C-C. An MutS homolog in mammals is called GTBP (G-T binding protein). Defects in PMS1, MLH1, MSH2 and [MSH3] enhances mutation rate in the yeast mitochondria, increases somatic mutability up to three orders of magnitude at certain loci (e.g. canavanine resistance), destabilizes $(GT)_n$ sequences. Mutations in the yeast Rth1 5'→3' exonuclease also increases mutability, particularly of plasmid-encoded genes. (See unscheduled DNA synthesis, DNA repair, mutator genes, microsatellite, homeologous recombination, hereditary nonpolyposis colorectal cancer, Lynch cancer families, Muir-Torre syndrome)

MISPARING: occurs when the homology between paired nucleotide sequences is imperfect and illicit binding takes place between nucleotides.

MISREADING: a triplet is translated into an amino acid different from its standard coding role. (See ambiguity in translation)

MISREPAIR: see DNA repair, SOS repair

MISREPLICATION: see error in replication

MISSENSE CODON: inserts in the polypeptide chain an amino acid different from that encoded by the wild type codon at the site. (See also nonsense codon)

MISSENSE MUTATION: change in DNA sequence that results in an amino acid substitution in contrast to mutations in synonymous codons where the change involves only the DNA but not the protein. (See also nonsense mutation)

MISSENSE SUPPRESSOR: enables the original amino acid to be inserted at a site of the peptide in the presence of a missense mutation. (See suppressor tRNA)

MISSPLICING: incorrect splicing. (See splicing)

MISTRANSLATION: see misreading, ambiguity in translation

MIT- : general designation of mitochondrial point mutations.

MITCHURIN: a 19-20th century Russian plant breeder who contributed a large number of improved varieties, mainly fruits, to the Russian and then to the Soviet agriculture. His career is often compared to that of the American Burbank. There was a very important difference, however, Burbank produced over 100 varieties to the USA agriculture without governmental pretensions that he made novel basic scientific discoveries. Mitchurin, on the other hand, published undigested and misunderstood theories and became the official forefather of lysenkoism, a state supported charlatanism. (See lysenkoism, Burbank)

MITE (mariner insect transposon-like element): an 1,457 bp DNA sequence in the vicinity of MLE containing a 24 bp tract with homology to Mos1 and supposedly responsible for the high frequency of unequal crossing over resulting in a duplication appearing as the Marie-Charcot-Tooth disease, and in the complementary deletion appearing as HNPP. Similar MITE elements occur also in rice, maize, sugarcane and other plants. (See Marie-Charcot-Tooth disease, HNPP, unequal crossing over, hot spot, Mos1, mariner, MLE)

MITOCHONDRIA: cellular organelles 1-10 μm long and 0.5-1 μm wide, surrounded by double membranes. The outer membrane has a diameter of 50-75 and the inner 75-100 Å. The latter forms the structures, called cristae. The outer membrane is associated monoaminoxidase and a NADH-cytochrome c reductase. Cardiolipin, phosphatidyl inositol, and cholesterol are major compounds associated with the inner membrane. Mitochondria have important role in generation of ATP and in electron transport. The ATP synthesis requires oxidative phosphorylation catalyzed by multiprotein complexes within the inner membranes of this organelle. In human mitochondria 13 out the 69 polypeptides involved in the process are coded for by mitochondrial DNA. The mitochondrial DNA and the prokaryotic type ribosomes in this organelle are capable of independent protein synthesis although they may not have all their necessary tRNA genes within the mt genome. The majority of the proteins within the mitochondria are synthesized in the cytosol and are imported. In yeast and some other organisms, under some circumstances, a single giant mitochondrion is formed that spreads all over the cell and can break up into smaller organelles. Mitochondria are usually transmitted only through the eggs. Exceptions exist, however. In a human cell about 100-1,000 mitochondria may be found, and each may contain 10 or more DNA molecules. During mammalian oogenesis the number of mitochondria may be 4,000 - 200,000. (See mtDNA, paternal leakage, compatibility of organelles with the nuclear genome, respiration, photorespiration, promitochondria, mitochondrial genetics, petite colony mutations, organelle sequence transfer, aspartate aminotransferase [glutamate oxaloacetate transaminase, GOT2], mitochondrial diseases in humans, cytoplasmic male sterility, mitochondrial abnormalities in plants, mitochondrial genetics, mitochondrial plasmids, mitochondrial import, *Neurospora* mitochondrial plasmids, mtTFA, sorting out, endosymbiont theory, evolution of organelles)

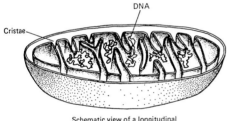

Schematic view of a longitudinal section of a mitochondrion

MITOCHONDRIAL ABNORMALITIES OF PLANTS: the cytoplasmic male sterility, wide-

Mitochondrial abnormalities of plants continued

spread in many species of plants and the *non-chromosomal striped* mutations of maize have proven mitochondrial defects. The latter ones have deletions either in the cytochrome oxidase (*cox2*) or in NADH-dehydrogenase (*ndh2*) mitochondrial genes. The phenotype (green and bleached stripes on the leaves, stunted growth, etc.) might suggest as if the plastids would have been involved. A similar mt-coded enzyme causes mitochondrial defects in animals. (See cytoplasmic male sterility, killer plasmids, senescence, mitochondrial disease in humans, mitochondrial genetic, mitochondrial plasmids)

MITOCHONDRIAL DISEASE IN HUMANS: up to the 1970s no human disease was attributed to mtDNA. The role of mitochondrial DNA was definitely proven only after molecular analysis became practical. The inheritance pattern is not always clear because of heteroplasmy. *Chloramphenicol resistance* in human cells is caused by several mutations in the mitochondrial DNA-encoded 16S ribosomal RNA genes. *Leber hereditary optic atrophy* (LHON), a progressive (most commonly single base pair substitution [Arg→His]) resulting in wasting away of the eyes (optic atrophy), abnormal heart beat accompanied by neurological anomalies characterizes this disease (involving 5 NADH dehydrogenases at different sites in the mtDNA) that affects predominantly the heteroplasmic males. It appears that a human X-chromosomal factor has a role in determining susceptibility to the mitochondrial defect. *Kearns-Sayre syndrome*, a progressive external ophtalmoplegia (paralysis of the eye muscles) is characterized by pigmentary degeneration of the retina and cardiomyopathy (inflammation of the heart muscles) and other less specific symptoms, is caused either by deletion or base substitution mutation in nucleotide 8,993 in the mtDNA. Deletions — in 30-50% of the afflicted — are flanked by the 5'-ACCTCCCTCACCA direct repeats and show the loss of nucleotides 8,468-13,446 and a less frequent deletion between the repeats 5'-CATCAACAACCG at positions 8,468-16,084. The larger deletions may affect simultaneously also other mitochondrial genes. One form of *myoclonic epilepsy*, MERRF (shock-like covulsions) is caused by a point mutation in the mtDNA-coded tRNA^{Lys}. *Mitochondrial myopathy* (eye degeneration), *encephalopathy* (brain degeneration), *lactic acidosis* (accumulation of lactic acid in the blood), and *stroke-like episodes* (*MELAS syndrome*) are due to a mtDNA encoded tRNA^{Leu} base substitution (A→G) at nucleotide 3,243. The same mutation occurs in about 20% of te patients with the recessive autosomal *Progressive external ophtalmoplegia* (PEO) and in some cases of diabetes mellitus. The *Pearson marrow-pancreas syndrome* is caused by mtDNA deletions affecting subunit 4 of NADH dehydrogenase, subunit 1 of cytochrome oxidase and subunit 1 of ATPase. *Oncocytoma*, responsible primarily for benign solid kidney tumors, loaded densely with mitochondria, has deletions in subunit 1 of cytochrome oxidase. In general, human males have more, or more are affected by mitochondrial mutations; 85% of the Leber's optic dystrophy cases found in males. Also Alzheimer disease, and Parkinson disease are associated with reduced sperm motility and fertility. *Leigh syndrome* (an ATP synthase defect at nucleotide 8527-9207) is progressive encephalopathy of children and it is accompanied by over 90% mutant mitochondria in the blood, muscle and nerve cells, although the inheritance appears autosomal. Several myopathies are associated with mutations also in various tRNAs of the mitochondria (tRNA^{Phe}, tRNA^{Ile} [cardiomyopathy], tRNA^{Glu} [cardiomyopathy], tRNA^{Met}, tRNA^{Ala}, tRN^{Asn}, tRNA^{Cys} [ophthalmoplegia], tRNA^{Tyr}, tRNA^{Ser}, tRNA^{Asp}, tRNA^{Lys} [MRRF syndrome], tRNA^{Gly}, tRNA^{Arg} [LHON], tRNA^{Ser}2, tRNA$^{Leu(CUN)}$ [skeletal myopathy], tRNA^{Glu}, tRNA^{Thr} [cytochrome b subunits]. A decrease in the number of mitochondria, caused by genetic factors or drugs may lead also to disease. *Aminoglycoside-sensitivity* (modest doses of streptomycin) may lead to hearing defects. In about 5×10^{-5} fraction of the human population is hypersensitive to chloramphenicol and may become anemic. Deletion of the COX (cytochrome oxidase) gene may result in *myoglobinuria*. *The autosomal dominant progressive external ophthalmoplegia* (adPEO) is a muscle weakness affecting primarily the eyes, is caused by a mutant gene in human chromosome10q23.3-q24.3 that causes multiple deletions. The mtDNA contains 37 genes most of them are transcribed

Mitochondrial diseases in humans continued

from the heavy chain but 9 are read in opposite direction from the light chain of the DNA and are thus somewhat overlapping. (See mitochondria, mtDNA, mitochondrial genetics, protoporphyria, aging, senescence, chondrome, optic atrophy, epilepsy, antibiotics, Alzheimer disease, Parkinson disease, Leigh's syndrome, diabetes mellitus, ophthalmoplegia, myoglobin, Wolfram syndrome)

MITOCHONDRIAL DNA: see mtDNA

MITOCHONDRIAL GENETICS: segregation of mitochondrial genes is followed by the cell phenotype because individual mitochondria are not amenable to direct genetic analysis. Despite the large number of mitochondria (>100; in the mouse oocytes ≈ 10^5) and mtDNA molecules (3-4/nucleoid), homoplasmic condition may be obtained at a much faster rate than expected on the basis of random sorting out. In mouse, it has been estimated that the number of segregating mitochondrial units is about 200 and the rapid sorting out was attributed to random genetic drift. Generally, 20 cell generations of the zygotes are used for appropriate mitochondrial typing followed by the use of selective media. The highest frequency of recombination is usually 20-25% because recombination between identical molecules also takes place and multiple "rounds of matings" occur within longer intervals. In *Schizosaccharomyces* the distribution of the mitochondria is mediated by microtubules.

Recombinational mapping is generally useful within short distances (≈ 1 kb). Genetic maps of mtDNA are generally constructed by physical mapping procedures. Also, the relative position of the genes may be ascertained by co-deletion — co-retention analyses. These methods resemble the deletion mapping of nuclear genes. Genes simultaneously lost by deletion, or retained in case of deletion of other sequences, must be linked. Mitochondrial genes have been mapped also by polarity. It was assumed that the yeast mtDNA carried some sex-factor-like elements, ω^+ and ω^-, respectively. The omega plus (ω^+) cells appeared to be preferentially recovered in polar crosses ([ω^+] x [ω^-]). In these so called *heterosexual crosses* the order of certain genes, single and double recombination and negative interference were observed. The exact mechanism of the recombination was not understood. The ω^+ factor turned out to be an intron within the 21S rRNA gene with a transposase-like function. Non-polar crosses have also been used to determine allelism. In the latter case all progeny was parental type and recombination was an indication of non-allelism. Linkage and recombination could be detected also in non-polar crosses but mapping was impractical. Complementation tests can be carried out in respiration deficient mutations on the basis that in "non-allelic" crosses respiration is almost completely restored within 5-8 hours whereas recombination may produce wild types only in 15-29 hours. Different loci are complementary and appear unlinked. Some loci may not complement each other.

Transformation of mitochondria requires the biolistic method. Usually co-transformation by nuclear and mitochondrial DNA sequences are employed. The nuclear gene (*URA3*) is used for selection of transformant cells. The ρ^- cells cannot be selected because they are defective in respiration. They are crossed then with mt⁻ cells that carry a lesion in the target gene and among the recombinants the transformants may be recovered. It is possible that the wild type *URA3* gene is transferred from the mitochondrion into the *ura3* mutant nucleus whereas the cytochrome oxidase gene (*cox2*) is maintained in the nucleoid.

The large subunit of the mitochondrial ribosomal RNA may house a mobile element, ω^+ This element (a Group I intron) may show polar transmission and can be used in a limited extent for recombination analysis as mentioned above. It encodes 235 amino acids and represents *Sce*I endonuclease. A *Chlamydomonas smithii* intron (from the *cytb* mitochondrial gene) can move into the same but intronless gene in *C. reinhardtii*. Several other Group I introns have similar mobility and the introns usually share the peptide LAGLI-DADG, a consensus conserved also in some maturases. Other Group I introns display the GIY-10/11aa-YIG pattern. The VAR-1 yeast gene (ca. 90% A+T) encoding the small mt ribosomal subunit is also an insertion element. Other mobile elements with characteristic G + C clusters are common features of

Mitochondrial genetics continued

these insertion elements. Group II introns present in various fungi are also mobile. Splicing/respiration deficient mutants become revertible if the defective intron is removed. Similar mobile elements can be found in the cpDNA of *Chlamydomonas* algae. Mitochondrial plasmids may be circular or linear. Some of the circular mt plasmids display structural similarities to Group I introns and their transcripts are reminiscent of reverse transcriptases (resembling protein encoded by Group II introns).

In yeast, nuclear genes *MDM1* (involved in the formation of non-tubulin cytoplasmic filaments), *MDM2* (encodes Δ9 fatty acid desaturase) and *MDM10* (determines mitochondrial budding) control the transmission of mitochondria to progeny cells. The subunit(s) of the mitochondrial transcriptase and the mtTFA transcription factor are coded in the yeast nucleus. Nuclear and mitochondrial promoters share cis elements. It seems that mitochondrial signals regulate nuclear genes controling mitochondrial functions. Partially reduced intermediates of oxygen metabolism, ROS, may cause mutations, oxidize proteins and may harm membranes. NADH dehydrogenase and coenzyme Q (ubiquinone) are hold responsible for the production of ROS. In the human diseases such as cancer, ischemic heart diseases, Parkinson disease, Alzheimer disease, and diabetes ROS products have been implicated. The mitochondrial ROS activates mammalian transcription factors NF-κB, AP and GLUT glucose transporters.

Mutation rates per mitochondrial D loop has been estimated to be 1/50 between mothers and offspring but other estimates used for evolutionary calculations are 1/300 per generation. (See physical mapping, mapping function, deletion mapping, rounds of matings, interference, petite colony mutations, mutations in cellular organelles, biolistic transformation, transformation of organelles, chloroplast genetics, mitochondria, mtDNA, mitochondrial import, introns, maturase, mitochondrial plasmids, mitochondrial diseases in humans, killer plasmids, senescence, MSF, mismatch repair, Parkinson disease, Alzheimer disease, ischemic, NF-κB, AP, GLUT, ROS, mitochondrial plasmids, mtTFA, bottleneck effects, heteroplasmy, Romanovs, sorting out, RNA editing, endosymbiont theory, RU maize, conplastic)

MITOCHONDRIAL HETEROSIS: it was claimed that the presence of different mitochondria may lead to complementation and increased vigor. (See hybrid vigor)

MITOCHONDRIAL IMPORT: transport into the mitochondria must pass two cooperating membrane layers. The outer membrane carries 4 receptors (Tom37, -70, -20, -22). The outer membrane import channel is built of at least six proteins; the inner membrane translocation system uses at least three proteins (Tom40, -6 -7). The inner membrane import channel proteins are Mas6p (MIM23), Sms1p (MIM 17) and Mpi1p (MIM44/ISP45). The mitochondrial heatshock 70 protein (Hsp70), the MIM44 complex and ATP play a central role in import. It appears that MIM44 first binds the incoming unfolded polypeptide chain as it is passing the entry site and it is then transfered to Hsp70 as the complex is dissociated by ATP. After this the polypeptide moves further and binds again to the complex and eventually traverses also the inner membrane. Although the mitochondria contain their own protein-synthetic machinery (ribosomes, tRNAs) the majority of the mitochondrial proteins is encoded by the nucleus and translated in, or imported from, the cytosol. The import system is also nuclearly encoded. The general import factors include the Hsp proteins, cyclophilin 20 and proteases. More specialized is the role of the imported assembly facilitator proteins. Yeast genes *SCO1, PET 117, PET191, PET100, OXA1, COX14, COX11, COX10* are such facilitators. The latter two are actually involved in heme biosynthesis. The COX10 product farnesylates protoheme b. (See mitochondria, mitochondrial genetics, mtDNA, chloroplast import, cyclophilin, heatshock proteins, farnesyl, heme)

MITOCHONDRIAL MAPPING: see mitochondrial genetics

MITOCHONDRIAL MYOPATHY: see mitochondrial disease in human

MITOCHONDRIAL PLASMIDS: circular or linear and occur in the mitochondria of some cytoplasmically male sterile lines of maize plants and relatives. Their sizes vary between 1.4 to 7.4 kb. The main types are S, R and D. S2, R2 and D2 are the same and R1 and D1 are apparently also identical with each other. S1 appears to have emerged as a recombinant between

Mitochondrial plasmids continued

R1 and R2. The *cms*-S plants carry the S1 and S2 plasmids. During the formation of S1 a terminal part of R1 (R*) was lost and it is inserted at two sites in the mtDNA. The S elements can be either integrated or free mitochondrial episomes. The S2 element encodes (*URF1*) a protein somewhat homologous to a viral RNA polymerase whereas another (*URF3*) appears to be homologous to a DNA polymerase. The *cms*-C and *cms*-T nucleoids are free from these plasmids whereas in the N-nucleoids the R1 and R2 sequences (from RU) are integrated. Another 2.3 kb plasmid (or a 2.15 kb derivative) is homologous to the tRNA^{Trp} and tRNA^{Pro} in the cpDNA, and also represents the only functional tRNA^{Trp} in the mitochondrion. (See cytoplasmic male sterility, mitochondrial genetics, mitochondria, tRNA, killer plasmids, senescence, episome, cytoplasmic male sterility, *Neurospora* mitochondrial plasmids)

MITOCHONDRION: the singular form of mitochondria. (See mitochondria, mitochondrial)

MITOCHONDRIOPATHIES: see mitochondrial diseases in humans

MITOGEN: collective name of substances stimulating mitosis and thus cell proliferation (such as growth factors).

MITOGENESIS: processes leading to cell proliferation.

MITOSIS: nuclear division leading to identical sets of chromosomes in the daughter cells. Mitosis

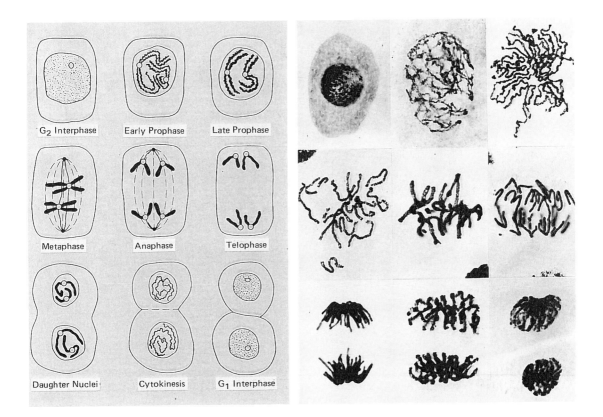

Left: THE MAJOR STEPS OF MITOSIS DIAGRAMED BY ONLY TWO CHROMOSOMES. Right: PHOTOMICROGRAPHS OF MITOSIS IN RYE, 2n = 14 (Courtesy of Dr. Gordon Kimber). Top Right to Left: INTERPHASE, EARLY PROPHASE, LATE PROPHASE, Middle; EARLY METAPHASE, METAPHASE, EARLY ANAPHASE, Bottom: TELOPHASE, LATE TELOPHASE, TWO DAUGHTER NUCLEI IN THE TWO PROGENY CELLS.

Mitosis continued
assures the genetic continuity of the ancestral cells in the daughter cells of the body. It involves one fully equational division. There is normally no (synapsis) pairing of the chromosomes. The centromeres split at metaphase separate at anaphase and the chromosomes relax after moving to the poles in telophase. The critical features of mitosis are diagramed above. Mitosis is different thus from meiosis where there is one reductional and one numerically equational division. During meiotic prophase the homologous chromosomes (bivalents) are synapsed and chiasmata may be observed. At the first meiotic division the centromeres do not split, and the undivided centromeres separate at anaphase I. Meiosis reduces the chromosome number to half, in contrast to mitosis which preserves the number of chromosomes in the daughter cells. A comparison between mitosis and meiosis is diagramed below. Mitosis and meiosis are the genetically most important processes of the eukaryotic cells (organisms). (See meiosis, cell cycle, nucleolus, nucleolar organizer, nucleolar reorganization, CENP)

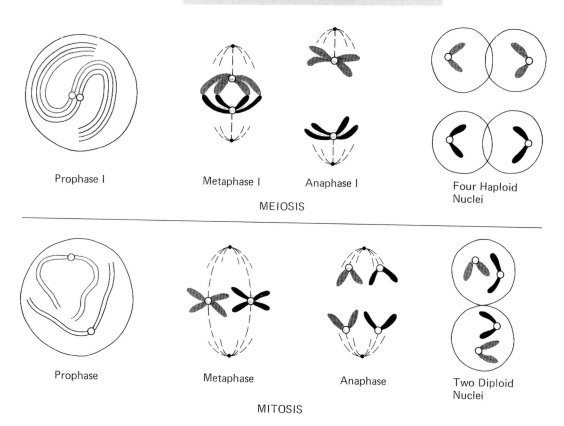

MITOSPORE: meiotic product of fungi ready for mitotic divisions.
MITOSTATIC: stopping or blocking the mitotic process. Many anti-cancer agents are mitostatic.
MITOTIC APPARATUS: the subcellular organelles involved in the nuclear divisions such as the spindle (microtubules), centromere (kinetochore), centriole, poles, CENP.
MITOTIC CENTER: see centrosome
MITOTIC CHROMOSOMES: are the chromosomes undergoing mitosis.
MITOTIC CROSSING OVER: recombination in somatic cells. Recombination is generally a

Mitotic crossing over continued

meiotic event (see crossing over) but in some organisms, the chromosomes may associate also during mitosis and this may be followed by genetic exchange of the linked markers. Although this may take place spontaneously, several agents capable of chromosomes breakage (radiation, chemicals) may enhance its frequency. For the detection of somatic recombination in higher eukaryotes, the markers must be cell and tissue autonomous, i.e. they should form sectors (spots) at the locations where such an event has taken place. In the first experiments with *Drosophila* the chromosomal construct diagramed below was used.

If exchange has taken place between *sn* and *y* then only a yellow sector (homozygous for *y*) was formed. *Twin spots* were observed only when in repulsion, the exchange took place between the proximal locus and the centromere. Somatic recombination takes place in the *Drosophila* male although this usually does not happen in meiosis. The characteristics of somatic recombination is that exchange is between two chromatids at the four-strand stage but instead of reductional division, as in meiosis, the centromeres are distributed equationally. Mitotic recombination is a relatively rare event in higher eukaryotes. Somatic recombination can be studied also in plants when the chromosomes are appropriately marked and from the sectors generative progeny can be isolated either from sectorial branches or by tissue culture techniques. In fungi (lower eukaryotes) where longer diploid phase exists (*Aspergillus*) mitotic recombination can be used even for chromosomal mapping because some of the parental and crossover products can selectively be isolated. The meiotic and mitotic maps are collinear but the relative recombination frequencies in the different intervals vary. Some research workers assumed that the mechanisms of the meiotic and mitotic recombination is the same. Several facts, however, cast some doubt about this view. In general, caution must be taken in interpreting genetic phenomena in somatic cells unless the assumptions can be confirmed also by classical progeny tests or by molecular data. (See diagram below, mitotic mapping, parasexual mechanisms, mitotic recombination as a bioassay in genetic toxicology)

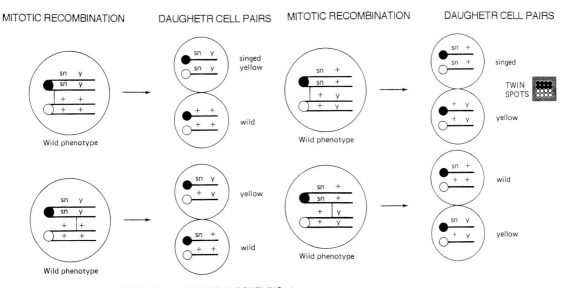

MITOTIC CROSSING OVER WITH MARKERS IN COUPLING ▲ AND REPULSION, AND THE CONSEQUENCES OF RECOMBINATION. (After the experiments of Stern, C. 1936 Genetics 21:625.

The wheat leaf shows white and dark green twin sectors on a pale green background because of nondisjunction in a heterozygote for the hemizygous ineffective Neatby's virescent gene.

MITOTIC INDEX: the fraction of cells in the process of mitosis at a particular time.

MITOTIC MAPPING: genetic recombination during mitosis is generally a very rare event in the majority of organisms. In some diploid fungi its frequency may reach 1 to 10 % that of meiosis. As a consequence of ionizing irradiation and certain chemicals the frequency of mitotic exchange may increase and on the basis of mitotic recombination genetic maps can be constructed. Although the order of genes are the same in mitotic and meiotic maps, the recombination frequencies may be quite different. (See mitotic crossing over, parasexual mechanisms, mitotic recombination as a bioassay in genetic toxicology)

MITOTIC NON-CONFORMITY: genetic alterations in fungi due to chromosomal translocations.

MITOTIC RECOMBINATION: see mitotic crossing over, mitotic mapping

MITOTIC RECOMBINATION AS A BIOASSAY IN GENETIC TOXICOLOGY: has been used in *Saccharomyces cerevisiae* yeast. In a diploid strain (D5) heterozygous for *ade2-40* and *ade2-119* alleles twin spots are visually detectable in case of mitotic crossing over. Homozygosity for *ade2-40* that has an absolute requirement for adenine involves formation of red color. Homozygosity for *ade2-119*, a leaky mutation, results in pink coloration. Since the two genes are complementary, the heterozygous cells do not require adenine and the colonies are white. Any environmental or chemical factor that promotes mitotic recombination is thus detectable. This procedure has not been extensively used in the genetic toxicology programs. (See mitotic crossing over)

MITOTIC SPINDLE: see spindle, spindle fibers

MITRAL PROLAPSE: is a buckling of the heart atrial leaflets as the heart contracts resulting in backward flow (regurgitation) of the blood. It is caused by an autosomal dominant allele and it is a common congenital heart anomaly affecting 4-8% of young adults, particularly females. It may accompany various other syndromes such as Marfan syndrome, Klinefelter syndrome, osteogenesis imperfecta, Ehlers-Danlos syndrome, fragile X syndrome, muscular dystrophies. In an autosomal recessive form ophtalmoplegia (paralysis of the eye muscles) is also involved (See heart disease and the other conditions named at separate entries)

MIXED-FUNCTION OXIDASES: generally flavoenzymes that oxidize NADH and NADPH and in the process may activate promutagens and procarcinogens. (See promutagen, procarcinogen, P-450, microsomes)

MIXOPLOID: the chromosome numbers in the different cells of the same organism vary.

MKP-1: dephosphorylates Thr^{183} and Tyr^{185} residues and thus regulates mitogen-activated MAP protein kinase involved in signal transduction. (See MAP, PAC)

Mle (maleless): *mle* is located in *Drosophila* chromosome 2-55.2, *mle3* in *Drosophila* chromosome 3.25.8, and similar genes are present also in the X-chromosomes. The common characteristics that homozygous females are viable but the homozygous males die. The underdeveloped imaginal discs of *mle3* may develop normally if transplanted into wild type larvae. MLE protein appears to be an RNA helicase and its amino and carboxyl termini may bind double-stranded RNA. Ribonuclease releases MLE from the chromosomes without affecting MSL-1 and -2 and RNA. MLE and MSL proteins appear to control dosage compensation. (See *Msl*, dosage compensation)

MLE (mariner-like element): transposable element in *Drosophila* that is present also in the human genome where it is responsible for a recombinational hot spot in chromosome 17. (See mariner, Charcot-Marie-Tooths disease, HNPP, MITE)

MLH (muscle enhancer factor, MLH2A): is a MADS box protein inducing muscle cell development in cooperation with basic helix-loop-helix proteins. (See MyoD, MADS box)

MLINK: is a computer program for linkage analysis.

MLL (mixed lineage leukemia): the protein encoded in human chromosome 11q23 has four homology domains, the A-T hook region, DNA methyltransferase homology, zinc fingers and the *Drosophila trithorax (trx*, 3-54.2) homology. (See leukemia, methyltransferase)

MLS (maximum likelihood lod score): see maximum likelihood, lod score, maximum likelihood method applied to recombination.

MLV: see Moloney mouse leukemia virus

MMP (microsatellite mutator phenotype): see microsatellite mutator

MMTV: murine mammary tumor virus

M-MuLV: Moloney murine leukemia virus.

MMTV: see mouse mammary tumor virus (MTV).

MN BLOOD GROUP (MN): the human chromosomal location is 4q28-q31. For the M blood type α-sialoglycoprotein (glycophorin A) and for the N type the δ peptide (glycophorin B) is responsible. Glycophorin deficient erythrocytes are resistant to *Plasmodium falciparum* (malaria). The En(a-) blood group variants are also lacking glycophorin A. The M and N alleles are closely linked to the S/s alleles and the complex is often mentioned as MNS blood group. The frequencies of the allelic combinations in England were MS (0.247172), Ms (0.283131, NS (0.08028) and Ns (0.389489). (See blood groups)

MNEMONS: Cuénot's historical (early 1900s) term for "genes".

MNNG: N-methyl-N'-nitro-N-nitrosoguanidine a monofunctional alkylating agent and a very potent mutagen and carcinogen (rapidly decomposing in light). It methylates O^6-position of guanine. Cell lines exist that are highly resistant to the cytotoxic effects of MNNG but they are even more sensitive to the mutagenic effects. (See mutagens, alkylating agents)

MNU (N-methyl-N-nitrosourea): is a monofunctional mutagen and carcinogen forming O^6-methylguanine. (See mutagens, alkylating agents, monofunctional)

MØ: see macrophage

MO15: is a CDK-activating kinase, related to CAK. Association of MO15 with cyclin H greatly increases its kinase activity toward Cdk2. (See cell cycle, cyclin, CDK, CAK, Cdk2)

MoAb: see monoclonal antibody

mob: bacterial gene facilitates the transfer of bacterial chromosome or plasmids into the recipient cell. In order to transfer the plasmids there is a need for the cis-acting *nick* site and a *bom* site. At the former the plasmid is opened up (nicked) and the *bacterial origin of mobilization* (*bom*) makes possible the conjugative transfer. (See Hfr, conjugation)

MOBILE GENETIC ELEMENTS: occur in practically all organisms and they represent different types of mechanisms and serve diverse purposes. Their general features are that they are capable of integration and excision from the genome (like the temperate bacteriophages), or move within the genome (like the insertion and transposable elements), fulfill general regulatory functions in normal cells (such as the switching of the mating type elements in yeast, phase variation in *Salmonella*), the generation of antibody diversity (by transposition of immunoglobulin genes in vertebrates), antigenic variation as a defense system in bacteria (*Borrelia*) and protozoa (*Trypanosomas*), parasitizing plant genomes by agrobacteria, etc. (See under separate entries, transposable elements, SDR, SINE, LINE, organelle sequence transfers, transposons, retroposons, retrotransposons)

MOBILIZATION: the binding of the ribosome to a mRNA initiates polysome formation; the process of conjugative tyransfer; also the release of a compound in the body for circulation.

MOBILIZATION OF PLASMIDS: transfer of conjugative plasmids to another cell. (See conjugation)

MODAL: adjective of mode. (See mode)

MODE: is the value of the variates (class of measurements) of a population that occurs at the highest frequency. (See mean, median)

MODEL: represents the essential features of a concept with minimum detail; model organism is a simple creature that shares the essential properties of some others and thus it facilitates understanding more complex systems.

MODELING: physical, mathematical and/or hypothetical construction for the exploration of the reality or mechanism of theoretical concepts. (See also simulation, Monte Carlo method)

MODEM (modulator/demodulator): links the computer to another computer through a telephone

Modem continued

line, e.g. fax modem (sends printed and graphic information) or e-mail modem sends and retrieves information from a mainframe computer to other computer operators through the information networks such as BITNET, INTERNET, and various online services.

MODIFICATION: most commonly in molecular biology means methylation. (See methylation of DNA, methylase)

MODIFIED BASES: occur primarily in tRNA and are formed mainly by post-transcriptional alterations. The most common modified nucleosides are ribothymidine, thiouridine, pseudouridine, isopentenyl adenosine, threonylcarbamoyl adenine, dihydrouridine, 7-methylguanosine, 3-methylcytidine, 5-methylcytidine, 6-methyladenosine, inosine, etc.

MODIFIER GENE: affects the expression of another gene. (See epistasis)

MODIFIED MENDELIAN RATIOS: are obtained when the product(s) of genes interact. These situations in case of two loci can be best represented by modified checkerboards using the zygotic rather than the gametic constitutions at the top and at the left side of the checkerboards. At the top left, a standard digenic situation is shown where for each locus the genotypes are $1AA$, $2Ab$, $1aa$ and $1BB$, $2Bb$, $1bb$, respectively, etc. Although in the boxes only the relative numbers of the phenotypes (genotypes) are shown (as a modification of the 9:3:3:1 Mendelian digenic ratio), their genetic constitution can be readily determined from the top left checkerboard. These schemes assume complete dominance. Additional variation in the phenotypic classes occurs in case of semidominance or codominance and the involvement or more than two alleic pairs. In common usage these modifications are mentioned as gene interactions but actually only the products of the genes do interact. (See Mendelian segregation, Punnett square, semidominance, codominance, phenotype, genotype)

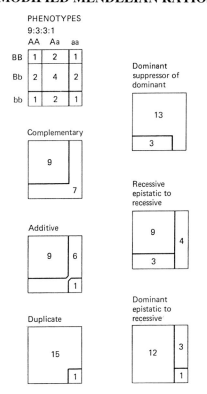

MODULATION: reversible alteration of a cellular function in response to intra- or extracellular factors.

MODY (maturity-onset diabetes of the young): are apparently dominant forms of familial diabetes with an onset at or after puberty but generally before age 25. Human chromosomal location is either 20q11.2 (MODY1) or 7p13 (MODY2) or 12q24 (MODY3) and it is a heterogeneous disease. MODYs are responsible for about 2-5% of the non-insulin-dependent diabetes. MODY1 is responsible for the coding of HNF-4α, a hepatocyte nuclear transcription factor, a member of steroid/thyroid hormone receptor family and an upstream regulator of HNF-1α. MODY2 encodes the glycolytic glycokinase which generates the signal for insulin secretion. MODY3 displays defects in the hepatocyte nuclear factor HNF-1a, a transcription factor which normally transactivates an insulin gene. (See diabetes mellitus, diabetes insipidus)

MOEBIUS SYNDROME: the major characteristic is facial paralysis of the sixth and seventh cranial nerves and often limb deformities and mental retardation. The dominant disorder was located to human chromosome 13q12.2-q13. The recurrence rate is below 1/50. (See neuromuscular diseases, mental retardation, periodic paralysis, limb defects)

MOHR SYNDROME: see orofacial digital syndrome II

MOHR-TRANEBJAERG SYNDROME (DDP): a recessive deafness encoded in Xq21.3-Xq22 involving poor muscle coordination, mental deterioration but without blindness. In the same chromosomal region there is another transcribed region with symptoms similar to DDP but with blindness, DFN-1. (See deafness)

MOI: see multiplicity of infection

MOLAR SOLUTION: 1 gram molecular weight compound dissolved in a final volume of 1 L.

MOLE: *Talpa europaea*, 2n = 34, an insect eating small underground mammal.

MOLE: see gram molecular weight; also fleshy (placental) neoplasia.

MOLECULAR BIOLOGY: the study of biological problems with physical and chemical techniques and interpretation of the functional phenomena on macromolecular bases.

MOLECULAR CHAPERONES: see chaperones, chaperonins

MOLECULAR CLOCK: see evolutionary clock

MOLECULAR CLONING: reproduction of multiple copies of DNA with the aid of a vector.

MOLECULAR COMPUTATION: could be called "reversed mathematics" because DNA or protein sequence information is used in an attempt to solve complex mathematical problems.

MOLECULAR DISEASE: the molecules involved have been identified, e.g., in sickle cell anemia a valine is replacing a glutamic acid residue in the hemoglobin β-chain. Since this first example numerous diseases have been explained in molecular terms.

MOLECULAR DRIVE: copies of redundant DNA sequences are rather well conserved in the genomes although one would have expected divergence by repeated mutations. The force behind this tendency for uniformity within species has been named molecular drive or concerted evolution. (See concerted evolution)

MOLECULAR EVOLUTION: studies the relationship of the structure and function of macromolecules (DNA, RNA, protein) among taxonomic groups. (See evolutionary distance, evolutionary tree, evolutionary clock, evolution of the genetic code, polymerase chain reaction. K_A/K_S, RNA world, origin of life)

MOLECULAR FARMING: use of transgenic animals or plants for the production of substances needed for the pharmaceutical industry or for other economic activities. (See transgenic)

MOLECULAR GENETICS: application of molecular biology to genetics; it is a somewhat unwarranted distinction because genetics as a basic science must always use the best integrated approaches.

MOLECULAR HYBRIDIZATION: annealing two different but complementary macromolecules (DNA with DNA, DNA with RNA, etc.). (See c_0t curve, probe)

MOLECULAR MARKERS: see RFLP, RAPD, VNTR, molecular weight, ladder

MOLECULAR MIMICS: may trigger autoimmune reaction because they have sequence homology to bacterial or viral pathogens. (See autoimmune diseases)

MOLECULAR MODELING: database and software information is available at <http://www.ncbi.nlm.nih.gov/Structure/>

MOLECULAR PLANT BREEDING: uses DNA markers to map agronomically desirable quantitative traits by employing the techniques of RFLP, RAPD, DAF, SCAR, SSCP, etc. The purpose is to incorporate advantageous traits into crops. (See separate entries, QTL)

MOLECULAR WEIGHT: the relative masses of the atoms of the elements are the atomic weights. The sum of the atomic weights of all atoms in a molecule determine the molecular weight of that molecule (MW). In the determination of the relative mass the mass of C^{12} is used that is approximately 12.01 (earlier the mass of the hydrogen was used, 1.08 but the calculations with it were more difficult). The relative molecular weight is abbreviated usually as M_r. The molecular weight of macromolecules is generally expressed in daltons, 1 Da = 1.661×10^{-24} gram. The molecular weights are determined by a variety of physico-chemical techniques. In gel electrophoresis of DNA fragments are compared with sequenced (known base number) restriction fragments of λ phage, generated by *HinD* III [125 to 23,130 bp], by *HinD* III - *Eco* RI double digests [12 to 21,226 bp], or *Eco* RI [3,530 to 21,226 bp] or pUC18 plasmid cleaved by *Sau* 3AI [36 to 955] or φX174 digested by *Hae* III [72 to 1,353] or commercially available synthetic *ladders* containing 100 bp incremental increases from 100

Molecular weight continued

to 1,600 or several other sizes, are most frequently used. For the large chromosome-size DNAs studied by pulsed-field gel electrophoresis, T7 (40 kb), T2 (166 kb), phage G (758 kb) or even larger constructs obtained by ligation are used. For protein electrophoresis bovine serum albumin (67,000 M_r), gamma globulin (53,000 and 25,000 M_r), ovalbumin (45,00 M_r), cytochrome C (12,400 M_r), and others can be employed.

MOLECULE: atoms covalently bound into a unit.

MOLONEY MOUSE LEUKEMIA ONCOGENE (MLV): integrates at several locations into the mouse genome; an integration site has been mapped to human chromosome 5p14. The Moloney leukemia virus-34 (Mov34) integration causes recessive lethal mutations in the mouse. Homolog to this locus is found in human chromosome 16q23-q24. The Mos oncogene encodes a protein serine/threonine kinase. It is a component of the cytostatic factor CSF and regulates MAPK. (See oncogenes, CSF, MAPK)

MOLONEY MOUSE SARCOMA VIRUS (MSV, MOS): the c-oncogene maps to the vicinity of the centromere of mouse chromosome 4. The human homolog MOS was assigned to chromosome 8q11-q12 in the vicinity of oncogene MYC (8-24). Break-point at human 8;21 translocations have been found to be associated with myeloblastic leukemia. It has been suspected that band 21q22, critical for the development of the Down syndrome, is responsible for the leukemia that frequently affects trisomic individuals for chromosome 21. The cellular mos protein (serine/threonine kinase) is required also for the meiotic maturation of frog oocytes. If the c-mos transcript is not polyadenylated maturation is prevented. Overexpression of c-mos causes precocious maturation, and after fertilization cleavage is prevented but disruption of the gene may lead to parthenogenetic development of mouse eggs. This protein is active primarily in the germline and may cause ovarian cysts and teratomas, it may cause oncogenic transformation also in somatic cells. (See cell cycle, teratoma, polyadenylation, leukemia, Down syndrome, c-oncogene)

MOLSCRIPT: a computer program for plotting protein structure. (See protein structure)

MOLTING: shedding the exoskeleton (shell) of insects during metamorphosis (developmental transition stages). (See *Drosophila*)

MOMENTS: are the expectations of different powers of a variable or its deviations from the mean, e.g., the first moment is $E(X) =$ the mean, the second moment is $E(X^2)$, the third $E(X^3)$. The first *moment about the mean*: $E(X - E[X])^2$, the second $E(X - E[X])^3$. The third moment about the mean (if it is not 0) indicates skewness, the fourth moment reveals kurtosis. The moment of the joint distribution of X and Y is the covariance. (See skewness, kurtosis, covariance, correlation)

MONGOLISM (mongoloid idiocy): now rejected name of Down's syndrome, human trisomy 21.

MONITOR: a video monitor receives and displays information directly received by a computer or a television monitor that accepts broadcast signals; monitoring: keeping track of something.

MONOALLIC EXPRESSION: in a diploid (or polysomic) cell or individual only one of the alleles is expressed such as the genes situated in one of the mammalian X chromosomes or in case of maternal or paternal imprinting. (See lyonization, imprinting)

MONOBRACHIAL CHROMOSOME: has only one arm. (See telocentric, chromosome morphology)

MONOCENTRIC CHROMOSOME: has a single centromere as it is most common.

MONOCHROMATIC LIGHT: literally light of a single color, practically light emission with a single peak within a very narrow wave-length band.

MONOCISTRONIC mRNA: codes for a single type of polypeptide. It is transcribed from a single separate cistron (not from an operon). Eukaryotic genes are usually monocistronic. The class I genes, 18, 5.8 and 28S rRNAs, including spacers, are transcribed as a single unit containing one of each gene's pre-rRNAs and are cleaved subsequently into mature rRNA. The 5S RNA genes are transcribed from separate promoters by polIII. (See ribosome)

MONOCLONAL ANTIBODY (Mab): is of a single type, produced by the descendants of one

Monoclonal antibody continued
cell and specific for a single type of antigen. MAbs are generated by injecting into mice purified antigens (immunization), than isolate spleen cells (splenocytes, lymphocytes) from the immunized animals and fuse these cells, in the presence of polyethylene glycol (a fusing facilitator), with bone marrow cancer cells (myelomas) that are deficient in thymidine kinase or (TK), or HGPRT. This process assures that the non-fused splenocytes rapidly senesce in culture and die. The unfused myeloma cells are also eliminated on HAT medium because they cannot synthesize nucleic acids, either by the *de novo* or by the salvage pathway. Some of the myelomas do not secrete their own immunoglobulins and thus hybrid immunoglobulins are not produced. Single hybridoma cells are cultured then in multi-well culture vessels and screened for the production of specific MAbs by the use of radioimmunoassays (RIA) or by enzyme-linked immunosorbent assay (ELISA). Most of the hybridomas will produce many types of cells and are of limited use, a few, however, may be more specific and these are rescreened for the specific type needed. E.g., animals immunized with melanoma cells produce HLA-DR (Human Leukocyte Antigen D-related) antibodies or melano-transferrin (a 95 kDa glycoprotein) or melanoma-associated chondroitin proteoglycan (heteropolysaccharides, glucosaminoglycans attached to extracellular proteins of the cartilage). Monoclonal antibodies have been used effectively for identification of tumor types in sera and histological assays. Some MAbs have been successfully applied for direct tumoricidal effect. Radioactively labeled monoclonal antibodies of Melanoma Associated Antigens (MAA) have been used for imaging (immunoscintigraphy) and detecting melanoma cells in the body when other clinical and laboratory methods failed. The problems of the "magic bullet" approach of combining specific monoclonal antibodies (prepared from individual cancer patients) against a specific cancer tissue with Yttrium^{90} isotope (emitting β rays of 2.24 MeV energy and 65 h half-life) or ricin (LD50 for mice by intravenous administration 3 ng/kg) or the deadly diphtheria toxin, or tumor necrosis protein (TNF) was expected to home on cancer cells and destroy them. This attractive scheme has not been proven successful in clinical trials so far. (See also abzymes). Another newer approach is to clone separately cDNAs of a variety of light and heavy chains of the antibody and allowing them to combine in all possible ways thus producing a combinatorial library of antibodies against all present and possible, emerging future epitopes. These antibodies can then be transformed into bacteria by using λ phage or filamentous phage such as M13. The phage plaques can then be screened with radioactive epitopes. The advantage of using filamentous phage is that they display the antibodies on their surface and permit screening in a liquid medium that enhances the efficiency by orders of magnitude. Although monoclonal antibodies did not completely fulfill all the (naive) therapeutic expectations, they still remain a power tool of biology. (See antibody, immunoglobulins, immune system, T cells, B cells, lymphocytes, somatic cell hybrids, Mab, TK, HGPRT, HAT medium, senescence, RIA, ELISA, hybridoma, melanoma, ricin, LD50, diphteria toxin, epitope, combinatorial library, quasi-monoclonal, bispecific monoclonal antibody)

MONOCOTYLEDONES: plants that form only one cotyledon such as the grasses (cereals).

MONOCYTES: mononuclear leukocytes that become macrophages when transported by the blood stream to lung and liver. (See microglia)

MONOECIOUS: separate male and female flowers on the same individual plant. (See autogamous, outcrossing, protandry, protogyny)

MONOFACTORIAL INHERITANCE: a single (dominant or recessive) gene (factor) determines the inheritance of a particular trait. (See Mendelian laws)

MONOFUNCTIONAL ALKYLATING AGENT: has only a single reactive group.

MONOGENIC HETEROSIS: same as overdominance, superdominance.

MONOGENIC INHERITANCE: the same as monofactorial inheritance

MONOGERM SEED: contains only a single embryo. The fruit of some plants (e.g. sugarbeet) is frequently used for propagation and usually it contains multiple seeds. This fruit may, however, be genetically modified to contain a single seed or mechanically fragmented to become monogerm. The agronomic advantage of the monogerm seed is that the emerging seedlings are not crowded and the labor-consuming thinning may be avoided or is at least facilitated.

MONOHYBRID: heterozygous for only one pair of alleles. (See also monofactorial inheritance)

MONOISODISOMIC: in wheat, 20"+i1", 2n=42, ["=disomic, i=isosomic]

MONOISOSOMIC: in wheat, 20"+i', 2n=41, ["=disomic, '=monosomic, i=isochromosome]

MONOISOTRISOMIC: in wheat, 20"+i2''', 2n=43, ["=disomic, i=isosomic, '''=trisomic]

MONOLAYER: non-cancerous animal cell cultures grow in a single layer in contact with a solid surface; single layer of lipid molecules. (See tissue culture, cell culture)

MONOMER: one unit of a molecule (which frequently has several in a complex); a subunit of a polymer (protein).

MONOMORPHIC LOCUS: in the population is represented by one type of allele.

MONOMORPHIC TRAIT: is represented by one phenotype in the population.

MONOOXYGENASES: introduce one atom of oxygen into a hydrogen donor, e.g. P450 cytochromes and have varied functions in cells in normal development and as detoxificants, and converting promutagens and procarcinogens into active compounds.

MONOPHYLETYIC: organisms have evolved from a single line of ancestry. (See also polyphyletic)

MONOPLOID: has only a single basic set of chromosomes. (See also haploid)

MONOPROTIC ACID: has only a single dissociable proton.

MONOSACHARIDE: a carbohydrate that is only one sugar of basic units $C_nH_{2n}O_n$ and thus can be diose, triose, pentose, hexose, etc., depending on the number of C atoms; they can be also aldose (e.g. glucose, ribose) or ketose sugars (e.g. fructose, deoxyribose)

MONOSODIUM GLUTAMATE ($HOOCCH(NH_2)CH_2CH_2COONa.H_2O$): used as a flavor enhancer (0.2-0.9%) in salted food or feed or for repressing the bitter taste of certain drugs or as a medication for hepatic coma. (See Chinese restaurant syndrome)

MONOSOMIC: the homologous chromosome(s) is represented only once in a cell or all cells of an individual. The gametes of diploids are monosomic for all chromosomes; a monosomic individual of an allopolyploid has 2n - 1 chromosomes. Medical cytogeneticists frequently call deletions as "partial monosomy" but this is a misnomer to be avoided because such a condition is hemizygosity for a particular locus or chromosomal region. Monosomics can produce both monosomic and nullisomic gametes. (See nullisomics, chromosome substitution, monosomic analysis)

MONOSOMIC ANALYSIS: is very efficient in Mendelian analysis of allopolyploids, such as hexaploid wheats. Monosomic individuals can produce monosomic and nullisomic gametes and their proportion is different in the males and in the females. The proportion also depends on the individuality of the particular chromosomes. On the average, the gametic output and the zygotic proportion of selfed monosomics of wheat is as shown in the body of the table below:

	SPERMS	
	MONOSOMIC (0.96)	NULLISOMIC (0.04)
EGGS ↓	ZYGOTES	
MONOSOMIC 0.25	DISOMIC ≈ **24%** (0.25 x 0.96)	MONOSOMIC ≈ **1%** (0.25 X 0.04)
NULLISOMIC 0.75	MONOSOMIC ≈ **72%** (0.75 x 0.96)	NULLISOMIC ≈ **3%** (0.75 x 0.04)

Very few nullisomic sperms are functional whereas the majority of the eggs are nullisomic because most of the monosomes (in the absence of a partner) remain and get lost in the metaphase plane and the eggs will receive only one representative from each chromosomes that have paired during prophase I. The monosomics can be used to assign genes to chromosomes and if the proper genetic constitution is used cytological test may not be required. On the basis

Monosomic analysis continued

```
    A                        a
  ————————      x        ————————
                              a
MONOSOMIC FEMALE         DISOMIC MALE
WITH DOMINANT ALLELE     WITH RECESSIVE ALLELE
```

In monosomic analysis the cross of the type, shown above, directly reveals the chromosomal location of the *a* gene in F_1.

of the Table above it is obvious that 75% of the F_1 will be of recessive phenotype if the genes are in the chromosome for what the female is monosomic. In case the recessive is in the monosomic female, only 3% will be recessive in the offspring.

(See nullisomy)

MONOSPERMY: fertilization by a single sperm.
MONOTELODISOMIC: in wheat, 20'+t2''', 2n=43, [''=disomic, t=telosomic, '''=trisomic]
MONOTELOMONOISOSOMIC: in wheat, 20''+t'+i', 2n=42, [''=disomic, '=monosomic, t=telosomic, i=isosomic].
MONOTELO-MONOISOTRISOMIC: in wheat, 20''+(t+i)1''', 2n=43, [''=disomic, '''=trisomic, t=telosomic, i=isosomic].
MONOTELOSOMIC: in wheat, 20''+t', 2n=41, [''=disomic, '=monosomic, t=telosomic].
MONOTOCOUS SPECIES: produces a single offspring by each gestation.
MONOTREME: belonging to the taxonomic order Monotremata, a primitive small mammalian group including the spiny anteater and the duck-billed platypus. They lay eggs and nurse their offspring within a small pouch through a nipple-less mammary gland. They show also other organizational similarities to marsupials. (See marsupials)
MONOTYPIC: a taxonomic category is represented by one subgroup, e.g. a monotypic genus includes a single species.
MONOZYGOTIC TWINS: develop from a single egg fertilized by a single sperm therefore they should be genetically identical, except mutations that occur subsequent to the separation of the zygote into two blastocytes after implantation into the wall of the uterus. Although MZ twins are generally identical genetically, their birthweight may be different because of differences in intrauteral nutrition due the path of blood circulation. Developmental malformations may affect only one of the MZ twins. The concordance in susceptibility to infectious diseases has been investigated but the information is not entirely unequivocal. Multifactorial diseases are more concordant in MZ twins than among dizygotic twins. (See twinning, dizygotic twins, heritability estimation in humans, zygosis, concordance, discordance, co-twin)
MONTE CARLO METHOD: is a computer-assisted randomization of large sets of tabulated numbers and can be used for testing against experimentally obtained data for determining whether their distribution is random or not. (See modeling, simulation)
MOORING SEQUENCE: is 11 nucleotide element anchored downstream to the base C to be RNA edited (CAAUUUGAUCAGUAUA). (See RNA editing)
MOOSE, North American: (*Alces alces*) 2n = 70.
MORALITY: the socially accepted principles and guidelines for the distinction of right from wrong in human behavior, usually based on customs and generally required by society to abide by. Some moral rules are present in all human societies, these rules may have, however, distinct variations. (See behavior genetics, ethology)
MORGAN: 100 units of recombination, 100 map units; usually the centiMorgan, 0.01 map unit is used. (See map unit)
MORPHACTINS: various plant growth regulators.
MORPHALLAXIS: see regeneration in animals, epimorphosis
MORPHINE: an opium type analgesic and addictive alkaloid; methylation converts it to the lower potency codein.
MORPHOALLELE: gene, involved in morphogenesis. (See morphogenesis)
MORPHOGEN: a compound that can affect differentiation and/or development, and can correct the morphogenetic pattern of a mutant that cannot produce it if it is supplied by an extract from

Morphogen continued

the wild type; the exact chemical nature of the morphogens is generally unknown, however some hormones and proteins have these properties. The morphogen is either a transcription factor or a type of transcriptional regulator. Their control involves concentration gradients over a threshold level. The activin signal (a member of the transforming growth factor-β family) spreads in a passive gradient about 300 μm or 10 cells diameter within a few hours in the vegetal cells of animals. (See also morphogenesis, vegetal pole, activin)

MORPHOGENESIS: is the process of development of form and structure of cells or tissues and eventually the entire body of an organism beginning from the zygote to embryonic and adult shape. Morphogenesis is mediated through morphogens in response to inner and outer factors. Morphogenesis takes place in three phases: determination, differentiation and development. Determination is a molecular change preparing the cell (or virus) to competence for differentiation. Differentiation is the realization of molecular and morphological structures that determine the differences among cells that are endowed with identical genetic potentials. The events of differentiation are coordinated in sequences of development. These steps generally occur in this order yet may be running in overlapping courses for different aspects of morphogenesis. The morphogenetic events vary in the different organisms because this is the basis of their identity yet some basic principles are common to all. The start point of morphogenesis is difficult to pinpoint because these events are running in cycles of the generations (what comes first the egg or the hen?). The life of an individual begins with fertilization of the egg by the sperm and the formation of the zygote. The zygote has all the genes that it will ever have (*totipotency*) yet many of these genes are not expressed at this stage. Also, there are *maternal effect* genes that control oogenesis and affect the zygote from outside without being expressed at this stage within the embryo's own gene repertory. Morphogenesis cannot be explained by one general set of theory, e.g. *gradients of morphogens* or the *signal relay* system. Apparently evidence for either one can be found in different morphogenetic pathways. (See also *Drosophila*, homeotic genes, morphogens, clonal analysis, developmental genetics, signal transduction, imaginal disks, oncogenes, morphogenesis in *Drosophila*, pattern formation, segregation asymmetric, organizer, morphogenetic furrow, development)

MORPHOGENESIS IN *DROSOPHILA*: has been studied the most extensively with genetic techniques. Oogenesis requires four cell divisions within the oocyst (see illustration at maternal effect genes) resulting in the formation of 16 cells, 1 oocyte and 15 nurse cells, the latter ones become polyploid and surrounded by a single layer of somatic (diploid) follicle cells. The nurse cells are in communication with the egg through cytoplasmic channels. Both the nurse cell genes (somatic maternal genes) and the egg (germline maternal genes) influence the fate of the zygote through morphogens. The oocyte itself is transcriptionally not active. These maternal effect genes determine the polarity of the zygote. The anterior - posterior gradients of the morphogens account for the future position of the head and tail, respectively. Genes *gurken* (*grk*, 2-30) and *torpedo* (2-10) play important role in anterior-posterior polarity and later during development also in the dorso-ventral determination. The dorso - ventral determination is responsible for the sites of the back and belly, respectively. The medio - lateral polarities are involved in the determination of the left and right sides of the body. The larvae and the adults develop from 12 compartments, one for the head, three for the thorax, and ten for the abdomen, formed already during the blastoderm stage of the embryo.

Mutation of the maternal genes mentioned in the box below usually cause recessive embryo lethality although the homozygous mothers are generally normal. The males are usually normal and fertile. The molecular bases of their action is known in a few cases. E.g., the amino terminal of the DL protein is homologous to the C-REL (avian reticuloendothelial viral oncogene homolog) protooncogene, that is present in human chromosome 2p13-2cen, and in mouse chromosome 11. The N terminus of *dorsal* is homologous to the product of gene *en* (*engrailed,* 2 -62) that is also expressed during gastrulation in stripe formation of the embryo. The carboxyl terminus of the *sna* gene product appears to contain five zinc-finger motifs

Morphogenesis in *Drosophila* continued
indicating a DNA binding mechanisms of transcription factors.

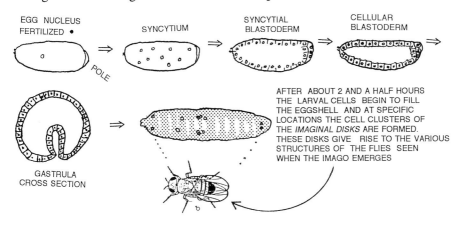

▲ EMBRYOGENESIS IIN *DROSOPHILA* ▲

ARISTA 1
ANTENNA 2
ORBITAL BRISTLES 3
OCELLUS 4
VIBRISSA 5
CARINA 7
PROBOSCIS 8
COMPOUND EYE 9
STERNOPLEURA 10
HUMERUS 11
HUMERAL BRISTLES 12
COXA 13
VERTICAL BRISTLES 14
POST-VERTICAL BRISTLES 15
PRESUTURAL BRISTLES 16
NOTOPLEURAL BRISTLES 17
SUPRA-ALARAL BRISTLES 18
DORSO-ALAR BRISTLES 19
POST-ALAR BRISTLES 20
SCUTELLAR BRISTLES 21
THORACIC SPIRACLES 22
MESOPLEURA 23
PTEROPLEURA 24
HYPOPLEURA 25
METANOTUM 26
WING 27
HALTERE 28

HUMERAL CROSSVEIN 29
COSTAL CELL 30
BASAL CELL 31
BASAL CELL 32
ALULA 33
AXILLARY CELL 34
ANAL CELL 35
MARGINAL CELL 36
LONGITUDINAL VEIN 37
SUBMARGINAL CELL 38
ANTERIOR CROSSVEIN 39
1st POSTERIOR CELL 40
DISTAL CELL 41
POSTERIOR CROSSVEIN 42
2nd POSTERIOR CELL 43
3rd POSTERIOR CELL 44

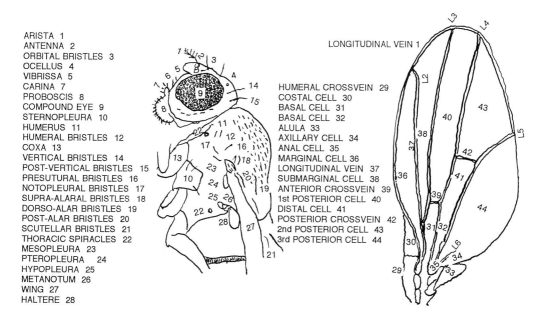

⇧ HEAD AND WING LANDMARKS OF THE *DROSOPHILA* BODY ⇧

> *Somatic maternal effect lethal genes*: in *Drosophila* {1} *tsl* (*torsolike*, chromosome 3-71) controls the anterior-posteriormost body structures (labrum, telson). Genes {2} *pip* (*pipe*, 3-47), {3} *dl* (*dorsal*, 2-52.9) eliminate ventral and lateral body elements, and their product is homologous to the *c-rel* proto-oncogene; a transcription factor homolog, NF-κB. {4} *ndl* (*nudel*, 3-17) exaggerates dorsal elements of the embryos, {5} *wbl* (*windbeutel*, 2-86) controls dorsal epidermis, and the {6} *sna* (*snail*, 2-51) strong alleles eliminate most mesodermal tissues. {7} The *gs* (*grand-childless*, 1-21), *gs(2)M* (chromosome 2) causes blockage of the embryos of normal-looking homozygous females before the blastoderm above 28.5 C°.

Morphogenesis in *Drosophila* continued

Maternal and germline genes (a very incomplete list):

{8} *ANTC* (*Antennapedia Complex*, 3-47.5) contains elements *lab* (*labial*), *pb* (*proboscipedia*) *Dfd* (*Deformed*), *Scr* (*Sex combs reduced*) and *Ant* affecting head structures in anterior-posterior relations. The *lab* and *pb* elements share cuticle protein genes. Elements (*ftz*) *fushi tarazu*, (*zen*) *zerknüllt* and *z* (*zen-2* [*zpr*]) affect segment numbers and (*bcd*) *bicoid* functions as a maternal effect gene eliminating anterior structures (head) and duplicating posterior elements (telsons). The product of *bcd* also binds RNA and acts as a translational suppressor of the *caudal* (*cad*) protein product. The *bcd* and *zen* genes are included in a 50 kb transcription unit called *Ama* (*Amalgam*). These latter four do not have homeotic functions. The segmental identity of the anterior thorax is regulated by *Scr* and the posterior part of the head is under the control of (*Scr*, *Dfd*, *pb* and *lab*). The entire complex encompasses 355 kb genomic DNA with multiple exons. Eight of the gene products are transcription factors and *Ant*, *Scr*, *Dfd*, *pb* and *lab* have homeotic functions. The *Antp* (*Antennapedia*) gene has both lethal and viable loss-of-function and gain-of-function recessive and dominant alleles controling structures anterior to the thorax and the thoracic segments. The first mutations observed converted the antennae of the adult into mesothoracic legs. The gene has two promoters and four transcripts that control differently the spatial expression, relying also on alternate splicing. The *bcd* gene is situated within *zen* and *Ama* is a maternal lethal and affects head and thorax development. The strong alleles in the females replace head and thorax of the embryos with duplicated telsons. Injection of *bcd*+ cytoplasm into the embryo (partially) remedies the topical alterations brought about by the mutant alleles. The RNA transcript is sticking to the anterior pole of the embryo and forming a steeply decreasing gradient in the posterior direction. This *bcd* gradient is regulated by genes *exu*, *swa*, and *stau* (see them below) and the gradient may be eliminated by mutations in these genes. In *bcd*- embryos the anterior activity of *hb* is eliminated and replaced by mirror image posterior *hb* stripes. The four exons are transcribed in either a long, complete RNA (2.6 kb) or a short one (1.6 kb) with exons 2 and 3 spliced out. The protein contains homologous tracts to the non-maternal effect genes *prd* (*paired*, 2-45, involved in the control of segmentation) and *opa* (*odd paired*, 3-48, that deletes alternate metasegments). Exon 3 contains the homeodomain with only 40% homology to other homeoboxes. The C-termini of the *bic* protein appear to be involved in transcription activation by binding to five high-affinity upstream sites of *hb* (TCTAATCCC). *Dfd* (*Deformed*) is a weak homeotic gene with recessive and dominant lethal alleles affecting the anterior ventral structures of the head and occasionally also thoracic bristles may appear on the dorsal part of the head. It is composed of five exons coding for a 586 residue protein. *ftz* (*fushi tarazu* [segment deficient in Japanese]) has both recessive late embryo lethal and dominant and viable regulatory alleles affecting genes in the *BXC* (*Bithorax complex*), *Ubx* (*Ultrabithorax*, 3-58.8), involved in the control of the posterior thorax and abdominal segments. The general characteristics of *ftz* is the pair-rule feature, i.e. in the mutants the even numbered abdominal and nerve cord segments are deleted (or fused). The striped pattern of the abdomen is controled within an one kb tract upstream of the beginning of transcription whereas a more distal upstream element regulates the central nervous system and an even more distal tract is required for the maintenance of the striped pattern. The homeobox is within the second of the two exons of this gene. The *ftz* ^{Rpl} mutations may transform the posterior halteres into posterior wing while the *ftz* ^{Ual} mutations convert patches of the first adult abdominal segment into a third abdominal segment like structure. (The latter two types are not embryonic lethal as the others). The *lab* (*labial*) mutations are embryonic lethal because of the failure of head structures. The protein product of the gene contains *opa* (*odd paired*, 3-48; deletes alternate metasegments) as well as a homeodomain although it does not display homeotic transformations. Gene *pb* (*proboscipedia*) may convert labial ("lip") portions into prothoracic leg structures or antennae. From the nine exons, #4 and #5 contain the homeobox and in exon

Morphogenesis in *Drosophila* continued

8 there are again *opa* sequences. The gene products (RNA and protein) are localized to the general area affected by the mutations. Null mutations in gene *Scr* (*Sex combs reduced*) are embryonic lethals. Homeotic transformations involve the labial and thoracic areas. Dominant mutation reduce the number of sex comb teeth. Gene *z2* (*zen-2*) has no detectable effects on development. Gene *zen* (*zerknüllt*) mutations may involve embryo lethality and the products may be required for post-embryonic development.

{9} *arm* (*armadillo*, 1-1.2): cell lethal at the imaginal disc stage because the posterior part of each segment is replaced by an anterior denticle belt. Transcripts have been found in all parts of the larva.

{10} *bcd* (*bicoid* : see *AntC*

{11} *bic* and *Bic* (*bicaudal*, 2-67, 2-52, 2-52.91) genes affect the anterior poles of the embryo by replacing these segments with posterior ones in opposite orientation. *Bic* apparently encodes a protein homologous to actin, part of the cytoskeletal system.

{12} *btd* (*buttonhead*, 1-31) mutations fail to differentiate the head.

{13} *BXC* (*Bithorax complex*, 3-58) is a cluster of genes that determine the morphogenetic fate of many of the thoracic and abdominal segments of the body. The second thoracic segment, which develops the second pair of legs and a pair of wings is the most basic part of the complex. The genetic map appears as follows:

abx bx Cbx **Ubx** bxd pbx   iab2 **abd-A** Hab iab3 iab4 Mcp iab5 iab6 iab7 **Abd-B** iab8 iab9

The entire complex is organized into three main integrated regions, *Ubx* (*Ultrabithorax*) is responsible for parasegments PS 5-6, *abd-A* (*abdominal-A*) defines the identity of PS7-13 and *Abd-B* (*Abdominal-B*) is expressed in PS10-14. In the *Ubx* domain *anterobithorax-bithorax* (*abx-bx*) region specifies PS5 and *bithorax-postbithorax* (*bxd-pbx*) defines PS6. In the *abd-A* region are *iab2, iab3* and *iab4* and in *Abd-B* are *iab5* to *iab9* elements. Mutations in *iab* (*infraabdominal*) tracts cause the homeotic transformation of an anterior segment to a more anterior abdominal (A) segment (e.g. A2→A1 or A3 [or more posterior ones]→A2, etc.) Mutations *abx* (*anterobithorax*) causes changes in thoracic (T) and abdominal (A) segments: T3 → T2, *bx* (*bithorax*): T3 → T2, *Cbx* (*Contrabithorax*): T2 → T3, *Ubx* (*Ultrabithorax*): A1+T3 → T2 and T2+T3→T1, *bxd* (*bithoraxoid*): A1 → T3, *pbx* (*postbithorax*): T3 → T2, *abd-A* (*abdominal-A*): A2 to A8 → A1, *Hab* (*Hyperabdominal*): A1+ T3 →A2, *Mcp* (*Miscadastral pigmentation*): A4 & A5 to an intermediate between A4 & A5, *Abd-B* (*Abdominal-B*): A5, A6, A7 may be weakly transformed into anterior forms. Most of these changes shown above involve only some structures in the segments but additional alterations may also occur.

{14} *cact* (*cactus*, 2-52) mutations reduce the dorsal elements and enhances ventral structures. This gene encodes a homolog of the Iκ-B protein, that forms a complex with product of *dl*, a NF-κB homolog transcription factor which is released from the complex upon phosphorylation of the *cact* product and after entering the cell nucleus it may participate in the activation of its cognate genes.

{15} *capu* (*cappucino*, 2-8) mutations may be lethal; causes somewhat similar alterations as *stau* in addition to making pointed appendages on the head.

(16) ci^D (*cubitus interruptus*-, Dominant, 4.0): the wing vein 4 is twice interrupted proximal and distal to anterior crossvein. In homozygotes the anterior portions of the denticle belts are duplicated in a mirror image manner in place of the posterior parts, and they are lethal.

{17} *dpp* (*decapentaplegic*, 2-4.0, [old name was *ho*]) is a complex locus with multiple developmental functions. The haploid-insufficiency *Hin*/+ condition is dominant embryonic lethal because of defects in gastrulation. The *Hin-Df* (deficiency), and *hin-r* (recessive) are also lethal. The *hin-emb* is an embryonic lethal mutation. It is complementary to the *shv* (*shortvein*) and *disk* (*imaginal disk*) region genes in the same complex. The *shv-lc* recessive larva-lethal mutants complement all *disk* alleles but mutant with *Hin* and *hin-r*. Mutants *shv-lnc* do not

Morphogensis in *Drosophila* continued

complement *disk*, *Hin* or *hin-r*, and *shv-p* alleles are viable and complement the disk mutants but not *Hin* or *hin-r*. Another mutation in the *shv* region, *Tg* (*Tegula*) causes the roof-like appearance of the wings. This mutation is complementary to all *dpp* genes. The *disk* group of mutations are either viable or lethal and may effect the eyes, wings, haltere, genitalia, head, imaginal disks, etc. The *dpp* gene apparently acts as regulator of mesodermal genes and it encodes a secreted protein of the TGF-β family transforming growth factor, involved in mammalian cancerous growth.

{18} *ea* (*easter*, 3-57) maternal effect lethal with loss- or gain-of-function effects on the dorsal, mesodermal or lateral structures, depending on the alleles. The mutants may be rescued by injection of normal cytoplasm.

{19} *ems* (*empty spiracles*, 3-53) the interior of the breathing orifices are partially missing; embryo lethal.

{20} *en* (*engrailed*, 2-62) some point mutations and chromosome breaks are viable. Others display "pair rule" defects, adjacent thoracic and abdominal segments fuse.

{21} *eo* (*extra organs*, 1-[66]) in the homozygous lethal embryos causes head defects and ventral hole.

{22} *eve* (*even skipped*, 2-58), lethal segmentation and head defects, "pair rule" effects. Its expression is reduced in *h* mutants and *en* segments do not appear. The expression of *Ubx* protein is high in the odd-numbered parasegments 7 to 13 rather than in every segment from 6 to 12; the *ftz* segments are disrupted.

{23} *exu* (*exuperentia*, 2-93) replaces the anterior part of the head with an inverted posterior midgut and anal pit (proctodeum).

{24} *fs(1)K10* (*female sterile*, 1-0.5) and a whole series of other *fs* genes in chromosome 1 may have both specific and overlapping effects. Expression may depend on cues from the oocyte. Eggs of homozygous females are rarely fertilized but if they are, the gastrulation is abnormal; the anterior ends are dorsalized.

{25} *fu* (*fused*, 1-59.5): veins L3 and L4 are fused beyond the anterior crossvein with the elimination of the latter. Heterozygous daughter of homozygous mothers have a temperature-sensitive segmentation problem that is not observed in reciprocal crosses.

{26} *ftz* (*fushi tarazu*) see *AntC* above (also in a separate entry)

{27} *gd* (*gastrulation defective*, 1-36.78) causes dorsal and ventral furrowing of the gastrula stage embryos.

{28} *gsb* (*goosberry*, 2-107.6) is homozygous lethal because the posterior part of segments are deleted and the anterior parts duplicated in mirror image fashion.

{29} *gt* (*giant*, 1-0.9) increased size larvae, pupae and imagos based on increased cell size. DNA metabolism is abnormal and both viable and lethal alleles are known affecting many ways the entire embryo. The protein product appears similar to that of *opa*.

{30} *h* (*hairy*, 3-26.5) displays extra microchaetae along the wing veins, membranes, scutellum and head. It is also a "pair rule" gene and affects the expression of *ftz*. It also regulates the expression of genes in the ASC (*achaete-scute* complex, 1.-0.0) involved in the control of hairs and bristles. The major gene products are located in the posterior and adjacent anterior parts of segment primordia.

{31} *hb* (*hunchback*, 3-48) alleles have different effects; the class I alleles among other effects lack thoracic and labial segments, class II mutants retain the prothoracic segment, class III alleles retain also the labial parts, class IV mutations prevent the formation of the mesothoracic segments, class V alleles also cause various gaps as well as segment transformations. These alleles are transcribed from two different promoters and produce up to five different transcripts. The products of this locus interacts with those of *kr* and *ftz* . Some of the *hb* alleles are activated by the product of *nos*.

{32} *hh* (*hedgehog*, 3-81): in homozygous embryos a posterior-ventral portion of each segment is

Morphogenesis in *Drosophila* continued

removed and the anterior denticle belt is substituted for in a mirror image. The embryos may not have demarcated segments. The gene has two activity peaks: during the first 3-6 hr and at 4-7 days of development (see more under separate entry).

{33} *kni* (*knirps*) [3-46]): these are lethal zygotic gap mutants. Its shorter transcript is expressed only until the blastoderm stage but the longer one is expressed even after gastrulation. The NH_2 end of the protein is homologous to one of the vertebrate nuclear hormone receptors. The *kni* box, a Zn-finger domain, is homologous to parts of the products of genes *knrl* (*knirps-related*, 3-[46]) and *egon* (*embryonic gonad*, 3-[47]).

{34} *Kr* (*Krüppel*, 2-107.6) mutants show gaps in thoracic and abdominal segments and other anomalies, and are lethal when homozygous. The protein product has similarity to the Zn-finger domain of transcription factor TFIIIA. It interacts with transcription factors TFIIB and TFIIEβ. In monomeric form it is an activator, in the dimeric form at high concentration it represses transcription. This protein binds to the AAGGGGTTAA motif upstream of *hb*. It also affects other maternal effect genes. The GLI oncogene is a homologue.

{35} *nkd* (*naked cuticle*, 3-47.3): denticle bands are partially missing; germ bands are shortened and thus lethal.

{36} *nos* (*nanos*, 3-66.2) is active in the pole cells and transport of its product in anterior direction is required for the normal abdominal pattern, and it is essential for the normal development of the germline. It is a maternal lethal gene. Its product represses that of *hb* in the posterior part of the embryo. Deficiency of both *hb* and *nos* is conducive, however, to normal development.

{37} *oc* (*ocelliless*, 1-23.1) some alleles are viable although the ocelli are eliminated, others (*otd*) involve lethality because of neuronal defects.

{38} *odd* (*odd skipped*, 2-8): embryonic lethal because posterior part of the denticle bands are replaced the anterior parts in mirror image fashion in T2, A1, A1, A3, A5 and A7.

{39} *opa* (*odd paired*, 3-48): alternate metasegments are genetically ablated. Denticle bands of T2, A1, A3, A5, A7 and naked cuticle of T3, A2, A4 and A6 are absent. The product of *en* is lost but that of *Ubx* increases in even-numbered parasegments.

{40} *osk* (*oskar*, 3-48): homozygous females and males are fertile but the embryos produced by homozygous females are defective in the pole cells and consequently also in abdominal segments; affects also *BicD* and *hb* expression.

{41} *phl* (*pole hole*, 1-0.5) blocks the formation of anterior - posterior end structures as well as the entire 8th abdominal segment. Phenotype is similar to that caused by *tso*. It is the *raf* oncogene in *Drosophila*.

{42} *pll* (*pelle*, 3-92) gene causes maternal embryo lethality by preventing the formation of ventral and lateral structural elements.

{43} *prd* (*paired*, 2-45): in strong mutants the anterior parts of T1, T3, A2, A4, A6, A8 and posterior parts of T2, AS1, A3, A5, A7 are absent.

{44} *run* (*runt*, 1-[65.8]; syn. *legless* [*leg*]): is a "pair rule" embryo lethal; eliminates the central mesothoracic and the uneven numbered abdominal denticle belts. The deletions are accompanied by duplication of the more anterior structures. The wild type allele appears to regulate the expression of *eve* and *ftz*..

{45} *slp* (*sloppy paired*, 2.8): parts of the naked cuticle are missing from T2, A1, A3, A5, A7 in an irregular way and it causes lethality.

{46} *sna* (*snail*, 2-51): embryonic lethals causes dorsalization and reducing or eliminating most of the mesodermal tissues. The C-terminus of the polypetide encoded has five Zn-finger motifs.

{47} *snk* (*snake*, 3-52.1) is a maternal lethal gene with dorsalizing effects. The encoded polypeptide contains elements homologous to serine proteases.

{48} *spi* (*spitz*, [*spire*], 2-54) is embryonic lethal blocking the development of anterior mesodermal tissues.

{49} *spz* (*spätzle*, 3-92): maternal lethal alleles accentuate dorsal structures.

Morphogenesis in *Drosophila* continued

{50} *stau* (*staufen*, 2-83.5) ablates pole cells and other anterior structures and some abdominal segments; causes the *grandchildless-knirps* syndrome.

{51} *swa* (*swallow*, 1-15.9) is expressed in the nurse cells and the product is transported to the oocyte and into the blastoderm until gastrulation leading to problems with nuclear divisions, head and abdominal defects. In *swa* homozygotes the *bcd+* products are disrupted.

{52} *tl* (*Toll*, 3-91) females heterozygous for the dominant or homozygous for the recessive *tl* mutant alleles produce lethal embryos that are defective in gastrulation and dorsal or ventral structures. The gene product is an integral membrane protein.

{53} *tll* (*tailles*, 3-102) deletes several posterior structures (Malpighian tubules, hindgut, telson) but brain and other anterior structures are also missing. Its expression is required for the manifestation of "pair rule" genes, *h* and *ftz* in the 7th abdominal segment and for site-specificities of *cad* (*caudal*, 2-[55] involved with head, thorax and abdominal structures and regulation of *ftz*), *hb* (see above) and *fkh* (*fork head*, 3-95, is involved with homeosis in both anterior and posterior structures of this non-maternal embryonic lethal). Gene *tll* has negative effects on genes *kni, Kr, ftz* and *tor*.

{54} *tor* (*torso*, 2-[57]) locus has both loss-of-function and gain-of-function alleles. The former type of alleles eliminate anterior-most head structures and segments posterior to the 7th abdominal segment. The latter type alleles are responsible for defects in the middle segments and enlargement of the most posterior parts of the body. The gene is expressed in the nurse cells, oocytes and early embryos. The expression of *ftz* is reduced in the gain-of-function *tor* mutants whereas *phl* mutations are epistatic to *tor*. With the exception of the NH_2 terminus, the protein is homologous to the growth factor receptor kinases of other organisms. It is concentrated in both pole cells and at the surface cells. Apparently, the product of this gene receives and transmits maternal information into the interior of the embryo.

{55} *trk* (*trunk*, 2-36) mutants lack anterior head structures as well as segments posterior to the 7th abdominal band.

{56} *Tub* (*tubulin* multigene families, scattered in chromosome 2 and 3). The *αTub* genes are responsible for the production of α-tubulin and are apparently active in the nurse cells, and the transcripts accumulate in the early embryos, and in the ovaries and control mitotic and meiotic spindle and cytoskeleton. The β-tubulin genes are expressed in the nurse cell, early embryos and various different structures and organs.

{57} *tud* (*tudor*, 2-[97]) the germline autonomous mutants display the "grandchildless-knirps" phenotype, lack pole cells yet about 30% of the embryos survive into sterile adults.

{58} *tuf* (*tufted*, 2-59): segment boundaries are duplicated in a mirror image and other parts deleted and neuronal pattern altered.

{59} *twi* (*twist*, 2-100) mutants are embryo lethals with defects in mesodermal differentiation. The embryos are twisted in the egg case. Mutations at the *dl, ea, pll* and *Tl* loci prevent the expression of *twi*. The *twi* polypeptides are homologous to DNA-binding myc proteins.

{60} *Ubx* (*Ultrabithorax*, 3-58.8) see *BXC*

{61} *vas* (*vasa*, 1-[64]): affects the pole region and segmentation (*grandchildless-knirp* syndrome). The protein product is homologous to murine peptide chain translation factor eIF-4A.

{62} *vls* (*valois*, 2-53): the phenotype is very similar to that caused by *vas*.

{63} *wg* (*wingless*, 2-30): visible viable and lethal alleles control segmentation pattern, imaginal disk pattern (wings and halteres). Its protein product is homologous to the mouse mammary oncogene *int-1* (INT1/Wnt). It appears that *frizzled* (*fz*) is the receptor of *wg*. The signal then may be transmitted to *Dsh* and *Arm* (β-catenin) and transcription of *En* is turned on with the assistance of other factors.

{64} *zen* (*zerknüllt*), a segment of *ANTC* {8} affecting segmentation and dorsal structrures of the early embryo

The maternal germline mutations are genetically identical in the unfertilized egg and their gene

Morphogenesis in *Drosophila* continued

products generally cause lethality in the egg or in the homozygous embryos. The heterozygous mothers are semi-sterile and the males are usually fairly normal in appearance and function. The molecular bases of a few such lethal genes is known. E.g., the *phl* gene appears to be homologous to the *v-raf* protooncogene and *phl* gene is the *Drosophila raf* gene that encodes a serine - threonine kinase protein also in humans and mice. The *snk* locus encodes a protein that appears to have a calcium binding site at the NH_2-end and with homology to several serine proteases at its C terminus. The product of the *ea* gene has some homology to an extracellular trypsin-like serine protease. Specific cytoplasm extracted from some (e.g., *osk, tor, ea, bcd*) unfertilized normal eggs or from normal embryos when injected into the mutant cytoplasm end may rescue the embryos that develop into sterile adults. The 923 amino acid protein encoded by *tor* has no homology to other known proteins in the NH_2-end but the rest of it is similar to a growth factor receptor tyrosine kinases and a hydrophobic segment appears to be associated with the cell surface membrane. The product of the *Tl* locus is an integral membrane protein with both cytoplasmic and extracytoplasmic domains containing 15 repeats of leucine-rich residues resembling yeast and human membrane proteins. The cytoplasmic domain is homologous to the interleukin-1 receptor (IL1R), the heterodimeric platelet glycoprotein 1b, coded in human chromosome 2q12 and mouse chromosome 1 near the centromere.

Although all of the mutants assigned to different chromosomal locations have different molecular functions, the phenotypic manifestation may not necessarily distinguishes that. The majority of the morphogenetic-developmental mutations are pleiotropic (e.g., affect the pole cells and cause the *grandchildless-fushi tarazu* syndrome). Many display epistasis, indicating the complex interactions of the regulatory processes involved (e.g., the expression of *twi* may be prevented by mutations at *dl, ea, pll and Tl*). The genes may be expressed at a particular position but their products form a diminishing gradient (e.g., *bcd, nos*). The DNA-binding protein encoded by *bcd*, depending on its quantitative level, regulates then qualitatively the transcription of e.g., *hb*. Some of the genes display the so called "gap" effect; they eliminate particular body segment, e.g., *kni, Kr, hb*. The so called "pair rule" genes may eliminate certain body segments and replace it with another (e.g. *ftz, eve*) or eliminate half of the segments and fuse them together in pairs (e.g., *en*). Several of the genes, particularly those in the huge *BTC* and *ATC* clusters may display homeotic effects. Although a great deal of information has been gathered on these genes primarily during the last decade, more specific knowledge is required for understanding the precise functions (especially the interacting circuits) of the morphogenetic processes. It appears that these genes are frequently, preferentially expressed at a particular position. Their mRNA or protein product is then spread in a gradient to the sites of the required action. In some instances only the RNA is spread and the protein is made locally. The position of morphogenetic function is controlled by a hierarchy of signals. These genes are expressed differently in different time frames and in coordinated sequence. The coordination is provided by the interaction of gene products. The same protein may turn on a set of genes and turn off others depending also on the local concentration of the products.

According to their **main** effects some of these genes may be classified into groups, others are more difficult to place in any of the groups because they all affect several stages and different structures. (The horizontal line stands for the embryo axis and the arrows or gaps illustrate the typical sites of action; the best representative genes are bracketed):

MATERNAL ANTERIOR GENES: →_____   1, 10 see [8] 23, 51,

MATERNAL POSTERIOR GENES: _____← 6, 11, [36], 40, 50, 57, 61, 62

MATERNAL END SEGMENT: ——— ← 1, 24, [54], 55,

MATERNAL DORSO-VENTRAL: ——↓↑— 2,[3] 4, 5, [14], 17, 18, 42, 46, 49, 52, 59, 63

ZYGOTIC GAP: __ ___ ____ __ ____ _   29, [31], [33], [34], 53

Morphogenesis in *Drosophila* continued

    ZYGOTIC PAIR RULE: — _ - _ — _ —   [22], 26, [30] , 38, 39, [14], 45

    ZYGOTIC SEGMENT POLARITY: —→←—→←—   9, 16, [20], 25, 28, 32, 35, 43, [63]

    HOMEOTIC GENES: genes within the *Antennapedia Complex* (8), and the *Bithorax Complex* (13)

The development of wing veins is controled by several genes. Locus *vn* (*vein*, 3-16.2, *Vein*, 3-19.6) disrupts longitudinal vein L4, posterior crossvein and sometimes L3, *ri* (*radius-incompletus* 3-46.8) interrupts L2, mutations in *px* (*plexus*, 2-100.5) produce extra veins. *N* (*Notch*, 1-3.0) complex (*Ax, Co, fa, l(1)N, N, nd, spl*) with a very large number of (dominant and recessive) alleles, are homo- and hemizygous lethal and remove small portions of the ends of the wings and affects also hairs and embryo morphogenesis, thickening of veins L3 and L5 and hypertrophied nervous system. *Notch* gene homologs are present also in vertebrates. The N complex codes for a protein with EGF-like repeats. The *Ax* (*Abruptex*) homozygotes reduce the length of longitudinal vein L5 and commonly L4, L2 and sometimes L3. The various *Ax* alleles are either positive or negative regulators of *Notch*. *Co* (*Confluens*) causes thickening of the veins. The *fa* alleles affect the eye facets and are non-complementing to the *spl* gene that also causes rough eyes and bristle anomalies. Gene *nd* (*notchoid facet*) is homozygous viable and displays some of the characteristics of the other genes within the complex. *E(spl)* (*Enhancer of split*, 3-89) mutation became known as exagerating the expression of *spl* (enhancer in this context does not correspond to the term enhancer as used in molecular biology). This gene produces 11 similar transcripts. They share homologies with *c-myc* oncogene (a helix-loop-helix) protein and also with the β-transducin G protein subunit, known to be involved in signal transduction. *Dl* (*Delta*, 3-66.2) causes thickening of the veins (and a number of other developmental defects) is responsible for a protein that has an extracellular element with nine repeats resembling the EGF, an apparent transmembrane and an intracellular domain with apparently five glycosylation residues. *Egfr* (*Epidermal growth factor receptor* [synonyms *top* {*Torpedo*}, *Elp*{*Ellipse*}, *fbl* {*faint little ball*}, 2-100) genes cause embryonic lethality and a number of other developmental effect including extra wing veins, eyes, etc. The genes *wg* (see above {63}) and *dpp* (see {17}) are involved in anterior-posterior specifications of the embryo and thus also in wing formation. Several other genes also affect wing and vein differentiation. Molecular evidence permits the assumptions that Dsh (dishevelled) a cytoplasmic protein is one of the receptors of the Wg protein signal. Dsh binds to N (Notch, a transmembrane protein) and inhibits its activity. When N binds to Delta it activates Su(H), suppressor of hairless (H). Su(H) then moves to the cell nucleus and operates as a transcription factor. The Wg signal can also activate the Shaggy-Zeste white (Sgg-Za3) serine- threonine kinase and the phosphorylation of Armadillo may lead Wg-dependent gene expression. The level of Armadillo may be elevated also by binding Wg to a member of the frizzled family (Dfz2). This *frizzled* (*fz*, 3-41.7) appears to be another receptor of Wg. Diffusible protein product of *optomotor-blind* (*omb*, 1-{7.5}) that was initially identified on the basis locomotor activity, is also required for the development of the distal parts of the wing within the *wg-dpp* system.

A general picture of neurogenesis is also emerging. (Modified after Campos-Ortega & Knust 1992, p. 347 in Development, Russo, V.E.A. *et al.*, eds, Springer Vlg., New York):

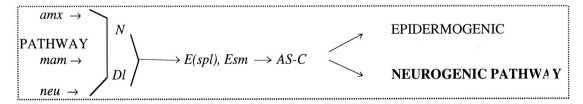

Morphogenesis in *Drosophila* continued

amx (*almondex*, 1-27.7) is a locus with multiple functions including eyes [some alleles complement the *lozenge* mutants]. Most relevantly, in the mutants there is hyperplasia of the central-peripheral nervous system and a concomitant reduction in epidermogenesis.

mam (*master mind*, 2-70.3) also affects the eyes but mutations lead to neural hyperplasia and epidermal hypoplasia.

neu (*neural*, 3-50) mutation also cause neural hyperplasia and epidermal hypoplasia.

N and *Dl* have been briefly described above. Both have protein products with epidermal growth factors (EGF)-like repeats. EGF has growth promoting signals and has ability to bind appropriate ligands.

E(spl): the wildtype alleles encode a protein with helix-loop-helix motifs, characteristic for transcription factors and displays similarities also to one subunit of the trimeric G proteins which have key roles in several signal transduction pathways.

ASC (*achaete-scute complex*, 1-0.0) controls sensilla and micro- and macrochaetae that are sensory organs of the flies and correspond to the peripheral nervous system. The ca. 100 kb region contains four major distinguishable areas *ac* (*achaeta*), *sc* (*scute*), *l(1)s* (*lethal scute*) and *ase* (*asense*). All four reduce and alter the pattern of the sensory organs. The dominant components, the *Hw* (*Hairy wing*) mutations increase hairynesss. Another regulatory mutation has been named *sis-b* (*sisterless b*). The complex includes nine transcription units. Four of them appear to be transcription factors because of the helix-loop-helix motifs.

svr (*silver*), *elav* (*embryonic lethal abnormal vision*), *vnd* (*ventral nervous system defective*) all are affected in the nervous system and located at 0 position of the X chromosome just like *ASC*. A large number of other genes at various locations are also involved in the nervous system. One must remember that even such a relatively simple organism as *Drosophila* 10,000 to 20,000 genes exist and their functions are too complex to be represented by simple models. (See also *Drosophila*, homeotic genes, morphogens, clonal analysis, developmental genetics, signal transduction, imaginal disks, oncogens, pattern formation, some genes are described with more details under separate entries)

MORPHOGENETIC FIELD: is an embryonal compartment capable of self-regulation.

MORPHOGENETIC FURROW: an embryonic tissue indentation marking the front line of the differentiation wave. Its progression is mediated by signal molecules. (See morphogenesis, morphogenesis in *Drosophila*, daughter of sevenless)

MORPHOLOGY: study of structure and forms.

MORPHOSIS: a phenocopy rather than being a mutation. (See phenocopy)

MORQUIO DISEASE: see gangliosidosis type I

MORTALITY: is the condition of being mortal, i.e. subject to death. The rate of mortality is computed as the average number of deaths per a particular (mid-year) population. An important factor of (particularly) infant death rate is the coefficient of inbreeding. First cousin marriages (inbreeding coefficient 1/16) approximately doubles infant mortality. The rate of human mortality calculated as the number of death per 1000 population may vary substantially in different parts of the world. In the years 1955-59, in the USA it was 9.4, in England 11.6, in Japan 7.8. In some other parts of the world it was double or higher. (See also inbreeding coefficient, age-specific birth and death rates, longevity, aging, juvenile mortality)

MORULA: a mass of blastomeres before implantation of the zygote onto the uterus.

Mos: see *mariner*

MOS (*mos*): see Moloney mouse sarcoma virus oncogene, MITE

MOSAIC: mixture of genetically different cells or tissues. Mosaicism is generally the result of somatic mutation, change in chromosome number, deletion or duplication of chromosomal segments, mitotic recombination, nondisjunction, sister chromatid exchange, sorting out of mitochondrial or plastid genetic elements, infectious heredity, intragenomic reorganization by the movement of transposons or insertion elements, lyonization, gene conversion, etc.

Mosaic continued

Introduction of foreign DNA into the cells by transformation or gene therapy may also be the cause of mosaic tissues. Somatic mutation may be of medical importance and may lead to oncogenic transformation and the expression of cancer. Somatic reversion of recessive genes causing disease, e.g. reversion of adenosine deaminase (ADA) or the tyrosinemia (FAH) genes, followed by selective proliferation of the normalized sector, may alleviate the disease. Loss of the extra chromosome in trisomics or nondisjunction in monosomics may restore the normal chromosome number. When a mutation is induced only in a single strand of the DNA, its expression may be delayed but in mice fur patches may occur in the heterozygotes and these animals are called *masked mosaics*. (See sex-chromosomal anomalies in humans, variegation)

MOSAIC GENES: contain exons and introns. (See introns, exons)

MOSOLOV MODEL: in the eukaryotic chromosome an extremely l ong DNA fiber is tightly packed into a very compact chromosome. One of the many existing models is shown here. Each line (←) represents an elementary chromosome fiber including the nucleosomal structure at various stages of folding. E: conceptualizes the chromomeres. The elementary fiber is about 25 Å in diameter and forms a tubular coil (solenoid). D: tightly coiled coils of the metaphase chromosomes. (See packing ratio, nucleosome, chromosome organization). (Diagram from Kushev, V.V. 1974 Mechanisms of Genetic Recombination. By permission of the Consultants Bureau, New York)

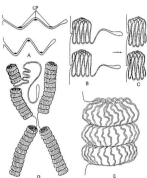

MOSQUITO: *Culex pipiens*, 2n = 6

MOTHEATEN: mutation in the SH2 domain of tyrosine phosphatase of mice leading to autoimmune anomalies. (See tyrosine phosphatase, autoimmune disease, SH2)

MOTIF: a structural domain or a sequence of amino acids or nucleotides present in different macromolecules. (See protein domains)

MOTOR PROTEINS: can move along filaments or macromolecules by deriving energy from the hydrolysis of energy-rich phosphates such as ATP; representatives include myosin, kinesin, dynein, helicases, bimC, etc. (See microtubule, anaphase)

MOUNTJACK (*Muntiacus muntjack*): the male is 2n= 7 and the female is 2n= 6 but the *Muntiacus reevesi* is 2n = 46.

MOUSE: of a computer is a pointer device.

MOUSE (*Mus musculus,* 2n = 40): rodent belonging to the subfamily Murinae, including about 300 species of mice and rats. They are extensively used for genetics and physiological studies because of the small size (25-40g), short lifecycle (10 weeks), gestation 19 days, 5-10 pups/litter and practically continuously breeding. The genome is 1.8×10^6 kDa in 2n = 40 chromosomes. In a m^2 laboratory space up to 3,000 individuals can be studied annually. Very detailed linkage information is available. According to the 1996 map the genetic length based 7,377 genetic markers including RFLP and other markers is 1,360.9 units. The average spacing between markers is now 400 kb. Transformation, gene targeting and other modern techniques of molecular genetic manipulations are well worked out. It is also very important that mouse is being used as a human genetic model for immunological, cancer and other human diseases. *Peromyscus* wild mice are 2n = 48. In mice Robertsonian translocations are common and therefore the chromosome number may vary in different populations and in different cultures (See databases, Encyclopedia of the Mouse Genome, Mouse Genome Database, Mouse Genome Informatics Group, Portable Dictionary of the Mouse Genome. Mapping information in *Nature* (Lond.) 380:149. Additional electronic information sources: <http://www.genome.wi.mit.edu> or by the single word 'help' to <genome_database @genome.wi.mit.edu>.

MOUSE GENOME: a news-letter by Journal Subscriptions Dept., Oxford Univ. Press, Walton Str., Oxford, OX2 6DP, UK

MOUSE GENOME DATABASE (MGD): integrates various types of information, mapping, molecular, phenotypes, etc. Contact: Mouse Genome Informatics, The Jackson Laboratory, 600 Main Str., Bar Harbor, ME 04609, USA, Phone: 207-288-3371, ext. 1900. Fax: 207-288-2516. INTERNET:<mgi-help@informatics.jax.org> for knockout: see TBASE in databases.

MOUSE GENOME INFORMATICS GROUP: an electronic bulletin board. Information: Mouse Genome Informatics User Support, Jackson Laboratory, 600 Main Str., Bar Harbor, ME 04609, USA. Phone: 207-288-3371, ext. 1900. Fax: 207-288-2516. INTERNET: <mgi-help@informatics.jax.org>.

MOUSE MAMMARY TUMOR VIRUS (MMTV): causes mammary adenocarcinomas. The virus is transmitted to the offspring by breast feeding. If the virus is transposed within 10 kb distance to the *Wnt-1* oncogene (homolog of the *Drosophila* locus *wingless* [*wg*]) the insertion may activate the oncogene because of the very strong enhancer in the viral terminal repeat. (See DNA-PK, pattern formation)

Mov34: see Moloney mouse leukemia virus oncogene

MOVABLE GENETIC ELEMENTS: see mobile genetic elements, transposons, transposable elements

MOVEMENT PROTEINS: their synthesis is directed by plant viruses in order to spread the infectious particles through the plasmodesmata with the aid of microtubules. (See microtubules, plasmodesma)

MOZ (monocytic leukemia zinc finger domain): see CREB, leukemia

M6P: mannose 6-phosphate. The M6P proteins are transmembrane proteins in the trans Golgi network. (See endocytosis, Golgi apparatus)

MPD: maximal permitted dose. (See radiation threshold, radiation hazards, radiation effects)

MPF: maturation protein factor, contains two subunits, a protein kinase (coded for by the $p34^{cdc2}$ gene) and B cyclin. Activation takes place (probably by phosphorylation at threonine 161 of the $p34^{cdc2}$ protein) during M phase and deactivation is mediated by degradation of cyclin subunit during the rest of the cell cycle. MPF is deactivated also by phosphorylation at Thr-14 and Tyr-15 amino acid residues; these sites are dephosphorylated probably by the product of gene $p80^{cdc25}$ or a homolog before the onset of mitosis. (See protein kinases, cyclin, signal transduction, $p34^{cdc2}$)

Mpl: is a regulator of megakaryocyte formation. (See megakaryocyte)

Mp1p: see mitochondrial import

MQM: marker-QTL-marker. (See QTL)

M_r (molecular weight): relative molecular mass of a molecule compared to that of the mass of a C^{12} carbon atom. This is different from the gram molecular weight, traditionally used in chemistry. (See also Dalton)

MR (Male recombination) factor of *Drosophila* (map location 2-54): is apparently a defective P element (see hybrid dysgenesis) that cannot move yet it can facilitate the movement of other P elements. Besides causing recombination in the male, it induces many of the symptoms of hybrid dysgenesis, including chromosomal aberrations, high mutation rate and mitotic exchange. (See male recombination, hybrid dysgenesis)

MRCA (most recent common ancestor): an evolutionary concept for divergence. (See evolutionary tree)

MRF: a member of the family of muscle proteins. (See MEF, myogenin, MYF5, MYOD)

mRNA: messenger RNA carries genetic information from DNA for the sequence of amino acids in protein. Its half-life in prokaryotes is about 2 min, in eukaryotes 6-24 hr or it may survive for decades in trees. In individual cells the life of the mRNA may vary by an order of magnitude. The unstable mRNAs even in eukaryotes may last only for a few minutes. The mRNA is produced by a DNA-dependent RNA polymerase enzyme, using the sense strand of the DNA as template and it is complementary to the template. The mRNA is derived from the primary transcript by processing, including the removal of introns (in eukaryotes). The size of

mRNA continued

the mRNA molecules varies a great deal because the size of the genes encoding the polypeptides is quite variable. Upstream of the coding sequences of the mRNA there are several regulatory sequences (G box, CAAT or AGGA box, etc.) which are important for transcription but they are not included in the mRNA. Transcription begins by the recognition and attachment of the transcriptase to a TATA box and the assembly of the transcription complex. The eukaryotic mRNA is capped with a methylated guanylic acid after transcription. Preceding the first amino acid codon (Met in eukaryotes and fMet prokaryotes) there is an untranslated leader sequence that helps the recognition of the ribosome. In prokaryotes the leader includes the Shine-Dalgarno sequence which assures a complementary sequence on the small ribosomal unit. In eukaryotes such a sequence is not known, however usually there is a $\boxed{\text{AG —CCAUGG}}$ prefered box around the first codon and the ribosomal attachment is relegated to a "scanning" for it in the leader. The structural genes then follow that in eukaryotes have been earlier spliced together from a highly variable number of exons. In the eukaryotic mRNA there are untranslated sequences also at the 3' end that include a polyadenylation signal (most commonly AAUAAA) to improve stability of the mRNA by post-transcriptional addition of over hundred adenylic residues. This signal is used also for discontinuing transcription. In prokaryotes either a rho protein-dependent palindrome or a rho-dependent GC-rich palindrome or a polyU sequence serves the same purpose. The number of mRNA molecules per cell varies according to the gene and the environment. In yeast cells, under good growing condition 7 mRNA molecules were detected with a half-life of about 11 minutes, indicating the formation of one mRNA molecule required about 140 seconds. The maximal transcription initiation rate for mRNA in yeast was found to be 6-8 second. (See introns, RNA polymerase II, transcription factors, transcription complex, open promoter complex, Hogness box, Pribnow box, UAS, up promoter, G-box, GC box, enhancer, leader sequence, cap, Shine-Dalgarno box, mRNA tail, transcription termination in eukaryotes, decapping, transcription termination in prokaryotes, regulation of gene activity, monocistronic mRNA, operon, regulon, mRNA degradation, aminoacyl-tRNA synthetase)

mRNA CAP: see cap

mRNA DEGRADATION: is not an incidental random process but it is under precise genetic control. The decay in eukaryotes may begin by shortening or removal of the poly(A) tail, that is followed by removal of the Cap and digestion by 5'→3' exoribonucleases (encoded by *XRN1*, *HKE1*). The decay by 3'→5' exonuclease is of minor importance. The poly(A) tail is removed by the PAN 3'→5' exonuclease but is requiring for activity the PABP (poly(A)-binding protein) and other proteins. The decapping is mediated by pyrophosphatases. The AU-rich elements (AURE) and other factors in the downstream regions regulate the decay process. Decay regulatory purine-rich elements (180-320 base) reside also within the coding region. The histone mRNAs lack poly(A) tail and their stability depends on 6-base double stranded stem and a 4- base loop, and their stability depends also on the 50-kDa SLBP (stem-loop-binding protein) and other similar proteins. The destabilization of these mRNAs depends on the process of translation, and probably on the association of the stem-and-loop proteins with the ribosomes, and it appears to be auto-regulated also by histone(s). The degradation of the mRNA is initiated by defects in the process of the translation or by encountering nonsense codons, wrong splicing, upstream open reading frames (uORFs), and transacting protein factors in the cytoplasm and the nucleus. Steroid hormones, growth factors, cytokines, calcium, iron also affect mRNA stability. Exo- and endoribonucleases, regulated by various factors may also be involved in processing and also in protecting mRNA. Viral infections may rapidly destabilize host mRNAs without affecting the rRNAs and tRNAs. (See mRNA, polyA mRNA, exonuclease, endonuclease, nonsense codon, histones, Cap, transcription termination, mRNA tail, URS, transcription factors)

mRNA LEADER: see leader sequence

mRNA TAIL: mature mRNAs of eukaryotes are generally tailed by ca. 200 adenylate residues.

mRNA tail continued

This is not the end of the primary transcript; transcription may continue by a thousand or more nucleotides beyond the end of the gene. Polyadenylation requires that the transcript be cut by an endonuclease and then a poly-A RNA polymerase attaches this poly-A tail which is probably required for stabilization of the mRNA. Several of the histone protein genes do not have, however, a poly-A tail. Other histone mRNAs, which are not involved with the mammalian cell cycle and histone mRNAs of yeast and *Tetrahymena,* are polyadenylated. The common post-transcriptional polyadenylation is signaled generally by the presence near the 3'-end an AATAAA sequence that is followed by two dozen bases downstream by a short GT-rich element. Polyadenylation takes place within the tract bound by these two elements. Within most genes tracts of AATAAA occur at more upstream locations but they are not used for poly-A tailing. Several genes may have alternative polyadenylation sites, however, and thus can be used for the translation of different molecules, e.g. for the membrane-bound or secreted immunoglobulin, respectively. The polyadenylation signal may also have a role in signaling the termination of transcription, no matter how much further downstream that takes place. In the nonpolyadenylated histone genes there is 6-base pair palindrome that forms a stem for a 4 base loop near the 3'-end of the mRNA and it is followed further downstream by a short polypurine sequence. The latter may pair with a U7 snRNP that facilitates termination. The U RNA transcripts of eukaryotic RNA polymerase II are not polyadenylated either. The formation of an appropriate 3'-tail requires that it would be transcribed from a proper U RNA promoter and the transcript would have the 5' trimethyl guanine cap. The 3'-end of U1 and U2 RNAs is formed by the signal sequence $GTTN_{0-3}AAAPU^{PUPU}_{PYPY}AGA$ [PU any purine, PY any pyrimidine] near the end. (See also transcription termination, decapping)

mRNAP: messenger ribonucleoprotein is a repressed mRNA.

MRP (mitochondrial RNA processing): is an RNase which cleaves the RNA transcribed on the H strand of mtDNA at the CSB elements. MRP designates frequently also mitochondrial ribosome proteins. (See CSB, DNA replication mitochondria, mtDNA, ribosomal RNA, ribosomal protein, ribonuclease P)

mrr : see methylation of DNA

MS2 PHAGE: a mainly single-stranded icosahedral RNA bacteriophage of about 3.6 kb with 4 genes, completely sequenced. (See bacteriophages)

MSAFP (maternal serum α-fetoprotein): analysis may detect prenatally chromosomal aneuploidy, (trisomy) and open neural-tube defects between the 15-20 weeks of pregnancy. The high level of this protein is generally accompanied also by oxidation products of estradiol and estrone (estriol) and chorionic gonadotropin. (See fetoprotein, gonadotropin, trisomy, Turner syndrome, Down syndrome)

msDNA (multicopy single-stranded DNA): is formed in bacteria which contain a reverse transcriptase and it is associated with this enzyme. This DNA has a length of 162 to 86 bases and it may be repeated a few hundred times. The 5' end of the msDNA is covalently linked by a 2'-5' phosphodiester bond to an internal G residue and thus forms a branched DNA-RNA copolymer of stem-and-loop structure. In some cases, because of processing, the msdRNA does not form a branched structure and it is only a single-stranded DNA. The 5' region of the msDNA is part of internal repeats within the copolymer. The system is transcribed from a single promoter and thus constitutes an operon. (See retron, reverse transcription)

msdRNA: see msDNA

MSF (mitochondrial import stimulating factor): it selectively binds mitochondrial precursor proteins and causes the hydrolysis of ATP. The MSF-bound precursor is made up of at least four proteins: Mas-20p,- 22p,-70p, -37p. (See mitochondria)

MSH: see melanocyte stimulating hormone

Msl (male-specific lethal): at least three of these genes exist that along with *Mle* (maleless) assure that a single X-chromosome in the male carries out all the functions at approximately the same level (dosage compensation) as two X-chromosomes in the female *Drosoophila. Msl-1,*

Ms1 continued

(2-53.3) and *Msl-2* (2-9.0) are located in the 2nd chromosome but similar genes are found also along the X-chromosome. Normally the male chromatin is highly enriched in a histone 4 monoacetylated at lysine-16 (H4Ac16). Mutation at this lysine alters the transcription of several genes. Mutation in *Msl* genes prevents the accumulation of H4Ac16 in the male X-chromosome. Msl-2 protein by containing a zinc-binding RING finger motif may specifically recognize X-chromosomal sequences and this way distinguishes between X and autosomes. (See dosage compensation, *Mle*, ring finger)

***Msp*I**: restriction endonuclease with recognition site C↓CGG.

Mss (mammalian suppressor of Sec4): is a guanosine-nucleotide exchange factor, regulating RAS GTPases. (See Sec4, Ypt1, Rab, GTPase, RAS)

MSV (*msv*): see Moloney mouse sarcoma virus

MTA α: yeast mating type α gene. (See mating type determination in yeast)

MTA *a*: yeast mating type *a* gene. (See mating type determination in yeast)

mtDNA (mitochondrial DNA): of animals consists of generally 5-6 small 16.5×10^3 bp mtDNA rings. The yeast mtDNA genome is circular and 17-101 kb. In *Paramecium* the mtDNA is linear and 40 kb. In plants it varies in the range of 200 to 2,500 kb, and occurs in variable size of mainly circular molecules. The human mtDNA contains 16,596 base pairs and it is similar to that of other mammals. The number of mtDNA rings in the eukaryotic cells may run into hundreds. The replication of the mouse mtDNA proceeds generally of 2 origins of replication and thus forming D loops. The buoyant density of the mtDNA is surprisingly uniform among many species; it varies between 1.705 to 1.707 g/mL. The heavy strand codes for 2 rRNAs, for 14 tRNAs and for about a dozen proteins (ATP synthase, cytochrome b, cytochrome oxidase, 7 subunits of NADH dehydrogenase). The plant mtDNA may encode 3 rRNAs and 15-20 tRNAs. The heavy chain of the mammalian mtDNA may be transcribed in a single polycistronic unit and subsequently cleaved into smaller functional units. There are also abundant shorter transcripts of the heavy chain of the mtDNA. The light DNA strands are transcribed into 8 tRNAs and into 1 NADH dehydrogenase subunit. The tRNA genes are scattered over the genome. The small mammalian mtDNA is almost entirely functional, and only a few (3 - 25) bases are between the genes without introns, and some genes overlap. The promoter sequences are very short. The upstream leader sequence is minimal and the typical eukaryotic cap is absent. In addition to the AUG initiator codon, AUA, AUU and AUC may start translation as methionine codons. The genetic code dictionary of mammalian and fungal (yeast) mitochondria further differs somewhat from the "universal code"; the UGA stop codon means tryptophan, the AUA isoleucine codon represent methionine in both groups, whereas the CUA leucine codon in yeast mtDNA spells threonine (but in *Neurospora* mtDNA it is still leucine) and the AGA and AGG arginine codons are stop codons in mammalian mtDNA. Other coding differences may still occur in other species. Some mtDNA genes have no stop codons at the end of the reading frame but a U or UA terminate the transcripts after processing that may become by polyadenylation a UAA stop signal. Since mitochondria have only 22 tRNAs, the anticodons must use an unusual wobbling mechanism. The two-codon-recognizing tRNAs can form also a G*U pairing and the four-amino-acid codon sets are base-paired either by two nucleotides only or the 5' terminal U of the anticodon is compatible with any other of the 3' bases of the codon. Nuclear mRNA may not be translated in the mitochondria because of the differences in coding. Mitochondrial translation does not use the Shine-Dalgarno sequence for ribosome recognition in contrast to prokaryotes. Rather it depends on translational activator proteins that connect the untranslated leader to the small ribosomal subunit. Plant mitochondria utilize the universal code.

The fission yeast and *Drosophila* mtDNAs are just slightly larger (19 kb) than that of mammals. The mtDNA of the budding yeast is about 80 kb yet its coding capacity is about the same as that of the mammalian mtDNA. The large mitochondrial DNA is rich in AT sequences and contain introns. Some yeast strains have the same gene in the long form (with introns)

mtDNA continued

and in other strains in short form (without intron). The introns may have maturase functions in processing the transcripts of the genes that harbor them (cytochrome b) or they may be active in processing the transcripts of other genes (e.g., cytochrome oxidase). The 1.1 kb intron of the 21S rDNA gene contains coding sequences for a site-specific endonuclease that facilitates its insertion into some genes lacking this intron as long as they contain the specific target site (5'-GATAACAG-3'). Thus, this intron is also an insertion element.

The *Saccharomyces* mtDNA genes are replicated from several (7 or more) scattered replicational origins, each containing 3 GC-rich segments separated by much longer AT tracts. Transcription is mediated by a mitochondrial RNA polymerase that is coded, however, within the nuclear DNA. The conserved 5'-$\overset{A}{T}$TTATAAGTAPuTA-3' promoter is positioned within 9 bases upstream from the transcription initiation site. Unlike the much smaller mammalian mtDNA genes, yeast genes have more or less usual upstream and 3' sequences. The much more compact mammalian or fission yeast genes do not use upstream regulatory sequences (UTRs). In the mtDNA a 13 residue sequence embedded in the $tRNA^{Leu(UUR)}$, a 34-kDa protein (mTERM) is bound and the complex is required for the termination of transcription. The rRNA genes, unlike in the nuclear genomes are separated by other coding sequences. Yeast mtDNA encodes at least one ribosomal protein; in mammals all mitochondrial ribosomal proteins are coded in the nucleus and imported into this organelle. The majority of the other proteins are also imported. Nuclear proteins mediate the translational control locus by specific or global manners.

The transport through the mitochondrial membrane takes place in several steps. The NH_2 terminus passes into the inner membrane through the protein import channel. The mitochondrial heatshock protein 70 (mtHsp70) stabilizes the translocation intermediate in an ATP-dependent process. The traffic may be in two ways, and it may be a passive transport.

The mtDNA of plants is much larger than that of other organisms, varying between 208 to 2,500 kbp. The mtDNA of *Arabidopsis* is 366,924 bp and contains 57 genes. The mtDNA of plants frequently exists in multiple size groups. This DNA is interspersed with large (several thousand kb) or smaller (200-300 kb) repeated sequences. Recombination between these direct repeats of a "master circle" may generate in stoichiometric proportion smaller, subgenomic DNA rings. In species without these repeats (e.g. *Brassica hirta*), subgenomic recombination does not occur. The very small mtDNA molecules are called plasmids. Plasmids, *S-1* and *S-2* share 1.4 kb termini. These plasmids may integrate into the main mtDNA.

The variety and the number of repeats may vary in the different species and some species (*Brassica hirta, Marchantia*) may be free of recombination repeats. The recombination repeats may contain one or more genes. The repeats display recombination within and between the mtDNA molecules. These events then generate chimeric sequences, deficiencies, duplications and a variety of rearrangements. Recombination between direct repeats tosses out DNA sequences between them and generate smaller circular and possibly linear molecules. The 570 kb maize mtDNA "master circles" may have 6 sets of repeats whereas in the T cytoplasm the repeats are quite complex. The size of the repeats may be 1 to 10 kb. Not all of them promote recombination and even the recombination repeats may vary in different species. Plant mitochondria may contain also single and double-stranded RNA plasmids. The latter may be up to 18 kb size. The single-stranded RNA plasmids may be replication intermediates of the double-stranded ones. Their base composition of some is different from that of the main mtDNA and appears to be of foreign (viral) origin.

It has been reported that passage of cells thorough *in vitro* culturing (tissue culture) may incite rearrangements in the mtDNA genome. This phenomenon may be the result, however, of different amplification of preexisting alterations. Formation of cybrids may also result in new combinations of the mtDNAs (see somaclonal variation). The mitochondrial protein complexes may be organized by chaperone-like mitochondrial proteases.

Some of the DNA sequences in the plant mitochondria are classified as "promiscuous" because they occur in the nuclei and in the chloroplasts too. These promiscuous DNA sequences may

mtDNA continued

contain tRNA (tRNA^{Pro}, tRNA^{Trp}) and 16S rRNA genes. Into the human nuclear DNA mtDNA has been inserted several times during evolution and is detectable mostly as pseudogenes. It has been estimated that mtDNA sequences are translocated into the nuclear chromosomes at the high frequency of 1×10^{-5} per cell generation. With the aid of biolistic transformation any type of DNA sequence can be incorporated into organellar genetic material, including mtDNA. The plant mitochondrial tRNAs may be encoded by mtDNA, nuclear DNA or plastid DNA directly or by plastid DNA inserted into the mtDNA. Plant mitochondria may code for some mitochondrial ribosome proteins. These ribosomal proteins may be different in different plant species. The liverwort mtDNA genome is very large and contains at least 94 genes.

Human mtDNA was also used for evolutionary studies to trace the origin of the present-day human population to a common female ancestor, the so called phylogenetic Eve. Restriction fragment length polymorphism carried out on the mtDNA of 147 people, representing African, Asian, Australian, Caucasian and New Guinean populations indicated an evolutionary tree by the maximal parsimony method. Accordingly, it appeared that Eve lived about 200,000 years ago in Africa because that population was the most homogeneous in modern times and other populations shared the most common mtDNA sequences with these samples. The advantage of the mtDNA for these studies was that mitochondria are transmitted through the egg and thus recombination with males would not alter its sequences. Later studies have shown, however that in mice (and probably in humans) a small number of mitochondria are transmitted also through the sperm. Therefore some of the conclusions of the original study regarding the time and population scales may require revision, the basic ideas about the human origin may be correct. *In vitro* in heteroplasmic cells the mutant mtDNA replication could be inhibited by peptide nucleic acid, complementary to the mutant sequences. (See mitochondria, introns, intron homing, spacers, kinetoplast, DNA replication mitochondrial, D loop, pan editing, RNA editing, mitochondrial import, cryptogene, mitochondrial disease, cytoplasmic male sterility, RFLP, evolution by base substitution, DNA fingerprinting, chondrome, plastid male transmission, paternal leakage, *Plasmodium*, Eve foremother, mitochondrial genetics, petite colony mutations, maximal parsimony, MSF, ancient DNA, DNA finger printing, heat shock proteins, chaperone, passive transport, heteroplasmy, peptide nucleic acid, mismatch repair, mtTFA, bottleneck effect, RU maize)

MTF (mouse tissue factor): a membrane protein initiating blood clotting.

mtTF: mitochondrial transcription factor (See transcription factors, transcription complex, mtDNA)

mtTFA (mitochondrial transcription factor): is nuclear gene product controling transcription in the mammalian mitochondria and possibly affecting transcription also in the nucleus. (See mtDNA, mitochondrial genetics)

MTHF: see photolyase

MTOC: microtubule organizing centers are formed early in the 16-cell stage within the cyst that gives rise to the primary oocyte and nurse cells. (See microtubule organizing center)

M-TROPIC (e.g. virus): homes on macrophages.

MTS1 (multiple tumor suppressor): gene encodes an inhibitor (p16) of the cyclin-dependent kinase 4 protein. When it is missing or inactivated by mutation, cell division may be out of control. MTS1 is implicated in melanoma, pancreatic adenocarcinoma. (See p16, melanoma, tumor suppressor, pancreatic adenocarcinoma)

MTT DYE REDUCTION ASSAY: measures cell viability by spectrophotometry.

Mu (Mutator): a transposable element system of maize increases the frequency of mutation of various loci by more than an order of magnitude. About 90% of the mutations induced carry an *Mu* element; their copy number in the genome may be 10-100. The element comes in various sizes but the longest ones are less than 2 kb. The shorter elements appear to be originated from the longer ones by internal deletions. There are relatively long (0.2) kb inverted repeats at the termini, and 9 bp direct repeats adjacent to them. The *Mu* elements appear to

Mu continued

transpose (in contrast to other maize transposable elements) by a replicative type of mechanism. The two best studied forms, *Mu1* (1.4 kb) and *Mu1.7* (1.745 kb) were identified also in circular extrachromosomal states. When *Mu* is completely methylated the mutations caused by it become stable, and less than complete methylation of the element and some other sequences (e.g., histone DNA) are associated with mutability. (See transposable elements, insertional mutation, controlling elements, hybrid dysgenesis)

Mu BACTERIOPHAGE: is a 37 kbp temperate bacteriophage. Its DNA is flanked by 5 bp direct repeats. Transcription of the phage genome during the lytic phase requires the *E. coli* RNA polymerase holoenzyme. During replication it may integrate into different target genes of the bacterial chromosome and thus may cause mutations (as the name indicates). Mu may integrate in more than one copy and if the orientation of the prophage is the same, it may cause deletion or if the orientation of the two Mu DNAs is in reverse, inversion may take place by recombination between the transposons. The phage carries three terminal elements at both left and right ends where recombination with the host DNA takes place. There is a transpositional enhancer (internal activation sequence) of about 100 bp at 950 bp from the left end. This left-enhancer-right complex is called LER. The 75 kDa transposase binds to the two ends and to the enhancer. The transposition is mediated by a complex of nucleoproteins, the transposome. In the presence of bacterial binding proteins (HU and IHF) a stable synaptic complex (SSC) is formed. Then at the 3' ends Mu is cleaved by the transposase and they form the Cleaved Donor Complex (CDC). Transesterification at the 3' OH places Mu into the host DNA. In the Strand Transfer Complex (STC) the 5' ends are still attached to the old flanking DNA but the 3'-ends are joined to the new target sequences and a cointegrate is generated by replication or nucleolytic cleavage separates Mu from the old flanks and then the gaps are repaired and the transposition is completed. (See mini Mu, transposons, mutator phage, cointegrate)

MUCOLIPIDOSES (ML): include a variety of recessive diseases connected to defects in lysomal enzymes. ML I is a neuroaminidase deficiency, ML II is an N-acetylglucosamine-1-phosphotoransferase deficiency (human chromosome 4q21-q23). This enzyme affects the targeting of several enzymes to the lysosomes. Congenital hip defects, chest abnormalities, hernia and overgrown gums but no excessive excretion of mucopolysaccharides are observed. ML III is apparently allelic to ML II although the mutant sites seem to be different. MP III (pseudo-Hurler polydystrophy) has also similarity to the Hurler syndrome (see mucopolysaccharidosis) although the basic defects are not identical. MP III A is a heparan sulfate sulfatase deficiency (see mucopolysaccharidosis, Sanfilippo syndrome A) whereas MP IIIB is basically the Sanfilippo syndrome B (see mucoplysaccharidosis). MP IV is a form of sialolipidosis with typical lamellar body inclusions in the endothelial cells that permit prenatal identification of this disease that may cause early death. (See mucopolysaccharidosis)

MUCOPOLYSACCHARIDOSIS (MPS): Hurler syndrome (*MPS I*) recessive 22pter-q11 deficiency of α-L-iduronidase (IDUA) resulting in stiff joints, regurgitation in the aorta, clouding of the cornea, etc. The IDUA protein (about 74 kDa) includes a 26-aminoacid signal peptide. The *MPS II (Hunter syndrome,* I-cell disease) is very similar to *MSP I* but it is located is human chromosome Xq27-q28. The symptoms in this iduronate sulfatase deficiency are somewhat milder and the clouding of the cornea is lacking. Dwarfism, distortion face, enlargement of the liver and spleen, deafness and excretion of chondroitin- and heparitin sulfate in the urine are additional characteristics. The following MPSs are controled by autosomal recessive genes and the deficiencies involve: *MPS IIIA* (Sanfilippo syndrome A; human chromosome 17q 25.3) heparan sulfate sulfatase, in *MPS IIIB* (Sanfilippo syndrome B, in human chromosome 17q21) N-acetyl-α-D-glucosaminidase, in *MPS IIIC* acetylCoA:α-glucosaminide-N-acetyl-transferase, *MPS IIID* (Sanfilippo syndrome D) N-acetylglucosamine-6-sulfate sulfatase, in *MPS IVA* (Marquio syndrome A) galactosamine-6-sulfatase, in *MPS IVB* (Marquio syndrome B) β-galactosidase, *in MPS VI* (Maroteaux-Lamy syndrome) N-acetylgalactosamine-4-sulfatase, in *MPS VII* (Sly syndrome) β-glucuronidase, in *MPS VIII* (DiFerrante syndrome) glucosamine-3-

6-sulfate sulfatase. (See mucolipidosis, iduronic acid, heparan sulfate, glucuronic acid, coronary heart disease, eye diseases, deafness)

MUCOSAL IMMUNITY: the mucosal membranes (in the gastrointestinal tract, nasal, respiratory passages and other surfaces that have lymphocytes) are supposed to trap 70% of the infectious agents. The mucosal cells rely on immunoglobulin A (IgA) rather than IgG used by the serum. In the gastrointestinal system are located the Peyer's patches that trap infectious agents and pass them to the antigen-presenting cells, T and B cells. The B cells generate the IgA antibodies. Concomitantly, the released antigen may stimulate the formation of IgG too. The mucosal immunity system can be triggered by providing attenuated forms of the pathogen (e.g. *Vibrio cholerae*) orally. Another approach is to deliver the antigens by bacteria that more effectively stimulate simultaneously both IgA and IgG production. Other delivery systems may use a biodegradable poly(DL-lactide-co-glycolide), PLG, or liposomes. The oral polio vaccine has been quite successful because it provides lasting protection but most of oral vaccines are short in this respect. (See Peyer's patch, antigen-presenting cell, T cell, B cell, immunoglobulins, antigen, immune system, vaccines)

Mud: a transposon modified (derived) from bacteriophage Mu. (See Mu bacteriophage)

MUIR-TORRE SYNDROME: is a familial autosomal dominant disease involving skin neoplasias apparently due to hereditary defects in the genetic (mismatch) repair system. (See mismatch repair, colorectal cancer, Gardner syndrome, polyposis hamartomatous, hereditary nonpolyposis colorectal cancer)

MULATTO: offspring of white and black parentage. (See miscegenation)

MULBERRY (*Morus* spp): fruit tree; x = 14. *M. alba* and *M. rubra* are 2n = 28 whereas the Asian *M. nigra* is 2n = 38.

MULBREY DWARFISM: see dwarfism [Mulbrey nanism]

MULE: see hinny

MULIBREY NANISM: an autosomal anomaly involving all or some of the characteristics: low birth weight, small liver, brain, eye, growth, triangular face, yellow eye dots, etc. (See stature in humans)

MULLER 5 TECHNIQUE: see *Basc*

MULLER'S RATCHET: genetic drift can lead to the accumulation of deleterious mutations, particularly in asexual populations. The expected time of losses of individuals with successive minimal number of mutations depends on the absolute number of individuals with minimal number of mutations: $N_m = q_m$ where q is their expected frequency and N = effective population size. Muller's ratchet may operate during the evolution of transposons and retroviruses and fix shorter sequences than the initial elements and establish elements that would depend on trans-acting elements for transposition. Conversely, the rapid mutations of retroposons may eliminate elements that would lose their transposase and thus can escape Muller's ratchet. (See also Y chromosome)

MÜLLERIAN DUCTS: gonadal cells begin to develop before the mouse embryo becomes two-weeks old. From the unspecialized primordial cells develop in the male the Wolffian ducts and in the female the Müllerian ducts. Initially, however both sexes develop both of these structures and appear bisexual but later, according to sex— one or the other— degenerates. The degeneration of the Müllerian ducts is caused by Müllerian inhibiting substance (MIS), a member TGF-β (transforming growth factor family of proteins) or by other name AMH (anti-Müllerian hormone). (See gonad, Wolffian ducts, SRY)

MULTICOMPONENT VIRUS: its genome is segmented, i.e. its genetic material is in several pieces similarly to the chromosomes in eukaryotic nuclei. (See bacteriophages)

MULTICOPY PLASMIDS: have several copies per cell. (See plasmid types).

MULTIDRUG RESISTANCE: is mediated by the multidrug transporter (MDT, 1,280 amino acids) phosphoglycoprotein that regulates the elimination (or uptake) of drugs from mammalian (cancer) cells in an ATP-dependent manner. In cancer cells the MDR gene is generally amplified. The MDR gene may be used for gene therapy by protecting the bone marrow from the effects of cytotoxic cancer drugs. The MDT gene controls a very broad base drug-resistance. (See amplification, multiple drug resistance)

MULTIFACTORIAL CROSS: the mating is between parents which differ at multiple gene loci.

MULTIFACTORIAL TRAIT: a quantitative trait based on control by several genes. (See polygenic inheritance, quantitative genetics, QT, complex inheritance)

MULTIFORKED CHROMOSOMES: in bacteria replication of the DNA may start again before the preceding cycles of DNA synthesis have been completed, and thus display multiple replication forks. (See replication fork, replication bidirectional, DNA replication)

MULTIGENE FAMILY: clusters of similar genes evolved through duplications and mutations and display structural and functional homologies. Some members of the families may be at different locations in the genome and some may be pseudogenes. (See pseudogenes)

MULTIGENIC: see polygenic inheritance

MULTILOCUS PROBE: in DNA fingerprinting simultaneously alleles of more than one locus are examined. (See DNA fingerprinting)

MULTIMAP: a computer program for linkage analysis using lod scores. (See lod score)

MULTIMERIC PROTEINS: have more than two polypeptide subunits.

MULTINOMIAL DISTRIBUTION: may be needed to predict the probability of proportions:

$P = \dfrac{n!}{r_1! r_2! ... r_z!} (a)^{r_1} (b)^{r_2} ... (z)^{r_z}$ where P is the probability $r_1, r_2 ... r_z$ stand for the expected numbers in case of a, b...z theoretical proportions and n = the total numbers. Example: in case of codominant inheritance and expected 0.25 AA, 0.50 AB, and 0.25 BB, the probability that in an AB x AB mating we would find in F_2 among 5 progeny 2 AA, 2AB and 1 BB is:

$P = \dfrac{5!}{2!2!1!}(0.25)^2 (0.50)^2 (0.25)^1 = \dfrac{120}{4}(0.0625)(0.25)(0.25) = 30(0.00391) \approx 0.117$. (See binomial distribution)

MULTIPARENTAL HYBRID: can be generated by fusion two or more different embryos of different parentage at the (generally) 4 - 8 cell stage and then reimplanted at (generally) the blastocyst stage into the uterus of pseudo-pregnant foster mothers. Such hybrids may appear chimeric if the parents were genetically different and can be used advantageously for study of development. (See allopheny, pseudo-pregnant)

MULTIPAROUS: an animal species that usually gives birth to multiple offspring by each delivery; a human female that gives birth to twins at least twice.

MULTIPATERNATE LITTER: is produced if the receptive multiparous female mates during estrus with several males. (See estrus, superfetation).

MULTIPLE ALLELES: more than two different alleles at a gene locus. The number of different combinations at a particular genetic locus can be determined by the formula $[n(n + 1)]/2$ where the number of different alleles at the locus is n. (See also allelic combinations)

MULTIPLE CLONING SITES (MCS): see polylinker

MULTIPLE CROSSOVERS: are genetically detectable only when the chromosomes are densely marked. In the absence of interference, the frequency of multiple crossovers is expected to be equal to the products of single crossovers. (See also coincidence, mapping, mapping function)

MULTIPLE DRUG RESISTANCE (MDR): is based on several mechanisms such as active detoxification system, improved DNA repair, altered target for the drugs, decreased uptake, increased efflux, inhibition of apoptosis. The MDR genes control multiple drug resistance in mammals. The 27 exon MDR1 locus (human chromosome 7q21.1) encodes a phosphoglycoprotein and controls drug transport and drug removal ('hydrophobic vacuum cleaner' effect). If tumors have high expression of this gene the prognosis for chemotherapy is not good. Upon drug treatment, the activity of the MDR protein may increase several fold. The MDR activity can be suppressed by some calcium channel blockers, some antibiotics, steroids, detergents, antisense oligonucleotides, anti-MDR ribozymes, etc. Non-P-glycoprotein-mediated MDR has also been restricted by MRP (a multiple-drug-resistance-related glycoprotein), encoded in human chromosome 16p13.1. (See cancer therapy, chemotherapy, multidrug resistance)

MULTIPLE ENDOCRINE NEOPLASIA (MEN): see endocrine neoplasia

MULTIPLE HAMARTOMA SYNDROME (MHAM): hamartomas are groups of proliferating, somewhat disorganized, mature cells — occurring under autosomal dominant control — on the skin, mucous membranes, gum but my be found as polyps in the colon or other intestines. The lesions may become malignant. The symptoms may be associated with a number of other defects. The Cowden disease was assigned to human chromosome 10q22-q23. Another form involving megalocephaly and epilepsy symptoms is the Lhermitte-Duclos disease. (See cancer)

MULTIPLE HIT: the mutagen causes mutation at more than one genomic site. (See kinetics)

MULTIPLE MYELOMA: is a bone marrow tumor causing anemia, decrease in immunoglobulin production and frequently accompanied by the secretion of the Bence-Jones proteins. Multiple myeloma may be genetically determined as it may appear in a familial manner. In some forms that have independent origin, within the same family or even within the same person, the protein may show some variations. It can be also induced in isogenic mice by injection of paraffin oil into the peritoneal cavity. Such animals may produce large quantities of the light chain immunoglobulin that can be subjected to molecular analysis. Such globulins are not necessarily monoclonal although from single transformation essentially similar molecules are expected. (See Bence-Jones proteins, immunoglobulin, monoclonal antibody)

MULTIPLE SCLEROSIS (MS): a disease caused by loss of myelin from the nerve sheath or even defects of the gray matter. It is called multiple because it frequently is a relapsing type of condition involving incoordination, weakness, abnormal touch sensation (paresthesia) expressed as a feeling of burning or prickling without an adequate cause. The exact genetic basis is not clear. It may be determined polygenically or it may be recessive; in both cases with much reduced penetrance. It had been suggested that it may be associated with defects of the HLA system but it has not been proven. There is molecular evidence that in some forms α-B-crystallin, a small heatshock protein is formed as an autoantigen. Viral initiation has also been considered. Among close relatives the incidence may be 20 times higher than in the general population. Females have about twice the risk as males. The prevalence in the USA is about 1/1,000 but it varies by geographical areas. The onset is between age 20 to 40. Some of the symptoms may be shared with other diseases, and conclusive identification requires laboratory analysis of myelin. Synonym: disseminated sclerosis. MS is under the control of several genes yet linkage to 5p14-p12 with a lod score of 3.4 was found. The Pelizaeus-Merzbacher disease, a late onset multiple sclerosis-like disorder with an autosomal dominant expression has been frequently confused with MS. Two principal regions in human chromosomes 17q22 and 6p21 (in the HLA area) control epistatically the susceptibility to MS. The concordance between monozygotic twins is about 30%. HLA DR2 carriers have a four fold increased relative risk. (See myelin, epilepsy, neuromuscular diseases, autoimmune disease, MHC, HLA, heatshock protein, Pelizaeus-Merzbacher syndrome, Addison-Schilder disease, HLA, concordance, twinning, lod score)

MULTIPLE TRANSLOCATIONS: see translocation complex

MULTIPLE TUMOR SUPPRESSOR: see MTS1

MULTIPLICATION RULE: see joint probability

MULTIPLICITY OF INFECTION: a single bacterial cell is infected by more than one phage particle. (See also double infection)

MULTIPLICITY REACTIVATION: when bacterial cells are infected with more than one phage particle, each, that have been inactivated by heavy doses of DNA-damaging mutagens, the progeny may contain viable viruses because replication and/or recombination restored functional DNA sequences. (See Weigle reactivation)

MULTIPOLAR SPINDLE: during mitosis more than two poles exist under exceptional conditions, e.g. aneuploidy caused by fertilization by more than one sperm. Such a condition may be the result of centrosome defects in animals. Multiple centrosomes may be formed in case the tumor suppressor gene p53 is not functioning. (See mitosis, centrosome, p53)

MULTI-REGIONAL ORIGIN: see out-of-Africa hypothesis

MULTISITE MUTATION: occurs generally when excessively large dose of a mutagen(s) is applied. Frequently deletions or other chromosomal aberrations are included. (See mutation, chromosomal aberration, deletion)

MULTISTRANDED CHROMOSOMES: contain more than two chromatids such as in the polytenic chromosomes produced by repeated replication without separation of the newly formed strands. At the early period of electronmicroscopic studies of ordinary chromosomes apparent multiple strands were observed, and it was assumed that each chromosome has many parallel strands of DNA. This assumption could not be validated by subsequent investigations. Each chromatid has only a single DNA double helix that is folded to assure proper packaging. (See packing ratio, polytenic chromosomes)

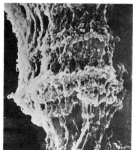

SCANNING ELECTRONMICROGRAPH OF A SEGEMENT OF POLYTENIC SALIVARY GLAND CHROMOSOME. (Courtesy of Dr. Tom Brady)

MULTIVALENT: association of more than two chromosomes in meiosis I. (See meiosis)

MULTIVARIATE ANALYSIS: has many uses in genetics when decision is needed in classifying syndromes with overlapping symptoms and the diagnosis requires quantitation. It may be also used to classify populations with similar traits. The assumption is that the variates $X_1...X_k$ are distributed according to a multivariate normal distribution. The variance of X_i, σ_{ii} and the covariance of X_i and X_j presumed to be identical in the two populations but σ_{ii} or σ_{ij} are not the same from one variate to another or one pair of variates to another pair, respectively. The difference between the two means for X_i is $\delta_I = \mu_{2i} - \mu_{1i}$. Hence the *linear discrimination function* $\Sigma L_i X_i$ provides the lowest probability of incorrect classification. The L_i values will be determined according to the procedure shown below. The δ/σ is maximalized for minimization of the chance of misclassification. The *generalized squared distance* is $\Delta^2 = (\Sigma L_i \delta_i)^2 / \Sigma\Sigma L_i L_j \sigma_{ij}$ and L_i is obtained by a set of equations: $\sigma_{11} L_1 + \sigma_{12} L_2 ... + \sigma_{1k} L_k = \delta_1 \quad \sigma_{k1} L_1 + \sigma_{k2} L_2 + ... + \sigma_{kk} L_k = \delta_k$. (See discriminant function, covariance [look up at correlation]

MUMMIES: dried animal or human bodies, preserved by chemicals and/or desiccation. They may contain protein (blood antigens), and DNA sequences to carry out limited molecular analysis using PCR technology. (See ice man, ancient DNA, ancient organisms)

MUNC18: a mammalian homolog of the Unc18 protein of *Caenorhabditis* and binds syntaxin. (See syntaxin)

MURASHIGE & SKOOG MEDIUM (MS1): for plant tissue culture is suitable for growing callus and different plant organs. Composition mg/L: NH_4NO_3 1650, KNO_3 1900, $CaCl_2.2H_2O$ 440, $MgSO_4.7H_2O$ 370, KH_2PO_4 170, KI 0.83, H_3BO_3 6.2, $MnSO_4.4H_2O$ 22.3, $ZnSO_4.7H_2O$ 8.6, $Na_2MoO_4.2H_2O$ 0.25, $CuSO_4.5H_2O$ 0.02, $CoCl_2.6H_2O$ 0.025, Ferric-EDTA 43, sucrose 3%, pH 5.7, inositol 100, nicotinic acid 0.5, pyridoxine.HCl 0.5, thiamin.HCl 0.4, indoleacetic acid (IAA) 1 - 30, kinetin 0.04 - 10. The microelements, vitamins and hormones may be prepared in a stock solution and added before use. For kinetin other cytokinins may be substituted such as 6-benzylamino purine (BAP) or isopentenyl adenine (or its nucleoside), for IAA, naphthalene acetic acid (NAA) or 2,4-D (dichlorophenoxy acetic acid) may be substituted or a combination of the hormones may be used in concentrations that is best suited for the plant and the purpose of the culture. For solid media use agar or gellan gum. Heat labile components are sterilized by filtering through 0.45 μm syringe filters. Variations of this medium are commercially available as a dry powder ready to dissolve but the pH needs to be adjusted. (See agar, gellan gum, syringe filter, Gamborg medium)

MURINE: pertaining to mice (or rats).

MUSCARINIC ACETYLCHOLINE RECEPTORS: are activated by the fungal alkaloid, muscarine, a highly toxic substance, causing excessive salivation (ptyalism, sialorrhea), lacrimation (shedding tears), nausea, vomiting, diarrhea, lower than 60 pulse rate, convulsions, etc. Antidote: atropine sulfate. (See signal transduction)

MUSCULAR ATROPHY: see Werdnig-Hoffmann disease, Kennedy disease, Charcot-Marie-Tooth disease, olivopontocerebellar atrophy, dentatorubral-pallidoluysian atrophy, Machado-Joseph syndrome, spinal muscular atrophy.

MUSCULAR DYSTROPHY: is a collection of anomalies involving primarily the muscle, controled by X and autosomal recessive or dominant genes. The most severe form is the (DMD) *Duchenne muscular dystrophy* in human chromosome Xp21. Its transmission is through the females because the males affected do not reach the reproductive stage. Similar defects were observed also in animals. The prevalence is about 3×10^{-4}. The frequency of female carriers is about twice as high as the affliction of males. The onset is around age 3 and within a few years the affected persons fail to walk and usually die by age 20. Mental retardation is common in this disease. The earliest diagnosis may be made by an abnormally high serum creatine phosphokinase (CPK) level at birth. Prenatal diagnosis is feasible and successful in about 90% of the cases. The dystrophin gene is one of the largest human gene involving about 2 megabase (almost half of the size of the entire genome of *E. coli*) may be involved. This gene includes about 60 introns with an average size of about 16 kb whereas the average exon is only about 50 kb. The gene in the muscles appears to use a promoter other than in the brain. The majority of the cases involve deletions and other chromosomal defects yet gene mutations are common too because of the large size of the gene. Intragenic recombination frequency was estimated to be 0.12 high. In DMD the dystrophin protein may be entirely missing whereas in the milder form of the diseases, the *Becker type* (BMD), the dystrophin protein is just shorter. The frequency of BMD is about 0.1 of that of DMD, and the patients may live until 35 and thus may have children. BMD and DMD are apparently coded by allelic genes. The *Emery-Dreyfuss dystrophy* affecting the shoulder muscles (scapulohumeral dystrophy) is also X-linked but at another location (Xq28). In the *limb-girdle muscular dystrophy* (LGMD) the defects are in the sarcoglycan subunits; α-sarcoglycan is encoded in chromosome 17q12-q21 the β subunit is coded in chromosome 4q12 whereas the γ subunit was localized to 13q12. Chromosomes 2p, 5q22.3-31.3 and 15q15 are also encoding LGMD subunits. The symptoms as the muscular dystrophies display considerable variation in onset and the severity of expression. The face and shoulder (fascioscapulohumeral) dystrophy is limited to the named body parts. The gene was assigned to human chromosome 4q35. The congenital dystrophy gene causes deficiency in the laminin, α2 chain (merosin) around the muscle fibers, was located to human chromosome 6q22-q23. Other types of muscular dystrophies may accompany symptoms of other human diseases. Dystrophin is a rod-like cytoskeletal protein, normally localized at the inner surface of the sarcolemma (the muscle fiber envelope). Dystrophin is attached within the cytoplasm to actin filaments and also to the membrane-passing β subunit of the dystroglycan (DG) protein and the α subunit of DG joins the basal lamina through the laminin protein (encoded in chromosome 6q22-q23). In the membrane, DG is associated with the β and γ subunits of sarcoglycan (SG); the α sarcoglycan is extracellularly attached to the other two subunits. In some LGMD patients there is also a defect in the muscle-specific protease calpain-3, encoded in chromosome 15q15. [The LGMD disease is also called severe childhood autosomal recessive muscular dystrophy or SCARMD]. In some of the SCARMDs the defect was associated with adhalin, a 50kDa sarcolemmal dystrophin-associated glycoprotein, encoded in human chromosome 17q. (See myotonic dystrophy, Charcot-Marie-Tooth disease, neuromuscular diseases, muscular atrophy, neuropathy, lamin, RFLP, apoptosis, emerin)

MUSHROOM BODY: is a central nerve complex in the brain.

MUSICAL TALENT, inheritance of: see Bach, dysmelodia

MUSTARD GAS (dichloroethyl sulfide, $[ClCH_2CH_2]_2S$): a radiomimetic, vesicant poisonous gas, used for warfare in World War I. (See nitrogen mustards, radiomimetic)

MUSTARDS (family of cruciferous plants): the taxonomy is not entirely clear. The white mustard (*Sinapis alba*) is 2n = 24. The black mustard (*Brassica nigra*) is 2n = 2x = 16. and supposedly the donor of its genome to the Ethiopian mustard (*B. carinata*) 2n = 34, (2 × [8 + 9]),

Mustards continued

an amphidiploid that received 9 chromosomes from *B. oleracea* (2n = 18). The brown mustard (*B. juncea*) 2n = 36 (2 x [8 + 10]) has one genome (x = 8) of the black mustard and another genome (x = 10) from *B. campestris* (2n = 20). The 38 (2 x (9 + 10)chromosomes of the rapes and swedes descended from *B. oleracea* (2n = 18) and *B. campestris* (n = 20). Sometimes *Arabidopsis* (2n = 10) is also called a mustard since mustard also means crucifer.

MUTABILITY: indicates how prone is a gene to mutate; it indicates a genetic instability.

MUTABLE GENE: has higher than usual rate of mutation. (See mutator genes, transposable elements, mismatch repair, DNA repair, mutator)

MUTAGEN: physical, chemical or biological agent capable of inducing mutation. (See also distal mutagen, promutagen, proximal mutagen, ultimate mutagen, activation of mutagen, physical mutagens, chemical mutagens, biological mutagens, effective mutagen, efficient mutagen, supermutagen, triple helix formation)

MUTAGEN ASSAYS: see bioassays in genetic toxicology, Ames test, ClB, autosomal dominant assays, autosomal recessive assays, Basc

MUTAGEN INFORMATION CENTER: see databases

MUTAGENESIS: see mutation induction, cassette mutagenesis, transposable elements, insertion elements, localized mutagenesis

MUTAGENESIS SITE-SELECTED: see localized mutagenesis, directed mutation

MUTAGENESIS, SITE-SPECIFIC: see localized mutagenesis, directed mutation, targeting genes, site-specific mutation

MUTAGENIC POTENCY: is difficult to determine for many agents that have low mutagenicity Therefore the genetic and carcinogenic hazard of many compounds is unknown and possibly harmless agents may have been found mutagenic in some studies when actually their harmful effect may not be verified by others. There are other agents that may be definitely mutagenic and/or carcinogenic but may have escaped attention. Also, there is no perfect way to quantitate mutagenic effectiveness, particularly for human hazards because of the different quantities humans may be exposed to. As an example the mutagenicity in the Ames *Salmonella* assay of a few compounds is listed (After McCann, J. *et al*. Proc. Natl. Acad. Sci. USA 72:5135):

COMPOUND	MUTATIONS/nmole	COMPOUND	MUTATIONS/nmole
Caffeine	0.002	Ethidium bromide	80
EDTA	0.002	Sodium azide	150
Sodium nitrite	0.010	Acridine ICR-170	260
Ethylmethane sulfonate	0.160	Nitrosoguanidine	1375
Captan (fungicide)	25.000	Aflatoxin B-1	7057
Proflavine	38.000	Nitrofuran (AF-2)	20800

Interesting to note that ethylmethane sulfonate, the probably most widely used mutagen, is only the ninth on this list and nitrofuran, an erstwhile food preservative, exceeds its effectiveness as a mutagen on molar basis more than three thousand fold. (See also environmental mutagens, Ames test, mutagen assays, bioassays in genetic toxicology)

MUTAGENIC SPECIFICITY: base analogs (5-bromouracil, 2-aminopurine) affect primarily the corresponding natural bases. The target of hydroxylamine is cytosine. Alkylating agents preferentially affect guanine. There are no simple chemicals, however, that would selectively recognize a particular gene or genes. With the techniques of molecular biology specifically altered genes can be produced by synthesis and introduced into the genetic material by transformation. Gene replacement by double crossing over, can substitute one allele for another. (See also synthetic genes, gene replacement, transformation, localized mutagenesis, TAB mutagenesis, Cre/Lox, knockout, knock in, targeting genes)

MUTAGENICITY and ACTIVE GENES: experimental data indicate that replicating or actively transcribed genes are prefered targets of mutagens probably because of the decondensation

GENETICS MANUAL

of the genetic material.

MUTAGENS-CARCINOGENS: a very large fraction of mutagens (genotoxic agents) are also carcinogens but not all carcinogens are mutagenic. (See bioassays in genetic toxicology)

MUTANT: an individual with mutation.

MUTANT ENRICHMENT: see screening, mutation detection, filtration enrichment

MUTANT FREQUENCY: the frequency of mutant individuals in a population, disregarding the time or event that produced the mutation and it is thus a concept different from mutation frequency. (See mutation rate)

MUTANT ISOLATION: see diagrams for the isolation of mutations below:

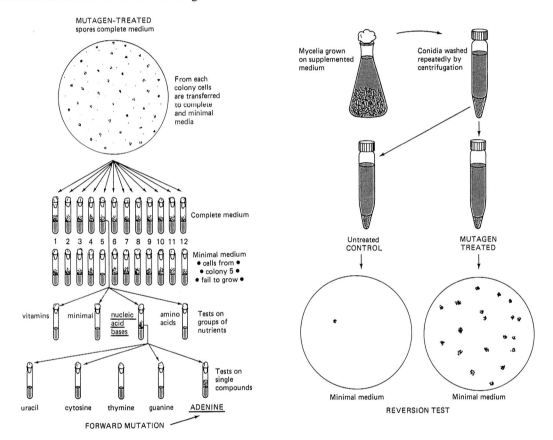

MUTANT ISOLATION IN MICROORGANISM

GENERAL SCHEME FOR THE NON-SELECTIVE ISOLATION OF MUTATIONS IN HAPLOID ORGANISMS. MUTAGEN-TREATED HAPLOID CELLS ARE PROPAGATED ON A COMPLETE MEDIUM (CONTAINING VITAMINS, AMINO ACIDS, NUCLEIC ACID BASES, ETC.) EACH COLONY ON THE PETRI PLATE (TOP LEFT) IS THE PROGENY OF A SINGLE CELL (BECAUSE OF THE GREAT DILUTION USED FOR PLATING). INOCULA ARE TRANSFERED TO BOTH COMPLETE AND MINIMAL MEDIA (WITHOUT ORGANIC SUPPLEMENT). AUXOTROPHIC MUTANTS FAIL TO GROW ON MINIMAL MEDIA BUT MAY GROW ON NUTRIENTS THAT SATISFY THEIR REQUIREMENT AND THIS WAY SPECIFIC FORWARD MUTATIONS CAN BE ISOLATED AND IDENTIFIED. REVERSION TEST ARE CARRIED OUT AS OUTLINED AT THE RIGHT PART OF THE DIAGRAM. ONLY THE REVERTANTS OF THE AUXOTROPHIC MUTANTS WILL GROW ON MINIMAL MEDIA

Basically similar procedures can be used in various microorganisms. The isolation of mutations is greatly facilitated if selective techniques are available. Revertants of auxotrophs

Mutant isolation continued

can grow on minimal media whereas the revertants do not need any special supplement for growth. Mutants resistant to certain chemicals (antibiotics, heavy metals, metabolite analogs, etc.) can be selectively isolated in the presence of the compound to what resistance is sought. Herbicide-resistant plants can be selected in the presence of the herbicide. (See mutation detection)

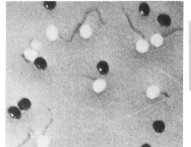

ISOLATION OF ALCOHOL DEHYDROGENASE MUTATIONS IN MAIZE POLLEN BY ALLYL ALCOHOL SELECTION. THE PICTURE DISPLAYS SEGREGATION OF BLACK (DEAD) ANDF WHITE, GERMINATING AND RESISTANT POLLEN. (Courtesy of Freeling, M. Nature 267:154)

Auxotrophic animal cells could be isolated in the presence of 5-bromodeoxyuridine because the wild type cells that incorporated that nucleoside analog after exposure to visible light — because of breakage of the analog-containing DNA—are inactivated. The non-growing mutant cells fail to incorporate the analogs and thus stay alive and after transfering them to supplemented media they may resume growth. Alcohol dehydrogenase mutations can be isolated by feeding the cells allyl alcohol. The wild type cells convert this substance to the very toxic acrylaldehyde and are killed, whereas the alcohol dehydrogenase inactive mutants (microorganisms, plants) cannot metabolize allyl alcohol and selectively survive. (See replica plating, fluctuation test, filtration enrichment, penicillin screen, selective medium, Ames test)

MUTAROTATION: a change in optical rotation of anomers (isomeric forms) until equilibrium is reached between the α and β forms.

MUTASE: enzyme mediating the transposition of functional groups.

MUTASOME: the enzyme complex (Rec A, UmuC-UmuD', pol III) involved in the error-prone DNA replication in translesion. (See DNA repair, translesion pathway)

MUTASYNTHESIS: a biochemically altered mutant of an microorganism may produce modified metabolites either directly or from a precursor analog; the new product (e.g. antibiotic) may be more useful as a drug.

MUTATION: heritable change in the genetic material; the process of genetic alteration.

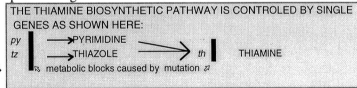

THE THIAMINE BIOSYNTHETIC PATHWAY IS CONTROLLED BY SINGLE GENES AS SHOWN HERE:
py ⟶ PYRIMIDINE
tz ⟶ THIAZOLE ⟶ th ⟶ THIAMINE
metabolic blocks caused by mutation

⇧ THE THIAZOLE REQUIRING MUTANTS OF *ARABIDOPSIS* FAIL TO GROW ON BASAL OR PYRIMIDINE MEDIA BUT DISPLAY NORMAL GROWTH ON THIAZOLE OR THIAMINE.. (From left to right: Basal, Thiazole, Pyrimidine, Thiamine media)

(See mutagen, spontaneous mutation, mutagenic potency, mutagenic specificity, mutant isolation, mutagen assays, mutation detection, mutant frequency, mutation rate, mutation in multicellular germline, somatic mutation, mutation in cellular organelles, mutator genes, mutation chromosomal, transposable elements, mutation beneficial, mutation neutral, mutation pressure, equilibrium mutations, mutation useful, mutation in human populations, genetic load, somatic hypermutation, auxotrophy)

MUTATION AND ALLELIC FREQUENCIES: see equilibrium mutations

MUTATION, BENEFICIAL: when a new mutant appears in the population with a reproductive success of 1.01 (selective advantage 0.01), the odds against its survival in the first generation $e^{-1.01} \cong 0.364$. Its chances for elimination by the 127th generation will be reduced to 0.973

Mutation beneficial continued

compared to the probability to a neutral mutation that has a 0.985 chance of extinction. Thus, ultimately, at a selective advantage of 0.01, its chance of survival will be 0.0197. According to the mathematical argument even mutations with twice as great fitness than the prevailing wild type, has a high (13%) probability of being lost, $e^{-2} \cong 0.1353$ during the first generation. Under normal conditions the mutants' selective advantage (s) is generally very small and the probability of its ultimate survival is $(y) = 2s$. The chance of its extinction is $(l) = 1 - 2s$. In order that a mutant would have a better than random chance (0.5) for survival the following conditions must prevail:

$$(1 - 2s)^n < 0.5 \text{ or } (1 - 2s)^n > 2, \text{ hence } - n \log_e (1 - 2s) > \log_e 2 \text{ or approximately}$$
$$-n(-2s) > \log_e 2, \text{ and therefore } n > \log_e 2/(2s), \text{ i.e., } 0.6931/(2s)$$

If $s = 0.01$ and n = number of mutations, n must be larger than $0.6931/(2 \times 0.01) = 34.655$.

In simple words, approximately 35 times must a mutation, with 0.01 selection coefficient, to occur to be ultimately accepted and fixed. If the spontaneous rate of mutation at a locus is 1×10^{-6}, a population of 35 million is required for providing an adequate chance for the survival of a mutant of that type. Since the proportion of advantageous mutations is probably no more than 1 per 10,000 mutations, very few new mutations have a real chance of survival. The cause of this may attributed to the fact that during the long history of evolution of the organisms, the majority of the possible mutations have been tried and the good ones were adopted by the species. The chances of new mutations is favored, however, when environmental conditions, such as agricultural practice, change in pests, pathogens and predators, etc. take place for which historical adaptation does not exist. (See mutation neutral, fitness, selective advantage, mutation rate)

MUTATION, CHROMOSOMAL: partial chromosomal losses (deletion, deficiency), duplications and rearrangements of the genetic material (inversion, translocation, transposition) are considered chromosomal mutations. The various types of aneuploids (nullisomy, monosomy, trisomy, polysomy, etc.) or increase in the number of genomic sets (polyploidy) can also be classified as chromosomal mutations. (See the specific entries for more details)

MUTATION, COST OF PRODUCTION: varies a great deal from organisms to organisms and also depends on the developmental stage when the mutagen is applied. Mutation induction at the gametic stage may be less expensive than at the multicellular (diploid) germline stage.

DETERMINATION OF THE COST OF MUTATION					
M_2 family size →	1	2	4	8	24
$M_1 + M_2$ size →	2+24	12+24	6+24	3=24	1+24
Mutant Expected→	6	5.244	4.098	2.697	1
Cost M_1 : M_2 ↓	Cost of 1 Mutation in Arbitrary Units Under the Conditions Shown Above and in the Left Side Column of this Tabulation				
1 : 1	8	**6.845**	7.321	10.011	25
1 : 2	12	**11.442**	13.177	18.910	49
1 : 3	**16**	16.018	19.034	27.809	73
2 : 1	12	9.153	**8.785**	11.123	26
3 : 1	16	11.442	**10.249**	12.236	27
4 : 1	20	13.783	**11.713**	13.348	28

In higher plants such as *Arabidopsis*, mutagens are generally applied to the mature seed and thus two generations are required for the isolation of mutants to be used for further experimental studies. The cost of raising these two generations may not be equal and both contribute to the final cost. Although in large M_2 families there is an increased probability for recovering the mutations induced, from the viewpoint of cost effectiveness large number of families with minimal size each, are desirable. In case of recessive mutations, depending on the (n) number of individuals in M_2 families the probability of recovery (P) of a mutant is at $n = 1$ (P = 0.25),

Mutation, cost of production continued

n = 2 (P = 0.437), n = 4 (P = 0.683), n = 8 (P = 0.899), n = 16 (P = 0.989), and n = 24 (P = 0.998) in case of heterozygosity of the M_1 generation, derived from a single cell. It is thus obvious that increasing the M_2 size from 1 to 24 involves the increase of probability of recovery only from 0.25 to 0.998. In the final cost, the cost of both generations must be included along with the effectiveness of recovery of the mutations induced. The table above indicates that M_2 family sizes larger than 4 generally increase the labor and the cost. The recovery of mutations is most effective if selective techniques are applicable. Unfortunately, in some cases (morphological mutants) this may not be possible.

MUTATION, DATING OF ORIGIN: G (number of generations since the emergence of the mutation) can be determined in a population using the information of linkage disequilibrium (δ) concerning closely linked genetic or molecular markers and their known recombination frequencies (θ):

$$G = \frac{\log[1 - Q/(1 - pN)]}{\log(1 - \theta)} \times \frac{\log \delta}{\log(1 - \theta)}$$

where $\delta = (pD - pN)/(1 - pN)$ and pD and pN represent the two different linkage phases of the homologous chromosomes. Q (probability) = $(1 - [1-\theta]^G)(1-pN)$

MUTATION DETECTION: depends on the general nature of the organisms. In haploid organisms (bacteria, algae, some fungi), in haploid cells (microspores, pollen, sperm) or in hemizygous cells (e.g., Chinese hamster ovary cell cultures), in heterozygotes for easily visible somatic markers, the mutations are readily detectable. In diploid or polyploid cells only the dominant mutations can immediately be observed and for the detection of recessive mutations more elaborate procedure is required (see mutation in the multicellular germline). The most effective methods involve selective screening. Molecular methods are also available but these are generally not suitable for large scale screening. When single stranded DNA is subjected to electrophoresis in non-denaturing gel, changed pattern is detectable by SSCP. Heteroduplexes can be resolved by instability in non-denaturing gradient gels (DGGE). Also, heteroduplexes move differently in non-denaturing gels. Cleavage of heteroduplexes by chemicals or enzymes may detect mutant sites. Polymorphism at a single locus can be detected by automated analysis using PCR and fluorescent dyes coupled with a quencher (see more under polymorphism). Mutation may be detected by chemical modification of the mismatches. Carbodiimide generates electrophoretically slow moving DNA containing mismatched deoxyguanylate or deoxythymidylates. Hydroxylamine modifies deoxycytidylate and osmium tetroxide modifies deoxycytidylate and deoxythymidylate in such a way that the DNA at these residues becomes liable to cleavage by strong base. Ribonuclease A cleaves at mismatches (depending on the context) of double-stranded nucleic acids. The Mut Y glycosylase of *E. coli* excises adenine from A-G and with somewhat less efficiency from A-C mispairs. In *E. coli* methyl-directed mismatch system is working primarily up to 3 nucleotide insertions or deletions but longer sequences are barely, if corrected at all, and C-C pairs are ignored. The MutH/L/S multi-enzyme complex misses only 1% of the G-T mispair-induced cuts at nearby GATC sequences. The repair is identified by PCR which also may be a source of replicational error. The latter errors can be estimated as $f = 2lna$, where f is the estimated fraction of mutations within the sequence, l = the length of the amplified sequence, n = the PCR cycles and a = the polymerase-specific error rate/nucleotides; (a for Taq polymerase is within the range of 10^{-6} to 10^{-7}). Direct sequencing may provide the most precise information at the highest investment of labor. (See also selective screening, replica plating, sex-linked lethal tests, specific-locus tests, muta-tion in human populations, Ames test, host-mediated assays, EMC, SSCP, DGGE, DNA sequencing, gel electrophoresis, polymorphism, PCR, footprinting genetic, mutant isolation, bioassays in genetic toxicology, substitution mutation)

MUTATION EQUILIBRIUM: see equilibrium mutations

MUTATION FREQUENCY: the same as mutation rate; see also mutant frequency

MUTATION IN CELLULAR ORGANELLES: cellular organelles mitochondria and plastids

Mutation in cellular organelles continued

are generally present in multiple copies per cell, and within individual organelles generally several copies of DNA molecules exist. Therefore, it generally is not expected that the mutations would be immediately revealed. For their visible manifestation they have to "sort out", i.e. the mutations may not be visible until single organelles become "homogeneous" regarding the mutation, and single cells would be "homoplasmonic" regarding that particular organelle. Although this process is frequently claimed to be stochastic, the direct observation does not seem to support the assumption. Organelles divide by fission, and the daughter organelles are expected to stay in the vicinity of the parental organelle within the viscous cytosol. Thus it is not mere chance where they are located, and the progeny organelles tend to remain clustered unless the plane of the cell division separates them. Therefore sorting out may be a relatively fast process. (See mitochondrial genetics, chloroplast genetics, mtDNA, chloroplasts)

MUTATION IN HUMAN POPULATIONS: is difficult to study directly because of the random or assorted mating system do not favor the identification of recessive mutations. Therefore human geneticists generally rely on the relative increase of "sentinel phenotypes" such as new dominant mutations, hemophilia, muscular dystrophy, cancer that may be used as "epidemiological indicators" for an increase of mutation. Also, sister chromatid exchange, chromosomal aberrations, sperm motility assays may reveal mutations. Molecular methods became more recently available such as RFLP, RAPD, PCR, and changed the previously held view that was based on the fact no transmitted genetic effects of radiation have been clearly detected by the traditional methods in human populations. Mutational hazards for humans is generally infered from mouse data which include induced mutations. The immature mouse oocytes are insensitive to radiation induced mutation but quite likely to be killed by 60 rads of neutrons and 400 R of X-rays or γ rays. Maturing or mature mouse oocytes on the other hand are very susceptible to mutation by acute radiation although the sensitivity to low-dose rate irradiation is 1/20 or less. In contrast the immature human oocytes are not susceptible to killing but their susceptibility to mutation is not amenable to testing. The estimated mutational hazard of mouse oocytes at various stages ranged from 0.17 to 0.44 times that in spermatogonia indicating the lower hazard to females than to males. In industrialized societies chemical mutagens constitute the major hazard, and because of their variety and potency, their effects are difficult to assess, especially at low levels of exposure. (See more under individual entries, mutation rate [undetected mutation], doubling dose, bioassays in genetic toxicology, specific locus mutations test, RBE, atomic radiations, base substitution mutation)

MUTATION IN MULTICELLULAR GERMLINE: reveals the number of cells in that germline at the time the mutation occurred. E.g., if a plant apical meristem is treated with a mutagen at the genetically effective 2 cell stage, in the second generation of this chimeric individual the segregation for a recessive allele will not be 3:1 but 7:1 because one of the cells will produce 4 homozygous wild type individuals, the other cell will yield 1 homozygous wild type + 2 heterozygotes (= 3 dominant phenotypes) and 1 homozygous recessive mutant. In case the germline contains 4 cells, the segregation ratio is expected to be 15:1. In case it consists of 8 cells the proportions are 31:1. (See mutation rate, GECN)

MUTATION INDUCTION: see mutagen, directed mutation, localized mutagenesis, cassette mutagenesis, chemical mutagens, physical mutagens, environmental mutagens

MUTATION, NEUTRAL: a neutral mutation is neither advantageous nor deleterious to the individual homo- or heterozygous for it. Its chance of immediate loss is determined by the first term of the Poisson series, $e^{-1} : \cong 0.368$. (See also evolution non-Darwinian). Consequently, its chance for survival is 1 - 0.362. The chance for its extinction for the second generation is $e^{-0.632} \cong 0.532$ and by the third generation it is $e^{-(1 - 0.532)} \cong 0.626$ ($e = 2.71828$). In general term the extinction of any neutral mutation is e^{x-1} where x is the probability of its loss in the preceding generation. According to R.A Fisher, by the 127th generation the odds against the survival of a neutral mutation is 0.985 and may eventually reach 100%. Several evolutionists, notably M. Kimura, have argued statistically against the conclusion of Fisher (and the Darwinian theory of the necessity of selective advantage). Accordingly, if the rate of mutation for a

Mutation neutral continued

locus is µ, the population size is N and the organism is diploid, the number of new mutations occurring per generation per gene are $2N\mu$. The chance for random fixation is supposed to be $(1/2)N$ because the new allele is represented only once among the total of 2N. The probability of fixation, accordingly, is expected to be $(2N\mu) \times [(1/2)N] = \mu$, indicating that the rate of incorporation of a new neutral mutation into the population is equal to the mutation rate. The average number of generations required for fixation of a neutral mutation (according to Kimura and Ohta) is expected to be $4N_e$ where N_e = the effective population size. Thus, if the effective population size = 1,000 individuals, it takes 4,000 generations for a neutral mutation to be fixed in the population. (See also effective population size, non-Darwinian evolution, mutation beneficial, mutation rate)

MUTATION PRESSURE: repeated occurrence of mutations in a population.

MUTATION PRESSURE OPPOSED BY SELECTION: mutations are frequently prevented from fixation by chance alone (see mutation neutral, mutation beneficial, non-Darwinian evolution). An equilibrium of selection pressure and mutation pressure may be required for the new mutations to have a chance for survival. Mathematically, the frequency of homozygous mutants is $q^2 = \mu/s$ where μ is the mutation rate and s is the coefficient of selection. (See also allelic fixation)

MUTATION RATE: frequency of mutation per locus (genome) per generation. This calculation is relatively simple if the cells population is haploid (prokaryotes, most of the fungi, or when the gametes are mutagenized). If mutation takes place in the multicellular germline of higher eukaryotes only indirect procedures can be used. First, the genetically effective cell number (GECN) in the germline must be determined (see genetically effective cell number). One must know also the level of ploidy. Thus mutation rate in these germline cells (R) is:

$$R = \frac{\text{number of independent mutational events}}{\text{survivors} \times \text{GECN} \times \text{ploidy}}$$

The standard error of mutation rates (s_m) can be computed as $\sqrt{\mu[1-\mu]/n}$ where μ = the mutation rate observed, n = the size of the population. In calculation of the mutation rate on the basis of survivors may pose problems if two different agents are compared. An example: if we use 10,000 haploid cells as a concurrent control and find 10 mutations, the spontaneous mutation rate would be $10/10,000 = 0.001$. If we expose another 10,000 cells to a mutagen that is lethal to 5,000 cells but again we obtain 10 mutations, the calculated apparent induced mutation rate is $10/5,000 = 0.002$ when actually the treatment may not have induced any mutation, it only reduced the survival. In case the mutagen is very potent and in each genome or family multiple mutations may occur, it may become very difficult to distinguish the multiple mutations from the single ones without further genetic analysis. Since the distribution of mutations follows the Poisson distribution, in such a case the average mutation rate may be better determined on the basis of the size of the fraction of the population that shows no mutation at all, this is the zero class of the Poisson series, $e^{-\mu}$. If e.g., the fraction of mutations of the population is 0.3, then zero class is $1 - 0.3 = 0.7 = e^{-\mu}$. Hence $-\mu = \ln 0.7 = -0.3566$ and $\mu \cong 0.36$. If no mutations are found at all in an experimental population that does not necessarily mean that the mutation rate is zero. An approximation to the possible mutation frequency may be made. If we assume perfect penetrance and "normal distribution" of undetected mutations, we may further assume that in a population of n genomes the frequency of mutations is $(1 - q)$ at a probability of P. In order to obtain an estimate of q we have to solve the equation:

$(1 - q)^n = (1 - P)/2$, hence $\hat{q} = 1 - \sqrt[n]{[1-P]/2}$. As an example, after arbitrary substitution of 10,000 for n and 0.99 for P, $\hat{q} \approx 1 - \sqrt[10000]{0.01/2} \approx 1 - 0.9995 \approx 5.3 \times 10^{-4}$. In simple words this means that if we observed no mutations at all, there is a high chance that actually more than 5 may have occurred under these conditions but we have missed them by chance. It is also possible — of course — that the rate of mutation in this case is much below of the 10^{-4}

Mutation rate continued

range or it may even be zero.

Mutation rate in human populations in case of dominance may be estimated: $\frac{number\ of\ sporadic\ cases}{2\ x\ number\ of\ individuals} = \mu$. Frequently it may be necessary to use a correction factor for penetrance or viability. Recessive X-linked mutations can be estimated on the basis of the afflicted males and the formula becomes $\mu = (1/3)s(n/N)$ where n = the number of new mutations, N = the size of the population examined and s = the relative selective value and/or penetrance. Estimation of the rate of recessive mutations is very difficult in human populations because the detection of homozygotes would require controled mating (inbreeding). This problem, intractable by classical techniques, may be solved by molecular methods. Human geneticists may have also great difficulty in identifying mutations which have symptoms overlapping several syndromes. For single cases multivariate statistics may be used but for many, this procedure may be arduous.

Mutation rate in bacteria has been estimated by various means.

Rate = $ln\ 2(M2 - M1)/(N2 - N1)$, alternatively $R = 2\ ln\ 2\left(\frac{M2}{N2} - \frac{M1}{N1}\right)/g$ where M1 and M2 are the number of mutant colonies at time 1 and 2, respectively, N1 and N2 are the corresponding bacterial counts, ln = natural logarithm, g = number of generations. In order to obtain reliable estimates on the rates, the culture must be started by large inocula to avoid bias due to mutation at the early generations when the population is still small, and the experiments must be maintained over several generations under conditions of exponential growth. (See mutation spontaneous, fluctuation test, multivariate analysis, discriminant function)

MUTATION RATE, INDUCED: using N-ethyl-N-nitrosourea, in mice induced at specific loci 1.5 x 10^{-3} mutation per locus. This rate may, however, be an overestimate for other eukaryotes, and the range appears to be about 10^{-4} or less. (See mutation rate spontaneous, supermutagens)

MUTATION RATE IN FITNESS: in *Drosophila* about 0.3/genome/generation was estimated as the frequency of deleterious mutations. In *E. coli* the rate of deleterious mutations per cell was estimated to be 0.0002, i.e. a more than 3 orders of magnitude difference. The *Drosophila* genome is about 50 fold larger than that of *E. coli*, and during a generation approximately 25 divisional cycles take place compared to 1 in the bacterium. If we take the liberty of making adjustments for these difference: 0.0002 x 50 x25 = 0.25, the deleterious mutation rate in the eukaryote and the prokaryote falls within the same range.

MUTATION SPECTRUM: indicates the range of mutations observed in a population under natural conditions or after exposure to different types of mutagens. The theoretical expectations would be that the common genetic material would mutate at the same rate at identical nucleotides in different organisms or under different conditions. This expectation is not met because some mutagenic agents have specificities for certain bases, others act by breaking the chromosomes, depending on their organization and the physical and biological factors present. Also, genes present is multiple copies per genome may suffer mutational alterations but this may not be observed by classical genetic analysis

A SAMPLE OF THE SPECTRUM OF MORPHOLOGICAL MUTATIONS IN ARABIDOPSIS EXPRESSED AT THE ROSETTE STAGE GROWN UNDER 9 HOURS DAILY ILLUMINATION. COLUMBIA WILD TYPE IS AT THE TOP RIGHT CORNER.

although molecular methods may detect them. In general, obligate auxotrophic mutations are very limited in higher plants when the screening uses selective culture media.

MUTATION, SPONTANEOUS: occurs at low frequency and its specific cause is unknown. The rate of spontaneous mutation per genome may vary; in bacteriophage DNA 7 x 10^{-5} to 1 x

Mutation spontaneous continued

10^{-11}, in bacteria 2×10^{-6} to 4×10^{-10}, in fungi 2×10^{-4} to 3×10^{-9}, in plants 1×10^{-5} to 1×10^{-6} in *Drosophila* 1×10^{-4} to 2×10^{-5}, in mice for seven standard loci is 6.6×10^{-6} per locus, in humans 1×10^{-5} to 2×10^{-6}. In RNA viruses 1×10^{-3} to 1×10^{-6} per base per replication has been estimated. The error rate of the reverse transcriptase is extremely high because the enzyme does not have editing function. The induced rate of mutation may be three order of magnitude higher but it varies a great deal depending on the locus, the nature of the mutagenic agent, the dose, etc. Mutation rates per mitochondrial D loop has been estimated to be 1/50 between mothers and offspring but other estimates used for evolutionary calculations are (1/300) per generation. (See mutation rate induced, mutation rate)

MUTATION, USEFUL: although the majority of the new mutations have reduced fitness and are rarely useful for agricultural or industrial applications, some have obvious economic value. Spontaneous and induced mutations have been incorporated into commercially grown crop varieties by improving disease and stress resistance, correcting amino acid composition of proteins, eliminating deleterious chemical components (erucic acid, alkaloids, etc.). Most of the natural variation in the species arose by mutation during their evolutionary history, and many have been added to the gene pool of animal herds and cultivated plants. Many of the floricultural novelties are induced mutants. Some of the animal stocks have accumulated single mutations, others such as the platinum fox is based on a single dominant mutation of rather recent occurrence. Genetic alterations in industrial microorganisms contributed very significantly to the production of antibiotics.

DARK PLATINUM FOX MUTANT
From Mohr & Tuff, J Heredity 30:227

MUTATIONAL DELAY: there is a time lag between the actual mutational event and the phenotypic expression because recessive mutations may show up only when they become homozygous. In organelles the mutations must sort out before becoming visible, etc. (See sorting out, pre-mutation)

MUTATIONAL DISTANCE: the number of amino acid or nucleotide substitutions between (among) macromolecules that may indicate their time of divergence on the basis of a molecular clock. (See evolutionary distance, molecular clock)

MUTATIONAL LOAD: see genetic load

MUTATIONAL SPECTRUM: the array and frequency of the different mutations that have been observed under certain conditions or in specific populations. The spectrum of auxotrophic mutations is much lower in most eukaryotes (except yeast and other fungi) than in lower and higher photoautotrophic organisms. (See mutation spectrum)

MUTATIONS UNDETECTED: see mutations rate

MUTATOR GENES: may be functioning on the basis of abnormal level of errors in DNA synthesis (error-prone DNA polymerase III) or abnormally low level of genetic repair due to defect of proofreading exonuclease in bacteria. The actual repair DNA polymerase in bacteria is pol I that also has $5' \rightarrow 3'$ polymerase and double and single-strand $3' \rightarrow 5'$ exonuclease capability. MutS protein recognizes DNA mismatches and MutL protein scans for nicks in the DNA and then through exonuclease action slices back the defective strand beyond the mismatch site and facilitates the replacement of the erroneous sequences with new and correct ones. In bacteria MutH protein recognizes mismatches not by the proofreading system of the exonucleases but by the distortion of the DNA molecules newly made. The recognition of the new strands depends on the not yet methylated A or C sites within GATC sequences in the new strands. Once the distortion is found the mismatched base can be selectively excised. If either of these repairs fail, mutator action is observed. Any defects in the bacterial gene *dam* (DNA methylase) may also result in high mutation rates because the correct base may be excised and replaced erroneously. Some of the so called mutator genes of the past were actually insertion

Mutator genes continued

elements that moved around in the genome and caused mutation by inactivation of genes through disrupting the coding or promoter sequences. The mutator genes *mutA* and *mutC* of *E. coli* cause the A•T→T•A and G•C→T•A transversions in the anticodon of two different copies of tRNA genes which normally recognize the GGU and GGC codons. As a consequence Gly is replaced by Asp at a rate of 1 to 2%. Defective methyl-directed mismatch repair (gene MutS) occurs in 1-3% of *E. coli* and *Salmonella* raising the mutation rate to antibiotic resistance (Rif [rifampicin], Spc [spectinomycin] and Nal [nalidixic acid]) up to hundreds of fold, depending on the strain of bacteria. High mutator activity may increase evolutionary adaptation in bacteria. (See insertional mutation, transposable elements, DNA replication prokaryotes, DNA repair, antimutators, mutations in cellular organelles, mismatch repair, mutation detection, RAD, DNA polymerases, anticodon, tRNA, transversion, amino acids)

MUTATOR PHAGE: the best characterized is the Mu bacteriophage. The Mu particle includes a 60 nm isosahedral head and a 100 nm tail (in the extended form) containing also base plates, spikes and fibers. The phage may infect enterobacteria (*E. coli, Citrobacterium freundii, Erwinia, Salmonella typhymurium*). Its DNA genetic material consists of a 33 kb (α) and a 1.7 kb (β) double-stranded sequences separated by a 3 kb essentially single-stranded G-loop (specifies host range). It has also variable length (1.7 kb) single-strand "split ends" (SE). Besides the coat protein genes, the *c* gene is its repressor (prevents lysis), *ner* (negative regulator of transcription), *A* (transposase), *B* (replicator), *cim* (controls superinfection [immunity]). *kil* (killer of host in the absence of replication), *gam* (protein protects its DNA from exonuclease V), *sot* (stimulates transfection), *arm* (amplifies replication), *lig* (ligase), *C* (positive regulator of the morphogenetic genes) and *lys* (lysis). Upon lysis 50 to 100 page particles are liberated. The Mu chromosome may exist in linear and circular forms. Mu can integrate at about 60 locations in the host chromosome with some preference. At the position of integration 5 bp target site duplications take place. The integration events cause insertional mutation in the host. Mu causes host chromosome deletions, duplications, inversions and transpositions. These functions require gene *A*, the intact termini of the phage and replication of the phage DNA. A related other phage is D108 has several DNA regions that are non-homologous. Its host range is the same as that of Mu. (See bacteriophages, temperate phage, Mu bacteriophage, insertion elements)

MUTON: as outdated term meaning the smallest unit of mutation; today we know it is a single nucleotide or nucleotide (pair)

MUTUAL EXCLUSIVENESS: e.g., alternative alleles at a particular genetic sites in a haploid cannot exist simultaneously.

MUTUALISM: a mutually beneficial or alternatively a selective situation for increased exploitative association of organisms. (See symbionts)

MVR (minisatellite repeat): see minisatellite

MX1: myxovirus (influenza) resistance; it is located at human chromosome 21q22.3.

Mxi1: a tumor suppressor gene. (See prostate cancer, MYC)

My: million years during the course of evolution; MYA means million years ago

MYASIS (myiasis): infestation of a live body with fly larvae (maggots) such as occurs as a consequence of screw worm, a common pest of southern livestock. (See genetic sterilization)

AN ADULT SCREWWORM FLY. IT HAS AN AVERAGE LIFE SPAN OF THREE WEEKS AND DURING THE PERIOD OF TIME IT MAY TRAVEL 100 TO 200 KILOMETERS BUT IT CAN SURVIVE ONLY DURING THE WARM WINTERS SUCH AS EXIST IN MEXICO OR IN THE SOUTHERN USA. (From Stefferud, A., ed. 1952 Insects. Yearbook of Agriculture, USDA, Washington, DC)

MYASTHENIA (gravis): generally involves muscular (eye, face, tongue, throat, neck) weakness, short of breath, fatigue and the development of antibodies against the acetylcholine receptors.

Myasthenia continued

The genetic determination is ambiguous, generally it appears autosomal recessive. The infantile form lacks the autoimmune feature, although it also includes a defect in the acetylcholine receptor. Anticholinesterase therapy and immunosuppressive drugs may be helpful. Newborns of afflicted mothers my be temporarily affected through placental transfer. (See neuromuscular diseases, epistasis indirect)

MYB ONCOGENE: is an avian myeloid leukemia (myeloblastosis) oncogene; its human homolog was assigned to chromosome 6q21-q23. MYB is actively transcribed in immature myeloid cells and its activity is substantially reduced as differentiation proceeds. Its product is a DNA-binding protein. The same gene is also called AMV (*v-amv*). (See oncogenes, AMV)

MYC: is an oncogene named after the myelocytomatosis retrovirus of birds from which its protein product was first isolated; it is a widely present DNA-binding nuclear proteins in eukaryotes. The MH-2 virus carries besides the *v-myc* gene also the *v-mil* gene which encodes serine/threonine kinase activity and invariably causes monocytic leukemia. Most of retroviruses in this group cause only the formation of non-immortalized macrophages requiring growth factors for proliferation. The cellular forms of the oncogenes are specially named after the type of cells they are found in. NMYC occurs in neuroblastomas (and retinoblastomas). Homologues of this gene were assigned to human chromosome 2p24 and to mouse chromosome 12 and 5. The LMYC genes are involved in human lung cancer where one was located to chromosome 1p32 and two to the mouse chromosomes 4 and 12. By alternative splicing and polyadenylation several distinct mRNAs are produced from a single gene. Metastasis is favored by the presence of a 6 kb restriction fragment of the DNA. In human chromosome 7 another form, the MYC-like 1 gene was detected with a 28 bp near perfect homology to the avian virus. In both man and mouse a translocation sequence was identified (PVT1) that has an activation role in Burkitt's lymphomas (caused by the Epstein-Barr virus) and plasmacytomas (a neoplasia). Myc/MAX heterodimer binds to the CDC25 gene and activates transcription after binding to the DNA sequence CACGTG. This binding site is recognized by a number of MYC-regulated genes. When cellular growth factors are depleted, Myc can induce apoptosis with the cooperation of CDC25. MAD protein holds MYC in check. MAD forms a heterodimer with MAX (MAD-MAX), and this successfully competes with MYC-MAX heterodimers for transcription and thus can control malignancy. The BIN1 protein interacts with MYC and serves as a tumor suppressor. (See oncogenes, PVT, CDC25, MAX, MAD, BIN1)

MYCELIUM: a mass of fungal hyphae. (See hypha)

MYCOBACTERIA: are Gram-positive bacteria and cause tuberculosis and leprosy, respectively. Their cell wall has low permeability and therefore they are rather resistant to therapeutic agents.

MYCOLOGY: the discipline of studying the entire range of mushrooms (fungi).

MYCOPHENOLIC ACID (MPA): see IMPDH

MYCOPLASMA: parasitic and/or pathogenic bacteria; they may contaminate also animal cell cultures. *Mycoplasma genitalium* was the first free-living organism to be completely sequenced by 1995, containing 580,070 bp DNA and 470 open reading frame; 31.8% of its known genes (101) is involved in translation.

MYCORRHIZA: a symbiotic association between plant roots and certain fungi.

MYELENCEPHALON: the part of the brain of the embryo that develops into the medulla oblongata (the connective part between the pons [brain stem]) and the spinal cord. (See brain)

MYELIN: a lipoprotein forming an insulating sheath around nerve tissues.

MYELOBLAST: a precursor (formed primarily in the bone marrow) of a promyelocyte and eventually a granular leukocyte. It contains usually multiple nucleoli. (See leukocyte)

MYELOCYTE: a precursor of the granulocytes, neutrophils, basophils and eosinophils.

MYELOCYTOMA: see myeloma

MYELODYSPLASIA: is a recessive form of leukemia, generally associated with monosomy of human chromosome 7q. The critical deletion apparently involves chromosomal segment 7q22-q34. (See leukemia)

MYELOMA: bone marrow cancer; myeloma cells are used to produce hybridomas. (See Bence Jones proteins, monoclonal antibody, hybridoma)

MYELOMENINGOCELE: see spina bifida

MYELOPATHY: diseases involving the spinal chord.

MYF-3 (myogenic factor): was located to human chromosome 11-p14 and its mouse homolog, *MyoD* is in chromosome 7. Controls muscle development, and may be subject to imprinting and may possibly be involved in the formation of embryonic tumors. The proteins is a helix-loop-helix transcription factor (MASH2); it is a mammalian member of the *achaete-scute* complex of *Drosophila*. (See imprinting, Mash-2, morphogenesis in *Drosophila*)

MYHRE SYNDROME: see dwarfism

MYLERAN (1,4-di[methanesulphonoxy] butane): is a clastogenic (causing chromosome breakage) alkylating mutagen. Also called busulfan.

MYOBLAST: muscle cell precursor. Myoblasts can be well cultured *in vitro* and reintroduced into animals where they fuse into myofibers. They can be used also as gene or drug delivery vehicles. (See vectors)

MYOCARDIAL INFARCTION: obstruction of blood circulation to the heart; may be accompanied by tissue damage. (See coronary heart disease)

MYOCLONIC EPILEPSY: occurs in different forms. The juvenile EJM gene is situated within the boundary of the HLA gene complex. It is characterized by generalized epilepsy with onset in early adolescence. Myoclonous epilepsy associated with ragged-red fibers is characterized also by epileptic convulsions, ataxia, myopathy (enlarged mitochondria with defects in respiration). The defect was attributed to a adenine → guanine transition mutation at position 8344 in the human mitochondrial DNA involving the tRNA^{Lys} gene. The autosomal recessive myoclonous epilepsy LaFlora shows up at about age 15 and results in death within ten years that after. The myoclonous epilepsy Unverricht and Lundborg is a 21q22.3 chromosome recessive with onset between age 6 to 13, beginning with convulsions turning within a few years into shock-like seizures (myoclonous). The latter disease is caused by mutation in the cystatin B (protease inhibitor) gene. A similar disease in cattle has an apparent defect in glycine-strychnine receptors. In mouse the *spastic* mutation in chromosome 3 affects the α-1 glycine receptors whereas the α-2 receptor defect is X-linked. One type of myoclonic epilepsy is a mitochondrial disease. (See epilepsy, mitochondrial disease in humans)

MYOCLONUS: sudden, involuntary contractions of the muscle(s) like in seizures of epilepsy or sometimes as a normal event during sleep. (See myoclonic epilepsy)

MyoD: is muscle-specific basic helix-loop-helix protein and it regulates muscle differentiation and the cessation of the cell cycle. Its DNA binding consensus is CANNTG in the promoter or enhancer of the genes controled. MyoD may activate p21 and p16 and promotes muscle-specific gene expression. When MyoD is phosphorylated by cyclinD1 kinase (Cdk) it may fail to transactivate muscle-specific genes. (See p21, p16, cell cycle, helix-loop-helix, Myf3, cyclin, Cdk, myogenin, MEF)

MYOFIBRIL: muscle fiber composed of actin, myosin and other proteins (bundle of myofilaments).

MYOGENESIS: muscle development.

MYOGENIN: a protein involved in muscle development. MyoD activates myogenin and overcoming the inhibitor of DNA-binding (helix-loop-helix) gene *Id*, and other muscle gene transcription and differentiation is set on course. (See MyoD, MYF5, MRF, MEF)

MYOGLOBIN: is a single polypeptide chain of 153 amino acids, attached to a heme group and functions in the muscle cells to transport oxygen for oxidation in the mitochondria. Recurrent myoglobinuria may result by a microdeletion in the mitochondrially encoded cytochrome C oxidase (COX) subunit III. (See hemoglobin, mitochondrial disease in humans)

MYO-INOSITOL: is an active form of inositol. (See inositol, embryogenesis somatic)

MYOKYMIA: hereditary spasmic disorder of the muscles, caused by potassium channel defects. (See ion channels)

MYOPATHY: a collection of diseases affecting muscle function, controlled by autosomal dominant, autosomal recessive, X-linked and mitochondrial DNA. Besides the weak skeletal muscles, ophthalmoplegia (paralysis of the eye muscles) usually accompanies it. The recessive homozygotes for myotonic myopathy are also dwarfs and have cartilage defects and myopia. Two types are known with carnitine palmitoyle transferase deficiency (CPT1; in 1pter-q12, CPT2) who show myoglobinuria, especially after exercise or fasting. Another recessive myopathy is based on succinate dehydrogenase and aconitase. Phosphoglycerate mutase deficiencies (human chromosome 10q25 and 7p13-p12) also involve myoglobinuria. X-linked forms display autophagy (cytoplasmic material is sequestered into lysosome associated vacuoles) or slow maturing of the muscle fibers, or swelling and hypertrophy in the quadriceps muscles, respectively. (See mitochondrial diseases in humans, cartilage, myopia, carnitine)

MYOPIA: is caused by the increased length of the eye lens in the front-to-back dimension, focusing the refracted light in front of the retina (nearsightedness). Most forms are under polygenic control with rather high heritability (around 0.6). Autosomal dominant, autosomal recessive (infantile) and X-linked forms have also been suggested. Some studies indicate significant positive correlation between myopia and intelligence. Myopia may be concomitant with various syndromes. (See eye diseases, human intelligence)

MYOSINS: a contractile proteins that form thick filaments in the cells, hydrolyze ATP and bind to actin and account for the mechanics of muscle function, organelle movement, phagocytosis, pinocytosis, etc. Members of the 11 families of myosin are represented in amoebas, insects, mammals and plants. (See filament, microfilament, myofibril, phagocytosis, pinocytosis, Usher syndrome)

MYOTONIA: the recessive myotonia congenita is apparently in human chromosome 7 and there is also a dominant form (Thomsen disease); both involve difficulties in relaxing the muscles. (See also periodic paralysis)

MYOTONIC DYSTROPHY: is a dominant human disorder expressed as wasting of head and neck muscles, eye lens defects, testicular dystrophy, speech defects, frontal balding and frequently heart problems. It is controlled by a dominant gene at the centromere of human chromosome 19. The prevalence is highly variable. In some isolated populations it may occur in 1/500 to 1/600 proportion while in other populations the occurrence is about 1/25,000. The manifestation of the symptoms is enhanced in subsequent generations, and more are found in children of affected mothers than affected fathers. This observation may be due, however, to anticipation. The condition is dominant and it is quite polymorphic. Molecular evidence indicates that the DNA sequences concerned are unstable in the 3' untranslated sequences downstream of the last exon of the protein kinase gene. The difficulties involving nucleosome assembly in DNA containing CTG triplet repeats have been implicated. On the basis of linkage with chromosome 19 centromeric genes, prenatal tests are feasible. Mutation frequency was estimated within the $1.1\text{-}0.8 \times 10^{-5}$ range. (See muscular dystrophy, Pompe's disease, Werdnig-Hoffmann disease, periodic paralysis, anticipation, prenatal diagnosis, mental retardation, imprinting, fragile sites, trinucleotide repeats)

MYOTUBES: aggregate of myoblasts into a multinucleated muscle cell. (See cadherin, integrin, fertilin, meltrin, acetylcholine)

MYRISTIC ACID (tetradecanoic acid, $CH_3(CH_2)_{12}COOH$): is a natural 14 carbon fatty acid without double bonds:. Myristoylation are common in oncoproteins, protein serine/threonine and tyrosine kinases, protein phosphatases, in the α-subunit of heterotrimeric G proteins, transport proteins, etc. (See fatty acids, kinase, G protein)

Myt1: see Cdc2

MYXOVIRUSES: a group of RNA viruses. (See RNA viruses)

MZ: monozygotic twin. (See twinning)

N

n: the gametic chromosome number; 2n = the zygotic chromosome number. (See x, genome, polyploid)

$N_2, N_3...N_n$: designate backcross generations. (See also B)

N51: a serum-inducible gene, identical to KC, MGSA and *gro*.

N NUCLEOTIDES: see immunoglobulins

NAA (α-naphthalene acetic acid): synthetic auxin. (See plant hormones, somatic embryogensis)

NAC (nascent-polypeptide associated complex): see protein synthesis

NAD: β-nicotinamide-adenine dinucleotide (diaphorase); a co-factor of dehydrogenation reactions and an important electron carrier in oxidative phosphorylation. DIA1 (cytochrome b5 reductase) is in human chromosome 22q13-qter, DIA2 in chromosome 7 and DIA4 in 16q22.1.

NADH: the reduced form of NAD. (See NAD)

NADP⁺ (nicotine adenine dinucleotide phosphate): an important coenzyme in many biosynthetic reactions. (See NAD, NADH)

NAIL-PATELLA SYNDROME (Turner-Kieser syndrome, Fong disease): malformation or absence of nails, poorly developed patella (a bone of the knee), defective elbows and other bone defects, kidney anomalies, collagen defects, etc. The dominant locus was assigned to human chromosome 9q34. (See collagen)

NAKED DNA: is not embedded in protein; pure DNA.

NALIDIXIC ACID (1-ethyl-1,4-dihydro-7-methyl-4-oxo-1,8-naphthydrine-3-carboxylic acid): an antibacterial agent (a DNA polymerase inhibitor); used also in veterinary medicine against kidney infections.

NANISM: see Mulbrey nanism, dwarfism

NAPHTHALENEACETIC ACID (naphthylacetic acid): see NAA →

NAPHTHYL ACETIC ACID: same as NAA

NARCOLEPSY: pathological frequent sleep, commonly associated with hallucinations and paralysis. (See autoimmune diseases)

NARINGENIN: a trihydroxyflavonone in various plants.

NARROW-SENSE HERITABILITY: see heritability

NarX: *E coli* kinase affecting nitrate reduction by regulator protein NarL. (See nitrate reductase, nitrogen fixation)

NASBA (nucleic acid sequence based amplification): an RNA strand in the presence of a primer is amplified by reverse transcriptase. Then the RNA strand is removed by RNase H and the resulting cDNA, using RNA primers, is amplified by (T7) RNA polymerase into sufficient quantities of RNA. (See RT-PCR, PCR, LCR)

NASCENT: just being born or synthesized; it is not combined yet with any molecule and therefore may be highly reactive.

NASOPHARYNGEAL CARCINOMA: see Epstein-Barr virus

NATIVE: original to a particular region; a natural form of a substance.

NATIVE CONFORMATION: the natural active structural form of a molecule.

NATURAL ANTIBODY: is produced "spontaneously" without deliberate immunization. (See antibody, immmunoglobulins, immune reaction)

NATURAL IMMUNITY: see immunity

NATURAL KILLER CELL (NK): see killer cell, T cell

NATURAL SELECTION: the action of forces in nature that maintain or choose the genetically fittest organisms in a habitat. The *fundamental theorem of natural selection* is "The rate of increase in fitness of any organism at any time is equal to its genetic variance in fitness at that time" or mathematically expressed $\Sigma\alpha \, dp = dt\Sigma\Sigma'(2pa\alpha) = Wdt$ where α = the average ef-

Natural selection continued

fect on fitness of introducing a gene, a = the excess over the average of any selected group, W = fitness, and $dt\Sigma'(2pa\alpha)$ is the sum of increase of average fitness due to the progress of all alleles considered [according to R.A. Fisher]. (See selection, fitness, cost of evolution)

NAVEL ORANGES: are commonly seedless but being "navel" may not rule out seed development since some navel citruses produce normal number of seeds. The Washington navel orange and the Satsuma mandarin are completely pollen sterile and that is the cause of the seedlessness when foreign pollen is excluded. The navel oranges (and grapefruits) have two or three whorls of carpels, resulting in a fruit-in-fruit appearance. This abnormal carpel formation may also interfere with pollination. (See seedless fruits, orange)

NBM PAPER: see diazotized paper

NBT: nitroblue tetrazolium used as a chromogen (0.5 g in 10 mL 70% dimethyl formamide).

NC (negative cofactors): interfere with TFIIB transcription factor binding to the preinitiation complex. (See transcription factors, TBP, PIC, preinitiation complex)

N-CAM (neural cell adhesion molecule): is a Ca^{++}-independent immunoglobulin-like protein which binds together cells by homophilic means (i.e. by being present on neighboring cells). (See also ICAM, neurogenesis)

NCAM (N-CAM): neural cell adhesion molecule. (See N-CAM)

N-DEGRON: the degradation signal of a protein, the NH_2-terminal amino acids and an internal lysine residue of the substrate protein. At that lysine several ubiquitin molecules form a multi-ubiquitin chain. The degradation may be mediated by a G-protein. (See ubiquitin, N-end rule, G proteins)

NEANDERTHAL PEOPLE: were hominids who lived about 30,000 to 200,000 years ago in France and Southern Germany and the Middle-East. They had large jaws and heavy bones and may have developed later (about 30,000 years ago) into the Cro-Magnon men that showed more similarity to present day humans. Their precise relationship to *Homo sapiens* is not known but DNA evidence indicates a dead end detour from human evolution. (See hominidae)

NEAREST-NEIGHBOR ANALYSIS: a technique to determine base sequences in oligonucleotides; a procedure of historical interest for DNA sequencing but gained new usefulness in designing DNA binding ligands for antisense therapy. (See antisense RNA)

NEB: see copia

NEBENKERN (paranucleus): a cellular body; generally a mitochondrial aggregate resembling the cell nucleus by microscopical appearance situated in the flagellum of the spermatozoon.

NEBULARINE ((9-β-D-ribofuranosyl-9H-purine): is a natural product of some fungi and streptomyces with an antineoplastic effect. It's subcutaneous LD50 for rodents varies from 220-15 mg/kg, depending on the species. (See LD50)

NEBULIN: is an actin-associated large protein in the skeletal muscles, built of repeating 35 residue units. (See titin)

NECROPHAGOUS: an adjective for organisms that thrive by eating dead tissues.

NECROSIS: death of an isolated group of cells or part of a tissue or tumor. It generally involves inflammation in animals because of the release of cell components toxic to other cells. Necrosis is different from apoptosis which involves elimination of cells no longer needed, and it is a normally programmed cell death. (See apoptosis, hypersensitive reaction)

NECROTIC: adjective for necrosis (See necrosis)

NECTAR: a sweet plant exudate, fed on by insects and small birds.

NEDD: ICE-related proteases. (See ICE, apoptosis)

NEGATIVE BINOMIAL: $(q - p)^{-k}$ after expansion $p_x = [(k + (x - 1)]! \dfrac{R^x}{q^k}$ where $p = m/k$, m = mean number of events, $q = 1 + p$, p_x = probability for each class, $R = p/q = m/(k + m)$, x = number of events/class and k must be determined by iterations using z_i scores and approximate k values until the z_i becomes practically zero. The detailed procedure cannot be shown

Negative binomial continued

here (see Rédei & Koncz 1992 in *Methods in Arabidopsis Research*, p.16 in Koncz. C, *et al.* eds., World Scientific, Singapore). The negative binomial distribution resembles that of the Poisson series but it may provide superior fit to data that are subject to more than one factor, each affecting the outcome according to the Poisson distribution, e.g. mutations occur according to the Poisson series but their recovery is following an independent Poisson distribution, etc. (See Poisson distribution)

NEGATIVE COMPLEMENTATION: intraallelic complementation when the polypeptide chain translated on one cistron of a locus interferes with the function of the normal polypeptide subunit(s) encoded by the same or another cistron of the multicistronic gene. Thus, actually there is an interference with the expression of the locus concerned. (See allelic complementation, complementation, hemizygous ineffective, cistron)

NEGATIVE CONTROL: prevention of gene activity by a repressor molecule. The gene can be turned on only if a ligand molecule (frequently the substrate of the enzyme that the gene controls) binds to the repressor resulting is moving the repressor from the operator site or from an equivalent position. (See *lac* operon, *tryptophan* operon, lambda phage, DNA binding proteins, DNA-binding domains)

NEGATIVE COOPERATIVITY: binding of a ligand or substrate to one subunit of a multimeric protein precludes the binding to another.

NEGATIVE DOMINANT: see dominant negative, negative complementation.

NEGATIVE FEEDBACK: see feedback inhibition

NEGATIVE INTERFERENCE: see coefficient of coincidence, interference, rounds of matings

NEGATIVE NUMBERS IN NUCLEOTIDE SEQUENCES: indicate position of bases upstream of the position (+1) where translation begins.

NEGATIVE REGULATOR: suppresses or reduces the transcription or translation in a direct or indirect manner. (See *Lac* operon, lambda phage)

NEGATIVE SELECTION OF LYMPHOCYTES: see positive selection of lymphocytes

NEGATIVE STAINING: is used for the study of macromolecules. The electronmicroscopic grid with the specimen on it is exposed to uranyl acetate or phosphotungstic acid that produce a thin film over it except where the macromolecule is situated. The electron beam "illuminates" the non-covered macromolecules while the metal-stained parts appear more dense. This gives then a negative image of viruses, ribosomes or other complex molecular structures. (See stains, electronmicroscopy)

NEGATIVE-STRAND VIRUS: see replicase

NEGATIVE SUPERCOIL: a double-stranded DNA molecule twisted in the opposite direction as the turn of the normal, right-handed double helix (e.g. in B DNA). (See also supercoiled DNA, Z DNA)

NEIGHBOR JOINING METHOD: is a relatively simple procedure for infering bifurcations of an evolutionary tree. Nucleotide sequence comparisons are made pairwise and the nearest neighbors are expected to display the smallest sum of branch lengths. (See evolutionary distance, evolutionary tree, least square methods, four-cluster analysis, unrooted evolutionary trees, transformed distance, Fitch-Margoliash test, DNA likelihood method, protein-likelihood method)

NEIGHBORLINESS: is a statistical method for the estimation of the position of two taxonomic entities in an evolutionary tree. (See evolutionary tree, transformed distance, Fitch-Margoliash test)

Neisseria gonorrhoeae: see antigenic variation

N-END RULE: the half-life of a protein is determined by the amino acids at the NH_2 end. (See N-degron)

NEOBIOGENESIS: the idea that living organisms arose from organic and inorganic material.

NEOCARZINOSTATIN (NCS): is a naturally occurring enediyne antibiotic. The NCS chromophore attacks specifically a single residue in a two-base DNA bulge. It attacks HIV type I RNA and other viruses. (See enediyne, bulge, HIV)

NEOCENTRIC: see neocentromere

NEOCENTROMERE: extra spindle-fiber attachment site in the chromosomes in eukaryotes of certain genotypes, different from the regular centromere position. (See centromere, centromere activation, holocentric, preferential segregation, knob)

NEO-DARWINIAN EVOLUTION: evolution is attributed to small random mutations accumulated by the force of natural selection and characters acquired as a direct adaptive response to external factors have no role in evolution. (See mutation beneficial, mutation neutral, selection, fitness, darwinism, directed mutation)

NEO-LAMARCKISM: see Lamarckism

NEOLITHIC AGE: about 7,000 years ago when humans turned to agricultural activity from hunting and gathering. (See also paleolithic, mesolithic)

NEOMORPHIC: a mutation that displays a new phenotype or any structural or other change that evolved recently.

NEOMYCIN (neo, $C_{23}H_{46}N_6O_{13}$): a group of aminoglycoside antibiotics. (See antibiotics, neomycin phosphotransferase, kanamycin)

NEOMYCIN PHOSPHOTRANSFERASE (NPTII): see kanamycin resistance, aminoglycoside phosphotransferase, aph(3')II, neo^r

NEOPLASIA: a newly formed abnormal tissue growth such as a tumor; it may be benign or cancerous. (See cancer, carcinogens, oncogenes)

neo^r: neomycin resistance gene encoding the APH(3')II enzyme. (See APH(3')II, aminoglycoside phosphotransferases)

NEO-SEXCHROMOSOME: is actually a translocation involving an autosome and the X or Y chromosome; some have binding sites for the MSL (male-specific lethal) proteins. (See MSL)

NEOTENY: the retention of some juvenile (or earlier stage) function in a more advanced stage.

NEO-X-CHROMOSOME: see neo-sexchromosome

NEO-Y-CHROMOSOME: see neo-sexchromosome

NEPHRITIS, FAMILIAL (nephropathy): is autosomal dominant and closely resembles the Alport syndrome. The kidney problems are preceded by elevated blood pressure, proteinuria and only microscopically detectable blood cells in the urine. In one form immunoglobulin G accumulates in the serum and the expression of the disease is promoted by dietary conditions (e.g. high gluten). (See Alport's disease, kidney disease)

NEPHROLITHIASIS: see kidney stones

NEPHRONOPHTHITIS: an autosomal recessive kidney disease characterized by anemia, passing of large amounts of urine (polyuria), excessive thirst (polydipsia), wasting of kidney tissues, etc. (See diabetes)

NEPHROSIALIDOSIS: autosomal recessive kidney inflammation caused by oligosaccharidosis (See lysosomal storage diseases, Hurler syndrome)

NEPHROSIS, CONGENITAL: is an autosomal recessive inflammation of the kidney. It can be detected prenatally by the accumulation of α-fetoprotein in the amniotic fluid. The basic defect is in the basement membrane structure of the glomerular membranes. Its incidence may be as high as 1.25×10^{-4} in populations of Finns or of Finnish descent. (See kidney diseases, fetoprotein, prenatal diagnosis, basement membrane)

NEPHROTOME: a part of the mesoderm that contributes to the formation of the urogenital tissues and organs.

NEPOTISM IN SELECTION: some disadvantageous individuals or groups may be favored by selection in case they promote the fitness of the reproducing groups that share genes with them. (See selection, indirect fitness)

NER (nucleotide exchange repair): see DNA repair, exchange repair

NERNST EQUATION: $\left(\frac{RT}{zF}\ln\frac{C_o}{C_i} = V\right)$ expresses the relation of electric potential across biomembranes to ionic concentration at both sides; V = equilibrium potential in volts, C_o and C_i are outside and inside ionic concentrations, $R = 2$ cal mol^{-1}°K^{-1}, T = temperature in K (Kalvin) degrees, $F = 2.3 \times 10^4$ cal V^{-1} mol^{-1}, z = valence, ln = natural logarithm.

NERVE CELL: see neuron

NERVE GROWTH FACTOR (NGF): stimulatory factor of growth and differentiation of nerve cells. It is present in small amounts in various body fluids. It is a hexamer composed of α, β and γ subunits but only the β subunit is active for the ganglions. The active component of the mouse submaxillary factor has a dimeric structure of two 118 amino acid residues (MW 13 kDa). The mouse gene for α is in chromosome 7, β is in chromosome 3. The human gene for β is in chromosome 1p13. It acts primarily on a tyrosine kinase receptor. (See signal transduction, growth factors, neurogenesis, neuropathy)

NES: see RNA export

NESTED GENES: partially overlapping genes, sharing a common promoter and the structural genes may be read in different registers. This is a very economical solution for extremely small genomes (φX174) for multiple utilization of the same DNA sequences within different reading frames. (See also overlapping genes, recoding, contiguous gene syndrome, knockout)

NESTED PRIMERS: the product of the first PCR amplification houses internally a second primer in order to minimize amplification of products by chance (See PCR, primers)

NETRINS: are secreted by neuronal target cells and assist homing in the proper nerve axons. (See also UNC-6, semaphorin, collapsin, axon, colorectal cancer [DCC])

NEU: synonymous with ERBB2, see ERBB1.

NEURAL CODE: represents the various stimuli as sensory experiences such as visual, olfactory, gustatory, mechanical, auditory qualities as well as intensities, frequencies and spatial relationships.

NEURAL CREST: ectodermal cells along the neural tube cleaved off from the ectoderm and migrate through the mesoderm to form the peripheral nervous system.

NEURAL PLATE: is a thickened notochordal overlay of nerve cell that develop into neural tubes.

NEURAL TUBE: the central nervous system of the embryo, derived from epithelial cells of the neural plate.

NEURAL-TUBE DEFECTS: see anencephaly, spina bifida, Meckel syndrome, microcephaly, hydrocephalus; such a condition may occur as part of a number syndromes including trisomy 18 and 13.

NEURAMINIC ACID: a pyruvate and mannosamin-derived 9-carbon aminosugar. Its derivatives such as sialic acid are biologically important. (See sialic acid, sialidoses, sialiduria, sialidase deficiency, neuroaminidase deficiency, formula on p. 977)

NEUREGULINS: are protein signals that activate acetylcholine receptor genes in synaptic nuclei. (See acetylcholine, agrin)

NEUREXINS: a great variety of nerve surface proteins generated by alternative splicing of three genes. They may function in cell-to-cell recognition and interact with synaptotagmin. (See splicing, synaptotagmin)

NEURITE: an extension from any type of neuron (axon or dendrite).

NEURITIS: nerve inflammation, increased sensitiveness or numbness, paralysis or reduced reflexes caused by one or more (polyneuritis) defects of nerves. (See neuropathy)

NEUROAMINIDASE DEFICIENCY: is a recessive autosomal lysosomal storage disease with multiple and variable characteristics. The basic common defect in the various forms is a deficiency of the sialidase enzyme. Sialidase cleaves the linkage between sialic acid (N-acetylneuraminic acid) and a hexose or hexosamine of glycoproteins, glycolipids or proteoglycans. The uncleaved molecules are then excreted in the urine. In sialurias rather free sialic acid is

Neuroaminidase deficiency continued
excreted. For the normal expression of the 76 kDa enzyme the integrity of the structural genes in human chromosome 10pter-q23, and a 32 kDa glycoprotein coded by chromosome 20 are required. The deficiency is characterized by cherry red muscle spots, progressive myoclonus (involuntary contraction of some muscles), loss of vision but generally normal intelligence. (See lysosomal storage diseases, neuraminic acid)

NEUROBLAST: a cell that develops into a neuron. (See neurogenesis)

NEUROBLASTOMA: a nerve cell tumor, located most frequently in the adrenal medulla (in the kidney). For its development a MYC oncogene seems to be responsible. Deletions of the short arm of human chromosome 1 inactivate the relevant tumor suppressor gene. (See cancer, MYC, ERBB1, tumor suppressor gene)

NEUROECTODERM: contributes to the formation of the nervous system. (See neurogenesis)

NEUROENDOCRINEIMMUNOLOGY: studies the interactions among the central-peripheral nervous system, endocrine hormones and the immune reactions.

NEUROEPITHELIOMA: see Ewing sarcoma

NEUROFIBROMATOSIS: autosomal dominant (NF-1), near the centromere of human chromosome 17q11.2, and another gene (NF-2) in the long arm of chromosome 22q12.2. They affect the developmental changes in the nervous system, bones and skin and cause light brown spots and soft tumors (associated with pigmentation) over the body, and in NF-2 particularly in the Schwann cells of the myelin sheath of neurons and involves mental retardation, etc. The literature distinguishes several forms of this syndrome. The loss of the NF1 protein activates the RAS signaling pathway through the granulocyte macrophage colony stimulating factor and makes the cells prone to develop juvenile chronic myelogenous leukemia. NF-1 also activates an adenylyl cyclase coupled to a G protein and it is deeply involved in the regulation of the overall growth of *Drosophila*. A pseudogene (NF1P1) was located in human chromosome 15. The frequency of NF1 is about 2×10^{-4}. Estimated mutation rate $1 - 0.5 \times 10^{-4}$. The neurofibromin protein is encoded in a 350 kb DNA tract with 59 exons. (See also epiloia, mental retardation, ataxia, hypertension, eye disease, Cushing syndrome, leukemia, RAS, GCSF, café-aux-lait spot, Lisch nodule, von Recklinghousen disease)

NEUROGENESIS: the nervous system has a great deal of similarity among all animals. The *neuroblasts* (the cells which generate the nerve cells) develop from the ectoderm. The *neural tube* (originating by ectodermal invagination) gives rise to the *central nervous system*, composed of *neurons* and the supportive *glial* cells. The differentiated neural cells do not divide again. The *neural crest* produces cells that eventually migrate all over the body and forms the *peripheral nervous system*. A neuron contains a dense cell body from which emanate the *dendrites* (reminding to the root system of plants). From the neurons extremely long *axons* may emanate which at the highly branched termini make contact with the target cells by *synapses*. The dendrites and axons are also called by the broader term *neurites*.

neuron with dendrites   axon   synaptic branches

Initially, the various components of the system develop at the points of migration and subsequently they are connected into a delicate network. The point of growth is named *growth cone* which manages a fast expansion, generating also some web-like structures (*microspikes or filopodia* organized into *lamellipodia*). The neurites are frequently found in *fascicles*, indicating that several growth cones travel the same track across tissues. This motion is mediated by cell adhesion molecules (N-CAM, cadherin, integrin). When the migrating growth cones reach their destination, they compete for the limited amount of *neutrophic factor* (NGF) released by the target cell and about a half of them starve to death. The nerve cells function at the target by

Neurogenesis continued

their inherent *neuronal specificity* rather than by their positional status. When the branches of several axons populate the same territory of control, they are trimmed back and some eliminated by a process of *activity-dependent synapse elimination*. Due to this process, the synaptic function becomes more specific. The migration of the motor neurons from the central nervous system toward the musculature begins already in the 10th hour of the embryo development in *Drosophila*. It has been found that the membrane-spanning receptor tyrosine phosphatases, DLAR, DPTP99A, DPT69D and others determine the "choice point" when the neuron heads toward the muscle fiber bundle. Apparently, after receiving the appropriate extracellular signal they dephosphorylate the relevant messenger molecules. Deleting these surface molecules, the axons lose their guidance system. When the neurons break out of the nerve fascicles, they are homing to the muscles by the attraction of the fasciclin III proteins in the muscle membranes. At this stage the muscles secrete also the chemorepellent semaphorin II and thus the neurons settle down to form eventual synapses. The slowing down of the growth cone and the formation of the synapse is mediated by the *Late Bloomer Protein* (LBP). The neurotransmitters are then released by the terminal arbors. The agrin protein secreted by the nerve cells binds to heparin and α-dystroglycan and causes the clustering of the acetylcholine receptors. Other molecules, still not identified, are likely to be involved.

Genetic study of neurogenesis is best defined in *Drosophila* and *Caenorhabditis*. In the latter, the nervous system contains only about 300 cells and a large array of mutant genes (many cloned) are available. In *Drosophila* the large *achaete-scute complex* (*AS-C*, in chromosome 1-0.0) and several other genes cooperate in neurogenesis. The differentiation and function of the systems appear to be operated by cell-to-cell contacts rather than by diffusing molecules. In most of the mutants this communication is disrupted. (See morphogenesis, nerve growth factor, N-CAM, cadherin, integrin, signal transduction, neuron-restrictive silencer factor)

NEUROGENETICS: the study of the genetic bases of nerve development, behavior and hereditary anomalies of the nervous system. (See <http://neurowww.cwru.edu/teach/teaching.htm>, <http://www_hbp.scripps.edu/Home.html>)

NEUROGENIC ECTODERM: a set of cells that separates from the epithelium and forms the neurons by moving into the interior of the developing embryo. (See gastrulation)

NEUROHORMONE: is secreted by neurons, such as vasopressin, gastrin. (See vasopressin, gastrin, gonadotropin releasing factor)

NEUROLEPTIC: changes affected by the administration of antipsychotic drugs.

NEOROMODULIN (GAP-43): a protein abundant at the nerve ends and may be involved in the release of neurotransmitters and a negative regulator of secretion at low levels of Ca^{2+}. (See neurotransmitter, synaptotagmin)

NEUROMUSCULAR DISEASES: include a variety of hereditary ailments with the common symptoms of muscle weakness and inability of normal control of movements. The specific and critical identification is often quite difficult because of the overlapping symptoms. (See lipidoses, gangliosidoses, sphingolipidoses, aminoacidurias, glutaric aciduria, muscular dystrophy, Werdnig-Hoffmann, syndrome, myotonic dystrophy, chromosome defects [Down's syndrome, trisomy 18, Cri-du-chat and Prader Willi syndrome], glycogen storage diseases, brain malformations, dystrophy, amyotrophic lateral sclerosis, Parkinson's disease, cerebral palsy, atrophy, multiple sclerosis, Kugelberg-Welander syndrome, Charcot-Marie-Tooth disease, abetalilipoproteinemia, Zellweger syndrome, Refsum diseases, diastematomyelia, Leigh's encelopathy, Moebius syndrome, ataxia, dystonia, palsy, myasthenia, aspartoacylase deficiency, carnosinemia, spinal muscular atrophy, Kennedy disease, Brody disease)

NEURON: a nerve cell capable of receiving and transmitting impulses. In an adult human body there are about 10^{12} neurons connected by a very complex network. (See neurogenesis)

NEURON-RESTRICTIVE SILENCER FACTOR (NRSF): is a protein that can bind to the neuron-restrictive silencer element (NRSE) by virtue of eight non-canonical zinc fingers.

NRSE sequences were detected in at least 17 genes expressed only in the nervous system. (See master molecule, silencer, Zn-finger)

NEUROPATHY: a non-inflammatory disease of the peripheral nervous system. (See Charcot-Marie-Tooth disease, Riley-Day syndrome, pain-insensitivity, sensory neuropathy 1, HNPP)

NEUROPEPTIDE: is a signaling molecule (peptide) secreted by nerve cells.

NEUROPEPTIDE Y (NPY): is supposed to modulate mood, cerebrocortical excitability, hypothalmic–pituitary signaling, cardiovascular physiology, sympathetic nerve function and feeding behavior. The latter claim is not supported by the evidence obtained with NPY mutant mice.

NEUROPHYSIN: a group of soluble carrier proteins (M_r about 10,000) of vasopressin and oxytocin and related hormones secreted by the hypothalmus. (See oxytocin, vasopressin, diabetes insipidus, hypothalmus, brain human)

NEUROSPORA CRASSA (red bread mold): an ascomycete (n = 7, DNA 4×10^7 bp) was introduced into genetic research more than half a century ago for the exact study of recombination in ordered tetrads. Its original home is in tropical and subtropical vegetations.

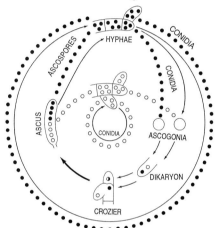

← THE LIFE CYCLE OF *NEUROSPORA*

The asexual life cycle is about one week whereas the sexual cycle requires about 3 weeks. The two mating types (*A* [5 kb] and *a* [3 kb]) are determined by a single locus. It was the first fungus to be used for centromere mapping on the basis of the frequency of second division segregation in the linear asci of 8 spores. The first auxotrophic mutants were induced in *Neurospora* and these experiments led to the formulation of the one-gene-one enzyme hypothesis that formed the cornerstone of biochemical genetics. Although it is a haploid organism, the availability of heterokaryons permits the study of dominance and allelic complementation. This was the first eukaryote where genetic transformation with DNA became feasible. Currently about 10^4 to 10^5 transformant can be obtained per μg DNA. *Neurospora* (and other ascomycetes) may display RIP (**r**epeat **i**nduced **p**ointmutation) observed at high frequency when duplicated elements are introduced into the nuclei (by transformation) resulting in GC→AT transitions and eviction of the duplications, resulting also in chromosomal rearrangements. Over 600 genes have been mapped and about 1/5th have been cloned. The genetic maps of *Neurospora* are generally shown in relative distances only because of the different *rec* genes present in the various mapping strains may alter recombination frequencies by an order of magnitude. Mitochondrial mutants are known. The respiration-deficient *poky* mutants are somewhat similar to the petite colony mutants of budding yeast. The *stopper* mutants are also 350-5,000 bp deletions and unable to make protoperithecia. Related species used for genetic studies are *N. sitophila* and *N. tetrasperma*. (See diagrammed life cycle, tetrad analysis, channeling, RIP, *Neurospora* mitochondrial plasmids)

NEUROSPORA MITOCHONDRIAL PLASMIDS: *Neurospora crassa* and *N. intermedia* strains may harbor 3.6-3.7 kb mitochondrial plasmids with a single open reading frame encoding a 81 kDA reverse transcriptase protein. The proteins appear similar to the class II intron-encoded reverse transcriptase-like elements. The Mauriceville and Varkud plasmid DNA apparently uses as primers a 3'-tRNA like structure. The latter two plasmids can integrate into the mitochondrial DNA at the 5'-end of the major plasmid transcript. *N. intermedia* contains also a smaller 'Varkud satellite plasmid' (VSP) that may be either linear or circular single-stranded RNA. The VSP plasmid bears resemblance to class I introns. (See mitochondrial plasmids, introns)

NEUROTOXIN: destroys nerve cells. (See also toxins)

NEUROTRANSMITTER: several types of signaling molecules (acetylcholine, glutamate, serotonine, γ-aminobutyric acid, catecholamines, neuropeptides, etc.) used for information transmission of neuronal signals. These signals can be excitatory (e.g. glutamate, acetylcholine, serotonin) or inhibitory such as gamma-aminobutyric acid and glycine and may open Cl⁻ channels. The secretion is activated by electric nerve impulses in response to extracellular signals. The signals are transmitted to other cells by chemical synapses across the synaptic cleft by attaching to the transmitter-gated ion channels. (See ion channels, AMPA, transmitter-gated ion channel, voltage-gated ion channel, signal transduction, synaps, calmodulin, GABA, synapse, synaptic cleft, see also other separate entries)

NEUROTROPHINS (NTs): are growth and nutritive factors for the nerves but may also potentiate nerve necrosis. Their receptor is either a glycoprotein (LNGFR) or trk. Neurturin, a structurally NT-related protein also activates the MAP kinase signaling pathway and promotes neural survival. (See NGF, MAP, trk, LNGFR, neurotrophin)

NEUROVIRULENCE: a viral infection attacking the nerves. (See virulence)

NEURTURIN: see neurotorphins

NEURULATION: the formation of the neural tube (the element that gives rise to the spinal chord and the brain) from the ectoderm during gastrulation.

NEUTERED: ovariectomized (ovaries surgically removed) female. (See spaying, castration, oophorectomy)

NEUTRAL MUTATION: see mutation neutral

NEUTRAL SUBSTITUTION: amino acid changes in the protein that have no effect on function.

NEUTRALIZING ANTIBODY: covers the viral surface and prevents binding of the virus to cell surface receptors and thus interferes with infection. (See antibodies)

NEUTRON: see physical mutagens

NEUTRON FLUX DETECTION: personal monitoring may be based on microscopic examination of proton recoil tracks in a film or dielectric materials (i.e., which transmit by induction rather than by conduction) such as plastic or glass. Precise measurement of neutron flux is more complex. (See radiation measurement)

NEUTROPENIA, CYCLIC: is a rare blood disease in humans and dogs. In humans it appears to be autosomal dominant while in dogs it is recessive. In 21 day cycles in humans (12 in dogs) fever, anemia, decrease in eosinophils, lymphocytes, platelets, monocytes recur and enhance the chances of infection. In some instances this childhood disease is outgrown by adulthood. (See endocardial fibroblastosis)

NEUTROPHIL: one of the specialized white blood cells (leukocyte). Generally they are distinguished by irregular-shaped, large nuclei and granulous internal structure (polymorphonuclear granulocyte). They contain lysosomes and secretory vesicles enabling them to engulf and destroy invading bodies (bacteria) and antigen-antibody complexes. (See also macrophage, eosinophils, basophils)

NEVIRAPINE: is a non-nucleoside inhibitor of HIV-1 reverse transcriptase. (See TIBO, acquired immunodeficiency syndrome, AZT)

NEVOID BASAL CELL CARCINOMA: is quite frequently caused by new dominant mutation in human chromosome 9q22.3. The patients generally have bifid ribs, bossing on the head, tooth and cranofacial defects, poly or syndactyly and reddish birthmarks that may become neoplastic. These cutanious neoplasias frequently become horny but are — in the majority of cases sporadic and benign basal cell carcinomas (BCC) — although they may become metastatic. The skin is very sensitive to ionizing radiation and in response develops numerous new spots (nevi). The more serious form is the basal cell nevoid syndrome (BCNS). This is the most common type of cancer in the USA and annually about 750,000 cases occur. It turned out that the major culprit in this disease is a homolog of the *Drosophila* gene *patch* (*ptc* or *tuf*

Nevoid basal cell carcinoma continued

[*tufted*] in chromosome 2-59). This segment polarity locus encodes a transmembrane protein and the wild type allele represses the transcription of members of the TGF-β (transforming growth factor). *Drosophila* gene *hh* (*hedgehog*, 3-81) has the opposite effect, i.e., it promotes the transcription of the TGF proteins. Homologs of these genes exist in other animals, including mouse. The human homolog has now been cloned by the use of a mouse probe of *ptc*. Nucleotide sequencing revealed that in one of the families, a 9 bp (3 amino acid) duplication occurred in the afflicted member of the family and in another kindred a 11 bp deletion was associated with expression of the BCNS. Other developmental genes may also contribute to the expression of these genes. (See cancer, TGF-β, morphogenesis in *Drosophila*)

NEVUS: an autosomal dominant red birthmark on infants that usually disappears in a few months or years. The autosomal dominant *basal cell nevus syndrome* is a carcinoma. It causes also bone anomalies of diverse types, eye defects, sensitivity to X-radiation, etc. (See vitilego, skin diseases, carcinoma, cancer)

NEWFOUNDLAND: is a rare blood group.

NEWT (*Triturus viridescens*): 2n = 2x = 22; its lampbrush chromosomes have been extensively studied. (See lampbrush chromosome)

NEWTON, ISAAC (1642-1727): famous mathematician, physicist, natural philosopher and afflicted by gout. (See gout)

NEXIN: protein connecting the microtubules with a cilium or flagellum.

NEXUS: a junction

NEZELOF SYNDROME: is an autosomal recessive T lymphocyte deficiency immune disease. The affected individuals lack cellular, while display humoral immunity. The development of the thymus is abnormal. (See immunodeficiency)

NF: nuclear factors required for replication. NF-1 (similar to transcription factor CTF) binds to nucleotides 19-39 of adenovirus DNA and stimulates replication initiation in the presence of DBP. NF-2 required for the elongation of replicating intermediates. NF-3 (similar to OTF-1 transcription factor) binds nucleotides 39-50 adjacent to the NF-1 binding site and stimulates replication initiation. (See CTF, OTF, DBP)

NF-1: human neurofibromatosis (neoplasia of the nerves) involves proteins homologous to GAP and IRA. NF is CCAAT binding protein. (See GAP, IRA1, IRA2, CAAT box)

NF-AT: is a family of transcription factors (NF-AT$_c$ and F-AT$_p$) that activate the interleukin (IL-2) promoter or the dominant negative forms block IL-2 activation in the lymphocytes. NF-AT is activated by the CD3 and CD2 antigen jointly but not separately and it is suppressed by cyclosporin (CsA). For the nuclear import of the cytoplasmic NF-AT, it has to be dephosphorylated. (See lymphocytes, interleukin, immune system, T cell, transcription factors, NF-κB, cyclosporin, cyclophyllin)

NF-κB (nuclear factor kappa binding): a transcription factor family dimeric with a REL oncoprotein and they are specific for the IκB (immunoglobulin kappa B lymphocytes) proteins; it is activated by interleukin 1 (IL-1R1) signaling cascade with the aid of TNF, hypoxia and viral proteins. NF-κB normally stays inactivated in the cytoplasm by being associated with inhibitor IκB. The activation dissociates the two proteins and IκB is degraded before NF-κB moves to the nucleus where it binds to the GGGACTTTCC consensus. The target genes are involved either with cellular defense or differentiation such as IL-2, IL-2 receptor, phytohemagglutinin (PHA), and phorbol ester (PMA) synthesis. Na-salicylate, aspirin, glucocorticoids, immunosuppressants (cyclosporin, rapamycin), nitric oxide, etc. inhibit this transcription factor involved with inflammation and infection processes. In arthritis, lupus erythromatosous, Alzheimer disease, HIV and influenza infection, in several types of cancer the level of NF-κB is elevated. The function REL subunit is necessary for the activation of TNF and TNF-α dependent genes and for the NF-κB protection against apoptosis. Inhibition of NF-κB translocation to the nucleus enhances apoptotic killing by ionizing radiation, some cancer drugs and TNF. Thus managing NF-κB may assist in drug therapy of cancer. (See REL, morphogen-

NF-κB continued

esis in *Drosophila* {3}, inerleukins, TNF, IκB, T cell, immunosuppressants, nitric oxide, glucocorticoid, autoimmune diseases, Alzheimer disease, HIV, apoptosis, phorbol-12-myristate-13-acetate)

NFAT: is a family of phosphoprotein transcription factors mediating the production of cell surface receptors and cytokines and regulate the immune response. (See T cell)

NFKB (nuclear factor kappa B, NF-kB, NF-κB): are transcription factors; NFKB1 in human chromosome 4q23 and NFKB2 in human chromosome 10q24. The former codes for a protein p105 and the latter for p49. Both are regulators or viral and cellular genes. They have homology also to the REL oncogene and the product of the *Drosophila* maternal gene *dl* and some regulatory proteins of plants. NF-κB is selectively activated by the nerve growth factor using the neurotrophin p75 receptor ($p75^{NTR}$), a member of the tyrosine kinase family, located in the Schwann cells. (See oncogenes, morphogenesis in *Drosophila* {3}, Schwann cell, NF-κB)

NF-E2: a transcription factor with a basic leucine zipper domain, and it regulates erythrocyte-specific genes. (See Maf, NRF)

NF-X, NF-Y: are transcription factors. (See major histocompatibility complex)

NG (nitrosoguanidine): see MNNG.

NGF: see nerve growth factor

NGFI-A: a mitogen-induced transcription factor (probably the same as egr-1, zif/268, Krox-24, and TIS8)

NGFI-B: see nur77 and TIS1

NICHE: a small depression on a surface; a special area of nature favorable for some species.

NICHOLAS II: tzar of Russia, distressed father of a hemophiliac son, Alexis. Some historians suggested that this monogenic disease (transmitted by mother Alexandra, granddaughter of Queen Victoria of England) had been one important factor in the disability of Nicholas II to deal with the social problems of his country that lead to his murder and to communist rule for about 3/4 of a century. A single gene altered then the history of about 1/4 of the world's population. (See hemophilia, antihemophilic factors, Romanov)

NICK: disruption of the phosphodiester bond in one of the chains of a double-stranded nucleic acid by an endonuclease. (See phosphodiester bond, endonuclease)

NICK TRANSLATION: DNA polymerase I enzyme of *E. coli* can attach to nicks and add (labeled nucleotides to the 3' end, while slicing off nucleotides from the 5' end, thus moving (translating) the nicks through this nucleotide labeling. This is a process of replacement replication. (See DNA replication)

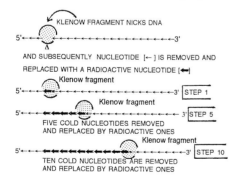

NICKED CIRCLE: a circular DNA with nicks. (See nick)

NICKING ENZYME: see nick

NICOTIANA (x = 12): genus in the *Solanaceae* family includes over 60 species. The majority are of new world origin although some are native in Australia and the South Pacific islands. About ten species have been used for smoking, and alkaloid production. Some tobaccos are ornamentals. Economically the most important is *N. tabacum* (2n = 48). The latter is an amphidiploid of presumably of *N. syvestris* (2n = 24) and either *N. tomentosiformis* (2n = 24) or *N. otophora* (2n =24). Synthetic amphidiploids with the latter resemble less *N. tabacum* than *N. tomentosiformis*. The fertility of *sylvestris* x *otophora* amphipolids are more fertile than those of *sylvestris x tomentosiformis*. *N. tabacum* has been widely used for cytogenetic analysis; monosomic and trisomic lines are available. This species has been extensively studied by *in vitro* culture techniques, including regeneration

Nicotiana continued

of fertile plants from single cells, because it is very easy to establish cell and protoplast cultures and carry out transformation by agrobacterial vectors. This was also the first higher plant species where foreign transposable elements (*Ac* of maize) could be expressed. For manipulations involving isolated cells most commonly the SR1 (streptomycin-resistant chloroplast mutation of Maliga) is used. This stock originated from the variety Petite Havana, that is relatively easy to handle because of its small size. Mendelian experiments are difficult to conduct with *N. tabacum* because of its allotetraploid nature. Single gene mutations or chromosomal markers are generally not available. Antibiotic resistant chloroplast mutations (streptomycin, lincomycin) have been produced in *Nicotiana* and served as tools to demonstrate recombination of the plastid genome. Transformation of chloroplast genes have been accomplished both by the biolistic methods and polyethylene glycol treatment of protoplasts. *N. plumbaginifolia,* one of the diploid species offers some advantages for Mendelian analysis but it is less suitable for cell cultures and regeneration.

NICOTINE (β-pyridyl-α-N-methylpyrrolidine): is a contact poison insecticide and an indispensable metabolite for nerve and other functions. Nicotine is considered as an addictive substance. (See following nicotine entries, smoking)

NICOTINE ADENINE DINUCLEOTIDE: see NAD^+

NICOTINE ADENINE DINUCLEOTIDE PHOSPHATE: see $NADP^+$

NICOTINIC ACETYLCHOLINE RECEPTORS: are ion-channel linked receptors in the skeletal muscles and presynaptic neurons. Thus nicotine and/or acetylcholine activate the nicotine-acetylcholine-regulated calcium channels resulting in the release of glutamate to its receptor in the postsynaptic cells of the hippocampus. This modulation of synaptic transmission may play an important role in learning, memory, arousal, attention, information processing. In Alzheimer diseases the nicotinic cholinergetic transmission degenerates. After smoking a cigarette about 0.5 μM nicotine may be delivered within 10 seconds to the brain and lung. (See signal transduction, nicotine, smoking)

NIDDM: non-insulin-dependent diabetes mellitus. (See diabetes mellitus)

NIEMANN-PICK DISEASE (sphingomyelin lipidosis): Types A, B, C, and D are known as autosomal recessive hereditary disorders of sphigomyelinase deficiency. The differences among these types involve the level of activity of the enzyme and the onset of the symptoms. The most common form, Type A is identified as a severe enlargement of the liver, degeneration of the nervous system and generally death by age 4. In the lysosomes phosphoryl choline ceramide (sphingomyelin) accumulates. Type B is also deficient in sphingomyelinase and since the nervous system is not affected the patients may live to adulthood. Type C is a milder form of Type A with low activity of the enzyme and the afflicted persons may survive up to age 20. In Type D the symptoms are similar to those in Type C, accumulate sphingomyelin yet the activity of the enzyme appears close to normal. The latter type apparently involves cholesterol transport from the lysosomes In Type E disease the nervous system remains normal and sphingomyelin accumulation is limited to some organs. Type E may not be directly determined genetically. A human acid sphingomyelinase gene has been mapped to chromosome 11p15.1-p15.4. (See sphingolipidoses, sphingolipids, epilepsy)

NIEUWKOOP CENTER: is a position of the early embryo for dorsal differentiation. (See Wnt, chordin, noggin)

nif: see nitrogen fixation

NIGERICIN (antibiotic of *Streptomyces*): facilitates K^+ and H^+ transport through membranes and activates ATPase in the presence of a lipopolysaccharide, and promotes the processing of interleukin-1β precursors. (See ionophore, interleukins, ATPase)

NIGHT BLINDNESS (nyctalopia): defective vision in dim light is caused by several choroid and retinal defects. The X-linked phenotype (Xp11.3) is distinguished from the autosomal type by frequent association with myopia (nearsightedness). The autosomal dominant Sorsby syndrome involve retinal degeneration caused by mutation in the tissue inhibitor metalloproteinase-3 gene (TIMP3). (See color blindness, day blindness, optic atrophy, Oguchi disease).

NIGHTSHADE (*Atropa belladonna*): solanaceus alkaloid-producing plant, 2n = 72. (See burdo)

NIH: National Institutes of Health, Bethesda, MD, USA

NIL (near isogenic line): see inbred

NIMA: a mitosis-specific protein kinase.

NINHYDRIN: triketohydrindene hydrate, a reagent for ammonia. Amino acids upon heating liberate ammonia that reduces ninhydrin and produces a blue color. This reagent is used for very sensitive colorimetric estimation of amino acids or as a spot test on paper or thin-layer chromatograms. (See paperchromatography, thin-layer chromatography)

NISH: non-isotopic *in situ* hybridization. (See also FISH, *in situ* hybridization)

NISHIMINE FACTOR: is a special hemostatic factor required for the generation of thromboplastic (blood clot forming) activity. (See antihemophilic factors)

NITRATE REDUCTASE: is used in prokaryotes to reduce nitrate to nitrite under the conditions of anaerobic respiration. The enzyme is associated with molybdenum and it is essential for normal function. Nitrate reductase is generally associated also with cytochrome b. Plants and fungi reduce nitrate before it can enter into the amino acid synthetic path. Nitrate reductase mutations are generally selected on chlorate media. (See chlorate, nitrogenase)

NITRIC OXIDE (NO): endothelial cells may make and release NO in response to liberation of acetylcholine by the nerves in the blood vessel walls. Nitric oxide activates also potassium channels through a cGMP-dependent protein kinase. Then the smooth muscles of the veins become relaxed and the blood flow is boosted. This mechanism is the basis of penal erection. NO is contributing toward the activation of macrophages and neutrophils in the body's defense reaction. NO may also cause cytostasis during differentiation. NO is called also a physiological messenger of the cardiovascular, immune and nervous system. Deficiency of NO may cause pyloric stenosis, depression, hypotension, inflammation, aggressive behavior, etc. (See depression, hypotension, pyloric stenosis, aggression, circadian rhythm, cGMP, ion channels)

NITRIFICATION: conversion of ammonium into nitrate.

NITROBLUE TETRAZOLIUM: see NBT, BCIP

NITROCELLULOSE FILTER: can be pure nitrocellulose or of cellulose acetate and cellulose nitrate mixtures. It can be used for immobilization of RNA in Northern blots, for Southern blotting of DNA, for replica plating and storage of λ phage or cosmid libraries, bacterial colonies, for Western blotting, for antibody purification, etc. Nucleic acids generally have superior binding to nylon filters. (See entries separately)

NITROGEN (N_2 = 28.02 MW): odor and colorless gas that becomes liquid at -195.8° C and solidifies at -210° C; 4/5 of the volume of air is nitrogen.

NITROGEN CYCLE: see nitrogen fixation

NITROGEN FIXATION: is the mechanisms of incorporation atmospheric nitrogen (N_2) into organic molecules with the aid of the nitrogenase enzyme complex, present in a few microorganisms (*Clostridia, Klebsiella, Cyanobacteria, Azotobacters, Rhizobia,* etc.). The first step in the process is the reduction of N_2 to NH_3 or NH_4. Atmospheric nitrogen can be converted into ammonia also industrially at high temperature and high atmospheric pressure (for the production of fertilizers and explosives). Many organisms are capable of using ammonia and can oxidize it into nitrite (NO_2^-) and nitrate (NO_3^-) by the process called *nitrification*.

Bacterial nitrogen fixation relies primarily on two enzymes : the tetrameric *dinitrogenase* (M_r 240,000), containing 2 molybdenum, 32 ferrum and 30 sulfur per tetramer and the dimeric *dinitrogenase reductase* (M_r 60,000) with a single Fe_4—S_4 redox center and two ATP binding sites. Nitrogen fixation begins by the reduced first enzyme. The reduction is the job of this second enzyme that hydrolyzes ATP. The role of ATP is donation of chemical and binding energy. The nitrogenase complex uses different means to protect itself from air (oxygen) that inactivates it. In the symbiosis between *Rhizobia* and legumes leghemoglobin, an oxygen-

Nitrogen fixation continued

binding protein is formed in the root nodules for protecting the nitrogen fixation system. In *Rhizobium leguminosarum* and related species the genes required for nitrogen fixation and nodulation are in the large conjugative plasmids. In *R. meliloti* these genes are within the bacterial chromosome. The fast growing (colony formation 4-5 days) *Bradyrhizobium* species are common in the tropical areas whereas in the moderate climates the various *Rhizobia* species are found on the legumes. These two main groups cannot utilize N_2 in culture but only in the nodules. The *Azorhizobia* can use N_2 without a symbiotic relation. The nodule formation requires a special interaction between the bacteria and the host. A series of nodulation factors have been identified, among others the *ENOD40*-encoded a 10 amino acid oligopeptide, a member of the TNF-R family. A homolog of this gene has been found also in non-legumes. The *Rhizobium leguminosarum* (*Vicia*) nodulation genetic system may be represented as:

```
  T  N M  L  EF     D      A B C        I J
  →  ←←← ← ←←       ←      →→→→→
```

The *nod* (nodulation) genes of the bacteria are located in a plasmid within the bacterial cell. The *nodD* expression is induced by root exudates (the flavonoids eryodyctyol, genistein).

RHIZOBIUM BACTERIA WITHIN A SOYBEAN ROOT NODULE. (Courtesy of Dr. W.J. Brill)

This is similar to the infection process of plants by *Agrobacteria* (related to *Rhizobia*) where the *virulence* gene cascade is induced by flavonoids (acetosyringone). As a consequence of turning on the *nod* system, signals are transmitted to the root hairs to curl up and develop into nodules inhabited then by bacteria. The different legume species produce different inducers and the bacteria may have different genes to respond, and this interaction specifies host and bacterial functions. Once within a root hair the infection spreads to neighboring cells. The bacteria are enveloped by plant membranes and form *bacteroids*. The nodule formation assures the right supply of oxygen to the bacteroids and the same time protects the nitrogenase system from oxygen that is detrimental to it. Nitrogen fixation is controled by the *nif* and *fix* genes, encoding the FixLJ proteins. The FixL protein is a transmembrane kinase that is activated by low oxygen tension. This kinase then phosphorylates the FixJ protein that in turn switches on *FixK* and *nifA* genes. The product of the latter interacts with an upstream element to promote the transcription of *fix* and *nif* operons. The nodule formation and nitrogen fixation depend then on complex circuits between bacterial and (and over 20) plant genes and environmental stimuli. The relationship between the host and bacteria is mutually advantageous and the symbiotic system has also enormous economic value in maintaining soil fertility and crop productivity. It would be desirable to transfer the genes required for nitrogen fixation into crop plants. Some of the homologs of the nodulin genes are already present in plants (leg-haemoglobin). The *nif* genes of *Klebsiella* had been transferred to plants but they failed to be expressed. To overcome this problem transformation of the (prokaryote-like) chloroplasts was considered but here it has not been possible to protect the O_2-sensitive nitrogenase from the oxygen evolution concomitant with photosynthesis. (See also *Agrobacteria*, bacteria, symbiosome)

NITROGEN MUSTARDS: are (radiomimetic) alkylating agents. The general formula where Al means alkyl groups; in this form it is trifunctional, has three chlorinated alkyl groups. When it is monofunctional it has one chlorinated alkyl group; the bifunctional has two such groups. Nitrogen mustards family includes antineoplastic and immunosuppressive compounds such as cyclophosphamide, uracil mustard, melphalan, chlorambucil, etc. (See sulfur mustards, mustards)

NITROGENASE: see nitrogen fixation

NITROGENOUS BASE: purine or pyrimidine of nucleic acids.

NITROSAMINES: see chemical mutagens

NITROSOMETHYL GUANIDINE (NNMG): is a very potent alkylating mutagen and carcinogen. It is light-sensitive and potentially explosive. (See chemical mutagens)

NITROUS ACID MUTAGENESIS: see chemical mutagens (ii)

NMDA (N-methyl-D-aspartate) **RECEPTOR**: is located in the post-synaptic membranes of the excitatory synapses of the brain. Activated NMDA receptors control the Ca^{2+} influx, excitatory nerve transmission. NMDA is regulated by tyrosine kinases and phosphatases. (See synapse, PTK, PTP)

NMDAR: see NITROSOGUANIDINE →

N-METHYL-N'-NITRO-N-NITROSOGUANIDINE: see MMNG

NMR: see nuclear magnetic resonanance spectroscopy

NMYC: see MYC

NOAH's ARK HYPOTHESIS: the low level of divergence among the different human populations indicates recent origin of the modern humans and the relatively fast replacement of the antecedent populations. (See Eve foremother, Y chromosome)

NOCICEPTIN (orphanin): see opiate

NOCICEPTOR: pain receptor involved in injuries. (See opiate)

NOCTURNAL ENURESIS (bedwetting): the dominant gene has been assigned to human chromosome 13q13-q14.3. The condition generally afflicts children under age 7 and then gradually improves and usually disappears later. Its cause is the improper regulation of the antidiuretic hormone and social maladjustment. (See antidiuretic hormone)

NOD: a kinesin-like protein on the surface of oocyte chromosomes of *Drosophila* that stabilizes the chromosomes to stay on the prometaphase spindle. A similar protein, Xklp1, is found in *Xenopus* oocytes. (See spindle)

NODE: a knot- or swelling-like structure; the widened part of the plant stem from where leaves, buds or branches emerge. Also a crossover site in the DNA. (See also internode)

NODULE: a small roundish structure; the root nodule of legumes harbors nitrogen-fixing bacteria. The rhizobial nodulation genes determine the formation of lipo-chitooligosaccharide signals that specify the Nod (nodulation) factors (NF). (See also nitrogen fixation, recombination nodule)

NODULIN: a plant protein in the root nodules of leguminous plants. (See nitrogen fixation)

NodV: *Rhizobium* kinase affecting nodulation regulator protein NodW. (See nitrogen fixation)

NOGGIN: see organizer

NOISE: random variations in a system, either biological or mechanical or electronic or disturbance and interference in any of the these systems.

NOMAD: specially constructed vectors built of fragment modules in a combinatorially rearranged manner. These vectors allow sequential and directional insertion of any number of modules in an arbitrary or predetermined way. They are useful for studying promoters, replicational origins, RNA processing signals, construction of chimeric proteins, etc. (See also Rebatchuk, D. *et al.*, 1996 Proc. Natl. Acad. Sci. USA 93:10891; <http://Lmb1.bios.uic.edu/NOMAD/NOMAD. html>)

NOMADIC GENES: dispersed repetitive chromosomal elements with high degree of transposition. (See also copia, insertion elements)

NOMARSKI DIFFERENTIAL INTERFERENCE CONTRAST MICROSCOPY: a phase shift is introduced artificially in unstained objects to cause a field contrast that is due to interference with diffracted light. As a result either dark field (phase contrast) or bright field contrast (Nomarski technique) can be obtained depending on the direction of the light by a

Nomarski differential interference contrast microscopy continued
quarter wave phase shift in a special microscopic equipment. In colored microscopic specimens the phase shift (and contrast of the image) is brought about by the differential staining by microtechnical dyes. (See also resolution optical, fluorescence microscopy, phase-contrast microscopy, confocal scanning, electronmicroscopy)

NOMENCLATURE: see gene symbols, gene nomenclature assistance, databases

NON-ALLELIC GENES: belong to different gene loci and are complementary. (See allelism).

NON-AUTONOMOUS CONTROLLING ELEMENT: a transposable element that lost the transposase function and can move only when this function is provided by a helper element. A non-autonomous controlling element is *Ds* or *dSpm* of maize whereas the helper element may be *Ac* and *Spm*, respectively. (See transposable elements, *Ac - Ds, Spm*)

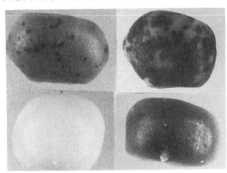

ACTIVE Spm TRANSPOSASE (TOP MAIZE KERNELS) AND INACTIVE TRANSPOSASE (BOTTOM ENDOSPERMS) IN EITHER SOLID DEEPLY COLORED (A) OR INACTIVATED (a-m¹) STATES. (Courtesy of Barbara McClintock)

NON-CHROMOSOMAL GENES: genes that are not located in the cell nucleus. It is more logical to call them extranuclear genes, such as the genes in mitochondria and plastids.

NON-CODING DNA: is not transcribed into any RNA.

NON-COMPETITIVE INHIBITION: enzyme inhibition is not relieved by increasing the concentration of substrate. (See inhibitor, repression, regulation of gene activity)

NON-CONJUGATIVE PLASMIDS: do not have transfer factors (*tra*) and are not transferred by conjugation. (See plasmid mobilization, plasmid conjugative, conjugation)

NON-COVALENT BOND: is not based on shared electrons; it is weak bond individually but many such bonds may result in substantial specific interactions.

NONCYCLIC ELECTRON FLOW: a light induced electron flow from water to NADP$^+$ as in photosystem I and II of photosynthesis. (See photosynthesis, Z scheme)

NON-DARWINIAN EVOLUTION: evolution by a random process rather than based on selective value . (See evolution non-darwinian, mutation neutral, Darwinian evolution)

NON-DIRECTIVENESS: the principle of genetic counseling that only the facts are disclosed but the decision is left to the persons seeking information. (See counseling genetic)

NONDISJUNCTION: one pair of chromosomes goes to the same pole in meiosis and the other pole then will have neither of them; in mitosis, the movement of both sister chromatids to the same pole. Left: NORMAL DISJUNCTION, Center: NORMAL METAPHASE, **Right: MITOTIC NONJUNCTION.** After Lewis K.R. & John, B. 1963 Chromosome Marker, Little Brown, Boston)

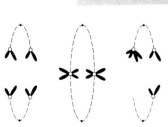

In meiosis nondisjunction can occur at meiosis I when the two non-disjoining chromosomes are the homologues (the bivalent) or at meiosis II when the non-disjoining elements are sister chromatids. Nondisjunction in a normal euploid cell is called *primary nondisjunction* and when the event takes place in a trisomic cell it is called *secondary nondisjunction*. Normal disjunction of meiotic chromosomes seems to have the prerequisite of chiasma (chiasmata). In the absence of chiasma the likelihood of nondisjunction increases. In cases when nondisjunction occurs after chiasma, the chiasma seems to be localized to distal position of that bivalent. The human X chromosome fails to disjoin in the achiasmatic case whereas the autosomes may be nondisjunctional even when distal exchanges take place. Meiosis II nondisjunction occurs when chiasmata take place in the near-centromeric region. In achiasmatic meiosis

Nondisjunction continued

the orderly disjunction depends on the centromeric heterochromatin. It is interesting to note that in human females nondisjunction increases by advancing age although the recombinational event determining chiasmata takes place prenatally. These facts require the assumption that the proteins (spindle and motor proteins) regulating normal chromosome segregation become less efficient during later phases of the development. (See aneuploidy, trisomy, monosomy, nullisomy, Down syndrome, him)

NONESSENTIAL AMINO ACIDS: not required in the diet of vertebrates because they can synthesize them from precursors (alanine, asparagine, aspartate, cysteine, glutamate, glutamine, glycine, proline, serine, tyrosine). (See amino acids, essential amino acids)

NONHEME IRON PROTEINS: contain iron but no heme group.

NON-HISTONE PROTEINS: are proteins in the eukaryotic chromosomes. In contrast to histones which are rich in basic amino acids, the non-histone proteins are acidic and are frequently phosphorylated. These proteins include enzymes, DNA- and histone-binding proteins and trascription factors. Since they have regulatory role(s), they are a heterogeneous group varying in tis-

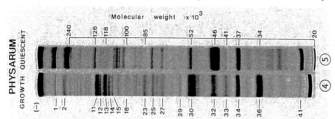

PHENOL-SOLUBLE ACIDIC PROTEIN PROFILES, SEPARATED BY ELECTROPHORESIS FROM *PHYSARUM* AT TWO DEVELOPMENTAL STAGES. (From Lestourgeon, A. *et al*. 1975 Arch. Biochem. Biophys. 159:861)

sue-specific and developmental stage-specific manner. Along with DNA and histones they are part of the chromatin. (See histones, chromosomin, chromatin)

NONHOMOLOGOUS: chromosomes do not pair in meiosis and their nucleotide sequences do not show substantial sequence similarities.

NON-HOMOLOGOUS RECOMBINATION: genetic exchange between chromosomes with less than the usual similarity in base sequences. (See illegitimate recombination)

NONIDET: a nonionic detergent used in electrophoresis.

NON-ISOTOPIC LABELING: see non-radioactive labeling, biotinylation, fluorochromes, FISH.

NON-LINEAR TETRAD: the ascus contains the spores in a random group of four. (See tetrad analysis, unordered tetrad)

NON-MENDELIAN INHERITANCE: extranuclear genetic elements are involved or gene conversion alters the allelic proportions. (See extranuclear genes, gene conversion, paramutation, mitochondrial genetics, chloroplast genetics)

NON-MHC-RESTRICTED CELL: see killer cells

NON-ORTHOLOGOUS GENE DISPLACEMENT: in evolutionarily related species a particular function is encoded by genes with dissimilar nucleotide sequence. (See orthologous, paralogous)

NON-PARAMETRIC TESTS: do not deal with parameters of the populations (such as means) and they estimate percentiles of a distribution without defining the shape of the distribution by the means of parameters. Non-parametric tests are simple and often they are the only ways of statistical estimations. The *Wilcoxon's signed-rank test* and the *Mann-Whitney test* substitute for a t-test by using paired samples. The *Spearman rank-correlation test* determines correlation between two variables. The *chi-square* goodness of fit test, the *association test* and many other frequently used tests in genetics are non-parametric. (See statistics, affected-sib-pair method, see also named above tests for description and use)

NONPARENTAL DITYPE (NPD): a tetrad or octad of spores in an ascus that display only recombinant chromosomes. (See tetrad analysis)

NON-PERMISSIVE CONDITION: is a regime where a conditional mutant may not thrive, the growth of the cell is not favored under these circumstances. (See temperature-sensitive mutation, conditional lethal)

NON-PERMISSIVE HOST: cell that does not favor autonomous replication of a virus; it may not prevent, however, the integration of the virus into the chromosome where the virus may initiate neoplasic transformation.

NON-PERMUTED REDUNDANCY: at the ends of T-even phage DNAs repeated sequences occur in the same manner in the entire population, e.g. |1234....1234| |1234....1234| (See also permuted redundancy)

NON-PLASMID CONJUGATION: some transposons (Tn1545, Tn916) are capable of conjugal transfer at frequencies about 10^{-6} to 10^{-8}. These larger transposons are very promiscuous and transfer DNA by a non-replicative process to different bacterial species. (See transposable elements, conjugation bacterial recombination, plasmids, restriction-modification)

NON-POLAR CROSSES (homosexual crosses): in mitochondrial crosses some genes appeared to segregate in a polar manner and others apparently failed to display such polarity. The polarity was attributed to a ω^+ and ω^- condition in yeast mitochondria. Actually ω is an intron within a 21S rRNA genes. (See mtDNA, mitochondrial genetics)

NON-POLAR MOLECULE: the molecule lacks dipole features, repels water and therefore has poor solubility (if any) in water-based solvents.

NON-PRODUCTIVE INFECTION: the virus DNA is inserted into the chromosome of the eukaryotic host and it is replicated as the host genes and no new virus particles are released. In such a situation the viral oncogene may initiate the development of cancer rather than the destruction of the cell as in productive infection. (See temperate phage, oncogenes)

NON-RADIOACTIVE LABELS: for DNA biotinylated probes are used most frequently. Biotin 11-dUTP contains a 11 atom linker between biotin and the 5'-position of deoxyuridine triphosphate. It is added to the probe by terminal deoxynucleotidyl transferase in place of thymidine. Then a signal-generating complex is added, containing streptavidin (protein with strong affinity for biotin) complex and peroxidase. After addition of peroxide and di-aminobenzidine tetrahydrochloride substrates a dark precipitate results if the probe is present. Other procedures use photobiotin® and other chemoluminiscent compounds. (See fluorochromes, FISH, chromosome painting, immunolabeling)

NONRANDOM MATING: mate selection is by choice, either by the organism or by the experimenter. (See assortative mating, inbreeding, autogamy, mating systems)

NON-RECIPROCAL RECOMBINATION: see gene conversion

NONRECURRENT PARENT: is not used for back-crossing of the progeny. (See recurrent parent, backcross)

NONREPETITIVE DNA: consists of unique sequences such as the euchromatin of the chromosomes that codes for the majority of genes and displays slow reassociation kinetics in annealing experiments. It does not involve redundancies such as SINE, LINE. (See unique DNA, euchromatin, SINE, LINE, c_0t curve, c_0t value)

NON-SECRETOR: see secretor, ABH antigens

NONSELECTIVE MEDIUM: any viable cell, irrespective of its genetic constitution, can use it in contrast to a selective medium containing, e.g. a specific antibiotic in which only the cells resistant to the antibiotic can survive. (See selective medium)

NONSENSE CODON: is stop signal for translation. (See nonsense mutation, code genetic)

NONSENSE MUTATION: converts an amino acid codon into a peptide-chain-terminator signal (UAA, ochre; UAG, amber; UGA, opal). (See genetic code, code genetic)

NONSENSE SUPPRESSOR: allows the insertion of an amino acid at a position of the peptide in spite of the presence of a nonsense codon at the collinear mRNA site. (See suppressor tRNA)

NONSPECIFIC PAIRING: see illegitimate pairing, chromosome pairing

NONTRANSCRIBED SPACER (NTS): nucleotide sequences in-between genes which are not transcribed into RNA, e.g., sequences between ribosomal RNA genes.

NON-VIRAL RETROELEMENTS: there is a class of *non-viral retrotransposable* elements in *Drosophila* and other organisms (*Ty* in yeast, *Cin* in maize, *Ta* and *Tag* in *Arabidopsis*, *Tnt* in tobacco). These transpose probably by RNA intermediates but lack terminal repeats although they encode reverse transcription type functions. They contain AT-rich sequences at the 3'-termini and frequently are truncated at the 5'-end. Some of these elements are no longer capable of movement because of their extensive (pseudogenic) modification. The *Drosophila*: *"D"* is a variable length element with up to 100 copies and variable number of target site duplications. *"F"* of variable length in 50 copies with 8 - 22 bp target site replication. The longest element of the *F* family is about 4.7 kb. The *"Fw"* element (3,542 bp) contains two long open reading frames (ORF), the longer of these bears substantial similarity to the polymerase domains of retroviral reverse transcriptases. The other transcript (ORF1) codes for a protein that has similarities to the DNA-binding sections of retroviral gag polypeptides. The *FB* family of transposons of *Drosophila* is not related to the *F* family mentioned here (see hybrid dysgenesis). *"G"* transposable element is of variable length up to 4 kb in 10 to 20 copies with 9 bp target site duplication. These elements are similar to the *F* family inasmuch as they have polyadenylation signals and poly-A sequences at the 3'-end. The *G* elements insert primarily into centromeric DNA. *"Doc"* is of variable length (up to 5 kb) with 6 to 13 bp target site duplication. The element has a variable 5'-terminus but a conserved 3'-terminus. *"Jockey"* of variable length up to 5 kb, in 50 copies. It has incomplete forms named *sancho 1, sancho 2* and *wallaby*. *"I"* is of variable length, up to 5.4 kb, with 0 to 10 complete copies plus about 30 incomplete elements with variable number of target site duplications (usually 12 bp). (See also copia, hybrid dysgenesis)

NOONAN SYNDROME (male Turner syndrome, female pseudo-Turner syndrome): is apparently caused by an autosomal dominant gene. The symptoms are variable and complex yet there is some resemblance to Turner syndrome females but this affects also males. It is frequently accompanied by mental retardation, heart and lung defect (stenosis), etc. but no visible chromosomal abnormality. (See Turner syndrome, stenosis, heart disease, face/heart defects)

NOPALINE: a dicarboxyethyl derivative of arginine is produced in plants infected by *Agrobacterium tumeciens* (strain C58). The nopaline synthase (*nos*) gene is located within the T-DNA of the Ti plasmid. The bacteria have also a gene for the catabolism of nopaline (*noc*). This and other opines, serve bacteria with a carbon and nitrogen source. The *nos* promoter and tailing have been extensively used in the construction of plant transformation vectors. (See *Agrobactrium*, opines, octopine, T-DNA)

NOR: see nucleolar organizer

NOPALINE →

$$HN=C\begin{matrix}NH_2\\NH-(CH_2)_3-CH-COOH\\|\\NH\\|\\HOOC-(CH_2)_2-CH-COOH\end{matrix}$$

NORADRENALINE: see norepinephrine

NOREPINEPHRINE (noradrenaline): see animal hormones

NORMAL DISTRIBUTION: is derived from the binomial and not

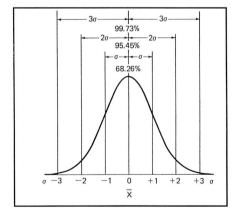

THE NORMAL DISTRIBUTION IS REPRESENTED BY A NORMAL CVURVE, CHARACTERIZED BY THE STANDARD DEVIATION σ AND WITHIN ± 1, ± 2 AND ± 3 σ 68.26, 95.45 ANS 99.73% OF THE POPULATION IS EXPECTED AROUND THE MEAN, e.g. IF THE POPULATION IS 100 AND THE MEAN IS 30 AND THE STANDARD DEVIATION = 5, APPROXIMATELY 68 INDIVIDUALS WILL BE WITHIN THE RANGE OF 30 ± 5.

from experimental data. The normal distribution is continuous yet it can be represented by the binomial distribution $(p + q)^k$ where k is an infinitely large number. The normal distribution takes the shape of a *bell curve*, however a variety of bell curves may exist depending on the two critical parameters μ (mean),

Normal distribution continued

and σ (standard deviation). The *mean* corresponds to the center of the bell curve, and σ measures the spread of the variates. The *normal distribution probability density function* is represented by the formula:

$$f(x) = Z = \frac{1}{\sqrt{2\pi\sigma^2}} e^{-(Y-\mu)^2/2\sigma^2} \quad \text{f } \mu = 0 \text{ and } \sigma = 1 \text{ then } f(z) = \frac{1}{\sqrt{2\pi}} e^{-z^2/2}$$

where Z indicates the height the ordinate of the curve, $\pi = 3.14159$, $e = 2.71828$, Y = variable.

Although in absolute sense perfect normal distribution is only a mathematical concept, experimental data may fit in a practically acceptable manner and can be used for the characterization of the available set of data. (See figure on preceding page, mean, standard deviation)

NORMAL SOLUTION: each liter (L) contains gram molecular weight equivalent quantity (quantities) of the solute thus it can be 1N or 2N...5N, etc. In other words in 1 L of the solvent there is a solute equivalent in gram(s) atom of hydrogen. Thus 1 normal HCl (MW = 36.465, analytical grade hydrochloric acid) requires 36.465 g HCl in 1L of water. [The commercially available HCl contains about 38% HCl (specific gravity about 1.19) thus the 1 N solution should have approximately 80.62 mL of the reagent in a total volume of 1 liter] The molecular weight of sulfuric acid (H_2SO_4) is 98.076 and thus requires for the 1 N solution 98.076/2 = 49.038 gram/L because H_2SO_4 has 2 hydrogen atom equivalence. The solutions are generally prepared in volumetric flasks. (See also mole)

NORRIE DISEASE: is an X-linked (Xp11.3-p11.2) retinal neoplastic disease (a pseudoglioma) often complicated by diverse other symptoms, cataracts, microcephalus, etc. (See eye diseases)

NORTHERN BLOTTING: electrophoresed RNA is transferred from gel to specially impregnated paper to which it binds. It is then hybridized to labeled (radioactive or biotinylated) DNA probes, followed by autoradiography or streptavidin-bound fluorochrome reaction for identification. (See autoradiography, biotinylation, aminobenzyloxymethyl paper, North-Western blotting, Southern blotting, Western blotting)

NORUM DISEASE: see lecithin:cholesterol acyltransferase deficiency

NOSOCOMIAL: a hospital originated condition, such as e.g. infection acquired in a hospital.

NOSOLOGY: the subject area of disease classification.

Notch (*N*): *Drosophila* gene locus, map location 1-3.00. Homozygotes and hemizygotes are lethal; the wing tips of heterozygotes "notched" (gapped), thoracic microchaetae are irregular. The expression is greatly variable, however, in the independently obtained mutants. Homozygotes may be kept alive in the presence of a duplication containing the normal DNA sequences. All of the mutants whether they are homozygotes or hemizygotes display aberrant differentiation in the ventral and anterior embryonic ectoderm. The nervous system is abnormal. The genetic basis of the mutations are deletions, rearrangements or insertions. Homologs of the *Notch* gene are found also in vertebrates and the transmembrane gene product is involved in cell fate determination. (See *Drosophila*, morphogenesis, CADASIL)

NOTOCHORD: rod-shape cell aggregate that defines the axis of development of the animal embryo giving rise to the somites that develop into the vertebral column in higher animals. (See organizer, somite)

NOTUM: the dorsal part of the body segments in arthropods. (See dorsal)

NPC (nuclear pore complex): is involved with the import of proteins into the nucleus. It is a 125 Mda complex, embedded into the nuclear envelope. It is associated with cytoplasmic proteins, importin α and β, Ran (small guanosine triphosphatase), Nup214/CAN, and nuclear transport factor 2 (NTF2). (See nuclear pore)

NPTII: see aph(3')II, neomycin phosphotransferase, amino glycoside phosphotransferase

NPXY: a protein amino acid sequence: Asn-Pro-X-Tyr. (See amino acid symbols in protein)

NRE1: see DNA-P

NRF (NF-E related factor): see NF-E
NSF: National Science Foundation, USA agency for supporting basic research and science.
NSF (N-ethylmaleimide-sensitive factor): is a component of the SNAP (soluble NSF attachment protein) in the fusion complex of vesicle. This then binds to the membrane receptor SNARE or tSNARE (target membrane SNARE). These play a role in the transport between the endoplasmic reticulum and the Golgi apparatus and neurotransmitter release. SNAPs are involved also in the regulation of transcription of snRNA by RNA polymerases II and III. (See SNARE, synaptic vessel, snRNA)
nsL-TP: non-specific lipid transfer proteins.
N-TERMINUS (amino end): the end of a polypetide chain where translation started.
NTP: nucleotidetriphosphates.
NtrB: a kinase in the bacterial nitrogen fixation system affecting NtrC. (See nitrogen fixation)
NtrC: protein regulates bacterial nitrogen assimilation signaling path by phosphorylation. (See nitrogen fixation, signal transduction)
NTS: see nontranscribed spacer
NU BODY: unit of a chromosome fiber containing 8 + 1 molecules of histones and about 200 nucleotide pairs of DNA. (See nucleosome)
Nu END: of the DNA enters first the phage capsid. (See packaging of DNA)
NUAGE: large cytoplasmic inclusions in the cells destined to become the germline.
NUCELLAR EMBRYO: develops from diploid maternal nucellus. (See nucellus, apomixis)
NUCELLUS: the maternal tissue of the ovule surrounding the embryo sac of plants. (See embryo sac, megagametophyte development)
NUCLEAR DIMORPHISM: in *Ciliates* there is the genetically active micronucleus and the much larger macronucleus, responsible for the metabolic functions of the cell but not for its inheritance. (See Paramecium)
NUCLEAR ENVELOPE: see nuclear membrane
NUCLEAR FAMILY: parents and children living in the same household. The *extended family* includes other relatives of the household. These are actually social terms. (See family)
NUCLEAR FISSION: splitting of heavy atoms to elements, e.g. uranium may thus produce barium, krypton, etc. resulting in the liberation of great amount of thermal energy causing the split nuclei to fly apart at great velocity. (See nuclear fusion, isotopes, nuclear reactor)
NUCLEAR FUSION: atomic energy can be liberated either by nuclear fission or by fusing lighter atomic nuclei (e.g. hydrogen) into heavier ones and in the process of thermonuclear reaction generating huge amount of energy. If this reaction could be made slower (rather than explosive), humanity could get access to vast amounts of inexpensive energy. (See nuclear fission)
NUCLEAR LAMINA: are three polypeptides forming a fibrous mesh within the cell nucleus attached to the inner nuclear membrane and participate in the formation of the nuclear pores. They anchor chromosomes to the membrane and control the dissolution of the nuclear membrane during mitosis. (See mitosis)
NUCLEAR LOCALIZATION SEQUENCES (NLS): direct the movement of proteins imported into the nucleus through ATP-gated pores. (See nuclear proteins, ion channels, importin, nuclear pore, Ran, footprinting, second cycle mutation)
NUCLEAR MAGNETIC RESONANCE SPECTROSCOPY (NMR): is a physical method for studying three-dimensional molecular structures. Electromagnetic radiation is pulsed at small 15-20 kDa proteins in a strong magnetic field. This results in a change in the orientation of the magnetic dipole of the atomic nuclei. When in an atomic nucleus the number of protons and neutrons is not equal they display a spin angular momentum. In the strong magnetic field the spin is aligned but it becomes misaligned in an excited state as a consequence of the radio frequencies of the electromagnetic radiation. When they return to the aligned state, electromagnetic radiation is emitted. This emission displays characteristics dependent on the neighbors of the atomic nuclei. Thus it is feasible to estimate, from the emission spectrum, the relative posi-

Nuclear magnetic resonance spectroscopy continued

tion of hydrogen nuclei in different amino acids of the protein. If the primary structure of the polypetide is known from amino acid sequence data, the three-dimensional arrangement of the molecules can be determined. The NMR technology — unlike X-ray crystallography —does not require that the material be in a crystalline state. NMR has found its use for the analysis of medical specimens and also in plant development and infection of tissue by pathogens. Relatively thick (500 µm) tissue slices can be studied by this non-destructive method. With the aid of color video processors the distribution of water content can be followed and photographed. (See X-ray crystallography, electromagnetic radiation, circular dichroism, Raman spectroscopy)

NUCLEAR MATRIX: is an actin-containing scaffold extended over the internal space within the nucleus participating in and supporting of various functions of the DNA, including replication, transcription, processing transcripts, receiving external signals, chromatin structure, etc. (See scaffold, actin)

NUCLEAR MEMBRANE: surrounds the cell nucleus of eukaryotes and the pores of this membrane facilitates the export and import of molecules. The nuclear membrane seems to disappear during the period of prophase - metaphase and is reformed again during anaphase and telophase. (See mitosis)

NUCLEAR PORES: are perforations in the nuclear membrane, defined on the inner side by the nuclear pore complex consisting of eight large protein granules in an octagonal pattern and it is associated also with nuclear lamina. The pores selectively control the in and out traffic through their about 9 nm channel. An active mammalian cell nucleus may have 3,000-4,000 nuclear pore complexes. The nuclear pore may have a mass of 125 kDa and is built of about 100 polypeptides. Small molecules may have passive passage but larger ones (up to 25 nm in diameter) require active transport. In a minute about 100 ribosomal proteins and 3 ribosomal subunits may pass through a pore. The transport requires appropriate import and export signals without clear consensus sequences. The major import factors are *importin* α and β; α picks up the molecule to be imported and β eases them through the pore by *Ran GTPase* and protein *pp15* provide the energy. The translocation may require effectors and other energy sources too. After delivery the import complex disengages and the two importin subunits quickly return to the cytoplasm. The export of RNA — generally following trimming and splicing — requires binding to special *nuclear export sequences* of export proteins. This process may be mediated by the *Rev factor* that has binding sequences to the HIV-1 transcripts and may allow through also unspliced RNA. Human cells have hRip (human Rev interacting protein) or Rab (Rev activation domain binding) proteins. Non-viral RNAs such as mRNA, tRNA, rRNA, snRNA may have each somewhat different export system variants. The general transcription factor TFIIIA has some features that may qualify it for the viral Rev functions. The capped (by m^{1}GppN-5') RNA ends are apparently joined in a cap-binding CBP80 and CBP20 protein complex (CBC) for export. The M9 region of the heterologous nuclear ribonucleoprotein (hnRNAP) also appears to be involved in the RNA trafficking through the nuclear pores. The Ca^{2+} content of the nucleus may regulate the traffic of intermediate size molecules (> 70 kDa) across the membrane. (See nucleus, cap, Rev, RCC1, transcription factors, hnRNA, RAN, RCC, RNA transport, NPC, nuclear localization signal)

NUCLEAR PROTEINS: function within the nucleus. (See nuclear localization signals)

NUCLEAR REACTOR (atomic reactor): splits uranium or plutonium nuclei, and the neutrons in chain reactions split additional atomic nuclei. The process is regulated by cadmium and boron rods that can absorb neutrons. For moderation (slowing down the neutron) heavy water and pure graphite is used. This makes the splitting more efficient. Only the uranium^{235} is fissionable but more than 99% of the naturally occurring uranium is U^{238} that makes enrichment mandatory. In some reactors U^{238} and thorium^{232}, by capturing one neutron and releasing two electrons, yield fissionable plutonium^{239} and U^{233}. The liberated thermal energy can thus be used to drive water vapor turbines to generate electric energy for peaceful purposes. In some countries (France, Hungary) very substantial part of the electric energy is provided by atomic power

Nuclear reactor continued

plants. Atomic reactors produce also radioactive tracers used in basic research and medicine. By combining U^{235} and plutonium (Pu^{239}) *atomic bombs* may be produced. In the *hydrogen bombs* uranium or plutonium ignites a fusion process between deuterium (heavy hydrogen isotope [D] of atomic weight 2.0141) and tritium (H^3) generating helium in the process millions of degrees of heat. All nuclear reactions generate some fallout of radioactive isotopes that may pose serious genetic danger to living organisms. The use of nuclear energy for peaceful purposes, under carefully shielded and monitored conditions may be justified as a compromise between environmental protection and sustainable industrial society. The use of thermonuclear weapons does not have any justification. (See radiation effects, radiation hazard assessment, radiation protection, fallout, atomic radiations, cosmic radiation, isotopes, plutonium)

NUCLEAR RECEPTORS: provide links between signaling molecules and the transcriptional system. They use several domains with different functions. The A/B domains are receptive to transactivation. The conserved C domains bind DNA (usually after dimerization). The D domain is a flexible hinge. The E domain binds ligands (hormones, vitamin D, etc.), dimerizes and controls transcription. (See also orphan receptors, transactivation, ligand, receptor)

NUCLEAR RNA: see hnRNA, transcript

NUCLEAR SIZE and RADIATION SENSITIVITY: are directly proportional except in polyploids where multiple copies of the genes may greatly delay the manifestation of the damage although the larger nucleus is a more vulnerable target for damaging agents, the multiple copies of the same gene may compensate for each other's function. (See physical mutagens, radiation effects)

NUCLEAR TRAFFIC: see nuclear pore, RNA transport, nuclear localization sequences

NUCLEAR TRANSPLANTATION: the nuclei of plant and animal cells can be inactivated by UV or X-radiation and then the nucleus can be replaced with another. Alternatively, the nucleus of one cell is evicted (enucleation, leading to the formation of a cytoplast) by cytochalasin (a fungal toxin) and the nucleus (karyoplast) is saved. Another cell is subjected to a similar procedure and the enucleated cytoplast saved (its karyoplast is discarded). Then by reconstitution, the nucleus of cell #1 is introduced into the cytoplast of cell #2. In successful cases the transplanted nucleus is functional and can direct the normal development of the cell into an organism (frog), indicating totipotency of nuclei harvested from different cell types. Similar transplantation procedures have been successfully employed for various mammals, including sheep where viable progeny has also been obtained when the oocytes with transplanted nuclei were implanted into ewes after several *in vitro* passages. The donor nuclei were obtained from still totipotent embryonic tissues. In 1997 a normal lamb was obtained by the transfer to enucleated metaphase II stage eggs a mammary cell nucleus of a 6 years old ewe. Before transfer the mammary cells were cultured at low concentration (0.5%) serum to force the cells to exit from the cell cycle into G_0 stage. This step was required to assure compatibility of the donor nucleus (reprograming) with the egg and avoiding DNA replication that may result in polyploidy and other types of chromosomal anomalies. It was assumed that the success of reprograming or remodeling of the donor nucleus was due to the presence of appropriate transcription factors and DNA-binding proteins. The uptake of the nucleus was facilitated by electrical pulses. The identity of the donor and recipient genotypes was verified by using microsatellite markers. The possibility of transplantation of nuclei from mature tissues (although it was unknown whether the successful cell was not in an embryonic stage) and raising viable offspring after reintroducing the reconstituted egg into incubator females, was the first example of cloning of an adult mammal. One must keep in mind that "perfect" cloning of males may not be feasible because the nuclear transfer takes place into an egg that may have different mitochondrial DNA. Also, if the eggs are collected from different females, even if the nuclei are obtained from a single individual, the offspring may not be entirely identical. This experiment foreshadowed the possibility of eventual cloning of other mammals, including humans. Therefore new political and ethical concerns surfaced regarding potential manipulation of the humanrace. Biologically, the clones are not different from monozygotic twins. The manipula-

Nuclear transplantation continued

tions, however, may cause abnormal development unless the procedure is perfected. Cloning of animals may contribute to the development of improved livestock and may facilitate medical research. In the past, homozygosity of animal herds and flocks was pursued by inbreeding that may lead also to reduced vigor (inbreeding depression) which is avoided by cloning in the short range but in the long range appropriate mating schemes are required for sexual reproduction. (See totipotency, differentiation, cytoplast, cytochalasin, cloning, transplantation of organelles cell cycle, G_0, inbreeding coefficient, mating systems, *Acetabularia*)

NUCLEASES: see endonuclease, exonuclease, restriction enzymes, ribonucleases, DNase

NUCLEASE-HYPERSENSITIVITY: see nuclease sensitive sites

NUCLEASE-SENSITIVE SITES: in the transcriptionally active chromatin some short DNA tracts are more easily digested by nuclease enzymes than the rest of the chromatin. These hypersensitive regions indicate that in order to be accessible to RNA polymerase and transcription factors, the DNA must be in a particular more open conformation. A well conserved 5'-CCGGNN-3' repeat sequence seems to exclude nucleosomes. The longer of the repeat the more effective is the nucleosome exclusion. Nuclease hypersensitive sites seem to be absent in regions where there is no potential transcription. (See nucleosomes, regulation of gene activity, chromatin)

NUCLEATION: in general usage it means the formation of an initial critical core in a process; during polymerization first a smaller number of monomers assemble in the proper manner that can be followed then by a rapid extension of the polymer.

NUCLEIC ACID BASES: in DNA adenine (A), guanine (G), thymine (T) and cytosine (C), in RNA uracil (U) is comparable to (T). Besides these main bases hydroxymethyl cytosine (in bacteriophages) and methyl cytosine (in eukaryotes) are regular components. To a lesser proportion other methylated bases also may occur in both DNA and RNA. In the transfer RNAs additional minor purine and pyrimidine bases may be found such as hypoxanthine, isopentenyl adenine, kinetin, zeatin, pseudouracil, 4-thiouracil, etc.

NUCLEIC ACID HOMOLOGY: is based on the complementarity of the bases in the nucleic acids. Complementary bases can anneal by the formation of hydrogen bonds, e.g. in two single stranded nucleic acid chains such as A-T-**A-G-C**-T-G and G-C-**T-C-G**-T-A only the bases represented by bold letters are complementary. (See hydrogen bond)

NUCLEIC ACID HYBRIDIZATION: annealing single strands of complementary DNAs or DNA with RNA. (See nucleic homology, DNA hybridization, c_0t curve, Southern blotting, *in situ* hybridization)

NUCLEIC ACID PROBE: see probe

NUCLEIC ACIDS: deoxyribonucleic acid (DNA) and ribonucleic acid (RNA); polymers of nucleotides. (See DNA, RNA)

NUCLEIN: a crude nucleic acid-containing preparation first obtained by Friedrich Miescher in 1868 and identified it as a common constituent of cell nuclei in pus cells and other tissue cells and yeast. (See nucleic acid)

NUCLEO — CYTOPLASMIC INTERACTIONS: chromosomal genes may affect the expression of organellar genes and organellar genes may influence the expression of nuclear genes. (See cytoplasmic male sterility, restorer genes, chloroplast genetics)

NUCLEOID: in a region at the prokaryotic cell membrane where the cell's DNA is condensed. This DNA is not surrounded, however, by membrane as is the case with the eukaryotic nucleus. The DNA in the mitochondria and chloroplasts is organized in a similar manner without a special envelope, however, the organellar genetic material is present in multiple copies. (See prokaryote, mtDNA, chloroplast, differentiation of plastid nucleoids)

NUCLEOLAR CHROMOSOME: has a nucleolar organizing region where the ribosomal genes are located. (See nucleolus, ribosomal RNA, nucleolar organizer, nucleolar reorganization)

NUCLEOLAR ORGANIZER: is the location of the highly repeated genes responsible for coding ribosomal RNA and it assembles also the products of their transcription until utilization.

Nucleolar organizer continued

Morphologically the nucleolar organizing regions (NOR) are identified by secondary constrictions. The number of nucleolar organizer regions per genome varies from one to several. The human genome has nucleolar organizers in five chromosomes (13, 14, 15, 21, 22). (See nucleolus, ribosome, rRNA, satellited chromosome)

NUCLEOLAR REORGANIZATION: the nucleolus becomes dispersed beginning with the prophase and it is invisible by the light microscope by metaphase and anaphase, and it starts reorganization by telophase, and finishes the reorganization by the end of the G1 phase. (See nucleolus, mitosis)

NUCLEOLIN: is a nucleolar protein, coating the ribosomal transcripts within the nucleolus. (See nucleolus)

NUCLEOLUS: a body (1-5 μm) associated with the nucleolus-organizer region of one or more eukaryotic chromosomes at the coding region of ribosomal RNAs. Besides RNA, the nucleolus contains various proteins and it is the site of the production of the ribosomal subunits. It has a larger variety of proteins than those incorporated into the ribosomes. By electronmicroscopy in the nucleolar organizer region paler fibrillar regions are distinguished where the DNA is not transcribed. The darker fibrillar elements indicate the rRNAs positions, and a granular background contains the ribosomal precursors. Although in some species there are several nucleolar organizer regions, the number of nucleoli may be smaller because of fusion. During nuclear division the nucleoli gradually disappear from view (as rRNA transcription tapers off) to be again reformed beginning telophase. Apparently, during the stage of no-show the ribosomal components are not destroyed but only dispersed to the surface of the chromatin bundle. (See cell structure, ribosome, rRNA, nucleolin, nucleolar organizer, nucleus, nucleolar chromosome, nucleolar reorganization, satellited chromosome)

NUCLEOLYTIC: the function of nucleases cutting phosphodiester bonds.

NUCLEOMORPH: a vestigial nucleus-like structure in some algae, enclosed within the *chloroplast endoplasmic reticulum*. The nucleomorph of the chlorachniophytes is only 380 kb, the smallest among eukaryotes. It contains three linear chromosomes, with subtelomeric rRNA genes at both ends. Two protein genes are co-transcribed. The 12 introns are very short (18-20 bp), and the spacers are too. Most of the genes are involved with the maintenance of the nucleomorph. In the cryptomonads the nucleomorph is somewhat larger (550-600 kb) yet it also has only three small chromosomes. In some organisms the chloroplast is enclosed by three or four layers of membranes indicating that they absorbed the nucleomorph(s) of the symbiont(s) and retained it as plastid(s). (See chloroplast)

NUCLEOPHILE: an electron rich group that donates electrons to electron-deficient carbon or phosphorus atom (electrophile).

NUCLEOPHILIC ATTACK: the reaction between a nucleophile and an electrophile. (See nucleophile)

NUCLEOPHOSMIN (NPM): see anaplastic lymphoma

NUCLEOPLASM: the solutes within the eukaryotic cell nucleus.

NUCLEOPLASMIN: acid-soluble proteins mediating the assembly of histones and DNA into chromatin; it is present in the nucleoplasm and in the nucleoli and carries out also chaperonin functions. It removes sperm-specific basic proteins from the pronucleus after fertilization and replaces them by H2A and H2B histones. (See chromatin, histones)

NUCLEOPORIN: see karyopherin

NUCLEOPROTEIN: nucleic acids associated with protein.

NUCLEOSIDE: a purine or pyrimidine covalently linked to ribose or deoxyribose or to another pentose.

NUCLEOSIDE DIPHOSPHATE KINASE: mediates the transfer of the terminal phosphate of a nucleoside 5'-triphosphate to a nucleoside 5'-diphosphate.

NUCLEOSIDE MONOPHOSPHATE KINASE: transfers the terminal phosphate of ATP to a nucleoside 5'-monophosphate.

NUCLEOSIDE PHOSPHORYLASE DEFICIENCY: is a dominant human chromosome 14q22 anomaly involving T cell immunodeficiency and neurological problems. (See T cell)

NUCLEOSOME: a histone octamer is wrapped around by about 1 and 3/4 times by DNA (core particle) and these nu bodies are connected with a 40-60 nucleotide long DNA linker with either histone 1 or histone 5 (see histones).

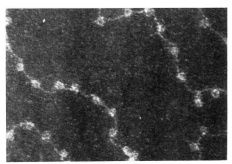

A DARK FIELD ELECTRONMICROGRAPH (260,000 X) OF A NUCLEOSOME STRING OF CHICKEN. (Courtesy of Drs. Olins, A.L. and Olins, D. E.)

The total size of the nucleosomes is different in different species and may vary between 150 to 250 nucleotide pairs. Organization similar to the nucleosome core is found in the $dTAF_{II}$ proteins associated with the TFIID transcription factors. Before transcription the nucleosomal structure is remodeled. The 2000 kDa SWI/SNF complex generates DNase hypersensitive sites by loosening the association of the DNA with histones and permitting the entry of the transcription factors. This function of SWI/SNF is only transient and requires ATP. There is another nucleosome remodeling factor (NURF) that remains associated with the nucleosomes. The compact yeast nucleosomes apparently lack histone 1. Recent studies indicate that H1 may be attached not only to the linker portion of the DNA but may be situated within the coiled section. Also, the role of the histones, considered earlier only to block transcription, in some cases may be actually slightly stimulative to transcription. The intact nucleosomal structure in place prevents the function of the DNA-dependent RNA polymerase, therefore the histone octamer is temporarily displaced and reestablished after the polymerase passed through. (See transcription factors, DNase hypersensitive site, histones, chromatosome, SWI, NURF, RAP, regulation of gene activity, chromatin remodeling, histone acetyl transferase, histone deacetylase)

NUCLEOSOME PHASING: nucleosome positions are not entirely random along the length of DNA but small variations exist and serve as controls of transcription and packaging of eukaryotic DNA. (See nucleosome)

NUCLEOTIDE: purine or pyrimidine nucleosides with 1 to 3 phosphate groups attached.

NUCLEOTIDE BIOSYNTHESIS : *purine biosynthetic path*: PHOSPHORIBOSYLAMINE -> GLYCIN-AMIDE RIBONUCLEOTIDE -> FORMYLGLYCINAMIDE RIBONUCLEOTIDE -> FORMYLGLYCINAMIDINE RIBO-NUCLEOTIDE -> 5-AMINOAIMIDAZOLE RIBONUCLEOTIDE -> 5-AMINO-IMIDAZOLE-4-CARBOXYLATE RIBONUCLEOTIDE -> 5-AMINOIMIDAZOLE-4-N-SUCCINOCARBOXAMIDE RIBONUCLEOTIDE -> 5-AMINO-IMIDAZOLE-4-CARBOXAMIDERIBONUCLEOTIDE -> 5-FORMAMIDOIMIDAZOLE-4-CARBOXAMIDE RIBONU-CLEOTIDE -> INOSINATE. From the latter ADENYLATE is formed through adenylosuccinate, and GUANYLATE is synthesized through xanthylate. Free purine bases may be formed by the hydrolytic degradation of nucleic acids and nucleotides.

The pyrimidine biosynthetic pathway: N-CARBAMOYLASPARTATE → DIHYDROOROTATE → OROTIDYLATE → URIDYLATE → CYTIDYLATE. From uridylate THYMIDYLATE is made by methylation. (See DNA replication, salvage pathway, orotic acid)

NUCLEOTIDE CHAIN GROWTH: at the initial position the 3' phosphates are retained, at subsequent sites nucleotide monophosphates are attached to the 3' position of the preceding ribose or deoxyribose after two phosphates of the triphosphonucleotides have been removed. The nucleotide chain always grows in the 5'→3' direction. (See diagram at DNA replication)

NUCLEOTIDE SEQUENCING: see DNA sequencing

NUCLEOTIDE SUBSTITUTION: see base substitution

NUCLEOTIDE TRIPLET REPEAT: see trinucleotide repeat

NUCLEOTIDYL TRANSFERASE: transfers nucleotides from one substance to another. (See terminal nucleotidyl transferase, DNA polymerase, RNA polymerase)

NUCLEUS: the genetically most important organelle (5-30 μm) in the eukaryotic cell surrounded by a double-layer membrane (ca. 25 nm) that encloses the chromosomes, proteins and RNA besides other solutes. The nuclear membrane is equipped with well organized pores for transsport of macromolecules in both directions. (See nuclear pore, nucleolus, chromosomes)

NUCLIDES: are atoms with characterized atomic number, mass and quantum. There are almost 1,000 nuclear species and about 40 are natural radioactive nuclides. By bombardment with radioactive energetic particles many additional ones have been generated in the laboratory. (See radionuclides, mass spectrometry)

NUDE MOUSE: genetically hairless; lacks thymus and thymic lymphocytes. Commonly used in immunogenetic research. (See lymphocytes, immune reaction, HLA)

NULL ALLELE: a non-expressed allele and it is commonly a deletion. (See deletion)

NULL HYPOTHESIS: assumes that the difference is null between the actually observed and the theoretically expected data. Statistical methods are then used to test the probability of this hypothesis. Obviously, if the data observed does not fit to the null hypothesis considered, it may be false to conclude that the null hypothesis is not valid; the only correct procedure is to determine the probability that the data would comply with the expectation. (See t-test, maximum likelihood, probability, significance level)

NULL MUTATION: entirely eliminates the function of a gene; deletions may appear as null mutations. (See null allele)

NULL PROMOTER: lacks TATA box and Initiator element and the transcription may begin at multiple start site sequences (MED-1). (See promoter, core promoter, base promoter)

NULLIPLEX: a polyploid or polysomic individual that at a particular locus has only recessive alleles. (See also simplex, duplex, triplex, quadruplex)

NULLISOMIC: cell or individual lacks both representatives of a pair of homologous chromosomes. Nullisomy is viable only in allopolyploids where the homoeologous chromosomes can compensate for the loss. Nullisomy may come about by selfing monosomics or by nondisjunction at meiosis I or II; in diploids it results, however in lethal gametes. Nullisomy is a normal condition for the Y chromosome in females (XX) whereas nullisomy for the X-chromosome is lethal. In allohexaploid wheat nullisomy has on an average only 4% transmission through the male whereas about 75% the eggs of monosomics are nullisomic. The cause of the high frequency of nullisomic eggs is that during meiosis I the univalent chromosome (of monosomics) fails to go to the pole and is thus lost. (See allopolyploid, monosomic, monosomic analysis, sex determination. nullisomic compensation, genome, *Triticum*)

THE COMPLETE SET OF THE 21 NULLISOMICS OF HEXAPLOID WHEAT, CHINESE SPRING. A, B AND D DENOTE THE GENOMES, AND THE NUMBERS INDICATE THE PARTICULAR CHROMOSOME WITHIN THE THREE SERIES OF 7. THE BOTTOM RIGHT EAR REPRESENTS THE NORMAL HEXAPLOID. (Courtesy of Professor E.R. Sears)

NULLISOMIC COMPENSATION: allopolyploids can survive as nullisomics but it is a deleterious condition. If, however, they are made tetrasomic for another homoeologous chromosome, their condition is ameliorated because of some degree of restoration of the genic balance. If, however, they are made tetrasomic for another chromosome, their condition is further aggravated. The response to added chromosome varies according the specific chromosome. The compensation may occur spontaneously by occasional nondisjunction. If such a nondisjunction takes place in the germline, the tissue receiving the compensating homoeologous chromosome will be at an advantage in producing gametes and there is a higher chance also for improved fertility. (See nullisomics, monosomic, nullisomic, homoeologous, tetrasomic, dosage effect)

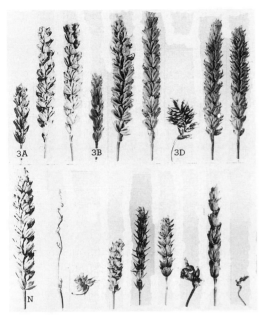

TOP ROW: NULLISOMIC 3A, NULLI 3A-TETRA 3B, NULLI 3A-TETRA 3D, NULLIISOMIC 3B, 3A, NULLI 3B-TETRA 3A, NULLI 3B-TETRA 3D, NULLISOMIC 3D, NULLI 3D-TETRA 3A, NULLI 3D-TETRA 3B
BOTTOM ROW: **NORMAL** HEXAPLOID (N), NULLI 2B-TETRA 4D, NULLI 4B-TETRA 5A, NULLI 5D-TETRA 4A, NULLI 6D-TETRA 1A, NULLI 7A-TETRA 1B, NULLI 7A-TETRA 4D, NULLI 7A-TETRA 6B, NULLI 3A-TRISOMIC 4A. OBVIOUSLY THE CORRESPONDING HOMOEOLOGOUS CHROMOSOMES COMPENSATED FOR THE ENTIRE LOSS OF THAT CHROMOSOME BUT THE NON-HOMOEOLOGOUS ADDITION EVEN AGGRAVATED THE CONDITION. (Courtesy of Professor E.R. Sears)

NULLIZYGOUS: a loss of both alleles in a diploid. (See nullisomic)
NULLOSOMIC: same as nullisomic (so used mainly by some human cytogeneticists)
NuMA (nuclear mitotic apparatus, centrophilin): is a nonhistone protein of about 250 kDa. It is present in the interphase nucleus and accumulates at the poles of the mitotic spindle until anaphase. Together with dynein and dynactin, NuMA tethers microtubules in the spindle pole and they assure the assembly and stabilization of the spindle pole. (See mitosis, nonhistone proteins, spindle, spindle fibers, dynein)
NUMERICAL APERTURE (NA): of a microscope lens determines the efficiency of the objective. The optical resolution of a dry lens with 0.75 NA at green light is about 0.5 μm and the depth of focus is about 1.3 μm whereas an oil immersion lens of 2 mm focal length and 1.3 NA has a resolution of 0.29 μm and a focal depth of 0.4 μm. (See resolution optical)
NUMERICAL TAXONOMY: classification of organisms into larger, distinct categories on the basis of quantitative measurements.
Nup214: see nuclear pore
Nup475: a transcription factor similar to TIS11 but differs in amino acid sequence at the NH_2 and COOH termini. (See TIS, transcription factors)
nur77: ligand-binding transcription factor including steroid and thyroid hormone receptors (similar to NGF-B and TIS1). Its level is high in apoptotic lymphocytes but not in growing T cells.
NURF: see histone
NURSE CELLS: in insect ovaries 15 (generally polyploid) nurse cells surround the oocyte within the follicles and their gene products affect, and have morphogenetic role, in the differentiation of the embryo at the early stages of development. (See morphogenesis)
NURTURE: nutritional (and also other environmental factors) that affect the manifestation of the hereditary properties (nature). In human genetics for the separation of the two components of

the phenotype twin studies are used. The differences between identical twins permits the quantitation of the extent of the influence of nurture. (See twinning)

nusA, nusB: are lambda bacteriophage genes involved in the regulation of RNA chain elongation. (See lambda becteriophage)

NUTMEG (*Myristica fragrans*): evergreen, dioecious spice plant; $2n = 6x = 42$.

NUTRITIONAL MUTANT: see auxotroph, mutation

NUTRITIONAL THERAPY: humans cannot synthesize the essential amino acids and depend on the diet for a steady supply. Smilarly, there is a dependence on an exogenous (dietary or medicinal) supply of vitamin C. Some epileptics may benefit from the administration of pyridoxin. Various hereditary defects are known in folic acid metabolism. Hereditary fructose intolerance, galactosemia, lactose intolerance can be kept in check by limiting the supply of these carbohydrates in the diet or infant formulas. Phenylketonurics must avoid phenylalanine consumption. (See epilepsy, fructose intolerance, galactosemia, disaccharide intolerance, phenylketonuria)

NYCTALOPIA: see night blindness

NYMPH (nympha): a sexually immature stage between larvae and adults of some arthropods such as ticks (*Ixodes*). (See Ixodiodia)

NYMPHA of Krause: the same as clitoris. (See clitoris)

NYMPHOMANIA: is an excessive sexual drive (abnormally long estrus) in the mammalian female based on hormonal disorders and usually accompanied by reduced fertility in mares and cows. The condition may have a clear hereditary component.

NYSTAGMUS: involuntary eye movement (displayed bysome albinos). This condition may be controlled by autosomal recessive, dominant or X-linked inheritance and may be associated as parts of some syndromes. (See eye diseases)

NYSTAGMUS-MYOCLONOUS: nystagmus accompanied by involuntary movement of other parts of the body. A rare congenital anomaly. (See nystagmus)

NYSTATIN: an antibiotic produced by a *Streptomyces* bacterium, and it is effective against fungal infections; it is also used as selective agent in mutant isolation of yeast and other fungi. It kills primarily the growing cells.

NZCYM BACTERIAL MEDIUM: H_2O 959 mL, casein hydrolysate (enzymatic, NZ amine) 10 g, NaCl 5 g, Bacto yeast extract 5 g, Casaminoacids 1 g, $MgSO_4$ 7 H_2O 2 g, pH adjusted to 7 with 5 N NaOH and filled up to 1 L. (See casamino acids, Bacto yeast extract)

NZM MEDIUM: the same as NCZYM but without casamino acids.

O

O: replicational origin. (See replication, replication fork, bidirectional replication)

Ω (omega): insertion element present in 0 to 1 copy per mitochondrion in yeast. (See mitochondria, mtDNA, insertion elements)

O ANTIGEN: see ABO blood group

O BLOOD GROUP: see ABO blood group

ω-AGATOXIN: ion channel blocking proteins from the *Agelenopsis* spiders.

OAK (*Quercus* spp): forest and ornamental trees with great morphological variety, 2n = 24.

OATS (*Avena* ssp.): a major cereal crop with somewhat reduced acreage since farm mechanization diminished the number of horses used in agriculture. The cultivated species (*A. sativa*) is an allohexaploid but diploid and tetraploid forms are also well known. Basic chromosome number, x = 7.

OBESITY: accumulation of excessive body weight (primarily fat) beyond the physiologically normal range. Differences in predisposition to obesity is long recognized by animal breeders and the different breeds of swine have large (over 100%) differences in fat content per body weight. Obesity is a health problem in humans because diabetes mellitus, hypertension, hyperlipidemia, heart disease and certain types of cancer appear to be associated with obesity. In mouse obesity is regulated by a gene *ob* in chromosome 6, sequenced in 1994. It has been suggested that the 167-amino acid protein product synthesized in the adipose (fat) tissues is secreted into the blood stream and regulates food intake through signaling to the hypothalamus. Reduction of this gene product or specific lesions to the ventromedial region of the hypothalamus stimulates food consumption and reduces energy expenditures. Some experimental data point to the *db* (*diabetes*) gene (mouse chromosome 4) receptor for the *ob*-encoded factor, leptin. The *tubby* gene of mouse (chromosome 7) also causes a maturity-onset obesity, insulin resistance, vision and hearing deficit. The *fat* mutation in mouse has a later onset than *ob*. In *ob/ob* mice the serum insulin level decreases with an increase of blood glucose level. In the *fat/fat* mice exogenously supplied insulin decreases serum glucose level. The fat mice store 70% of their insulin as proinsulin. Apparently, the *fat* gene causes deficiency in carboxypeptidase, an enzymes that normally processes proinsulin. In humans, obesity has been attributed to both dominant and recessive genetic factors with environmental (diet) factors contributing to about 40% of the variation in obesity. There was some indication of greater affect of human maternal than paternal body weight on the obesity of the progeny. Neuropeptide Y (NPY) appears to be a stimulant of food intake and an activator of a hypothalamic feeding receptor (Y5). A cAMP-dependent protein kinase (PKA) also plays an important role in obesity. This holoenzyme is a tetramer, containing two regulatory (R) and two catalytic (C) subunits. The catalytic function is phosphorylation of serine/threonine, and the regulatory units slow down the enzyme when the level of cAMP is low. A knockout of the RIIβ regulatory subunit leads to stimulation of energy expenditures in mice and they remained lean even on a diet that normally was conducive to obesity. In the inner membrane of the mammalian mitochondria body heat is generated by uncoupling oxidative phosphorylation. This process regulates energy balance and cold tolerance. Mice mutant in the uncoupling protein (ICP) have increased food intake but because of the increase in the rate of metabolism they do not become obese. (See leptin, diabetes, insulin, Prader-Willi syndrome, Alström syndrome, Bardet-Biedl syndrome, neuropeptide)

OBF (oct-binding factor; synonym BOB.1): regulates the lymphocyte specific oct sequence in the promoter of the transcription of the immunoglobulin genes; it is required for the development of the germinal centers. (See oct, immunoglobulins, germinal center)

OBJECTIVE LENS: microscope lens next to the object to be studied. (See light microscopy)

OBLIGATE: restricted to a condition, necessarily of a type, e.g. obligate parasite, obligate an-aerobe, etc.

OBLIQUE CROSSING OVER: in case of adjacent (tandem) duplications in homologous chromosomes pairing may take place in more than one register and crossing over may yield unequal products:

```
A A A A          A A A ↓A                    → A A A A A
A A A A          A A ↑A A        ⎤
normal           oblique          ⎦          → A A A
pairing          pairing and                   unequal
                 crossing over                  recombinant
                                                products
```

(See unequal crossing over)

OCA-B: same as OBF or Bob. (See OBF)

OCCAM's RAZOR (Ockham's razor): the philosophical precept of William Ockham [1280-1349], rebellious clergyman and venerabilis inceptor [= reverend innovator]: *"pluralites non est ponenda sine necessitate"*, meaning that multiple alternatives should not be offered in logical argumentation but the simplest yet adequate explanation should be chosen. (See maximal parsimony)

OCCLUSION: transcription from one promoter reduces transcription from a downstream promoter. (See downstream, promoter, transcription)

OCCUPATIONAL AND SAFETY AND HEALTH ADMINISTRATION: see OSHA

OCCURRENCE RISK: the chance that an offspring of a certain couple will express or become a carrier of a gene. (See genetic risk, recurrence risk)

OCELLUS (plural ocelli): the simple light sensor (eyelet) on the top of the head of insects, behind the compound eyes. (See compound eyes, ommatidium, rhabdomere, morphogenesis in *Drosophila*)

OCHRE: chain-terminator codon (UAA).

OCHRE SUPPRESSOR: mutation in the anticodon of tRNA that permits the insertion of an amino acid at the position of a normally chain terminating UAA RNA codon; ochre suppressors frequently suppress amber (UAG) mutations too. (See code genetic, nonsense codon, suppressor tRNA)

ω-CONOTOXINS: are *Conus* snail inhibitors of calcium ion channels. (See ion channels)

OCT-1: a mammalian gene regulatory protein with octa recognition sequence: $\begin{smallmatrix}\text{ATGCAAAT}\\\text{TACGTTTA}\end{smallmatrix}$. Oct-1 and Oct-2 regulate B-cell differentiation. The Oct-6 transcription factor regulates Schwann cell differentiation. (See Schwann cell, lymphocytes, OBF, B cell, immunoglobulins, Oct-2)

Oct-2: a lymphoid transcription factor, similar to Oct-1; both respond to BOB.1/OBF.1 activators.

OCTA: an 8-base sequence (ATTTGCAT) in the promoter of H2B histone gene, and some other genes. Several slightly different octa sequences are found in the promoter regions of different genes. (octo in Latin, οκτασ in Greek: number 8, Oct-1)

OCTAD: being of eight elements, e.g. the spores in an ascus if meiosis is followed by an immediate mitotic step as e.g. in *Neurospora, Ascobolus*, etc.

OCTAPLOID: a cell nucleus carrying eight genomes, 8x. (See polyploid)

OCTOPINE: a derivative of arginine, synthesized by a Ti plasmid gene (*ocs*) in *Agrobacterium* strain Ach 5. (See opines, *Agrobacterium*) →

OCULAR: microscope lens next to the viewer's eye. (See objective)

OCULAR ALBINISM: see albinism

OCULAR CICATRICIAL PEMPHIGOID: is an autosomal dominant autoimmune disease of the eye and possibly of other mucous membranes. It may be associated with defects in the HLA system. (See eye diseases, autoimmune disease, HLA)

OCULODENTODIGITAL DYSPLASIA: an autosomal dominant disorder involving defects in the eyes (microphthalmos) small teeth and polydactyly or syndactily. (See eye diseases, microphthalmos. polydactyly, syndactyly)

O.D.: optical density; indicates the absorption of light at a particular wavelength by a compound in a spectrophotometer. O.D. can be used to characterize a molecule, e.g. pure nucleic acids have maximal absorption at 260 nm but contamination and the solvent may alter the absorption pattern. (See also DNA measurements, extinction)

ODONTOBLAST: connective tissue cell, forming dentin and dental pulp of the teeth.

OENOTHERA: (*Onagraceae*, x = 7); several species, diploid and polyploid have been used for cytological study of multiple translocations and the nature and inheritance of plastid genes. (See gaudens, translocation, complex heterozygote, megaspore competition, zygotic lethal)

OESTROGEN (estrogen, estradiol): a steroid hormone. (See animal hormones)

OESTRUS (estrus): the periodically recurrent sexual receptivity, concomitant with sexual urge (heat) of mammals, except humans.

OFAGE (orthogonal-field alternation gel electrophoresis): the device used for the isolation of the small chromosomal size DNA of lower eukaryotes. (See pulsed field electrophoresis)

OFFERMANN HYPOTHESIS: was supposed to provide a mechanism for recombination within a short chromosomal region that would appear to be intragenic although the recombination just separated genes exerting position effect on their flanking neighbor. These flanking loci were not supposed to have any detectable phenotype themselves, except the position effect on the neighbor. This idea emerged in 1935, years before pseudoallelism has been discovered in 1940 by C.P. Oliver. (See pseudoallelism)

OFFSPRING-PARENT REGRESSION: see correlation

OGOD (one gene - one disorder): a hypothesis based on the analogy of the one gene - one polypeptide (one enzyme) theorem. The majority of the human (and animal) diseases cannot be reconciled with a single gene mutation, however, and the majority of the disease symptoms (syndromes) are under multigenic control although particular genes may have major effect. (See also behavior in humans, one gene—one enzyme theorem)

OGUCHI DISEASE: is a recessive human chromosome 2q mutation or deletion in the arrestin protein modulating light signal transduction to the eye or by defects in the rhodopsin kinase gene. Night vision is impaired but otherwise vision is normal. (See night blindness)

ohm (Ω = 1V/A): resistance of a circuit in which 1 volt electric potential difference produces a current of 1 ampere.

OHNO's LAW: the gene content of the X chromosome is basically the same in all mammals. Some exceptions are in humans, marsupials and in the monotreme, *Platypus* where genes in the short arm of the X chromosome may be of autosomal origin. There are other exceptions such as the chloride channel gene (*Cln4*) is autosomal in the mouse *Mus musculus* but it is X-linked in *Mus spretus*. Similarly the human steroid sulfatase (STS) gene is near the pseudo-autosomal region whereas in lower primates it is autosomal. The rationale of the Law is that translocations between autosomes and X-chromosome would upset sex-determinational gene balance. The X chromosomes of various mammals more or less conserved their sequence homologies; some rearrangements have, however, been detected by FISH probes. (See sex determination, FISH)

OIDIA: asexual fungal spores produced by fragmentation of hyphae into single spores.

OIL IMMERSION LENS: is the highest power objective lens of the light microscope. It is used with a special non-drying immersion oil, available at different viscosities with a refractive index of about 1.5150 for D line at 23° C, and it increases light-gathering power and improves resolution. (See light microscope, objective, resolution optical, numerical aperture)

OIL SPILLS: about 22 bacterial genera are known to have genetically determined ability to degrade petroleum hydrocarbons.

OK BLOOD GROUP: is encoded in human chromosome 19pter-p13. This antigen is present on the red cells of chimpanzees and gorillas but not in rhesus monkeys, baboons or marmosets. (See blood groups)

OKADAIC ACID: an inhibitor of PP1 and PP2a protein phosphatases.

OKAYAMA & BERG PROCEDURE: permits cloning of a full length mRNA genes. The

Okayama & Berg procedure continued

mRNA is extracted from post-polysomal supernatant of reticulocyte lysate of rabbits, made anemic by phenylhydrazine injection. The globin mRNA is recovered in the alcohol-precipitate of phenol extract or with the aid of a guanidinium thiocyanate method. The poly-A tailed mRNA is annealed to plasmid pBR322 that is equipped with a poly-T attached to a SV40 fragment inserted into the vector. In a following step an oligo-G linker is constructed, separated, and purified by agarose gel electrophoresis. Now cloning of the mRNA can be started. The poly-A tail is annealed to the poly-T end of the vector. Using reverse transcriptase, a DNA strand, complementary to the mRNA strand, already in the plasmid, is generated. Both to one strand of the plasmid vector and to the DNA strand of the RNA-DNA double strand, poly-C tails are added using terminal transferase enzyme. Now the oligo-G linker is added to the oligo-C ends and the plasmid is made circular by DNA ligase. Then the mRNA strand is removed by the use of RNase H and replaced by a complementary DNA strand, generated by DNA polymerase I and the construction is finished by ligation into a circular cloning vector. This new vector, containing the full-length cDNA, is transformed into *E. coli* cells for propagation. (See cloning vectors, ribonuclease H, linker, reverse transcriptase, terminal transferase, guanidinium thiocyanate, phenylhydrazine, RNA extraction)

OKAZAKI FRAGMENTS: are short (generally less than 1 kilobase in eukaryotes and about 2 kb in prokaryotes) DNA sequences formed during replication (of the lagging strand) and subsequently ligated into a continuous strand. (See also DNA replication, replication fork)

OKRA (*Abelmoschus esculenta*): an annual vegetable of the Malvaceae with about 30-40 species with variable chromosome numbers, generally higher than n = 34-36 have been reported.

OKT3: a monoclonal antibody capable of blocking interleukin production. (See also Oct)

OLFACTOGENETICS: is concerned with the genetically determined differences in body smell and the ability to recognize it. Scent is influenced by the chemical nature of secretions and to a great deal also by the diet and the microflora of the body. Human polymorphism in olfactory responses are of concern for the cosmetics (perfume) industry. It has been claimed that the ability to distinguish between various odors is determined by the *H2* locus of mice (an analog of the human *HLA* complex). Some people are incapable of smelling, anosmic for isobutyric acid or cyanide or urinary excretes of asparagus metabolites. The regulation of olfactory responses involve cAMP or phosphoinositide (IP_3)-regulated ion channels. (See bisexual, cAMP, phosphoinositide, IP_3)

OLFACTORY: concerned with smell sensation. (See olfactogenetics)

OLIGODENDROCYTE: non-neural cells that form the myelin sheath (neuroglia) of the central nervous system and they coil around the axons. (See neurogenesis)

OLIGOGENES: a small group of genes responsible for a particular trait; usually one has, however a major role. (See breast cancer, prostate cance, polygenes, QTL)

OLIGO-LABELING PROBES: short (10-20 nucleotides), commonly radioactively labeled, synthetic probes for the identification of genes for isolation, labeling in gel retardation assays, screening DNA libraries, etc. (See label, probe, gel retardation assay)

OLIGOMER: a polymer of relatively few units (amino acids, nucleotides, sugars, etc.)

OLIGONUCLEOTIDE-DIRECTED MUTAGENESIS: see site-specific mutagenesis

OLIGOPHRENIA PHENYLPYRUVICA: mental retardation due to phenylketonuria. (See phenylketonuria)

OLIGOSACCHARIDES: consist of sugar residues, such as glycans.

OLIGOSPERMIA: low sperm content in the semen. (See sperm, semen)

OLIVE (*Olea europea*): oil producing trees about 30 species. The cultivated forms are 2n = 2x = 46 although aneuploids have been identified.

OLIVOPONTOCEREBELLAR ATROPHY (OPCAI): the autosomal dominant or recessive, variable types of expressions involve ataxia, paralysis, incoordination, speech defects, and brain and spine degeneration. In some forms eye defects and other anomalies were also observed. Some cases displayed linkage to the HLA complex in human chromosome 6p21.3-

p21.2, in other instances such a linkage was not evident. Patients with this disease display 50% or less glutamate dehydrogenase activity. (See ataxia, palsy, glutamate dehydrogenase)

OMIM: an up-to-date catalog of autosomal dominant, autosomal recessive, X-linked and mitochodrial genes of humans available through the Internet (<http://www.ncbi.nlm.nih.gov/Omim/>). It is also available in book form: McKusick, Victor 1994 Mendelian Inheritance in Man. The Johns Hopkins University Press, Baltimore, MD, USA.

OMMATIDIUM (plural ommatidia): a self-sufficient element (facet) of the compound eye of arthropods, such as *Drosophila*. (See compound eye, rhabdomere)

OMMOCHROMES: insect eye pigments synthesized from tryptophan and the formation and condensation of hydroxykinurenine into xanthommatin and complexing with other components into, brown pigment granules. (See pteridines, *w* locus, pigmentation in animals)

OMNIPOTENT: see totipotent

omp A: outer membrane protein A of the bacterial cell.

OmpC, OmpF: bacterial outer membrane proteins are regulated by kinase EnvZ in response to osmolarity (at low osmolarity by C and at high by F); OmpR is an outer membrane regulatory protein. (See cell membrane, membrane proteins)

ONCOCYTOMA: see mitochondrial disease in humans, kidney diseases

ONCOGENES: of viruses (v oncogene) or similar genes in animal cells (c oncogene) are responsible for the first step in cancer initiation. The primary target of the majority of oncogenes is the cell cycle and they deregulate the function of genes that normally control the initiation or progression of the cell cycle. For details see ABL, AKT1, AMV, ARAF, ARG, BLYM, BMYC, CBL, DBL, ELK, EPH, ERBB, ERG, ETS, EVI, FES, FGR, FLT, FMS, FOS, GLI, HIS, HKR, HLM, HST, INT, JUN, KIT, LCA, LCK, LYT, MAS, MCF, MEL, MET, MIL, MYB, MYC, NGL, NMYC, OVC, PIM, PKS, PVT, RAF, RAS, REL, RET, RHO, RIG, ROS, SPI1, SEA, SIS, SK, SNO, SRC, TRK, VAV, YES, YUASA. (See cancer, carcinogens, retroviruses, non-productive infection, proto-oncogene, CATR1)

ONCOGENIC TRANFORMATION: development of a cancerous state. (See oncogenes, oncoproteins, oncogenic viruses)

ONCOGENIC VIRUSES: can integrate into mammalian cells and rather than destroying the host, they can induce cancerous proliferation of the target tissues. The oncogenic viruses may have double-stranded DNA genetic material such as the adenoviruses (genome size ca. 37 kbp), Epstein-Barr virus (ca. 160 kbp), human papilloma virus (ca. 8 kbp), polyoma virus (ca. 5-6 kbp) and the single-strand RNA viruses (retroviruses, 6-9 kb). (See adenoviruses, Epstein-Barr virus, papova viruses, retroviruses, acquired immunodeficiency, tumor viruses)

ONCOMOUSE: trade name for a mouse strain prone to breast cancer and suitable for this type of research. (See breast cancer)

ONCOPROTEIN: product of an oncogene, responsible for initiation and/or maintenance of hyperplasia and malignant cell proliferation. (See oncogenes)

ONCOSTATIN M: see APRF

ONE GENE-ONE ENZYME THEOREM: recognized that one gene is generally responsible for one particular biosynthetic step, mediated through an enzyme. Somewhat more precisely stated is the one gene (cistron)-one polypetide rule because some enzymatically active protein aggregates may be encoded by more than a single gene. There are also some other apparent exceptions, e.g., one mutation blocking the synthesis of homoserine may prevent the synthesis of threonine and methionine too because homoserine is a common precursor of these amino acids. Also, in the branched-chain amino acid (isoleucine-valine) pathway ketoacid decarboxylase and ketoacid transaminase enzymes control the pathways leading to both isoleucine and valine. (See homoserine, isoleucine-valine biosynthetic pathway, bifunctional enzymes)

ONE GENE-ONE POLYPEPTIDE: see one gene-one enzyme theorem

ONE-HYBRID BINDING ASSAY: is basically a gene fusion assay where a transcriptional activation domain is attached to a particular gene. The function of such a "hybrid" may then be assessed by the expression of an easily monitored reporter gene (e.g. luciferase), depending on the signal received from the particular gene. (See two-hybrid assay, gene fusion, luciferase, split-hybrid system, three-hybrid system)

ONE-STEP GROWTH: bacteriophages multiply within the bacterial cell until in one step, within less than 10 minutes, during the "rise" period, all the particles are released. The temperate phages may have a longer period preceding the rise after infection because the phage DNA may become integrated into the bacterial chromosome and then replicates synchronously with the bacterial genes. Upon induction the phage may switch to a lytic cycle which begins with autonomous replication followed by liberation of the phage particles. The number of phage particle released in then called the burst size. (See diagram on the left, bacteriophages, phage lifecycles)

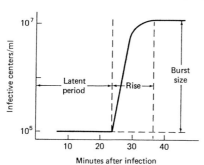

ONION (*Allium* spp): the Alloideae subfamily of the lily family includes 600 species. *A. cepa*, 2n = 2x = 16. Most of the North American species have x = 8. Polyploid species occur in different parts of the world with somatic chromosome numbers 24, 32, 40, 48.

ONOZUKA R-10: see cellulase

Allium, x = 8 →

ONSET: stage (time) of expression of a genetically determined trait.

ONTOGENY: the developmental course of an organism. (See also phylogeny)

ONTOS: *E. coli* computer data sets organized into object-oriented data base management system.

OOCYTE DONATION: is a means to overcome infertility of women who for some reason (older age, genetic risks, etc.) do not want or cannot conceive the normal manner but can serve as a recipient of either her own ovum obtained earlier and preserved or of an ovum from a donor. She can thus carry to term a normal baby. From genetic view point it is important that the ovum implanted should be carefully analyzed to be as risk free as possible. (See artificial insemination, surrogate mother, ART)

OOCYTE, PRIMARY: have the same chromosome number as other common body cells (2n, 4C) but upon meiotic division each gives rise to two haploid *secondary oocytes* (n, 2C). One of the two, the smaller, is called 1st polar body. By another division then the egg and three polar bodies (n, 1C) are formed The egg may become fertilized but the polar bodies do not contribute to the progeny and fade away. In human females the meiosis begins already in the four months old fetus and proceeds to diakinesis (dictyotene) stage until sexual maturity. After puberty in each four-week cycles one oocyte reaches the stage of the secondary oocyte and then after completing the equational phase of meiosis (meiosis II), an egg is released during ovulation. Each of the two human ovaries contain about 200,000 primary oocytes yet on the average only about 400 eggs are produced during the entire fertile period of the human female, spanning on the average about 30 to 40 years from the beginning at puberty and terminating with the onset of menopause. (See egg, spermatocyte, meiosis, C amount of DNA, menopause, gametogenesis, *Xenopus* oocyte culture, oogonium, oospore, oocyte translation system)

OOCYTE, SECONDARY: see oocyte primary

OOCYTE TRANSLATION SYSTEM: mRNAs injected into the amphibian oocyte (*Xenopus*) nucleus (germinal vesicle) may be transcribed, and in the cytoplasm these exogenous messengers are translated, the proteins may be correctly processed, assembled, glycosylated and phosphorylated, and delivered (targeted) to the proper location, etc. Similarly, injections of foreign DNA into the fertilized embryos may be replicated and inserted into the chromosomes. (See translation *in vitro*)

OOGAMY: the fertilization of (generally) a larger egg with a (smaller) sperm.

OOGENESIS: the formation of the egg. (See gametogenesis)

OOGONIUM: female sex organ of fungi fertilized by the male gametes.

OOGONIUM IN ANIMALS: the primordial female germ cell that is enclosed in a follicle by the term of birth and becomes the oocyte. (See also gametangium, gametogenesis)

OOPHORECTOMY: surgical removal of the ovary; same as ovariectomy, neutering, spaying. (See castration)

OOPLASM: egg cytoplasm.

OOSPORE: a fertilized egg in (fungi); it is either dikaryotic or diploid and frequently covered by a thick wall.

OPAL: chain-terminator codon (UGA). (See code genetic)

opaque (*o*): genes in maize (several loci); *o-2* gained particular attention because in its presence the prolamine and zein type proteins are reduced and the lysine content of the kernels increases. Thus the nutritional value improves significantly and has importance for some underdeveloped parts of the world where corn may be the main food staple. (See *floury*, kwashiorkor)

OPEN PROMOTER COMPLEX: a partially unwound promoter (the DNA strands separated) to facilitate the operation of the RNA polymerase. This separation is supposed to be the result of the attachment of the transcriptase to the promoter. The TATA box of the promoter is logical place for the attachment of the pol enzyme because there are only two hydrogen bonds between A and T in contrast to the three bonds between G and C and thus separation of the double helix is easier. This is followed by initiation of transcription. After the attachment of the RNA-elongation proteins, the σ subunit of the bacterial pol enzyme is evicted and transcription proceeds. (See also closed promoter complex, pol, Pribnow box, Hogness box, TBP, TAF, transcription factors, regulation of gene activity, RAD25, regulation of gene activity)

OPEN READING FRAME (ORF): a nucleotide sequence between an initiation and a terminator codon. In higher organisms most commonly one of the two DNA strands is transcribed into functional products although there are open reading frames in both strands.

OPEN SYSTEM: exchanges material and energy within its environment.

OPERAND: what is supposed to be operated (worked) on.

OPERATIONAL CONCEPTS: were frequently employed in genetics for providing an explanation when the underlying mechanism was not fully understood but from the visible behavior a conceptualization was possible in agreement with what was known. Examples: T.H. Morgan defined the gene as the unit of function, mutation and recombination before the nature of the genetic material was discovered. F.H.C. Crick concluded on the basis of frameshift mutagenesis — before the genetic code was experimentally determined— that the genetic code is most likely using nucleotide triplets.

OPERATIONAL TAXONOMIC UNIT (OTU): see character matrix

OPERATOR: recognition site of the regulatory protein in an operon or possibly in other systems such as suggested for controlling elements (transposable elements) of maize. (See transposable elements, operon, *Lac* operon, *Ara* operon)

OPERON: functionally coordinated group of genes producing polycistronic transcripts. Similar organization occurs also in the homeotic gene complexes of eukaryotic organisms, such as *ANTP-C* and *BX-C* of *Drosophila*. Many genes of the nematode (1/4), *Caenorhabditis* seem to be coordinately regulated and transcribed into polycistronic RNA that is processed into mRNA by transsplicing. (See *Lac* operon, *Arabinose* operon, *Tryptophan* operon, *His* operon, morphogenesis in *Drosophila,* homeotic genes, coordinate regulation, transsplicing, SL1, SL2)

OPHTHALMOPLEGIA: an autosomal dominant phenotypes involve defects in moving the eyes and the head. The autosomal recessive ophthalmoplegic sphingomyelin lipidosis appears to be allelic to the Niemann-Pick syndrome gene and it is associated with mitochondrial DNA mutations. (See mitochondrial diseases in humans, Kearns-Sayre syndrome, Niemann-Pick disease, eye disease, myopathy)

OPIATE: opium like substance. They regulate pain perception and pain signaling pathways and mood. *Endogenous opiates,* enkephalins and endorphins, were isolated from the brain and the pituitary gland, respectively. They contain a common four-aminoacid sequence and bind to the same cell surface receptors as morphine (and similar alkaloids). Nociceptin (orphanin) are

17 amino acid antagonists of the opioid receptor-like receptor. Opioids are opiate-like, but they are not derived from opium. (See enkephalin)

OPINES: are synthesized in crown-gall tumors of dicotyledonous plants under the direction of agrobacterial plasmid genes. The bacteria uses these opines as carbon and nitrogen sources. The octopine family of opines includes octopine, lysopine, histopine, methiopen, octopinic acid. The nopaline group includes nopaline and nopalinic acid. Agropines are agropine, agropinic acid, mannopine, mannopinic acid, agrocinopines. (See *Agrobacteria*, T-DNA octopine, nopaline)

OPIOCORTIN: is a prohormone (pro-opiocortin) translated as a precursor of several corticoid hormones and cut by proteases and processed into corticotropin, β-lipotropin, γ-lipotropin, α-MSH (melanin-stimulating hormone), β-MSH and β-endorphin as shown below: (See animal hormones)

PRO-OPIOCORTIN

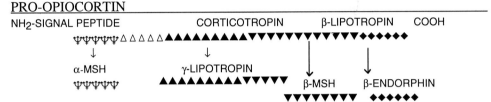

(See the individual peptide hormones under separate entries)

OPITZ SYNDROME (G syndrome, BBB syndrome): an apparently autosomal dominant anomaly with complex features such as hypertelorism (the paired organs are unusually distant), defects in the esophagus (the passage-way from the throat to the stomach), hypospadias (the urinary channel opens in the underside of the penis in the vicinity of the scrotum), etc. An autosomal recessive form polydactily, heart anomaly, triangular head, failure of the testes to descend into the scrotum and suspected deficiency of the mineralocorticoid receptor. The Opitz syndrome was assigned to chromosome 22q11.2. (See corticosteroid, head/face/brain defects, Opitz-Kaveggia syndrome)

OPITZ-KAVEGGIA SYNDROME: is an Xp22-linked phenotype involving large head, short stature, imperforate anus, heart defect, muscle weakness, defect in the white matter of the brain (corpus callosum), mental retardation. (See stature in humans, heart disease, mental retardation, head/face/brain defects, Opitz syndrome)

OPOSSUM (Marsupials): *Caluromys derbianus* 2n = 14; *Chironectes panamensis:* 2n = 22.

OPSINS: are photoreceptor proteins of the retina but non-visual photoinduction opsin exists in the pineal gland. (See pineal gland, rhodopsin)

OPSONINS: trigger phagocytosis by the scavenger macrophages and neutrophils. These substances bind to antigens associated with immunoglobulins IgG and IgM and facilitate the recognition of the antigen-antibody complexes by the defensive scavenger cells. Also, they may bind to the activated complement of the antibody and assist in the recognition of the cell surface antigens and thus mediate their destruction. (See antibody, complement, immunoglobulins, macrophage, neutrophil)

OPTIC ATROPHY: determined either by autosomal dominant, recessive (early onset types) or X-linked or mitochondrial defects of the eye, ear and other peripheral nerve anomalies. (See Behr syndrome, night blindness, Leber optic atrophy, Kearn-Sayre syndrome, Wolfram syndrome, mitochondrial disease in humans, myoclonic dystrophy, eye diseases, color blindness)

OPTICAL DENSITY: see O.D.

OPTICAL MAPPING: may be used for ordering restriction fragments of single DNA molecules. The fragments are stained by fluorochrome(s) and the restriction enzyme-generated gaps can be visualized. (See physical mapping, mapping genetic, FISH)

OPTICAL ROTATORY DISPERSION: a variation of optical rotation by wavelength of polarized light; it depends also on the difference in refractive index between left-handed and right-handed polarized light. The rotation is measured as an angle. It is very similar to circular dichroism. (See base stacking, circular dichroism)

OPTICAL SCANNER: is a device that generates signals from texts, diagrams, pictures, electrophoretic patterns, autoradiograms, etc. that can be then read or printed out with the assistance of a computer.

opus: see copia

ORANGE (*Citrus aurantium*, 2n = 18): is the familiar fruit tree. Botanically the peel of the fruit is the pericarp, containing at the lower face the fragrant oil glands. The juice sacs are enclosed by the carpels, containing the seeds. (See also navel orange)

ORC (origin recognition complex): is a six-subunit complex required before DNA replication can start in eukaryotic cell. Another protein ORC2 (Orp2 in fission yeast) apparently interacts with Cdc2, Cdc6 and Cdc18 proteins that regulate replication. ORC is also required for silencing of the HMRa locus of yeast, involved in mating type determination. The ORC homolog of *Drosophila* is DmORC2. (See replication, Cdc2, Cdc18, mating type determination in yeast, ARS, MCM, replication licensing factor)

ORCHARD GRASS (*Dactylis glomerata*): a shade and drought-tolerant forage crop, 2n = 28.

ORCHIDS (*Orchideaceae*, 2n = 20, 22, 34, 40): monocotyledonous tropical ornamentals.

ORDERED TETRAD: the spores in the ascus represent the first and second meiotic divisions in a linear sequence such as ⓪ ⓪ ⓪⓪ . (See tetrad analysis)

ORF: open reading frame, the nucleotide sequences between the translation initiator and the translation terminator codons.

ORGAN: a body structure destined to a function.

ORGAN CULTURE: growing organs *in vitro* to gain insight into function, differentiation and development.

ORGANELLE GENETICS: see mitochondrial genetics, chloroplast genetics, sorting out

ORGANELLE SEQUENCE TRANSFERS: during evolution apparently sequences homologous among the major organelles were transfered in the directions shown below:

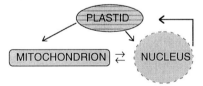

(See mobile genetic elements, mtDNA, chloroplasts)

ORGANELLES: membrane-enclosed cytoplasmic bodies such as the nucleus, mitochondrion, plastid, Golgi, lysosome. (See under separate entries)

ORGANIC: a carbon-containing compound; something associated with a metabolic function.

ORGANIC EVOLUTION: is the historical development of living beings of the past and present times. (See geological time periods)

ORGANISMAL GENETICS: studies inheritance in complete animals and plants and does not employ molecular methods or the tools of reversed genetics. (See reversed genetics, inter-organismal genetics)

ORGANIZER (Spemann's organizer): the dorsal lip of the blastopore (an invagination that encircles the vegetal pole of the embryo) and becomes a signaling center for differentiation. The formation of the organizer is preceded by induction of the mesoderm cell layer, resulting in the expression of organizer-specific homeobox genes and transcription of genes coding for signal molecules. This is followed by recruitment of the neighboring cells into axial mesoderm and neural tissues. Several of the nuclear genes responsible for the component of the organizer have been identified genetically and molecularly. The organization is controled by proteins such as Wnt that determines body axis, activin and fibroblast growth factor that are involved in

Organizer continued

signal reception to organize the formation of the mesoderm or by nogging that controls dorsal/ventral differentiation, chording affecting the development of the notochord, etc. (See vegetal pole, morphogenesis, epiblast, homeobox, homeotic genes, signal transduction, LIM)

ORGANOGENESIS: the development of organs from cells differentiated to special purposes.

ori: origin of replication. (See DNA replication, replication, plasmid ori_T, ori_V)

ORIGIN RECOGNITION ELEMENT (ORE): is the DNA site where proteins, that initiate replication, bind. (See replication fork)

ori_T: origin of transfer of the bacterial plasmid. Its A=T content is higher than the surrounding DNA. It has recognition sites for a number of conjugation proteins. Contains promoters for the *tra* genes that are localized in such a way that only after the complete transfer of the plasmid will all of them be transfered. The transfer of the single strand proceeds in 5'→3' direction in all known cases, including the transfer of T-DNA to plant chromosomes. The 3' end can accept added nucleotides. After the transfer has terminated, the plasmid re-circularizes. All these processes require specific proteins. (See bom, conjugation, T-DNA)

ori_V: origin of vegetative replication, used during cell proliferation of bacteria.

ORIENTATION SELECTIVITY: the majority of transposases work well only in one orientation. This orientation is chosen by accessory proteins associated with the transposases. It is synonymous with directionality of transposons. (See transposable elements, transposons)

ORIGIN of REPLICATION: starting point of replication during cell proliferation.

ORIGIN OF LIFE: prebiotic evolution laid down the foundations for the origin of life (see evolution prebiotic, evolution of the genetic code). The following steps might have been taken place: 1. generation of organic molecules, 2. polymerization of RNA, capable of selfreplication and heterocatalysis (cf. ribozymes), 3. peptide synthesis with the assistance of RNA, 4. evolution of translation and polypetide assisted replication and transcription, 5. reverse transcription of RNA into DNA, 6. development of DNA-RNA-Protein auto- and heterokatalytic systems, 7. sequestration of this complex into organic micellae formed by fatty acid - protein membrane like structures, 8. the appearance of first cellular organisms about 3 -4 billion years ago, 9. development of photosynthesis and autonomous metabolism. According to some estimates, the starter functions of life might have been carried out by as few as 20-100 proteins. The extensive duplications in all genomes may be an indication that the early enzymes did not have strong substrate specificity. The minimal cellular genome size might have been comparable to that of *Mycoplasma genitalium*, 580 kb long and coding 482 genes. The time required to develop from the 100 kb genome to a primitive heterotroph cyanobacterium with 7,000 genes might have taken about 7 million years. (See spontaneous generation unique or repeated, exobiology, geological time periods, evolution)

ORIGIN OF SPECIES: the title of a book by Charles Darwin, first published in 1859. This book is one of the most influential work on human cultural history. Darwin recognized the role of natural selection in evolution. Although he did not understand the principles of heredity (this book was published 7 years prior to the paper of Mendel). With the development of modern concepts of heredity, cytogenetics, population genetics and molecular biology, the seminal role of this book became generally recognized and appreciated even from a sesquicentennial distance. (See darwinism, evolution)

ORNITHINE ($NH_2[CH_2]_3CH[NH_2]COOH$): a non-essential amino acid for mammals. It is synthesized through the urea cycle. (See urea cycle)

ORNITHINE AMINOTRANSFERASE DEFICIENCY (hyperornithinemia, OAT): ornithine is derived either from arginine or N-acetyl glutamic semialdehyde or glutamic semialdehyde, and carbamoylphosphate synthetase converts it to citrulline. The decarboxylation of ornithine (a pyridoxalphosphate [vitamin B_6] requiring reaction) yields polyamines such as spermine and spermidine. OAT deficiency causes ornithinemia, 10-20 times increase of ornithine in blood plasma, urine and other body fluids. It also results in degeneration of eye tissue and tunnel vision and night blindness by late childhood. The OAT locus (21 kb, 11 exons) is in human

Ornithine aminotransferase deficiency continued
chromosome 10q23qter. Pseudogenes are at Xp11.3-p11.23 and at Xp11.22-p11.21. (See amino acid metabolism, ornithine transcarbamylase deficiency, ornithine decarboxylase, ornithine transcarbamylase deficiency, pseudogene)

ORNITHINE DECARBOXYLASE (ODC): the dominant allele (human chromosome 2p25, mouse chromosome 12, in *E. coli* 63 min) controls the ornithine→putrescine reaction and its activity is very sensitive to hormone levels. There is a second ODC locus in human chromosome 7 but the latter has much reduced function. (See amino acid metabolism, ornithine aminotransferase deficiency)

ORNITHINE TRANSCARBAMYLASE DEFICIENCY (OTC): This X-linked (Xp21.1) enzyme normally expressed primarily in the liver mitochondria, catalyzes the transfer of a carbamoyl group from carbamoyl phosphate to citrulline and ornithine is made while inorganic phosphate is released. The defect may lead to an accumulation of ammonia (hyperammonemia) because carbamoyl phosphate is generated from NH_4^+ and HCO_3^- (in the presence of 2 ATP). The high level of ammonia may cause emotional problems, irritability, lethargy, periodic vomiting, protein avoidance and other anomalies. The plasma ammonium level may be medicated by Na-benzoate, Na-phenylacetate and arginine. The incidence of OTC deficiency in Japan was found to be about 1.3×10^{-3}. In mice an OTC deficiency mutation is responsible for the *spf* (sparse fur) phenotype. (See amino acid metabolism, ornithine aminotransferase, ornithine decarboxylase, channelling)

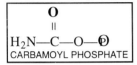

OROFACIAL-DIGITAL SYNDROME (OFD): OFD I is apparently a dominant X-linked or a recessive autosomal syndrome. The dominant form is lethal in males. OFD I shows hearing loss and polydactyly of the great toe, mental retardation, face and skull malformation, cleft palate, brachydactyly (short fingers and toes), etc. OFD II (Mohr syndrome) displays polysyndactyly, brachydactyly [short digits], and lobate tongue. OFD III and OFD IV include mental retardation, eye defects, teeth anomalies, incomplete cleft palate, hexadactyly, hunchback features, etc. OFD IV is distinguished on the basis of tibial dysplasia (shinbone defects). OFD V (Váradi-Papp syndrome), in addition, shows a nodule on the tongue and neural defects, etc. (See polydactyly, limb defects, mental retardation, Majewski syndrome)

OROSOMUCOID: there are two orosomucoid (serum glycoprotein) genes in humans ORM1 and ORM2 in 9q31-q32.

OROTIC ACID: is a precursor in the *de novo* pyrimidine synthesis:
N-CARBAMOYLASPARTATE $\xrightarrow{1}$ L-DIHYDROOROTATE $\xrightarrow{2}$ OROTATE $\xrightarrow{3}$ OROTIDYLATE $\xrightarrow{4}$ URIDYLATE (UMP) $\xrightarrow{5}$ URIDYLATE (UTP) $\xrightarrow{6}$ CYTIDYLATE (CTP). The enzymes mediating the reactions numbered are: (1) ASPARTATE TRANSCARBAMOYLASE, (2) DIHYDROOROTASE, (3) DIHYDROOROTATE DEHYDROGENASE, (4) OROTATE PHOSPHORIBOSYLTRANSFERASE, (5) OROTIDYLATE DECARBOXYLASE, (6) KINASES, (7) CYTIDYLATE SYNTHETASE. (See oroticaciduria)

OROTICACIDURIA: is a recessive deficiency of the enzyme orotidylate decarboxylase, encoded at human chromosome 3q13 or a deficiency of orotate phosphoribosyl transferase. The rare human diseases have counterparts in other higher eukaryotes. The clinical symptoms are anemia with large immature erythrocytes and urinary excretion of orotic acid. The symptoms are alleviated by supplying uridylate and cytidylate. (See orotic acid)

ORPHAN GENES: open reading frames in yeast (about 1/3 of the ORFs) that cannot be associated with known functions (FUN genes) and do not seem to have homologs in other organisms. Their number is expected to be smaller as information accrues. (See ORF, *Saccharomyces*)

ORPHAN RECEPTOR: has either no known ligand or it has the ligand-binding domain but lacks the conserved DNA binding site. (See nuclear receptors, receptor)

ORPHANIN: see opiate

ORPHON: a former member of a multigene family, but at a site separate from the cluster, and contains one coding region or they are pseudogenes. (See gene family, pseudogene)

ORTHOGENESIS: evolution in a straight line toward a goal or in a predetermined path.

ORTHOGONAL FUNCTIONS: can be employed for comparisons between observed data which have not affected each other.

ORTHOLOGOUS LOCI: genes in the direct line of evolutionary descent from an ancestral locus. (See paralogous loci, evolution of proteins, non-orthologous gene displacement)

ORTHOLOGY: common ancestry in evolution. (See orthologous loci, paralogy)

ORTHOSTITCHIES: see phyllotaxis

OSCILLIN: is a protein formed in the egg after the penetration by sperm, and it is involved with the oscillation of the level of Ca^{2+}, and it is apparently involved in triggering the early development of the embryo.

OSHA: Occupational Safety and Health Administration; in the USA. It provides information about and protects against potential hazards involved in laboratory and industrial operations.

OSLER-RENDU-WEBER SYNDROME: see telangiectasia hemorrhagic

OSMIOPHILIC: readily stains by osmium or osmium tetroxide.

OSMIUM TETROXIDE (OsO_4): a fixative for electronmicroscopic specimens. (See fixatives, microscopy)

OSMOSIS: diffusion of water through a semipermeable membrane toward another compartment (cell) where the concentration of solutes is higher. The osmosensing signal seems to be mediated by the Jnk protein kinase, a member of the mitogen-activated large family of proteins in mammals. In bacteria, a histidine kinase sensor (EnvZ) and a transcriptional regulator (OmpR) are involved. In yeast a similar mechanism is implicated, involving also HOG-1. (See HOG-1, histidine kinase)

OSMOTIC PRESSURE: in two compartments separated by a semipermeable membrane the solvent molecules pass toward the higher concentration of solutes. This flow may be prevented by applying high pressure to the compartment toward what the flow is directed.

OSMOTIN: protein in plants that may accumulate at high concentrations of salts.

OSP: see allele-specific probe for mutation

OSTEOARTHRITIS: a common human affliction involving degeneration of the cartilage, caused by the replacement of the [α1(II)] collagen by αII chains with reduced glycosylation. (See collagen, arthritis)

OSTEOBLAST: a cell involved in bone production, controled mainly by mouse gene CBFA-1.

OSTEOCHONDROMATOSIS (dyschondroplasia): an inhomogeneous autosomal dominant bone defect.

OSTEOGENESIS IMPERFECTA (OI): rare (prevalence 1/5,000 - 1/10,000) autosomal dominant or recessive disorders of collagen, resulting in abnormal bone formation due to mutation at numerous loci (ca. 17) that replace e.g. a glycine residue by a cysteine. Such a substitution disrupts the Gly - X - Pro tripeptide repeats in collagen that assures the helical structure of the molecules. Type I (chromosome 17) represents postnatal bone fragility, blue coloring of the eye white (sclera) and ear problems. Type II generally involves lethality about birth time due to a variety of bone defects. Most of the cases are considered new dominant mutations. Type III may result in prenatal bone deformities and the survivors are also crippled. Hearing loss may accompany the bone symptoms. This may be either recessive or dominant. Type IV has much milder expression of the similar symptoms as the other classes and may not prevent survival. (See collagen, hydrocephalus, mitral prolapse, connective tissue disorders)

OSTEOPETROSIS: occurs in different forms controlled by autosomal dominant or recessive genes in humans. The osteopetrosis with renal tubular acidosis is based on deficiencies in the carbonic anhydrase B or CA2 enzyme. The disease makes the calcified bones brittle. It generally involves mental retardation, visual problems, reduced growth and elevated serum acid phosphatase levels, and higher pH of the urine because of the excretion of bicarbonates and less acids in the urine. In the Albers-Schönberg disease, the osteopetrosis causes reduction in the size of the head, deafness, blindness, increased liver size and anemia. Defects in a colony stimulating factor (CSF) is infered from mouse experiments. A mild form of osteopetrosis is also known. A severe lethal form affects already the fetus by the 24th week and may cause stillbirth. (See carbonic anhydrase, osteoporosis, colony stimulating factor)

OSTEPONTIN: is a protein produced by the osteoblasts, and it is encoded in human chromosome 4. (See osteoblast)

OSTEOPOROSIS: abnormal thinning of the bone structure because of reduced activity of the osteoblasts to make bone matrix by using calcium and phosphates. This process is most frequently associated with advanced age and it is regulated by steroid and other hormones and vitamin D. The autosomal recessive juvenile form may be caused by defects in bone formation and the problems may be alleviated spontaneously by adolescence or may be treated successfully with steroid hormones. Another apparently autosomal recessive form in infancy involves also eye defects (pseudoglioma) and possibly other problems of the nervous system. The collagen genes COLIA1 and COLIA2 have been implicated. (See aging, hormone-receptor elements, Cbl, Src, collagen)

OSTEOSARCOMA: is an autosomal recessive, usually malignant, bone tumor with an onset in young adults. It is commonly part of a complex disease. People affected by bilateral (but not those afflicted by unilateral) retinoblastoma show a high tendency to develop also bone cancer. (See retinoblastoma, sarcoma)

OSTIOLE (ostium): a small (mouth-like) opening on various structures such as fungal fruiting bodies or internal organs.

OSTRICH (*Struthio camelus*): the zoological name reflects the ancient belief that this huge bird descended from a misalliance of the sparrow and the camel. It differs from other birds by having only two toes. Also, while other birds copulate by bringing together their cloacas (the combined opening for urine, faeces and reproductive cells), the male ostrich everts a penis-like structure from the cloaca during the process.

OTF-1, OTF-2: are binding proteins recognizing the consensus octamer 5'-ATGCAAAT-3' and facilitating transcription in several eukaryotic genes. OTF-1 is identical to NF-3 replication factor of adenovirus. (See NF, binding proteins)

OTOPALATODIGITAL SYNDROME (OPD): human chromosome Xq28 semidominant, variable expression deafness, cleft palate, broad thumbs and great toes. (See also cranioorodigital syndrome)

OTOSCLEROSIS: is an autosomal dominant hardening of the bony labyrinth of the ear resulting in lack of mobility of the structures and thus conductive hearing defect. It may be a progressive disease starting in childhood and fully expressed in adults. Its incidence as a hearing loss is about 0.003 among US whites and about an order of magnitude less frequent among blacks. (See deafness)

OTTER: *Amblonyx cinerea*, 2n = 38; *Enhydra lutris*, 2n = 38.

OTU (operational taxonomic unit): such as a particular population or a species, used in evolutionary tree construction. (See evolutionary tree, character matrix)

Ötzi: see ice man

OUABAIN (3[{6-deoxy-α-L-mannopyranosyl}oxy]1,5,11α,11,19-pentahydroxycard-20(22)-enolide): a cardiotonic steroid capable of blocking potassium and sodium transport through cell membranes; it is used as selective agent in animal cell cultures. (See steroids)

OUCHTERLONY ASSAY: see antibody detection

OUTGROUP: is at least two species that can be used to distinguish an ancestral from a derived species and for the rooting of phylogenetic trees. Characters (physical or molecular) shared between the ingroup and the outgroup are considered to be ancestral. (See evolutionary tree, PAUP)

OUT-OF-AFRICA: hypothesis suggests that in Africa was the origin of the human race. It is based on mtDNA composition showing that more differences exist in this respect in Africa than other parts of the world, supposedly because a few founders (highly conserved mtDNA) emigrated then spread and evolved relatively recently into the majority of the existing human racial groups. At the niche of origin, however, a longer evolutionary period permitted greater divergence. This hypothesis has been further supported on the basis linkage disequilibrium

Out-of-Africa continued

between an Alu deletion at the CD4 locus in chromosome 12 and short tandem repeat polymorphisms (STRP) used as nuclear chromosomal markers. The two markers are separated by only 9.8 kb. The mapping data seems to indicate that 1 cM of the human genome corresponds to about 800 kb. The Alu deletion was mainly associated with a single STRP in Northeast Africa and non-African populations sampled from 1600 individuals from European, Asian, Pacific and the Amerindian groups. In contrast in the sub-Saharan Africa a wide range of STRP markers were with the Alu deletion. These data also indicate that migration from Africa took place relatively recently, an estimated 102,000 to 313,000 years ago. This estimate is in relatively close agreement with the age of the first human fossil records in the Middle-East, dated to be 90,000-120,000 years old. *Homo sapiens* diverged from *H. erectus* an estimated 800,000 or more years ago. Some anthropologists do not accept this theory of human evolution and suggest multiregional origin for the modern humans. (See mtDNA, EVE foremother of mitochondrial DNA, founder principle, CD4, Alu, hominids)

OUTBREEDING: is the opposite of inbreeding; the mating is between unrelated individuals (crossbreeding, allogamy). (See allogamy, autogamy, inbreeding, protogyny, protandry)

OUTCROSSING: pollinating an (autogamous) plant with a different individual or strain or mating between animals of different genetic constitution. (See protandry, protogyny, incross)

OVALBUMIN: a nutritive protein of the eggwhite of chickens ($M_r \approx 4,500$); each molecule carries a carbohydrate chain. The gene is split by 7 introns into 8 exons and its transcription is induced by estrogen or progesterone. In the presence of the effector hormone the mRNA has a half-life of about a day but without it, its persistence may be reduced to 20%. The gene is part of a family of 3. (See oviduct, intron)

OVALOCYTOSIS: see elliptocytosis, acanthocytosis

OVARIECTOMY (oophorectomy): removal of the ovary. (See castration, spaying, neutering)

OVARY (ovarium): contains the ovules of plants; egg-producing organ of animals, a female gonad. (See gonad, gametogenesis)

OVC ONCOGENE: discovered in an ovarian cancer in a human chromosome 8 - 9 fusions. (See oncogenes)

OVERDOMINANCE: the *Aa* heterozygote surpasses both *AA* and *aa* homozygotes. (See hybrid vigor, superdominance, heterosis, overdominance and fitness, polar overdominance)

OVERDOMINANCE AND FITNESS: let us assume that in case of overdominance the selection

ALLELIC COMPLEMENTATION (= OVERDOMINANCE) OF TEMPERATURE-SENSITIVE PYRIMIDINE MUTANTS OF *ARABIDOPSIS*. TOP AND BOTTOM ROWS ARE HOMZYGOUS MUTANTS, IN THE MIDDLE ROW: THE F_1 HYBRIDS. (From Li, S.L. & Rédei, G.P.)

against the two homozygotes at a locus is 0.05. Thus, instead of the Hardy-Weinberg proportions, the population will be *AA* 95, *Aa* 200 and *aa* 95. The proportion of the surviving zygotes will be 390/400 = 0.975. This means a 2.5% reduction in the size of the population in a single reproductive phase. This may not be very serious in case of a single locus and large populations, but in case of 10, 100 and 500 loci it would be $0.975^{10} \cong 0.776$, $0.975^{100} \cong 0.0795$ and $0.975^{500} \cong 0.0000032$, respectively. Most likely such a great reduction in population size could not be tolerated. Thus overdominance may be advantageous only at 1 or very small number of loci. (See heterosis, fitness, hybrid vigor, allelic complementation, monogenic heterosis)

OVERDRIVE: a consensus of 5'-TAAPuTPyNCTGTPuTNTGTTTTGTTTG-3' in the vicinity of the T-DNA right border of some octopine plasmids facilitating the transfer of the T-DNA into the plant chromosomes. The VirC1 protein binds this region. (See *Agrobacterium*, T-DNA, virulence genes of *Agrobacterium*)

OVERHANG: a double-stranded nucleic acid has a protruding single-strand end: ‾‾‾‾‾‾

OVERLAPPING CODE: a historical idea about how neighboring codons may share nucleotides. (See overlapping genes)

OVERLAPPING GENES: occur in the small genomes of viruses. Sequences of the genetic material may be read in different registers; thus the same sequences may be representing two or more genes in a complete or partial overlap because they do not need equal amounts of proteins from the overlapping genes. The means for achieving this goal is either stop codon read-through or translational frame shifting on the ribosome. The murine leukemia virus (MLV) and the feline leukemia virus (FeLV) use the first alternative. In the HIV virus the RNA transcript makes a small loop and in the sequence 5' UUU|UUUA| GGGAAGAU-LOOP-GGAU an one base slippage (framed) takes place. This allows the normally UUA recognizing tRNA^{Leu} to pair with UUA (leucine) and next in place of GGG (glycine) an AGG (arginine) was inserted, and thus the stop codon was thrown out of frame and this permitted the synthesis of a fusion protein (*gag* + *pol*, i.e. envelop + reverse transcriptase). Such a ribosomal frame-shifting occurred only in a fraction (11%) of the ribosomes yet as a consequence instead of a 10-20 envelope:1 reverse transcriptase protein, the proportion changed to 8 envelope:1 reverse transcriptase (*gag* + *pol*) production. (See retroviruses, regulation of gene activity, nested genes, gag, pol, reverse transcription, frameshift, recoding, tRNA, deletion, knockout, contiguous gene syndrome, translational hopping, φX174)

OVERLAPPING INVERSIONS: part of one chromosomal inversion is included in another inversion. (See inversion, figure below)

TWO OVERLAPPING INVERSIONS IN A HETEROZYGOUS SALIVARY GLAND CELL OF *DROSOPHILA*. (Remember, the salivary gland chromosomes display somatic pairing).
From Dobzhansky, T. and Sturtevant, A.H. 1938 Genetics 23:28.

OVERWINDING: generates supercoiling in the DNA. (See supercoil)

OVIDUCT: the passage-way through which the externally laid eggs are released (e.g. in birds), the channel through which the egg travels from the ovary to the uterus (e.g. mammals). (See ovary, uterus, ovalbumin)

Ovis aries (sheep, 2n = 54): forms fertile hybrid with the wild mouflons but the sheep x goat (*Capra hircus*, 2n = 60) embryos rarely develop in a normal way although some hybrid animals had been obtained. (See nuclear transplantation)

OVIST: see preformation

OVOTESTIS: a gonad that abnormally contains both testicular and ovarian functions. (See hermaphrodite)

OVULATION: the release of the secondary oocyte from the follicle of the ovary that will be followed eventually by the formation of the egg and the second polar body. (See gametogenesis)

OVULE: megasporangium of plants in which a seed develops, the animal egg enclosed within the Graafian follicle. (See Graafian follicle)

OVUM (plural ova): an egg(s) ready for fertilization under normal conditions.

OX plural **OXEN**: castrated male cattle.

OX40: is a member of the tumor necrosis factor receptor family. (See TNF)

OXALOSIS: autosomal recessive forms involve excretion of large amounts of oxalate and/or glycolate through the urine caused either by a deficiency of 2-oxoglutarate (α-ketoglutarate):gly-

Oxalosis continued

oxylate carboligase or a failure of alanine:glyoxylate aminotransferase or serine:pyruvate aminotransferase. The accumulated crystals may result in kidney and liver disease. The glycerate dehydrogenase defects also inrease urinary oxalates and hydroxypyruvate. Hydroxypyruvate accumulates because the dehydrogenase does not convert it to phosphoenolpyruvate. It is also called peroxisomal alanine:glyoxylate aminotransferase deficiency (AGXT) located to human chromosome 2q36-q37. (See glycolysis)

oxi3: mitochondrial DNA gene responsible for cytochrome oxidase.

OXIDATION: loss of electrons from a molecule.

OXIDATION - REDUCTION: a reaction transfering electrons from a donor to a recipient.

OXIDATIVE DEAMINATION: e.g., an NH_2 group of cytosine is replaced by an O resulting into a conversion of C → Uracil. (See nitrous acid mutagenesis, chemical mutagens II)

OXIDATIVE DECARBOXYLATION: see pyruvate decarboxylation complex

OXIDATIVE DNA DAMAGES: occur when ionizing radiation or oxidative compounds hit cells. The major class of damages involves the formation of thymine glycol (isomers of 5,6-dihydroxy-5,6-dihydrothymine), 5-hydroxymethyluracil, 5,6-hydrated cytosine, 8 oxo-7,8-dihydroguanine, cross-linking between DNA and protein, elimination of the ribose. The oxidative damages usually repaired by exonucleases, endonucleases and glycosylases and other repair enzymes in prokaryotes and eukaryotes. (See DNA repair, oxidative deamination)

OXIDATIVE PHOSPHORYLATION: ATP is formed from ADP as electrons are transferred from NADH or $FADH_2$ to O_2 by a series of electron carriers. Such processes take place on the inner membrane of the mitochondria with the assistance of cytochromes as electron carriers. This is the main mechanism of ATP formation. (See mitochondria, ATP, NADH, FAD)

OXIDATIVE STRESS: is exerted by free radicals (reactive oxygen species, ROS) such as peroxides and hydroxyl species through the action of enzymes involved in mixed-function oxidation and autooxidation (P450 cytochrome complex, xanthine oxidase, phospholipase A_2). These reactions play important role in mutagenesis, aging, mitochondrial functions, etc., and considered to affect neuronal degeneration in Alzheimer and Parkinson diseases and in amyotrophic lateral sclerosis. (See diseases mentioned under separate entries)

OXIDIZING AGENT: an acceptor of electrons

8-OXODEOXYGUANINE: see oxidative DNA damage

OXPHOS (oxidative phosphorylation): a mitochondrial process by what molecular oxygen is combined with the electron carriers NADH and $FADH_2$ by the enzymes of the respiratory chain and mediate the $ADP + P_i$ → ATP conversion.

OXOPROLINURIA: see glutathione synthetase deficiency

OXYGEN EFFECT: in the presence of air or oxygen the frequency of chromosomal aberrations induced by ionizing radiation increases in comparison to conditions of anoxia (lack of air in the atmosphere). (See physical mutagens, radiation effects)

OXYGENASES: see mixed function oxidases

OXYNTIC CELLS: secrete acid (HCl) in the lining of the stomach.

OXYTOCIN: an octapeptide stored in the pituitary that regulates uterine contraction and lactation. It is synthesized in the hypothalmus with other associated proteins. They are assembled in the neurosecretory vesicles and transported to the nerve ends in the neurohypophysis (posterior lobe of the pituitary) and then may be secreted into the blood stream. The vasopressin and oxytocin (12 kb between them) and neurophysin genes are linked within the short arm of human chromosome 20pter-p12.21. They are transcribed from opposite strands of the DNA. It appears that the pre-proarginine - vasopressin - neurophysin are transcribed jointly and post-translationally separated by proteolysis. (See also vasopressin [antidiuretic hormone], neurophysin)

OZONE (O_3): a bluish, highly reactive form of oxygen. It is a disinfectant and its presence in the atmosphere protects the earth from the excessive ultraviolet radiation coming from the sun.

P

P: parental generation

P (Polyoma): regulatory DNA element in the viral basal promoter. (See Simian virus 40, Polyoma)

^{32}P: phosphorus isotope. (See isotopes)

P_1, P_2: designations of the parents, homozygous for different alleles, at the critical locus (loci) in a Mendelian cross. (See Mendelian laws, gametic arrays, genotypic segregation, allelic combinations)

p (petit): short arm of chromosomes, also denote frequencies. (See q)

ϕ (phi): symbol of some phages.

ψ (psi): see pseudouridine, formula ⇨. Also, packaging signal for virions.

PSEUDOURIDINE ⇨

p15^{INK4B}: is an inhibitor of CDK4 (encoded in human chromosome 9p21) appears to be an effector of TGF-β, a protein known to control the progression from G1 phase of the cell cycle to S phase. (See cell cycle, TGF, p16^{INK4}, p18, p19, cancer)

p16 (*MTSI* [multiple tumor suppressor], CDKN2): a cell cycle gene (in human chromosome 9p21) that has a major role in tumorigenesis. In 50% of the melanoma cells it is deleted and it is mutated in 25%. Over 70% of the bladder cancer cases are associated with deletions of 9p21 in both homologous chromosomes, in head and neck tumors 33%, in renal and other cells usually this tumor suppressor gene is lost. It normally restrains CDK4 and CDK6. (See melanoma, pancreatic adenocarcinoma, CDK4, CDK6, cell cycle)

p16^{INK4}: a protein inhibitory to CDK4 and thus appears to be a tumor suppressor because it inhibits the progression of the cell cycle. (See CDK4, tumor suppressor, PHO81, cell cycle, cancer, p18, p19)

p18^{INKC}: cell cycle inhibitors by blocking cell cycle kinases CDK4 and CDK6. (See cell cycle, cancer, p15, p16, p19)

p19^{INK4d}: cell cycle inhibitor of CDKs. (See p15, p16, p18, cell cycle, cancer)

p21: transforming protein of the Harvey murine sarcoma virus. A Ras-gene encoded 21 kDa protein binding GDP/GTP and hydrolyzing bound nucleotides and inorganic phosphate. This protein is involved in signal transduction, cell proliferation and differentiation and p21 controls the cyclin-dependent kinases (Cdk4, Cdk6, Cdk2), and binds to DNA polymerase δ processivity factor and it inhibits *in vitro* PCNA-dependent DNA replication but not DNA repair. In the absence of p21, cells with damaged DNA are arrested temporarily at the G2 phase and that is followed by S phases without mitoses. Consequently hyperploidy arises and apoptosis follows. Gene *p21* is under the control of p53 protein and the retinoblastoma tumor suppressor gene RB. Expression of the p21 Cdk inhibitors during differentiation of muscle cells and non-muscle cells is regulated also by MyoD, and the withdrawal from the cell cycle then does not require the participation of p53. p21 may not have an absolute requirement for induction by MyoD. Other antimitogenic signals are provided by the N terminal domains of p27 and p57. (See cell cycle, Cdk, mitosis, cancer, hyperploidy, apoptosis, p53, PCNA, p27, p57, MyoD, cell cycle, RAS)

p27: protein inactivates cyclin-dependent protein kinase 2. Its mutation results in increased body and organ size and neoplasia in mouse. (See Cdk2, p21, cell cycle, cancer, KIP)

p34^{cdc2-2}: gene coding for the catalytic subunit of MPF in *Schizosaccharomyces pombe* (counterparts, *CDC28* in *Saccharomyces cerevisiae*, *CDCHs* in humans have 63% identity with *cdc2*; these genes are present in all eukaryotes). The gene product is a serine/tyrosine kinase and its function is required for the entry into M phase of the cell cycle. If prematurely activated, it may cause apoptosis. (See cell cycle)

p38: a stress-activated protein kinase.

p52^{SHC}: is a RAS G-protein regulator protein; it is regulated through CTLA-4 - SYP associated phosphatase. (See CTLA-4, SYP, RAS)

p53: is a tumor suppressor gene when the wild type allele is present but single base substitutions may eliminate the suppressor activity and the tumorigenesis process may be initiated. Its product binds to specific DNA sequence, activates transcription from promoters with p53 protein binding sites; represses transcription from promoters lacking p53 binding sites, promotes annealing of DNAs, inhibits replication, causes G1 phase arrest, leads to apoptosis if DNA is damaged, interferes with tumorous growth, maintains genetic stability, reduces radiation hazards by its regulatory role in the cell cycle. This protein binds to a somewhat conserved consensus and it is phosphorylated at serine 315 residues by CDK proteins during S, G2 and M phases of the cell cycle but not at G1 although p53 controls an important G1 checkpoint. It also controls proteins p21, p27 and p57. p53 is a tetramer with separate domains for DNA binding, transactivation and tetramerization. Its transcriptional activation is mediated by coactivators TAFII40, TAF_{II} 60 and other TATA box binding factors. p53 is encoded in human chromosome 17p13. Protein p73 has functions similar to those of p53. (SCIENCE Magazine declared p53 the molecule of year 1993). (See tumor suppressor gene, annealing, apoptosis, TAF, TBP, transactivator, cancer, cell cycle, p21, p27, p57, substitution mutation)

p56^{chk1}: is a protein kinase and a checkpoint for mitotic arrest after mutagenic damage inflicted by UV, ionizing radiation or alkylating agents. The DNA damage results then in the phosphorylation of this protein in yeasts. The phosphorylation may prevent the mitotic arrest yet the cells may die later; p56 is not involved in DNA repair. Phosphorylation is required, however that other checkpoint genes to become/stay functional. (See cell cycle, DNA repair)

p57: is an antimitogenic protein; its carboxyl end assures nuclear localization and the amino end is involved in the inhibition of CDK proteins. (See CDK, p21, p27, cell cycle)

p70/p86: is the Ku autoantigen. (See DNA-PK, Ku)

p70^{s6k}: phosphorylates S6 ribosomal protein at serine/threonine residues before translation. Also called S6 kinase. (See translation initiation, p85^{s6k}, S6 kinase, signaling to translation)

p75: is a non-tyrosine kinase receptor protein.

p80^{sdc25}: is protein phosphatase that activates p34^{cdc2} - cyclin protein kinase complex by dephosphorylating Thr^{14} and Tyr^{15}. (See cell cycle)

p85^{s6k}: phosphorylates S6 ribosomal protein before translation at serine/threonine sites. Also called S6 kinase. (See translation initiation, p70^{s6k}, S6 kinase)

p95: is Fas (See also acrosomal process, Fas, APO1)

p107: a retinoblastoma protein-like regulator of the G1 restriction point of the cell cycle. (See restriction point, tumor suppressor, retinoblastoma)

p130: a retinoblastoma protein-like regulator of the G1 restriction point of the cell cycle. (See restriction point, tumor suppressor, retinoblastoma)

p300 (CBP): a cellular adaptor protein preventing the G_0/G_1 transition of the cell cycle, it may activate some enhancers and stimulate differentiation; it is also a target of the adenoviral E1A oncoprotein. Its amino acid sequences are related to CBP, a **C**REB-**b**inding **p**rotein. Nuclear hormone-receptors interact with CBP/p300 and participate in gene transactivation. PCAF is a p300/CBP associated factor in mammals, and it is the equivalent of the yeast Gcn5p (general controlled nonrepressed protein), an acetyltransferase working on histones 3, 4 (HAT A) and thus regulating gene expression. p300 functions also as a co-activator of NF-κB. (See adenovirus, CBP, CREB, NF-κB, histone acetyltransferase)

p350: is a DNA-dependent kinase and is a likely basic factor in severe combined immunodeficiency and it may be responsible also for DNA double-strand repair, radiosensitivity and the immunoglobulin V(D)J rearrangements. In association with the KU protein it forms a DNA-dependent protein kinase. (See severe combined immunodeficiency, kinase, KU, DNA-dependent protein kinase, immunoglobulins)

P450: family of genes coding for cytochrome enzymes involved in oxidative metabolism. They are widely present in eukaryotes and scattered around several chromosomes. All mammalian species have at least eight subfamilies. The homologies among the subfamilies is over 30% whereas the homologies among members of a subfamily may approach 70%. These cyto-

P450 continued
chromes posses monooxygenase, oxidative deaminase, hydroxylation, sulfoxide forming, etc. activities. The proteins are generally attached to the microsomal components of homogenized cells (endoplasmic reticulum [fragments]), often called also S9 fraction. Some of these enzymes (subfamily IIB) are inducible by phenobarbital. Their expression may be tissue-specific, predominant in the liver, kidney or intestinal cells. Mammalian P450 cytochrome fraction is generally added to the *Salmonella* assay media of the Ames test in order to activate promutagens. One member of the series is involved in the regulation of the synthesis of the 6th class of plant hormones, brassinosteroids. (See Ames test, cytochromes, hypoaldosteronism, steroid hormones, brassinosteroids)

P BLOOD GROUP: is controlled by two non-allelic loci; the non-polymorphic P blood group is located in human chromosome 6 and it is encoding globoside whereas the polymorphic P1 locus in human chromosome 22 encodes paragloboside. The frequency of the P gene in Sweden was found to be 0.5401 and that of P1 0.4599. According to other studies the frequency of P among caucasoids is about 0.75. The P1 blood type facilitates bacterial attachment to the epithelial cells of the urinary tract and kidney. Therefore infections are more common. Some P alleles raise the risk of abortions, and others may increase the chances of stomach carcinomas. Some of the literature calls P as P1 and P1 as P2. (See blood groups, globoside)

P1 CLONING VECTORS: have carrying capacity up 100 kbp DNA; thus it falls between Lambda and YAC vectors. (See vectors)

P CYTOTYPE: see hybrid dysgenesis

P ELEMENT: see hybrid dysgenesis

P GRANULE: are serologically definable elements in the cytoplasm of animal cells at fertilization that segregate to the posterior part of the embryo where stem cell determination takes place.

P NUCLEOTIDES: see immunoglobulins

P1 PHAGE: is an *E. coli* transducing phage and vector with near 100 kb carrying capacity.

P22 PHAGE: temperate bacteriophage of *Salmonella typhimurium*; its genome is about 41,800 bp.

P1 PLASMID: a cloning vector with a carrying capacity of about 100 kb. (See vectors, P1 phage)

P REGION OF GTP-BINDING PROTEINS: shares the G-X-X-X-X-G-K-(S/T) motif (see amino acid symbols) and are suspected to involve the hydrolytic process of GTP-binding and several nucleotide triphosphate-utilizing proteins. (See GTP binding protein superfamily)

P SITE: peptidyl site on the ribosome where the first aminoacylated tRNA moves before the second charged tRNA lands at the A site as the translation moves on. The binding of the tRNA to the 30S ribosomal subunit appears to be controlled by guanine residues at the 966, 1401 and 926 positions in the 16S rRNA. (See A site, protein synthesis, ribosome)

PABP: poly(A) binding protein (72 kDa) is the major protein that binds to the poly A tail of eukaryotic mRNA and converts it to mRNAP. (See binding proteins, mRNAP, mRNA tail, mRNA decay)

PAC: phage artificial chromosome; P1 phage PAC carries about 100-300 kb DNA segments. (See BAC, YAC)

pac: site in the phage genome where terminases bind and cut during maturation of the DNA before packing the DNA into the capsid. (See terminase, packaging of the DNA)

PAC-1: dephosphorylates Thr[183] and Tyr[185] residues and thus regulates mitogen-activated MAP protein kinase involved in signal transduction. (See MAP, MKP-1)

PACAP (pituitary adenylyl cyclase-activating polypeptide-like neuropeptide): is a neurotransmitter at the body-wall neuromuscular junction of *Drosophila* larvae. It mediates the cAMP-RAS signal transduction path. (See signal transduction, RAS, RAF)

PACHYNEMA: literally "thick thread" of chromosomes at early meiosis when the doublestranded structure of the chromosomes is not distinguishable by light microscopy because the chromatids are tightly appositioned. Also, the two homologous chromosomes are closely associated, unless structural differences prevent perfect synapsis. If a pair of chromosome is not comple-

Pachynema continued

ly synapsed by pachytene, they will not pair later either. In pachytene the chromosomal knobs and chromomeric structure is visible and can be used for identification of individual chromosomes. After pachytene the synaptonemal complex is dismounted and the chromosomes progressively condense. (See meiosis, pachytene analysis, synapsis, chiasma, chromomere)

PACHYONYCHIA: is a rare autosomal dominant keratosis of the nails and skin. (See keratosis)

PACHYTENE ANALYSIS: is the study of meiotic chromosomes at the pachynema stage when

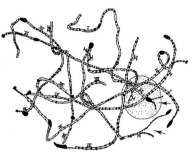

NATURALISTIC DRAWING OF THE 10 PACHYTENE CHROMOSOME PAIR OF A TEOSINTE X MAIZE HYBRID. NOTE (←) UNPAIRED ENDS OF CHROMOSOMES V, VII. (From Longley, A.E. 1937 J. Agric. Res. 54:835)

cytological landmarks, chromomeres, knobs are distinguishable by the light microscope and chromosomal aberrations (deletions, duplications, inversions, translocations, etc.) can cytologically be identified and correlated with genetic segregation information. The pachytene analysis of plants is analogous to the study of giant chromosomes in dipteran flies and other lower animals. The bands of the (somatic) salivary chromosomes are tightly appositioned chromomeres in these endomitotic chromosomes. (See meiosis, salivary gland, chromomere, endomitosis, recombination nodule)

PACHYTENE STAGE: the chromosomes form pachynema. (See pachynema)

PACKAGING of λ DNA: phage gene A recognizes the *cos* sites, gene D assists in filling the head (capsid) and genes W, F, V ILK and GMH assemble the phage from prefabricated elements and act in the processes shown diagrammatically:

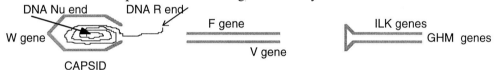

The DNA that first enters the phage capsid has the *Nu* end and the opposite end (the last) is the *R* end. (See lambda phage, development)

PACKING RATIO: the DNA molecule is much-much longer than the most extended chromosomes fibers. The packing ratio was then defined as the proportion of the DNA double helix and the length of the chromosome fibers. In the human chromosome complement the packing ratio was estimated to be more than 100:1 at metaphase. The length of the *Drosophila* genome at meiotic metaphase was estimated to be 7.8 μm and the length of a chain of 3,000 nucleotides is approximately 1 μm. The *Drosophila* genome contains about 8×10^7 bp, hence the total length of DNA within the *Drosophila* genome is about 26,667 μm and that would indicate a packing ratio of 3419:1. The packing ratio indicates some of the problems the eukaryotic chromosomes encounter in condensing to a small space an enormous length of DNA and still replicating, transcribing and recombining it in an orderly manner. To illustrate the problems in a trivial way: many eukaryotes have the same packing problem as folding a 2.5 km (1.6 mi) long thread into a 2.5 cm (1") skein. Prokaryotic type DNA — such as without nucleosomal structure — the excessive amount of plasmid DNA forms liquid crystalline molecular supercoils. (See Mosolov model, supercoiled DNA)

PADLOCK PROBE: contains two target-complementary segments connected by linker sequen-

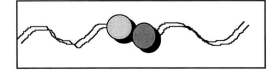

ces. Hybridization to target sequences brings the two ends close to each other and can be covalently ligated. The so circularized probes are thus catenated to the DNA (≈) like a padlock (OO). Such probes permit high-specificity detection, distinc-

tion among similar target sequences and can be manipulated without alterations or loss. (See probe)
PAGE: acronym for polyacrylamide gel electrophoresis.
PAGET DISEASE: two autosomal dominant forms have been described involving cancer of the bones or of the anogenital region (the region of the anus and genitalia).
PAH (polyaromatic hydrocarbon): the majority of them are carcinogenic.
PAI: plasminogen activation inhibitor. (See plasminogen activator)
PAIN-INSENSITYVITY: is controlled by defects causing hereditary sensory neuropathies. In the dominant form the dorsal ganglia are degenerated. In the recessive neuropathy the loss of myelinated A-fibers cause touch insensitivity. The congenital pain insensitivity with anhidrosis (CIPA) involves a defect of the nerve growth factor receptor (TRKA), and in the congenital insensitivity to pain without anhidrosis the small myelinated A-delta fibers are defective. (See neuropathy, Riley-Day syndrome, TRK, sensory neuropathy 1)
PAIR RULE GENES: determine the formation of alternating segments in the developing embryo. Similar segment pattern, although with variations, occurs in other insects too. (See morphogenesis, *Drosophila*, metamerism)
PAIRING (synapsis): the intimate association of the meiotic chromosomes. (See meiosis, zygotene, pachytene, somatic pairing, hydrogen pairing, base pair, tautomeric shift)
PAIRING ALKYLATED BASES: see alkylation
PAL: phenylalanine ammonia lyase.
PALEA: the inner, frequently translucent, bract around the grass flower.
PALEOLITHIC AGE: (old stone age) more than 20,000 years ago it marked the beginning of human tool formation and cave artistry by the Cro-Magnon humans. (See neolithic, mesolithic, geological time periods, Lascaux, Neanderthal)
PALEONTOLOGY: deals with the relics of past geological periods. Its methods and materials are used for the study of the evolution of biological forms. (See paleolithic age, geological time periods)
PALEOZOIC: geological period between about 225 to 570 million years ago; during the later part of this period land plants, amphibians and reptile appeared. (See geological time periods)
PALINDROME: region of a DNA strand where complementary bases are in opposite sequence, such as ATGCAC*GTGCAT. Palindromes may come about by inverted repeats of sections of the double-stranded DNA where these sequences of the opposite strand read the same forward and backward. Upon folding of these sequences in a single strand, they can assume structures with paired bases. Palindromic sequences in the DNA reassociate very rapidly because of the complementary bases are in close vicinity. A simple palindromic word is MADAM, it reads the same from left to right or from right to left. (See stem and loop, inverted repeats, insertion elements)
PALINGENESIS: regeneration of lost organs and parts or reappearance of evolutionarily ancestral traits during ontogeny. (According to Ernst Haeckel [1834-1914] the ontogeny recapitulates the phylogeny). (See ontogeny, phylogeny)
PALISADE CELLS: are oblong and arranged in a row; the large palisade parenchyma cells are below the upper epidermis of plant leaves and loaded with chloroplasts.
PALMPRINT: see fingerprint, Down's syndrome, simian crease.
PALOMINO: a horse with light tan fur color and flaxen mane.
PALSY (paralysis): cerebral palsy may be caused by physical injuries or may be part of the symptoms of diverse genetic syndromes. (See syndrome)
PAM: see evolutionary clock
PAMAM: see polyamidoamine dendrimers
PAN: the genus of chimpanzees.
PAN EDITING: adding U residues to the primary transcripts of mtDNA and thus causing extensive post-transcriptional changes in an RNA. (See kinetoplast, RNA editing)

PANCREATIC ADENOCARCINOMA: cancer of the pancreas, frequently associated with loss or defect of DCC, or p53 or MTS1 oncogene suppressors. (See DCC, p53, p16)

PANCREATITIS, HEREDITARY: autosomal dominant (7q35) gene (80% penetrance and variable expressivity) has an onset before the teen years, appearing as abdominal pain and other anomalies. The basic defect is in a cationic trypsinogen. (See trypsin)

PANDA (*Ailurus fulgens*): 2n = 36.

PANGENESIS: an ancient mis-concept about heredity, originated in the Aristotelian epoch and periodically revived during the centuries. Charles Darwin has also interpreted inheritance as pangenesis. Accordingly, all the information expressed during the life of the individuals is transported to the gametes from all parts of the body. Thus, pangenesis is the means of the inheritance of all, including the acquired characters. (See lysenkoism, acquired characters)

PANMICTIC INDEX: see fixation index, panmixis

PANMIXIS: random mating; in a population there is equal chance for each individual to mate with any other. (See Hardy-Weinberg theorem)

PANSPERMIA: the theory claiming that life has originated at several places in the universe and spread to earth by meteorites or by other means. (See origin of life)

PANTHER (*Panthera pardus*, leopard): 2n = 38: a feline species; in captivity may be crossed with lion but no mating is known to take place in the wild where conspecific sexual partners are available.

PANTROPIC: can affiliate with many different types of tissues.

PAP (Papanicolaou) **TEST**: is a cytological test for pre-malignant or malignant conditions (used primarily on smears obtained from the female urogenital tract). It detects also papilloma virus infections. (See malignant growth, papilloma virus)

PAPAIN: is a member of a family of proteolytic enzymes with an imidazole group near the nucleophilic SH group, and the former plays a role as a proton donor to the cleaved-off part. Papain cleaves immunoglobulin G into three near equal size fragments and this helped in clarifying the structure of antibodies. (See proteolytic, immunoglobulins, antibody)

PAPANICOLAOU TEST: see PAP test

PAPAYA (*Carica papaya*): a melon-like, edible fruit, latex-producing small tree with four genera and all 2n = 2x = 18; it is the source of the proteolytic enzyme papain. (See papain)

PAPER CHROMATOGRAPHY: a technique for the separation of (organic) molecules in filter paper by applying the mixture in a spot or band at the bottom of the paper and allowing an appropriate solvent to be sucked up and thus carry the components at different speed (to different height) so they can be separated. The components become visible by their natural color or by the application of specific reagents. A large variety of different modifications were worked out in one or two dimensions, in ascending and descending ways. Nowadays paper chromatography is not used very much. (See thin layer chromatography, Rf value, column chromatography, high performance liquid chromatography, affinity chromatography, ion exchange chromatography)

PAPILLARY THYROID CARCINOMA: caused by the RET oncogene. It accounts for about 80% of the thyroid cancers that have a prevalence in the 10^{-5} range. Its incidence is higher in females than males. (See RET)

PAPILLOMA VIRUS: is a double-stranded DNA ($\approx 5.3 \times 10^6$ Da or 8 kb) virus causing animal and human warts. It has been used as a genetic vector. (See papova viruses)

PAPOVA VIRUSES: a large class of (oncogenic) animal viruses of double-stranded, circular DNA includes the polyoma viruses, the bovine papilloma virus and simian virus 40 (SV40), etc. that have been used as genetic vectors for transformation of animal cells. Also, they have been extensively studied by molecular techniques to gain information on structure and function. (See Polyoma, Simian virus 40, Papilloma virus)

PAPS: 3'-phosphoadenosine-5'-phosphosulfate is a sulfate donor in several biochemical reactions, involving cerebrosides, glycosaminoglycans and steroids. It is generated by the pathway: ATP + sulfate → adenosine-3'-phosphosulfate (APS) + pyrophosphate, APS + ATP → PAPS + ADP.

PAR (pseudoautosomal region): where recombination takes place between the X and Y chromosomes. (See pseudoautosomal, X chromosome, Y chromosome)

PARABIOSIS: two animals joined together naturally such a Siamese twins or by surgical methods and can be used to study the interaction of hormones, transduction signals, etc. in-between two different individuals. Intrauterine parabiosis develops immune tolerance.

PARACENTRIC INVERSION: see inversion paracentric

Paracentrotus lividus: a sea urchin, extensively studied by embryologists. (See see urchins)

PARACRINE EFFECT: a ligand (e.g., hormone) is released by a gland and affects neighboring cells.

PARACRINE STIMULATION: one type of cell affects the function (such as proliferation) of another (nearby) cell.

PARADIGM: a model or an example to be followed.

PARAGANGLION: cells originating from the nerve ectoderm flanking the adrenal medulla, and darkly stained by chromium salts. These cells may form a type of phaechromocytoma tumors that secrete excessive amounts of epinephrine and norepinephrine. (See SHC oncogene)

PARAGENETIC: phenotypic alterations not involving hereditary mutation.

PARAHEMOPHILIA: is determined by homozygosity of semidominant autosomal genes. The symptoms involve bleeding similar to the conditions observed in hemophiliacs, bleeding from the uterus (menorrhagia) several days following child birth. The physiological basis is a deficiency of proaccelerin, a protein factor (V) involved in the stimulation of the synthesis of prothrombin. Therapy requires blood or plasma. (See antihemophilia factors, hemophilia, prothrombin deficiency, hemostasis)

PARAINFLUENZA VIRUSES: a group of immunologically related but distinguishable pathogens responsible for some respiratory diseases. (See Sendai virus)

PARALOCI: have the same properties as pseudoalleles. (See pseudoalleles)

PARALOGOUS LOCI: originated by duplication that was followed by divergence. (See orthologous loci, evolution of proteins, non-orthologous gene displacement)

PARALOGY: evolution by duplication of a locus. (See paralogous loci, orthology)

PARAMECIUM: unicellular Protozoon. Normally reproduces by binary fission, i.e. a single indi-

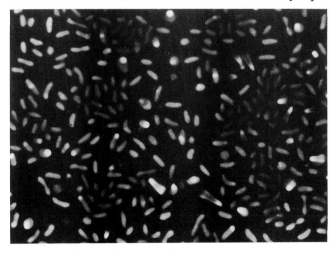

PARAMECIUM AURELIA (500 X) CELLS WITH BRIGHT AND NON-BRIGHT KAPPA PARTICLES SYMBIONTS (1,650X). THE SYMBIONTS ARE BACTERIA. THE BRIGHT PARTICLES CONTAIN THE SO CALLED R (REFRACTIVE) BODIES WHICH ARE BACTERIOPHAGES. THE NON-BRIGHT KAPPA CAN GIVE RISE TO BRIGHT INDICATING LYSOGENY. THE KAPPA-FREE (*kk*) PARAMECIA ARE SENSITIVE TO THE TOXIN PRODUCED BY THE BRIGHT PARTICLES AND MAY BE KILLED. THE *K*, KILLER STOCKS ARE IMMUNE TO THE TOXIN. (From Preer, J.R. et al. 1974 Bacteriol. Rev. 38:113 [Photo by C. Kung]; courtesy of Dr. J.R. Preer)

vidual splits into two. Each cell has two diploid micronuclei and a polyploid macronucleus. At fission the micronuclei divide by mitosis while the macronucleus is simply halved. These animals also have sexual processes (*conjugation*). Two of the slipper-shaped cells of opposite mating type attach to each other and proceed with meiosis of the micronuclei. Only one of the four products of meiosis survives in each of the conjugants. Each of these haploid cells divide

***Paramecium* continued**
into four cells (gametes). One of these gametes (male) is passed on into the other conjugating partner through a *conjugation bridge* and fuses with a haploid gamete (female). This is a reciprocal fertilization, resulting in diploid nuclei in the conjugants. Subsequently the pair separates into two *exconjugants*. The macronucleus disintegrates then in both. The diploid zygotic nuclei undergo two mitoses and form four diploid nuclei each. Two of the four nuclei function as separate micronuclei of the cells whereas the other two fuses into a macronucleus that becomes polyploid and that is responsible for all metabolic functions and for the phenotype. Besides this sexual reproduction (conjugation), *Paramecia* may practice selffertilization (*autogamy*). Meiosis takes place and the one surviving product divides twice by mitoses. Two of these identical cells then fuse and form two diploid, isogenic micronuclei. If the conjugation lasts longer, cytoplasmic particles may also be transfered through the conjugation bridge. Chromosome numbers may be 63-123. (See also killer strains, symbionts hereditary, *Ascaris*, macronucleus, chromosome diminution, internally eliminated sequences)

PARAMETER: a quantity that specifies a hypothetical population in some respect or a variable to which a constant value is attributed for a specific purpose or process.

PARAMETER ALPHA: see alpha parameter

PARAMETRIC METHODS IN STATISTICS: involve explicit assumptions about population distribution and parameters such as the mean, standard deviation of the normal distribution, the *p* parameter of the *Bernoulli process*, etc. (See Bernoulli process, normal distribution, non-parametric statistics)

PARAMUTATION: a *paramutable* allele becomes a *paramutant* one in response to a *paramuta-*

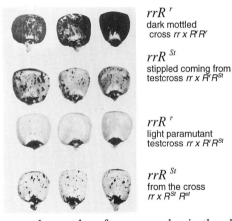

rrR^r
dark mottled
cross $rr \times R^rR^r$

rrR^{St}
stippled coming from
testcross $rr \times R^rR^{St}$

rrR^r
light paramutant
testcross $rr \times R^rR^{St}$

rrR^{St}
from the cross
$rr \times R^{St}R^{st}$

PARAMUTATION RESULTS IN REDUCED PIGMENTATION IN THE TRIPLOID ALEURONE OF MAIZE. THE R^R HOMOZYGOTES ARE FULLY COLORED (when all other color-determining alleles are present). THE r HOMOZYGOTES ARE COLORLESS. THE rrR^r GENOTYPE IS RESPONSIBLE FOR THE DARK MOTTLED ALEURONE. R^{st} CAUSES PARAMUTATION (STIPPLING) OF THE R^r PARAMUTABLE ALLELE THAT MAY BE MANIFESTED IN DIFFERENT GRADES. In the crosses the pistillate parents are shown first, left. (From Brink, R.A. 1956 Genetics 41:872)

genic allele if the two are in heterozygous condition; the alteration is similar but not identical to the paramutagenic allele. Both *paramutability* and paramutagenic functions are allele-specific. In contrast to gene conversion, paramutation may take place at low frequency also in the absence of a paramutagenic allele. At the *R* locus of maize partial reversion of the paramutant may happen but this has not been observed at the *B* locus of maize. The paramutant phenotype at the *R* locus may vary but at the *B* locus the phenotype appears to be uniform. The exact mechanism of this heritable alteration is not fully understood. It appears that the level of transcription is reduced at the paramutant allele compared to that in the paramutable one. Apparently at the *R* locus of maize hypermethylation is involved, at the *B* locus involvement of methylation has not been detected. Paramutation is not a general property of all genes although similar phenomena have been observed at a few other genes in maize and other plants. This phenomenon seems to violate the Mendelian principle that alleles segregate during meiosis independently and during the process no "contamination" takes place. (See gene conversion, copy choice, directed mutation, localized mutagenesis, pangenesis, blending inheritance, presence-absence hypothesis, graft hybrid, co-suppression, RIP, epigenesis, position effect)

PARAMYOTONIA: periodic paralysis (located in human chromosome 17). (See more there)

PARAMYXOVIRUS: single-stranded RNA viruses with a genome of 16-20 kb. Members of this group cause human mumps, respiratory diseases in human and other animals, including birds and reptiles. (See RNA viruses)

PARANEMIC COILS: the two components of the coil can be separated from each other without any entanglement as one can easily pull apart two spirals that were pushed together after they were wound separately, i.e. they are not interlocked. (See also plectonemic coils)

PARANEOPLASTIC NEURODEGENERATIVE SYNDROME: see autoimmune diseases

PARANOIA: is a psychological disorder in more (paranoia) or less severe (paranoid) state. The major characteristics are delusions of persecution (delusional jealousy, erotic delusions) or less frequently by feeling of grandiosity. In difference from schizophrenia, generally the rest of the personality and mental capacity may remain normal. Frequently, however, paranoid schizophrenia may occur. The precipitating factors are insecurity, frustration, physical illness, drug effects, etc. There is also an apparently undefined genetic component. (See also schizophrenia, affective disorders)

PARAOXONASE: see arylesterase, cholinesterase, pseudocholinesterase

PARAPATRIC SPECIATION: groups of organisms inhabiting an overlapping region become sexually isolated. (See allopatric, sympatric)

PARAPHYLETIC GROUP: does not include all descendents of the latest common ancestor.

PARAPLEGIA: paralysis of the lower part of the body; it may be hereditary.

PARAPROTEIN: is an abnormally secreted normal or abnormal protein, e.g, the Bence-Jones protein in myelogenous myeloma. It is also called M component.

PARAQUAT: an artificial electron acceptor of photosystem I. (See diquat, photosystem I)

PARASCARIS: a group of nematodes. (See *Ascaris*)

PARASEGMENT: unit of a metameric complex consisting of the posterior part of one segment and the anterior part of another in insect larval and subsequent stages. (See morphogenesis)

PARASELECTIVITY: an apparent (but not real) selectivity in pollination among plants.

PARASEXUAL MECHANISM OF REPRODUCTION: somatic-cell fusion and mitotic genetic recombination. The processes bear similarities to those common at sexual reproduction but they do not involve sexual mechanisms. (See mitotic recombination, cell fusion, somatic cell genetics)

PARASITEMIA: the blood contains parasites, e.g. *Plasmodium*. (See thalassemia, *Plasmodium*)

PARASITIC: lives on and takes advantage of another live organism. (See also biotrophic)

PARASITIC DNA: same as DNA selfish.

PARASTERILITY: caused by incompatibility between genotypes which may be fertile in other combinations. (See self-incompatibility alleles, Rh blood group)

PARASTITCHIES: the imaginary helical line in phyllotaxis. (See phyllotaxis)

PARATHORMONE (parathyroid hormone): is produced by the parathyroid gland next to the thyroid gland. It is a regulator of calcium and phosphate metabolism (mediated by cAMP) primarily in the bones, kidneys and the digestive tract. A recessive hypoparathyroidism was mapped to Xp27. (See hypercalcemia-hypocalciuria, hyperparathyroidism)

PARATHYROID HORMONE: regulates Ca^{2+} level in animals.

PARATOPE: the epitope-binding site of the antibody Fab domain. (See antibody, epitope)

PARENCHYMA: in plant biology it means storage cells, either near isodiametric, *spongy parenchyma*, closer to the lower surface of the leaves or the *palisade parenchyma* consisting of one or two layers of columnar cells with their long axis perpendicular to the upper epidermis. Both types of tissues contain conspicuous intercellular space. In zoology the parenchyma cells mean the functional units, rather than the network of an organ or tissue.

PARENS PATRIAE: the state or community right to intervene against individual rights or beliefs and protect the interest of a person against potentially serious or actually life-threatening conditions, e.g. compulsory immunization, genetic screening, prohibition of incest, etc.

PARENTAL DITYPE: see tetrad analysis

PARIETAL: situated on the wall or attached to the wall of a hollow organ.

PARIS CLASSIFICATION: of human chromosomes standardized (in 1971) the banding patterns and classified them by size groups; it is basically very similar as it is used today. Current maps show, however, only one of the two chromatids. (See human chromosomes, Denver classification, Chicago classification)

PARITY: in gene conversion the process can go equally frequently in the direction of one or the other allele. (See gene conversion). In human biology: the condition that a woman had borne offspring.

PARITY CHECK: in a digital system reveals whether the number of ones and zeros is odd or even.

PARKINSONISM: is a secondary symptom caused either by drugs, or inflammation of the brain (encephalitis), or Alzheimer's disease, or Wilson's disease, or Huntington's chorea, etc. (See Parkinson's disease and the conditions named above)

PARKINSON's DISEASE (PD): is a shaking palsy generally with late onset, however juvenile forms also exist. PD may include mental depression, dementia and deficiency of several different substances, notably dopamine, from the nervous system. The genetic determination of the heterogeneous symptoms is unclear; autosomal dominant, recessive, X-linked, polygenic and apparently only environmentally caused phenotypes have been observed. Its prevalence is about 0.001. Anticholinergic (blocking choline) and dopamine, glial-cell-derived neurotrophic factor (GDNF) treatments may be beneficial. An alternative approach may be using gene therapy by expressing transfected tyrosine hydroxylase in the striated muscle cells. This enzyme can produce dopamine. An early onset PD was located to human chromosome Xq28; an autosomal dominant form is in chromosome 22, and another locus encoding α-synuclein was assigned to human chromosome 4q21-q23. A dominant parkinsonism with dementia and pallido-pontonigral degeneration has been located to human chromosome 17q21. (See parkinsonism, neuromuscular diseases, dopamine, tyrosine hydroxylase)

PAROTID GLAND: is the salivary gland; the proline-rich parotid glycoprotein is encoded in human chromosome 12p13.2

PARP (poly[ADP-ribose] polymerase): an enzyme involved in surveillance and repair of DNA. It is cleaved by an ICE-like proteinase. (See ICE, apoptosis)

PARS (poly[ADP-ribose]) synthetase): attaches ADP-ribose units to histones and to other nuclear proteins. It is activated when DNA is damaged by nitric oxide.

PARSIMONY: see maximal parsimony, evolutionary tree

PARSLEY (*Petroselinum crispum*): roots are eaten and leaves are used for flavoring; $2n = 2x = 22$.

PARSNIP (*Pastinaca sativa*): a root vegetable; $2n = 2x = 22$.

PARTHENOCARPY: fruit development without fertilization. (See also parthenogenesis, apomixia)

PARTHENOGENESIS: embryo production from an egg without fertilization. Parthenogenesis may be induced in sea urchins by hypotonic media or in some amphibia by mechanical or electric stimulation of the egg. In some fishes, lizards and birds (turkey) it occurs spontaneously. Parthenogenesis in animals is most common among polyploid species. Parthenogenetic individuals produce only female offspring. On theoretical grounds parthenogenesis may be disadvantageous because it deprives the species of elimination of disadvantageous mutations on account of the lack of recombination available in bisexual reproduction. Parthenogenesis may cause embryonic lethality in mouse if the imprinted paternal genes are not expressed. Parthenogenesis is not known to occur in humans, however it may exist as a chimera when after fertilization the male pronucleus is displaced to one of the blastomeres and then the maternal chromosome set in the other blastomere is diploidized. Many plant species successfully survive by asexual reproduction as an evolutionary mechanism. Parthenoghenesis in plants is called apomis or apomixia. (See apomixia, gynogenesis, parthenocarpy)

PARTHENOTE: an individual developed from an egg stimulated to divide and develop in the absence of fertilization by sperm. (See parthenogenesis, apomixia)

PARTIAL DIGEST: the reaction is stopped before completion of nuclease action and thus the DNA is cut into various size fragments some of which may be relatively long because some of the recognition sites were not cleaved. (See restriction enzyme)

PARTIAL DIPLOID: see merozygote

PARTIAL DOMINANCE: is an incomplete dominance, semidominance.

PARTIAL LINKAGE: the genes are less than 50 map units apart in the chromosome and can recombine at a frequency proportional to their distance. (See crossing over, recombination)

PARTIAL TRISOMY: only part of a chromosome is present in triplicate. (See trisomy)

PARTICULATE INHERITANCE: the modern genetic theory that inheritance is based on discrete particulate material (written in nucleic acid sequences) that is transmitted conservatively rather than according to the pre-Mendelian theory of pangenesis, that claimed that the hereditary material is a miscible liquid subject to continuous changes under environmental effects. According to the particulate theory of genetics genes are discrete physical entities that are transmitted from parents to offspring without blending or any environmental influence, except when mutation or gene conversion or imprinting occurs. (See also pangenesis, gene conversion, mutation, imprinting)

PARTICULATE RADIATION: see physical mutagens

PARTITIONING: distribution of plasmids into dividing bacterial cells. Also, partition of variance in statistics among the identifiable experimental components to reduce the quantity of the residual or error variance. (See segregation, analysis of variance)

PARTURITION: the labor of child delivery.

PARVOVIRUS: icosahedral (18-25 nm), single-stranded DNA (5.5 kb) viruses.

PASA: is a special PCR procedure by what chosen allele(s) can be amplified if the primers match the end of that allele.

PASCALE TRIANGLE: represents the coefficients of individual terms of expanded binomials: $(p + q)^n$:

$$1p^n + \frac{n}{1!(n-1)!} p^{n-1}q + \frac{n!}{2!(n-2)!} p^{n-2}q^2 + ... + \frac{n!}{(n-1)!1!} p^{n-(n-1)} q^{n-1} + 1q^n$$

Since genetic segregation is expected to comply with the binomial distribution, the coefficients indicate the frequencies of the individual phenotypic (or in case of trinomial distribution) genotypic frequencies. (See binomial distribution, trinomial distribution, table on next page)

PASSENGER DNA: a DNA inserted into a genetic vector.

PASSIVE IMMUNITY: acquired by the transfer of antibodies or lymphocytes

PASSIVE TRANSPORT: does not require special energy donor for the process. (See active transport)

PASTEUR EFFECT: fast reduction of respiration (glycolysis) if O_2 is added to fermenting cells.

PÄTAU's SYNDROME: is caused by trisomy for human chromosome 13. This is one of the

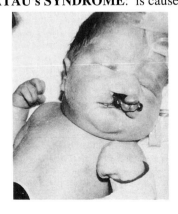

← PáTAU'S SYNDROME. (Courtesy of Dr. Judith Miles)

few (X, Y, 8, 18, 21, 22) trisomies that can be carried to term but it generally leads to death within 6 months because of severe defects in growth, heart, kidney and brain failures. It is accompanied also by face deformities (severe hare lip, cleft palate), polydactyly, club foot, defects of the genital systems, etc. Definite identification is carried out by cytological analysis, including FISH with the available chromosome-13-specific probes. An old designation of trisomy 13 was trisomy D because chromosome 13 belonged to the D group of human chromosomes. (See trisomy, trisomic analysis, aneuploidy)

Patch (*Ptc*): see sonic hedgehog, hedgehog

PATCH (patched duplex): resolution of a recombination intermediate (Holliday junction) without an exchange of the flanking markers (can be gene conversion). (See Holliday model L)

PATCH CLAMP TECHNIQUE: an electrode is tightly pressed against the plasma membrane so the flow of current through a voltage gated ion channel can be measured. It is used also for the sensitive *in situ* study of neurotransmitters. (See ion channels)

THE PASCAL TRIANGLE REPRESENTS THE COEFFICIENTS OF INDIVIDUAL TERMS OF EXPANDED BINOMIALS. THE EXPONENT OF THE BINOMIAL IS *n*. THE FIGURES DISPLAY A SYMMETRICAL HIERACHY. THE FREQUENCY OF A PARTICULAR CLASS CAN BE READILY CALCULATED BECAUSE OF THE SUM OF THE COEFFICIENTS IS DISPLAYED AT THE BOTTOM OF THE COLUMNS. MENDELIAN SEGREGATION FOLLOWS THE BINOMIAL DISTRIBUTION. (See text on preceding page)

$n \rightarrow$	1	2	3	4	5	6	7	8	9	10
	1	1	1	1	1	1	1	1	1	1
	1	2	3	4	5	6	7	8	9	10
		1	3	6	10	15	21	28	36	45
			1	4	10	20	35	56	84	126
				1	5	15	35	70	126	210
					1	6	21	56	126	252
						1	7	28	84	210
							1	8	36	120
								1	9	45
									1	10
										1
SUMS	2	4	8	16	32	64	128	256	512	1024

$n \rightarrow$	11	12	13	14	15	16	17	18	19	20
	1	1	1	1	1	1	1	1	1	1
	11	12	13	14	15	15	17	18	18	20
	55	66	78	91	105	120	136	153	171	190
	165	220	286	364	455	560	680	816	969	1140
	330	495	715	1001	1365	1820	2380	3060	3876	4845
	462	792	1287	2002	3003	4368	6188	8568	11682	15504
	462	904	1716	3003	5005	8008	12376	18564	27132	38760
	330	792	1716	3432	6435	11440	19448	31824	50388	77520
	165	495	1287	3003	6435	12870	24310	43758	75582	125970
	55	220	715	2002	5005	11440	24310	48620	92378	167960
	11	66	286	1001	3003	8008	19448	43758	92378	184756
	1	12	78	364	1365	4368	12376	31824	75582	167960
		1	13	91	455	1820	6188	18564	50388	125970
			1	14	105	560	2380	8568	27132	77520
				1	15	120	680	3060	11628	38760
					1	16	136	816	3876	15504
						1	17	153	969	4845
							1	18	171	1140
								1	19	190
									1	20
										1
SUMS	2048	4096	8192	16384	32768	65536	131072	262144	524288	1048576

PATENT: the so called gene patents do not protect the DNA sequence itself, rather the processs of manipulation is the object. The gene "ownership" only prevents the use or selling a particular sequence without permission. In general, according to US patent laws, the patent is pro-

Patent continued

tected for 17 years from date of issue. A requisite for patenting is that the subject of the patent application would be new and practically useful, e.g. be a probe for a gene. During the period of time of the patent, the patent-holder can prevent anybody from using it, including those who invented the same independently or even those who improved on the procedure to such an extent that the second invention meets the requirements for patenting. The inventor, however is obligated by law to disclose the invention in sufficient technical detail that anybody with the proper expertise could use it. The fact that an invention was arrived at under federal financial support does not exclude patentability but the inventor must report the patentable invention to the sponsoring agency. The intention of the government is that the invention would be used at maximum benefit to the public that can be achieved usually most effectively by commercial private enterprises. The patenting of the outcome of genetic research may be harmful to science if the investigators keep secret the ongoing work until it becomes patentable. If the discovery is published through the proper means of scientific communications prior to the patent application, it is disqualified from patenting. It is generally easier to patent a product than a process. Natural products (e.g. proteins) are usually not patentable unless they are modified in some way and are different from the natural product either in structure or function and these properties were not generally known. A DNA sequence, identified as the coding unit for a genetic disease or a genetic marker in its vicinity may be patentable but a cloned gene that may be used for translating a protein may not. DNA markers are patentable only if their direct use can be determined. The concept of patenting biological material raises several moral objections but it is defended by the biotechnology industry because it takes 100s of millions of dollars for the completion of such projects, and without the financial means, these investigations cannot be maintained. Between 1981 and 1995 a total of 1,175 human DNA sequences were patented. The patents laws vary in different countries, and also new legislation may take place any time. (See Cohen - Boyer patent on recombinant DNA, Eisenberg, R. S. 1992 in Gene Mapping, p. 226, Annas, G. J. & Elis, S., eds., Oxford Univ. Press, New York)

PATENT DUCTUS ARTERIOSUS: open blood vessels; is a disease caused by fetal rubella infection and also apparently by an autosomal dominant gene. Risk of recurrence in an affected family is about 1-2%. General incidence is less than 0.001. (See risk)

PATERNAL LEAKAGE: the transmission of mitochondrial DNA through the males. Generally mitochondria are not transmitted through the animal sperm. In mice, apparently the male transmission of mitochondria is within the range of 10^{-5}. In interspecific mouse crosses paternal mitochondria are transmitted but they are eliminated during early embryogenesis. The role of transmission of mitochondria in humans is not clear. Some cytological observations may indicate the incorporation of the midpiece of the sperm (containing mitochondria) into the egg. Genetic evidence for human paternal transmission of mitochondria is still needed. In some molluscans (mussel) there is a strong biparental inheritance of mtDNA. In *Mytilus* the paternal and maternal mtDNA displays 10 to 20% nucleotide divergence. The females transmit just one type of mitochondria whereas the males transmit also a second type of mtDNA genome. Biparental transmission of mtDNA may occur also in interspecific crosses of *Drosophila*. In *Paramecia,* through a cytoplasmic bridge, mitochondria may be transmitted. In fungi, the transfer is maternal although in some heterokaryonts cytoplasmic mixing may take place. In some slime molds mtDNA transmission is also mating type dependent. In *Physarum polycephalum* different *matA* alleles regulate the mtDNA transmission but a plasmid gene may also be involved and recombination can take place between mtDNAs. In *Chlamydomonas* algae several genes around the mating type factors were implicated. (See mtDNA, plastid male transmission, Eve foremother, mitochondrial disease in humans, mitochondrial genetics, plastid genetics)

PATERNITY EXCLOSURE: is based on genetic paternity tests. (See paternity testing, DNA fingerprinting)

PATERNITY TESTING: is frequently required in civil litigation suits but it may have significance for medical, immigration, archeological and other cases. The laboratory procedures are

Paternity testing continued

generally the same as used for DNA fingerprinting. Here as in DNA fingerprinting in general the exclusion of paternity is simple and straightforward but the determination of identity may pose more difficulties because in the multilocus tests more than 10% of the offspring may show one band difference and 1% may show two due to mutation. Therefore the following formula was recommended by Penas and Chakraborty (Trends in Genetics 10:204 [1994]):

$$PI = \frac{\binom{N}{U} \mu^U (1-\mu)^{N-U}}{\binom{n}{U} X^{n-U} (1-X)^U}$$

PI = paternity index, μ = mutation rate, X = band-sharing parameter, N = total number of bands per individual, n = number of test bands, U = number of bands not present in the alleged father. In rare instances (mistakes at maternity wards) similar test may be necessary to test maternity. The biological father of a child — even if the paternity can be accurately proven — cannot assert paternal rights against the will of the mother if she was/is married to another man [Hill J.L. 1991 New York University Law Review 66:353]. When "the child is born to a mother who is single or part of a lesbian couple, law does permit the biological father to assert his paternal rights, even if he clearly stated his intention prior to conception to have no relation" [Charo, R. A. 1994 in The Genetic Frontier. Ethics, Law and Policy. Frankel, M.S. & Teich, A., eds., Amer. Assoc. Adv. Sci., Washington, DC, USA]. (See forensic genetics, DNA fingerprinting, forensic index, utility index, surrogate mother)

PATH COEFFICIENT: method of Sewall Wright has been worked out for studying mathematically and by diagrams the paths of genes in populations, and genetic events determining multiple correlations. Here it is not possible to discuss meaningfully the mathematical foundations but one type of graphic application for determining some relations between offspring and parents can be found under *F* and inbreeding coefficient. (See inbreeding coefficient, correlation)

PATHOGEN IDENTIFICATION: the food industry may need rapid and highly sensitive methods for the detection of live pathogens in various products. In case of viable *E. coli* cells, this is feasible by infection with compatible bacteriophages carrying bacterial luciferase inserts. The genes in the phages are expressed only in live bacteria and if such are present, with a high-powered luminometer or by a microchannel plate enhanced image analyzer, even a single bacterial cell emitting light may be detected. (See luciferase bacterial)

PATHOGENESIS RELATED PROTEINS (PR): a variety of acidic or basic proteins synthesized in plants upon infection with pathogens. The chitinases and glucanases apparently act by damaging the cell wall of fungi, insects or even bacteria. (See host - pathogen relations)

PATHOGENIC: capable of causing disease.

PATHOGENICITY ISLAND: a group of genes in a pathogen involved in the determination and regulation of pathogenicity.

PATRISTIC DISTANCE: is the sum of the length of all branches connecting two species in an evolutionary tree. (See evolutionary tree)

PATROCLINOUS: inheritance through the male such as the Y chromosome, androgenesis, fertilization of a nullisomic female with a normal male, through non-disjunction the chromosome to be contributed by the female is eliminated, some of the gynandromorphs, sons of attached-X female *Drosophila*, etc. (See gynandromorph. nondisjunction, attached-X)

PATRONYMIC: a designation indicating the descent from a particular male ancestor, e.g. Johnson, son of John or O'Malley descendant of Malley. Common family names may assist in isolated populations to establish relationships. (See isonymy)

PATTERN FORMATION DURING DEVELOPMENT: developmental patterns may begin by intracellular differentiation (animal pole, vegetal pole, yolk), positional signals between cells and intracellular distribution of the receptors to various signals. Fibroblast growth factor and

Pattern formation during development continued

transforming growth factor-β have apparently major roles as mesoderm induction signals. The *Drosophila* gene *fringe* (*fng*) is involved with mesoderm induction and in wing embryonal disc formation. Juxta-position of *fng*-expressing and non-expressing genes seems to be required for establishing the dorsal ventral boundary of wing discs. The gene *fng* is expressed in the dorsal half of the wing disc whereas *wingless* (*wg*, 2-30) is limited to the dorsal-ventral boundary. Gene *hedgehog* (*hh*, 3-81) is expressed in the posterior half, and *decapentaplegic* (*dpp*, 2-40) is detected in the anterior-posterior boundary. Anterior - posterior patterning in *Drosophila* is affected by the *trithorax* (*trx*), *Polycom* (*Pc-G*) family members such as *extra sex combs* (*esc*). The homolog of the latter, *eed* (embryonic ectoderm development) controls anterior-posterior differentiation in mouse. Homologs to these genes have been identified in other animals and humans as well. In *Xenopus*, it appears that the FNG protein is translated with a signal peptide (indicating that it is secreted), in the proFNG peptide is terminated with a tetrabasic site for proteolytic cleavage and after this processing, it is ready for the normal function. *Wingless* of *Drosophila* is homologous to *Wnt1* mouse mammary tumor gene. Developmental pattern formation is under genetic control also in plants. The progress has been much slower, however, because the plant tissues and cells are more liable to dedifferentiation and redifferentiation. Mutants have been obtained with clear differences in morphogenesis but with the exception of flower differentiation and photomorphogenesis much less is known about the molecular mechanisms involved. (See morphogenesis, morphogenesis in *Drosophila,* flower differentiation, photomorphogenesis, MADS box, signal transduction, homeotic genes, cell lineages)

PAU GENES: see seripauperines

pauling: see evolutionary clock

PAUP (phylogenetic analysis using parsimony): is a computer program for the analysis of evolutionary descent on the basis of molecular data. (See evolutionary distance, evolutionary tree)

PAUSE: see attenuator region, σ

PAX (paired box homeodomain): several PAX proteins are known to be encoded in at least five different chromosomes and they mediate the development of the components of the eyes in insects (compound eyes) and humans, the central nervous system, the vertebrae and tumorigenesis. The 130 residue paired domain binds DNA and functions as a transcription factor for B cells, histones and thyroglobulin genes. (See Waardenburg syndrome, aniridia, Wilms tumor, rhabdosarcoma, animal models, B cell, histones, thyroglobulin, hox, homeotic genes)

PAXILLIN: is a protein that anchors actin filaments to fibronectin with the assistance of actinin, vinculin, talin and capping protein. (See CAM)

PBP: penicillin binding proteins are transpeptidases that inactivate cross-linking reactions of peptidoglycan synthesis (bacterial cell wall), and thus inactivate bacteria.

pBR322: non-conjugative plasmid (constructed by Bolivar & Rodriguez) of 4.3 kb, can be mobilized by helper plasmids because (although it lost its mobility gene) it retained the origin of conjugal transfer. It is one of the most versatile cloning vectors with completely known nucleotide sequence and over 30 cloning sites. It carries the selectable markers ampicillin resistance and tetracycline resistance. Insertion into these antibiotic resistance sites permit the detection of the success of insertion because of the inactivation of these target genes results in either ampicillin or tetracycline sensitivity. Although its direct use of this 20 years old plasmid has diminished during the last years, pBR322 components are present in many currently used vectors. (See plasmid, vectors, Amp, tetracyclin, see also diagram next page)

PBS: phosphate buffered saline.

PBX1, PBX2: are transcription factors involved in B cell leukemias, encoded in human chromosomes 1q23 and 3q222, respectively.

PCAF: is a human acetyltransferase of histones 3 and 4. (See p300, $TAF_{II}230/250$, histone acetyltransferases, nucleosome)

PCB: polychlorinated biphenyl is an obnoxious industrially employed carcinogen. *Pseudomonas* enzyme that may break it down. (See environmental carcinogens, sperm)

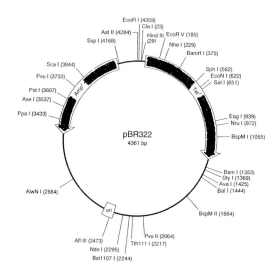

pBR332 plasmid cont. from preceding page.

ILLUSTRATION FROM PHARMACIA BIOTECH INC., BY PERMISSION.

PCL: putative cyclin. (See cyclin)

PCNA (**p**roliferating **c**ell **n**uclear **a**ntigen): is an auxiliary protein in pol δ and pol ε functions in eukaryotes; it has a similar role in DNA replication in general, and in the cell cycle and repair. Its function is similar to that of the β subunit of the prokaryotic pol III, it provides a 'sliding clamp' on the DNA to be replicated. PCNA replicative function may be inhibited by binding to p21. (See DNA replication, eukaryotes, cell cycle, DNA polymerases, p21, ABC excinuclease, excision repair)

PCR: see polymerase chain reaction

PCR-BASED MUTAGENESIS: any base difference between the amplification primer will be incorporated in the future template through polymerase chain reaction. Actually only half of the new DNA molecules would contain the alteration present in the original amplification primer unless a device is used, e.g. the undesired strand would be made unsuitable for amplification and therefore lost from the reaction mixture. The method may include multiple point mutations, small insertions or deletions too. It is possible that the amplification may result also in other nucleotide alterations as a result of the error-prone Taq polymerase. (See local mutagenesis, primer extension, polymerase chain reaction, DNA shuffling, VENT)

PCR TARGETING: see targeting genes

PC - TP: phosphatidylcholine transfer protein mediating transfer of phospholipids between organelles within cell.

PDECGF: platelet-derived endothelial cell growth factor.

pDelta: see γδ

PDGF: see platelet derived growth factor

pDUAL: see γδ

SEGREGATION FOR SMOOTH:WRINKLED WITHIN THE PEA POD.

PDZ DOMAINS: approximately 90 amino acid repeats involved in ion-channel and receptor clustering and linking effectors and receptors. (See ion channels, receptor, effector)

PEA (*Pisum* spp): several self-pollinating vegetable and feed crops: the Mendel's pea is *P. sativum*, and others are 2n = 2x = 14. (See Pisum, photograph above of segregation within pod)

PEA COMB: comb characteristic of poultry of *rrP(P/p)* genetic constitution. (See walnut comb)

PEACH (*Prunus persica*): x = 7, the true peaches are diploid.

PEACOCK'S TAIL: an evolutionary paradigm when a clear disadvantage (like the awkward tail) turns into a mating advantage because of the females' preference for the fancy trait and thus increasing the fitness of the males that display it. (See fitness, selection, sexual selection)

PEANUT: see groundnut

PEAR (*Pyrus* spp): about 15 species; x = 17 and mainly diploid, triploid or tetraploids. It is very difficult to hybridize it with apples but can be crossed with some *Sorbus*. (See apple)

PEARSON MARROW PANCREAS SYNDROME: see mitochondrial disease in humans

PECTIN: polygalacturonate sequences alternated by rhamnose and may contain galactose, arabinose, xylose and fucose side chains. Molecular weight varies from 20,000 to 400,000. Its role is intercellular cementing of plant cells. Acids and alkali may cause its depolymerization.

PEDICEL: the stalk of flowers in an inflorescence. (See peduncle)

PEDIGREE ANALYSIS: is generally carried out by examination of pedigree charts, used in human and animal genetics where the family sizes are frequently too small to conduct meaningful direct segregation studies. The pedigree chart displays the lines of descent among (close) relatives. Females are represented by circles, males by squares and if the sex is unknown a diamond (◊) is used. Abortion or still birth is indicated by the same but smaller symbols or by a vertical line over the symbol. Individuals expressing a particular trait are represented by a shaded or black symbol, and in case they are heterozygous for the trait half of the symbol is shaded. When an unaffected female is the carrier of a particular gene, there is a dot within the circle. In case of segregation for traits needs to be illustrated in the pedigree, the individual displaying both traits may be marked by a horizontal and vertical line within the symbol or only by a horizontal or vertical line, respectively. The parents are connected by horizontal lines and if the parents are close relatives the line is doubled The progeny is connected to the parental line with a vertical line and the subsequent generations are marked by Roman numerals at the left side of the chart, I (parents), II (children), III (grandchildren), and so on. Twins are connected to the same point of the generation line and if they are identical they are connected also to each other by a horizontal line. The order of birth of the offspring is from left to right and may be numbered accordingly below their symbol. An arrow to a particular symbol indicates the proband, the individual who first became known to the geneticist as expressing the trait. Males may also be represented by a *Mars shield* and females by a *Venus mirror* and interserxes by the sign. If a prospective offspring is considered at risk, the symbol is drawn by broken lines. Adoption is indicated by parenthesis around the symbol.

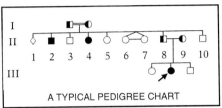

PEDOGENESIS: egg production by immature individuals such as larvae.

PEDUNCLE: the stalk of single-standing flowers. (See also pedicel)

PEG (polyethylene glycol): may be liquid or solid and comes in a range of different viscosities (200, 400, 600, 1500, etc.). It facilitates fusion of protoplasts, uptake of organelles, precipitation of bacteriophages, plasmids and DNA, promoting end-labeling, ligation of linkers, etc.

PELARGONIDINE: see anthocyanin

Pelargonium zonale (geranium): ornamental plant. Some variegated forms transmit the non-nuclear genes also through the sperm whereas in the majority of plants genes in the plastids are transmitted only through the egg. (See uniparental inheritance, chloroplasts, chloroplast genetics)

PELGER(-HUET) ANOMALY: an autosomal dominant condition in humans as well as rabbits, cats, etc. characterized by fewer (1.1-1.6) than normal (2.8) nuclear lobes in the granulocytic leukocytes. It may be a mild anomaly but it may be associated with other more serious ailments. The prevalence varies from 1×10^{-3} to 4×10^{-4}. Similar phenotypes were described also as autosomal recessive or X-linked.

PELIZAEUS-MERZBACHER DISEASE: is an X-chromosomal leukodystrophy that accumulates a proteolipid of the endoplasmic reticulum and the surface protein DM20. The clinical symptoms are defective myelination of the nerves and defective interaction between oligodendrocytes and neurons and pathogenesis of the central nervous system.

Pelle: a serine/threonine kinase, involved in dorsal signal transduction.

pelota: *Drosophila* gene involved in sperm function. (See azoospermia)

PEMPHIGUS: is a collection of skin diseases with the general features of developing smaller or larger vesicles of the skin that may or may not heal and in extreme cases may result in death. The autosomal dominant familial pemphigus vulgaris is an autoimmune disease of the skin and mucous membranes. In the majority of cases HLA-DR4 is involved. This anomaly is particularly common among Jews in Israel. (See Hailey-Hailey disease, HLA, skin diseases, desmosome, autoimmune disease)

PENDRED SYNDROME: is a recessive (probably human chromosome 7q21-34 region) thyroid anomaly and neurosensory deafness. For about 1-10% of the genetically determined hearing loss, this locus is responsible. (See deafness)

PENETRANCE: percentage of individuals that express a trait determined by gene(s) they con-tain. (See expressivity)

PENICILLIN: an antibiotic originally obtained from *Penicillium* fungi. (See antibiotics, *Penicillium*)

PENICILLIN ENRICHMENT: see penicillin screen

PENICILLIN SCREEN: was used for mass isolation of auxotrophic microbial mutations that failed to grow in basal media (in contrast to the wild type) and the presence of the antibiotic therefore did not lead to their death (in contrast to the wild type). After transfer to complete (or appropriately supplemented) media the auxotrophs grew and thus were selectively isolated. (See selective medium, replica plating, mutant isolation)

PENICILLINASE: see β-lactamase

PENICILLIN BINDING PROTEINS: see PBP

Penicillium notatum (fungus): x = 5

PENIS: is the male organ of urinary excretion and insemination (homologous to the female clitoris). It contains the corpus spongiosum through which the urethra passes. Above that are the corpora cavernosa that become extended when erection takes place due by enhanced blood supply to this elastic tissue as a consequence of NO (nitrogenmonoxide) gas flow to the muscles of the blood vessel wall, initiated by acetylcholine. The release of acetylcholine is controled by steroid hormones. (See animal hormones, acetylcholine, acetylcholine receptors, hypospadias, nitric oxide)

PENTAGLYCINES: see bacteria

PENTAPLOID: its cell nucleus contains five genomes (5x). Pentaploids are obtained when hexaploids (6x) are crossed with tetraploids (4x). The pentaploids are generally sterile or semifertile because the gametes generally have unbalanced number of chromosomes. (See polyploids, *Rosa canina*)

PENTOSE: sugar with 5 carbon atom backbone, such as ribose, deoxyribose, arabinose, xylose.

PENTOSE PHOSPHATE PATHWAY: glucose-6-phosphate + 2 NADP + H_2O → ribose-5-phosphate + 2 NADPH + 2 H+ + CO_2, i.e. the conversion of hexoses to pentoses generates NADPH, a molecule that serves as a hydrogen and electron donor in reductive biosynthesis. (See also Embden-Meyerhof pathway, Krebs-Szentgyörgyi cycle)

PENTOSE SHUNT: same as pentose phosphate pathway.

PENTOSURIA: an autosomal recessive non-debilitating condition characterized by excretion of increased amounts of L-xylulose (1-4 gm) in the urine because of a deficiency of the NADP-linked xylitol dehydrogenase enzyme. In Jewish and Lebanese populations the frequency of the gene was about 0.013- 0.03. (See gene frequency, allelic frequencies)

PEPCK: phosphoenolpyruvate carboxykinase, a regulator of energy metabolism.

PEPPER (*Capsicum* spp): they exist in a great variety of forms but all have 2n = 2x = 24 chromosomes. Some wild species are self-incompatible and the cultivated varieties also yield better if they have a chance for xenogamy. (See self-incompatibility, xenogamy)

PEPSIN: is an acid protease, formed from pepsinogens. It has preference for COOH side of phenylalanine and leucine amino acids.

PEPSTATIN: is a protease (pepsin, cathepsin D) inhibitor.

PEPTIDASE: hydrolyzes peptide bonds. In humans the peptidase gene PEPA is in chromosome

18q23, PEPB in 12q21, PEPC in 1q42, PEPD in 19cen-q13.11, PEPE in 17q23-qter, PEPS in 4p11-q12, and the tripeptidyl peptidase II (TPP2), a serine exopeptidase is in 13q32-q33.

PEPTIDE BOND: amino acids are joined into peptides by their amino and carboxyl ends (and they lose one molecule of water)

$$\underset{H}{\overset{|}{N}} = \overset{\overset{O}{\|}}{C} + H_2O$$
$\uparrow$ peptide bond

PEPTIDE ELONGATION: see protein synthesis, aminoacylation, aminoacyl-tRNA synthetase, elongation factors (eIF), ribosome

PEPTIDE INITIATION: see protein synthesis

PEPTIDE MAPPING: separation of (in) complete hydrolysates of proteins by two-dimensional paper chromatography or by two-dimensional gel electrophoresis for the purpose of characterization; the distribution pattern is the map or fingerprint, characteristic for each protein.

PEPTIDE NUCLEIC ACID: a nucleic acid base (generally thymine) is attached to the nitrogen of a glycine by a methylene carboxamide linkage in a backbone of aminoethylglycine units. Such a structure can displace one of the DNA strands and binds to the other strand. This highly stable complex has similar uses as the antisense RNA technology. It may also be used to screen for base mismatches and small deletions or base substitution mutations. Peptide nucleic acid complementary to mutant mtDNA selectively inhibits the replication of mutant mtDNA *in vitro*. (See antisense RNA, antisense DNA, mtDNA)

PEPTIDE PROCESSING: see post-translational processing

PEPTIDE TRANSPORTERS: TAP

PEPTIDE VACCINATION: synthetic polypeptides corresponding to CTL epitopes may result in cytotoxic T cell-mediated immunity but in some instances may enhance the elimination of anti-tumor CTL response. (See vaccination, CTL, epitope)

PEPTIDOGLYCAN: heteropolysaccharides cross-linked with peptides constituting the bulk of the bacterial cell wall.

PEPTIDYL SITE: see P site, ribosome, protein synthesis

PEPTIDYL TRANSFERASE: generates the peptide bond between the preceding amino acid carboxyl end (at the P ribosomal site) and the amino end of the incoming amino acid (at the A site of the ribosome). It is a ribozyme. (See ribosome, protein synthesis)

per (period) locus: in *Drosophila* (map location 1-1.4, 3B1-2) controls the circadian and ultradian rhythm and thus affecting eclosion, general locomotor activity, courtship, intercellular communication, etc. The mutations do not seem to affect the viability of the individuals involved, only the behavior is altered. When *per^s* (caused by a base substitution mutation in exon 5) in the brain is transplanted into *per^{01}* mutants (nonsense mutation in exon 4) causing short ultradian rhythm and multiple periods, some flies may be somewhat normalized. The locus has been cloned and sequenced and seems to code for a proteoglycan. The Per protein forms a heterodimeric complex with the Tim (*timeless* gene) protein and jointly autoregulate transcription. Tim is degraded in the morning in response to light and that results in the disintegration of the complex that is reformed again in dark in a circadian oscillation. (See circadian, ultradian, proteoglycan)

PERDURANCE: the persistence and expression of the product of the wild type gene even after the gene itself is no longer there.

PERENNIAL: lasts through more than one year.

PERFECT FLOWER: has both male and female sexual organs, i.e. it is hermaphroditic. (See hermaphrodite, flower differentiation)

PERFORIN: is pore-forming protein (homologous to component 9 of the complement) and establishes transmembrane channels; it is stored in vesicles within the $CD8^+$ cytotoxic T cells (CTL). These vesicles contain also serine proteases. Perforin mediates apoptosis by permitting killer substances slowly enter the cell. (See apoptosis, complement, T cells, fragmentin-2, granzymes)

PERFUSION: adding liquid to an organ through its internal vessels.

PERIANTH: designates both sepals and petals of the flowers. (see flower differentiation)

PERICARP: the fruit wall, developed from the ovary wall such as the pea pod, the outer layer of the wheat or maize kernels. The silique of *Arabidopsis*, the peel of the citruses, and the skin of apple, the shell of the nuts, etc. are also similar but exocarps. The outer layer of the common barley "seed" is not part of the fruit wall but it is a bract of the flower.

PERICENTRIC INVERSION: see inversion pericentric

PERICLINAL CHIMERA: contains genetically different tissues in different cell layers.

PERICYCLE: (root) tissue in-between the endodermis and the phloem. (See endodermis, phloem)

PERIDIUM: the covering of the hymenium or the hard cover of the sporangium of some fungi. (See hymenium, sporangium)

PERIODIC ACID-SCHIFF REAGENT (PAP): tests for glycogen, polysaccharides, mucins, and glycoproteins. It breaks C-C bonds by oxidizing near hydroxyl groups and forms dialdehydes and generates red or purple color.

PERIODIC PARALYSIS: a group of autosomal dominant human diseases manifested in periodically recurring weakness accompanied by low blood potassium level (hypokalemic periodic paralysis) or in other forms with high blood potassium level (hyperkalemic periodic paralysis). The latter types were attributed to base substitution mutations in a highly conserved region of the α subunit of a transmembrane sodium channel protein. In another type of the disease the blood potassium level appeared normal and the patients responded favorably to sodium chloride. (See ion channel, Moebius syndrome, myotonia, hyperkalemic periodic paralysis)

PERINATAL: around birth; the period after 28 weeks of human gestation and ending four weeks after birth.

PERIODICITY: the number of base pairs per turn of the DNA or the number of amino acids per turn of an α-helix of a polypetide chain. (See protein structure, Watson and Crick model)

PERIPHERAL NERVOUS SYSTEM: resides outside the brain and the spinal chord.

PERIPHERAL PROTEINS: are bound to the membrane surface by hydrogen bonds or electrostatic forces. (See membrane proteins)

PERIPLASMA: bacterial cell compartment between cell wall and cell membrane.

PERISTOME: a fringe of teeth at the opening of the sporangium of mosses or the buccal (mouth) area of ciliates.

PERITHECIUM: a fungal fruiting body of disk or flask shape with an opening (ostiole) for releasing the spores. A perithecium of *Neurospora* contains about 200 asci. Its primordium is called protoperithecium. (See ascogonium, apothecium, cleistothecium, gymnothecium)

PERIVENTRICULAR HETEROTOPIA: a human X-chromosomal mental retardation and seizures caused by anomalies of the brain cortex. (See also double cortex)

PERMAFROST: the soil layer in cold regions that remains permanently frozen even when the top may thaw.

PERMANENT HYBRID: see complex heterozygote

PERMEASE: enzymes involved in the transport of substances through cell membranes. (See membrane transport, membrane channels, membrane potential, ion channels)

PERMISSIVE CONDITION: at which a conditional mutant can survive or reproduce. (See conditional mutation)

PERMISSIVE HOST: a cell permits (viral) infection and/or development.

PERMUTATION: generating all possible orders of **n** numbers, and it can be obtained by the factorial: n! , e.g., the factorial of 4, $4! = 4 \times 3 \times 2 \times 1 = 24$

PERMUTED REDUNDANCY: at the termini of phage DNA a collection of redundant sequences occur in the phage population that start and end with permuted sequences of the same nucleotide sequence, e.g. 1234....1234, 2341....2341, 3412....3412, etc. This arrangement is characteristic for T-uneven phages, e.g. T1, T3, T5, etc. (See also non-permuted redundancy)

PERODICTUS: see Lorisidae

PEROXIDASE: heme protein enzymes which catalyze the oxidation of organic substances by peroxides. Glutathione peroxidase (and selenium) deficiency may cause hemolytic disease. Several peroxidase genes have been located in the human genome: GPX1 in 3q11-q12, GPX2 in 14q24.1, GPX3 in 5q32-q33, GPX4 (in testes) in chromosome 19, a rare eosinophil peroxidase (EPX) may compromise the immune system, a thyroid peroxidase deficiency (2p25) interferes with thyroid function.

PEROXIDASE AND PHOSPHOLIPID DEFICIENCY: an autosomal recessive anomaly of the eosinophils involving the enzyme defects named. (See microbody, Refsum disease, Zellweger syndrome)

PEROXIDES: display the $-O-O-$ linkage. Organic peroxides participate in activation and deactivation of promutagens, mutagens, procarcinogens and carcinogens and in many other physiological reactions. Peroxides are formed by the breakdown of amino acids and fatty material in the cell and may inflict thus serious damage. According to some views spontaneous mutation may be caused to a great extent by these regular components of the diet. Therefore eating rancy food may pose substantial risk. (See environmental mutagens, peroxidase, catalase peroxisomes, promutagen)

PEROXISOMAL 3-OXOACYLCOENZYME A THIOLASE DEFICIENCY (pseudo-Zellweger syndrome): autosomal recessive disease assigned to human chromosome 3p23-p22. (See microbody, Zellweger syndrome, adrenoleukodystrophy)

PEROXISOME: see microbodies

PERSONALITY can be characterized by five main groups of features: (1.) *extraversion* (being outgoing or the lack of it, ability to lead and sell their ideas versus reticent and avoiding company, [heritability about 0.71]; (2.) *neuroticism* (emotional versus stable, worrisome or self-assured, [heritability about 0.21]; (3.) *conscientiousness* (well-organized versus impulsive, responsible or irresponsible, reliable or undependable, [heritability 0.38-0.32]; (4.) *agreeableness* empathic or unfriendly, warm versus cold, cooperative versus quarrelsome, forgiving versus vindictive, [heritability about 0.49], and (5.) *openness* (insightful or lacking intelligence, imaginative versus imitative, inquisitive or superficial). These heritability estimates vary a great deal, however and may be very different in some populations. On the basis of twin studies several investigators concluded that overall close to 50% of the variance could be attributed to additive or non-additive genetic determination. (See behavior in humans, behavior genetics, human intelligence, affective disorders, heritability in humans)

PERTUSSIS TOXIN: is produced by *Bordatella* bacteria, responsible for whooping cough. The toxin stimulates ADP-ribosylation of the $G\alpha_1$ subunit of a G-protein in the presence of ARF and thus GDP stays bound to the G-protein and adenylate cyclase is not inhibited and K^+ ion channels do not open. As a consequence histamine hypersensitivity and reduction of blood glucose level follows. (See signal transduction, ARF, G-protein, ADP, GDP, cholera toxin, adenylate cyclase)

PEST: a carboxyl domain of IκB involved in stimulation of its proteolysis. (See IκB, NF-κB)

PEST ERADICATION BY GENETIC MEANS: see genetic sterilization, *Bacillus thüringiensis*, host - pathogen relations

PESTICIDE MUTAGENS: see environmental mutagens

PESTICIN: toxin of *Pasteurella* bacteria.

PESTILENCE: infectious epidemic of disease.

PET: see tomography

PETALS: generally the second whorl of modified leaves from the bottom of the flower. Frequently they are quite showy because of their anthocyanin or flavonoid pigmentation. The petal number is a taxonomic characteristic, although petal number my be altered by homeotic mutations converting the anthers and/or pistils into petals and appearing as sterile double flowers of floricultural advantage. (See morphogenesis of flowers, flower pigments, homeotic mutants)

PETIOLE: the stalk of leaves.

PETITE COLONY MUTANTS: of yeast form small colonies because they are deficient in respiration and lethal under aerobic conditions. The *vegetative petites* (ρ^-) are caused by (large) deletions in the mitochondrial DNA, the *segregational petites* are controled by nuclear genes at over 200 loci. The mitochondrial mutations occur at high (0.1 to 10%) frequency and using ethidium bromide as a mutagen their frequency may become as high as 100%. The mitochondrial petites fail to transmit this character in crosses with the wild type except one special group the *suppressive petites* that may be transmitted at a low frequency in outcrosses with the wild type. In yeast, the A+T content of the normal mitochondrial DNA is about 83%, in some of the mitochondrial mutants the A+T content may reach 96% because the coding sequences were lost and only the redundant A+T sequences were retained and amplified so the mtDNA content is not reduced. The *hypersuppressive petite* mutants have short (400-900 bp) repeats that share 300 bp (*ori* and *rep*) sequences with the wild type, necessary for replication. Yeast cells that have normal mitochondrial function make large colonies and are called grande. (See mitochondria, mtDNA)

PETRI PLATE: a (flat) glass or disposable plastic culture dish for microbes or eukaryotic cells.

Petunia hybrida (2n =28): *Solanaceae*; predominantly self-pollinating but allogamy also occurs. It has been used extensively for cell, protoplast and embryo culture, intergeneric and interspecific cell fusion, genetic transformation and the genetic control of pigment biosynthesis. It has a good number of related species.

PEUTZ-JEGHERS SYNDROME: see polyposis hamartomatous

PEV (position effect variegation): see position effect, heterochromatin, RPD3

PEYER's PATCHES: aggregated lymphatic nodes, plaques are instruments of mucosal immunity. (See mucosal immunity)

PFEIFFER SYNDROME: includes the autosomal dominant bone malformation affecting the head, thumbs and toes (acrocephalosyndactyly), the autosomal recessive head-bone (craniostenosis) and heart disease. The latter type seems to cosegregate with fibroblast growth factor receptor 1 in human chromosome 8p11. Another locus in chromosome 10q25-q26 represents also a fibroblast growth factor receptor, FGFR2. FGFR3 is located in chromosome 4p. The mutations in all three genes involve Pro→Arg replacements at identical sites, 253. This syndrome is allelic to the Crouzon and to the Jackson-Weiss syndromes. (See fibroblast growth factor, Alpert's syndrome, Crouzon syndrome, Jackson-Weiss syndrome, craniosynostosis syndromes)

PFGE (pulsed field gel electrophoresis): it separates very large nucleic acid fragments or even small chromosomes. The megabase size fragments can be used for physical mapping of large chromosomal domains (PFG mapping). (See also pulsed field gel electrophoresis)

pfu (p.f.u.): plaque forming unit. (See plaque)

PGA: phosphoglyceric acid, a 3-carbon product of photosynthesis. (See photosynthesis, Calvin cycle)

PGC: gynogenetic/parthenogenetic cell; the primordial cell of female gonads. (See gonad, gynogenesis, parthenogenesis)

PGK-neo: is commonly used transformation cassette for gene knockout where the neomycinphosphotransferase gene (*neo*) is fused to the phosphoglycerate kinase (PGK) promoter. (See knockout, vector cassette)

PGM: see phosphoglucomutase

PgtB: bacterial kinase that phosphorylates regulator protein PgtA. (See kinase, protein kinases)

pH: = - log (H_3O^+), the negative logarithm of the hydronium ion; pure water at 25° C contains 10^{-7} mole hydronium ions, whereas solutions of acids could contain 1 mole and solution of bases 10^{-13} moles per liter. The pH meters measure the electrical property of solutions which is proportional to pH; pH 7 is neutral and below it is acidic above it is alkalic (basic).

Ph[1] CHROMOSOME: see Philadephia chromosome

***Ph* GENE** (*pairing high*): controls selective pairing in hexaploid wheat. In its presence homeologous chromosomes do not pair. It is in chromosome 5B; plants nullisomic for this chromosome display multivalent associations in meiosis. A similar gene *Ph2* is in chromosome 3D. Chromosome pairing is regulated also by additional less powerful genes. (See *Triticum*)

PHA: phytohemagglutinin, a lectin of bean (*Phaseolus vulgaris*) plants; agglutinates erythrocytes and activates T lymphocytes. (See lectins, agglutination, erythrocyte, lymphocytes)

PHAEOCHROMOCYTOMA: is a bladder-kidney carcinoma, overproducing adrenaline and noradrenaline. (See animal hormones)

PHAEOMELANIN: a mammalian pigment. (See pigmentation of animals)

PHAGE (bacteriophage): virus of bacteria. (See bacteriophages, development, phage life cycle)

PHAGE CONVERSION: acquisition of new properties by the bacterial cell after infection by a temperate phage. (See temperate phage)

PHAGE CROSS: see round of matings

PHAGE DISPLAY: filamentous bacteriophages (M13, fd) have a few copies (3-5) of protein III gene at the end of the particles. This protein controls phage assembly and adsorption to the bacterial pilus. When short DNA sequences are inserted into gene III, the protein encoded is displayed on the surface of the particles. The peptides can be separated with antibody affinity chromatography. By the insertion of a large array of nucleotide sequences a huge combinatorial library of epitopes may be generated. This may be of applied significance for the pharmaceutical industry. (See filamentous phages, affinity chromatography, epitope screening, combinatorial library)

▼ AN OUTLINE OF A PHAGE LIFECYCLE ▼

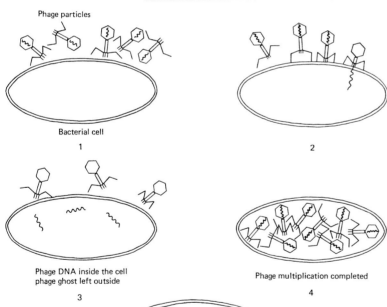

PHAGE GHOST: the empty protein shell of the virus.

PHAGE IMMUNITY: a lysogenic bacterium carrying a prophage cannot be infected by another phage of the same type. (See prophage)

PHAGE INDUCTION: stimulates the prophage to leave a site in the bacterial chromosome and become vegetative. Physical and chemical agents may be inducive (UV light, mutagens)

PHAGE LIFECYCLE: see diagram on the left.

(Redrawn after the illustration provided by Drs. Simon, L.D. and Anderson, T.F. Institute of Cancer Research, Philadelphia, PA, USA)

PHAGE MORPHOGENESIS: see one-step growth, development, phage life cycle.

PHAGEMIDS: are genetic vectors that generally contain the ColE1 origin of re-

Phagemids continued

plication and one or more selectable markers from a plasmid and a major intergenic copy of a filamentous phage (M13, fd1). When cells carrying such a combination is superinfected by a filamentous phage it triggers a rolling circle type replication of the vector DNA. This single-stranded product is used then for sequencing by the Sanger type DNA sequencing system, for oligonucleotide-directed mutagenesis and as strand-specific probes. The phagemids can carry up to 10 kb passenger DNA. Their replication is fast (in the presence of a helper), they can produce up to 10^{11} plaque-forming units (pfu)/mL bacterial culture. Their stability is comparable to conventional plasmids. They obviate subcloning the DNA fragments from plasmid to filamentous phage. The most widely used phagemids contain parts of phage M13 and pUC, πVX^c, and pBR322 vectors. (See plasmid, vectors, plasmovirus, vectors, pfu, pUC, DNA sequencing)

PHAGOCYTOSIS: a special cell (phagocyte) engulfes a foreign particle (microorganism, cell debris) and eventually exposes it to lysosomal enzymes for the purpose of destroying it. In lower animals this mechanism substitutes for the immune system. (See also pinocytosis)

PHAGOSOME: is a body (vesicle) surrounded by membrane within a phagocyte. (See phagocytosis)

PHALLOIDIN: is an amanotoxin, similar to but faster in action than amanitin. When labeled with fluorescent coumarin phenyl isothiocyanate it suitable to identify filamentous actin in the cells. It is extremely toxic. (See amatoxins, α-amanitin)

PHALLUS: the penis, also the symbol of generative power, also the fetal anlage of the penis and clitoris. (See penis, clitoris, anlage)

PHARMACOGENETICS: study of the reaction of individuals of different genetic constitution to various drugs and medicines.

PHARMING: production of pharmacologically useful compounds by transgenic organisms. (See transgenic)

PHAS-1: is a heat stable protein ($M_r \approx 12,400$); when it is not phosphorylated it binds to peptide initiation factor eIF-4E and inhibits protein synthesis. Its Ser^{64} site is readily phosphorylated by MAP and then no longer binds to eIF-4E and protein synthesis may be stimulated. (See eIF-4E, MAP)

PHASE VARIATION: is a programmed rearrangement in several genetic systems. The flagellin genes of the bacterium *Salmonella* display it at frequencies of 10^{-5} to 10^{-3} The flagellar protein has two forms, H1 and H2. The *H1* gene is a passive element. When *H2* is expressed no H1 protein is made. When *H2* is switched off the H1 antigen is made. The expression of *H2* is regulated by the expression of the *rh1* repressor (repressing the synthesis of the H1 protein) and the promoter of *H2*. This promoter is about 100 bp upstream from the gene and it is liable to inversion, and then *H2* and *rh1* are turned off. Such an event then switches on the synthesis of H1 protein. Reversing the inversion flip switches back to H2. The inversions are catalyzed by the Hin recombinase that is very similar to the invertases or recombinases of phage Mu or Cin from phage P1. They can functionally substitute for each other. Hin binds to the *hixL* and *hixR* recombination sites. Additional genes are also involved in the fine tuning. Somewhat similar mechanisms control the host-specificity genes of phage Mu and the mating type of budding yeast. (See the cassette model, regulation of gene activity, antigenic variation, mating type determination in yeasts, *Trypanosoma*, flagellin)

PHASE-CONTRAST MICROSCOPE: alters the phase of light passing through and around the objects and this permits its visualization without fixation and/or staining. (See also Nomarski, fluorescence microscopy, microscopy light, confocal microscopy, electronmicroscopy)

PHASEOLIN ($C_{20}H_{18}O_4$): an antifungal globulin in bean (*Phaseolus*).

PHASING CODON: initiates translation (such as AUG) and determines the reading frame. (See genetic code, reading frame)

PHASMID: is a plasmid vector equipped with the *att* site of the lambda phage and thus enables the plasmid to participate in a site-specific recombination with the λ genome resulting in incor-

Phasmid continued

poration of plasmid sequences into the phage (*lifting*). Because it contains both the λ and plasmid origins of replication it may be replicated either as a plasmid or as λ. (See lambda phage, phagemid, vectors)

PHENE: an observable trait that may or may not have direct genetic determination. (See gene)

PHENOCOPY: phenotypic change that mimics the expression of a mutation. (See phenotype)

PHENODEVIATE: an individual of unknown genetic constitution displaying a phenotype attributed to various genic combinations within the population.

PHENOGENETICS: attempts to correlate the function of genes with phenotypes.

PHENOGRAM: see character matrix

PHENOLICS: compounds containing a phenol ring such as acetosyringone, hydroxyacetosyringone, chalcone derivatives, phenylpropanoids, some phytoalexins. The mentioned compounds may excite or suppress the *vir* gene cascade of *Agrobacterium* and may affect the response to plant pathogenic agents. (See *Agrobacterium*, virulence genes of *Agrobacterium*, phytoalexins, chalcones, acetosyringone)

PHENOLOGY: the study of the effects of the environment on live organisms

PHENOME: group of organisms with shared phenotypes.

PHENOMENOLOGY: the concept that behavior depends on how a person interprets reality rather than what is the objective reality.

PHENOTYPE: appearance of an organism that may or may not represent the genetic constitution.

PHENOTYPIC LAG: a period of time may be required for gene expression after transformation or mutagenic treatment. (See transformation genetic, premutation)

PHENOTYPIC MIXING: mixed assembly of viral nucleic acids and proteins upon simultaneous infections by different types of viruses. Therefore the coat protein properties of the virions are not matching the viral genotype in serological or other behavior.

PHENOTYPIC REVERSION: is an apparent restoration of the normal expression of a mutant gene; it is not inherited, however. (See phenocopy)

PHENOTYPIC STABILITY FACTOR: is a measure of developmental homeostasis and it is calculated by the ratios of the quantitative expression of a parameter (gene) under two different environmental conditions. (See homeostasis, logarithmic stability factor)

PHENOTYPIC SEX: may not reflect the expectation based on the sex-chromosomal constitution. (See testicular feminization, hermaphrodite)

PHENOTYPIC SUPPRESSION: an apparently normal but non-hereditary phenotype brought about by translational error due to environmental effects and/or drugs. (See error in translation)

PHENOTYPIC VALUE: in quantitative genetics it is defined as the mean value of a population regarding the trait under study and it is generally represented as P value. (See breeding value)

PHENOTYPIC VARIANCE: see genetic variance

PHENYLALANINE: an essential water-soluble, aromatic amino acid (MW 165.19). Its biosynthetic path (*with enzymes involved in parenthesis*): Chorismate -> (*chorismate mutase*) > prephenate -> (*prephenate dihydratase*) -> Phenylpyruvate -> (*aminotransferase*, glutamate NH_3 donor) -> Phenylalanine. (See also chorismate, tyrosine, phenylketonuria)

PHENYLALANINE AMMONIA LYASE (PAL): deaminates phenylalanine into cinnamic acid and thus it is involved in the synthesis of plant phenolics.

PHENYLHYDRAZINE: is a hemolytic compound but it is used also as a reagent for sugars, aldehydes, ketones and a number of industrial purposes (stabilizing explosives, dyes, etc.). (See hemolysis)

PHENYLKETONURIA (PKU, PAH): gene was located to human chromosome 12q24.1. It is a recessive disorder that has a prevalence of about 1×10^{-4} (carrier frequency is about 0.02) in white populations. Thus an affected person has about 0.01 chance to have an affected child in case of a random mate but the recurrence rate in a family where one of the partners is affected and the other is a carrier it is nearly 0.5. Its incidence is substantially lower among Asian and

Phenylketonuria continued

black people (1/3 of that in whites). PKU is more frequent in European populations of celtic origin than in the other Europeans. This has been interpreted as the result of natural selection because PKU heterozygosity conveys some tolerance to the mycotoxin, ochratoxin A, produced by *Aspergillus* and *Penicillium* fungi, common in humid northern regions. Before the nature of this disorder and the method of treatment was identified about 0.5 to 1% of the patients in mental asylums were aflicted by PKU. The disease is caused by a deficiency of the enzyme *phenylalanine hydroxylase* and consequently the accumulation of phenyl pyruvic acid and a deficiency of tyrosine:

$$\text{PHENYL PYRUVIC ACID} \rightleftarrows \text{PHENYLALANINE} \Rightarrow \text{TYROSINE}$$

For the identification of the condition the Guthrie test was used: to cultures of *Bacillus subtilis* containing blood of the patients, β-2-thienylalanine was added. This phenylalanine analog is a competitive inhibitor of tyrosine synthesis. In the presence of excess amounts of phenylalanine, the bacterial growth does not stop however. Since in the different families the genetically determined defect in the enzyme varies, so does the severity of the clinical symptoms. The accumulation of phenyl pyruvic acid is apparently responsible for the mental retardation and the musty odor of the urine of the patients. The reduced amount of tyrosine prevents normal pigmentation (melanin) and thus results in pale color. The good aspect of this condition is that relative normalcy can be established if it is early diagnosed and dietary restrictions for phenylalanine are implemented. The restriction of phenylalanine must start as early as possible (before birth if feasible), and continue for the life to avoid mental disorders.

MENTALLY RETARDED HETEROZYGOUS CHILDREN OF A PHENYLKETONURIC MOTHER (indirect epistasis). Courtesy of Dr. C. Charlton Mabry; by permission of the New England Journal of Medicine [269:1404]

Phenylketonuria of the mother may damage the nervous system of genetically normal fetus through placental transfer (indirect epistasis). Because of the multiple metabolic pathways involving phenylpyruvic acid, besides the deficiency of phenylalanine hydroxylase, other genes and conditions may cause similar clinical symptoms. Phenylalanine hydroxylase activity requires the availability of the reduced form of the co-factor 5,6,7,8-tetrahydrobiopterin that is made by the enzyme *dihydrobiopterin reductase* from 7,8-dihydrobiopterin. The dehydrobiopterin reductase enzyme is coded in human chromosome 4p15.1-p16.1. Defect in this enzyme is also causing phenylketonuria symptoms but lowering the level phenylalanine in the diet does not alleviate the problems. Another form of phenylketonuria is based on a deficiency in dehydrobiopterin synthesis. Prenatal diagnosis can be carried out by several methods. (See epistasis, mental retardation, one gene-one enzyme theorem, genetic screening, Guthrie test, tyrosinemia, alkaptonuria, phenylalanine, amino acid metabolism, prenatal diagnosis)

PHENYLPROPANOID: see phenolics, phytoalexins

PHENYLTHIOCARBAMIDE TASTING (PTC): the major incompletely dominant gene appears to be in human chromosome 7. About 30% of North-American Whites and about 8-10% of Blacks cannot taste bitter this compound. Person affected by thyroid-deficiency (athyreotic) cretenism (mental deficiency) are non-tasters. (Phenylthiocarbamide (syn. phenylthiourea) has been used for classroom demonstration of human diversity but it should be kept in mind that it is a toxic compound (LD_{50} oral dose for rats 3 mg/kg and for mice 10 mg/kg). (See LD_{50})

PHEOCHROMOCYTOMA (phaeochromocytoma): an adrenal tumor induced by the SHC oncogene. (See paraganglion, SHC, adenomatosis endocrine multiple, endocrine neoplasia)

PHEROMONES: various chemical substances secreted by animals and cells for the purpose of signaling, generating certain responses by members of the species, such as sex-attractants - stimulants, territorial markers or other behavioral signals and cues. (See mating type determination in yeast).

PHIALIDE: fungal stem cells from which conidia are budded.

PHILADELPHIA CHROMOSOME: the short arm (p) of human chromosome 9, carrying the *c-abl* oncogene is translocated to the long arm (q) of chromosome 22 carrying site *bcr* (breakpoint cluster region). The *bcr - abl* gene fusion is then responsible for myelogenous (Abelson) and acute leukemia as a consequence of the translocation and fusion. (See ABL, BCR, leukemia)

PHLOEM: plant tissue involved in the transport of nutrients; it contains sieve tubes and companion cells, phloem parenchyma and fibers. (See sieve tube, parenchyma)

PHLORIZIN: a dichalcone in the bark of trees (Rosaceae); it blocks the reabsorption of glucose by the tubules of the kidney and thus causes glucosuria. (See disaccharide intolerance, chalcones)

PHO81: a yeast CDK inhibitor homologous to $p16^{INK4}$. (See CDK, $p16^{INK4}$)

PHO85: is a cyclin-dependent kinase of *Saccharomyces cerevisiae*. (See CDK, KIN28, CDC28, PHO81)

phoA: gene for alkaline phosphatase.

PHOCOMELIA: is the absence of some bones of the limbs proximal to the body; it may occur as a teratological effect of various recessive and dominant human genetic defects or as a consequence of teratogenic drugs, e.g. thalidomide use during pregnancy. (See limb defects in humans, Roberts syndrome, teratogen, thalidomide)

PHOGE (pulsed homogeneous orthogonal field): a type of pulsed field gel electrophoresis, within the range of 50 kb to 1 Mb DNA, permitting straight tracks of large number of samples. (See pulsed field gel electrophoresis)

PhoQ: *Salmonella* kinase, affecting regulator of virulence PhoP. (See virulence, *Salmonella*)

PhoR: phosphate assimilation is regulated by PhoR kinase upon phosphorylation regulator PhoB.

PHORBOL ESTERS: are facilitators of tumorous growth by activating protein kinase C. (See also TPA, protein kinases, procarcinogen, carcinogen, PMA, p. 802 for phorbol formula)

PHORBOL 12-MYRISTATE-13-ACETATE (PMA): see phorbol esters

PHOSPHATASES: in animals both acid and alkaline phosphatases are common, in plants acid phosphatases are found. Some of the phosphatases have high specificities and have indispensable role in energy release in the cells. A series of non-specific phosphatases carry out only digestive tasks. In humans the erythrocyte and fibroblast expressed acid phosphatase (ACP1) isozymes are coded in chromosomes 2 and 4. It has been suggested that these enzymes are splitting flavin mononucleotide phosphates. In megaloblastic anemia ACP1 level is increased. The tartrate-resistant acid phosphatase type 5 (TR-AP), is an iron-glycoprotein of 34 kDa (human chromosome 15q22-q26), and it is increased in the spleen in case of Gaucher disease. Lysosomal acid phosphatase (ACP2) is in human chromosome 11p12-p11. Alkaline phosphatase (ALPL) is present in the liver, bone, kidney and fibroblast, is often called the non-tissue-specific phosphatase (human chromosome 1p36-p34) is deficient in hypophosphatasias. The alkaline phosphatase ALPP is in the placenta (human chromosome 2q37) and several allelic forms have been identified. A similar alkaline phosphatase is present also in the testes and the thymus and the gene is also at the same chromosomal location but its expression is highly tissue-specific. (See also serine/threonine and tyrosine protein phosphatases, hypophosphatasia, hypophosphatemia, megaloblastic anemia, Gaucher's disease)

PHOSPHATIDYLINOSITOL (1,2-diacyl-sn-glycero-3-phospho[1-o-myoinositol]): is a cell membrane phospholipid. (See PIK, pleckstrin domain)

3'-PHOSPHOADENOSINE-5'-PHOSPHOSULFATE: see PAPS

PHOSPHODIESTER BOND: attaches the nucleotides into a chain by hooking up the incoming 5'-phosphate ends to the 3'-hydroxy tail of the preceding nucleotide: R^1 and R^2 represent

Phosphodiester bond continued
⇩

$$5'\text{-}R^1\text{-}3'\text{-}O\text{-}\overset{\overset{O}{\|}}{\underset{O^-}{P}}\text{-}O\text{-}5'\text{-}R^2\text{-}3'$$

nucleosides, O: oxygen, H: hydrogen, P: phosphorus. (See Watson-Crick model, formula at left)

PHOSPHODIESTERASES: are exonucleases. The snake venom phosphodiesterase starts at the 3'-OH ends of a nucleotide chain and splits off nucleoside-5'-phosphate. The 3'-phosphate terminus does not lend the nucleotide chain for its action. The spleen phosphodiesterase, on the other hand, generates nucleoside-3'-phosphate molecules by splitting on the other side of the nucleotides. Phosphodiesterase converts cyclic AMP into AMP. (See phosphodiester bond)

PHOSPHOFRUCTOSE KINASE 1 (PFK-1, phosphofructokinase): enzyme catalyzes the formation of fructose-1-6-bisphosphate from fructose-6-phosphate in the presence of ATP and Mg^{++}. PFKL (liver enzyme) is encoded in human chromosome 21q23.2.

PHOSPHOFRUCTOSE KINASE 2 (PFK-2): mediates the formation of fructose-2,6-bisphosphate formation from fructose-6-phosphate. It enhances the activity of fructosephosphate 1 enzyme by binding to it, and inhibits also fructose-2,6-bisphosphatase and therefore it enhances glycolysis. (See glycolysis)

PHOSPHOGLUCOMUTASE (PGM): is the enzyme that catalyzes the reaction:

GLUCOSE-1-PHOSPHATE ⇄ GLUCOSE-6-PHOSPHATE

Phosphoglucomutase proteins are homologous in structure through the animal kingdom. In humans there are several PGM enzymes and some with multiple allelic forms with characteristic patterns and are reasonably stable. Therefore PGM is used in forensic genetics for personal identification on samples up to 6 months old. Their human chromosomal locations are: PGM1 (1p31, PGM2 (4p14-q12), PGM3 (6q12), PGM5 (9p12-q12). (See forensic genetics)

PHOSPHOGLUCONATE OXIDATIVE PATHWAY: same as pentose phosphate pathway.

PHOSPHOGLYCERATEMUTASE DEFICIENCY: see myopathy, glycerophospholipid

PHOSPHOGLYCERIDE: see glycerophospholipid

PHOSPHOHEXOSE ISOMERASE (PHI): catalyzes the glucose-6-phosphate⇔fructose-6-phosphate conversions. It is encoded in human chromosome 19cen-q12. Its defects results in dominant hemolytic anemia. (See anemia, hemolytic anemia)

PHOSPHOINOSITIDE-3-KINASES: see phosphoinositides

PHOSPHOINOSITIDES: are inositol-containing phospholipids. They play an important role as second messenger and in phosphorylated/dephosphorylated forms they participate in the regulation of traffic through membranes, growth, differentiation, oncogenesis, neurotransmission, hormone action, sensory perception. The signals converge on phospholipase C (PLC, 20q12-q13.1) and it hydrolyzes phosphatidylinositol-4,5-bisphosphate ($PtdInsP_2$) into inositol trisphosphate ($InsP_3$) and diacylglycerol (DAG). $InsP_3$ regulates Ca^{2+} household and DAG activates PLC. $InsP_3$ levels regulate also pronuclear migration, nuclear envelope breakdown, metaphase-anaphase transitions and cytokinesis. Cytidine diphosphate-diacylglycerol synthase (CDS) is required for the regeneration of $PtdInsP_2$ from phosphatidic acid. CDS is a key regulator in G-protein-coupled phototransduction pathway. (See inositol, phospholipase, DAG, signal transduction)

PHOSPHOINOSITIDE-SPECIFIC PHOSPHOLIPASE Cδ: signal transducers and generate the second messengers inositol-1,4,5-triphosphate and diacylglycerol. (See signal transduction, second messenger)

PHOSPHOLIPASE (PL) **A, D, C**: each split specific bonds in phospholipids. PLC-β generates diacylglycerol and phosphatidylinositol 2,4,5-triphosphate from phosphatidylinositol 4,5-bisphosphate. These second messenger molecules play roles in signal transduction. PLC-γ is activated by receptor tyrosine kinases and one of its homologs is the SRC oncoprotein. PLA is present in mammalian inflammatory exudates. One form A2 is coded by human chromosome 12, the other PLA2B by chromosome 1. Phospholipase C enzymes are coded in human chromosomes at the following locations: PLCB3 (11q11), PLCB4 (20p12), PLCG2 (16q24.1). (See also serine/threonine phosphoprotein phosphatases, SRC, signal transduction)

PHOSPHOLIPID: a lipid with phosphate group(s). (See also liposome, lipids)

PHOSPHOMONOESTERASE: a phosphatase digesting phosphomonoesters, such as nucleotide chains. (See phosphodiester bond)

PHOSPHONITRICIN (Basta): see herbicides

PHOSPHORELAY: see two component regulatory system

PHOSPHORESCENCE: see fluorescence, luminescence

PHOSPHORIBOSYLGLYCINAMIDE FORMYLTRANSFERASE: autosomal dominant control of purine, pyrimidine biosynthesis and folate metabolism.

PHOSPHOROLYSIS: a glycosidic linkage holding two sugars together is attacked by inorganic phosphate and the terminal glucose is removed (from glycogen) as α-D-glucose-1-phosphate.

PHOSPHOROTHIOATES: are analogs of oligodeoxynucleotides and are used in antisense technology. Their attachment to 3'-end inhibits the activity of nucleases that attack RNA from that end. They can bind to proteins, but do not stimulate the activity of RNase H (phosphorodithioate modified heteroduplexes may stimulate RNase H), inhibit translation and relatively easily taken up by cells. Some of the effects of these molecules are not based on their antisense properties (e.g. binding to CD4, NF-κB, inhibition of cell adhesion, inhibition of receptors, etc.). Phosphorothioate-modified nucleotides (one of the oxygen attached to P is replaced by S) are used also *in vitro* mutagenesis to protect the template strand from nucleases while the strand to be modified is excised before resynthesis in a mutant form. (See antisense technologies, antisense RNA, ribonuclease H, CD4, NF-κB)

PHOSPHORYLASE b KINASE: is an enzyme that phosphorylates two specific serine residues in *phosphorylase b* and thus converting it into *phosphorylase a* upon the action of cAMP-dependent protein kinase (synonym protein kinase A). Phosphporylase b kinase mediates glycogen breakdown. This enzyme is a tetramer and for activation the two regulatory subunits (R) must be separated from the two catalytic subunits to be able to function. The dissociation is mediated by cAMP through A-kinase. The δ subunit is calmodulin. (See epinephrine, cAMP-dependent protein kinase, cAMP, A-kinases, calmodulin)

PHOSPHORYLASES (kinases): see serine/threonine kinases, tyrosine protein kinase, Jak kinase, phophorylase B, A-kinases, signal transduction, serine/threonine phosphoprotein phosphatases, phospholipase C. (See signal transduction, calmodulin, phosphorylase b kinase)

PHOSPHORYLATION: adding phosphate to a molecule. (See also oxidative phosphorylation, kinase, phosphorylases)

PHOSPHORYLATION POTENTIAL (Δg_p): the change in free energy within the cell after hydrolysis of ATP.

PHOSPHOSERINEPHOSPHATASE: hydrolyzes O-phosphoserine into serine; it is encoded in human chromosome 7p15.1-p15.1.

PHOTOAFFINITY TAGGING: the labels may be radioactive or fluorescent and bind to certain compounds by non-covalent bonds upon illumination.

PHOTOAGING: skin collagens and elastin are damaged by the ultraviolet light induced metalloproteinases and this results in wrinkling of the skin similar to what occurs during aging. These enzymes are upregulated by AP-1 and NF-κB transcription factors. (See aging, collagen, elastin, AP-1, NF-κB)

PHOTOALLERGY: immunological response to a substance activated by light.

PHOTOAUTOTROPH: an organism that can synthesize in light from inorganic compounds all its required organic substances and energy.

PHOTOCHEMICAL REACTION CENTER: the site of photon absorption and initiation of electron transfer in the photosynthetic system. (See photosynthesis)

PHOTODYNAMIC EFFECT: photosensitivation, photodestruction.

PHOTOELECTRIC EFFECT: has very wide applications of modern technology (television, computers and other electronic instruments). When light hits a suitable target electrons are emitted by the atoms. When X-rays hit a target very high energy photoelectrons may be generated.

PHOTOGENES: are chloroplast DNA encoded proteins involved in photosynthesis. One of the most studied is *photogene* 32 (*psbA*) codes for a 32 kDa thylakoid protein involved in electron transport in photosystem II. Also, it binds the herbicide atrazine and by removal or altering this binding site plants resistant to the weed killer can be obtained through molecular genetic manipulations. (See photosynthesis, herbicides)

PHOTOGRAPHY: in the laboratory has special requirements depending on the objects. Cell cultures in Petri plates can be best photographed through macrolenses (for extreme close ups use extension rings or teleconverter) and using highly sensitive color films comparable to Kodak Gold 400. To eliminate reflection, the blue photoflood lamps should be adjusted at about 45° angles. Agarose gels can be photographed with a polaroid camera mounted on a copying stand and using high speed (ASA 3000) films. Ultraviolet light sources of the longer wave length are less likely to damage the DNA. The contrast can be enhanced by the use of orange filters on the camera (such as Kodak Wratten 22A). Note that ultraviolet light is dangerous to the skin and particularly to the eyes. Use gloves, goggles and wear long-sleeve shirt. For photomicrography built in automatic exposure meters are very advantageous if frequently used. Otherwise numerous exposures, at the proper color temperature are necessary. For photocopying and editing halftone images computers with (color) scanners can be used. The resolution of the digital cameras may not be satisfactory for many biological applications.

PHOTOLABELING: is adding photoactivatable groups to proteins, membranes or other cellular constituents in order to detect their reaction path. The labels are generally small molecules, stable in dark and highly susceptible to light without photolytic damage to the target and stable enough to permit analytical manipulations of the sample. Synthetic peptides containing such as 4'-(trifluoromethyl-diazirinyl)-phenylalanine or 4'-benzoyl-phenylalanine, etc. have been used to analyze biological structures (membranes, proteins, etc.)

PHOTOLITOGRAPHY: is a modification of a more than a century old printing process. A solid plate is coated with a light-sensitive emulsion and overlaid by a photographic film and then it is illuminated. An image is formed after exposure of the plate to light. A similar principle has been adapted now to visualizing DNA sequences for the purpose of large scale mapping, fingerprinting and diagnostics. (See DNA chips)

PHOTOLYASE: is a repair enzyme (M_r 54,000) that splits pyrimidine dimers into monomers. In *E. coli* two chromophores assist the process; 5,10-methenyltetrahydrofolate absorbs the photoreactivating light (peak 380 nm) and 8-hydroxy-5-deazariboflavin (peak 380 nm) and the energy is then transferred to $FADH_2$, although the latter too absorbs some energy. The excited $FADH_2$* transfers then the energy to the dimer and while $FADH_2$ is regenerated the dimer splits up, the recipient member of the dimer breaks down and monomeric pyrimidines are formed. A second cofactor 5,10-methenyltetrahydrofolylpolyglutamate (MTHF) may be the light harvester. It is interesting that the blue light photoreceptor of plants bears substantial similarities to the bacterial photolyase and its cofactors are also the same yet the exact role of photolyases in plant DNA repair is unclear. (See DNA repair, direct repair, photoreactivation, pyrimidine dimer, cyclobutane ring)

PHOTOLYSIS: degradation of chemicals or cells by light.

PHOTOMIXOTROPHIC: an organism that can synthesize some of its organic requirements with the aid of light energy while for others it depends on supplied organic substances.

PHOTOMORPHOGENESIS: light-dependent morphogenesis. Light affects the growth and differentiation of plant meristems (photoperiodism), plastid differentiation, and directly or indirectly many processes of plant metabolism. Certain stages in photomorphogenesis can be reached at low intensity (fluence) illumination (or even in darkness) such as the formation of proplastids and etioplasts. Other steps such as the full differentiation of the thylakoid system and photosynthesis-dependent processes require high fluence rate and critical spectral regimes (red and blue). Several genes involved in the control of plastid development have been iden-

Photomorphogenesis continued

tified in *Arabidopsis* and other plants. The *lu* mutation is normal green at low light intensity but it is entirely bleached and dies at high light levels. Wild type plants can make etioplasts in the dark but the *de-etiolated* (*det1*), *constitutive photomorphogenesis* (*cop1* and *cop9*) mutants develop chloroplasts in darkness. The *gun* (*genome uncoupled*) mutations grow normally in the dark but do not allow the development of etioplasts into chloroplasts. Various pale *hy* (*high-hypocotyl*) mutants, deficient in phytochrome, make light green plastids indicating that phytochrome is not a requisite for plastid differentiation to an advanced stage. The *blu* (*blue light uninhibited*) class of mutants are inhibited in hypocotyl elongation by far red light. The *HY4* locus of *Arabidopsis* encodes a protein, homologous to photolyases and the recessive mutations are insensitive to blue light for hypocotyl elongation. Mutants were identified with no response to blue light and some with very high blue light requirement for curvature. Most of these light responses appear to be mediated by signal transduction pathways. The chlorophyll-b free, yellow green mutants (*ch*) display chloroplast structure appearing almost normal by electronmicroscopy. Several mutations defective in fatty acid biosynthesis and/or photosynthesis are rather normal in photomorphogenesis. Some mutants are resistant to high CO_2 atmosphere, and actually normal chloroplast differentiation requires high CO_2. Other mutants can be protected from bleaching only at 2% CO_2 atmosphere. The *Arabidopsis* nuclear mutants of the *im* (*immutans*) type display variegation under average greenhouse illumination but they are almost normal green under low light intensity and short daily light cycles whereas at high intensity continuous illumination they are almost entirely free of leaf pigments. Under the latter condition, by continuous feeding of the inhibitor and repressor of the *de novo* pyrimidine pathway, the leaf pigment content may increase twenty fold. In these variegated plants the green cells have entirely normal chloroplasts whereas the white cells lack thylakoid structure. The azauracil-treated plants display fully functional, although morphologically altered, thylakoids. An insertional mutation at the *ch-42 locus* (*cs*) identified a thylakoid protein, essential for normal greening of the plants without abolishing cell viability. The *PRF* (*pleiotropic regulatory factor*) locus, tagged by a T-DNA insertion, controls several loci involved in photomorphogenesis. The product of the gene is a subunit of the G-protein family. The *det2, cyp90, cop, fus, dim axr2* and the *cbb* dwarf mutations develop their characteristic phenotypes because of defects in the brassinosteroid pathway. The nuclear gene *chm* (*chloroplast mutator*) induces a wide variety of plastid morphological changes, due to extranuclear mutation. (See photoperiodism, phototropism, phytochrome, circadian, signal transduction, brassinosteroids, COP)

PHOTON: a quantum of electromagnetic radiation which has zero rest mass and an energy *h* times the frequency of the radiation. Photons are generated by collisions between atomic nuclei and electrons and other processes when electrically charged particles change momentum.

PHOTOPERIODISM: is the response of some species of plants to the relative length of the daily light and dark periods.

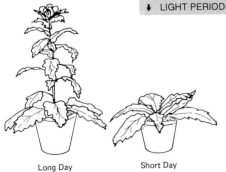

HENBANE (*Hyoscyanus niger*) LONG--DAY PLANTS FLOWER ONLY UNDER LONG DAILY ↓ LIGHT PERIODS (after appropriate cold treatment). Courtesy of Professor G. Melchers

Long Day   Short Day

Besides the length of the these cycles the spectral properties and the intensity of the light are also important. Responses of the plants include the onset of flowering, vegetative growth, elongation of the internodes, seed germination, leaf abscission, etc. *Short-day, long-day* and *day-neutral* plants are commonly distinguished on the basis of the critical daylength or in the latter category by the lack of it. The geographic distribution of plants is correlated with their photoperiodic response. In the near equatorial

Photoperiodism continued

regions short-day species predominate whereas in the regions extending toward the poles log-day plants are common. The onset of flowering of short-day plants is promoted by 15-16 hours of dark periods whereas in long-day plants the flowering is accelerated by continuous illumination or by longer light than dark daily cycles. The critical day-length is not an absolute term; it varies in the various species. Usually, there is a minimum number of cycles to evoke the photoperiodic response. The most important photoreceptor chromoprotein is *phytochrome*. The effect of phytochrome is affected by different plant hormones. Typical longday plants are henbane (*Hyoscyamus*), spinach, *Arabidopsis* (without a critical daylength), the majority of the grasses and cereal crops (wheat, barley, oats), lettuce, radish, etc. Typical short day plants are Biloxi soybean, cocklebur, aster, chrysanthemum, poinsettia, dahlia, etc. In the majority of species the photoperiodic response is controled by one or a few genes. (See phytochrome, photomorphogenesis, phototropism, vernalization)

PHOTOPHOSPHORYLATION: ATP formation from ADP in photosynthetic cells.

PHOTOREACTIVATION: elimination of the harmful effects of ultraviolet irradiation by subsequent exposure to visible light (that activates enzymes splitting up the pyrimidine dimers in the DNA). With a few exceptions, e.g., *Haemophilus influenzae,* most organisms possess light-activated repair enzymes. The majority of mammals do not have efficient photoreactivation system, except the marsupials. (See also light repair, photolyase dark repair, excision repair, glycosylases, error-prone repair, DNA repair)

PHOTORECEPTORS: see phytochrome, rhodopsin, phototropism, *sevenless*

PHOTOREDUCTION: in photosynthetic cells, light induced reduction of an electron acceptor.

PHOTORESPIRATION: oxygen consumption in illuminated plants used primarily for oxidation of the photosynthetic product phosphoglycolate; it also protects C3 plants from photooxidation. (See respiration, Calvin cycle, C3 plants)

PHOTOSYNTHESIS: using light energy for the conversion of CO_2 into carbohydrates with the assistance of a reducing agent such as water. (See photosystems, chlorophyll binding proteins)

PHOTOSYSTEMS: in photosynthesis there are photosystem I, excited by far red light ($\sim$ 700 nm) while photosystem II requires higher energy red light ($\sim$ 650-680 nm). Photosynthesis in bacteria that do not evolve oxygen and uses only photosystem I. Upon absorption of photons, photosystem I liberates electrons that are carried through a cascade of carriers to $NADP^+$ and that is reduced to NADPH. The departure of electrons generates a "void" in the P700 photoreaction center of photosystem I and that is filled then by electrons produced through splitting of water molecules in photosystem II. The overall reaction flow is:

$$2 H_2O + 2 NADP^+ + 8 \text{ photons} \rightarrow O_2 + 2 NADPH + 2 H^+$$

Mutants of *Chlamydomonas* alga lacking photosystem I survive as long as the actinic light (beyond violet) reaches 200 microeinsteins per m^2/second. (See also CAB, LHCP, antenna, chloroplast, Z scheme)

PHOTOTAXIS: is a movement of organisms (plants, animals and microbes) in response to light.

PHOTOTRANSDUCTION: the transmission of light signals mediating gene expression. (See signal transduction)

PHOTOTROPH: organism that uses light to generate energy and synthesizes its nutrients from inorganic compounds by using this energy.

PHOTOTROPISM: the reaction of an organ or organism to light involving apparently more than a single photoreceptor. It was suggested that one of the receptors is a membrane protein with autophosphorylating ability. (See photoreceptors, gravitropism)

PHRAGMOPALST: a hollow-looking ring- or barrel-like structure formed near the end of mitosis in the middle plane of plant cells before the *cell plate* appears, separating the two daughter cells. (See mitosis)

PHYCOBILINS: highly fluorescent photoreceptor pigments in blue-green, red and some other algae. They contain a linear tetrapyrrole prosthetic group for light harvesting. Contain also

Phycobilins continued

bile pigments and an apoprotein. This family of pigments includes the blue phycocyanins, the red phycoerythrins and the pale blue allocyanins. These pigments may form phycobilisome, attached to the photosynthetic membrane. Phytochromes are also related pigments. (See light-harvesting protein, phytochrome)

PHYCOCYANIN: the pigment of blue-green algae. (See phycobilins)

PHYCOERYTHRIN: the red pigment of red algae. (See phycobilins, phycocyanin)

PHYCOMYCETES: are fungi with some algal characteristics. *Ph. blakesleeanus* is easy to grow with 4 days asexual cycles and about 2 months sexual cycles. It forms heterokaryons (n = 14) and can be subjected to formal genetic analyses although the tetrads may be irregularly amplified. Transformation is feasible. Well suited for physiological and developmental studies.

PHYLETIC EVOLUTION: gradual emergence of species in a line of descent. The gaps in the fossil records is supposed to be due accidents in preservation of the intermediate forms.

PHYLLOQUINONE: is composed of a p-naphthokinone and a phytol radical and it catalyzes oxydation-reduction reactions in plants. (See vitamin K)

PHYLLOTAXY: consecutive leaves of plants do not occur above each other. Quite commonly single leaves are at opposite position (unless they occur in whorls). This arrangement makes sense for optimal utilization of light. In many plants the leaves may not alternate in 180° but they may be arranged in any other determined pattern. This pattern is called phyllotaxy. If the leaves are opposite to each other the phyllotaxy is ½. A common phyllotactic index is ⅖ (144°). This means that if the leaves are positioned by this index, leaf #1 will be followed by #2 at 144°, then #3 will take the place in a spiral at 288°, i.e., it will be above #1 (because 288:144 = 0.5 and 0.5 x 360 = 180), and so on. The arrangement of the fruits on the stem may also be caused by such an obliquity following either clockwise or counterclockwise directions. (See embryogenesis in plants) PHYLLOTAXY OF FRUITS →

PHYLOGENETIC TREE: graphically represents the phylogeny of organisms. (See evolutionary tree)

PHYLOGENETIC WEIGHTING: DNA sequence information from various taxa is included in the phylogenetic tree in decreasing order of relationship. Thus alignment from distant relatives should not precede alignment of closer relatives. This procedure prevents confounding similarity and descent. (See evolutionary tree, maximum parsimony, homology, DNA sequence alignment, homology)

PHYLOGENY: the evolutionary descent of a species or other taxonomic groups. (See evolution, ontogeny, speciation, information at Web site <http://phylogeny.harvard.edu/treebase>)

PHYLUM: the first main category of the plant and animal, and other kingdoms.

Physarum polycephalum: is a single-cell slime mold that displays physiological dioecy. The cell forms a plasmodium, i.e., the nuclei divide without cell division and thus the cell becomes multinucleate. In the early embryos only S and M phase of the cell cycle are detectable.

Physcomitrella patens: is a moss with a principal life phase as a haploid gametophyte. It can be used for the production of various mutants, for parasexual research, transformation, study of plant hormones on developmental processes and various tropisms.

PHYSICAL CONTAINMENT: see containment

PHYSICAL MAP: a map where the genome is ordered in DNA fragments or nucleotide sequences rather than in units of recombination. The first physical maps were constructed in bacteriophages with small genomes. The DNA of phage P4 was cleaved completely by restriction endonuclease EcoRI into four fragments which could be separated by electrophoresis according to size:

■   ■   ■   I
A   C   B   D

After incomplete digestion for 5 minutes larger fragments were also detected that contained fragments A + B + C, C + B and the combined size of C + D appeared but no fragment appeared with the size B + D. The cause of the absence of B + D must have been that B and D

Physical map continued

were not adjacent in the circular DNA. Therefore the sequence of the fragments in the chromosome could have been only: A - B - C - D.

The much larger polyoma genome was mapped by a different procedure. With a single EcoRI cut the circular DNA was linearized and that cut was designated as the zero coordinate of the map. HinDIII cut the circle into two fragments A - 55% and B - 45%. HpaII produced 8 fragments a = 27%, b = 21%, c = 17%, d = 13%, e = 8%, f = 7%, g = 5% and h = 2% of the total genome. When the DNA was cleaved by EcoRI and HpaII, fragment b (21%) was not detected by electrophoresis but instead two new fragments of 1% and 20% were found. Obviously, the EcoRI cut was 1% from one end and 20% from the other end of fragment b. In a following step the HinDIII generated A fragment was digested by HpaII and fragments c, d, e, g and h were found again (17 + 13 + 8 + 5 + 2 = 45) and two pieces of 3% and 7% were also obtained. When the HinDIII fragment of 45% length was exposed to HpaII fragment f remained intact but two other fragments of 18% and 20% were recovered. Therefore the fragments could be pieced together as follows:

HinDIII A: 7%   -   45%   -   3%
            part of a   part of b

   HinDIII B   18%   - 7% -   20%
                part of b   f   part of a

Incomplete digestion of A by HpaII, produced fragments a + c, c + e, e + d, h + g and g + b, therefore the polyoma DNA appeared as: b - f - a - c - e - d - h - g with the zero coordinate in b and g near the 100 coordinate.

Larger genomes such as *E coli*, yeast or of higher eukaryote's are generally pieced together by chromosome walking type of procedure, using overlapping fragments generated by several restriction endonucleases, e.g. :

                                    1   2   3
fragments generated by enzyme A: abcde fghijklmn oprstuvwz

                                      4   5
fragments generated by enzyme B: cdefghi   jklmnoprst

will be tied into the order 1, 2, 3 on the basis of hybridization of 4 with 1 and 2, and hybridization of 5 with 2 and 3 but not 5 with 1 or 4 with 3. In the initial steps generally YAC clones are used because they cover large segments of the genomes. This is usually followed by cosmid clones and eventually large continuities (contigs) are established without gaps. By the employment of anchors, fragments with genetically or functionally known sites, the physical map can be correlated with the genetic map determined by recombination frequencies and thus *integrated maps* are generated. The individual fragments can then be sequenced and thus maps of ultimate physical resolution can be obtained. (See RFLP, chromosome walking, integrated map, anchoring, contigs, cosmids, restriction enzymes, EcoRI, HinDIII, HpaII)

PHYSICAL MUTAGENS: the most widely used forms are *electromagnetic*, ionizing radiations such as X rays, γ rays emitted by radioisotopes. The most commonly used radiation sources for the induction of mutation by γ rays are cobalt^{60} (Co^{60}) and caesium^{137} (Cs^{137}). *Particulate radiations* such as produced by atomic fission are also ionizing. The ionization is the dislodging of orbital electrons of the atoms. Particulate (corpuscular) radiation source is uranium^{235} which releases neutrons, uncharged particles (slightly heavier than that the hydrogen atom) with very high penetrating power and the ability to release about 15 times as much energy along their path as the hard X rays (of short wave length and high energy). The *fast neutrons* have energies between 0.5 and 2.0 MeV (million electron volt). The *thermal neutrons* have much lower level of energy (about 0.025 eV) because they have been "moderated" by carbon and hydrogen atoms. Radioactive isotopes emit also β *particles* (electrons). Their level of energy and penetrating power depends a great deal on the source; H^3 (tritium) has very

Physical mutagens continued

short path (about 0.5 µm) and P^{32} is much more energetic (2,600 µm). Beta emitters are rarely used for mutation induction. They can, however, be incorporated directly into the genetic material by using radioactively labeled precursors or building blocks of nucleic acids and thus are capable of inducing localized damage; the degree of localization depends on the effective path length. Uranium^{238} emits *α particles* (helium nuclei) releasing thousands of times more energy per unit track than X rays. Because of the very low penetrating power it can be stopped by a couple sheet of cells in contrast to X rays and gamma rays which require heavy concrete or lead shielding. Alpha radiation, because of its high energy per short path, can very effectively destroy chromosomes. The most common genetic effect of all ionizing radiations are chromosome breakage and particularly deletions.

Another physical mutagen is *ultraviolet* (*UV*) radiation. The latter causes excitation in the biological material rather than ionization. Excitation may raise the orbital electrons to a higher level of energy from which they return to the ground state very shortly. UV radiation sources are commonly mercury or cadmium lamps (black light, germicidal and sun lamps). UV radiation is included in the natural sunlight, especially in clean air of the higher mountains. Near ultraviolet light, UV-B (290-400 nm) may be present in the emission of fluorescent light tubes and in the presence of sensitizers it may be genetically effective on a few layers of cells. The most common genetic effects of UV light is the production of pyrimidine dimers.

The effect of radiation on cells and organisms may be *direct*, i.e., the radiation actually hits the target molecules or it may be *indirect*, i.e., the radiation produces reactive molecules in the intra- or extra-cellular environment, and these in turn cause the genetic and/or physiological damage. Exposure to high temperature may enhance mutability. If radiation is received during DNA replication, damage is more likely than in the dormant state. Generally hydrated cells and tissues are more sensitive to ionizing radiation than dry or non-metabolizing cells. (See also X-rays, radioisotopes, radiation effects, ultraviolet light, chemical mutagens, maximal permissive dose, carcinogens, LET, chromosomal mutation, DNA repair, genetic sterilization, cosmic radiation, genomic subtraction, nuclear reactors, atomic radiations, electromagnetic radiation)

PHYSIOLOGY: the discipline dealing with the functions of living cells and organisms.

PHYTANIC ACID: 20-carbon, branched chain fatty acid is formed from the phytol alcohol ester of chlorophylls and it degraded by β-oxidation into propionyl-, acetyl- and isobutyryl- CoA. Deficiency of this oxidation leads to Refsum disease in humans. (See Refsum diseases, peroxisome)

PHYTOALEXINS: generally relatively low molecular weight yet diverse compounds synthesized through the phenylpropanoid pathway. They were attributed to defense systems against various plant pathogens. Current view is considering them mainly as consequences of infection rather than active defense molecules. (See host-pathogen relation)

PHYTOCHROMES: are regulatory proteins with alternating absorbance peaks in red and far-red light. Through their absorbance peaks (red [R] 666 nm and far-red [FR] 730) they control various photomorphogenic processes, such as short- and long-day onset of flowering, hypocotyl elongation, apical hooks, pigmentation, etc. These chromoproteins are homodimers of 124 kDa subunits and a tetrapyrrole complex, joined covalently through a cystein residue at about 1/3 distance from the NH_2 end. The molecule exists in two conformations corresponding to the R and FR absorption state. The interconversion between these states is mediated very rapidly by light of R and FR emission peaks. In etiolated plant tissue the inactive P_r conformation may constitute up to 0.5% of the protein. The transition from the P_r conformation into the active P_{fr} form also entails the degradation of this receptor. The apoprotein, coded by different genes (*PHYA* and *PHYB*) in *Arabidopsis* may have only about 50% homology in amino acid sequences although they bind the same chromophore. The specificity of PhyA (far-red) and PhyB (red) resides in the N-termini. Phytochrome can induce and silence the expression of genes in a specific selective manner. The transcription of the phytochrome genes is also light

Phytochromes continued

regulated; R light is reducing the transcription more effectively than FR. Phytochrome PHYC is a light-stable molecule. The phytochrome responses are under complex genetic regulatory systems involving light response elements, transcription factors and components of the signal transduction circuits. Although phytochrome is known as a ubiquitous plant product, the yeast *Pichia* also synthesizes phytochromobilin (PΦB), a precursor of this plant chromophore. Also, a phytochrome-like protein has been identified in prokaryotes. (See photoperiodism, photomorphogenesis, signal transduction, phycobilins, brassinosteroids) ↓ PHYTOCHROME CHROMOPHORE

PHYTOEXTRACTION: see bioremediation
PHYTOHEMAGGLUTININ: see PHA
PHYTOHORMONES: see plant hormones
PHYTOPLANKTON: aquatic, free-flowing plant.
PHYTOREMEDIATION: see bioremediation
PHYTOTRON: a plant growth chamber system with maximal physical regulation facilities.
Pi: inorganic phosphate
PI 3 KINASE: see phosphoinositide 3 kinase

(After Metzler D.E. Biochemistry, Acad. Press, New York)

PI VECTOR: contains packaging site (*pac*) and allows about 115 kb to be packaged, and it infects *E. coli* at a pair of *lox P* recombination sites at which the *Cre* recombinase circularizes DNA inside the host cell. (See vectors)
PIBIDS: see trichothiodystrophy

PIEBALD RAT →

PIC: preinitiation complex. Proteins associated with RNA polymerase before transcription. (See transcription factors, open promoter complex, TBP)
PIC: see polymorphic information content
PICORNAVIRUSES: their single-stranded RNA genomes of about 7.2 to 8.4 kb (ca. 2.5 to 2.9 x 10^6 Da) are transcribed into 4 major polypeptides. Their RNA transcript lacks the 5' cap in the mRNA, characteristic for other eukaryotic viruses. They include *enteroviruses* (a group of mostly unsymptomatic intestinal viruses but the paralytic *poliovirus* may also belong to this group), *cardioviruses* (responsible for myocarditis [causing inflammation of the heat muscles] and encephalomyelitis [inflammation of the brain and heart]), *rhinoviruses* (in over 100 variants responsible for the common cold and other respiratory problems in humans and animals), *aphtoviruses* (causing foot-and-mouth disease in cattle, sheep and pigs and occasionally infecing also people). The *hepatitis virus* may also be classified among the picornaviruses. (See papovaviruses, animal viruses)
PIEBALDISM: in animals is the result of hypomelanosis (low melanin), restricted to spots of the body; generally, white spots on black background. (See albinism, nevus, vitilego, melanin, Himalayan rabbit, pigmentation in animals, KIT oncogene, Hirschprung disease, figure above)
PIERRE-ROBIN SYNDROME: autosomal recessive defect involving the tongue (glossoptosis), small jaws (micrognathia) and sometimes cleft palate and syndactyly of toes. In an autosomal dominant form reduced digit number (oligodactyly) is also found. There is also an X-linked form involving clubfoot and heart defect. Another X-linked form increased the number of bones in the digits. (hyperphalangy).
PIG (*Sus crofa*): 2n =38. The domesticated breeds are the descendants of the crosses between the European wild boar and the Chinese pigs and they can still interbreed with the wild forms of similar chromosome number. The wild European pig is 2n = 36. The Caribbean pig-like peccaries (Tayassuidae) are 2n = 30. There are about 300 domesticated pig breeds. Sexual maturity begins by about 5-6 months and the gestation period is about 114 days. It is a multiparous species with a litter size of 4-12.
PIGEON: *Columbia livia*, 2n = 80.
PIGMENTATION DEFECTS: see albinism, hypomelanosis, incontinentia pigmenti, pigmentation in animals, LEOPARD syndrome, Fanconi anemia, hematochromatosis, neurofibromat-

osis, tuberous sclerosis, Waardenburg syndrome, polyposis hamartomatus, Addison disease, focal dermal hypoplasia erythermalgia. (See skin diseases)

PIGMENTATION OF ANIMALS: in mammals tyrosine is the primary precursor of the complex black pigment melanin. The enzyme tyrosinase (located in the melanosomes) hastens the oxydation of dihydroxyphenylalanine (DOPA) into dopaquinone that is changed by non-enzymatic process into leukodopachrome. Leukodopachrome is an indole-derivative that is oxydized also by tyrosinase into an intermediate of 5,6-dihydroxyindole. After another step of oxydation, indole-5,6-quinone is formed. Coupling the latter to 5,6-dihydroxyindole is the first step in additions of further dihydroxyindole units in the process of polymerization to melanin. When cysteine is combined with dopaquinone, through a series of steps, the reddish pigments of the hair and feathers are formed. The different pigments may have also other adducts at one or more positions to yield the various colors. In the formation of the eye color of insects tryptophan is a precursor to the formation of formylkynurenine - > kynurenine -> hydroxykynurenine - > ommin, ommatin. The catabolic pathway of amino acids contributes to the formation of pteridines that contribute to the coloration of insects, amphibians and fishes and serve also as a light receptor. Xanthopterin and leucopterin account for the yellow and white pigmentation of butterflies, sepiapterin is found in the eyes of *Drosophila* and biopterin is found in the urine and liver of mammals. The degradation of the heme group yields a linear tetrapyrrole from what the bile pigment biliverdin and ultimately bilirubin diglucuronide is synthesized. That is secreted into the intestines and may accumulate in the eyes and other organs causing jaundice when the liver does not function normally. Oxidized derivatives of bilirubin, urobilin and stercobilin color the urine. Mutations were detected already during the early years of genetics that block the biosynthetic paths of these pigments and thus contributed to understanding how genes affect the phenotype. The color of the skin in humans is determined by its melanin content. Phaeomelanin is a reddish pigment and eumelanin is black. The former is responsible for the light skin and red hair color and also it potentially generates free radicals and thus may make the individual susceptible to UV damage. Eumelanin provides protection against UV. The relative proportion of these two melanins is regulated by the melanocyte-stimulating hormone (MSH) and its receptor (MC1R). In mice, about 100 genes are known that control pigmentation. (See also chorismate, tryptophan, tyrosine, phenylalanine, albinism, melanin, eye color in humans, Himalayan rabbit, Siamese cat, pigmentation in plants, agouti, melanocyte-stimulating hormone, hair color, tanning)

P$_{II}$: proteins (are involved in bacterial glutamine synthesis) accelerate hydrolysis of NtrC in the presence of NtrB and ATP in limiting N supply and 2-ketoglutarate level. P$_{II}$ uridylylation increases permit the increase of NtrC-phosphate level and increased transcription from glnAp2 promoter. In excess N supply, P$_{II}$ is not altered resulting in no NtrC build-up and glnA2 activation ceases. (See *NtrB, NtrC, glnAp*)

PIK (phosphatidylinositol kinases, PI[3]K): preferentially phosphorylate the 3 and 4 position on the inositol ring. They participate in meiotic recombination, immunoglobulin V(D)J switches, chromosome maintenance and repair, progression of the cell cycle, etc. Their defect may lead to immunological disorders and cancer. PIK related kinases are TOR, FRAP, TEL, MEI, DNA-PK. Its inhibitor is wortmannin. (See phosphatidylinositol, immunoglobulins, DNA repair, cell cycle, wortmannin)

PILEUS: umbrella-shaped fleshy fungal head. Also, a membrane that may be present on the head of newborns.

PILIN: the protein material of the pilus. (See pilus)

PILUS: bacterial appendage which may be converted into a conjugation tube through which the entire or part of the replicated chromosome is transfered from a donor to a recipient cell. (See conjugation, conjugation mapping)

PIM ONCOGENE: is in human chromosome 6p21-p12, in mouse chromosome 17. The gene is highly expressed in blood-forming (hematopoietic) cells and myeloid cells and overexpressed in myeloid malignancies and some leukemias. The human protein is a serine/threonine kinase. (See oncogenes, serine/threonine kinases)

PIMENTO (*Pimento dioica*): also called allspice. Tropical dioecious spice tree; 2n = 2x = 22.

PIN1: is a peptidyl-prolyl cis/trans isomerase in human cells. It is important for protein folding assembly and/or transport. Its deficiency leads to mitotic arrest whereas its overproduction may block the cell cycle in G2 phase. It interacts with NIMA kinase. (See cell cycle, NIMA)

pIN: promoter of the transposase gene of a transposon. There are two GATC sites involved in *dam* methylation within pIN. (See RNA-IN, *dam*)

PINEAL GLAND: the site of melatonin synthesis and photoreception in the brain. (See melatonin, opsins)

PINEAPPLE (*Ananas comosus*): a monocotyledonous tropical or subtropical plant (2n = 50, 75, 100). The flowers and bracts are sitting on a central axis and form a fleshy fruits. The lack of seeds is caused by selfincompatibility of the commercial varieties but they develop seeds if allowed to cross pollinate by other varieties. (See seedless fruits)

PINES (Pinus spp): trees, all 94 species are 2n = 2x = 24. (See also spruce)

PINNA: the ear lobe; the lobe of a compound leaf or frond.

PINOCYTOSIS: formation of ingestion vesicles for fluids and solutes by the invagination of membranes of eukaryotic cells. (See also phagocytosis)

PINOSOME: a small cytoplasmic vesicle originating by invagination of the cell membrane.

PIP: phosphatidylinositol phosphate. (See phosphoinositides, PIP_2)

PIP_2: phosphatidylinositol bis-phosphate (is involved with PIP) in mediating the inositol phospholipid signaling pathway and in the activation of phospholipase C (PLC). (See phosphoinositides, PITP)

PIPES (piperazine-N,N'-bis(2-ethanesulfonic acid): a buffer within the pH range of 6.2-7.3.

PISTIL: a central structure of flowers (gynecium) consisting of the stigma, style and ovary. (See gametophyte female, gametophyte male, flower differentiation)

PISTILLATE: flower or plants that carries the female sexual organs. A female parent in plants.

Pisum sativum (pea): is a legume (2n = 14). It played an important role in establishing the Mendelian principles of heredity and it contributed further information on genetics. Curiously, the famous "wrinkled" gene of Mendel turned out to be an insertional mutation. (See Pea)

PIT: an indentation. Also, the stony endocarp of some fruits, e.g. plums.

PITCH: the length of a complete turn of a spiral (helix) and the translation per residue is the pitch divided by the number of the residues per turn; in a keratin alpha helix it is 0.54 nm/3.6 = 0.15 nm. Also, the physiological response of the ear to a sound depending on the frequency of vibration of the air. Also, a dark black residue after distillation.

PITH: the parenchyma tissue in the core of plant stems, e.g. elderberry (*Sambucus*).

PITHECIA (saki monkey): see Cepidae

PITP (phosphatidylinositol transfer protein): is required by for the hydrolysis of PIP_2 (phosphatidyl-inositol bis-phosphate) by PLC (phospholipase C). In a GTP-dependent signal pathway, PITP is required also by epidermal growth factor (EGF) signaling. (See PIP, PIP_2, EGF, GTP)

PI-TR: phosphatidylinositol transfer protein involved in transfer of lipids among organelles within cells.

PITUITARY (hypophysis): is located at the base of the brain and connected also to the hypothalamus (a ventrical part of the brain). The anterior part secretes the pituitary hormones and the posterior part stores and releases them. (See brain human)

PITUITARY DWARFISM: is due to recessive mutation, deletion or unequal crossing over in the gene cluster containing somatotropin and homologs in human chromosome 17q23-q24. Administration of somatotropin may restore growth. The defect may also be in the hormone receptor (human chromosome 5p13.1-p12, mouse chromosome 15) and in these cases the growth hormone level may be high (Laron types of dwarfisms). The level of somatomedin (insulin-like growth factors) may also be low. Somatomedin is a peptide facilitating the binding of binding proteins and in addition have insulin-like activity. In either case dwarfism may result. (See dwarfism, GH, insulin-like growth factor, hormone receptor, binding protein, stature in humans, pituitary gland, growth hormone pituitary)

PITUITARY TUMOR (GNAS1): caused by autosomal dominant mutations in the α chain of a G-protein (G_s). This protein is also called gsp oncoprotein. (See G-protein, McCune-Albright syndrome)

PIXEL: a picture element in the computer that represents a bit on the monitor screen or in the video memory.

pK$_a$: is the negative logarithm of the dissociation constant K_a; stronger acids have higher pK_a whereas weaker acids have lower. The dissociation of weaker acids is higher and that of stronger acids is lower.

PKA: protein kinase A (activated by cAMP). (See protein kinases)

PKB (protein kinase B): is a serine/threonine kinase; the same as Rac or Akt.

PKC: protein kinase C. (See protein kinases)

PKD: see polycystic kidney disease

PKR: a double-stranded RNA-dependent protein kinase, involved in NF-κB signaling. (See NFκB)

PKS ONCOGENES: are in human chromosomes Xp11.4 and 7p11-q11.2. These genes display very high homology to oncogene RAF1 and apparently encode protein serine/threonine kinases. (See *raf*, oncogenes)

PKU: see phenylketonuria

PLACEBO: a presumably inactive substance used in parallel but to different individuals in order to serve as a concurrent (unnamed) control for testing the effect of a drug. (See concurrent control)

PLACENTA: the maternal tissue within the uterus of animals that is in the most intimate contact with the fetus through the umbilical chord. Most commonly, the placenta is located on the side of the uterus; the placenta praevia is situated at lower part of the uterus. The latter situation may be correlated with the age of the mother. Also, the wall of the plant ovary to what the o-vules are attached.

PLACODE: a heavy embryonal plate of the ectoderm from what organs may develop. (See ectoderm, AER, ZPA, organizer)

PLAGUE: has been used to loosely defined wide-spread, devastating diseases. Strictly, the term applies today to infection by the *Pasteurella pestis* (*Yersinia pestis*) bacterium. The disease may occur in three main forms: *bubonic* plague (most important diagnostic features of is swelling lymph nodes, particularly in the groin area), *pneumonic* plague (attacking the respiratory system) and *septicemic* plague (causing general blood poisoning). Many of its symptoms are overlapping with other infectious diseases. It is used to be called also "black death" on account of the dark spots, appearing in largely symmetrical necrotic tissue with coagulated blood. The bacilli spread to human populations from rodents by fleas but infections occur also through cough drops of persons afflicted by pneumonic plague. Various animal diseases are also called plague (pestis) but, except those in rodents, are caused by other bacteria or viruses. Pasteurellosis can be effectively treated with antibiotics although some strains become resistant to a particular type of antibiotics (streptomycin, chloramphenicol). Eradication of rodent pests is the best measure of prevention. During the great epidemics in the 14th century an estimated 25 million victims were claimed by the disease. Sporadic occurrence is known even today in the underdeveloped areas of the world

PLAKOGLOBIN: 83 kDa protein localized to the cytoplasmic side of the desmosomes. (See desmosome, adhesion)

PLANCK CONSTANT (h): a constant of energy of a quantum of radiation and the frequency of the oscillator that emitted the radiation. $E = h\nu$ where E = energy, ν = its frequency; numerically 6.624×10^{-27} erg-sec

PLANKTON: collective name of many minute free-floating water plants and animals.

PLANT BREEDING: is an applied science involved in the development in high-yielding food, feed and fiber plants. It is concerned also with the production of lumber, renewable resources

Plant breeding continued

of fuel and many types of industrial raw products (such as latex, drugs, cosmetics, etc.). A major goal of plant breeding is to improve the nutritional value, safety and palatability of the crops. Plant breeding and technological improvements in agriculture resulted in near 10 fold increase in maize production and wheat yields doubled in this centrury. Plant breeding is based on population and quantitative genetics, and biotechnology.

PLANT DEFENSE: against herbivores is mediated by the signaling peptide *systemin* activating a lipid cascade. Membrane linolenic acid is released by the damage and converted into phytodienoic and jasmonic acids, structural analogs to the prostaglandins of animals. As a consequence, tomato plants produce several systemic *wound response proteins*, similar to those elicited by oligosaccharides upon pathogenic infections. Mutation in the octadecanoic (fatty acid) pathway blocks these defense responses. (See host-pathogen relations, jasmonic acid, prostaglandins, fatty acids, systemin)

PLANT DISEASE RESISTANCE: see host -pathogen relation, plant defense.

PLANT HORMONES: are auxins, gibberellins, cytokinins, abscisic acid and ethylene. The natural *auxin* in plants is indole-3-acetic acid (IAA) but a series of synthetic auxins are also known such as dichlorophenoxy acetic acid (2,4-D), naphthalene acetic acid (NAA), indolebutyric acid (IBA), etc. Auxins are involved in cell elongation, root development, apical dominance, gravi- and phototropism, respiration, maintenance of membrane potential, cell wall synthesis, regulation of transcription, etc. The bulk ($\approx$ 95%) of IAA in plants is conjugated through its carboxyl end to amino acids, peptides and carbohydrates. The conjugate regulates how much IAA is available for metabolic needs although some conjugates may be directly active as a hormone. Enzymes have been identified that hydrolyze the conjugates. Over the developing tissues auxins show concentration gradients, indicating its role in positional signaling similarly to animal morphogens. The conjugates may transport IAA within the plant. *Gibberellic acid* and gibberellins control stem elongation, germination and a variety of metabolic processes. *Cytokinins* occur also in wide variety of forms such as kinetin, benzylaminopurine (BAP), isopentenyl adenine (IPA), zeatin, etc. Their role is primarily in cell division but they regulate the activity of a series of enzymes. Regeneration of plants from dedifferentiated cells requires a balance of auxins and cytokinins. *Abscisic acid* terpenoids control abscission of leaves and fruits, dormancy and germination of seeds and a series of metabolic pathways. *Ethylene* was recognized as a *bona fide* plant hormone more recently. It is involved in the control of fruit ripening, senescence, elongation, sex determination, etc. The hormone type action of *brassinosteroids* in controlling elongation and light responses have been recognized by genetic evidence only in 1996. (See hormones, signal transduction, abscisic acid, ethylene, indole acetic acid, gibberellic acid, kinetin, brassinosteroids)

PLANT PATHOGENESIS: see host - pathogen relation

PLANT VIRUSES: vary a great deal in size, shape, genetic material and host-specificity. The majority of them have single-stranded RNA as genetic material and are either enveloped or not. The Reoviridae may have several double-stranded RNAs, and the Cryptovirus carries 2 double-stranded RNA. The Cauliflower (Caulimo) virus has double-stranded DNA whereas the Geminiviruses have single-stranded DNA genetic material. The size of their genome usually varies between 4 to 20 kb and their coding capacity is at least 4 proteins. The 5'-end may form methyl-guanine cap or it may have a small protein attached to it. The 3'-end may have a polyA tail or may resemble the OH end of the tRNA. Approximately 600-700 plant viruses have been described. (See viruses, cap, polyA tail, tRNA, viroid, TMV, CaMV, viroid)

PLANTIBODY: is an "antibody" produced in transgenic plants carrying the genetic sequences required for the recognition of the site of the viral coat protein involved in infection. (See antibody, host-pathogen relations)

PLAQUE: clear area formed on a bacterial culture plate (heavily seeded with cells) as a consequence of lysis of the cells by virus; turbid plaques indicate incomplete lysis. (See p. 797, lysis)

PLAQUE-FORMING UNIT: the number of plaques per mL bacterial culture. (See next page)

PLAQUE LIFT: on bacteriophage plates plaques are marked and overlaid by cellulose nitrate films. After denaturation and immobilization of the plaques on the filter, they are hybridized with probes to identify recombinants and return to the saved master plate for obtaining plugs of interest from the original plate. The procedure generally requires repetition in order to isolate unique single recombinants. (See colony hybridization)

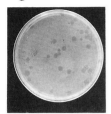

T3 BACTERIOPHAGE PLAQUES ON PETRI PLATE HEAVILY SEEDED BY BACTERIA. (Courtesy of Dr. C.S. Gowans)

PLASMA: the fluid component of the blood in which the particulate material is suspended. The blood plasma is free of blood cells but clotting has not been allowed during its isolation and it contains the platelets which harbor animal cell growth factors. (See PDGF, platelets, also serum, cytoplasm, cytosol)

PLASMABLAST: precursor of plasmacyte or precursor cell of the lymphocytes.

PLASMA CELL (plasmacyte): B lymphocytes can differentiate into either memory cells or plasma cells; the latter secrete immunoglobulins. (See lymphocytes, immunoglobulins, immune system)

PLASMA MEMBRANE: envelops all cells. (See cell membranes)

PLASMA PROTEINS: proteins in the blood plasma. The major components are serum albumin, globulins, fibrinogen, immunoglobulins, antihemophilic proteins, lipoproteins, α_1 antitrypsin, macroglobulin, haptoglobin and transfer proteins such as transferrin (iron), ceruloplasmin (copper), transcortin (steroid hormones), retinol-binding proteins (vitamin A) and cobalamin-binding proteins (vitamin B_{12}). The lipoproteins carry phospholipids, neutral lipids and cholesterol esters. In addition, there is a great variety of additional proteins present in the serum.

PLASMACYTOMA: cancer (myeloma) of antibody producing cells.

PLASMAGENE: non-nuclear genes (mitochondrial, plastidic or plasmid). (See mitochondrial genetics, chloroplast genetics)

PLASMALEMMA: a membrane around the cytoplasm or the membrane envelop of the fertilized egg.

PLASMID: dispensable genetic element that can propagate independently and can be maintained within the (bacterial) cell, and may be present in yeast and mitochondria of a number of organisms. The plasmids may be circular or linear double-stranded DNA. The conjugative plasmid posses mechanisms for transfer by conjugation from one cell to another. The non-conjugative plasmids lack this mechanism and therefore prefered for genetic engineering because they can be easier confined to the laboratory. (See also vectors, curing of plasmids, pBR322, pUC)

PLASMID ADDICTION: the loss of certain plasmids from the bacterial cells may lead to an apoptosis-like cell death, called post-segregational killing or plasmid addiction. (See apoptosis)

PLASMID, CHIMERIC: an engineered plasmid carrying foreign DNA.

PLASMID INCOMPATIBILITY: plasmids are compatible if they can coexist and replicate within the same bacterial cell. If the plasmids contain repressors effective for inhibiting the replication of other plasmids, they are incompatible. Generally closely related plasmids are incompatible and they thus belong to a different incompatibility group; the plasmids of enterobacteria belong to about two dozen incompatibility groups. Plasmids may be classified also according to the immunological relatedness of the pili they induce to form (such as F, F-like, I, etc.). The replication system of the plasmids defines both the pili and the incompatibility groups. Cells with F plasmids may form F sex pili, the R1 plasmids belong to FII pili group, etc. (See pilus, F^+, F plasmid, R plasmids, enterobacteria)

PLASMID INSTABILITY: indicates difficulties in maintenance caused by defect(s) in transmission, internal rearrangements and loss (deletion) of the DNA. (See also cointegration)

PLASMID, 2μm: is 6.3 kbp circular DNA plasmid of yeasts, present in 50-100 copies per haploid nucleus. It carries two 599 bp inverted repeats separating 2774 and 2346 bp tracts. Recombination between the repeats results in A and B type plasmids. Its recombination is controled by gene *FLP* and its maintenance requires the presence of the *REP* genes. (See yeast)

PLASMID MOBILIZATION: may take place by bacterial conjugation. Plasmid vectors use the gene *mob* (mobilization) if they do not have their own genes for conjugal transfer. Some plasmids may rely on *ColK* (colicin K, affecting cell membranes) that nicks plasmid pBR322 at the *nic* site, close to *bom* (basis of mobility). Mobilization proceeds from the nicked site (base 2254 in pBR322). Plasmids lacking the *nic/bom* system, e.g. pUC, cannot be mobilized.

PLASMID RESCUE: this procedure was designed originally for transformation with linearized plasmids of *Bacillus subtilis* that normally does not transform these bacteria. The linearized plasmid could be rescued for transformation in the presence of the *RecE* gene if recombination could take place. The linearized monomeric plasmid then could carry also any *in vitro* ligated passenger DNA into cells. If the host cells carry a larger number of plasmids (multimeric), special selection is necessary to find the needed one. Plasmid rescue has been used also for reisolation of inserts (plasmids) from the genome of transformed cells of plants.

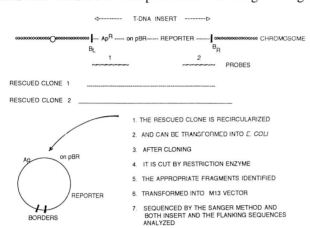

OUTLINE OF A PLASMID RESCUE PROCEDURE EXEMPLIFIED BY ISOLATING A T-DNA INSERT FROM *ARABIDOPSIS*. Ap = AMPICILLIN RESISTANCE GENE (Ap^R) OF THE pBR322 PLASMID, oripBR = ORIGIN OF REPLICATION OF THE pBR322 PLASMID PRESENT IN THE PLANT TRANSFORMING VECTOR, REPORTER IS HYGROMYCIN RESISTANCE, I LEFT (B$_L$) AND I RIGHT (B$_R$) BORDER SEQUENCES OF THE T-DNA. (After Koncz, C. *et al.* EMBO J. 86:8467)

The reisolation requires appropriate probes for (the termini) of the inserts to permit recognition, then it is recircularized and cloned in *E. coli* and they have at least one selectable marker, and an origin of replication compatible with the bacterium. The cloned DNA insert or fragments of it are inserted into M13 phage for nucleotide sequencing. This permits the identification of any changes that took place in the original transforming DNA and permits an analysis of the flanking sequences of the target sites as well. A number of different variations of the procedure have been adopted in prokaryotes, microbes, animals and plants. (See T-DNA, DNA sequencing, Rec)

PLASMID SHUFFLING: the general procedure in yeast first disrupts the particular gene in a diploid strain. After meiosis, the cells can be maintained only if the wild type allele is carried on a replicating plasmid (episome). Then mutant copies of that particular gene are introduced into the cell on a second episome and exchanged (shuffled) for the wild type allele. The phenotype of any of the mutant alleles can be studied in these cells that carry the disrupted (null) allele.

PLASMID TELOMERE: linear plasmids require exonuclease protection at the open ends. The problem may be resolved by capping with protein(s) or forming a lollipop type structure by fusing the ends of the single strands: ⇄ loops ⇄   ↑ proteins ↑

PLASMID VEHICLE: a recombinant plasmid that can mediate the transfer of genes from one cell (organism) to another. (See vectors)

PLASMIDS, AMPLIFIABLE: continues replication in the absence of protein synthesis (in the presence of protein synthesis inhibitor). (See amplification)

PLASMIDS, CONJUGATIVE: carries the *tra* gene, promoting bacterial conjugation and can be transferred to other cells by conjugation and can mobilize also the main genetic material of the bacterial cell. (See conjugation, F plasmid)

PLASMIDS, CRYPTIC: have no known phenotype.

PLASMIDS, MONOMERIC: present in a single copy per cell.

PLASMIDS, MULTIMERIC: have multiple copies a cell.

PLASMIDS, NON-CONJUGATIVE: lack the *tra* gene required for conjugative transfer but have the origin of replication and therefore when complemented by another plasmid for this function, they can be transfered. (See conjugation)

PLASMIDS, PROMISCUOUS: have conjugative transfer to more than one type of bacteria.

PLASMIDS, RECOMBINANT: is chimeric; it carries DNA sequences of more than one origin.

PLASDMIDS, RELAXED REPLICATION: may replicate to 1,000 or more copies per cell.

PLASMIDS, RUNAWAY REPLICATION: their replication is conditional, e.g. under permissive temperature regimes they may replicate almost out-of-control whereas under other conditions their number per cell may be quite limited.

PLASMIDS, SINGLE COPY: may have single or very few copies per cell.

PLASMIDS, STRINGENT MULTICOPY: may grow to 10 to 20 copies in a cell.

PLASMIN (fibrinolysin): proteolytic protein (serine endopeptidase) with specificity of dissolving blood clots, fibrin and other plasma proteins. For its activation urokinases (tissue plasminogen activator) are required. Plasmin may be used for therapeutic purposes to remove obstructions in the blood vessels. (See urokinase, plasminogen, CAM)

PLASMIN INHIBITOR DEFICIENCY (PLI, AAP): encoded in human chromosome 18p11-q11 as recessive gene, involved in the regulation of fibrinolysin. (See plasmin)

PLASMINOGEN: a precursor of plasmin. (See plasmin, plasminogen activator)

PLASMINOGEN ACTIVATOR (PLAT): cleaves plasminogen into plasmin; it is encoded in human chromosome 8q11-p11. The plasmin activator inhibitor (PLANH1) is encoded in human chromosome 7q21-q22 and PLANH2 at 18q21.1-q22. The plasminogen activator receptor was localized to 19q13.1-q13.2. (See plasminogen, plasmin)

PLASMODESMA (plural plasmodesmata): about 2µm or larger channels connecting neighboring plant cells, lined by extension of the endoplasmic reticulum. Functionally they correspond to the gap junctions of animal cells. (See gap junctions)

PLASMODIUM: a syncytium of the amoeboid stage of slime molds (such as in *Dictyostelium*).

PLASMODIUM: one of the several parasitic coccid protozoa causing malaria-type diseases in vertebrates, birds and reptiles. They invade the erythrocytes where destroying the host cells through the formation of merozoites (mitotic products) and spreading thus to other cells. The merozoites may develop into gametocytes (gamete forming cells) that infect blood-sucking mosquitos where they are transformed into sporozoites (the sexual generation) that are transmitted through insect bites to the higher animal host. The invaders first move to the liver where merozoites are formed and then go back to the erythrocytes and thus the cycle continues. *Plasmodium falciparum*: causes the falciparum malaria, *P. malariae* is responsible for the *quartan*, or 4th day recurring malaria. The protozoon contains two double-stranded extranuclear DNA molecules; the one of circular DNA resembles mitochondria whereas the second bears similarities to ctDNA. (See malaria, thalassemia, mtDNA, chloroplast)

PLASMOGAMY: fusion of the cytoplasm of two cells without fusion the two nuclei and thus resulting in dikaryosis. Plasmogamy is common in fungi but may occur in fused cultured cells of plants an animals. (See fungal life cycle, cell genetics)

PLASMOLEMMA (plasmalemma): plant cell membrane; the ectoplasm of the fertilized egg of animals.

PLASMOLYSIS: shrinkage of the plant cytoplasm caused by high concentration of solutes (salt) outside the cell resulting in loss of water. The cytoplasm separates from the cell wall.

PLASMON: sum of non-nuclear hereditary units such as exists in mitochondrial and plastid DNA. (See mtDNA, chloroplasts, plastome)

PLASMON-SENSITIVE GENE: see nucleo-cytoplasmic interaction.

PLASMOTOMY: fragmentation of multinucleate cells into smaller cells without nuclear division.

PLASMOVIRUS: bears some similarity to phagemids but in this case a retrovirus is combined with an independent vector cassette containing various elements. The envelope gene of the Moloney provirus is replaced by a transgene to prevent infective retroviral ability and it would not regain it by chance recombination with another retrovirus. Such a construct can express transgene(s) and can muliply within the target cells and provide a tool for cancer therapy. (See vectors, retrovirus, transgene, cancer therapy, viral vectors, phagemid)

PLASTID: cellular organelle of plants, containing DNA. It may differentiate into chloroplasts or etioplasts or amyloplasts or leucoplasts or chromoplasts. (See under separate names, plastid number per cell, plastid male transmission, ctDNA, chloroplast, chloroplast genetics)

PLASTID MALE TRANSMISSION: generally the genetic material in the plastids is transmitted only through the egg cytoplasm but in a few species of higher plants (*Pelargonium, Oenothera, Solanum, Antirrhinum, Phaseolus, Secale,* etc.) a variable degree of male transmission takes place. Biparental transmission of plastid genes (about 1%) may occur also in the alga *Chlamydomonas reinhardi*. The male transmission of these nucleoids is controled by one or two nuclear genes. The nucleoids of the plastid and mitochondria of the male are usually degraded or if they are included in the generative cells of the male, most commonly they fail to enter the sperm or not transmitted to the egg cytoplasm. In contrast to the angiosperms, conifers (pines, spruces, firs) the plastid DNA is usually transmitted through the males. In some interspecific hybrids exclusively paternal, exclusively maternal and biparental transmission was also observed. In other conifer crosses the mtDNA is transmitted maternally. In redwoods the transfer is paternal. It had been claimed that the destruction of the paternal ctDNA in the females is carried out by a restriction enzyme while the maternal ctDNA is protected by methylation. Others implicate a special nuclease C. (See chloroplast, genetics, ctDNA, mtDNA, paternal leakage)

PLASTID NUMBER PER CELL: in the giant cells of *Acetabularia* algae there may be one million chloroplasts but in the alga *Chlamydomonas* there is only one per cell. In higher plants the number of plastids vary according to the size of the cells, about 30-40 in the spongy parenchyma to about twice as many in the palisade parenchyma. (See plastid, ctDNA)

PLASTOCHRONE: the pattern of organ differentiation in time and space.

PLASTOME: sum of hereditary information in the plastids. (See ctDNA, chloroplast genetics)

PLASTOME MUTATION: mutation in the plastid (chloroplast) DNA. (See chloroplast genetics, mutation in cellular organelles,)

PLASTOQUINONE: an isoprenoid electron carrier during photosynthesis. (See isoprene)

PLATE: a Petri dish containing a nutrient medium for culturing microbial or plant cells; cell plate divides the two daughter cells after mitosis.

PLATE INCORPORATION TEST: is the most commonly used procedure for the Ames test when the *Salmonella* suspension (or other bacterial cultures), the S9 activating enzymes and the mutagen/carcinogen to be tested is poured over the bacterial nutrient plate in a 2 mL soft agar. After incubation for 2 days at 37 C° the number of revertant colonies is counted. (See Ames test, spot test)

PLATELET ABNORMALITIES: see Glanzmann's disease, thrombopathic purpura, thrombopathia, platelets.

PLATELET ACTIVATING FACTOR (PAF): is an inflammatory phospholipid. (See platelets)

PLATELET DERIVED GROWTH FACTOR (PDGF): a mitogen, secreted by the platelets, the 2-3 micrometer size elements in the mammalian blood, originated from the megakaryocytes of the bone marrow, and it is concerned with blood coagulation. PDGF controls the growth of fibroblasts, smooth muscle cells, nerve cells, etc. This protein bears substantial homologies to the oncogenic product of the simian sarcoma virus, the product of the KIT oncogene, the CSF1R (it activates also other oncogenes, such as c-fos). PDGF is required for the healing of vascular injuries and in these cases the expression induced Egr-1 (early growth response

Platelet derived growth factor continued
gene product) that may bind to the PDGF β chain promoter after displacing Sp1. PFGF- and insulin-dependent S6 kinase ($pp70^{S6k}$) are activated by phosphatidylinositol-3-OH kinase. Its receptor (PDGFR) is a tyrosine kinase. (See oncogenes, growth factors, signal transduction, platelets, Sp1, S6 kinase, phosphatidyl inositol)

PLATELETS: originate as cell fragments or "minicells" from megakaryocytes of the bone marrow. Their function is in blood clotting and repair of blood vessels; they secrete also mitogen(s). (See blood, platelet derived growth factor, blood serum)

PLATING EFFICIENCY: is the percentage of cells or protoplasts placed on a Petri plate that grow. The relative plating efficiency compares the fraction of growing cells in a treated series to that of an appropriate control.

PLATYSOME: the nucleosome core (when it was thought that it is a flat structure). (See nucleosome)

PLAYBACK: the number of non-repetitive sequences in a DNA can be determined by saturation of single-strand DNA with RNA of unique sequences. The kinetics of saturation, R_0t (by analogy to C_0t), is then determined. The annealed fraction is generally a small percent of the eukaryotic DNA that is highly redundant. To be sure that the RNA hybridized to only the unique DNA sequences, in the DNA-RNA hybrid molecules the RNA is degraded enzymatically and the remaining DNA is subjected to a reassociation test to determine its C_0t curve. This "playback" then reveals whether all the DNA so isolated has only genic DNA and is not redundant. Such studies may assist in estimating the number of housekeeping genes plus the genes that were transcribed when the RNA was collected. (See c_0t, housekeeping genes, gene number)

PLC: phospholipase C. (See phospholipase)

PLEATED SHEETS: relaxed β-configuration polypeptide chains hydrogen-bonded in a flat layer. (See protein structure)

PLECKSTRIN DOMAIN: occurs in many different proteins such as serine/threonine kinases, tyrosine kinases, and the substrates of these kinases, phospholipase C, small GTPase regulators and cytoskeletal proteins. Pleckstrin domains may participate in various signaling functions; they bind phosphatidylinositol 4,5-bisphosphate. (See also separate entries, SHC, SH2, SH3, WW, PTB, adaptor proteins, phosphatidylinositol)

PLECTIN: is a 500 kDa keratin of the cytoskeleton encoded in human chromosome 8q24. (See epidermolysis [bullosa simplex])

PLECTONEMIC COILS: the two coils were wound together, therefore they can be separated only by unwinding rather than simple pulling apart like the paranemic coils. The DNA double helix represents plectonemic coils. (See paranemic coils)

PLEIOMORPHIC: displays variable expression.

PLEIOMORPHIC ADENOMA: a salivary gland tumor caused by human chromosome breakage points, primarily at 8q12, 3p21 and 12q13-15. The translocation t(3;8)(p21;q12) results in swapping the promoters of PLAG1, a Zn-finger protein encoded in chromosome 8 and β-catenin (CTNNB1) and activation of the oncogene. (See zinc finger, β-catenin)

PLEIOTROPY: one gene affects more than one trait; mutation in various elements of the signal transduction pathways or in general transcription factors may have pleiotropic effects. (See signal transduction, transcription factors, epistasis)

PLEOMORPHISM: Nägeli's 19th century suggestion claiming lack of hard heredity in bacteria and they simply exist in a variety of pliable forms. This idea held back the development of bacterial genetics although physicians like Robert Koch and the taxonomist W. Migula sharply criticized it and stated that it ignored the facts known by the 1880s. (See *Hieracium*)

PLESIOMORPHIC: the trait in the more primitive state among several evolutionarily related species. (See apomorphic, symplesiomorphic, synapomorphic)

PLEURA: serous (moist) membrane lining the lung or insects' thoracic cavity.

PLOIDY: represents the number of basic chromosome sets in a nucleus. The haploids have one set (x), the diploids two (xx), autetraploids (xxxx), and so on. (See polyploidy)

PLTP (phospholipid transfer protein): mediates the exchange of HDL cholesteryl esters with very low density triglycerides and vice versa. (See HDL, cholesterol, CETP)

PLUM (*Prunus*): basic chromosome number x = 7 but a variety of polyploid forms exist.

PLUMULE: the plant embryonic shoot initial.

PLURIPOTENCY: same as totipotency.

PLUS AND MINUS METHOD: was an early version of DNA sequencing using dideoxy analogs of nucleosides (+ batch) during replication and after the analog was incorporated to a site, T4 exonuclease failed to continue degradation. In the minus (-) batch the synthesis stopped depending which single nucleotide was omitted (the precursor mixture containing only 3 deoxyribonucleotides). Thus nucleotide sequences of specific ends and length were generated and the fragments of different lengths were analyzed by electrophoresis. This method has been now replaced by the Sanger *et al.* method and its improvements. (See DNA sequencing)

PLUS END: the preferential growing end of microtubules and actin filaments. (See minus end)

PLUS STRAND: of the single stranded DNA or RNA of a virus is represented in the mature virion whereas the minus strand serves as a template for the transcription (replication) of the plus strand and the mRNA. (See replicative form, RNA replication)

PLUS STRAND → MINUS STRAND (MS) → PLUS STRAND → PLUS STRANDS

PLUTONIUM (Pu): a metallic fissile element (atomic number 94, atomic weight 242) produced by neutron bombardment of uranium (U^{238}) during the production of nuclear fuel and used for making nuclear weapons. Some of the heart pacemakers are powered by radioactive Pu. Thus the wearers as well as family members as well as the surgeons will be exposed to some radiation, generally below 1.28 Sv per person per year, a little more than the average natural background (the doses are additive, however). If Pu particles are inhaled (the most common type of ingestion), the element may affect the lung and may eventually be preferentially deposited in the skeletal system, causing bone cancer by the emission of X and γ rays. Pu^{239} has a halflife of 24.3×10^3 years and targets primarily the bone marrow. Other Pu isotopes have even much longer halflife. The level of Pu may be detected by radioactivity in the urine and by instruments placed on the body. Appropriate instruments can detect as low as 4 nCi (nanoCurie) values. (See atomic radiation, isotopes, radiation hazard assessment, Curie)

Plx1: is a kinase that phosphorylates the amino-terminal domain of Cdc25. (See Cdc25)

PLYMOUTH ROCK: a recessive white-feathered breed of chickens with the genetic constitution of *iicc*. The dominant *I* gene is a color inhibitor and *C* symbolizes color. (See White Wyan-dotte, Leghorn White)

PLZF: Zinc-finger protein encoded in human chromosome 11q23.

PMA: see phorbol 12-myristate-13-acetate, phorbolesters

pMB1: see ColE1

PML (promyelotic leukemia inducer): a zinc finger protein, encoded in human chromosome 15q21.

PNA: see peptide nucleic acid

Pneumococcus: see *Diplococcus pneumoniae*

PNPase: see polynucleotide phosphorylase

POCKET: is the motif of the retinoblastoma (RB) tumor-suppressor protein family that binds to viral DNA coded oncoproteins. Binding of RB to the E2F family of transcription factors blocks transcription, needed for the progression of the cell cycle. The pocket proteins share this retinoblastoma (RB) motif. (See E2F1, tumor suppressor, cell cycle, transcription factors)

Podospora anserina: n = 7, is an ascomycete.

pogo: see hybrid dysgenesis

POIKILODERMA ATROPHICANS (poikiloderma telangiectasia): see Rothmund-Thompson syndrome

POINT MUTATION: does not involve detectable structural alteration (loss or rearrangement of the chromosome), and are expected to involve base substitutions. (See substitution mutation)

POISSON DISTRIBUTION: basically an extreme form of the normal distribution that is found when in large populations rare events occur at random, such as e.g. mutation. The general formula is $e^{-m}(m^i/i!)$, and expanded $e^{-m}(m^0/0!, m^1/m!, m^2/2!...m^i/i!)$, where e = base of natural logarithm ($\cong 2.718$), m = mean number of events, i = the number by which a particular m is represented at a given frequency, ! = factorial (e.g. 3! = 3x2x1, but 0! = 1). (See figure, also negative binomial)

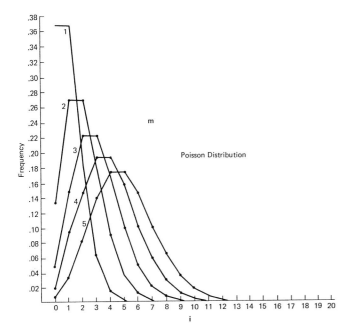

THE POISSON DISTRIBUTION. EACH CURVE CORRESPONDS TO A NUMBERED m VALUE. THE i CLASSES REPRESENT THE DISTRIBUTION OF EACH MEAN VALUE (m) WITH THE ORDINATE INDICATING THE FREQUENCIES.

poky (synonym: *mi-1*): a slow-growing and cyanide-sensitive respiration defective mitochondrial mutation in *Neurospora*. The basic defect appears to be a four-base deficiency of the 15 bp consensus at the 5'-end of the 19S rRNA of the mitochondria. Because of this defect, a further upstream promoter is used, making the transcript longer but during processing shorter RNAs are made. It is analogous to the petite colony mutations of budding yeast. (See petite colony mutation, stoppers, mtDNA)

pol (bacterial RNA polymerase): synthesizes all the bacterial and viral RNAs in the bacterial cells. Its subunits are $\alpha\alpha\beta\beta'$ and σ. The σ subunit identifies the promoter sequences and it is required for the initiation of transcription within the cell. After about a half dozen nucleotides were hooked up, it dissociates from the other subunits and further polymerization continues with the assistance of elongation protein factors. (See transcription, pol I, pol II, pol III eukaryotic RNA polymerases)

pol I: prokaryotic DNA polymerase where the polymerase (Klenow fragment) and exonuclease functions are located about 30 Å distance apart in a subunit, and editing (removal of wrong bases) follows the melt and slide model; it has the major role in prokaryotic repair and in the extension of the Okazaki fragments for the joining them into a contiguous strand by ligase. It adds 10-20 nucleotides/second to the chain and so it is much slower than pol III. (See melt and slide model, DNA replication, replication fork, Klenow fragment, pol III, DNA ligase)

pol I: RNA polymerase involved in the synthesis of ribosomal RNA (except 5S rRNA) in eukaryotes. By endonucleolytic cleavage it generates the 3' end of rRNA from longer transcripts. The upstream control regions for transcription initiation (binding proteins) vary from species to species. The human RNA pol I requires an activator UBF (upstream binding factor) and promoter selectivity factor SL1, including the TBF (TATA box binding protein) and associated subunits, TAF_I 110, TAF_I 63 and TAF_I 48. The former two keep contact with the promoter whereas TAF_I 48 interacts with UBF and prevents RNA pol II to use this promoter site. (See ribosome)

pol II: prokaryotic DNA polymerase, functions are not completely defined so far; it has known role in repair. (See DNA repair)

pol II: RNA polymerase transcribes messenger RNA and most of the snRNAs of eukaryotes with the assistance of different transcription factors. Nine or ten of its subunits are very similar to other polymerases; pol II has 4-5 smaller unique subunits. The two largest subunits are very similar in the three eukaryotic RNA polymerases and are similar also to prokaryotic subunits. It is most sensitive to α-amanitin inhibition (0.01 µg/mL). The site of sensitivity is in the largest, 220 kDa polypeptide. This large subunit is activated by phosphorylation. At the carboxy terminal there are 26 (yeast), 40 (*Drosophila*) or 52 (mouse) heptapeptide (Tyr-Ser-Pro-Thr-Ser-Pro-Ser) repeats. These repeats are essential for function. The Ser and Thr residues may be phosphorylated. Phosphorylation of the carboxy-terminal domain may affect the promoter-specificity of the enzyme. The C-terminus of (CTD) of the large subunit is instrumental also in the processing of the 3'-end of the transcript and the termination of transcription downstream of the polyA signal. CTD does not seem to affect initiation of transcription but it also mediates the response to enhancers. This enzyme is different from RNA pol I and RNA pol III inasmuch that it requires a hydrolyzable source of ATP for the initiation of transcription. RNA pol II is different from the other polymerases in the requirement for a large array of special transcription factors that modulate the transcription of the thousands of proteins. (See transcription factors, regulation of gene activity, α-amanitin)

pol III: prokaryotic DNA polymerase where the α subunit carries out the replication function and the ε-subunit is involved in editing (exonuclease) activity; it has the major role in replication of the leading and lagging strands. The replication has a speed of ≈1 kb/sec. There are only about 10-20 copies of the 10 subunit holoenzyme/cell (See DNA replication, replication fork, core polymerase, replisome)

pol III: RNA polymerase involved in the synthesis of transfer RNA, 5S rRNA, 7S rRNA in eukaryotes. Transcription of pol III is higher during S and G2 phases of the cell cycle than in G1. Many neoplastic cells display high pol III activity indicating that protein synthesis is demanded for tumorous growth. The retinoblastoma protein, RET appears to be a suppressor of increased pol III activity. (See tRNA, ribosomal RNA, ribosomes, La)

pol α: DNA polymerase, replicating the nuclear DNA (lagging strand) of eukaryotes. (See lagging strand, replication fork, DNA polymerases)

pol β: eukaryotic DNA repair polymerase. (See DNA polymerases)

pol δ: eukaryotic DNA polymerase (replicating the leading strand) of the nuclear chromosomes. (See replication fork, DNA polymerases)

pol $δ_2$: synonymous with pol ε. (See DNA polymerases)

pol ε: eukaryotic DNA polymerase with repair role. (See DNA polymerases)

pol γ: DNA polymerase replicating eukaryotic organelle DNA. (See θ type replication, DNA polymerases)

pol ζ: eukaryotic DNA polymerase without exonuclease activity. It is a repair enzyme in as much it can bypass pyrimidine dimers more efficiently than polα. It is insensitive to 200 µM aphidicolin (and in this respect it is similar to polβ and polγ) and it is insensitive dideoxynucleotide triphosphates (which inhibit polβ and polγ). It is moderately sensitive to 10 µM butylphenylguanosine triphosphates. It is relatively inactive with salmon sperm DNA or primed homopolymers. (See DNA polymerases)

POLAND SYNDROME: an autosomal dominant defect with low penetrance and the inheritance pattern is complicated by the teratogenic effects of diverse exogenous factors. It is characterized by fusion of fingers (syndactily), short fingers and anomalies of chest and sometimes other muscles. (See limb defects, syndactily)

POLAR: hydrophilic, i.e. soluble in water.

POLAR BODY: see gametogenesis in animals

POLAR BODY DIAGNOSIS: the genetic constitution of the polar body is tested by molecular techniques prenatally. (See prenatal diagnosis)

POLAR BOND: is covalent yet the electrons are more firmly tied to one of the two molecules and therefore the electric charge is polarized.

POLAR COORDINATE MODEL: of regeneration states that when cells are in non-adjacent positions, the process of growth restores all intermediate positions by the shortest numerical routes. The shortest intercalation mandates that small fragments may undergo duplication and large fragments may require regeneration The position of each cell on a collapsed cone (the idealized primordium) is specified by the radial distance from a central point at the tip of the cone and the circumferential position on the circle defined by the radius of the base. (See also distalization)

POLAR CYTOPLASM: is situated in the posterior (hind) portion of the fertilized egg cell. (See pole cells)

POLAR GRANULES: present in the posterior pole region of insect eggs and have maternal effect specification role during embryogenesis. (See animal pole, morphogenesis in *Drosophila*)

POLAR MOLECULE: is generally soluble in water; the distribution of the positive and negative charges are not even and thus resulting in a polarized effect.

POLAR MUTATION: may be a base substitution (nonsense mutation), insertion, frame shift or any chromosomal alteration that affects the expression of genes down-stream in the transcription — translation system. (See frame shift mutation)

POLAR NUCLEI: are in the embryosac of plants, formed at the third division of the megaspore. After they have fused (n + n) and fertilized by one sperm (n) they give rise to the triploid (3n) endosperm nucleus. (See megagametophyte, embryosac)

POLAR OVERDOMINANCE: is an unusual type of inheritance, i.e. mutants heterozygous for the dominant *callypige* gene of sheep (chromosome 18) transmitted the dominant (CLPG) allele only by the males but not by the females. (See imprinting, overdominance)

POLAR TRANSPORT: certain metabolites move only one direction in the plant body, e.g., the auxins under natural conditions are synthesized in the tissues over the ground and then move toward the roots.

POLARIMETER: measures the rotation of the plane of polarized light.

POLARITY, EMBRYONIC: is required for differentiation and requires asymmetric cell divisions. In *Caenorhabditi*s the PAR proteins (serine/threonine kinase) control embryonic polarity and a non-muscle type myosin II heavy chain protein (NMY-2) is a cofactor of this polarity. In *Drosophila* the major body axes, primarily the anterior-posterior polarity is controled by the gurken-torpedo gene products but other genes are also involved. (See morphogenesis in *Drosophila*, polar cytoplasm)

POLARITY MAPPING: see mapping mitochondrial genes

POLARIZATION: is the distortion of the electron distribution in one molecule caused by another.

POLARIZED DIFFERENTIATION: is the basis of morphogenesis, chemotactic response, response to pheromones, etc.

POLARIZED LIGHT: exhibits different properties in different directions at right angles to the line of propagation. Specific rotation is the power of liquids to rotate the plane of polarization.

POLARIZED RECOMBINATION: see polarized segregation

POLARIZED SEGREGATION: may be brought about by meiotic anomalies, e.g. in maize plants heterozygous for some knobbed chromosomes (and syntenic markers) are preferentially included into the basal megaspore. Polarized segregation has been observed as a result of gene conversion, e.g. in *Ascobolus immersus* alleles of the *pale* locus in the cross $\frac{188 \; w^+}{188^+ \; w}$ segregated in both cases 6:2 but in the first case the results were (4*[188]*+ 2 *[w+]*): 2 (*w*) whereas in the cross $\frac{w \; 137^+}{w^+ \; 137}$ the conversion asci were (4 *[w]*+ 2 *[137]*): 2 (*w+*). The genetic order of these alleles were <u>188 w 137</u>. Thus in the first cross *white* was in the minority class whereas in the second cross it was part of the majority class. (See also gene conversion, meiotic drive, map expansion)

POLARIZING MICROSCOPE: uses a *polarizer* (a polaroid screen) in front of the light beam and over the eyepiece an *analyzer* (permitting rotation). The anisotropic specimens (having difference in transmission or reflection depending on the angle of light) will display optical contrast. (See microscopy)

POLAROGRAPHY: electrochemical measurement of reducible elements.

POLAROID CAMERA: was developed in the 1940s and has found many uses in the biological laboratories because it can provide almost immediate negative or positive images for recording observations such as of electrophoretic gels. The combined developing and fixing solution is contained in between the exposed negative film and the receiving film or paper and when the storage "pod" bursts under pressure of pulling, the processing is carried out within the camera.

POLARON: part of a locus within which gene conversion (or recombination) is polarized. (See gene conversion, polarized segregation)

POLE CELLS: localized in the posterior-most part of the cellularized embryo and give rise eventually to the germline. (See germline)

POLIO VIRUSES: are icosahedral single-stranded RNA viruses with about 6.1 kb RNA in a total particle mass of about 6.8×10^6 Da; Type 1 was responsible for about 85% of the polyomyelitis (infantile paralysis) cases before successful vaccination (live oral, Sabin or inactivated, Salk) has been widely used in the developed countries. These small RNA viruses are highly mutable because their genetic material lacks repair systems. The three serotypes produce a cell surface receptor (PVR) by alternative splicing of its transcript. Susceptibility to polio virus was located to human chromosome 19q12-q13. Mice are very resistant to this virus because they lack the membrane receptor for the infection. (See picornaviruses, IRES)

POLLED: is a dominant gene in cattle, responsible for lack of horns.

POLLEN: is the male gametophyte of plants developing from the microspores by two postmeiotic divisions. The first division results in the formation of a vegetative and a generative cell. The round vegetative cell directs the elongation of the pollen tube growing through the pistil toward the ovule. The crescent-shaped generative cells may divide before or after the shedding of the pollen grains. One of them fertilizes the egg and gives thus rise to the diploid embryo, the other fuses with the diploid polar cell in the embryosac and thus contributes to the formation of the endosperm. The pollen tube elongates quite rapidly; it may grow 15 cm in just 5 to 15 hours. The pollen tube elongation is regulated by a protein that is glycosylated in that tissue. In allogamous species, a single individual may shed over 50,000,000 pollen grains whereas in autogamous species the number of pollen grains per anther may not exceed a couple of hundreds. Since the pollen grain is haploid and may be autonomous (gametophytic control), it may express its genetic constitution independently from the genotype of the anther tissues (e.g. waxy pollen, various color or sterility alleles), in some instances, however, the morphology of the pollen grain is under sporophytic control. Since the pollen is a more independent product than the megaspore, it is more likely to suffer from genetic defects for what the surrounding tissues cannot compensate, therefore pollen sterility is more common in plants than female sterility. Pollen sterility may not necessarily affects, however the fertility of the individuals because of the abundance of functional pollen grains in case of heterozygosity for the defects. (See microsporogenesis, gametogenesis, pollen tetrad, gametophyte, self-incompatibility)

POLLEN COMPETITION: see certation

POLLEN-KILLER: or spore-killer genes in wheat, tomato and tobacco render the pollen incapable to function effectively in fertilization and may cause segregation distortion. (See segregation distorter, pollentube competition, killer strains, killer plasmids)

POLLEN MOTHER CELL: microspore mother cell, microsporocyte. (See gametogenesis)

POLLEN STERILITY: is the inability of the pollen to function in fertilization. It can frequently be detected by the poor staining of the pollen grains with simple nuclear stains (acetocarmine,

Pollen sterility continued

acetoorcein, etc.). Deletions, translocations, and inversion heterozygosity generally result in pollen sterility. Mitochondrial plasmids may also be responsible for some types of male sterility. (See pollen, certation, gametophyte, cytoplasmic male sterility, fertility restorer genes)

POLLEN TETRAD: the four products of a single male meiosis. The components of the pollen tetrad may not stick together and may shed in a scrambled state. In some instances (*Salpiglossis, Elodea*, some orchids) the tetrads remain together, however in a way similar to the unordered tetrads of fungi. In *Arabidopsis* induced mutations (*qrt1, qrt2, quartet*) cause the four pollen grains to stay together because of the alteration of the outer membrane of the pollen mother cell. Each tetrad may then fertilize four ovules. (See tetrad analysis)

POLLEN TUBE: see pollen. (See for morphology page 701) POLLEN TETRAD ↗

POLLEN-TUBE COMPETITION: see certation

POLLINATION: transfer of the male gametophyte to the stigmatic surface of the style (ovary). (See gametophyte)

POLLINIUM: a mass of pollen sticking together and may be transported as such by the pollinator insects or birds.

POLLITT SYNDROME: see trochothiodystrophy

POLLUTION: spoiling the environment by the release of unnatural, impure, toxic, mutagenic, carcinogenic or any other undesirable and unesthetic material or to disturb nature by sound, odor, heat and light. Pollution may cause mutation, cancer and various other diseases.

POLY I-G: a DNA strand containing more cytosine is called heavy chain of a DNA double helix because it binds more of the polyI-G (inosine-guanosine) sequences. These annealed DNA double strands are separated by CsCl in the ultracentrifuge. (See ultracentrifuge, density gradient centrifugation, DNA heavy chain, inosine)

polyA mRNA: eukaryotic mRNAs post-transcriptionally polyadenylated at the 3' tail before leaving the nucleus, and subsequently in the cytoplasm the tail may be reduced to 50-70 residues or further extended to hundreds. Polyadenylation improves stability and efficiency of translation. PolyA tail is frequently added also to bacterial RNA. In *E. coli* the addition of polyA tail accelerates the decay of RNA I. In *Drosophila*, the length of the poly(A) tail may be correlated with the function in differentiation of the specific mRNA. The regulatory mechanism of polyadenylation is interchangeable between mouse and *Xenopus*. (See polyadenylation signal, mRNA tail, RNA I, PABP, mRNA degradation)

polyA POLYMERASE (*PAP*): adds the polyA tail post-transcriptionally to the eukaryotic mRNA and antisense RNA transcripts. In yeast, at least two other genes *RNA14* and *RNA15* are involved in the processing of the 3' end of the pre-mRNA. *E. coli* also encodes at least two PAP enzymes. (See mRNA tail, polyadenylation signal)

polyA TAIL: see polyadenylation signal, polyA mRNA

POLYACRYLAMIDE: see electrophoresis, gel electrophoresis

POLYADENYLATION SIGNAL: endonucleolytic processing of the primary transcript of the majority of eukaryotic genes is followed by post-transcriptional addition of adenylic residues downstream of the structural gene. The consensus signal for the process is AAUAAA in animals and fungi, about half of the plants use the same signal, the rest relies on diverse signals. In eukaryotes the number of added A residues may vary from 50 to 250. Polyadenylation is under the control of several genes. The RNA transcript of eukaryotes besides the poly(A) signal contains a CA element (PyA in yeast) and a GU-rich downstream element. To the AAUAAA *positioning element* binds the *polyadenylation specificity factor* (CPSF) that is a tetrameric protein consisting of 160, 73, 100 and 33 kDa subunits. The *cleavage stimulating factor* (CstF) is a trimeric protein of 64, 77 and 50 kDa subunits binds to GU-rich element of the RNA. The *polyadenylation polymerase* protein (PAP) binds downstream of the CPSF binding sites. The *cleavage factors* (CFI) and (CFII) are positioned upstream of the GU-rich element and they terminate the mRNA (p. 808). The polyadenylation complex of yeast is somewhat different.

Polyadenylation signal continued

```
    CPSF    PAP     CFs           CstF
____AAUAA_____CA↓_____GU-RICH_____RNA
```

In prokaryotes rarely a few adenine residues are also found at the mRNA 3'-terminus. Cordycepin (3'-deoxyadenosine) is an inhibitor of polyadenylation. (See mRNA tail, polyA polymerase)

POLY(ADP-Ribose) POLYMERASE: is DNA-binding enzyme but it appears to have no indispensable function.

POLYAMIDES: containing N-methylimidazole and N-methylpyrrole amino acids have high affinity for specific DNA sequences and may regulate the transcription similarly to DNA binding proteins. (See binding proteins)

POLYAMIDOAMINE DENDRIMERS (PAMAM): are highly branched, soluble, non-toxic molecules with amino groups on their surface. They are suitable for attaching to this surface antibodies, various pharmaceuticals and DNA. They are effective vehicles for transfection.

POLYAMINES: various molecules derived in part from arginine and present in cells in millimolar concentrations yet have important role in RNA and DNA transactions, replication, supercoiling, biosynthesis, etc. Typical polyamines are spermine, spermidine, putrescine, etc.

POLYANDRY: a form of polygamy involving multiple males for one female.

POLYAROMATIC COMPOUNDS: include various procarcinogens and promutagens such as benzo(a)pyrene, dibenzanthracene, methylcholanthrene etc.

POLYBRENE (hexadimethrine bromide): a polycation used for introduction of plasmid DNA into animal cells; it is also an anti-heparin agent. (See transformation genetic animal cells, heparin)

POLYCENTRIC CHROMOSOME: see neocentromeres

POLYCHLORINATED BIPHENYL (PCB): a highly carcinogenic compound and an inducer of the P-450 cytochrome group of monooxygenases. It had been used in electrical capacitors, transformers, fire retardants, hydraulic fluids, plasticizers, adhesives, pesticides, inks, copying papers, etc. (See microsomes, S-9)

POLYCHROMATIC: stainable by different dyes or displaying different shades when stained.

POLYCISTRONIC mRNA: is a contiguous transcript of adjacent genes, such as exist in an operon but it may be formed also in short genes of eukaryotes, e.g. oxytocin. The *Trypanosomas* produce multicistronic transcripts. (See oxytocin, *Trypanosoma*, *Caenorhabditis*)

POLYCLONAL ANTIBODIES: produced by a population of lymphocytes in response to antigens. These are not homogeneous as the monoclonal antibodies. (See monoclonal)

Polycomb (*Pc*, chromosome 3-47.1): *Drosophila* gene is a negative regulator of the *Bithorax* (*BXC*) and *Antennapedia* (*ANTC*) complexes. The homozygous mutants are lethal and the locus (and its homologs in vertebrates) is involved in the repression of homeotic genes controlling body segmentation. (See morphogenesis in *Drosophila*, *SWI*)

POLYCROSS: intercross among several selected lines to produce a "synthetic variety".

POLYCYCLIC AROMATIC HYDROCARBONS (PAH): generally carcinogenic and mutagenic compounds. They become more active during the process of the attempted detoxication by the microsomal enzyme complex. (See carcinogens, procarcinogens, mutagens, promutagens, benzo(a)pyrene, environmental mutagens)

POLYCYSTIC KIDNEY DISEASE: have two main types, and within each several variations exist. The adult type (ADPKD) dominant is controled apparently by the short arm of chromosome 16p13.31-p13.12. The autosomal recessive (ARPKD) generally has an early onset. Both forms occur at frequencies of 0.0025 to 0.001. Even the late onset type may be detectable early by tomography. The symptoms vary and involve kidney disease, cerebral vein aneurism (sac like dilatation), underdeveloped lungs, liver fibrosis and growth retardation, etc. The dominant type can be identified with high accuracy using chromosome 16p13 DNA probes but a less

Polycystic kidney disease continued

than 10% of the cases are due to genes not in chromosome 16. The autosomal recessive form is at an unknown location and it can be identified after the third trimester by ultrasonic methods because the kidneys are enlarged. The genetic transmission of the dominant and recessive diseases are very good. One polycystic kidney (PKHD1) locus was assigned to 6q21-p12, and sequences were found also in 2p25-p23 and 7q22-q31 that are homologous to polycystic kidney disease of the mouse. There is a PKD2 locus in 4q21-q23. The infantile type recessive PKD is also called Caroli disease. The ARPKD locus encodes a 968-amino acid protein which forms six transmembrane spans with intracellular amino and carboxyl ends It appears to be a voltage-activated Ca^{2+} (Na^+) channel protein. (See cardiovascular disease, hypertension, genetic screening, ion channels)

POLYCYSTIC OVARIAN DISEASE (Stein-Leventhal syndrome): generally involves enlarged ovaries, hirsuteness, obesity and lack or irregular menstruation and increased level of testosterone, high luteinizing hormone:follicle-stimulating hormone ratios and infertility. It appears to be due an autosomal factor yet 96% and 82% of the daughters of affected mothers and carrier fathers, respectively, developed the symptoms indicating a meiotic drive-like phenomenon. (See infertility, luteinization, Graafian follicle, meiotic drive)

POLYDACTYLY: presence of extra fingers or toes. In *postaxial* polydactyly (the most common type) the extra finger is in the area of the "little finger" and in *preaxial* cases this malformation is on the opposite side of the axis (thumb) of the palm or foot. The various types of polydactyly may be determined by autosomal recessive or dominant gene(s) and their expression is usually part of other syndromes. Crossed polydactyly indicates coexistence of postaxial and preaxial types with discrepancy between hands and feet. (See Ellis-van Creveld syndrome, Opitz syndrome, Meckel syndrome, Majewski syndrome, orofacial-digital syndromes, Pätau's syndrome, diastrophic dysplasia, syndactyly, Greig's cephalopolysyndactyly syndrome, Rubinstein-Taybi syndrome, focal dermal hypoplasia, ectrodactyl, adactyly, *hedgehog*)

POLYEMBRYONY: more than one cell of the embryo sac develops into an embryo in plants. (See also adventive embryos, embryo sac)

POLYETHYLENE GLYCOL (PEG): a viscous liquid or solid compound of low-toxicity, promoting fusion of all types of cells. PEG is widely used also in textile, cosmetics, paint and ceramics industry. (See also PEG)

POLYGALACTURONS: complex carbohydrates in the plant cell wall.

POLYGAMY: having more than one mating partner. In western human societies it is illegal but in others it is still acceptable that men have more than one wife simultaneously. Polyandry or polygyny is common practice in animal breeding but it may be objectionable to humans on moral grounds. In the USA the polygamy laws are applied to all citizens, irrespective of religious affiliation or cultural tradition.

POLYGENES: a number of genes involved in the control of a quantitative traits. (See gene number in quantitative traits, QTL)

POLYGENIC INHERITANCE: is determined by a number of non-allelic genes all involved in the expression of a single particular trait (such as height, weight, intelligence, etc.). Polygenic inheritance is characterized by counting and measurements and the segregating classes are not discrete but display continuous variation. (See quantitative genetics, QTL, complex inheritance, chaos)

POLYGENIC PLASMIDS: are obtained when two plasmids carrying identical genes cointegrate. Such plasmids may have merit in genetic engineering if the genes show positive dosage effect for anthropocentrically useful traits. (See cointegration)

POLYGENY: one male has more than a single female mate. (See polygamy)

POLYHAPLOID: has half the number of chromosomes of a polyploid. The gametes of polyploids are polyhaploid. (See polyploidy)

POLYHEDROSIS VIRUS, NUCLEAR (BmNPV): is an about 130 kbp DNA baculovirus of the silkworm (and other insects). It has been used (after size reduction) as a 30 kb cloning vector and it may propagate in a single silkworm larva about 50 µg DNA. (See baculoviruses, viral vectors)

POLYHYBRID: heterozygous for many gene loci.

POLYHYDROXYBUTYRATE (PHB): a polymer that can be manufactured by plants and it is biodegradable.

POLYKETENES: are polymers of $CH_2=C=O$ (ketene). Their biosynthesis is related to fatty acids. Several antibiotics (tetracycline, griseofulvin, etc.) contain ketenes. (See antibiotics)

POLYKINETIC CHROMOSOME: has centromeric activity at multiple sites. (See neocentromeres)

POLYLINKER: a DNA sequence with several restriction enzyme recognition sites (multiple cloning sites, MCS) used in construction of different cloning or transformation vehicles (plasmids). E.g TTCTAGAATTCT sequence has an overlapping XbaI (TCTAGA) and an EcoRI recognition sites (GAATTC) and thus linking it to the DNA may generate both types of cloning sites. (See vectors, restriction enzymes, cloning sites, pUC)

POLY-MARKER TEST: see DNA fingerprinting

POLYMER: a large molecule composed of a series of covalently linked subunits such as amino acids, nuclcotides, fatty acids, carbohydrates, etc. (See DNA, protein)

POLYMERASE: an enzyme that builds up large molecules from small units, such as the DNA and RNA polymerases generate from nucleotides DNA and RNA, respectively. (See pol)

POLYMERASE ACCESSORY PROTEIN (RF-C): is an essential part of the DNA replication unit in SV40. (See SV40)

POLYMERASE CHAIN REACTION: a method of rapid amplification of DNA fragments when short flanking sequences of the fragments to be copied are known. The reaction begins by denaturation of the target DNA, then primers are annealed to the complementary single strands. After adding a heat-stable DNA polymerase, such as Taq or Vent/Tli (originally the less thermostable Klenow fragment of polymerase I was used), chain elongation proceeds starting at the primers:

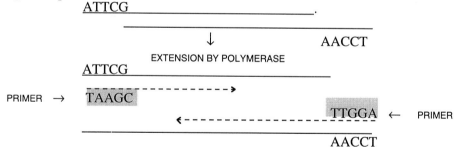

The cycles are repeated 20-30 times resulting in over a million fold (2^{20} = 1,048,567) replication of the target. The actual rate of replication may be less (80%) than the theoretically expected. The DNA amplified can be subjected to molecular analysis such as preimplantation analysis, genetic screening, prenatal analysis, sperm typing, gene identification, etc. The error frequency for the Klenow fragment is about 8×10^{-5}, for Taq 10^{-5} to 10^{-4}, for Tli 2 to 3×10^{-5}. All types of technical information and references are available at <http://apollo.co.uk/a/pcr>. (See RAPDS, DNA fingerprinting, vectorette, sperm typing, genetic screening, prenatal analysis, preimplantation genetics, tissue typing, primer extension, ancient DNA, molecular evolution, RT-PCR, *in situ* PCR, recursive PCR, methylation-specific PCR, PRINS)

POLYMERY: several genes cooperate in the expression of a trait. (See polygenes)

POLYMORPHIC: a trait occurs in several forms within a population. The polymorphism may be balanced and genetically determined. (See polymorphism, balaced polymorphism, RFLP)

POLYMORPHIC INFORMATION CONTENT (PIC): is used to identify and locate a hard-to-define marker locus. If the alleles of the marker locus are codominant then PIC is the fraction of the progeny (the informative offspring) that cosegregates by phenotype with an index locus.

Polymorphic information content continued

The index locus (which is used for the detection of linkage with marker alleles), has two alternative alleles, a wild type and a dominant (mutant) allele. The marker locus is polymorphic for dominant (genetic or physical [nucleotide sequences]) alleles. Only those progenies are informative where the index locus is homozygous in one of the parents and the other parent is heterozygous for the marker. The converse constitutions are not informative. In case both parents are heterozygous at the marker locus, only half of the offspring is informative.

$$\text{PIC} = 1 - \sum_{i=1}^{n} p_i^2 - \left(\sum_{i=1}^{n} p_i^2 \right)^2 + \sum_{i=1}^{n} p_i^4$$

where p_i = frequency of the index allele and i and n are the number of different alleles. The PIC values may vary theoretically from 0 to 1. A hypothetical example: four A alleles occur in a population with frequencies A^1: 0.2, A^2: 0.1, A^3 = 0.15 and A^4 = 0.55. After substitution PIC = 1- $(0.2^2 + 0.1^2 + 0.15^2 + 0.55^2) - (0.2^2 + 0.1^2 + 0.15^2 + 0.55^2)^2 + (0.2^4 + 0.1^4 + 0.15^4 + 0.55^4)$ thus PIC = 1 - 0.375 - 0.140625 + 0.0937125 ≈ 0.578, and in this case almost 58% of the progeny is informative. Usually PIC values of 0.7 or larger are required for showing good linkage. The larger the number of the marker alleles, the more informative is the PIC. (See microsatellite)

POLYMORPHISM: morphologically different chromosomes, or different alleles at a gene occur, or variable length restriction fragments are found within a population. Polymorphism can now be detected also by automated molecular techniques. During PCR amplification of a gene one or more fluorescent reporter probes are attached to the 5' end, and slightly downstream or at the 3' end a quencher substance(s) added. During amplification the quencher may be cleaved by the Taq polymerase if it hybridizes to an amplified segment. The cleavage of the quencher enhances the fluorescence of the reporter fluorochrome. The samples placed in a 96-well plate can be scanned at three wavelengths in about 5 minutes. The procedure may be sensitive enough to detect a single base difference. (See also balanced polymorphism, mutation detection, fluorochromes, PCR)

POLYMORPHONUCLEAR LEUKOCYTE (PMN): see granulocytes, leukocyte

POLYNEME: the linear structure includes more than one parallel strands, e.g. polytenic chromosomes (salivary gland chromosomes).

POLYNUCLEOTIDE: a nucleotide polymer hooked up through phosphodiester bonds.

POLYNUCLEOTIDE KINASE (PK): phosphorylates 5' positions of nucleotides in the presence of ATP such as ATP + XpYp $\xrightarrow{PK}$ p-5'XpYp + ADP (where X and Y are nucleotides).

POLYNUCLEOTIDE PHOSPORYLASE (PNPase): generates random RNA polymers $[(NMP)_n]$ — without a template — from ribonucleoside diphosphates (NDP) and releases inorganinic phosphate (P_i): $(NMP)_n + NDP \longrightarrow (NMP)_{n+1} + P_i$.

POLYOMA: is neoplasia induced by one of the polyomaviruses. The globoid (icosahedral) mouse polyoma viruses (a papova virus of 23.6×10^6 Da) contain double-stranded, circular DNA (4.5 kb). The BK and the JC viruses infect humans. (See Papova viruses)

POLYP: an outgrowth on mucous membranes such as may occur in the intestines, stomach or nose. They may be benign, precancerous or cancerous. (See, PAP, polyposis adenomatous, Gardner syndrome)

POLYPEPTIDE: a chain of amino acids hooked together by peptide bonds. (See protein synthesis, amino acids, peptide bond)

POLYPHYLETIC: an organism (cell) originated during evolution from more then one line of descent. A polyphyletic group may contain species that are classified into this group because of convergent evolution. (See convergence, divergence)

POLYPLOID CROP PLANTS: the most important polyploid crop plants include alfalfa (4x), apple (3x), banana (3x), birdsfoot trefoil (4x), white clover (4x), coffee (4x, 6x, 8x), upland cotton (4x), red fescue (6x, 8x, 10x), johnsongrass (8x), cultivated oats (6x), peanut (4x), European plum (6x), cultivated potatoes (4x), sugarcane *x), common tobacco (4x), bread wheat

(6x), macaroni wheat (4x). Most of these are apparently allopolyploids. (See allopolyploid)

POLYPLOIDY: having more than two genomes per cell. Definitive identification of polyploidy requires cytological analysis (chromosome counts) although many of the polyploids display broader leaves, larger stomata, larger flowers, etc.

← AUTOTETRAPLOID (top) AND DIPLOID (bottom) FLOWERS OF *Cardaminopsis petraea*.

(See autopolyploid, inbreeding autopolyploids, chromosome segregation, maximal equational segregation, alpha parameter, allopolyploid, tetrasomic, trisomy)

POLYPLOIDY IN ANIMALS: is rare and limited mainly to parthenogenetically reproducing species (e.g lizards). It occurs also in bees, silkworm and other species.

POLYPLOIDY IN EVOLUTION: is common in the plant kingdom but the majority of the polyploid species are allopolyploid. (See allopolyploid)

POLYPOSIS ADENOMATOUS INTESTINAL (APC): is controled by autosomal dominant genes responsible for intestinal, stomach (Gardner syndrome) or other types (kidney, thyroid, liver, nerve tissue, etc.) of benign or vicious cancerous tumors. The various forms are apparently controlled by mutations or deletions in the 5q21-q22 region of the human chromosome and represent allelic variations. Retinal lesions (CHRPE) are associated with truncations between codons 463-1387; truncations between codons 1403-1528 involve extracodonic effects, etc. In addition, it is conceivable that this is a *contiguous gene* region where adjacent mutations affect the expression of the polyposis. By the use of single strand conformation polymorphism technique, DNA analysis may permit the identification of aberrant alleles prenatally or during the presymptomatic phase of the condition. The situation is further complicated, however by the possibilities of somatic mutations. The *Min* gene of mouse appears to be homologous to the human APC and thus lending an animal model for molecular, physiological and clinical studies. The expression of *Min* is regulated also by the phospholipase-encoding gene *Mom1*, indicating also the involvement of lipids in the diet. Polyposis may affect a very large portion of the aging human populations, especially high is the risk for females. Certain forms of polyposis may affect already the young (juvenile polyposis). Regular monitoring by colorectal examination is necessary for those at risk. Bloody diarrhea and general weakness are usually too late symptoms for successful medical intervention. Molecular genetic information suggests that vertebrates use the same pathway of signal transduction as identified by *Drosophila* genes: *porcupine (porc*, 1.59)→*wingless* (*wg*, 2-30.0)→*disshevelled* (*dsh*, 1-34.5)→*zeste white3* (z^{w3}, 1.1.0)→*armadillo* (*arm*, 1-1.2)→cell nucleus. The normal human APC gene appears to be either a negative regulator (tumor suppressor) or an effector, acting between z^w and the nucleus. When it mutates, either it can no longer carry out suppression or it may become an effector. The product of *dsh* also appears to be a negative regulator of z^w. When the *zeste* product, glycogen synthase kinase (GSK3β) is inactive the *arm* product (catenin) is associated with the APC product and a signal for tumorigenesis is generated. Alternatively, when no signal is received, GSK phosphorylates and activates a second binding site on APC for catenin but that causes the degradation of catenin and thus no tumor signal is generated. The APC protein may act as a tumor gene also by docking at its COOH end with a human homolog of the *dlg1* (*disc large*, 1-34.82) of *Drosophila*). The *Dlg* product belongs to the *membrane associated guanylate kinase* protein family that are analogous to the proteins in vertebrates sealing adjacent cell membranes (tight junction). *Dlg* also considered to be a tumor gene. Although these molecular informations reveal a number of mechanisms of action, it is not clear which one is being used or multiple pathways are involved in polyposis. APC/FAP has a prevalence of about 1×10^{-4}. (See Gardner syndrome, Turcot syndrome, cancer, single-strand conformation, GSK3β, polymorphism, hereditary non-polyposis colorectal cancer, contiguous gene syndrome, animal models, tight junction, catenin, effector, polyposis hamartomatous)

POLYPOSIS HAMARTOMATOUS (Peutz-Jeghers syndrome): chromosome 19p dominant overgrowth of mucous membranes (polyp), especially in the small intestine (jejunum), but also in the esophagus (the canal from mouth to stomach), bladder, kidney, nose, etc. Melanin spots may develop on lips, inside the mouth and fingers. Ovarian and testicular cancer were also observed. (See pigmentation of the skin, cancer, Gardner syndrome, colorectal cancer Muir-Torre syndrome, polyposis adenomatous intestinal)

POLYPROTEIN: a contiguously translated long chain polypeptide that is processed subsequently into more than one protein.

POLYPURINE: a stretch of purine residues in nucleic acids.

POLYPYRIMIDINE: a sequence of multiple pyrimidines (mainly Us) in nucleic acids adjacent to the 3' splicing site. These are recognized by Py-tract-binding proteins (PTB) such as the essential splicing factor U2AF^{65}, the splicing regulator sex-lethal (Sxl), etc. (See splicing, introns, sex determination)

POLYRIBOSOME: same as polysome. (See also protein synthesis)

POLYSACCHARIDE: monosaccharides joined by glycosidic bonds (e.g. starch, glycogen)

POLYSOME: multiple ribosomes are held together by mRNA. (See ribosome, mRNA)

POLYSOMIC CELL: some chromosomes are present in more than the regular number of copies. The polyploids are polysomic for entire genomes. (See aneuploidy, polyploidy)

POLYSOMY: some of the chromosomes in a cell are present in more than the normal numbers. (See nondisjunction, polyploid, trisomy)

POLYSPERMIC FERTILIZATION: more than a single sperm enters the egg and because each may provide a centriole multipolar mitoses may take place resulting in aneuploidy and abnormal embryogenesis. (See fertilization)

POLYTENIC CHROMOSOMES: are composed of many chromatids (e. g., in salivary-gland cell nuclei) because DNA replication was not followed by chromatid separation. The polytenic chromosomes in the salivary glands nuclei of diptera may have undergone ten cycles of replication ($2^{10} = 1024$) without division and may have over 1000 strands. Also the polytenic chromosomes in the salivary glands are extremely long. A regular feature is the very close somatic pairing. Also, they all are attached at one point, at the chromocenter. The polytenic chromosomes have extensively exploited for analysis of deletions duplications, inversions and translocations. The characteristic banding pattern was used also as cytological landmarks for identification of the physical location of genes. Rarely, polyteny occurs in some specialized plant tissues (antipodals) too. (See salivary gland chromosomes, giant chromosomes)

POLYTOCOUS SPECIES: produce multiple offspring by each gestation. (See monotocous)

POLYTROPIC RETROVIRUS: see amphotropic retrovirus

POLYTYPIC: a species includes more than one variety or subtype.

POMGRANATE (*Punica granatum*): a Mediterranean fruit tree, 2n = 2x = 16 or 18.

POMPE's DISEASE: see glycogen storage diseases

PONGIDAE (anthropoid primates [hominoidea]: *Gorilla gorilla gorilla* 2n = 48; *Hyalobates concolor s* [gibbon] 2n = 52; *Hylobates lar* [gibbon] 2n = 44; *Pan paniscus* [pygmy chimpanzee] 2n = 48; *Pan troglodytes* [chimpanzee] 2n = 48; *Pongo pygmaeus* [orangutan] 2n = 48; *Symphalangus brachytanites* 2n = 50. (See primates)

PO-PS COPOLYMERS: phosphorothioate-phosphodiester copolymers are used for antisense technologies. (See antisense RNA)

POP': symbolizes the ends of the temperate transducing phage genome integrating into the bacterial host chromosome. The corresponding bacterial integration sites are BOB' and after integration (recombination) the sequence becomes: BOP' and POB', espectively (See attachment sites)

Pop1p: a protein component of ribonuclease P and MRP. (See ribonuclease P, MRP)

POPLAR (*Populus* spp): includes also cottonwood trees, 2n = 2x = 38.

POPPY (*Papaver somniferum*): its latex is a source of opium. The plant is grown for its oil-rich seed as a food and also for pharmaceutical purposes. Basic chromosome number x = 11; diploid and tetraploid forms are known.

POPULATION: a collection of individuals that may interbreed and freely trade genes (Mendelian population, deme) or it may be a closed population that is sexually isolated from other groups that share the same habitat. (See Hardy - Weinberg theorem, population equilibrium)

POPULATION CRITICAL SIZE: see critical population size

POPULATION DENSITY: the number of cells or individuals per volume or area.

POPULATION EFFECTIVE SIZE (N_e): the number of individuals in a group or within a defined area that actually transmit genes to the following reproductive cycles (offspring). Each breeding individual has 0.5 chance to contribute an allele to the next generation, and $0.5 \times 0.5 = 0.25$ is the probability to contribute two alleles. The probability that the same male contributes two alleles is $(1/N_m)0.25$ and for the same female it is $(1/N_f)0.25$ where N_m and N_f are the number of breeding males and females, respectively. The probability that any two alleles are derived from the same individual is $0.25N_m + 0.25N_f = 1/N_e$ and $N_e = 4N_mN_f/(N_m+N_f)$. (See founder principle, genetic drift, inbreeding and population size)

POPULATION EQUILIBRIUM: see Hardy - Weinberg theorem

POPULATION GENETICS: studies the factors involved in the fate of alleles in potentially interbreeding groups. The individuals within these groups (demes) may actually reproduce by random mating or selfing or by other within this range. Population genetics can be entirely theoretical and developing mathematical formulas for predicting the allelic frequencies and the effect of various factors that affect these frequencies and the historical paths of the genes and factors as they emerge, become established or disappear, form equilibria or remain unstable during microevolutionary periods. Experimental population genetics conducts biological studies in the sense of the theoretical framework. Population genetics thus deals with the consequences of mutation, migration, selection and breeding systems and is also one of the most important approaches to experimental (micro) evolution. It provides also the theory for many human genetics, animal and plant breeding research efforts.

POPULATION GROWTH, HUMAN: see age-specific birth and death rates, human population growth

POPULATION STRUCTURE: is endemic by subpopulation groups. The dispersal of the subdivisions reflect adaptive genetic differences, gene flow and natural selection pressure, and sometimes genetic drift. (See population genetics, endemic, natural selection, genetic drift)

POPULATION SUBDIVISIONS: smaller relatively separated breeding groups with restricted gene flow among them. (See gene flow, migration)

POPULATION WAVE: periodic changes in the effective population size. (See population size effective, random drift, founder principle, gene flow)

PORCUPINE MAN (ichthyosis histrix): is a dominant form of hyperkeratosis. (See keratosis, ichthyosis)

PORPHYRIA: is a collective name for a variety of genetic defects involved in heme biosynthesis resulting in under- and/or over-production of metabolites in the porphyrin-heme biosynthetic pathway. These diseases may be controled by recessive or dominant mutations. The affected individuals may be suffering from abdominal pain, psychological problems and photosensitivity. The autosomal dominant acute *intermittent porphyria* (human chromosome 11q23-ter) is caused by a periodic 40-60% reduction in porphobilinogen deaminase enzyme resulting in insufficient supplies of the tetrapyrrole hydroxymethyl bilane that is normally further processed by non-enzymatic way into uroporphyrinogen I. It was speculated that the famous Dutch paintter van Gogh was a victim of this rare disease. Prevalence is in the range of 10^{-4} to 10^{-5}. The periodic attacks are generally elicited by exogenous effects such as barbiturate, sulfonamide, alkylating and many other drugs, alcohol consumption, poor diet, various infections and hormonal changes. An *adult type* of (hepatocutaneous) porphyria, controlled by another human gene locus (1p34), involves light-sensitivity and liver damage by the accumulation of,

Porphyria continued

porphyrins caused by uroporphyrinogen decarboxylase deficiency. The general effect may be less severe than in the intermittent porphyria. The rare congenital *erythropoietic porphyria* is the result of a defect in the enzyme uroporphyrinogen III co-synthetase controled by a recessive mutation in human chromosome 10q25.2-q26.3. The laboratory identification is generally based on urine analysis for intermediates in the heme pathway. Porphyrias affect also various mammals. Defects in the porphyrin pathways are involved in several types of pigment deficiency mutations of plants. The *variegate porphyria* is caused by a defect of protoporphyrinogen oxidase (PPO) with symptoms basically similar to that of intermittent porphyria. This dominant disease has low penetrance. Its prevalence is very high (about 3×10^{-3}) in South-African populations of Dutch descent; it apparently represents founder effect. The mental problems of King George III of England was also attributed to variegate porphyria. (See porphyrin, heme, skin diseases, light-sensitivity diseases, founder effect, coproporphyria)

PORPHYRIN: four special pyrroles joined into a ring and generally with a central metal, like in the hemoglobin with iron or in chlorophylls with magnesium. (See porphyria, coproporphyria, haeme)

PORPHYRINURIA: see porphyria

PORPOISE: *Lagenorhynchus obliquidens*, 2n = 44. (See also dolphins)

PORTABLE DICTIONARY OF THE MOUSE GENOME: data on 12,000 genes and anonymous DNA loci of the mouse, homologs in other mammals, recombinant inbred strains, phenotypes, alleles, PCR primers, references, etc. Can be used on Macintosh, PC in File-Maker, Pro, Excel and text formats. Accessible through INTERNET (WWW, Gopher, FTP), CD-ROM or on floppy disk. Information: R.W. Williams, Center for Neuroscience, University of Tennessee, 875 Monroe Ave., Memphis, TN 36163. Phone: 901-448-7018. Fax: 901-448-7266.e-mail: <rwilliam@nb.utmem.edu>

PORTABLE PROMOTER: an isolated DNA fragment, including a sufficient promoter that can be carried by transformation to other cells, and may function in promoting transcription. (See promoter, transformation, gene fusion)

PORTABLE REGION OF HOMOLOGY: insertion and transposon elements may represent homologous DNA sequences and can recombine there. These recombinations may then generate deletions, cointegrates or insertion, inversions. These events can take place even in RecA- hosts. (See Tn*10*, cointegrate, deletion, inversion, targeting genes)

POSITION EFFECT: change in gene expression by a change in the vicinity of the gene. The new expression may be *stable* or variable (*variegation type position effect*). Stable position effect is observed when promoterless structural genes are introduced by transformation and the trans-gene is expressed with the assistance of a "trapped" promoter that is regulated differently than the gene's natural (original) promoter. Variegated position effect (PEV) is more difficult to interpret by molecular models. It has been assumed that heterochromatin affects the intensity of somatic pairing and the silencing is brought about by variations in somatic association and variations in cross-linking between the homologs by binding proteins. The *trithorax-like* gene of *Drosophila* encodes a GAGA-homology transcription factor that enhances variegation type position effect (PEV). The telomeric isochores have been also implicated in position effect (TPE). In *Drosophila* over 100 genes were found that affect variegation type position effect (PEV). It has been hypothesized that these genes control the packaging of the DNA. Many of the cancers develop after translocations or transpositions, indicating the significance of position effect on the regulation of growth. Transposable elements may also cause position effect. Position effects may be exerted even from long distances (2 Mb) and may make difficult to distinguish the position effect causing gene from mutation within the target gene. Such cases may complicate positional cloning. (See heterochromatin, LCR, Offermann hypothesis, regulation of gene activity, mating type determination in yeast, silencer, cancer, chromosomal rearrangements, chromosome breakage, isochores, transposable elements, epigenesis, paramutation, positional cloning, RPD3)

POSITIONAL CLONING: see chromosome walking, chromosome landing, map-based cloning

POSITIONAL INFORMATION: is provided by and to some cells in a multicellular organism and has important influence on differentiation and development. (See morphogenesis, differentiation)

POSITIONAL SENSING: provides information for specific differentiation functions. (See morphogenesis)

POSITIVE CONTROL: gene expression is enhanced by the presence of a regulatory protein (in contrast to negative control, where its action is reduced). The arabinose operon of *E. coli* is a classic example. The regulator gene *araC* produces a repressor (P_1) in the absence of the substrate arabinose. If arabinose is available P_1 is converted to P_2 (by a conformational change) which is an activator of transcription in the presence of cyclic adenosine monophosphate (cAMP). While the negative control (P_1) is correlated with low demand for expression, the activator (P_2) appears in response to the demand for high level of expression. In general cases the addition of an activator protein to the DNA makes possible normal transcription but adding a special ligand to the system removes the activator and the gene is turned off. (See *arabinose* operon, negative control, *lac* operon, autoregulation, catabolite activator protein, regulation of gene activity)

POSITIVE COOPERATIVITY: binding of ligand to one of the subunits of a protein facilitates the binding of the same to other subunits.

POSITIVE INTERFERENCE: see interference, coincidence

POSITIVE SELECTION OF LYMPHOCYTES: is a process of maturation of these cells into functional members of the immune system. In contrast, the negative selection eliminates, by apoptosis, early lymphocytes carrying autoreactive receptors.

POSITIVE-STRAND VIRUS: see replicase

POSITIVE SUPERCOILING: the overwinding follows the direction of the original coiling, i.e. It takes place right-ward. (See supercoiling)

POSTERIOR: pertaining to the hind part of the body or behind a structure toward the tail end.

POSTERIOR PROBABILITY: see Bayes theorem

POSTMEIOTIC SEGREGATION: takes place when the DNA was a heteroduplex at the end of meiosis. Among the octad spores of ascomycetes this may result in 5:3 and 3:5 or other types of aberrant ratios instead of the 4:4. (See tetrad analysis, gene conversion)

POSTREDUCTION: segregation of the alleles takes place at the second meiotic division. (See tetrad analysis, meiosis, prereduction)

POSTREPLICATIONAL REPAIR: see unscheduled DNA synthesis, DNA repair

PostScript: a computer application to handle text and graphics the same time. The PostScript code determines what the graphics look like when printed although may not be visible on the screen of the monitor.

POST-SEGREGATIONAL KILLING: see plasmid addiction

POSTTRANSCRIPTIONAL PROCESSING: the primary RNA transcript of a gene is cut and spliced before translation or before assembling into ribosomal subunits or functional tRNA; it includes removal of introns, modifying (methylating, etc.) bases, adding CCA to tRNA amino arm, polyadenylation of the 3' tail, etc. (See as an example opiotropin)

POSTTRANSLATIONAL MODIFICATION: enzymatic processing of the product of translation, the newly synthesized polypeptide chain. This may include proteolytic cleavage, glycosylation, conformational changes, assembly into quaternary structure, etc. (See protein synthesis, protein structure)

POSTZYGOTIC: see prezygotic

POTATO (*Solanum tuberosum*): it has 170 to 300 related species with basic a chromosome number $x = 12$. In nature, species with diploid, tetraploid and hexaploid chromosome numbers are found. The cultivated potatoes originated from *Solanum andigena* in Central America where they produce tubers under short-day conditions. The majority of the modern varieties are day-

Potato continued

neutral and develop tubers under long-day conditions. The cultivated potatoes are usually cross-pollinating species but many set seeds also by selfing. Generally the seed progeny is very heterogeneous genetically. Potatoes are rarely propagated by seed as a vegetable. The diploid relatives are usually self-incompatible whereas the polyploids may set seeds by themselves. Among the cultivated groups tuber color may vary from white to yellow to deep purple. Also the chemical composition of the tubers show a wide range, depending on the purpose of the market. Potato besides being a popular vegetable, is an important source of industrial starch. The related species carry genes of agronomic importance (disease, insect resistance, etc.) that have not been fully exploited yet for breeding improved varieties. Application of the molecular techniques of plant breeding seems promising.

POTATO BEETLE (*Leptinotarsa decemlineata*): is one of the most devastating pest of agricultural production of potatoes.

POTATO LEAF ROLL VIRUS: has double-stranded DNA genetic material.

POU: region with several transcriptional activators of 150-160 amino acids (including a homeo domain), involved with genes controling development. The acronym stands for a prolactin transcription factor (PIT), ubiquitous and lymphoid-specific octamer binding protein (OTF) and the *Caenorhabditis* neuronal development factor (Unc-86). (See homeodomain, transcription factor, octa, unc, *Caenorhabditis*, deafness)

pOUT: a strong promoter opposing pIN and directing transcription to the outside end of an insertion element. (See RNA-OUT, pIN)

POX VIRUS: is group of oblong double-stranded DNA viruses of 130-280 kbp. Some of them parasitize insects, others in the family are the chicken pox, cowpox (vaccinia) and smallpox viruses. Their transmission is by insect vectors or by dust or other particles. Engineered pox virus vectors that are not able to multiply in mammalian cells may have the ability to express passenger genes without the risk of disease. Due to the success of vaccination, small pox as a disease has now been eradicated and vaccination against is no longer necessary. Pox virus based vectors are being used orally to protect wild life (red fox) from rabies, for the protection of chickens against the Newcastle virus. Recombinant canarypox virus is employed for the protection of dogs and cats against the distemper, feline leukemia, equine influenza, etc. Highly attenuated derivatives, expressing rabies virus glycoprotein, Japanese encephalitis virus polyprotein or seven antigens of *Plasmodium falciparum* are used for safe and effective vaccination.

pp15: a protein factor required for nuclear import. (See membrane transport, RNA export)

pp125^{FAK}: see CAM

PPAR: peroxisome proliferator-activated receptor is a transcription factor in the adipogenic (fat synthetic) pathways.

pPCV: plasmid plant cloning vector, designation (with additional identification numbers and/or letters) of agrobacterial transformation vectors constructed by Csaba Koncz.

ppGpp: see discriminator region

ppm: parts per million.

PPRAα: see leukotrienes

PRADER-WILLI SYNDROME (Prader-Labhart-Willi syndrome): is a very rare (prevalence 1/25,000) dominant defect involving hypogonadism, obesity, short stature, small hands and feet, mental retardation, etc. by teen age development. The recurrence risk in affected families is about 1/1,000. This, and cytological evidence indicates that the condition is caused in about 60% of the cases by a chromosomal breakage in the long arm of human chromosome 15q11.2-q12. The same deletion when transmitted through the mother results in the Angelman syndrome. Molecular studies indicated in many cases the missing (uniparental disomy) or silencing (imprinting) of paternal DNA sequence in the patients. (See obesity, imprinting, disomic, Angelman syndrome, head/face/brain defects)

PRE-ADAPTIVE: trait or mutation that occurs before selection would favor it but it becomes important when the conditions become favorable for this genotype. (See adaptation)

PREBIOTIC: before life originated. (See evolution prebiotic)

PRECAMBRIAN: see Proterozoic, Cambrian, geological time periods

PRECISE EXCISION: the genetic vector or transposon leaves the target site without structural alterations and the initially disrupted gene or sequence can return to the original (wild type) form.

PREDETERMINATION: the phenotype of the embryo is influenced by the maternal genotypic constitution but the embryo itself does not carry the gene(s) that would be expressed in it at that particular stage. (See delayed inheritance)

PREDICTIVE VALUE: the true estimate of the number of individuals afflicted by a condition on the basis of the tests performed in the population.

PREDICTIVITY: of an assay system is e.g. the percentage of carcinogens correctly identified among carcinogens and non-carcinogens, by indirect carcinogenicity tests (based mainly on mutagenicity. (See accuracy, specificity, sensitivity, bioassays for environmental mutagens)

PREDISPOSITION: susceptibility to disease.

PREFERENTIAL REPAIR: transcriptionally active DNA is repaired preferentially. (See DNA repair)

PREFERENTIAL SEGREGATION: nonrandom distribution of homologous chromosomes toward the pole during anaphase I of meiosis. It may constitute a genetic load if harmful combination of genes (gene blocks) are preferentially included in the gametes. (See meiotic drive, neocentromere)

PREFORMATION: an absurd historical idea supposing that an embryo preexists in the sperm (spermists) or in the egg (ovists) of animals and plants, rather than developing by epigenesis from the fertilized egg. (See epigenesis)

PREGNANCY TEST: there are about 40 pregnancy tests known based on chemical study of blood and urine or other criteria. (See Aschheim-Zondek test)

PREIINITIATION COMPLEX: see PIC

PREIMMUNITY: see host-pathogen relation

PREIMPLANTATION GENETICS: detects genetic anomalies either in the oocyte or in the zygote before implantation takes place. This can be done by molecular and biochemical analyses, and cytogenetic techniques. The status of the egg — in some cases of heterozygosity for a recessive gene — may be determined prior to fertilization by examining the polar bodies. Since the first polar bodies are haploid products of meiosis, if they show the defect, then presumably the egg is free of it. The purpose of this test is to prevent transmission of identifiable familial disorders. (See gametogenesis, *in vitro* fertilization, ART, micromanipulation of the oocyte, polymerase chain reaction, sperm typing)

PRE-mRNA: primary transcript of the genomic DNA, containing exons and introns and other sequences. (See mRNA, RNA processing, introns, hnRNA, posttranscriptional processing, RNA editing)

PRE-MUTATION: a genetic lesion which potentially leads to mutation unless the DNA repair system remedies the defect before it is visually manifested. Pre-mutational lesions lead to delayed mutations. UV irradiation or chemical mutagens with indirect effects (that is the mutagen requires either activation or it induces the formation of mutagenic radicals, peroxides) frequently cause pre-mutations. Incomplete expansion of trinucleotide repeats may also be considered pre-mutational. (See chromosomal mutation, chromosome breakage, pointmutation, telo-mutation, trinucleotide repeats; see also Sherman paradox under mental retardation)

PRENATAL DIAGNOSIS: determines the health status or distinguishes among the possible nature of causes of a problem with a fetus before birth. The results of cytological or biochemical analysis permit the parents to prepare psychologically and medically to the expectations. Although chromosomal abnormalities cannot be remedied, for metabolic disorders (e.g. galactosemia) advance preparations can be made. Similarly fetal erythroblastosis may be prevented. In case of very severe hereditary diseases abortion may be an option if it is morally accept-

Prenatal diagnosis continued

able to the parents. Prenatal diagnosis is now available for more than hundred anomalies. Until recently, prenatal diagnosis required mainly amniocentesis or sampling of chorionic villi, now in some instances the maternal blood can be scanned for fetal blood cells and by the use of the polymerase chain reaction, the DNA of the fetus can be examined. (See amniocentesis, polymerase chain reaction, RFLP, DNA fingerprinting, PUBS, MSAFP, sonography, fetoscopy, echocardiography, hydrocephalus, galactosemias, genetic screening, chorionic villi, preimplantation genetics, ART)

PRENYLATION: is the attachment of a farnesyl alcohol in thioeter linkage with a cystein residue located near the carboxyl terminus of the polypeptide chain. The donor is frequently farnesyl pyrophosphate. Cytosolic proteins are frequently associated with the lipid bilayer of the membrane by prenyl lipid chains or through other fatty acid chains. Prenyl biogenesis begins by enzymatic isomerization of isopentenyl pyrophosphate ($CH_2=C[CH_3]CH_2CH_2OPP$) into dimethylallyl pyrophosphate ($[CH_3]_2C=CHCH_2OPP$). These then react to form geranyl pyrophosphate ($[CH_3]_2C=CHCH_2CH_2C[CH_3]=CHCH_2OPP$). Geranyl pyrophosphate is then converted into farnesyl pyrophosphate:

$$H_3C-\underset{H}{\overset{CH_3}{C}}=C-CH_2-CH_2-\underset{H}{\overset{CH_3}{C}}=C-CH_2-CH_2-\underset{H}{\overset{CH_3}{C}}=C-CH_2-O-\underset{O^-}{\overset{O}{\overset{\|}{P}}}-O-\underset{O^-}{\overset{O}{\overset{\|}{P}}}-O^-$$

2 farnesyl pyrophosphates are then converted to the 30 C squalene in the presence of NADPH

$$H_3C-\underset{H}{\overset{CH_3}{C}}=C-CH_2-\left[CH_2-\underset{H}{\overset{CH_3}{C}}=C-CH_2\right]_2-\left[CH_2-\underset{H}{\overset{CH_3}{C}}=C-CH_2\right]_2-CH_2-\underset{H}{\overset{CH_3}{C}}=C-CH_3 \quad \text{SQUALENE}$$

The $CH_2=\overset{CH_3}{\overset{|}{C}}-CH=CH_2$ units in the above molecules are called <u>isoprene</u> units. Members of the RAS family proteins, involved in signal transduction, cellular regulation, differentiation are prenylated at cystein residues of the COOH-terminus. Prenylation determines the cellular localization of these molecules. Cellular fusions are mediated by prenylated pheromones. The cytoskeletal lamins attaching to the cellular membranes are farnesylated. (See lipids, abscisic acid, lamin, RAS, cytoskeleton, pheromone)

PREPATENT: the period before an effect (e.g. infection) becomes evident.

PREPATTERN FORMATION: the distribution of morphogens precedes the appearance of the visible pattern of particular structures. (See morphogen)

PREPRIMING COMPLEX: a number of proteins at the replication fork of DNA involved in the initiation of DNA synthesis. (See DNA replication, replication fork)

PREPROTEIN: the molecule has not completed yet its differentiation (trimming and processing).

PREREDUCTION: the alleles of a locus separate during the first meiotic anaphase because there was no crossing over between the gene and the centromere. (See tetrad analysis, meiosis, postreduction)

PRE-rRNA: the unprocessed transcripts of ribosomal RNA genes; they are associated at this stage with ribosomal proteins and are methylated at specific sites. The cleavage of the cluster begins at the 5' terminus of the 5.8S unit and proceeds to the 18S and 28S units. (See rRNA, rrn, ribosomal RNA, ribosome)

PRESENCE-ABSENCE HYPOTHESIS: advocated by William Bateson during the first decades of the 20th century as an explanation for mutation. The recessive alleles were thought to be losses whereas the dominant alleles were supposed to be the presence of genetic determi-

Presence-absence hypothesis continued
nants. Similar views, in a modified form, have been maintained for decades and were debated in connection with the nature of induced mutations. (See null mutation, genomic subtraction)

PRESENILINS: proteins associated with precocious senility such the S182 (and like) proteins of the Alzheimer's disease. The presenilins are precursors of the amyloid-β proteins. Presenilin 2 accelerates apoptosis and thus may be the cause of the speedy aging in Alzheimer disease. (See Alzheimer disease, apoptosis)

PRESEQUENCE: is a generic name for signal peptides and transit peptides.

PRESETTING: a penchant of a transposable element to undergo reversible alteration in a new genetic milieu. It may be caused by methylation of the transposase gene. (See Spm, Ac-Ds)

PRESYMPTOTIC DIAGNOSIS: identification of the genetic constitution before the onset of the symptoms. (See prenatal diagnosis, genetic screening)

PRE-tRNA: see tRNA

PREVALENCE: the incidence of a genetic or non-genetic anomaly or disease in a particular human population. The (%) of hereditary diseases caused by presumably single nuclear genes in human populations: autosomal dominant 0.75, autosomal recessive 0.20, X-linked 0.05. Besides these, multifactorial abnormalities account for about 6% of the genetic anomalies. (See also mitochondrial diseases in humans)

PREVENTION OF CIRCULARIZATION OF PLASMIDS: see circularization

PREZYGOTIC: the DNA molecule in the prokaryotic cell before recombination (transduction or transformation); after integration it becomes postzygotic.

PriA: replication priming protein. (See replication fork, DNA replication)

PRIBNOW BOX (TATA box): 5'-TATAATG-3' (or similar) consensus preceding the prokaryotic transcription initiation sites by 5-7 nucleotides in the promoter region; its eukaryotic homolog is the Hogness box. (See Hogness box, open promoter complex)

PRIDE: a living and mating community of animals under the domination of a particular male(s).

PRIMARY CELLS: are taken directly from an organism rather than from a cell culture.

PRIMARY CONSTRICTION: the centromeric region of the eukaryotic chromosome.

PRIMARY NONDISJUNCTION: see nonsdisjunction

PRIMARY RESPONSE GENES: their induction occurs without the synthesis of new protein but require only pre-existing transcriptional modifiers such as hormones. (See signal transduction, secondary response genes)

PRIMARY SEX RATIO: ratio of males:females at conception. (See sex ratio)

PRIMARY SEXUAL CHARACTERS: the female and male gonad, respectively. (See secondary sexual characters)

PRIMARY STRUCTURE: the sequence of amino acid or nucleotide residues in a polymer.

PRIMARY TRANSCRIPT: the RNA transcript of the DNA before processing has been completed. (See processing)

PRIMASE: special RNA synthetase for the initiation of replication of the DNA lagging strand. In prokaryotes the primosome protein complex, in eukaryotes the pol α-associated primase fulfills the function. (See replication fork, DNA replication, PriA)

PRIMATES: the taxonomic group that includes humans, apes, monkeys and lemur. To the higher primates, called also anthropoidea or simians, belong the old world monkeys (cercopitheoidea) such as the Macaca, Cercopythecus, etc., hominoidea (chimpanzee [*Pan*], gorilla [*Gorilla*], orangutan [*Pogo*], and humans, also the now extinct early evolutionary forms. The anthropoidea includes also the new world monkeys (ceboidea). The lower primates or prosimians mean the genera of the lemur, galago, etc. According to data of D.E. Kohne *et al.* (1972 J. Hum Evol. 1:627) on the basis of thermal denaturation of hybridized DNA the number in million years of divergence (and the % of nucleotide difference) of various primates from man was estimated to be: chimpanzee 15 (2.4), gibbon, 30 (5.3), green monkey 46 (9.5), capuchin 65 (15.8), galago 80 (42.0). The taxonomic tree of primates can be outlined as:

Primates continued
PRIMATES: *I. Catarrhini.* IA1 Cercopithecidae (Old World Monkeys). IA1a Cercopithecinae, IA1b Colobinae. IA1c Cercopithecidae. IB. Hominidae (Gorilla, Homo, Pan, Pongo). IC. Hylobatidae (Gibbons). *II. Platyrrhini* (New World Monkeys): IIA. Callitrichidae (Marmoset and Tamarins). IIA1. Callimico. IIA2. Callithrix. IIA3. Cebuella. IIA4. Callicebinae. IIA5. Cebinae. IIA6. Pitheciinae. *III Strepsirhini* (Prosimians) IIIA Cheirogalidae. IIIA1 Cheirogaleus. IIIA2. Microcebus. IIIB. Daubentoniidae (Ayeayes). IIIB1. Daubentonia. IIIC Galagonidae (Galagos). IIIC1. Galago. IIIC2. Otolemur. IIID. Indridae. IIID1 Indri. IIID2. Propithecus (Sifakas). IIIE. Lemuridae (Lemurs). IIIE1. Eulemur. IIIE2. Hapalemur. IIIE3. Lemur. IIIE4. Varecia. IIIF. Loridae (Lorises). IIIF1. Loris. IIIF2. Nycticebus IIIF3. Perodicticus. IIIG Megalapidae. IIIG1. Lepilemur. *IV. Tarsii* (Tarsiers). IVA Tarsiidae (Tarpsiers). IVA1. Tarsius. (See human races, prosimii, Cebidae, Callithricidae, Cercopithecideae, Colobidae, Pongidae, *Homo sapiens*, Hominidae, evolutionary tree)

PRIMER: a short sequence of nucleotides (RNA or DNA) that assists in extending the complementary strand by providing 3'-OH ends for the DNA polymerase to start transcription. (See nested primers, primase, PCR)

PRIMER EXTENSION: an RNA (or single-strand DNA) is hybridized with a single strand DNA primer (30 - 40 base) which is 5'-end-labeled. Generally, the primers are complementary to base sequences within 100 nucleotides from the 5'-end of mRNA to avoid heterogeneous products of the reverse transcriptase which is prone to stop when it encounters tracts of secondary structure. After extension of the primer by reverse transcriptase, the length of resulting cDNA (measured in denaturing polyacrylamide gel electrophoresis) indicates the length of the RNA from the label to its 5'-end. When DNA (rather than RNA) is used as template DNA-DNA hybridization must be prevented. The purpose of the primer extension analysis to estimate the length of 5' ends of RNA transcripts and identify precursors of mRNA and processing intermediates. The cDNA so obtained can be directly sequenced by the Maxam - Gilbert method or also by the chain termination methods of Sanger if dideoxyribonucleoside triphosphates are included in the reaction vessels. (See DNA sequencing, primary transcript, post-transcriptional processing, chimeric proteins, PCR-based mutagenesis)

PRIMER WALKING: a method in DNA sequencing whereby a single piece of DNA is inserted into a large-capacity vector. After a shorter stretch had been sequenced a new primer is generated from the end of what had been already sequenced and the process is continued until sequencing of the entire insert is completed. (See DNA sequencing)

PRIMITIVE STREAK: the earliest visible sign of axial development of the vertebrate embryo when a pale line appears caudally at the embryonic disc as a result of migration of mesodermal cells. (See organizer, differentiation, morphogenesis, Hensen's node)

PRIMORDIUM: embryonic cell group that gives rise to a determined structure.

PRIMOSOME: complex of prepriming and priming proteins involved in replication of the Okazaki fragments and move along with the replication fork in the opposite direction to DNA synthesis. (See DNA replication, replication fork)

PRIMULA (Primrose): is an ornamental plant. *P. kewensis* ($2n = 36$) is an amphidiploid of *P. floribunda* ($2n = 18$) and *P. verticillata* ($2n = 18$).

PRINS (**pr**imed **in** **s**itu synthesis): is an *in situ* hybridization technique bearing some similarities to other methods of probing (e.g. FISH). The PRINS procedure uses small oligonucleotide (18-22 nucleotides) primers from the sequence of concern. After the primer is annealed to denatured DNA (chromosomal or other polynucleotides), a thermostable DNA polymerase is employed to incorporate biotin-dUTP or digoxygenin-dUTP. The procedure is very sensitive to mismatches (because the primer is short) and a mismatch at the 3'-end may prevent chain extension. The concentration of the primer $(C) = Ab_{260}/\varepsilon_{max} \times L$ where Ab_{260} = absorbance at 260 nm, ε_{max} = molar extinction coefficient (M^{-1}) and L = the path length of the cuvette of the spectrophotometer. The molar extinction coefficients are determined ε_{max} = (number of A x 15,200) + (number of T x 8,400) + (number of G x 12,010) + (number of C x 7,050) M^{-1}. (A =

Prins continued

adenine, T= thymine, G = guanine, C = cytosine). PRINS are useful for many purposes, including determination of aneuploidy, DNA synthesis, viral infection, etc. (See PCR, FISH, *in situ* hybridization, LISA, biotinylation, extinction)

PrintAlign: a computer program for graphical interpretation of fragment alignments in physical mapping of DNA. (See physical map)

PrintMap: computer program produces restriction map in PostScript code. (See PostScript)

PRION (PrP^C, PrP^{Sc}): infective glycoprotein particles, responsible for the degenerative brain diseases such as scrapie in sheep, BSE in cattle, kuru, Creutzfeldt-Jakob disease of man and may be also in Alzheimer disease. They appear like virus particles but are free of nucleic acid. It appears that a normal protein is structurally modified, the α helical structure is largely converted into β sheets, leading to the formation of these autonomous disease-causing proteins. In order to develop prion disease in mice, the organism must have PrPc, and if it is absent the animals become resistant to scrapie and show normal neuronal functions. Also microglia (cells that surround the nerves and phagocytize the waste material of the nerv tissue) must be present. If microglia are destroyed by L-leucine-methylester the neurotoxic PrP fragment, containing amino acids 106-126, does not harm the neurons. The transition from the normal PrP^C→ PrP^{Sc} (the scrapie prion) conformation involves changes in amino acid residues 121-231, involved in two antiparallel β-sheets and in three α-helices. It has been hypothesized—on the basis of experimental observations—that the toxicity of this protein is based on increased oxidative stress. The inactivation of the *PrP* gene in mice does not lead to an immediate deleterious condition but by the age of 70 weeks an extensive loss of the Purkinje cells (large neurons in the cerebellar cortex) takes place and the animals have problems with movement coordination (ataxia). The disrupted *PrP* genes makes them resistant to prions. In budding yeast two non-nuclear elements [*URE3*] and [*PSI*] appear to be the infectious prion forms of the Ure2p protein that is also a regulator of nitrogen catabolism. When Urep was overexpressed in wild type strains, the frequency of occurrence of the [*URE3*] increased 20-200 fold. If the overexpression of Urep was limited only to the amino ends of this protein, the frequency of occurrence of [*URE3*] increased 6,000-times. The carboxyl domain of Urep seemed to carry out nitrogen catabolism whereas the amino end induced the prion formation. Both [*URE3*] and [*PSI*] are the prion causing forms of nuclear genes *URE2* and *SUP35*, respectively. The *URE2* gene is involved in the control of utilization of ureidosuccinate as a nitrogen source, the *SUP35* nuclear gene encodes a subunit (eRF3, eukaryotic release factor) of the yeast translation termination complex. Mutations in both of the nuclear genes involve derepression of nitrogen catabolism that is normally repressed by nitrogen. The propagation of [*URE3*] and [*PSI*] depends on *URE2* and *SUP35* nuclear genes, respectively. *In vitro* the Sup35 protein may show prion-like properties. Structurally neither [*URE3*] nor [*PSI*] are similar to the mammalian PrP protein, indicating that there is more than one way for prions to arise. The infectious form of the normal Prp is also called Prp* and the PrP^{Sc} is designated also PrP^{res} (protease-resistant prion). PrP^C and PrP^{Sc} appear to be conformational isomers. The yeast prion [*PSI*$^+$] can be reversibly removed, "cured" to [*psi*$^-$] 100% in 7-8 generations when exposed to guanine hydrochloride or methanol. These denaturants induce the expression of chaperones, giving further support to the notion that the prion functions are based on conformational changes. Actually, the protein chaperones HSP104 and to a lesser extent HSP70, can affect the expression and transmission of [*PSI*$^+$] and its conversion to [*psi*$^-$]. When the *URE2* and *SUP35* genes or the N-terminal domain of their products are deleted the [*URE3*] and [*PSI*$^+$] elements permanently disappear. These yeast proteins are different from each other and from the prion proteins of higher eukaryotes, except the N-terminal region where homology exists. The vCJD (variant of Creutzfeldt-Jakob disease) prions appear to have either single amino acid differences or differences in glycosylation which may also be the cause or consequence of conformational differences. The differences in electrophoretic mobility of the protease digested prions is expected to shed light to the problems of tracing the transmission of prions from cattle to man or among different animal species. (See Creutzfeldt-Jakob disease, kuru, BSE, fatal familial insomnia, protein structure, curing plasmids, chaperones)

PRIVATE BLOOD GROUPS: is a collective name of various blood groups with low frequencies compared to *public blood* group systems that occur frequently.

PRL: see prolactin

PROACCELERIN: is a labile blood factor (V); its deficiency may lead to parahemophilia and excessive bleeding during menstruation or after surgery or bruising. (See antihemophilic factors)

PROBABILITY: is the statistical measure of chance on a scale between 0 to 1, inclusive. 0 means the lack of chance for an event to occur, 1 indicates a certainty that it will occur, and any values expressed as decimal fractions indicate the intermediate chances. The probability function indicates the value of a frequency predicted from the observations related to the parameter. The *simple probability* reveals the chance of a single event, the *compound probability* is the chance of multiple events. When two events are independent their *joint probability* is the product of their independent probabilities. *Alternate probability* exists in case of sex in dioecious species when an individual is either female or male and no intermediates are considered. One must keep in mind that probability does not absolutely prove or disprove a point, it simply indicates the chance of its occurrence. (See binomial probability, conditional probability, likelihood, maximum likelihood)

PROBAND: person(s) through which a family study of the inheritance of a human trait is initiated (also called propositus if male, or proposita if female). Determining the pattern of inheritance on the basis of families chosen by probands may display an excess of affected individuals relative to Mendelian expectations because of the bias in sampling of the population. (See ascertainment test)

PROBE: labeled nucleic acid fragment used for identifying or locating another segment by hybridization. Similarly, immunoprobes using primariliy monoclonal antibodies can also be employed. (See also synthetic DNA probes, heterologous probe, recombinational probe, immunoprobe, labeling, nick translation, padlock probe)

ProbeMaker: computer program converts DNA sequence files in FASTA format to digital restriction maps used for MapSearch Probes.

PROBOSCIS: tubular snout (nose-like emergence) on the head such as the feeding apparatus of *Drosophila*, trunk of the elephant, snout of tapirs, shrews, etc.

PROCAINE ANESTHETICS: are benzoic acid derivatives with local numbing of nerves or nerve receptors.

PROCAMBIUM: the primary meristem that gives rise to the cambium and the primary vascular tissue of plants. (See cambium, meristem)

PROCAPSID: empty capsid precursor of phage into what the DNA will be packaged. (See development, phage)

PROCARCINOGEN: requires chemical modification to become carcinogenic. (See carcinogen, phorbol esters)

PROCARYOTE: see prokaryote

PROCENTRIOLE: an immature centriole that upon maturing becomes the anchoring site of the spindle fibers, cilia and flagella. (See centriole)

PROCESSED GENES: are obtained by reverse transcriptase from mRNA and therefore are free of all elements (e.g. introns) removed during processing of the primary transcript. (See cDNA, intron, primary transcript)

PROCESSED PSEUDOGENE: is similar to a mRNA, lacks introns and may have polyA tail yet it is non-functional. It is assumed to have been inserted into the genome after faulty reverse transcription of a mRNA. (See reverse transcriptases, cDNA, processed genes)

PROCESSING: the trimming and modifying the primary transcripts of the DNA into functional RNAs or cutting and modifyng polypeptide chains preceding to become enzymes or structural proteins. (See primary transcripts, protein synthesis, post-translational processing)

PROCESSIVITY: defines the number of nucleotides added to the nascent DNA chain before the polymerase is dissociated from the template. The processivity for *E. coli* DNA polymerase I, II and III is 3 - 200, >10,000, >500,000, respectively. (See also error in aminoacylation)

PROCESSOR: a data processing hardware or a computer program (software) that compiles, assembles and translates information in a specific programming language.

PROCHIRAL MOLECULE: an enzyme substrate after attaching to the active site undergoes a structural modification and becomes chiral. (See chirality, active site)

PROCHLORON: see evolution of organelles

PRO-CHROMATIN: is a state of the chromatin that is conducive to transcription.

PROCHROMOSOME: heterochromatic blocks detected during interphase. (See heterochromatin, Barr body, mitosis)

PROCOLLAGEN: precursor of collagen. (See collagen)

PROCONVERTIN: antihemophilic factor VII and it deficiency may lead to excessive bleeding and hypoproconvertinemia. (See hypoproconvertinemia, antihemophilic factors)

PROCTODEUM: an invagination of the embryonal ectoderm where the anus is formed later.

PROCYCLIC: *Trypanosoma* is in the gut of the intermediate host (tsc-tsc fly) and at this stage is not infectious to higher animals. (See metacyclic *Trypanosoma*, *Trypanosoma*)

PRODUCT RATIO METHOD: see F_2 linkage estimation

PRODUCTIVE INFECTION: the virus is not inserted into the eukaryotic chromosome and can propagate independently from the host DNA and can destroy the cell while releasing progeny particles. (See lysis)

PROEMBRYO: the minimally differentiated fertilized egg.

PROFLAVIN: an acridine dye, capable of inducing frameshift mutations. (See acridine dye, frameshift mutation)

PROGENY TEST: a procedure for determining the pattern of inheritance. (See Mendelian segregation, Mendelian laws)

PROGERIA: see aging

PROGESTERONE: see also animal hormones ⟶

PROGRAM: is a set(s) of instructions in computer language (software) that permits the user to carry out specified tasks. In biology the development proceeds according to genetically determined pattern, realized by environmental effects.

PROGRAMMED CELL DEATH: see apoptosis

PROGRESSION: a process involved in oncogenic transformation after the initial mutation of a proto-oncogene an active oncogene. (See cancer, phorbol esters)

PROJECTIN: is a myosin-activated protein kinase. (See myosin)

PROJECTION FORMULA: modeling configurations of groups around chiral centers of molecules. (See chirality)

PROKARYON: same as prokaryote

PROKARYOTE: organism without membrane-enclosed (cell nucleus) genetic material (such as the case in bacteria). (See cell comparisons)

PROLACTIN (PRL): a mitogen, stimulating lactation.

PROLAMELLAR BODY: the crystalline-like, lipid-rich structure in the immature plastids that upon illumination develops into the internal lamellae of the proplastids and into the thylakoids of the chloroplasts. (See chloroplast)

PROLAMINE: see zein, high lysine corn

PROLIFERATING CELL NUCLEAR ANTIGEN: see PCNA

PROLINE BIOSYNTHESIS: proceeds from glutamate through enzymatic steps involving glutamate kinase, glutamate dehydrogenase, and finally Δ'-pyrroline-5-carboxylate is converted to proline by pyrroline carboxylate reductase. In some proteins, e.g. collagen, prolyl-4-hydroxylase generates 4-hydroxyproline from proline. The latter enzyme is coded in human chromosomes 10q21.3-q23.1 (α-subunit) and 17q15 (β-subunit). (See amino acid metabolism,

hyperprolinemia)

PROLOG: a database management and query system in physical mapping of DNA.

PROMASTIGOTE: see *Trypanosoma*

PROMETAPHASE: early metaphase. (See mitosis)

PROMISCUOUS DNA: homologous nucleotid sequences occurring in the various cell organelles (nucleus, mitochondrion, plastid) and they are assumed to owe their origin to ancestral insertions during evolution. (See insertion elements)

PROMISCUOUS PLASMIDS: see plasmids promiscuous

PROMITOCHONDRIA: organelles in anaerobically grown (yeast) cells that can differentiate into mitochondria in the presence of oxygen. (See mitochondria)

PROMOTER: site of binding of the transcriptase enzyme (RNA polymerase), transcription factor complexes and regulatory elements, including also the ribosome-binding untranslated sequences. The promoter is usually (basal promoter) situated in front of the genes although pol III may rely on both upstream and downstream promoters. The promoters of the 5S and tRNA genes are internal. The arrangement of the promoter used by pol II is outlined below:

enhancer - PROMOTER - leader - exons - introns - termination signal - polyadenylation signal - downstream regulators

TRANSCRIPTION FACTOR BINDING SITES, DNase HYPERSENSITIVE SITE, TATA BOX, TRANSCRIPTION START

Pol II enzymes generally have a TATA box both in prokaryotes and eukaryotes. The TATA box ca. 25 bp upstream from the initiation point of transcription is usually surrounded by GC-rich tracts. Near the transcription initiation site (-3 to +5) there is an initiator (In) with an average type of sequence: $(Pyrimidine)_2 CA(Pyrimidine)_5$. Some large eukaryotic genes utilize more than one promoter and the transcripts may vary. Some housekeeping and the RAS genes do not use TATA box. DNA-dependent RNA polymerase I synthesizes ribosomal RNAs; it has a core sequence adjacent to the transcription initiation site and upstream regulator binding sites (UCE). Pol III promoters facilitating the transcription of tRNA usually have split promoters with an A-box and a B-box about 40 base apart and situated inside the transcription unit 20 and 60 base downstream from the transcription initiation site. The pol III promoter of some U RNAs has, however a TATA box 30-60 base upstream from the transcription initiation site and further upstream a proximal sequence element (PSE) near the TATA box. Promoters (→) may be of different types and some genes may rely on multiple promoters:

TANDEM   OVERLAPPING   DIVERGENT PROMOTERS

(arrows symbolize promoters, boxes stand for structural genes)

The Promoter Scan II program identifies pol II promoters in genomic sequences and is available through Internet: <http://biosci.umn.edu/software/proscan/promoterscan.htm>. The Signal Scan can be used to find transcription factor binding sites by using TFD, TRANSFAC or IMD databases. (See also basal promoter, portable promoter, divergent dual promoter, divergent transcription, transcription complex, transcription factors, open promoter complex, closed promoter complex *Lac* operon, *Tryptophan* operon, *Arabinose* operon, pol I, pol II, pol III, regulation of gene activity, promoter clearance, promoter trapping, TATA box, TBP, TAF)

PROMOTER BUBBLE: see promoter clearance

PROMOTER CLEARANCE: the opened promoter (promoter bubble) starts moving forward as the first ribonucleotides are transcribed. (See also replication bubble)

PROMOTER SWAPPING: an exchange of promoter by e.g. reciprocal chromosome translocation. (See translocation, pleiomorphic adenoma)

PROMOTER TRAPPING: see trapping promoters, transcriptional gene fusion vectors, translational gene fusion vectors, gene fusion, promoter.

PROMUTAGEN: requires chemical modification (activation) to become a mutagen. (See mutagen, activation of mutagens)

PRONASE: a powerful general (non-specific) proteolytic enzyme isolated from *Streptomyces*.

PRONUCLEUS: the male and female gametic nucleus to be involved in the sexual union.

PROOFREADING: bacterial DNA polymerase I (and analogous eukaryotic enzymes) can recognize replicational errors and removes the inappropriate bases by its editing 3' - 5' exonuclease function. In bacteria proofreading is performed also by the *dnaQ* gene encoding the ε subunit (an exonuclease) of the DNA polymerase III holoenzyme. The base selection is carried out by the product of gene *dnaE*. Mismatches are repaired by the enzymes MutH, MutL and MutS and the corresponding homologs in higher organisms. The fidelity of replication due to the combined action of the sequentially acting bacterial genes was estimated to be in the range of 10^{-10} per base per replication. During the process of translation the EF-Tu•GTP → EF-Tu•GDP change releases a molecule of inorganic phosphate (P_i) and allows a time window to dissociate the wrong tRNA from the ribosome. A similar correction is made also by the aminoacyl synthetase enzyme by virtue of its active site, specialized for this function. (See DNA polymerase I, DNA polymerase III, DNA repair, error in replication, error in aminoacylation, ambiguity in translation, protein synthesis, DNA repair)

PROPELLER TWIST IN DNA: the surface angle formed between individual base-planes viewed along the C^6—C^8 line of a base pair.

PROPERDIN (Factor P): is a serum protein of 3-4 subunits (each ca. 56 kDa, encoded in human chromosome 6p21.3). It is an activator of the complement of the natural immunity system by stabilizing the convertase. (See convertase, complement, complement, immune system)

PROPHAGE: proviral phage is in an integrated state in the host cellular DNA and it is replicated in synchrony with the host chromosomal DNA until it is induced and thus becomes a vegatative virus. (See prophage induction, temperate phage, lysogeny, lambda phage)

PROPHAGE INDUCTION: treating the bacterial cells by physical or chemical agents that cause the moving of the phage into a vegetative life style resulting in asynchronous replication from the host and eventually lysis and liberation of phage. (See prophage, lysogeny)

PROPHAGE-MEDIATED CONVERSION: the integrated prophage causes genetic changes in the host bacterium, and it is expressed as altered antigenic property, etc.

PROPHASE: see meiosis, mitosis

PROPIONICACIDEMIA: see glycinemia ketonic

PROPIONYL-CoA-CARBOXYLASE DEFICIENCY: see glycinemia ketonic

PROPLASTID: young colorless plastid without the fully differentiated internal membrane structures; it may differentiate into chloroplast. (See etioplast, chloroplasts)

PROPORTIONAL COUNTERS: are used for measuring radiation-induced ionizations within a chamber. The voltage changes within are proportional to the energy released. It may be used for measuring neutron and α radiations with an efficiency of 35-50%. The equipment must be calibrated to the radiation source. (See radiation measurement, radiation hazard assessment)

PROPOSITUS (PROPOSITA): see proband

PROSIMII (prosimians): suborder of lower primates, including Galago, Lemur, Tarsius, Tupaia. Lorisidae. (See primates, Lemur, Tupaia, Lorisidae)

PROSOME: small ribonucleoprotein body.

PROST (pronuclear state embryo transfer): basically very similar to intrafallopian transfer of zygotes but the zygote is at a very early stage. (See intrafallopian transfer, ART)

PROSTAGLANDINS: long-chain fatty acids in different mammalian tissues with hormone-like muscle-regulating and inflammation-regulating functions; they exist in several forms. Prostaglandin synthesis is regulated by cyclooxigenases. (See animal hormones, cyclooxygenases)

PROSTATE CANCER: about 9-10% of USA males eventually develop this malignancy. The autosomal dominant gene has a high penetrance: about 88% of the carriers become afflicted by the age of 85. The high level of testosterone may increase the chances for this cancer. A metastasis suppressor gene, KAI1, in human chromosome 11p11.2 has been identified. The KaI1 protein appears to contain 267 amino acids with four transmembrane hydrophobic and one large hydrophilic domains. This glycoprotein is expressed in several human tissues and also in rats. A negative regulator of the MYC oncogene, MXI1 (encoded in human chromosome 10q24-q25) is frequently lost in prostate cancer. In chromosome 10pter-q11 region a prostate cancer suppressor gene, causing apoptosis of carcinoma, has been detected from loss of heterozygosity mutations (LOH). A major susceptibility locus was identified in human chromosome 1q24-q25. Candidate genes are expected in human chromosomes 3p, 7q, 8p and 8q, 9q, 10p, 13q, 16q, 17p, 18q and Y. (See PSA, cancer, tumor suppressor gene, MYC)

PROSTATES (prostata): gland in the animal male surrounding the base of the bladder and the urethra; upon ejaculation injects its content (acid phosphatase, citric acid, proteolytic enzymes, etc.) into the seminal fluid. (PSA, prostate cancer)

PROSTHESIS: any type of mechanical replacement of a body part, such as artificial limbs, false teeth, etc.

PROSTHETIC GROUP: a non-peptide group (iron or other inorganic or organic group) covalently bound (conjugated) to protein to assure activity.

PROT: Na^+/Cl^--dependent proline transporter that transports also glycine, GABA, betaine, taurine, creatine, norepinephrine, dopamine and serotonin in the brain. (See transporters)

PROTAMINE: a basic protein occurring in the sperm substituting for histones. (See histones)

PROTANDRY: in monoecious plants the pollen is shed before the stigma is receptive. (See monoecious, stigma, protogyny, self-sterility)

PROTANOPE: see color blindness

PROTEASE (proteinase): enzyme which hydrolyzes proteins at specific peptide bonds; for *protease 3* see antimicrobial peptides.

PROTEASOMES: ATP-dependent ubiqitinated proteins process intracellular antigens into short peptides that are then transported to the endoplasmic reticulum with the aid of TAP, and are responsible for MHC class I-restricted antigen presentation. Proteasomal polymorphism is determined, among others, by LMP2 and LMP7 genes encoded within the MHC class II region in the vicinity of TAPs that are upregulated by interferon γ. The 26S, (2000, kDa) proteasomes are hollow cylinders containing ubiquitins and degrade proteins with lysosomal proteases. The 20S (700 kDa) proteasomes contain multiple peptidases. Their active site is at the hydroxyl group of the N-terminal threonine in the β subunit. The PA proteins are **p**roteasome **a**ctivators. The 26S proteasome is associated with at least 18 ancillary and essential proteins (PSM proteins) and many of these are now genetically mapped to different human chromosomes. Chymostatin, calpain and leupeptin, etc. are inhibitors. Proteasomes have also ubiquitin-independent function such as the degradation of the excess amounts of ornithine decarboxylase, a key enzyme in polyamine biosynthesis. The proteasomes have important—although not fully understood—roles in differentiation and development by mediating protein turnover. (See ubiquitin, TAP, antigen presenting cell, MHC, antigen processing, immune system, polyamine)

PROTEGRIN: see antimicrobial peptides

PROTEIN: a large molecule (polymer) composed of one or more identical or different peptide chains. The distinction between protein and polypetide is somewhat uncertain; generally a protein has more amino acid residues (50-60) and therefore can fold. (See protein synthesis, protein structure, amino acid sequencing)

PROTEIN 14-3-3: a family of proteins named after their electrophoretic mobility, they occur in many forms in different organisms and have roles in signal transduction, exocytosis and regulation of the cell cycle, and oncogenes.

PROTEIN A: is isolated from *Staphylococcus aureus*; it binds the Fc domain of immunoglobulins without interacting with the antigen-binding site. It is used both in soluble and insoluble form for the purification of antibodies, antigens and immune complexes. (See antibody, immunoglobulins)

PROTEIN C: a vitamin K-dependent serine protease which selectively degrades antihemophilic factors VA and VIII:C, and it is thus an anticoagulant. (See protein C deficiency, antihemophilic factors, thrombin)

PROTEIN C DEFICIENCY (thrombotic disease): a human chromosome 2q13-q14 dominant and it may be a life-threatening cause of thrombosis. (See thrombosis, protein C)

PROTEIN CLOCK: see evolutionary clock

PROTEIN CONDUCTING CHANNEL: are membrane passageways for proteins that interact with the membrane protein and lipid components, (See protein targeting, SRP, translocon, translocase, TRAM, ABC transporters)

PROTEIN CONFORMATION: see conformation

PROTEIN DOMAINS: are generally formed by folding of 50-350 amino acid sequences for carrying out particular function(s). Small proteins may have only a single domain but larger complexes may have multiple modular units. The alternations of α helices and β sheets constitute a characteristic *motif*. The compact motifs are generally covered by polypeptide loops. Domain similarities among proteins from different organisms indicate possible functional relationship (homology) of those proteins. (See protein structure-β sheets, α helices, helix-turn-helix, helix-loop-helix, zinc finger, binding proteins, motif)

PROTEIN ENGINEERING: constructing proteins with amino acid replacements at particular domains and positions to explore their effect on function. (See directed mutation)

PROTEIN INTRON: see intein

PROTEIN ISOFORMS: closely related polypeptide chain family, encoded by a set of exons which share structurally identical or almost identical subset of exons. (See family of genes)

PROTEIN KINASE: phosphorylates one or more amino acids (frequently threonine, serine, tyrosine) at certain positions in a protein, and thus two negative charges are conveyed to these sites altering the conformation of the protein. This alteration then involves a change in the ligand-binding properties. The catalytic domain of this large family of enzymes is usually 250 amino acids. The amino acids outside the catalytic domains may vary substantially and specify the recognition abilities of the different kinases and serve in responding to regulatory signals. During the last 3-4 decades hundreds of protein kinases have been discovered that can be classified into serine/threonine (TGF-β, transforming growth factor), tyrosine (EGF [epidermal growth factor receptor], PDGF [platelet-derived growth factor receptor] protein kinases, SRC [Rous sarcoma oncogene product], Raf [product of the Moloney and MYC oncogenes], MAP kinase, cell cyclin-dependent kinase (Cdk), cell division cycle (Cdc), cyclic-AMP- and cyclic-GMP-dependent kinases, myosin light chain kinase, Ca^{2+}/calmodulin dependent kinases, etc. The consensus sequences for a few protein kinases are shown below:

Protein kinase A   (?)-Arg- (Arg/Lys)-(?)-(Ser/Thr)-(?)
Protein kinase G   (?)-{[Arg/Lys] 2x or 3x}-(?)-(Ser/Thr)-(?)
Protein kinase C   (?)-([Arg/Lys] 1-3x)-([?] 0-2x)-(Ser/Thr)-([?]0-2x)-(Se/[Thr]1-3x)-(?)
Ca^{++}/calmodulin kinase II   (?)-arg-(?)-(?)-(?)-(Ser-Thr)-(?)
Insulin receptor kinase   Thr-Arg-Asp-Ile-Tyr-Glu-Thr-Asp--Tyr-Tyr -Arg-Thr
EGF receptor kinase   Thr-Ala-Glu-Asn-Ala-Glu-Tyr-Leu-Arg-Val-Arg-Pro

(?) indicates any amino acid, the numbers after the amino acid with an "x" indicate how many times it may occur (See cAMP-dependent protein kinase, epinephrine, phosphorylase b kinase, signal transduction, obesity)

PROTEIN-LIKELIHOOD METHOD: is used to determine evolutionary distance when the organisms are not closely related and the non-synonymous base substitutions are higher than the synonymous ones. In such cases the protein method may provide more reliable information. (See evolutionary distance, evolutionary tree, least square methods, four-cluster analysis, unrooted evolutionary trees, transformed distance, Fitch-Margoliash test, DNA likelihood method)

PROTEIN PHOSPHATASES: remove phosphates from proteins. They include enzymes that reverse the action of protein kinases and have important role, together with the kinases, in signal transduction. (See protein kinases)

PROTEIN REPAIR: can be managed with assistance of chaperones. If the refolding is not feasible, proteolytic enzymes destroy them either directly or by the mediation of ubiquitins. Nascent polypetides, transcribed from truncated mRNAs without a stop codons acquire a C-terminal oligopeptide (Ala, Ala, Asn, Asp, Glu, Asn, Tyr, Ala, Leu, Ala, Ala or a variant), encoded by an *ssrA* transcript. The ssrA is a 362-nucleotide tRNA-like molecule that can be charged with alanine. The addition of the peptide tag takes place on the ribosome by cotranslational switching from the truncated mRNA to the ssrA RNA. The polypeptide chain so tagged is degraded in the *E. coli* cytoplasm or periplasm by carboxyl-terminal-specific proteases. (See amino acids, chaperone, ubiquitin, periplasm, protease, DNA repair)

PROTEIN-RNA RECOGNITION: almost all RNA functions involve RNA-protein interactions such as regulation of transcription, translation, processing, turnover, viral transactivation and gene regulatory proteins in general, tRNA aminoacylation, ribosomal proteins, transcription complexes, etc.

PROTEIN S (PROS): a human chromosome 3p11 vitamin K-dependent plasma proteins preventing blood coagulation and it is a cofactor for Protein C. Their deficiency results in thrombosis. (See protein C, thrombosis)

PROTEIN SEQUENCING: see amino acid sequencing

PROTEIN STRUCTURE: the *primary structure* means the sequence of the amino acids. The

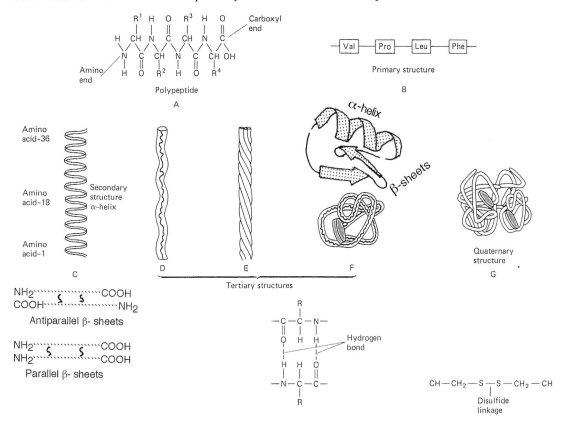

secondary structure is formed by the three-dimensional arrangement of the polypeptide chain such as the α helix (see diagram C) or the β conformation when hydrogen bonds are formed

Protein structure continued

between pleated sequences of the same chain. The latter type of conformation is frequently found in internal regions of enzyme proteins and in structural protein elements such as silk fibers and collagen. In diagramatic representation of protein molecules the α helixes are commonly represented by cylinders whereas the β sheets by ribbons frequently with arrows at their end. The *tertiary structure* involves a folding either to a globular (see F) or various types of rope kind structure (D, E) of the polypeptide chain that has already secondary structure (C). Mature proteins may be formed also from multiple identical or different polypeptide subunits and this type of association is the *quaternary structure* (G). For specialized biological functions more than one protein may be joined in *supramolecular complexes* such as the myosin and actin in the muscles, histones and various non-histone proteins in chromatin. The primary structure is genetically determined and all additional structural changes flow from this primary structure although trimming, processing, and association with prosthetic groups may be involved. Proteins with greater than 30% sequence homology assume generally the same basic structures. Protein structure information can be obtained through World Wide Web: < http://biotech.embl-heidelberg.de8400/> or <http://biotech.embl-ebi.ac.uk:8400/> or <http://pdb.bnl.gov:8400/> <http://www.ncbi.nlm.nih.gov/Structure>. (See MOLSCRIPT, protein synthesis, protein domains, molecular modeling, databases)

PROTEIN SYNTHESIS: has many basic requisites and a large number of essential regulatory elements. It intertwines with all cellular functions. The blueprint for protein synthesis in the vast majority of organisms (DNA viruses, prokaryotes and eukaryotes) is in the nucleotide sequences of the DNA code. In RNA viruses the genetic code is in RNA. The viruses do not have, however, their own machinery for the actual synthesis of protein, rather they exploit the host cell for this task. The genetic code specifies individual amino acids by nucleotide triplets, using one or several synonyms for each of the 20 natural amino acids. The triplet codons are in a linear sequence of the nucleic acid genes. In the organisms with DNA as the genetic material, the process of transcription produces a complementary RNA sequence from one or both strands of the antiparallel ($\rightleftarrows$) strands of the DNA. The double-strands unwind and the RNA polymerase(s) synthesize(s) a comple-

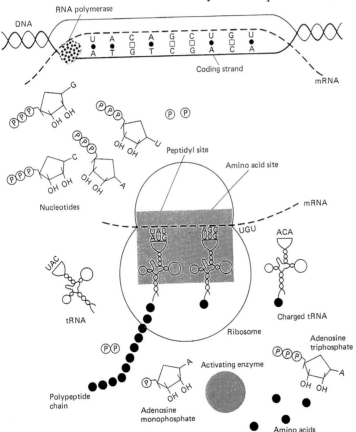

▲ AN OVERVIEW OF THE PROTEIN SYNTHESIZING MACHINERY ▲

plementary RNA copy of the *sense sequence* in the DNA. In the single stranded DNA and RNA viruses this DNA or RNA serves the purposes both of being the genetic material and also

Protein synthesis continued

the transcript for protein synthesis. In cellular organisms three main classes of RNAs are made, messenger RNA (mRNA), transfer RNA (tRNA) and ribosomal RNA (rRNA) and all three are indispensable for protein synthesis. In addition to these RNAs, a large number of proteins are required for the transcription process (transcription factors), for the organization of the ribosomes (50-80 ribosomal proteins), for the termination of transcription, for the activation of the tRNAs, etc. A broad overview (without details) are shown in the figure on the preceding page. ▼ A GENERAL OUTLINE OF GENETIC TRANSLATION ON THE RIBOSOMES ▼

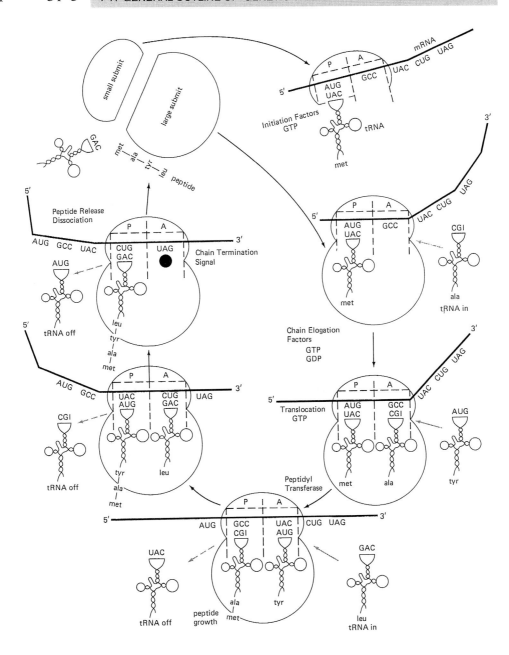

Protein synthesis continued

Some of the details of the transcriptional process are different in prokaryotes from that in eukaryotes. In the latter group, one DNA-dependent RNA polymerase is responsible for the synthesis of all RNAs. In eukaryotes pol I synthesizes rRNAs with the exception of the 5S and 7S rRNA, pol II transcribes mRNA and the small nuclear RNAs (snRNA) and pol III synthesizes tRNAs and 5S and 7S rRNA. The primary RNA transcripts must be processed to functional size molecules in all categories that may require splicing and other post-transcriptional modifications (capping, formylation, etc.). In prokaryotes the process of transcription and translation are *coupled*, i.e. as soon as the chain of mRNA unwinds from the DNA it is associated with the ribosomes and protein synthesis begins. See electronmicrograph on the left.

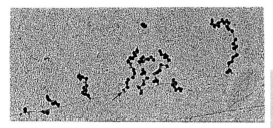

TRANSCRIPTION AND TRANSLATION COUPLED IN *E. coli*. THE THIN THREAD IS THE DNA, THE DARK ROUND STRUCTURES ARE POLYSOMES. THE TRANSCRIPTASE ATTACHMENT → IS INDICATED. (Courtesy of Drs. Miller, O.L. and Hamkalo, B.A. From Science 169:392 by permission)

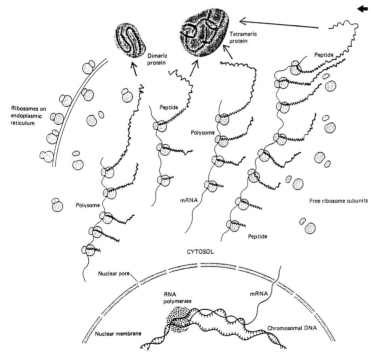

▲ OVERVIEW OF TRANSLATION IN EUKARYOTES ▲

← In eukaryotes when the mRNA is released from its DNA template it moves into the cytosol where protein synthesis takes place. The fate of the mRNA can be monitored by electronmicroscopy in both groups and these pictures show the elongation of RNA and protein strands. The first products of both display long strands and the short ones indicate the stage and place where they were started. The ribosomes are captured by the mRNA and form an association of multiple units in the form called *polysomes*. The prokaryotic mRNA is directed to the proper position in the 30S ribosomal subunit by the Shine-Dalgarno nucleotide sequence within a 8 to 13 base area upstream from the initiation codon. In eukaryotes such a sequence does not exist and the mRNA is simply scanned by the ribosome until the first methionine codon is found. These ribosomal units then slide from the 5'-end of the mRNA toward the 3'-end and thus the amino end of the polypeptide chain corresponds to the 5'-end of the mRNA. The ribosomes in both prokaryotes and eukaryotes are composed of a small and a large subunit. The size of these units is somewhat different in the two major taxonomic categories. The small and large subunits of the ribosomes jointly form two compartments, the so called P (peptidyl-tRNA binding site) and the A (aminoacyl-tRNA binding site). The ribosomes actually do not look like shown in these diagrams, because they are three-dimensional and have more elab-

Protein synthesis continued

orate structure. Before protein synthesis (translation) begins and the primary structure of the mRNA is translated from the nucleotide triplet codon words into the singular amino acid word language of the protein, the tRNA molecules must be charged with amino acids. This process is called also activation of tRNA. (See aminoacyl-tRNA synthetase)

The amino-acid-charged methione-tRNA (tRNA^{Met}) in eukaryotes and the formylated tRNA^{fMet} in prokaryotes seek out the cognate codon in the mRNA at the P site of the ribosome through the complementary anticodon. This event requires the presence of protein initiation factor(s) and GTP as energy source. The GTP is cleaved to GDP + inorganic phosphate (P^i) and thus liberates some of the needed energy. The elongation factor proteins and GTP and GDP complexes also police the system to prevent the wrong charged tRNA to go to an A site (proofreading function). Actually a similar correction mechanism is carried out earlier in the process by one of the active sites of the aminoacyl synthetase (activating) enzyme that usually dissociates the amino acid—tRNA link in case of a misalliance. With the double checks available, misincorporation of amino acids is approximately in the 10^{-4} range. Protein synthesis in the mitochondria and chloroplasts is essentially patterned after the prokaryotic systems.

The 5' base of the anticodon triplet may not be the exact and conventional base yet it functions normally (see wobble). The two subunits of the ribosomes are combined and the second charged tRNA can land now at the A ribosomal site. The carboxyl end of the methionine forms a peptide bond with the amino terminus of the next incoming amino acid at the A site. This process is mediated by the enzyme peptidyl transferase. For this transferase function not a protein but a 23S rRNA in the large subunit (a ribozyme) is responsible. Again energy donors and elongation protein factors are cooperating in the process of peptide chain growth (see initiation and elongation factors IF and eIF). When each peptide bond is completed the tRNA is released and recycled for another tour of duty. The *open reading frame* of the gene is terminated by a nonsense or chain-termination codon. When the ribosome slides to this point the mRNA is released from the ribosomes with the assistance of release factors (see transcription termination in eukaryotes, transcription termination in prokaryotes). Protein synthesis proceeds at a rather rapid rate; it has been estimated that in *E. coli* 50-200 amino acids may be incorporated into peptides in 5-10 seconds.

The dimeric NAC (nascent-polypeptide associated complex) interacts with the emerging polypeptide chains before 30 or fewer residue long chain is formed and protects the nascent chain from becoming associated with other cytosolic proteins until the signal peptide fully emerges and then the signal recognition particle (SRP) crosslinks to the polypeptide. The purpose of the NAC is to assure that the polypeptide would be oriented to the proper SRP and the endoplasmic reticulum. Alternatively if the protein does not carry a signal peptide, the nascent chain may be folded by chaperones such as heatshock proteins Hsp40 Hsp70 and TRiC. The completed amino acid sequences, the polypeptides, must be then converted to biologically active forms. This post-translational process may involve trimming (removal of some amino acids), proteolytic cleavage, folding to a tertiary structure, aggregation of different polypetide chains to form the quaternary structure, addition of prosthetic groups (such as heme, lipids, metals), and other non-amino-acid residues such as acyl, phosphate, methyl, isoprenyl and sugar groups. (See also code genetic, mRNA, tRNA, rRNA, ribosomes, aminoacyl-tRNA synthetase, wobble, cap, Shine-Dalgarno sequence, RNA polymerases, transcription factor, transcription initiation, elongation initiation factors, transcription termination, rho factor, transcription complex, signal sequences, transit peptide, signal peptides, regulation of gene activity, antibiotics, toxins, ambiguity in translation, chaperone, SRP, signaling to translation, translation initiation, introns, prenylation, TRiC, heatshock, E site)

PROTEIN SYNTHESIS INHIBITORS: see antibiotics, toxins, interferons

PROTEIN TARGETING: can be co-translational, i.e. newly synthesized proteins are delivered to specific sites (endoplasmic reticulum) in the cell before the chain is completed or post-translational when the transport takes place after the polypeptide is completed. (See signal hypothesis, signal sequence recognition particle, translocon, TRAM, protein conducting channel)

PROTEIN TRUNCATION TEST: may be used to detect the effects of several mutations that do not permit the completion of a polypeptide chain. The gene is transcribed by using polymerase chain reaction and the RNA is translated *in vitro* and the polypetide is analyzed in SDS minigels. (See PCR, SDS-polyacrylamide gel, rabbit reticulocyte *in vitro* translation)

PROTEIN-TYROSINE KINASES (PTK): phosphorylate tyrosine residues in some proteins. This function is frequently coded for by v-oncogenes of retroviruses but cellular oncogenes and other proteins may be involved and are controling signal transduction and other cellular processes such as cell proliferation and differentiation. Cytosolic tyrosine kinases preferntially phosphorylate their own SH2 domains or related SH2 domains with hydrophobic amino acids at key positions, e.g : Ile or Val at -1 and Glu, Gly or Ala at the +1 position. Receptor tyrosine kinases prefer Glu at -1 position. These preferences specify their signaling role. The RET oncogene's receptor tyrosine kinase product can shift substrate specficity and thereby cause multiple endocrine neoplasia. Quercetin, genistein, lavendustin A, erbstatin and herbimycin are all natural products and inhibitors of these enzymes. (See tyrosine kinase, protein kinases, SH2, endocrine neoplasia multiple)

PROTEIN TYROSINE PHOSPHATASE: see tyrosine phosphatase

PROTEINASE K: is a proteolytic enzyme, frequently used to remove nucleases during the extraction of DNA and RNA. With appropriate heat treatment any DNase associated with it can be safely removed. (See protease)

PROTEINOID: a polymerized mixture of amino acids formed during prebiotic stage of evolution (or simulated conditions in the laboratory). (See prebiotic)

PROTEOGLYCAN: heteropolysaccharides with a peptide chain attached through O-glycosidic linkage to a serine or threonine residue. Such molecules are enzymes, animal hormones, structural proteins, cellular lubricants (such as mucin), extracellular matrix proteins and the "antifreeze proteins" of antarctic fishes. (See antifreeze proteins)

PROTEOLYTIC: hydrolyzing peptide bonds of proteins.

PROTEOME: all the cellular proteins encoded by the cellular DNA in the cell. (See genome)

PROTEROZOIC (precambrian): the geological period 5 billion to 570 million years ago. Aquatic forms of living systems appeared during this era. (See geological time periods)

PROTEUS SYNDROME: involves gigantism of parts of the body probably caused by lipomatosis (abnormally large local fat accumulation). The genetic control is unclear.

PROTHALLIUM: the haploid gametophyte generation of ferns.

PROTHROMBIN DEFICIENCY: is caused by autosomal recessive, semidominant defects in the formation of anticoagulation factor VII, Stuart factor, Christmas factor and prothrombin. The human gene for prothrombin was assigned to chromosome sites 11p11-q12. These proteins have similar proteolytic properties and the synthesis of all four depends on the presence of vitamin K. The patients have a tendency of bleeding similarly to hemophiliacs. Hereditary deficiency of factor VII itself is rare but it may be fatal if bleeding affects the central nervous system. Stuart factor deficiency has symptoms similar to those in deficiency of factor VII. All of these conditions can be treated by transfusion with blood plasma. (See antihemophilia factors, hemophilia, vitamin K dependence, coumarin-like drug resistance)

PROTIST: a general term for single cell eukaryotic organisms. The *Monera* including bacteria, blue green algae, viruses are also sometimes called protists although these are prokaryotes.

PROTOCELL: abiotic ancestor of living cells under prebiotic conditions. (See origin of life)

PROTOCHLOROPHYLL: precursor of chlorophyll ($C_{55}H_{70}O_5N_4Mg$); if the magnesium is removed protophaeophytin results.

PROTOGYNY: in monoecious plants the stigma is receptive before the pollen is shed. (See protandry, monoecious, stigma, self-incompatibility)

PROTOMER: a polypeptide subunit of an oligomeric protein encoded by a cistron of a gene. (See cistron, oligomer)

PROTON: the positive nucleus of the hydrogen atom. The proton carries a positive charge equal to the negative charge of an electron but its mass is 1837 times larger.

PROTON ACCEPTOR: an anion capable of accepting protons. (See anion, proton)
PROTON DONOR: an acid
PROTON PUMP: mediates transport or exchange of protons across cellular membranes; energy is supplied usually by ATP or light. (See proton)
PROTONEMA: a filamentous stage in the formation of the gametophyte of mosses.
PROTONOMA: red color insensitive color blindness; an X-chromosomal anomaly. (See color blindness)
PROTO-ONCOGENES: are cellular genes which after genetic alteration(s) may initiate or predispose to cancerous transformation. They generally have their counterparts in oncogenic viruses (v-oncogenes). Also, they may be involved in processes of signal transduction in a variety of organisms in fungi, plants and animals. (See oncogenes, signal transduction, carcinogenesis, tumor suppressors, cell cycle)
PROTOPERITHECIUM: see ascogonia, perithecium
PROTOPLASIA: formation of a new tissue.
PROTOPLASM: the "live" content of a cell.
PROTOPLAST: a cell surrounded by the cell membrane but stripped of the cell wall, generally by a combination of pectin and cellulose digesting enzymes. Protoplasts under appropriate conditions may be regenerated into normal cells and intact plants. The bacterial protoplasts are generally called spheroplasts and may have some parts of the cell wall still attached. (See cellulase, macerozyme, pectinase)
PROTOPLAST FUSION: protoplasts may fuse in the presence of polyethylene glycol (and some other agents). The fusion may take place within sister cells or with the cells (protoplasts) of any taxonomically distant organisms such as mammalian and plant cells. These somatic hybrids,

PLANT PROTOPLAST →
From Durand, J. et al. 1973
Z. Pflanzenphysiol. 69:26

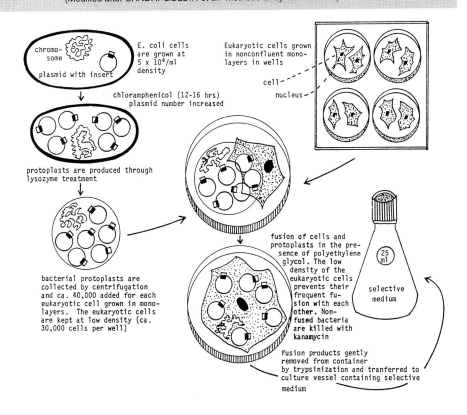

TRANSFORMATION BY PROTOPLAST FUSION BETWEEN BACTERIAL SPHEROPLASTS AND ANIMAL CELLS.
(Modified after SANDRI-GOLDIN et al. Methods Enzymol. 101:402)

Protoplast fusion continued

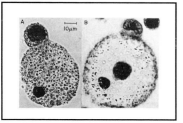

HUMAN HeLa CELLS ATTACHED TO TOBACCO PROTOPLAST (A)., THE HeLa NUCLEUS (larger) INSIDE THE TOBACCO CELL (B). (From Jones, C.W. et al. Science 193:401)

unlike the zygotes derived from the fusion of eggs and sperm, contain all the contents of the two cells, nuclei and cytoplasm, although some cytoplasmic organelles may be lost eventually. In certain rodent-human cell hybrids even the human chromosomes may be eliminated; similar observations are available for carrot and parsley cell hybrids. When the genetic differences between the fused protoplasts is large, the fused cells may not divide or may not divide continuously. Somatic hybrids between related species may, however, behave like allopolyploids and form fertile or sterile hybrids after regeneration. (See also cell fusion, polyethylene glycol)

PROTOPORPHYRIA: autosomal (human chromosome 18q21.3) dominant (or recessive) disease involving light sensitive itching, inflammation of the skin. The porphyrin level of the blood may increase by over 16 fold, to 1 g/100 mL. The excess protoporphyrin is deposited in the liver, causing potentially serious damage. The basic defect probably involves a deficiency of the mitochondrially located ferrochelatase. (See light-sensitivity defects, mitochondrial disease in humans, porphyria)

PROTOPORPHYRIN: the organic part of heme consisting of four pyrroles joined by methylene bridges. (See haeme)

PROTOTROPH: a genotype that has wild type nutritional requirement. (See also autotroph, auxotroph)

PROTOZOA: unicellular animals, mainly free-living (such as the *Paramecia*) some are, however, parasitic (such as the *Gaillardias* which frequently contaminate drinking water sources), the *Trypanosomas* and *Leishmanias* which cause potentially lethal infections in animals and humans. (See *Trypanosoma, Leischmania*)

PROVENANCE: the origin of a genetic stock.

PROVIRUS: a DNA sequence in the eukaryotic chromosomal DNA that is a reverse transcriptase product of a retroviral RNA. (See retroviruses, reverse transcription)

PROXIMAL: situated in the vicinity of a reference point; e.g. a gene near the centromere is proximal, versus another that is in the direction of the telomere, and thus called distal. In conjugational tranfer of bacteria the marker that is tranfered before another is the proximal. (See centromere, telomere, conjugation mapping)

PROXIMAL MUTAGEN: a chemical that has been activated into a mutagenic substance; it may not have reached yet its most reactive state. (See promutagen, activation of mutagens, ultimate mutagen, chemical mutagens, activation of mutagens)

Prp20p: is the yeast homolog of RCC1. (See RCC)

PRR: positive regulatory region. (See also negative regulation, *Arabinose* operon)

PRTF: pheromone receptor transcription factors, cooperating with GRM (general regulator mating factor) in the determination of mating type. (See pheromone, mating type determination in yeast, *Schizosaccharomyces pombe*)

PSA (prostate-specific antigen): a M_r 33,000 kallikrein type protease glycoprotein encoded at human chromosome 19q13. High levels of this protein in the serum may be an indication of prostatic carcinoma. The level of PSA varies a great deal and it is high after ejaculation and may provide false positive indication of cancer.

PSD-95: a family of membrane associated guanyl kinases; they also anchor K^+ channels by their PDZ domains. (See ion channels, GTP)

PSE: proximal sequence element. (See Hogness box)

PSE: pale soft exudative meat is controled in pigs by the *Halothane* gene.

PSEUDOACHONDROPLASIA: is a dominant human chromosome 19p12-p13.1 gene mutation controling the cartilage oligomeric matrix protein and it is responsible for short stature. (See achondroplasia)

PSEUDOALDOSTERTONISM (Liddle syndrome): is a human chromosome 4 hypertension associated with hypoaldosteronism, hypokalemia, reduced renin and angiotensin. (See aldosteronism, hypokalemia, renin, angiotensin)

PSEUDOALLELES: cluster of not fully complementing genes, separable by recombination. Pseudoalleles, e.g., a^1 and a^2 when heterozygous in trans position $a^1\ a^+//a^+\ a^2$ show mutant phenotype whereas in cis position $a^1\ a^2//a^+\ a^+$ are complementary (wild type), except when dominant alleles are involved. Since these alleles are closely linked, in order to be able to prove that recombination takes place (rather then mutation), the pseudoalleles must be genetically marked by flanking genes within preferably less than 10 m.u. apart of the locus. (See also complex locus, step allelomorphism, morphogenesis in *Drosophila*)

PSEUDOANEUPLOID: the chromosome number appears aneuploid but it is not truly the case only centromere fusion or misdivision of the centromeres have caused the changes in numbers. (See Robertsonian translocation, misdivision, B chromosomes)

PSEUDOAUTOSOMAL (PAR): are genes located in both telomeric regions of the X and Y chromosomes where recombination can take place and consequently, despite the sex-chromosomal location, sex-linkage is not obvious; e.g., a gene for schizophrenia was suggested to be pseudoautosomal. The SYBL1, encoding a synaptobrevin-like gene is present in both X and Y chromosomal PAR regions and it displays lyonization in the X chromosome and inactivation in the Y. (See autosome, differential segment, holandric genes, syntagmin)

PSEUDOCHOLINESTERASE DEFICIENCY (CH1, BCHE): is a dominant (human chromosome 3q26.1-q26.2) breathing difficulty (apnea) after treated with the muscle relaxant suxamethonium (succinylcholine chloride), a drug used for intubation, endoscopy, cesarean section, etc. as an adjuvant to anesthesia. Several allelic forms respond differently to drugs. Individuals with a defective enzyme may be particularly sensitive to cholinesterase inhibitor insecticides (parathion). The frequency of the gene varies a great deal in different populations. In Eskimos the frequency of the gene controlling the deficiency may be higher than 0.1; in other populations it may be less than 0.0002. The BCHE2 form was assigned to 2q33-35 and the same enzyme was suggested also to 16p11-q23.

PSEUDOBIVALENT: the chromosomes associated are not homologous. (See synapsis, illegitimate pairing)

PSEUDOBORDER: are DNA sequences in certain agrobacterial vectors or within the cloned foreign DNA and may cause deletions and rearrangements within the T-DNA inserts in the transgenic plants. (See T-DNA, transformation genetic)

PSEUDODOMINANCE: when a heterozygote loses the dominant allele, the recessive allele is uncovered (expressed) because of the lack of the dominant allele. Pseudodominance can be readily induced by treating heterozygotes with mutagens (e.g. ionizing radiation) that cause deletions. Before such experiments are conducted, it is advisable to place flanking genetic markers to the chromosome carrying the recessive markers to be able to rule out recombination and reversions. (See deletion)

PSEUDOEXTINCTION: disappearance of a species by evolution into another form.

PSEUDOGAMY: apomictic or parthenogenetic reproduction. (See apomixia, parthenogenesis)

PSEUDOGENE: has substantial homology with (clustered) functional genes of eukaryotes but it is inactive because of numerous mutations that prevent its expression and it is no longer available for transcription. The number of pseudogenes is variable in different species. Organisms with small genomes (e.g. *Drosophila*) have very few and it appears that some organisms eliminate from their genome the DNA sequences that are no longer functional. (See C-value paradox, gene relic)

PSEUDOHEMOPHILIA: is a bleeding disease, distinct from hemophilia; it is caused by some abnormalities of the platelets. (See hemophilia, platelet anomalies, hemostasis)

PSEUDOHERMAPHRODITISM: see hermaphrodite

PSEUDOHERMAPHRODITISM, MALE: is determined by a gene in human chromosome 17q11-q12. It is responsible for the deficiency of 17-ketosteroid reductase and consequently to feminization in prepubertal males and gynecomastia and virilization after puberty when usually the enzyme is expressed. The affected individuals may be surgically assisted to develop into sterile female phenotype (by removal of the hidden testes) or into male phenotype by reconfiguration of the external male genitalia. Infertility cannot be corrected, however. Recessive mutations in the luteinizig hormone receptor gene may also be responsible. (See gynecomastia, hermaphroditism, infertility, testicular feminization, luteinization)

PSEUDOHYPOPARATHYROIDISM: see Albright hereditary osteodystrophy

PSEUDOKNOT: is formed when a stem-and-loop RNA structure is bound at the base of the loop by a ligand resulting in a two-stem two-loop stacking. The actual configurations of the pseudoknots may vary. Pseudo-half-knots form only a single loop. Such structures may modulate RNA functions and can be exploited also in designing highly selective drugs. (See repeat inverted, antisense RNA)

PSEUDOLYSOGEN: lyses the bacterial cells so slow as if it would be lysogenic. (See lysogeny)

PSEUDOMONAS EXOTOXIN: kills by irreversible ribosylation of ADP and subsequent inactivation of translation elongation factor, EF-2. Its applied significance is the potential for cancer therapy. (See toxins)

PSEUDOMONAS TABACI: bacteria causing "wildfire" disease (necrotic spots) on tobacco leaves. The symptoms may be mimicked by methionine sulfoximine, a methionine analog.

PSEUDO-OVERDOMINANCE: in a population certain phenotype(s) may appear in excess of expectation because of the close linkage of the responsible gene to advantageous alleles. Also QTL loci may appear overdominant if they are relatively closely linked and display heterosis because the QTL mapping techniques cannot determine the map positions with great accuracy, and the molecular function of the genes involved is not known. (See overdominance, fitness, QTL, interval mapping)

PSEUDOPLASMODIUM: a migrating slug of cellular slime molds. (See *Dictyostelium*)

PSEUDOPODIUM: see amoeba

PSEUDOPREGNANT: female (mice) mated with vasectomized males and then implanted with blastocyst stage embryos derived from other matings. (See vasectomy, allopheny)

PSEUDOQUEEN: in social insects (bees, ants, termites) one worker (XX) may become fertile pseudoqueen after the loss of the queen of the colony.

PSEUDOSUBSTRATE: a molecule with similarity to an enzyme substrate but it is actually an inhibitor and special regulators are required for its removal so the enzyme is permitted to access its true substrate. (See substrate)

PSEUDOTEMPERATE PHAGE: it has a lysogenic cycle yet does not have a stable prophage state, e.g. the PBS1 transducing phage of *Bacillus subtilis*. (See lysogeny, prophage)

PSEUDO-TRISOMIC: is actually disomic but one of the chromosomes is represented by two telocentric chromosomes, each represent one and the other arm of the same chromosome, thus two telocentrics + one normal chromosome occurs. (See trisomy, telocentric chromosome)

PSEUDOTYPE: virus carrying foreign protein on his envelope and may expand the normal host range.

PSEUDOTYPING: if two types of viruses invade the same cell, genetic material of one may slip into the capsid of the other and this type of packaging permits the introduction of the viral genome into a host which normally would be incompatible with the normal virion. (See pseudovirus [pseudovirion])

PSEUDOURIDINE (ψ): a pyrimidine nucleoside (5-β-ribofuranosyluracil) occurs in the T arm of tRNA by post-transcriptional modification of a uracil residue. Pseudouridine has been found also in ribosomal RNAs and snRNAs. (See tRNA, ψ for formula)

PSEUDOVIRION (pseudovirus): contains non-viral DNA within the viral capsid and can thus be used to unload foreign DNA into a cell if a helper virus is provided. (See virion, capsid)

PSEUDOWILD TYPE: displays wild phenotype because a mutation at site different from the mutant locus that it masks but most commonly a duplicated segment masks the original and still present recessive mutation. In *Neurospora* it occurs at much higher frequency than expected by back mutation. It may be a suppressor mutation.

PSEUDOXANTHOMA ELASTICUM (PXE): autosomal recessive or dominant disorders caused by degenerative changes in the skin (peau d'orange = orange rind], neck, veins, eyes, intestines, etc. resulting in heart disease and hypertension. (See coronary heart disease, hypertension, skin diseases)

PSEUDO-ZELLWEGER SYNDROME: see peroxisomal 3-oxoacyl-coenzyme A thiolase deficiency

PSI (ψ): pseudouridine, and also the packaging signal in retrovirions. (See tRNA pseudouridine loop, retrovirus, retroviral vectors, pseudouridine formula on page 757)

PsnDNA: 150-300 bp pachytene DNA sequences flanking 800-3,000 bp internal chromosomal segments in eukaryotes, and the two short and the central DNA sequences are called PDNA (pachytene DNA). The PsnDNAs are supposed to be nicked by an endonuclease after homologous small nuclear RNA (snRNA) and a non-histone protein (PsnProtein) have opened the sequences to the action of the enzyme. These molecules appear only during late leptotene to pachytene and are assumed to be mediating recombination. (See crossing over, meiosis, snRNA, ZygDNA)

PSORALEN DYE: can combine with the DNA connecting nucleosomal core particles. After irradiation with near-ultraviolet light, cross-linking between the two DNA strands occurs. Psoralen-conjugated triple helix forming oligonucleotides have been used to induce site-specific mutations in mammalian cells. (See triple helix formation, site-specific mutation)

PSORIASIS: is a scaly type of skin defect determined either by dominant gene(s) of reduced penetrance or polygenic inheritance involving relatively few genes. Its incidence is common in Caucasian populations (1-3%) but it is much less frequent in Orientals (Eskimos, American Indians and Japanese). Recurrence rate may vary, depending on the type involved. If both parents are affected the recurrence among children may reach up to 75%. The psoriasis haplotype appears to include HLA-BW 17 and HLA-A 13 genes. Some observations indicate that psoriasis may be triggered by bacterial superantigens. One psoriasis susceptibility gene has been assigned to the end of the long arm of chromosome 17. (See HLA, keratosis, ichthyosis, skin diseases, Hirschprung disease)

PSYCHOMIMETIC: drugs affect the state of mind in a manner similar to psychoses. (See psychoses, psychotropic drugs, ergot)

PSYCHOSES: is a group of mental-nervous disorders with variable genetic and environmental components. (See autism, manic depression, schizophrenia, paranoia, affective disorders, attention deficit hyperactivity, Tourette's syndrome, IQ, dyslexia)

PSYCHOTROPIC DRUGS: affect the state of mind. They are used as medicine in various types of psychoses and may be very beneficial (e.g. lithium, valium, etc.) if applied under medical monitoring. Possible adverse side effects vary by the chemical nature of the drug and may include heart disease, birth defects, addiction, etc. (See psychoses, psychomimetic)

PTA DEFICIENCY DISEASE: is controlled by incompletely dominant (4q35) genes. Plasma thromboplastic antecedent protein deficiency is involved that results in unexpected bleeding after tooth extraction or various surgeries. Nose bleeding (epistaxis) is common but uterine bleeding (menorrhagia) or blood in the urine (hematuria) are rare. The carrier frequency in Ashkenazy Jewish populations is about 8.1%. (See antihemophilia factors, pseudohemophilia)

PTB (phosphotyrosine-binding domain): is present in proteins involved in signaling. (See SH2, SH3, WW, SCK, pleckstrin, signal transduction)

PTC: see phenylthiocarbamide; also papillary thyroid carcinoma; a variant of the RET oncogene caused neoplasia. (See RET)

PtdInsP$_2$: see phosphoinositides

PTERIDINES: purine derivatives, involved in coloring of insect eyes, wings, amphibian skin, etc. Pteridines may be light receptors. (See photoreceptors, rhodopsin, ommochromes)

PTG (protein targeting glycogen): forms complexes of phosphatases, kinases and glycogen synthase with glycogen. (See glycogen, kinase)

PTK: protein-tyrosine kinase involved in regulation of signal transduction and in growth and differentiation of cells. (See protein kinases)

PTP: see tyrosine phosphatase

PU.1: is a transcription factor in blood-forming cells regulating the differentiation of macrophages, B lymphocytes and monocytes; it belongs to the ETS family of oncogenes. (See ETS, monocytes, macrophages, lymphocytes)

PUBERTY: the time of sexual maturation, accompanied by the appearance of secondary sexual characteristics such as facial hair in males, enlargement of the breast in females, etc.

PUBERTY PRECOCIOUS: autosomal dominant disorders occur in two forms: isosexual, when sexual maturation in both males and females takes place before age 10 and 8.5, respectively but may be even much earlier, especially in females. Another form is male-limited. Testosterone production seems to be independent from gonadotropin releasing hormone production. The disorder is associated with a defect in the luteinizing hormone receptor. (See luteinizing hormone-releasing factor, animal hormones, hormonal effects on sex expression, G-proteins)

PUBLIC BLOOD SYSTEMS: see private blood groups

PUBS: percutaneous (through the skin) umbilical blood sampling, a method of prenatal biopsy for the identification of hereditary blood, cytological and other anomalies. (See also amniocentesis, prenatal diagnosis)

pUC VECTORS: are small (*pUC12/13* 1680 bp, *pUC18/19* 2686 bp) plasmids containing the replicational origin (*ori*) and the *Amp^r* gene of pBR322, and they carry the *LacZ'* fragment of the bacterial β-galactosidase. The Z' indicates that within this region there is multiple cloning site (MCS) for recognition by 13 restriction enzymes. The orientation of the MCS is in reverse in pUC18 relative to pUC19. Genes inserted into *Lac* may be expressed under the control of the *Lac* promoter as a fusion protein. Most commonly, the insertion inactivates the *Lac* gene and white colonies are formed in Xgal medium rather than blue when the gene is active. The pUC vectors can be used with JM105 and NM522 *E. coli* strains. (See vectors, *Xgal, Lac*). (The diagram is the courtesy of CLONTECH Laboratories Inc., Palo Alto, CA, USA)

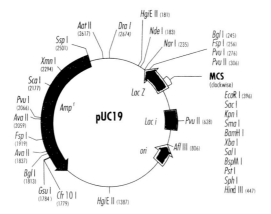

PUCCINIA GRAMINIS: see stem rust

PUFF: swollen area of polytenic chromosomes active in transcription. Puffing is induced by expression of transcription factor genes regulated by steroid hormones (ecdysone). Ecdysone formation comes in sequential pulses and thereby sequential activation of genes involved in metamorphosis of insects can be visualized at the level of the giant chromosomes. The puffs represent active transcription at particular genes and the pattern of puffing shifts along the salivary gland chromosomes during development and/or activation and the RNA extracted from the puffs reflect the differences in the base sequences of the genic DNA. Puffing has been desribed also in the rare polytenic chromosomes of some plant species, e.g. *Allium ursinum* or *Aconitum ranunculifolium*. These have been observed in specialized tissues of the chalaza or in the antipodal cells. (See giant chromosomes, ecdysone, drawing on the following page).

Puff continues

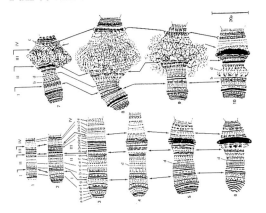

SELECTIVE ACTIVITY OF GENES DURING DEVELOPMENT OF THE DIPTERAN FLY *Rhynchosciara angelae* IS REFLECTED IN THE PUFFING PATTERN OF THE SALIVARY GLAND CHROMOSOMES. LOWER CASE LETTERS DESIGNATE BANDS. ROMAN NUMERALS INDICATE REGIONS OF THE CHROMOSOMES. (AFTER BREUER, M. E. AND PAVAN, C. 1954. By permission from Kühn, A. 1971 Lectures on Developmental Physiology. Springer-Vlg., New York)

PUFFERFISH, JAPANESE (*Fugu rubripes*): is a small vertebrate with about 400 Mbp DNA, i.e. only somewhat more than 1/10 of that of most mammals, and therefore it is suitable for structural and functional studies at the molecular level.

PULMONARY EMPHYSEMA: increase in size of the air space of the lung by dilation of the alveoli (small sac-like structures) or by destruction of their walls. Smoking may cause it.

PULMONARY STENOSIS: see stenosis

PULSE-CHASE ANALYSIS: expose cells for a period of time to a radioactive compound such as ^{3}H-thymidine (pulse) and examine the labeling of chromosomes in some cells. The culture is then transferred to non-radioactive thymidine and allowed to complete a division (chased to another stage). Study again the distribution of the label and determine its fate in the cell. The experiment diagramed permitted the conclusion that DNA replication is semi-conservative. (See radioactive tracer, radioactive label)

AUTORADIOGRAPHIC ANALYSIS OF THE REPLICATION OF THE DNA IN CHROMOSOMES BY THE PULSE CHASE PROCEDURE. (Drawn after Taylor, J.H. *et al.* 1957 Proc. Natl. Acad. USA 43:122)

PULSED FIELD ELECTROPHORESIS (PFGE): a procedure combining static electricity, alternating electric fields with gel electrophoresis for the separation of DNA of entire chromosomes of lower eukaryotes such as of yeast and *Tetrahymena* or large DNA fragments cloned in YAC vectors of any genome cut by rare-cutting restriction enzymes. (See CHEF, FIGE, OFAGE, YAC, PHOGE, TAFE, RGE)

PUMA (*Felis concolor*, *Puma concolor*): 2n = 38.

PUMP: various transmembrane proteins mediating active transport of ions and molecules through biological membranes. (See sodium pump)

PUNCTUATED EQUILIBRIUM: see punctuated evolution

PUNCTUATED EVOLUTION: a theory that evolution would follow alternating periods of rap-

Punctuated evolution continued
id changes and relatively stable intervals (punctuations). Natural selection of beneficial mutations appear after some intervals and spread over the population. (See speciation, beneficial mutation, neutral mutation, hopeful monster, shifting balance theory)

PUNCTUATION CODONS (UAA, UGA, UAG): terminate translation of the mRNA.

PUNNETT SQUARE: permits simple prediction of the expected pheno- and genotypic proportions. It is a checker board where on top and at the right column the male and female gametic output is represented and in the body of the table the genotypes are found. If, e.g., the heterozygote has the genetic constitution of *Aa, Bb* the gametes and genotypes will be:

MALE GAMETES →	AB	Ab	aB	ab
↓ FEMALE GAMETES				
AB	AB/AB	AB/Ab	AB/aB	AB/ab
Ab	Ab/AB	Ab/Ab	Ab/aB	Ab/ab
aB	aB/AB	aB/Ab	aB/aB	aB/ab
ab	ab/AB	ab/Ab	ab/aB	ab/ab

In case of linkage and recombination the frequency of each type of gamete must be used to obtain the correct genotypic proportions in the body of the checkerboard

(See also modified Mendelian ratios, Mendelian segregation)

PUPA: a stage in insect development between the larval stage and the emergence of the adult (imago). (See *Drosophila*, juvenile hormone)

PUPARIUM: the case in which the *Drosophila* (and other insect) pupa develops for about 4 days after hatching of the egg, and in another four days the imago emerges. (See *Drosophila*)

PURE CULTURE: involves only a single organism. (See also axenic culture)

PURE LINE: is genetically homogeneous (homozygous), and its progeny is expected to be identical with the parental line unless mutation occurs.

PURE-BREEDING: homozygous for the genes considered.

PURINE: a nitrogenous base composed of a fused pyrimidine and imidazole ring; the principal purines in the cells are adenine guanine, xanthine, hypoxanthine (but theobromine, caffeine and uric acid are also purines).

PURINE REPRESSOR (PurR): a member of the *Lac* repressor family of proteins regulating 10 operons involved in the biosynthesis of purine and affecting to some extent 4 genes controling *de novo* pyrimidine synthesis and salvage. Its ca. 60 amino acids, the NH_2 domain binds to DNA and its ca. 280 residue COOH domain binds effectors and it functions in oligomerization. (See *Lac* repressor, salvage pathway)

PURITY OF THE GAMETES: is one of the most important discoveries of Mendel. At anaphase I of meiosis of diploids the bivalent chromosomes segregate and at anaphase II the chromatids separate. Therefore in the gametes of diploids only a single allelic form of the parents is present with rare exceptions, e.g. nondisjunction and polyploids. (See meiosis, nondisjunction, gene conversion)

PURKINJE CELLS: in the cerebellum are large pear-shape cells and are connected to multibranched nerve cells traversing the cerebellar cortex. In the heart they are tightly appositioned cells transmitting impulses. (See cerebellum)

PUROMYCIN: an antibiotic, inhibits protein synthesis by binding to the large subunit of ribosomes; its structure resembles the 3'-end of a charged tRNA. Therefeore it can attach to the A site of the ribosome and forms a peptide bond but it cannot move to the P site and thus causes premature peptide chain termination. (See antibiotics, signaling to translation)

PUSHME-PULLYOU SELECTION: is a positive-negative selection system to isolate engineered chromosomes in somatic cell hybrids which have retained the segment positively selected for and lost the regions selected against.

PV16/18E6: a human papilloma virus oncoprotein. (See oncoprotein)

PVT1: is an oncogene (human chromosome 8q14) frequently present in human Burkitt lymphoma and murine lymphocytomas and may have activating role for MYC that is in the same chromosome. (See MYC, oncogenes, Burkitt lymphoma)

πVX: a microplasmid (902 bp) containing a polylinker and an amber suppressor for tyrosine tRNA. It can be used for cloning eukaryotic genes. (See recombinational probe)

PYCNIDIUM: a hollow spherical or pear-shaped fruiting structure of fungi producing the pycnidiospores which are released through the top opening, the ostiole. (See stem rust)

PYCNODYSOSTOSIS: a rare autosomal recessive (1q21) human malady characterized by defects in ossification (bone development) resulting in short stature, deformed skull with large fontanelles (soft, incompletely ossified spots of the skull common in fetuses and infants) and general fragility of the bones. The primary defects appears to be in cathepsin K, a major bone protease although interleukin-6 receptor has also been implicated. (See Toulouse-Lautrec, cleidocranial dysostosis, cathepsins)

THE FAMOUS FRENCH ARTIST HENRI TOULOUSE-LAUTREC MIGHT HAVE SUFFERED FROM THIS MALADY AND HIS SELF-PORTRAIT REVEALS SOME OF THE CHARACTERISTICS OF THE MALFORMATIONS.

PYCNOSIS: is a physiological effect of ionizing radiation on chromosomes expressed as clumping or stickiness. It is dose-dependent and the late prophase stage irradiation is most effective in causing it. Anaphase proceeds but the chromosomes have difficulties in separation, display chromatin bridges and may break up into fragments. (See bridge)

PYGMY: the Central African human tribe of about 100,000 has an average height of 142 cm. In comparison the average height of Swiss and Californian is 167-169 and 170-172 cm, respectively. The Pygmies do not respond to exogenous stomatotropin but the concentration of serum somatomedins in the adolescent Pygmies is about a third below that in non-Pygmies of comparable age. Although the shortness of Pygmies appears recessive, intermarriages indicate polygenic determination of height. (See dwarfism, stature in humans, nanism, somatomedin, somatotropin)

PYK2: links Src with G_i and G_q-coupled receptors with Grb2 and Sos proteins in the MAP kinase pathway of signal transduction. Its phosphorylation by Src is stimulated by lysophosphatidic acid (LPA) and bradykinin. Over-expressing mutants of Pyk or the protein tyrosine kinase Csk reduces the stimulation by LPA, bradykinin or over-expressed Grb2 and Sos. (See CAM, MAP, Src, G_i, G_q, lysophosphatidic acid, kininogen, Csk, Grb2, Sos)

PYLORIC STENOSIS: a smaller than normal opening of the pylorus, the lower gate of the stomach, that separates it from the small intestine (duodenum). It does not appear to have independent genetic control but it is part of some syndromes. If affects males five times as frequently as females; the overall incidence for both sexes is about 3/1,000 birth. About 20% of the sons of affected females displays this anomaly but only about 4% of the sons if the father has the malady. It may be caused by a deficiency of neuronal nitric acid synthase. (See sex-influenced, nitric oxide, imprinting)

PYO: personal years of observations.

PYOCIN: a bacteriotoxic protein produced by some strains *Pseudomonas aeruginosa* bacteria. (See bacteriocins)

PYRENOID: a dense, refringent protein structure in the chloroplast of algae and liverworts associated with starch deposition. (See chloroplast)

PYRETHRIN (pyrethroids, permethrin): insecticides are natural products of *Pyrethrum* (*Chrysanthemum cineraiaefolium*) plants (Compositae). They affect the voltage-gated Na^+ ion channels and humans may have severe allergic reactions to pyrethrins. (See ion channels)

PYRETHRUM (*Chrysanthemum* spp): a source of the natural insecticide pyrethrin with basic chromosome number $x = 9$. Some species are diploid or tetraploids or hexaploids. (See pyerthrin)

PYRIDINE NUCLEOTIDE: a coenzyme containing a nicotinamide derivative, NAD, NADP.

PYRIDOXIN (pyridoxal): vitamin B_6 is part of the pyridoxal phosphate coenzyme, instrumental in transamination reactions. An apparently autosomal recessive disorder involving seizures is caused by pyridoxin deficiency because of a deficit in glutamic acid decarboxylase (GAD) activity and consequently insufficiency of GABA, required for normal function of neurotransmitters. Administration of pyridoxin caused cessation of the seizures. The GAD gene was located in the long arm of human chromosome 2. An autosomal dominant regulatory pyridoxine kinase function has also been identified in humans. (See epilepsy)

PYRIDOXINE DEPENDENCY: may be manifested as autosomal recessive seizures with perinatal onset.

PYRIMIDINE: a heterocyclic nitrogenous base such as cytosine, thymine, uracil in nucleic acids but also the sedative and hypnotic analogs of uracil, barbiturate and derivatives.

PYRIMIDINE DIMER: cross-linked adjacent pyrimidines (thymidine or cytidine) in DNA causing a distortion in the strand involved and thus interfering with proper functions. It is induced by short-wavelength UV irradiation. The thymidine dimers may be split by visible light-inducible enzymatic repair (light repair) or by excision repair (dark repair). (See cyclobutane ring, physical mutagens, genetic repair, DNA repair, photolyase, CPD)

PYRIMIDINE DIMER N-GLYCOSYLASE: a DNA repair enzyme that creates an apyrimidinic site. Then the phosphodiester bond is severed and a 3'-OH group is formed on the terminal deoxyribose. Exonuclease 3'→5' activity of the DNA polymerase splits off the new 3'-OH end of the apyrimidinic site. After this, the replacement-replication—ligation process repairs the former thymine dimer defect. (See glycosylase, DNA repair, pyrimidine dimer)

PYRIMIDOPURINONES: a malondialdehyde-DNA adduct derived from deoxyguanosine. (See adduct)

PYRONIN: a histochemical red stain used for the identification of RNA.

PYRROLE: a saturated five-membered heterocyclic ring such as found in protoporphyrin. (See porphyrin, porphyria, heme)

PYRUVATE DEHYDROGENASE COMPLEX: contains three enzymes, pyruvate dehydrogenase, dihydrolipoyl transacetylase and dihydrolipoyl dehydrogenase and the function of the complex requires the coenzymes: thiamin pyrophosphate (TPP), flavine adenine dinucleotide (FAD), coenzyme A (CoA), nicotinamide adenine dinucleotide (NAD), and lipoate. The result of the reactions is oxidative decarboxylation whereby CO_2 and acetyl CoA are formed. (See oxidative decarboxylation)

PYRUVATE KINASE DEFICIENCY: is a recessive (human chromosome 1q21-q22) hemolytic anemia caused by actually two enzymes that are the products either of differential processing of the same transcript or chromosomal rearrangement. (See anemia, hemolytic anemia, glycolysis)

PYRUVIC ACID: a ketoacid ($CH_3COCOOH$) formed from glycogen, starch and glucose under aerobic conditions (under anaerobiosis is reduced to lactate and NAD^+ is formed). (See Embden-Meyerhof pathway, pentose phosphate shunt)

PyV: polyoma virus.

PZD: see micromanipulation of the oocyte.

Q

q: long arm of chromosomes. (See also p)

Q BANDING: chromosome staining with quinacrines that reveals cross bands. Because of the availability of newer microtechniques this procedure is no longer generally used. (See chromosome banding, quinacrine mustard)

Q-β: is an RNA bacteriophage of a molecular weight of about 1.5×10^6 Da. The Qβ replicase is an RNA-dependent RNA polymerase that synthesizes the single-stranded RNA genome of the phage without an endogenous primer. The replicase can use both the + and the − strand as a template and therefore it amplifies the genome rapidly. It is a heterotetramer consisting of a viral encoded and three host polypeptides: ribosomal protein S1, the EF-Tu (elongation factor), and Ts. (See replicase, plus strand)

QTL (quantitative trait loci): may be present in restriction fragments separated by electrophoresis. Their cosegregation is identified, and can be used for improving quantitative traits for plant breeding purposes, and for genetically defining behavioral traits and other polygenic characters. QTLs can be genetically mapped by several procedures, most commonly using the principles of maximum likelihood. In a backcross generation the phenotype (ϕ_t) and genotype (g_i) relations are expressed as: $\phi_t = \mu + bg_i + \varepsilon_i$ where g_i corresponds to the homozygous and heterozygous dominants of the QTL and its value may vary between 1 and 0. The mean of ε_i (a random variable) = 0 and its variance is σ^2. The values of μ, b and σ^2 are unknown. The genotypic value of Qq and other contributors to the quantitative trait is μ and b is the effect of a substitution of another allele at the quantitative trait loci. The statistical procedures shown below were adapted from Arús, P. & Moreno-González, J. 1993 Plant Breeding, pp.314, Hayward, M.D. et al. eds., Chapman & Hall, London, New York.

$$\text{The likelihood function } Lg_i(\mu, b, \sigma^2) = \frac{1}{\sqrt{2\pi\sigma^2}} e^{-\frac{(\phi_i - \mu - bg_i)^2}{2\sigma^2}}$$

and the likelihood that all individuals will be in the flanking parental marker classes (k) $M_1M_1M_2M_2$, $M_1M_1M_2m_2$, $M_1m_2M_2M_2$, and $M_1m_2M_2m_2$ will be $L_k(\mu, b, \sigma^2) = \Pi_t[P_i(1)L_i(1) + P_i(0)L_i(0)]$ where $P_i(1)$, and $P_i(0)$ are the probabilities that QQ and Qq quantitative genes will be in the recombinant classes, respective the flanking markers concerned. The maximum likelihood estimates for incomplete data can be also determined (Dempster et al. 1977. J. R. Stat. Soc. 39:1). The likelihood for all observations: $L(\mu, b, \sigma^2) = \Pi_k L_k(\mu, b, \sigma^2)$. For the determination of the LOD score, to ascertain that the information obtained is real rather than a false, spurious conlusion would be drawn, the following equation has to be resolved:

$$\text{LOD} = \frac{\log L(\mu, b, \sigma^2)}{L(\mu_0, b_0, \sigma_0^2)}$$

One must keep in mind that the estimates are as good as the data collected. Large populations and genes with greater quantitative effects improve the chances to find linkage. For estimating linkage information from multiple marker data, computer assistance is required; various programs are available. For mapping QTL in humans generally sib pairs are used. Statistical analysis indicates that choosing extreme discordant pairs makes the analysis more efficient. (See LOD score, mapping genetic, gene block, RFLP, liability, ASP analysis, interval mapping, complex inheritance)

QUADRANT: consists of four parts, e.g. a tetrad.

QUADRATIC CHECK: is used for testing two genes presumed to be required for phytopathogenic infection in the manner:

	Low Pathogenicity	High Pathogenicity
Plant Reaction Low →	NO INFECTION	INFECTION
Plant Reaction High →	INFECTION	INFECTION

Resistance in the plants is usually a dominant trait. (See Flor's model)

QUADRIVALENT: partially or completely identical four chromosomes in a polyploid that dis-

Quadrivalent continued
play pairing although of the four, at any particular position, only two can be synapsed. During meiosis they show quadrivalent association of four chromosomes. (See synapsis, meiosis, bivalent)

QUADRUPLETS: are fourfold twins and in the absence of the use of fertility increasing treatment, their expected frequency is about 1 in $(89)^3$ whereas the expectation for triplets and quintuplets is about $(89)^2$ and $(89)^4$, respectively. (See twinning)

QUADRUPLEX: a tetraploid or tetrasomic with four doses of the dominant alleles at a locus. (See autopolyploid, G quartet)

QUADRUPLICATE GENES: four genes conveying identical or similar phenotype but segregating independently in F_2 and display a dominant recessive proportion of 255:1.

QUANTITATIVE GENE NUMBERS: are difficult to determine because environmental effects obscure the impact of genes with minor effects. Several statistical procedures have been worked out for approximation. The simplest one is as follows:

$$N = \frac{R^2}{8(s_1^2 - s_2^2)}$$

where n = gene number, R is the difference between the parental means, s_1^2 = variance of the F_1, s_2^2 = variance of the F_2 generations. The most common view is that quantitative traits are determined by large number of genes and each of them contributes only little to the observed phenotype. The association between bristle number (a quantitative trait) and the *scabrous* locus of *Drosophila* indicated, however, that approximately 32% of the genetic variation in abdominal and 21% of the sternopleural bristle number was associated with DNA sequence polymorphism at this single locus. (See gene number, quantitative trait, QTL)

QUANTITATIVE GENETICS: studies genetic mechanisms involved with the expression of quantitative traits and its techniques involve those of population genetics and biometry. (See quantitative trait, QTL, population genetics, biometry, statistics)

QUANTITATIVE TRAIT: shows continuous variation of expression and can be characterized by measurement or by counting in contrast to qualitative traits which can be identified satisfactorily by simple description such as black or white. (See gene titration)

QUANTITATIVE TRAIT LOCI: see QTL

QUANTUM: unit to quantify energy (see photon).

QUANTUM SPECIATION: a rapid formation of a new species by selection and genetic drift. (See selection, genetic drift)

QUARANTENE: a state of isolation and observation without any external contact, especially from infection for a period of time.

QUARTET: a structure consisting of four elements. (See G quartet)

QUASI: in various combinations indicates almost, resembling or about of the notion that the following word specifies, e.g. quasi-species means that its difference from other form(s) may not qualify it for the status of a separate species with certainty.

QUASIDOMINANT: recessive inheritance is misclassified as dominant because the mating took place between a heterozygote and a homozygous recessive individual.

QUASI LINKAGE: see affinity

QUASI-MONOCLONAL ANTIBODY: was produced by mice heterozygous for the V(D)J IMimunoglobulin heavy chain (Ig) and the other allele being non-functional. Functional κ chain is also missing. When the heavy chain, specific for the hapten 4-hydroxy-3-nitrophenyl acetyl could join any λ chain, the antibody was monospecific but somatic mutation and secondary rearrangements changed the specificity of 20% of the B cell antigen receptors. Such a system can thus be used to study antibody diversity. (See antibody)

QUASI-SPECIES: small degree of genetic (nucleic acid) variation does not qualify it clearly for separate species status.

QUATERNARY STRUCTURE: the aggregate of multiple polypeptide subunits into a protein or by cross-linking DNA strands into a joint structure. (See cross linking, protein structure)

QUEEN: the reproductive female in cast insects such as bees, ants. (See pseudoqueen)

QUEEN VICTORIA: see hemophilias, Romanovs

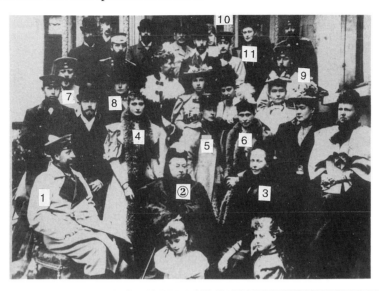

THE MOST FAMOUS FAMILY OF QUEEN VICTORIA OF ENGLAND, AFFECTED BY HEMOPHILIA, AT A REUNION ON APRIL 23, 1894. (1) KAISER WILHELM II, GRANDSON, **(2) QUEEN VICTORIA**, (3) DAUGHTER VICTORIA, (4) GRANDDAUGHTER TSARINA ALEXANDRA, (5) GRANDDAUGHTER IRENE, (6) GRANDDAUGHTER ALICE, (7) SON AND FUTURE KING OF ENGLAND EDWARD VII, (8) DAUGHTER BEATRICE, (9) SON ARTHUR, (10) GRANDDAUGHTER MARIE, (11) GRANDDAUGHTER ELIZABETH. FOR MOST LIKELY GENETIC CONSTITUTIONS REGARDING HEMOPHILIA SEE UNDER HEMOPHILIA. (COURTESY OF THE HUMANITIES RESEARCH CENTER, GERNSHEIM COLLECTION, UNIVERSITY OF TEXAS, AUSTIN, TX)

QUENCHING: suppression of fluorescence, transfer of electrons or suppression of an activator by blocking the binding site of the activator or binding it to another protein which prevents its binding to the activator binding site in the DNA.

QUICK-STOP: the temperature-sensitive DNA replication mutant *dna* of *E. coli* stops DNA replication immediately when the temperature rises to 42° C from the permissive 37 centigrade. (See temperature-sensitive mutation)

QUINACRINE MUSTARD (ICR 100): a light-sensitive, fluorescent compound used for chromosome staining. ICR 100 is strongly mutagenic; it is also a highly toxic antihelminthic drug. (See Q banding, acridine dyes)

QUINTUPLEX: see quadruplex.

QUORUM FACTORS: signaling molecules in autoinduction. (See autoinduction)

QUORUM-SENSING: a system of cell density-dependent expression of specific gene sets. (See autoinduction)

q.v. (quod vide): see it

R

R: (r, Röntgen, Roentgen) unit of ionizing radiation (1 electrostatic unit of charge in 1 cm³ dry air at 0° C and 760 mm pressure; about 93 ergs/living cells). (See also Rad, Rem, rep, Gy, Sv, cR)

ρ (rho): see buoyant density, petite colony mutants, transcription termination in prokaryotes

R1: methylation sites in the cytoplasmic region of *E. coli* chemotaxis transducer proteins.

R2: see hybrid dysgenesis I - R

rII: see rapid lysis mutants of bacteriophages

R-BANDED CHROMOSOME →

R BANDS: are heat-denaturation resistant chromosomal bands; half of them have telomeric sites. The bright field R bands usually show the reverse pattern to the G bands because the dark regions are euchromatic and the heterochromatic segments are bright. (See isochores, G bands)

R BODIES: refractive bodies are temperate bacteriophages within the κ particles of paramecia. (See symbionts hereditary. *Paramecium*)

R END: see packaging of λ DNA

R FACTORS: resistance factors in bacterial plasmids that may make host bacteria insensitive to antibiotics and to normally bacteriotoxic drugs. They are common among Gram-negative bacteria and are readily transmitted to other strains because the plasmids generally are endowed with transfer factors. These plasmids usually do not integrate into the bacterial host genome but can recombine with each other and generate new plasmids with multiple resistance factors. Because of the presence of multiple resistance factors, only simultaneous administration of multiple antibiotics may stop the multiplication of the bacteria. (See plasmid, plasmid mobilization, plasmid conjugative, antibiotics)

R GROUP (a radix): an abbreviation for an alkyl group or any other chemical substitutions.

R LOOP: DNA strand displaced by RNA in a double-stranded DNA-RNA heteroduplex; also the genomic DNA intron forms an R loop when the gene is hybridized with cDNA or mRNA.

R PLASMIDS: carry resistance factors (genes for antibiotic resistance and other agents).

R POINT (restriction point): before S phase, cells in G1 phase pause and may or may not continue with the cell cycle. Cancer cells bypass the restriction point and continue uncontroled divisions. Cultured cells may require serum or amino acids to escape from G1 toward S. (See cell cycle)

R UNIT: see r (Röntgen)

RAB: RAS oncogene homolog that regulates transport between intracellular vesicles, and controls endosome fusion. In human chromosome it was located to 19p13.2. Rab3A is functioning in the synaptic vessel. Basically they are GTPases. (See RAS oncogene, Sec, Ypt, Mss, synaptic vessel, GTPase, endosome)

RABBIT: *Oryctolagus cuniculus*, 2n = 44; *Sylvilagous floridanus*, 2n = 42.

RABBIT RETICULOCYTE *IN VITRO* TRANSLATION: mammalian mRNA (extracted from cells or transcribed *in vitro*) can be translated into protein under cell-free conditions using lysates of immature red blood cells of anemic rabbits. Anemia is induced by subcutaneous injection of the animals for 5 days by neutralized 1.2% acetylphenylhydrazine solutions (HEPES buffer). After the larger white blood cells are removed by centrifugation, the red blood cells are lysed at 0° C by sterile double-distilled water. Then the endogenous mRNA is destroyed by micrococcal nuclease in the presence of Ca^{2+}. Without calcium the nuclease does not work. The reaction is stopped by EGTA (ethylene glycol tetraacetic acid which chelates calcium). Hemin [$C_{34}H_{32}ClFeN_4O_4$], dissolved in KOH, is needed for suppressing an inhibitor of eukaryotic translation initiation factor eIF-2. The translation mixture must contain spermidine or RNasin ribonuclease inhibitors, creatine phosphate (an energy donor), dithiothreitol (a reducing agent to prevent the formation of sulfoxides from the S-labeled amino acids), all normal amino acids (except the one which will carry the radioactive label), buffer, the radioactive amino acid (e.g. [^{35}S]methionine), the reticulocyte lysate, tRNAs, KCl and magnesium acetate (to enhance translation), polyadenylate-tailed mRNA (to be translated into protein). All solu-

Rabbit reticulocyte *in vitro* translation continued

tions must be made up with RNase-free material and the vessels should be made RNase-free. Incubation is at 30° C for 30 to 60 minutes. Before precipitating (by 10% trichloroacetic acid) the protein synthesized, the ^{35}S-methionine-tRNA is destroyed either by 0.3 N NaOH or, in case SDS-polyacrylamide gels are used for subsequent analysis, by pancreatic ribonuclease. Immunoprecipitation may also be employed for the analysis of the translation product. The amount of protein synthesized can be measured by scintillation counting. Rabbit reticulocyte lysates are also available commercially. Alternatively wheat germ extract may also be used for *in vitro* translation. (See Wheat germ translation system, eIF-2, translation repressor proteins, polyA mRNA, SDS polyacrylamide gels, immunoprecipitation, scintillation counters)

RAC (same as Akt or PKB): is a serine/threonine kinase member of the RAS protein family and transmits signals from the cell surface membrane to the cytoskeleton. When activated it inhibits transferrin-receptor mediated endocytosis and regulates, with RHO, the formation of clathrin-coated vesicles and actin polymerization. It has an important role in RAS mediated oncogenic transformation. (See serine/threonine kinase, RAS, RHO, clathrin, signal transduction, cell membrane, cytoskeleton, PKB, endocytosis, transferrin)

RACCOON (*Procyon lotor*): 2n = 38.

RACE: group within a species, distinguished by several genetic characteristics. (See evolutionary distance, human races, racism)

RACE: rapid amplification of cDNA ends by PCR. (See polymerase chain reaction)

RACEMATE: a mixture of D and L optical stereoisomers (enantiomorphs); the mixture then becomes optically inactive. All naturally synthesized amino acid are in the L form but degradation generates the D enantiomorph. The degree of racemization of aspartic acid is faster than that of other amino acids and it has been used recently to determine the authenticity of ancient samples of DNA because the degradation of DNA and the racemization of amino acids, particularly Asp indicates whether the spurious DNA is really ancient or just contaminant in the archeological sample. In case the D/L Asp ratio exceeded 0.08 ancient DNA could not be retrieved. The degradation depends also on a number of factors, most notably the temperature what the specimen had been historically exposed to. The best preservation had taken place in insects enclosed in amber (although there is some controversy about these samples). In specimens where the D/L Asp ratio was about 0.05, up to 340 bp long DNA sequences could be detected using PCR technology. (See enantiomorph, carbon dating, evolutionary clock)

RACEME: inflorescence with elongated main stem and with flowers on near equal-size pedicels.

RACHIS: the axis of a spike (grass ear), and fern leaf (frond).

RACIAL DISTANCE: see evolutionary distance

RACISM: is the assumption of superiority of any particular ethnic group or groups and the consequently inferiority of some others. It advocates hatred and social discrimination on the basis of differences. The origin of racism can be traced back to prehistorical times for the purpose of exploitation of conquered or minority groups. Some forms of racism may be found in nearly all societies of Caucasians, Chinese, Japanese, Blacks, etc., even the Bible is not exempt from racist ideas, and it has been frequently used as a justification by bigots. In the 19th century the rise of the eugenics movement gave false scientific encouragement to racism, providing biological and ideological support for colonialism and social exploitation. Racist ideas were used to justify slavery. Racism culminated in the Third Reich of Hitler's Germany resulting not just in discrimination and suppression but physical mass elimination of "Non-Aryan" people and aiming to establish Rassenhygiene (racial hygiene). Racism cannot be justified on the basis of any scientific evidence and it is morally unacceptable by enlightened societies. Actually, the world most successful societies excelled because of their multiracial and multicultural composition. Ample biological evidence supports the superiority of hybrids of mammalian and plant species. (See human races, evolutionary distance, hybrid vigor, eugenics, miscegenation)

RAD: unit of ionizing **r**adiation **a**bsorbed **d**ose (100 ergs/wet tissue). (See r, rem, Gray, Sievert)

RAD: genes of yeast are involved in DNA repair and recombination. (See ABC excinucleases)

RAD1: a yeast gene involved in cutting of damaged DNA in association with RAD10 (ERRC1); its human homolog is XPF. (See DNA repair)

RAD2: a yeast gene involved in cutting DNA; its human homolog is XPG. (See DNA repair)

RAD3: a yeast DNA helicase and a component of transcription factor TFIIH; its human equivalent is XPD. (See DNA repair)

RAD14: yeast gene; its protein product binds to damaged DNA. The human homolog is XPA. (See DNA repair)

RAD25 (ERCC): a helicase subunit of the general transcription factor TFIIH and it is credited with promoter clearance for the beginning of transcription after ATP hydrolysis and after the open promoter complex is formed. It is also a DNA repair enzyme. (See transcription factors, open promoter complex, regulation of gene activity, helicase, DNA repair, promoter clearance)

RAD28: the yeast homolog of the gene of the Cockayne syndrome. (See Cockayne syndrome)

RAD50, RAD51, RAD52, RAD53, RAD54, RAD55, Rad56, RAD57: are yeast genes involved in radiation sensitivity DNA double-strand break, repair and recombination.

RAD51: gene of budding yeast regulates double-strand breaks and genetic recombination depending on ATP. The RAD51 protein bears similarity to the bacterial RecA and Rad1 homologs with similar functions occur also in mammals. Disruption of *RAD51* in mice has embryonic lethal effects. (See RecA, Dmc1)

RAD53: a yeast gene encoding pRAD53 signal transducer; it is called also SAD1, MEC2, SPK1.

RADIATION, ACUTE: the irradiation is delivered in a single dose at high rate, in contrast to *chronic radiation* when the same dose is administered during a prolonged period of time. (See radiation effects, physical mutagens)

RADIATION, ADAPTIVE: see radiation evolutionary

RADIATION CANCER: many of the different cancers are associated with chromosomal rearrangement(s), and ionizing radiation causes chromosomal breakage and rearrangements. Ultraviolet radiation may be responsible for the induction of skin cancer, especially if the body's genetic repair mechanism is weakened. According to estimates, 10 mSv may be responsible for 1 cancer death per 10,000 people. In the USA the permissible legal dose limit to the public is 0.25 mSv/year but it should be reduced to 0.20 mSv/year. (See Sievert, DNA repair, ultraviolet radiation, xeroderma pigmentosum, excision repair, physical mutagens, radiation hazard assessment, radiation safety hazards)

RADIATION CHIMERA: an antigenically different bone marrow transplant is harbored in a body after an extensive radiation treatment destroyed or substantially reduced the immune reaction of the recipient. Also, mutant sector(s) caused by radiation induced mutations or deletions.

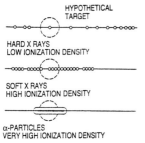

CHIMERIC DAHLIA FLOWER, THE CONSEQUENCE OF RADIATION EXPOSURE. (Photograph of Dr. Arnold Sparrow. Courtesy of the Brookhaven National Laboratory, Upton, NY)

RADIATION CHRONIC: see radiation acute

RADIATION DENSITY: is generally measured by LET (linear energy transfer) values, i.e., the average amount of energy released per unit length of the tract. In case the density is low the genetic damage is expected to be discrete. High LET radiation causes extensive damage along a very short path. (See physical mutagens, radiation effects)

THE PATTERN OF IONIZATION DENSITY ALONG THE TRACK OF HARD AND SOFT X-RAYS AND α-PARTICLES. (After Gray, L.H. from Wagner, R.P. & Mitchell, H.K. 1964 Genetics and Metabolism. Wiley, New York)

RADIATION DOUBLING DOSE: see doubling dose

RADIATION EFFECTS: ionizing radiation may cause gross chromosome breakage (deletions, duplications, inversions, reciprocal translocations, isochromatid breaks, transpositions, change in chromosome numbers if applied to the spindle apparatus) or minute changes including destruction of a single base in the nucleic acids or very short deletions involving only a few base pairs. These effects depend on the quality of the ionizing radiation and the status of the biological material involved. The physical effect of the radiation is frequently characterized by LET (linear energy transfer in keV/nm path), indicating the amount of energy released per unit tract as ionization and excitation in the biological target (see radiation density). Also, if the dose is delivered at a low rate, most of the damage may be repaired by the metabolic system of the cells. In prophase the chromatid breaks may remain open only for a few minutes but at interphase they may stay so substantially longer. The frequency of chromosome breakage is considerably increased in the presence of abundant oxygen, and anoxia has the opposite effect. Actively metabolizing cells are also more susceptible to radiation damage than dormant ones (germinating seeds versus dry ones). Chromosomal aberrations requiring two breaks (e.g., inversions, translocations) occur by second order kinetics whereas the induction rate of point mutations displays first order kinetics. Ultraviolet radiation causes excitation rather than ionization in the genetic molecules. The most prevalent damage is the formation of pyrimidine dimers although chromosome or chromatid breaks may also occur. Radiation-induced malignant growth appears to be mediated by protein tyrosine phosphorylation followed by activation of the RAF oncogene. Accumulation of radiation-induced mutations may be prevented by suppressing cell proliferation due to interferon regulatory factor (IRF) that arrests the cell cycle. Independently from IRF, p53 may have similar effect. Also, ionizing radiation may induce the transcription of p21, a cell cycle inhibitor, by a p53 and IRF dependent mechanism. (See also ionizing radiation hazards, doubling dose, radiation measurement, UV, ultraviolet light, DNA repair, RAF, signal transduction, p21, p53, interferon, radiation hazard assessment, X-ray caused chromosome breakage, kinetics)

RADIATION, EVOLUTIONARY: the spread of taxonomic categories as a consequence of adaptation and speciation mediated by forces of selection, mutation, migration and random drift.

RADIATION HAZARD ASSESSMENT: for human whole body exposure of X or γ radiation, the minimal biologically detectable dose in mSv (milliSievert): no symptoms 0.01-0.05, chromosomal defects detectable by cytological analysis 50-250, physiological symptoms at acute exposure 500-700, vomiting in 10% of people 750-1250, disability and hematological changes 1,500-2,000, median human lethal dose 3,000 but at doses above 4,500 the mortality is expected to be above 50%. Prolonged exposure below 1,000 Sv may cause leukemia and death. Single exposure of the spermatogonia to 0.5 Sv may block sperm formation. A single exposure of women to 3-4 Sv and 10-20 Sv over a longer time (2 weeks) may result in permanent sterility. A fetus in a pregnant woman should not be exposed to any radiation but in case of an emergency it should not exceed a total of 0.005 Sv and any person under age 18 should not receive an accumulated dose over 0.05 Sv/year. *Occupational exposure* maximal limit in mSv for whole body 50, for lens of the eye 150, and for other specific organs or tissues 500. For cumulative exposure the maximal limit should be below 10 mSv x age in years. For public and educational and training exposure recommended maximal limit in mSv/year: for whole body 1, for eye, skin and extremities 5.0. During the spring 1996 the European Union (EU) lowered the permissible dose limits. Members of the public can be exposed to a maximum of 1 mSv each year (earlier limit was 5 mSv). Radiation industry personnel's limit of exposure is now 100 mSv in five consecutive years and an average limit of 20 mSv/year (previous limit was 50 mSv). These guidelines must be implemented within four years by EU member states.

The exposure by routine *medical* X-ray examination is supposed to be no higher than 0.04 to 10 mSv, by fluoroscopy or X-ray movie no higher 25 mSv, by dental examination involving the entire jaws not above 30 mSv. By comparison working in a normally operated nuclear power plant the exposure may be from 3 to 30 mSv/year. The exposure by living within a gran-

Radiation hazard assessment continued

ite building may amount to 5 mSv/year and a transcontinental flight may involve an exposure of 0.03 mSv. A color television set may deliver 0.001 mSv/year to the viewer if he/she stays very close to the set. One should keep in mind that there may be no threshold below which radiation would have no effect. The current approximate incidence of mutation in liveborn human offspring and the estimated increase per rem per generation in parenthesis: autosomal dominant 0.0025 - 0.0075 (0.000005 -0.00002), recessive 0.0025, X-linked 0.0004 (0.000001), translocations 0.0006 (0.0025), trisomy 0.0008 (0.000001). Approximately 5 Gy (500 rem) is considered the human lethal dose, the bacterium *Deinococcus radiodurans* can recover from doses as high as 30,000 Gy. This high radiation resistance can be explained by the very efficient recombinational repair in this prokaryote. The chromosomes are just as well broken into pieces as other DNAs but its genetic material exist in pairs and in 12-24 hours repair by recombination at the Holliday junctures restores their integrity. Even small doses, such as delivered by therapeutic X-radiation, may increase by about 1/3 the number of broken chromosomes and radiation by isotope treatment has similar effect, depending on the dose and duration of the exposure. Eventually the broken chromosomes, or at least some of them, may be eliminated from the body. Radiation sensitivity is generally positively correlated with the size of the genetic material although in polyploids the damage may not be readily detectable because of the redundancy of the genes. (See atomic radiation, isotopes, radiation effects, radiation measurement, radiation effects, radiation protection, doubling dose, radiation threshold, DNA repair, Holliday model of recombination, X-ray chromosome breakage)

RADIATION HYBRID (RH): the human chromosomes are broken into several fragments with 8000 rad dose of X-rays. The irradiated cells are quickly fused to (with the aid of polyethylene glycol) into somatic cell hybrids with Chinese hamster cells and thus translocations and insertions into the hamster chromosomes are generated. The further apart are two human DNA markers, the higher the chance for a breakage. Thus estimating the frequency of breakage, an information is obtained about "recombination" in a manner analogous to classical genetic recombinational mapping. The recombination frequency in radiation hybrids varies between 0 and 1 (no recombination or the markers are always independent, respectively). In meiotic recombination the maximal value is 0.5 for independent segregation. The frequency of breakage is estimated by the formula below and the recombination frequencies expressed in centiRays (cR). At 65 Gy the estimated 1 cR ≈ 30 kb, at 90 Gy 1 cR ≈ 55 kb. $\theta = [(A^+B^-) + (A^-B^+)/[T(R_A + R_B - 2R_AR_B)]$ where (A^+B^-) are the hybrid clones retaining A but not B and (A^-B^+) retain B but not A, T = the total number of hybrids and R_A, R_B stand for the recombinant fractions. The fragments retained can be analyzed also by PCR procedures and they are expected to carry with them neighboring sequences and thus provide information on physical linkage for sequences of about 10 megabase. The recombination process is dose-dependent. In one study 50 Gy permitted the retention of an intact chromosome arm in 10% of the cases whereas 40% had fragments of 3-30 Mb and 50% 2-3 Mb. Using 250 Gy less than 6% of the hybrids involved larger than 3 Mb pieces. If the fragments generated are intended for positional cloning, usually higher doses are used. The retention of fragments varies; centromeric pieces are more likely to be retained. The fragmented DNA can be further analyzed by probing with the aid of Southern blots, polymerase chain reaction, using sequence tagged sites (STS), FISH, etc. The fragments are unstable unless they are fused with rodent chromosomes. The radiation hybrids may retain up to a dozen or slightly more fragments. (See somatic cell hybrid, mapping genetic, physical mapping, framework map, Rhalloc, Rhdb, recombination, Gy, STS, FISH, Southern blot, probe, positional cloning, θ [theta], WGRH, PRINS)

RADIATION HYBRID PANEL: a set of DNA samples containing radiation hybrid clones derived by fusion of human and rodent cells. (See radiation hybrid)

RADIATION INDIRECT EFFECTS: the radiation generates reactive radicals (e.g., peroxides) in the environment that in turn inflict biological damage. (See radiation effects, target theory)

RADIATION IONIZING: see ionizing radiation

RADIATION MEASUREMENT: see Geiger counter, scintillation counter, proportional counter,

RADIATION, NATURAL: see isotopes, cosmic radiation

RADIATION vs NUCLEAR SIZE: the harmful biological and genetic effects of ionizing radiations depend on the size of the cell nuclei, at the same dose. Larger nuclei present larger target and suffer more damage than smaller ones. Haploid nuclei are more sensitive because the genes are present generally in a single dose. Polyploids are relatively less sensitive because the multiple copies of the chromosomes. (See radiation effects, ionizing radiation, physical mutagens)

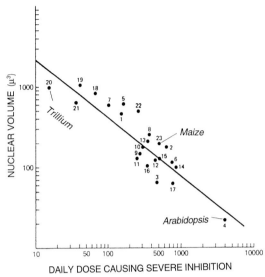

THE CORRELATION BETWEEN NUCLEAR SIZE AND RADIATION-SENSITIVITY IN PLANTS. 1. *Allium cepa*, 2. *Anethum graveolens*, 3. *Antirrhinum majus*, 4. *Arabidopsis thaliana*, 5. *Brodiaea bridgesi*, 6. *Graptopetalum bartramii*, 7. *Haworthia attenuata*, 8. *Helianthus annuus*, 9. *Impatiens sultanii*, 10. *Luzula pupurea*, 11. *Nicotiana glauca*, 12. *Oxalis stricta*, 13. *Pisum sativum*, 14. *Raphanus sativus*, 15. *Ricinus communis*, 16. *Saintpaulia ionantha*, 17. *Sedum orizyfolium*, 18. *Tradescantia ohiensis*, 19. *Tradescantia paludosum*, 20. *Trillium grandiflorum*, 21. *Tulbaghia violacea*, 22. *Vicia faba*, 23. *Zea mays*. (From Sparrow, A.H. et al. Radiation Bot. 1: 10)

RADIATION PARTICULATE: see physical mutagens

RADIATION PHYSIOLOGICAL FACTORS: the effect of radiation may be influenced by the species, age, developmental conditions, type of tissue and cells, metabolic state, genetic background, repair mechanisms, temperature, the chemical environment (presence of oxygen and other enhancing or protective compounds). (See physical mutagens, target theory, radiation hazard assessment)

RADIATION PROTECTION: using isotopes, ventilation should be by exhaustion via seamless ducts through the buildings and the particulate material must be trapped in appropriate filters or washed (scrubbed) before environmental discharge. The source of radiation must be shielded. The transmission of the shielding material depends on the peak voltage of the x-ray or on the energy of the emitting isotope and the thickness of the shield [attenuating material is characterized by half-value (HVL) or tenth-value layers (TVL)]. Examples :

kV	Lead millimeter		Concrete in centimeter	
X-ray	HVL	TVL	HVL	TVL
50	0.05	0.16	0.43	0.15
100	0.24	0.8	1.5	5.0
200	0.48	1.6	2.5	8.25
500	3.6	11.9	3.5	11.5
1,000	7.9	26.0	4.38	14.5
4,000	16.5	54.8	9.00	30.00
10,000	16.5	55.0	11.5	38.25
^{60}Cobalt	6.5	21.6	4.75	15.50

Cracks, seams, conduites, filters, ducts, etc. should be kept under continued surveillance for possible leaks. Protective clothing (aprons, gloves, etc.) affords very limited protection. Radioactive waste must be disposed of according to government and local standards whichever is more strict. Radiation areas should be identified by caution signs. "Radiation Area" is defined by the Occupational Safety and Health Standards of the USA as an area where a major portion of the body could receive in any hour a dose in excess of 5 millirem, or in any 5 con-

Radiation protection continued

secutive days a dose in excess of 100 millirem. "High Radiation Area" means any area, accessible to personnel, in which there exists radiation at such levels that a major portion of the body could receive in any one hour a dose in excess of 100 millirem. For non-ionizing electromagnetic radiation within the range of 10 MHz (megahertz, 10^7 cycles/sec) to 100 Ghz (gigahertz, 10^{11} cycles/sec) the energy density should not exceed 1 mW (milliwatt)/cm^2/0.1 hour. [Such radiations are within the realm of radio, microwave and radar range.] Emergency plans must be prepared for spillage, cleanup and fire. Never work with any hazardous material before sufficient training is obtained for safe storage, handling and emergencies by all personnel involved. (See isotopes, atomic radiation, ionizing radiation, radiation effects, radiation hazard assessment, radiation measurement, electromagnetic radiation)

RADIATION RESPONSE: deletions (single chromosomal breaks) and mutations occur with 1st order kinetics, whereas the majority of chromosomal rearrangement (inversions, translocations) are proportional to the square of the dose and thus follow 2nd or 3rd order kinetics. (See physical mutagens, radiation effects, kinetics, chromosomal aberrations,)

RADIATION SAFETY : see radiation protection, radiation hazard assessment, atomic radiation, cosmic radiation

RADIATION SAFETY STANDARDS: see radiation hazard assessment, radiation protection

RADIATION-SENSITIVITY: see DNA repair, chromosome breakage, aging, microcephaly [Nijmegen breakage syndrome], nevoid basal cell carcinoma, ataxia telengiectasia, xeroderma pigmentosum, radiation hazard assessment)

RADIATION SICKNESS: occurs when the human body or parts of it is exposed to ionizing radiation. The symptoms and hazards are dependent on dose. (See radiation hazard assessment)

RADIATION THERAPY: ionizing, electromagnetic radiation has anti-mitotic and destructive effects on live tissues and it is employed for suppress cancerous growth. Radiotherapy has been used to treat lymphnodes to suppress Hodgin's disease, radioactive isotopes can be applied by injection for localized radiation. Similar effects are expected by the use of magic bullets. Blood withdrawn from the body has been irradiated by UV light and returned to the system. (See radiation effects, magic bullet, Hodgin disease, lymphnode, radiation hazard assessment)

RADIATION THRESHOLD: the minimal harmful radiation dose is very difficult to determine because the visible physiological signs may not truly reflect the long-range mutagenic and carcinogenic effects. The maximal permissible doses for medical diagnostic or occupational radiation exposures reflect only conventional limits that have been revised many times as the sensitivity of physical and biological detection methods as well as the instrumentation of delivery have improved. It is conceivable that no low threshold exists. (See radiation response, cosmic radiation, mutation frequency-undetected mutations, radiation safety standards, radiation hazard assessment)

RADICAL: an atom or group with an unpaired electron, a free radical.

RADICAL SCAVENGER: may combine with free radicals and reduce the potential harm caused by the highly reactive molecules.

RADICLE: the seed or primary root of a plant embryo. The smallest branches of blood vessels and nerve cells.

RADIN BLOOD GROUP (Rd): is encoded in the short arm of human chromosome 1; its frequency is low.

RADIOACTIVE DECAY: see isotopes

RADIOACTIVE ISOTOPE: see isotopes, radioactive tracer, radioactive label

RADIOACTIVE LABEL: compounds (nucleotides, amino acids) containing radioactive isotopes are incorporated into molecules to detect their synthesis, fate or location (radioactive tracers) in the cells. For detection scintillation counters or autoradiography is used most frequently. Geiger counters may also detect qualitatively their presence. For cytological analysis those isotopes are used that have short path of radiation and display distinct, sharp marks on the film,

Radioactive label continued

e.g. tritium (^{3}H). For molecular biology most commonly ^{32}P (in DNA, RNA), or ^{14}C and ^{35}S (in proteins) or ^{125}I in immunoglobulins are used. (See also Southern blotting, Northern blotting, Western blotting, nick translation, non-radioactive labels, isotopes)

RADIOACTIVE TRACER: a radioactively labeled compound that permits tracing the biosynthetic transformations of the supplied chemical by determining when the radioactivity appears in certain metabolites after the supply. Also it reveals what part of a later metabolite has acquired the label from the supplied substance. The availability of $^{14}CO_2$ permitted tracing the path of photosynthesis and through the use of various isotopes, the metabolims of pharmaceuticals and role of hormones, etc. were determined. (See radioactive label, radioimmunoassay)

RADIOACTIVITY: emission of radiation (electromagnetic or particulate) by the disintegration of atomic nuclei.

RADIOACTIVITY DATING: see evolutionary clock

RADIOACTIVITY MEASUREMENT: see radiation measurements

RADIOAUTOGRAPHY: same as autoradiography

RADIOIMMUNOASSAY (RIA): the most commonly used isotope for radioimmunoassays, ^{125}I$^+$ is generated by oxidation of Na^{135}I by chloramine-T (N-chlorobenzene sulfonamide). This then labels tyrosine and histidine residues of the immunoglobulin without affecting the binding of the epitope and provides an extremely sensitive method for identifying minute quantities (1 pg) of the antigen in an experiment. Target proteins can be radiolabeled also by ^{35}S-methionine or ^{14}C. In the *Competition RIA* unlabeled target protein competes with a labeled antigen for binding sites on the antibody. The amounts of bound an unbound radioactivity are then quantified. In *Immobilized Antigen RIA* unlabeled antigen is attached to a solid support and exposed to radiolabeled antibody. The amount of radioactivity bound then measures the amount of specific antigen present in the sample. In the *Immobilized Antibody RIA* single antibody is bound to a solid support and exposed to labeled antigen. Again the amount of bound radioactivity indicates the amount of the antigen present. In the *Double-Antibody RIA* one antibody is bound to a solid support and exposed to an unlabeled antigen. After washing, the target antibody is quantitated by a second radiolabeled antibody (instead of the radiolabel biotinylation can also be used). This assay is very specific because it involves a step of purification. (See antibody, immune reaction, immunoglobulins, isotopes, Protein A)

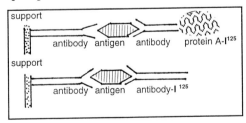

SUPPORT: diazobenzyoxymethyl cellulose or PVC
← IMMOBILIZED DOUBLE ANTIBODY RIA

← IMMOBILIZED ANTIGEN RIA

RADIOISOTOPE: see isotopes

RADIOISOTOPE DATING: see fossil records

RADIOMIMETIC: agents, primarily alkylating mutagens that may break single or both chromatids (isochromatids). Although it cannot be ruled out that the two chromatids break at the same place simultaneously, it is believed that isochromatid breaks are due to replicational events involving initially single chromatid breaks. They mimic the effects of ionizing radiation. (See alkylating agents, nitrogen mustards, sulfur mustards, epoxides, ionizing radiation)

RADIOMORPHOSES: morphological alterations in plants caused by ionizing irradiation during the life of the irradiated individuals; these effects may not be genetic.

RADIONUCLIDE: a nuclide that may disintegrate upon irradiation by corpuscular or electromagnetic radiation. (See nuclide)

RADIOTHERAPY: see radiation therapy

RADISH (*Raphanus sativus*): is a cruciferous vegetable crop; 2n = 2x = 18. (See *Raphanobrassica, Brassica oleracea*, mustards)

RADIX (root): is a multiplier of successive integral powers of a sequence of digits, e.g. if the radix is four, then 213.5 means 2 times 4 to the second power plus, 1 times 4 to the first power plus, 3 times 4 to the zero power plus 5 times 4 to the minus 1 power.

RADON (Rn): ^{219}Rn (An) is a member of the actinium series, ^{220}Rn (Tn) is a member of the thorium emanation group, ^{222}Rn is derivative of uranium, a heavy (generally accumulates in the basements), colorless radioactive noble gas and it is formed from uranium (radium emanation) contaminated rocks. It may pose health hazards in buildings at some locations and with poor ventilation.

RAF1: see *raf*

raf: v-oncogene (cytoplasmic product is protein-serine/threonine kinase). The cellular homolog RAF1 is closely related to ARAF oncogenes. The v-raf is homologous to the Moloney murine leukemia virus oncogene. The avian MYC oncogene is the equivalent to the murine RAF. RAF1 has been assigned also to human chromosome 3 and a pseudogene RAF2 is in chromosome 4. Human renal, stomach and laryngeal carcinoma cells revealed RAF1 sequences. Raf may be recruited to the cytoplasmic membrane by a carboxy-terminal anchor (RafCAAX) and its activation then is independent from Raf which is associated with the plasma membrane cytoskeletal elements and not with the lipid bilayer. (See ARA, MYC, Moloney, signal transduction, v-oncogene)

RAFT1: see TOR

RAFTK: see CAM

RAG1, RAG2 (recombination activating gene): are closely linked in mouse and encode the proteins of lymphocyte-specific recombination of the V(D)J sequences of the immunoglobulin genes. The functional part of RAG1 is the core sequence whereas other tracts can be deleted without affecting recombination. Mutation in RAG results in inability to form functional antigen receptors on B cells. (See antibody, immunoglobulins, junctional diversification, combinatorial diversification, RSS, V(J)D recombinase)

RAGE (receptor for advanced glycation endproduct): see Alzheimer disease

RAGWEED: mainly annual species of the genus *Ambrosia* (Compositae) widespread in North-America and Central-Europe and cause pollen allergy (hay fever) of the nose and eye, skin irritations and even asthma without hay fever. The susceptibility is genetically controled by a locus, *Ir*, within the HLA complex. The IgE antibody is elicited by the antigen E contained in the ragweed pollen. (See HLA, allergy, atopy, ragweed inflorescence and leaf above)

RAIDD: an adaptor protein that joins the ICE/CED-3 apoptosis effector molecules. (See apoptosis, ICE)

RALA: is RAS-like protein encoded in human chromosome 7. (See RAS oncogene)

RALB: is a RAS like protein, encoded in human chromosome 13. (See RAS oncogene)

RALOXIFENE: a lipid-like molecule that enters the cell nucleus and can bind to the special raloxifene-response element in the DNA and activates the tumor growth factor-β3 gene. Raloxifene is activated by 17-epiestriol an intermediate in the secretion of estrogen. (See hormone-response elements, tamoxifene, animal hormones)

RAM: see random-access memory; also rabbit-anti-mouse immunoglobulin.

ram: ribosomal ambiguity mutation causing high rate of translational error. (See ambiguity in translation)

RAMAN SPECTROSCOPY: is similar to infrared spectroscopy, it uses the 10,000 to 1,000 nm spectral regimes to detect different rotational and vibrational states of molecules. If two atoms are far apart, their interaction is negligible. If they are very close they may show repulsion. It is useful to obtain information on the physical state of nucleic acids. (See also FT-IR)

RAMET: clonal descendant of a clone.

RAMIE (*Boehmeria nivea*): subtropical-tropical, monoecious fiber plant; 2n = 2x = 14.

RAMUS (plural rami): a branch; it is used in various word combinations.

RAN (TC4): a guanosine triphosphatase (GTPase) required for import through nuclear pores; it participates also as a switch of GTP→GDP in DNA synthesis and cell cycle progression. The binding protein RanBP enhances the activity of RanGTP-activating protein Ran GAP. (See GTPase, RAS, signal transduction, cell cycle, RNA transport, RCC, GAP, nuclear pore)

RANA (genus of frogs): both the American leopard frog (*Rana pipiens*) and the European frog *R. temporaria* have 2n = 26 chromosomes. The frogs are generally more aquatic species than the toads. (See *Bufo, Xenopus*)

RANDOM ACCESS MEMORY (RAM): a storage of information that can be refered to at any order as long the computer is turned on.

RANDOM AMPLIFIED POLYMORPHIC DNA: see RAPD

RANDOM CHROMOSOME SEGREGATION: the gene under studied is (absolutely) or very closely linked to the centromere in polyploids. (See maximal equational segregation)

RANDOM FIXATION: see random genetic drift, founder principle

RANDOM GENETIC DRIFT: is change in gene frequency by chance. Such random changes are most likely to occur when the effective size of the population is reduced to relatively few individuals. Since random drift may occur repeatedly, eventually large changes may result in the genetic constitution of the population. Such changes are most likely when a few individuals migrate to a new (isolated) habitat. (See effective population size, founder principle)

RANDOM MATING: each individual in the population has an equal chance to mate with any other of the opposite sex (panmixis). In a population involving two different allelic pairs the frequency of the mating genotypes and the genotypic proportion of their progenies derived from the binomial distribution by expanding $(p + q)^4$:

MATES →	(A1A1)×(A1A1)	(A1A1)×(A1A2)	(A1A2)×(A1A2) (A1A1)×(A2A2)	(A1A2)×(A2A2)	(A2A2)×(A2A2)
FREQUENCY →	p^4	$4p^3q$	$6p^2q^2$	$4pq^3$	q^4
PROGENY		**A1A1**	**A1A2**	**A2A2**	
		p^4	$2p^3q$	p^2q^2	
		$2p^3q$	$4p^2q^2$	$2pq^3$	
		p^2q^2	$2p^3$	q^4	
		4	8	4	→ SUM = 16

(See Hardy-Weinberg theorem, mating systems, Pascal triangle)

RANDOM OLIGONUCLEOTIDE PRIMERS USED FOR SYNTHESIS OF RADIOACTIVE PROBES: heterogeneous oligonucleotides can anneal to different and many positions along a nucleic acid chain. They can also serve as primers for the initiation of DNA synthesis. If the precursors are one type of radioactive [α-^{32}P]-deoxyribonucleotide (dNTP), and cold dNTPs, highly radioactive probes can be obtained. Single stranded DNA templates can be copied by the aid of Klenow fragment of DNA polymerase I or in case of RNA template reverse transcriptase can be used. The primers are usually short (6 to 12 base) and can be generated either by DNA-ase digestion of commercially available DNA (from calf thymus or salmon sperm) or produced by an automatic DNA synthesizer. (See probe, nick translation)

RANDOM SAMPLE: is withdrawn from a collection without any selection.

RANTES: is a chemoattractant of cytokines for monocytes and T cells. The chemokine receptors appear to be seven membrane proteins, coupled to G proteins. It is also involved with a transient increase of cytosolic Ca^{2+} and also Ca^{2+} release. The opening of the calcium channel increases the expression of interleukin-2 receptor, cytokine release and T cell proliferation. Thus RANTES in addition to inducing chemotaxis, it can act as an antigen-independent activator of T cells *in vitro*. RANTES and MIP-1 chemokines along with the receptor CC CKR5 and fusin are believed to suppress replication of HIV. (See MIP-1a, fusin, chemotaxis, cytokine, CC CKR5, HIV, acquired immunodeficiency, T cell)

RAP1A, RAP1B, RAP2: are RAS related human proteins. The Rap1A is a suppressor protein of Ras-induced transformation; it has identical amino acid sequences with the effector region of Ras p21. RAP1 along with SIR3 are also transcriptional repressors of telomeric heterochromatin. RAP74 is a subunit of transcription factor TFIID and RAP74 is involved in binding to the serum response element. The RAP1 (repressor/activator protein) of yeast binds to upstream activator sequences (UAS) alone and in association with other proteins and activates many genes in addition to silencing the mating type and the telomerase functions. (See RAS oncogene, silencer, serum response element, transcription factors, mating type determination, telomerase, TRF1).

RAPAMYCIN: see FK506, TOR

RAPD: (pronounce rapid) markers are generated by random amplified polymorphic DNA sequences using (on average) 10 basepair primers and the PCR technique for the physical mapping of chromosomes on the basis of DNA polymorphism in the absence of "visible" genes. The map so generated may, however, be integrated into RFLP and classical genetic maps. (See polymerase chain reaction, physical mapping, sequence-tagged site, integrated map)

RAPE (*Brassica napus*): oil seed crop, 2n = 38. (See canola, erucic acid)

RAPHANOBRASSICA: is a man-made amphidiploid (2n = 36) of radish (*Raphanus sativus*, n = 9) and cabbage (*Brassica oleracea*, n = 9). (See *Brassica oleracea*, radish)

RAPHE: ridge on the seeds where the stalk of the ovule was attached; seam of tissues.

RAPHIDS: needle-like crystals within plant cells (often of oxaloacetic acid).

RAPHILIN: peripheral membrane protein. It may bind RAB proteins in a GTP-dependent manner and may be phosphorylated by various kinases and may bind Ca^{2+} and phospholipids. (See RAB)

RAPID LYSIS MUTANTS: *r* mutants of bacteriophage rapidly lyse the infected bacteria and therefore the size of the plaques are much larger than the ones made by wild type phage. (See lysis, plaque lift)

RAPID LYSIS PLAQUE

WILD TYPE PLAQUE

RAPSYN (M_r 43 K): peripheral membrane protein co-localized with the acetylcholine receptors at the neuromuscular synapse. (See acetyl choline)

RAR, RARE (retinoic acid receptor): and RXR-α, -β, -γ retinoid-X receptors are transducers of ligand-activated morphogenetic and homeostasis signals. RAR and RXR can form homodimers but usually found as heterodimers. These then bind to the cognate hormone response elements and increase the efficiency of transcription. RAR-α ligands can accomplish the binding of the RXR-RAR-α dimers to DNA causing RXR activation and initiating the transcriptional activity of RAR-α. It is encoded in human chromosome 17q21. (See hormone response elements, retinoic acid)

RARE (RecA-assisted restriction endonuclease): at the site of a restriction enzyme recognition in or near a locus a triplex structure, with an oligonucleotide, is generated with the assistance of enzyme RecA. The genome is then enzymatically methylated and RecA is removed. Only the site now unprotected will be cut by the restriction endonuclease. This procedure thus reduces the actual cleavage sites in the DNA. (See recA, restriction enzyme, DNA methylation)

RARE-MATING: cells of non-mating yeast strains are mixed with cells that are expected to mate under favorable conditions. Then low frequency of mating may take place and the progeny may be isolated by efficient selection. The rare-mating is presumed to be the result of mitotic recombination or non-disjunction of chromosome III in the non-mating strain. When normal mating fails, protoplast fusion may generate hybrids. The latter procedure results, however, in complex progeny. (See mating type determination, protoplast fusion)

RAS: is a protooncogene originally found in rat sarcoma virus; it codes for a monomeric GTP-binding protein in which point mutation in codons 12, 13 or 61 may lead to oncogenic transformation. The RAS proteins have important role in transmembrane signaling. This role may vary depending on cell type, from stimulation of adenylate cyclase to mating factor signal transduction and from proliferation to differentiation. The RAS protein becomes active only after prenylation by the 15 carbon farnesyl pyrophosphate. The prenylation is thioether forma-

RAS continued

tion with an amino acid, resulting in association of the protein with a membrane. RAS is one of the most important turnstile in signal transduction and one of the most common activated oncogen. The activation involves changing of the bound GDP into GTP. GAPs (GTPase activating proteins) inactivate RAS by hydrolysis of GTP. In contrast GNRPs (guanine nucleotide releasing proteins) mediate the replacement of bound GDP by GTP and the activation of RAS. Receptor tyrosine kinases activate RAS either by inactivating GAP or activating GNRP. The family RAS is represented in various human chromosomes: NRAS in 1p21, HRAS in 11p15, KRAS in 6p12-p11, RRAS in 19. In mammals on the basis of homology three groups may be classified: (i) RAS, RAL, RRAS, (ii) RHO, and (iii) RAB. In *Drosophila* there are *Ras1* in 3-(49), *Ras2* in 3-(15), *Ras3* to 3-1.4 (the latter has higher homology to *Rap1* and it is called now *Rap1*). In yeast the RAS homologs (*RAS1, RAS2*) are very closely related to the human protooncogenes and can be replaced by them. RAS protein is necessary also for the completion of mitosis, in association with other factors. Probably all eukaryotes carry RAS homologs. The p21 protein is also a mobile RAS protein. The RAS oncogene is generally active in tumorigenesis in the presence of the MYC or the E1A "immortalizing" oncogene products. The RAS promoter is high in GC and lacks a TATA box. The various RAS genes (human, mouse) may have over 50% difference at the nucleotide level but the amino acid composition is highly conserved. (See G-proteins, RAB, RALA, RALB, RHO, RAP, RASA, oncogenes, signal transduction, adenylate cyclase, farnesyl, prenylation, p21, GTPase)

RASA: is a guanosine triphosphate activating RAS protein (21 kDa [p21]) encoded by human chromosome 5q13.3 and in mouse chromosome 13. (See RAS oncogene)

RASPBERRY (*Rubus* spp.): the majority of raspberries are diploid (2n = 14) but loganberry is 2n = 42, and the blackberries exist with 2n = 28, 42 and 56 chromosomes. Some wild blackberries are probably allopolyploids with 2n = 35 and 2n = 84; the latter is dioecious.

RASMUSSEN's ENCEPHALITIS: see epilepsy

RASSENHYGENIE: the German term for negative eugenics [often so used]. Its goal was to protect the "purity" of the Aryan (German) race and enforced by the laws of the Nazi state. It resulted, between 1933-1945, in 350,000 forced sterilization, mass murders and ban on marriages between genetically fit and "unfit", and persons who's ancestry included more than 1/4 Jews, Gypsys or some other racial groups. (See eugenics)

RAT (*Rattus norvegicus*, 2n = 42): a genetic linkage map has been published by Jacob, H.J. *et al.* Nature Genetics (9:63 [1995]. The chromosome number of other rat species may be different.

RATE-LIMITING STEP: requires the highest amount of energy in a reaction chain or in a metabolic path the slowest step.

RATIONALE: the logical basis of an act, process or argument.

RATIONALIZE: an attempt to make something conform to reason. Sometimes apparent rationalization is applied in an effort to explain facts or ideas with inadequate justification.

RB: right border of the T-DNA. (See T-DNA)

Rb: see retinoblastoma

rBAT/4F2hc: are four-membrane-spanning proteins involved in membrane transport or regulation of transport of neutral and positively charged amino acids. (See cystinuria, transporters)

rbc: ribulose bisphosphate carboxylase/oxidase genes. (See chloroplast genetics)

RBE: relative biological effectiveness of radiation; it depends on a number of physical (type of radiation, wave length, dose rate, temperature, presence of oxygen, hydration, etc.), physiological (developmental stage) and biological factors (species, nuclear size and DNA content, level of ploidy, repair system, etc.). (See rem, radiation effects)

RBM (RNA-binding-motif): a gene family in the Y chromosome of mammals involved in male fertility. (See DAZ, boule)

R2Bm: is a silk worm retroelement without long terminal repeats. (See retrovirus)

RBTN (rhombotin): a cystin-rich oncoprotein family (encoded in human chromosome 11) containing a LIM domain. (See LIM domain)

RCC1: is a chromatin-bound guanine nucleotide release factor forming complexes with RAN (a G protein). Its deficiency interferes with the cell cycle progression, nuclear RNA export and protein import. It is a part of the nuclear pore complex. Its yeast homolog is Prp20p. (See RAN, cell cycle, nuclear pores, RNA transport)

RcsC: *E. coli* kinase affecting capsule synthesis regulator RcsB.

RDA (representational difference analysis): is genome scanning procedure for the detection and identification genetic markers representing disease genes. The cellular DNA is cut by one or more restriction endonucleases and the smaller fragments are amplified by PCR. Then DNA samples from affected and disease-free samples are denatured and the mixtures of the two samples are allowed to anneal. The sequences not matching fail to hybridize and those are expected to be responsible for the disease. The process is similar in principle to cascade hybridization. The relatively rapid mass screenings was expected to identify individuals liable to particular hereditary differences (diseases). The same procedure may be applicable also to non-disease genes of eukaryotes. RDA and GDRDA procedures can be used also to generate genetic maps in organisms with a paucity of chromosomal markers. (See also cascade hybridization, PCR, GMS, GDRDA, genetic screening, positional cloning, RNA fingerprinting, genomic subtraction)

rDNA: DNA complementary to ribosomal RNA. (See ribosome, rRNA)

REACTION, CHEMICAL: change in the atoms in or between molecules.

REACTION INTERMEDIATE: generally a short life chemical in a reaction path.

REACTION NORM: is the range of phenotypic potentials of expression of a gene or genotype. Usually, the genes do not absolutely determine the phenotype but they permit a range of expressions, depending on the genetic background, developmental and tissue-specificity conditions and the environment. (See genotype, phenotype, regulation of gene activity)

REACTIVATION: see multiplicity reactivation, Weigle reactivation, marker rescue)

READING DISABILITY: see dyslexia

READING FRAME: the triplets codons can be read in three different registers, starting with the first, second or third, however, only one may spell the correct protein. (See open reading frame, frame shift mutation)

READOUT: the DNA sequence recognition by proteins that may be *direct* (by hydrogen bonding) or *indirect* when the DNA conformation also plays a role.

READTHROUGH: ribosome continues translation downstream of a stop codon. (See translation termination, autogenous suppression)

READ-THROUGH PROTEIN: formed when a suppressor tRNA inserts an amino acid at a site where chain termination is normally expected because of a nonsense codon and thus produces a fusion protein from two different "in-frame" cistrons, separated by a nonsense codon. Read-through may be brought about by mutation in the anticodon of a tRNA or modification of the tRNA: e.g., selenocysteinyl-tRNA inserts selenocysteine into glutathione oxydase by recognizing the UGA (opal) stop codon. (See also gene fusion, transcriptional gene fusion, translational gene fusion, trapping promoters)

REAL TIME: means the actual time during which the physical process takes place.

REALIZED HERITABILITY: see gain

REANNEALING (reassociation): double-stranded DNA can be heat denatured (strands separated), and can be restored to double-stranded form, reannealed, when the temperature becomes lower than 60° C. (See c_0t curve)

REARRANGEMENTS: structural changes of the chromosome(s) (transclocation, inversion)

REASSOCIATION KINETICS: see c_0t curve

RecA PROTEIN: is 38.5 kDa polypeptide involved in homologous recombination by promoting pairing. It is a DNA-dependent ATPase, mediating strand exchange. RecA binds to single-stranded DNA and pre-synaptic nucleoprotein molecules mediate the pairing with the duplex DNA target. The paired DNA is then within a 25 Å hole. Within this cavity projecting toward

Rec A protein continued

the axis of the helix are mobile loops L1 and L2 representing the binding sites. The RecA protein expressed in transgenic plants substantially increases recombinational repair of DNA damage inflicted by mitomycin. (See recombination molecular mechanism prokaryotes, DNA repair, *RecA1*)

rec⁻: one or another type of recombination-deficient mutation.

recA1: a recombination deficient mutation of *E. coli* (map position 58 min) coding for a DNA-dependent ATP-ase, a 3522 amino acid residue enzyme. Plasmids carrying it remain monomeric and do not form multimeric circles. When M13 vectors carry it, the foreign passenger DNA has fewer deletions. The recA protein mediates the association of double stranded DNAs by synapsis mainly in the major grove but also in the minor grove of the DNA. The RecA-mediated pairing involves a triplex structure, i.e. along parts of the sequences double stranded DNA associates transiently with a single strand of the other DNA molecule. The pairing of the DNA molecules may be plectonemic (intertwined) and thus may not require stabilization by proteins. The paranemic coils are only juxtapositioned and require protein to keep them together. Experimental information indicates that ATP hydrolysis is not required for the exchange between paired strands rather the removal of RecA requires ATP hydrolysis. In case the homology between the DNAs is not perfect, ATP is needed for the exchange. RecA is also involved in branch migration but it is assisted in the process by RuvAB and RecG proteins. The extension of the DNA heteroduplex (at the rate of 2-10 bp/sec) in the 5'→3' direction needs ATP hydrolysis The length of the heteroduplex may become 7 kbp long. In both prokaryotes and eukaryotes besides RecA (or homologs), a stimulatory exchange protein, binding single strands of the DNA (SSB) is required. The SSB monomers (1/15 base in ssDNA) facilitate synapsis between the heterologous strands. After exchange RecA promotes DNA renaturation. The RecA homologs in yeast, mouse and humans are the RAD51 proteins, and Mei3 in *Neurospora*. (Recombination molecular mechanisms prokaryotes, branch migration, *recB* and other *Rec* genes below, DNA repair)

recB: *E. coli* gene (map position 60 min) encoding a subunit of exonuclease, controlling recombination and genetic repair. (See recombination molecular mechanism prokaryotes, recombination models, DNA repair)

RecBCD: an enzyme (ribozyme) complex functioning in recombination of prokaryotes. (See recombination molecular mechanisms prokaryotes, recombination models, DNA repair)

recC: *E. coli* gene (map position 60 min) encoding a subunit of exonuclease V, controlling recombination and genetic repair. (See recombination molecular mechanisms, DNA repair)

recE: locus of Rac prophage (map position 30 min), encoding exonuclease VIII and promotes homologous binding between single-stranded and double-stranded DNAs. (See RecA)

RecF: a single- and double-strand-binding recombination protein.

RecJ: a single-strand exonuclease, used in recombination of *E. coli*.

RecQ: *E. coli* DNA helicase.

RecR: mediates DNA renaturation during recombination of *E. coli*.

RecT: same as recE.

RECEPTACLE: the widened end of a flower stalk. Also, a container.

RECEPTOR: also called an operator, the site responding to the controlling element (transposase) as originally called by Barbara McClintock the components of the *Spm* transposable systems in maize. (See transposable elements of maize receptors)

RECEPTOR: proteins that bind to ligands with cellular signaling functions. The receptors may be located within the plasma membrane (transmembrane proteins) or intracellularly and bind ligands which penetrate cells by diffusion. (See signal transduction, hormone receptors, transmembrane proteins, serine/threonine kinase, receptor tyrosine kinase, receptor guanyl cyclase, receptor tyrosine phosphatase, receptors, adaptor proteins, T cell, nuclear receptor, orphan receptor, TCR)

RECEPTOR DOWN REGULATION: epidermal growth factor (EGF) binding receptors con-

Receptor down regulation continued

centrate in coated pits after binding with this growth factors. They go into the lysosomes where degradation of the receptor and EGF takes place. The cell surface will have a reduced number of them because of receptor down regulation. (See endocytosis, EGF)

RECEPTOR GUANYLYL CYCLASE: are transmembrane proteins associated at the cytosolic end with an enzyme that generates cyclic guanosine monophosphate (cGMP). cGMP then activates cGMP-dependent protein kinase (G-kinase) that phosphorylates serine/threonine residues in proteins. (See cGMP, serine/threonine kinase)

RECEPTOR SERINE/THREONINE KINASE: is the receptor of serine/threonine phosphorylating enzymes. (See serine/threonine kinase)

RECEPTOR TYROSINE KINASE (RTK): binds protein tyrosine kinase enzymes such as the receptors for the epidermal growth factor (EGF), insulin, insulin-like growth factor-1 (IGF-1), platelet-derived growth factor (PDGF), fibroblast growth factor (FGS), nerve growth factor (NGF), hepatocyte growth factor (HGF), vascular endothelial growth factor (VEGF), macrophage colony stimulating growth factor(M-CSF). These receptors are transmembrane proteins and when the receptor becomes associated with the cognate phosphorylase enzyme both the receptor and the target protein receive phosphate groups from ATP at certain tyrosine residues. The various regulatory proteins recognize different phosphorylated tyrosine residues in the receptor. Upon binding to their specific sites they may also be phosphorylated on their own tyrosine residues and become activated. A cascade of events may follow that activate entire signaling pathways. The different receptors and the associated proteins may control the separate or interacting signaling pathways. (See tyrosine kinase, signal transduction)

RECEPTOR TYROSINE PHOSPHATASE: binds protein tyrosine phosphates and splits off phosphate groups.

RECEPTOR-MEDIATED ENDOCYTOSIS: a very efficient delivery system of macromolecules (such as cholesterol), that adhere to coated pits, into cellular organelles. (See coated pits)

RECEPTORS: are proteins that bind other molecules (ligands). They can be *extracellular* receptors that respond to outside signals reaching the cells. They may be located on the *surface* of the cell membrane or *within the membrane* but their ligand-binding domains are exposed to the extracellular space. The *intracellular* receptors respond to ligands that diffuse into the cell. (See ligand, signal transduction, cargo receptors, receptor, adaptor proteins, T cell)

RECESSIVE: expression of a gene means that it is not visible in heterozygotes in the presence of the wild type or other dominant alleles of the locus. Recessivity is not necessarily an absolute lack of the expression of the gene (except in null alleles) because some very low level of transcription/translation may not be detectable by a particular type of study but may be observable by a finer analysis. (See dominance, semidominance)

RECESSIVE ALLELE: does not contribute to the phenotype in heterozygotes in the presence of the dominant allele.

RECESSIVE EPISTASIS: see epistasis, modified checkerboards

RECESSIVE LETHAL: dies when homozygous, and can be maintained only as heterozygote.

RECESSIVE LETHAL TESTS, *DROSOPHILA*: see *Basc, Cl* method, sex-linked recessive lethal, autosomal recessive lethal assay

RECESSIVE ONCOGENES: tumor-suppressor genes such as encoding p53. (See tumor sup pressors, p53)

recF: *E. coli gene* (map position 82 min) also called *uvrF*, controls recombination and radiation repair. (See recombination molecular mechanisms, DNA repair)

recG: *E. coli* gene (map position 82 min) controls recombination. (See recombination molecular mechanism)

RECIPIENT: bacterial cells of the F⁻ state receive genetic material from the donor F⁺ strains. Cell to what genetic material is transfered. (See conjugation, transformation)

RECIPIENT SITE: see donor site

RECIPROCAL CROSSES: for example A x B and B x A. In the majority of organisms, usually

Reciprocal crosses continued

the phenotype of the reciprocal hybrids is identical. In cases when between the two parents cytoplasmically determined differences exist, the F_1 offspring resembles closer the female parent that usually transmits the cytoplasm. These reciprocal differences may persist indefinitely in the advanced generations. Although the reciprocal differences usually are most obvious in plants, animal hybrids, e.g. the mule and the hinny are also easily distinguishable. (See mitchondrial genetics, chloroplast genetics)

RECIPROCAL HYBRIDS OF *Epilobium hirsutum* Essen and *Epilobium parviflorum* Tübingen. BOTH PARENTS ARE 2n = 36. IN THE CROSS AT THE LEFT AN *E. parviflorum* FEMALE WAS CROSSED BY AN *E. hirsutum* MALE. THE TWO PLANTS AT THE RIGHT REPRESENT THE RECIPROCAL CROSS WHEN THE *E. hirsutum* FEMALE PROVIDED THE CYTOPLASM.
(From Michelis, P. Umschau 1965 (4):106)

RECIPROCAL INTERCHANGE: the same as reciprocal translocation of chromosomes.

RECIPROCAL RECOMBINATION: is the most common exchange between homologous chromatids at the 4-strand stage of meiosis in eukaryotes. In case of single crossing over in an interval two parental types and two cross over strands are recovered. Exception is gene conversion where the exchange is non-reciprocal. In conjugational transfer in bacteria the reciprocal products of the event are not recovered and

| PARENTAL | AB and ab |
| RECIPROCAL RECOMBINANTS | Ab and aB |

their fate is unknown. In sexduction and specialized transduction reciprocal recombination may take place also in bacteria. (See recombination molecular mechanisms prokaryotes, conjugation, sexduction, specialized transduction)

RECIPROCAL SELECTION: see recurrent selection

RECIPROCAL TRANSLOCATION: segments of non homologous chromosomes are broken off and reattached to each other's place. As a consequence, generally 50% of the gametes of the translocation heterozygotes (formed by adjacent distributions) is defective because they do not have the correct amount of chromatin. (See more at translocation)

Rec-MUTANT: deficient in recombination and possibly altered in other functions of DNA.

RECODING: is a mechanism by what the same DNA sequence may be translated in more than one way. It is a common mechanism in viruses with overlapping genes. There are several other manners this can take place. Some genes utilize multiple promoters and depending on the choice of their utilization, the same RNA may code for more than one protein. Frameshifting may take place: e.g. the mRNA may show slippage on the ribosome, a $tRNA^{Leu}$ with an anticodon GAG may recognize CUUUGA in one frame and in a shift it inserts leucine (UUU) for 4 nucleotides: CUUU GA. Similar frameshifting cassettes may be determined by *E. coli* gene *SF2* and also in other prokaryotes. In the TY3 transposable element of yeast the GCG AGU U instead of the Ala (GCG) and Ser (AGU) it may read GCG A GUU Ala (GCG) and Val (GUU). The code words may be interpreted different ways and stop codons may specify selenocysteine, tryptophan and glutamine. The ribosome may also skip certain sequences, e.g. the T4 phage topoisomerase may bypass 50 contiguous nucleotides and after the long frame shift it continues translation. Variants of phage λ repressor and cytochrome b_{562} when translated from mRNA without a stop codon acquired an unusual COOH end by cotranslational switching of the ribosome reading from the defective mRNA to the tRNA-like ssrA transcript that is translated into Tyr-Ala-Leu-Ala-Ala (the normal carboxyl end would have been very similar Trp-Val-Ala-Ala-Ala). Recoding may be of importance in some human diseases, e.g. if in the cystic fibrosis transmembrane conductance regulator a glycine $codon_{542}$ or arginine $codon_{553}$ are replaced by an UGA (opal) stop codon, the disease symptoms are alleviated compared with some missense mutations because this opal codon permits some readthrough leakage. (See overlapping genes, frameshift, selenocysteine, topoisomerase, Ty, cystic fibrosis, set recoding)

RECODING SIGNAL: required for translational recoding. (See overlapping genes, recoding)

RECOGNITION SITE of RESTRICTION ENZYMES: see restriction enzyme

RECOIL: bounce back; electromagnetic radiation recoils from glass, metal. (See Compton effect)

RECOMBINAGENIC: may be involved in genetic recombination at increased frequency.

RECOMBINANT: individual with some of the parental alleles are reciprocally exchanged. (See reciprocal recombination)

RECOMBINANT CONGENIC: an outcross is followed by several generation of inbreeding in order to minimize the background genetic variations.

RECOMBINANT DNA: DNA that has been spliced *in vitro* from at least two sources with the techniques of molecular biology. (See vectors, cloning vectors, transformation genetic, genetic engineering, restriction enzymes, splicing)

RECOMBINANT DNA AND BIOHAZARDS: were feared by responsible scientists in the 1970s before the impact of the then new techniques could be fully assessed. Since then evidence has accumulated showing that some of the worries were not entirely justified, except by the wisdom of caution with a hitherto unknown and unused procedures. To avoid safety risks various levels of containments were made mandatory, depending on the organisms used to prevent accidental escape of the genetically engineered organisms. Certain types of gene transfers were entirely prohibited to avoid contagions and highly toxic products. Cloning vectors were constructed that would not survive outside the laboratory. Bacterial strain X^{1776} (so designated in honor of the then bicentennial anniversary of the USA national independence) had an absolute requirement for diaminopimelinic acid, an essential precursor of lysine and absent from the human gut. Cloning bacterial hosts were made deficient for excision repair (*uvrB*), auxotrophic for thymidine (indispensable for DNA synthesis), mutant for recombination (*rec*$^-$) and conjugational transfer of plasmids to other organisms. If reversion frequency for any of say 5 defects, each, would be in the range of 1×10^{-6}, then the joint probability for simultaneous reversion for all five would be $(10^{-6})^5 = 10^{-30}$. Since the mass of a single *E. coli* cell is about 10^{-12} g, in a mass of 10^{11} metric tons of bacteria could be expected to find such a five-fold mutation. Obviously, such a mass of bacteria would not likely to occur because the earth might not support them. For a comparison what this volume is, the wheat production of the world in 1980 was only 4.5×10^8 tons, and the estimated mass of the planet earth is 10^{20} tons. To avoid any problem, nevertheless, government authorization is required in all countries where this technology is used, for the release of any genetically engineered species (microbes, plants, animals) for the purpose of economic utilization. Objections to such carefully tested releases still occur, based not so much for public concerns but for personal or political reasons and most commonly because of ignorance. During the longer than 2 decades of practicing recombinant DNA technology, no major accident happened and with the guidelines available none is expected. (See also laboratory safety, containment, biohazards, recombinant DNA evolutionary potentials)

RECOMBINANT DNA, EVOLUTIONARY POTENTIALS: by molecular biology techniques genes can be transfered among organisms by means not routinely available in nature. However, it cannot be ruled out that during the process of natural evolution fragments of degraded DNA was taken up by direct transformation and exchanged between taxonomically unrelated species.

RECOMBINANT INBRED STRAIN PANELS: can be used for mapping in mouse. Information: Jackson Laboratory Animal Resources, 600 Main Str., Bar Harbor ME 04609, USA. Phone: 1-800-422 MICE or 207-288-3371. Fax: 207-288-3398.

RECOMBINANT INBREDS (RI): are generated for physical mapping of DNA by selfing F1 hybrid populations and selecting single seed or animal progenies for about 8 generations, until only $(0.5)^8$ (≈ 0.0039) fraction remains heterozygous for a particular marker. The parental lines are chosen on the basis of differences in their DNA sequences, and from the data the map position of these physical markers can be determined genetically by a combination of molecular

Recombinant inbreds continued

and progeny tests. In animals the calculation is as follows: R (the frequency of discordant individuals) is $R = (4r)/(1 + 6r)$ where r is the recombination in any single gamete. Because interference within the very short distances is practically complete, the distance in cM is about $d = 100r$. The recombination fraction ($\hat{r}$) in function of the size of the sample (N) is $\hat{r} = i/(4N - 6i)$ where i is the number of discordant strains and $\hat{d} = 100 \times \hat{r}$ in cM. In plants the frequency of recombinant monoploid gametes is calculated with basically the same formula $r = R/(2 - 2R)$ where R is the frequency of homozygous recombinant diploid individuals. (See RAPD, congenic resistant lines of mice, congenic strains)

RECOMBINANT JOINT: the site of connection of two molecules of DNA in a heteroduplex. (See heteroduplex)

RECOMBINANT PLASMID: generated either from two different DNAs by the techniques of molecular biology or by spontaneous or induced genetic recombination. (See plasmid)

RECOMBINANT VACCINE: is produced by *in vitro* modifications of the genes/proteins; it does not carry the full complement of the infectious agent. (See vaccine)

RECOMBINASE: enzyme, mediating recombination. (See *FLP/FRT, Cre/loxP, Rec*)

RECOMBINASE SYSTEM: see immunoglobulins

RECOMBINATION: a process by what the linkage phase (coupling or repulsion) of syntenic genes is altered. Recombination is most common during meiosis but mitotic recombination also takes place. The mechanism of the meiotic and mitotic events is not necessarily identical. (See also other recombination entries, linkage, repulsion, coupling, recombinational probe, flip-flop recombination, site-specific recombination, sex circle model of recombination, *Cre/loxP, FLP/FRT, rec*)

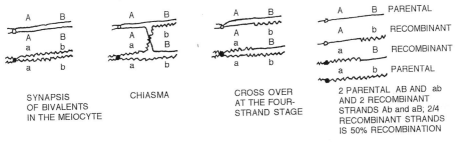

CYTOLOGICAL REPRESENTATION OF RECOMBINATION BETWEEN HOMOLOGOUS CHROMOSOMES IN COUPLING PHASE. EACH CHIASMA LEADS TO CROSSING OVER AND 50% RECOMBINATION. THE FREQUENCY OF RECOMBINATION DEPENDS, HOWEVER, ON THE DISTANCE BETWEEN THE TWO LOCI CONSIDERED. IF CROSSING OVER TAKES PLACE IN ALL MEIOCYTES BETWEEN THE BIVALENTS, THE FREQUENCY OF RECOMBINATION IS 50%. IF ONLY HALF OF THE BIVALENTS UNDERGO CROSSING OVER IN THAT PARTICULAR INTERVAL, THE RECOMBINATION FREQUENCY WILL BE 25% BECAUSE, SAY IN 4 MEIOCYTES WITH (16 CHROMATIDS) 4/16 = 0.25. IF ONLY TWO OF THE MEIOCYTES DISPLAY CROSSING OVER THEN 2/16 = 0.125 IS THE FREQUENCY OF RECOMBINATION. THE MAXIMAL RECOMBINATION FREQUENCY BY A SINGLE CROSSING OVER IS 50%; THE MINIMAL MAY BE EXTREMELY RARE IN CASE THE LINKAGE IS TIGHT.

RECOMBINATION BY REPLICATION: although at the beginning of the 20th century William Bateson suggested that recombination is basically associated with the process of replication. At that time neither of these phenomena were sufficiently understood or could even be hypothesized meaningfully. On the basis of cytological evidence (1930s) for marker exchange accompanied by chromosome exchange and later evidence that DNA exchange and phage gene exchange were correlated, the generally accepted view became that recombination does not require replication. The discovery of gene conversion remained, however a puzzling phenomenon although it was observed that about 50% of the gene conversion events involved flanking marker exchange. The Holliday and other molecular models of recombination permitted interpretation of classical crossing over and gene conversion without significant replication. Re-

Recombination by replication continued

cently it was found the mutation or loss of function of the PriA DNA replication protein blocked both replication and recombination in *E. coli*.

The SOS DNA repair activates a replication process that does not require the replicational *oriC* site or the normally functioning DnaA protein but it needs RecA and RecBCD activities. It was assumed and subsequently demonstrated that double-strand breaks may be assimilated into the DNA and result in double D loops in the presence of nearby chi elements. The *chi*s block nuclease activity and assist the initiation of replication. It appears that PriA and other proteins of the primosome generate a replication fork at the D loop, and relying on the DnaB helicase and the DnaG primase, replication and recombination can be turned on. First, apparently lagging strand synthesis begins by the replisome and the lagging strand then primes the synthesis of the leading strand. Defective PriA may however be compensated for by some elements of the primosome. The process of double-strand break repair and recombination appear the same with the exception that in repair only the defective region has to be corrected whereas in recombination the entire strand must be replicated in order to recover the recombinants. There are some observations that indicate joint events of replicational repair and recombination also in eukaryotes. (See reduplication theory, breakage and reunion, gene conversion, Holliday model, recombination molecular models, SOS repair, DNA repair, recA, recBCD, chi element, replication fork, D loop, replisome, lagging strand, leading strand)

RECOMBINATION IN AUTOTETRAPLOIDS: detection of linkage and recombination in autotetraploids is much more difficult than in diploids because of the multiplicity of the chromatids and alleles and because the segregation ratios are not simple to predict from genetic data without cytological information. The difficulties are practically almost insurmountable when the F_1 is a duplex or triplex and the genes concerned are far apart. Even in close linkage, generally large populations are required. In case of coupling test cross, in simplex individuals the procedure is very similar to that of a test cross in diploids as can be seen in an example (After deWinton, D. & Haldane, J.B.S. 1931, J. Genet. 24:121):

	PHENOTYPES OBSERVED (coupling)			
Parental	Recombinant	Recombinant	Parental	Total
SG	Sg	sG	sg	
336	215	210	353	1114

Recombination frequency $(p) = \dfrac{Sg + sG}{Total} = \dfrac{215 + 210}{1114} = \dfrac{425}{114} \cong 0.38$

In case of repulsion, the calculation presupposes a knowledge of the possible gametic series that can be derived as shown in the figure on next page and it is:

$(1)\dfrac{SB}{sb} : (p)\dfrac{SB}{sb} : (2-p)\dfrac{Sb}{sb} : (2-p)\dfrac{sB}{sb} : (1+p)\dfrac{sb}{sb}$ an example below

	PHENOTYPES OBSERVED (repulsion)			Total
SG (+Sg/sG)	Sg	sG	sg	
164	193	206	154	717

Recombination frequency $= \dfrac{SG + sg}{Total} = \dfrac{2(1+p)}{2(1+p) + 2(2-p)} = \dfrac{318}{717} \cong 0.44$

The manipulation of autetraploids with the techniques available for classical genetics is impractical despite the theoretical framework, and molecular analyses are also still lagging. (See also autopolyploids, alpha parameter, also derivation of gametic frequencies on next page)

RECOMBINATION, INTRAGENIC: see intragenic recombination

RECOMBINATION FREQUENCY: linkage is generally noticed in F_2 when independent segregation of the genes does not occur. Two genes in the homologous chromosomes can be at two different arrangements, repulsion (Ab/aB) or coupling (AB/ab). Some people call the repulsion trans and the coupling cis arrangement.

Recombinations most commonly calculated as the number of recombinants in percentage of the population in test crosses and this frequency is maximally 50% because at this value the

Recombination in autotetraploids continued from page 866

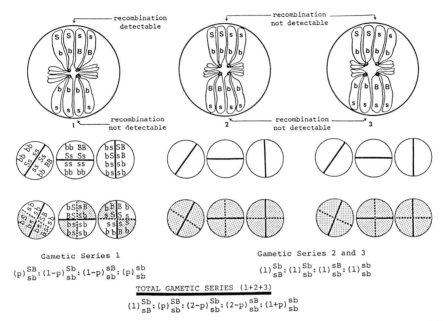

THE DERIVATION OF THE GAMETIC SERIES IN A SIMPLEX TETRASOMIC CASE IN REPULSION. RECOMBINATION IS CONSIDERED ONLY BETWEEN LOCI *S* AND *B*. IN MEIOSIS, THREE DIFFERENT QUADRIVALENT ASSOCIATIONS ARE POSSIBLE AS SHOWN AT THE TOP. RECOMBINATION IS DETECTABLE ONLY AMONG THE DESCENDANTS OF QUADRIVALENT 1. THE SECOND ROW REPRESENTS THE DIFFERENT TYPES OF DISJUNCTIONS AT ANAPHASE I, AND THE THIRD ROW OF CIRCLES SHOWS THE TYPES OF GAMETIC TETRADS FORMED. GAMETES CONTAINING RECOMBINANT STRANDS ARE OPEN, GAMETES WITH PARENTAL STRANDS ARE SHADED.

Recombination frequency continued from page 866

frequencies of recombinant and parental chromosomes is equal, i.e. the segregation is independent. Linkage is usually first observed in F_2 by deviation of the phenotypic proportions from the expectations for independent segregation. Example:

	PHENOTYPIC CLASSES EXPECTED			
	AB	Ab	aB	ab
Independent Segregation →	9/16	3/16	3/16	1/16
Linkage, Repulsion →	less	more	more	less
Linkage, Coupling →	more	less	less	more

PHENOTYPIC CLASSES IN TEST CROSSES IN TWO LINKAGE PHASES AND RECOMBINATION:
(A hypothetical case)

	AB	Ab	aB	ab
Repulsion cross (Ab/aB) x ab	5	45	45	5
Coupling cross (AB/ab) x ab	45	5	5	45

The linkage phase does not affect the frequency of recombination but it affects the frequency of the phenotypic classes. The frequency of recombination is the same in both cases $(5 + 5)/100 = 0.10 = 10\%$. In F_2 recombination frequencies cannot be calculated in such a simple way because in the heterozygotes the genetic constitution of the individual chromosome strands is concealed but may be revealed in F_3. Nevertheless, recombination frequencies can be calculated (see F_2 linkage estimation). Recombination takes place at the four-strand stage of

Recombination frequency continued

meiosis (see exception of mitotic recombination). The bivalents pair and at the simplest case two chromatids exchange segments The maximal frequency of recombination within a chromosomal interval is 50%. Recombination frequencies are generally converted to map units by multiplication with 100. The realistic conversion of recombination frequencies into map units generally requires the use of *mapping functions*. In physical measures one map unit has different meaning in different organisms, depending on the size of the genome in nucleotides (nucleotide pairs) and the genetic length of the genome. Thus, 1 map unit in the plant *Arabidopsis* appears to mean about 150 kbp, in maize the same is about 2,140 kbp and in humans about 1,100 kbp. The frequency of no recombination is a function (f) of the intensity of linkage and the population size; $f = (1 - r)^n$ where r is the recombination fraction and n is the number of testcross progeny. (See also mapping, mapping function, bacterial recombination, testcross, product ratio method, F_2 linkage estimation, F_3 linkage estimation, maximum likelihood method applied to recombination frequencies, recombination modification of, recombination variation of, sperm typing, chiasma)

RECOMBINATION FREQUENCIES IN BACTERIA: see bacterial recombination

RECOMBINATION, ILLEGITIMATE: see illegitimate recombination

RECOMBINATION, MECHANISMS, EUKARYOTES. YEAST: The Sep 1 (strand exchange protein) 132 kDa fragment of a 175 kDa protein of yeast initiates the transfer of one DNA strand from a duplex to a single-stranded circle with 5' to 3' polarity without an ATP requirement. It has also a 5' to 3' exonuclease activity and probably required for the preparation of 3'-end of single- and double-stranded DNA molecules for recombination. Mutation in Sep reduces mit-osis, sporulation, meiotic recombination and genetic repair. (The STPβ protein, encoded by gene *DST2/KEM1*, is probably identical to Sep 1). One monomer of Sep 1 binds to about 12 nucleotides of single-stranded DNA but this requirement is reduced by the presence of the 34 kDa SSB (single strand binding protein), product of yeast gene *RPA1* and by stimulatory factor SF1, a 33 kDA protein which at a concentration of 1 molecules per 20 nucleotides reduces the requirement for Sep 1 to about 1/100. The DPA protein (120 kDa) of yeast controls DNA pairing and promotes heteroduplex formation in a non-polar manner independently from ATP. It promotes single-strand transfer from double-strand DNA to single-stranded circular DNA if the former has single-strand tails. Protein STPα (38 kDa) increases 15 fold shortly before yeast cells are committed to recombination during meiosis. If the gene encoding it (*DST1*) mutates, meiotic recombination is greatly reduced without an effect on mitotic recombination. The *RAD50* gene product (130 kDa) has an ATP-binding domain and it binds stoichiometrically to duplex DNA. The *RAD51* gene product is homologous to the RecA protein of *E. coli* (see recombination, mechanism, prokaryotes) and binds single- and double-stranded DNA. The *DMC1* (*disrupted meiotic cDNA*) gene product appears during meiosis and along with the product of *RAD51*, performs functions similar to RecA in prokaryotes. Meiotic recombination in yeast is believed to involve double-strand break of the DNA. DROSOPHILA: protein Rrp 1 seems to promote exchanges between single-strand circular and linear duplex DNA. Its C-terminus has homology to *E. coli* exonuclease III and *Streptococcus pneumoniae* exonuclease A. MAMMALIAN CELLS: HPP-1 (human pairing protein with 5' to 3' exonuclease activity) binds to DNA and promotes strand exchange in 5' to 3' direction and it does not require ATP. Addition of the hRP-A (human single-strand binding) protein stimulates pairing about 70 fold and reduces the amount of HPP-1 requirement (cf. SF1 in yeast). The precise mechanism how the Holliday junction (see Holliday model, steps I to L) is resolved is not clear but endonuclease activity is postulated. BACTERIOPHAGE T4 gene 49 encodes endonuclease VII that under natural conditions cuts branched DNA structures. Similarly, bacteriophage T7 gene *3* product encodes endonuclease I that cleaves branched DNAs. In yeast, endonuclease XI (Endo XI, $\approx M_r$ 200,000 and other Endo proteins) was found in cells with mutations in the RAD genes and apparently cut cruciform DNA of the type expected by the Holliday juncture. (See recombination models, recombination molecular mechanisms in prokaryotes, recombination models, RAD, RAG, Sep 1, STPβ, synaptonemal complex, chiasma, sex circle model, gene conversion, databases)

RECOMBINATION MODELS: see Holliday model, Meselson - Radding model, Szostak et al. model

RECOMBINATION, MODIFICATION OF: the frequency of recombination may be altered by any means that affect genetic pairing such as chromosomal aberrations, by DNA inserts introduced through transformation, by temperature (either low or high), by physical mutagens, rarely by chemicals, by *rec⁻* genes, etc. In the heterogametic sex of *Drosophila* and silk worm meiotic recombination is usually absent although mitotic recombination occurs. In animals, recombination may be more frequent in females than in males and it is attributed to imprinting. In plants, when there is a sex difference in recombination its frequency is commonly lower in the megaspore mother cell. (See individual entries, recombination variation of, imprinting)

RECOMBINATION, MOLECULAR MECHANISMS, PROKARYOTES: the RecA protein (M_r 37,842) directs homologous pairing by forming a right-handed helix on the DNA and it catalyzes also the formation of DNA heteroduplexes. X-ray crystallography indicates that the DNA rests relaxed in the deep grove of this protein to facilitate the scanning for homologous sequences. The RecA protein is involved also in DNA repair functions (SOS repair). It digests the LexA bacterial repressor and instrumental indirectly in the derepression of over 20 genes involved in recombination and UV mutagenesis. The mechanism(s) of RecA activities can be studied by *in vitro* reactions. RecA can interact with 3 or 4 DNA strands by wrapping around the paired molecules. DNA-DNA pairing can take place between linear and circular DNA too. Strand exchange proceeds at a slow pace (2 to 10 base/sec) in a polar fashion (5' to 3'). The transfer begins at the 3' end of the duplex and is then transfered to a single strand DNA. Homology is a requisite for the RecA mediated reactions yet it tolerates some mismatches or insertions (up to even 1,000 bases or more) but these slow down the reactions. RecA can mediates pairing between two duplexes as long there are short single-strand stretches or gaps. Low pH, intercalating chemicals, Z- configuration and other structural changes of the DNA may alleviate the difficulties of binding two duplexes. The RecA protein is a low efficiency ATPase. ATP hydrolysis is not an absolute requirement, however, for RecA activities in recombination but to be more important for the repair reactions. In the presence of ATP the conformation of RecA is altered and in the nucleoprotein complex the DNA is substantially underwound (the spacing between bases extended from 3.4 Å to 5.1 Å). It is assumed that the paired DNA molecules are not just juxtapositioned but one molecule lays in the major groove of the other. The pairing may involve three or four strands.

DNA strand exchange requires that the RecA filament rotates along the longitudinal axis and the DNA molecules are "spooled" inside where they may form Holliday junction (see Holliday model). ATP is stabilizing the RecA - DNA association and when ATP is split into ADT, the heteroduplex is released and RecA is recycled. Besides the RecA protein, recombination requires the presence of a single-strand binding protein (SSB), DNA polymerase I, DNA ligase, DNA gyrase, DNA topoisomerase I and the products of genes *recB, recC, recD, recE, recF* (binding protein for single-strand DNA)*, recG, recJ* (an exonuclease acting on single-strand DNA), *recN, recO, recQ, RuvB* (helicases), *recR, ruvR, ruvB* and *ruvC*. RuvC nicks the DNA at the point of strand exchange. The RecBCD is a protein-RNA complex encoded by three genes (see above) that has the activities of (i) ATP-dependent double-strand exonuclease, (ii) ATP-dependent single-strand exonuclease, (iii) unidirectional DNA helicase, (iv) site-specific endonuclease to nick four to six nucleotides dowstream of chi, a recombinational hot spot (5'-GCTGGTGG-3'). It was suggested that RecBCD generates 3'-tails that are utilized by protein RecA for DNA strand exchange. RecB and RecC mutations can be suppressed by *sbcA* and *sbcB* mutations. Mutations in *sbcA* leads to the activation of product (exonuclease VIII) of *recE*. Mutation in *scbB* inactivates exonuclease I, an enzyme that digests single-strand DNA and its inactivation may assist the function of RecA in recombination. (See models of recombination). The precise mechanism how the Holliday junction (see Holliday model, steps I to L) is resolved is not clear but endonuclease (RuvC) activity is postulated.

Recombination, molecular mechanisms, prokaryotes continued

Bacteriophage T4 gene *49* encodes endonuclease VII that under natural conditions splits branched DNA structures. Similarly, bacteriophage gene *3* encodes endonuclease I, and cleaves branched DNAs. Some of the functions of the *ruv* operon of *E. coli* may be involved in the resolution of the Holliday junctions. *E. coli* has also *in vivo* systems where the molecular mechanism of resolution of recombination intermediates can be studied. Covalently closed plasmid DNA, DNA polymerase I and DNA ligase were transformed into *E. coli recA* mutants; both monomeric and dimeric plasmid progenies were found and the available markers permitted the conclusions that crossing over occurred in 50% of the progeny. Recombination is not limited to DNA but viral RNA molecules also can recombine. (See also recombination mechanisms eukaryotes, recombination models, recombinational probe, recombination by replication, illegitimate recombination, recombination RNA viruses)

RECOMBINATION NODULE: the suspected site of recombination seen through the electronmicroscope as a 100 nm in diameter densely stained structure adjacent to the synaptonemal complex. There are early nodules seen at the association sites of the paired meiotic chromosomes and also the late nodules are visible at pachytene when crossovers are juxtaposed. Non-crossovers do not show nodules after mid-pachytene. (See synaptonemal complex, chiasma, pachytene, association point, meiosis, recombination, crossing over, recombination RNA viruses)

RECOMBINATION PROFICIENT: do not have deficiencies involving enzymes mediating recombination. (See *Rec* and *rec* entries)

RECOMBINATION, RNA VIRUSES: takes place by template switching during replication thus its more like a copy choice than a breakage and reunion mechanism. The recombination can take place between homologous and non-homologous strands (illegitimate recombination). The latter mechanism may lead to deletions, duplications and insertions. Among picornaviruses the recombination frequency may be as high as 0.9 in case of high homology. Recombination in RNA viruses helps to eliminate disadvantageous sequences and can generate new variants. The estimated mutation rate per base is 6.3×10^{-4} and per genome about 5. The mutation rate is estimated as mutations per replication. (See copy choice, breakage and reunion, illegitimate recombination)

RECOMBINATION REPAIR: see DNA repair

RECOMBINATION, TARGETED: see *Cre/loxP*, *FLP/FRT*

RECOMBINATION, VARIATIONS OF: in the heterogametic sex of arthropods (male *Drosophila*, female silkworm) genetic recombination is usually absent or highly reduced. In the latter group of organisms mitotic recombination occurs, however, and these premeiotic exchanges may account for the observation of recombinants. The most common cause of variation is the presence of *rec-* genes. In the Abbott stock 4A x Lindegren's wild type crosses of *Neurospora*, post-reduction frequency was found to be 4.6 ± 1.2 whereas in the Lindegren's stock it was 13 ± 1.2, and in the Emerson's x Lindegren's crosses 27.6 ± 3.7. L.J Stadler, a pioneer of maize genetics, considered recombination as one of the most variable biological phenomena. (See recombination frequency, recombination modification of, tetrad analysis)

RECOMBINATIONAL HOT SPOT: see hot spot, chi site

RECOMBINATIONAL PROBE: one such short probe is inserted into the 902 bp πVX miniplasmid containing a polylinker and the *supF* suppressor gene. Lambda phage libraries containing also the miniplasmid construct are then propagated. If the phage carries a *supF* suppressible amber mutation, recombination between sequences homologous to the probe can be selectively recovered by forming plaques on an *E. coli* lawn. The recombination may take place even in the absence of perfect homology; less than ca. 8% divergence may be tolerated. The very large populations may reveal recombination within 60 base or longer probes very effectively. (See *rec*, *Rec*, miniplasmid, *supF*, πVX, lawn, see diagram on next page of the use of the πVX microplasmid for the selective isolation of eukaryotic genes by recombinational probes)

Recombinational probe continued

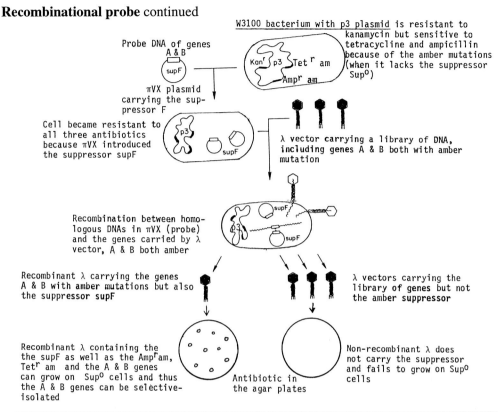

SELECTIVE ISOLATION OF SPECIFIC EUKARYOTIC GENES WITH THE AID OF THE πVX MICROPLASMID. SEVERAL OTHER PLASMIDS HAVE BEEN CONSTRUCTED FOR SIMILAR PURPOSES.

RECOMBINATIONAL REPAIR: see DNA repair

RECOMBINATOR: cis-acting chromosomal sites promoting homologous recombination. (See chi)

RECOMBINOGENIC: mutagenic agents that increase also recombination.

RECON: a historical term for the smallest recombinational unit; molecular biology has shown that recombination can take place between two nucleotides within a codon.

RECONSTITUTED CELL: was produced by fusing cytoplasts and karyoplasts. (See transplantation of organelles, cytoplast, karyoplast)

RECONSTITUTED VIRUS: into an empty viral capsid a complete genetic material is introduced e.g., into the coat of tobacco mosaic virus (TMV) the genome of the related Holmes ribgrass virus (HRV) was introduced and the new particle expressed the characteristic functions of the donor RNA. This classic experiment proved that the genetic material can be also RNA.

RECORD: a true document of an observation or of a hypothesis, in each case explicitly stated.

RECOVERIN: see rhodopsin

RECOVERY OF DNA FRAGMENTS FROM AGAROSE GEL: the fragment is driven by electrophoresis onto DEAE cellulose membrane by cutting a slit in front of the band and placing the DEAE sliver in the slit. Alternatively, electroelution can be used; or from low melting temperature agar the DNA can be extracted by phenol and precipitated by ammonium acetate in 2 volume ethanol and the DNA is collected by centrifugation. (See electrophoresis, DEAE cellulose)

RecQ: a family of helicase proteins. In case of mutation in he coding gene, chromosomal instability results in eukaryotes. In prokaryotes the *recQ* is involved in post-replicational repair. (See chromosomal rearrangements, helicase, RecA, RecB and other Rec genes)

RECQL: a RecQ-like protein in humans. (See RecQ)

RECRUITMENT: for the initiation of transcription some prokaryotic and eukaryotic genes require activators and the transcriptional complex will operate only if these activators are recruited to the transcriptional target. The GAL1 gene of yeast recruits the four units of the GAL4 activator at about 250 base upstream to begin transcription. The GAL4 units are blocked, however, unless galactose is available in the culture medium. The activators make contact with some sites of the RNA polymerase subunits. In bacteria, typical activators are the CAP (catabolite activator protein) and also the λ repressor may bind to the σ^{70} subunit of the polymerase. In yeast, the transcription complex includes more than 30 different proteins. (See transcription factors, transcription complex, activator proteins)

RECRUITMENT OF EXONS: evolving genes may acquire coding sequences for functional domains by borrowing exons through recombination. The recruited DNA sequences may occur in several protein genes of different function, e.g. the low-density lipoprotein (LDP) receptor (a cholesterol transport protein) has homology in 8 exons with the epidermal growth factor (EGF) peptide hormone gene. (See exon, LDP, EGF)

RECRUITMENT OF GENES: acquiring new genetic information through recombination or transfection (transformation). (See transformation, transfection, recruitment of exons)

RECTIFICATION, INWARD: through a voltage gated ion channel the current inward exceeds that of outward. In case of outward rectification the opposite holds. (See ion channel)

RECURRENCE RISK: the chance of having another child with the same defect by the same couple. (See risk, empirical risk, genetic risk, genotypic risk ratio, λ_s)

RECURRENT PARENT: (plant or animal) is mated with selected line(s) in several cycles for one or several back-crosses. (See recurrent selection)

RECURRENT SELECTION (reciprocal recurrent selection): is a variety of methods used for breeding superior hybrids of plants and animals of high productivity. The general procedure: lines (inbred or not) A and B are crossed in a reciprocal manner, i.e. A males are crossed to B females and B males are mated with A females. The initial lines are expected to be genetically different to assure the sampling of different gene pools. In this manner several lines are mated, not just A and B. The progenies are tested for performance and only the best parents are saved. The superior parents are then mated again to representatives of their own line, and on the basis of their progeny, the parents are reevaluated. The mating cycle is then repeated. Most commonly, each male is mated to several females of the other line in order to assure the availability of large enough population of offspring to be able to carry out statistically meaningful tests. The maintenance of the lines requires also that the females be mated to selected males within their own lines. This procedure results in inbreeding but enhances the chances for further selection. Therefore, the performance of the selected parental lines is expected to decrease but that of the hybrids will increase. An alternative simplified method involves selection for combining ability in only one set of lines. Thus line A is mated with a previously inbred tester which has an already known combining ability and the selection is restricted to within line A. This latter modification provides faster initial progress but the final gain may be limited. (See combining ability, hybrid vigor, heterosis, QTL, heritability, diallele analysis)

RECURSIVE PCR: is a method of DNA amplification. Synthetic oligonucleotide primers (50-90 bases) are used which have only terminal complementarity (17-20 bp). They are annealed at 52°C to 56°C. The heating cycles are 95°C, the cooling is at 56°C. The Vent polymerase is used at 72°C. This thermostable polymerase has capability not just for strand displacement but having exonuclease function and it carries out proofreading and therefore the fidelity of the amplification is very good. During the initial steps each 3' end is extended with the aid of the opposite strand as a template and duplex sections are thus generated. In further cycles one strand of the duplex is displaced by a primer oligonucleotide derived from one of the neighboring duplex. During the last step, high concentration of the terminal oligonucleotides assist in the amplification of the entire duplex. (See polymerase chain reaction, Vent)

red: see lambda phage, Charon vectors

RED613: is a fluorochrome, a conjugate of R-phycoerythrin and Texas Red. Excitation at 488 nm and emission at 613 nm. (See fluorochromes)

RED670: is a fluorochrome, a conjugate of R-phycoerythrin and a cyanine. Excitation at 488 nm from an argon-ion laser, emission at 670 nm. (See excitation)

RED BLOOD CELL: see erythrocyte, blood, sickle cell anemia

RED-GREEN COLOR BLINDNESS: see color blindness

RED QUEEN HYPOTHESIS: if a population does not continue to adapt at the same rate as its competitors, it will lose ecological niches where it can succeed, and if it stays put long enough it may become extinct. The name RQ was adapted from Lewis Carrol (pen name of C. L. Dodgson) 1872 fantasy story about a chess game, *Through the Looking Glass*. (See adaptation, extinction, beneficial mutation, equilibrium in populations, genetic homeostasis, treadmill evolution)

REDIFFERENTIATION: organ or organism formation from dedifferentiated cells, such as plants from callus. (See dedifferentiation, callus)

REDOX PAIR: an electron donor and the oxidized derivative.

REDOX REACTION: see oxidation-reduction

REDUCING SUGAR: its carbonyl carbon is not involved in glycosidic bond and can thus be oxidized. Glucose and other sugars can reduce ferric or cupric ions, and that property served for their analytical quantitation (Fehling reaction).

REDUCTANT: is an electron donor.

REDUCTION: gain of electrons.

REDUCTIONAL DIVISION: at meiotic anaphase I half of the chromosomes segregate to each pole and the two daughter cells have n number of chromosomes rather than $2n$ as the original meiocyte contained. In case of uneven numbers of crossing over between gene and centromere the numerical reduction of the chromosomes may not result in separation of the different pairs of alleles i.e. the reduction does not extend to the alleles. The reductional division at meiosis assures the constant chromosome numbers in the species and serves a basis for the Mendelian segregation. (See prereduction, postreduction, tetrad analysis, meiosis)

REDUCTIONAL SEPARATION: in meiotic anaphase I the parental chromosomes separate intact because there was no recombination between the bivalents. (See equational separation)

REDUCTIONISM: reducing ideas to simple forms or making efforts for explaining phenomena on the basis of the behavior of elementary units. This endeavor is frequently criticized because of the complexities of biological systems. One must also consider that without the analytical approach, science (molecular genetics) would not have progressed to the present stage.

REDUNDANCY: repeated occurrence of the same or similar base sequences in the DNA or multiple copies of genes. The repeated gene sequences are considered to have duplicational origin. About 38-45% of the sampled (ca. 1/3 of all) proteins in *E. coli* are expected to be duplicated and in the much smaller *Heamophilus influenzae* genome, completely sequenced by 1995, 30% appears to have evolved by processes involving duplications. In this small genome some gene families were represented by 10 to over 40 members whereas almost 60% of the genes appeared unique. In the large eukaryotic genomes the repetitious sequences are represented by much larger fractions. The number of protein kinases in higher eukaryotic cells may reach 2,000 and that of phosphatases about 1,000. Theoretically true redundancy should succomb to natural selection unless the rate of mutation is extremely low. In case the redundant genes, besides the shareds functions, have unique roles, they are maintainable. Redundant genes are saved also when the *developmental error* rate is high. (See also SINE, LINE)

REDUPLICATION HYPOTHESIS: at the down of Mendelism, William Bateson postulated that genetic recombination takes place by differential degree of replication and different associations of genes after they separated at interphase rather than by breakage and reunion of synapsed chromosomes. (See breakage and reunion, copy choice)

REELIN: a protein encoded by the *reeler* gene in mouse and expressed in the embryonic and postnatal periods. It is similar to extracellular matrix proteins involved in cell adhesion. Mutations in the gene impair coordination resulting in tremors and ataxia. (See ataxia, CAM)

REFERENCE LIBRARY DATABASE (RLDB): cosmid, YAC, P1 and cDNA libraries for public use on high-density filters. Information: Reference Library Database, Imperial Cancer Research Fund, Room A13, 44 Lincoln's Inn Fields, London WC2A 3PX, UK. Phone: 44-71269-3571. Fax: 44-71-269-3479. INTERNET: <genome@icrf.icnet.uk>, World Wide Web URL: <http://gea.lif.icnet.uk/>

REFRACTILE BODIES: *Paramecia* may contain bacterial symbionts and the bacteriophages associated with them may appear as bright (refractile) spots under the phase-contrast microscope. (See symbionts hereditary, killer strains, *Paramecium*)

REFSUM DISEASES: are autosomal recessive disorders occurring in adult and early onset forms. Both forms involve phytanic acid accumulation because of the deficiency of an oxidase enzyme residing in the peroxisomes. The symptoms include polyneuritis (inflammation of the peripheral nerves), cerebellar (hind part of the brain) anomalies and retinitis pigmentosa. The early onset form, in addition, displays facial anomalies, mental retardation, hearing problems enlargement of the liver, lower levels of cholesterols in the blood and accumulation of long-chain fatty acids and pipecolate (a lysine derivative). The symptoms of the infantile form overlap with those of the Zellweger syndrome. (See Zellweger syndrome, phytanic acid, microbodies, retinitis pigmentosa)

REGENERATION IN ANIMALS: is more limited than in plants where totipotency is preserved in most of the differentiated tissues. Regeneration can actually be classified into two main groups of functions: one is the regular replacement of cells (e.g. epithelia, hairs, nails, feathers, antlers, production of eggs and sperms, etc.) in a wide range of animals, and the other is the capacity to regenerate body parts lost by mechanical injuries. The latter type of regeneration may involve the formation of an entire animal from pieces of the body, such as it takes place by morphallaxis in e.g., sponges, Hydra, flatworms, annelids (preferentially from the posterior segments), echinoderms, etc. A more limited type of regeneration is found in the higher forms. Arthropods may replace lost appendages of the body. The vertebrate fishes can replace lost fins, gills or repair lower jaws. Some amphibians (salamanders, newts) readily regenerate lost limbs, tails and some internal organs. The reptile lizards can reproduce lost tails although the regenerated one is not entirely perfect. Regeneration of feathers and repair of beaks may take place in birds. In mammals lost blood cells may be replenished by activity of the bone marrow, or liver cells may regenerate new ones. More limited regeneration may occur in bone, muscle, skin and nerve cells but, unlike in plants, complete organisms cannot be regenerated from any part, except the embryonic stem cells or possibly from other cells after special treatments. (See regeneration of plants, transdetermination, homeotic genes, transplantation of nuclei, stem cells, transplantation of nuclei, nuclear transplantation)

REGENERATION OF PLANTS: see embryogenesis somatic, embryo culture, clone, vegetative reproduction, totipotency)

REGLOMERATE: see aggregulon

REGRESSION: the measure of dependence of one variate on another in actual quantitative terms in contrast to correlation which uses relative terms from 0 to 1. (See correlation for the calculation of regression coefficient, heritability, linear regression)

REGULATED GENE: the expression is conditional and affected by genetic and non-genetic factors. (See also housekeeping genes, constitutive genes, regulation of gene activity)

REGULATION OF ENZYME ACTIVITY: enzyme activity is characterized by various measures of enzyme kinetics (see Michaelis-Menten, Linweaver-Burk, Eadie-Hofstee). The reaction is controled by the quantity and/or activity of an enzyme. The quantity of the enzyme depends on protein synthesis/degradation controlled at the level of transcription, translation, processing of the protein, and its instability (see regulation of gene activity). The substrate of the enzyme may regulate the production of the enzyme protein (see enzyme induction, *lac* operon, catabolite repression, attenuation). *Feedback control* means that the accumulation

Regulation of enzyme activity continued

of the product of an enzyme may shut down the operation of a pathway at any step preceding the final product. Feedback control may be simple or multiple, i.e., more than one enzyme may be affected either simultaneously or sequentially or more than a single product of the pathway may act in a concerted manner (see feedback control). Feedback control may act either at the level of the synthesis (*feedback repression*) or by *inhibition* of the activity of a steady number of enzyme molecules. In general, the inhibitors are either *competitive* (bind to the enzyme and compete with the substrate for the active site) or may be *non-competitive* (the inhibitors act by attaching to the enzyme at a site other than the active site yet lower enzyme activity [by allosteric effect]). *Uncompetitve inhibitors* operate by binding to the enzyme-substrate complex. *Suicide inhibitors* are converted by the enzyme into an irreversibly binding molecule that permanently damages the enzyme. The inhibitors may affect simultaneously more than a single enzyme. *Mechanism-based inhibitors* are highly specific to a single enzyme and as such have great significance for medicinal chemistry. Among these are the *antisense inhibitors* (see antisense RNA). *Allosteric enzymes* may also be stimulated (*modulated*) by allosteric compounds. The modulator may be *homotropic*, i.e. essentially identical structurally with the substrate or *heterotropic* in case of non-identity with the substrate. The activity of an enzyme may require a proteolytic cleavage of the precursor protein, the *zymogen*. (See regulation of gene activity, protein synthesis, signaling, allostery, allosteric control)

REGULATION OF GENE ACTIVITY: the various types of cells and differentiated tissues of an organism generally contain the same genetic material (see totipotency, regeneration) yet their differences attest that the genes must function quite differently in order to bring about the variety of morphological and functional differences. Genetic regulation accounts for this variety. Many genes are expressed in every cell because they determine the metabolic functions essential for life. Another group of genes is responsible for such generally required structural elements as membranes, microtubules, chromosomal proteins, etc. (see *housekeeping*, *constitutive* genes). Other genes are not constitutive, i.e. they are regulated in response to external and internal control signals, in other words, they are expressed only when they are called up for a duty. The latter group of genes are responsible for the differences within an organism.

PRETRANSCRIPTIONAL REGULATION. The expression of genes is regulated by several means, including the structural organization of the eukaryotic chromosome. Although one time it was thought that DNA associated with histones is not or not efficiently transcribed. The nucleosomal organization of the DNA may not prevent transcription yet nucleosomal reorganization may be required for proper expression of the genes(see nucleosomes). It has been known since the early years of cytogenetics that, e.g., heterochromatic regions of the chromosomes were not associated with genes that could be mapped by recombinational analysis. It appears that these tightly condensed regions of the chromosome are not suitable for transcription in general. The coiling of the chromosomes is also genetically regulated. Position effect indicates that gene expression is altered or obliterated by transposition into heterochromatin. Similarly, lyonization of the mammalian X-chromosome involves heterochromatinization and silencing of genes (see silencer). Insertion of normal genes (by transformation) into the condensed telomeric region (about 10$_4$ bp length) interferes with their expression (see heterochromatin, position effect, lyonization, telomeres). Gene expression also depends in some way on the presence of nuclease-sensitive sites in the chromatin. At these nuclease hypersensitive sites, apparently the DNA is not wrapped around so tightly and is more accessible for transcription initiation (see nuclease sensitive sites). The effects of the chromatin locale on the expression of genes is shown by the large differences in the production of a specific mRNA in different transgenic animals and plants which carry a particular gene inserted at different chromosomal locations (see LCR). Also, in order to make the gene accessible to transcription or replication, in bacteria negative supercoils are formed and then must be relaxed. In eukaryotes, DNA in Z conformation may be preferentially available for initiation of transcription (see supercoiling, Z

Regulation of gene activity continued

DNA). Some genes are regulated by transposition; this mechanisms is common both in prokaryotes and eukaryotes to generate defense against the immune system of the host (see phase variation, antigenic variation) but it is used also for sex determination in yeast (see cassette model). At replication the four basic nucleotides are normally used, some nucleoside analog (e.g., 5-bromodeoxyuridine) may be incorporated into the DNA with some consequences on gene expression. In the T-even (T2, T4, T6) phages in place of cytosine 5-hydroxymethyl cytosine is found as a protection against most of the restriction enzymes. In eukaryotes 5 to 25% of the cytosine residues are 5-methylcytosine. Genes with methylated cytosine are generally not transcribed (see methylation of DNA, recruitement, SRB).

REGULATION OF TRANSCRIPTION AND TRANSCRIPTS: The cells have various options for the more direct regulation of transcription: (i) control of signal receptor and signal transmission circuits, (ii) construct or take apart assembly-lines geared to a particular function, (iii) transcriptional control, (iv) transcript processing and alternative splicing, (v) export of the mRNA to the cytosol in eukaryotes. In prokaryotes and cellular organelles the genetic material is not enclosed by a membrane and transcription and translation are coupled, (vi) selective degradation of mRNA or a carboxypeptidase may cleave the transcription factors.

Nucleotide sequences in the DNA (structural gene) specify the primary structure of the transcripts. Upstream cis elements (enhancers, promoters and other protein binding sequences) control the attachment and function of the DNA-dependent RNA polymerases (see pol I, pol II and pol III RNA polymerases). Some eukaryotic genes may have more than one promoter and the tissue or cell type and the physiological conditions select the promoter to be used. Transcript length is dependent on the promoter element used and the upstream, non-translated region contain binding sequences for further regulation of gene expression. The different upstream elements of the same gene may respond differently to cytokines, phorbol esters and hormones (see hormone receptors, hormone response elements). The enhancers may be positioned either upstream or downstream. Inducible genes receive cues through membrane receptors and transmitter cascades, generally regulated by kinases and phosphorylases (see signal transduction). Downstream DNA nucleotide sequences control the termination of transcription and in eukaryotes generally a polyA tail (exceptions are the histone genes) is added enzymatically without the use of a DNA template (see polyadenylation signals, transcription termination in eukaryotes and prokaryotes).

Gene expression begins by the initiation of transcription (see transcription, protein synthesis). The DNA displays some specific sequences in the major groves of the double helix that are recognized by DNA-binding proteins. (See *lac* operon, for the *E. coli lac* repressor binding site and the CAP site for binding of the catabolite activator protein. For phage λ the *cI* repressor binding element see lambda phage. For the consensus sequence of the budding yeast GAL4 upstream element see galactose utilization, for the mating type α2 consensus see mating type determination and for the transcription factor GCN4 see GCN4). In plants, the core sequence for a transcriptional activator protein is shown under G-box element. The binding proteins have a short α helix or a β sheet that fit into the major groove of the DNA at the specific sequence motif (see helix-turn-helix, zinc finger, leucine zipper, helix-loop-helix). Transcription may also be regulated also by specific activators. The activation may require a positive or negative control process (see arabinose operon, lac operon, CAT). Eukaryotic transcription requires for the initiation of transcription the presence of a general transcription factor protein complex (see open transcription complex). Additional specific transcription factors may modulate transcription (see transcription factors inducible). In the DNA there are also a number of *response elements* or regulatory sequences that bind a number of specific proteins and the proteins may bind additional modules (see response elements, hormone response elements). The inducible transcription factors in the eukaryotic nuclei help in the assembly of the transcription complex and activate or repress genes by assembling modules. The interacting elements are responsible for the fine tuning of metabolic pathways and regulating morphogenesis. These transcription factors may or may not be syntenic with the

Regulation of gene activity continued
genes they act on, and their number may vary depending on the gene concerned. The binding proteins may pile up in a specific way at the promoter after DNA looping brings them to that area. Also, the binding proteins may attract other molecules that act either in an activating or silencing manner. In an absolutely abstract form this may be visualized with a few computer symbols:

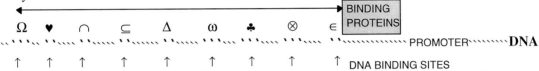

The bacterial DNA-dependent RNA polymerase (see pol) attaches to the double-stranded DNA, and generates an open promoter complex and proceeds with transcription (see open promoter complex). The bacterial RNA polymerase may rely on different σ subunits for transcribing different bacterial or viral genes. In some instances bacterial and eukaryotic genes also use activators of transcription to assist the RNA polymerase enzyme to generate the open promoter complex. These proteins may attach to the DNA at a region of some distance from the gene (enhancer) and looping may bring the protein to the promoter site (see looping of DNA):

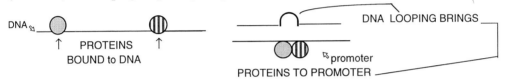

Actually the likelihood for association of two DNA sites by looping reaches an optimum at a distance of about 500 bp and it is much reduced when they are very close. Some of the enhancer DNA elements (binding sites for regulatory proteins) may be several thousands of nucleotides apart upstream or downstream of the structural gene (see enhancer). The various binding proteins (symbolized by: ∪, ♥, ∩, Ψ, Ω, ♠, ζ, •, ∇) may associate with the general transcription factors and with each other in different combinations and numbers either to activate or suppress, or modulate or silence the gene.

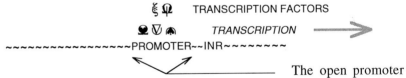

The open promoter complex includes the general transcription factors, RNA polymerase II, the TATA box and the transcription initiator (INR). These crude schematic figures cannot properly represent the interacting complexes that are required for turning on, turning off and modulating expression as needed for orchestration of intricate processes such as the temporal and topological control of morphogenesis (see morphogenesis). The transcription factors regulate these processes but the transcription proteins themselves are subject to regulation not only by transcription but by metabolic and environmental cues. These processes include conformational changes, combinatorial assembly of subunits, ligand binding, phosphorylation and dephosphorylation, presence of inhibitors and activators (see signal transduction). In eukaryotes there may be a need for chromatin remodeling in order that the activators and the TATA box binding protein could access the DNA (see nucleosome). For this process a histone acetylase or SWI/SNF complex may have to be recruited in preparation for transcription. In both prokaryotes and eukaryotes special control mechanisms have evolved for the termination of transcription (see transcription termination). The quantity of the transcripts is determined by regulation of the transcriptional process and the turnover of the transcripts. Many bacterial genes are organized into coordinated regulatory units employing negative, positive or a combination of these two

Regulation of gene activity continued

controls of transcription (see *lac* operon, *arabinose* operon). In these operons the genes are either exactly (see *tryptophan* operon) or with some modification (see *histidine* operon) are arranged according to the order of the biosynthetic pathway. The amino acid operons use, in addition, *attenuation* for the control of the quantity of the transcripts for maximal economy (see attenuator region, tryptophan operon). The operons are characterized by coordinated regulation of the transcription of several genes belonging to the same transcriptional unit and transcribe them into a polycistronic mRNA. Eukaryotes usually do not produce polycistronic mRNAs but the rRNA and tRNA transcripts are processed into functional units post-transcriptionally. Elements of a coordinated unit may not all be iuxta-positioned (see regulon, arabinose operon). The small phage (see φX174 [can be found under F]) and retroviral genomes may have over-lapping genes that specify more than one protein, depending on the register they are transcribed (see overlapping genes, recoding, retroviruses). The need for the protein products of these overlapping genes transcribed with the aid of the same promoter, may be not the same; some proteins, e.g., viral coat proteins may be needed in larger quantities than the replicase enzymes. Therefore, mechanisms evolved to skip internal stop signals and produce some fusion proteins that assist in achieving this goal (see overlapping genes, recoding). A still another means of regulation evolved in bacterial, plant and animal viruses for the regulation of gene activity at different steps by the use of antisense RNA. This mechanism is being explored now to develop specific drugs for the highly specific regulation of genes with minimal side effects or for the development of new, selective antimicrobial agents and more desirable crop plants without reshuffling the entire genome (see antisense technologies).

In prokaryotes a special short, transcribed stretch of nucleotides, the Shine - Dalgarno box controls the attachment of the mRNA to the small (30S) ribosomal subunit. For the same task, eukaryotes use "ribosome scanning", i.e. the mRNA tethers a 40S ribosomal subunit and by reeling locates the first initiator codon. Eukaryotic 40S ribosomal subunits can enter circular mRNAs if they contain internal ribosomal entry sites (IRS).

The primary transcripts are generally not suitable for translation into a protein or for an RNA product (rRNA, tRNA). The transcripts are processed to mRNA and/or other RNA units. Intrones are excised and the sequences corresponding to exons are spliced and may even be transspliced with the cooperation of spliceosomes (see intron, exon, spliceosome, alternative splicing, hnRNA, snRNA). The splicing itself may be genetically and organ-specifically regulated. Transposition of the P element of *Drosophila* is relatively rare in the soma but five times more common in the germline because one intron is not excised from the transposase transcripts in the somatic cells (see hybrid dysgenesis). Tissue-specificity and function-specificity of many proteins is controled partly by alternative splicing (see immunoglobulins, sex determination). Mitochondrial RNA transcripts may be modified also by replacing C residues with Us (see RNA editing).

The eukaryotic mRNAs are capped while still in the nucleus. The transcript is cut at the appropriate guanylic residue and it is then modified (see cap, capping enzymes). Capping increases the stability of the mRNA, facilitates its transport to the cytosol and assists in the initiation of translation by being recognized by initiation protein factors (eIF-4F, eIF-4B, etc.; see cap, eIF).

The tail of the eukaryotic mRNAs (with few exceptions, e.g., histones) are equipped with 50-250 adenylic units to increase their stability. Polyadenylation is controlled separately from transcription because these nucleotides are added by a special enzyme after processing of the transcript. Generally, the genes carry a short A-rich consensus (see polyadenylation signal) in the DNA that instructs the RNA polymerase to terminate transcription after the enzyme passes through the signal and also indicates the need for polyadenylation. Eventually, the poly-A tail is reduced to about 30 A units. In eukaryotes, the 3' tail may be specially regulated.

Some the transmembrane proteins have a hydrophobic amino acid sequence in the section that is going to be located within the membrane, whereas the cytosolic end contains a longer hydro-

Regulation of gene activity continued

philic carboxyl end. The transcript of the same coding sequences is differentially cut in such a manner as to assure such a terminus to be formed for the membrane-bound proteins whereas the otherwise identical circulating immunoglobulin molecule is terminated by a shorter hydrophilic end.

After these intricate preparatory processes, the eukaryotic mRNA is transported to the cytosol through the nuclear pores. Prokaryotes do not have membrane-enclosed nuclei but only nucleoids, anchored to the cell membrane, and there the translation goes *pari passu* with transcription.

POST-TRANSCRIPTIONAL REGULATION: The mRNA may be degraded before it could be translated into polypeptide chains. About half of the prokaryotic mRNAs may be degraded within 2-3 minutes after their synthesis. Eukaryotes have long-lived mRNAs which usually last for at least three times longer but sometimes in special dormant tissues of plants they may remain intact for many years. The degradation is mediated by special endonucleases that recognize mRNAs. Also A-U sequences in the non-translated downstream regions may remove the poly-A tails and thus in both cases stability is reduced.

Translation in eukaryotes begins with the transport of the capped mRNA outside the nucleus, into the cytosol. The mRNA tethers several ribosomes and the polysomal structures are formed. Some mRNAs are equipped with a signal coding sequence, coding for a special tract of 15 to 35 amino acids. That directs it toward the *signal sequence recognition particle* after only a few dozen amino acids are completed on the ribosome. The *signal peptide* then transports the nascent peptide chain into the lumen of the endoplasmic reticulum, Golgi vesicles, lysosomes and into mitochondria, plastids, etc. This mechanism facilitates the subcellular localization of the emerging proteins at places where they are most needed and from where they may be diffused in a gradient as required for embryonic differentiation (see signal sequence, signal peptide, signal sequence recognition particle, morphogenesis in *Drosophila*). Various control mechanisms have been involved in generation of the protein products of the genes: (i) translational control, (ii) post-translational modification of the polypetides, (iii) control of polypeptide assembly into proteins, (iv) regulation of protein conformation, (v) compartmentalization of the proteins, (vi) interaction of protein products and ribozymes, (vii) feedback controls at the level of protein synthesis and function (see induction, repression, attenuation, inhibition, silencers, etc.) may be required before, during and after the final protein products are made.

The state of phosphorylation of the eukaryotic initiation factor, eIF-2 is critical for the translation process. This protein may form a complex with guanosyl triphosphate (GTP) and it can assist then the attachment of the initiator tRNA^{Met} to the P site of the small subunit (40S) of the ribosome and scans the mRNA until it finds a methionine codon (AUG). This after, the large ribosomal (60S) subunit joins the small subunit to form the 80S ribosome and the same time one molecule of inorganic phosphate and the inactivated eIF-2 and GDP are released. Then eIF-2 can acquire another GTP and initiation goes on again (see protein synthesis).

Although all polypeptide chains are made by starting with a formylmethionine (prokaryotes) or methionine (eukaryotic), the final product is frequently truncated at both the amino and carboxyl termini. Many proteolytic enzymes are translated as large units and become activated only after cleaving off certain parts of the original protein. Insulin is made as a preproinsulin that must be tailored in steps into pre-, then pro-insulin and finally to insulin to become active. Several viral proteins, secreted hydrolytic proteins, peptide hormones and neuropeptides are made as polyprotein complexes and they have to be broken down into active units in the trans Golgi network, in secretory vesicles or even in the extracellular fluids to become fully functional. The formation of polyproteins appears to be justified as a protective measure against destruction in the cytosol until they can be sequestered and confined into some vesicles. The loaded vesicles then migrate to predetermined sites where upon receiving the cognate signals they release the active protein. The signals can be chemical, physical (electric potentials) or topological. Actually, the release of the members of the polyprote in group may be selective

Regulation of gene activity continued

regarding the site of release, different proteins can be released at different anatomical sites. Some proteins are synthesized in separate polypeptide chains but must be folded and/or assume a quaternary structure e.g, $\alpha\alpha\beta\beta$ and may even have to acquire a prosthetic group such as heme, a vitamin or other organic or inorganic group(s). The folding in prokaryotes begins after completion of the chain. In eukaryotes the folding may begin before finishing a polypeptide and thus higher complexity is generated in the large proteins The mRNA may be degraded before it could be translated into polypeptide chains.

Proteins are very commonly acetylated after translation, carbohydrate side chains are added (glycoproteins), prenylated, linked by covalent disulfide bonds, special amino acids (serine, threonine, tyrosine) are phosphorylated by kinase enzymes, lysine residues may be methylated, and extra carboxyl groups may be attached to aspartate and glutamate residues.

According to some estimates about 2,000 different protein kinases and 1,000 phosphatases may exist in a higher eukaryotic cell. They must be regulated in time, space, and for other specificities. This regulation is an extremely complex task and is expected to be mediated by associat-ions with modular, adaptor, scaffold and anchoring proteins working in a sequential cooperation through signal transduction pathways.

(See transcription, transcriptional activator, transcriptional modulation, protein synthesis, regulation of enzyme activity, chromatin, high mobility group of proteins translation initiation, translation, regulation of enzyme activity, signal transduction, serine/threonine phosphoprotein phosphatases, cell cycle, LCR, DNA looping, transcription complex, transcription factors, SL1, TBP, TAF, open promoter complex, RAD25, signaling to translation)

REGULATOR GENE: controls the function of other genes through transcription. (See also regulation of gene activity, enhancer, silencer)

REGULATORY ELEMENTS: generally upstream (enhancer) sequences located within 100 to 400 bp from the translation initiation nucleotide (+ 1) and control cell and developmental specificities. Some enhancers may be located at much more distant positions and also downstream. (See also basal promoter, regulation of gene activity)

REGULATORY ENZYME: allosteric or other modification alter its catalytic activity rate and thus affects other enzymes involved in the pathway.

REGULATORY SEQUENCE IN DNA: binds transcription factors, RNA polymerase and so regulates transcription. (See transcription factors, open transcription complex, enhancer, operon, attenuator site)

REGULON: non-contiguous set of genes under control of the same regulator gene. The different sections may communicate through looping of the DNA. (See looping of DNA, arabinose operon, regulation of gene activity)

REINITIATION: eukaryotic ribosomes can terminate an open reading frame and initiate another downstream (at low efficiency). Reinitiation takes place also when the translation of one reading frame is completed and the process moves on to the next cistron. (See regulation of gene activity, transcription, translation, cistron)

REITER SYNDROME: is a complex anomaly generally accompanied by overproduction of HLA-B27 histocompatibility antigen. The most common result is arthritis, inflammation of the eyes and the urethra (the canal that carries the urine from the bladder and in the males serves also as a genital duct). The inflammations may be related to sexually transmitted and to intestinal infections. (See HLA, rheumatic fever, arthritis, connective tisssue disorders)

REITERATED GENES: are present in more than one copy, possibly many times.

REJECTION: an immune reaction against foreign antigens such as may be present in transfused blood or grafted tissue. (See immune reaction, HLA)

REJOINING: see breakage and reunion, breakage-fusion-bridge cycles

rel (REL) **ONCOGENE**: turkey lymphatic leukemia oncogene, a transcription factor homologous with NF-κB. In its absence or inactivation the production of IL-3 and the granulocyte—macrophage colony-stimulating factor is impaired. (See also NF-κB, NFKB, IL-3, GMCSF, oncogenes, morphogenesis in *Drosophila* {3}).

RELATEDNESS, DEGREE OF: is term used in genetic counseling indicating the probability of sharing genes among family members. First-degree relatives such as a parent and child have half of their genes in common. Second-degree relatives such as a grandparent and grandchild have 1/4 of their genes identical. Population genetics prefers the use of the mathematically simpler terms such as the inbreeding coefficient, consanguinity and coefficient of coancestry. (See these concepts under separate entries, relationship coefficient)

RELATIONAL COILING: see chromosome coiling

RELATIONSHIP, COEFFICIENT OF: $r = 2F_{IR}/\sqrt{(1+F_I)(1+F_R)}$ where F_I and F_R the coefficients of inbreeding of I and R. If they are not inbred F_I and F_R equal to 0. (See coefficient of inbreeding, relatedness degree)

RELATIVE BIOLOGICAL EFFECTIVENESS: see RBE

RELATIVE FITNESS: see selection coefficient

RELATIVE MOLECULAR MASS (M_r): expresses molecular weight relative to ^{12}C isotope, (in $\frac{1}{12}$ units). It is comparable to molecular weight in daltons but it is not identical to molecular weight (MW) represented by the mass of the atoms involved. (See Dalton)

RELATIVE MUTATION RISK = 1/doubling dose. (See doubling dose, genetic risk)

RELATIVE SEXUALITY: the intensity of the sexual determination may have degrees in some organisms. In extreme cases a normally female gamete may behave as a male gamete toward a strong female gamete. (See isogamy)

RELAXED CIRCULAR DNA: is not supercoiled because of one or more nicks. (See nick, supercoiled DNA)

RELAXED CONTROL mutants (*relA*): lost stringent control and continue RNA synthesis during amino acid starvation of bacteria. (See stringent control, fusidic acid)

RELAXED REPLICATION CONTROL: the plasmids continue to replicate when the bacterial divisions stops. (See replication)

RELAXOSOME: is a DNA-protein structure mediating the initiation of conjugative transfer of bacterial plasmids. It contains a *nic* site at the origin of transfer (*oriT*). Relaxase catalyzes the nicking and it becomes covalently linked to the 5'-end through a tyrosyl residue. Then a single strand is transfered to the recipient by a rolling circle mechanism. (See conjugation, rolling circle, nick)

RELAY RACE MODEL OF TRANSLATION: a ribosome after passing a chain termination signal of an ORF does not completely disengage from the mRNA and may reinitiate protein synthesis if an AUG codon is within short distance downstream. (See translation, regulation of gene activity, reinitiation, ORF)

RELEASE FACTOR (Rf): when translation reaches a termination codon, the release factors let the polypeptide go free from the ribosome. In prokaryotes there are two direct release factors RF-1 and RF-2 and a third factor RF-3 stimulates the activity of RF 1 and 2. The eukaryotic release factors are named err. It has been suggested that all release factors are homologous to elongation factor G which mimics tRNA in its C-terminal domain and this would be the basis of the recognition of the Rfs of the ribosomal A site. (See transcription termination, protein synthesis, regulation of gene activity, EF-G)

RELEASING FACTORS: pituitary gland hormones are released under the influence of hypothalmic hormones. (See animal hormones)

RELICS: genes with major lesions (insertions and deletions) in one or more components; they are similar to pseudogenes. (See pseudogenes)

REM: Röntgen equivalent for man. It is the product of REB x rad. Generally 1 rem is considered to be 1 rad of 250 kV X-rays; 1 rem is 0.01 Sv (Sievert). (See R unit, rad, Gray, Sievert, REB)

REMI (restriction enzyme-mediated integration): an integrating vector is transformed into a cell in the presence of a restriction enzyme that facilitates insertion at the cleavage sites and may bring about insertional mutagenesis. (See insertional mutation, restriction enzyme)

RENAL CELL CARCINOMA (RCC): involves most commonly breakage in human chromosomes 1, 5 and 3p, each representing a different type. (See hypernephroma)

RENAL DYSPLASIA and LIMB DEFECTS: autosomal recessive underdevelopment of kidney and the urogenital system, accompanied by defects of the bones and genitalia. (See kidney disease, limb defects)

RENAL DYSPLASIA and RETINAL APLASIA: in the autosomal recessive condition kidney developmental anomaly is associated with eye defects. (See kidney disease, eye disease)

RENAL-HEPATIC-PANCREATIC DYSPLASIA (polycystic infantile kidney disease, ARPKD): autosomal recessive phenotypes include cystic (sac like structures) kidneys, liver and pancreas, sometimes associated with other anomalies such as blindness. The polycystic kidney disease of adult type dominant (ADPKD, human chromosome 16) is associated frequently with internal bleedings or arterial blood sacs (aneurysm). (See kidney disease)

RENAL TUBULAR ACIDOSIS: the autosomal dominant type I defect is primarily in the distal tubules with normal bicarbonate content in the serum. Type II is recessive and the defect is in the proximal tubules and there is a low level of bicarbonate in the urine. In another recessive form nerve deafness is present. A proximal type is X-linked recessive. The excretion of ammonium is reduced and the urine pH is usually above 6.5 in contrast to types I and II where it is around 5.5. Other variations have also been observed. (See kidney diseases)

RENATURATION: complementary single DNA strands reform double-strand structure by reannealing through hydrogen bonds. (See c_0t curve)

Renilla GFP: is green fluorescent protein with similarities to aequorin but only one absorbance and emission peak. (See aequorin)

RENIN (chymosin, rennet): protein hydrolase reacting with casein in cheese making; it is present in the kidneys and splits proangiotensin from α-globulin. (See angiotensin, pseudoaldosteronism)

RENNER COMPLEX: chromosomal translocation complex that is transmitted intact. (See translocation, translocation complex)

RENNER EFFECT: see megaspore competition

REOVIRUSES: RNA viruses causing respiratory and digestive tract diseases and arthritis-like symptoms in poultry and mammals. (See rotaviruses)

REP: repetitive extragenic consensus of 35 nucleotides, containing inverted sequences, in the bacterial chromosome. There are over 500 copies of it in *E. coli* in intergenic regions at 3' end of the genes. They are transcribed but not translated and appear to be the bacterial version of "selfish" DNA. (See selfish DNA)

Rep: an *E. coli* monomeric or dimeric binding protein and helicase. (See binding protein, helicase, monomer)

rep: röntgen equivalent physical; a rarely used unit of X- and γ radiation delivering the equivalent of 1 R hard ionizing radiation energy to water or soft tissues (≈ 93 ergs). (See R)

REPAIR GENETIC: see DNA repair, unscheduled DNA synthesis

REPAIROSOME: the protein complex mediating DNA repair. (See DNA repair)

REPEAT, DIRECT: (tandem) duplication of the same DNA sequence; may be present at the termini of transposable elements ABC-----------ABC. (See transposable element, transposon)

REPEAT-INDUCED GENE SILENCING: see co-suppression

REPEAT, INVERTED: the double-stranded DNA carries inverted repeats such as:

Such palindromic structures occur at the termini of prokaryotic and eukaryotic transposable elements. The single strands can fold back and form stem and loop structure such as:

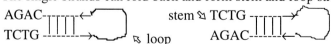

(See transposable elements)

REPEATS, TRINUCLEOTIDE: see fragile sites, trinucleotide repeats

REPERTOIRE, ANTIGENIC: the complete set of antigenic determinants of the lymphocytes.

REPERTOIRE SHIFT: after a secondary immunization with a hapten, following a primary immunization, the variable heavy/variable light (V_H/V_L) immunoglobulin genes display an altered spectrum of somatic mutations. (See immunoglobulins, hapten)

REPETITIVE DNA (repetitious DNA): similar nucleotide sequences occuring many times in eukaryotic DNA. (See SINE, LINE, redundancy, pseudogenes, α-satellite DNA)

REPLACEMENT VECTOR: has a pair of restriction enzyme recognition sites within the region of "non-essential" genes. Non-essential means that packaging and propagation in *E. coli* is not impaired by their removal and replacement by sequences of interest for the experimenter. (See vectors, stuffer)

REPLICA PLATING: has been designed to efficient selective isolation of haploid microbial mutants. Mutagen-treated cells were spread in greatly diluted suspension on the surface of complete medium and incubated to allow growth. Because of the dilution, each growing colony represent a single ori-ginal cell (clone). Then im-pressions are made of this master plate on minimal medium where only the wild type cells could grow. The ab-sence of growth on the minimal media plates indicated that auxotrophs exist at the spots where no growth was obtained. The impressions represented also a map of the colonies on the original, complete medium, master plate. Thus the experimenter could obtain cells from the original colonies and test them for nutritional requirement on differently supplemented media. This procedure thus permitted the isolation of the mutants and the identification of the nutrient requirement. (See outline on the left, mutant isolation)

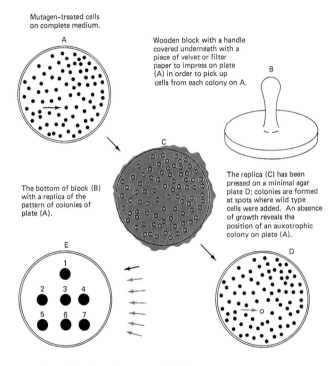

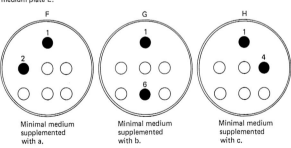

REPLICASE: is RNA-dependent RNA polymerase enzyme of viruses encoded by the viral RNA and packed to the progeny capsid so upon entry to a cell replication of the infective *negative-strand RNA* (influenza, stomatitisvirus [causing inflammation of mucous membranes]) forms the template for replication but does not code for

Replicase continued

viral proteins. Without the replicase this negative strand would not be able to function. The *positive-strand* RNA viruses (e.g. polio virus) are directly transcribed into protein, including the replicase, and this can be infectious in that form. The DNA-dependent DNA polymerase enzymes may also be called replicases. (See replication, RF, positive strand)

REPLICATING VECTOR: see transformation genetic, yeast

REPLICATION: see DNA replication, replication fork

REPLICATION, BIDIRECTIONAL: is the mode of replication in bacteria and also in the eukaryotic chromosome. Replication begins at an origin and proceeds in opposite direction on both old strands of the DNA double helix. Electron microscope reveals a θ (theta) resembling structure of the circular DNA whereas in the linear eukaryotic DNA bubble-like structures are visible. In prokaryotes this replication is mediated by DNA polymerase III, and in eukaryotes a DNA polymerase α type enzyme. Termination of replication in *E. coli* requires the 20 base long Ter elements and the associated protein Tus (termination utilization complex, M_r 36 K). The *TerA, D* and *E* stop the replication in the anticlockwise direction and *TerC, B* and *F* halt replication of the strand elongated clockwise. The Tus-Ter complex probably blocks the replication helicase. Similar mechanisms operate in most bacteria but replication fork arresting sites exist also in eukaryotes, including humans.

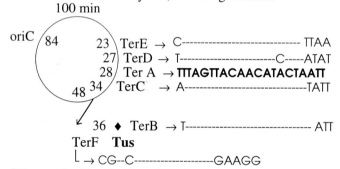

T7 RNA polymerase while replicating the template strand can by-pass up to 24 nucleotide gaps by making a copy of the deleted sequence using the corresponding non-template tract. (See DNA replication eukaryotes, DNA replication prokaryotes, θ replication, replication bubble, pol III, pol α, replication fork)

(Diagram after Kamada, K. *et al.* 1996 Nature 383:598)

REPLICATION BUBBLE (replication eye): indication of strand separation in a replicon. In an eukaryote nucleus an estimated 10^3 to 10^5 replication initiations occur during each cell cycle without any reinitiation per site. This assures te maintenance of gene number. (See replication bidirectional, DNA replication, replication fork, replicon, promoter bubble, ORC)

REPLICATION, CONSERVATIVE: a historical model of DNA replication, assuming that the two old (original) DNA strands produce two new copies which then anneal to each other. In other words, the double-stranded DNA is not composed of an old and a new strand as the current and experimentally demonstrated semiconservative replication mechanism shows. (See also semiconservative replication)

REPLICATION DEFECTIVE VIRUS: is mutant for the replication function or lost the genes required for producing infective particles. (See replicase)

REPLICATION, DISPERSIVE: an unproven old idea claiming that old and new double-strand DNA tracts alternate along the length of the molecule. (See replication, replication fork)

REPLICATION DURING THE CELL CYCLE: eukaryotic DNA replication takes place predominantly during the S phase of the cell cycle although some repair synthesis (unscheduled DNA synthesis may occur at other stages. In prokaryotes the replication in not limited to a particular stage and DNA synthesis may proceed without cellular fission. Such a phenomenon (endoreduplication) is exceptional in eukaryotes and most commonly limited to certain tissues only, e.g. to the salivary gland chromosomes of insects (*Drosophila, Sciara*) or a rare non-repeating process (endomitosis) that doubles the number of chromosomes. The replication in eukaryotes is an oscillatory process tied to the S phase of the cell cycle. The process of repli-

Replication during the cell cycle continued

cation shows some variations in even among the different eukaryotes and the process described below is modeled after that of *Saccharomyces cerevisiae* (the best known). During the G1 phase the pre-origin-of-replication complex (pre-ORC) the ORC is assembled after the cyclosome (APC) proteases degraded the cyclin B-cyclin-dependent kinase (Cyclin B-CDK). The cis-acting *replicator* element and the *initiator* proteins bind at each origin of replication (hundreds in eukaryotes). The replicator (0.5-1 kb) is a multimeric complex itself and its indispensable component is the A unit but B1, B2 and B3 are also used. The A, B1 and B2 form the core of the replicator and B3 is an enhancer that binds to the *autonomously replicating sequence* (ARS)-binding protein factor 1 (ABF1). The replicator (A + B1) hugs the ORC (origin recognition complex) composed of 6 subunits that form the hub of the replication process and attract other critical regulatory proteins. The site of the initiation in mammals may extend to 50 kb. At the origin, the nucleosomal structure is remodeled and during S, G2 and early M phase DNase hypersensitive sites are detectable that disappear before anaphase. It appears that protein Cdc7 is required for the remodeling. The CDC6 (or the homologous Cdc18 protein of fission yeast) is required in G1 or S phase for DNA synthesis (but the cells may proceed to an abortive mitosis and aneuploidy in its absence). CDC6 seems to be essential for the formation of the pre-ORC complex. Overexpression of this protein leads to polyploidy. The replication requires also a Replication Licensing Protein (RLF) and members of the MCM (minichromosome maintenance) proteins. Cyclin-dependent kinases (CLB5 and CLB6) are also required to establish the pre-initiation complex but after the assembly is completed some of them may be degraded. Some cyclin-dependent kinases also block the reinitiation of the complex until the cell passed through mitosis. Cyclin B5—cyclin-dependent kinase (Clb5-CDK) is inhibited by Sic1 (S phase inhibitory complex) that is removed by ubiquitin-mediated proteolysis at START point before the S phase is fired on. The ubiquitination is promoted by CDC34, CDC53, CDC4 SKP1, CLN1-Cdc28, CLN2-CDC28 and the APC proteins. The initiation of DNA replication may proceed also through another pathway, mediated by CDC7 and DNA-binding factor 4 (DBF4). (See cell cycle, replication fork, DNA replication eukaryotes, Replication protein A, DNA replication prokaryotes; DNA replication in mitochondria, rolling circle replication, θ (theta) replication, RNA replication, reverse transcription, bidirectional replication, and the other proteins and terms listed in alphabetical orders separately)

REPLICATION ERROR: is a source occurring when a nucleic acid base analog incorporated into the DNA at the structurally acceptable site but during the following replication, being only an analog, it may cause a replicational error that results in the replacement of the original base pair by another. E.g.: BrU—A base pair in the following replication is converted by error into BrU—G pair and finally resulting in the base substitution of C≡G at a site that was formerly T=A. Mispairing — in the absence of base analogs — can also occur, e.g. A=C, and the frequency of such errors may be within the range of 10^{-4} to 10^{-6}. (See base substitution, incorporation error, bromouracil, BUdR, hydrogen pairing, ambiguity in translation)

REPLICATION EYE: see replication bubble, DNA replication eukaryotes, replication bidirectional

REPLICATION FACTOR A: see RF-A, helix destabilizing protein

REPLICATION FACTOR C: see RF-C

REPLICATION FORK: represents the growing region of DNA where the strands are temporarily separated. (See DNA replication eukaryotes, DNA replication prokaryotes, GP32 protein, PriA, replication bubble, replication bidirectional, replication licensing, diagram on next page)

REPLICATION INTERMEDIATE: see, lagging strand, Okazaki fragment, replication fork

REPLICATION LICENSING FACTOR (RLF): the initiation of replication requires two competency signals: the binding of the RLF and an S-phase promoting factors. The sequential action of these two signals secure the accurate replication of the chromosomes. RLF has two elements RLF-M and RLF-B. RLF-M is a complex of MCM/P1. RLF-M protein binds to the chromatin early during the cell cycle but it is displaced after the S-phase. (See MCM1, MCM3, ORC, replication, cell cycle, replication bubble)

Replication fork continued from preceding page

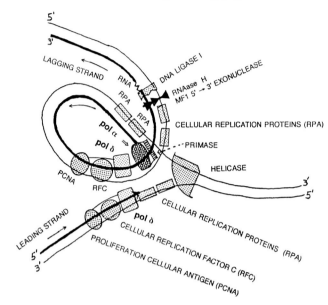

A MODEL OF THE EUKARYOTIC REPLICATION FORK BASED ON SV40 STUDIES. THIN LINES: OLD DNA STRANDS, HEAVY LINES: NEW STRANDS. THE REPLICATION FORK IS OPENED AT THE REPLICATIONAL ORIGIN BY HELICASES. THE *CELLULAR REPLICATION PROTEIN* (RPA) KEEPS THE FORK OPEN AND BRINGS THE pol α DNA POLY-MERASE COMPLEX TO THE RE-PLICATIONAL ORIGIN. AFTER A SHORT RNA PRIMER IS MADE (NOT SHOWN ON THE DIAGRAM) AT THE BEGINNING OF THE FIRST OKAZAKI FRAGMENT, *CELLULAR REPLICA-TION FACTOR C* (RFC) BINDS TO THE DNA AND DISPLACES POL α BY pol δ. THEN RFC, pol δ, AND PCNA (proliferating cell nuclear antigen) FORM A COMPLEX ON BOTH LEADING AND LAGGING STRANDS AND THE TWO NEW DNA STRANDS ARE REPLICATED IN CONCERT. THE SYNTHESIS OF THE LEADING STRAND IS STRAIGHTFORWARD. THE LAGGING STRAND IS MADE OF OKAZAKI FRAGMENTS (BY "BACK-STITCHING") BECAUSE THE DNA CAN BE ELONGATED ONLY BY ADDING NUCLEOTIDES AT THE 3'-OH ENDS. AFTER AN OKAZAKI FRAGMENT (100-200 BASES) IS COMPLETED, RNase H, MF1 EXONUCLEASE REMOVE THE RNA PRIMER (jagged line) AND DNA LIGASE I JOINS THE FRAGMENT(S) INTO A CONTINUOUS NEW STRAND. THE LONG ARROWS INDICATE THE DIRECTION OF GROWTH OF THE CHAIN. (Redrawn after Waga, S. & Stillman, B. 1994 Nature (Lond.) 369:707)

REPLICATION ORIGIN (o): the point in the genetic material where replication begins.

REPLICATION PROTEIN A (RPA): a complex of three different polypeptides (M_r 10,000, 34,000, 13,000), binds to a single-strand DNA, and may make the first step in DNA replication by participating in DNA unwinding. The p53 cancer suppressor gene interferes with its binding to the replication origin. RPA plays a role also in recombination and excision repair. It has affinity to xeroderma pigmentosum damage-recognition protein (XPA) and endonuclease XPG. (See also DNA replication eukaryotes, replication fork, xeroderma pigmentosum, endonuclease, DNA repair)

REPLICATION-RESTART: is a pathway of the SOS repair system of DNA by bypassing the mismatch and resulting in error-free replication. (See DNA repair)

REPLICATION SPEED: kbp/min: *E. coli* 45, yeast 3.6, *Drosophila* 2.6, toad 0.5, mouse 2.2.

REPLICATIONAL FIDELITY: DNA replication error

REPLICATIVE FORM (RF): double-stranded form of a single-stranded nucleic acid virus that generates the original complementary type of single-strand (+) nucleic acid. The necessity for the double-stranded replicative form is to generate a minus strand that is the template for the plus strand and is complementary to the "sense" molecule. (See DNA replication, RNA replication, plus strand)

REPLICATIVE INTERMEDIATE: see replicative form, replication intermediate, RNA replication

REPLICATIVE SEGREGATION: the newly formed cellular organelles display sorting out in the somatic cell lineages in case of mutation in organelle DNA. (See sorting out, cell lineage)

REPLICATIVE TRANSPOSITION: see transposition, transposable elements, cointegrate, Mu bacteriophage

REPLICATOR: the origin of replication in a replicon. (See replicon, ARS)

REPLICON: a replicating unit of DNA. The size of the replicational unit varies a great deal. In *E. coli:* ≈ 4.7 Mbp, *Saccharomyces cerevisiae*: (yeast): 40 kb, *Drosophila*: 40 kb, *Xenopus laevis* toad): 200 kb, mouse: 150 kb, broad bean (*Vicia faba*): 300 kb. (See also DNA replication, minireplicon)

REPLICON FUSION: see cointegration

REPLISOME: enzyme aggregate involved in the replication of DNA of *prokaryotes*. The DNA polymerase III holoenzyme consists of two functional enzyme units, one for the leading and the other for the lagging strand. The *polymerase core* contains one α subunit (for polymerization), the ε subunit (3'→5' exonuclease for editing repair) and the θ unit. It includes also a ring-like dimer of β *clamp* to hold on leash the DNA strands and a five-subunit *clamp loader* γ complex. There are two subunits that are organizing the two cores and the clamp loader into a pol III holoenzyme. The asymmetric replication of the leading and lagging strands is determined by the DnaC helicase, unwinding the double helix in front of the replisome. The helicase facilitates the hold of the complex onto the leading strand by the τ unit to make continuous extension of that strand. At the lagging strand the complex goes off and on, however as the Okazaki fragments are made. (See replitase, DNA replication in prokaryotes, replication fork in prokaryotes, GP32 protein, DNA polymerases)

REPLITASE: the replicational complex at the DNA fork in *eukaryotes*. (See replication fork in eukaryotes, replisome, DNA polymerases)

REPORTER GENE: is a structural gene with easily monitored expression (e.g. luciferase, β-glucuronidase) that will report the function as differentiation progresses, or any heterologous or modified promoter, polyadenylation or other signals attached to the gene by *in vitro* or *in vivo* gene fusion. (See luciferase, GUS, gene fusion)

REPORTER RING: according to the tracking concept of recombination between appropriate *res* (*resolvase*) points, recombination of two DNA molecules would retain during segregation "reporter rings" catenated to one of the two DNA strands of the DNA recombination substrate during synapse and after its resolution in the product. The "reporter rings" were expected to be limited to one of the catenated product molecule but the experimental data did not support this assumption. (See tracking, resolvase)

REPRESENTATIONAL DIFFERENCE ANALYSIS: see RDA

REPRESSIBLE: subject to potential repression. (See repression)

REPRESSION: control mechanism interfering with the synthesis (at the level of transcription) of a protein. A general type of repression is attributed to histones, tightly associated with DNA in the nucleosomal structure. When the histones are deacylated they reinforce repression by co-repressors (N-CoR, mSin3). Histone acetyl transferases acylate histones and permit the recruitement of transcription factors to the gene. (See feedback control, regulation of gene activity, regulation of enzyme activity, *Lac* repressor, *Arabinose* operon, repressor, *Tryptophan* operon)

REPRESSOR: protein product of the regulator gene that interferes with transcription of an operon. The majority of the DNA-binding proteins bind the DNA by their α helices but some of the repressors (*met, arc, mnt*) bind by β sheets or a combination of both (*trp*). Repression in eukaryotes have a variety of means for control. The short-range repressors are within 50-150 bp of the transcriptional activators. Since the promoters may be modular (repeating units), the short-range silencers may not affect other than the nearest activators. This organization assures the expression of genes controlling segmentation along the axis of the developing embryo. The long-range repressors may silence the activators from several kb distance. The short-range repressor proteins are monomeric whereas the long-range ones are multimeric. (See repression, *Lac* operon, *Lac* repressor, *Arabinose* operon, *Tryptophan* operon, corepressor, morphogenesis in *Drosophila*)

REPRODUCTION: see asexual, clonal, vegetative reproduction, dioecious, monoecious, auto-

Reproduction continued

gamy, apomixia, allogamy, parthenogenesis, hermaphroditism, cytoplasmic transfer, genetic systems, conjugation, conjugation *Paramecia*, life cycle, breeding system, fungal life cycle, social insects, vivipary, incompatibility

REPRODUCTIVE ISOLATION: prevents gene exchange between two populations by a hereditary mechanism. Most commonly reproductive isolation is caused by the inability of mating (pre-mating isolation) but in some instance inviability or sterility of the offspring is the barrier (post-mating isolation). In plants, chromosomal rearrangements most commonly lead to gametic disadvantage or sterility. In animals, translocated chromosomes are frequently transmitted at fertilization but the duplication-deficiency zygotes are inviable. (See incompatibility, sexual isolation, isolation genetic, speciation, founder principle, drift genetic, effective population size, translocation, inversion, infertility)

REPRODUCTIVE SUCCESS: see fitness, fertility, fecundity

REPRODUCTIVE TECHNOLOGIES: see ART

REPTATION: a theory about the movement of nucleic acid end-to-end in gels.

REPULSION: one recessive and one dominant allele are in the same member of a bivalent, such as $A\ b$ and $a\ B$. (See coupling, linkage)

RER: rough endoplasmic reticulum (endoplasmic reticulum with ribosomes sitting on it).

RESIDUE: the elements of polymeric molecules, such as nucleotides in nucleic acids, amino acids in proteins, and sugars in polysaccharides, fatty acids in lipids.

RESISTANCE TRANSFER FACTORS (RTF): plasmids that carry antibiotic or other drug resistance genes in a bacterial host and capable of conjugational transfer. (See plasmid(s))

RESOLUTION, OPTICAL: defines the ability of distinguishing between two objects irrespective of magnification. Magnification helps the human eye but does not improve the optical resolution. The power of resolution depends on the wave-length of the light and on the aperture of the lens used. (The achromatic lens is supposed to be free from color distortion, the apochromatic lens is free from color or optical distortions). The unaided human eye may discern details larger than 100 µm, the light microscope with a good oil-immersion objective lens may resolve 0.2 µm, the lowest limit for an ideal electronmicroscope is 0.1 nm, i.e. 1 Å. Under practical conditions the resolution of the electronmicroscope is about 2 nm. The resolution of the light microscope is generally defined as $\frac{0.61\lambda}{n \sin\theta}$ where λ is the wave-length of the light (for white light it may be about 530 nm), n stands for the refractive index of the immersion oil or air (when dry lens is used) and θ is the half of the angular width of the cone of the light beam focused on the specimen with the condensor of the microscope ($\sin\theta$ is maximally about 1). Then $\sin\theta$ is the *numerical aperture* of the lens and using oil-immersion its value may be 1.4. The immersion oil should be non-fluorescing, slow drying and of right viscosity for vertical or horizontal-inverted views. (See also fluorescence microscopy, phase-contrast microscopy, Nomarski, confocal microscopy, electronmicroscopy, oil immersion lens)

RESOLVASES: are endonucleases mediating site-specific recombination and instrumental in resolving cointegrates or concatenated DNA molecules; transposon Tn3 and γδ encoded proteins promoting site-specific recombination of supercoiled prokaryotic DNA containing replicon fusion, direct end repeates and internal *res* sites.

(See site-specific recombination, recombination molecular mechanism, recombination site-specific, reporter ring, TN3, concatenate, γδ element, EMC, phase variation, Tn*3*)

RESPIRATION: electrons are removed from the nutrients during catabolism and carried to oxygen through intermediaries in the respiratory chain. Oxygen is taken up and carbon dioxide is produced. (See also fermentation, Pasteur effect, chlororespiration, mitochondria)

RESPONDER (Rsp) : a component of the segregation distorter system in *Drosophila*, a repetitive

DNA at the heterochromatic site in chromosome 2-62. (See also hybrid dysgenesis, segregation distorter)

RESPONSE ELEMENTS: see hormone response elements, transcription factors inducible regulation of gene activity

RESPONSE REGULATOR: see two-component regulatory system

RESTITUTION CHROMOSOMAL: rejoining and healing of broken off chromosomes.

RESTITUTION NUCLEUS: unreduced product of meiosis.

RESTORER GENES: see cytoplasmic male sterility, fertility restorer genes

RESTRICTION ENDONUCLEASE: see restriction enzyme and table below

RESTRICTION ENZYME: endonucleases cut DNA at specific sites when the bases are not protected (modified, usually, by methylation). Bacteria synthesize restriction enzymes as a defense against invading foreign DNAs such as phages and foreign plasmids. Three major types have been recognized. Type II enzymes are used most widely for genetic engineering. *Type II* enzymes have separate endonuclease and methylase proteins. Their structure is simple, cleave at the recognition site(s); the recognition sites are short (4 - 8 bp) and frequently palindromic, they require Mg^{++} for cutting, the methylation donor is SAM. *Type III* enzymes carry out restriction and modification by two proteins with a shared polypeptide. They have 2 different subunits. Cleavage sites are generally 24-26 bp downstream from recognition site. The recognition sites are asymmetrical 5-7 bp. For restriction they require ATP and Mg^{++}. For methylation SAM, ATP, Mg^{++} are neded. *Type I* enzymes are single multifunctional proteins of three subunits, cleavage sites are random and at about 1 kb from specificity sites. The type I restriction enzymes are encoded in the *hsd* (host-specificity DNA) locus with three components *hsdS* (host sequence specificity), *hsdM* (methylation) and *hsdR* (endonuclease). Their recognition sites are bipartite and asymmetrical: TGA-N8-TGCT or AAC-N6-GTGC. For restriction and methylation SAM, ATP and Mg^{++} are required. Restriction enzymes may create a protruding and a receding ends or the two ends may be of equal length, e.g.:

EcoRI --------OH   PstI ---------------OH   AluI ---------OH
---------------p   ----------p   ----------p

The Type II enzymes may be of high specificity, ambiguous, or isoschizomeric, and may be prevented from action on methylated substrates or may be indifferent to methylation. Approximately 2,400 type II restriction enzymes with about 200 different specificities are known. Kilo- and Mega-base DNA substrates can rather precisely be cleaved by combining a DNA-cleaving moiety, e.g. copper:*o*-phenanthroline with a specific DNA binding protein (e.g. CAP). The complex thus cuts at the 5'-AAATGTGATCTAGATCACATTTT-3' DNA site of CAP recognition. The great specificity required that the cutting moiety would be attached to an amino acid in such a way that it would bend toward the selected target but not toward unspecific sequences. The IIS restriction endonucleases cleave DNA at a precise distance outside of their recognition site and produce complementary cohesive ends without disturbing their recognition site. (See also nucleases, isoschizomers, DNA methylation, restriction-modification, hsdR, CAP, antirestriction, RNA restriction enzyme)

RESTRICTION ENDONUCLEASES WITH RECOGNITION AND CUTTING SITES
(Isoschizomers: IS, Cutting site: ('), N: any base, ^{m}: Methylated base

Aal IS Stu I	Aos I IS: Avi II	Avr II IS: Bln I	Bse PI IS: Bss H II
Aat II GACGT'C	Apa I GGGCC'C	Bal I IS: Mlu NI	Bsi WI C'GTACG
Acc I GT'(A,C)(T,G)AC	Apo I IS: Acs I	Bam HI G'GATCC	Bsi YI CC(N)₅'NNGG
Acc II IS: Mro I	Apy I IS: Eco RII, Mva I	Ban I G'G(T,C)(A,G)CC	Bsm I GAATGCN'N
Acs I (A,G)'AATT(T,C)	Ase I: IS: Asn I	Ban II G(A,G)GC(T,C)'C	CTTAC'GNN
Acy I G(A,D)'CG(C,T)C	Asn I AT'TAAT	Bbr PI CAC'GTG	Bsp 12861 IS: Bmy I
Afl I IS: Ava II	Asp I GACN'NNGTC	Bbs I IS: Bpu AI	Bsp 14071 IS: Ssp BI
Afl II IS: Bfr I	Asp 700 GAANN'NNTTC	Bcl I T'GATCA	Bsp HI IS: Rca I
Afl III A'C(A,G)(T,C)GT	Asp 718 G'GTACC	Bfr I C'TTAAG	Bsp LU11I A'CATGT
Age IS: Pin AI	Asp EI GACNNN'NNGTC	Bgl I GCC(N)₄'NGGC	Bss HII C'CGCGC
Aha II IS: Acy I	Asp HI G(A,T)GC(T,A)'C	Bgl II A'GATCT	Bss GI IS: Bst XI
Aha III IS: Dra I	Asu II IS: Sfu I	Bln I C'CTAGG	Bst 1107 I GTA'TAC
Alu I AG'CT	Ava I G'(T,C)CG(A,G)(A,G)G	Bmy I G(G,A,T)GC(C,T,A)'C	Bst BI IS: Sfu I
Alw 44 I G'TGCAC	Ava II G'G(A,T)CC	Bpu AI GAAGAC(N)₂/₆	Bst EII G'GTNACC
Aoc I CC'TNAGG	Avi II TGC'GCA	Bse AI T'CCGGA	Bst NI IS: Mva I, Eco RII

Bst XI CCA(N)₅'NTGG	Hha I IS: Cfo I	Nar I GG'CGCC	Sca I AGT'ACT
Cel II GC'TNAGC	Hinc II IS: Hind II	Nci I CC'(G,C)GG	Scr FI CC'NGG
Cfo I GCG'C	Hind II GT(T,C)'(A,G)AC	Nco I C'CATGG	Sex AI A'CC(A,T)GGT
Cfr I IS:'Eae I	Hind III A'AGCTT	Nde I CA'TATG	Sfi I GGCC(N)₄'NGGCC
Cfr 10 I (A,G)'CCGG(T,C)	Hinf I G'ANTC	Nde II 'GATC	Sfu I TT'CGAA
Cla I AT'CGAT	Hpa I GTT'AAC	Nhe I G'CTAGC	Sgr AI C(A,G)'CCGG(T,C)G
Dde I C'TNAG	Hpa II C'CGG	Not I GC'GGCCGC	Sma I CCC'GGG
Dpn I Gᵐ A'TC	Ita I GC'NGC	Nru I TCG'CGA	Sna BI TAC'GTA
Dra I TTT'AAA	Kpn I GGTAC'C	Nsi I ATGCA'T	Sno I IS: Alw 44 I
Dra II (A,G)G'GNCC(T,C)	Ksp I CCGC'GG	Nsp I (A,G)CATG'(T,C)	Spe I A'CTAGT
Dra III CACNNN'GTG	Ksp 632 I CTCTTC(N)₁/₄	Nsp II IS: Bmy I	Sph I GCATG'C
Dsa I C'C(A,G)(C,T)GG	Mae II A'CGT	Nsp V IS: Sfu I	Ssp I AA'ATT
Eae I (T,C)'GGCC(A,G)	Mae III 'GTNAC	Pin AI AA'CCGGT	Ssp BI T'GTACA
Eag I IS: Ecl XI	Mam I GATNN'NNATC	Pma CI IS: Bbr PI	Sst I IS: Sac I
Eam 11051 IS: Asp EI	Mbo I IS: Nde II	Pml IS Bbr PI	Sst II IS: Ksp I
Ecl XI C'GGCCG	Mfe I IS: Mun I	Psp 1406 I AA'CGTT	Stu I AGG'CCT
Eco 47 III AGC'GCT	Mlu I A'CGCGT	Pst I CTGCA'G	Sty I C'C(A,T)(A,T)GG
Eco RI G'AATTC	Mlu NI TGG'CCA	Pvu I CGAT'CG	Taq I T'CGA
Eco RII 'CC(A,T)GG	Mro I T'CCGGA	Pvu II CAG'CTG	Tha I IS: Mvn I
Eco RV GAT'ATC	Msc I IS: Mlu NI	Rca I T'CATGA	Tru 9 I T'TAA
Esp I IS: Cel II	Mse I IS: Tru 91	Rsa I GT'AC	Tth 111 I IS: Asp I
Fnu DII IS: Mvn I	Msp I C'CGG	Rsr II CG'G(A,T)CCG	Van 91 I CCA(N)₄'NTGG
Fnu 4 HI IS: Ita I	Mst I IS: Avi II	Sac I GAGCT'C	Xba I T'CTAGA
Fok I GGATG(N)₉/₁₃	Mst II IS: Aoc I	Sac II IS: Ksp II	Xho I C'TCGAG
Fsp I IS. Avi II	Muᴜ I C'AATTG	Sal I G'TCGAC	Xho II (A,G)'GATC(T,C)
Hae II (A,G)GCGC'(T,C)	Mva I CC'(A,T)GG	Sau I IS: Aoc I	Xma III IS: Ecl XI
Hae III GG'CC	Mvn I CG'CG	Sau 3A 'GATC	Xmn I IS: Asp 700
Hgi AI IS: Asp HI	Nae I GCC'GGC	Sau 96 I G'GNCC	Xor II CGAT'CG

RESTRICTION FRAGMENT: a piece of DNA released after the digestion by a restriction endonuclease. The length of the fragments, depends on how many nucleotides are situated between the two cleavage sites. From the same genomic DNA, the same enzyme generates fragments of different lengths because the nucleotide sequence varies along the DNA length. (See restriction enzymes, restriction fragment number, RFLP)

RESTRICTION FRAGMENT NUMBER: can be predicted on the basis of the number of bases at the recognition sites. Since the polynucleotide chain has 4 bases (A, T, G, C), four cutters can have 4^4 = 256 bp average fragment length and six cutters 4^6 = 4096. The average frequency of these fragments is 0.25^4 = 0.0039 and 0.25^6 = 0.000244, respectively. These predictions would be valid, however, only if the distribution of the bases would be random, that is not the case in the coding sequences, e.g., in the ca. 49.5 kb λ DNA 12 EcoRI fragments would have been predicted but only 5 is observed. Four-, six-cutter indicates that the enzyme cleaves the substrate at a 4 or 6 nucleotide-specified sites. (See restriction enzymes, restriction fragment)

RESTRICTION FRAGEMENT LENGTH POLYMORPHISM: see RFLP

RESTRICTION MAP: see RFLP

RESTRICTION MEDIATED INTEGRATION: see REMI

RESTRICTION-MODIFICATION: the bacterial restriction enzymes are endonucleases and the modification enzymes are methyltransferases that recognize the same nucleotide sequence as the endonuclease and transfer a methyl group from S-adenosyl methionine either to C-5 of cytidine or to cytidine-N^4 or to adenosine-N^6. Examples: HpaII cuts C↓CGG and methylates C^mCGG, TaqI cuts T↓CGA and methylates TGC^{m}A. The biological purpose of this complex is to destroy invading nucleic acids (phages) by cleaving the foreign DNA with the aid of the restriction endonuclease(s), i.e. restrict the growth of the invader and the same time protect the bacterium's own genetic material by methylation. (See restriction enzymes, methylation of DNA, antirestriction)

RESTRICTION POINT: (See R point, cell cyle, cancer)

RESTRICTION SITE: the site where the restriction enzyme cleaves. (See cloning site)

RESTRICTIVE CONDITIONS: do not permit the growth or survival of some specific conditional mutants. (See conditional mutation, permissive conditions)

RESTRICTIVE TRANSDUCTION: see specialized transduction

RESVERATROL: see lipooxygenase

RET ONCOGENE: is in human chromosome 10q11.2. In *Drosophila* its homolog is *tor* (see morphogenesis in *Drosophila*). The protein product is a tyrosine kinase, essential for the de-

Ret oncogene continued

velopment of the nervous system; it is a signaling molecule for GDNF. Mutations at the RET locus may be responsible for familial medullary thyroid carcinoma (FMTC), multiple endocrine neoplasia (MEN2A and MEN2B) and Hirschprung disease. The RET gene has five important domains, cadherin binding, cystein-rich calcium-binding, transmembrane and two tyrosine kinase (TK) domains. The main course of the RET-activated signal pathway: RET receptor→Grb2→SOS→RAS→RAF→MAPKK→MAPK→NUCLEUS (transcription factors). (See oncogenes, tyrosine kinase, TCR, GDNF, endocrine neoplasia, Hirschprung disease, phaeochromocytoma, papillary thyroid carcinoma, Grb, SOS, RAS, RAF, MAPKK, MAPK)

RETARDATION: slower than normal growth and development. (See also mental retardation, gel retardation assay)

RETICULOCYTE: immature red blood cell displaying a reticulum (network) when stained with basic dyes. (See rabbit reticulocyte *in vitro* translation system)

RETICULOSIS: a complex autosomal recessive disease involving anemia, reduced platelet count, nervous disorders, immunodeficiency, etc. The symptoms may overlap different types of leukemias. The prevalence is about 5×10^{-5}. Bone marrow transplantation and chemotherapy have been beneficial in some cases. (See separate entries)

RETINA: the inner layer of the eyeball connected to the optic nerve.

RETINITIS PIGMENTOSA (RP): a group of human autosomal recessive (84%), X-linked recessive, (6%, Xp21.1) or autosomal dominant (10%, 13q14) diseases entailing visual defects and blindness with an onset during the first two decades of life with a prevalence in the 10^{-4} range. Only a small fraction of these pigmentary defect cases is associated with the rhodopsin receptor, a G-protein receptor. Several other diseases. e.g. some congenital deafness (Usher syndrome), hypogonadism—mental retardation, other neuropathies, mitochondrial deficiencies may also involve similar defects. Thus altogether about 14 genes may be involved in RP symptoms. A 'digenic retinitis pigmentosa' is due to mutations in the unlinked loci of peripherin (6p21-cen) and the ROM1 (retinal rod outer segment protein-1) in chromosome 11. Mutations in the α subunit of a cGMP phosphodiesterase (human chromosome 5q31.2-q34) gene, PDEA may also cause retinitis pigmentosa. (See Lawrence-Moon syndrome, rhodopsin, retinoblastoma, choroidoretinal degeneration)

RETINOBLASTOMA (RB): a tumor arising from the retinal germ cells, a glioma of the retina. The overall incidence is about $1 - 6 \times 10^{-5}$ per birth, expressed within the first two years. About 40% of the cases are genetically determined. The estimated mutation rate is within the range of 10^{-5} to 10^{-6}. The bilateral form is generally familial whereas the unilateral cases may be due to new mutations. A dominant RB gene was assigned to human chromosome 13q14. The mouse homolog Rb-1 is in mouse chromosome 14. Apparently RB is more frequently traced to the paternal chromosome 13 than to the maternal one. The incidence of sporadic RB is increasing with parental age. It appeared that in RB cells tumor growth factor B was absent. The retinoblastoma protein (or homologs) may play a general role in tumorigenesis upon phosphorylation. RB is frequently associated with small cell lung carcinoma (SCLC), osteosarcoma, other types of malignancies and esterase deficiency (in deletions). Retinoblastoma was the first type of cancer with recognized recessive inheritance in humans. About 5-10% of the retinoblastomas are associated with deletions at chromosome 13q14 and rearrangements involving that site. RB binding proteins (RBBP), with homology to the E7 transforming protein of a papillomavirus and to the large T antigen of SV40 have been identified. The normal allele of the retinoblastoma protein appears to be restraining abnormal proliferation by limiting the activity of pol III and pol I. The retinoblastoma protein stimulates the transcription of several genes primarily by activating the glucocorticoid receptors. RB has decisive role at the G1 restriction point decisions in the cell cycle regarding differentiation or continuation of cell divisions. Retinoblastoma may be unilateral (the majority of the non-hereditary cases) or bilateral (about two third) of the hereditary cases. (See eye disease, oncogenes, oncoprotein, pol III, deletion, papova virus, Simian virus 40, binding protein, sporadic, MDM2, glucocorticoid, restriction point, cell cycle, tumor suppressor)

RETINOIC ACID: is a carboxylic acid derivative of vitamin A; the aldehyde form is retinol. The 11-cis retinal is the light absorbing chromophore of the visual pigments (carotenoids). Retonic acids belong to a nuclear receptor family (RARE, RJR) and act as ligand-inducible transcription factors. Retinoids have a role in the anterior-posterior pattern of development of the body axis and limbs of vertebrates. (See transcription factors, RARE)

RETINOL: see retinoic acid

RETINOSCHISIS: are autosomal dominant, recessive or X-linked (Xp22.3-p22.1) degeneration of the retina involving splitting of that organ. (See eye diseases)

RETROELEMENT: see retroposon

RETROGENE: are pseudogenes with transcriptionally active promoter. (See pseudogenes, SINE)

RETROGRADE: backward. (See also anterograde)

RETRON: is responsible for the synthesis of msDNA. Retrons have several elements: the transcriptase *ret*, the coding regions of msDNA, msdRNA and requires also RNase H. The latter enzyme is required for the maintenance of the proper structure and the termination of the transcription. (See msDNA, RNase H, reverse transcriptase, retronphage)

RETRONPHAGE: are retrons that are parts of different proviruses, e.g. one in *E. coli* inserts within the selenocysteyl gene (*SelC*). (See retron, selenocyteine)

RETROPOSON: is a transposable element that is mobilized through the synthesis of RNA that is then converted to DNA again by reverse transcription before integration into the chromosome. The retroposon may be a viral element or it may have originated from an ancient viral element. Retroposons are the long interspersed elements (see LINE) and the short interspersed elements (see SINE), the copia elements in animals and several other of the hybrid dysgenesis factors, e.g. Ty in yeast, and they occur also in several species of plants. The majority of the plant retroposons have lost their ability to move. Retroposons may be distinguished from retrotransposons by the former not having long terminal repeats whereas the latter ones do. (See retroviruses, reverse transcriptase, copia, hybrid dysgenesis, transposable elements)

RETROPSEUDOGENE: see processed pseudogene

RETROREGULATION: RNase III may degrade mRNA from the 3' end; some mutations in temperate phages may prevent this degradation and thus permit the translation of the mRNA. This control is operating from the end (downstream) forward (upstream) and thus called retroregulation. (See retroviruses, regulation of gene activity, ribonuclease III)

RETROTRANSPOSON: retrovirus-like transposable elements with long terminal repeats and change position within the genome as retroviruses, however, they lack extracellular life style. Retrotransposons are common in the eukaryotic genomes and they are apparently not distributed random in the chromosomes. Of the five *Ty* elements of *Saccharomyces* are congregating in regions about 750 bp upstream of tRNA genes and *Ty5* is found at the telomeres. Usually the sites of insertion are methylated and not transcribed and that protects the genome from insertional mutations. In a 280 kb region flanking the maize alcoholdehydrogenase gene (*Adh1*, chromosome 1L-128) ten different retroelements were found crowded with repetitions and inserted within each other. The repetitive elements of the maize genome largely represent retrotransposons and constitute at least 50% of the total nuclear DNA. The size of the repeats vary from 10 to 200 kb and they are distributed throughout the genome. The *Arabidopsis* genome which is about 1/20th of that of maize also contains about 20 retrotransposons but only with 5-6 copy number. In *Vicia faba* plants with a genome size of 13.3 pg (about 1.3×10^{10} nucleotide pairs), 10% of the genome is retrotransposable elements but in related species their number is much smaller. In *Allium* plants the elements are mainly in the centromeric and telomeric heterochromatin regions whereas in other organisms they may be dispersed. The plant retroposons (with the exception of the Tnt1 of tobacco) fail to move because their transposase is pseudogenic. The major difference between retrotransposons and retroviruses is that the former do not produce envelope (Env) protein. (See retroviruses, insertion elements, retroposon, diagram next page transposable elements, transposons, hybrid dysgenesis, methylation of DNA, Ty, copia)

Retrotransposon continued

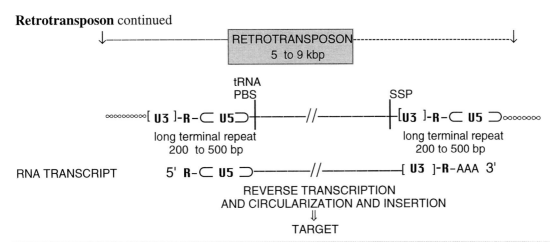

A DNA COPY OF A RETROTRANSPOSON IN THE CHROMOSOME IS REPLICATED THROUGH AN RNA TRANSCRIPT. THE REPLICATION BEGINS AT THE R SEGMENT OF THE 5' LONG TERMINAL REPEAT (LTR) AND PROCEEDS THROUGH SEGMENT U5 AND THE NON-REPETITIVE INTERNAL REGION OF THE ELEMENT (SHOWN WITHIN THE VERTICAL LINES). TOWARD THE 3' BOUNDARY OF THE R SEGMENT OF THE LONG TERMINAL REPEAT. THE FIRST STRAND OF THE DNA IS SYNTHESIZED BY THE REVERSE TRANSCRIPTASE, ENCODED WITHIN THE RETROTRANSPOSON AND IT IS PRIMED BY THE CCA-3'-OH END OF ONE OR ANOTHER KIND OF HOST tRNA THAT PAIRS BY COMPLEMENTARITY TO THE UPSTREAM LTR. THE REVERSE TRANSCRIPTION EMPLOYS ALSO A PROTEIN-tRNA COMPLEX THAT BINDS TO THE tRNA-PROTEIN SITE (tRNA-PPS). THE SECOND STRAND IS PRIMED AT THE SSP SITE (SECOND STRAND PRIMER). AT THE TARGET SITE, A DUPLICATION OCCURS:

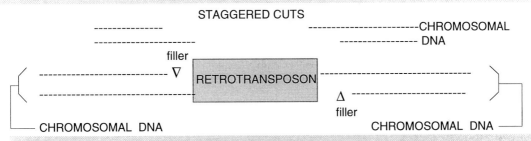

THE INTEGRASE MAKES STAGGERED CUTS AT THE TARGET AND THE SINGLE-STRAND OVERHANGS ARE FILLED IN BY COMPLEMENTARY BASES RESULTING IN TARGET SITE DUPLICATIONS

element	5' long terminal repeat	internal domain	3' long terminal repeat
copia	**TGTTGGA**...TAC**AACA** 276 bp	GGTTATGGGCCCAGTC...TTGAGGGGGCG 4190 bp	**TGTTGGA**...TAC**AACA** 276 bp
TY912	**TGTTGGA**...TTTCT**CA** 334 bp	TGGTAGCGCCCTGTGCT...TATGGGTGGTA 5250 bp	**TGTTGGA**...TTTCT**CA** 334 bp
Tag-3	**TGTTGGA**...GGTAA**CA** 514 bp	AGTGGTATCAGAGCCA.....AAGGTGGAGAT 4190 bp	**TGTTGGA**...GGTAA**CA** 514 bp

RETROTRANSPOSON AND RETROTRANSPOSON-LIKE ELEMENTS IN *DROSOPHILA* (COPIA), *SACCHAROMYCES CEREVISIAE* (TY 912), AND *ARABIDOPSIS* (TAG-3) ALL MAKE 5 bp TARGET SITE REPEATS. THEIR LONG DIRECT TERMINAL REPEATS ARE OF DIFFERENT LENGTH YET THEY HAVE HIGHLY CONSERVED SEQUENCES (ALIGNED IN BOLD). THEIR INTERNAL DOMAINS OF DIFFERENT LENGTH STILL HAS A FEW SIMILARITIES ALTHOUGH *COPIA* IS HIGHLY MOBILE, WHEREAS *TAG-3* IS NO LONGER MOVING BECAUSE ITS TRANSPOSASE GENE UNDERWENT TOO MANY CHANGES (PSEUDOGENIC). (AFTER VOYTAS & AUSUBEL 1988 NATURE (LOND.) 336:242

RETROVIRAL VECTORS: are capable of insertion into the chromosomes of a wide range of eukaryotic hosts. Generally the *gag, pol* and *env* genes are removed but all other elements required for integration and RNA synthesis are retained. Usually a selectable marker neomycin [*neo*], dehydrofolate-reductase [*dhfr*] resulting in methotrexate resistance (MTX) or bacterial hypoxanthine-phosphoribosyltransferase [*hprt*] or guanine-hypoxanthine-phosphoribosyltransferase [*gpt*] confering mycophenolic acid (MPA) resistance is inserted in such a way that its initiation codon (ATG) falls into the same place as the ATG of the group-specific antigen pro-

Retroviral vectors continued

tein (*gag*) gene. Transcription will be initiated at the 5' long terminal repeat (5'LTR) and translation will proceed from the ATG that is at the same place as that of *gag* :

----pBR322--- 5' LTR   PBS   SD   ψ   ATG   **SELECTABLE MARKER**   3' LTR---pBR322---

(pBR322: bacterial plasmid, 5' LTR: long terminal repeat, PBS: binding site for primer to initiate first-strand DNA synthesis, SD: splicing donor site, ψ : packaging signal for the virion, ATG the first translated codon (Met) of the selectable marker (e.g. antibiotic resistance), 3' LTR: long terminal repeat)

The generalized vector shown above is non-infectious and requires superinfection by a helper virus. After the DNA has been packaged into a virus particle the transfer of gene is almost fully efficient as long as the cell has the appropriate receptor. The recombinant DNA integrates into the host as a provirus. The provirus is generally present in a single copy per cell but this *producer cell* will proceed with production of recombinant retrovirions. The production of the recombinant virus is much enhanced if initially the cell is coinfected also by a wild type proviral plasmid. Such a system has the disadvantage of the presence of a helper virus which competes with a pseudovirion and it may also be hazardous by being pathogenic. The problems with the helper virus can be eliminated if it retains all the viral genetic sites except the packaging site (ψ) and thus it cannot produce infectious particles.

Broad host range (amphotropic) viral vectors can be constructed by replacing a narrow range (ecotropic) viral envelope protein with another, of an amphotropic virus. The envelope protein has the role of recognizing the cell surface receptors of the host. Caution must be taken because the vector or host cell DNA and the helper virus genetic material may recombine and give rise to infectious particles. By genetic engineering additional modifications have been introduced into the helper virus to prevent the formation of infectious single recombinants. This was achieved by replacing, e.g. the 3' LTR with a termination stretch of the SV40 eukaryotic virus DNA.

More useful are the *retroviral expression vectors*. The latter type not only prove that the viral vector is present in the cell but can propagate desirable genes (e.g. growth hormone genes, globin genes, etc.). Retroviral vectors may carry strong promoters and enhancers within the LTR region and thus can overexpress some genes. They may be employed also for insertional mutagenesis because they can insert at different chromosomal locations and can be used as tools to study animal differentiation and morphogenesis. (See vectors, retroviruses, viral vectors, epitope, SV40, mycophenolic acid)

RETROVIRUSES (Retroviridae) include onco-, lenti- and spumaviruses. In eukaryotes they contain generally dimeric single-stranded RNA as the genetic material that is replicated through a double-stranded DNA intermediate with aid of a reverse transcriptase enzyme. Retrovirus reverse transcription usually takes place after they infected the host. In the human foamy virus (spumavirus), the infectious particles already carry double-stranded DNA, indicating the reverse transcription precedes infection. Besides the polymerase gene (*pol*), they all carry group-specific antigen (*gag*) and envelope (*env*) protein genes. A provirus introduced into a cell by retroviral infection may become a retrotransposon (e.g. Ty element in yeast, copia in *Drosophila*, etc). They are characterized by long terminal repeats (LTR) of a few thousand nucleotide long. When the virus enters the host cell its RNA genetic material is converted into double-stranded DNA, and it may be covalently integrated into a host chromosome. The targets for integration spread all over the genome yet transcriptionally active regions appear to be favored and integration "hot spots" in chicken cells may be used by RSV a million fold more often than expected by chance alone, and some sequences in mouse cells (e.g. the HGPRT gene) may be avoided. After infection of a cell by a single virion, in a day, thousands of viral particle may be produced.

Retroviruses are considered to be diploid (dimeric) because they have a pair of genomes, two identical size RNAs of 7 to 9 kb. All retroviruses minimally encode a protease, a polymerase, ribonuclease H activity of the reverse transcriptase and an integrase function. A general

Retroviruses continued

ized structure of retroviruses is shown below (individual types may display variations of this scheme):

At the two ends of the viral genome are the <u>long terminal repeats</u> (LTR) of 2 to 8 kb. At the left and right termini of the LTRs are *attU3* and *attU5*, respectively for the attachment of the U3 (170 to 1,200 nucleotides) and U5 (80 to 120 nucleotides) direct repeats of the provirus in the host DNA. The *att* sequences at the 3'-end of U5 and at the 5'-end of U3' contain usually imperfect, inverted repeats where viral DNA joins the host DNA at 2 nucleotides from these ends. The terminal repeats represented by *R* (10 to 230 nucleotides) are used for the transfer of the DNA during reverse transcription. *E*: transcriptional enhancer, *P* promoter, *PA*: signal for RNA cleavage and polyadenylation. *PBS*: binding site for tRNA primer for first-strand DNA synthesis (different retroviruses use the 3'-OH end of different host tRNAs for initiation). *PPT* polypurine sequences which prime the synthesis of the second strand of DNA. *SD* splice donor (the site where *gag* and *pol* and *env* messages are spliced). *SA*: splice acceptor site (is the site where the second splice site joins to the first [donor] site). The three protein gag (group-specific antigen) a polyprotein, pol (polymerase, reverse transcriptase), env (envelope protein) are transcribed in different, overlapping reading frames and then for RNA packaging into virion. The genomic subunits are the same as the mRNA, i.e. they are (+) strands. The transcript RNA has a 7-methyl-guanylate group at the 5'-end, a 100 to 200 polyA tract at the 3'-end similarly to eukaryotic mRNAs. The different retroviruses code for different proteins with known or still unidentified functions. The gag-pol polyprotein complex contains information for proteolytic activities that generates a *protease* from the carboxyl end of gag and some other proteins. *Reverse transcriptase* (RT) and *integrase* (IN) and a protease are generated by proteolysis from the translated *pol* gene product. In the human foamy virus the pol protein is translated by splicing mRNA that does not include the gag domain. The reverse transcriptase varies among the various retroviruses. In the Rous sarcoma virus RT there is an RNA and DNA directed polymerase, an RNase H, it is also a tRNA-binding protein. It works with either RNA or DNA primers and synthesizes up to 10 kDa molecules from single RNA priming sites. Actinomycin D inhibits the replication on DNA but not on RNA template. Azidothymidine (AZT) inhibits polymerization and viral replication. The RNase H activity removes RNA in both 5' to 3' and 3' to 5' and digests the cap, tRNA and the polyA tail. All retroviruses have an integrase protein derived from the C end of the gag-pol polyprotein complex. Integrase (30 to 46 kDa protein with Zn-fingers) inserts the virus into the eukaryotic chromosomes:

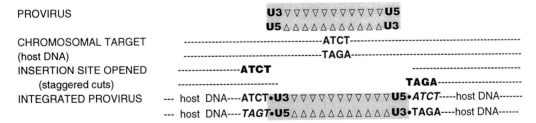

At the site of integration there is a target duplication as the recessed ends of the target filled in by complementary nucleotides (*in italics*).

The retroviral genome has a 10^5 times increased mutability compared with cellular genes. The retroviral genomes recombine with a 10 to 30% frequency during each cycle of replication. Integration may have profound effect on cellular genes, may inactivate suppressor genes control-

Retroviruses continued

ing cellular proliferation or may activate the transcription of cellular genes with the same effect and thus initiate carcinogenesis.

The major types of retroviruses: **1.** Bird's (avian) sarcoma and leukosis viruses such as Rous sarcoma virus (RSV), avian leukosis virus (ALV), Rous-associated viruses (RAV 1 and 2), **2.** Reticuloendotheliosis viruses (hyperplasia of the net-like and endothelial [tissues lining organ cavities]) , e.g. spleen necrosis virus (SNV), **3.** Mammalian leukemia and sarcoma viruses, e.g. Moloney murine sarcoma virus (Mo-MSV), Harvey murine sarcoma virus (Ha-MSV), Friend spleen focus-forming virus (FSFFV), feline leukemia virus (FLV), simian sarcoma-associated virus (SSAV), **4.** Mammary tumor viruses (MMTV), **5.** Primate-type D viruses, e.g. Mason-Pfizer monkey virus (MPMV), simian retrovirus (SRV-1), **6.** Human T-cell leukemia-related viruses (HTLV-1, HTLV-2), simian T cell leukemia virus (STLV) and bovine leukemia virus (BLV), **7.** Immunodeficiency and Lentiviruses, e.g human immunodeficency viruses (HIV1, HIV-2), Visna virus, simian immunodeficiency virus (SIV), caprine (goat) arthritis - encephalitis virus (CAEV), equine (horse) infectious anemia virus (EIAV). (Classification of Varmus, H. & Brown, P. 1989). About 1% of the human genome includes retroviral sequences (human endogenous retroviruses, HERV). These elements presumably inserted themselves during evolution and have been extensively modified. These HERV elements gave rise to the transposable elements or pseudogenic forms of them in the modern genome. The endogenous viruses synthesize the Gag and Env proteins and that enables them to become infective but may also prevent their transposition. The Env protein may also interfere with the viral receptors and thus limit reinfection. The HERV do not seem to have much significance for the genome at the present time, however retrotransposition may lead to mutation and loss of gene function, including loss of cancer gene suppression. (See also overlapping genes, retroposon, retrotransposon, retrogene, hybrid dysgenesis, copia elements, LINE, reverse transcription, retroviral vectors, oncogenes, tumor viruses, cancer, HTDV)

RETT's SYNDROME: see autism

Rev: is a splicing element originally identified in viruses; it assists exports through the nuclear pore. (See RNA export, nuclear pore)

REVERSAL OF DOMINANCE: see dominance reversal

REVERSE GENETICS: see reversed genetics

REVERSE LINKAGE: see affinity

REVERSE MUTATION: backmutation, change from mutant to wild type allele, $a \to A$.

REVERSE TRANSCRIPTASES: enzymes that transcribe DNA on RNA template. An outline of the function of the enzymes within the protein coat using the diploid template is shown on next page. A very similar process is followed in *in vitro* assays. Reverse transcriptases are commercially available from purified avian myeloblastosis (cancer of the bone marrow) cells or as cloned Moloney murine leukemia virus (Mo-MLV) gene product. The avian enzymes are dimeric and have strong reverse transcriptase and RNase H activities. The murine polymerase enzyme is monomeric and has only weak RNase H activity. Therefore the murine enzyme is the choice when mRNA is transcribed into cDNA. Also, RNase H can degrade DNA and may reduce the efficiency of cDNA synthesis. The temperature optimum of the avian enzyme is 42° C (pH 8.3) and at this temperature the murine enzyme is already degraded. The pH optimum for the murine enzyme is 7.6. Both enzymes have much lower activity slightly below or above the pH optima. Structurally complex RNAs are more efficiently transcribed by the avian enzyme. Reverse transcriptases are used for generating DNA from mRNA for vector construction or generating labeling probes, for primer extension, and for DNA sequencing by the dideoxy chain termination method. Reverse transcriptase does not have an editing (exonuclease) function therefore it may make errors at the rate of 5×10^{-3} to 1×10^{-6} per nucleotide. This rate is orders of magnitudes higher than the error rate of most eukaryotic replicases. The HIV-1 reverse transcriptase can use as template either RNA or DNA; in the latter case it makes double-stranded DNA. (See retroviruses, cDNA, central dogma, msDNA, error in repli-

Reverse transcriptases continued
cation. Reverse trancriptase action diagramed below)

```
1st retroviral RNA (-)                   R     U5   PBS    gag    pol    env
                        vvvvvvvvvvvvvvvvvvvvvvvvvvvvvvvvvvvvvvvvvvvvvvvvv
                           1st strand DNA       →→→→→→→→→→→→→→OH–tRNA primer
→→→→→→→→→→→→→→→→→→→→|  1st DNA synthesis continues after template switch to
vvvvvvvvvvvvvvvvvvvvvvvvvvvvvvvv   2nd retroviral RNA template
          U3      R
```

When the synthesis of the first strand reaches beyond the U3 site on the viral RNA template:

```
          5'- U3 - R - U5-----------------------------------U3 - R - U5- PBS - 3'
```

RNAase H generates a purine-rich sequence that primes the synthesis of the second strand of the DNA starting exactly at the point corresponding to the left (5') end of the LTR

```
               vvvvv←←←←←←←←←←←←←←←←←←←←←←←←←←←←←←←←  2nd DNA strand synthesis
RNA primer   ↑↑↑↑
generated by RNAase H
```

Synthesis then continues through U3-R-U5 and PBS with the tRNA still attached until it arrives to the first modified base in the tRNA (plus strand strong stop DNA). Then RNAase H removes the tRNA primer and after the two DNA strands become fully extended they anneal:

```
5'→→→→→→→→→→→→→→→→→→→→→→→→→→→→→→→→→→→→→→→→→→→→→→→→→→→→3'
3'←←←←←←←←←←←←←←←←←←←←←←←←←←←←←←←←←←←←←←←←←←←←←←←←←←←←5'
 U3 R U5                DOUBLE-STRANDED VIRAL DNA           U3 R U5
```

Reverse transcription of the retroviruses follows the generalized scheme. A single-stranded viral RNA (vvvv) serves as template for the synthesis of the first strand DNA (→→→). The synthesis is primed by a tRNA attached to the PBS (primer-binding site of the retroviral [-] strand). The host tRNA is base-paired by 18 nucleotides to a sequence next to U5. The first strand DNA (also called strong stop minus DNA) is extended at the rate of about 2 kb per h until the last part of the primer binding site is copied. When the synthesis is extended, the copying of the second DNA strand (←←←) begins. The process is practically the same in the cell and *in vitro* conditions.

REVERSE TRANSLATION: if the protein is sequenced and the amino acids of a short segment are coded by non-degenerate or moderately degenerate codons, one may synthesize a few RNAs on the basis of the presumed codon sequences, and one of them may be complementary to the DNA. This short RNA sequence can be hybridized to the DNA that is coding for this particular protein. Thus the gene may be isolated by reversing the translation and generating an appropriate probe. (See probe, synthetic probe)

REVERSED GENETICS: nucleic acids and proteins, etc. are first isolated and characterized *in vitro* by molecular techniques and subsequently their hereditary role is identified. Also, studying gene expression by introducing into cells (transformation) reporter genes with truncated upstream or downstream signals, *in vitro* generated mutations, etc., and thus determining the functional consequences of these alterations. Briefly, reversed genetics starts with molecular information and then deals with its biological role. Classical (forward) genetics recognizes genes when mutant forms become available and then studies their transmission, chromosomal location, mechanism of biochemical function, their fate in populations and evolution. Reversed genetics is sometimes called also surrogate genetics. (See genetics, inheritance, heredity)

REVERSION: backmutation either at the site where the original forward mutation took place or at another (tRNA) gene that may act as a suppressor tRNA or the reversion is caused by correcting frameshift. The possibility of reversion is frequently considered as evidence that the original forward mutation was not caused by a deletion. Suppressor mutation outside the mutant locus may restore, however, the non-mutant phenotype. Reversion may take place also when the mutation was caused by a duplication and a deletion evicted the duplicated sequence. (See backmutation, base substitution, suppressor, *sup*, suppressor tRNA, frameshift

REVERSION ASSAYS IN *SALMONELLA* AND *E. coli* IN GENETIC TOXICOLOGY: the *Salmonella* assay has been described by the Ames test. The most commonly used *E. coli*

Reversion assays in *Salmonella* and *E. coli* in genetic toxicology continued
test is employing strains WP2 and WP2$_{uvrA}$ that are deficient in genetic repair and are auxotrophic for tryptophan. They detect base substitution revertants but the assay does not respond to most frameshift mutagens (unlike some of the *Salmonella* strains TA97, TA98, TA2637 and derivatives). The *E. coli* systems does not offer any advantage over that of the *Salmonella* assay of Ames. (See Ames test, bioassays in genetic toxicology, mutation detection)

REX COLOR: of rodent (rabbit) hair appears in the presence the recessive fine fur gene, *r*, in certain combinations with black (*B/b*), agouti (*A/a*) and intensifier (*D/d*).

REYE SYNDROME: is a non-genetic inflammation of the brain of infants and may cause fever, vomiting, coma and eventually death.

RF (release factor): a protein which mediates the release of the peptide chain from the ribosome after it recognizes the stop codons. (See translation termination, protein synthesis, transcription termination in prokaryotes, transcription termination in eukaryotes)

RF: replicative form of single stranded nucleic acid viruses (DNA or RNA) where the original single strand makes a complementary copy that serves as template to synthesize replicas of the first (original) genomic nuclcic acid chain. (See replicase, plus strand)

Rf: fertility restorer genes in cytoplasmic male sterility. (See cytoplasmic male sterility)

Rf VALUE: in paper or thin-layer chromatography the distance from base line of the migrated compound divided by the distance of migration of the solvent (mixture). This Rf value which is always less then 1 is characteristic for a particular compound within a defined system of chromatography. (See paper chromatography, thin-layer chromatography)

$$Rf = \frac{distance - B}{distance - A}$$

< SUBSTANCE > (B)   > SOLVENT MIGRATION (A)

RF-A: replication factor A is a human single-strand-DNA binding protein, auxiliary to pol α and pol δ. (See pol, helix destabilizing protein, replication, replication fork eukaryotes)

RFA (replication factor A): the same as replication protein A (RPA). (See DNA replication eukaryotes)

RF-C: is DNA replication factor C, a primer/template binding protein with ATP-ase activity; it has a primary role in replicating the leading strand DNA in eukaryotes. RF-C loads PCNA on the DNA that tethers the DNA polymerase to the replication fork. RC-F is also called Activator I. (See replication fork, PCNA)

RFC (also RF-C): cellular replication factor. (See DNA replication eukaryotes)

RFLP (restriction fragment length polymorphism): restriction endonuclease enzymes cut DNA at specific sites and thus generate fragments of various sizes in their digest, depending on the distances between available recognition sites in the genome. When through mutation (during evolution) base changes occurred at the recognition sites, the length of fragments (within related strains) may have changed. After electrophoretic separation of a polymorphic pattern may be distinguished. These fragments may constitute co-dominant molecular markers for genetic mapping. Restriction fragment maps can be generated also by strictly physical means. If a small circular DNA is completely digested by a restriction enzyme yielding fragments, say A, B, C, D, E but incomplete digestion with the same enzyme produces ABD, DB, AD, BC and CE triple or double fragments, respectively but never AB, BE, DC or AC, the fragment sequence must be ADBCE because the double fragment must be neighbors. Another procedure is to digest by at least two enzymes and determine the overlaps by hybridization in a sequential manner. The overlapping fragments will indicate which fragments are next to each other e.g. ─── ─── ───   fragments by enzyme 1
                                                                                              ─── ───   fragments by enzyme 2

Restriction fragments can be used also in genetic linkage analysis. They represent "dominant" physical markers because the DNA fragments can be recognized in heterozygotes. RFLP markers are useful for following the inheritance of linked genetic markers which have variable expressivity and/or penetrance under unfavorable conditions. (See restriction enzymes, restriction fragment number, restriction fragment length, physical map)

RFLP MARKER: a restriction enzyme-generated DNA fragment which has been or can be mapped genetically to a chromosomal location and can be used for determining linkage to it; they are codominant and always expressed. Their inheritance, recombination can be determined in relatively small populations. (See restriction enzyme, RFLP, physical map, integrated map)

RFLP SUBTRACTION: a selective technique for the enrichment of particular polymorphic, eukaryotic genomic unique segments. Small restriction fragments are isolated and purified from one genome containing sequences that are in large fragments in another, related genome of mouse. By subtractive hybridization the segments with shared sequences by both genomes are removed. Thus small fragments unique to one or the other strain are obtained. These sequences are then mappable genetic markers. (See genomic subtraction)

RFLV: an RFLP variant. (See RFLP)

RFX: is a human DNA binding protein that promotes dimerization of MYC and MAX and thus stimulates transcription. A group of transcription factors for the major histocompatibility complex is also designated as RFX. (See MYC, MAX, MHC)

RGD: an amino acid sequence Arg-Gly-Asp in the extracellular matrix and in fibronectin is recognized by and bound to integrin. (See fibronectin, integrin, amino acid symbols in protein)

RGE (rotating gel electrophoresis): the gel is rotated 90° at switching the cycle of the electric pulses. (See pulsed field gel electrophoresis)

rGH (rat growth hormone): a thyroid hormone. (See animal hormones, hormone receptors, hormone-response elements)

RGS (regulator of G protein signaling): is actually the same as GAP (GTP-ase activating protein). The different RGS proteins have different specificities for the different $\alpha\beta$ and γ subunits of the trimeric G proteins. (See G proteins, GAP, signal transduction)

RH: see radiation hybrid

Rh BLOOD GROUP: the name comes from a misinterpretation of the early study, namely that this human antigen would have the same specificity as that of rhesus monkey red cells. It is now known that this was incorrect, the animal antigen is different but the name was not changed. Despite over a half century research the Rh antigen is not sufficiently characterized. The antigen may be controled by three closely linked chromosomal sites *C, D* and *E*, and on this basis 8 (2^3) different allelic combinations were conceivable; the triple recessive *cde* being a null combination. The 8 combinations are also designated as *R* or *r* with superscripts: *CDe* (R^1 or $R^{1,2,-3,-4,5}$), *cde* (r, or $R^{-1,-2,-3,4,5}$), *cDE* (R^2 or $R^{1,-2,3,4,-5}$), *cDe* (R^O or $R^{1,-2,-3,4,5}$), *cdE* (r'' or $R^{-1,-2,3,4,-5}$), *Cde* (r' or $R^{1,2,-3,-4,5}$), *CDE* (R^Z or $R^{1,2,3,-4,-5}$) and *CdE* (r^y or $R^{-1,2,3,-4,-5}$). The first 3 of these occur at frequencies about 0.42, 0.39 and 0.14, respectively in England and the others are quite rare. In some oriental populations R^1 (0.73%) and R^2 (0.19%) predominate and the recessives have a combined frequency of about 2%. This is in contrast to western populations where they occur in over 40%. Clinically the most important is the D antigen because 80% of the D⁻ individuals in response to large volume of D⁺ blood transfusion make anti-D antibodies. The *d* alleles are amorph. The Rh genes are in human chromosome 1p. Additional, regulatory loci have also been identified in chromosome 3. For phenotypic distinction antisera anti-D, anti-C, anti-E, anti-c and anti-e are used and on the basis of the serological reactions 18 phenotypes can be distinguished. Anti-D antibodies are usually immunoglobobulins of the G class (IgG) and develop only after immunization by Rh⁺ type blood. Anti-C antibodies are generally of IgM type and they occur along with IgG after an Rh⁻ person is immunized with Rh⁺ blood. Anti-E antibodies (IgG) are elicited in E negatives after exposure to E⁺ blood. Anti-c antibodies (IgG) occur in CDe/CDe individuals after transfusion with c⁺ erythrocytes. Anti-e antibodies are very rare (0.03). The major types of Rh antigens have several different variations. The Rh antigens are probably red blood cell membrane proteins. About 50 different Rh antigens have been identified. Rh deficiency may arise also by the activity of a special suppressor gene in human chromosome 6p11-p21.1 or by CD47 protein, encoded at

Rh blood group continued
human chromosome 3q13.1-q13.2. The RG gene is situated at 1p34.3-p36.1.
It is clinically very significant that about 15% of Western populations is *cde/cde*. In Oriental populations the frequency of this genotype is very low. This type of individuals — called Rh negatives — may respond with *erythroblastosis* when exposed to Rh positive blood. If an Rh negative female carries a fetus with Rh positive blood type, antibodies against the fetal blood may be produced by the mother. This may then cause severe anemia with a high chance of intrauterine death and abortion. Generally, during the first pregnancy this hemolytic reaction is absent but the chance in the following pregnancies, by when sufficient immunization has taken place by the fetal blood entering into the maternal bloodstream, the probabilty of erythroblastosis becomes high. Thus pregnancies of Rh⁻ females are monitored and appropriate serological treatment must be provided if antibody production is detected to prevent fetal erythroblastosis. Erythroblastosis may also occur if an Rh⁻ individual is transfused with Rh⁺ blood. The rodent antibodies responding to rhesus monkey red cells are called now LW blood group. (See erythroblastosis fetalis, blood groups, immunoglobulins, antibodies)

RHABDOMERE: rod-shaped element of the compound eye of insects. (See compound eye, ommatidium. *Drosophila*)

RHABDOMYOSARCOMA: a type of cancer involving chromosome breakage in the Pax-3 gene 2q35 and 13q14 (Rhabdomyosarcoma-2) or other translocations involving chromosome 3 and 11 (Rhabdomyosarcoma-1). These break points may also be related to the Beckwith-Wiedemann syndrome or WAGR syndrome. (See Pax, Beckwith-Wiedemann syndrome, Wilms tumor)

RHABDOVIRIDAE: oblong or rod-shape (130-380 x 70-85 nm) single-stranded RNA (13-16 kb) viruses with multiple genera and wide host ranges. (See CO_2-sensitivity in *Drosophila*)

Rhalloc: the sequences mapped by radiation hybrid methods by various mapping groups. (See radiation hybrids)

Rhdb: a database containing the mapping information obtained by radiation hybrids. (See radiation hybrids)

RHESUS BLOOD GROUP: see Rh blood group, LW blood group

RHESUS MONKEY (Macaque, *Macaca mulatta*, 2n = 42): is a representative of mainly South-East Asian and North-African species of long-tail monkeys. These intelligent animals have been used extensively for biological and behavioral studies. (See Rh blood group, LW blood group, cercopithecidae, primates)

RHEUMATIC FEVER (rheumatoid arthritis): are ailments affecting mainly the connective tissues and joints, but may cause also heart and nervous system anomalies. The disease is complex because environmental and susceptibility factors heavily confound the direct genetic determination. E.g., certain streptococcal infections can precipitate rheumatic fever. The familial forms are attributed to dominant genetic factor(s). The susceptibility has been attributed to recessive genes(s). Several antigens have been identified which appeared to be more predominant within affected kindreds. One monoclonal antibody D8/17 was present in 100% of the patients affected with the disease whereas two other monoclonal antibodies showed up between 70 to 90% coincidence and with 17 to 21% presence even among the unaffected people. (See also arthritis, rheumatoid, ankylosing spondylitis, HLA)

RHEUMATOID: means that it resembles rheumatic condition. (See rheumatic fever)

RHINOCEROS: *Ceratotherium simum*, 2n = 84; *Rhinoceros unicornis*, 2n = 82.

RHIZOBIUM: see nitrogen fixation

RHIZOFILTRATION: see bioremediation

RHIZOID: a structure resembling plant roots.

RHIZOME: an underground plant stem modified for storage of nutrients and propagation.

RHIZOMORPH: see hypha

RHKO (random homozygous knockout): see knockout

RhlB: a helicase of the DEADE-box family. (See helicase, DEAD-box, degradosome)

RHO: homolog of the RAS oncogene. It relays signals from cell-surface receptors to the actin cytoskeleton. It regulates myosin phosphatase and Rho-associated kinase. In yeast cells a RHO protein is involved in the stimulation of cell wall β (1→3)-D-glucan synthase and the regulation of protein kinase C, and in mediation of polarized growth and morphogenesis. Actually, RHO is a subunit of the glucan synthase enzyme complex. Serine-threonine protein kinase and protein kinase N (PKN) are apparently activated by RHO. RHO mediate also endocytosis. In human chromosomes, RHOs are designated as ARH6: 3pter-p12, ARH12: 3p21, ARH9: 5q 31-qter. The RHO family includes Rac, Cdc42, RhoG, RhoE, RhoL and TC10 proteins. (See RAS oncogene, cytoskeleton, receptor, receptors, RAC, Cdc42, endosome)

rho (ρ): a designation of density; high G+C content of DNA increases it, high A+T content decreases it. ($\rho = 1.660 + [0.098 \times \{G+C\}]$ fraction in DNA). The density is determined on the basis of ultracentrifugation in CsCl and refractometry of the bands. (See buoyant density)

rho FACTOR: protein involved in termination of transcription in (rho-dependent) prokaryotes is about 46 kDa, and its is a hexamer (~275 kDa) and for maximal efficiency it is present in about 10% of the molecular concentration of the RNA polymerase enzyme. It is basically an RNA-dependent ATPase. (See transcription termination in prokaryotes)

rho **GENE**: is responsible for the suppressive petite (mtDNA) condition in yeast. (See mtDNA)

rho⁻ MUTANTS: of yeast lost from their mitochondrial DNA most of the coding sequences and are very high in A+T content (the buoyant density of the DNA is low). (See mtDNA)

rho-DEPENDENT TRANSCRIPTION TERMINATION: actually none of the rho-dependent strains absolutely requires this protein factor for termination. (See rho factor, transcription termination, rho-independent)

RHODAMINE B: is a fluorochrome used for fluorescent microscopy; its reactive group forms a covalent bond with proteins (immunoglobulins) and other molecules. It is also a laser dye. Absorption maxima: 543 (355) nm. Caution: carcinogenic. (See fluorochromes)

RHODOPSIN: is a light-sensitive protein (opsin, $M_r \approx 28,600$) coupled with a chromophore, 11-cis retinal, which isomerizes to all-trans-retinal immediately upon the receipt of the first photon. It functions as the light receptor molecule in the disks of the photoreceptive membrane of the photoreceptor cells of the animal retina of the eye. Rhodopsin has seven short hydrophobic regions that pass through the endoplasmic reticulum (ER) membrane in seven turns. The amino end (with attached sugars) is within the ER lumen and the carboxyl end points out into the cytosol. In the rod shape photoreceptor cells rhodopsin is responsible for monochromatic light perception at low light intensities, and in the cone shape photoreceptor cells color vision is mediated by it in bright light. The photoreceptor cells transmit a chemical signal to the retinal nerves that initiate then the visual reaction series. When the receptor is activated, the level of cyclic guanylic monophosphate (cGMP) drops by the activity of *cGMP phosphodiesterase* and it is quickly replenished in dark by *guanylyl cyclase*. The activated opsin protein binds to transducin, an α_t G protein subunit, that activates cGMP phosphodiesterase. When one single photon of light hits rhodopsin, through an amplification cascade, 500,000 molecules of cGMP may be hydrolyzed, 250 Na⁺ channels may close and more than a million Na⁺ are turned back from entering the cell through the membrane within the time span of a second. In dark the sodium ion channels are kept open by cGMP, in light the channels are closed. The sodium-calcium channels being shut, in light, the intake of Ca^{++} is reduced and that leads to the restoration of the cGMP level through the action of the *recoverin* protein that cannot function well when it is bound to Ca^{++}. The rhodopsin gene has been assigned to human chromosome 3q21-qter and to mouse chromosome 6. *Drosophila* has three rhodopsin loci (*Rh2* [3-=65], *Rh3* [3-70], *Rh4* [3-45]). In the flies also the *nina* loci are involved in the synthesis of opsins affecting the ommatidia and ocelli (See also phytochrome, signal transduction, G-proteins, retinitis pigmentosa, retinoblastoma, color blindness, color vision, ommatidium, ocellus, opsins)

Rhoeo discolor: an ornamental plant with large chromosomes, 2n = 12; genome x = 14.5×10^9 bp.

rho-INDEPENDENT: the termination of transcription in some prokaryotic cells does not require rho factor. (See also rho factor, transcription termination prokaryote, rho-dependent)

RHUBARB (Rheum spp): about 50 species; 2n = 2x = 44. Its an accessory food plant and some species are used as medicinal herbs (cathartic [laxative]).

RHYNCHOSCIARAs: are dipteran flies with very clearly banded polytenic chromosomes in the salivary gland nuclei. (See *Sciara*, polytenic chromosomes)

RI: see recombinant inbreds

RI PARTICLES: are formed in cold *in vitro* during the 30S ribosomal subunit reconstitution experiment of rRNA and about 15 proteins. Upon heating to assume the proper conformation, they become RI* particles. (See ribosome, ribosomal RNA, ribosomal protein)

Ri PLASMID: root inciting plasmid of *Agrobacterium rhizogenes* can be used for genetic engineering similarly to the Ti plasmid of *Agrobacterium tumefaciens*. The bacterium is responsible for the hairy root disease of plants. Its T-DNA contains two segments. The right T-DNA (T_R) contains genes for the production of the opines mannopine and agropine and also for auxin. These auxin genes are higly homologous to the comparable genes in the Ti plasmid of *Agrobacterium tumefaciens*. The left portion of the T-DNA (T_L) includes 11 open reading frames with organization similar to eukaryotic genes but this segment is different from that of the Ti plasmid. (See Ti plasmid)

RIA: see radioimmunoassay

RIBOFLAVIN: is a vitamin precursor of flavin mononucleotide (FMN) and flavin adenine dinucleotide (FAD), oxidation coenzymes. Riboflavin is heat stable but rapidly decomposes in light.

RIBOFLAVIN-RETENTION DEFICIENCY: may prevent hatching of eggs in "leaky auxotrophic" chickens. The defect is not in absorption but the vitamin is rapidly excreted by a genetic default, and the *rd/rd* eggs have only about 10 μg of the vitamin rather than the normal level of about 70 μg. If 200 μg is injected into the eggs before incubation, hatching occurs.

RIBONUCLEASES: occur in a large number of specificities and they digest various types of ribonucleic acids. The bovine pancreatic ribonuclease is a small (124 amino acids) and very heat-stable enzyme. The pancreatic ribonuclease was the first enzyme chemically synthesized in the laboratory. An autoradiogram shown at left permits the distinction of the digestion patterns, separated in 20% polyacrylamide gel. NE: no enzyme control, T_1: G-specific enzyme, U_2: ribonuclease is A-specific, Phy M: *Physarum* enzyme M, specific for U + C, OH: random alkaline digest, B_c: *Bacillus cereus* enzyme with U + C specificity. On the left side guanosine positions are indicated from the 5'-end, on the right side the nucleotide sequences are shown as read from the gel. (Courtesy of P-L Biochemicals, Inc.). (See also RNases and ribonucleases below)

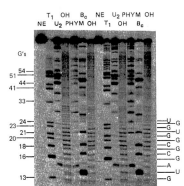

RIBONUCLEASE 1: degrades RNA I. (See RNA I)

RIBONUCLEASE II: is similar in action to Ribonuclease D; its role is not just processing but it can degrade the entire tRNA molecule. (See ribonuclease D)

RIBONUCLEASE III: is an endonuclease cutting RNA from the 3' or 5' end. It cleaves eukaryotic pre-rRNA at a U3 snoRNP dependent site. (See trimming, snoRNP)

RIBONUCLEASE A: is a family of RNA digesting enzymes, including pancreatic, brain ribonucleases as well the related eosinophil-derived neurotoxin (EDN), eosinophil cationic protein (ECP) and angiogenin, involved in defense functions. (See eosinophil, angiogenesis)

RIBONUCLEASE B: cuts at U+C sequences of RNA.

RIBONUCLEASE BN: is an exonuclease cutting tRNA. (See exonuclease)

RIBONUCLEASE D: processes tRNA primary transcripts at the 3' end into mature tRNA. (See tRNA, primary transcript)

RIBONUCLEASE E: cleaves RNAs with secondary structure within single-stranded regions rich in A and U nucleotides, such as RNA I. It is a 1061 amino acid 3'→5' exonuclease involved in RNA processing and degradation in *E. coli*. (See RNA I, *E. coli*, exonuclease)

RIBONUCLEASE H: digests RNA when paired with DNA but it does not cut single-strand RNA or double-strand RNA or double-strand DNA.

RIBONUCLEASE P: processes the 5' end of transfer RNA transcripts; its catalytic subunit is a ribozyme, a 377 nucleotide RNA that can do the processing even without the protein. (See ribozyme, external guide sequences, KH domain, RNase MRP)

RIBONUCLEASE T: exonuclease of tRNA cutting at the amino acid accepting end (CCA). (See tRNA)

RIBONUCLEASE T_1: is specific for G (guanine) linkages in RNA.

RIBONUCLEASE U2: is specific for A+U nucleotides in RNA.

RIBONUCLEIC ACID: see RNA

RIBONUCLEOPROTEIN (RNP): ribonucleic acid associated with protein. (See also RNP)

RIBONUCLEOTIDE: contains one of the four nitrogenous bases (A, U, G, C), ribose and phosphate. Building block of RNA. (See deoxyribonucleotide)

RIBONUCLEOTIDE REDUCTASE (RNR): converts ribonucleotide diphosphates into deoxyribonucleotide diphosphates. It is required for the completion of the cell cycle and malignancy. (See cell cycle, malignant)

RIBOSE: is an aldopentose sugar, present in ribonucleic acids with an OH group at both 2' and 3' positions. Its deoxyribose form lacks the O at the 2' position and it is present in DNA. (See aldose)

RIBOSOMAL DNA: codes for ribosomal RNAs. (See ribosome)

RIBOSOMAL FRAME SHIFTING (translational recoding): see overlapping genes, recoding

RIBOSOMAL GENES: see *rrn*, ribosomes, rRNA

RIBOSOMAL PROTEINS: are generally designated with an S or L indicating whether it is part of the small or large ribosomal subunit. The size of these 55 proteins in *E. coli* range from 6 kDa to 75 kDa. They bind to the RNAs at specific binding sites either directly or through their association. In *E. coli* the genes for these proteins are scattered among other genes in the chromosome. One of the bacterial ribosomal proteins is present in several copies whereas the other ones occur only once per ribosome. In eukaryotes, the more than 80 ribosomal protein genes generally occur in a single or a few copies. The ribosomal proteins assure the proper structural conditions on the ribosomes for translation. About 35% of the bacterial ribosomes is protein. The chloroplast ribosomal proteins are 2/3 imported from the cytoplasm and even larger fraction of the mitochondrial proteins are coded by the nucleus. The number of ribosomal proteins in organelles is higher than in prokaryotes. The number of proteins in the mammalian mitochondrial ribosomes is about 85 and nearly all are imported. The number of ribosomal proteins in the large mitochondria of higher plants is about 65. (See also nucleolus, ribosomes, ribosomal RNA, protein synthesis)

RIBOSOMAL RNA: about 65% of the bacterial ribosomes is RNA. The 16S bacterial rRNA (1.54 kb) has short double-stranded domains and single stranded loops and about ten of the bases near the 3' end are methylated. The 23S rRNA (3.2 kb) carries about 20 methylated bases. The 18S mammalian rRNA (1.9 kb) has more than 40 and the 28S (4.7 kb) more than 70 methylations. The ribosomal RNAs provide not just a niche for translation but they interact directly with translation initiation. The 16S rRNA cooperates with the anticodon at both the A and P sites and the 23S rRNA interacts with the CCA end of the tRNA. In the 23S rRNA two guanine sites are universally conserved (G2252, G2253), and the Cytosine 74 site of the acceptor end of the tRNA (CCA) is required for their functional interaction at the P site of the ribosome for protein synthesis. Methylated sequences in the rRNA mediate probably the joining of the small ribosomal subunit to the large subunit after translation is initiated, and hold on to the initiator tRNA^{fMet}. Some mutations in the 16S RNA may cause an override through

Ribosomal RNA continued

TRANSCRIPTION OF RIBOSOMAL RNA IN THE NUCLEOLAR ORGANIZER REGION OF THE CHROMOSOMES OF *ACETABULARIA*. THE GENES ARE SEPARATED BY NON-TRANSCRIBED INTERGENIC SPACERS. One transcription unit is about 1.7 μm. (Courtesy of Spring, H. *et al*. J. Microsc. Biol. Cell 25:107)

the stop codon and failure of termination of translation, and mutations in the 23S RNA may disturb the A and P ribosomal sites. Ribosomal ribozymes were shown to be involved in mediating peptidyl transferase functions. The prokaryotic 23S ribosomal RNA contains the pentapeptide coding minigene (GUGCGAAUGCUGACAUAAGUA) with canonical ribosome-binding site and appears to mediate resistance to the antibiotic erythromycin. (See also nucleolus, ribosome, ribosomal genes, class I genes, class III genes, rrn, introns, ribosomal proteins, protein synthesis, tRNA, gene size, RNase MRP)

RIBOSOME BINDING: see Shine-Dalgarno sequence, ribosome, mRNA, ribosome scanning

RIBOSOME-BINDING ASSAY: was used in the mid 1960's to identify several codons. RNA oligonucleotides bound to ribosomes attached only those charged tRNA molecules which had the specific anticodons and carried the appropriate amino acids. This way the relation between RNA codons and amino acids was revealed. (See genetic code, decoding)

RIBOSOME SCANNING: eukaryotic mRNAs do not have a Shine-Dalgarno consensus for ribosome binding; they are attached probably by the 5'-m^{7}G(5')pp. The (5')mRNA sequence reels on the ribosome until the initiator methionine codon is found. Circular viral or eukaryotic RNAs, if they contain internal ribosome entry sites (IRES), may be translated without a need for free 5' end. (Cap, Shine-Dalgarno, protein synthesis, IRES, dicistronic translation)

RIBOSOMES: provide the workshop and some of the tools for protein synthesis in all cellular organisms, including the subcellular organs, mitochondria and chloroplasts. The chloroplastic ribosome genes are situated in the characteristic inverted repeats, except in some *Fabaceae* and conifers. The prokaryotic and organellar ribosomes are similar and their approximate molecular weight is 2.5×10^6 Da with a sedimentation coefficient of 70S. The eukaryotic ribosomes, excluding the organellar ones, have a molecular weight of about 4.2×10^6 Da, and they are ≈ 80S. The ribosomes have a minor and a major subunit both are built of RNA and protein:

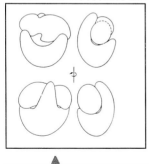

	PROKARYOTIC	EUKARYOTIC
SMALL SUBUNIT		
rRNA types	16S (1.54 kb)	18S (1.9 kb)
protein, kinds of molecules	21	~33
LARGE SUBUNIT	50S	60S
rRNA types	23S (3.2 kb)	28S (4.7 kb)
	5S (0.12 kb)	5 S (0.12 kb)
		5.8S (0.16 kb)
Protein, kinds of molecules	34	~49
RIBOSOMES, NUMBER/CELL	15,000-20,000 (*E. coli*)	more and variable
RIBOSOMAL GENE NUMBER	in 7 operons	200 genes in *Drosophila*/genome

(Small subunit row: 30S / 40S)

70S prokaryotic ribosome viewed from different angles. (Courtesy of Tischendorf, G.W. *et al*. Proc. Natl. Acad Sci. USA 72:4870)

The mitochondrial ribosomes do not contain 5S subunits but chloroplasts do have it. The number of ribosomes may greatly increase when protein synthesis is very rapid. During early embryogenesis in amphibia, the rRNA gene number may increase three order of magnitude by a process of amplification and the extra copies of the genes are sequestered into minichromosomes forming micronuclei. Their number in some higher plant regularly runs into thousands. The bacterial ribosomes are about 65% RNA and 35% protein. On the ribosomes several active centers can be distinguished. The A and P sites receive the tRNAs. This area extends to

Ribosomes continued
both small and large subunits. The tRNA after unloading of the amino acids leaves the ribosome at the exit site (E) of the large subunit. The translocation factor EF-G seems to occupy a space in between the two subunits. Elongation factor EF-TU may be located on the small subunit but it communicates with EF-G. Prokaryotic ribosomes are normally inhibited by chloramphenicol, erythromycin, lincomycin, streptomycin, spectinomycin, kanamycin, hygromycin, etc. Eukaryotic ribosomes are sensitive to cycloheximide, anisomycin, puromycin, tetracyclines, etc. (See nucleolus, protein synthesis, ribosomal proteins, E site, ribosomal RNAs, EF-G, EF-TU-GTP, antibiotics)

RIBOTHYMIDINE: a thymine in the tRNA attached to ribose rather than to deoxyribose as it occurs in the DNA. (See tRNA)

RIBOZYME: catalytic RNA, possessing enzymatic activity such as splicing RNA transcripts, cleavage of DNA, amide bond, polymerization and replication of RNA, etc. Thus these ribonucleic acid carry out functions very similar to those of protein enzymes. Ribozymes are metallo enzymes, generally using Mg^{++} for catalysis and stabilization. Most commonly they cleave phosphodiester bonds but they can also synthesize nucleotide chains. The ribozymes are generally large molecules yet the shortest ribozyme is only UUU and it acts on CAAA. Ribozymes commonly have an *internal guide* for substrate recognition near their 5' terminus and a *splice site* where they cleave and splice the molecules. Frequently ribozymes are classified into groups such as the hammerhead ribozymes (see diagram) that are used mainly by plant RNA viruses, the RNase P, the delta, group I and hairpin ribozymes. The hammerhead ribozymes cut at UCX sequence if the neighboring sequences are complementary and pair. The hairpin ribozymes must have at least 50 nucleotides in the catalytic domain and 14 in the substrate domain. The two domains pair in a two-stem form separated with a loop, most commonly containing an AGUC sequence and cleavage is usually between A and G. RNA-catalyzed RNA polymerization has also been identified. Ribozymes functioning as ligases, polynucleotide kinases, mRNA repair by targeted trans-splicing and isomerases as well as self-alkylating catalysts have also been isolated from large pools (10^{14}) of diverse RNAs. From evolutionary point of view these diverse ribozyme functions lend support to the ideas of the prebiotic RNA world. Ribozymes can be engineered to recognize specific mRNAs and by cleaving them the expression of a particular protein can be prevented. They have advantage over protein enzymes because it is less likely that they would incite an immune reaction. Because of their small size, their introduction into the cell is facilitated with the aid of transformation vectors. They can be propagated by small RNA transcription units that may accumulate up to 10^6 copies per cell. Transcription units for tRNA, U6 snRNA have been used. Although these units produce high ribozyme titer in the cell, polymerase II transcribed units target the ribozymes more effectively to the desired location. In such pol II units the ribozyme motif is inserted into the 5' untranslated sequences. Both hammerhead and hairpin ribozymes were introduced into human cells infected with HIV and the ribozymes reduced the level of the gag protein. (See introns, ribonuclease P, peptidyl transferase, CBP2, deoxyribozyme, ligase, kinase, alkylation, RNA world, RNA restriction enzyme, gene therapy, HIV)

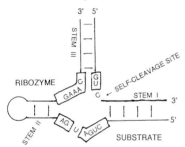

THE CRITICAL SEQUENCES IN THE HAMMERHEAD RIBOZYMES THE BOXED NUCLEOTIDES ARE CONSERVED

RIBULOSE 1,5-BISPHOSPHATE CARBOXYLASE - OXIDASE (Rubisco): a chloroplast located enzyme, generally the largest amount of protein in that organelle. Its large subunit is encoded and translated by the chloroplast system, the small subunit is coded for by a nuclear gene, translated in the cytosol and imported into the plastids. The abundance of the small subunit affects the translation of mRNA of the large subunit. Rubisco is involved in early steps of photosynthesis and makes through an intermediate 3-phosphoglycerol and it is involved in oxidative and reductive carboxylation. In dinoflagellates Rubisco is encoded by the nucleus. (See chloroplast genetics, photosynthesis)

RICE (*Oryza*): Gramineae, x = 12, genome size 4.5×10^8 bp; there are diploid or tetraploid species. Over 2,000 molecular markers are available in this small genome. Generation time 90-140 days. Database: <http://bioserver.myongji.ac.kr/ricemac.html>

RICIN: is an extremely toxic, ribosome-inactivating, dimeric toxin produced by the plant castor bean (*Ricinus*). It may have significance for cancer therapy research. (See magic bullet)

RICKETS: anomalies in bone development caused by defects of calcium and phosphorus absorbtion and/or vitamin D deficiency. Human autosomal recessive conditions may be caused by defects in the synthesis of calciferol (vitamin D) from sterols, and in such cases the dependency can be corrected by vitamin (D_3). In some forms the receptor is defective and the hereditary condition cannot be alleviated by the vitamin. Rickets may then have multiple phenotypic consequences such as alopecia, epilepsy, etc. (See hypophosphatasia, hypophosphatemia, vitamin D)

RICKETTSIA: small rod-shape or roundish gram-negative bacteria; they may carry typhus [typhoid fever, accompanied by eruptions, chills, headaches and high mortality], spotted fever, a tick-borne disease of cerebrospinal meningitis [brain inflammation] from animals to humans by infected arthropods (ticks, lice, fleas)

RIEGER SYNDROME: autosomal dominant eye, tooth and umbilical hernia syndrome. Chromosomal location (just as the Nazi discoverer of the disease) was controversial. Human chromosomes 21q22, 4q25, 13q14 and several others have been implicated. The basic cause is also unclear; epidermal growth factor, interleukin-2, alcohol dehydrogenase, fibroblast growth factor deficiencies appeared to be involved in chromosome 4. The chromosome 4 gene has now been cloned and it encodes a transcription factor with similarities to the *bicoid* gene of *Drosophila*. (See eye diseases, tooth agenesis, morphogenesis in *Drosophila* {8})

RIFAMPICIN: an antibiotic that combines with prokaryotic DNA polymerase (but not with mammalian DNA polymerase) and inhibits replication in *E. coli* and other prokaryotes. Rifamycin has similar effects. (See antibiotics)

RIGENS: is a translocation complex in *Oenothera muricata*. If during meiosis this complex goes to the top end of the megaspore tetrad, it may overcome its topological disadvantage and this megaspore may develop into an embryo sac because the other complex, *curvans* is not functional in the megaspores. (See megaspore competition, *Oenothera*, zygotic lethal)

RIG ONCOGENE: is probably required for all types of cellular growth and it is active in a very wide variety of cancers. (See oncogenes)

RIGS (repeat-induced gene silencing): an apparently epigenetic phenomenon caused by methylation of cytosine residues. Alternative process is that locally paired region of homologous sequences are flanked by unpaired heterologous sequences in transgenic plants. (See methylation of DNA, silencing, epigenesis)

RILEY-DAY SYNDROME (dysautonomia): human chromosome 9q31-q33 recessive neuropathy involving emotional instability, lack of tearing, unusual sweating, cold extremities, etc. The prevalence in Ashkenazic Jews is about $2-3 \times 10^{-4}$ but it is quite rare in other ethnic groups. Defects in the nerve growth factor receptor are suspected. (See neuropathy, nerve growth factor, pain-insensitivity)

RING BIVALENT: has terminalized chiasmata in both arms and thus in the early anaphase I the homologous chromosomes appear to be temporarily connected at the telomeric regions of the four chromatids. This is not a ring chromosome, however (See also translocation ring)

RING CANAL: are intercellular bridges during cytoblast differentiation. These transport mRNA and protein from nurse cells to oocytes. They are composed of actin, the hts (hui-li tai shao)

RING CHROMOSOME: a circular chromosome without free ends (○) such as the bacterial chromosome, as the ring DNAs in mitochondria and plastids. Ring chromosomes may result by different types of chromosome breakages. Simultaneous breaks across the centromere and the chromosome ends (telomeres) may result in fusion of the two broken termini generating one or two ring chromosomes. Also, crossing over between two ends of the same chromosome may give rise to a centric ring and acentric fragments. Sister chromatid exchange within a ring chromosome may result in a dicentric ring chromosome which at anaphase separation may break at various points and generate unequal size ring chromosomes and genetic instability. (See dicentric ring chromosome, also ring bivalent, translocation ring, sister strand exchange)

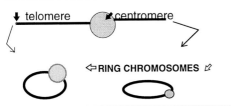

A SIMULTANEOUS BREAKAGE AT BOTH THE CENTROMERE AND AT THE TELOMERES MAY RESULT IN A FUSION BETWEEN THE BROKEN ENDS AND THE GENERATION OF RING CHROMOSOMES FROM BOTH OF THE ARMS OF THE NORMAL CHROMOSOME.

RING FINGER: is a cysteine-rich amino acid motif such as Cys-X2-Cys-X(9-27)-Cys-X(1-3)-His-X2-Cys-X2-Cys-X(4-48)-Cys-X2-Cys. X stands for any amino acids in the numbers shown in parenthesis. These are protein-protein, protein-membrane, protein-DNA interacting elements involved in the regulation of transcription, replication, recombination, restriction, development, cancerous growth, etc. The RING fingers may also bind Zinc and thus related to Zinc fingers. The name comes from the human gene RING—carrying such a motif— located in the vicinity of HLA. (See Zinc finger, DNA-binding protein domains)

RINGER SOLUTION: is prepared in somewhat different concentrations depending on type of tissues are used for as a sterilized physiological salt solution, in 100 mL water mg salts: NaCl 860, KCl 30, $CaCl_2$ 33; some formulations add also $NaHCO_3$ 20, NaH_2PO_4 and glucose 200.

RIP: term for "recombination induced premeiotically" and alternatively "repeat induced point mutation". When repeated DNA is introduced into fungi (*Neurospora* and others) by transformation, the duplications are ejected by a recombinational process that may generate high frequency of chromosomal aberrations and point mutations. (See also co-suppression)

RIP (ribosome-inactivating protein): are antiviral proteins in plants and animals with glycosylase activity and thus may depurinate RNA of susceptible ribosomes and thus block protein synthesis. (See depurination, glycosylases, saporins)

Rip1: see RNA export

RIPPING: the process generated by RIP mutations. (See RIP)

RISE: see one step growth

RISK: a combination of the degree of a hazard with its potential frequency of occurrence, e.g., a recessive gene causes a particular malformation in 30% of the fetuses, the probability of its homozygosity among the progeny of heterozygous parents is 0.25. Thus the risk of this malformation is 0.3 x 0.25 ≈ 0.075 i.e. 7.5% or if an individual is known to be heterozygous for a recessive lethal gene and marries a first cousin the risk that they will have a stillborn child may be as high as 0.5 x 0.25 = 0.125 = 1/8. If however, the carrier marries an unrelated person from the general population where the frequency of this gene is only 0.005, the risk will be 0.5 x 0.005 = 0.0025 = 1/400. In some cases the calculation of the genetic risk is not quite as simple. Let us assume that the penetrance of a dominant gene is 80% but non-transmissible factors (somatic mutation not included into the germline, environmental effects) may also evoke the same symptoms but in this case members of the family do not display the defect. Take an example, when new dominant germline mutations are responsible for 15% of the cases. In this instance the ancestors are not affected. The offspring of such a mutant individual

Risk continued

has, however, 0.8 (penetrance) x 0.5 (expected gametic transmission) = 0.4 chance for being affected. The risk of all their offspring for being not affected (1 - 0.15 = 0.85) and if not inherited (0 inheritance) the chance is 0.85 x 0. The probability of inherited (0.15) x penetrance (0.8) is ≅ 0.12 (12%). Further considerations are necessary for proper genetic counseling if the proband has already normal, not-affected offspring. We designate the hereditary status of the proband, the probability of being a hereditary case: P(H) = 0.15, as specified above. The probability that the first child being normal (not affected), despite the defective parental gene is P(N/H) = 1 - 0.4 = 0.6. The probability that the proband does not have this defective gene in the germline is P(H⁻) = 0.85. The conditional probability that an offspring would be normal is P(N/H⁻) ≅ 1. From Bayes' Theorem the probability that first child inherited but not expressed the trait is:

$$P(H/N) = \frac{P[H]P[N/H]}{P[H]P[N/H] + P[H^-]P[N/H^-]} = \frac{[0.15][0.6]}{[0.15][0.6] + [0.85][1]} = \frac{0.09}{0.94} \cong 0.096$$

With 2 not affected children P(N/H) = 0.6^2 = 0.36, and for n offspring 0.6^n is the probability of the parent being normal in phenotype although having the defective gene.

The second child being normal although carrying the defective gene is:
P(N/H) = 0.096 x 0.4 ≅ 0.0384 and after substitution into the Bayes' formula

$$P(H/N) = \frac{[0.15][0.0384]}{[0.15][0.0384] + [0.85][1]} = 0.0067$$

(See Bayes' theorem, genetic risk, recurrence risk, empirical risk, genetic hazards, genotypic risk ratio, λ_s, mutation in human populations, utility index for genetic counseling)

RK: rank of utility.

RK2 PLASMIDS: represent a family of broad host-range plasmids (56.4 kb), resistant to tetracycline, kanamycin and ampicillin. Size and selectability of other members of the family varies.

RLDB: see reference library database

RMSA-1 (regulator of mitotic spindle assembly): a protein which is phosphorylated only during mitosis and is substrate for Cdk2 kinase; it is required for the assembly of the spindle. (See spindle, Cdk)

RNA: ribonucleic acid is a polymer of ribonucleotides. There are three main classes of RNAs in the cell, the mRNA which provides the instructions for protein synthesis, the various ribosomal RNAs and the tRNAs. (See RNA I, mRNA, tRNA, rRNA, rrn)

RNA I: is an untranslated bacterial RNA controlling the maturation of RNA II that serves as a primer for plasmid DNA synthesis. RNA I and RNA II are synthesized on opposite DNA strands and RNA I binds to RNA II and thereby prevents its folding into a cloverleaf that is necessary for the formation of a stable DNA:RNA hybrid between RNA II and the plasmid DNA. This binding is promoted by the Rop protein (63 amino acid residues) coded for by 400 base downstream from the origin of replication. A single G → A transition mutation in Rop or upstream of it may contribute to plasmid amplification. RNA I and RNA II and Rop control also plasmid incompatibility.

```
                                          RNase H
         RNA II →    →  →  →  →  →  →              Replication
        _____|_____
        RNA I ←             ori                              ← Rop
```

RNase H cuts off the pre-primer section and prepares the primer for the actual DNA synthesis. RNA I may be polyadenylated and then its decay is hastened similarly to mRNA. RNA I and RNA II may interact initially by base pairing between their 7 nucleotide complementary loops. Rom/Rop protein may also bind to the transient complex and assures a more stable duplex of the two RNAs causing failure of replication initiation. (See polyadenylation, plasmid)

RNA II: see RNA I

RNA BINDING PROTEINS: can modify RNA structure locally or globally, may affect RNA

RNA binding proteins continued
trafficking, mRNA biosynthesis, translation, splicing, polyadenylation, differentiation and diseases. The length and composition of the binding domains may vary. A human RNA-binding protein is encoded in chromosome 8p11-p12 (RBP-MS). (See also DNA-binding protein domains)

RNA CAP: see cap, capping enzymes

RNA CODONS: see genetic code

RNA DEPENDENT RNA POLYMERASE: replicates the RNA genome of viruses.

RNA DRIVEN REACTION: in a DNA - RNA hybridization experiment, RNA is far in excess compared to single-stranded DNA. This assures that at all potential annealing sites hybridization will take place. (See DNA hybridization, nucleic acid hybridization)

RNA EDITING: is a means of post-transcriptional altering the RNA transcript. It is a very common process in the mitochondria of *Trypanosomes* (see *Trypanosoma brucei*). A separately transcribed 40-80 base "guide RNA" with homology to the 5'-end of the RNA to be modified pairs with the target. Then uracil residues from the 3' end tract of the guide are transfered into the target sequences. This editing thus changes the content of the message and the amino acids of the translated protein. In the mitochondria of this protozoon thousands of U nucleotides may be inserted into different pre-mRNAs. *Us* may also be removed at a 10 fold lower frequency. It has been hypothesized that the *U* replacements are provided by the 3'-end of the gRNA but recent experimental evidence indicates that they come from free UTPs. In the mitochondria of plants, in about 10% of the transcripts C may be replaced by U. RNA editing appears to be rare and limited in most higher animals. One C residue deamination in the apolipoprotein-B results in a U replacement and the creation of a stop codon and consequently a truncated protein. Another similar deamination in the middle of the transcripts alters the permeability of a Ca^{++} channel. Thus editing produces two different mRNAs from one. Although C→U is the common change, U→C may also occur exceptionally. The changes may also be brought about by simple deamination, addition or co-transcriptional errors, such as stuttering of the polymerase. RNA editing takes place also in different RNA viruses. Mammalian nuclear RNA editing may involve the deamination of adenosine into inosine in the double-stranded pre-mRNA of the glutamate-receptor subunits. The enzyme responsible for the process is dsRAD (double-stranded RNA adenosine deaminase, also called DRADA). In the $tRNA^{Asp}$ of marsupials the GCC anticodon is found that can recognize only glycine codon. In 50% of these tRNAs the middle base is edited to U, and thus the regular Asp anticodon is generated. The marsupial mitochondria have also the regular $tRNA^{Gly}$ with anticodon GGN which recognizes all four glycine codons but the edited codon can match up with only two of the glycine codons. It is a puzzling observation why among the only 22 tRNAs there are two for glycine (normal and edited). RNA editing occurs through the plant kingdom (with few exceptions), in mitochondria as well as in chloroplasts albeit at lower frequency in the latter. In the plastids there are about 25 editable sites whereas in the mitochondria their number may exceed 1,000. RNA editing may generate new initiation and termination codons in plant organelles and thus new reading frames. No U→C editing was observed in the mitochondria or chloroplasts of gymnosperms or in the chloroplasts of angiosperms. The site of editing apparently selected on the basis of the flanking sequences. RNA editing occurs also in nuclear genes and contributes to the regulation at an additional level. It appears that the neurofibromas are determined by editing in the neurofibromatosis gene (NF1). (See mtDNA, apolipoproteins, kinetosome, gRNA, genetic code, stuttering, wobble, DRADA, anticodon, mooring sequence)

RNA ENZYME: see ribozyme

RNA EXPORT: from the nucleus requires the presence of the nuclear export factor NES and a cellular cofactor Rip1. The NES function is part of the Gle1 yeast protein (M_r 62K). Gle1 interacts with Rip1 and nucleoporin (Nup 100) in the nuclear pore. The Rev splicing factor can substitute for the Gle1 function. (See RNA transport, REV, nuclear pore)

RNA EXTRACTION: essential requisite that RNase activity should be eliminated or prevented

RNA extraction continued

during all operations. The glassware can be made RNase free by baking for 8 hours or by chloroform washing. An 1% diethyl pyrocarbonate (DEPC [carcinogen!]) washing (2 hrs, 37 C°) may also be useful. RNase activity in the extraction media can be inhibited by vanadium or by the clay, Macaloid. These are subsequently eliminated by water-saturated phenol extraction. RNases can be blocked also by 4 M guanidium thiocyanate and β-mercaptoethanol. RNA is extracted from the tissues in a buffer containing a detergent (0.5% Nonidet) and a reducing agent (dithiothreitol). Proteins may be removed by proteinase K digestion. DNA is removed by RNase-free DNase. Finally RNA is precipitated by chilling in cold ethanol (containing Na-acetate). The RNA is taken up in TE, pH 7.6 buffer. Its quantity can be measured spectrophotometrically at 260 nm. Several variations of these general procedures are being used to isolate RNA. (See also DNA extraction, RNase-free Dnase. TE)

RNA FINGERPRINTING: the purpose is to identify the differential expression of the total array of genes that constitutes about 15% of all at a particular time in a mammalian genome. For this goal from a subset of mRNAs partial cDNAs are amplified by reverse transcription using PCR. The short sequences are then displayed on a sequencing gel (differential display). Pairs of primers are selected in such a way that each will amplify 50 to 100 mRNAs. One of the primers (5'-TCA) is anchored to the TG upstream of the poly(A) tail of the mRNA. This primer will recognize 1/12 (4!/2!) of the mRNAs with different combination of the last two 3' bases omitting T as the penultimate base. The primer will amplify then only this subpopulation. As 5' primers 6 to 7 bp arbitrary sequences are used. Such a procedure can be used not just for molecular analysis of development but eventually the genes producing the transcripts can be cloned. (See PCR, reverse transcription, DNA finger printing, fingerprinting of macromolecules, RDA)

RNA G8: contains about 300 nucleotides and it is associated with the ribosomes in *Tetrahymena thermophila*. It is transcribed by RNA polymerase III and conveys thermal tolerance to the cells. (See thermal tolerance)

RNA HELICASE: unwinds double-stranded RNA. (See eIF4A, helicase)

RNA, HETEROGENEOUS: see hnRNA

RNA LIGASE: catalyzes the joining of RNA termini, such as generated during the processing of tRNA transcripts, in a phosphodiester bond and may need ATP for the reaction. (See also ligase DNA, ligase RNA)

RNA MATURASES: ribosomal transcripts are processed to size by the U3, U8, U13 independently transcribed small RNAs (snoRNA, small nucleolar RNAs). The U13-U14 snoRNAs and E3 are encoded by the introns of protein-coding genes participating in the process of translation. They are co-transcribed with the pre-mRNA and removed during the processing of the genes. By the intron of the U22 host gene (UHG) seven U RNAs are transcribed. These U RNAs display 12-15 base complementary sequences to rRNAs. (See splicing, introns)

RNA MEDIATED RECOMBINATION: is thought to be involved in the exchange between the reverse transcript and the corresponding cellular allele. The Ty element-mediated recombination is supposed to involve RNA. (See Ty)

RNA MIMICRY: see translation termination

RNA MIMICS: 2'-modified oligodeoxynucleotides such as the 2'-O-methyl-modified ones that have enhanced binding affinity to complementary RNA and are resistant to some nucleases. (See antisense technology)

RNA PLASMIDS: exist in some mitochondria that are not homologous to the mtDNA. (See mtDNA)

RNA POLYMERASE: DNA-dependent RNA polymerases synthesize RNA on DNA template. In prokaryotes a single polymerase synthesizes all cellular RNAs. The prokaryotic pol enzyme contains four subunits αα, β, β' and σ. The large β subunits are evolutionarily highly conserved and they are the main instruments of polymerization with the other subunits. The prokaryotic RNA polymerases are of about 500 kDA in size. The α subunits recognize the promoter, the β'

RNA polymerase continued
subunit binds DNA and β is active in RNA polymerization. The σ is essential to start transcription in a specific way by opening the double helix for the action of the RNA polymerization. The eukaryotic organelles contain prokaryotic type RNA polymerases. The eukaryotic RNA polymerase II has about 10 subunits and is somewhat larger but highly homologous across wide phylogenetic ranges. The two largest subunits are similar to the β subunits of the prokaryotic enzyme. In eukaryotes there are three different DNA-dependent RNA polymerases (I, II, III). RNA polymerases replicate the genome of the RNA viruses. These enzymes — in contrast to the DNA polymerases — lack proof-reading function and consequently their error rate may be within the 10^{-3} to 10^{-4} range per nucleotide leading to the extreme diversity of the RNA viruses. (See figure below, pol, promoter, open promoter complex, transcription, transcription termination, error in replication, transcription factors)

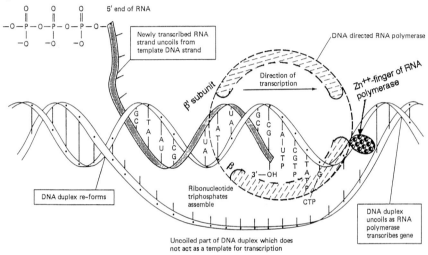

CONCEPTUALIZATION OF THE GENERAL PROCESS OF TRANSCRIPTION OF RNA ON DNA TEMPLATE FROM RIBONUCLEOSIDE 5'-TRIPHOSPHATES. THE FIRST 5'-NUCLEOTIDE RETAINS THE THREE PHOSPHATES, THE FOLLOWING ONES SPLIT OF A PYROPHOSPHATE AND ARE THEN HOOKED UP TO THE 3'-OH ENDS. (Modified after Page, D.S. Principles of Biological Chemistry, Willard Grant, Boston, MA)

RNA POLYMERASE HOLOENZYME: is a complex of RNA pol II, transcription factors, other regulatory proteins including Srb, a general transcriptional regulator of eukaryotes. (See transcription factors, co-activators, TBP, TAF)

RNA POLYMERIZATIONS: see pol

RNA PRIMER: see DNA replication

RNA PROCESSING: eukaryotic genes are "in pieces", although the DNA is continuous, in between the protein-coding nucleotide sequences (exons), non-translated or not translated together with exons, additional nucleotide sequences, introns occur. The long sequences are transcribed into a long RNA tract but the introns are removed and the exons are spliced to make the mRNA. Similar processing is carried out also with the rRNA and tRNA. The vast majority of prokaryotic genes do not require this processes. These cutoffs, the mRNA and the primary transcripts constitute the pool of the hnRNA (heterogeneous nuclear RNA). The primary RNA transcripts are coated with different proteins and thus form hnRNP, i.e. hnRNA-protein particles. These particles are instrumental in cutting the transcripts and splicing the exons into mRNA. The splicing is completed, the methylated guanylic cap and the polyA tail are added before the mRNA is exported into the cytosol. (See introns, hnRNA, spliceosome, post-transcriptional processing).

RNA REPLICATION: Some bacteriophages R17, MS2, fd2, f4, Qβ, M12, and some animal viruses such as the polio virus, the vesicular stomatitis virus, rhinoviruses (influenza viruses) and the vast majority of the plant viruses have RNA genetic material. This RNA is either single- or double-stranded. The best known is the Qβ replicase system. The tetramer (210 kDa) is encoded one viral (β subunit about 65 kDa) and three bacterial genes (translation factors). The replicase enzyme does not have an editing exonuclease function. These replicases do not replicate or transcribe host RNAs. The replication of the RNA is similar to the replication of DNA inasmuch as the strands are elongated at the 3'-OH ends. (The retrovirus replication is discussed under reverse transcription.) Most of the bacterial RNA viruses and many animal viruses have the same genetic material as their mRNA (+ strand viruses). They use a replicative intermediate (RI) for the synthesis of the first new (-) strand. The (-) strand can then generate as many (+) strands as needed. Some viruses are, however (-) strand viruses because their genetic material is not identical to the mRNA but complementary to it. The polio virus (+) strand is the genetic material that is associated with a protein at the 5'-end through a phosphodiester linkage to a tyrosine residue. The (-) strand lacks this feature. Apparently, the OH group of the tyrosine primes the synthesis of the (+) strand but the (-) strand can get by without it. The reoviruses of vertebrates contain 8 - 10 short double-stranded RNA molecules. When the virus enters the host cell the coat protein is shed and its RNA polymerase is activated that works conservatively and asymmetrically. A virion enzyme copies the (-) strand and a new (+) is released. The original RNA duplex is conserved. On the new (+) strand a (-) strand is made and the double-stranded RNA is reconstituted. (See also DNA replication, replicase, replicative intermediate, plus strand)

RNA RESTRICTION ENZYME: function exists in ribozymes. (See restriction enzyme, ribozyme)

RNA 10Sa: when a mRNA is truncated at the 3'-end and has no stop codon, the ribosome may switch from one RNA to another to terminate the translation. The switching results in the generation of a 11 amino residue at the carboxyl end that destines the protein for degradation. The last 10 amino acids are translated from a stable so called 10Sa RNA (363 nucleotides). So far these 10 amino acids were found on murine interleukin-6 translated in *E. coli* and also on λ phage *cI* and on cytochrome b-562 and polyphenylalanine translated on polyU. The 10Sa RNA is similar to a tRNA and it is charged with alanine. This alanine is found as the first amino acid of the tag. This saves the ribosome from the "unfinished" mRNA and permits its quasi normal operation. The mechanism carries out the same task as ubiquitin does in eukaryotes. (See translation non-contiguous, ubiquitin, lambda phage,)

RNA SEQUENCING: can be done by digesting RNAs with different ribonuclease enzymes that cut the phosphodiester linkages at the specific nucleotides and then separate the fragments by gel electrophoresis (ca. 20% polyacrylamide) in one or in two dimensions. The RNA is generally labeled by ^{32}P in order to autoradiograph the dried gels. T_1 ribonuclease (from *Aspergillus oryzae*) generates fragments with 3' guanosine monophosphate ends. U_2 ribonuclease (from *Ustilago sphaerogena)* cleaves at purine residues (intermediates 2',3'-cyclicphosphates). RNase CL 3 (from chicken liver) has about 16 fold higher activity for cytidylic than uridylic linkages. RNase B (from *Bacillus cereus*) cuts Up↓N and Ap↓N bonds. RNase Phy M (from *Physarum polycephalum*) has the specificity Ap↓N and Up↓N. When several of these enzymes are employed, from the separate electrophoretic patterns the sequences from the positions of the fragments with different termini can be deduced. In many ways sequencing of RNAs with the aid of enzymatic breakage is very similar in principle to the Maxam and Gilbert method of DNA sequencing. Recently RNA is sequenced by first converting it by reverse transcriptase into cDNA and then the much easier DNA sequencing methods are used to determine the RNA nucleotide sequences. (See DNA sequencing, ribonuclease)

RNA 7SL: an about 300 base long molecule forming the signal recognition particle (SRP) with six proteins of 10 to 75 kDa size.

RNA, SMALL: see snRNA

RNA SPLICING: the primary transcripts of RNA is much larger than required for particular functions, therefore during processing it is cut up and the transcripts corresponding to introns are removed. The remaining pieces are then reattached (spliced together) to form the mature RNA molecule. (See introns, RNA processing)

RNA TRANSCRIPT: an RNA copy of a segment of DNA.

RNA TRANSPORT: mRNA, snRNA, U3 RNA and ribosomal RNA (but not tRNA) are exported from the nucleus by RCC1 (yeast homolog is PRP20/MTR1). This nuclear protein is a guanine nucleotide exchange factor for the RAS-like guanosine triphosphatase (GTPase). (See RAN, RCC1, nuclear pore)

RNA TRIMMING: see trimming

RNA, UBIQUITOUS: see U RNA

RNA VIRUSES: see retroviruses, TMV, MS2, paramyxovirus, viral vectors, reovirus, togavirus

RNA WORLD: the pre-biotic era when auto- and heterocatalysis was carried out by RNA without DNA. (See origin of life, ribozyme)

RNA-IN: the leftward transcript of the bacterial transposase gene in transposable elements, transcribed from the pIN promoter. (See RNA-OUT, pIN)

RNA-OUT: transcript originating from the strong pOUT promoter of bacterial transposable elements. It opposes pIN and it directs transcription toward the outside end of the IS*10* element. (See RNA-IN, pIn, pOUT, Tn*10*)

RNAP: RNA polymerases; in prokaryotes there is only one DNA-dependent RNA polymerase whereas in eukaryotes RNA pol I, pol II and pol III are found. (See polymerases)

RNase: see ribonucleases

RNase-FREE DNase: heat at 100 C° for 15 min 10 mg RNase A per mL, 0.01 M Na-acetate (pH 5.2), then cool, adjusts pH to 7.4 (1 M Tris-HCl), store at -20 C°. (See DNase free of RNase)

RNase MRP: is a ribonuclease that cleaves rRNA transcripts upstream of the 5.8 rRNA. (See U RNPs)

RNASIN®: is a ribonuclease inhibitor.

RNKP-1: is a homolog of ICE and fragmentin-2. (See apoptosis, ICE, fragmentin-2)

RNP: ribonucleoprotein, any type of RNA associated with a protein particle such in the hnRNA or in the ribosomes. (See RNA, hnRNA, ribosomes)

RNR: see ribonucleotide reductase

ROAM MUTATION (**r**egulated **o**verproducing **a**lleles under **m**ating signals): activate yeast Ty elements by the influence of the MAT gene locus. Such a system may operate if the Ty insertion took place at the promoter of a gene and then the Ty enhancer may be required for the expression of that gene. These genes are expressed only in the *a* or *α* mating type cells but not in the diploid *a/α* cells. (See Ty, mating type determination in yeast)

ROAN: is a fur color (cattle, horse) with predominantly brown-red hairs interspersed with white ones. It is common sign of heterozygosity for the *R* and *r* alleles of a gene locus. (See co-dominance)

ROBERTS SYNDROME: actually overlaps with the *SC phocomelia* (= absence or extreme reduction of the bones of extremities located proximal to the trunk of the body) and with the *TAR syndrome* involving also *thrombocytopenia* (reduction in the number of blood platelets), mental retardation, cleft palate, etc. These three syndromes are all autosomal recessive and apparently basically the same and are caused by chromosomal instability. The prime suspect is a defect in the centromeric heterochromatin or the kinetochore itself. (See chromosome breakage, thrombocytopenia, Holt-Oram syndrome, Wiskott-Aldrich syndrome, mental retardation)

ROBERTSONIAN TRANSLOCATION (Robertsonian change): two non-homologous telocentric chromosomes fused at the centromere or more likely translocation between two non-homologous acrocentric chromosomes. The outcome is a replacement of two telo or acrocentric chromosomes with one clearly bi-armed chromosome. Robertsonian translocations are very common in mouse cell cultures but they occur also in wild natural populations, resulting in an

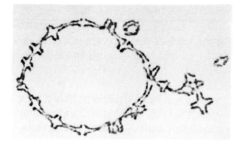

apparent change in chromosome morphology and numbers. If this translocation occurs, generally a minute piece of the acrocentric chromosome is lost but being genetically inert, it has no consequence for fitness. (See translocations, acrocentric, telocentric, fitness)

———————o- -o—— ➔ ————o———
two acrocentrics   Robertsonian translocation

LATE DIAKINESIS/EARLY METAPHASE I IN THE MOUSE, HETEROZYGOUS FOR DIFFERENT ROBERTSONIAN TRANSLOCATIONS; 15 METACENTRICS IN A SUPERCHAIN, 1 TRIVALENT, 2 BIVALENTS. THE SEX CHROMOSOMES ARE NOT INVOLVED IN ROBERTSONIAN CHANGES. (Courtesy of Capanna, E. *et al.* 1976 Chromosoma 58:341)

ROBINOW SYNDROME: an autosomal dominant phenotype involving, usually but not always, short stature, normal virilization but micropenis, hypertelorism of the face, etc. In the Robinow-Sorauf syndrome the main characteristic being the flattened and almost doubled big toe. An autosomal recessive Robinow syndrome included the same facial and genital features but in addition multiple ribs and abnormal vertebrae were present. (See limb defects, Chotzen syndrome, limb defects, stature in humans)

ROCKET ELECTROPHORESIS: a type of immunoelectrophoresis where antigens are partitioned against antisera. (See immunoelectrophoresis, antigen, antiserum, electrophoresis)

RODENTS (order Rodentia): a large number of species (mouse, rat, hamsters, rabbit) have been extensively used for genetic research because of the small size of these multiparous mammals and short generation time. Mice and rats reach their sexual maturity in 1 or 2 months, and their gestation period is 19 and 21 days, respectively. They have been exploited as laboratory models for the study of development, cancer, antibodies, population genetics, behavior genetics, radiation and mutational responses, etc. Rodents are carrier of several pathogens of humans (bubonic plague [*Pasteurella pestis*], tularemia [*Pasteurella tularensis*], etc.) Several inbred strains of mice contributed very significantly to the understanding of immunogenetics. See animal models, mouse)

ROENTGEN (röntgen): is a unit of ionizing radiation (X-rays). (See R unit, Röntgen machine)

ROGUE: an off-type of unknown (genetic) determination.

ROLLING CIRCLE: replication is common among circular DNAs (such as conjugative plasmids [such as the F plasmid], double-stranded [λ phage] and single-strand phages [M13, φX174], amplified rDNA minichromosomes in the amphibian oocytes). A protein nicks one of the DNA strands and remains attached to the 5'-end. The free 3'-OH terminus serves as the point of extension by DNA polymerase in such a way that the opened old strand is displaced from the circular DNA while the new strand is formed and immediately hydrogen bonded to the old template strand (see diagram). Thus the rolling circle remains intact and may generate new single stranded DNAs that may be doubled later. The displaced single strand may be formed just in a single unit length of the original duplex circle or it may become a single- or double-stranded concatamer or it may circularize in a single- or double-stranded form with assistance of a DNA ligase to join the open ends. (See illustration, conjugation, conjugation mapping, concatamer)

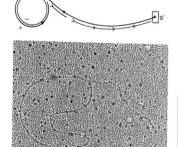

THE (-) STRAND IS THE TEMPLATE FOR THE (+) STRAND. THE → INDICATES THE DIRECTION OF GROWTH. THE 5' END OF THE (+) STRAND ATTACHES TO THE MEMBRANE (5). (Diagram is the courtesy of Gilbert, W. & Dressler, D. 1968 Cold Spring Harbor Symp. Quant. Biol. 33:473. Electronmicrograph is the courtesy of Nigel Godson)

ROM (RNA-one-Modulator): a protein of 63 amino acid residues affects the inhibitory activity of an antisense-RNA. A trans-acting inhibitor of plasmid replication is also named Rop/Rom.

(See RNA I, antisense technology)

ROMANOVs: the last Tsar of Russia, Nicholas II and his wife and three children were executed and buried near Ekaterinburg, Russia on 16 July 1918 after the bolshevik take-over of power. 75 years later the bodies were exhumed and from bone tissues the identities of the remains were determined on the basis of mitochondrial and nuclear DNAs. Tsarina Alexandra, was grand-daughter and Prince Philip (husband of the present Queen of England, Elizabeth) is great-grand-son of Queen Victoria. Their mitochondrial DNA should be identical, as well as those of the three daughters of Alexandra. Forensic DNA analysis confirmed the expectation. Sex determination from the bone samples was done by identifying the X-chromosomal amelogenin gene that is expressed in the enamel of the teeth and it is rich in GC (51%) and codes for a proline-rich protein (24%). The nuclear DNA samples confirmed the identity. The Tsar and his paternity and his mtDNA tied him to his brother (by the exhumed remains of Grand Duke Georgij) on the basis of heteroplasmy. The four other skeletons did not belong to royal family members but were that of the physician and of other people of the court. The two youngest children, Anastasia and Alexis were not found in the grave. The purported identity of Anastasia with Anna Anderson was not confirmed by the DNA analysis. (See DNA fingerprinting, heteroplasmy, Tsarevitch Alexis)

ROMK: is a potassium ion channel family member; it is encoded in human chromosome 11 and its isoforms are generated by alternative splicing. (See ion channels, isoform, splicing, Bartter syndrome)

Ron: a cell membrane tyrosine kinase, the receptor of the macrophage stimulating protein and both are located in human chromosome 3p21. (See macrophage-stimulating protein, HGF)

RÖNTGEN MACHINE: is an X-ray machine (invented by W.K. Röntgen), producing ionizing radiation (used for induction of mutation, mainly deletions) and medical examination of the body. The dose delivered is measured by R, Rad, Rem, rep, Sv, Gy. (See R, roentgen, radiation hazard, radiation effects, radiation threshold, radiation protection, radiation measurement)

roo: see copia

ROOT CAP: thin membrane-like protective shield at the tip of roots of plants.

ROOT NODULE: see nitrogen fixation

ROOT PRESSURE: guttation of wounded stem caused by osmosis in the roots of plants.

ROOTED EVOLUTIONARY TREE: indicates the origin of the initial split of divergence. (See evolutionary tree, unrooted evolutionary tree)

Rop PROTEIN: see RNA I

ROS: reactive oxygen species.

Rosa canina →

ROS ONCOGENE: is in human chromosome 6q22 and it is the c-homolog of the viral v-ros. It appears to be the same as MCF. (See oncogenes)

Rosa spp: ornamentals with 2n = 14, 21, 28. (See also *Rosa canina*)

Rosa canina (dog rose): a pentaploid species with 35 somatic chromosomes, and unlike other pentaploids it is fertile. In meiosis, the plants produce 7 bivalents and 21 univalents. The univalents are lost at gametogenesis in the male and so the sperms contain only 7 chromosomes, derived from the 7 bivalents. During formation of the megaspore all the 21 univalents and the 7 chromosomes from the 7 bivalents are incorporated into the embryo sac. The addition of the 7 male and the 28 female chromosomes to the zygote restores the 35 somatic chromosome number. The female contributes more chromosomes to the offspring, it is matroclinous. Recombination is limited to the 7 bivalents. Its breeding system is a unique mixture of generative and apomictic reproduction. (See pentaploids, apomixis, matroclinous)

ROSETTE: plant shoots with very much reduced internodes commonly found in dicots before the stem bolts after induction of flowering; any anatomical structure in animals arranged in a form resembling the petals of a rose.

r$_0$t: in a RNA-driven DNA - RNA hybridization reaction the concentration of RNA x the time of the reaction (analogous to c$_0$t in reassociation kinetic studies with DNA). (See c$_0$t, RNA driven reaction)

ROTAMASE: a group of enzymes catalyzing cis-trans isomerization. (See cyclophilin, immunophilins)

ROTAVIRUSES (*Reoviriodae*): their genomes consists of 10-12 double-stranded RNA and each particle carries a single copy of this genome. The terminal sequences control replication and packaging. Through internal deletions the RNAs may become aberrant, called DI RNA (defective interfering) resulting lower infectious capacity. The rotaviruses cause gastroenteritis (stomach and intestinal inflammation) and diarrhea in human babies and animals.

ROTATIONAL DIFFUSION: membrane proteins travel within the membranes by rotation perpendicular to the plane of the lipid bilayer. (See cell membrane)

ROTHMUND-THOMPSON SYNDROME: an autosomal recessive human disorder involving dermal (skin) lesions, dark pigmentation, light-sensitivity, early cataracts, bone and hair problems. May lead to squamous (scaly) carcinomas. Some similar types of diseases are determined by autosomal dominant genes. (See cancer, light-sensitivity diseases)

ROUGH ER: see endoplasmic reticulum rough, RER

ROUNDS OF MATINGS: after bacteriophages have been replicated within the host cell, the newly formed molecules of "vegetative DNAs" may recombine with each other several times. Because of the multiple exchanges, it appears that the maximal recombination between markers cannot exceed 30 - 40%. (See mapping function, negative interference, coefficient of coincidence)

ROUS SARCOMA: originally detected as a viral cancer in chickens. The protooncogene homolog was detected in rats and other mammals. (See RAS and homologs, oncogenes)

ROWLEY-ROSENBERG SYNDROME: is an autosomal recessive growth retardation, different from dwarfness, and characterized by aminoacidurias. (See dwarfism, aminoacidurias)

RPA: see replication protein A, see also DNA replication eukaryotes

RPB: subunits (differently numbered, each) of RNA polymerase II. (See RNA polymerase)

RPD3: a histone deacetylase that block the position effect exerted by centromeric and telomeric heterochromatin. (See position effect, PEV, histone)

R-PHYCOERYTHRIN: a phycobiliprotein fluorochrome isolated from algae. Maximal excitation at 545 and 565 nm but it is excited also at 480 nm. Maximal emission at 580 nm; hence the red color.

rpo: RNA polymerases such as A, B, C_1, C_2 in (organelles) plastids and mitochondria and resemble bacterial RNA polymerases. (See RNA polymerase)

r-PROTEINS: ribosomal proteins. (See ribosome)

RPTK: a sperm receptor protein tyrosine kinase that may bind to the ZP3 (zona pellucida) protein of the egg matrix. (See sperm, egg, fertilization)

RRAS: see RAS oncogene

rrn: in *E. coli* there are seven ribosomal transcription units, *rrn-A, -B, -C, -D, F, -G, -H* including the 16S - 23S - 5S RNAs, spacers and intercalated tRNA genes within the spacers. Maturation involves trimming of the co-transcript (cleavage by RNase III) and processing by other RNases, RNases P and D. (See diagram next page). The number of rRNA genes in eukaryotes is very variable and subject to amplification. At developmental stages of very active protein synthesis the number of ribosomal genes may be amplified to several thousands, and in e.g. the amphibian oocytes may be sequestered into mininuclei. Eukaryotic rRNA genes are transcribed in a ca. 45S precursor RNA, containing the 18S - 5.8S - 28S (in this order) and spacer sequences. The pathway of trimming may vary among the different species. The 18S rRNA is immediately methylated at about 40 sites and a few more methyl groups are added in the cytoplasm after maturation. The 28S rRNA methylated immediately after transcription at over 70 sites and these methylated sites are saved during the process of maturation. The cleavage takes place at the 5' side of the genes and between the spacers. The 5.8S rRNA eventually associates with the 28S rRNA by base pairing. (See diagram on next page, ribosomal RNA, ribosomal proteins, ribosomes)

a segment of a eukaryotic rRNA gene cluster in transcription →

rrn continued

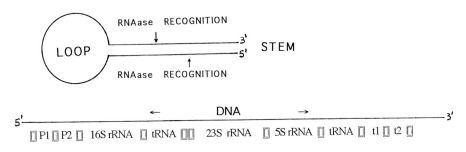

P1 and P2 : promoters, ☐ : spacers, t1 and t2: termination signals (not on scale)

THE DNA REGION OF THE rRNA AND tRNA GENE CLUSTER (rrn) in *E. coli* IS TRANSCRIBED INTO LONGER THAN 30S PRIMARY TRANSCRIPTS, INTERRUPTED BY SPACERS. THE RIBOSOMAL AND TRANSFER RNA GENES ARE CLUSTERED AND CO-TRANSCRIBED IN THE ORDER 5' - 26S - 23S - 5S - 3' AND WITHIN THE INTERGENIC SPACERS THE tRNA GENES ARE SITUATED. THE INDIVIDUAL GENE TRANSCRIPTS ARE TRIMMED BY RNase III AT THE DUPLEX STEMS (see above). THE DIFFERENT LOOPS CONTAIN CA. 1,600, 2,900 AND 120 NUCLEOTIDES, CORRESPONDING TO THE 16S, 23S AND 5S RIBOSOMAL RNAs, RESPECTIVELY. (THE 1,600 AND THE 2,900 BASE SEQUENCES ARE ALSO CALLED p16 AND p22, RESPECTIVELY). EACH OF THESE rRNA PRECURSORS ARE FURTHER TRIMMED BY RNase P, AND THE tRNA PRECURSORS ARE PROCESSED BY RNase P AT THE 5' AND BY RNase D AT THE 3' END.

rRNA: ribosomal RNA; structural component of the ribosomes; rRNAs may preferentially amplified in the oocytes of amphibians and other organisms. (See ribosome, ribosomal RNA)

RS DOMAIN: is rich in arginine (R) and serine (S); Rs proteins are parts of the spliceosomes. (See spliceosome)

RSF: see hybrid dysgenesis I - R system

RSK: a growth factor-regulated protein kinase; regulates the Coffin-Lowry syndrome. (See Coffin-Lowry syndrome)

RSS (recombinational signal sequence): V(J)D recombinase mediates recombination only in gene segments flanked by tripartite recombination signal sequences consisting of a highly conserved heptamer (7mer), an AT-rich nonamer (9mer) and 12 or 23 base long intervening nucleotides. (See immunoglobulin, V(J)D recombinase, RAG)

R^{St} (*R-st*): the stippled, paramutable allele of *R* locus of maize in the long arm of chromosome 10. (See paramutation)

r_0t **VALUE**: the measure of RNA - DNA or RNA - RNA hybridization; the product of the concentration of single-stranded RNA and time elapsed since the beginning of the reaction. (See also c_0t value)

RTF (resistance transfer factor): bacterial plasmids carrying various antibiotic and other resistance genes. (See conjugation in bacteria, plasmid mobilization)

RTK: see receptor tyrosine kinase

RTP (replication termination protein): functionally, but not structurally, similar to Tus. (See replication bidirectional)

RT-PCR: reverse transcription-polymerase chain reaction. The purpose of the procedure is similar to the PCR in general, in this instance to amplify the small amounts of RNA transcripts as cDNA. The reaction requires reverse transcriptase, mRNA, deoxyribonucleotides and primers that can be random DNA sequences, oligodeoxythymidine or antisense sequences. The method is very sensitive and the RNA of a single cell can be amplified and thus localized gene expression can be studied. Under well-controled conditions it can be semiquantitative. It is also used now for clinical diagnostic purposes. (See RNA fingerprinting, PCR, in situ PCR)

rtTA (reverse transactivator tetracycline): is basically tTA system containing a nuclear localization signal at the 5'-end and bind efficiently the *tetO* operator only in the presence of tetracycline derivatives such as doxycycline or anhydrotetracycline. (See tTA)

rtTA-nls: the same as rtTA

RU-486 (mifepristone): a pregnancy prevention drug. (See hormone receptors)

RU MAIZE: carry plasmid-like elements in their mitochondria yet they are not male sterile. (See cytoplasmic male sterility, mtDNA)

RUBELLA VIRUS (a toga virus): causes the disease of German measles. Infection during early pregnancy may cause intrauterine death of the human embryo and/or developmental anomalies in the newborn. (See teratogenesis)

RUBINSTEIN SYNDROME (Rubinstein-Taybi syndrome): defects in the heart valve of the pulmonary aorta, collagen scars on skin wounds, enlarged passageway between the skull and the vertebral column. This condition has very low recurrence risk (1%) and about 0.2-0.3% of the inmates of mental asylums are afflicted by it. It is determined by autosomal dominant inheritance. Haplo-insufficiency for the CBP transcription factor seems to be involved in the abnormal differentiation. Many afflicted individuals have breakpoints at chromosome 16p13, and this is the site of the human cyclic AMP response element binding protein (CBP/CREB). (See mental retardation, CREB, CBP, GLI3 oncogene)

RUBISCO: ribulose bisphosphate carboxylase - oxygenase (M_r 550,000) a chloroplast enzyme. The 8 large subunits (each M_r 56,000) are coded for by chloroplast DNA and the small subunits (each M 14,000) are under nuclear control. The carboxylase function catalyzes the covalent attachment of carbondioxide to ribulose-1,5-bisphosphate and then splitting into two molecules of 3-phosphoglycerate. The oxygenase function mediates the incorporation of O_2 into ribulose-1,5-bisphosphate and the resulting phosphoglycolate reenters the Calvin cycle. (See chloroplast, chloroplast genetics, photosynthesis)

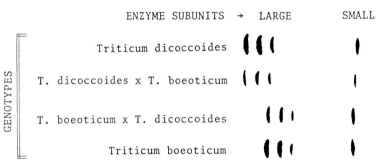

GENETIC AND ELECTROPHORETIC EVIDENCE THAT THE LARGE SUBUNITS OF RUBISCO ARE MATERNALLY INHERITED WHEREAS THE SMALL SUBUNIT IS TRANSMITTED BIPARENTALLY. MOLECULAR STUDIES MAPPED THE LARGE SUBUNITS IN THE CHLOROPLAST DNA. (After Chen, K., Gray, J.C. & Wildman, S.G. 1975 Science 190:1304

RULE 12/23: see immunoglobulins, V(D)J, RAG

Rumex hastatulus: a North-American herbaceus plant has variable chromosomal sexdetermination. The plants found in North Carolina have three pairs of autosomes, one X and two Y chromosomes. The form prevalent in Texas has four pairs of autosomes and one X and one Y chromosome:

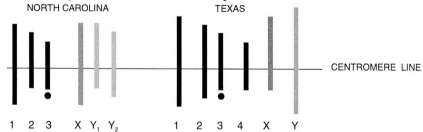

(Redrawn after Smith, B.W. 1964 Evolution 18:93)

RUNAWAY PLASMIDS: at lower temperature (30° C) they are present in relatively low number of copies and thus do not interfere with growth of the host cell. Above 35° C their copy

Runaway plasmids continued
number raises substantially and so does the DNA they contain but after about 2 hours (by the time they account for 50% of the DNA of the cell) they suppress cellular growth. (See vectors, copy-up mutation)

RUNAWAY REPLICATION: the replication is not restricted the normal manner.

RUN-OFF TRANSCRIPTION: the inducer of gene activity (e.g. light signal) is withdrawn and the tapering off of transcription is monitored by incorporation of labeled nucleotides.

RUN-ON TRANSCRIPTION: once the transcription is turned on in isolated nuclei it proceeds without further need for enhancers as measured by the incorporation of labeled nucleotides into mRNA. (See transcription, regulation of gene activity)

RUPERT: hemophiliac grandson of Leopold of Albany (a hemophiliac himself), son of Queen Victoria. (See hemophilias, Queen Victoria)

RUSSEL-SILVER SYNDROME (RSS): most commonly autosomal recessive, but X-linked or sporadic cases with low birth-weight dwarfism, frequently with asymmetric body and limbs, deformed fingers, relatively large skull; mental retardation are common. (See dwarfism)

RUST: disease of grasses (cereals) caused by *Puccinia* fungi. (See host - pathogen relations)

RUT: see oestrus

Rue: see branch migration, recombination molecular mechanisms in prokaryotes, recA

RVs: see BIN

RJR: see RARE

RYANODINE (ryanodol 3-(1H-pyrrole-2-carboxylate): a toxic extract (insecticide) from the new world tropical shrub *Ryania speciosa*. Regulates calcium ion channels in muscles. (See ion channels)

RYE (*Secale cereale*): 2n = 2x = 14 is an outbreeding crop plant used for the production of bread, biscuits, starch and alcohol. Its taxonomy is somewhat controversial. Rye can be crossed with a number of other cereals, among them the allopolyploid *Triticales* (wheat x rye hybrids, 2n = 42 and 2n = 56) are most notable. Addition lines, and transfer lines carrying rye chromosomes or chromosomal segments have been made. It belongs to those exceptional grain crops where autotetraploid varieties have agronomic value. Trisomic lines are known. Some varieties harbor a variable number of B chromosomes. Rather unusually, some of the plastids are transmitted also through the pollen. (See chromosome banding)

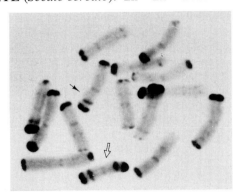

GIEMSA-STAINED RYE KARYOTYPE DISPLAYING AN ISOCHROMOSOME (⇨) AND THE CORRESPONDING ARM IN A NORMAL CHROMOSOME (→). (Courtesy of Dr. Goprdon Kimber)

← RYE EAR

(See *Triticale*, alien addition, chromosome substitution, alien substitution, transfer lines, holocentric)

RYEGRASS: *Lolium multiflorum* and *L. perenne*, both 2n = 14. ↓ RYEGRASS SPIKE

S

S: (such as in 5S RNA): see sedimentation coefficient

s: see selection coefficient

s: standard deviation of a set of experimental observations. (See also σ parametric)

σ: measure of superhelical density of DNA.

σ: a subunit of prokaryotic RNA polymerase enzyme, essential to start transcription in a specific way. This factor opens the double helix for the action of the RNA polymerase. Also, the σ^{70} is involved in pausing of transcription. The σ^{70} consists of two hexamers located at -10 and -35 positions from the transcription start point. In some algae the protein present in the chloroplast is encoded by the nucleus. In several plant species the same polypeptide is coded for by the chloroplast DNA. (See RNA polymerase, open promoter complex, chloroplast, chloroplast genetics)

σ: the parametric designation of standard deviation. (See standard deviation, standard error)

σ: yeast transposable element. (See Ty)

σ: viral infectious hereditary agent of *Drosophila*. (See CO_2 sensitivity, infectious heredity)

S9: see microsomes

S^{35} or ^{35}S: sulfur isotope. (See isotopes)

S49: a mouse lymphoma cell line

S ALLELES: control self-sterility in plants. (See self-incompatibility, incompatibility alleles)

S CYTOPLASM: present in some cytoplasmically male sterile lines. (See cms)

S FACTOR: mitochondrial plasmid-like element in male sterile plants. (See cms)

S6 KINASE: is actually the collective name for the cytosolic $p70^{s6k}$ and the nuclear $p85^{s6k}$ kinases that phosphorylate the S6 ribosomal protein before the initiation of translation. (See translation initiation, platelet-derived growth factor, phosphatidylinositol, S6 ribosomal protein)

S LOCUS: see selfsterility alleles

S1 MAPPING: when genomic DNA is hybridized with the corresponding cDNA or mRNA the introns cannot find partners to anneal with, and the single-stranded loops can be digested with S1 nuclease. The remaining DNAs that formed double-stranded structure can then be isolated by gel electrophoresis or their position and length can be determined by autoradiography if appropriately labeled material was used. Thus intron positions are revealed. (See introns, S1 nuclease, DNA hybridization, genomic DNA)

S_1 NUCLEASE: from *Aspergillus oryzae* cleaves single-stranded DNA (preferentially), and single stranded RNA. Double-stranded molecules and DNA-RNA hybrids are quite resistant to it unless used in very large excess. The enzyme produces 5'-phosphoryl mono- and oligonucleotides. S_1 has many applications in molecular biology: mapping of transcripts, removal of single-stranded overhangs from double-stranded molecules, analysis of the pairing of DNA-RNA hybrids, opening up "hairpin" structures. Its pH optimum is 4.5 and this may cause unwanted depurination. (See nucleases, S1 mapping)

S PHASE: of the cell cycle when regular DNA synthesis takes place. (See cell cycle)

S6 RIBOSOMAL PROTEIN: is phosphorylated at about 5 serine residues near its C terminus, and the phosphorylated state is correlated with the activation of protein synthesis on the ribosomes; the phosphorylation is stimulated by mitogens and growth factors. (See ribosome)

S-ADENOSYLMETHIONINE (SAM): a methyl donor for restriction-modification methylase enzymes, synonymous with Ado-met (See methylation of DNA)

SACCHARIN: a non-caloric sweetener; hundreds of times sweeter than sucrose. Oral LDLo for humans is 5 g/kg; suspected carcinogen and mutagen. (See LDLo, aspartame, fructose)

Saccharomyces cerevisiae: the eukaryotic budding yeast has chromosome number n = 16, its ge-

Saccharomyces cerevisiae continued

nome sizes is about 1.8×10^7 bp, approximately four time that of the prokaryotic *E. coli*. Recent sequencing and knockout information contradicted earlier estimates that less than 10% of its genome would be repetitive. In chromosome III of 55 open reading frames only 3 appeared indispensable for growth on a rich nutrient medium. Of 42 other genes only 21 displayed a phenotype. This information points to redundancy even in this small genome. This conclusion may be misleading, however, because some genes are called to duty only under specific circumstances. By 1996 its entire genome has been completely sequenced. The 5885 open reading frames are encoded by 12,068 kb. About 140 genes code for rRNA, 40 for snRNA and for 270 tRNA. About 11% of the total protein produced by the yeast cells (proteome) has metabolic function, 3% is involved in DNA replication and energy production, respectively; 7% is dedicated to transcription, 6% to translation and 3% (ca. 200) are different transcription factors. About 7% is concerned with transporting molecules. About 4% are structural proteins. Promoters, terminators, regulatory sequences and intergenic sequences with unknown functions occupy about 22% of the genome. Many proteins are involved with membranes. In rich nutrient media its doubling time is about one and half hour. ↓ **THE LIFE CYCLE OF SACCHAROMYCES**

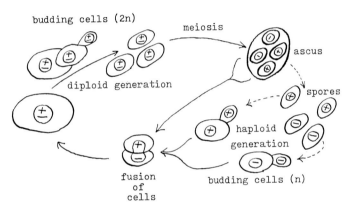

The organism has regular meiosis and mitosis. The vegetative multiplication is by budding (budding yeast), i.e. the new (daughter) cell is formed as a small protrusion (bud) on the surface of the mother cell. Haploid cells may fuse to generate diploidy and the diploid cells may undergo meiosis (sporulation) and the four haploid products are retained in an ascus as an unordered tetrad. The haploid cells may have *a* or *α* mating type. Although budding yeast is eukaryotic it can be cultured much like prokaryotes, and thus it combines many of the advantages of both groups of organisms. Approximately 25% of the human genes have significant homology with a yeast gene. The present genome might have evolved through extensive duplications. (See mating type determination in yeast, fungal life cycles, mtDNA, tetrad analysis, YAC, yeast vectors, yeast transformation, yeast transposable elements, gene replacement, *Schizosaccharomyces*, databases, sequencing <ftp://ftp.ebi.ac.uk/pub/databases/yeast> or <ftp://mips.embnet.org/yeast> or ftp://genome-ftp.stanford.edu/yeast/genome_seq>, protein coding sequences <ftp://ftp.ebi.ac.uk/pub/databases/lista>, YPD functions <ftp://isis.cshl.org/pub/yeast/YPD>, general database <http://genome.www.stanford.edu/ Saccharomyces>, Nature [Lond.] 1997; 387: issue no. 6632S, includes special World Wide Web addresses)

SAETHRE-CHOTZEN SYNDROME: see Chotzen syndrome

SAFETY: see laboratory safety, chemicals hazardous, recombinant DNA and biohazards, radiation hazard assessment, cosmic radiation, gloves, environmental mutagens

SAFFLOWER (*Carthamus tinctorius*): an oil crop of warmer climates, 2n = 24 but other related species have 2n = 20 or 2n = 44 chromosomes.

SAGE (Serial Analysis of Gene Expression): is a procedure that permits the analysis of the function of many genes by a sweeping procedure. The first step is to isolate all the mRNAs that are produced in a single organ, at a particular developmental stage. By reverse transcription they are converted into cDNA. The 3' end are then tagged by biotin. The cDNAs are digested by a restriction enzyme and the end fragments are trapped on streptavidin beads. After that, a second restriction enzyme is applied which cuts at least 9 bp from the fragments. In a following step each short (9 bp or longer) tags are amplified by PCR and the tagged pieces are linked into a

SAGE continued

single DNA molecule. Then each tag is sequenced by an automatic sequencer and the tags are counted. In this sweeping manner 20,000 genes can be monitored in a month. The same effort would require years if it would be conducted on separate genes. This procedure reveals not just the number of expressed genes in that organ but reveals also the level of their activity. Some may be expressed in a single copy others may be very active and are represented by multiple copies. (See genome project, biotin, streptavidin, DNA sequencing automated, electrospray MS, laser desorption MS, DNA chips)

SAGE: a computer program for the analysis of (human) genetic data.
SAIMIRI (squirrel monkey): see Cepidae
SAINFOIN (*Onobrychis vicifolia*): leguminous forage plant; 2n=14 or 28.
Sal **I**: restriction endonuclease with recognition site G↓TCGAC.
SALAMANDER: *Salamandra salamandra*, 2n = 24. SALICYLIC ACID →

SALICYLIC ACID (O-hydroxybenzoic acid): is a pain killer, keratolytic and fungicidal agent; it also mediates the expression of disease defense-related responses in plants. Its methyl salicylate derivative, a volatile compound, may carry out airborn signaling after infection of plants by pathogens. (See host - pathogen relation)

SALINE: water solution of NaCl; the "physiological saline" is 0.9% salt solution for humans.

SALIVARY GLAND CHROMOSOMES: are polytenic and because of their large size and clear landmarks, have been used extensively for cytogenetic analyses of *Drosophila* and other flies. The cultures to be used for this type of studies should be less-crowded and moist. Third instar larvae can be used as they crawl out of the medium before the cuticle hardens. The larvae are placed in aceto-orcein or into 7% aqueous NaCl on microscope slides. With a needle one holds the larva in place and with a second needle placed behind the mouth parts, larva is decapitated and the salivary gland is pulled out. In aceto-orcein on a clean slide the nuclei are stained in 5 to 10 min. The chromosomes may be spread by gentle pressure on the cover slide and after sealing at the edge with wax, they can be examined under a light microscope. (See polytenic chromosomes)

▲ SALIVARY GLAND CHROMOSOMES OF *DROSOPHILA* (Courtesy of Dr. H.K. Mitchell) UPPER LEFT (boxed) A REGULAR MITOTIC SET OF CHROMOSOMES AT ABOUT THE SAME SCALE. (Redrawn after Painter, T.S. 1934 J. Hered. 25:465)

SALIVARY GLAND CHROMOSOMES AND MAPPING: because the salivary chromosomes display clear topological markers (bands), they can be used to locate deletions duplications, inversions, translocations and can be used to associate mutant phenotypes with physical alterations. The genetic and cytological maps are collinear yet they are not exactly proportional by distance. (See coefficient of crossing over)

SALMON: *Salmo gardneri*, 2n = 58-65.

SALMONELLA (*typhimurium*): is a member of the enteric Gram-negative bacteria and as such related to *E. coli*, and its handling ease is similar to it. It is a human pathogen and even in the laboratory it requires some caution when manipulated. Several of the related species contaminate food and feed supplies and create health hazards. F^+, F' and Hfr strains are available. Its best known transducing phage is P22. (See histidine operon, Ames test, phase variation)

Salpiglossis variabilis: (Solanaceae) a plant species with the rather unusual characateristics that the four pollen grains, product of a single meiosis, stick together and therefore can be used for tetrad analysis and gene conversion in higher plants. (See tetrad analysis, gene conversion)

SALT BRIDGES: are non-covalent ionic bonds in multimeric proteins

SALTATION: the unproven evolutionary proposition that species (and even higher taxonomic categories) arise by non-Darwinian sudden, major alterations. (See hopeful monster)

SALTATORY REPLICATION: sudden amplification of DNA segments during evolution.

SALT-TOLERANCE: of plants is regulated by the HKT1(high affinity potassium [K^+] transporter). This protein is actually a Na^+ and K^+ cotransporter where at high K^+ level results in low level of Na^+ uptake thus resulting in some salt tolerance. (See ion channels)

SALVAGE PATHWAY: is a recycling pathway, in contrast to the *de novo* pathway, e.g., nucleotide synthesis from nucleosides after removal of the pentose followed by phosphorybosylation. (See phosphoribosyl transferase, HGPRT, HAT medium)

SAM: see S-adenosyl-L-methionine, see also substrate adhesion molecules

SAMPLING ERROR: can occur when the population is too small or when few individuals are chosen or whenever the selection is not random. (See drift genetic, genetic drift, effective population size, founder principle)

SANCHO: a non-viral retrotransposable element. (See transposable elements)

SANDHOFF'S DISEASE: is characterized either by the absence of both hexosaminidase A and B activity or by hexosaminidase B only. This autosomal recessive hereditary defect has very similar symptoms to those of the Tay-Sachs disease that is caused by hexosaminidase A deficiency. Hexosaminidase B defect blocks the degradation of β-galactosyl-N-acetyl-(1→3)-galactose-galactose-glucose-ceramide to β-(1→4)-galactosyl-galactose-glucose-ceramide and hexosaminidase A normally converts GM_2 ganglioside (neuraminic-N-acetyl-galactose-4-galactosyl-N-acetyl-glucose-ceramide) into GM_3 ganglioside (neuraminic-N-acetyl-galactose-glucose) ceramide. The genes for hexosaminidase A and B are in human chromosome 15q23-q24 and, 5q13, respectively. Their structural similarity indicates evolution by duplication. (See sphingolipidoses, sphingolipids, gangliosides, Tay-Sachs disease)

SANFILIPPO SYNDROME: see mucopolysaccharidosis

SANGER METHOD OF DNA SEQUENCING: see DNA sequencing

SANTA GERTRUDIS CATTLE: originated by a cross between the exotic-looking Brahman x shorthorn breeds (*Bos taurus*). The Brahman cattle descended from *Bos indicus*. (See cattle)

SAP-1 (stress activated protein): involved in the activation of MAP kinases in binding to their serum response factor (SRF). (See MADS box, MAP, SRF)

SAPK: a stress-activated protein kinase of the ERK family. SAPK can be activated by SEK a protein kinase, related to MAP kinase kinases. The signaling cascade involved is targeted to JUN. (See signal transduction, ERK, MAPKK, MEKK, JUN, JNK)

SAPONIFICATION: alkaline hydrolysis of triaglycerols to yield fatty acids. (See triaglyceride, fatty acids)

SAPORINS: plant glycosidases which remove adenine residues from RNA and DNA but not from ATP or dATP. (See RIP)

SAPROPHYTIC: lives on non-live organic material. (See biotrophic)

SAR (structure-activity relationship): is a field of study in carcinogen, mutagen and drug research. Structural modifications may or may not affect activity and it would be important to know what are the decisive factors.

SAR (systemic acquired resistance): may be induced in plants by pathogens, salicylic acid, etephon, a compound releasing ethylene. (See host - pathogen relationship, ethylene)

SAR: see scaffold

SAR1: a low activity GTPase, related to Arf. It regulates the traffic between the endoplasmic reticulum and the Golgi apparatus. (See endoplasmic reticulum, Golgi, GTPase, Arf, GTPase)

SAR by NMR (structure-activity relationship by nuclear magnetic resonance-based methods): is an essential and sophisticated method to synthetic drug design. Natural or synthetic molecules are screened and optimized analogs are synthesized to identify high-affinity ligands to develop effectively targetable drugs. (See SAR, NMR)

SARAN WRAP: a thin sheet of plastic that clings well to most any surface and suitable for covering laboratory dishes, gels, etc.

SARCOMA: a solid tumor tissue with tightly packed cells embedded in a fibrous or homogeneous substance; sarcomas are frequently malignant. (See Rous sarcoma, RAS, oncogenes)

SARCOMERE: muscle units of thick myosin, thin actin filaments between two plate-like Z discs. These units are repetitive. (See titin, nebulin)

SARCOPLASMIC RETICULUM: membrane network in the cytoplasm of muscle cells containing high concentration of calcium which is released when the muscle is excited.

SARCOSINEMIA: sarcosine (methylglycine, CH_3NHCH_2COOH) is normally converted to glycine (NH_2CH_2COOH) by the enzyme sarcosine dehydrogenase. A defect at this step increases the level of sarcosine in the blood and in the urine (hypersarcosinemia), and may result in neurological anomalies but it may have almost no effect at all. Glutaric aciduria and defects in folic acid metabolism may also cause hypersarcosinemia. (See glycine, folic acid)

SARKOSYL (N-lauroylsarcosine): biological detergent used for extraction of tissues.

SAS (switch-activating site): see *Schizosaccharomyces pombe*

SAT: see satellited chromosome

SATAC (satellite-based artificial chromosome): of mammals is expected to replicate independently from the genome and to express its gene content as a huge vector.

SATELLITE DNA: DNA fraction with higher or lower density (during ultracentrifugal preparations) than the bulk DNA; generally contains substantial repetitive DNA. (See repetitious DNA, ultracentrifugation, heterochromatin, α-satellite)

SATELLITE RNA: the transcript of satellite DNA: (See satellite DNA)

SATELLITE VIRUS: a defective virus co-existing with another (helper) virus to correct for its insufficiency.

SATELLITED CHROMOSOME: carries an appendage to one arm by a constriction. The bridge between the main body of the chromosome and appendage is the site of the nucleolar organizer. It was named originally SAT as an abbreviation for *sine acido thymonucleinico*, i.e., a place where there was — then — no detectable DNA (= thymonucleic acid, as it was called in the 1930s) in the chromosome. The appendages were also called trabants. (See chromosome morphology)   ← SATELLITE APPENDAGE →

SATSUMA MANDARIN: a pollen-sterile citrus variety that, if no foreign pollen can reach the flowers, produces seedless oranges. (See seedless fruits)

SATURATION OF MOLECULES: the carbon—carbon attachments of single covalent bonds (no double bonds when entirely saturated). (See saturated fatty acids)

SATURATED FATTY ACIDS: all their chemical affinities satisfied, have higher energy content than the unsaturated fatty acids that contain one or more double bonds. (See fatty acids, cholesterols)

SATURATION DENSITY: of a mammalian cell culture before contact inhibition takes place. (See cancer, malignant growth, contact inhibition)

SATURATION HYBRIDIZATION: one component in the nucleic acid annealing reaction mixture has excessive concentration to allow to find all possible sites of homology and hybridization. (See nucleic acid hybridization)

SATURATION MUTAGENESIS: induce mutations at all available sites to reveal the relative importance of these sites. (See mutagenesis, localized mutagenesis, linker scanning)

Sau **3A:** restriction endonuclease with recognition sequence ↓GATC; *Sau* 96 I: G↓GNCC

SAUR (small auxin-up RNA): RNAs encoded by several genes that are activated by auxins.

SBMA: see Kennedy disease

SC PHOCOMELIA: see Roberts syndrome

SC35: an SR protein. (See SR motif)

SCAFFOLD: the cytoskeleton of the cell or residual protein fibers left in the chromosome after the removal of histones. The bulk of the scaffold is of two proteins, Sc1 (a topoisomerase II) and Sc2. The scaffold is attached to SAR (scaffold attaching regions). (See cytoskeleton, topoisomerase, loop domains model)

SCAMP (secretory carrier membrane proteins): integral membrane proteins of secretory and transport vesicles, like the synaptic vessels, etc.

SCANNING ELECTRONMICROSCOPY (SEM): in contrast to transmission electron microscopy, the electron beam is reflected from the surface of the specimen, coated with a heavy metal vaporized in a vacuum, a process called *shadowing*. As the electron beam scans the specimen, secondary electrons are reflected according to the varying angles of the surface of the object and generating a *three-dimensional image* corresponding to the grade of reflections. The maximal resolution is 50-100 times less than with the transmission electronmicroscopy but the image obtained can be highly magnified. It is an important technique for developmental studies. (See also electronmicroscopy, stereomicroscopy)

SCANNING MUTAGENESIS: see linker scanning

SCANNING OF mRNA: the eukaryotic mRNA does not have a special consensus (such as the Shine-Dalgarno box in prokaryotes) for attachment to the ribosomes so the mRNA leader adhers by its methylated guanylic cap to the ribosomes and the ribosome runs on it until it finds an initiation codon (generally AUG) to start translation. A prefered sequence, however, may be around the AUG codon AG-----CCAUGG. The 3' terminus of the 18S rRNA of mammals bears some similarity to the prokaryotic 16S terminus:

$A^{Me\,2} A^{Me\,2}$ CCUGCGGAAGGAUGA-----UUA-3'-OH. (See Shine-Dalgarno sequence)

SCANNING TUNNELLING MICROSCOPE (STM): can resolve biological molecules at the atomic level; its use for DNA sequencing has been proposed. (See electrospray MS, laser desorption)

SCAP (SREBP cleavage-activating protein): regulates cholesterol metabolism by promoting the cleavage of transcription factors SREBP-1 and -2 (sterol regulatory element binding proteins). In low-sterol cells, discrete proteolysis cuts of the amino-terminal of SREBPs. As a consequence these proteins enter the nucleus and activate the LDL receptor and cholesterol and fatty acid biosynthetic enzymes. The system can be studied by mutant CHO cells that either cannot synthesize cholesterols or LDL receptors in response to sterol depletion or are sterol resistant and cannot terminate the synthesis of sterols or their LDL receptor. (See sterol, LDL, CHO)

SCAR (sequence-characterized amplified region): are physical markers obtained by polymerase chain reaction-amplified RAPD bands. (See amplification, RAPD, PCR)

SCARMD: see muscular dystrophy

SCATTER FACTOR (hepatocyte growth factor): its cellular responses are mediated by the Met tyrosine kinase receptor. It has multiple cell targets and it is probably involved in mesenchymalepithelial interactions and liver, kidney development, organ regeneration, metastasis, etc. (See hepatocyte growth factor, metastasis, macrophage-stimulating protein)

SCATTERING: deflection of electrons by collision(s). (See Compton effect)

SCD25: suppressor of gene *cdc25* mutations in yeast; increases the dissociation of Ras•GDP but does not affect Ras•GTP. (See cell cycle, *cdc25*)

SCE: see sister chromatid exchange

SCEUS (conserved evolutionary unit sequences): reveal evolutionary similarities of the DNA in the genomes across taxonomic boundaries. (See unified map)

SCHEIE SYNDROME: see Hurler syndrome

SCHIFF BASE: α-amino groups of amino acids may react reversibly with aldehydes and form a Schiff base; these are labile intermediates in enzymatic amino acid reactions.

SCHIFF'S REAGENT: retains a blue color in the presence of aldehydes. Aldehydes are exposed to a fuchsin solution (0.25 g/L H_2O) and decolorized by SO_2. (See aldehyde, fuchsin)

SCHINZEL SYNDROME: autosomal dominant ulnar-mammary syndrome shows complex symptoms including malformation of the hand and shoulder, mammary glands, delayed puberty, obesity, etc.

SCHINZEL-GEIDION SYNDROME: autosomal recessive malformation of the face and head, heart, growth retardation, telangiectasia, supernormal hair development. (See hypertrichosis, telangiectasis)

SCHISTOSOMIASIS: is the state of disease of animals and humans seized by one or another species of the parasitic flatworms *Schistosoma* (fluke). The parasites infect the blood vessels through contact with contaminated waters in warm climates of the world. In these species the male carries the female in a ventral sac (gynecophoral canal). Intermediate hosts are snails and molluscs. The *S. haematobium* is primarily a human parasite. The *S. japonicum* infects several animals as well. The symptoms of the disease may vary according to the various species and may include systemic irritations, cough, fever, eruptions, tenderness of the liver, diarrhea, etc. The infected intestines, liver, kidneys, brain etc. may be seriously damaged without medication. The therapeutic antimony derivatives are highly toxic to humans. The parasite is present in millions of people of the tropics including the ancient Egyptian mummies. A codominant locus in human chromosome 5q31-q33 provides some protection against *Schistosoma mansoni*. SCHISTOSOMA HAEMATOBIUM MALE CARRYING THE FEMALE ➡

SCHIZOCARP: a fruit where the carpels split apart in order to free the seeds.

SCHIZOPHRENIA (dementia precox): is a behavioral disorder. The afflicted individuals cannot distinguish well reality from dreams and imagination. Hallucinations and paranoid behavior, delusions, inappropriate emotional responses, lack of logical thought and concentration are common symptoms. The precise mechanism of inheritance is unknown yet in about 13% of the children of afflicted parents it reoccurs, and 45% of the identical twins are concordant in this respect versus only 15% of dizygotic twins. The general incidence in the population varies from 1 to 4%. Both autosomal, psedoautosomal single and multiple recessive and dominant loci were implicated. There were some indications that schizophrenia genes are in chromosome 6pter-p22, 8, 9, 20 and 22 and in the long arm of chromosome 5. The various manifestations of the disease have strong environmental components. MAO (monoamine oxidase) level is reduced in the individuals afflicted. MAO enzyme removes amino groups from neurotransmitters. The overproduction of dopamine (3,4-dihydroxyphenylethylamine), a precursor of neurotransmitters may also be suspected in causing it. Chloropromazine (a peripheral vasodilator, antiemetic drug) and reserpine (alkaloid) may alleviate the psychological symptoms by inhibiting dopamine receptors. In schizophrenia and other affective disorders frequently an expansion of trinucleotide repeats are detectable. (See also paranoia, psychoses, affective disorders, neurological disorders, pseudoautosome, concordant, twinning, trinucleotide repeats)

Schizosaccharomyces pombe: (fission yeast): the 3 chromosomes are of 5.7, 4.7 and 3.5 Mb size, respectively. The cells are 7 x 3 μm. This eukaryote, an ascomycete, has both asexual (by fission) and sexual life cycle (each meiosis producing 8 ascospores). Under good growing conditions it reproduces by mitotic divisions. At starvation (for any factors of growth) the plus (P) and minus (M) type cells fuse and meiosis follows. The mitotic cell cycle (2.5 h) has the typical G_1, S, G_2 and M phases. The G_2 phase takes 70% of the total time whereas the other phases equally share the rest. Under severe nutritional limitations instead of sexual development, the cells are blocked in either of the G phases and this dormant state is called "GO".

The mating type is determined by which of the *P* or *M* alleles is switched (transposed) from their silent position to the *mat1* locus where they are expressed. Actually both *P* and *M* genes have two alleles with different number of amino acids in their polypeptide products: *Pc* (118), *Pi* (159) and *Mc* (181) and *Mi* (42). The *c* alleles (required for meiosis and conjugation) are transcribed rightward from the centromere when nitrogen is available, and the *i* alleles (required only for meiosis) are transcribed in opposite direction in N starvation. The product of the *Pi* allele has a protein-binding domain, whereas that of *Mc* shows some homology to the *Drosophila Tdf* (testis-determining factor) and the mouse *Tdy*. Homothallic strains can switch between mating types but the heterothallic ones are either *P* or *M*. The *MAT* site is comparable to the disk drive of a computer (or the slot of a tape player) where either the *P* or the *M* floppy disk (or tape cassette) is plugged in and that determines then whether the mating type in the heterothallic strain will be *P* or *M*. The *P* (1113 bp) and *M* (1127 bp) sites are actually the storage sites for the *P* and *M* mating type information, respectively.

Schizosaccharomyces pombe continued

The mating type region in the right arm of chromosome II can be represented as shown:

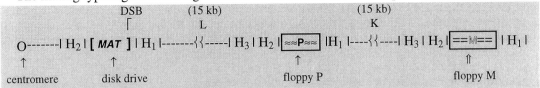

The *DSB* (double-strand break), near the *MAT* site — in about 25% of the cells — is probably required that the chromosome would permit the insertion of one or the other cassette. This breakage is probably transient and quickly restored so the continuity of the chromosome would not be compromised. The ≈15 kb *L* and *K* sequences are spacers where meiotic recombination is not observed. The H_1 (59 bp) and H_2 (135 bp) homology boxes are flanking the disk drive (*MAT*) and both floppy disks while the H_3 (57 bp) occurs at left of the *P* and *M* sequences only. It has been supposed that the reason why the *P* and *M* elements are silent at the storage sites is because of the H_3 presence there but not at the *MAT* site where they are expressed. The switching (transposition) is controled by *SAS1* and *SAS2* (switch-activating sites) right of *DSB* (within 200 kb). In addition, at least 11 other trans-acting (*swi*) loci regulate switching (transposition). Mating type determination in the budding yeast (*Saccharomyces cerevisiae*) is also controlled by transposition albeit in a different way, and the homology of the DNA sequences in the elements is also low. Fission yeast has contributed also to learning many aspects of the cell cycle control. (See *Saccharomyces cervisiae*, mating type determination, cell cycle, <http://www2.bio.uva.nl/pombe/>)

SCHWANN CELL: is a glial cell (forms myelin sheath for the peripheral nerves). Its differentiation requires the Oct-6 POU factor. (See Oct-1, POU)

SCIANNA BLOOD GROUP (Sc): is represented by antigenic groups Sc-1 and Sc-2, located in human chromosome 1. (See blood groups)

SCIARA: dipteran flies with polytenic, giant chromosomes in their salivary glands. The basic chromosome number in *Sciara coprophylla* is 3 autosomes and 1 X chromosome but there are also the heteropycnotic so called *limited chromosome*(s) which are present only in the germline. They are eliminated from the nuclei during the early cleavage divisions. The egg pronucleus contains three autosomes, an X chromosome and one or more limited chromosomes. The sperm contributes three autosomes, two X chromosomes and some limited chromosomes that are all of maternal origin. The first division of the spermatocyte is monocentric and separates the maternal chromosome set from the paternal one. The maternal chromosomes move to a single pole whereas the paternal set is positioned away from the pole and are never transmitted to the progeny. The single secondary spermatocyte displays an unusual unequal type division. The X chromosome divides longitudinally and both copies are included into the same cell and only this cell survives. From the cleavage nuclei first the limited chromosomes are eliminated and subsequently from the cells that become males one of the X chromosomes is also evicted thus the males become XO and the females are XX. (See also *Rhynchosciara*, polytenic chromosomes, salivary gland chromosomes, sex determination, chromosomal sex determination)

SCID (severe combined immunodeficiency): a heterogeneous group genetically determined diseases. It involves defective V(J)D recombination of immunoglobulin genes. Several frameshifts, point mutations or deletions in the IL-2Rγ chain (human chromosome Xq13) were found to be associated with SCID. The anomaly involves also the Jak/STAT and other signaling pathways. (See signal transduction, immunoglobulins, RAG)

SCIENCE: is a systematic study of natural phenomena with the explicit purpose to prove or negate a working hypothesis or hypotheses by experimental means. According to K.R. Popper for science the "criterion of potential satisfactoriness is thus testability, or improbability: only a highly testable or improbable theory is worth testing and is actually (and not merely potential-

Science continued

ly) satisfactory if it withstands severe tests — especially those tests to which we could point as crucial for the theory before they were ever undertaken...I refuse to accept the view that there are statements in science which we have, resignedly, accept as true merely because it does not seem possible...to test them." Applied science seeks to find economic use of the principles discovered by basic or pure science. (See experiment, genetics)

SCINTIGRAPHY: photographic location of radionuclides within the body after introduction of radioactive tracers. (See radiocative label, radioactive tracer)

SCINTILLATION COUNTERS: can be liquid scintillation counters or crystal scintillation-counters for solids. The counter is an electronic appliance where the sample is placed in a solution of organic compounds (cocktail). The radiation coming from the isotopes (even from the weak β–emitters) cause flashes in the fluorescing cocktail that are directed to a photoelectric cell. The cell then releases electrons that are amplified and registered (counted). Each flash corresponds to a disintegration of an atom of the isotope so the equipment displays (or prints out) the disintegrations per minute (dpm) or counts per minute (cpm), generally with background radiation subtracted. This information provides measures of the quantity of the label (or labeled compound). In the crystal scintillation counter the radiation (usually energetic γ-rays, X-rays or β-rays) emitted by the isotope hits a crystal of sodium iodide containing also traces of thallium iodide, and again the disintegrations are registered similarly to the liquid scintillation counter. (See radioisotopic tracers, radiolabeling, isotopes, dpm, radiation hazard assessment)

SCISSION: cuts in both strands of a DNA molecule at the same place. (See also nick)

SCISSORS GRIP: see Max

SCISSORS, MOLECULAR: see Cre/loxP, excision vector, targeting genes

SCK: is protein with phosphotyrosine binding domain but different from the SH2 domain of SRC. (See SRC, SH2, PTB)

SCLC: see small cell lung carcinoma

SCLERENCHYMA: plant tissues with tough, hard cell walls.

SCLERODERMA: a probably autosomal dominant disease involving scaly hardening of the skin and increased frequency of chromosome breakage. (See skin diseases)

SCLEROIDS: cells with unusually hardened walls.

SCLEROSIS: hardening caused by inflammation or hyperplasia of the connective tissue.

SCLEROTIA: see hypha

SCMRE: is a cis-acting element in the c-fos protooncogene, and it is responsible for induction by some mitogens. (See FOS, protooncogene, cis, mitogen)

scnDNA: single-copy nuclear DNA.

SCOLIOSIS: the spine is not straight but laterally curved; it may be due to polygenic causes or it can be part of skeletal syndromes. Its incidence is about 5-6% of the adult human populations.

SCOPOLAMINE: see *Datura*, alkaloids

SCOTOMORPHOGENESIS: differentiation of plants in the absence of light, e.g. etiolated growth. (See brassinosteroids)

SCOLIOTIC PIG ➡

SCP$_2$: sterol carrier protein 2; probably the same as nsl-TP. (See sterol)

SCRAPIE: see encephalopathies, Creutzfeldt-Jakob disease

SCREENING: selective classification of cell cultures for mutation or for special genes, conveying auxotrophy, antibiotic or other resistance, selecting antibodies by cognate antigens, plant populations for disease- or chemical-resistance, animal progenies for blood groups, etc.

SCREWWORM: see *Cochliomya hominivorax*, genetic sterilization, myasis

SCRIPTON (transcripton): unit of lambda phage transcription. (See)

scRNA: small cytoplasmic RNAs. (See snRNA)

scRNAP: small cytoplasmic ribonucleoprotein

SCROTUM: the pouch containing the testes and accessory sex organs of mammals.

SCUTELLUM: the single cotyledon of the grass embryo, S=scutellum, E=embryo➡

SD: see *Segregation distorter*, standard deviation

SDF-1 (stromal cell-derived factor): is a chemokine and natural ligand of fusin. (See fusin [LESTR], chemokines)

SDP: see short-day plants

SDR (short dispersed repeats): are organellar DNA sequences of 50-1000 bp that may occur in direct or inverted forms and may represent more than 20% of the chloroplast genomes of *Chlamydomonas reinhardtii* alga and various land plants. Similar or shorter redundant sequences occur also in the mitochondria of plants, animals and fungi. In *Saccharomyces cerevisiae* eight 200-300 bp *ori* and *rep* sequences and 200 G+C sequences of 20-50 bp may be clustered into several mtDNA gene families. Similar mobile G+C elements may occur also in other fungi. (See chloroplasts, mtDNA, mobile genetic elements, organelle sequence transfers)

SDR: see strain distribution pattern

SDS: sodium dodecyl sulfate; detergent, used for electrophoretic separation of protein and lipids.

SDS-PAGE: see SDS-polyacrylamide gel

SDS-POLYACRYLAMIDE GELS: electrophoretic gel containing sodium dodecyl sulfate (SDS, also called sodium lauryl sulfate [SLS], detergents) and polyacrylamide. This medium dissociates proteins into subunits and reduces aggregation. Generally the proteins are denatured before loading on the gel with heat and a reducing agent. The polypetides become negatively charged by binding to SDS and are separated in the gel according to size (rather than by charge). On the basis of the mobility, the molecular weight of the subunits can be estimated with the aid of appropriate molecular size markers (ladder) but caution is required because glycosylated proteins may not reflect the molecular mass of the protein. The concentration of the polyacrylamide determines the size of the polypeptides that can be separated. Polyacrylamides (bisacrylamide:acrylamide, 1:29) separarates [kDa proteins] as follows: 15% - [12 - 43], 10% [16 - 68], 7.5% [36 - 94], 5% [57 - 212]. (See also gel electrophoresis, electrophoresis)

SE: see standard error

SEA ONCOGENE: is an avian erythroblastosis virus oncogene. The human homolog is at human chromosome 11q13 in very close vicinity to INT2 and BCL1. (See oncogenes)

SEA URCHINS: *Strongylocentrotus purpuratus* and *Toxopneustus lividus*, both 2n = 36, and other echinodermata have been favorable objects of cell cycle studies, fertilization and embryogenesis. Their large size eggs can be easily collected and handled in the laboratory.

SEAL: *Callorhinus ursinus*, 2n = 36; *Zalophus californianus*, 2n = 36; *Crystophora crystata*, 2n = 34; *Erignatus barbatus* 2n = 34; *Helichoereus grypus* 2n = 32.

SecA PROTEIN: a peripheral membrane domain of the translocase enzyme and it is the primary receptor for the SecB/pre-protein complex by recognizing the leader domain of the pre-protein. Hydrolyzes ATP, GTP, promotes cycles of translocations and pre-protein release. (See membranes, translocase, Mss, Ypt, Rab, translocase, translocon, protein targeting, SRP, ARF)

SecB PROTEIN: a17-subunit chaperone involved in translocation of pre-proteins by complexing and keeping them in the right conformation and binding to membrane surface of the endoplasmic reticulum. Recognizes both the leader and mature protein domains. (See chaperones, membranes, endoplasmic reticulum, SRP, translocon, translocase, protein targeting)

SECKEL's DWARFISM (bird-headed dwarfism): a microcephalic autosomal recessive condition with reduced intelligence. (See dwarfism)

SECOND CYCLE MUTATION: is caused by the excision or movement of a transposable element and leaving behind a footprint which still causes some type of alteration in the expression of the gene. This type of alterations may be connected also with the defective nuclear localization of a transcription factor. (See transposon footprint, transposable elements, nuclear localization sequences)

SECOND DIVISION SEGREGATION: see tetrad analysis, postreduction

SECOND MESSENGER: molecules with key roles in signal transduction pathways such as cyclic-AMP, cyclic-GMP, and others. Animal physiologists call hormones first messengers. In animal and plant cells Ca^{++} is considered to play the role of second messenger. Inositol triphosphate is also a second messenger. (See cAMP, cGMP, mRNA, PIP, signal transduction, G proteins)

SECOND SITE REVERSION: it is actually a suppressor mutation at a site different from that of the original lesion but it is capable of restoring the normal reading of the mRNA. (See suppressor tRNA, suppressor gene, reversion)

SECOND STRAND SYNTHESIS: see reverse transcriptase

SECONDARY CONSTRICTION: see nucleolar organizer, satellited chromosome. (See SAT)

SECONDARY IMMUNE RESPONSE: an immune reaction conditioned by the memory cells when antigenic exposure occurs repeatedly. (See immunological memory)

SECONDARY LYMPHOID TISSUES: lymph nodes and spleen (in contrast to primary lymphoid tissues, the bone marrow and thymus). (See immune system)

SECONDARY METABOLISM: produces molecules that are not basic essentials for the cells and their products occurs only in specialized tissues, e.g. anthocyanin, hair pigments.

SECONDARY NONDISJUNCTION: see nondisjunction

SECONDARY RESPONSE GENES: their transcription is preceded by protein synthesis; probably primary response genes are involved in their induction. They are stimulated by mitogens alone without cycloheximide. (See mitogens, primary response genes, signal transduction)

SECONDARY SEX RATIO: the proportion of males to females at birth. (See age of parents, sex ratio, primary sex ratio)

SECONDARY SEXUAL CHARACTER: usually accompany the primary sexual characters but they are not integral part of the sexual mechanisms, e.g. facial hair in human males, red plumage of the male cardinal birds, increased size bosoms in females and higher pitch voice, etc. (See primary sexual characters)

SECONDARY STRUCTURE: the steric relations of residues that are next to each other in a linear sequence within a polymer such an α-helix, a pleated β-sheet of amino acids. (See protein structure)

SECONDARY TRISOMIC (isotrisomic): the third chromosome has two identical arm, originated by misdivision of the centromere or by the fusion of identical telochromosomes. (See trisomics, misdivision, telosome)

SECRETAGOGUE: a compound or factor that stimulates secretion.

SECRETION VECTOR: besides an expressed structural gene, it carries a secretion signal to direct the gene product to the appropriate site.

SECRETOR: secretes into the saliva the antigens of the *ABH* blood group. (See ABH antigen)

SECRETORY IMMUNOGLOBULIN A: an IgA dimer with a secretory component. (See immunoglobulins)

SECRETORY PROTEINS: mainly glycoproteins that are released by the cell after synthesis, such as hormones, antibodies and some enzymes.

SECRETORY VESICLE (secretory granule): releases stored molecules within the cell.

SECTIONING: is a generally required procedure for the preparation of biological specimens for histological examination. The material may need embedding before they are cut either by freehand or by microtomes. Some microtomes cut tissues frozen by CO_2. The 1-20 µm thin sections are subsequently placed on microscope slides and subjected to a series of manipulations (paraffin wax removal, dehydration, staining) before examination. (See microtome, embedding)

SECTORED-SPORE COLONIES: arise when the haploid spores carried heteroduplex DNA with different alleles in the heteroduplex region. (See heteroduplex)

SECTORIAL: displays sectors, e.g. mitotic recombinant, sorting-out of organelles, somatic mutation, etc. (See mosaic, chimeric)   ← SECTORIAL LEAF

SecY PROTEIN: integral membrane component of bacteria involved in chaperoning the assembly of membrane and some soluble proteins. (See membrane)

SecY/E PROTEIN: the membrane embedded domain of translocase enzyme consisting of SecY and SecE polypeptides. Stabilizes and activates SecA and facilitates membrane binding. (See membrane)

SEDIMENTATION COEFFICIENT: is the rate by what a molecule sediments in a solvent. It is characterized by the Svedberg unit (S) that is a constant of 1×10^{-13}, derived from the equation $s = (dx/dt)/\omega^2 x$, where x = the distance from the axis of rotation in the centrifuge, ω is the angular velocity in seconds ($\omega = \theta/t$, where θ is the angle of rotation and t is time). At a constant temperature (20 C°) in a solvent s depends on the weight, shape and hydration of a molecule. The S value is used for the characterization of macromolecules, e.g. 16S (= 16×10^{-13}).

SEED COAT: see integument

SEEDLESS FRUITS: may be the result of different genetic mechanisms. Aneuploids, triploids, self-incompatibility or gametic sterility genes may be most commonly responsible for this condition. (See seedless water melon, bananas, pineapple, naval orange, stenospermocarpy, seedless grapes)

SEEDLESS GRAPES: are the result of a gene that causes early embryo abortion although fertilization occurs normally (stenospermocarpy). (See seedless fruits)

SEEDLESS WATERMELONS: are triploids (produced by crossing tetraploids with diploids). They are more convenient to eat because the seeds do not have to be spit and thus may improve table manners. Their flavor and sweetness may make them superior to conventional varieties. (See seedless fruits)

SEGMENT POLARITY OF THE BODY GENES: in *Drosophila* mutations involved in the alteration of the characteristic body pattern and often accompanied by inverted repetition of the remaining structures. (See morphogenesis in *Drosophila*)

SEGMENTAL ANEUPLOID: contains an extra chromosomal fragment(s) in addition to the normal chromosome complement; it is a partial hyperploid. (See hyperploid, aneuploid)

SEGMENTATION GENES: control the polarity of body segments in animals. (See morphogenesis in *Drosophila*, metamerism)

SEGMENTED GENOMES: e.g., the DNA of T5 phage is in four linkage groups, the RNA genetic material of alfalfa mosaic virus is in four segments.

SEGREGATION: separation of the homologous chromosomes and chromatids at random to the opposite pole during meiosis and carrying in them the different alleles to the gametes. Independent segregation of non-syntenic genes (or ones which are more than 50 map units apart within a chromosome) is one of the basic Mendelian rules. In the haploid products of meiosis of diploids the 1:1 segregation of the alleles may be identified. (See Mendelian laws, autopolyploid, tetrad analysis, segregation distorter, preferential segregation, epistasis)

SEGREGATION, ASYMMETRIC: differentiation and morphogenesis require that the progeny cells would differ from the mother cell after cell division. This difference is brought about by the unequal distribution of cellular proteins. (See morphogenesis in *Drosophila*)

SEGREGATION DISTORTER: is a dominant mutation (*Sd*, map position 2-54 in *Drosophila*). When it is present, the homologous chromosomes (and the genes within) are not recovered in an equal proportion after meiosis. The second chromosomes that carry it are called SD and may be involved in chromosomal rearrangement and other (lethal) mutations. At the base of the left arm they may have also *E(SD)* [*enhancer of Sd*] or *Rsp* (*Responder of SD*, at the base of the right arm), and more distally *St(SD)*, *Stabilizer of SD* and other components of the system. *SD/+* males transmit the *SD* chromosome to about 99% of the sperm. When *Rsp* is in the homologous chromosome, *Sd* is preferentially recovered. Some of the other elements of the system act in a modifying manner and some may cause also recombination in the male. In many species of insects the infection of males by the bacterium *Wolbachia* kills the offspring of the infected males x uninfected females but the viability of the infected eggs is normal. (See also lethal factors, meiotic drive, transmission disequilibrium, certation, preferential segregation, polarized segregation, megaspore competition, tetrad analysis, gene conversion, epistasis, pollen-killer, infectious heredity)

SEGREGATION INDEX: the gene number in quantitative traits, also called effective number of loci. (See gene number in quantitative traits)

SEGREGATION LAG: the mutation or transformation is expressed only by third division of the bacteria until all chromosomes are sorted out.

SEGREGATION RATIO: the phenotypic (genotypic) proportions in the progeny of a heterozygous mating. (See segregation, Mendelian segregation, modified Mendelian ratios)

SEGREGATIONAL PETITE: see petite colony mutants, mtDNA

SEGREGATIONAL STERILITY: a heterozygote produces unbalanced gametes. (See inversion, translocation, gametophyte factor, hybrid dysgenesis, segregation distorter)

SEIZURE: a sudden attack precipitated by a defect in the function of the nervous system. (See epilepsy, double cortex, periventricular heterotopia)

Sek1: a tyrosine and threonine dual-specificity kinase involved in the activation of SAPK/JNK families of kinases and it also protects T cells from Fas and CD3-mediated apoptosis. (See SAPK, JNK, Fas, CD3, apoptosis, T cell)

SELECTABLE MARKER: permits the separation of individuals (cells), that carry, it from all other individuals, e.g. in an ampicillin or hygromycin medium (of critical concentration) only those cells (individuals) can survive that carry the respective resistance genes (selectable markers).

SELECTINS: are cell surface carbohydrate-binding, cytokine-inducible, transmembrane proteins. They bind also to endothelial cells in the small blood vessels along with integrins and enable white blood cells and neutrophils to ooze out at the sites of small lacerations to combat infection. The ESL-1 selectin ligand is a receptor of the fibroblast growth factor. The selectins contain an amino-terminal lectin domain, an element resembling epidermal growth factor and a variable number of complementary regulatory repeats and a cytoplasmic carboxyl end. The P and E selectins recruit T helper-1 but not T helper-2 cells to the site of inflammation. (See integrins, cell adhesion, cell migration, FGF, EGF, lectins, T cell)

SELECTION: unequal rate of reproduction of different genotypes in a population. (See also selection coefficient, allelic fixation, selection conditions, genetic load, fitness of hybrids, selection and population size, mutation pressure and selection, genetic drift, selection types, non-darwinian evolution)

SELECTION AND POPULATION SIZE: in very small population chance (random drift) may be more important than the forces of selection in determining allelic frequencies. When the selection pressure becomes very large, even in small populations there is a good chance for the favored allele to become fixed. (See also allelic fixation, random drift, genetic drift)

SELECTION COEFFICIENT: measure of fitness of individuals of a particular genetic constitution relative to the wild type or heterozygotes in a defined environment. If the fitness is 0 the selection coefficient 1. In other words, if an individual does not leave offspring (fitness is zero), the selection against it in a genetic sense is 1 (100%). The selection coefficient is generally denoted as (s) or (t), where the former indicates the selection coefficient of the recessive class and (t) indicates the selection coefficient of the homozygous dominants. The meaning and relation of fitness and selection coefficients can be illustrated best in a table form as shown below in a population complying with the Hardy - Weinberg theorem:

Genotypes	AA	Aa	aa	Total
zygotic frequencies	p^2	$2pq$	q^2	1
fitness	w_1	w_2	w^3	
gametes produced	$(w_1) \times (p^2)$	$(w_2) \times (2pq)$	$(w_3) \times (q^2)$	1

In case we take the fitness of one genotype as unity (in this case we choose the heterozygotes, e.g., $w_2 = 1$), then the *standardized fitness* becomes:

$$\frac{w_1}{w_2} = 1 - s \text{ and } s = 1 - \frac{w_1}{w_2} \text{ similarly } \frac{w_3}{w_2} = 1 - t \text{ and } t = 1 - \frac{w_3}{w_2}$$

(See fitness, allelic frequencies, Hardy - Weinberg theorem, advantageous mutation)

SELECTION CONDITIONS: selection may operate at different levels beginning in meiosis

Selection conditions continued

(distorters, gametic factors) or at any stage during the life of the individual beginning with the zygote through the entire reproductive period. The intensity of the selection depends on the genes, the overall genetic constitution of the individual and the environment, including the behavioral pattern of the population (e.g. protecting the young and infirm). The formulas representing the various means of selection were derived from the Hardy - Weinberg theorem $p^2 + 2pq + q^2 = 1$ (Table below) and their use is exemplified.

FORMULAS TO CALCULATE THE CHANGE IN ALLELIC FREQUENCIES PER GENERATION AT VARIOUS CONDITIONS OF SELECTION. IN CASE s OR q IS VERY SMALL, OMISSION OF THE sq PRODUCT FROM THE DENOMINATOR MAY BE OF VERY LITTLE CONSEQUENCE FOR THE OUTCOME.

Conditions of the selection	Change (Δq) in the allelic frequency per generation (+ : increase, - : decrease)
1. Selection in haploids	$-\dfrac{sq[1-q]}{1-sq}$
2. Differential selection in males and females (without sex linkage)	$q^2[1 - (1/2)(S_{male} + S_{female})]$
3. Selection at X-chromosomal linkage	In the heterogametic sex equation 1. holds; in the homogametic sex the rules of diploids apply
4. Selection against recessive lethals	$-\dfrac{q^2}{1+q}$
5. Selection against the allele without dominance	$-\dfrac{[1/2]sq[1-q]}{1-sq}$
6. Selection against dominant lethals	$+ (1 - q)$
7. Partial selection against homozygous recessives in case of complete dominance	$-\dfrac{sq^2[1-q]}{1-sq^2}$
8. Partial selection against completely dominant alleles	$+\dfrac{sq^2[1-q]}{1-\{s[1-q^2]\}}$
9. Selection against recessives in autotetraploids	$- spq^4$
10. Selection against intermediate heterozygotes	$-\dfrac{sq[1-q]}{1-2sq}$
11. Selection against heterozygotes	$+ 2spq(q - [1/2])$
12. Selection against both homozygotes (heterozygotes favored)	$+\dfrac{pq[s_1 p - s_2 q]}{1 - s_1 p^2 - s_2 q^2}$

The genotypic contributions to the *AA, Aa* and *aa* phenotypes are p^2, $2pq$ and $q^2(1-s)$,

Selection conditions continued

respectively, where s is the selection coefficient (see selection coefficient for derivation of s). The total contribution is now reduced from 1 to $1 - sq^2$ because sq^2 individuals are eliminated by selection. Thus the new frequency of the recessive alleles becomes $q_1 = \dfrac{q^2[1-s] + pq}{1 - sq^2}$ and the change in the frequency of q, is $\Delta q = \dfrac{q^2[1-s] + pq}{1 - sq^2} - q$ which reduces to: $-\dfrac{sq^2(1-q)}{1 - sq^2}$

The effectiveness of the selection depends also on the frequency of the allele involved. Selection against rare recessive alleles may be very ineffective because only the homozygotes are affected. Selection most frequently works at the level of the phenotype and the recessives may not influence the fitness of the heterozygotes. At low values of q the majority of the recessive alleles are in the heterozygotes and thus sheltered from the forces of selection. Dominant, semidominant and codominant alleles may be, however, very vulnerable if they lower the fitness of the individuals. As an example let us assume that the frequency of a recessive allele is $q = 0.04$ and by using formula 4 in the Table above the change in allelic frequency per generation is:

$\Delta q = -\dfrac{[0.04]^2}{1 + 0.04} \cong -0.00154$ and after $n = 25$ generations, the initial frequency of the gene, $q_0 = 0.04$ changes to: $q_n = \dfrac{q_0}{1 + nq_0} = \dfrac{0.04}{1 + [25 \times 0.04]} = 0.02$ meaning that complete elimination of all homozygotes ($s = 1$) for 25 generations reduces only to half the frequency of that recessives allele. The initial zygotic frequency of $(0.04)^2 = 0.0016$ (1/625) will thus change to $(0.02)^2 = 0.0004$ (1/2,500). If the same deleterious allele would be semidominant and it conveys a fitness of 0.5 relative to the homozygotes for the other allele, according to formula 5 in theTable above, the change of the frequency of this semidominant allele in a generation becomes

$\Delta q = -\dfrac{[1/2][0.5][0.04][0.96]}{1 - \{[0.5] \times [0.04]\}} \cong -0.0098$ Thus a semidominant allele with 0.5 selection coefficient will be selected against more than six times as effectively as a recessive lethal factor because $0.0098/0.00154 \cong 6.4$ in these examples.

The number of generations required to bring about a certain change in gene frequencies can be calculated in the simple case when the homozygous recessives are lethal, i.e. the selection coefficient is, $s = 1$:

$T_{generations} = \dfrac{q_0 - q_T}{q_0 q_T} = \dfrac{1}{q_T} - \dfrac{1}{q_0}$ where q_0 is the initial frequency of the allele and q_T is its frequency after T generations. If it is assumed that the genotypic frequency is $(q_0)^2 = 0.0001$ and $q_0 = 0.01$ then the number of generations required to reduce the initial frequency to $q_T = 0.005$ is $T = \dfrac{1}{0.005} - \dfrac{1}{0.01} = 100$. After 100 generations then the frequency of the recessive lethal allele becomes $(q_T)^2 = (0.005)^2 = 0.000025 = 1/40,000$ compared to the initial frequency of 1/10,000. The effectiveness of selection is much influenced by the heritability of the allele concerned. At complex cases more elaborate computations are required that cannot be illustrated here. (See also selection coefficient, balanced polymorphism, mutation pressure opposed by selection, selection and population size, allelic fixation, genetic load, fitness of hybrids, gametephyte, QTL, heritability, gain)

SELECTION DIFFERENTIAL: see gain

SELECTION INDEX IN BREEDING: see gain

SELECTION INTENSITY: see gain

SELECTION AND MEDICAL CARE: the progress of effective medical care saves increasing number of human lives. Some of the saved individuals will have a chance to transmit deleteri-

Selection and medical care continued

ous genes to their offspring so eventually some increase in detrimental alleles is expected. (See also selective abortion)

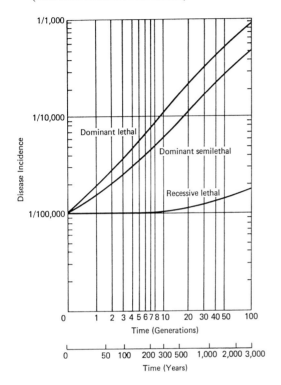

IF THE INITIAL FREQUENCY OF A DETRIMENTAL RECESSIVE ALLELE IS 0.001 OR LESS, THE CONSEQUENCES OF THE SELECTION MAY NOT BE EVIDENT FOR ABOUT 300 YEARS. THE FREQUENCY OF DOMINANT OR DOMINANT SEMI-LETHAL ALLELES MAY INCREASE MUCH FASTER. (From Bodmer, W.F. and Cavalli-Sforza, L.L. 1976 Genetics, Evolution, and Man. Freeman, San Francisco)

SELECTION, NATURAL: see natural selection

SELECTION PRESSURE: intensity of selection affecting the frequency of genes in a population.

SELECTION RESPONSE: (heritability) x (selection differential). (See heritability, gain)

SELECTION TYPES: (i) *Stabilizing selection* eliminates extreme types and favors the intermediate forms that have ability to survive under the most common but opposite conditions (such as cold and heat, draught and excessive precipitation). (ii) *Directional selection* shifts the mean values of a population either toward higher or lower values than the current mean. (iii) *Disruptive selection* breaks up the population into two or more subpopulations that each have adaptive advantage in particular niches of a larger habitat. (iv) *Frequency-dependent selection* favors an allele when it is relatively rare and may turn against it when it becomes abundant. Common examples are found in host - parasite, predator - prey relationships. When the number of predators increases beyond a point there will not be enough prey to maintain the predators and their number will decrease (See fitness, selection conditions, selection mechanisms)

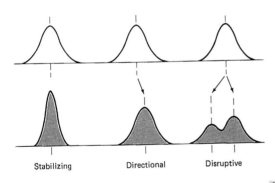

TOP: ORIGINAL FREQUENCY DISTRIBUTION OF POPULATIONS. BELOW: THE SHIFTS IN DISTRIBUTION AFTER SELECTION. (After Mother, K. 1953 Symp. Soc. Exp. Biol. 7:66)

SELECTION VALUE: see selection index, gain

SELECTIVE ABORTION: termination of a pregnancy by precocious removal of the fetus from the womb if the condition of the mother or of the fetus medically justifies it, and the legal system permits it. The genetic constitution or condition of the fetus may be tested with the aid of amniocentesis or sonography. From the viewpoint of genetics, selective abortion may pose biological problems. If all families would compensate for the abortions elected on the basis of genetic defects, the frequency of these defective genes may actually rise in the population. This may happen because this assists heterozygotes for genetic abnormalities to leave offspring that

Selective abortion continued

— although may not display the morbid trait — can again transmit the undesirable genes to future generations. If all carriers would refrain from reproduction, the frequency of the deleterious genes may sink to the level of new mutations. Selective abortion may involve also ethical, moral and political problems but these are beyond the scope of genetics. (See abortion spontaneous, amniocentesis, counseling genetic, selection and medical care)

SELECTIVE ADVANTAGE: in population genetics is expressed by the relative fitness of bearers of (two) genotypes. Generally the wild type has greater fitness (W_N) than a mutant type (W_M) and their relative fitness is (W_N)/(W_M). W (fitness) is the reproductive success. Usually, (W_N)/(W_M) = 1 - s, where s is the selection coefficient indicating the disadvantage of the mutant type. In case the fitness of a genotype exceeds 1 it has an advantage in survival. (See selection coefficient, fitness, beneficial mutation)

SELECTIVE FERTILIZATION: some sperms, because of their gene content, may be at a disadvantage in competition with other sperms for penetrating the egg, or where multiple eggs or megaspore cells are formed their success depends on their genetic constitution. Because of this, the genetic segregation may deviate from the standard Mendelian expectation. (See gamete competition, certation, megaspore competition, meiotic drive, sperm)

SELECTIVE MEDIUM: permits the propagation only of individuals or cells that carry a selectable marker, such as high or low temperature, antibiotic, drug resistance, etc. (See selectable marker)

SELECTIVE NEUTRALITY: assumes that random drift is responsible for the allelic frequencies in a particular population. (See random genetic drift, allelic frequencies)

SELECTIVE PEAK: is determined by the genetic homeostasis of a population, i.e. the gene frequencies are maintained at this optimum as long as catastrophic changes in the environment do not occur. (See homeostasis)

SELECTIVE SCREENING: see screening, selective medium, mutation detection

SELECTIVE VALUE: see fitness, selection coefficient

SELECTIVITY FACTOR (SF): is a human general transcription factor homologous to TFIIB of other eukaryotes. (See transcription factors)

SELECTOR GENES: are supposed to specify segmental differences during morphogenesis, such as e.g. anterior/posterior or dorsal/ventral. (See morphogenesis, morphogenesis in *Drosophila*)

SELENOCYSTEINE (SC, $C_6H_{12}N_2O_4Se_2$): a reactive, very toxic, oxygen-labile amino acid. Selenophosphate is the donor of selenium for the synthesis of selenocysteyl-tRNA which has the UCA anticodon and SC incorporation is directed by the UGA codon. Selenocysteine-containing proteins (glutathione peroxidase, 5'-deiodinases, formate and other dehydrogenases, glycine reductase) may have substantially higher or lower catalytic activity and may have theoretical and applied interest. (See code genetic)

SELEX (systematic evolution of ligands by exponential enrichment): is a method of ligand selection resulting in the isolation of aptamers. The technology may be applied to the search for new drugs and diagnostic compounds. (See aptamer)

SELF-ANTIGEN: stimulates autoimmune reaction by the T cells. (See T cells, autoantigen, MHC)

SELF-ASSEMBLY: a process of reconstituting a structure from components when there is enough information within the pieces to proceed without outside manipulation, e.g. the self-assembly of ribosomes from RNA and protein subunits.

SELF-COMPATIBILITY: self-fertilization can take place and the offspring is normal. (See also self-incompatibility, incompatibility alleles)

SELF-DESTRUCTIVE BEHAVIOR: see Lesh-Nyhan syndrome, Smith-Maganis syndrome

SELF-FERTILIZATION: can take place between the gametes in hermaphroditic individuals; autogamy is the strictest means of inbreeding. (See autogamy, inbreeding)

SELF-IMMUNITY: mediated by protein factors that bind to e.g. Moloney virus and thus prevent the integration into the sequences by another viral element. (See also cis-immunity, phage immunity)

SELF-INCOMPATIBILITY: failure of fusion of male and female gametes produced by the same individual, or lethality of the embryo formed by such fusion. The incompatibility for effective pollination is determined by secreted glycoproteins (SLG, on the surface of the stigma) and membrane-bound receptor protein kinases (SRK, attached to the stigma cells). These proteins are preferentially synthesized by the stigma cells and only low level of expression is found in the anthers of cruciferous plants. In several solanaceous plants (tomato, *Nicotiana alata, Petunia inflata*) a S (selfcompatibility) RNase may attack in the style the self pollen. In poppy plants, a receptor on the pollen recognizes a stylar protein. The binding between this ligand and receptor leads to release of calcium ions which in turn inhibit pollen tube growth. Self-incompatibility is usually determined by a large array of alleles in the populations and there are substantial differences among the species regarding the number of such alleles maintained depending on the effective population size. Heteromorphic self-incompatibility is based on mechanical barriers to self-fertilization. The stigma is located to a higher position in the flowers than the anthers, and the pollen usually does not reach the stigmatic surface. In such cases artificial pollination is successful. Sporophytic incompatibility means that the incompatibility factors operate not at the gametophyte (pollen - egg) level but the pistillar tissues prevent the growth of the pollen tube. (See incompatibility alleles, *S* alleles, histocompatibility, population effective size, gametogenesis, RNase, RNA I, ligand, gametophyte, pistil, pollen)

SELFING: is self-fertilization; it is symbolized by ⊗.

SELFISH DNA: an assumption for certain DNAs (introns, repetitive non-coding sequences) that they have no selective (adaptive, evolutionary) value for the carrier, therefore the presence of such sequences is of no advantage to the cells concerned, and are propagated only for selfish (parasitic) purposes. Some of the originally "selfish DNAs" (1979-80) turned out to have some functions, e.g. as maturases, and others represent transposable elements and continuously reshape the genome and thus are significant for mutation and evolution. The minisatellite DNAs and the trinucleotide repeats are implicated in an increasing number of hereditary diseases. The majority of The Y-chromosomal sequences of *Drosophila* do not seem to have any identifiable function yet male fertility may be impaired if they are deleted. (See also junk DNA, introns, ignorant DNA, copia, trinucleotide repeats, REP)

SELFISH REPLICON: small circular plasmids in eukaryotic nuclei (maize) without any apparent function beyond perpetuating themselves.

SELF-PRIMED SYNTHESIS: from single-strand DNA obtained through reverse transcription, one primer may be used at the 5' end to produce the second strand in a hairpin like structure. The synthesis by self priming is slow: → ⌒----------> 3'

SELF-REPLICATING DNA: is a replication mechanism for short double-stranded molecules that do not require assistance by proteins. Similar mechanisms may have operated during prebiotic evolution, and it can be reproduced in the laboratory. (See replication, self-assembly)

SELF-REPLICATING PEPTIDE: is an autocatalytic molecule capable of assembling amino acids into oligopeptides. The yeast transcription factor GCN4 leucine-zipper domain can promote its own synthesis of 15-17 amino acid residues. (See GCN4, leucine zipper)

SELFRESTRICTION: lymphocyte recognizes foreign antigen bound to self MHC molecule. (See lymphocyte, MHC)

SELVIN: a rarely used unit of absorbed radiation dose. (See Sievert)

SEM (scanning electronmicroscopy): see electronmicroscopy

SEM-5: homolog of *Grb2* in *Caenorhabditis* nematodes. (See *Grb2*)

SEMAPHORIN: is a family of membrane-associated, secreted protein factors, required for axonal pathfinding in neural development, and also regulate the development of the right ventricle and the right atrium of the heart as well as various cartilagineous and other tissues. Human semaphorins IV and V reside at the 3p21.3 chromosomal site; it is deleted in small cell lung carcinoma. (See axon, collapsin, netrin, small cell lung carcinoma)

SEMELPARITY: the organism reproduces only once during its lifetime, e.g. *Palingea longicauda* (Ephemeroptera).

SEMEN: the viscous fluid in the male ejaculate composed of the spermatozoa and secreted fluids of the prostate and other glands. The seminal fluid of *Drosophila* reduces the propensity of the females to mate with another male. Its higher quantity lowers the viability of the females. Thus the semen *per se* may have a role in fitness. (See testis, prostate)

SEMICONDUCTOR: materials of geranium, silicon and others are characterized by increased electric conductivity as temperature increases to room temperature. These materials are called semiconductors because their conductivity is much lower than that of metals. Also, in metals the increase in temperature lowers conductivity. In the semiconductor material the electronic motion is turned on through the crystal lattice structure. (The crystal lattice is a complex of atoms and molecules held together by electrons and atomic nuclei into an extremely large molecule-like structure). The energy states are in so-called bands. When all the sites in an energy band is completely occupied by electrons, there is no flowing electric current because none of the electrons can accept increased energy even if exposed to an electric field of ordinary magnitude. This is thus a non-conductor state. When the energy gap between two bands is small, the electrons can be thermally excited into a conduction band and the electrons under the influence of an external electric source can initiate an electric current. This state of the crystal is an *intrinsic semiconductor*. The carriers of the current are called positive holes. Transistors (electronic amplifying devices utilizing single-crystal semiconductivity) operate by the principles of conduction electrodes and mobile positive holes. Industrially used semiconductors are *extrinsic semiconductors*, meaning that into them small amounts of other material is introduced, and that results in enhanced conductive properties. These devices are essential components of electronic laboratory equipments and communication systems, computers and television sets.

SEMI-CONSERVATIVE REPLICATION: is the regular mode of DNA replication where one old strand serves as template for the synthesis of a complementary new strand and this then with an old strand becomes the daughter double helix. (See DNA replication, replication see figure at pulse-chase entry, Watson and Crick model)

SEMI-DOMINANT: the dominance is incomplete, and therefore such genes may be useful because the heterozygotes can be phenotypically recognized. (See incomplete dominance, codominance)

SEMIGAMY: occurs when the egg and sperm do not fuse rather they contribute separately to the formation of the embryo that may become thus a paternal-maternal chimera. (See also apomixis, androgenesis, parthenogenesis)

SEMI-LETHAL: genes reduce the viability of the individual, and may cause premature death. (See lethal equivalent, lethal factors, LD50, LDlo)

SEMINAL FLUID: see semen, sperm

SEMINAL ROOT: the root of the embryo in plant seeds.

SEMISTERILITY: indicates that in an individual some gametes or gametic combinations are not viable when others are normal. Semisterility is common after deletion and duplication in the offspring of inversion and translocation heterozygotes but it may be caused by self-incompatibility, incompatible non-allelic combinations, cytoplasmic factors, fungal or viral infections, adverse environmental conditions, etc. (See chromosomal aberrations, mtDNA)

SEMISYNTHETIC COMPOUNDS: are natural products but chemically modified.

SEMLIKI FOREST VIRUS: is a member of the alphavirus group. (See alphavirus)

SENDAI VIRUS: is a parainfluenza virus. In an ultraviolet light-inactivated form it has been used to promote fusion of cultured mammalian cells by its modifying effect on the lipids of the plasma membrane. (See cell genetics, cell fusion, cell membranes, polyethylene glycol, fusigenic liposome, alpha viruses)

senDNA: mitochondrial DNA (ca 2.5 kb), excised from the first intron of the *cox1* gene (cytochrome oxidase) and amplified in *Podospora anserina*. This and similar structures appear to be responsible for aging in vegetative cultures. (See senescence, aging, killer plasmids)

SENESCENCE: is the process of aging of organisms. At the cellular level it has a somewhat different meaning. Cell senescence indicates how many cell divisions are expected on the average from isolated mammalian cells. Generally, cell senescence is correlated with the age of the individual and organism from where it was explanted. Human fibroblast cells under normal conditions cease to proliferate after about ±50 divisions although individual lineages may vary. It has been suggested that the activity of the telomerase enzyme slows down and causes this phenomenon. Some rodent cell lines, after some somatic mutations, may not senesce. Also cancer cells and human fibroblast cells fused to cancer cells may divide indefinitely. Plant cells when provided with an appropriate regime of phytohormones may be maintained continuously and can even be regenerated into differentiated organisms. Senescence of plant cells is regulated by cytokinins, and the increase in cytokinin level inhibits the process of senescence. (See aging, monoclonal antibody, apoptosis, tissue culture, embryogenesis somatic, cell cycle, hybridoma, senDNA, killer plasmids, Hayflick's limit, telomerase, telomeres, plant hormones)

SENSE CODON: specifies an amino acid. (See genetic code)

SENSE STRAND: the DNA strand that carries the same nucleotide sequences as the mRNA, tRNA and rRNA (of course in the RNAs U stand in place of T). It does not carry an absolute meaning because of some cases both strands are transcribed although in context of a particular RNA it is correct. (See template strand, coding strand)

SENSE SUPPRESSION: see co-suppression

SENSILLUM: a cuticular sensory element; *sensilla campaniformia* are small circular structures along longitudinal veins of *Drosophila* wings. (See *Drosophila*)

SENSITIVITY: the percentage of correct identification of carcinogens on the basis of mutagenicity (or other rapid) assay system. (See accuracy, specificity of mutagen assays, predictability, bioassays for environmental mutagens)

SENSOR GENE: is supposed to be responsible for perceiving a signal. (See signal transduction)

SENSORYNEURAL: affecting the nerve mechanism of sensing.

SENSORY NEUROPATHY 1: is a dominant (human chromosome 9q221-q22.3) degenerative disorder of the sensory neurons and ulcerations and bone defects. (See neuropathy, pain-insensitivity)

SENTINEL PHENOTYPES: are used in human genetics to detect newly occurring mutations. These traits are supposed to be relatively easily detectable by direct appearance or clinical laboratory data can be obtained through routine examinations. Their frequencies are statistically evaluated to obtain epidemiological information regarding possible increase in mutagenicity/carcinogenicity in that environment. (See mutation in human populations, epidemiology)

Sep 1: a pleiotropic eukaryotic (yeast) strand exchange protein. (See recombination in eukaryotes, STPβ)

SEPAL: the whorl of (usually green) leaves below the petals in the flower. (See flower differentiation)

SEPHADEX: an ion exchanger or gel-filtration medium on cross-linked dextran matrix. (See dextran)

SEPHARDIC: Jews who moved to Spain after the Roman occupation of Israel and then in the middle ages to Western European countries. (See Ashkenazim, Jews and genetic diseases)

SEPHAROSE: anion exchanger of agarose matrix such as DEAE (diethylaminoethyl) sepharose.

SEPTAL: adjective for septum (dividing structure, a wall).

SEPTATE: separated by cross walls (septa).

SEPTIC SHOCK: is a bacterial lipopolysaccharide endotoxin-induced hypotension leading to inadequate blood supply to several organs and it is potentially fatal.

SEQUENASE ™: is a genetically engineered DNA polymerase. It combines the 85 kDa protein of phage T7 gene 5 and the 12 kDa *E. coli* thioredoxin protein (the latter keeps it associated with the template). The 3' → 5' exonuclease activity is suppressed. It synthesizes about 300 nucleotides per second, and it is used for DNA sequencing and oligolabeling. (See DNA sequencing, oligo-labeling probes)

SEQUENATOR: an automated equipment that breaks up a protein sequentially, starting at the NH_2 terminus, into amino acids, identifies them by chromatography and thus determines their sequence. (See amino acid sequencing)

SEQUENCE SPACE: the number of possible sequences of a particular length.

SEQUENCE-TAGGED CONNECTOR (STC): see genome project

SEQUENCE-TAGGED SITE: see sequenced tagged sites

SEQUENCED TAGGED SITES: single-copy DNA regions for which polymerase chain reaction (PCR) primer pairs are available and can be used for DNA mapping. (See PCR, primer)

SEQUENCING: see DNA sequencing, protein sequencing, RNA sequencing

SEQUESTER: lay away or separate (into a compartment).

SEQUIN: software tool for submitting nucleic acid sequence information to GenBank, EMBL, or DDBJ. It can be reached by <http://www.ncbi.nlm.nih.gov>. (See GenBank, EMBL, DBJ)

SER (smooth endoplasmic reticulum): an internal flat vesicle system in the cytoplasm involved in lipid synthesis. (See also RER)

SERCA: sarcoplasmic reticulum Ca^{2+} ATPase. (See Brody disease)

SERINE (Ser, S): an amino acid (β-oxy-α-amino-propionic acid, MW 105.09); soluble in water. RNA codons: UCU, UCC, UCA, UCG, AGU, AGC. Serine is derived from the glycolytic pathway:

3-PHOSPHOGLYCERATE -> 3-PHOSPHOHYDROXYPYRUVATE -> 3-PHOSPHOSERINE -> SERINE
| | |
phosphoglycerate dehydrogenase   phosphoserine aminotransferase   phosphoserine phosphatase

Serine dehydratase enzyme, with pyridoxalphosphate prosthetic group, degrades serine into pyruvate and NH_4^+. (See amino acids, amino acid metabolism, oxalosis)

SERINE KINASE: see MCF2 oncogene for serine phosphoprotein

SERINE PROTEASE: degrades proteins in extracellular matrix. (See matrix)

SERINE/THREONINE KINASE: phosphorylates serine and tyrosine residues in proteins. Their receptors are transmembrane proteins and they are attached to the cytosolic carboxyl end of the receptor. (See transforming growth factor β, PIM oncogene, PKS oncogene, activine bone morphogenetic protein, membrane proteins, protein kinases, receptor guanylyl cyclase, signal transduction)

SERINE/THREONINE PHOSPHOPROTEIN PHOSPHATASES: remove phosphate from serine and threonine residues of proteins. *Protein phosphatase-I* is inhibited by cAMP by promoting the phosphorylation of a *phosphatase inhibitor protein* through protein kinase A. *Protein phosphatase IIA* is the enzyme most widely involved in dephosphorylation of the products of serine/threonine kinases. *Protein phosphatase-IIB* (calcineurin) is most common in the brain where it is activated by Ca^{++}. *Protein phosphatase-IIC* has only minor role in the cells. The catalytic subunit of the first three are homologous but they contain also special regulatory subunits. *Phospholipase C* (PLC) may be coupled to G-proteins and upon its activation the level of Ca^{++} increases. This cation mediates numerous cellular reactions. (See serine/threonine kinases, phosphorylases, signal transduction, regulation of gene activity)

SERINE/THREONINE PROTEIN KINASE: see serine/threonine kinase

SERIPAUPERINES (PAU): a large group of proteins in eukaryotes, conspicuously low in serine and having amino-terminal signal sequence. Their function is still unknown. (See signal sequence)

SERODEME: a particular type of antigen produced by a clone. (See antigen)

SEROLOGY: deals with antibody levels and with the reactions of antigens. (See serum)

SEROTONIN (5-hydroxytryptamine): a tryptophan-derived neurotransmitter modulates sensory, motor and behavioral processes (including also feeding behavior) controlled by the nervous system. (See neurotransmitters, obesity, substance abuse, alcoholism)

SEROTYPE: distinguished from other cells by its special antigenic properties. (See antigen)

SERPINES (*se*rine-*p*rotease *in*hibitors): when mutated may be responsible for emphysema, thrombosis and angioedema. (See maspin, C1 inhibitor)

SERPROCIDIN: see antimicrobial peptides

SERTOLI CELLS: see Wolffian ducts

SERUM: a clear part of the blood from what the cells and the fibrinogen had been removed; the clear liquid that remains of the blood after clotting. The immune serum contains antibodies against specific infections. It differs from plasma which is the non-particulate portion of cells. (See plasma, antibody production, serology)

SERUM DEPENDENCE: animal cells may grow or differentiate only or preferentially in culture containing serum.

SERUM RESPONSE ELEMENT (SRE): is a DNA tract that assures transcriptional activation in response to growth factors in the serum. (See SRE, CArG box)

SESAME (*Sesamum indicum*): oil seed crop with about 37 related species; the cultivated form is $2n = 2x = 26$ but related species may have $x = 8$ and different levels of ploidy.

SESQUIDIPLOID: contains a diploid set of chromosomes derived from one parent and a higher-number set from the other. (See allopolyploid, allopolyploid segmental)

SESSILE: attached directly to a base without a stalk.

SET RECODING: see fuzzy inheritance

SETA: bristles.

SEVENLESS (*sev*): X-chromosomal gene (1-33.38) of *Drosophila* controlling the R7 rhabdomeres and thus altering photoreceptivity of the eye. The carboxy terminal of the protein product of the wild type allele shows homology to the tyrosine kinase receptor of *c-ras*, *v-src* and EGF. (See photoreceptor, rhodopsin, signal transduction, ommatidium, compound eye, daughter of sevenless)

SEVEN MEMBRANE PROTEINS (7tm): are integral parts of the plasma membrane and span the membrane by seven helices; they are important in signal receptor binding and in association with G-proteins. (See signal transduction, G-protein, transmembrane receptors)

SEVEN-PASS TRANSMEMBRANE PROTEINS: same as seven membrane protein.

SEVERE COMBINED IMMUNODEFICIENCY (SCID): is a less frequent (0.00001-0.00005) autosomal disease than the X-linked agammalobulinemia but it is generally lethal before age 2. The thymus is abnormally small and therefore there is a severe deficiency of the T and sometimes also the B lymphocytes. The afflicted infant cannot overcome infections. In some cases viral infection may severely damage the thymus, and this non-hereditary disease may closely mimic the symptoms of SCID. The DNA-dependent kinase (p350) encoded in human chromosome 8q11 is most likely responsible for SCID. (See agammaglobulinemia, hypogammaglobulinemia, immunodeficiency)

SEWALL WRIGHT EFFECT: same as drift genetic.

SEX: see also gender, sex cell, sex determination

SEX ALLOCATION: variation in sex ratio in favor of males or females due to non-chromosomal sex-determining mechanisms such as exists in social insects, caused by colony size, mating behavior and available resources. (See sex determination)

SEX BIVALENT: the X and Y chromosomes have homology only in the short common segment where they can pair and recombine. (See pseudoautosomal)

SEX CELL: is a gamete that can fuse with another sex cell of opposite mating type (sperm, egg) to form a zygote. (See zygote, mating type, gamete, isogamy)

SEX CHROMATIN: see Barr body

SEX CHROMOSOMAL ANOMALIES IN HUMANS: are of various types and they may occur at a frequency of 0.002 to 0.003 of the births. *Females* : X0, XXX, XXXX, XXXXX, X0/XX, X0/XXX, X0/XXX/XX, XX/XXX, X0/XYY, XXX/XXXX, XXX/XXXX/XX and

Sex chromosomal anomalies in humans continued

males : XX, XYY, XXY, XXYY, XXXY, XXXYY, XXY/XY, XYY/XYYY, X0/XXY/XY, XXYY/XY/XX, XXXY/XXXXY, and other even more complicated types have been reported. The most common mechanism by what these anomalies occur is nondisjunction in meiosis and mitosis. The more complex types mosaics (indicated by /) are the result of repeated nondisjunctional events. The X0 condition is called *Turner syndrome* , the XXX is *triplo-X,* XXY and other male conditions with multiple X and Y(s) are generally refered to as Klinefelter syndrome along with the XX males which have a Y-chromosome translocation to another chromosome. Similar sex-chromosomal anomalies have been identified in various other mammals. The X0 condition results in an abnormal female in humans and mice but it is a normal male in grasshopper or *Caenorhabditis*, and it is an abnormal male in *Drosophila*. (See trisomy, chromosomal sex determination, Turner syndrome, Klinefelter syndrome, triplo-X, XX males, gynandromorph, testicular feminization)

SEX CHROMOSOME: is unique in number and/or function to the sexes (such as X, Y or W, Z); see chromosomal sex determination. In the heterogametic sex, the X and Y chromosomes pair and may recombine in a relatively short terminal region although in the heterogametic sex in insects recombination is practically absent, except when transposable elements function.

SEX CIRCLE MODEL OF RECOMBINATION: basic tenets for fungi according to F.W. Stahl (1979): *1.* any marker can recombine either by reciprocal exchange or by gene conversion, *2.* close markers are more likely to recombine non-reciprocally, *3.* gene conversion observes the principle of parity, *4.* gene conversion is polar, *5.* in half of the cases gene conversion is accompanied by classical exchange of outside markers, *6.* reciprocal recombination is always accompanied by exchange of outside markers, *7.* conversion that does not involve outside exchange show no interference of flanking genes, *8.* gene conversion accompanied by outside marker exchange may involve also interference, *9.* conversion asci (5:3, 6:2) obey the principles listed under 1 to 8, *10.* all markers (except deletions, and a small fraction of conversion alleles) can segregate post-meiotically, *11.* the very rare aberrant 4:4 conversion asci may be the result of two events. (See recombination, gene conversion)

SEX COMB: special structures on the metatarsal region of the foreleg of *Drosophila* male. (See *Drosophila*)

SEX CONTROLED (sex influenced): the degree of expression of a gene is determined by the sex (e.g., baldness is more common in human males than females). (See Hirschprung's disease, Huntington's disease, imprinting)

SEX DETERMINATION: in dioecious animals and plants sex is usuallly determined by the presence of two X (female) and XY (male) chromosomal constitution, repectively. In other words the females are homogametic (i.e. the eggs all carry an X chromosome) and the males are heterogametic (i.e. they can produce sperm with either X or Y chromosome). In some species, e.g. birds, moths, the females are heterogametic and the males are homogametic. In the nematode *Caenorhabditis,* some grasshoppers, some fishes the females are XX and the males are of XO (single X) constitution. In *Drosophila* sex is determined by the proportion of the X-chromosome(s) and autosomes (A sets). Normally if the ratio is 1 X:2 sets of autosomes, the individual is male, if there are 2 Xs:2 sets of autosomes, the fly is female. All individuals with a sex ratio above 1 are also females and those with a ratio between 0.5 to 1 are intersexes. XO human and mouse individuals are females, however, and irrespective of the number of X chromosomes, as long as there is at least 1 Y chromosome, they appear male. In hermaphroditic plants the development of the gynoecia and androecea are determined by one or more gene loci. Actually, in *Drosophila* three major and some minor genes are known to control sex. *Sexlethal* (*Sxl*, 1-19.2) can mutate to recessive *loss-of-function* alleles that are deleterious to females but inconsequential to males. The dominant *gain-of-function* mutations do not affect appreciably the females but are deleterious to the males. The *Sxl* locus may produce ten different transcripts. Three transcripts (4.0, 3.1 and 1.7 kb) are expressed at the

Sex determination continued

blastoderm stage. Adult females have four transcripts (4.2, 3.3, 3.3, and 1.9 kb); the latter two are missing or reduced if the germline is defective. Adult males display three transcripts (4.4, 36 and 2.0 kb). The *Sxl* transcripts are alternatively spliced and functional in the female and are non-functional in the male. The *Sxl* gene product is apparently required for the maintenance of sexual determination and the processing of the downstream *tra* (*transformer*, 3-45) gene's product. The *Sxl* locus is regulated by other known genes: *da* (*daughterless*, 2-41.5) is a positive activator of *Sxl*, and it is suppressed by the gain-of-function mutations of the latter gene. The expression of *da*+ is necessary for the proper development of the gonads of the female in order to form viable eggs. In both sexes the product of *da*+ is required also for the development of the peripheral and central nervous system and the formation of the cells that determine the adult cuticle. Thus, the *da*+ gene has both maternal and embryonic influence. Females heterozygous for the *da^1* mutations produce sterile or intersex males and masculinize the exceptional daughters which are homozygous for *male* (*maleless*, 2-55.2); [*male* is lethal to single X males but has no affect on XX females]. The DA gene product is a helix-loop-helix protein with extensive homology to the human kE2 enhancer (human chromosome 19p13.3-p13.2) of the κ-chain family of immunoglobulins. Chromosomally female (XX) flies homozygous for the third-chromosome recessive *tra* mutations become sterile males. XXY *tra/tra* individuals are also sterile males but XY *tra/tra* males are normal males. A 0.9 kb transcript of the locus is female-specific and it is required in the female, and another 1.1 kb RNA is present in both sexes but no functions are known and is probably not essential. The splicing of the *tra* transcripts is controled by *Sxl* gene products. When the 0.9 transcript is expressed in XY flies, the body resembles that of females. Another *tra* locus (*transformer 2*, 2-70) regulates spermiogenesis and mating in normal males. Null mutations of *tra2*, when homozygous, transform XX females into sterile males. Actually, the *tra2* gene products seem to mediate the splicing of the *dsx* (*double sex*, 3-48.1) transcripts. Dominant mutations at the *dsx* locus when heterozygous with the wild type allele change XX individuals into sterile males but they have no effect on XY males. Null alleles of *dsx*, when homozygous, transform XX flies into intersexes. The recessive allele *dsx^{11}* transforms XY flies into intersexes, and the null alleles change both XX and XY flies into intersexes. Germline sexual differentiation is not affected by the normal allele of this gene but it is controlled by the X:autosome ratio. A 3.5 kb female-specific transcript is present in the larvae and adults. In the larvae a 3.8 and a 2.8 kb male-specific transcript is detectable and by adult stage, in addition, a 0.7 kb RNA also appears. The *ix* (*intersex*, 2-60.5) mutations when homozygous also change the XX flies into intersexes. Homozygous *ix* XY males appear normal morphologically but their courtship and mating behavior is altered. Thus sex determination in *Drosophila* appears to follow the cascade and *fru* regulates mating behavior and sexual orientation through the *tra* and *tra2* genes :

$$\textit{fru}$$
SEX CHROMOSOME : AUTOSOME RATIO → *Sxl* → *tra* → *tra2* → *dsx* → *ix*

In summary: the X:autosome ratio is the trigger mechanism for the alternate sex developmental pathways. In the males the *Sxl* and *tra* genes are expressed but their transcript is not spliced to functional forms. The critical male sex-determining function is attributed to locus *dsx* which in the wild type produces a protein blocking the genes required for female development. In the females, with a chromosomal constitution of 2X:2A sets, a functional *Sxl* product is made that mediates the female-specific splicing of its own transcripts. The *Sxl* protein mediates then the splicing of the *tra* transcripts, leading to the synthesis of a TRA protein which along with the TRA2 protein directs the female-specific splicing of the *dsx* transcript. The synthesized DSX protein blocks then all the genes with functions that would be conducive to male development. Sex determination in *Caenorhabditi*s is different from that in *Drosophila*, probably because the XX individuals are hermaphrodites and the nondisjunctional gametes lead to the develop-

Sex determination continued

ment of the rare XO males. The level of expression of the known sex-determination genes are:

XX	high	low	high	low	high	low	high	**FEMALE**

X:A→ ratio $xol-1$ → $sdc-1$ → $her-1$ → $tra-2$ → $fem-1$ → $tra-1$ →
　　　　　　$|sdc-1$　　　　　$|tra-3$　$|fem-2$
　　　　　　　　　　　　　　　　　$|fem-3$

XO	low	high	low	high	low	high	low	**MALE**

(Diagram after Kuwabara, P.E. and Kimble, J. 1992 Trends Genet. 8:164)

Sex determination in mammals is much more complex and the pathway is not entirely clear. For a number of years the H-Y antigen was thought to have a major role but this did not turn out to be correct. A major critical difference was found in an 11 amino acid segment of SMCX and SMCY proteins encoded within the X and Y homology region. Sex chromosomal anomalies in humans generally lead also to mental retardation. In several reptiles sex is determined by the temperature the eggs are exposed to during incubation in the sand they are laid in. The actual manifestation of sex may be deeply affected by endocrine hormones directly or indirectly through environmental pollutants. (See also mating type determination in yeast, freemartins, hormones in sex determination, gynandromorphs, intersex, hermaphrodite, testicular feminization, sex hormones, accessory sexual characters, sex phenotypic, mental retardation, chromosomal sex determination, mealy bug, dosage compensation, *Msl, Mle,* transsexual, H-Yantigen, SRY, mating type determination in yeast, *Schizosaccharomyces pombe,* F plasmid, sex plasmid, Hfr, sex-chromosomal anomalies in humans, social insects, *Sciara, schisotomiasis,* sex determination in plants, *Rumex*)

SEX DETERMINATION IN PLANTS: in dioecious plants sex determination is very similar to that in animals (see sex determination). In monoecious and hermaphroditic plants sex is controled without the presence of special chromosomes and a number of genes (nuclear, mitochondrial and plastidic) involved in morphogenesis, phytohormone synthesis and environmental responses determine the differentiation of the flowers and the oogenesis (female) and microsporogenesis (male) and so sexuality. Genes are known that are similar to those of sex reversal in animals and feminize or masculinize, respectively the monoecious or hermaphroditic flowers (e.g. *tassel seed, silkless* [in maize], *superman, gametophyte female* [in *Arabidopsis*], etc.). *Tasselseed 2* encodes a short-chain alcohol dehydrogenase that is involved in stage-specific floral organ abortion. Gibberellic acids, brassinosteroids, ethylene, chromosome-breaking agents, mutagens may also influence the expression of sexual development as well as temperature regimes and other environmental factors. (See gametophyte, gametophyte factors, vernalization, photoperiodism, self-incompatibility, flower differentiation, phytohormones)

SEX FACTOR: a transmissible plasmid in bacteria that carries the fertility factor(s) F. (See F^+, Hfr, F plasmid)

SEX HORMONES: have either estrogenic (female) or androgenic (male) influence. These are steroids of the ovaries and placenta (estradiol, progesterone) or of the testes (testosterone) or of the adrenal cortex, cortisol and aldosterol. Testosterone is required also in the females although in smaller amounts. Androgens control the reproductive organs but also affect hair growth (beard) and the early death of the hair follicles causing preferential male baldness. Androgens also promote bone and increased muscle growth. Some of the synthetic "anabolic hormones", without androgenic effects, are used (illegally) by athletes to boost performance. Testosterones are also precursors of estrogens. Estrogens are formed in the female-specific organs and their targets include the mammary glands, bones, and fat tissues. Estrogen synthesis is regulated by the follicle-stimulating hormone of the anterior pituitary. The pituitary luteinizing hormone mediates the release of the egg, and progesterone is required for the maintenance of pregnancy. The administration of exogenous estrogens and progestins inhibit ovulation and can be used as contraceptives. Other compounds act by prevention of the

Sex hormones continued

fusion of the sperm with the egg or implantation of the egg in the uterus, e.g. the drug RU486. Sperm production may be stopped by injection of progestin and androgen combinations or sperm maturation can be prevented by inhibiting epididymal functions or interfering with the release of enzymes required to break through the protective coat of the egg. Antiprogestins and antiestrogens or other inhibitors of steroid biosynthesis or by the use of non-peptide antigonadotropin-releasing hormone antagonists may serve as female contraceptives. The steroid hormone-controled sexual behavior is also mediated by the neuronal activity. (See animal hormones, hormone receptors, RU486, hyperlipoproteinemia, epididymis, transsexual)

SEX INFLUENCED: see sex-influenced, sex controled, imprinting

SEX LINKAGE: various genes in the sex chromosomes are inherited with the transmission of that chromosome. Sex linkage in females is generally partial because the two X-chromosomes may recombine. The recombination between the X and Y chromosomes is limited only to the homologous (pseudoautosomal) regions. In some insects (*Drosophila*, silkworm) recombination even between autosomes is usually absent in the heterogametic sex. (See recombination frequency, recombination mechanisms, crossing over, hemophilia, autosexing, genetic equilibrium, pseudoautosomal)

SEX MOSAIC: the sex-chromosomal constitution in the body cells may vary in a sectorial manner. Typical examples are the gynandromorphs in insects which have body sectors with both XX and XO constitution. Sex mosaicism occurs also in humans with variable numbers of X and Y chromosomal sectors. The mosaicism is generally the result of non-disjunction or chromosome elimination. (See sex determination, non-disjunction, gynandomorphs)

SEX, PHENOTYPIC: the phenotypic manifestation of the influence of the steroid sex hormones such as facial hair, increased phallic size in males, horns or special plumage in animals, and enlarged breast and mammary glands development in females. (See sex determination, animal hormones)

SEX PILUS: see pilus

SEX PLASMID: the bacterial F plasmid (See F element, F plasmid)

SEX PROPORTION: the proportion of male individuals in a population. (See sex ratio)

SEX RATIO: *primary sex ratio* is the number of male conceptuses relative to that of females. The *secondary sex ratio* indicates the number of females:males at birth. The *tertiary sex ratio* states the ratio among adult males and females. Since XX females are mated with XY males the proportion — just as in a testcross — should be 1:1. In the United States at birth the proportion is about 105-106 males : 100 females. In the West Indies the proportion is about 1:1 or slightly more females. In China and Korea, the newborns are about 115 males to 100 females. Generally the female : male ratio shifts in favor of females by progressing age. By about 21-22, in the USA, the female : male proportion becomes about 1:1, and because of mortality differential of the sexes, by age 65 there is about 145 females for 100 males. In dioecious plants the sex ratio may vary a great deal because of modifier genes and physiological factors (hormone supply). The problem whether infanticide alters the sex ratio by selection has been repeatedly considered by population geneticists since the 1930s. Infanticide generally biases the childhood ratio against females. This might have the consequence that the genes of families producing males would be favored, would have greater fitness and the secondary sex ratio would tend to be biased in favor of males. The problem is more complicated, however, in human societies because of the system of mating and the socio-economical conditions have a substantial influence. In some *Drosophila* stocks infection by filiform bacteria results in female offspring. Some of the gynandromorphs with very small XO sectors may also survive. Triploid intersexes live as well as females sex-transformed by *tra, ix, and dsx* genes (i.e., phenotypically males although XX). The sex ratio in the Seychelles warbler's eggs may vary according to the availability of food supply; in low-food territories they may have 77% male offspring whereas at high food supply the proportion of sons may be only 13%. This difference is not due to different viability of the eggs. (See also hermaphroditism, sex determination, sex-reversal, spirochete, infectious heredity, sex proportion, gynandromorph, age of parents and secondary sex ratio.)

SEX REALIZER: a substance that determines whether male or female gonads would develop.

SEXDUCTION: takes place when genes carried in the bacterial sex element (F' plasmid) recombine with the bacterial chromosome. (See also F' plasmid, Hfr, conjugational mapping, transduction)

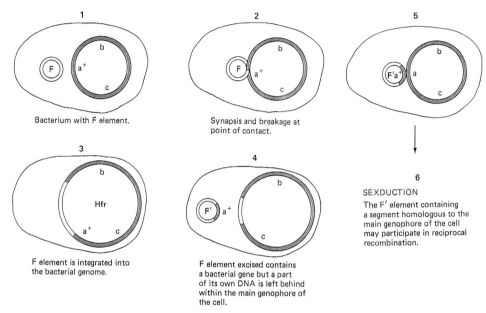

THE PROCESS LEADING TO SEXDUCTION

SEX-INFLUENCED: the degree of expression of the trait is different in male and female individuals, e.g., facial hairs in humans, color of plumage in birds, horns in deers, etc. (See hare lip, pyloric stenosis, Hirschprung's disease, lupus erythematosus, imprinting)

SEXING: distinguishing female from male forms of animals. This procedure may be difficult in young birds because the genitalia may appear ambiguous to those who do not have special expertise (see autosexing). Molecular sexing may make possible the identification of sex on the bases of DNA markers from any tissue sample. The avian males are homogametic (WW) and the females are heterogametic (ZW). On the Z chromosomes there are both the CHD-W and the CHD-NW genes whereas the W chromosome carry only the CHD-NW gene. The base sequences of these two genes are very similar, except in a short tract. When a restriction enzyme cuts within this segment of CHD-W, the females display three electrophoretic bands but the males show only one. (See sex determination, autosexing)

SEX-LIMITED: expression of a trait is limited to one sex (e.g., lactation to females, Wildervanck syndrome)

SEX-LINKED LETHAL MUTATIONS: in *Drosophila* served as the first laboratory test to quantitate mutation frequency and assess the mutagenic properties of physical and chemical agents. The old procedure was called the *Cl B* test (C stands for cross over exclusion brought about by the presence of usually three inversions, *l* is a recessive lethal gene and *B* indicates the dominant *Bar eye* mutation.) The principle of the techniques diagrammed at *ClB*. If any new recessive mutation (?) takes place in the X chromosome of a male, then in the F_2 only females may occur because the original recessive *l* gene present in the inverted *B l* chromosome will kill the hemizygous male progeny and if a new lethal mutation occurs that may kill

Sex-linked lethal mutations
(or much reduce the proportion) of the other type of males in F_2. Nowadays instead of the *ClB* method, generally the improved Basc chromosome is used (*B: Bar, a : apricot* eye color [*w* locus] and *sc: scute* inversions). Any recessive lethal mutation in the X-chromosome of the grandfather's sperm results in the death of one of the grandsons. Rarely, some exceptional females also found that are the result of unequal sister chromatid exchange in the inversion heterozygote mother. Somewhat similar manner autosomal recessive lethals can also be detected in *Cy L/Pm* stocks. (See bioassays in genetic toxicology, autosomal recessive lethal assay, *Basc*, *Clb* and diagrams there)

SEX-REVERSAL: the sex by karyotype does not always correspond to sex phenotype. By translocation, a testis determining factor (TDF) may move from the Y chromosome to an X chromsome, and thus 46 autosome XX males can occur. Also it has been claimed that an autosomal testis determining factor (TDFA) may be responsible for some of the intersexes and sex reversion cases. (See also intersexes, H-Y antigen, SRY, hermaphroditism, pseudohermaphroditism, testicular feminization, gonadal dysgenesis, sex determination, campomelic dysplasia, DSS)

SEXUAL DIFFERENTIATION: realization of sex-determination (gonads) and development of secondary sexual characters such as facial hair in men, differential plumage in birds, etc.

SEXUAL DIMORPHISM: the two sexes are morphologically distinguishable. In some species the differences appear only during later development or by the time of sexual maturity. The differences are not limited to morphology but also various functions may differ. In the males the language centers are localized in the left inferior frontal gyrus region of the brain, in females both left and right regions are active. (See human intelligence, autosexing, sex determination)

SEXUAL INCOMPATIBILITY: see incompatibility

SEXUAL ISOLATION: has significance in speciation by either preventing mating (isolation by life cycle, behavior or generative organs) between certain genotypes or because of gametic or zygotic death or inviability of their offspring. A very common form of sexual isolation is the inviability of the recombinants of chromosomal inversions. The cross-breeding at the incipient speciation may be prevented by pheromone gene(s) responsible for the production of differences in the female cuticular hydrocarbons in *Drosophila*. (See incompatibility, inversions, speciation, fertility, pheromone)

SEXUAL MATURITY: is the developmental stage by when reproductive ability is attained. It varies in different organisms (m: month, y: year): cat 6-12 m, cattle 6-12 m, chimpanzee 8-10 y, dog 9 m, elephant 8-16 y, horse 12 m, humans 12-14 y, mouse 1 m, rabbit 3-4 m, rat 2 m, sheep 6 m, swine 5-6 m. (See also gestation)

SEXUAL ORIENTATION: see homosexual

SEXUAL REPRODUCTION: production of offspring by mating of gametes of opposites sex or mating type.

SEXUAL SELECTION: competition among mates of the same sex or gametes or preferential choice of one or another type of mates. The general purpose of the sexual selection is to find mates with selective advantage for the offspring. The gynogen Sailfin Molly fish females reproduce clonally yet they rely on sperm of heterospecific males to initiate embryogenesis thus it appears that the males do not contribute the progeny. Yet the sexual forms of the females prefer those males which mate with the gynogens. Consequently the males exploited by the gynogens still benefit from the unusual sexual selection. Although sexual selection most frequently involves the selection of the males, in the sand lizard (*Lacerta agilis*) selection is achieved by the females. The females may copulate with several closely or distantly related males but it appears that the share of offspring sired by the remotely or unrelated males is higher in the same clutch. This was interpreted as an active selection mechanism by the female for the sperm of the more distantly related males. (See certation, megaspore selection, assortative mating, gynogenesis, heterospecific)

SEXUPAROUS: sexual production of offspring in species where parthenogenetic reproduction co-exists with sexual one. (See parthenogenesis)

SF: see hybrid dysgenesis I - R system. Also splicing factor protein (See introns)

SF1 (stimulatory factor): a yeast protein (33 kDa) reducing the binding requirement (at the concentration of 1 SF/20 nucleotides) of protein Sep 1 to DNA during recombination by two orders of magnitude. It probably has a role in DNA pairing. (See synapsis)

SF2/ASF: see SR protein, SR motif of binding proteins

SFF: a cell cycle regulating yeast protein. (See cell cycle)

SFM: serum-free medium.

Sfpi1: protooncogene; probably the same transcription factor as PU.1 and *Spi1*. (See *Spi1*, PU.1)

SGLT: sodium/glucose cotransporter; its defect results in Glucose-Galactose Malabsorption (GGM), a potentially fatal neonatal (human chromosome 22) recessive disorder unless the diet is sugar-free.

SH$_2$ (*src* homology domain): is an about 100 amino acid long binding site for tyrosine phosphoproteins. These phosphoproteins such as the SRC and ABL cellular oncogenes, phosphotyrosine phosphatases, GTPase activating protein, phospholipase C, and the Grb/Sem 5 adaptor protein have important role in signal transduction. (See SH3, SRC, WW, PTB, pleckstrin, signal transduction)

SH$_3$: (*src* homology domain): binding site for the proline-rich motif in an adaptor or mediator protein in the signal transduction pathway through RAS. By binding conformational and functional changes take place. The activity of the cellular SRC protein increases during normal and neoplastic mitoses. Protein p68 is, closely related to the GAP-associated p62, is bound to the SH3 domain of SRC. (See SH2, SRC, RAS, signal transduction, GAP)

SHADOWING: an electronmicroscopic preparatory procedure by what the surface of the specimen is coated with a vaporized metal, such as platinum. The shadowed objects display a three-dimensional effect in scanning but even in some cases in transmission electronmicroscopy. (See also scanning electronmicroscopy)

SHAH-WAARDENBURG SYNDROME: involves the endothelin-3 signaling pathway; it has the combined symptoms of the Hirschprung disease and the Waardenburg syndrome. (See Hirschprung disease, Waardenburg syndrome, endothelin, RET oncogene)

SHAKING (shak): alleles at several chromosomal locations in *Drosophila* cause shaking of the legs under anesthesia to a variable degree, depending on the locus and allele involved. Some mutants may display hyperactive behavior in a temperature-dependent manner. Some may be viable others are homozygous lethals. The *shakB* mutants may cause a defect in the synapse between the giant fiber neuron, postsynaptic interneuron and the dorsal longitudinal muscle and the nerves operating the tergotrochanter (back-neck) muscle (see illustrations at *Drosophila*). The shaking may be caused by a defect in a protein of a potassium ion channel.

SHASTA DAISY (*Chrysanthemum maximum*): ornamental plant; 2n ≈ 90.

SHC: is an adaptor protein involved in RAS-dependent MAP kinase activation after stimulation by insulin, epidermal (EGF), nerve (NGF), platelet derived growth factor (PDGF), interleukins (IL-2,-3,-5), erytropoietin and granulocyte/macrophage colony-stimulating growth factor (CSF), and lymphocyte antigen receptors. It binds to tyrosine-phosphorylated receptors and when phosphorylated at tyrosine it interacts with the SH2 domain of Grb2 which interacts with SOS in the RAS signal transduction pathaway. The phosphotyrosine-binding (PTB) domain also can recognize tyrosine-phosphorylated protein, and the latter is similar to the pleckstrin homology domain and most likely binds acidic phospholipids of the cell membrane. SHC is also an oncogene, involved in the development of pheochromocytoma neoplasias. (See signal transduction, phaeochromocytoma, adenomatosis endocrine multiple, neoplasia, insulin, EGF, NGF, PDGF, SOS, interleukins, CSF, pleckstrin domain)

SHEARING: cutting DNA to fragments by mechanical means, e.g. by rapid stirring or brusque pipetting.

SHEATS: any tube-like structure surrounding another; the part of a leaf that wraps the stem.

SHEEP, DOMESTICATED (*Ovis aries*): 2n = 54; some wild sheep has higher number of chromosomes.

SHEEP HYBRIDS: domesticated sheep (*Ovis aries,* 2n = 54) forms fertile hybrids with muflons but the goat x sheep hybrid embryos rarely develop normally. (See transplantation nuclear)

SHERMAN PARADOX: see mental retardation (caused by fragile X site)

SHIFT: an internal chromosomal segment, generated by two break points, translocated within the same or into another chromosome within a gap opened by a single break. A rare phenomenon. (See transposition, reciprocal interchange, translocation)

SHIFTING BALANCE THEORY of EVOLUTION: polymorphism in a population is determined by a dynamic interplay of the forces of pleiotropy, genotypic values, fitness and population structure. Alternative ideas would be the discredited neo-lamarckian internal drive or the well documented neutral mutation concepts. (See polymorphism, pleiotropy, genotype, fitness, neo-lamarckism, neutral mutation, population structure)

SHIGELLA: a group of gram-negative enterobacteria causing dysentery (intestinal inflammation and diarrhea) in humans and higher monkeys. (See primates)

SHIKIMIC ACID: is an intermediate in aromatic amino acid biosynthesis.

SHINE-DALGARNO SEQUENCE: nucleotide consensus (AGGAGG) in the non-translated 5' region of the prokaryotic mRNA (close to the translation initiation codon), complementary to the binding sites of the ribosomes. The terminus of the 16S ribosomal RNA is generally :
A^{Me2} A^{Me2} CCUGCGGUU GGAUGA<u>CCUCC</u>UUA-3'-0H. The eukaryotic mRNAs do not have this sequence because those messengers attach to the ribosomes by ribosomal scanning. Mitochondrial and ribosome mRNAs, generally but not always, have this or a modified Shine-Dalgarno. (See anti-Shine - Dalgarno, ribosome scanning)

SHOM (sequencing by hybridization to oligonucleotide microchips): is one of the automated (robotic) procedures developed for nucleotide sequence diagnostics. (See Yershov, G. *et al.* 1996 Proc. Natl. Acad. SciUSA 93:4913, see also DNA chips)

SHOOT: the plant part(s) above ground or a branch of a stem.

SHOPE PAPILLOMA: a viral disease of rabbits causing nodules under the tongue. The double-stranded DNA virus of about 8 kbp has 49 mole percent G+C content. (See bovine papilloma)

SHORT-DAY PLANTS: require generally less than 12-15 hr daily illumination for flowering; at longer light periods they usually remain vegetative. (See photoperiodism, long-day plants)

SHORT DISPERSED REPEATS: see SDR

SHORT PATCH REPAIR: is an excision repair removing then replacing about 20 nucleotides. (See DNA repair, excision repair)

SHORT SYNDROME: is an autosomal recessive phenotype characterized by the initials of the SHORT acronym: short **S**tature, **H**yperextensibility of joints and hernia, **O**cular depression, **R**ieger anomaly (partial absence of teeth, anal stenosis [narrow anus], hypertelorism [increased distance between organs or parts], mental and bone deficiencies, and **T**eething delay). (See stature in humans, dwarfism)

SHOTGUN CLONING: the DNA of an entire genome is cloned without aiming at particular sequences. From the cloned array of DNA fragments (library) the sequences of interest may be identified by appropriate genetic probes. (See cloning, DNA probe, DNA library)

SHOTGUN SEQUENCING: random samples of cloned DNA, e.g. the segments of a cosmid are sequenced at random. (See DNA sequencing)

SHP-1 (synonymous with SH-PTP1, PTP1C, HCP): tyrosine phosphatase; it contains the SRC homology domain SH2. Upon activation of T cells it binds to the kinase ZAP-70 resulting in increased phosphatase activity but in a decrease in ZAP-70 kinase activity. It is a negative regulator of the T cell antigen receptor. It is activated by radiation stress. (See T cell, ZAP-70)

SHPRINTZEN-GOLDBERG SYNDROME: see Marfanoid syndromes

SHREW: *Blarina brevicauda,* 2n = 50; *Cryptotis parva,* 2n = 52; *Neomys fodiens,* 2n = 52; *Notiosorex crawfordi,* 2n = 68; *Sorex caecutiens,* 2n = 42; *Suncus murinus,* 2n = 40.

SHUTTLE VECTOR: a "promiscuous" plasmid that can carry genes to more than one organism and can propagate the genes in the different cells, e.g. in *Agrobacterium*, *E. coli* and plant cells. (See vectors, cloning vectors, transformation genetic, promiscuous DNA)

SI: the unit of absorbed dose (1Joule/kg) of electromagnetic radiation and it is generally expressed in Gray (Gy) or Sievert (Sv = 1 rem) units. Earlier rad (=0.01 Gy) was used. (See r, rem, Gray, Sievert)

SIALIC ACID: an acidic sugar such as N-acetylneuraminate or N-glycolylneuroaminate; they are present in the gangliosides. Polysialic acid is involved in cell and tissue type differentiation, learning, memory and tumor biology. The synthesis is mediated by polysialyl transferase under the regulation of neural cell adhesion molecules. (See gangliosides, gangliosidoses, CAM, neuraminidase deficiency, fusogenic liposome)

SIALIDOSES: see neuraminidase deficiency

SIALIDASE DEFICIENCY: see neuraminidase deficiency

SIALURIA: is caused by an autosomal recessive gene defective in feedback-sensitivity of uridine diphosphate N-acetylglucosamine 2-epimerase enzyme by cytidine monophoshphate-neuro-aminic acid. The afflicted has defects in bone (dysostosis), psychomotor [movement and psychic activity]) development and infantile death may incur. (See neuroaminidase deficiency)

SIAMESE CAT: displays darker color at the extremities because of a temperature-sensitive gene slows down blood circulation and more pigment develops at specific locations of the body. (See pigmentation of animals, Himalayan rabbit, temperature-sensitive mutation, figure ↓).

SIB: same as a sibling.

SIBLING: natural children of the same parents. (See λ_s, risk, genetic risk, genotypic risk ratio)

SIBLING SPECIES: are morphologically very similar and frequently share habitat but they are reproductively isolated. (See species, speciation, fertility)

SIBPAL: a linkage analysis computer program for sib-pairs. (See sibling)

SIBSHIP: natural brothers and sisters. (See also kindred)   SIAMESE CAT →

SIC1: cell cycle S-phase cyclin-dependent kinase (CDK) inhibitor. (See CDC34)

SICKLE-CELL ANEMIA: a human hereditary disease caused by homozygosity of a recessive mutation(s) or deletions in the hemoglobin beta chain gene. Heterozygosity causes the sickle cell trait. Under low oxygen supply the red blood cells lose their plump appearance and partially collapse into sickle shape because the abnormal hemoglobin molecules aggregate.

HUMAN RED BLOOD CELLS FROM A SICKLE CELL ANEMIA PATIENT. LEFT: IN THE PRESENCE OF NORMAL OXYGEN SUPPLY, RIGHT: AT LOW OXYGEN SUPPLY. NORMAL RED BLOOD CELLS LOOK LIKE BICONCAVE DISCS VERY SIMILAR TO THAT AT LEFT HERE. THE SICKLING CELLS ARE UNABLE TO HOLD OXYGEN AND THAT CONDITION IS RESPONSIBLE FOR THE DISEASE. (Photographs are the courtesy of Cerami, A. & Manning, J.M. Proc. Natl. Acad Sci. USA 68:1180). CARRIERS OR SUFFERERS CAN BE UNAMBIGOUSLY IDENTIFIED BY ELECTROPHORETIC SEPARATION OF THE BLOOD PROTEINS ↓

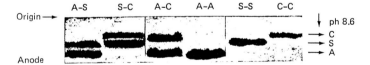

HOMOZYGOTES FOR THE NORMAL BLOOD PROTEIN (A-A), SICKLE CELL ANEMIA (S-S), HEMOGLOBIN C (C-C). IN THE HETEROZYGOTES (A-S, S-C AND A-C) — BECAUSE OF CODOMINANCE — BOTH TYPES OF PARENTAL PROTEINS ARE DETECTED. (From Edington, G.M. & Lehman, H. Trans. R. Soc. Trop. Med. Hyg. 48:332)

The disease is not absolutely fatal but crises may occur when the blood vessels are clogged. Complications may arise by poor blood circulation. In the classical form of the sickle cell disease, in the hemoglobin S a glutamine residue of the normal beta chain (hemoglobin A) was replaced by a valine residue. In the hemoglobin C at the same position a lysine replacement occurs, and this condition causes less severe clinical symptoms. Other forms, hemoglobin D and E, are less common. This disease provided the first molecular evidence that mutation leads

Sickle-cell anemia continued

to amino acid replacement. Sickle cell anemia affects more than 2 million persons world wide. About 10% the population of African descent in the United States of America is carrier (heterozygous) for this mutation and about 1/400 is afflicted with homozygosity at birth. In populations of European (except southern Europeans) descent, the frequency of this mutant gene is about 1/20 of that among Mediterreneans and Africans. The high frequency of the genetic condition in areas of the world with high infestation by malaria is correlated. The individuals without sickle cell anemia gene have about 2-3 times higher chance to be infected by *Plasmodium falciparum*. The mutation is selectively advantageous by protecting heterozygotes against malaria. The globin gene cluster has been located to human chromosome 11p15. The order is 5'—γG—γA—δ—β—3'. Correction of the defective allele is possible by transduction of the defective cells with retroviral or adenoviral vectors that can deliver the normal gene to the hematopoietic stem cells but these viral vectors may have deleterious consequences for the body. Another method is to introduce into the lymphoblastoid cells chimeric DNA-RNA oligonucleotides with a correction for the β^s allele mutation, brought about through gene conversion in the target cells. This procedure may eventually be clinically applicable. (See malaria, hemoglobin, thalassemia, sickle cell trait, genetic screening, viral vectors, gene therapy, gene conversion)

SICKLE CELL TRAIT: is due to heterozygosity of the recessive mutation in the gene controling the β-chain of hemoglobin. Normally, these heterozygotes do not suffer from this condition but under low oxygen supply, e.g. at high elevations, adverse consequences may arise. (See also sickle cell anemia, hemoglobin)

SIDE-ARM BRIDGE: an attachment of chromatids resembling chiasma but actually it is only an anomaly in mitosis or meiosis, usually arising when the division was disturbed by chemicals or radiation. (See bridge)

SIDEROCYTE ANEMIA (sideroblastic anemia): an anemia with erythrocytes containing non-hemoglobin iron; it may be controled by either autosomal recessive or dominant or X-linked genes. (See anemia)

SIEVE TUBE: a plant food transporting tube-shape, tapered, long cells; they may be connected by sieve plates that hold them together.

SIEVERT (Sv): the name for Sv (unit of absorbed dose equivalent [J/kg]) = 100 rem = 1 Sv. (See rem, Gray, rad, R)

SIGMA FACTOR: subunit of DNA-dependent bacterial RNA polymerase, required for the initiation of transcription and promoter selection. (See open promoter complex, RNA polymerase)

SIGMA REPLICATION: see rolling circle

SIGMA VIRUS of *Drosophila*: see CO_2 sensitivity

SIGN MUTATIONS: are frameshift mutations because an equal number base addition(s) (+) and deletion(s) (-) at the gene locus may restore the reading frame although may not always restore normal function. (See frameshift)

SIGNAL END: see immunoglobulins

SIGNAL HYPOTHESIS: postulated that the signal peptide of the nascent polypeptide chain guides it to the endoplasmic reticulum (and to other) membranes where the signal peptidase cleaves it off and subsequently the peptide chain is completed within the lumen of the membranes. This hypothesis has been validated for transport through bacterial cell membranes, mitochondria, plastids, peroxisomes, etc. (See signal peptides, transit peptide, protein targeting)

SIGNAL PEPTIDASE: see signal hypothesis

SIGNAL PEPTIDES (signal sequence): are 15 to 35 amino acid long sequences generally at the NH_2 terminus of the nascent polypeptide chains of proteins that have a destination for an intraorganellar or transmembrane location. They are made in eukaryotes and prokaryotes but not all the secreted proteins possess one. At the beginning of the sequence generally there are one or more positively charged amino acids, followed by a tract of hydrophobic amino acid

Signal peptides continued

residues that occupy about three fourth of the length of the chain. This hydrophobic region may be required to pass into the lipoprotein membrane. The amino acid sequences among the various signal peptides are not conserved, indicating that the secondary structure is critical for recognition by the signal peptide recognition particles and for the function within the membrane. The eukaryotic signal sequences are recognized by the prokaryotic transport systems and the prokaryotic signal peptides can function in eukaryotes. After the passage of the nascent peptide has started the signal peptides are split off by peptidases on the carboxyl end of (generally) glycine, alanine and serine. Consequently the majority of the proteins in the membrane or transported through the membranes, have the nearest downstream neighbor of one of these three amino acids at the amino end. One major characteristic of the signal peptide is the co-transcriptional targeting whereas the transit peptides are targeted postranslationally. The signal sequences may show polymorphism that might result in incorrect targeting. (See signal sequence recognition particle, transit peptide, leader peptide, endoplasmic reticulum)

SIGNAL RECOGNITION PARTICLE: see SRP

SIGNAL SEQUENCE: the amino terminal of some proteins signals the cellular destination of these proteins, such as the signal peptides. (See signal peptide)

SIGNAL SEQUENCE RECOGNITION PARTICLE (SRP): is a complex of six proteins and an RNA (7SL RNA) that recognizes the *SRP receptor protein* on the surface of the endoplasmic reticulum and the *signal peptides* of the nascent proteins translated on the ribosomes associated with the endoplasmic reticulum (rough endoplasmic reticulum) and facilitate the transport of these polypeptides into the lumen of the Golgi apparatus and lysosomes (co-translational transport). The SRP binds to the signal peptide after about a 70 amino acid chain is completed at the beginning of translation. The polypeptide chain elongation is somewhat relaxed until the SRP attaches to the SRP receptor. Then the SRP comes off and the amino acid chain and elongation resumes its normal rate and the entry of the chain through the membrane proceeds. In the meantime a peptidase inside the endoplasmic reticulum cuts off the 15 - 30 amino acid long signal peptide sequences. (See lysosomes, Golgi, signal peptides, protein synthesis, signal sequence particle, protein targeting, translocon, translocase, TRAM, protein conducting channel)

SIGNAL TRANSDUCTION: requires four major category of elements: signals, receptors, adaptors and effectors. **SIGNALS** — The extracellular signals interact with cell membrane receptors and make subsequently contacts with intracellular target molecules to stimulate a cascade of events leading to the formation of effector molecules that turn on and off genes and control cellular differentiation in structure and time by regulating transcription. The *signals* are proteins, peptides, nucleotides, steroids, retinoids, fatty acids, hormones, gases (ethylene, nitric oxide, carbon monoxide), inorganic compounds, light, etc. The target cells accept the signals by special sensors, *receptors*. The receptors are generally specific proteins with high binding specificities and positioned on the cell surface or within the plasma membranes and thus readily accept the signal *ligands*. The receptors may be also inside the cells, and the ligands may have to pass the cell membranes to reach them. As a consequence of the binding, a cascade of events is triggered and eventually the instruction reaches the cell nucleus and the relevant genes. The *paracrine signals* are restricted in movement to the proper, a generally nearby, target. The nerve cells are communicating by *synaptic signals*. The *endocrine hormone* signals may affect also distant targets in the entire body. The neurotransmitters are activated through long circuits of the nervous system by electric impulses emitted by neurons in response to the environment. The travel of the electric signals through the neurons is very fast, may pass through meters per second. The neurotransmitters have only a few nanometers to pass and the process takes only a few milliseconds. The local concentration of the endocrine hormones is extremely low. In contrast, the neurotransmitters, may be quite concentrated at very small target area. The neurotransmitters also may very rapidly, be removed either by reabsorption or by enzymatic hydrolysis. Generally, the hydrophobic signals persist longer in the

Signal transduction continued

cells than the hydrophilic ones. The membrane anchored growth factors and cell adhesion molecules are signaled through the *juxtacrine* mediators.

RECEPTORS. The target cells respond by *receptor proteins*. These receptors are endowed with specificities regarding the signal they respond to. Also, the same signal may have different receptors in differently specialized cells. In addition, the interpretation and use of the signal within similar cells may vary. The signals may act also in a combinatorial manner: several signals together may be involved in the cellular decisions and influence the length and quality of the effect of a signal received. The various receptors, despite substantial chemical differences of the signals (e.g., cortisol, estrogen, progesteron, thyroid hormones, retinoic acid, vitamin D) may bind ligands that control through the signal transduction path closely related and interchangeable upstream DNA consensus elements, involved in the regulation of transcription of different genes. Steroid hormone receptors, after bound to cognate hormones, may activate the transcription of the so called *primary response genes*. These proteins then repress the further transcription of the primary response genes and turn on the transcription of *secondary response genes* (see regulation of gene activity).

Receptors can be (i) *ion-channel* or (ii) *G-protein-* or (iii) *enzyme-linked* types. Group (i), also called transmitter-regulated ion channels, are involved in transmitting neuronal signals (see ion channels). Group (ii) receptors are transmembrane proteins of the so called seven-membrane type (see there) associated with guanosine phosphate-binding G proteins (see G-proteins). When GTP is bound to the G-protein, a cascade of enzymes or other proteins may be activated or an ion channel may become more permeable. G-protein linked receptors represent a large family of proteins with more than 100 of them have been already identified in a variety of eukaryotes. The receptors, generally monomeric and evolutionarily related proteins, respond to a variety of signals, such as hormones, mitogens, light, pheromones, etc. Activation of group (iii) receptors may directly or indirectly lead to the activation of enzymes. These three different types of signal transductions may not be entirely distinct because the function of the ion channels may interact with the pathways mediated through G-proteins and various kinases. Some of the receptors are *protein tyrosine phosphatases* (e.g., CD45 protein) and *serine/threonine phosphatases* residing within the membrane or in the cytosolic domain of transmembrane proteins or in the cytosol.

PATHWAYS may show a great variation depending on the signals and receptors involved. Ca^{++} is a general regulator. It may enter nerve cell terminals through voltage-gated Ca^{++} channels in the cell membranes and stimulates the secretion of neurotransmitters (see ion channels, voltage-gated ion channel). Alternatively, Ca^{++} may have a more general role by binding to G-protein linked receptors in the metabolism of inositol phospholipids, PIP (phosphoinositol phosphate) and PIP_2 (phosphoinositol bisphosphate). The specific trimeric G-protein, Gq is involved in the activation of *phospholipase C-β* which is specific for phosphoinositides, and splits PIP_2 into inositol triphosphates and diaglycerol. Hydrolysis of PIP_2 yields IP_3 (inositol 1,4,5-triphosphate). The latter sets free calcium from the endoplasmic reticulum through IP_3-gated channels, ryanodine receptors (see ryanodine). Upon further phosphorylation of IP_3 may give rise to IP_4 (inositol 1,3,4,5-tetrakisphosphate) that slowly yet steadily replenishes cytosolic calcium. The calcium level in the cytosol rises and subsides in very short bursts according to how the phosphatidyl-inositols regulate it (calcium oscillations). These oscillations still may assure increased secretion of second messenger and spares the cell from a constant level of the toxic Ca^{++} in the cytosol.

Besides pumping out Ca^{++} shielded in the endoplasmic reticulum, diaglycerol, and eicosanoids (arachidonic acid) may be produced. Diaglycerol and the latter lipid derivatives may activate the Ca^{++}-dependent enzyme, serine/threonine kinases which have key role in activating proteins which mediate signal transduction (see protein kinases). *Protein kinase C* may activate the cytosolic *MAPK* (mitogen activated protein kinase) by phosphorylation and may phosphorylate a cytoplasmic inhibitor complex such as IκB + NF-κB. Thus MAPK may phosphorylate

Signal transduction continued

DNA-binding proteins such as SRF (serum response factor) and Elk (member of the ETS oncoprotein family) sitting already at the upstream regulatory regions of a gene(s), such as the serum-response element (SRE). The phosphorylation then initiates transcription (see regulation of gene activity). The released protein factor NF-κB may migrate to the nucleus and by binding to its cognate DNA site, sets into motion transcription (if other factors are also present). The response of the genes to the traducing signals depends on the number of regulatory proteins responding to the signal. Monomolecular reactions display relatively slow response to the concentration of the signal molecules whereas if the number of effectors are multiple, the reaction to them may follow 3rd or multiple order kinetics. Similarly, prompt response is expected if the signal activates one reaction (e.g. phosphorylation) and the same time deactivates an inhibitor or suppressor (e.g. by phosphatase action).

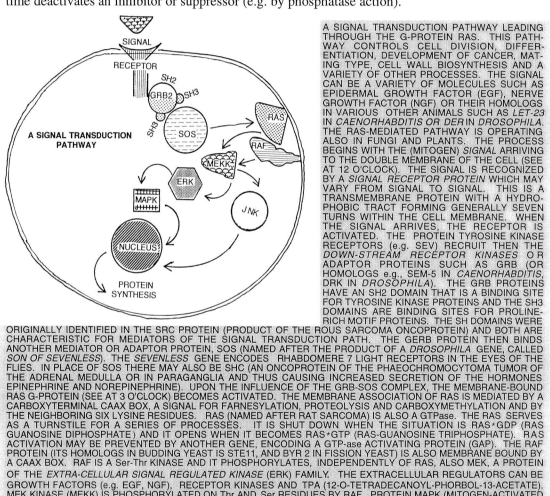

A SIGNAL TRANSDUCTION PATHWAY LEADING THROUGH THE G-PROTEIN RAS. THIS PATHWAY CONTROLS CELL DIVISION, DIFFERENTIATION, DEVELOPMENT OF CANCER, MATING TYPE, CELL WALL BIOSYNTHESIS AND A VARIETY OF OTHER PROCESSES. THE SIGNAL CAN BE A VARIETY OF MOLECULES SUCH AS EPIDERMAL GROWTH FACTOR (EGF), NERVE GROWTH FACTOR (NGF) OR THEIR HOMOLOGS IN VARIOUS OTHER ANIMALS SUCH AS *LET-23* IN *CAENORHABDITIS* OR *DER* IN *DROSOPHILA*. THE RAS-MEDIATED PATHWAY IS OPERATING ALSO IN FUNGI AND PLANTS. THE PROCESS BEGINS WITH THE (MITOGEN) *SIGNAL* ARRIVING TO THE DOUBLE MEMBRANE OF THE CELL (SEE AT 12 O'CLOCK). THE SIGNAL IS RECOGNIZED BY A *SIGNAL RECEPTOR PROTEIN* WHICH MAY VARY FROM SIGNAL TO SIGNAL. THIS IS A TRANSMEMBRANE PROTEIN WITH A HYDROPHOBIC TRACT FORMING GENERALLY SEVEN TURNS WITHIN THE CELL MEMBRANE. WHEN THE SIGNAL ARRIVES, THE RECEPTOR IS ACTIVATED. THE PROTEIN TYROSINE KINASE RECEPTORS (e.g. SEV) RECRUIT THEN THE *DOWN-STREAM RECEPTOR KINASES* OR ADAPTOR PROTEINS SUCH AS GRB (OR HOMOLOGS e.g., SEM-5 IN *CAENORHABDITIS*, DRK IN *DROSOPHILA*). THE GRB PROTEINS HAVE AN SH2 DOMAIN THAT IS A BINDING SITE FOR TYROSINE KINASE PROTEINS AND THE SH3 DOMAINS ARE BINDING SITES FOR PROLINE-RICH MOTIF PROTEINS. THE SH DOMAINS WERE ORIGINALLY IDENTIFIED IN THE SRC PROTEIN (PRODUCT OF THE ROUS SARCOMA ONCOPROTEIN) AND BOTH ARE CHARACTERISTIC FOR MEDIATORS OF THE SIGNAL TRANSDUCTION PATH. THE GERB PROTEIN THEN BINDS ANOTHER MEDIATOR OR ADAPTOR PROTEIN, SOS (NAMED AFTER THE PRODUCT OF A *DROSOPHILA* GENE, CALLED *SON OF SEVENLESS*). THE *SEVENLESS* GENE ENCODES RHABDOMERE 7 LIGHT RECEPTORS IN THE EYES OF THE FLIES. IN PLACE OF SOS THERE MAY ALSO BE SHC (AN ONCOPROTEIN OF THE PHAEOCHROMOCYTOMA TUMOR OF THE ADRENAL MEDULLA OR IN PARAGANGLIA AND THUS CAUSING INCREASED SECRETION OF THE HORMONES EPINEPHRINE AND NOREPINEPHRINE). UPON THE INFLUENCE OF THE GRB-SOS COMPLEX, THE MEMBRANE-BOUND RAS G-PROTEIN (SEE AT 3 O'CLOCK) BECOMES ACTIVATED. THE MEMBRANE ASSOCIATION OF RAS IS MEDIATED BY A CARBOXYTERMINAL CAAX BOX, A SIGNAL FOR FARNESYLATION, PROTEOLYSIS AND CARBOXYMETHYLATION AND BY THE NEIGHBORING SIX LYSINE RESIDUES. RAS (NAMED AFTER RAT SARCOMA) IS ALSO A GTPase. THE RAS SERVES AS A TURNSTILE FOR A SERIES OF PROCESSES. IT IS SHUT DOWN WHEN THE SITUATION IS RAS*GDP (RAS GUANOSINE DIPHOSPHATE) AND IT OPENS WHEN IT BECOMES RAS*GTP (RAS-GUANOSINE TRIPHOSPHATE). RAS ACTIVATION MAY BE PREVENTED BY ANOTHER GENE, ENCODING A GTP-ase ACTIVATING PROTEIN (GAP). THE RAF PROTEIN (ITS HOMOLOGS IN BUDDING YEAST IS STE11, AND BYR 2 IN FISSION YEAST) IS ALSO MEMBRANE BOUND BY A CAAX BOX. RAF IS A Ser-Thr KINASE AND IT PHOSPHORYLATES, INDEPENDENTLY OF RAS, ALSO MEK, A PROTEIN OF THE *EXTRA-CELLULAR SIGNAL REGULATED KINASE* (ERK) FAMILY. THE EXTRACELLULAR REGULATORS CAN BE GROWTH FACTORS (e.g. EGF, NGF), RECEPTOR KINASES AND TPA (12-O-TETRADECANOYL-PHORBOL-13-ACETATE). MEK KINASE (MEKK) IS PHOSPHORYLATED ON Thr AND Ser RESIDUES BY RAF. PROTEIN MAPK (MITOGEN-ACTIVATED PROTEIN KINASE) MAY BE CAPABLE OF AUTOPHOSPHORYLATION AND IT MAY BE PHOSPHORYLATED BY THE ERK FAMILY OF KINASES AT Tyr AND Thr RESIDUES. MAPK HOMOLOGS ARE KSS, HOG-1, FUS3, SLT-2, SPK-1, SAPK (A STRESS ACTIVATED KINASE), FRS (FOS REGULATING KINASE), etc. MAPK MAY THEN ACTIVATE THE FOS AND JUN ONCOGENE COMPLEX THAT IS ALSO KNOWN AS THE AP-1 HETERODIMERIC TRANSCRIPTION FACTOR OF MITOGEN-INDUCIBLE GENES. THE RAS TO MAPK ROUTE MAY BRANCH DOWNSTREAM THROUGH SEVERAL EFFECTOR PROTEINS AND MAY CONTROL THE TRANSCRIPTION OF SEVERAL DIFFERENT GENES. THE SPECIFICITY OF ACTIVATION DEPENDS ON COMBINATORIAL ARRANGMENTS OF THE EFFECTORS. (The size, shape or shading of the symbols were not intended to represent their structure.)

Signal transduction continued

In case of autophosphorylation or other reactions involving enzymes, which may bind their own products, with the increase of the number of product molecules, through a positive feedback, the activity of the enzyme may be increased in the course of time.

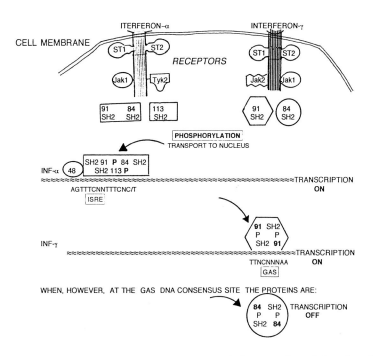

THE JAK-STAT SIGNAL TRANSDUCTION PATHWAY, BASED ON INTERFERON (IFN), MEDIATED RECEPTORS. *JAK* IS A FAMILY OF THE JANUS TYROSINE PROTEIN KINASE PROTEINS, AND *STAT* STANDS FOR 'SIGNAL TRANSDUCER AND ACTIVATOR OF TRANSCRIPTION'. WHEN THE LIGAND BINDS TO THE SIGNAL TRANSDUCING RECEPTORS (ST), THE RECEPTOR-ATTACHED *JAK KINASES* MODIFY THE *STAT PROTEINS*. AFTER DIMERIZATION (using SH2 domains), THEY ARE TRANSPORTED DIRECTLY TO THE *ISRE* (interferon-α reponse element) or to THE *GAS* (γ-interferon activation site) IN THE CHROMOSOMAL DNA. THESE TWO ELEMENTS VARY YET CONSENSUS SEQUENCES EXIST (as shown on the diagram). THE 84, 91, 113 AND 48 ARE PROTEINS (in kDa), BUT ADDITIONAL ONES MAY ALSO BE INVOLVED, DEPENDING ON THE NATURE OF THE SIGNALS RECEIVED. (After Darnell, J.E., Jr. *et al.* 1994 Science 264:14125 and Heim, M.H. *et al.* 1995 Science 267:1347)

The enzyme-linked signal receptors do not need G-proteins. The transmembrane receptor binds the ligand at the cell surface and the cytosolic domain functions as an enzyme or it associates with an enzyme and the transfer of the signal to the cell nucleus is more direct. An example is the cytokine-activated cell membrane receptor Jak *tyrosine kinases,* which when dimerized can combine with cytoplasmic STATs (signal transducers and activators of transcription) and chromosomal responsive elements. In response of interferon or other cytokine signals the Jaks phosphorylate tyrosine of the SH2 domains in a variety of STAT proteins. This may be followed by dimerization, and transfer of these proteins to the nucleus where they may turn on transcription of particular genes (see figure above). It has been estimated that about 1% of the human genes code for protein kinases. These kinases may be mainly either serine/threonine or tyrosine kinases. Some proteins may phosphorylate all three of these amino acids. The reliance on protein tyrosine kinases for signal transduction is rather general. Epidermal growth factor (EGF), nerve growth factor (NGF), fibroblast growth factor (FGF), hepatocyte growth factor (HGF), insulin, insulin-like growth factor (IGF), vascular endothelial growth factor (VEGF), platelet-derived growth factor (PDGF), macrophage colony stimulating factor (M-CSF), etc. function with the assistance of transmembrane *receptor tyrosine kinases.* Upon the arrival of the ligand (signal), the receptor is dimerized either by cross-linking two receptors by the dimeric ligand or by inducing autophosphorylation and linkage of two cytosolic domains of the receptors. The different phosphorylated sites may bind different cytoplasmic proteins. The insulin itself is a tetramer ($\alpha\alpha\beta\beta$) and thus does not need dimerization. After autophosphorylation it phosphorylates an insulin receptor (IRS-1) at tyrosine sites and that may bind to other proteins which also may become phosphorylated and may form different complexes and thus generating a variety of transcription factors. Alternatively, the *tyrosine kinase-associated receptors,* themselves are not tyrosine kinases but associate with proteins of this capability. Some of the enzyme-linked receptors are *serine/threonine kinases*

Signal transduction continued

with specificities for these two amino acids. The phosphorylated tyrosine residues are binding sites for proteins with SH2 domains. The *receptor tyrosine phosphatases* may activate or inhibit the signal pathways by the removal of phosphate from tyrosine residues. The receptor guanylate (guanylyl) cyclases operate in the cytosolic domain of the receptors and function by serine/threonine phosphorylation in association with trimeric G-proteins. The discoveries about signal transduction within the last decade changed the view about cellular functions and added a new dimensions to biology by integrating reversed and classical genetics. The signal transduction mechanisms have a large variety of means to regulate diverse functions of metabolism, differentiation and development.

PLANT SIGNALS: are somewhat different from the signals in animals. The plant hormones, similarly to animal hormones, are signaling molecules but most of them — except the brassinosteroids — are very different molecules. In plants, light and temperature signals (photoperiodism, vernalization, phytochrome) are very important for growth and differentiation. Salicylic acid is a signaling molecule for defense genes, etc. (See G-proteins, interferons, histidine kinase, regulation of gene activity, hormones, morphogenesis, AKAP79, T cells, integrin, adaptor proteins, photomorphogenesis, vernalization, photoperiodism, phytohormones, salicylic acid, host-pathogen relationship, signaling to translation, and the other terms mentioned)

SIGNAL TRANSFER PARTICLE: the PDGF associated with phospholipase C-γ, phosphatidyl inositol 3-kinase and the RAF protooncogene product regulates signaling. (See platelet derived growth factor, phosphatidyl inositol kinase, RAF)

SIGNALING: phosphorylation of enzymes mediated by second messengers leads to their activation through outfolding of the pseudosubstrate domains of the enzymes and thus opening the active sites for the true substrate. (See signal transduction)

SIGNALING MOLECULE: alerts cells to the behavior of other cells and environmental factors.

SIGNALING TO TRANSLATION: is directed toward the 5' untranslated region (UTR) and involve the translation of ribosomal proteins and elongation factors (eEF1A, eEF2). The target of the signals appear to be the 5' terminal oligopyrimidine sequences (5' TOP) and the UTR polypyrimidine tracts. Secondary structure formation with long UTRs may also be re-gulatory. Growth factors may phosphorylate the eukaryotic initiation factor eIF4E -binding protein, 4E-BP-1, and cause its dissociation from eIF4E. In order that translation would pro-ceede, the initiation factors eIF4A, -4B, -4G, -4E attach to the methylguanin cap and the secondary structure of the RNA is untwisted by the helicase action of eIF4A. Phosphorylation by the $p70^{S6k}$ kinase activates the S6 protein of the 40S ribosomal subunit, a process subject to enhancement by mitogens. This process is not a general requirement for translation, indicating that it affects only special genes with 5' TOP and polypyrimidine tracts in their UTRs. Cycloheximide and puromycin also are involved in the phosphorylation of S6. Both of these phosphorylations are inhibited by rapamycin. (See F506, $P70^{S6k}$, secondary structure, cycloheximide, puromycin, cap, regulation of gene activity, regulation of enzyme activity, protein synthesis, signal transduction)

SIGNIFICANCE LEVEL: indicates the probability of error by rejecting a null hypothesis that is valid or accepting one that is not correct. By convention 5% (*, significant), 1% (**, highly significant) and 0.1% (***, very highly significant) levels are used most commonly. These are not sacrosanct limits. In field experiments with crops the 5% level may be a satisfactory measure for comparative yields but even the 0.1% may not be acceptable for pharmaceutical tests because the chance of harming 1/1,000 persons is unacceptable. In general experimental practice, levels above 5% and below 0.1% are not considered meaningful although they may have relevance for pharmacology. (See goodness of fit, t-test, probability)

SILENCER: is a negative regulatory element reducing transcription of the region involving the target genes. Their action bears similarity to the heterochromatic chromosomal regions which reduce transcription of genes transposed to their vicinity. Silencing is mediated by a combin-

Silencer continued

ation of a protein and the site where silencing takes place. For example the *MATa* and *MATα* genes of yeast encode regulatory proteins that permit the expression of the *a* and *α* mating types, respectively, of the haploid cells and the non-mating phenotype of the sporulation-deficient *a/α* diploid cells. When these genes are at the *HMLa* and *HMRα* sites they are silenced until they are transposed to the *MAT* locus. The mating type switch is catalyzed by a cut mediated through HO endonuclease when the mating type alleles are at the MAT locus but not at the *HMLa* and *HMRα* locations. This indicates that the silencing is under the dual control of the silencer protein and a specific site. Inactivation of the *SIR2*, *SIR3* and *SIR4* derepresses *HML* and *HMR* and these genes affect the telomeric position effect of other genes as well. Mutations at the amino terminus of the *HISTONE 4* gene has also similar effect. Overexpression of *SIR2* causes hypoacetylation of this histone while *SIR3* mutations may alter the conformation of this histone bound to *HMR*. Loci *HML* and *HMR* both are flanked by *HML-E* and *HML-I* silencer elements. These silencers are similar to the autonomously replicating elements of yeast that are involved in DNA synthesis and apparently also in silencing. *HML-E* is capable of repression only in the presence of *HML-I* and the repressed domains appear to extend to 0.8 kb proximal to the centromere from *HML-E* and 0.4 kb distal from *HML-I*. (Based on Loo & Rine 1994 Science 264:1768). Silencer elements are present also in animal and plant systems. In plants, when multiple copies of a gene are introduced into the genome by transformation, all or most a gene copies are inactivated (trans-inactivation). The mechanism of this phenomenon is unclear. It has been suggested that when the level of a particular RNA is increased a degradative process is initiated. This have been attributed to a defense mechanism since the majority of plant viruses are RNA viruses. Some of the silencing appears, however, posttranscriptional. In fungi silencing have been attributed to premeiotic methylation when multiple copies are present in cis position. This view is supported by the long-standing knowledge that the repetitive sequences of the heterochromatin are not expressed. Position effect has also been known as a type of silencing. The reversible type of paramutation can also be considered as a trans-inactivation mechanism. (See targeting genes, enhancer, position effect, mating type determination in yeast, *Schizosaccharomyces pombe*, neuron-restrictive silencer factor, CREB, co-suppression, paramutation, heterochromatin, histone deacetylase)

SILENE: see *Melandrium*

SILENT MUTATION: base-pair substitution in DNA that does not involve amino acid replacement in protein and entail no change in function. (See mutation)

SILENT SITES: where mutations in the DNA base sequence have no consequence for function. (See synonymous codons)

SILICONIZATION: glassware is treated in a vacuum by dichlorodimethylsilane in order to prevent sticking DNA molecules to the vessel, and resulting in loss of recovery. (See DNA extraction)

SILIQUE: typical fruit of cruciferous plants; two carpels, dehiscing at the base at maturity, enclose the placentae which sit in one row on each of the opposite sides of the replum.

SILK FIBROIN: is a protein rich in glycine and alanine residues arranged largely in β sheets (β keratin). It is synthesized within the silk gland to protect the pupa. The pupa is called also chrysalis or coccoon. A fibroin gene is transcribed into about 10,000 long-life molecules of mRNA within a few days and they are translated several times into about a billion protein molecules and each gland manufactures about 10^{15} fibroin molecules (300 μg) in four days. Actually, the gland is a single cell but it contains polytenic chromosomes and thus the fibroin locus is amplified about a million fold ($10^9 \times 10^6 = 10^{15}$ fibroins). (See polytenic chromosomes, silk worm)

SILK WORM (*Bombyx mori*, 2n = 56): is one of the genetically best studied insects. Its genome contains a special type of transposable element, R2Bm that is present also in some other insects. R2Bm has no long terminal repeats. It is inserted in the 28S rRNA genes only and encodes an integrase and a reverse transcriptase function within one protein molecule. The R2

Silk worm continued

protein nicks one of the DNA strands and uses it also as a primer to transcribe its RNA genome which is then integrated as DNA-RNA heteroduplex. Subsequently a host polymerase synthesizes the second DNA strand. (See transposon, complete linkage, autosexing, tetraploidy, polyhedrosis virus, silk fibroin)

SIMIAN CREASE: see Down's syndrome for illustration. It can be rarely observed (1-4%) on the normal infants but it is characteristic for human trisomy 21, De Lange, Aarskoog and other syndromes.

SIMIAN SARCOMA VIRUS (SSAV): a gibbon/ape leukemia retrovirus with a homologous element in human chromosome 18q21. The long terminal repeat (535 bp) appears to contain transcriptional control and signal sequences. The human chronic lymphatic type leukemia seems to be associated with a break point of chromosome 18. (See leukemia)

SIMIAN VIRUS 40: is a eukaryotic virus of a molecular weight of 3.5×10^6 with double-stranded, supercoiled DNA genetic material of 5243 bp. The DNA is organized into a nucleosomal structure that does not have H1 histone. The DNA around the nucleosome cores is 187 ± 11 bp and these are separated by 42 ± 39 bp linkers. The viral particles are skewed icosahedral capsids and have 72 protein units. In primates, generally the virus follows a lytic lifestyle and the virions multiply in the cytoplasm, i.e. primates are *permissive hosts* for replication. Occasionally, in humans the viral DNA integrates into the chromosomes. Such an event may lead to cancerous transformation. Rodent cells are *non-permissive* hosts for viral replication and the viral DNA integrates into the chromosomes leading to cancerous tumor formation. The virus codes for early (t and T antigens) and late (VP1, 2, 3) viral proteins. The viral replication and transcription are bidirectional. In the non-permissive host only the early genes are expressed that are needed for replication of the genetic material before integration but there is no need for the coat proteins. The integration can take place at different sites, therefore it uses a mechanism of illegitimate recombination. The few integrated copies may be rearranged and may cause continued chromosomal rearrangement also in the host. The infectious cycle spans about 70 h. The joint replication and transcriptional origin (*ori*) area extends to about 300 bp and includes a rather sophisticated control system. The replication of the SV40 DNA begins at the 27 bp palindrome of the *ori* site that is adjacent to a region consisting of 17 A-T base pairs. Next to it, on the side of the late genes, there are three other units of 22, 21 and 21 GC-rich repeats that also promote replication although not absolutely essential to the process:

```
←─────────── LATE TRANSCRIPTS                              ───────────→ EARLY TRANSCRIPTS
72 bp         72 bp        21 bp     21 bp     22 bp    17 A-T box   ori
                                ◊         ◊         ◊
_____   _____   _____
      enhancer region              basal promoter region
                              ΔΔΔΔΔΔΔΔΔΔΔΔΔΔΔΔΔΔΔΔΔΔΔΔΔΔΔΔ ΔΔΔΔΔΔΔΔΔ ΔΔΔΔΔΔ
                                        T3                    T2      T1  binding sites
                                                                          for Tag
```

The 72 bp domain contains **GT**, **TC**, **SPH** and **P** (polyoma) elements in a sequence illustrated below by the staggering (although they are next to each other in linear order that cannot be represented here because of shortage of space):

GT II: 5'-GCTGTGGAATGT-3'
  GT I: 5' - GGTGTGGAAATG-3'   -
    TC I and **TC II**: 5'-TCCCCAG-3'
      SPH- II: 5'-AAGT<u>ATGCA</u>-3'
        SPH-I: 5'-<u>AAG</u>CATGCA-3'
          P: 5'-TTAGTCA-3'

> EACH Tag BINDING MOTIF HAS A MINIMUM OF TWO 5'-GAGGC-3' OF THIS PENTANUCLEOTIDE CONSENSUS

The SPH elements have an overlapping *octamer* that is present in other eukaryotic genes too and several other sequence motifs are similar in other promoters. The 72 elements include the

Simian virus 40 continued

47 bp B and the shorter 29 bp A domains that are parts of the essential enhancer region. A-T box is essentially a TATA box: 5'-TATTTAT-3'. For the start of replication the large T has to bind to the Tag binding sites Δ. At the initiation when low amounts of Tag are available, binding begins at the T1 site located at the pre-mRNA region (right of *ori*), and as more Tag will become available Tag binds to T2 and becomes an ATP-dependent helicase and with the cooperation of cellular proteins (DNA primase, polymerase, etc.) DNA replication proceeds. *Transcription*: For the function of *ori* in transcription the 17 bp A-T sequence is needed although this TATA box does not affect the rate of transcription. The 21◊21◊22◊ sequences promote transcription and the 2 hexamers 5'-GGGCGG-3' within these elements are essential for transcription. Their orientation and inversion does not interfere with transcription. The 72 bp element of SV40 is a capable enhancer also for mammalian, amphibian, plant and fission yeast genes. The natural host of SV40 is the rhesus monkey (*Macaca mulatta*), in the laboratory the kidney cell cultures of the African green monkey (*Cercopithecus aethiops*) are used primarily for its propagation. Single cells, each, may produce 100,000 viral genomes after lysis. The general assumption is that SV40 does not cause human cancer yet in many tumors its presence was detected by some laboratories and not by others. (See also SV40 vectors, COS cell)

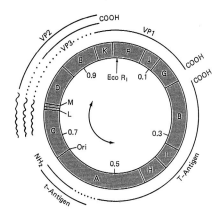

SimIBD: a computer procedure to assess affected-relative pair calculations.
SIMILARITY INDEX: see character index
SIMLINK: is a simulation based computer program for estimating linkage information.
SIMPLE PROTEIN: upon hydrolysis only amino acids are produced.
SIMPLE SEQUENCE LENGTH POLYMORPHISM (SSLP): variations in microsatellite sequences can be used for DNA mapping. PCR primers are designed for the unique flanking sequences and their length can thus be determined in the PCR products. (See microsatellite, PCR)
SIMPLESIOMORPHY: primitive features retained during evolution. It is not very useful to trace evolutionary development. (See synapomorphy, homology)
SIMPLEX: is a polyploid having only a single dominant allele at a particular gene locus; the other alleles are recessive at the locus. (See also duplex, triplex, quadruplex)
SIMPLEXVIRUS: member of the herpes family of viruses infecting humans and other mammals. (See herpes)
SIMPSON-GOLABI-BEHMEL SYNDROME (Simpson Dysmorphia): is a human Xq26 (500 kilobase stretch) syndrome, encoding the GPC3 gene responsible for the synthesis of glypican, a proteoglycan associated with the insulin-like growth factor (IGF2). Individuals afflicted by this condition are generally very tall, usually have facial anomalies, heart and kidney defects, cryptorchidism, hypospadia, hernias, bone anomalies and show susceptibility to cancer. Many of the symptoms involving overgrowth are shared with the Beckwith-Wiedeman syndrome. (See Beckwith-Wiedeman syndrome, IGF)
SIMULATION: representation of a biological system by a mathematical model generated frequently by a computer program. (See modeling, Monte Carlo method)
SINDBIS VIRUS: is a single-stranded RNA virus.
sine: the ratio of the side opposite (a) to an acute angle (A) of a right triangle and the hypotenuse (c): a/c, sine of angle A. (See arcsine, angular transformation)
SINE: short (≈ 0.5 kbp) interspersed repetitive DNA sequences that may occur over 100,000 times in the mammalian genomes. The B1 SINE of mouse of 130 to 150 bp in length and constitutes

SINE continued
nearly 1% of the genome; it is homologous to the human Alu sequences. The B2 SINE (~190 bp) has no human homolog. SINE type elements occur in all eukaryotes, including birds, fungi, insects and higher plants. These SINE sequences can also be used for fingerprinting. (See retroposons, transposable elements, LINE, Alu family, DNA fingerprinting)

SINGLE BURST EXPERIMENT: virus-infected bacterial population diluted and distributed into vessels in such a way that each vessel would contain a single infected bacterial cell.

SINGLE GENE TRAIT: it is controled by one gene locus, and shows monogenic inheritance.

SINGLE COPY PLASMIDS: see plasmids

SINGLE COPY SEQUENCE: DNA sequences containing non-redundant, genic portions.

SINGLE CROSS: see double cross

SINGLE STRAND ASSIMILATION: a single strand displaces another homologous strand and then takes its place during a recombinational event. (See recombination molecular models)

SINGLE STRAND BINDING PROTEIN: binds to both separated single stands of DNA and thus stabilize the open region to facilitate replication, repair and recombination. (See recombination molecular mechanisms, binding proteins)

SINGLE STRAND CONFORMATION POLYMORPHISM (SSCP): when small deletions or even single base substitutions take place in one of the DNA strands of a gene locus, this alteration may be detectable by the electrophoretic mobility of the DNA in denaturing polyacrylamide gels. The two strands, the normal and the affected may differ. If the individual is heterozygous for the amplified segment of the locus concerned, the electrophoretic analysis may indicate three or more band differences. In some cases even the homozygotes may show multiple bands. With this method nearly all of the alterations are detected in fragments of 200-300 bp. (See gel electrophoresis, polymerase chain reaction, gene isolation, DGGE, mutation detection, dideoxy fingerprinting)

SINGLETON: singly occurring whole-body mutations; the spontaneous frequency in mice for seven standard loci is 6.6×10^{-6} per locus. (See mutation rate)

SINGLETS: genes that occur only once in the genome.

SINK: storage of metabolites from where they can be mobilized on need.

SIPHONOGAMY: the immotile microgametes of higher plants are delivered to the archegonia through the elongating pollen tube. (See pollen tube, embryosac, zoidogamy)

SIPPLE SYNDROME: see phaeochromocytoma

SIR: see silencer

SIRE: male mammal; the term used primarily in animal breeding and applied animal genetics. (See also dam)

SIRENOMELIA: developmental malformation of fused legs and usually lack of feet.

SIS: simian sarcoma virus oncogene is in human chromosome 22q12.3-q13.1 and mouse chromosome 15. The SIS protein has high homology to the β chain of the platelet derived growth factor (PDGF), KIT oncogene, FOS oncogene and the colony stimulating factor. (See oncogenes, PDGF, colony stimulating factor)

SISTER CHROMATID COHESION: the juxtaposition of the sister chromatids until the end of metaphase in mitosis and until the end of metaphase II in meiosis. The physical basis of the cohesion is provided by the inner centromere proteins (INCENP) and the centromere-linking proteins (CLiP). The sister chromatids are closely juxtapositioned until anaphase indicating the presence of intersister connector structures. Dominant and recessive mutations have been identified in plants, animals and yeast that are defective in chromatid cohesion. (See mitosis, meiosis, synapsis, asynapsis, desynapsis, sister chromatids, sister chromatid exchange, chiasma)

SISTER CHROMATID EXCHANGE (SCE): sister chromatid exchanges are detectable in eukaryotic cells provided with 5-bromodeoxyuridine for (generally) one cycle of DNA replication. Subsequently at metaphase the chromosomes are stained with either the fluorescent compound Hoechst 33258 (harlequin staining) or according to a special Giemsa procedure. If sis-

Sister chromatid exchange continued

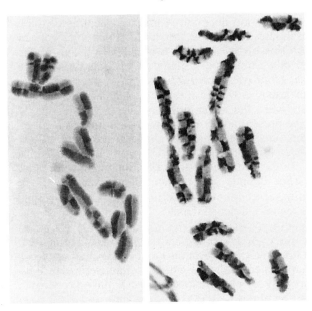

ter chromatids were reciprocally exchanged, sharp bands appear in mirror image-like fashion. The frequency of sister chromatid exchange is boosted by about a third by potential carcinogens and mutagens. This method has been successfully used in various animal and plant cells for identifying genotoxic agents. The data must be evaluated with care in comparison with the concurrent control because BrdU itself may break chromosomes under UV-B light. (See harlequin staining, Giemsa staining, bioassays in genetic toxicology, ring chromosomes, BrdU, ultraviolet light, genotoxic, sister chromatids, chiasma, crossing over)

SISTER CHROMATID EXCHANGE. LEFT: UNTREATED CONTROL. RIGHT: EXPOSED TO THE ALKYLATING COMPOUND THIOTEPA DURING DNA SYNTHESIS. (Courtesy of Professor B.A. Kihlman)

SISTER CHROMATIDS: are attached to the same side of the same centromere but they seem to be coiled in opposite directions. Their separation in mitosis requires the activation of a proteolytic enzyme encoded by the *Cut2* gene in *Schizosaccharomyces pombe*. (See chromatids)

SISTER-STRAND EXCHANGE: same as sister chromatid exchange

SIT: is a family of phosphatases

SITE-DIRECTED MUTAGENESIS: see directed mutation, localized mutagenesis, targeting

SITE-SPECIFIC CLEAVAGE: of nucleic acids is accomplished by restriction endonucleases, some special RNases and oligonucleotide-phenanthroline conjugates may cut both strands of the DNA in the presence of Cu^{2+} and a reducing agent. $EDTA-Fe^{2+}$ may do the same if tethered to triplex molecules albeit with low efficiency. In the presence of light, ellipticine attached to homopyrimidines may cleave a double helix within a triplex. (See restriction endonucleases, triplex, tethering)

SITE-SPECIFIC MUTATION: occur at particular nucleotides in the DNA and RNA, respectively. (See base substitution, localized mutagenesis, gene replacement, site-specific recombination, PCR-based mutagenesis, cassette mutagenesis. homolog-scanning mutagenesis, alanine-scanning mutagenesis, TAB mutagenesis, targeting vector)

SITE-SPECIFIC RECOMBINASES: are resolvases that attach at the two-base staggered cut sites. The enzyme is then covalently linked to the 5' ends and the PO_4 of the DNA is covalently linked to the OH group of the recombinase. Subsequently the broken DNA strand releases the deoxyribose hydroxyl group. The PO_4 is joined to another deoxyribose OH group and the DNA backbone is reconstituted. The members of the integrase group of enzymes attach at sites 6-8 base apart. The first breakage results in a Holliday juncture, that may lead to branch-migration and after a second strand exchange and rotation isomerization (see Holliday model steps H-J) the strands may be resolved either with an outside marker exchange (classical recombination) or in gene conversion (the constellation of the outside markers retained). It is conceivable that the broken ends are either reconstituted without any change or deletions may also take place or the position of the broken ends are inverted by 180 degrees resulting in what classical cytology called inversion. Resolvase and integrase reactions can be very specific for the sites and the reaction is secured by the assistance of additional proteins that bring into contact only the appropriate DNA stretches. These two enzymes act only on supercoiled DNA.

Site-specific recombinases continued

The integrase family of recombinase enzymes are more liberal in choice yet affected by various conditions. The Mu phage or the HIV integration does not require covalent association between the DNA and a protein. The phosphodiester bond of the donor DNA is hydrolized first to generate an OH group and between this group and a phosphodiester group of the receiving DNA the joining takes place and thus the strand is integrated. (See site-specific recombination, Holliday juncture, resolvase, *Cre/loxP, FLP/FRT*, integrase, phosphodiester linkage, transesterification, homing endonucleases)

SITE-SPECIFIC RECOMBINATION: occurs when the recombination is limited to a specific few nucleotide sequences. Homology may be present at the exchange region in both recombining molecules like at the integration - excision site of temperate phage or the specificity is limited to only one of the partners like at the 25 bp termini of the T-DNA or the direct and indirect repeats of the transposable elements. In the latter cases the recombinational target sites may have no or only minimal similarity. (See lambda phage, gene replacement, *Cre/Lox, FLP/FRT*, switching, site-specific recombinase, knockout, targeting genes, chromosomal rearrangement, ligand-activated site-specific recombination, T-DNA)

SITUS INVERSUS VISCERUM: a malformation of mammals, including humans, where the internal organs such as the heart is shifted to the right side of the chest (thorax). It is frequently accompanied by chronic dilation of the lung passages (bronchi) and inflammation of the sinus and the latter disorder is also called Kartagener syndrome which is characterized also by immotility of sperm and cilia. The anomaly may by either autosomal or X-linked recessive. Its incidence in the general population may be about 1/10,000. In the mouse the genes *iv* (chromosome 12) and the *inv* (chromosome 4) disturb left-right axis formation and cause 50 and 100% manifestation of situs inversus, respectively.

SIV (Simian immunodeficiency virus): is a relative of HIV. (See acquired immunodeficiency)

SJÖGREN-LARSSON SYNDROME: see ichthyosis

SJÖGREN (sicca = dry) **SYNDROME**: autosomal recessive autoimmune disease leading to the destruction of the salivary and lacrimal glands by the production of autoantibody against the SS-A (Ro RNA) and SS-B (La Sn RNA) particles. The autoantigens have been identified and purified; Ro autoantigen appears to be encoded in human chromosome 19pter-p13.2. The La autoantigen may be involved with RNA polymerase III. (See autoimmune disease)

SK ONCOGENE: probably regulates tumor progression; it was assigned to human chromosome 1q22-q24.

SKEWED DISTRIBUTION: the data are not symmetrical around the mean ; either one or the other extreme flank is predominant. (See normal distribution, kurtosis)

SKEWNESS: asymmetry in the distribution frequency of the data. (See also kurtosis, normal distribution, moments)

SKIN COLOR: see pigmentation of animals

SKIN DISEASES: see acne, epidermolysis, keratosis, ichthyosis, psoriasis, blisters, porphyria, pemphigus, acrodermatitis, familial hypercholesterolemia, Fabry disease, pseudoxanthoma elasticum, nevus, vitiligo, ectodermal dysplasia, focal dermal hypoplasia, scleroderma, lupus erythematosus, dermatitis, eczema, Gardner syndrome, cutis laxa, pigmentation defects, light-sensitivity, glomerulonephrotis, Rothmund-Thompson syndrome, Werner syndrome, epithelioma, dyskeratosis)

SKOTOMORPHOGENESIS: morphogenesis without dependence on light. (See also photomorphogenesis, de-etiolation)

SKP1: is an intrinsic kinetochore protein (22.3 kDa) widely conserved among species. It coordinates centromere and cell cycle events. (See kinetochore, cell cycle, CDC4)

SKUNK (*Mephitis mephitis*): 2n = 50; (*Spilogele putorius*), 2n = 64.

Sky: is a cellular kinase

SL1, SL2 (spliced leader): involved in transsplicing in *Caenorhabditis*. The 100 nucleotide leader donates its 5'-end 22 nucleotides to a splice acceptor site on the primary transcript. Trans-splic-

SL1, SL2 continued
ing is very common (70%) among the nematode's genes. This mechanism is used for the coordinately regulated gene clusters, transcribed in polycistronic RNA. The nematode operons use SL2 whereas other genes use SL1. (See transsplicing, coordinate regulation, operon)

SL1: is a transcription factor complex of RNA polymerase I. It is a complex of the TATA-box-binding protein (TBP) and the three TATA box associated factors (TAF). The TBP protein binds exclusively either SL1(RNA pol I) or TFIID (RNA pol II). In the case of RNA pol III TFIIIB is required for the recruitment of the polymerase to the promoter complex. (See pol I, pol II, TBP, TAF, transcription factors)

7SL RNA: is an RNA component in the signal recognition protein (SRP) complex. (See signal sequence recognition particle, Alu)

SLAM: a T cell receptor protein (M_r 70K) of the immunoglobulin family, constitutively and rapidly expressed on activated peripheral blood memory T cells, immature thymocytes and on some B cells. T cells carrying the $CD4^+$ antigens produce increased amounts of interferon γ without an increase of interleukins 4 or 5. SLAM function is independent of CD28. (See T cell, interferon, interleukin, CD28)

SLEEPING SICKNESS: a potentially fatal disease caused by *Trypanosomas*. The disease is spread by the tse-tse fly. (See *Trypanosomas*)

SLG: self-compatibility locus-secreted glycoprotein. (Self-incompatibility)

SLINK: is a computer program for estimating linkage information by a simulation approach.

SLIPPING: is a shifting of the translational reading frame. (See recoding, hopping)

SLITHERING: is a creeping type motion toward each other of the recombination sites and a recombinase enzyme within a supercoiled DNA molecule in case of site-specific recombination. (See site-specific recombination)

SLOT BLOT: binding cDNA onto slots on membrane filters for the analysis of transcription.

SLOW COMPONENT: during a reassociation reaction of single-stranded DNA the unique sequences anneal slowly. (See c_0t value, annealing)

SLOW STOP: bacterial mutant *dna* may complete slowly the replication underway but cannot start a new cycle at 42° C. (See replication)

SLS: sodium lauryl sulfate. (See SDS)

SLT: see specific locus mutations assay.

SLT-2: a protein kinase of the MAPK family. (See signal transduction, protein kinase)

SLUG: in general usage means a land mollusc but see also *Dictyostelium*.

Sma I: restriction endonuclease; recognition site CCC↓GGG. (See restriction enzymes)

SMAD: proteins when stimulated by TGF-β can suppress tumor formation. (See TGF)

SMALL CELL LUNG CARCINOMA (SCLC): is associated with a deletion of human chromosomal region 3p14.2; the susceptibility is dominant. It accounts for about 1/3 of all the lung cancers. Smoking may be the major cause of the development of this condition. Surgical remedy is usually not applicable because of the rapid metastasis but it generally responds to radiation and chemotherapy. Deregulation of the MYC oncogene is suspected cause. Recently a gene for fragile histidine triad, FHIT was associated with SCLC and with some of the non-small cell lung carcinomas (NSCLCs). The product of FHIT splits Ap_4A substrates asymmetrically into ATP and AMP. (See oncogenes, cancer, MYC, p53, ATP, AMP, semaphorin)

SMALL NUCLEAR RNA: see snRNA

SMALL t ANTIGEN: see SV40

SMART AMMUNITION: *Drosophila* P transposable element vectors with selectable markers to produce selectable (e.g. neomycin resistance) insertions. (See also hybrid dysgenesis, insertional mutation)

SMART CELLS: is a generalized concept that cells (genes) have the ability to sense internal and external cues and respond to them in a purposeful manner such as shown in signal transduction.

SMART LINKERS: synthetic oligonucleotides with multiple recognition sites for restriction enymes; they can be ligated to DNA ends to generate the desired types of cohesive ends. (See cloning vectors, cohesive ends, blunt end, blunt end ligation)

SMC: proteins involved in the structural maintenance of the chromosomes including condensation and segregation. (See also sex determination)

SMEAR: preparing a soft specimen for microscopic examination by gentle spreading directly on the microscope slide. (See also squash, sectioning, microscopy)

Smg p21: protein similar to Rap 1. (See Rap)

SMITH-LEMLI-OPITZ SYNDROME (SLO): is a high prevalence autosomal recessive anomaly involving microcephalus, mental retardation, abnormal male genitalia, polydactily, etc., encoded in human chromosome 7q32. (See mental retardation, dwarfness, head/face/brain defect)

SMITH-MAGENIS SYNDROME: involves head malformation, short brain, growth retardation, hearing loss, self-destructive behavior such as pulling off nails, putting foreign objects to ear, etc. Chromosome 17p11.2 region is generally deleted. (See mental retardation, self-destructive behavior)

SMOKING: is responsible for a wide variety of ailments such as heart disease, respiratory problems, cancer, etc. but it may decrease the risk of Parkinson's disease. The smoking habit is particularly prevalent in affective disorders. In the brain of smokers the level of monoamine oxidase B (MAOB) is 40% lower relative to that in non-smokers. MOAB degrades the neurotransmitter dopamine. The nicotinic acetylcholine receptors have, however, very important role in cognitive processes of the brain. (See Parkinson's disease, dopamine, affective disorders, nicotinic acetylcholine receptors, nicotine)

SMOOTH ENDOPLASMIC RETICULUM: has no ribosomes on its surface. (See SER)

SMOOTH MUSCLE: lacks sarcomeres; they are associated with arteries, intestines and other internal organs, except the heart. (See sarcomeres, striated muscles)

SMRT: is a silencing-mediator of retinoid and thyroid hormone receptors. (See retinoic acid, animal hormones)

SMUT: infection of grasses by *basidiomycete* fungi causing black carbon-like transformation of the inflorescence (by *Ustilago,* loose smut) or seed tissues (by *Tilletia,* covered smut).

SNAIL: *Helix pomatia univalens,* 2n = 24.

SNAKE VENOM PHOSPHODIESTERASE: releases 5'-nucleotides from the 3' end of nucleic acids.

SNAP: see NSF

SNAP-BACK: inverted repeat sequence in nucleic acids. (See repeat inverted)

SNAPDRAGON (Sippe 50 of Stubbe) ➔

SNAPDRAGON (*Antirrhinum majus*): 2n = 16, is a much employed dicotyledonous plant (*Scrophulariaceae*) for the study of mutation (transposable elements), flower pigments; it is also a popular ornamental. (See TAM)

SNAREs: are binding proteins attaching vesicles (v-SNAREs) to target membranes (T-SNAREs). The name probably comes from the surgeons wire tools by what polyps and tumors were removed by hooking these projections or from bird traps (snares). SNARE seems to be activated by Ypt1p. (See also NSF, Ypt, synaptobrevin)

SNO ONCOGENES: these are two SKI-related oncogenes. (See SKI, oncogene)

snoRNA (small nucleolar RNA): assist maturation of ribosomal RNAs in the nucleolus, folding of RNA, RNA cleavage, base methylation, assembly of pre-ribosomal subunits, export of RNP, etc. A family of snoRNAs of 10-21 nucleotides, complementary to the methylation sites of rRNA guides the methylation within the nucleolus. (See RNA maturase)

snRNA (small nuclear RNA): low molecular weight RNA in the eukaryotic nucleus, rich in uridylic residues. When associated with protein, mediate the splicing of primary RNA transcripts and freeing them from introns through assistance of the lariat. (See hnRNA, U1-RNA, RNP, introns, spliceosome, lariat,)

snRNP: small nuclear ribonucleoprotein, (pronounced also as "snurp"). (See KH domain)

SOB BACTERIAL MEDIUM: H_2O 950 mL, bacto tryptone 20 g, bacto yeast extract 5 g, NaCl 0.5 g plus 10 mL: of 250 mM KCl; pH is adjusted to 7 with 5 N NaOH and filled up to 1 L.

Just before use add 5 mL of 2 M MgCl$_2$.

SOC BACTERIAL MEDIUM: is the same as SOB but contains glucose (20 mM). (See SOB)

SOCIAL DARWINISM: unfortunate "application" of the Darwinian views to social order and to justify inequalities, harsh competition without adherence to ethics, aggression, imperialism, racism, unbridled "capitalism" as the necessity for the survival of the fittest. Social darwinism is no longer practised in the developed world. (See Darwinism, see also social engineering)

SOCIAL ENGINEERING: is the utopistic idea that the genetic determination of an individual is unimportant in defining the abilities and their realization but education, welfare, medical service, etc. may determine how a person will function in a society. Therefore the state and its institutions must actively control human life from cradle to death. Although, the importance of compassion, education, caring and many social safety nets are indispensable in a modern society, the significance of the individuality can not be ignored. (See social Darwinism)

SOCIAL INSECTS: like bees, wasps and ants, termites live in a colony and generally divide the various tasks among different casts such as workers (soldiers), queen and drones. The queen (gyne) and the workers and soldiers are diploid. The drones are different because they hatch from unfertilized eggs. The major differences between the queen and the workers is that the queens have predominantly 9-hydroxy-(E)2-decanoic acid and 9-keto-(E)2-decanoic acid in her mandibular glands. The workers have predominantly 10-hydroxy-(E)2-decanoic acid. These pheromones then determine their functional roles in the colony. The workers and soldiers have ovaries and under certain circumstances (especially the soldiers) may produce haploid eggs. Recently, the status of soldiers has been questioned as being genetically equivalent with the workers. The soldiers (not found in all species of social insects) role is protection of the colony. (See sex determination)

SOCIOBIOLOGY: study of the biology and behavior of social insects and other animal communities. (See social insects)

SOD (superoxide dismutase): see superoxide dismutase, amyotrophic lateral sclerosis

SODIUM AZIDE (N$_3$Na): is an inhibitor of respiration and it blocks the electron flow between cytochromes and O$_2$. It is also a potent mutagen for organisms that can activate it.

SODIUM CHANNEL: see ion channels

SODIUM DODECIL SULFATE: is a synonym for sodium lauryl sulfate; see dodecyl sulfate sodium salt.

SODIUM PUMP: a plasma membrane protein that moves Na$^+$ out and K$^+$ into the cells with the energy obtained by hydrolyzing ATP; it is called also Na$^+$— K$^+$ ATPase. (See ion channels)

SOFTWARE: a computer program that tells the hardware (the computer) what and how to carry out the instructions, the applications.

SOLENOID STRUCTURE: is a coiling electric conductor used for the generation of magnetic field; by analogy, the coiled nucleosomal DNA fiber is frequently described as a solenoid although it has no relation to electricity or magnetism; it only resembles those coils. The DNA solenoids are about 30 nm in diameter.

SOLID STATE CONTROL: is exercised by electric or magnetic means in solids, e.g. in a transistor. (See semiconductors)

SOLITARY LTR: long terminal repeats that lost their internal (transposase) sequences by recombination and excision. (See long terminal repeats, retrovirus, retroposon, retrotransposon)

SOLO ELEMENTS: terminal sequences (LTR) of retroposons that can exist in multiple copies without the coding sequences between the two direct repeats. (See Ty)

SOLUBLE RNA: a somewhat outdated term for transfer RNA. (See tRNA)

SOLUTE: any substance dissolved in a solvent.

SOLUTION HYBRIDIZATION: molecular hybridization in a liquid medium. (See nucleic acid hybridization)

SOMA: body cells distinguished from those which take part in sexual reproduction (germinal cells)

SOMACLONAL VARIATION: genetic variation occurring at a frequency higher than spontaneous mutation in cultured plant cells. The causative mechanism is poorly understood. It is conceivable that the asynchrony between nuclear and cell divisions are accompanied by chromosomal damage. Also, there is evidence that movement of endogenous transposable elements is involved. The mobility is attributed to the "stress" imposed by the culture.

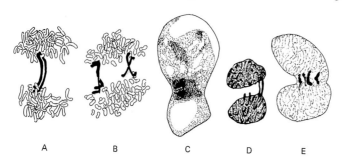

MITOTIC ANOMALIES IN A CLONE OF TOBACCO MAINTAINED IN CELL CULTURE. (A), (B) ANAPHASE BRIDGES, (C) TELOPHASE BRIDGE, (D) THE SAME AS (C) BUT ENLARGED NUCLEI, (E) EARLY INTERPHASE WITH CONJOINED NUCLEI. POLYPLOIDY IS ALSO OF COMMON OCCURRENCE. (From Cooper, L.S. et al. 1964 Amer. J. Bot. 51:284)

SOMATIC CELL: the majority of the (2n) body cells (reproduced by mitosis), including those of the germline but not the products of meiosis, the sex cells (gametes). (See germline, cell lineages, mitosis, cell genetics, parasexual mechanisms)

SOMATIC CELL HYBRIDS: are formed through fusion of different somatic cells of the same or different species. Somatic cell hybrids contain the nucleus of both cells and in addition all cytoplasmic organelles from both parents, in contrast to the generative hybrids where generally mitochondria and plastids are not transmitted through the male. For the fusion of cultured somatic cells the use of various special techniques are necessary. Most commonly protoplasts are used and the fusion medium is: polyethylene glycol [MW 1300-1600] 25 g, $CaCl_2 \cdot 2H_2O$ 10 mM, KH_2PO_4 0.7 mM, glucose 0.2 M in 100 mL H_2O, pH 5.5 for plants. The best media may vary according to species. The isolation of somatic animal cell hybrids was greatly facilitated by the use of selective media (see HAT medium, figure below).

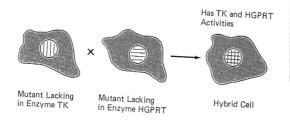

SELECTIVE ISOLATION OF ANIMAL SOMATIC CELL HYBRIDS DEFICIENT IN THYMIDINE KINASE AND HYPOXANTHINE-GUANINE PHOSPHORIBOSYLTRANSFERASE. ON 'HAT' MEDIUM ONLY THE COMPLEMENTARY HETEROKARYONS SURVIVE. (After Ephrussi, B. & Weiss, M.C. 1969 Sci. Amer. 220 [4]:26)

The fusion of animal cells is promoted by polyethylene glycol or by attenuated Sendai virus or by calcium salts at higher pH. Immediately after fusion, the somatic cells may become heterokaryotic but eventually the nuclei may also fuse. The availability of fused cells along with loss of one set of chromosomes or partial deletion of chromosomes made possible the localization of many human genes (see figure on next page). In human genetics most commonly mouse + human cells have been used. In such cultures the human chromosomes are gradually eliminated. If however one of the mouse chromosomes carry a defective gene, the human chromosome carrying a functional (wild type) allele must be retained in order that the culture could be viable in the absence of a particular supplement, required for normal function. The genetic constitution of the retained human chromosome can be verified also by enzyme assays, electrophoretic analysis of the proteins or immunological tests. For genetic analysis of higher mammals, and particularly in humans, where controled mating is not feasible, the availability of somatic hybrids opened a new very productive approach. In the somatic hybrid cells allelism and also synteny or linkage can be shown. If two or more human genes are consistently expressed in a particular chromosome retained, it is a safe conclusion that they are in the same chromosome. The prin-

Somatic cell hybrids continued

ciple of gene assignment is best explained by the diagram (p. 968, top). Genes can be mapped to particular chromosome bands by deletions (See diagram below). By somatic cell fusion hybrids can be obtained between taxonomically very distant species. All kinds of animal cells may be fused with each other, and in addition plant and animal cells can also be fused. Some of these exotic hybrids may have, however, difficulties of continuing cell divisions.

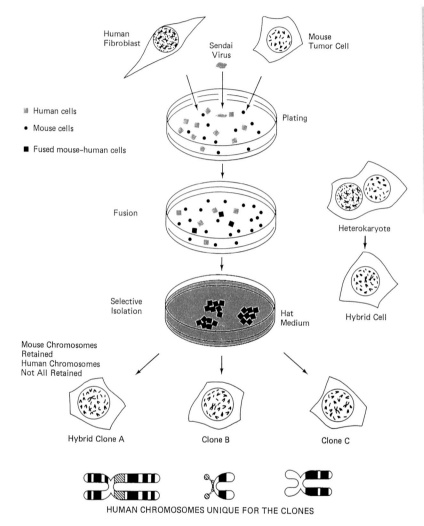

SELECTIVE ISOLATION OF MOUSE + HUMAN CELL HYBRIDS. THE FUSED CELL USUALLY RETAIN ALL MOUSE CHROMOSOMES BUT THE HUMAN FIBROBLAST CHROMOSOMES MAY BE PARTIALLY ELIMINATED AND IN SOME CLONES ONLY ONE OR ANOTHER HUMAN CHROMOSOME IS MAINTAINED. (Modified after Ruddle, F.H. and Kucherlapati, R. S. 1974. Sci. Amer. 231[1]:36)

Somatic cell fusion may make possible also to study recombination between various mitochondria or chloroplast DNAs in cells, where in generative hybrids, because of the uniparental (female) transmission of the organelles, such analyses are not feasible. Since plant cells generally retain totipotency in culture, the hybrid cells may be regenerated into intact organisms and can be further studied in favorable cases by the methods of classical progeny analysis. (See HAT medium, cell fusion, monoclonal antibody, human chromosome maps, *in situ* hybridization, mapping genetic, somatic embryogenesis, microfusion, radiation hybrids, see also diagrams on page 968)

SOMATIC CROSSING OVER: see mitotic crossing over

SOMATIC EMBRYOGENESIS: formation of embryos either directly, or first passing through a callus stage, from cultured adult plant cells or protoplasts. (See embryo culture, apomixis)

SOMATIC HYBRIDS: see somatic cell fusion, somatic cell hybrids, graft hybrids, microfusion

SOMATIC HYPERMUTATIONS: generally occur in a region of one to two kilobases around the rearranged V-J regions of the immunoglobulin genes and very rarely extend into the C (constant) sections. These mutations, usually transitions (*continued in the middle of p. 268*)

Somatic cell hybrids continued from preceding page

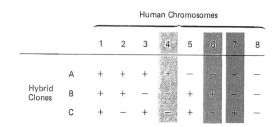

CHROMOSOME ASSIGNEMENT OF GENES ON THE BASIS OF INCOMPLETE CLONES OF MOUSE + HUMAN CELL HYBRIDS. A PANEL OF 3 CLONES, EACH CONTAINING A SET OF 4 OF 8 HUMAN CHROMOSOMES. IF A GENE IS EXPRESSED UNIQUELY IN CLONE (C) BUT NOT IN (A) OR (B), THE LOCUS MUST BE IN CHROMOSOME 7. BECAUSE ONLY C CARRIES CHROMOSOME 7. ADDITIONAL PANELS ARE NEEDED TO TEST GENES IN THE ENTIRE HUMAN GENOME. (After Ruddle & Kucherlapati 1974 Sci. Amer. 231[1]:36)

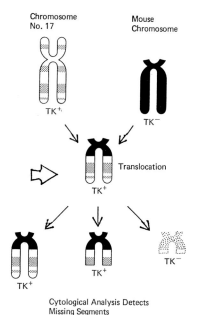

Cytological Analysis Detects Missing Segments

REGIONAL MAPPING OF THE THYMIDINE KINASE (TK) GENE TO A SEGMENT OF HUMAN CHROMOSOME 17. A TRANSLOCATION WAS OBTAINED BETWEEN HUMAN CHROMOSOME 17 AND A MOUSE CHROMOSOME THAT WAS DEFICIENT FOR TK. THE TRANSLOCATION CHROMOSOME WAS EXPOSED TO A CHROMOSOME-BREAKING AGENT THAT CLEAVED OFF SEGMENTS OF VARIOUS LENGTHS FROM THE END OF THE TRANSLOCATION. WHEN THE SEGMENT CONTAINING THE GENE (TK) WAS REMOVED, THE CELLS CARRYING THIS BROKEN CHROMOSOME NO LONGER DISPLAYED TK ACTIVITY AND THE GENE'S LOCATION WAS REVEALED AT 17q23.2-q25.3. (See human chromosome map)

Somatic hypermutations *continued from page 967*
and predominantly involving guanine, are most common in the complementarity determining region (CDR) and the events usually take place in the germinal centers. The prefered hot spots are purine-G-pyrimidine-(A/T) sequences although not all these sequences are hot spots for mutation. The serine codons AGC and AGT can represent a hot spot but the TCA, TCC, TCG and TCT are not. The codon usage in the CDER region appears to be evolutionarily determined to secure the maximum complementarity to antigens. It is not entirely clear what determines the targeting of the hypermutations to the special area but transcriptional enhancers appear to be involved. Hyper-mutations are limited to the coding strand and occur in a downstream polarity. (See immuno-globulins, CDR, germinal center, lymphocytes, transition, hot spot, enhancer, hypermutation, mosaic)

SOMATIC MUTATION: occurs in the body cells; it is undetectable if recessive, unless the individual is heterozygous. The mutation is manifested by sector formation. This procedure for testing mutability is particularly effective if multiple-marker heterozygotes are treated. It is not inherited to generative progeny unless it occurs or expands also to the germline. It has been successfully used for studies of mutation in the stamen hairs of the plant *Tradescantia*. In fungal cultures mutation in mitotic cells may appear as sectorial colonies. Somatic mutation can be studied in *in vitro* cell cultures. Recessive mutations are detectable for X-linked genes in hemizygous cells such as the male's. The procedure is effective if it is directed toward loci with selectable products. Molecular methods (PCR) may permit the identification of the mechanisms involved such point mutations, chromosomal deletions and rearrangement. The frequency of mutation may be affected by age, and various environmental factors and these can be analyzed. (See pseudodominance, bioassays in genetic toxicology, *Tradescantia* stamen hairs, somatic hypermutations, transposable elements, mosaic, PCR)

SOMATIC PAIRING: generally only the meiotic chromosomes pair during prophase (possibly late interphase) but in some tissues such as the salivary glands, the chromosomes are always tightly associated. Also, mitotic association of the chromosomes are a requisite for mitotic (somatic) crossing over in a few organisms where this phenomenon has been analyzed. (*Drosophila*, some fungi, *Arabidopsis*, etc.). (See mitotic recombination, parasexual mechanisms, pairing, intimate pairing)

SOMATIC RECOMBINATION: see mitotic crossing over.

SOMATIC REDUCTION: reduction of chromosome numbers during mitosis in polyploids.

SOMATIC SEGREGATION: unequal distribution of genetic elements during mitoses. (See mitotic crossing over, somatic mutation, sorting out, chloroplast genetics, mosaic)

SOMATOGAMY: fusion of sexually undifferentiated fungal hypha tips. (See fungal life cycles)

SOMATOMEDIN: is a second messenger-type polypeptide. In association with other binding proteins it is involved in the stimulation of several cellular functions. (See insulin-like growth factor, pituitary dwarfness, second messenger)

SOMATOPLASTIC STERILITY: in certain plant hybrids the nucellus may show excessive growth, and chokes the embryo to death. The embryo can be rescued if excised early and transferred to *in vitro* culture media. (See embryo culture)

SOMATOSTATIN: a 14-20 amino acid long hypthalmic neuropeptide inhibits the release of several hormones (somatotropin, thyrotropin, corticotropin, glucagon, insulin, gastrin) in contrast to the growth hormone-releasing factor that stimulates production of growth hormone. The somatostatin gene has been mapped to human chromosome 3q28 and to mouse chromosome 16. Two somatostatin receptors SSTR1 and 2 (391 and 369 amino acids, respectively) with very substantial homology have been identified. They are sevenpass-transmembrane proteins frequently bound to G-proteins and distributed all over the body, at particularly high levels in the stomach, brain and kidney. (See animal hormones, G-proteins, seven membrane proteins, signal transduction, GHRH)

SOMATOTROPIN: is the mammalian growth hormone (GH, $M_r \approx 21500$); it can correct some dwarfisms when its level is increased; also it stimulates milk production. Actually, there are 3 human growth hormones, all coded at chromosome 17q23-q24. Their mRNAs display about 90% homology and their amino acid sequences are shared as well. The placental lactogen protein is an even more effective growth hormone, also located nearby and it is highly homologous. The human growth hormone gene is transcribed only in the pituitary, whereas the homologs are expressed in the placental tissues. Human and other somatotropin genes have been cloned. Transformation of mice with rat growth hormone genes increased body size substantially. (See dwarfism, pituitary dwarfism, stature in humans)

SOMITES: paired mesoderm blocks along the longitudinal axis (notochord) of an embryo giving rise to the vertebral column and other segmented structures. After migration they may form the skeletal muscles.

SON OF SEVENLESS: see *SOS*

SONIC HEDGEHOG (*Shh*): a vertebrate gene that provides information in head - tail direction for development; it is a rather general signaling protein of animal differentiation. It is homologous to the *Drosophila hedgehog* (*hh*) gene; its receptors are *patched* (*ptc*) and *smoothened* (*smo*), signaling factor genes. (See *hedgehog*)

SONICATOR: an ultrasonic (≈ 20 kHz) equipment used for disrupting cells to extract contents.

SONOGRAPHY: is a method of ultrasonic prenatal analysis of possible structural and other defects in heart, kidney bone, sex organ, umbilical chord, body movement, for verifying pregnancy, etc. Presumably it entails no appreciable risk. (See ultrasonic, prenatal diagnosis, fetoscopy)

Sordaria **fimicola** (n = 7): an ascomycete with linear spore octads. It has been extensively used for genetic recombination. Large number of mutants are available. In the 1940 it was suspected that sexuality is relative in this fungus but the poor maters were just weak mutants.

SORGHUMS: arid, warm region crops. *S. bicolor* (and kaoliang), 2n = 2x = 20. *S. halepense*

(Johnson grass) is tetraploid.

SORI: see sorus

SORREL: a light chestnut fur color of horses determined by homozygosity for *d* gene; similar brownish color in other mammals. The sorrel plants *Rumex acetosella, R. scutatus* are used as tart vegetables; the sorrel tree is *Oxydendron* (See *Rumex*)

SORSBY SYNDROME: see night blindness

SORTING: the mechanism that ensures that molecules imported into organelles reach their destination. This process is mediated by transmembrane proteins and glycosylphosphatidylinositol-linked proteins. (See transit peptide, chloroplasts, mitochondria)

SORTING OUT: genetically different organelles (plastids, mitochondria) segregate into homo-

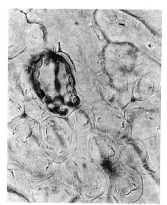
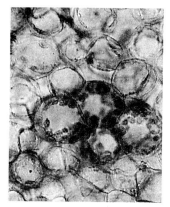

WHEN VARIEGATION IS CAUSED BY A MUTATION IN A NUCLEAR GENE ALL THE PLASTIDS WITHIN A CELL ARE EITHER GREEN OR COLORLESS (LEFT). MUTATION IN THE CHLOROPLAST GENETIC MATERIAL DISPLAY SORTING OUT AT THE BOUNDARY OF SECTORS, i.e., CELLS WITH BOTH GREEN AND COLORLESS PLASTIDS OCCUR (MIXED CELLS) AND IN A NON-STOCHASTIC PROCESS CELLS WITH ALL COLORLESS AND ALL GREEN PLASTIDS APPEAR IN LATER SECTORS (RIGHT)

geneous groups of cells or during embryogenesis cells of common origin reaggregate in order to form certain cell types and/or structures. The segregation of two different types, A and B, of mitochondria in a heteroplasmic cell line may be characterized by the formula of Solignac, M. *et al.* (Mol. Gen. Genet. 197:183): $V_n = \rho_0 (1-\rho_0)(1 - [1 - 1/N]^n)$ where V_n is the variance of ρ (the fraction of A within a cell) at the nth cell generation and ρ_0 = the fraction of A in the original cell line, N = the number of sorting out units. In *Schizosaccharomyces* the distribution of the mitochondria is mediated by microtubules. In humans the frequency of mutation in mtDNA is 10 times higher than in the nuclear DNA yet heteroplasmy is very rare except in some diseases. Despite the fact that the number of mitochondria in mammalian cell runs to the thousands per cell, usually the replication switches to one type. The number of the founder mtDNAs has been estimated within a wide range 1-6 and 20 to 200; in cattle and in *Drosophila* 370-740. These founders than undergo a restriction/amplification type of replication, i.e. they pass through a bottleneck and therefore heteroplasmy is very limited. (See ctDNA, plasmone mutation, plastid number, mtDNA, heteroplasmy, Romanovs)

SORUS (plural sori): a group of sporangia, such as found on lower surface of fern leaves.

SOS (son of sevenless): *Drosophila* gene that functions downstream from *sevenless*, encoding a receptor tyrosine kinase [RTK]) in the light signal transduction pathway. SOS is a gunanine nucleotide releasing protein (GNRP) and it interacts with RTK through the protein Drk receptor kinase, a homolog of the vertebrate Grb2, and SEM-5 in *Caenorhabditis*. These proteins function in a variety of signal transduction pathways involving EGF. SOS is frequently called also a mediator protein. (See BOSS, DRK, Grb2, receptor tyrosine kinase [RTK], rhodopsin, signal transduction, GNRP, daughter of sevenless)

SOS REPAIR: error-prone repair. (See DNA repair)

SOUTHERN BLOTTING: DNA fragments cut by restriction endonucleases are separated on agarose gel by electrophoresis, then transferred to membrane filters by blotting, to hybridize the pieces with radioactively (or fluorescent) labeled DNA or RNA and then identify the physical

Southern blotting continued

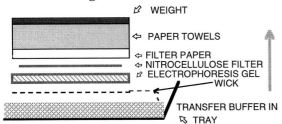

sites of restriction fragments and genes. The transfer to membranes may be achieved by capillary action of wicks and sucked through layers of filterpapers on top or by vacuum-driven devices (See restriction enzyme, RFLP, autoradiography, biotinylation, fluorochromes, nucleic acid hybridization, membrane filters)

AUTORADIOGRAPHY REVEALS THE PROBED SPOTS ON THE FILTER ON THE PHOTOGRAPHIC FILM ⟶

SOUTHERN HYBRIDIZATION: see Southern blotting

SOUTH-WESTERN METHOD: labeling simultaneously cDNA and binding proteins (transcription factors). The screening is carried out by hybridizing labeled probes of DNA to bind to polypeptides immobilized on nitrocellulose filters. (See Southern blot, Western blot)

SOVIET GENETICS: is not meant to define a special genetics because science does not have ideological, political, or ethnic attributes and it transcends all boundaries. A sad exemption is "soviet genetics", a misnomer because it was not genetics at all and it collapsed before that state that nurtured and enforced it did. Genetics in the Soviet Union had a very remarkable and successful beginning. In 1944 L. C. Dunn, Professor of Zoology at Columbia University noted: "There are today literally hundreds of trained genetical investigators in the U.S.S.R., certainly more than in any other country outside the U.S.A." (Science 99:2563). This outstanding research and teaching establishment was destroyed, however, in 1948 and geneticists suffered humiliation, persecution and almost total physical annihilation for a period of over 20 years by lysenkoism. (See lysenkoism, Mitchurin)

SOX (SRY type HMG box): mammalian genes encoding proteins with over 60% similarity to the HMG box of SRY. The SOX9 gene is critical for the differentiation of the Sertoli cells. Sox9 is apparently a transcription factor and capable of transactivation of genes involved in gonadal differentiation. The bacterial *Sox* genes are involved in superoxide responses. (See SRY, HMG, campomelic dysplasia, Wolffian duct)

SOYBEAN: *Glycine max*, 2n = 40; the basic chromosome number may be x = 10. Some related species are tetraploid. This crop is of great economic significance because of the high oil (20-23%) and high protein content (39-40%) of the seed.

Sp1: mammalian protein, general transcription factor for many genes recognizing the DNA sequence: $\substack{GGGCGG \\ CCCGCC}$ Sp1 elements also protect CpG islands of housekeeping genes from methylation. Sp3 and Sp4 are very similar; the latter is expressed in the brain. (See transcription factors inducible, transcription factors, DNA methylation)

SPACER DNA: non-transcribed nucleotide sequences between genes in a cell. In the rDNA there are spacers within (internal transcribed spacers, ITS) and between gene clusters (external transcribed spacers, ETS). These were thought of earlier as not-transcribed tracts but actually they represented very short transcripts. Animal mtDNA genes generally have very short (few bp) spacers. In fungi mtDNA spacers are common and variable in length because of recombination and slippage during replication; they may also be mobile. Plastid genes of higher plants are organized into operons without interruptions. The length of the spacers is determined by insertions, deletions and unequal recombinations. (See ribosomal RNA, rrn)

SPASTIC PARAPLEGIA (Strumpell disease): is a collection of paralytic diseases under a variety of genetic control, recessive, dominant autosomal and X-linked.

SPAYING: neutering of a female. (See castration)

SPEARMAN RANK-CORRELATION TEST (SPC): determines the relations between two variables without much elaborate calculations. It is a non-parametric test and can be used also

Spearman rank-correlation test continued

for comparison of traits that cannot be measured but can be classified subjectively (on the basis of their appearance). If the traits are quantified by measures, they are ranked according to their relative magnitude, i.e., assigning the highest rank to the largest. If an exact measurement is impractical (e.g., degree of susceptibility), they are simply ranked. In case of ties, an equal rank is assigned to both. The differences of the rank scores are squared and summed. Example:

Pairs	Trait X	Rank	Trait Y	Rank	Difference of Ranks (d)	d²
1	8	3	14	2	1	1
2	10	5	20	7	-2	4
3	12	7	17	5	2	4
4	9	4	15	3	1	1
5	6	1	13	1	0	0
6	11	6	18	6	0	0
7	7	2	16	4	-2	4
					Sum d = 0	Sum d² = 14

The SR correlation coefficient, $r = 1 - \dfrac{(n-1)\text{Sum}d^2}{n^3 - n}$, in this example $1 - \dfrac{6 \times 14}{7^3 - 7} = 0.75$

In case of ties, correction can be used, $T = \dfrac{m^3 - m}{12}$ where m stands for the number of measurements or classifications of identical values. If for example there are three measurements of 5, two measurements of 7, and two amounting to 8, the correction for ties among the X trait variables becomes: $\text{Sum } T_x = \dfrac{3^3 - 3}{12} + \dfrac{2^3 - 2}{12} + \dfrac{2^3 - 2}{12}$ and in a similar way the correction for ties can be determined in the Y series. Now we can obtain the terms:

$\text{Sum } X^2 = \dfrac{n^3 - n}{12} - \text{Sum}T_x$, and $\text{Sum } Y^2 = \dfrac{n^3 - n}{12} - \text{Sum } T_y$ and $r = \dfrac{\text{Sum } X^2 + \text{Sum}Y^2 - \text{Sum } d^2}{2\sqrt{\text{Sum}X^2 \text{Sum}Y^2}}$

The probabilities can be determined by tables constructed for various degrees of freedom. (See correlation, non-parametric tests, statistics)

SPECIALIZED TRANSDUCTION: the transducer temperate phage picks up a piece of the host DNA at the immediate vicinity of its established prophage site (and generally leaves behind a comparable length of its own). When this modified phage infects another bacterium, it may integrate into the host genome the gene that it picked up from the previous host. Lambda phage (λdgal; meaning a defective lambda that carries *gal*) is a typical specialized transducer with prefered site near 17 minutes on the map next to the *gal* locus, and it is a specialized transducer of this gene. Another well known specialized transducer phage is φ80trp that can carry the tryptophan operon (at about 27 and 1/2 min). Specialized transducer phage must be a temperate phages and they transduce (only) the special gene that is at their integration site. (See flowchart of specialized transduction on the following page, high-frequency lysate, double lysogenic, helper virus, bacterial recombination)

SPECIATION: is the process by what from an ancestral species new species diverge. Evolutionists may distinguish *conventional* methods of speciation when geographic isolation and accumulated mutations cause eventually reproductive isolation. According to the *quantum model* of speciation the divergence begins with spatial isolation followed by the survival of a few new type of individuals that give rise to a reproductively isolated new forms as mutations accumulate. *Saltational* speciation comes about by sudden major mutations. *Parapatric* (or stasipatric) speciation occurs without geographic separation and it is initiated from a relatively small number of individuals that are producing divergence under continued natural selection. *Sympatric* speciation occurs within the original area of dispersal, due to the emergence of a genetic isolation mechanisms. The mechanism of speciation in plants and animals differ because in animals, in contrast to plants, behavioral traits may also lead to speciation. In plants, sterility may not necessarily hinders propagation because some species reproduce pri-

Speciation continued

marily by vegatative means and they can practice both autogamy and cross-pollination. (See evolution, sibling species, phylogeny, co-speciation, theory of evolution, saltation, neo-Darwinism, neo-Lamarckism, species, fertility)

Specialized transduction continued from preceding page

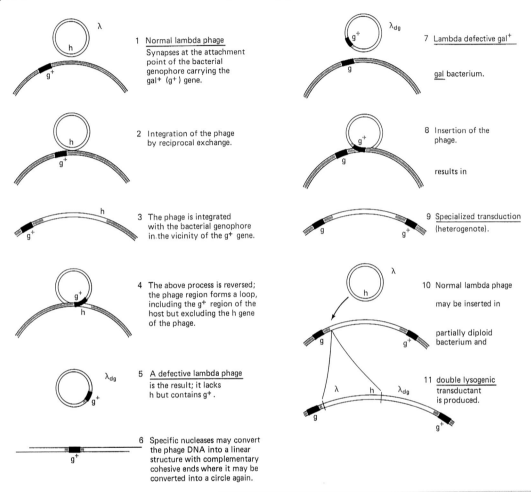

STEPS IN SPECIALIZED TRANSDUCTION EXAMPLIFIED BY THE LAMBDA PHAGE. (Modified after Stent, G.S. 1971 Molecuilar Genetics. Freeman, San Francisco, CA)

SPECIES: is a potentially interbreeding population which shows reproductive isolation from other species. The number of species is not known but about 10^6 have been described. The expectations are between 10 to 100 fold more that have not been discovered yet. (See speciation, sibling species, fertility, sexual isolation, species extant)

SPECIES EXTANT: with the increase of agricultural, industrial and changing land use the currently alive (A), % extinct (E), and % threatened (T) species indicate an alarming reduction of biodiversity: molluscs (A) 10^5, (E) 0.2, (T) 0.4; crustaceans: (A) $4\text{-}10^3$, (E) 0.01, (T) 3; insects (A) 10^6, (E) 0.005, (T) 0.07; vertebrates total (A) 4.7×10^4, (E) 0.5, (T) 5; mammals (A) 4.5×10^3, (E) 1, (T) 11; gymnosperms (A) 758, (E) 0.3, (T) 32; dicots (A) 1.9×10^5; (E) 0.2, (T) 9; monocots (A) 5.2×10^4, (E) 0.2, (T) 9. (Data from Smith, F.D.M. *et al.* 1993, NATURE 364:

Species extant continued

494). Within these larger categories some species are much more endangered that the above figures indicate. These numbers are not exactly valid because new species are being discovered; between 1980-90 more than 100 new mammals have been identified. The extinction percentages date from the year 1600 A.D. According to Dobson, A.P. *et al.* 1997 (Science 275: 550) in the United States of America the number of endangered species is 503 plants, 84 molluscs, 57 arthropods, 107 fish, 43 herptiles, 72 birds and 58 mammals. Important factors affecting the survival of the feral species is the annual increase of the human populations(1.7%), urbanization, industrial and livestock production, pollution, natural disasters. According to the World Conservation Union 1996 biennial report (http://www.iucn.org/themes/ssc/index.html), currently 5205 species are endangered and risk extinction. (See species, conservation genetics)

SPECIES SYNTHETIC: are produced in the laboratory by crossing putative ancestors of amphidiploid forms which are partially isolated genetically. The sterility of the offspring is prevented by doubling the chromosome number of the hybrids. The synthetic species permit a verification of the putative evolutionary path of existing polyploid species by studying the pairing behavior of chromosomes, the frequency of chiasmata, the number of univalents and multivalents formed, etc. These classical cytogenetic methods may be supplemented with nucleic acid hybridization, *in situ* hybridization, study of proteins, etc. Some of the hybrid species if ever occured in nature, were not maintained by natural selection. (See allopolyploids, *Triticale*, *Raphanobrassica*. (See speciation, evolution)

SPECIFIC ACTIVITY: the number of molecules of a substrate (μmol) acted on by enzymes (mg protein) per time (minutes) at standard temperature (25° C). Also, the relative amount of radioactive molecules in a chemical preparation.

SPECIFIC COMBINING ABILITY: see combining ability

SPECIFIC HEAT: joules or calories required to raise the temperature of 1 g substance by 1° C.

SPECIFIC LOCUS MUTATIONS ASSAYS (SLT): have been used to detect X-chromosomal mutations in hemizygous males (mouse), in autosomal heterozygotes for special fur color genes such as the Oak Ridge stock heterozygous for *a, b, p, c^{ch}, se, d, s* or the Harwell tester stock (HT) heterozygous for six loci (*a, bp, fz, ln, pa, pe*), or heterozygotes for thymidine kinase or hypoxanthine-guanine phosphoribosyltransferase genes in different mammalian cell cultures (mouse, Chinese hamster ovary cells). Mutation of the wild type allele is then immediately revealed in the first generation. Similar procedures are applicable also in plants and any other diploid systems. The animal assays are not as simple as the microbial assays (e.g., Ames test) but they are considered to be more relevant to human studies. (See bioassays in genetic toxicology, mutations in human populations, doubling dose)

SPECIFIC ROTATION (α): degrees of the plane of polarized light by an optically active compound of specific concentration at 25° C: $(\alpha) = \frac{rv}{nl}$, where the r = rotation in degrees, v = volume in cubic centimeter of the solution, l = length in centimeters of the light path.

SPECIFICITY: a measure of discrimination among compounds.

SPECIFICITY OF MUTAGEN ASSAYS: is the percentage of correct identification by mutagenesis presumably "non-carcinogenic" ("non-mutagenic") agents. (See sensitivity, accuracy, predictivity, bioassays in genetic toxicology)

SPECKLES INTRANUCLEAR: are storage areas for primary RNA transcript processing factors; the speckles are not the compartments of the eukaryotic splicing which takes place rather in association with transcription. (See splicing, introns)

SPECT: see tomography

SPECTINOMYCIN: see antibiotics

SPECTRAL KARYOTYPING: the chromosomes or chromosomal segments are hybridized with fluorescent dyes in a combinatorial manner. Five dyes (Cy2, Spectrum Green, Cy3, Texas Red and Cy5) provided enough combinatorial possibilities to "paint" each chromosome

Spectral karyotyping continued
with a different color or shade of it. Although the human eyes cannot distinguish these different hues, by the use of optical filters, Sagnac interferometer, a CCD camera and Fourier transformation, they were able to discriminate the special differences in standard classification colors using a computer program. The approach permitted classification of translocations that were not identifiable by other staining techniques. The lowest limit of differentiation by this technology is 500 to 1500 kbp. The procedure is applicable to clinical laboratory testing and evolutionary analyses. (See also Schröck, E. *et al.* 1996 SCIENCE 273:494, chromosome painting, FISH)

SPECTRIN: a filamentous protein (220 - 240 kDa) present in the red blood cell membranes and may constitute 30% of it. It is also involved in the formation of a network on the cytoplasmic surface in cooperation with actin, ankyrin and *band III* protein. (See ankryn, cytoskeleton, elliptocytosis, spherocytosis)

SPECTROPHOTOMETRY: estimating the quality and quantity of a substance in solution on the basis of absorption of monochromatic light passing through it.

SPEMANN'S ORGANIZER: the embryonic tissue site of signals that mediate the organization of the body. Its signaling center is at the blastopore of the gastrula and releases a variety of polypeptides controling neural and dorsal or ventral mesodermal differentiation. (See morphogenesis, organizer, gastrula, blastopore)

SPERM: it means the animal seminal fluid; geneticists generally understand it as the spermatozoon or plant male generative cells. In the first meaning the word has no plural, in the other sense both singular and plural are justified. The *Drosophila* sperm far exceeds in length that of any other animal's (up to 58 mm in *D. bifurcata* although that *of D. melanogaster* is about 1.91 mm; the human sperm is about 0.0045 mm). In a single normal human ejaculate the number of spermatozoa is within the range of 200-300 million. The human sperm count is showing a decreasing tendency during the last decades. The cause of this appears to be the increase of environmental pollutants with hormone-like effects, such as PCB (polychlorinated biphenyl) an industrial carcinogen and pesticide, dioxin (a solvent), phthalates (may be present in cosmetics), bisphenols (used in manufacturing resins and fungicides). The number of pollen grains (containing two sperms) released by a plant may vary between a few hundreds to tens of millions. (See also oocyte primary, environmental mutagens, fertilization, polyspermic fertilization, gametogenesis, RPTK, DDT, semen)

SPERM BANK: is a (human) sperm depository for the purpose of artificial insemination in case of sterility of the husband or other conditions which may warrant their use. Thousands of sperm banks (gene banks) exist in the world, most of the deposited samples are anonymous and the genetic constitution of the donors is not known completely. Since the 1930s H.J. Muller advocated the use of the sperm banks for positive eugenics purposes. Accordingly, the donors should be selected on the basis of superior talents, mental ability and physical constitution. This idea has not been widely accepted, however, because of moral objections and biological shortcomings. Most of the "superior phenotypes" cannot be evaluated by generally accepted criteria, and the phenotype may not fully represent the heritability of particular traits. Since artificial insemination of domestic animals became routine, sperm banks are exploited for animal breeding programs in order to produce maximal number of progeny of high-performance males. Even for animals this technology should be used with thorough consideration of population genetics principles in order to avoid narrowing the gene pool and inbreeding. (See *in vitro* fertilization, ART)

SPERM MORPHOLOGY ASSAYS IN GENETIC TOXICOLOGY: are based on the expectation that mutagens and carcinogens may interfere with normal spermiogenesis resulting in abnormal head shape, motility and viability of the treated or exposed sperm. This expectation is met with some agents but not with all. Thus sperm alterations may indicate mutagenic and/or carcinogenic properties but not all mutagens/carcinogens seem to affect these sperm parameters within non-lethal doses. (See bioassays in genetic toxicology)

SPERM RECEPTOR: see fertilization

SPERM STORAGE: insects commonly store sperm in the spermatheca for two weeks and fertilization may follow any time within this period. This is the cause why geneticists tend to use virgin females for controled matings. Sperm can be stored at the temperature of liquid nitrogen (-195.8° C) and retain their ability of fertilization. (See sperm bank)

SPERM TYPING: permit analysis of recombination in diploids using the recombinant gametes. The method of analysis requires the analysis of PCR-amplified gamete DNA sequences of known paternal types. This type analysis can resolve recombination even between single base pairs. The results must be scrupulously studied because the PCR method may have inherent errors. Single sperms can be separated with the aid of fluorescence activated cell sorters. If the sperms are subjected to 'primer extension preamplification' (PEP) before analysis, enough material can be obtained to carry out multipoint tests. This procedure is not practical with egg cells. (See recombination frequency, crossing over, maximum likelihood applied to recombination, polymerase chain reaction, cell sorter, prenatal diagnosis, genetic screening)

SPERMATHECA: is a storage facility of insect females for sperm to be used at a later fertilization following an initial mating.

SPERMATIA: male sperms produced at the tip of hyphae within the spermatogonia of rust fungi; they are comparable to the microconidia. (See conodia)

SPERMATID: a cell formed by the secondary spermatocyte and which differentiates into a spermatozoon. (See spermatozoon, sperm, gametogenesis)

SPERMATOCYTE (primary): is a diploid cell that by the first division of meiosis gives rise to the haploid secondary spermatocytes and those by cell division form the spermatids which differentiate into spermatozoa in animals. In plants, the spermatocytes function in a similar manner and thus producing the microspores which develop into pollen grains and within them the two sperms are formed either before or after the pollen tubes begins to elongate (See gonads, gametogenesis)

SPERMATOGENESIS: see gametogenesis, spermiogenesis

SPERMATOGONIA: are the primordials of the sperm cells; the secondary spermatogonia produce the primary spermatocytes. (See gametogenesis, spermatocyte)

SPERMATOPHYTE: seed-bearing plant. (See also cryptogamic plants)

SPERMATOZOON: the fully developed (differentiated) male germ cell, a sperm. (See sperm, spermatid)

SPERMIOGENESIS: is a postmeiotic process of differentiation of mature spermatozoa. It is supposed that CREM is involved in the control of genes required for the process because CREM-deficient mouse (obtained by recombination) cannot complete the first step of spermiogenesis and late spermatids are not observed. (See, sperm, CREM)

SPERMIST: see preformation

SPF: cell cycle S phase promoting factor, a kinase. (See cell cycle)

SPF CONDITION: specific pathogen-free condition of organisms maintained in a quarantine. (See quarantine)

Spfi-1: is an ETS family transcription factor. (See ETS oncogenes)

SPH: protein-binding DNA elements. (See Simian virus 40)

SPHEROCYTOSIS, HEREDITARY (HS): is the most common hemolytic anemia in Northern Europe. The basic defect involves both recessive and dominant mutations affecting ankyrin-1 and spectrin. The β-spectrin gene was assigned to human chromosome 14q22-q23. An autosomal dominant form was assigned to human chromosome 15q15. There is also an autosomal recessive type. (See ankyrin, spectrin, elliptocytosis, anemia)

SPHEROPLAST: spherical bacterial cell after (partial) removal of the cell wall. (See protoplast)

SPHINGOLIPIDOSES: are hereditary diseases involving the metabolism of sphingolipids with the enzyme defects indicated in parenthesis:

Farber's disease (ceramidase)   Lactosyl ceramidosis (β-galactosyl hydrolase)
Fabry's disease (α-galactosidase)   Metachromatic leukodystrophy (sulfatase)

Sphingolipidoses continued
Gaucher's disease (β-glucosidase)
Generalized gangliosidosis (β-galactosidase)
Krabbe's leukodystrophy (galactocerebrosidase)
Niemann-Pick disease (sphingomyelinase)
Sandhoff's disease (hexosaminidases A and B)
Tay-Sachs disease (hexosaminidase A)
(See these diseases under separate entries, see sphingolipids)

SPHINGOLIPIDS: are sphingosine-containing lipids with the general structure, and depending on the particular substitutions at X.
sphingosine — fatty acid (in neural cells most commonly stearic acid)
|
X

NEURAMINIC ACID →

hydrogen (ceramide)
glucose (glucosylcerebroside)   [neutral glycolipid]
glucose and galactose (lactosylceramide)   [neutral glycolipid]
complex of sialic acid, glucose, galactose, galactose amine (ganglioside G_{M2})
phosphocholine (sphingomyelin)
If sialic acid (acetyl neuramininc acid, glycolylneuraminic acid) is removed asiogangliosides result. (See also sphingosine, cerebrosides, gangliosides, sphingolipidoses, Sandhoff disease)

SPHINGOSINE: a solid fatty acid-like component of membranes. The principal naturally occurring sphingosine is: D(+) erythro-1,3-dihydroxy-2-amino-4-transoctadecene, $CH_3(CH_2)_{12}CH=CH(OH)-CH(NH_2)-CH_2OH$. In addition to the C_{18} sphingosines shown the molecules may have 14, 16, 17, 19 and 20 carbons. The molecules may also be branched or may contain an additional OH group.

Spi+: wild type λ phage is sensitive to phage P2 inhibition. Phage λ lacking the functions of *red* and *gam* can grow in P2 carrying-lysogens if it has *chi* (recombination sites for the *RecBC* system). (See lambda phage)

SPI1 ONCOGENE: spleen focus forming retrovirus homolog; it is in human chromosome 11p11.22. Spi1 is an ETS family transcription factor. (See oncogenes, ETS, PU.1)

SPIKE: an inflorescence which is alternatively called head or ear; a typical example is the spike of wheat or some other grasses. The spikelets or flowers are sessile on opposite sides of the axis that is called rachis. Also, short duration electrical variations along the nerve axon, a peak in electric potential.

SPIKE: the claw-like structures on the base plate of bacteriophages. (See development)

SPIKELET: a group of florets sitting on a common base in a spike. (See spike)

SPINA BIFIDA: a developmental disability in various mammals, including humans determined by autosomal dominant inheritance and reduced penetrance. The spinal column is incompletely closed and in some instances this involves no serious problem and detectable only by X-ray examination (spina bifida occulta). The more serious case is the spina bifida aperta when the spinal cord, the membranes (meninges), spinal cord and nerve ends are protruding (*myelomeningocoele*). This condition is frequently accompanied by hydrocephalus, incontinence, etc. The *meningocele* form involves only membrane extrusion and consequently it is a less severe defect. The overall frequency of these anomalies may be 0.2-0.3% in the general population. (See neural tube defects, prenatal diagnosis, genetic screening, MSAPF, mental retardation, hydrocephalus)

SPINACH (*Spinacia oleracia*): a dioecious plant, 2n = 12 (XX or XY, included). Many of the trisomics can be identified without cytological examinations. All 6 pairs of the chromosomes have distinct morphology.

SPINAL MUSCULAR ATROPHY (SMA): is a degeneration of the spinal muscles and it occurs in different forms and under different genetic controls. The adult type, proximal, is autosomal dominant and may be associated with different chromosomes. The juvenile (Kugelberg-Welander syndrome) affects primarily the proximal limb muscles and involves frequently twitching. This form was assigned to human chromosome 5q11.2-q13.3. The prevalence of this type is about 1/600 in newborns. The literature distinguishes also the Werdnig-

Spinal muscular atrophy continued

Hoffmann disease type, however, the various juvenile forms map to the same chromosomal segment although the expressions may differ. Deletions of this area is frequently the basis of the disease. The combined, estimated gene frequency is about 0.014. The spinal muscular dystrophy with microcephaly and mental retardation, the spinal muscular dystrophy distal, SMA proximal adult type and other variations of it appear autosomal recessive. (See also muscular atrophy, Kennedy disease, dystrophy, atrophy, muscular dystrophy, neuromuscular diseases, Kugelberg-Welander syndrome, Werdnig-Hoffmann disease)

SPINDLE: a system of microtubules ($\approx$ 20-30 nm) emanating from the poles (centrioles of the centrosomes in animals) during mitosis and meiosis and attaching to the kinetochores within the centromeres and pull the chromosomes toward the poles. In plants and also in some animal oocytes, there are no centrosomes and some of the centrosome functions are assumed by the chromosomes. Some of the microtubules reach from one pole to the other without attaching to the kinetochore. In *Drosophila* meiosis the spindles originate from each of the chromosomes and as the prophase progresses a bipolar spindle emerges. This stage uses a kinesin-like protein (NCD). The arrangement of the microtubules requires also the motor protein dynein. The meiotic pole is different from the mitotic centrosomes (DMAP60, DNAP 190 and γ-tubulin are apparently absent). In case of univalents, only monopolar spindle is formed. In some species the spindle origination from the chromosome is apparently suppressed by the centrosome. In some other species the chromosomes and the centrosomes cooperate in developing the meiotic spindle. The mitotic spindle may also need both chromosomes and the centrosomes (echinoderms). Microtubule assembly is not an exclusive property of the kinetochores as holocentric chromosomes indicate. Anaphase and cytokinesis, however, can take place in cells after the chromosomes have been removed. On the oocyte chromosomes of several species surface proteins (NOD, Xklp1) are found that stabilize premetaphase chromosomes, and in achiasmate meiosis substitute for the chiasmata. According to recent views the information for meiotic disjunction resides within the chromosomes and not in spindle apparatus. The kinetochore determines the transition from metaphase to anaphase. Tension of the kinetochore generates a checkpoint signal, and it is supposed that a phosporylated kinetochore protein attracts protein. The X chromosomes in XO cases of sex determination do not involve such a pause. (See nucleus, centromere, centrosome spindle fibers, tubulins, kinetochore, chromatid, mitosis, meiosis, univalent, holocentric chromosome, mitosis, meiosis, achiasmate, Swi6p, microtubule, RMSA-1, CENP)

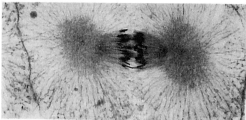
(Illustration by courtesy of Dr. P.C. Koller)

SPINDLE FIBERS: are microtubules clearly visible (by appropriate techniques) during mitotic and meiotic nuclear divisions. The microtubules originate at the spindle poles, the aster in animals. Three classes of fibers are emanating from the poles, the *astral microtubules* that radiate from the centrioles. The *polar microtubules* that meet at the divisional plane and appear to stabilize the spindle and the *kinetochore microtubules* that are anchored at the centromeres of the chromosomes, and at anaphase pull them toward the opposite poles. (See spindle, mitosis, meiosis, aster, centrioles, centromere, kinetochore, tubulins)

SPINDLE POISON: blocks the formation of spindle fibers and as a consequence polyploidy may result. (See polyploidy, colchicine)

SPINDLE POLE BODY: is the fungal equivalent of the centrosome. (See centrosome)

SPINOBULBAR MUSCULAR ATROPHY: see Kennedy disease

SPINOCEREBELLAR ATAXIA: see Huntington chorea, Kennedy disease, myotonic dystrophy, ataxia)

SPIRACLE: breathing hole on the insect body.

SPIRALIZATION: a pattern of winding of molecules or chromosomes. (See coiling)

SPIROCHETES (*Spirochaeta*): are filiform bacteria (5-6 µm) cause sex ratio distortion in *Drosophila* by the killing effect of its toxin on the male flies. (See sex ratio, symbionts hereditary).

SPIRONUCLEOSIDE (hydantocydin): is a herbicidal growth regulator of *Streptomyces hygroscopicus*. (See herbicide)

SPIROPLASMA: see symbionts hereditary, sex ratio, spirochetes

spitz (*Drosophila* chromosome 2-54): the locus encodes the Spitz protein, a ligand of DER. (See DER)

SPK-1: a protein kinase of the MAPK family. (See signal transduction)

SPLEEN: an upper left abdominal, oblong (ca. 125 mm) ductless gland. At the embryonal stage it participates in erythrocyte formation, in adults it makes also lymphocytes. By decomposing the erythrocytes, it provides to the liver the hemoglobin to form bile. The red pulp contains red blood cells and macrophages, the white pulp area carries the lymphocytes. It has a role in the defense mechanism of the body. (See asplenia)

SPLICE: resolution of a recombination intermediate (Holliday junction) resulting in an exchange of the flanking markers. (See Holliday model, introns, spliceosome)

SPLICEOSOME: is a protein-U snRNA (there are 5 main uridine-rich oligonucleotides) complex required for the folding of pre-mRNA into the proper conformation for removal of introns and splicing the transcripts of exons. The *majority* of eukaryotic primary transcripts contain introns with 5' GU and AG 3' splice sites, and the mammalian consensus is: (A/C)AG↓GUA/GAGU and in yeast (A/G)↓GUAUGU. In the excision the spliceosome complex U1, U2, U4-U6 and U5 snRNA and non snRNAs work together. Initially the U1 and U2 snRNA attaches by base pairing to the splice and to the branch sites, respectively. This pre-splicing complex is joined then by the U4-U6•U5 tri-snRNP complex. This is followed by snRNP-snRNP and mRNA—snRNP interactions. U6 base-pairs with U4. During spliceosome assembly U1 and U4 are displaced and U6 pairs with the 5' splice site and with U2 snRNA. The 5' splice site and the branch nucleotide then move to each other and the 2'-OH group of the latter serves as an electron donor for the first step of splicing. The excised 5'-exon and the lariat intron-3'-exon are the reaction intermediates. This first step is followed by a reaction between the electron donor nucleophile and the electron-deficient electrophile at the 3' splice junction by the 3'-OH of the freed 5'-exon, resulting in the ligation of the exons, and the intron lariat. A *minority* of the introns, the AT-AC introns, occur in some animal genes such as encoding PCNA. The 5' splice site of such introns has the consensus AT↓ACCTT and their branch site is TCCTTAAC. Their splicing complex includes U11 and U12 snRNP and one or more U5 snRNP variants. The PCNA branch site pairs with U12 and a loop of U5 aligns the exons of PCNA for ligation. U4 and U6 snRNAs are not used but the highly divergent U4atac and U6atac taking over their role. (See introns, exons, splicing, PCNA, snRNA)

SPLICING: joining of RNA with RNA or DNA with DNA at the sites of previous cuts. Constitutive splicing indicates that the exons are spliced in the same order as they occur in the primary RNA transcript, in contrast to alternative splicing when the exons may be joined in alternative manners and thus providing mRNAs for different proteins transcribed from the same gene. General scheme of pre-mRNA splicing:

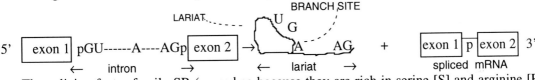

The splicing factor family SR (named so because they are rich in serine [S] and arginine [R]) contain an RNA recognition motif (RRM/RNP) and Ψ-RRM domain Ser-Trp-Gln-Asp-Leu-Lys-Asp, separated by a Gly rich tract. The Ser-Arg domains are well phosphorylated. These proteins select the splice sites and are active parts of the spliceosome complex. The ASF/SF2

Splicing continued
member of this family recruits the U1 snRNP to the RNA transcript. SR proteins communicate also between introns across exons, and affect alternative splicing. The SR proteins are subject to regulation. In-between the branch site and AG there are polypyrimidine sequences and these assist with the aid of various proteins to define the 3' splice site. Some of the many relevant proteins are PTB (polypyrimidine tract-binding protein) and PSF (PTB-associated splicing factor). Plants also have splicing factors but these are more variable than those of animals and this is the cause why animal introns are not spliced out in plants. (See restriction enzyme, vectors, introns, speckles intranuclear, lariat, spliceosome, U1 RNA, snRNA, alternative splicing)

SPLICING ENHANCER: repetitive GAA sequences associated with proteins.

SPLICING JUNCTURE (splice junction): sequences at the exon-intron boundaries (See introns, exons, spliceosome)

SPLIT GENE: discontinuous because of introns are intercalated between exons; the majority of the eukaryotic genes contain introns. (See introns, exons, splicing, spliceosome)

SPLIT-HYBRID SYSTEM: provides means for positive selection for molecules that disrupt protein-protein interactions. The genetically engineered construct is shown at left below:

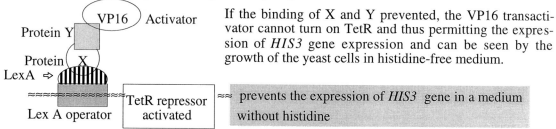

If the binding of X and Y prevented, the VP16 transactivator cannot turn on TetR and thus permitting the expression of *HIS3* gene expression and can be seen by the growth of the yeast cells in histidine-free medium.

~~ prevents the expression of *HIS3* gene in a medium without histidine

(See two-hybrid method, VP16, one-hybrid binding assays, split-hybrid system, three-hybrid system, tTA, rtTA)

Spm (*suppressor mutator*): a transposable element system of maize of an autonomous *Spm* and a non-autonomous (originally unnamed) element.

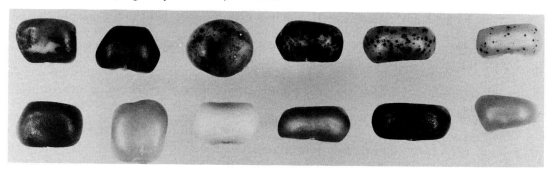

THE DOMINANT ALLELES OF MAIZE NORMALLY FORM DARK COLOR BUT THE a^{m-1} ALLELE POSSESSING THE Spm ELEMENT MAY DISPLAY ALTERNATIVE STATES. IN THE ABSENCE OF AN INACTIVE Spm ANY SHADE OF SOLID COLOR MAY APPEAR (bottom row) BUT IN THE PRESENCE OF AN ACTIVE Spm A VARIETY OF SECTORS ARE DISPLAYED (top row). (Courtesy of Barbara McClintock)

The non-autonomous element cannot insert or excise by its own power because it is defective in the transposase enzyme. This system is the same as the *En (Enhancer) -I (Inhibitor)* system. The non-autonomous component has also been called *dSpm (defective Spm)*. The original name represents the fact that insertional mutations caused by the non-autonomous *dSpm (I)*

Spm continued
element revert at high frequency only when *Spm* (*En*) is introduced into the genome because the latter has a functional transposase gene. The insertion does not always eliminate the function of the target gene and both a recessive mutation and the *Spm* element may be expressed. Such a case at the *A* (anthocyan) locus of maize (chromosome 3L-149) was designated as *a-m2* (*a* mutable) because of its frequent reversion to the dominant allele and displaying sectors in the presence of *Spm*. The *Spm-dependent* alleles harbor a non-autonomous (*dSpm* or *I*) insertion element that may jump out (and cause reversion) only when the functional *Spm* is introduced. The *Spm suppressible* alleles indicate the presence of a non-autonomous element that may or may not permit the expression of the gene but the presence of the active *Spm* allows the insert's removal. The terms *Spm-w* and *Spm-s* indicate weak and strong *Spm* transposase elements, respectively. *Inactivated Spm* has been called *Spm-i*, and elements that have alternating active and inactive phases were called *cycling Spm* (*Spm-c*) whereas very stably inactive forms were named *cryptic Spm* (*Spm-cr*). In addition a *Modifier* factor has been named that enhances the activity of *Spm*. The various forms of the *Spm* alleles were found to alter their activity in time, by developmental stage and extraneous factors which cause chromosome breakage (radiation, tissue culture, etc.). The term *presetting* was applied to the phenomenon that *Spm* determined the expression of a gene even after its removal from the genome but this effect later may fade away. Presetting was supposed to happen during meiosis and it was attributed to methylation. Indeed, the first exon of *Spm* is rich in cytidylic residues, the most commonly methylated base in DNA. Methylation may be instrumental in the variations and inactivation of the various *Spm* elements named above. A *Regulator* element is credited with the control of the extent of methylation. Although the various *Spm* alleles display some heritable qualities they appear to represent labile "changes in states" of the element, except the *dSpm* or other deletional forms. Transposition of the *Spm* elements is not tied to DNA replication and the transposition favors new insertion sites within the same chromosome although it may move to any other part of the genome. Integration of *Spm* re-results in a 3-bp duplication at the target site. The frequency of insertion and excision is influenced also by the base sequences flanking the target site. The *Spm* appears to be modified by the process of excision and involves various lengths of terminal deletions but the subterminal repetitious sequences may also be involved. Deletion of the terminal repeats abolishes transposition of the element. The abilities of *Spm* to transpose and to affect gene expression are inseparable.

The *Spm* element is transcribed into a 2.4 kb RNA with 11 exons; it includes two close open reading frames. One mRNA encodes 621 amino acids. The entire element (8.3 kb) is flanked by 13 bp inverted repeats (CACTACAAGAAAA and TTTTCTTGTAGTG). In a region 180 bp from the 5' end and 299 bp from its 3' end several copies of a receptive consensus CCGACATCTTA occur. The defective elements, *dSpm* have various length of internal deletions covering the entire length or part of the two open reading frames. Some partially deleted elements may still function as a weak *Spm-w* element. The function of an Spm element is regulated by sequences within the element and the location of the transposable element within the target genes. The function of the target genes depends on when, where and how the *Spm* element and the target gene's transcripts are spliced. (See illustration on page 980, transposable elements, hybrid dysgenesis, insertional mutation, Ac-Ds)

SPN (single nucleotide polymorphism): indicates single nucleotide difference between two nucleic acids.

SPONDYLOEPIPHYSEAL DSYPLASIA (SED): the *autosomal dominant* phenotype includes flattened vertebrae, short limbs, cleft palate, myopia (near-sightedness), muscle weakness, hernia and mental retardation. Collagen defects are incriminated in many cases. *Autosomal recessive* forms mimic arthritis-like symptoms (arthropathy) besides the short stature. The autosomal recessive forms may not involve flat vertebrae and the defect was attributed to a defici-

Spondyloepiphyseal dysplasia continued

ency of phosphoadenosine-5'-phosphosulfate and thus to undersulfated chondroitin. An *X-linked* SED was also described. (See achondroplasia, dwarfness, arthropathy-campylodactyly, chondroitin sulfate, collagen)

SPONTANEOUS GENERATION, CURRENT: until Louis Pasteur (1859-61) it was assumed by many scientists that microorganisms are formed from abiotic material even during the present geological period. The organisms found in broth and other rich nutrients grew — as he has demonstrated — only when the solutions were not heated to sufficiently high temperature and duration of time and exposed to unfiltered air. His discovery has been fundamental to modern microbiology and medicine and proved that spontaneous generation is not responsible for the current variations in microbial cultures. (See spontaneous generation unique or recurrent, lysenkoism, pleomorphism)

SPONTANEOUS GENERATION, UNIQUE or REPEATED: is an explanation for the abiotic origin of life. It is assumed that first, after the earth was formed, water, carbondioxide, ammonia and methane, all simple molecules were formed. Subsequently in a reducing atmosphere and energy sources being available (ultraviolet light), simple organic acids (acetic acid, formic acid) arose. In the presence of ammonia, methane, hydrogen, hydrogen cyanide and lightning energy amino acids could be formed. In following steps nucleotides could arise. Actually under simulated early earth conditions amino acids, polypeptides, carbohydrates and nucleic acids could be synthesized by chemists. These simple organic molecules might have aggregated into some sorts of micellae (bubbles), and after selfreplicating mechanisms came about the possibility opened up for the generation of an ancestral cell with a primitive RNA as the genetic material. It is not entirely clear when, where and how and how many times these events took place. Life is estimated to begin 3-4 billion years ago. It is not known whether on other planets, under similar conditions to that of the earth, living cells evolved. (See biopoesis, exobiology, spontaneous generation current, evolution prebiotic, origin of life, abiogenesis)

SPONTANEOUS MUTATION: occurs at relatively low frequency when no known mutagenic agent is or was present in the environment of the cell or organism; the cause of the mutation is thus unknown. The spontaneous frequency of mutation in mice for seven standard loci is 6.6×10^{-6} per locus. At some other loci and in other organisms it may be substantially higher or lower. (See mutation rate, mutation spontaneous)

SPOOLING: assumes that the triple- or quadruple-stranded naked DNA molecules are wound into the RecA protein filament for pairing and exchange. After the formation of heteroduplexes the DNAs are released. (See recombination, molecular mechanism)

SPORADIC: the rare occurrence of an off type does not show a clear familial pattern, and the etiology (cause or origin) is unknown. (See also epidemiology)

SPORANGIOPHORE: a sporangium-bearing branch. (See sporangium)

SPORANGIUM (plural sporangia): the spore-producing and containing structure in lower organisms (fungi, protozoa).

SPORE: a reproductive cell; generally the product of the eukaryotic *meiosis*, or it may arise *mitotically* as fungal conidiospores (conidia). The *bacterial spores* are metabolically dormant cells, surrounded by a heavy wall for protection under very unfavorable conditions.

SPORE MOTHER CELL: see sporocyte

SPORIDIUM: the "sexual" spore of basidiomycetes fungi. (See basidium)

SPOROCYTE: diploid cell that produces haploid spores as a result of meiosis. (See gametogenesis)

SPOROGENESIS: the mechanism or process of spore formation. (See meiosis, conidia)

SPOROPHORE: fruiting body capable of producing spores in fungi. (See spore)

SPOROPHYTE: the generation of the plant life cycle that produces by meiosis the (1n) gametophytes. The common form of plants (displaying leaves and flowers, etc.) is the (2n) sporophytic generation. (See gametophyte)

SPOROZOITE: the infective stage of protozoan life cycle; in malaria, they are formed within the mosquito. (See malaria, *Plasmodium*, *Anopheles*)

SPORULATION: in bacteria the process of formation of morphologically altered cells that can survive adverse conditions and assure the survival of the sporulating bacteria; also ascospore formation through meiosis.

SPOT TEST: a variation of the mutagen/carcinogen bioassays when the compound to be tested by bacterial reversions is added to the surface as a crystal or a drop after the Petri plates have been seeded by the bacterial suspension and the S9 microsomal fraction added. If the substance to tested is mutagenic, a ring of revertant cells should appear around the spot where it was added. This type of mutagenic assay is no longer much used. (See Ames test, plate incorporation test, reversion assays of *Salmonella*)

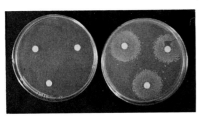

REVERSION ASSAY OF *SALMONELLA* BACTERIUM BY THE SPOT TEST. (Courtesy of Dr. G. Ficsor)

SPOTTING: see piebaldism, KIT oncogene, variegation

spp: the abbreviated plural of the word species.

spretus: *Mus spretus*, a species of mouse, commonly used for genetic analyses.

Sprekelia formosissima (Amaryllidecea): subtropical plant, n = ca. 60; genome sixe 1.8×10^{11} bp.

springer: see *copia*

SPRUCE (*Picea* spp): timber tree species with 2n = 2x = 24. (See pine, douglas fir)

SPT: a suppressor of transposition (e.g. of a Ty element)

SQUASH: *Cucurbita maxima* (winter squash) 2n = 24, 40; *C. pepo* (summer squash) 2n = 40; *C. moschata* (pumkin), 2n = 24, 40, 48.

SQUASH PREPARATION: for microscopic analysis of chromosomes in soft (softened) tissues there may be no need for sectioning but the fixed and stained material can be examined after smearing it directly on the microscope slide. In some cases the fixation is done by gentle heating on the slide followed by adding a small drop of aceto carmine or aceto orceine stain or even just placing the specimen into a drop of acetic stain and heating. These rapid procedures may permit an estimation of the stage of meiosis or mitosis. (See microscopy, stains)

SQUIRREL: *Tamiasciurus hudsonicus streatori*, 2n = 46; *Callospermophilus lateralis*, 2n = 42; *Ammospermophilus*, 2n = 32; *Citellus citellus*, 2n = 40.

SR1: a cytoplasmic mutant of tobacco resistant to streptomycin. (See *Nicotiana*)

SR MOTIF: serine/arginine-rich domains in RNA-binding proteins, involved in splicing pre-mRNA transcripts; they are required in the early steps of spliceosome assembly. One Sr protein (SC35) alone is sufficient to form a committed complex with human β–globin pre-mRNA. Different single SR proteins commit different pre-mRNAs to splicing and different sets of SR proteins may determine the alternative and tissue-specific splicing within an organism. (See splicing, processing, primary transcript, introns, tissue specificity)

SR PROTEINS: see SR motif

SRB: protein stabilizing RNA polymerase II binding to general transcription factors. They occur in eukaryotes from yeast to humans and have the forms, SRB 2, 4, 5, 6, 7. The *SRB* genes involve also mutation suppression in the carboxy-terminal of pol II, and they may encode kinase and cyclin-like proteins. (See transcription factors, transcription complex, kinase, cyclin, RNA polymerase holoenzyme, open promoter complex, regulation of gene activity)

SRBC: sheep red blood cell

SRC: Rous sarcoma virus oncogene of chicken. Its product is a protein-tyrosine kinase, a cellular signal transducer. Its homology domains SH2 and SH3 are present in several cytoplasmic mediator and adaptor proteins in the signal transduction pathways in different organisms. These domains bind phosphotyrosine or proline-rich residues. In humans SRC is in chromosome 20q12-q11. SRC may be involved both RAS-dependent and RAS-independent signaling

Src continued
pathways and may lead through either FOS or MYC to transcription factors. This family of nonreceptor kinases includes Src, Yes, Fgr, Fyn, Lck, Hck, Blk. Csk⁻ cells display increased Src, Fyn and Lyn activity. The Cbl oncogene acting downstream of Src is responsible for bone resorption in osteoporosis. (See oncogenes, SH2, SH3, signal transduction, Yes, Fgr, Fyn, Lck, Hck. Blk, Csk, Cbl, osteoporosis)

SRC-1 (steroid receptor coactivator-1): it enhances the stability of the transcription complex controlled by the progesteron receptor. (See transcription, progesteron)

SRE: a cis-acting enhancer element responding to serum induction: $CC(AT)_6GG$ (CarG box) is present in all serum response factor regulated gene. (see MADS box, TCF)

SREBP (sterol regulatory element binding proteins): see SCAP

SRF: serum response factor is a trans-acting regulatory protein binding to SRE and regulating serum-induced gene expression. (See trans-acting element, serum response element)

SRK: self-compatibility protein receptor kinase. (See self-incompatibility)

SRP: signal recognition particle is an element of polypeptide transport systems through the membranes of the endoplasmic reticulum. The subunit (Ffh) which recognizes the signal sequence as well as the α subunit (FtsY) of its receptor (SR) are GTPases. (See 7SL RNA, endoplasmic reticulum, signal recognition particle, signal peptide)

SRY (**s**ex-determining **r**egion **Y**): mammalian gene in the short arm of the Y chromosome responsible for testis determination and for the development of pro-B lymphocytes. The protein is a member of the high mobility group proteins. The expression of SRY initiates the formation of the Müllerian inhibiting substance (MIS) and the synthesis of testosterone. SRY contains the DNA minor groove binding domain, the HMG box that is conserved among mammals. Mutations affecting human sex-reversal are generally within the HMG box. A human chromosome 17q located gene SRA-1 (sex reversal autosomal) may also be controlled by SRY. In the short arm of the human X-chromosome there is the DSS (dosage-sensitive sex reversal) locus that also may be involved in sex reversal in case of its duplication. There is/are the SRYIF inhibitory factor(s) involved in gonadal differentiation. The voles (rodents) *Ellobius lutescens* 2n = 17, XO constitution in males and females and *E. tancrei* 2n = 32-54, XX in both males and females, as a normal condition do not have SRY whereas SRY is present in other rodents as well as in other eutherian and marsupial species. (See high mobility group proteins, SOX, TDF, Müllerian ducts, animal hormones, Wolffian ducts, campomelic dysplasia, eutherian)

σ^s (RpoS): is required for the expression of many growth phase and osmotically regulated prokaryotic genes. (See σ subunit of RNA polymerase)

Ss BLOOD GROUP: see MN blood group

SSB: single-strand (DNA) binding protein; in yeast it is encoded by gene *RPA1*. (See recombination molecular mechanism)

SSC: 1 x SSC is a solution of 0.15 M NaCl + 0.015 M sodium citrate, frequently used as a solvent for nucleic acids. (See DNA extraction)

SSCA (single-strand conformation analysis): see single-strand conformation polymorphism

SSCP: see single-strand conformation polymorphism

SSLP: see simple sequence length polymorphism

SSN6: yeast factor abolishing glucose repression of SUC2 invertase.

ssrA: see protein repair

SSV: simple sequence variation in DNA.

ST-1: is a single-stranded DNA phage, related to φX174 and G4.

STAB CULTURE: the microbial inoculum is introduced into agar medium by a stabbing motion of the inoculation needle or loop for the purpose of propagation.

STABILIZING SELECTION: see selection types

STABLE RNA: ribosomal and tRNA that persists long in the cell in comparison to mRNA that may be degraded in minutes. (See rRNA, tRNA, mRNA)

STAGGERED CUTS: after the cut of a double-stranded DNA, the length of the two polynucleotides is unequal such as: ──────── ────────

STAINS: for light microscopic examination of chromosomal specimens aceto carmine, aceto orcein or Feulgen stains are commonly used. Preparation: 0.5-1 g dry *carmine* powder is boiled for about half an hour under reflux in 100 mL 45% acetic or propionic acid. Orcein 1.1 g is dissolved in 45 mL glacial acetic acid or propionic acid and filled up to 100 mL by H_2O. Filter and store stoppered at about 5 C°. *Feulgen*: 1 g leuco-basic fuchsin is dissolved by pouring over 200 mL boiling H_2O, shake, cool to 50 C°, filter, add 30 mL 1/N HCl, then 3 g $K_2S_2O_5$, allow to bleach in dark for 24 hrs stoppered. Decolorize by 0.5 g carbon, shake 1 min then filter and store stoppered in refrigerator. For carmine or orcein staining, fix specimens in Carnoy and stain. For Feulgen, fix in Farmer's solution for a day, rinse with water, hydrolyze at 60 C° for 4-10 min. (the duration of hydrolyzation is critical and may need adjustment for each species). Rinse with stain for 1-3 hrs. Tease out tissue in 45% acetic acid, remove debris, flatten by coverslip and examine. May need overstaining with carmine if Feulgen staining is poor. For histological staining a variety of other stains may be used such as haematoxylin, methylene blue, ruthenium red, malachite green, sudan black, coomassie blue, fluorochromes, etc. may be employed. (See fixatives, sectioning, light microscopy, C-banding, G-banding, Q-banding, chromosome painting, harlequin stain, FISH, fluorochromes)

STAMEN: the male reproductive organs of plants composed of the anther which contains the pollen and the filament. (See anther, pollen)

STAMINA: capacity to endure, vigor.

STAMINATE FLOWER: male flower of monoecius or dioecious plants.

STAMMERING: is a most serious form of the speech defect, *stuttering*. The sufferers usually cannot talk fluently and after involuntary stops repeat syllables or entire words. The frequency of stammering is very frequent among Japanese and very unusual among American Indians. Autosomal dominant inheritance seems to be involved. (See stuttering)

STANDARD: an accepted point of reference, e.g. standard wild type. (See also control)

STANDARD DEVIATION (parametric symbol σ): is a measure of the variablity of members of the populations (s = $\sqrt{variance}$. (See variance, standard error)

STANDARD ERROR: measures the variation of the means of various samples of a population s/$\sqrt{m}$; (by some authors standard error and standard deviation are used as synonyms); standard error of proportions or fractions or frequencies = $\sqrt{\frac{p[1-p]}{n}}$ where *p* is the proportion or frequency and *n* is the population size (number of individuals or measurements).

STANDARD TYPE: generally it means the wild type used as a genetic reference. (See wild type)

STANDARDIZED FITNESS: see selection coefficient

STANFORD–BINET TEST: a modified Binet intelligence test. (See Binet test, human intelligence)

STANNIOCALCIN: calcium and phosphate, regulating hormones of vertebrates.

StAR (steroidogenic acute regulatory protein): enhances mitochondrial conversion of cholesterol into pregnenolone, an intermediate in steroid biosynthesis. Mutation in the coding gene results in deficiency of adrenal and gonadal steroidogenesis, and leads to congenital lipoid adrenal hyperplasia, an autosomal recessive disorder. (See cholesterol, steroid hormones, adrenal hyperplasia)

STARCH: see amylopectin

STARFISH: *Asterias forbesi*, 2n 36.

START: see cell cycle

START CODON: is AUG in RNA that specifies either formyl methionine (in prokaryotes) or methionine (in eukaryotic cells). Note that the Met codon in mitochondria varies in different organisms. (See genetic code)

START POINT: the position in the transcribed DNA where the first RNA nucleotide is incorporated. (See transcription)

STASIS: an equilibrium state without change.

STAT (signal transducers and activators of transcription): are cytoplasmic proteins which become activated by SRC mediated phosphorylation at tyrosine residues through the action of Jak kinases which are receptors for cytokine signals and enzymes at the cytosolic termini. Stat4 protein is essential for interleukin-12 mediated functions such as the induction of interferon-γ, mitogenesis and T lymphocyte killing and helper T lymphocyte differentiation. The latter process requires the action of the 42 kDa MAPK or ERK2. STAT recognizes and activates different genes. Cooperative binding of STAT makes possible the recognition of variations at the different binding sites. The sequence-selective recognition resides in their amino terminal domains. (See signal transduction, Jak-STAT pathway, signal transduction, SRC, lymphocytes, leptin)

STATHMOKINESIS: mitotic arrest. (See spindle poison)

STATIONARY PHASE: the population size is maintained without increase or decrease. (See growth curve)

STATISTIC: an estimate of a property of an observed set of data (e.g. mean) that bears the same relation to the data as the parameter does to the population. (See statistics)

STATISTICS: is a mathematical discipline that assists in collecting and analyzing data; guides to make conclusions and predictions on the basis of the analysis and reveals their trustworthiness by determining probability or likelihood. Statistics is used for collections of facts on demography, industrial productivity, trade, etc. such as in statistical yearbooks. (See probability, Bayes' theorem, likelihood, maximum likelihood, non-parametric tests)

STATOLITH: granules that are believed to be sensing gravity in cells. (See gravitropism)

STATURE IN HUMANS: is determined by environmental causes such as nutrition, disease, injuries, by the use of various medications and by simple or complex genetic factors. Prenatal anomalies of bone length may be determined by prenatal diagnosis. The most common types genetically determined human dwarfisms and other defects involving reduced stature are: achondroplasia, hypochondroplasia, achondrogenesis, osteochondromatosis, dyschondrosteosis, Russel-Silver syndrome, Smith-Lemli-Opitz syndrome, Opitz-Kaveggia syndrome, dwarfism, SHORT, Aarskog syndrome, Noonan syndrome, Hirschprung disease, Turner syndrome, Trisomy, Mulibrey nanism, hairy elbows, Pygmy. (See also growth hormone, growth retardation, limb defects, exostosis, regression)

STC (sequence-tagged connector): see genome project

STE II: mating factor receptors in budding yeast; in fission yeast the homolog is *Byr2* are extracellular signal regulated kinase homologs of MEK and ERK. (See signal transduction)

STEADY STATE: in a reaction enzyme-substrate concentration and other intermediates appear constant over time but input and output are in a flow.

STEEL FACTOR (stem cell factor): is a 40 - 50 kDa dimeric protein produced by the bone marrow and other cells and migrates to the hematopoietic stem cells. Its receptor is a transmembrane protein tyrosine kinase. (See cell migration, hematopoiesis, tyrosine kinase)

STEER: an emasculated bovine male animal. (See emasculation)

STELE: the core cylinder of vascular tissues in plant stems and roots.

STEM CELLS: of animals are not terminally differentiated, can divide without limit and when they divide the daughter cells can remain stem cells or can terminally differentiate in one or more ways. They occur in various tissues. Functionally they resemble the meristem in plants. (See meristem, regeneration in animals)

STEM - LOOP STRUCTURE: any DNA or RNA that may have non-paired single strand loop associated with a double-strand stem similar to a lollipop. (See palindrome, repeat inverted)

STEM RUST: caused by infection of the basidiomycete fungus *Puccinia graminis* on cereal plants. The haploid spores produced on the wheat plant germinate on the leaves on barberry shrubs and form pycnia (pycnidium). The pycniospores of different mating types undergo plasmogamy and form dikaryotic aecia (aecidia) on the lower surface of the barberry leaves. The aecisopores infect the wheat leaves and form the dark brown rust postules, called uredia. The dikaryotic uredospores reproduce then asexually and spread the disease. At the end of the growing season karyogamy takes place and the diploid teliospores are formed. The teliospores overwinter and eventually undergo meiosis and liberate the haploid basidiospores that germinate on barberry and restart the cycle. Stem rust may cause very substantial crop loss in wheat and other gramineae. The newer varieties are genetically more or less resistant to the fungus. *P.* spp have chromosome numbers 3-6. (See host - pathogen relationship)

STENOSIS: narrowing of a body canal or valve such in the aorta, heart valve, pulmonary artery, vertebral canal, etc.

STENOSPERMOCARPY: genetically determined abortion of the embryo soon after fertilization resulting in seedless but normal size berries, a desirable trait of table grapes. (See seedless fruits)

STEP ALLELOMORPHISM: is a historically important concept that paved the way to allelic complementation and to the study of gene structure. In the late 1920s Russian geneticists discovered that partial complementation among allelic genes may occur in a pattern and that was inconsistent with the then prevailing idea that the gene locus is the ultimate unit of function, mutation and recombination and that alleles are stereochemical modifications of an indivisible molecule. (See allelic complementation, Offermann hypothesis)

STEREOISOMERS: molecules with identical composition but with different spatial arrangement.

STEREOMICROSCOPY (dissecting microscopy): is used for visual analysis under relatively low magnification of natural specimens without sectioning. It has special advantage for dissecting structural elements with binocular viewing and top or side illumination without fixation and/or staining. (See confocal microscopy, scanning electronmicroscopy, microscopy)

STERIGMA: the small stalk at the tip of a basidium from which spores come off in some fungi.

STERILITY: either the male or the female or both types of gametes (haplontic) or the zygotes (diplontic) have reduced or no viability caused by lethal or semilethal genes, chromosomal defects, differences in chromosome numbers or incompatible cytoplasmic organelles. (See infertility, semisterility, somatoplastic sterility, cytoplasmic male sterility, incompatibility, self-incompatibility, azoospermia, infertility, deletion, inversion, translocation)

STERILIZATION, GENETIC: see genetic sterilization

STERILIZATION, HUMANS: had been practiced by various societies for different reasons. The eunuchs of the Chinese imperial courts and of the Osmanic harems served as guardians of the privileges of tyrannical social structures. The castrati of the Italian opera stages were exploited for singing female roles in an era when women were banned from performing arts. In the 1880s by the publications of Sir Francis Galton scientific justifications were sought for negative eugenics in order "to produce a highly gifted race of men by judicious marriages during consecutive generations." By 1890s sporadic sterilization of institutionalized, mentally retarded persons was initiated. Starting in 1907, in about 14 states (USA) laws were enacted for systematic sterilization of mentally retarded, blind, deaf, crippled or afflicted by tuberculosis, leprosy, syphilis, and chronic alcoholism. This "practical, merciful and inevitable solution" eventually degenerated into legal suggestions to eliminate criminal behavior, disease, insanity, weaklings and other defectives and "ultimately to worthless race types". Before strong moral objections gained noticeable ground in the year 1956 nearly 60,000 human individuals were legally sterilized. Interestingly, the Oklahoma law exempted from mandatory sterilization offenses against prohibition, tax evasion, embezzlement and political crimes. Several state laws advocated mandatory sterilization also as a guard against illegitimacy, particularly by unwed recipients of the Aid to Families with Dependent Children (AFDC). Until 1965, judicial approval could be obtained for a "good cause" for forced sterilization of mentally retarded indi-

Sterilization, humans continued

viduals whose family had hardship in supporting the offspring of promiscuous children. Although not all states rescinded yet the old laws, sterilization of humans is now practiced only voluntarily by ligation of the vas deferens (vasectomy) or tubal constriction or ovariectomy or by the use of various types of mechanical and hormonal contraceptives. Compulsory sterilization is objectionable on moral ground because reproduction is a basic human right. It is particularly reprehensible when advocated as a selective measure against certain human races. From genetic perspective it is controversial since 83% of the mentally retarded children are born to non-retarded parents. Also, selection against the majority of human defects is quite inefficient since the vast majority of the defective genes are in heterozygotes and many of the conditions are under polygenic control or are non-hereditary. Furthermore, there are no objective scientific or practical measures for the evaluation of most of the human traits. (See selection, eugenics, Genetics, Law, Social Policy by Reilly, P. 1977, Harvard Univ. Press)

STERNITES: the ventral epidermal structures of the abdomen. (See *Drosophila*)

STEROID HORMONES: are derived by the following pathway:

$$\text{[1] Cholesterol} \to \text{[2] Prognenolone} \to \text{[3] Progesterone} \to \begin{array}{l} \to \text{[4] Testosterone} \to \text{[5] Estradiole} \\ \to \text{[6] Corticosterone} \to \text{[7] Aldosterone} \\ \hspace{3.5cm} \to \text{[8] Cortisol} \end{array}$$

$$\downarrow$$
[9] Dexamethasone

[3] Is the principal hormone of the endocrine gland, corpus luteum, in the ovarian follicle after the release of the ovum. It also regulates the expression of the secondary sexual characters of females. [4] Is the main male sex hormone that is produced in 6-10 mg quantities daily in men and ca. 0.4 mg in women. It is responsible for the production of facial hair and baldness and the regulation of growth. [5] Is formed by oxidative removal of C-129 from its precursor; primarily a female hormone occuring in the ovaries and placenta and it is responsible for regulating — among other functions — bone growth, increased fat content and smoother skin of females compared to men. This hormone is present also in the testes. In cooperation with pro-, gesterone it regulates also the menstrual cycles. [6] and [7] are synthesized in the kidney cortex and regulate — among others — mineral (Na^+ Cl^-, HCO_3^-) reabsorption and are frequently called as mineralcorticoid hormones. [8] Is a glucocorticoid affecting protein, carbohydrate metabolism regulates the immune system, allergic reactions, inflammations, etc. [9] is also an anti-inflammatory glucocorticoid with a role in activating the glucocorticoid receptors. The number of steroid hormones is about 50 and they are present in practically every cell of the body — be-sides those mentioned — and they, along with thyroid hormones, have important roles in activation of genes. Up to 1966 the general assumption was that steroid hormones are not used by plants. It has been demonstrated that brassinolids (related to cholesterol, ecdysone) mediate several developmental processes in plants, such as elongation, light responses, etc. The steroid receptor superfamily includes receptors for estrogen, progesteron, glucocorticoid, mineralcorticoid, androgen, thyroid hormone, vitamin D, retinoic acid, 9-cis retinoic acid and ecdyson. The steroid hormone receptors stimulate the formation and then stabilize the pre-initiation complex of transcription. Most commonly the condition of their binding to the hormone response element is the binding to their appropriate ligands. Some, such as the thyroid hormone receptor, can bind to DNA in the absence of a ligand. In the absence of the ligand, they function as silencers through interaction with the TFIIB transcription factor. (See hormone-response elements, silencer, PIC, transcription factors, transcriptional activator, coactivator, regulation of gene activity, animal hormones, plant hormones, brassinosteroids)

STEROID SULFATASE DEFICIENCY: see ichthyosis

STEROIDS: contain a four-ring nucleus consisting of three six-membered rings and one five-membered ring. (See steroid hormones, brassinosteroids)

STEROLS: are lipids with a steroid nucleus. The concentration of free sterols determines the fluidity of the eukaryotic cell membranes. Esterification of sterols prevents their participation in

Sterols continued

membrane assembly. The process is mediated the ACAT complex (acy-CoA:cholesterol acyl transferase). Increase of ACAT activity my lead to hyperlipidemia and atherosclerosis. Sterol esterification may modify the LDL receptors and potentiates atherogenic processes. It may limit intestinal sterol absorption. (See hyperlipidemia, atherosclerosis, LDL, membranes)

stg (string, map position 3-99): *Drosophila* gene locus controlling the first ten embryonic divisions (similarly to gene *Cdc28* in *Shizosaccharomyces pombe);* it is a cyclin gene. (See cell cycle)

STICKINESS OF CHROMOSOMES: is observed as some sorts of adhesion between any chromosome within a cell. (See side-arm bridge)

STICKLER SYNDROME (arthroophthalmopathy, AOM): early and strong progressive myopia (nearsightedness). Retinal detachment may result in blindness caused probably by a dominant mutation in the collagen IIA1 gene (human chromosome 12q13.-q13.2). (See collagen, eye disease, connective tissue disorders, skin diseases)

STICKY ENDS: double-stranded DNA with a single-stranded over-hang to what complementary sequences are available and so they can stick by base pairing: →

STIGMA: the tip of the style that is normally receptive to the pollen of plants. In zoology, it means spot, such as a hemorrhagic small area on the body. (See gametophyte female, gametophyte male, protogyny, protandry)

STIGMASTEROL: a plant lipid derivative formed by methylation of ergosterol. For guinea pigs it is a vitamin necessary to avoid stiffness of the joints. (See also ergosterol, cholesterol)

STILL-BIRTH: birth of a dead offspring. It is caused by chromosomal defects in ca. 7% of the still-born or by other pathological conditions. (See chromosomal breakage)

STIPULE: a leaf like bract at the base of a leaf.

STM: see scanning tunneling microscope

STOCHASTIC: corresponds to a random process; a process of joint distribution of random variables. In a population — in contrast to a deterministic model — random drift and other chance events may determine the gene frequencies. Generally, deterministic and stochastic processes are running parallel and simultaneously. (See deterministic model)

STOCK: a genetically defined strain of organisms; also a root stock on what a scion is grafted.

STOCK, GARDEN (*Matthiola incana*): see *Matthiola*

STOLON: a horizontal underground stem such as the tuber-bearing structures of potatoes.

STOMA (plural stomata): a small hole on the leaf surface surrounded by two guard cells which control opening and closing. The stoma permits gas exchange (CO_2 uptake), and release of water vapors (transpiration). The opening of the stomata requires an increase in the turgor of the guard cells. It had been suggested that the process may be promoted by opening of K channels and the subsequent influx of K^+. The opening of the K channel is activated by a light controled proton pump. The closure of the stomata is controled by the hormone ABA and the influx of Ca^{2+} and the efflux of K^+. The calcium level is sensed by a cyclin-dependent protein kinase (CDPK). In the regulation, Ca^{2+}-dependent ATP-ases have major role These processes involve changes in the electric potentials (depolarization). (See ion channels, cell cycle, aequorin, calmodulin, cyclin, ATPase, proton pump)

STOMATIN: a cation conductance protein in the cell membrane.

STOP CODON: see nonsense codon, genetic code

STOP SIGNAL: see transcription termination in eukaryotes, transcription termination in prokaryotes, stop codon, release factor [RF]

STOPPERS: mitochondrial mutations in *Neurospora* displaying stop-start growth. (See poky)

STP: see signal transfer particle

STPβ (second strand transfer protein): see recombination mechanisms eukaryotes, Sep 1

STR: short tandem repeats, such as found in micro- and minisatellites. (See microsatellite, minisatellite)

STRABISMUS: is an anomaly of the eyes they may be either divergent or convergent or one di-

Strabimus continued

rected up, the other down because of the lack of coordination of the muscles concerned. Some persons display this anomaly only periodically. The pattern of inheritance is not entirely clear; most likely dominant factor(s) are involved. The recurrence among the offspring of convergent probands is higher than that among children of the divergent type. Its incidence in the general population is about 0.002. (See eye diseases)

STRAIN: an isolate of an organism with some identifiable difference from other similar groups. This term does not imply any stringent other criteria.

STRAND ASSIMILATION: The *exo* gene of lambda phage codes for a 5'-exonuclease (M_r 24,000) that can convert a branched DNA structure to an unbranched nicked duplex by the process called strand assimilation during recombination. A progressive incorporation of one DNA strand into another during recombination. (See recombination, lambda phage)

DOUBLE-STRANDED DNA

HETERODUPLEX AT STRAND ASSIMILATION

STRAND DISPLACEMENT: a type of viral replication involving the removal of the old strand before the new strand is completed. Similar mechanism is used by mtDNA, (See D loop)

STRAIN DISTRIBUTION PATTERN (SDR): the distribution of two alleles of a diploid among the progeny where linkage is studied either by a backcross or by recombinant inbred procedure. (See backcross, linkage, recombinant inbred)

STRAWBERRY (*Fragaria ananassa*): about 46 *Fragaria* species with x = 7; the wild Eurpean *F. vesca* is diploid (2n = 14), *F. moschata* (2n = 42), some east Asian species are tetraploid, the American strawberrys as well as the garden strawberries are 2n = 56.

STREAK (primitive streak): a sign on the early embryonal disc indicating the movement of cells and the beginning of the formation of the mesoderm and an embryonal axis. (See organizer)

STREPTAVIDIN: conjugated with rhodamine specifically binds to biotin (biotinylated nucleic acids, immunoglobulins) and permits their detection by fluorescence. The binding constant for biotin is $k_a = 10^{15}$ M^{-1}. (See also avidin)

STREPTAVIDIN-PEROXIDASE: identifies biotinylated antibodies in ELISA, in immunochemistry in general, and in protein blots. (See genomic subtraction, ELISA, biotinylation)

STREPTOCOCCUS A: is the common pathogenic bacterium causing pharyngitis (sore throat). About 5-10% of the infections may involve necrotic lesion of various severity. In rare extreme cases it may cause death. Some strains secrete substantial amount of a pyrogenic exotoxin A that stimulates the immune system as a superantigen. The excessive stimulation results in the overproduction of cytokines that may damage the lining of the blood vessels and thus causing fluid leakage, reduced blood flow and necrosis of the tissues because of the lack of oxygen. As a further consequence fasciitis (inflammation of the fibrous tissues) and myositis (inflammation of the voluntary muscles) may follow. The destruction of the tissues may result in death within a very short period of time after infection by the extremely virulent strain of "flesh eating bacteria". *Streptococcus pneumoniae* (*Pneumococcus*) provided the first information on genetic transformation in 1928. (See transformation genetic, necrosis, superantigen)

STREPTOLYGIDIN: an antibiotic that blocks the action of prokaryotic RNA polymerase.

STREPTOMYCES: is a group of Gram-negative bacteria of the actinomycete group, characterized by mycelia-like, septate colonies. On these mycelial colonies spore-bearing organs develop. These bacteria somewhat simulate a multicellular type of development.

STREPTOMYCIN: an antibiotic compound, precipitates nucleic acids, inhibits protein synthesis. Streptomycin resistant mutations in the ctDNA are maternally inherited; such mitochondrial DNA mutations may lead to hearing loss in humans. (See antibiotics, mitochondrial diseases in humans, mtDNA)

STRESS: a condition when living beings most cope with difficult mental or physiological conditions. In case of failure to successfully respond with some type of a homeostatic mechanism, death or substantial harm may result. Stress activates sphingomyelinase to generate ceramide

Stress continued

and the latter initiates apoptosis. Stress also activates heatshock proteins and glucose-regulated proteins (GRP). The GRP proteins are highly active during tumor progression. Their suppression may lead to apoptosis and rejection of the tumor cells. The stress signals may mediate the activation of genetic repair systems or in animals may proceed through three main pathways, the c-Abl or the JNK or the p53 routes. The first two are specific to different types of genotoxic agents, the p53 protein responds rather generally to various chemical stresses. The signal transducers eventually reach to the DNA by the activation of transcription factors. Ionizing or excitatory (UV) radiations may directly cause chromosome breakage resulting either in repair or apoptosis. In plants, stress stimulates the formation of elicitors and pathogenesis-related proteins. (See homeostasis, p38, GADD153, CAP, ceramides, sphingolipids, sphingolipidoses, SAP, SAPK, apoptosis, JNK cAbl, p53, pathogenesis-related proteins)

STRETCHING CHROMOSOMES: for more precise localization of FISH labels the chromosome can be extended 5-20 times their highly coiled length using hypotonically treated, unfixed metaphase chromosomes and centrifugation. (See FISH)

STRIATED MUSCLES: the heart and skeletal muscles are made of sarcomeres and thus striated transversely. (See also smooth muscles)

STRINGENT: rigidly controlled

STRINGENT CONTROL: in amino-acid-starved bacteria (auxotrophs), the product of the $relA^+$ gene shuts off ribosomal RNA synthesis as an economical device. In the presence of $relA$ amino acid synthesis is promoted (relaxed control) because ppGpp regulates the discriminator regions of the promoters. (See fusidic acid, relaxed control, discriminator region)

STRINGENT PLASMID: its low copy number is genetically controlled.

STRINGENT REPLICATION: limited replication of the low copy number plasmid DNA.

STRINGENT RESPONSE: under poor growth condition prokaryotic cells may shut down protein synthesis by limiting tRNA and ribosome formation. (See stringent control, magic spot)

STROKE: causes 150,000 death/year and afflicts three times as many in the USA. The major genetic factors involved are telangiectasia the Osler-Weber-Rendu syndrome, CADASIL, Ehlers-Danlos syndrome, polycystic kidney disease, Marfan syndrome, cardiovascular diseases, hypertension, melas syndrome. (See terms mentioned under separate entries)

STROMA: the aqueous solutes within an organelle. Also a pseudoparenchymatous association of fungal mycelia. The supportive tissue of an organ; stroma cells in the bone marrow may produce collagen and extracellular matrix.

STROMELYSIN: see transin

STRONG-STOP DNA: when reverse transcriptase initiates transcription of the first strand DNA — from the RNA— then a second strand is made on the 1st strand DNA template. The transcriptase pauses after the transcription of the R U5 segments (first strand) or the U5 R, U3 (second strand), and these "strong stop" DNA species accumulate (in the latter case including also small portions of the RNA primer) before transcription continues to completion of the first and then of the second strand, respectively. (See retroviruses, reverse transcriptase)

STRONTIUM: an earth metal with several isotopic forms, the ^{90}Sr, a β-emitter radioactive component of nuclear fallout is readily substituted for calcium and thus may be concentrated in the milk if the cows grazed on contaminated pastures. Its half-life is 28 years. After the bomb testings in the 1950s it became especially threatening to children whose bones accumulated 2.6 μμCi in contrast to adults (0.4 μμCi). (See radiation hazards, Ci, isotopes)

STRUCTURAL GENE: a primarily non-regulatory DNA sequence that codes for the amino acid sequence in a protein or for rRNA and tRNA.

STRUCTURAL HETEROZYGOSITY: involves normal and rearranged homologous chromosomes within cells. (See inversion, translocation, aberration chromosomal)

STRUMPELL DISEASE: see spastic paraplegia

STS: see sequenced tagged sites

STS-CONTENT MAPPING: in physical mapping the large sequences used (YAC clones) con-

tain STS tracts and their position can be thus mapped. (See sequenced tagged sites)
STUART FACTOR DEFICIENCY: see prothrombine deficiency, antihemophilic factors
STUDENT'S t DISTRIBUTION: a distribution that enables the statisticians to compute confidence limits for μ (the true mean of a population) when σ (the standard deviation of the true mean) is not known, and only the standard deviation of the sample s is available.

CRITICAL VALUES OF THE (TWO-TAILED) t-DISTRIBUTION. The calculated value at the determined degrees of freedom (df) must be identical or greater than the closest value found on the pertinent df line in order to qualify for the probability shown at the top of the columns. E.g., for df = 10, and t = 3.169 P = 0.01 but if the t would be only 3.168 P would be only 0.05 according to the Table and statistical conventions. More precise P values can be obtained by the use of t charts or linear interpolation using the logarithms of the two-tailed probability values (Simaika 1942 Biometrika 32:263). Remember that the t test indicates the probability of the null hypothesis that the two means would be identical.

df	P → 0.900	0.500	0.400	0.300	0.200	0.100	0.050	0.010	0.001
1	0158	1.000	1.376	1.963	3.078	6.314	12.706	63.654	636.620
2	0.142	0.816	1.061	1.386	1.886	2.920	4.303	9.925	31.599
3	0.137	0.765	0.978	1.250	1.638	2.353	3.182	5.841	12.924
4	0.134	0.741	0.941	1.190	1.533	2.132	2.776	4.604	8.610
5	0.132	0.727	0.920	1.156	1.476	2.015	2.571	4.032	6.869
6	0.131	0.718	0.906	1.134	1.440	1.943	2.447	3.707	5.959
7	0.130	0.711	0.896	1.119	1.415	1.895	2.365	3.500	5.408
8	0.130	0.706	0.889	1.108	1.397	1.860	2.306	3.355	5.041
9	0.129	0.703	0.883	1.100	1.383	1.833	2.262	3.250	4.781
10	0.129	0.700	0.879	1.093	1.372	1.812	2.228	3.169	4.587
11	0.129	0.697	0.876	1.088	1.363	1.796	2.201	3.106	4.437
12	0.128	0.696	0.873	1.083	1.356	1.782	2.179	3.054	4.318
13	0.128	0.694	0.870	1.080	1.350	1.771	2.160	3.012	4.221
14	0.128	0.690	0.868	1.076	1.345	1.761	2.145	2.977	4.140
15	0.128	0.691	0.866	1.074	1.341	1.753	2.131	2.947	4.073
16	0.128	0.690	0.865	1.071	1.337	1.746	2.120	2.921	4.015
17	0.128	0.689	0.863	1.069	1.333	1.740	2.110	2.898	3.965
18	0.127	0.688	0.862	1.067	1.330	1.734	2.101	2.878	3.922
19	0.127	0.688	0.861	1.066	1.328	1.729	2.093	2.861	3.883
20	0.127	0.687	0.860	1.064	1.325	1.725	2.086	2.845	3.850
21	0.127	0.688	0.859	1.063	1.323	1.721	2.080	2.831	3.819
22	0.127	0.686	0.858	1.061	1.321	1.717	2.074	2.819	3.792
23	0.127	0.685	0.858	1.060	1.320	1.714	2.069	2.807	3.768
24	0.127	0.685	0.857	1.059	1.318	1.711	2.064	2.797	3.745
25	0.127	0.684	0.856	1.058	1.316	1.708	2.060	2.787	3.725
26	0.127	0.684	0.856	1.058	1.315	1.706	2.056	2.779	3.707
27	0.127	0.684	0.855	1.057	1.314	1.703	2.052	2.771	3.690
28	0.127	0.683	0.855	1.056	1.312	1.701	2.048	2.763	3.674
29	0.127	0.683	0.854	1.055	1.311	1.699	2.045	2.756	3.659
30	0.127	0.683	0.854	1.055	1.310	1.697	2.042	2.750	3.646
40	0.126	0.681	0.851	1.050	1.303	1.684	2.021	2.704	3.551
60	0.126	0.679	0.848	1.046	1.296	1.671	2.000	2.660	3.460
120	0.126	0.678	0.845	1.041	1.289	1.658	1.980	2.617	3.373
	0.126	0.674	0.842	1.036	1.282	1.645	1.960	2.576	3.290

The quantity of t is determined by the equation: $t = \frac{\bar{x} - \mu}{s/\sqrt{n}}$ where $\bar{x}$ is the experimental mean and n is the population size. The critical t values are generally read from statistical tables after it had been quantified by the calculated t value, $t = \frac{d}{\sqrt{v}}$ where d is the difference between

Student's t distribution continued
means and V = variance. Under practical conditions the significance of the difference between two means is calculated by the formula $t = (\bar{x}_1 - \bar{x}_2)/\sqrt{[s_1]^2 + [s_2]^2}$ where the x values stand for the two means and s^2 values are the variances of the two populations. (See arithmetic mean, standard deviation, variance, t table above)

STUFFER DNA: part of the phage λ genome is not entirely essential for normal functions of the phage. Sequences between gene *J* and *att* representing about 1/4 of the genome can be removed and replaced (stuffed in) by genetic engineering without destroying viability of the phage. (See lambda phage, vectors)

sturt: is the unit of fate mapping. (See fate maps)

STUTTERING: a transcription termination phenomenon; poly U may easily break U-A associations. (See also stammering)

STYLUS (style): the slender structure leading from the stigma to the ovary of plants and through which the pollen tube grows to the embryosac. (See gametophyte female, gametophyte male)

su1 = supD, su2 = supE, su3 = supF, su4 = supC, su5 = supG and su7 = supU. (See at *su⁻*)

su⁻: the wild type allele of a suppressor mutation; the suppressor allele is *su⁺*.

su⁺: the suppressor allele at a locus in contrast to the wild type that is designated *su⁻*.

Su BLOOD TYPE: occurs in pigs and resembles the Rh blood type in humans. (See Rh blood type, erythroblastosis fetalis)

SUBCLONING: recloning a piece of DNA. (See cloning molecular, cloning vectors)

SUBCULTURING: transfering of a culture into a fresh medium.

SUBCUTANEOUS: beneath/under the skin.

SUBERIN: is a corky complex polymeric material (of fatty acids but no glycerol associated with it) on the surface and within plant cells. In many plants there is a subepidermal layer of suberin in air-filled cells and in various scar tissues. Suberin is frequently associated with cellulose, tannic acid, dark pigments (phlobaphenes) and inorganics. The commercial cork produced by the oak, *Quercus suber*, is suberin. (See also host-pathogen relation)

SUBLETHAL: only about 50% of the affected my live until sexual maturity.

SUBLIMON: sub-stoichiometric molecules of mtDNA that are supposed to be the products of recombination within short repeated sequences in this organelle. (see mtDNA)

SUBMETACENTRIC: chromosome with two arms clearly unequal in length. (See chromosome morphology)

SUBMISSION SIGNAL: in the majority of vertebrates aggressive behavior generally ends when the weaker partner in the conflict displays the submission signal, e.g. dogs lay on their back. The human race does not employ such definite signals and thus the conflicts frequently end in violence. (See aggression, behavior genetics, human behavior, ethology)

SUBSPECIES: a group of organisms within a species distinguishable by gene frequencies, chromosomal morphology and/or rearrangement(s), and may show some signs of reproductive isolation from the rest of the species. (See species)

SUBSTANCE ABUSE: the proclivity is genetically controlled. Morphine preference has at least three QTLs known, alcoholism has also several QTLs. Some of these conditions may be associated with variations in the serotonin transporter. A genetic factor for the latter is linked to the human ALPC2 locus, controling vulnerability to alcohol. The conditions leading to depression may generally affect substance abuse, some of the quantitative trait loci, mentioned do not appear, however, to act globally. (See alcoholism, serotonin, QTL)

SUBSTITUTION, DISOMIC: two homologous chromosomes replaced by two others. (See alien substitution, intervarietal substitution)

SUBSTITUTION LINE: one of its chromosomes (or pair) is derived from a donor variety or species. (See alien substitution, intervarietal substitution)

SUBSTITUTION, MONOSOMIC: one entire chromosome is substituted for another. (See substitution line)

SUBSTITUTION MUTATION: one base pair is replaced by another. (See transition and trans-, version, base substitution, point mutation, Li-Fraumeni syndrome, base sequences below)

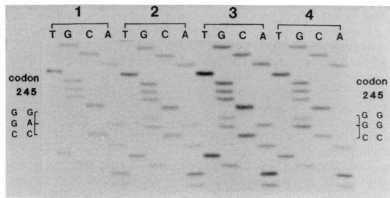

BASE SUBSTITUTION MUTATIONS IN THE p53 GENE IN THE NONCANCEROUS FIBROBLAST CELLS IN A FAMILY WITH THE LI - FRAUMENI SYNDROME. (1) PROBAND, (2) HIS BROTHER, (3) THEIR FATHER AND (4) A NORMAL CONTROL. IN CODON 245 GGC→GAC MUTATIONS ARE EVIDENT IN THE TWO GENERATIONS INVESTIGATED. (Courtesy of Professor Esther H. Chang, Nature 348: 747)

SUBSTRATE: the compound an enzyme can act on, or the culture medium for an organism, or a particular surface. (See pseudosubstrate)

SUBSTRATE ADHESION MOLECULES (SAM): bind as extracellular molecules to independent receptors on adhering cells.

SUBSTRATE CYCLE: see futile cycle

SUBSTRATE INDUCTION: enzyme synthesis is stimulated by the presence of the substrate of the enzyme. (See *lac* operon, substrate)

SUBTILIGASE: an enzyme capable of ligating esterified peptides in aqueous solutions.

SUBTILISIN: is a protease enzyme that cuts at serine in the context Gly-Th - **Ser** - Met-Ala-Ser; chymotrypsin also cuts at serine but within a different sequence.

SUBTRACTION GENOMIC: see genomic subtraction

SUBUNITS OF ENZYMES (protomers): the polypeptides that make up the oligomeric proteins.

SUBVITAL: has reduced viability yet has 50% chance to survive up to the reproductive period of the species. (See sublethal)

SUC2: yeast invertase gene

SUCROSE GRADIENT CENTRIFUGATION: see density gradient centrifugation

SUCROSE INTOLERANCE: see disaccharide intolerance

SUDAN BLACK: stains fatty tissues, wax, resins, cutins, etc. red in microscopic use.

SUDDEN INFANT DEATH SYNDROME (SID): unexpected death of healthy, normal infants within the first year of life during sleep. It appears to be associated with a deficiency in the binding of the muscarinic cholinergic receptors of the brain resulting in accumulation of carbon dioxide or lack of oxygen in the blood. (See muscarinic acetylcholine receptors)

SUG1: is an ATPase and activator of transcription; it can substitute in yeast for Trip1 and can interact with the transcriptional activation domain of GAL4 and herpes virus protein VP16. (See ATPase, transcriptional activator, Trip1, GAL4)

SUGAR BEET (*Beta vulgaris*): $2n = 18$. Is one of the greatest success stories of plant breeding. In the middle of the 18th century the average sugar content of the plant was about 2%; this was increased by the 20th century to about 20%, and the sugar yield per hectare increased to about 4 metric tons. Some of the modern varieties have numerous agronomically important features (disease resistance, monogermy), and this is a rare plant where triploid varieties (besides bananas) are grown commercially.

SUGARCANE: (*Saccharum*, $x = 10$); its diploid forms are unknown and the cultivated varieties have high and variable number of chromosomes ($2n = 80-173$), including polyploids and aneuploids. It is the most important source of saccharose or common sugar.

SUICIDE INHIBITOR: a molecule that inhibits enzyme action after the enzyme acted upon it.

Suicide inhibitor continued

In the original form it is only a weak inhibitor but after reacting with the enzyme it binds to it irreversibly and becomes a very potent inhibitor. Allopurinol, fluorouracil are such a molecules. (See allopurinol, regulation of enzyme activity)

SUICIDE MUTAGEN: uses a ^{32}P-labeled or other radioactive nucleotide that is incorporated into the genetic material and causes mutation by localized radiation. (See also magic bullet)

SUICIDE VECTOR: delivers a transposon into the host cells in which the vector itself cannot replicate but the transposon can be maintained and used for transposon mutagenesis. (See transposon mutagenesis

SULFHYDRYL GROUP: —SH, and when two are joined the linkage is a disulfide bond.

SULFONYLUREA: a group of compounds that stimulate insulin production by regulating insulin secretion and lower blood sugar level and it is used to treat patients with non-insulin dependent diabetes. These drugs interact with the sulfonylurea receptor of pancreatic β cells and inhibit the conductance of ATP-dependent K$^+$ ion channels. Reduction of potassium exit activates the inward rectifying Ca^{2+} channels and promotes exocytosis. Sulfonylureas are inhibitors of the acetolactate synthase enzyme and there are also sulfonylurea herbicides. (See ion channels, diabetes, herbicides)

SULFUR MUSTARD: S⟨Al - Cl / Al - Cl⟩   Al: alkyl group. If 2 alkyl groups are chlorinated it is bi-functional; if only 1 alkyl group is chlorinated it is mono-functional. (See also nitrogen mustard, alkylating agents)

SUM OF SQUARES: the sum of the squared deviations from their mean of the observations; it is used in statistical procedures to estimate differences. (See analysis of variance, intraclass correlation)

SUNBURN: in about 60% of the sunburn (UV) causes keratoses. The p53 gene may suffer mutation(s), mainly C→T transitions. These p53 mutant cells, after clonal propagation, may develop into squamous cell skin cancer (SCC); in over 90% of the SCC cells the p53 gene is mutant. The increased melanin production involved with ultraviolet light exposure may be correlated with the DNA repair system. (See p53, keratoses, DNA repair, ultraviolet light, pigmentation in animals, melanoma)

SUNFLOWER (*Helianthus*): an oil crop with about 70 xenogamous species, 2n = 2x = 34.

SUN-RED MAIZE: develops anthocyanin pigment when the tissues are exposed to sun-shine (e.g. through a stencil).

supC: a suppressor mutation for amber (5'-UAG-3') and ochre (5'-U3') chain-terminator codons. The mutation causes a base substitution in the anticodon of tyrosine tRNA (5'-GUA-3') and it changes to 5'-UUA-3' that can recognize both the ochre and amber codons as if they were tyrosine codons. Note: the pairing is 5' - 3' and 3' - 5'. For the recognition of the former, wobbling is required. (See suppressor, wobbling, anticodon, ochre, amber)

supD: an amber suppressor mutation that reads the amber (5'-UAG-3') chain termination codon as if it would be a serine (5'-UCG-3') codon because the normal serine tRNA anticodon (5'-CGA-3') mutates to (5'-CUA-3') therefore instead of terminating translation, a serine is inserted into the amino acid chain. (Remember, the pairing is antiparallel). (See suppressor, anticodon).

supE: an amber-suppressor mutation that reads the chain-termination codon 5'-UAG-3' (amber codon) in the mRNA as if it would be a glutamine codon (5'-CAG-3') because the glutamine tRNA anticodon mutates to 5'-CUA-3' from the 5'-CUG-3', and as a consequence the translation proceeds and a glutamine is inserted at the site where in the presence of an amber codon in the mRNA the translation would have terminated. (Note: the paring is antiparallel)

SUPERANTIGEN: are native bacterial and viral proteins that can bind directly (without breaking up into smaller peptides) to MHC class II molecules on antigen-presenting cells and to the variable regions of T cell receptor β chains and thus activate more T cells against e.g. enterotoxins of *Staphylococci* — causing toxic shock syndrome or food poisoning — than normal antigens. Cellular superantigens are supposed to be responsible for such diseases e.g. diabetes mellitus.

(See MHC, TCR, enterotoxin, toxic shock syndrome, diabetes mellitus, antigen)

SUPERCOILED DNA: may assume the *positive supercoiled* structure by twists in the same direction as the original generally (right handed) coiling of the double helix or it may be twisted in the opposite direction, *negative supercoiling*. Negative supercoiling (Z DNA) may be required for replication and transcription. The superhelical density expresses the superhelical turns per 10 bp and it is about 0.06 in cells as well as in virions. (See DNA replication prokaryotes, transcription, Z DNA, packing ratio)

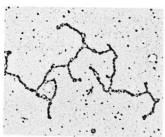

SUPERCOILED DNA PLASMID OF *Streptococcus lactis*. (Courtesy of Dr. Claude F. Garon)

SUPERDOMINANCE: same as overdominance or monogenic heterosis. (See hybrid vigor, overdominance)

SUPERFAMILY OF GENES: a group of genes that are structurally related and may have descended from common ancestors although their present function may be different.

SUPERFEMALE (metafemale): *Drosophila*, trisomic for the X-chromosome (XXX) but disomic for the autosomes; she is sterile. (See aneuploidy, supermale *Drosophila*)

SUPERFETATION: due to an apparently rare autosomal dominant gene ovulation may continue after implantation of the fertilized egg and an unusual type of twinning results. In animals when a female may mate repeatedly with several males, the same litter my become multipaternal. Human dizygotic twins may also be of different paternity when during the receptive period the female had intercourse with two different males. (See twinning, multipaternate litter)

SUPERGENES: are linked clusters of genes that are usually inherited as a block because inversion(s) prevent(s) the survival of the recombinants for the clusters and thus have evolutionary and applied significance in plant and animal breeding.

SUPERINFECTION: a bacterium is infected by another phage. This is generally not possible in a lysogenic bacterium because of immunity, i.e. the superinfecting phage cannot enter a vegetative cycle within the lysogen. (See immunity)

SUPERMALE *Asparagus*: can be obtained by regenerating and diploidizing plants obtained from Y chromosomal microspores or pollen by the techniques of cell culture. Thus their chromosomal constitution is 18 autosomes + YY. These plants are commercially advantageous because of the higher yield of the edible spears. (See YY *Asparagus*)

SUPERMALE *Drosophila*: has 1 X and 3 sets of autosomes i.e. the fly is monosomic for X and but he is trisomic for all the autosomes; he is sterile. (See superfemale, sex determination)

SUPERMAN: is a homeotic mutation in *Arabidopsis* resulting in excessive development of the androecium at the expense of other flower parts. (See flower differentiation)

SUPERMUTAGEN: an efficient mutagen that causes primarily point mutations without inducing frequent chromosomal defects. (See point mutation, ethylmethane sulfonate, nitrosoguanidine)

SUPERNATANT: the non-sedimented fraction after centrifugation of a suspension or in general any floating fraction derived of a mixture.

SUPERNUMERARY CHROMOSOME: see B chromosome, cat eye syndrome

SUPEROPERON: is a regulatory complex tying together e.g. photoreceptor pigment synthesis and photosynthesis. (See operon)

SUPEROVULATION: by injection of gonadotropic hormones (into mice) the number of eggs produced may increase several fold. Fertilization follows after about 13 hours and the eggs can surgically be collected to study *in vitro* the preimplantation development. Hormonal treatment may cause superovulation also in usually monoparous animals too. (See twinning)

SUPEROXIDE DISMUTASE (SOD): catalyzes the reaction $2O_2 + 2H^+ \leftrightarrows O_2 + H_2O_2$ and thus participates in the detoxification of the highly reactive (mutagenic) superoxide radical O_2^-. It is

Superoxide dismutase continued
encoded in human chromosome 22q21-q22. This enzyme is one of the main detoxificant of the mitochondrial free radicals. In its deficiency, cardiomyopathy and Lou Gehrig's disease may result. (See also amyotrophic lateral sclerosis, granulomatous disease, cardiomyopathies)

SUPER-REPRESSED: bacterial operon cannot respond to inducer. (See repression, inducer)

SUPERSHIFT: see gel retardation assays

SUPERSUPPRESSOR: a dominant suppressor acting on more than one allele or even on different gene loci. (See suppressor)

SUPERVITAL: its fitness exceeds that of the standard (wild) type.

supF: an amber-suppressor mutation in tRNA^{Tyr} that recognizes the chain-termination codon (5'-UAG-3') as if it would be a tyrosine codon (5'-UAC-3' or 5'-UAU-3') because a mutation at the anticodon sequence in the tyrosine tRNA changes the 5'-GUA-3' into a 5'-CUA-3' and thus the tyrosine tRNA inserts a tyrosine into the growing peptide chain where translation would have been terminated. (See suppressor tRNA, also πVX)

supG: suppressor mutation for both amber (5'-UAG-3') and ochre (5'-UAA-3') chain termination codons by a base substitution mutation in the anticodon of lysine tRNA from (5'-UUU-3') to 5'-UUA-3' that can recognize the amber and ochre codons in the mRNA as if they would be lysine (5'-AAA-3' or 5'-AAG-3') codons and thus inserting in the peptide chain a lysine rather than discontinuing translation. (Note: the pairing is antiparallel)

SUPPORTIVE COUNSELING: tends to alleviate the psychological problems involved with the discovery of hereditary disorders and birth defects in families. (See counseling genetic)

SUPPRESSOR, BACTERIAL: protein product of the bacterial regulator gene (such as e.g. *i* in the *Lac* operon) that when associated with the operator prevents transcription. (See *Lac* operon, see also *ara* operon)

SUPPRESSOR, EXTRAGENIC: the suppressor mutation is outside the boundary of the suppressed gene. (See suppressor tRNA)

SUPPRESSOR GENE: restores function lost by a mutation without causing a mutation at the site suppressed. The suppressor can be intragenic (within the cistron but also at another site) or at any other locus. (See suppressor extragenic, frameshift)

SUPPRESSOR, INTRAGENIC: the suppressor site is within the gene where suppression takes place. (See frameshift suppressor)

SUPPRESSOR MUTATION: see suppressor gene and suppressor tRNA, frameshift suppressor

SUPPRESSOR RNA: is preventing translation of a mRNA by partial base pairing with a specific sequence of the target. Two 22 and 40 nucleotide long tracts of the *lin-4* transcripts control the translation of gene *lin-14*. The latter is an early expressed nuclear protein in *Caenorhabditis* but its expression is blocked during later stages by the *lin-4* RNA that itself is not operating through a protein. Several other genes are subject to RNA suppressors in other organisms. The mechanism of this suppression seems to be different from that of the suppressor tRNA. (See also suppressor tRNA, suppressor gene, antisense RNA, antisense DNA)

SUPPRESSOR SELECTION GENE FUSION VECTOR: carries nonsense codon(s) in the structural gene and can be expressed only if the genome carries nonsense suppressors. (See vectors gene fusion, transcriptional gene fusion vector)

SUPPRESSOR T CELL: can suppress antigen-specific and allospecific T cell proliferation by competing for the surface of antigen-presenting cells.

SUPPRESSOR tRNA: makes possible the translation of nonsense or of missense codons in the original, normal sense because a mutation in the anticodon of the tRNA recognizes the complementary sequence in the codon but its specificity resides in the tRNA molecule. Exceptions are possible, however. In the anticodon 5'-CCA-3' of the tryptophan codon (5'-UGG-3') may be no mutation yet it may deliver to the opal position a tryptophan if at position 24 of the D loop of the tryptophan tRNA a guanine is replaced by an adenine. Similarly, if in the GGG codon of glycine another G base is inserted by a frameshift mutagen but subsequently in its anticodon (CCC) and extra C is inserted then the tRNA may read the four bases normally as a glycine codon rather than a chain terminator. (See tRNA, code genetic)

SUPRAVALVULAR AORTIC STENOSIS (SVAS): is a human chromosome 7 dominant mutation in an elastin gene with a prevalence of about 5×10^{-5} causing obstruction of the aortic blood vessels. The basic defect is due to a deficiency of elastin. It is pleiotropic and part of the Williams syndrome. (See coarctation of the aorta, cardiovascular diseases, Williams syndrome, elastin)

supU: is a suppressor of the opal (5'-UGA-3') chain terminator codon with an anticodon sequence of 5'-UCA-3'. The mutation changes the anticodon of the tryptophan tRNA from 5'-UCA-3' to 5'-CCA-3' and that permits the insertion of a tryptophan residue into the polypeptide chain where it would have been terminated without the suppressor mutation (codon-anticodon recognition is antiparallel). This suppressor mutation is unusual because it recognizes both the tryptophan codon (5'-UGG-3') and the opal suppressor codon, i.e. its action is ambivalent.

SURFACE ANTIGEN: generally glycoprotein molecules on the cell surface that determine the identity of the cells for immunological recognition. The display of surface antigens is generally regulated at the level of transcription. (See VSA, antigen, *Trypanosomas*, *Borrelia*)

SURROGATE GENETICS: see reversed genetics

SURROGATE MOTHER: a female who carries to term a baby for another couple. She may contribute actually the egg or she may just be a gestational carrier of a fertilized egg and has no genetic share in the offspring. In either case moral, ethical, psychological and legal problems must be pondered before this method of child bearing is chosen. Civil law generally keeps the maternal right of the gestational mother despite any contractual agreement contrary to natural parenthood. In case, however, the gestational mother is not the donor of the ovum, the maternal right pertains to the biological ovum donor. Obviously, there are serious ethical problems here beyond the principles of genetics. The society must protect the best interest of the child. (See oocyte donation, ART, paternity testing)

SURVIVAL: see neutral mutation, beneficial mutation, cost of evolution, genetic load, fitness

SURVIVAL OF THE FITTEST: see fitness, neo-Darwinism, social engineering, social Darwinism

SUSPENSION CULTURE: the cells are grown in liquid nutrient medium.

SUSPENSOR: is a line of cells through which the plant embryo is nourished by the maternal tissues. In general anatomy ligaments may be called suspensors.

SUTURE: a junction of various solid animal and plant tissues.

Sv: see Sievert

SV2s: glycosylated transmembrane proteins, homologous to prokaryotic and eukaryotic transporters. (See transporters)

SV40 Tag: simian virus 40 large T antigen. (See Simian virus 40)

SV40 VECTORS: SV40 plasmids (vectors) can be packaged only if their DNA is within the range of 3900 to 5300 bp. Since these small genomes do not have much dispensable DNA, it is almost impossible to construct a functional vector with any added genes to it. Fortunately, functions provided by helper DNA molecules may help to overcome these problems. Simian Virus 40 cannot replicate autonomously if the replicational origin (*ori*) is defective yet it can integrate into chromosomal locations of green monkey cells and can then be replicated along the chromosomal DNA (such a cell is COS [**c**ell **o**rigin **s**imian virus)]. Also, since the early genic region is normal (see SV40 for structure), it may produce the T antigen within the cell. If such a cell is transformed by another SV40 vector in what the viral early gene region was replaced by a foreign piece of DNA, the COS cell may act as a helper and replicate multiple copies of the second, the engineered SV40 DNA. Since the late gene region of this plasmid is normal, the viral coat proteins can be synthesized within the cell. The availability of the coat proteins permits the packaging of the engineered SV40 DNA into capsids. The virions so obtained can be used to infect other mammalian cells where the passenger DNA can be transcribed and translated and the foreign protein can be processed. The transformed cell thus can acquire a new function. Also, an SV40 plasmid can be constructed with insertion into the late

SV40 vectors continued

gene cluster a foreign gene with a desired function. This plasmid can then be used along with another SV40 plasmid with inactivated (deleted) early genes but with a good *ori site*. Upon coinfection a mammalian cell with these two plasmids, the SV40 plasmids can replicate to multiple copies and the inserted foreign gene can be expressed. Also, it is feasible to insert into a prokaryotic pBR322 plasmid the *ori* region of SV40 and another piece of DNA including all the necessary parts of a foreign gene. When this plasmid is transfected into a COS cell, the passenger gene can be transcribed, translated and processed thanks to the multiple copies replicated within this mammalian cell. With the assistance of SV40 based constructs, mammalian and other genes can be shuttled between mammalian and bacterial cells.

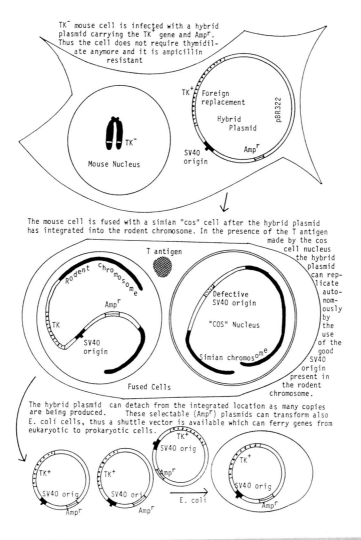

SV40 SHUTTLE VECTOR CAN FERRY GENES BETWEEN MAMMALIAN CELLS AND PROKARYOTES

The procedure: a thymidine kinase-deficient (tk^-) rodent cell is transformed by pBR322 bacterial plasmid carrying the *ori* of SV40 and the TK^+ (functional thymidine kinase), and an ampicillin resistance gene (amp^R) for bacterial selectability. Within the rodent cell the bacteri-

SV40 vectors continued

al plasmid is integrated into a chromosome of the rodent cell and this cell is then fused with a COS cell. The integrated pBR322 plasmid is replicated into many copies thanks to the presence of the COS nucleus. The hybrid plasmid, carrying a bacterial replicon, an SV40 *ori*, the TK^+ and amp^R can infect *E. coli* cells and can be selectively propagated there. Thus the shuttle function is achieved. Another SV40 and pBR322 based vector is the pSV plasmid. This contains, in a Pvu II and HinD III restriction enzymes generated fragment, the promoter signals and the mRNA initiation site. When any open reading frame is attached to it, transcription can proceed. In addition, an intron of the early region provides splicing sites for other genes. The region contains also the transcription termination and polyadenylation signals. Some other pBR322 parts may be equipped with additional specific, selectable markers. (See Simian virus 40, vectors, viral vectors, shuttle vector)

SVC: see carbondioxide (CO_2) sensitivity

SVEDBERG UNITS: see sedimentation

SVF-2: a member of the tumor necrosis factor receptor family. (See Fas, TNF)

SWEDE (*Brassica napus*): is a leafy fodder crop in Northern climates and it also an edible human vegetable; 2n = 38, AABB genomes). (See also rape)

SWEET CLOVER (*Melilotus officinalis*): a fragrant leguminous plant because of its coumarin content. The coumarin-free forms (*M. albus*) are used as hay, 2n = 16. (See coumarin)

SWEET PEA (*Lathyrus odoratus*), 2n = 14): is an ornamental that had been exploited for studies on the genetic determination of flower pigments. Another peculiarity is that the "long"/"disc" pollen shape (*L/l*) is determined by the genotype of the sporophytic (anther) tissue rather than by the gametophyte. Therefore delayed inheritance is observed. (See delayed inheritance)

SWEET POTATO (*Ipomoea batatas*): it is primarily a warm climate vegetable with about 25 species with basic chromosome number x = 15 and a single genome; the most common cultivated form is hexaploid although related species may be diploid or tetraploid.

SWEPT RADIUS: see linkage

SW1, SW2/SNF2, SWI3: are yeast genes encoding transcriptional activators. The homologous protein in *Drosophila* is BRAHMA. (See activator genes, co-activator, transcriptional activator, *Polycomb*)

Swi6: is a centromeric protein in *Schizosaccharomyces pombe*, essential for centromere function. (See centromere)

SWIMMING IN BACTERIA: movement by counterclockwise rotation of the flagella.

SWINE: most commonly used name for the female animals of the *Sus* species (hogs). (See pig)

SWI/SNF: see nucleosome

SwissProt DATABASE: (protein information): <http://expasy.hcuge.ch/sprot/sp-docu.html>

SWITCH, GENETIC: an individual cell may initially express immunoglobin gene Cμ but in its clonal progeny it may change to the expression of Cα as a result of somatic DNA rearrangement. Similar DNA switch may occur in the variable region of the light-chain genes. Switching occurs during the mating type determination of yeast and phase variation in prokaryotes. (See immunoglobulins, epigenesis, *Trypanosomas*, *Borrelia*, site-specific recombination, mating type determination in *Saccharomyces*, phase variation)

SWITCHING, PHENOTYPIC: may occur without any change in the genetic material and involves only altered regulation of transcription resulting in different phenotypes.

SWIVELASE (topoisomerase type I): after a single nick in a supercoiled DNA it permits the cut strand to make a turn around the intact one to relieve tension. (see DNA replication, prokaryotes, topoisomerase)

Sxl *(sex lethal)*: (chromosomal location 1-19.2) controls sexual dimorphism in *Drosophila*; it is required for female development. (See sex determination)

SYBASE: a computer program that links various databases for macromolecules such as GeneBank, EMBL. (See databases)

SYCAMORE (*Platanus* spp): large attractive monoecious trees, 2n = 42.

SYK (M_r 72 K): is a signal protein for the interleukin-2, granulocyte colony-stimulating factor and for other agonists. It is a protein tyrosine kinase of the SRC family and indispensable for B lymphocyte development. (See B cell, ZAP-70, agonist)

SYLLOGISM: a form of deductive reasoning using a *major premise*, e.g. mice show graft rejection, the rejection of transplants is based on the presence of the MHC system (*minor premise*), therefore mice must have a major histocompatibility system (*conclusion*). (See logic)

SYMBIONTS: mutually interdependent cohabiting organisms, such as the *Rhizobium* bacteria within the root nodules of leguminous plants or algae within green hydra animals.

SYMBIONTS, HEREDITARY: occur in a wide range of eukaryotic organisms and their maternal transmission simulates extranuclear inheritance. They may be more wide-spread than recognized. The temperate viruses of prokaryotes and the retroviruses may also be classified along these groups. In some strains of the unicellular protozoa, *Paramecium aurelia*, carrying the *K* gene, bacteria (e.g. *Caedobacter taenospiralis*) live in the cytoplasm and transmitted to the progeny. The first such infectious particles were named kappa (κ) particles before their bacterial nature was recognized and were supposed to be normal extranuclear, hereditary elements. Many of these kappa particles contain R (refractive) bodies that are bacteriophages. The κ particles with R bodies appear "bright" under the phase-contrast microscope. The non-bright cells may give rise to bright, indicating that the phages are in the free and infectious stage and the brights are in the integrated, proviral stage. The virus directs the synthesis of toxic protein ribbons that are responsible for killing kappa-free paramecia (κκ). The strains carrying the dominant *K* gene are immune to the toxin. Other strains have been discovered carrying different infectious particles lambda, sigma, mu, that also make toxins. The mu particles do not liberate free toxin and kill only the cells with which they mate (mate killer). Other symbionts delta, nu and alpha are not killer symbionts.

In *Drosophila* strains Rhabdovirus σ may be in the cytoplasm and responsible for CO_2-sensitivity. Normal flies can be anesthetized with the gas for shorter periods without any harm. Those which carry the virus may be paralyzed and killed by the same gas treatment. This virus is similar to the vesicular stomatitis virus of horses that cause fever and eruptions and inflammation in the mouth of horses, and to the fish rhabdovirus. In the non-stabilized strains only the females transmit the virus to part of the progeny (depending whether a particular egg contains or not the virus). Some of the non-stabilized may become stabilized. Stabilized strains transmit it through nearly 100% of the eggs and even some of the males transmit σ with some of the sperm yet the offspring of the male will not become stabilized. The *ref* mutants in chromosomes 1, 2 and 3 are refractory to infection. In some strains there are mutants of the virus that are either temperature-sensitive or constitutively unable to cause CO_2 sensitivity although they are transmissible. Different ribosomal picornaviruses can be harbored in *Drosophila* that may reduce the life and fertility of the infected females. Females with the sex ratio (SR) condition produce no viable sons and the transmission is only maternal. In their hemolymph (internal nutrient fluids) the females carry spiroplasmas, bacteria without cell wall. If the infection is limited to the XX sector of gynanders, they may survive but not if the infection is in the XO sector. Triploid intersexes or females, phenotypically sterile males because of the genes *tra* (*transformer*, chromosome 3-45), *ix* (*intersex* gene located at 2-60.5) or *dsx* (*double-sex* intersexes, chromosome 3-48.1) are not killed. The spiroplasmas may be destroyed by their special viruses. In plants (petunia, sugar beet), cytoplasmically inherited male sterility can be transmitted also by grafting. Some of the variegated tulips (broken tulips) are infected by viruses and had special ornamental value. During the 17th and 18th century "tulip mania" some rich Europeans paid for the bulbs of the most attractive varieties equal weights in gold to the mainly Turkish and Persian merchants. (See lysogeny, cytoplasmic male sterility, extranuclear inheritance, *Drosophila*, meiotic drive, segregation distorter, broken tulips, sex determination)

SYMBIOSOME: in legume root nodules 2 to 5 μm structures enclosing (by peribacteroid membrane) 2 to 10 bacteroids. The fixed nitrogen is released through this membrane to the plant and reduced carbon is received by the bacteroids from the plant. (See nitrogen fixation)

SYMBOLS: see gene symbols, *Drosophila*, pedigree analysis

SYMMETRIC HETERODUPLEX DNA: see Meselson - Radding model of recombination

SYMPATRIC: populations have overlapping habitats; it may be a beginning of speciation. (See speciation)

SYMPATRIC SPECIATION: species that live in the same, shared area for some reason become sexually isolated. (See: allopatric, parapatric)

SYMPETALY: fused petals in a flower.

SYMPLESIOMORPHIC: two or more species sharing a primitive evolutionary trait. (See plesiomorphic, apomorphic, synapomorphic)

SYMPLEX: a polyploid or triploid with one dominant allele at a locus. (See autopolyploid)

SYMPORT: cotransportation of different molecules through membranes in the same direction.

SYN: a prefix indicating union of tissues named after.

SYN: see ANTI

SYNAPOMORPHIC: species sharing an apomorphic trait. (See apomorphic, plesiomorphic, symplesiomorphic)

SYNAPOMORPHY: shared derived characters that can be used to advance phylogenetic hypotheses. (See simplesiomorphy)

SYNAPS (synapse): the site of connection between neural termini at which either a chemical or an electric signal is transmitted from one neuron to another (or to another type of cell). The neurotransmitter may diffuse across the synaps or the electric signal may be relayed from one cytoplasm to the other through a gap junction. The neurons must interpret the postsynaptic potentials (PSP) and integrate the excitatory (EPSP) and inhibitory (IPSP) paired pulse potentials. (See neurotransmitters, gap junction)

SYNAPSINS: bind to actin filaments, microtubules, annexing, SH3 domains, calmodulins and important regulators of synaptic vesicles. (See actin, annexin, SH3, calmodulin, synaptic vesicles, synaps)

SYNAPSIS: intimate chromosome pairing during meiosis between homologous chromosomes that may lead to crossing over and recombination. (See illegitimate pairing, crossing over, recombination, *res*, topological filter, tracking)

SYNAPTIC ADJUSTMENT: the degree of synapsis may change during meiosis in certain chromosomal regions. (See synapsis)

SYNAPTIC CLEFT: electric neuronal signals are transmitted from the presynaptic cell to the postsynaptic cell by the gap called the synaptic cleft. (See neuron)

SYNAPTIC VESICLES: originate from endosomes and store and release the neurotransmitters and other molecules, required for signal transmission between nerve and target cells. Their function is regulated by Ca^{2+}. (See neurotransmitters, syntaxin, RAB, synaptotagmins, synaptophysin, NSF, neuromodulin, neurogenesis)

SYNAPTINEMAL COMPLEX: see synaptonemal complex

SYNAPTOBREVINS (VAMPs): see syntaxin

SYNAPTOGAMIN: is an integral membrane protein, binding calcium and interacting with other membrane proteins in the synaptic vessels of the nerves. (See neurexin, synaptic vesicles)

SYNAPTOGYRIN: membrane proteins; may be phosphorylated on tyrosine.

SYNAPTOJANIN: a neuron-specific phosphatase (M_r 145,000) working on phosphatidylinositol and inositols and its putative role is in the recycling of synaptic vesicles. Its defect is responsible for Lowe's oculocerebrorenal syndrome. It binds to the SH3 domain of Grb2. (See also synaptotagmin, syntaxin, Lowe's oculocerebrorenal syndrome, Grb2, dynamin)

SYNAPTON: same as synaptonemal complex

SYNAPTONEMAL COMPLEX: proteinacious element between paired chromosomes in meiosis. They consist of two lateral and a central element. Denser spots within it are called recombination nodules and were supposed to have role in genetic recombination. It is supposed

Synaptonemal complex continued

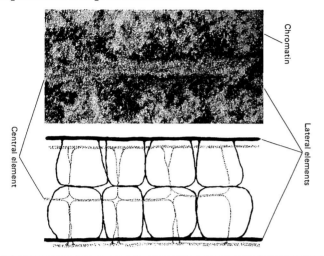

THE TRIPARTITE STRUCTURE MAY BE VISIBLE FROM INTERPHASE THROUGH DIPLOTENE. (Electronmicrogph by the courtesy of Dr. H.A. McQuade. Interpretative drawing after Comings, D.E. & Okada, T.A. Nature [Lond.] 227:451)

that the complex holds in place the recombination intermediates rather than actively promoting the process. Actually, in some fungi (*Saccharomyces*, *Aspergillus*) no synaptonemal complex is observed and concomitantly chromosome interference is absent. These observations led to the assumption that the complex is responsible for interference rather than recombination. Actually, recombination may precede the formation of the synaptonemal complex. (See synapsis, interference, recombination mechanism eukaryotes, association site, recombination nodule, crossing over)

SYNAPTOPHYSIN: are membrane-spanning proteins in the synaptic vessel involved in neurotransmitter release. (See synaptic vessel, neurotransmitter)

SYNAPTOTAGMINS: are proteins in the synaptic vesicles and have a role in Ca^{++} involved release of neurostransmitters, and in general in exo- and endocytosis. (See synaptojanin, syntaxin, synapse)

SYNCHRONOUS DIVISIONS: the cells are at the same stage of the cell cycle.

SYNCLINAL: see anticlinal

SYNCYTIAL BLASTODERM: an early stage of embryogenesis when the single layer of nucleated cytoplasmic aggregates do not have yet a cell membrane.

SYNCYTIUM: a collection of nuclei surrounded by cytoplasm without the formation of separate membranes around each, such as in early embryogenesis or among the progeny of a single spermatogonium or abnormal multinucleate cells or the plasmodia of slime molds. (See imaginal discs, blastoderm, *Dictyostelium*)

SYNDACTYLY: webbing or fusion between fingers and toes. In polysyndactyly, mutation in the polyalanine extension of the amino terminal of human homeotic gene (HOX13) is the responssible factor. (See Poland syndrome, limb defects, Rubinstein-Taybi syndrome, Greig's cephalopolysyndactyly, GLI3 oncogene, homeotic genes)

SYNDROME: a collection of symptoms, traits caused by a particular genetic constitution. The individual symptoms of different syndromes may, however, overlap among a large number of genetic and non-genetic disorders. Therefore the precise identification is often an extremely difficult task. More accurate identification will probably be possible when the genome projects will provide structural an topological evidence for all the loci concerned. (See genome projects, physical mapping)

SYNERGIDS: two haploid cells in the embryosac of plants flanking the egg. (See gametogenesis)

SYNERGISTIC ACTION: the participating elements enhance the reaction above the sum of the separate strength of their separate actions. (See interaction variance, epistasis)

SYNGAMY: the union of two gametes in fertilization leading potentially to the fusion of the two nuclei in the cell. (See plasmogamy, karyogamy, synkaryon, semigamy)

SYNGEN: a reproductively isolated group of ciliates.

SYNGENEIC: antigenically similar types of cells (in a chimera). (See antigen, immunoglobulins)

SYNKARYON: a cell (zygote, fused conidia or spores) with a nucleus originated by the union of two nuclei. (See karyogamy, heterokaryon, dikaryon)

SYNONYMOUS CODONS: have different bases in the triplet yet they specify the same amino acids. The 61 sense codons stand for 20 amino acids. Some amino acids have up to 6 codons. Thus mutation may not have any genetic consequence, except when exon splice site is involved. (See genetic code, splicing, exon, intron)

SYNTAXIN: is a synaptic membrane protein forming part of the nerve synaptic core complex along with the synaptosome associated protein and synaptobrevin (VAMP), a vesicle associated membrane protein. A synaptobrevin-like gene (SYBL1) was located to the pseudoautosomal region of the human X chromosome. It recombines with Y-chromosomal homolog and it displays lyonization in the X chromosome and it is inactivated in the Y chromosome. (See synaptotagmin, synaps, pseudoautosomal region, lyonization)

SYNTENIC GENES: are within in the same chromosome; they may, however, freely recombine if they are 50 or more map units apart. (See linkage, crossing over)

SYNTHASES: mediate condensation reactions without ATP. (See synthetases)

SYNTHETASES: mediate condensation reactions that require nucleoside triphosphates as energy source. (See synthases)

SYNTHETIC DNA PROBES: if the amino acid sequence in the protein is known but the gene was not yet isolated, a family of synthetic probes may be generated to tag the desired gene. This probe is generally no longer than 20 base because the difficulties involved in their synthesis. The genetic code dictionary reveals which triplets spell the amino acids. An amino acid sequence is selected that uses few synonymous codons. The possible combinations are generally chosen by a computer match. E.g. a probe for the His-Thr-Met peptide sequence would require the following 8 polynucleotide sequences to consider all possible sequences for a probe (see on the left).

His Thr Met
5' CACACUAUG 3'
CACACCAUG
CACACAAUG
CACACGAUG
CAUACUAUG
CAUACCAUG
CAUACAAUG
CAUACGAUG

The inclusion of **methionine** (having a single codon) is simplifying the task, **histidine** is relatively advantageous because it has only two synonymous codons, **threonine** with four codons make the work more difficult; leucine, serine and arginine containing parts of the proteins should be avoided (because they have six codons) but tryptophan, also with a single codon would be highly desired. (See probe)

SYNTHETIC ENHANCEMENT: is basically an epistatic process by increasing or reducing interaction between gene products by using crossing, knockouts, transformation, etc.

SYNTHETIC GENES: are produced *in vitro* by the methods of organic chemistry, by systematic ligation of synthetic oligonucleotides into functional units, including upstream and dowstream essential elements. The first successful synthesis involved the relatively short tyrosine suppressor tRNA gene. (See diagram on next page, suppressor tRNA)

SYNTHETIC LETHAL: is inviable except in certain genetic constitutions.

SYNTHETIC POLYNUCLEOTIDES: are nucleic acid oligomers or polymers generated in the laboratory by enzymatic or other synthetic methods. Some of them are produced by nucleic acid synthesizer machines.

SYNTHETIC SEED: somatic embryos encapsulated into a protective capsule (e.g. calcium alginate) and used for propagation in cases when regular seed is not available or homozygotes are difficult to obtain.

SYNTHETIC SPECIES: are amphidiploids of presumed progenitors of existing species obtained by crossing and diploidization. Some synthetic species have never existed in nature before as the *Raphanobrassica*, 2n = 36, an amphidiploid of radish (*Raphanus sativus*, n = 9) and cabbage (*Brassica oleracia*, n = 9). *Triticales* are similarly new amphidiploids either 2n = 48

Synthetic species continued
or 2n = 56, obtained by crossing tetraploid (2n = 28) hexaploid (2n = 42) wheat (*Triticum*) with diploid rye (*Secale cereale*, 2n = 14). Some synthetic species are only reconstructions of the evolutionary forms. e.g. *Nicotiana tabacum*, *Hylandra suecica*, *Primula kewensis*, etc.

Synthetic genes continued from page 1004

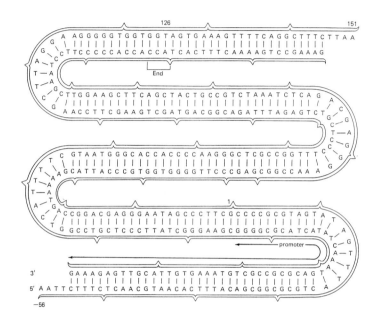

THE COMPLETE (including promoter and terminator) SYNTHETIC STRUCTURAL GENE OF THE TYROSINE SUPPRESSOR tRNA OF *E. coli*. THE PROJECTIONS ALONG THE SEQUENCE BRACKET THE SIZE OF THE FRAGMENTS LIGATED TOGETHER TO FORM THE COMPLETE GENE. (After Khorana, H.G. 1974 in Proc.Int. Symp. Macromol. p. 371. Mano, E.B., ed. Elsevier, Amsterdam, NL and Macaya, G. 1976 Recherche (Paris) 7:1080)

SYNTHETIC VARIETY: is composed over several selected lines which may reproduce by outcrossing within the group. (See polycross)

SYNTHON: a synthetically produced molecule, in the laboratory.

SYNTROPHIC: can be maintained (only) by cross-feeding. (See cross-feeding, channeling)

SYP: is a tyrosine phosphatase

SYRINGA: see lilac

SYRINGE FILTER: a syringe equipped with a commercially available sterilizing filter block (0.45 or 0.20 μm pores) and removes microbial contaminations instantly without heating.

SYRINGOMELIA: rare autosomal dominant or autosomal recessive cavitations (formation of cavities) in the spinal cord. It may be also due non-hereditary causes.

SYSTEMIC GENES: are cell autonomous versus genes which are regulated by intercellular communication.

SYSTEMIC AMYLOIDOSIS, INHERITED: is an extracellular deposition of fibrous proteins in the connective tissues under autosomal dominant control. (See amyloidosis)

SYSTEMIN: a 18 amino acid signaling peptide for plant defense mechanisms. (See plant defense)

SYSTEMS OF BREEDING: sexual reproduction may be allogamous, autogamous, inbreeding, assortative mating, hermaphroditic, monoecious and dioecious but reproduction may also be

asexual. (See breeding system)
SYT (synovial sarcoma): oncogene in human chromosome 18q11.2.
SZI: see micromanipulation of the oocyte
SZOSTAK MODEL OF RECOMBINATION: is a double break and repair model:

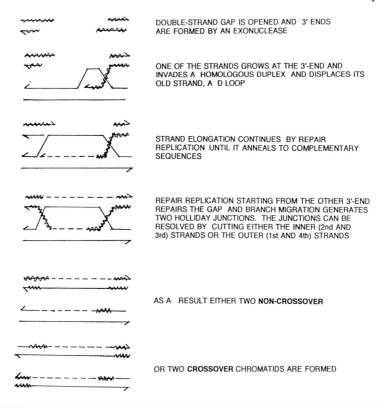

THE SZOSTAK *et al.* MODEL OF RECOMBINATION IS BASED ON DOUBLE-STRAND BREAKS IN CONTRAST TO THE HOLLIDAY OR THE MESELSON - RADDING MODELS THAT SUGGEST SINGLE-STRAND BREAKS IN THE DNA

It is applicable to transformational insertion of DNA molecules as well as it can account for gene conversion and/or conventional recombination by outside marker exchange. The version of the model shown in the diagram does not explain why in yeast 5:3 gene conversion does not occur. (*Saccharomyces cerevisiae* has 4 spores per ascus but *Schizosaccharomyces pombe* has 8 unordered ascospores.) Other modified models of Szostak can account for these types of experimental data. Genetic recombination in eukaryotes generally occurs by double-strand break. The frequency of double-strand breaks can be increased by radiation and various chemicals. Double-strand breaks can be increased in plant cells also by transformation of a restriction endonuclease into the cell. (See recombination molecular models, diagram above)

T: see thymine

t: time

τ: yeast retroelement. (See Ty)

T2: virulent bacteriophage. (See bacteriophages, T4)

T4: is a virulent (lytic) bacteriophage of *E. coli*; it has double-stranded DNA genetic material of 1.08×10^6 Da (about 166 kbp) with a total length of about 55 µm. Its cytosine exists in hydroxymethylated and glucosylated form. The linear DNA is terminally redundant and cyclically permuted. The redundancy occupies about 1% of the total DNA. It was an unexpected discovery that its thymidylate synthetase gene contains an intron. Introns are common in eukaryotes but exceptional in prokaryotes. The phage has over 80 genes involved in metabolism but only about a fourth of them are indispensable. The metabolic genes control replication, transcription and lysis. The other metabolic genes have functions overlapping those of the host. After infection, the phage turns off or modifies bacterial genes, degrades host macromolecules and dictates the transcription of its own genes and utilizes the host machinery for its own benefit. For the synthesis of its own DNA it relies on the nucleotides coming from the degradation of the host DNA. The cytidylic acid residues of the host are prevented from incorporation into phage DNA by phosphatases and a deaminase converts it to thymidylic acid or through a few steps to hydroxymethyl cytosine by a methylase enzyme. The hydroxymethylcytidylate is then glucosylated by a glucosyl transferase enzyme. The molar proportion of thymidylate is higher in T4 than in *E. coli*, presumably because of the conversion of cytidylate into thymidylate. An additional series of more than 50 genes are involved in morphogenesis. At least 40% of the genome is required for the synthesis and assembly of the viral particle. Head assembly is mediated by 24 genes, baseplate and tail requires at least 31 genes. (See development, one-step growth, bacteriophages)

2,4,5-T: see agent orange

***T* ALLELES**: see *Brachyury*

T ANTIGEN: of SV40 virus is a large multifunctional protein and it is an effector of DNA polymerase α function. It assists separating the DNA strands for replication and generates the replication bubble as the polymerase moves on. It has a special role in SV-induced tumors after tumor-suppressor proteins are eliminated or weakened. (See Simian virus 40)

t ANTIGEN: shares the same N-terminal sequences with the T antigen but the carboxyl end is different. The t antigen has homology to the G_t protein, an α subunit of the trimeric G-proteins involved in the activation of cGMP phosphodiesterase involved in photoreception and other processes. (See Simian virus 40, G-proteins)

T BOX: see MAR

T CELL RECEPTOR (TCR): T cell surface glycoproteins recognize antibody. The differentiation of the T cells begins with the differentiation of their receptors. At the beginning the TCR is double negative, i.e. they are CD4⁻ CD8⁻. The disulphide-linked heterodimers have α and β or γ and δ chains, containing variable and constant regions and are homologous to the corresponding antibodies. First, the TCRβ is rearranged, and that is followed by the rearrangement of the α chain. At this stage allelic exclusion may be signaled by the β chain and that means the end of rearrangements. After this the double positive stage follows, i.e. CD4⁺CD8⁺ TCR appears. Then a selection process results in an array of self MHC-restricted and self-tolerant TCRs. The TCR α chain is a transmembrane protein and the cytoplasmic (carboxyl) end has two potential phosphorylation sites and an Src homology 3 (SH3) domain. The TCRβ chain regulates the development of the T cells in the absence of the α chain. The αβ TCR generally recognizes antigens bound to the major histocompatibility (MHC) molecules. The TCR complex includes also the CD3 protein, required for signal transduction. It is made of the γ, δ, ε, and ζ chains. The ζ chain also plays a role in thymocyte development but it is not indispensable for signal transduction. The chromosomal site of the human α (14q11), β

T cell receptor continued

(6q35), γ (7p15-p14), δ (14q11.2) and ζ (1p22.1-q21.1) chains is shown in parenthesis.

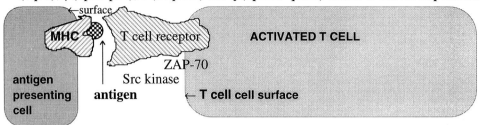

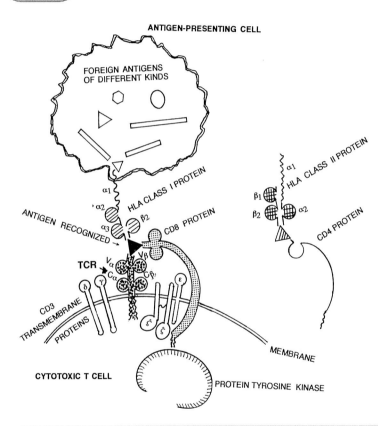

A GENERAL OUTLINE OF THE FUNCTIONS OF THE T CELL RECEPTOR (TCR) COMPLEX. ALTHOUGH DIAGRAMS ALWAYS GENERALIZE BEYOND REALITY, THEY MAY BE HELPFUL TO OBTAIN A BROAD UNDERSTANDING. THE FOREIGN ANTIGENS ARE PRESENTED TO THE T LYMPHOCYTES EITHER BY THE ANTIGEN-PRESENTING CELLS OR BY MACROPHAGES. THE MACROPHAGES ARE CAPABLE OF PARTIALLY DEGRADING THE LARGE MOLECULES OR THE INVADING CELLS. THESE CELLS ASSOCIATE THEN WITH EITHER CLASS I OR CLASS II MHC PROTEINS WHICH ARE ENCODED BY THE HLA GENES. THE MHC MOLECULES RECOGNIZE THE FOREIGN ANTIGENS AND BRING THEM TO THE TCR AND TO THE CD PROTEIN COMPLEX ASSOCIATED WITH THE T CELL SURFACE. THE CLASS I GENE PRODUCTS USE THE CD4 TRANS-MEMBRANE PROTEINS, WHEREAS THE CLASS II MOLECULES RELY ON CD8. THE COOH ENDS OF THE TCR CHAINS ARE INSIDE THE T CELL'S DOUBLE MEMBRANE AND THE NH_2 END IS INVOLVED IN THE RECOGNITION OF THE ANTIGEN IN ASSOCIATION WITH THE MHC AND CD ELEMENTS. THE TCR COMPLEX INCLUDES ALSO THE CD3 TRANSMEMBRANE PROTEINS. THE ζ SUBUNIT SERVES ALSO AS AN EFFECTOR IN SIGNAL TRANSDUCTION. PROTEIN TYROSINE KINASE IS AN IMPORTANT ELEMENT IN SEVERAL SIGNAL TRANSDUCTION PATHWAYS. THE RIGHT SIDE OF THE INCOMPLETE DIAGRAM SHOWS THE ASSOCIATION OF THE CLASS II AND CD4 PROTEINS WITH AN ANTIGEN. THE OTHER ELEMENTS OF THIS SYSTEM ARE VERY SIMILAR TO THAT SHOWN AT THE LEFT MAIN PART OF THE OUTLINE.

When the T cell receptor binds the MHC associated ligand on the antigen presenting cell activation is triggered. Co-stimulating signals may help. The CD3 component of the TCR may be altered, and protein tyrosine kinases (PTK) are activated. The PTKs may turn on the calcium-calcineurin, the RAS-MAP and the protein kinase signaling pathways. These path-

T cell receptor continued

ways then may activate the transcription factors NFAT, NF-κB, JUN, FOS (AP1) and ETS. They activate then new genes, some of them are specific transcription factors that facilitate the release of cytokines which in turn activate the clonal expansion of T cells. This is then followed by the production of antibodies by B cells or T cell cytotoxicity. Immune memory, immune tolerance, anergy and apoptosis are alternative functions to follow. Regulation of the development of TCR requires the cooperation of the protein tyrosine kinases (Src family), phospholipase C (PLCγ), CD5, CD28, VCP, ezrin, VAV, SHC, PtdIns, PIP2, PIP3. (See TCR genes, T cells, lymphocytes, immunoglobulins, Src, antibody, MHC, CD3, LCK, B lymphocytes, phosphoinositides, PIP, γδ T cells, immune system, signal transduction, thymus, and the other factors under separate entries)

T CELL REPLACING FACTOR: a lymphokine. (See lymphokines)

T CELLS: thymic lymphocytes control cell-mediated immune response (the foreign antigens are attached to them). The T lymphocytes originate in the bone marrow and then differentiate in the thymus and later migrate to the peripheral lymph nodes. While the early T cells are in the thymus a *negative selection* eliminates those T cells that react with self-antigens. At about the same stage of T cell development, a positive selection is taking place under the influence of the MHC complex, securing the survival of those T cells that can interact with antigens associated with one or another type of MHC molecules presented to these cells. These encounters lead to the differentiation and activation of the T cells. The cytotoxic T cells (CTL) are the major elements of the immune system; they cooperate with the natural killer cells (NK). The helper T_H and the T suppressor (T_S) cells mediate the humoral (secreted) immune responses. On the surface of the T cells are the T cell surface receptors (TCR). The T_H cells stimulate the proliferation of the B (bursa) cells when they recognize their cognate antigens. The joint action of T_H cell surface receptors (TCR) and the B cell's antigens bring about in the B cells the formation of growth and differentiation of proteins, named lymphokines that stimulate the propagation of B cells and the secretion of humoral antibody. The T cell surface receptors recognize foreign antigens only if they become associated with major histocompatibility (MHC) molecules carried to the TCR by the antigen presenting cells (APC) or macrophages. The TCR links to the various intracellular signaling pathways. The activation of the T cells requires phosphorylation by SRC tyrosine kinase of the CD3 immunoglobulin chains. For the full activation ZAP-70 protein-tyrosine kinase is also needed. The T cells are presented with a variety of antigen, including self-antigens that are carried by the APCs without discrimination. The self/foreign antigens are distinguished by the T cells. This ability begins to develop while the T cell are still in the thymus. The discrimination is a difficult task to achieve and the complex phosphorylations are required for surveying the very large array of ligands of varying degree of specificity (affinity). Alternatively, it is conceivable that first one peptide - MHC complex interacts first with one TCR. This TCR plays then the role in a contact cap and it detaches from the ligand and thus making possible for the ligand to bind another TCR. The process is repeated in a serial manner and assembles sufficient number of TCRs in the contact cap for productive signaling.

The activation of the T cell may be only partial in case there is subtle change (e.g. one amino acid replacement) in the peptide-MHC. The subtle variants may also inhibit the $CD4^+$ helper T cells to respond to the real antigen. Altered ligands or lack of co-stimulatory signals may cause anergy of the T cell, and cannot be stimulated by the TCR but may or may not proliferate under the influence of interleukin 2 (IL2). The ligands thus can be *agonists* that fully or partially activate the T cells, or altered ligands may be weakened agonists or even *antagonists* (reducing activation). The *null ligands* provoke no response. The weak agonists may not activate ZAP-70 and may have a different pattern of phosphorylation of the CD3 ζ chain. For the fully active immune reaction all the elements of the complex T cell activation must be in place. Some viral infections (HIV-1, hepatitis B, etc.) may lead to the production of antagonist ligands and then the cytotoxic T cells (CTL) cannot protect the body against the invader.

Some aspects of the T cell activation and regulation are outlined on page 1010. On the cell

T cells continued

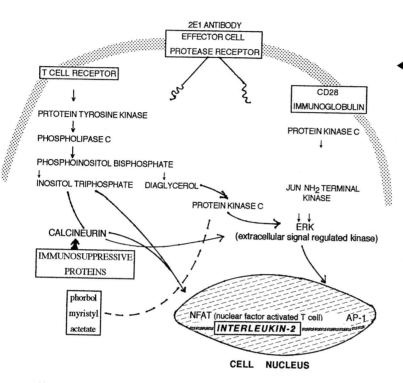

T CELL SIGNALING PATHWAYS. (Modified after Trucco, M. & Stassi, G. 1996 Nature (Lond.) 380:284)

membrane is the *T cell receptor* associated with CD3 and CD4 immuno globulins. This receptor TCR mediates the phosphorylation — with the aid of a *protein tyrosine kinase* (PTK) — *phospholipase C-γ1*. The so activated PLC-γ1 cleaves phosphatidylinositol-4,5 bisphosphate (PIP2) and generates *inositol trisphosphate* (IP3) and *diaglycerol* (DAG) that are second messengers. These second messengers activate *protein kinase C* and make available cytoplasmic Ca^{++} for *calcineurin* (also called protein phosphatase IIB). As a consequence the transcription factor NFAT (*nuclear factor activated T cell*) promotes the transcription of the *interleukin-2 gene*. Calcineurin also contributes to the activation of *ERK* (member of the mitogen-activated protein kinase family, MAPK). Co-stimulation is provided through the CD28 immunoglobulin system which activates protein tyrosine kinase C which — in collaboration with the RAS G protein — stimulates the *Jun NH$_2$-terminal kinase* (JNK). It is also known that the *FOS* and *JUN* oncogenes contribute to the formation of the *AP* group of transcription factors; probably by acting on protein kinase C (PKC). *Phorbol myristyl acetate* (PMA) is an adjuvant for the stimulation of the *IL-2* gene. These processes of the development of effector T cells may be down-regulated when the *2E1 antibody* attaches to the *effector cell protease receptor-1*. Further down-regulation may be caused by *immunosuppressive agents* such as FK506, cyclosporin and OKT3.

Another important player in the T cell response is CTLA (cytotoxic T lymphocyte antigen), a molecule with about 75% homology to CD28. While CD28 is a co-stimulator of T cell activation CTLA-4 is a negative regulator and protects against rampant lymphoproliferative disorders. The recognition of the role of CTLA offers an opportunity to neutralize its effect by specific antibodies and thereby accelerate the action of the antitumor interleukin production without serious side-effects. T cell differentiation is accompanied by the appearance of surface markers in an order: CD4, CD25, CD44, CD8, CD3. The commitment of T cell differentiation requires tumor necrosis factor-α (TNF-α) and interleukin-1α [IL-1α])

The understanding of these circuits may lead to designing better drugs against infections and the suppression of the rejection of tissue grafts and cancer. (See figure, see also, TCR, SLAM TIL, HLA, MHC, killer cells, proteasomes, TAP, vaccines, antibody, ZAP-70, CD3, CD4, CD8, CD28, Cd117, SRC, HIV, immune system, immunoglobulins, protein tyrosine kinase, phospholipase C, PIP, IP$_3$, EPR, calcineurin, NFAT, interleukin, ERK, MAPK, RAS, AP, phorbol esters, FK506, cyclosporin, OKT, LCK, RANTES, immunophilins, signal transduction, ICAM, AKAP79, TAP, RAD25, NF-AT, NF-κB, Fas, autoantigen, γδ T cells, thymus)

T COMPLEX: the products of the virD2 and virE2 agrobacterial virulence genes associated with the 5' end of the transfered strand. (See agrobacterial virulence genes, T-DNA)

t COMPLEX: see *brachyury*

T CYTOPLASM: Texas male sterile cytoplasm of maize; almost 100% of the pollen is incapable of fertilization (sporophytic control of male sterility). (See also cytoplasmic male sterility)

t DISTRIBUTION: see Student's t distribution, t value

t HAPLOTYPE: see brachyury

T LYMPHOCYTES: see T cells

T1 PHAGE: is a double-stranded DNA (48.5 kbp) virulent phage, a general transducer. Only 0.2% of its cytosines and 1.7% of the adenines are methylated. The terminal redundancy is about 2.8 kb. It infects *E. coli* and *Shigella* strains. (See bacteriophages)

T2, T4 and T6 PHAGES: are T-even virulent phages and are closely related. The 166 kbp linear genome of T4 contains glycosylated and hydroxymethylated cytosine and 3-5% of its DNA is terminal redundancy; it encodes about 130 genes with known function, and about 100 additional open reading frames have been revealed. (See bacteriophages, development)

T5 PHAGE (relatives BF23, PB, BG3, 29-α): Their genetic material is linear double-stranded DNA (≈121.3 kbp) with terminal repeats (≈10.1 kbp) but without methylated bases. Three internal tracts can be deleted without loss of viability. They are virulent phages with long tails. (See bacteriophages, development)

T7 PHAGE (T3 is related): a virulent phage; it does not tolerate superinfection (would be required for recombination). Its 39.9 kbp genetic material is enclosed in an icosahedral head with very short tail. It codes for about 55 genes. The T7 promoter is used in genetic vectors for *in vitro* transcription. Its replication requires DNA and RNA polymerase, helicase-primase complex, and single-strand-binding protein and endo- and exonuclease activity.

θ TYPE REPLICATION: occurs generally in small circular DNA molecules starting at a single origin and following a bidirectional course. (See replication, bidirectional replication)

t VALUE: the ratio of the observed deviation to its estimate of standard error: $t = \dfrac{d}{\sqrt{V}}$ it is used as a statistical device to estimate the probabilty of difference between two means. In its most commonly used form the t is calculated $t = \bar{x}_1 - \bar{x}_2 / \sqrt{[s_{\bar{x}_1}]^2 + [s_{\bar{x}_2}]^2}$ where $\bar{x}$ indicates the mean of the two sets of data and s stands for the standard deviation. (See standard deviation, Sudent's t distribution)

Ta: a copia-like element (5.2 kbp) in *Arabidopsis*, flanked by a 5 bp repeat but without transposase function. (See copia, retroposon, Tat1)

Tα: a non-viral retrotransposable element. (See retroposon, retrotransposon)

TAB1: is a human TAK1 kinase-binding protein. Overproduction of TAB1 enhanced the activity of the promoter of the inhibitor of the plasminogen activator gene which regulates TGF-β and increased the activity of TAK1 human kinase. (See TAK1, plasmin [plasminogen], TGF-β)

TAB MUTAGENESIS: (two amino acid Barany): a DNA segment containing two sense codons is introduced at a certain position into a gene *in vitro* and thus probing the effect of the two amino acid modification regarding the function of the protein product:

```
        Thr  Gly
        ACC  GGT         includes Hpa II recognition site C↓CGG
        T GG CCA


  AC           CGGT      after Hpa II cleavage
  TGCC         CA        pCGAATT   TAB single strand linker inserted
```

Tab mutagenesis continued

               Thr **Glu Phe** Gly   two **new** amino acids in the protein and a new
               ACC | GAA | TTC | GGT
               TGG **CTT AAG** CCA   *EcoR* I restriction site **CTTAA↓G** is generated

(See directed mutagenesis, cassette mutagenesis, localized, site-specific mutagenesis)

TABATZNIK SYNDROME: heart and hand disease II. (See Holt-Oram syndrome)

TAC: transcriptionally active complex. (See open promoter complex, transcription complex, transcription factors, transcription factors inducible)

TACE (tumor necrosis factor α converting enzyme): a member of the metalloproteinase family of proteins involved in inflammatory responses. (See metalloproteinases, TNF)

TACHYKININS: are peptides mediating secretion, muscle contraction and dilation of veins. (See also angiotensin)

TACHYTELIC EVOLUTION: see bradytelic evolution

TACTAAC BOX: a highly conserved consensus in mRNA introns of *Saccharomyces* yeast.

TAE (Tris-acetate-EDTA): see electrophoresis buffers

TAF: TATA box-associated factor. (See TBP)

TAF$_{II}$ (transcription activating factors): serve as coactivators of enhancer-binding proteins. TAF$_{II}$ appears to have sufficient homology to transcription factor TFIID that in the absence of TAF$_{II}$ transcription can still proceed in yeast. TFIID is actually a complex of TBP and TAFs. TAF causes a conformational change in transcription factor TFIIB. The acidic activator disrupts the amino and carboxy-terminal interactions within this molecule and this results in an exposure of the binding sites for the general transcription factors to enter into a preinitiation complex with TFIIB. TFIIB initiates the formation of an open promoter complex. (See regulation of gene expression, TF, open promoter complex, transactivator, co-activator)

TAF$_{II}$ 230/250: is a histone H3, H4 acetylating enzyme and it is part of the TFIID transcription complex. (See histone acetyltransferase)

TAFE (transverse alternating field electrophoresis): used for pulsed field gel electrophoresis when the current is pulsed across the thickness of the gel. (See pulsed field gel electrophoresis)

TAFFAZZIN: a fibroelastin group of proteins in the muscles. (See Barth syndrome)

Tag: large T antigen, an early transcribed gene of SV40. It functions as an ATP-dependent helicase in the replication of the DNA. The two Tag binding sites, each, include two consensus sequences 5'-GAGCC-3', separated by six or seven A=T base pairs. In order to start replication, Tag must bind to the replicational origin, *ori*, and to neighboring sequences of the virus. (See Simian virus 40)

tag: small t antigen of SV40, an early transcribed gene product.

TAGGING: identifying a gene by the insertion of a transposon, an insertion element, a transformation vector or by annealing with a DNA probe. These tags have known DNA sequences and can be detected on the basis of homology. When they are inserted within a structural gene or a promoter or other regulatory element of a gene, the expression of the gene may be modified or abolished and therefore their location may be detected by alteration in a specific function and may assist also in the isolation and cloning of the target DNA sequence. (See transposon tagging, transformation genetic, probe, chromosome painting)

TAILING HOMO-A: eukaryotic mRNA generally contain a post-transcriptionally added polyA tail. Poly-A or other homopolynucleotide sequences may be added to DNAs by terminal transferase. (See mRNA, terminal transferase)

TAIL-LESS: see *Brachyury, Manx*

TAK1: a human homolog of the kinase MAPKKK, is an activator of the TGF-β signal. (See TGF-β, TAB, MAP)

TALIN: a cytoskeletal protein binding integrin, vinculin and phospholipids. (See adhesion)

TAM: (transposable element *Antirrhinum majus*): are responsible for the high mutability of genes controlling the synthesis of flower pigments, known in this plant since pre-Mendelian

TAM continued

Photograph is by courtesy of B.J. Harrison and Rosemary Carpenter

times. There are several *TAM* elements. The termini of the *TAM1* and *TAM2* are homologous and their insertion results in 3 bp target site duplication. Their termini are almost identical to those of the *Spm/En* transposons of maize and somewhat homologous to the termini of the *Tgm1* transposon of soybean. The *TAM3* transposon is different from *TAM1* and *TAM2* and 7/11 bps of its terminal repeats are homologous to the *Ac* element of maize. Both *TAM3* and *Ac* may generate 8 bp target site duplications although *TAM3* may be flanked also by 5 bp repeats. *TAM* elements, similarly to the maize transposons, seem to move by excision and relocation. The excision is usually imprecise. The insertion within genes results in mutation and the resulting mutant phenotype depends on the site of the insertion. The excision results in more or less faithful restoration of the non-mutant phenotype depending on the extent of alteration left behind at the insertion site. (See transposable elements plants, transposons)

TAMARINS: new world monkeys. (See Callithricidae)

TAMOXIFEN (2-[4-{1,2-diphenyl-1-butenyl}phenoxy]-N,N-dimethylethanamine): is an antiestrogen drug used for treatment of breast cancer but it may cause endometrial cancer although the benefits may outweigh the risk. Tamoxifen binds to AP1 and to other estrogen response elements in the uterus. (See breast cancer, raloxifene, AP1)

TAN (translocation-7-9-associated *Notch* homolog): located in human chromosome 9q34. It is involved in T cell acute leukemia. (See morphogenesis in *Drosophila* [*Notch*], *Notch*, leukemia, T cell)

TANDEM DUPLICATIONS: see tandem repeats

TANDEM FUSION: elements associated head-to-tail, following each other in the same direction.

TANDEM REPEAT: adjacent direct repeats (such as ATG ATG ATG) of any size and number.

TANGIER DISEASE (HDL deficiency): has an autosomal recessive phenotype caused by a deficiency of the α-I component of the apolipoproteins. The afflicted individuals have enlarged orange tonsils, liver, spleen and lymph nodes and deficiency of the beneficial high-density lipoproteins. They accumulate cholesterol in their cells and prone to develop coronary heart disease. (See apolipoprotein, high-density lipoprotein, cardiovascular disease)

TANNING: the ability to produce darker skin color depends largely on the activity of the melanocyte stimulating hormone (MSH) and its receptor (MC1R). Red-hair individuals — low in these activities — do not tan easily and are susceptible to UV damage (skin cancer). (See pigmentation in animals, albinism, UV, melanoma, melanin, hair color)

TAP (transporter associated with antigen processing): deliver major histocompatibility class I molecule bound peptides with the cooperation of β_2 microglubulin to the endoplasmic reticulum. When these peptides exit from the endoplasmic reticulum they are carried to the T cell receptors (TCR). Mutation in TAP may prevent antigen presentation. (See HLA, major histocompatibilty antigen, microglubulin, immune system, TCR)

Tap: bacterial transducer protein responding to dipeptides.

TAPA-1: a membrane-associated protein that in association with CD19 protein and the complement receptor 2 (CR2) mediates early immune reaction by the B lymphocytes. (See CD19, complement, B lymphocyte, immunity,)

TAPETUM: the lining, nutritive tissue of the anther, sporangia or other plant or animal organs.

TAPIR: *Tapirus terrestris*, 2n = 80.

TAQ DNA POLYMERASE: is a single polypeptide chain, 94 kDa enzyme that extends DNA strands 5'→3'; it has also a 5'→3' exonuclease activity. The enzyme is obtained from the bacterium *Thermus aquaticus*. The commercially available, genetically engineered enzyme AmpliTaq™ has temperature optima of 75 to 80°C. The enzyme is used for DNA sequencing by the Sanger method, for cloning and for PCR procedures of DNA amplification. For the latter applications it is particularly useful because during the heat denaturation cycles it is not inactivated and it is not necessary to add new enzyme after each cycle. Phosphate buffers and EDTA are inhibitory to polymerization. (See DNA sequencing, polymerase chain reaction)

TAR (trans-activation responsive element): see transcription factors, hormone response elements. regulation of gene activity, DNA binding proteins

TAR: see also transformation activated recombination

Tar: bacterial chemotaxis transducer protein with aspartate and maltose being attractants and cobalt and nickel repellent.

TAR SYNDROME: see Robert's syndrome

TARGET: anything that is the place for an action, e.g., target cell, target organ, target for DNA insertion; the target of an X-ray machine is the surface hit by the electrons and then the electromagnetic radiation is emitted in the cathode tube. (See insertion element, transposons)

TARGET THEORY: interprets the effect of radiations by direct *hits* on sensitive cellular targets. It was recognized by physicists that the amount of energy delivered to living cells and causing biological (genetic) effects is extremely low and a comparable dose of heat energy would have no effect at all. Therefore there must be some special sensitive targets in the cells that respond highly to ionizing radiations. Studies with irradiated sperm and cytoplasm of *Drosophila* indicated that the target is the chromosomes and the genes. These experiments in the period between 1920s and 1940s paved the way to the physical inquiries into the nature of the genetic material. At this early period it was hoped that the different radiation sensitivities among genes will permit the estimation of the size of these genes. It turned out however that radiation-sensitivity of the same genes varied according to the physiological stage of the tissues (higher in imbibed seeds than in dry, dormant ones) and it was higher in spermatozoa than in spermatogonia. Furthermore, temperature, genetic background, irradiation of only the culture media of microorganisms affected radiation-sensitivity, indicating that radiation sensitivity is a more complex phenomenon and it does not precisely reveal the molecular nature of the gene. (See radiation effects, physical mutagens, radiation indirect effects)

TARGETED GENE TRANSFER: is used for "knockouts" (See targeting genes, knockout)

TARGETED RECOMBINATION: see *Cre/loxP, FLP/FRT,* targeting genes

TARGETING: aiming at or transporting to a site of some molecules. The homing of free cancer cells recognize endothelial surfaces by their peptide markers and permit organ selectivity. (See lymphocytes, metastasis, transit peptide, transit signal, site-specific mutagenesis)

TARGETING FREQUENCY: the number of insertions formed at homologous or quasi homologous sites in a genome by a transforming vector.

TARGETING GENES: can be accomplished either by insertional mutagenesis or gene replacement. *Inducible gene targeting* can be carried out by first introducing into an embryonic stem cell by homologous recombination the gene *loxP*, to a flanking position of the desired target gene; *lox* facilitates the recognition of the sites for the *Cre* recombinase of phage P1. Then the mouse is crossed with a transgenic line expressing the *Cre* recombinase under the control of an interferon-responsible cell type-specific promoter. The tissue-specific recombinase thus can remove from a particular type cell the targeted gene ("floxing"). The same procedure is applicable to other eukaryotic organisms using either the phage *Cre/loxP* or the yeast *FLP/FRT* system. Since the introduction of this site-specific alteration procedure, thousands of genes have been targeted, and in mouse alone several thousands of targeted stocks have been generated. Lately gene targeting became one of the most powerful tools in genetic analysis of eukaryotes. The general principles of the procedures are illustrated on next page.

Targeting genes continued

↓   homologous recombinantion   ↓

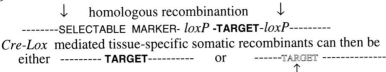

Cre-Lox mediated tissue-specific somatic recombinants can then be either   --------- **TARGET**----------   or   -----TARGET-------------
                                                                                             ↑
                                                                inactivated (or deleted)

By gene targeting through double crossover within the flanking chromosomal region different copies of the gene can be inserted (replaced) or the gene can be placed under the control of a specific endogenous or foreign promoter.

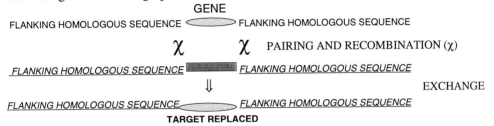

Targeting mammalian genes is feasible but the efficiency is fairly low (10^{-2} to 10^{-5}) compared to embryonic chicken stem cells (ES). Transfection of avian leukosis virus (ALV)-induced chicken pre-B cells the efficiency of recombination between the exogenous DNA and target locus may be as high as 10 to 100%. When a single mammalian chromosome is transfered to chicken cells by microcell fusion, in the somatic hybrid cell, the recombination proficiency of the mammalian chromosome at the selected locus may increase up to 10-15%. The recombined chromosome can then be shuttled back to mammalian cells for analysis.

Another targeting procedure takes advantage of the bacterial tetracycline repressor gene that attaches to the promoter of some genes and keeps them silent unless tetracycline is applied that binds to the repressor, and by inactivating it the genes are turned on. When this prokaryotic tetracycline repressor gene is inserted into a murine activator gene by transformation, the activator is incapacitated and the gene is silenced. Alternatively, inserting the tetracycline suppressor into a viral activator gene, all the genes of the transgenic mice that recognize the tetracycline suppressor—activator construct are turned on in the absence of tetracycline. Adding tetracycline to such a system, the antibiotic combines with the repressor—activator in the hybrid construct and the genes are now shut off because the suppressor—activator construct is removed from the activation position. Such a targeting construct can thus be used for on/off switching of particular genes. The success of targeting may be increased if the vector is an RNA-DNA hybrid molecule that pair more efficiently with the target. By the PCR targeting procedure a 20 bp DNA sequence tags may be generated using the photolithography procedure and DNA chips. The tag sequences are as different as possible yet possess hybridization properties to be identified simultaneously on high-density oligonucleotide arrays. Genomic DNA is isolated from a pool of deletions tagged and used as templates for amplification. For selectability a resistance gene (aminoglycoside phosphotransferase) may be used. The targeting sequence is amplified by PCR that have primers at the 3'-end homologous to the marker and at the 5'-end homologous to the target. This system is introduced into the cells by transformation. After homologous recombination at two flanks of the targeted open reading frame, the target is replaced by a construct including the 20 base tag, the selectable marker and the deletion mutation sequence. The large number of tagged deletion strains can then be pooled and tested under a variety of conditions to test how the deletion affected the function of the gene. The molecular tags are amplified and hybridized to a high-density array of known oligonucleotides, complementary to the tags. The relative intensity of hybridization reveals the relative proportion of the individual deletion strains in the pool and their fitness. Phenotypic meth-

Targeting genes continued

odological and other relevant information is contained in the TBASE <http://www.gdb.org/Dan/tbase/tbase.html>. (See insertional mutagenesis, local mutagenesis, gene replacement, knockout, homologous recombination, site-specific recombination, chromosomal rearrangement, homing endonucleases, chromosome uptake, IRES, photolithography, DNA chips)

TARGETING VECTOR: in a viral vector a section of the envelope protein gene is replaced by the coding sequences of e.g. 150 amino acids of erythropoietin (EPO) and thus it improves its ability to recognize the EPO receptor. Other approaches involve pseudotyping or attaching to the envelope special ligands. Liposomal vehicles may be conjugated with special antibodies for target recognition. (See liposome, pseudotyping, magic bullet)

TART: a telomere-specific retroposon of *Drosophila* with an about 5.1 kb 3' non-coding tract which is homologous to HeT-A; it encodes also a reverse transcriptase. (See telomere, Het-A, LINE, retroposon, reverse transcription)

TAS (termination associated sequences): are the signals for ending transcription. (See transcription)

Tassel-seed: mutations (*ts*) in maize result in kernels on the normally male inflorescence (tassel) as a result of the effemination. (See sex determination)

TASTE: is controled by a signal transducing G protein, gustducin. Both bitter and sweet tasting is mediated by gustducin. Ionic stimuli of salts and acids interact directly with ion channels and depolarize taste-receptors. Sugars, amino acids and most bitter stuff bind to specific receptors outside the cell membrane and these are connected then to G proteins. It is assumed that gustducin is involved with a phosphodiesterase. Phospholipase C also appears to have a role in the taste circuits. Gustducin receptors are present in the tongue and also in the stomach and in the intestines. (See ion channels, signal transduction)

Tat1: a transposon-like element (431 bp) in *Arabidopsis*, flanked by 13 bp inverted repeats and 5 bp target-site duplications but without any open readingframe and thus incapable of movement by its own power. (See *Arabidopsis*, transposons, open reading frame)

TAT-GARAT: the TAATGARAT enhancer motif of Herpes simplex virus. (See cigar)

TATA BOX: thymine (T) and adenine (A) containing binding sites for transcription factors and RNA polymerase complex. Many housekeeping genes and the RAS oncogene does not have this sequence. (See Pribnow box, Hogness box, transcription factors, transcription complex, open promoter complex, asparagine synthetase, promoter)

TATA BOX BINDING PROTEIN: see TBP

TATA FACTOR (TF): see transcription factors, TATA box

TATA Inr: core promoters may or may not contain these pyrimidine-rich transcription initiator elements. (See TATA box, open promoter complex, transcription factors)

TATSUMI FACTOR: is a blood clotting factor required for the activation of Christmas factor by activated PTA. It is controlled by an autosomal locus. (See antihemophilic factors, blood clotting pathways, PTA deficiency disease, hemostasis)

TAU: is a microtubule-associated protein.

TAUTOMERIC SHIFT: reversible change in the position of a proton in a molecule, affecting its chemical properties; it may trigger base substitution and thus mutation in DNA. (See hydrogen pairing, enol form, substitution mutation)

TAX: is a human T-cell leukemia virus protein that increases DNA binding of transcription factors containing a basic leucine zipper domain. (See leucine zipper. leukemia)

TAXOL: a spindle fiber blocking natural substance isolated from the yew *Taxus brevifolia*; a carcinostatic drug. (See spindle, carcinostasis)

TAXON (plural taxa): the collective name of taxonomic categories.

TAXONOMY: (biological) classification with a number of different systems. The bases of these classification is morphology, anatomy, genetics, biochemistry, physiology, cytology and macromolecular structure (DNA, RNA and proteins). Generally five broad categories are recognized: prokaryotes and viruses, protists, fungi, plants and animals. The taxonomic categ-

Taxonomy continued

ories of eukaryotes include Phylum, Class, Order, Family, Genus, Species and Subspecies (such as varieties, cultivars, breeds). In the past the classification was more rigid because the species was considered the mark of genetic isolation. Today, with somatic cell hybridization and transformation (transfection) there is no limit to genetic exchange between the various categories. There are over 300,000 plant and over 1,000,000 animal species named and classified by rules of nomenclature. For naming the binomial nomenclature is used. The genus is identified by capital first letter and then the species is designated in lower case letters. This is sometimes followed by the name of the first taxonomists who classified the organisms, e.g. *Arabidopsis thaliana (L.) Heynh.*, indicating *Arabidopsis* as the genus, *thaliana* as the species, L. stands for Linnaeus and Heynh. is the abbreviation for Heynhold who suggested the current name. For taxonomic relations see <http://www.ncbi.nlm.nih.gov>

TAY-SACHS DISEASE: is one of the most thoroughly studied biochemical diseases in human populations controled by an autosomal recessive gene. This defect occurs in all ethnic groups but it is particularly common among Ashkenazi (eastern European) Jews where the frequency of the gene is approximately 0.02 and the frequency of heterozygotes may be over 3%. The prevalence is about 1/2,500 to 1/5,000 per birth. Among the Sephardim Jews and other ethnic groups the frequency is about 1/1,000,000. Since the afflicted individuals generally die by age 3 to 4, the high frequency indicates that some heterozygote advantage must have existed for this gene. The onset is at age of a half year when general weakness, extension of the arms in response to sounds with a scared look, also muscular stiffness and retardation appears in an earlier apparently normal child. Then in rapid succession paralysis, reduction of mental abilities, vision problems leading to blindness become evident. One characteristic symptom is a cherry-red spot on the macula (grey opaque part of the cornea [eye]) caused by cell lesions. All these symptoms are the results of a deficiency of hexosaminidase A enzyme that controls the conversion of ganglioside G_{M2} into G_{M3}. As a consequence G_{M2} ganglioside accumulates leading to degeneration of myelin of the nervous system. Hexosaminidase A is composed of the α subunits (human chromosome 15q23-q24) and hexosaminidase B is a multimer of the β subunits (human chromosome 5q13). The Sandhoff's disease involves both hexosaminidase A and B deficiencies or only hexosaminidase B deficiency, and has somewhat similar symptoms with more rapid progression. Another (Type 3) milder form of G_{M2} gangliodisosis with some hexosaminidase A activity and may permit survival up to age 15. The older name of these ganglisidoses were called amaurotic familial idiocy. (See Sandhoff's disease, hexosaminidase, gangliosidoses, gangliosides, sphingolipids, lysosomal storage disease, genetic screening, Jews and genetic disease)

TAY SYNDROME: see trichothiodystrophy

T-BAM: the same as CD40 ligand.

T-BAND: telomeric regions of chromosomes with the highest concentrations of genes in the genome. (See band, chromosome banding, isochores)

TBP (TFIIτ): TATA box binding protein is a subunit of the general transcription factors, TFIID, SL1 and TFIIIB proteins that bind to DNA like a saddle in a two-fold symmetry. It has antiparallel β sheets at the concave area where it forms a reaction with the special A and T rich sequence of the DNA. The convex surface provides opportunities for interaction with other proteins: TBP → ⌒o⌐

⇑ DNA double helix in cross section

The TBP is highly conserved among different organisms from yeast to the plant *Arabidopsis* and to mammals. TBP is a complex of several TAF proteins. The TBP may also bind SL1 transcription factor of pol I but this binding is exclusively either with TFIID or SL1. The pol III TBP complex is called TFIIIB. The Dr1 repressor binds to TBP and selectively inhibits pol II and pol III action but pol I is not affected, however, because when pol II and III are repressed the relative output of pol I appears higher. The TFIID complex activity is restricted by the nucleosomal organization of the chromatin. (See transcription factors, TAF, pol I)

TC4: a small nuclear G protein. (See G proteins, RAN)

TC motif: see AP1, AP2 etc. transcription factors

TCF (ternary complex factors): are coactivators or transcription such as Elk, Sap-1a, Sap-1b, ERP-1 and other members of the ETS family of transcription factors and oncoproteins. Some are members of the high mobility group proteins. (See high mobility group proteins, ETS)

TCGF: see interleukin 2, IL-2

Tcl: see hybrid dysgenesis

TCLo: toxic concentration low; the lowest concentration of a substance in air that produced toxic, neoplastic, carcinogenic effects in mammals.

t-COMPLEX: of mouse consists of six complementation groups in chromosome 17, affecting tail development and viability. Homozygous mutants, of the same complementation group, are generally lethal. (See Brachyury)

TCP-1: a chaperonin coded for by the t-complex in mouse chromosome 17. A homolog of it occurs also in the pea leaf cytosol. (See chaperonins, chaperone)

TCR GENES: T cell receptors are glycoproteins and are similar to the antibody molecules. The TCR α chain has variable (V), diversity (D), junction (J) and constant (C) regions in the polypeptides that are generated with rearrangements of the gene clusters in human chromosome 14, in the proximity of the immunoglobulin heavy chain genes (IgH). These recombinations take place in the switching regions and the TCR genes also have the same hepta- and nanomeric sequences as the Ig genes. In the mouse TCRA (α) genes are in mouse chromosome 14 whereas the Ig heavy chains of the mouse are in chromosome 12. The TCR δ chain locus is situated within the human α gene between the V_α and J_α regions. The human β chain of TCR is encoded in chromosome 7q22 - 7qter. (In mouse, its homolog is in chromosome 6). The expression of the β genes also requires rearrangement of the VDJ genes next to the C genes. The γ_1 and γ_2 chain genes are in human chromosome 7p15-p14 area whereas their homologs are in mouse chromosome 6. The early lymphocytes carry TCR built of γ and δ polypeptides whereas later about 95% of the TCRs are built of α and β chains. The size of the αβ TCR is about 80 kDA, built of four subunits. After the αβ TCRs are formed, the γδ TCRs are eliminated. The TCR molecules are of a great variety, determined by the rearrangements just like in immunoglobulins. It appears that the TCR does not mutate somatically, in contrast to the Ig genes. The specificity of the TCR is augmented by their association either with Class I and Class II HLA gene products (MHC). The Class I HLA antigens are associated with the cytotoxic (killer) cells whereas the helper lymphocytes attach to the Class II antigens. In the function of the TCRs important role is played by the CD4 (in Class I) and CD8 (in Class II associations). Both types TCRs also require the CD3 transmembrane proteins, involved also in signal transduction. The CD peptides activate also some protein tyrosine kinases, important elements of several signal transduction pathways. Many forms of cancer are associated with chromosome breakage in the regions where the TCR proteins are coded. The human TCRβ locus consisting of 685 kb has been sequenced. This large family includes besides the TCR elements other genes too, such as a dopamine-hydroxylase-like gene, eight trypsinogen genes. The large locus involves besides the 46 functional genes 19 pseudogenes and 22 relics (genes with major lesions in one or more components). A portion of the locus is translocated from chromosome 7q22-7qter to 9. The V_β segments include promoters, the first exon as a signal peptide, with RNA splicing signals, the second exon is the V element and DNA rearrangment signal sequence. In some V_β families a conserved decamer interacts with binding proteins. (See T cell receptor, immunoglobulins, immune response, antibody, HLA, lymphocytes, CD4, CD8, SLAM, antigen-presenting cell, MHC)

TCV: see Turnip Crinkle Virus

TDF (testis determining factor): see sex determination, H - Y antigen

T-DNA (transfered DNA): of the Ti plasmid is bordered by 24-25 bp incomplete direct repeats:

T-DNA continued
TGGCAGGATATATTG X^G_A TTGTAAA for the left and TGGCAGGATATATTG X^G_A TTGTAAA for the right in the octopine plasmids of *Agrobacteria*. The border sequences of the nopaline plasmids are somewhat different. The left part of the sequences within the borders, T_L (14 kb) and right, T_R (7 kb) are distinguished. The left segment carries, among others, genes for plant oncogenicity (coding for the plant hormones indole acetic acid and the cytokinin, isopentenyl adenine) and either octopine or nopaline, the right segment contains genes for other opines and some others with unknown functions. The integration of the T-DNA into plant chromosomes is mediated by the virulence genes of the Ti plasmid and some chromosomal loci rather than by the T-DNA sequences. The T-DNA can integrate into the chromosome of plants by a process of illegitimate recombination at practically random locations. The integration involves most likely only one of the strands, called T-strand. Because of this transfer feature the T-DNA can be utilized as the most efficient plant transformation vector. The oncogenes or other sequences can be deleted and replaced by any desired DNA sequences (genes) and they are still inserted into the plant chromosome as long as the border sequences are retained. (See *Agrobacterium*, Ti plasmid transformation plants, virulence genes of *Agrobacterium*, binary vectors, cointegrate vectors, overdrive, opines)

TDT: see transmission disequilibrium test

TE: *Drosophila* transposable elements cytologically localized in chromosomes 1, 2 and 3.

TE BUFFER: contains Tris-EDTA, the pH range is 7.2 - 9.1. (See Tris-HCl buffer, EDTA)

TEC: a family of non-receptor tyrosine kinases required for signaling through the T cell receptor. (See TCR genes, T cell receptor)

TEL: phosphatidylinositol kinase of yeast. (See PIK)

TELANGIECTASIA, HEREDITARY HEMORRHAGIC (Osler-Rendu-Weber syndrome): a generally non-lethal bleeding disease (except when cerebral or pulmonary complications arise), is caused by lesions of the capillaries due to weakness of the connective tissues. The gene ORW1 in human chromosome 9q13 encodes a receptor (endoglin) for the transforming growth factor β expressed on vascular endothelium. ORW2 was mapped to chromosome 12, and it encodes an activin receptor-like kinase 1, a member of an endothelial serine/threonine kinase family. The prevalence of ORW syndrome in the USA is about 2×10^{-5}. (See hemostasis, Rothmund-Thompson syndrome, transforming growth factor β, endoglin, activin)

TELANGIECTASIS (telangiectasia): defective veins causing red spots of various sizes. (See poikiloderma telangiectasia, glomerulonephrotis)

TELEOLOGY: a dogma attributing a special vital force and ultimate purpose to natural processes beyond the material scientific evidence.

TELIOSPORES: are fungal spores protected by a thick wall and are either dikaryotic or diploid. (See stem rust, telium)

TELIUM: a fruiting structure (sorus) of fungi producing dikaryotic teliospores.

TELOCENTRIC CHROMOSOME: has terminal centromere (has one arm). Telochromosomes can be used to determine which genes are located in that particular single chromosome arm if a telotrisomic female is used according to the scheme below:

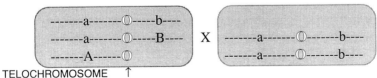
TELOCHROMOSOME

The segregation of *B:b* in both the 2n and 2n + telo progeny is expected to be 1:1 (testcross). Among the 2n offspring none or very few are expected to be *A* by phenotype because the telocentric egg can not remain functional because of dosage effect of the essential genes in the missing chromosome arm. The *A* phenotype is usually due to recombination between the telo

Telocentric chromosome continued

and the biarmed chromosomes. The 2n+telo offspring should be all *A* by phenotype and none *a* because the dominant *A* allele is in the trisomic arm. In allopolyploids the telochromosomes can be used to assign genes to the telochromosome and to determine recombination frequency between genes and the centromere. (See also centromere mapping in higher eukaryotes, misdivision of the centromere, Robertsonian translocation, tetrad analysis)

TELOCHROMOSOME: same as telocentric chromosome →

TELOISODISOMIC: in wheat, 20"+ti", 2n=42, ["=disomic, t=telosomic, i=isosomic]

TELOISOTRISOMIC: in wheat, 20"+(ti)1'", 2n=43, ["=disomic, '"=trisomic, t=telosomic, i=isosomic]

TELOMERASE: enzyme synthesizing telomeric DNA. This enzyme is different from other replicases inasmuch as the telomeric DNA is specified by an RNA template that is a part of the telomerase (ribozyme). The RNA has A and C repeats and therefore the telomeres are characterized by T and G repeats. The human telomerase template has 11 nucleotides: 5'-CUAA CCCUAAC and the telomere has the (5'-TTAGGG-3')$_n$ repeats. In budding yeast the *TLC1* gene is responsible for telomerase activity and gene *EST1* is also needed for the maintenance of the telomeres. The *TLC1* (telomerase component) gene is also required for the preservation of the RNA template and normal telomerase function. The telomeres are made mainly of double-stranded DNA repeats, however the far end has only single strand G repeats. The replication of the telomere takes place near the end of the cell cycle. Telomerase elongates only the G-rich strand and the C-rich is filled in later. Protein RCF binds also to the telomeric DNA and it is required for the replication of the leading strand by pol δ. Telomere-binding proteins (TBP) bind either to the single-strand terminal repeats (*Ostrichia* proteins) or to the double strand sequences (Rap1). Additional proteins bind to these TBPs. The E6 protein of the Human Papilloma Virus-16 (HPV16) activates telomerase but it does not immortalize the cells although it expands their life span. The newly replicated telomeres processed to size by proteins. Telomerase activity in tumor tissues is high but it is low in somatic cells or in benign neoplasias. The telomerase RNAs of *Tetrahymena*, *Euplotes* and *Oxytricha* seem to be transcribed by polymerase III and the human telomerase RNA may be the product of pol II because it is sensitive to α-amanitin. The telomere end of *Tetrahymena* where it was originally discovered is somewhat different. In the RNA template there are 5'-CAACCC-3' and in the telomere 5'-GGGTG-3' repeats occur. The Cdc13 protein of yeast mediates the access of the telomerase to the telomere. In the catalytic subunit of a telomerase of yeast (*EST2, ever shorter telomeres*) reverse transcriptase-like motifs were identified. (See telomeres, ribozyme, RAP, pol II, pol III, α-amanitin, *Tetrahymena*)

TELOMERE MAPPING: eukaryotic chromosome have characteristic telomeric repeats and can be used for mapping RFLPs relative to the telomeres. (See telomeres, RFLP)

TELOMERE TERMINAL TRANSFERASE: the ribozyme involved in replication telomeric DNA. (See telomerase)

TELOMERES: are special terminal structural elements of eukaryotic chromosomes rich in T and G bases. In human chromosomes and all vertebrates, *Trypanosomas*, fungi the many times repeated telomeric box is CCCTAA/TTAGGG. The telomeric region is highly conserved among diverse eukaryotes; in some species, however variations exists. In *Drosophila* the telomere is unusual inasmuch as rather than having the common repeats, it has a transposable element. This transposable element my assist the replication of the telomeric DNA. In most organisms proximal to the telomere middle-repetitive DNA is found (telomere associated DNA, TA) which may display some similarity to the *Drosophila* transposable sequences. The length of the telomeric sequences may vary among the organisms but variations exist during the life of the cells according to developmental stage. Mutants of shorter telomeres may function reasonably well although a transient pause may occur in the cell division. The non-nucleosomal special chromatin, including the repeats of yeast, is called telosome. In mammals the large telomeric DNA is nucleosomal, however. The major structural protein associated with

Telomeres continued
the yeast telomeres is Rap1. The same protein may also be present at other locations of the chromosomes and acts either as a repressor or activator of transcription. Telomeres are required for the proper replication of the linear eukaryotic chromosomes. This is probably the cause why among the diverse types of chromosomal aberrations cytologists failed to find attachments of internal chromosomal pieces to the telomeres or telomere to telomere fusions although chromosomes with broken ends may fuse into dicentric chromosomes after replication. The structure of the telomere may bear some similarity to the centromere and in some instances (e.g., rye) it may function as a neocentromere. It may be somewhat of a puzzle still how the normally fragmented somatic chromosomes of *Ascaris* produce their telomeres and how the germline chromosomes control the apparently multiple intercalary telomeres. In the polyploid ciliate macronucleus millions of centromeres may exist in some species. HIV-infected individuals appear to have shorter telomeres than healthy individuals of comparable age. Telomere loss or inactivation may play a role in senescence and telomerase activation may be a mechanism of cellular immortalization. The replication of the telomeric DNA is carried out by the ribozyme, telomerase. A specific telomere-binding human protein hTRF (60 kDa), recognizing the TTAGGG (and mammalian) sequences has been isolated. The corresponding repeat in *Tetrahymena* is TTGGGG and in yeast it is TG_{1-3}. One of its domains is similar to the Myb oncogene product. Telomeric sequences may be shortened in several types of cancer cells. In *Caenorhabditis* all chromosomes are capped by the same 4-9 kb tandem repeats of TTAGGC but the sequences next to it are different among the chromosomes. Chromosomes lacking telomeres can perform most functions but lack stability and undergo fusion, degradation, loss at a high rate and the chromatids may not be able to separate normally. (See telomerase, breakage-fusion-bridge cycle, neocentromere, chromosome diminution, immortalization, senescence, Myb, RAP, telomeric probe, HeT-A, TART, plasmid telomere)

TELOMERIC PROBE: are fluorochrome-labeled DNA sequences, complementary to the telomeric repeats TTAGGG and can be used for the cytological identification of short (cryptic) translocations. (See chromosome painting, telomere)

TELOMUTATION: a (dominant) premutation occurring in and transmitted through both sexes but expressed only in the offspring of the heterozygous female. (See premutation, delayed inheritance)

TELOPHASE: is the final major step of the nuclear divisions. The chromosomes have been pulled by the spindle fibers all the way to the poles. The microtubules attached to the kinetochores fade from view and the nuclear envelop appears again. The chromosomes relax and the nucleoli become visible again. (See mitosis, meiosis)

TELOSOME: a telocentric chromosome; the non-nucleosomal chromatin at the end of the yeast and ciliate chromosomes is also called telosomal. (See telomeres)

TELOTRISOMIC: is a trisomic having two bi-armed and one telochromosome. (See trisomy)

TELSON: the most posterior part of the arthropod body (opposite to the head). (See *Drosophila*)

TEM (transmission electronmicroscopy): see electronmicroscopy

TEM: triethylenemelamine, an alkylating agent. (See alkylating agent, alkylation)

TEMED: see acrylamide

TEMPERATE PHAGE: has both lysogenic and lytic life-styles. (See bacteriophage, lysogen, lysogeny, lambda phage, specialized transduction)

TEMPERATURE-SENSITIVE MUTATION (ts): the mutation causes such an alteration in the primary structure of the polypetide chain that its conformation varies according to temperature, and it is functional at either high or low temperature but it is non-functional at the other (non-permissive temperature). Temperature-sensitive conditional lethal mutants are very useful for various analyses because the biochemical/molecular basis of the genetic defect can be analyzed at the permissive temperature range when the cells or organisms can grow (normally). Ts condition is correlated with the buried hydrophobic residues in the protein. (See hyperthermia, cold hypersensitivity)

TEMPLATE: determines the shape or structure of a molecule because it serves as a "mold" for it (in someway similar to the mold to cast iron). The old DNA strands serve as templates for the new ones or one of the DNA strands may be a template for the mRNA and the other for another sense or nonsense RNA. Molecular biologists frequently call template (or antisense) the strand of the DNA that serves for the synthesis of mRNA by complementary base pairing. T7 RNA polymerase can bypass up to 24 nucleotide gaps in the template strand by copying a faithful sequence of the deletion using the non-template strand. (See semiconservative replication, replication fork, sense strand, coding strand)

TEMPLATE SWITCH: during replication the polymerase enzyme may jump to another DNA sequence and copy elements which were not present in the original DNA tract. Such a switch may occur when plasmid DNA (T-DNA) is inserted into a chromosomal target site. As a consequence deletions and rearrangements may follow. During primer extension the strand displaced may reanneal onto the template and the extended strand may be partially dissociated. A single-stranded sequence, attached to the 5' end of the displaced strand and complementary to the dissociated segment of the extending strand, can thus serve as an alternative template (See switch, T-DNA, primer extension)

TENASCIN: is a large glycoprotein complex with disulfide-linked peptide chains. It either promotes or interferes with cell adhesion depending on the type of cell and the different protein domains. It controls also cell migration. (See cell migration, cell adhesion)

TENDRIL: a plant organ that coils around objects and provides support.

TENSIN: cytoskeletal protein binding vinculin and actin; it contains an SH2 domain. (See actin, vinculin)

TENSION: the force(s) generated by pulling the mitotic/meiotic chromosomes to the opposite poles, opposed by the attachment between the homologs. (See spindle, kinetochore, pole)

TEOSINTE: see maize

ter **SITES**: see DNA replication prokaryotes

TERATOCARCINOMA: malignant tumor containing cells of embryonal nature; common in testes. The teratocarcinoma cells may differentiate into various types of tissues *in vitro* and have been used for studies of differentiation. (See cancer, teratoma)

TERATOGEN: agents causing malformation during differentiation and development.

TERATOGENESIS: malformation during differentiation and development. The inducing agents may be genetic defects, physical factors and injuries, infections, drugs and chemicals.

TERATOMA: is a mixed tissue group with cells of different potentials for development. It may be formed in various early or late animal tissue and may eventually become a malignant tumor. Teratomas are most common in the germinal tissues. In plants amorph, undifferentiated tumor tissue may be interspersed with differentiated elements giving rise to either shoots or roots. Teratomas may be formed in the tumors induced by *Agrobacterium rhizogenes*. (See *Agrobacterium*)

TERGITE: dorsal epidermis developed not from imaginal disks but from histoblasts. (See histoblast)

TERMINAL DIFFERENTIATION: is usually irreversible; in plants, however, under culture of high levels of phytohormones, dedifferentiation may be possible. (See dedifferentiation, differentiation)

TERMINAL NUCLEOTIDYL TRANSFERASE: an enzyme that can elongate DNA — at the 3-OH end — strands with any base that is present in the reaction mixture. It is used in genetic vector construction to generate homopolymeric cohesive tails for the purpose of splicing a passenger DNA to a plasmid. The passenger and the plasmid are equipped with complementary bases such as polyA and polyT, respectively, and can readily anneal and can be ligated. Terminal nucleotidyl transferase is also expressed in the adult bone marrow and generates immunoglobulin diversity. (See vectors, cloning vectors, immunoglobulins, T cell)

TERMINAL PROTEIN (TP): serine, threonine or tyrosine residues of the terminal proteins provide OH groups for the initiation of replication instead of 3' OH group of a nucleotide in a linear double-stranded viral DNA.

TERMINAL REDUNDANCY: repeated DNA sequences at the ends of phage DNA. (See permuted and non-permuted terminal redundancy)

TERMINAL TRANSFERASE: enzyme can add homopolymeric ends to DNA. It is useful in generating cohesive termini (in genetic vectors) if one (e.g., plasmid) of the reaction mixture contains adenylic acids and the other (e.g., passenger DNA) thymidylic acid (or G and C, respectively). (See terminal nucleotidyl transferase, cloning vectors)

TERMINALIZATION OF CHROMOSOMES: during meiosis the bivalents may display one or more chiasma(ta) which eventually (during separation) move toward the telomeric region in the process called terminalization. (See meiosis)

TERMINASE: phage enzyme binding to specific nucleotide sequences of the genome and cut in the vicinity of the binding at cohesive sites (*pac* or *cos*). (See lambda phage)

TERMINATION CODONS: see nonsense mutation, genetic code, terminator codons

TERMINATION FACTORS: proteins that release the polypeptide chains from the ribosomes.

TERMINATOR: the sequence at the end of a gene that signals for the termination of replication or transcription. (See transcription termination, protein synthesis, release factor)

TERMINATOR CODONS: UAA, UAG, and UGA in RNA (same as nonsense codons).

TERMISOME: nucleic acid terminal protein complexed with other proteins and DNA. (See terminator, protein synthesis)

TERNARY: has two meanings; either that it is made up of three elements or third in order.

TERNARY COMPLEX FACTORS: see TCF

TERPENES: hydrocarbons or derivatives with isoprene repeats; occur as animal pheromones and diverse types of plant fragrances. (See pheromones, fragrances, prenylation)

TERRESTRIAL RADIATION: coming from the unstable isotopes in the soil, such as uranium-containing rocks. (See cosmic radiation, radiation hazard assessment, radiation effects, ionizing radiation, isotopes)

TERRIFIC BROTH (TB): bacterial nutrient medium containing bacto-tryptone 12 g, bacto-yeast extract 24 g, glycerol 4 mL, H_2O 900 mL and buffered by 100 mL phosphate buffer.

TERTIARY STRUCTURE: the three-dimensional arrangement of the secondary structure of the polypeptide chain into layers, fibers or globular shape. A third order of complexity, folding or coiling the secondary structure once more. (See protein structure)

TESTA (seed coat): is maternal tissue in hybrids, therefore seed coat characters display expression delayed by one generation (e.g., to F_3 rather than in F_2 of recessive markers).

TESTCROSS: cross between a heterozygote and a homozygote for the recessive genes concerned, e.g., (*AB / ab*) x (*ab/ab*). (See recombination frequency)

TESTER: in a genetic cross is intended to reveal either the qualitative or quantitive gene content of the individual(s) tested. (See testcross, hybrid vigor, combining ability)

TESTES (sing. testis): the male gonads of animals, producing the male gametes. (See gonad, gamete)

TESTICLES: same as testes

TESTICULAR FEMINIZATION: is a developmental anomaly which occurs in humans and other mammals. The chromosomally XY individuals display female phenotype including the formation of a blind vagina (no uterus), female breasts, generally the absence of pubic hairs. Usually the individuals affected by this recessive disorder (human chromosome Xq11.1-q12) appear very feminine but sterile. Generally they develop abdominal or somewhat herniated small testes. About

TESTICULAR FEMINIZATION. (Courtesy of Dr. McL. Morris)

Testicular feminization continued

1.5×10^{-5} of the chromosomally males have this disorder. The condition is the result of a complete or partial deficiency or instability of the androsterone receptor protein (917 amino acids). The function of this androgen receptor protein may be either totally missing or only partial (Reifenstein syndrome). This receptor appears to be highly conserved among mammals. The gene extends to about 90 kb DNA. The protein binds to DNA by two domains encoded by exons 2 and 3 and five exons code for androgen binding while exon 1 has a regulatory function. (See chromosomal sex determination, hermaphrodite, hormone receptors, pseudohermaphroditism, dihydrotestosterone, figure on preceding page)

TESTOSTERONE: see animal hormones, hormonal effects on sex expression

TESTTUBE BABY: See *in vitro* fertilization, intrauterine insemination, intrafallopian transfer, ART

TESTWEIGHT: the mass of 1,000 seeds or kernels randomly withdrawn from a sample. (See absolute weight)

TET: tetracycline antibiotic, inhibits protein synthesis, genetic (selectable) marker in the pBR plasmids. (See pBR322)

TET (tubal embryo transfer): essentially the same as intrafallopian transfer of zygotes. (See intrafallopian transfer, ART)

Tet A: tetracycline resistance protein that controls the efflux of the antibiotic from the cells. (See tetracycline)

Tet R: tetracycline repressor (homodimer) regulates the expression of the antiporter (that exports the drug from the cells) at the level of transcription and it is activated by $[Mg - Tc]^+$ complex. (See tetracycline, antiport)

TETANY: a highly stimulated condition of the nervous and muscular system caused by low levels of calcium due to various diseases.

TETHERING: bringing together two distantly located nucleic acid sequences either by DNA looping, or catenanes or RNA lariats. (See introns, lariat RNA)

TETRACYCLINE: antibiotic that prevents the binding of amino acid-charged tRNAs to the A site of the ribosomes. (See antibiotics, → pBR322, ribosomes, protein synthesis, targeting genes, tTA, rtTA)

TETRAD, ABERRANT: the allelic proportions deviate from 2:2 because of polysomy, gene conversion, non-disjunction, suppression.

TETRAD ANALYSIS: the meiotic products of ascomycetes (occasionally some other organisms) stay together as the four products of single meiosis, as a *tetrad*. In some organisms the tetrad formation is followed by a post-meiotic mitosis within the ascus resulting in spore *octads*

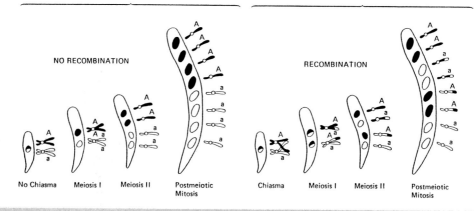

SPORE TETRADS AND OCTADS WITHOUT AND WITH RECOMBINATION BETWEEN A GENE AND THE CENTROMERE

Tetrad analysis continued

(see figure on preceding page). If the four spores are situated in the same linear order as produced by the two divisions of meiosis it is *ordered tetrad*. In the ordered tetrad, considering two genes *A* and *B*, three arrangements of the spores (parental ditype [PD], tetratype [TT], non-parental ditype [NPD]) can be distinguished as seen on the figure below:

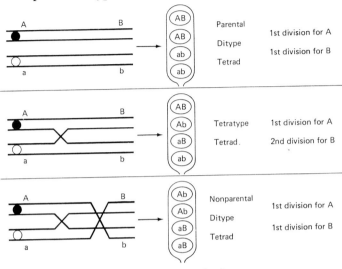

TRANSACTIONS AND RESULTS EXAMPLIFIED BY TWO GENE LOCI IN AN ORDERED TETRAD. GENE *A* IS SO CLOSE TO THE CENTROMERE THAT PRACTICALLY NO RECOMBINATION OCCURS BETWEEN THE TWO. (Diagram after Barratt, R.W. *et al*. 1954 Advances in Genetics 6:1)

The parental ditype (PD) indicates no crossing over, tetratype (TT) reveals one recombination between the two genes and also the second division segregation of the B/b alleles reveals recombination between the B/b gene and the centromere. The nonparental ditype (NPD) is an indication of double crossing over between the to gene loci.

The PD, TT and NPD may appear even is the genes are in separate chromosomes. An excess of PD over NPD is an indication of linkage. If the deviation from the 1:1 ratio between PD and NPD is small a chi square test may be used to test the probability of linkage by the formula.

$$\chi^2 = \frac{[PD - NPD]^2}{PD + NPD}$$

By counting the number of tetrads of the above three types, recombination frequency between the two loci can be calculated as shown in the second box at left. Recombination frequency between the B/b gene and the centromere can be calculated as shown by the box pointed to.

$$\frac{[1/2]TT + \text{all NPD}}{\text{all tetrads}}$$

$$\frac{TT[1/2]}{\text{all tetrads}}$$

The recombination frequencies (if they are under 0.15) multiplied by 100 provide the map distances in centimorgans. If the recombination frequencies are larger, mapping functions should be used. From the genetic constitution of the spores within the tetrads, a great deal of information can be derived regarding recombination (see figure on next page). When the four meiotic products are not in the order brought about by meiosis, the tetrad is *unordered*. For the estimation of gene - centromere distances from unordered tetrad data, one must rely on 3 markers, from that no more than 2 are linked, and algebraic solutions are required (e.g. Whitehouse 1950 Nature 165:893, see unordered tetrads). Tetrad analysis is most commonly used in ascomycetes (*Neurospora, Aspergillus, Ascobolus, Saccharomyces,* etc.) yet it can be applied to higher plants where the four products of male meiosis stick together (*Elodea, Salpiglossis*, orchids, *Arabidopsis* mutants). In *Drosophila* with attached X-chromosomes half-tetrad analysis is feasible. (See unordered tetrads, half-tetrad analysis, meiosis, mapping, linkage)

TETRAHYDROFOLATE (THF): the active (reduced) form of the vitamin folate, a carrier of one-carbon units in oxidation reactions, a pteridine derivative. (See hyperhomocysteinemia)

Tetrahymena pyriformis : (2n = 10) is a ciliated protozoon with linear mtDNA. Its rRNA transcripts are self-spliced. (See splicing, mtDNA, telomerase)

TETRALOOP: in structured RNAs duplex runs are connected by loops of 5'GNRA tetranucleotides (N = any nucleotide, R= purine). They are involved in long-range molecular interactions such as in hammerhead ribozymes, introns. (See ribozymes, introns)

Tetrad analysis continued

⇑ OCTADS OF NEUROSPORA, (Courtesy of Dr. D.R. Stadler, Genetics 41:528)

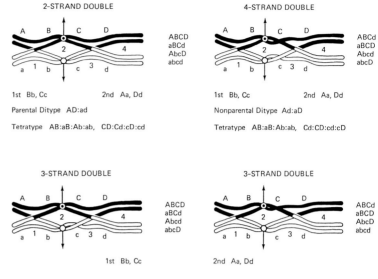

FOUR-POINT CROSS WITH GENES IN BOTH ARMS OF THE CHROMOSOMES. IT IS A FIVE-POINT CROSS IF WE CONSIDER THE CENTROMERE AS A GENETIC MARKER. FROM THE SPORE ORDER WE CAN DETERMINE EVEN IF THE CHROMATIDS ROTATED 180° AFTER THE EXCHANGE. (After Emerson, S. 1963 in Methodology in Basic Genetics, p. 167. Burdette, W.J., ed., Holden-Day, San Francisco)

TETRANUCLEOTIDE HYPOTHESIS: a historical assumption that nucleic acids are made of repeated units of equal numbers of the four bases (A, T/U, G and C), and because of this monotonous structure could not qualify for being the genetic material. During the period of 1909 to 1940 it was a rather widely accepted view, first proposed by Phoebus Levine.

TETRAPARENTAL OFFSPRING: results if *in vivo* fused blastulas of different matings are implanted together into the uterus. (See allophenic)

TETRAPLEX: consecutive guanine sequences may take four-stranded parallel or antiparallel conformations in the DNA or RNA. The tetraplex structure may have biological significance for the telomeres, specific recombination of the immunoglobulin genes, dimerization of the HIV genome, etc. (See telomere, immunoglobulins, HIV; see also tetrasomic)

TETRAPLOID: has four sets of genomes per nucleus. (See autopolyploidy, tetrasomic)

TETRASOMIC: one chromosome is present in four doses in the nuclei. Tetrasomy exists in tetraploids for all chromosomes and may not involve any serious anomaly although generally the fertility is reduced. Tetrasomy for individual chromosomes may have more serious consequences because of genic imbalance. Tetrasomy is usually not tolerated by animals. Tetrasomic mosaicism for human chromosome 12p leads to developmental anomalies, mental retardation, defects in the central nervous system and speech, etc. (See polyploidy, autopolyploid, trisomy, sex-chromosomal anomalies in humans, cat-eye syndrome)

TETRASOMIC - NULLISOMIC COMPENSATION: in allotetraploids the homoeologous chromosomes may compensate for each other and thus if two chromosomes are missing (nullisomy) and one type or their homoeologues is present in four copies (tetrasomy) the individuals may function rather well, depending on the particular chromosomes involved. When, however in the presence of nullisomy a non-homoeologous chromosome is substituted, the condition is worsened. (See also chromosome substitution, nullisomic compensation)

TETRATRICO SEQUENCES (TPR): are amphipathic α-helical amino acid tracts punctuated by proline-induced turns. Such motifs may control mitosis and RNA synthesis in the cells.

TETRATYPE: the meiotic products concerned with two genes show four types of combinations (e.g., AB Ab aB and ab). (See tetrad analysis)

TETRAZOLIUM BLUE: detects oxidation-reduction enzyme activity and thus identifies living cells and cancerous metabolism.

TETRODOTOXIN: see toxins

TEXAS RED: a fluorochrome with excitation at 580 nm and emission peak at 615 nm. (See fluorochromes)

TGF: see transforming growth factor

TF: transcription factors such as TF I, TF II or TF III, involved in the control of transcription by pol I, pol II or pol III, respectively; TFs assist transcription by cooperation with other binding proteins. (See transcription factors, open transcription complex)

TF: see transferrin

TFII: see transcription factors, TBP, PIC

TFIIτ: is synonymous with TBP.

TFM: see testicular feminization

TFO (triple-helix-forming oligonucleotide): see antisense RNA, antisense DNA, triple helix formation

TFR: transferrin receptor (see transferrin)

TGF: transforming growth factor is involved in cancerous growth. A member of the same family, Vg1, a factor localized to the vegetal pole of the animal embryo, may be involved in the induction of mesoderm formation in the embryo. The TGF-β family cytokines activate the receptors of the heterodimeric serine-threonine kinases. TGF-β binds directly to TGF receptor II, a kinase. The bound TGF-β then, binds by the TGF receptor I and becomes phosphorylated by it. TGFR I then transmits the cytokine signal along the signaling cascade. (See activin, bone morphogenetic protein, serine/threonine kinase, signal transduction, cytokines)

TGN: see trans-Golgi network

T$_H$ (or Th): T helper cell. (See T cell)

THALAMUS: a double-egg-shape area deep within the basal part of the brain involved in transmission of sensory impulses. (See fatal familial insomnia, brain human)

THALASSEMIA: hereditary defects of the regulation (or deletions) of hemoglobin genes causing anemia. In the thalassemias, generally the relative amounts of the various globins is affected because of deletions of the hemoglobin genes. *Thalassemia major* the most severe form of the disease in patients homozygous for a defect in the two β-chains and have an excessive amount of the F hemoglobin (HPFH: *hereditary persistence of fetal hemoglobin*). *Cooley's anemia* is also caused by β chain defects. *Thalassemia minor* is a relatively milder form with some hemoglobin A$_2$ present and usually slight elevation of the F hemoglobin is characteristic for the heterozygotes. In the *β-thalassemias* different sections of the β globin gene family from human chromosome 11p15.5 (5'- ϵ Gγ Gγ $\psi\beta$ δ β – 3') are missing. The deletions may involve only 600 bp from the 3' end as in in β^0 or about 50 kbp, eliminating most the family beginning from the 5' end and retaining only the β gene at the 3' end in human chromosome 11p. In the β chain gene about 100 point mutations and various deletions have been analyzed. In severe cases the symptoms of the β thalassemias are anemia, susceptibility to infections, bone deformations, enlargement of the liver and spleen, iron deposits, delayed sexual development, etc. may appear within a few months after birth. In the *α-thalassemias* (in human chromosome 16p13) various members of the α chain gene cluster (5'- ζ $\psi\zeta$ $\psi\alpha$ $\psi\alpha$ α2 α1 ∂ - 3') or even all 4 α chains are defective and the latter case results in the lethal hereditary disease, hydrops fetalis (Bart's hydrops), entailing accumulation of fluids in the body of the fetus and severe anemia, resulting in prenatal death. In α-thalassemia-1 both the α1 and the α2 genes are deleted, in thalassemia *α-thal-2* only either the left (α thal 2L) or right end (α thal 2R) of the α-gene(s) is lost. The *Hb Lepore hemoglobin* is the consequence of unequal crossing over within the α gene cluster resulting in N-terminal-$\delta\beta$-C protein fusion. The N-$\beta\delta$-C reciprocal protein fusion is called *Hb anti-Lepore* or *F thalassemia*. One of these types of hemoglobins were first re-

Thalassemia continued

ported in 1958 in the Lepore Italian family. Since that, a large number a Lepore type hemoglobins were discovered in various parts of the world and named Greece, Washington, Hollandia, etc. hemoglobin. Thalassemias have much higher incidence world-wide in areas where malaria is a common disease. Homozygotes for thalassemias rarely contribute to the gene pool, thus the heterozygotes appear to have selective advantage in the maintenance of this condition. Some recent studies, however, indicate that α thalassemia homozygotes have a higher incidence of uncomplicated childhood thalassemia and splenomegaly (enlargement of the spleen an indication of infection by *Plasmodium*). This increase of susceptibility to the relatively benign *P. vivax* which may provide some degree of immunization for later stages of life against the more severe disease caused by *P. falciparum* infection. This relatively high incidence of thalassemias in tropical and subtropical areas and the well known molecular genetics mechanisms involved, makes this disease a candidate for somatic gene therapy by either bone marrow transplantation or transformation. The prevalence of thalassemias varies from about $10 - 5 \times 10^{-4}$. Thalassemias are frequently associated with sickle cell anemias which are β globin defects and aggravating the conditions. Prenatal diagnosis is possible by the use of protein or DNA technologies. Fetal mutation may be identified after PCR amplification of chorionic samples (9-12 weeks) or in directly withdrawn cells from the amniotic fluid after the 3rd months of pregnancy. Hydrops can be identified by ultrasonic techniques during the second trimester. (See hemoglobin, methemoglobin, sickle cell anemia, plasma proteins, hemoglobin evolution, unequal crossing over, genetic screening, PCR, trimester, *Plasmodium*, Juberg-Marsidi syndrome)

THALIDOMIDE (N-phtaloyl glutamimide): a sedative and hypnotic agent of several trade names; it had been used experimentally also as an immunosuppressive and antiinflammatory drug. Its medical use as a tranquilizer during pregnancy caused one of the most tragic disasters of severe malformations of the embryos and newborns. (See teratogen)

THALLOPHYTE: a plant, fungus or algal body which is a thallus. (See thallus)

THALLUS: a relatively undifferentiated colony of plant cells without true roots, stem or leaves.

THANATOPHORIC DYSPLASIA: is a dominant lethal human dwarfism caused by deficiency of the fibroblast growth factor receptor 3, FGFR3. Its prevalence about 2×10^{-4}. (See dwarfism, fibroblast growth factor)

THEA (*Camellia sinensis*): 82 species, 2n = 2x = 30.

THEILERIOSIS: is an African parasitic infection of the cattle's immune system causing disease and death to the infected animals. In the T lymphocytes casein kinase II, a serine/threonine protein kinase is markedly higher. This protein is not identical with the common casein kinase found in lactating animals.

THELYOTOKY: parthenogenesis from eggs resulting in maternal females. (See arrhenotoky, deuterotoky, sex determination, chromosomal sex determination, sex determination)

THEOBROMINE: the principal alkaloid in cacao bean, containing 1.5 to 3% of the base. Present also in tea and cola nuts. The TDLo orally for humans is 125 mg/kg. It is a diuretic, smooth muscle relaxant, cardiac stimulant and a vasodilator (expands vein passages). (See TDLo, caffeine)

THERAPEUTIC RADIATION: see radiation hazard assessment

THERMAL CYCLER: is an automatic, programmable incubator use for the polymerase chain reaction. (See PCR)

THERMAL NEUTRONS: see physical mutagens

THERMAL TOLERANCE: is genetically determined and in plants may be affected by fatty acid biosynthesis and abscisic acid deficiency, etc. Heatshock proteins have also been credited with it. In *Tetrahymena thermophila* a small cytoplasmic RNA (G8 RNA) is responsible specifically for thermotolerance independently from a heat shock response. (See temperature-sensitive mutation, heat shock proteins, RNA G8, antifreeze protein)

THERMODYNAMIC VALUES: of chemicals are measured in calories; 1 calorie = 4.1840 absolute joule. (See joule)

THERMODYNAMICS, laws of: 1). When a mechanical work is transformed into heat, or the reverse, the amount of work is equivalent to the quantity of heat. 2). It is impossible to continuously transfer heat from a colder to a hotter body.

THERMOLUMINISCENT DETECTORS: for personnel monitoring device in a small crystalline detector (lithium fluoride, lithium borate, calcium fluoride or calcium sulfate, and metal ion traces as activators) radiation is absorbed in the crystalline body and upon heating it is released in the form of light. Useful range 0.003 - 10,000 rem. Radiophotoluminiscent (RPL) and thermally stimulated exoelectron emission (TSEE) detectors may be used instead of film badges in radiation areas. (See radiation measurements)

THERMOTOLERANCE: some mutants of plants display increased growth relatively to the wild type at higher temperature. Some of the temperature-sensitivity is associated with differences in fatty acid biosynthesis. (See also cold-regulated genes, temperature-sensitive mutation)

THF: see tetrahydrofolate

THETA REPLICATION: θ, is a stage of bidirectional form of replication in a circular DNA molecule. When the two replication forks reach about half way toward the termination sites the molecule resembles the Greek letter θ. (See DNA replication, θ replication, bidirectional replication, replication, prokaryotes)

THIAMIN (vitamin B_1): its absence from the diet causes the disease beri-beri, alcoholic neuritis and the Wernicke-Korsakoff syndrome. Iits deficiency was the first genetically determined auxotrophic mutations in *Neurospora,* and it is the most readily inducible auxotrophic mutation at several loci in lower and higher green plants such as algae and *Arabidopsis*. The vitamin is made of two moieties: 2-methyl-4-amino-5-aminomethyl pyrimidine and 4-methyl-5-β-hydroxyethyl thiazole. The pyrimidine requirement of the thiamin mutants cannot be met by the precursors of nucleic acids. (See thiamin pyrophosphate, Wernicke-Korsakoff syndrome, megaloblastic anemia)

THIAMIN PYROPHOSPHATE: the coenzyme of vitamin B_1. It is an essential cofactor for pyruvate decarboxylase (alcoholic fermentation), pyruvate dehydrogenase (synthesis of acetyl Co-A), α–ketoglutarate dehydrogenase (citric acid cycle), transketolase (photosynthetic carbon fixation), acatolactate synthetase (branched chain amino acid biosynthesis). (See thiamin)

THIN-LAYER CHROMATOGRAPHY (TLC): separation of (organic) mixtures within a very thin layer of cellulose or silica gel layer applied uniformly onto the surface of a glass plate or firm plastic sheet and used in a manner similar to paperchromatography. The material is applied at about 2 cm from the bottom, then the plate is dipped into an appropriate solvent (mixture). Generally, the substances are separated rapidly and with excellent resolution. Identification is made generally on the basis of natural color or with the aid of special color reagents. (See Rf)

THIOESTER: acyl carriers oligopeptide $\begin{array}{c} -C=O \\ | \\ SH \end{array}$ acyl groups covalently linked to reactive thiol; they are high energy in Coenzyme A; also thioester may assist the formation of without any cooperation by ribosomes. There are suggestions that thioethers are the relics of prebiotic conditions when sulfurous (volcanic) environment existed on the Earth and might have had a role in the origin of living cells by being energy carriers. (See sulfhydryl, disulfide)

THIOGUANINE (2-amino-1,7-dihydro-6H-purine-6-thione): is an antineoplastic/neoplastic agent. Its delayed cytotoxicity is attributed to postreplicative mismatch repair. After incorporation into the DNA it is methylated and pairs with either thymine or cytosine. The immunosuppressant, azathioprine (6-[1-methyl-4-nitroimidazol-5-yl]-thiopurine) used in organ transplantation can be converted to thioguanine and may be the cause of the increased incidence of cancer after transplantation. (See DNA repair, mismatch, base analogs, hydrogen pairing)

THIOL: see sulfhydryl

THIOREDOXIN: is a dithiol protein that mediates the reduction of disulfide bonds in proteins. Also, light regulates photosynthesis through reduced thioredoxin, linked to the electron transfer chain by ferredoxin. Light modulates also translation in the chloroplast by redox potential. (See photosynthesis, photosystems)

THIOTEPA: an alkylating agent, formerly used also as an antineoplastic drug; it induces a very high frequency of sister-chromatid exchange. (See alkylating agent, sister chromatid exchange)

THIRD MESSENGERS: propagate signals of the second messenger and thus activate or deactivate a series of genes. (See second messenger)

THOMSEN DISEASE: see myotonia

THORAX: the chest of mammals, the segment behind (posterior) the head in *Drosophila*.

THREE-HYBRID SYSTEM: is a construct that detects the role of RNA — protein interactions:

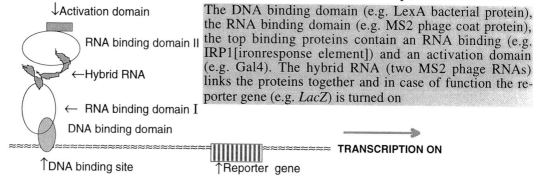

The DNA binding domain (e.g. LexA bacterial protein), the RNA binding domain (e.g. MS2 phage coat protein), the top binding proteins contain an RNA binding (e.g. IRP1[ironresponse element]) and an activation domain (e.g. Gal4). The hybrid RNA (two MS2 phage RNAs) links the proteins together and in case of function the reporter gene (e.g. *LacZ*) is turned on

(Modified after D.J.S. Gupta *et al.* 1996 Proc. Natl. Acad. Sci. USA 93:8496. (See two-hybrid system, one-hybrid binding assay, split-hybrid system)

THREE POINT TESTCROSS: it uses three genetic markers and thus permits the mapping of these genes also in a linear order. (See mapping genetic, testcross)

THREE-WAY CROSS: a single-cross (the F_1 hybrid of two inbred lines) is crossed with another inbred. The purpose is testing the performance of the single-cross. (See combining ability)

THREONINE: this amino acid is derived from oxaloacetate via aspartate and immediately from:

$(HO)_2(PO)OCH_2 CH (NH_2)COOH \rightarrow CH_3CH(OH)CH(NH_2)COOH$

O-phosphorylhomoserine *threonine synthase* threonine

Threonine dehydratase degrades threonine into α-ketobutyrate and NH_4^+; then five steps are required from α-ketobutyrate to produce isoleucine; the first of these steps is mediated by threonine deaminase. (See isoleucine - valine biosynthetic steps, threoninemia)

THREONYL tRNA SYNTHETASE (TARS): charges $tRNA^{Thr}$ by threonine. It is encoded in human chromosome 5 in the vicinity of leucyl tRNA synthetase gene. (See aminoacyl-tRNA synthetase)

THRESHOLD TRAITS: are expressed conditionally when the liability reaches a certain level. These characters are frequently under polygenic control yet they may fall into two recognizable classes, exhibiting it or not, although with special techniques more classes may be revealed. Many pathological syndromes may fall into this category and it makes very difficult to determine the genetic control mechanisms. (See syndrome, liability)

THROMBASTHENIA: autosomal dominant and autosomal recessive forms involve a platelet defect causing anemia and bleeding after injuries. The recessive form was assigned to human chromosome 17q21-q23. (See platelet, anemia)

THROMBIN (fibrinogenase): an enzyme that converts fibrinogen to fibrin and causes hemostasis (blood clotting). Thrombin has also another role as an anticoagulant by the proteolytic activation of protein C in the presence of Ca^{2+}. These seemingly conflicting functions are modulated by allosteric alteration brought about by Na^+ and the thrombomodulin protein, changing the substrate specificity of thrombin. The level of the activated protein C further increased by a compound named LY254603. (See fibrin, fibrin-stabilizing factor deficiency, hemostasis, antihemophilic factors, plasmin, protein C)

THROMBOCYTES: same as platelets

THROMBOCYTOPENIA (TAR syndrome): is an autosomal dominant bleeding disease caused by low platelet counts. An autosomal recessive form of it is associated with heart and kidney anomalies and absence of the radius (the thumb side arm bone), phocomelia, etc. Another thrombocytopenia is linked in human chromosome Xp11 and it is allelic to the Wiskott-Aldrich gene, WAS. (See Holt-Oram syndrome, Roberts syndrome, phocomelia, Wiskott-Aldrich syndrome, thrombopathic purpura)

THROMBOMODULIN: see thrombin

THROMBOPATHIA (essential athrombia): is a collection of blood clotting anomalies caused by problems in aggregation of the platelets. (See platelet abnormalities, hemophilias, hemostasis)

THROMBOPATHIC PURPURA (called also von Willebrand-Jürgen's syndrome): is caused by an autosomal recessive condition with symptoms resembling the bleeding and platelet abnormalities present in the Glanzmann's disease. The primary defect may involve the platelet membrane. Other symptoms may vary from case to case. Blood transfusion may shorten the time of bleeding. (See platelet abnormalities, hemophilias, hemostasis, thrombocytopenia, Glanzmann's disease)

THROMBOPLASTIN: see antihemophilic factors

THROMBOPOIETIN: regulates blood platelet formation; it is member of a cytokine receptor superfamily and it is similar to the erythropoietin and the granulocyte colony-stimulating factor receptors. (See megakaryocytes, platelet, erythropoietin, thrombocytopenia, G-CSF)

THROMBOSIS: a type of obstruction of the blood flow caused by aggregation of platelets, fibrin and blood cells. May be caused by mutation in serpin. (See serpin, protein C, protein C deficiency)

THROMBOSPONDIN: see apoptosis

THROMBOTIC DISEASE: see protein C deficiency

THROMBOXANES: induce the aggregation of platelets and act as vasoconstrictors. They are antagonists of prostacyclin G_2. (See cyclooxigenase)

Thy-1: is 19-25 kDa single chain glycoprotein expressed on mouse thymocytes but not on mature T cells of humans or rats.

THYLAKOID: flat sac-like internal chloroplast membranes; when stacked appear as grana. (See grana, chloroplast, girdle bands) electronmicrograph of thylakoids ⇗

THYMIDINE: nucleoside of thymine; thymine plus pentose (ribose or deoxyribose).

THYMIDINE KINASE: see TK

THYMIDYLATE SYNTHETASE: mediates the synthesis of thymidylic acid (dTMP) from deoxyuridine monophosphate (dUMP). Its gene has been assigned to human chromosome 18p11.32.

THYMIDYLIC ACID (nucleotide of thymine): thymine + pentose + phosphate

THYMINE: a pyrimidine base occurring almost exclusively in DNA, exceptionis the T-arm of tRNA. (See tRNA, pyrimidines)

THYMINE DIMER: UV light induced damage in DNA, covalently cross-linking adjacent thymine residues through their 5 and 6 C atoms. This cyclobutane structure interferes with replication and other functions of the DNA. Cytidine and uridine also can form similar dimers. The dimers can be eliminated by enzymatic excision and replacement replication (excision or dark repair). Alternatively, the dimers can be split by DNA photolyase, an enzyme that is activated by visible light maximally at 380 nm wave length (light repair). (See DNA repair, cyclobutane dimer)

THYMOMA: a cancer of the thymus. (See AKT oncogene)

THYMUS: is bilobal organ with three functionally important compartments, the subcapsular zone, the cortex and the medulla. It is invaded by the immature lymphoid cells coming from the bone marrow. The subcapsular space contains most of the immature CD4⁻ and CD8⁻ lymphoid stem cells. When they enter the cortex they begin to express the CD molecules and

Thymus continued
they rearrange the T cell receptors and form the different TCR αβ heterodimers. The epithelial cells in the cortex express MHC I and II molecules. The fate of the T cell will be determined here in response to the MHC molecules carried by the antigen presenting cells. The $CD4^+$ $CD8^-$ cells with TCR recognizing the MHC II complex or the $CD4^-$ $CD8^+$ with TCR specific for MHC I will leave the thymus and populate the secondary lymphoid tissue. Those cells which will not acquire self peptide—MHC specificity will be eliminated and also those which have strong avidity for self-peptide MHC will die by apoptosis because they would be autoreactive (non-autoimmune). (See T cells, T cell receptor, TCR genes, MHC, autoimmune, spleen)

THYROGLOBULIN: an iodine containing protein in the thyroid gland and has hormone-like action upon the influence of the pituitary hormone. (See animal hormones)

THYROID CARCINOMA: transforming sequence was localized to human chromosome 10q11-q12, and a tumor suppressor gene in human chromosome 3p may be involved.

THYROID HORMONE RESISTANCE: autosomal dominant mutations are known.

THYROID HORMONE RESPONSIVE ELEMENT: see TRE, hormones, hormone response elements, regulation of gene activity, goiter

THYROID HORMONE UNRESPONSIVENESS: is autosomal recessive in humans.

THYROID PEROXIDASE DEFICIENCY (TPO): is a group of recessive human chromosome 2p13 defects involving the incorporation of iodine into organic molecules.

THYRONINE (3p-[p(p-hydroxyphenoxy)-phenyl]-L-alanine): is a component of the thyroglobulin of the thyroid hormone. It generally occurs as 3,5,3'-triiodothyrosine. (See VDR)

THYROTROPIC: affecting (targeting) the thyroid gland.

THYROTROPIN: see goiter

THYROXIN: see goiter

TI ANTIGENS: stimulate antibody production independently of T cells and in the absence of MHC II molecules. The TI type 2 (TI-2) have polysaccharide antigens, are usually large molecules with repeating epitopes, activate the complement and are rather stable. The TI-1 type is mitogenic for mature and neonatal B cells. (See antigen, antibody, T cell, MHC)

Ti PLASMID: a large (about 200 kbp) tumor-inducing plasmid of *Agrobacterium tumefaciens* is

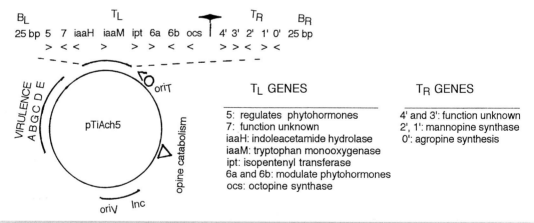

MAJOR LANDMARKS OF THE Ti OCTOPINE PLASMID, pTiACH5. ARROWHEADS INDICATE THE DIRECTION OF TRANSCRIPTION. ⊤ MARKS THE T_L - T_R BOUNDARY

reponsible for the crown gall disease of dicotyledonous plants. There are a few particularly important regions in this plasmid. The T-DNA is divided into the left (T_L) and right (T_R) segments and are flanked by the two border sequences (B_L and B_R) and in-between are several

Ti plasmid continued

genes. The virulence gene cascade, and its function is described under virulence genes of *Agro-bacterium*. The origin of vegetative replication (*oriV*) is functioning during proliferation of the cells, the nearby *Inc* (incompatibility) site determines host-specificity. The *oriT* is the origin of replication operated during conjugation. The latter is often called *bom* (base of mobilization) or CON (conjugation) because the synthesis of the transferred DNA begins there although the actual transfer begins when the mob site encoded protein (Mob, synonym Tra) attaches to a specific nick site, a single-strand cut, and one of the strands (the T-strand) is transported to the recipient bacterial cell through the cooperation also with some pore proteins. The oriT and Mob complex includes proteins TraI, TraJ and Tra H. TraJ attaches to a 19 bp sequence in the vicinity of the nick and binds also TraI which is a topoisomerase. These bindings are promoted by TraH. The process is a rolling circle type replication and transfer. The integrity of virulence genes *virD* and *virB* are absolutely required for productive conjugation. (See *Agrobacterium*, T-DNA, virulence genes of *Agrobacterium*, transformation, rolling circle, conjugation)

TIBO (O-Tibo and Cl-Tibo): are non-nucleoside inhibitors of HIV-1 reverse transcriptase. (See Neviorapine, acquired immunodeficiency syndrome, AZT)

Tie-1, Tie-2: receptor tyrosine kinases, expressed during endothelial cell growth and differentiation of the blood vessels. (See vascular endothelial growth factor, Flk-1, Flt-1, tyrosine kinase, transmembrane proteins, signal transduction)

TIER: an ordered arrangement by e.g. increasing stringency of a test.

TIF-1B: a transcription initiation factor of mouse binding to the core promoter of rRNA genes and controlling RNA polymerase I function.

TIGHT JUNCTION PROTEINS: form a seal between adjacent plasma membranes.

TIGER (*Panthera tigris*): 2n = 38. The tigert cat (*Leopardus tigrina*): 2n = 36.

TIL: see tumor infiltrating lymphocytes

TILLER: a lateral shoot of grasses arising at the base of the plant.

TIM (transfer inner membrane): a protein complex regulating the import of proteins into mitochondria. (See mitochondria, TOM)

TIME OF CROSSING OVER: appears to coincide with the meiotic prophase (late leptotene and early diplotene, probably at zygotene). Some experimental data indicate that treatments at S phase have an effect on the outcome; it is difficult to assess, however, whether these effects are direct or indirect. Meiosis is under the control of long series of genes acting sequentially and cooperatively and any of these may effect crossing over. (See crossing over, recombination, recombination mechanisms)

TIMELESS (Tim): protein of *Drosophila* is involved in the circadian clock. (See *per*)

TIP: tumor inducing principle. (See *Agrobacterium*)

TIS1: see nur77 and NGFI-B

TIS-8: a mitogen induced transcription factor. (See NGFI-A, egr-1)

TIS11: transcription factor, inducible by various hormones (similar to Nup475). (See Nup475)

TIS11b: murine homolog of cMG1. (See cMG1)

TIS11d: a transcription factor with 94% identity to 367 amino acids in TIS11b. (See TIS11b)

TISELIUS APPARATUS: is an early model of an electrophoretic separation equipment.

TISSUE CULTURE: *in vitro* culture of isolated cells of animals and plants. (See also cell culture, cell genetics, cell fusion, somatic cell fusion, embryogenesis somatic, axenic, aseptic)

TISSUE FACTOR (TF): is a cell surface glycoprotein mediating blood clotting after injuries by its interaction with blotting factor VII. (See antihemophilic factors, blood clotting pathways)

TISSUE PLASMINOGEN ACTIVATOR: cleaves plasminogen to plasmin and enhances fibrinolysis; it controls blood coagulation. (See blood clotting pathways)

TISSUE-SPECIFIC PROMOTER: facilitates gene expression limited to certain tissues or organs. (See promoter)

TISSUE-SPECIFICITY: the expression of a gene is limited to certain tissue(s).

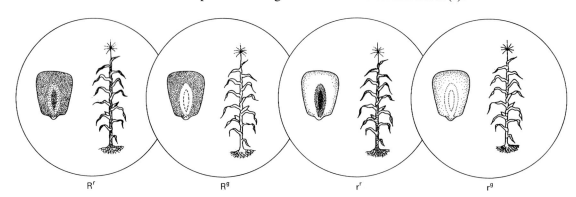

TISSUE-SPECIFICITY OF EXPRESSION (ANTHOCYANIN FORMATION) OF FOUR DIFFERENT *R* ALLELES OF MAIZE IN THE ALEURONE, EMBRYO, STALK, TASSEL AND LEAF TIPS.

The proportion of tissue specific genes have been estimated by transformation of *Arabidopsis* with transcriptional and translational gene fusion vectors. Of 200 transgenic plants about 10% displayed some degree of tissue-specific expression of the reporter gene, (aph[3']. (See also transcription illegitimate, housekeeping genes, aph)

TISSUE TYPING: determining the genetic constitution of potential graft before transplantation. It can be done by blood typing or even better by DNA analysis (RFLP) or using polymerase chain reaction. (See blood typing, DNA fingerprinting, polymerase chain reaction)

TITER: the amount of a reagent in titration required for a certain reaction. Also, phage titer the number of phage particles in a volume.

TITIN (connectin): is one of the largest protein molecules ($3 \times 10^6 M_r$) along with nebulin; it forms a network of fibers around actin and myosin filaments in the skeletal muscles. Titin keeps the myosin within the sarcomeres by being anchored to the Z discs (membrane bands in the striated muscles) and assures that the stretched muscles spring back. (See nebulin, glutenin, sarcomere, dystrophin)

TITRATION: adding measured amount of a solution of known concentration to a sample of another solution for the purpose of determining the concentration of the target solution on the basis of the appearance of a color, appearance of agglutination, etc. or determining the number of cells or phage particles in a series of dilutions. (See titer, gene titration)

TK: thymidine kinase phosphorylates thymidine. Its gene in humans encodes the cytosolic enzyme that contains 7 exons in the short arm of human chromosome 17q25-q25.3 but the mitochondrial TK gene is in chromosome 16; its sequences are apparently highly conserved in different species. (See HAT medium)

T_L: the left border of the T-DNA. (See T-DNA, Ti plasmid)

TLC: see thin layer chromatography

TLV (threshold limit value): the upper limit or time-weighted average concentration (TWA) of a substance people can be exposed to without adverse consequences.

T_m: see melting temperature

TM1, TM2: transmembrane amino acid domains of *E. coli* transducers, spanning the membrane layer inner space. TM1 is well conserved among the various transducers, TM2 is variable. (See transducer proteins)

TMF: TATA box modulatory factor. (See TATA box, promoter, transcription)

TMP: thymidine monophosphate

TMV (tobacco mosaic virus): is a single-stranded RNA virus of about 63900 bases and its cylindrical envelope contains 2130 molecules of a 158 amino acid protein. Its rod-shape particles

TMV continued

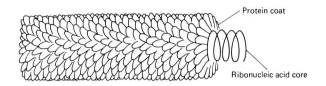

are about 3,000 Å long and 180 Å in diameter. The historical reconstitution experiments from coat protein and RNA demonstrated that RNA can be genetic material. Its mutagenesis by nitrous acid contributed substantially to the genetic confirmation of RNA codons.

Tn3 FAMILY OF TRANSPOSONS: are genetic elements that can move within a DNA molecule and from one DNA molecule to another and carry genes besides those required for transposition. The best known representative is the Tn3 element carrying ampicillin resistance (Ap^r). Tn3 is 4,957 bp long with 38 bp inverted terminal repeats and leaving behind— after moving at the transposition target site — 5 bp direct duplication. Although the terminal repeats of the various Tn elements vary, some sequences are well conserved, the GGGG sequence is generally present outside the repeats and ACGPyTAAG is common inside of one or both terminal repeats. In the Tn3 group an internal ACGAAAA is common. Normally transposition requires the presence of both terminal repeats, the presence of only one of them still may allow a lower frequency transposition. The sites of integration may vary yet AT-rich sequences are prefered and some homology between the terminal repeats and the insertional target my be needed. Some proteins of the host cell (IHF = integration host factor and FIS = factor for inversion stimulation) may facilitate the expression of the transposase. Other members of the family are Tn1, Tn2, Tn401, Tn801, Tn802, Tn901, Tn902, Tn1701, Tn2601, Tn2602, and Tn2660, all about 5 kb length with 39 bp terminal repeats and found in various Gram-negative bacteria. Related is the γδ (Tn1000) element in the F plasmid of 5.8 kb and with 36/37 bp terminal repeats and IS101 (insertion element 101), a cryptic element. Tn501 (8.2 kb) was the source of mercury resistance (Hg^r), Tn1721, Tn1771 (11.4 kb) for tetracycline resistance (Tc^r), Tn2603 (22 kb) for resistance to oxacillin (Ox^r), Hg^r, streptomycin (Sm^r)), and sulfonamide (Su^r). Tn21 (19.6 kb) carried resistance to Su^r, Hg^r, Ap^r, Sm^r, and Su^r. Tn4 (23.5 kb) was endowed with genes for Ap^r, Sm^r, and Su^r. Tn2501 (6.3 kb) was cryptic (i.e. expressed no genes besides the transposase). Tn551 and Tn917 (both 5.3 kb) from *Staphylococcus aureus* and *Streptococcus fecalis*, respectively, carried the erythromycin resistance (Ery^r) gene. The cryptic Tn4430 (4.1 kb) was isolated from *Bacillus thüringiensis*, R46 from enterobacteria, and pIP404 from *Clostridium perfringens* plasmids coded for the resolvase protein. The terminal repeats are generally within the range of 35-48 bp. The transposition is usually replicative and its frequency is about 10^{-5} to 10^{-7} per generation. The integration of the element requires the presence of a specific target site, called *res* site or IRS (internal resolution site), and genes *tpnA* (a transposase of about 110 MDa) and *tpnR*, encoding a resolvase protein (ca. 185 amino acids). The *res* site (about 120 bp) is where the resolvase binds and mediates site-specific recombination and protects the DNA against DNase I. Within the *res* site are located the promoters of *tnpR* and *tpnA* genes, functioning either in the same or in opposite direction, depending on the nature of the Tn element. The recombination between two DNA molecules requires the presence of at least two *res* sites in a negatively supercoiled DNA. The resolvase apparently has type I DNA topoisomerase function too. After synapses, mediated by resolvase and multiple *res* sites, strand exchange and integration may results. The recipient molecule thus acquires the donor transposon. The transposition event requires replication of the transposon DNA and then a fusion of the donor and the recipient replicons. This must be followed by a resolution of the cointegrate into a transposition product. (See transposable elements, gram-positive bacteria, transposon, cointegration, topoisomerase, antibiotics, resolvases)

Tn5: is a bacterial transposon of 5.8 kb of the following structure:

```
  O   IS50 L   I              2.8 kb                I   IS50 R   O
  <~~~~~~~~ ---------------------------------------- ~~~~~~~~>
     1.5 kb    p→     kan        ble        str          1.5 kb
                                                    ← transposase (tnp)
                                                    ← inhibitor (inh)
```

The inverted termini (~~) represent the IS50 insertion element that includes the *tnp* (58 kDa protein) and *inh* genes (product 54 kDa). The left (L) and right (R) IS50 elements are almost identical except that the L sequences contain an *ochre* stop codon in the *tnp* gene, rendering it non-functional, save when the bacteria carry an *ochre suppressor*. At the I (inside ends) site binds the IHF (integration host factor) protein. O is the outside end of the IS sequence. Within the repeat beginning at nucleotide 8 is the bacterial DnaA protein binding site TTAT^C_A CA^C_A A; DnaA product controls the initiation of DNA synthesis. Within the 2,750 bp central region are the genes for resistance to kanamycin (*kan*) and G418, bleomycin [phleomycin] (*ble*), and streptomycin (*str*) antibiotics; they are transcribed from the *p* promoter located within the Is L element at about 100 bp from the I end. These antibiotic resistance genes may not convey resistance in some cells, e.g., *str* may be cryptic in *E. coli*. The activation of this antibiotic resistance operon is contingent on the ochre mutation in the *tnp* gene. The inhibitor protein (product of *inh*) is apparently interacting with the terminal repeats rather than with the transcription or translation of the *tnp* gene.

Tn5 can insert one copy at many potential target sites within a genome. The IS terminal repeats are capable of transposition themselves without the internal 2.8 kb element. In *direct transposition* the complete Tn5 will occur in the same sequence as shown above. In *inverse transposition*, mediated by the I ends, the 2.8 kb central element is left behind, away from the termini. Inverse transposition occurs 2-3 order of magnitude less frequently than the direct one. In5 can form cointegrates with plasmids or bacteriophages and in these the IS elements and the entire Tn5 may occur and the orientation of the termini may be either:

direct: O←I kan I←O   or   indirect: O→I kan I←O

and the transposon may be present as a monomer or as a dimer. Transposition may not require special homologous target sequences yet some targets represent "hot spots" (displaying G•C or or C•G pairs next to the 9-bp target duplications) because they are being prefered for insertion. The less frequently used targets generally show G•C and A•T pairs at the ends. Tn5 insertions in general are almost random yet it appears that transcriptionally active promoters may present favorable targets. Insertion of the transposon into active genes usually results in inactivation because of the interruption of the coding sequences. Excision of the transposon may revert the gene to the original active state. This may occur at frequencies of 10^{-8} to 10^{-4} per cell divisions. Excision is not mediated by the transposase gene and it is independent from the bacterial *recA* gene but its depends on the structure of the inverted terminal repeats and re-plicational errors involving slippage of pairing between the new and template DNA strands. The presence of some sequences in the target may also promote Tn excision. The excision may involve also flanking DNA sequences and in this case the wild type function of the target gene is not restored. Several mutations in *E. coli* (*recB, recC, dam, mutH, mutS, mutD, ssb*) may promote excision and mutation to *drp* reduces excision. Tn5 — unlike Tn3 — does not have a resolvase function. Transposition may be affected by DNA gyrase, DNA polymerase I, DnaA protein, IHF and Lon (a protease cleaving effectively the SulA protein, a cell division inhibitor). Transposition of Tn5 is substantially increased in *dam* mutant strains that are deficient in methylating GATC sequences. The I ends contain GATC sequence and thus can be affected by methylation. The O ends do not have methylation substrate yet they are also affected by methylation in the I sequences (19 bp). (See transposon, Tn3, Tn7, Tn10, transposable elements bacteria, transposable elements, resolvase)

Tn7: is a 14 kb bacterial transposon of the following general structure:

LEFT |30 bp repeat| > dhfr aadA   tnsE tnsD tnsC tnsB tnsA < |30 bp repeat| RIGHT

dhfr = dehydrofolate reductase gene with much reduced sensitivity to trimethoprim, an inhibitor of the enzyme involved in the biosynthesis of both purines and pyrimidines and thus nucleic acids. *aadA* = adenylyl transferase gene, encoding an enzyme which inactivates the aminoglycoside antibiotics, streptomycin and spectinomycin and thus conveys resistance to the cells. The *tns* genes are responsible for transposition.

Tn7 element has high capacity to insert into the specific *att* Tn7 site of *E. coli* (25 kb counterclockwise from the origin of replication at map position 83) but when this site is not available it may transpose — at about two orders of magnitude lower frequency — to a *pseudo-att* Tn7 or to some other unrelated sites. At the *att* Tn7 site the right end is situated proximal to the bacterial *o* gene. Genes *tnsABC* mediate all transpositions but through different pathways; for transposition to *att* Tn7 and *pseudo-att* Tn7, in addition, the function of gene *D* is also required, whereas transposition to all other sites the expression of the *ABC + E* genes are needed. (The name *tns* abbreviates **tra**nspo**s**on **seven**.) The insertion at *att* Tn7 is also in a consistent orientation and it is within an intergenic region and thus does not harm the host. Insertion of Tn7 element protects the cell from an additional Tn7 insertion (immunity) yet under some conditions the immunity may not work. Integration results in 5-bp target site duplication; these duplications are different at *att* Tn7 and other sites. Several bacterial species besides *E. coli*, have specific *att* Tn7 sites in their genomes. For transposition both L and R terminal repeats are required. Elements with two L repeats do not move whereas two complete R termini can assure insertion to *att* Tn7 sites. Tn7, because of the site specificity, is not very useful for insertional mutagenesis but it has an advantage for inserting genes at the standard map position. The Tn7 family of transposons includes Tn73, Tn*1824*, and Tn*1527* with substantial similarity or even identity. Tn*1825* is a clearly distinct member. (See transposons bacterial, transposable elements)

Tn10: is a bacterial (*E. coli, Klebsiella, Proteus, Salmonella, Shigella, Pseudomonas*, etc.) transposable element of 9.3 kb. It may move within a bacterial genome, from the bacterial chromosome to temperate phage or plasmid and among different bacterial species. Its overall structure can be represented as:

L |1.3 kbp⇒| ←tetR →tetA   ←tetC →tetD→|←**1.3** kbp| R

The boxed left insertion element IS*10* (L) is a defective transposase. The boxed right IS*10* element (R) is a functional transposase. Thus Tn*10* is a composite transposable element. The *tetR, A, C* and *D* genes are involved in resistance against tetracycline. The *tetR* is a negative regulator and *tetA* encodes a membrane protein. The arrows indicate orientation and direction of transcription. The promoter of the Is*10* element may serve as promoter for adjacent outside genes (pOUT).

Insertion of Tn*10* within a gene, operon or upstream regulatory element may abolish the activity of the gene(s) or may modify their transcription. In the so called polar insertions transcription is initiated within the Tn element but it may be terminated when rho signals are encountered. In the non-polar insertions there is no rho sequences downstream to halt transcription. This type of *read-through* transcription may take place only in some Tn*10* derivatives but not in the wild type element. The rate of transposition for IS*10* is 10^{-4} and for Tn*10* 10^{-7} per cell cycle.

Both the IS*10* and the Tn*10* elements can cause chromosomal rearrangements (see upper diagram on next page). Insertion and transposon sequence may show portable region of homology and can undergo homologous recombination (see lower diagram on next page).

Tn*10* continued
These recombinations may generate deletions between or inversions in the regions in-between them at rates two orders of magnitude less frequent than transposition or lead to the formation of cointegrates.

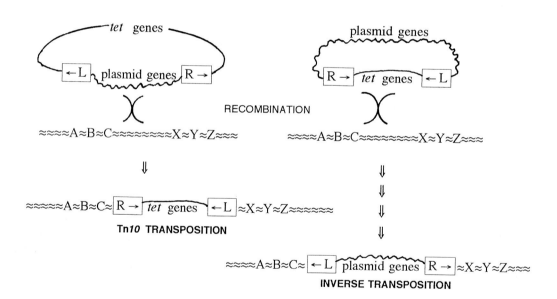

GENERATION OF DELETIONS AND INVERSIONS BY A PORTABLE REGION OF HOMOLOGY REPRESENTED BY TRANSPOSON Tn*10* (→ OR ←). GENES OR SITES ARE SHOWN BY CAPITAL LETTER IN ITALIC, χ INDICATES RECOMBINATION.

Tn*10* TRANSPOSITIONS. LEFT: THE NORMAL EVENT, AT RIGHT: THE INVERSE TRANSPOSITION (OR INSIDE-OUT) TRANSPOSITION. SOME OTHER EVENTS MAY LEAD ALSO TO DELETIONS OR DELETIONS AND INVERSIONS. THE L AND R BOXES REPRESENT Tn*10* TERMINI. THE THIN LINE STANDS FOR THE TRANSPOSON SEQUENCES WHEREAS THE SINGLE JAGGED LINE INDICATES THE SEQUENCES FLANKING THE TRANSPOSON IN THE ORIGINAL LOCATION OF THE PLASMID. THE ≈≈≈≈ SYMBOLIZES THE DNA SEQUENCES OF THE TARGET.

All DNA segments flanked by IS*10* can become transposable and thus may represent new composite transposons. IS*10* and Tn*10* may assist the fusion of different replicons and transfer information between bacterial chromosomes, plasmids and phages. The transpositions may also generate new units of regulated gene clusters by the movement of structural genes under

Tn*10* continued

the control of other regulatory sequences. The excision of the transposon may be "precise" if the nucleotide sequence is restored to its pre-insertion condition. Precise excision (average frequency 10^{-9}) may remove one copy of the 9-bp target site inverted duplications. The "near-precise excision" events involving removal of most of the internal sequences of Tn*10* occur at a frequency of about 10^{-6} and may later be followed by precise excision of the remaining sequences. These non-transposase mediated events are also independent from the host RecA recombination functions.

Tn*10* and IS*10* may transpose either by a non-replicative or a replicative mechanisms. In the former case the whole double-stranded element is lifted from its original position and transfered to another site. In the second case only one of the old strands of the transposon is integrated into the new position and the other strand represents the newly replicated one. At the target site apparently two staggered cuts are made 9 base apart. This causes then the 9-bp target duplications when the gap is filled and the protruding ends are used as templates:

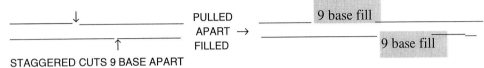

STAGGERED CUTS 9 BASE APART   PULLED APART → FILLED   TWO 9-BASEPAIR TARGET EXPANSIONS

The transposon (ca. 46 kDa) is coded within the IS*10* right terminus by about 1,313 nucleotides. The frequency of transposition may increase up to five orders of magnitude by increasing the expression of the transposase. The transposase action prefers being in the vicinity of the transposed sequences and its action is reduced by distance. Also, longer transposons are moved less efficiently than shorter ones; each kb increase in length involves about 40% reduction in movement within transposon sizes 3 to 9.5 kb. Transposition is regulated also by proteins IHF, HU proteins and DNA gyrase. The transposition frequency may be modified (within three orders of magnitude) also by the chromosomal context, i.e. cis-acting sequences. The integration hot spots appears to have some consensus sequences within the 9 bp target site in three bases:

1st	2nd	3rd position			
G A	C T	T	C	A	G
90%	98%	63%	23%	12%	2%, respectively.

Transposition seems to avoid actively transcribed and thus the most essential genes of the recipient. The transposase gene is transcribed from a very low efficiency pIN promoter originating near the O end of the right IS element. Its transcription is opposed also by the divergent pOUT promoter. Apparently, on the average each cell generation may not produce one molecule of transposase. Two *dam* (adenine) methylation sites (GATC) are located within pIN and the activity of this promoter is facilitated in the absence of methylation by the host *dam* methylase (located in *E. coli* at 74 min). Actually, the transposase promoter is usually hemimethylated. An increase in the number of Tn*10* copies per cell reduces transposase activity because the pOUT promoter generates an antisense RNA transcript that may pair to a 35 base complementary region of the pIN promoter region. The transposase activity is regulated also by foldback inhibition (FBI), hindering the attachment of the Shine-Dalgarno sequence of the transcript to the ribosome. The Tn*10* system is protected from transposase activation by inhibition of read-through.

Transposons inserted into a particular gene cause mutations because of the disruption of continuity of the coding sequences. Transposon mutagenesis has considerable advantage over chemical or radiation mutagenesis because it induces mutation only at the site of insertion whereas chemicals may affect simultaneously several genes. In addition, mutagenesis with

Tn*10* continued
    Tn*10* labels the gene also by tetracycline resistance, an easily selectable marker. Transposons may also label foreign genes cloned in *E. coli* or *Salmonella*. The tetracycline-resistance gene has been extensively used for monitoring insertions that result in the loss of tetracycline resistance but retain other (selectable) markers in the cloning vector. (See IHF, HU, gyrase, antisense RNA, FBI site, readthrough, transposable elements bacteria)

TNF (tumor necrosis factor): are proteins ($M_r \approx 17$ K) selectively cytotoxic or cytostatic to cancer cells of mammals (it is relatively harmless to normal cells); they are lymphokines. TNF was originally discovered in rodent cells infected by bovine *Mycobacterium* and then with endotoxin. The serum of these animals produced hemorrhagic necrosis and occasionally complete regression of transplanted tumors. The mature human TNFα consists of 157 amino acids after trimming off 73 amino acids from the pre-TNF. It has receptor sites on the surface of tumors and its action is synergistic with γ interferon. TNFα and TNBβ have similar functions and display ≈ 30% homology. These two genes, each, have 3 introns in their 3 kb sequence. In mice, it is localized within the *H-2* (histocompatibility cluster) and in humans in the homologous *MHC* region in chromosome 6p23 either between HLA-DR and HLA-A or proximal to the centromere. The genes have been cloned and sequenced in the mid-1980s. TNFα and IL-1 are the major inflammatory cytokines whereas IL-10 and TGFβ, IL-r and TNF-R are anti-inflammatory. TNF is produced mainly by macrophages. (See lymphokines, endotoxin, hemorrhage, necrosis, interferons, histocompatibility, HLA, TNF-R, TRAF, Crohn disease, arthritis, macrophage)

TNFR (tumor necrosis factor receptor): similar receptors in both animals and plants may be involved in processes of differentiation. TNFR-1 and -2 are distinguished, TNFR-1 mediates different effector functions through separate pathways. Recruitment of the Fas-associated protein with death domain (FADD) is involved in apoptosis. Tumor necrosis factor associated protein 2 (TRAF-2) and receptor interacting protein (RIP) are required for the activation of the Jun kinase (JNK) and nuclear factor kappa B (NF-κB). These two responses diverge downstream to TRAF-2. JNK is not participating in the apoptotic pathway and the activated NF-κB works against apoptosis. (See TNF, TRAF, NF-κB, nitrogen fixation [ENOD], apoptosis, Jun, Fas)

Tnp: transposase enzyme. (See Tn*5*)

Tnt: a non-viral retrotransposable element. (See transposable elements)

TNT: a solution containing 10 mM Tris.HCl buffer (pH 8.0), 150 mM NaCl, 0.05% Tween 20.

TOAD: *Bufo vulgaris*, 2n = 36; *Xenopus laevis*, 2n = 36.

TOBACCO (*Nicotiana* spp): the smoking tobacco is an allotetraploid (*N. sylvestris x N. tomentosiformis*) 2n = 48. The basic chromosome number is generally x = 12, however diploid forms with 2n = 20 (*N. plumbaginifolia*) and 2n = 18 (*N. langsdorfii*) and in the *Suavolens* group 2n = 36 (*N. benthamina*), 2n = 46 (*N. caviola*), 2n = 32 (*N. maritima*), 2n = 36 (*N. amplexicaulis*), 2n = 40 (*N. simulans*), 2n = 44 (*N. rotundifolia*) are also found. For genetic studies the diploid species are most useful (*N. plumbaginifolia*), for transformation *N. tabacum* is used most commonly because of the easy regeneration of plants from single cells. Antibiotic resistant chloroplast mutations are available. The S strain is a streptomycin resistant mutant of the variety Petite Havana. Mutation, recombination and transformation techniques are available for its plastid genome, using primarily antibiotic resistant cpDNA mutations.

TOBACCO MOSAIC VIRUS: see TMV

TOBACCO NECROSIS VIRUS (TNV): icosahedral single-stranded RNA virus of ≈ 4 kb, is a root plant-pathogen.

TOBACCO SATELLITE NECROSIS VIRUS (TSNV): a 17 nm in diameter single-strand RNA (≈ 1.2 kb) virus that depends on TNV for its replication. (See tobacco necrosis virus)

TOCOPHEROL (vitamin E): its deficiency results in nutritional muscular dystrophy and sterility, although, normally, humans with regular diet do not show a need for it; vitamin E may be beneficial by being an antioxidant and regulating nerve functions. (See vitamin E)

TOGAVIRUS: RNA plus strand viruses of about 12 kb. The capsid proteins are synthesized only after completion of replication. This viral family includes rubella, yellow fever and encephalitis viruses. (See RNA viruses)

TOILET: the medical meaning is cleansing, clearing.

TOLERANCE, IMMUNOLOGICAL: non-reactivity to an antigen that under other conditions would evoke an immune response. Tolerance can be induced by antigens provided to fetuses or neonates with immature immune system. In adults very high or very low dose of the antigen may cause tolerance. The tolerance is the result of either clonal elimination or inactivation of lymphocytes in the thymus. Liver transplantation may induce systemic immune tolerance for certain (kidney, heart) allografts. (See lymphocytes, immune response, antigen, allograft, immunosuppressants)

TOM (transfer outer membrane): a protein complex regulating transport through the outer layer of the mitochondrial membrane. (See mitochondria, TIM)

TOMATO (*Solanum lycopersicum*, 2n = 24): has about 8-10 related species. It is one of the cytologically and genetically best known autogamous plants and it is suitable for practically all modern genetic manipulations.

TOMATO BUSHY STUNT VIRUS: a single-strand RNA virus of about 4,000 bases enveloped by an icosahedral shell consisting of 180 copies of a 40 kDa polypeptide.

TOMOGRAPHY (body section radiography): is conducted by a tomograph in what a source of X-radiation moves in opposite direction to that of a film recording the image clearly only in one plane and blurring the rest of the images. In the *computerized axial tomography* (CAT scan) the scintillations produced by the radiation are recorded on a computer disk and the cross section of the body is analyzed electronically. The *positron emission tomography* (PET) involves the use of positron-labeled metabolites (e.g., γ-ray emitting glucose). Along the path of the radiation, positrons and electrons collide and the local concentration of the isotopes are recorded electronically. *Single-photon emission computed tomography* (SPECT) takes γ-ray photographs around the body and three-dimensional image is reconstructed by a computer, resulting in great resolution even of overlapping organs. *Ultrasonic tomography* uses ultrasound scanning. The radiation may not be without risk. (See X-rays, ultrasonic, sonography)

TONOPLAST: elastic membrane (≈ 8 nm) that surrounds the vacuoles. (See vacuoles)

TOOTH AGENESIS: is caused by an autosomal dominant (chromosome 4p) mutation in the homeodomain of transcription factor MSX1, and it involves failure to form the second premolars and the third molars. (See homeodomain, Rieger syndrome, amelogenesis imperfecta)

TOOTH MALPOSITION: apparently autosomal dominant gene controls the various misplacements or underdevelopment of the incisors and the canine teeth. (See also Hallermann-Streiff syndrome, Jackson-Lawler syndrome, amelogenesis imperfecta)

TOOTH SIZE: appears to be influenced by the human Y chromosome as indicated by observations on various sex-chromosomal dosage.

TOOTH-AND-NAIL DYSPLSIA: most commonly incisors, canine teeth and some of the molars are not or poorly developed. In some cases it is accompanied by abnormal toe nails in children. (See also hypodontia, Hallermann-Streiff syndrome, dental no-eruption, dentin dysplasia, denticle)

TOP AGAR: an 0.6% agar solution generally containing 0.5% NaCl and some organic supplements. About 2 mL of it, with suspended bacterial cells, is spread over the agar medium (30 mL/ 10 cm Petri plates) to initiate selective bacterial growth e.g. in Ames tests. (See Ames test)

TOP-DOWN ANALYSIS: starting with a mutant phenotype, the physiological or molecular mechanism responsible for the alteration is investigated. In contrast, the bottom-up analysis first studies the molecules and then the analysis is extended to their relations to the phenotype. Currently the major endeavors go beyond the role of individual molecules and major interest is being focused on the critical domains of the molecules. (See reversed genetics)

TOP-DOWN MAPPING: uses either traditional genetic recombinational analysis or radiation hybrid maps. (See mapping genetic, radiation hybrids, bottom-up map)

TOPOISOMERASE: enzymes which alter the tertiary structure of DNA without a change in the secondary or primary structure. Topoisomerase I nicks and closes single strands of DNA and change the linking number in one strand, topoisomerase II can cut and reattach both strands of the DNA and affects linking number in both strands. Topoisomerases disentangle DNA strands and have important role in DNA replication, transcription and recombination, suppression of mitotic recombination, stabilization of the genome (chromosome breakage), regulation of supercoiling, eukaryotic chromosome condensation, control of segregation of the chromosomes, regulation of the cell cycle, nuclear localization of imported molecules. Prokaryotic DNA topoisomerase III can function also as an RNA topoisomerase and can interconvert RNA circles and knots. The eukaryotic topoisomerase III is homologous with prokaryotic topoisomerase I has been located to human chromosome 17p11.2-p12. (See DNA replication, gyrase, linking number, linking number paradox)

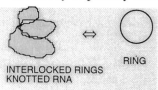

TOPOLOGICAL FILTER: is synonymous with two-step synapsis. The assumption is that synapsis requires an interaction between DNA *res* sites and the three subunits of a resolvase enzyme (step 1). After this initial step, the II and III subsites of the resolvase-*res* dimers pair in an antiparallel manner and the subunits interwrap. Then, the two I subsites of the resolvase dimer bind resulting in a *productive synaptic complex* (Step 2), capable to bring about DNA exchange. (See synapsis, *res*, resolvase)

TOPOLOGICAL ISOMERS of DNA: see linking number

TOR (target of rapamycin): phosphatidylinositol kinases of yeast (TOR1, TOR2). Also called RAFT1, FK506 and FRAP. (See PIK, FK506)

TORMOGEN: a component cell of the bristle which secretes the bristle socket. The other bristle cells are the trichogens which secrete the bristle shaft, also a neuron which contacts the shaft and through its axon connects with the nervous system.

TORTOISESHELL FUR COLOR: develops in female (cats) heterozygous for the X-chromosome-linked genes *black* and *yellow* colors because of selective, alternate inactivation of the two mammalian X-chromosomes. The tortoiseshell animals have mixed patches of black and yellow fur. This fur patterns occurs in the XX (female) or exceptionally in the XXY Klinefelter male cats. (See Lyon hypothesis, lyonization, calico cat)

TOTIPOTENCY: a characteristic of zygotic cells that permits differentiation into any type of cell or structure, including the whole organism. Plant cells maintain their totipotency in diverse adult tissues and after dedifferentiation may initiate other types of differentiations and somatic cells of entire organisms may be regenerated from cell cultures. Totipotency of animal cells is much more limited although lower animals such as hydra and earthworms may regenerate from differenetiated tissues. Embryonic stem cells (ES) of mouse comes close to totipotency/pluripotency inasmuch as they can be transferred to mouse embryos and can contribute to the formation of various cell types, including the germline. After special treatments differentiated animal cells may also revert to stem cell status. (See morphogenesis, somatic embryogenesis, redifferentiation, ES, nuclear transplantation)

TOUCH-AND-GO PAIRING: the end-to-end synapsis of the sex chromosomes in the heterogametic sex of some insects.

TOULOUSE-LAUTREC, HENRI: (1864-1901), one of the most remarkable painter of the end of the 19th century Paris life, a colleague of van Gogh and an influential precursor of modern art was affected by pycnodysostosis. His parents were first cousins. The diagnosis of his genetic malady has been questioned recently [Nature Genetics 11:363]. (See pycnodysostosis)

TOURETTE'S SYNDROME (Gilles de la Tourette disease, GTS): a human behavioral anomaly causing motor and vocal incoordination (tic, twitching), stuttering (echolalia), and the use of foul language (coprolalia). The onset is between 7 to 14 years of age and 3/4 of the affected are male. The genetic determination is apparently dominant with incomplete penetrance and

Tourette's syndrome continued

expressivity. Genes in several chromosomes (18, 7, 9, 3) have been implicated. The frequency of the defective gene(s) was estimated to be 0.4 to 0.9%, and a prevalence of 1% to 0.02% have been observed in different populations. Some of the mild cases are suspected to be responsible for male alcoholism and female obesity. (See affective disorders)

TOXIC SHOCK SYNDROME: is caused by infection of *Staphylococcus aureus* bacteria, affecting primarily menstruating women. It begins with sudden high fever, vomiting diarrhea, muscle pains (myalgia). Later rash, hypotension and potentially death may follow.

TOXINS: are organic poisons. The colicins are produced by a plasmid gene of *E. coli* and *Shigella* bacteria may affect sensitive bacteria in several ways. Cholera toxin is produced by the bacterium *Vibrio cholerae* interferes with the active transport through membranes and thus causes the loss of excessive amounts of fluids and electrolytes in the gastrointestinal system. The *Bordatella pertussis* toxin, pertussin contributes to high level of adenylate cyclase and thereby to the symptoms of whooping cough. Bungarotoxin (from *Bungarus multicinctus*) and cobrotoxin (from Formosan cobra) in snake venoms block the acetylcholine receptors or interfere with ion channel functions. Bungarotoxin has an LD_{50} value in mice of 0.15-0.21 µg/g. Other bacterial toxins include the tetanus toxin, diphtheria toxin, botulin, etc. The diphtheria toxin (of *Corynebacterium diphtheriae* if it carries a temperate phage with the *tox* gene) is one of the most dreaded human poison (sensitivity is encoded in human chromosome 5q23) with a minimal lethal dose of 160 µg/kg in guinea pigs. Mice and rats are, however, insensitive to this toxin. It inactivates eukaryotic initiation factor eIF-2 in translation on the ribosomes. The fungus *Amanita phalloides* toxin, amanitin (amatoxin) is a poison of RNA polymerase II. The about two dozen cytochalasins are synthesized by different fungi are composed of substituted hydrogenated isoindole rings fused with a macrocyclic ring (large organic compound). Colchicine is a plant alkaloid poison of the spindle fibers. The seed toxin of the plants *Strophantus* and *Acocanthera* are blocking agents for membrane transport and used as a selective agent in mammalian cell cultures. Abrin (from a legume) and ricin (from Castorbean seeds) are inhibitors of the attachment of aminoacylated t-RNAs to the ribosomes. The piscine toxin tetrodotoxin (from *Spheroides rubripes*; LD_{50} in mice 10 µg/kg) and the dinoflagellate, *Gonyalux* species saxitoxin (LD_{50} in mice 3.4 - 10 µg/kg) lock the sodium ion channels and block neurotransmission. The curare toxins were obtained originally as an arrow poison from the bark of the trees *Strychnos* and *Chondodendron*. The source of other curare toxins, bamboo curare, pot curare, gourd curare, etc. are members of the *Menispermaceae* family and are highly poisonous muscle relaxants; they block the acetylcholine receptors and some ion channels. Some of the curare toxins were used medically for treatment of tetanus shock and in surgery to alleviate muscle rigidity. (See for details colicins, amatoxin, diphtheria toxin cytochalasins, aflatoxins, colchicine, abrin, ricin, antibiotics, ion channel, acetylcholine receptors, neurotransmitters, LD_{50})

TOXOID: inactivated bacterial toxin which still may incite the formation of antitoxins.

TPA: is a phorbol ester (12-o-tetradecanoyl phorbol 13-acetate) that promotes neoplastic growth after induction has taken place. (See carcinogens, cancer)

TPA: inducer protein, a mitogen activating protein kinase C. (See protein kinases)

TPK1, TPK2, TPK3: catalytic subunits of A-kinase. (See protein kinases)

TPN: triphosphopyridine nucleotide; TPNH is the reduced form. They are synonymous with the analogs α-NADP and α-NADPH, respectively. Many enzymatic reactions require the β-NADP and β-NADPH (nicotinamide adenine dinucleotide phosphates). (See NAD, $NADP^+$)

TPR: see tetratrico sequences

tra (transformer): gene (chromosomal location 8-45) of *Drosophila* controls sterile male development in XX flies; XY *tra/tra* males are, however normal males. Its transcript occurs in both sexes but processed differently. The prokaryotic *tra* genes in conjugative plasmids control the conjugal transfer of DNA. (See sex determination)

***tra* GENES**: (more than 17) mediate the conjugal transfer of the F and other conjugative plas-mids. (See *tra*, conjugation, *ori*$_T$, relaxosome, plasmids)
TRABANT: a terminal chromosomal appendage. (See satellited chromosome)
TRACHEID: a long, lignified xyleme cell specialized for transport and support in plants.
TRACE ELEMENTS: are required only in minute amounts.
TRACER: a (radioactively) labeled molecule that permits the identification of the fate of the molecules into what it has been incorporated. (See isotopes)
TRACHEA: the duct leading from the throat (larynx) to the lungs of animals or the duct system of insects through what air is distributed into the tissues.
TRACHOPHYTE: vascular plant endowed with xylem, phloem and (pro)cambium in-between. (See xylem, phloem, cambium)
TRACKING: is a mechanism that would ensure that transposition would occur between two appropriate *res* (recombination sites). Experimental proofs for successful tracking (reporter rings) are not unequivocal. (See reporter ring)
TRACKING DYES: in electrophoresis loading buffers permit the visualization of the front migration toward the anode. Bromophenol blue in 0.5 x TBE buffer moves at the same rate as double-stranded linear DNA of 300 bp length. Xylene cyanol FF moves along with 4 kb linear double-strand DNA. (See electrophoresis, electrophoresis buffers)
TRADESCANTIA species: occur in the polyploidy range of 2x to 12x. The plants develop ca.
 100 stamen hairs in their inflorescence. Each hair represents a single cell line. When plants heterozygous for anthocyanin markers are exposed to mutagen, somatic mutations can be assessed as colored and colorless cell lines in ca. 150 flowers that single plants may form. Some of the species have been favorites for cytologists. (See also bioassays in genetic toxicology, somatic mutation)
TRADD (tumor necrosis factor receptor associated death domain): see tumor necrosis factor, death domain, apoptosis)
TRAF: tumor necrosis factor-associated receptor and a signal transducer for some interleukins. (See TNF, TNFR, CRAF, interleukins).
TRAILER SEQUENCE: follows the termination codon at the 3' end of the mRNA and it is not translated. (See polyA mRNA)
TRAIT: a distinguishable character of an organism that may or may not be inherited.
TRAM (translocating-chain associating membrane proteins): are membrane-spanning glycoproteins associated with the nascent peptide chain while the SRP is mediating its transfer to the endoplasmic reticulum. (See SRP, protein targeting, translocons, translocase, signal hypothesis)
TRANS: position indicates that two genetic markers are not on the same molecule or not on syntenic parts of the chromatids. (See also cis arrangement, synteny, chromatid)
TRANS ARRANGEMENT OF ALLELES: indicates that they are not in the same chromosome (DNA) strand (they are in repulsion) in contrast with the cis arrangement (coupling) when the two alleles are within the same strand. (See coupling, repulsion)
TRANSACETYLASE: a protein which transfers an acetyl group from acetyl coenzyme A (Acetyl-CoA) to another molecule. The third structural gene of the lactose operon (*lacA*) encodes a 275 amino acid polypeptide that forms a dimer of 60 kDa that is a transacetylase. (See *lac* operon)
TRANS-ACTING ELEMENTS: are proteins synthesized anywhere in the genome but regulating transcription by attachment to specific sites of a gene. (See also cis-acting element)
TRANSACTIVATION RESPONSIVE ELEMENT (TAR): in the HIV transcript the Tat proein binds to the TAR sequence near the 5'-end. This binding then mediates an increased expression of the viral genes and the synthesis of more mRNA. The Rev protein binds to a specific RNA site, the rev-responsive element (RRE), and facilitate the export of the unspliced

transcript to the cytoplasm where viral structural proteins and enzymes are made. (See acquired immunodeficiency, VP16, transactivator, VDR)

TRANSACTIVATOR: a protein domain attached to a specific inhibitory protein may prevent its blocking of transcription and may increase the transcription of the target gene(s) by several orders of magnitude. (See transactivation responsive element, VP16, p53)

TRANSAMINASES: see aminotransferases

TRANSCAPSIDATION (heteroencapsidation): the viral coat protein and the enclosed genetic material are of different origin (wolf in a sheep's skin).

TRANSCOBALAMIN (TCN1): is a ligand and transporter of the B-12 vitamin; it is encoded in human chromosome 11q11-q12.

TRANSCOBALAMIN DEFICIENCY (TC2): a recessive condition causing megaloblastic anemia was located to human chromosome 22q11-qter. (See megaloblastic anemia, anemia)

TRANSCONJUGANT: the bacterial genetic material was (partly) derived by recombination during conjugation. (See conjugation)

TRANSCORTIN DEFICIENCY (CBG): dominant, human chromosome 14q31-q32, reduction in a corticosteroid binding globulin. (See corticosteroid)

TRANSCRIBED SPACER: is a DNA element between genes that is transcribed but eliminated during the processing of the primary transcript. (See primary transcript)

TRANSCRIPT: an RNA copied on DNA and complementary to the template. (See transcription)

TRANSCRIPT CLEAVAGE FACTOR: induces cleavage and then release of the RNA transcript while it is still associated with the polymerase. In prokaryotes the *GreA* and *B*, in eukaryotes transcription factors carry out such functions.

TRANSCRIPT MAPPING: RNA transcripts are hybridized with specific DNA probes and subsequently the (not annealed) single strands are digested by S1 nuclease and thus the resistant (annealed) fragments represent the homologous tracts and demarcate the transcript. For the process of mapping an entire genome, sequence-tagged sites (STS) of cDNAs are used. By the PCR method the STS sequences can be amplified and their position can be mapped either by YAC clones or by radiation hybrid panels. (See nucleic acid hybridization, S1 nuclease, STS, radiation hybrid, YAC, PCR)

TRANSCRIPTASE: DNA dependent RNA polymerase or RNA dependent DNA polymerase or RNA dependent RNA polymerase enzyme. (See RNA polymerase, reverse transcription)

TRANSCRIPTION: synthesis of RNA complementary to a strand of a DNA molecule. In prokaryotes the majority of the transcriptionally active genes are located in the leading strand of replication and transcribed in the same direction as the DNA synthesis. In the absence of a functional DNA helicase, genes involved in the replication of the lagging strand are hampered by the transcription complex fork and stalled for many minutes, If, however, the DNA helicase is present the replication fork on the lagging strand can quickly pass the RNA polymerase complex. In prokaryotes, transcription and translation are coupled unlike in eukaryotes where the mRNA must be released through the nuclear pore complex into the cytoplasm. (See pol, class II and Class III genes of eukaryotes, transcription complex, transcription factors, transcription termination, regulation of gene activity, open transcription complex, signal transduction, mitochondria, mitochondrial genetics, chloroplasts, chloroplast genetics, replication fork, transcription complex, regulation of gene activity, inchworm model, antitermination, RNA polymerase)

TRANSCRIPTION COMPLEX: the TATA box-associated complex has the components TFIIA, TFIIB, TFIID, TFIIE, TFIIF, TFIIH, TFIIJ, TFIIK and RNA polymerase II. After the pol II enzyme moves downstream and away from the preinitiation complex and phosporylated, it can initiate transcription in the absence of other regulatory factors although generally specific transcription factors (transactivators) may boost and regulate its activity. (See transcription factors, regulation of gene activity, SRB, TBP, open promoter complex, transcription shortening)

TRANSCRIPTION-COUPLED REPAIR: see DNA repair, excision repair, colorectal cancer

TRANSCRIPTION FACTORS: a large number of different proteins that bind to either short upstream elements, terminator sequences of the gene or to the RNA-polymerases and modulate

Transcription factors continued

transcription. The *general transcription* factors have highly conserved sequences and are interchangeable even among such diverse organisms as mammals, *Drosophila*, yeast and plants. The transcription factors for RNA polymerase I, transcribing ribosomal RNA genes show a relatively simple organization.

The UBF (upstream binding factor) binds upstream in the promoter and regulates transcription by acting as an assembly factor for the transcription complex. TIF-1 or its vertebrate homolog SL1 are required for the attachment of Pol I to the promoter and the transcription is assisted by accessory proteins A and B.

The tRNAs, 5S rRNAs and some other small RNAs are transcribed by RNA polymerase III (PolIII). This enzymes has a requirement for protein factor TFIIIB and in case of transcribing the 5S rRNA it requires also TFIIIA and TFIIIC for the assembly of the transcription complex. Proteins TFIIIA and C, however, are detached after TFIIIB binds and only the latter stays on the DNA when PolIII lands and the transcription begins. For the transcription of the tRNAs TFIIIA is not required. The Pol III transcription units have also internal control regions A (box 5'-TG GCNNAGTGG-3') and B (box 5'-GGTCGANNC-3') or similar sequences. The transcription factors required for RNA polymerase II (Pol II) which transcribes protein-coding genes form the most elaborate complex.

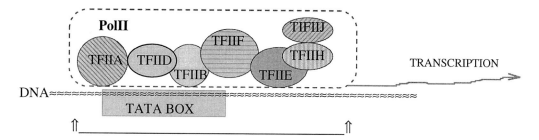

The *general transcription complex* is initiated by binding the TBP (TATA-box-binding protein) subunit of general transcription factor TFIID to the TATA box of eukaryotes (Hogness box). The TATA box is present in upstream regions of genes, coding for protein and transcribed by RNA polymerase II. TFIID attracts also TFIIB. After this, TFIIF, TFIIE and TFIIH proteins attach to RNA polymerase II (Pol II) to the TATA box. TFIIA is a co-activator and regulator of transcription and it binds 3' to the TATA box. The Pol II is inactive at this stage until TFIIH phosphorylates the bound Pol II using ATP as phosphate donor. The target of phosphorylation are several sites near the COOH-end of the largest subunit of pol II. Eukaryotic RNA polymerase II contains generally nine or ten subunits. The largest subunit is usually about 200 kDa. There are 26 (yeast) to 52 (mammals) repeats (Tyr-Ser-Pro-Thr-Ser-Pro-Ser) close to the carboxyl end. The phosphorylated Pol II then moves out of the complex and can now initiate transcription. The specific transcription factors are operative in special genes and at special tissues and time frames. The transcription factors may also bind a variety of other proteins before or during transcription and thus provide a great variety of fine tuning of gene expression. The TFIID TATA box-binding protein associated factors $dTAF_{II}42$ and $dTAF_{II}62$ form a heterotetramer resembling the heterotetrameric core of the histone octamer in the nucleosome. The general transcription factors are the basic instruments of transcription

Transcription factors continued

initiation but the modulation and regulation of transcription requires a large number of specific factors, activators, co-activators, suppressors and their interactions. Transcription factors with altered specificities can be generated in the laboratory using structure-based design and molecular technology. Besides the general transcription factors that are involved with almost all RNA polymerase II transcribed genes, specific and inducible transcription factors, activators and coactivators as well as chromatin reorganization factors also have important role. These molecules may be further explored for basic research and gene therapy. Viral repressors may prevent the association of RNA polymerase II with the transcriptional preinitiation complex. (See TF, TBP, mtTF, transcription complex, class II and class III genes, pol II [RNA polymerase II], regulation of gene activity, protein synthesis, gene therapy, ABC excinuclease, transcriptional activators, transcriptional modulation, mtTFA, CAAT box, enhancer, regulation of gene activity, chromatin, nucleosome, coactivator, transactivator, HMG, open promoter complex, gene number)

TRANSCRIPTION FACTORS, INDUCIBLE: are proteins that are synthesized within the cell in response to certain agents or metabolites. They bind to short upstream or dowstream DNA sequences and affect transcription, frequently by looping the bound DNA back to the promoter area and forming association with the other protein factors and the general transcription factors. Such transcription factors may be hormones, heatshock proteins (DNA binding site consensus: (**CNNGAANNTCCNNG**), phorbol esters (**TGACTCA**), serum response elements (**CCATATAGG**), etc. These protein factors exert their specificity in gene regulation not only by discriminative ability for individual genes (since their numbers must be lower than that of the genes) but by their modular assembly. (See transcription factors, hormone response elements, heatshock proteins, HSTF, regulation of gene activity)

TRANSCRIPTION FACTORS, INTERMEDIARY: do not associate directly with the promoter but either affect the conformation of the DNA or "adapt" other proteins to the transcription complex. (See open promoter complex, regulation of gene activity)

TRANSCRIPTION ILLEGITIMATE: takes place when very low level transcripts are detected in organs, tissues or developmental stages where these special transcripts are not expected to occur. (See also housekeeping genes, tissue-specificity)

TRANSCRIPTION SHORTENING: RNA polymerase II (RNAP II) hydrolyzes the 3' end of the transcript as part of the process of reading through pause signals and also secures fidelity of the transcription. It requires TFIIS protein.

TRANSCRIPTION TERMINATION IN EUKARYOTES: the ribosomal gene cluster is generally terminated much beyond the 28S rRNA gene and at about 200 bp upstream from the core promoter of the following pre-rRNA cluster. In the mouse the pol I termination signal contains a Sal I box (5'-AGGTCGACCAG[T/A][A/T]NTCCG-3') preceded by T-rich clusters. The actual termination is within the T-rich area and it is assisted by the Sal I box and the T-rich sequences around it. In humans the conserved repeats (5'-GACTTGACCA-3') terminate pre-tRNA transcription. In *Xenopus* (5'-GACTTGC-3') repeats and T-rich sequences in the spacer region bring about termination. Probably some proteins bind to the Sal I box. Recently, polypeptide chain release factors (RF) have been identified also in eukaryotes. Thus the termination mechanisms in different species vary substantially. In the mtDNA to a 13 residue sequence embedded in the tRNA$^{Leu(UUR)}$ a 34-kDa protein (mTERM) is bound and the complex is required for termination of transcription. (See also polyadenylation signal, transcription termination in prokaryotes, mRNA)

TRANSCRIPTION TERMINATION IN PROKARYOTES: ➔ can be rho-independent (intrinsic terminators exist in the RNA polymerase) and rho-dependent, i.e. the RNA polymerase requires the cofactor rho for termination of transcription. The terminator regions in

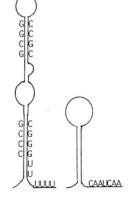

RHO-INDEPENDENT   RHO-DEPENDENT

GENETICS MANUAL

Transcription termination in prokaryotes continued

both systems have similar structures. They consist of palindromic sequences that can fold back into a hairpin. In the rho-independent terminator there are one or more G≡C rich sequences in the stem and at the base of the stem there are about six consecutive U residues. This structure mediates a pause in the movement of the RNA polymerase and thus causing a dissociation from the DNA template because the ribosyl-U of the transcript can make only weak hydrogen bonds with the deoxyribosyl-A in the DNA. In the rho-dependent termination, rho recognizes 50 to 90 bases before the hairpin facilitates termination. Polypeptide release factors (RF) may be used too in both prokaryotes and eukaryotes. (See diagram on page 1047, antitermination)

TRANSCRIPTION UNIT: the DNA sequences between the initiation and termination of transcription of a single or multiple (co-transcribed, multi-cistronic) gene(s).

TRANSCRIPTIONAL ACTIVATORS: are proteins that facilitate the activity of DNA-dependent RNA polymerase(s) in prokaryotes and eukaryotes. Transcription activators may have two independent domains: DNA-binding domain and transcriptional activating domain. These domains may not have to be covalently associated and their function may be carried out by dimeric association.

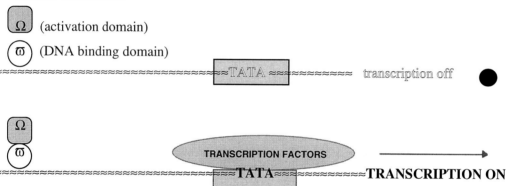

In this abstract sketch, the dimeric activators stimulate transcription by recruiting the components of the transcription complex to the TATA box, even when in their absence or in the presence of the monomers transcription may be hindered. (See positive control, negative control, transcription factors, DNA binding proteins, catabolite activator, regulation of gene activity, FK506, FKBP12, transcription complex, transcription termination, transcriptional modulation, SW, VDRI)

TRANSCRIPTIONAL ADAPTOR: is a histone acetyltransferase, acylating histone in the chromatin before activation of transcription. The yeast gene *GCN5* affects specifically histones 3 and 4. (See histone, nucleosome, transcription, transcription initiation)

TRANSCRIPTIONAL CO-ACTIVATOR: see TAF_{II}

TRANSCRIPTIONAL CONTROL: regulation of protein synthesis at the level of transcription. (See regulation of gene activity, signal transduction, transcription factors, operon, regulon)

TRANSCRIPTIONAL GENE FUSION VECTOR: carry transcription-termination codons (stop codons) in front of the promoterless structural gene, so when the structural gene fuses to a host promoter and thereby expressed (transcribed with the assistance of a host promoter), it will contain only the amino acid sequences specified by the inserted DNA. (See gene fusion, readthrough proteins, translational gene fusion, trapping promoters, diagram on next page)

TRANSCRIPTIONAL MODULATION: in many transcription factors proline- and glutamine-rich activation domains exist and modulating effects are attributed to them. (See transcriptional activators, transcription factor, transcriptional suppressor, WD-40)

TRANCRIPTIONAL SUPPRESSOR: can tightly bind to the operator or to other upstream elements of the DNA and thus prevents the initiation a transcription by elements (activators) of the transcription complex. Mot1 is an ATP-dependent inhibitor of the TATA box-binding pro-

Transcriptional suppressor continued

tein and the members of the NOT complex inhibit the transcription machinery by various ways (TBP, TAF, etc.). (See transcriptional activator, transcriptional co-activator, transcriptional modulation, nucleosome, mating type determination in yeast, silencer, MADS box)

Transcriptional gene fusion vector continued from page 1048

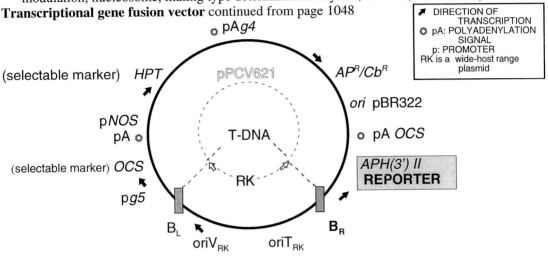

THE CRITICAL FEATURE OF THIS TRANSCRIPTIONAL GENE FUSION VECTOR IS THAT THE REPORTER (APH[3']II, OR LUCIFERASE OR GUS) HAS NO PROMOTER AND IT IS FUSED TO THE RIGHT BORDER OF THE T-DNA. THE STRUCTURAL GENE OF THE REPORTER CAN BE EXPRESSED ONLY IF IT INTEGRATES BEHIND A PLANT PROMOTER THAT CAN PROVIDE THE PROMOTER FUNCTION. IN FRONT THE STRUCTURAL GENE HERE, THERE ARE FOUR NONSENSE CODONS TO PREVENT THE FUSION OF THE PROTEIN WITH ANY PLANT PEPTIDES. BASICALLY, THE CONSTRUCT OF OTHER TRANSCRIPTIONAL FUSION VECTORS IS SIMILAR IN OTHER GROUPS OF ORGANISMS. For other symbols and abbreviation see translational gene fusion vectors. (Based on oral communications by Dr. Csaba Koncz)

TRANSCRIPTON: is a unit of genetic transcription.

TRANSDETERMINATION: a particular pathway of differentiation is overruled by genetic regulation thus e.g. at the *Antp* (*Antennapedia* locus 3-47.5) "gain of function mutants" in *Drosophila* the antenna is transformed into a mesothoracic leg or a wing may develop in the place of an eye, etc. These changes in the developmental pattern may be associated with chromosomal rearrangements. The breakpoints may be within the promoters. They may be altered as a cause of transcript heterogeneity and alternate splicing of the transcripts. Transdetermination occurs in plants (snapdragon, *Arabidopsis* and others) by transforming anthers and pistil into petals and

THE APICAL MERISTEM DEVELOPED INTO HORN-SHAPE STRUCTURES RATHER THAN INTO AN INFLORESCENCE IN *ARABIDOPSIS* MUTANTS

producing sterile full flowers, etc. (See morphogenesis, homeotic genes, imaginal disk)

TRANSDIFFERENTIATION: is a rare biological phenomenon of one type of differentiated cells are converted into another discrete type. (See transdetermination)

TRANSDUCER PROTEINS: in bacteria respond to effectors to relay information to cytoplasmic components of the excitation path to switch molecules. After the effector is diluted out, the adaptation pathway leads the restoration to the base condition.

TRANSDUCIANISM: see creationism

TRANSDUCIN: a G-protein, G_t, involved in transduction of light signals (RAS-related proteins) regulating cyclic GMP phosphodiesterase. It is activated by cholera toxin and inhibited by the pertussis toxin. (See G_t protein, cholera toxin, pertussis toxin, transduction)

TRANSDUCING PHAGE: see transduction

TRANSDUCTANT: a transduced cell. (See transduction)

TRANSDUCTION, ABORTIVE: ↓ transduced DNA is not integrated into the bacterial chromosome and therefore fails to replicate among the bacterial progenies and it is diluted out during the subsequent cell divisions. The non-replicating abortively transduced genetic material is transmitted in a unilinear fashion. (See diagram, transduction generalized, transduction specialized)

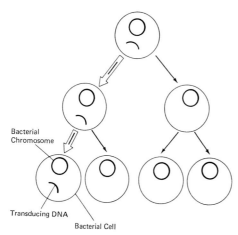

TRANSDUCTION, GENERALIZED: is a phage-mediated transfer of *unspecified* genes among bacteria (lysogenic or non-lysogenic). The virulent phage breaks down the host cell DNA into var-ious size fragments by the process called lysis. The transducing phage coat may then scoop up fragments of the DNA that fit into the capsule (head) that may not contain any phage genetic material. The DNA fragments enclosed in the phage head are picked up at random from the proper size group of the bacterial DNA without regard to the genes located in the fragments. For productive transduction the transferred fragments must be stably integrated into the chromosome of the recipient bacteria. (See also specialized transduction, transduction abortive, transduction mapping, pac site, marker effect)

TRANSDUCTION MAPPING: determines the map position of very closely linked bacterial genes on the basis of co-transduction frequencies; e.g. the donor DNA is $a^+ b^+$ the recipient bacterium is $a^- b^-$ then the recombination frequency is $[(a^+b^-) + (a^-b^+)] / [(a^+b^-) + (a^-b^+) + (a^+b^+)]$. (See transduction generalized)

TRANSDUCTION, SPECIALIZED (or restricted): is a temperate-phage-mediated transfer of *special* genes between bacteria. (See specialized transduction)

TRANSESTERIFICATION: is a replacement reaction catalyzed by an esterase enzyme. A nucleophile displaces an alcohol during the hydrolysis of an ester. A similar reaction occurs when in nucleic acid processing phosphodiester exchanges take place at the splice junctions of exons and introns. (See splicing, introns)

TRANSFECTION: originally, the term was coined for introduction of viral RNA or DNA into bacterial cells and the subsequent recovery of virus particles. Today, it is used for introduction of foreign DNA into animal cells where the genes carried (usually by viral vectors) may be expressed. (See transformation genetic).

TRANSFECTOMA: hybridoma cell producing a specific mouse/human chimeric antibody. (See hybridoma, antibody)

TRANSFER CLOCKWISE/COUNTERCLOCKWISE: the bacterial F plasmid may be integ-

Transfer clockwise/counterclockwise continued
ated in different orientations and at different locations in the bacterial chromosome and thus Hfr strains may be formed that transfer the chromosome either clockwise or counterclockwise during conjugation. (See Hfr, conjugation)

TRANSFER FACTORS: are bacterial plasmids capable of transfering information from one bacterial cell to another through conjugational mobilization. Some of the factors (e.g. ColE1) may not have genes for transfer yet they may be transfered to other cells by helper function of conjugative plasmids. These transfer factors may contain genes for resistance (transposable elements) and have great medical significance because of the transfer of antibiotic resistance and thus make the defense against pathogenic infection difficult. (See antibiotics, transposable elements bacterial, colicins)

TRANSFER LINE: a polyploid species carrying a relatively short foreign chromosomal segment in its genome. The transfer is generally made either by crossing over between homoeologous chromosomes, in the absence of a gene or chromosome (chromosome 5B in wheat) that would normally prevent homoeologous pairing. It can be obtained also by (X-ray) induced translocation. Construction of such lines may have agronomic importance for introducing disease resistance or any other gene(s) that are not available in the cultivated varieties or their close relatives (See alien addition, chromosome substitution, alien substitution, homoeologous chromosome)

TRANSFER RNA (tRNA): genes coding for tRNAs are clustered in both prokaryotes and eukaryotes. Some of the tRNA genes are located within the spacer regions of the ribosomal gene clusters. The majority of tRNA genes are clustered as a group in the DNA, and frequently occur in 2 - 3 copies. Some *E. coli* tRNA gene clusters include genes for proteins. The tRNA genes within the cluster are separated by intergenic sequences and they are transcribed as long pre-tRNA sequences. The primary transcript is processed at the 5'-end by RNase P and at the 3'-end by RNase D, BN, T, PH, RNase II and polynucleotide phosphorylase. The tRNAs are small (70 - 90 nucleotides) molecules. They assume a "clover leaf" secondary structure formed by single stranded loops and double-stranded sequences.

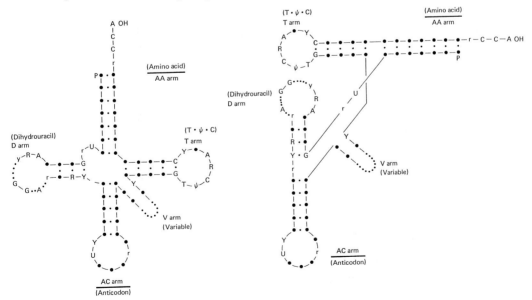

THE GENERAL STRUCTURAL FEATURES OF tRNAs. LEFT THE 'CLOVERLEAF', RIGHT THE L-SHAPED TERTIARY CONFORMATION. R STANDS FOR PURINE AND Y FOR PYRIMIDINE BASES IN ALL tRNAs; r AND y INDICATE THE OCCURRENCE OF THESE BASES IN MANY tRNAs, ψ IS PSEUDOURIDINE. (After KIM, S.H. Progr. Nucleic Acid Res. 17:181)

Transfer RNA continued

The functioning tRNAs assume an L-shape configuration. After charged with amino acids (aminoacylation), they haul the amino acids to the ribosomes for translation of the genetic code (protein synthesis). The amino acids are attached to the protruding C-C-A-(OH) amino acid arm and one of the C residues interact directly at the P site with the G2252 and G2253 of the 23S ribosomal subunit of prokaryotes. The anticodon loop contains a triplet complementary to the amino acid code word and thus it recognizes the code in the mRNA on the surface of the ribosome, the D-arm (dihydrouracil loop) is the recognition site for the aminoacyl-tRNA synthetase enzyme whereas the T-arm (a thymine-pseudouracil [ψ]-C consensus loop) recognizes the ribosomes. There is also a small variable loop (V arm). The presence of modified nucleotides is characteristic for tRNAs. These modifications take place right after transcription or during processing. The number of tRNAs in prokaryotes is higher then the number of genetic code words (64); in prokaryotes the number of different tRNA molecules may run into hundreds in the different species. Since there are only 20 amino acids, of the high number of tRNAs several deliver the same amino acid to the site of translation (isoaccepting tRNA). The vast majority of animal and fungal mitochondria synthesize their own tRNAs (about 22). Their anticodon reads the codons by the first two bases and thus do not need isoaccepting tRNAs; their structure; usually lack the pseudouridine loop, etc. The mitochondria of land plants, algae, *Paramecium*, *Tetrahymena* and *Trypanosomes* partially rely on tRNA import. Organelle tRNAs similarly to plants, add post-transcriptionally the 3'- CAA amino acid accepting terminus, in contrast to *E. coli*. The mitochondrial tRNAs in plants are generally quite variable and show similarities also to the chloroplast tRNAs. The chloroplast tRNAs of higher plants (about 30) may frequently be coded by more than one tract. The universal genetic code requires a minimum of 32 different tRNAs. Actually, however 26-24 tRNAs (with special wobbles) may suffice for protein synthesis. The mammalian and some fungal mitochondria do not code for all the tRNAs required and import these nuclearly coded molecules from the cytosol. Many fungi code for 25-27 tRNA genes, however. Prokaryotes and chloroplasts have a special tRNA$^{F\text{-Met}}$ for the initiation of translation and the same anticodon, CAU may recognize not only the Met codon (AUG) but three other initiator codons too. Nematodes initiate translation with UUG which is a leucine codon. (See also tRNA, *rrn* genes for tRNAs within ribosomal gene clusters, aminoacyl-tRNA synthase, isoaccepting tRNAs, code genetic, wobble, isoacceptor tRNA, ribosome, mitochondrial genetics, mtDNA, chloroplasts, chloroplast genetics, ribonuclease P)

TRANSFERASE ENZYMES: move a chemical group(s) or molecule(s) from a donor to an acceptor.

TRANSFERRIN (TF): a β–globulin (M_r ca. 75,000-76,000) that transports iron. The encoding gene is in human chromosome 3q21; the transferrin receptor gene was located nearby. Adenosine ribosylation factors affect the cellular redistribution of transferrin and endocytosis. (See endocytosis, BLYM, RAC)

TRANSFORMANT: a cell or organism that had been genetically transformed by the integration of exogenous DNA into its genetic material. (See transformation)

TRANSFORMATION ASSOCIATED RECOMBINATION (TAR): yeast cells are transformed simultaneously by a YAC vector (TAR) with terminal human genomic repeats such as Alu and a long piece of human genomic DNA containing interspersed repeats (Alu). Recombination between the YAC and the human genomic DNA within the homologous repeats yields large, stable circular YACs that can be used for mapping or cloning. If the TAR vector contains an *E. coli* F-factor cassette, the vector can be propagated also in bacterial cells. (See YAC, Alu, F factor, vector cassette)

TRANSFORMATION BY PROTOPLAST FUSION: see protoplast fusion

TRANSFORMATION, GENETIC: information transfer by naked DNA fragments or plasmid, obviating traditional sexual or asexual processes in prokaryotes and in eukaryotes. **BACTERIAL TRANSFORMATION** — Genetic transformation was discovered in bacteria in the late 1920s. It became a widely used genetic method only in the 1950. Originally, only naked bacterial DNA

Transformation, genetic continued

was used in fragments of 1/200 to 1/500 of the genome. This was provided to *competent* bacterial cells at a concentration of 5 - 10 µg/mL culture medium. The exogenous DNA can synapse with the bacterial genome and generally only one strand of the transforming DNA is integrated into the recipient although some bacteria (e.g., *Haemophilus influenzae*) preferentially takes up double-stranded DNA from its own species but integrates only one of the strands. Recognition of the homospecific DNA is mediated by uptake signal sequences (USS): 5'-AAGTGCGGT in the plus and 5'-ACCGCACTT in the minus strand. In the completely sequenced genome of 1,830,137 bp 1,465 such UUS were recognized. *Neisseria gonorrhoeae* also has USS elements (5'-GCCGTCTGAA).

Bacterial transformation generally may not involve an addition, rather it is a replacement of a part of the DNA of the recipient cell, except when plasmids are used. The non-integrated parts of the donor DNA are degraded and the rest replicates along the genes as a permanent integral part of the bacterial chromosome. The frequency of bacterial transformation may be in the range of 1% or as low as 10^{-3} to 10^{-5}, however, using bacterial protoplasts (spheroplasts) up to 80% transformation is attainable. For bacterial transformation most commonly various genetic vectors are used. Although different empirical procedures are employed in different laboratories, some general features of the methods are obvious. For transformation either high molecular weight, (DNase-free) DNA or plasmids (phage), dissolved in 1 x SSC, are used. Competence in recipient bacteria is induced by $CaCl_2$, $MnCl_2$, reducing agents and hexamminecobalt chloride or competent cells are purchased in a frozen state from commercial sources. Highly competent cells may yield ca. 10^7 to 10^9 colonies per 1 µg plasmid DNA. The success of transformation is improved by highly nutritious culture media and good aeration. The recognition of transformant cells is greatly facilitated by selectable markers. Transformation of bacteria by electroporation may be extremely efficient (10^{10} transformants/µg DNA). Cultures in mid-log phase are chilled and washed by centrifugation in low-salt buffer. The cells (3 x 10^{10}/mL) are suspended then in 10% glycerol and can be stored on dry ice or -70° C for up to half year. Thawed aliquots of the cells are mixed with properly prepared donor DNA and exposed to high voltage electric field in small volumes (20-40 µL). Gram-positive bacteria such as *Bacillus subtilis*, is more difficult to transform genetically than the gram-negative bacteria, e.g., *E. coli*. *B. subtilis* attracted interest for cloning because it is not pathogenic for humans. In the presence of the *B. subtilis recE* transformation is facilitated if the cell contains already a plasmid homologous to the vector. Also, spheroplasts in the presence of polyethylene glycol, take up exogenous DNA much easier. Vectors derived from *Staphylococcus aureus* containing tetracycline- (pT127) or chloramphenicol- (pC194) resistance have been successfully used for the development of vectors. *Staphylococcus aureus* is a serious pathogen. Shuttle vectors containing *E. coli* pBR322 and *S. aureus* plasmid elements were also used. Some of the antibiotic resistance genes, e.g., β-lactamase, have very different expression in different species of bacteria. *Streptomyces* were of substantial interest for transformation because of their efficient production of antibiotics. They can be transformed by a sex plasmid, liposomes and phage vectors. (See competence, SSC, DNA extraction, electroporation, vectors, cloning vectors, liposome, β-lactamase, antibiotics, sex plasmid).

FUNGAL TRANSFORMATION — is not entirely different from that in prokaryotes. Transformation of *Neurospora* started already during the early 1970 and caused genetic instabilities in the genome (see RIP). Transformation of budding yeast begun in the late 1970s and became very useful for various types of studies (cloning, YACs, gene replacement, etc.). Yeast cells are grown to about 10^7 density/mL and then suspended in a stabilizing buffer containing 1 M sorbitol. Subsequently the cell wall is removed by digestion with β-glucanase (an enzyme hydrolyzing glucan, the polysaccharide of the cell wall [yeast cellulose]). To the washed spheroplasts in sorbitol, in the presence of $CaCl_2$ and polyethylene glycol (PEG4000), the donor DNA is added. After about 10 min incubation of the mixture, the cells are gently embedded in 3% agar and layered over a selective medium in a Petri plate. The frequency of transformation depends

Transformation, genetic continued

a great deal on the type of vector used (between 1 to 10^6 colonies per μg DNA). Alternatively — although with lower yield — intact yeast cell have been treated with lithium salts before the DNA and polyethylene glycol is applied. This is followed by selection after spreading the cells onto the surface of selective media. This procedure does not require the production of spheroplasts and agar embedding. Both of these procedures may cause mutations. Similar methods of transformation have been used also in other fungi (*Neurospora, Aspergillus, Podospora*) and also in green algae. Shuttle vectors were advantageous for the transfer of genes between various fungi and between fungi and bacteria. (See YAC, gene replacement, episomal vector, integrating vector, replicating vector, centromeric vector, shuttle vector)

TRANSFORMATION OF ANIMAL CELLS — The most commonly used procedures involve precipitation of the donor DNA with calcium phosphate or DEAE-dextran. The precipitated granules may enter animal cells by phagocytosis and up to about 20% of the cells may integrate the donor DNA into the chromosomes. By precipitation, physically unlinked DNA molecules can also be transformed (cotransfected) into the cultured animal cells. The polycation Polybrene (Abbott Laboratories trade name for hexadimethrine bromide) is also used to facilitate the transformation by relatively low molecular weight DNA (plasmid vectors) when some other procedures are not working. Electroporation has also been successfully applied for stable or transient introduction of DNA into the cells. Bacterial (or even plant) protoplasts can also be used to bring about fusion of cell membrane (in the presence of polyethylene glycol) and this may be followed by transfer of plasmid DNA into animal cell nuclei. This procedure is less efficient than endocytosis mediated by calcium phosphate and the plasmids are frequently integrated in tandem into the chromosome(s) of vertebrates. The exogenous DNA may be introduced also by direct *microinjection* into (pro)nuclei or into embryonic stem cells (ES) and thus generate chimeras. In the latter case the transformed cells can be screened for insert copy number or the insert can be targeted to a specific site by homologous recombination. *Infection* of stem cells, bone marrow, zygotes, early embryos by vectors or by isolated chromosomes is also feasible (see microinjection). Gene replacement by homologous recombination (see targeting) is also an option. In the latter case sufficient information is required about the needs for critical cis-acting elements. The success of transformation of animals cells varies a great deal according to cell types used. Transformation of vertebrate cells became a very important tool of molecular biology and reversed genetics but unfortunately the transformed cells cannot be regenerated into complete individuals, except when germline or ES cells are transformed. In *Drosophila,* into the cloned *P* element an isolated gene can be inserted with the aid of genetic engineering and the element may be then microinjected into a young embryo where the DNA can integrate into the chromosome resulting in a stably transformed individual fly. The procedure is particularly effective if the P element vector is equipped also with a selectable marker. In mosquitos microinjection into the egg cells is feasible but was not very effective. A more successful approach was using viral vectors with the vesicular stomatitis virus glycoprotein envelope which binds to the cell membrane and delivers the foreign DNA. (See hybrid dysgenesis, ammunition, smart ammunition, SV40 vectors, Adenoma, Bovine Papilloma Virus vectors, retroviral vectors, DEAE-dextran, Polybrene, electroporation, polyethylene glycol, liposome, gene replacement, targeting genes, transgenic animals, gene therapy, ES, vesicular stomatitis virus, *Anopheles*).

TRANSFORMATION OF PLANTS — can be carried out by a variety of procedures. Most extensively used were the techniques of infecting leaf or root explants or protoplasts by agrobacteria, carrying genetically engineered plasmids. Practically all dicots can be readily transformed by agrobacteria but some monocots (*Dioscorea, Narcissus and Asparagus*) could also be transformed. The difficulty with monocot transformation by agrobacteria is apparently caused by the lack of secretion of substances needed for the activation of the virulence gene cascade or monocot cells fail to develop competence in response to infection by *Agrobacterium*. The vector plasmids are either cointegrate or binary. The cointegrate plasmids contain in cis also the virulence genes of the Ti plasmids whereas in the binary vectors the virulence genes are

Transformation, genetic continued

carried by a separate small helper plasmid. A common feature of all these vectors that the genes to be integrated into the plant chromosome are in-between the two 25-bp inverted repeats of the T-DNA. The virulence genes and all other DNA sequences are not integrated into the host genetic material. The left and right border sequences are important for successful transformation but only a few bp of the left border and either none or 1-3 bp or the right border are retained in the host (see *Agrobacterium*, Ti plasmid). The insertional target is not strictly specified in plants yet a few base similarity is frequently found. It appears that the border repeats scout for appropriate target sites and they are appositioned there. This process is directed also by the plasmid virulence genes and the bacterial chromosome has also some genes that assist the process. The target suffers initial staggered nicks, followed by degradation, and the DNA within the 25 bp borders of the T-DNA is integrated into the chromosome (see diagram below).

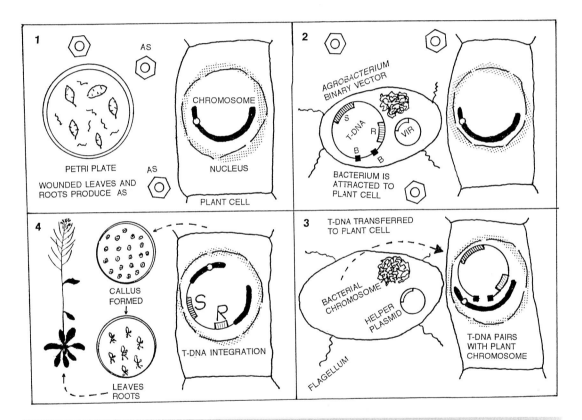

THE MAJOR STEPS OF TRANSFORMATION OF PLANTS BY THE USE OF AGROBACTERIAL BINARY VECTORS. AXENICALLY GROWN LEAVES, ROOTS OR STEM SEGMENTS ARE WOUNDED. THE WOUNDING STIMULATES THE PRODUCTION OF PHENOLICS SUCH AS ACETOSYRINGONE (AS) AND ATTRACTS THE BACTERIA CARRYING THE ENGINEERED PLASMIDS WITH SELECTABLE MARKERS (S) AND A REPORTER GENE (R), PLACED BETWEEN THE TWO BORDER SEQUENCES (B) OF THE T-DNA. A SMALL HELPER PLASMID CARRIES THE *VIR* GENES REQUIRED FOR TRANSFER AND INTEGRATION OF THE T-DNA. 2 DAYS AFTER INFECTION THE BACTERIAL GROWTH IS STOPPED BY ANTIBIOTICS. THE ISOLATED PLANT ORGANS DEVELOP CALLUS, ROOTS AND SHOOT AND EVENTUALLY COMPLETE PLANTS ON THE SELECTIVE MEDIA, ONLY IF THE TRANSFORMATION WAS SUCCESSFUL. THE TRANSFORMATION IS CONFIRMED ALSO BY THE EXPRESSION OF THE REPORTER GENE. THE TRANSFORMANTS PRODUCE SEED THAT DEVELOPS INTO HETEROZYGOUS PROGENY. THE ANTIBIOTIC MARKERS ARE DOMINANT IN THE PLANTS.

Transformation, genetic continued

In order to be expressed, the genes within the T-DNA generally carry appropriate (plant compatible) eukaryotic promoters and polyadenylation signals to be expressed in the plant cells. Some vectors may lack promoters and can be expressed only when fused (upon integration) *in vivo* with plant promoters. These may be translational or transcriptional fusion vectors. The translational fusion vectors lack the translation initiation methionine codon in order to facilitate the fusion of the structural gene in the T-DNA with some amino acid sequences of the plant host. The purpose of these types of transformation is to study the strength and tissue-specificity of different plant promoters and study the function of fusion proteins. The transcriptional fusion vectors carry one or more translational stop codons (nonsense codons) in the nucleotide tract preceding the ATG (translation initiation codon). Because of this, the structural transgene will be expressed if it will be driven by a genuine plant promoter and no fusion protein is obtained. These vectors are also shuttle vectors, they can be propagated in agrobacteria and *E. coli*. Being shuttle vectors greatly facilitates various manipulations. The vectors can be replicated in *E. coli* because they have the origin of replication of the pBR322 plasmid and outside the boundaries of the two T-DNA border sequences carry genes *oriV* (required for replication) and *oriT* (required for transfer) in *Agrobacterium*. The latter genes were derived from the promiscuous (wide host range) RK plasmid. The transformation procedure may be started by aseptically (axenically) grown plant tissues. The vectors cassettes generally carry selectable markers, most commonly for resistance against hygromycin B or kanamycin. The gene fusion reporter genes may be (bacterial or firefly) luciferase or GUS (β-glucuronidase) because of the easy monitoring, and the time and space of expression of the fused reporter gene. (See transcriptional gene fusion vector, [p. 1049], translational gene fusion vector [p. 1060])

The plant explants are generally wounded by scalpels as they are harvested and then dipped, drained then incubated for 2 days with a fresh bacterial suspension (grown to a density of about 10^6, washed and diluted to about half or less in plant nutrient solution) on the surfaces of an agar medium in Petri plates (see embryogenesis somatic). After 2 days the bacteria are removed either by claforan (syn. cefotaxime) or carbenicillin and the plant cells are grown further in media containing also hygromycin or kanamycin (G418) or other selective agent depending of the vector constructs. The regenerated plants may then be grown axenically to maturity in test tubes (in case of *Arabidopsis*) or in soil (in case of larger plants). The transgenes usually segregate as dominant alleles if the plant does not have a corresponding native locus that might mask their expression.

Alternatively, the agrobacterial infection can be applied to presoaked seeds of plants and eventually a small fraction of the embryos developing after meiosis will carry the transgene. Also, seedlings or plants can be infiltrated by agrobacterial suspensions and selection carried out at large scale in soil cultures treated with an appropriate herbicide (Basta) against what the vector carries resistance (*in planta* transformation). Plants can be transformed also by electroporation and by biolistic methods. Also cauliflower mosaic virus (CaMV) and geminivirus vectors have been developed but these are not widely used. (See *Agrobacterium*, electroporation, biolistic transformation, microinjection, gene transfer by microinjection, transformation of organelles, genetics of chloroplasts, genetics of mitochondria, gene trap vectors).

TRANSFORMATION MAPPING: a procedure in prokaryotes for determining gene order within the genetic region of integration of the homologous transforming DNA. It can be used as a three-point cross but the additivity of recombination is generally imperfect in transformation. The general principle is as follows: The recipient DNA is *abc*, the donor is *ABC*. Recombination frequencies are determined in three steps. First the recombination between *A* to *B* (I), then between *B* to *C*, (II) then between *A* to *C* (III) is determined. The frequency of recombination in interval I is calculated from the proportion of cells $[(Ab) + (aB)]/[(Ab) + (aB) + (AB)]$, in interval II $[(Bc) + (bC)]/[(Bc) + (bC) + (BC)]$ and in interval III by $[(Ac)+(aC)]/[(Ac) (aC) + (AC)]$. When the cells are classified, the third marker is left out of consideration, e.g. *Ab*

Transformation mapping continued

includes *AbC* and *Abc*. For the sake of simplicity the capital letters represent prototrophic and the lower case letters stand for auxotrophic markers. The recipient is always chosen as the auxotroph and the donor as the prototroph to facilitate selection of the classes. Example (the auxotrophic alleles are represented by (-), the prototrophic alleles by (+) signs):

Genes			Genotypes of Transformants				
a →	+	-	-	-	+	+	+
b →	+	+	-	+	-	-	+
c →	+	+	+	-	-	+	-
No.	12,000	3,400	700	400	2,500	100	1,200

Recombinants in the *A - B* interval = 3,400 + 400 + 2,500 + 100 = 6,400
Parental in the *A - B* interval = 12,000 + 1,200 = 13,200
Parental + Recombinant = 13,200 + 6,400 = 19,600
Recombinants in the *B - C* interval = 700 + 400 + 100 + 1,200 = 2,400
Parental in the *B - C* interval = 1,2000 + 3,400 = 15,400
Parental + recombinant = 2,400 + 15,400 = 17,800
Recombinants in the *A - C* interval = 3,400 + 700 + 400 + 2,500 + 100 + 1,200 = 8,300
Parental in the *A - C* interval = 12,000
Parental + Recombinant = 20,300

RECOMBINATION

in the *A - B* interval = $\frac{6400}{19600}$ = 0.327

in the *B - C* interval = $\frac{2400}{17800}$ = 0.135

in the *A - C* interval = $\frac{8300}{20300}$ = 0.409

and the map based on this three-point transformation is ≈: *A* 33 *B* 14 *C*. It is not entirely additive because the expectation for *A - C* might be 33 + 14 = 47 but the calculated is only ~ 41 yet we can see that the order is either *A B C* or *C B A*. (See recombination, mapping)

TRANSFORMATION OF ORGANELLES: mitochondria and plastids are also amenable to transformation by exposing protoplasts to appropriate vectors in the presence of polyethylene glycol (PEG) or even more effectively by employing the biolistic procedures. The efficiency of transformation is generally somewhat low. Transformation of organelles is still struggling with methodological difficulties yet it will have great potentials upon further improvements. The number of mitochondria (and plastids) within single cells may run into near hundred or even thousands and that would make possible the amplification of economically important proteins. (See chloroplast genetic, mitochondrial genetics, metabolite engineering, genetic engineering, protein engineering, biolistic transformation, transformation, human gene transfer, gene therapy)

TRANSFORMATION, ONCOGENIC: change to malignant (cancerous) cell growth (See cancer, carcinogen, gatekeeper, phorbol esters)

TRANSFORMATION, STABLE: produces cells which carry the transforming DNA in an integrated form and thus the acquired information is consistently transmitted to the progeny. (See transformation transient, transformation genetic)

TRANSFORMATION, TRANSIENT: the introduced DNA may be expressed only for a limited time (for 1 to 3 days) in the recipient cell because it is not integrated into the host genetic material. Electroporation is most commonly results in this kind of transformation. (See transformation stable, electroporation)

TRANSFORMATION VECTORS: see cloning vectors, vectors, transformation genetic

TRANSFORMED DISTANCE (TD): is an UPGMA procedure for species *i* and *j*:

$$d_{ij'} = (d_{ij} - d_{ir} - d_{jr})/2 + c$$

where *r* is a reference species within or outside the group and *c* is a constant to make $\bar{d}_{ij'}$ positive. This TD formula is unsuitable for the estimation of branch length in evolutionary trees. (See evolutionary tree, evolutionary distance, UPGMA, Fitch—Margoliash method for TD)

TRANSFORMING GROWTH FACTOR: see TGF

TRANSFORMING GROWTH FACTOR β: a superfamily of proteins that induces the change of undifferentiated tissues into specific types of tissues; TGFβ1 peptide factor causes reversible arrest in G1 phase of the cell cycle. TGFβ1 had been detected in cis position to several genes. These proteins are serine/threonine kinases. (See also activin, bone morphogenetic protein)

TRANSFORMING PRINCIPLE: a historical term used in the early bacterial transformation reports when it was not yet proven that DNA is the agent of transformation.

TRANSFORMYLASE: enzyme that adds a formic acid residue to the methionine-charged fMet tRNA (tRNA^{fMet}) in prokaryotes. (See protein synthesis)

TRANSGENE: a gene transfered to a cell or organism by isolated DNA rather than by sexual means. (See transformation genetic)

TRANSGENE MUTATION ASSAY: mouse transgenic for a prokaryotic reporter gene is exposed to mutagenic conditions (spontaneous or treated with an agent). The genomic DNA is isolated and rescued in phage lambda vector or used in a plasmid rescue system. The cloning bacteria are then plated and the number of mutant reporter genes compared with all the reporter genes analyzed to provide mutation frequency. Under experimental conditions spontaneous mutation rates (*lacZ*) were observed within the range about 6 to 80 x 10^{-6}, depending on the tissues from where the DNA was extracted. (See host-mediated assays, plasmid rescue, *lac* operon, β galactosidase, vectors, bioassays in genetic toxicology)

TRANSGENESIS: introducing a gene by genetic transformation. (See transformation genetic)

TRANSGENIC: carries gene(s) introduced into a cell or organism by transformation. Transgenic animals can potentially produce therapeutically needed proteins such as human tissue plasminogen activator (tPA) or $α_1$-antitrypsin (ATT), etc. Transgenic plants may have direct use in agriculture by virtue of their resistance to herbicides, pathogens or even by the production of various antigens that may be substituted for the standard type vaccines by eating them. (See transformation genetic)

TRANSGENOME: transformed eukaryotic cells by isolated whole or parts of chromosome(s) containing transgenes in their nuclei. (See transformation genetic [animal cells])

TRANSGENOSIS: an alternative term for transfer of genes by non-sexual means.

TRANS-GOLGI NETWORK (TGN): is the connection of the Golgi complex exit face to transport vesicles so the molecules coming from the endoplasmic reticulum will be transported to their proper destination. (See cis-Golgi, Golgi)

TRANSGRESSION: some segregants exceed both parents and the F1 hybrids.

TRANSIENT EXPRESSION DNA: see transformation transient

TRANSILIENCE, GENETIC: rapid changes in fitness by a multilocus complex in response changes in the genetic environment.

TRANSIN: cell-secreted metalloproteinase, a homolog of stromelysin.

TRANSINACTIVATION: see co-suppression

TRANSISTOR: see semiconductor

TRANSIT PEPTIDE is a dozen to five dozen amino acid residue leader sequence directing the import of proteins synthesized in the cytosol into mitochondria and chloroplasts. These peptides are generally rich in basic and almost free of acidic amino acids. Serine and threonine are usually very common. The transit peptide recognizes special membrane proteins, but itself is not transfered into the target organelle and it is cut off by a peptidase. The different transit-peptides do not appear to have conserved sequences. The transit peptide is targeted posttranslationally. Some mitochondrial and plastid proteins do not have these cleavable N-terminal sequences. The routing within the target seems to be influenced by the carboxy terminus or inner sequences of the proteins. The transit peptide engages the several proteins localized in the organelle membranes. Subsequently the protein inside the cell folds with the assistance of chloroplast chaperonin 60. (See signal peptide, chaperones)

TRANSITION MISMATCH: purine mispairs with a wrong pyrimidine. (See transversion mismatch, mismatch, transition mutation)

TRANSITION MUTATION: either a pyrimidine is replaced by another pyrimidine, or a purine by another purine in the genetic material leading to mutation. (See base substitutions, transversion)

TRANSLATION: converting the information contained in mRNA nucleotide sequences into amino acid sequences of polypeptides on the ribosomes. (See protein synthesis)

TRANSLATION INITIATION: is usually triggered by growth factors through signaling to the RAS/RAF G proteins and MEK/MAPK proteins. Upon phosphorylation 4E-BP1 releases the cap-binding protein eIF-4F and the mRNA cap associates with the 40S subunit of the eukaryotic ribosome. The phosphorylation of ribosomal protein S6 by protein kinase $p70^{s6k}$ is also required. Eukaryotic initiation factor eIF-2B is active when it is bound to GTP and ensures the supply of $tRNA^{Met}$. Insulin and other growth factors keep eIF2B attached to GTP whereas glycogen-synthase kinase (GSK) inactivates it because GSK inactivates insulin. In prokaryotes the initiation begins when the ribosomal binding site of the mRNA (including the Shine-Dalgarno sequence and AUG^{fMet} codon) binds to anti-Shine-Dalgarno sequence in the 16S RNA of the 30S ribosomal subunit. The AUG codon thus is directly placed into the P pocket of the ribosome and can interact with the formyl-methionine-charged fMet-tRNA. Before the 30S subunit combines with the 50S subunit a number of other interactions also take place. (See eIF, protein synthesis, DEAD box, cap, S6 kinase, ribosome, Shine-Dalgarno sequence)

TRANSLATION *IN VITRO*: (See wheat germ, rabbit reticulocyte)

TRANSLATION NON-CONTIGUOUS: generally the mRNA is translated in a collinear manner into amino acid sequences without skipping any parts of it. There are few exceptions to this continuity; 50 nucleotides in bacteriophage T4 *gene 60* are skipped during translation.

TRANSLATION REPRESSOR PROTEINS: may be attached to a site near the 5' end of the mRNA and preventing the function of the peptide chain initiation factors. (See protein synthesis, eIF-2, rabbit reticulocyte, aconitase, trinucleotide repeats)

TRANSLATION TERMINATION: takes place in the decoding A pocket of the ribosome where the polypetide release factors, RF1 recognizing prokaryotic stop codons UAG and UAA, and RF2 specific for UGA and UAA or RF3 without selectivity (and may cause misreading of all three stop signs) release the polypeptide chains. FR3 has homology to elongation factors EF-G and EF-Tu. This fact seems to indicate that termination and chain elongation processes bear similarities; in one case the stop codon is read, in the other the sense codons. The corresponding release factors in eukaryotes are eRFs. The Rfs may have additional homology domains, e.g. with the acceptor stem, the anticodon helix and T stem of tRNAs, called "tRNA mimicry". These homologies may assist their function. The yeast eRF3 is a prion-like element, psi^+. Interestingly the heatshock protein 104, a molecular chaperone can cure the cell from it. After the protein synthesis is terminated the termination complex and the ribosome are recycled. In the 16S rRNA of *E. coli*, mutation at nucleotide position C1054 causes translational suppression. Similarly at the corresponding site in the 18S eukaryotic rRNA, substitutions of A or G resulted in dominant nonsense suppression while the T substitution was a recessive antisuppressor and deletion of the site had a lethal effect. Although translation termination is mediated at the ribosomes, premature termination may result not just from nonsense codons but also by decay of the misspliced transcripts. (See translation initiation, release factor, stop codon, sense codon, autogenous suppression, readthrough, recoding, EF-G, EF-Tu-GTP, chaperone, prion, protein synthesis, ribosome)

TRANSLATIONAL CONTROL: protein synthesis is regulated during the process of translation on the ribosome; e.g. attenuation. (See attenuation, termination factors, translation termination, terminator codons, regulation of protein synthesis, translational termination, suppressor RNA)

TRANSLATIONAL ERROR: see ambiguity in translation, error in aminoacylation

TRANSLATIONAL GENE FUSION VECTORS: carry promoterless, 5'-truncated structural genes and when these are driven by the trapped host promoter, they direct the synthesis of fusion proteins containing amino acid residues coded for by both host and vector DNA sequences. (See diagram on next page, gene fusion, transcriptional gene fusion, read-through proteins, trapping promoters, transformation genetic)

Translational gene fusion vectors continued

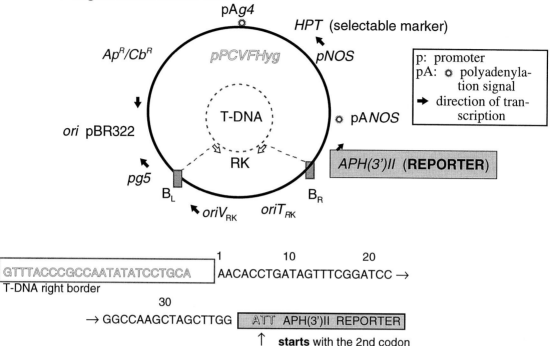

AGROBACTERIAL TRANSLATIONAL *IN VIVO* GENE FUSION VECTOR FOR PLANTS. THE CRITICAL FEATURE IS THAT THE REPORTER GENE IS FUSED TO THE RIGHT BORDER SEQUENCE OF THE T-DNA IN SUCH A MANNER THAT THE TRANSLATION INITIATOR CODON (AUG^{MET}) IS DELETED AND THE STRUCTURAL GENE OF THE REPORTER BEGINS WITH ITS SECOND CODON. IT DOES NOT HAVE A PROMOTER EITHER. THE REPORTER GENE CAN BE EXPRESSED ONLY WHEN IT IS INSERTED AND FUSED IN THE CORRECT REGISTER INTO AN ACTIVE PLANT PROMOTER. SINCE THE AUG CODON IS MISSING, THERE IS A GOOD CHANCE THAT THE REPORTER PROTEIN WILL BE FUSED WITH SOME PLANT (POLY-) PEPTIDES. THE USE OF SUCH A VECTOR PERMITS AN ANALYSIS OF THE EXPRESSION OF VARIOUS FUSION PROTEINS ON THE REPORTER. BESIDES STRUCTURE, SHOWN IN DETAIL AT THE LOWER PART OF THE DIAGRAM, THE TRANSFORMATION CASSETTE CONTAINS SELECTABLE MARKERS (e.g. HPT, PERMITTING SELECTIVE ISOLATION OF TRANSFORMANT ON HYGROMYCIN MEDIA), AMPICILLIN (Ap^R) AND CARBENICILLIN (Cb^R) RESISTANCE FOR SELECTABILITY IN BACTERIA AND ALSO THE REPLICATIONAL ORIGIN OF THE *E. coli* PLASMID pBR322. OUTSIDE THE BOUNDARIES OF THE T-DNA THERE ARE GENES FOR BOTH VEGETATIVE (oriV) AND CONJUGATIONAL TRANSFER (oriT) DERIVED FROM THE MULTIPLE HOST RANGE RK PLASMID. ONLY THE GENES BETWEEN THE TWO BORDER SEQUENCES (B$_L$ and B$_R$) ARE INSERTED INTO THE PLANT GENOME. THE BASIC PRINCIPLE OF THIS DIAGRAM HAVE BEEN EXPLOITED IN DESIGNING VECTORS ALSO FOR OTHER ORGANISMS. (After oral communication by Dr. Csaba Koncz)

TRANSLATIONAL HOPPING: occurs when a peptidyl-tRNA dissociates from its first codon and then reassociates with another dowstream. (See aminoacyl-tRNA synthetase, protein synthesis, overlapping genes, translational frameshift)

TRANSLATIONAL RECODING (same as ribosomal frame shift): see overlapping genes

TRANSLATIONAL RESTART: see reinitiation

TRANSLATIONAL TERMINATION: may take place by encountering stop codons, endonucleolytic cleavage, shortening the poly(A) tail, premature decapping of the mRNA. (See translational control)

TRANSLESION PATHWAY: is an SOS repair system of DNA; replication may lead to targeted

Translesion pathway continued
mutation at the site of mismatches such as at a thymine dimer, and at bases chemically modified or deleted by mutagens. The Rev proteins of yeast and the UmuC protein of *E. coli* are involved with translesion. The Rev polypeptides are subunits of DNA polymerase ζ. (See DNA repair, DNA polymerases)

TRANSLIN: a protein binding to GCAGA[A/T]C and CCCA[C/G]GAC sequences at the translocation breakpoint junctions in lymphoid malignancies and supposedly has a role in the rearrangement of immunoglobulin —T cell receptor. (See immunoglobulins, T cell, T cell receptor, TCR, lymphoma)

TRANSLOADING: modification of cancer vaccines by including non-self peptides so they would boost immunogenicity. (See cancer gene therapy)

TRANSLOCASE: a protein complex mediating transport of proteins through cell membranes. (See SecA, SecB, SecY/E, translocon, ABC transporters, protein targeting, signal hypothesis, SRP, ARF)

TRANSLOCATION: transfers codons of the mRNA on the ribosomes as the peptide chain elongates. In general any type of transfer of molecules from one location to another. (See protein synthesis)

TRANSLOCATION, CHROMOSOMAL: segment interchange between two nonhomologous chromosomes.

ORIGINAL CHROMOSOMES → RECIPROCAL TRANSLOCATION

Broken chromosomes do not stick however to telomeres; the interchange must involve internal regions. Fragments may be inserted in-between two ends of an internal breakpoint, and such an aberration is called *shift*. Translocations are detectable by the light microscope if the length of a chromosome arm is substantially altered. Translocations are usually reciprocal but during subsequent nuclear divisions one of the participant chromosomes which carries no essential genes may be lost. Heterozygotes for reciprocal translocations display cross-shaped configuration in meiotic prophase (see also diagram on next page). Translocation heterozygotes generally display 50% pollen sterility in plants because alternate and adjacent -1 distributions occur at about equal frequency in the absence of crossing over and adjacent-2 distributions, being non-disjunctional, are very rare. Note that inversion heterozygotes may also produce 50% male sterility but it occurs only if recombination within the inverted segment occurs freely (see inversion). In animals the gametes of translocation heterozygotes may succeed in fertilization but the zygotes or early embryos resulting from such a mating are generally aborted. Translocation homozygosity may not have any phenotypic consequence, however, it has been shown that many types of cancerous growth are associated with translocation breakpoints. Apparently, the DNA rearrangements in the vicinity of the genes interferes with the normal regulation of their activity as a kind of position effect. Translocation breakpoints reduce the frequency of crossing over between the breakpoint and the centromere (*interstitial segment*). Recombination in translocation homozygotes may be normal. Because the reciprocal interchange physically alters the synteny of genes, linkage groups may be reshuffled as a consequence of the exchange. Because translocations partially join two linkage groups, they can be exploited for assigning genes to chromosomes. The number of crosses to localize a gene to a chromosome may thus require fewer crosses. Also, the reduction of recombinations around the breakpoints may call attention to linkage over a somewhat larger chromosomal tract than a single marker. Furthermore, the association of certain genes with sterility may also be used as a chromosomal marker for the breakpoints. The sterility marker may not always be very useful because it can

TRANSLOCATED SALIVARY GLAND CHROMOSOME OF *Sciara coprophylla*. (Courtesy of Dr. Helen Crouse)

Translocation, chromosomal continued

be recognized only late during development (after sexual maturity). In some organisms translocation testers have been developed to expedite linkage analysis. A clear and early marker is translocated to several (tester) chromosomes. Then, the gene to be identified regarding its chromosomal position is crossed with all the translocation testers available. If according to previously obtained information with crosses involving non-translocated chromosomes, the marker in the tester showed independent segregation but crossed with the translocation testers it is linked to a single specific chromosome, its chromosomal position is revealed. The frequency of translocation may vary a great deal according to the species. In the *Oenothera* plants translocations are widespread and in *Oenothera lamarckiana* all the chromosomes are involved in translocations. According to some estimates there is about 0.004 chance that a human baby will carry a translocation. Many types of tumors carry translocations and the pattern involved is not a haphazard one. Potential oncogenes (MYC, RAS, SRC) are frequently translocated into the 14q11 region, the location of the T cell receptor (TCR) α and δ. MYC translocated to immunoglobulin genes is common in B cell neoplasia and Burkitt's lymphoma. The chain formation question is intriguing how these special translocations are controled. One interpretation may be that the gene fusions involved may lead to the creation of highly selective combinations to stimulate proliferation. (See also adjacent disjunction, position effect, synteny, multiple translocations, cancer, telomere, B chromosomes, trisomic tertiary, chromoso-

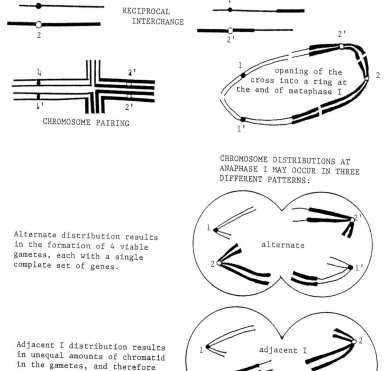

Alternate distribution results in the formation of 4 viable gametes, each with a single complete set of genes.

Adjacent I distribution results in unequal amounts of chromatid in the gametes, and therefore they are usually not viable.

Adjacent II distribution is rare because it is nondisjuctional (homologous centromeres go to the same pole). Because of duplications and deficiencies of the genetic material, they are usually non-viable.

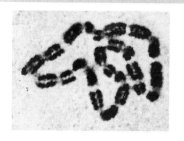

SIX RECIPROCAL TRANSLOCATIONS RESULTING IN A RING OF 12 IN *Rheo discolor* (Sax, K. J. Arnold Arboretum 16:216) ↓

mal rearrangement, oncogenes, Burkitt lymphoma, promoter swapping, unbalanced chromosomal constitution)

TRANSLOCATION COMPLEX: interchanged chromosomes of eukaryotes; members of the group are inherited as a complex that alone contributes viable gametes to the progeny.

TRANSLOCATION RING: multiple reciprocally translocated chromosomes, after terminalization, are attached end-to-end forming a ring of several chromosomes. (See illustration at the bottom of preceding page, ring bivalent, complex heterozygotes, terminalization)

TRANSLOCATION TEST, HERITABLE: see heritable translocation tests under bioassays in genetic toxicology

TRANSLOCON: the multiprotein complex (SecYp, SecGp in bacteria, Sec61p, Sbh1p, Ssh1p in yeast, sec61α, β, γ in mammals) involved in the transport of proteins through membranes. (See translocase, protein targeting, SRP, TRAM, ABC transporter, ARF)

TRANSMEMBRANE PROTEINS: they generally have three main domains; the amino terminus reaches into the cytoplasm where it usually associates with other cytosolic proteins, the hydrophobic domain generally makes seven turns within the cell membrane and the carboxylic end serves as a receptor for extracellular signals. (See cell membrane, membrane proteins, signal transduction, INT3 oncogene, KIT oncogene, MAS1 oncogene, seven membrane proteins, receptors)

TRANSMETHYLATION: see methylation of DNA

TRANSMISSION: indicates whether a particular gene or chromosome survives meiotic or post-meiotic selection and is recovered in the zygotes, embryos or adults. Transmission is generally reduced if the chromosomes have deficiencies, duplications or structural rearrangements. Monosomes and trisomes also have impaired transmission as well as all defective genes. Reduced transmission may be caused by segregation distorter genes and gametophyte factors. Genetic factors located within the cellular organelles and infectious heredity usually display uniparental (maternal) transmission. *Vertical transmission* indicates transfer by the gametes and *horizontal transmission* means spread of a condition through infectious agents without the involvement of the host genetic system. Transformation by foreign DNA gives the best documented case for horizontal transfer. Apparent homologies among genes carried by taxonomically different organisms may be interpreted as convergent evolution. The findings of transposable element (SINE) homologies in organisms widely separated taxonomically may be assumed to have origin in retroviral infections. Some of the homologous retrotransposons still carry the characteristic terminal repeats in e.g. *Vipera ammodytes* and the bovine genome. The transfers might have also been mediated by parasites such as ticks (*Ixodes*) common to a very wide variety of vertebrates (from reptiles to humans). (See megaspore competition, meiotic drive, preferential segregation, gene conversion, certation, self-incompatibility, infectious heredity, transformation, transposable elements, transposon, retroposon, retrotransposon)

TRANSMISSION DISEQUILIBRIUM TEST (TDT): is used to ascertain that a tentative association between two traits is or is not transmitted from the heterozygous parents. This test is not applicable meaningfully to a population where one of the alleles has very high frequency because in such a case the association will appear high although no causal relationship may be present. The TDT test is not a genetic linkage test. The TDT test can be used to estimate quantitatively the distribution of k offspring carriers of a mutation among r affected progeny by truncated binomials.

if $k = r$ $\quad t^k (1 - t^{s-r}) / \{1 - [t^s + (1 - t)^s]\}$

if $0 < k < r$ $\quad \binom{r}{k} t^k (1 - t)^{r-k} / \{1 - [t^s + (1 - t)^s]\}$

if $k = 0$ $\quad (1 - t)^r [1 - (1 - t)^{s-r}] / \{1 - [t^s + (1 - t)^s]\}$

where t = the segregation parameter, s = the size of the sibship.
Segregation distortion can also be determined on the basis of m carriers of the mutation among s genotyped progeny:

Transmission disequilibrium test continued

if parent is <u>typed</u>: $\binom{s}{m} t^m (1-t)^{s-m}$, if parent is <u>infered</u>: $\binom{s}{m} t^m (1-t)^{s-m} / \{1 - [t^s + (1-t)^s]\}$

(Formulas adopted from Hager, J. *et al.* 1995 Nature Genetics 9: 299)
(See association test, segregation distortion, binomial distribution, binomial probability)

TRANSMISSION GENETICS: it is actually a misnomer because genetics deals with inherited (transmitted) properties of organisms and in case there is no transmission there is no genetics. The term has been used to identify those aspects of genetics that deal only with the transmission of genes and chromosomes from parents to offspring involving also the study of segregation, recombination, mutation and other genetic phenomena without the use of biochemical and molecular analyses. It is used in the same sense as classical or Mendelian genetics. (See molecular genetics)

TRANSMITTER-GATED ION CHANNEL: converts chemical signals, received through neural synaptic gates, to electric signals. The channels in the postsynaptic cells receive the neurotransmitter. The process results in a temporary permeability change and a change in membrane potential, depending on the amount of the neurotransmitter. Subsequently, if the membrane potential is sufficient, voltage-gated cation channels may be opened. (See ion channels)

TRANSMOGRIFICATION: a complete change of living creatures such as the mythological chimeras, Satyrs, mermaids, etc. Genetic engineering and organ transplantation in medicine now brings into reality — in some way — the formerly imaginary beings; animals and plants expressing bacterial genes or vice versa. (See chimera, homeotic genes, gene fusion, allografts)

TRANSMUTATION: changing one species into another (an unproven idea); also changing one isotope into another by radioactive decay or changing the atomic number by nuclear bombardment.

TRANSOMIC: see transsomic

TRANSPEPTIDATION: transfer of an amino acid from the ribosomal A site to the P site. (See protein synthesis, aminoacyl-tRNA synthetase)

TRANSPIRATION: releasing water by evaporation through the stoma in plants, and through exhalation, through the skin, etc. in animals. (See stoma)

TRANSPLACEMENT: gene replacement with the aid of plasmid vectors. (See gene replacement vector, localized mutagenesis, targeting genes)

TRANSPLANTATION ANTIGENS: are proteins on the cell surface, encoded by the major histocompatibility (MHC) genes; they have major role in graft (allograft) rejection in mammals. (See HLA)

TRANSPLANTATION OF ORGANELLES: nuclei, isolated chromosomes, mitochondria and plastids can be transferred into other cells by cellular (protoplast) fusion and by microinjection. In case of nuclear transplantation, the resident nucleus is either destroyed (by radiation) or evicted by the use of the fungal toxins, cytochalasins. The enucleated cell is called *cytoplast* and the nucleus surrounded by small amount of cytoplasm is a *karyoplasts*. After introduction into the cytoplast another nucleus by fusion, a *reconstituted cell* is obtained. The individual components are labeled either genetically or by radioactivity or by staining or even mechanically (by 0.5 μm latex beads). The transfered organelles may express their genetic information and can be isolated efficiently and identified if selectable markers (e.g., antibiotic resistance) are used. (See cell fusion, transformation genetic, nuclear transplantation)

TRANSPORTERS: permease proteins that assist the transport of various molecules and ions through membranes. (See membranes, receptors, G-proteins, ABC transporters, CAT transporters, GLAST, PROT, GLYT, rbat/4F2hc, ASCT1)

TRANSPORTIN: a 90 kDa protein, distantly related to importin, and it mediates nuclear transport with the M9, 38 amino acid, transport signal by a mechanism different from that of the importin complex. (See importin, nuclear pore, RNA export)

TRANSPOSABLE ELEMENTS: occur in the majority of organisms; their major characteristic

Transposable elements continued

is that they are capable of changing their position within a genome or may move from one genome to another. The eukaryotic elements can be either retrotransposons (retrovirus-like) and have long direct terminal repeats (Class I.1) or do not have long terminal repeats (Class I.2) also called retroposons. Both types of Class I elements have also active or inactive reverse transcriptase. The Class II elements have inverted terminal repeats and code for transposase. (See transposable elements bacterial, transposable elements fungal, transposable elements animal, transposable elements plants, transposable elements viral, transposons, isochores, transposon footprint, second cycle mutation)

TRANSPOSABLE ELEMENTS ANIMAL: see copia, P element, LINE, SINE, hybrid dysgenesis, R2Bm, immunoglobulins

TRANSPOSABLE ELEMENTS, BACTERIAL: may be classified according to the Gram-negative host (Tn*3*, Tn*5*, Tn*7*, Tn*10*) or gram-positive host (Tn*554*, Tn*916*, Tn*1545*, Tn*551* and Tn*917*, Tn*4556*, Tn*4001*). The large transposon such as Tn*916* and Tn*1545* are capable of conjugative-like transfer to other cells. (See also insertion elements, non-plasmid conjugation)

TRANSPOSABLE ELEMENTS FUNGAL: see Ty, transposable elements yeast

TRANSPOSABLE ELEMENTS, PLANTS: see Ac-Ds, Spm (En), Dt , Mu, Tam, controling elements, somaclonal variation, retroposons, retrotransposons, transposons)

TRANSPOSABLE ELEMENTS, YEAST: see Ty [including δ, σ, τ], Ω, mating type determination, *Schizosaccharomyces pombe*

TRANSPOSASE: enzyme mediating the transfer of transposable genetic elements within the genome. The transposase function may be a part of the transposable element or it may be provided from trans position for elements that are defective in the enzyme. (See transposon)

TRANSPOSITION: transfer of a chromosomal segment to another position. The transposition may be *conservative* when the segment (transposon) is simply transferred to another location or it may be *replicative* when a newly synthesized copy is moved to another place while the original copy is still retained where it was. Transposition usually requires that both terminal repeats of the transposon would be intact. *One-sided transposition* — when one terminal repeat is lost — may still be feasible by replicative transposition. (See insertion element, transposons, transposable elements, Tn, hybrid dysgenesis, immunoglobulins, mating type determination in yeast, *Schizosaccharomyces pombe*)

TRANSPOSITION IMMUNITY: the transposable element does not move into a replicon which already carries another transposon or the inverted terminal repeats of a transposon. Transposition immunity is overcome by high expression of the transposase or defects in the terminal repeats of the resident transposon. Transposition of Tn*7* is inhibited by the presence of Tn*7* sequences within the same replicon. (See transposition, transposable elements, Tn*3*)

TRANSPOSON: see Tn, transposable elements, retroposon, retrotransposon

TRANSPOSON FOOTPRINT: short insertions left behind in the original target after the transposon exited from the sequences. These nucleotides may be the consequence of the genetic repair after excision, e.g. sequence before the insertion CTGGTGGC
                              after excision CTGGTGGC-TGGTGGC or
                                                 CTGGTGG**gc**TGGTGGC

(See transposable elements, second cycle mutation)

TRANSPOSON MUTAGENESIS: transposable and insertion elements can move in the genome (mobile genetic elements) and may insert within the boundary of genes. Such an insertion, by virtue of interrupting the normal reading frame, may eliminate or reduce or alter the expression of the gene and the event is recognized as a mutation. Recently, it has been shown that many of the insertions do not lead to observable change in the expression of the genes or their effect is minimal and their presence is revealed only by sequencing of the target loci. Mutations so generated have great advantage for genetic analysis because, the insertion serves as a tag on the gene permitting its isolation and molecular study. Many of the insertions are retrotransposons and in plants commonly located within introns. In animals the comparable elements are frequently within intergenic regions. (See also gene tagging, insertional mutation, transformation, labeling, gene isolation, plasmid rescue, suicide vector, retrotransposon, retroposon)

TRANSPOSON TAGGING: tagging a gene by the insertion of a transposon. The insertion disrupts the continuity of the gene causing a mutation and thereby the success of the tagging is identified by the phenotype. Subsequently, the gene can be isolated by using the labeled transposon as a probe. (See transposon mutagenesis, probe, gene isolation)

TRANSPOSONS, ANIMAL: see transposable elements animal

TRANSPOSONS, BACTERIAL: DNA segments which can insert into several sites of the genome and contain gene(s) besides those required for insertion; they are generally longer than 2 kilobases. It has been suggested that the introns of eukaryotic cells might have been introduced into the genes by broad host-range phages or transposons. (See Tn, insertion elements, accessory proteins)

TRANSPOSONS, FUNGAL; see Ty

TRANSPOSONS-CONTROLING ELEMENTS, PLANT: the major transposable elements in **maize** are *Ac-Ds, Spm, Dt, Mu*. Besides these there are much less well defined controling elements: *Bg (Bergamo), Fcu (Factor Cuna), Mr (Mutator of R), Mrh* (Mutator of *a1-m-rh*), *Mst (Modifier of allele R-st), Mut* (controling element of *bz1-m-rh*), *Cy* (regulatory element of *bz1-rcy*). (See controlling elements, Ac-Ds, Spm, Dt, Mu, Tam, Ta)

TRANS-SENSING: is an interaction between somatically "paired" homologous chromosomes affecting gene expression in diploids.

TRANSSEXUAL: has an innate desire to change her/his anatomical sex to the other form. The volume of the central subdivision of the bed nucleus of the strial terminals of the brain is larger in males than in females. In male-to-female transsexuals this particular area of the brain is female sized. Thus this anatomical condition may be a determining factor for transsexualism and sex hormon production. (See sex determination)

TRANSSOMIC LINE: carries microinjected chromosomal fragments in the cell nucleus.

TRANSSPLICING: splicing together exons that are not adjacent within the boundary of the gene but are remotely positioned and may be even in different chromosomes. (See introns, regulation of gene activity, SL1, SL2)

TRANS-TRANSLATION: may occur if the stop codon and preceding end of the mRNA is lost and the translation is completed by using another template RNA. (See recoding)

TRANS-VECTION: see cis-vection, trans-acting element, cis-trans effect

TRANSVECTION: a synapsis-dependent modification of activity in "pseudoalleles". In paired chromosomes genes in trans position may affect the expression of an allele. It has also been called trans-sensing. It has been interpreted as the result of interaction between DNA binding proteins attached to the two synapsed promoters. (See also co-suppression, RIP, pseudoalleles)

TRANSVERSION MISMATCH: is a mispairing involving either two purines or two pyrimidines, (See mismatch, transition mismatch)

TRANSVERSION MUTATION: substitution of a purine for a pyrimidine, or a pyrimidine for a purine in the genetic material. (See base substitutions, base substitution mutations)

TRAP: same as CD40 ligand

TRAP (tryptophan RNA-binding attenuation protein): in *Bacillus subtilis*. When activated by L-tryptophan, this protein binds to the mRNA leader causing a termination of transcription. This is in contrast to the situation in *E. coli* where the attenuation is brought about by an altered secondary structure of the nascent RNA transcript. Some sort of attenuation takes place also in eukaryotes but the mechanism of that is not entirely clear yet.

The TRAP protein containing 11 identical subunits is coded by the *mtrB* gene in *B. subtilis* and it binds single-stranded RNA. The β-sheet subunits form a wheel-like structure with a hole in the center and tryptophan is attached to the clefts between the β-sheets resulting in circularization of the RNA target in which eleven U/GAG repeats are bound to the surface of this ondecamer (11 subunit) protein modified by tryptophan. Similar mechanisms occur also in some other bacterial species. (See tryptophan operon, attenuation region)

TRAPPING PROMOTERS: when a promoterless structural gene is inserted into a host genome with the assistance of a transformation vector, the inserted sequences may become "in-frame" located within the host chromosome and a host promoter may drive the transcription of the foreign gene that in the vector had no promoter. Since the promoter and upstream regulatory elements control the transcription, directly or in association with transcription factors, the expression pattern (timing, tissue site) may be altered and the intensity of expression may be increased or decreased according to the nature of the promoter. (See also gene fusion, transcriptional gene fusion vectors, translational gene fusion vectors, read-through proteins)

TRAVELER'S DIARRHEA: is caused generally by bacterial (*E. coli, Salmonella*) infection.

TRCF (transcription repair coupling factor): a eukaryotic repair helicase corresponding to *UvrA* in *E. coli*; it is encoded by yeast gene *MFD* (mutation frequency decline).

TRE: thyroid hormone responsive element in the rat growth hormone gene with a consensus of AGGTCA....TGACCT. (See also ERBA, hormone response elements, regulation of gene activity)

TREACHER COLLINS SYNDROME: a dominant (human chromosome 5q32-q33) complex defect of the face.

TREADMILL EVOLUTION: see Red Queen hypothesis

TREADMILLING: is the addition of microtubule subunits to the growing plus end and loss of subunits at the minus end. (See microtubules)

TRIACYLGLYCEROLS (synonym triglycerides): are uncharged esters of glycerol and thus called also neutral fats. Triglicerides are energy storage compounds and contain four times as much energy in the human body than all the proteins combined. By lipase, they are hydrolized into glycerol and fatty acids. Lipolysis is controlled by cAMP in the adipose (fat) cells. Insulin inhibits lipolysis. (See epinephrine, norepinephrine, glucagon, adrenocorticotrophic hormone, fatty acids, triglycerol)

TRIBE: descendants of a female progenitor or a taxonomic group below a suborder or a group of primitive people with a common origin, culture and social system.

TRIBOLIUM: flour beetle.

TRiC: is a ring complex of eukaryotic chaperonin. (See chaperone, chaperonin)

TRICARBOXYLIC ACID CYCLE: see Krebs cycle

TRICHOGEN CELL: see tormogen

TRICHOGYNE: the hypha emanating from the protoperithecium, to what the conidia are attached prior to fertilization in some ascomycetes. (See hypha, conidia, ascomycete)

TRICHOME: hair or filament in plants, algae and animals; in some plants the hairs may be single filaments or they may have tripartite termini.

TRICHORHINOPHALANGEAL SYNDROME (TRPS1): dominant or recessive human chromosome 8q24 defect involving multiple exostoses (bone projections) and mental retardation. (See also Langer-Giedion syndrome)

TRICHOTHIODYSTROPHY: is a collective name for autosomal recessive human diseases involving low-sulfur abnormalities of the hair. The *Tay syndrome* involves also ichthyosiform erythroderma (scaly red skin), mental and growth retardation, etc. The *Pollitt syndrome* (trichorrhexis nodosa or trichothiodystrophy neurocutaneus) displays low cystin content of the hair, and the nails, and the head and the nervous system are also defective. Xeroderma pigmentosum IV includes trichothiodystrophy and sun- and UV-sensitivity. Also called PIBIDS. (See hair-brain syndrome, xeroderma pigmentosum, excision repair)

TRF: see T cell replacing factor, a lymphokine

TRF (thyrotropic release factor): see corticotropin

TRF1: 60 kDA telomeric TTAGGG repeat binding protein protects the telomere and facilitates its interaction with the telomerase enzyme. (See telomere, telomerase, RAP)

Trg: bacterial transducer protein with attraction to ribose and galactose.

TRICARBOXYLIC ACID CYCLE: see Krebs-Szentgyörgyi cycle

TRIGLYCERIDE: same as triaglycerol

TRIGLYCERIDEMIA: see familial hypertriglyceridemia

TRIHYBRID CROSS: the parental forms are homozygous altogether for 3 allelic pairs at (unlinked) loci, e.g., AABBdd x aabbDD and therefore in the F_2 8 phenotypic classes may be distinguished. (See gametic array, Mendelian segregation)

TRI-ISOSOMIC: in wheat, 20"+i"', 2n=43, ["=disomic, "'=trisomic, i=isosomic]

TRILLIUM : species have one of the largest normal chromosomes (2n = 10) in plants, about 50 times larger than *Arabidopsis*, a plant with one of the smallest chromosomes (2n = 10).

TRIMESTER: period of three months; the human pregnancy of 9 months includes 3 trimesters.

TRIMMING: the processing of the primary RNA transcripts to functional mRNA or ribosomal and tRNA. The cleavage of pre-rRNA transcripts by RNase III into 16S, 23S and 5S rRNA as well as into the tRNAs contained within the spacer sequences of the co-transcripts. The cleavage takes place at the duplex sequences forming the stem of the rRNA loops. (See post-transcriptional processing, *rrn* genes for diagram, introns)

TRINOMIAL DISTRIBUTION: $(1 + 2 + 1)^n$ can be expanded to predict the segregation of the genotypic classes (note that the quotients within parentheses must not be added!)

$$1(1+2)^n + \frac{n!}{1(n-1)!}(1+2)^{n-1} + ... + \frac{n!}{(n-1)!1}(1+2)^{n-(n-1)} + 1(1+2)^{n-n}$$

An example for three pairs of alleles:

$$1(1+2)^3 + + \frac{3!}{1(2!)}(1+2)^2 + 1 =$$

$$1 + (3 \times 2) + (3 \times 4) + 8 + 3 \times (1 + 2 + 2 + 4) + 3 \times (1 + 2) + 1$$

When rewritten in a symmetrical distribution:

1:2:1:2:4:2:1:2:1:2:4:2:4:8:4:2:4:2:1:2:1:2:4:2:1:2:1

the 27 terms indicate that triple heterozygotes are 8, double heterozygotes 4, and single heterozygotes are 2 in a distribution in compliance with Mendel's law. (See binomial, multinomials)

TRINUCLEOTIDE REPEATS: in >ten neurodegenerative diseases CAG (glutamine codons) are repeated many times. An unusual common feature of these diseases that in successive generations the symptoms appear earlier and with greater severity (anticipation) as gain-of-function mutations. The repeats form a hairpin structure and interfere with DNA replication. Also, the CpG sequences are likely to be methylated. The nature of the repeats may vary and may involve CGG, GCC, CAG, CTG sequences. These repeats may expand (from a few [5-50] in the normal to hundreds of copies) in an unstable manner in the 5' untranslated region of the FMR1 gene and cause translational suppression by stalling on the 40S ribosomal RNA. In *E. coli*, the larger expansions occur predominantly when the CTG trinucleotides are in the leading strands and deletions are mainly on the opposite lagging strands. (See anticipation, Huntington's chorea, Machado-Joseph disease, ataxia, Kennedy disease, dentatorubral-pallidolyusian atrophy, Jacobsen syndrome, fragile sites, fragile X, myotonic dystrophy, methylation of DNA, translation repressor proteins, human intelligence, FMR1 mutation, schizophrenia)

TRIOSE: sugar with 3 carbon backbone.

TRIOSEPHOSPHATE ISOMERASE DEFICIENCY (TPI1): is encoded in human chromosome 12p13 but pseudogenes seem to be present at other locations. The level of activity of the enzyme varies; null mutations are not expected to be viable since this is a key enzyme in the glycolytic pathway. The symptoms may be quite general weakness, neurological impairment, anemia, recurrent infections, etc. (See glycolysis)

Trip1 (thyroid hormone receptor): see Sug1

TRIPLE HELIX FORMATION: oligonucleotides may bind to polypurine-polypyrimidine tracts in the major groove of the DNA helix by Hoogsteen or reverse Hoogsteen bonding and prevent the access of transcription factors. This may block transcription and cleave the DNA but may

Triple helix formation continued

enhance repair DNA synthesis. In the triplex sequences, mutation rate in SV40 increased more than an order of magnitude in the suppressor gene, supFG1, employed as reporter, with 30 nucleotide long AG sequences (AG30). Shorter sequences or oligonucleotides of all four bases were either not or were much less effective. The triplex structure in xeroderma pigmentosum or in the Cockayne syndrome cells was not effective for mutation enhancement, indicating the requirement of excision repair for the events. (See DNA kinking, DNA repair, SV40, supF, xeroderma pigmentosum, Cockayne syndrome, Hoogsteen pairing, antisense technologies, psoralen dyes)

TRIPLET BINDING ASSAY: a historically important method to determine the meaning of genetic triplet codons. A single type of radioactively labeled amino acid, charged to the cognate tRNA was allowed to recognize and bind to ribosomes with mRNA attached. Each type of charged tRNA then recognized only their code words and the ribosomes were then trapped on the surface of a filter. Synthetic polynucleotides (mRNA) of known base composition retained only the cognate aminoacylated tRNA and thus providing the base composition and sequence of the true coding triplets. (See diagram at left, genetic code)

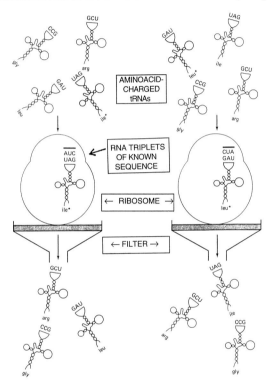

TRIPLET CODE: see genetic code

TRIPLEX: a three-stranded nucleic acid structure, e.g., an RNA oligonucleotide binding to double-stranded DNA and resulting in antisense effects. Triplex strands occur transiently in genetic recombination. (See antisense RNA, Hoogsteen pairing, peptide nucleic acid)

TRIPLEX: a polyploid with 3 dominant alleles at a gene locus. (See autopolyploid, trisomy)

TRIPLICATE GENES: convey identical or very similar phenotype and in a diploid, when segregate independently, display an F_2 phenotypic ratio of 63 dominant and 1 recessive.

TRIPLOID: a cell or organism with three identical genomes. Triploids (3x) are obtained when a (4x) is crossed with a diploid (2x). The majority of edible bananas, several cherry and apple varieties, many sterile ornamentals (chrysanthemums, hyacynths) are triploid. The seedles watermelons, produced by crossing tetraploids with diploids are triploids and have commercial value. Triploid sugar beets are in large scale agricultural production because of ≈10% or higher sugar yield per acre than the parental diploid varieties. (See also trisomic)

TRIPLO-X: females (XXX) occur in about 0.0008 of human births. Their phenotype is close to normal and they can conceive. Yet they are somewhat below average in physical and mental abilities although they tend to be somewhat tall. With an increasing number of X-chromosomes beyond 3 the adverse effects are further aggravated. (See sex determination, sex chromosomal anomalies in humans, trisomy, Turner syndrome)

Tripsacum: see maize

Tris-HCl BUFFER: contains tris[hydroxymethyl]aminomethane and hydrochloric acid; it is used in various dilutions within the pH range 7.2-9.1.

TRISKELION: three-legged proteins on the surface of vesicles built of three clathrin and three smaller proteins.

TRISOMIC: a cell or organism with one or more chromosomes (but not all) represented three times. (See trisomy, trisomic analysis, triploid)

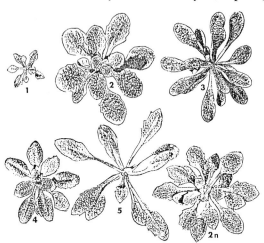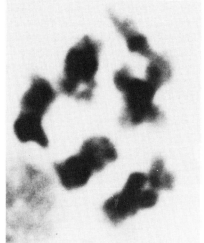

Left: THE FOUR PRIMARY TRISOMICS (1 - 4) AND A TELOTRISOMIC (5) OF *ARABIDOPSIS*. AT THE LOWER RIGHT CORNER IS A NORMAL DISOMIC INDIVIDUAL. ALL ARE IN COLUMBIA WILD TYPE BACKGROUND, GROWN UNDER SHORT DAILY LIGHT PERIODS. At right: THE CHROMOSOME COMPLEMENT OF A PRIMARY TRISOMIC. NOTE THE TRIVALENT ASSOCIATION AT THE UPPER RIGHT AREA. (The photomicrograph is the courtesy of Dr. Lotti Steinitz-Sears)

TRISOMIC ANALYSIS: trisomics have one or more chromosomes in triplicate (*AAA*) and thus produce both disomic (*AA*) and monosomic (*A*) gametes for gene loci in those chromosomes. Male transmission of the disomic gametes is usually poor, it rarely exceeds 1%. The transmission of disomic gametes through females varies according to the chromosome (genes) involved and depends also on environmental conditions; most commonly it is 1/4 or 1/3 of the normal (monosomic) gametes. The few viable human sex-chromosome trisomics (XXX, XXY, XYY) are either sterile or have very poor fertility because of the failure of normal development of the ovaries and testes (gonadal dysgenesis). From a single alternative allelic pair (*A/a*), disomics produce either *A* or *a* gametes. In contrast, trisomics potentially produce a maximum of five kinds of gametes, three disomic and two monosomic. The frequency of these five types of gametes depends also on the distance of the gene from the centromere, i.e. whether *chromosome segregation* (no crossing over between gene and centromere) or *maximal equational segregation* (random recombination between gene and centromere) occurs.

GAMETIC OUTPUT OF TRISOMICS

GENOTYPE	CHROMOSOME SEGREGATION					MAXIMAL EQUATIONAL SEGREGATION				
↓	AA	Aa	aa	A	a	AA	Aa	aa	A	a
AAa (duplex)	1	2	0	2	1	5	6	1	8	4
Aaa (simplex)	0	2	1	1	2	1	6	5	4	8

In order to obtain two identical recessive alleles from a duplex, the chromosomes must form trivalents and recombination must take place between the gene and the centromere and then at anaphase I distribution must move the two exchanged chromosomes toward the functional megaspore (equivalent to secondary oocyte in animals). At the most favorable coincidence of these events, in only half of the cases can we expect the two identical alleles to move into the same megaspore (equivalent to egg in animals). Thus the maximal chance of having a tetrad with a double recessive gamete (*aa*) will be $1/3 \times 1/2 = 1/6$. These fractions are based on 1 reductional and 2 equational disjunctions at meiosis I (1/3), and 2 alternative disjunctions at anaphase II (1/2) are the determining factors of the outcome.

Trisomic analysis continued

The phenotypic segregation can be derived from the gametic output by random combinations. One factor, transmission difference between the female and male gametes, is generally seriously alters the theoretically expected proportions.

PHENOTYPIC PROPORTIONS IN THE F_2 OF TRISOMICS (disomic and trisomic pooled)

TRANSMISSION‡ & GENOTYPE	CHROMOSOME SEGREGATION			MAXIMAL EQUATIONAL SEGREGATION		
	DOM.	REC.	aaa (%)	DOM.	REC.	aaa (%)
DUPLEX (*AAa*)						
MALE AND FEMALE	35	1	0	22.04	1	1.39¶
FEMALE ONLY	17	1	0	13.40	1	1.39
NONE	8	1	0	8	1	0.00
SIMPLEX (*Aaa*)						
MALE AND FEMALE	3	1	11§	2.41	1	13.9*
FEMALE ONLY	2	1	11	1.77	1	13.9
NONE	5	4	0	1.25	1	0

‡ refers to transmission of the disomic gametes   ¶ 1/576 tetrasomics (*aaaa*) are not included
§ 1/36 tetrasomic (*aaaa*) is not included   *25/576 tetrasomics (*aaaa*) are not included

PHENOTYPIC RATIOS IN TESTCROSS PROGENIES (trisomics + disomics)

GENOTYPES OF CROSS ↓	ONLY FEMALE TRANSMISSION OF DISOMIC GAMETES	NO TRANSMISSION AT ALL OF DISOMIC GAMETES
	Dom. : Rec.	Dom. : Rec.
AAa x aa	5 : 1	2 : 1
Aaa x aa	1 : 1	1 : 2

The genetic behavior of all chromosomes not present in triplicate in the trisomics is consistent with disomy. When the trisomic individuals are phenotypically distinguishable per se (do not require cytological identification of the chromosomal constitution), trisomy may be used very effectively to assign genes to chromosomes, irrespective of their linkage relationship. The segregation will not be 3:1 but it will vary (most commonly) between 17:1 to 8:1 as predicted by the 1st table above. In this case the gene *a* is in the chromosome which is triplicated. Genes in the disomic set of chromosomes is expected to display 3:1 segregation. In case telo trisomics are crossed with disomics (——o——*a*——, ——o——*a*——, o——*A*—— x normal disomic *a/a*) the dominant allele will be very rare among the diploid offspring but it may be expressed in every individual of 2n + telochromosome. Mapping can be done also by properly constructed isotrisomics, tertiary trisomics and compensating trisomics. (See disomic, monosomic, gonadal dysgenesis, chromosome segregation, maximal equational segregation, mapping by dosage effect, trisomy, Down syndrome, Pätau syndrome, Edwards syndrome)

TRISOMIC, COMPLEMENTING: can be of three types; one is an apparent trisomic only because there is a normal biarmed chromosome and the two other chromosomes are actually telosomes, representing the left and right arms of the intact, normal chromosome. The second complementing type also involves one biarmed normal chromosome but the two telosomes represent a pair of identical arms thus in essence this case is very similar to a secondary trisomic, having three identical arms and one single different arm. The third type is when there is a normal biarmed chromosome, the second chromosome has an arm identical to the first but its second arm is a translocated segment from a non-homologous chromosome, the third chromosome has a different translocated arm linked to the centromere and to an arm of a normal chromosome. (See trisomic analysis)

TRISOMIC, PRIMARY: chromosome number is 2n + 1, and the triplicate homologous chromosomes are structurally normal. (See trisomic analysis)

TRISOMIC, SECONDARY: the extra chromosome is an isochromosome, i.e. its two arms are identical; such a chromosome is an isochromosome. (See trisomic analysis, isochromosome)

TRISOMIC, TERTIARY: the extra chromosome is involved in a reciprocal translocation and is partly homologous with two standard chromosomes. (See trisomic analysis)

TRISOMY: involves one or more but not all chromosomes of a genome in triplicate; trisomics are aneuploids. They may arise by selfing triploids when some of the extra chromosomes are lost. Nondisjunction during meiosis may also lead to trisomic progeny. Trisomy may exist in various forms. The phenotype of the trisomics varies depending what genes are in the trisome. The phenotype may be very close, almost indistinguishable from that of normal disomics, and in other cases it may be lethal. Only a few of the autosomal human trisomies permit growth and development beyond infancy. Trisomy 9 allows near normal life expectancy although involves developmental and mental retardation. Individuals with trisomy 21 (Down's syndrome) also reach adult age but they remain mentally subnormal. Trisomy 22 may exist in mosaic form and the afflicted individuals are retarded in growth and mental abilities. About 3/4 of the trisomy 8 cases are mosaics, mentally retarded, and affected by head abnormalities to a variable degree, according to the extent of mosaicism. Trisomics for chromosome 11 have a variable extra long arm of chromosome 11 and accordingly with brain and other internal organ damage of variable extent. Trisomics for the long arm of chromosome 3 have very short life and variable abnormalities in internal organs and body development and severe mental retardation. Trisomy for the short arm of chromosome 4 is characterized by serious malformations of the brain, head, extremities, genitalia and usually die early. About 1/3 of the abortuses, due to autosomal trisomy, involves chromosome 16; these are never carried to term. Trisomy 13 (Pätau's syndrome) and trisomy 18 (Edwards syndrome) are live-born but die very early. All other autosomal human trisomy leads to abortion at various stages after conception. Sex-chromosomal trisomy generally does not compromise viability of humans yet they are usually sterile, display gonadal dysgenesis and a variety of physical and most commonly mental retardation as well. Trisomy is useful in plants for assigning genes to chromosomes. (See trisomics, trisomic analysis, Down syndrome, Edwards, Pätau's, Klinefelter syndromes, sex determination, sex-chromosomal anomalies in humans, cat-eye syndrome)

TRITANOPIA: see color blindness

TRITELOSOMIC: in wheat, 20"+t''', 2n=43, ["=disomic, '''=trisomic, t=telosomic]

TRITICALE: is a synthetic species produced by crossing wheat (*Triticum*) and rye (*Secale*) and doubling the chromosome number. The wheat parent may be tetraploid (2n = 28) or hexaploid (2n = 42) and the rye is generally diploid (2n = 14). Therefore the amphidiploids may have the chromosome number of either 2n = 42 (hexaploid) or 2n = 56 (octaploid). As a crop variety the former is used. On some soils suitable primarily for rye, the *Triticale* may provide grains with a milling quality approaching that of wheat. The new hybrids generally have shrunken kernels because some rye genes cause a poor development of the endosperm. These problems have been largely solved by breeding efforts and the commercial varieties have rather plump grains. (See *Triticum*, *Secale*, amphidiploid, chromosome doubling)

TRITICOLSECALE: see *Triticale*

TRITICUM (common wheat and relatives): form a series of allopolyploids. The common bread

WHEAT SPECIES: From right to left : *Triticum monococcum, T. turgidum (durum), T. timopheevi, T. aestivum* (Chinese Spring [most widely used for cytogenetic studies]), *T. compactum, T. spelta.* (Courtesy of Dr. Carlos Alonso-Arnedo)

wheat is *T. aestivum* (both winter and spring forms) and the *T. durum (T. turgidum)* are used mainly for various pastas (macaroni, spaghetti, etc.). The wheat kernel contains ca. 60-80% carbohydrates (starch), 8-15% protein that is usually low in lysine, tryptophan and methionine. Some of the wild relatives of the cultivated wheats produce kernels with higher amounts of protein than the commercial varie-

***Triticum* continued**

ties and these favorable traits can be tranfered by chromosomal engineering. Wheats are the staple food for about 1/3 of the human population.

DIPLOIDS (2n = 14)	genomic formula	former designation* and species, respectively
T. monococcum	A	T. boeoticum*, T. aegilopoides, T. thoudar. T. urartu
T. speltoides	S (= G?)	Aegilops speltoides*, Aegilops ligustica*
T. bicorne	S^b	Aegilops bicornis*
T. searsii	S^s	Aegilops searsii*
T. longissimum	S^l	Aegilops longissima* (Aegilops sharonensis ?)
T. tauschii	D	Aegilops squarrosa*, T. aegilops
TETRAPLOIDS (2n = 28)		
T. turgidum	AB	T. dicoccoides*, T. dicoccon, T. durum*, T. polonicum*, T. carthlicum*, T. persicum*
T. timopheevi	AG	T. araraticum, T. dicoccoides var. nudiglumis T. armeniacum
HEXAPLOIDS		
T. aestivum	ABD	T. vulgare*, T. spelta*, T. macha*, T. sphaerococcum*, T. vavilovii*

TRITIUM (H^3): see isotopes

TRITON X-100: is a non-ionic detergent. It used for the extraction of proteins and solubilization of biological materials. The reduced form is prefered when spectrophotometric measurement is required.

TRIVALENT: in a trisomic or polyploid individuals three chromosomes may associate during meiotic prophase. In the trivalent association at any particular place, only two homologous chromosomes can pair, however, at any particular time and site.

trk (TRK) **ONCOGENE**: its product in membranes is a receptor (TRKA [tyrosin kinase activity]) for NGF (nerve growth factor). It is localized to human chromosome 1q23-q24. (See neutrorophin)

tRNA (transfer RNA): the shortest RNA molecule in the cell (ca. 3.8S) consisting of about 76 to 86 nucleotides. They carry amino acids to the ribosomes during protein synthesis. The majority of cells have 40 to 60 types of tRNAs because most of the 61 sense codons have their own tRNA in the eukaryotic cytosol. The tRNAs which accept the same amino acid are *isoaccepting tRNAs*. In the human mitochondria there are only 22 different tRNAs and their number in the plant chloroplasts is about 30. It is called frequently an adaptor molecule because it adapts the genetic code for the formation of the primary structure of protein. Rarely (ca. 1/3000), a tRNA is charged with the wrong amino acid and in these cases the complex is usually disrupted and the tRNA is recycled. The tRNA genes may occur in between ribosomal RNA genes (promoter - 16S - -tRNA - 23S - 5S - tRNA - tRNA...) or they frequently form independent clusters of different tRNA genes (promoter - $tRNA^1$ - $tRNA^2$ - $tRNA^3$ - •). The different cognate tRNAs groups are generally identified by a superscript of the appropriate amino acid, e.g., $tRNA^{Met}$. Sometimes individual tRNA genes may be present in multiple copies. The tRNA gene clusters are transcribed into large primary molecules that require successive cleavage and trimming to form the mature tRNA. The endonuclease, RNase P (a ribozyme) recognizes the primary transcript whether it is a single tRNA sequence or a cluster of rRNA or tRNA genes and cleaves at the 5' terminus where the tRNA begins. For the tRNA to become functional, RNase D cuts at the 3'-end, and stops at a CCA sequence if any, or the 3'-end receives through post-transcriptional synthesis (by tRNA nucleotidyl synthetase) one, two or three bases and thus it is always finished by a 3'- CCA^{-OH} sequence. This 3'-OH end becomes the amino acid attachment site of the tRNA (see transfer RNA). During the process of maturation, through modification of the original bases, thiouridine (S4U), pseudouridine (ψ), ribothymidine, dihydrouridine (DHU), inosine (I), 1-methylguanine (m^1G), 1-dimethyl-

tRNA continued

guanine (m^1dG), N^6-isopentenyl adenosine (i^6A) may be formerd within the tRNA sequence. (See transfer RNA, amino acylation, aminoacyl-tRNA synthetase, protein synthesis, trimming, ribozyme, intron, pseudouridine, ribothymidine, dihydrouridine, isopentenyladenosine, hypoxanthine [for inosine])

tRNA DEACYLASE: cuts off the tRNA from the polypeptide chain after the complex reached the P site on the ribosome. (See protein synthesis, aminoacyl-tRNA synthetase)

tRNA MIMICRY: see translation termination

tRNA NUCLEOTIDYL TRANSFERASE: post-transcriptionally adds the C-C-A-3'-OH sequence to the amino acid accepting arm of tRNA. (See tRNA)

TROPHOBLAST: the surface cell layers of the blastocyst embryo.

TROPHOZOITE: growing and actively metabolizing cells of unicellular organisms vs. the cysts.

TROPIC: indicates after a word that something is aimed at it. e.g. T-tropic means that a virus targets T cells or directed at a site in some other way.

TROPIC HORMONE: stimulates the secretion of another hormone at another location.

TROPISM: growth of plants in the direction of some external factors.

TROPOMODULIN: maintains actin filament growth by capping the pointed ends of the actin filaments.

TROPOMYOSIN: see troponin

TROPONIN: Ca^{++}-binding, regulatory polypeptides in the muscle tissue. Troponin C binds four molecules of calcium. Troponin I has an inhibitory effect on myosin and actin, and troponin I binds tropomyosin, an accessory protein. In the relaxed muscles, troponin I binds to actin and moves tropomyosin to the position where actin and myosin would interact at muscle contraction. When the level of Ca^{2+} is high enough troponin I action is blocked so myosin can bind actin again allowing the muscle to contract. Troponin C is related in function to calmodulin. (See calmodulin, signal transduction)

Trp: see tryptophan operon, tryptophan

TRUE BREEDING: absence of segregation in the offspring.

TRUNCATION: is a cut-off point; e.g. in artificial selection individuals before or beyond an arbitrarily determined point are discarded or maintained, respectively. Also a cloned eukaryotic-gene without a polyadenylation signal or an incomplete upstream control element may be called truncated. (See selection)

TRYPANOSOMAS : are protozoa spread by the tse-tse fly (*Glossina*) and cause sleeping sickness. Various developmental stages are distinguished on the basis of the relative position of the flagella (basal flagella: trypomastigote, median: epimastigote, apical: promastigote, no flagella: amastigote). The *Trypanosomas* may reach a level of 10^9 -10^{10} individuals per mL blood of mammals. The chromosomes are small and variable in number because in addition to the stable chromosomes minichromosomes are also found. The ca. 100 minichromosomes (50 to 150 kb) contain open reading frames for the variable surface glycoproteins (VSG). These sequences are transcribed only when transposed to expression sites in the 0.2 to 6 megabase long 20 maxichromosomes. Since some of the genes may be present in more than single copy in different chromosomes, these organisms may resemble allodiploids. The genes do not have introns and some of the tandem array of genes are transcribed as long polycistronic pre-mRNA. At the 5' end each primary transcript is capped by a 39 nucleotide, spliced leader RNA (SLRNA). This cap itself is transcribed as a 139 base sequence but the 100 base sequence is not used in this trans-splicing reaction. The *Trypanosomas* have homologs of the mammalian U2, U4 and U5 small nuclear RNAs in the form of ribonucleoprotein particles (RNP). The mRNAs are polyadenylated. In *T. brucei* there is a family of about 1,000 genes involved in the recurring production of a great repertory of *variant antigen type* (VAT). At each flare-up of division of the parasite, it switches on the production of a different type of antigen (serodeme). The more than million molecules of antigens on the surface of its cells are the about 60 kDa *variable surface glycoproteins* (VSG). All the different VSGs have at least one

TRYPANOSOMAS continued

N-linked oligosaccharide and several cysteine residues near the N-end, and some similarities within the 50-100 amino acids at the C-terminus. Because of the rapid switches in production to new antigens, the vertebrate cell's immunological surveillance system cannot adapt rapidly enough to contain the infection. In chronic infections 50-100 different antigens may be produced. Although a particular *Trypanosoma* has only 1/million or less chance per cell division to switch to the production of a different antigen yet because of their immense number, the parasite has a good chance to escape the immune system of the mammalian host. At any one time only one of the VSG genes is expressed in the protozoon. The switching is not a direct response to the host antibody rather that acts only as a selecting mechanism. The switching of transcription may require a shut-off of a gene or a gene conversion type process takes place. The expressed surface antigen gene, the so called *expression-linked copy* (ELC), is always located in the telomeric regions of the chromosomes. Its promoter is located, however, 50 kbp upstream. In addition, non-VSG genes are also located in the expression region; these are called *expression site associated genes* (ESAG). The expression of these basic, silent genes requires a transposition into the activation region, in a manner similar to the mating type switching in yeast. The switching apparently depends on expression-linked copies (ELC) of the **m**ini **e**xon **d**ependent transcription into 140 base long eukaryotic type mRNA (called also medRNA). Because of this effective switching, there are serious problems in developing vaccines against the *Trypanosomas*. Mutants can be produced that turn on simultaneously more than one type surface glycoproteins. (See kinetosome, *Leishmania*, mating type determination in yeast, flagellar antigen) TRYPANOSOMA ➔

TRYPOMASTIGOTE: see *Trypanosoma*

TRYPSIN: a proteolytic enzyme synthesized as an enzymatically inactive zymogen, trypsinogen, that is activated by proteolytic cleavage. It is inactivated by the 6,000 M_r pancreatic trypsin inhibitor. Trypsin specifically cleaves polypetides at the carbonyl sides of Lys and Arg. Ser 195 and His 57 are at its active site.

TRYPSINOGEN DEFICIENCY (TRY1): is a 7q22-ter recessive hypoproteinemia and insufficient amino acid level. It is very similar to enterokinase deficiency. (See enterokinase deficiency)

TRYPTIC PEPTIDES: the products of digestion of a protein by trypsin. (See trypsin)

TRYPTOPHAN: is an essential aromatic amino acid (MW 204.22), soluble in dilute alkali, insoluble in acids and it is degraded when heated in acids. Its biosynthetic pathway (*with enzymes involved in parenthesis*): Chorismate-> (*anthranilate synthase*) -> Anthranilate -> (*anthranilate-phosphoribosyltransferase*) -> N-(5'-Phosphoribosyl)-anthranilate -> (*N-(5'-phosphoribosyl)-anthranilate isomerase*) -> Enol-1-*o*-carboxyphenylamino-1-deoxyribulose phosphate -> (*indole-3-glycerol phosphate synthase*) -> Indole-3-glycerol phosphate -> (*tryptophan synthase*) -> Tryptophan. In *E. coli* the first two enzymes shown constitute a single anthranilate synthase complex. Tryptophan is converted to formylkynurenine by *tryptophan dioxigenase* (tryptophan pyrrolase). Through the action of an *aminotransferase* tryptophan gives rise to indole-3-pyruvate which after *decarboxylation* forms indole-3-acetic acid, one of the most important plant hormone (auxin). Tryptophan and phenylalanine contribute to the formation of lignins, tannins, alkaloids (morphine), cinnamon oil, cloves, vanilla, nutmeg, etc. (See also chorismate, tyrosine, phenylalanine, melanin, plant hormones, pigments in animals, Hartnup disease, *tryptophan* operon, fragrances)

***TRYPTOPHAN* OPERON**: contains five structural genes, in *E. coli* they have been mapped (at 27 min) in exactly the same order as their sequence of action in the biosynthetic path (see tryptophan). This operon has a principal promoter and a secondary low-efficiency one. Between the operator and the proximal gene there is a 162 bp leader sequence (*trpL*) including an attenuator site *a*. (See temperature sensitive controling sequences of transcription). When there is sufficient amount of tryptophan in the cell and all the tRNA^{Trp} are charged with the amino acids, transcription of the leader sequence stops at base 140 (shown by the bar p. 1076). Thus, synthesis of the specific mRNA temporarily ceases when there is no immediate

Tryptophan operon continued

need for tryptophan. The primary sequence of the leader sequence is as shown below (the underlined sequences indicate ribosome binding, codons in outline):

```
                           Met                              Trp  Trp
pppAAGUUCACGUAAAAGGGUAUCGACAAUG AAAGCAAUUUUCGUACUGAAAGGUUGGUGG CGCACUUC
Opal           end of pause↓         nucleotides beyond the box are not assembled if attenuation works
UGA AACGGGCAGUGUAUUCACCAUGCGUAAAGCAAUCAGAUACCCAGCCCGCCUAAUGAGCGGGUUUUUUU
             Met starts trpE
UGAACAAAAUUAGAGAAUGCAAACACAAAAACCGACUCUCGAA    _   → →
```

The site of attenuation (*a*) is within the *trpL* (tryptophan leader) sequences. Before attenuation becomes effective, the RNA polymerase pauses at the *tp* site in the *trpL* (see attenuation). Expression of this operon is regulated by two main mechanisms, repression and attenuation. Repression prevents the initiation of transcription. The repressor is transcribed from the *trpR* gene (located at 100 min) and the tRNA^{Trp} is coded by gene *trpT* (84 min) whereas the amino-acylation of this tRNA is determined by gene *trpS* (74 min). The product of *trpR* is an aporepressor, i.e. it becomes active only when combined with tryptophan its corepressor (see tryptophan repressor). The tryptophan repressor alone may reduce transcription by a factor of 70 and attenuation may decrease it 8 to 10 fold but by the combination of these two controls transcription may be reduced by 8 x 70 (=560) or 10 x 70 (=700) fold. In *Neurospora* there are also 5 distinct genetic loci controlling tryptophan biosynthesis. These genes are derepressed coordinately with histidine, arginine and lysine loci and this phenomenon is called *cross-pathway regulation*. Humans do not have tryptophan biosynthetic genes and they depend on it (an essential amino acid) in the food. (See tryptophan, helix-turn-helix, *lac* operon, repressor, tryptophan repressor, essential amino acids, TRAP, antitermination)

TRYPTOPHAN REPRESSOR: consists of a repressor protein (using a helix-turn-helix motif) for binding to the operator site of the tryptophan operon and becomes active only when it is associated also with tryptophan. Illustrated with geometric symbols: ∪ (aporepressor protein), ♦ (co-repressor tryptophan), ⋓ (active tryptophan repressor). Actually the binding of tryptophan to the repressor protein facilitates the more intense binding of the complex due to a conformational change of the repressor protein. (See negative control, *tryptophan* operon, *Lac* operon, positive control, *arabinose* operon, helix-turn-helix motif, conformation)

TRYPTOPHANYL tRNA SYNTHASE (WARS): charges tRNA^{Trp} by the amino acid tryptophan. The gene WARS is in human chromosome 14. (See aminoacyl-tRNA synthetase)

ts: indicates temperature-sensitivity of an allele. (See temperature-sensitive mutation)

T$_S$: see T cell

TSAREVITCH ALEXIS: great-grandson of Queen Victoria who inherited her new mutation causing classic hemophilia. (See hemophilias)

tsBN2: is a temperature-sensitive baby hamster kidney (BHK) cell line.

TSE-TSE FLY: African fly of the genus *Glossina*, host of the parasitic *Trypanosomas*, causing sleeping sickness, a disease characterized by relapsing fever, enlargement of the lymph glands, anemia, severe emaciation and eventually death in humans and domestic animals. (See *Trypanosomas*)

TSHB: thyroid-stimulating hormone, thyrotropin. (See animal hormones)

Tsr: bacterial transducer protein recognizing serine as an attractant and leucine as a repellent. (See transducer proteins)

T-STRAND: a single-stranded intermediate of the T-DNA that is transfered from the Ti plasmid of *Agrobacterium* to the plant nucleus through the nuclear pores under the guidance of a virulence gene-encoded protein which is covalently attached to its 5' end. (See transformation, T-DNA, Ti plasmid)

tTA (tetracycline transactivator protein): is a fusion protein of the *tet* (tetracycline) repressor of *E. coli* and the transcriptional activation domain VP16 of herpes simplex virus. This system is generally driven by the *tetP* promoter which is actually a minimal immediate early cytomegalovirus (CMVIE) promoter, preceded by 7 copies of *tetO*, the tetracycline resistance operator of transposon 10 (Tn*10*). In the presence of tetracycline, this system is expressed at very low level and by removal of tetracycline the gene(s) under its control (e.g. luciferase, β-galactosidase) may be expressed at three orders of magnitude higher level. The system can be used also under the control of other promoters that suit best for regulating the expression pattern of the gene of interest. (See rtTA, tetracycline, split-hybrid system)

t-TEST: is used for the estimation of the statistical significance of the difference(s) between means. The t value is the ratio of the observed difference to the corresponding standard error: $t = (\bar{x} - m)/(s/\sqrt{n})$ where $\bar{x}$ and m are the two means, n = population size, s = standard deviation. More commonly the significance of the difference between two means is calculated as $t = \dfrac{m_1 - m_2}{\sqrt{[e_1]^2 - [e_2]^2}}$ where m_1 and m_2 are the two means ($\bar{x}$) and e_1 e_2 are the standard errors of the two means determined as $e = \dfrac{s}{\sqrt{n}}$ and $s = \sqrt{V}$ where V = variance = $\dfrac{\Sigma[(x - \bar{x})^2]}{n-1}$. When the t value is available the probability of the difference is determined with the aid of a "t table". (See Student's t distribution)

TTKs: are tubulin-associated kinases. (See tubulin)

TTP (tris-tetrapolin): cDNA shares 102 amino acid sequences in its product with TIS11 (insulin and serum-responsive transcription factor). (See TIS11, insulin, serum response element, transcription factors)

T-TROPIC: the T lymphocyte is targeted (e.g. by a virus).

TU: protein elongation factor (EF-TU) in prokaryotes binds aminoacyl-tRNA to the ribosomal A (acceptor) site. (See EF-TU.GTP, aminoacyl-tRNA, ribosome, protein synthesis)

TUBE NUCLEUS: it is in the vegetative cell at the tip of the growing pollen tube. It has only physiological and no genetic role because it does not enter the embryosac. (See gametophyte)

TUBER: an underground enlarged stem, specialized for food storage, e.g. in potato.

TUBEROUS SCLEROSIS: see epiloia

TUBOPLASTY: surgical repair of a defect on an internal tube such as the Fallopian tube.

TUBULINS: globular polypeptides of α and β subunits (50 kDa, each) are G-proteins concerned with signal transduction of nerve cells and are components of microtubules such as the spindle fibers and the cytoskeleton. (See spindle, cytoskeleton)

TULIP MANIA: see symbionts hereditary, tulips broken

TULIPS, BROKEN: variegation caused by infection with the tulip-breaking virus. The plants displaying this sectoring are valued by floriculturists. (See tulip mania, broken tulips)

TUMBLING: results when the bacterial flagella (singular: flagellum) rotate clockwise.

TUMOR: an abnormal clump of cells originated by benign or malignant growth. Malignant tumors may be invasive and show metastasis, common in many types of cancer. It may be caused by mutant genes, chromosome breakage and viral infections altering the normal regulation of cell proliferation. Tumorigenesis is generally a multi-phase process involving activation of cell cycle promoting genes and inactivation of tumor suppressors. Active MAPKK may stimulate tumorigenesis if transfected into mouse cells. Plant tumorous growth is often called callus that is never metastatic although the crown gall tumors of plants may be spread by new foci of infection of the inciting bacterium, *Agrobacterium tumefaciens*. Plant viruses and higher concentrations of phytohormones may also cause tumors. (See oncogenes, retroviruses, SV40, adenoviruses, *Agrobacterium*, genetic tumors, cancer, MAPKK, habituation, CATR1)

TUMOR ANTIGEN (tumor-specific antigen): tumor cell-surface antigens that can be presented by major histocompatibility molecules to T cells. (See MHC, T cell, antigen presenting cell)

TUMOR INFILTRATING LYMPHOCYTES (TIL): seek up tumors. TIL are isolated from solid tumors and cultured in single-cell suspension in a medium containing interleukin 2 (IL-2). Through genetic engineering they may be equipped with the gene of tumor necrosis factor through transformation by retroviral or other vectors. The transformed cells may selectively kill then cancerous tumor cells. (See tumor necrosis factor, retroviral vectors, lymphocytes)

TUMOR NECROSIS FACTOR: see TNF

TUMOR PROMOTER: see phorbol ester

TUMOR SUPPRESSOR GENE: its loss, inactivation or mutation permits neoplastic growth by deregulation. They most commonly have a role in the cell cycle or in the regulation of RNA polymerase II or III. Also, inhibition of peptide chain elongation may be a mechanism of tumor suppression. (Genes with cytostatic or cytotoxic effects are excluded from this category of tumor suppressors). Actually, the majority of cancer cells display deletions that may indicate the loss of a tumor suppressor gene; for the direct proof further evidence is required for the loss of a tumor suppressor. (See cancer, malignant growth, p53, p16, $p16^{INK4}$ p21, retinoblastoma, ELL, Elongin, DPC4, pol III, tumor suppressor factors, polycystic kidney disease, prostate cancer, oncogenes, cell cycle, apoptosis, colorectal cancer, tumor suppressor pathway)

TUMOR SUPPRESSOR FACTORS: the development of tumors follow multiple routes yet genes involved in the general control of differentiation, as revealed by studies of *Drosophila* morphogenesis, seem to be involved. Mammalian homologs of *Drosophila hedgehog (hh)*, sonic hedgehog (SHH), Indian hedgehog (IHH) and desert hedgehog (DHH) seem to be entailed in holoprosencephaly, a developmental anomaly connected to cancer. The receptor of *hedgehog* of *Drosophila* is *smoothened,* and its suppressor is *patched* which may be concerned with the development of basal cell carcinoma and defects in the central nervous and skeletal systems. The *cubitus interruptus* (*ci*) *Drosophila* gene appears to function as an effector of *hh* and similarly to the GLI genes of humans are responsible for a type of brain tumor and for the cephalopolysyndactyly syndrome. The decapentaplegic (*dpp*) *Drosophila* protein bears similarity to the mammalian tumor growth factor (TGF) and to the bone morphogenetic protein 4 (BMP4), causing defects in limb and gut formation. The effector of the mammalian TGF-β receptor (DPC4) involved in pancreatic tumors has its homolog in the *Drosophila mad* gene. The *Drosophila wingless* locus corresponds to the mammalian WNT loci controling mammary tumors. The *Drosophila zeste-white-3* gene codes for a signal molecule similar to a mammalian glycogen synthase kinase, and the *Drosophila armadillo* gene, encoding β-catenin, is also an oncoprotein, controlling intestinal tumors. (Based on Dean, M. 1996 Nature Genetics 14:245, see terms mentioned in the alphabetical list)

TUMOR VIRUSES: induce or participate in the formation of tumors (cancer). Their genetic material can be DNA (such as SV40, adenoviruses, papilloma virus, hepatitis-B virus, Epstein-Barr virus) or RNA (such as the retroviruses causing leukemia, lymphoma, AIDS, Kaposi's sarcoma, etc). The DNA viruses can integrate into the mammalian genetic material and activate cell replication by overwhelming the function of the tumor suppressor genes. The DNA tumor virus genetic material has no counterpart in the host genetic material. The RNA tumor viruses' genetic material is replicated by reverse transcription and produce a double-stranded counterpart of their genome. The viral RNA is then transcribed from the cellular DNA template. Most of the oncogenes in the cells (c-oncogenes) correspond to the v-oncogenes in the virus. (See SV40, adenovirus, Epstein-Barr virus, retroviruses, oncogenes)

TUMORIGENESIS: the formation of tumors. Fourier-transform infrared spectra (FT-IR) reveales structural modifications at many points in the DNA and marked differences between the primary and metastatic state. (See cancer, neoplasia, FT-IR)

TUMOROUS HYBRIDS: see genetic tumors

TUNICA: the cell layer in the plant apical meristem wrapping the inner corpus. (See meristem)

TUP1: a general suppressor of sugar metabolism of yeast; its product is a trimeric G-protein. It is similar to AAR1 and AAR2. TUP forms a repressor complex with yeast protein Cyc8. (See G proteins, glucose effect)

TUPAIA: a family of prosimii. *Tupaia*, 2n = 62; *Tupaia glis*, 2n = 60; *Tupaia montan*, 2n = 68. (See prosimii)

TURBID PLAQUE: see plaque

TURCOT SYNDROME: is an autosomal recessive malignant tumor of the central nervous system, associated with polyposis. (See polyposis adenomatous, Gardner syndrome)

TURGID: expanded because of water uptake.

TURGOR: the intracellular pressure caused by water absorption.

TURKEY: *Meleagris gallopavo*, 2n ≈ 80.

TURNER SYNDROME: is based on an X0 chromosomal constitution in some female mammals. Its incidence in humans is ≈ 0.0003 but the frequency in abortuses may be 0.01 to 0.02.

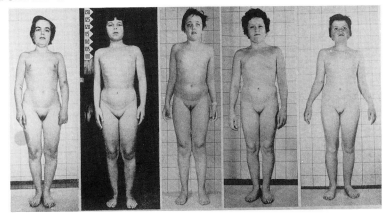

TURNER SYNDROME FEMALES OF DIFFERENT AGES. WITH THREE DOSES OF THE LONG ARM BUT ONLY ONE SHORT ARM OF THE X CHROMOSOME. DESPITE THE VARIATIONS IN APPEARANCE THE SIMILARITIES OF THE UNRELATED INDIVIDUALS IS OBVIOUS. (Courtesy of Lindsten J. *et al.* Ann. Hum. Genet. 26:383)

A NORMAL HUMAN **X** CHROMOSOME ➔

Turner females have usually short stature, webbed skin on a broad neck, underdeveloped genitalia and sterility, heart problems, prone to kidney disease, diabetes and hypertension but generally of normal or near normal intelligence. The retention of the paternal X usually results in better cognitive abilities (imprinting). In some instances they are also fertile; most of these cases are probably X0/XX mosaics. The symptoms are similar even when only the one of the short arms one of the X-chromosomes is missing. The single X chromosome is maternal in 75% of the cases. The underlying mechanism is puzzling because males normally have a single X and even in the females one of the X chromosomes is inactive, except during oogenesis and the first few weeks of embryogenesis. The majority of the studied female animals with single X (pigs, cattle, horse) are abnormal but the X0 female mice appears rather normal and fertile. In *Drosophila* the X0 individuals are sterile males. In Turner syndrome the most critical is the loss of the distal (pseudoautosomal) part of the short arm of the X chromosome. Some of the human developmental difficulties may be normalized by estrogen treatment. Artificial insemination or intrauterine implantation may permit reproduction. (See trisomy, sex determination, sex chromosomal anomalies in humans, human chromosome map, prenatal diagnosis, MSAFP, Noonan syndrome, imprinting, pseudoautosomal, ART)

TURNIP (*Brassica campestris*): is a polymorphic species with chromosome number 2n = 2x = 20, *Brassica napus* is 2n = 4x = 38. (See *Brassica oleracea*)

TURNIP CRINKLE VIRUS (TCV): is a plant RNA virus related to the Tomato Bushy Stunt Virus.

TURNOVER: is the depletion - repletion cycle of molecules.

TURNOVER NUMBER: the number of times an enzyme acts on a molecule in a unit of time at saturation.

TURNOVER RATE: the pace of decay and replacement of a molecule.

***Tus* GENE PRODUCT**: a protein that senses the *Ter* sequence signals in DNA replication in *E. coli*. (See DNA replication, replication bidirectional, prokaryotes, RTP)

TWEEN 20 (polyoxyethylene sorbitan, monolaurate): is an anionic biological detergent with about 50% lauric acid and the rest is myristic, palmitic and stearic acids.

TWIGDAM: Technical Working Group on DNA Analysis Methods. It is concerned with forensic application of DNA analytical and statistical techniques (See DNA fingerprinting)

TWIN HYBRIDS: see complex heterozygotes

TWIN SPOTS: are visible if (mitotic) somatic crossing over takes place between appropriately marked chromosomes or may be caused by nondisjunction. (See mitotic crossing over)

twine: *Drosophila* homolog of *Cdc25* meiosis-specific gene. (See cdc25, azoospermia)

TWINNING: is the phenomenon of developing two (or multiple) zygotes from a single impregnation in usually uniparous mammals. The frequency of twins is different in various ethnic groups. Among Nigerians it may be as high 4.5% and in some South American and Far-East populations it may be as low as 0.8%. In the USA its frequency among whites was about 0.89 to 0.94% and among blacks 1.37% before the wide-spread use of fertility promoting drugs. After *in vitro* fertilization more than half of the babies born are multiple.

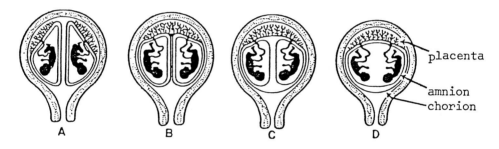

HUMAN TWINS CAN BE EITHER MONO- OR DIZYGOTIC AND THE TWO CONDITIONS CAN BE DISTINGUISHED ALSO ON ANATOMICAL BASIS OF THE DEVELOPING FETUS. MONOZYGOTIC TWINS ARE SURROUNDED BY A COMMON CHORION (C and D). (Modified after Stern, C. Principles of Human Genetics, Freeman. San Francisco)

Twins are either *monozygotic* (identical) or *dizygotic* (fraternal). The former ones are derived from a single fertilized egg and the latter ones develop from two separate eggs fertilized by different sperms. The monozygotic twins in the overwhelming majority of the cases are genetically identical whereas the dizygotic twins are comparable to any other siblings. It is possible that some dizygotic twins display higher similarity if one unfertilized egg gave rise to two blastomeres and then each was fertilized by two separate sperms. Another possibility is that one of the polar bodies (identical to the egg) becomes an egg due to a developmental mishap. The frequency of twinning has a genetically determined component as the studies of various ethnic groups indicate.

Monozygotic twins are expected to be of the same sex. Exceptionally, one of the XY male twin embryos may lose the Y chromosome and develops into an XO (possibly mosaic) female. The XO human cells develop into Turner syndrome females, unlike in *Drosophila* where XO individuals are sterile males. Sometimes it is difficult to distinguish between identical and non-identical twins on the basis of phenotypic similarities. DNA finger fingerprinting may resolve the problem. Dermatoglyphics may not always be conclusive due to developmental differences in digital, palmar or plantar (sole) ridge counts in the monozygotic twins. Mono- and dizygotic twins provide useful tools for the study of inheritance of polygenically determined human traits. Among women of families with dizygotic twins the rate of twinning is about double of that of the general population and it may be attributed to hereditary hormone levels. The inheritance of twinning through the male parents is lower than through the females.

Twinning continued

Monozygotic twinning may not have or may have very low genetic component. There are statistical methods for the discrimination between identical (MZ) and non-identical (DZ) twins based on concordance of alleles (DNA or any other type) Maynard Smith and Penrose (Ann. Human Genet. 19:273) worked out the following formula for the probability of concordance for DZ twins:

$$P = \left(1 + 2\sum_{i=1}^{n} p_i^2 + 2\left[\sum_{i=1}^{n} p_i^2\right]^2 - \sum_{i=1}^{n} p_i^4\right)/4$$

where i stands for the phenotype of the markers, n = number of alleles, p_i = allelic frequencies calculated on the basis of the binomial distribution for the various types of matings. (See DNA fingerprinting, fingerprints, forensic genetics, heritability estimation in humans, freemartin, superfetation, multipaternate litter, multiparous, zygosis, concordance, discordance, co-twin, quadruplex)

TWINS: see twinning

TWINTRON: see intron group II

TWISTING NUMBER: characterizes DNA supercoiling by indicating the number of contortions (writhing) and the number of twists, i.e. the number of nucleotides divided by the number of nucleotides per pitch. (See supercoiling, DNA)

TWITCHIN: is a myosin-activated protein kinase.

TWO-COMPONENT REGULATORY SYSTEMS: in bacteria pairs of proteins transduce environmental signals. In one of the proteins ca. 250 amino acids at the C-terminus and ca. 120 amino acids at the N-terminus are conserved. They control chemotaxis, virulence, nitrogen assimilation, dicarboxylic acid transport, sporulation, etc. They function by autophosphorylation of a histidine residue by the γ phosphate of ATP. The phosphate is then transfered to an aspartate, in the *response regulator*, that modifies regulatory activity of the C-terminal output domain. The system is also called a phosphorelay. A specific phosphatase may reset the system. Frequently more than two components are involved. (See signal transduction)

TWO-DIMENSIONAL GEL ELECTROPHORESIS: the (protein) mixture is first separated  by isoelectric focusing then separated by size using a slab of SDS polyacrylamide gel. Thus all proteins, except those rare molecules which have identical charge and molecular size, are distinguished. For the detection of small quantities of the molecules, they are labeled radioactively or by some non-radioactive means. A single two-dimensional gel slab permits the separation of hundreds of proteins at a time. (See electrophoresis, isoelectric focusing)

(Courtesy of ESA Inc. Chelmsford, MA 01824-4771)

TWO-HIT HYPOTHESIS: assumed that in order to develop cancer two genetic alterations must take place in succession. Some chromosomal aberrations also have called two-hit causes because two breaks are necessary to bring about an inversion or translocation. (See kinetics)

TWO-HYBRID METHOD: genetic construct facilitates the study of protein-protein interactions. The GAL4 protein is both an enzyme and an inducer. The native GAL4 protein contains an N-terminal UAS (upstream activator sequence)-DNA-binding region and a carboxyl-terminal transcription activating region. Both of these regions - in close vicinity - are required for the activation. Thus interaction between two proteins can be studied by fusing the N-terminal of say protein 𝒩 and the C terminal of protein 𝒞

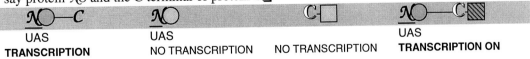

UAS	UAS		UAS
TRANSCRIPTION	NO TRANSCRIPTION	NO TRANSCRIPTION	**TRANSCRIPTION ON**

If the two proteins interact they reconstitute the link between the binding and activating domains and transcription of the reporter gene may proceed. Thus to show expression, the N

Two-hybrid method continued
binding domain must bind to the UAS element and contact is established with the C activation element of ❑ by joining the two components. The expression of the downstream reporter gene requires the interaction between the binding and activation domains. The advantage of the two-hybrid method that it can be used for testing protein interactions, determining the amino acid sequences critical for interactions and screening gene libraries for binding proteins or activators. The system can be applied for studies of the cell cycle and transcription factors, tumorigenesis, tumor suppression, etc. The *reverse two-hybrid* system detects mutations which are unable to bring the activation element to the DNA binding domain and thus cannot convert a potentially toxic compound to a toxic one or into a suicide inhibitor. (See galactose utilization, split-hybrid system, three-hybrid system, suicide inhibitor)

TWO-POINT CROSS: involves differences at two gene loci, e.g., *AB* x *ab*.

TWO-STEP SYNAPSIS: see topological filter

Ty: non-viral retroelements (retroposons) in the nucleus of budding yeast (*Saccharomyces cerevisiae*). Ty elements occur in several related forms and are designated Ty1 (25 - 30 copies), Ty2 (10 copies), Ty3 (2-4 copies), etc. The elements are flanked by long terminal direct repeats (334 - 371 bp) that are designated as δ for Ty1 and Ty2, σ for Ty3 and the open reading frame between the repeats (LTR) in Ty1 and Ty2 is identified as ε. Elements which resemble these terminal repeats, δ (ca. 100 copies), σ (20-30 copies) and τ (15-25 copies) are also found. The terminal repeats contain sequences identified as U3 (unique for 3'), R (repeated) and U5 (unique for 5' end) similarly to the designations in retroviruses. These LTRs contain the upstream gene activation sequence, TATA box, polyadenylation, transcription termination signals and the TG...CA bases involved in the integration of the reverse transcripts into the chromosomes. All the retroelements are generally flanked in the host by 5-bp target duplications. The open reading frames, distinguished as TYA and TYB, between the ends are also similar among them and resemble retroposons of other organisms. The TYA protein may be processed through proteolyis into several smaller proteins involved in the formation of the shell of the VLP (virus-like particle). TYB contains genes for protease (*pro*), integrase (*int*), reverse transcriptase (*rt*) and RNase H (*rnh*) but no *env* gene is present as would be expected for retroviruses. Although most of the Ty elements and the independently standing termini are intact, some contain deletions up to a few kb, and the LTR sequences may be truncated. There may be apparent insertions, duplications within the Ty elements or inversion(s) involving parts of a LTR. The heterogeneity of the coding regions is due to base substitutions.

Ty retrotransposons transpose by synthesizing an RNA that is reverse-transcribed into DNA for integration. The transcription is by RNA polymerase II and the transcripts are polyadenylated. Transcription may be prevented by mutations symbolized as *spt* (suppressor of Ty). The transcription of Ty elements may be induced by sex pheromones synthesized by the *MATa* or *MATα* genes but not in the *MATa/Mat α* diploids. *MAT* homozygotes do not affect transcription of Ty RNA. The Ty RNA can be packaged into virus-like particles (VLP). Proteolyis is involved in the processing of the TYA and TYB products required for the completion of VLPs. The reverse transcription appears to be primed by tRNA^{Met}. Degradation of the template RNA is due to RNase H. During reverse transcription, recombination, sequence modifications and deletions may occur at high frequency. Transcription of Ty is regulated by several *SPT* (suppressor of transposition), *ROC* (reducer of overproduction of transcripts) and *TYE* (Ty enhancer) genes. Transcription may be increased by exposure to UV light, ionizing radiation, chemical mutagens and the culture media and it is elevated 20 fold at low temperature. The frequency of transposition is measured by insertion into and activation of these particular genes by genetically tagged Ty elements.

The VLP is made in the cytosol but it must enter the nucleus for transposition to take place. Insertion of Ty into structural genes eliminates their function and reversion by excision is very rare because the target site is modified by the event. Insertion within non-coding sequences may activate or silence previously inserted elements. Ty1 and Ty2 elements appear to be distributed at random throughout the genome but the family of Ty3 elements are much more re-

Ty continued

stricted. The Ty3 element inserts at the transcription initiation sites of tRNA genes by pol III. The transcription factors TFIIIB and TFIIIC are required for this type of insertions. The Ty1 element has preference for integration into the 5' sequences of pol II-transcribed genes. Ty, δ, σ and τ are commonly targeted to the vicinity of tRNA genes (hot spots). The insertions are usually directed into the leader sequences (promoter sites) or near the 5'-end of the coding region. Among the abundant Ty elements recombination may occur and cause chromosomal rearrangements and deletions. Recombination is most common between LTR (δ) repeats. Gene conversion may also occur between these elements. Ty elements may incite deletions, inversions and translocations with all the phenotypic consequences. The mutation rate within Ty elements was estimated to be 0.15 per Ty per replication cycle, or about 1/15 of the retrotranspositions result in some type of mutation. This rate of mutation is approximately 4 to 6 orders of magnitude higher than that of the rest in the nuclear genome, but comparable to that in RNA viruses. This high mutation rate is attributable to the lack of proofreading ability of the reverse transcriptase. The mutations in Ty (about 30% of all) occur within the 7 bases of the primer-binding site (PBS) where the minus strand replication is primed by a tRNA. Some of the mutations are caused by imprecise cutting by the RNase H. Another error-prone event is the addition of nucleotides immediately adjacent to the tRNA primer binding site. After the replication reached the 5'-end nucleotide additions take place at the 3'-end. After the second-strand transfer, partial DNA•DNA duplexes are formed and the recessed 3'-ends prime the completion of the double strand synthesis. As a consequence of the addition of nucleotides, the 3'-end of the minus strand cannot anneal precisely with the plus-strand DNA template. In order to reconstitute the 5' LTR of the Ty, the mispaired 3'-primer ends are extended by reverse transcription. This process incorporates then the mispairs into the Ty element that will integrate into the yeast chromosome. The mismatches must then be fixed by DNA repair. Additional mutational mechanisms may also occur. These mutational processes are not unique features of the Ty elements but other retroelements use them as well. Retroviruses are somewhat different, however, because these show frameshift mutations and complex rearrangements. The Ty elements have been utilized as vectors (pGTy), as insertional mutagenic agents (transposon tagging), and for fusion of TyA proteins with certain proteins of interest to facilitate the purification of epitopes. Similar transposable elements occur also in other fungi (*Candida, Pichia, Hansenula*). (See retroposon, retrotransposon, reverse transcription, strong-stop DNA, mating type determination in yeast, protease, integrase, epitope, *Saccharomyces cerevisiae*, pol II, pol I, DNA repair, VLP)

Ty δ: is a terminal repeat of the Ty insertion element. (See Ty)

Tyk2: a non-receptor tyrosine kinase. (See Janus kinases)

TYLOSIS: formation of animal callus. (See keratosis)

TYPING: determination of the blood group antigens and HLA. General classification by type. Also, DNA typing by fingerprinting using restriction enzyme-generated fragment pattern. (See blood groups, HLA, DNA fingerprinting)

TYROSINASE: see albinism

TYROSINE: a non-essential aromatic amino acid (MW 181.19), soluble in dilute alkali. Its biosynthetic pathway (*with enzymes involved in parenthesis*): Chorismate -> (*chorismate mutase*) -> Prephenate -> (*prephenate dehydrogenase*) -> 4-Hydroxyphenyl pyruvate -> (*amino transferase* with glutamate NH_3 donor) -> Tyrosine. Tyrosine can come also from phenylalanine by dehydroxylation. Tyrosine is a precursor for norepinephrine, epinephrine, 3,4-dihydroxyphenylalanine (dopa), dopamine and catechol that form the catecholamine family of animal hormones. (See phenylalanine, chorismate, pigmentation of animals, alkaptonuria, tyrosinemia, tyrosine aminotransferase, tyrosine kinase, animal hormones, goiter, formula above)

TYROSINE AMINOTRANSFERASE (TAT): converts tyrosine into p-hydroxyphenylpyruvate. It is induced by glucocorticoid hormones. TAT deficiency occurs rarely in humans and it is controlled by a recessive gene (ca. 11 kb, 12 exons) in human chromosome 16q22.1-q22.3 (and

Tyrosine aminotransferase continued

mouse in chromosome 8). The condition involves elevated level of tyrosine and in some cases an increased urinary excretion of p-hydroxyphenylpyruvate and hydroxyphenylactate. The disease involves generally corneal ulcer, palm keratosis (callous skin), mental and physical retardation. The mitochondrial enzyme is under the control of another gene. A TAT regulator gene may be in the human X-chromosome and glutamic oxaloacetic transaminase (16q21) also regulates its activity. (See tyrosine, tyrosinemia)

TYROSINE HYDROXYLASE (TYH): encoded in human chromosome 11p15 (mouse chromosome 7) controls the synthesis of dopamine from phenylalanine. Dopamine is a hormone involved with adrenergic neurons, the sympathetic nerve fibers that liberate norepinephrine when an impulse passes the nerve synapse. This enzyme may play key role in fetal development and in manic depression. (See tyrosine, manic depression, dopa)

TYROSINE KINASE (protein tyrosine kinase): activity is essential for many processes of the signal transduction, tissue differentiation involving oncogenesis. The most frequently activated proteins are phospholipase C-γ, phosphatidylinositol 3-kinase, and GTPase activating kinase. (See also oncogenes ABL, ARG, Blk, Btk, FES, FGR, FLT, FHS, ERBB1, Fyn, RAF, SRC, KIT, LCA, Lyn, MET, RET, Syk, YES, morphogenesis in *Drosophila* [*tor*], *sevenless*, *SOS*, insulin [receptor β–chain], TCR [T cell receptor], EGFR, PDGF. MCSF, VEGF, Steel factor, hepatocyte growth factor, trk, neutrotrophin, FGF, receptor tyrosine kinase, signal transduction)

TYROSINE PHOSPHATASEs (protein tyrosine phosphatase, PTP): family of enzymes have at least 40 known members. Some have properties of transmembrane receptor proteins and the cytosolic forms have a characteristic ca. 240 amino acid catalytic domain. Each PTP has also a phosphate-binding, 11-residue motif containing catalytically active Cys and Arg. The proteins have critical role in signal transduction relevant to growth, proliferation and differentiation. (See signal transduction, PTK)

TYROSINE PROTEIN KINASE: see tyrosine kinase, phosphorylase

TYROSINE TRANSAMINASE: see tyrosine aminotransferase

TYROSINEMIA (FAH): the recessive gene responsible for the anomaly is in human chromosome 15q23-q25. Its prevalence is about 1/2,000 live births. FAH involves increased levels of tyrosine in the blood and a lack of p-hydroxyphenylpyruvate oxidase but the primary enzyme defect appears to be fumarylacetoacetase deficiency leading to accumulation of succinylacetone and succinylacetoacetate. Defects in porphyrinogenesis appears secondary. Because of the deterioration of the liver, as secondary symptoms, accumulation of methionine and other amino acids in the blood and urine is frequent. The particular odor of the body fluids may be due to α-keto-γ-methiolbutyric acid. Prenatal diagnosis of FAH relies on amniotic fluid analysis for succinylacetolactone or measurement of fumarylacetoacetase in cultured amniotic cells. Low tyrosine diet alleviates the symptoms but to avoid liver cancer, liver replacement before age 2 is advisable. Tyrosinemia Type III is due to a deficiency of 4-hydroxyphenylpyruvate dioxigenase. This disease does not usually involve dysfunction of the liver. The condition can be detected prenatally but carriers cannot be identified. (See tyrosine aminotransferase, genetic screening, methionine adenosyltransferase deficiency, methionine biosynthesis, amino acid metabolism, liver cancer, mosaic, gene therapy)

TYROSINOSIS: see tyrosine aminotransferase

U

U: uracil. (See pyrimidines)

U: uranium

U-937: human monocyte line. (See monocytes)

U PROTEIN (unwinding protein): unwinds the DNA strands at a distance from a nick.

U RNA (ubiquitous RN): a snRNA; their transcript has a 2,2,7-trimethyl guanosine cap and may have modified U residues and have no poly-A tail. (See snRNA, cap, polyA tail)

U1 RNA: has complementary sequences to the 5' consensus sequences of the splice sites and has presumably a role in mRNA generation from the primary transcript. (See hnRNA, RNA)

U2 RNA (snRNA): apparently recognizes the 3' end of introns at the lariat and seems to have a role in splicing. (See lariat, splicing, introns)

U3 snRNP: is the most abundant U RNAs ($\approx 10^6$ molecules/cell) processes near the 5'-end of the ribosomal RNA transcripts and stays on and generates a 5' knob characteristics for rRNA only. Binding U3 of the snRNP initiates the processing especially the transcripts of the 18S rRNA. (See rRNA, snRNP, ribosome)

U8 snRNP: is required for the upstream cleavage of the 5.8S and for cutting off of the 28S RNA ca. 500 nucleotides at the 3' end. (See rRNA, snRNP, ribosome)

U14 snRNP: is a maturase for the 18S rRNA. (See ribosome)

U22 snRNP: is essential for processing both ends of the 18S rRNA, separated by about 2,000 nucleotides.

UAA: ochre codon of translation termination. (See code genetic)

U2AF: a protein assisting U2 snRNA recognition. (See U2 RNA)

UAG: amber codon of translation termination. (See code genetic)

UAS: upstream activating sequences (regulate gene transcription). They behave similarly to enhancers. UAS encodes DNA-binding proteins, e.g., the GAL4 UAS codes for a 100 kDa protein and protects its 17 bp palindromic sequence against DNase I digestion or methylation. These proteins bind to special DNA sequences and to other proteins in the transcrpitional complex. Transcriptional activation and binding may rely on more than one tract of amino acids. (See galactose utilization, two-hybrid method)

UBC (E2): ubiquitin-conjugating enzymes. From UBC ubiquitin is transferred to a lysine residue of the target protein. This process may use also the ubiquitin ligase (E3). (See ubiquitin)

UBF (upstream binding factor): in association with protein SL-1 controls the transcription of rRNA genes. These protein factors differ even among closely related species; they are members of the high mobility group proteins. (See high mobility group proteins, transcription factors)

UBIQUITIN: an acidic polypeptide (76 amino acids), ubiquitously present in prokaryotic and eukaryotic cells. It may be associated with H2A histone and the conjugate is called UH2A. Its function involves binding to other proteins and the ubiquitinated proteins are then degraded by proteolytic enzymes. In the majority of the cases the degradation takes place in the cavity of the (20S-26S) *proteasomes*, cylindrical complexes of proteases, or the ligand-bound molecule can be internalized into vacuoles and degraded by lysosomal enzymes. Ubiquitination is involved in many aspects of cellular regulation, DNA repair, stress response, cell cycle progression, the formation of the synaptonemal complex, signal transduction, apoptosis. Ubiquitin is degraded by deubiquitinating enzymes (UBP). (See histones, proteolytic, lactacystin, lysosomes, proteasomes, UBC, UBP, CDC34 and the other entries listed separately)

UBIQUITOUS RNA: see URNA

UBP: is a group of ubiquitin degrading enzymes. (See ubiquitin)

UBR: an enzyme forms thioester with ubiquitin and UBC. (See UBC)

UCE (upstream control elements): regulate transcription. (See regulation of gene activity)

UDGase (uracil-DNA-glycosylase): is a repair enzyme capable of removing accidentally incorporated U or deaminated C, and the step is followed by removal of the apyrimidinic site with the aid of AP endonuclease. (See DNA repair, excision repair, AP endonuclease)

UDP: uridine diphosphate formed from uridine monophosphate and gives rise to uridine triphosphate by using ATP as the phosphate donor.

UEP: see unit evolutionary period.

UGA: opal stop codon. (See code genetic)

UH2A: see ubiquitin

UhpB: *E. coli* kinase involved in the regulation of sugar phosphate transport.

uidA: the gene for β-glucuronidase. (See β-glucuronidase)

ULCER: a corruption of the surface or deeper cell layers of the body, e.g. gastric ulcer, an ulceration inside the stomach or diabetic ulcers of the legs, or genital ulcers in case of sexually transmitted disease, etc.

ULLRICH-TURNER SYNDROME: same as Turner syndrome

ULTIMATE MUTAGEN: some chemical substances (promutagens) become mutagenic only after activation. In this process, first proximal mutagens are formed that subsequently may be converted to a form (ultimate mutagen) that is genetically most reactive with the DNA. (See activation of mutagens)

Ultrabar : see *Bar* mutation

ULTRACENTRIFUGE: a laboratory equipment suitable for the separation of cellular organelles and macromolecules by high (over 20,000 revolution/minute) centrifugal force in an evacuated and refrigerated chamber. At maximal speed it may exceed several hundred thousand times the gravitational force. The *analytical ultracentrifuge* monitors continuously or intermittently the boundary of movement of macromolecules in a solute (e.g., Cs_2SO_4). The more commonly used *preparative ultracentrifuge* fractionates organelles or macromolecules in sucrose or CsCl. The suspended material forms a band corresponding to its density and the density of the solute. Chloroplasts in sucrose gradients can be identified by color, DNA band in CsCl can be identified by staining with the dye ethidium bromide. The bands, containing homogeneous material, can be removed by careful suction from the centrifuge tubes or by drop-wise collection through a hole punctured at the bottom with a syringe needle.

ULTRADIAN: occurring periodically but less frequently than once a day. (See also circadian)

ULTRASONIC (ultrasound): radiation in excess of 2×10^4 hertz/second, generally 5×10^5 hertz/sec. It is used for breaking down cells, treatment of arthritis and by tomographic examination or sonography, etc. (See arthritis, tomography, sonography)

ULTRASTRUCTURE: fine structure beyond the resolution of the light microscope.

ULTRAVIOLET LIGHT (UV): emission below the wavelength of violet (400-424 nm). UV-A below 290 nm and UV-B between 290-400 nm wavelength. Nucleic acids have maximal absorption at about 260 nm. The absorption maximum may depend on base composition and pH. The maximal genetic effects of UV light coincides with the absorption maximum of nucleic acids. The major genetic effect of UV light is the generation of pyrimidine dimers and pyrimidinones and reactive peroxides. These compounds may damage the DNA and interfere with replication and transcription and can cause mutation and induce cancer. Oxidation of the membrane proteins may affect the signal transduction pathways and activate genes of the cellular defense systems. In prokaryotes the LexA repressor is inactivated leading to the upregulation of genes mediating mutation, recombination and DNA repair. In yeast, cell cycle genes and various kinases are activated. In animal cells, through the RAS signal transduction pathways AP and NF-κB transcription factors may be activated. In plant cells UV exposure may increase the synthesis of UV-absorbing flavonoids and phenylpropanoids and activates the octadecanoid defense pathway of fatty acids and generates less than 0.2 cyclobutane dimers per gene. Sometimes UV-C is distinguished as having the wavelength of 200-290 nm. (See DNA repair, genetic repair, physical mutagens, light response elements, light-sensitivity diseases, signal transduction, RAS, cyclobutane, pyrimidine dimer, flavonoids, phenylpropanoids, fatty acids, electromagnetic radiation)

ULTRAVIOLET SPECTROSCOPY: measuring the absorption of (≈ monochromatic) UV light and thus qualitatively or quantitatively identifies molecules such as DNA, RNA, and others.

UMP: uridine monophosphate. (See UDP)

UMU (UV mutagenesis): genes are involved in the repair of ultraviolet light damage to DNA. (See DNA repair [SOS repair])

UNBALANCED CHROMOSOMAL CONSTITUTION: parents heterozygous for inversions or translocations are functionally normal but may transmit to their offspring (unbalanced) duplication-deficiency gametes resulting in various physical and mental disabilities depending on the chromosomal region(s) involved. (See translocation, inversion)

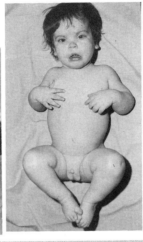
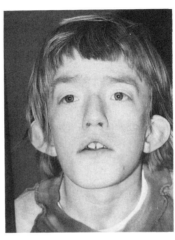

UNBALANCED CHROMOSOMAL CONSTITUTION INHERITED FROM PHENOTYPICALLY NORMAL CARRIER PARENTS. Left: DEFICIENCY FOR THE SHORT ARM OF CHROMOSOME 18. Middle: DUPLICATION OF THE TIP OF THE SHORT ARM OF CHROMOSOME 3. (Courtesy of Dr. Judith Miles). Right: DUPLICATION OF PART OF THE LONG ARM OF CHROMOSOME 4, RESULTING FROM A TRANSLOCATION HETEROZYGOSITY INVOLVING CHROMOSOMES 20 AND 4. (Courtesy of Dr. D.L. Rimoin)

UNC-6: a protein of the netrin family in *Caenorhabditis elegans* guiding ventral migration of the axon growth cone. (See netrins, semaphorins)

UNC-33: a *Caenorhabditis* protein regulating axon extension. (See CRMP, axon)

unc-86 (uncoordinated): *Caenorhabditis* gene with pivotal role in determining neural identities.

UNCHARGED tRNA: has no amino acid attached to it.

UNCOATING: during the initiation of infection the viral genome dissociates from the other constituents of the virus particle.

UNCOMPETITIVE INHIBITOR: see regulation of enzyme activity

UNCOUPLING AGENT: uncouples electron transfer from phosphorylation of ADP, e.g. dinitrophenol.

UNDERWINDING: characterizes negative supercoiling. (See supercoiling)

UNEQUAL CROSSING OVER: takes place between repeated sequences in the gene that may pair obliquely and therefore in the recombinant strands may have either more or less copies of the sequences than in the parental ones:

The extra DNA material may not be needed for the species and thus can be used for evolutionary experimentation and may contribute to the evolution of new function(s). (See duplication)

UNIDENTIFIED READING FRAME: see URF

UNIDIRECTIONAL REPLICATION: the replication fork moves only in one direction, left or right, from the origin. (See replication fork, bidirectional replication)

UNIFIED GENETIC MAP: represents similarities in the distribution of nucleotide sequences across phylogenetic groups and is expected to provide tools to evolutionary studies, and may assist in transfering economically advantageous genes to crops. (See SCEUS, evolution, comparative maps, CATS)

UNIFORMITY PRINCIPLE: see Mendelian laws

UNIMODAL DISTRIBUTION: has only one major peak.

UNINEME: a single strand (of e.g., DNA or of post-mitotic chromosome)

UNINFORMATIVE MATING: does not shed light on, e.g. linkage relationship.

UNIPARENTAL DISOMY: both homologous chromosomes are inherited from only one of the parents in a diploid. If these two chromosomes are identical the case is called *isodisomy*, if the are different: *heterodisomy*. (See nondisjunction)

UNIPARENTAL INHERITANCE: in the absence of male transmission of plastids and mitochondria, the genetic material of these organelles is transmitted to the progeny only through the egg. Also, telochromosomes and large deletions can generally be transmitted only through the female if transmitted at all. (See mtDNA, ctDNA, mitochondrial genetics, chloroplast genetics)

UNIQUE DNA: is present in a single copy per genome. (See singlet)

UNIT EVOLUTIONARY PERIOD (UEP): time in million years (MY) required for the fixation of 1% divergence in two initially identical nucleotide sequences.

UNIVALENT: eukaryotic chromosome without a pair. In case the chromosomes in a hybrid are not sufficiently homologous, they may form univalents and their distribution to the poles in meiosis may be disorderly. The frequency of univalents may permit conclusions regarding the lack of relatedness of the parental forms. (See Triticale)

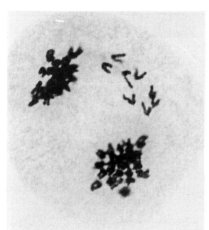

HYBRID OF AN OCTAPLOID (2n = 56) AND HYPOPLOID (2n = 41) HEXAPLOID TRITICALE. THE FEMALE GAMETE CONTRIBUTED 28, THE MALE ONLY 20 CHROMOSOMES. THEREFORE, THERE ARE 20 BIVALENTS AND 8 UNIVALENTS IN MEIOSIS. SEVEN OUT OF THE 8 UNIVALENTS REPRESENT THE *D* GENOME OF WHEAT. GENERALLY, THE LAGGING UNIVALENTS ARE NOT INCORPORATED INTO THE FUNCTIONAL GAMETES BECAUSE THEY FAIL TO REACH THE POLES. USUALLY, THEY DIVIDE BELATEDLY. (Courtesy of Kiss, Á. 1966 Z. Pflanzenzücht. 55:309)

UNIVERSAL BASES: 6-hydroxy- and 6-amino-5-azacytosine nucleosides or 1-(2'deoxy-β-D-ribofuranosyl)-3-nitropyrrole may serve for modified nucleotides in recognition of G, T and U with equal efficiency for Watson-Crick pairing. The latter maximizes stacking while minimizing hydrogen-pairing without sterically disrupting the double helix. These may be used to synthesize oligonucleotide probes and primers when the exact sequence needed cannot be infered because of the redundancy of the genetic code. (See hydrogen pairing, probe, primer)

UNIVERSAL CODE: the majority of DNAs across the entire phylogenetic range use DNA codons in the same sense. Notable exceptions exist in the mitochondria, and in a few species. (See genetic code)

UNIVERSAL DONOR: see ABO blood group

UNIVERSAL RECIPIENT: see ABO blood group

UNORDERED TETRADS: do not contain the spores in a linear order as generated in the first and second meiotic divisions. In contrast to ordered tetrads, unordered tetrads require the pre-

Unordered tetrads continued

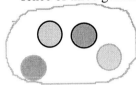
sence of three genetic markers and two of them must be in different chromosomes to be able to calculate gene - centromere distances. Again — like in ordered tetrads — we must determine the frequencies of tetratype tetrads for the, at least, three markers considered and we designate as p the tetratype frequency of say a and b, and q = tetratype frequency of b and c, similarly r = tetratype frequency for a and c. The exchange frequency between a and its centromere = x, between b and its centromere = y, and between c and its centromere = z.

Furthermore, we need the following three equations:

$$p = x + y - 3/2xy \qquad q = y + z - 3/2yz \qquad r = x + z - 3/2xz$$

The values, p, q and r being known, the unknown quantities, the recombination frequencies (x, y and z) between the three genes and their centromeres can be determined by resolving the three equations:

$$x = 2/3\left(1 \pm \sqrt{\frac{4 - 6p - 6r + 9pr}{4 - 6q}}\right), \quad y = 2/3\left(1 \pm \sqrt{\frac{4 - 6p - 6q + 9pq}{4 - 6r}}\right) \text{ and } z = 2/3\left(1 \pm \sqrt{\frac{4 - 6q - 6r + 9qr}{4 - 6p}}\right)$$

Once the recombination frequency between a gene and its centromere becomes available, the exchange frequencies between additional genes and their centromeres can also be calculated by the formula: $s = \frac{2(v - t)}{2 - 3t}$ where s = the unknown recombination frequency between marker d and its centromere, t = the known recombination frequency between gene e and its centromere, and v = the tetratype frequency for the unmapped gene d and the mapped gene e. (See also tetrad analysis)

UNROOTED EVOLUTIONARY TREES: do not indicate the initial split of the branching. (See evolutionary tree)

UNSATURATED FATTY ACID: contains one or more double bonds.

UNSCARE: United Nations Committee on Effects of Atomic Radiation.

UNSCHEDULED DNA SYNTHESIS: replication of DNA is outside normal S phase and this indicates a repair replication. The tests generally carried out on cultured hepatocytes or fibroblasts exposed to certain treatment(s) and then the incorporation of radioactive thymidine is monitored either by autoradiography or by scintillation counting. The data are compared with concurrent controls that were not exposed to any mutagen. Although the procedure appears attractive it is not very effective and practical for the identification of mutagens or carcinogens. (See bioassay in genetic toxicology)

UNSTABLE GENES: have higher than average mutation rate. Most commonly the instability is caused by the movement of insertion or transposable elements. Higher mutation rate may be due also to deficiency of genetic repair or defects in DNA replication. (See DNA repair, insertional mutation, error in replication, fractional mutation, variegation)

UNTRANSLATED REGIONS: are the leader sequences upstream from the first methionine codon and the downstream sequences beyond the stop codon of the mRNA. The upstream and the downstream regions include various control elements and at the 3'-end in eukaryotes the polyadenylation signal is situated. The untranslated regions of the mRNA do not code for any amino acid sequence. (See upstream, downstream, stop codon, polyadenylation)

UNUSUAL BASES: are modified forms of the normal DNA or RNA bases; they may be common in tRNA. Their incorporation into the DNA may lead to base substitution mutations. Methylation of C and A nucleotides may lead to imprinting, transient genetic variations in expression and alteration of RFLP. (See tRNA, RFLP, imprinting, methylation of DNA)

UNVERRICHT-LUNDBORG DISEASE: see myoclonic epilepsy

UNWINDING PROTEIN: facilitates unwinding of a DNA double helix and stabilizing single strands. (See DNA replication)

UP PROMOTER: boosts higher the frequency of transcription initiation. (See transcription factors, promoter)

UP-AND-DOWN: the structural arrangement of helical protein bundles comparable to the meander of β-sheets. (See meander, protein structure)

UPE: upstream promoter element. (See promoter)

UPGMA (unweighted pair group method with arithmetic means): are formulas for determining evolutionary distances. (See transformed distance, evolutionary distance)

UPSTREAM: in the direction of the 5' end of polynucleotides (DNA). (See downstream)

UPSTREAM ACTIVATION SEQUENCE: see UAS

UPSTREAM REGULATORY SEQUENCE (USR): regulatory element in the promoter region.

UPTAKE: eukaryotic cells may incorporate nuclei, plastids and mitochondria, pseudovirions, plasmids, liposomes and various other macromolecules besides smaller organic and inorganic molecules.

URACIL: a pyrimidine base in RNA; 2,4-dioxypyrimidine, MW 112.09, soluble in warm water but insoluble in ethanol. (See pyrimidines)

URE (ureidosuccinate utilization): cytoplasmic proteins involved in nitrogen metabolism in yeast are responsible for the production of a protein analogous to prion. (See prion)

UREA CYCLE: the formation of urea ($[NH_2]_2CO$) from amino acids and CO_2; ornithine is converted to citrulline and that to arginine; the hydrolytic cleavage of arginine produces urea and regenerates ornithine and thus the cycle is completed. (See arginine, ornithine, citrulline, carbamoylphosphate synthetase deficiency)

UREDIUM: a uredospore-producing sorus, a type of sporangium of fungi and protozoa. (See stem rust, sorus)

UREMIA: urine in the blood caused by a variety of factors.

UREOTELIC: excretes urea

URETHAN ($CH_3H_7NO_2$): is a toxic (lethal dose 2g/kg in rabbits) liquid (at 48-50 C°). It causes chromosomal breakage and mutation; it is also antineoplastic.

URF (unidentified reading frame): is capable of transcription (open) but the gene product has not been identified. (See reading frame)

URIC ACID: is a degradation product of xanthine and it is excreted in the urine, in particularly high amounts in hyperuricemic individuals under dominant and polygenic control and also in patients with glycogen storage diseases, in HPRT deficiency and gout. Birds and reptiles excrete high amounts normally. (See glycogen storage diseases, HPRT, Lesch-Nyhan syndrome, gout)

URICOTELIC: excretes uric acid.

URIDINE: uracil + ribose, an RNA nucleoside. (See pyrimidines)

URIDINE DIPHOSPHATE GLUCURONOSYL TRANSFERASE: is encoded in human chromosome 2; this group of enzymes glucuronate steroid hormones. (See Crigler-Najjar syndrome)

URIDYLATE: nucleotide of uracil; contains uracil + ribose + phosphate.

URL (uniform resource locator): the generic name for an Internet resource, such as WWW page, Gopher menu, a file transfer protocol server, etc. Information about the internet use is obtainable: <http://home.netscape.com/home/about-the-internet.html>. For automatic monitoring your interest on the Internet use 'The URL Minder': <http://www.netmind.com.URL-minder/URL-minder.html>. (See bookmarks)

U-RNA: uridine-rich nuclear RNA involved in transcript processing within the cell nucleus.

U-RNP: U-RNA associated with proteins involved in processing the primary transcripts of the genes within the nucleus.

UROCORTIN: is a neuropeptide similar to urotensin and corticotropin releasing factor. It evokes the synthesis adrenocorticotropic hormone and thus is involved stress-related endocrine and autonomic and behavioral responses. (See adrenocorticotropin, urotensin, corticotropin releasing factor)

UROGENITAL: involved with the system of urine secretion and the reproductive organs.

UROGENITAL ADYSPLASIA: autosomal dominant failure to develop one or both kidneys, frequently coupled with wide set eyes, low set ears, and other anomalies. The incidence among newborns may be as high as 4.5% and among adults 0.3%. (See kidney disease)

UROKINASE: is a plasminogen activator; the urokinase receptor is a glycosyl-phosphatidyl-inositol-linked cell surface protein that regulates cell adhesion. (See plasminogen, CAM)

UROPATHY: diseases of the urogenital system.

UROTENSIN: see urocortin

URS: upstream regulatory sequences, binding sites for various transcription factors. (See transcription factors, regulation of gene activity)

USE AND DISUSE: see lamarckism

USHER SYNDROME: is hereditary yet the pattern of inheritance is quite variable although in most cases probably it is autosomal recessive. Commonly it involves deafmutism, retinitis pigmentosa, mental disabilities and ataxia. The prevalence is about $4 - 5 \times 10^{-5}$. About 75% of the afflicted individuals have a severe defect in the USH1B gene in human chromosome 14q13.5, encoding myosin VIIA and USH2. The milder form, was assigned to 1q32. (See retinitis pigmentosa, ataxia, deafmutism, myosin)

USM (ubiquitous somatic mutations): are attributed to defects in DNA repair (mismatch repair) and in DNA replication.

U-snRNP: splicing factor of RNA transcript. (See splicing, spliceosome)

USP (chromosome-specific unique sequence probes): employs locus-specific fluorescent DNA sequences and it is suitable for identification of small deletions and duplications. (See FISH, chromosome painting, WCPP, telomeric probes)

Ustilago maydis (n = 2): is a basidiomycete which has been extensively used for meiotic and mitotic analysis of recombination, and for isolation of biochemical mutations, etc. This fungus causes the ear smut of maize. Several other *Ustilago* species are pathogenic to other Gramineae.

UTase (uridylyl transferase): catalyzes the transfer of uridyl group to the P_{II} regulatory subunit of adenylyl transferase (ATase), an enzyme which transfers an adenylyl group from ATP to a tyrosine-hydroxyl in glutamine synthetase. The complex ATase•P_{II}-uridylyl catalyzes phosphorolytic deadenylylation of glutamine synthetase. Glutamine synthetase is an enzyme involved in may functions. (See glutamine synthetase, glutamate dehydrogenase)

UTERUS: the female abdominal organ where the fertilized egg is embedded for the development of the embryo.

UTILITY INDEX FOR GENETIC COUNSELING: if the mother is heterozygous for a recessive disease allele d/D and other linked alleles (M_1/M_2), i.e. she is DM_1/dM_2 and the frequency of recombination between the two loci = r, half of her sons will be afflicted by the disease and 1 - r frequency of M_2 sons is expected to express the disease (as long as the penetrance and expressivity are high). A small *r* helps predictability. If her husband's genotype is DM_1, among her M_1M_2 daughters 1 - r will be carrier of the recessive disease allele *d* but one must know for sure which of the codominant *M* alleles is syntenic with *d*. In case of X linkage, for prediction on the basis of the *M* markers, the genetics counselor should know the genotype of the affected grandfather (dM_1 or dM_2). If the frequencies of the M_1 and M_2 markers is x_1 and x_2, respectively and the grandfather is dM_2, then the probability that the mother being informative for genetic counseling is $= 2x_1x_2$. Roychoudhury and Nei call this the *utility index of a polymorphic locus* for genetic counseling. If the grandfather has the recessive allele *m* and the grandmother is either MM or Mm, their heterozygous (Mm) daughter must have inherited the *M* allele maternally and the expected frequency of informative mothers is (1 - x)x. In case both *M* and *m* grandfathers are considered, the utility index of the mother becomes: $(1 - x^2)x$.

Genetic information regarding the mother can be obtained also from her children. In case the X-linked disease gene deemed to be in coupling with another marker, the Bayesian probability

Utility index for genetic counseling continued
for coupling is $(1-r)^2/[(1-r)^2 + r^2]$ and the probability for repulsion is $r^2/[(1-r)^2 + r^2]$ in case the mother has already two afflicted sons. In case the mother has both normal and afflicted sons: n_1 (DM_1), n_2 (DM_2), n_3 (dM_1) and n_4 (dM_2), the probability for coupling that she has 4 (= n) sons with the genotypes above is: $r^{n_2+n_3}(1-r)^{n_1+n_4}/2^n$; in case of repulsion the probability is: $r^{n_1+n_4}(1-r)^{n_2+n_3}/2^n$. The posterior probability that she is in coupling is $1/(1+\rho^\alpha)$ where $\rho = r/(1-r)$ and $\alpha = n_1 + n_4 - (n_2 + n_3)$, and in case $\alpha = 0$, the linked markers will not help in the prognosis. For repulsion the probability is $1/(1+\rho^\alpha)$. The probability regarding the genotype of the next offspring depends on the tightness of linkage.

In case of *autosomal dominant* disease in the presence of D, the disease is expected and both parents are informative. In case the mother is M_1M_2 and the father is M_1M_1 and the recombination frequency between the M and the D loci is r, the offspring of M_1M_1 genotype will have D with a probability of $1 - r$. DD homozygotes are expected very rarely because their occurrence depends on the product of the frequency of the D gene which is usually in the 10^{-5} range. In case there are multiple alleles at a locus, the total frequency of informative parents is $1 - \sum x_i^2 - (\sum x_i^2)^2 + \sum x_i^4$. In a mating of Dm/dM x dM/dm the offspring homozygous for m is expected to carry the dominant disease gene, D with a probability of $1 - r$ but the one with the dominant M phenotype is expected to be D at a frequency of $(1 + r)/3$. If the affected parent D is heterozygous and the other parent is dd, the frequency of the informative families is $2(1-x)x^4$. In case the mating is DdM_1M_2 x ddM_1M_1 and the children are n_1(DdM_1M_1), n_2 (DdM_1M_2), n_3 (ddM_1M_1) and n_4 (ddM_1M_2), respectively. The linkage phase can be estimated as shown above for X linkage [coupling: $1/(1+\rho^\alpha)$, repulsion $1/(1+\rho^\alpha)$]. All children of parent DdM_1M_2 will be informative, except when the spouse is ddM_1M_2 and all the progeny is heterozygous for the M locus but the probability that all children would be of such genetic constitution is $(0.5)^n$. The proportion of informative families when n>1 is: $2x_1x_2(1 - 0.5^{n-1}x_1x_2)$. If there are multiple markers, the proportion of the informative families (with x_i frequency of the ith allele): $2\sum_{i<j} x_i x_j (1 - 0.5^{n-1} x_i x_j)$.

In case of autosomal recessive diseases the calculation of the mathematical probabilities of informative families is more complex and biochemical or molecular (DNA) analyses are prefered. (See DNA fingerprinting, genetic counseling, counseling genetic, risk, paternity testing, see also Roychoudhury, A.K. & Nei, M. 1988 Human Polymorphic Genes. Oxford Univ. Press, New York)

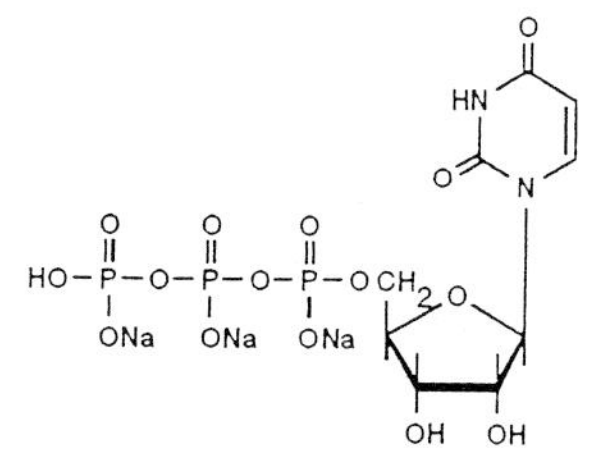

UTP: uridine triphosphate. (See UDP)
UTR: se untranslated region
UTROPHIN: see dystrophin (DRP2)
UV: see ultraviolet light
UVOMORULIN (UM): a transmembrane glycoprotein, also called E-cadherin. (See cadherins)
UvrABC: is an endonuclease complex of uvrA, uvrB and uvrC and these enzymes are involved in excision of ultraviolet light induced pyrimidine dimers. After the dimer is recognized, cuts are made on both sides of the dimer and thus excising the damaged area of about 12 nucleotides. (See DNA repair, ABC excinucleases)

V GENE: codes for the variable region of the antibody molecule.

V or V_{max}: the maximal velocity of the reaction when the enzyme is saturated with substrate.

v-mos: see Moloney mouse sarcoma

v-oncogene: see oncogenes

V-point: progression of the cell cycle beyond it ($\approx$ 6 hr before S phase) requires no insulin but only IGF-1. (See cell cycle, insulin, insulin-like growth factor)

VACCINES: a suspension of killed or attenuated pathogens for generating an immune defense system. The most successful vaccines (measles, mumps, rubella) are made of antigens generated against disease-causing microorganisms and injected into the blood stream to stimulate the development of circulating or serum antibodies (immunoglobulin G). More recently efforts are being made to develop vaccines that activate mucosal immunity. Membranes of the body, covering the gastrointestinal tract, the air-intake organs and the reproductive system are covered with mucosa. The purpose of this system is to trap infectious agents at the port of entry. The mucosa can develop sufficient quantities of immunoglobulin A. These new vaccines are expected to be delivered, without injection, orally. Some vaccines use live microorganisms that have been genetically engineered by removal part of their genome so they could not cause the disease yet they would promote the production of IgA and some also IgG. Other approaches would be introduction into the cells the genes of cytotoxic lymphocyte epitopes. Genetically engineered vaccines may be produced also in transgenic plants and these may eventually be edible. (See immune system, immunization, lymphocytes, mucosal immunity, CTL, epitope, immunization genetic, peptide vaccine)

VACCINIA VIRUS: is closely related to small pox (variola) and cowpox viruses. Apparently it does not occur in nature and can be found in the laboratory only where these two viruses are handled. Thus it appears to have derived somehow from variola and cowpox. Vaccinia vectors have been used to express antigens of unrelated pathogens (AIDS virus, hepatitis B) and employ them for immunization. The virus contains an about 190 kbp genome with about 260 potential open reading frames and about 200 bp telomeric sequences. (See immunization)

VACUOLES: are vesicles within plant and fungal cells, filled with various substances (nutrients, products of secondary metabolism, enzymes, crystals, solutes). Vacuoles may occupy minimal space in meristematic cells whereas in older cells they take up to 90% of the cell inner volume. The vacuoles are enclosed by the elastic tonoplast membrane that permits the change of their size. They may regulate the osmotic pressure of the cytosol by releasing smaller molecules or polymerizing them as needed to maintain a constant value in the cytoplasm. Vacuoles also regulate pH by a similar balancing action. The vacuoles supply the cells with storage nutrients, hydrolytic enzymes, anthocyanin pigments, and in some cases various toxic substances such as tannins, phenolics, alkaloids, etc. (See lysosomes, cells)

VAGILITY: is the ability of organisms to disperse in a natural habitat and it is thus a factor of speciation and survival.

VAGINA: the female organ of copulation beginning at the vulva and extending to the cervix of the uterus. Vaginismus is an involuntary painful contraction of the vaginal muscles and may cause severe pain by intercourse. An autosomal recessive condition of vaginal atresia (absence of vagina) is known. In general, the term describes an anatomical sheath. (See clitoris, Fallopian tube, vaginismus, ovary, egg)

VAGINAL PLUG: after successful copulation of mice, part of the male ejaculate forms a vaginal plug that closes the vagina for 16-24 hours and in some strains even for a few days. The presence of the plug reveals to the breeder the success of the mating.

VALIDATION: confirmation of experimental results or working hypotheses by repeated tests.

VALINE BIOSYNTHESIS: see isoleucine-valine biosynthetic pathway

VALINEMIA: see hypervalinemia

VALYL tRNA SYNTHETASE (VARS): is encoded in human chromosome 9. VARS charging tRNA^{Val} by the amino acid valine. (See aminoacyl-tRNA synthetase)

VAMP (synaptobrevin): a synaptic protein. (See syntaxin)

VAN der WAALS FORCE: weak, short-range attraction between non-polar (hydrophobic) molecules.

VAN der WOUDE SYNDROME: dominant, human chromosome 1q32-q41 located cleft lip and palate syndrome. (See cleft palate, harelip)

VAN GOGH, VINCENT: 1853-1890, the famous Dutch painter might have been a sufferer of intermittent porphyria. (See porphyria)

VANCOMYCIN ($C_{66}H_{75}Cl_2N_9O_{24}$) : is a very potent glycopeptide antibiotic.

VANILLA (*Vanilla planifolia*): tropical spice tree; 2n = 2x =32.

VÁRADI-PAPP SYNDROME: see orofacial-digital syndrome V

VARIABILITY: the condition of being able or apt to vary.

VARIABLE NUMBER TANDEM REPEATS: see VNTR

VARIABLE REGIONS: of the antibody are situated at the amino end of both the light and heavy chains, and this region determines antibody specificity and antigen binding. (See immunoglobulins, antibody)

VARIABLE SURFACE GLYCOPROTEIN: see *Trypanosoma*

VARIANCE: is the mean of the squared deviations of the variates from the mean of the variates:

$$V = \Sigma[(x - \bar{x})^2]/n - 1$$

where x are the variates and $\bar{x}$ is the mean of the variates and n is the number of variates (individuals). (See variate, invariance, standard deviation, standard error, analysis of variance, intraclass correlation, genetic variance)

VARIANT: a cell or individual different from the standard type.

VARIATE: a variable quantity measured in a sample of a population.

VARIATION: may be *continuous* and the individual measurements do not fall into discrete classes, e.g., the traits determined by polygenic systems. In case of *discontinuous* variation the measurements can be classified into distinct classes such as the qualitative traits (black and white [and no gray]) in a segregating population. (See variance, genetic variance, continuous variation, discontinuous variation)

VARIEGATION: sector formation or mosaicism of the somatic cells due a number of different mechanisms such as nondisjunction, somatic mutation, segregation of organelles (chloroplasts) deletion, disease, etc. (See also uniparental inheritance, lyonization, tulips broken, transposable elements, position effect, piebaldism, mitotic recombination, nondisjunction)

VARIETY: organism of a distinct form or function. (See also cultivar, cultigen)

VARIOGRAM: a plot of genetic distance relative to geographic distance. (See genetic distance)

VARKUD PLASMID: see *Neurospora* mitochondrial plasmids

VAS DEFERENS: see CBAVD

VASCULAR DISEASES: see cardiovascular diseases

VASCULAR ENDOTHELIAL GROWTH FACTOR (VEGF): is required for angiogenesis and it is particularly active in some tumor tissues to provide the necessary blood supply for proliferation. Its receptor is a tyrosine kinase. (See angiogenesis, Flk-1, Flt-1, tyrsosine kinase)

VASCULAR TISSUE: of plants includes the xylem, phloem, the (pro)cambium and the surrounding fibrous parenchyma; in animals the blood vessels are the primary vascular tissue.

VASECTOMY: surgical removal of the vas deferens (ductus deferens), the excretory channel of the semen. It is a method of fertility control for males. (See birth control drugs)

VASODILATOR: causes expansion of (blood) vessels.

VASOPRESSIN: see antidiuretic hormone, oxytocin

VAT: variant antigen type. (See *Trypanosoma*)

```
      ┌─── S ─── S ───┐
  Cys-Tyr-Phe-Gln-Asn-Cys-Pro-Arg-GlyNH₂
           HUMAN VASOPRESSIN
```

VAV ONCOGENE: it is in human chromosome 19p13.2-p12, and appears to be a transcription factor requiring tyrosine phosphorylation. It regulates lymphocyte development and activation. (See oncogenes, B lymphocytes, T cell receptor)

vBNS (very high-speed Backbone Network Service): is a computer network linking five supercomputer centers to facilitate fast scientific communication and remote control by the use of special equipment.

vCJD: variant of the Creutzfeldt-Jakob disease; a possible contagious form of the mad cow disease. (See Creutzfeldt-Jakob disease, encephalopathies)

VCP (vasoline containing protein): is involved in lipid metabolism. (See T cell receptor)

V(D)J (variable[diversity]juncture): sequences in immunoglobulins where antibody diversity is generated by recombination. (See immunoglubulins, T cell receptor)

VDR (vitamin D3 receptors): they control homeostasis, growth and differentiation. Preferentially bind to response elements of direct repeats, palindromes and inverted palindromes of hexameric core-binding domains, particularly well when they are spaced by three nucleotides. They can dimerize with 3,5,3'-triiodothyronine, a thyroid hormone receptor that can direct sensitivity of ligands for transactivation. (See vitamin D, transactivator, hormone response elements)

VECTOR: generally an insect or other organisms transmitting parasites and/or pathogens. (See also vectors, vector cassette)

VECTOR, ALGEBRAIC: see matrix algebra.

VECTOR CASSETTE: is a transformation construct carrying all essential elements (including reporter genes, selectable marker, replicator, etc.), and it can be used for insertion of different DNA sequences. (See vectors, reporter gene, transformation, knockout, targeting genes)

VECTORETTE: is a short DNA sequence serving as a specific linker-primer for PCR amplification. It generally contains an inner, non-complementary sequence (bubble), flanked by two short pieces of duplex DNA. The 5'-end may be either blunt or complementary to a restriction site, depending on the restriction enzyme used to digest the DNA. Ligation of the 3'-end is prevented by a two A residue overhang. (See polymerase chain reaction, amplification, linker, primer, ligase, blunt end, overhang, restriction enzyme)

VECTORS: molecular genetic constructs, generally a circular plasmids that can introduce exogenous genetic material into prokaryotic and eukaryotic cells. They may be *cloning vectors* that only replicate the DNA according to the plasmid replicon. The *expression vectors* carry genes complete with all the elements required for expression (promoter, structural gene, termination signals, etc.). *Gene fusion vectors* do not have promoters and the expression of the transferred gene is contingent on fusion with a host cell promoter. *Shuttle vectors* can carry DNA among different hosts (such as *E. coli* and COS cells, *Agrobacterium* and plants). All vectors must have as a minimum a replicator site, selectable marker(s) and mechanisms for introduction of parts of their sequences into the host genetic material. (See also cloning vectors, cosmids, phagemids, ColE1, plasmovirus, yeast vectors, viral vectors, excision vectors, plasmids, transposable elements, NOMAD, BAC, BIBAC, YAC, HAC, PAC)

VEGETAL POLE: is the lower end of the animal egg where the yolk is concentrated. The opposite end of the egg is called animal pole. After fertilization the yolk moves to the central position and becomes the starting site of the differentiation of axes (anterior-posterior, dorsal-ventral, median-lateral) of the embryo. (See morphogenesis *Drosophila*, pole cell)

VEGETATIVE CELL: is involved in metabolism but not in sexual reproduction. (See also gametogenesis in plants)

VEGETATIVE HYBRIDS: see graft hybrids

VEGETATIV INCOMPATIBILITY: see fungal incompatibility

VEGETATIVE NUCLEUS (macronucleus): see *Paramecium*, tube nucleus

VEGETATIVE PETITE: see petite colony mutants

VEGETATIVE REPRODUCTION: is common practice in many species of plants (grafting, rooting), and lower organisms that use fission for propagation. The advantage of this type of

Vegetative reproduction continued
reproduction is that the progeny forms a genetically homogeneous clone unless or until mutation takes place. (See also somatic embryogenesis, grafting, regeneration, clone, tissue culture)

VEGETATIVE STATE: asexual, unconscious, non-replicating, non-infectious, etc. depending on context.

VEGF: vascular endothelial growth factor is synthesized by smooth muscle cells. It is somewhat related to PDGF. The VEGF gene is divided among 8 exons and by alternative splicing three different proteins are produced. (See signal transduction, PDGF)

VEHICLE: see vectors

VELANS: see *gaudens*

VELOCARDIOFACIAL SYNDROME: a heart and face, kidney, parathyroid and thymus defect caused by deletion in human chromosome 22q11. (See DiGeorge syndrome, face/heart defects)

VENA (plural venae, adj. venous): a vein that is carrying blood toward the heart.

VENT: a DNA polymerase extracted from *Thermococcus litoralis*. (See recursive PCR)

VENTRAL (from the Latin venter meaning abdomen): position at the side opposite to the back.

VENTRICULAR: adjective for belonging to a ventriculus (cavity such as in the heart).

VENTURE : are a variable number repeating elements (40 to >150) of 14-15 nucleotides, rich in guanine. The shorter VENTR alleles are associated with susceptibility to insulin-dependent diabetes mellitus (IDDM) but one of the long repeats (14 x 50 = 700 nucleotides) appears to be protective against IDDM. (See diabetes mellitus, imprinting, VNTR)

VENULES: small vessels that collect blood from the capillary veins.

VERMICULITE ™: a commercial silicate medium to grow plants under greenhouse conditions.

VERMILION EYE COLOR of INSECTS: it is controled by the *v* locus of *Drosophila* encoding tryptophan pyrrolase, an enzyme that converts tryptophan to formylkynurenin. The recessive vermilion eye color is actually bright scarlet because the brown ommochrome is not formed. The *v* eye discs when transplanted into normal tissues develop wild-type eye color. (See animal pigments, ommochromes, tryptophan)

VERNALIZATION: some biannual or winterannual species of plants have a low temperature requirement for the induction and completion of the bolting and flowering stage of development. This need can be satisfied in spring planting by exposing the germinating seeds to near freezing temperature for a genetically required and variable period. It is called also yarowization (in Russian yarowie kchleba means spring cereal). (See photoperiodism, photomorphogenesis)

VERSENES (EDTA): are widely used laboratory chelating agents that at higher concentrations may cause chromosome breakage. (See EDTA)

VERTICAL TRANSMISSION: see transmission

VERY LOW DENSITY LIPOPROTEIN: see VLDL

VESICLES: membrane surrounded sacs in the cell, generally with storage and transport functions.

VESICULAR STOMATITIS VIRUS (VSV): negative-strand RNA virus of 11,161 nucleotides, enclosed by a nucleocapsid (N). N is a 35-turn helix within the membrane-surrounded oval particles. There is a transmembrane G protein on the surface of the virion for binding cell surface receptors, required for infection. VSV viruses can be engineered into useful genetic vectors. (See CO_2 sensitivity, viral vectors)

VETO CELL: recognizes T cells and inactivates them. (See immune suppression)

Vg-1: a protein that sends signals in animals to develop head and other nearby organs.

VIABILITY: the ability to survive; this is a property of organisms depending on genetic, developmental and environmental factors. A normal human fetus may become viable outside the womb after it reached a weight of about 500 g and about 20 weeks after gestation. The viability of a mutant is often expressed as the survival rate relative to the wild type.

Vicia faba: see broad bean, 2x = 2n = 12; its large chromosomes well suited for cytological study.

VILLUS: vesicular projections on a membrane. The amniotic villi, near the end of te umbilical cord are sampled for genetic examination during prenatal amniocentesis. (See amniocentesis)

VIMENTIN: is a constituent of the filament network extending through the cytoplasm of eukaryotic cells. (See intermediate filaments)

VINBLASTIN and vincristine: are antineoplastic alkaloids from the shrub *Vinca rosea*. (See ↓)

VINCULIN: a cytoskeletal protein binding α-actinin, talin, paxillin, tensin, actin filaments and phospholipids. (See mentioned items under separate entries, adhesion)

VIOLENT BEHAVIOR: in humans impulsive aggression has been attributed to reduced levels of 5-hydroxyindole-3-acetic acid in the cerebrospinal fluid and a nonsense mutation in MAOA enzyme resulted in aggressive behavior in a kindred. (See MAOA, behavior in humans)

VIR GENES: see virulence genes *Agrobacterium*

VirA: agrobacterial kinase phosphorylating the product of virulence gene *VirG*.

VIRAL CANCER: see cancer, oncogene

VIRAL ENVELOPE: protein-lipid coat of viruses. (See viruses)

VIRAL GHOST: are empty viral capsids, without their own genetic material but can be filled with DNA and become a genetic vector. (See transformation genetic; see also generalized transduction)

VIRAL ONCOGENE: see v-oncogene

Vinca rosea ↗

VIRAL VECTORS: *in vitro* genetically modified viral DNA (e.g. adenovirus, SV40, bovine papilloma, Epstein-Barr, BK viruses, Baculovirus, Polyhedrosis virus, etc.), containing non-viral genes to be introduced into eukaryotic cells. *Autonomous stable viral vectors* have also been constructed that replicate in the cytoplasm. In order to prevent killing the cells, their copy number is limited by introducing copy number regulators. From the *Bovine Papilloma Virus* (BPV) autonomous (episomal) and shuttle vectors have been constructed that maintain a low (10-30) copies in the cytoplasm. The *shuttle vectors* can be rescued from the mammalian cells and can propagate various protein genes depending on a number of intrinsic and extrinsic factors. The *Epstein—Barr Virus* (EBV) vector can be propagated in the cytoplasm of various types of mammalian cells at low copy number (2-4) and suitable for the study of gene expression, regulatory proteins, etc. The vector has up to 35 kb carrying capacity and can be rescued. The *BK* (baby kidney) *Virus* has been advantageously used for human cells. HIV and other lentivirus vectors may have special advantages because they can integrate into non-dividing cells such as hepatocytes, hematopoietic stem cells and neurons, in contrast to the most widely used mouse leukemia virus vectors which require DNA replication for integration. For human applications *replication-deficient retroviral* vectors can accommodate 9 kb exogenous DNA and they are generally used in *ex vivo* studies. *Adenovirus vector* can carry 7.5 kb DNA and can be taken up by the cell by a specific virus receptor and the $\alpha_V\beta_3$ or $\alpha_V\beta_5$ surface integrins. The adenoviral vector is not integrated into the human genome and thus does not lead to permanent genetic change and the treatment has to be reapplied periodically (in weeks or months). Also, the current vectors may cause inflammation because of antivector cellular immunity. The *Baculovirus* and *Polyhedrosis Virus* are used as vectors for insect cells. The targeting of viral vectors to specific tissues can be increased by genetically engineering into the envelope protein a special receptor for a target ligands e.g. into the Moloney murine leukemia virus the erythropoietin was inserted or into the avian retrovirus envelope an integrin sequence was added. (See vectors genetic, retroviral vectors, shuttle vector, HIV, episomal vector, adenovirus, plasmid rescue, gene therapy, erythropoietin, integrin)

VIRGIN: has not been mated or did not have prior sexual intercourse. In *Drosophila* the females can store the sperm received by prior mating and therefore the paternal identity can be secured only if virgin females are used.

VIRGIN T CELL: see immune response

VIRILE: has the characteristics of an (adult) male, masculine.
VIRION: a complete virus particle (coat and genetic material).
VIROID: non-encapsidated RNA (ca. 1.2×10^5 daltons), capable of autonomous replication and (plant) pathogenesis, such as the potato spindle tuber viroid.
VIRULENCE: determines or indicates the infectivity or pathogenicity of an organism. Recently it was discovered that several bacterial species of different structure and functions acquired a shared mechanism for virulence. About 15-20 protein genes with relatively low G-C contents (below 40%) are assembled in a "pathogenicity island" of either the bacterial chromosome or in a plasmid. These genes encode the molecular machinery (Type III virulence) to produce and transmit to their target the bacterial toxins. (See also neurovirulence)
VIRULENCE GENES OF *AGROBACTERIUM*: the Ti plasmid carries — in an about 35 kb DNA — major (*A, B, C, D, E*) and minor (*F, H*) virulence genes that mediate the process of infection and T-DNA transfer. Gene *VirA* codes for a single protein that is a transmembrane receptor. Its N-terminal periplasmic region responds to sugars and pH whereas the periplasmic loop between the 2 membrane layers responds to phenolic compounds (e.g., acetosyringone, secreted by the wounded plant tissues) which play important role in the induction of the cascade of all *Vir* genes although *VirA* itself is constitutive yet it is modulated by several factors. The C-terminus of VirA protein is autophosphorylated. *VirG* also codes for a single protein, a transcriptional regulator. For expression, it requires phosphorylation by VirA. *VirG* regulates by feedback the phosphate metabolism, mediated by *VirA*, and these two genes together are involved in conjugational transfer. The *VirB* operon encoding 11 proteins is also a conjugational mediator. VirB1 is a lysozyme-like protein. *VirC* determines host range and C1 protein binds to the *overdrive* repeats near the right border of some octopine plasmids. The *VirD* genes are responsible for 4 polypeptides; VirD1 is a topoisomerase, VirD2 is an endonuclease; in addition, this locus codes for a binding protein and a pilot protein guiding the T-DNA to the plant chromosome. *VirE2* codes also for a binding protein which mediates the transfer of DNA single strand into the plant nucleus. *VirF* probably encodes an extracellular protein that regulates *Vir* functions. *VirH* product may metabolize plant phenolics. Nopaline plasmids contain also the gene *tzs* (trans-zeatin secretion) and *pin* (plant inducible) *F* loci. The chromosomal virulence loci (*ChvA, ChvB, Chv, psc*) are involved in the production of bacterial surface polysaccharides. Chromosomal genes *cbg, pgl*, when present, may enhance virulence. (See *Agrobacterium*, Ti plasmid, T-DNA, overdrive, transformation genetic, crown gall)
VIRULENT: in general the poisonous form of prokaryotes; the virulent bacteriophages do not have the prophage life style and after reproduction they destroy the host bacteria by lysis. (See prophage, lysis)
VIRUS: a small particle with either DNA or RNA as the genetic material. Viruses are the ultimate parasites because they lack any element of the metabolic machinery and absolutely depend on the host for assistance to express their genes. They are generally so small in size (15-200 nm) that light microscopy cannot reveal them, except the pox viruses which may be up to 450 nm in length. Bacterial, animal and plant viruses exist. The majority of plant viruses have single-stranded RNA genetic material but a few have double-stranded DNA, e.g. CaMO. (See animal viruses, plant viruses, bacteriophages, oncogenic viruses, retroviruses, cauliflower mosaic virus, TMV)
VIRUS, COMPUTER: a deliberately generated destructive program that can be spread through borrowed computer disks as well as network services. The damage to the files can be usually prevented by the use of continuously updated virus monitoring and eliminating programs.
VIRUS HYBRID: when the *2b* gene of cucumber mosaic RNA virus is replaced by its homolog of tomato, virulence of the interspecific hybrid virus increased. (See vaccinia virus)
VIRUS MORPHOLOGY: see development, T4, lambda phage, retroviruses
VIRUS RECONSTITUTION: see reconstituted virus
VIRUS RESISTANCE: the Fv and Rfv genes of mouse convey resistance against the mouse leu-

Virus resistance continued
kemia (Friend virus) by encoding a retroviral envelope protein or the viral gag (group specific antigen). In the latter case the Fv gene product blocks the entry of the virus into the nucleus or these proteins may exercise a dominant negative effect on the virus that happened to be there. Introduction into the plant genome tobacco mosaic virus coat protein genes restricts the virulence of the virus in the normally susceptible plants.

VIRUS-FREE PLANTS: plants regenerated under axenic conditions from apical meristems are generally free of virus disease until they are reinfected. Plants so obtained are commercially useful for the production of virus-free seed stocks.

VIRUSOID: 300-400 nucleotide long RNAs, pathogenic to plants and accompany other plant viruses. (See viroid)

VISIBLE MUTATION: can be identified by the phenotype seen.

VITAL STAIN: colors living cells without serious damage to viability.

VITAL STATISTICS: involves birth marriage and death registrations. The information may assist in constructing human pedigrees and important information on family histories, congenital and hereditary disease, longevity, etc.

VITALISM: postulates that living beeings are controled not only by physical and chemical mechanisms but life is associated with a transcendental vital force. (See genetic essentialism)

VITAMIN A: also called retinol is synthesized from carotenoids. Its deficiency result in visual and skin anomalies. Excessive amounts may be harmful.

VITAMIN B$_1$: see thiamin →

VITAMIN B$_2$: see riboflavin, riboflavin retention deficiency

VITAMIN B$_6$: see pyridoxine

VITAMIN B COMPLEX: includes thiamin, riboflavin, nicotinic acid (amide), pathotenic acid, pyridoxin and vitamin B$_{12}$.

VITAMIN B$_{12}$ DEFECTS: vitamin K or its coenzyme form has a MW of about 1,355. It is composed of a core ring with Cobalt (Co^{3+}) at its center and to it, through isopropanol, a dimethyl benzimidazole ribonucleotide is joined; it contains also a 5'-deoxyadenosine. It is usually isolated as a cyanocobalamin because during process of purification a cyano group may be attached to the cobalt at the place where the 5'-deoxyadenosyl group is positioned in the coenzyme. B$_{12}$ is not synthesized in plants and animals thus it is not usually present in the diet. Intestinal microorganism make it from the meat consumed and it is absorbed when the so called "intrinsic factor", a glycoprotein, is available in satisfactory amounts. In case less than 3 µg/day is accessible, pernicious anemia develops in humans. B$_{12}$ deficiency may result in case an autosomal recessive mutation prevents the release of the lysosomally stored vitamin. Cysteinuria is often called cobalamin F (cbl F) disease. In methylmalonicacidemia combined with homocystinuria, methylmalonyl-CoA mutase and homocysteine methyltetrahydrofolate methyltransferase (cbl C) deficiencies are involved. (See methylmalonicaciduria, cysteinuria, amino acid metabolism)

VITAMIN C: see ascorbic acid, formula above

VITAMIN D: is an antirachitic fat soluble vitamin and its deficiency leads to ricketsia and defects in bone development and maintenance. Vitamin D$_2$ (ergocalciferol) is formed upon irradiation of ergosterol and vitamin D$_3$ (cholecalciferol) from 7-dehydrocholesterol. Children need about 20 µg/day in their diet. Autosomal recessive human defects in vitamin D receptors do not respond favorably to vitamin D fortified diet. (See also Williams syndrome, VDR, hormone-response elements)

VITAMIN E (tocopherols): vitamin E is an antioxidants (in several different forms) normally present in satisfactory amounts in a balanced diet. In an autosomal recessive condition vitamin E malabsorption was observed, causing

Vitamin E continued

anomalies and accumulation of cholesterol. The severe symptoms could be alleviated by 400-1,200 international units (IU) of vitamin E. It appeared that the affected individuals lacked an α-tocopherol binding protein, required to build it into very low density lipoproteins. The familial vitamin E deficiency is under autosomal recessive control in humans. (See tocopherols, VLDL)

VITAMIN K (phylloquinone): a plant lipid cofactor of blood coagulation (vitamin K_1) and a related substance, menaquinone (vitamin K_2) is synthesized by intestinal bacteria of animals, the synthetic menadione (vitamin K_3) has also some vitamin K activity. (See Vitamin K-dependent blood clotting factors)

VITAMIN K-DEPENDENT BLOOD CLOTTING FACTORS: some autosomal recessive bleeding diseases respond favorably to the administration of vitamin K. It is apparently required for posttranslational modification of at least six proteins involving the conversion of the NH_2-end of glutamic acid into γ-carboxyglutamic acid. A deficiency of this process may occur also as a consequence of treatment by coumarin drugs — such as warfarin — used as anticoagulants. (See prothrombin deficiency, resistance to coumarin-like drugs, antihemophilic factors, warfarin, γ-glutamyl carboxylase)

VITAMINS: minor dietary supplements, without measurable caloric value. They serve generally the role of coenzymes.

VITELLINE LAYER: a yolk (heavier) layer around the eggs; in mammals it is (a thinner) zona pellucida. (See egg)

VITELLOGENIN: yolk protein.

VITILIGO: autosomal dominant halo skin spots (may be identical on opposite side of the body) present after birth that may spread or regress. (See piebaldism, nevus, skin diseases)

VIVIPARY: giving birth to live offspring. In plants the seed germinates before shedding from the fruits.

V(J)D RECOMBINASE: assembles immunoglobulins and the T cell receptors. (See immunoglobulins, T cell receptors, RAG, RSS, accesibility, SCID)

VLDL (very low density lipoprotein): see low-density lipoproteins, familial hypercholesterolemia, sterol, Alzheimer disease, hypertension)

VLP: virus-like particle, such as e.g., some transposable elements (retroposons). (See Ty, retroposons)

V_{max}: the maximal level of enzyme activity (maximal velocity reaction).

v-mil: see MIL, MYC

VNTR (variable number tandem repeats): loci are used in forensic DNA fingerprinting. These repeats may display hundreds of alleles per single loci and are extremely polymorphic in restriction fragments. They are useful as physical markers for mapping. Also, because matching patterns occur by a chance of 10^{-7} to 10^{-8} only, they are well suited for a criminal or other personal, legal identification on the basis of minute amounts of DNA extracted from drops of body fluids, blood or semen. VNTR can also be used for taxonomic and evolutionary studies in animals and plants . (See DNA fingerprinting, diabetes mellitus)

VOLE: *Microtus agrestis*, 2n = 50; *Microtus arvalis*, 2n = 46; *Microtus montanus*, 2n = 24.

VOLT (V): the unit of electric potential. In case the resistance is 1 Ω (Ohm = 1 V/A = 1 m^2 kg sec^{-3} A^{-2}) and the electric current is 1 Ampere, the voltage is 1. (See Ampere, Watt)

VOLTAGE-GATED ION CHANNELS: are opened/closed for the transport of ions in response to a change in voltage across cell membranes. (See ion channels, signal transduction)

v-oncogene: viral oncogene, homologous to a c-oncogene, but carried by oncogenic viruses, capable of causing cancerous growth. (See oncogenes, cancer, c-oncogene, retrovirus)

VON GIERKE DISEASE: see glycogen storage disease

Von HIPPEL-LINDAU SYNDROME (VHL): dominant phenotype (human chromosome 3p26-p25) involving tumorous growth primarily of the blood vessels of the eye (hemangioma) and the brain (hemangioblastoma). The kidneys (phaeochromocytoma) and the pancreas may also become tumorous. The incidence estimates vary between 0.00002 to 0.00003, and the mutation rate appears to be about $2\text{-}4 \times 10^{-6}$. The primary cause of the tumor is the inactivation of the VHL tumor suppressor gene. (See ELONGIN, eye diseases, kidney diseases, phaeochromocytoma, mutation rate, renal cell carcinoma, hypernephroma)

VON RECKLINGHOUSEN DISEASE (NF1): see neurofibromatosis

VON WILLEBRAND'S DISEASE (VWD): is a complex hereditary bleeding condition based on the deficiency of an antihemophilic factor VII. It is different from hemophilias inasmuch as the bleeding from the gastrointestinal, urinary system and the uterus is prolonged. Several forms of this disease have been distinguished based on which component of the large VIII antigen is affected. The bleeding can be readily stopped upon supplying normal blood to the patients. The dominant gene which causes the reduction of factor VIII, was assigned to human chromosome 12pter-p12. In the recessive forms qualitative alteration of the protein is involved and one type is encoded by the X-chromosome. The type III form causes very severe symptoms and it is encoded by an autosome; this is the least common type of VWD. The frequency of heterozygotes has been estimated from 1.4 to 5%, and thus VWD appears quite frequent. Recurrence risk for the dominant form, in case of one of the parents is affected, is about 50%. Both heterozygotes and homozygotes express VWD. The dominant gene (human chromosome 12p-p13.3) was estimated to be 178 kb with 52 exons. Percutaneous umbilical blood sampling (PUBS) may permit prenatal diagnosis. (See hemophilias, antihemophilic factors, hemostasis, prenatal diagnosis)

VON WILLEBRAND-JÜRGEN's SYNDROME: see thrombopathic purpura

VP16: is a herpes simplex virus transcription activation domain that can facilitate the expression of other genes by a factor of 10^5. (See transactivator)

Vpg: genome-linked viral protein attaches to the 5' end of RNA viruses and assist as a primer in the replication of the nucleic acid.

VRE (ventral response element): participates in the regulation of ventral development by preventing switching on an activator that operates the dorsal development.

VSG: variable surface glycoprotein. (See *Trypanosoma*)

VSP (very short patch repair): a prokaryotic repair involving T-G mismatches by restoring the original base pairs.

VSV: see CO_2 sensitivity

V-TYPE ATPases: are responsible for the acidification of cellular organelles (vacuoles, lysosomes, Golgi complex) by the maintenance of the vacuolar type ATPase — proton pump in plant and animal cells.

V-TYPE POSITION EFFECT: is a variegated expression of genes transposed into the vicinity of heterochromatin. This phenomenon is common in *Drosophila* but only few cases have been demonstrated in plants. One of the most common cause of variegation is the movement of insertion or transposable elements or sorting out of plastid-DNA encoded mutations. (See position effect, heterochromatin, variegation, chloroplast genetics, transposable elements)

VULVA: the outer region of the external female genital organ, the vaginal orifice and organs associated with it.

VULVOVAGINITIS: is an autosomal dominant allergy to the semen resulting in vaginal inflammation lasting from a couple to several hours after coitus.

W CHROMOSOME: corresponds to the Y chromosome in the heterogametic females, WZ, in birds and butterflies. (See chromosomal sex determination)

w **LOCUS**: the first mutation discovered in *Drosophila* by Morgan is involved in the control, production and distribution of brown (ommochrome) and red (pteridine) pigments of the eyes and ocelli and some other anatomical structures. The gene (at 1-1.5) apparently encodes an ATP-binding membrane transport protein for the precursors of the pigments. More than 200 alleles have been identified within an 0.03 centimorgan region which has been mapped by intragenic recombination into 7 domains. The wild type allele is incompletely dominant over many mutant alleles. The alleles do not show partial complementation with the exception of the w^{sp} (white spotted) allele that displays allelic complementation with the majority of other alleles in the presence of the z^a *(zeste)*, a regulatory gene at 1-10 location. (See map unit, recombination)

W MUTAGENESIS: tendency of increased mutation after Weigle reactivation. (See Weigle)

W POINT: a stage just before S phase when cells still have serum growth factor requirement to enter S phase. (See cell cycle)

W REACTIVATION: same as Weigle reactivation.

WAARDENBURG SYNDROME: autosomal dominant forms may be distinguished on the basis of displacement (type I) and without displacement (type II) of the eyelids. Variegation in the color of the iris, white forelock, eyebrows and eyelashes, syndactyly, heart defects may occur in an autosomal recessive type. The Waardenburg syndrome type 2 gene (MITF [microphthalmia-associated transcription factor]) converts fibroblasts into melanocyte-like cells. Human chromosomal locations: type I 2q35, type II 3p12-p14. (See PAX, eye defects, polydactily, microphtalmos)

WAGR: see Wilms tumor

WAHLUND'S PRINCIPLE: when two populations, each with different allelic frequencies and both in Hardy - Weinberg equilibrium are mixed by migration, there will be an overall decrease of heterozygotes: $\bar{H} = 2\bar{p}\bar{q}\ [1 - (\sigma^2/\bar{p}\bar{q})]$. The decrease of overall heterozygosity indicates the degree of heterogeneity between the two populations and $(\sigma^2/\bar{p}\bar{q})$ is the Wahlund's variance of gene frequencies. (See allelic frequencies, migration)

WALDEMAR OF PRUSSIA: one of the hemophiliac great-grandson of Queen Victoria. (See hemophilia A)

WALKER BOX: an ATP-binding protein domain. (See ATP)

WALKER-WAGNER SYNDROME: an autosomal recessive hydrocephalus (accumulation of fluid in the enlarged head) generally associated with retinal detachment, congenital muscular dystrophy and lissencephaly. (See Miller-Dieker syndrome, hydrocephalus, prenatal diagnosis, head/face/brain defects)

WALKING: see chromosome walking

WALLABY: *Wallabia bicolor,* 2n = 11 in males and 10 in females; *Wallabia eugenii* 2n =16.

WALLABY: non-viral retrotransposable element named after the jumping small Australian kangaroo. (See retroposon)

WALNUT (*Juglans* spp): both wild and cultivated forms; 2n = 2x = 32.

WALNUT COMB: in poultry determined by the genetic constitution *RrPp* and as a result of epistasis occurs in 9/16 frequency in F_2 after brother sister mating of the same double heterozygotes. The other phenotypes in the segregating F_2 are rose *Rr/RR, pp* (3), pea *rr, PP/Pp* (3), and single *pp, rr* (1). FROM LEFT TO RIGHT: WALNUT, ROSE, PEA, SINGLE COMB

WALRUS: *Odobenus rosmarus*, 2n = 32

WALTZING MOUSE: a chromosome deletion causing involuntary movements.

WANDERING SPOTS: an older method of sequencing short oligonucleotides.

WARD-ROMANO SYNDROME (WRS): an autosomal dominant or recessive LQT disease involving anomalous heart muscle fibrillations, fainting (syncope) and possibly sudden death. (See LQT, electrocardiography, ion channels, HERG, Jarvell and Lange-Nielson syndrome)

WARFARIN ($C_{19}H_{16}O_4$): a slightly bitter, water and alcohol soluble compound. Depresses the formation of prothrombin, necessary for blood clotting and may cause fragility of the capillary veins leading to hemorrhages. It is used in certain surgeries and treatment of diseases that block arteries by blood clots. It is also a rodent poison. A single ingestion may not be necessarily very hazardous to humans but the rats or mice eating it repeatedly in baits suffer internal bleeding and die. The antidote of warfarin is vitamin K. Mutations in rodents may make them resistant to warfarin. (See also anticoagulation factors, prothrombin deficiency, coumarin-like drug resistance, vitamin K-dependent clotting factors, chondrodysplasia)

WASP: *Habrobracon* spp. 2n = 20 for female, 2n = 10 for the male.

WATERCRESS (*Rorippa nasturtium-aquaticum*): is a northern European vegetable with x = 16 but the species may be diploid, sterile triploid or tetraploid.

WATERMELON (*Citrullus vulgaris*): annual fruit, 2n = 22. Triploids are grown commercially.

WATSON and CRICK MODEL: as described in by J.D. Watson and F.H.C. Crick. 1953 in Nature (Lond.): 171:964) became perhaps the world's most famous biological model ever conceived.

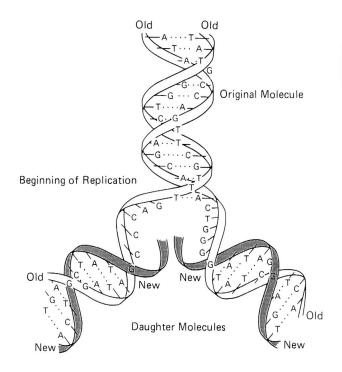

THE DOUBLE-STRANDED DNA MOLECULE IS JOINED THROUGH 2 AND 3 HYDROGEN BONDS BETWEEN THE A=T AND G≡C NUCLEOTIDES, RESPECTIVELY. THE STAIRCASE-LIKE RIBBONS REPRESENT THE SUGAR - PHOSPHATE BACKBONE OF THE DOUBLE HELIX. DURING REPLICATION THE OLD PLECTONEMIC STRANDS UNWIND AND EACH OLD SEPARATED STRAND SERVES AS TEMPLATE FOR THE FORMATION OF THE NEW MOLECULES THAT ARE COMPOSED FROM ONE OLD AND ONE NEW SINGLE STRAND (See DNA replication, replication fork). THIS IS CALLED THEREFORE THE SEMI-CONSERVATIVE MODE OF REPLICATION. THE SIGNI-FICANCE OF THE MODEL IS THAT IS CONSISTENT WITH ALL GENETIC PHEN-OMENA (MUTATION, RECOMBINATION, GENE EXPRESSION, ETC.). SINCE THE MODEL WAS ORIGINALLY PROPOSED, THE DETAILS OF THE MECHANISMS OF THE DNA TRANSACTIONS HAVE BEEN WORKED OUT IN GREATER DETAIL BUT BASICALLY NONE OF THE ESSENTIAL FEATURES HAD TO BE REVISED. THIS MODEL SERVED AS BASIS FOR THE *CENTRAL DOGMA OF GENETICS* INDICATING THAT THE FLOW OF INFORMATION IS FROM DNA TO RNA AND TO PROTEIN. DURING THE 1960s IT WAS DISCOVERED THAT THROUGH REVERSE TRANSCRIPTASE INFORMATION CAN BE DIRECTED BY REVERSE TRANSCRIPTION FROM RNA TO DNA BUT NOT FROM PROTEIN TO RNA AND DNA. THE DISCOVERY OF PRIONS MAKES THE ROLE OF PROTEINS IN HEREDITY SOMEWHAT AMBIGUOUS, HOWEVER. ALL OTHER PROPOSALS CONCERNING HEREDITARY MOLECULES (BESIDES DNA AND RNA) FADED NOW INTO OBLIVION. THE WATSON AND CRICK MODEL IS SHOWN IN ANOTHER FORM ON NEXT PAGE.

Watson and Crick model continued

THE WATSON & CRICK MODEL AS REPRESENTED BY JOSSE, J., KAISER, A.D., and KORNBERG, A. 1961 J. Biol. Chem. 236:864 ↓ (See DNA types)

WATT (W): is the product of Volts and Amperes in case of direct current. 1 W = 1 joule/sec = 0.293 calories/sec = 1/735 HP (horse power). In other words 1 W power is generated by the electric potential between two points of 1 Volt and 1 Ampere current. (See Volt, Ampere)

WCPP (whole chromosome painting probe): contains a mixture of many probes, specific for a single chromosome and thus may label with color its entire length. The multicolor labeling probes may permit the differentiation of all chromosomes in a single karyotype. (See chromosome painting, USP, FISH)

WD-40: repeat (N) motif of tryptophan (W) - aspartic acid (D) in several eukaryotic regulatory proteins (absent in prokaryotes):

$$(X_{6-94} \text{———} [\text{Gly-His} \text{———} X_{23-41} \text{———} \text{Trp-Asp}])^{N_{4-6}}$$

↑ VARIABLE LENGTH ↑ CONSTANT CORE

WD repeats are involved in signal transduction, RNA processing, developmental regulation, cell cycle, vesicular traffic, etc. (See signal transduction)

W-DNA: a left-handed zig-zag duplex with the same directions as B-DNA but other characteristics match that of the Z-DNA. (See DNA types)

WEASEL: *Mustela erminea*, 2n = 44; *Mustela frenata*, 2n = 44.

WEBB (WB): a very rare blood group involves an altered glycosylation of glycophorin. (See glycophorin, Gerbich, En, MN, blood groups)

WEE1: a protein kinase which inactivates *cdc2* gene through phosphorylation of the tyrosine-15 residue. (See kinase, cell cycle, *cdc*)

WEIGHTED MEAN: is the calculated mean multiplied by the pertinent frequency, e.g. in a po-

Weighted mean continued

pulation the mean value of the homo- and heterozygous dominants (*AA and Aa*) is 250 and that of the homozygous recessives (*aa*) is 200, and the respective frequencies are 0.8 and 0.2, then the weighted mean of the population is (250 x 0.8) + (200 x 0.2) = 240. (See mean)

WEIGLE MUTAGENESIS: increase in mutation of phage by mutagenic treatment of the host.

WEIGLE REACTIVATION: increase in phage survival when mixed with host cells exposed to low doses of UV light. (See DNA repair, marker rescue, multiplicity reactivation)

WEISMANNISM: inheritance takes place by the transmission of the genetic determinants through the germline (Keimbahn) and environmentally induced phenotypic variations are not inherited. (August Weismann 1885)

WERDNIG-HOFFMANN DISEASE: is a recessive (5q12.2-q13) infantile muscular dystrophy affecting primarily the spinal cord muscles. The prevalence is approximately 1×10^{-4}. The gene frequency was estimated to be about 0.014 and the frequency of heterozygotes about 0.02. It encodes a protein that shows homology to dystrophin. The survival rate varies depending on the severity of symptoms. A subunit of transcription factor TFIIH is usually suffers deletions. (See dystrophin, muscular dystrophy, neuromuscular diseases, spinal muscular atrophy, transcription factors)

WERNER SYNDROME (WRN): involves premature aging, hardening of the skin, cataracts, arteriosclerosis, diabetes mellitus, etc. Cultured cells have higher chromosome breakage and mutability. Neoplastic growth develops frequently. Ulceration around the ankles and soft tissue calcification are symptoms of this condition, unrelated to aging. The recessive defect appeared to be controlled by human chromosome Xp12-p112 but current linkage information including complete sequencing on the basis of positional cloning, confirms its location to 8p12. The infered protein contains 1,432 amino acids and resembles helicases. The defects in the helicase (frameshift, nonsense mutation) explains well most of the symptoms on the basis of a defect in DNA metabolism. (See aging, progeria, helicase)

WERNICKE-KORSAKOFF SYNDROME: is an autosomal recessive disorder involving transketolase deficiency caused by the reduced ability of the enzyme to bind thiamin pyrophosphate. (See thiamin pyrophosphate)

WEST SYNDROME: involves an X-linked central nervous defect resulting in infantile spasms and mental retardation. (See epilepsy)

WESTERN BLOT: identification of polypeptides separated electrophoretically on SDS polyacrylamide gels, then transfered to nitrocellulose filter and labeled by immunoprobes such as radioactive or biotinylated antibodies. (See immunoblotting, gel electrophoresis)

WGRH (whole genome radiation hybrid): is made when the donor is a diploid cell line for the generation of radiation hybrids. (See radiation hybrid)

WHALE: *Baleonoptera* species are 2n = 44; *Kogia breviceps*, 2n = 42.

WHEAT: see *Triticum*

WHEAT GERM *IN VITRO* TRANSLATION: cell-free wheat germ extracts can be used for the translation of viral, prokaryotic and eukaryotic mRNAs into protein. The supernatant of the extract must be chromatographically purified from inhibitory endogenous amino acids and pigments before translation. Such an extract contains tRNA, rRNA, and other factors required for protein synthesis. Phosphocreatine and phosphocreatine kinase additions are needed for supplying the energy. Spermidine is added to stimulate the translation efficiency and prevent premature termination of the polypeptide chain. Magnesium acetate and potassium acetate, mRNA (to be translated) and amino acids (including one in radioactively labeled form) are also necessary. Incubation is at 25° C for 1 to 2 hours. In general, the procedure is very similar to the rabbit reticulocyte system. (See rabbit reticulocyte *in vitro* translation system)

WHITE BLOOD CELL: see leukocyte

WHITE FORELOCK: see forelock white

WHITE LEGHORN (chicken): has the constitution for genes controlling color of the plumage *CC, OO, II* and the color is white. *C* and *O* both needed for pigmentation but *I* is an inhibitor of color. (See white silkie, white wyandotte)

WHITE SILKIE (chickens): are of the constitution *cc, OO, ii* have white feathers because only one of the 2 dominant genes *O*, required for pigmentation is present. (See White Leghorn, White Wyandotte)

WHITE WYANDOTTE (chickens): have the genes *CC, oo, ii* and are white because only one, *C* of the 2 dominant genes is present but not the other, *O*. (See White Leghorn, White Silkie)

WHORL: a circular or spiral arrangement of structures, such as the various parts of the flowers or the dermal ridges in a human fingerprint. (See fingerprinting, flower differentiation)

WIDOW's PEAK: an autosomal dominant pointed hairline in humans.

WILCOXON's SIGNED-RANK TEST: is a non-parametric substitute for the t-test for paired samples. The desirable minimal number of paired samples is 10 and it is expected that the population would have a median, be continuous and symmetrical. The differences between the variates is tabulated and ranked; the largest receives the highest rank. In case of ties, each should be assigned to a shared rank. The smaller group of signed-rank values are then summed as the T value. This T is then compared with figures in a statistical table and if the figure obtained is smaller than that in the body of the table under probabilty and on the line corresponding to the number of pairs tested, then the null hypothesis is rejected and the conclusion is justified that the two samples are different. Example:

Pairs	Difference	Signed Ranks +	Signed Ranks −
1	+6	7	
2	+5	6	
3	+10	10	
4	−3		4
5	+4	5	
6	+7	8	
7	−2		3
8	−1		0.5
9	+9	9	
10	−1		0.5
			T = 8.0

	Probability	
n	0.05	0.01
10	10	5
11	13	7
12	17	10
14	25	16
16	35	23
18	47	33
20	60	43
22	75	55
24	91	69
26	110	84

Since at T being 8.0 according to the first line of the table the difference between the two sets of data is significant at the 0.05 probability but not at 0.01.

A more general procedure for determining the probabilities relies on determining the Z value either for threshold probabilities or more precisely by using a table of the cumulative normal variates (such as the Biometrika Tables for Statisticians, Vol. 1, Pearson, E.S. & Hartley, H.O., eds.). Z values larger than 1.960, 2.326 and 3.291 correspond to P 0.05, 0.01 and 0.001, respectively. These probabilities rule out the null hypothesis.

$$Z = \frac{\mu - T - 0.5}{\sigma} \quad \text{and} \quad \mu = \frac{n(n+1)}{4} \quad \text{and} \quad \sigma = \sqrt{\frac{[2n+1]\mu}{6}}$$

where n = the number of paired data. (See non-parametric statistics, t test)

WILD TYPE: the standard genotype (that is most common in wild population).

WILDERVANCK SYNDROME: appears to be an X-linked dominant deafness, frequently associated with other disorders. Approximately 1% of the deaf females are affected by it. It does not occur in males, presumably because it is lethal when homo- or hemizygous. It has been suggested that it is polygenically determined with still unknown mechanism of male exclusion. (See deafness, sex-limited, imprinting)

WILDFIRE DISEASE OF PLANTS: is caused by the toxin (methionine analog) of the bacterium *Pseudomonas tabaci*, and it leads to necrotic spots on the leaves.

WILLIAMS SYNDROME (Williams-Beuren syndrome, WMS): is an autosomal dominant condition involving stenosis (narrowing) of the aorta, of the arteries, of the lung, elfin face (elfins are diminutive mythological creatures), malformation of teeth and stature, mental defi-

Williams syndrome continued
ciency and excessive amounts of calcium in the blood (hypercalcemia) and in some tissues. Various types of deletions in different chromosomes (15, 4, 6) have been suspected. Chromosome-specific probes indicated a microdeletion of 7q11.23. Lower calcium in the diet may alleviate some of the symptoms. Vitamin D2 anomaly is suspected. The function of the Williams factor is unrelated. Recent findings indicate the involvement of LIMK protein kinases. (See supravalvular aortic stenosis, cardiovascular diseases, dwarfism, vitamin D, Williams factor, LIM domain, face/heart defects)

WILLIAMS FACTOR (Flaujeac factor deficiency): autosomal recessive mutation in human chromosome 3q26-qter causing deficiency of a high molecular weight kininogen, a precursor of a blood clotting factor. (See kininogen, blood clotting pathways, antihemophilic factors)

WILMS TUMOR: is usually associated with a deletion in the short arm of human chromosome 11 extending from 11p13 to 11p15.5 spanning apparently several genes and frequently designated also as the WAGR syndrome (**W**ilms tumor-**a**niridia-**g**enitourinary anomalies and **R**AS oncogene-like function). The cases usually involve symptoms of all or parts of the functions implied by WAGR. Wilms tumor is caused by mutation in a cancer-suppressor transcription factor with 4 Zn finger domains. The Wilms tumor (WT) is extremely complex and additional WT genes may be in chromosomes 16q, and 1p, 4p, 8p, 14p, 17p and q, 18q were also implicated. The transcript displays alternative splicings. The Wilms tumor suppressor gene is expressed only in the maternally transmitted allele. Prevalence is about 10^{-4} during the first five years of children. The latest evidence indicates that WT is in the 17q12-q21 region. (See aniridia, RAS, deletion, hypertension, kidney diseases, Zinc finger, breast cancer, imprinting, Rhabdosarcoma, Denys-Drash syndrome)

WILSON DISEASE (WD): is a recessive disease encoded in human chromosome 13q14-q21. It affects primarily persons of age 30 and over although juvenile forms have also been described. The major symptom is cirrhosis of the liver and psychological ailments, caused by a deficiency in ceruloplasmin resulting in copper accumulation. The basic defect is in a copper transporting ATPase. The prevalence of WD in the USA is about 3×10^{-5}. Similar anomaly has been observed in Long-Evans Cinnamon rats that may serve as an animal model for the study of the disease. Prenatal diagnosis of offspring of carrier parents may be possible by using linkage with DNA markers. (See also Menke's disease, acrodermatitis, hemochromatosis)

WISKOTT-ALDRICH SYNDROME (WAS): is an X-chromosomal immunodeficiency disease causing eczema, reduced platelet size, bloody diarrhea, high susceptibility to infections, lymphocyte malignancies and usually death before age 10. The prevalence is about 4×10^{-6}. The affected individuals are deficient in a 115 kDa lymphocyte membrane protein and the platelets are abnormally low in a glycoprotein (sialophorin). Carriers may be identified by linkage, by lymphocyte analysis and nonrandom inactivation of the X-chromosomes. WAS is allelic to the human Xp11.23 thrombocytopenia gene. The involvement of CDC42 signaling defect is likely. (See immunodeficiency, thrombocytopenia, thrombopathic purpura, cancer, CDFC42)

WNT1: see INT1 oncogene (*wingless* gene product in *Drosophila*), morphogenesis in *Drosophila* {63}, pattern formation, organizer

WOBBLE: the 5' base of the anticodon can recognize more than one kind of base at the 3' position of the codon, e.g both U or C in the mRNA may pair with G, and both G and A may pair with U, or A or U, or C may recognize I (inosinic acid at the 5' position in the anticodon). According to the classical or universal genetic code of 61 sense and 3 missense codons, a minimum of 32 tRNAs would be required to recognize all the amino acids, Further simplifications permit protein synthesis, however, by 22-24 tRNAs. (See genetic code, tRNA, anticodon, isoacceptor tRNA, mtDNA)

WOLF: (*Canis lupus*, 2n =78) can form fertile hybrids with the domesticated dogs (*Canis familiaris*, 2n = 78) as well as with the coyote (*C. latrans*, 2n = 78) but not with the foxes (*Vulpes vulpes*, 2n = 36). The domesticated dogs appear to be much closer evolutionarily to wolves (on the basis of mtDNA) than to the coyotes. (See fox).

WOLF-HIRSCHHORN SYNDROME: involves a deletion of one of the short arms of human

Wolf-Hirschorn syndrome continued
chromosome 4p16.1 (usually the paternal) resulting in severe growth, mental, face and genitalia defects, etc. The deletion generally eliminates HOX7 (homeobox 7), responsible for normal development in humans and mice. (See deletion, homeobox, homeotic genes)

WOLFFIAN DUCTS: develop as a precursor of the male gonads of vertebrates. This development is enhanced under influence of testosterone hormone. The male gonad (testes) then formed by the Sertoli cells which eventually surround the spermatogonia. The Leydig cells secrete the steroid testosterone and the Sertoli cells produce the anti-Müllerian hormone (human chromosomal location 12q13) which causes the regression of the uterus and the Fallopian tubes. (See also Müllerian ducts)

WOLFRAM SYNDROME (DIDMOAD): this large mitochondrial deletion extending over several coding sequences (7.6 kb) involves **d**iabetes **i**nsipidus, **d**iabetes **m**ellitus, **o**ptic **a**trophy and **d**eafness. (See diabetes, optic atrophy, deafness, mitochondrial diseases in humans)

WOLMAN DISEASE (lysosomal acid lipase deficiency): due to autosomal recessive genes in the long arm of human chromosome 10 and in mouse chromosome 19. The early onset forms are deficient in this enzyme (cholesteryl ester hydrolase) involve liver and spleen enlargement, failure to feed normally and death by 2 to 4 months. The accumulation of cholesterol esters is caused by mutant alleles of the same locus. Some forms permit survival to the teens. (See cholesterol, lysosomes)

WOODCHUCK: see *Marmota monax*

WOODRATS: several species mainly with 2n = 52.

WOODS' LIGHT: is a ultraviolet light source with nickeloxide filter and with a maximal transmission at about 365 nm while most other spectral regions are blocked. (See ultraviolet light)

WOOLLY HAIR: may be black (and autosomal dominant) or blond (and autosomal recessive). The hair in both cases is short and tightly curled.

WORKING HYPOTHESIS: is an experimentally testable assumption regarding a problem.

WORM GENETICS: an informal reference to *Caenorhabditis elegans*. (See *Caenorhabditis*)

WOERTMANNIN: protein from *Penicillium fumiculosum*; is a stimulator of neutrophils and inhibitor of PIK. (See neutrophil, PIK)

WRAPPING CHOICE: generalized transducing phage "chooses" to scoop up host rather than viral DNA and "wraps" it into the phage capsid. (See transduction)

WRIGHT BLOOD GROUP: is very rare; the frequency of the Wr(a) antigen is about 3×10^{-4} in Europe. (See blood groups)

WRINKLED/SMOOTH: gene locus of pea, immortalized by Mendel's discovery of monogenic inheritance. The recessive "wrinkled allele" turned out to be an insertional mutation.

WRITHING NUMBER: indicates the contortion of a DNA double helix in a supercoiled state, i.e. the number of times the double helix crosses its own axis.

WRONGFUL BIRTH: potential responsibility of physician or genetic counselor for negligence in informing or prenatal care of prospective parent(s) about risk involving childbirth.

WRONGFUL LIFE: a potentail responsibility of parents, physicians and genetic counselors for not preventing the birth of a child with serious hereditary disease or for illegitimacy which may carry a social stigma. The affected offspring may sue. (See counseling genetic, genetic privacy, confidentiality, paternity test)

WW DOMAIN: is a two-tyrosine motif of signaling proteins and a binding site for proline-rich peptides. (See SH2, SH3, pleckstrin, PTB, signal transduction)

www (world wide web): the system available through Internet with the aid of a browser program which makes possible the search through a computer linked to the system. (See Internet)

wx **GENE** (*waxy*): occurs in various cereal plants and used as a chromosome marker (in the short arm of chromosome 9-56 of maize). In its presence starch (stained blue by iodine stain) is replaced by amylopectin (stained red-brown); it is defective in NDP-starch glucosyltransferase. (See iodine stain, amylopectin)

X

x: basic chromosome number. (See polyploids, n)

$\bar{x}$: arithmetic mean of the sample.

χ: symbol for crossing over or chiasma.

χ^2: see chi square

X^{1776}: (bicentennial) bacterial strain with an absolute requirement for diaminopimelic acid (a lysine precursor) and needed for the growth of viable bacteria, and thus cannot survive outside the laboratory. It was so named in the US bicentennial year at the Asilomar Conference in 1976. (See Asilomar Conference)

χ ELEMENT: see chi elements

Xa: blood coagulation factor. Also a plasmid factor that specifically cleaves protein after Arg of the tetrapeptide Ile-Glu-Gly-Arg that connects the 31 amino terminal of phage λ cII protein. (See clotting, lambda phage)

XANTHINE: is a purine derived from either adenine through hypoxanthine by xanthine oxidase or from guanine by deamination. By xanthine oxidase it is converted to uric acid. (See uric acid)

XANTHOMATOSIS: see cerebral cholesterinosis

XANTHOPHYLL: yellow carotenoid pigments that play an accessory role in light absorption.

XBP: encodes the 89 kDa subunit of human transcription factor TFIIH, corresponding to yeast *SSL2* encoding a 105 kDa polypeptide. (See transcription factors)

X-CHROMOSOME: one of the sex chromosomes generally present in two doses in the females and in one in the males. In some species the females are XY and the males are either XX or XO or other chromosomal doses may be found. In species where the female is heterogametic, her sex-chromosomal constitution is often designated as WZ and the male's as ZZ. (See Y chromosome, chromosomal sex determination, Barr body, lyonization, Ohno's law)

X-CHROMOSOMAL INACTIVATION: see lyonization, Barr body

XCID: X-chromosome-linked severe combined immunodeficiency. (See SCID, immunodeficiency)

XENIA: expression of the gene(s) of the male in the endosperm following fertilization; the expression in the embryo may not yet be visible. (See also metaxenia)

XENOBIOTCS: compounds that naturally do not occur in living cells.

XENOGAMY: fertilized by different neighboring plants.

XENOGENEIC: transplantation from another species (xenotransplantation).

XENOGENETICS: the study of the effect of environmental factors on conditions under polygenic control. (See polygenic)

XENOGRAFT: transplantation of tissue from another species (e.g. animal→human); it poses problems of rejection and the possibilities of viral infection. The rejection is mediated by the complement cascade of the immune system. If swine, transgenic for the human complement, was used as heart donor to baboons, the function of the heart was prolonged by several hours. (See complement, immune system, transgenic, epitope)

XENOLOGY: a study of the mixture of original and foreign genetic sequences within an organism or group of organisms caused by horizontal transfer and transformation. (See homology, transmission, transformation, infectious heredity)

Xenopus laevis (South African clawed toad): 2n = 4x = 36, a species of frog. Because of being tetraploid and long generation time it is not a favorite object of genetic manipulations although it is an excellent object of embryological studies. Pioneering research was conducted with nuclear transplantation. Transformation of embryonic tissues was not as successful as expected because too few cells expressed the transgenes. Now techniques have been developed for transformation of sperms, and this will permit the study of dominant transgenes on various developmental and physiological processes. (See toad, frog, transformation, transgene,

Xenopus oocyte culture)

XENOPUS OOCYTE CULTURES: *Xenopus* frog oocytes are of about 1 μL in volume; they contain large amounts of DNA (12 pg in the nucleus, 25 pg in the nucleoli, 4 ng in mitochondria). They can synthesize daily 20 ng of RNA and 400 ng of protein. About 10^6 to 10^7 bp DNA can directly be injected into the *germinal vesicle* (the nucleus). The injected DNA is packaged into nucleosomal structure, and it is replicated and translated by the cellular machinery and the products can be ready for analysis within a few hours. Recombinant DNA can be introduced and studied the same way. Exogenous DNA is replicated according to the cell cycle of the oocytes and no reinitiation of the foreign DNA replication occurs. All foreign DNAs are replicated according to the oocyte cell cycles and all the foreign replicational signals are overridden. The replication in the oocytes is also extremely rapid because transcription is apparently halted during replication. *Xenopus* oocytes have been very successfully used for the structural, morphological analysis of vertebrate embryogenesis. They proved to be extremely useful for the molecular analysis of transcription, regulation, cell-to-cell communication and early morphogenesis. (See oocyte, protein synthesis, morphogenesis)

XENOTRANSPLANTATION: is transplantation of organs/tissues among different species. (See xenogeneic)

XENOTROPIC RETROVIRUSES: are replicated only in the cells of species other than the species from which the virus originated. (See also ecotropic and polytropic virus)

XERODERMA PIGMENTOSUM: the first symptoms may arise during the first year of life as very intense freckles induced by sunshine become proliferative and appear as skin cancer. In some cases the central and peripheral nervous system are also affected. Usually A (9q22.2-q34.3), B (2q21), C (3), D (19), E, F (15) G (13). and H complementation groups are distinguished.

A MILD FORM OF XERODERMA PIGMENTOSUM. THE INITIAL SIGNS ARE USUALLY VERY HEAVY FRECKLES WHICH GRADUALLY TURN INTO DIFFERENT TYPES OF SKIN CANCER. THE PROGRESS OF THE DISEASE IS ENHANCED BY SUNSHINE. A VARIETY OF AILMENTS MAY ACCOMPANY THE MAIN SYMPTOMS AND THE AFFLICTED PERSONS RARELY REACH ADULTHOOD. (Courtesy of the March of Dimes—Birth Defects Foundation)

Complementation groups B and D encode helicases. Type F is defective in a DNA repair endonuclease (homologous to yeast Rad1). Also, another form with much milder symptoms is known with an apparently dominant inheritance. In some forms the defect is not in the excision repair but a post-replicational anomaly is causing the light-sensitivity. Similar gene(s) occur(s) in mouse and five RAD genes of yeast have also defects in excision repair. (See DNA repair, excision repair, Bloom syndrome, ataxia telangiectasia, Fanconi syndrome, Cockayne syndrome, *RAD*, light-sensitivity diseases, thrichothiodystrphy, helicase, complementation groups)

XEROPHYTE: draught-tolerant plant, thrives at low-precipitation regions.

XG (Xg[a]): is a blood group antigen determining dominant factor in the short arm of the human X-chromosome. It appears that this end of the X-chromosome can recombine with the Y chromosome, and it does not undergo lyonization as the majority of the X-chromosomal genes. (See lyonization, blood groups, ichthyosis)

Xgal (5-bromo-4-chloro-3-indolyl-β-D-galactopyranoside): when β-galactosidase enzyme hydrolyzes this chromogenic substrate blue color is formed in a bacterial culture plate. (See also β–

galactosidase)

XID: sex-chromosome-linked immunodeficiency.

Xist (chromosome inactivation center): see lyonization

Xklp1: see NOD

XLA: X-linked agammaglobulinemia. (See agammaglobulinemia)

X-LINKED: the gene is within the X-chromosome. (See Z linkage)

XLNO 38: a protein with homologies to nucleoplasmin. Although it is not part of the mature ribosomes, it is associated with both small and large subunits of ribosomes. Its role appears to be chaperoning the assembly of basic proteins on ribosomal RNA precursors. (See chaperone)

X-NUMERATOR: the number of X chromosomes relative to autosomes (X:A) in the determination of sex. (See chromosomal sex determination)

XO: having a single X chromosome in a diploid cell. (See sex determination, Turner syndrome)

XPA, XPD, XPF: see ABC excinucleases

XPB: DNA helicase. Equivalent Ss12. (See DNA repair, excision repair)

XPG: xeroderma pigmentosum endonuclease. (See xeroderma pigmentosum, endonuclease, DNA repair)

X-RAY CAUSED CHROMOSOME BREAKAGE: the major effects of ionizing radiations on living cells is chromosome breakage. Chromosome breakage has been extensively utilized for genetical analyses for knocking out genes, for the production of radiation hybrids, deletion mapping, genomic subtraction, etc. The destructive effects of the radiation depends on the nature of radiation. Soft X-rays having higher density of linear energy transfer, break the chromosome more effectively than the shorter wave-length and high energy hard X-rays. The destructive effect depends also on the species exposed, the type of tissues irradiated, the physiological conditions during the delivery of the radiation, the repair system, etc. In *Drosophila* and locusts usually 1000 - 5000 R doses are employed to cause chromosome breakage in adults and embryos. In case of gonadal radiation, 100 to 500 R may be effective. In the testis cells of *Macaca mulatta* monkey an increase of radiation dose from 25 R to 400 R resulted in close to exponential increase of chromosome breakage reaching about 1.5% of the chromosomes at the highest dose. In the large nuclei and chromosomes of *Trillium* and *Vicia faba* generally 100 to 500 R break the root tip chromosomes and lower doses may be sufficient if applied to pollen mother cells. For the smaller nuclei of maize 800 to 1500 R may be chosen. The very small somatic nuclei of *Arabidopsis* may tolerate 5-10 fold higher doses of irradiation than maize. (See radiation sensitivity, radiation hazard assessment, radiation effects, LET)

X-RAY CRYSTALLOGRAPHY: see X-ray diffraction analysis

X-RAY DIFFRACTION ANALYSIS: is used for the analysis of the structure of molecules by determining the angles of electron scattering upon exposure to X-rays. When a large number of a particular type of molecules in an array are irradiated they will scatter the incident electrons. Where the scattered beams cancel each other, no bright image is formed. Where however the scattered electrons reinforce each other because they diffracted by a certain common arrangement of crystals or molecules, a bright image is formed on the screen. Thus the intensity of the spots on the screen provides a basis for calculating and determining the internal, three-dimensional structure of the object. (See also nuclear magnetic resonance spectrography, X-rays)

X-RAY HAZARD: see X-rays, radiation hazard assessment, radiation protection, radiation effects, X-ray chromosome breakage

X-RAY REPAIR: involves excision repair mechanisms. A human X-ray repair gene locus was isolated through complementation of excision repair-deficient Chinese hamster ovary cells by human DNA. The locus XRCC1 was assigned to human chromosome 19q13.2-q13.3. XRCC4 locus (human chromosome 5) is involved in the determination of X-ray sensitivity in he the G1 phase of the cell cycle but its response in the S phase appears normal, indicating that the defective mutants (in Chinese hamster ovary cells) are deficient in the repair of DNA double strands. (See radiation sensitivity, excision repair, DNA repair, physical mutagens, X-rays)

X-RAY SENSITIVITY: see X-ray repair, X-ray chromosome breakage

X-RAY THERAPY: see radiation therapy

X-RAYS: ionizing electromagnetic radiation emitted by the cathode tubes of Röntgen machines within the range of 10^{-11} and 10^{-8} m wavelength. The shorter ones (hard rays) have greater penetration and lower ionization density while the longer wavelength (soft rays) have reduced penetration and denser dissipation of the energy. Hard rays cause more discrete lesions to the genetic material, the soft rays are expected to cause more chromosomal breakage. The spectrum of the radiation may be controled by filters. The effectiveness of the filters depends on the attenuation of the radiation by the nature of the filter. The fluence of the radiation can be defined as $I = (I_0)e^{-\mu x}$ where I = fluence at a certain depth x; Io = the fluence rate at the surface and μ = the specific attenuation coefficient and μ/p is mass attenuation coefficient (p = density of the absorbing material). At photon energy MeV = 0.1, the mass attenuation coefficients (cm²/g) are for aluminum (0.171), iron (0.370), lead (5.400), water (0.171) and concrete (0.179). (See ionizing radiation, X-ray repair, Compton effect, radiation hazard assessment, radiation protection, radiation efffects)

THICKNESS OF THE PROTECTIVE SHIELD NEEDED IN mm USING LEAD, DEPENDING ON VOLTAGE AND AMPERAGE OF THE X-RADIATION.

Kilo Volt	milliAmpere		
	≤5	5 - 10	30
50	0.5	0.6	0.7
125	1.5	3.0	3.5
250	6.0	7.0	8.0
400	16.0	18.0	21.0

XRCC: see X-ray repair

XREF: cross-referencing model organism genes with human disease and other mammalian phenotypes. (See databases)

XX MALES: occur as a normal condition in birds and some other species, it is a rare (0.000,05) condition of male mammalian (human) births. The recurrence risk is, however, about 25%. Actually most of them (90%) have an X-chromosome—Y-chromosome short arm translocation. Their phenotype and infertility resembles to that of Klinefelter syndrome individuals although the individuals tend to be shorter in stature. Their distinction from hermaphrodites requires the use of appropriate, fluorochrome-labeled cytogenetic probe for the critical Y-chromosomal segment. (See sex determination, sex chromosomal anomalies in humans, Klinefelter syndrome, hermaphrodite)

XXX: see triplo-X, metafemale, sex-chromosomal anomalies in humans

XXY: see Klinefelter syndrome, sex-chromosomal anomalies in humans

XYLEM: transporting tracheid vessels of plants carrying nutrients and water from the roots toward the leaves.

XYLENE ($C_6H_4[CH_3]_2$): synonymous with xylol; a higly flammable and irritant, narcotic liquid used as solvent in microtechnique. The permissible threshold of vapors in the air is 100 ppm.

XYLENE CYANOLE FF: see tracking dyes

XYLOSE (wood sugar, $C_5H_{10}O_5$): an epimer of aldopentoses. It used in the tanning industry, as a diabetic carbohydrate food and in clinical tests of intestinal absorption. (See epimer)

XYLULOSE ($C_5H_{10}O_5$): is an intermediate in the pentose phosphate pathway and accumulates in the urine of pentosuriac patients. (See pentose phosphate pathway, pentosuria)

XYY: see sex-chromosomal anomalies in humans

Y

Y: symbol of pyrimidines in nucleic acid sequences

Y CHROMOSOME: one of the sex chromosomes, present in males (XY) generally. However in some species (birds, insects, fishes) the females are XY and the males may be XX or XO. In organisms with homogametic males their chromosomal constitution is commonly identified as WW. In the majority of the species the Y chromosome has only genes for sex differentiation or sex determination. The major part of the Y chromosomes thus lacks homology to the X chromosome, except in the short common segment. Evolutionist suggested the general low gene number is the consequence of the absence of recombination. R.A. Fisher assumed that the absence of recombination and genes is due to the fact that recombination would mess up sex determination system and would lead to intersexes. The diminution of the gene content of the Y chromosome has been attributed also to Muller's ratchet. Molecular markers for the Y chromosomes are increasing and it is becoming possible to trace paternal lines of evolutionary descent on the basis of the variations in these chromosomes in a manner similar to the mtDNA was used for the development of the (mitochondrial) Eve's origin. (See sex determination, SRY, recombination variations of, Muller's ratchet, Eve foremother of molecular mtDNA, holandric genes, azoospermia)

YAC: **y**east **a**rtificial **c**hromosome vectors equipped with a yeast centromere and some (*Tetra-*

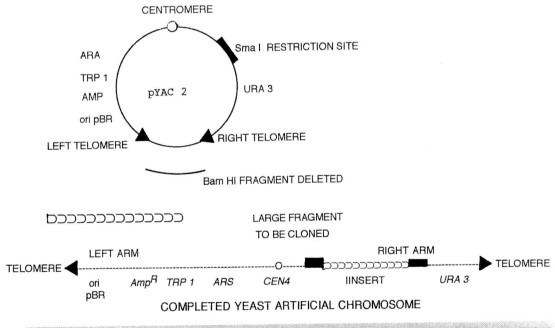

Top: A CIRCULAR YEAST PLASMID WITH *TETRAHYMENA* TELOMERES, A SEGMENT OF THE pBR322 PROKARYOTIC PLASMID, CONTAINING THE REPLICATIONAL ORIGIN AND THE Amp^R GENE (SELECTABLE IN *E. coli* FOR AMPICILLIN RESISTANCE), *ARS* (autonomous replication sequence), CLONED CENTROMERE OF YEAST CHROMOSOME 4 (CEN 4), RESTRICTION ENZYME RECOGNITION SITES WHERE THE PLASMID CAN BE OPENED FOR INSERTION AND LATER LIGATION. The *TRP1* (tryptophan) AND *URA 3* (uracil) GENES OF YEAST SERVE TO ASCERTAIN THAT BOTH ARMS ARE PRESENT AND THE TRANSFORMANT CAN SYNTHESIZE TRYPTOPHAN AND URACIL. THE DELETION OF THE PIECE OF THE PLASMID BETWEEN TWO Bam SITES DELETED ALSO THE YEAST *HIS* (histidine) GENE. THIS VECTOR, BECAUSE OF THE pBR322 SEQUENCES, REPLICATES WELL IN *E. coli* AS WELL AS IN EUKARYOTIC CELLS. ALL DETAILS ARE NOT SHOWN AND THE DIAGRAM IS NOT ON SCALE. OTHER YAC VECTORS ARE SIMILAR BUT NOT IDENTICAL WITH THIS MODEL. Bottom: COMPLETED LINEAR YAC. (Modified after Burke, D.T. *et al.* Science 236:806)

YAC continued

hymena) telomeres in a linear plasmid containing selectable markers, ARS (autonomously replicating sequence) for maintenance and propagation of eukaryotic DNA inserts in cloning of even larger than 200 kb size sequences.

YACs have important role for identifying contigs in physical mapping of larger genomes, for in *situ* hybridization, map based gene isolation, etc. The rate of instability of YAC was estimated to be about 2%. In mitotic yeast cells YACs behave like other chromosomes; meiosis can be analyzed by tetrads although recombination appears to be reduced. YACs may be maintained through some cell divisions in the mouse cytoplasm and behave like *double minute* chromosomes or they may be integrated into the mouse chromosomes. (See also anchoring, contigs, YAC library, *in situ* hybridization, chromosome walking, pulsed field gel electrophoresis, DM)

YAC LIBRARY: contain large restriction fragments of genomic DNA cloned in YAC vectors and separated by pulsed field gel electrophoresis. A YAC library contains generally 100 to 250 kb (or larger) size DNA fragments in multiple (at least five) copies if possible, covering the entire genome of an organism. Selecting the appropriate YAC clone for the purpose of finding the region of interest, the YAC library in yeast colony filter hybridization experiments is probed by RFLPs, PCR, inverse polymerase chain reaction probes, plasmid rescue or by other means. (See also YAC, contig, pulsed field gel electrophoresis, colony hybridization, RFLP, PCR, inverse polymerase chain reaction, plasmid rescue, probe, genome project)

YAK (*Bos grunniens*): 2n = 60.

YAM (*Dioscorea* spp): tropical food crops. The Asian and African species are x = 10 but the Americans are x = 9. The actual chromosome numbers vary from diploid up to decaploid.

YAMA: a Ced-3-like protease, also called CPP32β. (See apoptosis)

YANG CYCLE: the pathway of biosynthesis of ethylene from 2-keto-4-methylthiobutyrate (KMB)→ S-adenosyl-L-methionine (AdoMet)→5'-methylthioadenosine (MTA)→5'-methylthioribose (MTR)→5'-methylthioribose-1-phosphate (MTR-1-P)→KMB. AdoMet is then converted to 1-amino-cyclopropane-1-carboxylic acid (ACC) by ACC synthase and ACC oxidase generates then ethylene. (See ethylene, plant hormones, AdoMet)

YAROWIZATION: see vernalization

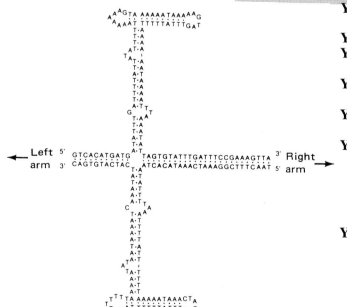

YEAST CENTROMERIC REGION 3. From Clark. L. *et al.* Stadler Symp.13:9

Y-CHROMOSOMAL LINKAGE: see holandric genes,

Ycp: see yeast centromeric vectors

YEAST (budding yeast): see *Saccharomyces cervisiae*

YEAST (fission yeast): see *Schizosaccharomyces pombe*

YEAST ARTIFICIAL CHROMOSOMES: see YAC

YEAST CENTROMERIC VECTORS (Ycp): carry centromeres and telomeres, and ARS, can be linear or circular; their stability improves with increased length. (See also YAC, vectors)

YEAST EPISOMAL VECTOR: carries the origin of replication of the 2 μm yeast plasmid. (See *Saccharomyces cerevisiae*, yeast plasmid 2μm and can be propagated independently in the cytoplasm or integrated into a chromosome. (See vectors)

YEAST INTEGRATING VECTOR (YI): replicated only by the host, it behaves like a gene in the chromosome, can show homologous recombination, duplications and substitutions like bacterial episomes. Although it can generate chromosomal rearrangements its stability is extremely high but its ability to transform is extremely low. (See vectors)

YEAST PLASMID 2 µm: is a circular duplex DNA plasmid of 6318 bp contains genes for replication and thus can maintain a copy number of about 50/cell but it lacks any selectablemarker in native form. By attaching it to the pBR322 *E. coli* plasmid shuttle vectors (ca. 8.5 kbp) have been generated that carry also the bacterial histidine operon (as selectable marker), and it is also expressed in bacteria as well as in budding yeast because a segment of this plasmid can serve as a promoter for the operon. (See yeast vectors, *Saccharomyces cerevisiae*, histidine operon, selectable marker)

YEAST REPLICATING VECTORS (Yrp): carry autonomously replicating sequences (ARS), have moderate stability and relatively low copy number, they may be integrated into chromosomes. (See vectors)

YEAST TRANSFORMATION: may involve a number of different changes in the yeast chromosome at the site of transformation. (See figure below, transformation genetic fungal transformation)

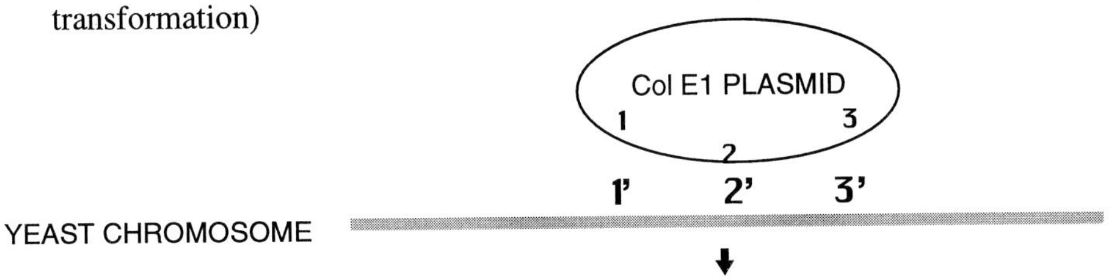

EXCHANGE REGION	HOMOLOGOUS RECOMBINATION							TYPE
1	------------ 1	2	3	Col E1	1'	2'	3' --------	addition
2	------------ 1'	2'- 2	3	Col E1	1	2- 2'	3' --------	transformation
3	-------------1'	2'	3'	Col E1	1	2	3 --------	single crossover
1 - 3	--------------------------1'		2		3' --------------			substitution
2 - 3	--------------------------1'		2' - 2		3' --------------			transformation or
1 - 2	--------------------------1'		2 - 2'		3' --------------			double cross over or gene conversion

The various mechanisms of transformation and recombination between plasmid and chromosomal DNA in yeast. Gene **2'** is an auxotrophic marker gene 2 is a prototrophic homolog; **1'** and **3'** are flanking chromosomal sequences homologous to 1 and 2 in the plasmid with a ColE1 or other replicator. (Modified after Hinnen, Botstein and Davis in *Molecular Biology of the Yeast Saccharomyces*, Strathern *et al.*, eds, Cold Spring Harbor Lab. Press., Cold Spring Harbor, NY.

YEAST TRANSPOSABLE ELEMENTS: see Ty, Ω
Yep: see yeast episomal vector
YES: yeast - *E. coli* shuttle vector. (See shuttle vector)
YES1 ONCOGENE: is in human chromosome 18q21. Its protein product is homologous to that of Roux' sarcoma virus (SRC) and it is also a protein tyrosine kinase. (See tyrosine kinases). (See oncogenes)

YI: see yeast integrating vector
Y-LINKED: see holandric gene
YOLK: the complex nutrients embedding the animal egg. (See egg)
YPD: nutrient medium containing g/L yeast extract 10, glucose 20, Bacto-Peptone 10.
YPGE: a nutrient medium containing yeast extract (10 g), Bacto-Peptone (10 g), glycerol (20 g) and ethanol (10 g) per liter H_2O.
YPT: a homolog of Sec, RAB (See Sec, RAB)
Yrp: see yeast replicating vectors
Yt BLOOD GROUP: is coded in human chromosome 7q.
YT BACTERIAL MEDIUM: H_2O 900 mL, bacto-tryptone 16 g, bacto-yeast extract 10 g, NaCl 5 g, pH 7.0 (adjusted by 5N NaOH), filled up to 1 L and diluted to half before use.
Yttrium (Y): a very rare metal; the ^{90}Y has a half-life of 64 hours and has been used for internally exposing cancerous tissues to β-radiation. (See ionizing radiation, isotopes, magic bullet)
YUASA ONCOGENE: its human homolog is in chromosome 7.
YY ASPARAGUS: by regeneration of plants from microspores and chromosome doubling, plants can be obtained with this chromosomal constitution. The practical advantage of this vegetatively reproducible vegetable is the approximately 30% higher yield of edible spears. (See diagram below, embryogenesis somatic)

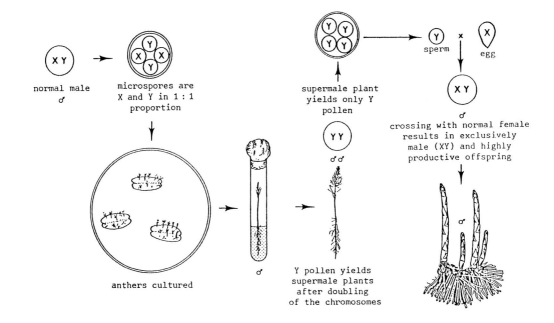

THE PRINCIPLE OF PRODUCTION OF (YY) ALL MALE ASPARAGUS.

Z

Z (or z): is the standard normal probability density function calculated by the formula:

$$Z = \frac{1}{\sigma\sqrt{2\pi}} e^{-(Y-\mu)^2/2\sigma^2}$$

where Y is the normal variate, $\pi \cong 3.14159$ and $e \cong 2.71828$, μ is the mean and σ is the standard deviation. The z values are generally read from statistical tables. Z indicates the height of the ordinate of the curve and thereby the density of the items. (See standard deviation, normal distribution, confidence intervals)

Z BUFFER: $Na_2HPO_4.7H_2O$ 0.06 M, NaH_2PO_4 0.04 M, KCl 0.01 M, $MgSO_4.7H_2O$ 0.001 M, β-mercapto ethanol 0.05 M, pH 7. Not to be autoclaved.

Z CHROMOSOME: the sex-chromosome present in both sexes of species with heterogametic females (comparable thus to the X chromosome). The males are ZZ and the females are WZ. (See also W chromosome, sex determination)

Z DISC: see sarcomere

Z DISTRIBUTION: is a statistical device for testing the significance of the differences between correlation coefficients in case the null hypothesis is not r = 0. The relation of z to r was elaborated by R.A. Fisher as $z = (1/2)[ln(1 + r) - ln(1 - r)]$ and its standard error as $\sigma_z = \frac{1}{\sqrt{(n-3)}}$.

For routine calculations usually tables are available in statistical textbooks. (See covariance for correlation coefficient)

Z DNA: is a relatively rare left-handed double helix that may be formed in short (8-62 bp) regions of alternating purines and pyrimidines (ATGTGTGT, GCATGCAT). By altering DNA structure transcription may be facilitated. (See DNA types)

Z LINKAGE: genes syntenic in the Z chromosome. (See chromosomal sex determination, X-linked)

Z RNA: may occur in double-stranded RNA molecules and it is similar to Z DNA. (See Z DNA)

Z SCHEME: an additional (zig-zag) scheme to generate enough ATP by photosynthsis. By two steps (in photosystem I and II) an electron passes from water but there is not enough energy in a single quantum of light to energize the electron directly and efficiently all the way from PS II to top of PS I and make $NADP^+$ The leftover energy makes pumping H^+ possible across the membranes to capture some light energy for synthesizing ATP. (See photosystems)

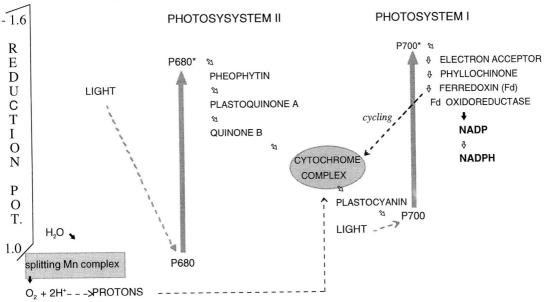

ZAP-70 (zeta-associated protein 70): is a cytosolic protein tyrosine kinase expressed only in T cells and natural killer cells. By binding to phosphorylated ζ-chains of the CD3 T-cell antigen-receptor complex, it assists in the activation of T cells. Its defect may lead to severe combined immunodeficiency. (See SHP-1, CD3 T cell, killer cell, immunodeficiency)

Zea mays (L.) : see maize

ZEATIN (6[4-hydroxy-3-methyl-cis-2-butenylamino]purine): a cytokinine plant hormone. In the *Agrobacterium* T-DNA-located *ipt* gene encodes an isopentenyl transferase that mediates the synthesis of trans-zeatin and isopentenyl adenosine. In plant tissue culture media it is commonly used alternative to kinetin, benzylamino purine or isopentenyl adenosine. (See plant hormones)

ZEBRA: *Equus quagga*, 2n = 44; it can form hybrids with both horses (*Equus caballus*, 2n = 64) and donkeys (2n = 62). *Equus grevyi*, 2n = 46; *Equus zebra hartmanniae*, 2n = 32. The Zebra duiker (*Cephalophus zebra*, 2n =58) is not a member of the Equidae family but it is the male of a bovine species.

ZEBRA (Zta): a non-acidic activator protein of the lytic cycle of the Epstein-Barr virus; it promotes the assembly of the DA complex (TFIID-TFIIA) of transcription factors. (See transcription factors, open transcription complex)

ZEBRAFISH (*Brachydanio rerio*, 2n = 50): a 3-4 cm tropical fresh water fish; the genome size is about 2×10^9 bp. It is easy to raise, sexually mature in 2-3 months (4 generations/year). Embryogenetic pattern is laid down in 12 hrs, and it is well suited for the analysis of developmental pathways and cell lineages in this small vertebrate. The hundreds of eggs laid externally. The embryos are transparent to permit viewing of gastrulation, development of the brain and heart, etc. Haploids survive for several hours and mutants are available. About 500 genes involved in its development have been identified. (See EP, <http://zebra.sc.edu>)

ZEBU (*Bos indicus*): 2n = 60. (See Santa Gertrudis cattle)

ZEIN: is a prolamine protein in maize (homologous to gliadin in wheat) may make up to 50% of the grain proteins. It is a protein of low nutritional value (and thus undesirable) because of the low lysine and tryptophan content. It is deposited in zein bodies at the place of synthesis. In the high-lysine maize varieties (*opaque, floury*) prolamines are very low and non-prolamine proteins increase. (See high-lysine corn, glutenin)

ZELLWEGER SYNDROME: is a brain-liver-kidney (cerebrohepatorenal) recessive disease involving human chromosome 7q11.12-q11.13 but a dozen of other loci may have similar effects. The basic defect is due to peroxisome anomalies. (See microbodies, neuromuscular diseases, chondrodysplasia punctata, cataract, pseudo-Zellweger syndrome)

ZERO, ABSOLUTE: the minimum lowest temperature; Kelvin 0° = Celsius - 273.15°.

ZERO TIME BINDING: is the status of reassociation of two single-strand palindromic DNAs at the beginning of an annealing kinetics experiment. These are the fastest reassociating fractions because they are repeats and are close to each other. (See c_0t curve)

zeste: see *w* locus

ZFY (zinc finger Y): a sequence in the Y chromosomes assumed to be involved in the maturation of testes or sperm. A 729 bp intron located immediately upstream of the zinc finger exon shows no or possibly very little sequence variation in world wide human samples. (See Zinc finger)

ZIG-ZAG INHERITANCE: see criss-cross inheritance

zif: see NGFI-A

ZINC FINGERS: are binding mechanisms of transcription factors and other regulatory proteins containing tandemly repeated cysteine and histidine molecules, and these fold in a "finger-like" fashion cross-linked to Zn. Other highly conserved amino acids in it are phenylalanine (F), leucine (L) and tyrosine (Y). Some Zinc finger proteins bind also RNA to DNA. (See hormone receptor, DNA-binding protein domains, acrodermatitis enteropathica, ZFY, RING finger, binuclear zinc cluster, Cys_4 receptor)

ZINC RING FINGER: see RING finger

ZIPPER DOMAIN: part of dimeric DNA-binding proteins where the two subunits are held together by repeating amino acid residues; in the other parts the subunits are separated from each other. (See leucine zipper)

ZOIDOGAMY: fertilization takes place by motile antherozoids. (See also siphonogamy)

ZOLLINGER-ELLISON SYNDROME (Wermer syndrome MEN 1): is a multiple endocrine adenomatosis (neoplasia) encoded in human chromosome 11q13. (See adenomatosis multiple endocrine)

ZONA PELLUCIDA: a yolky layer around the mammalian egg. (See fertilization [animals], vitelline layer, egg)

ZONULA OCCLUDENS: tight cell junction associated with a 210-255 kDa membrane-tied guanylate kinase.

ZOO BLOT: Southern hybridization experiments with probes derived from a variety of different species to test potential homologies.

ZOOSPORE: motile (swimming) spore.

ZP1, ZP2, ZP3: zona pellucida glycoproteins. (See fertilization, zona pellucida)

ZPA (zone of polarizing activity): determines the anterior/posterior differentiation of the limbs and it is located behind AER in the limb bud. (See AER, limb bud)

ZPR: zinc finger protein binding to epidermal growth factor receptor. (See zinc fingers, EGFR)

Zta: see ZEBRA

ZWISCHENFERMENT: see G6PD

ZWITTERION: dipolar ion with separated positive and negative poles.

ZygDNA: it has been reported that 0.1 to 0.2% of the eukaryotic DNA may not be replicated until late leptotene–zygotene. This delayed replication involves dispersed 4-10 kb stretches and it was assumed that these segments (ZygDNA) code for genes with products aiding chromosome pairing. The delayed replication was suspected to be caused by a leptotene (L) lipoprotein. (See meiosis)

ZYGOMORPHIC: a structure of bilateral symmetry, like e.g. a snapdragon flower.

ZYGONEMA: the chromosome at the zygotene stage. (See meiosis, zygotene stage)

ZYGOSIS: twins can be identical (monozygotic, MZ) or non-identical (dizygotic, DZ). The distinction is not always simple because dizygotic twins (like any sibs) may show several identical features, depending on the genetic constitution (consanguinity) of the parents. If we designate the probability of monozygosis between twins with identity in a genetic marker as $P(A_1/B)$ where A_1 and B are different markers and the probability of the twins being either dizygotic or erroneously assumed to be monozygotic is: $1 - P(A_1/B)$. The calculation may be based on the formula:

$P(A_1/B) = \dfrac{1}{1 + [Q \times L]}$ where Q is the dizygotic:monozygotic proportion in the population (DZ/MZ), and L = likelihood ratio of the conditional probabilities for DZ and MZ twins would be identical for a particular genetic condition. The conditional probabilities that if one of the twins is of a particular dominant type, the second would also be of the same type at dizygosis is 0.5 - 1.0, and at monozygosis 1.0. In case of recessive markers, these probabilities are 0.25 and 1.0, respectively. The probabilities depend also on the genetic constitution of the parents. Recessive markers may be expressed only if both parents carry that particular allele. If either of the parents is homozygous for a dominant marker, then both twins must carry that marker, irrespective of zygosity. L can be computed as $L = \dfrac{P[DZ]}{P[MZ]} L_1 \times L_2 ... L_n$ where $\dfrac{P[DZ]}{P[MZ]}$ is the empirical probability of the DZ:MZ proportions in the general population, and $L_1 \times L_2 ... L_n$ are the conditional probabilities for the genetic markers 1 to *n* used, either 0.25 or 1.0. Further complications may arise if either the penetrance or expressivity of the markers vary. The DZ:MZ proportions may vary generally from 0.65:0.35 to 0.70:0.30 but it is somewhat different in various ethnic groups, the age of the mother and the use of fertility drugs and artificial insemination, etc. DNA markers, either by RFLP or the PCR method of typing, can better re-

Zygosis continued

solve the problem than using blood types or other genetic analyses. (See conditional probability, likelihood, RFLP, PCR, DNA fingerprinting, twinning, monozygotic, dizygotic, concordance, discordance)

ZYGOSPORE: is formed by the fusion of two spores or two multinucleate gametangia. (See heterothallism, homothallism, gametangia)

ZYGOTE: cell resulting of the union of two gametes of opposite sexes. (See gamete)

ZYGOTENE STAGE: of meiosis when the homologous chromosomes begin to synapse. The pairing usually begins at the termini and proceeds toward the centromeric region. At this stage the synaptonemal complex is detectable by electronmicroscopy. The intimate bivalent pairing appears to be a requisite for chiasma formation and presumably for genetic recombination. (See meiosis, chiasma, synaptonemal complex, association point)

ZYGOTIC COMBINATIONS: see Punnett square, allelic combinations

ZYGOTIC GENE: during embryo development it participates in the early control of differentiation in contrast to the maternal effect genes which are transcribed from the maternal genome and their product is transfused to the embryo. (See maternal effect genes, morphogenesis)

ZYGOTIC INDUCTION: usually the integrated λ prophage is inherited as an integral gene of the bacterial chromosome. When, however, a lysogenic Hfr cell, carrying the λ is crossed to a *nonlysogenic* F⁻ recipient, the prophage leaves the chromosomal position (induction) and becomes an infectious vegetative phage after replicating to about 100-200 particles. After zygotic induction, only the markers transmitted before the position of the prophage has changed are recovered in the recombinants. Zygotic induction does not occur if the *F⁻ cells are lysogenic*, irrespective wheather the Hfr is lysogenic or not. The F⁻ recipient is immune to superinfection by free λ phage and it cannot support the vegetative development of the chromosomal λ phage transmitted by the Hfr donor. This indicates that the F⁻ cytoplasm carries an immunity substance or a repressor. (See lysogeny, Hfr, F⁻, lysogenic repressor, lambda phage)

ZYGOTIC LETHAL: genetic factor that permits the function of the gametes but kills the zygote.

GAMETIC LETHALITY

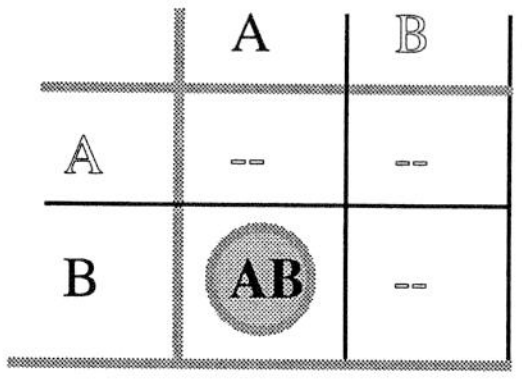

	A	B
A	--	--
B	AB	--

THE GAMETIC LETHAL FACTORS A AND B ARE UNABLE TO FORM SELFED ZYGOTES, AND ONLY THE HETEROZYGOUS EMBRYOS ARE VIABLE. ONLY THE FEMALE CAN CONTRIBUTE A VIABLE *B* GAMETE AND ONLY THE MALE PRODUCES VIABLE *A* GAMETES

ZYGOTIC LETHALITY

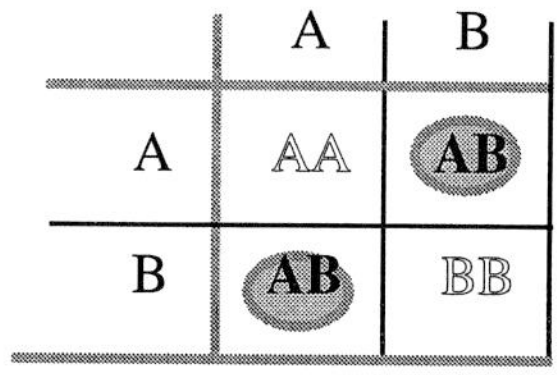

	A	B
A	AA	AB
B	AB	BB

IN ZYGOTIC LETHALITY ONLY THE HETEROZYGOTES ARE VIABLE AND BOTH TYPES OF HOMOZYGOTES ARE LETHAL

The two systems shown above occur in *Oenothera* plants which carry translocation complexes. The inviable gametes and zygotes are printed in outline.

(See translocation, complex heterozygote, translocation [in animals], *Oenothera*)

ZYMOGEN: inactive enzyme precursor. (See regulation of enzyme activity)

ZYMOGRAM: electrophoretically separated isozymes are identified in the electrophoretic medium (starch, agarose) by supplying a chromogenic substrate *in situ* and this reveals the position of the functionally active enzyme bands. (See electrophoresis, isozyme)

ZYMOLASE: hydrolyzes 1→3 glucose linkages such as those existing in yeast cell walls; it is not an exactly defined mixture of proteins extracted from *Athrobacter luteus*.

ZYMOTYPE: electrophoretically determined pattern of enzymes (proteins), characteristic for individuals or group of individuals. (See isozymes, electrophoresis, zymogram)

GENERAL REFERENCES

A list of books, published mainly during the last few years, and selected on the basis of relevance to the topics covered in this volume. The grouping is somewhat arbitrary because of the overlaps in scope and contents.

AGING *(see also molecular biology, cancer, diseases)*

Bellamy, D. 1995 Aging: A Biomedical Perspective. Wiley, New York
Goate, A. and Ashall, F. 1995 Pathobiology of Alzheimer's Disease. Acad. Press, San Diego, CA, USA
Holbrok, N.J., Martin, G.M. and Lockshin, R.A. 1995 Cellular Aging and Cell Death. Wiley, New York
Holliday, R. 1995 Understanding Aging. Cambridge Univ. Press, New York
Kanungo, M.S. 1994 Genes and Aging. Cambridge Univ. Press., New York
Medina, J.J. 1996 The Clock of Ages: Why We Age — How We Age — Winding Back the Clock. Cambridge University Press, New York
Rattan, S.I.S. and Toussaint, O., eds. 1996 Molecular Gerontology. Plenum Press, New York
Rose, M.R. and Finch, C.E. 1994 Genetics and Evolution of Aging. Kluwer, Norwell, MA, USA
Smith, D.W.E. 1993 Human Longevity. Oxford Univ. Press, New York
Tomel, D.L. and Cope, F.O., eds. 1994 Apoptosis II. Molecular Basis of Apoptosis in Disease. Cold Spring Harbor Laboratory Press, Cold Spring Harbor, NY, USA

BEHAVIOR *(see also neurobiology)*

Begleiter, H. and Kissin, B. 1995 The Genetics of Alcoholism. Oxford Univ. Press, New York
Boake, C.B., ed. 1994 Quantitative Genetic Studies of Behavioral Evolution. Univ. Chicago Press, Chicago, IL, USA
Bock, G.R. and Goode, J.A., eds. 1996 Genetics of Criminal and Antisocial Behavior. Wiley, New York
Bouchard, T.J. and Propping, P. 1993 Twins as a Tool of Behavioral Genetics. Wiley, New York
Plomin, R., DeFries, J.C. and McClearn, G.E. and Rutter, M. 1997 Behavioral Genetics. Freeman, New York
Quiatt, D. and Reynolds, V. 1995 Primate Behavior. Cambridge Univ. Press, New York
Rutter, M. ed. 1995 Genetics of Criminal and Antisocial Behavior. Wiley, New York
Sternberg, R.J. and Ruzgis, P., eds. 1994 Personality and Intelligence. Cambridge Univ. Press, New York
Turner, J.R., Cardon, L.R. and Hewitt, J.K., eds. 1995 Behavior Genetic Approaches in Behavioral Medicine. Plenum, New York
de Waal, F. 1996 Good Natured: The Origins of Right and Wrong in Humans and Other Animals. Harvard Univ. Press, Boston, MA, USA

BIOCHEMISTRY *(see also molecular biology, laboratory technology, biotechnology)*

Aidley, D. J. and Stanfield, P.R 1996 Ion Channels: Molecules in Action. Cambridge Univ. Press, New York
Blackburn, G.M. and Gait, M.J., eds. 1996 Nucleic Acids in Chemistry and Biology. Oxford Univ. Press, New York
Bodansky, M. 1993 Principles of Peptide Synthesis. Springer-Vlg., New York
Branden, C. and Tooze, J. 1991 Introduction to Protein Structure. Garland, New York
Burrell, M.M., ed. 1993 Enzymes of Molecular Biology. Humana, Totowa, NJ, USA
Campbell, J.L., ed. 1995 DNA Replication. Methods in Enzymology, Vol. 262. Acad Press, San Diego, CA, USA
Cohen, N.C., ed. 1966 Guidebook on Molecular Modeling in Drug Design. Acad. Press, San

Biochemistry continued
Diego, CA, USA
Conly, E.C. 1995 Ion Channel FactsBook. Extracellular Ligand-Gated Ion Channels. Acad. Press, San Diego, CA, USA
Elliott, W.H. and Elliott, D.C. 1997 Biochemistry and Molecular Biology. Oxford Univ. Press, New York
D'Alessio, G. and Riordan, J.F., eds. 1997 Ribonucleases. Structures and Functions. Acad.Press, San Diego, CA, USA
Devlin, T.M., ed. 1997 Textbook of Biochemistry with Clinical Correlations. Wiley, New York
Ellis, R.J., ed. 1996 The Chaperonins. Acad. Press, San Diego, CA, USA
Hardie, G.D. and Hanks, S. 1995 The Protein Kinase FactsBooks. Acad. Press,San Diego, CA USA
Hecht, S.M., ed. 1996 Bioorganic Chemistry. Nucleic Acids, Oxford Univ. Press, New York
Hurtley, S., ed. 1996 Protein Targeting. IRL Press, New York
Juo, P.S., ed. 1996 Concise Dictionary of Biomedicine and Molecular Biology. CRC Press, Boca Raton, FL, USA
Kornberg, A. and Baker, T.A. 1992 DNA Replication. Freeman, New York
Kricka, L.J., ed. 1995 Nonisotopic Probing, Blotting and Sequencing. Acad. Press, San Diego, CA, USA
Kyte, J. 1995 Structure in Protein Chemistry, Garland, New York
Metzler, D.E. 1977 Biochemistry. The Chemical Reactions of Living Cells. Acad Press, New York
Pain, R.H. ed. 1994 Mechanism of Protein Folding. ORL Press, New York
Papa, S. and Tager, J.M., eds. 1995 Biochemistry of Cell membranes. A Compendium of Selected Topics. Birkhäuser, Cambridge, MA, USA
Soyfer, V.N. and Potaman, V.N. 1996 Triple-Helical Nucleic Acids. Springer-Vlg., New York
Sperelakis, N., ed. 1995 Cell Physiology Source Book. Acad. Press, San Diego, CA, USA
Stryer, L. 1995 Biochemistry, Freeman, New York
Voet, D. and Voet, J. 1995 Biochemistry, Wiley, New York
Wallsgrove, R.M., ed. 1995 Amino Acids and their Derivatives in Higher Plants. Cambridge Univ. Press, New York
Webb, E.C. 1992 Enzyme Nomenclature. Acad. Press, San Diego, CA, USA
Woodgett, J.R., ed. 1995 Protein Kinases. IRL, Oxford Univ. Press, New York
Zubay, G.L., Parson, W.W., and Vance, D.E. 1995 Principles of Biochemistry. Brown, Dubuque, IA, USA

BIOLOGY (see also molecular biology, microbiology)

Alberts, B., Bray, D., Lewis, J., Raff, M., Roberts, K., and Watson, J.D. 1994 Molecular Biology of the Cell. Garland, New York
Calisher, C.H. and Fauqquet, C.N., eds. 1992 Stedman's/ICTV VirusWorlds. Williams & Williams, Baltimore, MD. USA
Cheresh, D.A. and Mecham, R.P., eds. 1994 Integrins. Molecular and Biological Responses to the Extracellular Matrix. Acad. Press, San Diego, CA, USA
Clemente, C.D., ed. 1985 Gray's Anatomy. Lea & Febiger, Philadelphia, PA, USA
Cross, P. C. and Mercer, K.L. 1993 Cell and Tissue Ultrastructure. A Functional Perspective. Freeman, New York
Desjardins, C. and Ewing, L.L., eds. 1993 Cell and Molecular Biology of the Testis. Oxford Univ.Press, New York
Devor, E.J., ed. 1993 Molecular Application in Biological Anthropology. Cambridge Univ. Press, New York
Dulbecco, R., ed. 1997 Encyclopedia of Human Biology. Acad. Press. San Diego, CA, USA

Biology continued

Epstein, H.F. and Shakes, D.C., eds. 1995 *Ceanorhabditis elegans*. Modern Biological Analysis of an Organism. Acad. Press, San Diego, CA, USA
Ferraris, J.D. and Palumbi, S.R., eds. 1966 Molecular Zoology. Advances, Strategies and Protocols. Wiley-Liss, New York
Graham, P.H., Sadowsky, M.J. and Vance, C.P. 1994 Symbiotic Nitrogen Fixation. Kluwer, Norwell, MA
Grundzinskas, J.G. and Yovich, J.L., eds. 1994 Gametes. The Oocyte. Cambridge Univ. Press, New York
Grundzinskas, J.G. and Yovich, J.L., eds. 1995 Gametes. The Spermatozoon. Cambridge Univ. Press, New York
Hall, B.K. ed. 1994 Homology. The Hierarchial Basis of Comparative Biology. Acad. Press, San Diego, Ca, USA
Holt, J.G., Bruns, M.A., Caldwell, B.J. and Pease, C.D., eds. 1993 Bergey's Manual of Determinative Bacteriology. Williams & Williams, Baltimore, MD, USA
Hutchison, C. and Glover, D.M., eds. 1995 Cell Cycle Control. IRL, New York
Jong, S.-C., Birmingham, J.M. and Ma, G., eds. 1993 Stedman's/ATCC Fungus Names. Williams & Williams, Baltimore, MD, USA
Kohen, E., Santus, R. and Hirschberg, J.G. 1995 Photobiology. Acad. Press, Sandiego, CA, USA
Krstíc, R.V. 1994 Human Microscopic Anatomy. An Atlas for Students of Medicine and Biology. Springer-Vlg, New York
Larsen, W.J. 1993 Human Embryology. Churchill Livingstone, New York
Lodish, H., Baltimore, D., Berk, A., Zipursky, S.L., Matsudaira, P. and Darnell, J. 1995 Molecular Cell Biology. Freeman, New York
Mabberly, D.J. 1997 The Plant-Book. A Portable Dictionary of the Vascular Plants. Cambridge Univ. Press, New York
Mayr, E. 1997 This is Biology: The Science of the Living World. Harvard Univ. Press. Cambridge, MA, USA
Merz, K.M. and Roux, B., eds. 1966 Biological Membranes. A Molecular Perspective from Computation and Experiment. Birkhäuser Vlg., Boston, MA, USA
Peracchia, C., ed. 1994 Handbook of Membrane Channels. Acad. Press, San Diego, CA, USA
Raven, P.H. and Johnson, G.B. 1992 Biology. Mosby, St. Louis, MO, USA
Scott, T.A., ed. & Translator, 1966 Concise Encyclopedia. Biology. Walter de Gruyter, New York
Thorpe, T.A. ed. 1995 In Vitro Embryogenesis in Plants, Kluwer, Norwell, MA, USA
Weiss, K.M. 1993 Genetic Variation and Human Diseases. Principles and Evolutionary Approches. Cambridge Univ. Press, New York
Wood, J.W. and de Gruyter, A. 1954 Dynamics of Human Reproduction. Biology, Biometry, Demography. Hawthorne, New York
Yu, H.-S. 1994 Human Reproductive Biology. CRC Press, Boca Raton, FL, USA

BIOMETRY

Bailey, N.T.J. 1995 Statistical Methods for Biologists. Cambridge Univ. Press, New York
Elands-Johnson, R.C. 1971 Probability Models and Statistical Methods in Genetics. Wiley, New York
Emery, A.E.H. 1986 Methodology in Medical Genetics. An Introduction to Statistical Methods. Churchill Livingstone, Edinburgh, UK
Forthofer, R.N. anf Lee, E.S. 1995 Introduction to Biostatistics. A Guide to Design, Analysis, and Discovery. Acad. Press, San Diego, CA, USA
Hays, W.L. and Winkler, R.L. 1971 Statistics: Probability, Inference and Decision. Holt, Rinehart and Winston, New York

Biometry continued

Howson, C. and Urbach, P. 1993 Scientific Reasoning. The Bayesian Approach. Open Court, La Salle, IL, USA
Kemptorne, O. 1957 An Introdution to Genetic Statistics. Wiley. New York
Lange, K. 1997 Mathematical and Statistical Methods for Genetic Analysis. Springer, New York
Lindley, D.V. 1965 Introduction to Probability and Statistics from a Bayesian Viewpoint. Cambridge Univ. Press, New York
Malécot, G. 1969 The Mathematics of Heredity. Freeman, San Francisco, CA, USA
Mather, K. 1965 Statistical Analysis in Biology. Methuen, London, UK
Mitchell, M. 1996 An Introduction to Genetic Algorithms. MIT Prees, Cambridge, MA, USA
Norton, N.E., Dabeeru, C.R. and LaLuel, J-M. 1983 Methods in Genetic Epidemiology. Karger, New York
Paterson, A.H., ed. 1997 Molecular Dissection of Complex Traits. CRC Press, Boca Raton, FL, USA
Shoukri, M.M. and Edge, V.L. 1995 Statistical Methods for Health Sciences. CRC Press, Boca Raton, FL, USA
Snedecor, G.W. and Cochran, W.G. 1967 Statistical Methods. Iowa State Univ. Press, Ames, IA, USA
Sokal, R.R. and Rohlf, F.J. 1969 Biometry. Freeman, San Francisco, CA, USA
Young, I.D. 1991 Introduction to Risk Calculation in Genetic Counselling. Oxford Univ. Press, Nerw York
Zwillinger, D. 1996 CRC Standard Mathematical Tables and Formulae. CRC Press, Boca Raton, FL, USA

BIOTECHNOLOGY *(see also laboratory technology, molecular biology, biochemistry)*

Borrebaeck, C.A.K., ed. 1995 Antibody Engineering. Oxford Unov. Press, New York
Burden, D.W. and Whitney, D.B. 1995 Biotechnology. Proteins to PCR. A Course in Strategies and LabTechniques. Birkhäuser Boston, Cambridge, MA, USA
Crooke, S.T. and Lebleu, B., eds. 1993 Antisense Research and Applications. CRC Press, Boca Raton, FL, USA
Endreß, R. 1994 Plant Cell Biotechnology. Springer-Vlg., New York
Glazer, A.N. and Nikaido, H. 1995 Microbial Biotechnology. Fundamentals of Applied Microbiology. Freeman, New York
Gresshoff, P.M., ed. 1997 Technology Transfer of Plant Biotechnology. CRC Press, Boca Raton, FL, USA
Heikki, M.T. and Lynch, J.M., eds. 1995 Biological Control. Cambridge Univ. Press, New York
Howe, C. 1995 Gene Cloning and Manipulation. Cambridge Univ. Press, New York
Kingsman, S.M and Kingsman, A.J. 1988 Genetic Engineering. An Introduction to Gene Analysis and Exploitation in Eukaryotes. Blackwell Scientific. Oxford, UK.
Maclean, N., ed. 1995 Animals with Novel Genes. Cambridge Univ. Press, New York
Meyers, R.A., ed. 1995 Molecular Biology and Biotechnology. A Comprehensive Desk Reference. VCH, New York
Monsatersky, G.M. and Robl, J.M., eds. 1995 Strategies in Transgenic Animal Science, ASM Press, Washington, DC
Old, R.W. and Primrose, S.B. 1994 Principles of Gene Manipulation. An Introduction to Genetic Engineering, Blackwell Scientific, Cambridge, MA, USA
Pinkert, C.A., ed. 1994 Transgenic Animal Technology. Acad. Press, San Diego, CA, USA
Potrykus, I. and Spangenberg, G., eds. 1995 Gene Transfer to Plants, Springer Vlg., New York
Shuler, M.L. *et al.*, eds. 1994 Baculovirus Expression System and Biopesticides. Wiley-Liss, New York
Smith, J.E. 1996 Biotechnology. Cambridge Univ. Press, New York
Strauss, M, and Barranger, J.A., eds. 1997 Concepts in Gene Therapy. W. De Gruyter, New York

Biotechnology continued
Sullivan, N.F. 1995 Technology Transfer: Making Most of Your Intellectual Property. Cambridge Univ. Press, New York
Swindell, S.R., Miller, R.R.and Myers, G.S.A., eds. 1996 Internet for the Molecular Biologist. Horizon Scientific, Portland, OR, USA
Tzotzos, G.T., ed. 1995 Genetically Modified Organisms. A Guide to Biosafety. CAB Internat. Oxford, UK
Vining, L.C. and Stuttrad, C., eds. 1994 Genetics and Biochemistry of Antibiotic Production. Butterworth-Heinemann, Stoneham, MA, USA
Wang, K., Herrera-Estrella, A. and van Montagu, M., eds 1995 Transformation of Plants and Soil Microorganisms. Cambridge Univ. Press, New York

CANCER *(see also diseases, aging, viruses, molecular biology, cytology and cytogenetics)*

Bertino, J.R., ed. 1996 Encyclopedia of Cancer, Acad. Press, San Diego, CA, USA
Eeles, R. Ponder, B., Easston, D. and Horwich, A. 1996 Genetic Predisposition to Cancer. Chapman and Hall, New York
Hesketh, R.T. 1995 The Oncogene Factsbook. Acad. Press, San Diego, CA, USA
Lindahl, T., ed. 1996 Genetic Instability in Cancer. Cold Spring Harbor Lab. Press, Cold Spring Harbor, NY, USA
Marcus, A.I. 1994 Cancer from Beef. DES, Federal Regulation, and Consumer Confidence. Johns Hopkins Univ. Press, Baltimore, MD, USA
Mendelsohn, J., Howley, P.M., Isreal, M.A. and Liotta, L.A. eds. 1994 The Molecular Basis of Cancer. Saunders, Philadephia, USA
Minson, A.C., Neil. J.C. and McCrae, M.A., eds. 1954 Viruses and Cancer. Cambridge Univ. Press, New York
Mitelman, F. 1997 Catalog of Chromosomal Aberrations in Cancer. Wiley, New York
Peters, G. and Vousden, K. 1997 Oncogenes and Tumour Suppressor. IRL/Oxford Univ. Press, New York
Pettit, G.R., Person, F.H. and Herald, C.L. 1994 Anticancer Drugs from Animals, Plants and Microorganisms. Wiley, New York
Ponder, B.A.J., Cavenee, W.K. and Solomon, E., eds. 1995 Genetics and Cancer. A Second Look. Cold Spring Harbor Laboratory Press, Cold Spring Harbor, NY, USA
Vogt, P.K. and Verma, I.M., eds. 1995 Oncogene Techniques. Acad. Press, San Diego, CA, USA
Weinberg, R.A. 1996 Racing to the Beginning of the Road. The Search for the Origin of Cancer. Crown, New York

CONSERVATION *(see also evolution, population genetics)*

Avise, J.C. and Hamrick, J.L., eds. 1996 Conservation Genetics: Case Histories from Nature. Chapman and Hall. New York
Caughly, G. and Gunn, A. 1995 Conservation Biology in Theory and Practice, Blackwell Science, Cambridge, MA
Clark, T.W. 1997 Averting Extinction. Reconstructing Endangered Species Recovery. Yale Univ. Press, New Haven, CN, USA
Ehrenfeld, D., ed. 1995 Plant Conservation. Blackwell Science, Cambridge, MA
Frankel, O.H., Brown, A.D. and Burdon, J.J. 1995 The Conservation of Plant Biodiversity. Cambridge Univ. Press, New York
Hunter, M.L., Jr. 1995 Fundamentals of Conservation Biology. Blackwell Science, Cambridge MA, USA
Huston, M.A. 1994 Biological Diversity. The Coexistence of Species in Changing Landscapes. Cambridge Univ. Press, New York.

Conservation continued

Juo, A.S.R. and Freed, R.D., eds. 1993 Agriculture and Environment. Bridging Food Production and Environmental Protection in Developing Countries. Amer. Soc. Agron, Madison, WI, USA
Lawton, J.H. and May, R.M., eds. 1995 Extinction Rates. Oxford Univ. Press, New York
Meffe, G.K. and Carrol, C.R. 1995 Principles of Conservation Biology. Sinauer, Sunderland, MA, USA
Paehlke, R., ed. 1995 Conservation and Environmentalism. An Encyclopedia. Garland, New York
Reaka-Kudle, M.L., Wilson, D.E. and Wilson, E.O., eds. 1996 Biodiversity II. Understanding and Protecting our Biological Resources. Joseph Henry Press (Natl. Acad. Sci. USA), Washingon, DC, USA
Samways, M.J. 1994 Insect Conservation Biology. Chapman and Hall, New York
Shaw, I.C. and Chadwick, J. 1996 Principles of Environmental Toxicology. Taylor & Francis, London,UK.

CYTOLOGY and CYTOGENETICS (*see also molecular biology, evolution, cancer*)

Berezney, R. and Riordan, J.F., eds. 1997 Nuclear Matrix. Structural and Functional Organization. Acad. Press, San Diego, CA, USA
Blackburn, E. H. and Greider, C.W., eds. 1996 Telomeres. Cold Spring Harbor Lab. Press, Cold Spring Harbor, NY, USA
Borgaonkar, D. 1994 Chromosomal Variation in Man. A Catalog of Chromosomal Variants and Anomalies. Wiley-Liss, New York
Choo, K.H. 1997 The Centromere. Oxford Univ. Press, New York
Darlington, C.D. and Wylie, A.P. 1955 Chromosome Atlas of Flowering Plants. Allen & Unwin, London, UK
Davis, K.E. and Warren, S.T., eds., 1993 Genome Rearrangement and Stability. Cold Spring Harbor Laboratory Press, Cold Spring Harbor, NY, USA
DeGrouchy, J. and Turleau, C. 1984 Clinical Atlas of Human Chromosomes. Wiley, New York
Drlica, K. and Riley, M., eds. 1990 The Bacterial Chromosome. Amer. Soc. Microbiol., Washington, DC, USA
Elgin, S. C.R., ed. 1995 Chromatin Structure and Gene Expression, IRL Press, New York
Fukui, K. and Nakyama, S., eds. 1996 Plant Chromosomes. Laboratory Methods. CRC Press. Boca Raton, FL, USA
Goodhew, P., Keyse, R.J. and Lorimer, J.W. 1997 Introduction to Scanning Transmission Electron Microscopy. Springer, New York
Gosden, J.R. ed. 1994 Chromosome Analysis Protocols. Humana, Totowa, NJ, USA
Gustafson, J.P. and Flavell, R.B., eds. 1996 Genomes of Plants and Animals. 21st Stadler Genetics Symposium. Plenum, New York
Heim, S. and Mitelman, F. 1995 Cancer Cytogenetics. Wiley-Liss, New York
Henriquez-Gil, N., Parker, J.U.S. and Puertas, M., eds. 1997 Chromosomes Today. Chapman and Hall, New York
Hsu, T. C. and Benirschke, K. 1967-1977 An Atlas of Mammalian Chromosomes. Springer-Vlg., New York
Hutchison, C. and Glover, D.M., eds. 1995 Cell Cycle Control. Oxford Univ. Press, New York
Hyams, J.S. and Lloyd, C.W. 1994 Microtubules. Wiley-Liss. New York
Jauhar, P.P., ed. 1996 Methods of Genome Analysis of Plants. CRC Press, Boca Raton. FL, USA
Javois, L.C., ed. 1995 Immunocytochemical Methods and Protocols. Humana, Totowa, NJ, USA
John, B. and Lewis, K.R. 1968 The Chromosome Complement. Spriger Vlg, New York
Kipling, D. 1995 The Telomere. Oxford Univ. Press, New York
Macgregor, H.C. 1993 An Introduction to Animal Cytogenetics. Chapman and Hall, New York

Cytology and cytogenetics continued

Moore, R.J. 1973 Index to Plant Chromosome Numbers 1967-1971. Oosthoek, Utrecht, Netherlands

Obe, G. and Natarayan, A.T., eds. 1994 Chromosomal Aberrations. Origin and Significance. Springer-Vlg., New York

Polak, J. and Van Noorden, S. 1997 Introduction to Immunocytochemistry. Springer. New York

Rooney, D.E. and Czepulkowski, B.H. 1997 Human Chromosome Preparations: Essential Techniques. Wiley, New York

Shapiro, H.M. 1994 Practical Flow Cytometry. Wiley-Liss, New York

Sharma, A.K. and Sharma, A. 1994 Chromosome Techniques. Harwood, Langhorne, PA, USA

Sheppard, C. and Shotton, D. 1997 Confocal Laser Scanning Microscopy. Springer, New York

Therman, E. 1993 Human Chromosomes: Structure, Behavior, Effects. Springer-Vlg., New York

Van Driel, R. and Otte, A. 1997 Nuclear Organization, Chromatin Structure and Gene Expression. Oxford Univ. Press, New York

Wagner, R.P., Maguire, M.P. and Stallings, R.L. 1993 Chromosomes. A Synthesis. Wiley-Riss, New York

Wang, X. F. and Herman, B., eds. 1996 Fluorescence Imaging Spectroscopy and Microscopy. Wiley, New York

Wolffe, A. 1995 Chromatin. Structure and Function. Acad. Press, San Diego, CA, USA

DEVELOPMENT *(see also molecular biology, neurology)*

Bate, M. and Martinez Arias, A., eds. 1994 The Development of *Drosophila melanogaster*. Cold Spring Harbor Lab. Press, Cold Spring Harbor, NY, USA

Bowman, J. ed. 1994 Arabidopsis. An Atlas of Morphology and Development. Springer-Vlg., New York

Campos-Ortega, J.A. and Hartenstein, V. 1997 The Embryonic Development of *Drosophila melanogaster*. Springer, New York

Duboule, D., ed. 1994 Guidebook to the Homeobox Genes. Oxford Univ. Press, New York

Findley, J.K. ed. 1994 Molecular Biology of the Female Reproductive System. Acad. Press, San Diego, CA, USA

Fosket, D.E. 1994 Plant Growth and Development: A Molecular Approach. Acad. Press, San Diego, CA, USA

Gilbert, S.F. 1994 Developmental Biology. Sinauer, Sunderland, MA, USA

Gilbert, L.I., Tata, J.R. and Atkinson, B.G., eds. 1996 Metamorphosis. Postembryonic Reprogramming of Gene Expression in Amphibian and Insect Cells. Acad. Press, San Diego, CA., USA

Gu, J. 1997 Analytical Morphology. Theory, Applications and Protocols. Springer. New York

Harrison, L.G. 1993 Kinetic Theory of Living Patterns. Cambridge Univ. Press, New York

Hunter, R.H.F. 1995 Sex Determination, Differentiation and Intersexuality in Placental Mammals. Cambridge Univ. Press, New York

Kalthoff, K. 1995 Analysis of Biological Development. McGraw-Hill, New York

Kaufman, M.H. 1992 The Atlas of Mouse Development. Acad. Press, San Diego, CA, USA

Lawrence, P.A. 1992 The Making of a Fly: The Genetics of Animal Design. Blackwell Scientific, Cambridge, MA, USA

Martini Neri, M.E., Neri, G., and Opitz, J.M., eds. 1996 Gene Regulation and Fetal Development. Wiley-Liss, New York

Müller, W.A. 1997 Developmental Biology. Springer, New York

Ohlsson, R., Hall, K. and Ritzen, M., eds. 1995 Genomic Imprinting - Causes and Consequences. Cambridge Univ. Press, New York

Potten, C. ed., 1996 Stem Cells. Acad. Press. San Diego, CA, USA

Reik, W. and Surani, A. 1997 Genomic Imprinting. Frontiers in Molecular Biology. IRL/Oxford Univ. Press, New York

Development continued

Russo, V.E.A., Brody, S., Cove, D. and Ottolenghi, S. 1992 Development. The Molecular Genetic Approach. Springer-Vlg, New York
Russo, V.E.A., Martienssen, R.A. and Riggs, A.D., ed. 1996 Epigenetic Mechanisms of Gene Regulation. Cold Spring Harbor Laboratory Press, Cold Spring Harbor, NY, USA
Solari, A.J. 1994 Sex Chromosomes and Sex Determination in Vertebrates. CRC Press, Boca Raton, FL, USA
Wachtel, S.S. 1993 Molecular Genetics of Sex Determination. Acad. Press, San Diego, CA, USA
Wilkins, A.S. 1993 Genetic Analysis of Animal Development, Wiley-Liss, New York
Williams, E.G. et al., eds. 1994 Genetic Control of Self-Incompatibility and Reproductive Development in Flowering Plants. Kluwer, Norwell, MA

DISEASES *(see also cancer, aging, molecular biology, immunology)*

Baker, H.F. and Ridley, R.M., eds. 1996 Prion Diseases. Humana, Totowa, NJ., USA
Becker, R. 1996 Alzheimer Disease. From Molecular Biology to Therapy. Springer, New York
Brioni, J.D. and Decker, M.W., eds. 1997 Pharmacological Treatment of Alzheimer's Disease. Molecular and Neurobiological Foundations. Wiley, New York
Clarke, J.T.R. 1996 A Clinical Guide to Inherited Metabolic Diseases. Cambridge Univ. Press, Cambridge, UK
Collinge, J. and Palmer, M.S. 1997 Prion Disease. Oxford Univ. Press. New York
Elles, R. 1996 Molecular Diagnosis of Genetic Diseases, Humana Press, Totowa, NJ, USA
Edwards, R.G., ed. 1993 Preconception and Preimplantation Diagnosis of Human Genetic Diseases. Cambridge Univ. Press, New York
Evans, D.A.P. 1994 Genetic Factors in Drug Therapy. Clinical and Molecular Pharmacogenetics. Cambridge Univ. Press, New York
Ewald, P.W. 1994 Evolution of Infectious Diseases. Oxford Univ. Press, New York
Giraldo, G., Bologesi, D.P., Salvatore, M. and Beth-Giraldo, E., eds., 1996 Development and Applications of Vaccines and Gene Therapy in AIDS. Karger, Basel
Gupta, S., ed., 1996 Immunology of HIV Infection. Plenum, New York
Hall, L.L., ed. 1996 Genetics and Mental Illness: Evolving Issues for Research and Society. Plenum, New York
Harper, P.S. 1993 Practical Genetic Counseling. Butterworth, Oxford, UK
Jeffery, S., Booth, J. and Butcher, P. 1997 Nucleic Acid-Based Diagnosis. Springer. New York
McCance, K.L. and Huether, S.E. 1994 Pathophysiology: The Biologic Basis of Disease in Adults and Children. Mosby, St. Louis, MO, USA
Prusiner, S.B. 1966 Prions, Prions, Prions. Springer Vlg. New York
Robinson, A. and Linden, M.G. 1993 Clinical Genetics Handbook. Blackwell Scientific, Oxford, UK
Roland, P.E. 1997 Brain Activation. Wiley, New York
Ross, D.W. 1997 Introduction to Molecular Medicine. Springer. New York
Scriver, C.R. et al., eds., 1995 The Metabolic and Molecular Bases of Inherited Disease. McGraw-Hill, San Francisco, CA, USA
Stamatoyannopoulos, G., Nienhuis, A.W., Majerus, P.W. and Varmus, H. 1994 The Molecular Basis of Blood Diseases. Saunders. Philadelphia, PA, USA
Stine, G.J. 1994 Acquired Immunodeficiency Syndrome. Biological, Medical, Social, and Legal Issues. Prentice Hall. Englewood Cliff, NJ.
Terry, R.D., Katzman, R. and Bick, K.L., eds. 1994 Alzheimer Disease. Raven. New York
Thoene, J.G., ed. 1996 Physician's Guide to Rare Diseases. Dowden, Montvale, NJ, USA
Vile, R.G. 1997 Understanding Gene Therapy. Springer, New York
Vos, J-M. H. 1995 DNA Repair Mechanisms. Impact on Human Diseases and Cancer. Springer, New York

EVOLUTION *(see also population genetics, cytology and cytogenetics)*

Anderson, E. 1949 Introgressive Hybridization. Wiley, New York
Andersson, M. 1994 Sexual Selection. Priceton Univ. Press. Princeton, NJ, USA
Avise, J.C. 1994 Molecular Markers, Natural History and Evolution. Chapman & Hall, New York
Bell, G. 1997 Selection: The Mechanism of Evolution. Chapman and Hall, New York
Cavalli-Sforza, L.L. and Cavalli-Sforza, F. 1995 The Great Human Diasporas: The History of Diversity and Evolution. Addison-Wesley, Boston, Ma, USA
Cavalli-Sforza, L.L., Menozzi, P. and Piazza, A. 1994 The History and Geography of Human Genes. Princeton Univ. Press, Princeton, NJ, USA
Crozier, R.H. and Pamilo, P. 1996 Evolution of Social Insect Colonies. Oxford Univ. Press, New York
Dayhoff, M.O., ed. 1972 Atlas of Protein Sequences and Structure. Vol. 5. Natl. Biomed Res. Fund. Washington, DC, USA
Dobzhansky, Th. *et al.* 1977 Evolution. Freeman, San Francisco, CA, USA
Fleagle, J.G. and Kay, R.F., eds. 1994 Anthropoid Origins. Plenum. New York
Ford, E.B. 1971 Ecological Genetics. Chapman and Hall, London, UK
Gerhart, J. and Kirschner, M. 1997 Cells, Embryos and Evolution: Developmental Understanding of Phenotypic Variation and Evolutionary Adaptability. Blackwell Science. Cambridge, MA, USA
Gesteland, R.F. and Atkins, J.F., eds. 1993 The RNA World. Cold Spring Harbor Laboratory Press, Cold Spring Harbor, NY, USA
Gibbs, A.J., Calisher, C.H. and Garcia-Arenal, F., eds. 1995 Molecular Basis of Virus Evolution. Cambridge Univ. Press, New York
Gillespie, J.H. 1992 The Causes of Molecular Evolution. Oxford Univ. Press, New York
Hall, B.K, ed. 1994 Homology. The Hierarchial Basis of Comparative Biology. Acad. Press, San Diego, CA
Herrmann, B. and Hummel, S., eds. 1994 Ancient DNA. Recovery and Analysis of Genetic Material from Paleontological, Archeological, Museum, Medical and Forensic Specimens. Springer-Vlg, New York
Hillis, D.M., Moritz, C. and Mable, B.K., eds. 1996 Molecular Systematics. Sinauer, Sunderland, MA, USA
Jablonka, E. and Lamb, M.J. 1995 Epigenetic Inheritance and Evolution: The Lamarckian Dimension. Oxford Univ. Press, New York
Jones, S. *et al.*, eds. 1994 The Cambridge Encyclopedia of Human Evolution. Cambridge Univ. Press, New York
Lawton, J.H. and May, R.M., eds. 1995 Extinction Rates. Oxford Univ. Press, New York
Lerner, I.M. and Libby, W.J. 1976 Heredity, Evolution and Societey. Freeman. San Francisco, CA, USA
Lewin, R. 1997 Patterns in Evolution: The New Molecular View. Freeman, New York
Li, W-H. 1997 Molecular Evolution. Sinauer. Sunderland, MA, USA
Margulis, L. 1993 Symbiosis in Cell Evolution. Freeman. New York
Maynard-Smith, J. and Szathmáry, E. 1995 The Major Transitions in Evolution. Freeman, New York
Mayr, E. 1963 Animal Species and Evolution. Harvard Univ. Press, Cambridge, MA, USA
Myamoto, M.E. and Cracraft, J., eds. 1991 Phylogenetic Analysis of DNA Sequences. Oxford Univ. Press, New York
Nei, M. 1987 Molecular Evolutionary Genetics. Columbia Univ. Press, New York
Ohno, S. 1970 Evolution by Gene Duplication. Springer-Vlg., New York
Osawa, S. 1995 Evolution of the Genetic Code. Oxford Univ. Press, New York
Raff, R.A. 1996 The Shape of Life. Genes, Development, and the Evolution of Animal Form. Univ. Chicago Press, Chicago, IL, USA
Robert, D.MCL *et al.*, eds. 1996 Evolution of Microbial Life. Cambridge Univ. Press, New York

Evolution continued
Roff, D.A. 1997 Evolutionary Quantitative Genetics. Chapman and Hall, New York
Rose, M.R. and Lauder, G.V., eds. 1996 Adaptation. Acad. Press, San Diego, CA, USA
Ruse, M. 1997 Monad to Man: The Concept of Progress in Evolutionary Biology. Harvard Univ. Press, Cambridge, MA, USA
Simmonds, N.W., ed. 1976 Evolution of Crop Plants. Longman, London, UK
Stebbins, L.G. 1950 Variation and Evolution in Plants. Columbia Univ. Press, New York
Strickberger, M.W. 1995 Evolution. Jones and Bartlett, Boston, MA, USA
Zeuner, F.E. 1963 A History of Domesticated Animals. Harper & Row, New York

GENETICS MONOGRAPHS (*see also genetics textbooks, molecular biology, cytology, transposable elements, mapping, recombination, mutation, radiation, organelles, evolution*)

Annual Reviews of Genetics. Acad. Press. San Diego, CA, USA
Bowling, A.T. 1996 Horse Genetics. CAB Internat/Oxford Univ. Press, New York.,
Bridge, P.J. 1994 The Calculation of Genetic Risks. Worked Examples of DNA Diagnostics. Johns Hopkins Univ. Press. Baltimore, MD, USA
Bulmer, M.G. 1980 The Mathematical Theory of Quantitative Genetics. Oxford Univ. Press, New York
Chaudharty, B.R. and Agarwal, S.B., eds., 1996 Cytology, Genetics, and Molecular Biology of Algae. SPB Academic, Amsterdam, The Netherlands
Epstein, H.F. and Shakes, D.C., eds. 1995 *Caenorhabditis elegans*: Modern Biological Analysis of an Organism. Acad. Press, San Diego, CA, USA
Esser, K. and Kuenen, R. 1967 Genetics of Fungi. Springer Vlg. New York
Falconer, D.S. and Mackay, T.E.C. 1996 Introduction to Quantitative Genetics. Longman/ Addison Wesley, White Plains, NY, USA
Freeling, M. and Walbot, V., eds. 1994 The Maize Handbook. Springer-Vlg., New York
Goldstein, L.S.B. and Fryberg, E.A., eds. 1994 *Drosophila melanogaster*. Practical Uses in Cell and Molecular Biology. AP Professional, Cambridge, MA, USA
Kearsey, M.J. and Pooni, H.S. 1996 The Genetical Analysis of Quantitative Traits. Chapman and Hall, New York
King, R.C., ed. 1974 Handbook of Genetics. Plenum, New York
Koncz, C., Chua, N.-H. and Schell, J., eds. 1993 Methods in Arabidopsis Research. World Scientific, Singapore
Lindsley, D.L. and Zimm, G.G. 1992 The Genome of *Drosophila melanogaster*. Acad Press, San Diego, CA, USA
Lyon, M.F., Rastan, S. and Brown, S.D.M. 1996 Genetic Variants and Strains of the Laboratory Mouse. Oxford Univ. Press, New York
Martinelli, S.D. and Kinghorn, J.R. eds. 1994 Aspergillus: 50 Years On. Elsevier, Amsterdam, The Netherlands
Mather, K. and Jinks, J.L. 1977 Introduction to Biometrical Genetics. Cornell Univ. Press, Ithaca, NY, USA
McKusick, V.A., ed. 1978 Medical Genetic Studies of the Amish. Johns Hopkins Univ. Press, Baltimore,MD, USA
McKusick, V.A. 1994 Mendelian Inheritance in Man: A Catalog of Human Genes and Genetic Disorders. Johns Hopkins Univ. Press, Baltimore, MD, USA
Meyerowitz, E.M. and Somerville, C.R., eds. 1994 Arabidopsis. Cold Spring Harbor Laboratory Press, Cold Spring Harbor, NY, USA
Neale, M.C. and Cardon, L.R. 1992 Methodology for Genetic Studies of Twins and Families. Kluwer, Dordrecht, The Netherlands
Nicholas, F.W. 1987 Veterinary Genetics. Clarendon Press, Oxford, UK
Nicholas, F.W. 1996 Introduction to Veterinary Genetics. Oxford University Press, New York
Ohlsson, R., Hall, K. and Ritzen, M., eds. 1995 Genomic Imprinting. Causes and Consequences.

Genetics monographs continued
   Cambridge Univ. Press, New York
Piper, L. and Ruvinsky, A., eds. 1997 The Genetics of the Sheep. CAB/Oxford Univ. Press, New York
Powell, K.A., Renwick, A. and Peberdy, J.F., eds. 1994 The Genus Aspergillus: From Taxonomy and Genetics to Industrial Applications. Plenum, New York
Riddle, D.E., *et al.* eds. 1997 *C. elegans* II. Cold Spring Harbor Laboratory Press, Cold Spring Harbor, NY, USA
Rimoin, D.L., Connor, J.M. and Peyeritz, R.E., eds. 1996 Emery and Rimoin's Principles and Practice of Medical Genetics. Churchill Livingstone, New York
Silver, L.M. 1995 Mouse Genetics. Concepts and Applications. Oxford Univ. Press, New York
Singer, M. and Berg, P., eds. 1997 Exploring Genetic Mechanisms. University Science Books. Sausalito, CA, USA
Stevens, L. 1991 Genetics and Evolution of the Domestic Fowl. Cambridge Univ. Press, New York
Strachan, T, and Read, A,P. 1996 Human Molecular Genetics. Bios Scientific/Coronet Books. Philadelphia, PA, USA
Stubbe, H. 1966 Genetik und Zytologie von *Antirrhinum* L. Sect. *Antirrhinum*. Fischer, Jena, Germany
Weir, B.S 1996 Genetic Data Analysis II. Sinauer, Sunderland, MA, USA
Wheals, A.E., Rose, A.H. and Harrison, S.J., eds. 1995 The Yeasts. Vol. 6. Yeast Genetics. Acad. Press, San Diego, USA
Young, I.D. 1991 Introduction to Risk Calculation in Genetic Counselling. Oxford Univ. Press, New York

GENETICS TEXTBOOKS *(see also genetics monographs, history of genetics, molecular biology, cytology)*

Birge, E.A. 1994 Bacterial and Bacteriophage Genetics. Springer-Vlg, New York
Emery, A.E.H. and Mueller, R.F. 1992 Elements of Medical Genetics. Churchill Livingstone, New York
Emery, A.E.H. and Malcolm, S. 1995 An Introduction to Recombinant DNA in Medicine. Wiley, New York
Fincham, J.R.S. 1994 Genetic Analysis: Principles, Scope and Objectives. Blackwell Science, Boston, MA, USA
Fincham, J.R.S., Day, P.R. and Radford, A. 1979 Fungal Genetics. Blackwell, Oxford, UK
Griffiths, A.J.F., *et al.* 1996 An Introduction to Genetic Analysis. Freeman, New York
Hartl, D.L. 1996 Essential Genetics. Jones and Bartlett, Boston, MA, USA
Jorde, L.B., Carey, J.C. and White, R.L. 1997 Medical Genetics. Mosby, St. Louis, MO, USA
Lewin, B. 1997 Genes VI. Oxford Univ. Press, New York
Old, H.W and Primrose, S.B. 1995 Principles of Gene Manipulation. Blackwell Science, Cambridge,MA, USA
Rédei, G.P. 1982 Genetics, Macmillan, New York
Schleif, R. 1993 Genetics and Molecular Biology. Johns Hopkins Univ. Press, Baltimore, MD, USA
Russel, P.J. 1996 Genetics. HarperCollins, New York
Serra, J.A. 1965 Modern Genetics. Acad. Press. New York
Singer, M. and Berg, P. 1991 Genes & Genomes. A Changing Perspective. University Science Books, Mill Valley, CA, USA
Strickberger, M.W. 1985 Genetics. Macmillan, New York
Thompson, J.N. *et al.* 1997 Primer of Genetic Analysis. Cambridge Univ. Press. New York
Vogel, F. and Motulsky, A.G. 1996 Human Genetics. Springer. New York
Whitehouse, H.L.K. 1972 Towards an Understanding of the Mechanism of Heredity. St. Martin's

Press, New York

HISTORY OF OF GENETICS *(see also genetics monographs)*

Allen, G.E. 1978 Thomas Hunt Morgan. The Man and his Science. Princeton Univ. Press, Princeton, NJ, USA
Bearn, A.G. 1996 Archibald Garrod and the Individuality of Man. Oxford Univ. Press, New York
Cairns, J., Stent, G. and Watson, J.D. 1966 Phage and the Origins of Molecular Biology. Cold Spring Harbor Lab. Press. Cold Spring Harbor, NY, USA
Cambrosio, A. and Keating, P. 1996 Exquisite Specificity: The Monoclonal Antibody Revolution. Oxford Univ. Press, New York
Carlson, E.A. 1966 The Gene: A Critical History. Saunders, Philadelphia, PA, USA
Carlson, E.O. 1981 Genes, Radiation, and Society. The Life and Work of H.J. Muller. Cornell Univ. Press, Ithaca, NY, USA
Dunn, L.C. 1965 A Short History of Genetics. The Development of Some of the Main Lines of Thought: 1864—1939. McGraw-Hill, New York.
Emery, A.E.H. and Emery, M.L.H. 1995 The History of a Genetic Disease: Duchenne Muscular Dystrophy or Meryon Disease. Roy. Soc. Medicine Press, London, UK
Fedoroff, N. and Botstein, D., eds. 1992 Barbara McClintock's Ideas in the Century of Genetics. Cold Spring Harbor Laboratory Press, Cold Spring Harbor, NY, USA
Fisher, E.P. and Lipson, C. 1988 Thinking About Science. Max Delbrück and the Origins of Molecular Biology. W.W. Norton, New York
Focke, W.O. 1881 Die Pflanzenmischlinge. Borntraeger, Berlin, Germany
Friedberg, E.C. 1997 Correcting the Blueprint of Life. An Historical Account of the Discovery of DNA Repair Mechanisms. Cold Spring Harbor Lab. Press, Cold Spring Harbor, NY, USA
Goldschmidt, R.B. 1956 Portraits from Memory. Recollections of a Zoologist. Univ. of Washington Press, Seattle, WA, USA
Hall, M.N. and Linder, P., eds. 1993 The Early Days of Yeast Genetics. Cold Spring Harbor Laboratory Press, Cold Spring Harbor, NY, USA
Harwood, J. 1993 Styles of Scientific Thought —The German Genetics Community 1900-1933, Univ. Chicago Press. Chicago, IL, USA
Jacob, F. 1997 La Souris, la Mouche et l'Homme. Odile Jacob, Paris, France
Johannsen, W. 1909 Elemente der Exakten Erblichkeitslehre. Fisher, Jena, Germany
Judson, F. 1996 The Eighth Day of Creation. Makers of the Revolution in Biology. Simon and Schuster, New York
Lerner, I.M. 1950 Genetics in the U.S.S.R. An Obituary. Univ. British Columbia Press, Vancouver, Canada
Magner, L.N. 1993 A History of the Life Sciences. Dekker, New York
McCarty, M. 1985 The Transforming Principle: Discovering that Genes are Made of DNA. Norton, New York
Medvedev, Zs. 1969 The Fall and Rise of T.D. Lysenko. Columbia Univ. Press, New York
Morgan, T.H., Sturtevant, A.H., Muller, H.J. and Bridges, C.B. 1915 The Mechanism of Mendelian Heredity. Holt, New York
Morgan, T.H. 1919 The Physical Basis of Heredity. Lippincott, Philadelphia, PA, USA
Müller-Hill, B. 1996 The *lac* Operon. A Short History of a Genetic Paradigm. De Gruyter, New York
Neel, J.V. 1994 Physician to the Gene Pool: Genetic Lessons and Other Stories. Wiley, New York
Olby, R.C. 1966 Origins of Mendelism. Schocken Books, New York
Olby, R. 1974 The Path to the Double Helix. University of Washington Press, Seattle, WA, USA
Orel, V. 1996 Gregor Mendel The First Geneticist. Oxford Univ. Press, New York
Portugal, F.H. and Cohen, J.S. 1977 A Century of DNA. MIT Press, Cambridge, MA, USA
Potts, D.M and Potts, W.T.W. 1995 Queen Victoria's Gene: Hemophilia and the Royal Family.

History of genetics continued
   Alan Sutton, Dover, NH, USA
Provine, W.B. 1971 The Origins of Theoretical Population Genetics. Univ. Chicago Press, Chicago, IL, USA
Rabinow, P. 1996 Making PCR. A Story of Biotechnology. Univ. Chicago Press, Chicago, IL, USA
Roberts, H.F. 1965 Plant Hybridization Before Mendel. Hafner, New York
Rushton, A.R. 1994 Genetics and Medicine in the United States, 1800-1922. Johns Hopkins Univ. Press, Baltimore, MD, USA
Sarkar, S., ed. 1992 The Founders of Evolutionary Genetics: A Centenary Reappraisal. Kluwer Academic, Dordrecht, NL/Boston, MA
Shnoll, S.E. 1997 Heroes, Martyrs and Villains in Russian Life Sciences. Springer, New York
Sinsheimer, R.L. 1994 The Strands of Life: The Science of DNA and the Art of Education. Univ. California Press, Berkeley, CA, USA
Sirks, M.J. and Zirkle, C. 1964 The Evolution of Biology. Ronald Press, New York
Stent, G.S., ed. 1981 The Double Helix: Text, Commentary, Reviews, Original Papers. Norton, New York
Stubbe, H. 1972 History of Genetics from Prehistoric Times to the Rediscovery of Mendel's Laws. MIT Press, Cambridge, MA, USA
Sturtevant, A.H. 1965 A History of Genetics. Harper and Row, New York
Watson, J.D. 1997 The Double Helix. Weidenfeld & Nicholson, New York
Watson, J.D. and Tooze, J. 1981 The DNA STORY: A Documentary History of Gene Cloning. Freeman, San Francisco, CA, USA
Weir, R.F., Lawrence, S.C. and Fales, E., eds. 1994 Genes and Human Self-Knowledge. Historical and Philosophical Reflections on Modern Genetics. Univ. Iowa Press. Iowa City, IA, USA
Wilson, E.B. 1925 The Cell in Development and Heredity. Macmillan. New York
Zirkle, C. 1935 The Beginnings of Plant Hybridization. Univ. Pennsylvania Press, Philadephia, PA, USA

IMMUNOLOGY (*see also molecular biology, biology*)

Austyn, J.M. and Wood, K.J. 1993 Principles of Cellular and Molecular Immunology. Oxford Univ. Press, New York
Benjamini, E., Sunshine, G. and Leskowitz, S. 1996 Immunology. A Short Course. Wiley, New York
Birch, J.R. and Lennox, E.S., eds. 1995 Monoclonal Antibodies. Wiley-Liss, New York
Bona, A. and Bonilla, F.A. 1996 Textbook of Immunology. Gordon and Break & Harwood Academic, Toronto, CAN
Fernandez, N. and Butcher, G. 1997 MHC1. A Practical Approach. Oxford Univ. Press, New York
Gregory, C.D., ed. 1995 Apoptosis and the Immune Response. Wiley-Liss, New York
Herbert, W.J., Wilkinson, P.C. and Sott, D.I., eds. 1995 The Dictionary of Immunology. Acad. Press, San Diego, CA, USA
Honjo, T. and Alt, F.W., eds. 1995 Immunoglobulin Genes. Acad. Press, San Diego, CA, USA
Leffell, M.S., Donnenberg, A.D. and Rose, N.R. 1997 Handbook of Human Immunology. CRC Press, Boca Raton, FL, USA
Law, S.K.A. and Reid, K.I.M 1995 Complement. IRL/Oxford Univ. Press, New York
Liu, M.A., Hilleman, R.A. and Kurth, R., eds. 1995 DNA Vaccines. New York Acad. Sci. Annals. 772
McCafferty, J., Hoogenboom, H.R. and Chiswell, D.J. 1996 Antibody Engineering. A Practical Approach. IRL/Oxford Univ Press, New York
Paul, W.E. 1993 Fundamental Immunology. Raven, New York
Pillai, S. 1997 Lymphocyte Development. Springer, New York
Reid, M.E. and Lomas-Francis, C. 1997 The Blood Group Antigen. Acad. Press, San Diego, CA,

Immunology continued
 USA
Ritter, M.A. and Ladyman, H.M., eds. 1995 Monoclonal Antibodies. Production, Engineering and Clinical Application. Cambridge Univ. Press, New York
Roitt, I. 1991 Essential Immunology. Blackwell Scientific, Oxford, UK
Roitt, I.M. and Delves, P.J. 1992 Encyclopedia of Immunology. Acad. Press, San Diego, CA, USA
Stites, D.P., Terr, A.I. and Parslow, T.G., eds. 1994 Basic & Clinical Immunology. Appleton & Lange, Norwalk, CN., USA
Tilney, N.L., Strom, T.B. and Paul L.C., eds. 1996 Transplantation Biology. Cellular and Molecular Aspects. Lippincott-Raven, Philadelphia, PA, USA
Zanetti, M. and Capra, J.D., eds. 1995 The Antibodies. Harwood, Langhorne, PA, USA

LABORATORY TECHNOLOGY (see also molecular biology, biotechnology, cytology)

Alfa, C. et al.., eds. 1993 Experiments with Fission Yeast. A Laboratory Course Manual. Cold Spring Harbor Laboratory Press, Cold Spring Harbor. NY, USA
Amdur, M.O., Doull, J, and Klaassen, C.D., eds. 1991 Casaretts and Doull's Toxicology. The Basic Sciences of Poisons. Pergamon Press, Elmsford, NY, USA
Ansorge, W., Voss, H. and Zimmermann, J., eds. 1996 DNA Sequencing Strategies. Automated and Advanced Approaches. Wiley, New York
Armour, M-A. 1996 Hazardous Laboratory Chemicals Disposal Guide. CRC Press, Boca Raton, FL, USA
Ausubel, F.M., et al., eds. 1987- Current Protocols in Molecular Biology. Wiley, New York
Bernstam, V.A. 1992 Handbook of Gene Level Diagnostics in Clinical Practice. CRC Press, Boca Raton, FL USA
Bhojwani, S.S. and Razdan, M.K. 1996 Plant Tissue Culture: Theory and Practice. Elsevier Science, New York
Birren, B. and Lai, E. 1993 Pulsed Field Gel Electrophoresis: A Practical Guide. Acad Press, San Diego, CA, USA
Birren, B. and Lai, E., eds. 1996 Nonmammalian Genomic Analysis. A Practical Guide. Acad. Press, San Diego, CA, USA
Birren, B. et al., eds. 1996 Genome Analysis: A Laboratory Manual. Cold Spring Harbor Laboratory Press, Cold Spring Harbor, NY, USA
Bishop, M. and Rawlings, C. 1997 DNA and Protein Sequence Analysis. A Practical Approach. Oxford Univ. Press, New York
Bozzola, J.J. and Russel, L.D. 1992 Electron Microscopy. Principles and Techniques for Biologists. Jones & Bartlett, Boston, MA, USA
Bretherick, L. 1990 Bretherick's Handbook of Reactive Chemical Hazards. Butterworth, London, UK
Butler, M. 1996 Animal Cell Culture and Technology. IRL/Oxford Univ. Press, New York
Celis, J., ed. 1997 Cell Biology. A Laboratory Handbook. Acad Press, San Diego, CA, USA
Chrispeels, M.J. and Sadava, D.E. 1994 Plants, Genes and Agriculture. Jones and Bartlett, Boston, MA, USA
Clark, M. 1997 Plant Molecular Biology. A Laboratory Manual. Springer, New York
Coligan, J.E. et al., eds. 1991 Current Protocols in Immunology. Wiley, New York
Crowther, J.R. 1995 ELISA. Humana. Totowa, NJ, USA
Dangler, C.A., ed. 1996 Nucleic Acid Analysis. Principles and Bioapplications. Wiley, New York
Davis, L.G., Kuehl, W.M. and Battey, J.F. 1994 Basic Methods in Molecular Biology. Appleton and Lange. Norwalk, CN, USA
deBoer, A.G. and Sutanto, W., eds. 1997 Drug Transport Across the Blood-Brain Barrier. In Vitro and In Vivo Techniques. Gordon & Breach, Amsterdam, NL
Dieffenbach, C.W. and Dveksler, G.S., eds. 1995 PCR Primer. A Laboratory Manual. Cold

Laboratory technology continued
   Spring Harbor Laboratory Press. Cold Spring Harbor, NY, USA
Dodds, J.H. and Roberts, L.W. 1995 Experiments in Plant Tissue Culture. Cambridge Univ. Press, New York
Dracapoliu, N.C. et al., eds. 1994 Current Protocols in Human Genetics. Wiley, New York
Echalier, G. 1997 *Drosophila* Cells in Culture. Acad. Press, San Diego, CA, USA
Edwards, S. and Collin, H.A. 1997 Plant Cell Culture. Springer, New York
Evans, I.H., ed. 1966 Yeast Protocols. Methods in Cell and Molecular Biology. Humana Press. Totowa, NJ. USA
Ferré, F. 1997 Gene Quantification. Springer, New York
Freshney, R.I., ed. 1992 Animal Cell Culture. IRL Press, New York
Furr, A.K., ed. 1989 CRC Handbook of Laboratory Safety. CRC Press, Boca Ratoon, FL, USA
Garman, A. 1997 Non-Radioactive Labelling. Acad. Press, San Diego, CA, USA
Glasel, J.A. and Deutscher, M.P., eds. 1995 Introduction to Biophysical Methods for Protein and Nucleic Acid Research. Acad. Press, San Diego, CA, USA
Glover, D. and Hames, B.D., eds. 1995 DNA Cloning: A Practical Approach. Oxford Univ. Press, New York
Gold, L.S. and Zeigler, E., eds. 1997 Handbook of Carcinogenic Potency and Genotoxicity Databases. CRC Press, Boca Raton, FL, USA
Goldstein, L.S.B. and Fryberg, E.A., eds. 1994 *Drosophila melanogaster*. Practical Uses in Cell and Molecular Biology. Acad. Press, San Diego, CA, USA
Gosden, J.R., ed. 1996 Prins and *In Situ* PCR Protocols. Humana Press, Totowa, NJ, USA
Grout, B. 1995 Genetic Preservation of Plant Cells in Vitro. Springer, New York
Hames, B.D. and Higgins, S.J., eds. 1996 Gene Probes 2. Oxford Univ. Press, New York
Hames, B.D. and Rickwood, D. 1990 Gel Electrophoresis of Proteins: A Practical Approach. IRL Press, New York
Harrison, M.A. and Rae, I.F. 1997 General Techniques of Cell Culture. Cambridge Univ. Press, New York
Harwood, A.J. ed. 1994 Protocols for Gene Analysis. Humana, Totowa, NJ, USA
Hogan B. et al. 1994 Manipulating the Mouse Embryo. A Laboratory Manual. Cold Spring Harbor Laboratory Press, Cold Spring Harbor, NY, USA
Hames, B.D. and Higgins, S.J., eds. 1996 Gene Probes 2. A Practical Approach. IRL/Oxford Univ. Press, New York
Inman, K. and Rudin, N. 1997 Introduction to Forensic DNA Analysis. CRC Press, Boca Raton, FL, USA
Isaac, P. G., ed. 1994 Protocols for Nucleic Acid Analysis by Nonradioactive Probes. Humana, Totowa, NJ, USA
Johnston, J.R. 1994 Molecular Genetics of Yeast. A Practical Approach. IRL Press, New York
Jones, G.E., ed. 1996 Human Cell Culture Protocols. Humana Press, Totowa, NJ, USA
Joyner, A.L. 1994 Gene Targeting: A Practical Approach. Oxford Univ. Press, New York
Kendall, D.A. and Hill, S.J., eds. 1995 Signal Transduction Protocols. Humana, Totowa, NJ., USA
Kneale, G.G., ed. 1994 DNA—Protein Interactions. Principles and Protocols. Humana, Totowa, NJ, USA
Kobayashi, T., Kitagawa, Y. and Okumura, S., eds. 1994 Animal Cell Technology. Basic and Applied. Kluwer, Norwell, MA, USA
Krieg, P. A., ed. 1996 A Laboratory Guide to RNA. Isolation, Analysis, and Synthesis. Wiley-Liss, New York
Landegren, U. 1996. Laboratory Protocols for Mutation Detection. Oxford Univ. Press. New York
Larrick, J.W. and Siebert, P.D., eds. 1995 Reverse Transcriptase PCR. Ellis Horwood, London, UK
Lasick, D.D. 1997 Liposomes in Gene Delivery. CRC Press, Boca Raton, FL, USA

Laboratory technology continued

Lowestein, P.R. and Enquist, L.W., eds. 1996 Protocols for Gene Transfer in Neuroscience: Towards Gene Therapy of Neurological Disorders. Wiley, New York

Maliga, P., *et al.*, eds. 1995 Methods in Plant Molecular Biology. A Laboratory Course Manual, Cold Spring Harbor Lab. Press, Cold Spring Harbor, NY, USA

Malik, V.S. and Lillehoj, eds. 1994 Antibody Techniques. Acad. Press, San Diego, CA, USA

Markie, D., ed. 1995 YAC Protocols. Humana, Totowa, NJ, USA

Meier, T. and Fahrenholz, F. 1997 A Laboratory Guide to Biotin-Labeling in Biomolecule Analysis. Springer, New York

Milligan, G., ed. 1992 Signal Transduction: A Practical Approach. IRL Press, New York

METHODS IN ENZYMOLOGY: continuous series including wide areas of biology. Acad. Press, San Diego, CA, USA

METHODS IN MOLECULAR BIOLOGY: continuous volumes. Humana Press. Totowa, NJ, USA

METHODS IN NEUROSCIENCES: continuous volumes. Acad. Press. San Diego, CA, USA

Montesano, R. Bartsch, H., Boyland, Della Porta, G., Fishbein, L., Griesemer, R.A., Swan, A.B. and Tomatis L., eds. 1982 Handling chemical carcinogens in the laboratory - Problems of safety. Biol. Zbl. 101:653-70.

New, R.R.C., ed. 1990 Liposomes: A Practical Approach. Oxford Univ. Press, New York

Nuovo, G.J. 1994 PCR in Situ Hybridization. Raven, New York

Ormerod, M.G., ed. 1990 Flow Cytometry: A Practical Approach. Oxford Univ. Press, New York

Poirier, J., ed. 1997 Apoptosis Techniques and Protocols. Humana Press, Totowa, NJ, USA

Rapley, R. ed. 1996 PCR Sequencing Protocols, Humana Press, Totowa, NJ, USA

Reinert, J. and Bajaj, Y.P.S. 1977 Plant Cell, Tissue, and Organ Culture. Springer-Vlg., New York

Rickwood, D. and Hames, B.D. 1990 Gel Electrophoresis of Nucleic Acids: A Practical Approach. IRL Press, New York

Ridley, D.D. 1996 Online Searching. A Scientist's Perspective. A Guide for the Chemical and Life Sciences. Wiley, New York

Sambrook, J, Fritsch, E.F. and Maniatis T. 1989 Molecular Cloning: A Laboratory Manual. Cold Spring Harbor Lab. Press, Cold Spring Harbor, NY, USA

Sansone, E.B. and Tewari, Y.B. 1978 The permeability of laboratory gloves to selected. nitroamines, Pp. 517-43. In Environmental Aspects of N-Nitroso Compounds, Walker, E.A., Griciute, L., Castenegro, M. and Lyle R.E., eds., Int. Agency Res. Cancer, Sci. Publ. 19, Lyon, France

Schaefer, B.C. 1997 Gene Clonig and Analysis: Current Innovations. Horizon, Wymondham, UK

Seeley, H.W., VanDemark, P.J. and Li, J.J. 1991 Microbes in Action: A Laboratory Manual of Microbiology. Freeman, New York

Smith, B.J., ed. 1996 Protein Sequencing Strategies. Humana, Totowa, NJ, USA

Trower, M.K. 1996 In Vitro Mutagenesis Protocols. Humana Press, Totowa, NJ, USA

Tymmis, M.J., ed. 1995 In Vitro Transcription and Translation Protocols. Humana, Totowa, NJ, USA

Vasil, I.K., ed. 1984 Cell Culture and Somatic Cell Genetics of Plants. Acad. Press, San Diego, CA, USA

Walker, J.M., ed. 1996 The Protein Protocols Handbook. Humana Press, Totowa, NJ, USA

Wasserman, P.A. and DePamphilis, M.L., eds. 1993 Guide to Techniques in Mouse Development. Methods in Enzymology 225. Acad. Press, San Diego, CA, USA

Weising, K. *et al.* 1995 DNA Fingerprinting in Plants and Fungi. CRC Press, Boca Raton, FL, USA

White, B.A., ed. 1996 PCR Cloning Protocols. Humana Press, Totowa, NJ, USA

Wilkinson, D.G., ed. 1992 *In Situ* Hybridization: A Practical Guide. IRL Press, New York

Wu, W. *et al.* 1997 Methods in Gene Biotechnology. CRC Press, Boca Raton, FL, USA

MAPPING *(see also recombination)*

Boultwood, J., ed. 1996 Gene Isolation and Mapping Protocols. Humana Press, Totowa, NJ, USA
Liu, B.H. 1997 Statistical Genomics. Linkage, Mapping, and QTL Analysis. CRC Press, Boca Raton, FL, USA
Mather, K. 1957 The Measurement of Linkage in Heredity. Methuen. London, UK
O'Brien, S.J., ed., 1993 Genetic Maps. Locus Maps of Complex Genomes. Cold Spring Harbor Laboratory Press, Cold Spring Harbor, NY, USA
Paterson, A.H., ed. 1996 Genome Mapping in Plants. Acad. Press, San Diego, CA, USA
Primrose, S.B. 1995 Principles of Genome Analysis. A Guide to Mapping and Sequencing DNA from Different Organisms. Blackwell Science, Cambridge, MA, USA
Schook, L.B., Lewin, H.A. and McLaren. D.G., eds., 1991 Gene-Mapping Techniques and Applications. Marcel Dekker, New York
Speed, T. and Waterman, M.S. 1996 Genetic Mapping and DNA Sequencing. Springer, New York

MICROBIOLOGY *(see also molecular biology, biology, viruses, laboratory technology)*

Adolph, K.W., ed. 1995 Microbial Gene Techniques. Acad. Press, San Diego, CA, USA
Atlas, R.M. 1993 Handbook of Microbiological Media. CRC Press, Boca Raton, FL, USA
Baltz, R.H., Hageman, G.D. and Skatrud, P.L., eds. 1993 Industrial Microorganisms. Basic and Applied Molecular Genetics. Amer. Soc. Microbiol, Washington, DC, USA
Baron, E.J. *et al.*, eds. 1994 Medical Microbiology. Wiley, New York
Baumberg, S., Young, J.P.W., Wellington, E.M.H. and Saunders, J.R., eds. 1995 Population Genetics of Bacteria. Cambridge Univ. Press, New York
Boyd, R.F. 1995 Basic Medical Microbiology, Little Brown, New York
Bruijn de, F.J., Lupski, J.R. and Weinstock, G.M. 1998 Bacterial Genomes: Physical Structure and Analysis. Chapman and Hall, New York
Bryant, D.A. 1994 The Molecular Biology of Cyanobacteria. Kluwer, Norwell, MA, USA
Clewell, D.B., ed. 1993 Bacterial Conjugation. Plenum, New York
Cooper, G.M., Temin, R.G. and Sugden, R., eds. 1995 The DNA Provirus. Amer. Soc. Microbiol. Washington, DC, USA
Dale, J.W. 1994 Molecular Genetics of Bacteria. Wiley, New York
Dangl, J.L., ed. 1994 Bacterial Pathogenesis of Plants and Animals. Molecular and Cellular Mechanisms. Springer-Vlg., New York
Dorman, C.J. 1994 Genetics of Bacterial Virulence. Blackwell Scientific, Cambridge, MA, USA
Elliott, C.G. 1993 Reproduction of Fungi: Genetical and Physiological Aspects. Chapman and Hall, New York
Funell, B.E. 1996 The Role of Bacterial Membrane in Chromosome Replication and Partitition. Chapman and Hall, New York
Gartland, K.M.A. and Davey, M. 1995 Agrobacterium Protocols. Humana Press, Totowa, NJ, USA
Goset, F. and Guespin-Michel, J. 1994 Prokaryotic Genetics: Genome Organization, Transfer and Plasticity. Blackwell Scientific, Cambridge, MA, USA
Gow, N.A.R. and Gadd, G.M., eds. 1994 The Growing Fungus. Chapman & Hall, New York
Griffin, D.H. 1996 Fungal Physiology. Wiley, New York
Hawkey, P.M. and Lewis, D.A., eds. 1989 Medical Bacteriology: A Practical Approach. IRL Press, New York
Holt, J.G., ed. 1993 Bergey's Manual of Determinative Bacteriology. William & Wilkins, Baltimore, MD, USA
Kwon-Chung, K.J. and Bennett, J.E. 1992 Medical Mycology. Lea & Febiger. Philadelphia, PA, USA
Levine, J. *et al.*, eds. 1994 Bacterial Endotoxins. Wiley-Liss, New York

Microbiology continued

Lin, E.C.C. and Lynch, A.S., eds. 1996 Regulation of Gene Expression in *Escherichia coli*. Chapman and Hall, New York
McKane, L. and Kandel, J. 1995 Microbiology. Essentials and Applications. McGraw-Hill, New York
Miller, V.L., Kaper, J.B, Portnoy, D.A. and Isberg, R.R., eds. 1994 Molecular Genetics of Bacterial Pathogenesis. Amer. Soc. Microbiol. Press, Washington, DC, USA
Moat, A.G. and Foster, J.W. 1995 Microbial Physiology. Wiley, New York
Murray, P.R. *et al*., eds., 1995 Manual of Clinical Microbiology. ASM Press. New York
Neidhardt, F.C. *et al*., eds. 1966 *Escherichia coli* and *Salmonella*. Amer. Soc. Microbiol, Washington, DC, USA
Nickoloff, J.A., ed. 1995 Electroporation Protocols for Microorganisms. Humana, Totowa, NJ, USA
Singleton, P. 1995 Bacteria in Biology, Biotechnology and Medicine. Wiley, New York
Somasegaran, P. and Hoben, H.J. 1994 Handbook of Rhizobia. Methods in Legume-Rhizobia Technology. Springer-Vlg., New York
Sonenshein, A.L., Hoch, J.A. and Losick, R., eds. 1993 *Bacillus subtilis* and other Gram-Positive Bacteria. Biochemistry, Physiology, and Molecular Genetics. Amer. Soc. Microbiol., Washington, DC, USA
Streips, U.N. and Yasbin, R.E., eds. 1991 Modern Microbial Genetics. Wiley-Liss, New York
Summers, D.K. 1996 The Biology of Plasmids. Blackwell Science, Cambridge, MA, USA
White, D. 1995 The Physiology and Biochemistry of Prokaryotes. Oxford Univ. Press, New York

MOLECULAR BIOLOGY *(see also laboratory technology, biotechnology, biochemistry)*

Abelson, J.N., ed. 1996 Combinatorial Chemistry. Methods in Enzymology 267. Academic. Press, San Diego, CA, USA
Agrawal, S. 1996 Antisense Therapeutics. Humana. Press, Totowa, NJ, USA
Alphey, L. 1997 DNA Sequencing. Springer, New York
Ashley, R.H. 1966 Ion Channels: A Practical Approach. Oxford Univ. Press, New York
Bauerle, P.A., ed. 1995 Inducible Gene Expression. Hormonal Signals. Birkhäuser, Cambridge, MA, USA
Beugelsdijk, T.J., ed. 1997 Automation Technologies in Genome Characterization. Wiley, New York
Bishop, J.E. and Waldholz, M. 1990 Genome. Simon and Schuster, New York
Blow, J., ed. 1996 Eukaryotic DNA Replication. IRL/Oxford Univ. Press, New York
Calladine, C.R. and Drew, H.R. 1997 Understanding DNA: The Molecule and How it Works. Acad. Press, San Diego, CA, USA
Callards, R. and Gearing, A. 1994 The Cytokine FactsBook. Acad. Press, San Diego, CA, USA
Clapp, J.P., ed. 1995 Species Diagnostics Protocols, PCR and other Nucleic Acid Methods. Humana, Totova, NJ. USA
Conley, E.C 1995-96 The Ion Channel FactsBook. Vols. 1-4 Acad. Press, San Diego, CA, USA
Cortese, R., ed. 1995 Combinatorial Libraries. Synthesis, Screening and Application Potentials. De Gruyter, New York
DePamphilis, M.L., ed. 1996 DNA Replication in Eukaryotic Cells. Cold Spring Harbor Laboratory Press, Cold Spring Harbor, NY, USA
Docherty, K., ed. 1996 Gene Transcription: DNA Binding Proteins. Wiley, New York
Docherty, K., ed., 1996 Gene Transcription: RNA Analysis. Wiley, New York
Doolittle, R.F., ed. 1996 Computer Methods for Macromolecular Sequence Analysis. Methods in Enzymology 266. Acad. Press, San Diego, CA, USA
Eckstein, F. and Lilley, D.M.J. 1997 Catalytic RNA. Springer, New York
Ellis, R.J., ed. 1966 The Chaperonins. Acad. Press, San Diego, CA, USA

Molecular biology continued
Eun, H-M. 1996 Enzymology Primer for Recombinant DNA Technology. Acad. Press, San Diego, CA, USA
Farrel, R.E. and Leppertt, G. 1997 rDNA Manual. Springer, New York
Gelvin, S.B. and Schilperoort, R.A. 1994 Plant Molecular Biology Manual. Kluwer, Norwell, MA, USA
Goodbourn, S., ed. 1996 Eukaryotic Gene Transcription. Oxford Univ. Press, New York
Goodfellow, J.M., ed. 1995 Computer Modeling in Molecular Biology. VCH, New York
Gribskov, M. and Devereux, J. 1991 Sequence Analysis Primer. Freeman, New York
Griffin, A.M. and Griffin, H.G., eds. 1994 Computer Analysis of Sequence Data. Humana, Totowa, NJ, USA
Harris, D.A. 1997 Molecular and Cellular Biology. Horizon. Wymondham, UK
Harris, D.A., ed. 1997 Prions: Molecular and Cellular Biology. Horizon, Wymondham, UK
Hawkins, J.D. 1996 Gene Structure and Expression. Cambridge Univ. Press, New York
Hershey, J.W.B., Mathews, M.B. and Sonenberg, N., eds. 1995 Translational Control. Cold Spring Harbor Lab. Press, Cold Spring Harbor, NY, USA
Hideaki, H. *et al*. eds. 1996 Intracellular Signal Transduction. Acad. Press, San Diego, CA, USA
Hills, D. and Moritz, C., eds. 1996 Molecular Systematics. Sinauer, Sunderland, MA, USA
Horton, R.M., ed. 1997 Genetic Engineering with PCR. Horizon, Wymondham, UK
Houdebine, L.M., ed. 1997 Transgenic Animals. Generation and Use. Harwood, Newark, NJ, USA
Hoy, M.A. 1994 Insect Molecular Genetics. An Introduction to Principles and Applications. Acad. Press, San Diego, CA, USA
Hughes, M.A. 1996 Plant Molecular Genetics. Addison Wesley, Boston, MA, USA
Innis, M.A., Gelfand, D.H. and Sninsky, J.J., eds. 1995 PCR Strategies. Acad. Press, San Diego, CA, USA
Jost, J-P. and Saluz, H-P, eds., 1993 DNA Methylation: Molecular Biology and Biological Significance. Birkhäuser Vlg., New York
Kaplitt, M.G. and Loewy, A.D., eds. 1995 Viral Vectors. Gene Therapy and Neuroscience Applications. Acad. Press, San Diego, CA, USA
Kirby, L.T. 1990 DNA Fingerprinting. An Introduction. Stockton Press, New York
Kneale, G.G., ed. 1994 DNA-Protein Interactions. Principles and Protocols. Humana, Totowa, NJ, USA
Krainer, A.R., ed. 1997 Eukaryotic mRNA Processing. Oxford Univ. Press, New York
Lachtman, D.S. 1995 Eukaryotic Transcription Factors. Acad. Press, San Diego, CA, USA
Levy, S.B., ed. 1997 Antibiotic Resistance. Wiley, New York
Lilley, D.M.J., ed. 1995 DNA - Protein. Structural Interactions. IRL Press. New York
Litvak, S. 1996 Retroviral Reverse Transcriptases. Chapman and Hall, New York
McDonald, C.J., ed. 1997 Enzymes in Molecular Biology. Essential Data. Wiley, New York
Moley, J.F. and Kim, S.H. 1994 Molecular Genetics of Surgical Oncology. CRC Press, Boca Raton, FL, USA
Morimoto, R.I., Tissières, A. and Georgopoulos, C., eds. 1994 The Biology of Heat Shock Proteins and Molecular Chaperones. Cold Spring Harbor Laboratory Press, Cold Spring Harbor, NY, USA
Mullis, K.B., Ferré, F. and Gibbs, eds. 1994 The Polymerase Chain Reaction. Birkhäuser, Boston, MA, USA
Nagai, K. and Mattaj, I.W., eds. 1995 RNA—Protein Interactions, IRL Press, New York
Newton, C. and Graham, A. 1997 PCR. Springer, New York
Nicola, N.A., ed. 1995 Guidebook to Cytokines and Their Receptors. Oxford Univ. Press, New York
North, R.A., ed. 1994 Ligand- and Voltage-Gated Ion Channels. CRC Press, Boca Raton. FL, USA

Molecular biology continued

Saluz, H.P. and Wiebauer, K. 1995 DNA and Nucleoprotein Structure In Vivo. CRC Press, Boca Raton, FL, USA
Schenkel, J. 1997 RNP Particles. Springer, New York
Sealfon, S.C. 1994 Receptor Molecular Biology. Acad. Press, San Diego, CA
Setlow, J.K. and Hollaender, A., eds. 1979-1996 Genetic Engineering: Principles and Methods. Plenum, New York
Shinitzky, M., ed. 1995 Biomembranes. Signal Transduction Across Membranes. VCH, New York
Sinden, R.R. 1994 DNA Structure and Function. Acad. Press, San Diego, CA, USA
Skalka, A.M. and Goff, S.P., eds. 1993 Reverse Transcriptase. Cold Spring Harbor Laboratory Press, Cold Spring Harbor, NY, USA
Söll, D. and Rajbahandary, U.L., eds. 1994 tRNA. Structure, Biosynthesis and Function. AMS Press, Washington, DC, USA
Tait, R.C. 1997 An Introduction to Molecular Biology. Horizon, Wymondham, UK
Tomiuk, J., Wöhrmann, K. and Sentker, A. 1996 Transgenic Organisms. Springer, New York
Vaddi, K., Keller, M. and Newton, R.C. 1997 The Chemokine FactsBook. Acad. Press, San Diego, CA, USA
Vega, M.A., ed. 1995 Gene Targeting. CRC Press. Boca Ratoon, FL, USA
Watson, J.D. *et al.* 1987 Molecular Biology of the Gene. Benjamin/Cummings, Menlo Park, CA, USA
Wilks, A.E. and Harpur, A.G. 1996 Intracellular Signal Transduction: The Jak-STAT Pathway. Chapman and Hall, New York
Wingender, E. 1993 Gene Regulation in Eukaryotes. VHC, New York
Woodgett, J.R., ed. 1995 Protein Kinases. Oxford Univ. Press, New York

MUTATION *(see also radiation)*

Cooper, D.N. and Krawczak, M. 1993 Human Gene Mutation. Bios Scientific, Oxford, UK
Cotton, R.G.H. 1997 Mutation Detection. Oxford Univ. Press, New York
Friedberg, E.C., Walker, G.C. and Siede, W. 1995 DNA Repair and Mutagenesis. Amer. Soc. Microbiol. Press, Washington, DC, USA
Gold, L.S. and Zeiger, E., eds. 1996 Handbook of Carcinogenic Potency and Genotoxicity Databases. CRC Press, Boca Raton, FL, USA
Hollaender, A. ed. 1971- . Chemical Mutagens. Plenum, New York
Klaassen, C.D., ed. 1996 Casarett and Doull's Toxicology. The Basic Science of Poisons. McGraw-Hill, New York
Landegren, U., ed. 1996 Laboratory Protocols for Mutation Detection. Oxford Univ. Press, New York
Loveless, A. 1966 Genetic and Allied Effects of Alkylating Agents, Pennsylvania State Univ Press, University Park, PA, USA
McPherson, M.J. 1991 Directed Mutagenesis. IRL Press, New York
Ross, W.C.J. 1962 Biological Alkylating Agents. Butterworth, London, UK
Pfeifer, G.P., ed. 1966 Technologies for Detection of DNA Damage and Mutation. Plenum, New York
Taylor, G.R., ed. 1997 Laboratory Methods for the Detection of Mutations and Polymorphisms in DNA. CRC Press, Boca Raton, FL, USA
Vogel, F. and Röhrborn, G., eds. 1970 Chemical Mutagenesis in Mammals and Man. Springer Vlg., New York

NEUROBIOLOGY *(see also molecular biology, behavior)*

Ancill, R.J., Holliday, S. and Higgenbottam, J., eds. 1995 Schizophrenia. Exploring the

Neurobiology continued
Spectrum of Psychosis. Wiley, New York
Blum, K. and Noble, E.P., eds. 1997 Handbook of Psychiatric Genetics. CRC Press, Boca Raton, FL, USA
Burrows, M. 1996 The Neurobiology of the Insect Brain. Oxford Univ. Press, New York
Butler, A.B. and Hodos, W. 1996 Comparative Vertebrate Neuroanatomy. Evolution and Adaptation. Wiley, New York
Cervós-Navarro, J. and Ulrich, H. 1995 Metabolic and Degenerative Diseases of the Nervous System. Pathology, Biochemistry and Genetics. Acad. Press, San Diego, CA, USA
Crick, F. 1994 The Astonishing Hypothesis: The Scientific Search for the Soul. Simon and Schuster, Old Tappan, NJ, USA
Galaburda, A.M., ed. 1993 Dyslexia and Development. Neurobiological Aspects of Extra-Ordinary Brains. Harvard Univ. Press, Cambridge, MA, USA
Gershon, E.S. and Cloninger, C.R., eds. 1994 Genetic Approcahes to Mental Disorders. Amer. Psychiatric Press, Washington, DC, USA
Honavar, V. and Uhr, L., eds. 1994 Artificial Intelligence and Neural Networks. Steps Toward Principled Integration. Acad. Press, San Diego, CA, USA
Johnston, D. and Wu, S.M.-S. 1994 Foundations of Cellular Neurophysiology. MIT Press, Cambridge, MA, USA
Latchman, D., ed. 1995 Genetic Manipulation of the Nervous System. Acad. Press, San Diego, CA, USA
Lyon, G., Adams, R.D. and Kolodny, E.H. 1996 Neurology of Hereditary Metabolic Diseases of Children. McGraw-Hill, New York
McGaugh, J.L., ed. 1995 Brain and Memory. Modulation and Mediation of Neuroplasticity. Oxford Univ. Press, New York
Merchant, K.M. 1996 Pharmacological Regulation of Gene Expression in the CNS. Towards an Understanding of Basal Ganglion Functions. CRC Press, Boca Raton, FL, USA
Micevych, P.E. and Hammer, R.P., Jr., eds. 1995 Neurobiological Effects of Sex Steroid Hormones. Cambridge Univ. Press, New York
Phillips, M.I. and Evans, D., eds. 1994 Neuroimmunology. Acad. Press, San Diego, CA, USA
Revest. P. and Longstaff, A. 1997 Molecular Neuroscience. Springer, New York
Schachter, D.L. 1996 Searching for Memory: The Brain, the Mind, and the Past. Basic Books, New York
Smith, C.U.M. 1996 Elements of Molecular Neurobiology. Wiley, New York
Wess, J. 1995 Molecular Mechanism of Muscarinic Acetylcholine Receptor Function. Springer, New York
Wheal, H. and Thomson, A., eds. 1995 Excitatory Amino Acids and Synaptic Transmission. Acad. Press, San Diego, CA, USA

ORGANELLES (see also genetics monographs)

Attardi, G.M. and Chomyn, A., eds. 1995 Mitochondrial Biogenesis and Genetics. Methods in Enzymology. Vol. 260, Acad. Press, San Diego, CA, USA
Beale, G. and Knowles, J. 1978 Extranuclear Genetics. Arnold, London, UK
Darley-Usmar, V, and Schapira, A.H.V., eds. 1994 Mitochondria, DNA, Proteins and Disease. Portland Press, Chapel Hill, NC, USA
Di Mauro, S. and Wallace, D.C., eds. 1992 Mitochondrial DNA in Human Pathology. Raven Press, New York
Grun, P. 1976 Cytoplasmic Genetics and Evolution. Columbia Univ. Press, New York
Gillham, N.W. 1994 Organelle Genes and Genomes. Oxford University Press, New York
Herrmann, R.G., ed. 1992 Cell Organelles. Springer, New York
Kirk, J.T.O. and Tilney Bassett, R.A.E. 1978 The Plastids. Elsevier. Amsterdam, Netherlands
Margulis, L. 1970 Origin of Eukaryotic Cells. Yale Univ. Press, New Haven, CN, USA

Organelles continued

Sager, R. 1972 Cytoplasmic Genes and Organelles. Acad. Press, New York

PHILOSOPHY *(see also social and ethical issues, biotechnology, evolution)*

Bayertz, K., ed. 1995 GenEthics: Technological Intervention in Human Reproduction as a Philosophical Problem. Cambridge Univ. Press, New York
Holton, G. 1993 Science and Antiscience. Harvard Univ. Press, Cambridge, MA, USA
Horgan, J. 1996 The End of Science. Facing the Limits of Knowledge in the Twighlight of the Scientific Age. Helix (Addison-Wesley), New York
Kitcher, P. 1996 The Lives to Come: The Genetic Revolution and Human Possibilities. Penguin, New York
Murphy, M.P. and O'Neill, L.A.J., eds. 1997 What is Life: The Next 50 Years. Speculations on the Future of Biology. Cambridge Univ. Press, New York
Popper, K.R. 1959 The Logic of Scientific Discovery. Basic Books. New York
Popper, K.R. 1983 Realism and the Aim of Science. Rowman and Littlefield. Totowas, NJ, USA
Walters, L. and Palmer, J.G. 1997 The Ethics of Human Gene Therapy. Oxford Univ. Press, New York

PLANT and ANIMAL BREEDING *(see also biometry, biotechnology, cytology and cytogenetics, population genetics, plant pathology)*

Bahl, P.N. 1996 Genetics, Cytogenetics and Breeding of Crop Plants. Science Pubs. Bio-Oxford & IBH, Lebanon, NH, USA
Elsevier's Dictionary of Plant Genetic Resources 1995 Elsevier Science, New York
Evans, L.T. 1996 Crop Evolution, Adaptation and Yield. Cambridge Univ. Press, New York
Hayward, M.D., Bosemark, N.O. and Romagosa, eds. 1993 Plant Breeding. Principles and Prospects. Chapman and Hall, London, UK
Kang, M.S. and Gauch, H.G., Jr. 1996 Genotype-by-Environment Interaction. CRC Press, Boca Raton, FL, USA
Le Roy, H.L. 1966 Elemente der Tierzucht: Genetik, Mathematik, Populationsgenetik. Bayerischer Landwirthschafts-Vlg., München, Germany
Mason, I.L. 1996 A World Dictionary of Livestock Breeds. CAB Internat/Oxford Univ. Press, New York
Mrode, R.A. 1996 Linear Models for the Production of Animal Breeding Values. CAB International/Oxford Univ. Press, New York
Purdom, C.E. 1993 Genetics and Fish Breeding. Chapman and Hall, New York
Simmonds, N.W. 1979 Principles of Crop Improvement. Longman. Harlow, Essex, UK
Sobral, B.W. 1966 The Impact of Plant Molecular Genetics. Birkhäuser, Boston, MA, USA
Stoskopf, N.C., Tomes, D.T. and Christie, B.R. 1993 Plant Breeding. Theory and Practice. Westview, Boulder, CO, USA
Sybenga, J. 1992 Cytogenetics in Plant Breeding. Springer-Vlg., Berlin, Germany
Tsuchya, T. and Gupta, P.K., eds., 1991 Chromosome Engineering in Plants. Genetics, Breeding, Evolution. Elsevier Science, New York
Van Vleck, L.D., Pollak, E.J. and Oltenacu, E.A.B. 1987 Genetics for Animal Sciences. Freeman, New York

PLANT PATHOLOGY *(see also plant physiology, molecular biology)*

Agrios, G.N. 1997 Plant Pathology. Acad. Press, San Diego, CA, USA
Boller, T. and Meins, F., Jr., eds. 1992 Plant Gene Research: Genes Involved in Plant Defense. Springer-Vlg., New York
Daniel, M. and Purkayastha, R.P., eds. 1994 Handbook of Phytoalexin Metabolism and Action.

Plant pathology continued
 Dekker, New York
 Entwistle, P.F. et al., eds., 1994 Bacillus thuriengesis. An Environmental Biopesticide. Theory and Practice. Wiley, New York
 Goodman, R.N. and Novacky, A.J. 1994 The Hypersensitive Reaction in Plants to Pathogens. APS Press, The American Phytopathological Society, St. Paul, MN, USA
 Gurr. S.J., McPherson, M.J. and Bowles, D.J. 1992 Molecular Plant Pathology. IRL Press. New York
 Matthews, R. and Burnie, J.M. 1995 Heatschock Proteins in Fungal Infections. Springer, New York
 Scandalios, J.G., ed. 1997 Oxidative Stress and Molecular Biology of Antioxidant Defenses. Cold Spring Harbor Laboratory Press. Plainview, NY, USA
 Sigee, D.C. 1993 Bacterial Plant Pathology. Cell and Molecular Aspects. Cambridge Univ. Press, New York

PLANT PHYSIOLOGY *(see also biochemistry, molecular biology, biotechnology)*

Basra, A.S., ed., 1994 Stress-Induced Gene Expression in Plants. Harwood, Langhorne, PA, USA
Fosket, D.E. 1994 Plant Growth and Development. A Molecular Approach. Acad. Press, San Diego, CA, USA
Dashek, W.V., ed. 1997 Methods in Plant Biochemistry and Molecular Biology. CRC Press, Boca Raton, FL, USA
Dey, P. M. and Harbourne, J.B., eds. 1997 Plant Biochemistry. Acad. Press, San Diego. CA, USA
Hall, D.O. and Rao, K.K. 1994 Photosynthesis. Cambridge Univ. Press, New York
Mohr, H. and Schopfer, P. 1995 Plant Physiology. Springer, New York
Pirson, A., ed. 1993 Encyclopedia of Plant Physiology. Springer, New York
Taiz, L. and Zeiger, E. 1991 Plant Physiology. Benjamin/Cummings, Redwood City, CA, USA

POPULATION GENETICS *(see also biometry, evolution, cytology and cytogenetics)*

Baumberg, S. et al., eds. 1995 Population Genetics of Bacteria. Cambridge Univ. Press, New York
Bonne-Tamir, B. and Adam, A. eds. 1992 Genetic Diversity Among Jews: Diseases and Markers at the DNA Level. Oxford Univ. Press, New York
Cavalli-Sforza, L.L. and Bodmer, W.F. 1971 The Genetics of Human Populations. Freeman, San Francisco, CA, USA
Cavalli-Sforza, L.L., Menozzi, P. and Piazzo, A. 1994 The History and Geography of Human Genes. Princeton Univ. Press, Princeton, NJ, USA
Chadwick, D. and Cardew, G., eds. 1996 Variation in the Human Genome. Wiley, New York
Crow, J.F. and Kimura, M. 1970 An Introduction to Population Genetics Theory. Harper and Row, New York
Fisher, R.A. 1958 The Genetical Theory of Natural Selection. Dover Publications, New York
Fisher, R.A. 1965 The Theory of Inbreeding. Acad. Press, New York
Goodman, R.M and Motulsky, A.G., eds. 1979 Genetic Diseases Among Ashkenazi Jews. Raven, New York
Haldane, J.B.S. 1935 The Causes of Evolution. Longmans Green, New York
Hartl, D.L. 1980 Principles of Population Genetics. Sinauer, Sunderland, MA, USA
Hoelzel, A.R., ed. 1992 Molecular Genetic Analysis of Populations. IRLPress, New York
Lerner, I.M. 1954 Genetic Homeostasis. Oliver & Boyd, Edinburgh, UK
Levin, S.A., ed. 1994 Frontiers of Mathematical Biology. Springer, New York
Li, C.C. 1976 First Course in Population Genetics. Boxwood Press, Pacific Grove, CA, USA
Mettler, L.E., Gregg, T.G. and Schaffer, H.E. 1988 Population Genetics and Evolution. Prentice-Hall, Englewood Cliffs, NJ, USA

GENETICS MANUAL

Population genetics continued

Mourant, A.E., Kopec, A. C. and Domaniewska-Sobczak, K. 1978 Genetics of the Jews. Clarendon Press. Oxford, UK
Nei, M. 1975 Molecular Population Genetics and Evolution. American Elsevier, New York
Real, L.A., ed. 1994 Ecological Genetics. Princeton Univ. Press, Princeton, NJ, USA
Roychoudhury, A.K. and Nei, M. 1988 Human Polymorphic Genes: World Distribution. Oxford Univ. Press, New York
Spiess, E.B. 1977 Genes in Populations. Wiley, New York
Weir, B., ed. 1995 Human Identification. The Use of DNA Markers. Kluwer Academic, Boston, MA, USA
Weiss, K., ed. 1996 Variation in the Human Genome. Wiley, New York
Wilmsen Thornhill, N., ed. 1993 The Natural History of Inbreeding and Outbreeding. Theoretical and Empirical Perspectives. Univ. Chicago Press, Chicago, IL, USA
Wright, S. 1968-76 Evolution and the Genetics of Populations. Univ. Chicago Press, Chicago, IL, USA

RADIATION *(see also mutation)*

Dunlap, R.R., Kraft, M.E. and Rosa, E.A., eds. 1993 Public Reactions to Nuclear Waste. Duke Univ. Press. Durham, NC, USA
Heald, M.A. and Marion, J.B. 1995 Classical Electromagnetic Radiation. Saunders, Philadelphia, PA, USA
Hendry, J.H. and Lord, B.L. 1995 Radiation Toxicology. Bone Marrow and Leukemia. Taylor & Francis, London, UK
Lindee, S. 1995 Suffering Made Real: American Science and the Survivors at Hiroshima. Univ. Chicago Press, Chicago, IL, USA
Prasad, K.N. 1995 Handbook of Radiobiology. CRC Press, Boca Raton, FL, USA
Shigematsu, I. *et al.*, eds. 1995 Effects of A-Bomb Radiation on the Human Body. Harwood, Lanhorne, PA, USA (translation)
Shull, W.J. 1995 Effects of Atomic Radiation. Wiley - Liss, New York
Turner, J.E. 1995 Atoms, Radiation, and Radiation Protection. Wiley, New York
Wilkening, G.M. 1991 Ionizing radiation, pp. 599-655, in Patty's Industrial Hygiene and Toxicology, Clayton, G.D. and Clayton, F.E., eds. Wiley, New York

RECOMBINATION *(see also mapping)*

Bailey, N.T.J. 1961 Introduction to the Mathematical Theory of Genetic Linkage. Clarendon Press, Oxford, U.K.
Bishop, M. 1994 Guide to Human Genome Mapping. Acad. Press, San Diego, CA, USA
Broach, J.R., Pringle, J.R. and Jones, E.W., eds. 1991 Recombination in Yeast. Cold Spring Harbor Laboratory Press, Cold Spring Harbor, NY, USA
Leach, D.R.F. 1996 Genetic Recombination. Blackwell Science, Cambridge, MA, USA
Mather, K. 1957 The Measurement of Linkage in Heredity. Methuen, London, UK
Nelson, D.L. and Brownstein, B.H., eds., 1993 YAC Libraries. Oxford Univ. Press, New York
Paszkowski, J., ed. 1994 Homologous Recombination and Gene Silencing in Plants. Kluwer, Norwell, MA, USA
Stahl, F.W. 1979 Genetic Recombination. Thinking About it in Phage and Fungi. Freeman, San Francisco, CA, USA
Terwilliger, J.D. and Ott, J. 1994 Handbook of Human Genetic Linkage. Johns Hopkins Univ. Press, Baltimore, MD, USA
Whitehouse, H.L.K. 1982 Genetic Recombination. Understanding the Mechanisms. Wiley, New York

SOCIAL and ETHICAL ISSUES *(see also philosophy)*

Annas, G. J. and Elias, S., eds. 1992 Gene Mapping. Using Law and Ethics as Guides. Oxford Univ. Press. New York
Ballantyne, J., Sensabaugh, G. and Witkowski, J. eds. 1989 DNA Technology and Forensic Science. Cold Spring Harbor Laboratory Press, Cold Spring Harbor, NY, USA
Frankel, M.S. and Teich, A.H., eds. 1994 The Genetic Frontier: Ethics, Law and Policy. AAAS, Washington, DC, USA
Kitcher, P. 1996 The Lives to Come. Simon and Schuster, New York
Reilly, P. 1977 Genetics, Law and Social Policy. Harvard Univ. Press, Cambridge, MA, USA
Reiss, M.J. and Straughan, R. 1996 Improving Nature? Cambridge Univ. Press, New York
Thompson, P. ed. 1994 Issues in Evolutionary Ethics. State Univ. New York Press, Albany, NY, USA
Walters, L. and Palmer, J.G. 1996 The Ethics of Human Gene Therapy. Oxford Univ. Press, New York

TRANSPOSABLE ELEMENTS *(genetics monographs)*

Arkhipova, I.R., Lyubomirskaya, N.V. and Ilyin, N.V. 1995 Drosophila Retrotransposons. CRC Press, Boca Raton, FL, USA
Berg, D.E. and Howe, M.M., eds. 1989 Mobile DNA. Amer. Soc. Microbiol., Washington, DC, USA
Sherratt, D.J., ed. 1995 Mobile Genetic Elements. IRL Press, New York

VIRUSES *(see also microbiology, molecular genetics)*

Adolph, K.W., ed. 1995 Viral Gene Techniques. Acad. Press, San Diego, CA, USA
Calendar, R., ed. 1988 The Bacteriophages. Plenum, New York
Cann, A.J. 1997 Principles of Molecular Virology. Acad. Press, San Diego, CA, USA
Cullen, B., ed. 1993 Human Retroviruses. IRL/Oxford Univ. Press, New York
Hendrix, R.W., Roberts, J. W., Stahl, F. W. and Weisberg, R.A., eds. 1983 *Lambda II*. Cold Spring Harbor Laboratory Press, Cold Spring Harbor, NY, USA
Kaplitt, G. and Loewy, A.D., eds. 1995 Viral Vectors. Gene Therapy and Neuroscience Applications. Acad.Press, San Diego, CA, USA
Levy, J.A., ed. 1995 The Retroviridae. Plenum, New York
Levy, J.A., Fraenkel-Conrat, H. and Owens, R.A. 1994 Virology. Prentice-Hall, Englewood Cliffs, NJ, USA
Mahy, B. 1996 A Dictionary of Virology. Acad. Press, San Diego, CA, USA
Morse, S.S., ed. 1993 The Evolutionary Biology of Viruses. Raven, New York
Murphy, F.A. *et al.* 1995 Virus Taxonomy. Springer, New York
Webster, R.G. and Granoff, A., eds. 1994 Encyclopedia of Virology. Acad. Press, San Diego, CA, USA
White, D.O. and Fenner, F.J. 1994 Medical Virology. Acad. Press, San Diego, CA, USA
Wimmer, E., ed. 1994 Cellular Receptors for Animal Viruses. Cold Spring Harbor Laboratory Press, Cold Spring Harbor, NY, USA

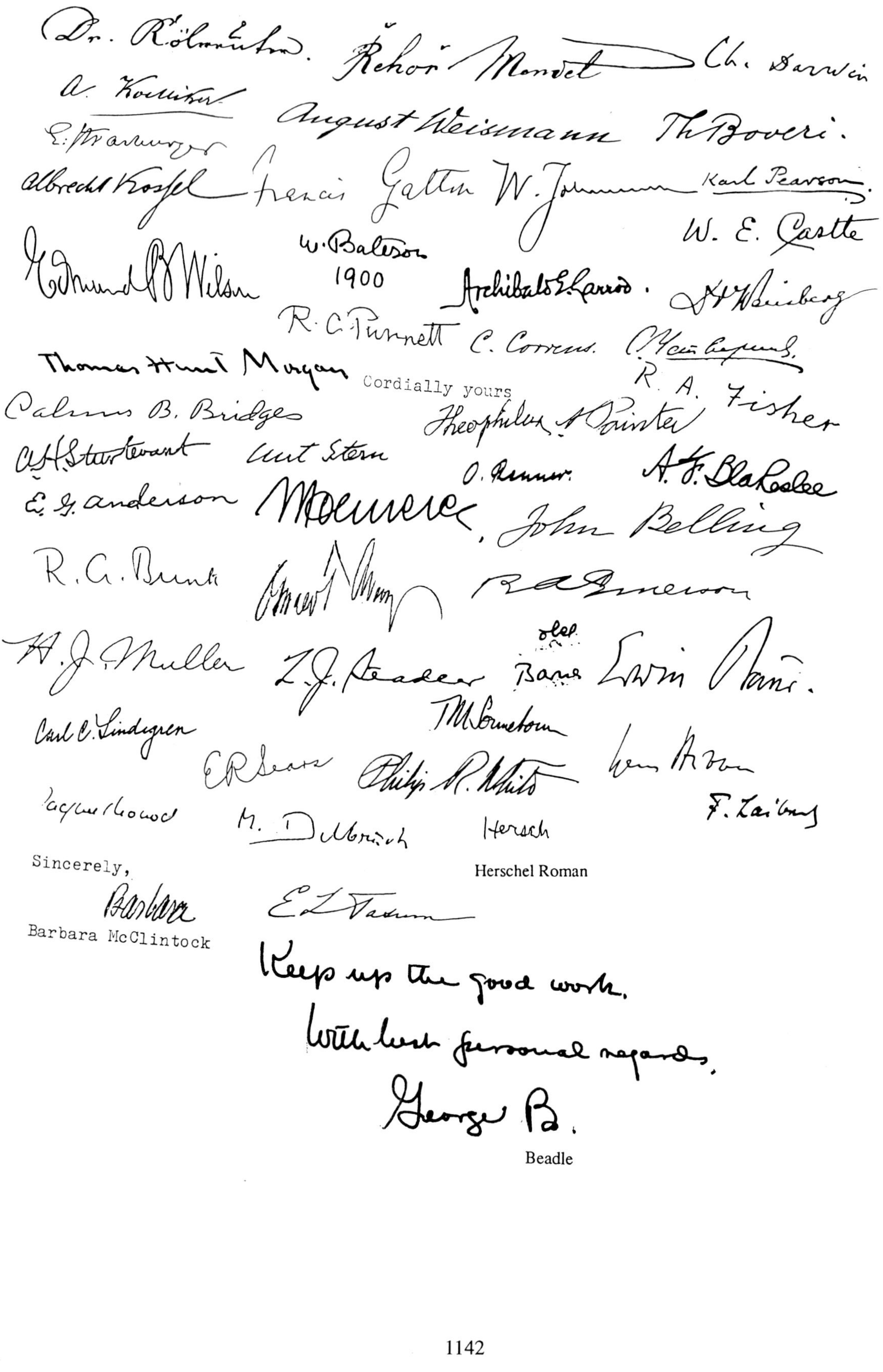

GENETICS MANUAL

[Page of signatures including:]

Dr. Kölreuter, Rehor Mendel, Ch. Darwin, A. Kölliker, E. Strasburger, August Weismann, Th. Boveri, Albrecht Kossel, Francis Galton, W. Johannsen, Karl Pearson, W. E. Castle, Edmund B. Wilson, W. Bateson 1900, Archibald E. Garrod, H. Weinberg, R. C. Punnett, C. Correns, (von Tschermak), Thomas Hunt Morgan, R. A. Fisher, Calvin B. Bridges, Theophilus S. Painter, A. H. Sturtevant, Curt Stern, E. G. Anderson, O. Renner, A. F. Blakeslee, M. Oenerre, John Belling, R. A. Brink, Ernst (Caspari?), R. A. Emerson, H. J. Muller, L. J. Stadler, Barbara (old), Erwin Bauer, Carl C. Lindegren, T. M. Sonneborn, E. R. Sears, Philip R. White, F. Zaibing, Jacques Monod, M. Delbrück, Hersch

Sincerely,
Barbara McClintock

Herschel Roman

E. L. Tatum

Keep up the good work. With best personal regards,
George B.
Beadle